Teacher Edition

Printed in the U.S.A.

ISBN 978-0-358-05531-0

4 5 6 7 8 9 10 2331 28 27 26 25 24 23 22 21

4500825686 C D E F G

Authors

Edward B. Burger, PhD, is a mathematician who is also the president of Southwestern University in Georgetown, Texas. He is a former Francis Christopher Oakley Third Century Professor of Mathematics at Williams College, and a former vice provost at Baylor University. He has authored or coauthored numerous articles, books, and video series; delivered many addresses and workshops throughout the world; and made many radio and television appearances. He has earned many national honors, including the Robert Foster Cherry Award for Great Teaching. In 2013, he was inducted as one of the first fellows of the American Mathematical Society.

Juli K. Dixon, PhD, is a professor of mathematics education at the University of Central Florida (UCF). She has taught mathematics in urban schools at the elementary, middle, secondary, and post-secondary levels. She is a prolific writer who has published books, textbooks, book chapters, and articles. A sought-after speaker, Dr. Dixon has delivered keynotes and other presentations throughout the United States. Key areas of focus are deepening teachers' content knowledge and communicating and justifying mathematical ideas. She is a past chair of the National Council of Teachers of Mathematics Student Explorations in Mathematics Editorial Panel and a member of the board of directors for the Association of Mathematics Teacher Educators. You can find her on social media at @TheStrokeOfLuck.

Timothy D. Kanold, PhD, is an award-winning international educator, author, and consultant. He is a former superintendent and director of mathematics and science at Adlai E. Stevenson High School District 125 in Lincolnshire, Illinois. He is a past president of the National Council of Supervisors of Mathematics (NCSM) and the Council for the Presidential Awardees of Mathematics (CPAM). He has served on several writing and leadership commissions for the National Council of Teachers of Mathematics during the past two decades, including the *Teaching Performance Standards* task force. He presents motivational professional development seminars worldwide with a focus on developing professional learning communities (PLCs) to improve teaching, assessing, and learning of *all* students. He has recently authored nationally recognized articles, books, and textbooks for mathematics education and school leadership, including *What Every Principal Needs to Know about the Teaching and Learning of Mathematics* and *HEART: Fully Forming Your Professional Life as a Teacher and Leader.* You can find him on social media at @tkanold.

Robert Kaplinsky, MEd, is a mathematics educator with over fifteen years of experience as a classroom teacher and teacher specialist for Downey Unified School District, instructor for the University of California, Los Angeles (UCLA), and presenter at conferences around the world. He created the #ObserveMe moment, where educators open their classroom doors so that they can learn from each other. He's also the co-founder of the website OpenMiddle.com as well as the Southern California Math Teacher Specialist Network, a group that includes over 200 math teacher specialists from more than 5 counties. His work has been published by the American Educational Research Association (AERA), Education Week, and many others. You can find him on social media at @robertkaplinsky.

Matthew R. Larson, PhD, is a past president of the National Council of Teachers of Mathematics (NCTM). Prior to serving as president of NCTM, he was the K–12 mathematics curriculum specialist for Lincoln Public Schools (Nebraska), where he currently serves as Director of Elementary Education. A prolific speaker and writer, he is the coauthor of more than a dozen professional books. He was a member of the writing teams for the major publications *Principles to Actions: Ensuring Mathematical Success for All* (2014) and *Catalyzing Change in High School Mathematics: Initiating Critical Conversations* (2018). Key areas of focus include access and equity and effective stakeholder communication. He has taught mathematics at the secondary and college levels and held an appointment as an honorary visiting associate professor at Teachers College, Columbia University. You can find him on social media at @mlarson_math.

Steven J. Leinwand is a principal research analyst at the American Institutes for Research (AIR) in Washington, DC, and has nearly 40 years in leadership positions in mathematics education. He is a past president of the National Council of Supervisors of Mathematics and served on the National Council of Teachers of Mathematics Board of Directors. He is the author of numerous articles, books, and textbooks and has made countless presentations with topics including student achievement, reasoning, effective assessment, and successful implementation of standards. You can find him on social media at @steve_leinwand.

Program Consultants

English Language Development Consultant

Harold Asturias is the director for the Center for Mathematics Excellence and Equity at the Lawrence Hall of Science, University of California. He specializes in connecting mathematics and English language development as well as equity in mathematics education.

Blended Learning Consultant

Weston Kieschnick, ICLE Senior Fellow, a former teacher, principal, instructional development coordinator, and dean of education, has driven change and improved student learning in multiple capacities throughout his educational career. Now, as an experienced instructional coach and senior fellow with ICLE, Kieschnick shares his expertise with teachers to transform learning through online and blended models.

Program Consultant

David Dockterman, EdD, operates at the intersection of research and practice. A member of the faculty at the Harvard Graduate School of Education, he provides expertise in curriculum development, adaptive learning, professional development, and growth mindset.

Open Middle™ Consultant

Nanette Johnson, MEd, is a Secondary Math Teacher Specialist at Downey Unified School District. She has presented at conferences including the California Mathematics Council (CMC) – South in Palm Springs, CA, CMC-North in Asilomar, CA, and the National Council for Teachers of Mathematics (NCTM) conferences across the United States.

STEM Consultants

Michael Despezio has authored many HMH instructional programs for science and mathematics. He has also authored numerous trade books and multimedia programs on various topics and hosted dozens of studio and location broadcasts for various organizations in the US and worldwide. Recently, he has been working with educators to provide strategies for implementing the Next Generation Science Standards.

Bernadine Okoro is a chemical engineer by training and a playwright, novelist, director, and actress by nature. Okoro went from working with patents and biotechnology to teaching in K–12 classrooms. She is a 12-year science educator, Albert Einstein Distinguished Fellow, original author of NGSS, and a member of the Diversity and Equity Team. Okoro currently works as a STEM learning advocate and consultant.

Marjorie Frank An educator and linguist by training, a writer and poet by nature, Marjorie Frank has authored and designed a generation of instructional materials in all subject areas. Her other credits include authoring science issues of an award-winning children's magazine, writing game-based digital assessments, developing blended learning materials, and serving as instructional designer and coauthor of school-to-work software. She has also served on the adjunct faculty of Hunter, Manhattan, and Brooklyn Colleges.

Cary I. Sneider, PhD While studying astrophysics at Harvard, Cary Sneider volunteered to teach in an Upward Bound program and discovered his real calling as a science teacher. After teaching middle and high school science, he settled for nearly three decades at Lawrence Hall of Science in Berkeley, California, where he developed skills in curriculum development and teacher education. Over his career, Cary directed more than 20 federal, state, and foundation grant projects and was a writing team leader for the Next Generation Science Standards.

Math Solutions® Program Consultants

Deepa Bharath, MEd
Professional Learning Specialist
Math Solutions
Jupiter, Florida

Nicole Bridge, MEd
Professional Learning Specialist
Math Solutions
Attleboro, Massachusetts

Treve Brinkman
Director of Professional Learning
Math Solutions
Denver, Colorado

Lisa K. Bush, MEd
Sr. Director, Professional Development
Math Solutions
Glendale, Arizona

Carol Di Biase
Professional Learning Specialist
Math Solutions
Melbourne, Florida

Stephanie J. Elizondo, MEd
Professional Learning Specialist
Math Solutions
Ocala, Florida

Christine Esch, MEd
Professional Learning Specialist
Math Solutions
Phoenix, Arizona

Le'Vada Gray, MEd
Director of Professional Learning
Math Solutions
Country Club Hills, Illinois

Connie J. Horgan, MEd
Professional Learning Specialist
Math Solutions
Jerome, Idaho

Monica H. Kendall, EdD
Professional Learning Specialist
Math Solutions
Houston, Texas

Lori Ramsey, MEd
Professional Learning Specialist
Math Solutions
Justin, Texas

Lisa Rogers
Professional Learning Specialist
Math Solutions
Cape Coral, Florida

Derek Staves, EdD
Professional Learning Specialist
Math Solutions
Greeley, Colorado

Sheila Yates, MEd
Professional Learning Specialist
Math Solutions
Sioux Falls, South Dakota

Classroom Advisors

Rebecca Boden
Grant County Board of Education
Grant County Schools
Williamstown, Kentucky

Tara Brandt
Director of Mathematics
Holyoke Public Schools
Holyoke, Massachusetts

Bob Cloud
East Samford School
Auburn City Schools
Auburn, Alabama

Eric Creel-Flores
Valley High School
Chambers County Schools
Valley, Alabama

Michelle Demney
East Samford School
Auburn City Schools
Auburn, Alabama

Kimberley Hoffman
Estrella Foothills High School
Buckeye Union High School District
Goodyear, Arizona

Vicki S. Ramseyer Morrow
Bolingbrook High School
Valley View 365U
Community USD
Bolingbrook, Illinois

Todd Taylor
Vestavia Hills High School
Vestavia Hills, Alabama

Unit 1

Essentials of Geometry

MODULE 1 Geometry in the Plane

LEARNING ARC FOCUS

Each lesson includes a full learning arc: building concepts prior to applying and practicing procedures. The key below helps you interpret the learning arc focus of each lesson as well as see a progression across lessons within a module.

Build Conceptual Understanding | Connect Concepts and Skills | Apply and Practice

MODULE 2 Tools for Reasoning and Proof

Unit 2

Parallel and Perpendicular Lines

MODULE 3 Lines and Transversals

MODULE 4 Lines on the Coordinate Plane

Build Conceptual Understanding | Connect Concepts and Skills | Apply and Practice

Unit 3

Transformations

MODULE 5 Transformations that Preserve Size and Shape

MODULE 6 Transformations that Change Size and Shape

Unit 4

Triangle Congruence

MODULE 7 Congruent Triangles and Polygons

Build Conceptual Understanding   Connect Concepts and Skills   Apply and Practice

MODULE 8 Triangle Congruence Criteria

(t) ©Dmitry Melnikov/Alamy; (b) ©Houghton Mifflin Harcourt

Unit 5

Relationships Within Triangles

MODULE 9 Properties of Triangles

MODULE 10 Triangle Inequalities

Build Conceptual Understanding | Connect Concepts and Skills | Apply and Practice

Unit 6

Quadrilaterals, Polygons, and Triangle Similarity

MODULE 11 Quadrilaterals and Polygons

MODULE 12 Similarity

(l) ©davorana/Shutterstock; (r) ©Cameris/Shutterstock

Build Conceptual Understanding Connect Concepts and Skills Apply and Practice

Unit 7

Right Triangle Trigonometry

MODULE 13 Trigonometry with Right Triangles

MODULE 14 Trigonometry with All Triangles

Unit 8

Properties of Circles

MODULE 15 Angles and Segments in Circles

Build Conceptual Understanding Connect Concepts and Skills Apply and Practice

MODULE 16 Relationships in Circles

MODULE 17 Circumference and Area of a Circle

Unit 9

Surface Area and Volume

MODULE 18 Surface Area

MODULE 19 Volume

Build Conceptual Understanding   Connect Concepts and Skills   Apply and Practice

Unit 10

Probability

MODULE 20 Probability of Multiple Events

MODULE 21 Conditional Probability and Independence of Events

Build Conceptual Understanding   Connect Concepts and Skills   Apply and Practice

TEACHER RESOURCES

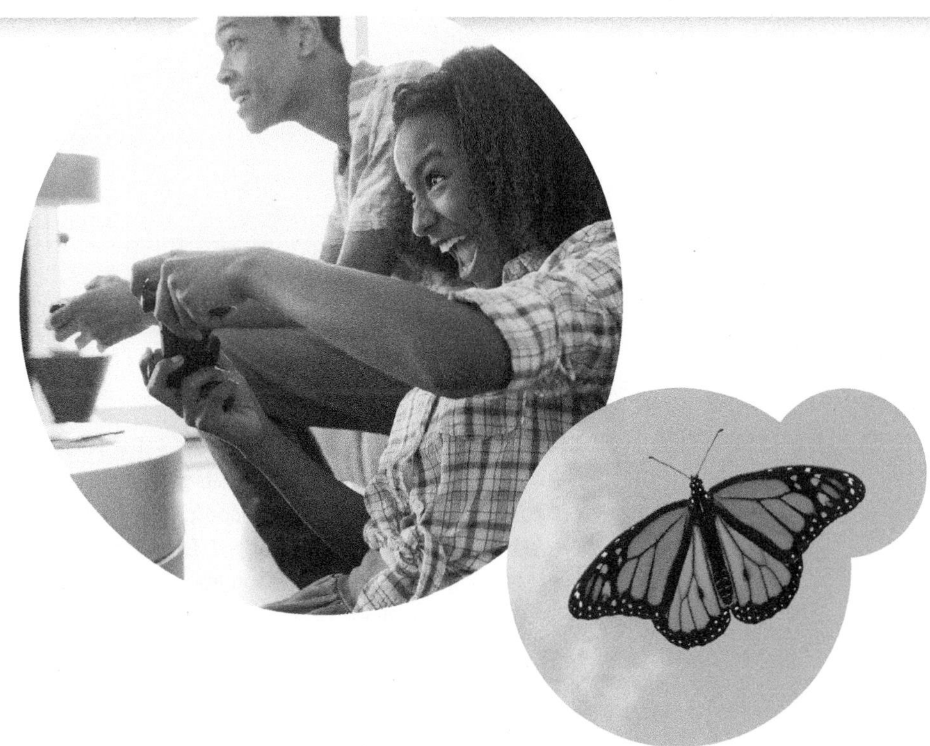

UNIT 1

THE ESSENTIALS OF GEOMETRY

Module 1: Geometry in the Plane

Module 2: Tools for Reasoning and Proof

Wildlife Conservationist 

- **Say:** *Think about animals and plants that are classified as endangered species. What steps can environmentalists and scientists take to conserve these species?*
- Explain that a wildlife conservationist uses data to evaluate the status of the population of an endangered species. Then introduce the STEM task.

STEM Task

Ask students if they have any previous knowledge of how to represent and analyze data.

- Ask students whether the data about the population of right whales could be made more meaningful if represented on a graph.
- Have students create a line graph to display the data.
- Have students state a hypothesis about the status of the population of right whales and use the data to support their hypothesis.
- Ask students to propose a public policy that could help conserve the population of right whales.

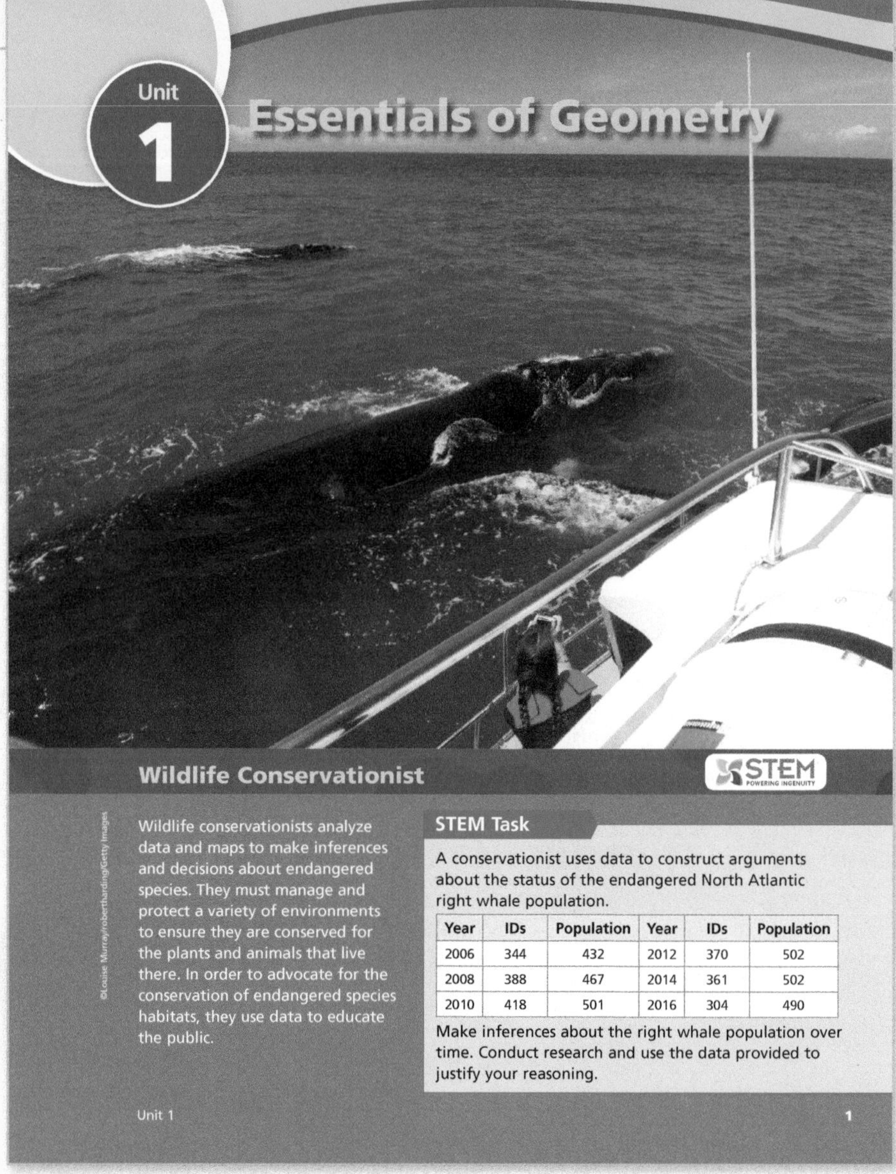

Unit 1

Essentials of Geometry

Wildlife Conservationist

STEM POWERING INGENUITY

Wildlife conservationists analyze data and maps to make inferences and decisions about endangered species. They must manage and protect a variety of environments to ensure they are conserved for the plants and animals that live there. In order to advocate for the conservation of endangered species habitats, they use data to educate the public.

STEM Task

A conservationist uses data to construct arguments about the status of the endangered North Atlantic right whale population.

Year	IDs	Population	Year	IDs	Population
2006	344	432	2012	370	502
2008	388	467	2014	361	502
2010	418	501	2016	304	490

Make inferences about the right whale population over time. Conduct research and use the data provided to justify your reasoning.

Unit 1 Project The Birds and the Trees

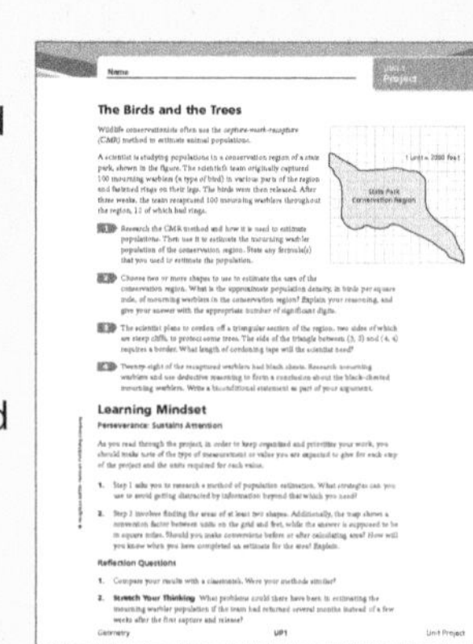

Overview: In this project students calculate the area and perimeter of a polygon in the coordinate plane that represents a conservation region. They use deductive reasoning to form a conclusion about black-chested mourning warblers.

Materials: paper and pencil or a computer used for word processing

Assessing Student Performance: Students' reports should include:

- a correct estimate of the mourning warbler population using the CMR method **(Lesson 1.3 or earlier)**
- a correct area and population density **(Lesson 1.3)**
- a correct triangle perimeter **(Lesson 1.4)**
- an accurate conclusion using deductive reasoning and a biconditional statement **(Lesson 2.1 and Lesson 2.3)**

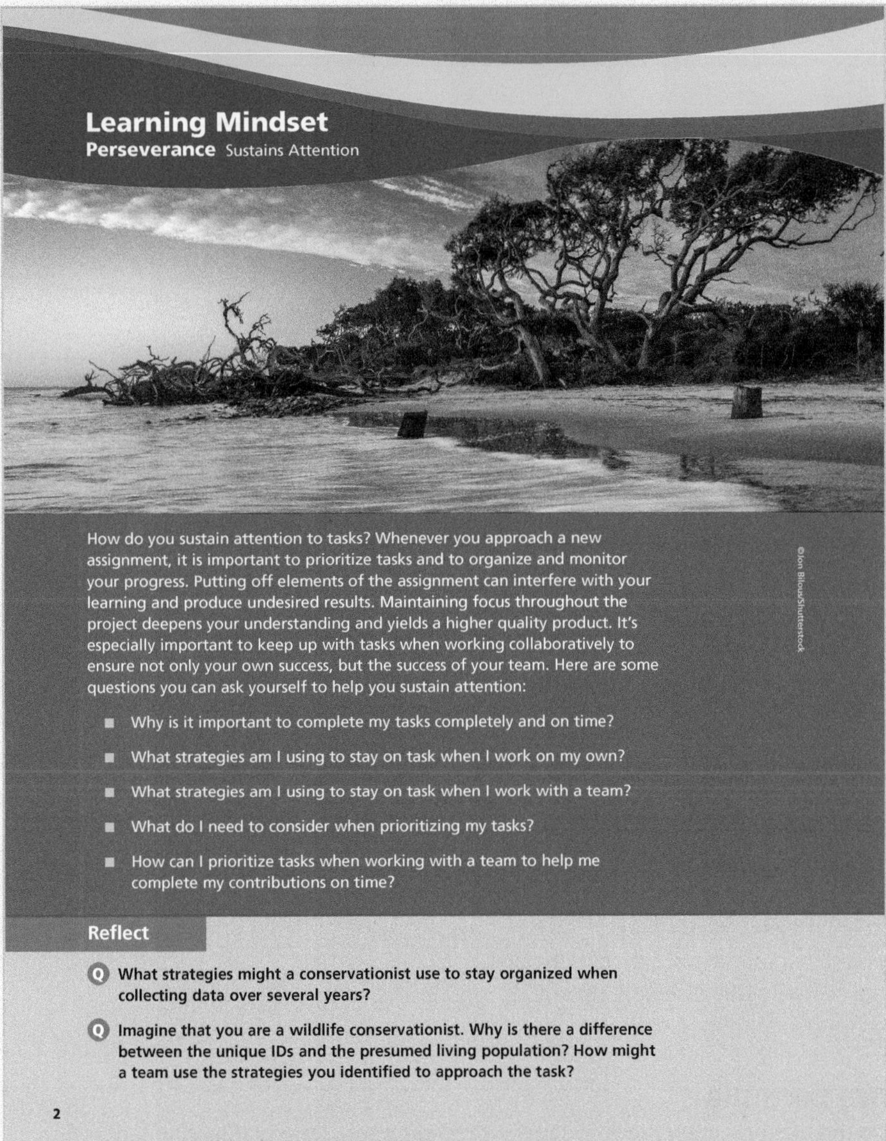

Learning Mindset
Perseverance Sustains Attention

How do you sustain attention to tasks? Whenever you approach a new assignment, it is important to prioritize tasks and to organize and monitor your progress. Putting off elements of the assignment can interfere with your learning and produce undesired results. Maintaining focus throughout the project deepens your understanding and yields a higher quality product. It's especially important to keep up with tasks when working collaboratively to ensure not only your own success, but the success of your team. Here are some questions you can ask yourself to help you sustain attention:

- Why is it important to complete my tasks completely and on time?
- What strategies am I using to stay on task when I work on my own?
- What strategies am I using to stay on task when I work with a team?
- What do I need to consider when prioritizing my tasks?
- How can I prioritize tasks when working with a team to help me complete my contributions on time?

©Jon Bilous/Shutterstock

Reflect

Q What strategies might a conservationist use to stay organized when collecting data over several years?

Q Imagine that you are a wildlife conservationist. Why is there a difference between the unique IDs and the presumed living population? How might a team use the strategies you identified to approach the task?

2

Perseverance

Learning Mindset

Monitors Knowledge and Skills

The learning-mindset focus in this unit is *perseverance*, which refers to a person's ability to persist in moving forward even after experiencing setbacks. Examples of learning setbacks in mathematics include an inability to solve a problem or performing poorly on tests and quizzes.

Mindset Beliefs

Students who react to learning setbacks often say, "Just tell me how to do this." or "I'm never going to get this." They tend to believe that their inability to learn a particular concept or skill is predetermined and permanent. They demonstrate a fixed mindset.

A student who experiences a learning setback may feel frustrated and disheartened, in contrast to one who exercises resilience and sees a learning setback as an opportunity. Gaps in learning mathematics often require students to review prior content, practice their earlier skills, and/or try alternative strategies. Students who are perseverant should be encouraged to think about what they might do differently when stuck on a problem.

Have a discussion with students about what it means to persevere. Do they understand that struggling with or failing at a learning task may be useful? Help students see that although their initial responses to a new learning task may be frustrating, they can eventually overcome these feelings and become more confident.

Mindset Behaviors

Encourage students to become more perseverant by asking themselves why they are having difficulty. Self-monitoring can take the form of asking the following questions:

When reading mathematical text:

- How can I use the book to help me learn?
- What are the important ideas I need to learn?

When practicing a mathematical skill:

- Why is this skill important?
- Is this skill related to something I already know?

When solving a mathematical problem:

- Have I checked each step I used when solving the problem?
- Can I solve this problem another way?

As students become more proficient at monitoring their learning, they can become better at learning new math concepts and skills. Students may experience fewer learning setbacks as they learn how to recognize they need to start over when their thinking is leading them astray. Perseverant students will recognize that overcoming learning setbacks can lead to a deeper understanding and appreciation of mathematics.

What to Watch For

Watch for students who are struggling with math despite their best efforts and hard work. Help them become perseverant by

- praising them for staying on task,
- posing questions that they can use to self-monitor their thinking,
- suggesting that they try a different approach or method, and
- encouraging them to keep trying.

Watch for students who become so discouraged by a learning setback that they give up trying. Help them move from a fixed mindset by

- acknowledging that feeling frustrated is natural when solving a problem is difficult,
- emphasizing that it is important to remove distractions, and
- assuring them that keeping on task can eventually lead to success.

"Keep your eye on the task, not on yourself. The task matters, and you are a servant."

—Peter F. Drucker, American-Austrian educator

GEOMETRY IN THE PLANE

Introduce and Check for Readiness

- Module Performance Task • Are You Ready?

Lesson 1.1—2 Days

Points, Lines, and Planes

Learning Objective: Understand precise geometric notation, bisect a segment using a compass and straightedge, and apply the Midpoint Formula to solve problems in the coordinate plane involving distance.

Review Vocabulary: line, plane, point

New Vocabulary: bisect, collinear, congruent, coplanar, distance, endpoint, line segment, midpoint, postulate, ray, undefined term

Lesson 1.2—2 Days

Define and Measure Angles

Learning Objective: Name and classify angles, measure and draw angles using a protractor, construct an angle bisector using a compass, and write and solve equations to solve mathematical problems involving angle relationships.

Review Vocabulary: acute angle, complementary angles, degree, obtuse angle, ray, reflex angle, right angle, straight angle, supplementary angles

New Vocabulary: adjacent angles, angle, angle bisector, vertex

Lesson 1.3—2 Days

Polygons and Other Figures in the Plane

Learning Objective: Find the perimeter and area of polygons.

Review Vocabulary: area, nonpolygon, polygon, rectangle, regular polygon, trapezoid, triangle

New Vocabulary: *n*-gon

Lesson 1.4—2 Days

Apply the Distance Formula

Learning Objective: Find the perimeter and the area of a figure on the coordinate plane using the Distance Formula, and model irregular figures with simple polygons to estimate perimeter and area.

New Vocabulary: Distance Formula

Assessment

- Module 1 Test (Forms A and B)
- Unit 1 Test (Forms A and B)

LEARNING ARC FOCUS

 Build Conceptual Understanding  Connect Concepts and Skills  Apply and Practice

TEACHING FOR SUCCESS

TEACHING FOR DEPTH: Geometry in the Plane

Meaning of Distance on the Coordinate Plane. It is important for students to understand how to use the Distance Formula to calculate distances in the coordinate plane. The Distance Formula $d = \sqrt{(x_2 - x_1)^2 + (y_2 - y_1)^2}$ gives the distance d between two points whose coordinates are (x_1, y_1) and (x_2, y_2). Students can use the Distance Formula to solve the following types of problems:

- The coordinates of the vertices of a triangle are (0, 6), (0, 0), and (8, 0). Use the Distance Formula to show that the triangle is a right triangle.
- What is the perimeter and area of a quadrilateral whose vertices have coordinates of (0, 0), (2, 3), (6, 3), and (4, 0)?
- Is the quadrilateral whose vertices are located at (1, 2), (2, 4), (4, 3), and (3, 1) a rectangle? A square?

Students will use the Midpoint and Distance Formulas and properties of segments and angles to classify polygons, prove lines are parallel and perpendicular, prove figures are congruent or similar, and transform figures plotted in the coordinate plane.

Mathematical Progressions

Prior Learning	Current Development	Future Connections
Students: • wrote and solved linear, multi-step equations in one-variable. • used terms such as ray, angle, and vertex. • showed that the product of a number and its multiplicative inverse (reciprocal) is 1. • applied the Pythagorean Theorem to calculate the length of one side of a triangle given the lengths of the two other sides.	**Students:** • measure and construct segments. • name and classify angles. • use a protractor to measure and draw angles. • bisect angles and line segments. • write and solve equations about segments and angles. • find the perimeters and area of polygons. • find areas and perimeters of irregular shapes. • apply the Distance Formula to find the distance between points in the coordinate plane.	**Students:** • will prove theorems involving segments. • will prove theorems involving angles. • will prove theorems about parallel and perpendicular lines. • will prove triangles are congruent or similar. • will identify polygons based on the properties of their segments and angles. • will prove statements about polygons and circles plotted in the coordinate plane. • will transform geometric figures in the coordinate plane.

TEACHER TO TEACHER

From the Classroom

Implement tasks that promote reasoning and problem solving. When learning how to approximate the perimeter and area of an irregular shape, my students often struggle with finding a model that best represents a shape for the task at hand.

As an example, draw an irregular shape on the board—one that is not polygonal—and ask students how they might estimate its perimeter and area. (A good shape might be a map of the state in which you live.) Have students suggest one or more polygons that they might use to enclose the shape, such as a rectangle, two rectangles, or a rectangle and a triangle. Point out that the figures they suggest do not have to be the same.

Then ask students how they might use the properties of the polygon(s) to estimate the perimeter and area of the irregular shape. Remind them that this method entails applying both the Segment Addition Postulate and the Area Addition Postulate, and that the measures obtained are always approximations of reality.

This method is also an interesting way to introduce the idea of fractal geometry. For example, how could you measure a coastline? If you placed a ruler around a map of its edge, you would get a certain value. And as you got closer and closer to it, you would get a better approximation, but ultimately, you would find that its length is infinite.

LANGUAGE DEVELOPMENT • Planning for Instruction

By giving all students regular exposure to language routines in context, you will provide opportunities for students to **listen, speak, read,** and **write** about mathematical situations and develop both mathematical language and conceptual understanding at the same time.

Using Language Routines to Develop Understanding

Use the Professional Cards for the following routines to plan for effective instruction.

Co-Craft Questions Lessons 1.1 and 1.3

Students think of natural questions to ask about a given situation or problems similar to a given task and answer the questions they have developed or problems they have created.

Information Gap Lesson 1.2

Students recognize when information given in a problem situation is incomplete, and they pose questions and share knowledge with others to discover any missing facts or relationships and work together to solve the problem.

Three Reads Lessons 1.3 and 1.4

Students read a problem three times with a specific focus each time.

1st Read What is the situation about?
2nd Read What are the quantities in the situation?
3rd Read What are the possible mathematical questions that we could ask for the situation?

Critique, Correct, and Clarify Lesson 1.4

Students working in pairs correct a flawed explanation, argument, or solution and share their thoughts and refine the sample work.

Connecting Language to Geometry in the Plane

Watch for students' use of the review and new terms listed below as they explain their reasoning and make connections with new concepts.

Key Academic Vocabulary

Prior Learning and Current Development • Review and New Vocabulary

line a straight path of points in a plane; has no thickness; continues forever in both directions

supplementary angles two angles whose measures have a sum of 180°

plane a flat surface; has no thickness; extends forever in all directions

point a geometric figure that names a location; has no dimension

angle a geometric figure formed by two line segments or rays that share the same endpoint

bisect to divide into two congruent parts

Distance Formula distance between two points (x_1, y_1) and (x_2, y_2) on the coordinate plane is $\sqrt{(x_2 - x_1)^2 + (y_2 - y_1)^2}$, where (x_1, y_1) and (x_2, y_2) are the endpoints of the segment

midpoint the point that divides a segment into two congruent segments

postulate a statement that is accepted as true without proof

Linguistic Note

Listen for how students distinguish among the terms *line*, *point*, and *plane*. Some English Language Learners may already be familiar with these three words in everyday conversations. However, these geometric terms are probably new to them and ironically are given as undefined. To help students better understand their significance in geometry, stress the dimensionality of each term using models such as a dot, string, and a board.

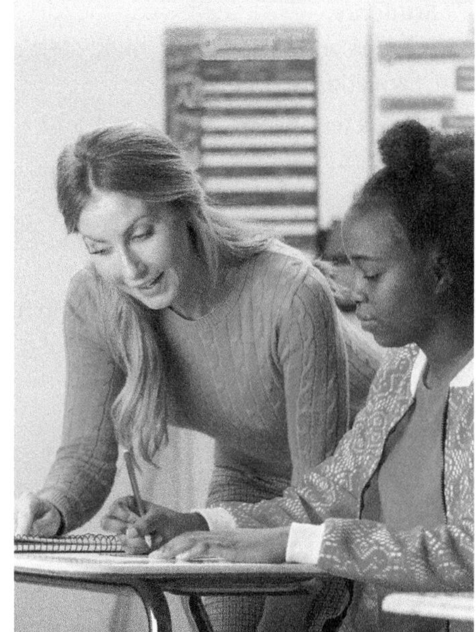

Module 1

Geometry in the Plane

Module Performance Task: *Spies and Analysts*™

FIGHT WILDFIRES

How can we figure out what percentage of the fire is contained?

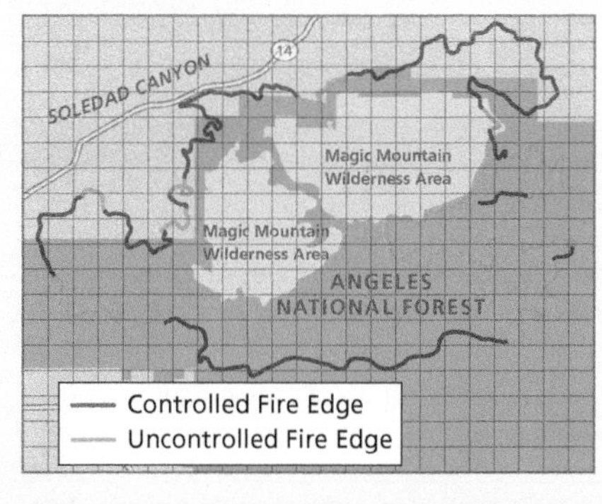

Module 1 3

Connections to This Module

One sample solution might involve creating a model of an irregular polygon to represent the fire edge.

- Draw a series of connected line segments to form an *n*-gon around the perimeter of the enclosed area. **(1.3)** The greater the number of segments in the model, the closer the perimeter of the model will be to the perimeter of the irregularly-shaped fire edges.
- Find the length of each segment using a coordinate grid and the Distance Formula. **(1.4)** Then use The Segment Addition Postulate and add the lengths of the segments in the model to find the perimeter of the enclosed area. **(1.1)**
- Form the ratio of the estimated perimeter of the controlled fire edge to the perimeter of the entire fire edge. Then divide and multiply the quotient by 100 to approximate the percent of the fire that is contained.

FIGHT WILDFIRES

Overview

This problem requires students to make assumptions to determine how to estimate the percentage of a contained wildfire. Examples of assumptions:

- An irregular polygon is a good model of the size of the fire.
- The information about the controlled and uncontrolled fire edges is accurate.

Be a *Spy*

Students should first decide what information about the situation they need to know.

- What does the word *contained* mean when talking about wildfires?
- How can you use the shape of the region to calculate the percent contained?
- How can you approximate the perimeter of an irregular shape?

Students should recognize that the shape formed by the fire edges is irregular. Therefore, to estimate the percent of the fire contained they can create a mathematical model of the perimeters.

Be an *Analyst*

Help students understand that the percent of the contained fire can be approximated by using the ratio of the perimeter of the controlled fire edge to the perimeter of the entire fire edge. Students are to use geometry to find a strategy to approximate the perimeter of each area.

Alternative Approaches

In their analysis, students might:

- use a coordinate grid and trace the edge of each area to approximate its perimeter.
- use a coordinate grid and apply the Distance Formula to approximate distances.
- use a ruler to measure the perimeters.
- use string to measure the perimeters.

ONLINE

Assign the Digital Are You Ready? to power actionable reports including

- proficiency by standards
- item analysis

Are You Ready?

Diagnostic Assessment

data checkpoint

- Diagnose prerequisite mastery.
- Identify intervention needs.
- Modify or set up leveled groups.

Have students complete the *Are You Ready?* assessment on their own. Items test the prerequisites required to succeed with the new learning in this module.

Areas of Triangles Students will apply a previously learned formula for the area of a triangle to approximate areas of irregular shapes.

Areas of Composite Figures Students will apply previously learned formulas for finding the area of triangles and quadrilaterals to approximate areas of irregular shapes.

Angle Relationships Students will apply definitions of types of angles to establish relationships among them, including adjacent angles, complementary angles, and supplementary angles.

ONLINE 

Are You Ready?

Complete these problems to review prior concepts and skills you will need for this module.

Areas of Triangles

Find the area of the triangle with the given base and height.

1. $b = 2$ cm and $h = 7.5$ cm
   7.5 cm²
2. $b = 6\frac{3}{5}$ ft and $h = 7$ ft
   $23\frac{1}{10}$ ft²
3. $b = 3$ m and $h = 10\frac{1}{2}$ m
   $15\frac{3}{4}$ m²

Areas of Composite Figures

Find the area. Round to the nearest tenth.

4.
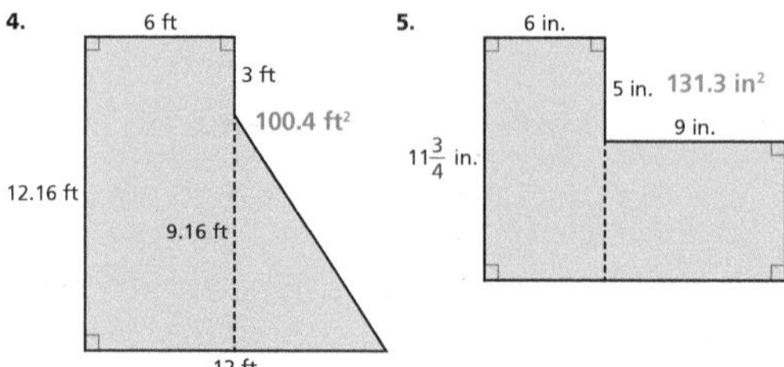

5.

Types of Angle Pairs

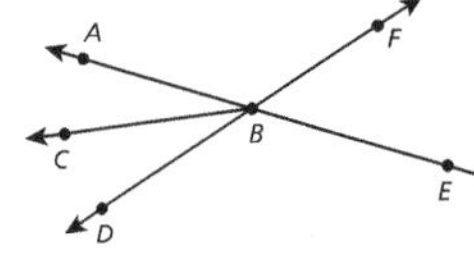

6. Name a pair of vertical angles.
   $\angle ABD$ and $\angle FBE$ or $\angle ABF$ and $\angle DBE$
7. Name a pair of adjacent angles.
   Possible answer: $\angle ABD$ and $\angle ABF$
8. If $m\angle ABD = 83°$, $m\angle ABC = (x + 10)°$, and $m\angle DBC = (4x - 12)°$, what is the measure of each angle?
   $m\angle ABC = 27°$ and $m\angle DBC = 56°$

Connecting Past and Present Learning

Previously, you learned:

- to use the area formula to find the area of triangles,
- to write the area formulas to find the area of composite figures, and
- to use angle relationships to find the measurement of angles.

In this module, you will learn:

- to understand the difference between undefined and defined terms,
- to use the distance formula to find the length of a line, and
- to use area formulas to find the area of a figure on the coordinate plane.

4

DATA-DRIVEN INTERVENTION

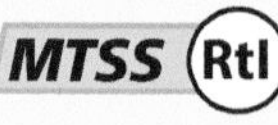

Concept/Skill	Objective	Prior Learning *	Intervene With
Areas of Triangles	Find the area of a triangle given the triangle's base and height.	Grade 6, Lesson 12.2	• Tier 3 Skill 1 • Reteach, Grade 6 Lesson 12.2
Areas of Composite Figures	Find areas of figures composed of triangles and quadrilaterals.	Grade 7, Lesson 10.4	• Tier 2 Skill 1 • Reteach, Grade 7 Lesson 10.4
Angle Relationships	Solve problems involving complementary, supplementary, vertical, and adjacent angles.	Grade 7, Lesson 7.5	• Tier 2 Skill 2 • Reteach, Grade 7 Lesson 7.5

* Your digital materials include access to resources from Grade 6–Algebra 2. The lessons referenced here contain a variety of resources you can use with students who need support with this content.

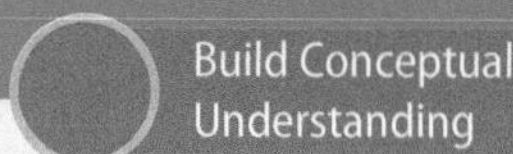

1.1 Points, Lines, and Planes

LESSON FOCUS AND COHERENCE

Mathematics Standards

- Make formal geometric constructions with a variety of tools and methods (compass and straightedge, string, reflective devices, paper folding, dynamic geometric software, etc.). Copying a segment; copying an angle; bisecting a segment; bisecting an angle; constructing perpendicular lines, including the perpendicular bisector of a line segment; and constructing a line parallel to a given line through a point not on the line.

Mathematical Practices and Processes 

- Attend to precision.
- Use appropriate tools strategically.
- Reason abstractly and quantitatively.

I Can Objective 

I can copy and add segments.

Learning Objective

Understand precise geometric notation, bisect a segment using a compass and straightedge, and apply the Midpoint Formula to solve problems in the coordinate plane involving distance.

Language Objective

Explain the steps for performing a construction with a compass and straightedge.

Vocabulary

Review: line, plane, point

New: bisect, collinear, congruent, coplanar, distance, endpoint, line segment, midpoint, postulate, ray, undefined term

Lesson Materials: compass, straightedge, toothpicks

Mathematical Progressions

Prior Learning	Current Development	Future Connections
Students: • graph polygons on the coordinate plane. **(Gr6, 4.2)** • draw geometric shapes using rulers. **(Gr7, 4.1 and 4.2)**	**Students:** • understand the meaning of undefined and defined geometric terms. • learn geometric notation. • use a compass and straightedge to copy, add, and bisect segments. • use the Midpoint Formula to solve problems.	**Students:** • will expand their geometric vocabulary and use geometric terms to reason about properties. **(1.2, 2.3, 4.3)** • perform more sophisticated constructions. **(1.2, 9.1–9.3)**

UNPACKING MATH STANDARDS

Make formal geometric constructions with a variety of tools and methods (compass and straightedge, string, reflective devices, paper folding, dynamic geometric software, etc.). Copying a segment; copying an angle; bisecting a segment; bisecting an angle; constructing perpendicular lines, including the perpendicular bisector of a line segment; and constructing a line parallel to a given line through a point not on the line.

What It Means to You

Students will use a compass and straightedge to create a segment equal to the length of two smaller segments, and to bisect a segment in order to find the midpoint. Students may have experience using drawings to facilitate their reasoning, but using constructions may be a new idea. The emphasis for this standard should be on using constructions to create precise segment lengths and to find the precise midpoint of a segment.

WARM-UP OPTIONS

ONLINE

PROJECTABLE & PRINTABLE 

ACTIVATE PRIOR KNOWLEDGE • Reason about Lines

Use these activities to quickly assess and activate prior knowledge as needed.

Problem of the Day 

How could you determine if two lines, not on the coordinate plane, are parallel? Possible answer: I could measure the shortest distance between the lines at two different points. If the distances are the same, the lines are parallel.

Quick Check for Homework

As part of your daily routine, you may want to display the Teacher Solution Key to have students check their homework.

Make Connections

Based on students' responses to the Problem of the Day, choose one of the following:

1. Project the Interactive Reteach, Grade 7, Lesson 9.4.

2. Complete the Prerequisite Skills Activity:

Have students work in pairs. Have them start with 2 toothpicks and arrange the toothpicks to find the smallest number of intersections. Then, have the students rearrange the toothpicks to find a different number of intersections. Students should repeat the process for 3 and 4 toothpicks and record their answers, making sure to find all possible answers.

- *If you have two parallel lines and a third line intersecting one of the lines, how many intersection points would you have?* I would have two intersection points because lines go on forever and the third line will eventually cross both parallel lines.
- *Why can't you have 2 intersection points with 4 parallel lines?* If I have 4 parallel lines and I tilt one of them to intersect, that line will automatically intersect all 3 parallel lines, creating 3 intersection points.

SHARPEN SKILLS

If time permits, use this on-level activity to build fluency and practice basic skills.

Vocabulary Review

Objective: Students demonstrate an understanding of basic geometric terms.
Materials: index cards

Have students work in pairs. Prepare a matching activity. Write the terms used in the lesson such as *point, line, line segment, ray, plane, collinear, bisect, midpoint, etc.* on one set of index cards. Write the definitions of the terms on another set of index cards. On a set of index cards, sketch drawings illustrating the terms. Ask students to match the term, the definition, and the illustration that all refer to the same concept.

Encourage students to explain in their own words how they matched the cards.

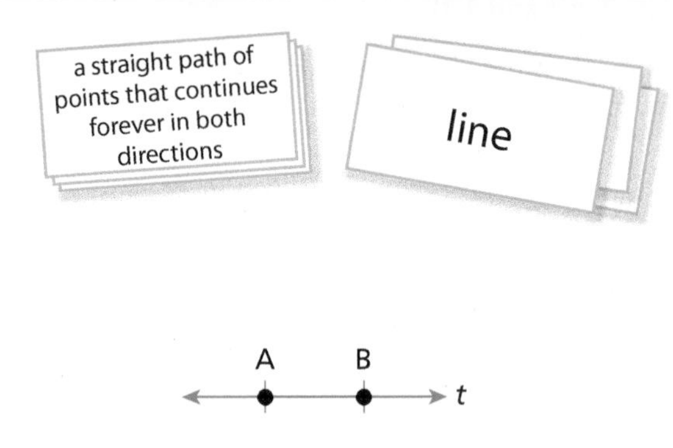

PLAN FOR DIFFERENTIATED INSTRUCTION

Small-Group Options

Use these teacher-guided activities with pulled small groups.

On Track

Materials: computer

Have students investigate the free app GeoGebra and create lines and segments, bisect segments, and find distances on a coordinate plane.

Almost There

Materials: compass and straightedge

If some students have difficulty using a compass, have them work in a small group to share the construction steps. Make sure their compasses are not too loose to make a good arc, and check that their compasses do not slip while performing their constructions.

Ready for More

Materials: patty paper

Have students use the construction for bisecting a segment to create a triangle with a bisected angle. Students should understand that bisecting creates two equal parts: a bisected line segment produces two equal line segments, and a bisected angle produces two equal angles. Students can use patty paper to verify that their triangle contains a bisected angle.

Math Center Options

Use these student self-directed activities at centers or stations.

Key: ● Print Resources ● Online Resources

On Track

- ● Interactive Digital Lesson
- ●● Journal and Practice Workbook
- ● Interactive Glossary (printable): **line**, **plane**, **point**, **bisect**, **collinear**, **congruent**, **coplanar**, **distance**, **endpoint**, **line segment**, **midpoint**, **postulate**, **ray**, **undefined term**

Almost There

- ● Reteach 1.1 (printable)
- ● Interactive Reteach 1.1

Ready for More

- ● Challenge 1.1 (printable)
- ● Interactive Challenge 1.1

ONLINE 

View data-driven grouping recommendations and assign differentiation resources.

During the *Spark Your Learning,* listen and watch for examples students use. See samples of student work on this page.

Use Mathematical Terms with Precision

Strategy 1

The English meaning of *segment* is a part or section of something that is divided. This English definition is helpful for understanding the mathematical meaning of a line segment. A segment of a line is just a part or section of a line with two endpoints.

If students . . . use mathematical terms with precision and connect meanings to English definitions, they are demonstrating an excellent understanding of mathematical terms.

Have these students . . . explain how someone may use the word segment in real life. **Ask:**

- Q Do you think we can use the word segment in a way not connected to geometry?
- Q Why do some mathematical terms have different meanings in everyday life?

Use Some Mathematical Terms

Strategy 2

The English meaning of *line* is a path or drawn mark. This English definition is helpful for visualizing lines.

If students . . . identify mathematical terms, they are demonstrating a good understanding of the situation. However, they may not make strong connections between English and mathematical meanings.

Activate prior knowledge . . . by having students explain what a line is in their own words and give an example of how to use the term in everyday speech. **Ask:**

- Q What do we call part of a line that has a beginning and end?
- Q How are the English and mathematical meanings of *line* and *segment* related?

COMMON ERROR: Uses General Terms

I can see lines connecting the bridge to the ground.

If students . . . do not relate mathematical terms to their meanings in English, they may not understand the necessity for using precise language in mathematics.

Then intervene . . . by asking students to list all the geometric objects they see in the photo. **Ask:**

- Q Do any of the terms you listed have meanings outside of mathematics?
- Q How do the English meanings of the words relate to the mathematical meanings?

1.1

Points, Lines, and Planes

I Can copy and add segments.

Spark Your Learning

Geometry is the study of shapes, and it requires careful use of precise language. Many words that have a specific meaning in geometry have additional and sometimes related meanings that we use in everyday speech.

©Philip Mugridge/Alamy

Complete Part A as a whole class. Then complete Parts B–D in small groups.

A. What mathematical terms are suggested by the photo and the situation it represents? Construct a definition for each term and use the picture to explain your reasoning.

A. Possible answer: Mathematical terms that are suggested by the photo are point, line segment, and arc. See Additional Answers for definitions.

B. Do the English meanings of these terms help you understand the mathematical meanings? What strategy and tool would you use to give an example where the English meaning of a term is helpful? See Strategies 1 and 2 on the facing page.

C. Give an example where the English meaning is not helpful.

C, D. See Additional Answers.

D. Does your answer depend on the context in which you are using the term? Explain your reasoning.

Turn and Talk
- Discuss with your partner other words that have meanings that depend on the situation.
- Discuss how the meaning of *plane* is different in Geometry and English. See margin.

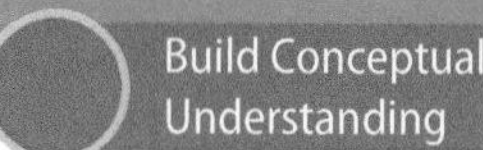

Build Conceptual Understanding

1 Spark Your Learning

▶ MOTIVATE

- Have students look at the photo in their books and read the information contained in the photo. Then complete Part A as a whole-class discussion.
- Give the class the additional information they need to solve the problem. This information is available online as a printable and projectable page in the Teacher Resources.
- Have students work in small groups to complete Parts B–D.

▶ PERSEVERE

If students need support, guide them by asking:

Q **Advancing • Use Tools** Which tool could you use to solve the problem? Why choose that tool and not some other? Students' choices of tools and reasons for choosing them will vary.

Q **Assessing** How can you find English meanings of words? I could look them up in a dictionary or online.

Q **Advancing** How would you define the meaning of a point in geometry and the meaning of a point in everyday life? Possible answer: In everyday life, one could say, "the point of this conversation," which means the focus of the conversation, while in Geometry a point shows a place on a coordinate plane.

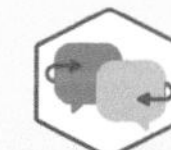

Turn and Talk Students should consider why the word *plane* in English can have one meaning in everyday use but a different meaning in a math classroom. In everyday language, people may use the word *plane* to refer to an airplane or any flat surface. In geometry, a plane is a specific two-dimensional flat surface that extends in all directions forever.

▶ BUILD SHARED UNDERSTANDING

Select groups of students who used various strategies and tools to share with the class how they solved the problem. As they present their solutions, have each group discuss why they chose a specific strategy and tool.

CULTIVATE CONVERSATION • Co-Craft Questions

If students have difficulty identifying geometric objects and naming them, have them name several shapes and their components. What are some natural questions to ask about the shapes?

Work together to craft the following questions:

- How would you describe the sides of the shapes you drew?
- How do the shapes you drew compare to real-world objects, such as a box, ball, or wedge?

Then have students think about how to answer these questions. **Ask:**

- Do you see any of the shapes you drew in the photo?
- Are any components of the bridge straight? Are any curved?

2 Learn Together

Build Understanding

Task 1 **MP** **Attend to Precision** Students make sense of definitions of basic geometric terms, different ways to name them, and how to write them using specific mathematical notation.

CONNECT TO VOCABULARY

Have students use the **Interactive Glossary** to record their understanding of the vocabulary in this task.

Sample Guided Discussion:

Q How would you interpret the statement that a line segment has no thickness but it has length? A line has no thickness, and since a line segment is part of a line, it also has no thickness. A line segment does have length, which is the distance between the two endpoints of the line segment.

Turn and Talk By giving each other directions, students will understand the importance of precise language. Draw a 4-sided figure where two adjacent sides have the same length and the other two sides have a different length.

Build Understanding

Use Geometry Vocabulary

Most words, or terms, used in geometry are defined, but some concepts are so basic that it is impossible to write a definition that does not refer to the term being defined. These are called **undefined terms**.

Undefined Terms		
Term	**Figure**	**Ways to Name Figure**
A **point** names a location. A point has no dimension.	C, D, E (points)	point *C*, point *D*, point *E*
A **line** names a straight path of points in a plane. A line has no width or thickness, and it continues forever in one dimension.	 A B ℓ	line ℓ, $\overleftrightarrow{AB}$ and $\overleftrightarrow{BA}$
A **plane** is a flat surface. A plane has no thickness and it extends forever in two dimensions.	A, B, C, $\mathcal{R}$ (plane)	plane $\mathcal{R}$, plane *ABC*

Defined Terms		
Term	**Figure**	**Ways to Name Figure**
An **endpoint** is a point at an end of a segment or the starting point of a ray.	A, B (segment); F (ray)	The endpoints are *A* and *B* or *F*.
A **ray** is a part of a line that starts at an endpoint and extends forever in one direction.	 H I	ray *HI* or $\overrightarrow{HI}$
A **line segment** is part of a line consisting of two endpoints and all points between them.	A B	$\overline{AB}$ or $\overline{BA}$

1 A. Draw a plane and label three points in the plane. Name the plane in two ways. How many points do you need to use to name a plane? Does the order of the points matter? A–C. See Additional Answers.

B. Draw a ray with *P* as an endpoint. Write a name for the ray. How many points do you need to use to name a ray? Does the order of the points matter? Explain.

C. Draw a line segment and a ray that lie on the same line and that share an endpoint. Explain how the rules for naming the two figures are different.

Turn and Talk Describe a figure to your partner and have them draw the figure from your description. Did you get the expected results? Trade roles and do it again. See margin.

6

LEVELED QUESTIONS

Depth of Knowledge (DOK)	Leveled Questions	What Does This Tell You?
Level 1 **Recall**	Can we name a plane using two letters? no; A plane can be named using one letter, like plane $\mathcal{R}$, or named used three points in the plane, like plane *ABC*.	Students' answers will indicate whether they understand the different ways to name a plane.
Level 2 **Basic Application of Skills & Concepts**	How can you determine the endpoint of ray *AB*? If the ray is called $\overrightarrow{AB}$, then the endpoint will be the first letter, point *A*.	Students' answers will demonstrate whether they can distinguish the parts of a ray based on the name of the ray.
Level 3 **Strategic Thinking & Complex Reasoning**	Lines can intersect at a point. Can planes intersect? If so, what would the intersection of two planes look like? Possible answer: Yes, two planes can intersect; The intersection will be a line.	Students' answers will demonstrate their level of readiness to reason abstractly about how different geometric figures intersect.

Measure and Add Segments

The **distance** between two points is a measure of the length of the shortest line segment that would connect them. Distance is not defined until a unit of measure has been chosen. Points are **collinear** if they lie on the same line. Points and lines are **coplanar** if they lie in the same plane.

A **postulate** is a statement that is accepted as true without proof. Congruency allows you to say segments have the same length without knowing the length.

Segment Addition Postulate

Assume that A, B, and C are collinear points. If B is between A and C, then $AB + BC = AC$.

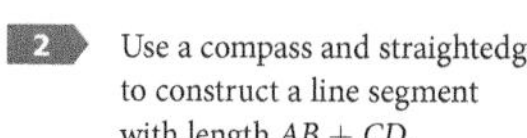

2 Use a compass and straightedge to construct a line segment with length $AB + CD$.

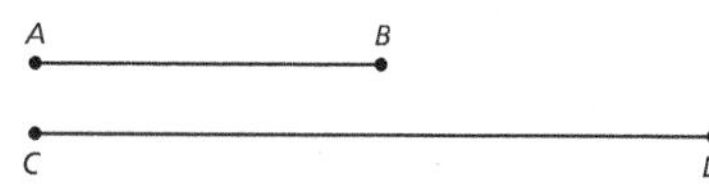

A. Use a straightedge to draw a line segment that is longer than $AB + CD$. Label point M near the left end of the segment. Open the compass to length AB.

B. Place the compass on point M and draw an arc. Label the intersection N. Compare AB and MN. How do you know this relationship is true?

B–D. See Additional Answers.

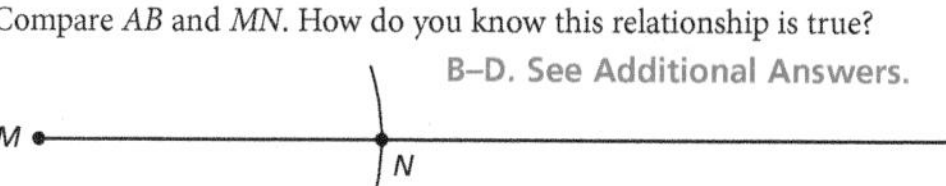

C. Use the information in Parts A and B to explain how to add $\overline{CD}$ to $\overline{MN}$. Explain why the length of the resulting segment is equal to the sum of the lengths of $\overline{CD}$ and $\overline{MN}$.

D. Would the length of the resulting segment be the same or different if you started with a copy of $\overline{CD}$ instead of with a copy of $\overline{AB}$? Explain your reasoning.

Turn and Talk Discuss with your partner how you could use a ruler to draw a segment with length $AB + CD$. Rank the accuracy of the two methods: using a ruler and using a compass and straightedge. Which method is better? Explain your reasoning. See margin.

Task 2 (MP) **Use Tools** Encourage students to use mathematical language correctly when referring to the geometric terms.

By following the directions for the construction, students will demonstrate their understanding of mathematical language.

CONNECT TO VOCABULARY

Have students use the **Interactive Glossary** to record their understanding of the vocabulary in this task.

Sample Guided Discussion:

Q Could you construct a line segment with length $AB + CD$ by copying $\overline{CD}$ first, then copying $\overline{AB}$? Explain your answer. yes; The order in which the segment lengths are copied does not matter.

Turn and Talk Students may want to construct the line segment by using a ruler to measure the smaller segments. Help students understand the need to use a compass and straightedge to construct a line segment equal to the sum of two shorter line segments. Using a straightedge and compass is more accurate because you copy the length without the need for rounding or estimating. Using a straightedge and compass is a precise way to construct the line segment.

EL PROFICIENCY LEVEL

Beginning

Focus on the vocabulary words *bisect, midpoint, arc,* and *intersect*. Have students illustrate and label the words in their notes.

Intermediate

Have students work in groups or pairs. Encourage students to verbalize each step they take in the construction. Say, "Tell me what you are doing now?"

Advanced

Have students use precise terms to explain the reasons for each step as they bisect a segment to find the midpoint

Task 3 **(MP) Use Tools** Encourage students to repeat the process of constructing a segment bisector several times to become more comfortable using the compass. Using tools with ease will help students focus on the meaning of the construction.

CONNECT TO VOCABULARY

Have students use the **Interactive Glossary** to record their understanding of the vocabulary in this task.

Sample Guided Discussion:

Q Why do we draw the arcs with a compass open to more than a half of the length of the line segment? If the compass is open to less than a half of the segment, the arcs will not meet. If it is open to more than a half, then you know the arcs will intersect twice.

Turn and Talk You may wish to have students try the activity using folded paper. I will draw a line segment on the paper. Then, I will fold the paper so that the two end points meet exactly. That will fold half of the segment on top of the other half thus splitting it in two equal parts. The fold line will go through the midpoint of the segment.

Bisect a Segment to Find the Midpoint

Two segments that have the same length are **congruent**, and you can write a congruence statement using the symbol $\cong$. So if the segments $\overline{AB}$ and $\overline{BC}$ are the same length, you can write $\overline{AB} \cong \overline{BC}$.

A line segment can be bisected using a compass and straightedge to find the midpoint. To **bisect** a figure is to divide it into two congruent parts. The **midpoint** of a segment is the point that bisects the segment into two congruent segments.

3 Use a compass and straightedge to bisect a segment. A–C. Check students' work.

A. Draw $\overline{AB}$ on a piece a paper. Place your compass on the endpoint A. Open the compass to about $\frac{2}{3}$ the length of the segment. Draw an arc that starts above the line and continues below the line to form about half a circle.

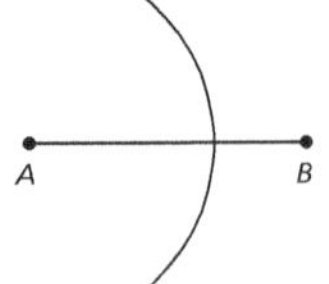

B. Do not change the compass setting. Place the compass on endpoint B. Draw an arc that intersects your first arc both above and below the line segment.

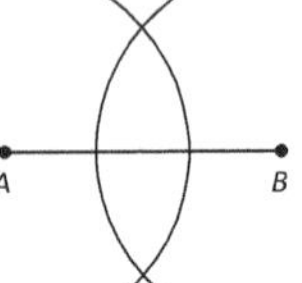

C. Connect the two points where the two arcs intersect each other. Label the point M where this segment intersects the original segment.

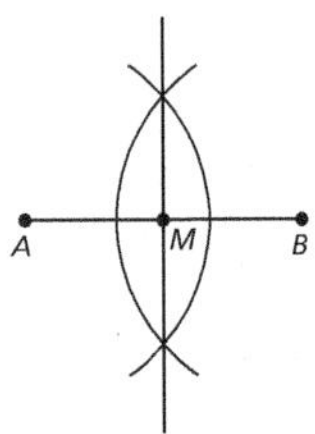

D. How can you verify that $\overline{MA} \cong \overline{MB}$? Explain your reasoning. See Additional Answers.

Turn and Talk Describe a process to bisect a segment using paper folding. See margin.

Step It Out

Identify Points and Segments on the Coordinate Plane

Remember that points on the coordinate plane can be represented using an ordered pair (x, y) where x is the x-coordinate and y is the y-coordinate. To find the midpoint of a segment on the coordinate plane, you can use the Midpoint Formula.

Midpoint Formula

The midpoint M of $\overline{AB}$ with endpoints $A(x_1, y_1)$ and $B(x_2, y_2)$ is given by $M\left(\frac{x_1+x_2}{2}, \frac{y_1+y_2}{2}\right)$.

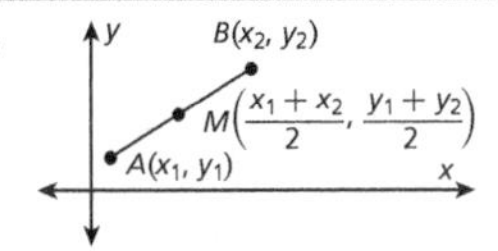

4 Given the points $A(-3, 2)$, $B(5, 2)$, $C(1, 1)$, and $D(1, 7)$, find the midpoint of $\overline{AB}$, $M_{\overline{AB}}$, and the midpoint of $\overline{CD}$, $M_{\overline{CD}}$. Do the segments intersect at one of the midpoints? Explain your reasoning.

Use the Midpoint Formula to calculate the coordinates of $M_{\overline{AB}}$.

$$M_{\overline{AB}} = \left(\frac{x_a + x_b}{2}, \frac{y_a + y_b}{2}\right) = \left(\frac{-3+5}{2}, \frac{2+2}{2}\right) = (1, 2)$$

Use the Midpoint Formula to calculate the coordinates of $M_{\overline{CD}}$.

$$M_{\overline{CD}} = \left(\frac{x_c + x_d}{2}, \frac{y_c + y_d}{2}\right) = \left(\frac{1+1}{2}, \frac{1+7}{2}\right) = (1, 4)$$

A. Why is the y-coordinate of the midpoint the same as the y-coordinate of both endpoints?

B. Why is the x-coordinate of the midpoint the same as the x-coordinate of both endpoints?

C. You can graph the segments to find the point where they intersect. Do the segments intersect at one of the midpoints? Explain your reasoning.

A–C. See Additional Answers.

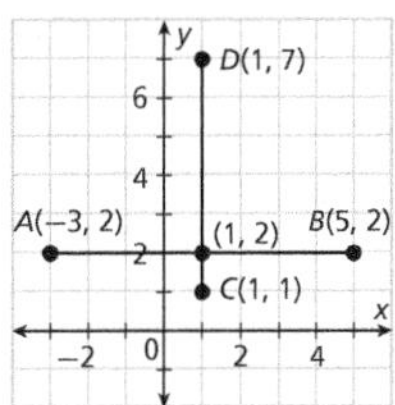

Turn and Talk If $\overline{CD}$ has an endpoint $C(4, -6)$ and the midpoint is $M(-2, 6)$, what are the coordinates of the endpoint D? Explain how you found the endpoint D. See margin.

Step It Out

Task 4 (MP) **Reason** Students use a formula to find the midpoint of a segment on a coordinate plane.

Point out to students that to find the x-coordinate of the midpoint they need to find the average, or the mean, of the x-coordinates of the endpoints. To find the y-coordinate of the midpoint, they need to find the average, or the mean, of the y-coordinates of the endpoints.

Sample Guided Discussion:

Q Is there another way to find the x-coordinate of the midpoint M of $\overline{AB}$? Find the difference between 5 and −3, which is 8. Then divide 8 by 2, which is 4. Then add 4 to the lesser endpoint x-coordinate, −3, to get the x-coordinate for the midpoint, 1.

Q If one of the x-coordinates is in the second quadrant, will that change the way we find the coordinates of the midpoint? Explain your answer. No, it should not change the formula used to find the midpoint; For example, if the x-coordinates are −1 and 1, we know that the midpoint is 0. If we apply the formula, $\frac{-1+1}{2} = 0$.

Turn and Talk So far students have used the Midpoint Formula already knowing the coordinates of both endpoints. This task asks them to find the coordinates of one of the endpoints knowing the coordinates of the other endpoint and the midpoint. You can find the coordinates of the unknown endpoint by substituting the known information in the Midpoint Formula: $\left(\frac{4+x_2}{2}, \frac{-6+y_2}{2}\right) = (-2, 6)$. Solve the equations $\frac{4+x_2}{2} = -2$ and $\frac{-6+y_2}{2} = 6$ to find the coordinates of endpoint D, $(-8, 18)$.

ONLINE

Assign the Digital On Your Own for
- built-in student supports
- Actionable Item Reports
- Standards Analysis Reports

On Your Own

Assignment Guide

The chart below indicates which problems in the On Your Own are associated with each task in the Learn Together. Assign daily homework for tasks completed.

Learn Together Tasks	On Your Own Problems
Task 1, p. 6	Problems 5–8
Task 2, p. 7	Problems 9–13 and 15
Task 3, p. 8	Problem 14
Task 4, p. 9	Problems 16–37

Check Understanding

1. Draw points A and B. Draw a line through them. What is a name of the line?
2. Given two segments, describe a method of drawing a segment with length equal to the sum of the lengths of the given segments. 1–3. See Additional Answers.
3. Explain how to find the midpoint of a segment using a compass.
4. Suppose $\overline{AB}$ on the coordinate plane is horizontal and has length 4 units. If the coordinates of A are $(4, 6)$ and B is to its right, give the coordinates of B and of the midpoint of $\overline{AB}$. $B(8, 6)$; midpoint $(6, 6)$

On Your Own

5. Tell whether each term is *defined* or *undefined*.
   - **A.** point undefined
   - **B.** line undefined
   - **C.** segment defined
   - **D.** ray defined
   - **E.** plane undefined
   - **F.** endpoint defined

Use the campus map to answer Problems 6–8.

Campus Shuttle Map

to stadium

to airport

to town

6. Name three different rays in the figure. $\overrightarrow{EF}$, $\overrightarrow{FG}$, $\overrightarrow{EC}$
7. Name three different points in the figure. point A, point B, point C
8. Name three different ways to name the plane. plane CEF, plane ABC, plane ABD
9. Use a compass and straight edge to construct a segment with length $AB + CD$. Check students' drawings.

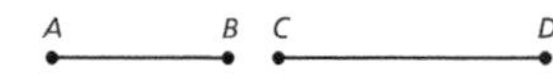

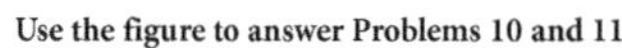

Use the figure to answer Problems 10 and 11.

10. If $AB = 49$ and $BC = 22$, what is the length AC? $AC = 71$
11. If $AC = 62$ and $BC = 27$, what is the length AB? $AB = 35$
12. Use the figure to solve for x. Find AB and BC. $x = 11$, $AB = 20$, $BC = 16$

A $2x - 2$ B $x + 5$ C

36

13. The length of $\overline{DF}$ is $(4x + 2)$ inches, with E as the midpoint. The length of $\overline{DE}$ is 17 inches. What is the value of x? $x = 8$
14. Use a ruler to draw a 4-inch long segment $\overline{MN}$. Use a compass and straightedge to locate the midpoint of $\overline{MN}$. Check students' drawings.
15. Point B lies along $\overline{AC}$ between points A and C. The length of $\overline{AC}$ is 38 centimeters. If $AB = 7x - 1$ and $BC = 4x + 6$, is B the midpoint of $\overline{AC}$? Explain your reasoning. See Additional Answers.

10

data checkpoint

3 Check Understanding

Formative Assessment

Use formative assessment to determine if your students are successful with this lesson's learning objective.

Students who successfully complete the Check Understanding can continue to the On Your Own practice.

For with students who miss 1 problem or more, work in a pulled small group using the Almost There small-group activity on page 5C.

ONLINE

Assign the Digital Check Understanding to determine
- success with the learning objective
- items to review
- grouping and differentiation resources

4 Differentiation Options

Differentiate instruction for all students using small-group activities and math center activities on page 5C.

Reteach

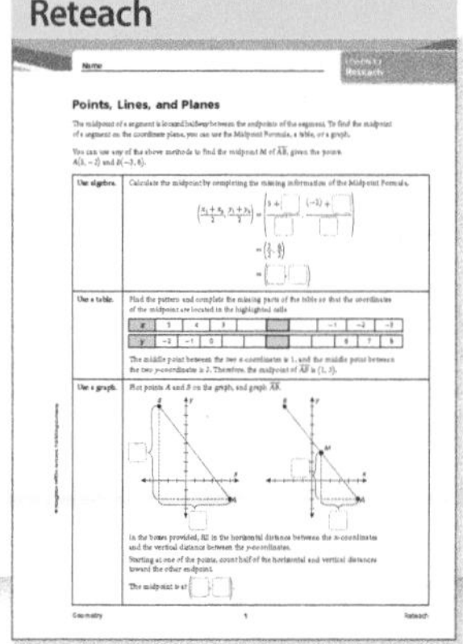

Challenge

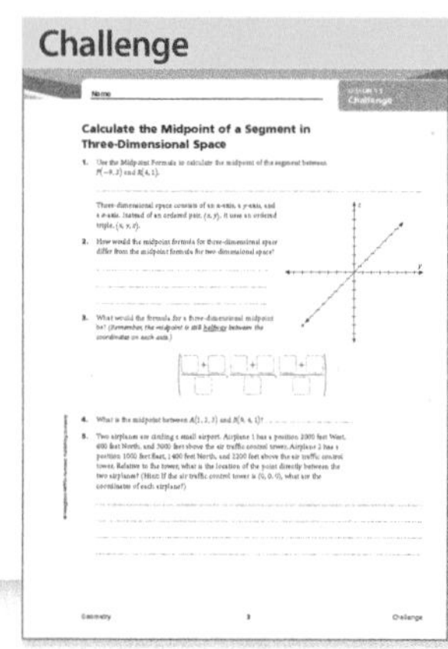

Use the figure to answer Problems 16–21. Assume integer coordinates.

16. Find the midpoint of $\overline{AB}$. $(-1, 2)$

17. Find the midpoint of $\overline{HI}$. $(-3, -2)$

18. Find the midpoint of $\overline{CD}$. $(3, -1)$

19. Do the segments $\overline{AB}$ and $\overline{CD}$ have the same length? Yes, they are both 6 units long.

20. Do the segments $\overline{HI}$ and $\overline{CD}$ have the same length? No, HI is 4 units long and CD is 6 units long.

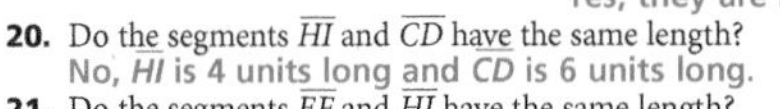

21. Do the segments $\overline{EF}$ and $\overline{HI}$ have the same length? Yes, they are both 4 units long.

Find the midpoint *M* of a segment with the given endpoints.

22. $A(-4, 6)$ and $B(2, 8)$ $(-1, 7)$

23. $C(-4, 0)$ and $D(0, -10)$ $(-2, -5)$

24. $E(-5, -3)$ and $F(3, 7)$ $(-1, 2)$

25. $G(-9, 11)$ and $H(-13, 1)$ $(-11, 6)$

26. $J(-7, 3)$ and $K(2, -8)$ $(-2.5, -2.5)$

27. $L(-8, 11)$ and $N(-3, 12)$ $(-5.5, 11.5)$

28. The lacrosse player shown passes the ball to a team member who is directly in front of the center of the net and is directly below her in the diagram. If the team member catches the ball and then throws it to the center of the net, what is the total distance of the two throws? 50 yards

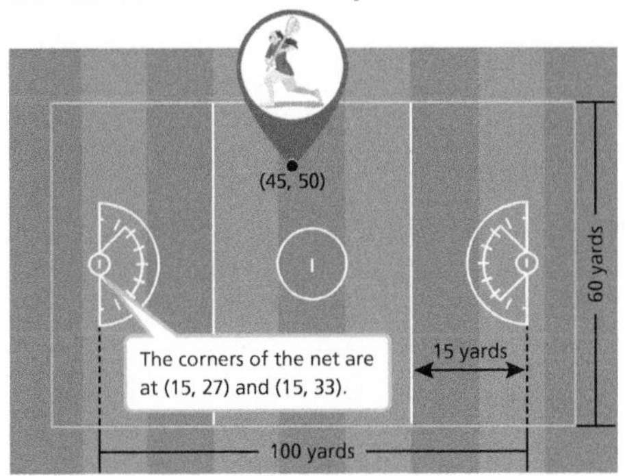

29. Two vertices of rectangle *ABCD* are located at $A(-4, -2)$ and $B(5, -2)$. If the midpoint of $\overline{AD}$ is located 5 units down from *A*, what are the coordinates of the vertices *C* and *D*? $C(5, -12)$ and $D(-4, -12)$

30. **Open Ended** Suppose $\overline{AB}$ has length 7 units and the coordinates of *A* are (4, 6). Give coordinates for three possible locations for *B* and the midpoint of each of those possible segments. See Additional Answers.

31. (MP) **Use Structure** If $\overline{BC}$ has endpoints $B(-6, -2)$ and $C(3, -2)$, which quadrant does the midpoint lie in? Quadrant III

32. If the midpoint of a segment on the coordinate plane is the origin, can the endpoints be located in the same quadrant? Explain.

32, 33. See Additional Answers.

33. (MP) **Critique Reasoning** A teacher asked students to find the midpoint of $\overline{AB}$, with $A(-8, 7)$ and $B(2, -1)$. Describe and correct the error John made in finding the midpoint.

$$M_{\overline{AB}} = \left(\frac{-8-2}{2}, \frac{7-(-1)}{2}\right)$$
$$= (-5, 4) \times$$

Questioning Strategies

Problem 31 Can you explain the difference between the quadrants on the Cartesian coordinate plane in terms of positive and negative values of the coordinates? Both *x*- and *y*-values are positive in Quadrant I and negative in Quadrant III. In Quadrant II, *x*-values are negative but *y*-values are positive. In Quadrant IV, the *x*-values are positive but the *y*-values are negative.

Watch for Common Errors

Problem 16 Some students may think that each grid line represents 1 unit instead of 2 units. This is likely the case if students give a midpoint of $(-0.5, 1)$. Remind students to determine the scale of the coordinate grid before identifying points.

5 Wrap-Up

Summarize learning with your class. Consider using the Exit Ticket, Put It in Writing, or I Can scale.

Exit Ticket 

$\overline{AB}$ is 6 units long and on the x-axis. Point A is located at $(7, 0)$. Determine the coordinates for two different points that could be the midpoint of $\overline{AB}$. Explain your answer.

$(4, 0)$ or $(10, 0)$; Since $\overline{AB}$ is 6 units long and on the x-axis, point B can be at either $(13, 0)$ or $(1, 0)$. Substitute the coordinates of the endpoints in the Midpoint Formula to find the coordinates of the midpoint.

Put It in Writing 

Describe how you can use a compass and straightedge to construct a segment that represents the difference of the lengths of two given segments.

I Can 

The scale below can help you and your students understand their progress on a learning goal.

4	I can copy and add segments, and I can explain my steps to others.
3	I can copy and add segments.
2	I can draw and name lines, line segments, and planes.
1	I can identify defined and undefined geometric terms.

Spiral Review • Assessment Readiness

These questions will help determine if students have retained information taught in the past and can also prepare them for high-stakes assessments. Students must write an solve an equation **(Alg1, 2.2)**, identify the y-intercept of a linear equation in standard form **(Alg1, 3.1)**, recall function notation **(Alg1, 4.1)**, and reflect a shape over an axis **(Gr8, 1.3)**.

34. A city is planning to include a skate park in the renovation of a recreational facility. The skate park will include a bowl that runs the length of one end of the park. What are the coordinates of A and B, the top and bottom of the bowl? $(-5, 5)$ and $(-6, -4)$

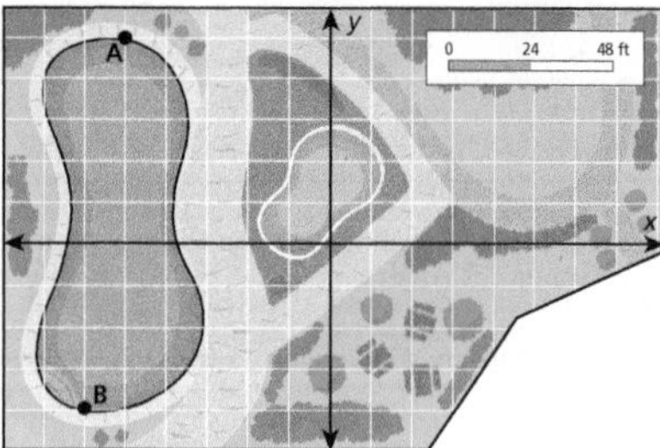

35. Open Ended Suppose the coordinates of one endpoint of $\overline{CD}$ are $C(1, 6)$. If the midpoint is 2 units from C, plot 4 possible locations for point D. Explain how you found those points. See Additional Answers.

36. Harrison's house is halfway between the library and town hall. If Harrison's house is at $(6, -2)$ on a coordinate plane and the library is at $(4, -9)$. What is the location of town hall? $(8, 5)$

37. Open Middle™ Using the digits 1 to 9, at most one time each, fill in the boxes to create a line segment's two endpoint and midpoint.

Endpoints $(\square, \square)$ and $(\square, \square)$ Midpoint $(\square, \square)$

Possible answer: endpoints: $(1, 4)$ and $(3, 6)$; midpoint: $(2, 5)$

Spiral Review • Assessment Readiness

38. Liam goes to the movies with 5 friends. The movie costs \$5.25 per person. They also bought 5 sodas at \$2.75 each. How much did they spend at the movies?

Ⓐ \$5.75
Ⓑ \$13.75
Ⓒ \$26.25
Ⓓ \$40.00

39. What is the y-intercept of the graph of the equation $x + 2y = 2$?

Ⓐ $(0, 1)$
Ⓑ $(0, 2)$
Ⓒ $(1, 0)$
Ⓓ $(2, 0)$

40. Given the function $f(x) = 6(3x - 5)$, if $f(x) = 42$, what is x?

Ⓐ 3
Ⓑ 4
Ⓒ 5
Ⓓ 6

41. If a rectangle is drawn in the first quadrant and reflected over the y-axis, what quadrant is the new image in?

Ⓐ I
Ⓑ II
Ⓒ III
Ⓓ IV

 **I'm in a Learning Mindset!**

Did I manage my time effectively when constructing segments and determining their midpoints? What steps did I take to manage my time?

Keep Going to Journal and Practice Workbook

Learning Mindset

Perseverance Collects and Tries Multiple Strategies

Point out that managing time effectively when solving problems is an important skill. Encourage students to develop a roadmap of the solution process to avoid spending too much time on a part of the solution that may not be important. *How can creating a roadmap to the solution help you avoid unnecessary steps in a solution process? Why is time management important? What can you do if one solution strategy is taking a very long time?*

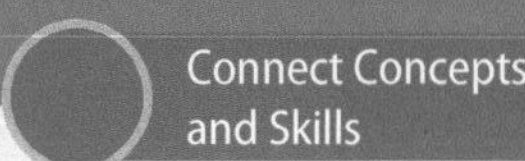

1.2 Define and Measure Angles

LESSON FOCUS AND COHERENCE

Mathematics Standards

- Know precise definitions of angle, circle, perpendicular line, parallel line, and line segment, based on the undefined notions of point, line, distance along a line, and distance around a circular arc.
- Make formal geometric constructions with a variety of tools and methods (compass and straightedge, string, reflective devices, paper folding, dynamic geometric software, etc.). Copying a segment; copying an angle; bisecting a segment; bisecting an angle; constructing perpendicular lines, including the perpendicular bisector of a line segment; and constructing a line parallel to a given line through a point not on the line.

Mathematical Practices and Processes 

- Attend to precision

I Can Objective 

I can copy and measure angles.

Learning Objective

Name and classify angles, measure and draw angles using a protractor, construct an angle bisector using a compass, and write and solve equations to solve mathematical problems involving angle relationships.

Language Objective

Explain the steps needed to construct an angle bisector.

Vocabulary

Review: acute angle, **complementary angles**, **obtuse angle**, **ray**, **reflex angle**, **right angle**, **straight angle**, **supplementary angles**

New: adjacent angles, **angle**, **angle bisector**, **vertex**

Lesson Materials: ruler, compass, protractor

Mathematical Progressions

Prior Learning	Current Development	Future Connections
Students: • wrote and solved linear, multi-step equations in one-variable. **(Alg1, 2.2)** • use terms such as ray, angle, and vertex. **(Gr4, 13.1)**	**Students:** • name, classify, and measure angles. • construct congruent angles and angle bisectors. • use linear equations to represent and solve problems involving angle relationships.	**Students:** • will prove theorems involving angles. **(2.4)** • will work with angle bisectors as they relate to triangles. **(9.3)**

PROFESSIONAL LEARNING

Using Mathematical Practices

This lesson provides an opportunity to address Mathematical Practice MP.2, which calls for students to make sense of quantities and their relationships to problem situations. They abstract a given situation and represent it symbolically, manipulate the representing symbols, and pause as needed during the manipulation process in order to probe into the referents for the symbols involved. Students use quantitative reasoning to create coherent representations of the problem at hand; consider the units involved; attend to the meaning of quantities, not just how to compute them; and know and flexibly use different properties of operations and objects.

WARM-UP OPTIONS

ONLINE

PROJECTABLE & PRINTABLE 

ACTIVATE PRIOR KNOWLEDGE • Use Terms about Angles

Use these activities to quickly assess and activate prior knowledge as needed.

Problem of the Day 

Donte and Jada leave school to walk home. Donte must walk northeast to get home and Jada must walk directly east to get home. Use the letters S, D, and J to represent the school, Donte's house, and Jada's house. The streets Donte and Jada walk on to get home continue in the same direction after they arrive. Draw a diagram on your paper to represent this situation, labeling all three points. Answer the questions below using your diagram.

- What do you call the figure created by the story?
- What do you call point S?
- What is the name of the figure drawn from school to Donte's house? From school to Jada's house?
- Estimate the number of degrees in the angle you drew.
- If Donte's house is northwest from the school instead of northeast, do any of your answers change?

angle; vertex; ray SD, ray SJ; measure is less than 90°; Yes, the answer to the 4th bullet changes. The measure is greater than 90° but less than 180°.

Quick Check for Homework

As part of your daily routine, you may want to display the Teacher Solution Key to have students check their homework.

Make Connections

Based on students' responses to the Problem of the Day, choose one of the following:

1. Project the Interactive Reteach, Grade 4, Lesson 13.1
2. Complete the Prerequisite Skills Activity:

Have students work in pairs. Ask each student to draw an angle and its name on their paper. For example, a student draws an angle and names it ∠*ABC*. Ask students to switch papers and name the vertex and the rays of the angle drawn using the letters given in the angle name. Students verify each other's work. Next, ask students to complete the same steps, only have them draw two intersecting lines on their paper, naming each line and the intersection point. Lastly, ask students to draw a polygon, name the polygon using letters, and repeat the steps.

- *When do two different rays create an angle?* when the two rays meet at a common vertex
- *When can you use one letter to name an angle as opposed to needing three letters to name an angle?* You can use one letter to name an angle if no other angles have the same vertex. You must use 3 letters to name an angle when the angle shares a vertex with another angle.
- *Can you name any rays in your third drawing? Why or why not?* You cannot name any rays in the 3rd drawing because the figure is a polygon, so is made up of line segments, not rays.

SHARPEN SKILLS

If time permits, use this on-level activity to build fluency and practice basic skills.

Vocabulary Review

Objective: Students use a graphic organizer to define ray, angle, and vertex.
Materials: Word Description (Teacher Resource Masters)

Have students work in pairs. Have pairs work together to write each term in the center of the organizer, and then fill in each box with information about the three terms: angle, ray, and vertex.

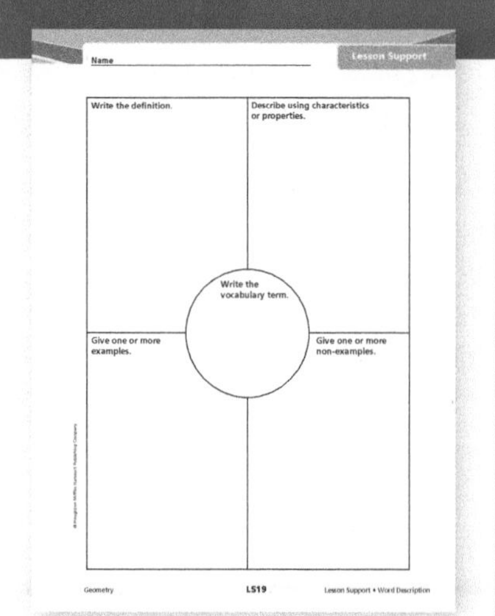
Name
Lesson Support
Write the definition.
Describe using characteristics or properties.
Write the vocabulary term.
Give one or more examples.
Give one or more non-examples.
Geometry
LS19
Lesson Support • Word Description

PLAN FOR DIFFERENTIATED INSTRUCTION

Small-Group Options

Use these teacher-guided activities with pulled small groups.

On Track

Instruct each group member to write a word problem that involves supplementary or complementary angles. The measure of each angle should not be given, but rather the word problem should include a sentence that allows a reader to write algebraic expressions to represent each angle. Ask students to solve the word problem they wrote to create an answer key. Once every group member has written a word problem, take turns sharing the word problems. Instruct students to consider the following when writing word problems:

- The angle measures that result from the word problem should be whole numbers.
- The word problem should include if the angles are complementary or supplementary.

Almost There

Materials: worksheet

Give each group member a vocabulary table. Have students fill in the vocabulary column with the vocabulary from the lesson (there are 13 terms). Next, ask students to define each and draw a picture of each. At the bottom of the page, students can create a problem for their group members to solve that practices using one of the terms mathematically.

Term	Definition	Picture
1		
2		
3		
4		
5		
6		
7		
8		
9		
10		
11		
12		
13		
Example		

Ready for More

Materials: protractor, compass

Have groups arrange themselves in a circle. Ask each group member to draw either an acute or an obtuse angle on a piece of paper. Each student then passes their paper to the student to their right. That student must use a compass to construct the angle bisector for the angle they received. When done, have students pass their paper back to the person who drew the angle. That student will verify the angle bisector is correct by using a protractor to measure each angle created to verify the two angles are congruent.

Math Center Options

Use these student self-directed activities at centers or stations.

Key: ● Print Resources ● Online Resources

On Track

- ● Interactive Digital Lesson
- ● Interactive Glossary (printable): **acute angle**, **complementary angles**, **obtuse angle**, **ray**, **reflex angle**, **right angle**, **straight angle**, **supplementary angles**, **adjacent angle**, **angle**, **angle bisector**, **vertex**
- ●● Journal and Practice Workbook

Almost There

- ● Reteach 1.2 (printable)
- ● Interactive Reteach 1.2

Ready for More

- ● Challenge 1.2 (printable)
- ● Interactive Challenge 1.2
- ● Illustrative Mathematics: Bisecting an Angle

ONLINE 

View data-driven grouping recommendations and assign differentiation resources.

During the *Spark Your Learning*, listen and watch for strategies students use. See samples of student work on this page.

Bisect an Angle

Strategy 1

I can find the angle formed between the goal posts $(80° - 52° = 28°)$ and then bisect that angle $\left(\frac{28°}{2} = 14°\right)$ to find the middle of the angle. She should kick the ball 14° degrees from either goal post.

If students . . . bisect the angle between Kimora's lines of sight to the goalposts according to the diagram, they have demonstrated an exemplary understanding of how angles are used in real-world situations.

Then, have these students . . . explain how they determined their expressions and why they bisected the angle. **Ask:**

- Q How did you use the given information and relationship between the angle measures to set up your work?
- Q Why did you use the measure of the angles made by bisecting instead of the total angle?

Draw a Diagram

Strategy 2

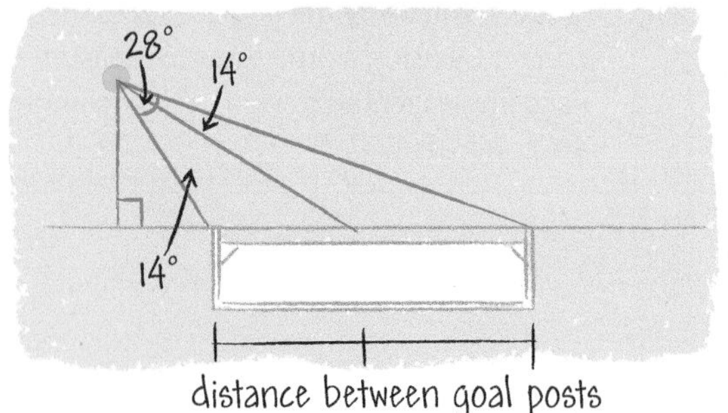

The angle formed between the goal posts is 28°. If Kimora kicks right in the middle of the two posts, she will kick it at a 14° angle.

If students . . . draw and label a diagram to find the angle of the kick, they understand that the angle at which the ball should be kicked is formed by the bisector of the greater angle but may not be able to show this in an expression.

Activate prior knowledge . . . by asking students how they can take what they know about the diagram shown to them and write an expression to find the same information without using an additional diagram. **Ask:**

- Q Where would you be looking if you were in Kimora's position according to the photo and diagram?
- Q What is the ideal angle at which Kimora should kick the ball to have an equal amount of space on either side of the angle for errors?

COMMON ERROR: Uses the Complete Angle

The angle between Kimora's lines of sight to the goalposts as $80° - 52° = 28°$. Kimora can kick the ball at an angle of 28° to make it into the goal.

If students . . . find the angle between the lines of sight but do not bisect the angle, they may not understand that the total measure of the angle is not the angle at which the ball should be kicked.

Then intervene . . . by pointing out that the angle they found refers to the total angle between the goal posts, not the angle at which the ball should be kicked. **Ask:**

- Q If the ball is kicked at an angle of exactly 28° will the ball enter the net of the goal? How do you know?
- Q If Kimora were to kick the ball in the middle of the angle formed between the goal posts, what angle would she kick the ball at?

1.2

Define and Measure Angles

I Can copy and measure angles.

Spark Your Learning

Kimora is practicing for an in-game scenario. A teammate crosses the ball beyond the far post, leaving the goalie out of position and an empty net for Kimora.

Complete Part A as a whole class. Then complete Parts B–D in small groups.

A. What is a mathematical question you can ask about this situation? What information would you need to know to answer your question?

A. Where should the ball be kicked?; where the ball is in relation to the goal

B. What would give you the most room for error? Should you consider lengths or angles? Explain your reasoning. See Additional Answers.

C. To answer your question, what strategy and tool would you use along with all the information you have? What answer do you get? See Strategies 1 and 2 on the facing page.

D. How does precision in language help you organize your thinking? See Additional Answers.

Turn and Talk Predict how your answer would change for each of the following changes in the situation: See margin.

- Kimora was in the center of the field.
- Kimora was closer to the end-line.

CULTIVATE CONVERSATION • Information Gap

Ask students questions to help them decide what missing information they need to answer the question, "Where should the ball be kicked?"

1 Do you have enough information to determine where the ball should be kicked? Explain. no; I know that Kimora is not directly in front of the goal, but I don't know exactly where on the field she is.

2 Does the projected image give you additional information to help determine where the ball should be kicked? Explain. yes; The projected image shows the angles from Kimora's position to both goal posts, so I can tell where she needs to kick the ball in order to make a goal.

3 In addition to the angles, what information might be helpful? Possible answer: It might be helpful to know how far Kimora is from the goal and whether it is windy. If the wind is strong and she is far enough from the goal, she may need to adjust where she kicks the ball.

1 Spark Your Learning

▶ MOTIVATE 

- Have students look at the photo in their books and read the information contained in the photo. Then complete Part A as a whole-class discussion.
- Give the class the additional information they need to solve the problem. This information is available online as a printable and projectable page in the Teacher Resources.
- Have students work in small groups to complete Parts B–D.

▶ PERSEVERE

If students need support, guide them by asking:

Q **Advancing • Use Tools** Which tool could you use to solve the problem? Why choose that tool and not some other? Students' choices of tools and reasons for choosing them will vary.

Q **Assessing** Where should Kimora aim her kick to have the best chance of scoring? She should aim at the center of the goal.

Q **Assessing** If Kimora moves further to the left of the goal, how does that affect where she should aim her kick? If Kimora moves further to the left, her kick should be aimed further to the right and closer to the goal post on the right side.

Turn and Talk Help students understand that they can use an angle to describe the direction in which Kimora kicks the ball. An angle is formed by two rays, so students also need a reference direction to compare with the direction the ball is kicked. Assuming Kimora is running towards the end line, the projectable gives angles relative to the direction in which she is running. Students might also give angles relative to the lines of sight, since that is where she is looking. If Kimora is in the center of the field, the angle will be as large as possible and the location will match the midpoint of the goal-line. If Kimora is closer to the end line, the angle will be smaller and the location will be furthest from the midpoint of the goal line.

▶ BUILD SHARED UNDERSTANDING

Select groups of students who used various strategies and tools to share with the class how they solved the problem. As they present their solutions, have each group discuss why they chose a specific strategy and tool.

2 Learn Together

Build Understanding

Task 1 (MP) **Attend to Precision** Students draw figures including angles from verbal and written descriptions.

CONNECT TO VOCABULARY

Have students use the **Interactive Glossary** to record their understanding of the vocabulary in this task.

Sample Guided Discussion:

Q **In Parts B and C, why can't an angle be named using just the letter of its vertex?** When you have two adjacent angles, or two angles that are not adjacent but share a vertex, you must use three letters to name the angle. If you use only the letter that names the vertex, Angle B, it is impossible to know which angle you are referring to.

Turn and Talk Have students work in pairs. Ask students to complete the task a few times, taking turns being the student who describes and the student who draws. Ask them to include terms such as ray, vertex, adjacent, and non-adjacent when describing the figure. Have pairs practice labeling rays and vertices so all angles in the drawing can be named. A student can ask their partner to draw a figure with three angles that are adjacent and share a common vertex. Then together they can label the rays and vertex with letters, and give all the possible names of all angles in the drawing.

Build Understanding

Draw and Name Angles

You can use the undefined terms point and line to define terms used with angles.

Defined Terms

Term	Figure	Names
A **ray** is a part of a line that starts at an endpoint and extends forever in one direction.	A B	$\overrightarrow{AB}$ or ray AB
An **angle** is formed by two line segments or rays that share the same endpoint.	A B C	$\angle ABC$ or $\angle B$
A **vertex** of an angle is the common endpoint of the two rays that form the angle.	A B C	The vertex is at B.
Adjacent angles are two angles in the same plane with a common vertex and a common side, but no common interior points.	A C B D	$\angle ABC$ and $\angle CBD$

1 A. Draw two rays that form an angle. Draw two rays that do not form an angle. A–C. See Additional Answers.

B. Draw and label adjacent angles ABC and CBD. Write all possible ways to name all three angles using the point names.

C. Draw and label angles ABC and DBE so that they share the vertex B but are not adjacent angles.

Turn and Talk Describe a figure to your partner that has at least one angle and have them draw the figure from your description. Did you get the expected results? Trade roles and do it again. See margin.

14

LEVELED QUESTIONS

Depth of Knowledge (DOK)	Leveled Questions	What Does This Tell You?
Level 1 **Recall**	How many rays are necessary to form an angle? two	Students' answers will indicate whether they understand the definitions of ray and angle.
Level 2 **Basic Application of Skills & Concepts**	Compare the number of rays required to draw two angles that are adjacent to the number of rays required to draw two non-adjacent angles that share a common vertex. 3 rays; 4 rays	Students' answers will demonstrate whether they understand the difference between adjacent and non-adjacent angles.
Level 3 **Strategic Thinking & Complex Reasoning**	How is the geometric shape of a circle related to the concept of adjacent angles? Possible answer: The degree measure of a circle can be thought of as a set of adjacent angles all around a common vertex.	Students' answers will reflect whether students can recognize and generalize a pattern and develop logical arguments for a concept.

Measure and Classify Angles

The measure of a segment is a description of its length. The measure of an angle is a description of the distance around a circular arc. A common measure for angles is degrees. One degree, written 1°, is $\frac{1}{360}$ of the way around the circular arc.

Angle Classification by Measure		
Name	**Measure**	**Example**
acute	$0° < m\angle A < 90°$	
right	$m\angle A = 90°$	
obtuse	$90° < m\angle A < 180°$	
straight	$m\angle A = 180°$	
reflex	The measure of the reflex angle of a given acute, right, or obtuse angle will be greater than 180° and less than 360°.	

 A. Classify each angle as acute, right, obtuse, or straight. A–C. See Additional Answers.

$\angle ABG$	$\angle DBC$
$\angle ABD$	$\angle LHC$
$\angle GBD$	$\angle ABF$
$\angle DBF$	$\angle LHK$
$\angle KHC$	$\angle ABC$

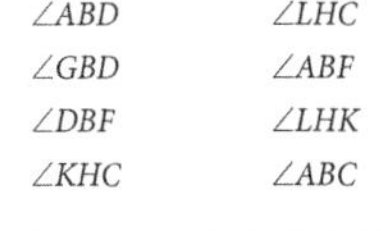

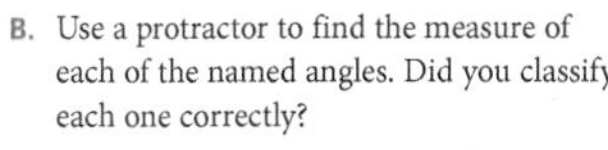

B. Use a protractor to find the measure of each of the named angles. Did you classify each one correctly?

C. When can estimating an angle measure be useful in solving a problem?

 **Turn and Talk** Reflex angles are defined in terms of a given angle; they are considered the "other half" of an acute, right, or obtuse angle. Together, an angle and its reflex angle form a complete rotation, or a full angle of 360°.

- What is the measure of the reflex angle of $\angle LHC$ in the rug shown?
- If the measure of an angle is $x°$, what is the measure of its reflex angle? See margin.

Task 2 (MP) **Attend to Precision** Students recognize situations where estimating the measure of an angle is sufficient, and understand how to use a protractor.

CONNECT TO VOCABULARY

Have students use the **Interactive Glossary** to record their understanding of the vocabulary in this task.

Sample Guided Discussion:

Q Can $\overrightarrow{AB}$ be named as the ray of an obtuse angle? Yes, $\overrightarrow{AB}$ is part of $\angle ABD$, which is greater than 90° and less than 180°, so it is obtuse.

Q What is a second way to name $\angle ABG$? $\angle ABG$ can also be named $\angle GBA$, as long as the vertex, point B, is named in the middle.

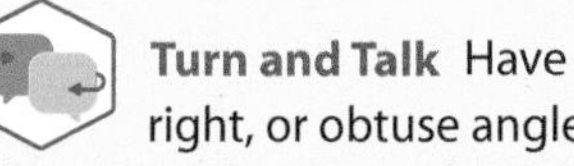

Turn and Talk Have students draw an acute, right, or obtuse angle, and label it A. Then instruct students to draw a curve from one side of the angle, through the exterior of the angle, to the other side. The curve they draw represents the reflex angle for angle A. Because $m\angle LHC = 90°$, then the measure of its reflex angle is $360° - 90° = 270°$; $360° - x°$

Step It Out

Task 3 (MP) **Attend to Precision** Students understand that, when used with precision, geometric construction tools can be used to draw precise geometric figures.

CONNECT TO VOCABULARY

Have students use the **Interactive Glossary** to record their understanding of the vocabulary in this task.

Sample Guided Discussion:

Q **What must be true about the two arcs that are swung on each ray of the original angle?** The arcs that are swung on each ray of the original angle must be of equal distance from the vertex of that angle.

Q **How are you sure that the arcs that you swing will create a center ray that is equidistant from each ray of the original angle?** The center ray constructed is equidistant from the two rays of the original angle because the radius used on the compass, when swinging the arcs, remains constant throughout the construction.

Turn and Talk Help students realize that they can check their construction of an angle bisector by using this paper folding technique. Use a ruler to draw an angle on a piece of paper. Fold the paper so that the two rays of the angle drawn lie on top of each other. The crease created by folding the paper is the angle bisector of the original angle drawn.

Step It Out

Bisect an Angle

You can use a compass and straightedge to bisect an angle. You can write a congruence statement about angles using the same symbol you used for segments: $\angle A \cong \angle B$.

Connect to Vocabulary

To bisect a figure is to divide it into two congruent parts. The **angle bisector** of an angle bisects the angle.

The steps you can use to construct an angle bisector using a compass and straightedge are shown. Put the steps in order. E, C, A, B, F, D

A. Place the point of the compass on X and draw an arc.

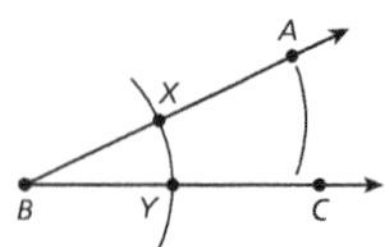

B. Place the point of the compass on Y and draw an arc.

C. Place the point of the compass on vertex B. Draw an arc that intersects both sides of the angle. Label the intersections X and Y.

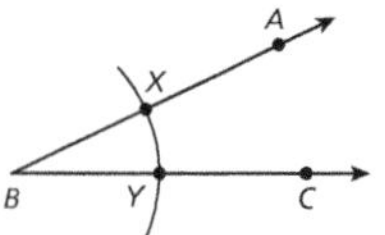

D. Measure each angle with a protractor to verify that $\angle ABD \cong \angle DBC$.

E. Draw an angle and label it $\angle ABC$.

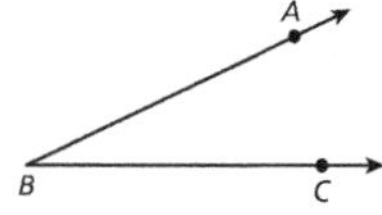

F. Label the intersection of the arcs D. Use a straightedge to draw $\overrightarrow{BD}$.

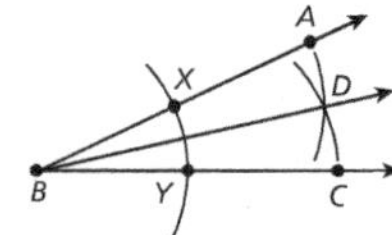

Turn and Talk Describe a process to bisect an angle using paper folding. See margin.

EL PROFICIENCY LEVEL

Beginning

Have students work in pairs. Give each pair a sheet that shows several angles, some that are bisected into two congruent angles and some that are divided into two non-congruent angles. Ask students to share with their partner whether each angle on the sheet has been bisected or not.

Intermediate

Provide pairs of students with an illustration of the construction of an angle bisector. Ask one student to describe the steps of creating this construction while the second student traces the steps using a compass. This will help students memorize the steps of constructing an angle bisector using a compass.

Advanced

Have students tell how they can test that the construction of an angle bisector has been performed correctly.

Analyze Angle Relationships

Two angles are **supplementary** if the sum of their angle measures is equal to 180°. Two angles are **complementary** if the sum of their angle measures is equal to 90°.

Angle Addition Postulate

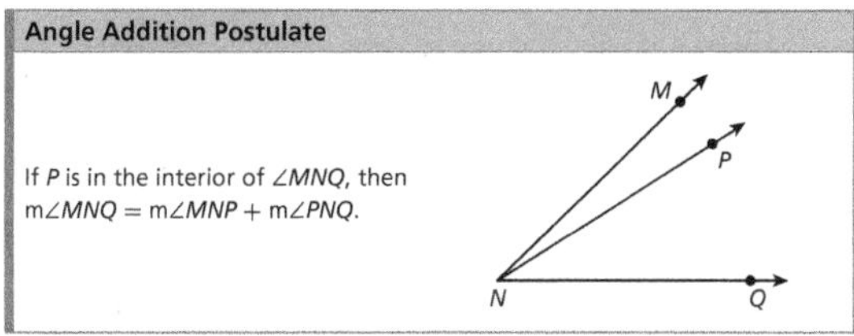

If P is in the interior of $\angle MNQ$, then $m\angle MNQ = m\angle MNP + m\angle PNQ$.

4 In the image $m\angle ABC = 156°$. Find $m\angle ABD$ and $m\angle CBD$.

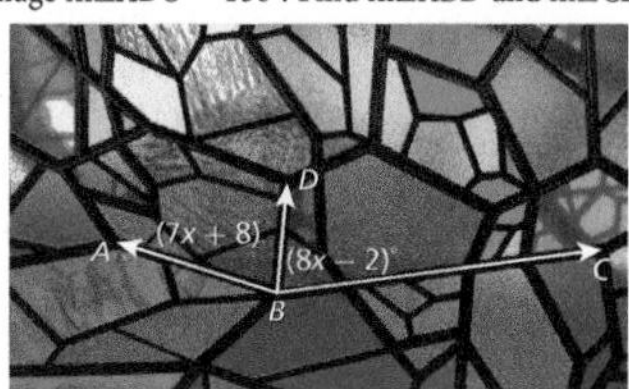

$m\angle ABC = m\angle ABD + m\angle CBD$ — A. What postulate can be used to find $m\angle ABC$? Angle Addition Postulate

$156 = (7x + 8) + (8x - 2)$ — B. What property justifies this step? Substitution Property of Equality

$156 = 15x + 6$

$150 = 15x$

$10 = x$

So, $m\angle ABD = 78°$ and $m\angle CBD = 78°$. — C. How were these measures determined? $m\angle ABD = 7(10) + 8 = 78°$; $m\angle CBD = 8(10) - 2 = 78°$

Turn and Talk Suppose $\angle ABC$ and $\angle CBD$ are adjacent complementary angles and $m\angle ABC = (3x + 11)°$ and $m\angle CBD = (6x - 2)°$.
What is the value of x? What are the measures of the angles?
Does the shared ray bisect the angle? See margin.

Task 4 (MP) **Attend to Precision** Students use algebra and angle addition to determine the measures of angles given a figure and algebraic expressions for angle measures.

CONNECT TO VOCABULARY

Have students use the **Interactive Glossary** to record their understanding of the vocabulary in this task.

Point out to students that when solving an equation, the goal is to get the variable on one side of the equal sign and the value of that variable on the opposite side. In this example, the variable only appears on the right side of the equation. Use properties of numbers to simplify the right side of the equation and get the variable by itself.

Sample Guided Discussion:

Q What properties were used to rewrite the expression $(7x + 8) + (8x - 2)$ as the expression $15x + 6$? Possible answer: Associative Property, then Commutative Property, then Associative Property, then Distributive Property: $(7x + 8) + (8x - 2) = 7x + (8 + 8x) - 2 = 7x + (8x + 8) - 2 = (7x + 8x) + (8 - 2) = (7 + 8)x + 6 = 15x + 6$

Q How is the third equation different than the equation beneath it? How does this suggest the property that was used to obtain the fourth equation? The number 6 was subtracted from 156 to make 150 on the left side of the equation; This suggests that the Subtraction Property of Equality was used to isolate the variable term on the right side of the equation.

Turn and Talk The examples students have experienced so far involve finding the value of x that makes an equation true. In this problem, students must go a step further by evaluating each expression to determine the degree measure of each angle. Solving the equation, $3x + 11 + 6x - 2 = 90$, the value of x is 9. The first angle has a degree measure of $3(9) + 11$, or 38°, and the second angle has a degree measure of $6(9) - 2$, or 52°. So, the shared ray is not a bisector because the two angles are not congruent.

ONLINE

Assign the Digital On Your Own for
- built-in student supports
- Actionable Item Reports
- Standards Analysis Reports

On Your Own

Assignment Guide

The chart below indicates which problems in the On Your Own are associated with each task in the Learn Together. Assign daily homework for tasks completed.

Learn Together Tasks	On Your Own Problems
Task 1, p. 14	Problems 5–11
Task 2, p. 15	Problems 12–14, 17–22, and 28
Task 3, p. 16	Problems 15 and 16
Task 4, p. 17	Problems 23–27 and 29–31

Check Understanding

1. Draw each figure. A, B. Check students' work.
   - **A.** ray *CA*
   - **B.** angle *ABC*; Identify the vertex.
2. Classify the angles below without using a protractor.

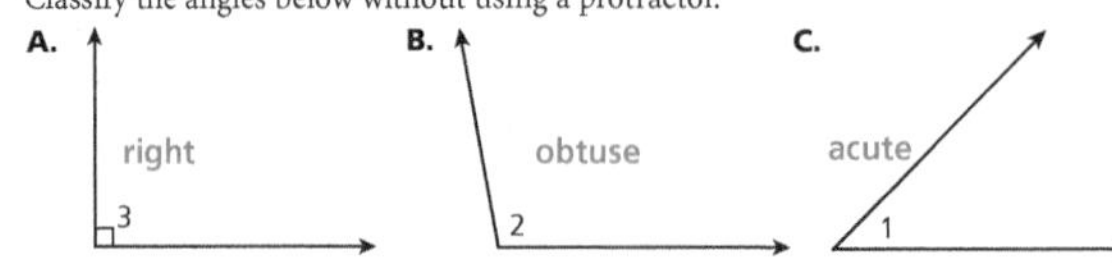

3. Use a protractor to draw a 42° angle. Label the angle *ABC*. Use a compass and straightedge to construct the angle bisector of $\angle ABC$. See Additional Answers.
4. $\angle CDE$ and $\angle EDF$ are supplementary angles. If $m\angle CDE = (2x + 6)^\circ$ and $m\angle FDE = (x - 9)^\circ$ what is the value of *x*? $x = 61$

On Your Own

Name each angle in three different ways.

5. 6. 7.

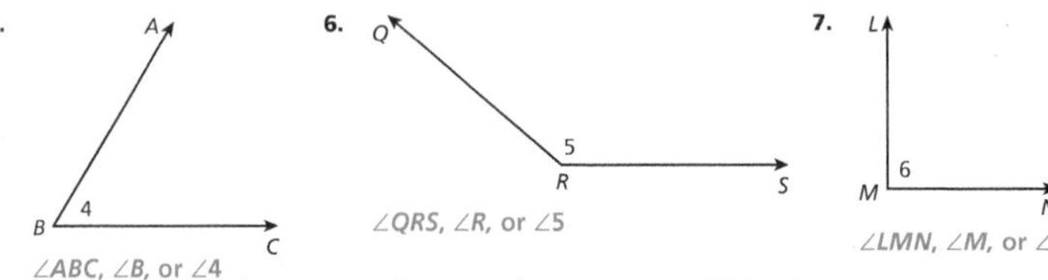

8. Draw two segments that intersect to form an angle. 8–16. See Additional Answers.
9. Draw two rays that do not form an angle.
10. Draw and label adjacent angles *ABC* and *CBD*.
11. Draw and label angles *PQR* and *SQT* so that they share the vertex *Q* but are not adjacent angles. Name two other angles formed in your drawing.

Use a protractor to draw an angle with the given measure.

12. 79° 13. 125° 14. 185°

Use a compass and straightedge to construct an angle bisector.

15. 16.

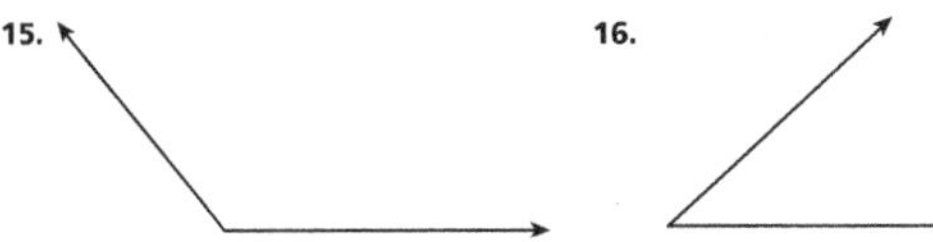

18

③ Check Understanding

Formative Assessment

Use formative assessment to determine if your students are successful with this lesson's learning objective.

Students who successfully complete the Check Understanding can continue to the On Your Own practice.

For students who miss 1 problem or more, work in a pulled small group using the Almost There small-group activity on page 13C.

ONLINE

Assign the Digital Check Understanding to determine
- success with the learning objective
- items to review
- grouping and differentiation resources

④ Differentiation Options

Differentiate instruction for all students using small-group activities and math center activities on page 13C.

Reteach

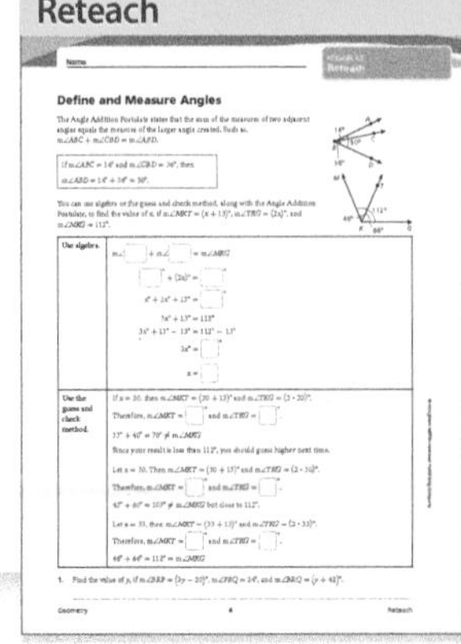

Challenge

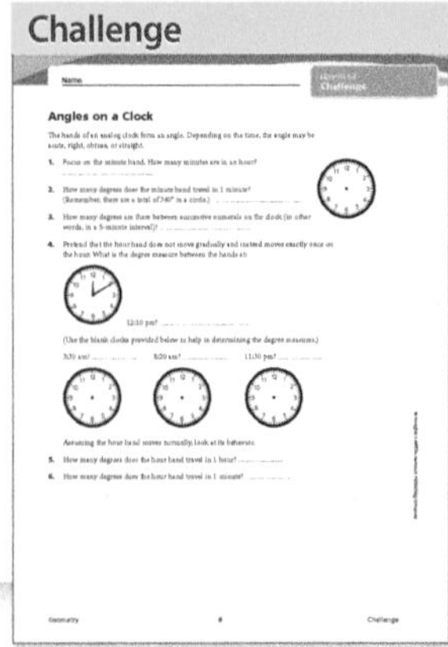

Use the protractor photo to find the measure of each of the following angles.

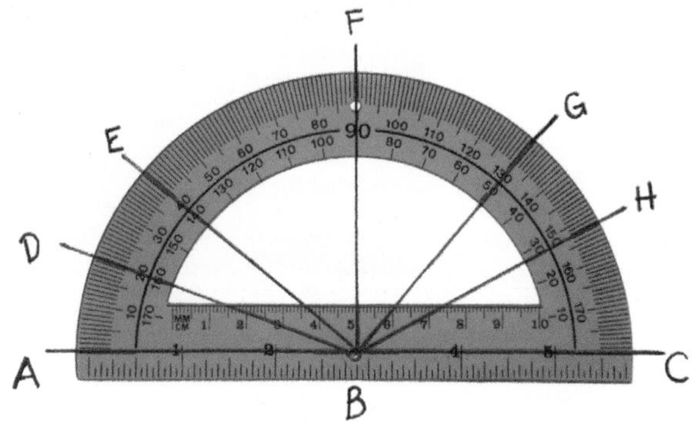

17. $\angle ABD$ 20°

18. $\angle EBF$ 50°

19. $\angle GBC$ 50°

20. $\angle DBF$ 70°

21. $\angle EBG$ 90°

22. $\angle DBC$ 160°

23. Find the measure of $\angle ABD$ and $\angle DBC$ given $m\angle ABC = 77°$.

A
D
4x − 2
3x + 2
B
C

$m\angle ABD = 42°$ and $m\angle DBC = 35°$

24. Find the measure of $\angle ABD$ and $\angle DBC$ given $m\angle ABC = 140°$.

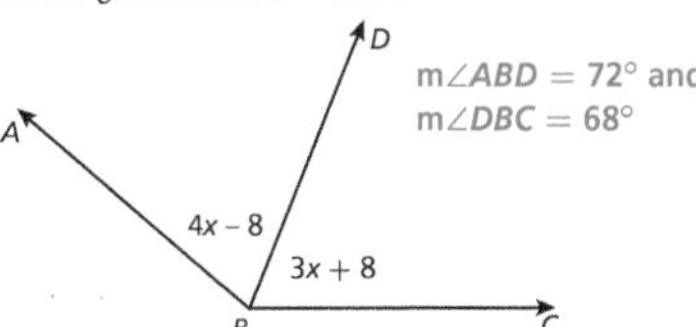

$m\angle ABD = 72°$ and $m\angle DBC = 68°$

25. If $\angle MNO$ and $\angle QRS$ are complementary angles and $m\angle MNO$ is 63°, what is $m\angle QRS$? 27°

26. If $\angle MNO$ and $\angle QRS$ are supplementary angles and $m\angle MNO$ is 74°, what is $m\angle QRS$? 106°

27. (MP) **Use Structure** Angles P and Q are supplementary angles. If $m\angle P$ is 3 times $m\angle Q$ minus 4, what are the measures of the two angles? 134° and 46°

28. (MP) **Reason** Use a compass to create a new angle with measure equal to $m\angle A - m\angle B$. Check students' work. The new angle should be 80°.

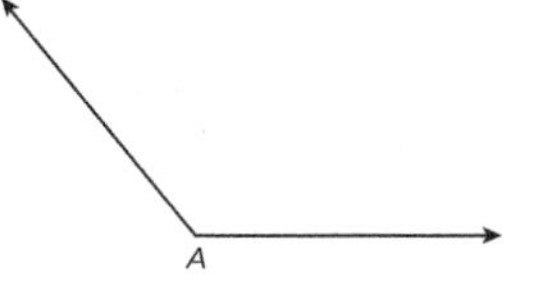

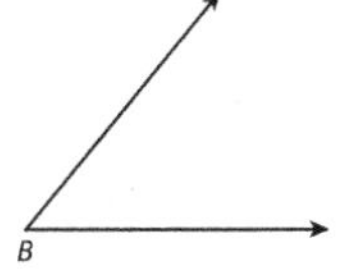

Watch for Common Errors

Problem 17–22 A common error when using a protractor is that students read the measurement using the wrong set of numbers that are on the protractor. Be sure to remind students where to place the vertex of the angle and the initial side of an angle. Depending on the placement of the initial side, count up from zero to reach the reading of the terminal side.

Questioning Strategies

Problem 23 What properties might you use to find the value of x? Possible answer: Angle Addition Postulate, Transitive Property of Equality, Division Property of Equality

5 Wrap-Up

Summarize learning with your class. Consider using the Exit Ticket, Put It in Writing, or I Can scale.

Exit Ticket 

$\angle ABC$ is a straight angle. $\overrightarrow{BD}$ is drawn so that two adjacent angles are created. $m\angle ABD = 22x - 9$ and $m\angle CBD = 4x + 7$. Does $\overrightarrow{BD}$ bisect $\angle ABC$? Show your work.

$4x + 7 + 22x - 9 = 180$

$26x - 2 = 180$

$26x = 182$

$x = 7$

$m\angle ABD = 22(7) - 9 = 145$

$m\angle CBD = 4(7) + 7 = 35$

Therefore, $\overrightarrow{BD}$ is not an angle bisector because the two angles are not congruent.

Put It in Writing 

Describe some strategies you can use to construct two adjacent, congruent angles.

I Can 

The scale below can help you and your students understand their progress on a learning goal.

4	I can copy and measure angles, find missing angles using angle relationships, and explain how to perform these skills to others.
3	I can copy and measure angles.
2	I can find the measure of a missing angle using concepts of angle relationships.
1	I can measure, draw, and name angles that meet certain given criteria.

Spiral Review • Assessment Readiness

These questions will help determine if students have retained information taught in the past and can also prepare them for high-stakes assessments. Here, students must solve one-variable, multi-step, linear equations **(Alg1, 2.2)**, calculate the area of a triangle **(Gr6, 12.2)**, and find the midpoint given a segment's end points **(1.1)**.

29. The Fan Bridge (or Merchant Square Bridge) is designed to open to allow boats to travel through the canal that it crosses. When it opens, each of the angles created by two adjacent sections is congruent. When $m\angle AFD = 60°$, what are the measures of each of the other angles formed by adjacent sections? What is the measure of $m\angle AFE$? $20°$; $80°$

30. The measure of $\angle ABC$ is $174°$ and $\overrightarrow{BD}$ bisects the angle into $\angle ABD$ and $\angle DBC$. If $\angle ABD$ measures $(7x - 10)°$, what is the value of x? $x \approx 13.9$

31. Main St. is a straight road that runs through the center of town. Sycamore Street and Rosewood Street both intersect Main Street at the same angle. Sycamore intersects Main Street at an obtuse angle with measure $(9x - 5)°$. Rosewood Street intersects Main Street at an acute angle with measure $(4x + 3)°$. Sketch the roads. What are the measures of the given angles these streets make with Main Street?

See Additional Answers for art. Sycamore Street: $121°$; Rosewood Street: is $59°$

Spiral Review • Assessment Readiness

32. Solve the equation $x - 7 = 4(x + 5)$.

Ⓐ $x = -9$

Ⓑ $x = -7$

Ⓒ $x = 7$

Ⓓ $x = 9$

33. A proposed structure shaped as a right triangle will have side lengths of 50 yards and 120 yards. The hypotenuse length will be 130 yards. What is the area of the triangle?

Ⓐ $3000\ yd^2$

Ⓑ $3250\ yd^2$

Ⓒ $6000\ yd^2$

Ⓓ $6500\ yd^2$

34. Match the segment described on the left with its midpoint on the right.

Segment Endpoints		Midpoint
A. $G(4, 7)$ and $H(3, -9)$	4	**1.** $(2.5, -2)$
B. $J(4, 7)$ and $K(-3, 9)$	2	**2.** $(0.5, 8)$
C. $L(-4, -7)$ and $M(9, 3)$	1	**3.** $(-0.5, 1)$
D. $N(-4, -7)$ and $P(3, 9)$	3	**4.** $(3.5, -1)$

 **I'm in a Learning Mindset!**

When I analyze angle relationships, what strategies do I use to persevere through difficulties when the diagrams get more complex?

Keep Going to Journal and Practice Workbook

Learning Mindset

Perseverance Sustains Focus

Point out the activities in the lesson when students were dealing with complex diagrams. Have students isolate specific angles within the diagrams. *How can you determine which parts of a complex diagram are important? How can you simplify a complex diagram? What presents the greatest challenge when identifying information from a complex diagram?*

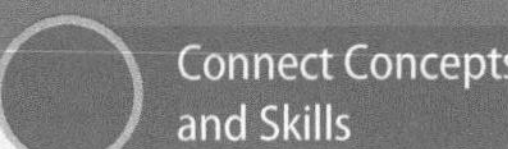

1.3 Polygons and Other Figures in the Plane

LESSON FOCUS AND COHERENCE

Mathematics Standards

- Apply concepts of density based on area and volume in modeling situations (e.g., persons per square mile, BTUs per cubic foot).
- Use geometric shapes, their measures, and their properties to describe objects (e.g., modeling a tree trunk or a human torso as a cylinder).
- Know precise definitions of angle, circle, perpendicular line, parallel line, and line segment, based on the undefined notions of point, line, distance along a line, and distance around a circular arc.

Mathematical Practices and Processes 

- Attend to precision.
- Look for and make use of structure.
- Use appropriate tools strategically.
- Model with mathematics.
- Reason abstractly and quantitatively.

I Can Objective 

I can identify and measure a polygon.

Learning Objective

Find the perimeter and area of polygons.

Language Objective

Explain how to use polygons to model real-world shapes, and apply formulas and the Area Addition Postulate to estimate perimeter and area.

Vocabulary

Review: **area**, **nonpolygon**, **polygon**, **rectangle**, **regular polygon**, **trapezoid**, **triangle**

New: ***n*-gon**

Lesson Materials: ruler, drawing compass

Mathematical Progressions

Prior Learning	Current Development	Future Connections
Students: • examined points and lines and defined a plane as a two-dimensional space. **(1.1)** • defined and classified angles as acute, obtuse, right, straight, or reflex. **(1.2)**	**Students:** • define polygons and classify two-dimensional closed figures as polygons or nonpolygons. • define and construct regular polygons. • use polygons to model complex figures, estimate their perimeters and areas, and calculate their population density.	**Students:** • will find area and perimeter of plane figures in the coordinate plane. **(1.4)** • will identify congruent triangles and polygons. **(7.1)**

PROFESSIONAL LEARNING

Visualizing the Math

Students may find it helpful to make a table similar to the one here. It includes the name and sketch of several polygons. Notice that not all polygons have to be convex polygons. If any sides of a polygon form reflex angles, the polygon is concave.

Name	Triangle	Quadrilateral	Pentagon	Hexagon	9-gon
Polygon					

PROJECTABLE & PRINTABLE

WARM-UP OPTIONS

ACTIVATE PRIOR KNOWLEDGE • Find Perimeter and Area

Use these activities to quickly assess and activate prior knowledge as needed.

Problem of the Day 

Grayson is replacing the floor in his kitchen. Before he buys the necessary materials, he needs to know the perimeter and area of the room. The figure below is a floor plan of the kitchen with its dimensions given in feet. What are the perimeter and area of the kitchen floor? perimeter: 40 ft, area: 92 ft^2

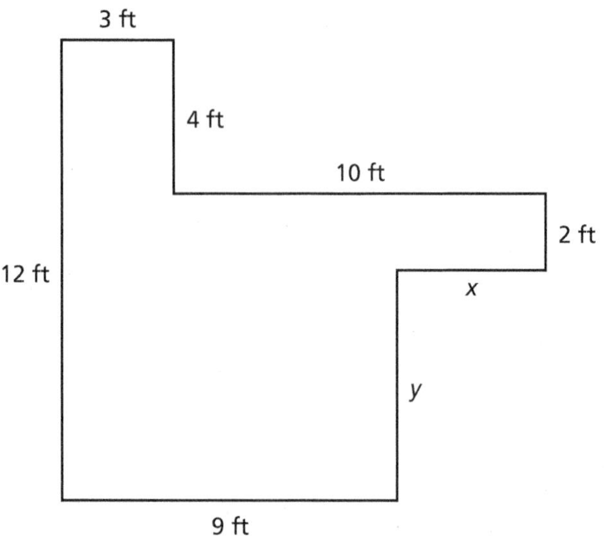

Quick Check for Homework

As part of your daily routine, you may want to display the Teacher Solution Key to have students check their homework.

Make Connections

Based on students' responses to the Problem of the Day, choose one of the following:

 Project the Interactive Reteach, Grade 7, Lesson 10.4.

 Complete the Prerequisite Skills Activity:

Have students work in groups of 3 or 4. Each student draws a rectangle with the side lengths labeled in inches. The students then cut out their shapes and form a composite shape using the four figures without overlap. The group should determine the area and perimeter of their composite figure. The group should then form a different composite figure with their rectangles and repeat the process.

- *What is the easiest way to find the area of the composite figure?* Possible answer: Find the area of each rectangle and then the add the individual areas.
- *Can you add the perimeters of all the rectangles together to find the perimeter of the composite figure? Why or why not?* no; Not all sides of the rectangles are used to form the sides of the composite figure.
- *Will the area of the second composite figure that you form have the same area as the first composite figure? Why or why not?* yes; The rectangles do not overlap so the area of the composite figure will still be the sum of the areas of the rectangles.
- *Will the perimeter of the second composite figure that you form have the same perimeter as the first composite figure?* Not necessarily. It depends how the composite shape was formed.

If students continue to struggle, use Tier 2 Skill 1.

SHARPEN SKILLS

If time permits, use this on-level activity to build fluency and practice basic skills.

Mental Math

Students use mental math to find the perimeter of a figure.

Figure *ABCD* is a rectangle, and diagonal $\overline{AC}$ divides the rectangle into two identical right triangles. Ask students what they need to find to calculate the perimeter of triangle *ADC* and how they can find it. Ask students to mentally compute the squares of sides $\overline{AD}$ and $\overline{DC}$. Ask students for a strategy for finding the sum of the squares of the sides. Ask students to mentally compute the square root of 25. Finally, ask students to mentally compute the perimeter of triangle *ADC* and to share their solution strategies with the class.

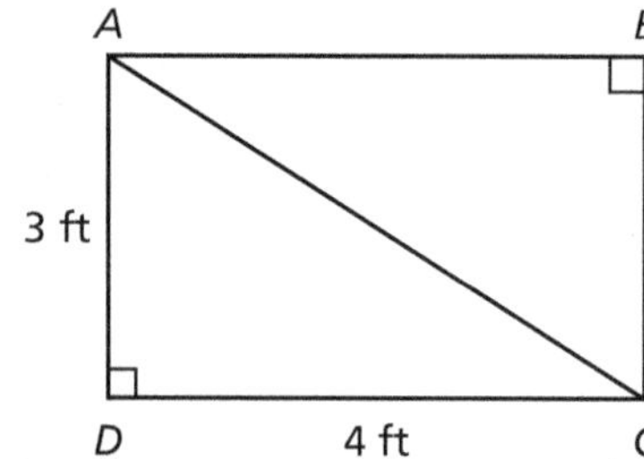

PLAN FOR DIFFERENTIATED INSTRUCTION

Small-Group Options

Use these teacher-guided activities with pulled small groups.

On Track

The continent of Africa measures about 4600 miles from West to East at its farthest points and about 5000 miles from North to South. Its population in 2018 was about 1.298 billion people. Have students use a polygon to model and estimate the area of Africa from a map and then find its population density per square mile.

Almost There

Display the following figure:

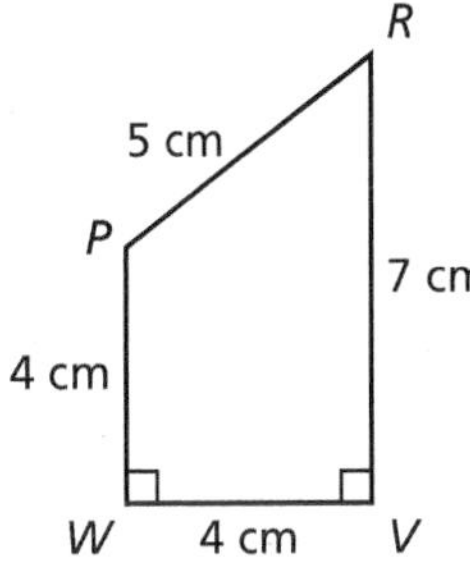

Have students do the following:

- Explain why the figure is a polygon using the definition.
- Classify the polygon.
- Name the polygon using the labels on its vertices.
- Identify two polygons they can use to calculate the area of the polygon.

Ready for More

A naturalist studying the American monarch butterfly reported that the population density of butterflies in a field was about 680 butterflies per acre. Using this information, what would be the approximate population of butterflies in 40 acres?

Math Center Options

Use these student self-directed activities at centers or stations. Key: ● Print Resources ● Online Resources

On Track

- ● Interactive Digital Lesson
- ●● Journal and Practice Workbook
- ● Interactive Glossary (printable): **area**, **nonpolygon**, **polygon**, **rectangle**, **regular polygon**, **trapezoid**, **triangle**, ***n*-gon**
- ● Module Performance Task

Almost There

- ● Reteach 1.3 (printable)
- ● Interactive Reteach 1.3
- ● RtI Tier 2 Skill 1: Areas of Composite Figures
- ● Desmos: Area v. Perimeter

Ready for More

- ● Challenge 1.3 (printable)
- ● Interactive Challenge 1.3
- ● Illustrative Mathematics: Hexagonal Pattern of Beehives

Unit Project Check students' progress by asking to see how they divided up the conservation region in the coordinate plane using polygons.

View data-driven grouping recommendations and assign differentiation resources.

During the *Spark Your Learning,* listen and watch for strategies students use. See samples of student work on this page.

Use Two Polygons to Model Areas

Strategy 1

For Staten Island, I drew a triangle whose base is 13.9 miles and whose height is 7.3 miles.

Its area is $A = \frac{1}{2}\ bh = \frac{1}{2}\ (13.9)(7.3) = 50.735\ mi^2$.

I divided to find the population density.

479,000/50.735 ≈ 9441 people per square mile

For Manhattan, I drew a rectangle with base of 2.3 miles and height of 13.4 miles. Its area is $bh = (13.4)(2.3) = 30.82\ mi^2$.

I divided to find its population density.

1,665,000/30.82 ≈ 54,023 people per square mile.

If students . . . use a triangle to model Staten Island and a rectangle to model Manhattan, they are employing an efficient method and demonstrating an exemplary understanding of drawing and analyzing shapes to solve problems from Grade 7.

Have these students . . . explain why they chose the polygons they used and how they found the corresponding areas and population densities. **Ask:**

- Q What polygon did you use to represent each borough?
- Q How did you use population and areas to find the population density of each borough?

Use a Ratio to Compare Areas

Strategy 2

I used a rectangle whose base is 2.3 miles and height is 13.4 miles to model the area of Manhattan. Its area is $bh = (2.3)(13.4) = 30.82\ mi^2$.

I used another rectangle to model the area of Staten Island.

The bases of the rectangles are approximately equal (13.4 and 13.9), and the height of the rectangle for Staten Island is about 3 times the height of the rectangle for Manhattan: 7.3/2.3 ≈ 3.2,

So, I reasoned that the area of the rectangle that models Staten Island is about 3 times the area of the rectangle that models Manhattan, so its value is about 3 x 30.82, or about $92.46\ mi^2$.

I divided to find the population density of Manhattan as 1,665,000/30.82 ≈ 54,023 people per square mile, and the population density of Staten Island as 479,000/92.46 ≈ 5181 people per square mile.

If students . . . estimate the area of Staten Island using a rectangular model, they may not understand that the more closely a shape resembles the shape of a region, the better the estimates of area and population density will be.

Activate prior knowledge . . . by having students explain why an estimate in a real-world situation such as this can never be exact. **Ask:**

- Q How could the estimate of the area of Staten Island be improved?
- Q Regardless of the accuracy of these estimates, do they still tell you something important about each borough?

COMMON ERROR: Divides Area by Population to Find Population Density

The area of the triangle that models Staten Island equals $\frac{1}{2}bh = \frac{1}{2}(13.9)(7.3) = 50.735\ mi^2$, and the area of the rectangle that models Manhattan equals $bh = (13.4)(2.3) = 30.82\ mi^2$.

So, the population density of Staten Island equals 50.375 ÷ 479,000, or about 0.00011 people per square mile. The population density of Manhattan equals 30.82 ÷ 1,665,000, or about 0.000019 people per square mile.

If students . . . do not use the correct order of division, they may not understand the definition of population density; that is, population density is the ratio between population and unit area.

Then intervene . . . by pointing out that because they were to find out how many people live within a square mile of each borough, they needed to divide the given population by the area of the region, not the other way around. **Ask:**

- Q Does it make sense that the population densities that you found are nearly zero?
- Q Do the units of the divisor and dividend in each calculation match the units in your answers?

1.3

Polygons and Other Figures in the Plane

I Can identify and measure a polygon.

Spark Your Learning

Researchers who study demographics often use census data about the United States when studying an area. One important factor of a location is *population density*, or the number of people per unit of area. Staten Island and Manhattan, two boroughs of New York City, are compared below.

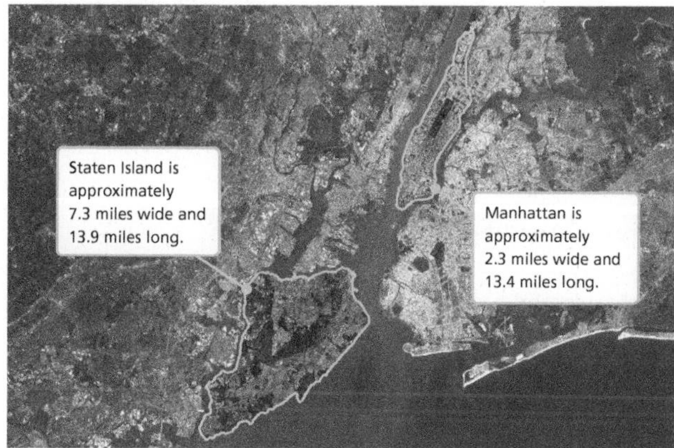
Staten Island is approximately 7.3 miles wide and 13.9 miles long.
Manhattan is approximately 2.3 miles wide and 13.4 miles long.

Complete Part A as a whole class. Then complete Parts B–D in small groups.

A. What is a mathematical question you can ask about this situation? What information would you need to know to answer your question? A. What is the population density of each borough?; the population and area of each borough

B. How could geometric shapes play a role in this situation? How could someone comparing the two boroughs use geometric shapes? B, D. See Additional Answers.

C. To answer your question, what strategy and tool would you use along with all the information you have? What answer do you get? See Strategies 1 and 2 on the facing page.

D. How could the information you found be applied to a real-world scenario about the two regions? Why is this an important comparison?

Turn and Talk How do you think your answer would change if you chose different geometric shapes to model the boroughs? Is there only one correct way to create a model? How could you make your estimation more accurate? See margin.

CULTIVATE CONVERSATION • Co-Craft Questions

If students have difficulty formulating a mathematical question about the situation in the Spark Your Learning, ask them to imagine they are a researcher who has been asked to find the population density of each borough. What are some natural questions to ask about this situation?

Work together to craft the following questions:

- What information do you need to find the population density of each borough?
- How can you use shapes to find the areas of the boroughs?
- What mathematical operation can you use to calculate the population density of each borough?

Then have students think about what additional information, if any, they would need to answer these questions. **Ask:**

- Can you determine the exact area of each borough? Explain.
- Can you determine the exact population density of each borough? Explain.

1 Spark Your Learning

▶ MOTIVATE 

- Have students look at the photo in their books and read the information contained in the photo. Then complete Part A as a whole-class discussion.
- Give the class the additional information they need to solve the problem. This information is available as a projectable page in the Module Resources.
- Have students work in small groups to complete Parts B–D.

▶ PERSEVERE

If students need support, guide them by asking:

Q **Advancing • Use Tools** Which tool could you use to solve the problem? Why choose that tool and not some other? Students' choices of tools and reasons for choosing them will vary.

Q **Assessing** What geometric shape does each borough most resemble? Staten Island resembles a triangle, and Manhattan resembles a rectangle.

Q **Assessing** How can you find the areas of triangles and rectangles? triangle: $A = \frac{1}{2}bh$, rectangle: $A = lw$

Q **Advancing** How is population density related to population and area? I can divide the population of each borough by the area of the shape that represents the borough. This gives the population per square mile, also known as population density.

Turn and Talk When predicting how their answers would change, students should consider using different shapes to model the areas. Possible answers: The final calculations for population density would vary, but not by a large amount if you chose shapes that fit the land area well; There are many ways to model the regions; To make the calculation more accurate, Manhattan could be a rectangle with a triangle at either end rather than a single rectangle.

▶ BUILD SHARED UNDERSTANDING

Select groups of students who used various strategies and tools to share with the class how they solved the problem. As they present their solutions, have each group discuss why they chose a specific strategy and tool.

2 Learn Together

Build Understanding

Task 1 MP **Attend to Precision** Students determine whether given figures are polygons by making explicit use of the definition to justify their conclusions.

CONNECT TO VOCABULARY

Have students use the **Interactive Glossary** to record their understanding of the vocabulary in this task.

Sample Guided Discussion:

Q **In Part A, what should you look for to determine whether each figure is a polygon or not?** To determine if a figure is a polygon, I should look for whether the figure is a closed figure with at least three line segments that intersect exactly two other line segments at their endpoints.

Task 2 MP **Use Structure** Students construct polygons using descriptions of characteristics of the polygon.

Q **Could you draw a square for Part B?** No, because the lengths of the sides of a square have the same length.

Turn and Talk Guide students who are describing the figure they drew to concentrate on the general characteristics of the figure, such as the number of sides and lengths. Possible answer: No, both polygons do not look the same; There are many different ways to draw geometric figures.

Build Understanding

Understand the Definition of a Polygon

The definition of a polygon varies across the field of mathematics. In this course, we define a **polygon** as a closed plane figure formed by three or more line segments such that each segment intersects exactly two other segments only at their endpoints; no two segments with a common endpoint are collinear. Other definitions of a polygon may allow lines to intersect or allow two polygons to be called a polygon. A **nonpolygon** is a geometric object that does not meet the definition of a polygon.

1 A. Determine whether each figure is a polygon.

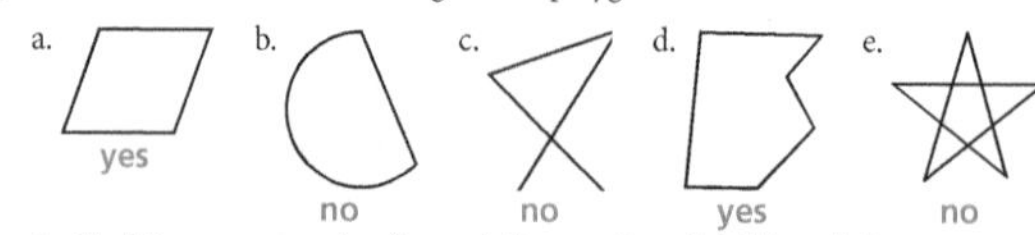

B. Explain your reasoning for each figure. See Additional Answers.

Classify Polygons by the Number of Sides

Polygons can be classified according to the number of sides in the figure.

Name	Triangle	Quadrilateral	Pentagon	Hexagon	*n*-gon
Number of Sides	3	4	5	6	*n*

You may also see a polygon described as an ***n*-gon**, where *n* is the number of sides in the figure. For example, a polygon with nine sides may be called a 9-gon.

Polygons are named by listing the vertices in order moving either clockwise or counterclockwise around the figure. The sample quadrilateral can be named many different ways, including *ABCD*, *BCDA*, and *ADCB*, but it cannot be named *ABDC*.

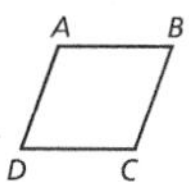

2 Draw the polygons described below. Then classify the polygon and name it using the vertices. A–C. See Additional Answers.

A. a polygon with five vertices

B. a polygon with four sides of different lengths

C. a polygon with six sides of equal length

Turn and Talk Secretly draw a polygon, then describe and ask your partner to draw the polygon. Does each person's drawing look the same? Why or why not? Trade roles and try the exercise again. See margin.

22

LEVELED QUESTIONS

Depth of Knowledge (DOK)	Leveled Questions	What Does This Tell You?
Level 1 **Recall**	Why is a polygon a closed plane figure? because it is made up of three or more line segments that intersect only at their endpoints	Students' answers will indicate whether they understand the definition of a polygon.
Level 2 **Basic Application of Skills & Concepts**	How can you determine whether this figure is a polygon? I can use the definition of a polygon and see that the segments that form the sides of the figure intersect only at their endpoints.	Students' answers will demonstrate whether they can use the definition of a polygon to determine whether a given figure is a polygon.
Level 3 **Strategic Thinking & Complex Reasoning**	Why is a circle not a polygon? because a polygon is made up of line segments and a circle is a closed curve	Students' answers will reflect whether they understand the definition of a polygon and can reason strategically about the parts of a figure that make it a polygon.

Step It Out

Construct Regular Polygons

A polygon is a **regular polygon** when all the sides and angles of the polygon are congruent.

The steps for constructing a regular hexagon using a ruler and compass are shown.

3

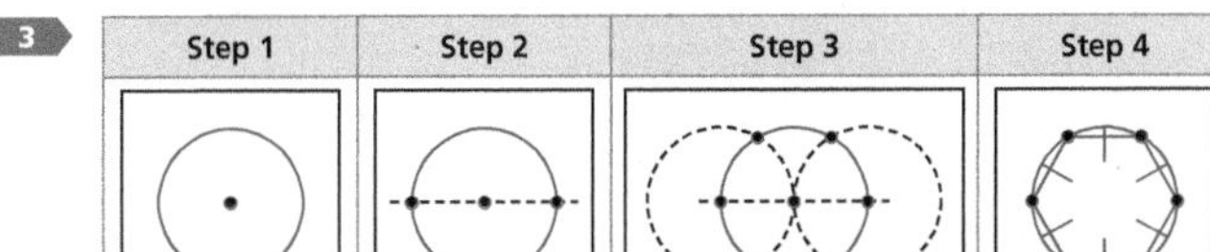

A. Describe what happens in each step. A–C. See Additional Answers.

B. What is the purpose of this construction? Do you achieve the same result if the points are not evenly spaced?

C. What is the purpose of the hash marks on each side of the hexagon in Step 4?

Turn and Talk How could you change the construction to create a regular triangle inscribed in the circle? See margin.

Model to Estimate Area and Perimeter

The **area** of a geometric figure is the surface contained within the boundaries of a two-dimensional object such as a **triangle**, **rectangle**, or **trapezoid**. The perimeter of a two-dimensional shape is the distance all the way around the figure, found by adding all the side lengths. The table below reviews common area formulas.

Polygon	Triangle	Rectangle	Trapezoid
Figure	h, b	w, ℓ	b_1, h, b_2
Area Formula	$A = \frac{1}{2}bh$	$A = lw$	$A = h\left(\frac{b_1 + b_2}{2}\right)$

If a figure on a plane is more complex, it can often be divided into the basic shapes shown in the table. You can then calculate the area of each individual shape.

Area Addition Postulate
If a figure is composed of two or more shapes, the area of the figure is the sum of the areas of the individual shapes.

Step It Out

Task 3 MP **Use Tools** Students use a ruler and a drawing compass to construct a regular hexagon.

CONNECT TO VOCABULARY

Have students use the **Interactive Glossary** to record their understanding of the vocabulary in this task.

Sample Guided Discussion:

Q **What property of the original circle determines the distances between the vertices of the hexagon?** The distance between the vertices is the length of the radius of the congruent circles.

Q **If you draw segments from each vertex of the hexagon to its center, what are the resulting figures?** six equilateral triangles

Turn and Talk Point out that a regular triangle has three equal sides and three equal angles and is called an equilateral triangle. Connect every other point rather than all six.

EL PROFICIENCY LEVEL

Beginning

Give each student an index card. Have them write the word "polygon" on one side and "nonpolygon" on the other side. Then have them draw a triangle on the side of the card where they think it belongs. Repeat the activity with several shapes, both polygons and nonpolygons (e.g., a circle, an open six-sided figure, etc.).

Intermediate

Have students work in groups. Give each group the description of a polygon as in Task 2. The groups should draw the polygon and classify it. Have each group share its figure with other groups.

Advanced

Draw an *n*-gon on the board, where *n* is any positive integer. Then have students complete the statement, "This figure is an *n*-gon because ______." Then have students explain why the polygon is regular or irregular.

Task 4 (MP) **Model with Mathematics** Students see how to use a grid and geometric shapes to model the area of a given region and apply the Area Addition Postulate to approximate its value. They use the area and the given population of birds to calculate the approximate population density of birds on the island.

Encourage students to use more than one shape to model the area of the island. Point out that using multiple shapes may allow them to better cover the shape of the island and find a better approximation of its area.

CONNECT TO VOCABULARY

Have students use the **Interactive Glossary** to record their understanding of the vocabulary in this task.

Sample Guided Discussion:

Q Why is it helpful to use a grid to estimate the area of the island? A grid allows you to better approximate the distances represented by the dimensions of the polygons used and to approximate the area. The smaller the grid, the better the approximation.

Turn and Talk Remind students that the perimeter of the island is the distance around its edges. Possible answer: no; I would choose shapes that followed the shape of the island more exactly if I were estimating the perimeter.

Population density is the measurement of population over a certain area. You can calculate population density by dividing the population by the area of the region.

4 Scientists tracking an endangered bird species need to estimate the area of a small island that will be a sanctuary. The current bird population is 1,058. To determine if the population is increasing, they need to relate the current bird population to the area of the island.

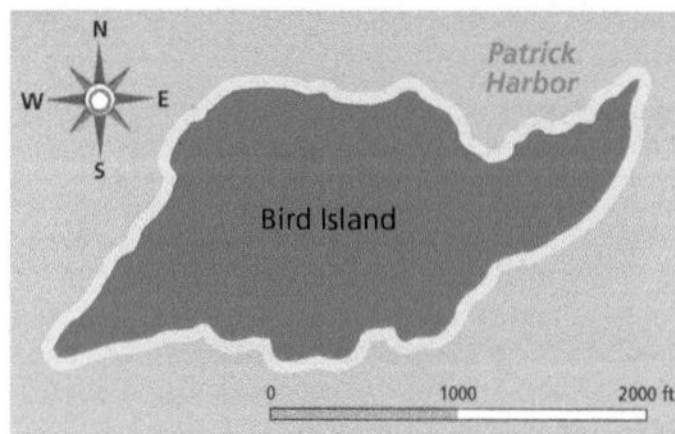

A. Choose one or more shapes to model the island. A–E. See Additional Answers.

B. Find the area of each individual shape. Then, using the Area Addition Postulate, find the total approximate area of the island.

C. Compare your solution to the one shown. Which estimate appears more accurate? Why?

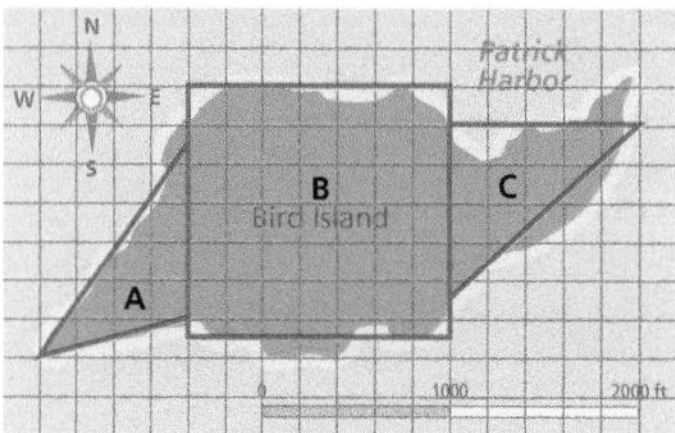

D. The population density of endangered birds on the island gives the number of birds per square foot. What do the units of the population density indicate about how to calculate it?

E. If the population of endangered birds on the island is 1058, use the area you calculated in Part B to find the approximate population density of endangered birds on the island.

Turn and Talk Would you use the same shapes to approximate the island if you were measuring perimeter instead of area? If not, what changes would you make? See margin.

Prove the Pythagorean Theorem

Recall that the Pythagorean Theorem, $a^2 + b^2 = c^2$, can be used to find unknown side lengths or the hypotenuse length of any right triangle.

Pythagorean Theorem

In a right triangle with legs *a* and *b* and hypotenuse *c*, the square of the length of the hypotenuse is equal to the sum of the squares of the lengths of the legs.

$a^2 + b^2 = c^2$

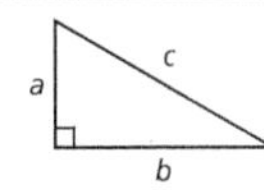

5 The figure shows right triangle *ABC*, which has been copied four times to make the shape of a square with vertices *A*, *B*, *F*, and *E*.

Notice the placement of the four triangles also creates a centrally located square.

Follow the steps below to examine why the Pythagorean Theorem applies to any right triangle.

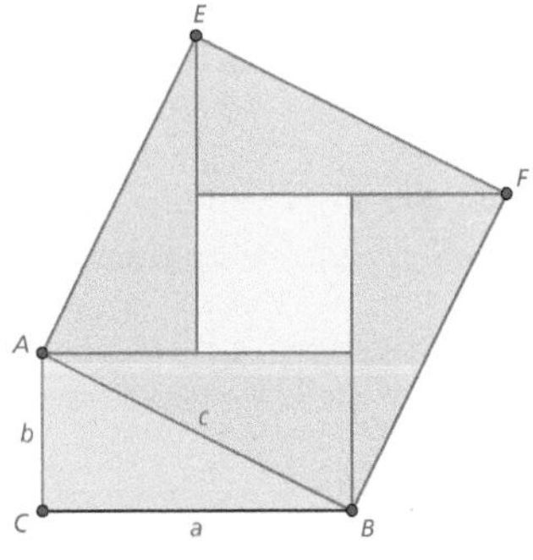

Write an equation to describe the area of square *ABFE*. A–C. See Additional Answers.

$A = lw$

$A = c \cdot c = c^2$

A. Why does this describe the area of square *ABFE*?

We can break the area of the square into the area of the individual shapes that make up the square.

Area of the square = 4(area of triangle *ABC*) + area of central square

$c^2 = 4\left(\frac{1}{2}bh\right)_{\text{triangle}} + (lw)_{\text{square}}$

$c^2 = 4\left(\frac{1}{2}ab\right) + (a - b)(a - b)$

B. Why does this describe the area of the smaller central square?

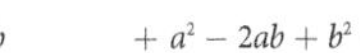

$c^2 = 2ab + a^2 - 2ab + b^2$

C. Explain how you obtain this trinomial from the product $(a - b)(a - b)$.

Simplify the equation to $c^2 = a^2 + b^2$.

Turn and Talk What formula results if $b > a$? What formula results if $b = a$? How do you know the four triangles meet at right angles to form square *ABFE*. See margin.

Task 5 (MP) **Reason Abstractly and Quantitatively**
Students examine a diagram made up of congruent right triangles and squares to establish the Pythagorean Theorem. Encourage students to label the corresponding sides of the copied triangles in terms of *a*, *b*, and *c*.

SUPPORT SENSE-MAKING Three Reads

Have students read the problem three times. Use the questions below for a different focus each time.

1. What is the situation about?
2. What are the quantities in the situation?
3. What are the possible mathematical questions that you could ask about this situation?

Sample Guided Discussion:

Q **What is the length of the sides of the central square in terms of *a* and *b*?** You can use the grid to express the lengths of the sides of the central square. The four right triangles within square *AEFB* have side lengths of *a*, *b*, and *c*, where *c* is the hypotenuse in each triangle. The length of the sides of the central square are each $a - b$.

Q **How can you find the product of $(a - b)$ and $(a - b)$?** Apply the Distributive Property: $(a - b) \times a - (a - b) \times b$, which equals $a^2 - ab - ab + b^2$ or $a^2 - 2ab + b^2$.

Q **Why is c^2 equal to $a^2 + b^2$?** The value of c^2 is equal to the sum of the area of the four right triangles and the area of the square. The area of the four right triangles is $4\left(\frac{1}{2}ab\right)$, which equals $2ab$. The area of the square is the product of $(a - b)(a - b)$ which is $a^2 - 2ab + b^2$. The sum of these two expressions results in the addition of opposites, $2ab$ and $-2ab$, which is 0. Therefore, c^2 is equal to $a^2 + b^2$.

Turn and Talk If students are not sure what happens if $b > a$ or $b = a$, encourage them to sketch two right triangles with hypotenuses labeled *c*, and whose sides have each of those properties. Then use tools to create the square on the hypotenuse and the copies of the triangle. Possible answer: If $b > a$ then $(a - b)$ changes to $(b - a)$ in the solution, but the resulting formula is the same; If $b = a$, then the formula can simplify to $2a^2 = c2$; The triangles meet at right angles because the sum of the measures of the two acute angles in a right triangle is 90°.

ONLINE

Assign the Digital On Your Own for
- built-in student supports
- Actionable Item Reports
- Standards Analysis Reports

On Your Own

Assignment Guide

The chart below indicates which problems in the On Your Own are associated with each task in the Learn Together. Assign daily homework for tasks completed.

Learn Together Tasks	On Your Own Problems
Task 1, p. 22	Problem 5
Task 2, p. 22	Problems 13–15
Task 3, p. 23	Problems 6, 10
Task 4, p. 24	Problems 7, 8, 11, 12, 16–21
Task 5, p. 25	Problem 9

Check Understanding

1. (MP) **Construct Arguments** Draw a figure that is a polygon, a figure that is not a polygon, and a figure that is a polygon under some but not all definitions of polygon. Explain your reasoning. 1–3. See Additional Answers.
2. Classify and name the polygon. Is it a regular polygon? Justify your answer.

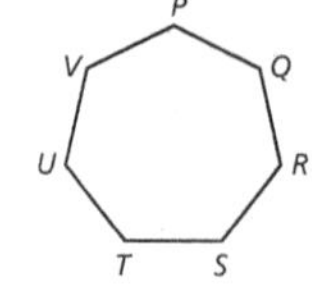

3. Draw a regular polygon with 9 vertices. Label the vertices and classify the polygon.
4. The population density of a square platform must be less than 4 people per square yard to meet safety requirements. The expected attendance at the event is 150 people. What is the minimum side length, in whole yards, of the platform that meets safety requirements? 7 yards

On Your Own 5–7. See Additional Answers.

5. Give an example of a two-dimensional figure that is not a polygon. Justify your answer.
6. Kayla examined the given figure and determined it was a regular hexagon because all the angles are congruent. Did Kayla classify the figure correctly? Why?

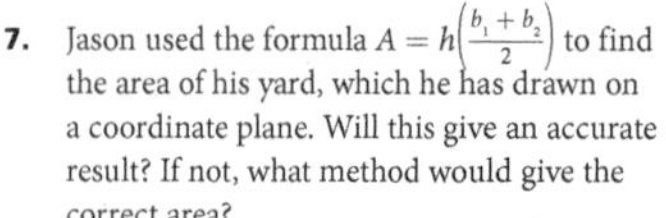

7. Jason used the formula $A = h\left(\frac{b_1 + b_2}{2}\right)$ to find the area of his yard, which he has drawn on a coordinate plane. Will this give an accurate result? If not, what method would give the correct area?

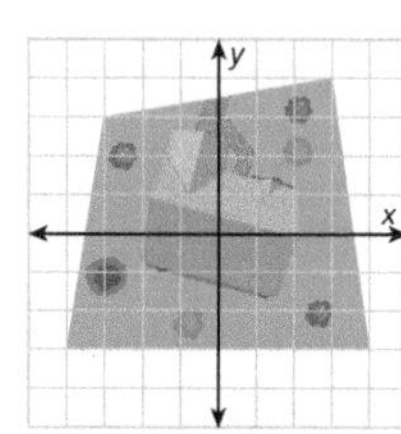

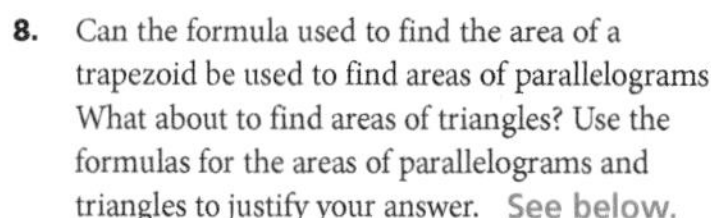

8. Can the formula used to find the area of a trapezoid be used to find areas of parallelograms? What about to find areas of triangles? Use the formulas for the areas of parallelograms and triangles to justify your answer. See below.

Determine if each statement provides enough information to find the value. If yes, find the value.

9. the area of a triangle with a hypotenuse of 5 ft no
10. the perimeter of a regular pentagon with a side length of 3 m yes; 15 m
11. the width of a rectangle with an area of 24 in^2 and a length of 6 in. yes; 4 in.
12. the side length of a square with an area of 64 ft^2 yes; 8 ft

8. Yes; The average of the base lengths multiplied by height will give the area of a parallelogram because the bases are equal in length in a parallelogram. The average of the base lengths (in which one base length is zero) multiplied by the height will give the area of a triangle.

26

3 Check Understanding

Formative Assessment

Use formative assessment to determine if your students are successful for this lesson's learning objective.

Students who successfully complete the Check Understanding can continue to the On Your Own practice.

For students who miss 1 problem or more, work in a pulled small group using the Almost There small-group activity on page 21C.

ONLINE

Assign the Digital Check Understanding to determine
- success with the learning objective
- items to review
- grouping and differentiation resources

4 Differentiation Options

Differentiate instruction for all students using small-group activities and math center activities on page 21C.

Reteach

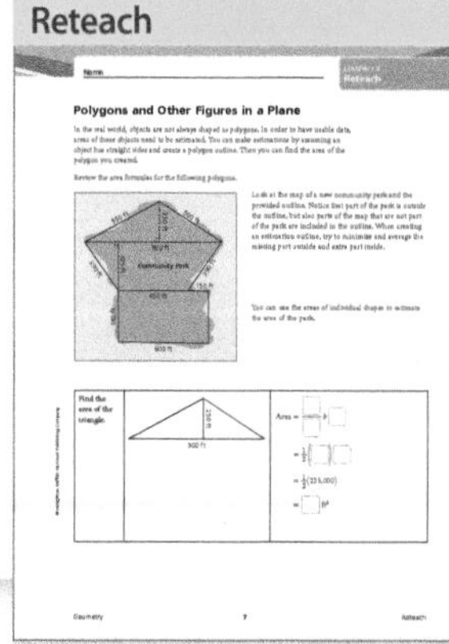

Challenge

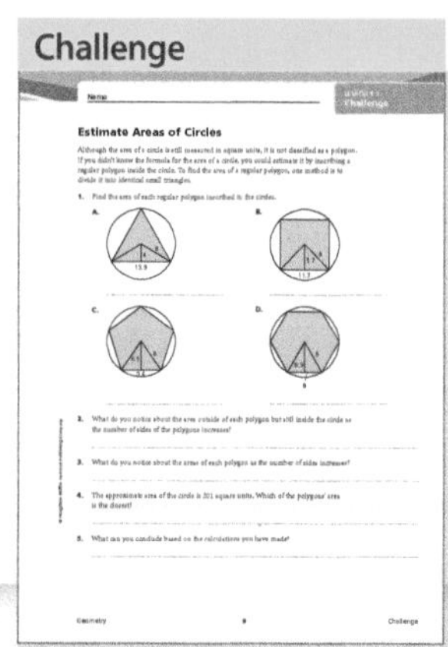

Name each polygon three different ways. 13–16. See Additional Answers.

13.

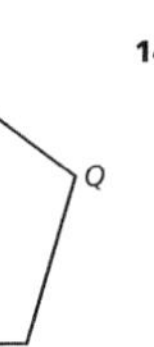

14.

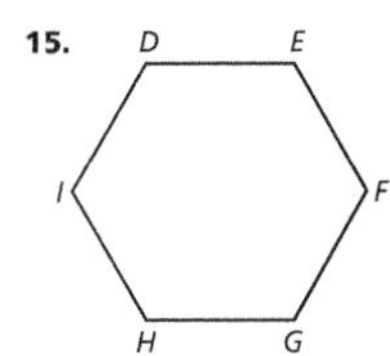

15.

16. The equation used to find the area of a geometric figure is shown. Draw a figure that matches the formula.

$A = 5 \cdot \left(\frac{1}{2}\right)(6)(4)$

17. (MP) **Model with Mathematics** Quinn is coating the sides of her tent in a water repellant spray to prepare for an upcoming camping trip. Each can of spray lists the approximate square footage of fabric it will cover, but she isn't sure of the area of the tent's surface. Approximate the surface area by modeling each side using geometric figures. Assume that the tent has a square base. about 112 ft²

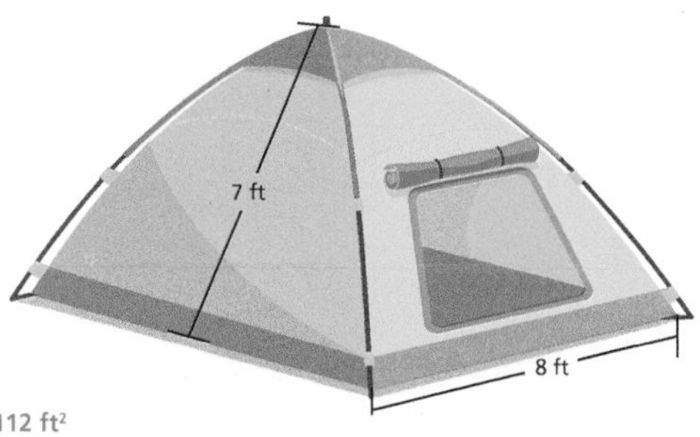

18. A rectangular pool that is fifteen feet wide and twenty feet long is surrounded by a deck that is four feet wide. What is the area of the surface of the deck? 344 ft²

19. (MP) **Use Repeated Reasoning** There are two homeroom classes at Whiteford High School. The plans below show the layouts for Ms. Chang's and Mr. Edwards's classrooms.

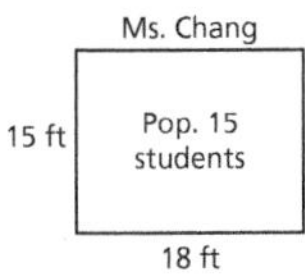

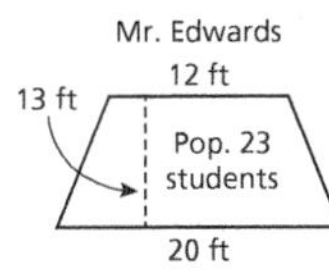

A. Which classroom has a greater population density? Explain. See Additional Answers.

B. Use the population and area to determine how many square feet of classroom space each student has in Ms. Chang's room. 18 ft²

C. How many students should be added to the classroom with a lower population density in order to make the population density of the two classrooms approximately equal? 15 students

Questioning Strategies

Problem 17 Suppose that the can of water repellant spray that Quinn buys covers 150 ft². Would she have enough spray to cover the sides and the bottom of the tent? Explain. yes; There are four sides to Quinn's tent, and the area of each side can be approximated with a triangle and the given dimensions. The area formula for a triangle is $A = \frac{1}{2}(8)(7) = 28$ ft², so the approximate area of the four sides of the tent equals $4(28) = 112$ ft², which is less than 150 ft².

Watch for Common Errors

Problem 18 Some students may add 4 feet to the given dimensions of the pool, so that when they find the area of the rectangle that includes the pool and the deck, the product is (19 ft)(24 ft) = 456 ft². However, have the students draw a model of the pool and surrounding deck. Point out that the deck extends out 4 feet around all four sides of the pool, and that the length and width of the resulting rectangle are each increased by 8 feet, not 4 feet. The area of the pool and surrounding deck is (23 ft)(28 ft) = 644 ft². The area of the pool is equal to the difference between 644 ft² and 300 ft², which equals 344 ft².

5 Wrap-Up

Summarize learning with your class. Consider using the Exit Ticket, Put It in Writing, or I Can scale.

Exit Ticket 

The state of Massachusetts is about 150 miles long and 50 miles wide, and its area can be approximated by a rectangle. The population of Massachusetts in 2018 was about 6,742,000.

What was the population density in square miles of Massachusetts in 2018? The population density of a region is the ratio of the population and the area of the region. Using the given dimensions, the total area of Massachusetts is about 150 mi $\times$ 50 mi $=$ 7500 mi^2. Therefore, in 2018 the population density in the state was about $\frac{6{,}742{,}000}{7500} = 899$ people per square mile.

Put It in Writing 

Describe how you can find the area of geometric figures composed of several shapes.

I Can 

The scale below can help you and your students understand their progress on a learning goal.

4	I can use polygons to model shapes in real-world situations and use formulas to estimate regional areas and perimeters.
3	I can identify and measure a polygon.
2	I can identify and draw polygons and nonpolygons.
1	I can identify a closed figure in the plane.

Spiral Review • Assessment Readiness

These questions will help determine if students have retained information taught in the past and can also prepare them for high-stakes assessments. Here, students must classify an angle **(1.2)**, determine the midpoint of a segment determined by a set of ordered pairs **(Gr6, 11.3)**, identify angle pairs that are supplementary **(1.2)**, and use the Pythagorean Theorem **(Gr8, 11.3)**.

20. Science A wildlife preserve aims to keep the population density of large mammals at a natural balance. In order to help determine what the relative and total population density of a preserve should be, a researcher is looking at populations in a large wild area covering about 9000 square miles. Compare the population densities of the different mammals. What conjectures do you have about why some populations are denser than others? See Additional Answers.

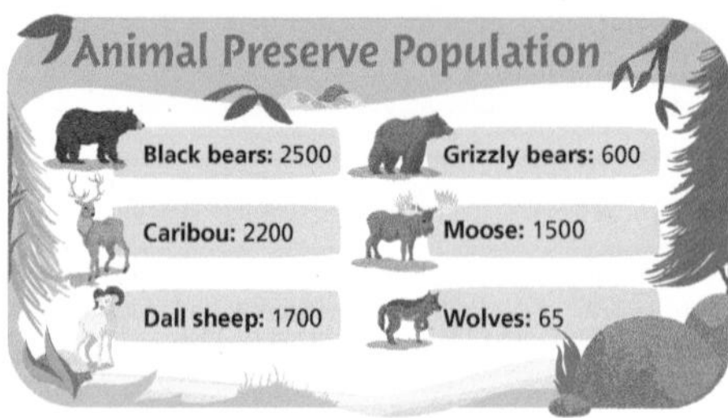

21. Open Ended Create a population density problem that can be solved by modeling an area. Show two different methods of choosing shapes to model the area. Then calculate the area from each model. Compare the two solutions. Check students' work.

Spiral Review • Assessment Readiness

22. Which figure represents $\angle AEF$?

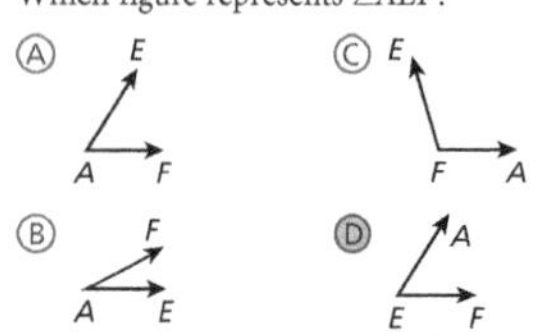

Ⓐ Ⓑ Ⓒ Ⓓ

23. Point (0.5, 3) is the midpoint of which set of ordered pairs?

Ⓐ $(-4, -4), (5, -3.5)$
Ⓑ $(1, 3.5), (-2, 4)$
Ⓒ $(2, 9), (-1, -3)$
Ⓓ $(8, 0.5), (7, -3.5)$

24. Which sets of angles are supplementary angles? Select all that apply.

Ⓐ 60°, 30°
Ⓑ 30°, 150°
Ⓒ 60°, 60°
Ⓓ 50°, 130°
Ⓔ 50°, 40°
Ⓕ 130°, 130°

25. A sketch of a section of tempered glass to be used for a shelf is shown. Which expression represents the length of x?

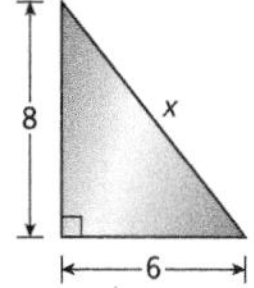

Ⓐ $(6 + 8)^2$
Ⓑ $8^2 + 6^2$
Ⓒ $6^2 - 8^2$
Ⓓ $\sqrt{6^2 + 8^2}$

 **I'm in a Learning Mindset!**

Do my methods for estimating area give me results that have acceptable levels of accuracy? What evidence supports that claim?

28 Keep Going to Journal and Practice Workbook

Learning Mindset

Perseverance Checks for Understanding

Point out that estimation is an important skill for many areas of mathematics. Encourage students to develop the habit of checking for accuracy of their estimations of area by comparing their estimates with other students' estimates. They can also research to find information about the place whose area they are estimating. Also remind students that acceptable levels of accuracy with estimation will vary depending on the factor being studied. *How does using polygons to model real-world situations help you understand that reasonable answers to problems are not always exact? Can you think of other situations in which modeling a real-world situation might be useful?*

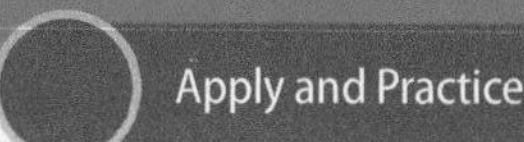

1.4 Apply the Distance Formula

LESSON FOCUS AND COHERENCE

Mathematics Standards

- Use coordinates to compute perimeters of polygons and areas of triangles and rectangles, e.g., using the Distance Formula.
- Use geometric shapes, their measures, and their properties to describe objects (e.g., modeling a tree trunk or a human torso as a cylinder).

Mathematical Practices and Processes 

- Look for and express regularity in repeated reasoning.
- Look for and make use of structure.
- Model with mathematics.

I Can Objective 

I can measure the distance between two points on the coordinate plane.

Learning Objective

Find the perimeter and the area of a figure on the coordinate plane using the Distance Formula, and model irregular figures with simple polygons to estimate perimeter and area.

Language Objective

Explain how the Distance Formula is used to find the perimeter and the area of a figure on the coordinate plane.

Vocabulary

New: Distance Formula

Lesson Materials: geometric drawing tool, graph paper

Mathematical Progressions

Prior Learning	Current Development	Future Connections
Students: • found distances in the coordinate plane using the Pythagorean Theorem. **(G8, 11.4)**	**Students:** • discover that translating one side of a parallelogram along the line containing it without changing its length results in another parallelogram with the same area but with a different perimeter. • use knowledge of the Pythagorean Theorem to justify the Distance Formula. • estimate the area of an irregular shape on the coordinate plane.	**Students:** • will use coordinates to prove statements about segments. **(2.4)** • will use coordinates to perform transformations. **(5.1)** • will use coordinates to prove statements about polygons and circles. **(11.1–11.5 and 16.1)**

UNPACKING MATH STANDARDS

Use coordinates to compute perimeters of polygons and areas of triangles and rectangles, e.g., using the Distance Formula.

What It Means To You

Students demonstrate an understanding of finding lengths using the Distance Formula and using the lengths to compute the perimeters and the areas of polygons on the coordinate plane. This understanding develops from Grade 8, where they found distances on the coordinate plane using the Pythagorean Theorem.

The emphasis for this standard is on using models to estimate perimeters and areas and using the Distance Formula to calculate lengths used in the estimates. In Lesson 5.1, students will use the Distance Formula to define transformations that preserve size.

WARM-UP OPTIONS

ONLINE

PROJECTABLE & PRINTABLE

ACTIVATE PRIOR KNOWLEDGE • Find Perimeter and Area in the Coordinate Plane

Use these activities to quickly assess and activate prior knowledge as needed.

Problem of the Day

Find the length of each segment.

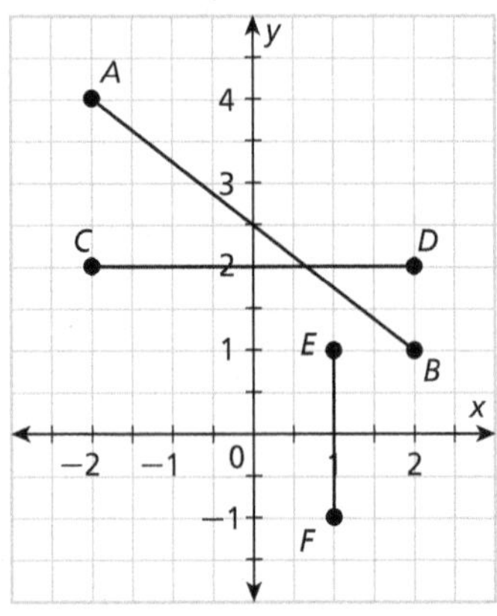

$AB = 5$, $CD = 4$, $EF = 2$

Quick Check for Homework

As part of your daily routine, you may want to display the Teacher Solution Key to have students check their homework.

Make Connections

Based on students' responses to the Problem of the Day, choose one of the following:

1. Project the Interactive Reteach, Grade 8, Lesson 11.4.
2. Complete the Prerequisite Skills Activity:

Materials: graph paper

Have students work in pairs. One student should draw a slanted segment, a horizontal segment, and a vertical segment on a coordinate grid. The other student should find the length of each segment. Then have students switch roles to create and find the lengths of new segments.

- *How do you find the length of a horizontal segment?* Calculate the absolute value of the difference of the *x*-coordinates.
- *How do you find the length of a vertical segment?* Calculate the absolute value of the difference of the *y*-coordinates.
- *How do you find the length of a slanted segment?* Draw a horizontal line through one endpoint of the segment and a vertical line through the other to create a right triangle. The vertex of the right angle is the intersection of the horizontal and vertical lines. Use the Pythagorean Theorem to find the hypotenuse (length of the slanted segment).

If students continue to struggle, use Tier 2, Skill 4.

SHARPEN SKILLS

If time permits, use this on-level activity to build fluency and practice basic skills.

Quantitative Comparison

Objective: Students make a comparison between two quantities.

Write the following problem on the board. Ask students to choose the letter representing the correct answer and to explain their reasoning.

Quantity A	**Quantity B**
area of a right triangle with sides 6, 8, and 10	area of a rectangle with length 12 and width 2

A. Quantity A is greater.
B. Quantity B is greater.
C. The two quantities are equal. C; Quantity A is 24 square units and Quantity B is 24 square units.
D. The relationship cannot be determined from the information given.

PLAN FOR DIFFERENTIATED INSTRUCTION

Small-Group Options

Use these teacher-guided activities with pulled small groups.

On Track

Materials: index cards

Give each student a card with a triangle or a rectangle on a coordinate grid. Have each student calculate the perimeter and the area of the figure on the given card. Then have students pair up and switch cards to calculate the perimeter and the area of the figure on their partner's card. Each student should pair up with as many other students as time allows.

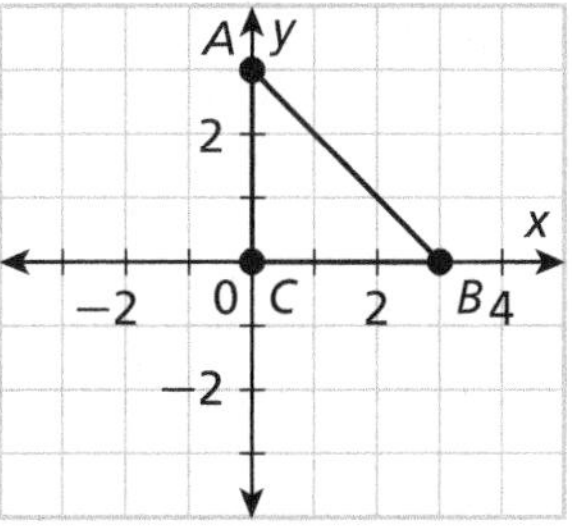

Almost There

Materials: graph paper

Draw an irregular shape on a transparency that can easily be modeled with a combination of triangles and rectangles. Have students do the following:

- Model the area of the shape with triangles and rectangles.
- Find the area of the model.
- Find the perimeter of the model.
- Have students pair up and critique their solutions.

Ready for More

Materials: index cards

Give each student a card with an irregular shape on a coordinate grid. Have each student calculate the perimeter and the area of the figure on the given card. Then have students pair up and switch cards to calculate the perimeter and the area of the figure on their partner's card. Each student should pair up with as many other students as time allows.

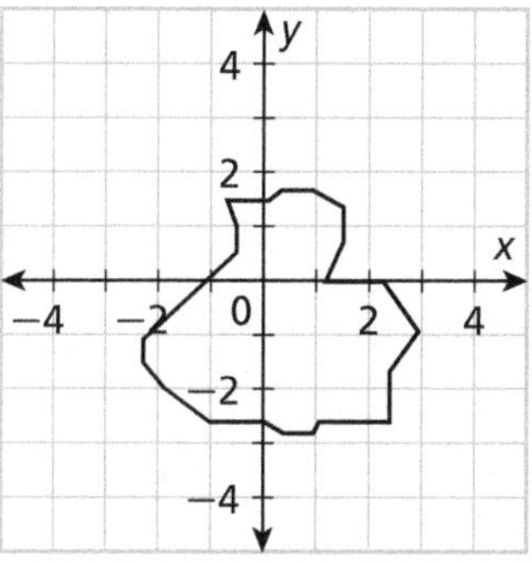

Math Center Options

Use these student self-directed activities at centers or stations. Key: ● Print Resources ● Online Resources

On Track

- ● Interactive Digital Lesson
- ●● Journal and Practice Workbook
- ● Interactive Glossary (printable): **ordered pair**, **Distance Formula**

Almost There

- ● Reteach 1.4 (printable)
- ● Interactive Reteach 1.4
- ● RtI Tier 2 Skill 4: The Pythagorean Theorem and Its Converse
- ● Illustrative Mathematics: Squares on a Coordinate Grid

Ready for More

- ● Challenge 1.4 (printable)
- ● Interactive Challenge 1.4
- ● Illustrative Mathematics: Triangle Perimeters
- ● Desmos: Move any side of a triangle along a line that contains it. How is the perimeter and area affected?

Unit Project Check students' progress by asking what formula they are using to find the side lengths of the triangle.

View data-driven grouping recommendations and assign differentiation resources.

During the *Spark Your Learning*, listen and watch for strategies students use. See samples of student work on this page.

Use a Model

Strategy 1

The area of the entire city is about $25 \times 15 = 375$ square units.
So, each of the 7 regions should have an area of about $\frac{375}{7} = 53.6$ square units.

Because the park land and water are not populated areas, they should be divided evenly between the regions or included in regions with greater area.

If students . . . divide the area into simple shapes, they are employing an efficient method and demonstrating an exemplary understanding of using a model to find the area.

Have these students . . . explain how they chose their model. **Ask:**

- Q How did you decide how to divide the area?
- Q What simple polygons did you use?

Count Squares

Strategy 2

I counted squares to find the area. There are about 350 squares.
So, each of the 7 regions should have an area of about $\frac{350}{7} = 50$ square units.

If students . . . count squares to find the area, they understand area but may not know how to model the problem geometrically.

Activate prior knowledge . . . by having students write formulas for the area of triangles, rectangles, and squares. **Ask:**

- Q How do you find the area of triangles, rectangles, and squares?
- Q Can you model the area by dividing the irregular shape into these simpler shapes?

COMMON ERROR: Uses a Poor Model

I used a rectangle model to find the area. The rectangle is 25 units by 16 units, so the area of the entire city is about 400 square units. Each of the 7 regions should have an area of about $\frac{400}{7} = 57.1$ square units.

If students . . . use a poor model to find the area, they do not understand how to represent the irregular shape with a realistic model.

Then intervene . . . by pointing out that the square is not a realistic model for the shape **Ask:**

- Q How much of the area inside your model is *not* part of the irregular shape?
- Q How close do you think the area of your model is to the area of the irregular shape?
- Q How can you make a better model?

1.4

Apply the Distance Formula

I Can measure the distance between two points on the coordinate plane.

Spark Your Learning

A city has recently redrawn their city map. There are city councilors that represent different areas within the city. The new city map is shown below.

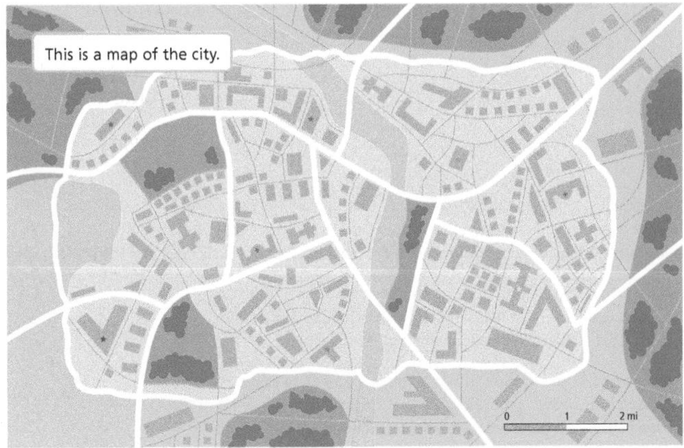

Complete Part A as a whole class. Then complete Parts B–D in small groups.

A. What is a mathematical question you can ask about this situation? What information would you need to know to answer your question? A. Where should the borders be drawn?; the number of councilors and what resources should be divided

B. How do you ensure the entire city is evenly represented? See Additional Answers.

C. What strategy and tool would you use to determine how to divide the city? In what ways could you divide the city? Why? See Strategies 1 and 2 on the facing page.

D. How can you compute the area of each region you drew within the city? See Additional Answers.

Turn and Talk Compare the perimeters of each region of the city. Are the perimeters all the same? Does it make sense for the perimeters to be equal as well as the areas? See margin.

SUPPORT SENSE-MAKING • Three Reads

Tell students to read the information in the photo three times and prompt them with a different question each time.

- What is the situation about? The situation is about a city that has six councilors representing approximately the same area. The city needs to be divided into 6 equal areas.
- What are the quantities in this situation? How are those quantities related? The quantities are the area of the city and the number of councilors assigned to the city; The areas covered by each councilor should be approximately equal.
- What are possible questions you could ask about this situation? Possible answer: What is the area of the city? How can I divide the city into six regions of equal area?

1 Spark Your Learning

▶ MOTIVATE

- Have students look at the diagram in their books and read the information contained in the diagram. Then complete Part A as a whole-class discussion.
- Have students work in small groups to complete Parts B–D.

▶ PERSEVERE

If students need support, guide them by asking:

Q **Advancing • Use Tools** Which tool could you use to solve the problem? Why choose that tool and not some other? Students' choices of tools and reasons for choosing them will vary.

Q **Assessing** How can a coordinate grid help you find area? The area of a region is the number of squares in the region.

Q **Assessing** If a rectangle were drawn around the city, what would the approximate area be? 90 square units

Q **Assessing** How can you approximate the area of the city? I can approximate the number of squares in the rectangle around the city that are not in the city and subtract that number from 90.

Q **Advancing** How does knowing the approximate area of the city help you solve the problem? I can divide the area of the city by six to find the approximate area of each region. Then, I can divide the city into six regions, each with approximately the same area.

Turn and Talk When finding perimeter, students will be adding horizontal and vertical lengths if they divide the area shown on the map into rectangles. Possible answer: No, it does not make sense for the perimeters to be equal because the region is divided into rectangles with different dimensions. The perimeters of these rectangles can vary while the area remains the same.

▶ BUILD SHARED UNDERSTANDING

Select groups of students who used various strategies and tools to share with the class how they solved the problem. As they present their solutions, have each group discuss why they chose a specific strategy and tool.

② Learn Together

Build Understanding

Task 1 (MP) **Use Repeated Reasoning** Students divide a parallelogram into two triangles. They then use the formula for the area of a triangle to find the areas and add them together to find the area of the parallelogram.

CONNECT TO VOCABULARY

Have students use the **Interactive Glossary** to record their understanding of the vocabulary in this task.

Sample Guided Discussion:

Q **In Part B, how can you divide the parallelogram into two triangles so that for each triangle, the base is horizontal and the height is vertical?** Make the base of each triangle the diagonal of the parallelogram on the x-axis.

Turn and Talk Help students understand that moving one vertex of a parallelogram changes the shape of the quadrilateral. The figure will no longer have opposite sides parallel. Have students use a geometric drawing tool to replicate the situation. The height of the bottom triangle changes, but the base length does not change and the top triangle does not change; Use the same process: divide the figure into two triangles, find the area of each, and then find the sum.

Build Understanding

Find Area on the Coordinate Plane

When you calculate the area of a composite figure on a coordinate plane, you can often split the figure into simple shapes with horizontal and vertical sides to make it easier to calculate the individual areas.

1 Find the area of parallelogram $PQRS$.

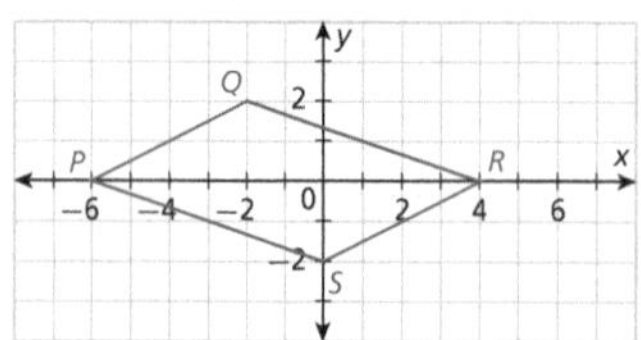

A. How can you use the coordinates of the vertices to find the horizontal distance between P and R? How can you use the coordinates to find the vertical distance between Q and S? What are those distances? A–E. See Additional Answers.

B. How can you divide the parallelogram into shapes that allow you to calculate the total area of the parallelogram using the distances you found in Part A? Describe your reasoning. Find the area.

C. Copy the figure. Draw a line through points P and Q. Suppose $\overline{PQ}$ moves along that line. Draw a parallelogram that shows a possible result of this transformation. Find the area. Compare the two areas and describe your findings.

D. Copy the figure. Draw a line through points Q and R. Suppose $\overline{QR}$ moves along that line. Do you think the result would be the same as when you moved $\overline{PQ}$? Why or why not?

E. The parallelogram in the figure below has a similar shape to parallelogram $PQRS$. Can you use the same strategy to calculate the area of this parallelogram as you did for parallelogram $PQRS$? What makes this orientation easier or more difficult to calculate the area?

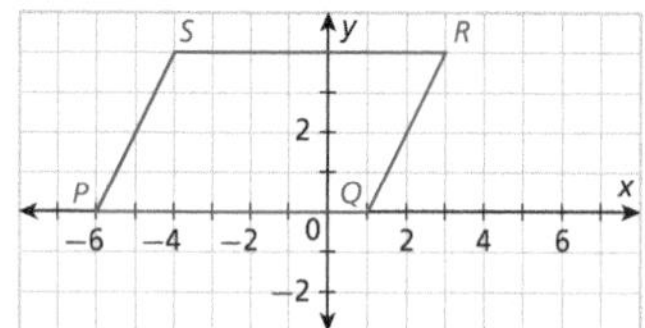

Turn and Talk Suppose S is moved from $(0, -2)$ to $(0, -4)$. What changes? What process can you use to find the area? See margin.

30

LEVELED QUESTIONS

Depth of Knowledge (DOK)	Leveled Questions	What Does This Tell You?
Level 1 **Recall**	What is the formula for the area A of a triangle with base b and height h? $A = \frac{1}{2}bh$	Students' answers will indicate whether they know the formula for the area of a triangle.
Level 2 **Basic Application of Skills & Concepts**	In Part B, give coordinates for the endpoints used to calculate the height of the upper triangle. $(-2, 0)$ and $(-2, 2)$	Students' answers will demonstrate whether they can identify endpoints of segments in the coordinate plane.
Level 3 **Strategic Thinking & Complex Reasoning**	Explain how to find the area of a parallelogram if one of the diagonals is on the y-axis. Possible answer: Divide the parallelogram into two triangles that share the diagonal as a common base. Find the area of each triangle and add.	Students' answers will reflect whether they can reason strategically about how to determine the area of a parallelogram by dividing it into two triangles.

Step It Out

Find Length on the Coordinate Plane

Previously you've used the Pythagorean Theorem to calculate the distance between two points on the coordinate plane. The **Distance Formula** is a variation of the Pythagorean Theorem that can be easily applied when working with coordinates to calculate the distance between points on a coordinate plane.

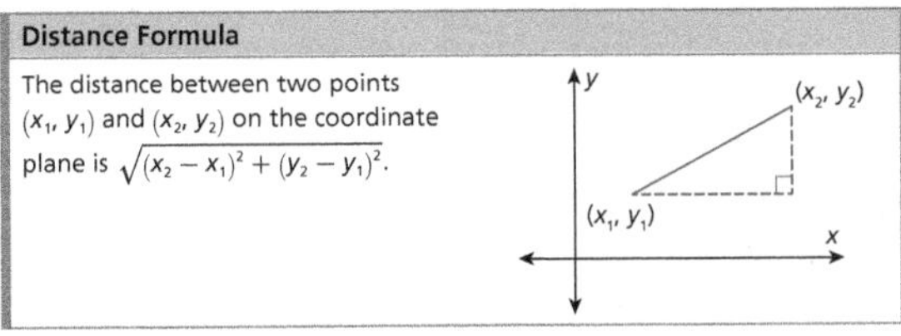

Distance Formula

The distance between two points (x_1, y_1) and (x_2, y_2) on the coordinate plane is $\sqrt{(x_2 - x_1)^2 + (y_2 - y_1)^2}$.

The Distance Formula can be derived using a right triangle where the hypotenuse represents the segment and vertical and horizontal legs are drawn. Substitute expressions for the lengths of the legs into the Pythagorean Theorem.

$a^2 + b^2 = c^2$	Pythagorean Theorem
$(x_2 - x_1)^2 + (y_2 - y_1)^2 = c^2$	Substitute coordinates.
$\sqrt{(x_2 - x_1)^2 + (y_2 - y_1)^2} = c$	Distance Formula

 2 Calculate the perimeter of the triangle, rounded to the nearest hundredth.

A. Identify the coordinates of the vertices. **A. $A(0, 5)$, $B(5, -5)$, $C(-5, -5)$**

B. The calculation for AB is shown. Calculate the length of the each side.

$A(0, 5)$ and $B(5, -5)$	Points
$\sqrt{(x_2 - x_1)^2 + (y_2 - y_1)^2}$	Distance Formula
$= \sqrt{(5 - 0)^2 + (-5 - 5)^2}$	Substitute coordinates for A and B.
$= \sqrt{(5)^2 + (-10)^2}$	Simplify.
$= \sqrt{125} \approx 11.18$	Simplify.

B. $BC = 10$; $AC = 11.18$

C. Calculate the sum of the lengths to find the perimeter. **32.36**

Turn and Talk What happens to the perimeter if you move the figure on the coordinate plane? What happens to the perimeter if you move one side along the line that contains it? How does that compare to your findings about the area from the previous task? See margin.

PROFICIENCY LEVEL

Beginning

Write the Distance Formula and the points $A(5, 3)$ and $B(2, 1)$. Say, "The Distance Formula can be used to find the distance between points A and B." Then write (x_1, y_1) and (x_2, y_2) below the coordinates of A and B and ask students to rewrite the formula, replacing the variables with the appropriate numbers.

Intermediate

Have students work in groups. Give each group an index card. Each card should show two points and their coordinates, such as $A(4, 2)$ and $B(1, 3)$. Ask students to explain how to place the coordinates into the Distance Formula, and then have them calculate the distance between the two points. Have groups switch cards and repeat.

Advanced

Have students explain how to use the Distance Formula to determine the perimeter of a triangle.

Step It Out

Task 2 (MP) **Use Structure** Students use the structure of the coordinate plane to find the perimeter of a triangle using the Distance Formula.

CONNECT TO VOCABULARY

Have students use the **Interactive Glossary** to record their understanding of the vocabulary in this task.

Sample Guided Discussion:

Q How do you find the coordinates? Use the graph to find the missing coordinates.

Q When substituting the coordinates of $\overline{AB}$ into the Distance Formula, which point is (x_1, y_1) and which point is (x_2, y_2)? Possible answer: Point A is (x_1, y_1) and point B is (x_2, y_2).

Q Do you have to use the Distance Formula to find the length of all three sides of the triangle? Explain. no; $\overline{CB}$ is horizonal and the horizontal distance between points (x_1, y_1) and (x_2, y_2) is $|x_2 - x_1|$.

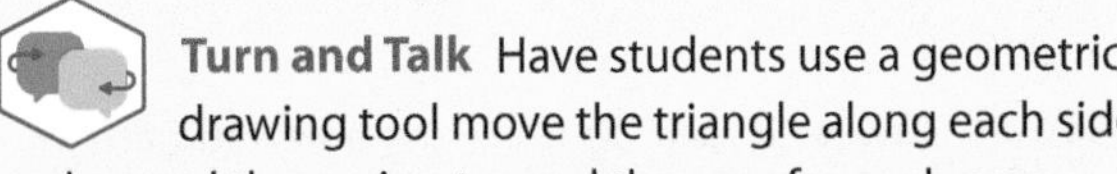

Turn and Talk Have students use a geometric drawing tool move the triangle along each side and record the perimeter and the area for each case. If you move the figure in the coordinate plane without changing the shape, the perimeter does not change; If you move one side along the line that contains it, the perimeter changes, unlike the area, which stays the same.

Task 3 **(MP) Model with Mathematics** Students use one or more shapes to model the irregular shape of a lake and estimate its area.

Sample Guided Discussion:

Q **What kinds of geometric shapes could you use to model different regions of the lake?** Possible answer: triangles, squares, rectangles, trapezoids

Q **Would modeling the lake by using one large rectangle and finding its area give you an accurate estimate? Why or why not?** Possible answer: no; If one rectangle were used to model the lake, the estimated area would be larger than the actual area of the lake because there would be a lot of white space in the rectangle that is not part of the lake.

(EL) OPTIMIZE OUTPUT • Critique, Correct, and Clarify

Have students work with a partner to discuss how the estimate described in Part C of Task 3 is different from their own estimate. Encourage students to use the vocabulary terms *area* and *estimate* in their discussions. Students should revise their estimates if necessary after talking about the problem with their partner.

Turn and Talk To make the area estimate more accurate, students need to reduce the amount of white space within the shapes they use. This can be done by increasing the number of shapes. Remind students that area is a square measure and perimeter is a linear measure. To estimate perimeter, the model should closely resemble the outer boundaries of the lake. Possible answer: I could divide the area into more shapes to make the area estimate more accurate; I would use shapes that match the outline of the lake more closely if I were estimating perimeter.

ANSWERS

C. Possible answer: I used more shapes that more closely match the outline, so my solution is more precise; There is more than one way to estimate the area, but the solutions should all provide answers that are reasonably close.

D. Possible answer: I could divide the area into more/smaller shapes to make the area estimate more accurate. The challenge is choosing and calculating the areas of more and more shapes. The increase in accuracy may not be worth the extra time to make the calculations depending on the needed precision.

Model Area on the Coordinate Plane

3 A group of scientists are observing a lake in order to assess the effects of recent high temperatures on the water level. To do this, they can compare two aerial photos and estimate the loss of lake area. The most recent photo of the lake is shown.

A. Choose one or more shapes to model the area of the lake. A–D. See Additional Answers.

B. Count or use the distance formula to calculate the lengths needed to find the areas of each shape you chose to model the lake. Then add the areas to estimate the total area.

C. Compare your solution to the one below. Which estimate do you think is more accurate? Is there only one solution to estimate an area? Justify your answer.

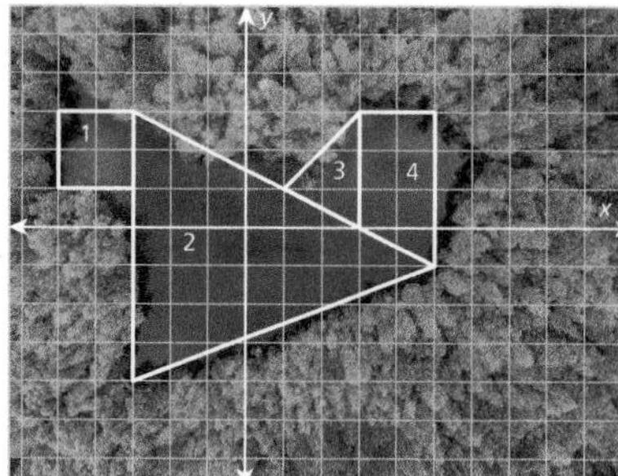

1: $A = lw = (2)(2) = 4$

2: $A = \frac{1}{2}bh = \frac{1}{2}(7)(8) = 28$

3: $A = \frac{1}{2}bh = \frac{1}{2}(3)(2) = 3$

4: $A = \frac{1}{2}(b_1 + b_2)h = \frac{1}{2}(3 + 4)(2) = 7$

Total: $4 + 28 + 3 + 7 = 42$

One estimate of the area is 42 square units.

D. How could you revise your model to make the area estimate more accurate? Discuss the challenges of obtaining a more accurate estimate.

Turn and Talk What shapes would you use to estimate the perimeter of the lake? Can you use the sum of the perimeters of the individual shapes as an estimate of the perimeter of the lake? See margin.

Check Understanding

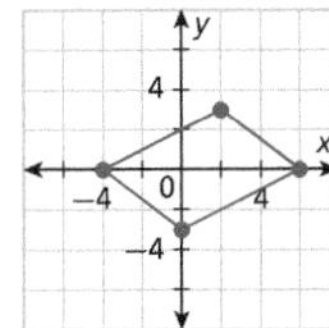

1. Find the area of the parallelogram. $A = 30$ square units
2. Calculate the perimeter of the parallelogram. Round to the nearest hundredth. about 23.42 units
3. Kyle states he can find the distance between two points on the coordinate plane by creating a triangle and using the Pythagorean Theorem. Ava states she can find the distance between the two points using only the coordinates of each point and the Distance Formula. Which student is correct? Explain your reasoning. See Additional Answers.
4. (MP) **Model with Mathematics** Meg walks to school and work each day and wants to track how far she walks each day. In the morning, Meg walks 7 blocks due east to school. After school, she walks 2 blocks north then 4 blocks west to reach work. She walks straight home from work. How far does she walk in all? about 16.6 blocks

On Your Own

Are the areas of the two parallelograms equal? Explain. 5, 6. See Additional Answers.

5. 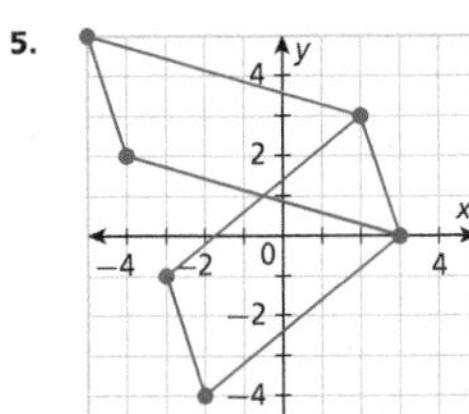

6. 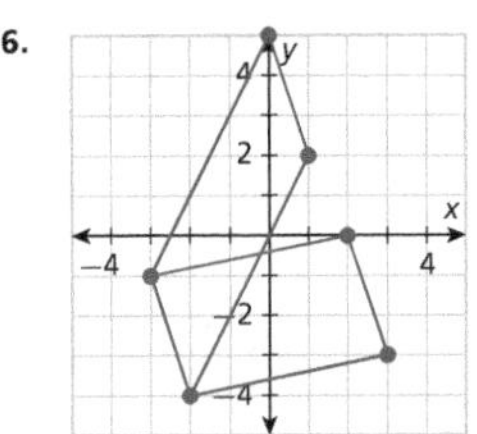

7. Describe a change you could make to a geometric figure so that the perimeter would change but the area would remain the same. See Additional Answers.
8. Use the figure shown.
   A. Which segments are congruent? $\overline{AB} \cong \overline{CD}$ and $\overline{EF} \cong \overline{IJ}$
   B. What are the lengths of the congruent segments?
   $AB = CD = \sqrt{73} \approx 8.54$ units and $EF = IJ = \sqrt{50} \approx 7.07$ units

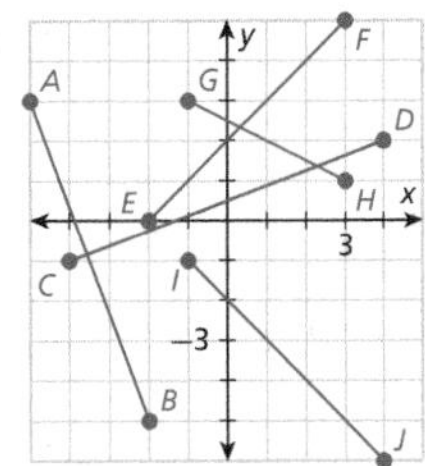

ONLINE

Assign the Digital On Your Own for
- built-in student supports
- Actionable Item Reports
- Standards Analysis Reports

On Your Own

Assignment Guide

The chart below indicates which problems in the On Your Own are associated with each task in the Learn Together. Assign daily homework for tasks completed.

Learn Together Tasks	On Your Own Problems
Task 1, p. 30	Problems 5–7, 15–18, and 25
Task 2, p. 31	Problems 8–14 and 20–23
Task 3, p. 32	Problems 19 and 24

3 Check Understanding

Formative Assessment

Use formative assessment to determine if your students are successful with this lesson's learning objective.

Students who successfully complete the Check Understanding can continue to the On Your Own practice.

For students who miss 1 problem or more, work in a pulled small group using the Almost There small-group activity on page 29C.

ONLINE

Assign the Digital Check Understanding to determine
- success with the learning objective
- items to review
- grouping and differentiation resources

4 Differentiation Options

Differentiate instruction for all students using small-group activities and math center activities on page 29C.

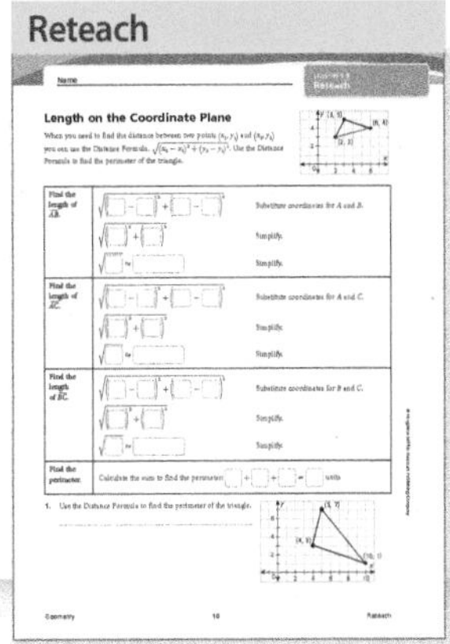

Questioning Strategies

Problem 9 What is the perimeter of an isosceles right triangle if the vertices of the hypotenuse are (0, 0) and (5, 0) and the other vertex is in the 1st quadrant? The other vertex is (2.5, 2.5). The lengths of the legs are each 3.54 units and the length of the hypotenuse is 5 units. The perimeter is 12.08 units.

Watch for Common Errors

Problem 13 Some students may connect the points in the wrong order. Point out that the vertices should be connected in order in which they are listed. Suggest that students label the points, in order, from A to E, so that the figure is pentagon $ABCDE$.

Draw a figure on the coordinate plane that matches the given description. Then calculate the perimeter. 9, 10. Check students' drawings.

9. an isosceles right triangle with endpoints of one leg at (0, 0) and (5, 0) about 17.1 units

10. a triangle that contains the vertices (−2, −1), (4, 3), and (−3, 3). about 18.3 units

Find the perimeter and area of the described figure.

11. a polygon with vertices (−5, −3.5), (0, −2), (2.5, 4), and (−1, 6) $P \approx 26.1$ units; $A = 33.75$ units²

12. a polygon with vertices (−10, −4), (−2, 1), and (−5, 4) $P = 23.1$ units; $A = 19.5$ units²

13. a polygon with vertices (−2, −6), (0, −2), (1, 3), (−1, 3) and (−2, 1) $P = 20.8$ units; $A = 15.5$ units²

14. MP **Critique Reasoning** Braden is trying to find the length of the given line segment to the nearest tenth of a unit. Is his work accurate? If not, identify and explain the error. Then show the correct work and result.

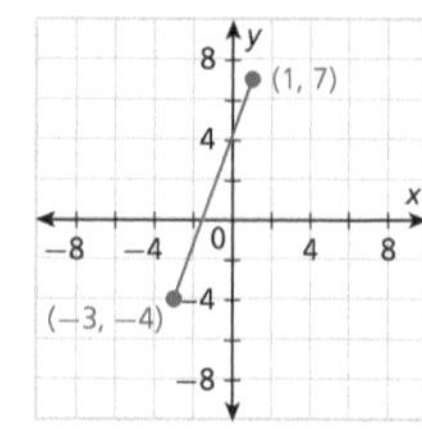

Distance Formula $\sqrt{(x_2 - x_1)^2 + (y_2 - y_1)^2}$

Step 1 $\sqrt{(-3 - (-4))^2 + (1 - 7)^2}$

Step 2 $\sqrt{(1)^2 + (-6)^2}$

Step 3 $\sqrt{1 + 36}$

Step 4 $\sqrt{37} \approx 6.1$

See Additional Answers.

Find the area of each figure.

15.

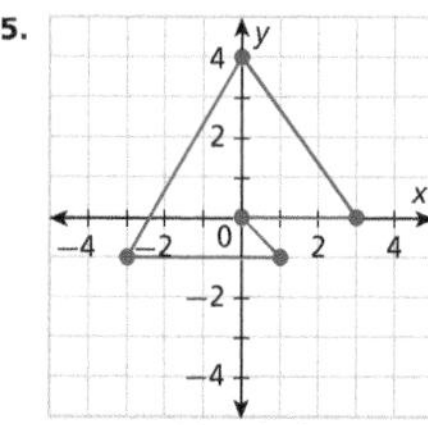

14 units²

16.

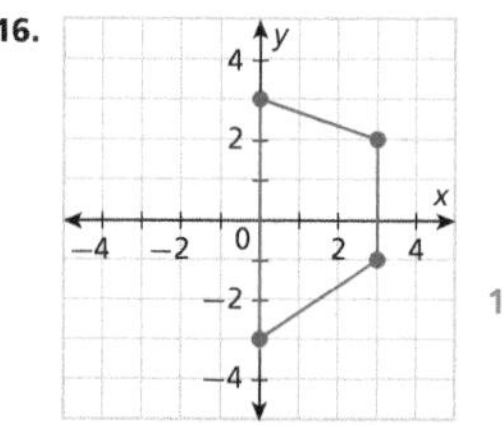

13.5 units²

17.

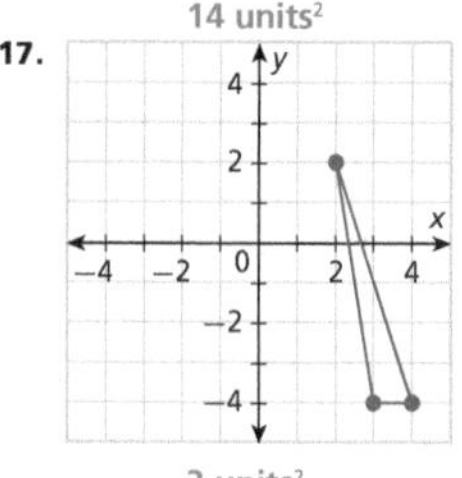

3 units²

18.

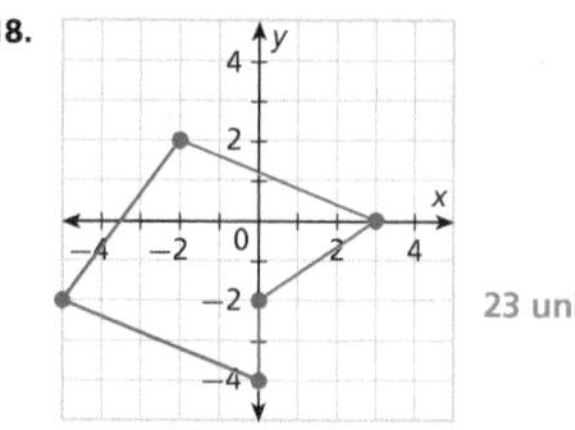

23 units²

19. Sam is landscaping the area around a backyard pool. He has 150 feet of fencing to enclose the area and will be using landscape rock to cover the enclosed area around the pool. Sam's sketch of the situation is shown. Each unit represents 10 feet.

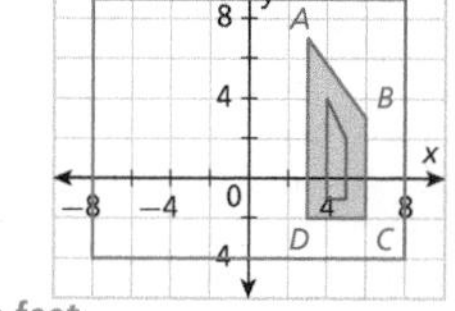

A. Does Sam have the right amount of fencing? How much extra does he have or how much more does he need? no; he needs 70 more feet.

B. How many square feet of rock should he purchase? 1700 ft^2

C. Sam's client is considering moving B and C 20 feet to the right. What are the new coordinates of B and C? What is the perimeter of the expanded fenced area $ABCD$? $B(8, 3)$, $C(8, -2)$; $P \approx 254$ feet

Find the perimeter of each figure.

20. 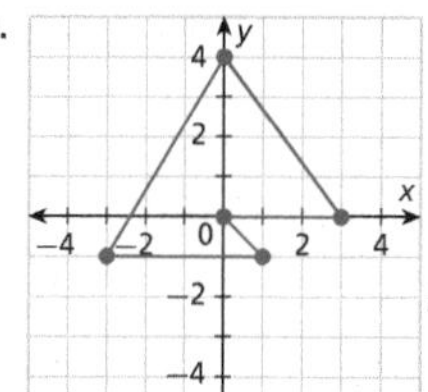

$P \approx 19.2$ units

21. 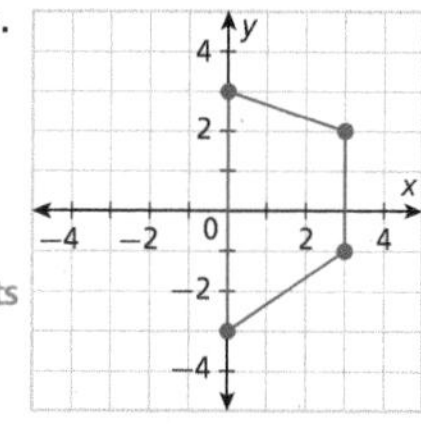

$P \approx 15.8$ units

22. A city planner designs a trail through a new green space. On the plan, the trail begins at $(-4, -0.5)$ and ends at $(5, 1.75)$. The trail has 3 straight segments with endpoints at $(-4, -0.5)$, $(0, 0)$, $(3, 1)$, and $(5, 1.75)$. The city would like to use the trail for an upcoming 10K race. If each unit on the plan represents 1 km, will the trail qualify for a 10K race? Justify your reasoning. See Additional Answers.

23. Aliya is viewing a sign located at the Ferris wheel. She sees the locations of three picnic areas labeled on the map. She wants to eat lunch at the closest picnic area. Which area should she choose?
Area C is the closest.

Watch for Common Errors

Problem 22 Some students may think that the points are vertices of a polygon and find the perimeter. Elicit from students that the trail starts at $(-4, -0.5)$ and ends at $(5, 1.75)$.

Questioning Strategies

Problem 23 Which picnic areas are greater than 5.5 units from Aliya's location? Explain. Area A is 5.83 units from Aliya's location, and Area B is 6.40 units from her location.

5 Wrap-Up

Summarize learning with your class. Consider using the Exit Ticket, Put It in Writing, or I Can scale.

Exit Ticket 

Padma draws a graph of a home plate on a baseball diamond. The graph is a polygon with vertices $A(8.5, 8.5)$, $B(8.5, 0)$, $C(0, -8.5)$, $D(-8.5, 0)$, and $E(-8.5, 8.5)$. What are the perimeter to the nearest whole number and the area to the nearest hundredth, of the home plate? perimeter ≈ 58 units, area ≈ 216.75 square units

Put It in Writing 

Describe some strategies you can use to model area in the coordinate plane.

I Can 

The scale below can help you and your students understand their progress on a learning goal.

4	I can measure the distance between two points on the coordinate plane and explain my solution steps to others.
3	I can measure the distance between two points on the coordinate plane.
2	I can estimate the area of irregular shapes on a coordinate grid.
1	I can decompose figures into smaller shapes to find the area of a figure.

Spiral Review • Assessment Readiness

These questions will help determine if students have retained information taught in the past and can also prepare them for high-stakes assessments. Here, students must recognize complementary angles **(1.2)**, recognize properties of equality **(Alg1, 2.2)**, and recognize if figures are polygons or nonpolygons **(1.3)**.

24. An orchard needs to estimate their crops for next spring. Approximate the area of the region containing orange trees. The roads separate the orchards and each unit represents 50 feet. See Additional Answers.

25. Open Middle™ Using the digits 1 to 9, at most one time each, fill in the boxes to make coordinates for a triangle's vertices with the greatest possible area.

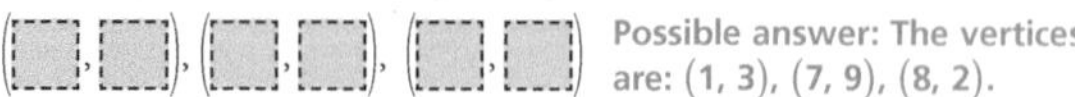

Possible answer: The vertices are: (1, 3), (7, 9), (8, 2).

Spiral Review • Assessment Readiness

26. Which sets of angles are complementary angles? Select all that apply.

Ⓐ 10°, 170°
Ⓑ 55°, 45°
Ⓒ 65°, 25°
Ⓓ 70°, 20°
Ⓔ 40°, 50°
Ⓕ 120°, 60°

27. To start to solve the equation $9x + 3 = 3x$, Mohammed writes $9x - 3x + 3 = 3x - 3x$. Which property of equality did he use?

Ⓐ Subtraction
Ⓑ Division
Ⓒ Multiplication
Ⓓ Addition

28. Determine whether each figure is a polygon.

	A.	B.	C.	D.
Polygon	?	?	?	?
Nonpolygon	?	?	?	?

 I'm in a Learning Mindset!

How did an initial failure with estimating area lead to learning growth when determining more exact areas of figures on the coordinate plane?

Learning Mindset

Perseverance Learns Effectively

Point out that estimating perimeter and area of an irregular shape on the coordinate plane involves modeling the shape with polygons, such as triangles, rectangles, and squares. If students made poor models, have them reflect on what they did wrong and how they could do better. Encourage students to think about the mistakes they made and how they can correct those mistakes. *When you first made a model to estimate the area of an irregular shape, how good was the model? What mistakes did you make? What did you learn from your mistakes? How can you improve the model? How could you apply what you learned to make a model to find the perimeter of an irregular shape? How can you refine the model to make it better?*

Module 1 Review

Segment Addition

Segment Addition Postulate: If points A, B, and C are collinear and B is between A and C, then $AB + BC = AC$.

$$AB + BC = AC$$
$$2x - 5 + 3x + 6 = 46$$
$$5x + 1 = 46$$
$$5x = 45$$
$$x = 9$$
$$AB = 2 \cdot 9 - 5 = 13 \text{ and } BC = 3 \cdot 9 + 6 = 33.$$

Angle Addition

Angle Addition Postulate: If D is in the interior of $\angle ABC$, then $m\angle ABC = m\angle ABD + m\angle CBD$.

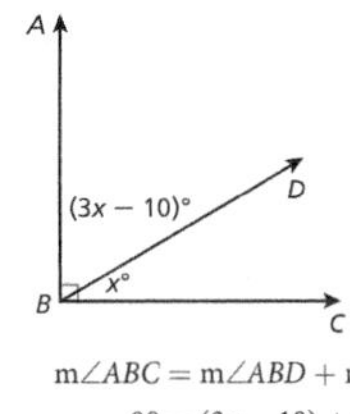

$$m\angle ABC = m\angle ABD + m\angle CBD$$
$$90 = (3x - 10) + x$$
$$90 = 4x - 10$$
$$25 = x$$
$$m\angle ABD = 3 \cdot 25 - 10 = 65° \text{ and } m\angle CBD = 25°.$$

Area

Area Addition Postulate: If a figure is formed by two or more shapes that do not overlap, the area of the figure is the sum of the areas of the individual shapes.

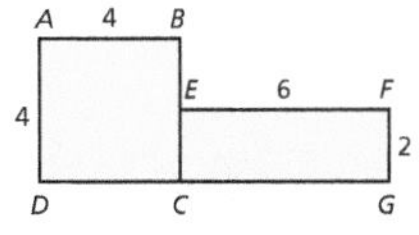

$$\text{Area} = ABCD + EFGC$$
$$= 4 \cdot 4 + 6 \cdot 2$$
$$= 16 + 12$$
$$= 28 \text{ square units}$$

Perimeter

The Distance Formula can be used to find lengths on the coordinate plane.

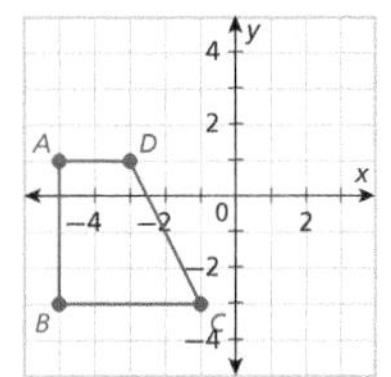

$$DC = \sqrt{(x_2 - x_1)^2 + (y_2 - y_1)^2}$$
$$= \sqrt{(-1 - (-3))^2 + ((-3) - 1)^2}$$
$$= \sqrt{20} \approx 4.5$$
$$\text{Perimeter} \approx 2 + 4 + 4 + 4.5 = 14.5 \text{ units}$$

Module 1 37

ONLINE

Assign the Digital Module Review for

- built-in student supports
- Actionable Item Reports
- Standards Analysis Reports

MODULE 1 REVIEW

Module Review

Use the first page of the Module Review to summarize and connect the main ideas of the module. Use the second page to assess students' understanding of the vocabulary, concepts, and skills presented in the module.

Sample Guided Discussion:

Q Segment Addition Why is it important that *A*, *B*, and *C* are collinear? Possible answer: If the points are not collinear, then $AB + BC$ would be greater than AC because the direct path from A to C is shorter than passing through B.

Q Angle Addition How are $\angle ABD$ and $\angle DBC$ related? Possible answer: They are adjacent complementary angles. **If $\angle ABD$ and $\angle DBC$ were supplementary, how would the solution process change?** I would substitute 180 for the measure of $\angle ABC$ instead of 90.

Q Area What specific kinds of polygons are *ABCD* and *EFGC*, and which has the greater area? How much greater? *ABCD* is a square and *EFGC* is a rectangle. The area of *ABCD* is 16 square units, and the area of *EFGC* is 12 square units; The square is 4 square units greater in area.

Q Distance Formula Explain how you can use the Area Addition Postulate to find the area of trapezoid *ABCD*? Possible answer: Draw $\overline{BD}$. Then add the areas of the resulting triangles. area $ABCD$ = area $\triangle ABD$ + area $\triangle BDC$

Module Review continued

Possible Scoring Guide

Items	Points	Description
1–6	2 each	identifies the correct term(s)
7	2	correctly finds the midpoint of a segment
8	2	correctly finds the measure of two supplementary angles
9	2	correctly finds the population density
10, 11	2 each	correctly finds the area of the polygon
12	2	correctly finds the perimeter of the rectangle
Total points possible = 24 points		

The Unit 1 Test in the Assessment Guide assesses content from Modules 1 and 2.

Vocabulary

Choose the correct term from the box to complete each sentence.

Vocabulary
bisector collinear midpoint *n*-gon point undefined term vertex

1. A(n) __?__ is a(n) __?__; it is a basic figure that is not defined in terms of other figures. point; undefined term
2. The __?__ of a segment divides the segment into two congruent segments. midpoint
3. A polygon with *n* sides is called a(n) __?__. *n*-gon
4. A(n) angle __?__ divides an angle into two congruent parts. bisector
5. Points that lie on the same line are __?__. collinear
6. The common endpoint of the sides of an angle is the __?__. vertex

Concepts and Skills

7. Sam lives exactly halfway between his friends, Danny and Leo. Danny's house is located at the point $(-4, 4)$ and Leo lives at the point $(9, -8)$. At what point is Sam's house located? $\left(\frac{5}{2}, -2\right)$
8. Two angles are supplementary. The measure of $\angle ABD$ is $(2x - 9)°$ and the measure of $\angle CBD$ is $(4x + 12)°$. What are the measurements of the two angles? $m\angle ABD = 50°$ and $m\angle CBD = 130°$
9. On July 4, 2017, the United States of America had a population of 325.7 million people. The total land area of the United States is 3,535,932 square miles. What was the population density of the United States on July 4, 2017? 92.1 people per square mile

Find the area of the figure.

10. 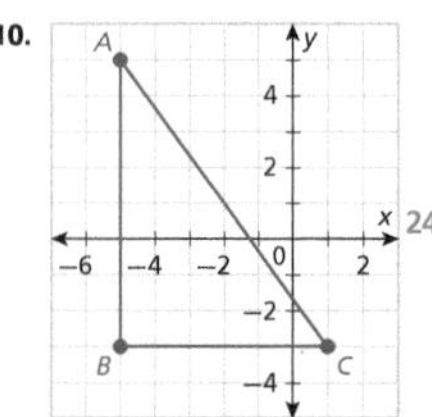

24 square units

11. 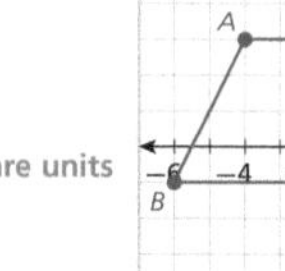

30 square units

12. (MP) **Use Tools** Liam is putting up fence around a garden. He has poles located at $A(7, 7)$, $B(16, 7)$, $C(2, 2)$, and $D(16, 2)$. Each unit on his coordinate grid represents 1 foot. How many feet of fencing does he need to fence in the garden? Round to the nearest foot. State what strategy and tool you will use to answer the question, explain your choice, and then find the answer. 35 feet

38

DATA-DRIVEN INSTRUCTION

Before moving on to the Module Test, use the Module Review results to intervene based on the table below.

MTSS (RtI)

Items	Lesson	DOK	Content Focus	Intervention
7	1.1	3	Find the midpoint of a segment in the coordinate plane.	Reteach 1.1
8	1.2	3	Find the measure of two supplementary angles.	Reteach 1.2
9	1.3	3	Find population density.	Reteach 1.3
10, 11	1.4	3	Find the areas of polygons in the coordinate plane.	Reteach 1.4
12	1.4	3	Find the perimeter of a rectangle using the Distance Formula.	Reteach 1.4

Module Test

The Module Test is available in alternative versions in your Assessment Guide. All items are presented in standardized test formats.

ONLINE

Assign the Digital Module Test to power actionable reports including

- proficiency by standards
- item analysis

Form A

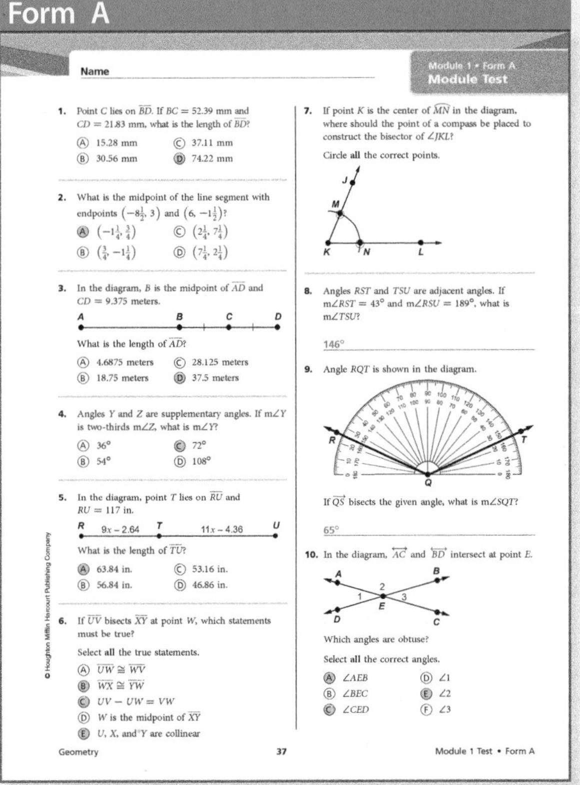

Name

Module 1 • Form A
Module Test

1. Point C lies on $\overline{BD}$. If $BC = 52.39$ mm and $CD = 21.83$ mm, what is the length of $\overline{BD}$?
   (A) 15.28 mm (B) 30.56 mm (C) 37.11 mm (D) 74.22 mm

2. What is the midpoint of the line segment with endpoints $\left(-8\frac{1}{2}, 3\right)$ and $\left(6, -1\frac{1}{2}\right)$?
   (A) $\left(-1\frac{1}{4}, \frac{3}{4}\right)$ (B) $\left(\frac{3}{4}, -1\frac{1}{4}\right)$ (C) $\left(2\frac{1}{4}, 7\frac{1}{4}\right)$ (D) $\left(7\frac{1}{4}, 2\frac{1}{4}\right)$

3. In the diagram, B is the midpoint of $\overline{AD}$ and $CD = 9.375$ meters.
   A B C D
   What is the length of $\overline{AD}$?
   (A) 4.6875 meters (B) 18.75 meters (C) 28.125 meters (D) 37.5 meters

4. Angles Y and Z are supplementary angles. If $m\angle Y$ is two-thirds $m\angle Z$, what is $m\angle Y$?
   (A) 36° (B) 54° (C) 72° (D) 108°

5. In the diagram, point T lies on $\overline{RU}$ and $RU = 117$ in.
   R $9x - 2.64$ T $11x - 4.36$ U
   What is the length of $\overline{TU}$?
   (A) 63.84 in. (B) 56.84 in. (C) 53.16 in. (D) 46.86 in.

6. If $\overline{UV}$ bisects $\overline{XY}$ at point W, which statements must be true?
   Select **all** the true statements.
   (A) $\overline{UW} \cong \overline{WV}$
   (B) $\overline{WX} \cong \overline{YW}$
   (C) $UV - UW = VW$
   (D) W is the midpoint of $\overline{XY}$
   (E) U, X, and Y are collinear

7. If point K is the center of $\widehat{MN}$ in the diagram, where should the point of a compass be placed to construct the bisector of $\angle JKL$?
   Circle **all** the correct points.
   J M K N L

8. Angles RST and TSU are adjacent angles. If $m\angle RST = 43°$ and $m\angle RSU = 189°$, what is $m\angle TSU$?
   146°

9. Angle RQT is shown in the diagram.
   R T Q
   If $\overrightarrow{QS}$ bisects the given angle, what is $m\angle SQT$?
   65°

10. In the diagram, $\overleftrightarrow{AC}$ and $\overleftrightarrow{BD}$ intersect at point E.
   A B 2 1 3 E D C
   Which angles are obtuse?
   Select **all** the correct angles.
   (A) $\angle AEB$ (B) $\angle BEC$ (C) $\angle CED$ (D) $\angle 1$ (E) $\angle 2$ (F) $\angle 3$

© Houghton Mifflin Harcourt Publishing Company

Geometry 37 Module 1 Test • Form A

Form A

Module 1 • Form A
Module Test

Name

11. Natasha built a deck as shown in the diagram. The deck is engineered to support 244 kilograms per square meter. Each unit on the grid represents 1.5 meters.
   How many kilograms can the deck support?
   21,960

12. The diameter of a circle has endpoints (5, 21) and (29, 3).
   Part A
   What is the radius of the circle in units?
   15
   Part B
   What are the coordinates of the center of the circle?
   (17,12)

13. A polygon has vertices $A(5, -1)$, $B(21, 11)$, $C(26, -1)$, and $D(2, -8)$.
   Part A
   What is the perimeter of $ABCD$ in units?
   Possible answer: $58 + \sqrt{58}$
   Part B
   What is the area of $ABCD$ in square units?
   Possible answer: 199.5

14. A rainwater collection system collects water into a holding tank. For every 1 cm of rainfall on a 100 m^2 roof, 1,000 L of water can be collected.
   Part A
   How many liters of water can be collected from 1 cm of rainfall on a 120 m^2 roof?
   1,200
   Part B
   How many centimeters of rainfall are needed to fill a 40,000 L holding tank from a 120 m^2 roof?
   Possible answer: 33.33

15. A farmer is preparing to plant a rectangular plot of land. A body of water divides the land into two regions with the dimensions shown in the diagram.
   625 ft 750 ft 500 ft 250 ft 500 ft
   Part A
   Which measurement is the BEST estimate of the area of the shaded regions?
   (A) 218,750 ft^2 (B) 312,500 ft^2 (C) 531,250 ft^2 (D) 687,500 ft^2
   Part B
   The land will be planted with corn at a density of 0.4 pound per 1,000 ft^2. How many pounds of corn are needed to plant the shaded regions?
   Possible answer: 212.5

© Houghton Mifflin Harcourt Publishing Company

Geometry 38 Module 1 Test • Form A

Form B

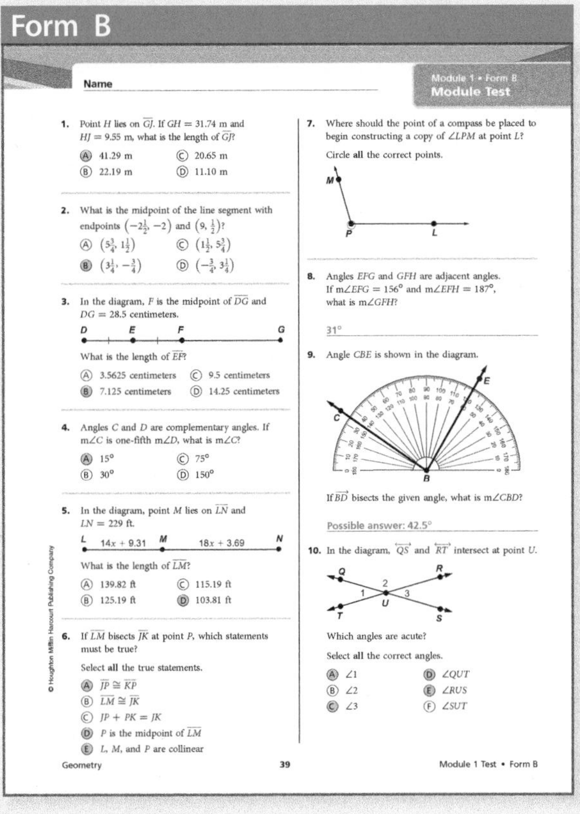

Name

Module 1 • Form B
Module Test

1. Point H lies on $\overline{GJ}$. If $GH = 31.74$ m and $HJ = 9.55$ m, what is the length of $\overline{GJ}$?
   (A) 41.29 m (B) 22.19 m (C) 20.65 m (D) 11.10 m

2. What is the midpoint of the line segment with endpoints $\left(-2\frac{1}{2}, -2\right)$ and $\left(9, \frac{1}{2}\right)$?
   (A) $\left(5\frac{3}{4}, 1\frac{1}{2}\right)$ (B) $\left(3\frac{1}{4}, -\frac{3}{4}\right)$ (C) $\left(1\frac{1}{2}, 5\frac{3}{4}\right)$ (D) $\left(-\frac{3}{4}, 3\frac{1}{4}\right)$

3. In the diagram, F is the midpoint of $\overline{DG}$ and $DG = 28.5$ centimeters.
   D E F G
   What is the length of $\overline{EF}$?
   (A) 3.5625 centimeters (B) 7.125 centimeters (C) 9.5 centimeters (D) 14.25 centimeters

4. Angles C and D are complementary angles. If $m\angle C$ is one-fifth $m\angle D$, what is $m\angle C$?
   (A) 15° (B) 30° (C) 75° (D) 150°

5. In the diagram, point M lies on $\overline{LN}$ and $LN = 229$ ft.
   L $14x + 9.31$ M $18x + 3.69$ N
   What is the length of $\overline{LM}$?
   (A) 139.82 ft (B) 125.19 ft (C) 115.19 ft (D) 103.81 ft

6. If $\overline{LM}$ bisects $\overline{JK}$ at point P, which statements must be true?
   Select **all** the true statements.
   (A) $\overline{JP} \cong \overline{KP}$
   (B) $\overline{LM} \cong \overline{JK}$
   (C) $JP + PK = JK$
   (D) P is the midpoint of $\overline{LM}$
   (E) L, M, and P are collinear

7. Where should the point of a compass be placed to begin constructing a copy of $\angle LPM$ at point L?
   Circle **all** the correct points.
   M P L

8. Angles EFG and GFH are adjacent angles. If $m\angle EFG = 156°$ and $m\angle EFH = 187°$, what is $m\angle GFH$?
   31°

9. Angle CBE is shown in the diagram.
   E C B
   If $\overrightarrow{BD}$ bisects the given angle, what is $m\angle CBD$?
   Possible answer: 42.5°

10. In the diagram, $\overleftrightarrow{QS}$ and $\overleftrightarrow{RT}$ intersect at point U.
   Q R 2 1 3 U T S
   Which angles are acute?
   Select **all** the correct angles.
   (A) $\angle 1$ (B) $\angle 2$ (C) $\angle 3$ (D) $\angle QUT$ (E) $\angle RUS$ (F) $\angle SUT$

© Houghton Mifflin Harcourt Publishing Company

Geometry 39 Module 1 Test • Form B

Form B

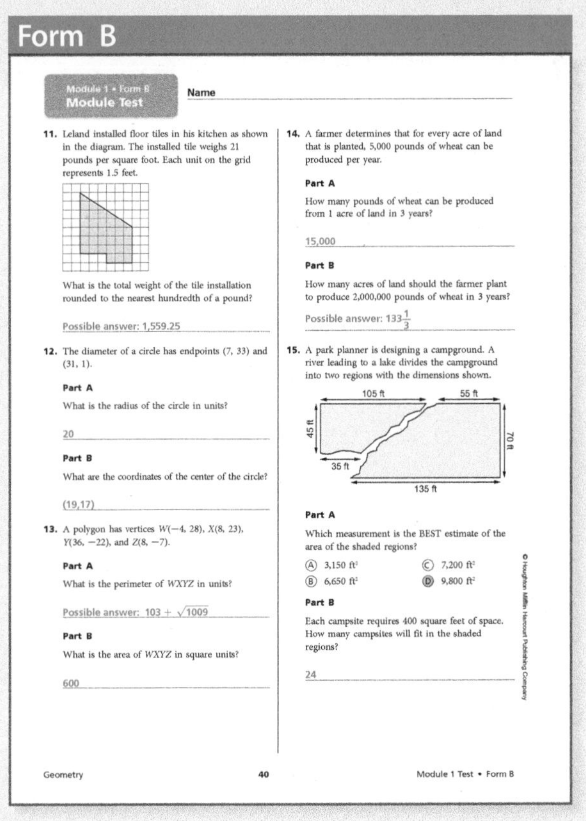

Module 1 • Form B
Module Test

Name

11. Leland installed floor tiles in his kitchen as shown in the diagram. The installed tile weighs 21 pounds per square foot. Each unit on the grid represents 1.5 feet.
   What is the total weight of the tile installation rounded to the nearest hundredth of a pound?
   Possible answer: 1,559.25

12. The diameter of a circle has endpoints (7, 33) and (31, 1).
   Part A
   What is the radius of the circle in units?
   20
   Part B
   What are the coordinates of the center of the circle?
   (19,17)

13. A polygon has vertices $W(-4, 28)$, $X(8, 23)$, $Y(36, -22)$, and $Z(8, -7)$.
   Part A
   What is the perimeter of $WXYZ$ in units?
   Possible answer: $103 + \sqrt{1009}$
   Part B
   What is the area of $WXYZ$ in square units?
   600

14. A farmer determines that for every acre of land that is planted, 5,000 pounds of wheat can be produced per year.
   Part A
   How many pounds of wheat can be produced from 1 acre of land in 3 years?
   15,000
   Part B
   How many acres of land should the farmer plant to produce 2,000,000 pounds of wheat in 3 years?
   Possible answer: $133\frac{1}{3}$

15. A park planner is designing a campground. A river leading to a lake divides the campground into two regions with the dimensions shown.
   105 ft 55 ft 45 ft 70 ft 35 ft 135 ft
   Part A
   Which measurement is the BEST estimate of the area of the shaded regions?
   (A) 3,150 ft^2 (B) 6,650 ft^2 (C) 7,200 ft^2 (D) 9,800 ft^2
   Part B
   Each campsite requires 400 square feet of space. How many campsites will fit in the shaded regions?
   24

© Houghton Mifflin Harcourt Publishing Company

Geometry 40 Module 1 Test • Form B

TOOLS FOR REASONING AND PROOF

Introduce and Check for Readiness

- Module Performance Task • Are You Ready?

Lesson 2.1—2 Days

Write Conditional Statements

Learning Objective: Write a conditional statement and related conditional statements and determine whether the statements are true.

New Vocabulary: biconditional statement, conditional statement, conjecture, contrapositive, converse, counterexample, definition, inverse

Lesson 2.2—2 Days

Use Inductive and Deductive Reasoning

Learning Objective: Differentiate between inductive and deductive reasoning and apply deductive reasoning in the context of geometric proofs.

New Vocabulary: deductive reasoning, inductive reasoning, proof, theorem

Lesson 2.3—2 Days

Write Proofs about Segments

Learning Objective: Use congruence and the Segment Addition Postulate to complete proofs about segments.

New Vocabulary: symbolic notation

Review Vocabulary: congruent, line segment

Lesson 2.4—2 Days

Write Proofs about Angles

Learning Objective: Apply proof concepts to situations and theorems involving angles.

Review Vocabulary: linear pair, vertical angles

Assessment

- Module 2 Test (Forms A and B)
- Unit 1 Test (Forms A and B)

LEARNING ARC FOCUS

Build Conceptual Understanding 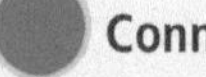 Connect Concepts and Skills  Apply and Practice

TEACHING FOR SUCCESS

TEACHING FOR DEPTH: Tools for Reasoning and Proof

Make Connections. It is important for students to recognize that mathematics involves more than just mastering and applying procedural skills. In this module, students extend their mathematical experiences by learning about conditional statements and viable arguments.

Unlike procedural skill problems, proving conjectures in Euclidean geometry requires logic and reason. To that end, students learn how to apply known definitions and theorems to prove conjectures about mathematical objects: that is, given an hypothesis, they learn how to prove that a conclusion is true.

Students will begin the module by learning about conditional, or if-then, statements. Starting with a conditional statement, students learn how to write its converse, inverse, and contrapositive and verify whether each is true or false. These concepts lead into learning the distinction between inductive and deductive reasoning and how each can be used to establish a desired conclusion. In the last two lessons, students focus on definitions related to segments and angles and learn how to apply deductive reasoning to complete formal proofs of theorems about them.

Mathematical Progressions

Prior Learning	Current Development	Future Connections
Students: • solved basic linear equations and literal equations. • used linear equations and inequalities in one variable to model and solve real-world problems. • solved inequalities in one variable and graphed the solutions. • graphed linear equations and linear inequalities in two variables. • solved systems of linear equations. • solved quadratic equations in one variable.	**Students:** • define and write conditional and biconditional statements. • use conditional statements to establish whether statements are true or false. • define and contrast inductive and deductive reasoning. • use deductive reasoning to write and understand proofs. • apply the properties of congruence to relationships among segments and angles. • prove theorems about segments and angles.	**Students:** • will apply and prove theorems about angles to prove lines in the plane are parallel or perpendicular. • will apply theorems about segments and angles to prove triangles are congruent or similar. • will apply theorems about segments and angles to establish polygonal properties and theorems. • apply relationships between segments and angles in right triangles to establish the trigonometric ratios.

TEACHER TO TEACHER

From the Classroom

Pose purposeful questions. Many students who are enrolled in a formal plane geometry course are not yet ready for deductive proofs. Some students don't see the need to prove anything, and even if they do, some just "don't get it," as hard as they may try.

One way to approach how to prove a theorem deductively is to start by asking students what they have to prove. Ask "What conditions must be true for this statement to be true?" "Is there a figure we can draw to model the situation?" "What do we know about the given situation?" For example, to prove that vertical angles are congruent, draw two intersecting lines and label the pairs of vertical angles 1 and 3, and 2 and 4. Looking at angles 1 and 2, ask what they know about these angles. Looking at angles 2 and 3, ask what they know about these angles. "What do you know about linear pairs and supplementary angles?" Emphasize that angles supplementary to the same angle are congruent so vertical angles are therefore, congruent.

Another way to approach proving a theorem deductively is to prove it indirectly. Start by assuming the desired conclusion is false. For example, if the conclusion is that vertical angles are congruent, ask: "What is the opposite of this conclusion?" Answer: "Vertical angles are not congruent." Then show that this assumption leads to the contradiction of a known fact.

LANGUAGE DEVELOPMENT • Planning for Instruction

By giving all students regular exposure to language routines in context, you will provide opportunities for students to **listen**, **speak**, **read**, and **write** about mathematical situations and develop both mathematical language and conceptual understanding at the same time.

Using Language Routines to Develop Understanding

Use the Professional Cards for the following routines to plan for effective instruction.

Co-Craft Questions Lesson 2.2

Students think of natural questions to ask about a given situation or problems similar to a given task and answer the questions they have developed or problems they have created.

Three Reads Lessons 2.1, 2.3, and 2.4

Students read a problem three times with a specific focus each time.

1st Read What is the situation about?
2nd Read What are the quantities in the situation?
3rd Read What are the possible mathematical questions that we could ask for the situation?

Information Gap Lesson 2.3

Students recognize when information given in a problem situation is incomplete, and they pose questions and share knowledge with others to discover any missing facts or relationships and work together to solve the problem.

Critique, Correct, and Clarify Lesson 2.2

Students correct the work in a flawed explanation, argument, or solution method and share with a partner and refine the sample work.

Connecting Language to Tools for Reasoning and Proof

Watch for students' use of the review and new terms listed below as they explain their reasoning and make connections with new concepts.

Linguistic Note

Listen for the vocabulary that students use when working in this module. Many mathematical terms are new and unfamiliar, such as *conditional statements* and *deductive reasoning*, but are at the heart of geometry and proof.

Key Academic Vocabulary

Current Development • Review and New Vocabulary

linear pair a pair of adjacent angles whose noncommon sides are opposite rays

vertical angles nonadjacent angles formed by two intersecting lines

conditional statement a statement that can be written in the form "if p, then q," where p is the hypothesis of the statement, and q is the conclusion

contrapositive a statement formed by both exchanging and negating the hypothesis and conclusion of a conditional statement; if not q, then not p.

converse a statement formed by exchanging the hypothesis and conclusion of a conditional statement; if q, then p.

deductive reasoning the use of facts, definitions, and logic to prove a statement is true

Inductive reasoning showing that a statement is true by looking at a specific case or cases

inverse a statement formed by negating the hypothesis and conclusion of a conditional statement; if not p, then not q.

postulate a statement that is accepted as true and does not need to be proven

proof an argument that uses factual statements and logic to arrive at a conclusion

theorem a statement that can be proved to be true

Module

2

Tools for Reasoning and Proof

Module Performance Task: Focus on STEM

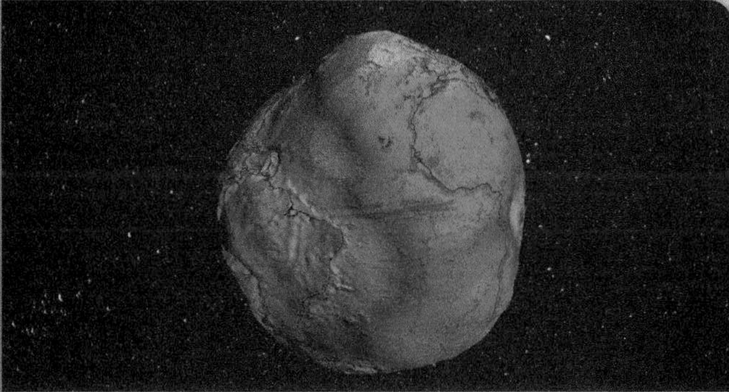

(bg) ©gutesk/Shutterstock; (inset) ©European Space Agency/HPF/DLR/Science Source

Scientific Reasoning

The image above shows a geoid model of Earth, which depicts gravitational anomalies at the surface. Scientists often display quantitative data through pseudocolor images, where information is visualized by mapping a functions magnitude to a specific color.

The graph below shows the relationship between mass (in kilograms) and weight (in newtons) as measured at the Amundson-Scott South Pole Station.

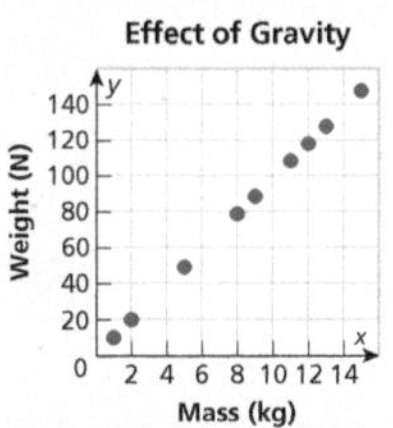

A–D. See margin.

A. Form a conjecture from the data. Use evidence to explain your reasoning.

B. Write a conditional statement, an if-then statement, to test your conjecture.

C. If additional data supports your conjecture, are you proven correct? Explain.

D. A measurement taken at Mount Nevado Huascaran in Peru shows a mass of 10 kilograms has a weight of 97.6 N. Does this prove your conjecture wrong? How can you refine your conjecture to account for this data?

Module 2 39

Mathematical Connections

Task Part	Prerequisite Skills
Part A	Students should be familiar with the equation of a linear function: $y = mx + b$ **(Algebra 1, 2.4)**.
Parts B and C	Students should be able to examine a set of ordered pairs (x, y) and write a linear equation that describes their relationship, such as $y = mx + b$. Students should also recognize that if the coordinates of a point do not lie on the graph of a function, then the coordinates do not satisfy the equation of the function **(Algebra 1, 2.4)**.
Part D	Students should understand that if data from a new observation do not satisfy a given algebraic relationship, then either the equation is not correct or the data are suspect or incorrect **(Algebra 1, 2.4)**.

Scientific Reasoning

Overview

Scientific reasoning involves the formation of theories that explain observed phenomena. Scientists in all fields begin by stating a hypothesis that describes an observed phenomenon and collecting evidence by performing repeated observations and/or experiments that either establish the validity of the hypothesis or refute it.

Career Connections

Suggest that students research a career in *geodesy*, the branch of applied mathematics that determines size, shape and coordinates on the surface of the Earth and that also includes the study of Earth's gravitational field and rotation.

Answers

A. Possible answer: The weight of an object in newtons is about 9.8 times the mass in kilograms. The y-values on the graph increase by 9.8 N for every 1 kg increase in the x-value.

B. Possible answer: If the mass of an object is x kilograms, then the weight is $9.8x$ newtons.

C. Possible answer: No, a conjecture cannot be proven correct by a finite number of specific cases. A conjecture can be proven true by a logical argument, and proven wrong by a counterexample or a logical argument.

D. Possible answer: Yes, if a counterexample can be found, then the conjecture is not true; I can refine my conjecture to include the specific location where measurements were taken.

ONLINE

Assign the Digital Are You Ready? to power actionable reports including

- proficiency by standards
- item analysis

Are You Ready?

Diagnostic Assessment

- Diagnose prerequisite mastery.
- Identify intervention needs.
- Modify or set up leveled groups.

Have students complete the *Are You Ready?* assessment on their own. Items test the prerequisites required to succeed with the new learning in this module.

Justify Steps for Solving Equations Students will apply the properties of equality to solve linear equations in one variable.

The Pythagorean Theorem and its Converse Students will apply the Pythagorean Theorem, $a^2 + b^2 = c^2$, to find the length of a side of a right triangle given the lengths of the other two sides.

Angle Relationships in Triangles Students will find the measure of the third angle in a triangle given the measures of the other two angles or solve for x, given algebraic expressions in terms of x for the measures of the three angles.

ONLINE Ed

Are You Ready?

Complete these problems to review prior concepts and skills you will need for this module.

Justify Steps for Solving Equations

Solve the equation. Justify each step of the solution. 1–4. See Additional Answers.

1. $4x + 7 = 39$
2. $\frac{5}{8}t = 2\frac{1}{2}$
3. $6m - 11 = 2m + 13$
4. $0.4(c - 2) = -1.6$

The Pythagorean Theorem and Its Converse

Find the missing side length of the right triangle.

5. $a = 5, b = 12, c = ?$ $c = 13$
6. $a = 24, b = 7, c = ?$ $c = 25$
7. $a = 6, b = ?, c = 10$ $b = 8$
8. $a = ?, b = 15, c = 17$ $a = 8$

Use the Converse of the Pythagorean Theorem to determine if the triangle is a right triangle.

9. $a = 8, b = 6, c = 10$ right triangle
10. $a = 2, b = 7, c = 8$ not a right triangle

Angle Relationships in Triangles

11. A triangle has angles that measure 40° and 80°. What is the measure of the third angle of the triangle? 60°
12. The three angle measures of a triangle are $(2x)°$, $(x + 5)°$, and $(5x - 25)°$. Solve for x. $x = 25$
13. The three angle measures of a triangle are $(x - 5)°$, $(x + 30)°$, and $(2x - 5)°$. Solve for x. $x = 40$
14. A triangle has angles that measure 35° and 95°. What is the measure of the opposite exterior angle? 130°

Connecting Past and Present Learning

Previously, you learned:

- to determine horizontal and vertical lengths on the coordinate plane,
- to measure and classify angles, and
- to classify polygons by the number of sides.

In this module, you will learn:

- to add the lengths of segments and show segments are congruent,
- to show that angles are congruent, and
- to prove that statements are true.

40

DATA-DRIVEN INTERVENTION

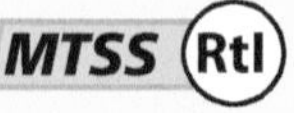

Concept/Skill	Objective	Prior Learning *	Intervene With
Justify Steps for Solving Equations	Use properties to justify each step when solving a linear equation.	Algebra 1, Lesson 2.2	• Tier 2 Skill 3 • Reteach, Algebra 1, Lesson 2.2
The Pythagorean Theorem and its Converse	Use the Pythagorean Theorem and its converse to solve problems.	Grade 8, Lessons 11.1–11.3	• Tier 2 Skill 4 • Reteach, Grade 8 Lessons 11.1–11.3
Angles Relationships in Triangles	Use the Triangle Sum Theorem and the Exterior Angle Theorem to solve problems.	Grade 8, Lesson 4.1	• Tier 2 Skill 5 • Reteach, Grade 8 Lesson 4.1

* Your digital materials include access to resources from Grade 6–Algebra 2. The lessons referenced here contain a variety of resources you can use with students who need support with this content.

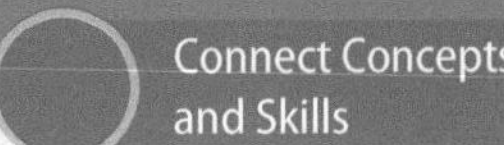

2.1 Write Conditional Statements

LESSON FOCUS AND COHERENCE

Mathematics Standards

- Know precise definitions of angle, circle, perpendicular line, parallel line, and line segment, based on the undefined notions of point, line, distance along a line, and distance around a circular arc.

Mathematical Practices and Processes (MP)

- Reason abstractly and quantitatively.
- Construct viable arguments and critique the reasoning of others.

I Can Objective 

I can write conditional statements and related conditional statements.

Learning Objective

Write a conditional statement and related conditional statements, and determine whether the statements are true.

Language Objective

Write a biconditional statement from a mathematical definition.

Vocabulary

New: biconditional statement, **conditional statement**, **conjecture**, **contrapositive**, **converse**, **counterexample**, **definition**, **inverse**

Mathematical Progressions

Prior Learning	Current Development	Future Connections
Students: • proved the Pythagorean Theorem. **(Gr 8, 11.1)**	**Students:** • understand basic postulates used to study geometry. • write conditional statements in the form "if *p*, then *q*." • write the converse, inverse, and contrapositive of conditional statements and determine if the statements are true. • rewrite definitions as biconditional statements.	**Students:** • will write proofs about triangles and quadrilaterals. **(12.1, 12.3)**

PROFESSIONAL LEARNING

Math Background

An important goal of this lesson is to help students make sense of conditional statements. Much of geometry involves proofs and theorems. To understand how proofs and theorems work, there must be an understanding of the relationship between the hypothesis and the conclusion. The use of an if-then relationship is a common form of argument that is used not just in geometry but many other areas of math, science, and everyday life. For example, "Given the equation $y = 2x + 3$, if $x = 3$, then $y = 9$" or "If you don't study, you will not do well on the exam."

WARM-UP OPTIONS

ONLINE

PROJECTABLE & PRINTABLE 

ACTIVATE PRIOR KNOWLEDGE • Prove the Pythagorean Theorem

Use these activities to quickly assess and activate prior knowledge as needed.

Problem of the Day 

Consider the figure. Find the area of the four triangles, the inner square, and the entire square. Use the areas to prove the Pythagorean Theorem: $a^2 + b^2 = c^2$.

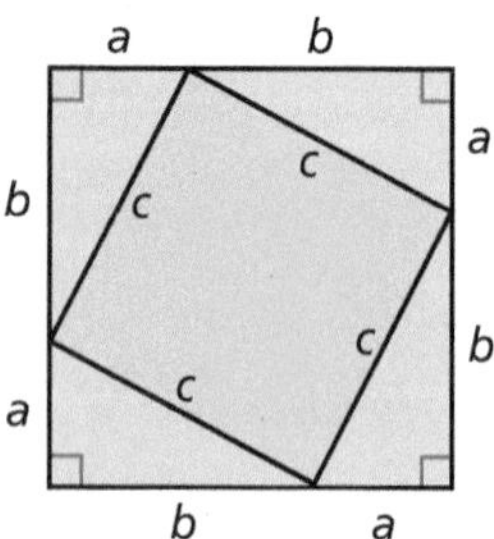

Four triangles: $4\left(\frac{1}{2} \cdot b \cdot a\right) = 2ab$

Inner square: c^2

Outer square: $(a + b)(a + b) = a^2 + 2ab + b^2$

Proof: $2ab + c^2 = a^2 + 2ab + b^2$

$2ab + c^2 - 2ab = a^2 + 2ab + b^2 - 2ab$

$c^2 = a^2 + b^2$

Quick Check for Homework

As part of your daily routine, you may want to display the Teacher Solution Key to have students check their homework.

Make Connections

Based on students' responses to the Problem of the Day, choose one of the following:

1. Project the Interactive Reteach, Grade 8, Lesson 11.1.
2. Complete the Prerequisite Skills Activity:

Have students examine the right triangle in the figure.

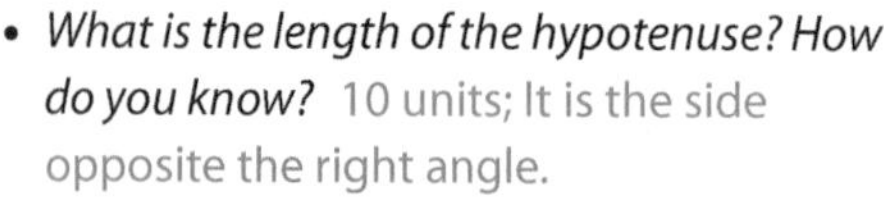

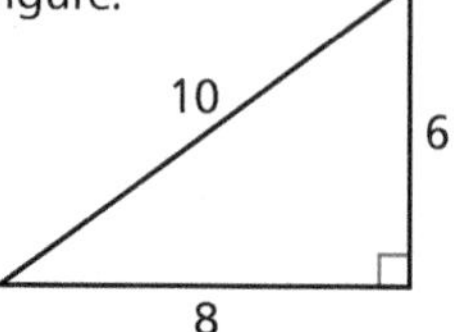

- *What is the length of the hypotenuse? How do you know?* 10 units; It is the side opposite the right angle.
- *What are the lengths of the legs of the triangle?* The legs have lengths of 8 units and 6 units.
- What do the variables in the Pythagorean Theorem $a^2 + b^2 = c^2$ represent? The a and b represent the legs of the triangle and c represents the hypotenuse.
- Substitute the values from the triangle into the Pythagorean Theorem to confirm the theorem is true for this triangle.

$a^2 + b^2 = c^2$

$8^2 + 6^2 = 10^2$

$64 + 36 = 100$

$100 = 100$

If students continue to struggle, use Tier 2 Skill 4.

SHARPEN SKILLS

If time permits, use this on-level activity to build fluency and practice basic skills.

Vocabulary Review

Objective: Students use a graphic organizer to review the definitions of point, line, and plane.
Materials: Word Description (Teacher Resource Masters)

Have students work in groups of three. Provide each student with a graphic organizer. Assign each student one of the three vocabulary review words and have them complete the graphic organizer for the term they were assigned.

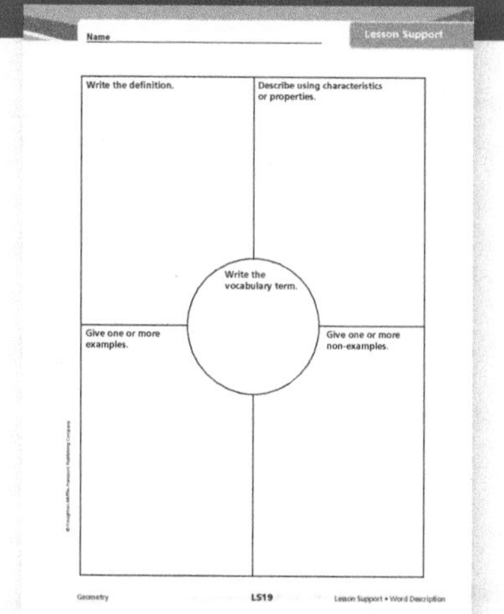

PLAN FOR DIFFERENTIATED INSTRUCTION

Small-Group Options

Use these teacher-guided activities with pulled small groups.

On Track

Give students this slogan for a restaurant chain: "Your meal is free when it's your birthday." Have students write the conditional statement, and its converse, inverse, and contrapositive statement. Students should discuss whether each statement is true or false, and which statement would be the best for the restaurant to use.

Almost There

Materials: index cards

Write the statement "If a shape has three sides, then the shape is a triangle." Give students four index cards and have them do the following:

- Determine the hypothesis of the statement and write in on an index card. Write *p* on the other side of the card.
- Determine the conclusion of the statement and write in on an index card. Write *q* on the other side of the card.
- Write the word *not* on the other two index cards.
- Use the cards to help write the converse, inverse, and contrapositive of the conditional statement.

Ready for More

Write the beginning of the conditional statement "If a shape has four sides, then ________." Have students write a conclusion for which the following are true:

- The conditional statement is true.
- The inverse is false.
- The converse is true but the contrapositive is false.
- The converse is true but the inverse is false.
- The biconditional statement is true.

Math Center Options

Use these student self-directed activities at centers or stations. Key: ● Print Resources ● Online Resources

On Track

- ● Interactive Digital Lesson
- ● Interactive Glossary (printable): **counterexample**, **conjecture**, **conditional statement**, **converse**, **inverse**, **contrapositive**, **biconditional statement**, **definition**
- ●● Journal and Practice Workbook

Almost There

- ● Reteach 2.1 (printable)
- ● Interactive Reteach 2.1
- ● RtI Tier 2 Skill 4: The Pythagorean Theorem and Its Converse

Ready for More

- ● Challenge 2.1 (printable)
- ● Interactive Challenge 2.1

Unit Project Check students' progress by asking to see their biconditional statement.

View data-driven grouping recommendations and assign differentiation resources.

Analyze Given Information

Strategy 1

The if-then statements below are true.

If the group is using fireworks, then the group is assessed a fine of $300.

If the group is littering, smoking, or fishing, then the group is assessed a fine of $100.

If only the $300 fine is known, then the group could have been using fireworks or they could have been littering, smoking, and fishing, but it is impossible to know without more information.

If students . . . indicate that there is uncertainty in knowing which rule was broken, they are employing an analytical method and demonstrating attention to detail and precision.

Have these students . . . explain how they determined their solution. **Ask:**

- Q How did you know there was not one solution?
- Q What would be one way to know which rule(s) the group broke?

Scan Given Information

Strategy 2

The if-then statement below is true.

If the group is using fireworks, then the group is assessed a fine of $300.

This means that if the group is assessed a fine of $300, then the group was using fireworks.

If students . . . choose the first activity that goes with a $300 fine, they understand the relationship between the fines and the prohibited activity, but they fail to acknowledge that there is another reason a $300 fine might be assessed.

Activate prior knowledge . . . by asking students to write their solution using an if-then statement. **Ask:**

- Q What information is given? What is the *if* part of the statement?
- Q Could you draw any other conclusions using just the *if* part of your statement?

COMMON ERROR: Reads Only Top of Sign

The group was fined for being in the park after hours.

If students . . . choose the top rule of the sign, they have not read the entire sign.

Then intervene . . . by pointing out that there is not a fine associated with being in the park after hours and that only specific activities have fines associated with them. **Ask:**

- Q What are the prohibited activities? What is the consequence of engaging in a prohibited activity?
- Q What fine is associated with being in the park after hours?

2.1 Write Conditional Statements

I Can write conditional statements and related conditional statements.

Spark Your Learning

Clara meets with friends at a public park. They notice another group being issued a fine and check the posted rules and regulations.

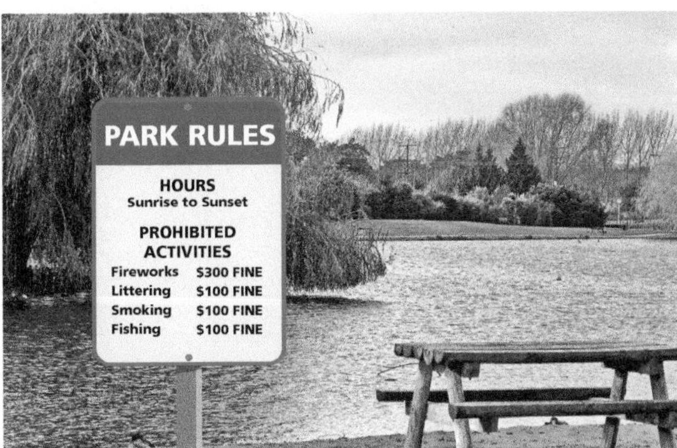

PARK RULES
HOURS
Sunrise to Sunset
PROHIBITED
ACTIVITIES
Fireworks $300 FINE
Littering $100 FINE
Smoking $100 FINE
Fishing $100 FINE

©Kenneth William Caleno/Shutterstock

Complete Part A as a whole class. Then complete Parts B–D in small groups.

A. What is a mathematical question you can ask about this situation? What information would you need to know to answer your question? A. What rule did the group break?; amount of fine

B. How can you rewrite each rule as an if-then statement? Is each statement still true if you swap the words in the "if" part with the words in the "then" part? Explain your reasoning. B. See Additional Answers.

C. To answer your question, what strategy and tool would you use along with all the information you have? What answer do you get? C. See Strategies 1 and 2 on the facing page.

D. How could you rewrite one of the statements using the word "not" either once or twice to make a new true statement? Explain your reasoning. D. See Additional Answers.

 **Turn and Talk** How would your answer change if the group was fined $100? See margin.

CULTIVATE CONVERSATION • Three Reads

Tell students to read the information in the photo three times and prompt them with a different question each time.

1. What is the situation about? The situation is about the rules and regulations of a park and the consequences for being caught breaking them.
2. What are the components in this situation? How are those components related? The components are the prohibited activities and the fines; Each activity has a specific fine that must be paid if you are caught breaking it.
3. What are possible questions you could ask about this situation? Possible answer: Can a group be fined for breaking more than one rule? Why are there different consequences for breaking different rules?

1 Spark Your Learning

▶ MOTIVATE 

- Have students look at the photo in their books and read the information contained in the photo. Then complete Part A as a whole-class discussion.
- Give the class the additional information they need to solve the problem. This information is available online as a printable and projectable page in the Teacher Resources.
- Have students work in small groups to complete Parts B–D.

▶ PERSEVERE

If students need support, guide them by asking:

Q **Advancing • Use Tools** Which tool could you use to solve the problem? Why choose that tool and not some other? Students' choices of tools and reasons for choosing them will vary.

Q **Assessing** What happens if you get caught breaking a park rule? You will pay a fine.

Q **Assessing** What do you know happened if a person paid a fine of $300 for breaking a single rule? If someone paid a fine of $300 for breaking a single rule, then they were caught using fireworks in the park.

Q **Advancing** If you switch the "if" and "then" parts of an if-then statement, is the resulting statement true? not in general; Possible answer: For example, the statement "If it is noon, then I eat." may be true, but the related statement "If I eat, then it is noon." is not true since I eat at other times of the day.

 **Turn and Talk** When determining their responses, allow students to make different assumptions about the number of rules that could have been broken. Answers will vary. Assume that you can only be fined for breaking a single rule. There are three different ways to get a $100 fine: by fishing, by littering or by smoking. If you only know that the amount of the fine is $100, you cannot determine which rule the group broke.

▶ BUILD SHARED UNDERSTANDING

Select groups of students who used various strategies and tools to share with the class how they solved the problem. As they present their solutions, have each group discuss why they chose a specific strategy and tool.

(2) Learn Together

Build Understanding

Task 1 (MP) **Reason** Students use sketches to make conclusions about mathematical statements.

CONNECT TO VOCABULARY

Have students use the **Interactive Glossary** to record their understanding of the vocabulary in this task.

Sample Guided Discussion:

(Q) **In Part B, did you create a sketch that is a counterexample? If so, how do you think your sketch proves the statement is false?** I cannot draw a sketch that is a counterexample.

(Q) **What type of everyday objects can you use to visualize a plane?** Answers will vary. Possible answer: I can use a piece of paper to visualize a plane.

Turn and Talk Emphasize that only one counterexample is needed for a statement to not always be true. Possible answer: The difference of two positive integers is positive. Counterexample: $5 - 8 = -3$.

Build Understanding

Make Sketches From Descriptions

In this course, you will use geometric reasoning to write proofs. As a starting point, you will use some postulates that are so basic that you cannot prove them.

Often, you will see the phrases "there exists" and "exactly one," such as "There exists an acute triangle that is isosceles" or "There is exactly one solution to the equation $x + 4 = 9$." "There exists" means that there is at least one, but there may be many more than one. "Exactly one" means there is only one.

Connect to Vocabulary

A postulate is a statement that is accepted as true without proof. A **counterexample** is an example that proves that a conjecture or statement is false.

Postulates of Geometry

- Through any two points, there is exactly one line.
- Through any three noncollinear points, there is exactly one plane containing them.
- If two points lie in a plane, the line containing them also lies in the plane.
- If two lines intersect, they intersect in exactly one point.
- If two planes intersect, then they intersect in exactly one line.

1 A. Draw a sketch that demonstrates that through any two points, there is exactly one line. A–I. See Additional Answers.

B. Can you find a counterexample of the statement? If so, draw a sketch of the counterexample.

C. What do your findings in Parts A and B suggest about the statement?

D. Draw a sketch that demonstrates that if two lines intersect, they intersect in exactly one point.

E. Can you find a counterexample of the statement? If so, draw a sketch of the counterexample.

F. What do your findings in Parts D and E suggest about the statements?

G. Draw sketches that demonstrate that through any three noncollinear points, there is exactly one plane containing them, if two points lie in a plane, the line containing them also lies in the plane, and if two planes intersect, then they intersect in exactly one line.

H. Can you find a counterexample of each statement in Part G? If so, draw a sketch of the counterexample.

I. What do your findings in Parts G and H suggest about the statements?

Turn and Talk Think of a mathematical statement about numbers that is not always true. Show that it is not true by finding a counterexample. See margin.

42

LEVELED QUESTIONS

Depth of Knowledge (DOK)	Leveled Questions	What Does This Tell You?
Level 1 **Recall**	If two lines intersect, at how many points do they intersect? exactly one	Students' answers will indicate whether they understand the basic premise of one of the postulates.
Level 2 **Basic Application of Skills & Concepts**	If a counterexample does not exist for a conjecture, what does that say about the statement? The statement is true.	Students' answers will demonstrate whether they understand the use of counterexamples to disprove a conjecture.
Level 3 **Strategic Thinking & Complex Reasoning**	How can you show that the statement "Through any three points, there is exactly one plane" is false? I can make a sketch of three collinear points through which there is more than one plane.	Students' answers will help them reflect on the importance of key terms in a statement and reason strategically about how the omission or addition of terms can affect the truth value of the statement.

Write Related Conditional Statements

A **conjecture** is a statement that is believed to be true. A **conditional statement** is a statement that can be written in the form "if p, then q," where p is the hypothesis of the statement and q is the conclusion.

There are three statements that are related to a conditional statement. These are the **converse**, **inverse**, and **contrapositive** of a conditional statement.

Conditional Statements		
Statement	**Definition**	**Symbols**
Conditional Statement	A statement that can be written in the form "if p, then q," where p is the hypothesis and q is the conclusion.	If p, then q.
Converse	A statement formed by exchanging the hypothesis and conclusion of a conditional statement.	If q, then p.
Inverse	A statement formed by negating the hypothesis and conclusion of a conditional statement.	If not p, then not q.
Contrapositive	A statement formed by both exchanging and negating the hypothesis and conclusion of a conditional statement.	If not q, then not p.

A conditional statement and its contrapositive are either both true or both false. The converse of a conditional statement and the inverse of a conditional statement are also either both true or both false.

 Use the following conjecture for this task. A–E. See Additional Answers.

> The product of two real numbers is a negative number when exactly one of the factors is a negative number.

A. Identify the hypothesis and conclusion in the conjecture.

B. Write the conjecture in the form "if p, then q."

C. Write the converse of the statement. Is the converse a true statement? Explain.

D. Write the inverse of the statement. Is the inverse a true statement?

E. Write the contrapositive of the statement. Is the contrapositive a true statement?

Turn and Talk Write a true conditional statement. Show that the contrapositive of the statement is also true. See margin.

PROFICIENCY LEVEL

Beginning

Focus on the vocabulary terms *conditional statement, converse, inverse,* and *contrapositive*. Write "If it rains, then the game will be cancelled". Ask students "What is the converse?" Then ask them for the inverse and the contrapositive.

Intermediate

Have students work in groups of four. One student comes up with a conditional statement. The other students come up with the converse, inverse, and contrapositive. The groups make a presentation to the class with each student reading their statement and identifying it as the conditional statement, the converse, the inverse, or the contrapositive.

Advanced

Have students explain how to form the converse, inverse, and contrapositive from a conditional statement. Their explanation should include words and symbols.

Task 2 (MP) **Critique Reasoning** Students write and determine the truth value of conditional statements and explain their conclusions to others.

CONNECT TO VOCABULARY

Have students use the **Interactive Glossary** to record their understanding of the vocabulary in this task.

Sample Guided Discussion:

Q What term or terms in the statement help identify the hypothesis? Why? The term "when" is related to the word "if"; The statement that follows is the hypothesis.

Q For Part D, determine a numerical answer that represents the inverse. Can you think of a counterexample? Possible answer: $(-3)(-4) = 12$; There are no counterexamples.

Turn and Talk Have students think of a true conditional statement. Have them find the contrapositive of the statement and verify if it is also true. Possible answer: If an angle measures 90 degrees, then it is a right angle. The following contrapositive of the statement is also true: If an angle is not a right angle, then it does not measure 90 degrees.

Step It Out

Task 3 (MP) **Construct Arguments** Students write biconditional statements from definitions, and determine if it matters which statement in the definition is used for p and which is used for q.

Encourage students to build their arguments using logical reasoning and stated assumptions.

CONNECT TO VOCABULARY

Have students use the **Interactive Glossary** to record their understanding of the vocabulary in this task.

Sample Guided Discussion:

Q In Part C, why are there two different ways to write the definition of an acute angle as a biconditional statement? There are two different ways because both the conditional statement "If the measure of an angle is greater than 0 degrees and less than 90 degrees then it is an acute angle" and its converse "If an angle is an acute angle, then the measure of the angle is greater than 0 degrees and less than 90 degrees" are true.

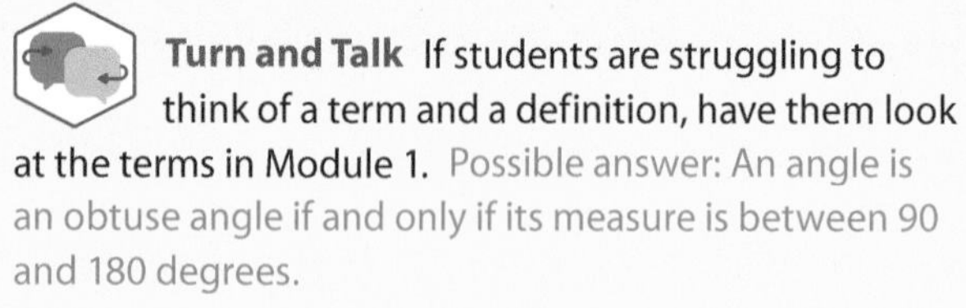

Turn and Talk If students are struggling to think of a term and a definition, have them look at the terms in Module 1. Possible answer: An angle is an obtuse angle if and only if its measure is between 90 and 180 degrees.

Task 4 (MP) **Construct Arguments** Students analyze a conditional statement and identify its converse, inverse, and contrapositive.

Sample Guided Discussion:

Q What part of the conditional statement is the hypothesis? What is the conclusion? The hypothesis is "you bought it from us". The conclusion is "you bought the best".

Q Is the conditional statement in the slogan a true statement? Explain. Possible answer: not necessarily; Advertisers create slogans to try and sway the consumer. The product may not be the best product a consumer can buy.

Turn and Talk If students are struggling to think of a reason, have them look at the contrapositive of the slogan and compare it to the original slogan and the converse and inverse. Possible answer: The contrapositive states the conclusion first.

Step It Out

Write Definitions as Biconditional Statements

If a conditional statement and its converse are both true, they can be combined into a single **biconditional statement**. A biconditional statement is a statement that can be written in the form "p if and only if q," where p is the hypothesis and q is the conclusion. Biconditional statements can be used to write **definitions**.

Connect to Vocabulary

You have used definitions to understand the meanings of new words. In mathematics, a **definition** is a statement that describes a mathematical object and can be written as a true biconditional statement.

3 Write the given definition as a biconditional statement. A–C. See Additional Answers.

Midpoint of a line segment: the point that divides the segment into two congruent segments

Let p be "a point is a midpoint of a line segment." Let q be "the point divides a segment into two congruent segments."

A. Does it matter which statement is used for p and for q in a biconditional statement?

The definition of a midpoint of a line segment, written as a biconditional statement, is: A point is a midpoint of a line segment if and only if the point divides the segment into two congruent segments.

B. What other way can you write the biconditional statement?

C. Write the definition of an acute angle as a biconditional statement.

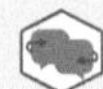

Turn and Talk Write the definition of a mathematical term you know as a biconditional statement. See margin.

Apply Conditional Statements in the Real World

4 A store creates a banner with its new advertising slogan.

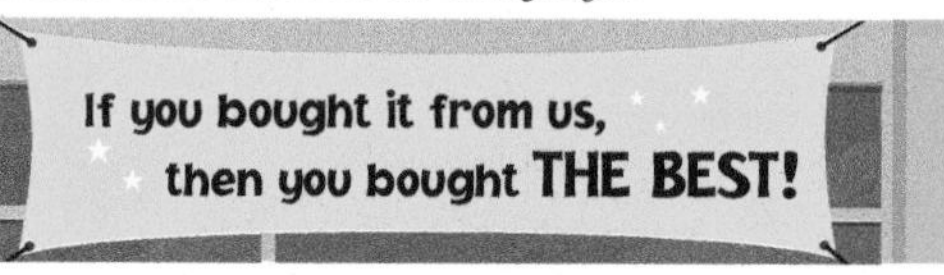
If you bought it from us,
then you bought THE BEST!

Match each related statement to the type of statement it is.

	Type	Statement
3	**A.** Converse	**1.** If you did not buy it from us, then you did not buy the best.
1	**B.** Inverse	**2.** If you did not buy the best, then you did not buy it from us.
2	**C.** Contrapositive	**3.** If you bought the best, then you bought it from us.

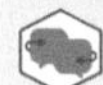

Turn and Talk Why would an advertiser choose to use the contrapositive of a conditional statement? Explain. See margin.

44

Check Understanding

1. (MP) Reason Draw a sketch that demonstrates this statement. Is it possible to find a counterexample? If so, draw a sketch of the counterexample. 1–4. See Additional Answers.

   Two distinct lines that intersect will intersect at exactly one point.

2. Write the statement "Yesterday was Monday, so today is Tuesday." in if-then form. Write the converse, inverse, and contrapositive of the statement.
3. Write the definition of a right angle as a biconditional statement.
4. A pizza chain has a slogan "Your pizza will be made right, or your next one is free." Write this slogan in if-then form.

On Your Own

Draw a sketch that demonstrates each statement. Determine if it is possible to find a counterexample. If possible, draw a sketch of the counterexample. 5, 6. See Additional Answers.

5. There is only one plane that exists between three noncollinear points.
6. Lines in the same plane that do not intersect are parallel. Through a point not on a line, there is exactly one line that is parallel to the given line.

Write the conditional statement in if-then form.

7. $4x + 7 = 15$ when $x = 2$.
   If $x = 2$, then $4x + 7 = 15$.
8. Today is Thursday, so tomorrow is Friday.
   If today is Thursday, then tomorrow is Friday.
9. The sum of the measures of two supplementary angles is 180 degrees.
   If two angles are supplementary angles, then the sum of their measures is 180 degrees.

Write the conditional statement, the converse, the inverse, and the contrapositive of the statement.

10. $2t + 3 = 13$ when $t = 5$. 10–12. See Additional Answers.
11. The dog gets a treat when it performs the trick.
12. An isosceles triangle has two sides with the same length.

Write the definition as a biconditional statement. 13–16. See Additional Answers.

13. A scalene triangle is a triangle with three sides with different lengths.
14. A square is a rectangle with four sides that are the same length.
15. Perpendicular lines are lines that intersect at a 90° angle.
16. (MP) Reason Write the statement as an if-then statement. Write the converse of this statement. Is the converse a true statement? Explain.

    The sum of two positive numbers is a positive number.

ONLINE

Assign the Digital On Your Own for
- built-in student supports
- Actionable Item Reports
- Standards Analysis Reports

On Your Own

Assignment Guide

The chart below indicates which problems in the On Your Own are associated with each task in the Learn Together. Assign daily homework for tasks completed.

Learn Together Tasks	On Your Own Problems
Task 1, p. 38	Problems 5, 6, 17, and 18
Task 2, p. 39	Problems 7–12, 16, 19, and 22
Task 3, p. 40	Problems 13–15 and 20
Task 4, p. 40	Problem 21

Watch for Common Errors

Problem 11 Some students may write the conditional in the order of the problem statement, "If the dog gets a treat, then it performs the trick." Point out that they need to determine the conclusion that naturally follows the hypothesis.

3 Check Understanding

Formative Assessment

Use formative assessment to determine if your students are successful with this lesson's learning objective.

Students who successfully complete the Check Understanding can continue to the On Your Own practice.

For students who miss 1 problem or more, work in a pulled small group using the Almost There small-group activity on page 37C.

ONLINE

Assign the Digital Check Understanding to determine
- success with the learning objective
- items to review
- grouping and differentiation resources

4 Differentiation Options

Differentiate instruction for all students using small-group activities and math center activities on page 37C.

Reteach

Write Biconditional Statements

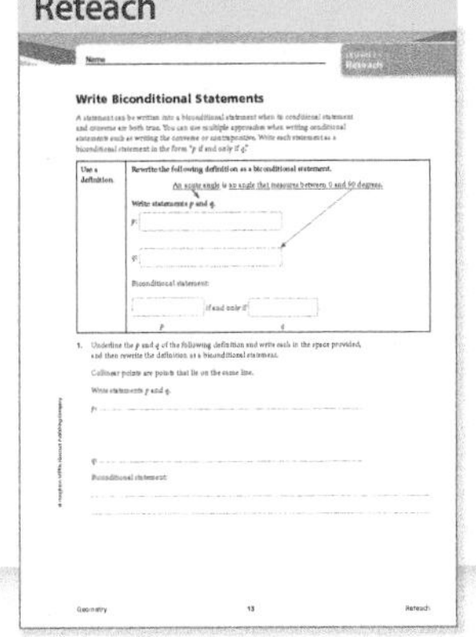

Challenge

Write Conditional Statements

5 Wrap-Up

Summarize learning with your class. Consider using the Exit Ticket, Put It in Writing, or I Can scale.

Exit Ticket 

Consider the statement "If you multiply two negative integers, then the product is positive." What is the converse? Is it true or false? If the product of two integers is positive, then the integers are negative. The converse is false.

Put It in Writing 

Describe some strategies you can use to write a conditional statement and its converse, inverse, and contrapositive.

I Can 

The scale below can help you and your students understand their progress on a learning goal.

4	I can write conditional statements and related conditional statements, and I can explain the process of constructing the statements to others.
3	I can write conditional statements and related conditional statements.
2	I can identify the hypothesis and the conclusion of a statement.
1	I can use counterexamples to prove a statement is false.

Spiral Review • Assessment Readiness

These questions will help determine if students have retained information taught in the past and can also prepare them for high-stakes assessments. Here, students must find the distance between two points on the coordinate plane **(1.4)**, identify a polygon based on its characteristics **(1.3)**, and identify justifications for steps used to solve a linear equation in one variable **(Gr 6, 9.4)**.

Give a counterexample for each conclusion.

17. The difference of two negative numbers is always negative. $-5-(-8)=3$

18. A number times a greater number is always greater than the number. $-3(5)=-15$

19. Open Ended Write a conditional statement that is not true, but the inverse of the statement is true. Answers will vary.

20. Can the statement be written as a biconditional statement? If so, write the statement as a biconditional statement. If not, explain why not. See Additional Answers.

The y-intercept of a line is the point where the line intersects the y-axis.

21. A painting company has its slogan on a van. Write this statement as an if-then statement, then find the converse, inverse, and contrapositive of the statement. See Additional Answers.

22. Consider the statement "If you bought a ticket to a concert, then you attended the concert." Find the converse, inverse, and contrapositive of the statement. Is the converse of the statement true? Explain your reasoning. See Additional Answers.

Spiral Review • Assessment Readiness

23. The tip of a toy arrowhead shaped as an isosceles right triangle is sketched on a graph. The endpoints of one leg are (0, 0) and (0, 10). What is the length of the longest side of the tip?

Ⓐ $10\sqrt{2}$ units
Ⓑ 100 units
Ⓒ $20\sqrt{2}$ units
Ⓓ 200 units

24. A polygon has five sides. What type of polygon is it?

Ⓐ quadrilateral
Ⓑ pentagon
Ⓒ hexagon
Ⓓ octagon

25. For $8(-2x+1)-10=-6$, match each step in the solution with the its justification.

Step	Justification	
A. $8(-2x+1)-10=-6$	**1.** Combine constants.	A. 3
B. $-16x+8-10=-6$	**2.** Addition Property of Equality	B. 5
C. $-16x-2=-6$	**3.** Given equation	C. 1
D. $-16x=-4$	**4.** Division Property of Equality	D. 2
E. $x=0.25$	**5.** Distributive Property	E. 4

 **I'm in a Learning Mindset!**

What strategies can I use to concentrate when writing the related statements of conditional statements?

46 Keep Going Journal and Practice Workbook

Learning Mindset

Perseverance Sustains Attention

Point out the importance of identifying the hypothesis p and the conclusion q before students write conditional statements. Encourage students to follow the guidelines to write the related converse, inverse, and contrapositive statements. Remind students that when they are trying to determine the truth of the conditional statement and the related statements, they are looking for counterexamples that disprove the statement. *How can you identify the hypothesis and conclusion in a conditional statement? What does it mean to negate a statement?*

2.2 Use Inductive and Deductive Reasoning

LESSON FOCUS AND COHERENCE

Mathematics Standards

- Prove theorems about lines and angles. Theorems include: vertical angles are congruent; when a transversal crosses parallel lines, alternate interior angles are congruent and corresponding angles are congruent; points on a perpendicular bisector of a line segment are exactly those equidistant from the segment's endpoints.
- Prove theorems about triangles. Theorems include: measures of interior angles of a triangle sum to 180°; base angles of isosceles triangles are congruent; the segment joining midpoints of two sides of a triangle is parallel to the third side and half the length; the medians of a triangle meet at a point.

Mathematical Practices and Processes 

- Construct viable arguments and critique the reasoning of others.
- Reason abstractly and quantitatively.

I Can Objective 

I can apply deductive reasoning in a mathematical context.

Learning Objective

Differentiate between inductive and deductive reasoning and apply deductive reasoning in the context of geometric proofs.

Language Objective

Explain the logic behind the two-column proof format.

Vocabulary

New: deductive reasoning, **inductive reasoning**, **proof**, **theorem**

Lesson Materials: spreadsheet software

Mathematical Progressions

Prior Learning	Current Development	Future Connections
Students: • learned about points, lines, planes, angles, and polygons. **(1.1, 1.2, 1.3)** • learned about the structure of logical, conditional statements. **(2.1)**	**Students:** • understand the concepts of inductive and deductive reasoning. • differentiate between examples of inductive and deductive reasoning. • apply deductive reasoning in a mathematical context.	**Students:** • explain and justify each step required to construct a formal geometric proof. **(2.3, 2.4, 3.1, 3.2, 3.3)**

PROFESSIONAL LEARNING

Using Mathematical Practices

Construct viable arguments and critique the reasoning of others.

This lesson provides an opportunity for students to master the concept of deductive reasoning. Students will first evaluate the reasoning of others and determine whether it is deductive reasoning. Students then learn to use deductive reasoning to construct their own arguments. Then they learn to apply those skills in a mathematical context.

WARM-UP OPTIONS

ONLINE

PROJECTABLE & PRINTABLE 

ACTIVATE PRIOR KNOWLEDGE • Interpret Conditional Statements

Use these activities to quickly assess and activate prior knowledge as needed.

Problem of the Day 

Are these statements true? If you think they are not, find a counterexample.

1. If a line goes through both point A and point B, then the points are collinear.
2. If two rays are drawn, then the rays form an angle.
3. If $x^2 = 16$, then $x = 4$

1. correct.

2. incorrect; Two rays do not have to form an angle.

3. incorrect; $x = \pm 4$

Quick Check for Homework

As part of your daily routine, you may want to display the Teacher Solution Key to have students check their homework.

Make Connections

Based on students' responses to the Problem of the Day, choose one of the following:

 Project the Interactive Reteach, Geometry, Lesson 2.1.

 Complete the Prerequisite Skills Activity:

Have students work in pairs. Each student should write a conditional statement based on definitions introduced or reviewed in Module 1. Then, have students switch and interpret whether the statements are true or false. Have students discuss each statement.

- *Think again about the statement. Can you find a counterexample?* yes; I can find counterexamples for all false statements.
- *When you wrote the statement, did you think you wrote a correct or incorrect statement?* I thought my statement was correct.

SHARPEN SKILLS

If time permits, use this on-level activity to build fluency and practice basic skills.

Mental Math

Objective: Students demonstrate an understanding of deductive reasoning.
Materials: index cards

Have students work in small groups. Have each group write two statements, one statement per index card. One statement should be an example of inductive reasoning, and one statement should be an example of deductive reasoning. Then each group will take turns presenting their statements to the rest of the class, and their classmates will vote on which statement exemplifies each type of reasoning.

PLAN FOR DIFFERENTIATED INSTRUCTION

Small-Group Options

Use these teacher-guided activities with pulled small groups.

On Track

Materials: prepared index cards

Give each student two index cards, one with an example of inductive reasoning and one with an example of deductive reasoning, or both cards with the same type of reasoning. Ask students to mark each card with an I or a D to indicate which type of reasoning it is. Collect all the cards and discuss a few of the cards with the whole class. Have students explain whether the I or D was indicated correctly.

Almost There

Give each student a linear equation in one variable and tell them to solve it. Tell them to justify each step with one of the properties of equality. Remind students to only do one computation in each step and to only use one property of equality in each step. Remind students that doing multiple computations in one step increases the likelihood of making mistakes. Encourage students to use a two-column format to solve the equation and write the justifications.

Ready for More

Draw a two-column table for a geometric proof on the board. Draw a line on the board, with four points labeled on the line: A, B, C, and D. Place A and B equidistant from each other toward the left end of the line. Place C and D equidistant from each other toward the right end of the line. Ask students to help you prove that if $AC = BD$, then $AB = CD$.

Statements	Reasons
1.) $AC = BD$	1.) Given
2.) $AB + BC = AC$ $BC + CD = BD$	2.) Segment Addition Postulate
3.) $AB + BC = BD$	3.) Substitution
4.) $AB + BC = BC + CD$	4.) Substitution
5.) $AB = CD$	5.) Subtraction Property of Equality

Math Center Options

Use these student self-directed activities at centers or stations. Key: ● Print Resources ● Online Resources

On Track

- ● Interactive Digital Lesson
- ● Interactive Glossary (printable): **inductive reasoning**, **deductive reasoning**, **proof**, **theorem**
- ●● Journal and Practice Workbook

Almost There

- ● Reteach 2.2 (printable)
- ● Interactive Reteach 2.2

Ready for More

- ● Challenge 2.2 (printable)
- ● Interactive Challenge 2.2

View data-driven grouping recommendations and assign differentiation resources.

Make a List

Strategy 1

1. Felicia needs to know how big her booth is compared to last year's size.
2. She needs to figure out whether she should pack for the two days (if there is a place to store her products), or bring more items for the second day.
3. If the booth is placed outside, Felicia should look at the weather forecast.
4. Felicia should look at the types of items she sold most the previous year and pack more of those.

She should record the profit she made after the fair and reflect upon her predictions.

If students . . . use an organization method, like a list or table, to make sure they are considering a wide range of factors, and justify their reasoning, they are employing an efficient method and demonstrating an excellent understanding of inductive and deductive reasoning from Grade 8.

Then, have these students . . . explain how they determined the factors they chose. **Ask:**

- Q How did using a list help you answer the question?
- Q Did you start with all of the items on the list, or did the list help you add more things you did not consider at first?

Use a Narrative

Strategy 2

Felicia wakes up on the day of the fair. She packs items to sell and a notebook to keep track of sales. She checks the weather forecast. It is sunny, so she leaves her umbrella at home. When she gets to the fair, she checks her booth. She lays out the items in the booth with the best-selling items from last year in front.

If students . . . use a narrative to verbalize their thinking they can create a good chronology of the problem, but they may miss out on events that happen or could happen simultaneously.

Activate prior knowledge . . . by having students organize their information in a more methodical way. **Ask:**

- Q What are all of the things Felicia needs to do before the fair?
- Q How could the knowledge from last year's fair help with her projections?

COMMON ERROR: Writes Incomplete Answer

Felicia needs to remember to smile when selling items. Customer service is important for sales. It's good to have items in a lot of different colors. Her booth is bigger this year so she packs more items to be on the safe side. Felicia remembers to bring change for money. She also brings her lucky bandana.

If students . . . spend too much time analyzing the characters to create a story, they might needlessly complicate the problem and overlook important information.

Then intervene . . . by pointing out that there might be other factors that are critical for the solution of the problem. **Ask:**

- Q Felicia was at the fair last year as well. What, besides the size, could she know that will help her sell well this year?
- Q Have you thought about what Felicia will need for both days of the fair?

2.2

Use Inductive and Deductive Reasoning

I Can use inductive and deductive reasoning to justify conjectures.

Spark Your Learning

Felicia is packing a variety of items to show in a craft fair booth.

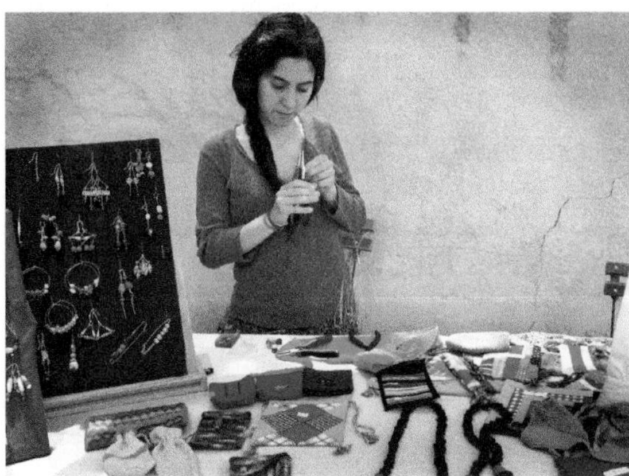

Complete Part A as a whole class. Then complete Parts B–C in small groups.

A. What is a mathematical question you can ask about this situation? What information would you need to know to answer your question?
A. How can Felicia decide what to bring?; size of the booth, size of each item, past sales records, and audience

B. To answer your question, what strategy and tool would you use along with all the information you have? What answer do you get?
B. See Strategies 1 and 2 on the facing page.

C. After the fair, how will Felicia know whether she prepared well for the fair? What might Felicia learn that could affect her preparation for the next fair?
See Additional Answers.

Turn and Talk Suppose that the craft fair went very well and that Felicia plans on returning to the same craft fair in three months. What considerations would cause Felicia to prepare differently? See margin.

CULTIVATE CONVERSATION • Co-Craft Questions

If students have difficulty formulating a mathematical question about the situation in the Spark Your Learning, ask them to visualize themselves organizing a shed or the trunk of a car. What are some natural questions to ask about this situation?

Work together to craft the following questions:

- How can Felicia utilize the space she has?
- How much time should she spend preparing the booth compared to the amount of time she will spend selling her items to her customers?

Then have students think about what additional information, if any, they would need to answer these questions. **Ask:**

- Can you determine the exact outcome of Felicia's endeavors? Why or why not?

1 Spark Your Learning

▶ MOTIVATE

- Have students look at the photo in their books and read the information contained in the photo. Then complete Part A as a whole-class discussion.
- Give the class the additional information they need to solve the problem. This information is available online as a printable and projectable page in the Teacher Resources.
- Have students work in small groups to complete Parts B and C.

▶ PERSEVERE

If students need support, guide them by asking:

Q **Advancing • Use Tools** Which tool could you use to solve the problem? Why choose that tool and not some other? Students' choices of tools and reasons for choosing them will vary.

Q **Assessing** When Felicia is preparing to pack for the craft show, what is one of the most fundamental things that she should consider? Possible answer: One of the most important things for Felicia to take into consideration is the geometric dimensions of her booth.

Q **Advancing** What types of reasoning will Felicia need to use? Possible answer: Felicia probably needs to use a combination of inductive and deductive reasoning.

Turn and Talk Encourage students to think about external factors like the weather, and whether or not school will be in session. Possible answer: Answers will vary. The craft fair will be in a different season and the products brought may no longer be popular during the new season. The product may not be something customers repeatedly buy, or maybe the customer base has been exhausted.

▶ BUILD SHARED UNDERSTANDING

Select groups of students who used various strategies and tools to share with the class how they solved the problem. As they present their solutions, have each group discuss why they chose a specific strategy and tool.

② Learn Together

Build Understanding

Task 1 (MP) **Reason Abstractly** Students learn to differentiate between examples of deductive and inductive reasoning. Students learn to write their own examples of each type of reasoning.

CONNECT TO VOCABULARY

Have students use the **Interactive Glossary** to record their understanding of the vocabulary in this task.

Sample Guided Discussion:

Q **How can you tell which reviews are examples of deductive reasoning?** The reviews that exemplify deductive reasoning are based on objective facts like size and battery life.

Q **How can you tell which reviews are examples of inductive reasoning?** The reviews that exemplify inductive reasoning are based on opinions or experiences of customers.

Turn and Talk Encourage students to keep in mind which reviews are based on deductive reasoning. Possible answer: yes; There are more positive reviews than negative reviews.

Build Understanding

Compare Inductive and Deductive Reasoning

Inductive reasoning and deductive reasoning are two processes that are used to determine if a statement is true.

Inductive reasoning is the process of reasoning that a rule or a statement may be true by looking at specific cases.

Deductive reasoning is the process of using logic to draw conclusions. Deductive reasoning uses facts, definitions, postulates, theorems, and logic to *prove* a statement is true for all cases.

1 Rebecca is shopping for a new phone. She is reading reviews of a brand of phone on a website. A–D. See Additional Answers.

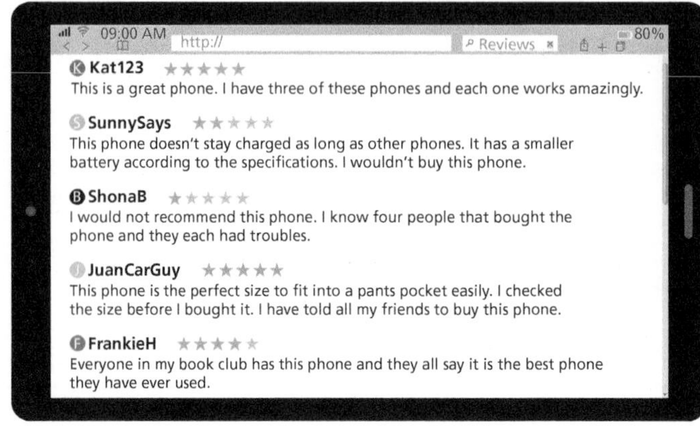

A. Which reviews use inductive reasoning? Which reviews use deductive reasoning? Explain your reasoning.

B. Does the use of inductive reasoning or deductive reasoning indicate a positive review or a negative review? Explain your reasoning.

C. Write your own review of a phone using inductive reasoning. Explain why your review uses inductive reasoning.

D. Write your own review of a phone using deductive reasoning. Explain why your review uses deductive reasoning.

Turn and Talk Do you think Rebecca should buy this phone based on the reviews? Explain your reasoning. See margin.

48

LEVELED QUESTIONS

Depth of Knowledge (DOK)	Leveled Questions	What Does This Tell You?
Level 1 **Recall**	How can you tell the difference between examples of inductive and deductive reasoning? Deductive reasoning uses facts and logic. Inductive reasoning uses specific cases.	Students' answers will indicate whether they understand the definitions of deductive and inductive reasoning.
Level 2 **Basic Application of Skills & Concepts**	How can you be sure that a review is an example of inductive reasoning? A review is an example of inductive reasoning if it is based primarily on a specific user experience.	Students' answers will demonstrate whether they can use the definitions to analyze examples.
Level 3 **Strategic Thinking & Complex Reasoning**	How can you write a review in a way that exemplifies deductive reasoning? Possible answer: I can make sure that my review is based on objective facts.	Students' answers will reflect whether they understand the meaning of deductive reasoning well enough to write their own examples.

Step It Out

Apply Properties of Equality

When you solve an algebraic equation, you are using deductive reasoning. As you solve an equation, you can state the rule or property that justifies each of the steps in the solution process.

Properties of equality are rules that describe ways that you can change both sides of an equation in the same way and have the equality remain true. These properties allow you to add, subtract, multiply, or divide both sides of an equation by the same quantity in the process of solving an equation.

Properties of Equality	
Property	**Symbols**
Addition Property of Equality	If $a = b$, then $a + c = b + c$.
Subtraction Property of Equality	If $a = b$, then $a - c = b - c$.
Multiplication Property of Equality	If $a = b$, then $ac = bc$.
Division Property of Equality	If $a = b$, and $c \neq 0$, then $\frac{a}{c} = \frac{b}{c}$.

2 Use algebra and deductive reasoning to justify the statement. A–C. See Additional Answers.

If $3x + 8 = -x$, then $x = -2$.

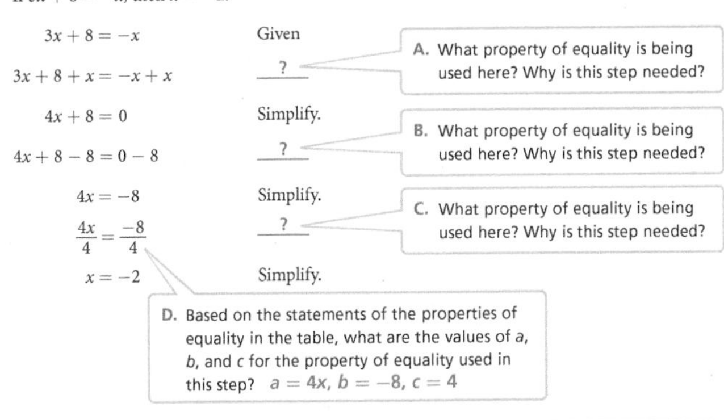

$3x + 8 = -x$ — Given

$3x + 8 + x = -x + x$ — ?

$4x + 8 = 0$ — Simplify.

$4x + 8 - 8 = 0 - 8$ — ?

$4x = -8$ — Simplify.

$\frac{4x}{4} = \frac{-8}{4}$ — ?

$x = -2$ — Simplify.

A. What property of equality is being used here? Why is this step needed?

B. What property of equality is being used here? Why is this step needed?

C. What property of equality is being used here? Why is this step needed?

D. Based on the statements of the properties of equality in the table, what are the values of a, b, and c for the property of equality used in this step? $a = 4x$, $b = -8$, $c = 4$

Turn and Talk

- Use algebra and deductive reasoning to justify the statement. Be sure to name a property of equality when you use one.

  If $\frac{1}{4}z - \frac{3}{4} = 1$, then $z = 7$. See margin.
- Write an if-then statement for the equation $\frac{1}{4}z + \frac{3}{4} = 1$. What steps and justifications change in the solution?

PROFICIENCY LEVEL

Beginning

Write examples/symbolic forms of each property of equality on the board and ask volunteers to name them.

Intermediate

Ask students to explain, in their own words, the meaning of equality as it pertains to mathematics.

Advanced

Have students explain why an equation remains true when one of the properties of equality is applied. For example, they could say that the same operation is performed on both sides of the equation.

Step It Out

Task 2 (MP) **Reason Quantitatively** Students identify which properties of equality are used in the process of solving a linear equation in one variable.

Sample Guided Discussion:

Q How can you determine which property of equality, if any, was used in each step? I can tell based on which operation was used to simplify both sides of the equation.

Q How does this example relate to the topic of geometry? This example uses deductive reasoning. Geometry also uses deductive reasoning to prove theorems based on postulates and other proven theorems.

Turn and Talk Students may not immediately recognize which properties apply to which steps. Encourage them to take their time and look closely at the steps in their justification.

- $\frac{1}{4}z - \frac{3}{4} = 1$ — Given

  $\frac{1}{4}z - \frac{3}{4} + \frac{3}{4} = 1 + \frac{3}{4}$ — Addition Property of Equality

  $\frac{1}{4}z = \frac{7}{4}$ — Simplify.

  $4\left(\frac{1}{4}z\right) = 4\left(\frac{7}{4}\right)$ — Multiplication Property of Equality

  $z = 7$ — Simplify.
- If $\frac{1}{4}z + \frac{3}{4} = 1$, then $z = 1$; In the second step, you would use the Subtraction Property of Equality, but the other justifications would remain the same.

Task 3 (MP) **Construct Arguments** Students use a two-column format to justify the steps in the simplification of a linear equation in one variable.

OPTIMIZE OUTPUT Critique, Correct, and Clarify

Have students discuss the statements written in the left column of the proof to justify the need for each statement, or present an argument in favor of changing the order of statements or of eliminating a statement.

Sample Guided Discussion:

Q What is the purpose of the two-column format for a proof? The justification for each statement is given in the same row, and the next row follows logically from the previous. In this way, the deductive reasoning in the proof can be clearly analyzed by anyone.

Q Why do you not need specific values to complete the proof? The purpose of proofs is to use deductive reasoning to establish a general case. The use of specific values in a proof would be more closely aligned with the concept of inductive reasoning.

Turn and Talk Direct students to review the definition of deductive reasoning. Possible answer: It has statements and reasons headers and given and prove statements, but the reasoning is the same; If decimals were used, the fourth step would probably use the Division Property of Equality instead.

Write a Two-Column Proof

A **theorem** is a statement you can prove is true, using a series of logical steps. A **proof** is an argument that uses true statements and logic to arrive at a conclusion. Once you prove that a statement or theorem is true, you can use it in later proofs.

There are different formats for proofs. A common format is a two-column proof listing statements in the first column and reasons in the second column. The left column contains numbered mathematical statements about the given information. It also contains the results of applying known properties to statements that have already been made. The right column gives the reason that the corresponding statement is true.

In Geometry, you will write proofs of theorems which are intended to prove that something is true in every case. You will also write proofs, as in the task below, where you are proving something about a specific situation.

3 Write a two-column proof of the statement. A, B. See Additional Answers.

If $\frac{1}{4}z - \frac{3}{4} = 1$, then $z = 7$.

Given: $\frac{1}{4}z - \frac{3}{4} = 1$

Prove: $z = 7$

Statements	Reasons
$\frac{1}{4}z - \frac{3}{4} = 1$	Given
$\frac{1}{4}z - \frac{3}{4} + \frac{3}{4} = 1 + \frac{3}{4}$	___?___ Property of Equality
$\frac{1}{4}z = \frac{7}{4}$	Simplify.
$4\left(\frac{1}{4}z\right) = 4\left(\frac{7}{4}\right)$	___?___ Property of Equality
$z = 7$	Simplify.

A. Is this the Addition or Subtraction Property of Equality? How do you know?

B. Is this the Multiplication or Division Property of Equality? How do you know?

Turn and Talk

- How is this two-column proof different from the algebraic solution in Task 2?
- Suppose decimals were used instead of fractions in the given equation in Task 3. Would this change the properties of equality you would write as reasons when you write the proof? Explain your reasoning. See margin.

50

Check Understanding 1, 2. See Additional Answers.

1. **(MP) Reason** Explain why the given conclusion uses inductive reasoning.
   For the sequence 5, 10, 15, 20, ... , the next term is 25 because the previous terms are multiples of 5.
2. Is proving the Midpoint Formula an example of inductive reasoning or deductive reasoning? Explain.
3. You are given that $\overline{ST}$ has length 23. Prove that $t = 7$ by copying and completing the two-column proof.

Given: $ST = 23$

Prove: $t = 7$

$4t - 5$, S, T

Statements	Reasons
$4t - 5 = 23$	Given
$4t - 5 + 5 = 23 + 5$	? Addition Property of Equality
$4t = 28$	Simplify.
$\frac{4t}{4} = \frac{28}{4}$	? Division Property of Equality
$t = 7$	Simplify.

On Your Own 4–9. See Additional Answers.

Explain why the given conclusion uses inductive reasoning.

4. It always snows on my birthday.
5. The next term in the pattern 4, 8, 12, ... is 16.
6. Justin wears a white shirt four days in a row, so Justin only has white shirts.
7. Antonia made a list of her soccer team's game results. Antonia's soccer team wins when they score 3 or more goals.

8. $15 + 17 = 32$, $8 + 4 = 12$, and $2 + 7 = 9$, so the sum of two positive integers is a positive integer.
9. Janet is working on a puzzle that includes the following.

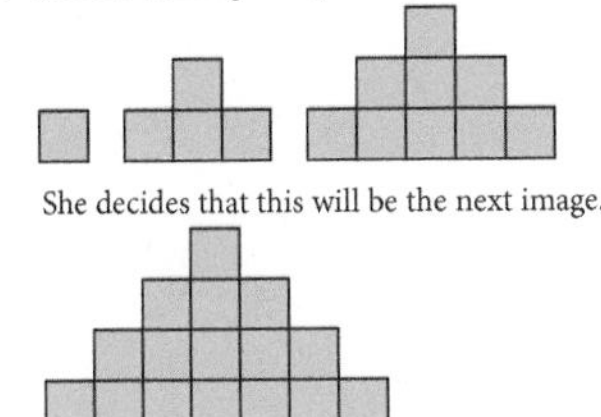

She decides that this will be the next image.

ONLINE

Assign the Digital On Your Own for
- built-in student supports
- Actionable Item Reports
- Standards Analysis Reports

On Your Own

Assignment Guide

The chart below indicates which problems in the On Your Own are associated with each task in the Learn Together. Assign daily homework for tasks completed.

Learn Together Tasks	On Your Own Problems
Task 1, p. 44	Problems 4–16
Task 2, p. 45	Problems 17–37
Task 3, p. 46	Problems 38–52

3 Check Understanding

Formative Assessment

Use formative assessment to determine if your students are successful with this lesson's learning objective.

Students who successfully complete the Check Understanding can continue to the On Your Own practice.

For students who miss 1 problem or more, work in a pulled small group using the Almost There small-group activity on page 43C.

ONLINE

Assign the Digital Check Understanding to determine
- success with the learning objective
- items to review
- grouping and differentiation resources

4 Differentiation Options

Differentiate instruction for all students using small-group activities and math center activities on page 43C.

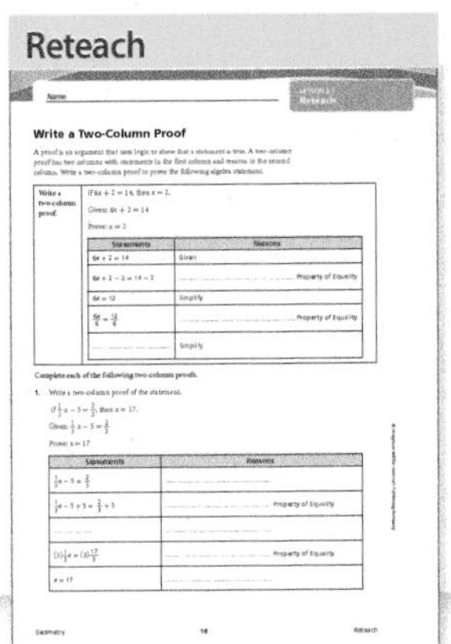

Challenge

Write a Two-Column Proof

Questioning Strategies

Problem 15 Suppose the next time Tammy goes bowling her score is only 80 points. How does that support your conclusion about the type of reasoning used? The fact that her next score could be any number less than or equal to 300 (the maximum score in bowling) supports the idea that the given prediction is not based on objective facts and deductive reasoning. It is based on a specific case and inductive reasoning.

In Problems 10–15, decide if inductive reasoning or deductive reasoning is used to make the conjecture. Explain your reasoning. 10–15. See Additional Answers.

10. Each time Jose went to the store this week, he bought a bag of apples. If Jose is going to the store today, you can conclude that he will buy a bag of apples.

11. All right angles measure 90°. $\angle ABC$ is a right angle, so $m\angle ABC$ is 90°.

12. All mammals are warm-blooded animals. If an otter is a mammal, you can say that it is a warm-blooded animal.

An otter can hold its breath for up to 8 minutes.

13. Cindy rolls a number cube six times and the result is an even number each time. You can conclude that she will roll an even number the next time she rolls the number cube.

14. A scalene triangle has three sides with different lengths. If triangle *XYZ* is a scalene triangle, it has three sides with different lengths.

15. Tammy goes bowling and her scores were 122, 138, and 117. When Tammy goes bowling the next time, she will get a score of over 100 points.

16. Consider the following pattern. See Additional Answers.

A. Assume that the pattern continues using just these two shapes. Describe how the pattern might continue. Did you use inductive or deductive reasoning?

B. Describe another way that the pattern might continue.

Use deductive reasoning to write a conclusion.

17. If two distinct lines intersect, they intersect at one point. Lines *m* and line *n* intersect. Line *m* and line *n* intersect at one point.

18. If a person is over 48 inches tall, they can ride the roller coaster. Emma is over 48 inches tall. Emma can ride the roller coaster.

19. If an integer is not divisible by 2, then it is an odd number. 17 is not divisible by 2. 17 is an odd number.

20. If the measure of an angle is greater than 90°, then it is an obtuse angle. The measure of an angle is 120°. The angle is an obtuse angle.

21. If a number is divisible by 6, then it is divisible by 2. 54 is divisible by 6. 54 is divisible by 2.

22. If Victoria has less than $50, she does not go to the movies. Victoria has $35. Victoria does not go to the movies.

Write the statement as an if-then statement.

23. $x - 7 = 22; x = 29$
If $x - 7 = 22$, then $x = 29$.

24. $t + 4 = 17; t = 13$
If $t + 4 = 17$, then $t = 13$.

25. $3a + 4 = 10; a = 2$
If $3a + 4 = 10$, then $a = 2$.

26. $2m - 7 = 25; m = 16$
If $2m - 7 = 25$, then $m = 16$.

27. $x = 7; 2x - 13 = 1$
If $x = 7$, then $2x - 13 = 1$.

28. $r = -2; 6 - 2r = 10$
If $r = -2$, then $6 - 2r = 10$.

29. $n = 12; n - 20 = -8$
If $n = 12$, then $n - 20 = -8$.

30. $b = 0; 4b - 9 = -9$
If $b = 0$, then $4b - 9 = -9$.

MP Reason **For Problems 31–37, select the word that makes the statement true.**

31. If a triangle has an obtuse angle it (must, may, cannot) be an acute triangle. cannot

32. If one endpoint of a segment is on a line so that two angles are formed, then the two angles formed (must, may, cannot) be supplementary. must

33. If the sun is shining this morning, then it (must, may, cannot) rain this afternoon. may

34. Acadia National Park is in Maine. Joshua lives in Maine. So Joshua (must have, may have, never) visited Acadia National Park. may have

35. If a number is even, then it (must, may, cannot) be a whole number. may

36. A quadrilateral is (always, sometimes, never) a parallelogram. sometimes

37. If a is 0, then ab is (always, sometimes, never) equal to 0. always

Prove that the solution to the equation is true. State each property of equality that you use. 38–43. See Additional Answers.

38. $y - 23 = 7; y = 30$

39. $n + 14 = 19; n = 5$

40. $3a + 7 = 16; a = 3$

41. $2r - 11 = 1; r = 6$

Write a two-column proof of the statement. Include Given and Prove statements.

42. If $6b + 7 = b + 2$, then $b = -1$.

43. If $9k - 5 = 7k + 3$, then $k = 4$.

In Problems 44–47, determine whether the statement is true based on the true statements.

- If Jin goes to the library, she will borrow a book.
- If Jayden goes to the library, he will borrow a movie.
- If Jin does not have a softball game, she goes to the library.
- Jin and Jayden are at the library.

44. Jin borrowed a book. true

45. Jayden borrowed a book. false

46. Jin had a softball game. false

47. Jayden borrowed a movie. true

48. MP **Construct Arguments** Jerome states that if a line segment is 23 units long, and part of the segment is 9 units long, then the rest of the segment is 14 units long. What postulates allow Jerome to make this conclusion? See Additional Answers.

Watch for Common Errors

Problem 32 Some students may confuse statements that are true in certain circumstances with statements that are always true. Encourage students to use test cases to determine if the statement is ever true or ever false.

5 Wrap-Up

Summarize learning with your class. Consider using the Exit Ticket, Put It in Writing, or I Can scale.

Exit Ticket 

Thomas lives in a town where all the avenues are parallel to each other and all the streets are parallel to each other. However, the avenues do not meet the streets at right angles. Thomas wants to know why any given pair of angles formed by the intersections in the town are either congruent or supplementary. Explain why. In the scenario described, any given pair of angles would fall into one of the following categories: vertical, linear pair, alternate interior, alternate exterior, same-side interior, same-side exterior, or corresponding; All of those categories of angle pairs are either congruent or supplementary.

Put It in Writing 

Describe some strategies you can use to differentiate between deductive and inductive reasoning.

I Can 

The scale below can help you and your students understand their progress on a learning goal.

4	I can use deductive reasoning to construct a two-column geometric proof.
3	I can apply deductive reasoning in a mathematical context.
2	I can use the definitions of deductive and inductive reasoning to differentiate between examples of each.
1	I can understand the definitions of deductive and inductive reasoning.

Spiral Review • Assessment Readiness

These questions will help determine if students have retained information taught in the past and prepare them for high-stakes assessments. Here, students must calculate distance in the coordinate plane **(1.4)**, recognize attributes of polygons **(1.3)**, identify a conditional statement **(2.1)**, and measure segments **(1.1)**.

49. Marita claims that the sum of three consecutive counting numbers is three times the second number. Prove that this conjecture is true.

A. Let the second number be n. Write expressions for the first and third numbers using n. $n-1$, $n+1$

B. Find the sum of the three numbers. $(n-1)+n+(n+1)=3n$

C. Explain why your result proves the conjecture. The sum of the three numbers is 3 times n which is the middle number.

D. Is Marita's claim true for negative numbers? Explain your reasoning. yes; n can be any integer.

50. Open Ended Write a true statement. Use deductive reasoning to explain why the statement is true. Answers will vary.

51. A rectangle with four congruent sides is a square. What conclusion can you make about rectangle $JKLM$ based on the sides? rectangle $WXYZ$? Rectangle $JKLM$ is a square. Rectangle $WXYZ$ is not a square.

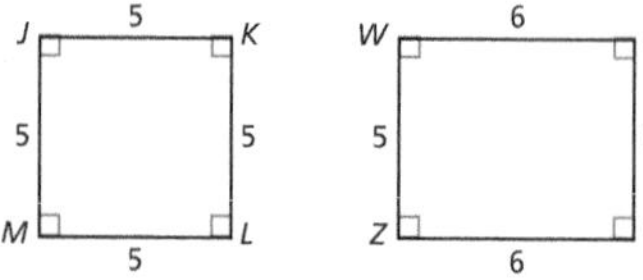

52. Open Ended Write a math problem that uses the Segment Addition Postulate, the Addition Property of Equality, and the Division Property of Equality to find the length of a segment. Check student's work; solutions should involve an equation of the form $ax-b=c$.

Spiral Review • Assessment Readiness

53. What is the length of $\overline{BD}$?

Ⓐ $\sqrt{14}$ units
Ⓑ $\sqrt{34}$ units
Ⓒ 8 units
Ⓓ 15 units

54. A fenced-in area for riding horses has six sides. What is the name of this shape?

Ⓐ quadrilateral
Ⓑ pentagon
Ⓒ hexagon
Ⓓ octagon

55. The ____ of a statement is formed by negating the hypothesis and conclusion of a conditional statement.

Ⓐ inverse
Ⓑ converse
Ⓒ contrapositive
Ⓓ biconditional

56. On a number line, point X is located at 8 and point Z is located at 44. Point Y is between X and Z where $XY = 14$ units. What is YZ?

Ⓐ 32 units
Ⓑ 22 units
Ⓒ 20 units
Ⓓ 18 units

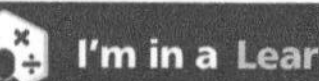

When a proof requires many steps, what strategies can I use to stay focused when writing each step and reason in the correct, logical sequence?

54

Keep Going ▶ Journal and Practice Workbook

Learning Mindset

Perseverance Sustains Focus

Point out that making sure a given step in solving a problem follows logically from the previous step is essential in all areas of mathematics, not just when solving equations. Encourage students to develop the habit of justifying the steps of their solutions when solving any mathematical problem. Also let students know that although they used deductive reasoning and the properties of equality to solve various types of problems in this lesson, in the next lessons they will focus on traditional geometric proofs. *How does justifying each step when solving a mathematical problem help you have a more confident mindset? When you can't think of a good justification for one of your solution steps, what might that suggest about your approach to solving the problem?*

2.3 Write Proofs about Segments

LESSON FOCUS AND COHERENCE

Mathematics Standards

- Prove theorems about lines and angles. *Theorems include: vertical angles are congruent; when a transversal crosses parallel lines, alternate interior angles are congruent and corresponding angles are congruent; points on a perpendicular bisector of a line segment are exactly those equidistant from the segment's endpoints.*

Mathematical Practices and Processes 

- Use appropriate tools strategically.
- Attend to precision.
- Look for and express regularity in repeated reasoning.
- Reason abstractly and quantitatively.

I Can Objective 

I can use properties of segments to show congruence.

Learning Objective

Use congruence and the Segment Addition Postulate to complete proofs about segments.

Language Objective

Explain the meaning of the properties of congruent segments and the Segment Addition Postulate.

Vocabulary

New: symbolic notation

Mathematical Progressions

Prior Learning	Current Development	Future Connections
Students: • defined line segments. **(1.1)** • solved a simple equation in one variable. **(Alg1, 2.2)**	**Students:** • prove theorems about line segments. • use the Segment Addition Postulate to prove line segment congruence. • solve equations in one variable as part of a proof of segment congruence.	**Students:** • will prove theorems about angles. **(2.4)** • will prove theorems about parallel and perpendicular lines. **(3.2–3.3)** • will show that two triangles are congruent using corresponding pairs and angles. **(7.1–7.3)**

UNPACKING MATH STANDARDS

Prove theorems about lines and angles. *Theorems include: vertical angles are congruent; when a transversal crosses parallel lines, alternate interior angles are congruent and corresponding angles are congruent; points on a perpendicular bisector of a line segment are exactly those equidistant from the segment's endpoints.*

What It Means to You

Students prove segment congruence by writing a proof using the Reflexive, Symmetric, and Transitive Properties, as well as the Segment Addition Postulate. This skill was introduced in the previous lesson, but for many students justifying their reasoning in this way may seem difficult. Mathematical proofs are used to show that a statement is true. Students will use mathematical vocabulary and grammar to demonstrate deductive reasoning mathematically. The process of logical reasoning and deduction are important to understanding geometry concepts and mathematics as a whole. In this course, students will write proofs about angles, segments, parallel and perpendicular lines, and triangles.

WARM-UP OPTIONS

ONLINE

PROJECTABLE & PRINTABLE

ACTIVATE PRIOR KNOWLEDGE • Determine Lengths of Segments Using Equations

Use these activities to quickly assess and activate prior knowledge as needed.

Problem of the Day 

On a map, a straight street has 3 points labeled with locations *A*, *B*, and *C*, where *B* is between points *A* and *C*. The distance from point *A* to point *C* is 52 meters. The distance from *A* to *B* is represented by the expression $4x + 12$ and from *B* to *C* by $6x - 10$. Write and solve an equation to determine how far it is from point *A* to *B* and from point *B* to *C*. $4x + 12 + 6x - 10 = 52$; $x = 5$, so the distance from *A* to *B* is 32 meters and the distance from *B* to *C* is 20 meters.

Quick Check for Homework

As part of your daily routine, you may want to display the Teacher Solution Key to have students check their homework.

Make Connections

Based on students' responses to the Problem of the Day, choose one of the following:

 Project the Interactive Reteach, Geometry Lesson 1.1.

 Complete the Prerequisite Skills Activity:

Have students work in pairs. Have one student construct a line segment that is 20 centimeters in length, $\overline{LN}$, with a midpoint *M*. Provide the students with a notecard with the expressions $8x + 12$ and $4x + 6$. Have students work together to determine which expression represents the length of $\overline{LN}$ and which represents the length of $\overline{MN}$. Then have students determine the value of *x* and the values of *LM* and *MN*.

- *Look at the expressions. If the value of x is the same in both expressions, which expression represents the length of the whole segment? How do you know?* If the expression $4x + 6$ is half of the segment or the length of $\overline{LM}$, then 2 of these $4x + 6 + 4x + 6 = 8x + 12$ is the length of $\overline{LN}$.
- *How can you determine the value of x? What is the value of x?* The length of $\overline{LN}$ is 20 centimeters, so set the expression $8x + 12$ equal to 20 and solve for *x*; $x = 1$
- *What is the next step in finding the length of $\overline{LN}$?* Substitute $x = 1$ into the expression $4x + 6$.
- *What are the lengths of $\overline{LM}$ and $\overline{MN}$?* 10 centimeters

SHARPEN SKILLS

If time permits, use this on-level activity to build fluency and practice basic skills.

Mental Math

Objective: Students demonstrate an understanding of equivalent expressions.
Materials: prepared index cards

Have students work in pairs. On half of the index cards, write an expression that is the sum of 2 two-step expressions (i.e., $3x + 1 + 2x - 2$). On the other cards, write an expression, such that expressions are equivalent to one of the cards from the first half of the set. Give each group a set of each type of card. Have students take turns selecting a card from the first set and mentally evaluating the expression to find the matching equivalent expression from the second set.

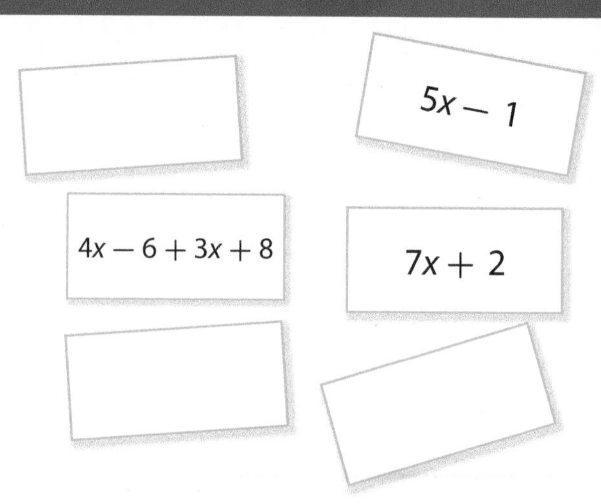

PLAN FOR DIFFERENTIATED INSTRUCTION

Small-Group Options

Use these teacher-guided activities with pulled small groups.

On Track

Materials: index cards

Give each student a card showing a labeled line segment, where a point between the two endpoints (does not have to be the midpoint) is also labeled. Provide them with the length of each section of the segment, written as an expression, and the total length of the line segment as a numeric value ($RS = 5x$, $ST = 7x$, $RT = 72$) written on it. Have students do the following:

- Use the Segment Addition Postulate to solve for x.
- Find the length of each segment.
- Determine if the point in between the two endpoints is the midpoint of the line segment.

Almost There

Materials: interlocking cubes

Have students connect 7 cubes to make a "train" of cubes. Display a line segment $\overline{DF}$ with point E along the line. Label $\overline{DE}$ with $4x$ and $\overline{EF}$ with $3x$. Have students do the following:

- Explain how their "train" model represents the line segment.
- Find the total length of the line segment, ignoring the color difference in the cubes.
- Find the value of each cube or x, if $\overline{DF}$ is 35 units long.
- Write an algebraic equation that could be used to find the value of x.
- Solve for x without using the colored cubes.

Ready for More

Write the following problem. Have students find the unknown distance.

Malachi hiked a 5-mile trail at the local park. At the trail marker T he is 2 miles from the trail head. The trail marker at point U is the midpoint between the T and V. How far is trail marker U from the end of the trail, V?

Math Center Options

Use these student self-directed activities at centers or stations. Key: ● Print Resources ● Online Resources

On Track

- ● Interactive Digital Lesson
- ●● Journal and Practice Workbook
- ● Interactive Glossary (printable): **symbolic notation**
- ● Module Performance Task

Almost There

- ● Reteach 2.3 (printable)
- ● Interactive Reteach 2.3

Ready for More

- ● Challenge 2.3 (printable)
- ● Interactive Challenge 2.3

Unit Project Check students' progress by asking to see their conclusion and deductive reasoning steps.

View data-driven grouping recommendations and assign differentiation resources.

During the *Spark Your Learning*, listen and watch for strategies students use. See samples of student work on this page.

Using Reasoning

Strategy 1

School to Soccer Field = Soccer Field to Home

School to Library = Library to Soccer Field

$\frac{3}{8}$ mile = Library to Soccer Field

School to Library = $\frac{3}{8}$ mile

School to Soccer Field = $\frac{3}{8} + \frac{3}{8} = \frac{6}{8}$ mile

Soccer Field to Home = $\frac{6}{8}$ mile

School to Home = $\frac{6}{8} + \frac{6}{8} = \frac{12}{8} = 1\frac{1}{2}$ miles

If students . . . use reasoning about midpoints and congruent segments, they are demonstrating an exemplary understanding of congruent segments to solve problems.

Have these students . . . explain how they determined the relationships between each distance. **Ask:**

- Q How did you use the given information, the relationship between the segments, and the midpoints to determine the distances between the stops?
- Q What is the relationship of the distance between the school and the soccer field to the distance between the soccer field and home?

Use a Diagram

Strategy 2

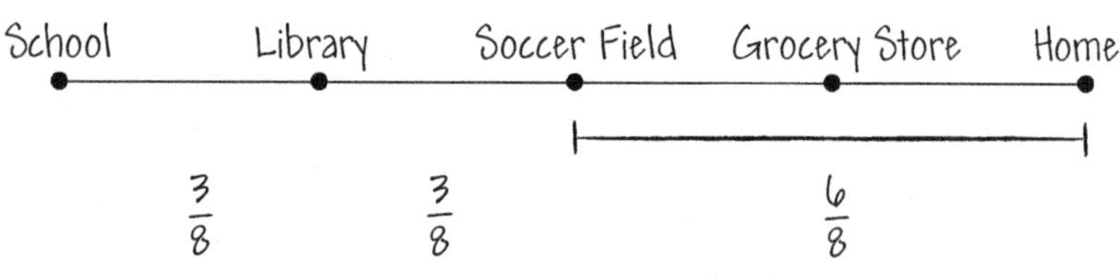

$\frac{3}{8} + \frac{3}{8} + \frac{6}{8} = \frac{12}{8} = 1\frac{1}{2}$

The distance from school to home is $1\frac{1}{2}$ miles.

If students . . . use a diagram to solve the problem, they understand the meaning of midpoint and how segments are related but may not know how to explain their reasoning without a model.

Activate prior knowledge . . . by having students write expressions to represent the relationships between the segments. **Ask:**

- Q How can you write statements of equality relating the segments?
- Q Can you find the individual distances between the soccer field and grocery store and between the grocery store and home? If not, what can you determine?

COMMON ERROR: Uses Midpoint Incorrectly

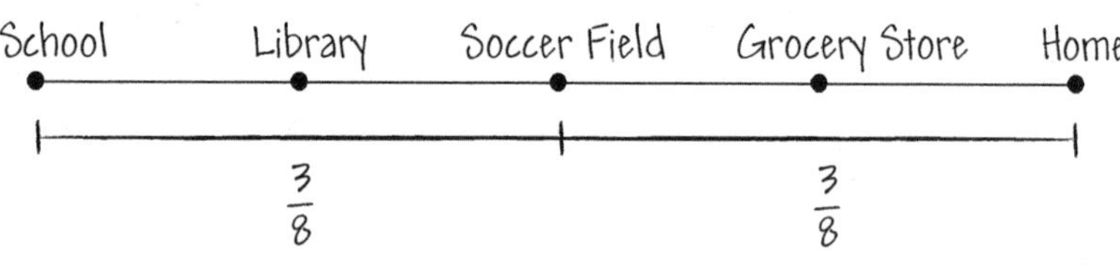

$\frac{3}{8} + \frac{3}{8} = \frac{6}{8}$

The distance from school to home is $\frac{6}{8}$ mile.

If students . . . use the measure of $\frac{3}{8}$ mile as the distance between the school and soccer field, they may not understand that the library being at the midpoint means that the distance from the library to the soccer field is $\frac{3}{8}$ mile.

Then intervene . . . by having them reread the information provided and explain which distance is described as $\frac{3}{8}$ mile. **Ask:**

- Q If the midpoint between the school and the soccer field is the library, then how far is the school from the library?
- Q How can you use this information and that the midpoint between the school and home is the soccer field to find the distance from the school to home?

2.3

Write Proofs about Segments

I Can use properties of segments to show congruence.

Spark Your Learning

Brianna and Carl make several stops on their walk home from school. They use an app to keep track of their distance for only one portion of their trip.

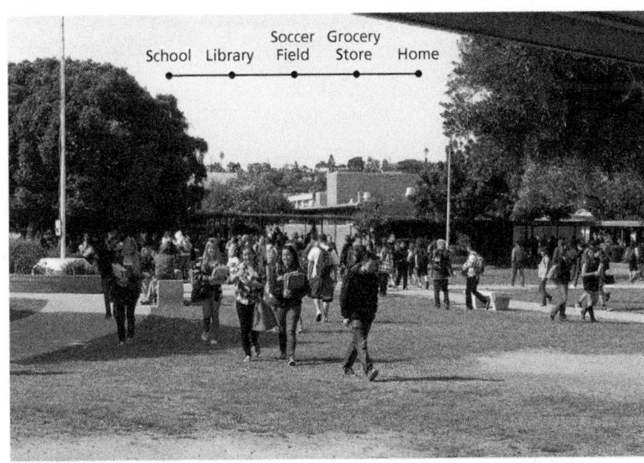

©Blake Young/Alamy

Complete Part A as a whole class. Then complete Parts B–D in small groups.

A. What is a mathematical question you can ask about this situation? What information would you need to know to answer your question?
A. How long is their walk home?; the distance between each stop

B. What relationship(s) are involved in this situation? What postulate(s) or theorem(s) are involved in this situation? How do you know?
B. See Additional Answers.

C. To answer your question, what strategy and tool would you use along with all the information you have? What answer do you get?
C. See Strategies 1 and 2 on the facing page.

D. Does your answer make sense in the context of the situation? Justify your reasoning with relationship(s), postulate(s), or theorem(s).
D. See Additional Answers.

Turn and Talk Is the grocery store the midpoint between the soccer field and their home? Explain your reasoning. See margin.

CULTIVATE CONVERSATION • Information Gap

Ask students questions to help them decide what missing information they need to answer the question, "What is the distance from the school to home?"

1. Do you have enough information to find the distance Brianna and Carl travel from school to the soccer field? Explain. yes; Since the library is the midpoint between the school and the soccer field, and the library is $\frac{3}{8}$ mile from the soccer field, then the distance from the school to the soccer field is also $\frac{3}{8}$ mile.

2. Do you have enough information to find the distance Brianna and Carl travel from the soccer field to home? yes; Since the soccer field is the midpoint between the school and home, then the distance from the soccer field to home is the same as the distance from the school to the soccer field.

3. What information would you need to determine the distance between the soccer field and the grocery store or the grocery store and home? Possible answer: I need to know if the grocery store is the midpoint between the soccer field and home.

1 Spark Your Learning

▶ MOTIVATE

- Have students look at the image in their book and read the information contained in the photo. Then complete Part A as a whole-class discussion.
- Give the class the additional information they need to solve the problem. This information is available online as a printable and projectable page in the Teacher Resources.
- Have students work in small groups to complete Parts B–D.

▶ PERSEVERE

If students need support, guide them by asking:

Q **Advancing • Use Tools** Which tool could you use to solve the problem? Why choose that tool and not some other? Students' choices of tools and reasons for choosing them will vary.

Q **Assessing** What must be true of the distances from the school to the library and from the library to the soccer field? The distances are equal, because the library is the midpoint between the school and the soccer field.

Q **Advancing** If the distance from the grocery store to home is half that of the distance from the soccer field to the grocery store, what is the distance from the grocery store to home? Explain. If the distance from the soccer field to home is $\frac{6}{8}$ mile, then the distance from the grocery store to home is $\frac{2}{8}$ mile, because it is half of $\frac{4}{8}$ mile (the distance from the soccer field to the grocery store).

Turn and Talk When determining if the grocery store is the midpoint between the soccer field and home, have students draw and label the diagram with all of the information that they have been provided to accurately answer the question. There is not enough information to determine if the grocery store is the midpoint between the soccer field and home.

▶ BUILD SHARED UNDERSTANDING

Select groups of students who used various strategies and tools to share with the class how they solved the problem. As they present their solutions, have each group discuss why they chose a specific strategy and tool.

② Learn Together

Build Understanding

Task 1 (MP) **Use Tools** Students determine the congruence of line segments using statements about the properties of segment congruence and the reflective, symmetric, and transitive properties.

CONNECT TO VOCABULARY

Have students use the **Interactive Glossary** to record their understanding of the vocabulary in this task.

Sample Guided Discussion:

Q **In Part C, assign a length to $\overline{GH}$. How can you use this to determine which property can be used to determine congruence?** If $\overline{GH}$ is 2 inches long, then $\overline{JK}$ is 2 inches long. If $\overline{JK}$ is 2 inches long, then $\overline{LM}$ is also 2 inches long. Each segment is 2 inches long, so $\overline{GH} \cong \overline{LM}$. This can also be determined using the Transitive Property.

Turn and Talk Remind students that they know the relationships between $\overline{GH}$, $\overline{JK}$, and $\overline{LM}$, so they can also determine the relationship between these segments and some other rope segment $\overline{NP}$ using the same reasoning they used in Part B. Use the Transitive Property and the fact that $\overline{JK} \cong \overline{LM}$ to show that $\overline{JK} \cong \overline{NP}$. Use the Transitive Property and the fact that $\overline{GH} \cong \overline{LM}$ to show that $\overline{GH} \cong \overline{NP}$.

Build Understanding

Investigate Properties of Congruence

Remember that two segments are congruent if the lengths of each segment are equal.

Connect to Vocabulary

Symbolic notation includes symbols used to represent figures and geometric relationships. Lines, segments, rays, and angles can be written using symbolic notation.

The table below shows how the properties of congruence are applied to segments.

Properties of Segment Congruence	
Property	**Example**
Reflexive	$\overline{AB} \cong \overline{AB}$
Symmetric	If $\overline{AB} \cong \overline{CD}$, then $\overline{CD} \cong \overline{AB}$.
Transitive	If $\overline{AB} \cong \overline{CD}$ and $\overline{CD} \cong \overline{EF}$, then $\overline{AB} \cong \overline{EF}$.

1 The climbing structure shown is made using rope segments. A–D. See Additional Answers.

©Olaf Speier/Shutterstock

A. Why would you expect some rope segments to be congruent?

B. How can you determine if $\overline{TV}$ is congruent to $\overline{WX}$ if you know that $\overline{WX} \cong \overline{TV}$?

C. How can you determine if $\overline{GH}$ is congruent to $\overline{LM}$ if you know that $\overline{GH} \cong \overline{JK}$ and $\overline{JK} \cong \overline{LM}$?

D. What does the Reflexive Property of Congruence tell you about each rope segment?

Turn and Talk If a rope segment $\overline{NP}$ is congruent to $\overline{LM}$ in Part C, how can you show that $\overline{NP}$ is congruent to $\overline{JK}$ or $\overline{GH}$? See margin.

LEVELED QUESTIONS

Depth of Knowledge (DOK)	Leveled Questions	What Does This Tell You?
Level 1 **Recall**	What property states that if a segment is congruent to another then the reverse is also true? Symmetric Property of Segment Congruence	Students' answers will indicate whether they understand the meaning of one of the properties of segment congruence.
Level 2 **Basic Application of Skills & Concepts**	If $\overline{1}$ is congruent to $\overline{2}$ and $\overline{2}$ is congruent to $\overline{3}$, what property can you use to show that $\overline{3}$ is congruent to $\overline{1}$? I can use the Transitive Property of Segment Congruence. If $\overline{1} \cong \overline{2}$ and $\overline{2} \cong \overline{3}$, then $\overline{1} \cong \overline{3}$.	Students' answers will demonstrate whether they can apply the properties of congruence to show that two segments are congruent.
Level 3 **Strategic Thinking & Complex Reasoning**	If you know that three segments are congruent and a fourth line segment is congruent to 1 of those, is it possible for the fourth line segment to *not* be congruent to any of the other segments? By the Transitive Property of Segment Congruence, all four segments are congruent.	Students' answers will reflect whether they understand the meaning of segment congruence and the relationship between congruent segments.

Step It Out

Use Segment Congruence in a Real-World Problem

2 Jessica is constructing a bookshelf for her bedroom. The instructions show a diagram of the completed bookshelf. The boards used to construct the bookshelf are represented by $\overline{AB}$, $\overline{BE}$, $\overline{EG}$, $\overline{AG}$, $\overline{CH}$, $\overline{DK}$, and $\overline{FJ}$.

A. Jessica wants to organize the boards into groups of congruent boards before she starts working. She sets aside the boards represented by $\overline{AB}$, $\overline{AG}$, and $\overline{DK}$. Which boards appear to be congruent to these segments?

$\overline{AB}$: $\overline{EG}$
$\overline{AG}$: $\overline{BE}$
$\overline{DK}$: none

Turn and Talk Why might Jessica want to work on groups of congruent boards at the same time? See margin.

Use the Segment Addition Postulate

3 The length of $\overline{AC}$ is 24. Prove that $\overline{AB} \cong \overline{BC}$.
A–C. See Additional Answers.

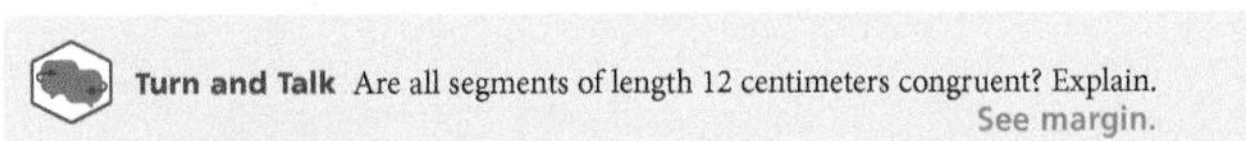

Statements	Reasons
1. $AC = 24$, $AB = 12$	1. Given
2. $AB + BC = AC$	2. ?
3. $12 + BC = 24$	3. Substitution
4. $BC = 24 - 12$	4. ?
5. $BC = 12$	5. Simplify.
6. $\overline{AB} \cong \overline{BC}$	6. Definition of congruent segments

A. What allows you to write $AB + BC = AC$?

B. What allows you to write $BC = 24 - 12$?

C. Which statements allow you to use this reason?

Turn and Talk Are all segments of length 12 centimeters congruent? Explain. See margin.

PROFICIENCY LEVEL

Beginning
Draw two congruent segments measuring 10 inches labeled $\overline{AB}$ and $\overline{CD}$. Say, "These are congruent segments because each segment is 10 inches." Then draw a segment measuring 5 inches and labeled $\overline{PQ}$ and ask students to draw a segment that is congruent to it, labeled $\overline{RS}$.

Intermediate
Provide students with a segment $\overline{DF}$ that is 16 centimeters in length where point E is the midpoint of the segment. Give them the following statements $DE = 8$ cm and $DF = 16$ cm. Ask students to show that $\overline{DE} \cong \overline{EF}$ by measuring the segments and by using mathematical reasoning.

Advanced
Have students use the properties of segment congruence to show that $\overline{MN} \cong \overline{NO}$ when the length of $\overline{MO}$ is 20 cm, point N lies on the segment, and the length of $\overline{NO}$ is 10 cm. Have students explain the steps they used.

Step It Out

Task 2 (MP) **Attend to Precision** Students use clear definitions to explain which segments in a diagram are congruent.

Have students look at the bookshelf and its parts as rectangles. They should use what they know about rectangles and their sides to determine which pairs of sides are congruent, and be able to relay their reasoning accurately using segment names and the congruency symbol.

Sample Guided Discussion:

Q **Look at segment AB. Which segment is congruent to $\overline{AB}$?** Segment EG

Q ***JCDK* and *DEFK* seem to provide the same amount of storage space. What must be true, in terms of midpoints, for the space to be the same?** D must be the midpoint of $\overline{CE}$ and K must be the midpoint of $\overline{JF}$.

Turn and Talk For students that are unfamiliar with building furniture using diagrams, point out that gathering parts that are the same is helpful for determining which parts go where. Possible answer: The boards are all cut to the same length.

Task 3 (MP) **Use Repeated Reasoning** Students complete a proof by reasoning about the relationships of two line segments as they relate to the larger segment of which they compose, using patterns of segment addition.

Sample Guided Discussion:

Q **In Part A, which two segments make up $\overline{AC}$? What property is shown by the statement $AB + BC = AC$?** $\overline{AB}$ and $\overline{BC}$; Segment Addition Postulate

Q **For Part C, look for the two statements in the proof that show the length of $\overline{AB}$ and $\overline{BC}$. How can those statements be used to show that the two segments are congruent?** If two segments have the same length, then the segments are congruent.

Turn and Talk Students may be able to understand that segments of 12 centimeters are congruent, but may have difficulty explaining this fact. Have them use the definition of congruent segments to explain. yes; Possible answer: By the definition of congruent segments, any segments that are the same length are congruent.

Task 4 (MP) **Reason** Students use properties of segment congruence in squares to organize a proof.

Point out to students that the first statement(s) in a proof should be taken from the given information in the problem and that those given statements should be written in the same order as they are written in the problem.

Sample Guided Discussion:

Q What do you know about the lengths of $\overline{AB}$ and $\overline{DE}$? Which statement and reason match this? The lengths are the same because all sides of a square are equal; $AB = DE$; All sides of a square are equal length.

Q What should then follow about the congruence of segments $\overline{AB}$ and $\overline{BC}$? The segments are congruent because their lengths are congruent, so $\overline{AB} \cong \overline{DE}$ by the definition of congruent segments.

Q What can you use as a reason for $\overline{DE} \cong \overline{BC}$? If $\overline{AB} \cong \overline{BC}$ and $AB = DE$ then $\overline{DE} \cong \overline{BC}$ by the Transitive Property of Congruence.

Turn and Talk Have students draw rectangle *ABED* with the associated change. Then have them look at the statements and reasons to determine which would need to change. Possible answer: You can use the definition of a rectangle to show that $\overline{AB} \cong \overline{DE}$. Then use the same steps as with a square.

Task 5 (MP) **Attend to Precision** Students use algebraic reasoning and accurate calculations to determine the value of a variable in an equation representing congruent segment lengths.

SUPPORT SENSE-MAKING Three Reads

Have students read the problem three times. Use the questions below for a different focus each read.

1. What is the situation about?
2. What are the quantities in this situation?
3. What are the possible mathematical questions that you could ask for the situation?

Sample Guided Discussion:

Q If the two segments are congruent, then what must also be true about the value of x? It is the same for both expressions.

Turn and Talk Have students determine which property can be used to show that the length of $\overline{LN}$ is the same as the sum of the two parts, *LM* and *MN*. Add the expression for *LM* and *MN*.

Prove Segment Congruence

4 **Given:** *ABED* is a square.

$\overline{AB} \cong \overline{BC}$

Prove: $\overline{DE} \cong \overline{BC}$

A, B. See Additional Answers.

A. Write the statements in the correct order.

B. Write the reasons in the correct order.

Statements	Reasons
$\overline{AB} \cong \overline{DE}$	Given
$\overline{DE} \cong \overline{BC}$	All sides of a square are equal in length.
$AB = DE$	Transitive Property of Congruence
ABED is a square.	Given
$\overline{AB} \cong \overline{BC}$	Definition of congruent segments

Turn and Talk What parts of the proof would need to change if *ABED* were a rectangle instead of a square? Explain. See margin.

Apply Algebra to Ensure Segment Congruence

5 What value of x will make $\overline{LM} \cong \overline{MN}$?

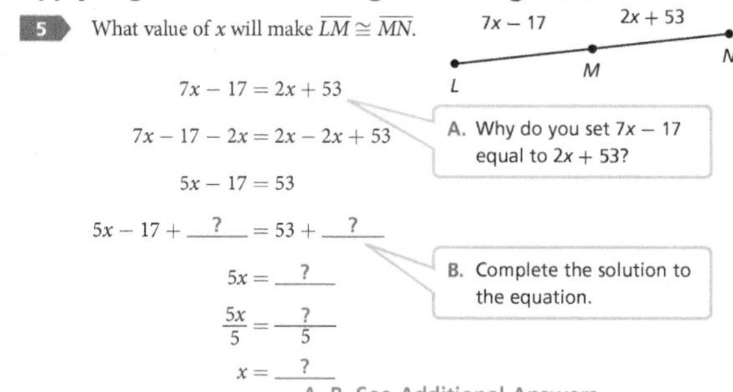

7x − 17
2x + 53
L
M
N
A. Why do you set 7x − 17 equal to 2x + 53?
B. Complete the solution to the equation.

$7x - 17 = 2x + 53$

$7x - 17 - 2x = 2x - 2x + 53$

$5x - 17 = 53$

$5x - 17 + \underline{\ ?\ } = 53 + \underline{\ ?\ }$

$5x = \underline{\ ?\ }$

$\frac{5x}{5} = \frac{?}{5}$

$x = \underline{\ ?\ }$

A, B. See Additional Answers.

Turn and Talk How would you write an expression for the length of $\overline{LN}$? See margin.

Check Understanding

1. You know that $\overline{PQ} \cong \overline{ST}$ and that $\overline{XY} \cong \overline{ST}$. What can you conclude about $\overline{PQ}$ and $\overline{XY}$? They are congruent by the Transitive Property.

2. The floor design for a new hotel is shown in the figure. The design involves a rectangle, $DEGF$, joined with an isosceles triangle, ABC. The triangular portion of the floor design is centered where it meets the rectangular portion so that points B and C are the midpoints of $\overline{DC}$ and $\overline{BE}$, respectively. Which outside edges of the floor design are congruent? See Additional Answers.

3. Suppose $VW = 4$ and $WX = 5$. What must be true about V, W, and X to conclude that $VX = 9$? V, W, and X must be on the same segment with W between V and X.

A D B C E F G

4. In the figure, $DEGF$ is a rectangle, $\overline{DB} \cong \overline{CE}$, and $FG = 3BC$. Write the correct statements to complete the proof to show that $\overline{BC} \cong \overline{DB}$. See Additional Answers.

Statements	Reasons
1. $\overline{DB} \cong \overline{CE}$	1. Given
2. ?	2. Definition of congruent segments
3. $DE = DB + BC + CE$	3. Segment Addition Postulate
4. ?	4. Substitution
5. ?	5. Opposite sides of a rectangle have equal length.
6. $FG = DB + BC + DB$	6. Substitution
7. ?	7. Given
8. $3BC = DB + BC + DB$	8. Substitution
9. $BC = DB$	9. Simplify.
10. ?	10. Definition of congruent segments

5. If $AB = 3x - 12$ and $CD = 44 - x$, what value of x will make $\overline{AB} \cong \overline{CD}$? 14

On Your Own

Match the statement with the property of congruence that makes the statement true.

A. Reflexive Property **B.** Symmetric Property **C.** Transitive Property

6. $\overline{BC} \cong \overline{FG}$, so $\overline{FG} \cong \overline{BC}$. B
7. $\overline{PQ} \cong \overline{PQ}$ A
8. $\overline{AB} \cong \overline{CD}$ and $\overline{CD} \cong \overline{EF}$, so $\overline{AB} \cong \overline{EF}$. C
9. $\overline{CD} \cong \overline{CD}$ A
10. $\overline{LM} \cong \overline{NO}$ and $\overline{NO} \cong \overline{PQ}$, so $\overline{LM} \cong \overline{PQ}$. C
11. $\overline{WX} \cong \overline{YZ}$, so $\overline{YZ} \cong \overline{WX}$. B

Use the Segment Addition Postulate to find the missing length if the given points are collinear and appear in alphabetical order.

12. $AB = 7$, $AC = 15$, $BC = ?$ 8
13. $XY = 11$, $XZ = 13$, $YZ = ?$ 2
14. $JK = 8$, $KL = 10$, $JL = ?$ 18
15. $FG = 15$, $GH = 12$, $FH = ?$ 27

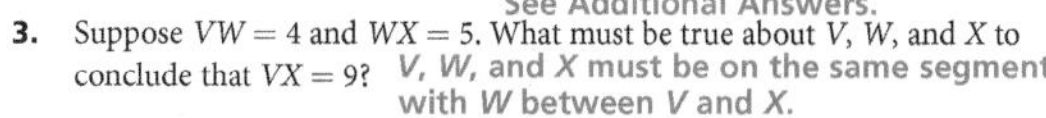

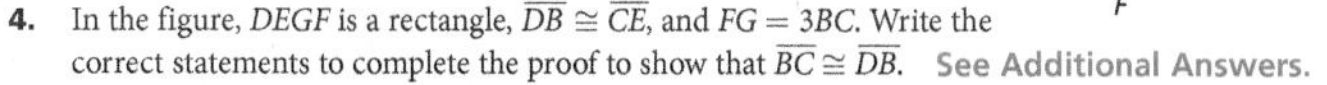

ONLINE

Assign the Digital On Your Own for
- built-in student supports
- Actionable Item Reports
- Standards Analysis Reports

On Your Own

Assignment Guide

The chart below indicates which problems in the On Your Own are associated with each task in the Learn Together. Assign daily homework for tasks completed.

Learn Together Tasks	On Your Own Problems
Task 1, p. 52	Problems 6–11 and 36
Task 2, p. 53	Problems 20–22
Task 3, p. 53	Problems 12–19, 25, and 32–35
Task 4, p. 54	Problems 23 and 24
Task 5, p. 54	Problems 26–31, 37, and 38

3 Check Understanding

Formative Assessment

Use formative assessment to determine if your students are successful with this lesson's learning objective.

Students who successfully complete the Check Understanding can continue to the On Your Own practice.

For students who miss 1 problem or more, work in a pulled small group using the Almost There small-group activity on page 51C.

ONLINE

Assign the Digital Check Understanding to determine
- success with the learning objective
- items to review
- grouping and differentiation resources

4 Differentiation Options

Differentiate instruction for all students using small-group activities and math center activities on page 51C.

Reteach

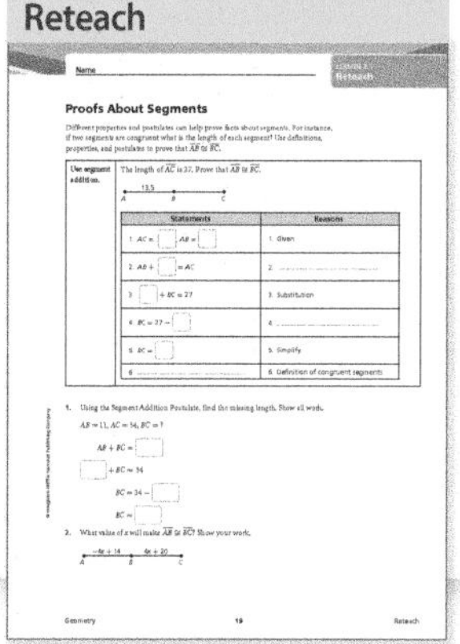

Challenge

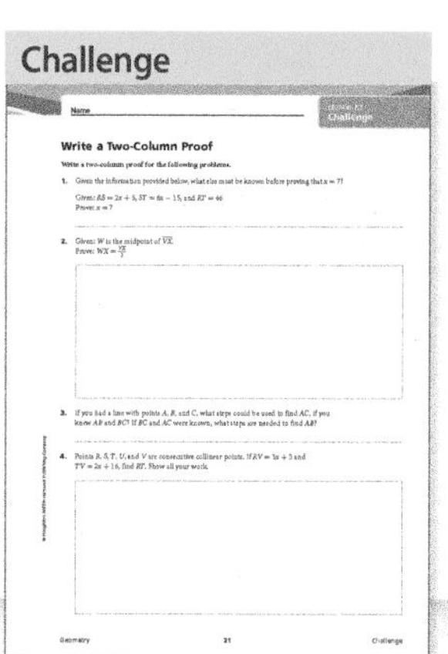

Questioning Strategies

Problem 22 If a triangle *DEF* is equilateral, what is the relationship between the sides? The sides of an equilateral triangle are congruent.

Problem 23 If $\overline{AD}$ consists of $\overline{AB}$, $\overline{BC}$, and $\overline{CD}$, and $AB = CD$, what expression could you write to show that $AB = BD$? $AB + BC = CD + BC$

A, B, and C are collinear and B is between A and C. Use the Segment Addition Postulate to determine if $\overline{AB} \cong \overline{BC}$.

16. $AB = 8$, $AC = 18$ no

17. $AC = 18$, $BC = 9$ yes

18. $AB = 12$, $AC = 24$ yes

19. $BC = 15$, $AC = 20$ no

List the groups of segments in the figure that appear to be congruent.

20.

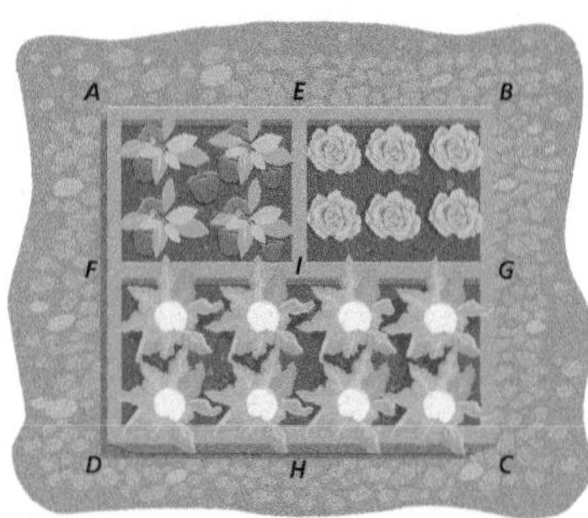

21.

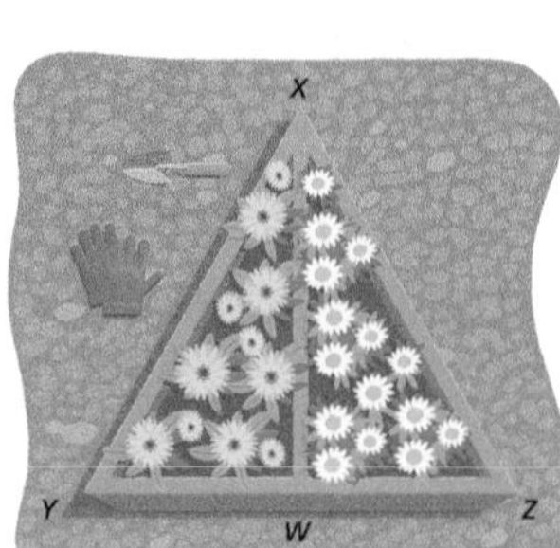

20–25. See Additional Answers.

22. Point *S* lies on $\overline{RT}$. The length of $\overline{RS}$ is 25 units. The length of $\overline{RT}$ is 50 units. Is $\overline{RS}$ congruent to $\overline{ST}$? Explain.

23. Triangle *DEF* is an equilateral triangle. $\overline{FG} \cong \overline{DE}$. Write a two-column proof that proves that $\overline{EF} \cong \overline{FG}$.

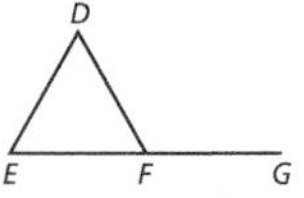

24. Write a two-column proof.
Given: $AB = CD$
Prove: $AC = BD$

A B C D

25. (MP) **Reason** Raul is building a table from the kit. The instructions include a diagram of the finished table. Which pieces of the frame appear to be congruent?
first group: $\overline{AB}$, $\overline{CD}$; second group: $\overline{AC}$, $\overline{BD}$, $\overline{CE}$, $\overline{DF}$

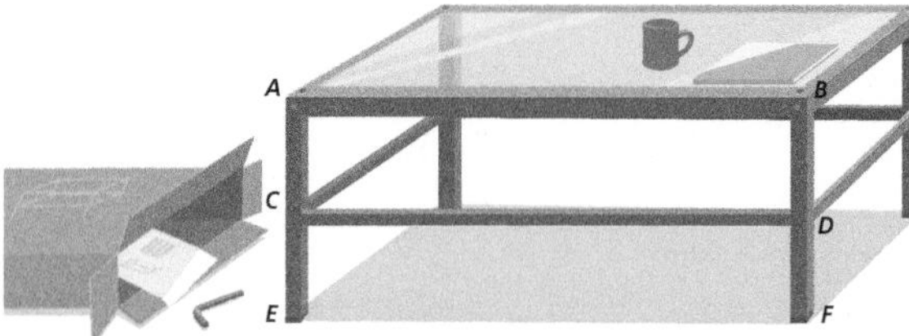

Solve for x.

26. $\overline{BD}$ has a point C between points B and D. $BC = 7x - 4$, $CD = 3x + 1$, and $BD = 9x$. $x = 3$

27. $\overline{XZ}$ has a point Y between points X and Z. $XY = 17$, $YZ = 4x + 9$, and $XZ = 11x - 9$. $x = 5$

28. $\overline{LN}$ has a point M between points L and N. $LM = x + 13$, $NM = 2x + 1$, and $LN = 6x - 7$. $x = 7$

29. $\overline{FH}$ has a point G between points F and H. $\overline{FG} \cong \overline{GH}$, $FG = 8x - 7$, and $FH = 13x - 2$. $x = 4$

30. $\overline{JL}$ has a point K between points J and L. $\overline{JK} \cong \overline{KL}$, $KL = 2x - 1$, and $JL = 5x - 12$. $x = 10$

31. $\overline{PR}$ has a point Q between points P and R. $\overline{PQ} \cong \overline{QR}$, $PQ = x + 1$, and $PR = 3x - 13$. $x = 15$

32. Find the length of segment GI, if $GJ = 64$ centimeters. 47 cm

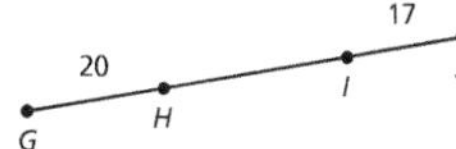

33. Find the length of segment WY, if $WZ = 128$ feet and segment WX is congruent to segment YZ. 91 ft

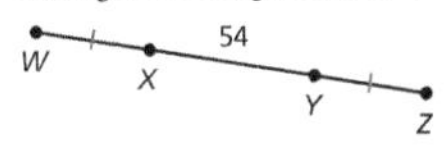

34. Open Ended Write a problem with three points on a segment that uses the Segment Addition Postulate to solve. See below.

35. (MP) **Model with Mathematics** The distance from the convention center to the airport is 16 miles. A restaurant lies on a line segment drawn from the convention center to the airport. The restaurant is 4 miles from the airport. Sketch the situation. How far is the restaurant from the convention center?
12 miles; See Additional Answers for sketch.

36. STEM An engineering student creates a model of a bridge for a project. The photo shows the student's model. The student intends that $\overline{AB} \cong \overline{CD}$ and $\overline{DF} \cong \overline{GH}$. What additional information would you need to know to determine that $\overline{AB} \cong \overline{DF}$?
Possible answer: $\overline{AB} \cong \overline{GH}$

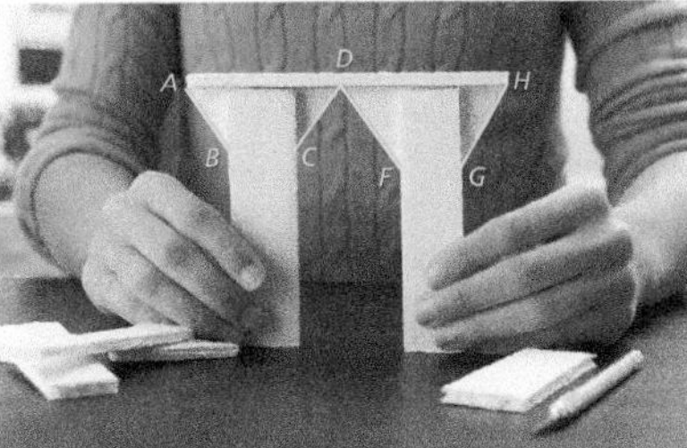

34. Point B lies on $\overline{AC}$ between points A and C. $AB = 12$ and $BC = 18$. Find AC.

Watch for Common Errors

Problems 26–28 Some students may assume that the given point is the midpoint of the described line segment. They should add the lengths of the sections and set them equal to the length of the entire segment to solve for x. Have these students reread the problem and draw what is described in the problem.

Questioning Strategies

Problem 35 How can the Segment Addition Postulate help you to answer this question? The postulate tells us that the distance from the airport to the convention center is the same as the sum of the distance from the airport to the restaurant and the distance from the restaurant to the convention center. So you can use subtraction to find the missing distance.

Wrap-Up

Summarize learning with your class. Consider using the Exit Ticket, Put It in Writing, or I Can scale.

Exit Ticket 

Triangle *ABC* is a right triangle. $\overline{AB} \cong \overline{CD}$. Solve for *x*. Is triangle *ABC* an isosceles right triangle? Explain how you found your answer.

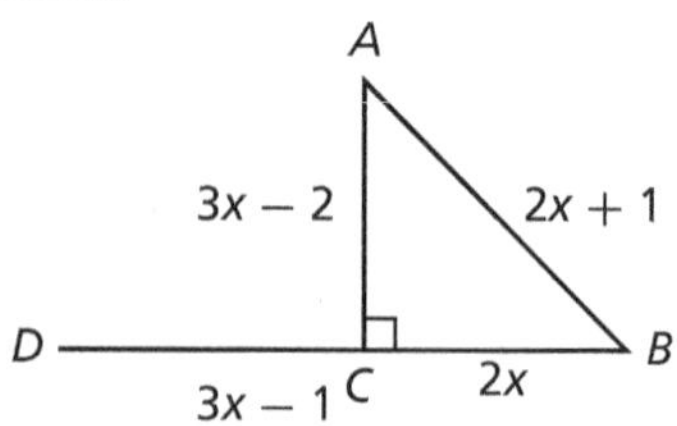

$x = 2$; yes; Possible answer: I wrote the equation that sets the expression for *AB* equal to the expression for *CD*, and solved for *x*. Then I substituted this value into the expressions for the lengths of $\overline{BC}$ and $\overline{AC}$. Triangle *ABC* is an isosceles right triangle with two sides congruent where $AC = 4$ units and $BC = 4$ units.

Put It in Writing 

Describe some strategies you can use to prove that two line segments are congruent.

I Can 

The scale below can help you and your students understand their progress on a learning goal.

4	I can use the properties of segments to show congruence, and I can explain my solution steps to others.
3	I can use properties of segments to show congruence.
2	I can recognize the properties of segments to show congruence.
1	I can recognize segment congruence in diagrams.

Spiral Review • Assessment Readiness

These questions will help determine if students have retained information taught in the past and can also prepare them for high-stakes assessments. Here, students must use properties of equality to complete a proof **(2.2)**, measure angles using a protractor **(1.2)**, and identify the converse, inverse, and contrapositive of a statement **(2.1)**.

37. Find the value of *a* that makes $\overline{XY} \cong \overline{YZ}$. What are the lengths of $\overline{XY}$, $\overline{YZ}$, and $\overline{XZ}$?

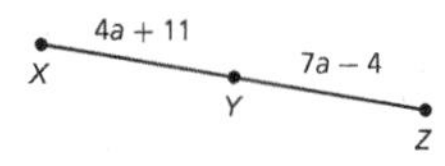

$a = 5$; $XY = 31$, $YZ = 31$, $XZ = 62$

38. Find the value of *n* when $AC = 48$. What are the lengths of $\overline{AB}$ and $\overline{BC}$? Is $\overline{AB} \cong \overline{BC}$?

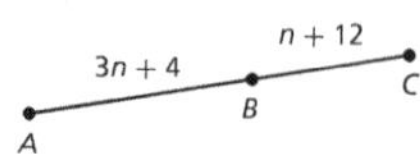

$n = 8$; $AB = 28$, $BC = 20$; $\overline{AB}$ is not congruent to $\overline{BC}$

Spiral Review • Assessment Readiness

39. What is the missing operation that completes the proof?
If $6x + 8 = 74$, then $x = 11$.

Statements	Reasons
$6x + 8 = 74$	Given
$6x = 66$	___?___ Property of Equality
$x = 11$	Division Property of Equality

Ⓐ Addition
Ⓑ Multiplication
Ⓒ Subtraction
Ⓓ Division

40. Find the measure of $\angle FGI$ and $\angle IGH$ given $\angle FGH = 128°$.

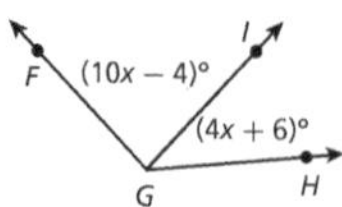

Ⓐ 86°, 42°
Ⓑ $74\frac{6}{7}°$, $53\frac{1}{7}°$
Ⓒ 94°, 34°
Ⓓ $76\frac{1}{7}°$, $51\frac{6}{7}°$

41. Consider the conditional statement: "*If a member of the track team is a sprinter, then they are a fast runner.*" Identify the type of related statement with the conditional.

Related Statement	Converse	Inverse	Contrapositive
A. If a member of the track team is not a fast runner, then they are not a sprinter.	?	?	**?** (shaded)
B. If a member of the track team is a fast runner, then they are a sprinter.	**?** (shaded)	?	?
C. If a member of the track team is not a sprinter, then they are not a fast runner.	?	**?** (shaded)	?

I'm in a Learning Mindset!

Is my understanding of proofs using segments improving? Are there any adjustments I need to make to enhance my learning?

Keep Going ▶ Journal and Practice Workbook

Learning Mindset

Perseverance Checks for Understanding

Have students think about what they know about segments and how to use proofs to show this understanding. Let students think about the way they completed proofs and any struggles they may have had in writing them. *How does knowing the properties of congruent segments help to complete a proof? Is there something you struggled with in completing the proofs about segments? What could you do to help complete additional proofs like these?*

2.4 Write Proofs About Angles

LESSON FOCUS AND COHERENCE

Mathematics Standards

- Prove theorems about lines and angles. Theorems include: vertical angles are congruent.

Mathematical Practices and Processes 

- Reason abstractly and quantitatively.
- Construct viable arguments and critique the reasoning of others.

I Can Objective 

I can use definitions and relations between lines and angles to prove theorems involving lines and angles.

Learning Objective

Apply proof concepts to situations and theorems involving angles.

Language Objective

Explain why the given reasons for the steps in a proof are logical and make sense.

Vocabulary

Review: linear pair, vertical angles

Mathematical Progressions

Prior Learning	Current Development	Future Connections
Students: • classified logical statements as examples of inductive or deductive reasoning. **(2.2)** • applied the properties of equality to solve linear equations. **(Alg1, 2.2)** • applied the properties of congruence to justify logical statements about segments. **(2.3)**	**Students:** • apply the properties of congruence to justify logical statements about angles. • apply definitions and theorems to find angle measures. • use reason to justify each step in a deductive proof about angles.	**Students:** • will identify the relationships between angles formed by parallel lines crossed by a transversal. **(3.1)** • will find measures of angles formed by parallel lines crossed by a transversal. **(3.1)** • will apply theorems to deductively prove whether lines are parallel. **(3.2)**

PROFESSIONAL LEARNING

Visualizing the Math

Students need to be able to identify relationships between pairs of angles when expressed within a task and/or represented in figures. For example, two intersecting lines form linear pairs of angles, supplementary angles, and vertical angles. Angles in a linear pair are adjacent and supplementary. The sum of the measures of supplementary angles, whether adjacent or not, equals 180°. The vertical angles created when two lines intersect are congruent.

In this figure, $\overline{AC}$ and $\overline{BD}$ intersect to form 4 linear pairs, and two pairs of vertical angles that are congruent.

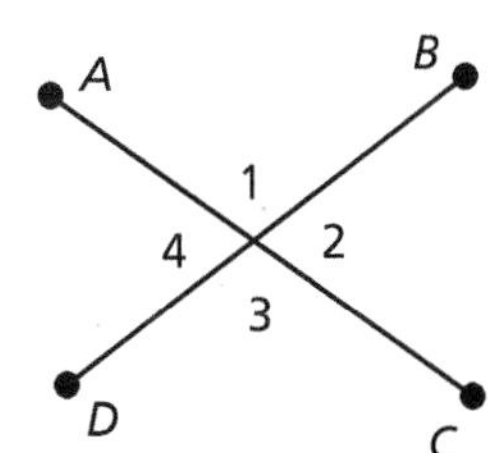

WARM-UP OPTIONS

ONLINE

PROJECTABLE & PRINTABLE 

ACTIVATE PRIOR KNOWLEDGE • Apply Congruence Properties

Use these activities to quickly assess and activate prior knowledge as needed.

Problem of the Day 

What congruence property makes each of the statements true?

- $\overline{MN} \cong \overline{PQ}$, so $\overline{PQ} \cong \overline{MN}$. Symmetric
- $\overline{PQ} \cong \overline{PQ}$ Reflexive
- $\overline{PQ} \cong \overline{MN}$ and $\overline{MN} \cong \overline{RS}$, so $\overline{PQ} \cong \overline{RS}$. Transitive

Make Connections

Based on students' responses to the Problem of the Day, choose one of the following:

 Project the Interactive Reteach, Geometry, Lesson 2.3.

 Complete the Prerequisite Skills Activity:

Have students work in pairs. One student should name a property of congruence and write the hypothesis of an if-then statement. The other student should write the conclusion of the statement. Then have the partners switch roles and choose a different property. One student should write the hypothesis and the other student should write its conclusion.

- *What kind of statement is an if-then statement?* a conditional statement
- *What is the hypothesis in the statement "If AB = 5, then 5 = AB?"* $AB = 5$
- *Given $\overline{XL} \cong \overline{YZ}$ and $\overline{YZ} \cong \overline{WQ}$, what is the conclusion?* $\overline{XL} \cong \overline{WQ}$

SHARPEN SKILLS

If time permits, use this on-level activity to build fluency and practice basic skills.

Quantitative Comparison

Objective: Students make a comparison between two quantities.

Write the following problem on the board. Ask students to choose the letter representing the correct answer and to explain their reasoning.

Quantity A	**Quantity B**
the measure of the complement of 65°	the measure of the angle vertical to an angle whose measure is 115°

A. Quantity A is greater.
B. Quantity B is greater. B; Quantity A is 25°, and Quantity B is 115°.
C. The two quantities are equal.
D. The relationship cannot be determined from the information given.

PLAN FOR DIFFERENTIATED INSTRUCTION

Small-Group Options

Use these teacher-guided activities with pulled small groups.

On Track

Materials: ruler, protractor

Have students draw figures to represent each of the following situations. Have them include labels and the measures of all angles in each figure.

- a pair of non-adjacent, non-congruent supplementary angles
- a pair of obtuse vertical angles
- a pair of adjacent complementary angles
- two congruent supplementary angles

Almost There

Tell whether each of the following statements are sometimes, always, or never true.

- Supplementary angles are ________ congruent.
- Angles of a linear pair are ________ complementary.
- Vertical angles are ________ formed by two intersecting segments.
- Congruent angles ___________ have equal measures.

Ready for More

Have students work to prove that if a pair of vertical angles are supplementary, then the angles are right angles.

Math Center Options

Use these student self-directed activities at centers or stations.

Key: ● Print Resources ● Online Resources

On Track

- ● Interactive Digital Lesson
- ●● Journal and Practice Workbook
- ● Interactive Glossary (printable): **linear pair**, **vertical angles**
- ● Module Performance Task

Almost There

- ● Reteach 2.4 (printable)
- ● Interactive Reteach 2.4

Ready for More

- ● Challenge 2.4 (printable)
- ● Interactive Challenge 2.4
- ● Illustrative Mathematics: Congruent angles made by parallel lines and a transversal
- ● Desmos: Lines, Transversals, and Angles

ONLINE 

View data-driven grouping recommendations and assign differentiation resources.

Spark Your Learning • Student Samples

During the *Spark Your Learning,* listen and watch for strategies students use. See samples of student work on this page.

Apply Reason

Strategy 1

The scissor legs have the same length and do not change as they move up and down. The measures of the pairs of angles between the legs do change. As the legs move upwards, the measures of angles 1 and 3 decrease, getting smaller and smaller. As the legs move apart and become more horizontal, the measures of angles 1 and 3 increase, getting close to 180°.

If students . . . use reason to decide whether the movement of the scissor legs affects the measures of the angles between them, they are demonstrating a solid understanding of the situation.

Have these students . . . explain how they determined whether the measures of the angles changed as the legs moved. **Ask:**

- **Q** How could you model this situation?
- **Q** What pairs of angles form supplementary pairs? What pairs of angles appear to be congruent?

Model the Situation

Strategy 2

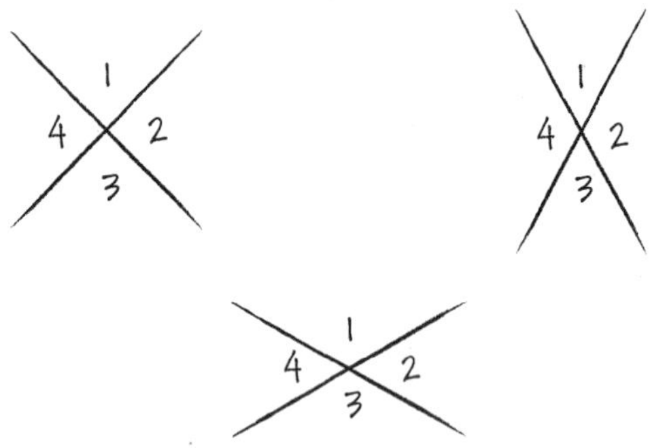

Angles 1 and 3 are always congruent, and angles 2 and 4 are always congruent. The lengths of the bars are always equal.

If students . . . draw correct models of the situation, they understand that the two line segments representing the bars remain congruent, and that although the measures of the angles between them change, they are equal.

Activate prior knowledge . . . by having students recall what they know about angle relationships. **Ask:**

- **Q** Which of your models shows the platform at a higher position than the others?
- **Q** What angles in the figures form linear pairs?

COMMON ERROR: Reason Incorrectly

The lengths of the bars and the measures of the angles shrink as the legs move upwards and increase as the legs move downwards.

If students . . . do not recognize that the four rigid bars have a constant length that doesn't change as they lift or lower the platform, they may not be visualizing the scenario correctly.

Then intervene . . . by arranging two pencils with the same length into the shape of an X, and rotating them about their intersection point so that they become nearly vertical or horizontal. **Ask:**

- **Q** What do you notice about the lengths of the pencils as they are rotated about their intersection?
- **Q** How do the measures of the angles between the pencils change as the pencils are moved closer together or farther apart?

2.4

Write Proofs about Angles

I Can prove theorems about angles.

Spark Your Learning

Heavy cargo is raised by a hydraulic scissor lift table to be loaded into an airplane. The scissor legs move along the base track from a horizontal to a vertical position, raising the platform to the desired height.

©mrclausen/iStock/Getty Images Plus/Getty Images

Complete Part A as a whole class. Then complete Parts B–D in small groups.

A. What is a mathematical question you can ask about the scissor legs as they move? What information would you need to know to answer your question?

A. How do the angles made by the scissor legs change as the platform moves up and down? how the scissor legs move as the platform height changes

B. What reasonable assumptions can you make? What shapes can you use to model the scissor legs? What do you know about the parts of these shapes? B. See Additional Answers.

C. To answer your question, what strategy and tool would you use along with all the information you have? What answer do you get? C. See Strategies 1 and 2 on the facing page.

D. Does your answer make sense in the context of the situation? Explain why or why not. D. See Additional Answers.

Turn and Talk Describe what would happen in the following scenarios.

- The scissor legs were not straight.
- The pair of scissor legs on one side of the platform were not in the same position as the pair on the other side. See margin.

SUPPORT SENSE MAKING • Three Reads

Tell students to read the information in the problem three times and prompt them with a different question each time.

1. What is the situation about? It is about what changes and what stays the same when the diagonal bars in a scissor lift raise a car.
2. What parts of a scissor lift change when a car is raised? the measures (size) of the angles between each pair of diagonal bars
3. What are possible mathematical questions you could ask about this situation? Possible answers: How does the measure of the angles between the diagonal bars change when a car is lifted? How does it change when the car is lowered?

1 Spark Your Learning

▶ MOTIVATE

- Have students look at the photo in their books and read the information contained in the photo. Then complete Part A as a whole-class discussion.
- Give the class the additional information they need to solve the problem. This information is available online as a printable and projectable page in the Teacher Resources.
- Have students work in small groups to complete Parts B–D.

▶ PERSEVERE

If students need support, guide them by asking:

Q **Advancing • Use Tools** Which tool could you use to solve the problem? Why choose that tool and not some other? Students' choices of tools and reasons for choosing them will vary.

Q **Assessing** What appears to be true about the four diagonal bars in the scissor lift? They have the same length.

Q **Advancing** How would you describe the angles between each pair of diagonal bars as the car is lifted? The measures of the angles change.

Turn and Talk Have students imagine or sketch what the scissor lift would look like if the bars were not straight. Then, have them imagine how the platform would be affected if the angles at the intersection of each pair of bars had different measures.

- The relationship between the angles might change. The angle pairs 1 and 3, and 2 and 4, would no longer be vertical angles and so might not be congruent.
- One side of the platform will be higher than the other, which might cause the cargo to fall off the lift table.

▶ BUILD SHARED UNDERSTANDING

Select groups of students who used various strategies and tools to share with the class how they solved the problem. As they present their solutions, have each group discuss why they chose a specific strategy and tool.

② Learn Together

Build Understanding

Task 1 (MP) **Reason** Students deduce the relationship between angle congruence and angle measures to prove the Right Angle Congruence Theorem.

Sample Guided Discussion:

Q **The Transitive Property is used to justify Step 3. Explain why this is correct.** Possible answer: The Symmetric Property of Congruence states that if $m\angle B = 90°$, then $90° = m\angle B$. Because $m\angle A = 90°$, the Transitive Property implies that $m\angle A = m\angle B$.

Q **How does the notation $m\angle A$ differ from the notation $\angle A$?** The notation $m\angle A$ is a number that is the degree measure of $\angle A$, while the notation $\angle A$ represents a geometric figure—angle A.

Turn and Talk Help students understand that Euclidean plane geometry consists of logical sets of undefined and defined terms, postulates, and theorems. The Right Angle Congruence Theorem is a theorem that relies on the definition of a right angle and two congruence postulates. Once a theorem has been proven, you can use it without proof to prove a new, and possibly more complicated theorem. Possible answer: Once you have proven this theorem, you can just state two right angles are congruent by right angle congruence.

Build Understanding

Analyze Congruence and Equal Measure in a Proof

Two angles are congruent if and only if the angles have the same measure. Recall the properties of congruence for segments that you learned previously. These properties of congruence are also true for angles.

Properties of Angle Congruence	
Property	**Words**
Reflexive	$\angle A \cong \angle A$
Symmetric	If $\angle A \cong \angle B$, then $\angle B \cong \angle A$.
Transitive	If $\angle A \cong \angle B$ and $\angle B \cong \angle C$, then $\angle A \cong \angle C$.

Right Angle Congruence Theorem
All right angles are congruent.

1 Consider the following proof of the Right Angle Congruence Theorem.

Given: $\angle A$ and $\angle B$ are right angles.

Prove: $\angle A \cong \angle B$

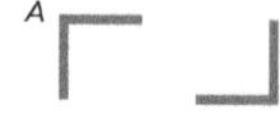

Statements	Reasons
1. $\angle A$ and $\angle B$ are right angles.	1. Given
2. $m\angle A = 90°$ $m\angle B = 90°$	2. Definition of Right Angle
3. $m\angle A = m\angle B$	3. ___?___
4. $\angle A \cong \angle B$	4. Definition of Congruent Angles

A. In Step 3 of the proof, the measures of $\angle A$ and $\angle B$ are shown to be equal. What is the property that justifies this step? A–D. See Additional Answers.

B. Why would the definition of right angles be used in Step 2?

C. Why is it correct to apply the definition of congruent angles in Step 4? Explain your reasoning.

D. What is a reason for using the angle names, then switching to angle measures, and then switching back to the angle names?

Turn and Talk You proved the Right Angle Congruence Theorem in Task 1. Describe why you no longer have to state the definition and angle measures of right angles when proving any theorems that contain right angles, such as any theorem involving rectangles or right triangles. See margin.

64

LEVELED QUESTIONS

Depth of Knowledge (DOK)	Leveled Questions	What Does This Tell You?
Level 1 **Recall**	What property allows you to say that the measures of $\angle A$ and $\angle B$ equal 90°? All right angles have a measure of 90 degrees.	Students' answers will indicate whether they know that the measure of a right angle is 90 degrees.
Level 2 **Basic Application of Skills & Concepts**	Given that $m\angle B = 90°$, what property of congruence can you use to conclude that $90° = m\angle B$? I can use the Symmetric Property of Congruence.	Students' answers will demonstrate whether they can use the properties of congruence to complete steps in a proof.
Level 3 **Strategic Thinking & Complex Reasoning**	If $\angle A$ and $\angle B$ are both right angles in a figure with 4 right angles, what must be true about the remaining 2 angles in the figure? Possible answer: The remaining 2 angles are congruent to $\angle A$ and $\angle B$.	Students' answers will reflect whether they understand that if two angles are right, then angles congruent to them are also right angles.

Step It Out

Justify Each Step in a Solution A–C. See Additional Answers.

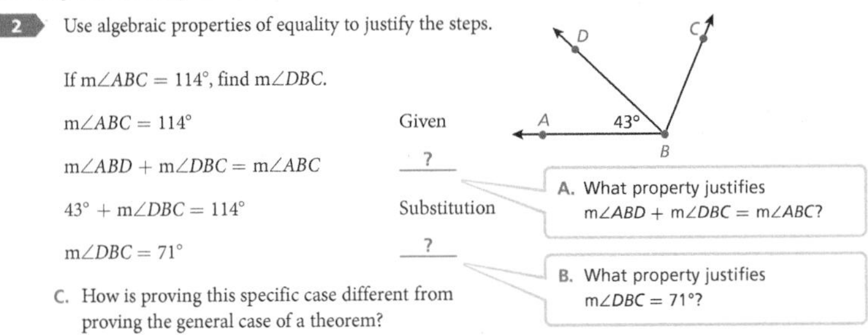

2 Use algebraic properties of equality to justify the steps.

If $m\angle ABC = 114°$, find $m\angle DBC$.

$m\angle ABC = 114°$	Given
$m\angle ABD + m\angle DBC = m\angle ABC$	?
$43° + m\angle DBC = 114°$	Substitution
$m\angle DBC = 71°$	?

A. What property justifies $m\angle ABD + m\angle DBC = m\angle ABC$?

B. What property justifies $m\angle DBC = 71°$?

C. How is proving this specific case different from proving the general case of a theorem?

Turn and Talk Write the problem in this task as an if-then statement. See margin.

Prove the Congruent Supplements Theorem

Congruent Supplements Theorem
If two angles are supplements of the same angle, or congruent angles, then the two angles are congruent.

3 Prove the Congruent Supplements Theorem.

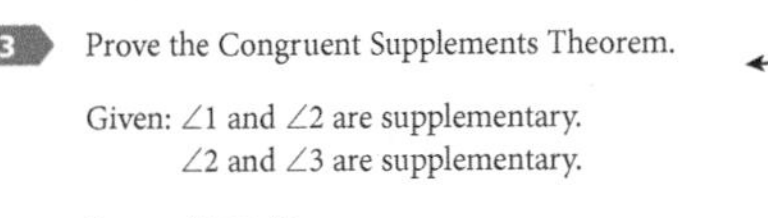

Given: $\angle 1$ and $\angle 2$ are supplementary.
$\angle 2$ and $\angle 3$ are supplementary.

Prove: $\angle 1 \cong \angle 3$

A–C. See Additional Answers.

Statements	Reasons
1. $\angle 1$ and $\angle 2$ are supplementary. $\angle 2$ and $\angle 3$ are supplementary.	1. Given
2. $m\angle 1 + m\angle 2 = 180°$ $m\angle 2 + m\angle 3 = 180°$	2. ?
3. $m\angle 1 + m\angle 2 = m\angle 2 + m\angle 3$	3. ?
4. $m\angle 1 = m\angle 3$	4. Subtraction Property of Equality
5. $\angle 1 \cong \angle 3$	5. ?

A. What reason justifies $m\angle 1 + m\angle 2 = 180°$ and $m\angle 2 + m\angle 3 = 180°$?

B. What reason justifies $m\angle 1 + m\angle 2 = m\angle 2 + m\angle 3$?

C. What reason justifies $\angle 1 \cong \angle 3$?

Turn and Talk Describe how to use the proof of the Congruent Supplements Theorem as a model for a proof of the Congruent Complements Theorem. See margin.

PROFICIENCY LEVEL

Beginning
Write the word "supplementary" on the board and explain how it states that the sum of the measures of two angles is 180°.

Intermediate
Have students decide whether the following statement is always, sometimes, or never true. "If two angles are supplementary, then one angle is acute and the other angle is obtuse." Then have students rewrite the statement depending on their answers.

Advanced
Have students explain or prove that if two supplementary angles are congruent, each is a right angle.

Step It Out

Task 2 (MP) **Construct Arguments** Students justify the steps in a simple, formal proof.

Sample Guided Discussion:

Q What is the relationship between $\angle ABD$ and $\angle DBC$? They are adjacent angles because they have the same vertex and share a common side, $\overline{BD}$.

Q Why is the equation $m\angle DBC + m\angle ABD = m\angle ABC$ also correct? Possible answer: Because addition is commutative.

Turn and Talk Students may incorrectly apply the Angle Addition Postulate and find the sum of the two given measures rather than their difference. If they are struggling, have them think of the equation in terms of x, such that $43 + x = 114$. If $m\angle ABC = 114°$ and $m\angle ABD = 43°$, then $m\angle DBC = 71°$.

Task 3 (MP) **Critique Reasoning** Students use critical thinking to determine the correct reasons for the statements given in a proof.

Sample Guided Discussion:

Q How would you classify $\angle 1$, $\angle 2$, and $\angle 3$ based on their appearance? $\angle 1$ and $\angle 3$ appear to be acute angles, and $\angle 2$ appears to be an obtuse angle.

Q Why is it important to prove that $\angle 1 \cong \angle 3$, instead of using the angle diagrams that are shown to identify congruence? The way figures are drawn can suggest a relationship, but additional information must be given to draw a logical conclusion.

Turn and Talk Have students recall the definition of complementary angles and redraw $\angle 2$ to reflect that it would have to be an acute angle. Change the sum of the angles from 180° to 90°.

Task 4 (MP) **Construct Arguments** Students use the definition of linear pairs and supplementary angles to prove the Linear Pairs Theorem.

CONNECT TO VOCABULARY

Have students use the **Interactive Glossary** to record their understanding of the vocabulary in this task.

Sample Guided Discussion:

Q Why are $\angle 1$ and $\angle 2$ adjacent angles? They have the same vertex, B, and share a common side, $\overline{BD}$.

Q What geometric figure is formed by the noncommon sides of a linear pair of angles? The noncommon sides are opposite rays that form a line.

Task 5 (MP) **Construct Arguments** Students use the definition of vertical angles and the Angle Addition Postulate to prove the Vertical Angles Theorem.

Sample Guided Discussion:

Q What property states that if two angles are congruent to a third angle, then they are congruent to each other? the Transitive Property

Q How can you use what you know about algebraic equations to justify that $m\angle 1 + m\angle 2 = m\angle 2 + m\angle 3$ is equivalent to $m\angle 1 = m\angle 3$? The Subtraction Property of Equality says that if we subtract the same value from both sides of an equation, the equation remains true.

SUPPORT SENSE-MAKING Three Reads

Have students read the problem three times. Use the questions below for a different focus each read.

1. What is the situation about?
2. What are the quantities in the situation?
3. What are the possible mathematical questions that you could ask for the situation?

Turn and Talk Remind students to review the Given and Prove statements to make sure that they finish their proofs with the final conclusion that was asked for. The theorem itself and the Prove line in the proof ask for a relationship between angles not angle measures. The proof steps themselves involved determining equal angle measures, but at the end of the proof the equal measures need to be converted into a statement about congruent angles.

Prove Theorems about Angles

Linear Pairs Theorem

If two angles form a linear pair, then they are supplementary.
$\angle 1$ and $\angle 2$ are a linear pair, so $m\angle 1 + m\angle 2 = 180°$.

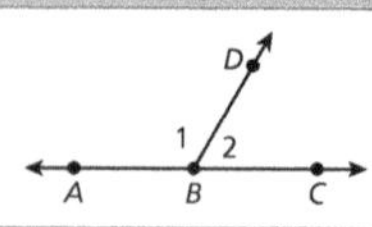

Connect to Vocabulary

A **linear pair** is a pair of adjacent angles whose noncommon sides are opposite rays. **Vertical angles** are the nonadjacent angles formed by two intersecting lines.

Vertical Angles Theorem

Vertical angles are always congruent.
$\angle 1$ and $\angle 3$ are vertical angles so $\angle 1 \cong \angle 3$.
$\angle 2$ and $\angle 4$ are vertical angles so $\angle 2 \cong \angle 4$.

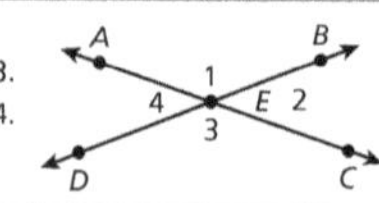

4 Prove the Linear Pairs Theorem.

Given: $\angle 1$ and $\angle 2$ form a linear pair.

Prove: $\angle 1$ and $\angle 2$ are supplementary.

A, B. See Additional Answers.

Statements	Reasons
1. $\angle 1$ and $\angle 2$ form a linear pair.	1. Given
2. $\overrightarrow{BA}$ and $\overrightarrow{BC}$ are opposite rays.	2. ?
3. $m\angle ABC = 180°$	3. Definition of straight angle
4. $m\angle 1 + m\angle 2 = m\angle ABC$	4. ?
5. $m\angle 1 + m\angle 2 = 180°$	5. Transitive Property
6. $\angle 1$ and $\angle 2$ are supplementary.	5. Definition of supplementary angles

A. What property justifies $\overrightarrow{BA}$ and $\overrightarrow{BC}$ being opposite rays?

B. What property justifies $m\angle 1 + m\angle 2 = m\angle ABC$?

5 Prove the Vertical Angles Theorem.

Given: $\angle 1$ and $\angle 3$ are vertical angles.

Prove: $\angle 1 \cong \angle 3$

A, B. See Additional Answers.

Statements	Reasons
1. $\angle 1$ and $\angle 3$ are vertical angles.	1. Given
2. $m\angle AEC = 180°$; $m\angle DEB = 180°$	2. Definition of straight angle
3. $m\angle 1 + m\angle 2 = 180°$ $m\angle 2 + m\angle 3 = 180°$	3. Angle Addition Postulate
4. $m\angle 1 + m\angle 2 = m\angle 2 + m\angle 3$	4. ?
5. $m\angle 1 = m\angle 3$	5. ?
6. $\angle 1 \cong \angle 3$	6. Definition of congruent angles

A. What property justifies $m\angle 1 + m\angle 2 = m\angle 2 + m\angle 3$?

B. What property justifies $m\angle 1 = m\angle 3$?

Turn and Talk In Task 5, why is the final step of the proof needed? See margin.

66

Check Understanding

1. **MP Attend to Precision** Are ∠1 and ∠2 congruent? Explain your reasoning. See Additional Answers.

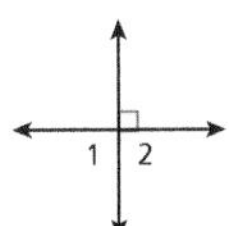

2. Find m∠ABC, if m∠ABD = 120°. 50°

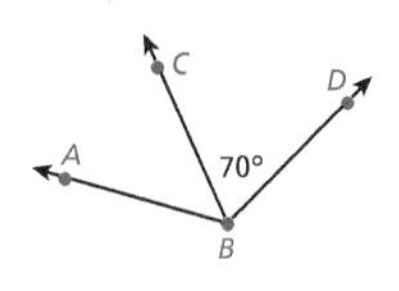

3. If ∠X ≅ ∠Y and ∠Y ≅ ∠Z, what can you conclude about the relationship between ∠X and ∠Z? The angles are congruent by the Transitive Property.

On Your Own

Match each statement with the property of congruence that makes the statement true.

A. Reflexive Property
B. Symmetric Property
C. Transitive Property

4. ∠C ≅ ∠D, so ∠D ≅ ∠C. B
5. ∠L ≅ ∠M and ∠M ≅ ∠N, so ∠L ≅ ∠N. C
6. ∠BCD ≅ ∠BCD A
7. ∠XYZ ≅ ∠RST and ∠RST ≅ ∠ABC, so ∠XYZ ≅ ∠ABC. C
8. ∠Q ≅ ∠Q A
9. ∠JKL ≅ ∠FGH, so ∠FGH ≅ ∠JKL. B

Find the measure of each angle.

10. ∠1 and ∠2 form a linear pair and m∠2 = 52°. m∠1 = 128°, m∠2 = 52°
11. ∠1 and ∠2 are supplementary, ∠2 and ∠3 are supplementary, and m∠2 = 32°. m∠1 = 148°, m∠2 = 32°, m∠3 = 148°
12. ∠1 and ∠2 are vertical angles and m∠1 = 44°. m∠1 = 44°, m∠2 = 44°
13. ∠1 and ∠2 form a linear pair and m∠1 = 132°. m∠1 = 132°, m∠2 = 48°
14. ∠1 and ∠2 are supplementary, ∠2 and ∠3 are supplementary, and m∠2 = 85°. m∠1 = 95°, m∠2 = 85°, m∠3 = 95°
15. ∠1 and ∠2 are vertical angles and m∠2 = 103°. m∠1 = 103°, m∠2 = 103°
16. A section of fencing has four angles that are labeled as shown. Which angle is a vertical angle to ∠4? Which angles form linear pairs with ∠4? ∠2; ∠1 and ∠3

ONLINE

Assign the Digital On Your Own for
- built-in student supports
- Actionable Item Reports
- Standards Analysis Reports

On Your Own

Assignment Guide

The chart below indicates which problems in the On Your Own are associated with each task in the Learn Together. Assign daily homework for tasks completed.

Learn Together Tasks	On Your Own Problems
Task 1 p. 60	Problems 4–9
Task 2, p. 61	Problems 19, 32, and 36
Task 3, p. 61	Problems 11, 14, 22, 24, 29, 30, 35, and 36
Task 4, p. 62	Problems 10, 13, 16, 18, 21 25, 27, 30, and 34
Task 5, p. 62	Problems 12, 15, 16, 17, 20, 23, 26, 28–31, and 33–35

3 Check Understanding

Formative Assessment

Use formative assessment to determine if your students are successful with this lesson's learning objective.

Students who successfully complete the Check Understanding can continue to the On Your Own practice.

For students who miss 1 problem or more, work in a pulled small group using the Almost There small-group activity on page 59C.

ONLINE

Assign the Digital Check Understanding to determine
- success with the learning objective
- items to review
- grouping and differentiation resources

4 Differentiation Options

Differentiate instruction for all students using small-group activities and math center activities on page 59C.

Reteach

Challenge

Questioning Strategies

Problem 18 What is the converse of the Linear Pair Theorem? The Linear Pairs Theorem is "If two angles form a linear pair, then they are supplementary." The converse is "If two angles are supplementary, then they form a linear pair." The converse is not true because not all supplementary angles form a linear pair. By definition, if the sum of the measures of two angles is 180°, the angles are supplementary.

Problem 24 What equations can you write to find $m\angle 2$? $m\angle 2 = 180 - (12x + 17)$; $m\angle 2 = 180 - (23x - 5)$

17. (MP) **Construct Arguments** Prove the measure of angle AEC is equal to twice the measure of angle FEB.

17–19. See Additional Answers.

18. (MP) **Reason** What is the converse of the Linear Pair Theorem? Is this a true statement? Explain your reasoning.

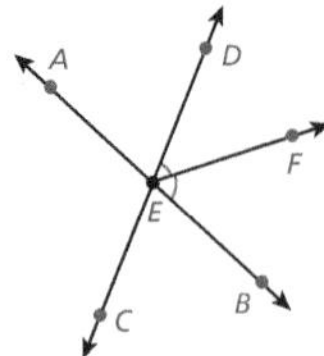

19. Write a two-column proof.

Given: $\angle 1 \cong \angle 2$; $\angle 2$ is a complement of $\angle 3$.

Prove: $\angle 1$ is a complement of $\angle 3$.

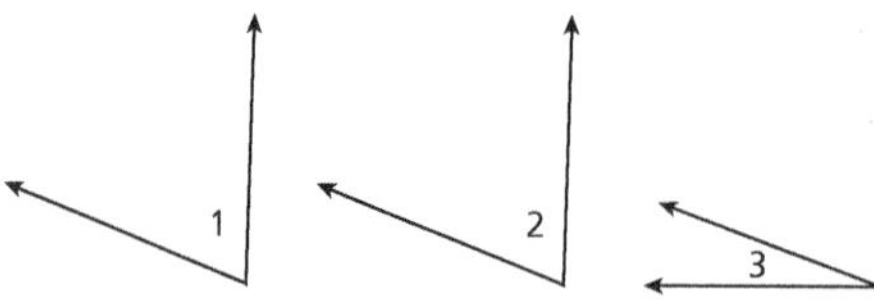

In Problems 20–28, find the value of x.

20. $\angle 1$ and $\angle 2$ are vertical angles. $m\angle 1 = 40°$ and $m\angle 2 = (7x + 5)°$. $x = 5$

21. $\angle 1$ and $\angle 2$ are a linear pair. $m\angle 1 = (11x + 6)°$ and $m\angle 2 = 75°$. $x = 9$

22. $\angle 1$ and $\angle 2$ are supplementary. $\angle 2$ and $\angle 3$ are supplementary. $m\angle 1 = (4x)°$ and $m\angle 3 = (x + 48)°$. $x = 16$

23. $\angle 1$ and $\angle 2$ are vertical angles. $m\angle 1 = (7x - 8)°$ and $m\angle 2 = (5x + 34)°$. $x = 21$

24. $\angle 1$ and $\angle 2$ are supplementary. $\angle 2$ and $\angle 3$ are supplementary. $m\angle 1 = (12x + 17)°$ and $m\angle 3 = (23x - 5)°$. $x = 2$

25. $\angle 1$ and $\angle 2$ are a linear pair. $m\angle 1 = (4x + 17)°$ and $m\angle 2 = (11x - 17)°$. $x = 12$

26. $\angle 1$ and $\angle 2$ are vertical angles. $m\angle 1 = (9x - 13)°$ and $m\angle 2 = (6x + 38)°$. $x = 17$

27. $\angle 1$ and $\angle 2$ are a linear pair. $m\angle 1 = (18x + 43)°$ and $m\angle 2 = (31x - 10)°$. $x = 3$

28. $\angle 1$ and $\angle 2$ are vertical angles. $\angle 2$ and $\angle 3$ are supplementary. $m\angle 1 = (2x - 13)°$ and $m\angle 3 = (4x + 1)°$. $x = 32$

29. Which angles are congruent to each other in the figure if you know that angle 1 is congruent to angle 5? Explain your reasoning.
$\angle 2 \cong \angle 3 \cong \angle 6 \cong \angle 7$; $\angle 1 \cong \angle 4 \cong \angle 5 \cong \angle 8$

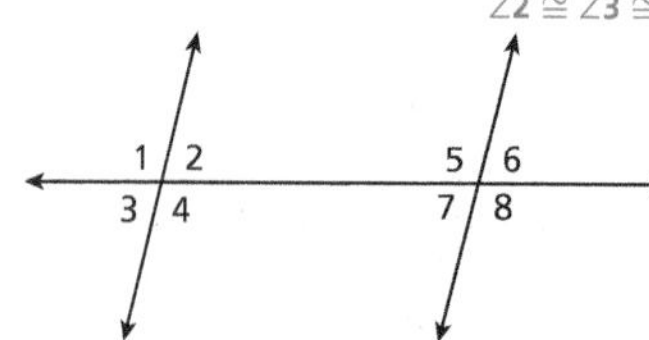

30. STEM Engineers design structures to support the weight of the structure itself and other forces that might be applied to the structure. The diagram shows the internal support structure of the Statue of Liberty. Four angles are labeled in the diagram. What is the measure of ∠1, ∠2, and ∠3? Justify your reasoning for each measure.
See Additional Answers.

31. How many pairs of vertical angles are shown in the drawing? List the pairs of angles. 3; ∠1 and ∠4, ∠2 and ∠5, ∠3 and ∠6

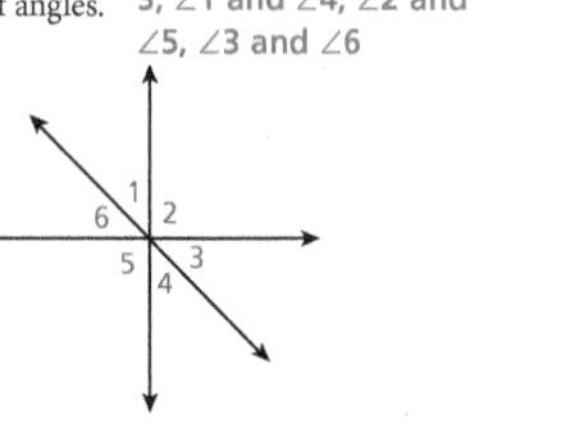

32. ∠*HEF* and ∠*DEH* are complementary angles. What is the measure of ∠*HEG*? 96°

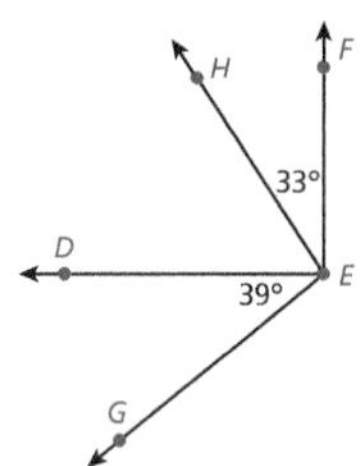

33. The lines through $\overline{AD}$ and $\overline{BE}$ intersect at point *F*. Complete the proof that m∠*AFE* = m∠*BFC* + m∠*CFD*.

Statements	Reasons
1. ∠*AFE* and ∠*BFD* are vertical angles.	1. ∠*AFE* and ∠*BFD* are opposite angles sharing a vertex at the intersection of two lines.
2. ∠*AFE* ≅ ∠*BFD*	2. Vertical Angles Theorem
3. m∠*AFE* = m∠*BFD*	3. Definition of congruent angles
4. m∠*BFD* = m∠*BFC* + m∠*CFD*	4. Angle Addition Postulate
5. m∠*AFE* = m∠*BFC* + m∠*CFD*	5. Transitive Property

34. Find the measures of ∠1, ∠2, and ∠3. Justify your reasoning for the measure of each angle.
See Additional Answers.

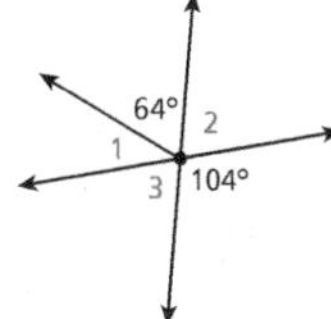

Watch for Common Errors

Problem 32 Have students trace the angles that are complementary, ∠*HEF* and ∠*DEH*, to correctly identify the angles whose measures have a sum of 90°. Have them describe what they are asked to find in the problem, m∠*HEG*, and how they could find that measure before using mathematical means to solve. This may eliminate the possibility that they incorrectly find m∠*DEH* by subtracting m∠*HEF* from 90°.

Questioning Strategies

Problem 33 What is the relationship between the types of and measures of ∠*AFE* and ∠*BFC* and ∠*CFD*? They are vertical angles, and vertical angles are congruent.

5 Wrap-Up

Summarize learning with your class. Consider using the Exit Ticket, Put It in Writing, or I Can scale.

Exit Ticket 

In the figure below, $\angle 1$ and $\angle 2$ are vertical angles, and $\angle 1$ and $\angle 2$ are supplementary. What is $m\angle 1$?

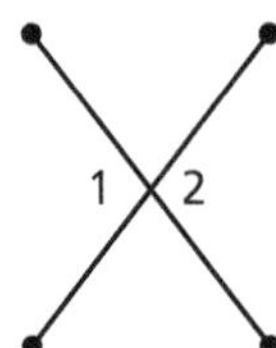

Because $\angle 1$ and $\angle 2$ are vertical angles, $m\angle 1 = m\angle 2$. Because $\angle 1$ and $\angle 2$ are supplementary, the sum of their measures is equal to $180°$. Therefore, if $x = m\angle 1$ and $x = m\angle 2$, $2x = 180°$, and $x = 90°$. So, $m\angle 1 = 90°$.

Put It in Writing 

Describe some strategies you can use to prove theorems about angles.

I Can 

The scale below can help you and your students understand their progress on a learning goal.

4	I can use definitions and relations between lines and angles to prove theorems involving lines and angles, and I can explain my proofs to others.
3	I can use definitions and relations between lines and angles to prove theorems involving lines and angles.
2	I can use definitions and relations between lines and angles to solve problems involving missing measurements.
1	I can recognize definitions and relations between lines and angles to solve problems with help.

Spiral Review • Assessment Readiness

These questions will help determine if students have retained information taught in the past and can also prepare them for high-stakes assessments. Here, students must be able to recognize a pair of vertical angles **(Gr7, 7.5)**, restate a given equation and solution as an if-then statement **(2.2)**, and recognize the properties of congruence **(2.3)**.

35. The figure shows a map of five streets that meet at Winthrop Circle. The measure of the angle formed by Mea Road and Hawk Lane is 130°. The measure of the angle formed by East Avenue and Hawk Lane is 148°. Tulip Street bisects the angle formed by Mea Road and Hawk Lane. Summer Drive bisects the angle formed by East Avenue and Hawk Lane. What is the measure of the angle formed by Tulip Street and Summer Drive? Explain your reasoning.
139°; The angle is the sum of half of each of the two given angles.

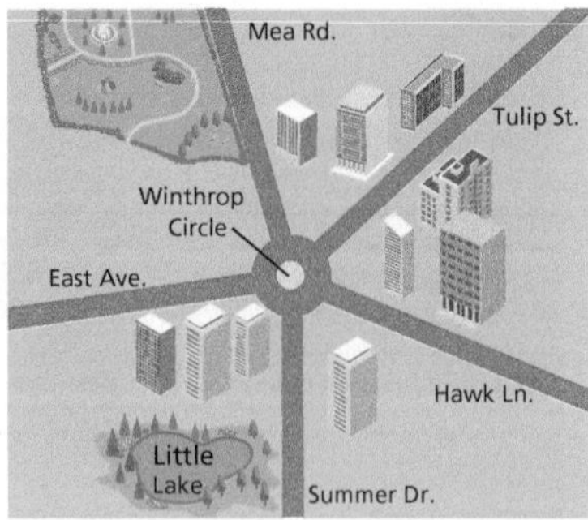

36. Open Ended Write a math problem that can be solved using the Congruent Supplements Theorem. Answers will vary.

37. A rhombus is a quadrilateral with four sides of equal length. Draw a rhombus. Use a compass and straightedge to bisect one of the angles in the rhombus. What do you notice. Confirm your conjecture using at least two additional rhombuses.
See Additional Answers.

Spiral Review • Assessment Readiness

38. Two braces support a table top. Which pair(s) of angles are vertical angles? Select all that apply.

Ⓐ $\angle 1, \angle 3$
Ⓑ $\angle 3, \angle 4$
Ⓒ $\angle 2, \angle 3$
Ⓓ $\angle 2, \angle 4$

39. Write the statement as an if-then statement.
$5x - 7 = 8; x - 3$

Ⓐ If $x = 3$, then $5x = 15$.
Ⓑ If $x = 3$, then $5x - 7 = 8$.
Ⓒ If $5x - 7 = 8$, then $5x = 1$.
Ⓓ If $5x - 7 = 8$, then $x = 3$.

40. Match each statement with the property of congruence that makes it true.

Statement	Property
1. $\overline{RS} \cong \overline{TU}$, so $\overline{TU} \cong \overline{RS}$. B	**A.** Reflexive Property
2. $\overline{UV} \cong \overline{WX}$ and $\overline{WX} \cong \overline{YZ}$, so $\overline{UV} \cong \overline{YZ}$. C	**B.** Symmetric Property
3. $\overline{CD} \cong \overline{CD}$ A	**C.** Transitive Property

I'm in a Learning Mindset!

When I prove and utilize theorems about angles in group settings, what strategies do I use to stay on task while working with my peers?

Keep Going ▶ Journal and Practice Workbook

Learning Mindset

Perseverance Sustains Attention

Point out that proving theorems about angles requires an understanding of the definition of each type of angle, the relationship between congruence and measures of angles, and/or the special relationships between complementary, supplementary, vertical angles, and linear pairs of angles. When proving theorems about angles in group settings, encourage students to work together to identify all the given relationships and list any justifications that might be needed. Ask them to consider the logical progression of steps. *How does justifying each step in proving a theorem help you develop a more confident mindset? How can having multiple people work on the same proof provide an advantage? How can you effectively communicate ideas to other group members?*

Module 2

Review

Conditional Statements

A rectangle that has four equal sides is a square.

Conditional Statement If a rectangle has four equal sides, then it is a square.

Converse If a rectangle is a square, then it has four equal sides.

Inverse If a rectangle does not have four equal sides, then it is not a square.

Contrapositive If a rectangle is not a square, then it does not have four equal sides.

Inductive and Deductive Reasoning

Inductive Reasoning Helen measured the sides of five squares and noticed that all sides are the same length. Helen concludes that all squares have four sides that are the same length.

Deductive Reasoning A rectangle with four equal sides is a square.

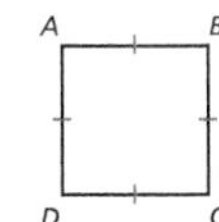

Rectangle *ABCD* is a square.

Segments

$\overline{JK}$ has a midpoint at *L*.

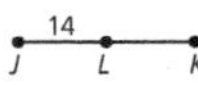

Because *L* is the midpoint of $\overline{JK}$, $JL = LK$.

Because the lengths of the segments are equal, the segments are congruent.

So, $\overline{JL} \cong \overline{LK}$.

Figure *ABCD* is a square.

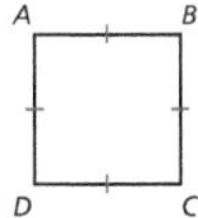

Because the figure is a square, all sides have the same length. So, $\overline{AB} \cong \overline{BC} \cong \overline{CD} \cong \overline{DA}$.

Angles

Figure *ABCD* is a square.

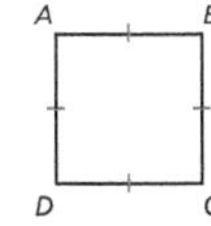

Each angle in square *ABCD* is a right angle, so each angle measures 90°. Each angle has the same measure, so each angle is congruent.

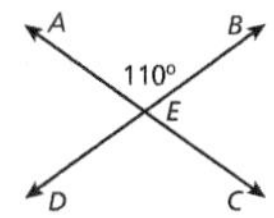

$\angle AEB$ and $\angle DEC$ are vertical angles, so they are congruent. So, $m\angle DEC = 110°$.

$\angle AEB$ and $\angle AED$ form a linear pair, so they are supplementary. So, $m\angle AED = 70°$.

$\angle AED$ and $\angle BEC$ are vertical angles, so they are congruent. So, $m\angle BEC = 70°$.

ONLINE

Assign the Digital Module Review for

- built-in student supports
- Actionable Item Reports
- Standards Analysis Reports

Module Review

Use the first page of the Module Review to summarize and connect the main ideas of the module. Use the second page to assess students' understanding of the vocabulary, concepts, and skills presented in the module.

Sample Guided Discussion:

Q Conditional Statements The expression $p \rightarrow q$ is a conditional statement, where p is the hypothesis and q is the conclusion. How can you write the statement "A rectangle that has four equal sides is a square." and its converse, inverse, and contrapositive in terms of p and q? The conditional statement is $p \rightarrow q$. The converse is $q \rightarrow p$, the inverse is not $p \rightarrow$ not q, and the contrapositive is not $q \rightarrow$ not p.

Q Inductive and Deductive Reasoning Explain the difference between inductive and deductive reasoning as given in this section. Possible answer: Inductive reasoning: In the first example, Helen measured the sides of one square and concluded that the four sides of all squares are equal. Deductive reasoning: In the second example, no measures were involved. Rather, geometric facts and theorems were applied to prove that a rectangle with four equal sides is a square.

Q Segments If point L is between J and K as shown and $JL \neq LK$, then what conclusion could you draw? Possible answer: Point L is not the midpoint of $\overline{JK}$.

Q Angles A square is a rectangle with four equal sides. What can you conclude about the angles of a rectangle? Possible answer: Each angle in a rectangle is a right angle.

Module Review continued

Possible Scoring Guide

Items	Points	Description
1–4	2 each	identifies the correct term(s)
5, 6	2 each	correctly writes the definition of the term as a biconditional statement
7	2	correctly writes each type of statement
8	2	correctly uses deductive reasoning to write a conclusion
9, 10	2 each	correctly decides whether inductive or deductive reasoning was used
11	2	correctly uses the Segment Addition Postulate to solve for x
12	2	correctly uses the Segment Addition Postulate and deductive reasoning to establish whether two segments are congruent
13, 14	2 each	correctly finds the measure of each angle
Total points possible = 28 points		

The Unit 1 Test in the Assessment Guide assesses content from Modules 1 and 2.

Vocabulary

Choose the correct term from the box to complete each sentence.

Vocabulary
conditional statement
contrapositive
deductive reasoning
linear pair

1. A two-column proof is an example of ___?___. deductive reasoning
2. A ___?___ is a pair of adjacent angles whose noncommon sides are opposite rays. linear pair
3. The statement "If a number is divisible by 2, then it is even" is an example of a ___?___. conditional statement
4. A statement formed by both exchanging and negating the hypothesis and conclusion of a conditional statement is the ___?___. contrapositive

Concepts and Skills 5–7. See Additional Answers.

Write the definition of the term as a biconditional statement.

5. obtuse angle
6. pentagon
7. Write the statement "Monday is a weekday" in if-then form. Then write the converse, inverse, and contrapositive of the statement.
8. A triangle with three acute angles is an acute triangle. Triangle RST has three acute angles. Use deductive reasoning to write a conclusion. $\triangle RST$ is an acute triangle.

In Problems 9 and 10, decide if inductive reasoning or deductive reasoning is used to reach the conclusion. Explain your reasoning.

9. All acute angles measure less than 90°. See below.
   $\angle LMN$ is an acute angle, so $\angle LMN$ has a measure less than 90°.
10. (MP) **Use Tools** Every day last week, Sally had a turkey sandwich for lunch. Sally will have a turkey sandwich for lunch today. State what strategy and tool you will use to determine which type of reasoning is used, explain your choice, and then find the answer. inductive reasoning; The conclusion is based on observing Sally's lunch each day last week.
11. Segment AC has a point B between points A and C. $AB = 4x + 2$, $BC = 3x - 1$, and $AC = 9x - 11$. Solve for x. $x = 6$
12. $\overline{PR}$ had a point Q between points P and R. The length of $\overline{QR}$ is 15 units. The length of $\overline{PR}$ is 32 units. Is $\overline{PQ} \cong \overline{QR}$? Explain. no; If $PR = 32$ and $QR = 15$, then $PQ = 17$.

Find the measure of each angle.

13. $\angle 1$ and $\angle 2$ are vertical angles. $m\angle 1 = 66°$. $m\angle 1 = 66°$, $m\angle 2 = 66°$
14. $\angle 1$ and $\angle 2$ are supplementary. $\angle 2$ and $\angle 3$ are supplementary. $m\angle 1 = 133°$. $m\angle 1 = 133°$, $m\angle 2 = 47°$, $m\angle 3 = 133°$

9. deductive reasoning; The definition of an acute angle is used to draw the conclusion.

72

DATA-DRIVEN INSTRUCTION

Before moving on to the Module Test, use the Module Review results to intervene based on the table below.

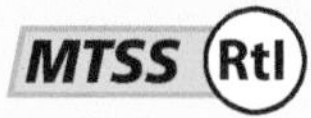

Items	Lesson	DOK	Content Focus	Intervention
5, 6	2.1	2	Write biconditional statements.	Reteach 2.1
7	2.1	3	Write conditional statements.	Reteach 2.1
8	2.2	3	Write a conclusion using deductive reasoning.	Reteach 2.2
9, 10	2.2	3	Recognize inductive and deductive reasoning.	Reteach 2.2
11	2.3	2	Use the Segment Addition Postulate and the definition of the midpoint of a segment.	Reteach 2.3
12	2.3	3	Use the Segment Addition Postulate and deduction to show that two segments are not congruent.	Reteach 2.3
13, 14	2.4	2	Find the measures of vertical and supplementary angles.	Reteach 2.4

Module Test

The Module Test is available in alternative versions in your Assessment Guide. All items are presented in standardized test formats.

ONLINE

Ed

Assign the Digital Module Test to power actionable reports including

- proficiency by standards
- item analysis

MODULE 2

TEST

Form A

Name

Module 2 • Form A
Module Test

1. Which equation or inequality is a counterexample to the statement?
   If A, B, and C are collinear points, AB + BC = AC.
   Ⓐ $BC = AB$
   Ⓑ $BC < AC$
   Ⓒ $AB > AC$
   Ⓓ $AB > BC$

2. Statement 1: *If a quadrilateral has four right angles, then it is a rectangle.*
   Statement 2: *If a quadrilateral is not a rectangle, then it does not have four right angles.*
   What is the relationship between the statements?
   Ⓐ Statement 2 is the inverse of statement 1.
   Ⓑ Statement 2 is the converse of statement 1.
   Ⓒ Statement 2 is the biconditional of statement 1.
   Ⓓ Statement 2 is the contrapositive of statement 1.

3. If $\overline{AB} \cong \overline{BC}$ and $\overline{BC} \cong \overline{CD}$, which property shows $\overline{AB} \cong \overline{CD}$?
   Ⓐ Addition Property
   Ⓑ Reflexive Property
   Ⓒ Symmetric Property
   Ⓓ Transitive Property

4. Given $\overline{QR} \cong \overline{ST}$. If $QR = \frac{2}{3}x$ in. and $ST = 0.4$ in., what is the value of x?
   Possible answer: $\frac{3}{5}$

5. Point G lies on $\overline{DE}$. If $DE = 21$ m, $DG = 9x - 7$ m, and $GE = 6x - 2$ m, what is the length of $\overline{DG}$ in meters?
   11

6. The diagram shows four rays that share endpoint V, and $\angle WVX \cong \angle YVZ$.
   W X Y V Z
   Prove: $m\angle WVY = m\angle XVZ$

Statements	Reasons
1. $\angle WVX \cong \angle YVZ$	1.
2. $m\angle WVX = m\angle YVZ$	2.
3. $m\angle XVY = m\angle XVY$	3.
4. $m\angle WVX + m\angle XVY = m\angle YVZ + m\angle XVY$	4.
5. $m\angle WVX + m\angle XVY = m\angle WVY$; $m\angle YVZ + m\angle XVY = m\angle XVZ$	5.
6. $m\angle WVY = m\angle XVZ$	6.

   Which reasons correctly complete the proof for the given statement?
   Select all the correct reasons.
   Ⓐ 1. Given
   Ⓑ 2. Definition of Congruent Angles
   Ⓒ 3. Transitive Property
   Ⓓ 4. Reflexive Property
   Ⓔ 5. Angle Addition Postulate
   Ⓕ 6. Symmetric Property

7. If $\angle S$ and $\angle T$ are a linear pair and $m\angle S = 87°$, what is $m\angle T$ in degrees?
   93°

Form A

Module 2 • Form A
Module Test

Name

8. Place an X in the table to show whether each statement is always true, never true, or sometimes true.

	Always True	Never True	Sometimes True
A triangle has 2 obtuse angles.		X	
Points X, Y, and Z are collinear.			X
Line MN is longer than segment MN.	X		
If ∠F and ∠G are supplementary angles, then m∠F = 90°.			X

9. The top of a three-story building is 80.5 feet from the ground. The bottom of the building's second story is 25.75 feet from the ground. What is the distance in feet from the bottom of the second story to the top of the building?
   Possible answer: 54.75

10. Angle A and angle B are supplementary. If $m\angle A = (16x - 7)°$ and $m\angle B = (21x + 2)°$, what is the measure of angle A in degrees?
    73°

11. Three lines intersect at a point to form the angles shown.
    X Y Z
    If $m\angle Y = 33°$ and $m\angle Z = 87°$, what is $m\angle X$ in degrees?
    60°

12. Line EF intersects $\overline{AB}$ and $\overline{DC}$ as shown, and $AB = DC$.
    A E B D F C
    Part A
    Which statements about the diagram are true?
    Select all the true statements.
    Ⓐ $\overline{AD} \cong \overline{BC}$
    Ⓑ $\overline{AE} \cong \overline{EF}$
    Ⓒ $\overline{BC} \cong \overline{EF}$
    Ⓓ $\overline{EB} \cong \overline{DF}$
    Ⓔ $\overline{EB} \cong \overline{EF}$
    Part B
    If $AB = 27.25$ m and $DF = 16.4$ m, what is the length in meters of $\overline{AE}$?
    Possible answer: 10.85

13. Lines CD and EF intersect at point X, such that $m\angle CXE = \frac{3}{5}m\angle FXE$.
    Part A
    What is $m\angle EXD$ in degrees?
    Possible answer: 67.5°
    Part B
    What is $m\angle DXF$ in degrees?
    Possible answer: 112.5°

Form B

Name

Module 2 • Form B
Module Test

1. Which statement is a counterexample to the statement?
   If ∠A and ∠B are supplementary angles, then m∠A ≠ m∠B.
   Ⓐ Angle A is an obtuse angle.
   Ⓑ Angle A is a straight angle.
   Ⓒ Angle B is an acute angle.
   Ⓓ Angle B is a right angle.

2. Statement 1: *If a quadrilateral is a rhombus, then it has four congruent sides.*
   Statement 2: *If a quadrilateral is not a rhombus, then it does not have four congruent sides.*
   What is the relationship between the statements?
   Ⓐ Statement 2 is the inverse of statement 1.
   Ⓑ Statement 2 is the converse of statement 1.
   Ⓒ Statement 2 is the biconditional of statement 1.
   Ⓓ Statement 2 is the contrapositive of statement 1.

3. Which property shows that if $\overline{MN} \cong \overline{PQ}$, then $\overline{PQ} \cong \overline{MN}$?
   Ⓐ Reflexive Property
   Ⓑ Subtraction Property
   Ⓒ Symmetric Property
   Ⓓ Transitive Property

4. Given $\overline{CD} \cong \overline{GH}$. If $CD = \frac{1}{2}x - 3$ units and $GH = 5 - 3x$ units, what is the value of x?
   Possible answer: $\frac{16}{7}$

5. Point S lies on $\overline{AB}$. If $AB = 34$ units, $AS = 5 + 2x$ units, and $SB = x + 2$ units, what is the length of $\overline{AS}$ in units?
   23

6. In the diagram, $\overline{JK} \cong \overline{LM}$.
   J K L M
   Prove: $JL = KM$

Statements	Reasons
1. $\overline{JK} \cong \overline{LM}$	1.
2. $JK = LM$	2.
3. $KL = KL$	3.
4. $JK + KL = LM + KL$	4.
5. $JK + KL = JL$; $LM + KL = KM$	5.
6. $JL = KM$	6.

   Which reasons correctly complete the proof for the given statement?
   Select all the correct reasons.
   Ⓐ 1. Given
   Ⓑ 2. Pythagorean Theorem
   Ⓒ 3. Transitive Property
   Ⓓ 4. Multiplication Property of Equality
   Ⓔ 5. Segment Addition Postulate
   Ⓕ 6. Substitution

7. If $\angle Q$ and $\angle R$ are a linear pair and $m\angle R = 136°$, what is $m\angle Q$ in degrees?
   44°

Form B

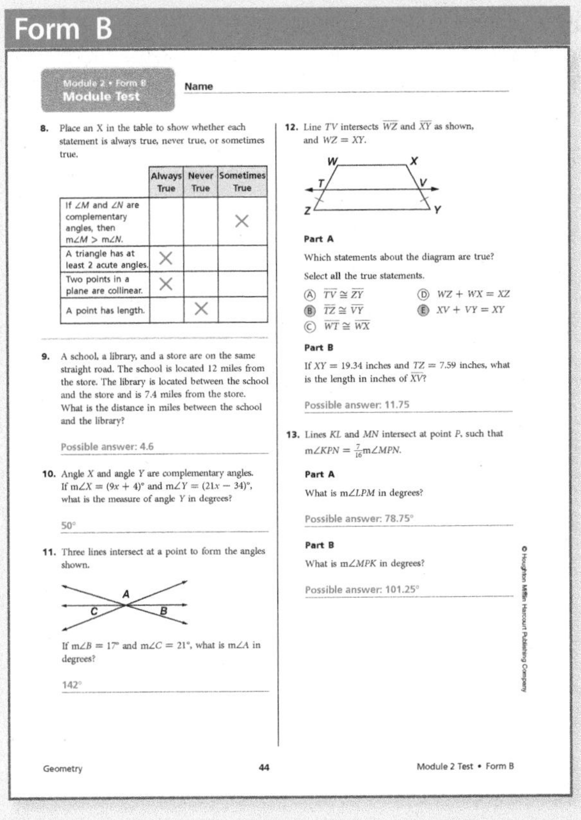

Module 2 • Form B
Module Test

Name

8. Place an X in the table to show whether each statement is always true, never true, or sometimes true.

	Always True	Never True	Sometimes True
If ∠M and ∠N are complementary angles, then m∠M > m∠N.			X
A triangle has at least 2 acute angles.	X		
Two points in a plane are collinear.	X		
A point has length.		X	

9. A school, a library, and a store are on the same straight road. The school is located 12 miles from the store. The library is located between the school and the store and is 7.4 miles from the store. What is the distance in miles between the school and the library?
   Possible answer: 4.6

10. Angle X and angle Y are complementary angles. If $m\angle X = (9x + 4)°$ and $m\angle Y = (21x - 34)°$, what is the measure of angle Y in degrees?
    50°

11. Three lines intersect at a point to form the angles shown.
    A C B
    If $m\angle B = 17°$ and $m\angle C = 21°$, what is $m\angle A$ in degrees?
    142°

12. Line TV intersects $\overline{WZ}$ and $\overline{XY}$ as shown, and $WZ = XY$.
    W X T V Z Y
    Part A
    Which statements about the diagram are true?
    Select all the true statements.
    Ⓐ $\overline{TV} \cong \overline{ZY}$
    Ⓑ $\overline{TZ} \cong \overline{VY}$
    Ⓒ $\overline{WT} \cong \overline{WX}$
    Ⓓ $WZ + WX = XZ$
    Ⓔ $XV + VY = XY$
    Part B
    If $XY = 19.34$ inches and $TZ = 7.59$ inches, what is the length in inches of $\overline{XV}$?
    Possible answer: 11.75

13. Lines KL and MN intersect at point P, such that $m\angle KPN = \frac{7}{16}m\angle MPN$.
    Part A
    What is $m\angle LPM$ in degrees?
    Possible answer: 78.75°
    Part B
    What is $m\angle MPK$ in degrees?
    Possible answer: 101.25°

Module 3: Lines and Transversals

Module 4: Lines on the Coordinate Plane

Textile Engineer STEM

- **Say:** *Think about the textiles or fabrics you see around you every day. These materials were developed and designed by people in the textile industry, including textile engineers. These materials are finely honed from either natural fibers, like cotton and silk, or from man-made materials, like nylon, rayon, and vinyl.*
- Explain that textile engineers use mathematics along with concepts from science to design and work with or invent equipment that makes fabric, yarn, and fiber. Then introduce the STEM task.

STEM Task

Ask students what they may know about the terms *textile*, *engineering*, and *carbon fiber*.

- Have students look around the classroom and point out different types of textiles and patterns they see. Ask if these textiles come from natural sources or are man-made. How can they tell? What forms do the designs on some of the textiles take? Are they regular or irregular designs? Are they made up of parallel and/or perpendicular lines?
- Discuss different types of engineering professions, such as civil, electrical, and mechanical engineering. What is the focus of each of these professions?
- Explain that carbon is a metallic element that occurs naturally in different forms, such as coal and diamonds. Carbon fiber is a polymer, a long chain of many carbon filaments. This material can be thinner than a strand of hair, but is five times stronger and two times stiffer than steel.
- In the STEM task, the area density of the carbon fiber used is 75 kg/m^2. To answer the question posed, students first need to find an area and express it in square meters.

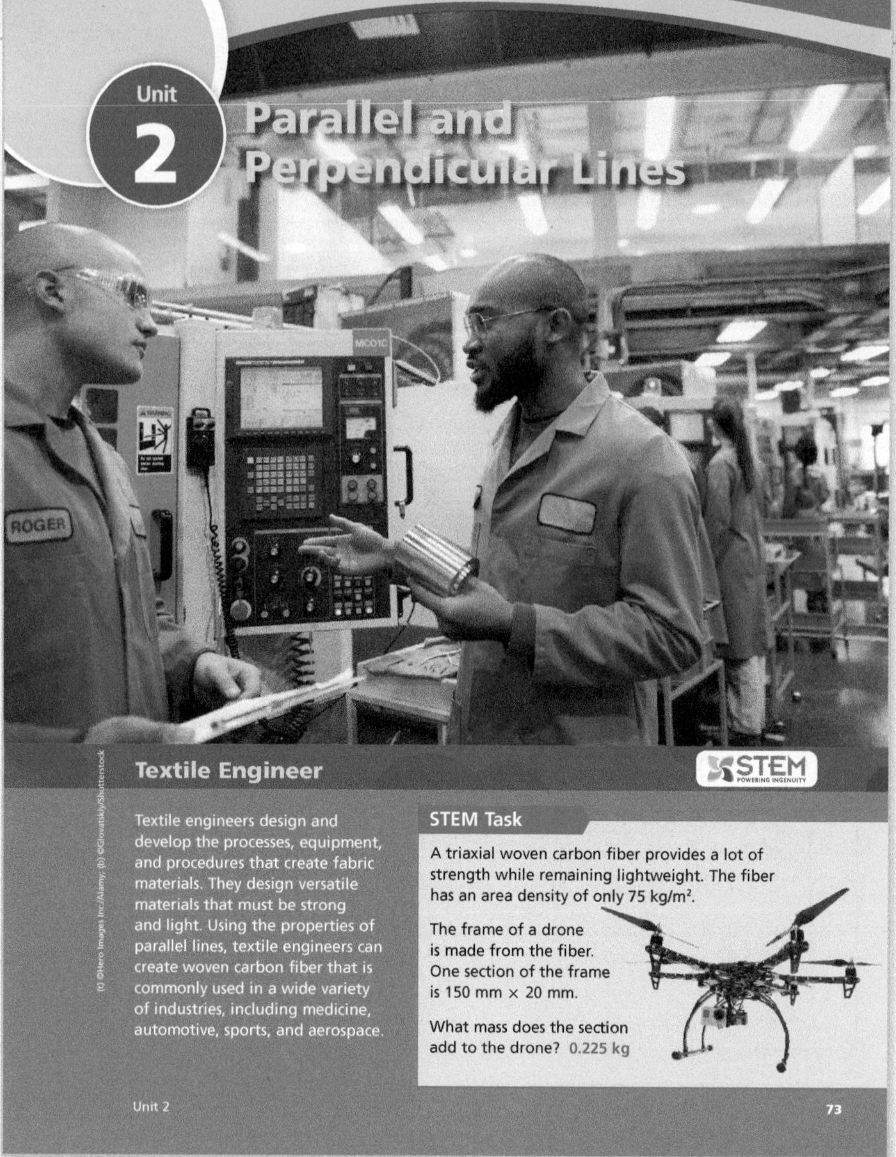

Unit 2

Parallel and Perpendicular Lines

Textile Engineer STEM

Textile engineers design and develop the processes, equipment, and procedures that create fabric materials. They design versatile materials that must be strong and light. Using the properties of parallel lines, textile engineers can create woven carbon fiber that is commonly used in a wide variety of industries, including medicine, automotive, sports, and aerospace.

STEM Task

A triaxial woven carbon fiber provides a lot of strength while remaining lightweight. The fiber has an area density of only 75 kg/m^2.

The frame of a drone is made from the fiber. One section of the frame is 150 mm × 20 mm.

What mass does the section add to the drone? 0.225 kg

(t) ©Hero Images Inc/Alamy; (b) ©Glovatskiy/Shutterstock

Unit 2 73

Unit 2 Project Fiber for Flight

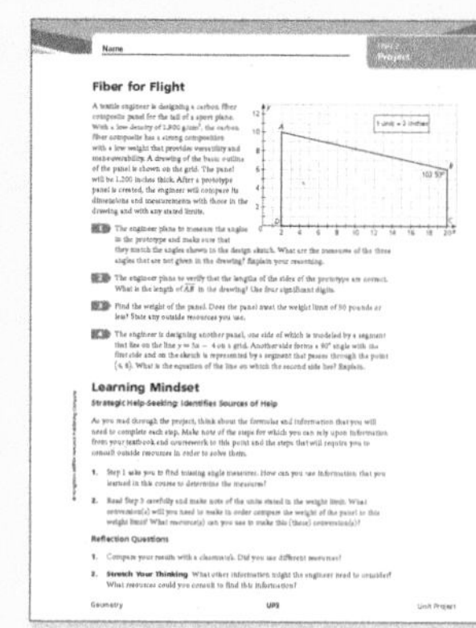

Overview: In this project students use properties of parallel and perpendicular lines to determine the measures of the sides and angles of a carbon fiber composite panel for the tail of a sport plane.

Materials: display or poster board for presentations

Assessing Student Performance: Students' presentations should include:

- three correct angle measures using the properties of parallel and perpendicular lines **(Lesson 3.1)**
- a correct distance *AB* **(Lesson 4.2 or earlier)**
- a correct panel weight and determination that the panel meets the weight limits **(Lesson 4.2 or earlier)**
- a correct equation representing the line of the second side of the panel **(Lesson 4.2)**

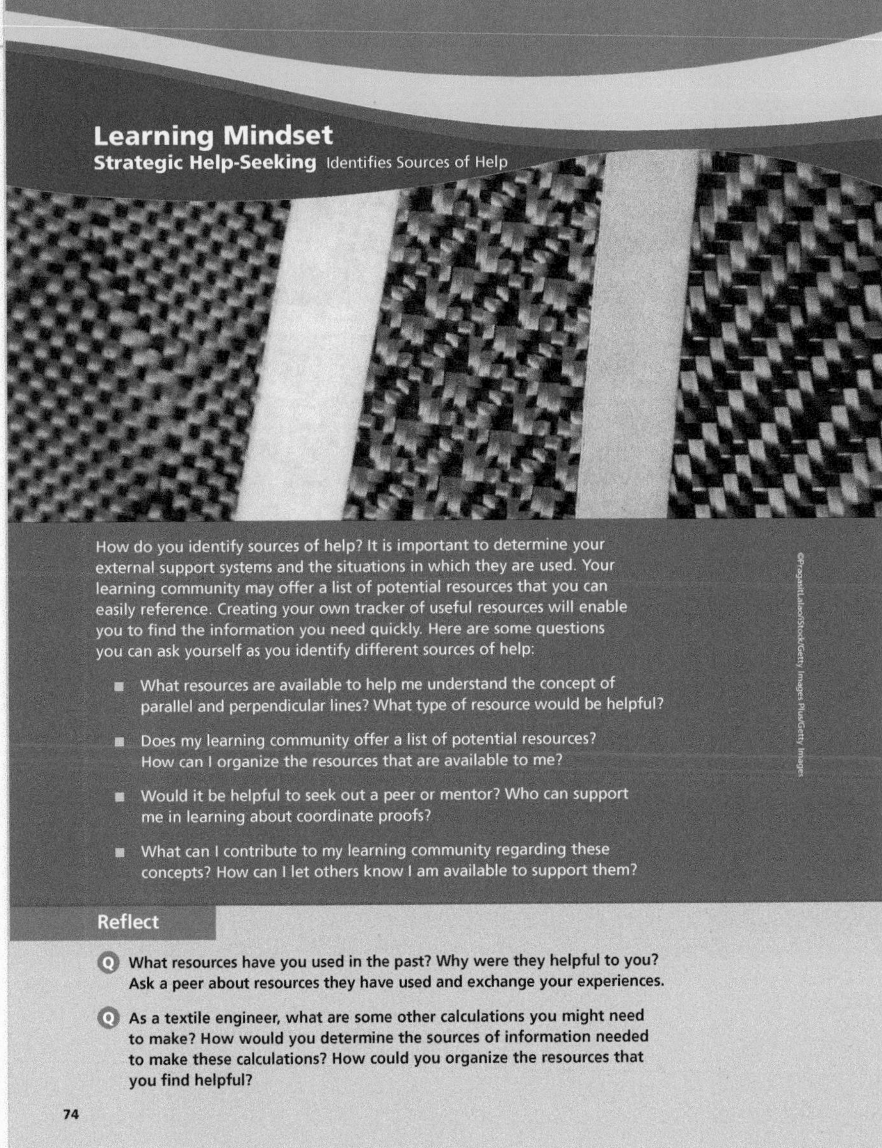

Learning Mindset

Strategic Help-Seeking Identifies Sources of Help

How do you identify sources of help? It is important to determine your external support systems and the situations in which they are used. Your learning community may offer a list of potential resources that you can easily reference. Creating your own tracker of useful resources will enable you to find the information you need quickly. Here are some questions you can ask yourself as you identify different sources of help:

- What resources are available to help me understand the concept of parallel and perpendicular lines? What type of resource would be helpful?
- Does my learning community offer a list of potential resources? How can I organize the resources that are available to me?
- Would it be helpful to seek out a peer or mentor? Who can support me in learning about coordinate proofs?
- What can I contribute to my learning community regarding these concepts? How can I let others know I am available to support them?

Reflect

Q What resources have you used in the past? Why were they helpful to you? Ask a peer about resources they have used and exchange your experiences.

Q As a textile engineer, what are some other calculations you might need to make? How would you determine the sources of information needed to make these calculations? How could you organize the resources that you find helpful?

74

What to Watch For

Watch for students who are struggling with the material and do not ask for help. Help them consult with others for support by

- asking leading questions related to each step in solving a problem and
- encouraging them to explain or write down what they know about a concept or skill.

Watch for students who resist asking for help, either in class or outside of class. Encourage them by

- suggesting websites that provide mathematical support about whatever skill or concept they don't understand.

> "The only mistake you can make is not asking for help."
>
> —Sandeep Jauhar, doctor and author

Strategic Help-Seeking

Learning Mindset

Identifies Sources of Help

The learning-mindset focus in this unit is *strategic help-seeking*, which refers to a person's ability to describe his or her external support systems and situations and how they can help them in their learning.

Mindset Beliefs

Students who have difficulty learning a new concept or skill may be reluctant to look to others for help. Working with peers provides a preferred support system for some, not only because some peers may have a better grasp of the material to be learned, but also because there's less risk of being embarrassed admitting that one doesn't understand something.

Mindset Behaviors

Encourage students to describe the problems they are having mastering the objectives of this unit.

When asked to write an equation of a line, ask:

- How can I express the properties of a line algebraically?
- Do I understand the connection between the terms in an equation of a line and its geometric properties?

When calculating distances or midpoint coordinates ask:

- Does it matter which point I start with?
- In either formula, do I add or subtract corresponding x- and y-coordinates?

Where can I get help on understanding the material?

- Where can I go online to access help?
- Is there someone in my class or in my family who might tutor me?

As students become more comfortable reaching out to others, they can realize greater success mastering the mathematics involved in their lessons. They might find an alternate strategy that can help them or another way to think about a problem. It's true in mathematics, especially, that there is often more than one way to analyze and solve a problem, including asking others for help.

LINES AND TRANSVERSALS

Introduce and Check for Readiness

- Module Performance Task • Are You Ready?

Lesson 3.1—2 Days

Parallel Lines Crossed by a Transversal

Learning Objective: Identify, explain, and prove the relationships formed when a transversal crosses parallel lines.

New Vocabulary: flow proof

Review Vocabulary: alternate exterior angles, alternate interior angles, consecutive exterior angles, consecutive interior angles, corresponding angles, parallel lines, transversal

Lesson 3.2—2 Days

Prove Lines are Parallel

Learning Objective: Students will be able to prove whether or not two lines are parallel.

Lesson 3.3—2 Days

Prove Lines are Perpendicular

Learning Objective: Students will be able to define and construct the perpendicular bisector of a line segment as the set of points that are equidistant from its endpoints.

New Vocabulary: perpendicular bisector, perpendicular lines

Assessment

- Module 3 Test (Forms A and B)
- Unit 2 Test (Forms A and B)

LEARNING ARC FOCUS

 Build Conceptual Understanding

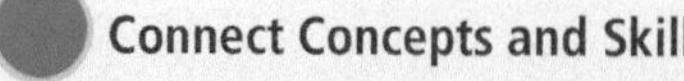

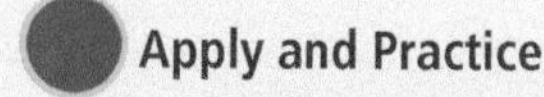

TEACHING FOR SUCCESS

TEACHING FOR DEPTH: Lines and Transversals

Represent and Explain. More than 2000 years ago, Euclid defined a line as *a "breadthless length."* Today, a line is defined mathematically as a one-dimensional figure having no thickness and extending infinitely in both directions.

There are two possibilities for the orientation of two lines in a plane. They can intersect at exactly one point or not intersect at all. In Lesson 1, students review the definition of parallel lines, which in the words of Euclid, *"…do not meet one another in either direction."* If a third line—a transversal—intersects the lines, it creates four pairs of congruent angles by virtue of their position. Students prove theorems about these angles in this lesson.

It is important to point out to students that a construction is not a proof. Rather, students must apply appropriate definitions, postulates, and theorems to support their constructions. For example, when constructing the line perpendicular to a given line through a point not on the line, students construct equal arcs whose corresponding equal chords form the sides of two isosceles triangles. Students can then establish the validity of their construction by proving that the triangles are congruent.

Mathematical Progressions

Prior Learning	Current Development	Future Connections
Students: • established facts about the angles created when parallel lines are cut by a transversal. • used postulates to prove theorems about lines and angles (Angle Addition Postulate, Supplementary Angles, Linear Pair Postulate, and Vertical Angles Theorem). • used a straightedge and compass to copy an angle. • used a straightedge and compass to bisect a segment and angle. • explained a proof of the Pythagorean Theorem and its converse.	**Students:** • construct and identify angle pairs formed by transversals. • identify, explain, and prove the relationship between angle pairs formed when a transversal crosses parallel lines. • prove whether two lines are parallel. • define and construct the perpendicular bisector of a line segment as the set of points that are equidistant from its endpoints.	**Students:** • will use properties of lines and angles to justify geometric constructions. • will use properties of angles formed when a transversal crosses parallel lines to prove theorems about triangles. • will use properties of parallel lines to represent translations. • will use properties of perpendicular lines to represent reflections. • will use properties of perpendicular lines to prove theorems about triangles. • will construct a circle that circumscribes a triangle.

TEACHER TO TEACHER

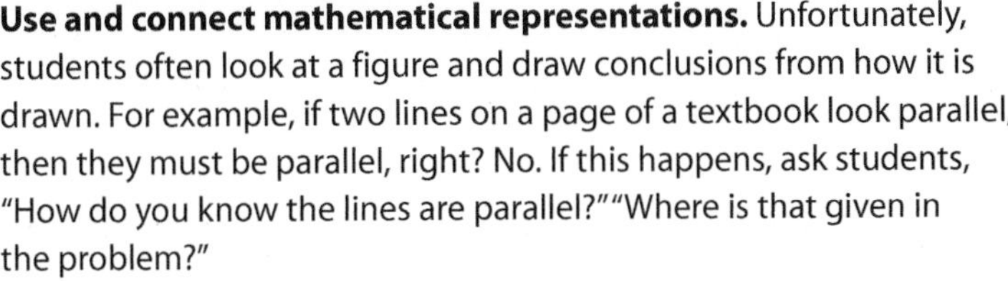

From the Classroom

Use and connect mathematical representations. Unfortunately, students often look at a figure and draw conclusions from how it is drawn. For example, if two lines on a page of a textbook look parallel, then they must be parallel, right? No. If this happens, ask students, "How do you know the lines are parallel?" "Where is that given in the problem?"

This module focuses on relationships among parallel and perpendicular lines and the angles formed by them. Help students develop strategies to analyze mathematical representations. For example, given two parallel lines and a transversal, how can students identify the types of angles formed? One way to help is to suggest tracing the sides of the angles. If a trace forms the letter Z or N, then the angles are alternate interior angles. If the trace is an F or a backwards-F, the angles are corresponding angles. Remember, a picture is worth a thousand words, but in geometry, can lead to a false conclusion if you make unsubstantiated assumptions.

LANGUAGE DEVELOPMENT • Planning for Instruction

By giving all students regular exposure to language routines in context, you will provide opportunities for students to **listen, speak, read,** and **write** about mathematical situations and develop both mathematical language and conceptual understanding at the same time.

Using Language Routines to Develop Understanding

Use the **Professional Cards** for the following routines to plan for effective instruction.

Co-Craft Questions and Problems Lesson 3.1

Students think of natural questions to ask about a given situation or problems similar to a given task and answer the questions they have developed or problems they have created.

Three Reads Lessons 3.1 and 3.3

Students read a problem three times with a specific focus each time.

1st Read What is the situation about?
2nd Read What are the quantities in the situation?
3rd Read What are the possible mathematical questions that we could ask for the situation?

Information Gap Lesson 3.2

Students recognize when information given in a problem situation is incomplete, and they pose questions and share knowledge with others to discover any missing facts or relationships and work together to solve the problem.

Critique, Correct, and Clarify Lesson 3.2

Students correct the work in a flawed explanation, argument, or solution method and share with a partner and refine the sample work.

Connecting Language to Lines and Transversals

Watch for students' use of the review and new terms listed below as they explain their reasoning and make connections with new concepts.

Key Academic Vocabulary

Prior Learning and Current Development • Review and New Vocabulary

transversal a line that intersects two or more coplanar lines at different points

parallel lines lines in the same plane that do not intersect

flow proof uses boxes and arrows to show the structure of a logical argument

perpendicular bisector a line perpendicular to a segment at the midpoint of the segment

perpendicular lines lines that intersect at 90° angles

Linguistic Note

Listen for vocabulary and terminology that students use when writing and talking about this lesson. To correctly identify pairs of angles formed between parallel lines and transversals requires understanding the words *alternate, corresponding, interior,* and *exterior*. Each have meanings outside of geometry, but here, it is vital that students understand how to use these terms to identify pairs of angles and the relationships between them based on their position.

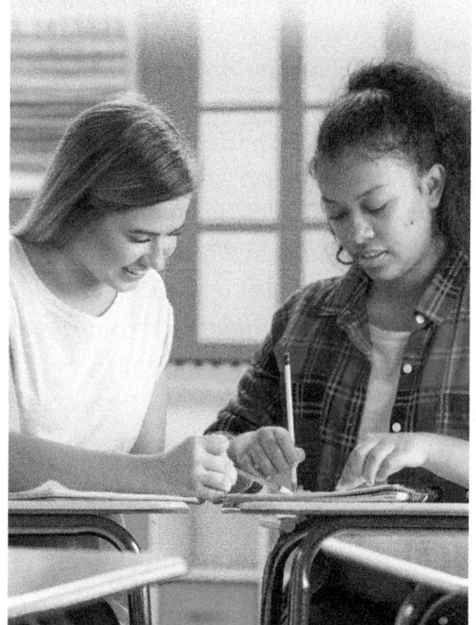

Module 3

Lines and Transversals

Module Performance Task: Focus on STEM

Geometry of Truss Bridges

Truss bridges are one of the oldest types of modern bridges. The bridges are composed of trusses, which are beams, called struts, which connect together to form triangular shapes for strength. The triangular shapes contract to handle heavy weights and expand during extreme weather conditions.

A. Identify two parallel struts in the diagram. How can an engineer ensure that the struts are parallel during construction? Explain your reasoning.

B. Identify two perpendicular struts in the diagram. How can an engineer ensure that the struts are perpendicular during construction? Explain your reasoning.

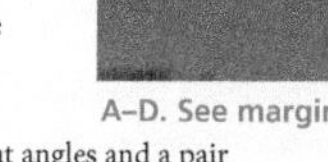

A–D. See margin

C. Use a transversal to identify a pair of congruent angles and a pair of supplementary angles. Explain how you determined your answer.

D. Would an engineer use parallel lines to ensure angles are congruent, or congruent angles to ensure lines are parallel? Explain your reasoning.

Mathematical Connections

Task Part	Prerequisite Skills
Part A	Students should know that if a transversal intersects two lines so that pairs of corresponding angles are congruent, the lines are parallel **(3.2)**.
Parts B	Students should know that lines intersecting at right angles are perpendicular **(3.2, 3,3)**.
Part C	Students should know that when parallel lines are cut by a transversal, pairs of alternate interior and alternate exterior angles are congruent and corresponding angles are congruent **(3.1)**.
Part D	Students should realize that the placement of the struts depends on accurately measuring the angles between them. **(3.2)**.

Geometry of Truss Bridges

Overview

A bridge is a construction that that lets us cross over an obstacle, like a river or a road. There are six types of bridges. One of the simplest and most useful is the truss bridge, a structure of connected elements arranged in triangular units, as the photo shows. Some triangular elements themselves are parallel to one another. Others are perpendicular. Some elements can be considered transversals that intersect both parallel and perpendicular elements.

Career Connections

Suggest that students research a career in civil engineering. Civil engineers have responsibilities for many tasks, including the conception, design, construction, and maintenance of bridges, roads, airports, tunnels, and many other structures or systems in both the public and private sectors.

Answers

A. Possible answer: $\overline{AG}$ is parallel to $\overline{CE}$. An engineer can ensure that two alternate interior angles are congruent, such as $\angle EAG$ and $\angle AEC$.

B. Possible answer: $\overline{AC}$ is perpendicular to $\overline{AG}$. An engineer can ensure that adjacent angles are congruent and supplementary.

C. Possible answer: $\angle EAG \cong \angle AEC$ because they are alternate interior angles of parallel lines cut by a transversal. $\angle BHE$ is supplementary to $\angle DEH$ because they are consecutive interior angles of parallel lines cut by a transversal.

D. Possible answer: Engineers would use congruent angles to ensure that lines are parallel. They can control the measure of angles, which would allow them to reach their goal of creating parallel lines.

ONLINE

Assign the Digital Are You Ready? to power actionable reports including

- proficiency by standards
- item analysis

Are You Ready?

Diagnostic Assessment

data checkpoint

- Diagnose prerequisite mastery.
- Identify intervention needs.
- Modify or set up leveled groups.

Have students complete the *Are You Ready?* assessment on their own. Items test the prerequisites required to succeed with the new learning in this module.

Solve Multi-Step Equations Students will apply their previous knowledge of solving simple linear equations to solving equations involving more than one step.

Types of Angle Pairs Students will apply theorems about adjacent, complementary, supplementary, and vertical angles and their relationships to find measures of three angles.

Parallel Lines Cut by a Transversal Students will apply theorems about vertical and supplementary angles to find the measures of angles formed between two parallel lines cut by a transversal.

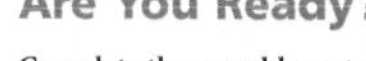

ONLINE Ed

Are You Ready?

Complete these problems to review prior concepts and skills you will need for this module.

Solve Multi-Step Equations

Solve each equation.

1. $5t - 7 = -23 - 3t$ $t = -2$
2. $6(z + 4) = 10z + 12$ $z = 3$
3. $8m + 1 = 3m - 4$ $m = -1$
4. $5(w - 7) = w + 9$ $w = 11$
5. $-2(3x + 8) + 4 = x + 9$ $x = -3$
6. $4r + 6 = 6r - 6$ $r = 6$
7. $5(2n + 3) = 15 - n$ $n = 0$
8. $-3(4 - y) + 5 = y + 1$ $y = 4$

Types of Angle Pairs

Given that $\angle 1 = 25°$, find each missing measure.

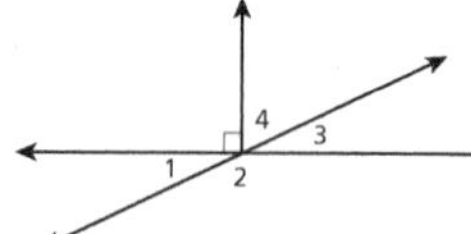

9. $\angle 2$ 155°
10. $\angle 3$ 25°
11. $\angle 4$ 65°

Parallel Lines Cut by a Transversal

Two parallel lines are cut by a transversal. Find the measure of each angle.

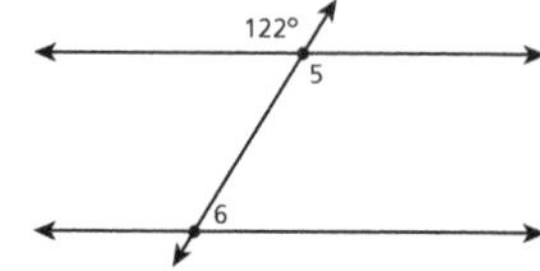

12. $\angle 5$ 122°
13. $\angle 6$ 58°

Connecting Past and Present Learning

Previously, you learned:

- to find missing side lengths in right triangles,
- to determine if two angles are congruent, and
- to determine if two angles are supplementary.

In this module, you will learn:

- to determine if angles formed by a transversal are congruent or supplementary,
- to construct parallel and perpendicular lines, and
- to prove or disprove if two lines are parallel or perpendicular.

76

DATA-DRIVEN INTERVENTION

Concept/Skill	Objective	Prior Learning *	Intervene With
Solve Multi-Step Equations	Solve linear equations with grouping symbols and/or the variable on both sides.	Grade 8, Lesson 3.1	• Tier 3 Skill 2 • Reteach, Grade 8 Lesson 3.1
Types of Angle Pairs	Solve problems involving complementary, supplementary, vertical, and adjacent angles.	Grade 7, Lesson 7.5	• Tier 2 Skill 2 • Reteach, Grade 7 Lesson 7.5
Parallel Lines Cut by a Transversal	Solve problems about angle measures when parallel lines are cut by a transversal.	Grade 8, Lesson 4.3	• Tier 2 Skill 6 • Reteach, Grade 8 Lesson 4.3

* Your digital materials include access to resources from Grade 6–Algebra 2. The lessons referenced here contain a variety of resources you can use with students who need support with this content.

3.1 Parallel Lines Crossed by a Transversal

LESSON FOCUS AND COHERENCE

Mathematics Standards

- Know precise definitions of angle, circle, perpendicular line, and line segment, based on the undefined notions of point, line, distance along a line, and distance around a circular arc.
- Prove theorems about lines and angles.

Mathematical Practices and Processes 

- Attend to precision.
- Use appropriate tools strategically.
- Construct viable arguments and critique the reasoning of others.

I Can Objective 

I can determine the relationship between angle pairs formed by a transversal crossing parallel lines.

Learning Objective

Identify, explain, and prove the relationships formed when a transversal crosses parallel lines.

Language Objective

Explain how angle pairs formed by a transversal intersecting two lines change when the lines are parallel or not parallel.

Vocabulary

Review: **alternate interior angles**, **alternate exterior angles**, **consecutive interior angles**, **consecutive exterior angles**, **corresponding angles**, **transversal**, **parallel lines**

New: flow proof

Lesson Materials: tracing paper

Mathematical Progressions

Prior Learning	Current Development	Future Connections
Students: • established facts about the angles created when parallel lines are cut by a transversal. **(Gr8, 4.3)** • used postulates to prove theorems about lines and angles (Angle Addition Postulate, Supplementary Angles, Linear Pair Postulate and Vertical Angles Theorem). **(2.4)**	**Students:** • identify angle pairs formed by a transversal crossing parallel lines and categorize their relationship as congruent or supplementary. • prove the relationship between angle pairs formed by a transversal crossing parallel lines. • determine unknown angle measures using parallel lines and transversals and provide proof. • prove the relationship among interior angles in a trapezoid using parallel lines and transversals.	**Students:** • will use properties of lines and angles to justify geometric constructions. **(3.2)** • will use properties of angles formed when a transversal crosses parallel lines to prove theorems about triangles. **(9.1)**

PROFESSIONAL LEARNING

Using Mathematical Practices and Processes

Construct viable arguments and critique the reasoning of others.

This lesson provides an opportunity to address this Mathematical Practice Standard, which requires students to "construct viable arguments". Students identify pairs of angles formed when a transversal intersects two lines. Then they explore the relationships between the angles when the intersected lines are parallel. Students use theorems and postulates about defined angle pairs to prove relationships between other angles in a diagram.

ONLINE

PROJECTABLE & PRINTABLE 

WARM-UP OPTIONS

ACTIVATE PRIOR KNOWLEDGE • Use Theorems about Vertical Angles and Linear Pairs

Use these activities to quickly assess and activate prior knowledge as needed.

Problem of the Day 

A carpenter connects two beams as shown in the diagram. If the top angle measures 48°, how can the carpenter use vertical angles and linear pairs to find the measurements of angles *m* and *p*? The 48° angle and angle *m* form a linear pair. This means the sum of the two angles is 180°. Subtracting $180 - 48 = 132$ finds that angle *m* is 132°. The 48° angle and angle *p* are vertical angles which have the same measure, so angle *p* is 48°.

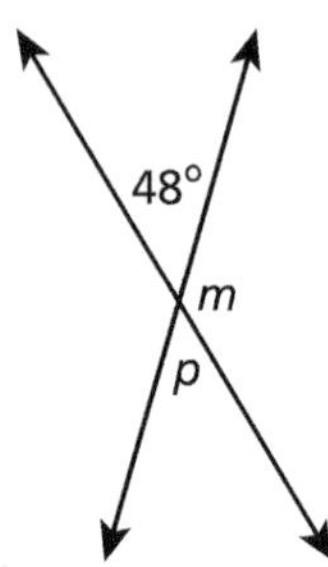

Quick Check for Homework

As part of your daily routine, you may want to display the Teacher Solution Key to have students check their homework.

Make Connections

Based on students' responses to the Problem of the Day, choose one of the following:

1. Project the Interactive Reteach, Geometry, Lesson 2.4.
2. Complete the Prerequisite Skills Activity:

Have students work in pairs. Ask the students to draw a pair of intersecting lines. Have students work to identify all pairs of vertical angles. Then ask students to find all linear pairs shown in their drawing. Have students review theorems and postulates related to vertical angles and linear pairs.

- *How many pairs of vertical angles are in your drawing?* 2
- *What is true about the measures of the angles in a pair of vertical angles?* They have the same measure by the Vertical Angle Theorem.
- *How many linear pairs are in your drawing?* Any two angles adjacent to each other that form a straight line on one side are a linear pair. There are four pairs of adjacent angles that form a straight line on one side.
- *What is the sum of two angles in a linear pair?* 180°

If students continue to struggle, use Tier 2 Skill 2.

SHARPEN SKILLS

If time permits, use this on-level activity to build fluency and practice basic skills.

Vocabulary Review

Objective: Students review the definition and properties of vertical angles.
Materials: Bubble Map (Teacher Resource Masters)

Have students use the Bubble Map. Write the term "Vertical Angles" in the center. In two bubbles, work with students to sketch a drawing of two different examples of vertical angles. Then use one bubble to write a definition of vertical angles. In the last bubble, describe how the measures of a pair of vertical angles are related.

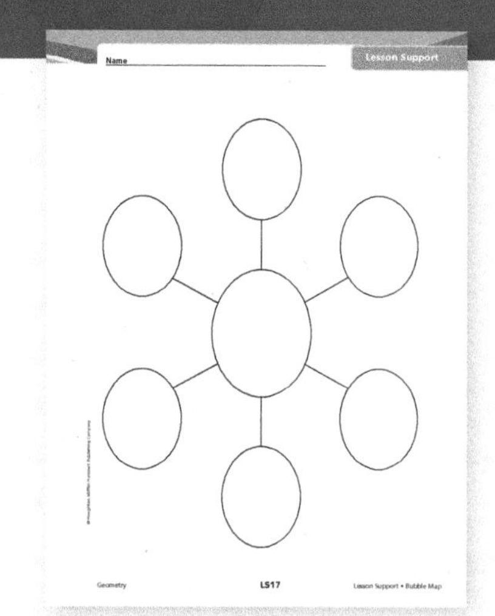

PLAN FOR DIFFERENTIATED INSTRUCTION

Small-Group Options

Use these teacher-guided activities with pulled small groups.

On Track

Draw the diagram on the board.

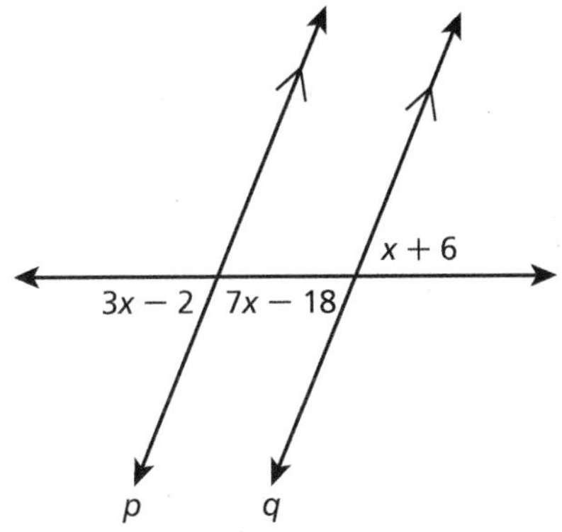

Have students work in small groups to do the following:

- Solve for *x*, showing all algebraic steps and reasoning.
- Ask students to work together to find another way to solve for *x*. What theorems are you using each time you solve?

Almost There

Materials: index cards

Give each student an index card. Assign each student a type of pair of angles from this lesson. Ask the student to draw the pair of angles formed by a transversal intersecting two lines. On the back, have the student write the name of the type of angles shown.

Have students walk around the room and share their angle pair and name with 5 different students.

front of card | back of card

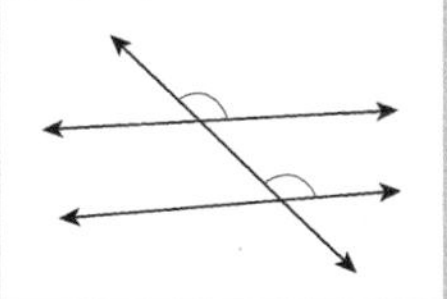

Corresponding Angles

Ready for More

Draw the diagram on the board.

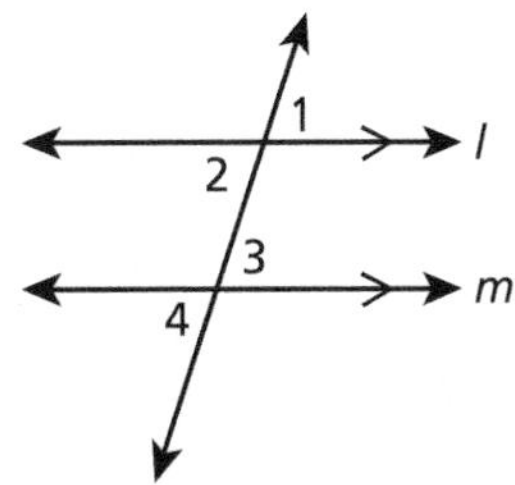

Have students work in small groups to do the following:

- Identify the pair of angles formed by $\angle 1$ and $\angle 4$.
- Ask the students to prove the angles are congruent using a flow proof.
- Then have students try to write a second proof using a two-column proof.

Math Center Options

Use these student self-directed activities at centers or stations. **Key:** ● Print Resources ● Online Resources

On Track

- Interactive Digital Lesson
- Journal and Practice Workbook
- Interactive Glossary (printable): **alternate interior angles**, **alternate exterior angles**, **consecutive interior angles**, **consecutive exterior angles**, **corresponding angles**, **parallel lines**, **transversal**, **flow proof**
- Module Performance Task
- Desmos: Lines, Transversals, and Angles

Almost There

- Reteach 3.1 (printable)
- Interactive Reteach 3.1
- RtI Tier 2 Skill 2: Types of Angle Pairs
- Illustrative Mathematics: Street Intersections

Ready for More

- Challenge 3.1 (printable)
- Interactive Challenge 3.1
- Illustrative Mathematics: Congruent Angles Made by Parallel Lines and a Transverse

Unit Project Check students' progress by asking them to describe the relationship between each of the sides and the other sides.

View data-driven grouping recommendations and assign differentiation resources.

During the *Spark Your Learning*, listen and watch for strategies students use. See samples of student work on this page.

Use an Drawing

Strategy 1

I drew a picture of the shelf unit so the angle the side support makes with the bottom shelf is the same as the angle it makes with the top shelf.

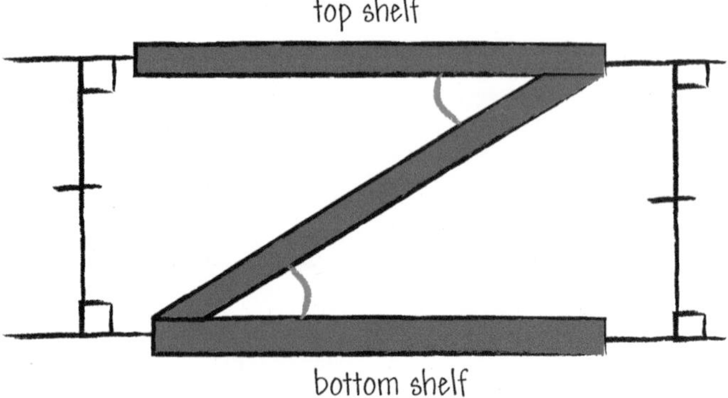

Using these angles means the top and bottom shelves are parallel, because the distance between them is always equal.

If students . . . use a drawing to solve the problem, they are using their knowledge of parallel lines and transversals from Grade 8.

Have these students . . . explain how they used their drawing to determine if the shelves are parallel. **Ask:**

- Q What did you draw?
- Q How did your drawing help you determine if the shelves are parallel?
- Q What should be true about the planes represented by the shelves if they are parallel?

Explain in Words

Strategy 2

The top shelf is at the same angle as the bottom one, so the shelves must be parallel. The side supports are positioned the way they are so the top shelf is level.

If students . . . explain in words why they think the shelves are parallel, they understand that the shelves should be parallel, but they might not see how the angles made with the side support are related to the parallel shelves.

Activate prior knowledge . . . by having students review the definition of parallel lines and transversals. **Ask:**

- Q What must be true about two lines or two planes for them to be parallel?
- Q What kind of measurements could you make to tell if the desktop is flat?

COMMON ERROR: Makes an Incorrect Conclusion

The side support is at an angle, not straight up and down. This means the top shelf must also be at an angle, not level. So, the shelves cannot be parallel.

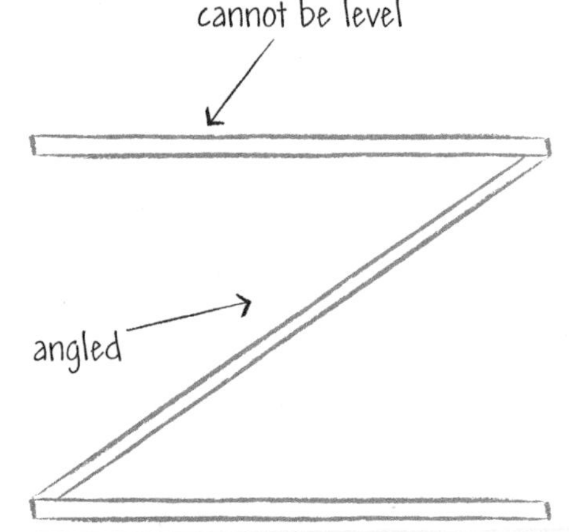

If students . . . think the shelves cannot be parallel because the side support is not perpendicular, they may not understand how the angles and sides are related.

Then intervene . . . by pointing out that the angled side support can be designed so the shelves are parallel. **Ask:**

- Q How could the side support be cut so the shelf could lie flat on it?
- Q What would happen if the side support was placed vertically?

Parallel Lines Crossed by a Transversal

I Can determine the relationship between angle pairs formed by a transversal crossing parallel lines.

Spark Your Learning

Leon has a Z-shaped shelf unit.

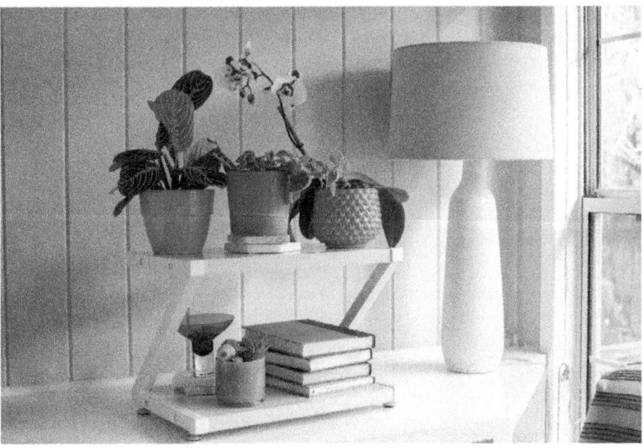

Complete Part A as a whole class. Then complete Parts B–D in small groups.

A. What question can you ask about the design of the shelf unit?

B. To answer your question, what strategy and tool would you use along with all the information you have? What answer do you get? See Strategies 1 and 2 on the facing page.

C. Why do you think the sides of the shelf unit are designed this way?

D. What is another design for the side supports of the shelf that will not affect the position of the shelves in the image? Another design would be sides that are perpendicular to the shelves.

A. How do you know the two shelves are parallel?

C. Possible answer: Using a Z-shape like this for the sides ensures that the shelves are parallel to each other.

Turn and Talk What happens if the design of the shelf unit changes as described below?

- The acute angles between the shelves and the side supports remain congruent, but become smaller?
- One of the acute angles between the shelves and the side supports becomes larger than the other? See margin.

CULTIVATE CONVERSATION • Co-Craft Questions

If students have difficulty formulating a mathematical question about the situation in the Spark Your Learning, ask them to consider the shape formed by the shelves and the side supports. What are some natural questions to ask about this situation?

Work together to craft the following questions:

- Why do you think the designers used a Z shape for the shelf unit?
- What other shape could the three pieces of wood form besides a Z shape?
- What appears to be true about the top and bottom horizontal portions of the side supports?

Then have students think about what additional information, if any, they would need to answer these questions. **Ask:**

- Can you determine if the shelves are parallel? Why or why not?
- If the shelves are not parallel, how could the side supports be adjusted to make them parallel?

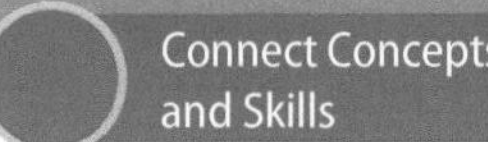

Connect Concepts and Skills

1 Spark Your Learning

▶ MOTIVATE 

- Have students look at the photo in their books and read the information contained in the photo. Then complete Part A as a whole-class discussion.
- Give the class the additional information they need to solve the problem. This information is available online as a printable and projectable page in the Teacher Resources.
- Have students work in small groups to complete Parts B–D.

▶ PERSEVERE

If students need support, guide them by asking:

Q **Advancing • Use Tools** Which tool could you use to solve the problem? Why choose that tool and not some other? Students' choices of tools and reasons for choosing them will vary.

Q **Assessing** If the front part of a shelf is lower than the back part, what could happen to your materials? They could roll or slide off.

Q **Advancing** How would the cuts made to the side supports need to change if you wanted more space between the shelves? The angles would remain congruent to each other, but the angle would be closer to 90°.

Turn and Talk Have students sketch a picture of the shelf unit that shows each of the given changes. Have students think about whether the shelves remain parallel. The shelves move closer together; the shelves would no longer be parallel.

▶ BUILD SHARED UNDERSTANDING

Select groups of students who used various strategies and tools to share with the class how they solved the problem. As they present their solutions, have each group discuss why they chose a specific strategy and tool.

2 Learn Together

Build Understanding

Task 1 (MP) **Attend to Precision** Students will identify and locate pairs of angles formed by intersecting and transversal lines.

CONNECT TO VOCABULARY

Have students use the **Interactive Glossary** to record their understanding of the vocabulary in this task.

Sample Guided Discussion:

Q **How can you remember how to find pairs of corresponding angles?** Each angle in a pair will be on the same side of the transversal and either both above or both below the intersected lines.

Q **In Part C, how can you show that two angles are congruent?** Possible answer: You could trace one of the angles using tracing paper and then move it over other angles to see if they are congruent.

Turn and Talk Sketch a pair of non-parallel lines and a transversal on the board. Point out the relative measurements of angle pairs. Then, sketch a pair of parallel lines and a transversal. Point out the relative measurements of angle pairs. The angles in the same positions relative to the transversal will have the same measure.

Build Understanding

Explore Angle Pairs Formed by Transversals

A **transversal** is a line that intersects two or more coplanar lines at different points. In the diagram, line t is the transversal.

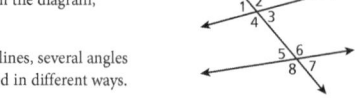

When a transversal intersects two lines, several angles are formed. These angles are related in different ways.

Angles that are on the same side of the transversal are consecutive angles. Angles that are on opposite side of the transversal are alternate angles.

Angles that are on the outside of two lines are exterior angles. Angles that are on the inside of the two lines are interior angles.

Angle Pair Relationships When Lines are Cut by a Transversal

Term	Examples
Corresponding angles lie on the same side of the transversal and on the same sides of the intersecting lines.	∠1 and ∠5, ∠2 and ∠6 ∠3 and ∠7, ∠4 and ∠8
Alternate interior angles are nonadjacent angles that lie on opposite sides of the transversal between the intersected lines.	∠3 and ∠5 ∠4 and ∠6
Consecutive interior angles lie on the same side of the transversal between the intersected lines.	∠3 and ∠6 ∠4 and ∠5
Alternate exterior angles lie on opposite sides of the transversal outside the intersected lines.	∠1 and ∠7 ∠2 and ∠8
Consecutive exterior angles lie on the same side of the transversal outside the intersected lines.	∠1 and ∠8 ∠2 and ∠7

1 A. Draw two lines intersected by a transversal. Label the lines and number the angles formed. A–C. See Additional Answers.

B. Give one example for each of the following types of angle pairs: corresponding angles, alternate exterior angles, alternate interior angles, consecutive exterior angles, and consecutive interior angles.

C. Are any pairs of angles in your diagram congruent? Are any pairs of angles supplementary? Explain your reasoning.

Turn and Talk What do you think will happen with the angle relationships if the two lines crossed by the transversal are parallel? See margin.

LEVELED QUESTIONS

Depth of Knowledge (DOK)	Leveled Questions	What Does This Tell You?
Level 1 **Recall**	What is the definition of a pair of alternate interior angles? a pair of angles that are on opposite sides of the transversal, but are in between the intersected lines	Students' answers will indicate whether or not they understand where to find alternate interior angles.
Level 2 **Basic Application of Skills & Concepts**	In the diagram, explain why ∠2 and ∠5 are not alternate interior angles? Although the angles are on opposite sides of the transversal, ∠2 is an exterior angle and ∠5 is an interior angle.	Students' answers will indicate if they can determine the difference between interior and exterior angles.
Level 3 **Strategic Thinking & Complex Reasoning**	How can you make sure you have found all pairs of corresponding angles? There are 4 pairs of corresponding angles. In each pair, the angles should have the same location relative to the transversal and the intersected lines.	Students' answers will indicate they understand there are multiple pairs of corresponding angles that appear on either side of the transversal and either above or below the intersected lines.

Investigate Transversals and Parallel Lines

Parallel lines are any lines in the same plane that do not intersect.

Transversals intersecting parallel lines occur in real-world situations. For example, consider the paths across this university campus.

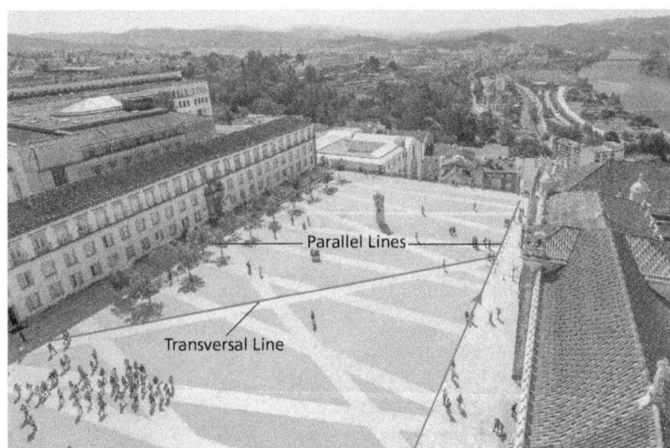

When a transversal intersects two parallel lines, there are relationships between the pairs of angles formed.

2 Two parallel lines m and n are intersected by a transversal t.

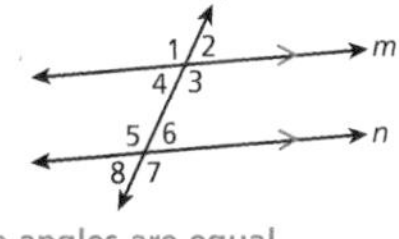

A. Trace the diagram onto tracing paper. Slide the tracing paper so that $\angle 5$ on the tracing paper is positioned over $\angle 1$. What do you notice? The measures of the angles are equal.

B. Use the tracing-paper method to compare the angle measures of all of the corresponding angle pairs. What appears to be true about the corresponding angles?
The measures of pairs of corresponding angles are equal.

C. Use tracing paper to compare the measures of alternate interior angle pairs and alternate exterior angle pairs. What do you notice about the measures of the pairs of angles?
C. The measures of pairs of alternate interior angles and alternate exterior angles are equal.

D. Use tracing paper to compare the measures of consecutive interior angle pairs and consecutive exterior angle pairs. What do you notice about the measures of the pairs of angles? The measures of pairs of consecutive interior angles and consecutive exterior angles are supplementary.

Turn and Talk Compare your drawings in Task 1 and Task 2. Do you think these angle pair relationships will change if the lines are not parallel? Explain. See margin.

Build Understanding

Task 2 (MP) **Use Tools** Students will use tracing paper to compare the measurements of angles in a diagram.

CONNECT TO VOCABULARY

Have students use the **Interactive Glossary** to record their understanding of the vocabulary in this task.

Sample Guided Discussion:

Q Before tracing your diagram, what predictions can you make about the relationships between the angle pairs? It appears that some angle pairs will be congruent, while other pairs may not be congruent.

Q In Part D, how do the relationships between the consecutive interior and consecutive exterior angles differ from the other angle pairs? The consecutive exterior and interior angle pairs are not congruent, but the alternate interior and exterior and corresponding angles pairs are congruent.

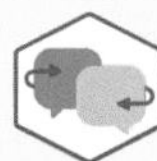

Turn and Talk Have students look at the same pair of corresponding angles from each diagram. Encourage students to compare the size of the angles in the two diagrams. Yes, because if the lines aren't parallel, then the angles formed by the lines and the transversal will not be the same.

EL PROFICIENCY LEVEL

Beginning

Have students take turns describing the relationships between angle pairs while pointing at the angle pairs in the diagram.

Intermediate

Assign students a partner. Have one student write a sentence describing the angle pairs that are congruent. Have the other student write a sentence describe the angle pairs that are not congruent. Then have students compare answers.

Advanced

Have students write a paragraph describing how to identify different pairs of angles formed by two lines and a transversal. Have the students highlight the pairs that have congruent angles. Have students use a different color to highlight the pairs that have non-congruent angles.

Task 3 (MP) **Construct Arguments** Students will use theorems and postulates to prove a pair of angles is congruent.

Sample Guided Discussion:

Q What type of angles are ∠1 and ∠2? vertical angles

Q What do you know about ∠2 and ∠3? They are corresponding angles and are congruent because the lines are parallel.

Q In Part B, look at the flow proof. How can you connect the congruent angle pairs to come to the conclusion? Use the Transitive Property of Congruence because ∠2 is common in two angle pairs.

Turn and Talk Point out the diagram at the top of the page. Pose the following question to students: Could you use the Alternate Exterior Angle Theorem to prove something about ∠7 and ∠8? Encourage students to brainstorm what other properties or theorems they could use to help. Possible answer: Have the alternate interior angles be the postulate. You can use it and linear pairs to prove all the other theorems.

Prove Relationships Between Angle Pairs Formed by Parallel Lines and a Transversal

When two parallel lines are intersected by a transversal, every angle formed is either congruent or supplementary to a given angle. Many of these relationships are described in the following table.

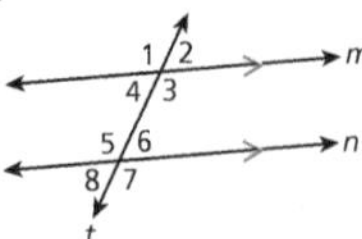

Angle Pair Relationships When Parallel Lines are Cut by a Transversal	
Postulate or Theorem	**Examples**
Corresponding Angles Postulate If two parallel lines are cut by a transversal, then the resulting corresponding angles are congruent.	∠1 ≅ ∠5 ∠2 ≅ ∠6 ∠3 ≅ ∠7 ∠4 ≅ ∠8
Alternate Interior Angles Theorem If two parallel lines are cut by a transversal, then the pairs of alternate interior angles are congruent.	∠3 ≅ ∠5 ∠4 ≅ ∠6
Consecutive Interior Angles Theorem If two parallel lines are cut by a transversal, then the pairs of consecutive interior angles are supplementary.	$m\angle 3 + m\angle 6 = 180°$ $m\angle 4 + m\angle 5 = 180°$
Alternate Exterior Angles Theorem If two parallel lines are cut by a transversal, then the pairs of alternate exterior angles are congruent.	∠1 ≅ ∠7 ∠2 ≅ ∠8
Consecutive Exterior Angles Theorem If two parallel lines are cut by a transversal, then the pairs of consecutive exterior angles are supplementary.	$m\angle 1 + m\angle 8 = 180°$ $m\angle 2 + m\angle 7 = 180°$

A **flow proof** uses boxes and arrows to show the structure of a logical argument. The justification for each step is written below the box.

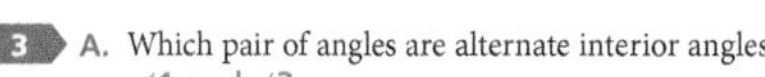

3 A. Which pair of angles are alternate interior angles?
∠1 and ∠3

B. What are the justifications for each step in the flow proof below?

Prove the Alternate Interior Angles Theorem.

Given: $a \parallel b$

Prove: ∠1 ≅ ∠3

Given, Vertical Angles Theorem, Corresponding Angles Postulate, Transitive Property of Congruence

Turn and Talk Suppose any one of the theorems in the list above could instead be a postulate used to prove the other theorems. Which angle pair relationship should be a postulate? Explain your reasoning. See margin.

80

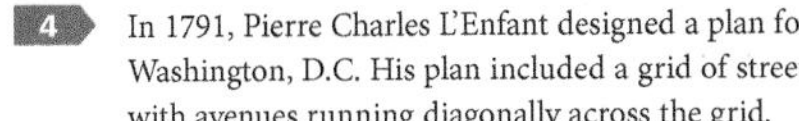

Step It Out

Use Parallel Lines to Determine Angle Measures

4 In 1791, Pierre Charles L'Enfant designed a plan for Washington, D.C. His plan included a grid of streets with avenues running diagonally across the grid.

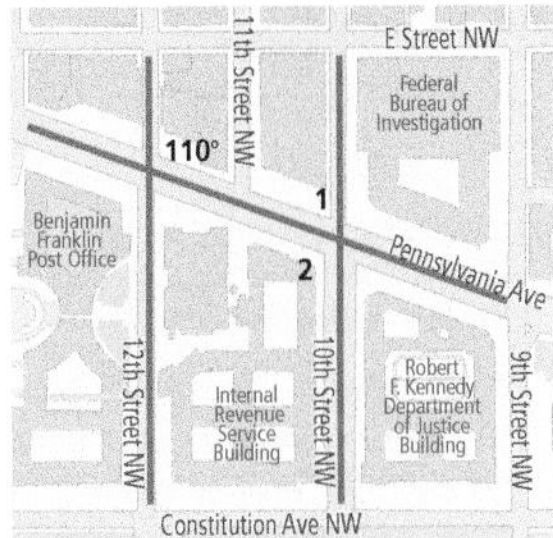

Look at the map of a section of Washington, D.C. 12th St. NW and 10th St. are parallel. They are intersected by Pennsylvania Avenue.
See Additional Answers
Find the measures of $\angle 1$ and $\angle 2$.

$m\angle 1 = 70^\circ$ — Explain how you can find this measure.

$m\angle 2 = 110^\circ$

5 The highlighted streets in Washington, D.C. form two trapezoids with a common side. The vertical streets are parallel and the horizontal streets are parallel.

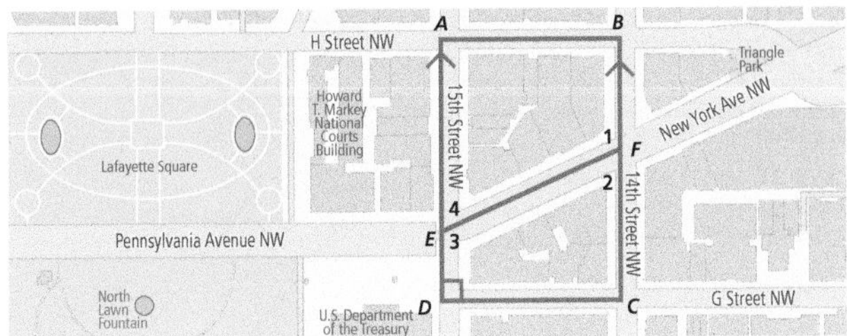

Prove that the sum of the angle measures of the trapezoid is 360°.

Given: $\overline{AD} \parallel \overline{BC}$, $\angle D$ is a right angle.

Prove: $m\angle 2 + m\angle 3 + m\angle C + m\angle D = 360^\circ$

A. Consecutive Interior Angles Theorem
B. yes; also Consecutive Interior Angles Theorem

Statements	Reasons
1. $\overline{AD} \parallel \overline{BC}$	1. Given
2. $m\angle 2 + m\angle 3 = 180^\circ$	2. ?
3. $m\angle C + m\angle D = 180^\circ$	3. ?
4. $m\angle 2 + m\angle 3 + m\angle C + m\angle D = 180^\circ + 180^\circ$	4. Substitution Property of Equality
5. $m\angle 2 + m\angle 3 + m\angle C + m\angle D = 360^\circ$	5. Simplify.

A. What is the reason Statement 2 is true?

B. Is Statement 3 true for the same reason?

Turn and Talk If $m\angle 1 = (7n - 2)^\circ$ and $m\angle 4 = (4n + 6)^\circ$, find the values of n, $m\angle 1$, and $m\angle 4$. See margin.

Task 4 (MP) **Construct Arguments** Students will use a diagram and theorems to find missing angle measures.

SUPPORT SENSE-MAKING Three Reads

Have students read the problem three times. Use the questions below for a different focus each read.

1. What is the situation about?
2. What are the quantities in the situation?
3. What are the possible mathematical questions that you could ask for the situation?

Sample Guided Discussion:

Q What is the name of the angle pair made of 110° and $\angle 2$? These are alternate interior angles, which are congruent when the lines are parallel.

Q $\angle 1$ and $\angle 2$ are a linear pair. What do you know about two angles in a linear pair? A linear pair is two angles that are supplementary or have a sum of 180°.

Task 5 (MP) **Construct Arguments** Students will make logical arguments to prove properties in a real-world situation.

Sample Guided Discussion:

Q What type of angles are $\angle 2$ and $\angle 3$? consecutive interior angles

Q In Part B, how might you be able to see the relationship between $\angle C$ and $\angle D$? Possible answer: You can extend the lines to see that the angles are on the same side of the transversal formed by a segment between points C and D.

Turn and Talk Point out to students that $\angle 1$ and $\angle 4$ are consecutive interior angles. This means they have a sum of 180° because they are supplementary. Students can write the equation $(7n - 2) + (4n + 6) = 180$ and solve. $n = 16$, $m\angle 1 = 110^\circ$, $m\angle 4 = 70^\circ$

ONLINE

Assign the Digital On Your Own for
- built-in student supports
- Actionable Item Reports
- Standards Analysis Reports

On Your Own

Assignment Guide

The chart below indicates which problems in the On Your Own are associated with each task in the Learn Together. Assign daily homework for tasks completed.

Learn Together Tasks	On Your Own Problems
Task 1, p. 78	Problems 5 and 12
Task 2, p. 79	Problems 6, 7, and 23
Task 3, p. 80	Problems 8–11, 14, 15, 22, 24, and 27
Task 4, p. 81	Problems 13, 16–21, and 23
Task 5, p. 81	Problems 25 and 26

Check Understanding

Use the diagram for Problems 1–3.

1–4. See Additional Answers.

1. Give one example for each of the following types of angle pairs: corresponding angles, alternate exterior angles, alternate interior angles, consecutive exterior angles, and consecutive interior angles.
2. Which angles are congruent to $\angle 3$?
3. Use the diagram to explain how you could use the Consecutive Interior Angles Theorem and a linear pair of angles to prove the Alternate Interior Angles Theorem.
4. Megan is making a wood model of each letter of the alphabet. She drew a sketch of the letter N. The vertical sides of the N are parallel. What is the value of x?

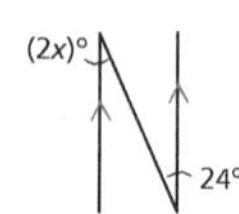

On Your Own

5. **Attend to Precision** List all of the pairs of corresponding angles, alternate interior angles, alternate exterior angle, consecutive interior angles, and consecutive exterior angles.
See Additional Answers.

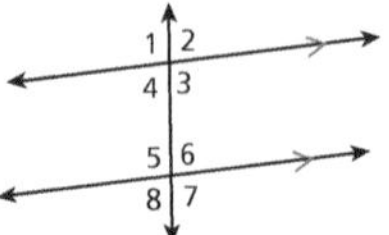

For Problems 6–11, use the diagram shown.

6–14. See Additional Answers.

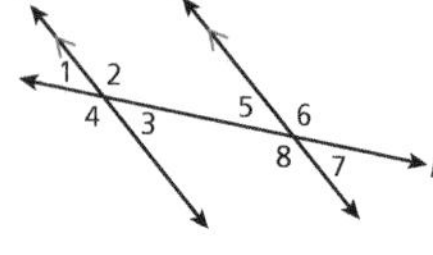

6. Which angles are congruent to $\angle 2$?
7. Which angles are supplementary to $\angle 7$?
8. Which postulate or theorem justifies that $\angle 1$ is congruent to $\angle 5$?
9. Which postulate or theorem justifies that $\angle 4$ is supplementary to $\angle 7$?
10. Which postulate or theorem justifies that $\angle 3$ is congruent to $\angle 5$?
11. Which postulate or theorem justifies that $\angle 2$ is supplementary to $\angle 5$?
12. **Open Ended** Draw a pair of parallel lines intersected by a transversal. Label the angles. Choose one of the angles and list the angles congruent and supplementary to that angle.
13. **Science** Venation is the pattern of the veins in a leaf. Some leaves have parallel venation. Find the measure of the angles in the photo of the leaf.
14. **Construct Arguments** Prove the Consecutive Exterior Angles Theorem.

82

data checkpoint

3 Check Understanding

Formative Assessment

Use formative assessment to determine if your students are successful with this lesson's learning objective.

Students who successfully complete the Check Understanding can continue to the On Your Own practice.

For students who miss 1 problem or more, work in a pulled small group using the Almost There small-group activity on page 77C.

ONLINE

Assign the Digital Check Understanding to determine
- success with the learning objective
- items to review
- grouping and differentiation resources

4 Differentiation Options

Differentiate instruction for all students using small-group activities and math center activities on page 77C.

Reteach

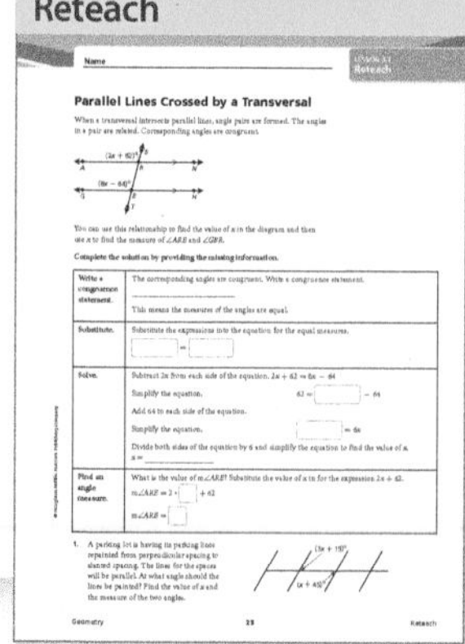

Parallel Lines Crossed by a Transversal

Challenge

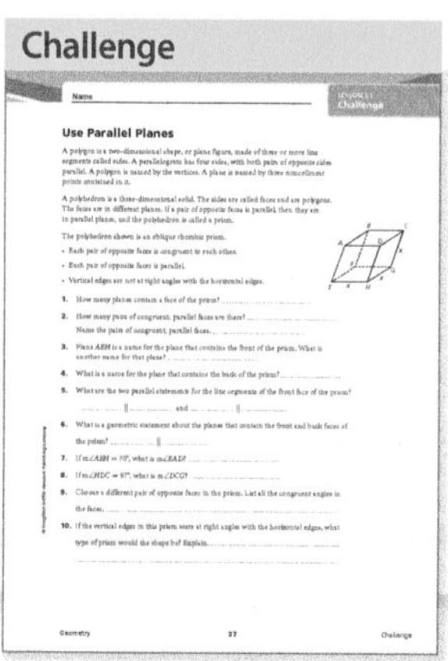

Use Parallel Planes

15. **Reason** Let the fact that consecutive interior angles are supplementary be a postulate. Use this postulate to prove the relationship between corresponding angles as a Corresponding Angles Theorem.

Find the value of x. 15–24. See Additional Answers.

16.

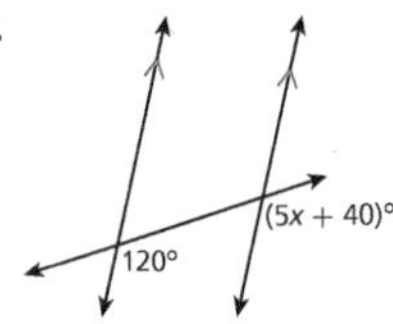

17.

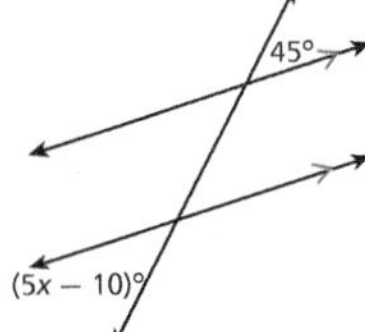

18.

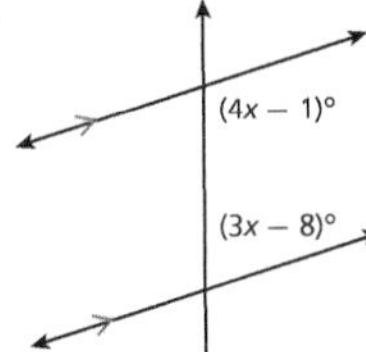

19.

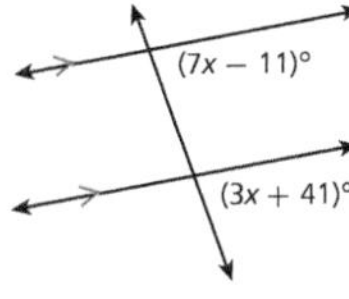

20.

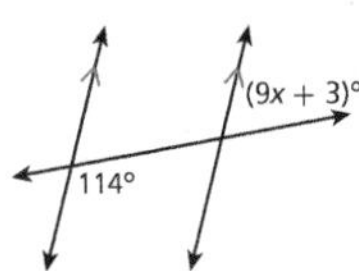

21.

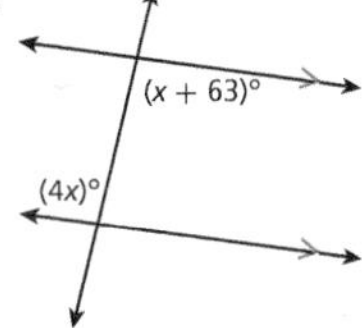

22. **Construct Arguments** Prove the Alternate Exterior Angles Theorem.

23. **STEM** An engineer is designing a railing for a staircase.

A. Which angle is congruent is to $\angle 4$?

B. Which postulate or theorem did you use to determine the congruent angle?

C. Suppose that the measure of $\angle 4$ is 75°. What are the measures of the other angles?

24. **Construct Arguments** Prove the Consecutive Interior Angles Theorem.

Questioning Strategies

Problem 18 Ask students to identify the pair of angles labeled in the diagram. What is the relationship between this type of pair of angles? How can this help you write an equation? The angles are consecutive interior angles which are supplementary; You can write and solve an equation where the sum of the expressions is equal to 180°, $(4x - 1) + (3x - 8) = 180$.

Watch for Common Errors

Problem 20 Some students might initially think they can use a theorem directly, but show students that the labeled angles are not a defined pair of angles. Ask students to use the 114° angle to find the value of another angle, such as the one directly above it. Then, the student can use the Corresponding Angles Postulate to write and solve an equation.

5 Wrap-Up

Summarize learning with your class. Consider using the Exit Ticket, Put It in Writing, or I Can scale.

Exit Ticket 

For the diagram below, a student says that $x + y = 180$ because the angles are consecutive interior angles. Is the student correct? Explain. The student is incorrect; The angles are consecutive interior angles, but they are not supplementary because the intersected lines are not parallel.

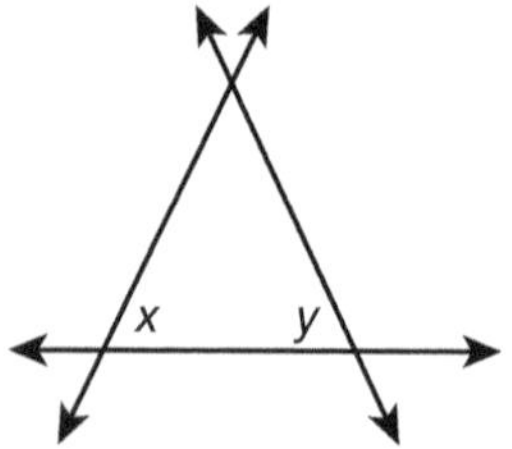

Put It in Writing 

Explain how to identify the different types of pairs of angles formed by a transversal which intersects two lines.

I Can 

The scale below can help you and your students understand their progress on a learning goal.

4	I can determine the relationship between angle pairs formed by a transversal crossing parallel lines and explain to others how to use this information in a proof.
3	I can determine the relationship between angle pairs formed by a transversal crossing parallel lines.
2	I can identify all pairs of angles formed by a transversal crossing parallel lines.
1	I can identify some pairs of angles formed by a transversal crossing parallel lines.

Spiral Review • Assessment Readiness

These questions will help determine if students have retained information taught in the past and can also prepare them for high-stakes assessments. Here, students must use segment addition properties to solve problems **(2.3)**, find the slope between two points **(Alg1, 3.2)**, use properties of vertical angles **(2.4)**, and analyze conditional statements **(2.1)**.

Use the diagram for Problems 25 and 26.

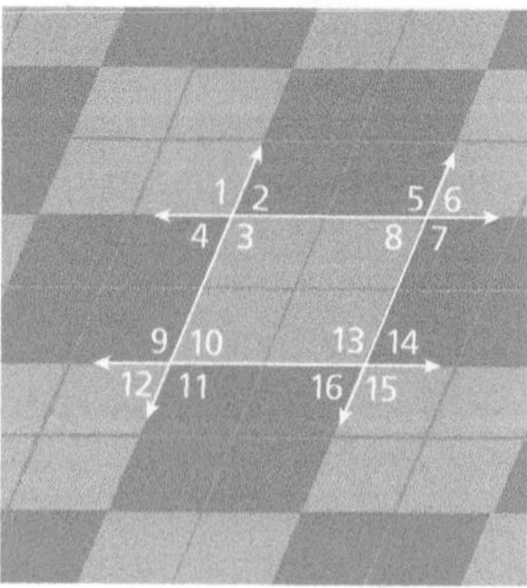

25. Two parallel lines are intersected by a transversal. $\angle 3$ and $\angle 9$ are alternate interior angles. The measure of $\angle 3$ is $(4m + 1)°$. The measure of $\angle 9$ is $(187 - 2m)°$. What is the measure of each angle?

25–27. See Additional Answers.

26. Two parallel lines are intersected by a transversal. $\angle 12$ and $\angle 15$ are consecutive exterior angles. The measure of $\angle 12$ is $(14a + 17)°$ and the measure of $\angle 15$ is $(26a + 43)°$. What is the measure of each angle?

27. Line a is parallel to line b. Line x is parallel to line y. Prove that $\angle 2$ and $\angle 13$ are supplementary angles.

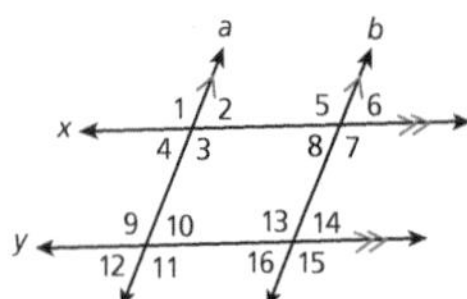

Spiral Review • Assessment Readiness

28. The length of a piece of wood represented by $\overline{RT}$ is 65 inches. Suppose you cut the wood at some point S so that $RS = 4b + 7$ and $ST = 7b - 8$. What is RS?

Ⓐ 6 inches
Ⓑ 31 inches
Ⓒ 34 inches
Ⓓ 65 inches

29. Segment AB has endpoints at $A(4, 1)$ and $B(6, -2)$. What is the slope of the segment?

Ⓐ $-\frac{3}{2}$
Ⓑ $-\frac{2}{3}$
Ⓒ $\frac{2}{3}$
Ⓓ $\frac{3}{2}$

30. $\angle 1$ and $\angle 2$ are vertical angles. $m\angle 2 = 122°$ and $m\angle 1 = (4z - 10)°$. What is z?

Ⓐ 10
Ⓑ 33
Ⓒ 122
Ⓓ 132

31. If a conditional statement is given by "If a number is a whole number, then it is an integer," then its ________ statement is "If a number is an integer, then it is a whole number."

Ⓐ inverse
Ⓑ definition
Ⓒ contrapositive
Ⓓ converse

 **I'm in a Learning Mindset!**

What resources are available for me to understand the relationship between angle pairs formed by transversals and parallel lines?

Keep Going Journal and Practice Workbook

Learning Mindset

Strategic Help Seeking Identifies Sources of Help

Encourage students that it will take time to learn and be confident with each of many new definitions, theorems, and postulates. Remind students to ask for clarification, when needed, from classmates, teachers, and to look up definitions in the textbook. *Which of the angle pairs is the most difficult for you to remember? Can you think of anything to help you remember the relationship between those pairs of angles? When you get stuck working on your homework, where can you go for help?*

3.2 Prove Lines Are Parallel

LESSON FOCUS AND COHERENCE

Mathematics Standards

- Prove theorems about lines and angles.
- Make formal geometric constructions with a variety of tools and methods (compass and straightedge, string, reflective devices, paper folding, dynamic geometric software, etc.).
- Know precise definitions of angle, circle, perpendicular line, parallel line, and line segment, based on the undefined notions of point, line, distance along a line, and distance around a circular arc.

Mathematical Practices and Processes 

- Use appropriate tools strategically.
- Construct viable arguments and critique the reasoning of others.

I Can Objective 

I can ensure that two lines are parallel by construction.

Learning Objective

Students will be able to prove whether or not two lines are parallel.

Language Objective

Use language related to parallel lines cut by transversal postulates, postulate converses, and angle pairs.

Lesson Materials: compass, straightedge, translucent paper, index cards

Mathematical Progressions

Prior Learning	Current Development	Future Connections
Students: • proved theorems about angles formed when a transversal crosses parallel lines. **(3.1)** • used a straightedge and compass to copy an angle. **(1.2)**	**Students:** • construct a line parallel to a given line through a point. • use converses of parallel line theorems to prove whether or not lines are parallel. • prove the transitive property of parallel lines.	**Students:** • will use properties of parallel lines to represent translations. **(5.1)** • will use properties of lines and angles to justify geometric constructions. **(9.1–9.3)** • will use properties of angles formed when a transversal crosses parallel lines to prove theorems about triangles. **(9.1 and 9.2)**

PROFESSIONAL LEARNING

Math Background

The goal of the lesson is for students to be able to construct parallel lines as well as evaluate whether lines are parallel by checking for congruent angle pairs. Later in the course, students will apply their knowledge of parallel lines when they explore properties of quadrilaterals and in modeling 3D shapes. Parallel line postulates and their converses are integral to a variety of proofs in Euclidean geometry.

WARM-UP OPTIONS

ONLINE

PROJECTABLE & PRINTABLE 

ACTIVATE PRIOR KNOWLEDGE • Angle Constructions

Use these activities to quickly assess and activate prior knowledge as needed.

Problem of the Day 

Given angle A, construct angle B such that $m\angle B = 2m\angle A$ Possible answer: Copy $\angle A$ so that it shares a ray with the original angle.

Quick Check for Homework

As part of your daily routine, you may want to display the Teacher Solution Key to have students check their homework.

Make Connections

Based on students' responses to the Problem of the Day, choose one of the following:

 Project the Interactive Reteach, Geometry Lesson 1.2.

 Complete the Prerequisite Skills Activity:

Discuss the construction steps needed in copying an angle using a straightedge and compass.

- *What is the first step in copying an angle?* Open the compass and make an arc on the angle that needs to be copied. Then, keeping the compass open to the same radius, make the same arc where you need to copy the angle.
- *Is that enough to copy the angle?* No, it is not. We need to copy the opening of that arc, measuring with compass the distance between the two intersection points of the arc with the angle rays.

Have all students copy angles to become comfortable using the compass.

SHARPEN SKILLS

If time permits, use this on-level activity to build fluency and practice basic skills.

Mental Math

Objective: Students identify pairs of angles given two parallel lines cut by a transversal.
Materials: index cards or pictures with traced parallel lies cut by a transversal

Have students work in pairs. Have all index cards in a pile. Students take turns picking an index card. The first person who correctly identifies the marked angles keeps the index card. The person with most index cards wins the game.

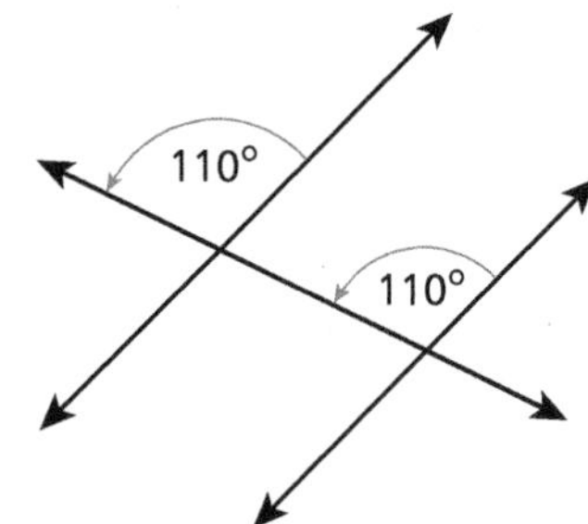

PLAN FOR DIFFERENTIATED INSTRUCTION

MTSS 

Small-Group Options

Use these teacher-guided activities with pulled small groups.

On Track

Materials: transparent paper and straightedge

Have students work in small groups to draw a line on the transparent paper and discuss how folding can be used to make the following:

- a line perpendicular to the first line
- a line parallel to the first line
- a transversal cutting the two parallel lines

Have students describe the angle pairs formed by the parallel lines cut by a transversal.

Almost There (RtI)

Materials: sheets of paper

Have students make a fold in a sheet of paper. The fold should not be parallel to an edge. Have them cut the sheet along the fold. Have students mark a pair of angles and rotate the cut-off sheet to prove the converses of angle postulates.

Ready for More

Have students discuss in small groups whether it is possible to find the measures of all angles in a figure where two parallel lines are cut by two parallel lines, given the measure of only 1 of the 16 angles.

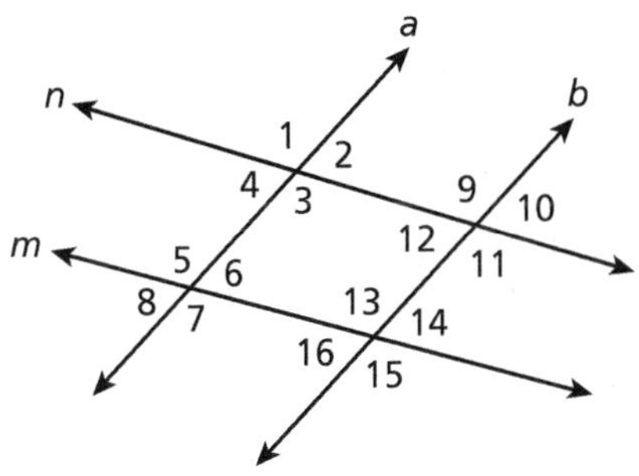

Math Center Options

Use these student self-directed activities at centers or stations. Key: ● Print Resources ● Online Resources

On Track

- ● Interactive Digital Lesson
- ●● Journal and Practice Workbook
- ● Module Performance Task

Almost There

- ● Reteach 3.2 (printable)
- ● Interactive Reteach 3.2

Ready for More

- ● Challenge 3.2 (printable)
- ● Interactive Challenge 3.2
- ● Desmos: Lines, Transversals and Angles

ONLINE  View data-driven grouping recommendations and assign differentiation resources.

During the *Spark Your Learning*, listen and watch for strategies students use. See samples of student work on this page.

Use Logical Reasoning

Strategy 1

I looked at the top two rows and saw that all sides of the rectangles were on the first orange line. Since the rectangles all have same height, that means the second orange line, that goes through the bases of the rectangles, is parallel to the first line since opposite sides of rectangles are parallel.

If students . . . use logical reasoning on the basis of the provided information to solve the problem, they are demonstrating an exemplary understating of the concept of parallel lines.

Have these students . . . explain how they determined that the rest of the orange lines are also parallel. **Ask:**

- Q How do you know that the third orange line is also parallel to the first?
- Q What can you conclude for the rest of the orange lines?

Use a Tool to Verify

Strategy 2

I knew the lines have to be parallel since the rectangles have the same height but I wanted to verify that. I used a ruler to measure the first and last rectangle in each row. They had equal heights.

If students . . . use a tool to verify that the lines are parallel by measuring the heights of the rectangles, they may not see how the rectangles are related to the lines since the lines are placed on the opposite sides of the parallel sides of the rectangles.

Activate prior knowledge . . . by having students think about sides of rectangles. **Ask:**

- Q What do you know about the opposite sides of each rectangle?
- Q If two lines go through opposite sides of rectangles with the same height in a row, what can you conclude about those lines?

COMMON ERROR: Uses a Tool to Verify

I understood that the first two rows have rectangles with the same heights and that the first two orange lines are parallel. However, there was no information about the rest of the rectangles. So, I used a ruler to measure the heights of several rectangles below. It turned out they had equal heights as well.

If students . . . do not continue the logic of the problem to encompass and explain the whole picture, students may not understand that if two lines go through opposite sides of rectangles with the same height in a row, then the two lines must be parallel.

Then intervene . . . by pointing out that the opposite sides of rectangles are parallel. **Ask:**

- Q Can you have two non-parallel lines going through two opposite sides of a rectangle?
- Q If you have two parallel lines, and the lines are connected to form rectangles, what can you tell about the heights of the rectangles?

3.2

Prove Lines Are Parallel

I Can ensure that two lines are parallel by construction.

Spark Your Learning

Look at the optical illusion, known as the Café Wall Illusion.

©consinternational.Com/Alamy

Complete Part A as a whole class. Then complete Parts B–D in small groups.

A. What is a question you can ask about the optical illusion? What information would you need to know to answer your question?

A. Are the orange lines between the rows parallel?; if the lines are straight and if they will eventually intersect

B. Focus on two or three rows of the illusion. How does this help you answer your question? See Additional Answers.

C. To answer your question, what strategy and tool would you use along with all the information you have? What answer do you get? See Strategies 1 and 2 on the facing page.

D. How can you use tools to verify your answer? See Additional Answers.

Turn and Talk

- Suppose all of the black squares are changed to white. How do you think changing the color would change the illusion? Explain your reasoning.
- Does turning the image so that the stripes are vertical change the illusion? What difference do you see, if any? See margin.

EL CULTIVATE CONVERSATION • Information Gap

Have students think about what additional information, if any, they would need to answer the question "Are the orange lines between the rows parallel?"

1. Can you determine whether the first two lines are parallel if you know that the rectangles in the first row are not congruent? Why or why not? yes; If the rectangles are not congruent but have heights that are congruent, the distance between the two lines is constant. This means the lines are parallel.

2. Can you determine whether the first two lines are parallel if you know that the rectangles in the first row are congruent? Explain. yes; If the rectangles are congruent, then the height of each of the rectangles between the top two orange lines are congruent, so the distance between the two lines is constant. This means the lines are parallel.

1 Spark Your Learning

▶ MOTIVATE

- Have students look at the photo in their books and read the information contained in the photo. Then complete Part A as a whole-class discussion.
- Give the class the additional information they need to solve the problem. This information is available online as a printable and projectable page in the Teacher Resources.
- Have students work in small groups to complete Parts B–D.

▶ PERSEVERE

If students need support, guide them by asking:

Q **Advancing • Use Tools** Which tool could you use to solve the problem? Why choose that tool and not some other? Students' choices of tools and reasons for choosing them will vary.

Q **Assessing** How can a geometric tool, for example, a ruler, help you figure out whether the lines are parallel? I can measure the distance between the lines. Parallel lines should have a constant distance between them.

Q **Assessing** If you look at the top row knowing that the height of the rectangles is the same, can you make an argument that the top two orange lines are parallel? The two orange lines lie on the top and bottom of the rectangles which all have the same height. Therefore, the lines must be parallel.

Q **Advancing** How is the optical illusion achieved? It may be because the rectangles are not directly under each other but moved slightly to the left and to the right.

Turn and Talk If needed, have students color in the white rectangles. Possible answer: The optical illusion would no longer be visible and it would be obvious that the horizontal lines are parallel. If you turn the image, it will still be an optical illusion.

▶ BUILD SHARED UNDERSTANDING

Select groups of students who used various strategies and tools to share with the class how they solved the problem. As they present their solutions, have each group discuss why they chose a specific strategy and tool.

② Learn Together

Build Understanding

Task 1 **(MP) Use Tools** Students construct a line parallel to a given line through a point using a compass and straightedge.

CONNECT TO VOCABULARY

Have students use the **Interactive Glossary** to record their understanding of the vocabulary in this task.

Sample Guided Discussion:

Q Why do we need to draw $\overrightarrow{XP}$? By drawing $\overrightarrow{XP}$, we create $\angle PXY$, which we then copy.

Q When we copied $\angle PXY$, we drew an arc with the same center at point X and the same radius. What will happen if the radii are not the same? If the radii are not the same, then the distances between the points where the arcs touch, $\overrightarrow{XP}$ and $\overleftrightarrow{XY}$, and $\overrightarrow{XP}$ and $\overleftrightarrow{PZ}$, respectively, are not going to be the same. That means that the corresponding angles are not going to be congruent and the lines are not going to be parallel.

Turn and Talk Have students perform the construction if needed. The steps would be the same no matter which point you choose to start on the ray.

Build Understanding

Construct Parallel Lines

You can use the Parallel Postulate to construct a pair of parallel lines.

Parallel Postulate

For point P not on line l, there is exactly one line parallel to l through point P.

1 Use a compass and a straightedge to construct parallel lines. This construction involves copying an angle.

A. Draw a line and a point P not on the line. Draw and label points X and Y on the line. Use a straightedge to draw $\overrightarrow{XP}$. Explain why P must not lie on $\overleftrightarrow{XY}$.
If point P is on $\overleftrightarrow{XY}$, then a line parallel to point P will coincide with $\overleftrightarrow{XY}$.

B. Use a compass to copy $\angle PXY$ at point P. Draw point Z. Draw a straight line through points P and Z. Why is $\overleftrightarrow{XY}$ parallel to $\overleftrightarrow{PZ}$?
The two lines do not intersect.

C. What is true about the corresponding angles? What would be true about $\overleftrightarrow{XY}$ and $\overleftrightarrow{PZ}$ if the corresponding angles did not have this relationship? The corresponding angles are congruent; The lines would not be parallel.

D. How is the Parallel Postulate used in this construction? The Parallel Postulate guarantees that, for any line, you can construct a parallel line through a point that is not on the line.

Turn and Talk How would the construction steps be different, if at all, if you drew $\overrightarrow{YP}$ instead of $\overrightarrow{XP}$? See margin.

Converses of the Parallel Lines Theorems

Recall that the converse of a conditional statement is formed by exchanging the hypothesis and conclusion of the statement. The converse of a statement in the form "If p, then q" is "If q, then p."

Conditional statement: If $x + 4 = 6$, then $x = 2$.

Converse: If $x = 2$, then $x + 4 = 6$.

LEVELED QUESTIONS

Depth of Knowledge (DOK)	Leveled Questions	What Does This Tell You?
Level 1 **Recall**	How can we use semi-transparent paper to make sure that we have constructed parallel lines? We can trace one of the angles and move the picture on top of the other angle to check if the angles are the same.	Students' answers will indicate whether they understand that when two lines are parallel, the corresponding angles at the transversal line are congruent.
Level 2 **Basic Application of Skills & Concepts**	When two parallel lines are cut by a transversal, what is true of the corresponding angles? Corresponding angles are congruent.	Students' answers will indicate whether they can use parallel lines and a transversal to define congruent angles.
Level 3 **Strategic Thinking & Complex Reasoning**	How would you copy $\angle X$ so that you construct an alternate interior angle to $\angle X$ through P? I would follow the same process but rather than starting above point P, I will copy the angle below P so that point Z will be located on the left side of P.	Students' answers will indicate whether they understand they could construct parallel lines using other congruent angles.

The parallel lines postulates and theorems you learned previously have true converses. These converses are listed below along with an example from the diagram.

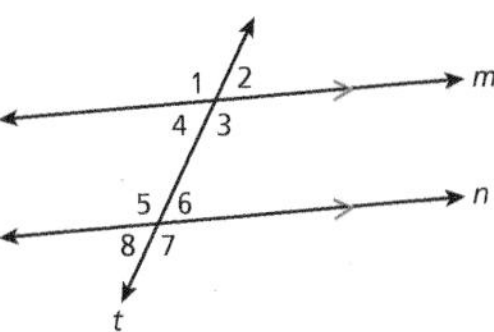

Converses of Parallel Lines Postulates and Theorems	
Postulate or Theorem	**Example**
Converse of the Corresponding Angles Postulate If two lines are cut by a transversal so that corresponding angles are congruent, then the lines are parallel.	If $\angle 1 \cong \angle 5$, then $m \parallel n$.
Converse of the Alternate Interior Angles Theorem If two lines are cut by a transversal so that alternate interior angles are congruent, then the lines are parallel.	If $\angle 4 \cong \angle 6$, then $m \parallel n$.
Converse of the Alternate Exterior Angles Theorem If two lines are cut by a transversal so that alternate exterior angles are congruent, then the lines are parallel.	If $\angle 1 \cong \angle 7$, then $m \parallel n$.
Converse of the Consecutive Interior Angles Theorem If two lines are cut by a transversal so the consecutive interior angles are supplementary, then the lines are parallel.	If $\angle 3$ and $\angle 6$ are supplementary, then $m \parallel n$.
Converse of the Consecutive Exterior Angles Theorem If two lines are cut by a transversal so the consecutive exterior angles are supplementary, then the lines are parallel.	If $\angle 1$ and $\angle 8$ are supplementary, then $m \parallel n$.

2 In the photo, $\angle 2 \cong \angle 6$.

A. What theorem or postulate can you use to show that line j and line k are parallel?
Converse of the Corresponding Angles Postulate

B. How can you use a different theorem or postulate to show that line j and line k are parallel?
See Additional Answers.

Turn and Talk Are converses of conditional statements always true? Give an example that supports your answer. See margin.

Task 2 (MP) **Construct Arguments** Students will be able to understand the converses of the parallel line theorems.

Sample Guided Discussion:

Q If you think of copying angles 5, 6, 7 and 8 on paper and moving the paper up to place it over angles 1, 2, 3, and 4, which angle will be the corresponding angle to $\angle 6$? $\angle 2$

Q If you think of copying angles 5, 6, 7 and 8 on paper and moving the paper up to place it over angles 1, 2, 3, and 4, which angle will be the corresponding (or congruent) to $\angle 4$? $\angle 8$

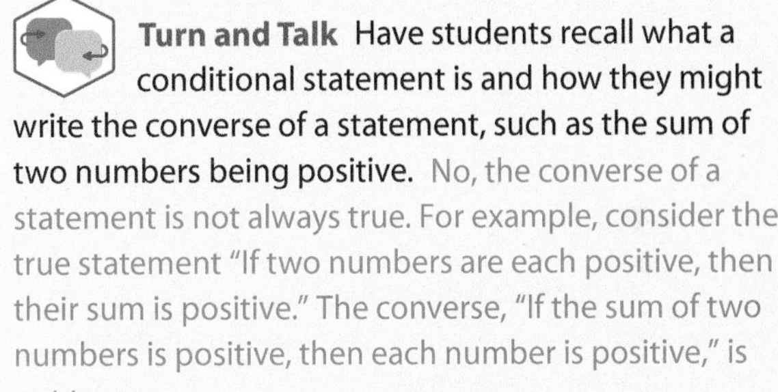

Turn and Talk Have students recall what a conditional statement is and how they might write the converse of a statement, such as the sum of two numbers being positive. No, the converse of a statement is not always true. For example, consider the true statement "If two numbers are each positive, then their sum is positive." The converse, "If the sum of two numbers is positive, then each number is positive," is not true.

PROFICIENCY LEVEL

Beginning
Have students use the term *converse* in a sentence structure. Make statements such as, "If it's snowing, then it must be cold." Ask the students whether the converse is true. Encourage them to say, "The converse is true/not true."

Intermediate
Have students practice if-then statements by varying the angle pairs given. Encourage students to make complete sentences such as, "If the corresponding angles are congruent, then the lines are parallel."

Advanced
Have students explain the difference between a Parallel Lines Postulate and a Converse to a Parallel Lines Postulate.

Step It Out

Task 3 (MP) **Construct Arguments** Students prove whether or not lines are parallel. They discuss their reasoning to complete a two-column proof.

Sample Guided Discussion:

Q **Other than ∠1, what other angle is congruent to ∠2?** ∠4

Q **What do you need to prove in order to claim that the two lines are parallel?** Corresponding angles ∠1 and ∠4 are congruent.

Turn and Talk Have students look at the converses of parallel lines postulates and theorems to determine which might apply to this situation. Converse of the Alternate Interior Angles Theorem

Task 4 (MP) **Construct Arguments** Students prove whether or not lines are parallel. They consider different theorems to justify statements.

Sample Guided Discussion:

Q **What can we determine about the relationship between Water St. and Pearl St.?** Angle C is 109° because the 71° angle and angle C are supplementary and have a sum of 180°. Angle B is 109° because vertical angles are congruent. So, Pearl St. and Water St. are parallel.

Q **What is the measure of angle A and how can you use this information to determine if Oak Ave. is parallel to Pearl Street?** 108°; Corresponding angles of Oak Ave. and Pearl St. are not congruent and therefore the streets are not parallel.

Q **If you have two parallel lines a and b and a third line c not parallel to a, can c still be parallel to b?** No, line c will eventually intersect both a and b.

Turn and Talk Have students talk about how roads are similar to and different from lines or line segments. Possible answer: Roads are not always flat. Roads have hills. Roads have different widths in different places. You are trying to model a three-dimensional object with an item that is one-dimensional.

Step It Out

Prove Whether or Not Two Lines Are Parallel

You can use the converses of the parallel line postulates and theorems to prove whether or not two lines are parallel.

Belinda is tiling a floor using quadrilateral tiles. Regardless of the color, the central tiles are identical. She knows that ∠1 ≅ ∠2. She wants to prove $l \parallel m$.

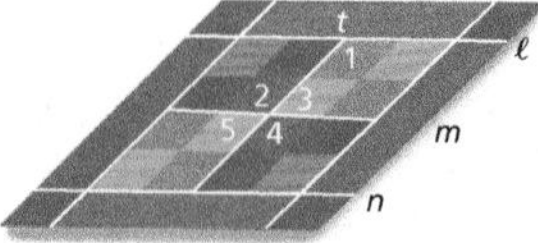

Given: ∠1 ≅ ∠2

Prove: $l \parallel m$

Statements	Reasons
1. ∠1 ≅ ∠2	1. Given
2. ∠2 ≅ ∠4	2. ?
3. ∠1 ≅ ∠4	3. ?
4. $l \parallel m$	4. ?

A. What reason justifies ∠2 ≅ ∠4? Vertical Angles Theorem

B. What reason justifies ∠1 ≅ ∠4? Transitive Property of Congruence

C. What reason justifies $l \parallel m$? Converse of the Corresponding Angles Postulate

Turn and Talk What theorem can you use to prove $l \parallel m$ using fewer steps? See margin.

Are Oak Avenue and Pearl Street parallel? Write a proof that justifies your conclusion.

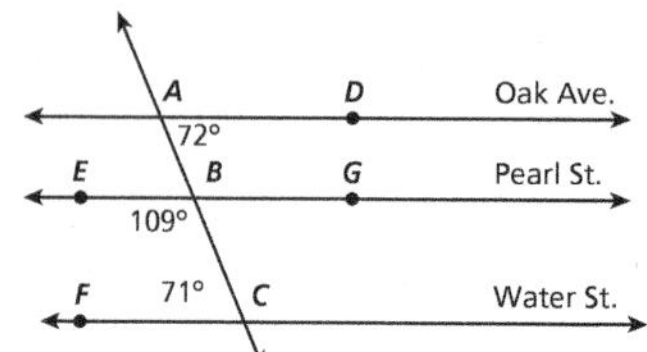

A. What information are you given in the diagram? Use this to write the Given statement. See Additional Answers.

B. What is the Prove statement? $\overleftrightarrow{AD} \nparallel \overleftrightarrow{EG}$

Statements	Reasons
1. ?	1. Given
2. m∠ABE = 71°	2. ?
3. ?	3. ?

C. What theorem justifies m∠ABE = 71°? Linear Pairs Theorem

D. What reason completes the proof? Alternate Interior angles are not congruent.

E. What conclusion can you make? Oak Avenue is not parallel to Pearl Street.

Turn and Talk Why is it not always valid to model a road with a line or a line segment? See margin.

88

Transitive Property of Parallel Lines

The Transitive Property can be applied to parallel lines.

Transitive Property of Parallel Lines	
If two lines are parallel to the same line, then they are parallel to each other. If $a \parallel b$ and $b \parallel c$, then $a \parallel c$.	a, b, c (three parallel lines)

5 A drawing of a subway system shows the different levels used to reach the subway platforms. Level 1 is line p, level 2 is line q, and level 3 is line r. Prove that Level 2 and level 3 are parallel.

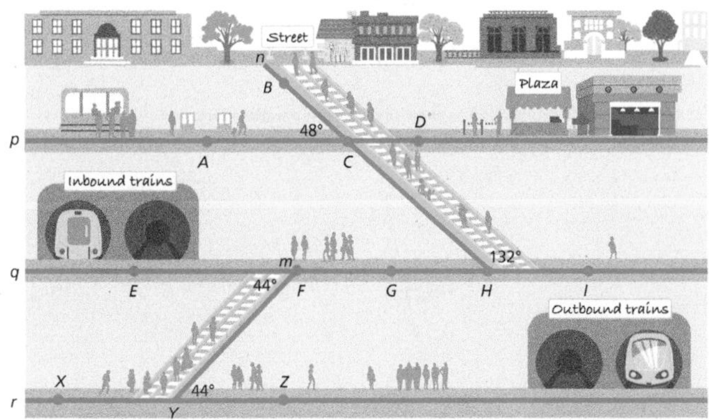

A–D. See Additional Answers.

Statements	Reasons
1. $m\angle ACB = 48°$, $m\angle CHI = 132°$	1. Given
2. $m\angle DCH = 48°$	2. ? A. What property justifies $m\angle DCH = 48°$?
3. $48° + 132° = 180°$	3. Addition
4. $m\angle DCH + m\angle CHI = 180°$	4. Substitution Property of Equality
5. $\angle DCH$ and $\angle CHI$ are supplementary.	5. Definition of supplementary angles
6. $p \parallel q$	6. ? B. What property justifies $p \parallel q$?
7. $m\angle EFY = 44°$, $m\angle FYZ = 44°$	7. Given
8. $\angle EFY \cong \angle FYZ$	8. Definition of congruent angles
9. $p \parallel r$	9. ? C. What property justifies $p \parallel r$?
10. $q \parallel r$	10. ? D. What allows you to write $q \parallel r$?

Task 5 (MP) **Construct Arguments** Students prove the transitive property for parallel lines.

OPTIMIZE OUTPUT Critique, Correct, and Clarify

Have students work with a partner to discuss the missing parts of the double-column proof. Encourage students to use the correct vocabulary when talking about the postulates and theorems.

Sample Guided Discussion:

Q In Statement 2, what are the possible types of angles to choose from? complementary, supplementary, and vertical

Q In Statement 9, how can you tell whether the reason is a postulate or a converse to a postulate? If we are making a conclusion that two lines are parallel based on congruent angles, then it is a converse.

ONLINE

Assign the Digital On Your Own for
- built-in student supports
- Actionable Item Reports
- Standards Analysis Reports

On Your Own

Assignment Guide

The chart below indicates which problems in the On Your Own are associated with each task in the Learn Together. Assign daily homework for tasks completed.

Learn Together Tasks	On Your Own Problems
Task 1, p. 86	Problems 18–19
Task 2, p. 87	Problems 7–10
Task 3, p. 88	Problems 11–15
Task 4, p. 88	Problems 16 and 17
Task 5, p. 89	Problems 20 and 21

Check Understanding

1. Draw line k and point P that is not on line k. How many lines can be drawn through P that are parallel to line k? How do you know?

1. One line can be drawn through P that is parallel to k by the Parallel Postulate. See Additional Answers for drawing.

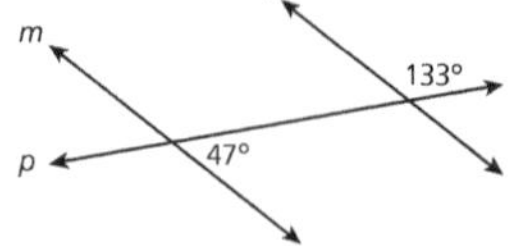

In Problems 2–5, name a postulate or theorem that can be used with the given information to prove that the lines are parallel.

2. $\angle 3 \cong \angle 7$ — 2. Converse of the Corresponding Angles Postulate
3. $\angle 3 \cong \angle 5$ — 3. Converse of the Alternate Interior Angles Theorem
4. $\angle 2$ and $\angle 5$ are supplementary — 4. Converse of the Consecutive Interior Angles Theorem
5. $\angle 1 \cong \angle 7$ — 5. Converse of the Alternate Exterior Angles Theorem

6. Is line m parallel to line n? Explain your reasoning.

6. The vertical angle of the angle measured 133° and the angle measured 47° are supplementary consecutive interior angles. So, the lines are parallel by the Converse of the Consecutive Interior Angles Theorem.

m n p 133° 47°

On Your Own

Determine if there is enough information to show that line a is parallel to line b. If there is, tell which postulate or theorem you would use.

8–9. See Additional Answers.

7. no

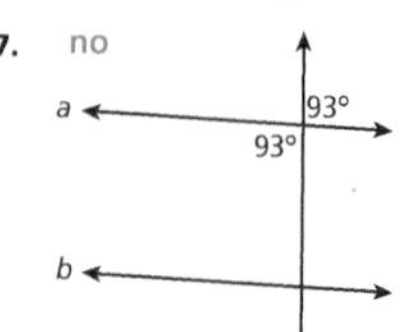

8.

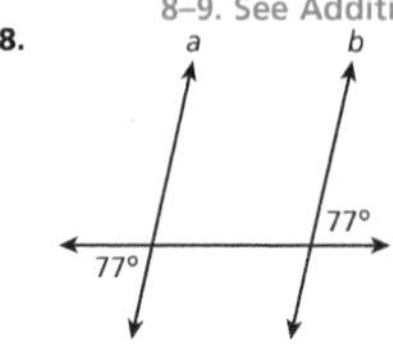

9.

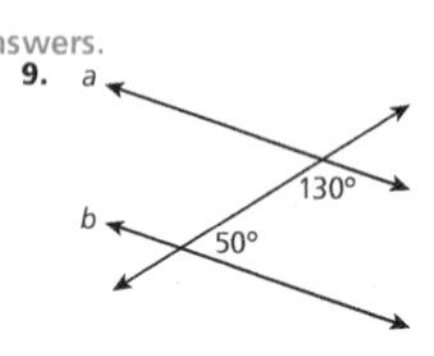

10. Which sides of $QRST$ are parallel if you know that $\angle 1 \cong \angle 2$? $\overline{QT}$ and $\overline{RS}$

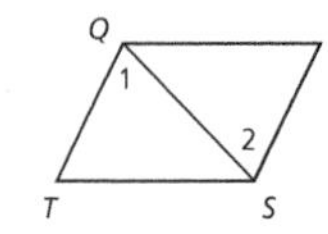

11. (MP) **Construct Arguments** Fill in the missing steps of the proof of the Converse of the Alternate Interior Angles Theorem.

Given: $\angle 1 \cong \angle 2$ — $\angle 1 \cong \angle 3$; Converse of the Corresponding Angles Postulate

Prove: $a \parallel b$

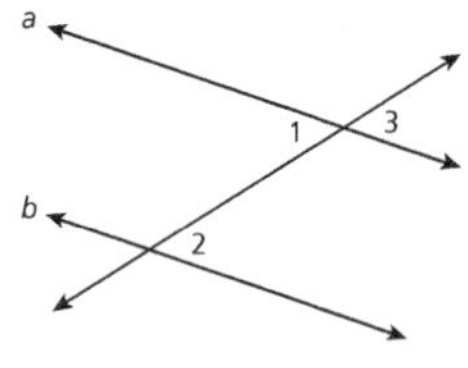

Statements	Reasons
1. $\angle 1 \cong \angle 2$	1. Given
2. ___?___	2. Vertical Angles Theorem
3. $a \parallel b$	3. ___?___

90

3 Check Understanding

Formative Assessment

Use formative assessment to determine if your students are successful with this lesson's learning objective.

Students who successfully complete the Check Understanding can continue to the On Your Own practice.

For students who miss 1 problem or more, work in a pulled small group using the Almost There small-group activity on page 85C.

ONLINE

Assign the Digital Check Understanding to determine
- success with the learning objective
- items to review
- grouping and differentiation resources

4 Differentiation Options

Differentiate instruction for all students using small-group activities and math center activities on page 85C.

Reteach

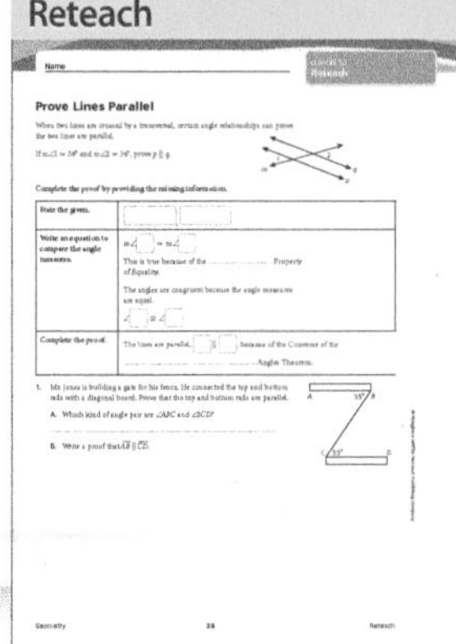

Challenge

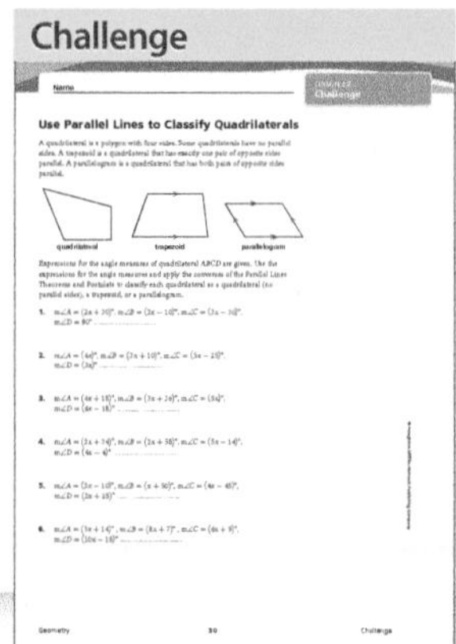

Find the value of x that makes $p \parallel q$.

12. $x = 15$

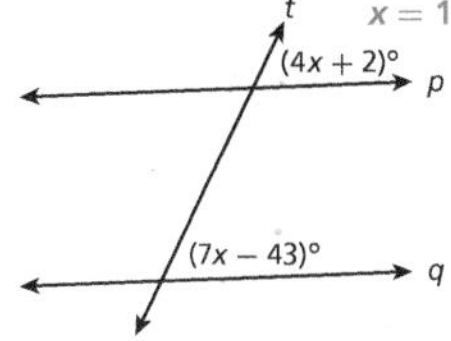

13. $x = 7$

(17x + 8)°
(10x + 57)°

14. $x = 27$

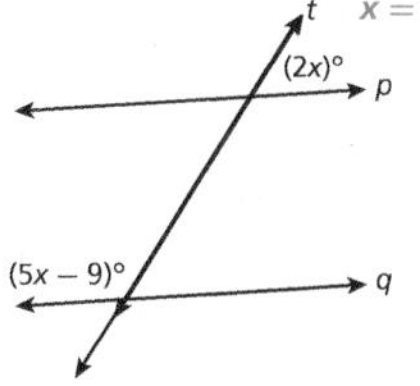

15. $x = 9$

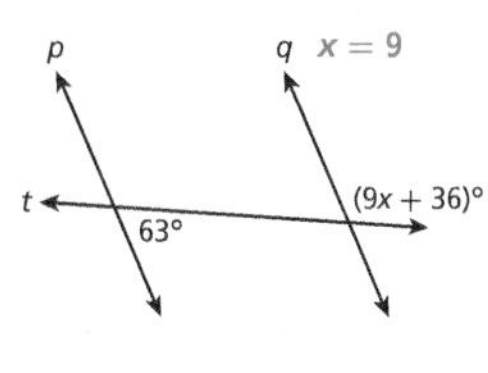

Use the diagram to write each two column proof.
16–20. See Additional Answers.

16. Prove the Converse of the Alternate Exterior Angles Theorem.

Given: $\angle 1 \cong \angle 7$

Prove: $a \parallel b$

17. Prove the Converse of the Consecutive Interior Angles Theorem.

Given: $\angle 4$ and $\angle 5$ are supplementary.

Prove: $a \parallel b$

18. Each lane in the swimming pool is parallel to the lane to its right. Explain why the left side of the pool is parallel to the right side of the pool.

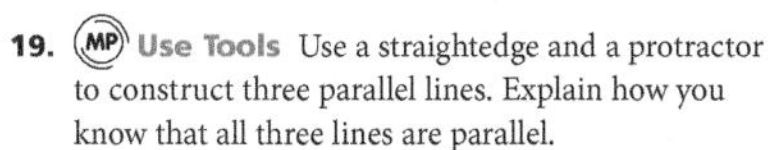

19. MP **Use Tools** Use a straightedge and a protractor to construct three parallel lines. Explain how you know that all three lines are parallel.

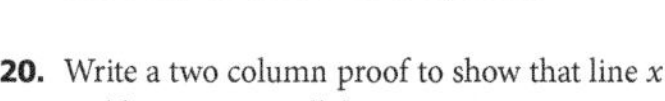

20. Write a two column proof to show that line x and line z are parallel.

Given: $\angle 1$ and $\angle 2$ are supplementary. $\angle 3 \cong \angle 4$

Prove: $x \parallel z$

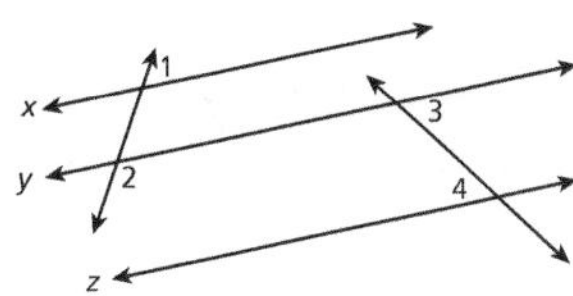

Questioning Strategies

Problem 16 Suppose $\angle 1$ is congruent to $\angle 2$. How does that change the problem? If the angles are the same, that means that they are both right angles and line t is perpendicular to both lines a and b. Lines a and b are still going to be parallel.

Watch for Common Errors

Problem 21 Some students may be misled by a drawing where lines look parallel but they are in fact not parallel. Suggest to students that as soon as they prove two lines a parallel to mark them on the diagram.

(5) Wrap-Up

Summarize learning with your class. Consider using the Exit Ticket, Put It in Writing, or I Can scale.

Exit Ticket 

Draw a line *l* and a point *D* not on the line. Using a compass and a straightedge, construct a parallelogram such that one of its sides is on *l* and point *D* is one of its vertices. Explain in writing how you know that the shape is a parallelogram. See students' constructions. Possible answer: I drew line *k* through *D* to intersect *l* at point *A*. I drew point *B* on *l*. I copied $\angle DAB$ at point *D*, thus creating two congruent corresponding angles. I drew line *t* parallel to line *l*. I copied $\angle A$ at point *B* to the right of point *B*, thus creating corresponding congruent angles. I drew line *q* parallel to line *k*. The shape is a parallelogram by construction.

Put It in Writing 

Describe some strategies you can use to recognize that two lines are parallel.

I Can 

The scale below can help you and your students understand their progress on a learning goal.

4	I can ensure that two lines are parallel by construction, and I can explain my steps to others.
3	I can ensure that two lines are parallel by construction.
2	I can construct two parallel lines using corresponding angles.
1	I can perform some of the steps needed to construct two parallel lines.

Spiral Review • Assessment Readiness

These questions will help determine if students have retained information taught in the past and can also prepare them for high-stakes assessments. Here, students must use the midpoint to determine the length of a segment **(2.3)**, find the measure of an angle in a linear pair **(2.4)**, and match pairs of angles with their correct name **(3.1)**.

21. A marching band is performing at halftime of a football game. A section of their formation is shaped as shown. See Additional Answers.

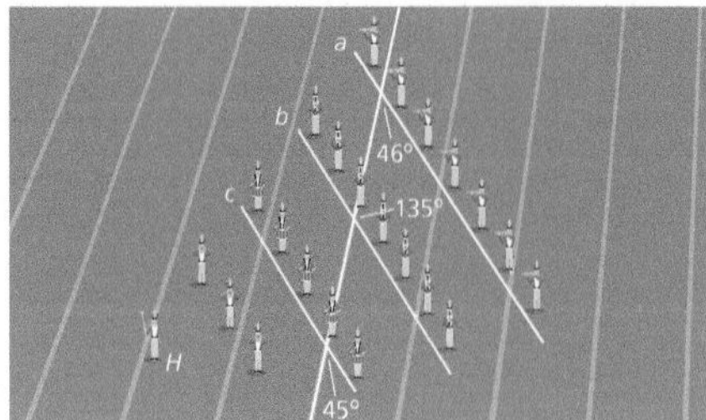

A. Which pairs of lines of marchers are parallel? Explain how you know these lines are parallel.

B. Are there any lines that are not parallel with the others? Explain your reasoning.

Spiral Review • Assessment Readiness

22. $\overline{AC}$ represents the width of a banner. The length of $\overline{AC}$ is 72 inches. Point *B* is the midpoint of $\overline{AC}$. What is the length of $\overline{AB}$ in inches?

Ⓐ 36 inches
Ⓑ 48 inches
Ⓒ 72 inches
Ⓓ 144 inches

23. $\angle 1$ and $\angle 2$ form a linear pair. $m\angle 1 = (5x + 7)^\circ$ and $m\angle 2 = (8x + 4)^\circ$. What is $m\angle 1$?

Ⓐ 72°
Ⓑ 108°
Ⓒ 130°
Ⓓ 167°

24. Match each pair of angles on the left with the correct description on the right.

Angle pair	Description	
A. $\angle 1$ and $\angle 7$	**1.** Consecutive interior angles	A. 2
B. $\angle 3$ and $\angle 6$	**2.** Alternate exterior angles	B. 1
C. $\angle 4$ and $\angle 8$	**3.** Alternate interior Angles	C. 4
D. $\angle 4$ and $\angle 6$	**4.** Corresponding angles	D. 3

1 2 4 3 5 6 8 7

 **I'm in a Learning Mindset!**

Who can support me while learning about proving that lines are parallel?

Keep Going ▶ Journal and Practice Workbook

Learning Mindset

Strategic Help-Seeking Identifies Sources of Help

Point out that writing two-column proofs is a work in progress. Assure students that they will continue to develop their understanding with logical explanations and proofs. They should also identify resources that are available to help them. *What helped you understand the logic of the two-column proofs? How does discussing the "statements" part of the two-column proof with peers help you understand the logic behind it? What other resources are available to help you with proofs?*

3.3 Prove Lines Are Perpendicular

LESSON FOCUS AND COHERENCE

Mathematics Standards

- Prove theorems about lines and angles.
- Make formal geometric constructions with a variety of tools (compass and straightedge, string, reflective devices, paper folding, dynamic geometric software, etc.)
- Know precise definitions of angle, circle, perpendicular line, parallel line, and line segment, based on the undefined notions of point, line, distance along a line, and distance around a circular arc.

Mathematical Practices and Processes 

- Use appropriate tools strategically.
- Construct viable arguments and critique the reasoning of others.

I Can Objective 

I can ensure that a line is a perpendicular bisector of a segment by construction.

Learning Objective

Students will be able to define and construct the perpendicular bisector of a line segment as the set of points that are equidistant from its endpoints.

Language Objective

Explain how you can ensure that a line is a perpendicular bisector of a segment.

Vocabulary

New: perpendicular, perpendicular bisector

Lesson Materials: compass and straightedge, geometry software

Mathematical Progressions

Prior Learning	Current Development	Future Connections
Students: • used a straightedge and compass to bisect a segment and angle. **(1.1 and 1.2)** • explained a proof of the Pythagorean Theorem and its converse. **(Gr8, 11.1–11.3)**	**Students:** • construct a perpendicular bisector. • prove the Perpendicular Bisector Theorem. • prove the Converse of the Perpendicular Bisector Theorem. • construct perpendicular lines. • prove theorems about right angles.	**Students:** • will use properties of perpendicular lines to represent reflections. **(5.3)** • will use properties of lines and angles to justify geometric constructions. **(9.2 and 9.3)** • will use properties of perpendicular lines to prove theorems about triangles. **(9.2)** • will construct a circle that circumscribes a triangle. **(8.1)**

PROFESSIONAL LEARNING

Visualizing the Math

Have students create an equilateral triangle using geometry software. Have them construct the perpendicular bisector of each side. Tell them that the point of intersection of the perpendicular bisectors of the sides of a triangle is the center of a circle that will go through all three vertices. With the radius equal to the distance from the point of intersection to any vertex and the point of intersection as the center, draw a circle. Explain how the image displays a reflection. Students can fold their papers on any of the perpendicular bisectors to see the symmetry of the figure.

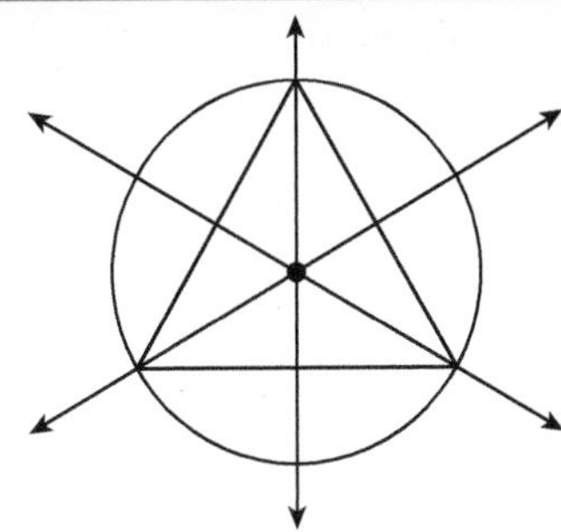

WARM-UP OPTIONS

ONLINE

PROJECTABLE & PRINTABLE 

ACTIVATE PRIOR KNOWLEDGE • Apply the Pythagorean Theorem

Use these activities to quickly assess and activate prior knowledge as needed.

Problem of the Day 

Padma is hiking. She walks 5 kilometers south and continues west until she is 13 kilometers from her starting point. How many kilometers west did Padma walk? Let $w =$ number of kilometers west that Padma walks. Then $5^2 + w^2 = 13^2$, so $w = \sqrt{13^2 - 5^2} = \sqrt{144} = 12$. Padma walks 12 kilometers west.

Quick Check for Homework

As part of your daily routine, you may want to display the Teacher Solution Key to have students check their homework.

Make Connections

Based on students' responses to the Problem of the Day, choose one of the following:

1. Project the Interactive Reteach, Grade 8, Lesson 11.1.
2. Complete the Prerequisite Skills Activity:

Have students work in pairs. One student writes either hypotenuse x or leg x, where x is an integer between 1 and 20. The other student writes leg y, where y is an integer less than x. Have the students work together to determine the missing hypotenuse or leg.

- *If the hypotenuse of a right triangle is x and a leg is y, what is the missing leg?* $\sqrt{x^2 - y^2}$.
- *If a leg of a right triangle is x and the other leg is y, what is the hypotenuse?* $\sqrt{x^2 + y^2}$

If students continue to struggle, use Tier 2 Skill 4.

SHARPEN SKILLS

If time permits, use this on-level activity to build fluency and practice basic skills.

Quantitative Comparison

Objective: Students make a comparison between two quantities.

Write the following problem on the board. Ask students to choose the letter representing the correct answer and to explain their reasoning.

Quantity A	**Quantity B**
the hypotenuse of a right triangle with sides that measure 3 cm and 4 cm	the unknown side of a right triangle with one side that measures 2 cm and a hypotenuse of $\sqrt{29}$ cm

A. Quantity A is greater.

B. Quantity B is greater.

C. The two quantities are equal. C; Quantity A is 5 and Quantity B is 5.

D. The relationship cannot be determined from the information given.

PLAN FOR DIFFERENTIATED INSTRUCTION

Small-Group Options

Use these teacher-guided activities with pulled small groups.

On Track

Materials: worksheet with 10 isosceles triangles each with altitude to the base

Have students work in pairs. Each student writes numbers or variables for each triangle as in Problems 3 and 4 of the Check Understanding. They represent at least 2 of each of the following:

- A triangle that uses the Perpendicular Bisector Theorem to find the variable.
- A triangle that uses the Converse of the Perpendicular Bisector Theorem to find the variable.
- A triangle for which neither theorem can be used and the value of the variable cannot be found.

Each student enters the answers on the other side of the paper.

Students switch papers and solve the triangles.

Almost There

Draw the triangle.

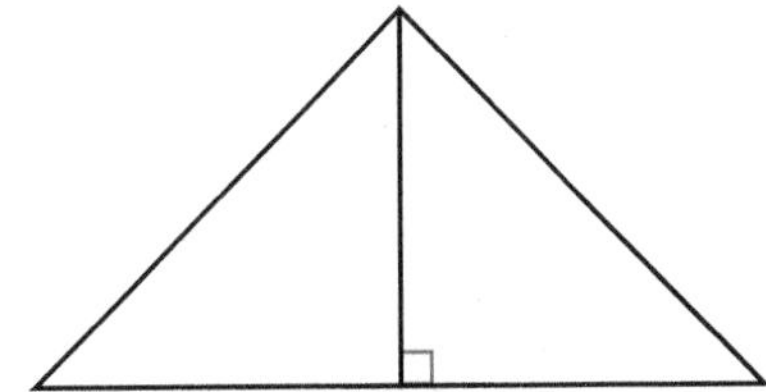

- Label the slanted sides 10. Write $3x + 4$ and 22 to indicate the lengths of the two horizontal segments.
- Ask if the vertical segment is the perpendicular bisector of the horizontal segment. Why or why not? If so, what is the value of x?
- Repeat by changing the values in the figure to side: $2x + 6$, side: 20, base segment: 12, and base segment: 12.
- Repeat by changing the values in the figure to side: 8, side: 7, base segment: 8, and base segment: $2x + 1$.

Ready for More

Have students draw an acute scalene triangle and then do the following:

- Construct the perpendicular bisector of each side.
- Set the compass to the distance between the point of intersection of the perpendicular bisectors of the sides.
- Use the point of intersection of the perpendicular bisectors of the sides as the center and then circumscribe a circle on the triangle.

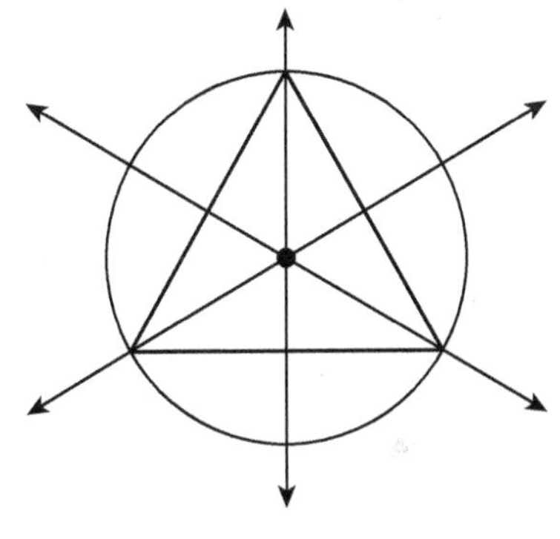

Math Center Options

Use these student self-directed activities at centers or stations. Key: ● Print Resources ● Online Resources

On Track

- ● Interactive Digital Lesson
- ●● Journal and Practice Workbook
- ● Interactive Glossary (printable): **perpendicular**, **perpendicular bisector**
- ● Module Performance Task
- ●● Standards Practice: Prove Theorems About Lines and Angles

Almost There

- ● Reteach 3.3 (printable)
- ● Interactive Reteach 3.3
- ● Illustrative Mathematics: Construction of perpendicular bisector
- ● RtI Tier 2 Skill 4: The Pythagorean Theorem and Its Converse

Ready for More

- ● Challenge 3.3 (printable)
- ● Interactive Challenge 3.3
- ● Illustrative Mathematics: Reflected Triangles

View data-driven grouping recommendations and assign differentiation resources.

Spark Your Learning • Student Samples

During the *Spark Your Learning*, listen and watch for strategies students use. See samples of student work on this page.

Use the Converse of the Pythagorean Theorem Strategy 1

I measured all sides of the upper-left triangle and the side lengths satisfied the Pythagorean Theorem. So, one of the angles at the fountain is a right angle. Because of vertical angle and linear pairs, all the angles are right angles.

If students . . . use the Pythagorean Theorem to solve the problem, they are employing an efficient method and demonstrating an exemplary understanding of the Pythagorean Theorem and its converse from Grade 8.

Have these students . . . explain how they determined to use the converse of the Pythagorean Theorem and how they solved it. **Ask:**

- Q Why did you use the converse of the Pythagorean Theorem?
- Q How did you determine that the triangle was a right triangle?

Use Paper Folding Strategy 2

I traced the boundaries of the park and the paths through the center of the park. I folded the trace on one of the paths and then on the other. The result of the fold was all the angles at the fountain overlap in four square corners. So, the paths are perpendicular.

If students . . . use paper folding to solve the problem, they understand the visual dynamics of the problem but may not know how model the problem geometrically.

Activate prior knowledge . . . by having students recall the Pythagorean Theorem. **Ask:**

- Q How can you show that a triangle is a right triangle?
- Q What is the relationship of the sides of a right triangle?

COMMON ERROR: Misuses the Pythagorean Theorem

I measured the three sides of the triangle and got 4, 5, and 3. Because $4^2 \neq 5^2 + 3^2$, the triangle is not a right triangle. The paths are not perpendicular.

If students . . . misuse the Pythagorean Theorem, then they need to recall the conditions for it.

Then intervene . . . by pointing out the relationship of the sides of a right triangle. **Ask:**

- Q What is the name of the longest side of a right triangle?
- Q What are the other two sides called?
- Q What is the relationship between the three sides?

3.3

Prove Lines Are Perpendicular

I Can ensure that a line is a perpendicular bisector of a segment by construction.

Spark Your Learning

A city is designing a park that is shaped like a triangle.

Complete Part A as an entire class. Then complete Parts B–D in small groups.

A. What geometric question can you ask about the design of the park? What information do you need to answer your question?

A. Are the paths that intersect in the center of the park perpendicular?; length and angle measures

B. What information about relationships between geometric figures in the park does the diagram *not* give? Possible answer: congruent angles

C. To answer your question, what strategy and tool would you use along with all the information you have? What answer do you get? See Strategies 1 and 2 on the facing page.

D. What other relationships do you see among the paths in the park? Explain. See Additional Answers.

Turn and Talk

- Predict whether the paths would remain perpendicular if the path parallel to a side of the park were moved closer to that side.
- Would you be able to determine anything about whether the paths were perpendicular if you knew that the paths intersecting sides of the park each intersected the midpoint of the side? See margin.

SUPPORT SENSE-MAKING • Three Reads

Tell students to read the information in the illustration three times and prompt them with a different question each time.

1. What is the situation about? The situation is about a triangular-shaped park being designed that contains two crossing paths.
2. What are the quantities in this situation? How are those quantities related? The quantities are the lengths of sides of the park and the paths and the angle measures of the intersections of the paths and the sides; One of the paths is parallel to a side of the park. Two sides of the park are congruent.
3. What are possible questions you could ask about this situation? Possible answer: Are the two paths perpendicular? Is the path from a vertex to the opposite side perpendicular to the opposite side?

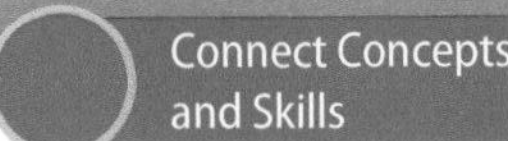

Connect Concepts and Skills

1 Spark Your Learning

▶ MOTIVATE 

- Have students look at the photo in their books and read the information contained in the photo. Then complete Part A as a whole-class discussion.
- Give the class the additional information they need to solve the problem. This information is available online as a printable and projectable page in the Teacher Resources.
- Have students work in small groups to complete Parts B–D.

▶ PERSEVERE

If students need support, guide them by asking:

Q **Advancing • Use Tools** Which tool could you use to solve the problem? Why choose that tool and not some other? Students' choices of tools and reasons for choosing them will vary.

Q **Assessing** If two paths are parallel, what does that tell you about angles formed? Possible answer: Alternate interior angles are congruent and corresponding angles are congruent.

Q **Advancing** How can you show that the paths that intersect at the center of the circular path are perpendicular? Possible answer: You need to show that all the angles of intersection are congruent. You can measure them with a protractor.

Turn and Talk Ask students to think about what happens to the corresponding angles if the parallel line moves closer to the longest side. Ask if there would be any change in their measure. Possible answer:

- The paths would remain perpendicular because if you are moving the parallel path closer to the longest side, you are not changing angle measures.
- Yes, as long as the segments marked congruent in the figure remain congruent, the actual lengths do not matter.

▶ BUILD SHARED UNDERSTANDING

Select groups of students who used various strategies and tools to share with the class how they solved the problem. As they present their solutions, have each group discuss why they chose a specific strategy and tool.

② Learn Together

Build Understanding

Task 1 **(MP) Use Tools** Students use definitions, properties, and theorems to develop a proof of the Perpendicular Bisector Theorem.

CONNECT TO VOCABULARY

Have students use the **Interactive Glossary** to record their understanding of the vocabulary in this task.

Sample Guided Discussion:

Q What is a reflection of a line segment? Possible answer: A transformation where every point on a line segment appears an equal distance on the other side of a given line of reflection.

Q How would the proof change if point C were reflected across $\overline{AB}$? Possible answer: There would be no changes.

Q Would the proof be valid if $\overline{CD}$ were changed to $\overleftrightarrow{CD}$ in the Given statement? Explain. yes; Possible answer: $\overline{CD}$ would be contained in $\overleftrightarrow{CD}$.

Turn and Talk Have students describe the triangle by the side lengths and angle measures to determine the type of triangle. $\triangle ACB$ is an isosceles triangle because $AC = BC$.

Build Understanding

Prove the Perpendicular Bisector Theorem

Perpendicular lines intersect at 90° angles. A **perpendicular bisector** of a segment is a line perpendicular to the segment at the midpoint of the segment.

Perpendicular Bisector Theorem

In a plane, if a point is on the perpendicular bisector of a segment, then it is equidistant from the endpoints of the segment.

In the diagram, $\overleftrightarrow{CD}$ is a perpendicular bisector of $\overline{AB}$. Each triangle is formed by the reflection of a segment that is part of $\overleftrightarrow{CD}$, so $AE = BE$, $AF = BF$, and $AG = BG$. Each point on $\overleftrightarrow{CD}$ is the same distance from the endpoints A and B.

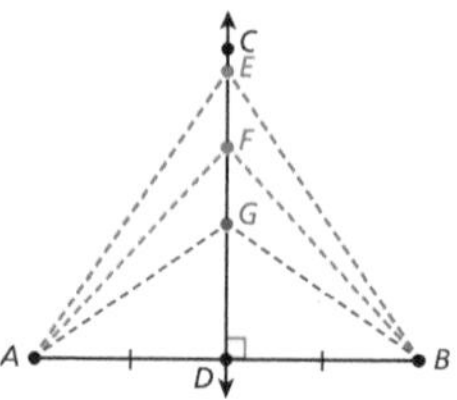

1 Prove the Perpendicular Bisector Theorem.

Given: $\overleftrightarrow{CD}$ is a perpendicular bisector of $\overline{AB}$.
Prove: $AC = BC$

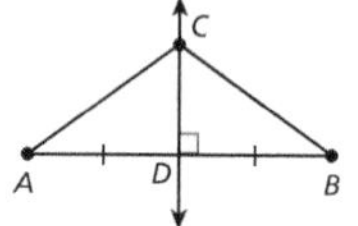

Statements	Reasons
1. $\overleftrightarrow{CD}$ is a perpendicular bisector of $\overline{AB}$.	1. Given
2. $AD = BD$	2. Definition of perpendicular bisector
3. $m\angle ADC = m\angle BDC = 90°$	3. Definition of perpendicular bisector
4. $CD = CD$	4. ? (A. What property justifies Step 4? Reflexive Property)
5. $\triangle ADC$ and $\triangle BDC$ are right triangles.	5. Definition of right triangle
6. $(AC)^2 = (AD)^2 + (CD)^2$ $(BC)^2 = (DB)^2 + (CD)^2$	6. ? (B. What theorem justifies Step 6? Pythagorean Theorem)
7. $(BC)^2 = (AD)^2 + (CD)^2$ (C. Which steps combined to create Statement 7? Steps 2 and 6)	7. Substitution Property
8. $(AC)^2 = (BC)^2$	8. ? (D. What property justifies Step 8? Transitive Property)
9. $AC = BC$	9. If $x^2 = y^2$ and x and y are both nonnegative, then $x = y$.

Turn and Talk What type of triangle is $\triangle ACB$ in the proof of the Perpendicular Bisector Theorem? Explain your reasoning. See margin.

94

LEVELED QUESTIONS

Depth of Knowledge (DOK)	Leveled Questions	What Does This Tell You?
Level 1 **Recall**	What property allows you to say that a number is equal to itself? Reflexive Property of Equality	Students' answers will indicate whether they understand the meaning of one of the properties of equality that can be used in the proof.
Level 2 **Basic Application of Skills & Concepts**	Is a triangle with side lengths 7, 25, and 24 a right triangle? Explain. yes; The triangle is a right triangle because $25^2 = 7^2 + 24^2$.	Students' answers will demonstrate if they know how to apply the Pythagorean Theorem.
Level 3 **Strategic Thinking & Complex Reasoning**	If the left side of the equation in Step 7 were changed to $(AC)^2$, what would the right side be and how would that affect Steps 8 and 9? $(DB)^2 + (CD)^2$; Possible answer: Steps 8 and 9 would remain unchanged.	Students' answers will reflect whether they understand proofs and can reason strategically about how to provide an alternate proof.

Prove the Converse of the Perpendicular Bisector Theorem

The Converse of the Perpendicular Bisector Theorem is true.

Converse of the Perpendicular Bisector Theorem

If a point is equidistant from the endpoints of a segment, then it lies on the perpendicular bisector of the segment.

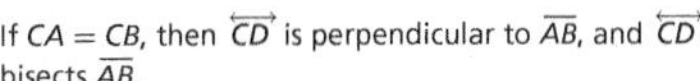

If $CA = CB$, then $\overleftrightarrow{CD}$ is perpendicular to $\overline{AB}$, and $\overleftrightarrow{CD}$ bisects $\overline{AB}$.

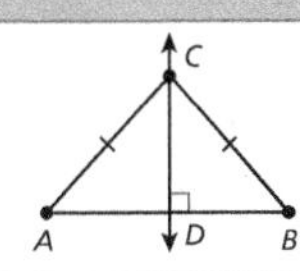

To prove this Converse, you may need to add a perpendicular line or segment into a given figure. You can do this because of the Perpendicular Postulate.

Perpendicular Postulate

If there is a line and a point that is not on the line, there is exactly one line through the point that is perpendicular to the given line.

2 Prove the Converse of the Perpendicular Bisector Theorem.

Given: $CA = CB$

Prove: C lies on the perpendicular bisector of $\overline{AB}$.

A figure for the given information would have $\overline{CA}$ and $\overline{CB}$ drawn and labeled congruent. Draw $\overline{AB}$ and draw a line through C that crosses $\overline{AB}$ at a point D such that $\overleftrightarrow{CD}$ is perpendicular to $\overline{AB}$. (See the Perpendicular Postulate above.)

A. What do you know about $\triangle ACD$ and $\triangle BCD$?
The triangles are right triangles.

B. How can you use the Pythagorean Theorem to describe the relationship of the side lengths of $\triangle ACD$ and $\triangle BCD$?
$(AD)^2 + (CD)^2 = (CA)^2$ and $(BD)^2 + (CD)^2 = (CB)^2$

C. What is true about $(CA)^2$ and $(CB)^2$? How do you know? $(CA)^2 = (CB)^2$; because the lengths CA and CB are given as equal, then the squares of those lengths will also be equal.

D. What equation can you write using the information that $(CA)^2 = (CB)^2$? How do you know this is true? $(AD)^2 + (CD)^2 = (BD)^2 + (CD)^2$; the Transitive Property

E. When you simplify the equation from Part D, you find that $AD = BD$. Explain how that relationship completes the proof. Because $AD = BD$, then $\overleftrightarrow{CD}$ bisects $\overline{AB}$. $\overleftrightarrow{CD}$ is also perpendicular to $\overline{AB}$, so C lies on the perpendicular bisector of $\overline{AB}$.

Turn and Talk Why can you draw the extra lines on the diagram as part of the proof? See margin.

Task 2 (MP) **Construct Arguments** Students use the Pythagorean Theorem to prove the Converse of the Perpendicular Bisector Theorem.

Sample Guided Discussion:

Q What two relationships do you have to know to show that a line is a perpendicular bisector? Which relationship is drawn to be true as a given in this proof and which relationship is then proven? A line that is a perpendicular bisector must be both perpendicular to the segment and also must divide it into two congruent segments. In the proof, $\overleftrightarrow{CD}$ is drawn to be perpendicular to $\overline{AB}$ as a given, which produces right triangles. Then the Pythagorean Theorem is used to prove that the segment has been bisected.

Q Why is the Pythagorean Theorem used in the proof? Possible answer: The Pythagorean Theorem was used to create equations with lengths of segments involved in the proof.

Turn and Talk Explain to students that in order to introduce extra lines in a proof, there must be reasons to justify their existence. Possible answer: You can draw the extra lines because between every two points there exists a line and because there is a line perpendicular to a line through any given point.

PROFICIENCY LEVEL

Beginning

Have students work in pairs to look up and write the definitions of perpendicular and bisector. Then have them explain how the two definitions combine to make the definition of a perpendicular bisector.

Intermediate

Have students work in groups. Each group receives a set of index cards showing diagrams for situations representing either the Perpendicular Bisector Theorem or its Converse.

Have students identify the information that indicates whether the diagram would be used in a proof using the Theorem or using the Converse.

Advanced

Students draw diagrams and write given and prove statements on index cards, indicating if their situation represents the Perpendicular Bisector Theorem or represents its Converse. Then, they have another student confirm if their decision (Theorem or Converse) was correct. These cards may also be used for the intermediate activity.

Step It Out

Task 3 (MP) **Use Tools** Students use a compass and straightedge to construct the perpendicular bisector of a segment.

Sample Guided Discussion:

Q **In the first step of the construction, is there a minimum distance for the compass setting? Explain.** yes; Possible answer: The compass setting must be greater than one half the length of the given segment. If the setting were less than one half the length of the given segment, the arcs centered at each end point would not intersect. If the setting were one half the length of the given segment, the arcs would intersect at the midpoint of the segment.

Q **Is there a maximum distance for the compass setting in the first step of the construction? Explain.** no; Possible answer: For any compass setting greater than one half the length of the given segment, the arcs would intersect in two points, one above the segment and one below.

Task 4 (MP) **Use Tools** Students use a compass and straightedge to construct a line perpendicular to a line through a point.

Sample Guided Discussion:

Q **Is there a minimum distance for the compass setting in the second step of the construction? Explain.** yes; Possible answer: The compass setting must be greater than one half the length of $\overline{AB}$ because if the setting were less than one half the length of $\overline{AB}$, the arcs centered at each end point would not intersect. If the setting were one half the length of $\overline{AB}$, the arcs would intersect at the midpoint of the segment.

Q **Is there a maximum distance for the compass setting in the second step of the construction? Explain.** no; Possible answer: For any compass setting greater than one half the length of $\overline{AB}$, the arcs would intersect in two points, one above the segment and one below.

Step It Out

Construct Perpendicular Bisectors and Lines

3 The steps below show the construction of the perpendicular bisector of a segment.

Use a compass to draw an arc centered at X such that the radius of the arc is greater than $\frac{1}{2}XY$.

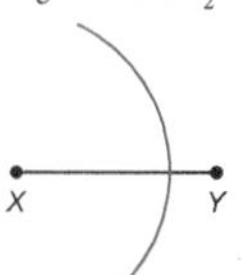

Without changing the compass setting, draw an arc centered at Y.

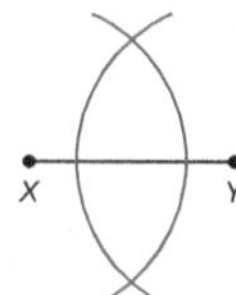

Use a straightedge to draw a line through the two points where the arcs intersect.

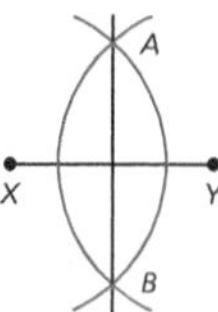

A. Justify the construction using the Perpendicular Bisector Theorem or its Converse. See Additional Answers.

B. These construction steps are also used to construct the midpoint of a segment. Explain why the midpoint is also on the perpendicular bisector of the segment. See Additional Answers.

4 Use a compass and a straightedge to construct a line perpendicular to a line through a point.

Place the compass on P. Draw an arc that intersects line m at two points. Label the intersections A and B.

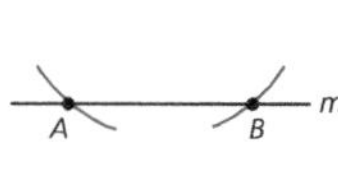

Draw an arc centered at A below line m. Use the same compass setting to draw an arc centered at B that intersects the arc centered at A.

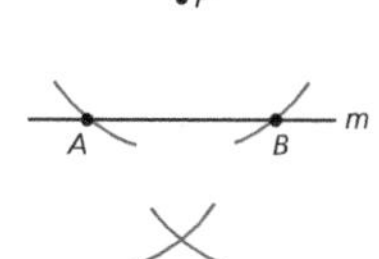

Use a straightedge to draw a line through point P and the point below m where the arcs intersect.

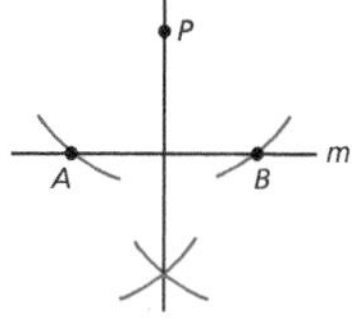

A. Are points A and B equidistant from point P? yes

B. Justify the construction using the Perpendicular Bisector Theorem or its Converse. See Additional Answers.

96

Use Theorems About Right Angles

Perpendicular Transversal Theorem

If a line is perpendicular to one of two parallel lines, then it is perpendicular to the other line as well.

If $m \parallel n$ and $t \perp m$, then $t \perp n$.

5 Find the length of the bridge supports indicated by $\overline{PQ}$ and $\overline{RQ}$. The measurements are in feet.

A. How do you know that line t is perpendicular to line n?

Since $m \parallel n$ and $m \perp t$, the Perpendicular Transversal Theorem tells you that $n \perp t$.

Given: $PS = RS$

Find the value of x.

$PQ = RQ$

B. Why can you write this equation?

$8x + 31 = 12x - 5$

$36 = 4x$

$9 = x$

Point Q is equidistant from P and R because Q is on line t, the perpendicular bisector of $\overline{PQ}$.

Determine PQ and RQ.

$PQ = 8x + 31$

$PQ = 8(9) + 31$

$PQ = 72 + 31 = 103$

C. Why do you only need to calculate PQ and not also RQ?

Since you know the lengths are equal, you only need to find the value once.

The bridge supports are each 103 feet long.

Turn and Talk How would your answer change if you calculated RQ instead of PQ? See margin.

Task 5 (MP) **Construct Arguments** Students construct a proof of the Perpendicular Transversal Theorem.

SUPPORT SENSE-MAKING Three Reads

Have students read the problem three times. Use the questions below for a different focus each read.

1. What is the situation about?
2. What are the quantities in the situation?
3. What are the possible mathematical questions that you could ask for the situation?

Sample Guided Discussion:

Q **What is the Given for the Perpendicular Transversal Theorem?** If $m \parallel n$ and $t \perp m$, then $t \perp n$.

Q **If $m \parallel n$, how are the corresponding angles formed by transversal t related?** The corresponding angles are congruent.

Q **If $t \perp m$, what kind of angle is $\angle QSR$?** It is a right angle.

Turn and Talk Before they calculate RQ, ask students what they expect the length to be. The result would not change, but the expression used would. $RQ = 12x - 5 = 12(9) - 5 = 108 - 5 = 103$; $RQ = PQ = 103$

ONLINE

Assign the Digital On Your Own for
- built-in student supports
- Actionable Item Reports
- Standards Analysis Reports

On Your Own

Assignment Guide

The chart below indicates which problems in the On Your Own are associated with each task in the Learn Together. Assign daily homework for tasks completed.

Learn Together Tasks	On Your Own Problems
Task 1, p. 94	Problems 5, 6, 9, and 16
Task 2, p. 95	Problems 5, 6, 8, 10, 11, 13, 14, and 25
Task 3, p. 96	Problems 7 and 18
Task 4, p. 96	Problem 13
Task 5, p. 97	Problems 12, 15, 17, 19–24, and 26–28

Check Understanding

1. How can you construct a perpendicular bisector of a segment RS by folding a piece of paper? See Additional Answers.
2. In the figure, is $\overline{BD}$ perpendicular to $\overline{AC}$? Explain why or why not. See Additional Answers.

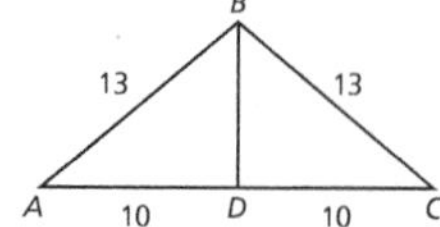

Find the value of the variable. Explain your reasoning.

3. $b = 6$ by the Perpendicular Bisector Theorem.

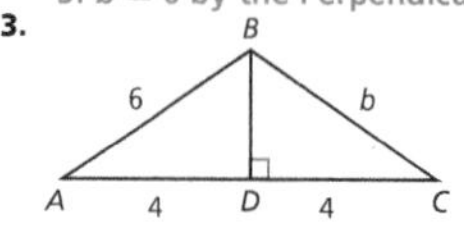

4.

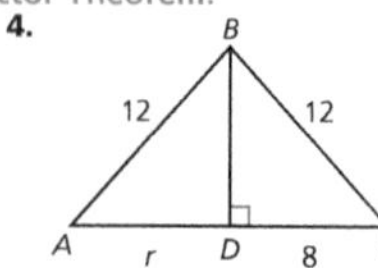

4. $r = 8$ by the Converse of the Perpendicular Bisector Theorem.

On Your Own

Decide if there is enough information to conclude that P lies on the perpendicular bisector of $\overline{QR}$. 5, 6. See Additional Answers.

5.

6.

7. MP Use Tools Draw a segment. Construct the perpendicular bisector. See Additional Answers.

Find the value of the variable.

8. $x = 24$

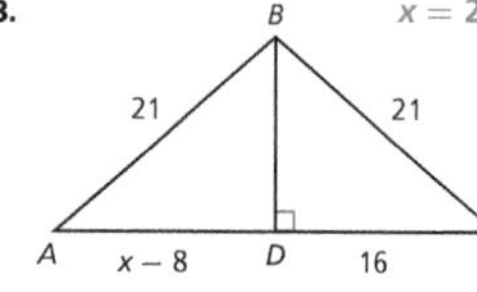

9. $a = 4$

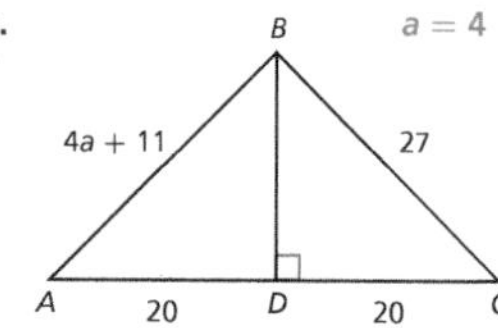

10. $s = 3$

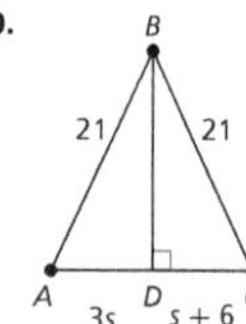

11. $d = 7$

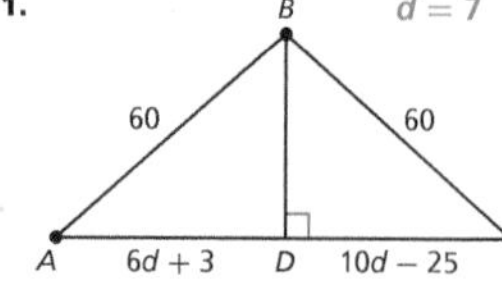

98

(l) ©Positiffy/Shutterstock; (r) ©Datsenko Maryna/Shutterstock

data checkpoint

3 Check Understanding

Formative Assessment

Use formative assessment to determine if your students are successful with this lesson's learning objective.

Students who successfully complete the Check Understanding can continue to the On Your Own practice.

For students who miss 1 problem or more, work in a pulled small group using the Almost There small-group activity on page 93C.

ONLINE

Assign the Digital Check Understanding to determine
- success with the learning objective
- items to review
- grouping and differentiation resources

4 Differentiation Options

Differentiate instruction for all students using small-group activities and math center activities on page 93C.

Reteach

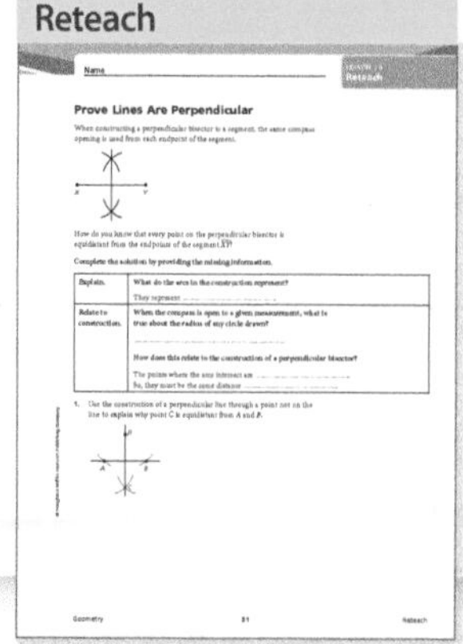

Challenge

Use Parallel and Perpendicular Lines to Classify Quadrilaterals

12. Is line t perpendicular to line j? Explain how you know.
yes; You know that $j \parallel k$ and $t \perp k$. By the Perpendicular Transversal Theorem, you know that $t \perp j$

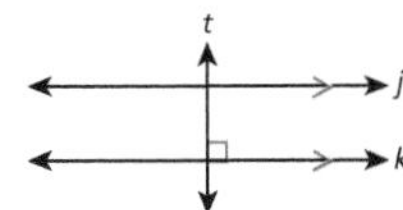

13. (MP) **Construct Arguments** If $AC = BC$ and $AD = BD$, prove $\overline{CD} \perp \overline{AB}$.

Given: $AC = BC$ and $AD = BD$

Prove: $\overline{CD} \perp \overline{AB}$

See Additional Answers.

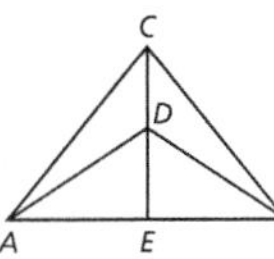

Use the diagram to find the lengths. $\overline{BF}$ is the perpendicular bisector of $\overline{AC}$. $\overline{EC}$ is the perpendicular bisector of $\overline{BD}$.

14. Suppose $AC = 24$. What is the length of $\overline{FC}$? 12

15. Suppose $BD = 30$. What is the length of $\overline{AB}$? 15

16. Suppose $ED = 11$. What is the length of $\overline{BE}$? 11

17. Suppose $AB = 14$. What is the length of $\overline{BD}$? 28

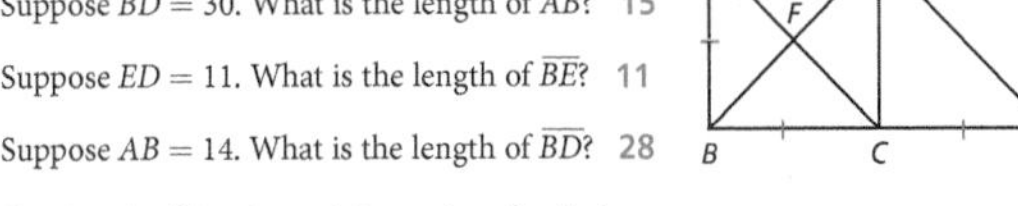

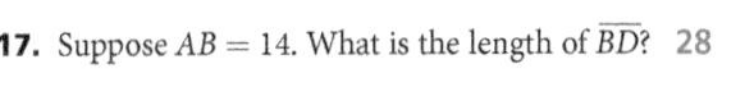

18. Construct a diameter and the center of a circle.
A–D. See Additional Answers.

A. Draw a circle. Use a straightedge to draw a segment that intersects the circle in two points. Label the intersections with the circle A and B.

B. Use a compass to draw a perpendicular bisector of $\overline{AB}$. How do you know that the perpendicular bisector of $\overline{AB}$ is a diameter of the circle?

C. Use a compass to draw a perpendicular bisector of the diameter of the circle. How do you know that the point of intersection of this perpendicular bisector and the diameter of the circle is the center of the circle?

D. Is it possible for a point to be the center of a circle, but not be on the perpendicular bisector of a chord? Explain your reasoning.

19. Prove the Perpendicular Transversal Theorem.

Given: $a \perp t$ and $a \parallel b$

Prove: $b \perp t$ See Additional Answers.

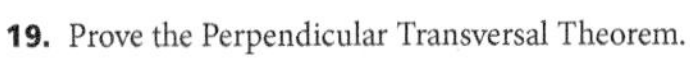

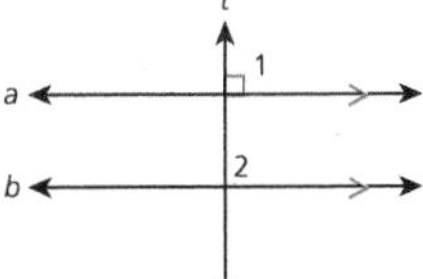

In 20–24, consider the possible design variations in the pattern shown. In the figure, $XY = ZY$ and $XW = ZW$.

20. Suppose $m\angle 3 = 48°$. What is $m\angle 1$? 42°

21. Suppose $WZ = 5$ and $XZ = 8$. What is WY? 3

22. Find $m\angle 1 + m\angle 2$. 90°

23. Suppose $WY = 15$ and $XZ = 16$. What is WZ? 17

24. Suppose $m\angle 4 = 51°$. What is $m\angle 1$? 51°

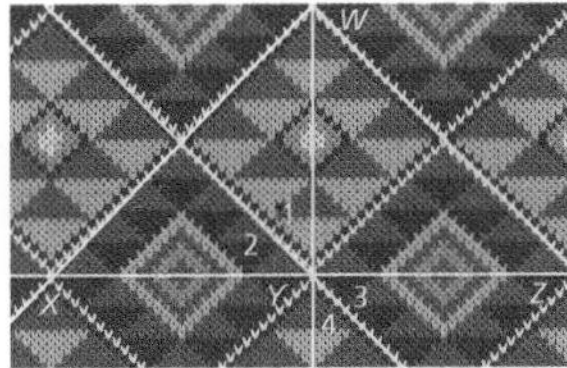

Watch for Common Errors

Problem 18A If the segment a student draws is a diameter, the student may become confused. The perpendicular bisector drawn in Part B will also be a diameter. Then in Part C, the perpendicular bisector drawn to the diameter from Part B will be the diameter from Part A. To avoid confusion, you may want to ask students to draw a segment (in Part A) that is obviously not a diameter.

Questioning Strategies

Problem 18C The perpendicular bisector of the diameter of the circle and the diameter intersect the circle in 4 points. If you connect these 4 points in clockwise order, what type of quadrilateral is formed? Explain. Possible answer: The perpendicular bisector of the diameter of the circle is also a diameter. Four identical isosceles right triangles are formed with legs that are radii of the circle. The 4 hypotenuses are congruent and meet at right angles (formed by the two 45° angles of two isosceles right triangles). So, the quadrilateral is a square.

5 Wrap-Up

Summarize learning with your class. Consider using the Exit Ticket, Put It in Writing, or I Can scale.

Exit Ticket 

The construction of a perpendicular bisector is shown.

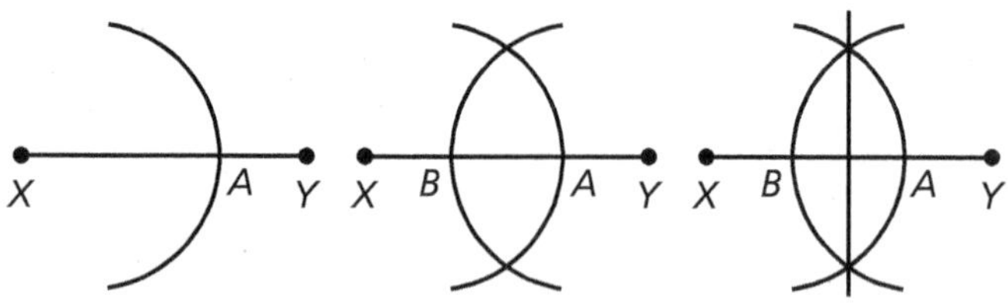

Explain why the line through the two intersecting arcs is the perpendicular bisector of $\overline{XY}$? Possible answer: The arc lengths are the same, so the distances from X to each intersection and from Y to each intersection must be equal. So, the intersections lie on the perpendicular bisector of $\overline{XY}$ by the Converse of the Perpendicular Bisector Theorem.

Put It in Writing 

Describe the perpendicular bisector of a segment and the measurements and angles associated with a bisector.

I Can 

The scale below can help you and your students understand their progress on a learning goal.

4	I can ensure that a line is a perpendicular bisector of a segment by construction, and I can explain my steps to others.
3	I can ensure that a line is a perpendicular bisector of a segment by construction.
2	I can identify and construct a perpendicular bisector.
1	I can identify a perpendicular bisector.

Spiral Review • Assessment Readiness

These questions will help determine if students have retained information taught in the past and can also prepare them for high-stakes assessments. Here, students must find the slope of a line **(Gr8, 1.1)**, determine congruent angles of parallel lines cut by a transversal **(3.1)**, define angles formed by two lines cut by a transversal **(3.2)**, and solve a supplementary angles problem **(2.4)**.

25. STEM A water molecule has two hydrogen atoms and one oxygen atom. The distance between each hydrogen atom and the oxygen atom is the same. Do you have enough information to determine if the oxygen atom lies on the perpendicular bisector of a segment between the two hydrogen atoms? Explain your reasoning.

25–28. See Additional Answers.

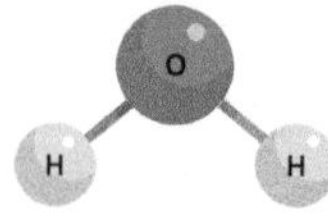

26. If $m \perp p$ and $n \perp p$, show that $m \parallel n$.

Given: $m \perp p$ and $n \perp p$

Prove: $m \parallel n$

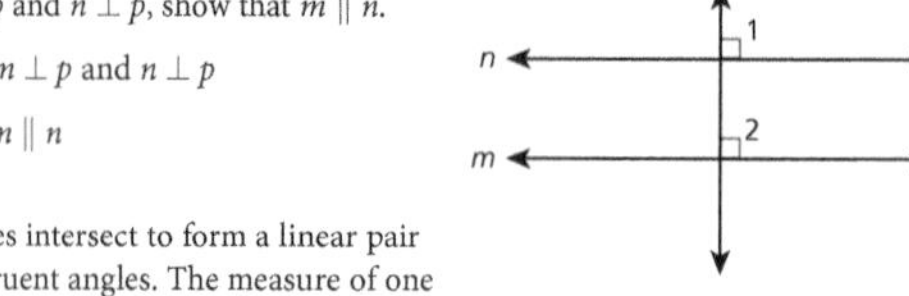

27. Two lines intersect to form a linear pair of congruent angles. The measure of one angle is $(15w + 15)°$. The measure of the other angle is $\left(\frac{25v}{2}\right)°$. Find the values of w and v. Explain your reasoning.

28. Music The valve pistons on a trumpet are all perpendicular to the lead pipe. Explain why the valve pistons must be parallel to each other.

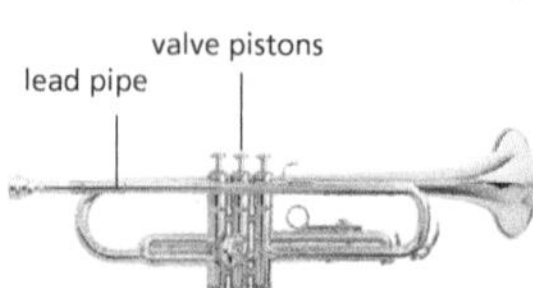

Spiral Review • Assessment Readiness

29. What is the slope of a line that passes through $(-4, 7)$ and $(2, -5)$?

Ⓐ 2
Ⓑ $\frac{1}{2}$
Ⓒ $-\frac{1}{2}$
Ⓓ -2

30. Which angle(s) are congruent to $\angle 3$? Select all that apply.

Ⓐ $\angle 1$
Ⓑ $\angle 2$
Ⓒ $\angle 4$
Ⓓ $\angle 5$
Ⓔ $\angle 6$
Ⓕ $\angle 7$

31. ___?___ angles are nonadjacent angles that lie on opposite sides of the transversal between the intersected lines.

Ⓐ Alternate exterior
Ⓑ Alternate interior
Ⓒ Consecutive interior
Ⓓ Consecutive exterior

32. $\angle 1$ and $\angle 2$ are supplementary angles. $m\angle 1 = (6y + 7)°$ and $m\angle 2 = (9y - 7)°$. What is y?

Ⓐ 12
Ⓑ 16
Ⓒ 79
Ⓓ 101

 **I'm in a Learning Mindset!**

What can I contribute to my learning community to prove that intersecting lines are perpendicular?

Keep Going Journal and Practice Workbook

Learning Mindset

Strategic Help-Seeking Identifies Sources of Help

Point out that for intersecting lines to be perpendicular, they must meet to form right angles. Remind students that if two lines form four congruent angles, the line are perpendicular. Point out that congruent angles that form a linear pair are right angles. Let students know that the perpendicular bisector of a segment is perpendicular to the line containing the segment. *Where can I find what it takes to prove two intersecting lines are perpendicular? What resources are available to help me prove two lines are perpendicular? Who could I ask to help me prove that two intersecting lines are perpendicular?*

Module

3 Review

Angle Pairs Formed by Parallel Lines and a Transversal

Two paths at a park are parallel. These paths are crossed by another path as shown.

Using the Corresponding Angles Postulate, Alternate Interior Angles Theorem, Alternate Exterior Angles Theorem, Consecutive Interior Angles Theorem, and Consecutive Exterior Angles Theorem, we know the following relationships:

$\angle 1$ is congruent to $\angle 4$, $\angle 5$, and $\angle 8$.

$\angle 1$ is supplementary to $\angle 2$, $\angle 3$, $\angle 6$, and $\angle 7$.

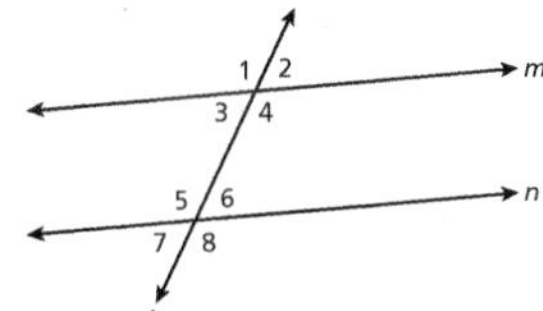

Parallel Lines

Andrea studied another section of the park. She wanted to know if the North Trail and South Trail are parallel. She measured the angles formed by a trail that crosses them.

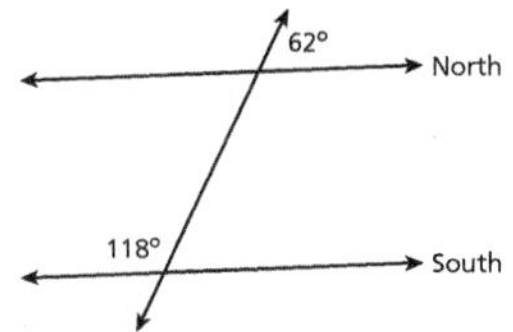

She uses the fact that the vertical angle of the 62° angle also measures 62°.

Because $62° + 118° = 180°$, she can use the Converse of the Consecutive Interior Angles Theorem to show that the North Trail and South Trail are parallel.

Perpendicular Lines

The park plans to build another trail to connect to the East Trail. The trail should be a perpendicular bisector of the East Trail.

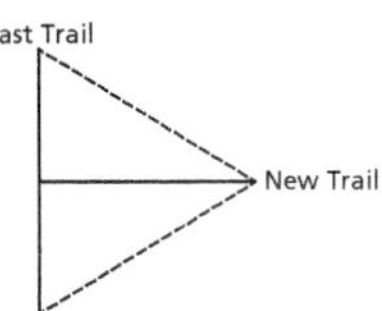

The park measures the distances from the end of New Trail to the ends of East Trail and find they are equal. New Trail also bisects East Trail.

The park knows New Trail is a perpendicular bisector by Converse of the Perpendicular Bisector Theorem.

ONLINE

Assign the Digital Module Review for

- built-in student supports
- Actionable Item Reports
- Standards Analysis Reports

Module Review

Use the first page of the Module Review to summarize and connect the main ideas of the module. Use the second page to assess students' understanding of the vocabulary, concepts, and skills presented in the module.

Sample Guided Discussion:

Q Angle Pairs Formed by Parallel Lines and a Transversal What is the relationship between the congruent angles $\angle 1$ and $\angle 5$ and why? They are corresponding angles because they lie on the same side of the transversal and above each of the parallel lines.

Q Parallel Lines What is another theorem you can use to determine that the two trails are parallel? The measure of the angle adjacent to the 62° angle at the North Trail is supplementary to the 62° angle and its measure is 118°. The measure of the interior angles on either side of the transversal are both equal, so by the Converse of the Alternate Interior Angles Theorem, the trails are parallel.

Q Perpendicular Lines Can you draw another line from a point not on the New Trail that is also the perpendicular bisector of the East Trail? No. As given in Lesson 3, the Perpendicular Postulate states that from a point not on a line, there is one and only one line that is perpendicular to the line. The New Trail contains all the points that lie on the perpendicular bisector of the East Trail.

Module Review continued

Possible Scoring Guide

Items	Points	Description
1–3	2 each	identifies the correct term
4–7	2 each	identifies pairs of angles formed by a transversal and explains if they are congruent or supplementary
8	2	finds the measure of two consecutive interior angles whose measures are expressed algebraically
9	2	given three lines and a transversal, uses the measures of three angles to determine which sets of lines are parallel
10	2	finds the value of *z* that determines the lengths of the segments of the base
11	2	finds the value of *x* that will make alternate interior angles congruent
12	2	finds the length from a point on the perpendicular bisector of a segment to one of its endpoints
13	2	finds the value of a variable of a point on the perpendicular bisector of a line segment
Total points possible = 26 points		

The Unit 2 Test in the Assessment Guide assesses content from Modules 3 and 4.

Vocabulary

Choose the correct term from the box to complete each sentence.

Vocabulary
parallel
perpendicular
transversal

1. A(n) __?__ is a line that intersects two coplanar lines at two distinct points. transversal
2. __?__ lines are lines in the same plane that do not intersect. parallel
3. Two lines are __?__ if they intersect to form 90° angles. perpendicular

Concepts and Skills

Given that $m \parallel n$, identify an example of each angle pair and explain if they are supplementary or congruent.

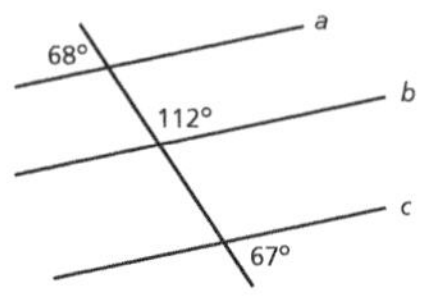

4–7. See Additional Answers.

4. Corresponding angles
5. Consecutive interior angles
6. Alternate interior angles
7. Alternate exterior angles
8. Two parallel lines are intersected by a transversal. Two angles, $\angle 1$ and $\angle 2$, are consecutive interior angles. The measure of $\angle 1$ is $(7t - 4)°$. The measure of $\angle 2$ is $(t + 24)°$. What is the measure of each angle? $m\angle 1 = 136°$, $m\angle 2 = 44°$
9. Determine if each set of lines is parallel or not. Justify your reasoning. See Additional Answers.

Find the value of the variable.

10. $z = 12$

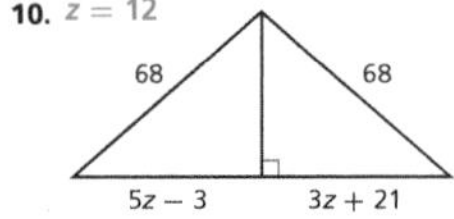

11. $x = 18.5$

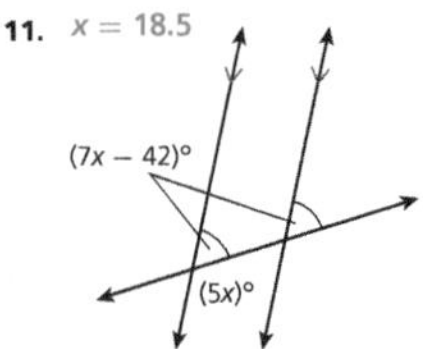

12. $\overline{LM}$ is a perpendicular bisector of $\overline{NP}$. The length of $\overline{LN}$ is $12w + 7$, and the length of $\overline{LP}$ is $15w - 5$. What is the length of $\overline{LN}$? 55
13. MP **Use Tools** $\overline{AB}$ is a perpendicular bisector of $\overline{CD}$. The length of $\overline{AD}$ is $7h + 11$, and the length of $\overline{AC}$ is $15h + 3$. What is the value of h? State what strategy and tool you will use to answer the question, explain your choice, and then find the answer. 1

102

DATA-DRIVEN INSTRUCTION

Before moving on to the Module Test, use the Module Review results to intervene based on the table below.

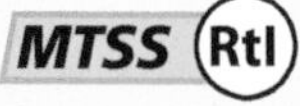

Items	Lesson	DOK	Content Focus	Intervention
4–7	3.1	2	Identify pairs of angles when a transversal intersects two parallel lines.	Reteach 3.1
8	3.1	3	Find the measures of two consecutive interior angles formed by a transversal.	Reteach 3.1
9	3.2	3	Determine which sets of three given lines are parallel.	Reteach 3.2
10, 12, 13	3.3	3	Use the Perpendicular Bisector Theorem to find the lengths of segments.	Reteach 3.3
11	3.1	3	Find the value of *x* that will make two alternate interior angles equal.	Reteach 3.1

Module Test

The Module Test is available in alternative versions in your Assessment Guide. All items are presented in standardized test formats.

ONLINE

Assign the Digital Module Test to power actionable reports including

- proficiency by standards
- item analysis

Form A

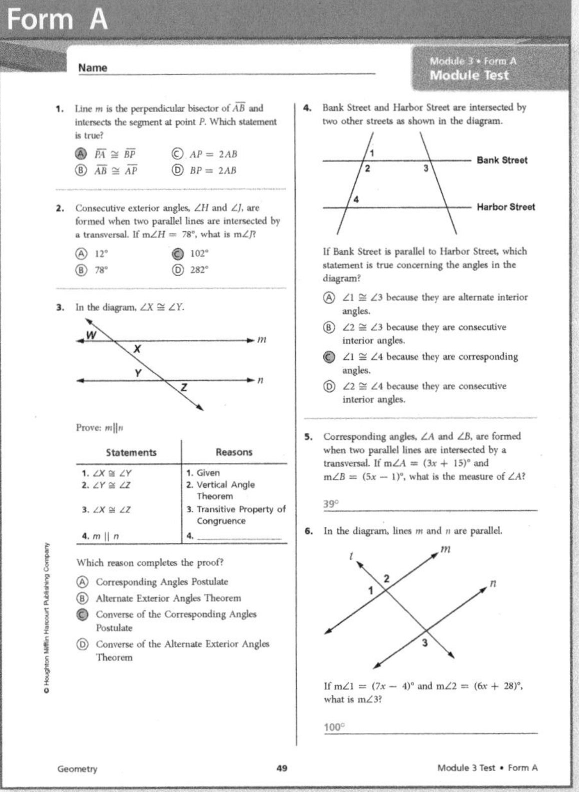

Name ____________________ Module 3 • Form A **Module Test**

1. Line m is the perpendicular bisector of $\overline{AB}$ and intersects the segment at point P. Which statement is true?

Ⓐ $\overline{PA} \cong \overline{BP}$ Ⓒ $AP = 2AB$
Ⓑ $\overline{AB} \cong \overline{AP}$ Ⓓ $BP = 2AB$

2. Consecutive exterior angles, $\angle H$ and $\angle J$, are formed when two parallel lines are intersected by a transversal. If $m\angle H = 78°$, what is $m\angle J$?

Ⓐ 12° Ⓒ 102°
Ⓑ 78° Ⓓ 282°

3. In the diagram, $\angle X \cong \angle Y$.

Prove: $m \parallel n$

Statements	Reasons
1. $\angle X \cong \angle Y$	1. Given
2. $\angle Y \cong \angle Z$	2. Vertical Angle Theorem
3. $\angle X \cong \angle Z$	3. Transitive Property of Congruence
4. $m \parallel n$	4. ________

Which reason completes the proof?

Ⓐ Corresponding Angles Postulate
Ⓑ Alternate Exterior Angles Theorem
Ⓒ Converse of the Corresponding Angles Postulate
Ⓓ Converse of the Alternate Exterior Angles Theorem

4. Bank Street and Harbor Street are intersected by two other streets as shown in the diagram.

Bank Street
Harbor Street

If Bank Street is parallel to Harbor Street, which statement is true concerning the angles in the diagram?

Ⓐ $\angle 1 \cong \angle 3$ because they are alternate interior angles.
Ⓑ $\angle 2 \cong \angle 3$ because they are consecutive interior angles.
Ⓒ $\angle 1 \cong \angle 4$ because they are corresponding angles.
Ⓓ $\angle 2 \cong \angle 4$ because they are consecutive interior angles.

5. Corresponding angles, $\angle A$ and $\angle B$, are formed when two parallel lines are intersected by a transversal. If $m\angle A = (3x + 15)°$ and $m\angle B = (5x - 1)°$, what is the measure of $\angle A$?

39°

6. In the diagram, lines m and n are parallel.

If $m\angle 1 = (7x - 4)°$ and $m\angle 2 = (6x + 28)°$, what is $m\angle 3$?

100°

© Houghton Mifflin Harcourt Publishing Company

Geometry 49 Module 3 Test • Form A

Form A

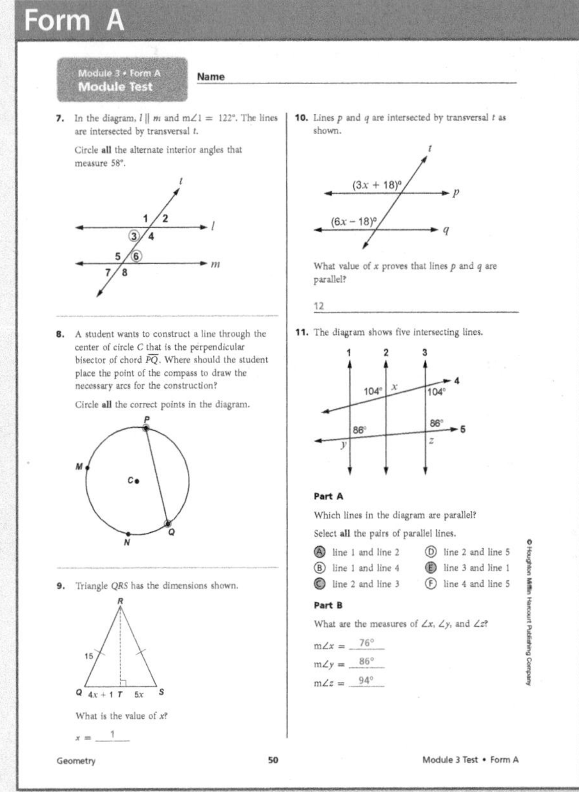

Module 3 • Form A **Module Test** Name ____________________

7. In the diagram, $l \parallel m$ and $m\angle 1 = 122°$. The lines are intersected by transversal t.

Circle **all** the alternate interior angles that measure 58°.

8. A student wants to construct a line through the center of circle C that is the perpendicular bisector of chord $\overline{PQ}$. Where should the student place the point of the compass to draw the necessary arcs for the construction?

Circle **all** the correct points in the diagram.

9. Triangle QRS has the dimensions shown.

What is the value of x?

$x =$ 1

10. Lines p and q are intersected by transversal t as shown.

What value of x proves that lines p and q are parallel?

12

11. The diagram shows five intersecting lines.

Part A

Which lines in the diagram are parallel?

Select **all** the pairs of parallel lines.

Ⓐ line 1 and line 2 Ⓓ line 2 and line 5
Ⓑ line 1 and line 4 Ⓔ line 3 and line 1
Ⓒ line 2 and line 3 Ⓕ line 4 and line 5

Part B

What are the measures of $\angle x$, $\angle y$, and $\angle z$?

$m\angle x =$ 76°
$m\angle y =$ 86°
$m\angle z =$ 94°

© Houghton Mifflin Harcourt Publishing Company

Geometry 50 Module 3 Test • Form A

Form B

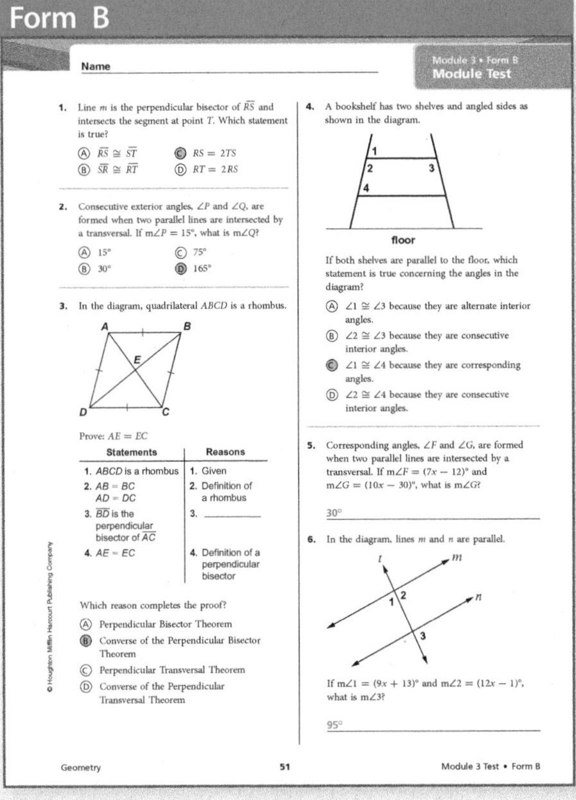

Name ____________________ Module 3 • Form B **Module Test**

1. Line m is the perpendicular bisector of $\overline{RS}$ and intersects the segment at point T. Which statement is true?

Ⓐ $\overline{RS} \cong \overline{ST}$ Ⓒ $RS = 2TS$
Ⓑ $\overline{SR} \cong \overline{RT}$ Ⓓ $RT = 2RS$

2. Consecutive exterior angles, $\angle P$ and $\angle Q$, are formed when two parallel lines are intersected by a transversal. If $m\angle P = 15°$, what is $m\angle Q$?

Ⓐ 15° Ⓒ 75°
Ⓑ 30° Ⓓ 165°

3. In the diagram, quadrilateral $ABCD$ is a rhombus.

Prove: $AE = EC$

Statements	Reasons
1. $ABCD$ is a rhombus	1. Given
2. $AB = BC$, $AD = DC$	2. Definition of a rhombus
3. $\overline{BD}$ is the perpendicular bisector of $\overline{AC}$	3. ________
4. $AE = EC$	4. Definition of a perpendicular bisector

Which reason completes the proof?

Ⓐ Perpendicular Bisector Theorem
Ⓑ Converse of the Perpendicular Bisector Theorem
Ⓒ Perpendicular Transversal Theorem
Ⓓ Converse of the Perpendicular Transversal Theorem

4. A bookshelf has two shelves and angled sides as shown in the diagram.

floor

If both shelves are parallel to the floor, which statement is true concerning the angles in the diagram?

Ⓐ $\angle 1 \cong \angle 3$ because they are alternate interior angles.
Ⓑ $\angle 2 \cong \angle 3$ because they are consecutive interior angles.
Ⓒ $\angle 1 \cong \angle 4$ because they are corresponding angles.
Ⓓ $\angle 2 \cong \angle 4$ because they are consecutive interior angles.

5. Corresponding angles, $\angle F$ and $\angle G$, are formed when two parallel lines are intersected by a transversal. If $m\angle F = (7x - 12)°$ and $m\angle G = (10x - 30)°$, what is $m\angle G$?

30°

6. In the diagram, lines m and n are parallel.

If $m\angle 1 = (9x + 13)°$ and $m\angle 2 = (12x - 1)°$, what is $m\angle 3$?

95°

© Houghton Mifflin Harcourt Publishing Company

Geometry 51 Module 3 Test • Form B

Form B

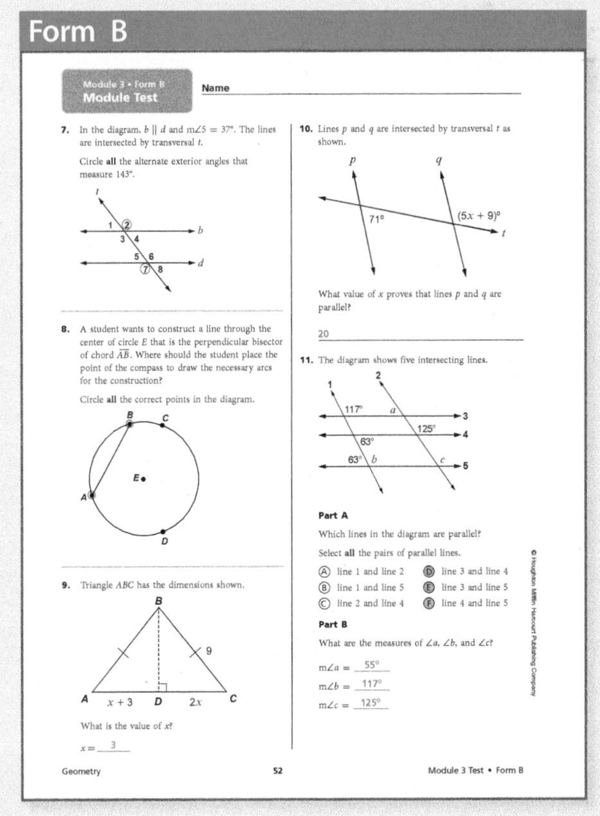

Module 3 • Form B **Module Test** Name ____________________

7. In the diagram, $b \parallel d$ and $m\angle 5 = 37°$. The lines are intersected by transversal t.

Circle **all** the alternate exterior angles that measure 143°.

8. A student wants to construct a line through the center of circle E that is the perpendicular bisector of chord $\overline{AB}$. Where should the student place the point of the compass to draw the necessary arcs for the construction?

Circle **all** the correct points in the diagram.

9. Triangle ABC has the dimensions shown.

What is the value of x?

$x =$ 3

10. Lines p and q are intersected by transversal t as shown.

What value of x proves that lines p and q are parallel?

20

11. The diagram shows five intersecting lines.

Part A

Which lines in the diagram are parallel?

Select **all** the pairs of parallel lines.

Ⓐ line 1 and line 2 Ⓓ line 3 and line 4
Ⓑ line 1 and line 5 Ⓔ line 3 and line 5
Ⓒ line 2 and line 4 Ⓕ line 4 and line 5

Part B

What are the measures of $\angle a$, $\angle b$, and $\angle c$?

$m\angle a =$ 55°
$m\angle b =$ 117°
$m\angle c =$ 125°

© Houghton Mifflin Harcourt Publishing Company

Geometry 52 Module 3 Test • Form B

LINES ON THE COORDINATE PLANE

Introduce and Check for Readiness

- Module Performance Task • Are You Ready?

Lesson 4.1 – 2 Days

Slope and Equations of Parallel Lines

Learning Objective: Use slope to identify, write, and use equations of parallel lines.
Review Vocabulary: coordinate proof, slope

Lesson 4.2 – 2 Days

Slope and Equations of Perpendicular Lines

Learning Objective: Write the equation of a line that is perpendicular to a given line.
Review Vocabulary: perpendicular lines

Lesson 4.3 – 2 Days

Write a Coordinate Proof

Learning Objective: Use coordinates to prove simple geometric theorems algebraically.

Assessment

- Module 4 Test (Forms A and B)
- Unit 2 Test (Forms A and B)

LEARNING ARC FOCUS

 Build Conceptual Understanding  Connect Concepts and Skills  Apply and Practice

TEACHING FOR SUCCESS

TEACHING FOR DEPTH: Lines on the Coordinate Plane

Meaning of Slope. The informal meaning of the word *slope* suggests steepness, or the rise or fall of a surface. In coordinate geometry, *slope* represented by the letter m is a number that represents the direction and steepness of a nonvertical line in the plane. The direction of the line is given by the sign of m and the steepness by the absolute value of m.

The slope of a nonverical line is defined to be the ratio between the *rise* and the *run* between two points in the coordinate plane. Point out to students that the rise is the difference between the y-coordinates of the two points, taken in either order, and the run is the corresponding difference between the x-coordinates.

Finally, the slope of a line describes its orientation in the plane. Remind students of the following facts:

- If $m = 0$, the line is horizontal.
- If $m < 0$, the line falls from left to right.
- If $m > 0$, the line rises from left to right.
- If $|m| < 1$, the line rises or falls less steeply than the line $y = x$.
- If $|m| > 1$, the line rises or falls more steeply than the line $y = x$.

Mathematical Progressions

Prior Learning	Current Development	Future Connections
Students: • explained why the slope is the same between any two distinct points on a nonvertical line. • understood that lines with the same slope are parallel. • wrote equations of lines in slope-intercept form. • proved the Pythagorean Theorem. • wrote an equation of a line given two points.	**Students:** • use slope to identify, write, and use equations of parallel and perpendicular lines. • prove the Distance Formula and use it to prove congruence of segments on the coordinate plane. • determine the coordinates of the midpoint of a segment in the coordinate plane. • apply the Distance Formula to find the length of a segment.	**Students:** • will investigate properties of circles graphed in the coordinate plane. • will investigate transformations in the coordinate plane. • will write coordinate proofs about triangle relationships. • will write coordinate proofs about parallelograms.

TEACHER TO TEACHER

From the Classroom

Establish mathematics goals to focus learning. In this module, students need to be able to choose the correct formulas to calculate the slope of a line or segment, the distance between points or the length of a segment between two points, and the coordinates of the midpoint of a segment. Each formula involves knowing when to find a difference, a sum, and/or the square or square root of a number.

To help students distinguish among these three goals, choose and label the coordinates of two points. (Be sure that the points do not determine either a vertical or horizontal line.) For example, choose coordinates such as $A(-3, 3)$ and $B(2, 5)$. Have students work in pairs and plot the points in the coordinate plane and draw the line through the points. Then have them calculate the following three values for the given coordinates. Be sure to emphasize the correct symbolism in each one.

- $m_{AB} = \dfrac{5-3}{2-(-3)} = \dfrac{2}{5}$
- $d_{\overline{AB}} = \sqrt{(2-(-3))^2 + (5-3)^2} = \sqrt{29} \approx 5.4$
- $\text{midpoint}_{\overline{AB}} = \left(\dfrac{-3+2}{2}, \dfrac{3+5}{2}\right) = \left(-\dfrac{1}{2}, 4\right)$

When everyone is done and agrees with their answers, have students explain the meaning of each value with respect to the line and segment.

LANGUAGE DEVELOPMENT • Planning for Instruction

By giving all students regular exposure to language routines in context, you will provide opportunities for students to **listen, speak, read,** and **write** about mathematical situations and develop both mathematical language and conceptual understanding at the same time.

Using Language Routines to Develop Understanding

Use the **Professional Cards** for the following routines to plan for effective instruction.

Co-Craft Questions and Problems Lesson 4.3

Students think of natural questions to ask about a given situation or problems similar to a given task and answer the questions they have developed or problems they have created.

Three Reads Lesson 4.2

Students read a problem three times with a specific focus each time.

1st Read What is the situation about?
2nd Read What are the quantities in the situation?
3rd Read What are the possible mathematical questions that we could ask for the situation?

Information Gap Lesson 4.1

Students recognize when information given in a problem situation is incomplete, and they pose questions and share knowledge with others to discover any missing facts or relationships and work together to solve the problem.

Critique, Correct, and Clarify Lesson 4.3

Students correct the work in a flawed explanation, argument, or solution method and share with a partner and refine the sample work.

Connecting Language to Lines on the Coordinate Plane

Watch for students' use of the review terms listed below as they explain their reasoning and make connections with new concepts.

Key Academic Vocabulary

Prior Learning • Review Vocabulary

coordinate proof a style of proof where generalized coordinates are used to prove geometric theorems

perpendicular lines two lines in the same plane that intersect to form 90° angles

slope of a nonvertical line the ratio m of the vertical change (the *rise*) to the horizontal change (the *run*) between any two points on a line

Linguistic Note

Listen for how students use the specialized terms in this lesson. The word *slope*, for example, has a very common meaning, but in mathematics it is a number. *Vertical* and *nonvertical lines* have special meanings in a coordinate grid. A vertical line stands straight up and down, but doesn't have a slope. A nonvertical line always has a slope in the conventional sense, although here, it is also a number. A horizontal line can be graphed on a coordinate grid, but has a slope of 0. This is like a flat stretch of road that has no steepness.

Finally, it will take several examples to define *opposite reciprocals*. The word *opposite* may be clear—as in positive and negative—but the word *reciprocal* is far less familiar. One meaning is that it describes when two people agree to help one another in a similar way. However, in this context, the word implies a product equal is 1.

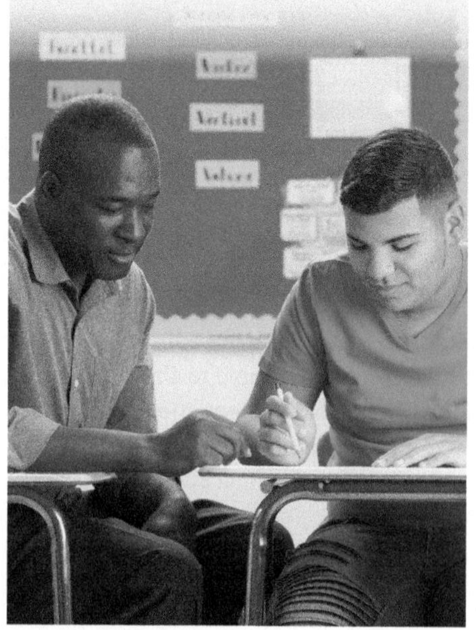

Module 4

Lines on the Coordinate Plane

Module Performance Task: Focus on STEM

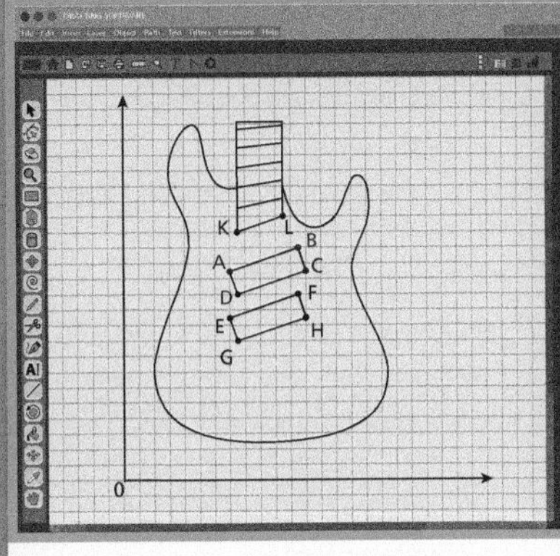

Pickup Placement in Acoustic Design

Electric guitars contain pickups, which create a magnetic field. The movement of the strings create deviations in the magnetic field, and these deviations are converted to reproduce the sound of the strings.

The placement of the pickups affects the tone of the guitar, so the holes for the pickups in the body must be planned into the design of the guitar. The designer wants the pickups to be parallel to the last fret. The body is designed using Computer Aided Drafting software and designates the coordinates of the cutout $ABCD$ at $A(14, 27)$, $B(23, 30)$, $C(24, 27)$, and $D(15, 24)$. The end points of the last fret are $K(15, 32)$ and $L(21, 34)$. A–D. See margin.

A. Prove that the last fret of the guitar is parallel to the cutout of the pickup $ABCD$.

B. What relationships do you notice between the other segments of the cutout and the last fret of the guitar? Justify your statement.

C. The cutout $EFGH$ needs to be in the same orientation as $ABCD$. Point E must be located 6 cm directly below A. What are the coordinates of cutout $EFGH$, given that 1 unit is equivalent to 1 cm?

D. Another fret has end points at $M(15, 40)$ and $N(21, 41)$. Is this fret parallel to the cutout of pickup $ABCD$? Explain your reasoning.

Module 4 103

Mathematical Connections

Task Part	Prerequisite Skills
Part A	Students should know how to calculate the slope of a nonvertical line or segment **(4.1)**.
Parts B and C	Students should know that if the slopes of two lines or segments are opposite reciprocals, the lines or segments are perpendicular **(4.2)**.
Part C	Students should understand that the distance between two points on a vertical line is the difference between their *y*-coordinates **(1.1 and 1.4)**.
Part D	Students should know that if the slopes of two lines or segments are not equal or are not opposite reciprocals, the lines or segments are neither parallel nor perpendicular **(4.1 and 4.2)**.

Pickup Placement in Acoustic Design

Overview

The word *acoustic* refers to the science of sound. An acoustic guitar is one that does not employ electronic amplification. In an acoustic guitar, the strings are plucked or strummed with the fingers or with a pick. The body of the acoustic guitar amplifies the sound. Guitar strings made of steel produce a very twangy and bright sound. A classical guitar uses nylon strings, which produce softer but more subtle sounds.

Career Connections

Suggest that students research becoming a luthier. A luthier makes and repairs stringed instruments, including acoustic and electric guitars, violins, and cellos. To become a luthier, one can take courses and/or apprentice with a master luthier.

Answers

A. slope of $\overline{KL} = \frac{34-32}{21-15} = \frac{2}{6} = \frac{1}{3}$;

slope of $\overline{AB} = \frac{30-27}{23-14} = \frac{3}{9} = \frac{1}{3}$

The slopes of the segments are equal, so the segments are parallel.

B. slope of $\overline{DC} = \frac{27-24}{24-15} = \frac{3}{9} = \frac{1}{3}$

slope of $\overline{AD} = \frac{24-27}{15-14} = -3$

slope of $\overline{BC} = \frac{27-30}{24-23} = -3$

$\overline{DC}$ is parallel to the last fret because the slopes of the segments are equal.
$\overline{AD}$ and $\overline{BC}$ are perpendicular to the last fret because the slopes of the segments are opposite reciprocals.

C. $E(14, 21)$, $F(23, 24)$, $G(24, 21)$, $H(15, 18)$

D. slope of $\overline{MN} = \frac{41-40}{21-15} = \frac{1}{6}$

The slopes are not equal, so the segments are not parallel.

ONLINE

Assign the Digital Are You Ready? to power actionable reports including

- proficiency by standards
- item analysis

Are You Ready?

Diagnostic Assessment

- Diagnose prerequisite mastery.
- Identify intervention needs.
- Modify or set up leveled groups.

Have students complete the *Are You Ready?* assessment on their own. Items test the prerequisites required to succeed with the new learning in this module.

Slopes of Lines Students will apply the slope formula to determine the slopes of two or more lines in the coordinate plane and establish whether they are parallel or perpendicular.

Point-Slope Form Students will write equations of lines in slope-intercept form that are parallel or perpendicular to a given line in the coordinate plane.

Distance and Midpoint Formulas Students will apply the Pythagorean Theorem to prove the Distance Formula and use the formula to prove segments in the coordinate plane are congruent.

ONLINE

Are You Ready?

Complete these problems to review prior concepts and skills you will need for this module.

Slopes of Lines

Find the slope of each line given two points on the line.

1. $(2, 3)$ and $(4, 9)$ 3
2. $(-4, -1)$ and $(4, 1)$ $\frac{1}{4}$
3. $(0, 2)$ and $(12, -1)$ $-\frac{1}{4}$
4. $(3, 7)$ and $(8, 7)$ 0
5. $(1, 4)$ and $(5, 8)$ 1
6. $(-3, -7)$ and $(-1, -15)$ -4

Point-Slope Form

Write an equation in point-slope form for each line given its slope and a point on the line.

7. -2 and $(0, 1)$ $y - 1 = -2x$
8. $\frac{4}{3}$ and $(3, -1)$ $y + 1 = \frac{4}{3}(x - 3)$
9. $\frac{1}{6}$ and $(2, 1)$ $y - 1 = \frac{1}{6}(x - 2)$
10. $-\frac{3}{2}$ and $(-2, -2)$ $y + 2 = -\frac{3}{2}(x + 2)$
11. 5 and $(-6, 2)$ $y - 2 = 5(x + 6)$
12. -4 and $(7, -1)$ $y + 1 = -4(x - 7)$

Distance and Midpoint Formulas

Find the length and the midpoint of a segment with the endpoints A and B. Round to the nearest hundredth.

13. $A(0, 0)$ $B(6, 10)$ distance: $\sqrt{136} = 2\sqrt{34} \approx 11.66$ midpoint: $(3, 5)$
14. $A(-4, -6)$ $B(5, 12)$ distance: $\sqrt{405} = 9\sqrt{5} \approx 20.12$ midpoint: $(0.5, 3)$
15. $A(1, 5)$ $B(7, 12)$ distance: $\sqrt{85} \approx 9.22$ midpoint: $(4, 8.5)$
16. $A(-11, -9)$ $B(-2, -5)$ distance: $\sqrt{97} \approx 9.85$ midpoint: $(-6.5, -7)$

Connecting Past and Present Learning

Previously, you learned:

- to find the slope of a line given two points,
- to prove if lines are parallel or perpendicular using angle pairs, and
- to write equations for lines on the coordinate plane.

In this module, you will learn:

- to write equations for parallel and perpendicular lines,
- to determine if two lines are parallel or perpendicular based on their slope, and
- to write coordinate proofs.

104

DATA-DRIVEN INTERVENTION

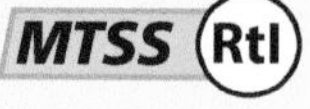

Concept/Skill	Objective	Prior Learning *	Intervene With
Slopes of Lines	Find the slope of a line given two points on the line.	Grade 8, Lesson 5.1	• Tier 3 Skill 3 • Reteach, Grade 8 Lesson 5.1
Point-Slope Form	Use point-slope form to write an equation of a line.	Algebra 1, Lesson 4.4	• Tier 2 Skill 7 • Reteach, Algebra 1 Lesson 4.4
Distance and Midpoint Formulas	Find lengths and midpoints of line segments in the coordinate plane.	Geometry, Lessons 1.1 and 1.4	• Tier 2 Skill 8 • Reteach, Geometry Lessons 1.1 and 1.4

* Your digital materials include access to resources from Grade 6–Algebra 2. The lessons referenced here contain a variety of resources you can use with students who need support with this content.

4.1 Slope and Equations of Parallel Lines

LESSON FOCUS AND COHERENCE

Mathematics Standards

- Prove the slope criteria for parallel and perpendicular lines and use them to solve geometric problems (e.g., find the equation of a line parallel or perpendicular to a given line that passes through a given point).

Mathematical Practices and Processes 

- Look for and make use of structure.
- Reason abstractly and quantitatively.
- Construct viable arguments and critique the reasoning of others.
- Model with mathematics.

I Can Objective 

I can find the equation of a line that is parallel to a given line.

Learning Objective

Use slope to identify, write, and use equations of parallel lines.

Language Objective

Explain the steps need to write an equation of a line parallel to a given line that passes through a given point.

Vocabulary

Review: coordinate proof, slope

Lesson Materials: ruler

Mathematical Progressions

Prior Learning	Current Development	Future Connections
Students: • explained why the slope is the same between any two distinct points on a nonvertical line in the coordinate plane **(Gr8, 5.1)** • understood that lines with the same slope are parallel. **(Gr8, 6.2)** • wrote equations of lines in slope-intercept form. **(Alg1, 3.2)**	**Students:** • identify and investigate properties of parallel lines. • investigate the properties of parallel segments. • use the Slope of Parallel Lines Postulate to write equations of parallel lines. • use the Slope of Parallel Lines Postulate in a real-world context.	**Students:** • will write coordinate proofs about triangle relationships **(9.4 and 9.5)** • will write coordinate proofs about parallelograms. **(11.1–11.5)**

UNPACKING MATH STANDARDS

Prove the slope criteria for parallel and perpendicular lines and use them to solve geometric problems (e.g., find the equation of a line parallel or perpendicular to a given line that passes through a given point).

What It Means to You

Students demonstrate an understanding of the slope criteria of parallel lines and parallel segments. This understanding develops from investigating and identifying properties of parallel lines. A coordinate proof of the slope criteria is written. The slope criteria is applied to write equations of nonvertical, vertical, and horizontal parallel lines both in and out of a real-world context.

In the next lesson, students will use slope to identify, write, and use equations of perpendicular lines.

WARM-UP OPTIONS

ONLINE

PROJECTABLE & PRINTABLE 

ACTIVATE PRIOR KNOWLEDGE • Write Equations for Lines in Slope-Intercept Form

Use these activities to quickly assess and activate prior knowledge as needed.

Problem of the Day 

Write the equation of a line that passes through (2, −1) and (−2, 3) in slope-intercept form. $y = -x + 1$

Quick Check for Homework

As part of your daily routine, you may want to display the Teacher Solution Key to have students check their homework.

Make Connections

Based on students' responses to the Problem of the Day, choose one of the following:

 1 Project the Interactive Reteach, Algebra 1, Lesson 3.2.

 2 Complete the Prerequisite Skills Activity:

Students work in pairs. One student writes a number between −3 and 3, inclusive, to represent the slope. The other student writes a pair of integer coordinates, between −5 and 5, inclusive, to represent a point on the line. The pair writes an equation of the line in slope-intercept form with the given slope that passes through the given point, justifying each step.

- *What is the equation of a line in slope-intercept form?* $y = mx + b$
- *What is m?* the slope of the line
- *What is b?* the y-intercept of the line
- *How do you find b if you know m?* Possible answer: Substitute the given point into the equation $y = mx + b$ and solve for b.

If students continue to struggle, use Tier 3 Skill 3.

SHARPEN SKILLS

If time permits, use this on-level activity to build fluency and practice basic skills.

Vocabulary Review

Objective: Students use a graphic organizer to review the definitions of nonvertical parallel lines, vertical parallel lines, and parallel segments.
Materials: Word Description (Teacher Resource Masters)

Have students work in pairs. Provide each pair with a word description graphic organizer. Have students take turns writing a definition and drawing an example of nonvertical parallel lines, vertical parallel lines, and parallel segments.

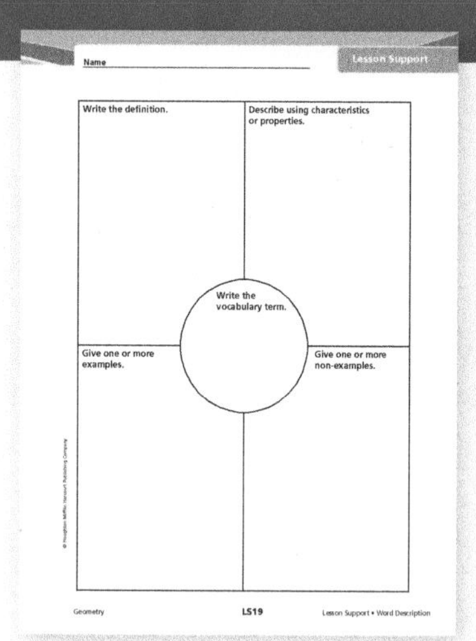
Name
Lesson Support
Write the definition.
Describe using characteristics or properties.
Write the vocabulary term.
Give one or more examples.
Give one or more non-examples.
Geometry LS19 Lesson Support • Word Description

PLAN FOR DIFFERENTIATED INSTRUCTION

MTSS (RtI)

Small-Group Options

Use these teacher-guided activities with pulled small groups.

On Track

Materials: index cards with equations; index cards with points

Students work in pairs. One student picks an equation, the other a point. They work together to find the equation of the line parallel to the given line that passes through the given point.

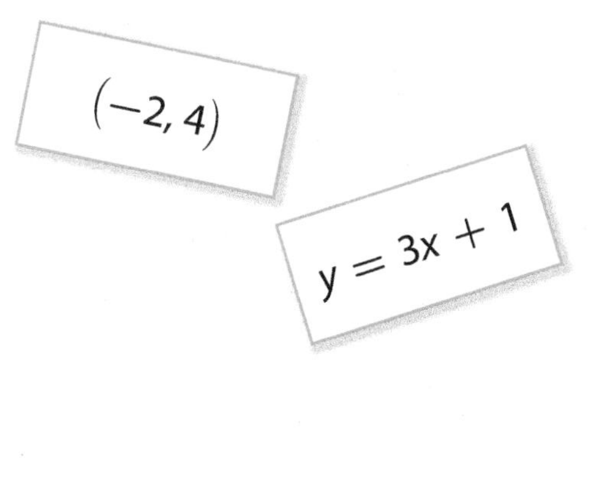

Almost There

Write the following equations.

$y = 2x - 1$	$3x + y = -6$
$y = -3x - 4$	$y = 2x - 1$
$-2x + y = 3$	$-2x = 7 - y$

- Have students find the lines that are parallel and explain why.

Write the equation $y = 2x - 5$ and the point $(4, -3)$.

- Lead students to find the equation of a line parallel to the given line through the given point by having them list reasons for each step in their work.

Ready for More

Materials: index cards with point on them

Students work in pairs. Pick three points from the cards. Find the equation of the line through two of the points. Then, find the equation of the line parallel to that line passing through the other point.

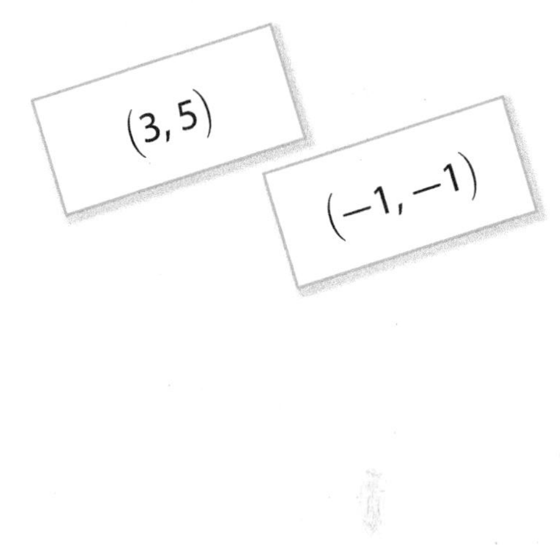

Math Center Options

Use these student self-directed activities at centers or stations. Key: ● Print Resources ● Online Resources

On Track

- ● Interactive Digital Lesson
- ●● Journal and Practice Workbook
- ● Interactive Glossary (printable): **coordinate proof**, **slope**
- ● Illustrative Mathematics: Slope Criterion for Perpendicular Lines

Almost There

- ● Reteach 4.1 (printable)
- ● Interactive Reteach 4.1
- ● RtI Tier 3 Skill 3: Slopes of Lines
- ● Illustrative Mathematics: Parallel Lines

Ready for More

- ● Challenge 4.1 (printable)
- ● Interactive Challenge 4.1
- ● Illustrative Mathematics: Slope Criterion for Perpendicular Lines

ONLINE  View data-driven grouping recommendations and assign differentiation resources.

During the *Spark Your Learning*, listen and watch for strategies students use. See samples of student work on this page.

Use a Diagram and a Definition — Strategy 1

Use a diagram and the definition of slope.
The slope of the equation $y = x - 2$ is 1.
The horizontal and vertical distance between the lines is c, so

slope $= \frac{\text{rise}}{\text{run}} = \frac{c}{c}$.

The y-intercept of the line directly above the given line is $-2 + c$. The intercept directly below is $-2 - c$, etc. The equations of the lines are $y = x - 2 + nc$, where c is the horizontal and vertical distance between the lines and n is an integer between −5 and 1, inclusive.

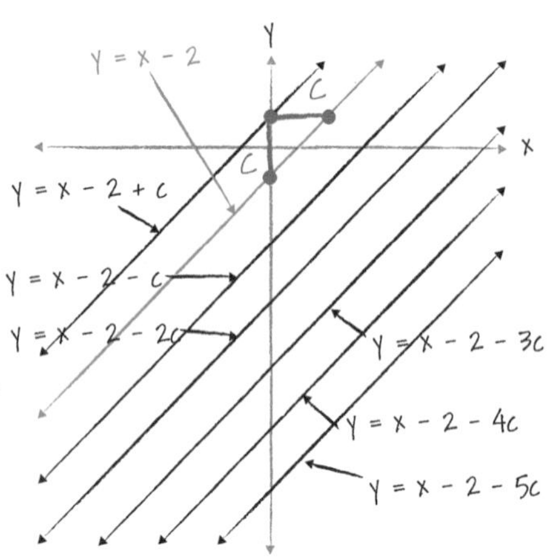

If students . . . use a diagram and the definition of slope to solve the problem, they are employing an efficient method and demonstrating an exemplary understanding of writing equations of lines from Algebra 1.

Have these students . . . explain how they determined their equation. **Ask:**

- Q How did you use the given information to create your diagram?
- Q How did the diagram help you write your equation?

Use a Definition — Strategy 2

The slope of the equation $y = x - 2$ is 1. The horizontal and vertical distance between the parking lines is c, so slope $= \frac{\text{rise}}{\text{run}} = \frac{c}{c}$. The y-intercept of the parking space directly above the given space is c. The next is $2c$, etc. The equations of the other lines are $y = x - 2 + nc$, where c is the horizontal and vertical distance between the parking lines and n is an integer.

If students . . . use a definition and not a diagram, they understand slope of 1 could be written as c divided by c.

Activate prior knowledge . . . by having students evaluate the situation. **Ask:**

- Q How many parking places line up parallel to $y = x - 2$?
- Q Should there be restrictions on n?

COMMON ERROR: Uses Wrong Slope

Some students may think the slope of the parallel lines in the parking lot is c and that the equations are $y = cx$.

If students . . . chose the wrong slope, then they are having trouble visualizing the situation.

Then intervene . . . by pointing out that there are 7 lines that are parallel. **Ask:**

- Q What is the slope of the given line?
- Q What is the slope of any line parallel to it?

4.1

Slope and Equations of Parallel Lines

I Can find the equation of a line that is parallel to a given line.

Spark Your Learning

A city is building a parking lot and needs to plan the spacing of each parking space.

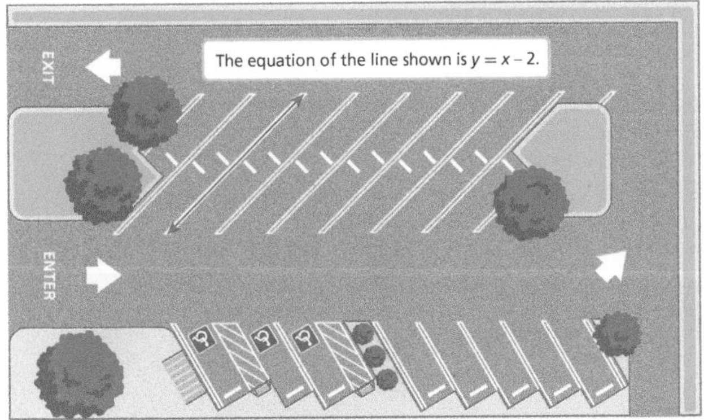

Complete Part A as a whole class. Then complete Parts B–D in small groups.

A. What is a mathematical question that you can ask about this situation? What information would you need to know to answer your question?

A. What are the equations of the other lines?; the given equation of one of the parking space lines and more information about the line spacing

B. What mathematical facts about the parking space lines do you already know? What form for an equation of a line on a coordinate plane could be helpful? See Additional Answers.

C. To answer your question, what strategy and tool would you use along with all the information you have? What answer do you get? See Strategies 1 and 2 on the facing page.

D. Suppose you extend the lines to the x-axis to find the x-intercepts. How would this help you check your answer for Part C?

Turn and Talk Could you determine the perpendicular distance between two of these lines if you knew the horizontal distance between two lines? What might those steps look like if so? See margin.

D. Possible answer: Each line's x-intercept and y-intercept will be the same (but opposite sign). I can't always see the y-intercept on the graph but I can read the x-intercepts to check the answers in Part C.

CULTIVATE CONVERSATION • Information Gap

Ask students questions to help them decide what missing information they need to answer the question, "What are the equations of the other lines?"

1. Do you have enough information to find the equations of the other lines? Explain. no; I know the equation of one of the lines but I don't know how that line relates to the other lines.

2. What information do you need? I need more information about the line spacing and about the slopes of the lines.

3. What other information would help? Possible answer: It would help if I knew the relationship, if any, between the given line and the other lines.

1 Spark Your Learning

▶ MOTIVATE

- Have students look at the diagram in their books and read the information contained in the diagram. Then complete Part A as a whole-class discussion.
- Give the class the additional information they need to solve the problem. This information is available online as a printable and projectable page in the Teacher Resources.
- Have students work in small groups to complete Parts B–D.

▶ PERSEVERE

If students need support, guide them by asking:

Q **Advancing • Use Tools** Which tool could you use to solve the problem? Why choose that tool and not some other? Students' choices of tools and reasons for choosing them will vary.

Q **Assessing** What is the slope and y-intercept of the given line? slope $= 1$, y-intercept $= -1$

Q **Assessing** What do you know about the slopes of parallel lines? The slopes of parallel lines are equal.

Q **Advancing** If 3 lines are parallel and the distance between the top line and the middle line is d, what is the distance between the top line and the bottom line? $2d$

Turn and Talk Elicit from students that slope is equal to rise over run. *Rise* is a vertical distance and *run* is a horizontal distance. And if the slope of a line is 1, the rise and run between two points are equal. Possible answer: Yes, since the slope of the line is 1, this means the vertical distance is the same as the horizontal distance between the lines. I can then create a right triangle using the horizontal and vertical distances between the lines. I can calculate length of the hypotenuse using the Pythagorean Theorem. I can create another right triangle using the horizontal (or vertical distance) and half the hypotenuse. I can then use the Pythagorean Theorem again to solve for the perpendicular distance.

▶ BUILD SHARED UNDERSTANDING

Select groups of students who used various strategies and tools to share with the class how they solved the problem. As they present their solutions, have each group discuss why they chose a specific strategy and tool.

② Learn Together

Build Understanding

Task 1 (MP) **Use Structure** Students use the structure of the slope formula with subscript notation to show an instance of two parallel lines having the same slope.

Sample Guided Discussion:

Q Graphically describe lines with negative slopes, zero slope, and positive slopes? Possible answer: If the slope is negative, the line goes downward from left to right. If the slope is 0, the line is horizontal. If the slope is positive, the line goes upward from left to right.

Q What would be another way to describe the slope of a horizontal line? Possible answer: The slope of a horizontal line is undefined because the change in *x*, the run, is 0, and division by 0 is undefined.

Turn and Talk Remind students that the slope is a ratio of the change in *y* to the change in *x*. no; Possible answer: The calculation for the slope would still result in the same answer because the *x*- and *y*-values would change by the same amount.

Build Understanding

Investigate Properties of Parallel Lines

The **slope** m of a nonvertical line is the ratio of the vertical change (the *rise*) to the horizontal change (the *run*) between any two points on the line. For the line shown, $m = \frac{\text{rise}}{\text{run}} = \frac{y_2 - y_1}{x_2 - x_1}$, where (x_1, y_1) and (x_2, y_2) are two points on the line. The subscript notation indicates whether each value in the ratio is the *x*- or *y*-coordinate of the first point or the second point.

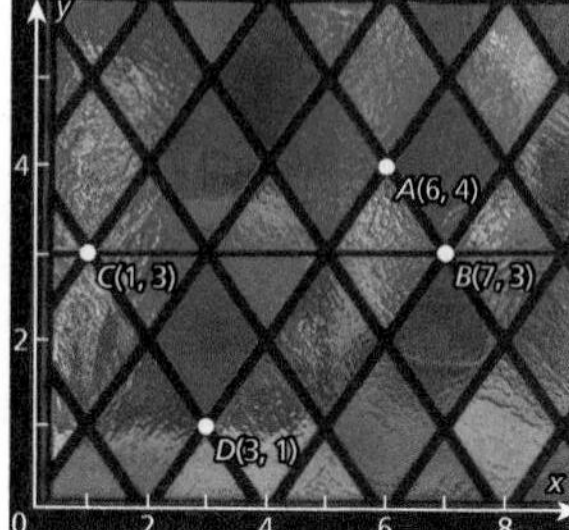

Recall that parallel lines are two lines in the same plane that do not intersect. There is a special relationship between the slopes of any two parallel lines.

1 The coordinates of two points on each line are given. Indicate which of the points you chose as the first point and which is the second point.

$A(6, 4)$, $B(7, 3)$, where A is (x_1, y_1) and B is (x_2, y_2)

$C(1, 3)$, $D(3, 1)$, where C is (x_1, y_1) and D is (x_2, y_2)

Find the slope of $\overleftrightarrow{AB}$ and $\overleftrightarrow{CD}$.

$\overleftrightarrow{AB}$: $m = \frac{y_2 - y_1}{x_2 - x_1}$, $m = \frac{3-4}{7-6} = -1$

$\overleftrightarrow{CD}$: $m = \frac{y_2 - y_1}{x_2 - x_1}$, $m = \frac{1-3}{3-1} = \frac{-2}{2} = -1$

A. Does it matter which point is chosen as the first point and which is the second point? If the assignment of the first point and second point are interchanged, does it change the slope of $\overleftrightarrow{AB}$? Explain. See Additional Answers.

B. How can you graphically confirm the slope should be negative? The lines slope downward from left to right.

C. What do you observe about the slopes of the parallel lines? The slopes of the parallel lines are the same.

D. What do you predict the outcome would be if we highlighted a third parallel line in the figure and then calculated the slope of that line? Why? See Additional Answers.

Turn and Talk Would the results of Task 1 be different if different points from each line had been used in the slope equation? Why or why not? See margin.

LEVELED QUESTIONS

Depth of Knowledge (DOK)	Leveled Questions	What Does This Tell You?
Level 1 **Recall**	What is the slope of the line that passes through (1, 2) and (3, 8)? 3	Students' answers will indicate if they know the formula to find the slope of a line.
Level 2 **Basic Application of Skills & Concepts**	Is the line passing through *A*(3, 1) and *B*(2, 2) parallel to the line passing through *C*(1, −1) and *D*(−1, 1)? Explain yes; The slope of each line is −1.	Students' answers will demonstrate if they can use slope to determine if lines are parallel.
Level 3 **Strategic Thinking & Complex Reasoning**	If the line passing through *A*(2, −3) and *B*(3, 2) is parallel to the line passing through *C*(0, 1) and *D*(−1, *y*), what is the value of *y*? −4	Students' answers will reflect whether they understand that parallel lines have the same slope and can reason strategically about how to determine that two lines are parallel.

Identify Parallel Segments

Slope Criteria of Parallel Lines
Two nonvertical lines are parallel if and only if they have the same slope.

A line segment is a section of a line that is contained within the line. The equation of the line segment is the same equation of the line that contains it, except limitations on the domain are given for the segment. The limitations describe the endpoints of the segments.

Parallel segments, like parallel lines, have equal slopes and do not intersect.

2 A parallelogram is a four-sided shape (or quadrilateral) with two pairs of parallel, opposite sides. You can determine if a shape shown on a coordinate plane is a parallelogram by examining the slopes of each of its sides.

The coordinates of the vertices of the figure shown are. $A(4, 10)$, $B(7, 5)$, $C(12, 15)$, $D(8, 18)$

A. Why do you need the coordinate points to determine the slopes? 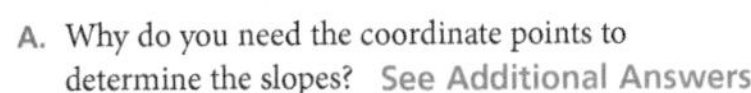 See Additional Answers.

B. What should be true about the slopes of any parallel sides? 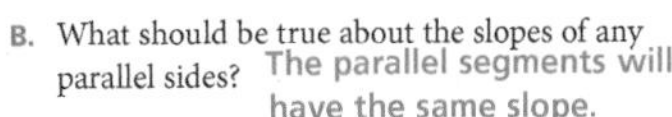 The parallel segments will have the same slope.

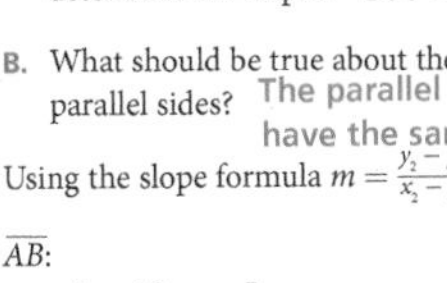

Using the slope formula $m = \frac{y_2 - y_1}{x_2 - y_1}$, the slopes of each segment are:

$\overline{AB}$:

$$m = \frac{5 - 10}{7 - 4} = \frac{-5}{3}$$

$$m = -\frac{5}{3}$$

$\overline{BC}$:

$$m = \frac{15 - 5}{12 - 7} = \frac{10}{5}$$

$$m = 2$$

$\overline{CD}$:

$$m = \frac{18 - 15}{8 - 12} = \frac{3}{-4}$$

$$m = -\frac{3}{4}$$

$\overline{DA}$:

$$m = \frac{10 - 18}{4 - 8} = \frac{-8}{-4}$$

$$m = 2$$

C. Which segments are parallel? How can you classify the shape using this information? C, D. See Additional answers.

D. Examine the lengths of $\overline{DA}$ and $\overline{BC}$. Is it necessary that the two segments be of equal length or be located near each other on the coordinate plane to be parallel?

Turn and Talk How can you change one point on the figure to create a parallelogram? See margin.

Task 2 MP **Reason** Students reason abstractly and quantitatively to determine if a quadrilateral on a coordinate plane is a parallelogram.

Sample Guided Discussion:

Q Does the figure look like a parallelogram? Explain no; Only one pair of opposite sides appears to be parallel.

Q After you find the slopes of all four sides, how can you tell if the figure is a parallelogram? Possible answer: If both pairs of opposite sides have equal slopes, the figure is a parallelogram.

Turn and Talk Have students identify the non-parallel sides. Ask "Which point can I change to make a parallelogram?" Possible answer: You could change point *D* to (9, 20) to create a parallelogram.

PROFICIENCY LEVEL

Beginning

Draw a parallelogram and a quadrilateral that is not a parallelogram. Say "Which quadrilateral is a parallelogram?" Have students write "parallelogram" and its definition along with a sketch.

Intermediate

Have students work in groups. Give each group a set of index cards. Each card should show a quadrilateral in a coordinate plane. Have students find the slopes of the sides and write whether the quadrilateral is a parallelogram, trapezoid, or neither explaining why in terms of slope.

Advanced

Have students work in pairs. Each pair draws a parallelogram on a coordinate plane and writes a paragraph describing how they drew the parallelogram and why the figure is a parallelogram.

Task 3 (MP) **Construct Arguments** Students construct a coordinate proof for the slope criteria for parallel lines.

CONNECT TO VOCABULARY

Have students use the **Interactive Glossary** to record their understanding of the vocabulary in this task.

Sample Guided Discussion:

Q What is the image of the translation of a line 3 units left and 2 units down? a line parallel to the given line 3 units left and 2 units down from the given line

Q What is the image of a translation of any line? a parallel line

Q What does slope tell you about a line? Possible answer: Slope tells the steepness and direction of a line.

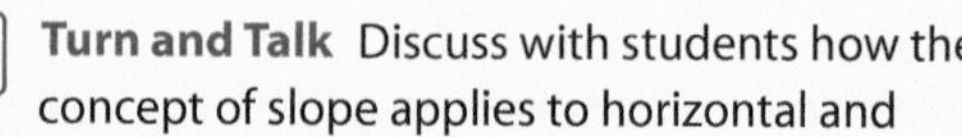

Turn and Talk Discuss with students how the concept of slope applies to horizontal and vertical lines. Possible answer: A vertical line would not have a value for "run" or any change in the x-direction to fill into the slope equation. A horizontal line would not have a value for "rise" or change in the y-direction to fill into the slope equation.

Prove the Slope Criteria of Parallel Lines

A **coordinate proof** is a style of proof where generalized coordinates are used to prove a geometric theorem. You can assign variables to one or more points and then prove the geometric theorem applies to any points on the coordinate plane.

3 **Given:** $p \parallel q$, neither line is vertical or horizontal

Prove: $m_p = m_q$

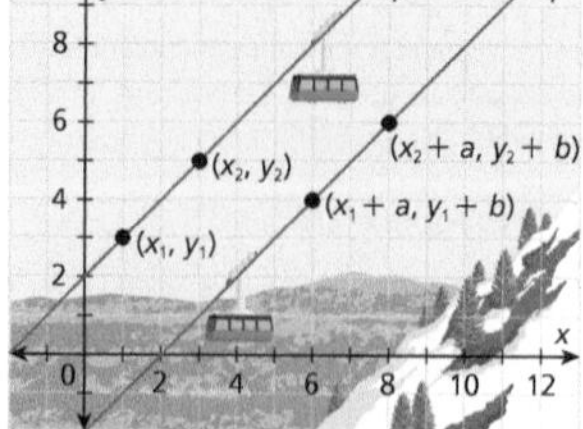

Choose two points from each line.
The points identified on the parallel lines are labeled to represent that the points on line q are a translation of the points on line p. Recall that if any lines are parallel, one can be mapped to the other by a translation.

Find the slopes.
The slopes of line p and line q are calculated using the slope formula, $m = \frac{y_2 - y_1}{x_2 - x_1}$, and the x-and y-coordinates of each point.

Line p:

$$m_p = \frac{y_2 - y_1}{x_2 - x_1}$$

Line q:

$$m_q = \frac{(y_2 + b) - (y_1 + b)}{(x_2 + a) - (x_1 + a)}$$

$$m_q = \frac{y_2 + b - y_1 - b}{x_2 + a - x_1 - a}$$

$$m_q = \frac{y_2 - y_1}{x_2 - x_1}$$

Substitute.
Using substitution, we can see that $m_p = m_q$ for any pair of parallel lines.

A. For line q, explain the meaning of a and b. What do they represent? See Additional Answers.

B. In your own words, how does the fact that any line parallel to another line is a translation of that line help you prove the slopes of the two lines are the same? Possible answer: Since you know one line is a translation of the other, there is a concrete way each point from one line relates to a point on the other line.

C. Would the outcome of the proof change if we had used lines with a negative slope to choose generic points? No, the outcome is the same.

Turn and Talk Why did the given information specify that the lines were not vertical or horizontal? What would happen to the equations used in the roof in both cases if the lines were either vertical or horizontal? See margin.

Step It Out

Write Equations of Parallel Lines

When given the equation of a line, you can find the slope by rearranging the equation into slope-intercept form, $y = mx + b$, where m is the slope and b is the y-intercept.

4 Write an equation for a line parallel to $y = 0.25x - 5$ that passes through the point $(8, 5)$.

First, collect the given information.

$m = 0.25, x = 8, y = 5$ — A. Where do you find the values of x and y? *from the given coordinate point through which the line must pass*

The only unknown piece of information to write an equation in the form $y = mx + b$ is the y-intercept of the parallel line.

Use the slope-intercept form and substitute the given information.

$y = mx + b$

$5 = (0.25)8 + b$

$5 = 2 + b$

$b = 3$ — B. What does this value of b tell you? *The value of b gives the location where the line crosses the y-axis.*

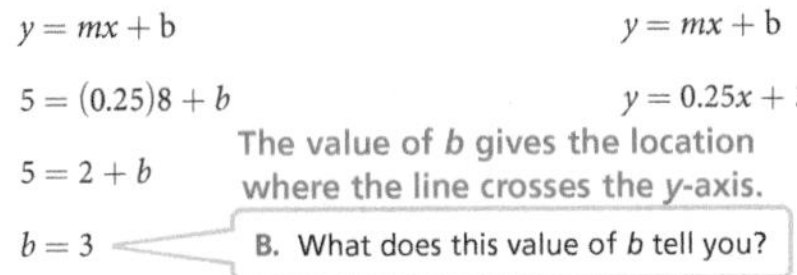

Replace the values for m and b back into the general slope-intercept form.

$y = mx + b$

$y = 0.25x + 3$ — C. Why do we substitute the values back into the slope-intercept form? *to find the general form for the equation that describes any point on the line*

The equation for a line parallel to $y = 0.25x - 5$ that passes through the point $(8, 5)$ is $y = 0.25x + 3$.

Write Equations of Parallel, Horizontal, and Vertical Lines

Remember that horizontal lines have a slope of 0, and vertical lines have an undefined slope.

The equation $y = 3$ can be written in slope-intercept form as $y = 0x + 3$ since the slope of a horizontal line is 0. However, $x = -1$ cannot be written in slope-intercept form since the slope is undefined.

The equation for a horizontal line that passes through the point (a, b) is $y = b$.

The equation for a vertical line that passes through the point (a, b) is $x = a$.

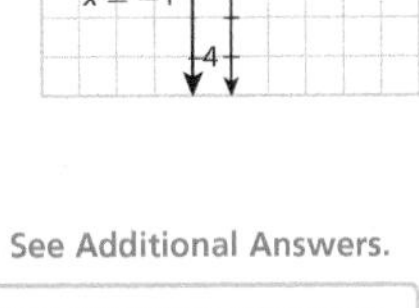

5 Write an equation for a line parallel to $y = -5$ that passes through $(3, 6)$ and an equation for a line parallel to $x = 4$ that passes through the point $(-1, 2)$.

A, B. See Additional Answers.

Parallel to:	Passes through:	Equation
$y = -5$	$(3, 6)$	$y = 6$
$x = 4$	$(-1, 2)$	$x = -1$

A. How do you know these two lines have the same slope?

B. Why are the slopes of $x = 4$ and $x = -1$ undefined?

Step It Out

Task 4 MP **Use Structure** Students use the structure of an equation in slope-intercept form to write equations of parallel lines.

SUPPORT SENSE-MAKING Three Reads

Have students read the problem three times. Use the questions below for a different focus each read.

1. What is the situation about?
2. What are the quantities in the situation?
3. What are the possible mathematical questions that you could ask for the situation?

Sample Guided Discussion:

Q What do you need to know to write an equation of a line in slope-intercept form? the slope and y-intercept

Q If two lines are parallel, what is true about their slopes? The slopes are equal.

Task 5 MP **Use Structure** Students use a chart to write equations of lines parallel to a horizontal line and a vertical line based on their knowledge of slope in these special cases.

Sample Guided Discussion:

Q What is the slope of a horizontal line going through the point (c, d)? The slope of a horizontal line is 0.

Q What is the equation of a horizontal line going through the point (c, d)? $y = d$

Q What is the slope of a vertical line going through the point (c, d)? The slope of a vertical line is undefined.

Q What is the equation of a vertical line going through the point (c, d)? $x = c$

 (MP) **Model with Mathematics** Students use a linear equation as a model of a parallel path to a fountain in a garden.

Sample Guided Discussion:

Q What do you know about the slope of parallel lines? The slopes are equal.

Q What is the slope-intercept equation of a line? $y = mx + b$

Q What do you need to determine the equation of a line in slope-intercept form? You need to know the slope and y-intercept of the line.

Turn and Talk Elicit from students the standard form of an equation of a line. Have students make a plan to transform the equation from standard form to slope-intercept form. Possible answer: Rewrite the equation in slope-intercept form. $y = -\frac{Ax}{B} + \frac{C}{B}$; $m = -\frac{A}{B}$

Apply the Slope Criteria for Parallel Lines

When solving a problem involving equations of parallel lines, first determine how to use the given information to find the answer. For example, you may need to find the slope of a given line, and then use the slope to write an equation for a parallel line through a given point.

6 A landscape designer is planning parallel walking paths through a large rectangular garden. The figure shows the centerline of two paths. A fountain on the other side of the garden will be located along the center of a third path. What equation represents the centerline of the third path?

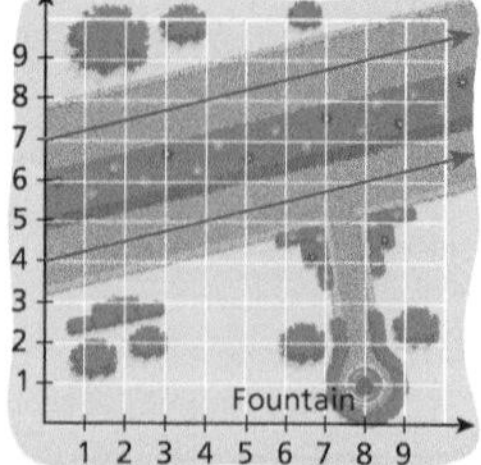

Find the slope of a given line.

The given parallel lines have the same slope. Choose either to determine the slope for the unknown equation.

$m = \frac{\text{rise}}{\text{run}} = \frac{1}{4}$

A. Why might you count rise and run here versus using the slope formula? The rise and run are easily visible on the figure.

Determine the point through which the line must pass.

The fountain is located at $(8, 1)$.

Find the y-intercept of the unknown line.

$y = mx + b$

$1 = \frac{1}{4}(8) + b$

$1 = 2 + b$

$b = -1$

B. How can finding the y-intercept help to graph the line on the coordinate plane? The y-intercept gives a coordinate point from which you can start drawing the line using the slope.

Write the equation of the line.

Write the equation for the third path using the same slope as the given parallel lines and the y-intercept calculated in the previous step.

C. What is the equation for the path parallel to the paths shown in the figure that passes through the point $(8, 1)$? $y = \frac{1}{4}x - 1$

Turn and Talk How would you find the slope of a line given in standard form? Write the slope of $Ax + By = C$ in terms of A, B, and C. See margin.

Check Understanding

1. Can you write an equation for a line parallel to a given line that passes through exactly one point on the given line? Explain. See Additional Answers.
2. Which two equations could describe two parallel sides of a quadrilateral? Explain your reasoning. lines *a* and *b*; Both lines have a slope of −3.
   - **A.** $3x + y = -5$ from $x = 0$ to $x = 5$
   - **B.** $y = \frac{1}{3}(21 - 9x)$ from $x = -1$ to $x = 2$
   - **C.** $y = 3x + 2.5$ from $x = -10$ to $x = -3$
3. Points (x_1, y_1) and (x_2, y_2) lie on line *u*. Points $(x_1 + 2, y_1 - 1)$ and $(x_2 + 2, y_2 - 1)$ lie on line *v*. Are lines *u* and *v* parallel? Justify your reasoning. See Additional Answers.
4. Write an equation for a line that passes through $(0, 0)$ and is parallel to a line that passes through $(-5, 3)$ and $(2, 1)$. $y = -\frac{2}{7}x$
5. Are the lines $y = \frac{1}{2}$ and $x = \frac{1}{2}$ parallel? Justify your answer. no; One is horizontal and one is vertical.
6. On a map, Padma's house is located at $(0, 2)$ on a street modeled by the equation $y = 2x + 2$. Each unit on the map represents one block. Padma leaves her house, walks four blocks east, then turns and walks one block north. She continues walking parallel to her street. What equation models her new path? $y = 2x - 5$

On Your Own

In Problems 7–9, use the graph to find the unknown slope.

7. A line is parallel to the line with a *y*-intercept of −2. What is its slope? $m = \frac{3}{5}$
8. A line is parallel to the line with a *y*-intercept of 4. What is its slope? $m = -\frac{2}{3}$
9. A line is parallel to the line with a *y*-intercept of −1. What is its slope? $m = -4$

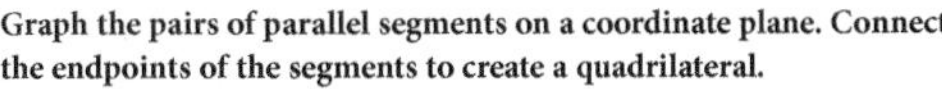

Graph the pairs of parallel segments on a coordinate plane. Connect the endpoints of the segments to create a quadrilateral.

10. $y = \frac{4}{5}x - 1.5$ from $x = 2$ to $x = 6$ $y = \frac{4}{5}x + 3$ from $x = 0$ to $x = 3$ See Additional Answers.
11. $y = 3x$ from $x = -1.5$ to $x = 3$ $y = 3x - 5$ from $x = 1$ to $x = 3$ See Additional Answers.

For Problems 12–15, determine if the two lines are parallel.

12. Line *m* passes through points $(1, 5)$ and $(3.5, 7)$. Line *n* passes through points $(4, 7)$ and $(7.5, 10)$. not parallel
13. Line *a* passes through points $(2, -3)$ and $(6, 9)$. Line *b* passes through points $(0, 9)$ and $(2, 3)$. not parallel
14. Line *p* passes through points $(4, 1)$ and $(8, 3)$. Line *q* passes through points $(-8, 1)$ and $(4, 7)$. parallel
15. Line *j* passes through points $(-3, 0)$ and $(2, 4)$. Line *k* passes through points $(1, -9)$ and $(9, 1)$. not parallel

ONLINE

Assign the Digital On Your Own for
- built-in student supports
- Actionable Item Reports
- Standards Analysis Reports

On Your Own

Assignment Guide

The chart below indicates which problems in the On Your Own are associated with each task in the Learn Together. Assign daily homework for tasks completed.

Learn Together Tasks	On Your Own Problems
Task 1, p. 106	Problems 7–9 and 12–15
Task 2, p. 107	Problems 10, 11, and 16
Task 3, p. 108	Problem 17
Task 4, p. 109	Problems 18–27
Task 5, p. 109	Problems 28–33 and 36
Task 6, p. 110	Problems 34, 35, 37, and 38

3 Check Understanding

Formative Assessment

Use formative assessment to determine if your students are successful with this lesson's learning objective.

Students who successfully complete the Check Understanding can continue to the On Your Own practice.

For students who miss 1 problem or more, work in a pulled small group using the Almost There small-group activity on page 105C.

ONLINE

Assign the Digital Check Understanding to determine
- success with the learning objective
- items to review
- grouping and differentiation resources

4 Differentiation Options

Differentiate instruction for all students using small-group activities and math center activities on page 105C.

Reteach

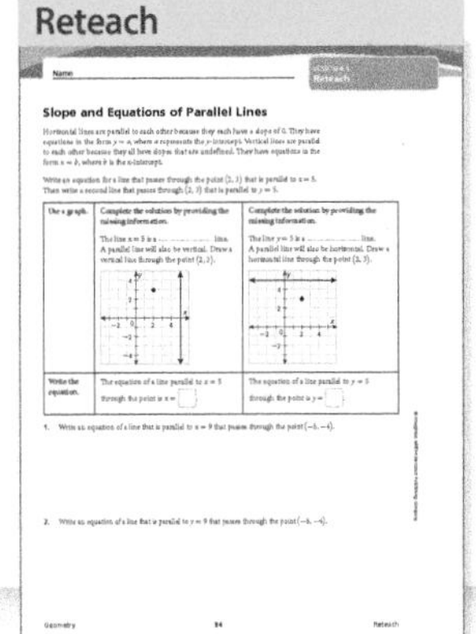

Challenge

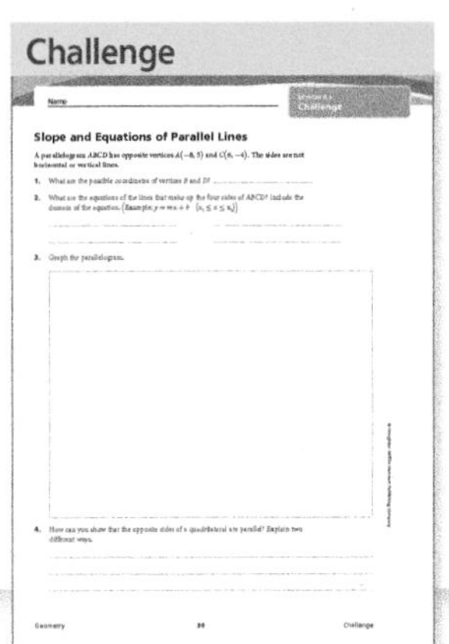

Questioning Strategies

Problem 16 Eva is asked to create a blueprint of a square tabletop centered at (0, 0) that has a side length of 2 units. Write equations with domain restrictions that Eva uses in her blueprint.

$y = -1$; from $x = -1$ to $x = 1$
$y = 1$; from $x = -1$ to $x = 1$
$x = -1$; from $y = -1$ to $y = 1$
$x = 1$; from $y = -1$ to $y = 1$

Watch for Common Errors

Problem 18 Some students may get confused with which points to use. Have them label the first point (the one the line passes through) P and label the points the line passes through A and B. Students find the line passing through P and parallel to $\overleftrightarrow{AB}$.

16. (MP) **Use Repeated Reasoning** Eva is asked to create a blueprint of a tabletop before the prototype is produced. The tabletop is in the shape of a parallelogram, which is a quadrilateral with two pairs of parallel sides. She graphs the following segments.

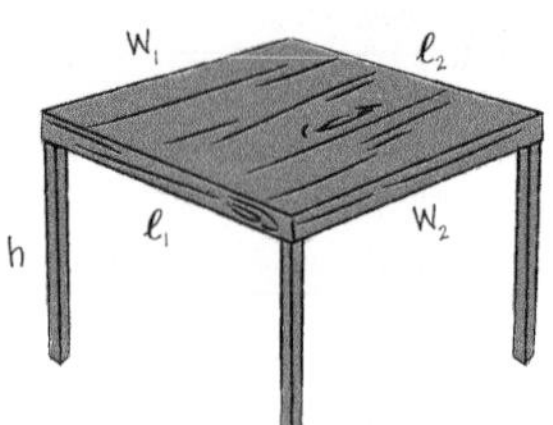

See Additional Answers.

Length 1: $y = 5$ from $x = -1$ to $x = 4$

Length 2: $y = 10$ from $x = -1$ to $x = 4$

Width 1: $x = -1$ from $y = 5$ to $y = 10$

Width 2: $x = 4$ from $y = 5$ to $y = 10$

Is Eva's graph a parallelogram? Discuss the slopes in your justification.

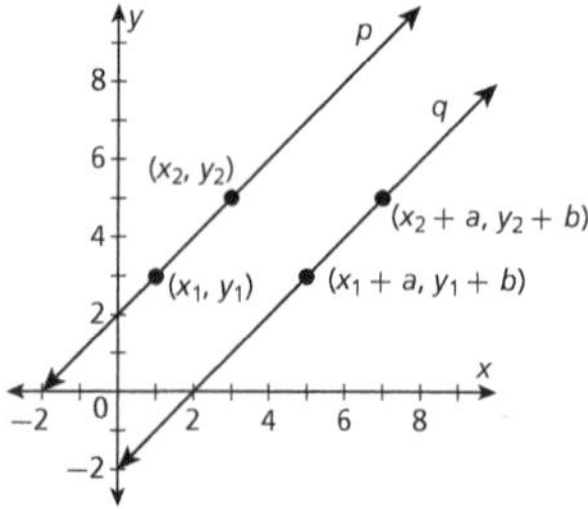

17. Put the reasons of the proof that parallel lines have the same slope in the correct order.

Given: $p \parallel q$ 1. C, 2. A, 3. B, 4. E, 5. D

Prove: $m_p = m_q$

Statements	Reasons
1. Lines p and q are parallel.	?
2. If a point on line p is (x, y), any point on line q is $(x + a, y + b)$, where a and b are constants.	?
3. $m_p = \frac{y_2 - y_1}{x_2 - x_1}$, $m_q = \frac{(y_2 + b) - (y_1 + b)}{(x_2 + a) - (x_1 + a)}$	?
4. $m_q = \frac{(y_2 + b) - (y_1 + b)}{(x_2 + a) - (x_1 + a)} = \frac{y_2 - y_1}{x_2 - x_1}$	?
5. $m_p = m_q$	?

A. Any parallel lines can be mapped to each other by a translation.

B. slope $m = \frac{y_2 - y_1}{x_2 - x_1}$

C. Given

D. Substitution Property of Equality

E. Combine like terms.

Find the equation for each line.

18. Line n passes through $(-4, -1.5)$ and is parallel to a line that passes through $(0, 2)$ and $(5, 4)$. $y = \frac{2}{5}x + \frac{1}{10}$

19. Line f passes through $\left(\frac{1}{4}, 0\right)$ and is parallel to a line that passes through $(7, -3)$ and $(-10, -2)$. $y = -\frac{1}{17}x + \frac{1}{68}$

20. Line j passes through $(15, 13)$ and is parallel to a line that passes through $(-0.5, 1)$ and $(11.5, 9)$. $y = \frac{2}{3}x + 3$

Find the equation for the line parallel to the given line and passing through the given point.

21. $y = -\frac{1}{3}x + 7$; $(-2, 5.5)$ $y = -\frac{1}{3}x + \frac{29}{6}$

22. $y = \frac{4}{5}x - 1$; $(0, 6)$ $y = \frac{4}{5}x + 6$

23. $-2x + y = -8$; $(11, 0)$ $y = 2x - 22$

24. $3y = -12x + 4$; $(1, 1)$ $y = -4x + 5$

25. $x = y$; $(1.25, 3)$ $y = x + 1.75$

26. $y = mx + b$; (x_1, y_1) $y = mx + (y_1 - mx_1)$

27. Open Ended A line passes through $(0, -1)$ and $(5, 10)$. Write the equation of any line parallel to that line, and provide a point other than the y-intercept through which the line passes. Show your work and explain the steps used to find a point along the parallel line. Answers will vary. Check students' work.

Identify the slope of each given line. Write an equation for the line parallel to the given line passing through the given point.

28. $x = -3$; $(2, 0)$ undefined; $x = 2$

29. $x = 5$; $(1.4, 8)$ undefined; $x = 1.4$

30. $y = -10$; $(0, 0)$ $m = 0$; $y = 0$

31. $y = \frac{2}{3}$; $(-4, 3)$ $m = 0$; $y = 3$

32. $x = 0$; $(20, 14)$ undefined; $x = 20$

33. $y = 8.5$; $\left(1, \frac{3}{5}\right)$ $m = 0$; $y = \frac{3}{5}$

34. STEM An engineer has a sketch of a U-shaped bracket to be manufactured. Before manufacturing, the part is modeled using computer-aided design (CAD) software, which then tells a machine how to cut the part. One leg and the bottom of the bracket have been modeled using line segments. The legs must be parallel and the same length. Write the equation and domain limitation of the segment that represents the missing leg, and draw the segment. See Additional Answers.

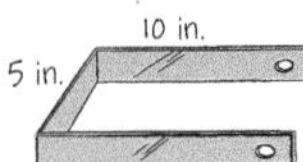

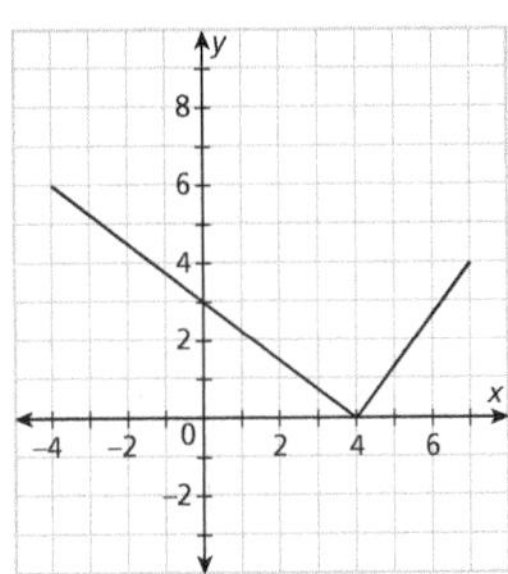

35. Use the figure to answer Parts A–C.

A. Are lines a and b parallel? Use the slopes to justify your answer.

B. If so, write the equation to a third parallel line. If not, find the equation for line a.

C. Suppose line b passed through $(0, -2)$ and $(6, -4)$. How would your answer to Part A change?

A–C. See Additional Answers.

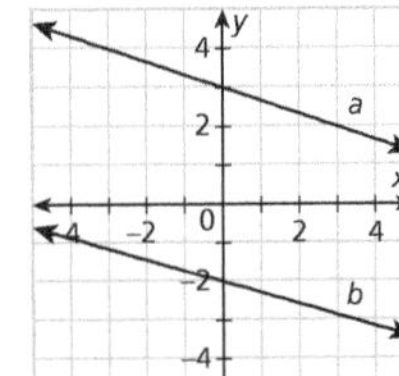

Watch for Common Errors

Problem 28 Students are asked to find the slope of the given line. For the line $x = -3$, they may use the coefficient of x as the slope. Remind them that vertical lines have no slope.

5 Wrap-Up

Summarize learning with your class. Consider using the Exit Ticket, Put It in Writing, or I Can scale.

Exit Ticket 

Find the equation for the line parallel to $3x + 2y = 6$ that passes through the point $(-4, 1)$.

Possible answer: The equation $3x + 2y = 6$ in slope-intercept form is $y = -\frac{3}{2}x + 3$. The slope is $-\frac{3}{2}$. The equation of the parallel line has the form $y = -\frac{3}{2}x + b$. Substitute $(-4, 1)$ for x and y to find that $b = -5$. The equation of the parallel line is $y = -\frac{3}{2}x - 5$.

Put It in Writing 

Explain how to determine if $ax + by = c$ and $dx + ey = f$ are parallel.

I Can 

The scale below can help you and your students understand their progress on a learning goal.

4	I can find the equation of a line that is parallel to a given line, and I can explain how I found the line to others.
3	I can find the equation of a line that is parallel to a given line.
2	I can determine if two lines are parallel using the slope criteria.
1	I can determine the slope of a line given the equation.

Spiral Review • Assessment Readiness

These questions will help determine if students have retained information taught in the past and can also prepare them for high-stakes assessments. Here, students must find a value that will make corresponding angles of parallel lines equal **(3.2)**, use the Pythagorean Theorem to find a leg of a right triangle **(3.3)**, and describe parallel and perpendicular concepts by matching **(3.1–3.3)**.

36. (MP) **Critique Reasoning** Consider the statement "Two lines in a coordinate plane are parallel if and only if the slopes of the lines are m and m, where m is any real number." Is this statement true or false? Explain your reasoning.

36–38. See Additional Answers.

37. A city is planning to add a light rail track and a community walking path alongside an existing train line track. The current rail bed and track are shown in the diagram. The new light rail track will be placed in the middle of the space between the existing track and the farther edge of the track bed. The new walking path will be located 15 feet into the existing green space on the other side. What lines do the centers of the light rail track and the community walking path follow?

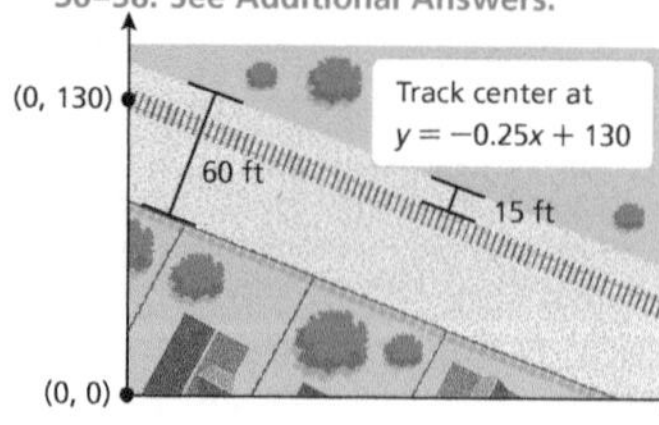

38. (Open Middle™) Using the digits 1 to 9, at most one time each, fill in the boxes to create linear equations for two lines that are parallel.

$y = \frac{\square}{\square}x + \square$; $\square x - \square y = \square$

Spiral Review • Assessment Readiness

39. What value of x makes lines a and b parallel?

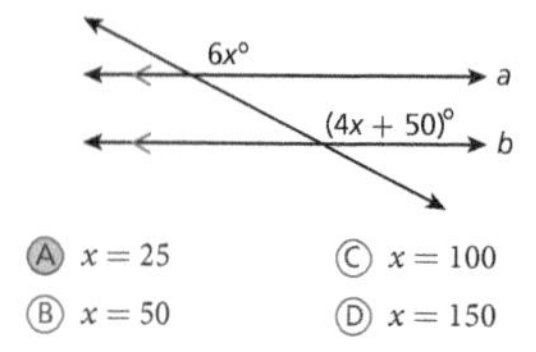

Ⓐ $x = 25$
Ⓑ $x = 50$
Ⓒ $x = 100$
Ⓓ $x = 150$

40. $\overline{AB}$ is 13 inches long. $\overline{BD}$ is 12 inches long. What is the length of $\overline{AD}$?

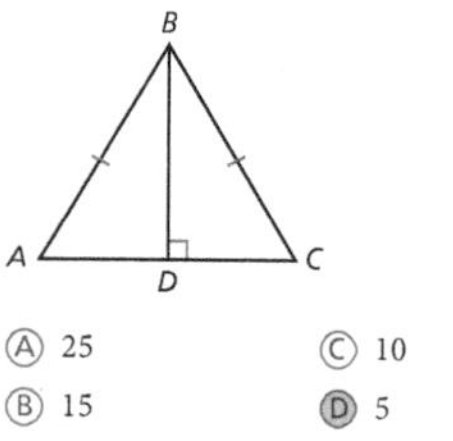

Ⓐ 25
Ⓑ 15
Ⓒ 10
Ⓓ 5

41. Match the name on the left with its correct description on the right. A. 2; B. 1; C. 3; D. 4

A. parallel lines
B. perpendicular lines
C. transversal
D. perpendicular bisector

1. intersect at a 90° angle
2. do not intersect
3. intersects two lines at two different points
4. perpendicular to a segment at the midpoint

 **I'm in a Learning Mindset!**

What can I contribute to my learning community regarding how to use parallel lines to solve real-world problems?

114

Keep Going ▶ Journal and Practice Workbook

Learning Mindset

Strategic Help-Seeking **Identifies Sources of Help**

Making the transition between understanding how to solve a mathematical exercise and understanding how to apply the same principles to real-world problems can present a learning hurdle for some students. Sometimes it is simply a matter of understanding how mathematical terms can manifest in the real world. Have students look around the classroom and point out examples that relate to this lesson. Then extend those same concepts to things they can visualize from the outside world. *Look at your textbook. What parts of the front cover could be thought of as parallel lines? (the edges) If the edges are parallel lines, what is the plane that includes both the top and bottom edges? (the front cover) Think of the ramp at the front entrance to the school. What measurements would you have to make to determine its slope? How can you prove that the painted lines between parking spaces are parallel? How can you use parallel lines on a football field to tell how close the ball is to the end zone?*

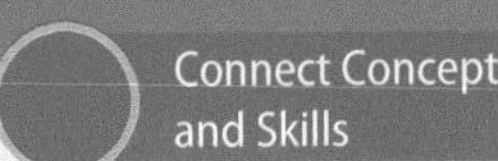

4.2 Slope and Equations of Perpendicular Lines

LESSON FOCUS AND COHERENCE

Mathematics Standards

- Prove the slope criteria for parallel and perpendicular lines and use them to solve geometric problems (e.g., find the equation of a line parallel or perpendicular to a given line that passes through a given point).

Mathematical Practices and Processes 

- Reason abstractly and quantitatively.
- Construct viable arguments and critique the reasoning of others.
- Look for and make use of structure.
- Model with mathematics.

I Can Objective 

I can use slope to write the equation of a line that is perpendicular to a given line.

Learning Objective

Write the equation of a line that is perpendicular to a given line.

Language Objective

Explain the generalization that can be made about the slopes of perpendicular lines.

Vocabulary

Review: perpendicular lines

Mathematical Progressions

Prior Learning	Current Development	Future Connections
Students: • explained why the slope is the same between any two distinct points on a nonvertical line in the coordinate plane. **(Gr8, 5.1)** • wrote equations of lines in slope-intercept form. **(Alg1, 3.2)**	**Students:** • recognize slopes of perpendicular lines as being opposite reciprocals. • use slope to determine if lines are horizontal or vertical. • use slope to write the equation of a line that is perpendicular to a given line.	**Students:** • will write coordinate proofs about triangle relationships. **(9.4 and 9.5)** • will write coordinate proofs about parallelograms. **(11.1–11.5)**

PROFESSIONAL LEARNING

Visualizing the Math

The more students are able to see the graphs of perpendicular lines, find their slopes, and compare them, the better chance they will have of retaining the fact that the slopes are opposite reciprocals. The graph to the right shows two perpendicular lines. The slopes of the lines are 2 and $-\frac{1}{2}$, which are opposite reciprocals.

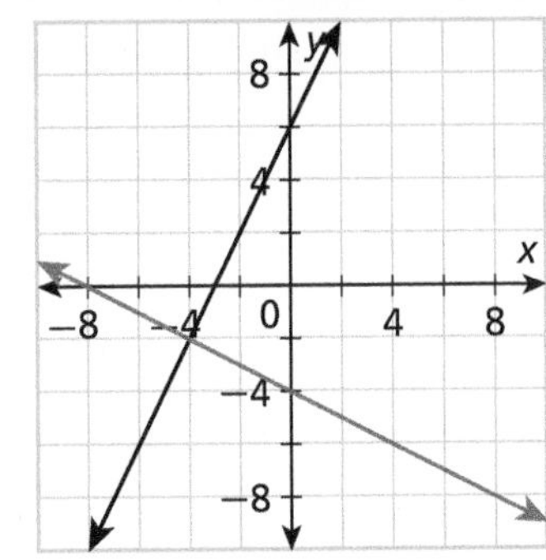

WARM-UP OPTIONS

ONLINE

PROJECTABLE & PRINTABLE

ACTIVATE PRIOR KNOWLEDGE • Write Equations in Slope-Intercept Form

Use these activities to quickly assess and activate prior knowledge as needed.

Problem of the Day 

The cost of admission to a concert is $65 for 3 children and 5 adults. If x is the cost of 1 child and y is the cost of 1 adult, write a linear equation in slope-intercept form to represent this situation. The admission cost for 3 children and 5 adults is $65. Symbolically, this can be written as $3x + 5y = 65$. To get the equation in slope intercept form, you must solve for y;

$y = -\frac{3}{5}x + 13$

Quick Check for Homework

As part of your daily routine, you may want to display the Teacher Solution Key to have students check their homework.

Make Connections

Based on students' responses to the Problem of the Day, choose one of the following:

1. Project the Interactive Reteach, Algebra 1, Lesson 3.2.
2. Complete the Prerequisite Skills Activity:

Have students work in pairs. Give each pair a list of linear equations written in the form $ax + by = c$, such as $3x + 4y = 16$. Ask students to practice writing each equation in slope-intercept form. Once they have all of the equations in $y = mx + b$ form, ask them to use inverse operations to change the equations back into standard form.

- *What do you have to do to change an equation written in standard form into an equation written in slope-intercept form?* Solve the equation for y.
- *What is the first step? What property do you use?* First use inverse operations to get the x-term to the other side; the Addition or Subtraction Property of Equality
- *What is the next step? What property do you use?* Divide all three terms of the equation by the coefficient of y; the Division Property of Equality

SHARPEN SKILLS

If time permits, use this on-level activity to build fluency and practice basic skills.

Mental Math

Objective: Students use mental math to determine the slope of a line.
Materials: index cards

Stand in front of the room with a set of index cards. Each card has a linear equation written on it in standard form. Display the first card and ask for volunteers to state the slope of the line without doing any calculations on paper. Some students may verbally give the steps that must be performed in order to get y by itself, which is fine. But the objective is to perform the inverse operations mentally to determine the slope of the line.

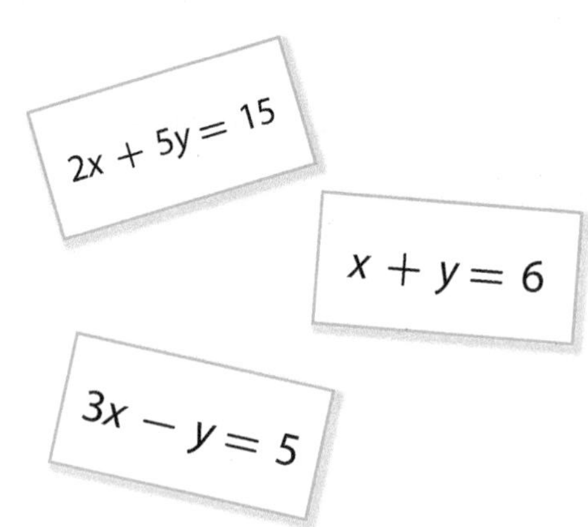

PLAN FOR DIFFERENTIATED INSTRUCTION

Small-Group Options

Use these teacher-guided activities with pulled small groups.

On Track

Materials: index cards

Give each student a card that shows either a linear equation or an ordered pair. Ask students to pair up with another student so that each pair has an equation and an ordered pair to work with. Ask pairs to work together to write the equation of a line that is perpendicular to the line shown on the index card but that passes through the ordered pair shown on the second index card.

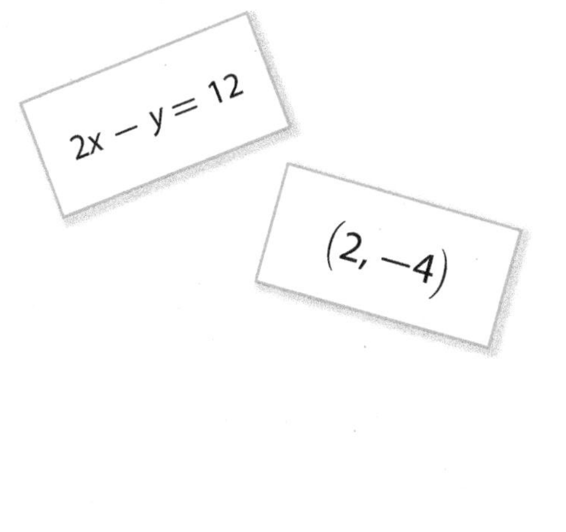

Almost There

Materials: 0.5-centimeter grid paper (Teacher Resource Masters)

Have students work in pairs. Give each pair a sheet of grid paper. Have students do the following:

- One student draws a line on the grid whose *y*-intercept is an integer.
- That same student also plots a single point, not on the line, somewhere on the grid.
- The second student writes the equation of a line perpendicular to the line drawn that passes through the single point plotted.
- Reverse roles.

Ready for More

Write the equation of the given line: $ax - by = 24$. Have students find the values of a and b for which the following are true:

- $2y + 6 = -x$ is perpendicular to the given line.
- The given line passes through the point (0, 4).

Math Center Options

Use these student self-directed activities at centers or stations. Key: ● Print Resources ● Online Resources

On Track

- ● Interactive Digital Lesson
- ●● Journal and Practice Workbook
- ● Interactive Glossary (printable): **perpendicular lines**
- ● Module Performance Task

Almost There

- ● Reteach 4.2 (printable)
- ● Interactive Reteach 4.2
- ● Illustrative Mathematics: Slope Criterion for Perpendicular Lines

Ready for More

- ● Challenge 4.2 (printable)
- ● Interactive Challenge 4.2
- ● Illustrative Mathematics: When are two lines perpendicular?

Unit Project Check students' progress by asking what slope they are using for the equation they are to write.

View data-driven grouping recommendations and assign differentiation resources.

During the *Spark Your Learning*, listen and watch for strategies students use. See samples of student work on this page.

Use Coordinate Grid

Strategy 1

$y = \frac{2}{3}x$ is the given equation of one of the windmill blades.
The line representing the perpendicular blade goes through the points (0, 0) and (−2, 3). Drawing a vertical line from (−2, 3) to the x-axis, the rise over the run is $\frac{3}{2}$. But since this blade is going downhill from left to right, it must have a negative slope. So, the perpendicular blade has an equation with a slope of $-\frac{3}{2}$ and a y-intercept of 0. $y = -\frac{3}{2}x$

If students . . . use the coordinate grid to solve the problem, they are employing an exemplary understanding of how to write the equation of a line when given its graph from Grade 8.

Have these students . . . explain how they determined their equation and how they solved it. **Ask:**

- Q How did you know the relationship of the lines representing the windmill blades?
- Q What did you do to find the *y*-intercept of the perpendicular blade?

Use Estimation

Strategy 2

$y = \frac{2}{3}x$ is the given equation of one of the windmill blades.
The line representing the perpendicular blade goes through the point (0, 0), so its equation has a y-intercept of 0. It also goes downhill so must have a negative slope. I can estimate the slope to be about $-\frac{3}{2}$. $y = -\frac{3}{2}x$

If students . . . use estimation to determine the slope of the line representing the perpendicular blade, they understand that slope can be found by putting the rise over the run but may not know how to use two points on the line to find it.

Activate prior knowledge . . . by having students name a point on the line other than (0, 0). **Ask:**

- Q Can you draw a right triangle where the perpendicular blade is the hypotenuse?
- Q What is the ratio of the height of the triangle to its base?

COMMON ERROR: Uses Wrong Slope

$y = \frac{2}{3}x$ is the given equation of one of the windmill blades.
The line representing the perpendicular blade goes through the points (0, 0) and (−2, 3). Drawing a vertical line from (−2, 3) to the x-axis, the rise over the run is $\frac{3}{2}$. So, the perpendicular blade has an equation with a slope of $\frac{3}{2}$ and a y-intercept of 0. $y = \frac{3}{2}x$

If students . . . use a positive slope in their equation instead of a negative slope, then they do not understand, mathematically or visually, that perpendicular lines must have slopes that are opposite signs.

Then intervene . . . by pointing out that a line that rises to the right has a positive slope and a line that falls to the right has a negative slope. **Ask:**

- Q When looking at the perpendicular blade from left to right, does it rise to the right or fall to the right?
- Q What must you place in front of the slope if the line is decreasing?

4.2

Slope and Equations of Perpendicular Lines

I Can use slope to write the equation of a line that is perpendicular to a given line.

Spark Your Learning

Windmills have evenly-spaced blades to maximize capturing wind energy and to maintain balance while the blades are turning.

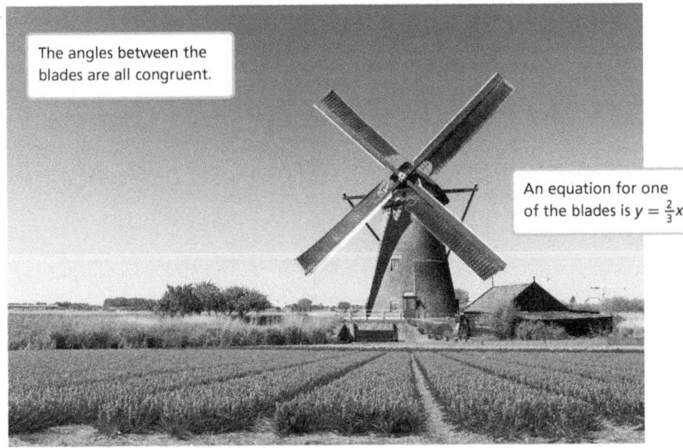

The angles between the blades are all congruent.
An equation for one of the blades is $y = \frac{2}{3}x$.

©Neirfy/Shutterstock

Complete Part A as a whole class. Then complete Parts B–D in small groups.

A. What is a mathematical question you can ask about this windmill? What information would you need to know to investigate your question?

B. Lines representing the blades would have what kind of relationship? Explain how you know. See Additional Answers.

C. Would the same relationship between the lines be true for a windmill with six blades? How do you know? See Additional Answers.

D. To answer your question, what strategy and tool would you use along with all the information you have? What answer do you get? See Strategies 1 and 2 on the facing page.

A. What is an equation for the other blade?; angle measures and the relationship between the lines representing the blades

Turn and Talk Suppose you know that each of the four blades is 32 feet long. What can you determine about the center point of all four blades by using the Perpendicular Bisector Theorem or its converse? See margin.

CULTIVATE CONVERSATION • Three Reads

Tell students to read the information given in the photo three times and prompt them with a different question each time.

1. What is the situation about? This situation involves determining the relationship that exists between two lines created by two windmill blades.
2. What important piece of information is given in the photo that can be used to help you solve the problem? The photo gives the equation of one of the lines representing one of the windmill blades and that the angles formed by the lines are congruent.
3. What prior knowledge do you have that could help solve the problem? We know that a linear pair of angles that are congruent must each have a measure of 90 degrees and that lines that meet at a right angle are called perpendicular lines.

1 Spark Your Learning

▶ MOTIVATE

- Have students look at the photo in their books and read the information contained in the photo. Then complete Part A as a whole-class discussion.
- Give the class the additional information they need to solve the problem. This information is available online as a printable and projectable page in the Teacher Resources.
- Have students work in small groups to complete Parts B–D.

▶ PERSEVERE

If students need support, guide them by asking:

Q **Advancing • Use Tools** Which tool could you use to solve the problem? Why choose that tool and not some other? Students' choices of tools and reasons for choosing them will vary.

Q **Assessing** What type of angles are formed by the two lines represented by the blades? The angles are right angles.

Q **Assessing** What do you call two lines that meet at a right angle? perpendicular lines

Q **Advancing** What generalization can be made about the slopes of perpendicular lines? Perpendicular lines have slopes where one is positive and the other is negative.

Turn and Talk Students should review what a perpendicular bisector and a midpoint are when they attempt to reason and respond about what can be determined using the given information. Possible answer: For each line segment formed by two continuous blades, the center point of the windmill is equidistant from both endpoints. So, the center point lies on the perpendicular bisector of each segment. The line segments formed by two continuous blades are then perpendicular bisectors of each other.

▶ BUILD SHARED UNDERSTANDING

Select groups of students who used various strategies and tools to share with the class how they solved the problem. As they present their solutions, have each group discuss why they chose a specific strategy and tool.

② Learn Together

Build Understanding

Task 1 (MP) **Reason** Students make sense of the parameters of an equation, break apart the equation into pieces that have meaning, and interpret a situation given visually in order to represent the situation symbolically.

CONNECT TO VOCABULARY

Have students use the **Interactive Glossary** to record their understanding of the vocabulary in this task.

Sample Guided Discussion:

Q How do you know if the slope of a line is positive or negative? An increasing line has a positive slope and a decreasing line has a negative slope.

Turn and Talk Students should review the difference between horizontal and vertical lines, recalling that when an equation starts with x and is set equal to a constant, it is a vertical line that crosses the x-axis at that constant. When an equation starts with y and is set equal to a constant, it is a horizontal line that crosses the y-axis at that constant. Possible answer: $y = 2$; An infinite number of lines fit this description. It is possible if I know all vertical lines have an undefined slope, and all horizontal lines have a slope of 0. All vertical lines are perpendicular to all horizontal lines, so any horizontal line will be perpendicular to the given line.

Build Understanding

Investigate Properties of Perpendicular Lines and Segments

Perpendicular lines are two lines in the same plane that intersect to form 90° angles. When given the equations of two lines, you can determine if they are perpendicular by examining the slopes of the lines.

The figure shows a pair of perpendicular lines, along with the equation of each line. The slopes of the lines are 2 and $-\frac{1}{2}$. Do those slopes appear to have any special relationship? You will examine another case in Task 1 below.

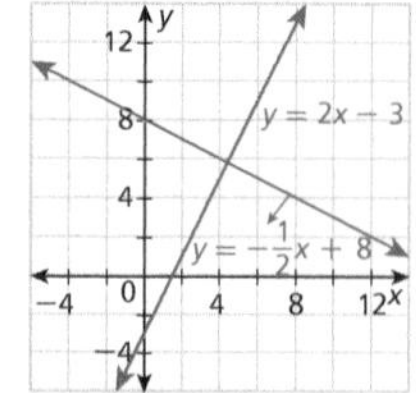

1 A. Examine the figure on the coordinate plane. Copy the table and write the slope for each line or segment.
A–D. See Additional Answers.

Name	Slope
$\overleftrightarrow{AD}$	?
$\overline{AB}$	?
$\overline{BC}$	?
$\overleftrightarrow{CD}$	?

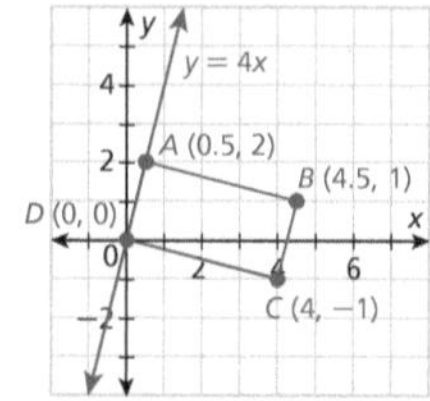

B. What do you notice about the slopes of $\overleftrightarrow{AD}$ and $\overline{BC}$? What do you notice about the slopes of $\overline{AB}$ and $\overleftrightarrow{CD}$? Is there a general relationship between the two slope values?

C. Think about the general relationship you found between the slopes of the perpendicular lines. How could you use this relationship to determine if two lines are perpendicular when presented with the equations of the lines?

D. Suppose you have the set of lines $y = 3x - 0.5$, $x + 3y = 6$, and $y = x + 3$. Which lines, if any, are perpendicular? How do you know? If any are perpendicular, graph them on a coordinate plane.

Turn and Talk What is an equation of a line perpendicular to $x = -4$? How many lines fit this description? How can you use the slopes of the lines to support your answer? See margin.

LEVELED QUESTIONS

Depth of Knowledge (DOK)	Leveled Questions	What Does This Tell You?
Level 1 **Recall**	A line has the equation $y = \frac{4}{7}x - 10$. What is the slope of any line perpendicular to this line? $-\frac{7}{4}$	Students' answers will indicate whether they understand slopes of perpendicular lines.
Level 2 **Basic Application of Skills & Concepts**	What is the equation of a line that is perpendicular to $2x + 4y = 12$ and passes through the point $(0, 5)$? $y = 2x + 5$	Students' answers will indicate whether they understand equations of perpendicular lines.
Level 3 **Strategic Thinking & Complex Reasoning**	A student was given the equation $x - 3y = -6$ and writes the equation of a line perpendicular to it that passes through the point $(4, 3)$. The student writes the equation $y = 3x - 9$. What mistake did they make? The student used the reciprocal of the slope instead of the opposite reciprocal.	Students' answers will indicate whether they understand the relationship between perpendicular lines and their slopes and if they can draw valid conclusions and explain their reasoning.

Prove the Slope Criteria for Perpendicular Lines Theorem

In Task 1, you discovered lines with a slope of 4 are perpendicular to lines with a slope of $-\frac{1}{4}$. In general, the slopes of perpendicular lines are opposite reciprocals. Notice that the product of any pair of opposite reciprocals is -1. So, $4 \cdot \left(-\frac{1}{4}\right) = -1$.

Slope Criteria for Perpendicular Lines
The product of the slopes m of any two nonvertical perpendicular lines is -1. $m_1 \cdot m_2 = -1$

 **Given:** Nonvertical lines p and q are perpendicular.

Prove: The product of their slopes is -1.

The quilt shown includes triangles you can use to verify the slope criteria. Mark the point (a, b) along line p. Draw a line from the point to the x-axis to create a right triangle.

For line p, the slope is $\frac{b}{a}$.

The quilt shows a rotation of the triangle by 90° about the origin. The image is a right triangle with the hypotenuse on top of line q. The coordinate where the triangle intersects line q is $(-b, a)$. The slope of line q is $\frac{a}{-b} = -\frac{a}{b}$.

Show that the product of the slopes of lines p and q is -1.

$$(\text{slope of line } p) \cdot (\text{slope of line } q) = -1$$

$$\frac{b}{a} \cdot \left(-\frac{a}{b}\right) \stackrel{?}{=} -1$$

$$-\frac{ab}{ab} = -1 \checkmark$$

A–C. See Additional Answers.

So, if two lines are perpendicular, the product of their slopes must be -1.

A. How do you know the triangle created by the image after the rotation is a right triangle with the same side lengths as the preimage?

B. Is the following statement always true? "If two lines are perpendicular and one line has a negative slope, the other must have a positive slope." Why?

C. Does the theorem apply when one of the lines is vertical? Why or why not?

 **Turn and Talk** In the proof, what could change if you started with a line that has a negative slope? Will the proof still be valid? See margin.

PROFICIENCY LEVEL

Beginning

Give students a pair of numbers, such as 6 and −6, or $\frac{1}{2}$ and −2, and ask them to classify those numbers as opposites, reciprocals, opposite reciprocals, or none of these. Be sure to give several examples of each.

Intermediate

Have students work in groups. Give each group a set of index cards that show a number on each. Ask groups to place the cards face up in front of them. Students take turns choosing two cards and explaining why the numbers displayed on the cards either are or are not opposite reciprocals.

Advanced

Have students explain in words how to write the equation of a line that is perpendicular to $x - y = 3$ and passes through the point $(0, -3)$.

Step It Out

Task 2 (MP) **Construct Arguments** Students understand and use stated assumptions to construct valid arguments. They justify their conclusions, communicate them to others, and respond to the arguments of others.

Sample Guided Discussion:

Q **What sign does a product have when one factor is positive and the other is negative?** The product is negative.

Q **What does the Inverse Property of Multiplication tell us?** When you multiply a number by its reciprocal, the product is always 1.

Turn and Talk Students should review the concept of rotations by checking to see if the proof works for a variety of rotations, both clockwise and counterclockwise. Possible answer: We could rotate the preimage 90° clockwise rather than counterclockwise to create the image. The proof would still be valid.

Task 3 **MP** **Use Structure** Students will use what they know about the structure of a linear equation and its parameters to algebraically write equations of lines that are perpendicular to a given line.

Sample Guided Discussion:

Q **How do you find the slope of a line when given two points on the line?** Divide the change in *y* by the change in *x*.

Q **If you know the slope of a line and a point that lies on the line, how can you solve for the missing *y*-intercept?** You can replace the variables *y*, *m*, and *x* in the equation $y = mx + b$ with the values given for each and then solve the equation for *b*.

Turn and Talk Have students consider both the slope and *y*-intercept when predicting how the equation would change. The *y*-intercept would change.

$y = mx + b$
$2 = 4(2) + b$
$2 = 8 + b$
$b = -6$

$y = mx + b$
$y = 4x - 6$

Task 4 **MP** **Use Structure** Students will use what they know about the characteristics of functions and the equations of vertical and horizontal lines to make generalizations about special cases of perpendicular lines.

Sample Guided Discussion:

Q **Why is the slope of a vertical line undefined?** The change in *y* divided by the change in *x* results in a denominator of 0, which means the slope is undefined.

Q **On the horizontal line $y = 4$, what is the *y*-coordinate of every point that lies on the line?** 4

Q **On the vertical line $x = 4$, what is the *x*-coordinate of every point that lies on the line?** 4

Step It Out

Write Equations of Perpendicular Lines

3 Write an equation of the line perpendicular to $\overline{AB}$ that passes through $(-1, -1)$.

A (−4, 2)
B (2, 0.5)
(0, 1)
(−1, −1)

First, identify two points on the given segment.

$(-4, 2)$ and $(0, 1)$

Find the slope of $\overline{AB}$.

$m = \frac{\text{rise}}{\text{run}} = -\frac{1}{4}$

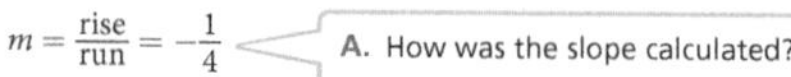
A. How was the slope calculated?

Find the slope of the perpendicular line segment.

The slope of $\overline{AB}$ is $-\frac{1}{4}$, so the slope of a perpendicular line is 4.

B. Why is the slope of the perpendicular line 4?

Write the equation of the perpendicular line segment.

Use $(-1, -1)$ and the slope-intercept form to find b.

$y = mx + b$

$-1 = 4(-1) + b$

$-1 = -4 + b$, or $b = 3$

Finally, substitute the slope and the calculated value of b back into slope-intercept form.

$y = mx + b$

$y = 4x + 3$

C. Why do you substitute these values back into the general slope-intercept form of a line?

The equation of the line is $y = 4x + 3$.

A–C. See Additional Answers.

Turn and Talk Explain how the equation would change if the perpendicular line passed through the point $(2, 2)$ instead. See margin.

4 Vertical and horizontal lines are a special case of perpendicular lines. Write an equation for the line perpendicular to $y = -3$ that passes through the point $(5, 7)$.

Two methods of finding the equation of the line are shown.

$y = -3$, or $y = 0x - 3$, so $m = 0$	$y = -3$
The opposite reciprocal slope of $m = 0$ is $m = -\frac{1}{0}$, which is undefined. Therefore, the answer must be a vertical line. $x = a$ is the vertical line that passes through (a, b), so the line perpendicular to $y = -3$ that passes through the point $(5, 7)$ is $x = 5$.	I know the line $y = -3$ is a horizontal line. I know that vertical lines are perpendicular to horizontal lines, so the answer must be a vertical line. The *x*-coordinate given in the problem is 5. Therefore the line perpendicular to $y = -3$ that passes through point $(5, 7)$ is the vertical line $x = 5$.

A, B. See Additional Answers.

A. What are the coordinates of the point where the two lines intersect?

B. Explain why the vertical line passes through the *x*-coordinate given in the problem.

118

Apply the Criteria for Slopes of Perpendicular Lines

When solving a problem involving equations of perpendicular lines, first determine how to use the given information to find the answer. For example, you may need to find the slope of a given line, and then use the slope to write an equation for a perpendicular line through a given point.

5 A ship in distress is located at the point shown on the grid. The quickest way to reach the shoreline is to direct the ship along a line perpendicular to the shore. What is the equation of the line the navigator must use to sail the ship perpendicular to the shoreline from their current location?

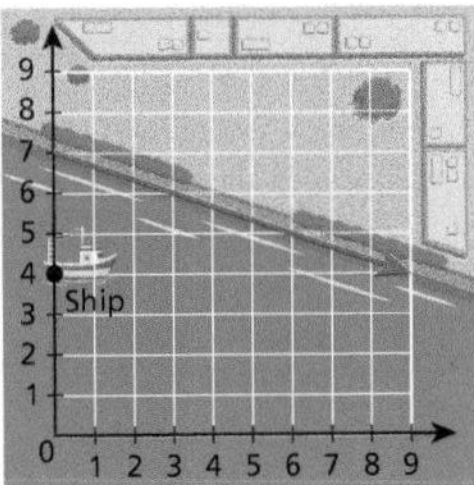

Write an equation for the shoreline.

The line has a slope of $-\frac{1}{3}$, and the y-intercept is 7.

$y = -\frac{1}{3}x + 7$

A. How do you know this is the slope?

Find the slope of the line the captain will follow.

The line the captain will follow has a slope of 3, the opposite reciprocal of $-\frac{1}{3}$.

B. Why must you find the slope before you can write the equation?

Write the equation of the perpendicular line the ship will follow.

The ship is located at $(0, 4)$, so the perpendicular line should pass through this point. This point lies on the y-axis, so the y-intercept is 4. The equation of the line the ship will follow is $y = 3x + 4$.

C. How would this step be different if the given point was not the y-intercept?

A–C. See Additional Answers.

Turn and Talk At what point would the two lines intersect? What are two different methods you can use to find the solution? See margin.

Task 5 (MP) **Model with Mathematics** Students will apply mathematics they know to solve problems arising in everyday life and identify important quantities in a practical situation in order to arrive at a valid mathematical conclusion.

Sample Guided Discussion:

Q How do you find the slope of a line when given a graph? Choose two points on the line and divide the difference in the y-coordinates by the difference in the x-coordinates.

Q Why is it easier to write the equations of perpendicular lines when you are given the y-intercept? It is easier because you do not have to use algebra to solve for the y-intercept because you already know both parameters, slope and y-intercept.

Turn and Talk Encourage students to first try and find the intersection point by using the equations and not the graph. Students should recall how to solve a system of linear equations algebraically, a skill they practiced in Algebra 1. The lines intersect at (1.5, 8.5). I can determine this using the figure or by solving a system of equations using the equation of each line.

ONLINE

Assign the Digital On Your Own for
- built-in student supports
- Actionable Item Reports
- Standards Analysis Reports

On Your Own

Assignment Guide

The chart below indicates which problems in the On Your Own are associated with each task in the Learn Together. Assign daily homework for tasks completed.

Learn Together Tasks	On Your Own Problems
Task 1, p. 116	Problems 13, 19c, and 21
Task 2, p. 117	Problems 12 and 23
Task 3, p. 118	Problems 7–10, 19a–b, 20, 22, and 27
Task 4, p. 118	Problems 6, 11, 14, and 15
Task 5, p. 119	Problems 24–26

Check Understanding

1. Find the slopes of the line and a line perpendicular to $y = -x + 20$. slope of the given line: -1; slope of a perpendicular line: 1
2. If k is a constant, is the line $y = mx + k$ perpendicular to the line $y = -mx + k$? Explain your reasoning. no; m and $-m$ are opposites, not opposite reciprocals.
3. What is the equation of a line perpendicular to $y = \frac{1}{3}x - 9$ that passes through the point $(0, 1)$? $y = -3x + 1$
4. Write an equation for a line perpendicular to $y = 7$ that passes through the point $(2, 13)$. $x = 2$
5. On a map, Peach Street is modeled by the equation $4x - y = 7$. Apple Street is perpendicular to Peach Street and passes through the point $(12, 2)$. Find the equation that models Apple Street. $y = -\frac{1}{4}x + 5$

On Your Own

Write the equation of a line perpendicular to the line that passes through the given point.

6. $y = 2$ $(1, -3)$ $x = 1$
7. $y = -x + 20$ $(-4, 2)$ $y = x + 6$
8. $y = -5x - \frac{4}{3}$ $(6, -4)$ $y = \frac{1}{5}x - \frac{26}{5}$
9. $3x - 2y = 10$ $(7, 5)$ $y = -\frac{2}{3}x + \frac{29}{3}$
10. $\frac{1}{2}y - 8x + 3 = 0$ $(-8, 0)$ $y = -\frac{1}{16}x - \frac{1}{2}$
11. $12x = 33$ $(0, 0)$ $y = 0$

For Problems 12—18, determine whether each statement is *always, sometimes,* or *never* true.

12. A line with slope $\frac{a}{b}$ and a line with slope $-\frac{b}{a}$, where $a \neq 0$ and $b \neq 0$, are perpendicular. always
13. Two perpendicular lines in the same plane intersect just once. always
14. The line $y = c$, where c is a constant, is perpendicular to the line $y = x$. never
15. A line perpendicular to another line has an undefined slope. sometimes
16. The sum of the slopes of perpendicular lines is 0. sometimes
17. Two lines are both parallel and perpendicular. never
18. If two lines are perpendicular, exactly one of the lines has a negative slope. sometimes
19. The endpoints of one side of square $PQRS$ are at $P(1, 1)$ and $Q(3, 4)$.
    A. What is the equation for the line that contains $\overline{QR}$? $y = -\frac{2}{3}x + 6$
    B. What is the equation for the line that contains $\overline{SP}$? $y = -\frac{2}{3}x + \frac{5}{3}$
    C. Are $\overline{PQ}$ and $\overline{QR}$ parallel or perpendicular? Write an equation using their slopes that proves your choice.
    perpendicular; $\frac{3}{2}\left(-\frac{2}{3}\right) = -\frac{6}{6} = -1$

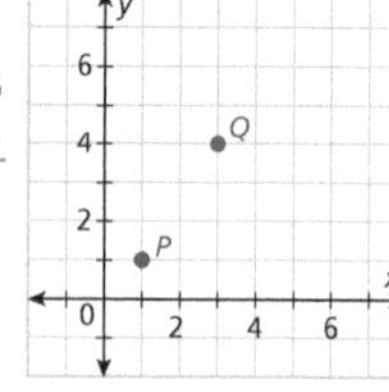

120

data checkpoint

3 Check Understanding

Formative Assessment

Use formative assessment to determine if your students are successful with this lesson's learning objective.

Students who successfully complete the Check Understanding can continue to the On Your Own practice.

For students who miss 1 problem or more, work in a pulled small group using the Almost There small-group activity on page 115C.

ONLINE

Assign the Digital Check Understanding to determine
- success with the learning objective
- items to review
- grouping and differentiation resources

4 Differentiation Options

Differentiate instruction for all students using small-group activities and math center activities on page 115C.

Reteach

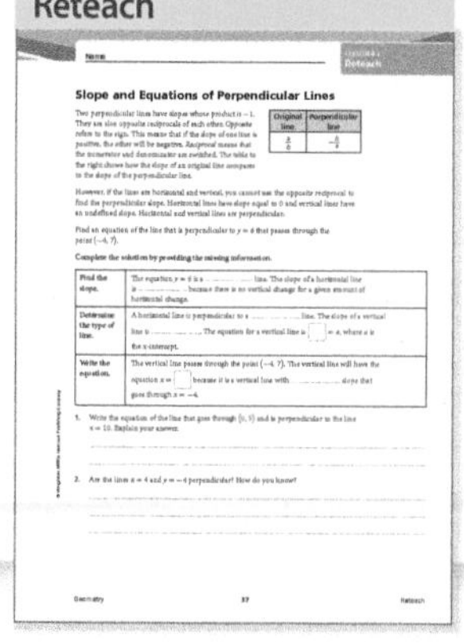

Challenge

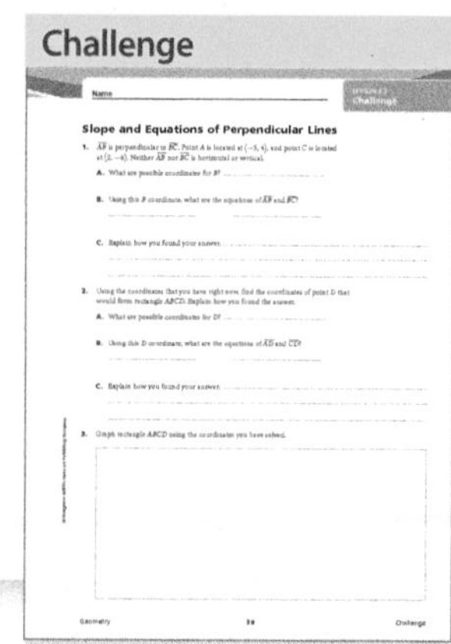

20–23. See Additional Answers.

20. Consider the line segments in the graph.

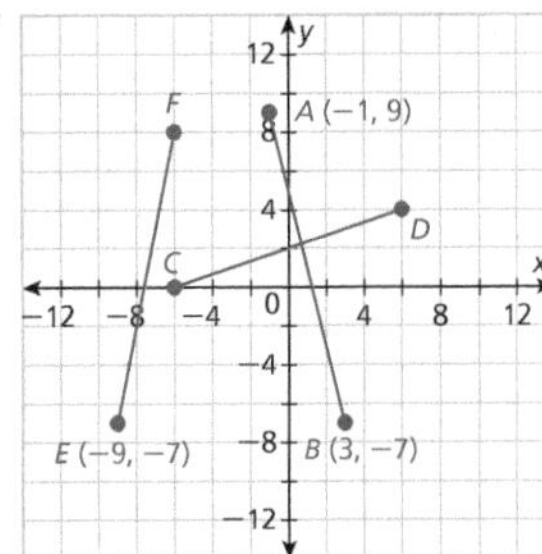

A. Are any of the line segments perpendicular? Justify your answer.

B. Write equations for $\overline{AB}$ and a line perpendicular to $\overline{AB}$ that passes through point $(8, -27)$.

C. Write equations for $\overline{CD}$ and a line perpendicular to $\overline{CD}$ that passes through point $(-9, -10)$.

D. Write equations for $\overline{EF}$ and a line perpendicular to $\overline{EF}$ that passes through point $(15, 0)$.

E. Where could you place point G so that $\overline{FG}$ is perpendicular to $\overline{CD}$?

21. (MP) **Use Repeated Reasoning** Is it possible for segments along the lines $y = 4x - 7$, $y = -\frac{1}{4}x + 2$, $y = -4x + 2$, and $y = -\frac{1}{4}x + 2$ to form a rectangle? Justify your answer.

22. (MP) **Construct Arguments** Check Michael's work for errors. He wants to write an equation of the line perpendicular to $6x + 8y = -4$ that passes through the point $(-2, 6)$ but isn't sure if he's made any mistakes.

Step 1	Step 2	Step 3
$-4 = 6x + 8y$	$y = mx + b$	$y = mx + b$
$y = -\frac{3}{4}x - \frac{1}{2}$	$6 = -\frac{4}{3}(-2) + b$	$y = -\frac{4}{3}x + \frac{10}{3}$
	$b = \frac{10}{3}$	

A. In which step(s) does Michael make an error, if any? Describe the error(s).

B. If there are any errors, correct them and write the correct equation.

23. The converse of the Slope Criteria of Perpendicular Lines Theorem states, "If the slopes of two lines are opposite reciprocals, the two lines are perpendicular." Write a proof to represent the converse of the Slope Criteria for Perpendicular Lines Theorem.

24. A helicopter landing pad is often a large circle marked with an "H". In a scale drawing of the ship's landing pad, the segment of $x = -5$ from $y = -2$ to $y = 6$ represents the vertical left leg of the H. The horizontal leg bisects both vertical legs and is 6 units long. What segment equations represent the middle leg and the right leg of the H?
middle leg: $y = 2$ from $x = -5$ to $x = 1$;
vertical right leg: $x = 1$ from $y = -2$ to $y = 6$

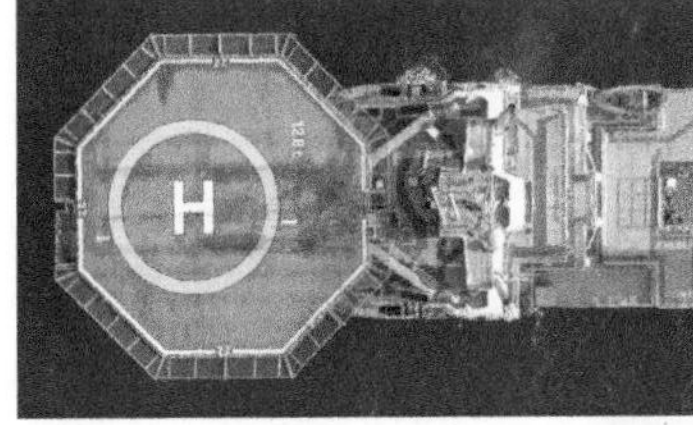

Questioning Strategies

Problem 22 Suppose Michael's work was not provided. What steps would you take to find the answer? I would first put $6x + 8y = -4$ in the form $y = mx + b$ to see what the slope of the line is. Then, I would use the opposite reciprocal of that number and the x- and y-values of the given point to find the y-intercept of the line perpendicular to the given line.

Watch for Common Errors

Problem 20 Sometimes when students are writing the equation of a line that is perpendicular to a given line, they forget that they need to solve for the y-intercept and use the same y-intercept as the one in the given equation, or the y-coordinate of the given point. For example, in Problem 17a, segment $\overline{AB}$ has a slope of -4, so the slope of a line perpendicular to segment $\overline{AB}$ would have a slope of $\frac{1}{4}$. When writing the equation of the perpendicular line, they use a slope of $\frac{1}{4}$, but then write the y-coordinate of the point on the line, -27, as the y-intercept instead of using the new slope and the given point on the line to solve for the y-intercept.

5 Wrap-Up

Summarize learning with your class. Consider using the Exit Ticket, Put It in Writing, or I Can scale.

Exit Ticket 

Line m passes through the points (2, 2) and (−1, 3). Find the slope of this line and write an equation of a line perpendicular to line m but that passes through the point (6, −4).

$$m = \frac{2-3}{2-(-1)} = -\frac{1}{3}$$

$$y = mx + b$$
$$-4 = 3(6) + b$$
$$-4 = 18 + b$$
$$b = -22$$
$$y = 3x - 22.$$

Put It in Writing 

Describe a strategy you can use to prove that two line segments on a grid are perpendicular.

I Can 

The scale below can help you and your students understand their progress on a learning goal.

4	I can use slope to write the equation of a line that is perpendicular to a given line, and I can explain my solutions steps to others.
3	I can use slope to write the equation of a line that is perpendicular to a given line.
2	I can write the slope of a line that is perpendicular to a line with a given equation.
1	I can recognize when two equations represent perpendicular lines by comparing their slopes.

Spiral Review • Assessment Readiness

These questions will help determine if students have retained information taught in the past and can also prepare them for high-stakes assessments. Here, students must prove lines are parallel **(3.2)**, prove lines are perpendicular **(3.3)**, recognize equations of parallel lines **(4.1)**, and use coordinate geometry to find the length of a line segment when given its endpoints **(4.3)**.

25. (MP) **Use Repeated Reasoning** Ivis walks home from school, but today she takes a detour due to construction. Her school is located on a horizontal street at the point (0, 2). Each unit on the coordinate plane represents one block. She walks 3 blocks due south, 2 blocks east, 1 block north, and 2 more blocks east. Her house is located on a vertical street that passes through this point. Find the equation of the line that models her street. $x = 4$

26. A trail through a national park is being extended to include side paths to different natural features. The planners would like to add a perpendicular path that leads to a banyan tree, a unique type of tree that grows new roots from the canopy down to the ground. The trail and tree are modeled on the coordinate plane.

Banyan Tree

A. What are the slopes of the current trail and the new perpendicular path? See Additional Answers.

B. What is the equation for the line that contains the new path to the tree? $y = 3x + 1$

C. A trail marker will be placed where the trail and path intersect. What is the location of the marker? (1.2, 4.6)

D. The planners want to add another side path leading to a small waterfall. This path will be parallel to the path to the banyan tree. Is the path to the waterfall perpendicular to the trail as well? Explain. See Additional Answers.

27. Open Middle™ Using the digits 1 to 9, at most one time each, fill in the boxes to create equations for two lines that are perpendicular. See Additional Answers.

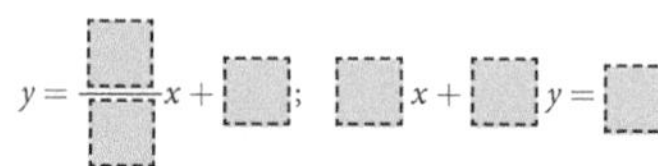

Spiral Review • Assessment Readiness

28. When two lines are cut by a transversal, and corresponding angles are congruent, the two lines are __?__.

Ⓐ perpendicular
Ⓑ parallel
Ⓒ congruent
Ⓓ complementary

29. Which lines are parallel to $5y = -x + 10$? Select all that apply.

Ⓐ $y = -5x + 2$
Ⓑ $-x - 3 = 5y$
Ⓒ $2y = 10x - 9$
Ⓓ $-5y - 5x = 2$
Ⓔ $2x = -10y + 2$
Ⓕ $y = 2x + 5$

30. Any point on a perpendicular bisector of a segment is __?__ from the endpoints of that segment.

Ⓐ equilateral
Ⓑ parallel
Ⓒ perpendicular
Ⓓ equidistant

31. What is the length of the line segment with endpoints $A(-1, 2)$ and $B(-7, 4)$?

Ⓐ $3\sqrt{5}$
Ⓑ 6.5
Ⓒ $2\sqrt{10}$
Ⓓ 5.5

 I'm in a Learning Mindset!

How did I use input from my group members to solve the Spark Your Learning Task from the beginning of the lesson?

122 Keep Going Journal and Practice Workbook

Learning Mindset

Strategic Help-Seeking Identifies Sources of Help

Have a discussion with students about the opportunity they had to collaborate with peers while working on the Spark Your Learning exercise. Discuss how that experience was helpful, as opposed to if they had done the activity by themselves. Collaboration with peers allows a student to learn different strategies that they may not have thought of themselves, but that are an accurate way of obtaining an answer. For example, some students may have used visualization to answer some of questions, where others may have resorted to using a graphing calculator to see a mathematical representation of the windmills. Collaboration should be encouraged not only in the classroom but outside of class as well, on their own time. Point out that peer tutoring is a proven method of increasing one's chances of retaining content that is learned. *Does your stress level increase when you are asked to complete a mathematical task on your own compared to being asked to complete a task with a partner? Were there parts of the Spark Your Learning exercise where a peer had a different strategy than you did? How can working with a partner help you retain new information?*

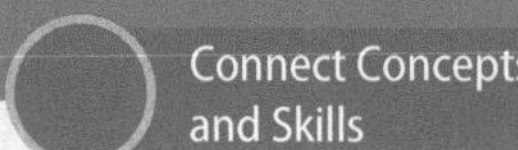

4.3 Write a Coordinate Proof

LESSON FOCUS AND COHERENCE

Mathematics Standards

- Use coordinates to prove simple geometric theorems algebraically.

Mathematical Practices and Processes 

- Construct viable arguments and critique the reasoning of others.
- Attend to precision.
- Look for and express regularity in repeated reasoning.
- Reason abstractly and quantitatively.

I Can Objective 

I can use the Distance Formula to show congruence on the coordinate plane.

Learning Objective

Students will use coordinates to prove simple geometric theorems algebraically.

Language Objective

Given two points, students should be able to explain the process of how to find the distance between the points using the Distance Formula and the Pythagorean Theorem and how to find the midpoint of the line connecting the points using the Midpoint Formula.

Mathematical Progressions

Prior Learning	Current Development	Future Connections
Students: • proved the Pythagorean Theorem. **(Gr8, 11.1)** • wrote an equation of a line given two points. **(Alg1, 4.4)**	**Students:** • explore the concept of distance in the coordinate plane. • apply known information to write a formal proof of a known concept. • use the Distance Formula to determine lengths of segments in the coordinate plane. • use the Distance Formula to determine congruency of sides of a polygon. • use the Distance Formula to solve a real-world problem.	**Students:** • will write coordinate proofs about triangle relationships. **(9.4 and 9.5)** • will write coordinate proofs about parallelograms **(11.1–11.5)**

UNPACKING MATH STANDARDS

Use coordinates to prove simple geometric theorems algebraically.

What It Means to You

Students use coordinates and equations to discover geometric facts. They will establish relationships based on coordinates, such as the distance between two points or the midpoint of the line connecting two points, by coordinate proofs. Because coordinate proofs employ both algebra and geometry, students will begin to make connections between the two disciplines and discover that geometry can be seen through an algebraic lens.

WARM-UP OPTIONS

ONLINE

PROJECTABLE & PRINTABLE 

ACTIVATE PRIOR KNOWLEDGE • Write an Equation of a Line Given Two Points

Use these activities to quickly assess and activate prior knowledge as needed.

Problem of the Day 

Write the equation of the line that passes through the points $(-4, 3)$ and $(5, 12)$. $y = x + 7$; $m = \frac{12 - 3}{5 - (-4)} = \frac{9}{9} = 1$;

$y = mx + b$; $12 = 1(5) + b$; $7 = b$

Quick Check for Homework

As part of your daily routine, you may want to display the Teacher Solution Key to have students check their homework.

Make Connections

Based on students' responses to the Problem of the Day, choose one of the following:

1. Project the Interactive Reteach, Algebra I, Lesson 4.4.

2. Complete the Prerequisite Skills Activity:

Have students work in pairs. Each student writes an ordered pair, and the students work together to find the slope of the line that connects those pairs. Then, each student uses the slope and their ordered pair to solve the equation $y = mx + b$ for b. They write the equation of line that passes through the two points and compare their answers to make sure they are the same.

- *What is the formula to find the slope of a line?* $m = \frac{y_2 - y_1}{x_2 - x_1}$
- *What is the slope-intercept form of a line?* $y = mx + b$
- *What will happen if the slope turns out to be zero or undefined?* If the slope is zero, the line will be a horizontal line in the form of y equals some number. If the slope is undefined, the line will be a vertical line in the form of x equals some number.
- *Why doesn't it matter which point is used to solve for b and write the equation of the line?* Both points are on the same line which only has one equation to describe it.

If students continue to struggle, use Tier 2 Skill 7.

SHARPEN SKILLS

If time permits, use this on-level activity to build fluency and practice basic skills.

Quantitative Comparison

Objective: Students make a comparison between two quantities.

Write the following problem on the board. Ask students to choose the letter representing the correct answer and to explain their reasoning.

Quantity A	**Quantity B**
the distance between $(-5, 12)$ and $(4, 13)$	the distance between $(-4, 3)$ and $(-12, 3)$

A. Quantity A is greater.

B. Quantity B is greater.

C. The two quantities are equal.

D. The relationship cannot be determined from the information given.

A; Quantity A $= \sqrt{82}$ and Quantity B $= 8$.

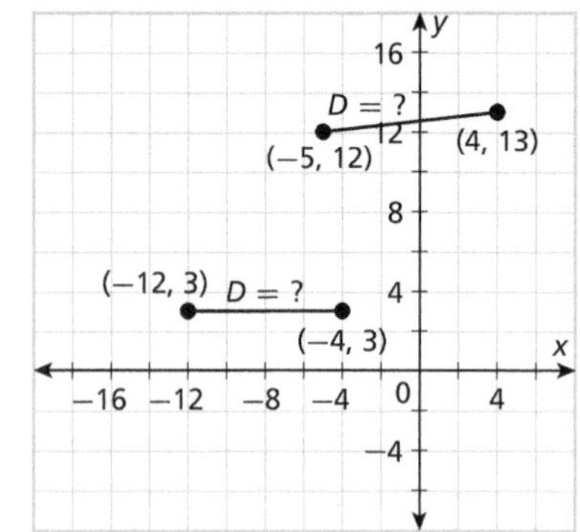

PLAN FOR DIFFERENTIATED INSTRUCTION

Small-Group Options

Use these teacher-guided activities with pulled small groups.

On Track

Materials: index cards

Give each student a card with an ordered pair such as $(3, 4)$ or $(-5, -2)$. Have students pair up and calculate the distance between the points and the midpoint of the line that connects the two points.

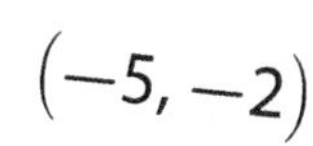

Almost There (RtI)

Write the Distance and Midpoint Formulas and the ordered pairs $(5, -12)$ and $(3, 6)$.

- Have students label the ordered pairs (x_1, y_1) and (x_2, y_2).
- Check that students have plugged correct values into the formulas.
- Have students calculate the distance between the points and the midpoint of the line that connects the two points using the formulas.

Ready for More

One point of a line segment is $(3, 5)$. The midpoint of the line segment is $(7, 12)$.

- What is the other point of the line segment?
- How long is the line segment?
- What is the other coordinate and midpoint of the line segment that has the same length and one coordinate of $(1, 4)$?

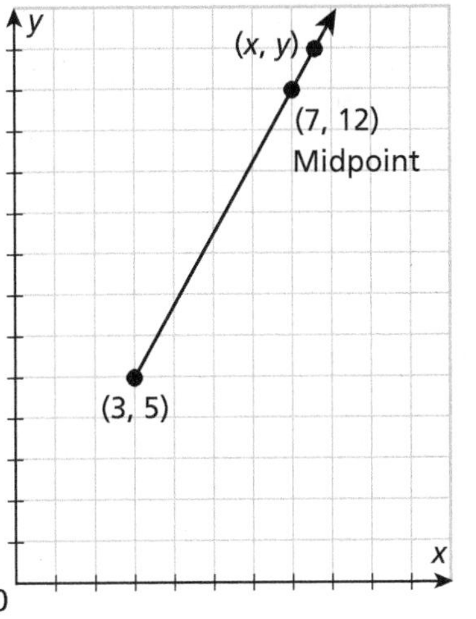

Math Center Options

Use these student self-directed activities at centers or stations. Key: ● Print Resources ● Online Resources

On Track

- ● Interactive Digital Lesson
- ●● Journal and Practice Workbook
- ● Module Performance Task

Almost There

- ● Reteach 4.3 (printable)
- ● Interactive Reteach 4.3
- ● RtI Tier 2 Skill 7: Point-Slope Form

Ready for More

- ● Challenge 4.3 (printable)
- ● Interactive Challenge 4.3
- ● Illustrative Mathematics: A Midpoint Miracle

ONLINE Ed

View data-driven grouping recommendations and assign differentiation resources.

Spark Your Learning • Student Samples

During the *Spark Your Learning*, listen and watch for strategies students use. See samples of student work on this page.

Break Complex Problems into Steps

Strategy 1

Both paths are complex, with many twists and turns. It would be hard to create a single equation to apply the Distance Formula to either entire path. I would break each path into smaller segments. By counting the grid units in each segment, I could apply the Distance Formula to each segment. Then I would add each set of segments to determine the length of the two paths.

If students . . . break the trails into smaller segments, they are demonstrating that they understand that breaking the complex paths into smaller segments with fewer curves will allow their estimations of each segment to be more accurate.

Have these students . . . explain how they chose the smaller segments. **Ask:**

- Q What formula could you use to estimate the length of these segments?
- Q If you forgot the formula, how could you use the information from the coordinate plane to calculate the distance between endpoints of each segment?

Use Tools to Estimate

Strategy 2

Use string to measure the trails.

Place pieces of string along each trail and cut them to be the same length as the trail.

Compare the strings to determine which trail is shorter.

If students . . . use a flexible tool to measure the trails, they are demonstrating that they understand that complex paths are more difficult to measure than straight ones. But, they may not be comfortable with breaking complex problems into smaller steps (paths into segments) or they may not be comfortable using the Distance Formula.

Activate prior knowledge . . . by having students write the Distance Formula for segments of the path. **Ask:**

- Q What shapes would you be trying to create with your segment selections?
- Q How would the Distance Formula apply to each shape?

COMMON ERROR: Oversimplifies the Problem

I decided to estimate the length of each trail by drawing straight lines between each trail's beginning and end and then measuring the line with a ruler.

If students . . . simply make their calculations using the endpoints for each path, they are oversimplifying. They are not taking into account the twists and turns in either path. "As the crow flies" may indicate that Path A is shorter, but hikers may have to go around many more obstacles than if they had taken Path B. The walking distance for Path A may actually be far longer than Path B.

Then intervene . . . by explaining that no estimation can be undertaken without looking at the problem as a whole. **Ask:**

- Q Looking at the map, does either path have more twists and turns that might impact its length?
- Q What other kinds of geographic features might impact the number of steps a hiker might have to take to walk a path from beginning to end?

4.3

Write a Coordinate Proof

I Can use the Distance Formula to show congruence on the coordinate plane.

Spark Your Learning

A state park has two hiking trails to an extinct volcano dome. One trail travels around the north side of the volcano dome, and the other trail travels around the south side.

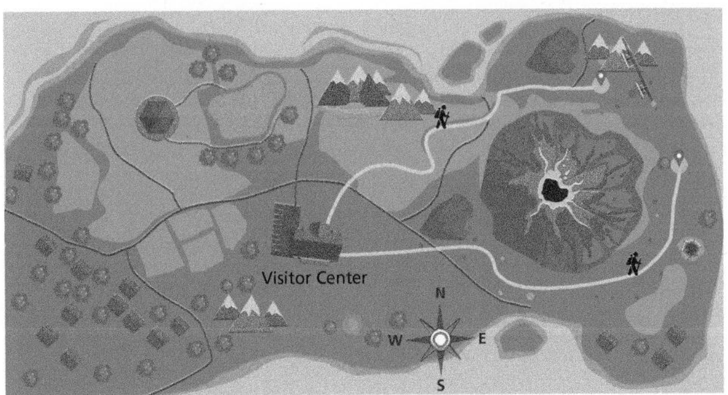

Complete Part A as a whole class. Then complete Parts B–D in small groups.

A. What is a mathematical question you can ask about this situation? What information would you need to know to answer your question? A. Which trail is shorter?; the length of each trail

B. What tools can you use to answer the question? How can you use them to make an estimate? See Additional Answers.

C. How would you use the information in the photo and the additional information your teacher gave you to answer your question? What is the answer? See Strategies 1 and 2 on the facing page.

D. Can you think of any ways to find a more precise answer? D. Possible answer: Estimate the length of smaller sections of the trails one at a time.

Turn and Talk Predict how your answer would change for each of the following changes in the situation: See margin.

- The trails are straight lines.
- The trails have hills or other elevation changes not mapped.
- The scale is 5 miles per unit.

CULTIVATE CONVERSATION • Co-Craft Questions

If students have difficulty formulating a mathematical question about the situation in the Spark Your Learning, ask them to think about how to measure the trail using manipulatives. What are some natural questions to ask about this situation?

Work together to craft the following questions:

- How can a piece of string be used to determine which trail is shorter?
- How can a ruler be used to determine which trail is shorter?
- Why are straight trails easier to measure than curved ones?

Then have students think about what additional information, if any, they would need to answer these questions. **Ask:**

- Can you determine an exact measure of each curved trail? Why or why not?
- Can you determine an exact measure if each trail were straight? Explain.

Connect Concepts and Skills

1 Spark Your Learning

▶ MOTIVATE

- Have students look at the map in their books and read the information contained in the map. Then complete Part A as a whole-class discussion.
- Give the class the additional information they need to solve the problem. This information is available online as a printable and projectable page in the Teacher Resources.
- Have students work in small groups to complete Parts B–D.

▶ PERSEVERE

If students need support, guide them by asking:

Q **Advancing • Use Tools** Which tool could you use to solve the problem? Why choose that tool and not some other? Students' choices of tools and reasons for choosing them will vary.

Q **Assessing** By reading the grid and using the horizontal label first and then the vertical label, where is the volcano dome located? 14H

Turn and Talk Tell students to visualize walking on a trail and encountering a hill. Climbing and descending the hill would involve extra steps, making the distance of the trail longer.

- The new estimate would be more precise.
- Possible answer: The new estimate would be longer because the current estimate did not take into account change in elevation, etc.
- The new estimate would be 10 times as long as the current estimate.

▶ BUILD SHARED UNDERSTANDING

Select groups of students who used various strategies and tools to share with the class how they solved the problem. As they present their solutions, have each group discuss why they chose a specific strategy and tool.

② Learn Together

Build Understanding

Task 1 (MP) **Construct Arguments** Students learn to use stated assumptions and previously established rules to prove the Distance Formula.

Because the proof relies on the construction of a right triangle, show students that it is impossible to draw a right triangle if either $x_1 = x_2$ or $y_1 = y_2$ and thus the assumptions that $x_1 \neq x_2$ and $y_1 \neq y_2$ are needed.

Sample Guided Discussion:

Q **What ordered pair, in terms of the ordered pairs used for *P* and *Q*, represents the third vertex of the triangle?** (x_2, y_1)

Turn and Talk Encourage students to draw a picture. Using visualization is a great first step toward understanding the problem. A sketch of the two points can help students see the resulting right triangle and prompt students to think about the Pythagorean Theorem. Possible answer: Draw a right triangle and use the Pythagorean Theorem to calculate the length of the hypotenuse or distance between the points.

Build Understanding

Prove the Distance Formula

The Pythagorean Theorem states that $c^2 = a^2 + b^2$, where a and b are the lengths of the legs of a right triangle and c is the length of the hypotenuse. You can use the Distance Formula to apply the Pythagorean Theorem to find the distance between points on the coordinate plane.

Distance Formula
The distance between two points (x_1, y_1) and (x_2, y_2) on the coordinate plane is given by $\sqrt{(x_2 - x_1)^2 + (y_2 - y_1)^2}$.

A coordinate proof is one that uses both coordinate geometry and algebra. You can use a coordinate proof to place a figure on the coordinate plane to prove the Distance Formula.

1 **Given:** $P(x_1, y_1)$ and $Q(x_2, y_2)$, $x_1 \neq x_2$ and $y_1 \neq y_2$

Prove: $PQ = \sqrt{(x_2 - x_1)^2 + (y_2 - y_1)^2}$

Begin by finding the length of each side of the right triangle by using the coordinates of each vertex.

The length of $\overline{PR}$ is the difference of the x-coordinates of the two endpoints: $PR = x_2 - x_1$.

The length of $\overline{QR}$ is the difference of the x-coordinates of the two endpoints: $QR = y_2 - y_1$.

The length of the hypotenuse of the triangle is PQ.

Write the Pythagorean Theorem and then substitute the lengths in terms of the coordinates.

$$c^2 = a^2 + b^2$$
$$PQ^2 = PR^2 + QR^2$$
$$PQ^2 = (x_2 - x_1)^2 + (y_2 - y_1)^2$$
$$PQ = \sqrt{(x_2 - x_1)^2 + (y_2 - y_1)^2}$$

Therefore, for any two points on the coordinate plane, the distance between the points is given by $\sqrt{(x_2 - x_1)^2 + (y_2 - y_1)^2}$.

A. How do you know $\triangle PQR$ is a right triangle? A, B. See Additional Answers.

B. Suppose the triangle is drawn in a different quadrant, and $x_2 - x_1$ is a negative number. Is the proof still valid? Why or why not?

Turn and Talk How can you find the distance between two points on the coordinate plane if you forget the Distance Formula? See margin.

LEVELED QUESTIONS

Depth of Knowledge (DOK)	Leveled Questions	What Does This Tell You?
Level 1 **Recall**	What kind of proof uses both coordinate geometry and algebra? a coordinate proof	Students' answers will indicate whether they understand the definition of a coordinate proof.
Level 2 **Basic Application of Skills & Concepts**	What is the distance between $(-2, 5)$ and $(5, 8)$? $\sqrt{(8-5)^2 + (5-(2))^2} = \sqrt{3^2 + 7^2} = \sqrt{9 + 49} = \sqrt{58}$	Students' answers will indicate whether they can apply the Distance Formula when given two points.
Level 3 **Strategic Thinking & Complex Reasoning**	Given the distance between two points and the coordinates of one point, how many ordered pairs satisfy the coordinates of the other point? Expain. an infinite number; Draw a point. Then, extend the distance out in all directions from the point. The distances become the radii of a circle, the points of whose circumference satisfy the Distance Formula.	Students' answers will indicate whether they can visualize a problem that has more than one correct answer.

Prove the Midpoint Formula

Recall that the coordinates of the midpoint of a segment are the averages of the x-coordinates and of the y-coordinates of the endpoints.

The coordinates of the midpoint of $\overline{AB}$, shown in the figure, are found using the Midpoint Formula.

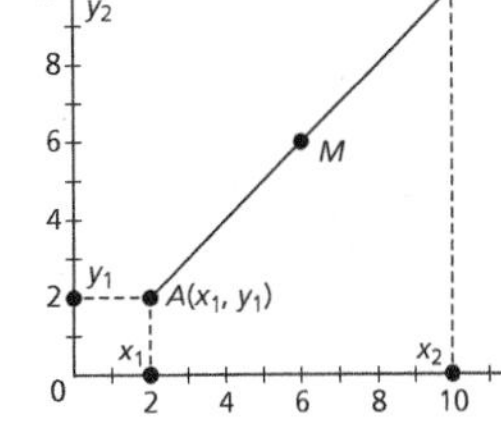

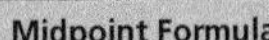

Midpoint Formula

The midpoint of the segment between any points (x_1, y_1) and (x_2, y_2) is $M\left(\frac{x_2+x_1}{2}, \frac{y_2+y_1}{2}\right)$.

2 Prove the Midpoint Formula.

Given: $A(x_1, y_1)$ and $B(x_2, y_2)$

Prove: The midpoint of $\overline{AB}$ is $M\left(\frac{x_2+x_1}{2}, \frac{y_2+y_1}{2}\right)$.

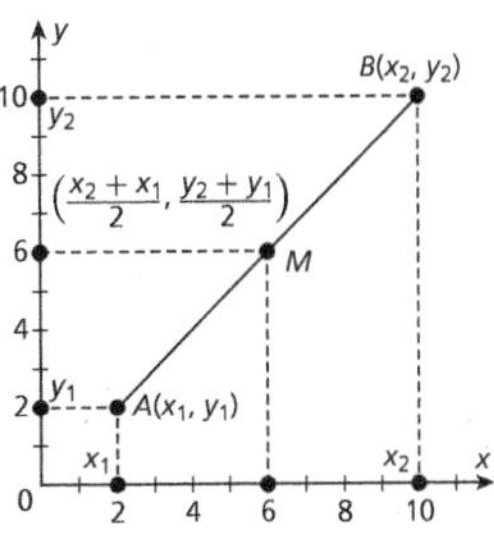

The horizontal distance from A to B is $x_2 - x_1$.

If M is halfway between A and B, the distance from A to M is half the distance from A to B. The horizontal distance between A and M is $\frac{x_2 - x_1}{2}$.

To find the x-coordinate of point M, add the horizontal distance from the origin to A to the horizontal distance from A to M.

x-coordinate of M: $x_1 + \frac{x_2 - x_1}{2}$	The horizontal distance from the origin to A added to the horizontal distance from A to M
$\frac{2x_1}{2} + \frac{x_2 - x_1}{2}$	Find a common denominator.
$\frac{2x_1 + x_2 - x_1}{2}$	Add the fractions.
$\frac{x_1 + x_2}{2}$	Combine like terms.

A. Show the steps to find the y-coordinate of M. A–C. See Additional Answers.

B. Suppose $x_1 = x_2$ or $y_1 = y_2$. Would the Midpoint Formula still apply? What would change?

C. Could you find the x-coordinate of M by subtracting the horizontal distance between A to M from the x-coordinate of B, rather than adding it to the x-coordinate of A? Why?

Turn and Talk Suppose you used the Midpoint Formula to calculate the distance between $(-2, 5)$ and $(5, -8)$, and then measured the distance with a ruler to verify. Would that be sufficient to prove the Midpoint Formula? Why or why not? See margin.

Task 2 **MP** **Attend to Precision** Students learn to use concrete references such as a drawing to prove the Midpoint Formula.

Students may have trouble recognizing that if M is halfway between A and B, the distance from A to M is half the distance from A to B. Consider showing this to students by using a diagram with specified values.

Sample Guided Discussion:

Q Why does the Midpoint Formula include a fraction with a denominator of 2? The midpoint is an average of the two x-coordinates and the two y-coordinates. An average is found by adding the given values and dividing by the number of given values, which in this case is 2.

Q In Part A, what does $\frac{y_2 - y_1}{2}$ represent? It represents the vertical distance between A and M.

Turn and Talk Remind students that this is an example of inductive reasoning. Inductive reasoning uses examples to reach a conclusion that is likely, but not certain. no; Possible answer: Proving one case by measuring does not prove the formula will work in every case.

PROFICIENCY LEVEL

Beginning

Give students several simple real-world scenarios involving the coordinates of two points. For each scenario, ask a question that can be solved using either the Distance Formula or the Midpoint Formula. Students should respond with "Distance Formula" or "Midpoint Formula" to indicate which should be used to solve the problem.

Intermediate

Have students explain how to calculate the distance between the points $(1, 2)$ and $(5, 5)$ by using the Distance Formula.

Advanced

Have students graph the ordered pairs $(1, 1)$, $(1, 5)$ and $(4, 1)$. Have them explain how to calculate the distance between each pair of points, and then explain how to verify that the triangle is a right triangle by using the Pythagorean Theorem with the distances found.

Step It Out

Task 3 (MP) **Use Repeated Reasoning** Students recognize patterns in solving problems involving the Distance Formula or the Midpoint Formula.

Suggest to students that the first step of any problem using ordered pairs should be to label the ordered pairs (x_1, y_1) and (x_2, y_2). This will help them to substitute the correct values into the Distance Formula.

OPTIMIZE OUTPUT Critique, Correct, and Clarify

Have students work with a partner to discuss the importance of matching the found value for *y* with the correct *x*-value in Task 3. Prompt them to correct each other's mistakes and help clarify any misconceptions while using the formula to find the length of the line segments.

Sample Guided Discussion:

Q How can you find the length of the segment of a line using the equation of the line and the *y*-coordinates of the endpoints? Substitute the given *y*-values into the equation and solve for *x*.

Q In Part C, what property justifies that the sum of two endpoints is the same despite the order in which they are added? the Commutative Property of Addition

Turn and Talk Some students will want to use a calculator to prove that $7\sqrt{10} \neq 10\sqrt{5}$. Show students that another way to disprove the equality is to square both sides. $\left(7\sqrt{10}\right)^2 = 49(10) = 490$

$\left(10\sqrt{5}\right)^2 = 100(5) = 500$

No; $7\sqrt{10} \neq 10\sqrt{5}$

Step It Out

Use the Distance Formula to Find Segment Length and Prove Congruence

Recall that a line segment is a portion of a line bounded by two endpoints. You can use the Distance Formula to determine the length of a line segment on the coordinate plane.

Sometimes a line segment will be the side of a figure. So, you find the distance between the vertices of the figure. The triangle with vertices $A(-2, -2)$, $B(3, 6)$, and $C(8, -2)$ has two congruent sides.

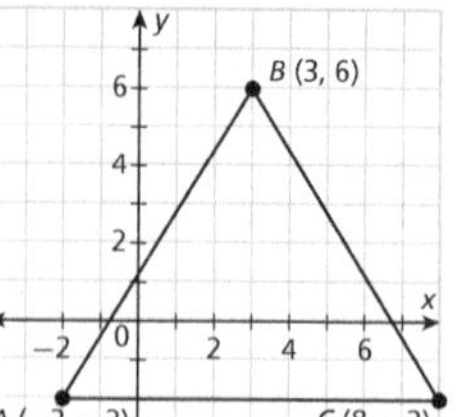

AB: $\sqrt{(3-(-2))^2 + (6-(-2))^2} = \sqrt{89}$

BC: $\sqrt{(8-3)^2 + (-2-6)^2} = \sqrt{89}$

CA: $\sqrt{(8-(-2))^2 + (-2-(-2))^2} = \sqrt{100}$

You can also find the length of a segment of a line using the equation of the line and the *x*-coordinates of the endpoints.

3 Find the length of the segment of the line $y = 3x - 6$ from $x = -3$ to $x = 4$.

Find the endpoints of the line segment by substituting the *x* values into the equation.

Line	Endpoint 1 (when $x = -3$)	Endpoint 2 (when $x = 4$)
$y = 3x - 6$	$y = 3(-3) - 6 = -15$ Endpoint 1: $(-3, -15)$	$y = 3(4) - 6 = 6$ Endpoint 2: $(4, 6)$

Use the Distance Formula to find the distance between the endpoints.

$$\begin{aligned}\text{Distance between Points 1 and 2} &= \sqrt{(x_2 - x_1)^2 + (y_2 - y_1)^2} \\ &= \sqrt{(4-(-3))^2 + (6-(-15))^2} \\ &= \sqrt{(7)^2 + (21)^2} = \sqrt{49 + 441} \\ &= \sqrt{490} = 7\sqrt{10}\end{aligned}$$

The length of the segment of the line $y = 3x - 6$ from $x = -3$ to $x = 4$ is $7\sqrt{10}$.

A. Find the length of the segment of the line $y = 2x + 8$ from $x = -8$ to $x = 2$.
A–C. See Additional Answers.

B. In the first step, why do you substitute the given *x*-values into the equation?

C. In the Distance Formula, is the answer different if you use $(y_2 - y_1)^2 + (x_2 - x_1)^2$ under the radical instead?

Turn and Talk Is the length of the segment of the line $y = 3x - 6$ from $x = -3$ to $x = 4$ the same as the length of the segment of the line $y = 2x + 8$ from $x = -8$ to $x = 2$? Why or why not? See margin.

126

Apply the Distance Formula to a Real-World Problem

 **4** A ship is traveling along the line $y = 1.5x + 20$ from a port located at $x = 0$ to an anchor point along the same line 30 nautical miles horizontally from the port. One unit represents one nautical mile. Approximately how far does the ship travel? What point represents half the distance traveled?

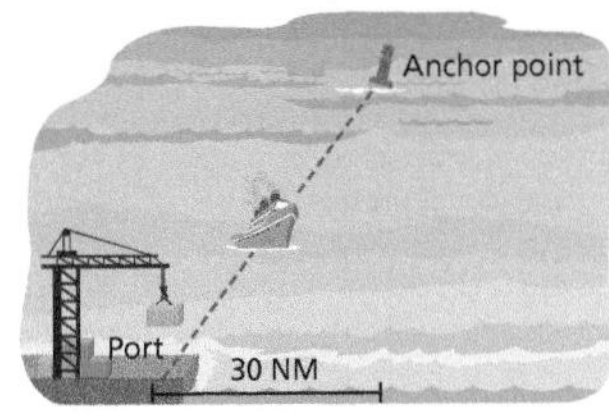

Determine the coordinates of the line segment described in the problem.

Starting Point	**Ending Point**
$y = 1.5x + 20; x = 0$	$y = 1.5x + 20; x = 30$
$y = 1.5(0) + 20 = 20$	$y = 1.5(30) + 20 = 65$
$(0, 20)$	$(30, 65)$

A. What clue tells us the ending point is at $x = 30$?

A, B. See Additional Answers.

Determine the distance between the coordinates using the Distance Formula.

$\sqrt{(x_2 - x_1)^2 + (y_2 - y_1)^2}$

$\sqrt{(30 - 0)^2 + (65 - 20)^2}$

$\sqrt{(30)^2 + (45)^2}$

$\sqrt{900 + 2025} = \sqrt{2925} \approx 54.1$

The boat travels approximately 54.1 nautical miles.

B. How can an approximate answer be helpful in real-world problems?

Determine the halfway point between the coordinates using the Midpoint Formula.

$M\left(\frac{x_2 + x_1}{2}, \frac{y_2 + y_1}{2}\right)$

$M\left(\frac{30 + 0}{2}, \frac{65 + 20}{2}\right) = \left(\frac{30}{2}, \frac{85}{2}\right)$

$M(15, 42.5)$

The midpoint is $(15, 42.5)$.

C. How could you find the location when the boat has traveled one-fourth the total distance?

Possible answer: I could find the midpoint between the starting point and the halfway point.

 **Turn and Talk** A second ship travels along the line $y = 2x + 25$ from $x = 5$ to $x = 32$. Which ship travels a longer distance? Explain your reasoning. See margin.

Task 4 (MP) **Reason** Students use information from a diagram to solve a real-world problem.

The diagram breaks apart the problem in order to find the solution. Some students may be overwhelmed when a problem contains many numbers. Using a diagram may help students see how each given number plays a role in the solution to the problem.

Sample Guided Discussion:

Q How would the process of solving the problem change if the ship traveled 30 nautical miles vertically from the port? The 30 would represent the y-value, and the equation would be solved for x.

Q In Part B, what additional information is needed to determine the rate at which the ship travels? the time it takes to travel 54.1 nautical miles

 **Turn and Talk** Remind students that the second ship is not starting from a port, so neither x-coordinate will equal 0. The second ship travels about 60.4 NM, which is a longer distance than the first ship travels.

$y = 2(8) + 25 = 35 \rightarrow (5, 35)$

$y = 2(32) + 25 = 89 \rightarrow (32, 89)$

ONLINE

Assign the Digital On Your Own for

- built-in student supports
- Actionable Item Reports
- Standards Analysis Reports

On Your Own

Assignment Guide

The chart below indicates which problems in the On Your Own are associated with each task in the Learn Together. Assign daily homework for tasks completed.

Learn Together Tasks	On Your Own Problems
Task 1, p.124	Problem 6
Task 2, p. 125	Problem 7
Task 3, p. 126	Problems 8–19
Task 4, p. 127	Problems 20–25

Check Understanding

1. How is the Distance Formula related to the Pythagorean Theorem? 1–3. See Additional Answers.
2. In your own words, explain how averaging the x-coordinates and averaging the y-coordinates of the endpoints of a line segment provide the coordinates of the midpoint of the segment.
3. What is the length of the segment $y = 2.5x + 10$ from $x = 0$ to $x = 16$? $8\sqrt{29}$
4. A triangle has vertices $(2, -3)$, $(-2, 1)$, and $(1, 2)$. Are any of the sides congruent? No, they all have different lengths.
5. A straight jogging path runs from $(-6, -9)$ to $(5, 4)$ on the map of the city park. Each unit on the map represents half of a mile. Approximately how long is the jogging path? approximately 8.5 miles

On Your Own

6. (MP) **Reason** When using the Pythagorean Theorem to prove that the distance between two points on the coordinate plane is given by $\sqrt{(x_2 - x_1)^2 + (y_2 - y_1)^2}$, why do you assume $x_1 \neq x_2$ and $y_1 \neq y_2$? 6–7. See Additional Answers.
7. (MP) **Construct Arguments** Given that endpoints of a segment are (x_1, y_1) and (x_2, y_2), explain why the x-coordinate of the midpoint of the segment is $\left(\frac{x_1 + x_2}{2}\right)$ rather than $\left(\frac{x_2 - x_1}{2}\right)$.

Determine whether $\overline{AB}$ is congruent to $\overline{CD}$. Justify your answer. 8–15. See Additional Answers.

8. $A(-2, -6), B(7, 12)$
   $C(-9, -12), D(2, 3)$
9. $A(-4.5, 2.5), B(10.5, 22.5)$
   $C(0, 0), D(15, 20)$
10. $A(20, 3), B(18, 7)$
    $C(-19, -4), D(-15, -6)$
11. $A(4, 11), B(1, 1)$
    $C(-3, -3), D(-5, -12)$
12. $A(1, 7), B(3, -5)$
    $C(2, 2), D(14, 0)$
13. $A(-8, -2), B(-5, 1)$
    $C(11, 3), D(14, 6)$
14. 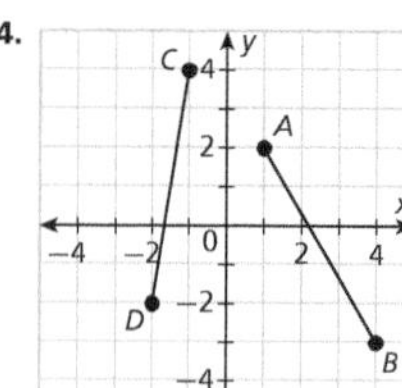

15. 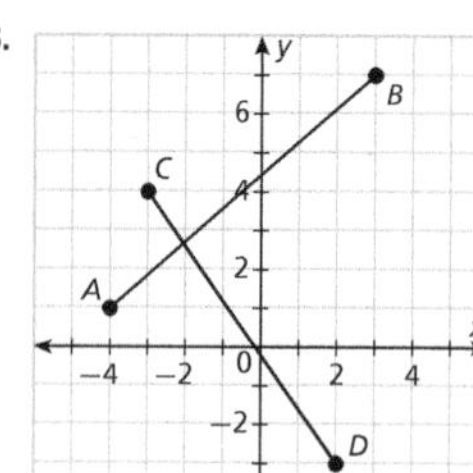

data checkpoint

3 Check Understanding

Formative Assessment

Use formative assessment to determine if your students are successful with this lesson's learning objective.

Students who successfully complete the Check Understanding can continue to the On Your Own practice.

For students who miss 1 problem or more, work in a pulled small group using the Almost There small-group activity on page 123C.

ONLINE

Assign the Digital Check Understanding to determine

- success with the learning objective
- items to review
- grouping and differentiation resources

4 Differentiation Options

Differentiate instruction for all students using small-group activities and math center activities on page 123C.

Reteach

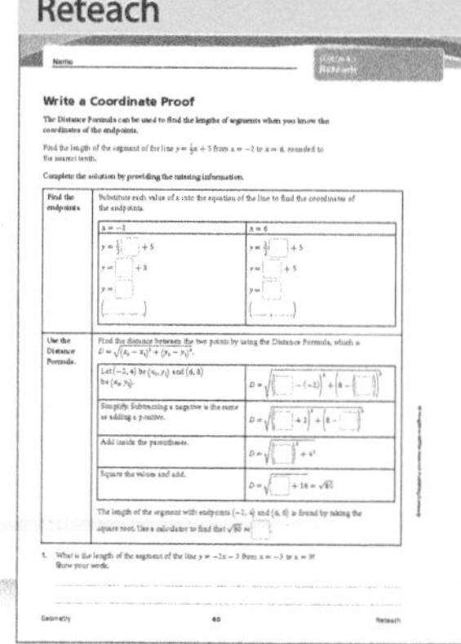

Challenge

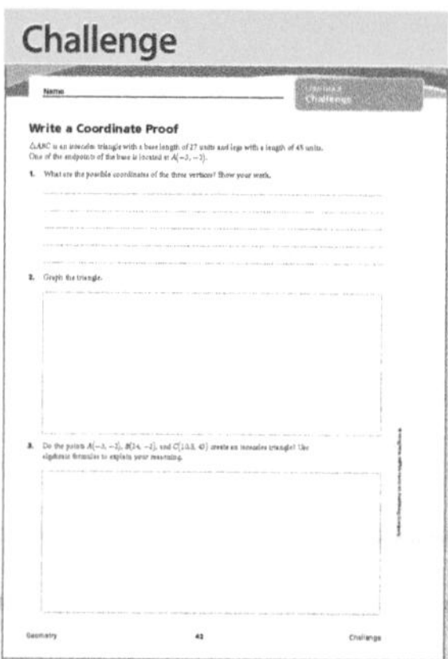

Use the figure for Problems 16–19.

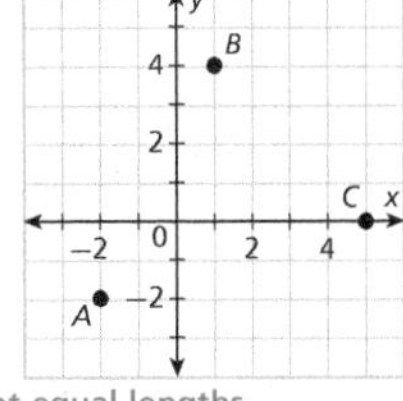

16. Find the length and midpoint of $\overline{AB}$.
16–18. See Additional Answers.

17. Find the length and midpoint of $\overline{BC}$.

18. Find the length and midpoint of $\overline{CA}$.

19. An *equilateral triangle* has three sides of equal length. If you form a triangle with vertices *A*, *B*, and *C*, is it an equilateral triangle? Explain your reasoning.
No, it is not an equilateral triangle; The legs are not equal lengths.

20. Open Ended Write a real-world problem that can be solved using the Midpoint Formula. Then solve the problem and justify your results. Answers will vary. Check students' work.

21. Health and Fitness Amy is training for a race by running laps around her school gym. To estimate the distance of each lap, she calculates the perimeter of the gym, which has vertices of $W(-4, -3)$, $X(-2, 1)$, $Y(5, 5)$, and $Z(3, -1)$ on the school map.

A. Approximately how many feet are in one lap around the gym if each unit on the map represents 50 feet? Round to the nearest foot. approximately 1307 feet per lap

B. There are 5280 feet in a mile. Approximately how many full laps must she run to train for 3 miles? 12 laps

22. A triangle has the vertices $(-4, -3)$, $(-1, 0)$, and $(-3, 2)$. Is it the same size as a triangle with the vertices $(4, 1)$, $(9, 2)$, and $(7, 4)$? Explain.

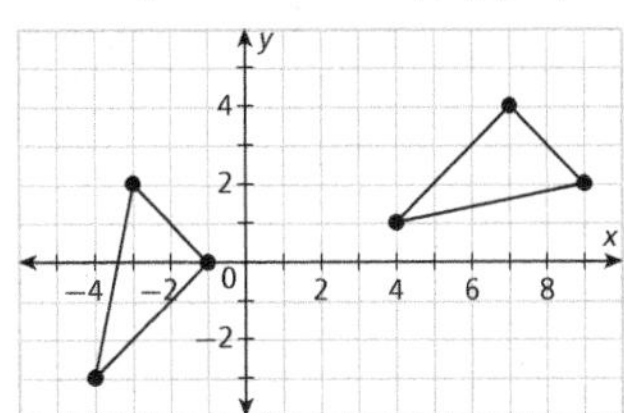

yes; The three sides of the triangles have equal lengths.

23. A trolley around a botanical garden needs a new schedule for the tourist season. A transportation planner plots the main station at the origin, and the first and second stops on the map, where each unit represents 0.5 mile.

A. Determine the distance the trolley travels between the station and the first stop. If it leaves the station at 9:15 a.m. and travels at a rate of 15 mi/h, what time will the trolley reach the first stop? 6.5 miles; 9:41 a.m.

B. How far will the trolley travel from the first stop to the second stop? about 4 miles

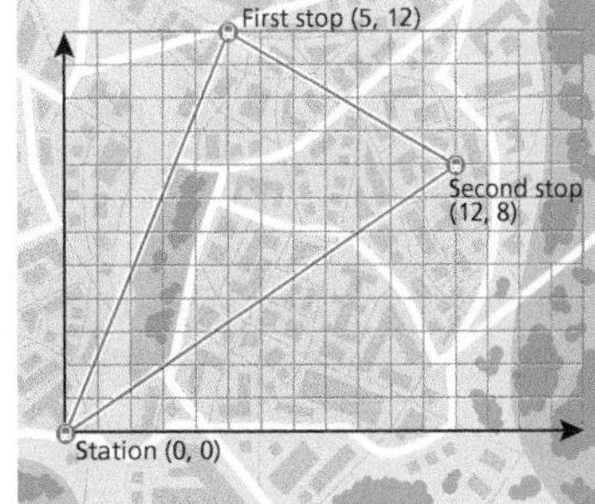

Watch for Common Errors

Problem 16 Students may inadvertently switch values for *x* and *y* when substituting the ordered pairs into the Midpoint Formula. Encourage students to get in the habit of labeling the ordered pairs (x_1, y_1) and (x_2, y_2) before beginning the problem so that they use the correct values in the formula.

Questioning Strategies

Problem 19 An isosceles triangle has two sides of equal length. If you form a triangle with vertices *A*, *B*, and *C*, is it an isosceles triangle? Explain your reasoning. No, it is not an isosceles triangle; No two sides have the same lengths.

5 Wrap-Up

Summarize learning with your class. Consider using the Exit Ticket, Put It in Writing, or I Can scale.

Exit Ticket 

Find the distance between $(-4, 1)$ and $(1, 13)$ and the midpoint of the line that connects them.

$$D = \sqrt{(1-(-4))^2 + (13-1)^2} = \sqrt{5^2 + 12^2}$$

$$= \sqrt{25 + 144} = \sqrt{169} = 13;$$

$$MP = \left(\frac{-4+1}{2}, \frac{1+13}{2}\right) = \left(-\frac{3}{2}, 7\right)$$

Put It in Writing 

Given two ordered pairs, describe the process of finding the distance between them using the Distance Formula and the Pythagorean Theorem.

I Can 

The scale below can help you and your students understand their progress on a learning goal.

4	I can use the Distance Formula to show congruence on the coordinate plane, and I can explain my solution steps to others.
3	I can use the Distance Formula to show congruence on the coordinate plane.
2	I can verify the distance between two points I found by using the Distance Formula with the Pythagorean Theorem.
1	I can apply the Distance Formula if given two points.

Spiral Review • Assessment Readiness

These questions will help determine if students have retained information taught in the past and can also prepare them for high-stakes assessments. Here, students must write the equation of a line perpendicular to a given line through a specific point **(4.2)**, use transformations to write the coordinates of an ordered pair **(5.1)**, and determine if equations of lines are parallel, perpendicular, or neither **(4.1 and 4.2)**.

24. **Model with Mathematics** A car travels 26 miles down a straight road, starting at the origin. The final location of the car is the $(10, y_2)$.

A. Use the Distance Formula to write and solve an equation to find the final y-coordinate of the car. Show your work. See Additional Answers.

B. What is the midpoint of the path of the car? $(5, 12)$

25. A wooden bridge has a triangular support system. One support has the vertices $P(-1, 3)$, $Q(1, -2)$, and $R(-3, -2)$.

A. Which legs of the triangle are congruent? $\overline{PQ}$ and $\overline{PR}$ are congruent.

B. If each unit on the coordinate plane represents 2 ft, how many whole triangular supports can be constructed with 100 ft of lumber? 3 whole supports

Spiral Review • Assessment Readiness

26. Which line is perpendicular to $x = -5$ and passes through $(1, 6)$?

Ⓐ $x = 1$
Ⓑ $y = 6$
Ⓒ $x = 5$
Ⓓ $y = -5$

27. If (x, y) represents the point $(2, 4)$, what represents the point $(5, 6)$?

Ⓐ $(x - 3, y + 2)$
Ⓑ $(x - 2, y - 3)$
Ⓒ $(x + 2, y + 3)$
Ⓓ $(x + 3, y + 2)$

28. For each set of linear equations, determine if the lines are parallel, perpendicular, or neither.

Equations	Parallel	Perpendicular	Neither
A. $y = \frac{2}{5}x + 3$ $y = \frac{5}{2}x - 3$	?	?	?
B. $y = x - 2.5$ $y = -x + 9$	?	?	?
C. $y = 2x$ $y = 2x - 10$	?	?	?

 **I'm in a Learning Mindset!**

What resources are available to help me understand writing a coordinate proof?

Learning Mindset

Strategic Help-Seeking **Identifies Sources for Help**

Point out that by definition, coordinate proofs involve both coordinate geometry and algebra. This lesson repeatedly showed that placing a figure on the coordinate plane can help students prove various formulas. Remind students that this kind of visualization can also be used to help them understand the meaning of word problems. Students can also verify answers found using the Distance Formula by using the Pythagorean Theorem and answers found by the Midpoint Formula by using the Distance Formula. *How can the answer found using the Distance Formula be verified by using the Pythagorean Theorem? How can the answer found using the Midpoint Formula be verified by using the Distance Formula?*

Module

4 Review

The Distance Formula

Two races through the city are shown. Race A follows the path $y = 2x - 4$ from $x = -1$ to $x = 4$. Race B follows the path $y = -0.25x - 2$ from $x = -6$ to $x = 5$. Are the races the same length?

Find the coordinates of the endpoints.

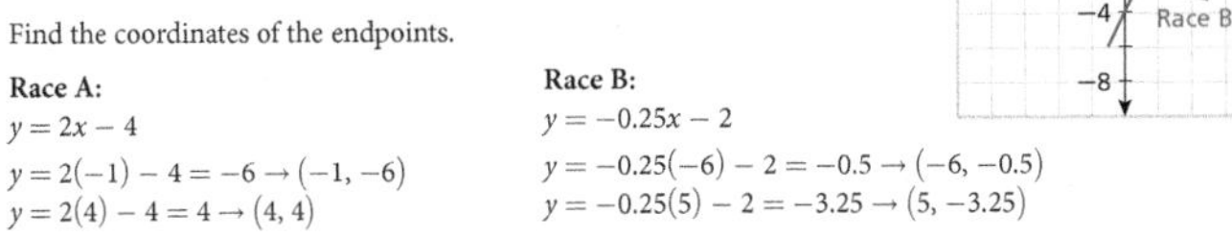

Race A:

$y = 2x - 4$

$y = 2(-1) - 4 = -6 \rightarrow (-1, -6)$

$y = 2(4) - 4 = 4 \rightarrow (4, 4)$

Race B:

$y = -0.25x - 2$

$y = -0.25(-6) - 2 = -0.5 \rightarrow (-6, -0.5)$

$y = -0.25(5) - 2 = -3.25 \rightarrow (5, -3.25)$

Use the Distance Formula with the endpoint coordinates of each segment to determine the length of each race.

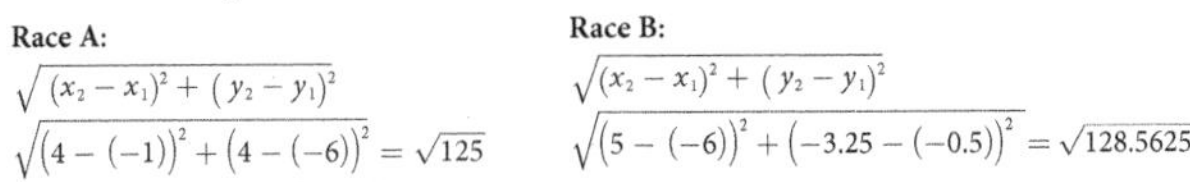

Race A:

$\sqrt{(x_2 - x_1)^2 + (y_2 - y_1)^2}$

$\sqrt{(4 - (-1))^2 + (4 - (-6))^2} = \sqrt{125}$

Race B:

$\sqrt{(x_2 - x_1)^2 + (y_2 - y_1)^2}$

$\sqrt{(5 - (-6))^2 + (-3.25 - (-0.5))^2} = \sqrt{128.5625}$

The segments are not congruent. Race B is the longer race because $\sqrt{128.5625} > \sqrt{125}$.

Parallel Lines

A major roadway runs parallel to Race A and passes through the point $(1, 1)$. What is the equation of the parallel road?

Parallel lines have equal slopes.

Race A: $y = 2x - 4$

Parallel Road: $y = 2x + b$

$y = mx + b$, where m is the slope

Use slope-intercept form and the point through which the road passes to find the missing y-intercept.

$y = mx + b$

$1 = 2(1) + b$

$-1 = b$

Write the equation of the parallel line using the slope and y-intercept. The equation that describes the parallel road is $y = 2x - 1$.

Perpendicular Lines

A city walking path is perpendicular to Race A and passes through the point $(4, 2)$. What is the equation of the path?

Perpendicular lines have opposite reciprocal slopes.

Race A: $y = 2x - 4$.

Path: $y = -\frac{1}{2}x + b$

m and $-\frac{1}{m}$ are opposite reciprocals.

Use slope-intercept form and the point through which the path passes to find the missing y-intercept.

$y = mx + b$

$2 = -\frac{1}{2}(2) + b$

$3 = b$

Write the equation of the perpendicular path using the slope and y-intercept. The equation that models the path is $y = -\frac{1}{2}x + 3$.

ONLINE

Assign the Digital Module Review for

- built-in student supports
- Actionable Item Reports
- Standards Analysis Reports

Module Review

Use the first page of the Module Review to summarize and connect the main ideas of the module. Use the second page to assess students' understanding of the vocabulary, concepts, and skills presented in the module.

Sample Guided Discussion:

Q The Distance Formula Why is it valid to apply the Distance Formula to find the length of the races? The races are represented as segments that lie along two straight lines in the coordinate plane. The coordinates of the endpoints of each segment are given, so you can substitute these values into the Distance Formula to find and compare the lengths of each race.

Q Parallel Lines Suppose you know the equation of line *L* in the coordinate plane and the coordinates of a point not on *L*. What information would you need to write an equation of the line parallel to *L* through the point not on *L*? I know that parallel lines have the same slope, so first I would need to find the slope m of line L from its equation. Then I could substitute the values for m and the coordinates of the point into the slope-intercept equation $y = mx + b$ to find b.

Q Perpendicular Lines Now suppose you know the equation of line *Q* in the coordinate plane and the coordinates of a point not on *Q*. What information would you need to write an equation of the line perpendicular to *Q* through the point not on *Q*? Lines in the coordinate plane that are perpendicular have slopes that are opposite reciprocal. So, first, I would need to find the slope m of line Q from its equation and determine $-\frac{1}{m}$. Then I could substitute the values of $-\frac{1}{m}$ and the coordinates of the point into the slope-intercept equation to find b.

Module Review continued

Possible Scoring Guide

Items	Points	Description
1–4	2 each	identifies the correct term
5–10	2 each	writes equations of lines parallel and perpendicular to a given line
11–14	2 each	writes equations of lines parallel and perpendicular to a given line that passes through a given point
15	2	writes an equation for a line perpendicular to a given line through a point not on the line
16	2	justifies steps of a proof for the Distance Formula
17	2	determines whether two segments are congruent and explains using the Distance Formula
18	2	classifies a triangle and justifies the answer
Total points possible = 36 points		

The Unit 2 Test in the Assessment Guide assesses content from Modules 3 and 4.

Vocabulary

Choose the correct term from the box to complete each sentence.

Vocabulary
coordinate proof
parallel lines
perpendicular lines
subscript notation

1. __?__ never intersect and have equal slopes. parallel lines
2. Formulas with __?__ use small numbers on the lower right corner of variables to help identify different parts of the equation. subscript notation
3. __?__ intersect at 90° angles and have opposite reciprocal slopes. perpendicular lines
4. A __?__ uses geometry and algebra to show each step of a mathematical conclusion. coordinate proof

Concepts and Skills

Write an equation for a line parallel to and a line perpendicular to the given line. 5–10. See Additional Answers.

5. $y = 4x - 5$
6. $3x - y = 9$
7. $y = -2x + 5$
8. $4x + 5y = 2$
9. $y = \frac{3}{4}x - 8$
10. $-6x + 2y = -7$

Write an equation for a line parallel and a line perpendicular to the given line that passes through the given point. 11–14. See Additional Answers.

11. $2x + 3y = 6; (-2, -6)$
12. $-4y + x + 12 = 0; (5, 3)$
13. $y = -4x + 6; (1, 1)$
14. $y = \frac{1}{3}x - 5; (-4, 2)$

15. (MP) Use Tools A landscape architect is designing a water feature for a botanical garden. The current path follows the line $y = -3x + 1$. The water feature will be located at $(7, -4)$. Write an equation for a path that will run to the water feature and is perpendicular to the current path. State what strategy and tool you will use to answer the question, explain your choice, and then find the answer. $y = \frac{1}{3}x - \frac{19}{3}$

16. **Justify each step of the proof for the Distance Formula.**

A.	$\triangle PQR$ is a right triangle.
B.	$c^2 = a^2 + b^2$
C.	$PQ^2 = (x_2 - x_1)^2 + (y_2 - y_1)^2$
D.	$PQ = \sqrt{(x_2 - x_1)^2 + (y_2 - y_1)^2}$

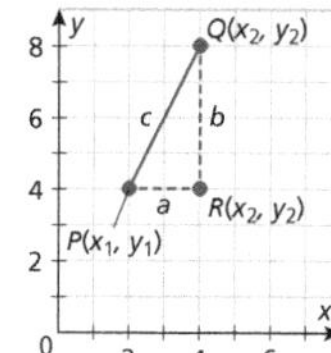

16–18. See Additional Answers.

17. Are the segments $y = 4x + 9$ from $x = 0$ to $x = 5$ and $y = -0.5x - 2$ from $x = -3$ to $x = 1$ congruent segments? Explain using the Distance Formula.
18. A triangle has vertices $A(10, -2)$, $B(14, 1)$, and $C(14, -4)$. Classify the triangle and justify your answer.

132

DATA-DRIVEN INSTRUCTION

Before moving on to the Module Test, use the Module Review results to intervene based on the table below.

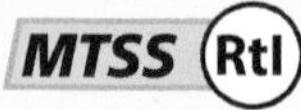

Items	Lesson	DOK	Content Focus	Intervention
5–10	4.1, 4.2	2	Write an equation for a line parallel to and for a line perpendicular to a given line.	Reteach 4.1, 4.2
11–14	4.1, 4.2	2	Write an equation for a line parallel to and for a line perpendicular to a given line through a given point not on the line.	Reteach 4.1,4.2
15	4.2	2	Write an equation for the line perpendicular to a given line through a point not on the line.	Reteach 4.2
16	4.3	3	Prove the Distance Formula.	Reteach 4.3
17	4.3	2	Determine the lengths of segments in the coordinate plane.	Reteach 4.3
18	4.3	2	Classify a triangle by finding the length of the sides.	Reteach 4.3

Module Test

The Module Test is available in alternative versions in your Assessment Guide. All items are presented in standardized test formats.

ONLINE

Ed

Assign the Digital Module Test to power actionable reports including

- proficiency by standards
- item analysis

Form A

Name

Module 4 • Form A
Module Test

1. What is the midpoint of the line segment with endpoints (1, 2) and (11, 6)?

Ⓐ (1.5, 8.5) **Ⓒ (6, 4)**
Ⓑ (5, 2) Ⓓ (12, 8)

2. What is the length of the line segment with endpoints (12, −21) and (15, −9)?

Ⓐ $3\sqrt{13}$ units Ⓒ $12\sqrt{15}$ units
Ⓑ $3\sqrt{17}$ units Ⓓ $12\sqrt{29}$ units

3. Which equation represents a line that is parallel to $y = 3x - 4$?

Ⓐ $y = -\frac{1}{3}x + 2$ **Ⓒ $y = 3x + 1$**
Ⓑ $y = -3x + 7$ Ⓓ $y = \frac{1}{3}x - 4$

4. A line contains points (−3, 19) and (1, 18). Which equation represents the line that is parallel to the given line and passes through point (−4, −9)?

Ⓐ $y = \frac{1}{4}x - 8$ Ⓒ $y = -\frac{1}{4}x - 8$
Ⓑ $y = \frac{1}{4}x - 10$ **Ⓓ $y = -\frac{1}{4}x - 10$**

5. Which pair of linear equations could describe the opposite sides of a parallelogram?

Ⓐ $5x - 5y = 30$ and $y = x + 16$
Ⓑ $2x + y = 24$ and $y = 2x + 13$
Ⓒ $-9x - 9y = 18$ and $y = x - 4$
Ⓓ $-x + 2y = 16$ and $y = 2x + 7$

6. Which equation represents a line that is perpendicular to $y = \frac{1}{2}x + 12$?

Ⓐ $y = -2x + 3$
Ⓑ $y = -\frac{1}{2}x + 4$
Ⓒ $y = \frac{1}{2}x - 9$
Ⓓ $y = 2x$

7. The graph of a line is shown.

Which equation represents the line that is perpendicular to the given line and has a y-intercept at (0, 4)?

Ⓐ $y = \frac{1}{2}x + 4$ Ⓒ $y = -2x - 4$
Ⓑ $y = \frac{1}{2}x - 4$ Ⓓ $y = -2x + 4$

8. A line contains points (1, −1) and (−2, −10). Which equation represents the line that is perpendicular to the given line and passes through point (6, −4)?

Ⓐ $y = 3x - 1$ **Ⓒ $y = -\frac{1}{3}x - 2$**
Ⓑ $y = 3x - 22$ Ⓓ $y = -\frac{1}{3}x + \frac{14}{3}$

9. What is the slope of a line that is parallel to $14x - 18y = 72$?

$\frac{7}{9}$

10. Write an equation of the line that is parallel to $6x - 8y = 56$ and passes through point (22, −8.5).

Possible answer: $y = \frac{3}{4}x - 25$

11. Write an equation of the line that is parallel to $y = 16$ and passes through point (−7, 43).

$y = 43$

12. What is the slope of a line that is perpendicular to $-4x + \frac{1}{2}y = 7$?

$-\frac{1}{8}$

Geometry 53 Module 4 Test • Form A

Form A

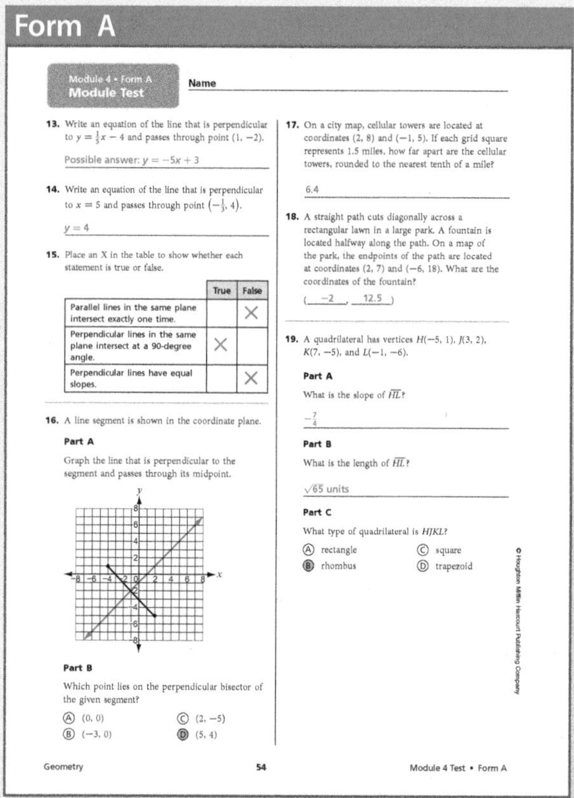

Module 4 • Form A
Module Test

Name

13. Write an equation of the line that is perpendicular to $y = \frac{1}{5}x - 4$ and passes through point (1, −2).

Possible answer: $y = -5x + 3$

14. Write an equation of the line that is perpendicular to $x = 5$ and passes through point $\left(-\frac{1}{3}, 4\right)$.

$y = 4$

15. Place an X in the table to show whether each statement is true or false.

	True	False
Parallel lines in the same plane intersect exactly one time.		X
Perpendicular lines in the same plane intersect at a 90-degree angle.	X	
Perpendicular lines have equal slopes.		X

16. A line segment is shown in the coordinate plane.

Part A

Graph the line that is perpendicular to the segment and passes through its midpoint.

Part B

Which point lies on the perpendicular bisector of the given segment?

Ⓐ (0, 0) Ⓒ (2, −5)
Ⓑ (−3, 0) **Ⓓ (5, 4)**

17. On a city map, cellular towers are located at coordinates (2, 8) and (−1, 5). If each grid square represents 1.5 miles, how far apart are the cellular towers, rounded to the nearest tenth of a mile?

6.4

18. A straight path cuts diagonally across a rectangular lawn in a large park. A fountain is located halfway along the path. On a map of the park, the endpoints of the path are located at coordinates (2, 7) and (−6, 18). What are the coordinates of the fountain?

(−2, 12.5)

19. A quadrilateral has vertices $H(-5, 1)$, $J(3, 2)$, $K(7, -5)$, and $L(-1, -6)$.

Part A

What is the slope of $\overline{HL}$?

$-\frac{7}{4}$

Part B

What is the length of $\overline{HL}$?

$\sqrt{65}$ units

Part C

What type of quadrilateral is $HJKL$?

Ⓐ rectangle Ⓒ square
Ⓑ rhombus Ⓓ trapezoid

Geometry 54 Module 4 Test • Form A

Form B

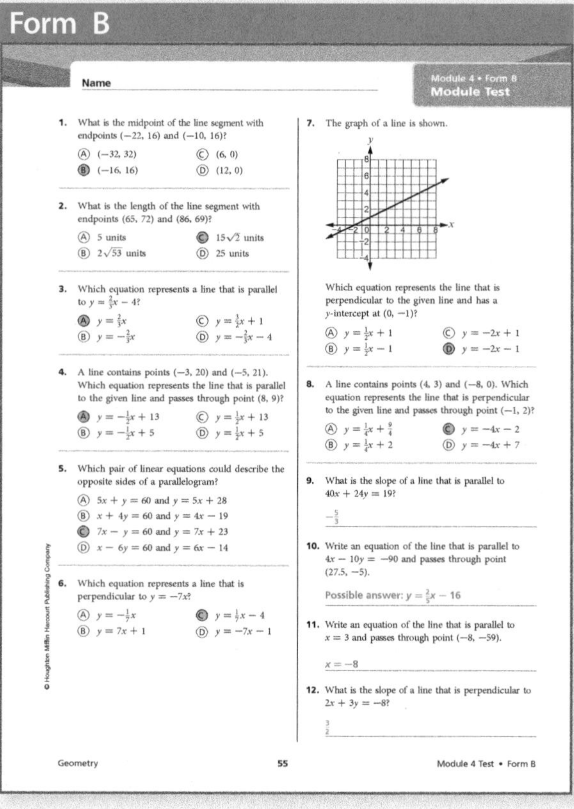

Name

Module 4 • Form B
Module Test

1. What is the midpoint of the line segment with endpoints (−22, 16) and (−10, 16)?

Ⓐ (−32, 32) Ⓒ (6, 0)
Ⓑ (−16, 16) Ⓓ (12, 0)

2. What is the length of the line segment with endpoints (65, 72) and (86, 69)?

Ⓐ 5 units **Ⓒ $15\sqrt{2}$ units**
Ⓑ $2\sqrt{53}$ units Ⓓ 25 units

3. Which equation represents a line that is parallel to $y = \frac{2}{3}x - 4$?

Ⓐ $y = \frac{2}{3}x$ Ⓒ $y = \frac{3}{2}x + 1$
Ⓑ $y = -\frac{2}{3}x$ Ⓓ $y = -\frac{2}{3}x - 4$

4. A line contains points (−3, 20) and (−5, 21). Which equation represents the line that is parallel to the given line and passes through point (8, 9)?

Ⓐ $y = -\frac{1}{2}x + 13$ Ⓒ $y = \frac{1}{2}x + 13$
Ⓑ $y = -\frac{1}{2}x + 5$ Ⓓ $y = \frac{1}{2}x + 5$

5. Which pair of linear equations could describe the opposite sides of a parallelogram?

Ⓐ $5x + y = 60$ and $y = 5x + 28$
Ⓑ $x + 4y = 60$ and $y = 4x - 19$
Ⓒ $7x - y = 60$ and $y = 7x + 23$
Ⓓ $x - 6y = 60$ and $y = 6x - 14$

6. Which equation represents a line that is perpendicular to $y = -7x$?

Ⓐ $y = -\frac{1}{7}x$ **Ⓒ $y = \frac{1}{7}x - 4$**
Ⓑ $y = 7x + 1$ Ⓓ $y = -7x - 1$

7. The graph of a line is shown.

Which equation represents the line that is perpendicular to the given line and has a y-intercept at (0, −1)?

Ⓐ $y = \frac{1}{2}x + 1$ Ⓒ $y = -2x + 1$
Ⓑ $y = \frac{1}{2}x - 1$ **Ⓓ $y = -2x - 1$**

8. A line contains points (4, 3) and (−8, 0). Which equation represents the line that is perpendicular to the given line and passes through point (−1, 2)?

Ⓐ $y = \frac{1}{4}x + \frac{9}{4}$ **Ⓒ $y = -4x - 2$**
Ⓑ $y = \frac{1}{4}x + 2$ Ⓓ $y = -4x + 7$

9. What is the slope of a line that is parallel to $40x + 24y = 19$?

$-\frac{5}{3}$

10. Write an equation of the line that is parallel to $4x - 10y = -90$ and passes through point (27.5, −5).

Possible answer: $y = \frac{2}{5}x - 16$

11. Write an equation of the line that is parallel to $x = 3$ and passes through point (−8, −59).

$x = -8$

12. What is the slope of a line that is perpendicular to $2x + 3y = -8$?

$\frac{3}{2}$

Geometry 55 Module 4 Test • Form B

Form B

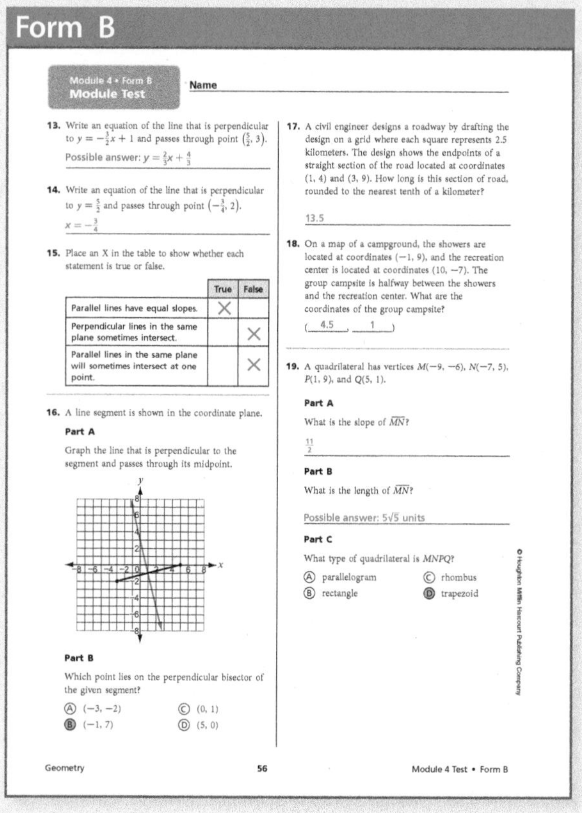

Module 4 • Form B
Module Test

Name

13. Write an equation of the line that is perpendicular to $y = -\frac{3}{2}x + 1$ and passes through point $\left(\frac{5}{2}, 3\right)$.

Possible answer: $y = \frac{2}{3}x + \frac{4}{3}$

14. Write an equation of the line that is perpendicular to $y = \frac{5}{2}$ and passes through point $\left(-\frac{3}{4}, 2\right)$.

$x = -\frac{3}{4}$

15. Place an X in the table to show whether each statement is true or false.

	True	False
Parallel lines have equal slopes.	X	
Perpendicular lines in the same plane sometimes intersect.		X
Parallel lines in the same plane will sometimes intersect at one point.		X

16. A line segment is shown in the coordinate plane.

Part A

Graph the line that is perpendicular to the segment and passes through its midpoint.

Part B

Which point lies on the perpendicular bisector of the given segment?

Ⓐ (−3, −2) Ⓒ (0, 1)
Ⓑ (−1, 7) Ⓓ (5, 0)

17. A civil engineer designs a roadway by drafting the design on a grid where each square represents 2.5 kilometers. The design shows the endpoints of a straight section of the road located at coordinates (1, 4) and (3, 9). How long is this section of road, rounded to the nearest tenth of a kilometer?

13.5

18. On a map of a campground, the showers are located at coordinates (−1, 9), and the recreation center is located at coordinates (10, −7). The group campsite is halfway between the showers and the recreation center. What are the coordinates of the group campsite?

(4.5, 1)

19. A quadrilateral has vertices $M(-9, -6)$, $N(-7, 5)$, $P(1, 9)$, and $Q(5, 1)$.

Part A

What is the slope of $\overline{MN}$?

$\frac{11}{2}$

Part B

What is the length of $\overline{MN}$?

Possible answer: $5\sqrt{5}$ units

Part C

What type of quadrilateral is $MNPQ$?

Ⓐ parallelogram Ⓒ rhombus
Ⓑ rectangle **Ⓓ trapezoid**

Geometry 56 Module 4 Test • Form B

Module 5: Transformations that Preserve Size and Shape

Module 6: Transformations that Change Size and Shape

Pulmonologist STEM

- **Say:** *What is the science of pulmonology? Pulmonologists are doctors who diagnose and treat conditions affecting the respiratory system. They treat respiratory disorders such as lung cancer, obstruction of the airway, lung diseases marked by scarring and inflammation, airway disorders, and sleep disordered breathing.*
- Explain that a pulmonologist uses various tools and mathematical models to understand how human lungs develop and function. Then introduce the mathematics of the STEM task.

STEM Task

Ask students if they have any previous knowledge of fractals.

- Have students research fractals and self-similarity.
- Have students find pictures of human lung growth.
- Discuss the meaning of *morphogenesis* and specifically, lung morphogenesis with students.
- Have students create a bulletin board that contains pictures of fractals and lung growth.
- Elicit from students the relationship between self-similar fractals and lung growth.

Unit 3 Transformations

Pulmonologist STEM

A pulmonologist is a physician that specializes in diagnosing and treating illnesses of the lungs. Some lung diseases and conditions are asthma, pneumonia, tuberculosis, and emphysema. Pulmonologists use various tools and models to understand how the lungs develop and operate.

STEM Task

Lung morphogenesis is an iterative process of sequential branching. Each branch is divided into two smaller parts, compartmentalizing the lung into self-similar sections.

Describe how the lung structure develops using transformations. Explain your reasoning.

(t) ©Helen King/Corbis/Getty Images; (b) ©Science Picture Co/Science Source

Unit 3 133

Unit 3 Project Making Some Breathing Room

Overview: In this project students use transformations to map out a basic 2-dimensional representation of the lungs.

Materials: paper and pencil or a computer used for word processing

Assessing Student Performance: Students' reports should include:

- an explanation of the nature of the main vertical branch at the top of the model and a correct stage of human development for the start of branch B **(Lesson 5.4 or earlier)**
- a correct identification of the symmetry of the two branches and a correct equation for the line of symmetry **(Lesson 5.4)**
- the correct coordinates for R' **(Lesson 5.4)**
- a correct value for the y-coordinate of point R **(Lesson 6.1)**

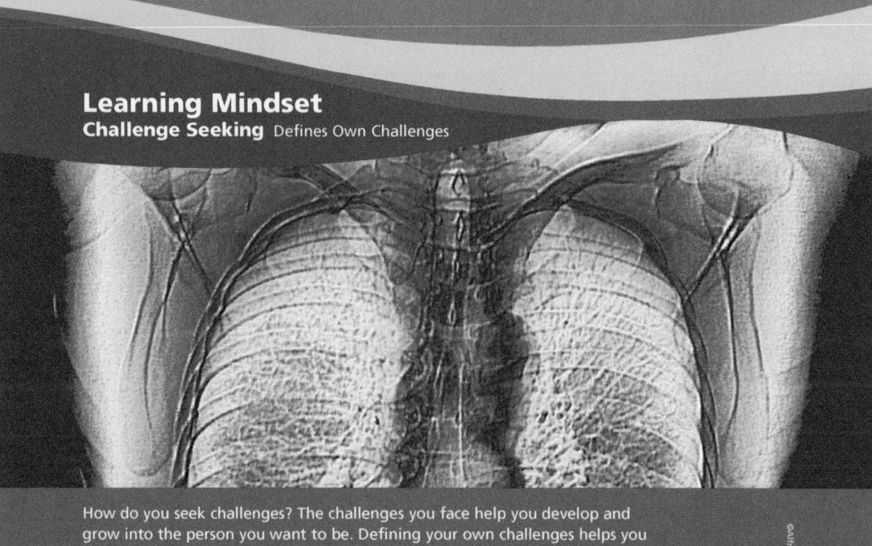

How do you seek challenges? The challenges you face help you develop and grow into the person you want to be. Defining your own challenges helps you to take ownership of your own development. The goals we set should stretch our capabilities, but still be within reach. Here are some questions you can ask yourself when thinking about challenging yourself:

- What do I want to be like? How can I set goals for myself to grow in this direction?
- What goals have I set for myself? Will my goals cause me to stretch my abilities and grow?
- What can I do if a challenge is too simple or difficult? How can I redefine my goals to make them more manageable?
- How do I proactively seek out additional learning opportunities? What resources do I use to seek out additional learning opportunities?
- What benefit is there to creating my own learning path? Why is it important to take ownership of my own development?

Reflect

Q **Think of a time when have you defined your own challenge. How did you benefit from that experience? How was it different from having somebody else set your goals?**

Q **As a pulmonologist, why would you continue to challenge yourself to learn more? How would you benefit, and how would your community benefit?**

134

Challenge Seeking
Learning Mindset

Defines Own Challenges

Monitors Knowledge and Skills

The learning-mindset focus in this unit is *challenge seeking*, which refers to a person's ability to seek appropriate new challenges and learning. In mathematics, students should discover the pleasure of setting and meeting their own goals and challenges.

Mindset Beliefs

Some students who are satisfied to get by learning the bare minimum might say, "Just show me how to do this." or "Is this all I need to know?" Or "why do I have to learn this?" Other students may believe that their ability to learn a concept or skill is easy and straightforward and requires little effort.

Students who are challenge-seeking go beyond what is expected or straightforward. They might ask themselves, "What does this concept or skill lead to?" "How can I use this information to learn new things? "What are its applications?" Questions such as these demonstrate a growth mindset.

Discuss with students how they can discover the satisfaction of setting individual goals and challenges. Create new challenges for them if they are eager to learn more but may not know how to proceed.

Mindset Behaviors

Encourage students to become more challenge-seeking by asking themselves about the importance of defining their own goals. Self-monitoring can take the form of the following questions:

When reading a mathematical textbook:

- Why is it important that I study my textbook?
- What else can I learn from this skill or concept?

When practicing a mathematical skill:

- What else can I learn by practicing this skill?
- Where and how can I use this skill?

When successfully solving a mathematical problem:

- Can I solve a similar, but more difficult, problem?
- How can I use this solution method to solve more difficult problems?

As students become more proficient at monitoring their own knowledge and skills, it is likely that they will be more amenable to accept and welcome further challenges. Students will also be more likely to achieve their personal goals.

What to Watch For

Watch for students who are struggling with learning mathematics despite their best efforts and hard work. Help them become challenge-seeking by

- encouraging them to be flexible and open-minded,
- pointing out that it may be necessary to adjust their goals, and
- encouraging them to seek advice and assistance when needed.

Watch for students who are satisfied with the status quo. Help them define their own challenges by asking them these questions.

- Are you satisfied with what you already know?
- Do you seek appropriate new challenges when they arise?
- How can you seek out new challenges and new learning opportunities?

"If you aren't in over your head, how do you know how tall you are?"

— T.S. Eliot

TRANSFORMATIONS THAT PRESERVE SIZE AND SHAPE

Introduce and Check for Readiness

- Module Performance Task • Are You Ready?

Lesson 5.1—2 Days

Define and Apply Translations

Learning Objective: Develop a definition of translation as a function that preserves measures of segments and angles and draw the image of a figure under such a transformation.

New Vocabulary: component form, **image**, **isometry**, **preimage**, **rigid motion**, **transformation**, **translation**, **vector**

Lesson 5.2—2 Days

Define and Apply Rotations

Learning Objective: Develop a definition of rotation as a function that preserves measures of segments and angles and draw the image of a figure under such a transformation.

New Vocabulary: angle of rotation, **center of rotation**, **rotation**

Lesson 5.3—2 Days

Define and Apply Reflections

Learning Objective: Develop a definition of reflection as a function that preserves measures of segments and angles and draw the image of a figure under such a transformation.

New Vocabulary: line of reflection, **reflection**

Lesson 5.4—2 Days

Define and Apply Symmetry

Learning Objective: Explain how to determine the number of lines of symmetry and the angle of rotational symmetry for any regular polygon.

New Vocabulary: angle of rotational symmetry, **line of symmetry**, **line symmetry**, **rotational symmetry, symmetry**

Assessment

- Module 5 Test (Forms A and B)
- Unit 5 Test (Forms A and B)

LEARNING ARC FOCUS

Build Conceptual Understanding   Connect Concepts and Skills   Apply and Practice

TEACHING FOR SUCCESS

TEACHING FOR DEPTH: Transformations that Preserve Size and Shape

Meaning of Translation. A translation is a transformation in the plane that maps every point of a figure or graph onto an image so that every point on the image is moved the same distance from points on the preimage and in the same direction. A translation is an isometry, or rigid motion, that preserves distances, angle measures, collinearity, and betweenness of points.

A translation in the coordinate plane can be expressed as a vector using the component form $\langle a, b \rangle$, where a is the horizontal change and b is the vertical change. The horizontal change a and the vertical change b can be used to write a rule for the translation as

$(x, y) \rightarrow (x + a, y + b)$.

In this module, students gain experience with translations in the following ways.

- They can translate a figure in the plane using a vector.
- They can translate a figure in the coordinate plane using the coordinates of a preimage.
- They can identify the translation vector given the coordinates of a preimage and its image.

Students will use translations in future mathematics courses and even physics courses. When a line or segment is translated, the image is a parallel line or segment. The definition of congruence states that figures are congruent if they can be made to coincide. Therefore, one way of making figures coincide, and hence, congruent, is by a translation.

Mathematical Progressions

Prior Learning	Current Development	Future Connections
Students: • verified the properties of reflections experimentally. • learned that a two-dimensional figure is congruent to another if the second can be obtained from the first by a sequence of rotations, reflections, and translations. • described the effect of reflections on two-dimensional figures using coordinates.	**Students:** • measure and construct segments in the coordinate plane. • define a translation, rotation, and reflection as a function that preserves measures of segments and angles. • draw the image of a translated, rotated, or reflected figure. • describe rotations and reflections that carry a given figure onto itself. • examine the properties of symmetry in the plane.	**Students:** • will define dilations, stretches, and skews. • will apply sequences of transformations. • will identify congruent triangles and polygons. • will prove triangles are congruent if and only if pairs of corresponding sides and angles are congruent. • will apply the ASA, SAS, and SSS theorems to prove triangle congruence. • will use HL and AAS to prove right triangle congruence.

TEACHER TO TEACHER

From the Classroom

Implement tasks that promote reasoning and problem solving. I like to use the symmetry of regular polygons to allow my students opportunities to use their reasoning and problem solving skills.

I draw an equilateral triangle, a square, a regular pentagon, a regular hexagon, and a regular octagon on the board or project them using a dynamic geometry application. I have my students identify the type(s) of symmetry for each figure, making them draw the lines of symmetry and to justify their findings.

I encourage students use dynamic geometric software to create their own symmetric figures and have them explain each step in their constructions. I find that my students enjoy using geometry software. It helps them visualize the geometry.

LANGUAGE DEVELOPMENT • Planning for Instruction

By giving all students regular exposure to language routines in context, you will provide opportunities for students to **listen, speak, read,** and **write** about mathematical situations and develop both mathematical language and conceptual understanding at the same time.

Using Language Routines to Develop Understanding

Use the **Professional Cards** for the following routines to plan for effective instruction.

Co-Craft Questions and Problems Lesson 3

Students think of natural questions to ask about a given situation or problems similar to a given task and answer the questions they have developed or problems they have created.

Three Reads Lessons 1, 2, and 4

Students read a problem three times with a specific focus each time.

1st Read What is the situation about?
2nd Read What are the quantities in the situation?
3rd Read What are the possible mathematical questions that we could ask for the situation?

Information Gap Lesson 2

Students recognize when information given in a problem situation is incomplete, and they pose questions and share knowledge with others to discover any missing facts or relationships and work together to solve the problem.

Critique, Correct, and Clarify Lessons 1 and 4

Students correct the work in a flawed explanation, argument, or solution method and share with a partner and refine the sample work.

Linguistic Note

Listen for how students use the terms *symmetry, line symmetry, line of symmetry, rotational symmetry,* and *angle of rotational symmetry*. Although some English learners might have encountered the word *symmetry* in everyday life, they are probably not familiar with these mathematical terms.

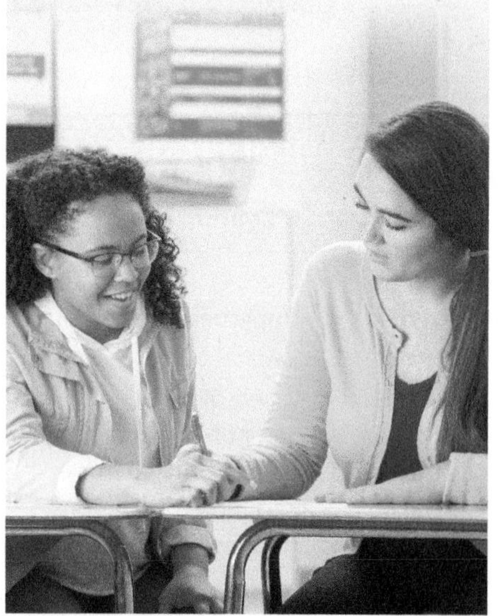

Connecting Language to Transformations that Preserve Size and Shape

Watch for students' use of the new terms listed below as they explain their reasoning and make connections with new concepts.

Key Academic Vocabulary

Current Development • New Vocabulary

angle of rotational symmetry the smallest angle of rotation that maps a figure to itself

image the corresponding points of a figure after a transformation of a preimage

isometry a rigid motion

line of symmetry a line that divides a plane figure into two congruent reflected halves

rigid motion a transformation that does not change the size or shape of a figure

rotation a rigid motion that turns a figure about a point P, such that each point and its image are the same distance from P

rotational symmetry a rotation of a figure about its center by an angle of 180° or less so that the image coincides with the preimage

transformation a function that changes the position, size, or shape of a figure or graph

vector a quantity that has both direction and magnitude

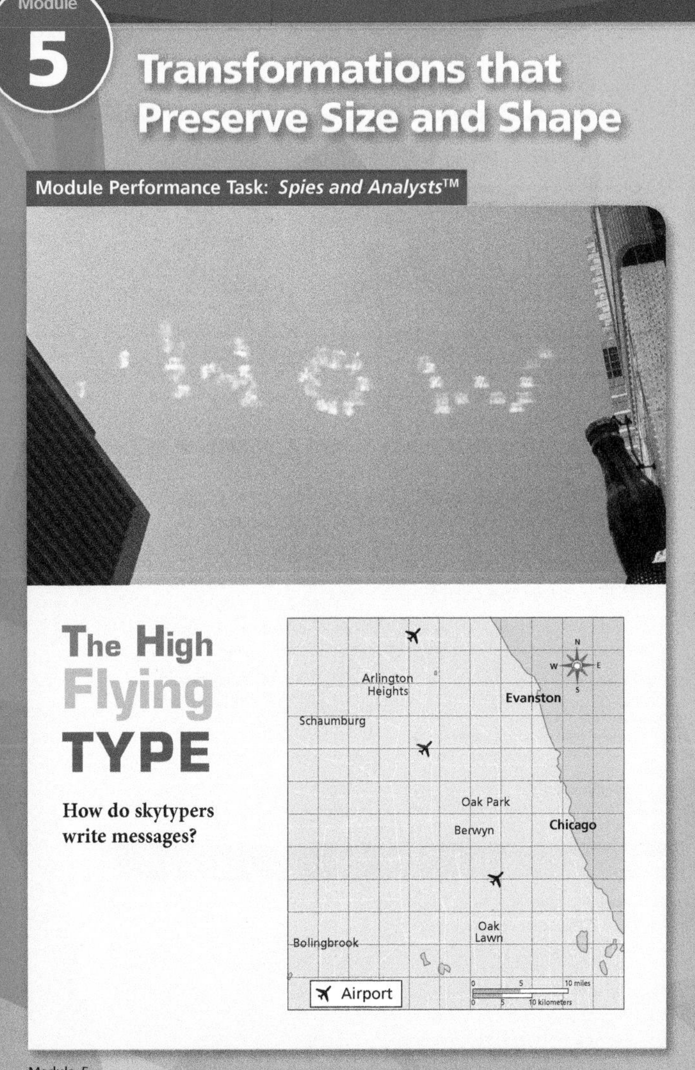

Module 5
Transformations that Preserve Size and Shape
Module Performance Task: Spies and Analysts™
The High Flying TYPE
How do skytypers write messages?
Arlington Heights
Schaumburg
Evanston
Oak Park
Berwyn
Chicago
Oak Lawn
Bolingbrook
Airport
©Werner Anans/Getty Images
Module 5
135

Connections to This Module

One sample solution might involve using the following transformations to fix skytyping problems.

- Use a reflection to create a mirror image. **(5.1)**
- Use a translation to move to a new location. **(5.1)**
- Use a rotation to change direction. **(5.2)**
- Use a dilation to change size. **(6.1)**

The High Flying TYPE

Overview

In this problem, students view a scenario about skytypers. Skytyping involves using a small plane that expels special smoke that creates a message during flight. The patterns that the plane makes while flying creates a readable message for someone on the ground. This scenario is based upon the problem described here: https://robertkaplinsky.com/work/skytypers/

Be a *Spy*

First, students must determine what information they need to know, including the following:

- What is the skytyping message?
- How can you write the message on a coordinate grid so a skytyper can know precisely where to write the message?
- How can you correct any mistakes?

Students should understand that they will be making a message using "dots" in the sky the same way that a dot matrix printer prints letters. These dots need to be printed in exact locations to clearly produce a message either in print or in the sky.

Be an *Analyst*

Help students understand that if they have made a mistake in recording their message, they can either replot the points and redraw the message by hand or use a computer to make the necessary correction(s).

Alternative Approaches

For alternate approaches to solving the problem, students could reorder the sequence of transformations they originally used to solve the problem. For example, if they performed a translation, a reflection, and then a rotation to solve the problem, they may be able to obtain the same solution by performing.

ONLINE

Assign the Digital Are You Ready? to power actionable reports including

- proficiency by standards
- item analysis

Are You Ready?

Diagnostic Assessment

- Diagnose prerequisite mastery.
- Identify intervention needs.
- Modify or set up leveled groups.

Have students complete the *Are You Ready?* assessment on their own. Items test the prerequisites required to succeed with the new learning in this module.

Translate Figures in the Coordinate Plane Students will apply previous knowledge of translations to develop a definition of a translation as a function that preserves measures of segments and angles.

Rotate Figures in the Coordinate Plane Students will apply previous knowledge of rotations to develop a definition of a rotation as a function that preserves measures of segments and angles.

Reflect Figures in the Coordinate Plane Students will apply previous knowledge of reflections to develop a definition of a reflection as a function that preserves measures of segments and angles.

ONLINE 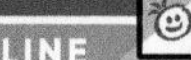

Are You Ready?

Complete these problems to review prior concepts and skills you will need for this module.

Translate Figures in the Coordinate Plane

The coordinates of a triangle are $(2, 7)$, $(5, -3)$, and $(-1, -1)$. Find the coordinates of the image after translating by the rule.

1. $(x, y) \rightarrow (x + 1, y - 2)$
$(3, 5)$, $(6, -5)$, and $(0, -3)$

2. $(x, y) \rightarrow (x - 3, y - 1)$
$(-1, 6)$, $(2, -4)$, and $(-4, -2)$

3. $(x, y) \rightarrow (x - 4, y + 2)$
$(-2, 9)$, $(1, -1)$, and $(-5, 1)$

4. $(x, y) \rightarrow (x + 3, y + 6)$
$(5, 13)$, $(8, 3)$, and $(2, 5)$

Rotate Figures in the Coordinate Plane

5. The coordinates of a triangle are $(-2, -1)$, $(1, 6)$, and $(3, -2)$. Find the coordinates of the image after rotating the triangle 180° about the origin.
$(2, 1)$, $(-1, -6)$, and $(-3, 2)$

6. The coordinates of a quadrilateral are $(-4, 0)$, $(1, 5)$, $(2, -3)$, and $(-3, -1)$. Find the coordinates of the image after rotating the quadrilateral 270° counterclockwise about the origin.
$(0, 4)$, $(5, -1)$, $(-3, -2)$, and $(-1, 3)$

7. The coordinates of a quadrilateral are $(2, 5)$, $(6, 1)$, $(1, -3)$, and $(-1, -1)$. Find the coordinates of the image after rotating the quadrilateral 90° counterclockwise about the origin.
$(-5, 2)$, $(-1, 6)$, $(3, 1)$, and $(1, -1)$

Reflect Figures in the Coordinate Plane

8. The coordinates of a triangle are $(6, -4)$, $(2, 0)$, and $(4, 4)$. Find the coordinates of the image after reflecting the triangle across the y-axis. $(-6, -4)$, $(-2, 0)$, and $(-4, 4)$

9. The coordinates of a triangle are $(-3, -1)$, $(2, -7)$, and $(5, -3)$. Find the coordinates of the image after reflecting the triangle across the x-axis. $(-3, 1)$, $(2, 7)$, and $(5, 3)$

10. The coordinates of a rectangle are $(-2, 6)$, $(-2, -3)$, $(3, 6)$, and $(3, -3)$. Find the coordinates of the image after reflecting the rectangle across the y-axis. $(2, 6)$, $(2, -3)$, $(-3, 6)$, and $(-3, -3)$

11. The coordinates of a rectangle are $(-5, 1)$, $(-5, -7)$, $(2, 1)$, and $(2, -7)$. Find the coordinates of the image after reflecting the rectangle across the x-axis.
$(-5, -1)$, $(-5, 7)$, $(2, -1)$, and $(2, 7)$

Connecting Past and Present Learning

Previously, you learned:

- to graph points in the coordinate plane and
- to transform figures in the coordinate plane.

In this module, you will learn:

- to define transformations as functions and
- to describe symmetries of figures based on transformations.

136

DATA-DRIVEN INTERVENTION

Concept/Skill	Objective	Prior Learning *	Intervene With
Translate Figures in the Coordinate Plane	Find the image of a figure that is translated in the coordinate plane.	Grade 8, Lesson 1.2	• Tier 2 Skill 9 • Reteach, Grade 8 Lesson 1.2
Rotate Figures in the Coordinate Plane	Find the image of a figure that is rotated in the coordinate plane.	Grade 8, Lesson 1.4	• Tier 2 Skill 10 • Reteach, Grade 8 Lesson 1.4
Reflect Figures in the Coordinate Plane	Find the image of a figure that is reflected in the coordinate plane.	Grade 8, Lesson 1.3	• Tier 2 Skill 11 • Reteach, Grade 8 Lesson 1.3

* Your digital materials include access to resources from Grade 6– Algebra 2. The lessons referenced here contain a variety of resources you can use with students who need support with this content.

5.1 Define and Apply Translations

LESSON FOCUS AND COHERENCE

Mathematics Standards

- Represent transformations in the plane using, e.g., transparencies and geometry; describe transformations as functions that take points in the plane as inputs and give other points as outputs. Compare transformations that preserve distance and angle to those that do not (e.g. translation versus horizontal stretch).
- Develop definitions of rotations, reflections, and translations in terms of angles, circles, perpendicular lines, parallel lines, and line segments.
- Given a geometric figure and a rotation, reflection, or translation, draw the transformed figure using, e.g., graph paper, tracing paper, or geometry software. Specify a sequence of transformations that will carry a given figure onto another.

Mathematical Practices and Processes 

- Use appropriate tools strategically.
- Attend to precision.

I Can Objective 

I can translate figures in the plane.

Language Objective

Develop a definition of translation as a function that preserves measures of segments and angles and draw the image of a figure under such a transformation.

Language Objective

Explain how to determine the location of the image of a translation relative to the preimage when given the translation vector in component form.

Vocabulary

New: component form, image, isometry, preimage, rigid motion, transformation, translation, vector

Lesson Materials: compass, straightedge

Mathematical Progressions

Prior Learning	Current Development	Future Connections
Students: • verified the properties of translations experimentally. **(Gr8, 1.2)** • understood that a two-dimensional figure is congruent to another if the second can be obtained from the first by a sequence of rotations, reflections, and translations. **(Gr8, 1.5)**	**Students:** • describe translations using translation vectors. • define translations as rigid motions. • use a compass and a straightedge to construct the image of a polygon along a translation vector.	**Students:** • will specify a sequence of transformations that will carry a given figure onto another. **(6.2)** • will use the definition of congruence in terms of rigid motions to decide if two figures are congruent. **(7.1)**

UNPACKING MATH STANDARDS

Represent transformations in the plane using, e.g., transparencies and geometry; describe transformations as functions that take points in the plane as inputs and give other points as outputs.

What It Means to You

Students have already experimented with transformations in the plane in Grade 8, but they will now investigate transformations in a more formal sense. In understanding a transformation as a function with domain and range of both the set of points in a plane, students connect concrete, geometric concepts with the more abstract concept of a function. This is emphasized by using notation such as $T(A) = A'$ to indicate that A' is the image of A under the transformation T. Students then work with a definition of rigid motions in terms of this formal definition of transformations, furthering their growth and experience with mathematical definitions.

WARM-UP OPTIONS

ONLINE

PROJECTABLE & PRINTABLE 

ACTIVATE PRIOR KNOWLEDGE • Graph Polygons on a Coordinate Plane

Use these activities to quickly assess and activate prior knowledge as needed.

Problem of the Day 

Polly lives in a city where all streets run north to south or east to west and are spaced evenly. She has drawn a city map on a coordinate plane with her home at the origin so that each 1×1 square represents a city block. Her school is 2 blocks east and 3 blocks north of her home, and the library is 1 block west and 5 blocks north of her home. Sketch the triangle with vertices at Polly's home, her school, and the library.

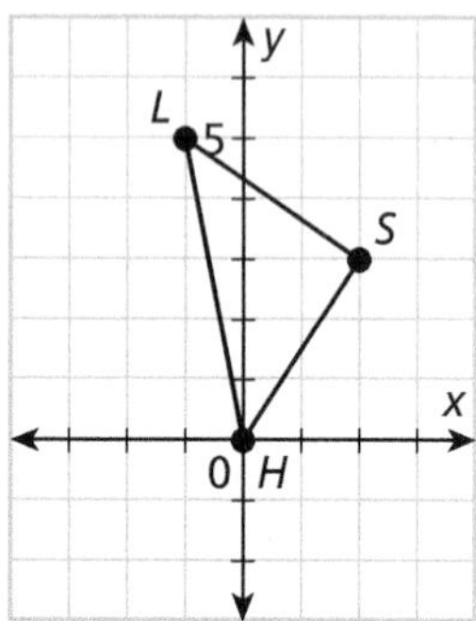

Quick Check for Homework

As part of your daily routine, you may want to display the Teacher Solution Key to have students check their homework.

Make Connections

Based on students' responses to the Problem of the Day, choose one of the following:

1. Project the Interactive Reteach, Grade 6, Lesson 4.2.
2. Complete the Prerequisite Skills Activity:

Have students work in pairs. Have them take turns sketching a triangle or quadrilateral on a coordinate plane without showing their partner, giving their partner the coordinates of the vertices, and then checking to see whether the polygon their partner sketches matches theirs.

- *How can we interpret the x-coordinate of a point?* It is the directed distance we go along the x-axis from the origin to graph the point. A positive x-coordinate means we go to the right from the origin, while a negative x-coordinate means we go to the left.
- *How can we interpret the y-coordinate of a point?* It is the directed distance we go along the y-axis from the origin to graph the point. A positive y-coordinate means we go up from the origin, while a negative y-coordinate means we go down.

SHARPEN SKILLS

If time permits, use this on-level activity to build fluency and practice basic skills.

Vocabulary Review

Objective: Students demonstrate an understanding of the definition of transformation.
Materials: Bubble Map (Teacher Resource Masters)

Have students work in small groups. Each group should work together to build a Bubble Map for the term "transformation," clarifying that we are referring to the transformations of two-dimensional figures that they explored in previous classes. If a group gets stuck, prompt them with an informal term, such as "slide" or "flip." Don't be too concerned with the formality of their language, as in this module students will formalize their definitions and understandings of transformations in the plane.

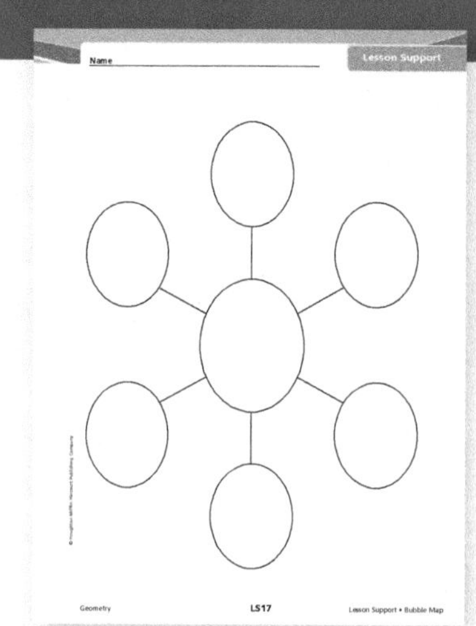

PLAN FOR DIFFERENTIATED INSTRUCTION

Small-Group Options

Use these teacher-guided activities with pulled small groups.

On Track

Materials: coordinate planes (Teacher Resource Masters), index cards

Have students work in small groups. Give each group coordinate planes and a set of index cards, some with the coordinates of a convex polygon and the others with a vector in component form. Have students do the following:

- Draw one of each type of card at a time.
- Graph the preimage and image of the described polygon along the transformation.
- Label vertices of preimages and images accordingly.

Almost There

Materials: tracing paper or transparencies, index cards

Have students work in pairs. Give each pair of students tracing paper or a transparency and a set of index cards, some with a polygon with labeled vertices and the others with an arrow representing a vector. Have students do the following:

- Take turns drawing a polygon card and a vector card.
- Sketch the translation of the polygon along the vector while their partner coaches them.

When finding the direction of the vector, students should position the cards with adjacent sides (this is to keep students from rotating the vector card to always give a horizontal translation).

- Label vertices of preimages and images accordingly.

Ready for More

Materials: coordinate planes (Teacher Resource Masters), index cards

Have students work in small groups. Give each group coordinate planes and a set of index cards, some with the coordinates of a convex polygon and the others with a vector in component form. Have students do the following:

- Draw one of each type of card at a time.
- Graph the preimage and image of the described polygon along the transformation.
- Find a pair of translation vectors that result in a transformation equivalent to that of the given vector when they are composed.
- Label vertices of preimages and images accordingly.

Math Center Options

Use these student self-directed activities at centers or stations. Key: ● Print Resources ● Online Resources

On Track

- ● Interactive Digital Lesson
- ●● Journal and Practice Workbook
- ● Interactive Glossary (printable): **component form**, **image**, **isometry**, **preimage**, **rigid motion**, **transformation**, **translation**, **vector**
- ● Module Performance Task
- ● Desmos: Translations with Coordinates

Almost There

- ● Reteach 5.1 (printable)
- ● Interactive Reteach 5.1
- ● Illustrative Mathematics: Identifying Translations

Ready for More

- ● Challenge 5.1 (printable)
- ● Interactive Challenge 5.1

View data-driven grouping recommendations and assign differentiation resources.

During the *Spark Your Learning*, listen and watch for strategies students use. See samples of student work on this page.

Use Reasoning

Strategy 1

The pattern shown has no gaps or overlaps until it reaches the edge of a wall. There are no straight parts of the pattern, so there will be gaps at the edges of the wall.

If students . . . reason about the shape of the tile, they have demonstrated that they understand that they can predict the result of repeated translations of the tile.

Have these students . . . explain how they determined their strategy and how they carried it through. **Ask:**

- Q How did you know to use the shape of the tiles in finding your answer?
- Q Can the tiles be arranged in a different way to avoid gaps at the edges of the wall?

Draw a Diagram

Strategy 2

I can draw what the kitchen tiling might look like and trace tiles from the photo to cover the surface. There are gaps left at the edges of the wall.

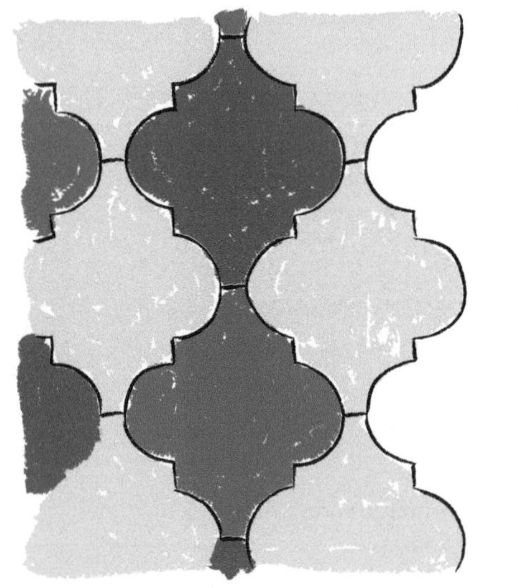

If students . . . draw tiles to cover the surface, they understand that the surface must be covered with no gaps but may not be able to predict the result of repeated translations of the tile.

Activate prior knowledge . . . by having students reason about the shape of the tile. **Ask:**

- Q Why doesn't it matter where we start tiling?
- Q Can you use the shape of the tile to predict what will happen when we come to a wall?

COMMON ERROR: Overlaps Tiles

I can draw what the kitchen floor might look like and trace tiles from the photo to cover the floor.

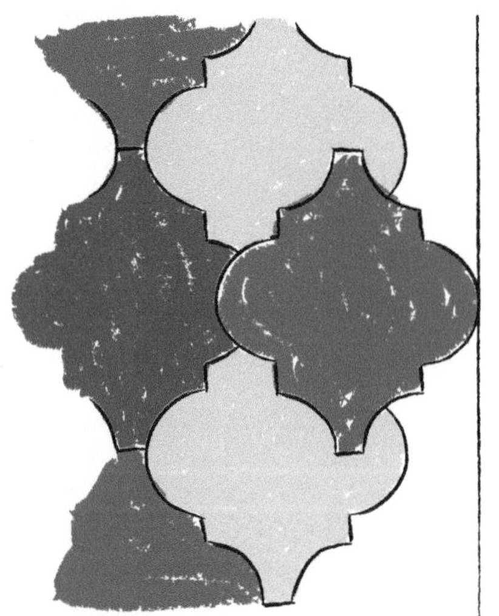

If students . . . overlap tiles, then they may not understand that a tessellation is repeated translations with no overlaps.

Then intervene . . . by pointing out that there should not be overlapping tiles. **Ask:**

- Q How should tiles be arranged if we were covering a floor or wall?
- Q What problems might we have if a surface had overlapping tiles?

Define and Apply Translations

I Can translate figures in the plane.

Spark Your Learning

Ya'ara is designing a kitchen, and her client has chosen the tiles shown.

Complete Part A as a whole class. Then complete Parts B–D in small groups.

A. What is a geometric question you can ask about the situation? What information do you need in order to answer your question?

A. How can this pattern be described with math?; how the parts of a pattern relate mathematically and how they can be changed to form new mathematical patterns

B. Once the pattern is started, describe how the position of the next tile is determined. See Additional Answers.

C. To answer your question, what strategy and tool would you use along with all the information you have? What answer do you get? See Strategies 1 and 2 on the facing page.

D. This pattern is being used on a surface in a kitchen. Why is it important for all of the tiles to have the same orientation? See Additional Answers.

Turn and Talk The pattern created by the tiles is called a tessellation. Where are tesselating patterns commonly used? What makes them useful in these situations? See margin.

SUPPORT SENSE-MAKING • Three Reads

Tell students to read the information in the photo three times and prompt them with a different question each time.

1. What is the situation about?
   The situation is about covering a surface with tiles of a certain shape.
2. What are the quantities in the situation? How are those quantities related?
   The quantities are the sizes or dimensions of the tiles and the distances between tiles; The sizes of the tiles are all the same, and the distances between tiles next to each other are all zero, since there are no gaps between tiles touching each other. We could also describe how tiles are turned or rotated with angles, with each angle having a measure of 0°.
3. What are the possible mathematical questions that you could ask for the situation?
   Possible answer: What is the area covered by one tile? Can we cover the entire surface with these tiles? How can this pattern be described mathematically?

1 Spark Your Learning

▶ MOTIVATE

- Have students look at the photo in their books and read the information contained in the photo. Then complete Part A as a whole-class discussion.
- Give the class the additional information they need to solve the problem. This information is available online as a printable and projectable page in the Teacher Resources.
- Have students work in small groups to complete Parts B–D.

▶ PERSEVERE

If students need support, guide them by asking:

Advancing • Use Tools Which tool could you use to solve the problem? Why choose that tool and not some other? Students' choices of tools and reasons for choosing them will vary.

Assessing How would you lay the tiles? Possible answer: I would start by laying one tile, then lay tiles one by one, making sure they lay next to each other and fit together like puzzle pieces.

Assessing What do you notice about the shapes and sizes of the tiles? The shapes and sizes of the tiles are all the same. All of the tiles are congruent.

Advancing Could the tiles be rotated and still make the pattern as shown? All tiles are going in the same direction, but the tiles could be rotated 180° and still result in the same pattern.

Turn and Talk Help students understand that when they describe something as a tessellation, they are only formalizing patterns with which they are already likely familiar. Help them make the distinction between tessellations and patterns in which not all figures are congruent or figures have different orientations. Possible answer: Tesselating patterns are used to cover surfaces with tiles of various sorts on floors and walls. You want the surface to be smooth, so you don't want there to be any gaps in the pattern.

▶ BUILD SHARED UNDERSTANDING

Select groups of students who used various strategies and tools to share with the class how they solved the problem. As they present their solutions, have each group discuss why they chose a specific strategy and tool.

② Learn Together

Build Understanding

Task 1 (MP) **Use Tools** Students identify translations as rigid motions and investigate their properties. Make sure students understand that a transformation is a function that acts on points in a plane.

CONNECT TO VOCABULARY

Have students use the **Interactive Glossary** to record their understanding of the vocabulary in this task.

Sample Guided Discussion:

Q **Why do you think we use notation such as *A* and *A′* when referring to this pair of points?** We want to indicate that point *A′* is the image of point *A*.

Turn and Talk Help students understand that the vector *v* indicates both the distance and direction of the translation, but the location of the vector is not important. It may be helpful to give them tracing paper or transparencies to better visualize the points of *A′B′C′* moving along the vector *v*. If you translated the preimage *A′B′C′* by the vector *v* it would move the preimage the same amount and direction as the initial translation of *ABC*, creating a new image, *A″B″C″*.

Build Understanding

Properties of Translations

A **transformation** is a function that changes the position, size, or shape of a figure or graph. A transformation maps a **preimage** determined by set of points in the plane to a corresponding set of points called the **image**.

An **isometry** or **rigid motion** is a transformation that does not change the size or shape of a figure. The properties of rigid motions are listed below.

Properties of Rigid Motions
• Rigid motions preserve distance.
• Rigid motions preserve angle measures.
• Rigid motions preserve collinearity.
• Rigid motions preserve betweenness.

One type of rigid motion is a translation, or slide. A **translation** is a transformation that maps every point of a figure or graph the same distance in the same direction. A translation is a rigid motion.

A **vector** is a quantity that has both direction and magnitude. A vector has an initial point and a terminal point. You can use a vector to describe the distance and direction of a translation.

You can use the notation $T_{\vec{v}}(A) = A'$ to identify a translation of point A to A' along the vector v or $\vec{v}$ as shown in the figures.

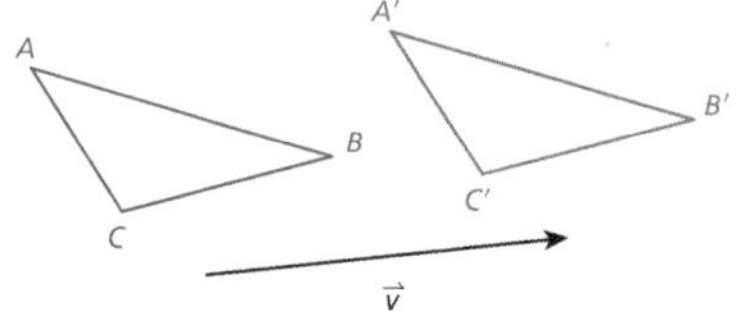

1 Confirm the properties of rigid motions under translations.

A. Copy $\triangle ABC$ and $\triangle A'B'C'$ above. What is the relationship between the lengths of corresponding sides of the triangles? A–D. See Additional Answers.

B. What is the relationship between the measures of corresponding angles in the triangles?

C. Are the corresponding sides of the triangles parallel? Justify your answer.

D. What is the distance between corresponding vertices?

 **Turn and Talk** What would happen if you translated $A'B'C'$ by the vector v? See margin.

138

LEVELED QUESTIONS

Depth of Knowledge (DOK)	Leveled Questions	What Does This Tell You?
Level 1 **Recall**	Is a translation a rigid motion? Yes, a translation only slides the preimage and does not change its size or shape.	Students' answers will indicate whether they understand the definition of rigid motions and translations.
Level 2 **Basic Application of Skills & Concepts**	Triangle *ABC* is translated along vector *v* of length 5 units. What is the distance between *A* and *A′*? 5 units	Students' answers will indicate whether they understand translation vectors.
Level 3 **Strategic Thinking & Complex Reasoning**	Describe the result of repeatedly translating images of a figure along the same vector? You would have multiple copies of the original preimage lined up and evenly spaced from each other.	Students' answers will indicate whether they can predict the result of repeated composition of a translation.

Construct a Translation

In the context of transformations, the orientation of a figure means the order of the vertices around the figure. A translation preserves the orientation of the shape because it does not change the order of the vertices.

In the figures below, ABC and $A'B'C'$ have the same orientation, but ABC and $A''B''C''$ do not have the same orientation.

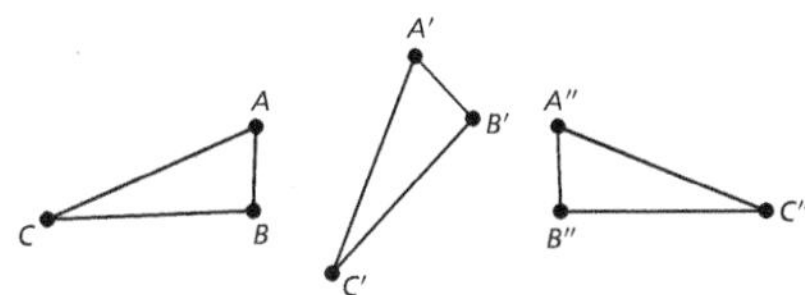

 Construct a translation of quadrilateral $CDEF$ using vector XY.

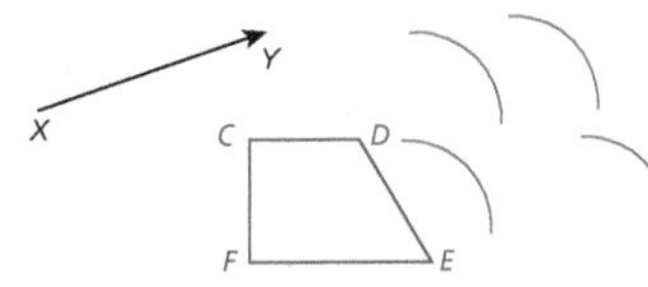

Step 1 Set a compass to the distance from X to Y.

Step 2 Place the point of the compass on C and make an arc. Repeat this step for each of the other vertices.

A. In Step 2, how do you know to make an arc above and to the right of each vertex?

A–C. See Additional Answers.

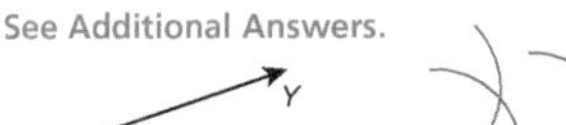

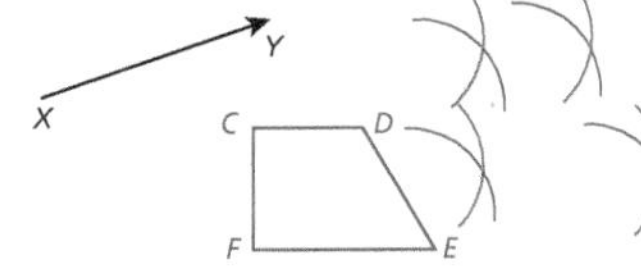

Step 3 Open your compass to the distance from X to C. Slide the compass point up the vector until the compass point is at point Y. Draw an arc from point Y so that it intersects the first arc.

Step 4 Repeat Step 3 for the other vertices.

Step 5 Use a straightedge to connect the points where the arcs intersect.

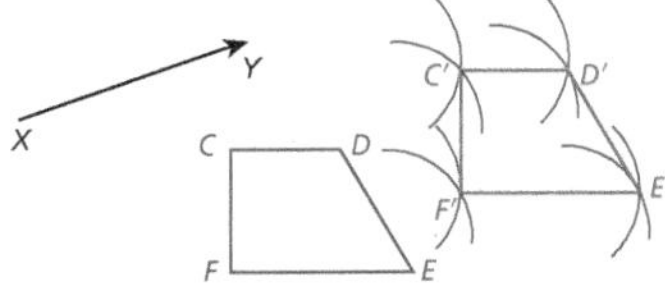

B. Are $\overline{CC'}$, $\overline{DD'}$, $\overline{EE'}$, and $\overline{FF'}$ parallel, perpendicular, or neither. Describe how you can use lined paper to check that your answer is reasonable.

C. How do you know that the distances CC', DD', EE', and FF' are all equal to XY?

 **Turn and Talk** Identify parallel lines cut by a transversal within your figure. Then, identify the corresponding angles. Use translations to justify the Corresponding Angles Postulate. See margin.

Task 2 MP **Use Tools** Students use a compass and a straightedge to construct the image of a quadrilateral along a translation vector.

Students should understand that they are constructing four segments parallel to vector XY and of the same length as the vector. These segments can be used to define vectors that are all equal to XY. Students may find it confusing to hear that two vectors with different locations are actually the same vector, but emphasize that only direction and length are needed to define a vector.

Sample Guided Discussion:

Q **What is true about any point on the arc you drew with the compass point at point C?** The distance from any point on the arc to point C is equal to the length of vector XY.

Q **What is true about the intersection point (call it Z) of the arcs you drew with the compass point at points C and Y?** The distance from any point on the arc to point Y is equal to the distance from point X to point C. This point and points X, Y, and C are the vertices of a parallelogram.

Q **What is true about vectors XY and CZ?** They are parallel and have the same length, so they are the same vector.

Turn and Talk Help students understand that while the image and preimage of quadrilaterals are composed of segments, the segments can be extended to form lines. They may want to extend the sides of the quadrilaterals beyond the vertices to make this more apparent. See students' work. Students should identify parallel lines, such as $\overline{CD}$ and $\overline{EF}$, which are cut by the transversals $\overline{CF}$ and $\overline{DE}$. Corresponding angles, such as $\angle CDE$ and $\angle C'D'E'$, are also identified. Identical translations of all angles of one figure can result in the other figure.

Step It Out

Task 3 (MP) **Use Tools** Students translate a quadrilateral in a coordinate plane along a vector described in component form.

Make sure students understand that the translation vector describes functions on the x- and y-coordinates of points, so this translation can be thought of as the composition of vertical and horizontal translations.

CONNECT TO VOCABULARY

Have students use the **Interactive Glossary** to record their understanding of the vocabulary in this task.

Sample Guided Discussion:

Q How can we think of this translation as the result of two translations in order? We can first translate quadrilateral *ABCD* 4 units to the right and then translate the resulting image 2 units down. We could also perform the vertical translation first followed by the horizontal translation and get the same result.

Q Is ⟨4, −2⟩ a point? Explain no; The brackets indicate that ⟨4, −2⟩ is a vector. We could, however, draw the vector by drawing a segment with endpoints (0, 0) and (4, −2) and putting an arrowhead at (4, −2).

Q How would a vector describing a horizontal translation be written? a vertical translation? ⟨*a*, 0⟩ and ⟨−*a*, 0⟩ describe horizontal translations, while ⟨0, *b*⟩ and ⟨0, −*b*⟩ describe vertical translations.

Turn and Talk Help students understand that while rays and vectors are drawn in the same way, our drawings are only representations so we can visualize these objects. We would not be able to see an actual ray, vector, or line segment since they have no thickness. A vector has a direction and a distinct length. A ray has a direction and a starting point, and thus, it has infinite length A line segment consists of two endpoints and all the points in between them. It has a distinct length but no direction Yes, it is possible to decompose a vector into distinct components by breaking it down to the x- and y-coordinates; You would decompose a vector into components when you take a preimage and redraw an image.

Step It Out

Translate in a Coordinate Plane

When translating a figure in a coordinate plane, you can write the translation vector in **component form** ⟨*a*, *b*⟩ where *a* is the horizontal change and *b* is the vertical change. The component from of $\overrightarrow{ST}$ in the diagram is ⟨4, −3⟩.

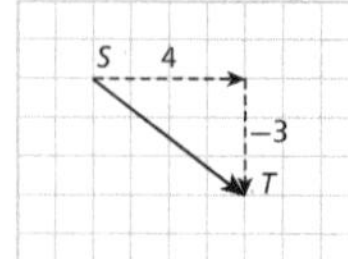

The horizontal change *a* and the vertical change *b* can be used to write a rule for a translation in the coordinate plane.

Rules for Translations on a Coordinate Plane	
Translation *a* units to the right, or positive direction	$(x, y) \to (x + a, y)$
Translation *a* units to the left, or negative direction	$(x, y) \to (x - a, y)$
Translation *b* units up, or positive direction	$(x, y) \to (x, y + b)$
Translation *b* units down, or negative direction	$(x, y) \to (x, y - b)$

The rules for translations can be combined. For example, when a figure is translated *a* units to the left and *b* units down, the rule for the translation is $(x, y) \to (x - a, y - b)$.

3 Draw the preimage and image of the polygon with vertices $A(0, 6)$, $B(4, 8)$, $C(4, 4)$, and $D(0, 2)$ translated using the vector ⟨4, −2⟩.

Preimage	Image
$A(0, 6)$	$A'(4, 4)$
$B(4, 8)$	$B'(8, 6)$
$C(4, 4)$	$C'(8, 2)$
$D(0, 2)$	$D'(4, 0)$

What coordinate rule is used to determine the image coordinates?

$(x, y) \to (x + 4, y - 2)$

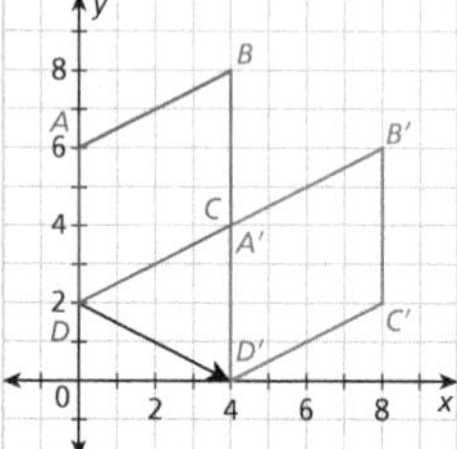

Turn and Talk How is a vector different from a ray? How is a vector different from a line segment? Is it always possible to write a vector using its horizontal and vertical components? When and why would you want to use the component form of a vector? See margin.

EL PROFICIENCY LEVEL

Beginning
Show students a quadrilateral on a coordinate plane with its image along the vector ⟨−2, 3⟩. Say, "translate left two, up three." Have students say out loud the translations described by other pairs of images and preimages under translations.

Intermediate
Have students work in pairs. For each pair, give one student a polygon on a coordinate plane and a transformation vector, and the other student a blank coordinate plane. Have the student with the preimage and vector give their partner directions while their partner graphs on their coordinate plane the image of the polygon. Make sure only the student giving the directions can see the preimage and vector. After they have completed the transformation, have them switch and try another.

Advanced
Have students explain how to determine the translation vector for a given image and preimage on a coordinate plane.

Identify a Translation Vector

4 A. Match the correct image with the correct table. Justify your answer. See Additional Answers.

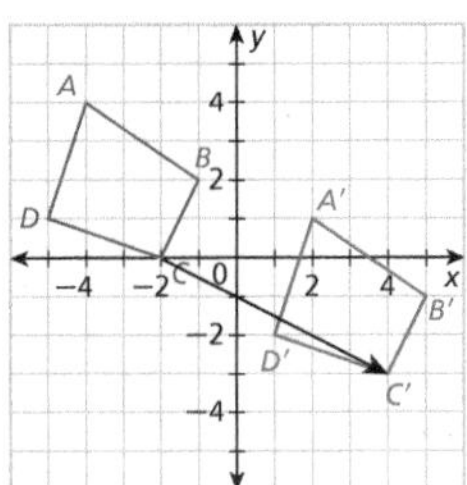

Preimage coordinates (x, y)	Image $\langle 6, -3 \rangle$
$A(-4, 4)$	$A'(2, 1)$
$B(-1, 2)$	$B'(5, -1)$
$C(-2, 0)$	$C'(4, -3)$
$D(-5, 1)$	$D'(1, -2)$

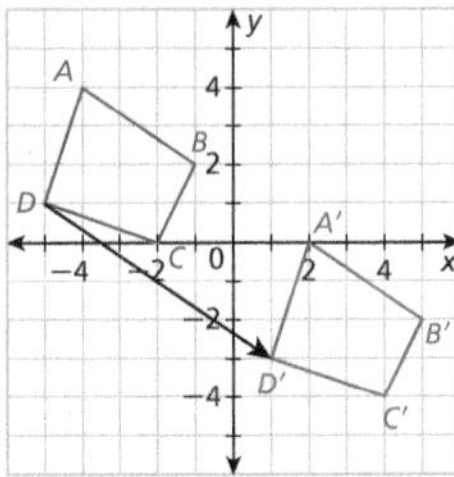

Preimage coordinates (x, y)	Image $\langle 6, -4 \rangle$
$A(-4, 4)$	$A'(2, 0)$
$B(-1, 2)$	$B'(5, -2)$
$C(-2, 0)$	$C'(4, -4)$
$D(-5, 1)$	$D'(1, -3)$

The table shows a translation of *ABCD* to *A′B′C′D′*.

Preimage coordinates (x, y)	Image $\langle ?, ? \rangle$
$A(-4, 4)$	$A'(-6, 9)$
$B(-1, 2)$	$B'(-3, 7)$
$C(-2, 0)$	$C'(-4, 5)$
$D(-5, 1)$	$D'(-7, 6)$

B. What vector maps *ABCD* to *A′B′C′D′*?

$\langle -2, 5 \rangle$

Turn and Talk Can you invert a translation? How can you identify the translation? Is it a translation if every point is mapped to itself? See margin.

Task 4 (MP) **Attend to Precision** Students identify a translation vector from a preimage and image.

Emphasize that we can think of a translation as two changes on the coordinates of points. Since a translation induces the same change on all *x*-coordinates of points and the same change on all *y*-coordinates, an interesting discussion could begin by asking how the slope of segments connecting corresponding points relates to the translation vector.

OPTIMIZE OUTPUT Critique, Correct, and Clarify

Have students work with a partner and discuss which calculations, if any, are correct for finding the translation vector for the first pair of preimage and image: $\langle -4-2, 4-1 \rangle$, $\langle 2-(-4), 4-1 \rangle$, $\langle -4-2, 1-4 \rangle$, or $\langle 2-(-4), 1-4 \rangle$. Encourage students to use the terms *translate*, *image*, and *preimage* in their discussions.

Sample Guided Discussion:

Q If we know for a fact that a pair of figures are the preimage and image of a translation, how many corresponding points do we need to find the translation vector? We only need one pair since a translation moves all points the same distance in the same direction. However, it may be helpful to confirm with additional pairs to make sure we didn't make an arithmetic error.

Q Which arithmetic operation could we use to find the translation vector from a pair of corresponding points? We could use subtraction since we want to find the change in *x*- and *y*-coordinates.

Turn and Talk Help students understand that a translation is a function on points, specifically, an invertible function. Students may need to be reminded that to invert means to reverse. Through discussion of the last question, students should reinforce their concept of an identity function. It may be helpful to prompt with, "What happens if we translate along the vector $\langle 0, 0 \rangle$?" Possible answer: Yes, you can invert a translation by reversing the process used to find the image of an original translation. It is still a translation; Yes, by definition, a translation maps every point of the preimage to the image.

ONLINE

Assign the Digital On Your Own for
- built-in student supports
- Actionable Item Reports
- Standards Analysis Reports

On Your Own

Assignment Guide

The chart below indicates which problems in the On Your Own are associated with each task in the Learn Together. Assign daily homework for tasks completed.

Learn Together Tasks	On Your Own Problems
Task 1, p. 138	Problems 20 and 21
Task 2, p. 139	Problems 6–8
Task 3, p. 140	Problems 10–16, 19, and 24
Task 4, p. 141	Problems 9, 17, 18, 22, and 23

Check Understanding

1. Quadrilateral *DEFG* is translated using $\overrightarrow{AB}$. The magnitude of $\overrightarrow{AB}$ is 26 mm. What is the length of $DD' + EE' + FF' + GG'$? 104 mm

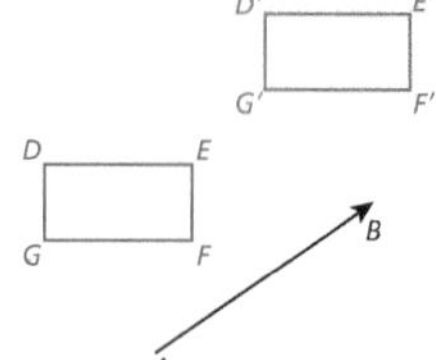

2. Copy $\triangle ABC$ shown below. Then translate the triangle using $\overrightarrow{XY}$. See Additional Answers.

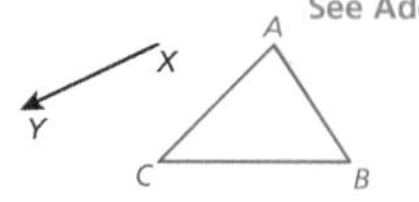

3. Draw a polygon with vertices $A(6, 7)$, $B(8, 4)$, $C(5, 2)$, and $D(2, 3)$. Then draw its image after a translation by the vector $\langle -4, -2 \rangle$. Write a coordinate rule for the translation. See Additional Answers.

Each table gives the coordinates of the vertices of a figure and its image after a translation. Give the component form of a vector that maps $\triangle PQR$ to $\triangle P'Q'R'$.

4.

$\triangle PQR$	$\triangle P'Q'R'$
$P(-3, 2)$	$P'(3, -1)$
$Q(1, 3)$	$Q'(7, 0)$
$R(-2, -1)$	$R'(4, -4)$

$\langle 6, -3 \rangle$

5.

$\triangle PQR$	$\triangle P'Q'R'$
$P(1, 2)$	$P'(-4, 3)$
$Q(5, 2)$	$Q'(0, 3)$
$R(8, -1)$	$R'(3, 0)$

$\langle -5, 1 \rangle$

On Your Own

$F'G'H'J'$ is the image of $FGHJ$ after a translation along $\overrightarrow{ST}$. Determine whether each statement is *always*, *sometimes*, or *never* true.

6. $\angle F \cong \angle G$ sometimes
7. $HH' = GG'$ always
8. $HJ = H'J'$ always

9. $A'B'C'$ is the image of ABC after a translation in the coordinate plane. Write a coordinate rule for the translation.
$(x, y) \rightarrow (x - 2, y + 8)$

Preimage coordinates	Image coordinates
$A(-6, 4)$	$A'(-8, 12)$
$B(5, 9)$	$B'(3, 17)$
$C(3, 0)$	$C'(1, 8)$

Draw the figure with the given vertices. Then draw its image after a translation by the given vector.

10. $A(-1, 5)$, $B(-1, -1)$, $C(-6, 2)$; vector $\langle 4, 2 \rangle$ 10–12. See Additional Answers.
11. $A(2, 7)$, $B(4, 6)$, $C(4, 3)$, $D(-1, 5)$; vector $\langle 5, -3 \rangle$
12. $A(-1, 5)$, $B(5, 5)$, $C(5, 3)$, $D(-1, 3)$; vector $\langle 1, 4 \rangle$

142

3 Check Understanding

Formative Assessment

Use formative assessment to determine if your students are successful with this lesson's learning objective.

Students who successfully complete the Check Understanding can continue to the On Your Own practice.

For students who miss 1 problem or more, work in a pulled small group using the Almost There small-group activity on page 137C.

ONLINE

Assign the Digital Check Understanding to determine
- success with the learning objective
- items to review
- grouping and differentiation resources

4 Differentiation Options

Differentiate instruction for all students using small-group activities and math center activities on page 137C.

Reteach

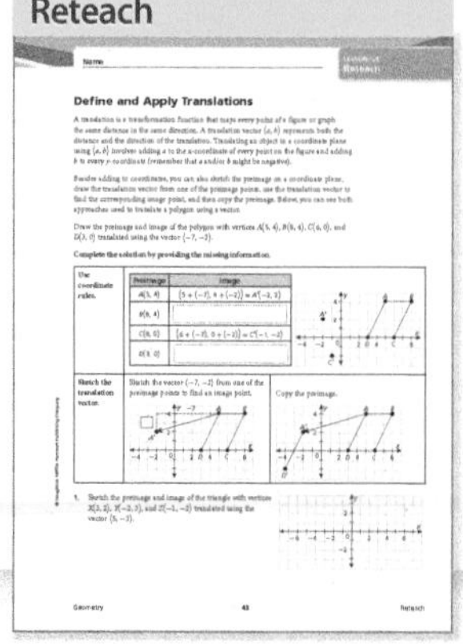

Challenge

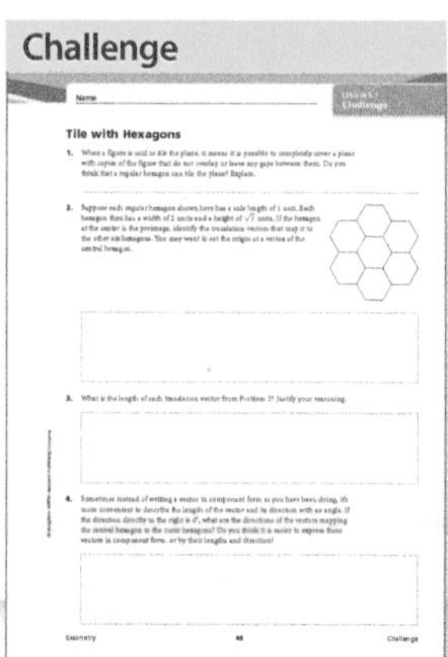

Match each set of coordinates for a preimage with the coordinates of its image after applying the vector $\langle -4, 4\rangle$.

13. $X(-6, 2), Y(-1, -1), Z(-1, 5)$ C

14. $W(9, 5), X(4, 7), Y(3, 2), Z(1, 7)$ D

15. $X(-4, -2), Y(-1, -2), Z(-1, 5)$ B

16. $W(4, 9), X(3, 2), Y(6, 9), Z(6, 2)$ A

A. $W'(0, 13), X'(-1, 6), Y'(2, 13), Z'(2, 6)$

B. $X'(-8, 2), Y'(-5, 2), Z'(-5, 9)$

C. $X'(-10, 6), Y'(-5, 3), Z'(-5, 9)$

D. $W'(5, 9), X'(0, 11), Y'(-1, 6), Z'(-3, 11)$

Specify the component form of the vector that maps each figure to its image.

17.

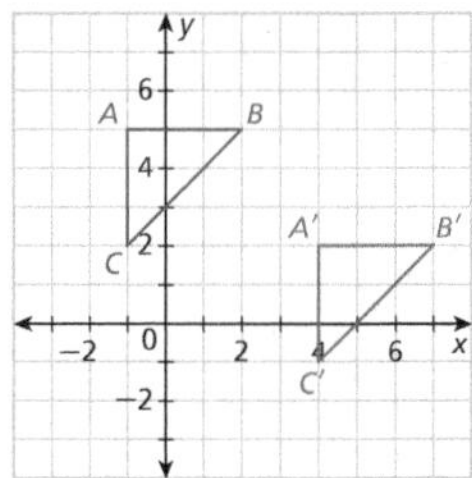

$\langle 5, -3\rangle$

18.

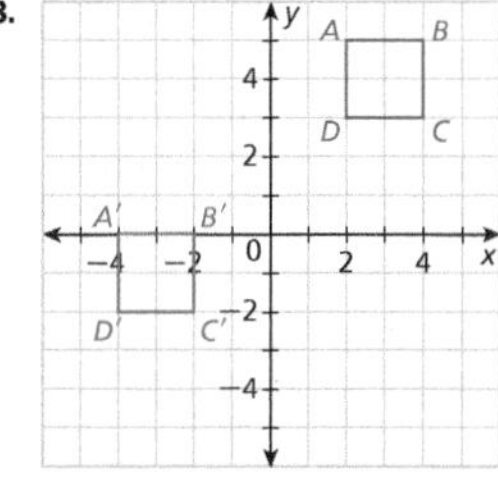

$\langle -6, -5\rangle$

19. Part of a fabric pattern is shown. When the pattern is translated so that the left edge touches the right edge or so that the top edge touches the bottom edge, the pattern continues. Describe the vectors that can be used to map A to each of the points B, C, and D. Explain your reasoning. See Additional Answers.

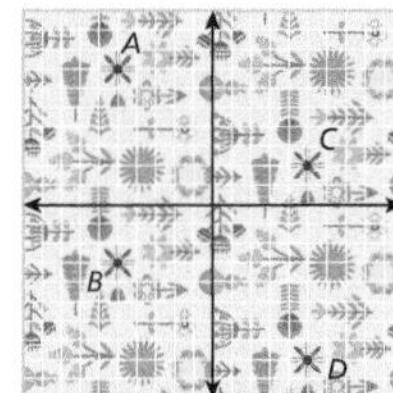

20. MP **Reason** Mark, Josh, and Will are standing in a classroom in different spots. They each draw a map of the classroom on a coordinate plane. Each student marks where the other students are standing, and places himself at the origin. Point M represents Mark, point J represents Josh, and point W represents Will. Does a translation map graph A onto graph B? graph A onto graph C? Two of the graphs are correct. Which graph is incorrect? How do you know? See Additional Answers.

A

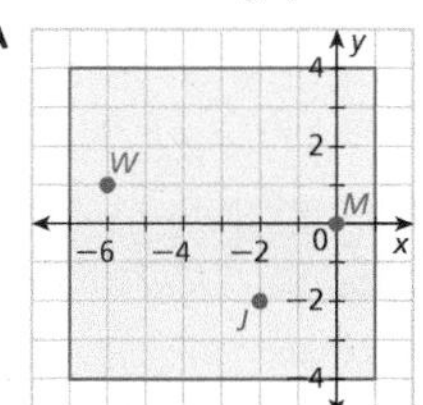

B

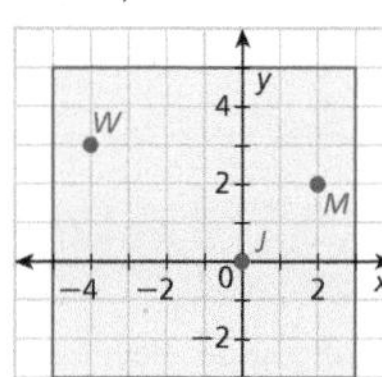

C

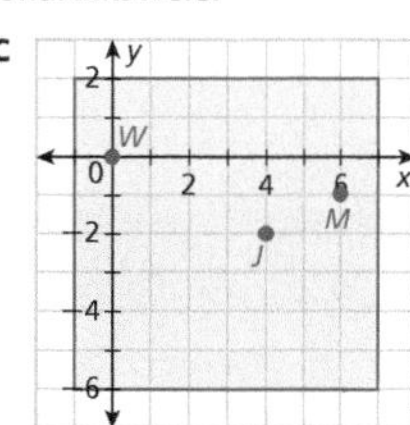

Watch for Common Errors

Problem 18 Students may be so used to going left to right on a coordinate plane that they may confuse quadrilateral $A'B'C'D'$ with the preimage and give a vector with positive components. Encourage them to carefully read notation instead of assuming the direction of translation.

Questioning Strategies

Problem 20 Suppose Mark and Josh make graphs representing maps of the classroom in a coordinate plane again (this time correctly), but with the positive y-axis in the direction they are facing. When will a translation map Josh's graph onto Mark's? This will happen when Mark and Josh are facing in the same direction. If they are not facing in the same direction, a translation will not map Josh's graph onto Mark's since the translation would have to rotate the graph. A translation does not rotate the preimage.

5 Wrap-Up

Summarize learning with your class. Consider using the Exit Ticket, Put It in Writing, or I Can scale.

Exit Ticket 

Triangle DEF has vertices $D(0, 3)$, $E(-1, -4)$, and $F(5, 1)$. What is the image when DEF is translated along the vector $v = \langle 2, -4 \rangle$? In which directions was the preimage shifted and how far? Triangle $D'E'F'$ has vertices $D'(2, -1)$, $E'(1, -8)$, and $F'(7, -3)$; The preimage was moved 2 units to the right and 4 units down.

Put It in Writing 

How can you tell the direction of a translation from the component form of its translation vector?

I Can 

The scale below can help you and your students understand their progress on a learning goal.

4	I can translate figures in the plane and explain to others how to determine a translation vector from a preimage and image in a coordinate plane.
3	I can translate figures in the plane.
2	I can describe what a translation vector does to the coordinates of a figure in a coordinate plane.
1	I can tell why a translation is a rigid motion.

Spiral Review • Assessment Readiness

These questions will help determine if students have retained information taught in the past and can also prepare them for high-stakes assessments. Here, students must find the distance between two points **(4.3)**, find the equation of a line through a point and perpendicular to another line **(4.2)**, and classify an angle by its measure **(5.2)**.

21. Mr. Smith wants to tile a wall in his bathroom. The wall is 5 feet long × 6 feet tall. He is using black and white tiles to create a checkerboard pattern. Each tile measures 6 inches by 3 inches. Will he use the same number of black and white tiles to make a checkerboard pattern? If so how many of each will he use? yes; He will use 120 black tiles and 120 white tiles.

22. The cyclist is riding her bicycle through a city. She starts by riding 6 blocks south, then 3 blocks west, then 2 blocks south, and lastly 4 blocks west. What vector describes the position of the cyclist from her starting position to her final destination? Let 1 unit represent 1 block. $\langle -7, -8 \rangle$

23. (MP) **Critique Reasoning** A student is trying to identify the vector that maps the preimage $P(1, 4)$, $Q(4, 1)$, and $R(-2, -1)$ to the image $P'(4, 1)$, $Q'(7, -2)$, and $R'(1, -4)$. The student says that it is vector $\langle -3, -3 \rangle$. Explain the error. See Additional Answers.

24. (MP) **Use Structure** Find the vertices of the preimage of a figure if the image coordinates are $A'(2, 6)$, $B'(8, 8)$, $C'(8, 4)$ and $D'(4, 2)$ and the preimage is translated using the vector $\langle 4, -4 \rangle$. $A(-2, 10)$, $B(4, 12,)$, $C(4, 8)$ and $D(0, 6)$

Spiral Review • Assessment Readiness

25. Find the distance between $(4, 6)$ and $(3, 2)$.

Ⓐ 4.1
Ⓑ 7.1
Ⓒ 8.2
Ⓓ 16.3

26. Find the equation of a line that is perpendicular to the line $y = -6x + 5$ that passes through $(-3, 2)$.

Ⓐ $y = -6x - 16$
Ⓑ $y = -\frac{1}{6}x + \frac{5}{2}$
Ⓒ $y = \frac{1}{6}x + \frac{5}{2}$
Ⓓ $y = 6x + 5$

27. For each angle, identify its classification.

Angle	Acute	Right	Obtuse	Straight
A. $m\angle C = 90°$	?	?	?	?
B. $90° < m\angle D < 180°$	?	?	?	?
C. $m\angle F < 90°$	?	?	?	?

 **I'm in a Learning Mindset!**

How can I apply what I have learned about translations to careers in the STEM field?

Keep Going to Journal and Practice Workbook

Learning Mindset

Challenge-Seeking Defines Own Challenges

Point out how challenging and interesting content can be a good motivator for exploration of careers that apply concepts related to the content. Translations have many real-world applications, especially in STEM fields. One example of a field that makes heavy use of translations is graphics programming, where images and other displayed elements are often translated to different positions. Encourage students to research careers related to the content; a simple internet search can often yield surprising results! *How does learning about how classroom concepts are used in interesting careers give you a feeling of control over your learning? Where can you go when you'd like to see how classroom concepts are used?*

5.2 Define and Apply Rotations

LESSON FOCUS AND COHERENCE

Mathematics Standards

- Represent transformations in the plane using, e.g. transparencies and geometry software; describe transformations as functions that take points in the plane as inputs and give other points as outputs. Compare transformations that preserve distance and angle to those that do not.
- Develop definitions of rotations, reflections, and translations in terms of angles, circles, perpendicular lines, parallel lines, and line segments.

Mathematical Practices and Processes 

- Use appropriate tools strategically.
- Attend to precision.

I Can Objective 

I can rotate figures in the plane.

Learning Objective

Develop a definition of rotation as a function that preserves measures of segments and angles and draw the image of a figure under such a transformation.

Language Objective

Explain how the properties of circles are used when finding the image of a figure rotated about a center by an angle.

Vocabulary

New: angle or rotation, center of rotation, rotation

Lesson Materials: compasses, geometry software, protractors, straightedges, tracing paper/transparencies

Mathematical Progressions

Prior Learning	Current Development	Future Connections
Students: • verified the properties of rotations experimentally. **(Gr8, 1.4)** • understood that a two-dimensional figure is congruent to another if the second can be obtained from the first by a sequence of rotations, reflections, and translations. **(Gr8, 1.4)** • described the effect of rotations on two-dimensional figures using coordinates. **(Gr8, 1.4)**	**Students:** • define rotations as rigid motions. • use a compass and a straightedge to construct the image of a figure given a center of rotation and an angle of rotation. • identify the coordinates of the image of a figure rotated a multiple of 90° about the origin. • identify angles mapping a regular polygon to itself when it is rotated by those angles.	**Students:** • will describe the rotations and reflections that carry a rectangle, parallelogram, or regular polygon onto itself. **(15.1)** • will specify a sequence of transformations that will carry a given figure onto another. **(6.2)**

PROFESSIONAL LEARNING

Math Background

In this lesson, students find the coordinates of images of points rotated by a multiple of 90° about the origin. For example, in Task 4 students determine which quadrilateral is the image of another quadrilateral rotated 180° about the origin by applying the rule that a point (x, y) is mapped to $(-x, -y)$ under such a rotation. Students will revisit this concept in Algebra 2 when they analyze functions' graphs for evidence of even or odd behavior, since the graph of an odd function has 180° symmetry about the origin.

WARM-UP OPTIONS

ONLINE

PROJECTABLE & PRINTABLE 

ACTIVATE PRIOR KNOWLEDGE • Construct a Perpendicular Bisector

Use these activities to quickly assess and activate prior knowledge as needed.

Problem of the Day 

Maria and Mark are marking off a volleyball court on the beach and have already measured one sideline with a measuring tape. They want to mark off a center line next. How can they do it with only their measuring tape, knowing that the center line is perpendicular to the sideline and intersects its midpoint?

Possible answer: They can construct a perpendicular bisector of the sideline. One person holds the end of the measuring tape at one end of the sideline while the other person pulls a fixed length of tape and marks an arc in the sand while pulling the tape tight, and then marks an arc in the same way from the other end of the sideline so the arcs intersect. Doing the same on the other side of the sideline gives two intersection points that lie on a perpendicular bisector of the sideline.

Make Connections

Based on students' responses to the Problem of the Day, choose one of the following:

1. Project the Interactive Reteach, Geometry, Lesson 3.3.
2. Complete the Prerequisite Skills Activity:

Have students work in pairs, giving each student a compass and a straightedge. Try to pair students who struggled with the Problem of the Day with students who appear more proficient with the prior knowledge addressed. Have them work together as they review the construction of a perpendicular bisector of a line segment.

- *What are we doing when we mark off arcs with the compass point at either end of the segment?* Since we are keeping the compass at the same setting for both, we are marking arcs that are parts of circles of the same radius.
- *How is the intersection point of one pair of the arcs related to the endpoints of the line segment?* Since it is on two circles of the same radius centered at the endpoints, the intersection point is equidistant from the endpoints.

Quick Check for Homework

As part of your daily routine, you may want to display the Teacher Solution Key to have students check their homework.

SHARPEN SKILLS

If time permits, use this on-level activity to build fluency and practice basic skills.

Mental Math

Objective: Students find angles coterminal with quadrantal angles.
Materials: index cards

Give each group a set of index cards. On one side of each card, write a multiple of 90° somewhere from 360° to 900°, and on the other side write the coterminal angle of the angle on the front of the card with a nonnegative measure less than 360°. Make sure the cards are distributed with the larger angle measure facing up.

Have students take turns drawing a card and finding the measure of the angle on the other side of the card.

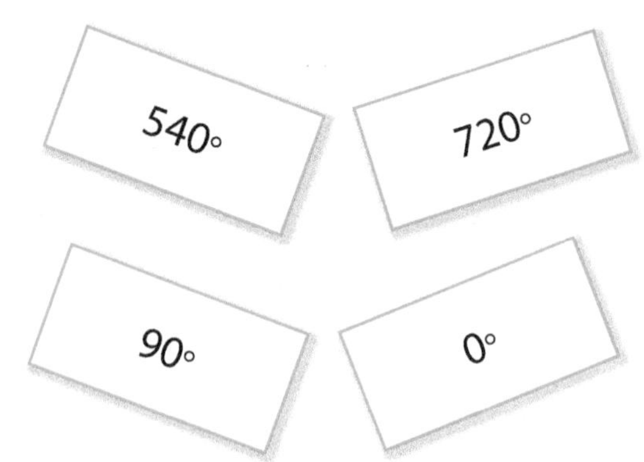

PLAN FOR DIFFERENTIATED INSTRUCTION

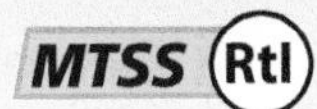

Small-Group Options

Use these teacher-guided activities with pulled small groups.

On Track

Materials: coordinate planes (Teacher Resource Masters), index cards

Have students work in small groups. Give each group coordinate planes and a set of index cards. Some of the index cards should have the coordinates of a convex polygon and the other cards should each have a nonzero multiple of 90°. Have students do the following:

- Draw one of each type of card at a time.
- Find the coordinates of the image point when the polygon is rotated counterclockwise about the origin by the angle.

Almost There

Materials: index cards, protractors, rulers, tracing paper or transparencies

Have students work in pairs. Make index cards with a polygon with labeled vertices and a labeled center of rotation not on or inside the polygon and other cards with a multiple of 10° between 0 and 180°. Have students do the following:

- Take turns drawing a polygon card and an angle card.
- Trace the polygon and center of rotation.
- Use a protractor and ruler to sketch the image.

Their partner should coach them. Encourage students to label vertices of preimages and images accordingly and check their image by rotating the transparency.

Ready for More

Materials: compasses, geometry software, protractors, straightedges, tracing paper/transparencies

Have students work in small groups. Say, "We've seen how some transformations you saw before are called rigid motions, and we can perform multiple rigid motions on a figure. Is there any way we can represent rotation of a figure about a center as a combination of other rigid motions?"

Have students experiment with whichever materials they think will help. Since students have not yet formally worked with reflections in this module, you may want to prompt them with questions about "flips" if they get stuck.

Math Center Options

Use these student self-directed activities at centers or stations.

Key: ● Print Resources ● Online Resources

On Track

- ● Interactive Digital Lesson
- ●● Journal and Practice Workbook
- ● Interactive Glossary (printable): **angle of rotation**, **center of rotation**, **rotation**
- ● Module Performance Task

Almost There

- ● Reteach 5.2 (printable)
- ● Interactive Reteach 5.2
- ● Illustrative Mathematics: Identifying Rotations

Ready for More

- ● Challenge 5.2 (printable)
- ● Interactive Challenge 5.2
- ● Illustrative Mathematics: Defining Rotations
- ● Desmos: Transformation Golf: Rigid Motion

ONLINE 

View data-driven grouping recommendations and assign differentiation resources.

During the *Spark Your Learning*, listen and watch for strategies students use. See samples of student work on this page.

Reason about Orientations of Figures

Strategy 1

While it looks like every country is upside down compared with maps I'm used to, the orientations of the countries are the same. The map is a rotation of the map I'm used to.

If students . . . reason about the orientation of figures on the map representing countries, they have demonstrated an understanding of rigid motions.

Have these students . . . explain how they determined their approach and how they came to their conclusion. **Ask:**

- Q How did you know to look at the orientations of countries on the map?
- Q How did you use the orientations of countries on the map to come to a conclusion?

Use a Physical Model

Strategy 2

I can hold a map I'm used to next to the map on the projected page and try to move it around the map on the projected page so they line up and lie directly on top of each other. If I flip the projected page map it looks like Africa is west of South America, but if I turn the map the two continents are where they're supposed to be. I can get the map on the projected page by turning the other map.

If students . . . use a physical model, they have an understanding of rigid motions but may not be visualizing them yet without visual aids.

Activate prior knowledge . . . by having students predict the result of a transformation. **Ask:**

- Q How can you visualize how a map would look if you flipped it?
- Q How can you visualize how a map would look if you turned it upside down?

COMMON ERROR: Disregards Orientation

Since the map is upside down, I can tell that you can get this map by flipping a map that we're more used to.

If students . . . disregard the orientation of figures, they may be confusing a rotation with a reflection.

Then intervene . . . by pointing out that the orientations of figures in a map would be reversed in a reflection. **Ask:**

- Q If you flipped the map on the projectable, north and south are where we are used to having them, but how about east and west?
- Q How else can you get north and south where we are used to having them?

5.2

Define and Apply Rotations

I Can rotate figures in the plane.

Spark Your Learning

Here is an unfamiliar map of the world.

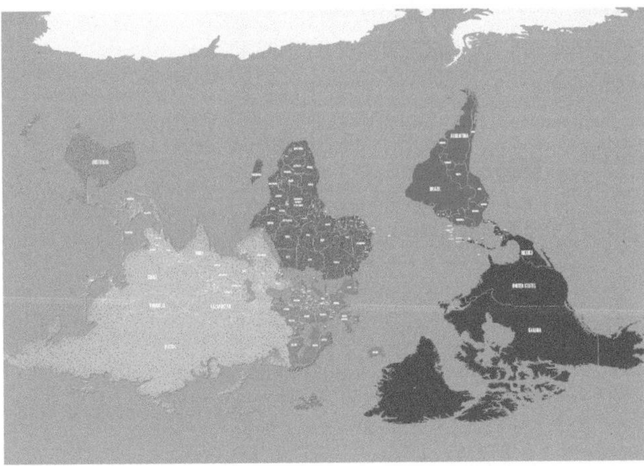

©Pyty/Shutterstock

Complete Part A as a whole class. Then complete Parts B–D in small groups.

A. What is a geometric question you can ask about this map? Is there additional information you would need to know in order to answer your question?

A. Does turning a map affect the size or shape of the countries on the map?; why the map looked like this

B. What are some reasons that you might use a map that looks like this? See Additional Answers.

C. If you took a map that looks more familiar, and tried to match the country positions, what kind of transformation could you use? To answer your question in Part A, what strategy and tool would you use along with all the information you have? What answer do you get? See Strategies 1 and 2 on the facing page.

D. What are the clues that make it clear that this map is shown in the intended position? All text, symbols, and keys are written in a readable way in the position shown.

Turn and Talk Why do you think that someone might want to use the map that has been turned in this way? See margin.

SUPPORT SENSE-MAKING • Information Gap

Ask students questions to help them decide what missing information they need to answer the question, "If you took a map that looks more familiar and tried to match the country positions, what kind of transformation could you use?"

1. Do the picture in your textbook and the projected page give you enough information to conclude the map is the result of a transformation of a more familiar map? Explain. yes; The positions of countries on the map are different from how they appear on more familiar maps.

2. Do the picture in your textbook and the projected page give you enough information to conclude the map is the result of a rigid motion of a more familiar map? Explain. yes; The shapes and sizes of objects on the map seem to be the same as they are on more familiar maps.

3. Will a more familiar map help you find what kind of transformation you could use to get the map in your textbook? Explain. yes; We can compare the positions of countries on a more familiar map with the positions of the same countries on the map in the textbook. We can also compare their orientations.

1 Spark Your Learning

▶ MOTIVATE

- Have students look at the photo in their books and read the information contained in the photo. Then complete Part A as a whole-class discussion.
- Give the class the additional information they need to solve the problem. This information is available online as a printable and projectable page in the Teacher Resources.
- Have students work in small groups to complete Parts B–D.

▶ PERSEVERE

If students need support, guide them by asking:

Q **Advancing • Use Tools** Which tool could you use to solve the problem? Why choose that tool and not some other? Students' choices of tools and reasons for choosing them will vary.

Q **Assessing** The map is unfamiliar, but is it still usable? yes; We just need to remember that the upward direction is south.

Q **Assessing** Which directions on the map represent east and west? East is now to the left, and west is to the right.

Q **Advancing** Is this map the result of a rigid motion of a map we're more used to seeing? Explain. yes; All distances appear to be unchanged, as do the shapes, sizes, and orientations of land masses and bodies of water.

Turn and Talk Help students think about the change of perspective that the rotated map shows. Possible answer: To show the countries in the southern hemisphere at the top; to make people consider the globe from a new perspective by orienting the map toward the southern pole as opposed to the northern pole.

▶ BUILD SHARED UNDERSTANDING

Select groups of students who used various strategies and tools to share with the class how they solved the problem. As they present their solutions, have each group discuss why they chose a specific strategy and tool.

2 Learn Together

Build Understanding

Task 1 **MP** **Use Tools** Students identify rotations as rigid motions and investigate their properties with geometry software.

Students have seen rotations before and may call them "turns," but encourage them to use more formal language.

Sample Guided Discussion:

Q **How are translations and rotations similar?** Both are rigid motions, so they preserve shape, size, and orientation.

Q **How are translations and rotations different?** Possible answers: A translation does not "turn" the preimage, but a rotation does. If we rotate by a multiple of 360° then every point will be mapped to itself, while the translation along ⟨0, 0⟩ is the only translation mapping every point to itself.

Turn and Talk When sketching a rotation using tracing paper or a transparency, we turn the sheet while holding it stationary about a point, so students may think a rotation rotates the plane. Help them understand that the sheet is only a model for the plane and geometrically, the plane is fixed. Possible answer: When you rotate the copy, you rotate just the figure. The vertices of the image define the endpoints of the boundary lines along which all points on the figure lie.

Build Understanding

Explore Rotations as Rigid Motions

In order to be included as a type of rigid motion, a rotation (or turn) must meet all of the criteria for rigid motions described in the previous lesson.

1 Use a geometry drawing tool to investigate a rotation.
Draw $\triangle DEF$ and point C not on the triangle. Mark C as the center. Select $\triangle DEF$ and rotate it 75° about point C. Label the image $\triangle D'E'F'$.

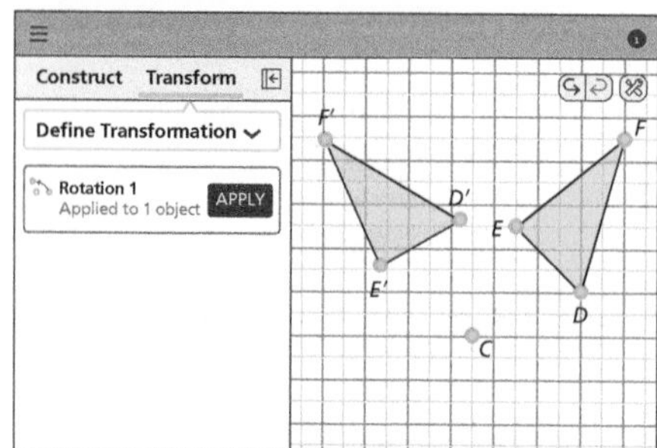

A. Does this transformation meet all of the properties of a rigid motion? Does this transformation preserve orientation? Explain your reasoning.
A, B. See Additional Answers.

B. Measure $\angle DCD'$, $\angle ECE'$, and $\angle FCF'$. Explain why these angles have the same measure.

C. Measure the distance from C to D and from C to D'. What do you notice? Does this relationship remain true as you move point C? The distance form C to D and C to D' are the same; yes

Turn and Talk Suppose you rotate a copy of the triangle, are you rotating the whole plane or just the figure? Why is it enough to find the image of the vertices and then claim that the entire figure has been rotated? See margin.

Construct a Rotation

A **rotation** is a rigid motion that turns a figure through an **angle of rotation** about a point P, such that each point and its image are the same distance from P. All the angles with vertex P formed by a point and its image are congruent. The point P is called the **center of rotation**.

A rotation is a function that takes points in the plane as inputs. The function notation $R_{C\theta}(P) = P'$ can be used for a rotation by angle θ with center C where point P' is the image of P.

When rotating about a given point, a figure can rotate counterclockwise or clockwise. When no direction is specified, you can assume a counterclockwise rotation. Also, a counterclockwise rotation of $x°$, is the same as a clockwise rotation of $(360 - x)°$.

146

LEVELED QUESTIONS

Depth of Knowledge (DOK)	Leveled Questions	What Does This Tell You?
Level 1 **Recall**	Is a rotation a rigid motion? Yes, a rotation only rotates or turns a figure and does not change its size or shape.	Students' answers will indicate whether they understand rotations as rigid motions.
Level 2 **Basic Application of Skills & Concepts**	What would happen if you rotated triangle ABC about point C by 180°, then rotated the image $A'B'C'$ about point C by 180°? The first rotation leaves ABC on the opposite side of point C from the preimage $A'B'C'$, and the second rotation maps every point of $A'B'C'$ back to its corresponding point of ABC.	Students' answers will indicate whether they understand that transformations can be composed, and that compositions of rotations can be equivalent to the identity rotation.
Level 3 **Strategic Thinking & Complex Reasoning**	Can we get translations from rotations? Yes, if I rotate a figure 180° about a point and then rotate the image 180° around another point, the final image will be a translation of the original image.	Students' answers will indicate whether they understand rotations with different centers can be composed and can reason strategically to compose reflections to obtain a desired result.

2 Rotate $\triangle ABC$ about point P by the reference angle K.

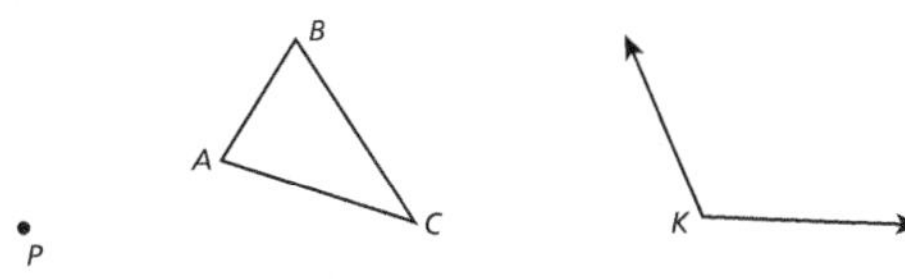

A. Draw a segment from point P to each vertex. Then construct an angle congruent to $\angle K$ with P as the vertex and C on one ray of the angle. What point will lie on the segment drawn? vertex C'

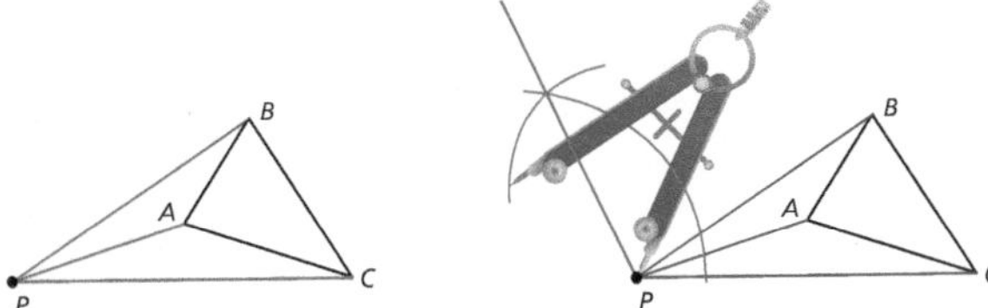

B. Use a compass to find the distance from point P to each vertex. Mark each distance on the corresponding ray. Then connect the images of the vertices. Explain how you know that $\triangle ABC \cong \triangle A'B'C'$.

A rotation is a rigid motion. Since $\triangle A'B'C'$ is a rotation of $\triangle ABC$, $\triangle ABC' \cong \triangle A'B'C'$.

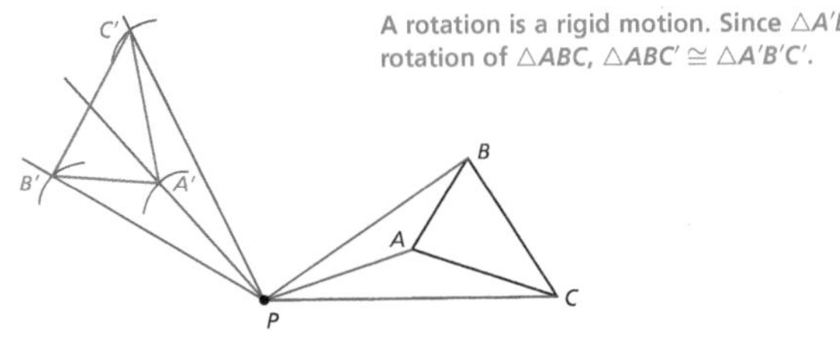

C. Construct circles with center P and with radius AP, radius BP, and radius CP. What do you notice about the circles? C–E. See Additional Answers.

D. How can tracing paper be used to check the construction?

E. Suppose you do not have a compass. Explain how to rotate $\triangle ABC$ about point P by the reference angle K using only a protractor and a ruler.

Turn and Talk How does the construction of a triangle use the properties of a rotation to produce an accurate image? See margin.

Task 2 (MP) **Use Tools** Students use a compass and a straightedge to construct the rotation of a triangle about a given center and angle.

Draw attention to the properties of circles that are used in the construction. Emphasize the fact that a rotation has two geometric parameters, a fixed point (the center of rotation) and an angle, in contrast to the single geometric parameter of a translation vector. Point out that the positions of points change as they are rotated about a fixed point but their distances to the fixed point remain unchanged.

CONNECT TO VOCABULARY

Have students use the **Interactive Glossary** to record their understanding of the vocabulary in this task.

Sample Guided Discussion:

Q **You can geometrically describe a translation with a vector. How many geometric objects do you need to describe a rotation?** two, a center and an angle

Q **Why do you construct a circle with center P and with radius $\overline{AP}$?** The point on the image A' needs to be equidistant from P with the point on the preimage A. The circle with center P and with radius $\overline{AP}$ is the set of all points equidistant with A from P, so A' will be somewhere on the circle.

Turn and Talk Help students use the properties of rotations as rigid motions to understand and explain why their image was accurate, if they were careful, and to further generalize to other triangles. It may be helpful to have them refer back to Lesson 5.1 and review the properties of rigid motions. Possible answer: A triangle uses the properties of rotation as the distance from the center of rotation, and it has equal angles to produce an accurate rotation.

Step It Out

Task 3 (MP) **Use Tools** Students use a compass and a straightedge to locate the center of a rotation given the preimage and image, and then identify the angle of the rotation.

Emphasize the idea that points move along concentric circles as they are rotated about the center, so we can use properties of circles to draw conclusions.

Sample Guided Discussion:

Q How does the perpendicular bisector of segment $\overline{BB'}$ relate to any circle through the points *B* and *B'*? The perpendicular bisector of $\overline{BB'}$ is the set of all points equidistant from *B* and *B'*, so any circle through points *B* and *B'* has its center on the perpendicular bisector.

Q Why do we construct the perpendicular bisectors of segments $\overline{BB'}$ and $\overline{CC'}$, or another pair of segments with corresponding points for endpoints? The intersection of the perpendicular bisectors is the center of rotation.

Q How can you determine which point would also lie on the circle with radius $\overline{AD}$? The preimage of triangle *ABC* has point *A* along the circle, so the rotation would result in the corresponding point lying on the circle. The image of *A'B'C'* has point *A'* along the circle as well.

Step It Out

Identify Parameters of a Rotation

When a figure is rotated, every point is moved along a circle centered at the center of rotation. This means that the points are equidistant from the center, and you can use this fact to find the center of rotation given the preimage and image of a rotation.

Find the center of rotation and the angle of rotation that maps $\triangle ABC$ onto $\triangle A'B'C'$.

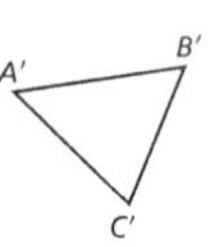

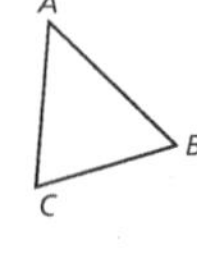

Step 1 Draw a line segment between two sets of corresponding points. Then draw the perpendicular bisectors of these segments.

Step 2 Locate the intersection of the two perpendicular bisectors. This is the center of rotation.

Step 3 Connect two corresponding vertices to the center of rotation. Use your protractor to measure this angle, which is the angle of rotation.

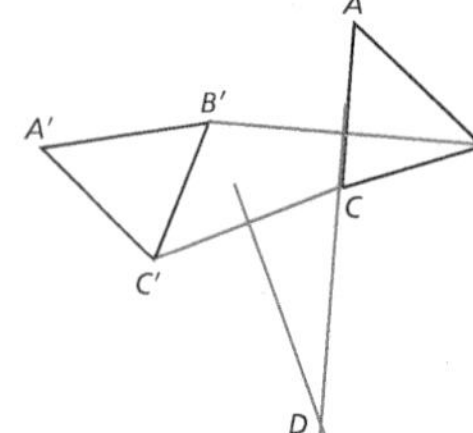

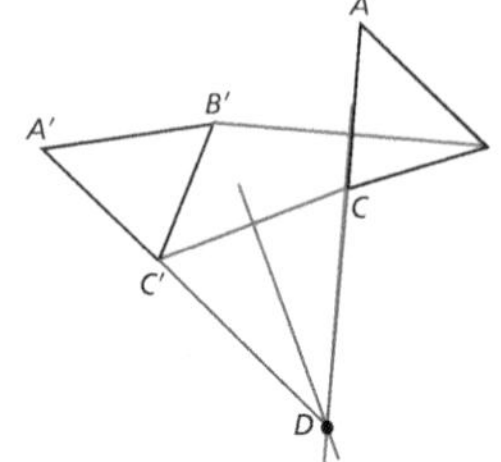

A. What point represents the center of rotation? What angle represents the angle of rotation? point *D*; $\angle CDC'$

B. Suppose a circle with radius *AD* is drawn with the compass point on *D*. Name another point that lies on the circle. point *A'*

Rotations in a Coordinate Plane

For certain rotations in the coordinate plane, there are simple rules you can use to calculate the coordinates of the image for a given preimage point.

The table summarizes the rules for counterclockwise rotations in the coordinate plane.

Rules for Counterclockwise Rotations About the Origin	
90° rotation	$(x, y) \rightarrow (-y, x)$
180° rotation	$(x, y) \rightarrow (-x, -y)$
270° rotation	$(x, y) \rightarrow (y, -x)$
360° rotation	$(x, y) \rightarrow (x, y)$

148

PROFICIENCY LEVEL

Beginning

Show students a scalene triangle *ABC* along with a point *D* not on or in the triangle, and the image *A'B'C'*. Say, "rotate triangle *ABC* 100 degrees about *D*." While doing so, you may want to emphasize the rotation with a tracing of the triangle on tracing paper or a transparency. Have students describe out loud the rotations shown by other similar diagrams.

Intermediate

Have students work in pairs and have them take turns giving their partner directions as they use a compass and a straightedge to find the parameters of a given rotation. Encourage students to use vocabulary such as "center," "image," "preimage," and "perpendicular bisector."

Advanced

Have students explain how to find the center and angle of rotation for a given image and preimage.

4 Which graph shows a 180° rotation of *ABCD*? Explain your reasoning. See Additional Answers.

A.

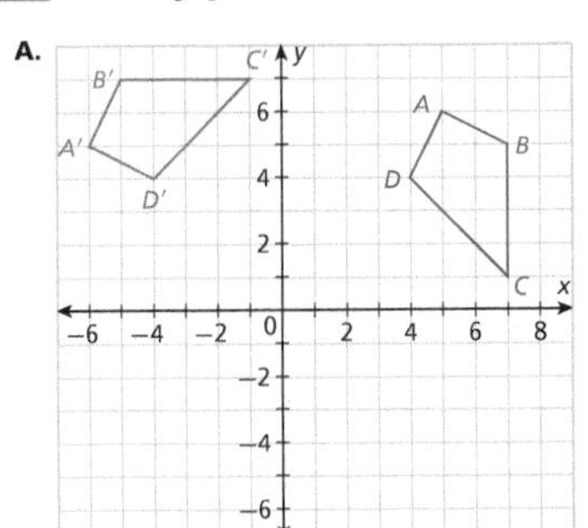

B.

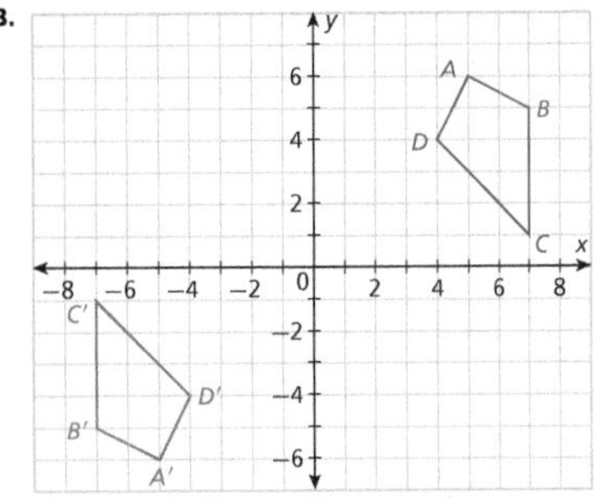

Turn and Talk How would the rules for the rotation of an image change if the center of rotation is no longer (0, 0)? See margin.

Rotate a Figure Onto Itself

Regular polygons can be rotated so that the image of the figure looks exactly like the preimage. The angles of rotation that map a regular polygon to itself depend on the number of sides of the polygon. A rotation of 360° will always map a figure to itself. The other angles of rotation that will map a regular n-gon to itself are multiples of $\frac{360}{n}$.

5 What is the smallest angle of rotation less than or equal to 360° that will map the glass tile onto itself.

A. Equilateral triangle

$\frac{360°}{3} = 120°$

A. What are all the rotations that map the triangle onto itself?

B. Square

$\frac{360°}{4} = 90°$ A, B. See margin.

B. What are all the rotations that map the square onto itself?

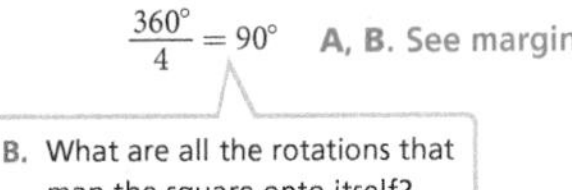

Turn and Talk Draw a polygon that is not regular that can be rotated so that it maps to itself. Identify the angles of rotation that will map the figure to itself. See margin.

ANSWERS

Task 5

A. 120°, 240°, 360°, and multiples of 120° that are greater than 360° will map the triangle onto itself.

B. 90°, 180°, 270°, 360°, and multiples of 90° that are greater than 360° will map the square onto itself.

Task 4 (MP) **Use Tools** Students calculate the coordinates of image points for rotations using coordinate rules about rotations by quadrantal angles about the origin.

SUPPORT SENSE-MAKING Three Reads

Have students read the problem three times. Use the questions below for a different focus each read.

1. What is the situation about?
2. What are the quantities in the situation?
3. What are the possible mathematical questions that you could ask for the situation?

Sample Guided Discussion:

Q How can you use the rules for counterclockwise rotations around the origin to determine which graph correctly represents the rotation of *ABCD*? Pick a point on figure *ABCD*, such as point $A(5, 6)$. A 180° rotation has a rule of $(x, y) \rightarrow (-x, -y)$, so the image $A'B'C'D'$ would have point A' at $(-5, -6)$. Then, select the graph that also has that same point.

Turn and Talk Help students understand that the axes and origin of a coordinate plane are arbitrary geometric objects used for reference. We can often make computations much easier by translating axes so a convenient reference point is our new origin, and then translating the axes back to their original locations. Possible answer: If the preimage is not rotated around the origin, translate the point of rotation to the origin, and translate the preimage using the same rule. Rotate the preimage around the origin. Then translate the preimage, image, and the point of rotation so that the preimage and point of rotation are both in the original positions.

Task 5 (MP) **Attend to Precision** Students rotate regular polygons about themselves and identify angles of rotation that map the polygons onto themselves.

Make it clear that a center is needed to define a rotation, with this center possibly in the interior of the rotated figure.

Turn and Talk Help students to see how irregular figures may only map to themselves when rotated fully. Possible answer: A parallelogram can be rotated 360°.

ONLINE

Assign the Digital On Your Own for
- built-in student supports
- Actionable Item Reports
- Standards Analysis Reports

On Your Own

Assignment Guide

The chart below indicates which problems in the On Your Own are associated with each task in the Learn Together. Assign daily homework for tasks completed.

Learn Together Tasks	On Your Own Problems
Task 1, p. 146	Problems 7–9, 11, 12, and 24–26
Task 2, p. 147	Problem 10
Task 3, p. 148	Problems 12–14
Task 4, p. 149	Problems 15–19
Task 5, p. 149	Problems 20–23

Check Understanding

In Problems 1–3, use $\triangle PQR$, point M, and $\angle Z$.

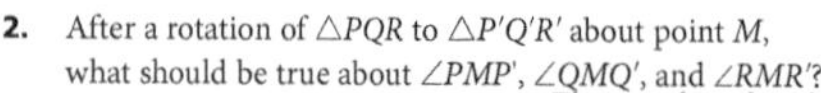

1. Copy $\triangle PQR$ and rotate it about point M by the reference angle Z. See Additional Answers.
2. After a rotation of $\triangle PQR$ to $\triangle P'Q'R'$ about point M, what should be true about $\angle PMP'$, $\angle QMQ'$, and $\angle RMR'$? The angles should be congruent.
3. After a rotation of $\triangle PQR$ to $\triangle P'Q'R'$ about point M, where will the perpendicular bisectors of $\overline{RR'}$ and $\overline{QQ'}$ intersect? The segments will intersect at point M.

Draw the image of the figure under the given rotation. 4, 5. See Additional Answers.

4. counterclockwise 90°

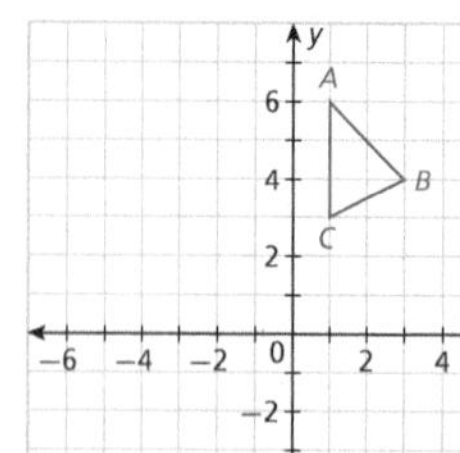

5. counterclockwise 270°

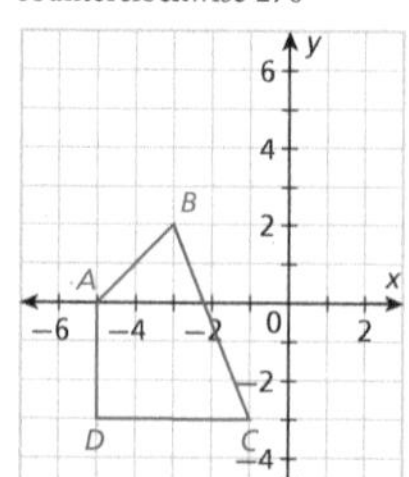

6. Sketch a polygon that will map to itself after a rotation of 60°. Check students' polygons. Possible answer: regular hexagon

On Your Own

A rotation about point N maps $\triangle FGH$ to $\triangle F'G'H'$. Tell whether the statement about the rotation is *always*, *sometimes*, or *never* true. Explain your reasoning. 7–11. See Additional Answers.

7. $\triangle FGH \cong \triangle F'G'H'$
8. $NH = NH'$
9. $\angle GHF \cong \angle HFG$
10. Copy $WXYZ$ and point Q. Rotate $WXYZ$ about point Q using the measure of $\angle WZY$ in the quadrilateral as the angle of rotation.

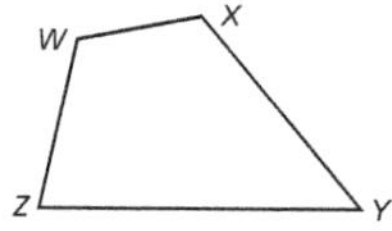

11. (MP) **Critique Reasoning** Rocco drew the image of $\triangle L'M'N'$ after a rotation of 104°. His work is shown at the right. Did Rocco rotate $\triangle L'M'N'$ correctly? Explain why or why not.

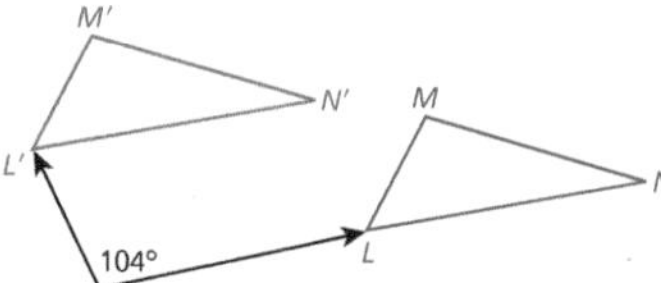

150

3 Check Understanding

Formative Assessment

Use formative assessment to determine if your students are successful with this lesson's learning objective.

Students who successfully complete the Check Understanding can continue to the On Your Own practice.

For students who miss 1 problem or more, work in a pulled small group using the Almost There small-group activity on page 145C.

ONLINE

Assign the Digital Check Understanding to determine
- success with the learning objective
- items to review
- grouping and differentiation resources

4 Differentiation Options

Differentiate instruction for all students using small-group activities and math center activities on page 145C.

Reteach

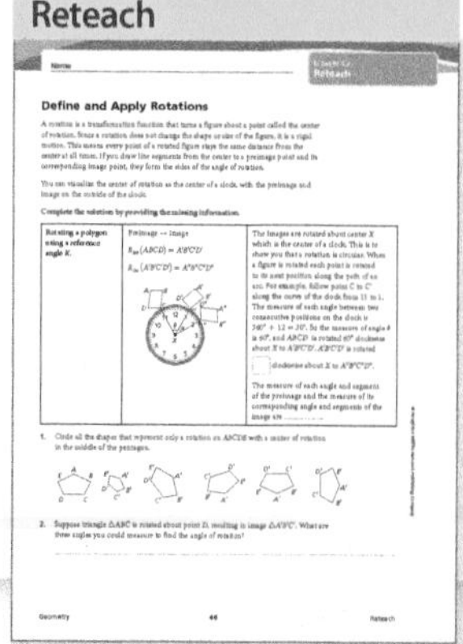

Challenge

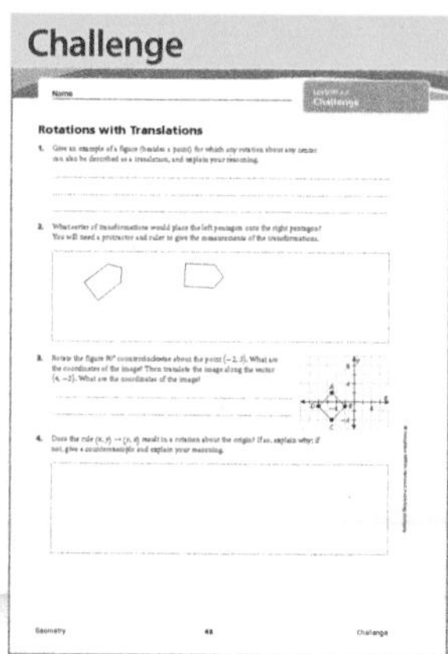

In Problems 12 and 13, copy each set of figures which show the image after a rotation and its preimage. Find the angle of rotation and the center of rotation using a compass, a straightedge, and a protractor. 12–14. See Additional Answers.

12.

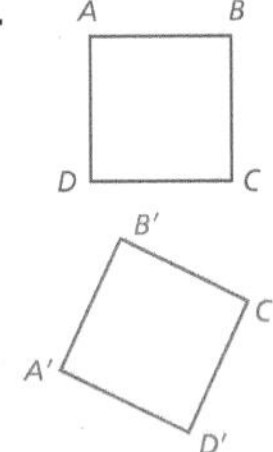

13.

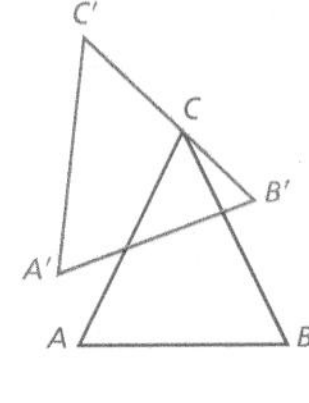

14. MP **Critique Reasoning** Mehta said that since all points turn about the center of rotation by the same angle, all points move the same distance under a rotation. Do you agree with Mehta's statement? Why or why not?

In Problems 15–17, draw the preimage and image of each polygon under the given rotation. 15–18. See Additional Answers.

15. Polygon $A(2, 5)$, $B(5, 4)$, $C(5, 2)$, $D(1, 2)$; counterclockwise 180°

16. Polygon $A(2, 7)$, $B(4, 6)$, $C(4, 3)$, $D(-1, 5)$; counterclockwise 270°

17. Rectangle $A(-1, 5)$, $B(5, 5)$, $C(5, 3)$, $D(-1, 3)$; clockwise 90°

18. Triangle ABC has vertices $A(-2, 2)$, $B(-7, 2)$, and $C(-6, 5)$. The triangle is rotated 270° about the origin. Will the coordinates of the vertices of the image of the triangle have positive or negative x-coordinates? Explain your reasoning.

19. MP **Use Repeated Reasoning** Given the preimage of a triangle with vertices at $A(7, -5)$, $B(6, -9)$, and $C(3, -6)$, rotate the image counterclockwise rotation of 1350°. What are the coordinates of the image? $A'(-5, -7)$, $B'(-9, -6)$ and $C'(-6, -3)$

Describe any rotations less than or equal to 360° that map the polygon onto itself.

20. Regular pentagon 72°, 144°, 216°, 288°, 360°

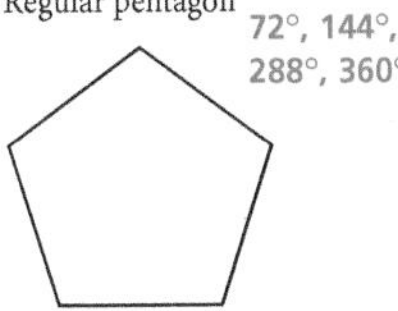

21. Regular octagon 45°, 90°, 135°, 180°, 225°, 270°, 315°, 360°

22. **History** The first Ferris wheel was invented in 1893. It had 36 cars that were equally spaced around the circumference of the wheel. The wheel rotates so that the car at the bottom of the ride is replaced by the next car. By how many degrees does the wheel rotate between consecutive cars? 10°

Watch for Common Errors

Problem 12 The regularity of the square makes several rotations possible that result in an image in the same position as the given image. Students may assume a rotation that would take A to B'. They need to carefully read the vertex labels of the preimage and image in order to correctly understand the indicated rotation.

Questioning Strategies

Problem 19 Suppose instead that triangle ABC is rotated 1350° clockwise about the origin instead. What are the coordinates of the image? First, we divide 1350° by 90°, and see that $\frac{1350°}{90°} = 15$. The remainder when 15 is divided by 4 is 3, so a clockwise rotation of 1350° is equivalent to a clockwise rotation of $3 \cdot 90° = 270°$. We then know that a clockwise rotation of 270° is equivalent to a counterclockwise rotation of 90°. So, we use the rule $(x, y) \rightarrow (-y, x)$, which gives us the image points $A'(5, 7)$, $B'(9, 6)$, and $C'(6, 3)$.

5 Wrap-Up

Summarize learning with your class. Consider using the Exit Ticket, Put It in Writing, or I Can scale.

Exit Ticket 

Triangle XYZ has vertices $X(2, 5)$, $Y(-1, 4)$, and $Z(-2, 1)$. Rotate triangle XYZ counterclockwise 270° about the origin and give the coordinates of the points of its image.

Triangle $X'Y'Z'$ has vertices $X'(5, -2)$, $Y'(4, 1)$, and $Z'(1, 2)$.

Put It in Writing 

How can you rotate a point about a fixed center with a compass and a straightedge?

I Can 

The scale below can help you and your students understand their progress on a learning goal.

4	I can rotate figures in the plane, and I can explain my steps to others.
3	I can rotate figures in the plane.
2	I can describe what a rotation about the origin by a multiple of 90° does to the coordinates of a figure in a coordinate plane.
1	I can tell why a rotation is a rigid motion.

Spiral Review • Assessment Readiness

These questions will help determine if students have retained information taught in the past and can also prepare them for high-stakes assessments. Here, students must find the distance between two points **(4.3)**, reflect a point across an axis in a coordinate plane **(5.3)**, find the equation of a line that is perpendicular to a line through two points **(4.2)**, and translate a point in a coordinate plane along a vector in component form **(5.1)**.

23. The 12-sided polygon shown is regular. Suppose the polygon is rotated counterclockwise about its center so that $\overline{AB}$ maps onto $\overline{FG}$. What is the angle or rotation? 210°

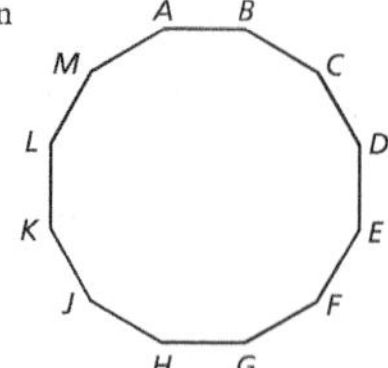

24. MP **Use Repeated Reasoning** Oliver was walking from his house to the grocery store. He started walking at 4:12 pm and arrived at the grocery store at 4:25 pm. Through what angle of rotation did the minute hand turn? 78°

The Skylon Tower, in Niagara Falls, Canada, has a revolving restaurant 775 feet above the falls. The restaurant makes a complete revolution once every hour.

25. While a visitor was at the tower, the restaurant rotated through 135°. How long was the visitor in the tower? 22.5 minutes

26. A visitor was in the tower for 40 minutes. How many degrees did the restaurant rotate? 240°

Spiral Review • Assessment Readiness

27. Find the distance between (6, 7) and (3, −12).

Ⓐ 9.25
Ⓑ 10.8
Ⓒ 12.6
Ⓓ 19.2

28. If you reflect the point (6, 2) across the x-axis, in which quadrant is the new point located?

Ⓐ I
Ⓑ II
Ⓒ III
Ⓓ IV

29. Find the equation of a line that is perpendicular to the line that passes through points (4, 6) and (−2, 3).

Ⓐ $y = -\frac{1}{2}x - 8$
Ⓑ $y = \frac{1}{2}x + 8$
Ⓒ $y = -2x - 8$
Ⓓ $y = 2x + 8$

30. Find the image of (3, −2) after you translate the point using the vector ⟨4, −2⟩.

Ⓐ (7, −4)
Ⓑ (−7, −4)
Ⓒ (7, 4)
Ⓓ (−7, 4)

 **I'm in a Learning Mindset!**

How do I know that the tasks about rotations are the right level of challenge for me?

Keep Going to Journal and Practice Workbook

Learning Mindset

Challenge-Seeking Defines Own Challenges

Point out how monitoring your own level of challenge is an important aspect of learning and that getting some questions incorrect does not necessarily mean the level is too challenging. In fact, if a student was able to rotate figures about given points without any trouble, it is probably a good sign that they are not being challenged. A good level of challenge is one where learners are initially incorrect about some things but are able to recognize and correct their errors. *How does being adequately challenged help put you in a learning mindset? What should you do if it seems you're answering all of the questions incorrectly or all of the questions correctly?*

5.3 Define and Apply Reflections

LESSON FOCUS AND COHERENCE

Mathematics Standards

- Represent transformations in the plane using, e.g., transparencies and geometry software; describe transformations as functions that take points in the plane as inputs and give other points as outputs. Compare transformations that preserve distance and angle to those that do not (e.g., translation versus horizontal stretch).
- Given a geometric figure and a rotation, reflection, or translation, draw the transformed figure using, e.g., graph paper, tracing paper, or geometry software. Specify a sequence of transformations that will carry a given figure onto another.

Mathematical Practices and Processes 

- Use appropriate tools strategically.
- Construct viable arguments and critique the reasoning of others.

I Can Objective 

I can reflect figures in a plane.

Learning Objective

Develop a definition of reflection as a function that preserves measures of segments and angles and draw the image of a figure under such a transformation.

Language Objective

Describe properties of reflection and the steps for reflecting figures on the coordinate plane using mathematical language.

Vocabulary

Review: line of reflection, reflection

Lesson Materials: tracing paper, compass, straightedge, mirrors, patterns blocks

Mathematical Progressions

Prior Learning	Current Development	Future Connections
Students: • verified the properties of reflections experimentally. **(Gr8, 1.3)** • understood that a two-dimensional figure is congruent to another if the second can be obtained from the first by a sequence of rotations, reflections, and translations. **(Gr8, 1.5)** • described the effect of a reflections on two-dimensional figures using coordinates. **(Gr8, 1.5)**	**Students:** • develop a definition of reflection as a function that preserves measures of segments and angles. • construct the image of a figure using compass and straightedge. • identify multiple transformations.	**Students:** • will describe the rotations and reflections that carry a rectangle, parallelogram, or regular polygon onto itself. **(5.4)** • will specify a sequence of transformations that will carry a given figure onto another. **(6.2)** • will use the definition of congruence in terms of rigid motions to decide if two figures are congruent. **(7.3)**

UNPACKING MATH STANDARDS

Given a geometric figure and a rotation, reflection, or translation, draw the transformed figure using, e.g., graph paper, tracing paper, or geometry software. Specify a sequence of transformations that will carry a given figure onto another.

What It Means to You

Students demonstrate understanding of identifying transformations, identifying the line of reflection, and evaluating whether a reflection has been performed correctly. They explain how to construct images using appropriate tools, capitalizing on their skills in constructing parallel and perpendicular lines. Students then construct arguments to explain how multiple transformations are applied one after another in a sequence.

WARM-UP OPTIONS

ONLINE

PROJECTABLE & PRINTABLE 

ACTIVATE PRIOR KNOWLEDGE • Transformations of Figures

Use these activities to quickly assess and activate prior knowledge as needed.

Problem of the Day 

Describe what kind of transformation will be required to move figure A to each of the figures (1–3).

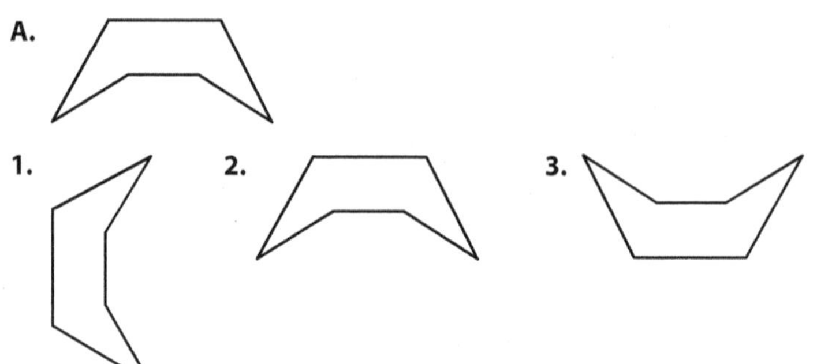

Figure 1 is obtained by rotating figure A counterclockwise; Figure 2 is translated; Figure 3 is reflected.

Quick Check for Homework

As part of your daily routine, you may want to display the Teacher Solution Key to have students check their homework.

Make Connections

Based on students' responses to the Problem of the Day, choose one of the following:

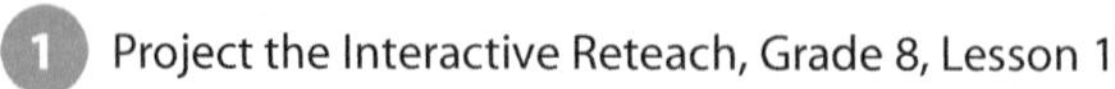

1. Project the Interactive Reteach, Grade 8, Lesson 1.

2. Complete the Prerequisite Skills Activity:

Provide students with rectangular tracing paper and have them work in pairs. Ask students to draw an irregular image and use the tracing paper to create reflections, translations, and rotations.

- *Think about the tracing paper as a rectangle. When you turn the rectangle 90°, what happens to the figure?* The figure also turns 90°.
- *If you slide your figure to the right and then down, what transformations did you apply to the figure?* a horizontal translation and then vertical translation
- *If you drew a line on the desk and flipped your paper over the line, what transformation did you apply to the figure?* a reflection

If students continue to struggle, use Tier 2 Skill 13.

SHARPEN SKILLS

If time permits, use this on-level activity to build fluency and practice basic skills.

Quantitative Comparison

Objective: Students make a comparison between two quantities. Write the following problem on the board. Ask students to discuss the problem in pairs and explain their reasoning algebraically.

Quantity A	**Quantity B**
the area of circle *A*	the area of semicircle *B* whose diameter is twice the diameter of circle *A*

A. Quantity A is greater.

B. Quantity B is greater. B; The area of circle *A* is πr^2. The area of semicircle *B* is $\frac{1}{2}\pi(2r)^2 = 2\pi r^2$.

C. The two quantities are equal.

D. The relationship cannot be determined from the information given.

PLAN FOR DIFFERENTIATED INSTRUCTION

Small-Group Options

Use these teacher-guided activities with pulled small groups.

On Track

Materials: coordinate plane (Teacher Resource Master)

Have students work in pairs. Have one student from the pair draw a design in one of the quadrants on the coordinate plane, such as shown here. Then, have the other student reflect the design in each of the other 3 quadrants.

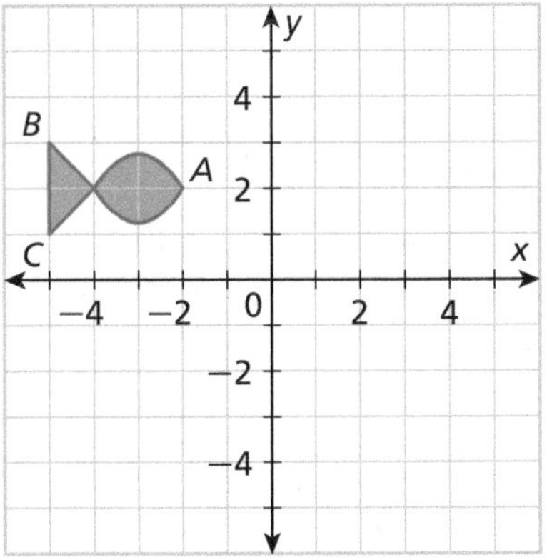

Almost There

Materials: pattern blocks, coordinate plane (Teacher Resource Master)

Have students work in pairs to do the following:

- Divide a sheet of paper into four parts.
- One student places a block in one of the quadrants.
- The other student places the same block in all other quadrants by reflecting across the axes.
- Repeat the same activity on a coordinate plane.

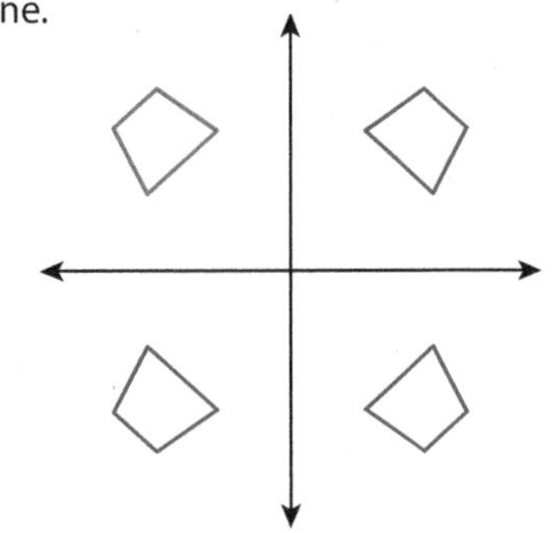

Ready for More

Materials: coordinate plane (Teacher Resource Master)

Have students work in pairs. One student draws a figure in one of the quadrants in such a way that the design falls in at least two quadrants. Then, the other student reflects the design.

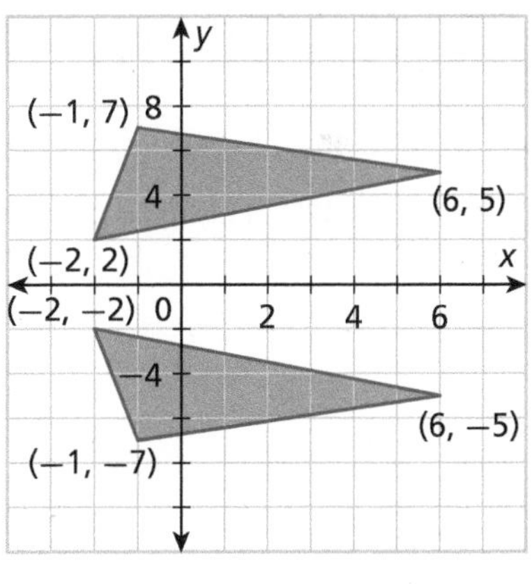

Math Center Options

Use these student self-directed activities at centers or stations. Key: ● Print Resources ● Online Resources

On Track

- ● Interactive Digital Lesson
- ●● Journal and Practice Workbook
- ● Interactive Glossary (printable): **line of reflection**, **reflection**
- ● Module Performance Task

Almost There

- ● Reteach 5.3 (printable)
- ● Interactive Reteach 5.3
- ● RtI Tier 2 Skill 13: Sequences of Transformations

Ready for More

- ● Challenge 5.3 (printable)
- ● Interactive Challenge 5.3
- ● Desmos: Transformation Golf: Rigid Motion

ONLINE 

View data-driven grouping recommendations and assign differentiation resources.

During the *Spark Your Learning*, listen and watch for strategies students use. See samples of student work on this page.

Use Spatial Reasoning

Strategy 1

I know that half of the image is a reflection of the other across a vertical line (the middle of the figure). The figure is exactly the same size and shape on both sides. Rotations and translations would not result in the same figure.

If students . . . use spatial reasoning to explain the image in terms of reflections, they are demonstrating exemplary understanding of transformations from Grade 8.

Have these students . . . explain how they determined that the image half is not translated or rotated. **Ask:**

- Q How do you know that half of the image has not been translated or rotated?
- Q If half of the figure is reflected over the vertical midline, what is true of the reflected image?

Use Folding to Model

Strategy 2

The image has two halves that are identical in size and shape. If I fold it in half, the two parts will cover each other exactly.

If students . . . use spatial reasoning by visualizing folding, they understand the meaning of reflection, but may not have well developed spatial reasoning skills.

Activate prior knowledge . . . by having students use transformation terms. **Ask:**

- Q If the left half of the figure was reflected onto the right half, what would be true of the reflected image?
- Q What is true about the two halves of the figure?

COMMON ERROR: Uses Visual Comparison

The image looks like it was drawn or painted to look the same on both sides. The two halves look the same because the artist made them to look that way.

If students . . . do not relate the image to geometric transformations, they may not fully understand they need to relate their explanation to transformations.

Then intervene . . . by pointing out that the image was painted and a transformation was used. **Ask:**

- Q How can you get the other half of the image if you only have one half of the painted halves?
- Q How would the image look if half of the image has been rotated?

5.3

Define and Apply Reflections

I Can reflect figures in the plane.

Spark Your Learning

Magid was working in the art studio and created the image below.

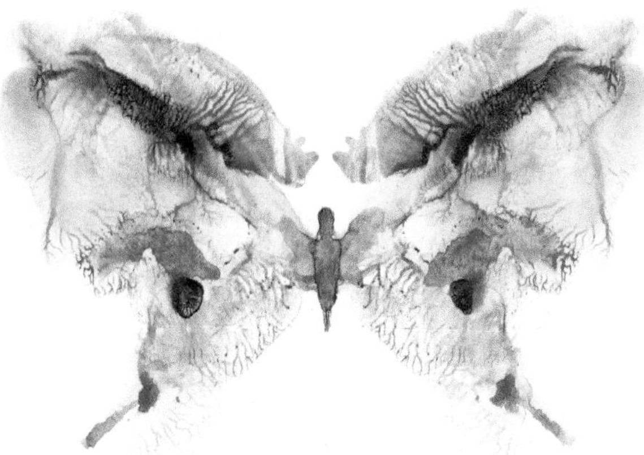

©Pobytov/DigitalVision Vectors/Getty Images

Complete Part A as a whole class. Then complete Parts B–D in small groups.

A. What is a geometric question you can ask about this image? What information would you need to know to answer your question?

A. How are the two halves of the picture related?; need to know how the image was created

B. How do you think the image was created? See Additional Answers.

C. To answer your question, what strategy and tool would you use along with all the information you have? What answer do you get? See Strategies 1 and 2 on the facing page.

D. How do you know that a translation or a rotation was not used to create the image shown? See Additional Answers.

Turn and Talk How can you use a mirror to create an image and its reflection in the same plane? How is this different from the reflection you see when you look at yourself in the mirror? See margin.

CULTIVATE CONVERSATION • Co-Craft Questions

If students have difficulty formulating a mathematical question about the situation in the Spark Your Learning, ask them to consider folding the image. What are some natural questions to ask about this situation?

Work together to craft the following questions:

- Can you use paper folding to create this picture?
- Can this image be obtained by any student?
- Can this image be obtained using a computer software?

Then have students think about what additional information, if any, they would need to answer these questions. **Ask:**

- Can you determine whether this is a complete image? Why or why not?
- Can you determine whether the image was done by hand? Explain.

1 Spark Your Learning

▶ MOTIVATE

- Have students look at the photo in their books and read the information contained in the photo. Then complete Part A as a whole-class discussion.
- Give the class the additional information they need to solve the problem. This information is available online as a printable and projectable page in the Teacher Resources.
- Have students work in small groups to complete Parts B–D.

▶ PERSEVERE

If students need support, guide them by asking:

Q **Advancing • Use Tools** Which tool could you use to solve the problem? Why choose that tool and not some other? Students' choices of tools and reasons for choosing them will vary.

Q **Assessing** If you have only part of the picture, would you be able to recreate it? The picture seems symmetrical vertically, so if I have the left half or the right half, I would be able to recreate it.

Q **Advancing** If the right half of the image is placed in Quadrant 1 on a coordinate graph, how would you obtain the other half? You have to reflect the picture over the y-axis, although they will not be connected unless the midline of the image was directly on the y-axis.

Turn and Talk Help students consider how the position of the mirror affects the reflected image. Possible answer: The mirror has to be perpendicular to the plane that the image appears in. When you look at yourself in the mirror, you and the image are in two different planes.

▶ BUILD SHARED UNDERSTANDING

Select groups of students who used various strategies and tools to share with the class how they solved the problem. As they present their solutions, have each group discuss why they chose a specific strategy and tool.

2 Learn Together

Build Understanding

Task 1 **MP Use Tools** Students use geometric tools such as tracing paper, a protractor, and a ruler to reflect an image and identify the properties of reflections.

Sample Guided Discussion:

Q If you did not have tracing paper, what other material would have helped you perform the transformation? Possible answer: I could have used graphing paper.

Q In Part A, how can you tell that you might have not performed the transformation correctly? Possible answer: if the image is larger or smaller or if point D is not on the line of reflection

Turn and Talk Remind students about a previous lesson where they had to construct parallel lines cut by a transversal, given a line and a point not on the line, using a compass and straightedge. Ask students to recall that they copied an angle. Possible answer: Consider the preimage and reflect it across the line again. Yes, you can copy it using a compass and straightedge.

Build Understanding

Explore Reflections as Rigid Motions

In order for a transformation to be considered a rigid motion, it must meet all of the criteria for rigid motions.

1 Use tracing paper to reflect an image across a line.

Draw and label a line p on tracing paper. Then draw and label a quadrilateral $DEFG$ with vertex D on line p. Fold the tracing paper along line p. Trace the quadrilateral. Then unfold the paper and label the new image $D'E'F'G'$.

A. Does the transformation meet all of the properties of a rigid motion? Does the transformation preserve orientation? Explain your reasoning. See Additional Answers.

B. Draw line segments to connect the corresponding vertices of the two quadrilaterals. Use a protractor to measure the angles formed by each line segment and line p. What do you notice? Each line segment forms a right angle to line p.

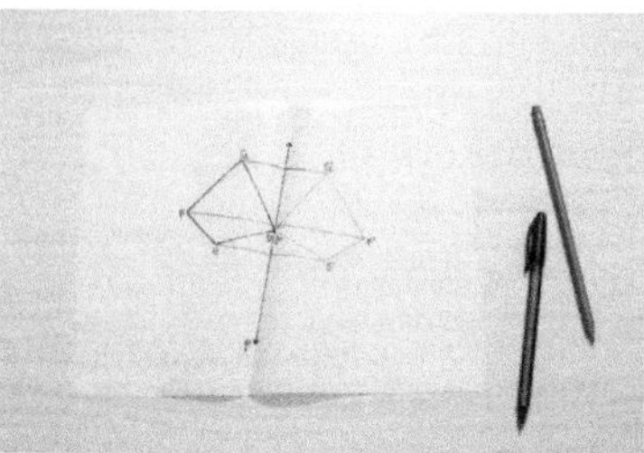

C. Using a ruler, measure each segment that connects corresponding vertices. Then measure the part of the segment that connects each vertex to line p. How is line p related to $\overline{EE'}$, $\overline{FF'}$, and $\overline{GG'}$? Line p is a perpendicular bisector of the connecting line segments.

Turn and Talk How would you map the image back to the preimage? Is it possible to do this without using a reflection? See margin.

LEVELED QUESTIONS

Depth of Knowledge (DOK)	Leveled Questions	What Does This Tell You?
Level 1 **Recall**	How can you tell that a performed transformation is a rigid motion? The corresponding segments and angles are congruent.	Students' answers will indicate whether they understand the meaning of rigid motion and will be able to identify whether it occurred.
Level 2 **Basic Application of Skills & Concepts**	What else can you use instead of tracing paper to perform the transformation across line p? I can use a compass and a straightedge to copy the preimage or I can use dynamic software.	Students' answers will indicate whether they can identify tools to use for performing transformations.
Level 3 **Strategic Thinking & Complex Reasoning**	How can you use a reflection to obtain an image in the same orientation? I will have to reflect the image again.	Students' answers will indicate whether they understand the meaning of transformations and if they can apply sequences of transformations.

Construct a Reflection

A **reflection** is a transformation across a line, called the **line of reflection**, such that the line of reflection is the perpendicular bisector of each segment joining each point and its image.

The function notation $R_m(P) = P'$ can be used for a reflection across line m where point P' is the image of P.

2 Use a compass and straightedge to reflect $\triangle ABC$ across line p.

Construct a perpendicular line from vertex A to line p. Use a compass to measure the distance from vertex A to line p along this line. Copy this segment onto the perpendicular line so that one endpoint is on line p. Label the intersection as A'.

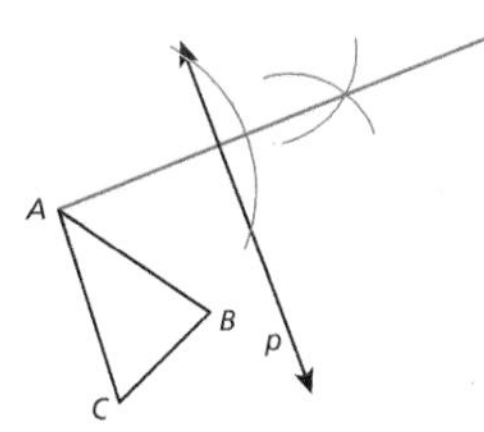

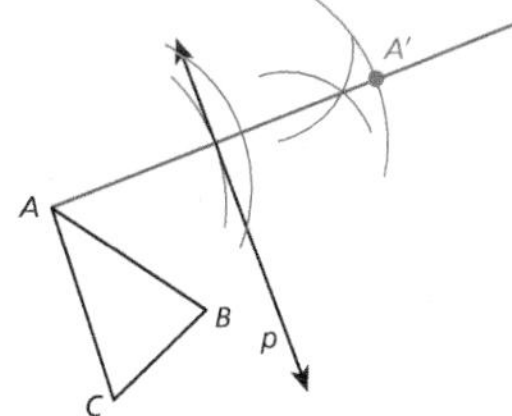

A–C. See Additional Answers.

A. Explain how you know that point A' is a reflection of point A.

B. Repeat the process to find reflections of vertex B and vertex C. Use a straightedge to connect A', B', and C'. Is $\triangle ABC \cong \triangle A'B'C'$? Explain why or why not.

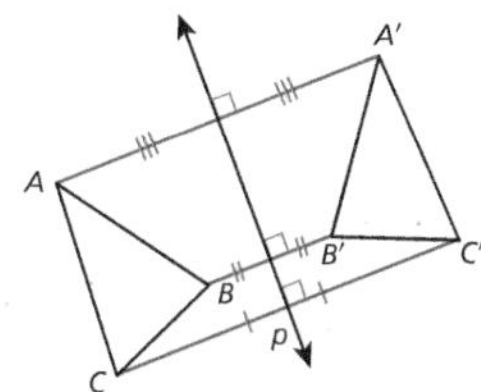

C. Suppose vertex C is reflected first instead of vertex A. Will the resulting image be the same? Explain.

Turn and Talk Explain how the construction results in the point and its image being the same distance from the line of reflection. See margin.

PROFICIENCY LEVEL

Beginning

Have students use mirrors to understand the meaning of reflections. With the help of the mirrors, have students reflect triangle *ABC*. Encourage students to use the words *reflection, preimage,* and *image* during the activity.

Intermediate

Have students explain in writing the similarity and differences between the meanings of *perpendicular* and *perpendicular bisector.*

Advanced

Have students explain how they could use the distance from each vertex to the line of reflection to construct the image, triangle $A'B'C'$.

Step It Out

Task 2 (MP) **Use Tools** Students use a compass and a straightedge to reflect a triangle across a line.

CONNECT TO VOCABULARY

Have students use the **Interactive Glossary** to record their understanding of the vocabulary in this task.

Sample Guided Discussion:

Q Can you use tracing paper to achieve the same result? Yes, you can fold the paper along line p and trace the preimage on the other half.

Q In Part B, instead of constructing another perpendicular line to line p, can you construct a parallel line to segment AA' through vertex B? Explain. yes; If I construct a parallel line to AA' through B, the line will be perpendicular to line p since AA' is perpendicular to line p.

Q Why does the order in which you reflect the vertices not affect the resulting image? If each vertex is reflected over line p correctly and the vertices are connected with line segments, then the image will be an accurate reflection of the preimage.

Turn and Talk Remind students that they copied the distance from each vertex to line p and from line p to the image along a perpendicular line. Possible answer: A compass is used to copy the distance from each vertex to the line of reflection. The image and preimage of the segment are along a line that is perpendicular to the line of reflection, so the two segments represent the distance from the vertices to the line of reflection. Copied segments have the same measure, so the point and its image are the same distance from the line of reflection.

Task 3  **Use Tools** Students explain the process for drawing a line of reflection using a compass and straightedge.

Sample Guided Discussion:

Q Can you use paper folding to find the line of reflection? If we fold the paper in such a way that the preimage and the image of the bird overlap, then the crease is the line of reflection.

Q In Step C, how are the intersecting arcs drawn? A compass was set to the same radius and centered at both A and A' to make the two intersecting arcs. The two intersection points are equidistant from both points A and A'.

Turn and Talk Remind students that if the line of reflection is a perpendicular bisector of segment AA' that connects the corresponding vertices, then the line of reflection goes through the midpoints of AA'. Possible answer: She can connect each pair of corresponding points and use a ruler to measure and find the midpoint of each segment. Then she can use the ruler to draw the line of reflection through the two midpoints.

Step It Out

Identify a Line of Reflection

If you are given a preimage and its image after a reflection, you can locate the line of reflection by finding the perpendicular bisector of segments connecting corresponding vertices.

3 The photo shows the reflection of a bird. The steps for drawing the line of reflection are out of order. Place them in the correct order. Explain each step. A–D. See Additional Answers.

A.

B.

C.

D.

 **Turn and Talk** Suppose Ellie has ruler, but she does not have compass. Explain how she can find the line of reflection in the photo using two pairs of corresponding points on the preimage and image. See margin.

(all) ©Getty Images/PhotoDisc

Reflections in a Coordinate Plane

For certain lines in a coordinate plane, there are simple rules you can use to calculate the coordinates of the image for a given preimage point.

The table summarizes the rules for reflections in the coordinate plane.

Rules for Reflection in a Coordinate Plane	
Reflection across the x-axis	$(x, y) \to (x, -y)$
Reflection across the y-axis	$(x, y) \to (-x, y)$
Reflection across the line $y = x$	$(x, y) \to (y, x)$
Reflection across the line $y = -x$	$(x, y) \to (-y, -x)$

156

4 Which graph shows the correct image of $\triangle ABC$ reflected across the line $y = -x$? Explain how you know. See Additional Answers.

A.

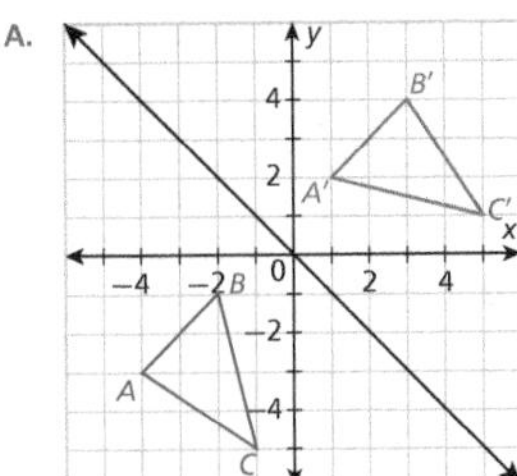

B.

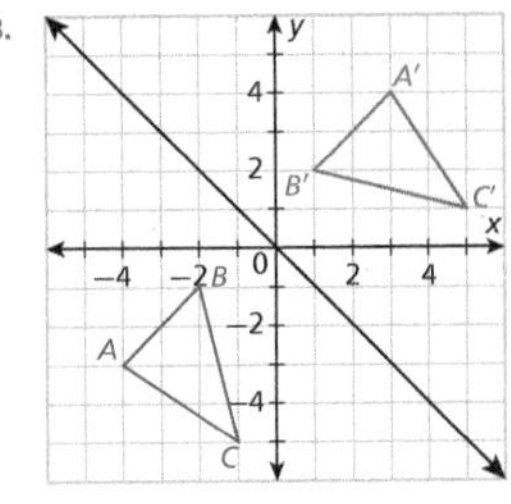

Turn and Talk How would the rules for reflecting a figure in a coordinate plane change if the line of reflection did not pass through the origin but instead was parallel to one of the four types of lines in the table of reflection rules? See margin.

Perform Multiple Transformations

Transformations can be applied one after another in a sequence where you use the image of the first transformation as the preimage for the next transformation.

Pairs of reflections can be used to generate translations and rotations.

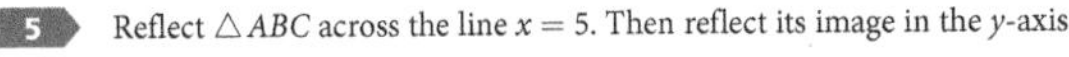

5 Reflect $\triangle ABC$ across the line $x = 5$. Then reflect its image in the y-axis.

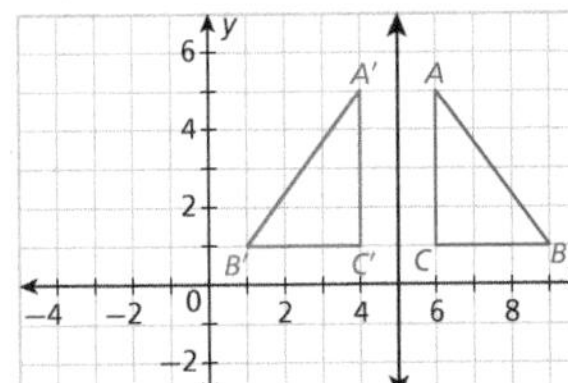

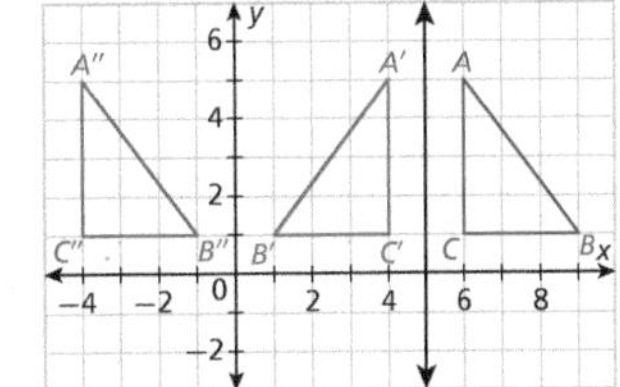

A. Describe one transformation that maps $\triangle ABC$ to $\triangle A''B''C''$. Translate $\triangle ABC$ using the vector $\langle -10, 0 \rangle$.

B. What kind of transformation is the result of reflecting a figure across a line and then reflecting its image across a line parallel to the first line of reflection? a translation

Turn and Talk Draw a triangle in a coordinate plane. Reflect the triangle across the x-axis, and then reflect its image across the y-axis. Describe one transformation that maps the resulting image to the original triangle. See margin.

Task 4 (MP) **Use Tools** Students apply their knowledge about coordinate planes and linear functions to identify coordinates of an image on a coordinate plane.

Sample Guided Discussion:

Q Could you identify the correct image if the vertices of the image were not labeled? No, I would not be able to because the two figures are identical.

Q How can you use the rule from the table to determine the coordinate pairs for each vertex of the image? The rule $(x, y) \to (-y, -x)$ can be applied to the coordinates of each vertex to find the coordinates of the image vertices. So, $A(-4, -3)$ reflected is $A'(3, 4)$.

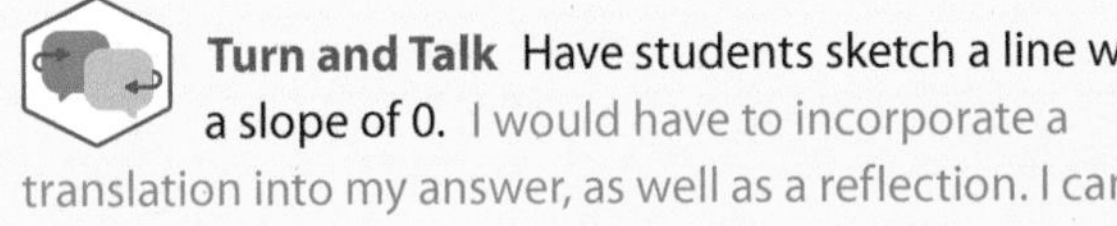

Turn and Talk Have students sketch a line with a slope of 0. I would have to incorporate a translation into my answer, as well as a reflection. I can follow these rules.

- Temporarily translate the line of reflection so that it does pass through the origin. Also translate the figure to be reflected in the same way.
- Perform the reflection according to the appropriate rule from the table of reflection rules.
- Identify the reverse of the original translation and apply it to the translated line of reflection, the translated original figure, and the image of the translated original figure. The translated image of the translated original figure is now the image of the original figure after reflection across the original line.

Task 5 (MP) **Construct Arguments** Students will explain how to reflect a triangle across a vertical line and then across the y-axis.

Sample Guided Discussion:

Q In Part A, how can you describe the changes in the coordinates of each vertex? The x-coordinates of each vertex move 10 units to the left but the y-coordinates stay the same.

Q If triangle ABC were not reflected, what transformation could you use to result in triangle $A''B''C''$? a slide or translation

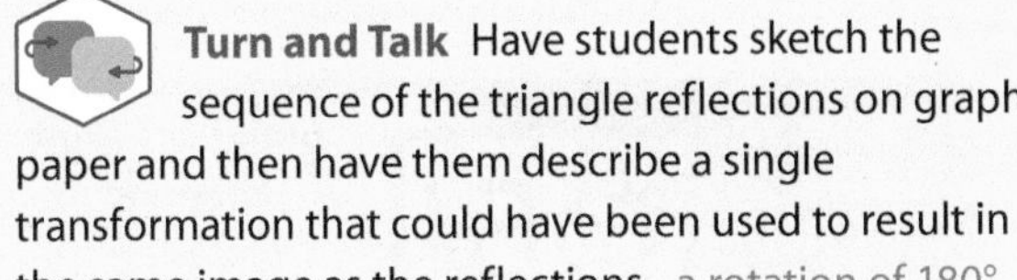

Turn and Talk Have students sketch the sequence of the triangle reflections on graph paper and then have them describe a single transformation that could have been used to result in the same image as the reflections. a rotation of 180°

ONLINE

Assign the Digital On Your Own for

- built-in student supports
- Actionable Item Reports
- Standards Analysis Reports

On Your Own

Assignment Guide

The chart below indicates which problems in the On Your Own are associated with each task in the Learn Together. Assign daily homework for tasks completed.

Learn Together Tasks	On Your Own Problems
Task 1, p. 154	Problem 6
Task 2, p. 155	Problems 7–9
Task 3, p. 156	Problems 12, 13, 19
Task 4, p. 157	Problems 10, 11, 14, 15, and 20
Task 5, p. 157	Problems 16–18 and 21

Check Understanding

1. Use the figure and the line of reflection.
   A. Copy the figure and the line of reflection. Use a compass and a straightedge to construct the reflection of the figure across the line.
   B. Use a straightedge to connect each vertex of the preimage with its corresponding image. What should be true about the segments?
   A, B. See Additional Answers.

In each diagram, $\triangle A'B'C'$ is the image of $\triangle ABC$ after a reflection. Copy the triangles and draw the line of reflection.

2. 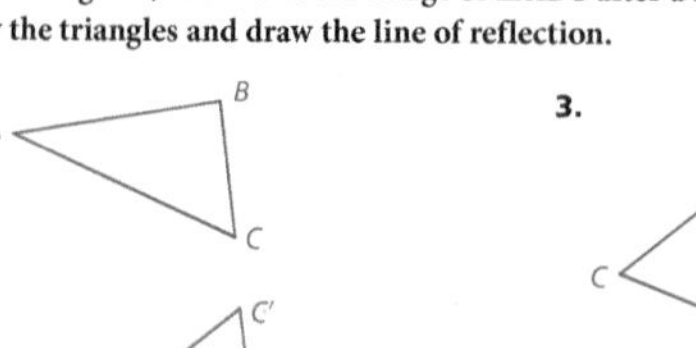

3. 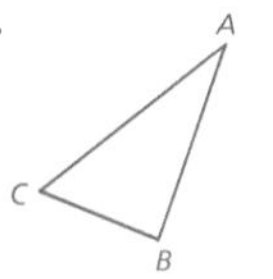 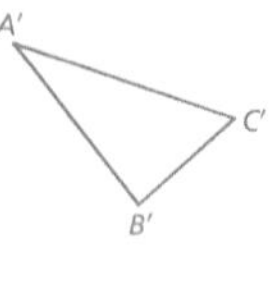

2–5. See Additional Answers.

In a coordinate plane, quadrilateral *PQRS* has vertices $P(0, 7)$, $Q(4, 6)$, $R(2, 3)$, and $S(-1, 3)$. Find the coordinates of the vertices of the image after each reflection.

4. Reflection across the x-axis
5. Reflection across the line $y = x$

On Your Own

6. Quadrilateral $K'L'M'N'$ is the image of *KLMN* after a reflection across line q. If vertex M lies on line q, then what is true about point M and its image M'? Point M and point M' are the same point.

Copy each figure and the line of reflection. Use a compass and a straightedge to construct the reflection across the line. Label the vertices of the image. 7, 8. See Additional Answers.

7. 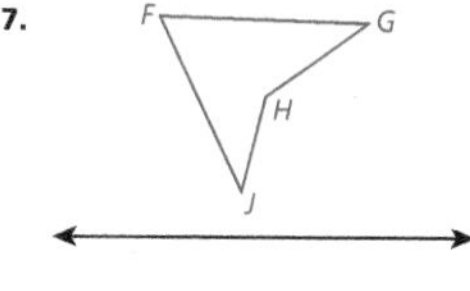

8. 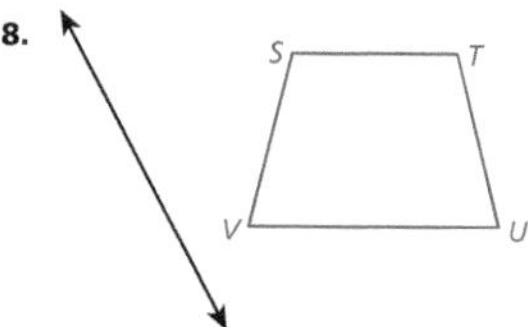

9. A trail designer is planning two trails that connect campsites A and B to a point on the river. She wants the total length of the trails to be as short as possible.
   A. Copy the points and the line. Find the image of point B after a reflection across the line. Label the point B'.
   B. Draw $\overline{AB'}$. Find the intersection of $\overline{AB'}$ and the line of reflection. Label this point X. Then draw $\overline{BX}$.
   C. What is the shortest distance between point A and point B'? Use your answer, along with the fact that $\overline{BX} \cong \overline{B'X}$, to explain why $AX + BX$ is least when X is in this position. A–C. See Additional Answers.

River

158

data checkpoint

3 Check Understanding

Formative Assessment

Use formative assessment to determine if your students are successful with this lesson's learning objective.

Students who successfully complete the Check Understanding can continue to the On Your Own practice.

For students who miss 1 problem or more, work in a pulled small group using the Almost There small-group activity on page 153C.

ONLINE

Assign the Digital Check Understanding to determine

- success with the learning objective
- items to review
- grouping and differentiation resources

4 Differentiation Options

Differentiate instruction for all students using small-group activities and math center activities on page 153C.

Reteach

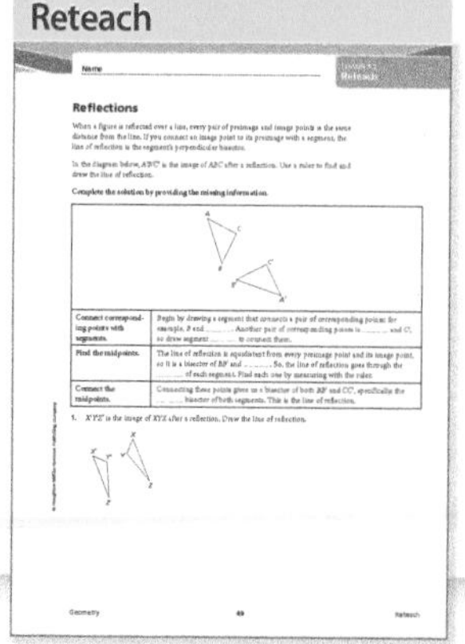

Challenge

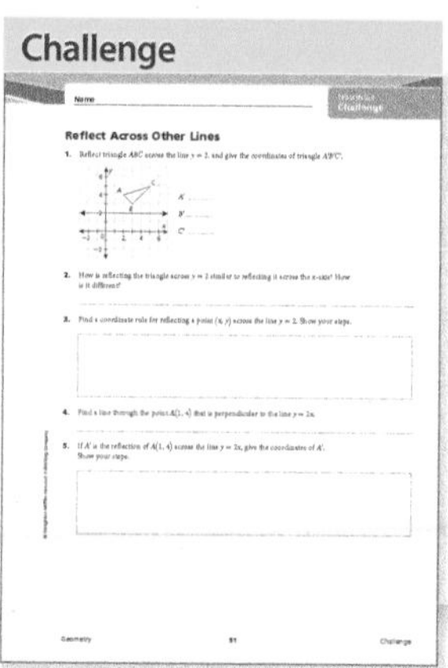

Each diagram of a hole at a miniature golf course shows the starting position for a ball and the hole. Draw a diagram for the path of a ball that will reach the hole in one shot.
10, 11. See Additional Answers.

10.

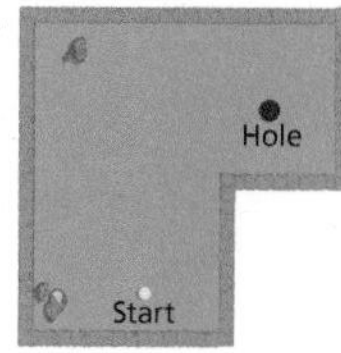

11.

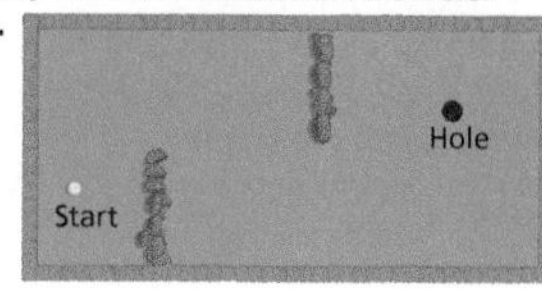

In each diagram, the red figure is the image of the blue figure after a reflection. Copy the figures and draw the line of reflection. 12, 13. See Additional Answers.

12.

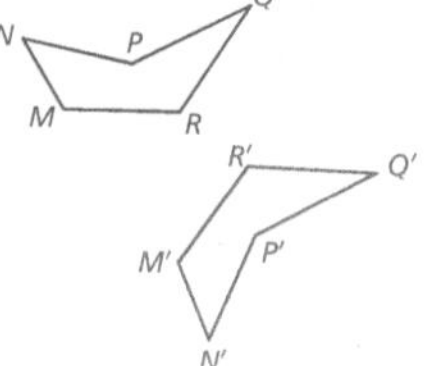

13.

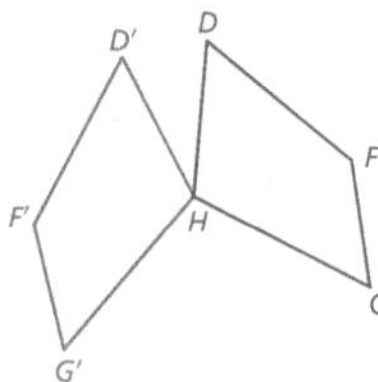

Find the images of the points in the pattern after the given reflection.
14, 15. See Additional Answers.

14. $A\left(\frac{5}{2}, \frac{5}{2}\right)$, $B(4, 0)$, $C\left(\frac{1}{2}, 3\right)$ reflected across $y = -x$

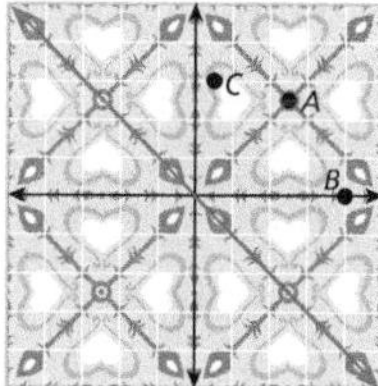

15. $A(-1, 1)$, $B(2, 3)$, $C(-2, -2)$ reflected across the y-axis

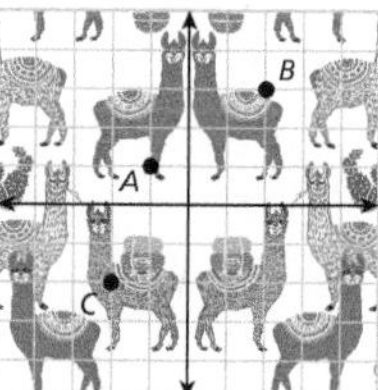

In the diagram, $\triangle ABC \cong \triangle PQR$.
16–18. See Additional Answers.

16. Describe one transformation that maps $\triangle ABC$ to $\triangle PQR$.

17. Describe two reflections that map $\triangle ABC$ to $\triangle PQR$.

18. Reflect $\triangle ABC$ across the line $y = x$. Describe how to map the image to $\triangle PQR$.

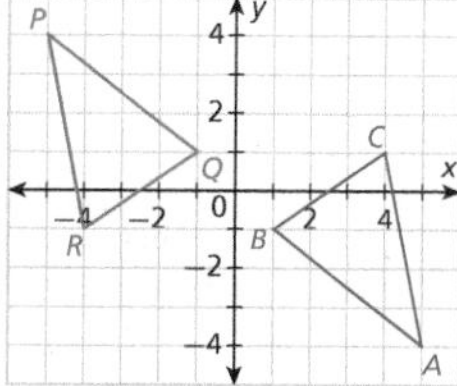

Questioning Strategies

Problem 13 How would you find the line of reflection using tracing paper? I would fold the paper through point H so that D and D' would overlap. The crease will be the reflection line.

Watch for Common Errors

Problem 14 Point out to students the rules of reflection. Remind them that reflections across the lines $y = x$ and $y = -x$ result in swapped coordinates.

5 Wrap-Up

Summarize learning with your class. Consider using the Exit Ticket, Put It in Writing, or I Can scale.

Exit Ticket 

Describe how to reflect rectangle *ABCD* across line *p* given that vertex *D* lies on line *p* and side $\overline{CD}$ is perpendicular to line *p*. *If point D is on line p and $\overline{CD}$ is perpendicular to line p, then side $\overline{AD}$ and its image $A'D'$ coincide and lie on line p. I would not need to draw perpendicular lines since $\overline{AD}$ and $\overline{CD}$ are already perpendicular. I will extend $\overline{AB}$ and $\overline{CD}$ and use a compass to copy the distances AB and CD to mark B′ and C′. Then I will connect B′ and C′.*

Put It in Writing 

Explain in writing how you would translate an image on a coordinate plane using the rule $(x, y) \rightarrow (x + 7, y - 1)$.

I Can 

The scale below can help you and your students understand their progress on a learning goal.

4	I can reflect figures in a plane, and I can explain my steps to others.
3	I can reflect figures in a plane.
2	I can perform most of the steps to reflect a figure using a straightedge and compass.
1	I can reflect figures using tracing paper.

Spiral Review • Assessment Readiness

These questions will help determine if students have retained information taught in the past and can also prepare them for high-stakes assessments. Here, students must translate a point given a vector, calculate distances between points in a coordinate plane **(5.1)**, determine the distance between two pairs of points **(4.3)**, identify an angle of rotation of a two-dimensional shape given coordinates **(5.2)**, and identify the angle of rotation about its center to map a square onto itself **(5.4)**.

19. MP **Reason** In the diagram, $\triangle A'B'C'$ is the image of $\triangle ABC$ after a reflection.

A. Write an equation for the line of reflection.

B. Explain how you can use the equation of the line of reflection to determine the slope of $\overline{AA'}$. Then find the slope of $\overline{AA'}$.

A, B. See Additional Answers.

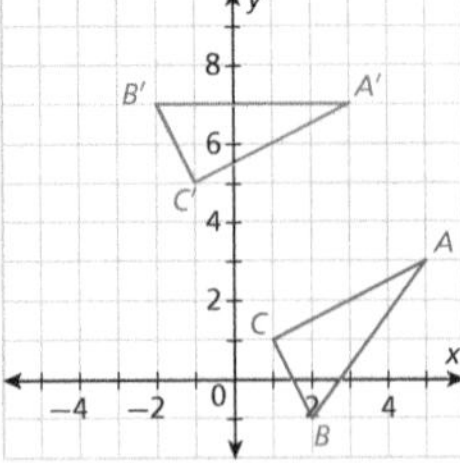

20. MP **Critique Reasoning** A student reflects triangle $A(-2, 6)$, $B(5, 5)$, $C(5, 3)$ across $y = -2$ and says the image is $A'(-2, -6)$, $B'(5, -5)$, $C'(5, -3)$. Explain the student's error. Then give the correct coordinates for $\triangle A'B'C'$. See Additional Answers.

21. Open Middle™ Using the digits 1 to 9, no more than one time each, fill in the blanks to make a true statement.

__?__ reflections across the *x*-axis and/or the *y*-axis is the same as __?__ 90° counterclockwise rotations about the origin. Possible answer: 2 and 4

Spiral Review • Assessment Readiness

22. Point $A(9, 4)$ is translated using the vector $\langle -6, -3 \rangle$. What is the image of the translated point?

Ⓐ (3, 1) Ⓑ (3, 7) Ⓒ (7, 3) Ⓓ (12, 4)

23. Which of the following have a distance greater than or equal to 5.9? Select all that apply.

Ⓐ $A(-3, 4)$ and $B(8, -3)$

Ⓑ $C(3, 4)$ and $D(1, 1)$

Ⓒ $C(3, 4)$ and $E(1, -4)$

Ⓓ $G(9, 4)$ and $H(8, -3)$

Ⓔ $I(-9, 3)$ and $J(0, 4)$

24. A rotation maps (4, 9) onto (4, 9). What is the angle of rotation?

Ⓐ 90° Ⓑ 180° Ⓒ 270° Ⓓ 360°

25. The square is rotated about its center. Which of the following rotations will not map the square onto itself?

Ⓐ 90° Ⓑ 135° Ⓒ 270° Ⓓ 360°

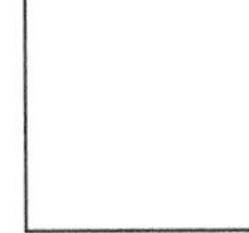

 **I'm in a Learning Mindset!**

How can I use reflections in my career of choice?

Keep Going ▶ Journal and Practice Workbook

Learning Mindset

Challenge Seeking Defines Own Challenges

Point out to students that understanding how reflections are created makes us more comfortable with the mathematics and allows us not only to perform them but to challenge ourselves with more complicated tasks. *How comfortable are you in reflecting figures? Do you feel you understand the process? Do you think you can challenge yourself by doing more constructions or attempt to perform multiple transformations? How could reflections be used in a career of your choosing?*

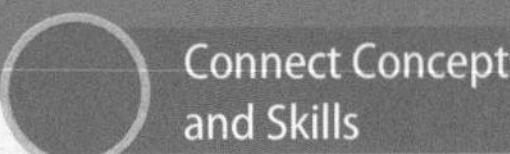

5.4 Define and Apply Symmetry

LESSON FOCUS AND COHERENCE

Mathematics Standards

- Given a rectangle, parallelogram, trapezoid, or regular polygon, describe the rotations and reflections that carry it onto itself.

Mathematical Practices and Processes (MP)

- Look for and make use of structure.
- Look for and express regularity with repeated reasoning.

I Can Objective

I can identify symmetry in figures.

Learning Objective

Describe the rotations and reflections that carry a given figure onto itself.

Language Objective

Explain how to determine the number of lines of symmetry and the angle of rotational symmetry for any regular polygon.

Vocabulary

New: angle of rotational symmetry, line of symmetry, line symmetry, rotational symmetry, symmetry

Mathematical Progressions

Prior Learning	Current Development	Future Connections
Students: • developed definitions of rotations, reflections, and translations in terms of angles, circles, perpendicular lines, parallel lines, and line segments. **(5.1–5.3)** • drew transformed figures, given a geometric figure and a rotation, reflection, or translation. **(5.1–5.3)**	**Students:** • determine whether a figure has symmetry. • find the lines of symmetry and the angle of rotational symmetry for a figure. • create figures with symmetry in the coordinate plane.	**Students:** • will prove theorems about triangles. **(9.1–9.5)** • will prove theorems about parallelograms. **(11.1 and 11.2)** • will recognize even and odd functions from their graphs and algebraic expressions for them. **(Alg2, 3.1)**

PROFESSIONAL LEARNING

Visualizing the Math

Students may find it helpful to use paper models to visualize a line of symmetry dividing a shape into two mirror images. Students can trace a figure shape onto a piece of paper and cut the figure out. By finding all the ways that the shape can be folded in half such that the edges of the shape line up perfectly, students will have found the lines of symmetry for the figure.

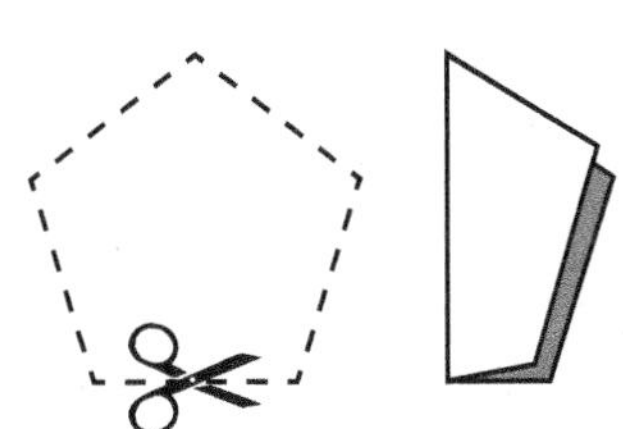

WARM-UP OPTIONS

ONLINE

PROJECTABLE & PRINTABLE

ACTIVATE PRIOR KNOWLEDGE • Reflections in the Coordinate Plane

Use these activities to quickly assess and activate prior knowledge as needed.

Problem of the Day 

The preimage *ABCD* is shown below. If *ABCD* is reflected across the *x*-axis, what will be the coordinates of the image?

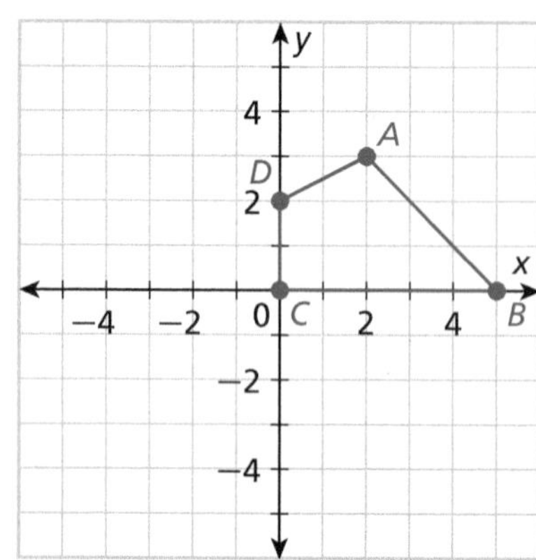

$A'(2, -3)$
$B'(5, 0)$
$C'(0, 0)$
$D'(0, -2)$

Quick Check for Homework

As part of your daily routine, you may want to display the Teacher Solution Key to have students check their homework.

Make Connections

Based on students' responses to the Problem of the Day, choose one of the following:

1. Project the Interactive Reteach, Geometry, Lesson 5.4.
2. Complete the Prerequisite Skills Activity:

Have students work in pairs. Each student plots and labels three points on the coordinate plane. Have students trade graphs and find the coordinates of the image reflected across the *y*-axis. After finding the coordinates, each student should graph the image on the coordinate plane.

- *Consider the point $F(5, 2)$. How far is the point from the y-axis?* 5 units
- *If the point $F(5, 2)$ is reflected over the y-axis, how far from the y-axis will the reflected point be?* 5 units
- *If the point $F(5, 2)$ is reflected over the y-axis, what will be the coordinates of the reflected point?* $(-5, 2)$
- *Compare the coordinates of the point F and its reflection. Which values are the same and which are different?* The *y*-coordinates are the same. The values of the *x*-coordinates are opposites.

If students continue to struggle, use Tier 2, Skill 11.

SHARPEN SKILLS

If time permits, use this on-level activity to build fluency and practice basic skills.

Vocabulary Review

Objective: Students demonstrate an understanding of lines of symmetry.
Materials: index cards, number cubes

Have students work in small groups. Students will draw one figure on each card that has between 1 and 6 lines of symmetry. Once all the index cards have been completed, arrange the cards so that every group member can see the figures on each card. One student rolls a number cube, and each of the other team members picks up one card that shows a figure with the number of lines of symmetry shown on the number cube. Repeat with a different student rolling the number cube until all cards have been picked up.

PLAN FOR DIFFERENTIATED INSTRUCTION

Small-Group Options

Use these teacher-guided activities with pulled small groups.

On Track

Materials: index cards

Give each student an index card with a geometric shape that resembles a star with 4, 5, or more points. Students should use their understanding of symmetry to draw all the lines of symmetry for the star and to determine the angle of rotational symmetry.

Almost There

Materials: coordinate grids, scissors, protractors

Have students trace and cut out a regular polygon. Have students do the following:

- Place a dot near one vertex of the regular polygon to keep track of the rotations.
- Place the polygon such that the center of the polygon is at the origin and the vertex with the dot is on the *x*-axis.
- Rotate the polygon counterclockwise about the origin until the next vertex is on the *x*-axis.
- Place a dot on the coordinate grid where the vertex with the dot is, then remove the polygon.
- Draw a line from the origin to the point. Measure the angle formed by the line and the *x*-axis to determine the angle of rotational symmetry.

Ready for More

Challenge students to find a regular polygon whose number of lines of symmetry is as close as possible to the measure of its angle of rotational symmetry. Students should be able to explain how they determined their answer and how many sides the polygon has.

Math Center Options

Use these student self-directed activities at centers or stations. **Key:** ● Print Resources ● Online Resources

On Track

- ● Interactive Digital Lesson
- ●● Journal and Practice Workbook
- ● Interactive Glossary (printable): **angle of rotational symmetry**, **line of symmetry**, **line symmetry**, **rotational symmetry**, **symmetry**
- ● Illustrative Mathematics: Symmetries of Rectangles
- ● Module Performance Task

Almost There

- ● Reteach 5.4 (printable)
- ● Interactive Reteach 5.4
- ● RtI Tier 2 Skill 11: Reflect Figures in the Coordinate Plane

Ready for More

- ● Challenge 5.4 (printable)
- ● Interactive Challenge 5.4
- ● Illustrative Mathematics: Symmetries of a Quadrilateral I

Unit Project Check students' progress by asking to see their line of symmetry before they start finding the equation. Then ask to see their equation.

View data-driven grouping recommendations and assign differentiation resources.

During the *Spark Your Learning*, listen and watch for strategies students use. See samples of student work on this page.

Use a Compass and a Protractor

Strategy 1

The circles in the piece of art all have the same center, so I know they're concentric. To make a design like this, I could use a compass to create concentric circles. Some of the triangles are repeated in a pattern around the circle, so I can use the protractor to draw multiple triangles that are the same size.

If students . . . use a compass and a protractor to solve the problem, they are drawing circles and constructing triangles with given side lengths and angle measures from Grade 7.

Have these students . . . explain how they determined what tools to use to create their design. **Ask:**

- Q What parts of the piece of art made you decide to use a protractor to complete the design?
- Q What parts of the piece of art made you decide to use a compass to complete the design?

Fold Paper to Create Patterns

Strategy 2

I see some complex circles that all have the same center. To make a design like this, I cut out circles and fold each in half twice to find the center. Then I can arrange the circles so that the centers are all at the same point and trace all the circles onto one piece of paper. The triangles repeat in a pattern around the circle, so I can draw one triangle and then fold the paper in half to determine where to draw triangles on the other side of the circle. I can complete the pattern by repeating this many times.

If students . . . fold paper to help create their designs, they understand how to recognize the patterns in the art but they may not know the most efficient method to recreate the patterns.

Activate prior knowledge . . . by having students think about tools they could use to construct the art piece. **Ask:**

- Q What tool can you use to help you make circles that all have the same center? Describe how to make concentric circles using this tool.
- Q What tool can you use to help you make triangles that are all the same size? Describe how to make similar triangles using this tool.

COMMON ERROR: Ignores Patterns in Design

The piece of art has many circles of different sizes, so I should make sure the design has many circles of different sizes. The piece of art also has lots of similar triangles, so I should include lots of triangles that are the same size in the design.

If students . . . create their own pattern using the same shapes that are in the piece of art but without considering how the piece of art is designed, they may not understand that the design is made of repeating patterns of triangles and concentric circles.

Then intervene . . . by pointing out that the problem asks them to recreate the piece of art as closely as possible. **Ask:**

- Q Does every design that has circles of many sizes look alike?
- Q Does every design that has congruent repeating triangles look alike?
- Q What tools can you use to make sure your design looks like the piece of art?

5.4

Define and Apply Symmetry

I Can identify symmetry in figures.

Spark Your Learning

While visiting a museum, Rose saw this piece of art. It inspired her to create a piece of art just like it.

Complete Part A as a whole class. Then complete Parts B–D in small groups.

A. What is a geometric question you can ask about this situation? What information would you need to know to answer your question? A. How was such a complex pattern created?; need to know the geometric techniques used by the artist

B. To answer your question, what strategy and tool would you use along with all the information you have? What answer do you get? B. See Strategies 1 and 2 on the facing page.

C. What transformations do you see in the art? Describe them. C, D. See Additional Answers.

D. What tools do you think would be useful in creating this type of art?

Turn and Talk How can you create your own circle art? How would you make sure it is symmetric? See margin.

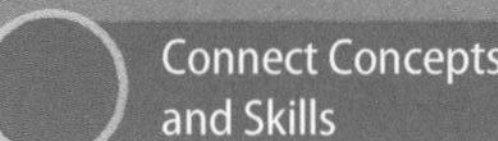

1 Spark Your Learning

▶ MOTIVATE

- Have students look at the photo in their books and read the information contained in the photo. Then complete Part A as a whole-class discussion.
- Give the class the additional information they need to solve the problem. This information is available online as a printable and projectable page in the Teacher Resources.
- Have students work in small groups to complete Parts B–D.

▶ PERSEVERE

If students need support, guide them by asking:

Q **Advancing • Use Tools** Which tool could you use to solve the problem? Why choose that tool and not some other? Students' choices of tools and reasons for choosing them will vary.

Q **Assessing** What congruent shapes do you see in the art piece? congruent triangles

Q **Advancing** Imagine a line drawn on the piece of art that divides the art into two pieces that are mirror images of each other. What must be true about such a line? Possible answer: The line will pass through the center of the concentric circles.

Turn and Talk Students with an interest in art may want to describe a way to draw geometric art by hand. Remind students that circles and other geometric shapes should be drawn using mathematic tools. Possible answer: I would use a compass to draw the circles and a protractor to divide the circle into equal parts. Then I would use a ruler to draw the triangles.

▶ BUILD SHARED UNDERSTANDING

Select groups of students who used various strategies and tools to share with the class how they solved the problem. As they present their solutions, have each group discuss why they chose a specific strategy and tool.

EL CULTIVATE CONVERSATION • Three Reads

Tell students to read the information in the photo three times and prompt them with a different question each time.

1. What is the situation about? The situation is about a piece of art that Rose saw in a museum and how she could make her own piece of art that looks similar.
2. What are the shapes in the figure? How are those shapes related? The figure shows congruent triangles and similar circles; The circles are arranged so they all have the same center, and the triangles are arranged in a repetitive pattern.
3. What are possible questions you could ask about the situation? Possible answer: What steps could you take to recreate this piece of art? How would you make sure the triangles are congruent? How would you make sure the circles all have the same center?

② Learn Together

Build Understanding

Task 1 (MP) **Use Structure** Students will use the positions of congruent sides and angles in geometric figures to find lines of symmetry.

CONNECT TO VOCABULARY

Have students use the **Interactive Glossary** to record their understanding of the vocabulary in this task.

Sample Guided Discussion:

Q How does a line of symmetry pass through a polygon with an even number of sides? Possible answer: For polygons with an even number of sides, the lines of symmetry can pass through the midpoints of opposite sides; or they can pass through opposite angles.

Q Describe a polygon that has no lines of symmetry. Possible answer: a scalene triangle

Turn and Talk Remind students that a rectangle is any quadrilateral with 4 right angles. Not all rectangles have 2 different side lengths. The rectangle would have to be a square. Two lines of symmetry pass through opposite vertices and two lines of symmetry pass through the midpoints of opposite sides.

Build Understanding

Explore Line Symmetry

A figure has **symmetry** if there is a transformation that maps the figure to itself.

A figure has **line symmetry** if it can be reflected over a line so that the image coincides with the preimage. That line is called a **line of symmetry**. A line of symmetry divides a plane figure into two congruent, reflected halves. Two examples are shown.

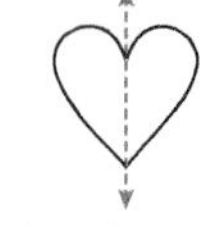
1 line of symmetry

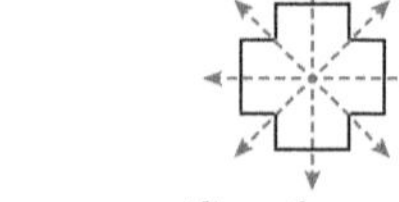
4 lines of symmetry

If a figure can be folded along a straight line so that one half of the figure exactly matches the other, the figure has line symmetry. The crease is the line of symmetry.

1 A. How many lines of symmetry does each figure have? Draw other polygons with four sides or five sides that have the same number of lines of symmetry.
A–C. See Additional Answers.

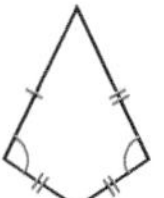
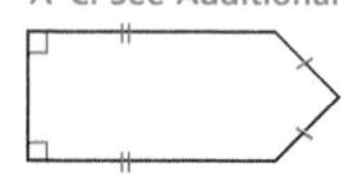

B. How many lines of symmetry does each figure have? Draw a polygon with an odd number of sides and a polygon with an even number of sides. Determine the number of lines of symmetry of each polygon.

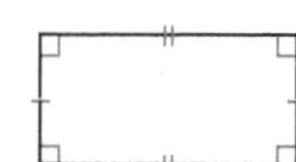

C. How does a line of symmetry intersect a polygon with an odd number of sides?

D. Draw a figure with many lines of symmetry. How many lines of symmetry does your figure have? Possible answer: I drew a regular hexagon with 6 lines of symmetry.

E. Does a figure exist that has an infinite number of lines of symmetry?
Yes, a circle has an infinite number of lines of symmetry.

Turn and Talk You determined that a rectangle has only 2 lines of symmetry. Is it possible for a rectangle to have 4 lines of symmetry? Explain why or why not. Use diagrams to support your answer. See margin.

162

LEVELED QUESTIONS

Depth of Knowledge (DOK)	Leveled Questions	What Does This Tell You?
Level 1 **Recall**	What is a line of symmetry? A line of symmetry is a line that divides a plane figure into two congruent reflected halves.	Students' answers will indicate whether they understand the definition of a line of symmetry.
Level 2 **Basic Application of Skills & Concepts**	How many lines of symmetry does an isosceles triangle have? 1	Students' answers will indicate whether they can determine the number of lines of symmetry for a specific figure.
Level 3 **Strategic Thinking & Complex Reasoning**	Does every rhombus have the same number of lines of symmetry? Explain. no; Possible answer: If the rhombus is a square, it has 4 lines of symmetry, but if the rhombus is not a square, it has 2 lines of symmetry.	Students' answers will indicate whether they can reason about whether all figures of a certain type will have the same number of lines of symmetry.

Explore Rotational Symmetry

A figure that can be rotated about its center by an angle of 180° or less so that the image coincides with the preimage has **rotational symmetry**. The **angle of rotational symmetry** is the smallest angle of rotation that maps a figure to itself.

A figure with rotational symmetry does not necessarily have line symmetry, but if it does, the intersection of the lines of symmetry is the center of rotational symmetry.

This flower has 3 lines of symmetry. It can be rotated twice to show rotational symmetry.

3 lines of symmetry
$360° \div 3 = 120°$

Rotate the flower 120°.

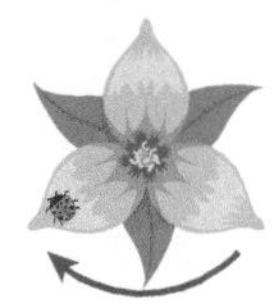

Rotate the flower again.
$120° + 120° = 240°$

To find the smallest angle of rotation, divide 360° by the number of lines of symmetry. Any additional angles of rotation will be multiples of the smallest angle.

2 Draw a square. Trace it onto tracing paper. Hold the center of the traced figure against the original figure with your pencil. Rotate the traced figure counterclockwise until it coincides again with the original figure beneath it.

A. By how many degrees did you rotate the figure? What are all of the angles of rotation up to and including 360°?
90°; 90°, 180°, 270°, 360°.

B. Draw the lines of symmetry of the original figure. How many lines of symmetry does it have? Describe how the lines of symmetry can be used to determine the angle of rotational symmetry.
4; Divide 360° by the number of lines of symmetry.

C. Repeat the procedure to find all of the angles of rotation up to and including 360° of the figure at the right. Are lines of symmetry helpful in this case? Explain.
90°, 180°, 270°, 360°; no; The figure does not have any lines of symmetry.

Turn and Talk Why can't an angle of rotational symmetry be 0°? Why can't it be greater than 180°? See margin.

Task 2 (MP) **Use Repeated Reasoning** Students use their understanding of angle measurements to find angles of rotation for figures with rotational symmetry.

CONNECT TO VOCABULARY

Have students use the **Interactive Glossary** to record their understanding of the vocabulary in this task.

Sample Guided Discussion:

Q The flower is rotated clockwise to illustrate its angle of rotation. Could the angle of rotation be illustrated by rotating the flower counterclockwise? Explain. yes; You can rotate the flower either direction to show the angle of rotation.

Q Describe a figure that has both rotational symmetry and line symmetry. What is the angle of rotational symmetry for your figure? Possible answer: an equilateral triangle; 120°

Turn and Talk Students should understand that repeatedly adding the angle of rotation should eventually produce a sum of 360°. Possible answer: The angle of rotational symmetry must be greater than 0° because the image will not have moved if it is equal to 0°. If it is less than or equal to 180° it has rotational symmetry. If it is greater than 180°, it cannot have rotational symmetry because if it did, there would be a smaller angle of rotational symmetry. For instance, suppose a figure can be rotated clockwise $x°$, where $180 < x < 360$, so that the figure coincides with itself. Then, it can also be rotated counterclockwise $(360 - x)°$, where $0 < 360 - x < 180$.

Step It Out

Task 3 (MP) **Use Repeated Reasoning** Students will determine a relationship between the number of lines of symmetry and the angle of rotational symmetry for a regular polygon.

Sample Guided Discussion:

Q The lines of symmetry for a regular polygon intersect to form congruent angles inside the polygon. Is the measure of the congruent angles formed the same as the measure of the angle of rotational symmetry? Explain. no; Possible answer: For a regular polygon, the measure of the angles formed by the intersection of the lines of symmetry will be half the measure of the angle of rotational symmetry.

Q The angle of rotational symmetry for a regular polygon is 30°. How many lines of symmetry does the polygon have? How do you know? 12; $360° \div 30° = 12$

Turn and Talk In a square, some of the right triangles created by lines of symmetry of regular polygons can be rotated onto their adjacent right triangles. Encourage students to determine whether a regular polygon exists such that every right triangle created by the lines of symmetry can be rotated onto the adjacent triangle on either side. Possible answer: I am decomposing the regular polygons into right triangles; no; Possible answer: Each triangle can be reflected onto its adjacent triangle, since the orientation of the triangle needs to change. Rotating the triangle will not change its orientation.

Step It Out

Describe Symmetry in Regular Polygons

Recall that a regular polygon is a polygon that is both equilateral and equiangular.

The table shows the types of symmetry in some regular polygons.

Regular polygon	Sides	Lines of symmetry	Angle of rotational symmetry	Image
Equilateral triangle	3	3	120°	
Square	4	4	90°	
Pentagon	5	5	72°	

3 What are the lines of symmetry and the angle of rotational symmetry for a regular hexagon?

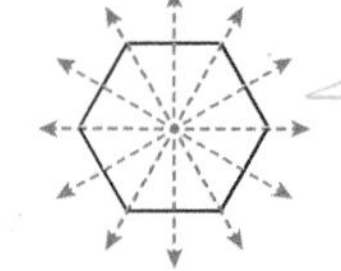

A. How many lines of symmetry does the hexagon have? 6

B. Explain why the angle of rotational symmetry is 60°. See Additional Answers.

C. Consider a regular decagon. How many lines of symmetry does it have? What is its angle of rotational symmetry? 10; 36°

D. Consider a regular polygon with n sides. How many lines of symmetry does it have? Write an expression for its angle of rotational symmetry in terms of n. n lines; $\frac{360°}{n}$

Turn and Talk When you draw lines of symmetry of regular polygons, what kind of shapes are you creating? Can each piece be rotated onto its adjacent piece? Explain why or why not. See margin.

164

PROFICIENCY LEVEL

Beginning

Show students a square that includes the lines of symmetry and an equilateral triangle that does not include the lines of symmetry. Point to all the lines of symmetry and say "The square has four lines of symmetry." Then ask students to determine the number of lines of symmetry for the equilateral triangle.

Intermediate

Have students work in groups. Give each group a set of index cards. Each card should show a figure that has rotational symmetry or a figure that does not have rotational symmetry. Ask students to explain whether each figure has rotational symmetry and how they know.

Advanced

Have students write a paragraph explaining how to determine the angle of rotational symmetry for any regular polygon.

Use Symmetry in a Coordinate Plane

 The x-axis and the y-axis are lines of symmetry of an image. Use the lines of symmetry to draw the entire image.

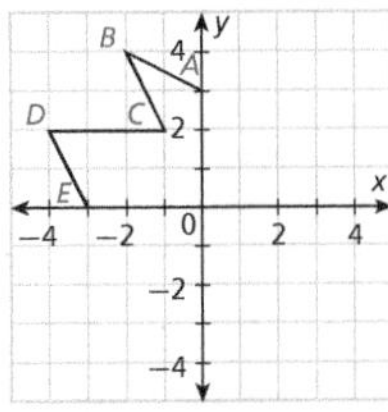

Reflect the part of the image shown across the x-axis.

$(x, y) \rightarrow (x, -y)$

$A(0, 3) \rightarrow J(0, -3)$

$B(-2, 4) \rightarrow H(-2, -4)$

$C(-1, 2) \rightarrow G(-1, -2)$

$D(-4, 2) \rightarrow F(-4, -2)$

A. Why is point E not included in the list of reflected points?

See Additional Answers.

Reflect the new image across the y-axis.

$B(-2, 4) \rightarrow B'(2, 4)$

$C(-1, 2) \rightarrow C'(1, 2)$

$D(-4, 2) \rightarrow D'(4, 2)$

$E(-3, 0) \rightarrow E'(3, 0)$

$F(-4, -2) \rightarrow F'(4, -2)$

$G(-1, -2) \rightarrow G'(1, -2)$

$H(-2, -4) \rightarrow H'(2, -4)$

B. How do the coordinates change when a point is reflected across the y-axis?

Take the opposite of the x-coordinate, and keep the y-coordinate the same.

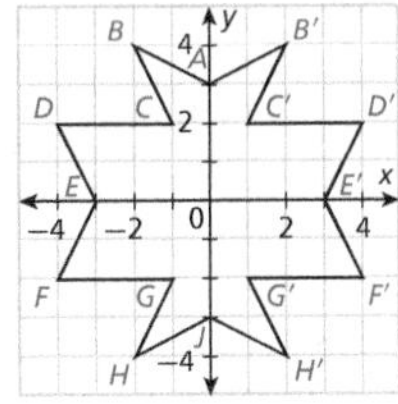

C. Suppose the original image was reflected across the y-axis first, and then the image was reflected across the x-axis. Would the resulting image be the same? Explain why or why not. yes; The order of the reflections does not affect the final image.

D. Describe any rotational symmetry in the final image. The final image maps onto itself by a rotation of 180°.

E. Explain why the final image could not be created using only rotations of the original image. See Additional Answers.

 **Turn and Talk** Change the location of one vertex in the original image so that the final image can be obtained using only rotations. See margin.

Task 4 (MP) **Use Repeated Reasoning** Students use the properties of a coordinate plane to create a symmetric figure by repeatedly reflecting portions of the figure across an axis.

OPTIMIZE OUTPUT Critique, Correct and Clarify

Have students work with a partner to discuss how to use the original image to create a figure that has a smaller angle of rotational symmetry. Students should be able to determine a possible angle of rotational symmetry and use that angle to create their figures. After creating their figure, students should determine whether any other figures with rotational symmetry can be made from the original figure.

Sample Guided Discussion:

Q Does the final image have any other lines of symmetry beside the x-axis and the y-axis? Explain your reasoning. no; Possible answer: Besides the x-axis and y-axis, no other lines divide the shape into two halves that are mirror images of each other.

Turn and Talk In order to create a figure with both rotational symmetry and lines of symmetry, students should understand that they are changing the location of one vertex such that a rotation of 90° looks the same as a rotation across an axis. Possible answer: Change vertex C to $(-2, 2)$.

ONLINE

Assign the Digital On Your Own for

- built-in student supports
- Actionable Item Reports
- Standards Analysis Reports

On Your Own

Assignment Guide

The chart below indicates which problems in the On Your Own are associated with each task in the Learn Together. Assign daily homework for tasks completed.

Learn Together Tasks	On Your Own Problems
Task 1, p. 162	Problems 7–9, 10, 17, 18, 23, 25, and 29–31
Task 2, p. 163	Problems 11–13, 15–16, 18, 23, 28, and 31
Task 3, p. 164	Problems 14, 17, 19–22, and 26
Task 4, p. 165	Problems 27 and 32

Watch for Common Errors

Problem 9 Students might mistake the figure for a regular polygon because the shape has 4 congruent sides. Remind students that the figure must also be equiangular to be a regular polygon, and that regular polygons have symmetries that non-regular polygons may not have.

Check Understanding

Determine whether each image has line symmetry. If so, how many lines of symmetry?

1. 
yes; 1 vertical line of symmetry

2. 
yes; 5 lines of symmetry

Determine whether each square nautical flag has rotational symmetry. If so, describe the rotations up to and including 360° that map the flag to itself.

3. yes; 180°, 360°

4. yes; 180°, 360°

5. How many lines of symmetry does a regular nonagon have? What is its angle of rotational symmetry? 9 lines; 40°

6. In a coordinate plane, polygon $ABCDEF$ has one line of symmetry which is $y = x$. Given the coordinates of the vertices $A(3, 3)$, $B(3, 0)$, $C(1, -3)$, $D(-3, -3)$, what are the coordinates of vertices E and F? $E(-3, 1)$, $F(0, 3)$

On Your Own

In Problems 7–9, determine whether each figure has line symmetry. If so, copy the shape and draw all the lines of symmetry.

7. 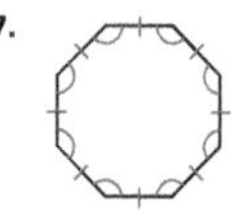
See Additional Answers.

8. no

9. See Additional Answers.

10. MP **Reason** Why is the diagonal of a rectangle not a line of symmetry? See Additional Answers.

Determine whether each figure has rotational symmetry. If so, describe the rotations up to and including 360° that map the figure to itself.

11. 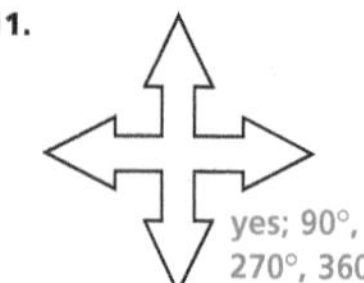
yes; 90°, 180°, 270°, 360°

12. 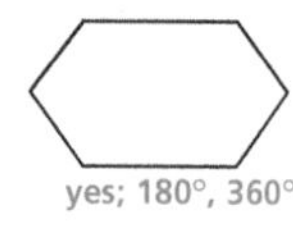
yes; 180°, 360°

13. 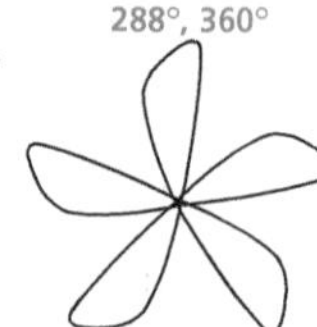
yes; 72°, 144°, 216°, 288°, 360°

166

data checkpoint

3 Check Understanding

Formative Assessment

Use formative assessment to determine if your students are successful with this lesson's learning objective.

Students who successfully complete the Check Understanding can continue to the On Your Own practice.

For students who miss 1 problem or more, work in a pulled small group using the Almost There small-group activity on page 161C.

ONLINE Ed

Assign the Digital Check Understanding to determine

- success with the learning objective
- items to review
- grouping and differentiation resources

4 Differentiation Options

Differentiate instruction for all students using small-group activities and math center activities on page 161C.

Reteach

Challenge

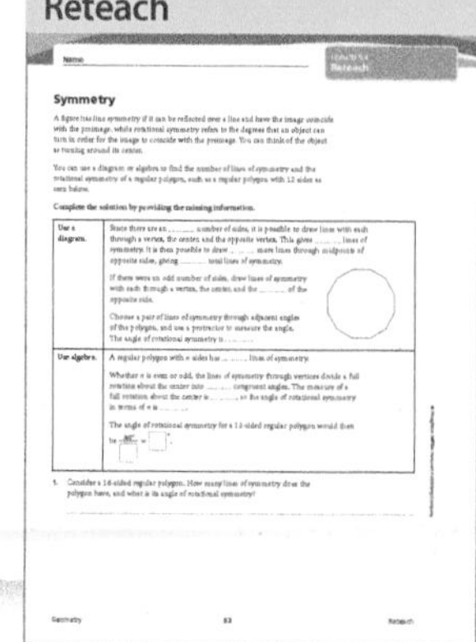

14. (MP) **Critique Reasoning** Donna correctly states that the measure of an interior angle of a square is 90°, and the angle of rotational symmetry of the square is 90°. Then she states that the angle of rotational symmetry of an equilateral triangle is 60° because the measure of an interior angle is 60°. Explain her error.
14–18. See Additional Answers.

15. What must be true for the snowflake shown at the right to have rotational symmetry?

16. (MP) **Use Structure** Point symmetry is when every part of a figure has a matching part and is the same distance from the central point but in the opposite direction. Name a polygon that has point symmetry.

17. **Open Ended** Draw a polygon with more than four sides that has exactly two lines of symmetry and an angle of rotation of 180°.

18. Compare the line symmetry and rotational symmetry of a rectangle and an ellipse.

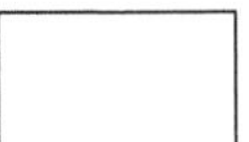

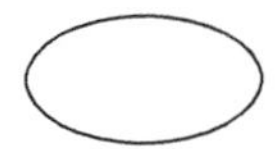

Use the given information about a regular polygon to determine how many sides the polygon has.

19. This polygon has exactly 7 lines of symmetry. 7 sides

20. This polygon has exactly 20 lines of symmetry. 20 sides

21. The angle of rotational symmetry for this polygon is 30°. 12 sides

22. The angle of rotational symmetry for this polygon is 18°. 20 sides

In Problems 23–26, octagon *ABCDEFGH* is a regular octagon with its center at point *P*. Determine whether each statement about the octagon is true or false.

23. $\overleftrightarrow{HD}$ is a line of symmetry. true

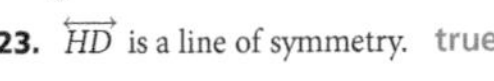

24. $\overline{BC}$ maps onto $\overline{EF}$ by a clockwise rotation about point *P* of 90°. false

25. $\overleftrightarrow{FC}$ is a line of symmetry. false

26. A clockwise rotation of 315° about point *P* maps the octagon onto itself. true

27. A figure has a 90° angle of rotational symmetry about the origin. Part of the image is shown at the right.
    A. Draw the entire figure. See Additional Answers.
    B. Does the entire figure have line symmetry? no
    C. Can any parts of the entire figure be determined using only reflections? Explain.
    See Additional Answers.

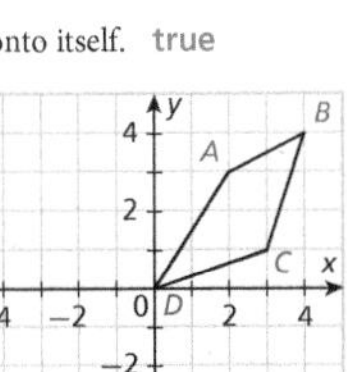

Questioning Strategies

Problem 17 Do any regular polygons with more than four sides have exactly two lines of symmetry and an angle of rotation of 180°? Explain your reasoning. no; Possible answer: Regular polygons all have more than 2 lines of symmetry, and regular polygons always have an angle of rotation that is 120° or less.

Watch for Common Errors

Problem 20 Students who draw a polygon as a solution method could find the wrong number of lines of symmetry by counting each line of symmetry twice as they count lines of symmetry in order around the polygon. Remind students that for regular polygons, the number of sides equals the number of lines of symmetry.

5 Wrap-Up

Summarize learning with your class. Consider using the Exit Ticket, Put It in Writing, or I Can scale.

Exit Ticket 

Draw a regular polygon and a non-regular polygon that each have exactly 4 lines of symmetry and an angle of rotation of 90°. Possible answer:

 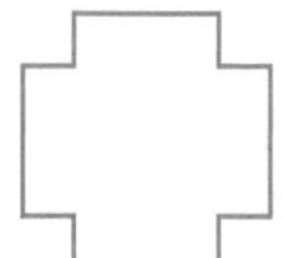

Put It in Writing 

Write a paragraph describing how to determine the lines of symmetry and the angle of rotation for a non-regular polygon.

I Can 

The scale below can help you and your students understand their progress on a learning goal.

4	I can identify symmetry in figures, and I can explain how to identify lines of symmetry and angles of rotational symmetry to others.
3	I can identify symmetry in figures.
2	I can create a figure that has symmetry.
1	I can differentiate between line symmetry and rotational symmetry.

Spiral Review • Assessment Readiness

These questions will help determine if students have retained information taught in the past and can also prepare them for high-stakes assessments. Here, students must determine the coordinates of a reflected image given the coordinates of the preimage **(5.3)**, recall the fundamental property of rigid motions **(5.1)**, and match coordinate notations to descriptions of transformations **(5.2)**.

Describe any symmetry you see in each object found in nature.

28. 

rotational symmetry

29. 

1 vertical line of reflection down the center

30. 1 vertical line of reflection down the center

31. many of lines of symmetry and rotational symmetry

32. (Open Middle™) Using the digits 1 to 9, at most one time each, create 4 ordered pairs to represent the vertices of a quadrilateral that has rotational symmetry and has 4 lines of symmetry. Possible answer: $(1, 6)$, $(3, 5)$, $(4, 7)$ and $(2, 8)$

Spiral Review • Assessment Readiness

33. Given the preimage $A(3, 3)$, $B(3, -1)$, and $C(6, 3)$, reflect the image across $y = -x$. What are the coordinates of the image?

Ⓐ $A'(1, -3)$, $B'(-3, -3)$, $C'(-3, 6)$
Ⓑ $A'(-3, 3)$, $B'(-3, -1)$, $C'(-6, -3)$
Ⓒ $A'(-3, -3)$, $B'(1, -3)$, $C'(-3, -6)$
Ⓓ $A'(-3, 6)$, $B'(1, -3)$, $C'(-3, -3)$

34. Which of the following is not a property of a rigid motion?

Ⓐ preserves angle measures
Ⓑ preserves orientation
Ⓒ preserves distance
Ⓓ preserves collinearity

35. Match the coordinate notation with the description of the transformation in the coordinate plane.

Coordinate notation	Description of transformation
A. $(x, y) \rightarrow (x, y - 1)$	**1.** a rotation of 90° about the origin
B. $(x, y) \rightarrow (-y, x)$	**2.** a translation of 1 unit to the left
C. $(x, y) \rightarrow (x, -y)$	**3.** a translation 1 unit down
D. $(x, y) \rightarrow (x - 1, y)$	**4.** a reflection across the x-axis

A. 3; B. 1; C. 4; D. 2

 **I'm in a Learning Mindset!**

What have I learned about symmetry and how can I make sure that the tasks in this lesson are challenging enough for me?

Keep Going Journal and Practice Workbook

Learning Mindset

Challenge Seeking Defines Own Challenges

Remind students that each person likely has a unique set of topics that they can master easily, as well as a unique set of topics that they will find challenging. Students that have easily mastered symmetry can examine connections between symmetry and previous lessons on reflections and rotations to develop more challenging problems. *Did you find the concept of symmetry challenging? Can you develop more challenging symmetry questions involving rotations and reflections?*

Assign the Digital Module Review for
- built-in student supports
- Actionable Item Reports
- Standards Analysis Reports

MODULE 5 REVIEW

Module 5

Review

Translations

Ralph is redesigning his living room. He moved a table as shown.

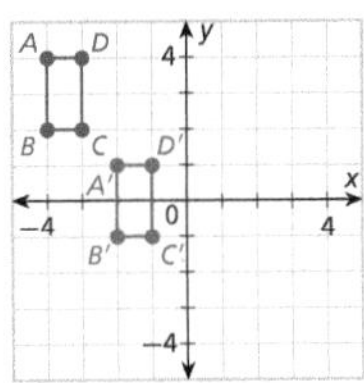

The horizontal change from $(-4, 4)$ to $(-2, 1)$ is $-2-(-4) = 2$.

The vertical change from $(-4, 4)$ to the $(-2, 1)$ is $1 - 4 = -3$.

The vector that represents the movement of the table is $\langle 2, -3 \rangle$.

Rotations

Ralph wants to rotate a triangular table 90° counterclockwise around the center of the room.

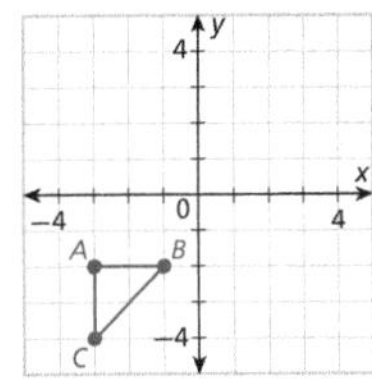

The rotated image of (x, y) is $(-y, x)$.

The new coordinates of the table are below.

$A(-3, -2) \rightarrow A'(2, -3)$

$B(-1, -2) \rightarrow B'(2, -1)$

$C(-3, -4) \rightarrow C'(4, -3)$

Reflections

Ralph has a cabinet against a wall that he wants to reflect to the other side of the room.

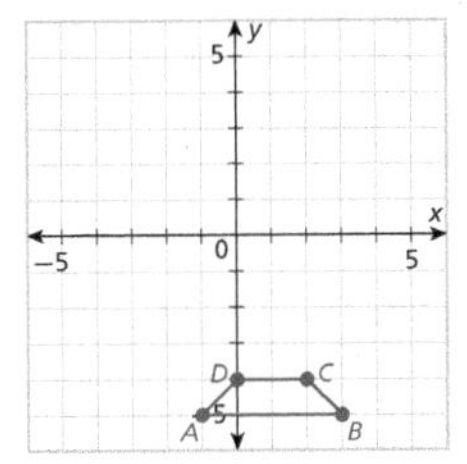

Reflect the cabinet across the x-axis. The new coordinates of the cabinet are below.

$A(-1, -5) \rightarrow A'(-1, 5)$

$B(3, -5) \rightarrow B'(3, 5)$

$C(2, -4) \rightarrow C'(2, 4)$

$D(0, -4) \rightarrow D'(0, 4)$

Symmetry

Ralph has an octagonal table. The shape of the table is a regular octagon.

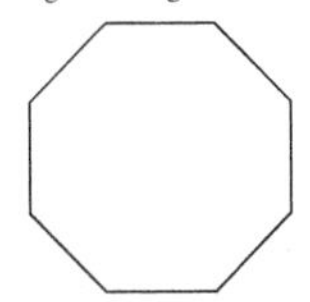

He wants to know what angles he can rotate the table and have it match his original position.

He finds that the angle of rotational symmetry of the table is 45°.

He can rotate the table 45°, 90°, 135°, 180°, 225°, 270°, 315°, and 360°.

Module 5 169

Module Review

Use the first page of the Module Review to summarize and connect the main ideas of the module. Use the second page to assess students' understanding of the vocabulary, concepts, and skills presented in the module.

Sample Guided Discussion:

Q Translations What vector represents the movement of the table back to its original position? $\langle -2, 3 \rangle$

Q Rotations How could you use a series of reflections to perform an equivalent transformation of the triangle? Possible answer: Reflect the triangle over the x-axis: $(x, y) \rightarrow (x, -y)$, then reflect the image over the line $y = x$: $(x, -y) \rightarrow (-y, x)$. So, $(x, y) \rightarrow (-y, x)$

Q Reflections How could you replace the reflection with an equivalent rotation? Possible answer: Rotate $ABCD$ 180° counterclockwise about the point $P(1, 0)$.

Q Symmetry Explain how you can use lines of symmetry in an octagon to find the center and radius of the circumscribed circle. Possible answer: Draw the four lines of symmetry though four pairs of opposite vertices. The intersection of the lines of symmetry is the center of the circumscribed circle. The radius is the distance between the center and one of the vertices.

Module Review continued

Possible Scoring Guide

Items	Points	Description
1–4	2 each	identifies the correct term
5	2	rotates a triangle 180° in the coordinate plane
6	2	reflects a triangle in the coordinate plane across the x-axis
7	2	translates a triangle in the coordinate plane according to a given rule
8	2	reflects a triangle across the y-axis
9–12	1 each	identifies the lines of symmetry for four figures
9–12	2 each	lists the measures of the angles of rotations for figures that have rotational symmetry
Total points possible = 24 points		

The Unit 3 Test in the Assessment Guide assesses content from Modules 5 and 6.

Vocabulary

Choose the correct term from the box to complete each sentence.

Vocabulary
reflection
rotation
rotational symmetry
transformation

1. A transformation about a point P such that each point and its image are the same distance from P is a __?__. rotation
2. A __?__ is a change is the position, size, or shape of a figure or graph. transformation
3. A figure has __?__ if it can be rotated about a point by an angle less than 360° so that the image coincides with the preimage. rotational symmetry
4. A __?__ is a transformation across a line such that the line is the perpendicular bisector of each segment joining each point and its image. reflection

Concepts and Skills

Draw the image of the figure under the given transformation.
5–8. See Additional Answers.

5. 180° rotation
6. reflection across the x-axis
7. $(x, y) \rightarrow (x - 4, y + 1)$
8. reflection across the y-axis

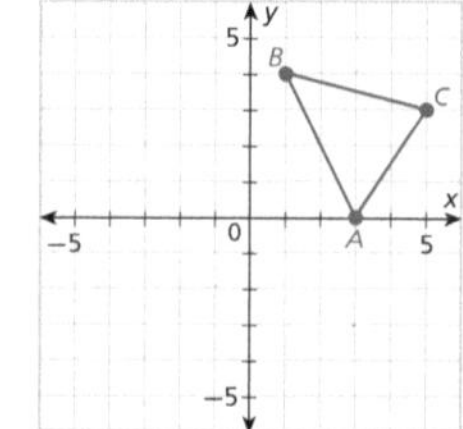

Copy the figure. Identify and describe the number of lines of symmetry for the figure. List the angles of rotation for the figures that have rotational symmetry.

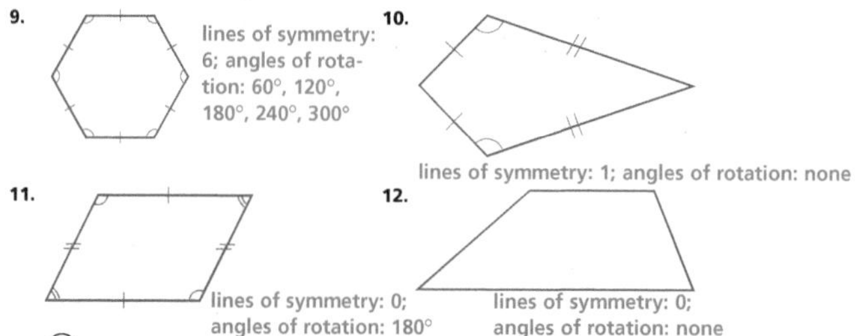

9. lines of symmetry: 6; angles of rotation: 60°, 120°, 180°, 240°, 300°
10. lines of symmetry: 1; angles of rotation: none
11. lines of symmetry: 0; angles of rotation: 180°
12. lines of symmetry: 0; angles of rotation: none

13. (MP) **Use Tools** A carnival ride operates by rotating cars around a central axis. The coordinates of one of the cars at the start of the ride are $A(3, -8)$, $B(5, -12)$, and $C(1, -12)$. What are the coordinates of the car after it has rotated counterclockwise 180°? State what strategy and tool you will use to answer the question, explain your choice, and then find the answer. $A'(-3, 8)$, $B'(-5, 12)$, $C'(-1, 12)$

170

DATA-DRIVEN INSTRUCTION

Before moving on to the Module Test, use the Module Review results to intervene based on the table below.

MTSS (RtI)

Items	Lesson	DOK	Content Focus	Intervention
5	5.2	2	Rotate a figure 180°.	Reteach 5.2
6	5.3	2	Reflect a figure across the x-axis.	Reteach 5.3
7	5.1	2	Translate a figure according to a rule.	Reteach 5.1
8	5.3	2	Reflect a figure across the y-axis.	Reteach 5.3
9–12	5.4	3	Identify and describe the number of lines of symmetry for a figure. List the measures of the angles of rotation for the figures that have rotational symmetry.	Reteach 5.4

Module Test

The Module Test is available in alternative versions in your Assessment Guide. All items are presented in standardized test formats.

ONLINE

Ed

Assign the Digital Module Test to power actionable reports including

- proficiency by standards
- item analysis

Form A

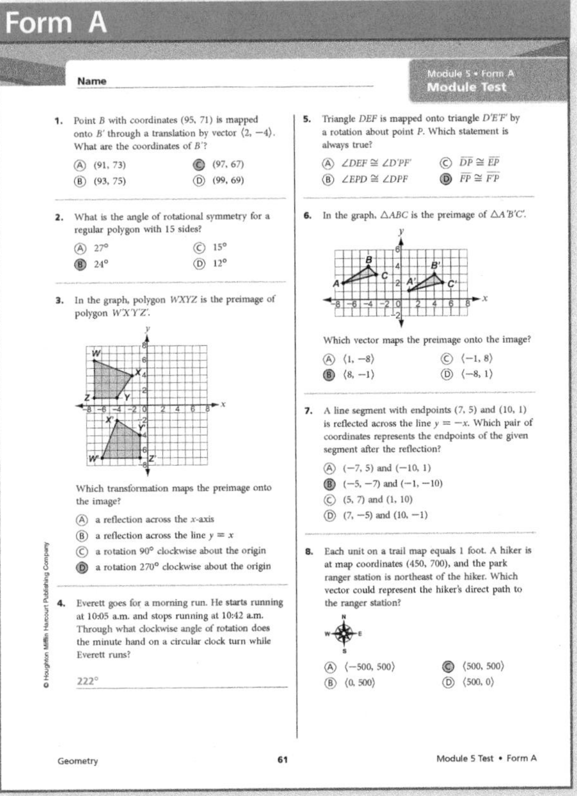

Name

Module 5 • Form A
Module Test

1. Point B with coordinates $(95, 71)$ is mapped onto B' through a translation by vector $\langle 2, -4 \rangle$. What are the coordinates of B'?

   (A) $(91, 73)$ (B) $(93, 75)$ (C) $(97, 67)$ (D) $(99, 69)$

2. What is the angle of rotational symmetry for a regular polygon with 15 sides?

   (A) 27° (B) 24° (C) 15° (D) 12°

3. In the graph, polygon $WXYZ$ is the preimage of polygon $W'X'Y'Z'$.

   Which transformation maps the preimage onto the image?

   (A) a reflection across the x-axis
   (B) a reflection across the line $y = x$
   (C) a rotation 90° clockwise about the origin
   (D) a rotation 270° clockwise about the origin

4. Everett goes for a morning run. He starts running at 10:05 a.m. and stops running at 10:42 a.m. Through what clockwise angle of rotation does the minute hand on a circular clock turn while Everett runs?

   222°

5. Triangle DEF is mapped onto triangle $D'E'F'$ by a rotation about point P. Which statement is always true?

   (A) $\angle DEF \cong \angle D'PF'$ (B) $\angle EPD \cong \angle DPF$ (C) $\overline{DP} \cong \overline{EP}$ (D) $\overline{FP} \cong \overline{F'P}$

6. In the graph, $\triangle ABC$ is the preimage of $\triangle A'B'C'$.

   Which vector maps the preimage onto the image?

   (A) $\langle 1, -8 \rangle$ (B) $\langle 8, -1 \rangle$ (C) $\langle -1, 8 \rangle$ (D) $\langle -8, 1 \rangle$

7. A line segment with endpoints $(7, 5)$ and $(10, 1)$ is reflected across the line $y = -x$. Which pair of coordinates represents the endpoints of the given segment after the reflection?

   (A) $(-7, 5)$ and $(-10, 1)$
   (B) $(-5, -7)$ and $(-1, -10)$
   (C) $(5, 7)$ and $(1, 10)$
   (D) $(7, -5)$ and $(10, -1)$

8. Each unit on a trail map equals 1 foot. A hiker is at map coordinates $(450, 700)$, and the park ranger station is northeast of the hiker. Which vector could represent the hiker's direct path to the ranger station?

   (A) $\langle -500, 500 \rangle$ (B) $\langle 0, 500 \rangle$ (C) $\langle 500, 500 \rangle$ (D) $\langle 500, 0 \rangle$

Form A

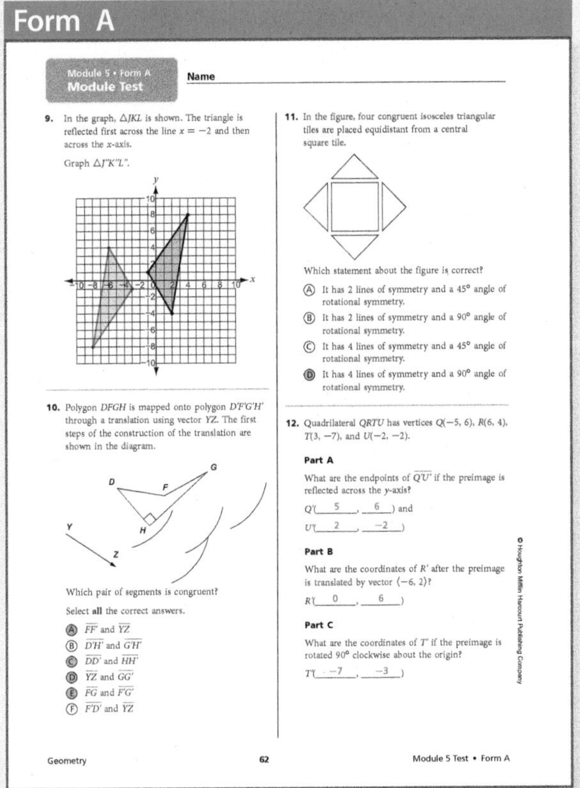

Module 5 • Form A
Module Test

Name

9. In the graph, $\triangle JKL$ is shown. The triangle is reflected first across the line $x = -2$ and then across the x-axis.

   Graph $\triangle J''K''L''$.

10. Polygon $DFGH$ is mapped onto polygon $D'F'G'H'$ through a translation using vector YZ. The first steps of the construction of the translation are shown in the diagram.

    Which pair of segments is congruent?

    Select **all** the correct answers.

    (A) $\overline{FF'}$ and $\overline{YZ}$
    (B) $\overline{D'H'}$ and $\overline{G'H'}$
    (C) $\overline{DD'}$ and $\overline{HH'}$
    (D) $\overline{YZ}$ and $\overline{GG'}$
    (E) $\overline{FG}$ and $\overline{F'G'}$
    (F) $\overline{F'D'}$ and $\overline{YZ}$

11. In the figure, four congruent isosceles triangular tiles are placed equidistant from a central square tile.

    Which statement about the figure is correct?

    (A) It has 2 lines of symmetry and a 45° angle of rotational symmetry.
    (B) It has 2 lines of symmetry and a 90° angle of rotational symmetry.
    (C) It has 4 lines of symmetry and a 45° angle of rotational symmetry.
    (D) It has 4 lines of symmetry and a 90° angle of rotational symmetry.

12. Quadrilateral $QRTU$ has vertices $Q(-5, 6)$, $R(6, 4)$, $T(3, -7)$, and $U(-2, -2)$.

    Part A

    What are the endpoints of $\overline{Q'U'}$ if the preimage is reflected across the y-axis?

    $Q'($ 5 , 6 $)$ and

    $U'($ 2 , −2 $)$

    Part B

    What are the coordinates of R' after the preimage is translated by vector $\langle -6, 2 \rangle$?

    $R'($ 0 , 6 $)$

    Part C

    What are the coordinates of T' if the preimage is rotated 90° clockwise about the origin?

    $T'($ −7 , −3 $)$

Form B

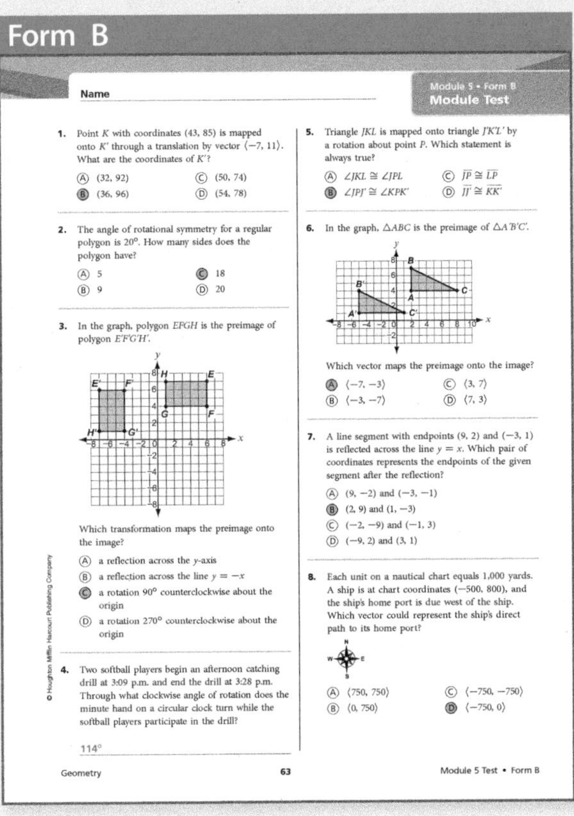

Name

Module 5 • Form B
Module Test

1. Point K with coordinates $(43, 85)$ is mapped onto K' through a translation by vector $\langle -7, 11 \rangle$. What are the coordinates of K'?

   (A) $(32, 92)$ (B) $(36, 96)$ (C) $(50, 74)$ (D) $(54, 78)$

2. The angle of rotational symmetry for a regular polygon is 20°. How many sides does the polygon have?

   (A) 5 (B) 9 (C) 18 (D) 20

3. In the graph, polygon $EFGH$ is the preimage of polygon $E'F'G'H'$.

   Which transformation maps the preimage onto the image?

   (A) a reflection across the y-axis
   (B) a reflection across the line $y = -x$
   (C) a rotation 90° counterclockwise about the origin
   (D) a rotation 270° counterclockwise about the origin

4. Two softball players begin an afternoon catching drill at 3:09 p.m. and end the drill at 3:28 p.m. Through what clockwise angle of rotation does the minute hand on a circular clock turn while the softball players participate in the drill?

   114°

5. Triangle JKL is mapped onto triangle $J'K'L'$ by a rotation about point P. Which statement is always true?

   (A) $\angle JKL \cong \angle JPL$ (B) $\angle JPJ' \cong \angle KPK'$ (C) $\overline{JP} \cong \overline{LP}$ (D) $\overline{JJ'} \cong \overline{KK'}$

6. In the graph, $\triangle ABC$ is the preimage of $\triangle A'B'C'$.

   Which vector maps the preimage onto the image?

   (A) $\langle -7, -3 \rangle$ (B) $\langle -3, -7 \rangle$ (C) $\langle 3, 7 \rangle$ (D) $\langle 7, 3 \rangle$

7. A line segment with endpoints $(9, 2)$ and $(-3, 1)$ is reflected across the line $y = x$. Which pair of coordinates represents the endpoints of the given segment after the reflection?

   (A) $(9, -2)$ and $(-3, -1)$
   (B) $(2, 9)$ and $(1, -3)$
   (C) $(-2, -9)$ and $(-1, 3)$
   (D) $(-9, 2)$ and $(3, 1)$

8. Each unit on a nautical chart equals 1,000 yards. A ship is at chart coordinates $(-500, 800)$, and the ship's home port is due west of the ship. Which vector could represent the ship's direct path to its home port?

   (A) $\langle 750, 750 \rangle$ (B) $\langle 0, 750 \rangle$ (C) $\langle -750, -750 \rangle$ (D) $\langle -750, 0 \rangle$

Form B

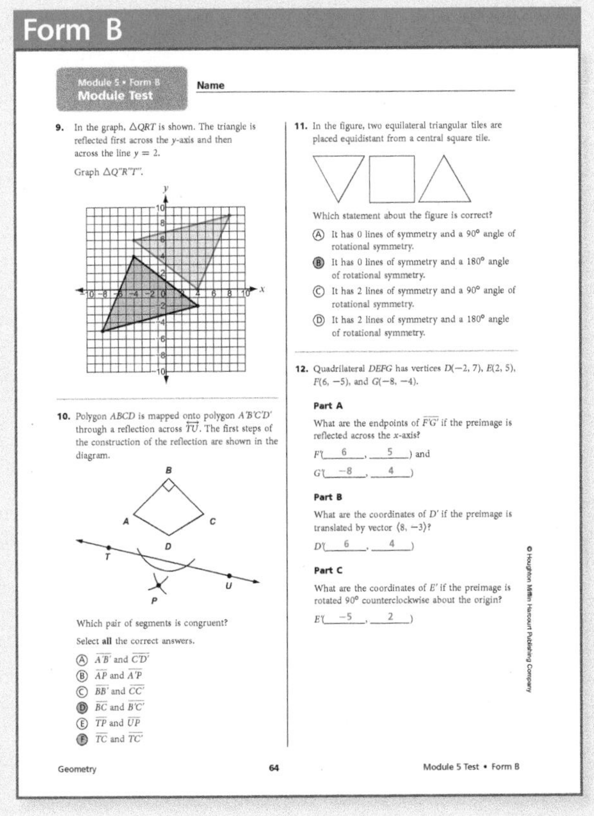

Module 5 • Form B
Module Test

Name

9. In the graph, $\triangle QRT$ is shown. The triangle is reflected first across the y-axis and then across the line $y = 2$.

   Graph $\triangle Q''R''T''$.

10. Polygon $ABCD$ is mapped onto polygon $A'B'C'D'$ through a reflection across $\overleftrightarrow{TU}$. The first steps of the construction of the reflection are shown in the diagram.

    Which pair of segments is congruent?

    Select **all** the correct answers.

    (A) $\overline{A'B'}$ and $\overline{C'D'}$
    (B) $\overline{AP}$ and $\overline{A'P}$
    (C) $\overline{BB'}$ and $\overline{CC'}$
    (D) $\overline{BC}$ and $\overline{B'C'}$
    (E) $\overline{TP}$ and $\overline{UP}$
    (F) $\overline{TC}$ and $\overline{TC'}$

11. In the figure, two equilateral triangular tiles are placed equidistant from a central square tile.

    Which statement about the figure is correct?

    (A) It has 0 lines of symmetry and a 90° angle of rotational symmetry.
    (B) It has 0 lines of symmetry and a 180° angle of rotational symmetry.
    (C) It has 2 lines of symmetry and a 90° angle of rotational symmetry.
    (D) It has 2 lines of symmetry and a 180° angle of rotational symmetry.

12. Quadrilateral $DEFG$ has vertices $D(-2, 7)$, $E(2, 5)$, $F(6, -5)$, and $G(-8, -4)$.

    Part A

    What are the endpoints of $\overline{F'G'}$ if the preimage is reflected across the x-axis?

    $F'($ 6 , 5 $)$ and

    $G'($ −8 , 4 $)$

    Part B

    What are the coordinates of D' if the preimage is translated by vector $\langle 8, -3 \rangle$?

    $D'($ 6 , 4 $)$

    Part C

    What are the coordinates of E' if the preimage is rotated 90° counterclockwise about the origin?

    $E'($ −5 , 2 $)$

MODULE 6

PLANNING

TRANSFORMATIONS THAT CHANGE SIZE AND SHAPE

Introduce and Check for Readiness

- Module Performance Task • Are You Ready?

Lesson 6.1—2 Days

Define and Apply Dilations and Stretches

Learning Objective: Perform and analyze transformations to include dilations, stretches, and compressions. Use coordinate rules and geometric drawing tools to investigate the effect of multiplication on the points in a figure.

Review Vocabulary: dilation

New Vocabulary: center of dilation, compression, scale factor, stretch

Lesson 6.2—2 Days

Apply Sequences of Transformations

Learning Objective: Students will apply sequences of transformations to figures, specify sequences that map a given preimage to a given image, and make predictions about the results of applying a sequence of transformations.

New Vocabulary: composition

Assessment

- Module 6 Test (Forms A and B)
- Unit 6 Test (Forms A and B)

LEARNING ARC FOCUS

 Build Conceptual Understanding

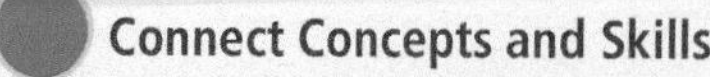 Connect Concepts and Skills

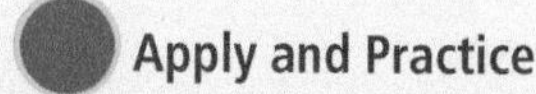 Apply and Practice

TEACHING FOR SUCCESS

TEACHING FOR DEPTH: Transformations that Change Size and Shape

Make Connections. Transformations occur in many aspects of the real world. Mathematically, transformations are functions that use points in a plane as inputs and produce other points as outputs. Transformations include translations, rotations, reflections, and dilations.

A rigid motion is a kind of transformation that changes an object's location but preserves its shape and size. Translations (slides), rotations (turns), and reflections (flips) each preserve distances, angle measures, collinearity, and betweenness of points. A dilation is a transformation, but it is not a rigid motion. Dilations alter distance and produce an image that has the same shape as its preimage, but has a different size by either stretching or shrinking the original figure.

Where can transformations be found in everyday life? Actually, they are everywhere. Here are some common examples.

- Translations can be found in the arrangement of seats in auditoriums and movie theaters, in repetitive floor tiles, and in the movements of vehicles.
- Reflections can be found in mirrors, on the surface of lakes, and on the glass windows of buildings.
- Rotations can be found in wheels of all types, ceiling fans, and analog clocks.
- And dilations can be found at the grocery store (in different size cans and boxes), at the zoo (adult and baby elephants), dress and shoe sizes, and athletic fields.

Mathematical Progressions

Prior Learning	Current Development	Future Connections
Students have: • learned the definitions of rotations, reflections, and translations in the plane. • described the effects of translations, rotations, and reflections on figures in the *x* coordinate plane. • learned that two-dimensional figures are similar if the second is the image of the first through a rotation, reflection, or translation. • represented and described transformations in the plane as functions that take points as inputs and yield points as outputs. • compared transformations that preserve distances and angles to those that do not.	**Students:** • extend transformations to include dilations and stretches. • compare rigid and nonrigid transformations. • apply sequences of transformations to move figures in the plane. • specify a transformational sequence that maps a preimage to an image. • predict the result of applying a sequence of transformations.	**Students will:** • will use similarity transformations to determine whether two figures are similar. • will apply similarity transformations to establish the AA Theorem and prove that two triangles are similar. • will use similarity to define the trigonometric ratios.

TEACHER TO TEACHER

From the Classroom

Use and connect mathematical representations. I use a blueprint as a mathematical representation of a dilation. I divide my students into groups and tell them that they are going to create a blueprint of the classroom. I assign a section of the room to each group and have them create a scale drawing of the space. I provide students with measuring tapes and either yardsticks or meter sticks. I have the class decide on the scale: e.g., 1 in. = 1 ft.

After each group has completed its drawing, I have them come together as a class and put the drawings together to create a blueprint for the entire room.

If existing blueprints of your school are available, use them to discuss the scale factors and describe how each represents the dimensions of the part of the school represented by the blueprint.

LANGUAGE DEVELOPMENT • Planning for Instruction

By giving all students regular exposure to language routines in context, you will provide opportunities for students to **listen, speak, read,** and **write** about mathematical situations and develop both mathematical language and conceptual understanding at the same time.

Using Language Routines to Develop Understanding

Use the **Professional Cards** for the following routines to plan for effective instruction.

Co-Craft Questions and Problems Lesson 6.1

Students think of natural questions to ask about a given situation or problems similar to a given task and answer the questions they have developed or problems they have created.

Three Reads Lesson 6.1

Students read a problem three times with a specific focus each time.

1st Read What is the situation about?
2nd Read What are the quantities in the situation?
3rd Read What are the possible mathematical questions that we could ask for the situation?

Information Gap Lesson 6.2

Students recognize when information given in a problem situation is incomplete, and they pose questions and share knowledge with others to discover any missing facts or relationships and work together to solve the problem.

Critique, Correct, and Clarify Lesson 6.2

Students correct the work in a flawed explanation, argument, or solution method and share with a partner and refine the sample work.

Connecting Language to Transformations that Change Size and Shape

Watch for students' use of the review and new terms listed below as they explain their reasoning and make connections with new concepts.

Key Academic Vocabulary

Prior Learning and Current Development • Review and New Vocabulary

dilation a transformation that changes the size of a figure by the same amount in all directions

center of dilation a fixed point in the plane that does not change when a dilation is applied

composition a transformation that directly maps a preimage to the final image after each image is used as a preimage in the next transformation.

compression a transformation that changes the shape of a figure in one direction by a factor greater than 0 and less than 1

scale factor the ratio of the length of a segment on the image to the length of the corresponding segment on the preimage

stretch a transformation that changes the shape of a figure by a factor greater than 1 in one direction

Linguistic Note

Listen for students' use of the terms *dilation, stretch*, and *compression*. Students are probably familiar with the word stretch, but the other two terms may be less common. Explain that when a doctor examines your eyes, he or she *dilates* them. That is, the pupils in your eyes are made larger so the doctor can examine them more closely. A *compression* is the opposite of a dilation. Here, things keep their shape, but are made smaller. Then point out that these meanings are the same in mathematics, only they apply to geometric figures.

Module 6

Transformations that Change Size and Shape

Module Performance Task: Focus on STEM

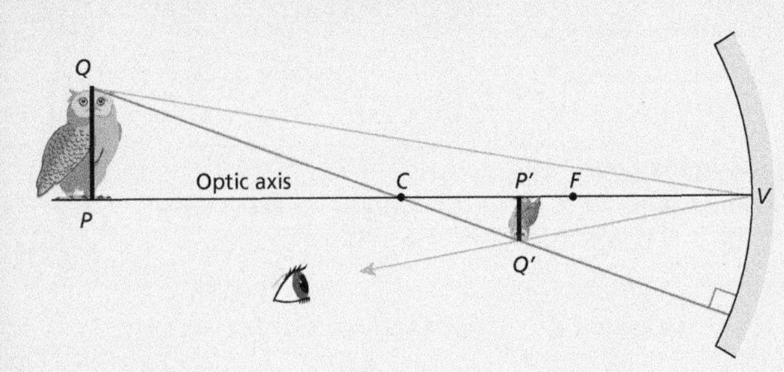

Geometric Optics

Geometric optics is one of two major modeling techniques for describing the propagation of light. The model is based on the underlying assumption that light travels in a straight line through a homogenous medium as a ray.

The diagram above shows a concave, spherical mirror with center C and an object represented by segment QP. The focal point of the mirror is located at point F and the vertex of the mirror, V, is located at the intersection of the optic axis and the mirror.

The location of the reflected image can be located using a process called ray tracing. The law of reflection states that the measure of the angle formed by the ray and the optic axis is equal to the measure of the angle formed by the reflected ray and the optic axis. Using this law, and the fact that a ray from the object through the center C is reflected back onto itself, you can determine the location of the image.

A. Describe how to locate the image $\overline{Q'P'}$ using a sequence of transformations. See margin.

B. Use a compass and straight edge to construct the image $\overline{Q'P'}$ when the location of $\overline{QP}$ is beyond C (as shown here), at C, between C and F, at F, and between F and V. See Additional Answers.

C. Compare the the images created from each of the locations of $\overline{QP}$ in Part B. See margin.

D. Repeat for a convex, spherical mirror. Explain any differences or similarities between the two mirrors. See margin.

Module 6 171

Mathematical Connections

Task Part	Prerequisite Skills
Part A	Students should be familiar with dilations **(6.1)**.
Parts B–D	Students should be familiar with dilations and sequences of transformations **(6.1 and 6.2)**.

Geometric Optics

Overview

Students examine sequences of transformations used to describe the propagation of light rays and the creation of images.

Students use a schematic ray diagram to identify the properties of images formed when a light ray is reflected off a concave spherical mirror. Students construct the images created when a light ray passes through one of several locations and identify how the sequence of transformations affects the orientation and size of the image of an object.

Career Connections

Optics is the scientific study of light and the behavior of light. Suggest that students research a career in *geometry optics*, a branch of optics that deals with light rays and the study of how lenses and mirrors reflect light rays. Such effects are captured by various optical instruments. Persons working in geometric optics include electro-optical engineers, laser engineers, optical designers, optical engineers, optical scientists, and optical technicians

Answers

A. $\overline{QP}$ is reflected across the optic axis $\overline{PV}$ and then dilated by scale factor of $\frac{Q'P'}{QP}$. The center of the dilation is V.

B. See Additional Answers for possible student constructions.

C. (1) For the object beyond C, the image is inverted and reduced in size.
(2) For the object at C, the image is inverted and the same size.
(3) For the object between C and F, the image is inverted and is enlarged.
(4) For the object at F, there is no image because the rays are parallel.
(5) For the object between F and V, the image is upright and enlarged and appears on the other side of the mirror (so it is not shown).

D. For a convex mirror, the image is always upright and reduced in size.

ONLINE

Assign the Digital Are You Ready? to power actionable reports including

- proficiency by standards
- item analysis

Are You Ready?

Diagnostic Assessment

- Diagnose prerequisite mastery.
- Identify intervention needs.
- Modify or set up leveled groups.

Have students complete the *Are You Ready?* assessment on their own. Items test the prerequisites required to succeed with the new learning in this module.

Scale Drawings Students will apply previous knowledge of scale drawings to use scaled factors of dilations in compressions and stretches.

Dilate Figures in the Coordinate Plane Students will apply previous knowledge of dilations to explore the properties of dilations and to write transformation rules.

Sequences of Transformations Students will apply previous knowledge of transformations to apply and analyze sequences of transformations and to write compositions of functions.

ONLINE

Are You Ready?

Complete these problems to review prior concepts and skills you will need for this module.

Scale Drawings

Consider the scale drawing.

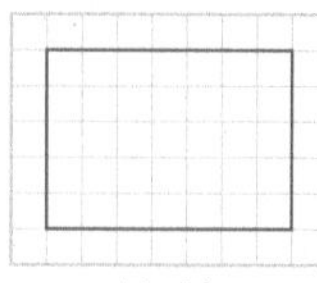

1. Find the actual dimensions of the rectangle. 30 feet by 42 feet
2. Find the actual area of the rectangle. 1260 square feet

Dilate Figures in the Coordinate Plane

The vertices of a triangle are $(8, -4)$, $(0, 12)$, and $(-4, 4)$. Find the vertices of the image after a dilation using $(0, 0)$ as the center of dilation for each scale factor.

3. $\frac{1}{2}$ $(4, -2)$, $(0, 6)$, and $(-2, 2)$
4. 2 $(16, -8)$, $(0, 24)$, and $(-8, 8)$
5. $\frac{3}{2}$ $(12, -6)$, $(0, 18)$, and $(-6, 6)$
6. $\frac{1}{4}$ $(2, -1)$, $(0, 3)$, and $(-1, 1)$

Sequences of Transformations

7. A quadrilateral has vertices $(7, -2)$, $(1, 1)$, $(4, -3)$, and $(6, -4)$. Find the vertices of the image after reflecting the quadrilateral across the y-axis and then translating 2 units up and 3 units left. $(-10, 0)$, $(-4, 3)$, $(-7, -1)$, and $(-9, -2)$
8. A triangle has vertices $(4, -8)$, $(5, 1)$, and $(-1, 2)$. Find the vertices of the image after rotating the triangle 90° clockwise about the origin and then dilating by a scale factor of 2. $(-16, -8)$, $(2, -10)$, and $(4, 2)$
9. A line segment has endpoints $(-4, 9)$ and $(3, -2)$. Find the endpoints of the image after reflecting the segment across the x-axis, translating 4 units down, and then reflecting across the line $x = 4$. $(12, -13)$ and $(5, -2)$

Connecting Past and Present Learning

Previously, you learned:

- to apply rigid transformations,
- to create scale drawings, and
- to describe symmetries of figures.

In this module, you will learn:

- to dilate figures in the coordinate plane and
- to apply more than one transformation on a figure.

172

DATA-DRIVEN INTERVENTION

Concept/Skill	Objective	Prior Learning *	Intervene With
Scale Drawings	Compute actual lengths and areas from a scale drawing.	Grade 7, Lesson 1.6	• Tier 3 Skill 4 • Reteach, Grade 7 Lesson 1.6
Dilate Figures in the Coordinate Plane	Find the image of a figure that is dilated in the coordinate plane.	Grade 8, Lesson 2.2	• Tier 2 Skill 12 • Reteach, Grade 8 Lesson 2.2
Sequences of Transformations	Find the image of a figure after a sequence of transformations in the coordinate plane.	Grade 8, Lessons 1.5 and 2.3	• Tier 2 Skill 13 • Reteach, Grade 8 Lesson 1.5 and 2.3

* Your digital materials include access to resources from Grade 6–Algebra 2. The lessons referenced here contain a variety of resources you can use with students who need support with this content.

6.1 Define and Apply Dilations, Stretches, and Compressions

LESSON FOCUS AND COHERENCE

Mathematics Standards

- Represent transformations in the plane using, e.g., transparencies and geometry software; describe transformations as functions that take points in the plane as inputs and give other points as outputs. Compare transformations that preserve distance and angle to those that do not (e.g., translation versus horizontal stretch).
- A dilation takes a line not passing through the center of the dilation to a parallel line, and leaves a line passing through the center unchanged.
- The dilation of a line segment is longer or shorter in the ratio given by the scale factor.
- Use geometric descriptions of rigid motions to transform figures and to predict the effect of a given rigid motion on a given figure; given two figures, use the definition of congruence in terms of rigid motions to decide if they are congruent.

Mathematical Practices and Processes 

- Use appropriate tools strategically.
- Attend to precision.

I Can Objective 

I can dilate and stretch a figure and determine how a figure has been transformed.

Learning Objective

Perform and analyze transformations to include dilations, stretches, and compressions. Use coordinate rules and geometric drawing tools to investigate the effect of multiplication on the points in a figure.

Language Objective

Compare and contrast a dilation with a stretch or a compression.

Vocabulary

Review: dilation

New: center of dilation, compression, scale factor, stretch

Lesson Materials: compass, geometric drawing tool, straightedge

Mathematical Progressions

Prior Learning	Current Development	Future Connections
Students: • developed definitions of rotations, reflections, and translations in terms of angles, circles, perpendicular lines, parallel lines, and line segments. **(5.1–5.3)** • described the effect of dilations, translations, rotations, and reflections on two-dimensional figures using coordinates. **(5.1–5.3)**	**Students:** • discern between rigid and nonrigid transformations. • identify properties of a dilation. • determine the center and scale factor of a dilation given a preimage and its image. • perform dilations on the coordinate plane. • predict the effect of a given transformation.	**Students:** • given two figures, will use the definition of similarity in terms of similarity transformations to decide if they are similar; will explain using similarity transformations the meaning of similarity for triangles as the equality of all corresponding pairs of angles and the proportionality of all corresponding pairs of sides. **(12.1)**

PROFESSIONAL LEARNING

Using Mathematical Practices and Processes

Attend to precision.

This lesson provides an opportunity to address this Mathematical Practice Standard. They apply this standard when they make careful measurements to draw dilations, determine scale factors, and decide the type of transformation that has occurred. Students pay close attention to how scale factors applied to preimages affect the resulting images.

WARM-UP OPTIONS

ONLINE

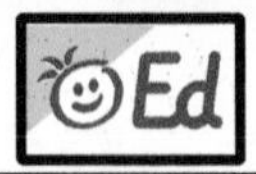

PROJECTABLE & PRINTABLE

ACTIVATE PRIOR KNOWLEDGE • Describe Translations, Reflections, and Rotations

Use these activities to quickly assess and activate prior knowledge as needed.

Problem of the Day 

The plans for a garden include a triangle-shaped rock on each side of a walkway. The plans are drawn on a coordinate plane where the walkway is the y-axis and the coordinates of one rock are $X(-3, 1)$, $Y(-1, 1)$, and $Z(-3, 4)$. If the second rock is the reflection of the first rock over the walkway, what are the coordinates for the second rock?
$X'(3, 1), Y'(1, 1), Z'(3, 4)$

Quick Check for Homework

As part of your daily routine, you may want to display the Teacher Solution Key to have students check their homework.

Make Connections

Based on students' responses to the Problem of the Day, choose one of the following:

1. Project the Interactive Reteach, Geometry, Lesson 5.3.

2. Complete the Prerequisite Skills Activity:

Write the following words on the board: *translation*, *reflection*, *rotation*. Have students work in pairs to describe each term in words and sketch an example of each type of transformation. Then have pairs of students share their descriptions and sketches with other groups.

- *What happens when a figure undergoes a translation?* The figure moves horizontally or vertically or both.
- *What happens when a figure undergoes a reflection?* The figure is flipped over a line of reflection.
- *What happens when a figure undergoes a rotation?* The figure turns around a center point.
- *What happens to the size and shape of the figure when these types of transformations occur?* The size and shape stay the same for each transformation. The location changes for all three transformations. In the reflection and the rotation, the orientation also changes.

If students continue to struggle, use Tier 2 Skill 9.

SHARPEN SKILLS

If time permits, use this on-level activity to build fluency and practice basic skills.

Vocabulary Review

Objective: Students demonstrate understanding of dilations.
Materials: Word Description (Teacher Resource Masters)

Have students use the word description graphic organizer and write the word *dilation* in the center. Work together as a class to write the definition and characteristics in the top two boxes. Then have students sketch examples and non-examples of dilations in the bottom two boxes. Have students share their drawings with the class.

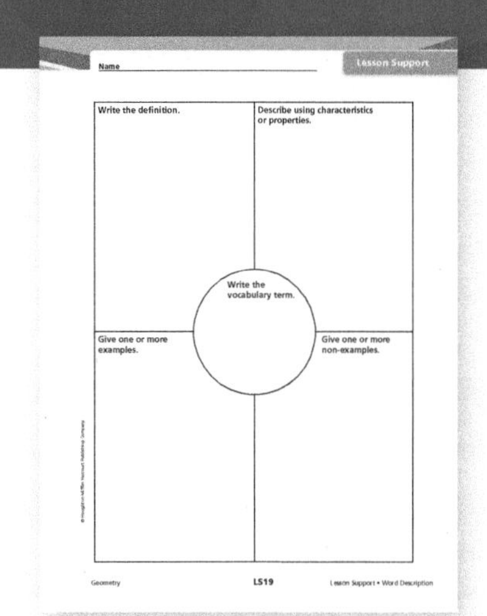

Name

Lesson Support

Write the definition.

Describe using characteristics or properties.

Write the vocabulary term.

Give one or more examples.

Give one or more non-examples.

Geometry LS19 Lesson Support • Word Description

PLAN FOR DIFFERENTIATED INSTRUCTION

Small-Group Options

Use these teacher-guided activities with pulled small groups.

On Track

Materials: coordinate planes (Teacher Resource Masters)

Have students dilate a triangle by drawing a triangle, choosing a random scale factor, and determining the coordinates of the image.

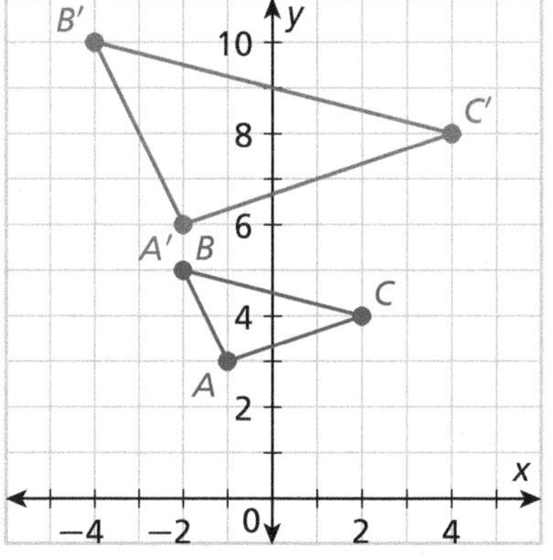

Almost There

Materials: index cards

Make sets of three cards. On one card, write an ordered pair that represents a preimage point. On a second card, write a scale factor. On the third card, write the ordered pair that represents the image point after the dilation. Make enough sets so that each student can have one card.

- Shuffle the cards and pass out one to each student.
- Have students walk around the room to find the two cards that correspond to their own card.
- Go over the card sets together as a class, identifying which points are the preimage and image for each set.

Ready for More

Write the rules on the board.

$(x, y) \rightarrow (2x, 2y)$

$(x, y) \rightarrow (x, 2y)$

$(x, y) \rightarrow (2x, y)$

$(x, y) \rightarrow (x + 2, 2y)$

$(x, y) \rightarrow (2x, x + 2)$

Then have students do the following.

- Discuss in small groups which coordinate rules, if any, show dilations, stretches, or compressions.
- What other transformations are present?
- If needed, have students sketch examples using each rule to verify their answers.

Math Center Options

Use these student self-directed activities at centers or stations.

Key: ● Print Resources ● Online Resources

On Track

- ● Interactive Digital Lesson
- ●● Journal and Practice Workbook
- ● Interactive Glossary (printable): **dilation**, **center of dilation**, **compression**, **scale factor**, **stretch**
- ● Illustrative Mathematics: Dilations and Distances
- ● Module Performance Task
- ● Coordinate planes (Teacher Resource Masters)

Almost There

- ● Reteach 6.1 (printable)
- ● Interactive Reteach 6.1
- ● RtI Tier 2 Skill 9: Translate Figures in the Coordinate Plane

Ready for More

- ● Challenge 6.1 (printable)
- ● Interactive Challenge 6.1
- ● Illustrative Mathematics: Dilating a Line

Unit Project Check students' progress by asking to see the length of QR that they are working with.

View data-driven grouping recommendations and assign differentiation resources.

During the *Spark Your Learning*, listen and watch for strategies students use. See samples of student work on this page.

Analyze the data in the tables

Strategy 1

I added the number of responses in the first data table and noticed that there were 150 data points.

48 + 34 + 23+ 35 + 10 = 150

I added the number of responses in the second data table and noticed there were 365 responses. 80 + 53+ 67 + 125 + 40 = 365

This means the adults must have been able to choose more than one "subjects still studied".

The bars in the graph seem to have the correct heights but the widths vary in the first graph. They should all have the same width.

If students . . . analyze the data in the tables, they are comparing graphs and data tables from Grade 6.

Have these students . . . explain how they compared the data in the tables to the representations in the graph. **Ask:**

- Q What kind of information did you gain from the data in the tables?
- Q What differences do you notice between the graphs and the tables?

Compare the graphs

Strategy 2

In the first graph, the tallest triangle is for English Literature. This is the most popular subject. However, the triangles should have the same width. When some are wider than others, this implies there may have been more answers for that subject.

In the second graph, most people are still studying science because the bar is the highest. The bar graph is more accurately constructed because the widths are all the same.

If students . . . compare only the graphs, they know how the size of the bars represents the data from the survey, but may not know how to evaluate the graphs using the data.

Activate prior knowledge . . . by having students recall how to display data using a bar graph. **Ask:**

- Q What should be true about the widths and the heights of the bars?
- Q What does the height of the Science bar in each graph tell you about the participants in the survey?
- Q When you look at the data in the tables, what information can you learn about the participants in the survey?

COMMON ERROR: Focuses on the size of the icons

The most favorite subject is English Literature because it has the largest triangle.

Most students are still studying Science because it has the tallest bar.

If students . . . look only at the overall size of the bars, they may not understand the widths of the icons should all be the same, and that the data in the tables gives them more information.

Then intervene . . . by pointing out that the heights match the corresponding numbers in the tables. **Ask:**

- Q What does the wider width of some of the triangles seem to indicate?
- Q Based on the data in the table, do the wider widths accurately represent the data?
- Q If you add up the numbers in each table, what do you notice?

6.1

Define and Apply Dilations, Stretches, and Compressions

I Can dilate and stretch a figure and determine how a figure has been transformed.

Spark Your Learning

Jocelyn is researching what subjects adult learners choose to study. She found the charts below in a report about a survey of a group of adults taking classes at a university.

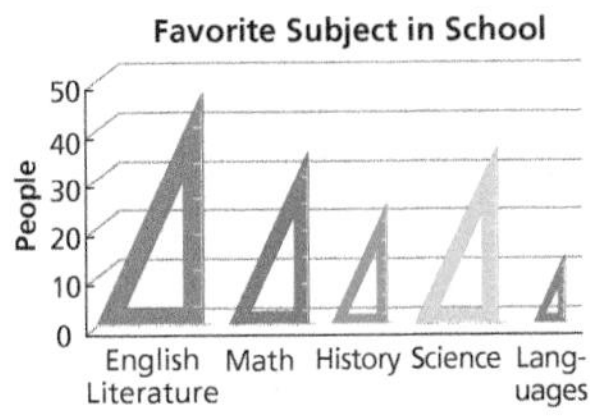

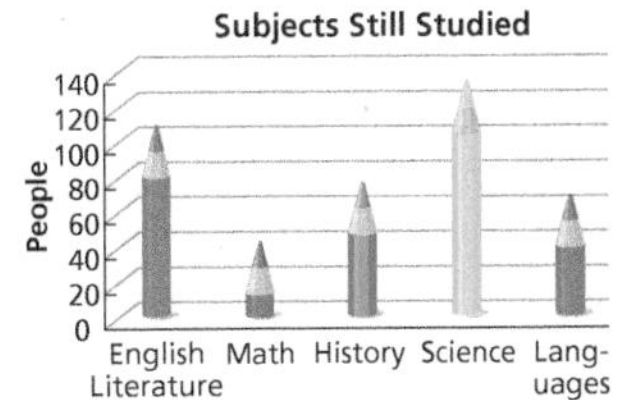

Complete Part A as a whole class. Then complete Parts B and C in small groups.

A. What is a mathematical question you can ask about the charts and how they were made? What information would you need to know to answer your question?

A. Do the displays accurately represent the information?; need to know the data used to create the displays

B. To answer your question, what strategy and tool would you use along with all the information you have? What answer do you get? See Strategies 1 and 2 on the facing page.

C. Do these charts appear to accurately represent the survey data? See Additional Answers.

Turn and Talk Think about ways the data display might change.

- Suggest a different way to present the data in one of the charts. Explain why your display would make the information easier to understand.
- The totals in the two displays are different. What must be true if these represent a survey of the same group of adults? See margin.

CULTIVATE CONVERSATION • Co-Craft Questions

If students have difficulty formulating a mathematical question about the situation in the Spark Your Learning, ask them to examine and compare the triangles and rectangles used to represent the data. What are some natural questions to ask about this situation?

Work together to craft the following questions:

- How are the heights related to the data points in each graph?
- How are the widths related to the data points in each graph?
- How many adult learners were surveyed to collect this data?

Then have students think about what additional information, if any, they would need to answer these questions. **Ask:**

- Can you determine the answers a specific adult learner chose for each survey question? Why or why not?
- Which graph represents more data points? Explain.

1 Spark Your Learning

▶ MOTIVATE

- Have students look at the graphs in their books and read the information contained in the graphs. Then complete Part A as a whole-class discussion.
- Give the class the additional information they need to solve the problem. This information is available online as a printable and projectable page in the Teacher Resources.
- Have students work in small groups to complete Parts B and C.

▶ PERSEVERE

If students need support, guide them by asking:

Q **Advancing • Use Tools** Which tool could you use to solve the problem? Why choose that tool and not some other? Students' choices of tools and reasons for choosing them will vary.

Q **Assessing** What do you notice about the size and shape of the figures used to represent the data in the two graphs? In the first graph, the triangles are different heights and different widths. In the second graph, the bars are all the same width, but different heights.

Turn and Talk When deciding how to change the graphs, encourage students to see that the different widths of the triangles in the first graph do not accurately represent the data. Use pencil icons for both graphs and treat the data in the same way. Then it would be easier to compare the data so that you can think about whether people changed their opinions about what subjects are interesting to learn.
Explain to students that the total is found by adding the data in the tables. Point out that there are 48 people whose favorite subject is English Literature, but 80 people are still studying it. Ask students to brainstorm how that might be possible. The people were allowed to give more than one answer to the Subjects Still Studied question.

▶ BUILD SHARED UNDERSTANDING

Select groups of students who used various strategies and tools to share with the class how they solved the problem. As they present their solutions, have each group discuss why they chose a specific strategy and tool.

② Learn Together

Build Understanding

Task 1 (MP) **Use Tools** Students use a geometric drawing tool to identify differences in types of transformations.

CONNECT TO VOCABULARY

Have students use the **Interactive Glossary** to record their understanding of the vocabulary in this task.

Sample Guided Discussion:

Q **In Part A, why do you think the measures of the sides of the triangles stay the same after the transformation?** Each x-coordinate is moved horizontally in the same direction and each y-coordinate is moved vertically in the same direction. The side lengths do not change.

Turn and Talk Encourage students to think about times where a real-life object stays the same size and shape. Then have students think about times when an object changes size or shape. Sample answer: A would happen when an object is moved. B would happen when a bigger copy of an object is made, for instance a larger size of clothes. C would happen when a smaller copy of an object is made, for instance a scale model. D would happen when an object is squished vertically, for instance a sandwich is sat upon.

Build Understanding

Investigate Transformations

Some transformations change the size and shape of a figure. Two nonrigid transformations are dilations and stretches.

A **dilation** changes the size of a figure by the same amount in all directions.

preimage

dilation by a factor greater than 1

dilation by a factor less than 1

A **stretch** changes the shape of a figure by a factor greater than 1 in one direction. A **compression** changes the shape of a figure by a factor greater than 0 and less than 1 in one direction.

preimage

vertical compression

horizontal stretch

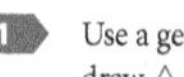

1 Use a geometric drawing tool to draw $\triangle ABC$ with vertices $A(-2, 2)$, $B(2, 5)$, and $C(2, 2)$.
A–D. See Additional Answers.

A. Transform $\triangle ABC$ using the rule $(x, y) \to (x - 2, y + 1)$. What kind of transformation is this?

B. Transform $\triangle ABC$ using the rule $(x, y) \to (3x, 3y)$. How does the figure change? Is this a rigid transformation?

C. Repeat Step B using the rule $(x, y) \to \left(\frac{1}{2}x, \frac{1}{2}y\right)$.

D. Repeat Step B using the rule $(x, y) \to (x, 2y)$.

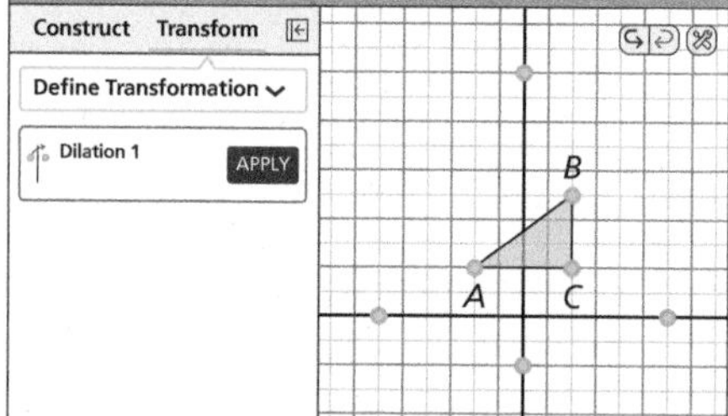

©Valentina Razumova/Shutterstock

Turn and Talk Identify a real-world situation when each transformation would arise.
See margin.

174

LEVELED QUESTIONS

Depth of Knowledge (DOK)	Leveled Questions	What Does This Tell You?
Level 1 **Recall**	Two squares are plotted in a coordinate plane. One square has side lengths of 3 and the other square has side lengths of 6. What type of transformation can be used to map the smaller square to the larger square? a dilation	Students' answers will indicate whether they can identify a dilation when given a verbal description.
Level 2 **Basic Application of Skills & Concepts**	The rule $(x, y) \to (5x, 5y)$ transforms a figure. What happens to the size and shape of the figure? The shape stays the same, but the size becomes 5 times larger in every direction.	Students' answers will indicate whether they can apply a rule and analyze the result of the transformation.
Level 3 **Strategic Thinking & Complex Reasoning**	How can you write a rule to create only a stretch or a compression, but not both? Multiply only one of the coordinates by a constant factor.	Students' answers will indicate whether they understand that a compression or a stretch happens for one direction, either horizontally or vertically.

Explore Dilations

The **center of dilation** is the fixed point in the plane that does not change when the dilation is applied. The center can be on the image or elsewhere on the plane. Rays drawn from the center of dilation to the preimage will intersect corresponding points on the image.

The **scale factor** k of a dilation is the ratio of the length of a segment on the image to the length of the corresponding segment on the preimage. For example, $k = \frac{S'T'}{ST}$.

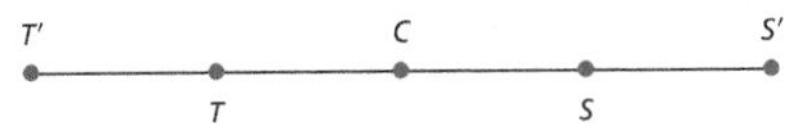

2 A. Use a compass and straightedge to construct a dilation with scale factor of 2.
Check students' constructions.

Draw triangle PQR and a center of dilation C outside the triangle.

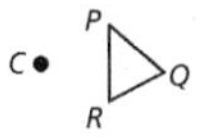

Draw a ray from the center of dilation through each vertex.

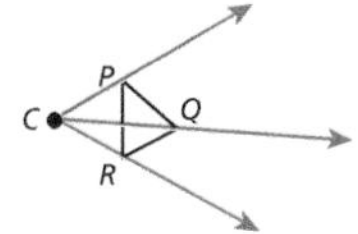

Set the compass to the distance CP. Then mark this distance along the ray CP from point P. Repeat for all vertices.

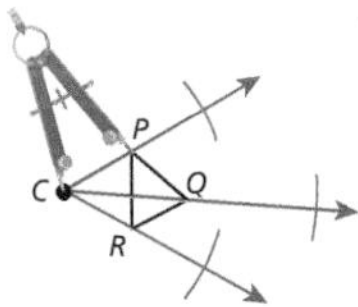

The intersections of the arcs and rays are the vertices of the image. Draw and label the image.

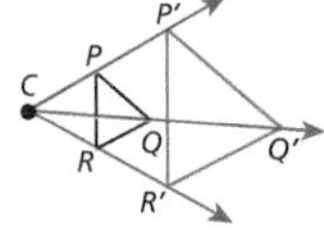

B. What is the scale factor of the dilation? How do you know?
The scale factor is 2; Each vertex is twice as far from the center of dilation.

C. How are the ratios of the lengths of corresponding sides of the preimage and image related? C, D. See Additional Answers.

D. Do $\overline{PR}$ and $\overline{P'R'}$ appear to be parallel? How can you check?

Turn and Talk

- Make conjectures about the relationship between segments in the preimage and image when the center of dilation is on the preimage and when it is not on the preimage. See margin.
- A horizontal stretch originates from a vertical line in the same way that a dilation originates from a center point. Describe how you could construct a horizontal stretch by a factor of 2.

Step It Out

Task 2 (MP) **Use Tools** Students use a compass and a straightedge to construct a dilation.

CONNECT TO VOCABULARY

Have students use the **Interactive Glossary** to record their understanding of the vocabulary in this task.

Sample Guided Discussion:

Q Why do you use the measurements of the distance from the center to each vertex? This keeps the ratio of the side lengths the same from the preimage to the image.

Q In Part B, what tools can you use to verify the scale factor? I know the compass lengths were used twice. I could also use a ruler to measure and compare the side lengths of the preimage and image.

Turn and Talk Have students repeat the steps in Part A, but this time place C on point P of the triangle. If the center of dilation is on a segment of the preimage, then the image of the segment will be collinear with its preimage. If the center of dilation is not on a segment of the preimage, then the image of the segment will be parallel to its preimage. **Have students draw a vertical line and a triangle off the line. Then ask students how the vertical line could be related to the vertices of the triangle.** Instead of drawing rays that extend to vertices from a center point as a dilation does, draw separate rays that are perpendicular to the fixed line through each vertex of the figure. Do the third and fourth steps of the construction in Task 2.

PROFICIENCY LEVEL

Beginning
Give a verbal description of a dilation, stretch, or compression aloud. Have the students turn to a partner and give the word that best represents the description. For example, say "double the size of a photograph" and students should say "dilation".

Intermediate
Assign each student a value for a scale factor k. Ask students to write what would happen to a preimage when that scale factor is applied in a dilation. Have students share their examples with the class.

Advanced
Ask students to write a short paragraph describing the how the value of the scale factor affects the result of the image in a dilation.

Task 3 (MP) **Use Tools** Students examine and analyze a one-point perspective drawing using measurements.

Sample Guided Discussion:

Q What part of the drawing appears to be the center of dilation? There are several lines that all meet near the back of the picture, where the road seems to disappear. The point where these lines meet appears to be the center of dilation.

Q In Part B, how can you use the given measurements to find the scale factor? I can write the ratio of the measurements of the image compared to the preimage. The ratio should always be the same: $\frac{(BC)'}{BC} = \frac{\frac{2}{3}}{2} = \frac{1}{3}$ and $\frac{(AB)'}{AB} = \frac{1}{3}$.

Turn and Talk Help students by reminding them that to find the scale factor, they divided measurements for the image by the preimage. Ask students to consider what would happen if they divide the measurements for the preimage by the image. Then have students compare the resulting value. A scale factor that is the reciprocal of the scale factor used to create the image would turn the image back into the preimage. Scale factors between 0 and 1 reduce, and scale factors greater than 1 enlarge.

Step It Out

Use Properties of Dilations

Dilations share some of the properties with rigid motions, but they change lengths of corresponding segments.

Properties of Dilations
• preserve angle measure
• preserve collinearity
• preserve orientation
• map a segment to another segment whose length is the product of the scale factor and the length of the preimage
• map a line not passing through the center of the dilation to a parallel line and leave a line passing through the center unchanged

3 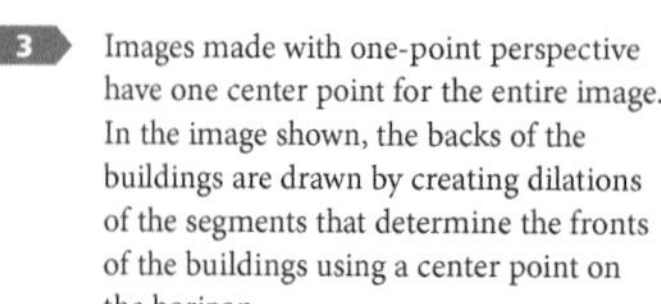
Images made with one-point perspective have one center point for the entire image. In the image shown, the backs of the buildings are drawn by creating dilations of the segments that determine the fronts of the buildings using a center point on the horizon.

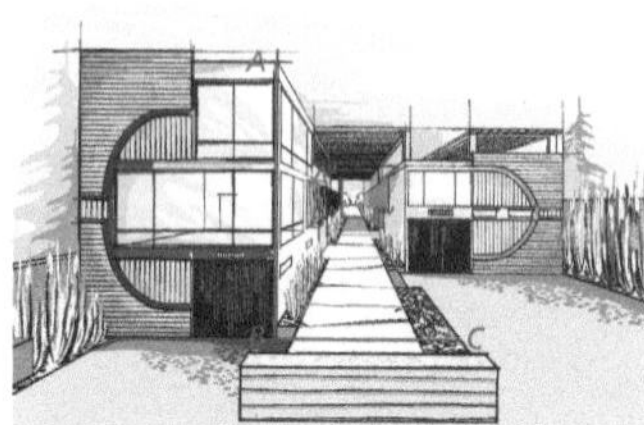

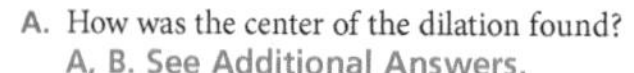
A. How was the center of the dilation found?
A, B. See Additional Answers.

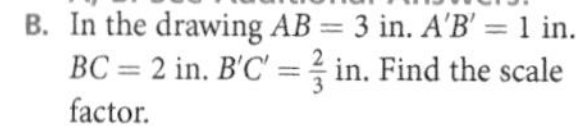
B. In the drawing $AB = 3$ in. $A'B' = 1$ in. $BC = 2$ in. $B'C' = \frac{2}{3}$ in. Find the scale factor.

Turn and Talk If $ABCD$ is the preimage and $A'B'C'D'$ is the image, what would it take to turn $A'B'C'D'$ back into $ABCD$? Compare scale factors that will reduce an image to those that will enlarge an image. See margin.

Dilations, Stretches, and Compressions on the Coordinate Plane

The coordinate notation for dilations, stretches, and compressions are related. Each involves multiplying at least one of the coordinates by a scale factor k.

Transformation	Center at Origin
Dilation	$(x, y) \rightarrow (kx, ky)$
Vertical Stretch ($k > 1$)	$(x, y) \rightarrow (x, ky)$
Horizontal Stretch ($k > 1$)	$(x, y) \rightarrow (kx, y)$
Vertical Compression ($0 < k < 1$)	$(x, y) \rightarrow (x, ky)$
Horizontal Compression ($0 < k < 1$)	$(x, y) \rightarrow (kx, y)$

4 Draw a dilation with scale factor $\frac{1}{2}$ of quadrilateral $ABCD$ with vertices $A(-4, -2)$, $B(-6, -4)$, $C(-4, -6)$, and $D(-2, -4)$ centered at the origin.

Use the coordinate rule $(x, y) \to \left(\frac{1}{2}x, \frac{1}{2}y\right)$.

A. How do you know whether the image will be larger or smaller than the preimage?

It will be smaller because the scale factor is between 0 and 1.

Preimage	Image
$A(-4, -2)$	$(-2, -1)$
$B(-6, -4)$	$(-3, -2)$
$C(-4, -6)$	$(-2, -3)$
$D(-2, -4)$	$(-1, -2)$

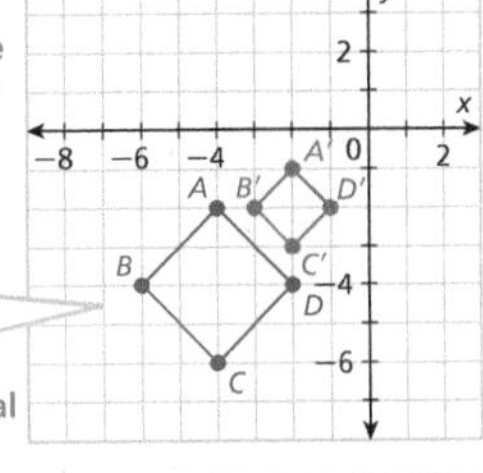

Plot and connect points to draw the dilation. Be sure to use prime notation for the image.

B. Which quadrilateral is the image?

The smaller quadrilateral is the image.

Turn and Talk How can you use the graph to check the dilation? See margin.

Predict the Effect of a Transformation Rule

5 Look at the coordinate points and predict what kind of transformation will happen in each case. Match the transformation in column A with the rule in column B.
1-C, 2-F, 3-B, 4-E, 5-A, 6-D

A. Match the transformation in the left column with the rule in the right column.

1. Translation	A. $(x, y) \to (x, 3y)$
2. Dilation	B. $(x, y) \to (3x, y)$
3. Horizontal stretch	C. $(x, y) \to (x + 3, y)$
4. 270° Rotation	D. $(x, y) \to (x, 0.3y)$
5. Vertical stretch	E. $(x, y) \to (y, -x)$
6. Vertical compression	F. $(x, y) \to (3x, 3y)$

B. Is each transformation rigid or nonrigid?
2-F, 3-B, 5-A, and 6-D are nonrigid; 1-C and 4-E are rigid.

Turn and Talk

- Compare the transformation rules that result in rigid transformations and nonrigid transformations.
- Compare the results of applying rigid transformations and nonrigid transformations to a figure. See margin.

Task 4 (MP) **Attend to Precision** Students will use a given rule to dilate a figure on a coordinate plane.

SUPPORT SENSE-MAKING Three Reads

Have students read the problem three times. Use the questions below for a different focus each read.

1. What is the situation about?
2. What are the quantities in the situation?
3. What are the possible mathematical questions that you could ask for the situation?

Sample Guided Discussion:

Q How can you describe the scale factor used in the rule? The scale factor is $\frac{1}{2}$, which is between 0 and 1.

Q In Part B, how can you tell the difference between the image and preimage when you consider the scale factor? The scale factor is between 0 and 1, which means the image will be smaller than the preimage. The smaller quadrilateral will be the image.

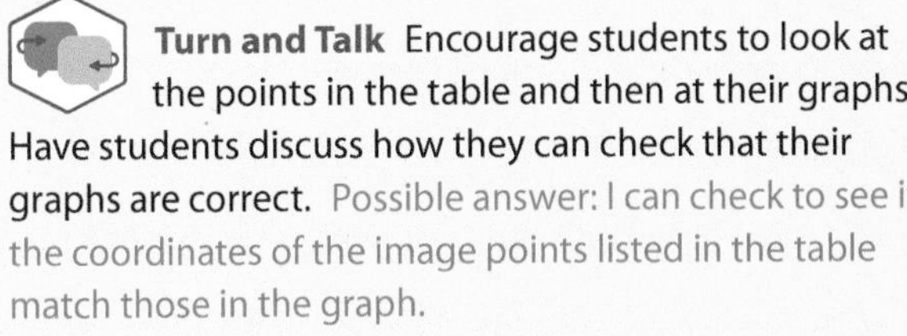

Turn and Talk Encourage students to look at the points in the table and then at their graphs. Have students discuss how they can check that their graphs are correct. Possible answer: I can check to see if the coordinates of the image points listed in the table match those in the graph.

Task 5 (MP) **Attend to Precision** Students apply coordinate rules and examine how the rules affect a preimage.

Sample Guided Discussion:

Q How could you verify the effect of a coordinate rule? I could choose points for a preimage, apply the rule, and then compare the graphs of the preimage and image.

Turn and Talk Suggest that students study all the types of transformations given in Task 5 and think about which are rigid and which are non-rigid.

- Transformation rules that result in rigid motions add values to the coordinates, swap coordinates, or multiply one or both coordinates by −1. Transformations that result in nonrigid motions multiply one or both coordinates by a value other than −1.
- Rigid transformations do not change a figure's side lengths or angle measures. Nonrigid transformations change lengths if both coordinates are multiplied by the same value, and they change the lengths and angles if the coordinates are multiplied by different values.

ONLINE

Assign the Digital On Your Own for

- built-in student supports
- Actionable Item Reports
- Standards Analysis Reports

On Your Own

Assignment Guide

The chart below indicates which problems in the On Your Own are associated with each task in the Learn Together. Assign daily homework for tasks completed.

Learn Together Tasks	On Your Own Problems
Task 1, p. 174	Problems 6 and 7
Task 2, p. 175	Problems 8 and 9
Task 3, p. 176	Problems 18, 20, 21, 23, and 24
Task 4, p. 177	Problems 19, 22, and 25
Task 5, p. 177	Problems 10–17

Check Understanding

1. What type of transformation is the result of applying the rule $(x, y) \rightarrow (2(x, y))$? dilation

Use the rule to transform the triangle with vertices $A(1, 4)$, $B(0, 1)$, and $C(-4, 0)$. Graph and label the image and the preimage.

2. $(x, y) \rightarrow (x, 3y)$
3. $(x, y) \rightarrow (2x, y)$ 2, 3. See Additional Answers.

Name the transformation used to map the blue figure to the red figure.

4. 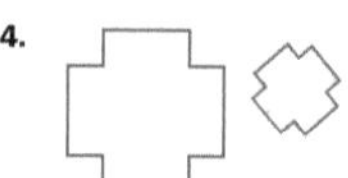
dilation and rotation

5. 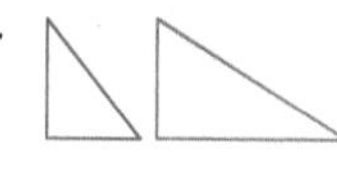
horizontal stretch

On Your Own

6. (MP) **Critique Reasoning** Richard says that the rule $(x, y) \rightarrow (0.2x, y)$ describes a horizontal stretch because only the x-coordinates are affected by the rule. Is Richard correct? Why or why not? incorrect; Since the scale factor is between 0 and 1, it is actually a horizontal compression.

7. Maya wants to figure out what sort of transformation was used to map the first picture to the second picture.

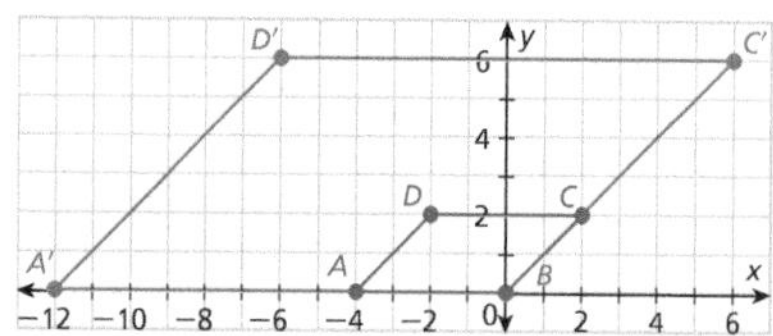

   A. Describe a method Maya could use to solve the problem.
   B. Why are the pictures given misleading?
   C. What kind of transformation was used?

A–C. See Additional Answers.

8. Figure $A'B'C'D'$ is a dilation of $ABCD$. What was the scale factor? Write a rule for the dilation. scale factor = 3; $(x, y) \rightarrow (3x, 3y)$

9. (MP) **Critique Reasoning** Bridget was constructing a dilation with scale $\frac{1}{2}$. She drew rays from the center through each vertex. Then she used a compass to find the distance from the center to each vertex, and then marked off that distance again along her rays. Then she connected the corresponding vertices. Did Bridget correctly construct the dilation? Why or why not?
no; Bridget dilated by a scale factor of 2 instead of $\frac{1}{2}$.

178

data checkpoint

3 Check Understanding

Formative Assessment

Use formative assessment to determine if your students are successful with this lesson's learning objective.

Students who successfully complete the Check Understanding can continue to the On Your Own practice.

For students who miss 1 problem or more, work in a pulled small group using the Almost There small-group activity on page 173C.

ONLINE

Assign the Digital Check Understanding to determine

- success with the learning objective
- items to review
- grouping and differentiation resources

4 Differentiation Options

Differentiate instruction for all students using small-group activities and math center activities on page 173C.

Reteach

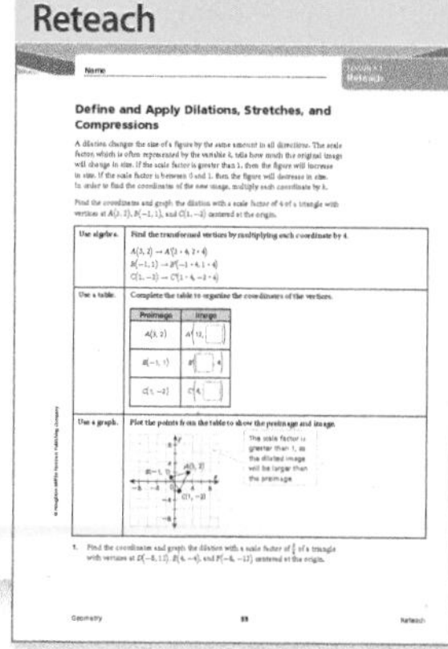

Challenge

Other Transformation Rules

In Problems 10–17, name the type of transformation that will result from applying the given rule.

10. $(x, y) \rightarrow (2x, 2y)$
dilation by scale factor of 2

11. $(x, y) \rightarrow (x + 2, y)$
translation to the right 2 units

12. $(x, y) \rightarrow \left(\frac{1}{4}x, \frac{1}{4}y\right)$
dilation by scale factor of $\frac{1}{4}$

13. $(x, y) \rightarrow (7x, y)$
horizontal stretch with scale factor 7

14. $(x, y) \rightarrow (-x, -y)$
rotation of 180°

15. $(x, y) \rightarrow \left(x + 2, \frac{1}{2}y\right)$

16. $(x, y) \rightarrow (x, 1 - y)$

17. $(x, y) \rightarrow \left(2(x + 1), 2(y - 2)\right)$

15–17. See Additional Answers.

18. Art Danica is studying perspective drawing in her art class. What kind of transformation did she use to create her drawing?
dilation

19. Open Ended Draw a triangle on the coordinate plane. Dilate the triangle. Describe the steps you took.
Check students' work.

20. (MP) **Critique Reasoning** Joshua attempted to dilate a right triangle but his new triangle is not a right triangle. Could this be correct? Explain why or why not.
See Additional Answers.

21. STEM Jeremy is creating an app that shrinks and zooms pictures to fit. What scale factor would create a figure three-fifths the size of the original? scale factor of $\frac{3}{5}$

22. (MP) **Use Structure** You can use the steps in Parts A–D to write a coordinate rule to dilate a figure using center (a, b) not at the origin. A–D. See Additional Answers.

A. Write a rule that will translate the figure (p, q) with center (a, b) so that the center is the origin $(0, 0)$.

B. Starting with your rule from Part A, write a rule that will dilate the figure by scale factor k.

C. Starting with your rule from Part B, write a rule that will translate the figure with the center at the origin so that the center is mapped back to its original location (a, b).

D. Draw a triangle and choose a center of dilation that is not the origin. Using scale factor 2, apply your rule from Part C to your triangle. Does your rule correctly dilate your figure?

23. Health and Fitness AJ wants to design a running track with two lanes. The lanes should be transformations of each other with the smaller one inside the larger one. Can AJ use a translation to draw one of the lanes from the other? Why or why not?
See Additional Answers.

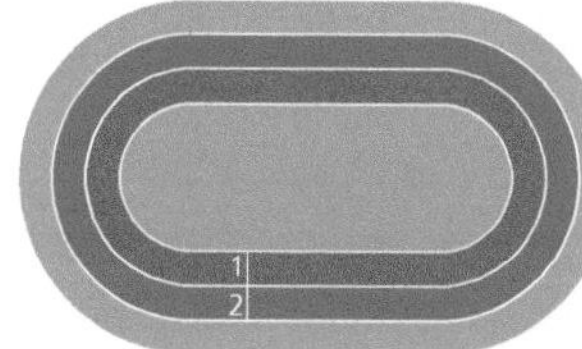

Questioning Strategies

Problem 20 Joshua likely used a coordinate rule when attempting to dilate the right triangle. In using the rule, he probably multiplied each x- and y-coordinate by a number. What must be true about the numbers used to multiply the coordinates when performing a dilation? The multiplier must be the same number. For example, using the rule, $(x, y) \rightarrow (kx, ky)$ for a positive k value would be a dilation because the coordinates are both multiplied by k.

Watch for Common Errors

Problem 23 A student might think that a translation can be used because the horizontal parts of the smaller track can be lined up with the horizontal parts of the original track by translations. Remind students that if a translation moves vertically, all parts of the figure will move vertically in the same manner.

5 Wrap-Up

Summarize learning with your class. Consider using the Exit Ticket, Put It in Writing, or I Can scale.

Exit Ticket 

Two students perform a transformation on the same preimage using the rules below.

$(x, y) \rightarrow (2x, y)$
$(x, y) \rightarrow (x, 2y)$

Are the students' images the same? Explain.

The images are not the same; When the first student multiplied the x-coordinate by 2, a horizontal stretch was performed. When the other student multiplied the y-coordinate by 2, a vertical stretch was performed. The first student made a wider image and the second student made a taller image.

Put It in Writing 

Describe the similarities and differences of dilations and stretches or compressions.

I Can 

The scale below can help you and your students understand their progress on a learning goal.

4	I can dilate and stretch a figure and determine how a figure has been transformed, and I can explain my reasoning to others.
3	I can dilate and stretch a figure and determine how a figure has been transformed.
2	I can write a coordinate rule to dilate or stretch a figure.
1	I can examine a coordinate rule to determine if a figure has undergone a transformation.

Spiral Review • Assessment Readiness

These questions will help determine if students have retained information taught in the past and can also prepare them for high-stakes assessments. Here, students must identify a rotated figure **(5.2)**, find lines of symmetry **(5.4)**, identify nonrigid transformations **(6.2)**, and identify reflections **(5.3)**.

24. Kevin has a set of nesting dolls. He wants to draw a picture of them, and he wonders if he can use the same scale factor to reduce his first sketch to create the second as he would use to reduce the second sketch to create the third, etc. The heights of the dolls are 6 in., 4.5 in., 3.2 in., 2 in., and 1.25 in. Can Kevin use the same rule for each dilation?

24, 25. See Additional Answers.

25. Open Middle™ Using the integers −9 to 9, at most one time each, fill in the boxes two separate times so that the triangle with vertices at $(1, -3)$, $(2, 3)$, and $(-1, -2)$ has been dilated.

Image vertices: $(\square, \square)$, $(\square, \square)$, and $(\square, \square)$;

Dilation point: $(\square, \square)$; Scale factor: $\square$

Spiral Review • Assessment Readiness

26. Which of the following is a rotation of *ABCD* by less than 360°?

(A) (B) (C) (D)

27. How many lines of symmetry does an equilateral triangle have?

(A) 0
(B) 1
(C) 2
(D) 3

28. Which of the following are nonrigid transformations? Select all that apply.

(A) translation
(B) compression
(C) dilation
(D) rotation
(E) reflection
(F) stretch

29. Decide whether each pair of figures show a reflection.

	Pentagons	Triangles	Rectangles	Triangles
Reflection	?	?	?	?
No reflection	?	?	?	?

I'm in a Learning Mindset!

A photogrammetrist is someone who uses the science of making reliable measurements by the use of photographs and especially aerial photographs (as in surveying). If I were a photogrammetrist how would this lesson be helpful?

Keep Going to Journal and Practice Workbook

Learning Mindset

Challenge-Seeking Defines Own Challenges

Point out that the aerial photographs used by photogrammetrists may be taken from drones, airplanes, or even satellites. Each photograph is a scale image of the ground below and gives information about a large area in a small, manageable form. When the goal is to analyze very large areas, a lot of planning goes into making sure the entire area is imaged. Gaps between the photos can easily introduce error into the final product. Taken from directly overhead, the images change the size of their target area by the same amount in all directions. *Given that last statement, these images involve what kind of transformation? How could the photogrammetrist find the scale factor of the photograph? What measurements would the photogrammetrist need to make sure the scale factor is correct? How can the scale factor be used on the photograph to find measurements on the ground below? If this is a new method of measuring for the photogrammetrist, what type of mathematical concepts might be important to learn or review?*

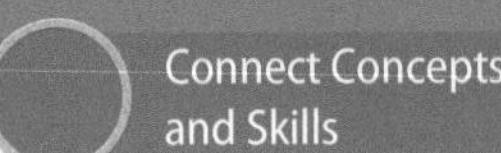

6.2 Apply Sequences of Transformations

LESSON FOCUS AND COHERENCE

Mathematics Standards

- Develop definitions of rotations, reflections, and translations in terms of angles, circles, perpendicular lines, parallel lines, and line segments.
- Given a geometric figure and a rotation, reflection, or translation, draw the transformed figure using, e.g., graph paper, tracing paper, or geometry software. Specify a sequence of transformations that will carry a given figure onto another.

Mathematical Practices and Processes 

- Use appropriate tools strategically.
- Attend to precision.
- Look for and make use of structure.

I Can Objective 

I can determine the effects of a sequence of transformations on a figure.

Learning Objective

Apply sequences of transformations to figures, specify sequences that map a given preimage to a given image, and make predictions about the result of applying a sequence of transformations.

Language Objective

Understand and use the language of transformations and compositions of transformations.

Lesson Materials: index cards

Vocabulary

New: composition

Mathematical Progressions

Prior Learning	Current Development	Future Connections
Students: • represented transformations in the plane; described transformations as functions that take points in the plane as inputs and give other points as outputs. **(5.1–5.3)** • developed definitions of rotations, reflections, and translations in terms of angles, circles, perpendicular lines, parallel lines, and line segments. **(5.4)**	**Students:** • represent transformations graphically and algebraically. • understand and use both sequences of transformations and compositions of transformations. • find the inverse of a sequence or combination of transformations.	**Students:** • given two figures, will use the definition of similarity in terms of similarity transformations to decide if they are similar; will explain using similarity transformations the meaning of similarity for triangles as the equality of all corresponding pairs of angles and the proportionality of all corresponding pairs of sides. **(12.1)**

UNPACKING MATH STANDARDS

Use geometric descriptions of rigid motions to transform figures and to predict the effect of a given rigid motion on a given figure; given two figures, use the definition of congruence in terms of rigid motions to decide if they are congruent.

What It Means to You

Students demonstrate an understanding of multiple rigid transformations applied to geometric figures and predict the effect of a given rigid transformation on a figure. This understanding develops from the lessons in Module 5, where they transformed figures using a single transformation applied to a geometric figure. The emphasis for this standard is on multiple transformations applied to a figure and predicting what the effect of those transformations will do. This will lead them to write compositions of transformations and use them to predict the effects on a figure.

WARM-UP OPTIONS

ONLINE

PROJECTABLE & PRINTABLE

ACTIVATE PRIOR KNOWLEDGE • Identify Transformations

Use these activities to quickly assess and activate prior knowledge as needed.

Problem of the Day 

Given a triangle with vertices (0, 0),(−3, 0) and (0, −4), determine whether it is congruent with another triangle with vertices (2, 2), (2, 5) and (6, 2). If the two are congruent, can the first triangle be translated such that it will occupy the position of the second triangle? The two figures are congruent right triangles, but the first cannot be translated into the position of the second because their orientation is different.

Quick Check for Homework

As part of your daily routine, you may want to display the Teacher Solution Key to have students check their homework.

Make Connections

Based on students' responses to the Problem of the Day, choose one of the following:

1. Project the Interactive Reteach, Geometry, Lesson 6.2.

2. Complete the Prerequisite Skills Activity:

Have students work in pairs. Give the example of a square *ABCD* with vertices at *A*(1, 1), *B*(1, 3), *C*(3, 3), and *D*(3, 1). Have students work together to solve the following problems.

- *What rotation(s) would transform the square so that the side $\overline{BC}$ is perpendicular to the x-axis?* A clockwise or counterclockwise rotation of 90 degrees would result in side $\overline{BC}$ being perpendicular to the *x*-axis.
- *What type of transformation would position the square so that it is centered at the origin?* a translation down 2 units and left 2 units
- *How many 45° rotations would rotate a square back to its original position if only two were clockwise and more than two were counterclockwise?* Twelve rotations would be required. If two are clockwise, then an additional two rotations would be needed to return the figure to its original position. From that point, it would take eight counterclockwise rotations to return the figure to its original position.

If students continue to struggle, use Tier 2 Skill 14.

SHARPEN SKILLS

If time permits, use this on-level activity to build fluency and practice basic skills.

Mental Math

Objective: Students perform simple transformations of figures shown on index cards.
Materials: index cards

Have students work in pairs. Give each pair two sets of cards that have corresponding information, one with a transformation indicated on each card and the other with a figure or figures. Ask students to take turns, with one showing a card while the other names a transformation that would fit what is shown on the card.

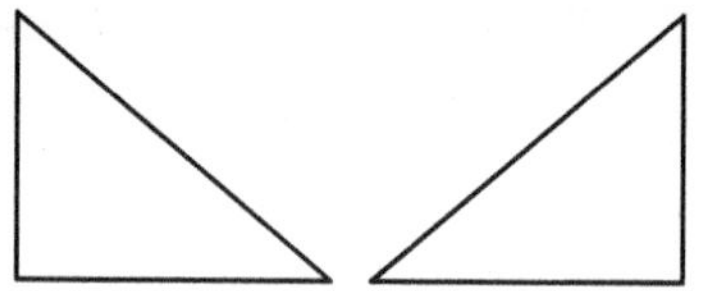

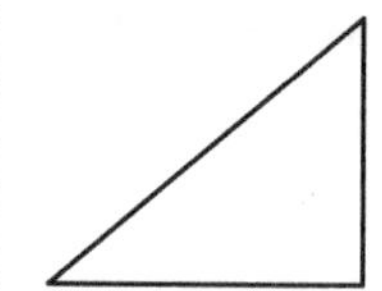

90° rotation	translation 1 unit to the right

PLAN FOR DIFFERENTIATED INSTRUCTION

Small-Group Options

Use these teacher-guided activities with pulled small groups.

On Track

Materials: index cards

Have students work in small groups. Distribute the index cards, which show a sequence of transformations and a final result. Have students put the cards in a logical order so that they lead to the final result. Each group gets identical cards. Have a volunteer report to the class.

Almost There

Have students work in small groups to visualize transformations of the following:

- a right triangle with its right angle at the origin
- a parallelogram with its base on the *x*-axis and a 60° angle from the origin, lying entirely in Quadrant I

Next, have students sketch their transformations on graph paper, along with the figures in their original positions. Ask volunteers to present their transformations to the class.

Ready for More

Materials: index cards from the On Track activity

Have students write the transformation from each card to the next. Then have them turn the cards over and write the inverse transformations.

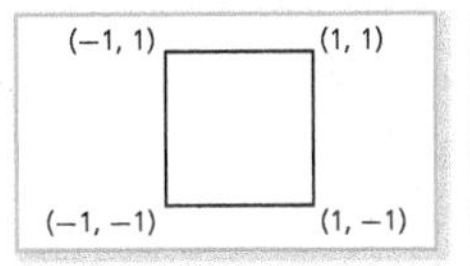

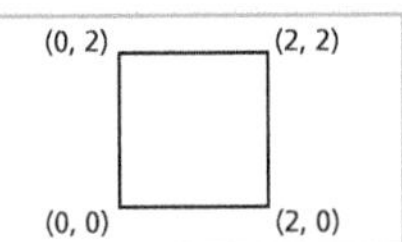

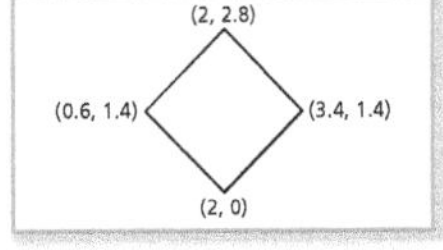

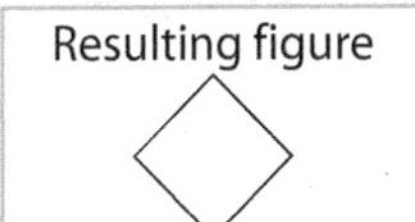

Math Center Options

Use these student self-directed activities at centers or stations. Key: ● Print Resources ● Online Resources

On Track

- ● Interactive Digital Lesson
- ●● Journal and Practice Workbook
- ● Interactive Glossary (printable): **composition**
- ● Module Performance Task
- ● Illustrative Mathematics: Identifying Rotations

Standards Practice:

- ●● Use Transformations that Do and Do Not Preserve Measurement
- ●● Develop Definitions of Transformations in Terms of Geometric Shapes
- ●● Draw a Transformed Figure and Specify the Transformations Used

Almost There

- ● Reteach 6.2 (printable)
- ● Interactive Reteach 6.2
- ● RtI Tier 2 Skill 14: Congruent Figures
- ● Illustrative Mathematics: Reflecting Reflections

Ready for More

- ● Challenge 6.2 (printable)
- ● Interactive Challenge 6.2
- ● Illustrative Mathematics: Showing a Triangle Congruence: The General Case

View data-driven grouping recommendations and assign differentiation resources.

Spark Your Learning • Student Samples

During the *Spark Your Learning*, listen and watch for strategies students use. See samples of student work on this page.

Describe Transformations

Strategy 1

The animal paws are symmetric to each other, like mirror images, but one is above the other. They have an opposite orientation along a vertical line of symmetry. The transformation is a reflection across the y-axis and a translation up by a certain number of units.

If students . . . think about the problem in terms of symmetry and general transformations, they are displaying a conceptual understanding of transformations.

Have these students . . . explain how they determined the steps in the transformation. **Ask:**

- Q How did you use your knowledge of transformations to approach the problem?
- Q What made you realize the paws were reflected?

Use a Coordinate Plane

Strategy 2

I copied the left paw onto a coordinate plane, where it lies to the left of the y-axis. Then I copied the right paw onto the same graph paper, just to the right of the y-axis, where it is aligned vertically with the other paw. Then I applied a translation to the right paw. Finally, I compared my drawing to the paw prints.

If students . . . think of a transformation on a coordinate plane, then they understand how the figures are related but may not yet know how to apply transformations more broadly.

Activate prior knowledge . . . by having students graph simple transformations of a rectangle on graph paper. **Ask:**

- Q How can you apply a reflection of a rectangle in Quadrant 2 so that it ends up in Quadrant 1?
- Q What transformation would you use to move the image of the rectangle upward so that it is similar to the paw prints?

COMMON ERROR: Uses Wrong Transformation

Using a translation of the left paw a certain number of units to the right and up would land the left paw at the right paw's position.

If students . . . use the wrong transformation, such as a translation instead of a reflection, they may not know how to apply multiple transformations to a figure.

Then intervene . . . by pointing out there are many possible transformations. **Ask:**

- Q What different motions can you think of that can affect a figure?
- Q How can you describe a transformation that yields the mirror image of another figure?

6.2

Apply Sequences of Transformations

I Can determine the effects of a sequence of transformations on a figure.

Spark Your Learning

Bryce has found animal tracks while hiking.

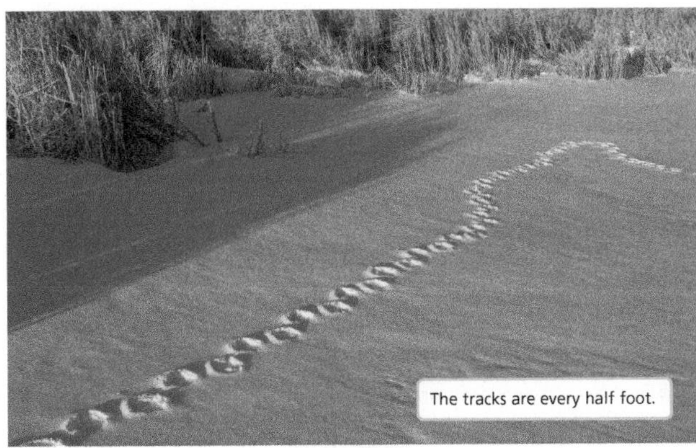
The tracks are every half foot.

Complete Part A as a whole class. Then complete Parts B–D in small groups.

A. What is a mathematical question you can ask about this situation? What information would you need to know to answer your question? A. What transformations can be used to describe the pattern made by the tracks?; the locations and orientations of the tracks

B. What transformation(s) are involved in this situation? How would you describe each transformation? See Additional Answers.

C. To answer your question, what strategy and tool would you use along with all the information you have? What answer do you get? See Strategies 1 and 2 on the facing page.

D. Does your answer make sense in the context of the situation? How do you know? See Additional Answers.

 **Turn and Talk** Will animal tracks always involve a reflection and a translation? See margin.

CULTIVATE CONVERSATION • Information Gap

Ask students questions to help them decide what missing information they need to answer the question, "What transformation(s) are involved in this situation?"

- Do you have enough information to conclude what kinds of transformations the prints represent? Explain. yes; We can tell that the prints are reflections of each other and involve a translation.
- Suppose that, out of a set of about 6 paw prints, one of the prints appeared much broader than the others. How would you explain that? It would suggest that the animal's paw slipped or slid while on the ground.
- What makes you think there is a second transformation involved? If the transformation were a simple reflection, then the tops and bottoms of both prints would be aligned; i.e., they would both fit inside the same upper horizontal line and lower horizontal line.

1 Spark Your Learning

▶ MOTIVATE 

- Have students look at the photo in their books and read the information contained in the photo. Then complete Part A as a whole-class discussion.
- Give the class the additional information they need to solve the problem. This information is available online as a printable and projectable page in the Teacher Resources.
- Have students work in small groups to complete Parts B–D.

▶ PERSEVERE

If students need support, guide them by asking:

Q **Advancing • Use Tools** Which tool could you use to solve the problem? Why choose that tool and not some other? Students' choices of tools and reasons for choosing them will vary.

Q **Assessing** In what way is the left paw print different from the right paw print? Possible answer: The two paw prints appear identical. The right paw print is the result of reflecting the left paw and then translating upward.

Q **Advancing** Using the two given prints, how could you duplicate the left print to form the right print with only one type of transformation? Possible answer: I could reflect the right print over a horizontal line of reflection that is just above the claws of the left print and then reflect over a vertical line of reflection that is equidistant from each paw.

 **Turn and Talk** Help students visualize the mirrored symmetry of animal tracks. Have them think of tracks they have seen in the snow or mud. Possible answer: Any animal with feet will have tracks with reflection. If the animal is moving forward (and not, for example, a bird that landed and flew away without taking any steps) then the tracks will involve a translation.

▶ BUILD SHARED UNDERSTANDING

Select groups of students who used various strategies and tools to share with the class how they solved the problem. As they present their solutions, have each group discuss why they chose a specific strategy and tool.

2 Learn Together

Build Understanding

Task 1 (MP) **Use Tools** Students use the coordinate plane to determine the sequence of transformations applied to a triangle.

Sample Guided Discussion:

Q How can you determine the center of rotation for triangle *ABC* to map it to *A′B′C′*? Connect the corresponding vertices of each figure with a line segment, so *AA′*, *BB′*, and *CC′*. Then draw perpendicular bisectors of each segment. The intersection of the perpendicular bisectors is the center of the rotation, in this case, the origin.

Q What examples can you find of transformations for which the order of applying them does and does not affect the resulting image? The order is unimportant for transformations of the same type—for example, two reflections, two translations, or two rotations. When rotations follow reflections, however, the order does affect the resulting image.

Turn and Talk Help students understand that a sequence of rigid transformations involving two or more of the same type of transformation (whether two translations, two rotations, and so on) has only one possible final result. no; The three translations can be written as a single translation; no; The reflections can be written as a single reflection; The rotations can be written as a single rotation.

Build Understanding

Apply Two Rigid Motions to a Figure

You can apply a sequence of two or more transformation to a figure. When you do this, the image of the first transformation is the preimage for the second transformation, and so on.

1 Suppose you have two transformations to apply to a figure, but the order in which the transformations should be applied is not specified. It can be helpful to understand when the order in which you apply the transformations will affect the final image.

A. Describe the sequence of transformations used to map *ABC* to *A′B′C′* and to map *A′B′C′* to *A″B″C″*. Apply the translations in the other order. Does the order of applying the translations affect the final image? Explain your reasoning.
A–E. See Additional Answers.

B. Make a conjecture: Can the order in which you apply two translations ever affect the final image? Explain your reasoning. Write a single transformation to justify your conjecture.

C. Describe the sequence of transformations used to map *DEF* to *D′E′F′* and to map *D′E′F′* to *D″E″F″*. Apply the transformations in the other order. Does the order of applying the transformations affect the final image? Explain your reasoning.

D. What are the possible sequences of two transformations including translations, rotations, and reflections that you might apply to a figure? Make and test predictions about when the order of two rigid transformations applied in a sequence affects the final image. Include translations, rotations, and reflections in your reasoning. You may want to use a geometric drawing tool to investigate the possibilities.

E. Make a conjecture about when the order of two rigid transformations applied in a sequence affects the final image. Include translations, rotations, and reflections in your reasoning.

Turn and Talk

- If you apply a sequence of three translations to a figure, will the order of the transformations matter? Explain your reasoning. See margin.
- If you apply a sequence of three reflections or three rotations with the same center, will the order of the transformations matter? Explain your reasoning.

182

LEVELED QUESTIONS

Depth of Knowledge (DOK)	Leveled Questions	What Does This Tell You?
Level 1 **Recall**	What is the difference between a reflection and a rotation? A reflection is the flip of a figure across a major line of symmetry. A rotation is the turn of a figure by a certain number of degrees.	Students' answers will indicate an understanding of the difference between reflections and rotations.
Level 2 **Basic Application of Skills & Concepts**	How many 45° rotations in a sequence would leave a figure unchanged? Why? Eight rotations would leave a figure unchanged, because $45° \times 8 = 360°$, which is a full turn.	Students' answers will indicate whether they are able to understand how rotations function and the degree of turn that equates to a full turn or return to the original position.
Level 3 **Strategic Thinking & Complex Reasoning**	Can you think of a sequence of transformations that would change the final image if the order is changed? A sequence containing both a reflection and a rotation could not be done in reverse as the resulting images would have different orientations/locations.	Students' answers will indicate how well they are able to think about and visualize transformations in sequence and the effects of changing the order of the sequence.

Apply Transformations to Map an Image Back to Its Preimage

Sometimes, you may need to map an image back to its preimage. While it can be helpful to have the sequence that was used in the original mapping, you may find a sequence using different transformations that accomplishes what you need.

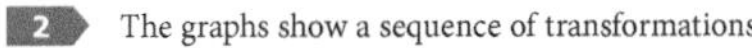

2 The graphs show a sequence of transformations.

A. Describe the transformations used to map $ABCD$ to $A'B'C'D'$ and to map $A'B'C'D'$ to $A''B''C''D''$. Describe two different sequences of transformations you could you use to map $A''B''C''D''$ to $ABCD$.
A–E. See Additional Answers.

B. Suppose the transformations in Part A included a horizontal stretch instead of a dilation. Predict how this would change the transformations used to map $A''B''C''D''$ to $ABCD$.

C. Make a conjecture to describe a general rule for reversing the effect of a dilation, stretch, or compression.

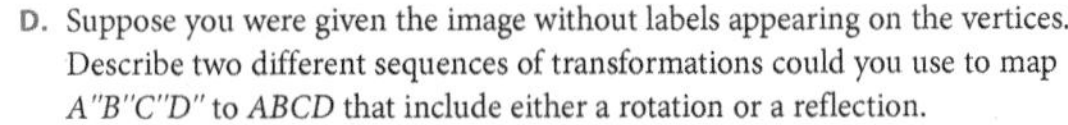

D. Suppose you were given the image without labels appearing on the vertices. Describe two different sequences of transformations could you use to map $A''B''C''D''$ to $ABCD$ that include either a rotation or a reflection.

E. In the image below, the blue polygons are all congruent. Describe the transformations that could have been used to map A to B. Can you use the same transformations to map B to C and C to D? Predict what happens if you reverse the order of the transformations.

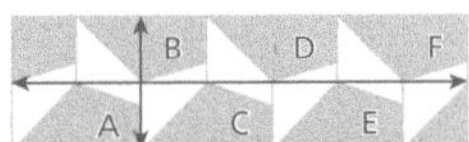

F. The pattern shown with Part E is called a glide reflection. How is this name related to the transformations used to create the pattern?
The pattern involves a translation (or glide) and a reflection.

Turn and Talk Make a conjecture about whether you can use a horizontal stretch to undo a vertical compression. Provide evidence to justify your conjecture. See margin.

Task 2 (MP) **Use Tools** Students determine the sequence of transformations applied to a figure to map it onto its image and work backward to describe the transformation in reverse on a coordinate plane.

Sample Guided Discussion:

Q How many different types of transformations are required to map $ABCD$ to $A''B''C''D''$? Two types are required: translation and dilation.

Q In what way is a stretch different from a dilation? A stretch involves stretching a figure along one line of symmetry, such as the x-axis. A dilation scales a figure by a factor that affects all sides of that figure.

Q In a vertical stretch or compression of a figure, which coordinates of coordinate pairs are affected? a horizontal stretch or compression? In a vertical stretch or compression, the y-coordinates are affected. In a horizontal stretch or compression, the x-coordinates are affected.

Turn and Talk Have students think about the meaning of the words *stretch* and *compression*. Possible answer: no; A horizontal stretch affects the x-coordinates while a vertical stretch affects the y-coordinates, so they cannot undo each other.

Step It Out

Task 3 (MP) **Attend to Precision** Students specify a transformation that yields the image in the illustration using precise coordinate notation for applying each transformation.

OPTIMIZE OUTPUT Critique, Correct, and Clarify

Have students work with a partner to discuss which graph shows a dilation and which does not. Encourage students to use the terms *dilation, reflection*, and *translation* in their discussions. Students should revise their answers, if necessary, after talking about the problem with their partner.

Sample Guided Discussion:

Q What happens to the circle in the preimage in the first set of transformations? Judging by the circle alone, the circle is translated from the second quadrant to the first.

Q What transformation was applied to the preimage to result in the first image? a reflection across the y-axis

Q How can you determine if the first image is dilated, stretched, or compressed to form the second image? Possible answer: I can calculate the length of each of the sides of the first image and then the length of each side of the second image. If the corresponding sides are proportional, then the first image was dilated.

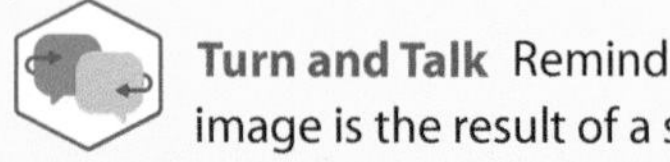

Turn and Talk Remind students that the final image is the result of a sequence of transformations at various stages. Possible answer: The final transformation is a reflection and a dilation by a factor of $\frac{1}{2}$. Yes, the image on the shirt can be described as a single transformation of the original: $(x, y) \rightarrow (-0.5x, 0.5y)$

Step It Out

Specify a Sequence of Transformations

3 Greg wants to create a T-shirt using an image he created for a poster. He plans to print the image on transfer paper and then iron it onto a shirt. The image on the shirt will be about one half the size of the original image. Greg uses the rough sketch below to figure out how to transform the figure to create the shirt.

Find a sequence of rigid motions that maps the original figure to the final image for the transfer. Give coordinate notation for the transformations you use.

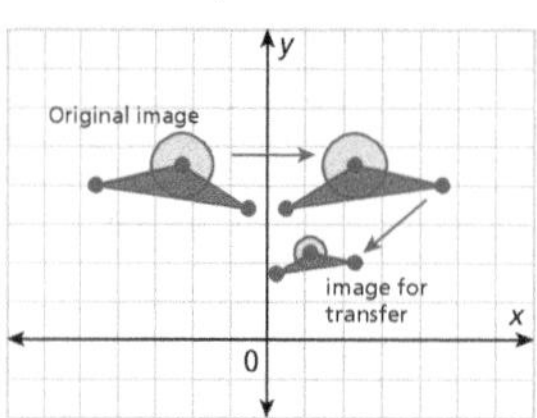

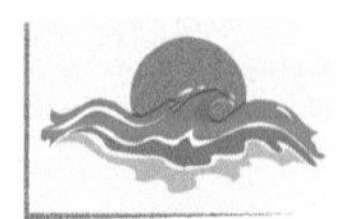

A. How do you know Greg used a reflection instead of a rotation or translation? Write a transformation to map the original image to the second image. A–B. See Additional Answers.

B. Is the third image a dilation, a stretch, or a compression? How do you know? What is the center of this transformation? What is the factor used for the transformation?

Turn and Talk What is the final transformation that happens when the image is transferred to the shirt? Can the relationship between the original image and the image on the shirt be described using a single transformation? See margin.

Write a Composition of Functions

You can write a sequence of transformations as a function. The **composition** of several transformations is a new transformation that directly maps a preimage to the final image after each image is used as a preimage in the next transformation.

If each of the transformations included in a composition is a rigid motion, then the composition is also a rigid motion. If any nonrigid motion is included, then the composition may not be a rigid motion.

For example, consider the transformation rules $(x, y) \rightarrow (2x, 2y)$ and $(x, y) \rightarrow (y, x)$. The composition of these transformations is $(x, y) \rightarrow (2y, 2x)$.

184

PROFICIENCY LEVEL

Beginning
Have students work in pairs to model transformations, using pattern blocks and grid paper, that are translations, reflections, rotations, and dilations. Partners should choose specific transformations and tell each other what they are.

Intermediate
Have students work in pairs to sketch sequences of transformations provided in written form, such as "a translation 2 units up". Have students explain the sequences to each other. Ask volunteers to report to the class.

Advanced
Have students work in groups to find and discuss a sequence of transformations, written in coordinate notation, that yields the original figure in its original place as the final result. An extra challenge would be to include an even number of dilations in the sequence. Ask volunteers to report to the class.

4 A figure is transformed by a dilation centered at the origin with a scale factor of 2, and then a translation of 4 units to the right and 3 units down.

A. Write the sequence of transformations using coordinate notation for each transformation. $(x, y) \to (2x, 2y)$; $(x, y) \to (x + 4, y - 3)$

B. Which of the two transformations below represents the composition of transformations? Explain your reasoning. See Additional Answers.

$(x, y) \to (2x + 4, 2y - 3)$ $\quad$ $(x, y) \to \left(2(x + 4), 2(y - 3)\right)$

Turn and Talk On slips of paper, write coordinate rules for three different transformations including nonrigid transformations. Combine your rules with ones that your partner has written. Select two of the rules and write a composition of them. Then have your partner determine the rules you used in the composition. See margin.

In a glide reflection, a figure is translated then reflected in a line parallel to the translation. A glide reflection is a composition of a translation and reflection, and it is a rigid transformation.

Glide reflections are commutative: In a glide reflection, the resulting image is the same regardless of the order in which the transformations are performed.

5 The figure shows a glide reflection.

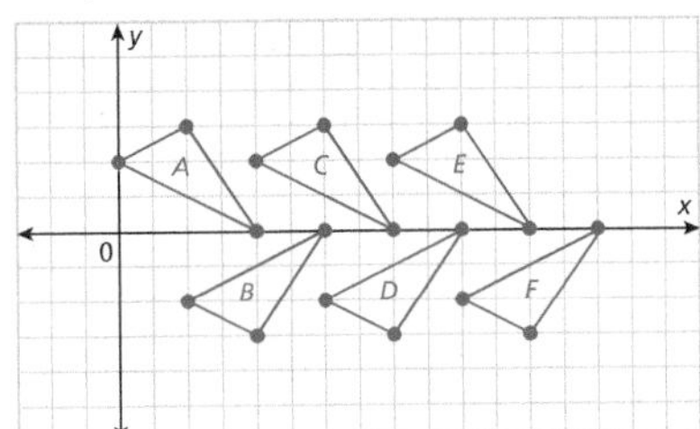

A. The functions that are used to map A to B are a translation by 2 units right and a reflection across the x-axis. Does the order matter for these transformations? no

B. The glide reflection can be written as a composition of $(x, y) \to (x + 2, y)$ and $(x, y) \to (x, -y)$. The composition of these functions is $(x, y) \to (x + 2, -y)$. Notice that each of the transformations only affects one of the coordinates. Use this to justify your answer to Part A. See Additional Answers.

Turn and Talk Name another pair of transformations that are commutative, meaning that the order in which you apply the transformations does not matter. See margin.

Task 4 (MP) **Use Structure** Students use coordinate notation to write a composition of transformations.

Sample Guided Discussion:

Q In what way is a composition of transformations different from a sequence of transformations? A sequence of transformations is one transformation followed by another or by several other transformations. A composition of transformations yields the final result of a sequence of transformations but is written so that the composition contains all the transformations.

Q What is a necessary condition for a composition of transformations? It must preserve the original transformations without altering them in any way. Every composition should be able to be reversed. If the original transformations are not preserved, then reversing the process will not yield the original figure.

Turn and Talk Remind students again that the final image is the result of a sequence of transformations at various stages and add that a transformation composition must be equivalent to the sequence in the final result. See students' work.

Task 5 (MP) **Attend to Precision** Students perform a glide reflection on the coordinate plane and use precise language to describe the effect of changing the order of the translation and reflection.

CONNECT TO VOCABULARY

Have students use the **Interactive Glossary** to record their understanding of the vocabulary in this task.

Sample Guided Discussion:

Q What is the difference is between a reflection and a glide reflection? A reflection is one transformation whereas a glide reflection is two, a translation and a reflection together. In a glide reflection, the reflection is across a line parallel to the translation.

Turn and Talk Have students model each of the types of rigid transformations on grid paper to help them recall which types are commutative. (Sequences of the same type of rigid transformation are commutative.) Because the reflection only affects the y-coordinates and the translation only affects the x-coordinates, the order does not matter because neither transformation affects the other.

ONLINE

Assign the Digital On Your Own for
- built-in student supports
- Actionable Item Reports
- Standards Analysis Reports

On Your Own

Assignment Guide

The chart below indicates which problems in the On Your Own are associated with each task in the Learn Together. Assign daily homework for tasks completed.

Learn Together Tasks	On Your Own Problems
Task 1, p. 182	Problems 5 and 6, 17, 22, and 26
Task 2, p. 183	Problems 7–9 and 23
Task 3, p. 184	Problems 12 and 15–16
Task 4, p. 185	Problems 10–11, 13–14, 18–21, and 25
Task 5, p. 185	Problem 24

Check Understanding

Use the figure to the right for Problems 1–4.

1. Which figures are the result of a rotation being applied to triangle A? triangles B and D
2. Write a sequence of transformations to map triangle A to triangle E. See Additional Answers.
3. Write a sequence of transformations to map triangle D to triangle A. See Additional Answers.
4. Write a composition of functions that maps triangle A to triangle C. $(x, y) \to \left(2(x-2), \frac{4}{3}\left(y-\frac{1}{2}\right)-4\right)$ or $(x, y) \to \left(2x-4, \frac{4}{3}y-\frac{14}{3}\right)$

On Your Own

The figures given are congruent. Find a sequence of rigid motions that maps one figure to the other.

5. $BCDE$: $(6, 2)$, $(8, 2)$, $(6, 6)$, $(-2, 6)$ to $FGHI$: $(2, -4)$, $(4, -4)$, $(2, -8)$, $(-6, -8)$ reflect over x axis, translate left 4 units and down 2 units
6. $JKLM$: $(1, 8)$, $(4, 4)$, $(-2, -2)$, $(-2, 6)$ to $OPQR$: $(5, -1)$, $(1, -4)$, $(-5, 2)$, $(3, 2)$ See Additional Answers.

Predict whether the order matters in the given transformations on $BCDE$. Check your prediction. 7–9. See Additional Answers.

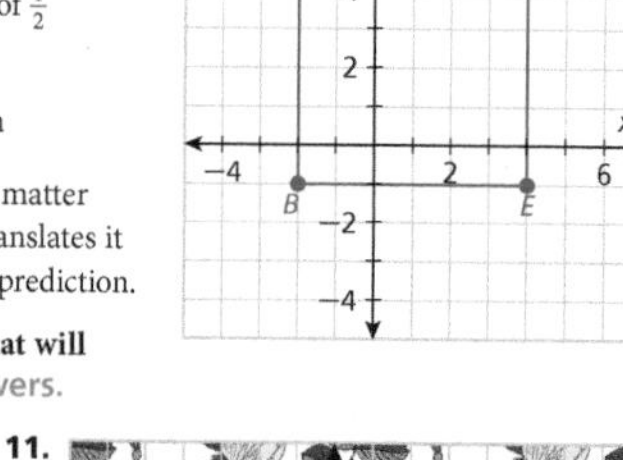

7. translate left 2 units and down 4 units; dilate by a factor of $\frac{1}{2}$ with center at the origin
8. dilate by factor of $\frac{1}{3}$; rotate 90° clockwise about the origin
9. (MP) **Critique Reasoning** Brett predicts that it doesn't matter whether he dilates rhombus $ABCD$ first or rotates and translates it first and then dilates it. Provide evidence for or against his prediction.

In Problems 10 and 11, write a composition of functions that will map Figure A to Figure B. 10, 11. See Additional Answers.

10.

11.

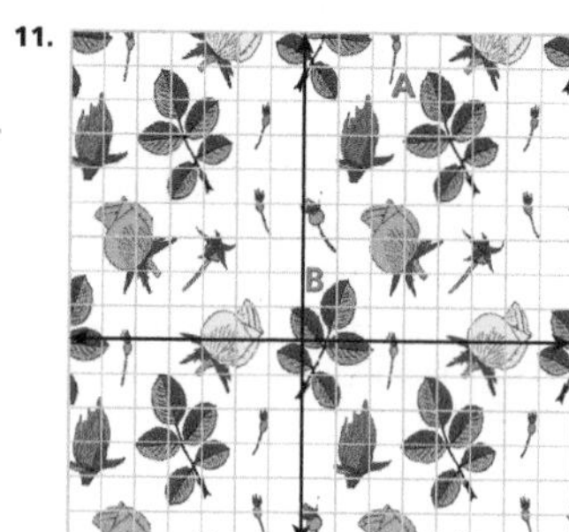

186

(l) ©Kilchytska Maryana/Shutterstock; (r) ©Natalya Danko/Alamy

3 Check Understanding

Formative Assessment

Use formative assessment to determine if your students are successful with this lesson's learning objective.

Students who successfully complete the Check Understanding can continue to the On Your Own practice.

For students who miss 1 problem or more, work in a pulled small group using the Almost There small-group activity on page 181C.

ONLINE

Assign the Digital Check Understanding to determine
- success with the learning objective
- items to review
- grouping and differentiation resources

4 Differentiation Options

Differentiate instruction for all students using small-group activities and math center activities on page 181C.

Reteach

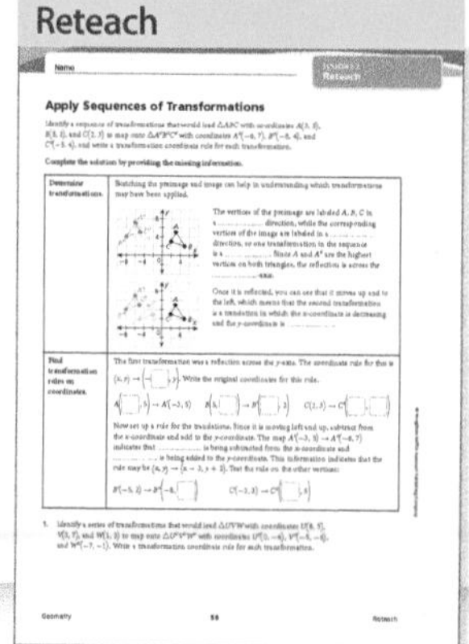

Apply Sequences of Transformations

Challenge

Apply Sequences of Transformations

12. Which sequence of transformations maps rectangle $ABCD$ into quadrant 2? B

A. Reflect over x-axis; translate 8 units left; rotate clockwise 90°.

B. Rotate clockwise 270°; reflect over y-axis; translate 6 units up.

Apply the composition of transformations to triangle ABC with coordinates $A(-4, -2)$, $B(0, 3)$, and $C(-4, 3)$. Write the coordinates of the image. 13–16. See Additional Answers.

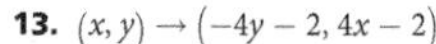

13. $(x, y) \rightarrow (-4y - 2, 4x - 2)$

14. $(x, y) \rightarrow (-x, -3y)$

Write a sequence of transformations to map Figure 1 to Figure 2. Then write a sequence of transformations to map Figure 2 to Figure 1.

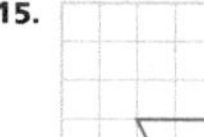

15.

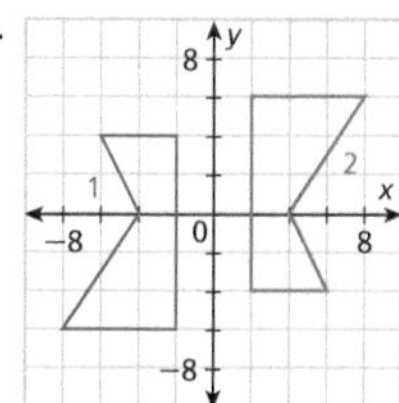

16.

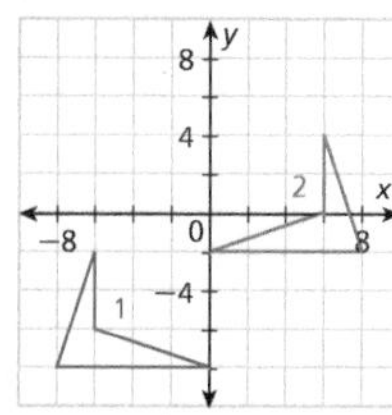

17. **Open Ended** Deshaun has written a computer program that translates an image when a key is pressed. He wants to apply three transformations to move a triangle around a coordinate grid. He wants it to start in quadrant I and end up quadrant IV. Write a sequence of transformations that he could use. Possible answer: Rotate the figure 90° counterclockwise each time.

In Problems 18–21, write a composition of transformations using the order of the sequence given. 18–20. See Additional Answers.

18. $ABCD$, centered at the origin, is rotated 180° about the origin, and dilated by a factor of $\frac{1}{3}$.

19. EFG, centered at the origin, is reflected over the x-axis, rotated clockwise 90° about the origin, and translated down 4 units and to the right 1 unit.

20. HIJ is dilated by a factor of 4 with the origin as the center of dilation, reflected over the y-axis, and translated right 3 units and up 5 units.

21. $LMNO$ is translated right 5 units and up 3 units, rotated counterclockwise 90°, and dilated by a factor of 0.5. $(x, y) \rightarrow (-0.5y - 1.5, 0.5x + 2.5)$

22. **STEM** Nakia is coding a computer game. She needs a character to jump up 2 units and right 1 unit, go down to the ground to duck under overhead objects, and then pivot around to face enemies. What transformations could she use to do this? See Additional Answers.

23. The coordinates of $\triangle ABC$ are $A(1, 1)$, $B(3, -6)$, and $C(5, 1)$. A–D. See Additional Answers.

A. Is $\triangle ABC$ an equilateral triangle, an isosceles triangle, or a scalene triangle? Explain your reasoning.

B. Apply a translation 6 units left and a dilation with the origin as the center of dilation, by a factor of $\frac{1}{2}$ to $\triangle ABC$. What are the coordinates of the vertices of the image?

C. Is $\triangle A'B'C'$ the same type of triangle as you determined in Part A?

D. Give an example of a sequence of transformations for which $\triangle ABC$ and its image $\triangle A'B'C'$ are different types of triangles.

Questioning Strategies

Problem 12 Suppose option A was that rectangle *MPQR* would be translated 8 units up instead of 8 units left, with the other transformations remaining the same. Would the figure lie in Quadrant 2? A portion of the figure would lie in Quadrant 2, but most of the figure would lie in Quadrant 1.

Watch for Common Errors

Problem 15 Students may get stuck on this problem if they attempt to use translations and rotations to describe the sequence of transformations. Point out that students can solve this problem with a sequence of two of the same type of transformations.

5 Wrap-Up

Summarize learning with your class. Consider using the Exit Ticket, Put It in Writing, or I Can scale.

Exit Ticket 

Suppose you are standing on the southeast corner of the intersection of West 19th Street and North 8th Avenue. West 19th Street runs east and west, and North 8th Avenue runs north and south.

You need to go to the opposite side of North 8th Avenue, but traffic is blocked on the south side of West 19th because of street repairs. Write a sequence of transformations that will get you to the opposite side of North 8th Avenue, directly opposite from where you are started. The street represents four units when you cross it and the avenue represents 8 units when you cross. Possible answer: $(x, y) \rightarrow (x, y + 4)$; $(x, y) \rightarrow (x - 8, y)$; $(x, y) \rightarrow (x, y - 4)$

Put It in Writing 

Write a summary of the different transformations you know and how to apply them to a figure, such as a right triangle.

I Can 

The scale below can help you and your students understand their progress on a learning goal.

4	I can determine the effects of a sequence of transformations on a figure, and I can explain my steps to others.
3	I can determine the effects of a sequence of transformations on a figure.
2	I can use sequences of transformations to create the image of a figure in a quadrant different from that of its preimage.
1	I can understand and perform rigid motions on figures in the coordinate plane.

Spiral Review • Assessment Readiness

These questions will help determine if students have retained information taught in the past and can also prepare them for high-stakes assessments. Here, students must identify a scale factor for dilations in coordinate notation **(6.1)**, identify lines of reflection for a square **(5.4)**, identify the relationship between congruence and transformations **(5.3)**, and understand and state the nature of reflections **(5.1)**.

24. DESIGN In the rug design, the shapes A and B are congruent.

A. Find two sequences of rigid motions that map A to B using different types of transformations.

B. Is the orientation of the figure affected by either sequence of transformations? Explain. A, B. See Additional Answers.

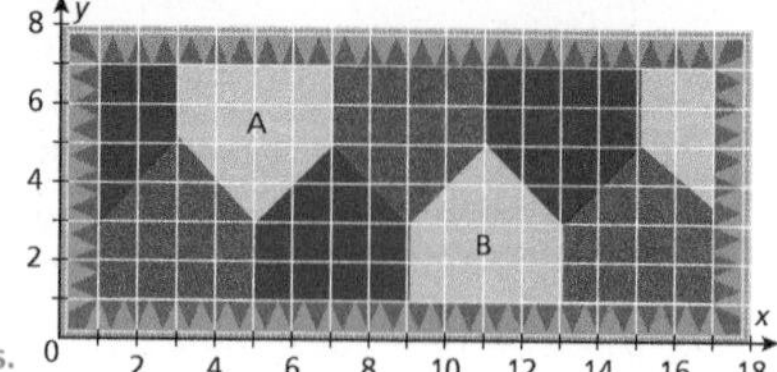

25. MP **Use Structure** Figure $L'M'N'$ has been transformed with function $(x, y) \rightarrow \left(\frac{1}{2}(-y + 4), \frac{1}{2}(x + 20)\right)$. What transformation will map it to the preimage LMN? $(x, y) \rightarrow (2y - 20, -2x + 4)$

26. Open Middle™ List three sequences of transformations that take preimage $ABCD$ to image $A'B'C'D'$. reflection, rotation, translation

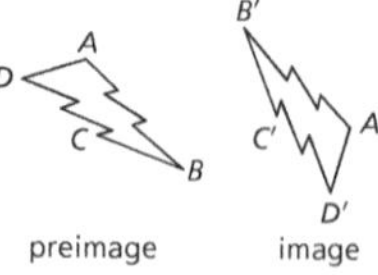

Spiral Review • Assessment Readiness

27. Which of the following will produce a dilation such that the preimage is 3 times bigger than the image?

Ⓐ $(x, y) \rightarrow \frac{1}{3}(x, y)$

Ⓑ $(x, y) \rightarrow (x + 3, y + 3)$

Ⓒ $(x, y) \rightarrow 3(x, y)$

Ⓓ $(x, y) \rightarrow (x - 3, y - 3)$

28. How many lines of reflection does the figure have?

Ⓐ 2

Ⓑ 3

Ⓒ 4

Ⓓ 8

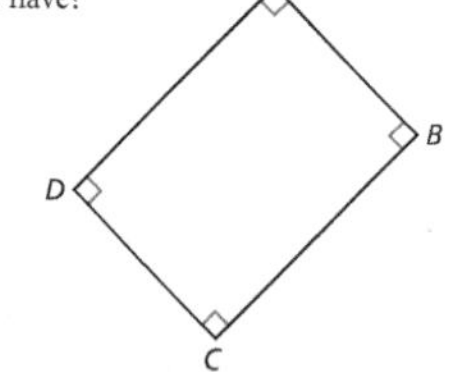

29. If two polygons are congruent, which of the following is *sometimes* true?

Ⓐ They can be transformed into each other with a single nonrigid transformation.

Ⓑ They are the same shape.

Ⓒ They can be transformed into each other with a sequence of nonrigid transformations.

Ⓓ They are the same size.

30. In a reflection, which of the following is true?

Ⓐ The *y*-coordinate is always the opposite of the *y*-coordinate in the preimage.

Ⓑ A preimage is reflected across the perpendicular bisector to get the image.

Ⓒ Each point will be equally far away from the line of reflection.

Ⓓ The image and preimage are the same size and shape but a different orientation.

I'm in a Learning Mindset!

How do I know whether a task is challenging enough for me?

188 Keep Going Journal and Practice Workbook

Learning Mindset

Challenge Seeking Defines Own Challenges

Composition of transformations is a crucial area of mathematics for students to grasp because of its many applications in engineering, design, and architecture. Students might define their challenge as gaining an understanding of how a sequence of transformations can be written as a composition of functions. Have students look for real-world applications of sequences of transformations, such as in bridge building and skyscraper design. *How do you find the motivation to take on challenges? Is it possible to find new motivation in school subjects? When you define your own challenge and find yourself doing well, does that help build your confidence?*

Review

Dilations

Kimora drew a scale drawing and made sure of the following information.

- Angle measures are preserved.
- Side lengths are proportional.

Roche wanted to find the center and the scale factor.

- Identify corresponding points and connect them with a line. All three lines intersect at $(-3, 0)$.
- Compare the length of corresponding sides. The preimage to the image is related by the scale factor of $\frac{1}{3}$.

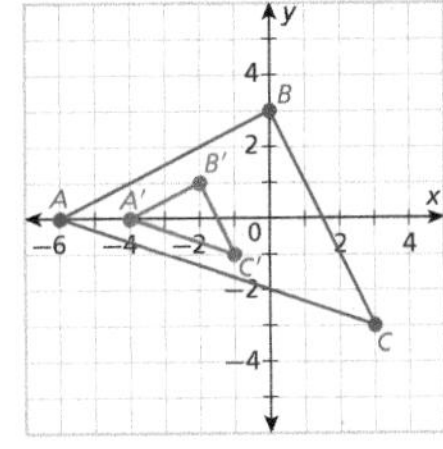

Sequences of Transformations

Write a sequence of transformation to that maps $\triangle ABC$ to $\triangle A'B'C'$.

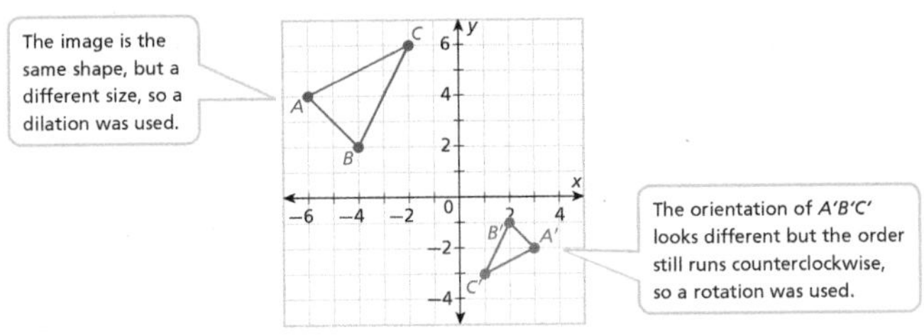

Map $\triangle ABC$ to $\triangle A'B'C'$ with a clockwise rotation of 180° about the origin, followed by a dilation with a scale factor of $\frac{1}{2}$.

ONLINE

Assign the Digital Module Review for

- built-in student supports
- Actionable Item Reports
- Standards Analysis Reports

MODULE 6 REVIEW

Module Review

Use the first page of the Module Review to summarize and connect the main ideas of the module. Use the second page to assess students' understanding of the vocabulary, concepts, and skills presented in the module.

Sample Guided Discussion:

Q Dilations Suppose that a vertex of a triangle is at the origin. Why is the vertex of the corresponding image under a dilation also at the origin? Possible answer: A dilation in the coordinate plane can be expressed as $(x, y) \rightarrow (kx, ky)$, where k is the scale factor. At the origin, $x = 0$, and $y = 0$. Therefore, the image of the vertex is $\left(k(0), k(0)\right)$ or $(0, 0)$, the origin.

Q Sequences of Transformations
A sequence of transformations is performed on a triangle. The image is a triangle whose sides have the same lengths as the corresponding sides of the preimage. Could the sequence of transformations have included nonrigid transformations? Explain. Possible answer: yes; It could have included two counteracting dilations. For example, a dilation with a scale factor $\frac{1}{2}$ and a dilation with scale factor 2 would result in an image with the same lengths.

Module Review continued

Possible Scoring Guide

Items	Points	Description
1–4	2 each	identifies the correct term
5	2	constructs a dilation
6	1	explains why the figure is a dilation
6	2	explains why a figure is a dilation, determines the coordinates of the center of the dilation, and determines two possible scale factors
7 and 8	2 each	finds a sequence of transformations that maps a preimage onto an image
Total points possible = 16 points		

The Unit 3 Test in the Assessment Guide assesses content from Modules 5 and 6.

Vocabulary

Choose the correct term from the box to complete each sentence.

Vocabulary
center of dilation
dilation
rigid motion
scale factor

1. A __?__ is a transformation that does not change the size or shape of a figure. rigid motion
2. The __?__ is the fixed point about which all other points are transformed in a dilation. center of dilation
3. A __?__ is a transformation that changes the size of a figure but does not change the shape. dilation
4. The ratio of the lengths of corresponding sides in the image and the preimage of a dilation is the __?__. scale factor

Concepts and Skills

5. (MP) **Use Tools** Describe how to construct the image of $\triangle ABC$ under a dilation with center C and scale factor 3. State what strategy and tool you will use to answer the question, explain your choice, and then find the answer. See Additional Answers.

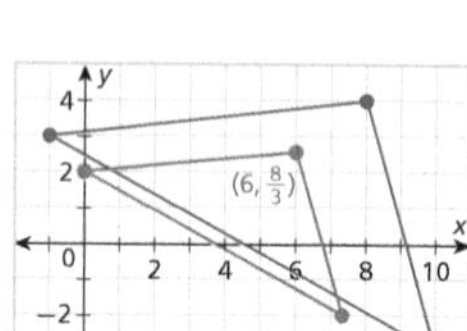

6. A dilation on the coordinate plane is shown.
   A. Explain why it is a dilation. See Additional Answers.
   B. Determine the coordinates of the center of dilation. (2, 0)
   C. Determine 2 possible scale factors. $\frac{2}{3}$ or $\frac{3}{2}$

Find a sequence of transformations for the indicated mapping.

7.

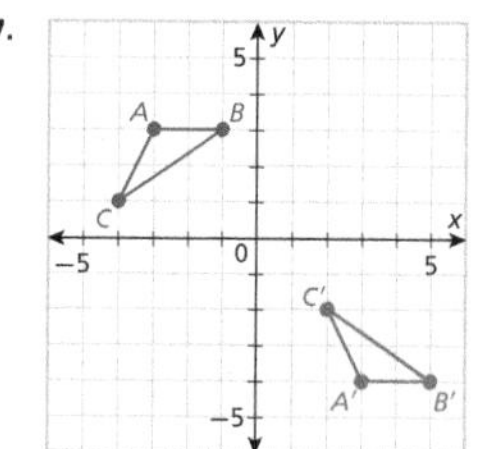

Translate 6 units right and 1 unit up and then reflect across the x-axis.

8.

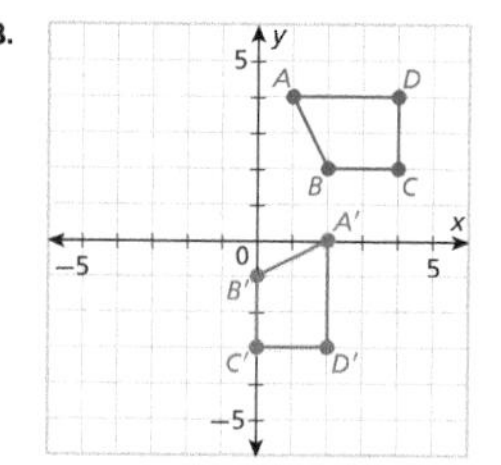

Rotate 270° counterclockwise and then translate 2 units left and 1 unit up.

190

DATA-DRIVEN INSTRUCTION

Before moving on to the Module Test, use the Module Review results to intervene based on the table below.

Items	Lesson	DOK	Content Focus	Intervention
5	6.1	3	Construct a dilation.	Reteach 6.1
6	6.1	2	Explain why a figure is a dilation, find the center, and identify two possible scale factors.	Reteach 6.1
7 and 8	6.2	2	Identify and describe a sequence of transformations that maps a preimage to an image.	Reteach 6.2

Module Test

data checkpoint

The Module Test is available in alternative versions in your Assessment Guide. All items are presented in standardized test formats.

ONLINE

Ed

Assign the Digital Module Test to power actionable reports including

- proficiency by standards
- item analysis

Form A

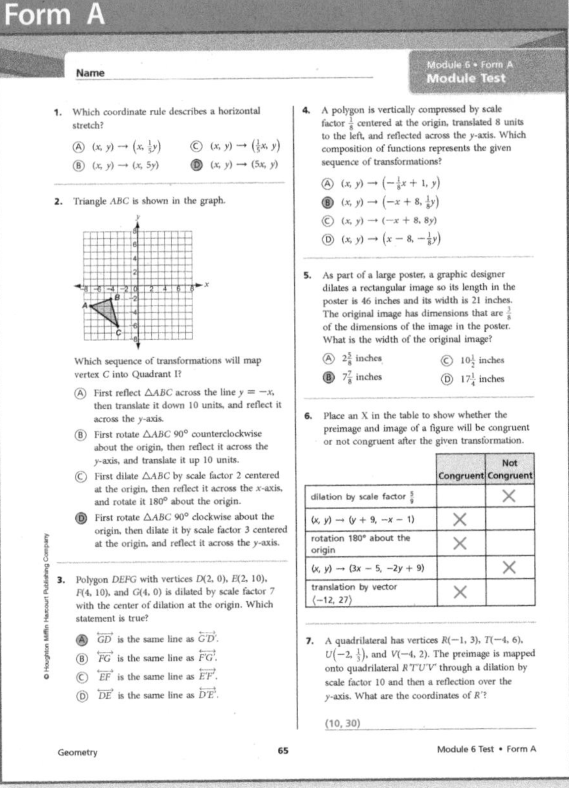

Form A

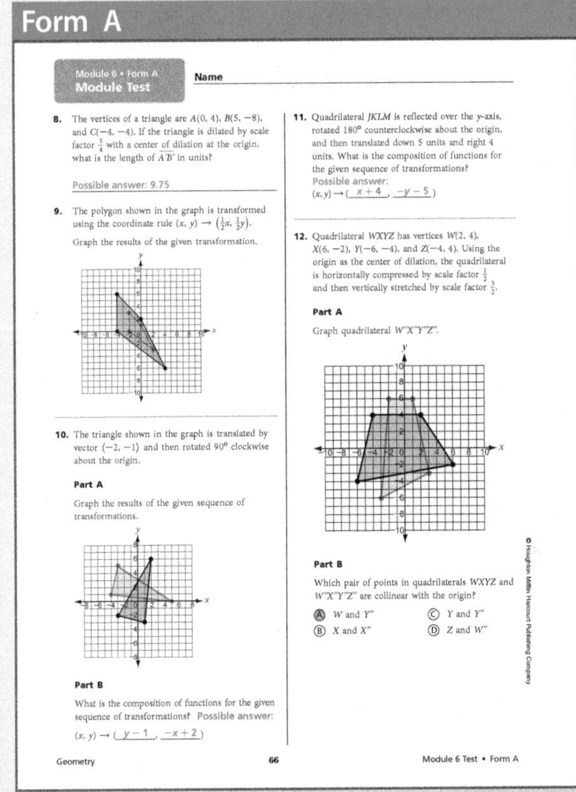

Form B

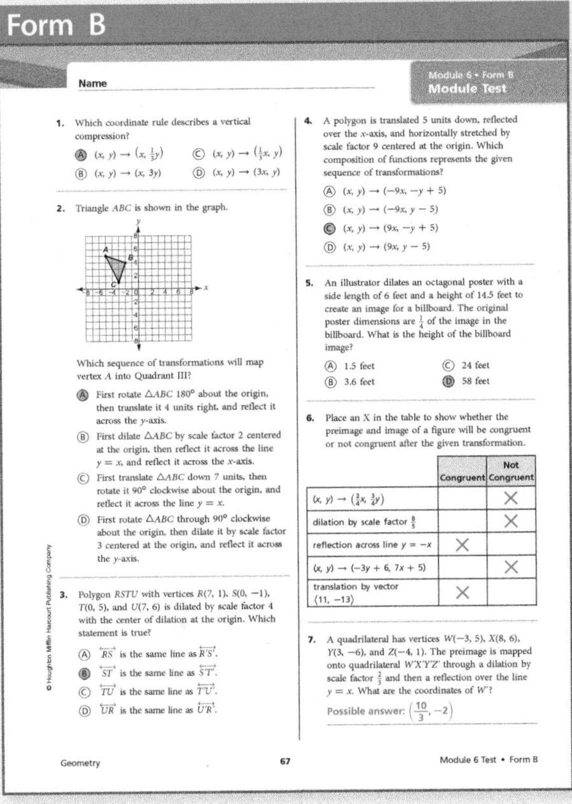

Form B

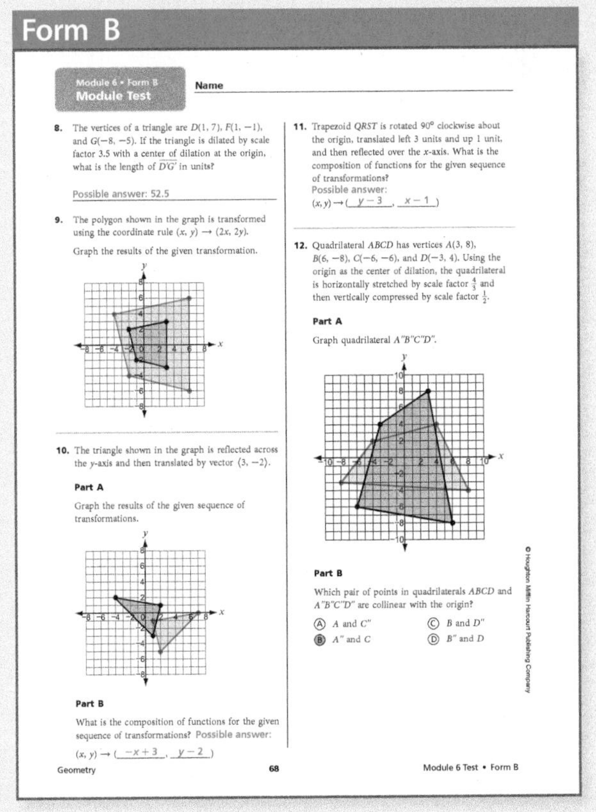

Module 7: Congruent Triangles and Polygons

Module 8: Triangle Congruence Criteria

Architect STEM

- **Say:** *Think of all the housing developments, office buildings, high-rise buildings, and shopping malls you have seen. An architect oversees the planning and designing of those structures.*
- Architects manage the design and construction of a structure. They must have knowledge of technical and environmental issues and an understanding of business. They lead a design team of structural, electrical, and mechanical engineers that makes sure the structure is built to its design specifications.

Then introduce the mathematics of the STEM task.

STEM Task

In preparation for this lesson

- have students research diagrids.
- have students create a bulletin board that contains pictures of Hearst Tower in New York City.

During the lesson

- discuss diagrids and their role in architecture.
- have students identify the types of triangular shapes used for the windows.
- use the bulletin board to complete the task.

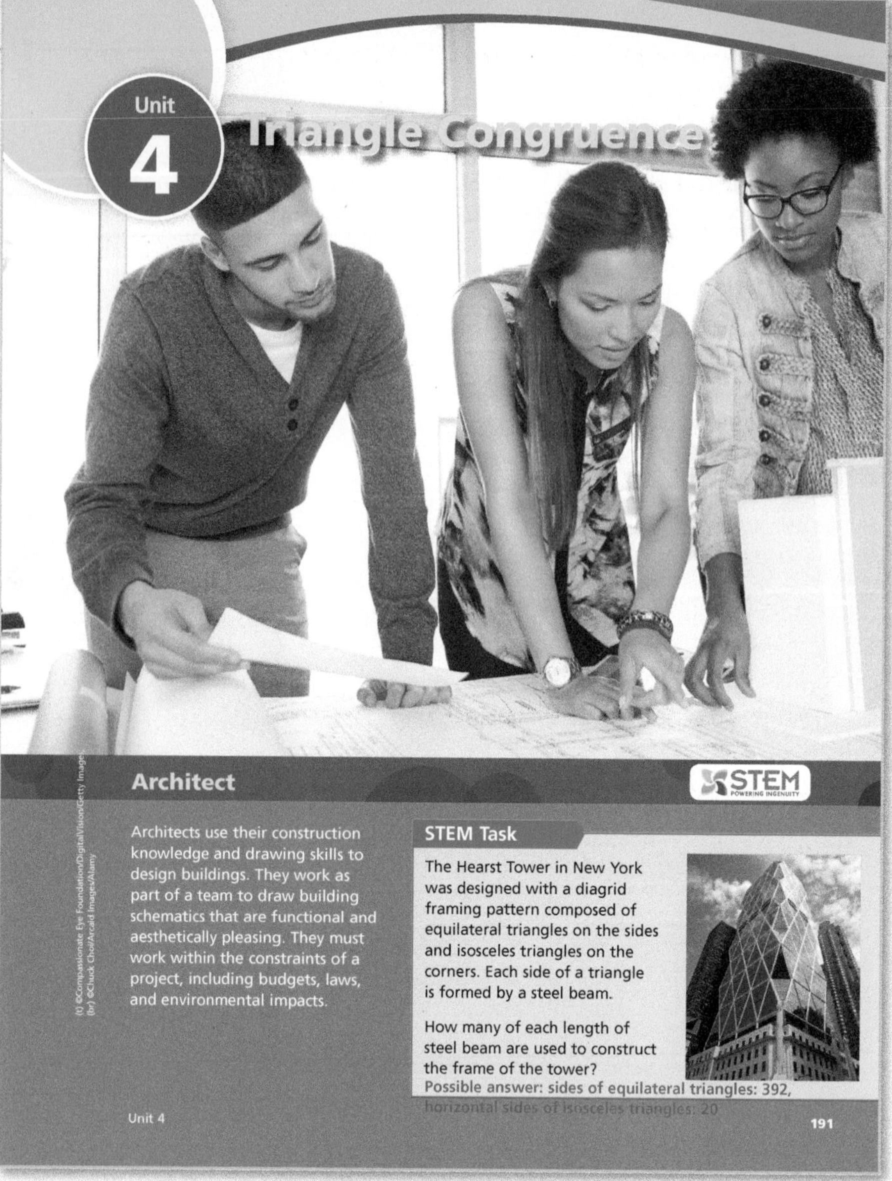
Unit 4
Triangle Congruence

Architect
STEM POWERING INGENUITY

Architects use their construction knowledge and drawing skills to design buildings. They work as part of a team to draw building schematics that are functional and aesthetically pleasing. They must work within the constraints of a project, including budgets, laws, and environmental impacts.

STEM Task

The Hearst Tower in New York was designed with a diagrid framing pattern composed of equilateral triangles on the sides and isosceles triangles on the corners. Each side of a triangle is formed by a steel beam.

How many of each length of steel beam are used to construct the frame of the tower?
Possible answer: sides of equilateral triangles: 392, horizontal sides of isosceles triangles: 20

(t) ©Compassionate Eye Foundation/DigitalVision/Getty Images; (br) ©Chuck Choi/Arcaid Images/Alamy

Unit 4 191

Unit 4 Project Coordinating Congruencies

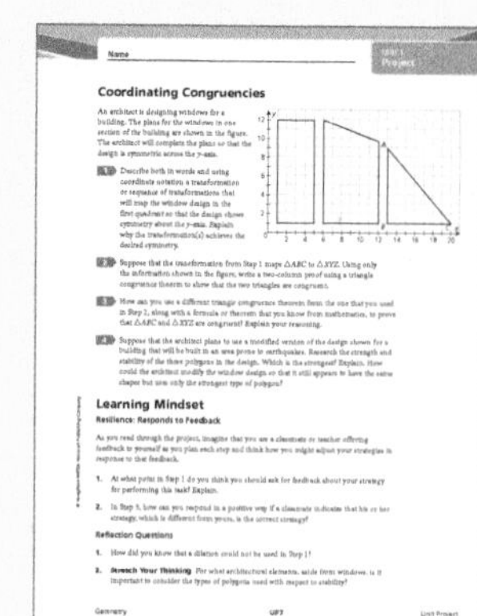
Coordinating Congruencies

Learning Mindset

Overview: In this project students use triangle congruence theorems to determine symmetry in a window designed by an architect.

Materials: display or poster board for presentations

Assessing Student Performance: Students' presentations should include:

- a correct series of transformations that create a figure such that the y-axis is a line of symmetry **(Lesson 7.3)**
- a correct proof using SAS that the two triangles are congruent **(Lesson 8.2)**
- a correct proof using SSS that the two triangles are congruent **(Lesson 8.3)**
- a correct description of triangles being the strongest and a reasonable modification to the design to increase its strength **(Lesson 8.3 or later)**

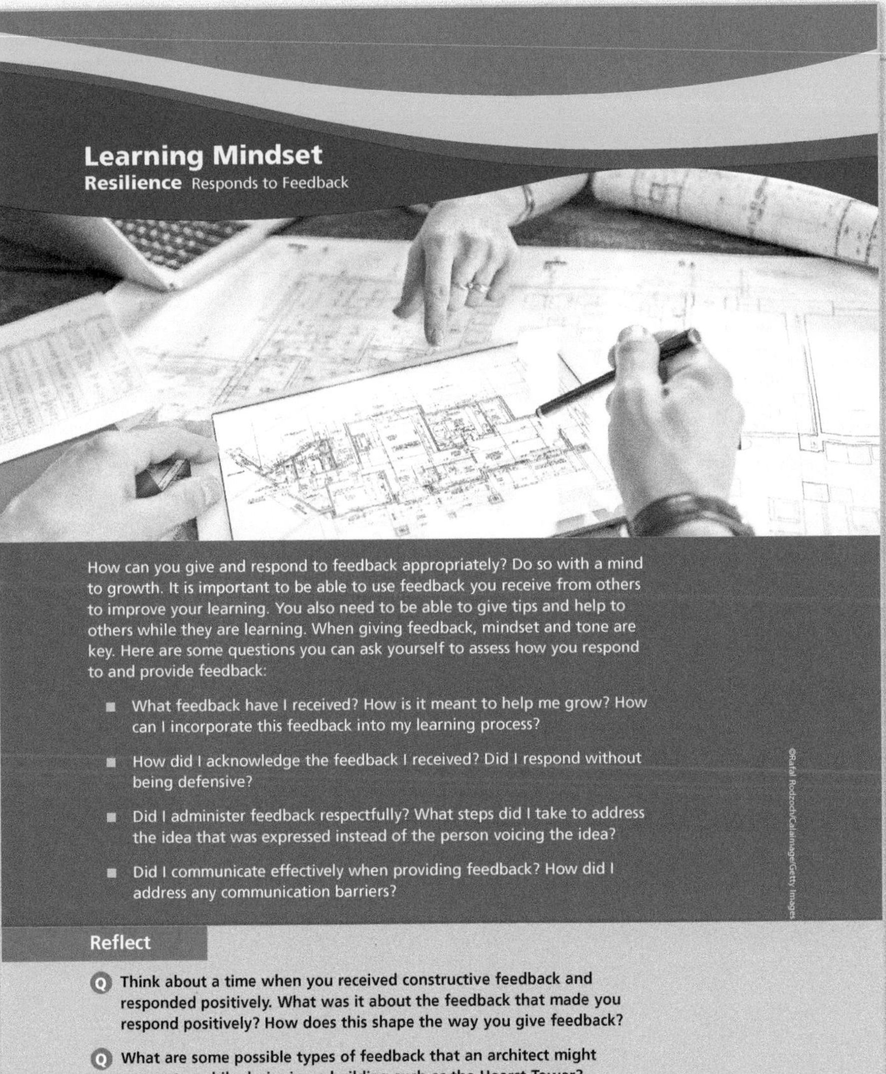

Learning Mindset
Resilience Responds to Feedback

How can you give and respond to feedback appropriately? Do so with a mind to growth. It is important to be able to use feedback you receive from others to improve your learning. You also need to be able to give tips and help to others while they are learning. When giving feedback, mindset and tone are key. Here are some questions you can ask yourself to assess how you respond to and provide feedback:

- What feedback have I received? How is it meant to help me grow? How can I incorporate this feedback into my learning process?
- How did I acknowledge the feedback I received? Did I respond without being defensive?
- Did I administer feedback respectfully? What steps did I take to address the idea that was expressed instead of the person voicing the idea?
- Did I communicate effectively when providing feedback? How did I address any communication barriers?

©Rafal Rodzoch/Caiaimage/Getty Images

Reflect

- **Think about a time when you received constructive feedback and responded positively. What was it about the feedback that made you respond positively? How does this shape the way you give feedback?**
- **What are some possible types of feedback that an architect might encounter while designing a building such as the Hearst Tower? How would you respond to ensure the project was successful?**

192

Learning Mindset

Resilience

Responds to Feedback

Monitors Knowledge and Skills

The learning-mindset focus in this unit is *responds to feedback*, which refers to how a person gives and responds to feedback. In mathematics, students should possess a growth mindset that enables them to respond positively to improve and increase their learning.

Mindset Beliefs

Some students who receive negative feedback might become defensive and shut down saying, "I did the best I could" or "I've always done it this way!" Or they may physically and mentally withdraw from engaging in a discussion by saying, "I'll never get this right!" Such responses prevent students from developing a growth mindset.

Discuss with students how they can view receiving or providing positive and negative feedback as learning opportunities. Create appropriate challenges for students who seem eager to learn more. Or alternatively, encourage them to go back and review previously learned ideas to develop greater self-confidence and overcome any negative feedback.

Mindset Behaviors

Encourage students to become more responsive to feedback by asking themselves about its importance. Self-monitoring can take the form of asking these questions:

When studying in a group:

- Did I communicate effectively? How did I overcome any communication barriers?
- How did I listen to opposing points of view about a specific task or concept? Did I respect others?

When practicing a mathematical skill:

- How did I assess my own understandings or gaps in what I need to know?
- How did I respond to negative feedback I received or gave to others? Was I objective?

When solving a mathematical problem successfully:

- How can I best communicate to others a way to solve a problem?
- To whom can I turn to help me find another way to solve a problem?

As students become more proficient at responding to and providing feedback to others, they may become more comfortable communicating with you and other students to achieve personal goals.

What to Watch For

Watch for students who are struggling with learning mathematics despite their best efforts and hard work. Help them become more responsive to feedback by asking:

- How can you assert your own needs and viewpoints in nonaggressive or defensive ways?
- How do you respond to positive feedback?

Watch for students who resist receiving feedback about their own work or who are reluctant to provide feedback to others by asking them these questions.

- How might communicating with others help you be more successful?
- How can you provide helpful feedback to others?

"We all need people who will give us feedback. That's how we improve."

—Bill Gates, entrepreneur

MODULE 7

PLANNING

CONGRUENT TRIANGLES AND POLYGONS

Introduce and Check for Readiness

- Module Performance Task
- Are You Ready?

Lesson 7.1—2 Days

Understand Congruent Figures

Learning Objective: Use rigid motions to show figures are congruent.

New Vocabulary: corresponding angles, corresponding sides

Review Vocabulary: congruent

Lesson 7.2—2 Days

Corresponding Parts of Congruent Figures

Learning Objective: Use congruency of corresponding parts to prove triangles are congruent.

Lesson 7.3—2 Days

Use Rigid Motions to Prove Figures Are Congruent

Learning Objective: Write proofs involving congruent figures.

Assessment

- Module 7 Test (Forms A and B)
- Unit 4 Test (Forms A and B)

LEARNING ARC FOCUS

 Build Conceptual Understanding

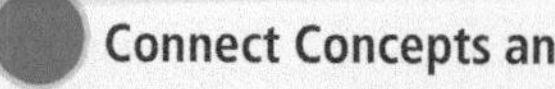

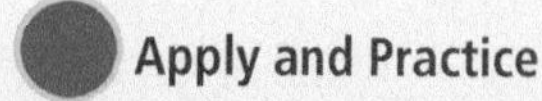

TEACHING FOR SUCCESS

TEACHING FOR DEPTH: Congruent Triangles and Polygons

Making Connections. Congruent figures have the same number of sides, and the measures of their corresponding sides and angles are equal. In other words, two figures are congruent if they are identical in every way except for their location in the plane. Alternately, two figures are congruent if one can be made to coincide exactly with the other. Triangles are congruent if three pairs of corresponding sides and three pairs of corresponding angles are congruent.

The rigid motion transformations in the plane are translations, reflections, and rotations. Each transformation or sequences of these transformations preserve lengths, angle measures, and betweenness of points. Dilations are not rigid transformations because they do not preserve lengths.

Rigid motion transformations can be used to define congruence. Two figures are congruent if and only if one can be transformed into the other by a sequence of rigid motions. If one figure is transformed into the other, the preimage and its image coincide. The measures of their corresponding sides and angles are equal. They are identical regardless of their location in a plane.

Using rigid motions to define congruence provides a connection between Euclidean geometry and transformational geometry. Rigid motions can be defined algebraically and results can be plotted in the coordinate plane.

Mathematical Progressions

Prior Learning	Current Development	Future Connections
Students: • understood that 2-dimensional figures are congruent if the 2nd can be obtained from the 1st by a sequence of rigid motions. • represented transformations in the plane. • developed definitions of rotations, reflections, and translations. • specified transformations that map a given figure onto another. • used rigid motions to transform figures.	**Students:** • use rigid motions to show figures are congruent. • use congruency of corresponding parts to prove triangles are congruent. • write proofs involving congruent figures.	**Students:** • will write coordinate proofs about triangle relationships. • will write coordinate proofs about parallelograms. • will identify and describe relationships among lines and angles of a circle, including the relationships among central, inscribed, and circumscribed angles. • will construct inscribed and circumscribed circles of a triangle, and prove properties of angles for a quadrilateral inscribed in a circle.

TEACHER TO TEACHER

From the Classroom

Build procedural fluency from conceptual understanding. When I see students struggling with understanding what congruent figures are and proving them to be congruent, the source of their confusion often traces back to a shallow understanding of congruency, such as having the same shape and size. Students who struggle with "Are these figures congruent?" also struggle with the question, "What parts of the figures are congruent?" I believe that conceptual understanding is key to supporting procedural fluency. Students who are fluent thinkers in mathematics understand what they read and hear and share independent thoughts that can be communicated to others with clarity. I want my students to have the conceptual understandings that foster procedural fluency.

LANGUAGE DEVELOPMENT • PLANNING FOR INSTRUCTION

By giving all students regular exposure to language routines in context, you will provide opportunities for students to **listen, speak, read,** and **write** about mathematical situations and develop both mathematical language and conceptual understanding at the same time.

Using Language Routines to Develop Understanding

Use the **Professional Cards** for the following routines to plan for effective instruction.

Co-Craft Questions and Problems Lesson 7.1

Students think of natural questions to ask about a given situation or problems similar to a given task and answer the questions they have developed or problems they have created.

Three Reads Lessons 7.1–7.3

Students read a problem three times with a specific focus each time.

1st Read What is the situation about?
2nd Read What are the quantities in the situation?
3rd Read What are the possible mathematical questions that you could ask for the situation?

Information Gap Lesson 7.3

Students recognize when information given in a problem situation is incomplete, and they pose questions and share knowledge with others to discover any missing facts or relationships and work together to solve the problem.

Connecting Language to Congruent Triangles and Polygons

Watch for students' use of the review and new terms listed below as they explain their reasoning and make connections with new concepts.

Key Academic Vocabulary

Prior Learning and Current Development • Review and New Vocabulary

congruent two figures such that one can be obtained from the other by a sequence of rigid motions	**corresponding angles** angles in the same position in polygons with an equal number of sides **corresponding sides** sides in the same position in polygons with an equal number of sides

Linguistic Note

Listen for students' ability to repeat and apply the phrase *Corresponding Parts of Congruent Figures are Congruent* (CPCFC). Some English learners might be overwhelmed by all the words that start with "c" and the repeated use of the word *congruent*. Help these students better understand the meaning of these individual words and the phrase by drawing two congruent triangles and displaying measures of their corresponding sides and angles. Have the students use their fingers to trace the six pairs of corresponding congruent parts while repeating the "*CPCFC*" phrase along with each trace.

Module

7

Congruent Triangles and Polygons

Module Performance Task: Focus on STEM

Image Stitching

Image stitching is the process of combining several overlapping images to create a larger image. It is used to create high-resolution satellite images and to stabilize shaky video recordings. The process identifies key features in a scene to align the photos before stitching them into one seamless image.

A. Select a group of images that would be ideal to stitch together. Explain your reasoning in terms of transformations and in terms of measures. A–D. See margin.

B. Describe an algorithm or procedure that can be used to select images that would be ideal to stitch together.

C. For the images you did not choose, why would they be less than ideal to stitch together? Explain your reasoning.

D. Describe how you would alter your algorithm or procedure to incorporate images that you identified as less than ideal.

Module 7 193

Mathematical Connections

Task Part	Prerequisite Skills
Parts A and C	Students can identify congruent figures and justify why figures are congruent using corresponding parts **(7.1, 7.2)**.
Parts B and D	Students can describe rigid transformations on a coordinate grid and how an algorithm can be used to identify congruent figures in object recognition **(7.3)**.

Image Stitching

Overview

Stunning photographic mosaics can be created using *image stitching*, a technique that creates a photographic panorama from a set of overlapping images. Image stitching computer programs can make high-resolution photo-mosaics and create satellite photos and digital maps. Most digital cameras and camcorders use image-stitching technology to produce spectacular panoramas.

In this STEM activity, students will identify images that have congruent parts that can be stitched together.

Career Connections

Students with an interest in art and photography can consider careers that involve creating original panoramas using image stitching. Alternately, they may work as computer programmers and write image-stitching software or as technicians who can use computers to manipulate and create image-stitching images.

Answers

A. images 1, 3, 5, and 6 all have congruent, overlapping features and can be translated so the features with the same size can be on top of each other. Also, together they include the most possible extension of scenery in all directions.

B. Possible answer: Identify important features in the scene. Identify images that contain these features. Match images with congruent measures. Align orientation and position of the features. Stitch together.

C. Possible answer: Image 2 would not work because the houses are larger than in the other images. Image 4 would not work because there is no information that cannot be obtained from the other images, and it is not taken from quite the same perspective.

D. Possible answer: The algorithm can be broadened by allowing for similar figures in addition to congruent figures. If key features contain the same angle measures, they can be scaled to the appropriate size before being stitched.

ONLINE

Assign the Digital Are You Ready? to power actionable reports including
- proficiency by standards
- item analysis

Are You Ready?

data checkpoint

Diagnostic Assessment

- Diagnose prerequisite mastery.
- Identify intervention needs.
- Modify or set up leveled groups.

Have students complete the *Are You Ready?* assessment on their own. Items test the prerequisites required to succeed with the new learning in this module.

Translate Figures in the Coordinate Plane Students will extend previous knowledge of translating figures in the coordinate plane to prove figures are congruent by rigid motions.

Rotate Figures in the Coordinate Plane Students will extend previous knowledge of rotating figures in the coordinate plane to prove figures are congruent by rigid motions.

Reflect Figures in the Coordinate Plane Students will extend previous knowledge of reflecting figures in the coordinate plane to prove figures are congruent by rigid motions.

ONLINE 

Are You Ready?

Complete these problems to review prior concepts and skills you will need for this module.

Translate Figures in the Coordinate Plane

The coordinates of a quadrilateral are $(-1, 2)$, $(2, 4)$, $(4, 3)$, and $(1, -1)$. Find the coordinates of the image after translating by the given rule.

1. $(x, y) \rightarrow (x - 4, y + 1)$ $(-5, 3), (-2, 5), (0, 4), (-3, 0)$
2. $(x, y) \rightarrow (x + 2, y - 5)$ $(1, -3), (4, -1), (6, -2), (3, -6)$
3. $(x, y) \rightarrow (x, y - 3)$ $(-1, -1), (2, 1), (4, 0), (1, -4)$
4. $(x, y) \rightarrow (x - 3, y - 2)$ $(-4, 0), (-1, 2), (1, 1), (-2, -3)$

Rotate Figures in the Coordinate Plane

The coordinates of a triangle are $(-1, 1)$, $(3, -1)$, and $(-2, -4)$. Find the coordinates of the image after performing the rotation about the origin.

5. 90° counterclockwise $(-1, -1), (1, 3), (4, -2)$
6. 270° counterclockwise $(1, 1), (-1, -3), (-4, 2)$
7. 180° $(1, -1), (-3, 1), (2, 4)$

Reflect Figures in the Coordinate Plane

The coordinates of a quadrilateral are $(-2, 3)$, $(0, 4)$, $(4, 2)$, and $(-1, -1)$. Find the coordinates of the image after performing the reflection across the given line.

8. *x*-axis $(-2, -3), (0, -4), (4, -2), (-1, 1)$
9. *y*-axis $(2, 3), (0, 4), (-4, 2), (1, -1)$

Connecting Past and Present Learning

Previously, you learned:
- to transform figures in the coordinate plane,
- to determine unknown values using angle pairs, and
- to write proofs about segments and angles.

In this module, you will learn:
- to use transformations to prove if figures are congruent, and
- to use corresponding parts of congruent figures to determine unknown values.

194

DATA-DRIVEN INTERVENTION

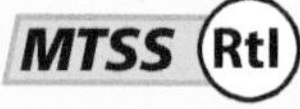

Concept/Skill	Objective	Prior Learning *	Intervene With
Translate Figures in the Coordinate Plane	Find the image of a figure that is translated in the coordinate plane.	Grade 8, Lesson 1.2	• Tier 2 Skill 9 • Reteach, Grade 8 Lesson 1.2
Rotate Figures in the Coordinate Plane	Find the image of a figure that is rotated in the coordinate plane.	Grade 8, Lesson 1.4	• Tier 2 Skill 10 • Reteach, Grade 8 Lesson 1.4
Reflect Figures in the Coordinate Plane	Find the image of a figure that is reflected in the coordinate plane.	Grade 8, Lesson 1.3	• Tier 2 Skill 11 • Reteach, Grade 8 Lesson 1.3

* Your digital materials include access to resources from Grade 6–Algebra 2. The lessons referenced here contain a variety of resources you can use with students who need support with this content.

7.1 Understand Congruent Figures

LESSON FOCUS AND COHERENCE

Mathematics Standards

- Use the definition of congruence in terms of rigid motions to show that two triangles are congruent if and only if corresponding pairs of sides and corresponding pairs of angles are congruent.

Mathematical Practices and Processes (MP)

- Use appropriate tools strategically.
- Look for and make use of structure.
- Attend to precision.
- Model with mathematics.

I Can Objective 

I can determine whether figures are congruent.

Learning Objective

Use rigid motions to show figures are congruent and find unknown measures in congruent figures.

Language Objective

Explain how to determine if two figures are congruent using rigid motions or corresponding angles and sides.

Vocabulary

Review: congruent

New: corresponding angles, corresponding sides

Lesson Materials: tracing paper, ruler, protractor

Mathematical Progressions

Prior Learning	Current Development	Future Connections
Students: • used geometric descriptions of rigid motions to transform figures and to predict the effect of a given rigid motion on a given figure; given two figures, used the definition of congruence in terms of rigid motions to decide if they are congruent. **(5.1–5.3)**	**Students:** • understand that if one figure is obtained from the other by a sequence of rigid transformations, then the figures are congruent. • use congruence of corresponding parts to determine whether figures are congruent. • use rules for rigid motions in the coordinate plane to prove figures are congruent. • use congruency and given information to determine unknown lengths.	**Students:** • will write coordinate proofs about triangle relationships. **(9.4, 9.5)** • will write coordinate proofs about parallelograms. **(11.1–11.5)**

UNPACKING MATH STANDARDS

Use the definition of congruence in terms of rigid motions to show that two triangles are congruent if and only if corresponding pairs of sides and corresponding pairs of angles are congruent.

What It Means to You

Students demonstrate an understanding of congruence in terms of rigid motions by using tracing paper. When rigid motions are used to map one triangle onto another, students will see that corresponding pairs of sides and angles are congruent for congruent triangles.

The emphasis for this standard is on making connections between rigid motions and congruent figures and their corresponding parts. In future lessons, students will use knowledge of congruent corresponding parts to write formal triangle congruence proofs.

WARM-UP OPTIONS

ONLINE

PROJECTABLE
& PRINTABLE

ACTIVATE PRIOR KNOWLEDGE • Rotate Triangles

Use these activities to quickly assess and activate prior knowledge as needed.

Problem of the Day 

A painter has painted the wall around a triangular shaped piece of art, so that the art did not need to be removed. After the paint dries, the painter rotates the art work 180° around the center point. When the art work is rotated, will part of the wall be unpainted? Explain your answer.

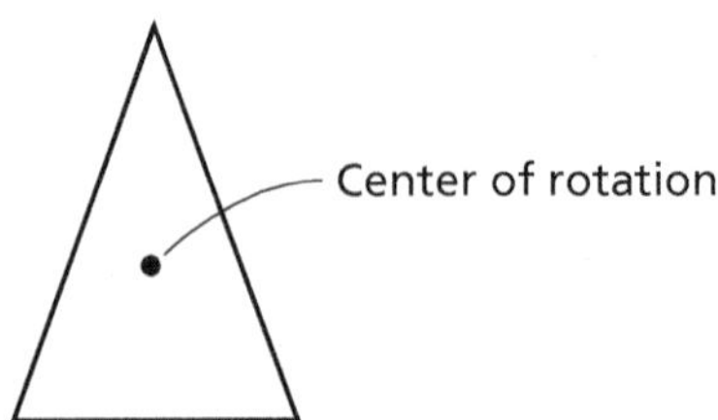

Part of the wall will be unpainted; The rotation produces a congruent triangle, but in a different orientation. There will be unpainted parts of the wall that show where the bottom two vertices of the triangle were originally.

Quick Check for Homework

As part of your daily routine, you may want to display the Teacher Solution Key to have students check their homework.

Make Connections

Based on students' responses to the Problem of the Day, choose one of the following:

1. Project the Interactive Reteach, Geometry, Lesson 5.2.

2. Complete the Prerequisite Skills Activity:

Have students use a coordinate plane to draw a triangle with vertices (−2, 0), (2, 0), and (0, 4). Then ask the following questions.

- *Apply the coordinate rule $(x, y) \rightarrow (-x, -y)$ to your original triangle and draw the new triangle. What type of transformation has occurred?* a 180° rotation
- *Apply the coordinate rule $(x, y) \rightarrow (-y, x)$ to your original triangle and draw this new triangle. What type of transformation has occurred?* a rotation 90° counterclockwise about the origin
- *How can you determine if either new triangle is congruent to the original triangle?* A rotation is a rigid motion that preserves side lengths and angle measures. The original triangle maps onto the new triangles using a rotation about the origin. You could also measure the sides and angles to confirm the congruency.

If students continue to struggle, use Tier 2 Skill 14.

SHARPEN SKILLS

If time permits, use this on-level activity to build fluency and practice basic skills.

Vocabulary Review

Objective: Students review the definition and characteristics of congruent figures, citing examples and non-examples.
Materials: Word Description (Teacher Resource Masters)

Have students write the words "Congruent Figures" in the center box. Work together as a class to write a definition. Then have students work in pairs to draw examples and non-examples of congruent figures. Finally, come back together as a class to list some characteristics of congruent figures.

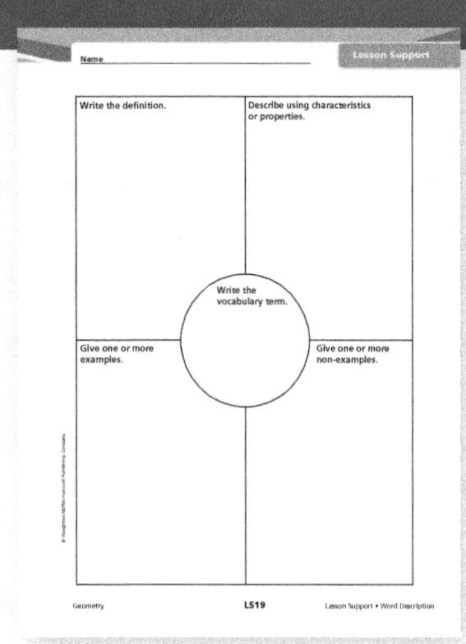

Name

Lesson Support

Write the definition.

Describe using characteristics or properties.

Write the vocabulary term.

Give one or more examples.

Give one or more non-examples.

Geometry LS19 Lesson Support • Word Description

PLAN FOR DIFFERENTIATED INSTRUCTION

Small-Group Options

Use these teacher-guided activities with pulled small groups.

On Track

Materials: coordinate plane, tracing paper

Have students work with partners to do the following:

- One student draws a triangle on a coordinate plane.
- The second student make a list of three rigid transformations.
- Together have the students apply the transformations to the triangle.
- Use the tracing paper to trace the original triangle and perform the transformations to see that the final triangle is congruent.

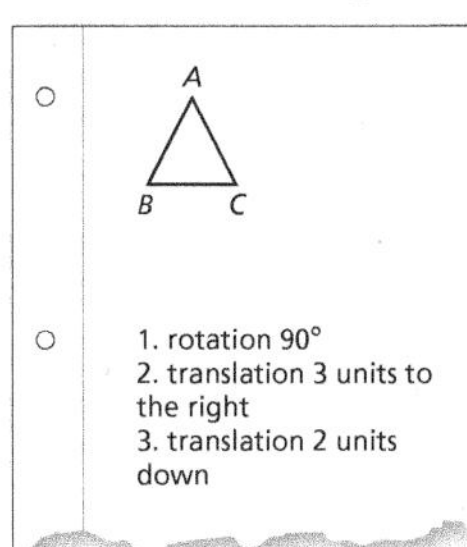

Almost There

Materials: index cards

Make sets of three cards where two of the cards show congruent figures and the third is not congruent to the other two.

- Have each pair examine the three cards and identify the figure that is most likely not congruent.
- Then have students write 1–2 sentences describing what they would need to know to determine if the other two figures are congruent
- Redistribute the sets of cards so each group has another set of three cards and repeat the activity.

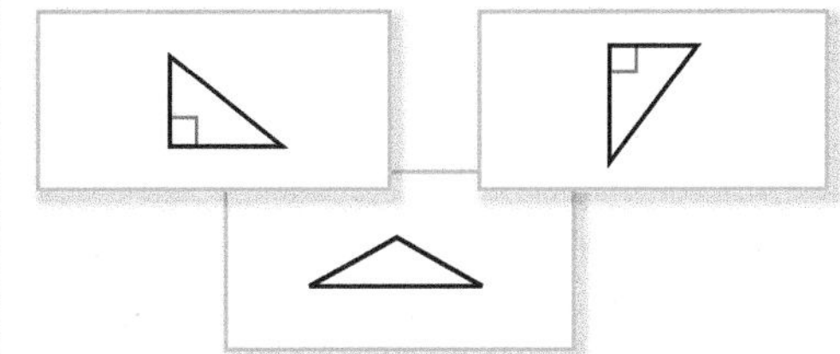

Ready for More

- Display these triangles on the board.

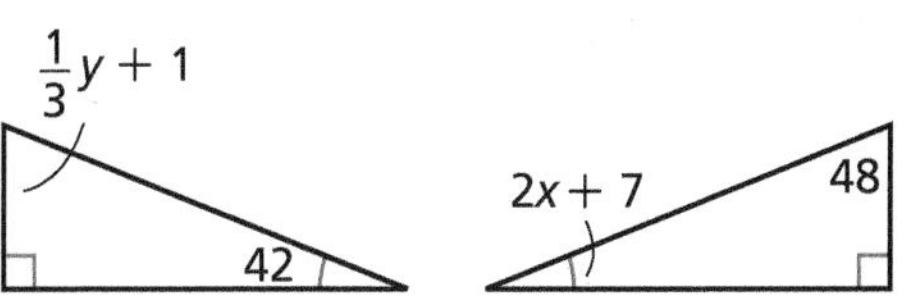

- Have students work individually to solve for the variables in the diagram.
- Go over the answers together as a class. Ask a volunteer to show their work and explain their reasoning.
- Have students work in pairs to discuss whether the triangles are congruent. Ask the questions: *Do you have enough information to tell if the triangles are congruent? If yes, explain. If no, what other information do you need?*

Math Center Options

Use these student self-directed activities at centers or stations. Key: ● Print Resources ● Online Resources

On Track

- ● Interactive Digital Lesson
- ●● Journal and Practice Workbook
- ● Interactive Glossary (printable): **congruent**, **corresponding angles**, **corresponding sides**
- ● Module Performance Task

Almost There

- ● Reteach 7.1 (printable)
- ● Interactive Reteach 7.1
- ● RtI Tier 2 Skill 14: Congruent Figures
- ● Illustrative Mathematics: Properties of Congruent Triangles

Ready for More

- ● Challenge 7.1 (printable)
- ● Interactive Challenge 7.1
- ● Illustrative Mathematics: Building a Tile Pattern by Reflecting Octagons

View data-driven grouping recommendations and assign differentiation resources.

During the *Spark Your Learning*, listen and watch for strategies students use. See samples of student work on this page.

Use a Ruler and a Protractor to Measure

Strategy 1

I measured the side lengths of one triangle and then compared them to the corresponding side lengths of a second triangle.

I also measured the angles in one triangle and compared the measures to the corresponding angles in a second triangle.

Then I compared these measurements to a third triangle to see that the windows are the same size and shape.

If students . . . use measurements to solve the problem, they are showing a good understaing of congruent figures from Grade 8.

Have these students . . . explain why they measured the sides and angles. **Ask:**

- Q How did you use measurements of side lengths and angles to determine if the triangles are congruent?
- Q Is it enough to measure only the sides or only the angles? Explain.
- Q How might the triangles be congruent in real-life even if they aren't in the picture?

Use Transformations

Strategy 2

I traced one of the triangles on tracing paper. Then I moved the triangle using translations, rotations and/or reflections to see if it has the same size and shape of another triangle.

It looks like the triangles are not all the same size in the photo, but they might be in the actual building.

If students . . . use tracing paper, they are applying their knowledge of rigid transformations and congruent figures but are not using a more advanced way of looking at it mathematically.

Activate prior knowledge . . . by having students think about how the drawing may not be the most accurate depiction of the triangles. **Ask:**

- Q Does the photograph show the triangles straight on or is the image a little distorted because of the way the photo was taken?
- Q How could you increase the accuracy of your triangle comparisons?

COMMON ERROR: Makes an Assumption Without Evidence

The triangles all appear to be equilateral which means that they all have the same size and shape.

If students . . . make an assumption about the figures in the photo, they may not recall that side lengths cannot be assumed from a diagram.

Then intervene . . . by pointing out that the photo does not indicate any lengths on the triangles, so an assumption that the triangles are equilateral cannot be made just by viewing the photo, but there are ways to determine if the triangles are the same size. **Ask:**

- Q How could you use a transformation such as a translation, reflection, or rotation to determine if the triangles are the same size?
- Q How could you use actual measurements to determine congruency?
- Q Suppose it turns out that the triangles are all equilateral. Does that automatically mean they are the same size?

7.1

Understand Congruent Figures

I Can determine whether figures are congruent.

Spark Your Learning

The architect chose to incorporate a wall of triangles as part of the design of this building.

©Chris Hellier/age fotostock/Getty Images

Complete Part A as a whole class. Then complete Parts B–D in small groups.

A. What is a mathematical question you can ask about the exterior shapes of the building? A. Are triangles *ABC*, *DEF*, and *GHJ* the same size and shape?

B. Describe a method you could use to answer your question. See Additional Answers.

C. To answer your question, what strategy and tool would you use along with all the information you have? What answer do you get? See Strategies 1 and 2 on the facing page.

D. Explain how the answer to your question is important in the design of a building like the one shown. Possible answer: Since the design must be repeated, the triangles should be alike or congruent.

Turn and Talk How can the figures used in Parts B–D be transformed to map one onto the other? Describe each transformation and state whether each transformation is a rigid motion. See margin.

CULTIVATE CONVERSATION • Co-Craft Questions

If students have difficulty formulating a mathematical question about the situation in the Spark Your Learning, ask them to imagine they are a company replacing a broken window. What are some natural questions to ask about this situation?

Work together to craft the following questions:

- Can I use the original measurements of the windows when making the replacement window?
- What would I need to know about the size and shape of the current windows?
- If the broken window is at the top of the building, how could I use the measurements of a window closer to the ground to help design the replacement?

Then have students think about what additional information, if any, they would need to answer these questions. **Ask:**

- Can you determine if the windows are the same size just by measuring the side lengths? Why or why not?
- Can you determine if the windows are the same size just by measuring the angles? Why or why not?

1 Spark Your Learning

▶ MOTIVATE 

- Have students look at the photo in their books and read the information contained in the photo. Then complete Part A as a whole-class discussion.
- Give the class the additional information they need to solve the problem. This information is available online as a printable and projectable page in the Teacher Resources.
- Have students work in small groups to complete Parts B–D.

▶ PERSEVERE

If students need support, guide them by asking:

Q **Advancing • Use Tools** Which tool could you use to solve the problem? Why choose that tool and not some other? Students' choices of tools and reasons for choosing them will vary.

Q **Assessing** What can you tell about the shapes of the windows? They look like they are all equilateral triangles.

Q **Assessing** If the triangles were the same size and shape, what would you expect to be true about *AB*, *DE*, and *GH*? They should have the same length.

Turn and Talk Remind students that a rigid transformation changes the orientation or location of the figure but not the size or the shape. A non-rigid transformation changes the size or shape of a figure. Possible answer: The triangles in the windows can be reflected or just translated; It depends on where they are being moved in the design. The transformations of the windows are rigid transformations.

▶ BUILD SHARED UNDERSTANDING

Select groups of students who used various strategies and tools to share with the class how they solved the problem. As they present their solutions, have each group discuss why they chose a specific strategy and tool.

② Learn Together

Build Understanding

Task 1 (MP) **Use Tools** Students use tracing paper to map one triangle to another in order to compare side lengths and angle measures of two congruent figures.

CONNECT TO VOCABULARY

Have students use the **Interactive Glossary** to record their understanding of the vocabulary in this task.

Sample Guided Discussion:

Q **What are three types of rigid transformations where the size and shape of a figure do not change?** translation, reflection, and rotation

Q **In Part B, what type of translation is used to map point *A* to point *D*?** a vertical and a horizontal translation

Q **In Part D, over which line segment is point *C* reflected over to become point *F*?** It is reflected over $\overline{AB}$.

Turn and Talk Have students brainstorm different dilations in the real-world like photo enlargements or a sweater shrinking in the dryer. Have students review the definition of a dilation that multiplies the coordinates of a figure by a given scale factor. No, they are not congruent. A dilation changes the size of the figure. Congruent figures are not different sizes.

Build Understanding

Transform Figures with Congruent Corresponding Parts

Two figures are **congruent** if and only if one can be obtained from the other by a sequence of rigid motions.

The definition of congruent figures above is written as a biconditional statement because it contains the phrase "if and only if." A true biconditional statement can be rewritten as a true conditional statement and its true converse.

Conditional statement: If a figure is obtained from another figure by a sequence of rigid motions, then the two figures are congruent.

Converse: If two figures are congruent, then one can be obtained from the other by a sequence of rigid motions.

1 A. In the figures *ABC* and *DEF*, all corresponding side lengths and angles are congruent. Trace *ABC* on a piece of tracing paper and trace *DEF* on a separate piece of tracing paper. How can you show that the figures are congruent according to the definition of congruent figures?

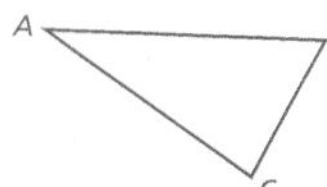

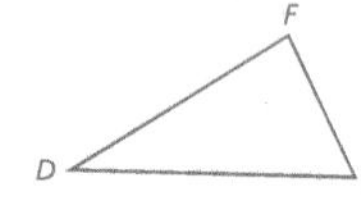

You must show that a series of rigid motions will map *ABC* to *DEF*.

B. Arrange the two figures on a desk as shown. Describe the transformation you can use to map point *A* to point *D*. a translation

C. Describe the transformation you can use to map point *B* to point *E*.

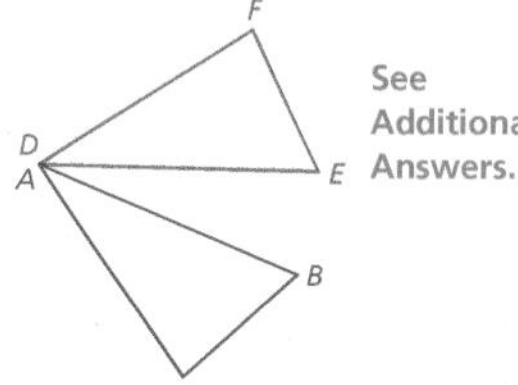

See Additional Answers.

D. Describe the transformation you can use to map point *C* to point *F*.

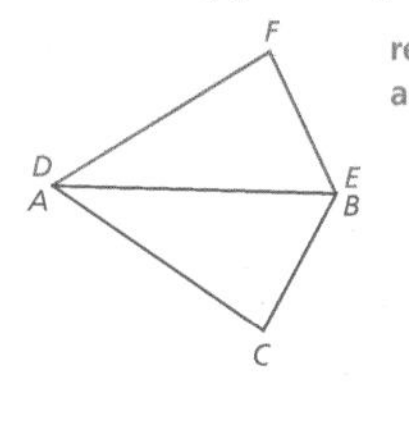

reflection across $\overline{ED}$

E. Complete the statements to verify that the image of $\angle B$ is $\angle E$:

If two triangles are congruent, then their corresponding sides and angles map to each other.

Steps B–D showed that __?__ through a sequence of rigid motions. $\triangle ABC \cong \triangle DEF$

Based on $\triangle ABC \cong \triangle DEF$, the image of $\angle B$ is __?__. $\angle E$

Turn and Talk Suppose a dilation is used to map a figure onto another figure. Are those figures congruent? See margin.

196

LEVELED QUESTIONS

Depth of Knowledge (DOK)	Leveled Questions	What Does This Tell You?
Level 1 **Recall**	Which transformations change the location and the orientation of a figure? rotations and reflections	Students' answers will indicate whether they know a reflection and a rotation change both the location and the orientation of a figure.
Level 2 **Basic Application of Skills & Concepts**	Why does a rigid transformation keep the size and shape of the figure the same? The rigid transformations only change the location or orientation of a figure. The rigid transformations do not change the length of the sides or the measure of the angles.	Students' answers will indicate whether they know applying a rigid transformation to a figure preserves the figures size and shape.
Level 3 **Strategic Thinking & Complex Reasoning**	If rigid transformations are performed in a different order, is the resulting figure the same as in the original order of transformations? The size and shape of the figure does not change, but the orientation and location may be different than the original order of transformations.	Students' answers will indicate whether they understand that a rotation, reflection, and translation will produce a congruent figure, and the order of transformations does not affect the size or shape but can affect the resulting orientation and location.

Use Corresponding Parts to Show Figures Are Congruent

Corresponding angles and **corresponding sides** are located in the same position for polygons with an equal number of sides. You can write a congruence statement for two figures by matching the congruent corresponding parts. In the statement $\triangle LMN \cong \triangle PQR$, $\overline{LM}$ corresponds and is congruent to $\overline{PQ}$, $\angle L$ corresponds and is congruent to $\angle P$, and so on.

Corresponding Parts of Congruent Figures are Congruent (CPCFC)
Two figures are congruent if and only if corresponding pairs of sides and corresponding pairs of angles are congruent.

This biconditional statement can be rewritten as a conditional statement and its converse.

Conditional statement: If all pairs of corresponding sides and all pairs of corresponding angles of two figures are congruent, then the figures are congruent.

Converse: If two figures are congruent, then all pairs of corresponding sides and all pairs of corresponding angles are congruent.

2 Determine whether the indicated triangles are congruent.

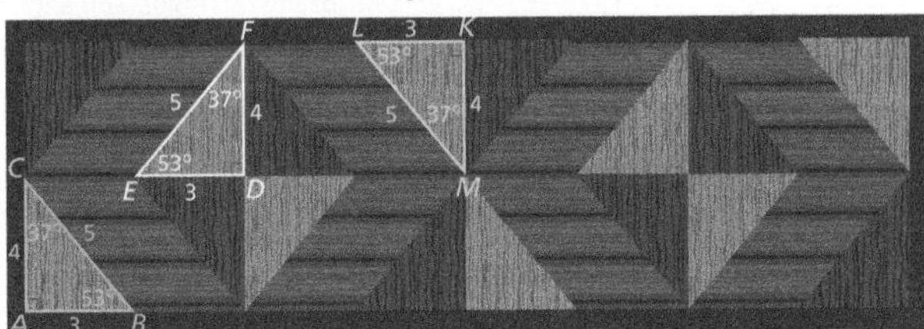

Use a table of corresponding parts.

$\triangle ABC$	$\triangle DEF$	$\triangle KLM$
$\overline{AB}$	$\overline{DE}$	$\overline{KL}$
$\overline{BC}$	$\overline{EF}$	$\overline{ML}$
$\overline{AC}$	$\overline{FD}$	$\overline{KM}$
$\angle A$	$\angle D$	$\angle K$
$\angle B$	$\angle E$	$\angle L$
$\angle C$	$\angle F$	$\angle M$

A. What information are you given about the lengths of the corresponding sides? The corresponding side lengths are congruent.

B. What information are you given about the measures of the corresponding angles? The corresponding angles are congruent.

C. Which triangles are congruent? How do you know? All three triangles are congruent because all corresponding sides and all corresponding angles are congruent.

D. Write a congruence statement for the congruent triangles. Is there more than one way to write the congruence statement? Explain. $\triangle ABC \cong \triangle DEF \cong \triangle KLM$; Yes; Corresponding parts need to be in corresponding order in the congruence statement.

E. Describe the transformations that map one of the highlighted triangles to the other two. Answers will vary but must be a sequence of rigid motions.

Turn and Talk Is it possible to determine congruent corresponding parts of two figures given only a congruence statement, such as $GHJK \cong WXYZ$? Explain. See margin.

Task 2 (MP) **Use Structure** Students examine corresponding sides angles of triangles to determine congruence using CPCFC.

CONNECT TO VOCABULARY

Have students use the **Interactive Glossary** to record their understanding of the vocabulary in this task.

Sample Guided Discussion:

Q **How many pairs of corresponding sides and angles are there between two congruent triangles?** There are three pairs of corresponding sides and three pairs of corresponding angles.

Q **In Part D, what must be true about the order of vertices of the triangles when writing congruence statements?** The corresponding angles must be in the same order for both triangles. For example, if $\angle A$ corresponds to $\angle D$, they must be listed in the same location in the congruence statement, i.e. both first, both second, or both last.

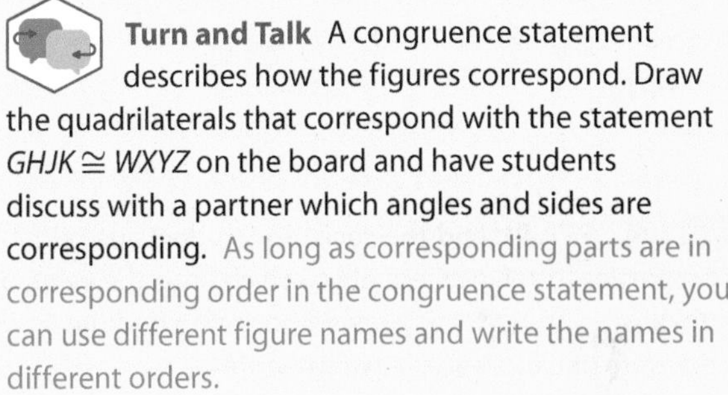
Turn and Talk A congruence statement describes how the figures correspond. Draw the quadrilaterals that correspond with the statement $GHJK \cong WXYZ$ on the board and have students discuss with a partner which angles and sides are corresponding. As long as corresponding parts are in corresponding order in the congruence statement, you can use different figure names and write the names in different orders.

PROFICIENCY LEVEL

Beginning
Have students write in words what the letters CPCFC represent. Then have them draw and label a diagram to illustrate.

Intermediate
Write the congruence statement $\triangle PQR \cong \triangle STU$ and draw the triangles on the board. Have students explain to each other which angles and which sides are congruent given the congruence statement.

Advanced
Have students write a paragraph explaining how to tell which pairs of angles and which pairs of sides are congruent when given a congruence statement.

Step It Out

Task 3 (MP) **Attend to Precision** Students use the Third Angles Theorem by examining corresponding and congruent angle pairs.

Sample Guided Discussion:

Q What is true about the sum of the angles in each of the triangles? The total sum of the angles in the triangles is 180°.

Q Why do the two pairs of congruent angles indicate that $\angle L$ and $\angle P$ are also congruent? Since two pairs of angles are congruent, you can subtract this amount from 180°. The difference of 180° and the sum of the two known angles is the measure of the third angle in each triangle.

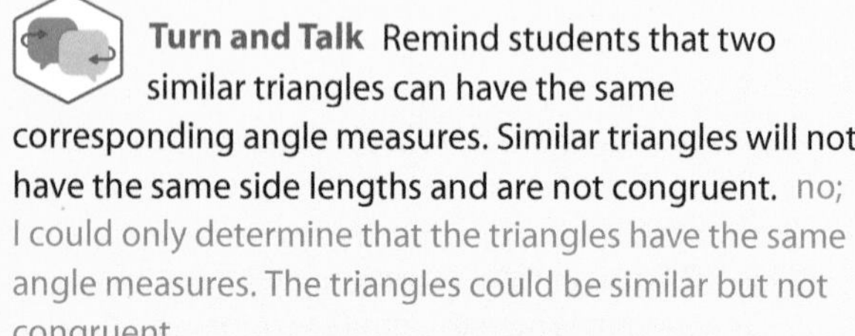

Turn and Talk Remind students that two similar triangles can have the same corresponding angle measures. Similar triangles will not have the same side lengths and are not congruent. no; I could only determine that the triangles have the same angle measures. The triangles could be similar but not congruent.

Task 4 (MP) **Attend to Precision** Students use algebraic properties to solve an equation which finds the value of an unknown variable. This allows students to find the measure of two angles in two triangles.

Sample Guided Discussion:

Q If two angles are congruent, what does this mean about their measures and how they can be found when their measures are given as expressions? The measures are the same, so the expressions can be written equal to each other and solved for the variable.

Q Why is 10 not the measure of the angle? 10 is the value of x. To find the measure of the angles, substitute $x = 10$ into the expressions for each angle.

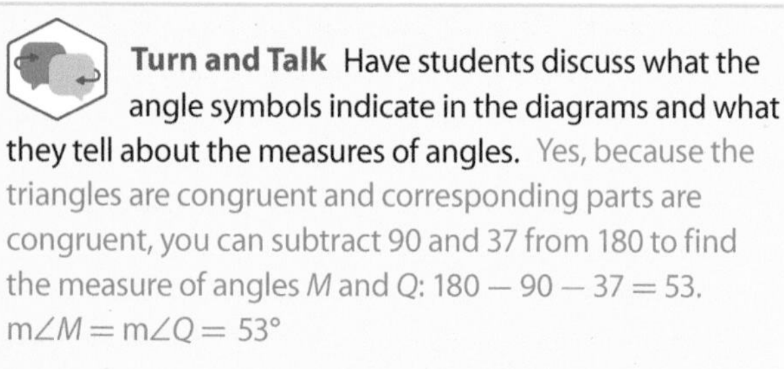

Turn and Talk Have students discuss what the angle symbols indicate in the diagrams and what they tell about the measures of angles. Yes, because the triangles are congruent and corresponding parts are congruent, you can subtract 90 and 37 from 180 to find the measure of angles M and Q: $180 - 90 - 37 = 53$. $m\angle M = m\angle Q = 53°$

Step It Out

Apply the Third Angles Theorem

Third Angles Theorem

If two angles of one triangle are congruent to two angles of another triangle, then the third pair of angles are congruent.
If $\angle A \cong \angle D$ and $\angle C \cong \angle F$, then $\angle B \cong \angle E$.

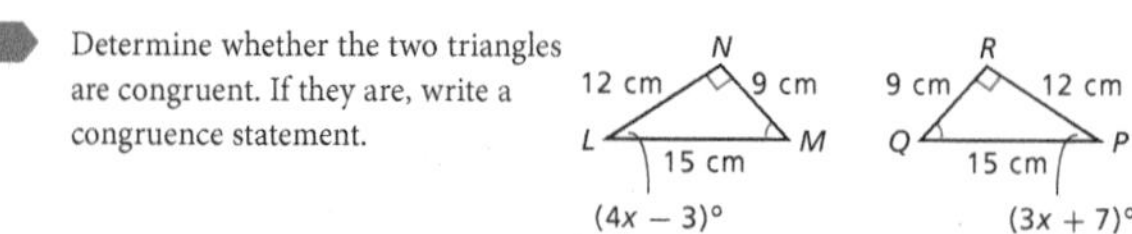

3 Determine whether the two triangles are congruent. If they are, write a congruence statement.

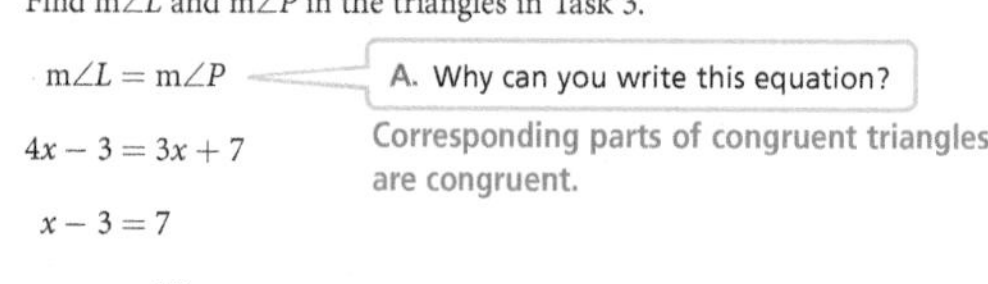

The markings on the figure show that the following corresponding parts are congruent:

Sides: $\overline{LN} \cong \overline{PR}$, $\overline{NM} \cong \overline{RQ}$, $\overline{LM} \cong \overline{PQ}$

Angles: $\angle N \cong \angle R$, $\angle M \cong \angle Q$

By the Third Angles Theorem, $\angle L \cong \angle P$.

A. Why can the Third Angles Theorem be used here?

Two pairs of the corresponding angles are congruent. This means the third set of corresponding angles is congruent also.

All pairs of corresponding sides are congruent, and all pairs of corresponding angles are congruent, so $\triangle LMN \cong \triangle PQR$.

Turn and Talk Suppose the side lengths of the triangles are not given. Can you still determine whether the triangles are congruent? Explain why or why not. See margin.

4 Find $m\angle L$ and $m\angle P$ in the triangles in Task 3.

$m\angle L = m\angle P$

A. Why can you write this equation?

Corresponding parts of congruent triangles are congruent.

$4x - 3 = 3x + 7$

$x - 3 = 7$

$x = 10$

Find the measure of each angle.

$m\angle L = (4x - 3)° = (4(10) - 3)° = 37°$

So, $m\angle L = 37°$ and $m\angle P = 37°$.

B. How do you know that $m\angle P = 37°$?

$\angle P$ and $\angle L$ are congruent.

Turn and Talk Using the given information, can you determine the measures of the other angles in the figures? Explain your reasoning. See margin.

198

Apply Properties of Congruent Figures

Patterns made of congruent figures are common in nature. These patterns can sometimes be adapted for industrial or safety uses.

5 Scientists have been able to use a difference between the skin of whales and sharks to develop a method that prevents bacteria growth. While whales can have barnacles, algae, and other sea creatures growing on their skin, sharks are not affected by such parasites because shark skin has a pattern of repeating congruent hexagons that does not allow parasites to stick.

Scientists copied this pattern to print an adhesive film that can be used on surfaces in hospitals and public restrooms to prevent bacteria from growing. Use the given information to determine the side lengths in the hexagons.

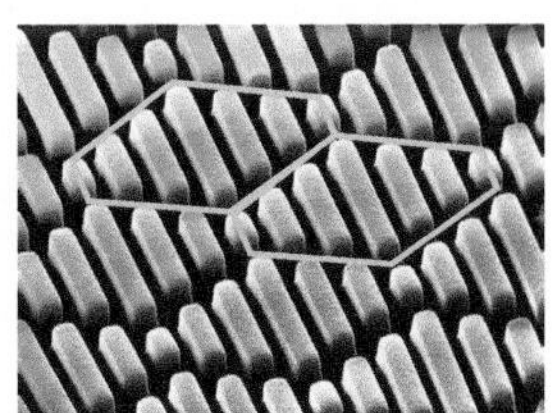

A, B. See Additional Answers.

B, C, A, D, K, F, E, J, G, H
$2x + 5$, $4z - 8$, 7, 4, $2y + 3$

A. How do you know that the corresponding angles in each figure are congruent?

B. How do you know that the corresponding sides of the figures are congruent?

What values of the variables result in congruent figures?

Solve for x.	Solve for y.
$AB = ED$	$HJ = ED$
$2x + 5 = 7$	$2y + 3 = 7$
$2x + 5 - 5 = 7 - 5$	$2y + 3 - 3 = 7 - 3$
$2x = 2$	$2y = 4$
$\frac{2x}{2} = \frac{2}{2}$	$\frac{2y}{2} = \frac{4}{2}$
$x = 1$	$y = 2$

C. Use a similar strategy to find the value of z.

$z = 3$

Turn and Talk How do you know there must be a rigid motion that maps *ABCDEF* onto *EDKJHG*? See margin.

Task 5 (MP) **Model with Mathematics** Students will use properties of congruent figures and their corresponding congruent parts to solve for a variable in a real-world situation.

SUPPORT SENSE-MAKING Three Reads

Have students read the problem three times. Use the questions below for a different focus each read.

1. What is the situation about?
2. What are the quantities in the situation?
3. What are the possible mathematical questions that you could ask for the situation?

Sample Guided Discussion:

Q What do you know about the size and shape of hexagon *ABCDEF* and *EDKJHG*? They have the same size and shape because they are congruent figures.

Q How could you use the values of the variables to find the lengths of the sides? You can substitute the values of the variables into the expressions for each side length.

Turn and Talk Have students review the transformations that are considered rigid motions. What happens to the size and shape of a figure when it undergoes a translation, reflection, or a rotation. The definition of congruence states that figures are congruent if and only if there is a sequence of rigid motions that map one to the other.

ONLINE

Assign the Digital On Your Own for

- built-in student supports
- Actionable Item Reports
- Standards Analysis Reports

On Your Own

Assignment Guide

The chart below indicates which problems in the On Your Own are associated with each task in the Learn Together. Assign daily homework for tasks completed.

Learn Together Tasks	On Your Own Problems
Task 1, p. 196	Problems 5, 6, 9, 12, 15, and 16
Task 2, p. 197	Problems 13 and 14
Task 3, p. 198	Problems 7 and 8
Task 4, p. 198	Problems 7 and 8
Task 5, p. 199	Problem 17

Check Understanding

1. (MP) **Construct Arguments** How can you show two figures congruent?
   Use rigid transformations.
2. Suppose $PQRS \cong TUVW$.
   A–C. See Additional Answers.
   A. Write a biconditional statement that describes the congruence relationship between $PQRS$ and $TUVW$.
   B. Describe a sequence of rigid motions that maps $PQRS$ to $TUVW$.
   C. List the corresponding congruent parts.

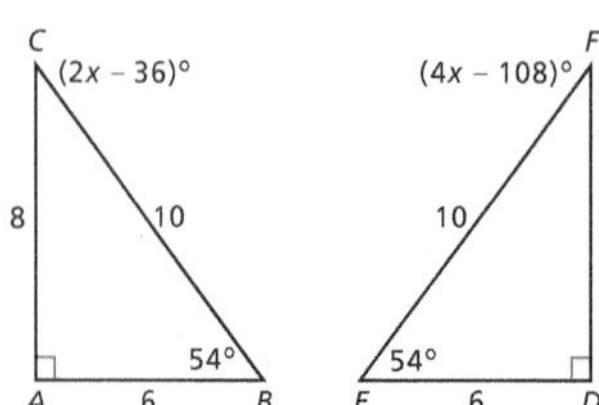

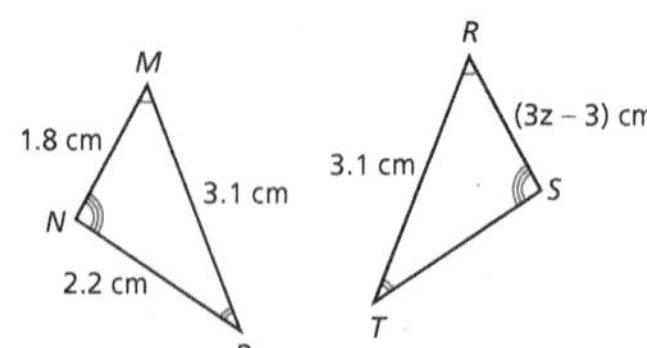

3. Find the measures of $\angle C$ and $\angle F$.
   $m\angle C = m\angle F = 36°$

4. Find the value of the variable that results in congruent triangles.
   $z = 1.6$

On Your Own

5. (MP) **Critique Reasoning** Chelsea says that the concept of rigid motion means two triangles cannot be congruent if any pair of corresponding parts is not congruent. Is Chelsea correct? Explain why or why not.
   yes; Corresponding parts of congruent figures are congruent.
6. Describe a sequence of rigid motions that maps $\triangle MNP$ onto $\triangle MQR$ to show $\triangle MNP \cong \triangle MQR$.
   Possible answer: reflection across a line through M that bisects $\angle NMQ$

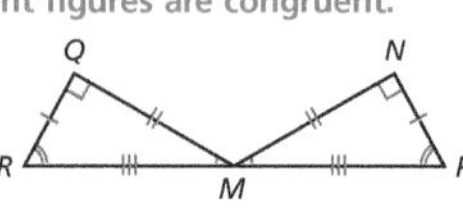

In Problems 7 and 8, determine whether the triangles are congruent. Explain your reasoning.

7.

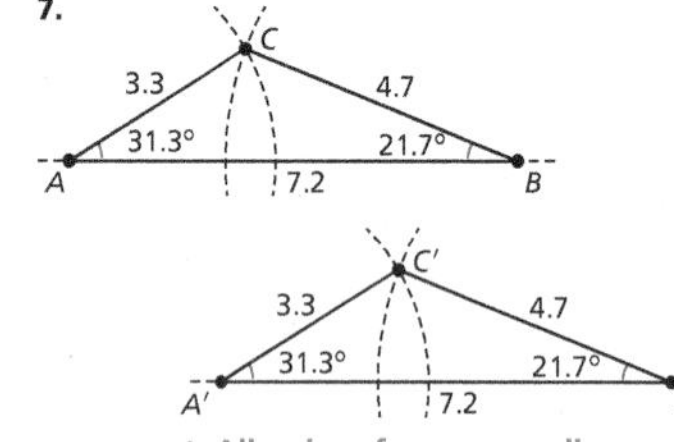

congruent; All pairs of corresponding parts are congruent.

8.

not congruent; Not all pairs of corresponding parts are congruent.

200

data checkpoint

3 Check Understanding

Formative Assessment

Use formative assessment to determine if your students are successful with this lesson's learning objective.

Students who successfully complete the Check Understanding can continue to the On Your Own practice.

For students who miss 1 problem or more, work in a pulled small group using the Almost There small-group activity on page 195C.

ONLINE

Assign the Digital Check Understanding to determine

- success with the learning objective
- items to review
- grouping and differentiation resources

4 Differentiation Options

Differentiate instruction for all students using small-group activities and math center activities on page 195C.

Reteach

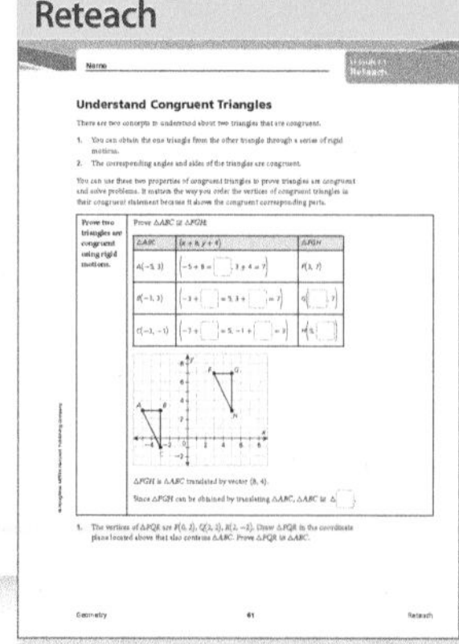

Challenge

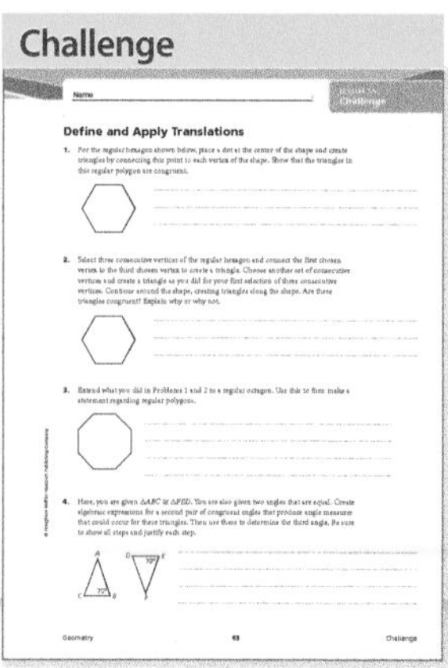

In Problems 9–12, describe a sequence of transformations you could use to prove that the figures are congruent. If there is no such sequence, explain why not. See margin.

9. Triangles *KLM* and *PQR*

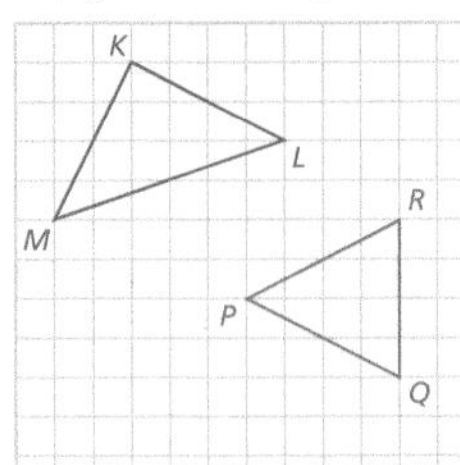

10. Pentagons *GHIJK* and *LMNOP*

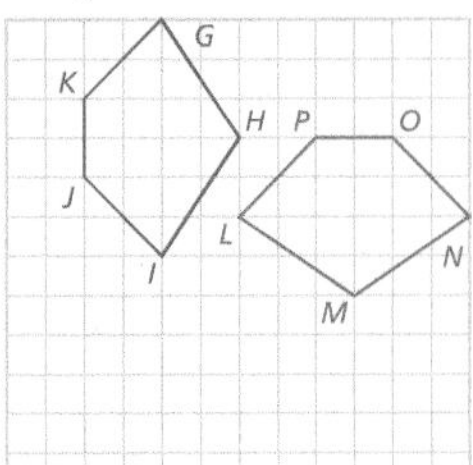

11. Squares *NOPQ* and *RSTU*

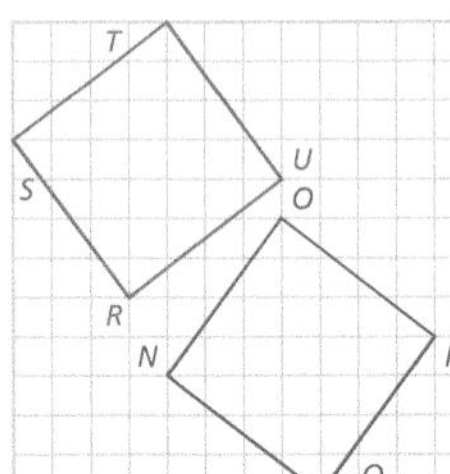

12. Hexagons *ABCDEF* and *UVWXYZ*

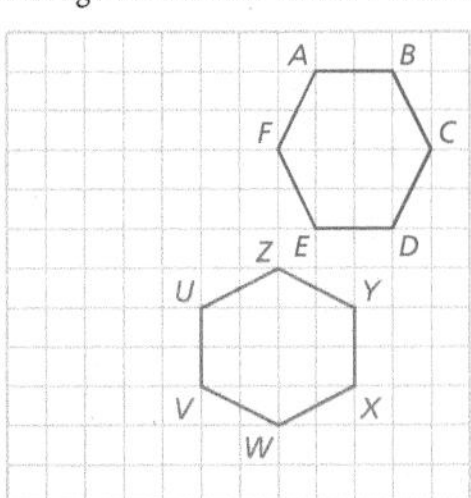

In Problems 13 and 14, use a ruler and protractor to measure the figures. Is each pair congruent? Why or why not?

13.

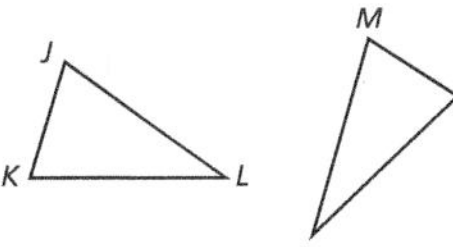

14.

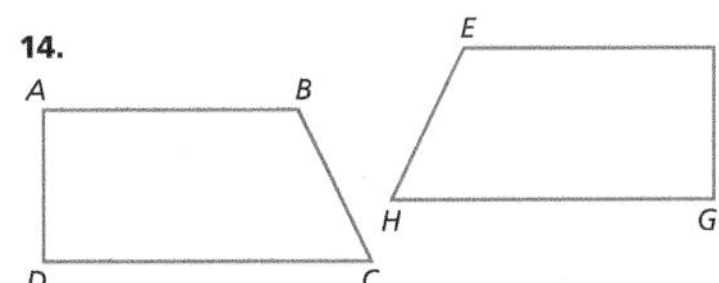

13, 14. See Additional Answers.

15. (MP) **Use Structure** Shamara is replacing the outlined pieces of glass. She needs to know whether the two shapes are congruent. How can she prove the shapes are congruent? She could use a sequence of rigid motions to map one shape to the other.

A

B

16. (MP) **Critique Reasoning** Dan says two figures are congruent if one figure can be transformed into the other by any series of transformations. Provide a counterexample to show why his definition is not complete. Answers will vary and should include a non-rigid transformation such as a dilation.

Watch for Common Errors

Problem 13 Students might measure the sides and angles of the figures correctly but then not compare corresponding pairs of sides and angles. Encourage students to make a list of sides with their corresponding measurements in order from smallest to largest for each triangle. Then have them compare the corresponding sides from smallest to largest. Have students measure the angles and then compare them from smallest to largest to help identify corresponding angle pairs.

Questioning Strategies

Problem 16 Suppose Dan uses a dilation as a part of the series of transformations. Is a dilation a rigid transformation? Suppose Dan uses two dilations so that the figure returns to its original size. What must be true about the series of transformations to prove two figures are congruent? A dilation is not a rigid transformation; Even though the figure may return to its original size, it does not follow the definition of congruent figures if one can't be obtained using a sequence of rigid motions.

ANSWERS

9. no; The figures are not the same size and shape. *Possible answer:* You can translate *K* to *P* (which also translates *L* to *Q* because $\overline{KL} \parallel \overline{PQ}$) and then reflect over $\overline{PQ}$, but *M* does not map to *R*.

10. yes; *Possible answer:* You can translate *G* to *L*, rotate counterclockwise about *L* by m∠*HLM* to map *H* to *M*, and then reflect over $\overline{LM}$ to map *I* to *N*, *J* to *O*, and *K* to *P*.

11. yes; *Possible answer:* You can translate *N* to *R* and then rotate counterclockwise about *R* by m∠*ORS* to map *O* to *S* (which also maps *P* to *T* and *Q* to *U*).

12. yes; *Possible answer:* You can translate *A* to *U*, rotate clockwise about *U* by m∠*BUV* (90°) to map *B* to *V*, and then reflect over $\overline{UV}$ to map *C* to *W*, *D* to *X*, *E* to *Y*, and *F* to *Z*.

5 Wrap-Up

Summarize learning with your class. Consider using the Exit Ticket, Put It in Writing, or I Can scale.

Exit Ticket 

Two triangles in a tile design are congruent so that $\triangle ABC \cong \triangle DEF$. Describe a sequence of two rigid transformations that map the tile formed by $\triangle ABC$ onto the tile formed by $\triangle DEF$.

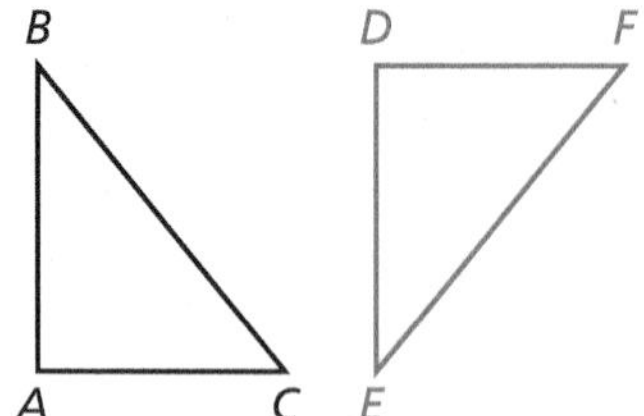

Possible answer: You can use a translation to the right and then a reflection. You can use a reflection and then a translation.

Put It in Writing 

A series of rigid transformations or finding measurements of sides and angles can be used to show two figures are congruent or not congruent. Write a paragraph describing how each method can be used to show that two figures are congruent.

I Can 

The scale below can help you and your students understand their progress on a learning goal.

4	I can determine whether figures are congruent and explain my reasoning to others.
3	I can determine whether figures are congruent.
2	I can measure and compare side lengths and angle measures in figures.
1	I can identify some pairs of congruent corresponding sides or congruent corresponding angles.

Spiral Review • Assessment Readiness

These questions will help determine if students have retained information taught in the past and can also prepare them for high-stakes assessments. Here, students must use compositions of transformations to identify figures **(6.1)**, determine if two triangles are congruent **(7.2)**, and identify rigid and non-rigid motions **(6.2)**.

17. The illustration shows a square from a "Yankee Puzzle" quilt.

A. Use the idea of congruent shapes to describe the design of the quilt square. A–C. See Additional Answers.

B. Explain how the triangle with base $\overline{AB}$ can be transformed to the position of the triangle with base $\overline{CD}$.

C. Explain how you know $CD = AB$.

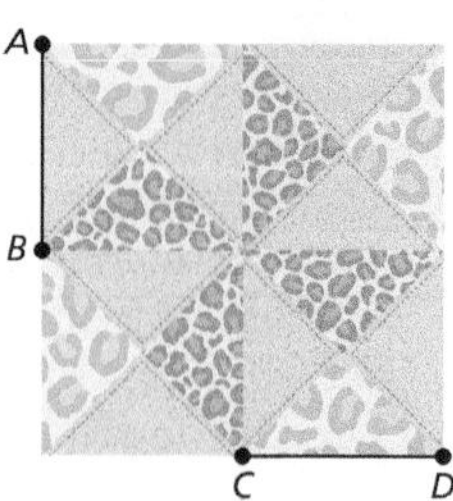

Spiral Review • Assessment Readiness

18. Which transformation will produce a dilation such that the original is 3 times smaller than the new image?

Ⓐ $(x, y) \rightarrow \frac{1}{3}(x, y)$

Ⓑ $(x, y) \rightarrow (x + 3, y + 3)$

Ⓒ $(x, y) \rightarrow (3x, 3y)$

Ⓓ $(x, y) \rightarrow (x - 3, y - 3)$

19. Are the given figures congruent? Explain.

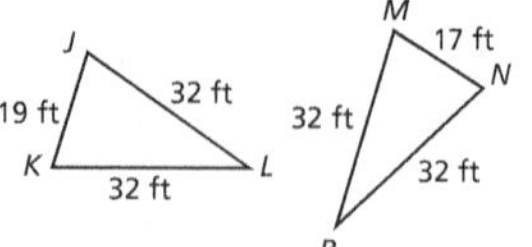

Ⓐ yes; by CPCFC

Ⓑ yes; The triangles have congruent angles.

Ⓒ no; $\overline{JK}$ and $\overline{MN}$ are not congruent.

Ⓓ no; $\overline{KL}$ and $\overline{MP}$ are not congruent.

20. For each type of transformation, identify whether it is a rigid motion or a nonrigid motion.

Transformation	Rigid Motion	Nonrigid Motion
A. Translation	**?**	?
B. Compression	?	**?**
C. Dilation	?	**?**
D. Rotation	**?**	?
E. Reflection	**?**	?
F. Stretch	?	**?**

I'm in a Learning Mindset!

What steps am I taking to direct my own learning to understand congruent figures?

202

Keep Going to Journal and Practice Workbook

Learning Mindset

Resilience Managing the Learning Process

Students may have difficulty thinking of different ways to use rigid transformations to prove congruence. Encourage students to try transformations in different sequences if their first attempt does not produce a congruent figure. *What other transformation could you try first? What would happen if you tried the transformations in a different order? What happens if you use the same transformation more than once in a sequence; for example, a translation, reflection, and then a second translation?*

7.2 Corresponding Parts of Congruent Figures

LESSON FOCUS AND COHERENCE

Mathematics Standards

- Use the definition of congruence in terms of rigid motions to show that two triangles are congruent if and only if corresponding pairs of sides and corresponding pairs of angles are congruent.

Mathematical Practices and Processes (MP)

- Use appropriate tools strategically.
- Construct viable arguments and critique the reasoning of others.
- Look for and make use of structure.

I Can Objective 

I can use congruent figures to solve problems.

Learning Objective

Use congruent figures to identify congruent parts of figures, solve for unknown measures, and prove geometric statements.

Language Objective

Explain how knowing two figures are congruent can help us solve for unknown measures of parts of the figures.

Lesson Materials: rulers, protractor, scissors, geometry tiles, GeoGebra

Mathematical Progressions

Prior Learning	Current Development	Future Connections
Students: • compared transformations that preserve distance and angle to those that do not (e.g., translation versus horizontal stretch). **(5.1–5.3)** • given two figures, used the definition of congruence in terms of rigid motions to decide if they are congruent. **(5.1–5.3)**	**Students:** • use the converse of Corresponding Parts of Congruent Figures are Congruent (CPCFC) to identify congruent sides and angles between congruent figures. • use the converse of CPCFC to solve for unknown measures. • use the converse of CPCFC to prove geometric statements.	**Students:** • will write coordinate proofs about triangle relationships. **(9.4–9.5)** • will write coordinate proofs about parallelograms. **(11.1–11.5)**

PROFESSIONAL LEARNING

Using Mathematical Practices and Processes

Use appropriate tools strategically.

This lesson gives an opportunity to address this Mathematical Practice Standard. According to the standard, mathematically proficient students "consider the available tools when solving a mathematical problem." In this lesson, students focus on the converse of Corresponding Parts of Congruent Figures are Congruent (CPCFC). While students should eventually progress to more abstract reasoning, using physical models of two congruent figures can help draw attention to corresponding parts being mapped to each other through transformations. This can also be done with dynamic geometry software such as GeoGebra. Students can also use rulers and protractors to confirm congruence of corresponding parts.

WARM-UP OPTIONS

ONLINE

PROJECTABLE & PRINTABLE 

ACTIVATE PRIOR KNOWLEDGE • Predict the Effect of a Rigid Motion on a Figure

Use these activities to quickly assess and activate prior knowledge as needed.

Problem of the Day 

Triangle $\triangle ABC$ has vertices $A(-2, 2)$, $B(-4, 3)$, and $C(-4, 4)$. If $\triangle ABC$ is reflected across $\overline{AC}$, what will be the coordinates of B'?

We know that the line $\overline{AC}$ goes through the origin and $\overline{BC}$ is a vertical segment with C positioned 1 unit above B, so $\overline{B'C'}$ will be a horizontal segment with C' positioned 1 unit to the left of B'. C is mapped to itself by the reflection, so B' is 1 unit to the right of C at $(-4 + 1, 4) = (-3, 4)$.

Quick Check for Homework

As part of your daily routine, you may want to display the Teacher Solution Key to have students check their homework.

Make Connections

Based on students' responses to the Problem of the Day, choose one of the following:

 Project the Interactive Reteach, Geometry, Lesson 5.3.

 Complete the Prerequisite Skills Activity:

Have students work in small groups and give each group a set of index cards. Some cards should have three or four named ordered pairs; for example, $A(1, 2)$, $B(-4, 0)$, $C(-3, 1)$, while the other cards should have the description of a reflection on the coordinate plane. Have students predict the coordinates of the images of the points on one card with the reflection described on another card, and then graph and confirm their predictions.

- *What does a reflection across the x- or y-axis do to the coordinates of a point?* It changes the sign of a coordinate.
- *What does a reflection across the line $y = x$ do to the coordinates of a point?* It switches the x- and y-coordinates.

If students continue to struggle, use Tier 2 Skill 11.

SHARPEN SKILLS

If time permits, use this on-level activity to build fluency and practice basic skills.

Quantitative Comparison

Objective: Students make a comparison between two quantities.

Write the following problem on the board. Ask students to choose the letter representing the correct answer and to explain their reasoning.

Quantity A	**Quantity B**
the distance between the points $A(3, -1)$, $B(2, 5)$	the distance between the points $C(-2, 4)$, $D(2, 8)$

A. Quantity A is greater. A; Quantity A is $\sqrt{(3-2)^2 + (-1-5)^2} = \sqrt{1+36} = \sqrt{37}$, while
B. Quantity B is greater. Quantity B is $\sqrt{(-2-2)^2 + (4-8)^2} = \sqrt{16+16} = \sqrt{32}$.
C. The two quantities are equal.
D. The relationship cannot be determined from the information given.

PLAN FOR DIFFERENTIATED INSTRUCTION

Small-Group Options

Use these teacher-guided activities with pulled small groups.

On Track

Materials: index cards

Have students work in small groups. Give each pair a set of index cards. Each card should have two congruent figures with the measures of corresponding parts given as linear expressions. Point out to students that the figures are congruent. Use the same variable for each pair of corresponding measures. Have students work together to solve for the values of the variables.

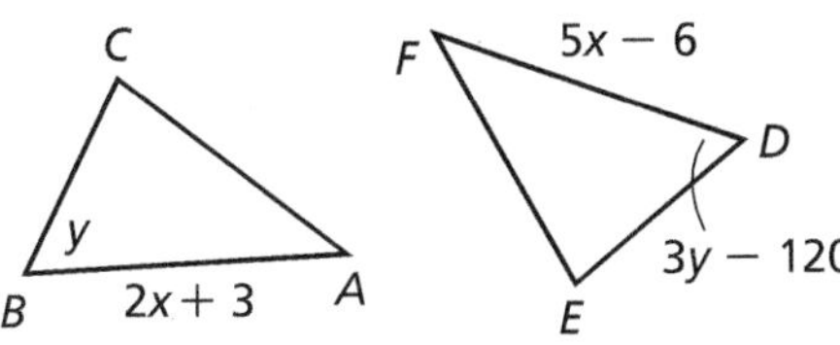

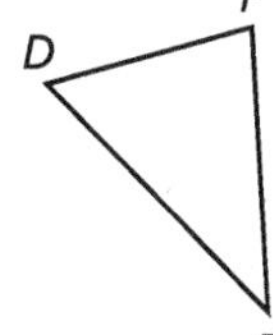

Almost There

Materials: index cards

Give each pair a set of index cards with two congruent figures with labeled vertices. Use a variety of transformations for the image of the figure on each card, a rotation, a reflection, or a translation. Point out to students that the figures are congruent, and have them take turns naming pairs of congruent angles and sides.

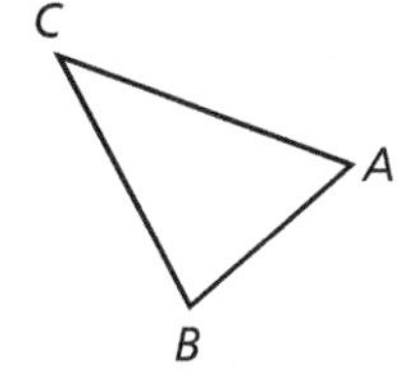

Ready for More

Materials: index cards

Give each group a set of index cards. Each card should have two similar figures with some measures given and others missing. If the figures are similar but not congruent, make sure a pair of corresponding side lengths are given. Point out to students that the figures are similar, and have them work together to determine unknown measures or explain why they cannot be found.

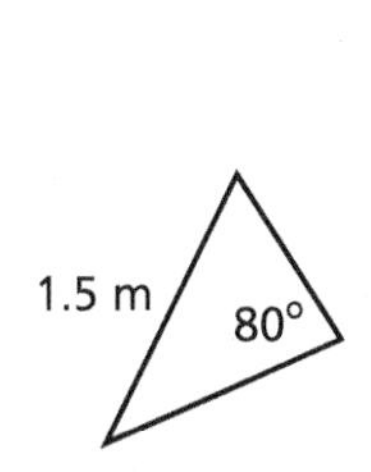

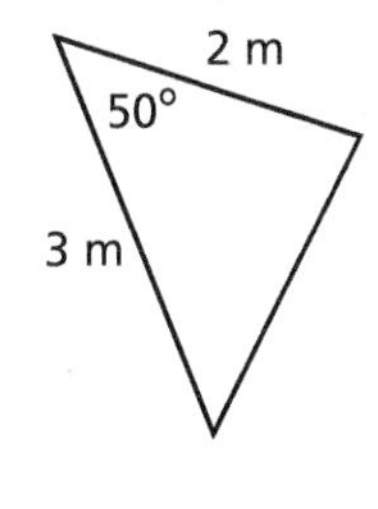

Math Center Options

Use these student self-directed activities at centers or stations. Key: ● Print Resources ● Online Resources

On Track

- ● Interactive Digital Lesson
- ●● Journal and Practice Workbook
- ● Module Performance Task

Almost There

- ● Reteach 7.2 (printable)
- ● Interactive Reteach 7.2
- ● RtI Tier 2 Skill 11: Reflect Figures in the Coordinate Plane
- ● Illustrative Mathematics: Congruent Triangles

Ready for More

- ● Challenge 7.2 (printable)
- ● Interactive Challenge 7.2
- ● Illustrative Mathematics: Is This a Parallelogram?

View data-driven grouping recommendations and assign differentiation resources.

During the *Spark Your Learning,* listen and watch for strategies students use. See samples of student work on this page.

Reason with Congruent Triangles

Strategy 1

The triangles are congruent, so the corresponding parts are congruent. AB and DE are equal, so the distance across the river is 400 ft.

If students . . . reason with congruent triangles, they are demonstrating an understanding of Congruent Parts of Congruent Figures are Congruent from a previous lesson.

Have these students . . . explain how they determined their equation and how they solved it. **Ask:**

- Q How did you use the relationships between the parts of the triangles to conclude $AB = DE$?
- Q How do you know that $AB \neq CE$?

Measure Side Lengths

Strategy 2

I measured the side lengths of the triangles in the picture in my book and found that AB and DE are equal. The distance across the river is 400 ft.

If students . . . measure side lengths in the diagram of their textbook, they likely have an understanding of the concept of corresponding parts but may not understand that lengths in a diagram may not be to scale.

Activate prior knowledge . . . by having students reason about lengths without measuring. **Ask:**

- Q Can you always assume that drawings or diagrams are to scale and accurate?
- Q Since we know the triangles are congruent, can you relate side lengths another way?

COMMON ERROR: Uses the Wrong Corresponding Sides

The triangles are congruent, so the corresponding parts are congruent. Sides $\overline{AC}$ and $\overline{CE}$ are corresponding sides. AC and CE are then equal, so the distance across the river is 475 ft.

If students . . . use the incorrect corresponding sides, then they may not understand that the question is asking them to find the length of $\overline{DE}$.

Then intervene . . . by pointing out that the bridge will be located along $\overline{DE}$, because it will be the shortest length across the river. **Ask:**

- Q What is the shortest side of triangle CDE that still crosses the river?
- Q Which side of triangle ABC is the image of $\overline{DE}$?

7.2

Corresponding Parts of Congruent Figures

I Can use congruent figures to solve problems.

Spark Your Learning

As part of a river festival, organizers plan to build a temporary bridge across a river.

Complete Part A as a whole class. Then complete Parts B–D in small groups.

A. What is a mathematical question you can ask about this situation? What information would you need to know to answer your question?
A. What is the distance across the river?; need to know the lengths and angle measures and whether the triangles are congruent

B. What measurements are involved in this situation? What do you know?
the side lengths of the triangles; The distances and the triangles are congruent.

C. To answer your question, what strategy and tool would you use along with all the information you have? What answer do you get?
See Strategies 1 and 2 on the facing page.

D. Does your answer make sense in the context of the situation? How do you know?
yes; The corresponding parts of the triangles are congruent.

Turn and Talk Predict how your answer would change for each of the following changes in the situation:

- At a different point in the river, similar measurements are made, but angles B and D are obtuse and still congruent. If AC and BC do not change, what happens to AB?
- At a different point in the river, similar measurements are made, but angles B and D are acute and still congruent. If AC and BC do not change, what happens to AB? See margin.

SUPPORT SENSE-MAKING • Three Reads

Tell students to read the information in the photo three times and prompt them with a different question each time.

1. What is the situation about? The situation is about a bridge being built across a river. There are two congruent triangles, with the bridge as a side of one of the triangles.
2. What are the quantities in this situation? How are those quantities related? The quantities are the side lengths of the triangles and the measures of the triangles' angles; The triangles each have a right angle, so the measures of those angles are equal.
3. What are possible questions you could ask about this situation? Possible answer: How far is it across the river? Are the triangles congruent?

1 Spark Your Learning

▶ MOTIVATE

- Have students look at the photo in their books and read the information contained in the photo. Then complete Part A as a whole-class discussion.
- Give the class the additional information they need to solve the problem. This information is available online as a printable and projectable page in the Teacher Resources.
- Have students work in small groups to complete Parts B–D.

▶ PERSEVERE

If students need support, guide them by asking:

Q **Advancing • Use Tools** Which tool could you use to solve the problem? Why choose that tool and not some other? Students' choices of tools and reasons for choosing them will vary.

Q **Assessing** What do we know about the shapes and sizes of the triangles? The triangles are congruent, so they have the same shape and size.

Q **Assessing** Does it look like one triangle is the preimage and the other triangle the image of a transformation? If so, what is the transformation? yes; a rotation

Q **Advancing** How should the side and angle measures of the corresponding parts of the preimage and image of a rotation compare? The measures of all corresponding parts should be equal.

Turn and Talk Help students understand that when the triangles change as described, they may not maintain their relationship of congruence. Students may want to sketch the alternate scenarios or use geometry software to gain a greater understanding. It gets shorter; It gets longer.

▶ BUILD SHARED UNDERSTANDING

Select groups of students who used various strategies and tools to share with the class how they solved the problem. As they present their solutions, have each group discuss why they chose a specific strategy and tool.

2 Learn Together

Build Understanding

Task 1 (MP) **Use Tools** Students use cut-out representations of congruent quadrilaterals to explore the converse of Congruent Parts of Congruent Figures are Congruent.

Make sure students understand that the cut-out shapes are physical models of figures in the plane, and in moving the shapes around, they are modeling the actions of rigid motions on the figures.

Sample Guided Discussion:

- Q **Why can we think of one quadrilateral as a transformation of the other?** The cut-out congruent quadrilaterals represent figures in the plane, with one quadrilateral moved to become the other.
- Q **Is the transformation the result of at least one rigid motion? Explain.** yes; The shape and size of the first does not change when it is mapped to the other.
- Q **What do we know about the distance between the mappings of two points by a rigid motion?** The distance between the mappings of the points is equal to the distance between the preimage and the image.

Turn and Talk Help students find a way to describe why the two figures have the same perimeters and area by describing how to calculate each. Possible answer: Yes, because the corresponding side lengths are all equal, their sums will be equal; Possible answer: Yes, because the corresponding side lengths and heights are equal, the product of corresponding bases and heights will be equal.

Build Understanding

Identify the Corresponding Congruent Parts of Congruent Figures

In this lesson you focus on the converse of Corresponding Parts of Congruent Figures are Congruent. If you know two figures are congruent, you can use CPCFC to conclude the sides and the angles of the figures are congruent.

Converse of CPCFC: If two figures are congruent, then all pairs of corresponding sides and all pairs of corresponding angles are congruent.

1 Use the steps below to create and investigate congruent figures.
I can map one to the other using a rotation and a translation.

A. Fold a sheet of paper in half. Use a straightedge to draw a quadrilateral on the folded sheet. Then cut out the quadrilateral, cutting through both layers of paper. Label the quadrilaterals *ABCD* and *EFGH*. Use transformations to explain why the quadrilaterals are congruent.

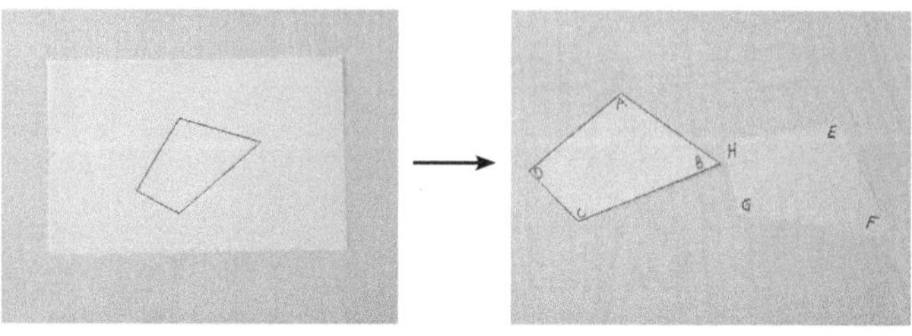

B. Write a congruence statement. Why is it better to start with a quadrilateral that has no symmetry or congruent sides? B–D. See Additional Answers.

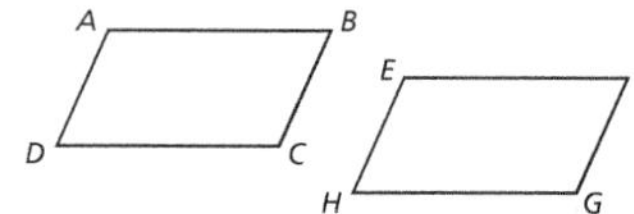

C. List the corresponding sides of the quadrilaterals above. What do you know about the corresponding side lengths? Justify your reasoning.

D. List the corresponding angles of the quadrilaterals above. What do you know about the corresponding angle measures? Justify your reasoning.

E. What can you conclude about the diagonals of the quadrilaterals?

E. The diagonals are congruent because they are corresponding lengths in the congruent figures.

Turn and Talk

- Can you conclude that the quadrilaterals have the same perimeter? Explain.
- Can you conclude that the quadrilaterals have the same area? Explain. See margin.

204

LEVELED QUESTIONS

Depth of Knowledge (DOK)	Leveled Questions	What Does This Tell You?
Level 1 **Recall**	If one figure was flipped, could you still identify corresponding parts? yes; A flip doesn't change the size or shape.	Students' answers will indicate whether they understand properties of a rigid motion.
Level 2 **Basic Application of Skills & Concepts**	If you rotate parallelogram *EFGH* about the center so that angle *G* is where angle *E* is located, what would be the corresponding angle and sides? $\overline{AB} \cong \overline{GH}$, $\overline{BC} \cong \overline{HE}$, $\overline{CD} \cong \overline{FE}$, $\overline{DA} \cong \overline{FG}$, $\angle A \cong \angle G$, $\angle B \cong \angle H$, $\angle C \cong \angle E$, $\angle D \cong \angle F$	Students' answers will indicate whether they understand they can apply CPCFC when figures are known to be congruent and can correctly pair congruent parts.
Level 3 **Strategic Thinking & Complex Reasoning**	If you reflected parallelogram *ABCD* across $\overline{BC}$ to get parallelogram *BADC*, would $ABCD \cong BADC$ be a correct congruence statement? No, since $\angle A$ is obtuse and $\angle B$ is acute, the vertices do not correspond in that order.	Students' answers will indicate whether they can reason strategically to determine pairs of corresponding parts resulting from transformations.

Step It Out

Use Congruent Corresponding Parts

If you know that figures are congruent, you can use that information to solve problems using corresponding lengths and angles within the figures.

Use the congruence statement to determine the corresponding parts of the triangles.

2 Find $m\angle S$ given that $\triangle HJK \cong \triangle RST$.

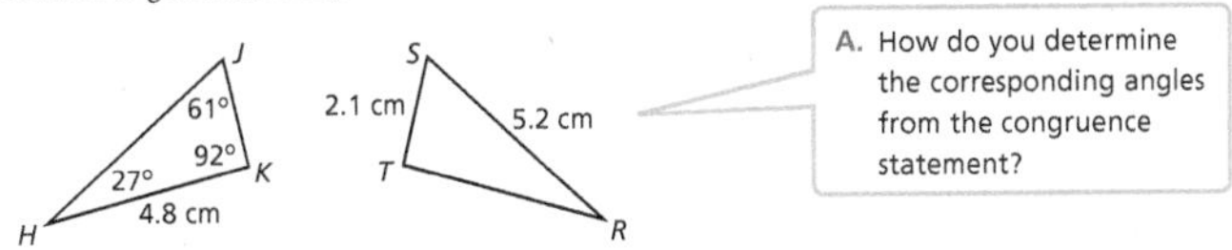

A. How do you determine the corresponding angles from the congruence statement?

Since $\triangle HJK \cong \triangle RST$, $\angle S \cong \angle J$. It is given that $m\angle J = 61°$, so $m\angle S = 61°$.

Turn and Talk For the triangles shown in Task 2, find $m\angle T$ and RT. See margin.

Prove a Geometric Relationship

When you need to prove relationships between two figures, one common strategy is to look for congruent figures and then use that congruence to prove corresponding segments or lengths are congruent.

3 Write a two-column proof.

Given: $\triangle WXY \cong \triangle YZW$

Prove: $\overline{XW} \parallel \overline{ZY}$

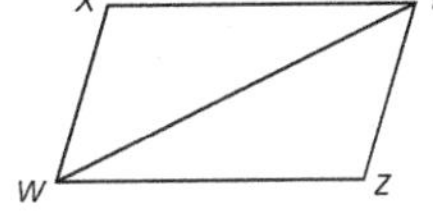

Statements	Reasons
1. $\triangle WXY \cong \triangle YZW$	1. Given
2. $\angle WXY \cong \angle YZW$	2. Corresponding parts of congruent figures are congruent.
3. $\overline{XW} \parallel \overline{ZY}$	3. Converse of the Alternate Interior Angles Theorem

A. Explain how you know $\angle XWY$ and $\angle ZYW$ are corresponding angles.
Use the congruence statements to determine the corresponding parts.

B. What is the Converse of the Alternate Interior Angles Theorem?
See Additional Answers.

Turn and Talk Use the proof in Task 3 as a template to prove $\overline{XY} \parallel \overline{ZW}$. See margin.

PROFICIENCY LEVEL

Beginning
Show students a pair of congruent triangles $\triangle ABC \cong \triangle DEF$ and say, "Angle *A* is congruent to angle *D*," pointing at each. One by one, name other parts of the triangles and have students say the names of congruent parts.

Intermediate
Have students work in pairs. Provide each pair with a pair of congruent triangles with some measures given and other measures missing. Have students take turns explaining how to determine missing measures.

Advanced
Have students explain how they can reason with transformations to determine congruent parts of congruent figures.

Step It Out

Task 2 (MP) **Construct Arguments** Students use congruence of triangles to determine unknown measures.

Sample Guided Discussion:

Q What kind of transformation of $\triangle HJK$ results in $\triangle RST$? a reflection

Q Is $\triangle HJK$ also the image of a rigid motion applied to $\triangle RST$? Explain. yes; a reflection across the same line

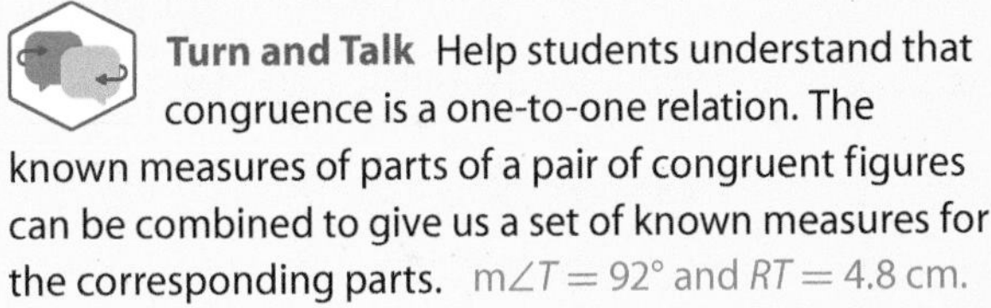

Turn and Talk Help students understand that congruence is a one-to-one relation. The known measures of parts of a pair of congruent figures can be combined to give us a set of known measures for the corresponding parts. $m\angle T = 92°$ and $RT = 4.8$ cm.

Task 3 (MP) **Use Structure** Students use the converse of CPCTC to prove a relationship between parts of triangles and apply their reasoning to a parallelogram.

Make sure students understand that triangles and other figures are often parts of an overall geometric structure and can be used to draw conclusions about the overall structure.

Sample Guided Discussion:

Q How can you use a transformation to determine corresponding pairs of sides and angles? One triangle is a reflection of the other, so we can visualize where points would go if we folded the quadrilateral on $\overline{WY}$.

Q What is meant by the converse of a theorem? We switch the "if" and "then" of the statement.

Turn and Talk Help students use similar reasoning to show that the quadrilateral is a parallelogram. Suggest they draw a new diagram so the diagonal $\overline{WY}$ is not there to distract from the parallelogram. Yes, because if the alternate interior angles are congruent, then the lines are parallel.

ONLINE

Assign the Digital On Your Own for
- built-in student supports
- Actionable Item Reports
- Standards Analysis Reports

On Your Own

Assignment Guide

The chart below indicates which problems in the On Your Own are associated with each task in the Learn Together. Assign daily homework for tasks completed.

Learn Together Tasks	On Your Own Problems
Task 1, p. 204	Problems 5, 6, 13, and 17
Task 2, p. 205	Problems 9–12 and 14
Task 3, p. 205	Problems 4, 7, 8, 15, and 16

Check Understanding

1. **MP Construct Arguments** Reggie claims the two figures shown are not congruent. Prove Reggie right or wrong.
See Additional Answers.

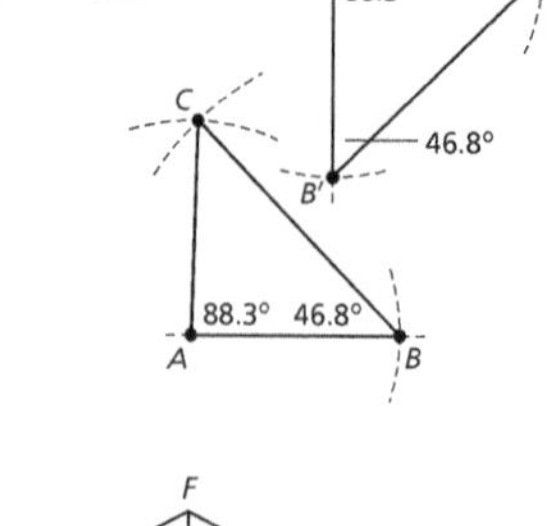

In Problems 2 and 3, write the indicated proof.

2, 3. See Additional Answers.

2. **Given:** $PQTU \cong QRST$

Prove: $\overline{QT}$ bisects $\overline{PR}$.

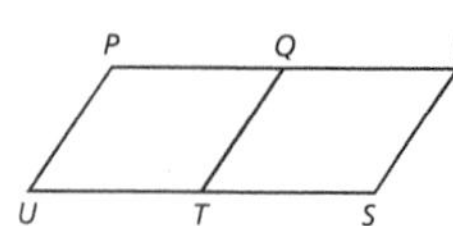

3. **Given:** $\triangle AFG \cong \triangle CAB$ and $\triangle DCG \cong \triangle CAB$

Prove: $\overline{FG} \cong \overline{CG}$

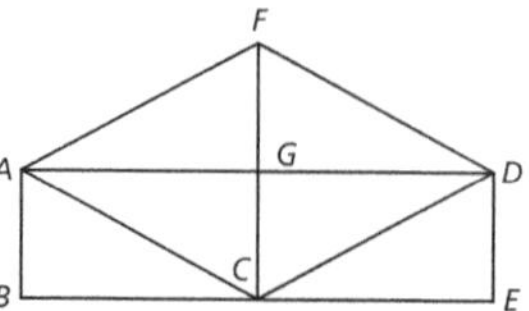

On Your Own

4. Write the indicated proof.
Given: $\triangle ABC \cong \triangle ADC$
Prove: $\overline{AC}$ bisects $\angle BAD$ and $\overline{AC}$ bisects $\angle BCD$

4–8. See Additional Answers.

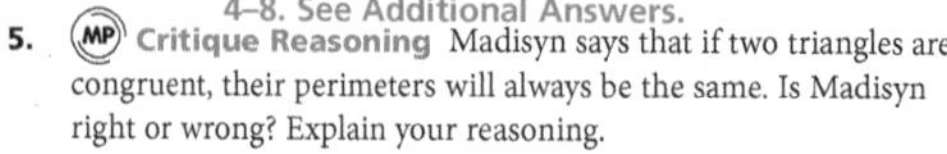

5. **MP Critique Reasoning** Madisyn says that if two triangles are congruent, their perimeters will always be the same. Is Madisyn right or wrong? Explain your reasoning.

6. If $\triangle ABC$ is a right triangle and $\triangle ABC \cong \triangle DEF$, is it possible for $\triangle DEF$ to not be a right triangle? Explain your reasoning.

7. $ABCD$ and $WXYZ$ are quadrilaterals.
A. Use transformations to prove $ABCD \cong XWZY$.
B. Use transformations to prove $\overline{AD} \cong \overline{XY}$.

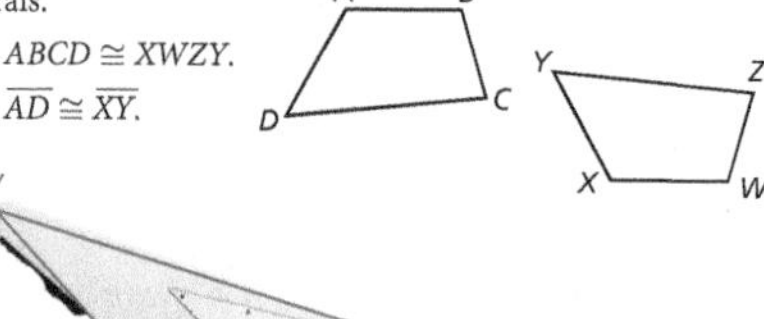

8. Write the indicated proof.
Given: $\triangle SVT \cong \triangle SWT$
Prove: $\overline{ST}$ bisects $\angle VSW$.

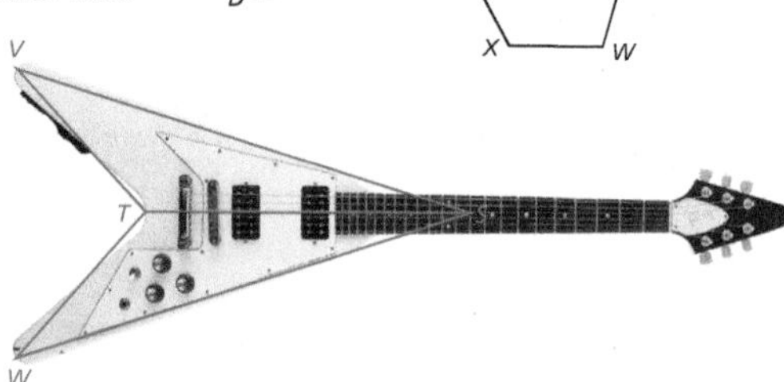

3 Check Understanding

Formative Assessment

Use formative assessment to determine if your students are successful with this lesson's learning objective.

Students who successfully complete the Check Understanding can continue to the On Your Own practice.

For students who miss 1 problem or more, work in a pulled small group using the Almost There small-group activity on page 203C.

ONLINE

Assign the Digital Check Understanding to determine
- success with the learning objective
- items to review
- grouping and differentiation resources

4 Differentiation Options

Differentiate instruction for all students using small-group activities and math center activities on page 203C.

Reteach

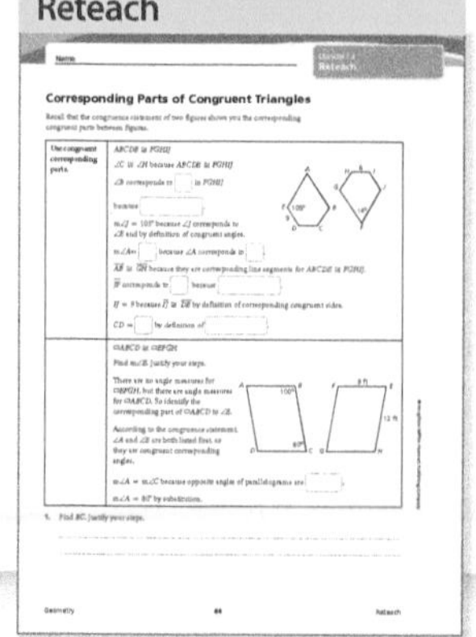

Challenge

In Problems 9–12, explain how you know whether the figures are congruent. Then find the indicated measure, if possible. 9–12. See Additional Answers.

9. w

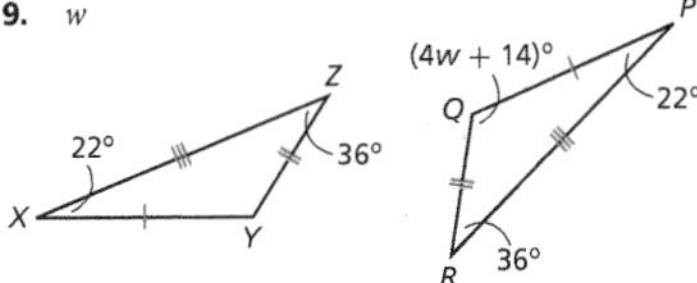

10. y

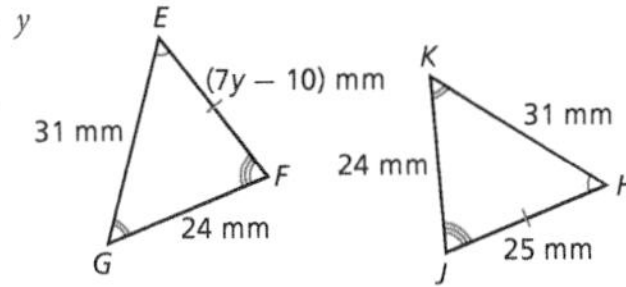

11. $m\angle V$

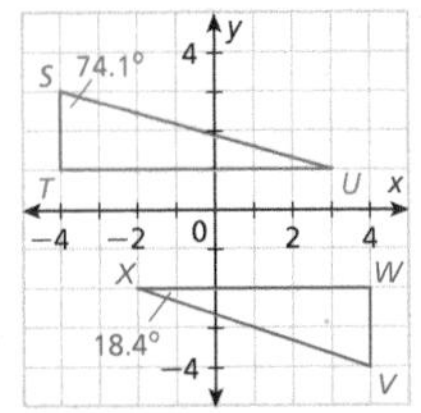

12. $m\angle C$

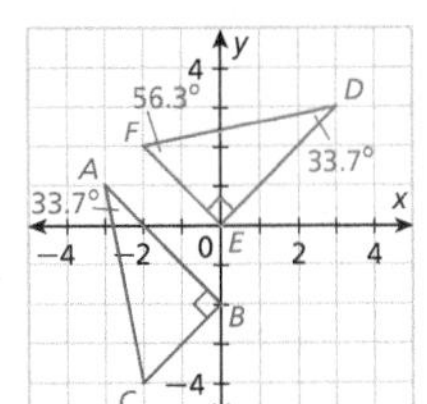

13. Which of the puzzle pieces could fit in the empty space? Explain how you know.
13, 14. See Additional Answers.

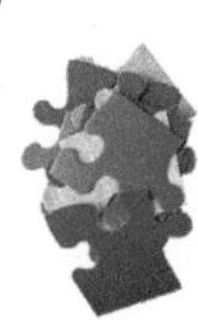

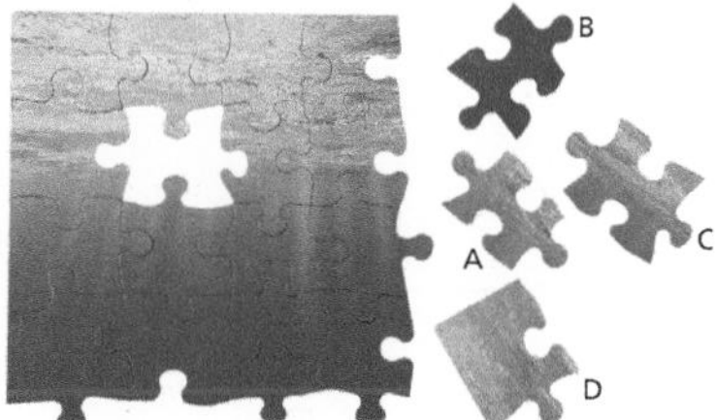

14. In the figure, $\triangle ABC \cong \triangle DEF$.

A. Find $m\angle D$.

B. Find AB.

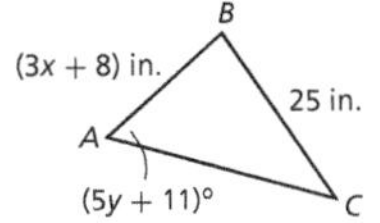

(6y + 2)°
F
D
(5x) in.
83°
E

In Problems 15 and 16, write the indicated proof.
15, 16. See Additional Answers.

15. Given: $\triangle STU \cong \triangle VTU$; $\overline{ST} \cong \overline{SV}$

Prove: $\triangle STV$ is equilateral.

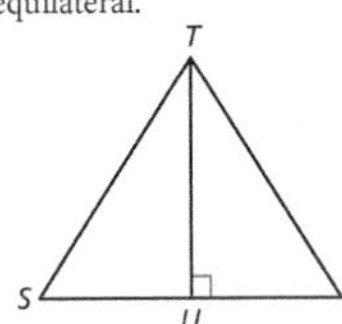

16. Given: $\triangle MNO$ and $\triangle QPR$ as marked

Prove: $\triangle MNO \cong \triangle QPR$

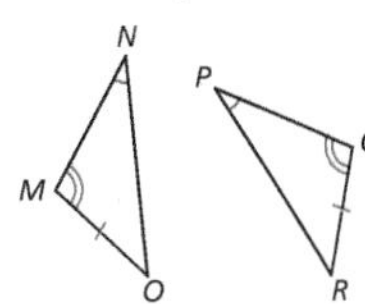

Watch for Common Errors

Problem 10 Students may set $7y - 10 = 31$ and solve for y since the sides $\overline{EF}$ and $\overline{KH}$ seem to be roughly in the same position on their respective triangles. Students should instead identify a pair of corresponding sides by seeing which pair of angles are at the vertices of $\overline{EF}$, and then matching those angles with their congruent corresponding angles in the other triangle.

Questioning Strategies

Problem 14 Suppose instead that $m\angle D = 5x$ and $DE = 6y + 2$. Is it possible to solve for $m\angle D$ and AB? Explain. yes; We would set $5y + 11 = 5x$ and $3x + 8 = 6y + 2$. Instead of a pair of linear equations with each in one variable, this is a system of linear equations in two variables:

$$\begin{cases} 5y + 11 = 5x \\ 6y + 2 = 3x + 8 \end{cases}$$

We can solve the second equation for x and then substitute for x in the other equation:

This gives us $AB = DE = 6\left(\frac{21}{5}\right) + 2 = \frac{126}{5} + 2 = \frac{136}{5}$.

We can then substitute for y in the first equation to solve for $5x$, which is the value of $m\angle D: 32° = 5x = m\angle D$.

5 Wrap-Up

Summarize learning with your class. Consider using the Exit Ticket, Put It in Writing, or I Can scale.

Exit Ticket 

Suppose we can fold $\triangle ABC$ in half along a segment through C and another point D that lies on $\overline{AB}$ so points A and B lie on top of each other. Prove that $\angle ACD \cong \angle BCD$. Since we can fold $\triangle ABC$ in this way, the triangles $\triangle ACD$ and $\triangle BCD$ are reflections of each other, so $\triangle ACD \cong \triangle BCD$. This means corresponding parts of the triangles are congruent, so $\angle ACD \cong \angle BCD$.

Put It in Writing 

Describe how to use transformations to find corresponding parts of congruent figures.

I Can 

The scale below can help you and your students understand their progress on a learning goal.

4	I can use congruent figures to solve problems, and I can explain the process to others.
3	I can use congruent figures to solve problems.
2	I can write a congruence statement for two figures and tell which parts of the figures are congruent.
1	I can list corresponding parts of congruent figures.

Spiral Review • Assessment Readiness

These questions will help determine if students have retained information taught in the past and can also prepare them for high-stakes assessments. Here, students must find the value of a variable in expressions for corresponding angle measures that makes two figures congruent **(7.1)**, solve for a variable in expressions for corresponding side lengths when figures are assumed congruent **(7.3)**, and identify transformations by a composition of transformation **(6.2)**.

17. Gardening Marissa's design has multiple sections that should be identical, but she's not sure whether they are. See Additional Answers.

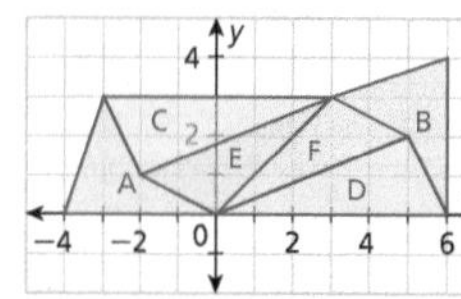

A. Prove that the figures A and B are congruent by mapping one to the other using transformations.

B. Will the same transformations as you used in Part A map figure C to figure D? If not, describe a different set of transformations you can use to prove C is congruent to D.

Spiral Review • Assessment Readiness

18. What value of the variable makes the figures congruent?

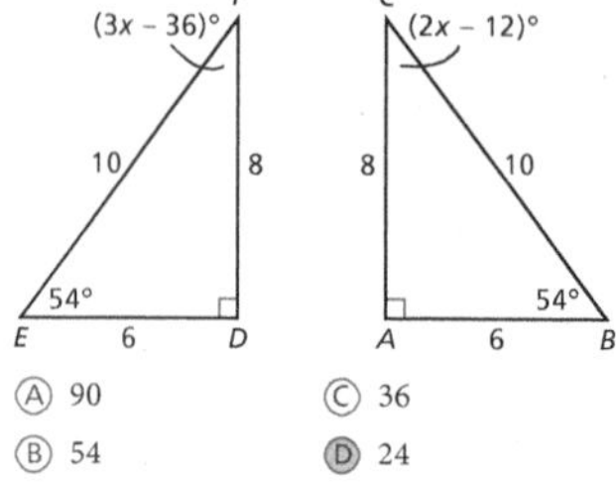

Ⓐ 90
Ⓑ 54
Ⓒ 36
Ⓓ 24

19. What is the length of $\overline{LM}$ if $GHJK \cong LMNP$?

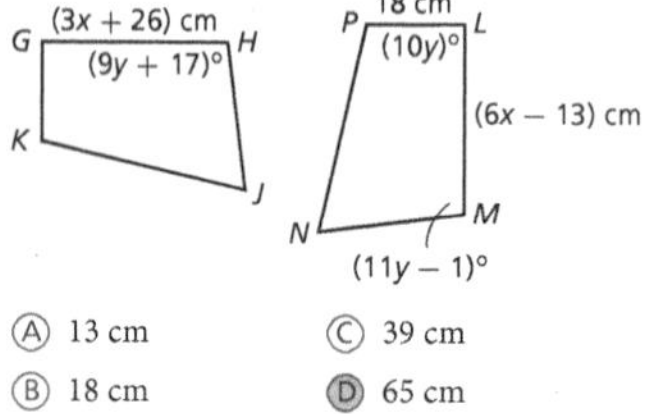

Ⓐ 13 cm
Ⓑ 18 cm
Ⓒ 39 cm
Ⓓ 65 cm

20. Match the rule on the left with the type of transformation on the right.

A. $(x, y) \rightarrow \left(\frac{1}{3}x, \frac{1}{3}y\right)$ — **1.** Dilation — A. 1

B. $(x, y) \rightarrow (x + 3, y + 3)$ — **2.** Reflection — B. 4

C. $(x, y) \rightarrow (y, -x)$ — **3.** Rotation — C. 3

D. $(x, y) \rightarrow (-x, y)$ — **4.** Translation — D. 2

 **I'm in a Learning Mindset!**

How can I give and receive feedback about the meaning of congruent figures?

208

Keep Going to Journal and Practice Workbook

Learning Mindset

Resilience Responds to Feedback

Point out that good listening skills are critical in both giving and receiving feedback. When discussing congruent figures, students should be specific instead of vague when referring to sides and angles. At the same time, it's important for students to first listen carefully when receiving feedback instead of blurting out more questions if something isn't clear. Careful listening can help us slow down and think more deeply about congruency questions. *Can taking care to be specific when giving feedback help you understand congruent figures more deeply? How does slowing down and thinking carefully about feedback you've received put you in a learning mindset?*

7.3 Use Rigid Motions to Prove Figures Are Congruent

LESSON FOCUS AND COHERENCE

Mathematics Standards

- Use geometric descriptions of rigid motions to transform figures and to predict the effect of a given rigid motion on a given figure; given two figures, use the definition of congruence in terms of rigid motions to decide if they are congruent.
- Use the definition of congruence in terms of rigid motions to show that two triangles are congruent if and only if corresponding pairs of sides and corresponding pairs of angles are congruent.

Mathematical Practices and Processes 

- Construct viable arguments or critique reasoning of others.
- Look for and make use of structure.

I Can Objective 

I can use rigid motions to show that figures are congruent.

Learning Objective

Use the definition of congruence in terms of rigid motions to determine if two given figures are congruent.

Language Objective

Explain how you know if an image is congruent to its preimage and describe transformations that resulted in the image.

Mathematical Progressions

Prior Learning	Current Development	Future Connections
Students: • represented and described transformations in the plane as functions that take points in the plane as inputs and give other points as outputs, compared transformations that preserve distance and angle to those that do not. **(5.1 and 5.2)** • developed definitions of rotations, reflections, and translations. **(5.1 and 5.2)** • used geometric descriptions of rigid motions to transform figures and predicted the effect of a given rigid motion on a given figure. **(5.1 and 5.2)**	**Students:** • prove congruence using rigid transformations. • use transformation notation to describe a transformation.	**Students:** • will write coordinate proofs about triangle relationships. **(9.4 and 9.5)** • will write coordinate proofs about parallelograms. **(11.1–11.5)**

PROFESSIONAL LEARNING

Visualizing the Math

Students may benefit from determining congruence by inspection, such as the problem in Task 1 on page 210. When provided several triangles, they can predict which are congruent just by looking at them and use tracing paper to verify congruence.

WARM-UP OPTIONS

ONLINE

PROJECTABLE & PRINTABLE

ACTIVATE PRIOR KNOWLEDGE • Sequences of Transformations

Use these activities to quickly assess and activate prior knowledge as needed.

Problem of the Day 

A triangle is drawn on a coordinate plane in Quadrant 1. After a sequence of two transformations, the triangle's image lies in Quadrant 3 and is congruent to the preimage. What could have been the sequence of transformations performed on the original triangle? Answers will vary. Possible answer: The triangle was rotated 90° clockwise and then reflected over the *y*-axis.

Quick Check for Homework

As part of your daily routine, you may want to display the Teacher Solution Key to have students check their homework.

Make Connections

Based on students' responses to the Problem of the Day, choose one of the following:

 Project the Interactive Reteach, Grade 8, Lesson 1.5.

 Complete the Prerequisite Skills Activity:

Give each student a sheet of grid paper. On the top half, ask students to draw $\triangle ABC$ and $\triangle A'B'C'$, the image of $\triangle ABC$ after a transformation of their choice. Ask students to describe the transformation they performed using precise mathematical language or symbols and then cut the paper to separate the image and preimage. Mix up the figures and hand one to each student. Ask students to find the student with the matching figure, reminding them that they may not be congruent any longer.

- *Is the image the same size?* Answers will vary.
- *Which transformations result in images that are the same size?* rotations, reflections, and translations
- *Did the image get smaller or larger?* Answers will vary.
- *Which transformations result in an image that is bigger or smaller?* dilations

If students continue to struggle, use Tier 2 Skill 13.

SHARPEN SKILLS

If time permits, use this on-level activity to build fluency and practice basic skills.

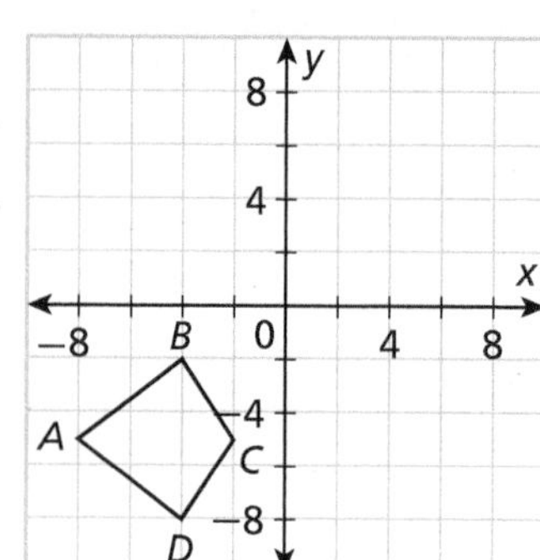

Mental Math

Objective: Students demonstrate an understanding of congruent figures.
Materials: index cards

Have students work in small groups. Display a figure on a coordinate grid and the description of the following series of transformations: a rotation of 180°, followed by a reflection over the *x*-axis, followed by a translation 4 units down and 3 units up.

Have students discuss the characteristics of the final image without actually drawing the image. Ask questions such as: Where does the figure land after the 1st transformation? after the 2nd? after the 3rd? Is the final image congruent? How do you know?

PLAN FOR DIFFERENTIATED INSTRUCTION

MTSS (RtI)

Small-Group Options

Use these teacher-guided activities with pulled small groups.

On Track

Materials: index cards and geometry software or coordinate planes

Give students an index card that shows the coordinates of two congruent figures. Have each student write down a sequence of two rigid transformations that map the first figure onto the second figure. Have them use geometry software or coordinate planes to perform each sequence to verify that a congruent image is created.

Figure 1:	$J(6, 2)$ $K(8, 12)$ $L(6, 6)$ $M(-2, 6)$
Figure 2:	$W(2, -4)$ $X(4, -4)$ $Y(2, -8)$ $Z(-6, -8)$

Almost There (RtI)

Materials: worksheet and colored pencils or markers

Give each group a worksheet that displays two congruent figures. Have students take turns circling corresponding angles in the same color, using a different color for each pair. Have them do the same for corresponding sides by tracing corresponding pairs of side lengths. Then, have the group come up with rigid motion(s) that mapped one figure to the other.

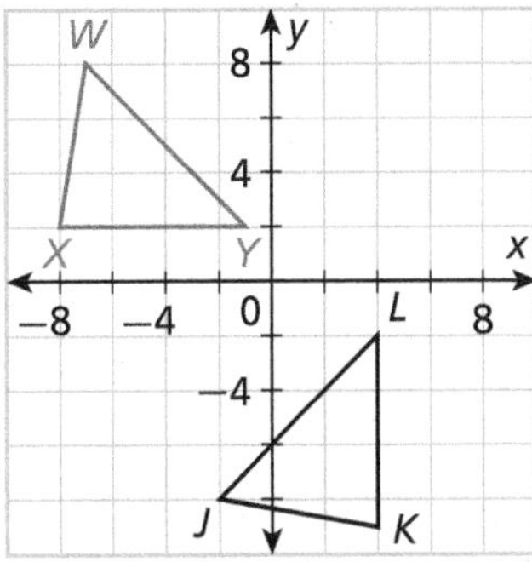

Ready for More

Materials: grid paper and markers

Provide groups with notation describing a sequence of two or more transformations. Have the group create a diagram of two figures, *A* and *B*, where *B* is the image of *A* after performing the sequence of transformations described. Have groups present their work to the class, including an explanation of what each notation means and how each is shown on their paper, as well as why the sequence is an example of a rigid motion.

1. $(x, y) \rightarrow (-x, -y)$

2. $(x, y) \rightarrow (x-4, y)$

3. $(x, y) \rightarrow (-x, y)$

Math Center Options

Use these student self-directed activities at centers or stations. Key: ● Print Resources ● Online Resources

On Track

- ● Interactive Digital Lesson
- ●● Journal and Practice Workbook

Standards Practice:

- ●● Determine if Two Figures, One a Transformation of the Other, are Congruent
- ●● Show that Two Triangles are Congruent
- ● Module Performance Task

Almost There

- Reteach 7.3 (printable)
- Interactive Reteach 7.3
- RtI Tier 2 Skill 13: Sequences of Transformations
- Illustrative Mathematics: Triangle Congruence with Coordinates

Ready for More

- Challenge 7.3 (printable)
- Interactive Challenge 7.3
- Illustrative Mathematics: Building a Tile Pattern by Reflecting Hexagons
- Desmos: Transformation Golf: Rigid Motion

Unit Project Check students' progress by asking for the coordinates of the vertices for each new piece of the window.

View data-driven grouping recommendations and assign differentiation resources.

During the *Spark Your Learning*, listen and watch for strategies students use. See samples of student work on this page.

Use a Rotation

Strategy 1

Because there will need to be 1 complete heart in each identical slice and there are 12 hearts on the pie top, then the greatest number of slices will be 12 slices. Each slice will be $\frac{1}{12}$ of the full pie, and $\frac{1}{12} \cdot 360° = 30°$ per slice. Rotating a slice multiples of 30° about the center of the pie shows that the slice exactly matches every other slice, which means that the slices will all be identical.

If students . . . use rotations to solve the problem, they are showing a solid understanding of the transformations that create congruent images from Module 5.

Have these students . . . explain how they determined that the figures are congruent. **Ask:**

- Q How did you use the given information and what you know about transformations to solve the problem?
- Q What transformation did you use to prove congruence?

Match Parts

Strategy 2

Because there will need to be 1 complete heart in each identical slice and there are 12 hearts on the pie top, then the greatest number of slices will be 12 slices. Every slice will have two edges that are identical radii of the circular pie. Each slice will have 1 complete heart. Each slice will have a 30° angle at its point $\left(\frac{1}{12} \cdot 360° = 30°\right)$. Each slice will have a curved edge that is the same fractional part of the circumference of the circle. Since all parts of one slice can be matched to corresponding parts of every other slice, the slices will all be identical.

If students . . . match parts to solve the problem, they understand how to show a shape is congruent to another shape but may not have a solid understanding of the transformations that create congruent figures.

Activate prior knowledge . . . by having students recall prior knowledge about transformations. **Ask:**

- Q What type of transformations can result in a congruent figure?
- Q What transformation moves a figure around a point, such as the center of the pie?

COMMON ERROR: Does Not Use Enough Information

In order to know if two figures are congruent, you must know the side lengths and angle measures. Since we are not given the dimensions of the triangles in the photo, we cannot assume that they are congruent. So, there is not enough information to answer this question.

If students . . . believe there is not enough information given to solve the problem, then they do not recall that concepts about transformations can be used to solve the problem and are depending on dimensions alone.

Then intervene . . . by pointing out that you can solve this problem without knowing the actual angle measures and side lengths. **Ask:**

- Q What types of transformations result in congruent figures?
- Q How can we use transformations to help us get to an answer in this problem?

7.3

Use Rigid Motions to Prove Figures Are Congruent

(I Can) use rigid motions to show that figures are congruent.

Spark Your Learning

Phyl is teaching a workshop on making a fancy pie crust with a repeated design of hearts for the top of a cherry pie. Each serving of the pie should be identical.

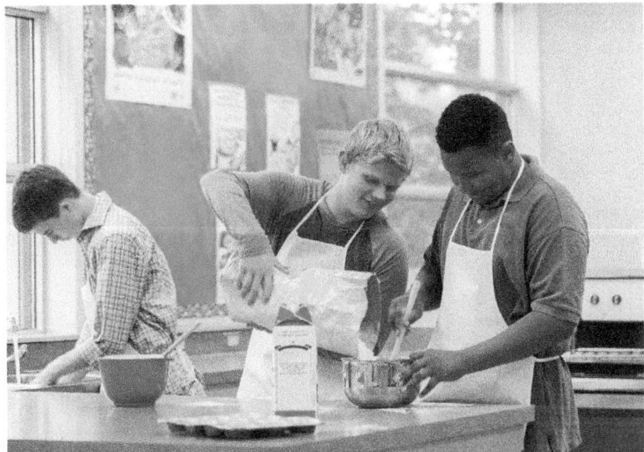

Complete Part A as a whole class. Then complete Parts B–D in small groups.

A. What is a mathematical question you can ask about this situation? What information would you need to know to answer your question?

A. How can you cut the greatest number of identical servings from the pie?; need to know what the design looks like

B. What assumptions must you make about the hearts to answer the question? See Additional Answers.

C. To answer your question, what strategy and tool would you use along with all the information you have? What answer do you get? See Strategies 1 and 2 on the facing page.

D. Does your answer make sense in the context of the situation? How do you know? See Additional Answers.

Turn and Talk What if there had been a heart shape placed at the center of the pie top in addition to the shapes around the edge? How many identical servings could you make then? See margin.

CULTIVATE CONVERSATION • Information Gap

Ask students questions to help them decide what information they need to answer the question, "How can you cut the greatest number of identical servings from the pie?"

Do you have enough information to determine how many identical servings can be made from a pie in the workshop? no; I would need to know how many hearts are on the pie and how they are to be laid out on the top of the pie.

If one serving of pie included one heart and another serving had no hearts, could those servings be considered identical? no; The design on the top of each serving of pie must look identical for the servings to be considered identical.

What is the shape of a full pie? a circle What polygon does a slice of pie resemble? a triangle What would be true about each of the straight sides of a serving? They would each be a radius of the circular pie, so all congruent.

1 Spark Your Learning

▶ MOTIVATE

- Have students look at the photo in their books and read the information contained in the photo. Then complete Part A as a whole-class discussion.
- Give the class the additional information they need to solve the problem. This information is available online as a printable and projectable page in the Teacher Resources.
- Have students work in small groups to complete Parts B–D.

▶ PERSEVERE

If students need support, guide them by asking:

Advancing • Use Tools Which tool could you use to solve the problem? Why choose that tool and not some other? Students' choices of tools and reasons for choosing them will vary.

Assessing What does it mean when two shapes are congruent? Corresponding sides have equal lengths, and corresponding angles have equal angle measures.

Assessing What transformations result in images that are congruent to preimages? reflections, translations and rotations

Advancing What is the center of rotation for the pie? The triangles are rotated about the center of the pie.

Turn and Talk When considering the response to this question, students should discuss how an additional design element affects the transformations needed to match up slices. one; Possible answer: No cuts are possible. The heart shape does not have rotational symmetry, so placing a heart shape at the center of the pie makes it impossible to cut the pie and have slices that exactly match when rotated. So, the pie will have one slice — the whole pie.

▶ BUILD SHARED UNDERSTANDING

Select groups of students who used various strategies and tools to share with the class how they solved the problem. As they present their solutions, have each group discuss why they chose a specific strategy and tool.

② Learn Together

Build Understanding

Task 1 (MP) **Construct Arguments** Students use prior knowledge of transformations to make conjectures and build a logical progression of statements to justify their conjectures.

Sample Guided Discussion:

Q What four transformations have you already learned about? reflections, rotations, translations, and dilations

Q In Part C, what transformation is being performed to create the image? a dilation

Turn and Talk Help students recall what is meant by orientation and that it refers to the position of the angles of a figure and how the figure is placed in the space that it occupies. They should be remembering which transformations are rigid motions. No, changing the orientation happens when a figure is reflected. Reflections are rigid motions and rigid motions do not change the size of a figure.

Build Understanding

Determine Whether or Not Figures Are Congruent

Recall that two figures are congruent if and only if one can be obtained from the other by a sequence of rigid motions that may include reflections, translations, and rotations.

A–C. See Additional Answers.

1 A. Use tracing paper to trace *ABCD*. Then move the tracing paper so that *ABCD* is mapped onto *EFGH*. Explain the translation used to map *ABCD* onto *EFGH*. How do you know that *ABCD* is congruent to *EFGH*?

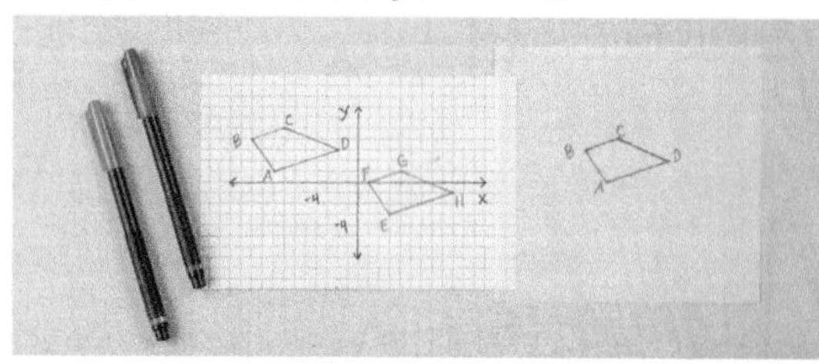

B. Trace *JKLM* and *NPQR*. Fold the paper so that *JKLM* is mapped onto *NPQR*. What transformation is used? Is *JKLM* congruent to *NPQR*? How do you know?

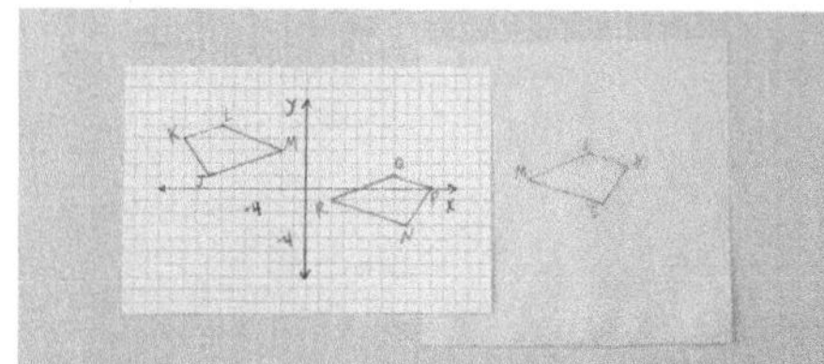

C. Is there a rigid transformation that maps *STUV* to *WXYZ*? Is *STUV* congruent to *WXYZ*? How do you know?

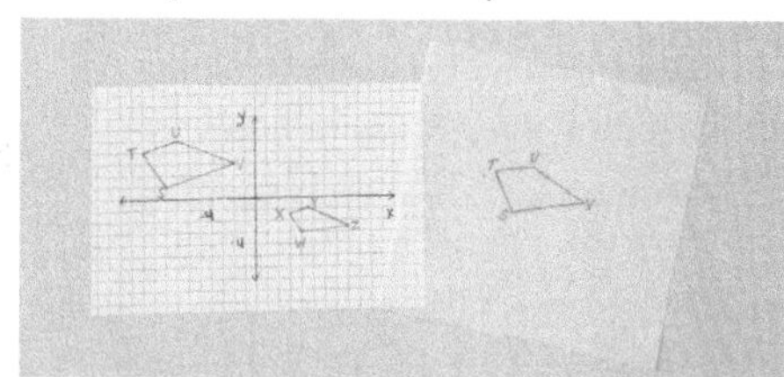

D. How do the sizes of the pairs of figures help determine if they are congruent? If both figures are not the same size, they are not congruent.

 **Turn and Talk** Does changing the orientation of a figure affect its size and shape? Use the figures above to support your answer. See margin.

210

LEVELED QUESTIONS

Depth of Knowledge (DOK)	Leveled Question	What Does This Tell You?
Level 1 **Recall**	A figure is reflected over the *y*-axis. Does the size of the image change? No, a reflection preserves side length.	Students' answers will indicate whether they understand the result of a reflection in terms of congruence.
Level 2 **Basic Application of Skills & Concepts**	A figure is reflected over the *x*-axis and then dilated by a factor of 2. Is this sequence of transformations considered to be that of rigid motions? No, a dilation changes the size of a figure, so this sequence is not considered rigid motions.	Students' answers will indicate whether they understand the characteristics of rigid transformations.
Level 3 **Strategic Thinking & Complex Reasoning**	If a series of rigid motions includes a rotation, reflection, translation, and a dilation, is it possible for the image to be mapped to the preimage? This is only possible if the dilation has a scale factor of 1, but all other transformations in the series could potentially result in the image being mapped to the preimage.	Students' answers will indicate if they understand the difference between transformations that are rigid motions and those that are not, as well as being able to explain when a series of transformation results in a figure mapping to itself.

Step It Out

Find a Sequence of Rigid Motions

2 In the diagram, $JKLM \cong WXYZ$. Find a sequence of rigid motions that maps one figure onto the other.

Make a table of coordinates to look for a pattern.

$J(4, 1)$	$W(2, -2)$
$K(3, 1)$	$X(1, -2)$
$L(-1, 3)$	$Y(-3, -4)$
$M(3, 3)$	$Z(1, -4)$

A. What type of transformation is suggested by the signs of the *y*-coordinates?
A. See Additional Answers.

B. What is suggested by the differences between each pair of *x*- and *y*-coordinates?

Map *JKLM* to *WXYZ* with a reflection across the *x*-axis, followed by a translation.

B. The *x*-coordinates all change such that $x \to x - 2$ and the *y*-coordinates all change such that $y \to -y - 1$. This suggests a reflection across the *x*-axis, followed by a translation left 2 and down 1.

Reflection: $(x, y) \to (x, -y)$

Translation: $(x, y) \to (x - 2, -y - 1)$

Turn and Talk What translation would you need to use if you wanted to translate *JKLM* before applying the reflection? See margin.

3 In the diagram, $\triangle KLH \cong \triangle DFG$. Find a sequence of rigid motions that maps $\triangle DFG$ to $\triangle KLH$.

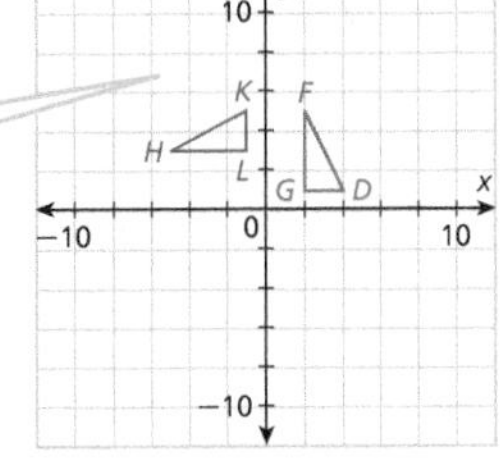

A. Based on the diagram, what transformations should be performed?
a rotation and a translation

Two transformations that map *DFG* to *KLH* are shown.

Transformation 1: 90° rotation counterclockwise around the origin

Transformation 2: translation up 1 unit

B. The coordinate notation for each transformation is shown. Which notation describes Transformation 1? Which notation describes Transformation 2?

$(x, y) \to (x, y + 1)$ Transformation 1: $(x, y) \to (-y, x)$

$(x, y) \to (-y, x)$ Transformation 2: $(x, y) \to (x, y + 1)$

Turn and Talk Describe a sequence of transformations that map *KLH* to *DFG*. See margin.

PROFICIENCY LEVEL

Beginning
Have students give examples of congruent figures in the classroom. Students might mention rectangular desk tops or floor tiles that have the same size and shape.

Intermediate
Have students work in pairs and display a figure with vertices labeled. Pairs copy the figure onto their graph paper. Have one student state a sequence of transformations and the other verbally predict where the image will lie and how it will be oriented. Then they can perform the sequence to check. Switch roles.

Advanced
Have students describe a sequence of transformations that will result in an image that is the same as the preimage no matter in which order the transformations are performed.

Step It Out

Task 2 (MP) **Use Structure** Students look closely at described transformations to determine a pattern or structure and have the facility to shift perspectives.

Sample Guided Discussion:

Q What happens to the *x*- and *y*-coordinates when a reflection is performed on a figure over the *x*-axis or *y*-axis? If it is a reflection over the *x*-axis, then the *y*-coordinate changes to its opposite. If it is a reflection over the *y*-axis, then the *x*-coordinate changes to its opposite.

Q In Part B, how can you tell a translation was performed? A translation adds or subtracts the same number of units from either the *x*- and *y*-coordinates depending on the direction of the translation.

Turn and Talk Students may not understand how to express transformations using mapping rules. Suggest students try finding the correct sequence of transformations on graph paper first. left 2 and up 1

Task 3 (MP) **Use Structure** Students will utilize the properties of transformations to determine how images are created and understand how to name those transformations using expressions.

Sample Guided Discussion:

Q Does it look like the figure flipped, turned, slid, enlarged, or shrunk? It looks like the figure was turned and slid.

Q In Part B, how is the notation for a rotation different than the notation for a translation? The notation for a translation adds or subtracts a number from the *x*- and *y*-coordinates, and the notation for a rotation totally changes the *x*- and *y*-coordinates of each vertex.

Turn and Talk Discussing how changing the order of the transformations in a sequence either changes or does not change the image is important. Allow students time to try changing the order using graph paper so they can visually identify the changes. Possible answer: No, instead of translating left one unit you would need to translate down one unit. Translate the figure down 1 unit, then rotate 90° clockwise about the origin.

ONLINE

Assign the Digital On Your Own for
- built-in student supports
- Actionable Item Reports
- Standards Analysis Reports

On Your Own

Assignment Guide

The chart below indicates which problems in the On Your Own are associated with each task in the Learn Together. Assign daily homework for tasks completed.

Learn Together Tasks	On Your Own Problems
Task 1, p. 210	Problems 3, 12–13, and 15–17
Task 2, p. 211	Problems 8–11 and 14
Task 3, p. 211	Problems 4–7

Questioning Strategies

Problem 3 Suppose you are given a statement like the one Marta is given. How does the congruence of angles and segments relate to the congruence of two figures? Since rigid motions preserve angle measure and distance, verifying that corresponding angles and segments have the same measure determines whether the two figures are congruent.

Check Understanding

1, 2. See Additional Answers.

1. Suppose you are given this statement: If *ABCD* and *EFGH* are congruent, then *ABCD* can be mapped onto *EFGH* using a rotation and a translation. Determine whether or not this statement is true or false. Then explain your reasoning.

2. The figures shown are congruent. Find a sequence of transformations to map *PQRST* onto *DEFGH*. Give coordinate notation for the transformations you use.

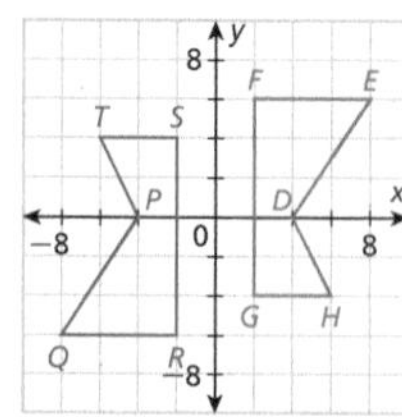

On Your Own

3. 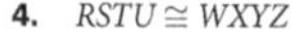 **Critique Reasoning** Marta says that given the statement $ABCD \cong EFGH$, she can write all the side and angle congruence statements for the quadrilaterals without a diagram. Is Marta correct? Explain your answer. yes; The congruence statement determines the corresponding parts of the congruent figures.

In Problems 4–7, write a sequence of rigid motions that maps the first figure in the congruence statement to the second. Use coordinate notation to write the transformations. 4–7. See Additional Answers.

4. $RSTU \cong WXYZ$

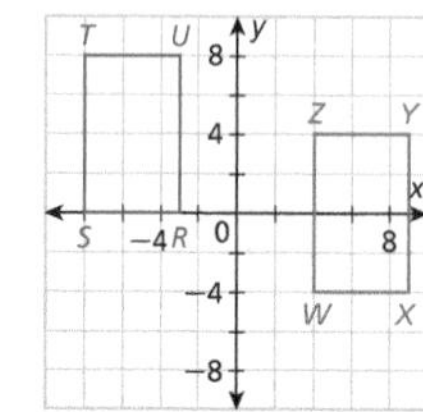

5. $\triangle ABC \cong \triangle DEF$

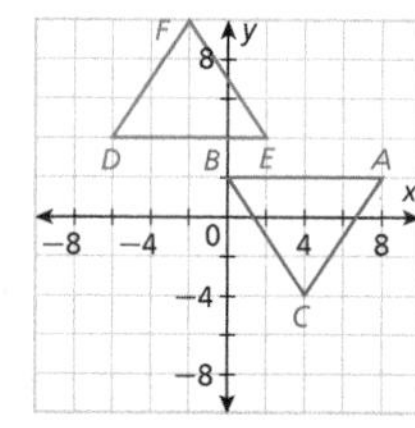

6. $DEFGH \cong PQRST$

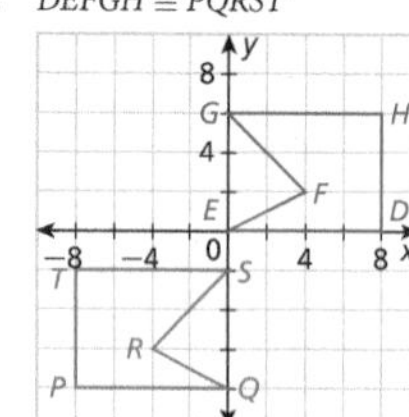

7. $\triangle WXY \cong \triangle CED$

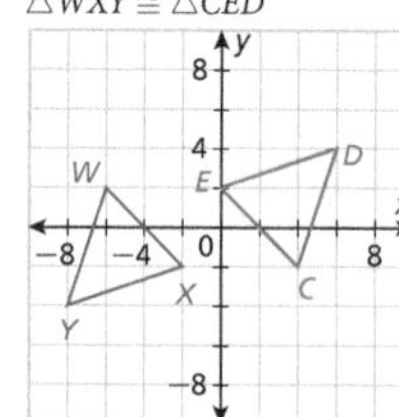

212

data checkpoint

3 Check Understanding

Formative Assessment

Use formative assessment to determine if your students are successful with this lesson's learning objective.

Students who successfully complete the Check Understanding can continue to the On Your Own practice.

For students who miss 1 problem or more, work in a pulled small group using the Almost There small-group activity on page 209C.

ONLINE

Assign the Digital Check Understanding to determine
- success with the learning objective
- items to review
- grouping and differentiation resources

4 Differentiation Options

Differentiate instruction for all students using small-group activities and math center activities on page 209C.

Reteach

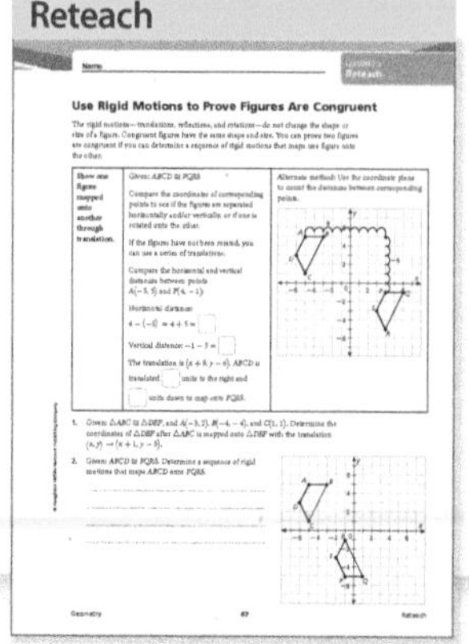

Challenge

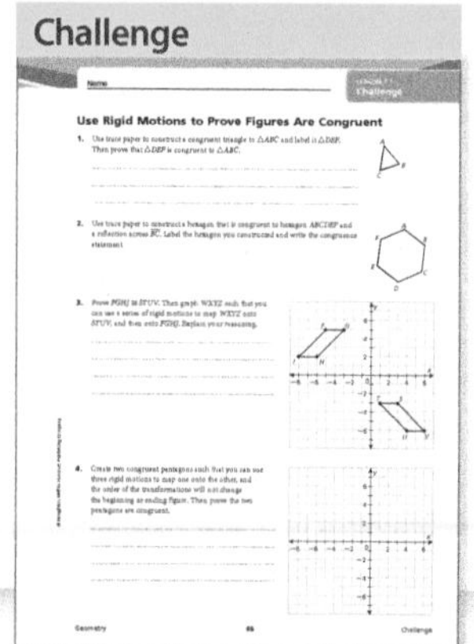

In Problems 8–11, use the definition of congruence to decide whether or not the two figures are congruent. Explain your answer using coordinate notation for any transformations you use. 8–11. See Additional Answers.

8.

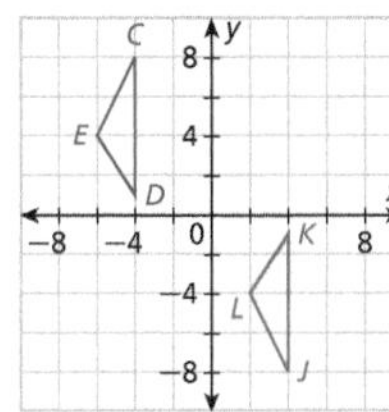

9.

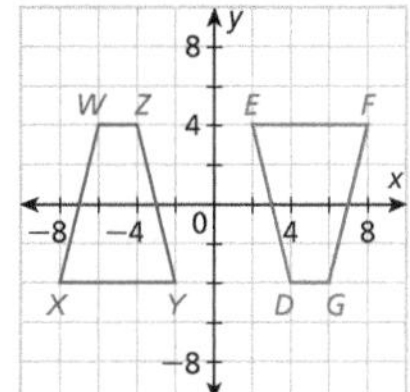

10.

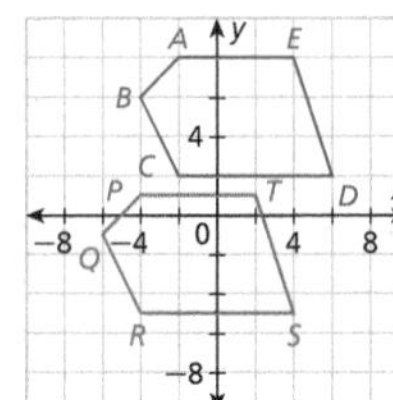

11.

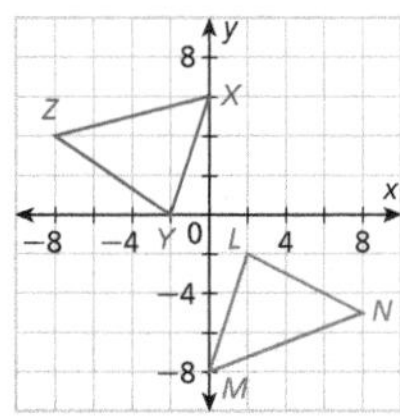

In Problems 12 and 13, describe the transformations that can be used to show that the shapes in each logo are congruent. 12, 13. See Additional Answers.

12.

13.

14. **Justify Reasoning** Two students are each trying to show the two figures are congruent. The first student maps *ABCDE* onto *FGHIJ* using a rotation of 180° about the origin, followed by a translation 4 units up. The second student uses the rule $(x, y) \to (-x, -y + 4)$. Are both students correct? If not, explain which student made an error and how to fix their work.
Both students are correct.

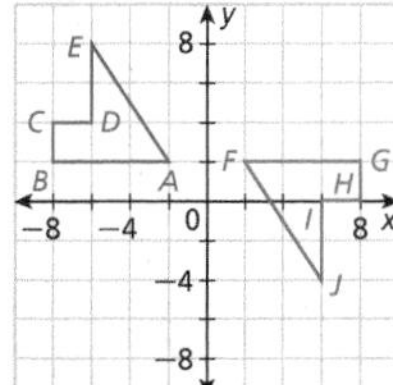

Watch for Common Errors

Problem 8 Some students may think that if two figures are congruent, then there is one type of rigid motion that maps one figure to the other. Explain that sometimes more than one type of sequence of rigid motions can be used map one figure to its congruent image.

5 Wrap-Up

Summarize learning with your class. Consider using the Exit Ticket, Put It in Writing, or I Can scale.

Exit Ticket

Explain in words what the following sequence of transformations does to a preimage and if the resulting image is congruent.

1. $(x, y) \rightarrow (-x, y)$ 2. $(x, y) \rightarrow (x - 1, y + 2)$

The first reflects the figure over the x-axis. The second translates the reflected image 1 unit left and 2 units up. The image is congruent to the preimage because both reflections and translations are rigid motions.

Put It in Writing

Have students explain how they know if two figures are congruent using the definition of rigid motion. Prompt students to provide an example and a non-example of transformations that create congruent images.

I Can

The scale below can help you and your students understand their progress on a learning goal.

4	I can use rigid motions to show that figures are congruent, and I can explain my solution steps to others.
3	I can use rigid motions to show that figures are congruent.
2	I can determine which rigid motion is performed when given a diagram of a preimage and its image.
1	I can examine figures shown a coordinate grid to and determine if they are congruent.

Spiral Review • Assessment Readiness

These questions will help determine if students have retained information taught in the past and can also prepare them for high-stakes assessments. Here, students must identify steps to a triangle congruence proof **(7.2)**, identify which transformation is performed when given the preimage and image **(7.3)**, and show an understanding of triangle congruence theorems **(Gr8, 1.5)**.

In Problems 15 and 16, find a sequence of transformations for the indicated mapping. Give coordinate notation for the transformations you use.

15, 16. See Additional Answers.

15. Map *PQRSTU* to *ABCDEF*.

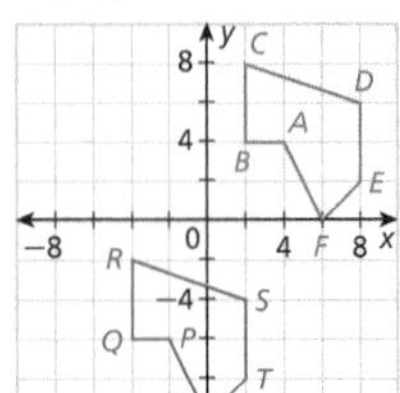

16. Map *DEF* to *KLM*.

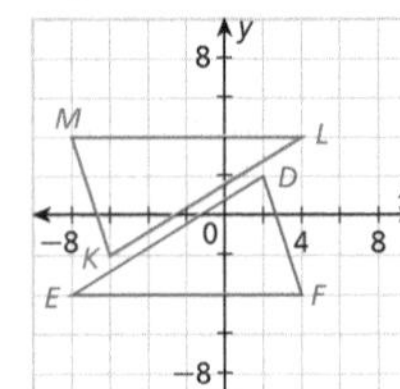

17. (Open Middle™) What is the fewest number of transformations needed to take preimage *ABCD* to image *A′B′C′D′*? Check students' work.

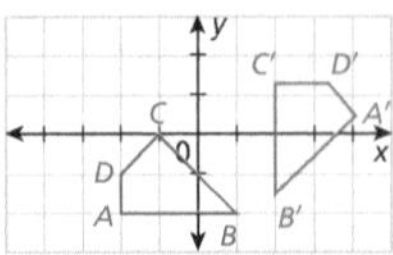

Spiral Review • Assessment Readiness

18. Given $\triangle ABC \cong \triangle EDC$, which reason proves that $\angle A \cong \angle E$?

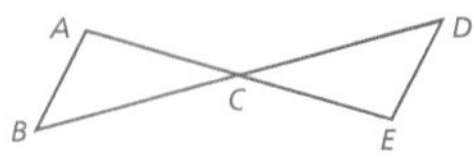

Ⓐ CPCFC
Ⓑ Alternate Interior Angles Theorem
Ⓒ Converse of the Alternate Interior Angles Theorem
Ⓓ Third Angles Theorem

19. What transformation will map *KLM* to *PQR*?

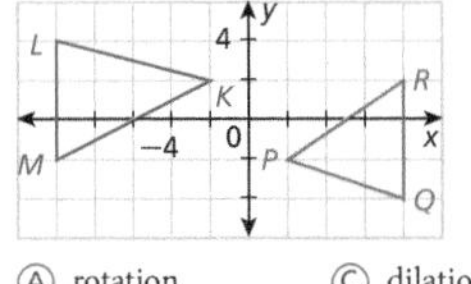

Ⓐ rotation
Ⓑ reflection
Ⓒ dilation
Ⓓ none of these

20. Which combinations of angles and sides can be used to prove two triangles are congruent? Choose all that apply.

Ⓐ Angle-Side-Angle
Ⓑ Angle-Angle-Angle
Ⓒ Side-Angle-Side
Ⓓ Angle-Side-Side
Ⓔ Angle-Angle-Side
Ⓕ Side-Angle-Side

I'm in a Learning Mindset!

How can I make sure I understand every step of a proof and can give a reason for each step?

Keep Going to Journal and Practice Workbook

Learning Mindset

Resilience Monitors Knowledge and Skills

Point out that identifying the skills that are required for a task is essential in all areas of mathematical problem solving. Encourage students to develop the habit of using prior knowledge of transforming functions and transfer that knowledge to this new content of congruence. *What have you learned in the past that relates to what we are learning now about congruence? What transformations result in congruent figures and which do not? When you cannot think of a good reason for a step in a proof, what might that suggest about your knowledge of this content?*

Module 7 Review

Congruent Polygons

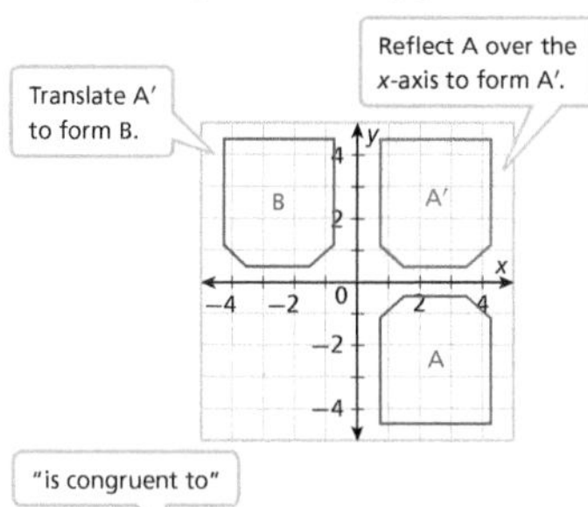

"is congruent to"

Polygon A ≅ Polygon B because a series of rigid transformations can map Polygon A onto Polygon B.

Corresponding Parts of Congruent Triangles Are Congruent

The order gives the correspondence of points.

If $\triangle ABC \cong \triangle DEF$, then all corresponding parts are congruent.

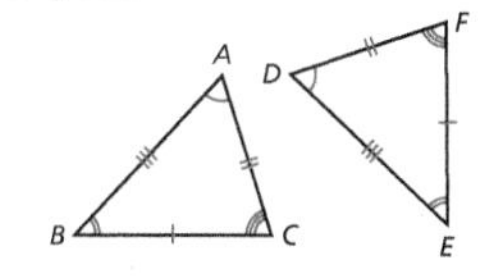

$\angle A \cong \angle D$	$\overline{BC} \cong \overline{EF}$
$\angle B \cong \angle E$	$\overline{CA} \cong \overline{FD}$
$\angle C \cong \angle F$	$\overline{AB} \cong \overline{DE}$

Using Congruence

Congruence can be used to find angle or side measures of figures. Given $\triangle ABC \cong \triangle DEF$, determine $m\angle B$.

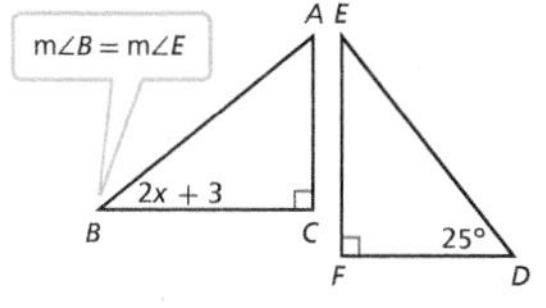

Triangle Sum Theorem

$$m\angle D + m\angle E + m\angle F = 180$$
$$m\angle D + m\angle B + m\angle F = 180$$
$$25 + 2x + 3 + 90 = 180$$
$$2x = 62$$
$$x = 31$$

Substitution

$m\angle B = 2x + 3 = 2(31) + 3 = 65$

Proofs

Congruence can be used to prove geometric relationships.

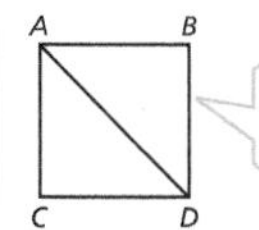

Reflect over $\overline{AD}$ or rotate around midpoint of $\overline{AD}$.

Given: Square $ABDC$; $\overline{AD}$ bisects $\angle A$ and $\angle D$.
Prove $\triangle ABD \cong \triangle ACD$

Statement	Reason
Square $ABDC$; $\overline{AD}$ bisects $\angle A$ and $\angle D$.	Given
$\overline{AB} \cong \overline{CD} \cong \overline{AC} \cong \overline{BD}$	Definition of square
$\angle A \cong \angle B \cong \angle C \cong \angle D$	Right angles
$\angle BAD \cong \angle CDA \cong \angle CAD \cong \angle BDA$	Definition of bisection
$\triangle ABD \cong \triangle ACD$	CPCTC

ONLINE

Assign the Digital Module Review for

- built-in student supports
- Actionable Item Reports
- Standards Analysis Reports

Module Review

Use the first page of the Module Review to summarize and connect the main ideas of the module. Use the second page to assess students' understanding of the vocabulary, concepts, and skills presented in the module.

Sample Guided Discussion:

Q Congruent Polygons What is another sequence of rigid transformations you can use to map *A* onto *B*? Possible answer: Reflect *A* over the *y*-axis and then reflect its image over the *x*-axis.

Q Corresponding Parts of Congruent Triangles Are Congruent What are two other ways to indicate the correspondence between the congruent triangles *ABC* and *DEF*? Possible answer: $\triangle BCA \cong \triangle EFD$ and $\triangle CAB \cong \triangle FDE$.

Q Using Congruence If the two triangles in the figure are congruent, what are the three pairs of corresponding sides? Possible answer: $\overline{DF} \cong \overline{AC}$, $\overline{EF} \cong \overline{BC}$, and $\overline{DE} \cong \overline{AB}$.

Q Proofs

Given: Square *ABCD*; $\overline{AD}$ bisects $\angle A$ and $\angle D$.

Prove: $\angle BAD \cong \angle ADC$

Possible answer: $\overline{AB} \parallel \overline{CD}$ because opposite sides of a square are parallel. $\overline{AD}$ is a transversal by definition. So, $\angle BAD \cong \angle ADC$ because they are alternate interior angles formed by the intersection of $\overline{AD}$ and the parallel sides.

Module Review continued

Possible Scoring Guide

Items	Points	Description
1–4	2 each	identifies the correct term
5	2	correctly writes congruency statements for corresponding parts
6	2	correctly explains whether the figures are congruent
7	1	correctly describes a sequence of transformations to map $\triangle XYZ$ onto $\triangle X'Y'Z'$
7	2	correctly describes a sequence of transformations to map $\triangle XYZ$ onto $\triangle X'Y'Z'$ and two sequences to map $\triangle X'Y'Z'$ onto $\triangle XYZ$
8	2	correctly writes a proof to show that the two triangles are congruent
9	1	correctly writes a congruence statement, describes the rigid motion, and finds PQ
9	2	correctly writes a congruence statement, describes the rigid motion, identifies parts corresponding to the diagonal, writes a proof, and finds PQ
Total points possible = 18 points		

The Unit 4 Test in the Assessment Guide assesses content from Modules 7 and 8.

Vocabulary

Choose the correct term from the box to complete each sentence.

Vocabulary
biconditional statement
conditional statement
congruent
rigid motion

1. A __?__ is a transformation that does not change the size or shape of a figure. rigid motion
2. Two figures that have the same size and shape are __?__. congruent
3. A __?__ is a statement that can be written in the form "if p, then q." conditional statement
4. A __?__ is a statement that can be written in the form "p if and only if q." biconditional statement

Concepts and Skills

5. Given $\triangle JKL \cong \triangle PQR$, write congruency statements for all corresponding parts. See Additional Answers.
6. A triangle is translated, dilated, and rotated to map onto an image. Are the two figures congruent? Explain your reasoning. no; A dilation is not a rigid motion.
7. You can show that $\triangle XYZ \cong \triangle X'Y'Z'$.
   A. Explain how to map $\triangle XYZ$ onto $\triangle X'Y'Z'$.
   B. Explain how to map $\triangle X'Y'Z$ onto $\triangle XYZ$.
   C. Explain how to map $\triangle X'Y'Z$ onto $\triangle XYZ$ with a different sequence of transformations.
   A–C. See Additional Answers.
8. Write a proof.
   Given: $\triangle ABC \cong \triangle ADC$
   Prove: $BD = 2BC$
   See Additional Answers.

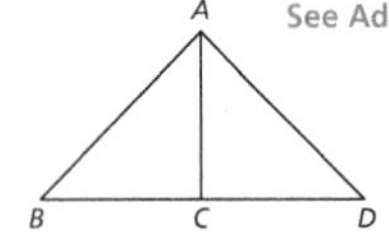

9. Parallelogram $PQRS$ has 2 pairs of opposite sides that are congruent and a diagonal.
   A. Identify 2 congruent triangles and write a congruence statement. $\triangle PQR \cong \triangle RSP$
   B. Describe a rigid motion that demonstrates the congruency. a rotation of 180 degrees about the midpoint of $\overline{PR}$
   C. What parts correspond to the diagonal? What property allows you to state that they are congruent? See Additional Answers.
   D. Write a formal proof to prove the congruency. See Additional Answers.
   E. (MP) Use Tools Determine PQ. State what strategy and tool you will use to answer the question, explain your choice, and then find the answer. $PQ = 22$

216

DATA-DRIVEN INSTRUCTION

Before moving on to the Module Test, use the Module Review results to intervene based on the table below.

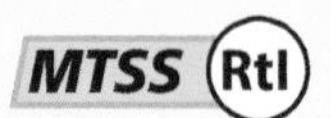

Items	Lesson	DOK	Content Focus	Intervention
7 and 9B	7.3	2	Use rigid motions to show figures are congruent.	Reteach 7.3
6	7.3	2	Use rigid motions to determine whether figures are congruent.	Reteach 7.3
5, 9A, 9C, and 9E	7.2	2	Write congruence statements about all corresponding parts.	Reteach 7.2
8 and 9D	7.1	3	Write proofs involving congruent figures.	Reteach 7.1

Module Test

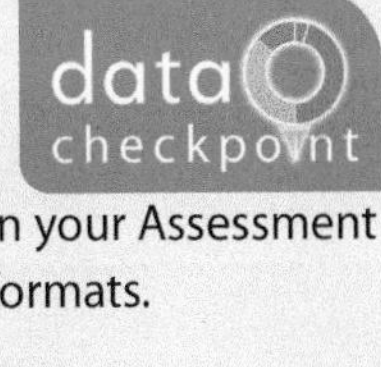

The Module Test is available in alternative versions in your Assessment Guide. All items are presented in standardized test formats.

ONLINE

Assign the Digital Module Test to power actionable reports including

- proficiency by standards
- item analysis

Form A

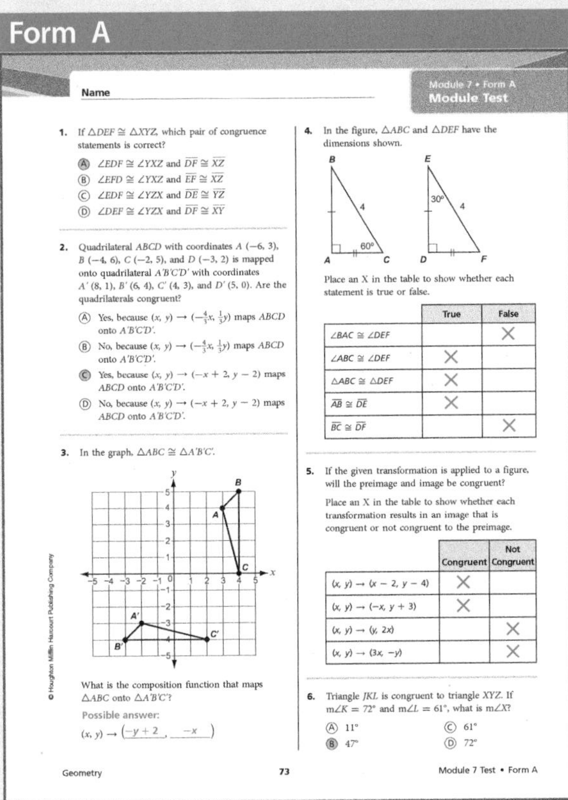

Name

Module 7 • Form A
Module Test

1. If $\triangle DEF \cong \triangle XYZ$, which pair of congruence statements is correct?
   Ⓐ $\angle EDF \cong \angle YXZ$ and $\overline{DF} \cong \overline{XZ}$
   Ⓑ $\angle EFD \cong \angle YXZ$ and $\overline{EF} \cong \overline{XZ}$
   Ⓒ $\angle EDF \cong \angle YZX$ and $\overline{DE} \cong \overline{YZ}$
   Ⓓ $\angle DEF \cong \angle YZX$ and $\overline{DF} \cong \overline{XY}$

2. Quadrilateral $ABCD$ with coordinates $A(-6, 3)$, $B(-4, 6)$, $C(-2, 5)$, and $D(-3, 2)$ is mapped onto quadrilateral $A'B'C'D'$ with coordinates $A'(8, 1)$, $B'(6, 4)$, $C'(4, 3)$, and $D'(5, 0)$. Are the quadrilaterals congruent?
   Ⓐ Yes, because $(x, y) \to \left(-\frac{4}{3}x, \frac{1}{3}y\right)$ maps $ABCD$ onto $A'B'C'D'$.
   Ⓑ No, because $(x, y) \to \left(-\frac{4}{3}x, \frac{1}{3}y\right)$ maps $ABCD$ onto $A'B'C'D'$.
   Ⓒ Yes, because $(x, y) \to (-x + 2, y - 2)$ maps $ABCD$ onto $A'B'C'D'$.
   Ⓓ No, because $(x, y) \to (-x + 2, y - 2)$ maps $ABCD$ onto $A'B'C'D'$.

3. In the graph, $\triangle ABC \cong \triangle A'B'C'$.

   What is the composition function that maps $\triangle ABC$ onto $\triangle A'B'C'$?

   Possible answer:
   $(x, y) \to (-y + 2, -x)$

4. In the figure, $\triangle ABC$ and $\triangle DEF$ have the dimensions shown.

   Place an X in the table to show whether each statement is true or false.

	True	False
$\angle BAC \cong \angle DEF$		X
$\angle ABC \cong \angle DEF$	X	
$\triangle ABC \cong \triangle DEF$	X	
$\overline{AB} \cong \overline{DE}$	X	
$\overline{BC} \cong \overline{DF}$		X

5. If the given transformation is applied to a figure, will the preimage and image be congruent?

   Place an X in the table to show whether each transformation results in an image that is congruent or not congruent to the preimage.

	Congruent	Not Congruent
$(x, y) \to (x - 2, y - 4)$	X	
$(x, y) \to (-x, y + 3)$	X	
$(x, y) \to (y, 2x)$		X
$(x, y) \to (3x, -y)$		X

6. Triangle JKL is congruent to triangle XYZ. If $m\angle K = 72°$ and $m\angle L = 61°$, what is $m\angle X$?
   Ⓐ 11° Ⓒ 61°
   Ⓑ 47° Ⓓ 72°

© Houghton Mifflin Harcourt Publishing Company

Geometry 73 Module 7 Test • Form A

Form A

Module 7 • Form A
Module Test

Name

7. In the figure, polygon $QRST$ is shown.
   Given: $\triangle QRT \cong \triangle STR$
   Prove: $m\angle TQR + m\angle QRT + m\angle TRS = 180°$

Statements	Reasons
1.	1. Given
2.	2. Triangle Sum Theorem
3.	3. Congruent Parts of Congruent Triangles are Congruent
4.	4. Triangle Sum Theorem

   What is the correct statement for the third step of the proof?
   Ⓐ $\angle TQR \cong \angle RST$ Ⓒ $\overline{TQ} \cong \overline{RS}$
   Ⓑ $\angle QRT \cong \angle STR$ Ⓓ $\overline{QR} \cong \overline{ST}$

8. In the graph, polygons $JKLM$ and $YXWZ$ are congruent.

   Which sequence of transformations will map polygon $JKLM$ onto polygon $YXWZ$?
   Ⓐ a rotation 90° clockwise about the origin, then a reflection across the x-axis and a translation by $\langle 5, 6\rangle$
   Ⓑ a translation by $\langle 5, 6\rangle$, then a rotation 90° clockwise about the origin and a reflection across the x-axis
   Ⓒ a rotation 90° counterclockwise about the origin, then a reflection across the y-axis and a translation by $\langle -6, -5\rangle$
   Ⓓ a translation by $\langle -6, -5\rangle$, then a rotation 90° counterclockwise about the origin and a reflection across the y-axis

9. In the figure, $\triangle ABC \cong \triangle EDF$ as shown.

   Part A
   What is $m\angle ABC$ in degrees?
   52°

   Part B
   What is the length of $\overline{DE}$ in inches?
   Possible answer: 34.2

© Houghton Mifflin Harcourt Publishing Company

Geometry 74 Module 7 Test • Form A

Form B

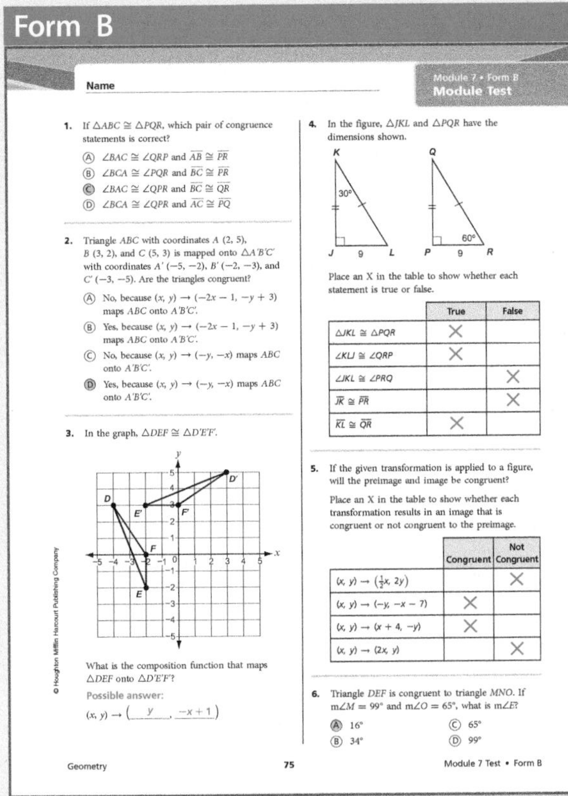

Name

Module 7 • Form B
Module Test

1. If $\triangle ABC \cong \triangle PQR$, which pair of congruence statements is correct?
   Ⓐ $\angle BAC \cong \angle QRP$ and $\overline{AB} \cong \overline{PR}$
   Ⓑ $\angle BCA \cong \angle PQR$ and $\overline{BC} \cong \overline{PR}$
   Ⓒ $\angle BAC \cong \angle QPR$ and $\overline{BC} \cong \overline{QR}$
   Ⓓ $\angle BCA \cong \angle QPR$ and $\overline{AC} \cong \overline{PQ}$

2. Triangle ABC with coordinates $A(2, 5)$, $B(3, 2)$, and $C(5, 3)$ is mapped onto $\triangle A'B'C'$ with coordinates $A'(-5, -2)$, $B'(-2, -3)$, and $C'(-3, -5)$. Are the triangles congruent?
   Ⓐ No, because $(x, y) \to (-2x - 1, -y + 3)$ maps ABC onto $A'B'C'$.
   Ⓑ Yes, because $(x, y) \to (-2x - 1, -y + 3)$ maps ABC onto $A'B'C'$.
   Ⓒ No, because $(x, y) \to (-y, -x)$ maps ABC onto $A'B'C'$.
   Ⓓ Yes, because $(x, y) \to (-y, -x)$ maps ABC onto $A'B'C'$.

3. In the graph, $\triangle DEF \cong \triangle D'E'F'$.

   What is the composition function that maps $\triangle DEF$ onto $\triangle D'E'F'$?

   Possible answer:
   $(x, y) \to (y, -x + 1)$

4. In the figure, $\triangle JKL$ and $\triangle PQR$ have the dimensions shown.

   Place an X in the table to show whether each statement is true or false.

	True	False
$\triangle JKL \cong \triangle PQR$	X	
$\angle KLJ \cong \angle QRP$	X	
$\angle JKL \cong \angle PRQ$		X
$\overline{JK} \cong \overline{PR}$		X
$\overline{KL} \cong \overline{QR}$	X	

5. If the given transformation is applied to a figure, will the preimage and image be congruent?

   Place an X in the table to show whether each transformation results in an image that is congruent or not congruent to the preimage.

	Congruent	Not Congruent
$(x, y) \to \left(\frac{1}{2}x, 2y\right)$		X
$(x, y) \to (-y, -x - 7)$	X	
$(x, y) \to (x + 4, -y)$	X	
$(x, y) \to (2x, y)$		X

6. Triangle DEF is congruent to triangle MNO. If $m\angle M = 99°$ and $m\angle O = 65°$, what is $m\angle E$?
   Ⓐ 16° Ⓒ 65°
   Ⓑ 34° Ⓓ 99°

© Houghton Mifflin Harcourt Publishing Company

Geometry 75 Module 7 Test • Form B

Form B

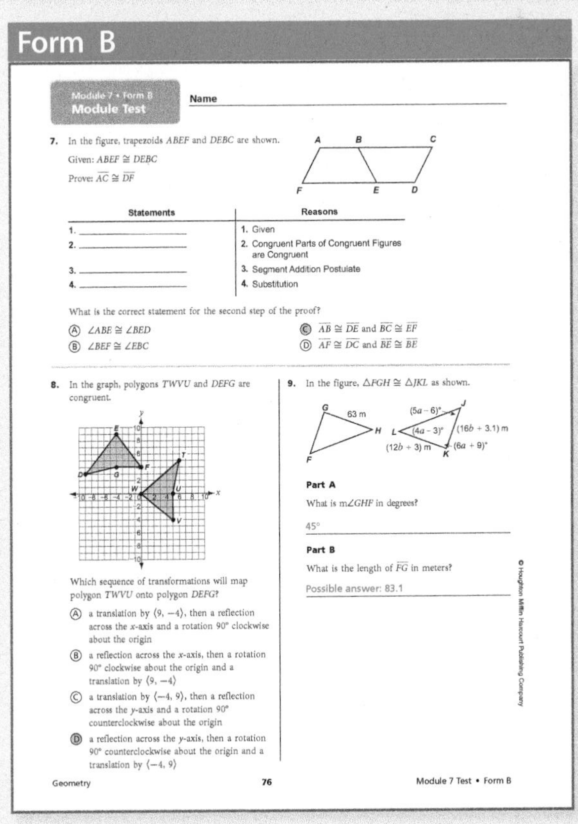

Module 7 • Form B
Module Test

Name

7. In the figure, trapezoids $ABEF$ and $DEBC$ are shown.
   Given: $ABEF \cong DEBC$
   Prove: $\overline{AC} \cong \overline{DF}$

Statements	Reasons
1.	1. Given
2.	2. Congruent Parts of Congruent Figures are Congruent
3.	3. Segment Addition Postulate
4.	4. Substitution

   What is the correct statement for the second step of the proof?
   Ⓐ $\angle ABE \cong \angle BED$ Ⓒ $\overline{AB} \cong \overline{DE}$ and $\overline{BC} \cong \overline{EF}$
   Ⓑ $\angle BEF \cong \angle EBC$ Ⓓ $\overline{AF} \cong \overline{DC}$ and $\overline{BE} \cong \overline{BE}$

8. In the graph, polygons $TWVU$ and $DEFG$ are congruent.

   Which sequence of transformations will map polygon $TWVU$ onto polygon $DEFG$?
   Ⓐ a translation by $\langle 9, -4\rangle$, then a reflection across the x-axis and a rotation 90° clockwise about the origin
   Ⓑ a reflection across the x-axis, then a rotation 90° clockwise about the origin and a translation by $\langle 9, -4\rangle$
   Ⓒ a translation by $\langle -4, 9\rangle$, then a reflection across the y-axis and a rotation 90° counterclockwise about the origin
   Ⓓ a reflection across the y-axis, then a rotation 90° counterclockwise about the origin and a translation by $\langle -4, 9\rangle$

9. In the figure, $\triangle FGH \cong \triangle JKL$ as shown.

   Part A
   What is $m\angle GHF$ in degrees?
   45°

   Part B
   What is the length of $\overline{FG}$ in meters?
   Possible answer: 83.1

© Houghton Mifflin Harcourt Publishing Company

Geometry 76 Module 7 Test • Form B

TRIANGLE CONGRUENCE CRITERIA

Introduce and Check for Readiness

- Module Performance Task • Are You Ready?

Lesson 8.1—2 Days

Develop ASA Triangle Congruence

Learning Objective: Students use ASA congruence criteria to prove that two triangles are congruent.
New Vocabulary: included side

Lesson 8.2—2 Days

Develop SAS Triangle Congruence

Learning Objective: Students use SAS congruence criteria to prove that two triangles are congruent.
New Vocabulary: included angle

Lesson 8.3—2 Days

Develop SSS Triangle Congruence

Learning Objective: Students use SSS congruence criteria to prove that two triangles are congruent.

Lesson 8.4—2 Days

Develop AAS and HL Triangle Congruence

Learning Objective: Students use AAS and HL congruence criteria to determine if triangles are congruent.

Assessment

- Module 8 Test (Forms A and B)
- Unit 4 Test (Forms A and B)

LEARNING ARC FOCUS

 Build Conceptual Understanding
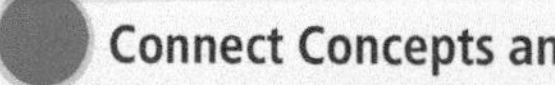

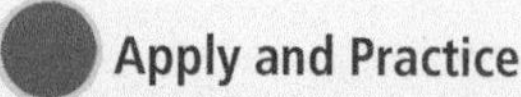

TEACHING FOR SUCCESS

TEACHING FOR DEPTH: Triangle Congruence Criteria

Making Connections. There is a connection between congruent triangles and pairs of corresponding angles and corresponding sides. To prove triangles are congruent, however, it is not necessary to prove that the three pairs of sides and three pairs of angles are congruent. Rather, there are certain combinations of corresponding parts—sides and/or angles—that ensure that triangles are congruent. As a result, the remaining corresponding parts are also congruent.

Students have learned that the sum of the measures of the angles of a triangle is 180°. If three angles of one triangle are congruent to three angles of another triangle, the triangles will have the same shape, but they won't necessarily be congruent. The lengths of at least one pair of sides need to be given to show congruence. Triangles that have equal angles but no equal sides are *similar*. Establishing congruence between triangles requires one of the four following criteria:

- Three sides of one triangle are congruent to three side of another other, triangle (SSS).
- Two angles and the included side of one triangle are congruent to two angles and the included side of another triangle (ASA).
- Two sides and the included angle of one triangle are congruent to two sides and the included angle of another triangle (SAS).
- Two angles and a side of one triangle are congruent to two angles and a side of another triangle, which is a special case of SAS (AAS).
- The hypotenuse and one leg of a right triangle are congruent to the hypotenuse and leg of another right triangle (HL).

Mathematical Progressions

Prior Learning	Current Development	Future Connections
Students: • represented transformations in the plane and described transformations as functions. • developed definitions of rotations, reflections, and translations. • used rigid motions to transform figures. • used definitions of rigid motions to determine congruency. • proved triangle congruence by rigid motions.	**Students:** • use rigid motions to show figures are congruent. • use congruency of corresponding parts to prove triangles are congruent. • use ASA, SSS, SAS, AAS, and HL congruence criteria to prove that two triangles are congruent.	**Students:** • will write coordinate proofs about triangle relationships. • will write coordinate proofs about parallelograms. • will use the properties of similarity transformations to establish the AA criterion for two triangles to be similar. • will construct an equilateral triangle, a square, and a regular hexagon inscribed in a circle.

TEACHER TO TEACHER

From the Classroom

Facilitate meaningful mathematical discourse. When I assign the first triangle congruence proof to students, I divide them into small groups, where each group has access to a geometry software application if they choose to use it. I explain that I want each group to write one proof of the congruence theorem using the given criteria and a second proof involving a transformation. This gives students an opportunity to work together and discuss how they will go about creating the steps in each proof. When everyone is ready, I bring the class together to review their proofs and discuss the pros and cons of different points of view.

LANGUAGE DEVELOPMENT • Planning for Instruction

By giving all students regular exposure to language routines in context, you will provide opportunities for students to **listen, speak, read,** and **write** about mathematical situations and develop both mathematical language and conceptual understanding at the same time.

Using Language Routines to Develop Understanding

Use the **Professional Cards** for the following routines to plan for effective instruction.

Co-Craft Questions and Problems Lesson 8.2

Students think of natural questions to ask about a given situation or problems similar to a given task and answer the questions they have developed or problems they have created.

Three Reads Lessons 8.1–8.3

Students read a problem three times with a specific focus each time.

1st Read What is the situation about?
2nd Read What are the quantities in the situation?
3rd Read What are the possible mathematical questions that you could ask for the situation?

Information Gap Lesson 8.4

Students recognize when information given in a problem situation is incomplete, and they pose questions and share knowledge with others to discover any missing facts or relationships and work together to solve the problem.

Critique, Correct, and Clarify Lessons 8.1 and 8.4

Students correct the work in a flawed explanation, argument, or solution method and share with a partner and refine the sample work.

Connecting Language to Triangle Congruence Criteria

Watch for students' use of the new terms listed below as they explain their reasoning and make connections with new concepts.

Key Academic Vocabulary

Current Development • New Vocabulary	
included angle an angle formed by two sides of a triangle	**included side** a side connecting the vertices of two angles

Linguistic Note

Listen for students' use of the terms *included sides* and *included angles.* Some English learners may have used the word *included* in other contexts. For example, "My lunch included dessert" or "The cost of this item included the sale tax." Here, *included* means *contained.*

Then explain that *included* in geometry means *between.* So, an included side is between two angles, and an included angle is between two sides. Have students draw and label different geometric figures and identify included sides and angles and explain their choices.

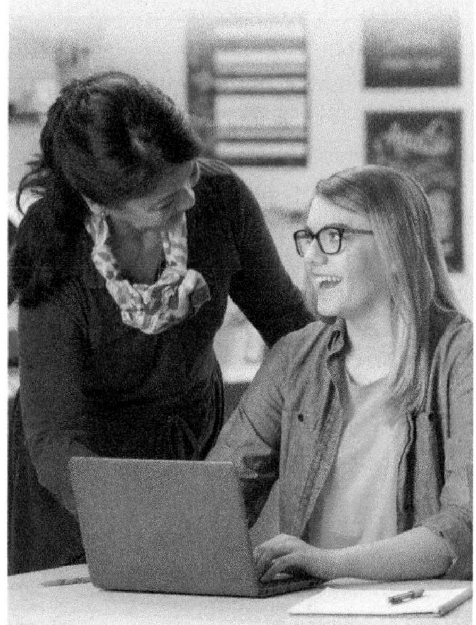

Module 8

Triangle Congruence Criteria

Module Performance Task: Focus on STEM

Lunar Laser Ranging

During the Apollo missions to the moon, astronauts placed arrays of Cube Corner Retroreflectors (CCRs) on the lunar surface to use in future Lunar Laser Ranging (LLR) experiments.

A CCR is created by cutting three perpendicular surfaces, similar to a corner you might see where two walls meet each other and the floor, within a circular pupil.

A, B. See Additional Answers.

A. For the Apollo 15 mission, the CCRs were arranged in large rectangular panels. If the diameter of each CCR is 38 mm, estimate the surface area of a rectangular portion of the panel containing a 12×9 array of CCRs. Explain your reasoning.

B. A new retroreflector is scheduled to launch to the moon. Given that the diameter of this CCR is 100 mm, what is the total surface area of a similar 12×9 array of the new CCRs? How does this compare to the surface area of the array of 38 mm CCRs?

C. In the LLR experiment, the laser beam reflects off each surface of the corner before returning to Earth. Given the speed of light c and the time taken for the signal to travel to and return from the moon t, write an equation that represents the distance d between Earth and the moon. $d = \frac{ct}{2}$

D. Estimate the distance from Earth to the moon given $c \approx 3 \times 10^8$ m/s and $t \approx 2.6$ s. Then, research and compare your estimate to more exact values. Determine if your estimate was too high or too low. What might contribute to the difference in estimates? See margin.

Module 8 217

Mathematical Connections

Task Part	Prerequisite Skills
Parts A and B	Students can use triangle congruence criteria to determine congruent triangles **(8.1–8.4)**.
Parts C and D	Students can use formulas involving the speed of light to find the distance from Earth to the moon **(Algebra 1, 2.2)**.

Lunar Laser Ranging

Overview

Students use congruence criteria and rigid motions to prove that triangles in an applied problem in lunar science are congruent.

Lunar Laser Ranging experiments involve the use of Corner Cube Retroreflectors on the surface of the Moon to determine its distance from the surface of Earth. Since there have been so few missions to the moon, there are only a few retroflectors on the moon.

A retroreflector is an optical device that reflects light back to its source with minimal scattering. A Cube Corner Retroreflector is made up of three planes that form the inside corner of a cube. When a ray of light hits the first side of the plane, it is reflected to the second side and then to the third side, and then reflected back to its source.

Career Connections

Students who are interested in Lunar Laser Ranging can consider a career as a lunar and planetary geodesist, someone who studies the rotation rates, masses, and gravity fields of the moon and planets, or as a cartographer who produces maps of the location and elevation of surface features of these bodies. Related career goals include becoming a physicist, an astronomer, an optical engineer, or even an astronaut.

Answer

D. The estimated distance to the moon is about 390,000,000 meters.

$d = \frac{ct}{2} = \frac{3 \times 10^8(2.6)}{2} = 3.9 \times 10^8$ meters

Possible answer: A more exact value is given by 384,400,000 meters. Variations in measurement could be due to rounding the time, variations in the diameter of Earth at a laser location, or the approximation of the speed of light.

ONLINE

Assign the Digital Are You Ready? to power actionable reports including

- proficiency by standards
- item analysis

Are You Ready?

Diagnostic Assessment

data checkpoint

- Diagnose prerequisite mastery.
- Identify intervention needs.
- Modify or set up leveled groups.

Have students complete the *Are You Ready?* assessment on their own. Items test the prerequisites required to succeed with the new learning in this module.

Types of Angle Pairs Students extend previous knowledge of types of angle pairs to establish triangle congruence criteria by analyzing angle relationships.

The Pythagorean Theorem and Its Converse Students use previous knowledge of the Pythagorean Theorem to establish congruence criteria for right triangles.

Congruent Figures Students extend previous knowledge of congruent figures to establish and prove triangle congruence criteria.

ONLINE 

Are You Ready?

Complete these problems to review concepts and skills you will need for this module.

Types of Angle Pairs

Use the given angle measure to find the missing value.

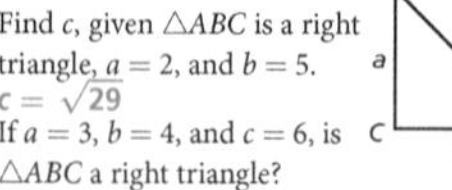

1. $m\angle 1 = 100°$ $m\angle 8 =$? ° 80
2. $m\angle 2 = 55°$ $m\angle 3 =$? ° 125
3. $m\angle 6 = 42°$ $m\angle 4 =$? ° 42
4. $m\angle 5 = 115°$ $m\angle 2 =$? ° 65
5. $m\angle 7 = 120°$ $m\angle 5 =$? ° 120
6. $m\angle 3 = 118°$ $m\angle 7 =$? ° 118

The Pythagorean Theorem and Its Converse

Use the given diagram with the given information.

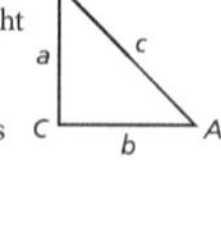

7. Find a, given $\triangle ABC$ is a right triangle, $c = 15$, and $b = 12$. $a = 9$
8. Find c, given $\triangle ABC$ is a right triangle, $a = 2$, and $b = 5$. $c = \sqrt{29}$
9. If $a = 10$, $b = 24$, and $c = 26$, is $\triangle ABC$ a right triangle? yes
10. If $a = 3$, $b = 4$, and $c = 6$, is $\triangle ABC$ a right triangle? no

Congruent Figures

Find each side length or angle measure, given $\triangle PQR \cong \triangle XYZ$.

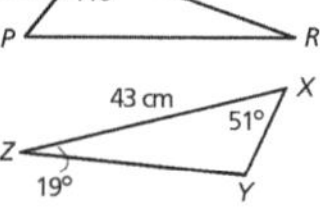

11. PR $PR = 43$ cm
12. $m\angle ZYX$ $m\angle ZYX = 110°$
13. $m\angle QRP$ $m\angle QRP = 19°$
14. XY $XY = 15$ cm

Connecting Past and Present Learning

Previously, you learned:

- to use angle pairs to determine unknown values,
- to find missing side lengths of right triangles using the Pythagorean Theorem, and
- to use corresponding parts of congruent figures to determine unknown values.

In this module, you will learn:

- to develop triangle congruency criteria from rigid transformations,
- to use triangle congruency criteria to solve problems, and
- to use triangle congruency criteria to prove relationships in geometric figures.

218

DATA-DRIVEN INTERVENTION

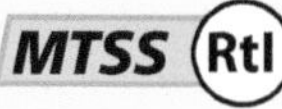

Concept/Skill	Objective	Prior Learning *	Intervene With
Types of Angle Pairs	Solve problems involving complementary, supplementary, vertical, and adjacent angles.	Grade 7, Lesson 7.5	• Tier 2 Skill 2 • Reteach, Grade 7 Lesson 7.5
The Pythagorean Theorem and Its Converse	Use the Pythagorean Theorem and its converse to solve problems.	Grade 8, Lesson 11.1–11.3	• Tier 2 Skill 4 • Reteach, Grade 8 Lesson 11.1–11.3
Congruent Figures	Use corresponding parts of congruent figures to determine unknown values.	Grade 8, Lesson 1.5	• Tier 2 Skill 14 • Reteach, Grade 8 Lesson 1.5

* Your digital materials include access to resources from Grade 6–Algebra 2. The lessons referenced here contain a variety of resources you can use with students who need support with this content.

8.1 Develop ASA Triangle Congruence

LESSON FOCUS AND COHERENCE

Mathematics Standards

- Explain how the criteria for triangle congruence (ASA, SAS, and SSS) follow from the definition of congruence in terms of rigid motions.

Mathematical Practices and Processes

- Use appropriate tools strategically.
- Look for and make use of structure.
- Construct viable arguments and critique the reasoning of others.
- Model with mathematics.

I Can Objective

I can use ASA congruence criteria to prove that two triangles are congruent.

Learning Objective

Students use ASA congruence criteria to prove that two triangles are congruent.

Language Objective

Explain the steps needed to prove triangles congruent using ASA congruence criteria.

Vocabulary

New: included side

Lesson Materials: ruler, protractor, geometry software

Mathematical Progressions

Prior Learning	Current Development	Future Connections
Students: • given two figures, used the definition of congruence in terms of rigid motions to decide if they are congruent. **(7.3)** • used the definition of congruence in terms of rigid motions to show that two triangles are congruent if and only if corresponding pairs of sides and corresponding pairs of angles are congruent. **(7.1–7.3)**	**Students:** • prove that two triangles are congruent showing that the one triangle can be obtained from the other by a sequence of rigid motions. • prove that two triangles are congruent showing that all pairs of corresponding angles and corresponding sides are congruent. • prove that two triangles are congruent using ASA congruence criteria.	**Students:** • will write coordinate proofs about triangle relationships. **(9.4 and 9.5)** • will write coordinate proofs about parallelograms. **(11.1–11.5)** • will use the properties of similarity transformations to establish the AA criterion for two triangles to be similar. **(12.3)**

UNPACKING MATH STANDARDS

Explain how the criteria for triangle congruence (ASA, SAS, and SSS) follow from the definition of congruence in terms of rigid motions.

What It Means to You

Students demonstrate an understanding of using ASA Triangle Congruence criteria to prove two triangles are congruent. This understanding develops from the last module, where they used corresponding parts and/or rigid motions to prove congruence.

In this lesson, the ASA Triangle Congruence Theorem uses three congruent pairs of corresponding parts to prove congruence, an angle, an included side, and a second angle. Students use rigid motions to determine congruence and use proofs to show that two triangle are congruent. In future lessons, students will continue to develop other types of triangle congruence criteria, such as SAS and SSS.

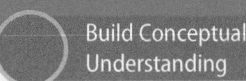

ONLINE

PROJECTABLE & PRINTABLE

WARM-UP OPTIONS

ACTIVATE PRIOR KNOWLEDGE • Use Rigid Motions to Show Triangle Congruence

Use these activities to quickly assess and activate prior knowledge as needed.

Problem of the Day

A series of transformations maps $\triangle ABC$ into $\triangle GHI$. Is $\triangle ABC \cong \triangle GHI$? Explain why or why not. Possible answer: $\triangle ABC$ is transformed into $\triangle GHI$ by a reflection across the x-axis. Then it is translated 2 units left onto $\triangle GHI$. So, $\triangle ABC \cong \triangle GHI$.

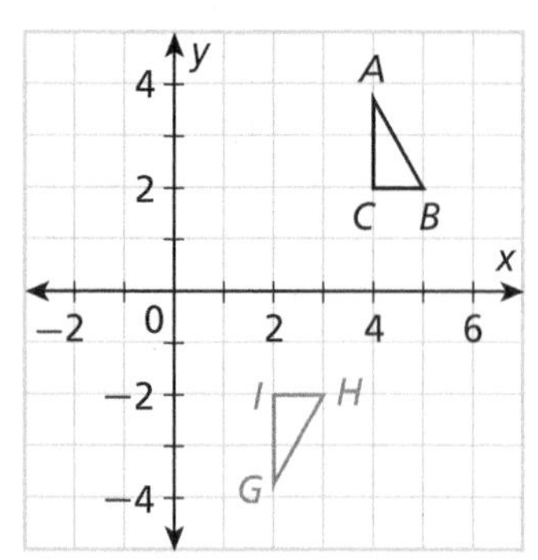

Quick Check for Homework

As part of your daily routine, you may want to display the Teacher Solution Key to have students check their homework.

Make Connections

Based on students' responses to the Problem of the Day, choose one of the following:

1. Project the Interactive Reteach, Grade 8, Lesson 1.5.
2. Complete the Prerequisite Skills Activity:

Have students work in pairs. One student draws two different but similar looking triangles on a coordinate plane or geometry software. A third triangle should also be drawn such that it is congruent to one of the first 2 triangles, which can be proven via a transformation (translation, reflection, or rotation). The other student identifies the congruent triangles and the transformation.

- *What are 3 rigid motions?* Possible answer: Reflections, rotations, and translations are rigid motions.
- *Which two triangles are congruent?* Possible answer: the ones in the first and fourth quadrants
- *Which transformation produced the congruent triangles?* Possible answer: a reflection across the x-axis

If students continue to struggle, use Tier 2 Skill 14.

SHARPEN SKILLS

If time permits, use this on-level activity to build fluency and practice basic skills.

Mental Math

Objective: Students demonstrate their knowledge about the angles of a triangle.
Materials: index cards

Have students work in groups of three. Give each group of students a set of index cards. The cards show the measure of two angles of a triangle on the front and the measure of the third angle on the back. One student draws a card and shows the two angles to one partner. The partner must give the measure of the third angle and tell if the triangle is acute, right, or obtuse. If the partner gets both right, he/she receives 1 point. If the partner is wrong, the third group member can get the point with both correct answers. Rotate the roles so that each student gets the same number of chances to score in a certain amount of time. The student who receives the most points is declared the winner.

50° 105°

PLAN FOR DIFFERENTIATED INSTRUCTION

Small-Group Options

Use these teacher-guided activities with pulled small groups.

On Track

Materials: compass, straightedge, protractor, blank transparencies, transparency markers

Give each group a transparency and some markers. Assign each group a rigid motion and a triangle. Each group will be responsible for making a class presentation showing how to use the rigid motion to create a congruent triangle and then to verify the congruency by measuring corresponding parts.

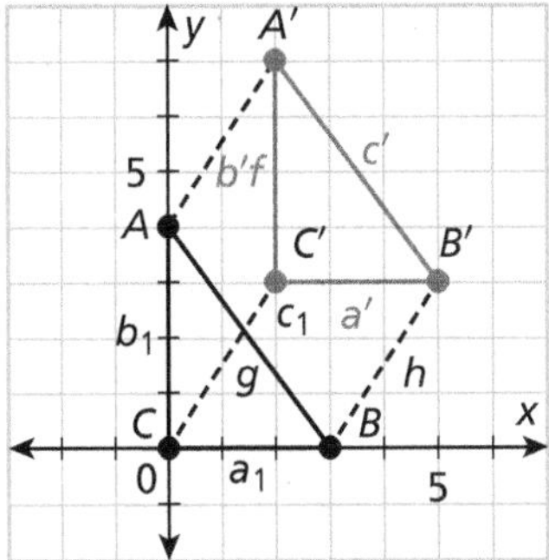

Almost There

Materials: ruler, protractor

Provide students with a set of 4 triangles that look similar, but with only 1 pair of congruent triangles in different orientations. Have students do the following:

- Determine corresponding parts of the triangles.
- Measure and label the sides and angles of each triangle.
- Determine which pair of triangles is congruent.

Ready for More

Materials: geometry software or coordinate plane

Have students create a worksheet with 3 problems on pairs of congruent triangles where one triangle in each pair is the result of 2 transformations of the other triangle. Have them include the following as part of their worksheet:

- Each problem should have a drawing showing the triangles after each transformation.
- Each problem will ask for a congruence statement and a list of the transformations used to create the congruency.
- Each problem will come with an answer key.

Students work on each other's worksheets.

Math Center Options

Use these student self-directed activities at centers or stations.

Key: ● Print Resources ● Online Resources

On Track

- ● Interactive Digital Lesson
- ●● Journal and Practice Workbook
- ● Interactive Glossary (printable): **included side**
- ● Module Performance Task

Almost There

- ● Reteach 8.1 (printable)
- ● Interactive Reteach 8.1
- ● RtI Tier 2 Skill 14: Congruent Figures
- ● Illustrative Mathematics: Why does ASA Work?

Ready for More

- ● Challenge 8.1 (printable)
- ● Interactive Challenge 8.1
- ● Illustrative Mathematics: Properties of Congruent Triangles

View data-driven grouping recommendations and assign differentiation resources.

During the *Spark Your Learning,* listen and watch for strategies students use. See samples of student work on this page.

Use Corresponding Parts

Strategy 1

The angles in both triangles are 60° and 60° with a side of 2 feet between the two angles. The corresponding sides and the corresponding angles of the triangles are congruent. So, the triangles are must be congruent.

If students . . . use corresponding parts to prove the triangles congruent, they are demonstrating an exemplary understanding of corresponding parts of triangles from Module 7.

Have these students . . . explain how they determined they could use corresponding parts to determine congruency. **Ask:**

- Q Can you determine that the third angle in both triangles is congruent?
- Q If the third angle is opposite a length that you know, what do you think is true of the two lengths that you don't know?
- Q How can you use CPCFC to conclude that the triangles are congruent?

Use a Translation

Strategy 2

Make a sketch of the 2 signs. Draw the height of △ABC using a dashed line. Translate △ABC straight down the same distance as the height of the triangle. This translation of △ABC coincides with △DEF. So, △ABC ≅ △DEF.

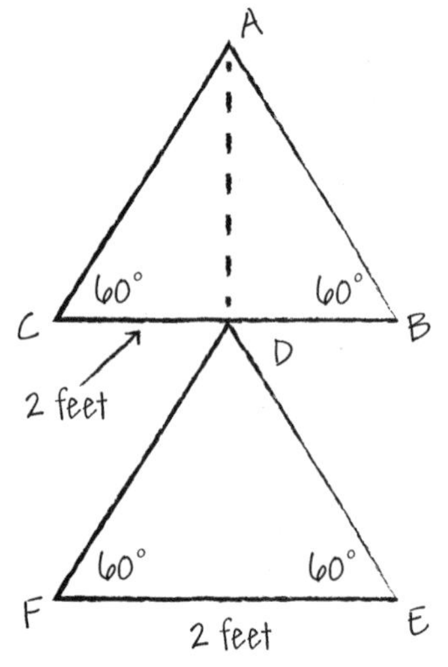

If students . . . use a translation to solve the problem, they understand that transformations can be used to prove congruence but may not know how to use previous concepts without models to show congruence.

Activate prior knowledge . . . by having students recall methods of proving congruence. **Ask:**

- Q What two methods can you use to prove two figures congruent?
- Q Which method is based on previous understanding of corresponding parts of congruent figures?

COMMON ERROR: Uses Wrong Assumptions

The triangles have 3 pairs of corresponding angles, 60°, 60°, 60°, that are congruent. This means the triangles are congruent.

If students . . . use an assumption about congruent angles to determine the triangles are congruent, then they may not understand that angles do not determine side length, such as in similar triangles with the same angle measures.

Then intervene . . . by pointing out that similar triangles can have congruent angles and different side lengths. **Ask:**

- Q If two triangles have congruent angles and at least 1 pair of corresponding congruent sides, what can you reason about the triangle?
- Q What do you know about corresponding sides of congruent figures?

Develop ASA Triangle Congruence

I Can use ASA congruence criteria to prove that two triangles are congruent.

Spark Your Learning

Triangular road signs are used to alert drivers of upcoming roadway hazards.

©Martin Ludlam/Shutterstock

Complete Part A as a whole class. Then complete Parts B–D in small groups. A. Are the two triangles congruent?; the angle measures and side lengths

A. What mathematical question can you ask about the two triangular signs? What information would you need to know to answer your question?

B. To answer your question, what strategy and tool would you use along with all the information you have? What answer do you get?
See Strategies 1 and 2 on the facing page.

C. What can you conclude about the angles of each triangle? What can you conclude about the side lengths of each triangle?
C, D. See Additional Answers.

D. How do you think your conclusion would apply to different triangles?

Turn and Talk Think about what you already know about congruent triangles. What parts must be congruent? Can you think of various ways to check for congruency besides confirming all six corresponding parts of both triangles are congruent? See margin.

SUPPORT SENSE MAKING • Three Reads

Tell students to read the information in the photo three times and prompt them with a different question each time.

1. What is the situation about? The situation is about comparing the size and shape of two triangular signs.
2. What are the quantities in this situation? How are those quantities related? Possible answer: The quantities are the measures of the angles of the triangles and the lengths of the sides; The bottom two angles of each triangle are 60° and the bottom sides are each 2 feet long.
3. What are possible questions you could ask about this situation? Possible answer: Are the triangles congruent?

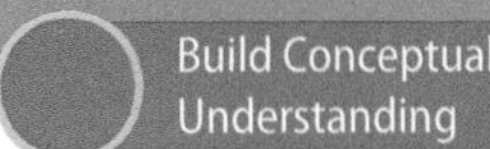

1 Spark Your Learning

▶ MOTIVATE

- Have students look at the photo in their books and read the information contained in the photo. Then complete Part A as a whole-class discussion.
- Give the class the additional information they need to solve the problem. This information is available online as a printable and projectable page in the Teacher Resources.
- Have students work in small groups to complete Parts B–D.

▶ PERSEVERE

If students need support, guide them by asking:

Q **Advancing • Use Tools** Which tool could you use to solve the problem? Why choose that tool and not some other? Students' choices of tools and reasons for choosing them will vary.

Q **Assessing** What are the labeled congruent measures of the two triangular signs? They each have two angles that measure 60° and a side that measures 2 feet.

Q **Assessing** How do you prove two figures are congruent? Possible answer: Show that the figure can be obtained from the other by a sequence of rigid motions or show that all pairs of corresponding angles and corresponding sides are congruent.

Q **Advancing** If 3 of the angles and/or sides are congruent, do you think the other 3 angle and/or sides would be congruent? Possible answer: If the 3 angles of one triangle were congruent to the three angles of another triangle, the sides of the triangles could have different lengths. If 3 sides are congruent then, the two triangles are congruent.

Turn and Talk Ask students to think about their answers to Parts B through D and how those answers relate to the two triangles. Possible answer: All parts of congruent triangles must be congruent; I could measure two triangles to prove congruency. I could use a series of rigid transformations to prove congruency.

▶ BUILD SHARED UNDERSTANDING

Select groups of students who used various strategies and tools to share with the class how they solved the problem. As they present their solutions, have each group discuss why they chose a specific strategy and tool.

② Learn Together

Build Understanding

Task 1 (MP) **Use Tools** Students use a ruler and protractor to draw a triangle with two angles, 40° and 50°, with an included side of 3 inches. Then they compare the triangle they drew to the given triangle in the text.

Sample Guided Discussion:

Q **When you use the protractor to draw the 40° angle, do you use the inside scale or the outside?** inside

Q **When you use the protractor to draw the 50° angle, do you use the inside scale or the outside?** outside

Q **What is the measure of ∠*PRQ* and why?** 90° because the sum of the measures of the angles of a triangle is 180°.

Turn and Talk Have students think about the relationship between the angles and the side measures of a triangle. Possible answer: If two angles and the side between the angles on one triangle are congruent to two angles and the side between the angles on another triangle, then the triangles are congruent.

Build Understanding

Draw Triangles Given Two Angles and a Side

You have learned two ways to show that two figures are congruent.

- Show that the figure can be obtained from the other by a sequence of rigid motions.
- Show that all pairs of corresponding angles and corresponding sides are congruent.

However, there are special theorems you can apply to triangles to check for congruence without using these methods. These theorems depend on measures of certain sides and angles of the triangles. When applying these theorems, accurate measures must be used.

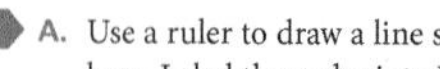

A–C. See Additional Answers.

1 A. Use a ruler to draw a line segment 8 inches long. Label the endpoints *P* and *Q*.

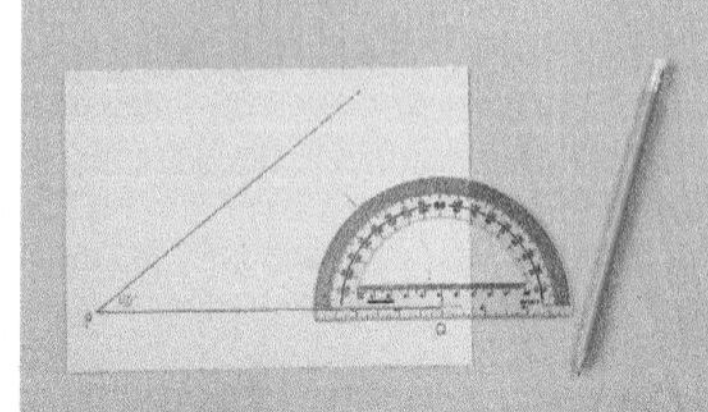

B. Use a protractor to draw a 40° angle with segment *PQ* as one side and the vertex at point *P*.

C. Use a protractor to draw a 50° angle with segment *PQ* as one side and the vertex at point *Q*. Label the point where the two segments intersect as point *R*.

D. Find a partner and compare your triangles. Did you draw the same triangle? If not, is there a set of rigid transformations that maps your triangle to your partner's triangle? What can you conclude about the two triangles? Possible answer: Yes, we drew the same triangle; The triangles are congruent because all corresponding parts are congruent.

Turn and Talk Based on your results, how can you decide whether two triangles are congruent without checking that all six pairs of corresponding sides and corresponding angles are congruent? See margin.

Justify ASA Triangle Congruence Using Transformations

In a triangle, the side connecting the vertices of two consecutive angles is called their **included side**. Given two side angle measures and the length of the included side, you can draw only one triangle. So, triangles with those measures are congruent.

Angle-Side-Angle (ASA) Triangle Congruence Theorem

If two angles and the included side of one triangle are congruent to two angles and the included side of another triangle, then the triangles are congruent.

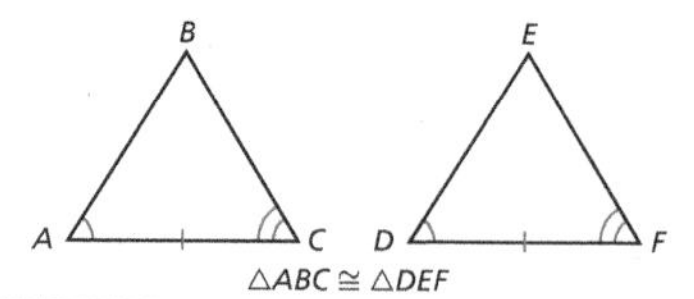

LEVELED QUESTIONS

Depth of Knowledge (DOK)	Leveled Questions	What Does This Tell You?
Level 1 **Recall**	How can you find the measure of the third angle of the triangle? Possible answer: Add the two given angles and subtract the result from 180.	Students' answers will indicate whether they understand the Triangle Sum Theorem.
Level 2 **Basic Application of Skills & Concepts**	Which side of the triangle is the longest side and why? Possible answer: The longest side is $\overline{PQ}$ because it is opposite the largest angle.	Students' answers will indicate whether they understand the relationship between the sides and angles of a triangle.
Level 3 **Strategic Thinking & Complex Reasoning**	If ∠*P* were changed to 90° to form a new triangle, how do the sides and angles of the new and old triangles compare? Possible answer: The angles of both triangles would be 40°, 50°, and 90°. The old triangle would have a 3-inch hypotenuse and the new triangle would have a 3-inch leg.	Students' answers will demonstrate that they understand the concept of ASA Congruence.

 A. Use tracing paper to create two copies of your triangle from Task 1 as shown. Label the angles and side you know to be congruent. What can you do to show that these triangles are congruent? Possible answer: Use a series of rigid transformations to map $\triangle PQR$ to $\triangle TUV$.

B. Start with a transformation that maps point P to point T. What transformation did you use? Is it a rigid transformation? a translation; yes

C. Now use a rotation to map point Q to point U. What point is the center of rotation? What angle of rotation did you use? point P or T; $\angle UPQ$ or $\angle UTQ$

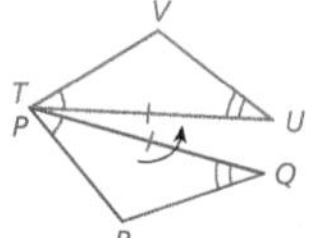

D. How do you know the image of point Q is point U? Point P and point T have been mapped to each other, and $PQ = TU$. Therefore, point Q maps to point U.

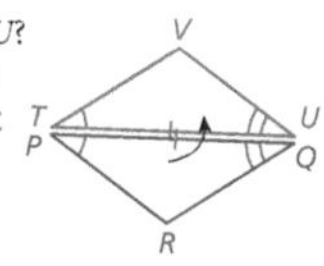

E. Finally, what rigid motion will map point R to point V? a reflection over $\overline{TU}$

F. Since $\angle P \cong \angle T$ and $\angle Q \cong \angle U$, the image of $\overline{PR}$ lies on $\overline{TV}$, and the image of $\overline{QR}$ lies along $\overline{VU}$. Point V is the only point that lies on both $\overline{TV}$ and $\overline{VU}$, and the image of point R must also lie on both $\overline{TV}$ and $\overline{VU}$. You can conclude that point V is the image of point R. What else does this reasoning allow you to conclude? What is the sequence of rigid motions that maps $\triangle PQR$ to $\triangle TUV$. $\triangle PQR \cong \triangle TUV$; a translation that maps P to T, a rotation that maps point Q to point U, and a reflection over $\overline{TU}$ that maps point V to point R

 **Turn and Talk** Aliyah comments that this process works for any two triangles with two congruent angles, not necessarily only those with a congruent included side as well. Do you agree? Explain. See margin.

Task 2 MP **Use Tools** Students transform a given triangle that maps to another in order to prove congruence.

CONNECT TO VOCABULARY

Have students use the **Interactive Glossary** to record their understanding of the vocabulary in this task.

Sample Guided Discussion:

Q **What transformation can you use to "slide" point *P* to point *T*?** a translation

Q **How can you measure the angle of rotation to map point *Q* to point *U*?** Possible answer: Triangle *PQR* has to be rotated so that point *Q* moves up to map it to point *U*. The angle between the two points is the angle of rotation, or angle *UPQ*.

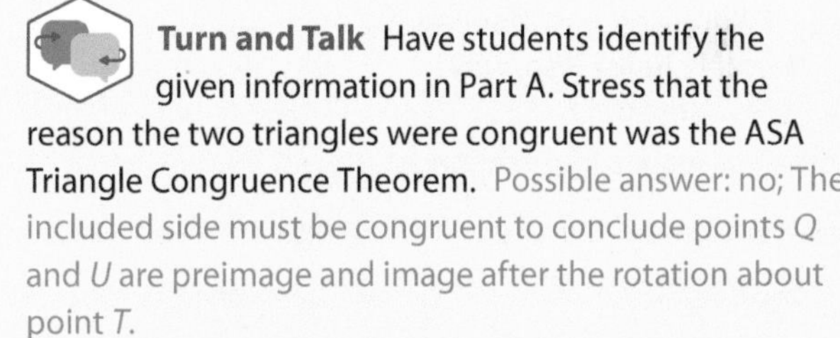

Turn and Talk Have students identify the given information in Part A. Stress that the reason the two triangles were congruent was the ASA Triangle Congruence Theorem. Possible answer: no; The included side must be congruent to conclude points *Q* and *U* are preimage and image after the rotation about point *T*.

PROFICIENCY LEVEL

Beginning

Have three students role play as two angles and the included side. Students stand in a row. The left person says, "Jess and I are the two angles and Tom is the included side." The middle person says he or she is the included side and names the angles. The right person states their status. Students change positions so that each one is the included side.

Intermediate

Give pairs of students a set of index cards with labeled triangles. Have students draw a card and use a complete sentence to name 2 angles and the included side.

Advanced

Students are given blank index cards. On the front they draw a triangle with 2 angles and the included side marked. On the back, they name the side and angles. Then have students draw a counterexample of a triangle with the same angles and a side that is congruent, but not included. Have students prove why these two triangles are not congruent.

Task 3 (MP) **Use Structure** Students use the ASA Triangle Congruence Theorem to determine if two triangles are congruent.

Sample Guided Discussion:

Q **What is the measure of $\angle C$? $\angle F$?** 50°; 50°

Q **If you only know that the corresponding angles of two triangles are congruent can you prove the triangles are congruent? Explain.** no; Possible answer: If corresponding angles are congruent, the triangle will have the same shape, but the lengths of the sides could be different.

Task 4 (MP) **Construct Arguments** Students use what they know about triangles to prove that two overlapping triangles are congruent.

Sample Guided Discussion:

Q **Would drawing and labeling the overlapping triangles separately help? If so, why?** yes; Possible answer: It would clearly show the two triangles and it would be easier to mark congruent parts.

Q **Can you prove another pair of triangles in the figure are congruent by ASA? Why?** yes; Possible answer: If a point is labeled at the intersection of sides $\overline{VW}$ and $\overline{XY}$, as point Z, then, by ASA Congruence Theorem, $\triangle XVZ \cong \triangle WYZ$.

Turn and Talk Point out to students that the corresponding vertices of the triangles in a congruence statement must match. Possible answer: The order is important because it indicates which vertices are corresponding; There are many cases where there is more than one possible order; One example would be a pair of congruent equilateral triangles in which any vertex corresponds to any other vertex.

Determine Whether or Not Triangles Are Congruent Using ASA Triangle Congruence

3 Use the ASA Triangle Congruence Theorem to determine whether the triangles are congruent.

Compare angle measures.

$m\angle A = m\angle D = 30°$

$m\angle B = m\angle E = 100°$

So, $\angle A \cong \angle D$ and $\angle B \cong \angle E$.

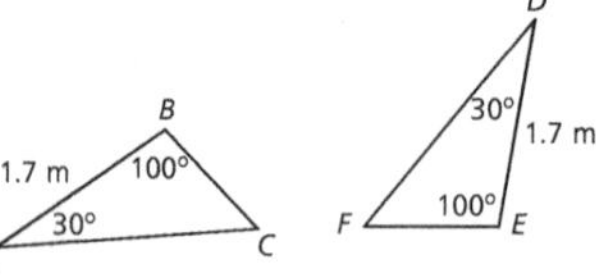

A. What other information is needed to prove that the triangles are congruent using the ASA Congruence Theorem? The included sides must be proved congruent.

B. Is $\overline{AB} \cong \overline{DE}$? How do you know? yes; Each side is 1.7 m long.

C. Notice that the triangles have two pairs of congruent angles, and the included sides between each pair are also congruent. Can you conclude that $\triangle ABC \cong \triangle DEF$? Explain your reasoning. yes; The criteria to use ASA Triangle Congruence have been met.

Prove Triangles Are Congruent

4 Write a proof that shows $\triangle XYV \cong \triangle WVY$.

Given: $\angle XVY \cong \angle WYV$ and $\angle XYV \cong \angle WVY$.

Prove: $\triangle XYV \cong \triangle WVY$

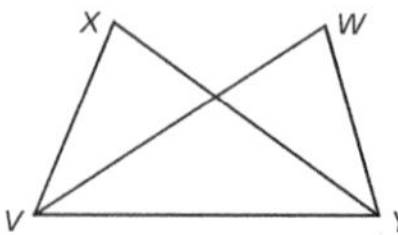

A. Can the ASA Triangle Congruence Theorem be used to show that the triangles are congruent? Explain why or why not. yes; The triangles have two congruent angles and share the included side.

B. The proof is shown below. Why does Step 3, $\overline{VY} \cong \overline{VY}$, need to be included in the proof? to prove the included sides between the congruent angles are congruent

Statements	Reasons
1. $\angle XVY \cong \angle WYV$	1. Given
2. $\angle XYV \cong \angle WVY$	2. Given
3. $\overline{VY} \cong \overline{VY}$	3. Reflexive Property of Congruence
4. $\triangle XYV \cong \triangle WVY$	4. ASA Triangle Congruence Theorem

C. Suppose the two triangles don't share side VY. What additional piece of information would you need to prove the triangles are congruent? See Additional Answers.

Turn and Talk Why is the order in a congruence statement so important? Is there ever a case where there is more than one possible order? Explain. See margin.

222

Step It Out

Apply the ASA Triangle Congruence Theorem

5 Kaylee is designing a kite to enter in a contest at the local park. Can she use the given information to conclude that the leading edges have the same length and the trailing edges have the same length?

Determine the given information and what needs to be proved.

The diagram shows that $\angle QPS \cong \angle RPS$ and $\angle QSP \cong \angle RSP$.

It needs to be shown that the leading edges have the same length and the trailing edges have the same length. If the triangles are congruent, all corresponding sides have the same length.

Write a proof.

Given: $\angle QPS \cong \angle RPS$ and $\angle QSP \cong \angle RSP$.

Prove: $PQ = PR$, $QS = RS$

A. How can you use triangle congruence to prove this statement?

Statements	Reasons
1. $\angle QPS \cong \angle RPS$	1. Given
2. $\angle QSP \cong \angle RSP$	2. Given
3. $\overline{PS} \cong \overline{PS}$	3. Reflexive Property of Congruence
4. $\triangle PQS \cong \triangle PRS$	4. ASA Triangle Congruence Theorem
5. $\overline{PQ} \cong \overline{PR}$, $\overline{QS} \cong \overline{RS}$	5. Corresponding parts of congruent figures are congruent.
6. $PQ = PR$, $QS = RS$	6. Definition of congruent segments

B. In Statement 4, can a different congruence statement be written? If so, give an example. yes; Possible answer: $\triangle PSQ \cong \triangle PSR$

C. In Statement 5, how do you know that $\overline{PQ}$ corresponds to $\overline{PR}$ and $\overline{QS}$ corresponds to $\overline{RS}$? The congruence statement is written so congruent angles are corresponding. We can use the congruence statement to identify corresponding sides.

Turn and Talk Could you still prove $\triangle PQS \cong \triangle PRS$ if the congruent angles are not marked in the figure, but instead you know $\overline{PS}$ bisects $\angle QPS$ and $\angle QSR$? See margin.

Step It Out

Task 5  **Model with Mathematics** Students use the ASA Triangle Congruence Theorem to prove that a kite has two pairs of congruent sides.

OPTIMIZE OUTPUT Critique, Correct, and Clarify

Have students work with a partner to discuss all pairs of triangles in the kite that are congruent and to justify the congruence. It may be helpful to label the intersection of $\overline{PS}$ and $\overline{QR}$ as point T and explain that they are perpendicular. Encourage students to use the vocabulary term *included side* and the *ASA Triangle Congruence Theorem* in their discussions. Students should revise their answers, if necessary, after talking about the problem with their partner.

Sample Guided Discussion:

Q **Which lengths represent the leading edges?** *PQ and PR*

Q **Which lengths represent the leading edges?** *QS and RS*

Q **What do you know about side $\overline{PS}$?** Side $\overline{PS}$ is congruent for both triangles because it is one length.

Turn and Talk Have students identify the corresponding sides and corresponding angles of the congruent triangles. yes; If you know segment $\overline{PS}$ bisects both angles, then you know both halves of the angles are congruent by the definition of an angle bisector.

ONLINE

Assign the Digital On Your Own for
- built-in student supports
- Actionable Item Reports
- Standards Analysis Reports

On Your Own

Assignment Guide

The chart below indicates which problems in the On Your Own are associated with each task in the Learn Together. Assign daily homework for tasks completed.

Learn Together Tasks	On Your Own Problems
Task 1, p. 220	Problem 6
Task 2, p. 221	Problems 6 and 7
Task 3, p. 222	Problems 8–11 and 17–21
Task 4, p. 222	Problems 12–16
Task 5, p. 223	Problem 22

Check Understanding

1. What is one way you can determine if two triangles are congruent without checking all six pairs of corresponding angles and sides? See Additional Answers.
2. If two triangles have two pairs of congruent angles and a pair of congruent included sides, will a series of rigid transformations always map one triangle to the other? Explain. yes; Congruent figures can always be mapped to one another using rigid transformations.
3. Triangles ABC and DEF are being compared. In the triangles, $m\angle A = 45°$, $m\angle B = 45°$, $m\angle D = 45°$, $m\angle F = 90°$, and $AB = DE = 1.5$ cm. Are the two triangles congruent? yes
4. What must be true in order to use the ASA Triangle Congruence Theorem to prove that two triangles are congruent? See Additional Answers.
5. Two triangular pieces of artwork are hanging side by side. One is an equilateral triangle with a side length of 3 ft. The second is an equilateral triangle with a side length of 4 ft. Is it possible to prove the triangles are congruent using the ASA Triangle Congruence Theorem? Why or why not? no; The two triangles have no corresponding congruent sides.

On Your Own

6. (MP) **Use Tools** Sara draws a triangle by first creating segment $\overline{AB}$ that is 2 in. long. She uses a protractor to draw a 20° angle with $\overline{AB}$ as one side and B as the vertex. Next, she draws a 50° angle with $\overline{AB}$ as one side and A as the vertex. Follow Sara's steps to create a triangle, labeling each vertex and angle. Is there a series of rigid transformations that would map your triangle to Sara's triangle? Justify your reasoning.

6–11. See Additional Answers.

7. Describe a series of transformations that maps $\triangle LMN$ to $\triangle XYZ$.

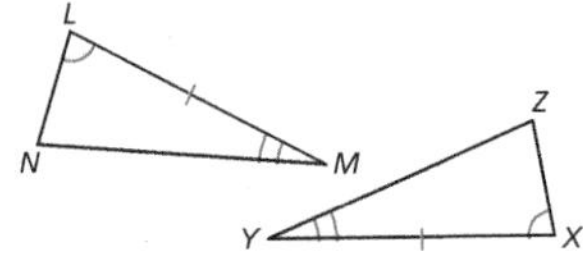

Determine if the triangles are congruent. Explain your reasoning.

8.
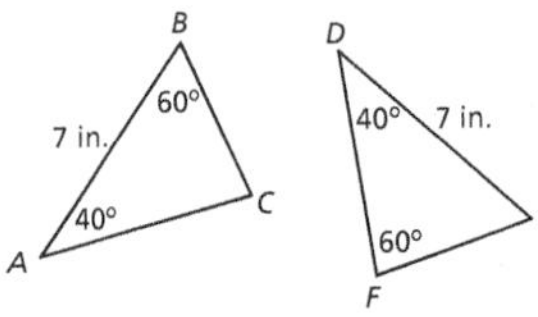

9.
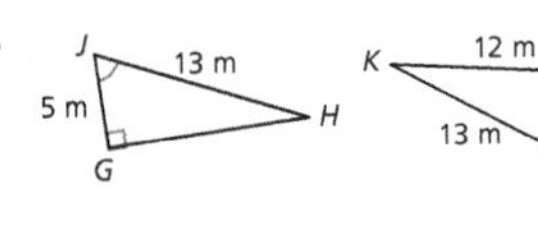

10.
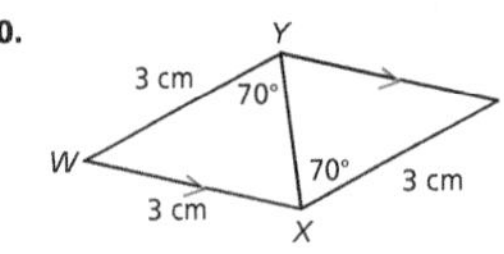

11.
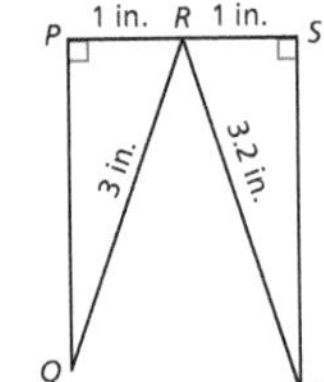

224

3 Check Understanding

Formative Assessment

Use formative assessment to determine if your students are successful with this lesson's learning objective.

Students who successfully complete the Check Understanding can continue to the On Your Own practice.

For students who miss 1 or more problems, work in a pulled small group using the Almost There small-group activity on page 119C.

ONLINE

Assign the Digital Check Understanding to determine
- success with the learning objective
- items to review
- grouping and differentiation resources

4 Differentiation Options

Differentiate instruction for all students using small-group activities and math center activities on page 119C.

Reteach

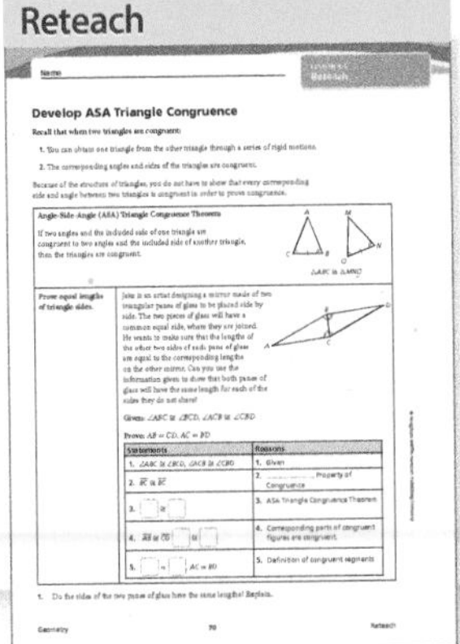

Challenge

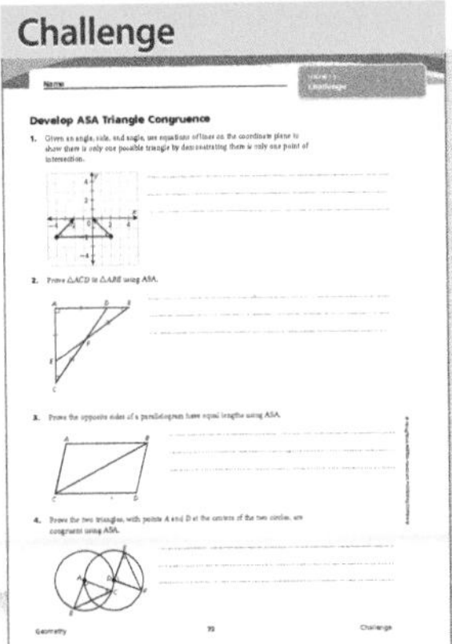

12. The figure shows figure $HJKL$. What additional information do you need to prove $\triangle HJL \cong \triangle KLJ$ by the ASA Triangle Congruence Theorem?

12–16. See Additional Answers.

13. Copy and complete the proof.

Given: $\overline{WX}$ bisects $\angle YWZ$ and $\angle YXZ$.

Prove: $\triangle YWX \cong \triangle ZWX$

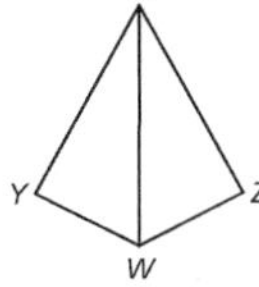

Statements	Reasons
1. $\overline{WX}$ bisects $\angle YWZ$ and ___?___.	1. Given
2. $\angle YWX \cong \angle ZWX$	2. Definition of angle bisector
3. $\angle YXW \cong \angle ZXW$	3. Definition of angle bisector
4. $\overline{XW} \cong \overline{XW}$	4. ___?___
5. ___?___	5. ___?___

Write each proof.

14. **Given:** $\overline{LM} \parallel \overline{PQ}$ and $\overline{LP} \parallel \overline{MQ}$.

Prove: $\triangle LMQ \cong \triangle QPL$

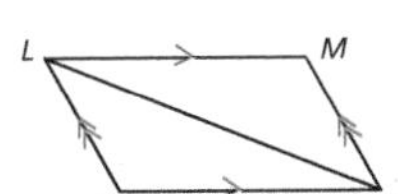

15. **Given:** $\overline{BD}$ is perpendicular to $\overline{AC}$, D is the midpoint of $\overline{AC}$, and $\angle A \cong \angle C$.

Prove: $\triangle ADB \cong \triangle CDB$

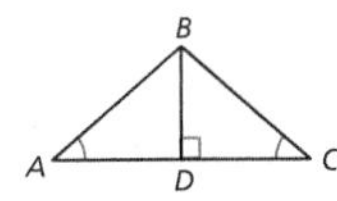

16. **Given:** $\overline{SQ}$ bisects $\overline{PT}$ and $\angle T \cong \angle P$.

Prove: $\triangle SRT \cong \triangle QRP$

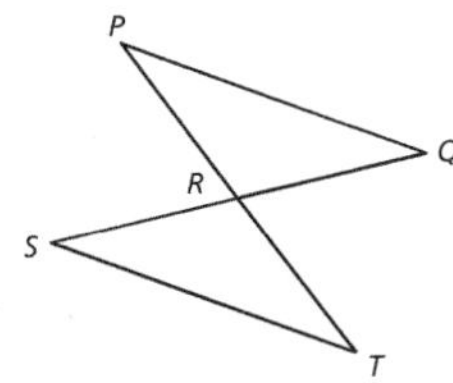

17. **Use Repeated Reasoning** For what values of x and y is $\triangle ABD$ congruent to $\triangle ACD$?

$x = 10$
$y = 4$

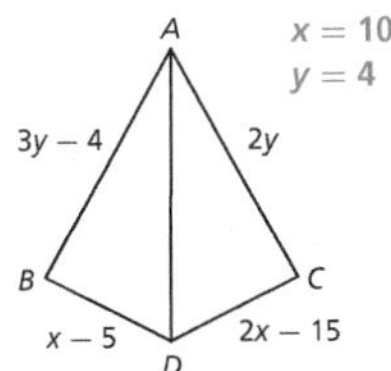

Watch for Common Errors

Problem 12 Students may not understand that they need to know $\overline{JH} \parallel \overline{KL}$ to accurately determine that the second angle comprising angles J and L are congruent. Have students think of $\overline{JL}$ as a transversal that intersects segments $\overline{JH}$ and $\overline{HL}$. Ask what they would need to know in order to prove that the alternate interior angles are congruent.

Questioning Strategies

Problem 15 Suppose that $\overline{BD}$ is the perpendicular bisector of $\overline{AC}$. Can you still prove the triangles are congruent by the ASA Triangle Congruence Theorem? Explain. no; Possible answer: You need to know that $\angle A \cong \angle C$ to use the ASA Triangle Congruence Theorem.

5 Wrap-Up

Summarize learning with your class. Consider using the Exit Ticket, Put It in Writing, or I Can scale.

Exit Ticket 

Prove two of the triangles congruent using the ASA Triangle Congruence Theorem showing steps and reasons. Are the two triangles congruent? Explain.

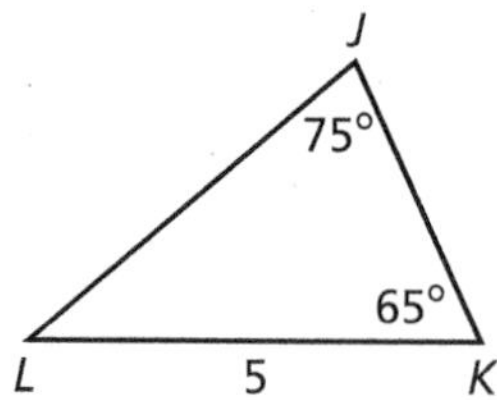

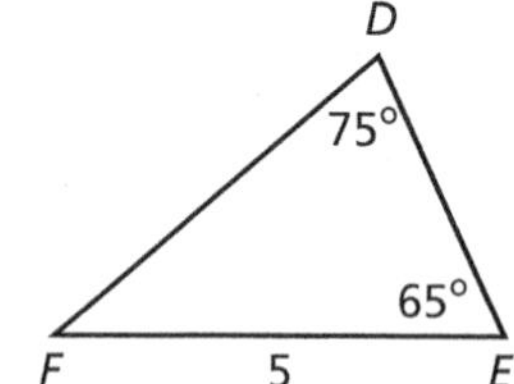

yes; By the Third Angles Theorem, $\angle L \cong \angle F$. Then by the ASA Triangle Congruence Theorem, the triangles are congruent.

Put It in Writing 

Explain why the ASA Triangle Congruence Theorem works.

I Can 

The scale below can help you and your students understand their progress on a learning goal.

4	I can use ASA congruence criteria to prove that two triangles are congruent, and I can explain my proof to others.
3	I can use ASA congruence criteria to prove that two triangles are congruent.
2	I can recognize when to use ASA congruence criteria but cannot apply it.
1	I can recognize some congruent parts of two triangles.

Spiral Review • Assessment Readiness

These questions will help determine if students have retained information taught in the past and can also prepare them for high-stakes assessments. Here, students must identify a method to prove figures are congruent **(7.1)**, select corresponding parts of congruent triangles **(7.2)**, and select a congruent transformation from compositions of transformations **(7.3)**.

Determine if enough information is given to prove $\triangle LMN \cong \triangle LKN$ using the ASA Triangle Congruence Theorem. Explain your reasoning.

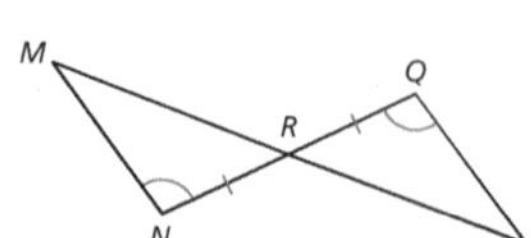

18. $\overline{LN} \perp \overline{MK}$

19. $\angle M \cong \angle K$

20. $\overline{LN}$ bisects $\overline{MK}$, $\overline{LN} \perp \overline{MK}$

21. $\angle MNL \cong \angle KNL$, $\overline{LN}$ bisects $\angle MLK$

18–22. See Additional Answers.

22. (MP) **Use Structure** Consider the diagram shown. Sometimes you can use triangle congruence to prove other relationships.

A. Explain how you know the triangles are congruent by the ASA Congruence Theorem.

B. Write a flow proof to show that $\overline{MN} \cong \overline{PQ}$.

M Q R N P

Spiral Review • Assessment Readiness

23. Which method can be used to prove two figures are congruent?

Ⓐ finding equivalent perimeters
Ⓑ rigid transformations
Ⓒ finding equivalent areas
Ⓓ counting vertices

24. If $\triangle PQR \cong \triangle XYZ$, which statements must be true? Select all that apply.

Ⓐ $\overline{RP} \cong \overline{ZX}$
Ⓑ $\angle QRP \cong \angle ZXY$
Ⓒ $\angle R \cong \angle Z$
Ⓓ $\overline{RP} \cong \overline{QR}$
Ⓔ $\overline{YX} \cong \overline{PQ}$
Ⓕ $\angle P \cong \angle Y$

25. Which function represents a congruent transformation?

Ⓐ $(x, y) \to (-x, -y)$
Ⓑ $(x, y) \to (2x, 2y)$
Ⓒ $(x, y) \to (x + y, y + x)$
Ⓓ $(x, y) \to \left(\frac{x}{2}, \frac{y}{2}\right)$

26. Is it necessary to prove all corresponding angles and sides between two triangles are congruent to prove the triangles are congruent?

Ⓐ Yes, you must check all six parts.
Ⓑ Yes, all parts of congruent triangles must be congruent.
Ⓒ No, triangle congruence theorems can use less information than all six parts.
Ⓓ No, some corresponding parts may not be congruent in congruent triangles.

 **I'm in a Learning Mindset!**

How did I evaluate opposing points of view when trying to determine triangle congruency as a group?

Keep Going to Journal and Practice Workbook

Learning Mindset

Resilience Responds to Feedback

Point out that two figures can be proven congruent by using a sequence of rigid motions or by showing all pairs of corresponding sides and all pairs of corresponding angles are congruent. A pair of triangles has six pairs of corresponding parts, 3 corresponding sides and 3 corresponding angles. Because of the relationship of the parts of a triangle, congruence can be proven by showing 3 pairs of corresponding parts are congruent. The ASA Triangle Congruence is one method of proof that can be used to evaluate opposing views of triangle congruency. *When determining triangle congruency as a group, how well do you communicate? How do you evaluate other points of view? When trying to resolve differences in methods or in content, are you positive and encouraging? Do you play an active part in trying to come to solutions to problems?*

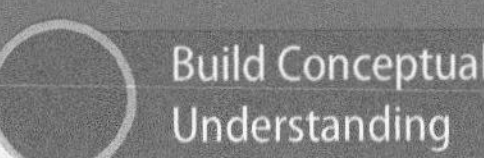

8.2 Develop SAS Triangle Congruence

LESSON FOCUS AND COHERENCE

Mathematics Standards

- Explain how the criteria for triangle congruence (ASA, SAS, and SSS) follow from the definition of congruence in terms of rigid motions.

Mathematical Practices and Processes 

- Use appropriate tools strategically.
- Look for and express regularity in repeated reasoning.
- Construct viable arguments and critique the reasoning of others.

I Can Objective 

I can use SAS congruence criteria to prove that two triangles are congruent.

Learning Objective

Define and identify SAS triangle congruence criteria and prove they are sufficient to assume triangle congruence and use SAS congruence criteria to prove triangles are congruent.

Language Objective

Use corresponding sides and angles and SAS congruence to define congruent triangles.

Vocabulary

New: included angle

Lesson Materials: compass, protractor, scissors, ruler

Mathematical Progressions

Prior Learning	Current Development	Future Connections
Students: • showed that two triangles are congruent if and only if corresponding pairs of sides and corresponding pairs of angles are congruent. **(7.1–7.3)**	**Students:** • define and identify triangle measures in an SAS configuration. • prove SAS congruence criteria are sufficient to assume triangles are congruent. • use SAS congruence criteria to prove triangles are congruent.	**Students:** • will write coordinate proofs about triangle relationships. **(9.4 and 9.5)** • will write coordinate proofs about parallelograms. **(11.1–11.5)** • will construct an equilateral triangle, a square, and a regular hexagon inscribed in a circle. **(15.2)**

PROFESSIONAL LEARNING

Using Mathematical Practices and Processes

Look for and express regularity in repeated reasoning.

This lesson provides an opportunity for students to see how this Mathematical Practice Standard might be applied incorrectly. According to the standard, mathematically proficient students "notice if calculations are repeated, and look both for general methods and for shortcuts." What may seem to be a valid shortcut, though, can turn out to lead to incorrect results. For example, in Task 3 students may identify two pairs of congruent sides and a pair of congruent angles and then incorrectly assume the conclusion of the SAS Triangle Congruence Theorem. While the comparing of measures was a repeated procedure, assuming the conclusion of the theorem requires the additional step of confirming their reasoning that the angle pair is included between the corresponding sides.

WARM-UP OPTIONS

ONLINE

PROJECTABLE & PRINTABLE

ACTIVATE PRIOR KNOWLEDGE • Describe a Sequence of Rigid Motions

Use these activities to quickly assess and activate prior knowledge as needed.

Problem of the Day

Give a sequence of rigid motions that shows $\triangle ABC \cong \triangle DEF$.

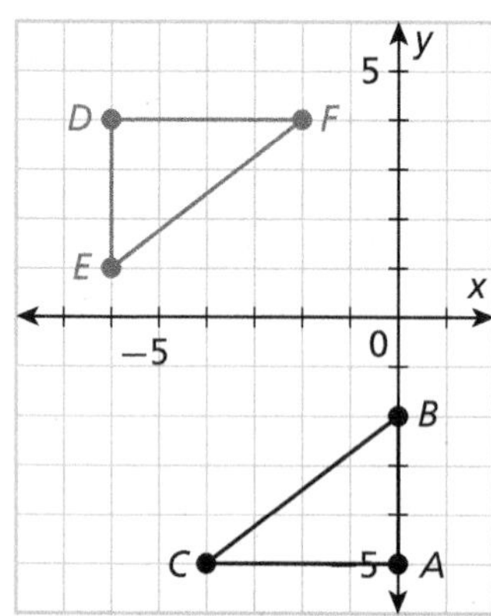

Possible answer: Translate $\triangle ABC$ by the rule $(x + 6, y + 1)$ then rotate $\triangle A'B'C'$ by 180° about the origin.

Quick Check for Homework

As part of your daily routine, you may want to display the Teacher Solution Key to have students check their homework.

Make Connections

Based on students' responses to the Problem of the Day, choose one of the following:

1. Project the Interactive Reteach, Geometry, Lesson 7.3.
2. Complete the Prerequisite Skills Activity:

Have students work in pairs. Give each pair a transparency and a coordinate plane (Teacher Resource Masters). One student should graph three non-collinear points of a triangle with integer coordinates in one quadrant and label them. Then while the other student is not watching, use the transparency to compose one or two rigid motions on the triangle and then plot the final image with integer coordinates on the coordinate plane. The other student can then try to determine the rigid motions that map the preimage onto the image.

- *How can we tell a translation has been applied?* The orientations of the preimage and image are the same, and the image does not seem to have been turned.
- *How can we tell a rotation has been applied?* The image seems to have been turned.
- *How can we tell a reflection has been applied?* The orientations of the preimage and image are flipped.

If students continue to struggle, use Tier 2 Skill 13.

SHARPEN SKILLS

If time permits, use this on-level activity to build fluency and practice basic skills.

Vocabulary Review

Objective: Students demonstrate an understanding of terms related to a transversal.
Materials: Word Description (Teacher Resource Masters)

Have students work in pairs. Each pair should work together to build a word description for the term "included side" without looking anything up. Encourage students to work together to recall as much information as possible and to not worry too much about being exact at this point.

When all pairs have had ample time to work, have a class discussion and create a word description on the board.

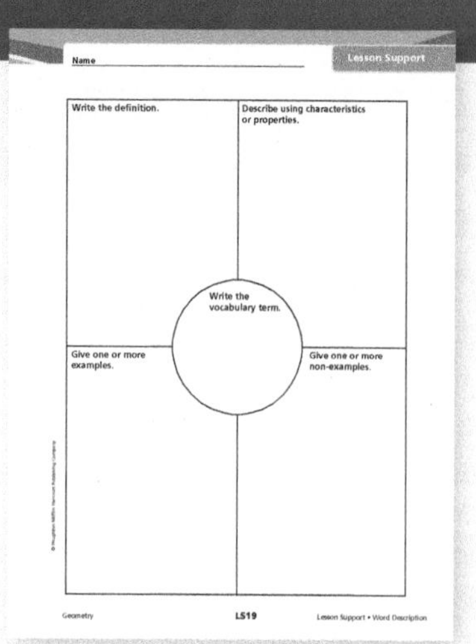

PLAN FOR DIFFERENTIATED INSTRUCTION

Small-Group Options

Use these teacher-guided activities with pulled small groups.

On Track

Have students work in pairs. Ask,

"Suppose $\triangle ABC$ is isosceles with $\overline{AB} \cong \overline{AC}$." Have students do the following:

- Show that the bisector of $\angle A$ is a perpendicular bisector of $\overline{BC}$.
- Show that $\angle B \cong \angle C$.
- Draw a triangle that is congruent to $\triangle ABC$ with one angle that measures 45°, using SAS as their reasoning for congruence.

Almost There

Materials: index cards, protractor, ruler

Give each pair a set of six index cards labeled with one of the following measures: 3 cm, 4 cm, 5 cm, 40°, 60°, or 80° written on it. Have groups do the following:

- Draw two side length cards and one angle measure card.
- Attempt to create a triangle with those measures or explain why it is not possible.
- If it was possible to create a triangle with the given measures, attempt to create another or explain why the first triangle is unique.

Ready for More

Materials: index cards, protractor, ruler

Give each pair a set of six index cards labeled with one of the following measures: 3 cm, 4 cm, 5 cm, 40°, 60°, or 80° written on it. Have groups do the following:

- Draw three cards.
- Attempt to create a triangle with those measures or explain why it is not possible.
- If it was possible to create a triangle with the 3 given measures, attempt to create another or explain why the first triangle is unique.

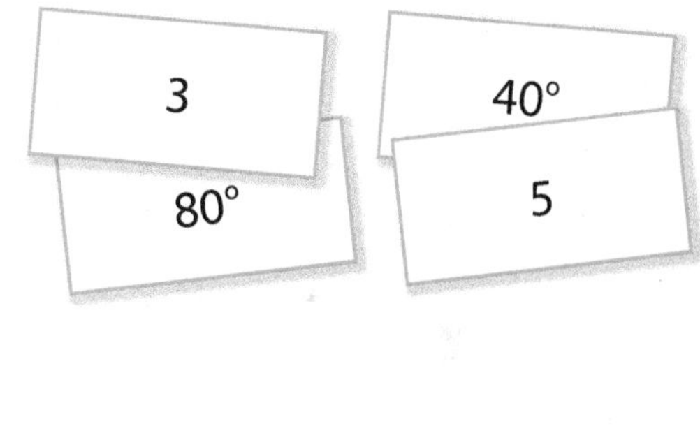

Math Center Options

Use these student self-directed activities at centers or stations.

Key: ● Print Resources ● Online Resources

On Track

- ● Interactive Digital Lesson
- ●● Journal and Practice Workbook
- ● Interactive Glossary (printable): **included angle**
- ● Module Performance Task

Almost There

- ● Reteach 8.2 (printable)
- ● Interactive Reteach 8.2
- ● RtI Tier 2 Skill 13: Sequence of Transformations
- ● Illustrative Mathematics: Why Does SAS Work?

Ready for More

- ● Challenge 8.2 (printable)
- ● Interactive Challenge 8.2
- ● Illustrative Mathematics: When Does SSA Work to Determine Triangle Congruence?

Unit Project Check students' progress by asking what sides and angles they are using in their proof.

View data-driven grouping recommendations and assign differentiation resources.

During the *Spark Your Learning*, listen and watch for strategies students use. See samples of student work on this page.

Use Side Lengths and Angle Measure — Strategy 1

The triangles share one side, so the triangles have 2 pairs of congruent sides and a congruent angle that is between the two congruent sides, or included. I know only one triangle can be created when given two side lengths and the angle between those two sides.

If students . . . use the given measures and the Reflexive Property of Congruence to identify the shared side as a third congruent measure, they are demonstrating an exemplary understanding of triangle relationships.

Have these students . . . explain how they determined their criteria and how they knew it makes a triangle unique. **Ask:**

- Q How did you use the given measurements to describe the triangle?
- Q What if the measured parts were arranged differently?

Use Rigid Motions — Strategy 2

After tracing one triangle, I reflected the traced triangle and translated it so it matches up with the other triangle. The triangles are congruent.

If students . . . use rigid motions to map one triangle to the other, they understand congruence as a result of rigid motions but may not understand how some measurement criteria make triangles unique.

Activate prior kowledge . . . by having students reason with the given measurements. **Ask:**

- Q What if the measures of two angles and the side between them was given instead?
- Q How many different triangles can you create that have corresponding parts congruent to one of the given triangles?

COMMON ERROR: Does Not Use the Common Side

I only know the triangles have one pair of corresponding congruent sides and one pair of corresponding congruent angles, which is not enough information to tell whether the triangles are congruent.

If students . . . do not use the common side of the triangles in their reasoning, they may not understand the Reflexive Property of Congruence and triangle relationships.

Then intervene . . . by pointing out that the triangles share another pair of corresponding congruent sides through the common side. **Ask:**

- Q Can you trace both triangles so they are not touching?
- Q Why can you reuse the side they share for each of the triangles?

8.2

Develop SAS Triangle Congruence

I Can use SAS congruence criteria to prove that two triangles are congruent.

Spark Your Learning

A triangular window must have precisely cut panes in order to fit together properly. In this window, the builder needed to cut all panes the same size and shape.

Complete Part A as a whole class. Then complete Parts B–D in small groups.

A. What mathematical question can you ask about the triangles in the window? What kinds of information would you need to know to answer your question?

A. Are the triangles congruent?; the angle measures or side lengths

B. State the facts you already know about triangle congruence. See Additional Answers.

C. To answer your question, what strategy and tool would you use along with all the information you have? What answer do you get? See Strategies 1 and 2 on the facing page.

D. How do you think your investigation and conclusion relates to other triangles? See Additional Answers.

Turn and Talk Think about what you already know about triangle congruence. How can you combine what you already know with the information presented above to draw new conclusions about triangle congruence? Consider the given angle measures. How do you think that relates to your conclusion? See margin.

CULTIVATE CONVERSATION • Co-Craft Questions

If students have difficulty formulating a mathematical question about the situation in the Spark Your Learning, ask them to imagine they are designing the window. What are some natural questions to ask about this situation?

Work together to craft the following questions:

- How should the panes be cut so they fit together correctly?
- Are the triangles congruent?
- Would the triangles fit correctly if they were not congruent?

Then have students think about what additional information, if any, they would need to answer these questions. **Ask:**

- Can you determine whether the triangles are congruent if you only have the measures of some of their angles? Why or why not?
- Can you determine whether the triangles are congruent if you only have the lengths of some of their sides? Why or why not?

1 Spark Your Learning

▶ MOTIVATE

- Have students look at the photo in their books and read the information contained in the photo. Then complete Part A as a whole-class discussion.
- Give the class the additional information they need to solve the problem. This information is available online as a printable and projectable page in the Teacher Resources.
- Have students work in small groups to complete Parts B–D.

▶ PERSEVERE

If students need support, guide them by asking:

Q **Advancing • Use Tools** Which tool could you use to solve the problem? Why choose that tool and not some other? Students' choices of tools and reasons for choosing them will vary.

Q **Assessing** How many pairs of corresponding congruent angles do the triangles have? Explain. one pair; They each have an angle measuring 50°.

Q **Assessing** How many pairs of corresponding congruent sides do the triangles have? Explain. two pairs; They each have a side measuring 6 inches and they share a side.

Turn and Talk Help students determine whether the given congruent measures are sufficient to conclude the triangles are congruent. Encourage them to think about how given parts are arranged as they did in the last lesson. Students should see that they are given an SAS configuration of measures for each triangle, which is similar to the ASA configuration they worked with in the last lesson. Possible answer: Two triangles can be proven to be congruent if they have two pairs of congruent angles and the pair of included sides are congruent; I also know all parts of two congruent triangles must be congruent. I can conclude that SAS also proves congruence because the angle provided in the figure is the included angle, similar to how the side was included in the previous lesson.

▶ BUILD SHARED UNDERSTANDING

Select groups of students who used various strategies and tools to share with the class how they solved the problem. As they present their solutions, have each group discuss why they chose a specific strategy and tool.

2 Learn Together

Build Understanding

Task 1  **Use Tools** Students use a protractor and ruler to draw a fixed angle and side length and model another fixed side length with a strip of paper. They investigate whether these fixed side-side-angle measures are sufficient to define a triangle and then whether the same set of measures in a side-angle-side configuration are sufficient.

Sample Guided Discussion:

Q If you can draw more than one triangle with a given set of measures, can you conclude that two triangles sharing that set of measures are congruent? Explain. no; The two triangles might not be congruent because the measures not provided for the two different triangles may not be the same.

Turn and Talk Have students look at the triangles they made in Part A and the order of the relationship between the two sides and the angle. Point out that the order of these measures affects the congruency of two triangles. Possible answer: The triangles must be congruent if the corresponding congruent angles are the included angles between the pairs of corresponding congruent sides.

Build Understanding

Draw Triangles Given Two Sides and an Angle

In the previous lesson, you learned that two triangles are congruent if two pairs of corresponding angles and their included sides are congruent. The following steps explore a second way to check for triangle congruency.

1 Use a ruler to draw a horizontal line. Use a protractor and a ruler to draw a 3-inch line segment that meets the horizontal line at a 45° angle.

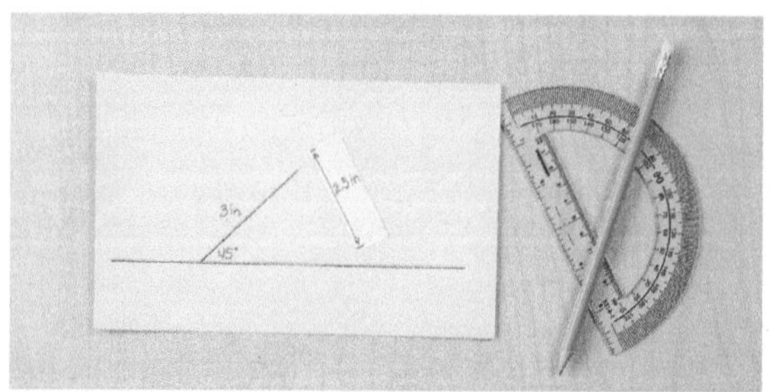

A. Use a thin strip of paper that is 2.5 inches long to complete the triangle. How many different triangles can you form? Draw each option you discover. **A, B. See Additional Answers.**

B. Now place the paper strip so that the angle between the 3-inch segment and 2.5-inch paper strip has a measure of 45°. With this arrangement, is there more than one way to complete the triangle? Why or why not?

 **Turn and Talk** If two triangles have two pairs of congruent corresponding sides and one pair of congruent corresponding angles, under what conditions can you conclude that the triangles must be congruent? Explain. See margin.

Justify SAS Triangle Congruence Using Transformations

Two sides of a triangle form an angle called the **included angle**. Given two side lengths and the measure of the included angle, you can draw only one triangle. So, two triangles with those measures are congruent.

Side-Angle-Side (SAS) Triangle Congruence Theorem

If two sides and the included angle of one triangle are congruent to two sides and the included angle of another triangle, then the triangles are congruent.

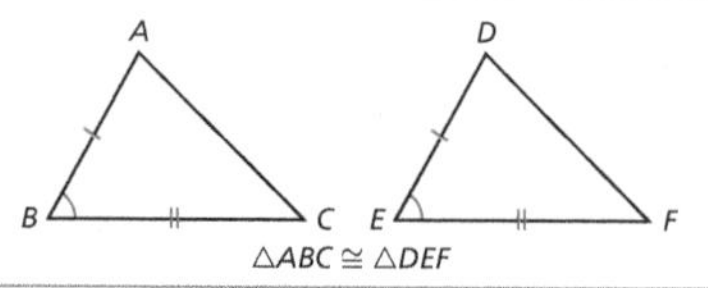

228

LEVELED QUESTIONS

Depth of Knowledge (DOK)	Leveled Questions	What Does This Tell You?
Level 1 **Recall**	For which part of the task was the 45° angle included between the two fixed side lengths? Part B	Students' answers will indicate whether they understand the meaning of parts of a triangle.
Level 2 **Basic Application of Skills & Concepts**	How many triangles, $\triangle ABC$, are there with $m\angle BCA = 50°$, $AC = 10$ cm, $BC = 8$ cm? How many with $m\angle BAC = 50°$, $AC = 10$ cm, and $BC = 8$ cm? one; two	Students' answers will indicate whether they understand the possible number of triangles in the SSA case.
Level 3 **Strategic Thinking & Complex Reasoning**	Suppose for Part A of the task you can use paper strips of lengths other than 2.5 inches. Are there lengths that make it possible to form only one triangle? Are there lengths that make it impossible? A strip about 2.1 inches long or longer than 3 inches will only form one triangle; A strip shorter than 2.1 inches will not form a triangle.	Students' answers will indicate whether they can reason why a given side-side-angle set of measures is insufficient in general to assume uniqueness of a triangle and can reason further to determine cases where such a set of measures would be unique to a triangle or not possible.

Because congruent figures can be mapped to one another using rigid transformations, we can justify SAS Triangle Congruence Theorem using transformations. Remember only rigid transformations such as reflections, translations, and rotations preserve the angle measurements and side lengths between the preimage and image.

2 A. Construct $\triangle LMN$ by copying $\angle C$, side $\overline{CA}$, and side $\overline{CB}$. How do the vertices of $\triangle LMN$ correspond to the vertices of $\triangle ABC$?
L corresponds to *A*, *M* corresponds to *B*, *N* corresponds to *C*.

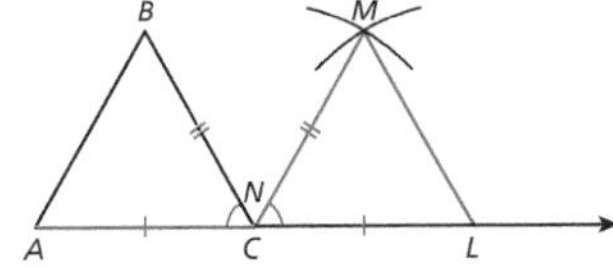

B. What transformation maps $\triangle LMN$ to $\triangle ABC$? Is the transformation a rigid motion?
Possible answer: A reflection over a vertical line through points *N* and *C*; Yes, it is a rigid transformation.

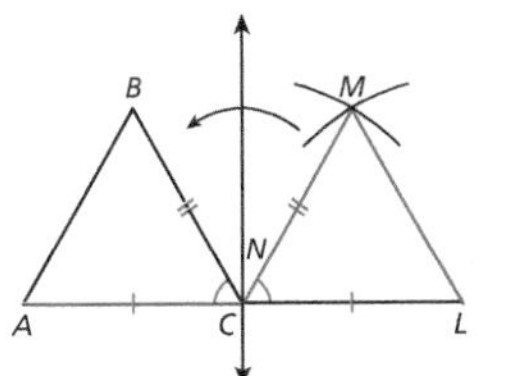

C. Can you conclude that $\triangle LMN \cong \triangle ABC$? Explain your reasoning.
yes; The two triangles are congruent because a sequence of rigid transformations maps one triangle to the other.

Turn and Talk Trace $\triangle LMN$ and $\triangle ABC$ on a separate piece of paper. Cut out each triangle and place them on a desktop so $\overline{AC}$ and $\overline{LN}$ do not lie on the same line. Is it possible to use only one transformation to map $\triangle LMN$ to $\triangle ABC$? If not, describe the sequence of transformations needed to map $\triangle LMN$ to $\triangle ABC$. See margin.

Determine Whether or Not Triangles Are Congruent Using SAS Triangle Congruence

3 Use the SAS Triangle Congruence Theorem to determine whether the triangles are congruent.

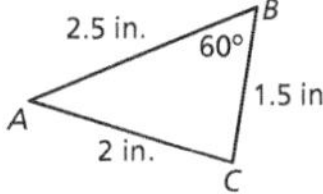

Compare side lengths.

$\overline{AB}$ corresponds to $\overline{DF}$ because $AB = DF = 2.5$ in.

$\overline{BC}$ corresponds to $\overline{FE}$ because $BC = FE = 1.5$ in.

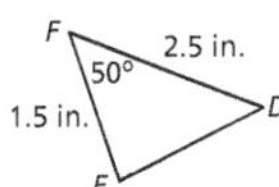

A. Why can you conclude that $\overline{AB} \cong \overline{DF}$ and $\overline{BC} \cong \overline{FE}$?
A–C. See Additional Answers.

B. $\angle B$ corresponds to $\angle F$ because they are the included angles between the corresponding sides. Is $\angle B \cong \angle F$?

C. Can the SAS Triangle Congruence Theorem be used to determine whether these triangles are congruent? Explain why or why not.

PROFICIENCY LEVEL

Beginning

Show students a triangle $\triangle ABC$ and say, "Angle *ABC* is included between sides *AB* and *BC*" while writing, "$\angle ABC$ is included between $\overline{AB}$ and $\overline{BC}$." Prompt students with an angle or side of the triangle and have them name the order in which the given sides and angle are shown in the triangle, such as sides $\overline{BC}$ and $\overline{CA}$ and angle *BCA*. Students would respond angle *BCA* is included between sides $\overline{BC}$ and $\overline{CA}$.

Intermediate

Show students a triangle $\triangle ABC$ and say, "Angle *ABC* is included between sides $\overline{AB}$ and $\overline{BC}$" while writing, "$\angle ABC$ is included between $\overline{AB}$ and $\overline{BC}$." Have students write similar statements for other included angles and sides of the triangle, making sure to use correct notation.

Advanced

Have students write an explanation of how to determine which angles are included between which sides from how a polygon is named, such as hexagon *GHIJKL*.

Task 2 (MP) **Use Tools** Students are formally introduced to the SAS Congruence Theorem. They use a compass and straightedge to copy two sides and an included angle of a triangle and justify SAS Triangle Congruence with transformations.

Make sure students understand that this was only one special case of SAS congruence, but the reasoning used can be extended to the general case.

CONNECT TO VOCABULARY

Have students use the **Interactive Glossary** to record their understanding of the vocabulary in this task.

Sample Guided Discussion:

Q How do you know a rotation does not map $\triangle ABC$ to $\triangle LMN$? The orientations of the triangles are flipped and not turned.

Q In Part B, how can you construct a reflection line between $\triangle ABC$ and $\triangle LMN$? I would construct a line perpendicular to $\overleftrightarrow{AL}$ through point *C* (which is the same point as *N*).

Turn and Talk Make sure students label the vertices of their triangles so they can identify image points when prompted. Encourage them to be creative with the placements of their triangles. Have students specify whether the orientation of the preimage triangle has been preserved through the one transformation or a sequence of transformations. Check students' answers. Each student will arrange the triangles differently.

Task 3 (MP) **Use Repeated Reasoning** Students use SAS Triangle Congruence to determine whether triangles are congruent. While matching part measurements is a repeated process, students still need to take into account where corresponding parts lie on their respective triangles.

Sample Guided Discussion:

Q Are two triangles always congruent if they have two pairs of congruent sides and a pair of congruent angles? Explain. no; If the corresponding angles are not both included between the corresponding sides, then the triangles might not be congruent.

Q If $m\angle E = 60°$, can you say that $\angle B$ and $\angle E$ are corresponding angles? Not if $\overline{AB}$ corresponds to $\overline{DF}$ and $\overline{BC}$ corresponds to $\overline{FE}$, since $\angle B$ is included between $\overline{AB}$ and $\overline{DF}$ but $\angle E$ is not included between $\overline{DF}$ and $\overline{FE}$.

Step It Out

Task 4 (MP) **Construct Arguments** Students analyze a real-world problem involving triangles and interpret an SAS Triangle Congruence from the situation, applying the Reflexive Property of Congruence as an intermediate step. They then use CPCTC in their reasoning to argue that corresponding parts are congruent.

SUPPORT SENSE-MAKING **Three Reads**

Have students read the problem three times. Use the questions below for a different focus each read.

1. What is the situation about?
2. What are the quantities in the situation?
3. What are the possible mathematical questions that you could ask for the situation?

Sample Guided Discussion:

Q What kind of strategy can you use to show $\overline{MQ}$ and $\overline{RN}$ are congruent? I can find two triangles that have $\overline{MN}$ and $\overline{RQ}$ as sides, prove the triangles are congruent, and then use CPCTC.

Q Which triangles have $\overline{NP}$ and $\overline{MP}$ as sides, respectively? $\overline{NP}$ is a side of $\triangle RPN$, while $\overline{MP}$ is a side of $\triangle MPQ$.

Q How can you more easily see the sides that include $\angle P$ on each triangle? I can sketch $\triangle RPN$ and $\triangle MPQ$ separately so they do not overlap or share sides or angles.

Turn and Talk Help students see that ASA congruence requires two pairs of congruent angles should make it clear that they do not initially have enough information to use the criterium. Note that while it may be possible to solve for the additional needed information, a more direct method of solution is available in this case. Possible answer: ASA criteria is not a good choice for this problem because we are not provided initial information about any angles. We are given information about sides, so using criteria involving more sides than angles is a better choice.

Step It Out

Apply SAS Triangle Congruence in a Real-World Context

You can use the SAS Triangle Congruence Theorem to prove triangles are congruent in real-world problems. You may need to determine if the given information meets the criteria for the SAS Triangle Congruence Theorem.

4 An engineer is building a prototype for a clothing rack. The cross bar is positioned so that $\overline{NP} \cong \overline{QP}$ along the congruent sides $\overline{MP}$ and $\overline{RP}$. Prove that the slanted supports $\overline{MQ}$ and $\overline{RN}$ are congruent.

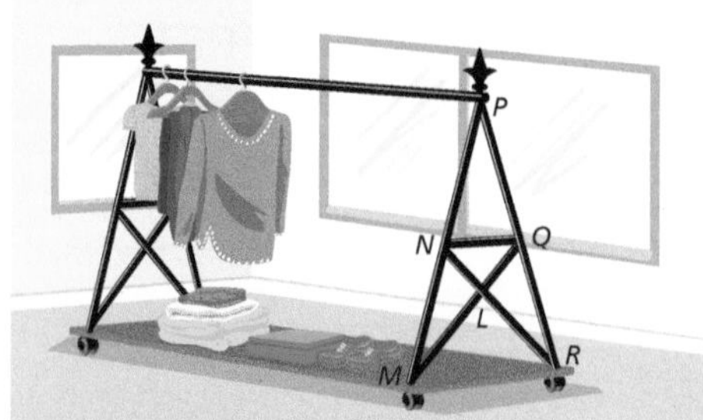

Collect the given information.

You know $\overline{MP} \cong \overline{RP}$ and $\overline{NP} \cong \overline{QP}$.

A. Why should the SAS Triangle Congruence Theorem be used to show congruent triangles?

Possible answer: Two pairs of corresponding congruent sides are given, and we can find information about the included angle.

Prove the unknown information.

The given information tells you about two sides of $\triangle MPQ$ and $\triangle RPN$, specifically that two sets of corresponding sides are congruent.

In both triangles, the included angle between those sides is $\angle P$. By the Reflexive Property of Congruence, $\angle P \cong \angle P$.

B. What is a congruence statement for the two triangles?

$\triangle MPQ \cong \triangle RPN$

The triangles are congruent by the SAS Triangle Congruence Theorem.

Because all corresponding parts of congruent figures are congruent, $\overline{MQ} \cong \overline{RN}$.

C. What other information do you know about the triangles now that you know they are congruent?

$\angle PMQ \cong \angle PRN$, $\angle MQP \cong \angle RNP$

Turn and Talk Why is the ASA Triangle Congruence Theorem not the best choice for this problem? See margin.

Check Understanding

1. Kim draws a triangle with one 4 mm side, one 6 mm side, and a 25° included angle between those sides. If you follow the same steps as Kim to draw a triangle, is your triangle congruent to Kim's triangle? Explain. yes; It is only possible to construct one triangle when given two sides and an included angle.
2. Two triangles have two pairs of congruent sides and a congruent pair of included angles. Can you map one triangle to the other? 2, 3. See Additional Answers.
3. Explain the differences between the situations in which you would use the SAS Triangle Congruence Theorem versus the ASA Triangle Congruence Theorem.
4. $\overline{QS}$ is perpendicular to $\overline{RP}$, and $\overline{QS}$ is an angle bisector of $\angle PQR$. Can you prove $\triangle PQS \cong \triangle RQS$? Justify your answer. yes, by SAS criteria; $\overline{PQ} \cong \overline{QR}$, $\angle PQS \cong \angle RQS$ and shared side $\overline{QS}$ is congruent to itself.

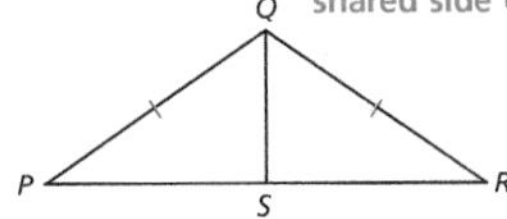

On Your Own

Determine if the triangles are congruent, not congruent, or if there is not enough information to determine. Explain your reasoning. 5–10. See Additional Answers.

5.

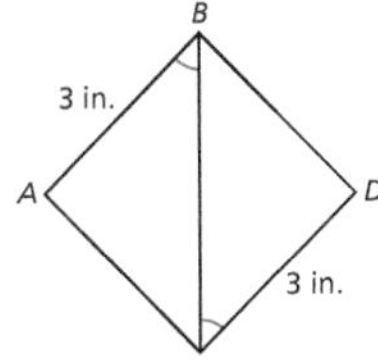

6.

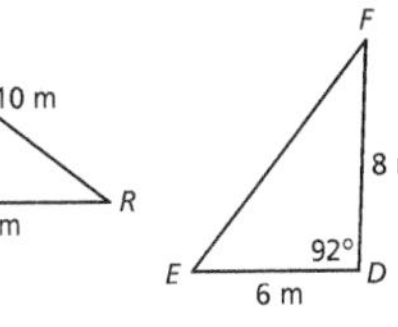

7.

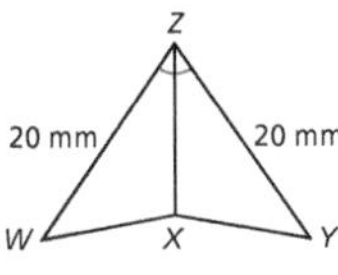

8.

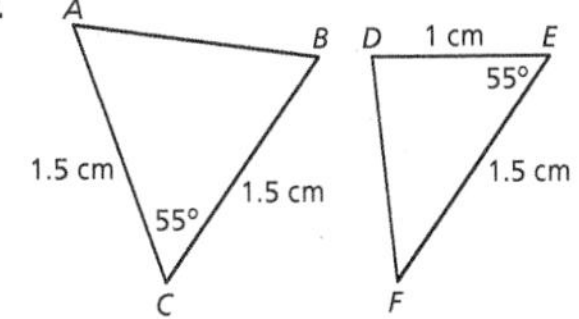

9. Two triangular road signs are mounted on the same post. They each have two pairs of 4-inch sides, and both have an included angle of 30°. Is it possible to prove the triangles are congruent? Explain.
10. **Open Ended** Two congruent triangles have one side that is twice the length of a second side, and a third side that is equal in length to one of the other sides. One angle is not congruent to the other two. Draw two congruent triangles that meet the description.

ONLINE

Assign the Digital On Your Own for
- built-in student supports
- Actionable Item Reports
- Standards Analysis Reports

On Your Own

Assignment Guide

The chart below indicates which problems in the On Your Own are associated with each task in the Learn Together. Assign daily homework for tasks completed.

Learn Together Tasks	On Your Own Problems
Task 1, p. 228	Problems 10, 21, and 22
Task 2, p. 229	Problems 5, 7, and 18
Task 3, p. 229	Problems 5–9, 15–17, 19, 20, and 23
Task 4, p. 230	Problems 11–14, 19, and 24

3 Check Understanding

Formative Assessment

Use formative assessment to determine if your students are successful with this lesson's learning objective.

Students who successfully complete the Check Understanding can continue to the On Your Own practice.

For students who miss 1 problem or more, work in a pulled small group using the Almost There small-group activity on page 227C.

ONLINE

Assign the Digital Check Understanding to determine
- success with the learning objective
- items to review
- grouping and differentiation resources

4 Differentiation Options

Differentiate instruction for all students using small-group activities and math center activities on page 227C.

Reteach

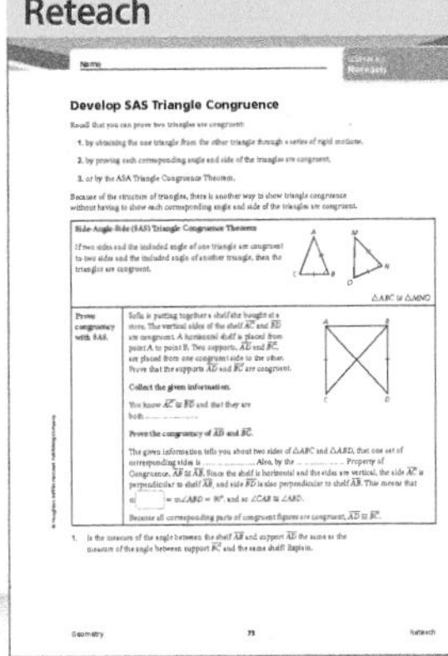

Challenge

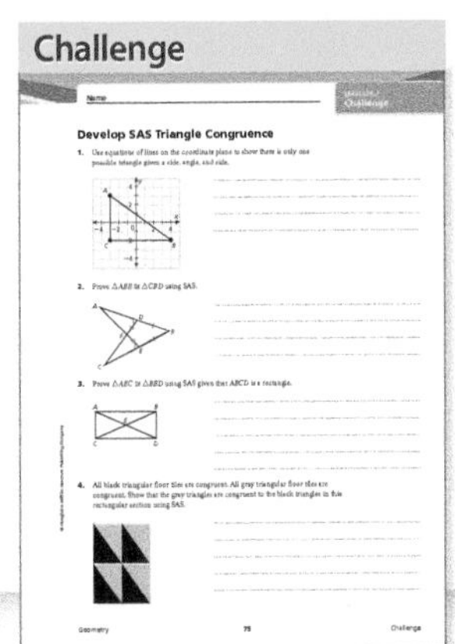

Questioning Strategies

Problem 11 Suppose Zach created the following proof instead. Is his reasoning correct? Explain.

Statements	Reasons
1. $\overline{AD} \parallel \overline{BE}$	1. Given
2. $\overline{DB}$ bisects $\overline{AE}$.	2. Given
3. $\overline{AC} \cong \overline{CE}$	3. Definition of a Bisector
4. $\overline{BC} \cong \overline{CD}$	4. Definition of a Bisector
5. $\angle ACD \cong \angle ECB$	5. Vertical Angles Theorem
6. $\triangle ACD \cong \triangle ECB$	6. SAS Triangle Congruence Theorem

Zach's reasoning is incorrect; We do not know that $\overline{BC} \cong \overline{CD}$ since we are not explicitly told that $\overline{AE}$ bisects $\overline{DB}$. So, Zach does not have enough information to show SAS congruence.

Watch for Common Errors

Problem 13 Students may make the following error in their reasoning:

Statements	Reasons
1. $\overline{PQ} \parallel \overline{RS}$, $PQ = RS$	1. Given
2. $PQ = RS$	2. Given
3. $\angle PSQ \cong \angle RQS$	3. Alternate Interior Angles Theorem
4. $\overline{SQ} \cong \overline{SQ}$	4. Reflexive Property of Congruence
5. $\triangle PQS \cong \triangle RSQ$	5. SAS Triangle Congruence Theorem

Step 5 does not follow since the angles in Step 3 are not included between corresponding sides. Encourage students to make sure corresponding angles are included between corresponding sides before assuming the conclusion of the SAS Triangle Congruence Theorem.

11. MP **Critique Reasoning** Zach creates the following proof to show $\triangle ACD$ is congruent to $\triangle ECB$. Where did he make a mistake? Explain and correct his error. **11–14. See Additional Answers.**

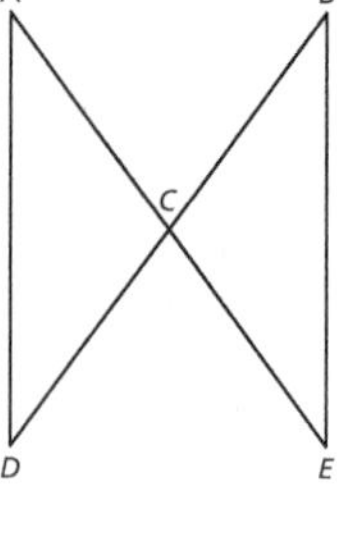

Given: $\overline{AD} \parallel \overline{BE}$ and $\overline{DB}$ bisects $\overline{AE}$.

Prove: $\triangle ACD \cong \triangle ECB$

Statements	Reasons
1. $\overline{AD} \parallel \overline{BE}$	1. Given
2. $\overline{DB}$ bisects $\overline{AE}$.	2. Given
3. $\overline{AC} \cong \overline{CE}$	3. Definition of a bisector
4. $\angle ACD \cong \angle ECB$	4. Vertical Angles Theorem
5. $\triangle ACD \cong \triangle ECB$	5. ASA Triangle Congruence Theorem

12. In a regular polygon, all the sides have the same length and all the angles have the same measure. Copy and complete the proof.

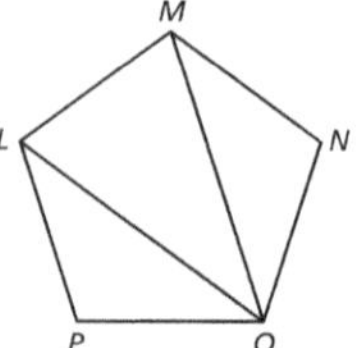

Given: Polygon $LMNOP$ is a regular pentagon.

Prove: $\overline{LO} \cong \overline{MO}$

Statements	Reasons
1. $LMNOP$ is a regular pentagon.	1. ?
2. $MN = LP$, $NO = PO$	2. Definition of regular polygon
3. $\overline{MN} \cong \overline{LP}$, $\overline{NO} \cong \overline{PO}$	3. Definition of congruence
4. ?	4. Definition of regular polygon
5. $\angle P \cong \angle N$	5. ?
6. $\triangle LPO \cong \triangle MNO$	6. ?
7. ?	7. Corresponding parts of congruent figures are congruent.

Write each proof using the SAS Triangle Congruence Theorem.

13. Given: $\overline{PQ} \parallel \overline{RS}$, $PQ = RS$

Prove: $\triangle PQS \cong \triangle RSQ$

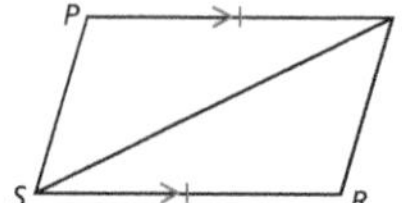

14. Given: $\overline{AC} \cong \overline{DE}$, $\overline{BC} \cong \overline{BD}$, $\angle BCD \cong \angle BDC$

Prove: $\triangle ABC \cong \triangle EBD$

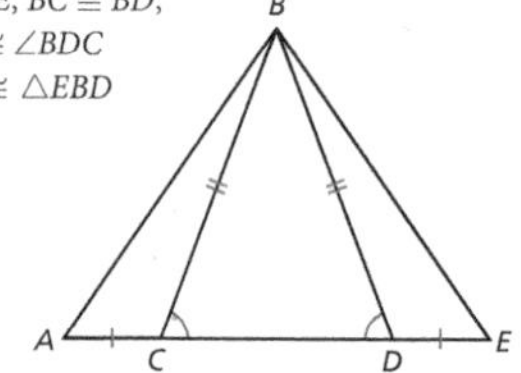

Find the value of the variable that results in congruent triangles.

15. $x = -8$

16.

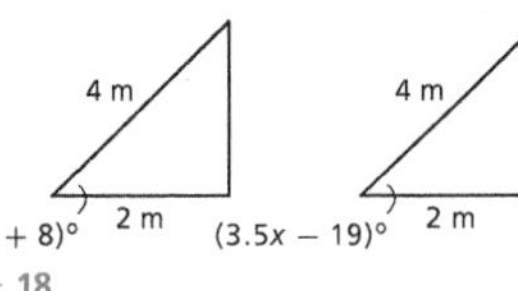

$x = 18$

17. (MP) **Critique Reasoning** Ava claims she can use the SAS Congruence Theorem to prove two right triangles are congruent. Both triangles contain a 40° angle and a side that is 9 inches long. Is Ava correct? Is there a different theorem she could use if not? Justify your answer. **See Additional Answers.**

18. Refer to the figure to answer each question.

A. Explain why the triangles are congruent using any triangle congruence theorem.

A, B. See Additional Answers.

B. Describe a sequence of transformations that maps one triangle to the other.

C. Write a congruence statement to relate the two triangles. $\triangle ABC \cong \triangle XYZ$

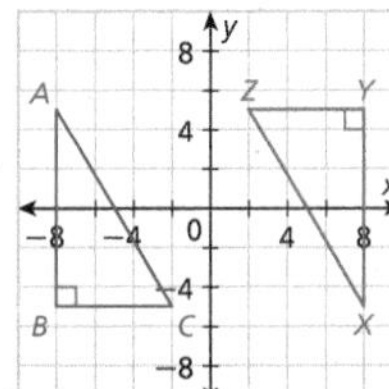

19. For a class project, students are sewing a quilt made of triangular pieces. Each piece must be congruent for the quilt pattern to fit together. To check the pattern pieces, a student carefully measures two different triangles. Triangle 1 has a longest side that is twice as long as the shortest side, with an included angle of 20°. Triangle 2 has an 8-inch side, a 4-inch side, and an included angle of 20°. Are the two triangles congruent? If not, what additional information would the student need to record to prove the triangles are congruent?

19, 20. See Additional Answers.

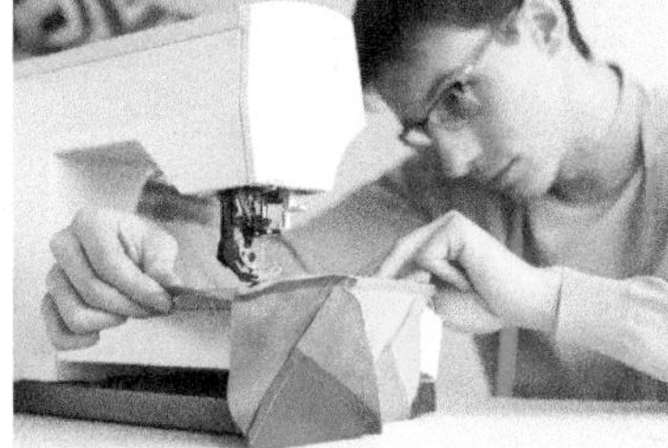

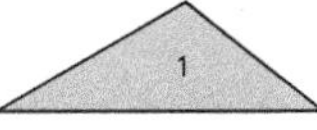

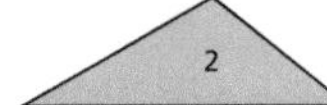

20. Anna and Sharon both construct a triangle. Anna begins by drawing a segment with a length of 3 inches. Starting from one vertex, she draws another segment with a length of 5 inches so that the two segments have an included angle of 35°. Sharon constructs her triangle by drawing a segment with a length of 5 inches, measuring a 35° angle, and drawing a ray from one vertex. Then she measures and terminates the ray at 3 inches. Do Anna and Sharon create congruent triangles? Explain.

Questioning Strategies

Problem 17 Suppose instead that Ava knows the legs of both right triangles are 9 inches and 6 inches long. Can Ava use the SAS Congruence Theorem to prove the triangles are congruent? Explain. yes; Since the hypotenuse of a right triangle is always opposite from its right angle, the right angle is included between the legs. Ava now knows a pair of congruent corresponding angles that are included between corresponding congruent sides.

5 Wrap-Up

Summarize learning with your class. Consider using the Exit Ticket, Put It in Writing, or I Can scale.

Exit Ticket

Suppose for triangles $\triangle ABC$ and $\triangle DEF$, you know $\overline{AB} \cong \overline{DE}$, $\overline{AC} \cong \overline{DF}$, and $\angle C \cong \angle F$. Either tell why you can say $\triangle ABC \cong \triangle DEF$, or give one more pair of congruent parts that would let us say the triangles are congruent. I cannot say the triangles are congruent since the corresponding angles are not included between the corresponding sides. If I knew $\angle A \cong \angle D$, I could say the triangles were congruent by ASA, or if I knew $\overline{BC} \cong \overline{EF}$, I could say the triangles were congruent by SAS.

Put It in Writing

Explain how you know there is only one possible triangle $\triangle ABC$ with $AB = 5$ cm, $BC = 6$ cm, and $m\angle B = 50°$.

I Can

The scale below can help you and your students understand their progress on a learning goal.

4	I can use SAS congruence criteria to prove that two triangles are congruent, and I can explain my solution steps to others.
3	I can use SAS congruence criteria to prove that two triangles are congruent.
2	I can use the SAS Triangle Congruence Theorem to find missing measures of parts of triangles.
1	I can find a pair of corresponding angles included between two pairs of corresponding sides.

Spiral Review • Assessment Readiness

These questions will help determine if students have retained information taught in the past and can also prepare them for high-stakes assessments. Here, students must use CPCFC to solve for missing measures **(7.2)**, determine how many triangles exist with three given side lengths **(8.2)**, use ASA congruence criteria two prove two triangles are congruent **(8.1)**, and identify a sequence of rigid motions on a figure **(7.3)**.

21. (MP) **Construct Arguments** Explain in your own words why there is no SSA Triangle Congruence Theorem. It is possible to create two different triangles when given two sides and an angle that is not included between those two sides. Two triangles with the same SSA criteria would not necessarily be congruent.

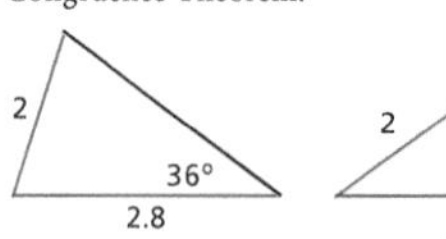

22. Explain what additional information you could use to prove two triangles are congruent when dealing with the ambiguous SSA case.

22–24. See Additional Answers.

23. Assume that $\triangle ABC$ is an isosceles triangle with $AB = AC$. Use the SAS Triangle Congruence Theorem to show that the angle bisector of $\angle A$ separates $\triangle ABC$ into two congruent triangles.

24. If both diagonals of the rectangle shown are the same length and bisect each other, explain how you know that $\overline{FG} \cong \overline{JH}$ and $\overline{FJ} \cong \overline{GH}$.

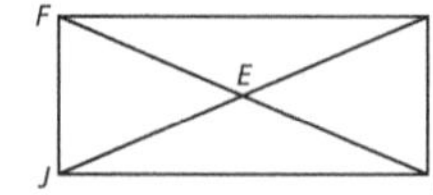

Spiral Review • Assessment Readiness

25. Triangle ABC is congruent to triangle PQR, where $AB = 3.5$ cm and $BC = 4.2$ cm. The measure of the included angle is 50°. The measure of $\angle Q$ is $(3x - 10)°$. What is the value of x?

Ⓐ 20
Ⓑ 50
Ⓒ 60
Ⓓ 130

26. Is $\triangle LMN \cong \triangle XMY$?

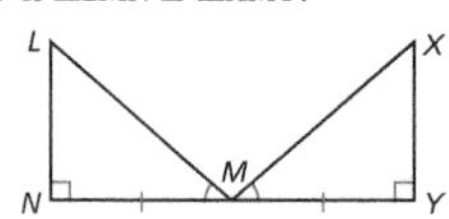

Ⓐ Yes, all right triangles are congruent.
Ⓑ Yes, by ASA Triangle Congruence.
Ⓒ No, all three angles are not congruent.
Ⓓ No, all three sides are not congruent.

27. Two triangles have three pairs of corresponding sides of lengths 3, 4, and 5 inches. How many different triangles can you construct that meet these criteria?

Ⓐ exactly one
Ⓑ exactly two
Ⓒ exactly three
Ⓓ infinitely many

28. A figure is transformed according to the following function. Which movements describe the transformations? Select all that apply.

$(x, y) \to (x + 2, -(y - 2))$

Ⓐ two units to the left
Ⓑ two units to the right
Ⓒ two units up
Ⓓ two units down
Ⓔ reflection over the x-axis
Ⓕ reflection over the y-axis

I'm in a Learning Mindset!

Did I administer feedback respectfully? How did I address the idea that was presented instead of the person who voiced the idea?

Keep Going ▶ Journal and Practice Workbook

Learning Mindset

Resilience Responds to Feedback

Point out how we can sometimes get excited or emotional when we're sure somebody else's arguments are incorrect, but we need to be respectful when trying to convince the other person. Logical arguments are not refuted by questioning the knowledge of the person who made them, and doing so can make the other person less willing to listen. Other people will be more willing to listen to our arguments when we discuss ideas instead of the people who made them. For example, instead of saying, "you chose an angle that is not included between corresponding sides," it would be better to say, "the angle needs to be included between corresponding sides." *Would you rather hear someone say, "the correct reasoning is…" instead of "you're wrong because…"? How can being respectful in giving feedback help make other people more willing to listen?*

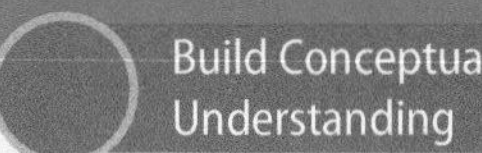

8.3 Develop SSS Triangle Congruence

LESSON FOCUS AND COHERENCE

Mathematics Standards

- Explain how the criteria for triangle congruence (ASA, SAS, and SSS) follow from the definition of congruence in terms of rigid motions.

Mathematical Practices and Processes 

- Use appropriate tools strategically.
- Construct viable arguments and critique the reasoning of others.
- Look for and make use of structure.

I Can Objective 

I can use SSS congruence criteria to prove that two triangles are congruent.

Learning Objective

Analyze triangle congruency using the SSS criteria through constructions, rigid transformations, formal proofs, and inspection of diagrams. Solve problems related to triangles that are congruent by SSS.

Language Objective

Describe when two triangles are congruent by identifying three pairs of corresponding pairs of congruent sides.

Lesson Materials: compass, ruler, strips of paper

Mathematical Progressions

Prior Learning	Current Development	Future Connections
Students: • used geometric descriptions of rigid motions to transform figures and to predict the effect of a given rigid motion on a given figure and decide if two transformed figures are congruent. **(5.1–5.3, 7.3)** • used the definition of congruence in terms of rigid motions to show that the two triangles are congruent if and only if corresponding pairs of sides and corresponding pairs of angles are congruent. **(7.1–7.3)**	**Students:** • understand that if three sides of a triangle are congruent to three sides of another triangle, then the two triangles are congruent. • recognize that the SSS criterium follows from the definition of congruence. • use transformations to justify SSS Triangle Congruence. • use SSS Triangle Congruency to prove that triangles are congruent. • use congruency justified by SSS Triangle Congruence to solve a real-world problem.	**Students:** • will write coordinate proofs about triangle relationships. **(9.4, 9.5)** • will write coordinate proofs about parallelograms. **(11.1–11.5)** • will construct an equilateral triangle, a square, and a regular hexagon inscribed in a circle. **(15.2)**

PROFESSIONAL LEARNING

Math Background

The Side-Side-Side congruence theorem adds to the student's growing list of ways to prove triangles are congruent. They already know that a pair of triangles with Angle-Side-Angle or Side-Angle-Side criteria proves two triangles congruent. In the next lessons, students will examine congruency specific to right triangles using the Hypotenuse-Leg theorem. They will also see how having two pairs of congruent angles and a congruent pair of sides proves congruency in an Angle-Angle-Side situation. When students have a full toolbox of triangle congruency criteria, they can assess diagrams and real-world situations to determine when triangles are congruent.

ONLINE

PROJECTABLE & PRINTABLE 

WARM-UP OPTIONS

ACTIVATE PRIOR KNOWLEDGE • Analyze a Figure Given a Sequence of Transformations

Use these activities to quickly assess and activate prior knowledge as needed.

Problem of the Day 

A knitted sweater design calls for two triangles reflected over a center line as shown in the diagram. What are the missing side lengths in the design? Explain your answer.

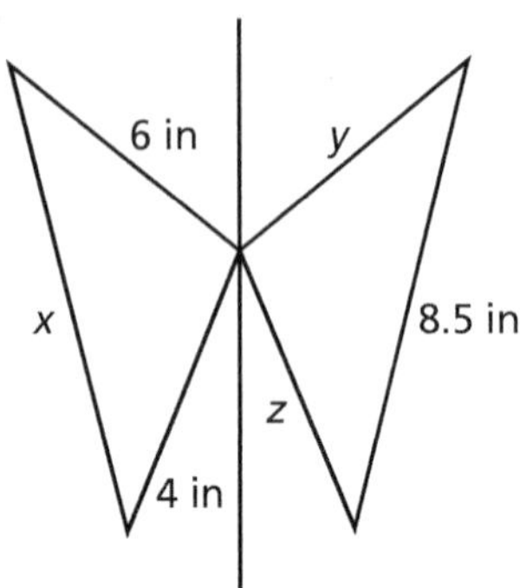

$x = 8.5$, $y = 6$, $z = 4$; A reflection is a rigid transformation which preserves side lengths.

Quick Check for Homework

As part of your daily routine, you may want to display the Teacher Solution Key to have students check their homework.

Make Connections

Based on students' responses to the Problem of the Day, choose one of the following:

1. Project the Interactive Reteach, Geometry, Lesson 5.3.
2. Complete the Prerequisite Skills Activity:

Have students draw the triangle formed by the coordinates $A(-3, 1)$, $B(-1, 1)$, and $C(-1, 5)$. Then have students reflect the triangle over the y-axis to form $\triangle A'B'C'$. Then have students reflect $\triangle A'B'C'$ over the x-axis to form $\triangle A''B''C''$. Students should label the coordinates of their points.

- *How do you know the first triangle is congruent to $\triangle A'B'C'$?* A reflection is a rigid transformation which preserves the side lengths and angle measures.
- *How do you know $\triangle A'B'C'$ is congruent to $\triangle A''B''C''$?* A reflection is a rigid transformation which preserves the side lengths and angle measures.
- *How do you know the first triangle is congruent to $\triangle A''B''C''$?* Two reflections were performed which are both rigid transformations. Also, the Transitive Property of Congruence shows that $\triangle ABC \cong \triangle A''B''C''$.

If students continue to struggle, use Tier 2 Skill 11.

SHARPEN SKILLS

If time permits, use this on-level activity to build fluency and practice basic skills.

Quantitative Comparison

Objective: Students make a comparison between two quantities.

Sketch the diagram and problem on the board. Ask students to choose the letter representing the correct answer and to explain their reasoning.

$2x + 5$ $3x + 2$

Both quantities are equal.

The two triangles are congruent.

Quantity A	Quantity B
The value of x.	3

A. Quantity A is greater.
B. Quantity B is greater.
C. The two quantities are equal.
D. The relationship cannot be determined from the information given.

C. Since the triangles are congruent, the corresponding sides are congruent, so

$2x + 5 = 3x + 2$
$5 = 3x - 2x + 2$
$5 = x + 2$
$5 - 2 = x$
$3 = x.$

PLAN FOR DIFFERENTIATED INSTRUCTION

Small-Group Options

Use these teacher-guided activities with pulled small groups.

On Track

Materials: compass, ruler, protractor

- In pairs, have one student use a ruler to draw a triangle.
- Then have the students work together using the compass and ruler to construct a congruent triangle.
- Have students use the ruler and protractor to confirm that the corresponding side lengths and corresponding angles have equal measures.
- Have students explain how they would know the triangles are congruent even if they hadn't measured the angles.

Almost There

Materials: index cards

Draw two triangles on several index cards. Have some of the cards show two congruent triangles by SSS and some of the cards show two triangles that are not congruent.

- Hand a card to each student.
- Have the student decide if their triangles are congruent and think of a reason that explains their conclusion.
- Have students share their card and their reasoning with a partner.

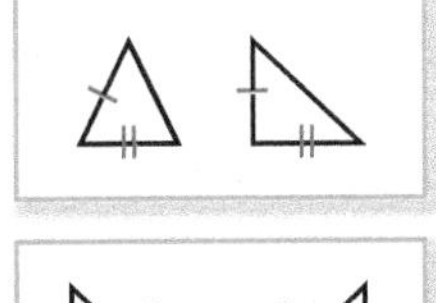

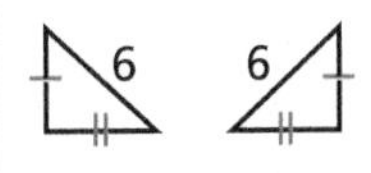

Ready for More

Draw the diagram on the board. Tell students the sum of the side lengths in one triangle is 49.3.

In pairs, have students work to answer the questions.

- Are the triangles congruent using the SSS congruence criteria? Why or why not?
- Which variable can you solve for first?
- What is the value of x and y?

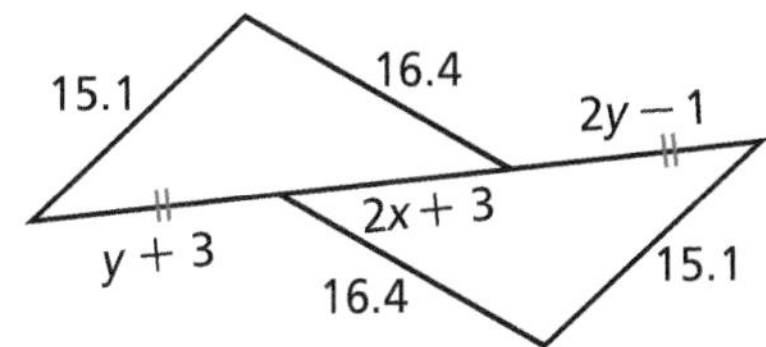

Math Center Options

Use these student self-directed activities at centers or stations.

Key: ● Print Resources ● Online Resources

On Track

- ● Interactive Digital Lesson
- ●● Journal and Practice Workbook
- ● Illustrative Mathematics: Why does SSS work?
- ● Module Performance Task

Almost There

- ● Reteach 8.3 (printable)
- ● Interactive Reteach 8.3
- ● RtI Tier 2 Skill 11: Reflect Figures in the Coordinate Plane

Ready for More

- ● Challenge 8.3 (printable)
- ● Interactive Challenge 8.3
- ● Illustrative Mathematics: SSS Congruence Criterion

Unit Project Check students' progress by asking what triangle congruence theorem they are using in their proof.

View data-driven grouping recommendations and assign differentiation resources.

During the *Spark Your Learning*, listen and watch for strategies students use. See samples of student work on this page.

Use Rigid Motions with a Model

Strategy 1

My classmate and I each made a triangle with three strips of paper with lengths 2 in., 4 in., and 5 in. Then we used rotations, reflections, and translations, and we could see that our triangles mapped onto each other.

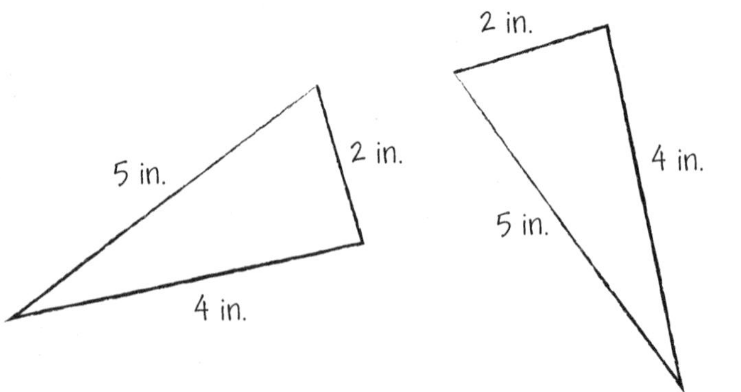

If students . . . construct a model and use rigid motions to solve the problem, they are demonstrating an exemplary understanding of rigid motions to prove congruency from previous geometry lessons.

Have these students . . . explain how they determined their equation and how they solved it. **Ask:**

- Q How did you compare the two triangles made with the same lengths?
- Q How do you know the triangles are congruent?

Measure the Side Lengths

Strategy 2

My classmate and I each drew a triangle using the side lengths 2 in., 4 in., and 5 in. Then we used a ruler to confirm our measurements were the same. Since we both have triangles with the same side lengths, the triangles are congruent. The triangles look like they are the same size and the same shape.

If students . . . compare two triangles using the measurements of the sides, they are showing that they understand the relationship between corresponding parts, but may not understand how to show congruence using rigid motions.

Activate prior knowledge . . . by having students use tracing paper to map one triangle onto the other using rigid transformations. **Ask:**

- Q If you trace one triangle onto tracing paper, what transformations could you use to map it onto the second triangle?
- Q How could you use tracing paper to compare both the corresponding side lengths and the corresponding angle measures?

COMMON ERROR: Assumes Angles Must Be Known

I drew a triangle with side lengths of 2 in., 4 in., and 5 in. Other triangles with these measures may look different because I don't know the angles of the triangle.

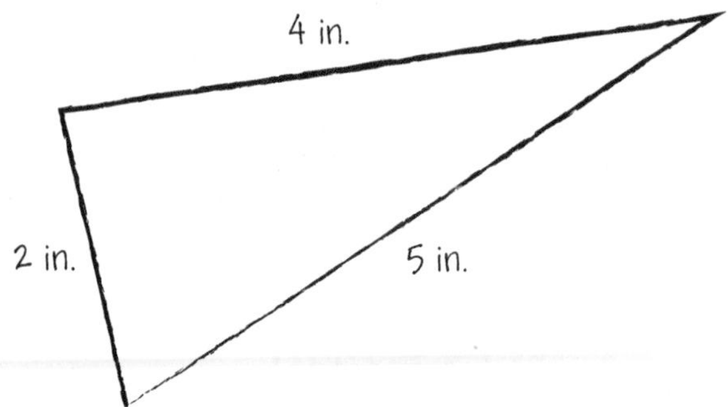

If students . . . assumed the measures could be used to draw multiple triangle because they do not know the angle measures.

Then intervene . . . by pointing out that they do not need to know the angles if they have all three side measures. Only one triangle can be drawn, where another triangle with the same lengths would be congruent. **Ask:**

- Q If you draw a second triangle with the same side lengths, how could you compare the triangles to see if they are congruent?
- Q How do you know only one triangle can be drawn with these three side lengths?

8.3

Develop SSS Triangle Congruence

I Can use SSS congruence criteria to prove that two triangles are congruent.

Spark Your Learning

A welder is making a wall sculpture of triangles made using standard lengths of metal pieces.

The welder chooses from metal pieces in five standard lengths.
The sculpture is to be made of triangular shapes. Each shape uses three of the metal pieces.

Complete Part A as a whole class. Then complete Parts B–D in small groups.

A. Suppose the welder uses the same three lengths for more than one triangle. What is a mathematical question you can ask about the situation? **A. If two triangles are made from the same metal-piece lengths, will the two triangles be congruent?**

B. Use tools and the additional information your teacher gives you to cut out strips of paper to match the metal pieces the welder uses. Build a triangle with those metal-piece lengths. Compare with your classmates' triangles. Does everyone's triangle look the same or different? Does this prove the answer to your question? See Strategies 1 and 2 on the facing page.

C. How does the construction of your triangle compare and contrast with the constructions of triangles from the previous lessons? See Additional Answers.

D. What do you know about congruent triangles and transformations you could use to check your theory? See Additional Answers.

 **Turn and Talk** Could you determine if two triangles are congruent knowing only two side lengths? Support your answer. See margin.

SUPPORT SENSE-MAKING • Three Reads

Tell students to read the information in the photo three times and prompt them with a different question each time.

1. What is the situation about? The situation is about a welder that makes triangles with three different lengths of metal that will be used to make a sculpture.
2. What are the quantities in this situation? How are these quantities related? The quantities are the lengths of the metal pieces; Three of the lengths will be used to make a triangle, 2 inches, 4 inches, and 5 inches.
3. What are possible questions you could ask about this situation? Possible answer: What type of triangle is formed using the three lengths of metal? Can only one triangle be formed using three given lengths?

1 Spark Your Learning

▶ MOTIVATE 

- Have students look at the photo in their books and read the information contained in the photo. Then complete Part A as a whole-class discussion.
- Give the class the additional information they need to solve the problem. This information is available online as a printable and projectable page in the Teacher Resources.
- Have students work in small groups to complete Parts B–D.

▶ PERSEVERE

If students need support, guide them by asking:

Q **Advancing • Use Tools** Which tool could you use to solve the problem? Why choose that tool and not some other? Students' choices of tools and reasons for choosing them will vary.

Q **Assessing** How many possible triangles with side lengths 2 inches, 4 inches, and 5 inches are there? Only 1 triangle can have those side lengths.

Q **Assessing** What type of transformations could the welder use to see if two triangles made from the same side lengths are congruent? The welder could use any rigid transformation such as a translation, a reflection, or a rotation.

Q **Advancing** What happens to the angles in the triangle when the welder chooses the three side lengths for two triangles? The angles will be the same for the two triangles. The given side lengths can form the triangle in only one way.

 **Turn and Talk** Ask students to think about the angle between those two side lengths and if it would have the same measure for both triangles. Encourage students to use two paper strips of different lengths to form two sides of a triangle. Have them determine if there is more than one possible triangle. no; There would not be enough information to prove two triangles are congruent. More than one triangle could be drawn that matched those conditions.

▶ BUILD SHARED UNDERSTANDING

Select groups of students who used various strategies and tools to share with the class how they solved the problem. As they present their solutions, have each group discuss why they chose a specific strategy and tool.

2 Learn Together

Build Understanding

Task 1 (MP) **Use Tools** Students use a compass and ruler to construct and compare triangles with three given lengths.

Sample Guided Discussion:

Q **What is the relationship between the arc and the point when you set your compass to 3 inches?** The arc is part of a circle with a center of point *A* and radius 3 inches. This means the distance from point *A* to anywhere on the arc is 3 inches.

Q **In Part E, how can you show the triangle is congruent to the first triangle you drew?** The second triangle can be transformed using rigid motions so that it maps onto the first triangle.

Turn and Talk Have students compare their first three triangles. Encourage students to think about how the constructions have the same steps, but are done in a different order. yes; Possible answer: The triangle may be oriented differently but would still have side lengths of 2 inches, 3 inches, and 4 inches, therefore it would still be congruent to the triangles constructed in Parts D, E, and F; Two triangles with three congruent sides are congruent.

Build Understanding

Draw Triangles Given Three Side Lengths

Remember that two triangles are congruent only if a series of rigid transformations maps one triangle onto the other.

1 A. Use a ruler to draw a horizontal line that is 2 inches long. Label the endpoints *A* and *B*. A–C. Check students' constructions.

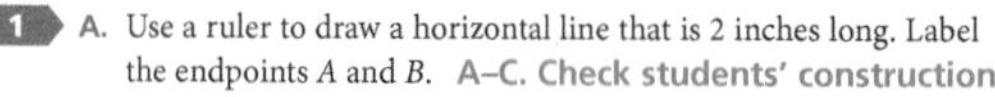

B. Set a compass to 3 inches. Place the point on *A*, and draw an arc above the horizontal line.

C. Next, set the compass to 4 inches. Place the point on *B*, and draw an arc that it intersects the first arc.

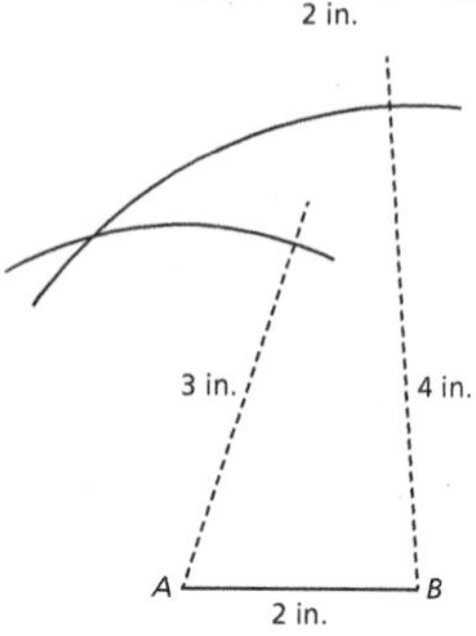

D. Label the intersection of the two arcs *C*. Draw $\triangle ABC$. How do you know that the result is a triangle with side lengths of 2, 3, and 4 inches? D–F. See Additional Answers.

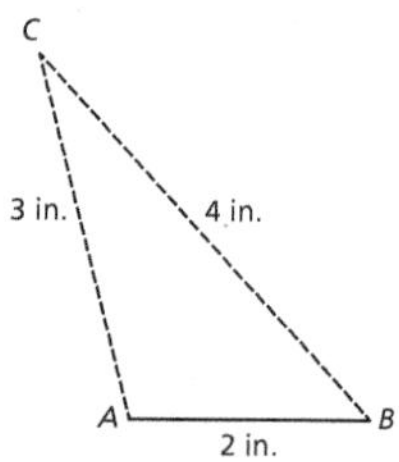

E. Draw a second horizontal 2 inch segment on your paper. Repeat steps B–D by drawing the arcs below the line. Is the resulting triangle congruent to the triangle you drew in part D?

F. Create a second triangle with side lengths 2, 3, and 4 inches. This time, begin with the 4-inch segment as the horizontal line. What is the relationship between the triangles? How do you know?

Turn and Talk

- Do you expect that you would get another congruent triangle if you started the construction with the 3-inch side? Explain.
- What can you say about two triangles with 3 pairs of congruent sides if you do not know anything about the angles? See margin.

236

LEVELED QUESTIONS

Depth of Knowledge (DOK)	Leveled Questions	What Does This Tell You?
Level 1 **Recall**	How do you draw a length of a given distance from a point using a compass? You open the compass width to a given length using a ruler, then place the compass on the point and draw an arc and connect the point to the arc.	Students' answers will indicate whether they know how to construct a segment or an arc of a given length.
Level 2 **Basic Application of Skills & Concepts**	How can you compare two triangles to determine if they are congruent using rigid motions? You can use translations, rotations, or reflections to see if one figure maps onto the other figure.	Students' answers will indicate whether they understand how congruent figures are related using rigid motions.
Level 3 **Strategic Thinking & Complex Reasoning**	To draw a triangle when you are given three lengths, does it not matter which side you draw first? It does not matter which side you draw first because you construct all three lengths and all triangles with those three lengths will be congruent.	Students' answers will indicate whether they understand the order of the construction does not change the size or the shape of the resulting triangle with three given side lengths.

Justify and Use SSS Triangle Congruence

You can use three pairs of congruent sides to determine that two triangles are congruent.

Side-Side-Side (SSS) Triangle Congruence Theorem

If three sides of a triangle are congruent to three sides of another triangle, then the triangles are congruent.

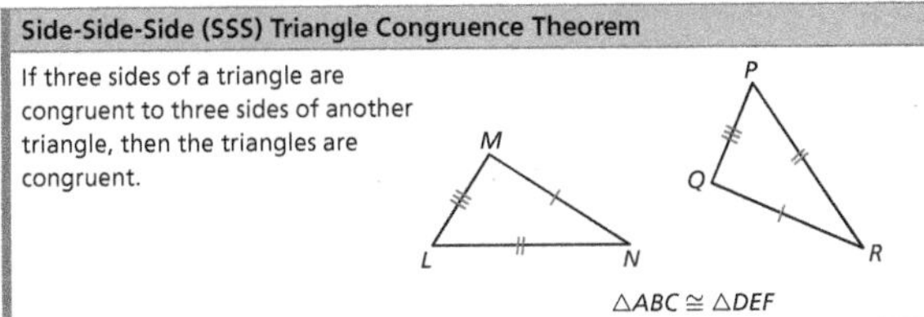

$\triangle ABC \cong \triangle DEF$

2 Use transformations to prove the SSS Triangle Congruence Theorem.

Let triangles LMN and PQR have congruent sides.

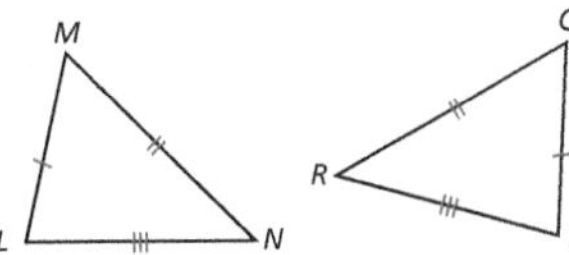

A. Translate $\triangle PQR$ along $\overrightarrow{PL}$ so that P lies on L. Then rotate $\triangle PQR$ about L with a rotation angle of $\angle RPN$. Check students' constructions.

B. Reflect $\triangle PQR$ across $\overline{LN}$. Does this map Q to M? Why or why not? Did the series of transformations prove the two triangles are congruent? See Additional Answers.

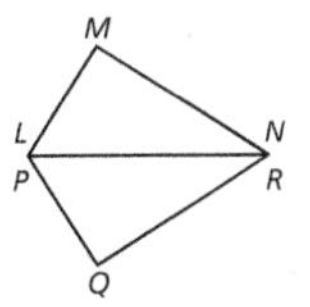

3 In the triangles shown, you know that $\overline{AB} \perp \overline{BC}$ and $\overline{DE} \perp \overline{EF}$. See Additional Answers.

A. Use the Pythagorean Theorem to find the lengths BC and DF. What allows you to use the Pythagorean Theorem to find these lengths?

B. Complete the statement: Three pairs of corresponding sides are congruent, so by the SSS Triangle Congruence Theorem, __?__. $\triangle ABC \cong \triangle DEF$

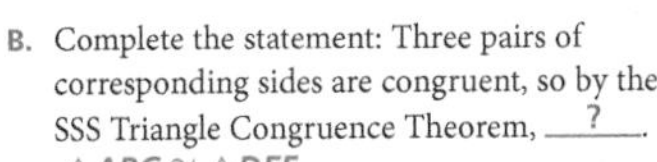

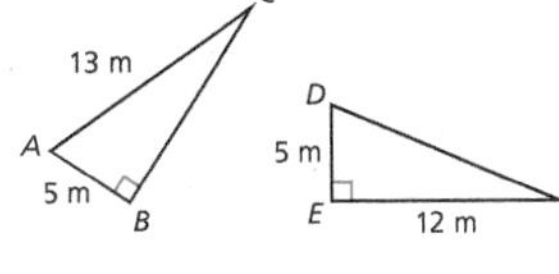

C. Why was the SSS Triangle Congruence Theorem used in this case? Could one of the other congruence theorems been applied instead? See Additional Answers.

Turn and Talk The side lengths of one triangle are 5 inches, 3 inches, and 7 inches. The side lengths of another triangle are 5 cm, 3 cm, and 7 cm. Are the two triangles congruent? See margin.

PROFICIENCY LEVEL

Beginning

Have students write in words what SSS stands for in the theorem.

Intermediate

Ask students to briefly discuss in pairs how the SSS Triangle Congruence Theorem is different from the SAS and ASA Triangle Congruence Theorems.

Advanced

Have students write a short paragraph describing two situations when the SSS Triangle Congruence Theorem can be used to prove two triangles are congruent.

Task 2 (MP) **Construct Arguments** Students perform rigid motions to the SSS Triangle Congruence Theorem.

Sample Guided Discussion:

Q Based on the markings in the diagram, which pairs of sides are corresponding? $\overline{MN}$ and $\overline{QR}$, $\overline{ML}$ and $\overline{QP}$, and $\overline{LN}$ and $\overline{PR}$

Q What other way could you use a similar sequence of transformations to map $\triangle LMN$ onto $\triangle PQR$? Possible answer: You could translate along $\overline{NR}$, so N lies on R. Then you could perform a rotation and then a reflection.

Task 3 (MP) **Construct Arguments** Students use the Pythagorean Theorem and the SSS Triangle Congruence Theorem to prove two triangles are congruent.

Sample Guided Discussion:

Q In Part A, what equations can you use to find the missing side lengths? $5^2 + x^2 = 13^2$ and $5^2 + 12^2 = x^2$.

Q How can you use the side lengths to help you write the congruence statement in Part B? Possible answer: Follow the vertices around the triangle from smallest length to longest length for both triangles.

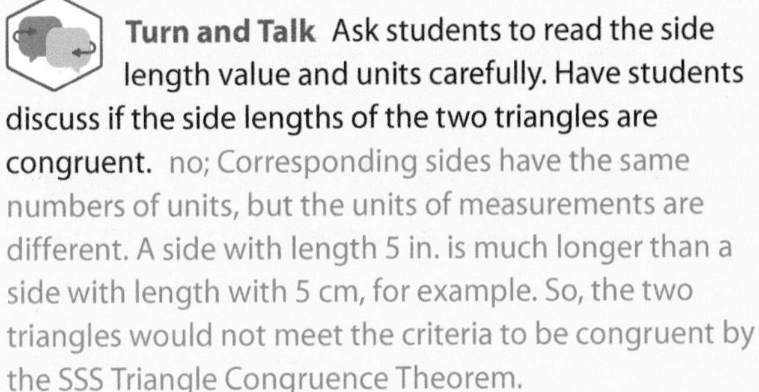

Turn and Talk Ask students to read the side length value and units carefully. Have students discuss if the side lengths of the two triangles are congruent. no; Corresponding sides have the same numbers of units, but the units of measurements are different. A side with length 5 in. is much longer than a side with length with 5 cm, for example. So, the two triangles would not meet the criteria to be congruent by the SSS Triangle Congruence Theorem.

Step It Out

Task 4 (MP) **Use Structure** Students use an algebraic equation to find the missing length in two congruent triangles with SSS congruence.

Sample Guided Discussion:

Q It looks like one of the corresponding angle pairs in the triangles is a right angle. Why is this not enough to use a different congruence theorem? There are no angle measures marked in the figure. Even if the angles look like right angles, they must be marked.

Q In Part C, why is 3 not the length of the third side of the triangles? The number 3 is the value of x. The side lengths have values of $5x - 7$ and $2x + 2$ when $x = 3$.

Turn and Talk Encourage students to discuss what possible information could allow them to use SAS triangle congruence. Possible answer: You would need to know that any pair of corresponding angles are congruent or have the same angle measure. For example, you would need to know that the measures of the included angles between the sides labeled 10 and 12, in both triangles, are congruent.

Step It Out

Apply SSS Triangle Congruence in a Real-World Context

You can use the SSS Triangle Congruence Theorem to prove triangles are congruent in real-world problems.

4 Micah designs a pair of triangular beaded earrings for an art project. The numbers of beads on each side of the triangles are given. For what value of x can you use the SSS Triangle Congruence Theorem to prove the triangles are congruent?

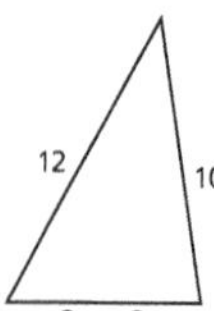

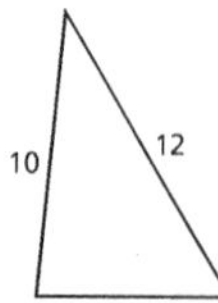

Choose a strategy.
Each triangle has one side with length 10 cm and one side with length 12 cm. So two pairs of corresponding sides are congruent. In order for the triangles to be congruent by the SSS Triangle Congruence Theorem, the third pair of corresponding sides must also be congruent.

A. Why choose the SSS Triangle Congruence Theorem to prove the triangles are congruent? See Additional Answers.

Write an equation.
If the third side of both triangles must be congruent, we can write an equation that shows the lengths of both sides are equal.

$5x - 7 = 2x + 2$

B. Why are you able to write this equation? See Additional Answers.

Solve the equation.
Solve the equation to determine the value of x.

$5x - 7 = 2x + 2$

$3x - 7 = 2$

$3x = 9$

$x = 3$

C. What is the unknown side length in each triangle? 8 cm

Answer the question.
You can use the SSS Triangle Congruence Theorem to prove the two triangles are congruent when the value of x is 3.

Turn and Talk What additional information would you have needed to prove the triangles are congruent by the SAS Triangle Congruence Theorem? See margin.

Check Understanding

1–3. See Additional Answers.

1. Two triangles have side lengths of 3 inches, 4 inches, and 5 inches. Is it possible to draw two different triangles that meet this criteria? Justify your answer.

2. Describe a set of transformations that maps $\triangle ABC$ to $\triangle DEF$.

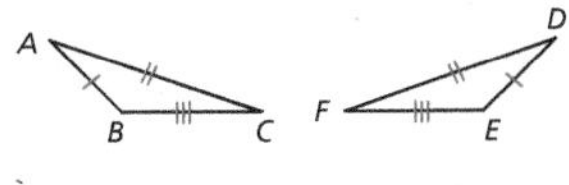

3. What additional information is needed to use the SSS Triangle Congruence Theorem to provethat the triangles are congruent?

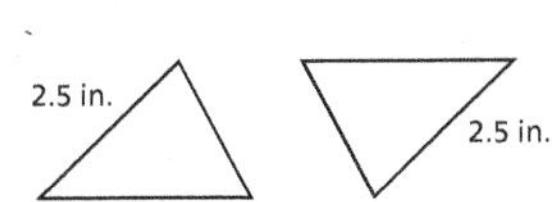

4. Triangles PQR and XYZ are equilateral triangles. The sides of $\triangle PQR$ are 5 inches long. One side of $\triangle XYZ$ is $(2x - 3)$ inches. What value of x makes the triangles congruent? $x = 4$

On Your Own

Use a compass and a ruler to construct two triangles with the given side lengths, and determine if they are congruent. Label each side.

5, 6. See Additional Answers.

5. 2 in., 3.5 in., 5 in.

6. 9.5 cm, 6 cm, 12 cm

In Problems 7–12, determine whether the triangles are *congruent*, *not congruent*, or if there is *not enough information*. Which triangle congruence theorem can be used? Explain your reasoning.

7–12. See Additional Answers.

7.

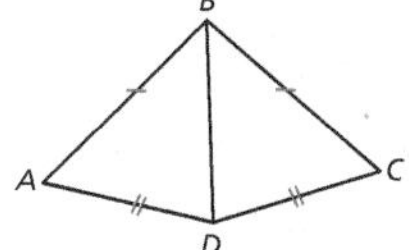

8.

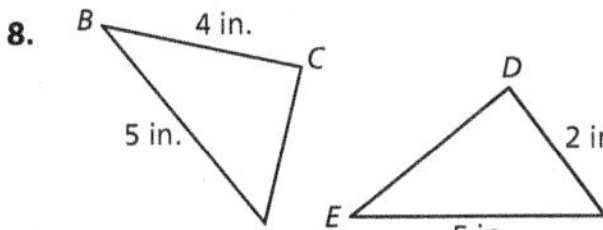

9.

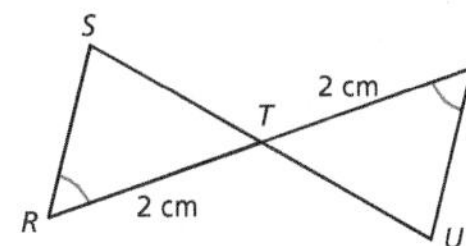

10.

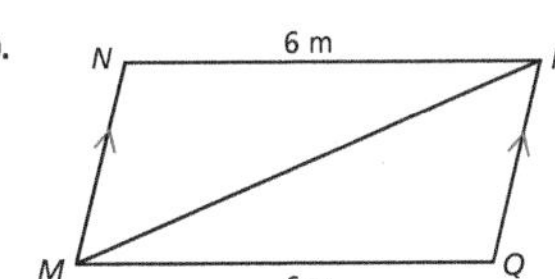

11.

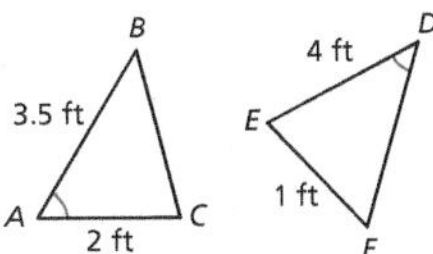

12.

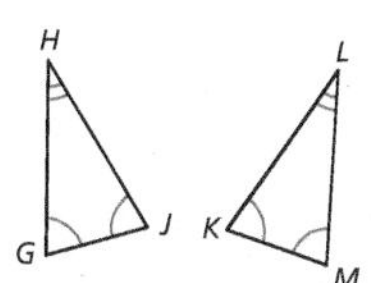

ONLINE

Assign the Digital On Your Own for

- built-in student supports
- Actionable Item Reports
- Standards Analysis Reports

On Your Own

Assignment Guide

The chart below indicates which problems in the On Your Own are associated with each task in the Learn Together. Assign daily homework for tasks completed.

Learn Together Tasks	On Your Own Problems
Task 1, p. 236	Problems 5, 6, 23, and 26
Task 2, p. 237	Problems 13, 14, and 21
Task 3, p. 237	Problems 7–12, 24, 25, 27, and 28
Task 4, p. 238	Problems 15–20 and 22

3 Check Understanding

Formative Assessment

Use formative assessment to determine if your students are successful with this lesson's learning objective.

Students who successfully complete the Check Understanding can continue to the On Your Own practice.

For students who miss 1 problem or more, work in a pulled small group using the Almost There small-group activity on page 235C.

ONLINE

Assign the Digital Check Understanding to determine

- success with the learning objective
- items to review
- grouping and differentiation resources

4 Differentiation Options

Differentiate instruction for all students using small-group activities and math center activities on page 235C.

Reteach

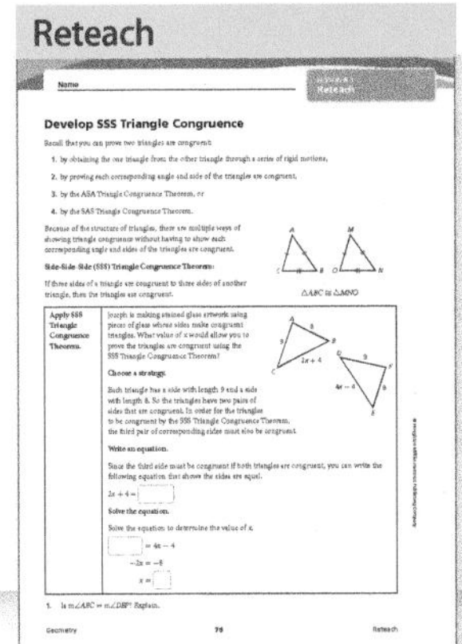

Challenge

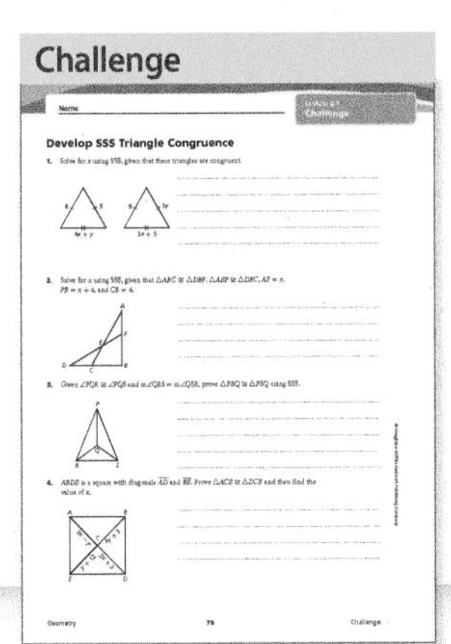

Watch for Common Errors

Problem 22 Draw attention to the sentence indicating that corresponding sides are using the same variable. Make sure students are setting the correct same variable expressions equal to each other. Remind students that the values of the variables can be substituted back into the expressions to find the lengths of each side.

Identify a series of rigid transformations that maps $\triangle ABC$ onto $\triangle DEF$.

13.

B A C

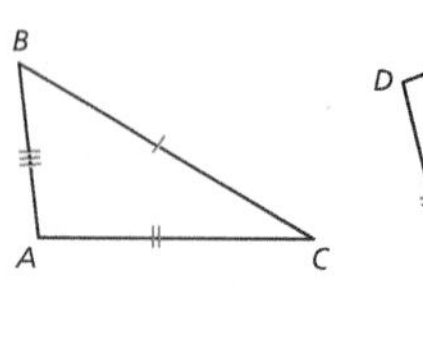

14.

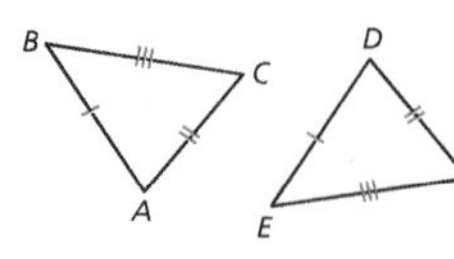

14. Possible answer: Translate $\triangle ABC$ along $\overrightarrow{BE}$ to map B to E. Rotate $\triangle ABC$ counterclockwise about E with an angle of rotation of $\angle CEF$ to map C to F. Reflect $\triangle ABC$ across $\overline{EF}$ to map A to D.

Find the values of x and y for which $\triangle ABC$ and $\triangle XYZ$ are congruent. Some problems may have more than one solution. 15–21. See Additional Answers.

15. $\triangle ABC$: side lengths of 5, 8, and x

$\triangle XYZ$: side lengths of 5, $2x$, and 4

16. $\triangle ABC$: side lengths of 10, 10, and $2x - 1$

$\triangle XYZ$: side lengths of 10, 10, and $4x - 4$

17. $\triangle ABC$: side lengths of 2, $x + 3$, and $y - 3$

$\triangle XYZ$: side lengths of $3x - 6$, 6, and 2

18. $\triangle ABC$: side lengths of 16, 20, and $4x - 8$

$\triangle XYZ$: side lengths of 20, 12, and $30 - 2y$

19. $\triangle ABC$: side lengths of 8, $6y$, and x

$\triangle XYZ$: side lengths of 18, 12, and 8

20. $\triangle ABC$: side lengths of 3, $x + y$, and x

$\triangle XYZ$: side lengths of 2, 3, and $3x - y$

21. The Great Pyraminds of Giza are a set of ancient Egyptian structures large enough to be visible from space. Can you prove the front faces of the two pyramids are congruent triangles? Why or why not?

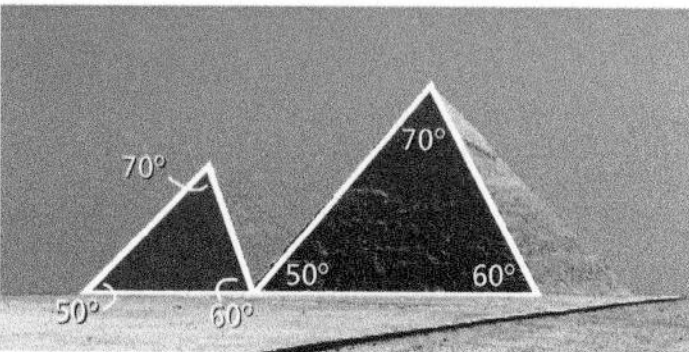

22. Sam is designing triangular metal braces to use while building a pergola for a patio. What value of the variables allows you to use the SSS Triangle Congruence Theorem to prove the two triangles are congruent? The sides using the same variable are corresponding. $x = 3$, $y = 5$, $z = 3$

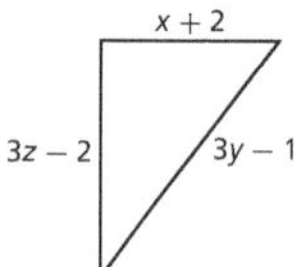

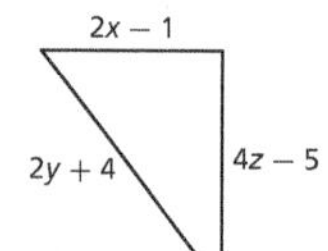

23. (MP) **Use Tools** Draw a 2-inch line segment labeled $\overline{DE}$ on your paper. Use a ruler and a compass to construct an equilateral triangle with $\overline{DE}$ as one of the sides. Then explain your construction. 23–25. See Additional Answers.

24. Complete the proof.

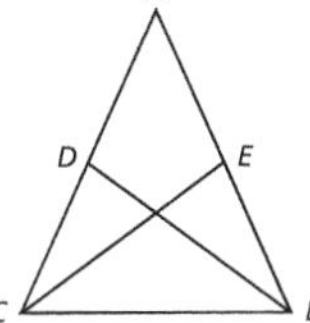

Given: $\overline{AC} \cong \overline{AB}$, $\overline{AD} \cong \overline{AE}$, $\overline{BD} \cong \overline{CE}$

Prove: $\triangle BCD \cong \triangle CBE$

Statements	Reasons
1. $\overline{AC} \cong \overline{AB}$, $\overline{AD} \cong \overline{AE}$	1. Given
2. $AC = AB$, $AD = AE$	2. Definition of congruence
3. $AC = AD + DC$, $AB = AE + EB$	3. ?
4. $AD + DC = AE + EB$	4. ? Property of Equality
5. $DC = EB$	5. ? Property of Equality
6. ?	6. Definition of congruence
7. ?	7. Given
8. $\overline{BC} \cong \overline{CB}$	8. Reflexive Property of Congruence
9. $\triangle BCD \cong \triangle CBE$	9. SSS Triangle Congruence Theorem

25. **STEM** Any geostationary satellite orbits approximately 35,700 km above the surface of Earth at the equator and travels at the same rate as Earth's spin. Because of this, it will appear to be fixed in one spot in the sky to a viewer on the ground. Usually this type of satellite is used for communications, due to the constant signal available to the locations within the transmission range. Using the given information and the radius r of Earth, show that the distance the signal travels in each direction is congruent.

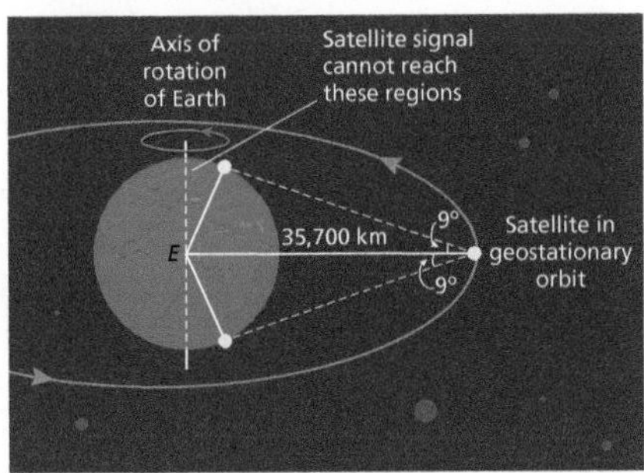

Questioning Strategies

Problem 23 Suppose you open your compass to the width of $\overline{DE}$ and draw an arc from point D. How big of an arc do you need to draw? Why do you need more than one arc to complete the equilateral triangle? The arc from point D needs to be long enough to intersect a second arc drawn from point E; The arc drawn from point E will have also have the compass width of $\overline{DE}$.

Questioning Strategies

Problem 24 To prove the triangles are congruent, consider the fact that they share one side. Which side do they share? What else would you need to prove to use the SSS criteria for triangle congruence? The triangles share side $\overline{BC}$ so this is one pair of congruent sides; You need to also prove the other two pairs of corresponding sides are congruent.

Watch for Common Errors

Problem 25 Students might try to use SSS to prove the triangles are congruent. Encourage students to use the fact that the radius from the center of the Earth is always equal. Also, the segment from the satellite to the Earth is tangent to the circle, forming a 90° angle. This allows students to find the third angle of the triangle and use SAS to prove the triangles congruent.

5 Wrap-Up

Summarize learning with your class. Consider using the Exit Ticket, Put It in Writing, or I Can scale.

Exit Ticket

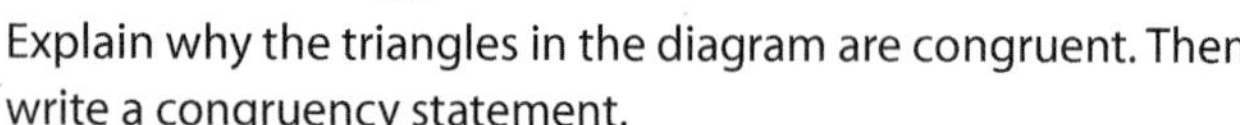

Explain why the triangles in the diagram are congruent. Then write a congruency statement.

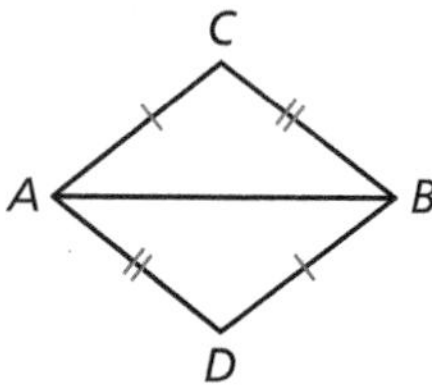

The triangles have two marked pairs of congruent sides. The third pair of corresponding sides is the side that is shared in the diagram, $\overline{AB}$. $\Delta ACB \cong \Delta BDA$

Put It in Writing 

Explain why Side-Side-Side congruence proves two triangles are congruent, but Angle-Angle-Angle does not.

I Can 

The scale below can help you and your students understand their progress on a learning goal.

4	I can use SSS congruence criteria to prove that two triangles are congruent, and I can explain my reasoning to others.
3	I can use SSS congruence criteria to prove that two triangles are congruent.
2	I can identify when SSS congruence can be used by analyzing diagrams with markings, numbers, and variables.
1	I can identify when SSS congruence can be used when the sides of the triangles are labeled with values.

Spiral Review • Assessment Readiness

These questions will help determine if students have retained information taught in the past and can also prepare them for high-stakes assessments. Here, students must identify theorems for proving triangle congruency **(8.1–8.3)**, use the Pythagorean Theorem with right triangles **(Gr8, 11.3)**, and prove two triangles are congruent with SAS **(8.2)**.

26. MP **Attend to Precision** Draw a 1.5-inch diameter circle with center O. Use a ruler and compass to construct a regular hexagon. Explain your construction.

26–28. See Additional Answers.

27. MP **Critique Reasoning** Kevin draws two right triangles with angles of 45°, 45°, and 90°. He then creates a proof to show the two triangles are congruent using AAA criteria. Can you recreate the proof? Show the proof or describe why not.

28. Using the SSS Triangle Congruence Theorem and the Distance Formula, show that $\triangle PQR \cong \triangle XYZ$.

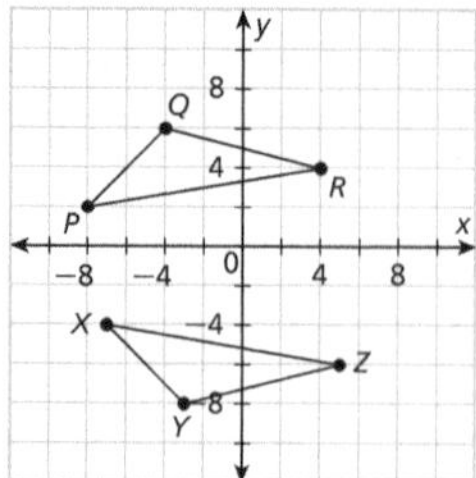

Spiral Review • Assessment Readiness

29. Which set of criteria cannot be used to prove any two triangles are congruent?

Ⓐ Side-Angle-Side
Ⓑ Angle-Side-Angle
Ⓒ Side-Side-Angle
Ⓓ Side-Side-Side

30. Two right triangles each have a side that is 3 inches long and a hypotenuse that is 5 inches long. Can you prove the triangles are congruent?

Ⓐ Yes, all right triangles are congruent.
Ⓑ Yes, the Pythagorean theorem shows the pair of other legs are congruent.
Ⓒ No, the unknown leg could be any length.
Ⓓ No, not enough information is given to prove congruency.

31. Match the statement with the correct reason to show that $\triangle ABC \cong \triangle DCB$.

A. $\overline{AC} \cong \overline{BD}$	**1.** Given	A. 1
B. $\overline{AB} \cong \overline{CD}$	**2.** SAS Triangle Congruence Theorem	B. 4
C. $\angle ABC \cong \angle DCB$	**3.** Alternate Interior Angles Theorem	C. 3
D. $\triangle ABC \cong \triangle DCB$	**4.** Given	D. 2

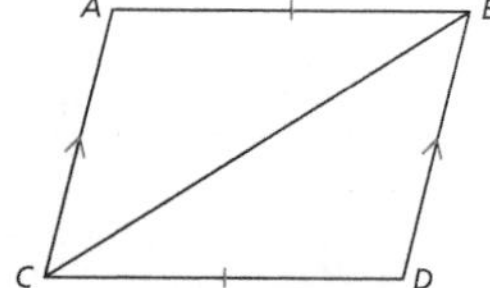

 I'm in a Learning Mindset!

How did I assert my own viewpoints and opinions when working with a group?

Keep Going to Journal and Practice Workbook

Learning Mindset

Resilience Responds to Feedback

There is often more than one way to construct a figure, solve a problem, or write a proof. In this lesson, a triangle with given side lengths is constructed in more than one way, but the final result was a congruent triangle. Be confident in explaining your methods for solving problems, but also be open to other approaches as you work. *What are the advantages or disadvantages to the method you used? How does your classmate's work to solve the problem differ from your work? Is there more than one way to approach this problem?*

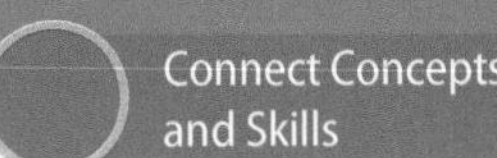

8.4 Develop AAS and HL Triangle Congruence

LESSON FOCUS AND COHERENCE

Mathematics Standards

- Explain how the criteria for triangle congruence (ASA, SAS, and SSS) follow from the definition of congruence in terms of rigid motions.

Mathematical Practices and Processes 

- Construct viable arguments and critique the reasoning of others.
- Use appropriate tools strategically.
- Look for and make use of structure.

I Can Objective 

I can use AAS and HL congruence criteria to determine if triangles are congruent.

Learning Objective

Use HL and AAS congruence criteria to determine if triangles are congruent.

Language Objective

Explain the difference between HL congruence criteria and the congruence criteria required for other triangle congruency theorems.

Lesson Materials: protractor, compass, ruler, index cards

Mathematical Progressions

Prior Learning	Current Development	Future Connections
Students: • given a geometric figure and a rotation, reflection, or translation, drew the transformed figure and specified a sequence of transformations that will carry a given figure onto another. **(5.1 and 5.2)** • used geometric descriptions of rigid motions to transform figures and predicted the effect of a given rigid motion on a given figure. **(5.1, 5.2, and 7.3)** • used the definition of congruence in terms of rigid motions to show that two triangles are congruent. **(7.1–7.3)**	**Students:** • use Triangle Congruency Theorems to show triangles are congruent. • write proofs about congruent triangles using properties and theorems about triangles. • understand the HL and AAS Triangle Congruency Theorems and their criteria. • use the Pythagorean Theorem to solve problems.	**Students:** • will write coordinate proofs about triangle relationships. **(9.4 and 9.5)** • will write coordinate proofs about parallelograms. **(11.1–11.5)** • will construct an equilateral triangle, a square, and a regular hexagon inscribed in a circle. **(15.2)**

PROFESSIONAL LEARNING

Visualizing the Math

Students may find it helpful to create a visual aid showing the criteria for each Triangle Congruency Theorem to refer to throughout the lesson or when working on independent practice. Visualizing the criteria for the theorems will help students retain those characteristics so they able to recall those criteria when writing proofs.

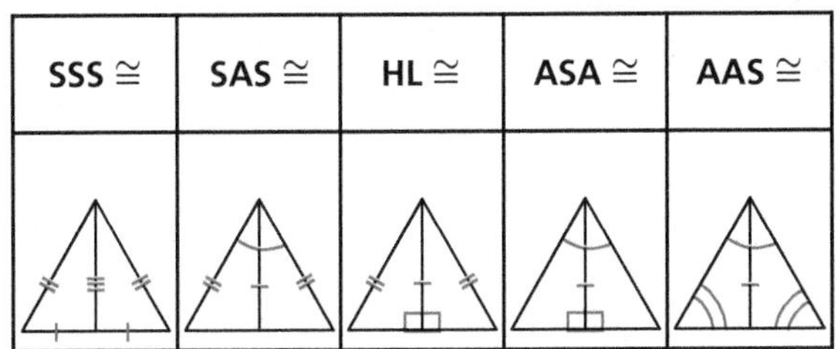

WARM-UP OPTIONS

ONLINE

PROJECTABLE & PRINTABLE

ACTIVATE PRIOR KNOWLEDGE • Use Rigid Transformations to Create Congruent Triangles

Use these activities to quickly assess and activate prior knowledge as needed.

Problem of the Day 

Triangle *ABC* has coordinates at $A(0, 0)$, $B(-4, 4)$, and $C(-4, -4)$. Reflect *ABC* over the *y*-axis and label the image *A'B'C'*. Give a sequence of 2 rigid motions that would also map *ABC* onto *A'B'C'*. What do you know is true about the two triangles? Explain. A reflection over the *x*-axis followed by a rotation of 180° about the origin would also map *ABC* to *A'B'C'*; I know the two triangles are congruent because a reflection and a rotation are rigid transformations, which preserve distance and angle measure.

Quick Check for Homework

As part of your daily routine, you may want to display the Teacher Solution Key to have students check their homework.

Make Connections

Based on students' responses to the Problem of the Day, choose one of the following:

 Project the Interactive Reteach, Grade 8, Lesson 1.5.

 Complete the Prerequisite Skills Activity:

Have students work in pairs. Ask one student to draw a shape on a coordinate grid, labeling each vertex with a letter. Have the second student describe a sequence of two transformations, while the first student performs them on the shape they drew. The first student must then state whether the two figures are congruent, explaining why or why not using precise mathematical language. Have students switch roles and repeat the steps.

- *How can the sequence of transformations your partner described be shown symbolically?* Answers will vary.
- *Did the image of your original figure get smaller or larger?* Answers will vary.
- *If an image gets smaller or larger, what transformation must have been performed?* a dilation
- *What transformations lead to congruent figures?* reflections, rotations, and translations

If students continue to struggle, use Tier 2 Skill 14.

SHARPEN SKILLS

If time permits, use this on-level activity to build fluency and practice basic skills.

Mental Math

Objective: Students demonstrate an understanding of congruent figures.
Have students work in small groups and display pentagon *ABCDE*. Ask group members to take turns describing a transformation while the other group members predict the coordinates of the image using mental math and whether the image will be congruent or not. Ask the group members to share their ideas and work together to come to a consensus on the coordinates of the image. Then, the next group member describes a different transformation for the group to work with. Each transformation described should be different than the previous transformation.

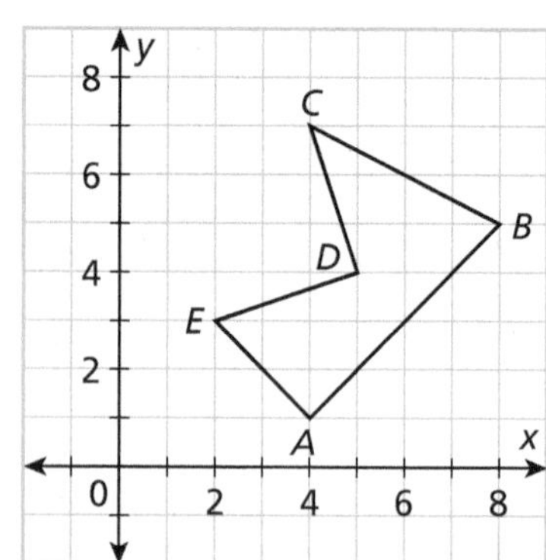

PLAN FOR DIFFERENTIATED INSTRUCTION

Small-Group Options

Use these teacher-guided activities with pulled small groups.

On Track

Materials: index cards, protractors, rulers

Give each group a set of index cards listing the abbreviations for each Triangle Congruency Theorem, including the ones they proved to be invalid in this lesson. Have groups work together to come up with a diagram of two accurate triangles, using the tools provided, to support each theorem, labeling congruent sides and angles with degree measures. Share examples with the class. Students should identify the theorem(s) that are invalid and explain why drawing a pair of triangles is not possible.

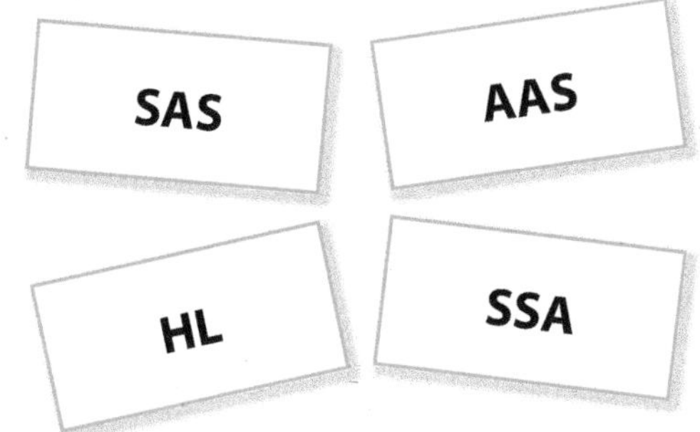

Almost There

Materials: index cards

Provide each group with a set of 10 index cards, each showing 2 triangles that are labeled with congruency marks such that one of the Triangle Congruency Theorems could be used to prove the triangles are congruent. Be sure to include a card for each theorem learned, with some appearing more than once. Have the group come up with the list of ten theorems that can be used to prove congruency before they begin analyzing the triangles. Have groups share their answers and discuss discrepancies.

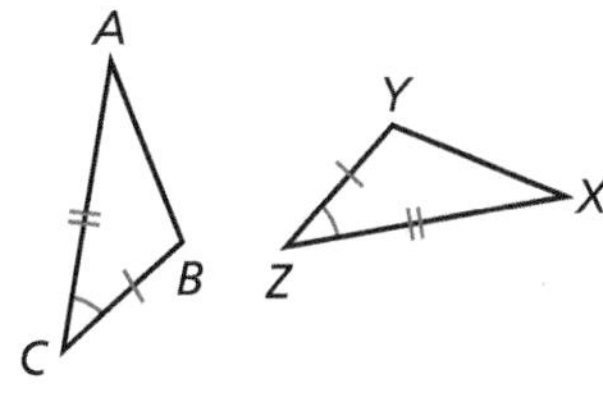

Ready for More

Display figure *KITE* and ask groups to work together to answer the following questions:

- Can you prove that $\triangle KEI \cong \triangle TEI$ using the HL Congruency Theorem? Explain.
- What would you need to know about the relationship between $\overline{IE}$ and $\overline{KT}$ in order to use HL Triangle Congruency Theorem to prove that $\triangle KOI \cong \triangle TOI$ and $\triangle EOK \cong \triangle EOT$?
- What Triangle Congruency Theorem can you use to prove $\triangle KEI \cong \triangle TEI$ without any additional information being given?

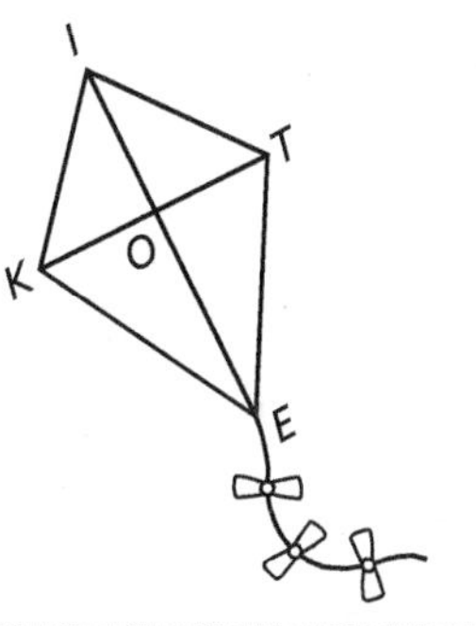

Math Center Options

Use these student self-directed activities at centers or stations.

Key: ● Print Resources ● Online Resources

On Track

- ● Interactive Digital Lesson
- ●● Journal and Practice Workbook
- ●● Standards Practice: Use ASA, SAS, and SSS
- ● Module Performance Task

Almost There

- ● Reteach 8.4 (printable)
- ● Interactive Reteach 8.4
- ● RtI Tier 2 Skill 14: Congruent Figures
- ● Illustrative Mathematics: Are The Triangles Congruent

Ready for More

- ● Challenge 8.4 (printable)
- ● Interactive Challenge 8.4
- ● Illustrative Mathematics: Right Triangles Inscribed in Circles I

View data-driven grouping recommendations and assign differentiation resources.

During the *Spark Your Learning*, listen and watch for strategies students use. See samples of student work on this page.

Use Rigid Motions

Strategy 1

To determine if $\triangle ABC \cong \triangle BDC$, I would have to rotate $\triangle BDC$ so that $\angle BDC$ maps onto $\angle ABC$. When I do this, the sides and angles that are congruent do not map onto each other, which means the triangles cannot be congruent.

If students . . . use rigid motions to solve the problem, they are recalling that rigid transformations map one figure to another, which shows the figures are congruent, from Lesson 7.3.

Have these students . . . explain how they determined the triangles are not congruent. **Ask:**

- Q How did you use the given information and the concept of rigid transformations to solve the SYL problem?
- Q What facts allowed you to determine the triangles were not congruent?

Use Angle Measures

Strategy 2

$\triangle ABC$ contains one angle that is 40° and two other angles that are both acute, so $\triangle ABC$ is an acute triangle. $\triangle BDC$ contains one angle that is 40°, a 2nd angle that is acute, and a 3rd angle that is obtuse, so $\triangle BDC$ is an obtuse triangle. Therefore, the two triangles cannot be congruent.

If students . . . use angle measure/size, they understand that an acute triangle cannot map onto an obtuse triangle by performing a rigid motion but may not understand the criteria involved in proving triangles are congruent.

Activate prior knowledge . . . by having students think about the triangle congruency theorems they have already learned. **Ask:**

- Q How did you know if two triangles were congruent using SAS, ASA, or SSS?
- Q What parts of the triangles are important when proving triangles are congruent?

COMMON ERROR: Uses SAS

$\triangle ABC \cong \triangle BDC$ because the two triangles have one congruent angle and two congruent sides. Therefore, the two triangles are congruent using the SAS theorem of triangle congruency.

If students . . . prove the triangles are congruent using SAS, they do not understand that the corresponding congruent angle of the two triangles must be included in the two corresponding congruent sides.

Then intervene . . . by reviewing the triangle congruency theorems they have already learned. **Ask:**

- Q Is the 40° angle included within the two congruent corresponding side lengths?
- Q What order do the congruent parts go in?

8.4

Develop AAS and HL Triangle Congruence

I Can use AAS and HL congruence criteria to determine if triangles are congruent.

Spark Your Learning

As part of a design challenge, a group is planning to paint a pattern of triangles on the wall.

Complete Part A as a whole class. Then complete Parts B–D in small groups.

A. What is a mathematical question you can ask about the image? What additional information would you need to answer your question? A. Are the two triangles congruent?; the measurements of sides or of sides and angles

B. What methods do you know to check for triangle congruence? Consider theorems and modeling. See Additional Answers.

C. To answer your question, what strategy and tool would you use along with all the information you have? What answer do you get? See Strategies 1 and 2 on the facing page.

D. Does your answer make sense in the context of the problem? Explain your reasoning. Possible answer: yes; If I copy both triangles onto my paper, I cannot map one to the other.

Turn and Talk How does the information given in this case compare to the congruence criteria cases you have studied so far? See margin.

CULTIVATE CONVERSATION • Information Gap

Ask students questions to help them decide what missing information they need to answer the question, "Are the two triangles congruent?"

1. Do you have enough information to conclude that the triangles are congruent due to any of the congruency theorems you have learned about so far? Explain. no; We are only given two congruent pairs of characteristics instead of 3.

2. Would you be able to prove the triangles are congruent if the pair of corresponding sides adjacent to the 40° angle were shown to be congruent? Explain. no; The 40° angle is not included in the two congruent sides, and there is not a pair of congruent sides included between two congruent pairs of angles.

3. To prove congruency using one of the theorems you have already learned, what information must be given? We would need to know that the other two pairs of corresponding sides are congruent to use SSS, that the two pairs of corresponding sides that include the given angle measure are congruent to use SAS, or that the pair of corresponding angles adjacent to the given angle are congruent to use ASA.

1 Spark Your Learning

▶ MOTIVATE

- Have students look at the photo in their books and read the information contained in the photo. Then complete Part A as a whole-class discussion.
- Give the class the additional information they need to solve the problem. This information is available online as a printable and projectable page in the Teacher Resources.
- Have students work in small groups to complete Parts B–D.

▶ PERSEVERE

If students need support, guide them by asking:

Q **Advancing • Use Tools** Which tool could you use to solve the problem? Why choose that tool and not some other? Students' choices of tools and reasons for choosing them will vary.

Q **Assessing** What needs to be true for two triangles to be congruent? Corresponding pairs of angles and sides must be the same measure and length.

Q **Assessing** What information about congruence have you been given? We know that one pair of corresponding angles and one pair of corresponding sides are congruent.

Q **Advancing** Why doesn't having one congruent pair of corresponding angles lead you to conclude the triangles are congruent? Similar shapes have congruent pairs of corresponding angles, and similar triangles are not congruent. They are the same shape but different sizes.

Turn and Talk Have students recall all of the congruence criteria they have already learned when determining congruency between triangles. Possible answer: So far we have studied triangle congruence cases in which two of the same type of value is given and an included value for the other type of value is given (S-A-S or A-S-A), or three side lengths were given (S-S-S). Triangles *ABC* and *DBC* in the case here show two pairs of congruent sides and an angle that is not included.

▶ BUILD SHARED UNDERSTANDING

Select groups of students who used various strategies and tools to share with the class how they solved the problem. As they present their solutions, have each group discuss why they chose a specific strategy and tool.

② Learn Together

Build Understanding

Task 1 (MP) **Construct Arguments** Students are given the opportunity to use and show understanding of previously learned congruence theorems and convey their mathematical analysis and thinking using relevant facts and symbols to prove the AAS Triangle Congruence Theorem.

Sample Guided Discussion:

Q Is the given information enough to prove that the two triangles are congruent? Why or why not? No, the given information is not enough; The congruent pair of sides is not included between the two congruent pairs of angles.

Q What is the Third Angles Theorem? It states that if two angles of a triangle are congruent to two angles in a second triangle, then the third pair of angles is also congruent.

Turn and Talk Encourage students to think back to before they learned any triangle congruence theorems and consider how they proved two triangles were congruent then. Possible answers: I could find a series of rigid transformations to map one triangle to the other; The process is similar in that we know all parts of congruent triangles are congruent, so we either prove all parts are congruent, or prove the criteria exists to use a triangle congruence theorem we have already proven.

Build Understanding

Prove the AAS Triangle Congruence Theorem

The Angle-Angle-Side Triangle Congruence Theorem describes another set of three congruent corresponding pairs that you can use to prove two triangles are congruent.

Angle-Angle-Side (AAS) Triangle Congruence Theorem

If two angles and a non-included side of one triangle are congruent to the corresponding angles and non-included side of another triangle, then the triangles are congruent.

A. Look at the Given and Prove statements. What theorems do you know that can be used to prove the AAS Triangle Congruence Theorem?

Given: $\angle A \cong \angle X$, $\angle B \cong \angle Y$, $\overline{AC} \cong \overline{XZ}$

Prove: $\triangle ABC \cong \triangle XYZ$

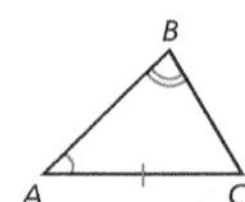

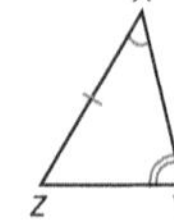

Statements	Reasons
1. $\angle A \cong \angle X$ and $\angle B \cong \angle Y$	1. Given
2. $\overline{AC} \cong \overline{XZ}$	2. Given
3. $\angle C \cong \angle Z$	3. Third Angles Theorem
4. $\triangle ABC \cong \triangle XYZ$	4. ASA Triangle Congruence Theorem

A. Possible answer: If I know the third angles are congruent, I could use ASA Triangle Congruence Theorem. If I could prove $\overline{AB}$ is congruent to $\overline{XY}$, I could use the SAS Triangle Congruence Theorem.

B. Why is the final step proving the triangles are congruent using the ASA Triangle Congruence Theorem and not the AAS Triangle Congruence Theorem?
You cannot use the theorem you're trying to prove in its own proof.

C. What previous information allows you to use the Third Angles Theorem in Line 2? What previous information allows you to use the ASA Triangle Congruence Theorem in Line 3? C, D. See Additional Answers.

D. Why doesn't the final step show the triangles are congruent by AAA criteria?

Turn and Talk

- What is another method you could use to prove the AAS Triangle Congruence Theorem?
- How is the proof of this theorem related to the proofs of the other congruence criteria theorems? See margin.

244

LEVELED QUESTIONS

Depth of Knowledge (DOK)	Leveled Questions	What Does This Tell You?
Level 1 **Recall**	Which triangle congruency theorems could be used to prove triangle are congruent given three angles and a side? AAS and ASA	Students' answers will indicate whether they can identify AAS and ASA cases.
Level 2 **Basic Application of Skills & Concepts**	Suppose triangles *ABC* and *DEF* are proven congruent using the SAS Triangle Congruence Theorem. What possible pairs of congruent sides and angle measures could have been given? Possible answer: $\overline{AB} \cong \overline{DE}$, $\angle B \cong \angle E$, $\overline{BC} \cong \overline{EF}$	Students' answers will indicate whether they understand the concept of corresponding sides and angles and the characteristics of the SAS Triangle Congruence Theorem.
Level 3 **Strategic Thinking & Complex Reasoning**	How are the triangle congruence theorems ASA and AAS different? ASA requires that the included sides are congruent. AAS requires that a pair of non-included sides are congruent.	Students' answers will indicate whether they understand the difference between AAS and ASA.

Explore SSA Triangle Congruence

The table shows a summary of the triangle congruence theorems you have studied.

Some Methods for Proving Triangles are Congruent		
Theorem	**Statement**	**Example**
ASA Triangle Congruence Theorem	If two angles and the included side of one triangle are congruent to two angles and the included side of another triangle, then the triangles are congruent.	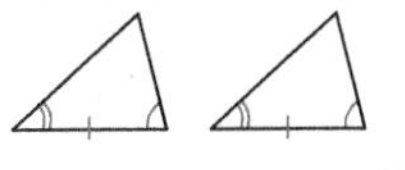
SAS Triangle Congruence Theorem	If two sides and the included angle of one triangle are congruent to two sides and the included angle of another triangle, then the triangles are congruent.	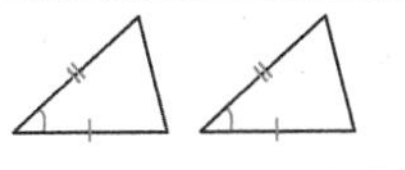
SSS Triangle Congruence Theorem	If three sides of a triangle are congruent to three sides of another triangle, then the triangles are congruent.	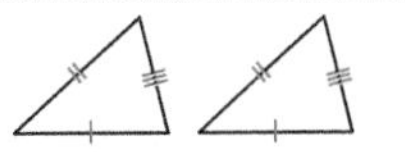
AAS Triangle Congruence Theorem	If two angles and a non-included side of one triangle are congruent to the corresponding angles and non-included side of another triangle, then the triangles are congruent.	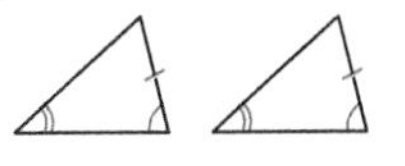

2 Another possible set of three congruent corresponding parts that might be used to prove triangles are congruent are adjacent sides and a non-included angle. Is there a Side-Side-Angle Triangle Congruence Theorem?

A. Draw a horizontal line. Use a ruler and protractor to draw a 3-inch line segment that meets your horizontal line at a 30° angle. Label the point where they meet as point *A* and the other endpoint as *B*.
A, B. Check students' constructions.

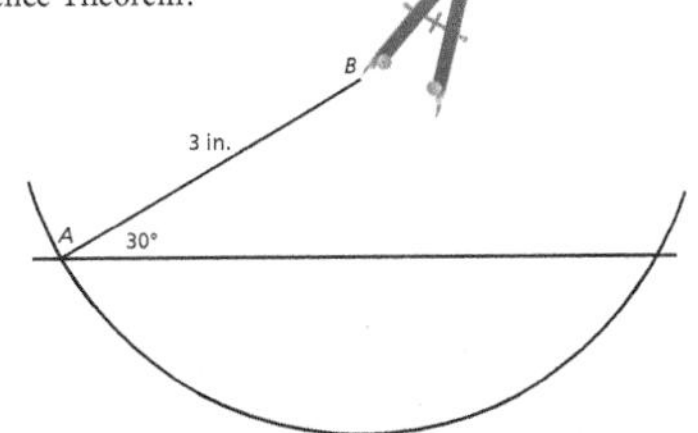

B. Set a compass for 2 inches, and place the point of the compass on *B*. Draw an arc with a radius of 2 inches from point *B* through the horizontal line.

C. Label the points where it meets the horizontal line as point *C*, and draw a segment from *B* to each point to complete the triangles. Why is it possible to create more than one triangle?
C, D. See Additional Answers.

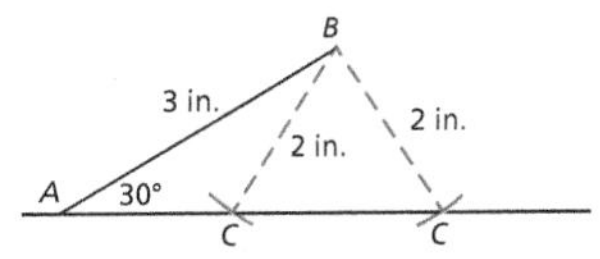

D. This construction creates two possible triangles. Can you use Side-Side-Angle criteria to prove two triangles are congruent? Why or why not?

Turn and Talk How many triangles can you draw that have sides of length 2 inches and 3 inches and a non-included angle of 90°? Explain. See margin.

Task 2 (MP) **Use Tools** Students use rulers, protractors, and compasses to draw a counterexample to verify that a Side-Side-Angle Triangle Congruence Theorem is invalid.

Sample Guided Discussion:

Q In Part A, how many congruent line segments can be drawn from point *B* to the horizontal line you drew? two congruent line segments

Q Would the outcome have changed if you changed the degree measure of the acute angle drawn? Why? no; No matter how big or small the angle drawn is, there will be two congruent line segments drawn to the line, one that hits to the left of *B* and one that hits to the right. So, there will still be two noncongruent triangles formed.

Turn and Talk If students have trouble visualizing how to form a triangle with the given measures, ask them to try drawing a right triangle in two ways: with the 3-inch side used as the hypotenuse (that is, drawn opposite the 90° angle) and with the 2-inch side used as the hypotenuse. 1 triangle; You can draw a single right triangle provided the 3-inch side is the hypotenuse. If the 2-inch side is used as the hypotenuse, no triangle is possible.

Step It Out

Task 3 (MP) **Construct Arguments** Students use prior knowledge of theorems about right triangles to prove that the Hypotenuse-Leg Theorem is valid, justify their conclusions, and communicate them to others.

Sample Guided Discussion:

Q What relationship can you recall about the three sides of a right triangle? The sum of the squares of the two legs of a right triangle is equal to the square of its hypotenuse.

Q Why doesn't the given information allow us to use SAS to prove the triangles are congruent? The given pair of congruent corresponding angles is not included in the two given pairs of congruent corresponding sides.

Turn and Talk Help students realize that sides of triangles can be named in two ways, either using the end points of each side to name the segment, or by using a lower case letter that is the same letter as the labeled angle opposite that side. If the triangles were not right triangles, we could not use the Pythagorean Theorem to write the equations we used to show a is equal to x; Because the triangles are right triangles, we have a pair of congruent corresponding angles. Using the same methods as shown in Task 3, we can show that $b = y$ and $a = x$, so $\overline{AC} \cong \overline{XZ}$ and $\overline{BC} \cong \overline{YZ}$. Then the triangles are congruent by the SAS Triangle Congruence Theorem.

Step It Out

Prove the HL Triangle Congruence Theorem

In Task 2, you showed that Side-Side-Angle criteria are not sufficient to prove triangle congruence. Right triangles are a special case in which you can prove two right triangles are congruent using the hypotenuse and a leg of the triangle.

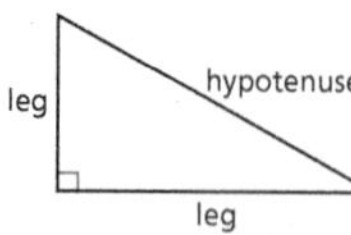

Hypotenuse-Leg (HL) Triangle Congruence Theorem

If the hypotenuse and a leg of a right triangle are congruent to the hypotenuse and a leg of another right triangle, then the right triangles are congruent.

3 Prove the HL Triangle Congruence Theorem.

Given: $\angle C$ and $\angle Z$ are right angles. $\overline{AB} \cong \overline{XY}$ and $\overline{AC} \cong \overline{XZ}$.

Prove: $\triangle ABC \cong \triangle XYZ$

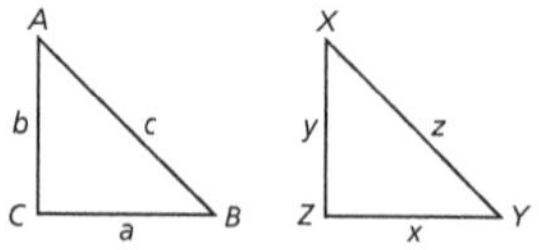

(Triangle ABC with sides a, b, c; triangle XYZ with sides x, y, z.)

Since $\overline{AB} \cong \overline{XY}$, you know that $c = z$ and $c^2 = z^2$.

Since $\overline{AC} \cong \overline{XZ}$, you know that $b = y$ and $b^2 = y^2$.

The triangles are right triangles, so $a^2 + b^2 = c^2$ and $x^2 + y^2 = z^2$.

A. Why can you use the Pythagorean Theorem? A. You can use the Pythagorean Theorem because the given information proves both triangles are right triangles.

$c^2 = z^2$

$a^2 + b^2 = x^2 + y^2$

$a^2 + b^2 = x^2 + b^2$

$a^2 = x^2$

$a = x$

B. Why can you write this equation? What allows you to make the substitution in the second line? B. See Additional Answers.

All three pairs of corresponding sides are congruent, so by the SSS Triangle Congruence Theorem, $\triangle ABC \cong \triangle XYZ$.

C. Why can't you use the HL Triangle Congruence Theorem in the proof? You cannot use the theorem you're trying to prove in its own proof.

Turn and Talk

- What parts of the proof would not be valid if the triangles were not right triangles?
- Explain how you could use the SAS Triangle Congruence Theorem instead of the SSS Triangle Congruence Theorem in the proof. See margin.

246

EL PROFICIENCY LEVEL

Beginning

Have students label the parts of a right triangle, identifying the hypotenuse, the legs, and the right angle. Then ask them to label the other two angles with appropriate degree measures.

Intermediate

Have students work in groups and provide them with a sheet showing several triangles, some right triangles and some not. Ask them to describe which ones are right triangles and explain how they know.

Advanced

Have students write a short paragraph to explain the relationship that exists between the three sides of a right triangle.

Apply AAS and HL Triangle Congruence

4 Shar is planning her running route and needs to know if two triangular blocks are congruent. Determine whether the two blocks are congruent.

Examine the given information.
Notice the two blocks have corresponding 50° angles and congruent vertical angles.

A. Explain why the intersection contains congruent angles.

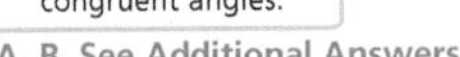

A, B. See Additional Answers.

Determine a possible congruence theorem.
There are two congruent angles, so consider AAS and ASA. The positions of the sides that are marked congruent mean that you should use the AAS Triangle Congruence Theorem.

B. How does the position of the sides marked congruent tell you which theorem to use?

Answer the question.
The blocks have two pairs of corresponding congruent angles and a corresponding side that is not included between the two angles. The blocks are congruent by the AAS Triangle Congruence Theorem.

5 Shar is considering another possible running route. Are the routes shown congruent?

A–D. See Additional Answers.

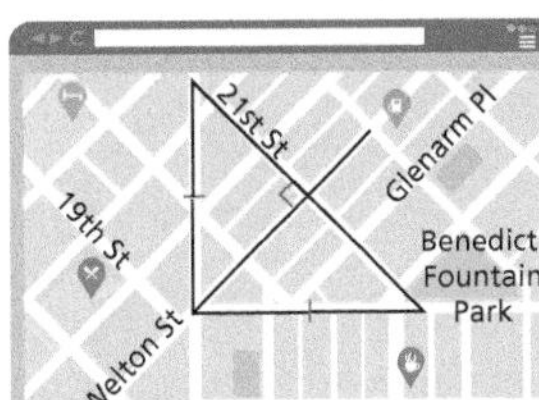

Examine the given information.
By the given information, we know the intersection of Welton Street and 21st Street creates two right angles.

A. How do you know the measure of these angles?

Determine possible congruent sides.
The hypotenuses of the triangles are congruent.

B. Why are these sides hypotenuses?

C. What side is shared by the two triangles? Is this shared side congruent?

Answer the question.
The triangles are congruent by the HL Triangle Congruence Theorem.

D. What three congruent pairs are used to make this conclusion?

Turn and Talk How are the HL criteria and the SSS criteria related? See margin.

Task 4 (MP) **Use Structure** Students are shown a pair of triangles that have two pairs of corresponding congruent angles and one pair of non-included sides. They utilize properties of intersecting lines to find congruent angles that are not given.

Sample Guided Discussion:

Q Why can't we use the HL Triangle Congruence Theorem to prove the two triangular routes are congruent? There is no information telling us that one hypotenuse is congruent to the other.

Task 5 (MP) **Use Structure** Students recognize congruent angles and sides in a diagram of two triangles with a shared side that is perpendicular to corresponding sides of the two triangles. They use these relationships to determine that two triangles are congruent by the HL Triangle Congruence Theorem.

Sample Guided Discussion:

Q How is this diagram different than the diagram you saw in Task 4? In Task 4, the right triangles shared a common vertex, so it had a pair of vertical angles. In Task 5, the two right triangles share a common side.

Q What property tells us that a segment is congruent to itself? Reflexive Property of Congruence

Turn and Talk Encourage students to discuss the difference between HL Theorem and the other triangle congruency theorems they have used, pointing out that HL Theorem requires only two pairs of congruent sides, where the other theorems require three pairs of congruent sides or angles. Possible answer: The HL criteria and SSS criteria are related because the HL criteria is a shortcut of the SSS criteria used for right triangles only. If a triangle is a right triangle, the three sides are related by the Pythagorean Theorem. So, we know if the HL criteria are met, the SSS criteria is met by default.

OPTIMIZE OUTPUT Critique, Correct, and Clarify

Have students work with a partner to discuss when it is appropriate to use the HL Triangle Theorem of Congruence or when using AAS or ASA is more appropriate. Have them each draw diagrams to justify their claims. Encourage students to use vocabulary from the lesson in their discussions.

ONLINE

Assign the Digital On Your Own for
- built-in student supports
- Actionable Item Reports
- Standards Analysis Reports

On Your Own

Assignment Guide

The chart below indicates which problems in the On Your Own are associated with each task in the Learn Together. Assign daily homework for tasks completed.

Learn Together Tasks	On Your Own Problems
Task 1, p. 244	Problems 6–8, 11, and 12
Task 2, p. 245	Problems 8 and 11
Task 3, p. 246	Problem 13
Task 4, p. 247	Problems 14–15
Task 5, p. 247	Problems 9, 10, 16, 17, and 19–21

Check Understanding

1. Compare the AAS Triangle Congruence Theorem to the ASA Triangle Congruence Theorem. 1–3. See Additional Answers.

2. A triangle has side lengths of 2 inches and 3 inches and a non-included angle of 20° adjacent to the 3-inch side. Use a ruler and a protractor to draw the triangle or triangles that meet these criteria. Justify your answer.

3. In your own words, explain why the SSA criteria does prove congruence when the angle used in the theorem is a right angle.

4. What additional information must be known to prove the triangles are congruent using AAS Triangle Congruence? Either $\overline{BC} \cong \overline{ZX}$ or $\overline{AC} \cong \overline{YX}$

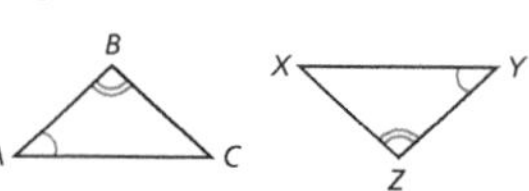

5. Can you prove that the two triangles are congruent? If so, explain how. See Additional Answers.

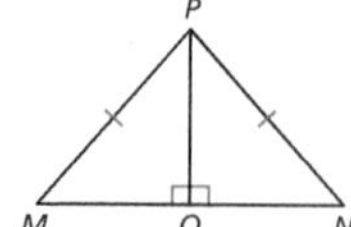

On Your Own

Is enough information given to prove the two triangles are congruent? If they are congruent, explain what triangle congruence criteria should be used.

6. 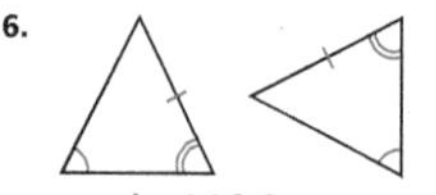 

yes; by AAS Congruence

7. 

yes; by AAS Congruence

8. 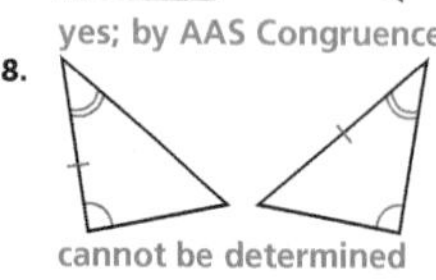

cannot be determined

9. 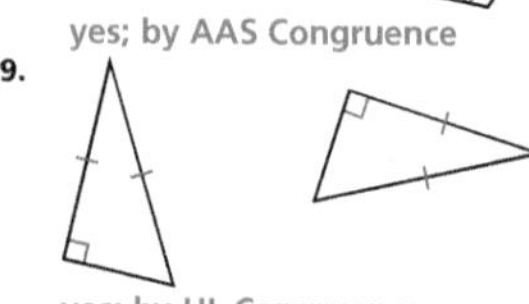

yes; by HL Congruence

10. 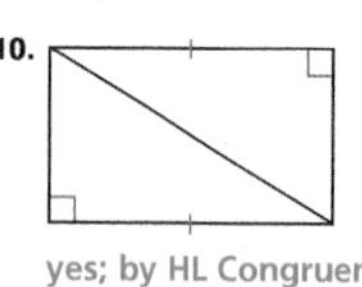

yes; by HL Congruence

11. 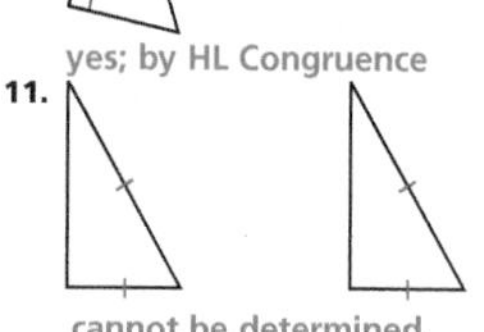

cannot be determined.

What value of x will make each pair of triangles congruent? Explain.

12. 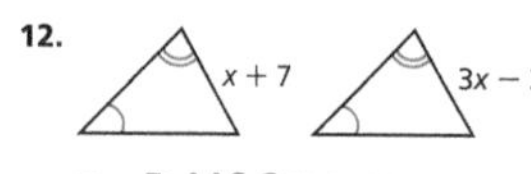

$x = 5$; AAS Congruence

13. 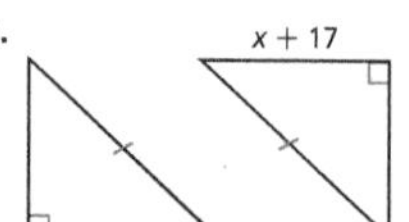

$x = 3$; HL Congruence

248

data checkpoint

3 Check Understanding

Formative Assessment

Use formative assessment to determine if your students are successful with this lesson's learning objective.

Students who successfully complete the Check Understanding can continue to the On Your Own practice.

For students who miss 1 problem or more, work in a pulled small group using the Almost There small-group activity on page 243C.

ONLINE

Assign the Digital Check Understanding to determine
- success with the learning objective
- items to review
- grouping and differentiation resources

4 Differentiation Options

Differentiate instruction for all students using small-group activities and math center activities on page 243C.

Reteach

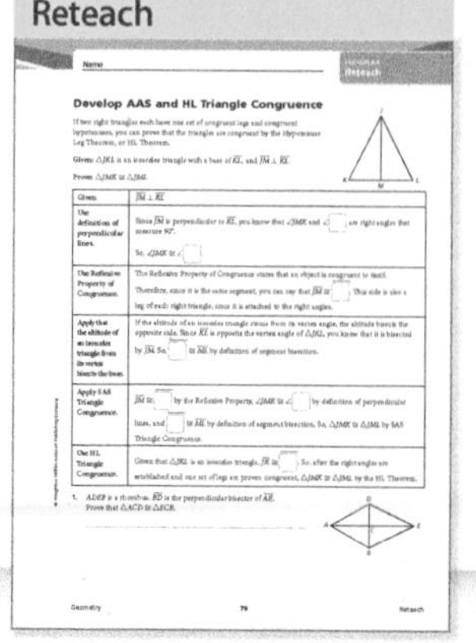

Challenge

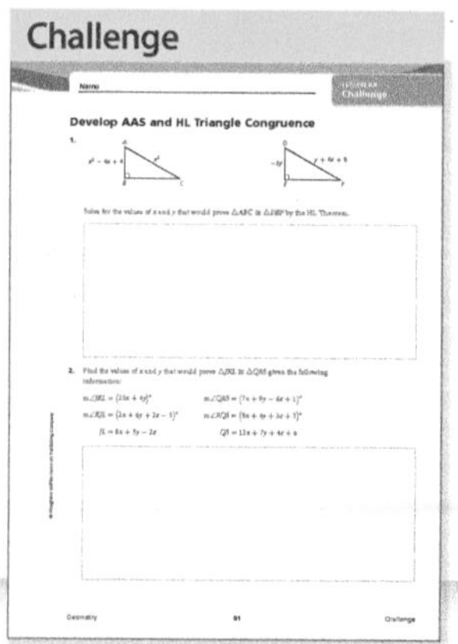

Write a two-column proof to show the triangles are congruent.

14–19. See Additional Answers.

14. **Given:** $\overline{JK} \parallel \overline{MN}$, and $\overline{JN}$ bisects $\overline{KM}$.

Prove: $\triangle JKL \cong \triangle NML$

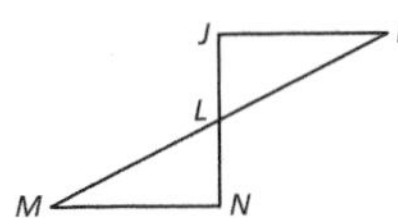

15. **Given:** $\angle A \cong \angle C$, and $\overline{DB}$ bisects $\angle ABC$.

Prove: $\triangle ABD \cong \triangle CBD$

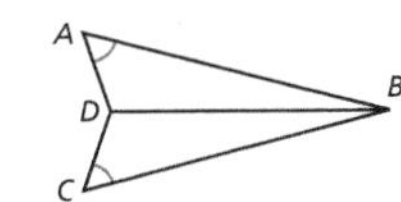

16. **Given:** $\overline{PT} \perp \overline{SQ}$, $\overline{ST} \cong \overline{PQ}$, and R is the midpoint of $\overline{PT}$.

Prove: $\triangle PQR \cong \triangle TSR$

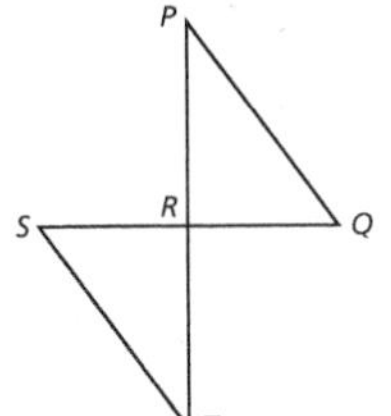

17. **Given:** $\angle E$ and $\angle C$ are right angles. $\overline{AE} \cong \overline{BC}$

Prove: $\triangle ABC \cong \triangle BAE$

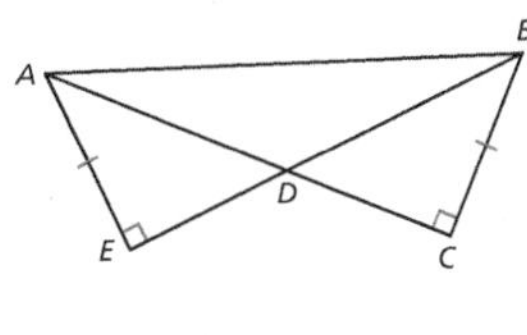

18. **Open Ended** Write a proof problem that would use the AAS Triangle Congruence Theorem. Provide a diagram with labels and list the Given and Prove statements. Explain how the AAS Theorem would be used in the proof.

19. An inspector needs to find the dimensions of a roof truss. The length of the bottom chord is 2.5 times the height of the truss. The length of the top chord is 21 ft, which includes a 5 ft overhang. The bottom chord and the king post are perpendicular. What are the lengths of the bottom chord and the king post? Round your final answers to the nearest foot.

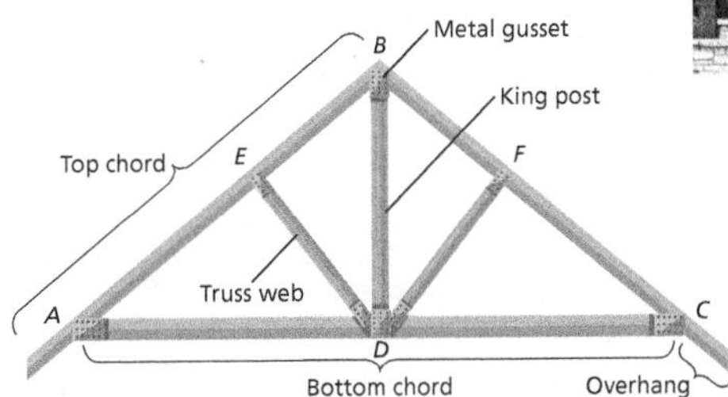

Questioning Strategies

Problem 14 Suppose that the statement $\overline{JN}$ bisects $\overline{KM}$ was not given but the statement $\overline{ML} \cong \overline{LK}$ was given. Would you have enough information to prove the triangles are congruent? No because there is no information telling us that angles *N* and *J* are right angles, so we cannot use HL. We would only have one congruent pair of sides and one congruent pair of angles so we could not use any of the other theorems either.

Watch for Common Errors

Problem 16 Some students may neglect to mark up a diagram based on the given statements, which creates a difficult task in proving what needs to be proved, in this case, that $PQR \cong TSR$. In this example, once the diagram is marked with the given information, there are no other facts needed to get to the proof statement. Also, students should know that the reasons are most often the names of the theorems themselves.

5 Wrap-Up

Summarize learning with your class. Consider using the Exit Ticket, Put It in Writing, or I Can scale.

Exit Ticket 

What value of x makes right triangles JKL and JKM congruent?

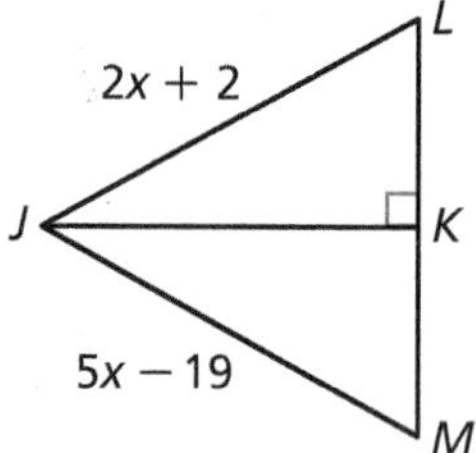

$x = 7$

Put It in Writing 

Describe how the HL Congruence Theorem is different than the other congruence theorems we have studied.

I Can 

The scale below can help you and your students understand their progress on a learning goal.

4	I can use AAS and HL congruence criteria to determine if triangles are congruent, and I can explain my reasoning to others.
3	I can use AAS and HL congruence criteria to determine if triangles are congruent.
2	I can state the characteristics of AAS and HL Congruence Theorems and draw diagrams to show those characteristics.
1	I can state the characteristics of AAS and HL Congruence Theorems.

Spiral Review • Assessment Readiness

These questions will help determine if students have retained information taught in the past and can also prepare them for high-stakes assessments. Here, students must understand the characteristics of AAS Triangle Congruence Theorem **(8.1)**, understand the characteristics of SAS Triangle Congruence Theorem **(8.2)**, understand whether SSS or AAA criteria can be used to determine congruency **(8.3)**, and know the sum of the interior angles of a triangle **(Gr8, 4.1)**.

20. (MP) **Construct Arguments** Maya claims that if two right triangles share a hypotenuse, they must be congruent. Is Maya correct? Explain. **20, 21. See Additional Answers.**

21. John is designing the base for a new sculpture to be placed in a city plaza. To determine how much material is needed for the border of the base, John needs to approximate the perimeter of the sculpture to plan its placement in the park. The figure represents the overhead view of the sculpture. Find the perimeter of the sculpture. Explain your steps.

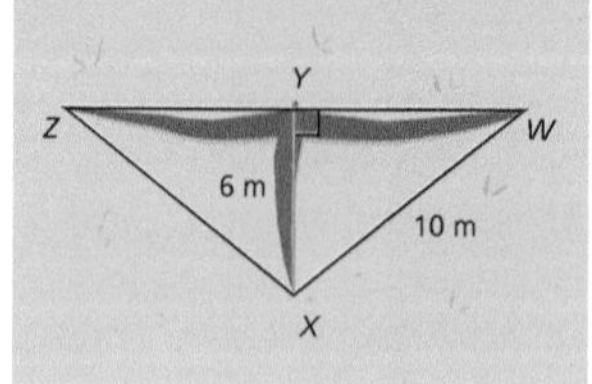

Spiral Review • Assessment Readiness

22. Two triangles with two pairs of congruent angles and any one corresponding congruent side, included or not, are congruent.

Ⓐ always true
Ⓑ sometimes true
Ⓒ never true
Ⓓ can not be determined

23. The angles used to prove congruency in the SAS Triangle Congruence Theorem must be __?__.

Ⓐ right angles
Ⓑ opposite a side being used in the theorem
Ⓒ acute
Ⓓ included

24. If two triangles are congruent by the SSS Triangle Congruence Theorem, you can also prove they are congruent using only AAA criteria.

Ⓐ Yes, because corresponding parts are congruent.
Ⓑ No, three congruent angles don't prove congruency.
Ⓒ Yes, because the AAA Theorem and SSS Theorem are equivalent.
Ⓓ No, triangles congruent by the SSS Theorem may not have three congruent angles.

25. An equilateral triangle contains an interior angle of 60°. What is the sum of all three angles within the triangle?

Ⓐ 60°
Ⓑ 120°
Ⓒ 180°
Ⓓ 360°

 **I'm in a Learning Mindset!**

Did I communicate effectively? Did I respond non-defensively when questions were asked about the ideas I communicated?

Keep Going to Journal and Practice Workbook

Learning Mindset

Resilience Responds to Feedback

Point out that comparing results to problems with group members or classmates is essential in all areas of mathematics, not just when working with triangle congruence theorems in this lesson. Therefore, it is important to be able to communicate effectively when analyzing the steps in a proof, the given information, or the theorems that are appropriate to solve a problem. Respond to the viewpoints of others non-defensively. On the contrary, students should welcome feedback from others as a means of correcting mistakes and misconceptions. *When given the opportunity to work in groups, are you accepting of the feedback you receive from peers when comparing answers? Are you willing to listen to the viewpoints of others in order to expand your knowledge and discover new strategies? Do you take feedback given from teachers or peers personally?*

Module 8 Review

ASA and AAS Triangle Congruence

Any two triangles with two pairs of congruent angles and a congruent included side are congruent by the Angle-Side-Angle (ASA) Triangle Congruence Theorem.

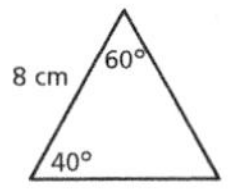

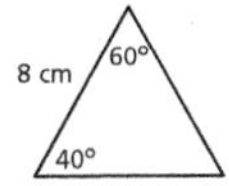

We also know by the Third Angles Theorem, that the missing angles are also congruent, so we could also say they are congruent by the Angle-Angle-Side (AAS) Triangle Congruence Theorem.

SAS Triangle Congruence

Any two triangles with two pairs of congruent sides and a congruent included angle are congruent by the Side-Angle-Side (SAS) Triangle Congruence Theorem.

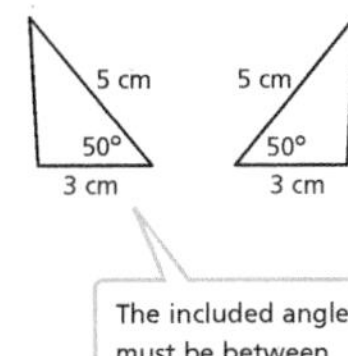

SSS Triangle Congruence

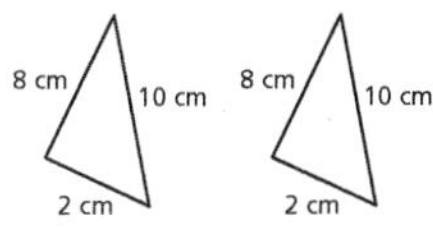

Only one possible triangle can be created when given all three side lengths. Therefore, any two triangles with three pairs of congruent sides must be congruent by the Side-Side-Side (SSS) Triangle Congruence Theorem.

HL Triangle Congruence

In right triangles, the third side is determined by the Pythagorean Theorem. Thus, any two right triangles with a congruent hypotenuse and leg are congruent by SSS or SAS.

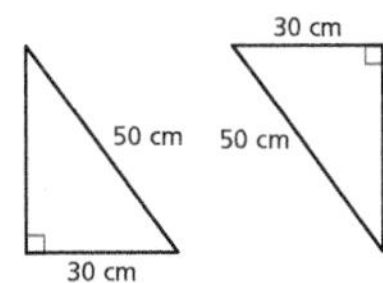

The two right triangles above have a congruent hypotenuse and a congruent leg. Therefore, they are congruent by the Hypotenuse-Leg (HL) Triangle Congruence Theorem.

Module 8 251

ONLINE

Assign the Digital Module Review for

- built-in student supports
- Actionable Item Reports
- Standards Analysis Reports

MODULE 8 REVIEW

Module Review

Use the first page of the Module Review to summarize and connect the main ideas of the module. Use the second page to assess students' understanding of the vocabulary, concepts, and skills presented in the module.

Sample Guided Discussion:

Q ASA and AAS Triangle Congruence What is the measure of the third angle in each triangle? Explain. The sum of the measures of the angles in a triangle is 180°; So, $40° + 60° + x = 180°$, and $x = 80°$.

Q SAS Triangle Congruence How can you tell whether an angle in a triangle is included between two sides? Possible answer: Each side of the angle is a side of the triangle. The third side is opposite the angle.

Q SSS Triangle Congruence What rigid transformation could you use to establish that the two triangles in this figure might be congruent? Possible answer: a horizontal translation because the orientation of the triangles and the measures of the sides are preserved

Q HL Triangle Congruence Given the lengths of leg *a* and hypotenuse *c* in a right triangle, what expression can you write for the length of leg *b*? $\sqrt{c^2 - a^2}$

Module Review continued

Possible Scoring Guide

Items	Points	Description
1–5	2 each	identifies the correct term
6	2	correctly lists all triangle congruence theorems and examples of each
7	2	correctly identifies the theorem relating ASA and AAS congruence theorems
8	2	correctly relates the HL congruence theorem to other congruence theorems
9	2	correctly explains why SSA is not a congruence theorem
10–11	2 each	correctly finds values of *x* that make two triangles congruent
12–15	2 each	correctly proves the two triangles are congruent
Total points possible = 30 points		

The Unit 4 Test in the Assessment Guide assesses content from Modules 7 and 8.

Vocabulary

Choose the correct term from the box to complete each sentence.

Vocabulary
- congruent triangles
- corresponding angles
- corresponding sides
- included angle
- included side

1. When a transversal crosses two parallel lines, angles in the same position in each intersection are called __?__ and are congruent. corresponding angles
2. A(n) __?__ is formed by two adjacent sides of a triangle. included angle
3. All corresponding parts of __?__ are congruent. congruent triangles
4. A(n) __?__ connects two consecutive angles in a triangle. included side
5. Two sides in the same position of two congruent triangles are called __?__. corresponding sides

Concepts and Skills

6. List all triangle congruence theorems and give an example of each. See Additional Answers.
7. What theorem relates the ASA and AAS triangle congruence theorems? Third Angles Theorem
8. The HL Congruence Theorem is a specific case of what other theorem? Explain your reasoning. SSS or SAS; They are related by the Pythagorean Theorem.
9. (MP) Use Tools Explain why there is no Side-Side-Angle Triangle Congruence Theorem. State what strategy and tool you will use to answer the question, explain your choice, and then find the answer. See Additional Answers.

What value of x will make each pair of triangles congruent.

10. $x = 10$

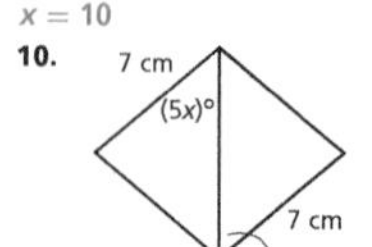

11. $x = -2.5$

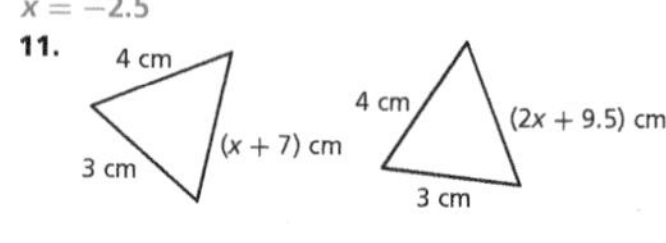

Write a proof to prove that the triangles in each pair are congruent.
12–15. See Additional Answers.

12.

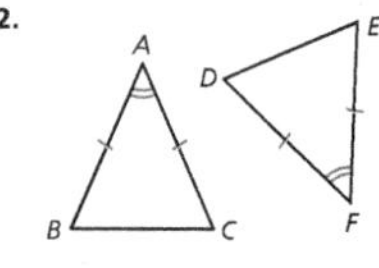

13.

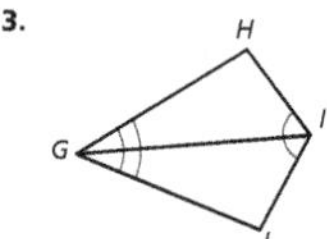

14.

15.

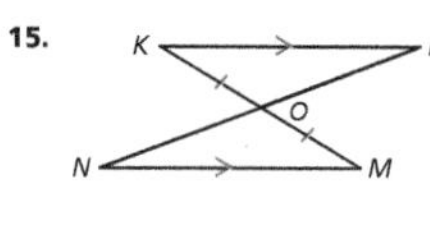

252

DATA-DRIVEN INSTRUCTION

Before moving on to the Module Test, use the Module Review results to intervene based on the table below.

MTSS (RtI)

Items	Lesson	DOK	Content Focus	Intervention
6	8.1–8.4	2	List triangle congruence theorems with examples.	Reteach 8.1–8.4
7	8.4	2	Identify an application of the Third Angles Theorem.	Reteach 8.4
8	8.4	3	Explain why HL is a specific case of SSA.	Reteach 8.4
9	8.4	3	Explain why SSA is not a triangle congruence theorem.	Reteach 8.4
10–11	8.2	2	Find values of *x* that make triangles congruent.	Reteach 8.2
12–15	8.1–8.4	3	Prove triangles are congruent.	Reteach 8.1–8.4

Module Test

The Module Test is available in alternative versions in your Assessment Guide. All items are presented in standardized test formats.

ONLINE

Assign the Digital Module Test to power actionable reports including

- proficiency by standards
- item analysis

Form A

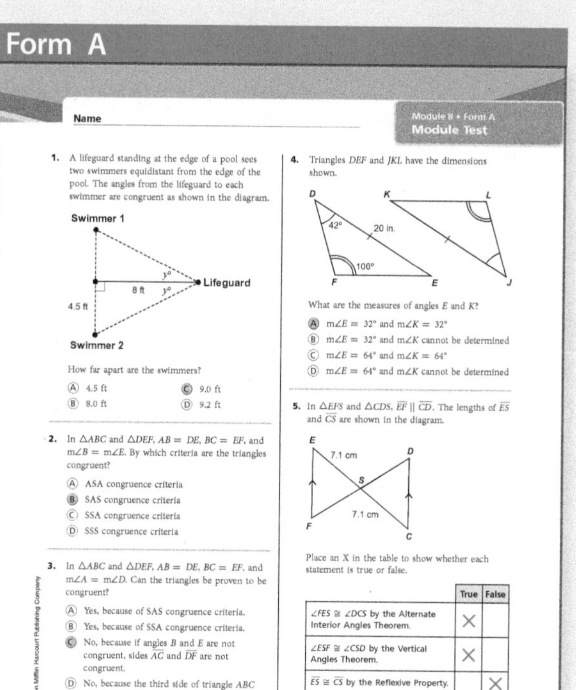

Form A

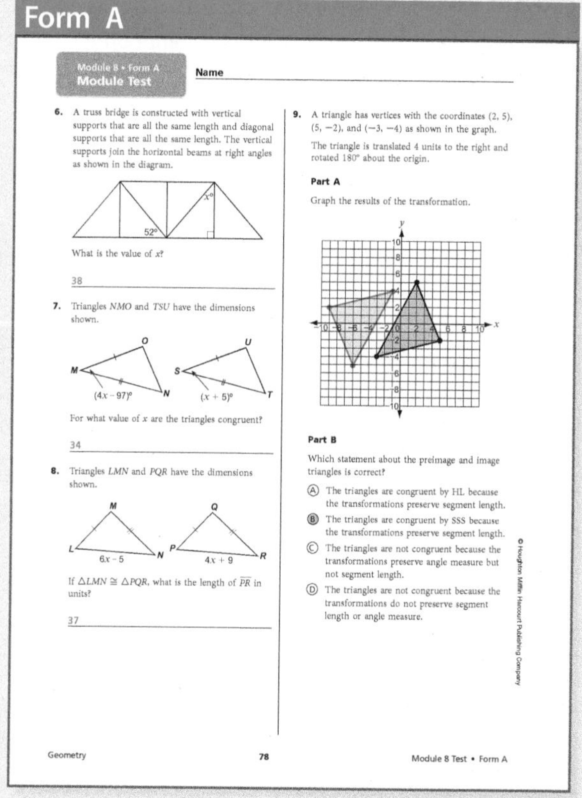

Form B

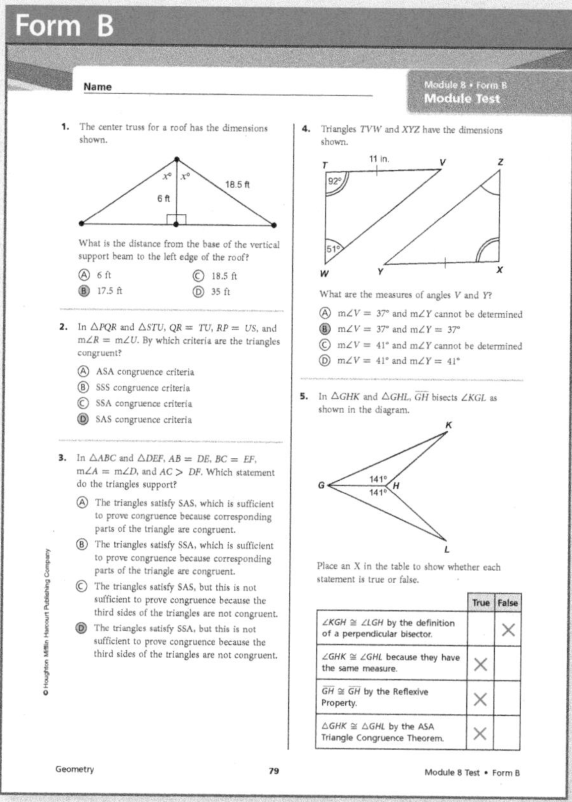

Form B

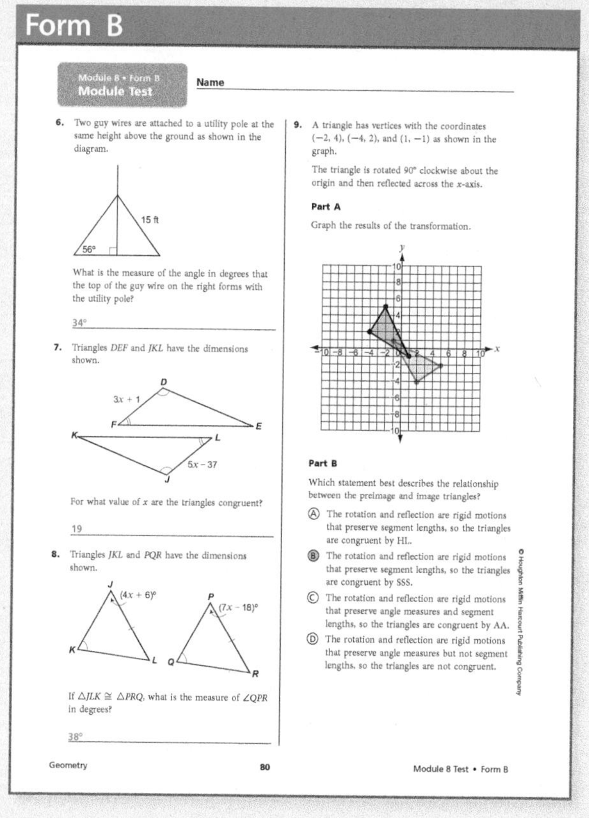

Module 9: Properties of Triangles

Module 10: Triangle Inequalities

Environmental Chemist STEM

- **Say:** *Chemists deal with the elements and compounds that make matter. Environmental chemists study how chemicals affect living things and their habitats. They research how manmade chemicals get into the environment, how they change or contaminate the environment, what can be done to prevent the release of harmful chemicals, and how to remove them where they already exist.*
- Explain that environmental chemists study natural reactions that occur in water, in air, and on land, especially those that affect living organisms. This STEM task is focused on studying the concentration of toxic bacteria, called *cyanobacteria*, caused by the presence of excess phosphorus in the Maumee River that flows into Lake Erie, one of the five Great Lakes.

STEM Task

Ask students what they know about algae, aquatic organisms capable of photosynthesis. Then discuss the situation represented by the Maumee River data.

- Review with students that algae are oxygen producers that live in fresh or marine waters. A harmful algal bloom (HAB), such as the one described here, results from the rapid increase in the population of algae in an aquatic system.
- The Maumee River, the source of the data on the graph, empties into Lake Erie. Have students look at a map of the Great Lakes region and locate the Maumee River. Have them trace it from its source to Lake Erie and describe the land and cities through which the river runs.
- Have students investigate the element phosphorous and its compounds. What are its properties? Does it occur naturally? Why is phosphorus essential for life? What might be the source of the excess phosphorous in the river?
- What environmental conditions encourage the growth of HABs and what are the harmful effects of HABs?
- How might climate change affect the growth of HABs?

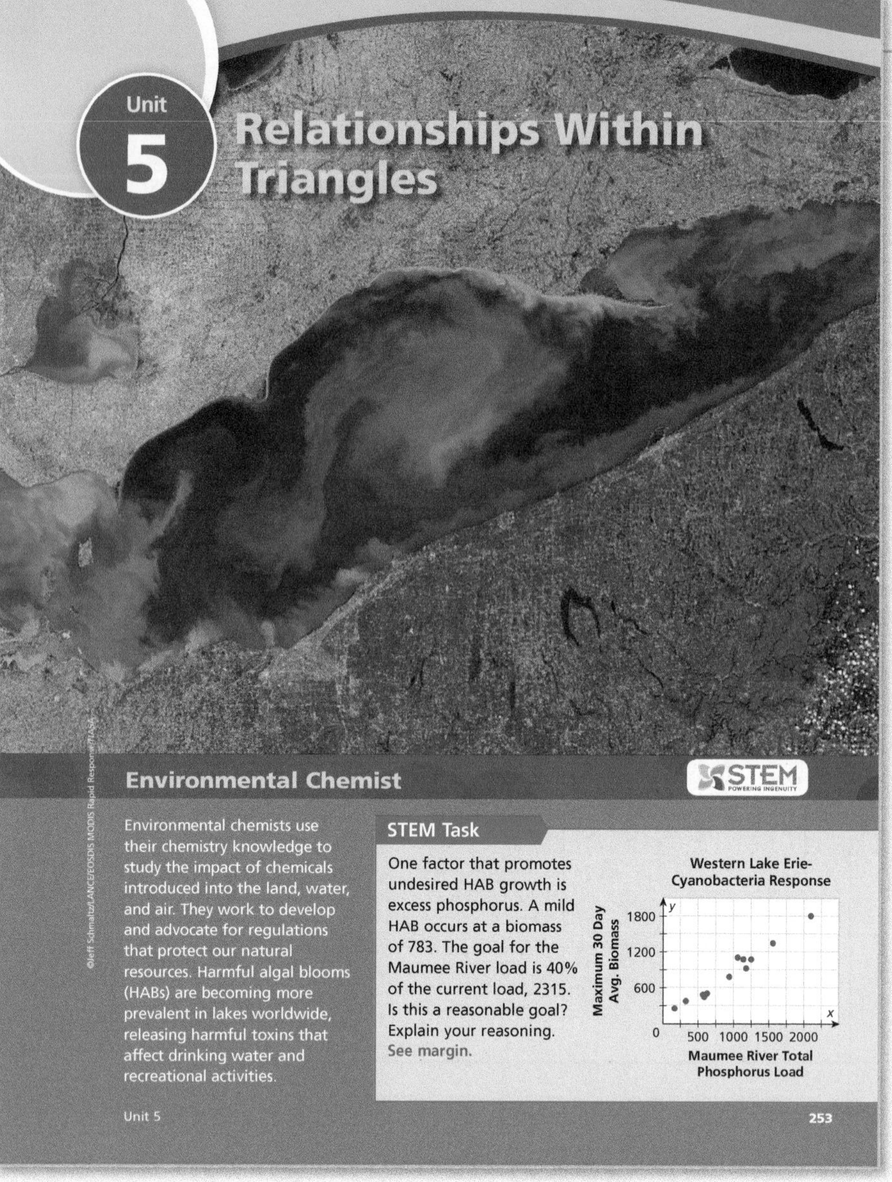

Unit 5

Relationships Within Triangles

©Jeff Schmaltz/LANCE/EOSDIS MODIS Rapid Response/NASA

Environmental Chemist STEM

Environmental chemists use their chemistry knowledge to study the impact of chemicals introduced into the land, water, and air. They work to develop and advocate for regulations that protect our natural resources. Harmful algal blooms (HABs) are becoming more prevalent in lakes worldwide, releasing harmful toxins that affect drinking water and recreational activities.

STEM Task

One factor that promotes undesired HAB growth is excess phosphorus. A mild HAB occurs at a biomass of 783. The goal for the Maumee River load is 40% of the current load, 2315. Is this a reasonable goal? Explain your reasoning. See margin.

Unit 5 253

Unit 5 Project Room for Bloom

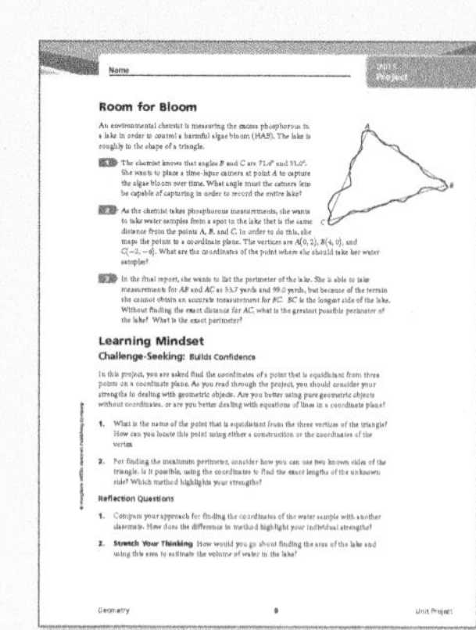

Overview: In this project students use the properties of triangles to determine the camera angle, the location of a water sample, and the perimeter of a lake.

Materials: paper and pencil or a computer used for word processing

Assessing Student Performance: Students' reports should include:

- a correct determination of the missing angle in the triangle **(Lesson 9.1)**
- correct coordinates for the circumcenter of the triangle **(Lesson 9.2)**
- a correct upper bound for the perimeter using the Triangle Inequality and a correct exact value of the perimeter **(Lesson 10.1)**

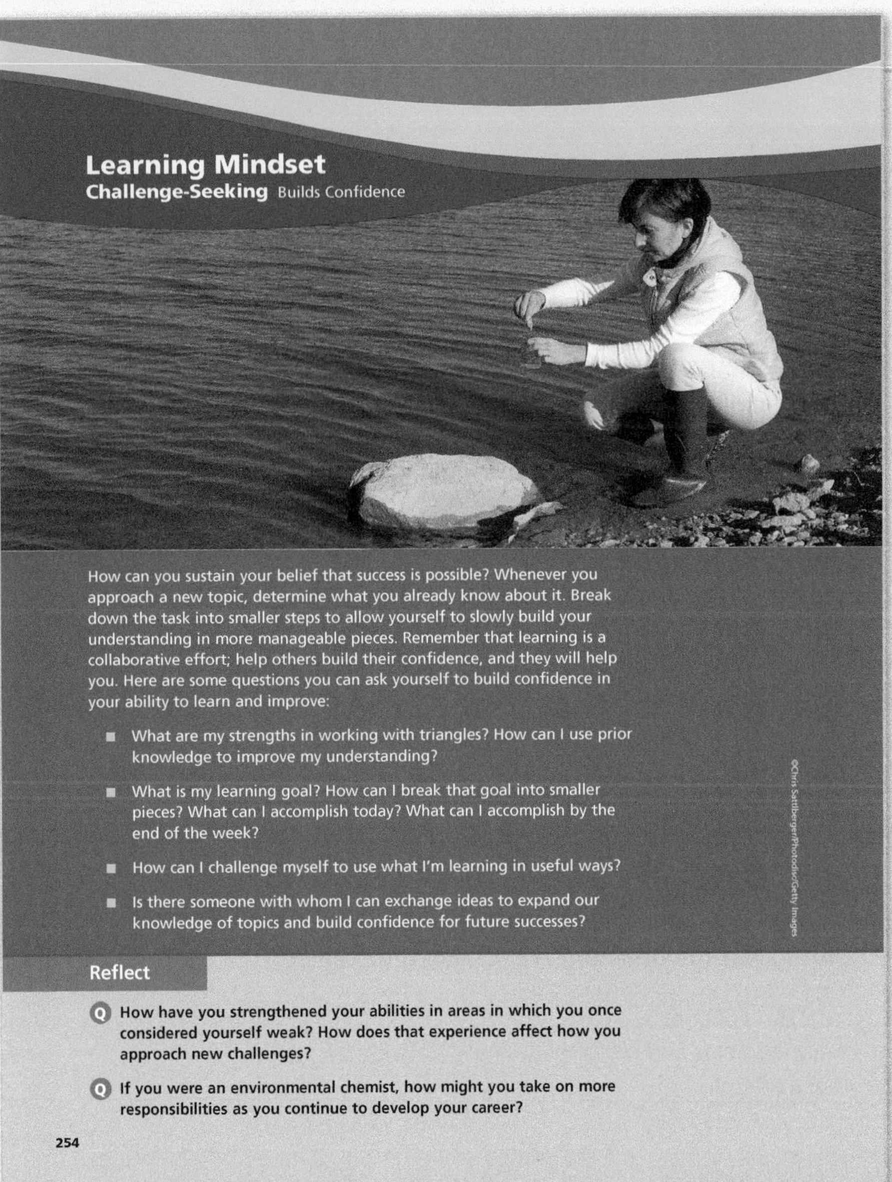

Learning Mindset
Challenge-Seeking Builds Confidence

©Chris Sattlberger/Photodisc/Getty Images

How can you sustain your belief that success is possible? Whenever you approach a new topic, determine what you already know about it. Break down the task into smaller steps to allow yourself to slowly build your understanding in more manageable pieces. Remember that learning is a collaborative effort; help others build their confidence, and they will help you. Here are some questions you can ask yourself to build confidence in your ability to learn and improve:

- What are my strengths in working with triangles? How can I use prior knowledge to improve my understanding?
- What is my learning goal? How can I break that goal into smaller pieces? What can I accomplish today? What can I accomplish by the end of the week?
- How can I challenge myself to use what I'm learning in useful ways?
- Is there someone with whom I can exchange ideas to expand our knowledge of topics and build confidence for future successes?

Reflect

Q **How have you strengthened your abilities in areas in which you once considered yourself weak? How does that experience affect how you approach new challenges?**

Q **If you were an environmental chemist, how might you take on more responsibilities as you continue to develop your career?**

254

What to Watch For

Watch for students who accept learning challenges. Help them become more self-confident by

- evaluating what they understand about a particular skill or concept,
- asking them why they enjoy a learning challenge, and
- applauding their efforts.

Watch for students who are very self-confident. Encourage this attitude toward learning by

- reminding them that being self-confident involves preparation and mastering of underlying skills and concepts and
- acknowledging their enthusiasm.

"One important key to success is self-confidence. An important key to self-confidence is preparation."

—Arthur Ashe

Learning Mindset

Challenge Seeking

Builds Confidence

The learning-mindset focus in this unit is *challenge seeking*, which refers to a person's desire to go beyond what's expected: to learn new things and derive personal satisfaction from being able to solve complex problems.

Mindset Beliefs

Students who react to challenges enjoy learning. In mathematics, this attitude often takes the form of choosing more difficult problems to solve or by looking at the table of contents of a math textbook to ask "What comes next?" Students who seek challenges usually exhibit an above average level of self-confidence. They are not likely to step into the shadows when confronted by a new and difficult situation. They optimistically persevere in acquiring more information and finding solutions for new challenges.

A person who seeks challenges demonstrates unique strengths. In mathematics, students who enjoy challenges push ahead. If they reach a stumbling block, they are willing to go back and review where they might have gone wrong. They ask, "What can I do differently to get past this hurdle?" or "Is there another way to solve this?" They demonstrate both resilience and perseverance.

Discuss with students what they do when they are confronted with a difficult situation, or in math, a confusing problem. Do they try different methods to solve the problem? Do they check with others? Do they recognize that it is sometimes necessary to start over? Are they confident about their ability to be successful?

Mindset Behaviors

Encourage students to become willing to accept learning challenges by asking themselves questions about how they approach learning. This self-monitoring can take the form of questions like the following:

When reading a mathematical textbook:

- Do I understand the conditions of the problem or situation?
- Am I able to recall previously learned skills to use in solving this new problem or situation?

When looking for a challenge:

- Can I connect a given problem or situation to similar problems I've seen in the past?
- Am I confident in my ability to be successful in addressing this challenge?

When addressing a new challenge:

- How confident am I in my solution to the problem?
- How can I verify that my solution to a challenge is valid?

Students who continue to accept and resolve learning challenges become more self-confident.

MODULE 9

PLANNING

PROPERTIES OF TRIANGLES

Introduce and Check for Readiness

- Module Performance Task • Are You Ready?

Lesson 9.1—2 Days

Angle Relationships in Triangles

Learning Objective: Prove theorems about triangle angles.
Review Vocabulary: isosceles triangle
New Vocabulary: auxiliary lines, corollary, exterior angle of a triangle, interior angle of a triangle, remote interior angle

Lesson 9.2—2 Days

Perpendicular Bisectors in Triangles

Learning Objective: Construct perpendicular bisectors and use the point of concurrency to circumscribe triangles with circles.
New Vocabulary: circumcenter of a triangle, circumcircle, circumscribed, concurrent lines, perpendicular bisector of a side of a triangle, point of concurrency

Lesson 9.3—2 Days

Angle Bisectors in Triangles

Learning Objective: Prove that angle bisectors are concurrent and inscribe circles in triangles.
New Vocabulary: angle bisector of a triangle, incenter

Lesson 9.4—2 Days

Medians and Altitudes in Triangles

Learning Objective: Construct medians and altitudes to find centroids and orthocenters.
New Vocabulary: altitude of a triangle, centroid, median of a triangle, orthocenter

Lesson 9.5—2 Days

The Triangle Midsegment Theorem

Learning Objective: Construct midsegments and prove the Triangle Midsegment Theorem.
New Vocabulary: midsegment of a triangle

Assessment

- Module 9 Test (Forms A and B)
- Unit 5 Test (Forms A and B)

data checkpoint

LEARNING ARC FOCUS

Build Conceptual Understanding

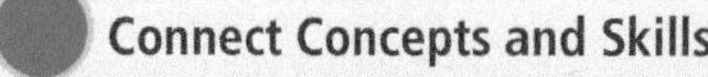

Connect Concepts and Skills

Apply and Practice

TEACHING FOR SUCCESS

TEACHING FOR DEPTH: Properties of Triangles

Make Connections. In this module, students focus on the special relationships within triangles. Although a triangle is the simplest of all the polygons, there are unique relationships and theorems about the lines that can be drawn, or constructed, within. Students continue their study of triangles by applying what they have learned about bisecting angles and segments to learn what happens when the angles of a triangle are bisected and when its medians are drawn. These lines create points of concurrency—the incenter and centroid, respectively.

Students also learn that lines that are perpendicular to the sides—the altitudes and perpendicular bisectors—also create unique points of concurrency—the orthocenter and circumcenter, respectively.

Each type of center and radius creates a circle, and students construct and prove related theorems that establish the formal relationships between a triangle and those circles.

Mathematical Progressions

Prior Learning	Current Development	Future Connections
Students: • proved triangle congruence criteria. • proved lines are perpendicular and constructed perpendicular bisectors of segments. • proved theorems about perpendicular bisectors.	**Students:** • prove theorems about perpendicular bisectors and angle bisectors of triangles. • construct angle bisectors. • use constructions to find the incenter and circumcenter of a triangle. • prove that the altitudes of a triangle meet at a point. • prove that the medians of a triangle meet at a point. • prove and use the Triangle Midsegment Theorem.	**Students:** • will prove theorems about triangle inequalities. • will prove theorems about quadrilaterals.

TEACHER TO TEACHER

From the Classroom

Support productive struggle in learning mathematics. My students find it hard to not confuse the different points of concurrency in a triangle and how they are created. They know that the angle bisectors, perpendicular bisectors, altitudes, and medians each intersect at a point, but which center goes with which type of lines?

I think it's helpful to have students draw or sketch different types of triangles to see which lines create which centers (and circles). For example, have them draw acute, right, and obtuse triangles and sketch two angle bisectors. Ask them to decide where the point of concurrency appears to always lie. If it's in the interior, then this point is the *incenter* and equidistant from each side of the triangle. On the other hand, in a right triangle, the perpendicular bisectors of the sides always intersect on the hypotenuse, equidistant from its endpoints. So, the hypotenuse can be considered a diameter of the circle that *circumscribes* the triangle.

This leaves the altitudes and medians. Again, drawing each type of segment in different types of triangles should suggest where these points of concurrency lie. Then, because the prefix *ortho-* means *right*, this point of concurrency is associated with the intersection of perpendicular lines.

As for the medians of a triangle, the easiest way my students can remember their point of concurrency is to draw a triangle on a firm sheet of paper or cardboard, construct its medians, and then cut out the triangle and see if it balances when a pencil is inserted at the point of concurrency. (It may wobble a little!)

LANGUAGE DEVELOPMENT • Planning for Instruction

By giving all students regular exposure to language routines in context, you will provide opportunities for students to **listen, speak, read,** and **write** about mathematical situations and develop both mathematical language and conceptual understanding at the same time.

Using Language Routines to Develop Understanding

Use the **Professional Cards** for the following routines to plan for effective instruction.

Co-Craft Questions and Problems Lessons 9.1 and 9.2

Students think of natural questions to ask about a given situation or problems similar to a given task and answer the questions they have developed or problems they have created.

Three Reads Lessons 9.2, 9.3, and 9.5

Students read a problem three times with a specific focus each time.

1st Read What is the situation about?
2nd Read What are the quantities in the situation?
3rd Read What are the possible mathematical questions that you could ask for the situation?

Information Gap Lesson 9.4

Students recognize when information given in a problem situation is incomplete, and they pose questions and share knowledge with others to discover any missing facts or relationships and work together to solve the problem.

Critique, Correct, and Clarify Lesson 9.4

Students correct the work in a flawed explanation, argument, or solution method and share with a partner and refine the sample work.

Connecting Language to Properties of Triangles

Watch for students' use of the review and new terms listed below as they explain their reasoning and make connections with new concepts.

Key Academic Vocabulary

Prior Learning and Current Development • Review and New Vocabulary

isosceles triangle a triangle with at least two congruent sides

altitude of a triangle the perpendicular segment from a vertex to the opposite side or to a line that contains the opposite side

centroid the point of concurrency of the medians of a triangle

circumcenter of a triangle the point of concurrency of the perpendicular bisectors

incenter the point of concurrency of the angle bisectors of a triangle

median of a triangle the segment whose endpoints are a vertex of the triangle and the midpoint of the opposite side

midsegment of a triangle the segment that joins the midpoints of two sides of a triangle

orthocenter the point of concurrency of the altitudes of the triangle

perpendicular bisector of a side of a triangle the segment that is perpendicular to and bisects a side of a triangle

Linguistic Note

Listen to see whether students recognize and are able to use the 18 new mathematical terms in this module. Most students will not be familiar with these terms and may be overwhelmed by them. They may not fully understand them, and/or be able to discriminate between them: for example, *incenter* and *circumcenter*. As students go through each lesson, have them write each new term (and any old ones) on one side of an index card and its meaning (or sketch) on the reverse side as appropriate. Have students then pair up and alternate reading a term aloud to each other, describe what it means, and check its description on the other side of the card to see if they were correct. Have them continue adding new terms to their stack of cards and repeat this exercise several times until students are comfortable speaking and using the terminology.

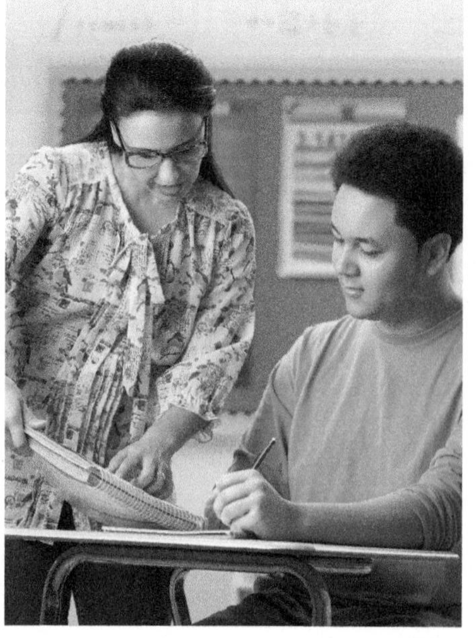

Module 9

Properties of Triangles

Module Performance Task: *Spies and Analysts*™

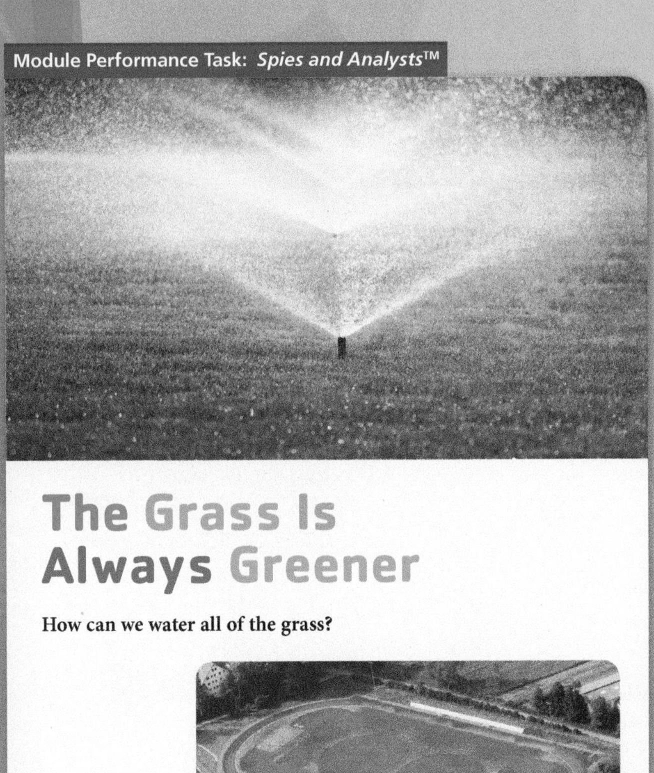

The Grass Is Always Greener

How can we water all of the grass?

Connections to This Module

One way to approach solving this problem is to assume that the circles all have the same radius. Then make a sketch showing rows of adjacent circles that represent how far each spray can reach. Draw the circles so that they are tangent to one another. Connecting the centers of three adjacent circles determines a triangle.

- Use a reasonable scale and the area formula for a circle to estimate the radius of each spray.
- The triangle formed is equilateral. Use the radii of the circles to determine the distances between each circle. **(9.1)**
- Draw the altitude in the triangle and estimate the area of the triangle. **(9.4)**
- Use the approximate dimensions of the field to determine how many spray circles (and triangles) you will need. **(9.4)**

Decide how to use the information above to place the sprinklers so you can most effectively water the grassy field.

The Grass Is Always Greener

Overview

This problem requires students to decide where to place the sprinklers in a large field so that all of the grass in a field is watered. Examples of assumptions:

- The shape of the field is rectangular.
- There are multiple circles in the field that are the same size.

Be a *Spy*

First, students must determine what information it is necessary to know. This should include:

- the size of the field
- the sizes of the circles

Students should realize that each sprinkler waters a circular area and that the radii of the circles can be fixed or changed. It is also important that the arrangement of the sprinklers should be such that most of the water is confined to the grassy area and not wasted along the edges and areas next to the field.

Be an *Analyst*

Help students understand that they need to know how to determine the number of sprinklers needed and where the sprinklers need to be located to most efficiently water all of the grassy field.

Alternative Approaches

In their analysis, students might:

- sketch the layout of the field and the irrigation circles to create a visual model.
- use algebraic formulas involving circles and triangles to determine the number of sprinklers needed, as shown on the left.
- relate the total area of the field to the area of each circular spray area to estimate the number of sprinklers.

ONLINE

Assign the Digital Are You Ready? to power actionable reports including

- proficiency by standards
- item analysis

Are You Ready?

Diagnostic Assessment

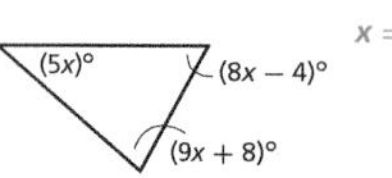

- Diagnose prerequisite mastery.
- Identify intervention needs.
- Modify or set up leveled groups.

Have students complete the *Are You Ready?* assessment on their own. Items test the prerequisites required to succeed with the new learning in this module.

Slopes of Lines Students will apply previous knowledge of slopes of lines to identify and prove properties relating to altitudes and midsegments.

Angle Relationships in Triangles Students will apply previous knowledge of angle relationships in triangles to locate points of concurrency within triangles and to prove theorems about triangles.

Distance and Midpoint Formula Students will apply previous knowledge of the Distance and Midpoint Formula to determine the lengths of special segments within triangles, to locate points of concurrency within triangles, and to prove theorems about triangles.

ONLINE 

Are You Ready?

Complete these problems to review prior concepts and skills you will need for this module.

Slopes of Lines

Find the slope of the line that passes through each pair of points.

1. $(5, 1), (2, -1)$ $\frac{2}{3}$
2. $(-4, 3), (-2, -3)$ -3
3. $(1, -6), (5, -1)$ $\frac{5}{4}$
4. $(3, 8), (7, 6)$ $-\frac{1}{2}$

Angle Relationships in Triangles

Find the value of x.

5. $x = 8$

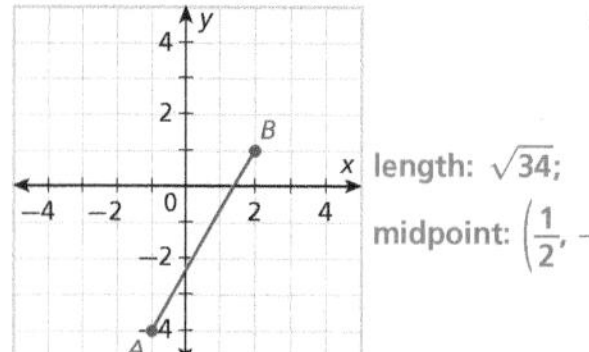

6. $x = 25$

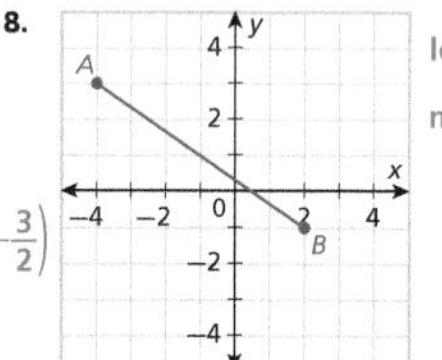

Distance and Midpoint Formulas

Find the length and midpoint of each line segment.

7. length: $\sqrt{34}$; midpoint: $\left(\frac{1}{2}, -\frac{3}{2}\right)$

8. length: $2\sqrt{13}$; midpoint: $(-1, 1)$

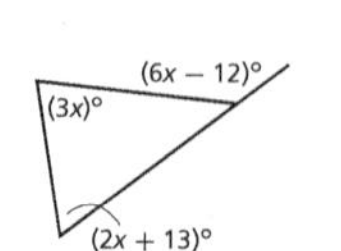

Connecting Past and Present Learning

Previously, you learned:

- to determine the midpoint of a line segment,
- to calculate unknown angle measures in triangles, and
- to construct perpendicular bisectors and angle bisectors.

In this module, you will learn:

- to determine the lengths of special segments within triangles,
- to locate points of concurrency within triangles, and
- to prove theorems about triangles.

256

DATA-DRIVEN INTERVENTION

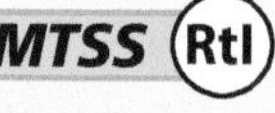

Concept/Skill	Objective	Prior Learning *	Intervene With
Slopes of Lines	Find the slope of a line given two points on the line.	Grade 8, Lesson 5.1	• Tier 3 Skill 3 • Reteach, Grade 8 Lesson 5.1
Angle Relationships in Triangles	Use the Triangle Sum Theorem and the Exterior Angle Theorem to solve problems.	Grade 8, Lesson 4.1	• Tier 2 Skill 5 • Reteach, Grade 8 Lesson 4.1
Distance and Midpoint Formulas	Find lengths and midpoints of line segments in the coordinate plane.	Geometry, Lessons 1.1 and 1.4	• Tier 2 Skill 8 • Reteach, Geometry Lessons 1.1 and 1.4

* Your digital materials include access to resources from Grade 6–Algebra 2. The lessons referenced here contain a variety of resources you can use with students who need support with this content.

9.1 Angle Relationships in Triangles

LESSON FOCUS AND COHERENCE

Mathematics Standards

- Prove theorems about triangles. Theorems include: measures of interior angles of a triangle sum to 180°; base angles of isosceles triangles are congruent; the segment joining midpoints of two sides of a triangle is parallel to the third side and half the length; the medians of a triangle meet at a point.
- Construct an equilateral triangle, a square, and a regular hexagon inscribed in a circle.

Mathematical Practices and Processes 

- Construct viable arguments and critique the reasoning of others.
- Attend to precision.

I Can Objective 

I can prove theorems about triangle angles.

Learning Objective

Students prove theorems about triangle angles and apply the theorems in solving problems.

Language Objective

Students construct arguments and prove theorems about angle relationships using precise language.

Vocabulary

Review: **isosceles triangle**

New: **auxiliary line**, **corollary**, **exterior angle**, **interior angle**, **remote interior angle**

Lesson Materials: protractor, Bubble Map (Teacher Resource Masters), Index Cards, straws, tape

Mathematical Progressions

Prior Learning	Current Development	Future Connections
Students: • proved triangle congruence criteria. **(8.1–8.4)** • defined and measured angles. **(1.2)**	**Students:** • prove theorems about triangle angles. • apply theorems about triangle angles to solve problems.	**Students:** • prove theorems about other properties of triangles. **(9.2–9.5)** • prove theorems about quadrilaterals. **(11.1–11.5)**

UNPACKING MATH STANDARDS

Prove theorems about triangles. Theorems include: measures of interior angles of a triangle sum to 180°; base angles of isosceles triangles are congruent.

What it Means to You

Students formalize their understanding about the sum of the interior angles of a triangle by proving the Triangle Sum Theorem and its corollary. They also extend their understanding of isosceles triangles by proving the Isosceles Triangle Theorem. They apply their understanding of the theorems in solving problems where they find unknown angles.

WARM-UP OPTIONS

ONLINE

PROJECTABLE & PRINTABLE

ACTIVATE PRIOR KNOWLEDGE • Understand Complementary and Supplementary Angles

Use these activities to quickly assess and activate prior knowledge as needed.

Problem of the Day

$\angle MNQ$ and $\angle QNP$ are supplementary.

$m\angle MNQ = (2x + 55)°$

$m\angle QNP = (x + 17)°$

What is the measure of each angle?

Solution:

$2x + 55 + x + 17 = 180$

$3x + 72 = 180$

$3x = 108$

$x = 36°$

$m\angle MNQ = 2(36) + 55 = 127°$

$m\angle QNP = 36 + 17 = 53°$

Quick Check for Homework

As part of your daily routine, you may want to display the Teacher Solution Key to have students check their homework.

Make Connections

Based on students' responses to the Problem of the Day, choose one of the following:

1. Project the Interactive Reteach, Geometry, Lesson 1.2.
2. Complete the Prerequisite Skills Activity:

Have students work in pairs to define and illustrate the following terms:

a) complementary angles
b) supplementary angles
c) angle bisector

- *What term is used when the sum of angles forms a straight line?* The angles are supplementary.
- *If a right angle is bisected, what are the two angles obtained called?* complementary

If students continue to struggle, use Tier 2 Skill 5.

SHARPEN SKILLS

If time permits, use this on-level activity to build fluency and practice basic skills.

Vocabulary Review

Objective: Students demonstrate understanding of triangle congruency postulates.
Materials: Bubble Map (Teacher Resource Masters), index cards

Have students work in small groups. Each pair of students picks a card and needs to explain to their peers which triangle congruence postulate the card exemplifies and why. Then the students glue the cards on the bubble map.

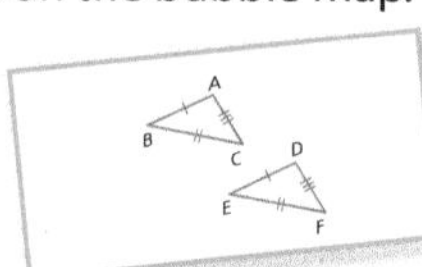

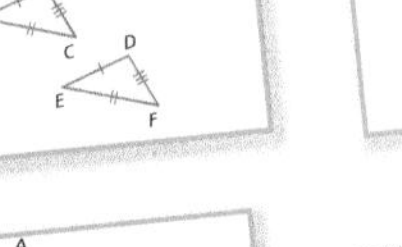

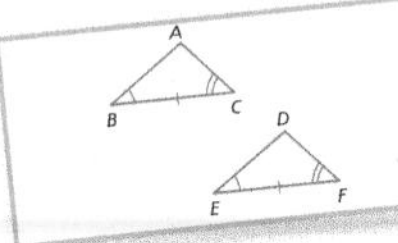

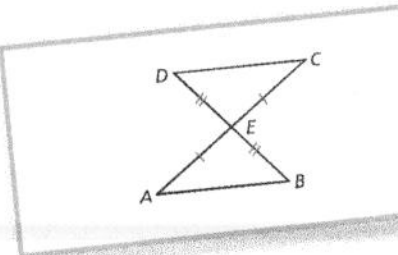

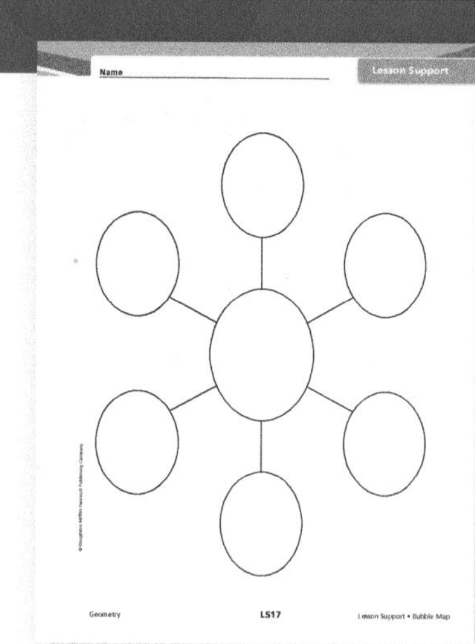

PLAN FOR DIFFERENTIATED INSTRUCTION

Small-Group Options

Use these teacher-guided activities with pulled small groups.

On Track

Materials: paper

Have students work in pairs. Direct each pair to construct two different isosceles triangles which have the same base length. Have the students attach the triangles at their bases, obtaining a kite. Encourage the students to prove that the base angles of the triangles form two congruent angles of the kite.

Example:

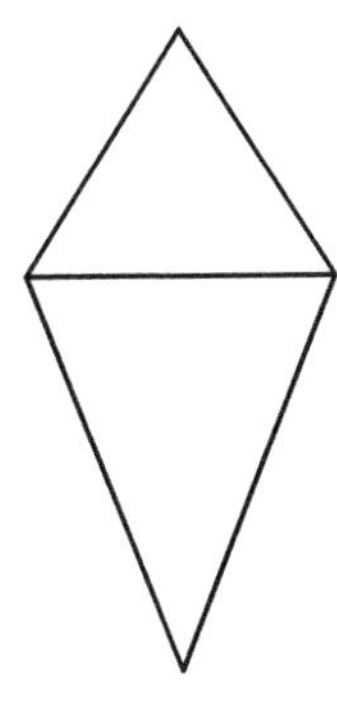

Almost There

Materials: protractor, straws, tape

Have students work in pairs. Ask students to demonstrate the Exterior Angle Theorem by designing a model where they could change the exterior angle and measure the change of the remote interior angles.

Example:

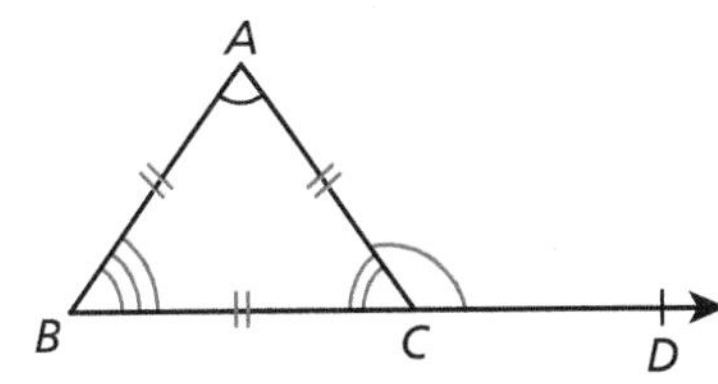

Ready for More

Have students discuss the angles of an isosceles trapezoid. Prompt them to think about the trapezoid as composed of a rectangle and two triangles. Encourage them to prove that the two triangles are congruent in order to prove that two base angles of each base are congruent.

Math Center Options

Use these student self-directed activities at centers or stations. Key: ● Print Resources ● Online Resources

On Track

- Interactive Digital Lesson
- Journal and Practice Workbook
- Interactive Glossary (printable): **isosceles triangle**, **auxiliary line**, **corollary**, **exterior angle**, **interior angle**, **remote interior angle**

Almost There

- Reteach 9.1 (printable)
- Interactive Reteach 9.1
- RtI Tier 2 Skill 5: Angle Relationships in Triangles

Ready for More

- Challenge 9.1 (printable)
- Interactive Challenge 9.1
- Illustrative Mathematics: Sum of Angles in a Triangle

Unit Project Check students' progress by asking what the sum of the angles in a triangle must be.

View data-driven grouping recommendations and assign differentiation resources.

During the *Spark Your Learning,* listen and watch for strategies students use. See samples of student work on this page.

Write a Proof in a Narrative

Strategy 1

If the chain is parallel to the ground, the two triangles are similar. They share an angle and they have two pairs of congruent corresponding angles formed by the parallel lines cut by the transversals (the two sides of the ramp).

The measures of the interior angles will not change as the ramp is adjusted because the sum of the interior angles in any triangle is always 180°.

If students . . . use a proof to solve the problem, they are showing an exemplary understanding of constructing an argument by stating the reasons for their claims.

Have these students . . . consider how the ramp changes as the chain is shortened. **Ask:**

- Q How would the relationship between the two triangles change if the ramp becomes steeper but the chain still stays parallel to the ground?
- Q What would you add to the design of the ramp to collect data for the relationship?

Explore the Relationship by Creating a Model

Strategy 2

I think that if the chain is parallel to the ground, the two triangles are similar, as you can see in the picture below. I used a protractor and measured the angles to prove that corresponding angles are congruent. The interior angles will change as the ramp changes, but the sum of the angles will not change; it will stay 180°.

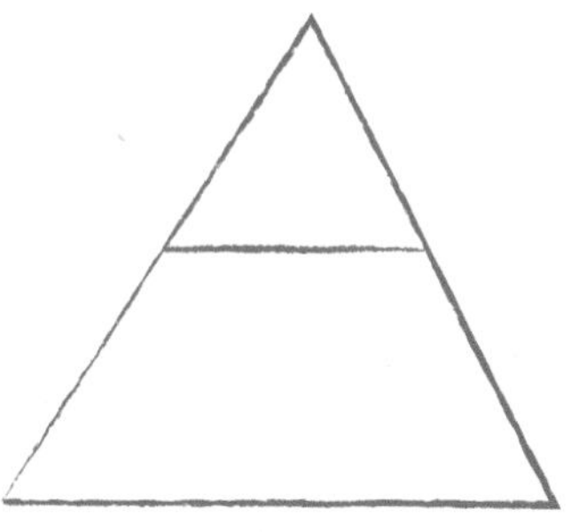

If students . . . use a model and measure the angles, they show an excellent understanding of the properties embedded in the model, but may not recall results about parallel lines and a transversal.

Activate prior knowledge . . . by having students recall similar triangles. **Ask:**

- Q Can you point to a pair of lines cut by a transversal?
- Q What congruent angle pairs are formed by the parallel lines and the transversal?

COMMON ERROR: Focuses on Details but Misses the Big Picture

I will use a protractor to measure the angles of both triangles. As the ramp is adjusted, the angles will also change in both triangles.

If students . . . do not consider that the sum of the interior angles will stay the same, they may not recall that the sum of the interior angles in any triangle is 180°.

Then intervene . . . by pointing out that individual angles will change but their total number of degrees will not. **Ask:**

- Q How can you write the sum of the three angles algebraically?
- Q What could some angle measurements be?

9.1

Angle Relationships in Triangles

I Can prove theorems about triangle angles.

Spark Your Learning

As part of a dog agility competition, a dog must run up and down the ramp shown. The height of the ramp can be adjusted by changing the length of the chain attached to both sides of the ramp.

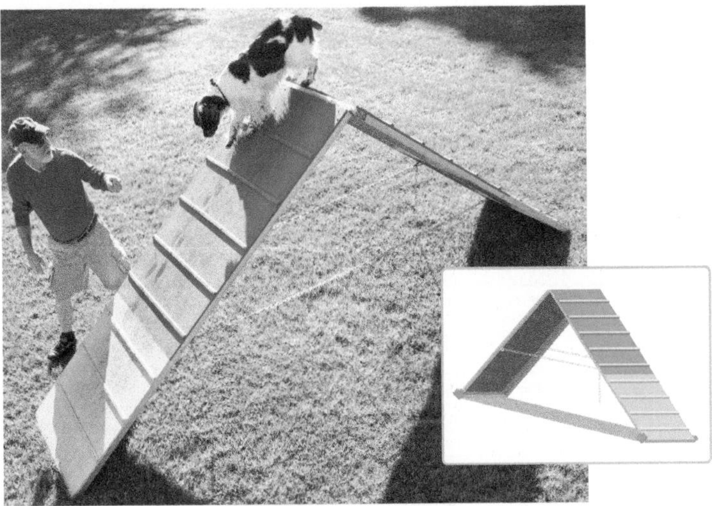

Complete Part A as a whole class. Then complete Parts B and C in small groups.

A. What is a geometric question you can ask about this ramp? A. What happens to the measures of the interior angles of the triangle formed by the chain as the ramp is adjusted?

B. To answer your question, what strategy and tool would you use along with all the information you have? What answer do you get? See Strategies 1 and 2 on the facing page.

C. Does your answer make sense in the context of this situation? How do you know? Yes, it makes sense because the total measure of all angles in a triangle is 180°, and the right angle measures 90°.

Turn and Talk Suppose the chain is parallel to the ground. How is the triangle formed by the sides of the ramp and the ground related to the smaller triangle formed by the chain and the sides of the ramp? See margin.

EL SUPPORT SENSE-MAKING • Co-Craft Questions

If students have difficulty formulating a mathematical question about the situation in the Spark Your Learning, ask them to visualize themselves adjusting the ramp. What are some natural questions to ask about this situation?

Work together to craft the following questions:

- What can you tell about the geometric shape that you see?
- If you are the judge of the show, would you want to know some measurements of the ramp?
- If you need to make the ramp steeper, would you make the chain longer or shorter?

Then have students think about what additional information, if any, they would need to answer these questions. **Ask:**

- Can you determine the type of triangle the ramp forms? Why or why not?
- Can you determine whether the length of the chain can be measured? Explain.

1 Spark Your Learning

▶ MOTIVATE

- Have students look at the photo in their books and read the information contained in the photo. Then complete Part A as a whole-class discussion.
- Give the class the additional information they need to solve the problem. This information is available online as a printable and projectable page in the Teacher Resources.
- Have students work in small groups to complete Parts B–D.

▶ PERSEVERE

If students need support, guide them by asking:

Q **Advancing • Use Tools** Which tool could you use to solve the problem? Why choose that tool and not some other? Students' choices of tools and reasons for choosing them will vary.

Q **Assessing** How would the sum of the interior angles change if the ramp becomes steeper? The sum of the interior angles of any triangle is 180° and that would not change no matter the steepness of the ramp.

Q **Assessing** How would the measures of the interior angles change if the ramp becomes steeper? The top angle would become smaller and therefore the angles at the base would compensate and become larger.

Q **Advancing** If the triangle formed by the ramp is obtuse, would that make the ramp steeper or not compared to an acute triangle ramp? An obtuse triangle ramp (top angle is the obtuse angle) would be less steep.

Turn and Talk Remind students that we can consider the ground as a straight line. The two triangles have the same interior angle measures. They share one angle, and the other two angles are base angles of isosceles triangles. The angle formed by the ramp and the ground is congruent to the angle formed by the side of the ramp and the chain because they are corresponding angles of two parallel lines cut by a transversal.

▶ BUILD SHARED UNDERSTANDING

Select groups of students who used various strategies and tools to share with the class how they solved the problem. As they present their solutions, have each group discuss why they chose a specific strategy and tool.

② Learn Together

Build Understanding

Task 1 (MP) **Attend to Precision** Students prove the Triangle Sum Theorem and use mathematical language with precision.

CONNECT TO VOCABULARY

Have students use the **Interactive Glossary** to record their understanding of the vocabulary in this task.

Sample Guided Discussion:

Q If line l Is parallel to $\overline{AC}$, what does that make $\overline{AB}$? It is a transversal.

Q In Part A, how can you use the measure of a straight angle? On line l, the sum of angles 4, 2, and 5 add to a straight angle, which is 180°.

Turn and Talk Point out to students that although they have informally used the fact that the sum of the measures of the interior angles of a triangle is 180°, they now have written a formal proof of the theorem. If you know measures of two angles of a triangle, you can use the fact that the sum of the angles of a triangle is 180° to find the measure of the other angle.

Build Understanding

Prove the Triangle Sum Theorem

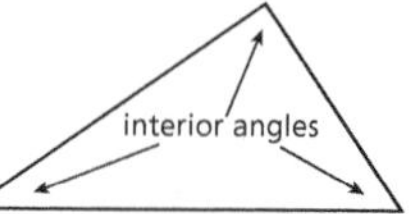

An **interior angle** is an angle formed by two sides of a polygon with a common vertex.

An **auxiliary line** is a line drawn in a figure to aid in a proof. The word auxiliary means something that is helpful or gives assistance.

Triangle Sum Theorem
The sum of the measures of the interior angles of a triangle is 180°.

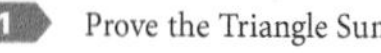

1 Prove the Triangle Sum Theorem.

A. Look at the diagram below. How is the auxiliary line l used in the proof? What do you know about each of the angles in the diagram? See Additional Answers.

B. What reason should be given for Step 2? Alternate Interior Angles Theorem

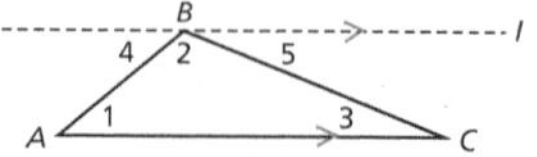

Given: $\triangle ABC$

Prove: $m\angle 1 + m\angle 2 + m\angle 3 = 180°$

Statements	Reasons
1. Draw line l through point B parallel to $\overline{AC}$.	1. Parallel Postulate
2. $\angle 1 \cong \angle 4$, $\angle 3 \cong \angle 5$	2. ?
3. $m\angle 1 = m\angle 4$, $m\angle 3 = m\angle 5$	3. Definition of congruent angles
4. $m\angle 4 + m\angle 2 + m\angle 5 = 180°$	4. Angle Addition Postulate and definition of a straight angle
5. $m\angle 1 + m\angle 2 + m\angle 3 = 180°$	5. Substitution Property of Equality

Turn and Talk How can the Triangle Sum Theorem help you solve problems? See margin.

Prove the Exterior Angle Theorem

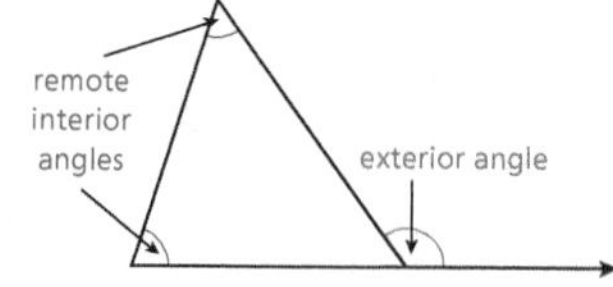

An **exterior angle** is an angle formed by one side of a polygon and the extension of an adjacent side.

A **remote interior angle** is an interior angle that is not adjacent to the exterior angle.

You can find the relationship between exterior angles and interior angles as well.

Exterior Angle Theorem
The measure of an exterior angle of a triangle is equal to the sum of the measures of its remote interior angles.

LEVELED QUESTIONS

Depth of Knowledge (DOK)	Leveled Questions	What Does This Tell You?
Level 1 **Recall**	Recall the congruent angle pairs when we have two parallel lines cut by a transversal. Which angle pairs do you see? $\angle 1$ and $\angle 4$ and $\angle 3$ and $\angle 5$ are alternate interior angles.	Students' answers will indicate that they have the prior knowledge needed to understand the reason for the construction of the auxiliary line, which is the foundation of the proof.
Level 2 **Basic Application of Skills & Concepts**	How is the sum of the interior angles of the triangle related to line l? Line l forms a straight angle of 180°. The angle is also the sum of the interior angles of the triangle or also 180°.	Students' answers will indicate that they understand the steps in the proof.
Level 3 **Strategic Thinking & Complex Reasoning**	How does the fact that we have proven the Triangle Sum Theorem help us in proving other triangle properties? Now that we have formally proved the theorem, we can use it as a reason for statements in other proofs.	Students' answers will indicate that they understand the importance of formal proofs in geometry.

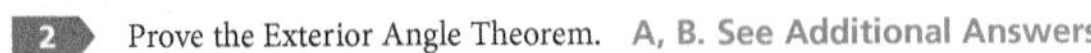

2 Prove the Exterior Angle Theorem. A, B. See Additional Answers.

A. What do you know about the relationship between ∠3 and ∠4 from the diagram? What theorem(s) allow you to reach your conclusion?

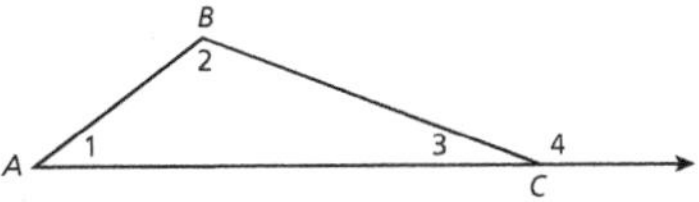

Given: ∠4 is an exterior angle of $\triangle ABC$.

Prove: $m\angle 4 = m\angle 1 + m\angle 2$

Statements	Reasons
1. ∠4 and ∠3 are supplementary.	1. Linear Pairs Theorem
2. $m\angle 4 + m\angle 3 = 180°$	2. Definition of supplementary angles
3. $m\angle 1 + m\angle 2 + m\angle 3 = 180°$	3. Triangle Sum Theorem
4. $m\angle 4 + m\angle 3 = m\angle 1 + m\angle 2 + m\angle 3$	4. Substitution Property of Equality
5. $m\angle 4 = m\angle 1 + m\angle 2$	5. Subtraction Property of Equality

B. Describe how the Triangle Sum Theorem is being used to prove the Exterior Angle Theorem.

Prove the Isosceles Triangle Theorem

Recall that an **isosceles triangle** is a triangle with a least two congruent sides.

Isosceles Triangle Theorem
If two sides of a triangle are congruent, then the two angles opposite the congruent sides are congruent.

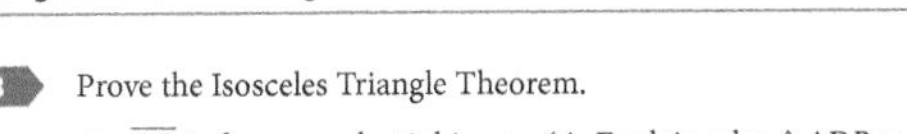

3 Prove the Isosceles Triangle Theorem.

A. $\overline{AD}$ is drawn so that it bisects ∠A. Explain why $\triangle ADB \cong \triangle ADC$.
A, B. See Additional Answers.

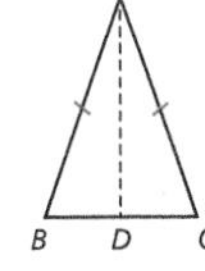

Given: $\overline{AB} \cong \overline{AC}$

Prove: $\angle B \cong \angle C$

Statements	Reasons
1. $\overline{AB} \cong \overline{AC}$	1. Given
2. Draw $\overline{AD}$ so that it bisects ∠A.	2. An angle has one angle bisector.
3. $\angle DAB \cong \angle DAC$	3. Definition of angle bisector
4. $\overline{AD} \cong \overline{AD}$	4. Reflexive Property of Congruence
5. $\triangle ADB \cong \triangle ADC$	5. SAS Triangle Congruence Theorem
6. $\angle B \cong \angle C$	6. CPCFC

B. The Isosceles Triangle Theorem can also be proven by drawing $\overline{AD}$ so that D is the midpoint of $\overline{BC}$. Explain the steps of this proof.

PROFICIENCY LEVEL

Beginning

Have students draw and demonstrate how their drawings explain the terms isosceles, scalene, and equilateral.

Intermediate

Have students work with a partner. Encourage the students to explain how the idea of drawing an auxiliary line has helped in the proofs.

Advanced

Have students discuss the proofs of the theorems informally. Have them prove the theorems as a list of logical statements rather than a double-column proof.

Build Understanding

Task 2 (MP) **Construct Arguments** Students attempt to prove the Exterior Angle Theorem and reason through the formally developed proof.

CONNECT TO VOCABULARY

Have students use the **Interactive Glossary** to record their understanding of the vocabulary in this task.

Sample Guided Discussion:

Q Do you see a straight angle in the diagram? Angles 3 and 4 form a straight angle of 180°.

Q In Part A, if we assign to angle 3 the measure of 50°, what would be the measure of angle 4? Angle 4 would be 180° − 50° = 30°.

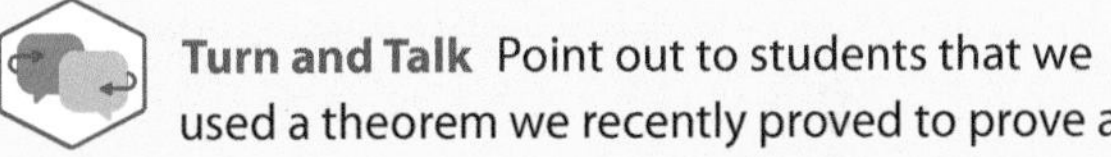

Turn and Talk Point out to students that we used a theorem we recently proved to prove a new theorem. Both theorems use the sums of the measures of the angles of a triangle, which equals 180°.

Task 3 (MP) **Construct Arguments** Students use two different methods to prove that the base angles of isosceles triangles are congruent.

Sample Guided Discussion:

Q How would proving that $\triangle ADB \cong \triangle ADC$ allow us to conclude that $\angle B \cong \angle C$? When two triangles are congruent, then all of their corresponding parts are also congruent.

Q In Part A, how does the auxiliary line $\overline{AD}$ help us in the proof? The auxiliary line bisects ∠A in two equal parts, which lets us use SAS to prove that the triangles are congruent.

Step It Out

Task 4 (MP) **Construct Arguments.** Students justify arguments to prove that the base angles of isosceles triangles are congruent.

CONNECT TO VOCABULARY

Have students use the **Interactive Glossary** to record their understanding of the vocabulary in this task.

Sample Guided Discussion:

Q What makes the proof a corollary? The fact that we proved that the sum of the interior angles in a triangle is 180° allows us to infer that if one of the angles is 90°, then the sum of the other two angles is also 90° $(90° + 90° = 180°)$ or that the other two angles are complementary.

Q In statement 4, how is the Substitution Property demonstrated? Since in 2, $m\angle L = 90°$, therefore 90° can be substituted for $\angle L$.

Task 5 (MP) **Construct Arguments.** Students reason about possible theorems they could use to justify presented solutions.

Sample Guided Discussion:

Q Looking at the picture only, how do you know you can find the unknown angle? I know that the sum of the interior angles of every triangle is 180°, so it is only a matter of computation.

Q What do the red tick marks mean in the first triangle in the Turn and Talk task? They indicate that the two marked sides are congruent.

Turn and Talk Have students describe the given information in each picture to help them determine which theorem they need to apply. Triangle Sum Theorem and Isosceles Triangle Theorem; Exterior Angle Theorem

Step It Out

Apply Triangle Theorems

A **corollary** is a theorem whose proof follows directly from another theorem.

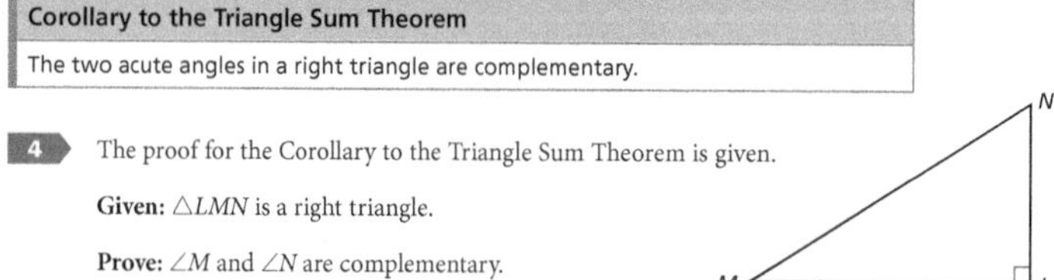

Corollary to the Triangle Sum Theorem

The two acute angles in a right triangle are complementary.

4 The proof for the Corollary to the Triangle Sum Theorem is given.

Given: $\triangle LMN$ is a right triangle.

Prove: $\angle M$ and $\angle N$ are complementary.

Statements	Reasons
1. $\triangle LMN$ is a right triangle.	1. Given
2. $m\angle L = 90°$	2. Definition of a right angle
3. $m\angle L + m\angle M + m\angle N = 180°$	3. ?
4. $90° + m\angle M + m\angle N = 180°$	4. Substitution Property of Equality
5. $m\angle M + m\angle N = 90°$	5. ?
6. $\angle M$ and $\angle N$ are complementary.	6. Definition of complementary angles

A. What justifies the third statement? Triangle Sum Theorem

B. What justifies the fifth statement? Subtraction Property of Equality

5 Find the value of the unknown angle measure.

$72° + 25° + x° = 180°$

$97° + x° = 180°$

$x° = 83°$

What theorem was used to write this equation? Triangle Sum Theorem

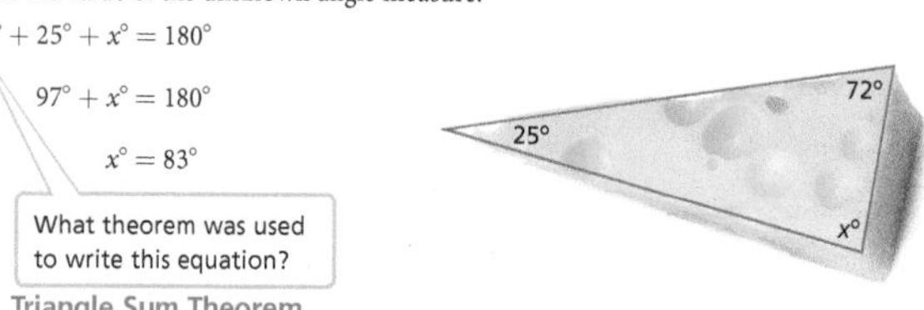

Turn and Talk For each triangle, which theorem(s) do you apply to write an equation to find the value of the varaible? See margin.

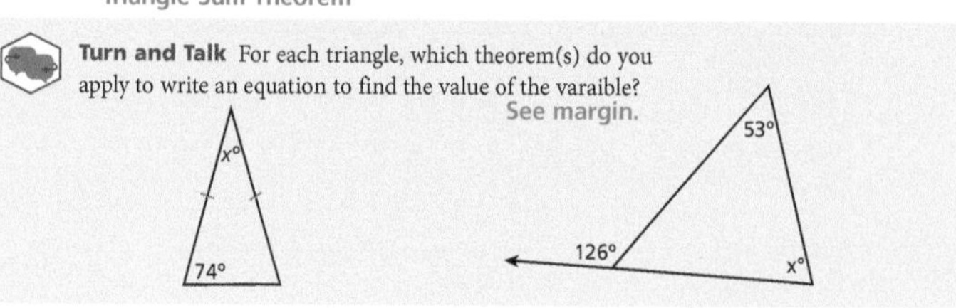

Check Understanding

1. If the interior angles of a triangle are congruent, what is the measure of each angle? 60°
2. The measure of an exterior angle of a triangle is 54°. The measure of one of the remote interior angles is 33°. What is the measure of the other remote interior angle? What is the measure of the other angle of the triangle? 21°; 126°
3. If an isosceles triangle has a right angle, what are the measures of the other interior angles? 45°, 45°

Find the measure of the unknown interior angle.

4. 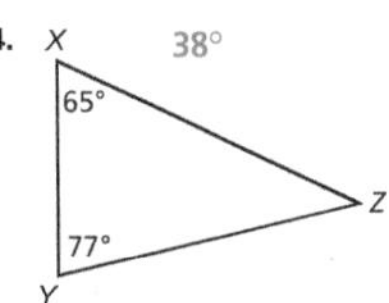

5. 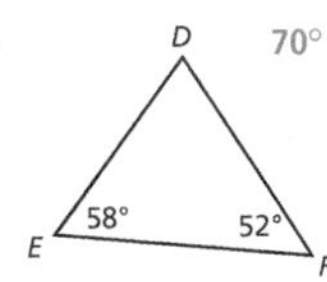

On Your Own

(MP) Reason **In Problems 6–8, answer each question and explain your reasoning.** 6–11. See Additional Answers.

6. Is it possible for a triangle to have two obtuse angles?
7. Is it possible for an exterior angle of a triangle to be a right angle?
8. An isosceles triangle has an angle that measures 100°. Do you have enough information to determine the other two angles of the triangle?
9. Write a two-column proof of the converse of the Isosceles Triangle Theorem. If two angles of a triangle are congruent, the two sides opposite the angles are congruent.

   Given: $\angle B \cong \angle C$

   Prove: $\overline{AB} \cong \overline{AC}$

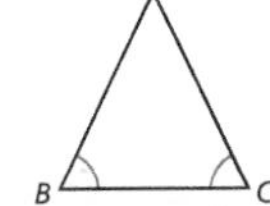

An asterism is a group of stars that is easier to recognize than a constellation. Find the value of x in each asterism.

10. 

The Summer Triangle

11. 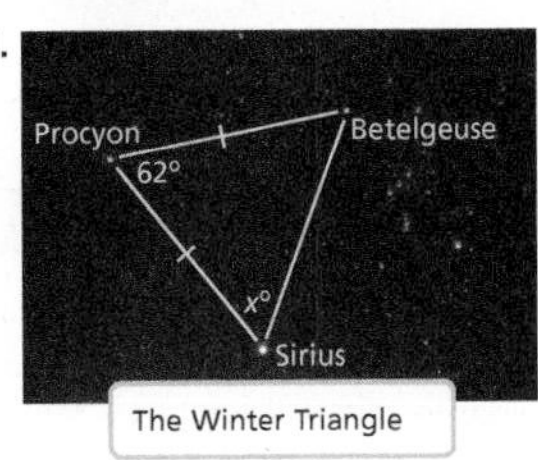

The Winter Triangle

ONLINE

Assign the Digital On Your Own for
- built-in student supports
- Actionable Item Reports
- Standards Analysis Reports

On Your Own

Assignment Guide

The chart below indicates which problems in the On Your Own are associated with each task in the Learn Together. Assign daily homework for tasks completed.

Learn Together Tasks	On Your Own Problems
Task 1, p. 258	Problems 6, 25, 26, 33, and 34
Task 2, p. 259	Problems 7, 18, 24, 26, and 34
Task 3, p. 259	Problems 8, 9, 28–30, 31, and 32
Task 4, p. 260	Problems 6–8
Task 5, p. 260	Problems 10–17, 19–23, 25, and 27–30

3 Check Understanding

Formative Assessment

Use formative assessment to determine if your students are successful with this lesson's learning objective.

Students who successfully complete the Check Understanding can continue to the On Your Own practice.

For students who miss 1 problem or more, work in a pulled small group using the Almost There small-group activity on page 257C.

ONLINE Ed

Assign the Digital Check Understanding to determine
- success with the learning objective
- items to review
- grouping and differentiation resources

4 Differentiation Options

Differentiate instruction for all students using small-group activities and math center activities on page 257C.

Reteach

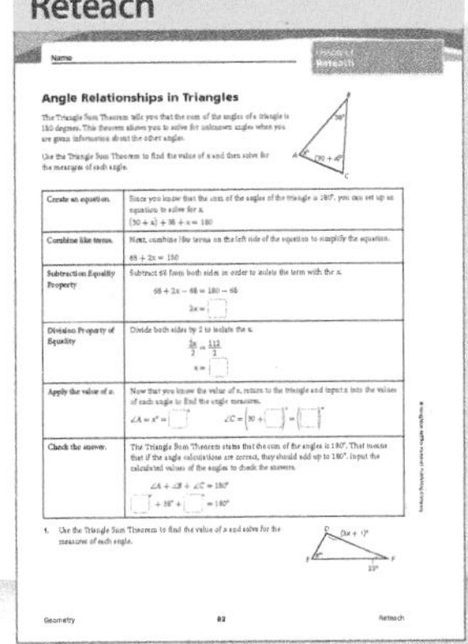

Challenge

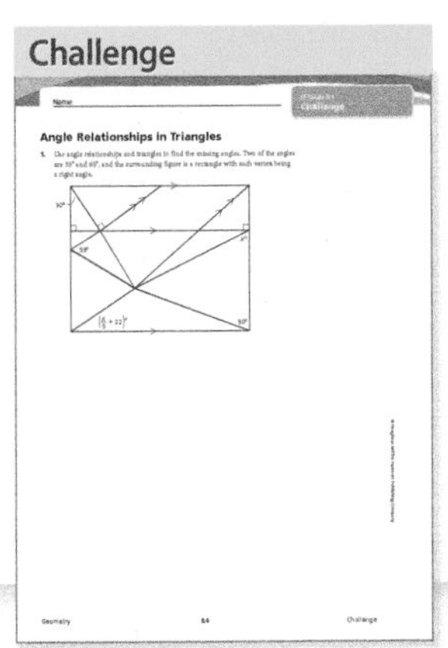

Watch for Common Errors

Problem 14 Point out that a picture may be misleading. Just because a triangle looks equilateral, we cannot assume that it is equilateral unless the information is provided or we prove that this is the case.

Find the value of x.

12. $x°$, 80°, 150° $x = 70$

13.

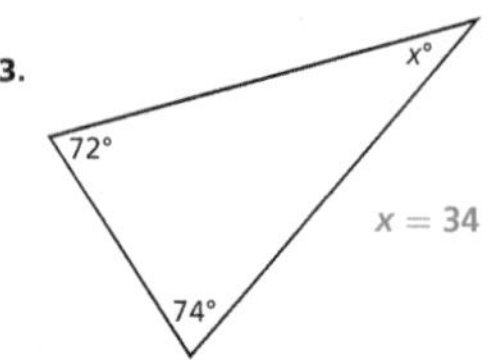

$x = 34$

14. $x°$, 57° $x = 66$

15.

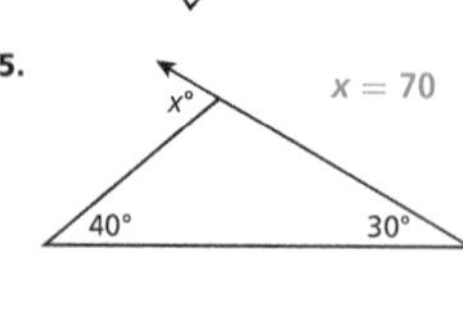

$x = 70$

16. 74°, $x°$ $x = 74$

17.

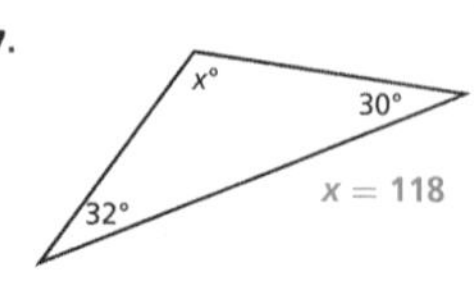

$x = 118$

18. MP **Critique Reasoning** A triangle has two angles that measure 82° and 44°. Jessica states that the measure of the exterior angle of the unknown angle is 126°. Tristan states that the measure of the exterior angle is 54°. Who is correct? Explain your reasoning. 18, 19. See Additional Answers.

19. Open Ended Write a problem that uses the Triangle Sum Theorem to find the value of a variable.

20. Two spotlights illuminate a stage as shown. Find the value of x, y, and z. $x = 30$, $y = 130$, $z = 20$

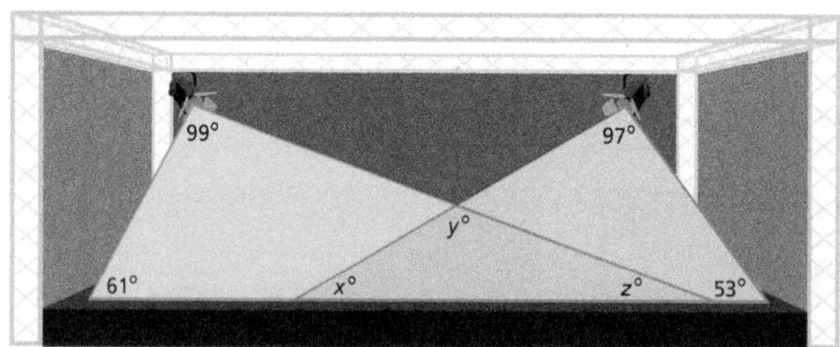

21. Lily is cutting triangular shapes from fabric for a quilt she is making. One piece she cuts has an angle that measures 54°. The measure of one of the unknown angles is twice the measure of the other unknown angle. What are the measures of the unknown angles? 42°, 84°

262

22. A person at T is observing a drone flying along $\overleftrightarrow{AC}$. The drone's path is parallel to a horizontal line at the person's eye level. At point A, the angle of elevation to the drone is 35°. After the drone has traveled 240 feet to point B, the angle of elevation is 70°. What is BT? Explain your reasoning.
See Additional Answers.

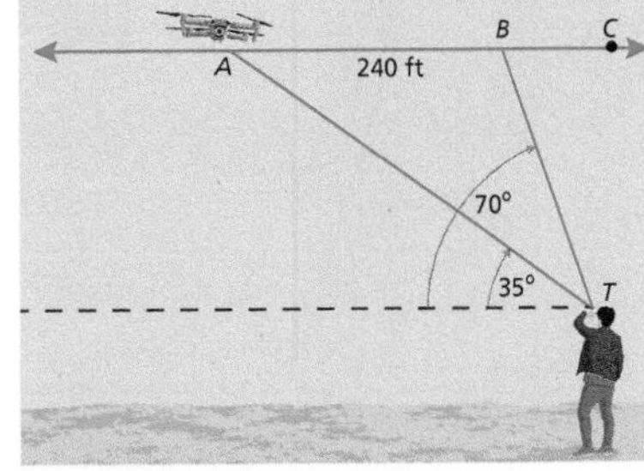

23. (MP) **Reason** If you know the measure of an exterior angle of a triangle, can you determine the measures of the remote interior angles? Explain your reasoning.
See Additional Answers.

24. Consider $\triangle XYZ$ shown. If $m\angle X$ is five times as great as $m\angle Y$, find $m\angle X$ and $m\angle Y$.
$m\angle X = 100°$, $m\angle Y = 20°$

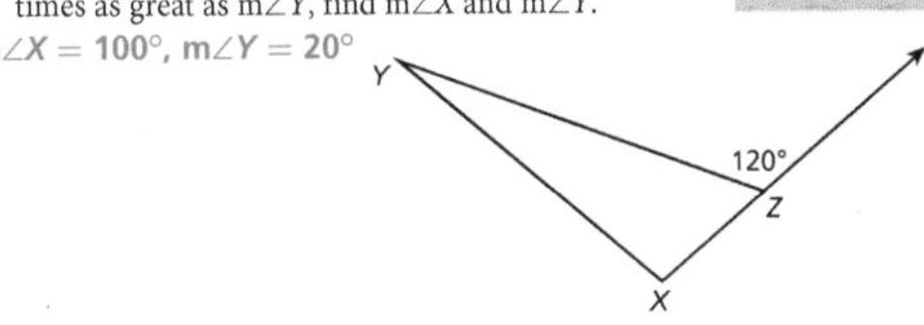

Find the measure of each interior angle of the triangle.

25. A $(5x - 7)°$, B $(x + 16)°$, C $(8x + 3)°$
$m\angle A = 53°$, $m\angle B = 28°$, $m\angle C = 99°$

26. B $(4m + 3)°$, A $(6m + 22)°$, C $(3m - 3)°$
$m\angle BAC = 26°$, $m\angle B = 91°$, $m\angle C = 63°$

27. A $(17x + 1)°$, B $(20x + 8)°$, C $(24x - 12)°$
$m\angle A = 52°$, $m\angle B = 68°$, $m\angle C = 60°$

28. A, B $(13t - 16)°$, C $(4t + 2)°$
$m\angle A = 75°$, $m\angle B = 75°$, $m\angle C = 30°$

29.

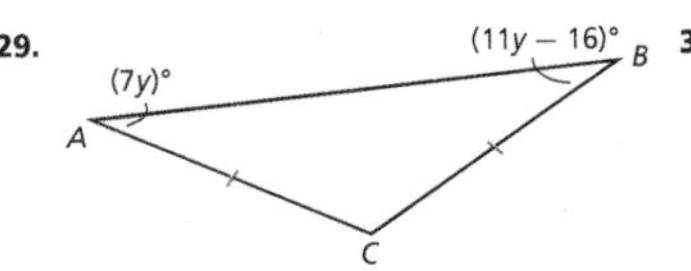

$m\angle A = 28°$, $m\angle B = 28°$, $m\angle C = 124°$

30. B $(a + 9)°$, A $(6a - 25)°$, C $(3a)°$
$m\angle CAB = 103°$, $m\angle B = 26°$, $m\angle C = 51°$

Questioning Strategies

Problem 26 How are the three given angle measures related? How do you know? The exterior angle measure is equal to the sum of the measures of the interior angles; the Exterior Angles Theorem

5 Wrap-Up

Summarize learning with your class. Consider using the Exit Ticket, Put It in Writing, or I Can scale.

Exit Ticket 

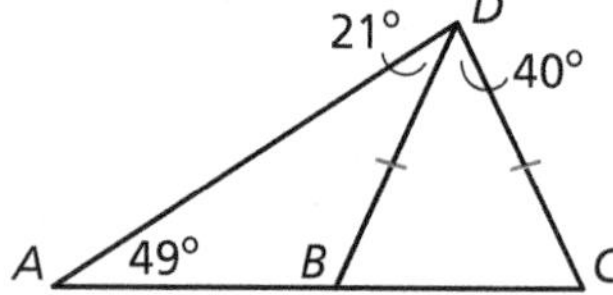

Consider triangle ADC. What is the measure of angle BCD?

Possible Solution:

$m\angle DBC = m\angle ADB + m\angle DAB = 70°$ (Exterior Angle Theorem)

$\angle DBC = \angle DCB = 70°$(Isosceles Triangle Theorem)

Put It in Writing 

Explain in writing how you would know a triangle is not isosceles if given the angles of the triangle.

I Can 

The scale below can help you and your students understand their progress on a learning goal.

4	I can prove theorems about triangle angles, and I can explain my reasoning to others.
3	I can prove theorems about triangle angles.
2	I can apply triangle theorems to solve problems.
1	I can apply the Triangle Sum Theorem to solve problems.

Spiral Review • Assessment Readiness

These questions will help determine if students have retained information taught in the past and can also prepare them for high-stakes assessments. Here, students must apply their knowledge of Triangle Congruence Theorems **(8.1–4)**, apply their knowledge about slopes of perpendicular lines **(4.2)**, and connect the lesson to future applications such as properties of parallelograms **(11.1–2)**.

In Problems 31 and 32, find MN.

31. 35

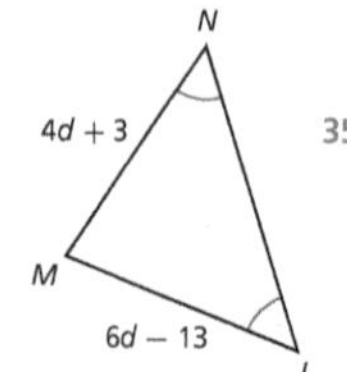

32. 51

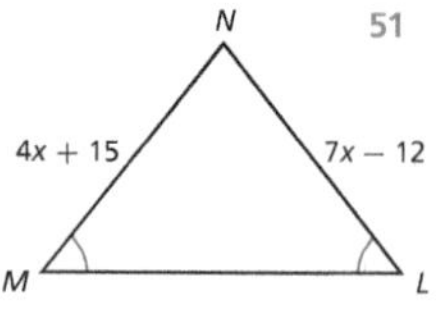

33. (MP) **Construct Arguments** In isosceles $\triangle ABC$, $\angle B$ and $\angle C$ are the base angles. The measures of the interior angles are integers. How do you know that the measure of the vertex angle, $\angle A$, is an even number? 33, 34. See Additional Answers.

34. (Open Middle™) Using the integers from 1 to 9 at most one time each, fill in the boxes to give possible measures for the angles in the triangle.

$m\angle KAB = \square\square\square°$

$m\angle ABC = \square\square°$

$m\angle BCA = \square\square°$

Spiral Review • Assessment Readiness

35. What value of x makes $\triangle EDG \cong \triangle FDG$?

Ⓐ 3
Ⓑ 5
Ⓒ 9
Ⓓ 26

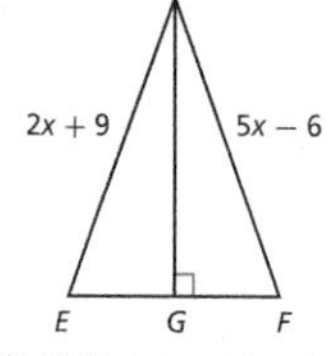

36. Rectangle $PQRS$ is divided into two triangles by diagonal $\overline{PR}$. Which congruence statement is correct?

Ⓐ $\triangle PRS \cong \triangle RPQ$
Ⓑ $\triangle RSP \cong \triangle QRP$
Ⓒ $\triangle PRS \cong \triangle QRP$
Ⓓ $\triangle SPR \cong \triangle PQR$

37. Line m and line n are perpendicular. The slope of line m is 3. What is the slope of line n?

Ⓐ 3
Ⓑ $\frac{1}{3}$
Ⓒ $\frac{-1}{3}$
Ⓓ −3

38. Which theorem can be used most directly to show that $\triangle ABC \cong \triangle DEF$?

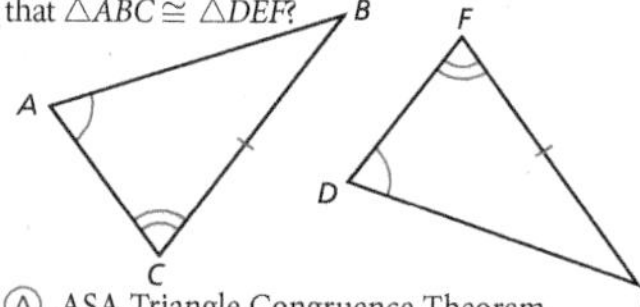

Ⓐ ASA Triangle Congruence Theorem
Ⓑ AAS Triangle Congruence Theorem
Ⓒ SAS Triangle Congruence Theorem
Ⓓ SSS Triangle Congruence Theorem

I'm in a Learning Mindset!

How do I proactively seek out additional learning opportunities?

Keep Going ▶ Journal and Practice Workbook

Learning Mindset

Challenge Seeking Builds Confidence

Encourage students to talk to each other about their confidence in applying the theorems discussed in the lesson. Point out that most of the information presented today was not radically new for them. Rather, they have seen it informally in previous years when exploring geometric ideas. *How comfortable are you with the theorems and problems we learned today? Did you find yourself confident in most of the tasks? Were there any tasks you found challenging at first but now you feel more confidents about?*

9.2 Perpendicular Bisectors in Triangles

LESSON FOCUS AND COHERENCE

Mathematics Standards

- Construct the inscribed and circumscribed circles of a triangle, and prove properties of angles for a quadrilateral inscribed in a circle.
- Make formal geometric constructions with a variety of tools and methods (compass and straightedge, string, reflective devices, paper folding, dynamic geometric software, etc.). Copying a segment; copying an angle; bisecting a segment; bisecting an angle; constructing perpendicular lines, including the perpendicular bisector of a line segment; and constructing a line parallel to a given line through a point not on the line.
- Prove theorems about triangles. Theorems include: measures of interior angels of a triangle sum to 180°, base angles of isosceles triangles are congruent; the segment joining midpoints of two sides of a triangle is parallel to the third side and half the length; the medians of a triangle meet at a point.

Mathematical Practices and Processes 

- Use appropriate tools strategically.
- Construct viable arguments and critique the reasoning of others.
- Attend to precision.

I Can Objective 

I can construct perpendicular bisectors and use the point of concurrency to circumscribe triangles with circles.

Learning Objective

Use the Circumcenter Theorem to solve problems in the coordinate plane, and use the Circumcenter Theorem to determine the lengths of the segments that connect the circumcenter to the vertices of a triangle.

Language Objective

Describe the steps needed to show that a right triangle in the coordinate plane has perpendicular bisectors that intersect on the triangle.

Vocabulary

New: circumcenter of a triangle, circumcircle, circumscribed, concurrent lines, perpendicular bisector, point of concurrency

Lesson Materials: geometric drawing software

Mathematical Progressions

Prior Learning	Current Development	Future Connections
Students: • proved lines are perpendicular and constructed perpendicular bisectors of segments. **(3.3)**	**Students:** • use the circumcenter of a triangle to solve problems in the coordinate plane. • use the Circumcenter Theorem to find the lengths of segments that connect the circumcenter to a vertex of a triangle.	**Students:** • will construct angle bisectors and prove that angle bisectors are concurrent. **(9.3)**

PROFESSIONAL LEARNING

Visualizing the Math

Students may find it helpful to explore where the point of concurrency is located for the perpendicular bisectors of different kinds of triangles by performing their own constructions using pencil and paper. The triangle at the right shows that the perpendicular bisectors of an obtuse triangle will have a point of concurrency that is outside the triangle.

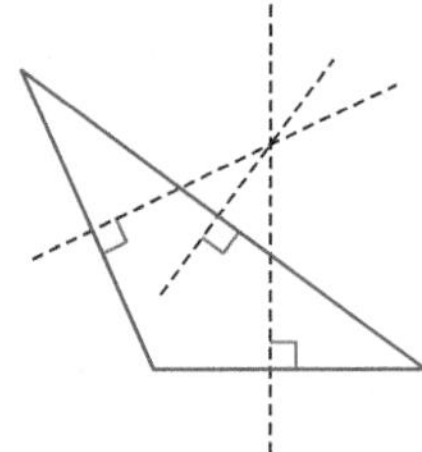

WARM-UP OPTIONS

ONLINE

PROJECTABLE & PRINTABLE

ACTIVATE PRIOR KNOWLEDGE • Find the Midpoint of a Line Segment

Use these activities to quickly assess and activate prior knowledge as needed.

Problem of the Day

Determine the equation of line $\overleftrightarrow{AB}$ and the equation for a line perpendicular to $\overleftrightarrow{AB}$.

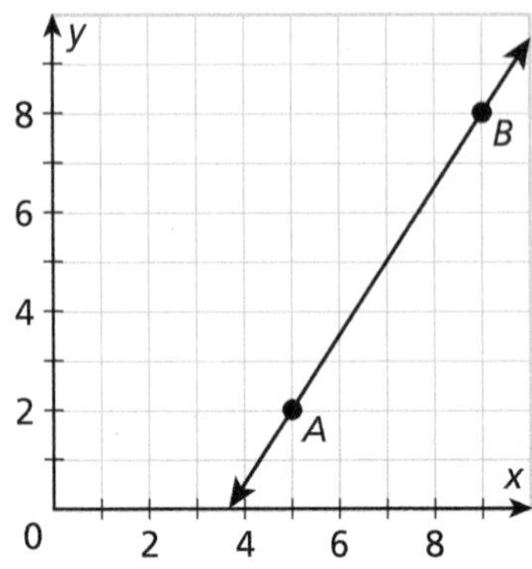

$y = \frac{3}{2}x - \frac{11}{2}$; $y = -\frac{2}{3}x + 6$

Quick Check for Homework

As part of your daily routine, you may want to display the Teacher Solution Key to have students check their homework.

Make Connections

Based on students' responses to the Problem of the Day, choose one of the following:

1. Project the Interactive Reteach, Geometry, Lesson 4.2.
2. Complete the Prerequisite Skills Activity:

Have students work in pairs. Each student should choose the coordinates for one of two points in the coordinate plane, such as $C(4, 8)$ and $D(6, 9)$. Have the students work together to determine the equation for the line that contains their points and a line that is perpendicular to the line containing their points.

- *If you know two points on the line, what equation can you use to find the slope of the line?* $m = \frac{y_2 - y_1}{x_2 - x_1}$
- *What is the slope of the line containing the points $C(4, 8)$ and $D(6, 9)$?* $m = \frac{y_2 - y_1}{x_2 - x_1} = \frac{9 - 8}{6 - 4} = \frac{1}{2}$
- *Write the equation for $\overleftrightarrow{CD}$ in point-slope form.*
  $y = \frac{1}{2}x + b$
  $8 = \frac{1}{2}(4) + b$
  $b = 6$
  $y = \frac{1}{2}x + 6$
- *What will the slope be for a line that is perpendicular to line $\overleftrightarrow{CD}$?* -2
- Write the equation of a line that is perpendicular to line $\overleftrightarrow{CD}$.
  Possible answer: $y = -2x + 5$

If students continue to struggle, use Tier 3 Skill 3.

SHARPEN SKILLS

If time permits, use this on-level activity to build fluency and practice basic skills.

Mental Math

Objective: Students determine the value of *x* which makes two segments the same length.
Materials: index cards, drawing of triangle labeled as shown

Have students work in small groups. Each index card shows an expression that represents the length of both $\overline{DE}$ and $\overline{DF}$, which are both equal to 10 units according to the Circumcenter Theorem. Students take turns drawing a card and mentally determining the value for *x* which makes their expression equal to 10.

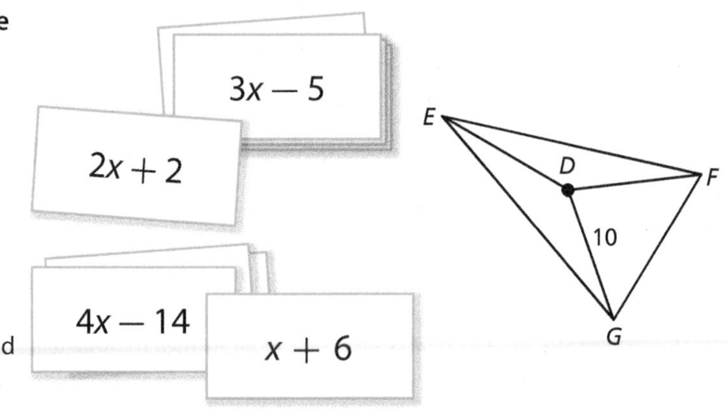

PLAN FOR DIFFERENTIATED INSTRUCTION

Small-Group Options

Use these teacher-guided activities with pulled small groups.

On Track

Materials: index cards

Have each student select a card showing the coordinates of a point in the coordinate plane. Have students form groups of three so that their group has the coordinates for three different points. Students can work together to find the coordinates for the circumcenter of the triangle formed by their points.

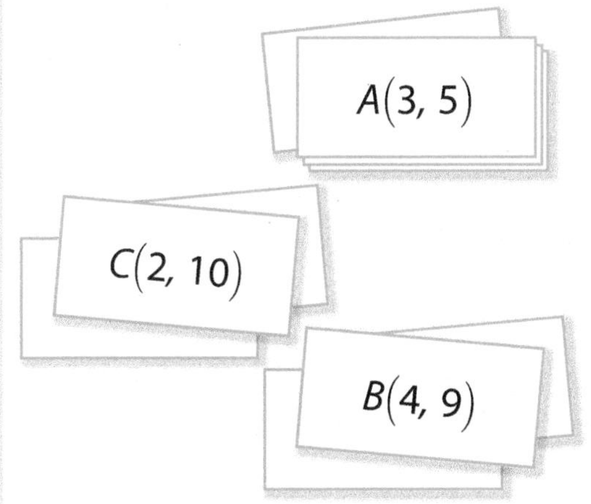

Almost There

Materials: straight edge

Have students use a straight edge to draw a large triangle ABC. Have students do the following:

- Fold their paper so that $\overline{AB}$ is folded exactly in half, with point A lining up exactly with point B.
- Fold their paper so that $\overline{AC}$ is folded exactly in half, with point A lining up exactly with point C.
- Fold their paper so that $\overline{BC}$ is folded exactly in half, with point B lining up exactly with point C.
- Each of the three folds is the perpendicular bisector for one the sides of the triangle ABC. Have students find the point of concurrency where all three perpendicular bisectors meet.

Ready for More

Give students the coordinates for two vertices of a triangle, $A(7, 9)$ and $B(6, 4)$, as well as the coordinates of the circumcenter for the triangle, $D(4, 7)$. Have students find the coordinates for two points which could be the third vertex of the triangle.

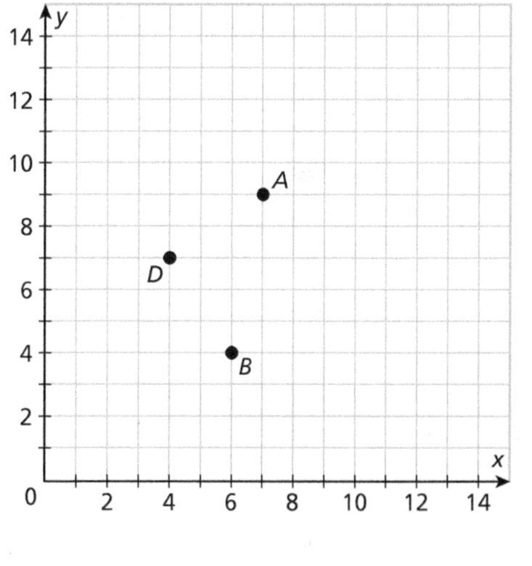

Math Center Options

Use these student self-directed activities at centers or stations. Key: ● Print Resources ● Online Resources

On Track

- ● Interactive Digital Lesson
- ●● Journal and Practice Workbook
- ● Interactive Glossary (printable): **circumcenter of a triangle**, **circumcircle**, **circumscribed**, **concurrent lines**, **perpendicular bisector**, **point of concurrency**
- ● Illustrative Mathematics: Circumcenter of a Triangle

Almost There

- ● Reteach 9.2 (printable)
- ● Interactive Reteach 9.2
- ● RtI Tier 3 Skill 3: Slopes of Lines
- ● Illustrative Mathematics: Placing a Fire Hydrant

Ready for More

- ● Challenge 9.2 (printable)
- ● Interactive Challenge 9.2
- ● Illustrative Mathematics: Construction of a Perpendicular Bisector

Unit Project Check students' progress by asking how the circumcenter of a triangle is constructed.

View data-driven grouping recommendations and assign differentiation resources.

During the *Spark Your Learning,* listen and watch for strategies students use. See samples of student work on this page.

Use the Distance Formula

Strategy 1

Since (11, 5) is the midpoint of $\overline{BC}$, the central location will be on the line containing (1, 5) and (11, 5). Let (x, 5) be the coordinates of the central location. Use the Distance Formula and the coordinates for point A and point B to determine the value of x.

$$\sqrt{(x-1)^2+(5-5)^2}=\sqrt{(x-11)^2+(5-9)^2}$$
$$\sqrt{(x^2-2x+1)}=\sqrt{(x^2-22x+121)+16}$$
$$x^2-2x+1=x^2-22x+137$$
$$-2x+1=-22x+137$$
$$20x+1=137$$
$$20x=136$$
$$x=6.8$$

The coordinates of the central location are (6.8, 5).

If students . . . use the Distance Formula to solve the problem, they are employing an efficient method and are demonstrating an understanding of finding distances on the coordinate plane from Grade 8.

Have these students . . . explain how they determined their equation and how they solved it. **Ask:**

- Q How did you use the given information to determine the *y*-coordinate for the central location?
- Q What properties did you use to solve the equation and find the *x*-coordinate for the central location?

Use a Compass

Strategy 2

The central point should be the same distance from each of the three food trucks. Open a compass to an estimate for the distance from the central location to the food truck.

Place the point of the compass at the points A, B, and C and draw a line in the middle of the three points. If the three lines you drew do not meet at a single point, change your compass to a different length and repeat the process until you find a point that is equidistant from each of the points A, B, and C.

The central location should be close to (7, 5).

If students . . . use or compass to solve the problem, they understand how to find a point that is the same distance from the points *A*, *B*, and *C* but may not know how to use the given coordinates to find the central location.

Activate prior knowledge . . . by having students examine what must be true about the coordinates of the central location. **Ask:**

- Q How can you use the coordinates of the given points to determine one of the coordinates for the central location?
- Q What formulas do you know that can be used to determine distances in the coordinate plane?

COMMON ERROR: Uses Central Location that Is Not the Same Distance From All Points

The central location should be the same distance from B (11, 9) as it is from C (11, 1). The point (11, 5) is the same distance from B as it is from C, so the central location should be at (11, 5).

If students . . . only consider 2 points when determining the central location, they may not understand that the central location must be the same distance from each of the three food trucks.

Then intervene . . . by pointing out that they can check their answer to see if the point they chose as the central location is the same distance from each food truck. **Ask:**

- Q Is the distance from (11, 5) to point *A* the same as the distance from (11, 5) to point *B*?

9.2

Perpendicular Bisectors in Triangles

I Can construct perpendicular bisectors and use the point of concurrency to circumscribe triangles with circles.

Spark Your Learning

A catering company that has food trucks in three locations wants to relocate its main office to a central location.

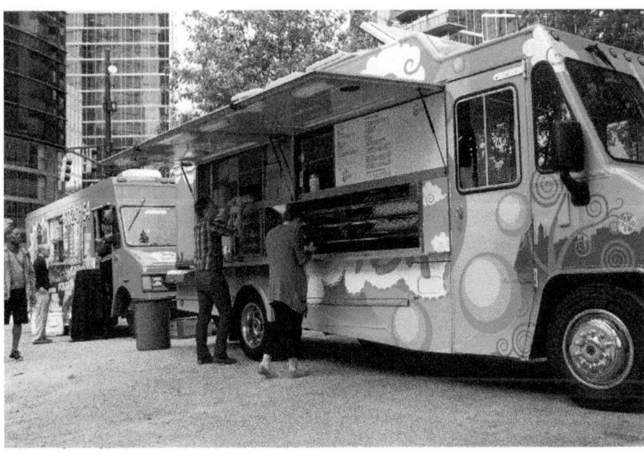

©Blulz60/Shutterstock

Complete Part A as a whole class. Then complete Parts B–D in small groups.

A. What is a mathematical question you can ask about this situation? What information would you need to know to answer your question?

A. What location is an equal distance from each truck?; need to know the locations of the food trucks

B. How could you answer your question visually? See Additional Answers.

C. To answer your question, what strategy and tool would you use along with all the information you have? What answer do you get? See Strategies 1 and 2 on the facing page.

D. What is the answer to your question? A location that is an equal distance from all three food trucks is around point (7, 5).

Turn and Talk Can you think of any reasons that the new main office could not be built at the exact central location? See margin.

CULTIVATE CONVERSATION • Co-Craft Questions

If students have difficulty formulating a mathematical question about the situation in the Spark Your Learning, ask them to imagine that three seats represent the food trucks as they try to determine the location of the main office. What are some natural questions to ask about this situation?

Work together to craft the following questions:

- How can you find the midpoint of the line segment connecting two of the food trucks?
- How can you find the set of all points which are the same distance from two of the food trucks?

Then have students think about what additional information, if any, they would need to answer these questions. **Ask:**

- Can you determine the location of the main office if you know the coordinates of the points representing two of the food trucks? Why or why not?
- Can you determine the location of the main office if you know the coordinates of the points representing all three food trucks? Explain.

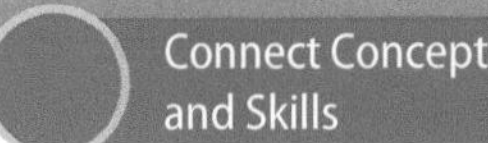

Connect Concepts and Skills

1 Spark Your Learning

▶ MOTIVATE

- Have students look at the photo in their books and read the information contained in the photo. Then complete Part A as a whole-class discussion.
- Give the class the additional information they need to solve the problem. This information is available online as a printable and projectable page in the Teacher Resources.
- Have students work in small groups to complete Parts B–D.

▶ PERSEVERE

If students need support, guide them by asking:

Q **Advancing • Use Tools** Which tool could you use to solve the problem? Why choose that tool and not some other? Students' choices of tools and reasons for choosing them will vary.

Q **Assessing** How do you know the point $(11, 5)$ is the same distance from point B as it is from point C? Possible answer: You can use the Distance Formula to show that $(11, 5)$ is 4 units from both point B and point C.

Q **Assessing** How can you show that all three food trucks are about the same distance from the main office at point $(7, 5)$? Possible answer: Use a compass to draw a circle with center at $(7, 5)$ whose radius is equal to the distance from the main office to point $(1, 5)$. The locations of all three food trucks should be close to being on the circle.

Turn and Talk Remind students that their coordinate grids show the location of the three food trucks and the main office, but the locations of other structures and landmarks are not shown on their grids. The central location could be located on a river, in a park, or where another business is already located.

▶ BUILD SHARED UNDERSTANDING

Select groups of students who used various strategies and tools to share with the class how they solved the problem. As they present their solutions, have each group discuss why they chose a specific strategy and tool.

② Learn Together

Build Understanding

Task 1 (MP) **Use Tools** Students will use geometry drawing programs to examine the intersections of the perpendicular bisectors of the sides of a triangle.

CONNECT TO VOCABULARY

Have students use the **Interactive Glossary** to record their understanding of the vocabulary in this task.

Sample Guided Discussion:

Q How can you tell that a line bisects a line segment? A line bisects a line segment when it divides the line segment into two equal segments.

Task 2 (MP) **Construct Arguments** Students prove that the perpendicular bisectors of a triangle always intersect at a single point.

CONNECT TO VOCABULARY

Have students use the **Interactive Glossary** to record their understanding of the vocabulary in this task.

Sample Guided Discussion:

Q What does the Perpendicular Bisector Theorem state? If a point is on the perpendicular bisector of a segment, then it is equidistant from the segment's endpoints.

Build Understanding

Investigate Perpendicular Bisectors

A **perpendicular bisector** of a side of a triangle is a segment that is perpendicular to and bisects a side of a triangle.

1 Use a geometric drawing tool to draw a triangle. Construct the midpoint of each side of the triangle and a line perpendicular to each side through the midpoint.

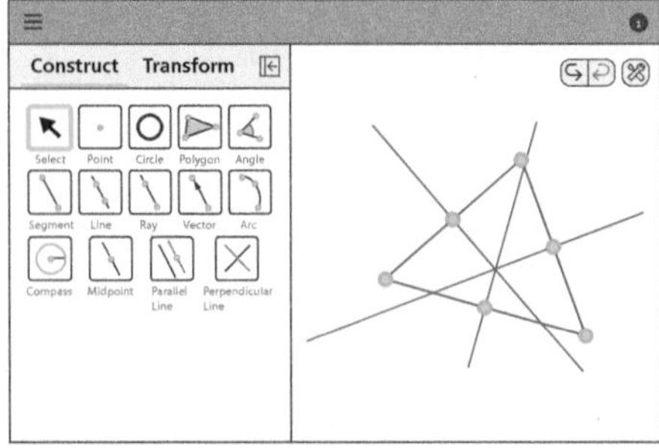

A. Move the vertices of the triangle. What is true about their perpendicular bisectors? They appear to intersect at a point.

B. In what type of triangle is the intersection of the perpendicular bisectors inside the triangle? outside the triangle? on the triangle? an acute triangle, an obtuse triangle; a right triangle

Prove that Perpendicular Bisectors Are Concurrent

In Task 1, the perpendicular bisectors of the sides of a triangle intersect in a point. **Concurrent lines** are three or more lines that intersect at one point. The **point of concurrency** is the point where concurrent lines intersect.

2 Prove that the perpendicular bisectors of a triangle are concurrent.

Given: Lines ℓ, m, and n are perpendicular bisectors of $\triangle ABC$.

Prove: Lines ℓ, m, and n are concurrent.

A. Why do you know line ℓ and line m intersect? A, B. See Additional Answers.

B. Explain how you know that lines ℓ, m, and n are concurrent in Step 6.

Statements	Reasons
1. Lines ℓ, m, and n are perpendicular bisectors of $\triangle ABC$.	1. Given
2. Let P be the intersection of m and ℓ.	2. Definition of intersecting lines
3. $PB = PC$; $PB = PA$	3. Perpendicular Bisector Theorem
4. $PA = PC$	4. Transitive Property of Equality
5. P is on line n.	5. Converse of Perpendicular Bisector Theorem
6. Lines ℓ, m, and n are concurrent.	6. ___?___

LEVELED QUESTIONS

Depth of Knowledge (DOK)	Leveled Questions	What Does This Tell You?
Level 1 **Recall**	What is a perpendicular bisector of a side of a triangle? A line that bisects a side of a triangle and is also perpendicular to that side of the triangle.	Students' answers will indicate whether they understand the characteristics of a perpendicular bisector of a side of a triangle.
Level 2 **Basic Application of Skills & Concepts**	Will the perpendicular bisector of a side of a triangle always pass through the midpoint for that side? Explain. yes; The midpoint divides the side into two equal segments.	Students' answers will indicate whether they understand the relationship between the midpoint of a side of a triangle and the perpendicular bisector for that side.
Level 3 **Strategic Thinking & Complex Reasoning**	What must be true for two of the perpendicular bisectors of the sides of a triangle to be perpendicular to each other? The triangle must be a right triangle.	Students' answers will indicate whether they can reason logically about when the perpendicular bisectors of the sides of a triangle are perpendicular to each other.

Prove the Circumcenter Theorem

The point of concurrency of the perpendicular bisectors of a triangle is called the **circumcenter of a triangle**.

A circle that contains all the vertices of a polygon is **circumscribed** about the polygon. A circle circumscribed about a polygon is called the **circumcircle** of the polygon.

Circumcenter Theorem

The perpendicular bisectors of the sides of a triangle intersect at a point that is equidistant from the vertices of the triangle.
$PA = PB = PC$

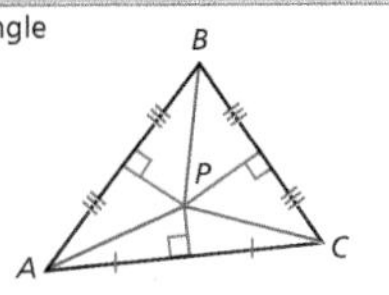

3

Show that the distance from the circumcenter of the triangle to each vertex of the triangle is equal.

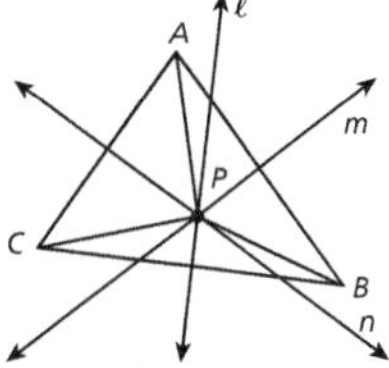

Given: Lines ℓ, m, and n are perpendicular bisectors of $\triangle ABC$.
P is the circumcenter of $\triangle ABC$.

Prove: $PA = PB = PC$

Statements	Reasons
1. Lines ℓ, m, and n are perpendicular bisectors of $\triangle ABC$.	1. Given
2. P is the circumcenter of $\triangle ABC$.	2. Given
3. P is on the perpendicular bisector of $\overline{AB}$, so $PA = PB$.	3. ___?___
4. P is on the perpendicular bisector of $\overline{BC}$, so $PB = PC$.	4. ___?___
5. $PA = PB = PC$	5. ___?___

A. What reasons should be given in Steps 3, 4, and 5? Should a step be included stating that P is on the perpendicular bisector of $\overline{AC}$? Why or why not?
See Additional Answers.

B. Draw a triangle and construct the circumcenter. Using the circumcenter as the center and $\overline{AP}$ as the radius, construct a circle. Why is the point of concurrency of the perpendicular bisectors of a triangle called the circumcenter of the triangle? Possible answer: The circumcenter of the triangle is the center of the circumscribed circle.

C. The circumcenter of a triangle is used to draw a circle circumscribed about the triangle. If one side of the triangle is a diameter of the circumcircle, then how can you classify the triangle by its angles? right triangle

Turn and Talk Why is it only necessary to construct two perpendicular bisectors to circumscribe a triangle with a circle? See margin.

PROFICIENCY LEVEL

Beginning

Show students pictures which include a circle and a triangle. Some of the pictures should show circles that are circumscribed about the triangle and others will show circles that are not. Have students state whether each picture shows a circumcircle or does not show a circumcircle.

Intermediate

Have students work in small groups. Give each group a picture showing the intersection of the perpendicular bisectors of a triangle. Ask students to explain how to find the circumcenter of a triangle.

Advanced

Have students write a short paragraph describing the steps they would use to circumscribe a circle about a triangle.

Task 3 (MP) **Attend to Precision** Students use their knowledge of perpendicular bisectors of sides of triangles to prove the Circumcenter Theorem.

CONNECT TO VOCABULARY

Have students use the **Interactive Glossary** to record their understanding of the vocabulary in this task.

Sample Guided Discussion:

Q In part C, where is the circumcenter in relationship to the triangle? The circumcenter is the midpoint of the hypotenuse of the right triangle.

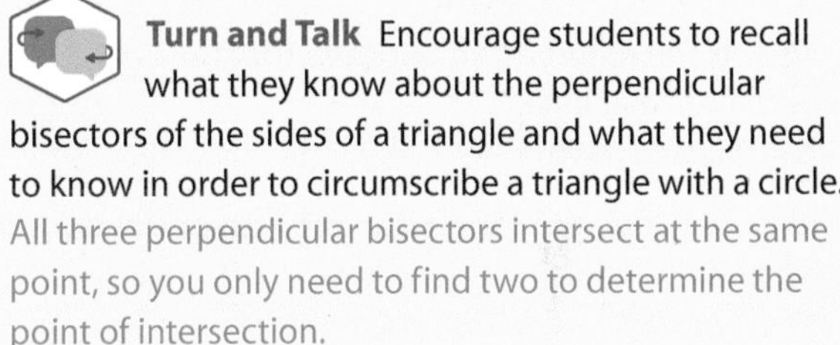

Turn and Talk Encourage students to recall what they know about the perpendicular bisectors of the sides of a triangle and what they need to know in order to circumscribe a triangle with a circle. All three perpendicular bisectors intersect at the same point, so you only need to find two to determine the point of intersection.

Step It Out

Task 4 **(MP) Use Tools** Students use the coordinate grid to find perpendicular bisectors of sides of a triangle.

SUPPORT SENSE-MAKING **Three Reads**

Have students read the problem three times. Use the questions below for a different focus each read.

1. What is the situation about?
2. What are the quantities in the situation?
3. What are the possible mathematical questions that you could ask for the situation?

Sample Guided Discussion:

Q What is the relationship between the slopes of perpendicular lines? The slopes of perpendicular lines are opposite reciprocals.

Q Why do the midpoint of $\overline{AB}$ and the circumcenter have the same x-value? Possible answer: Side $\overline{AB}$ is parallel to the x-axis, so the perpendicular bisector of $\overline{AB}$ is a vertical line. All points on a vertical line share the same x-value.

Turn and Talk Encourage students to examine how the coordinates of the vertices of the triangle in a coordinate plane can be used to determine the equations of the perpendicular bisectors for each side. Possible answer: Draw the perpendicular bisectors of each side of the triangle and find the circumcenter. Draw a circle with the center at the circumcenter that passes through the points of the triangle. When working on a coordinate plane, you can find the equations of the perpendicular bisectors and determine the point of intersection of the two lines.

Task 5 **(MP) Attend to Precision** Students write and solve equations to find the length of a line segment connecting a vertex of a triangle to its circumcenter.

Sample Guided Discussion:

Q What is the length of side $\overline{PB}$? 71 units

Q Could the length of side $\overline{PA}$ be $9x + 1$? Explain. no; The value of x is 8, so $9x + 1 = 9(8) + 1 = 72$. The distance from the circumcenter to each vertex must be the same.

Step It Out

Use Circumcenters to Solve Problems

4 The diagram shows a bass player (point A), a drummer (point B), and a guitar player (point C) in a recording studio. Where should an overhead microphone be placed so that it is the same distance from each musician?

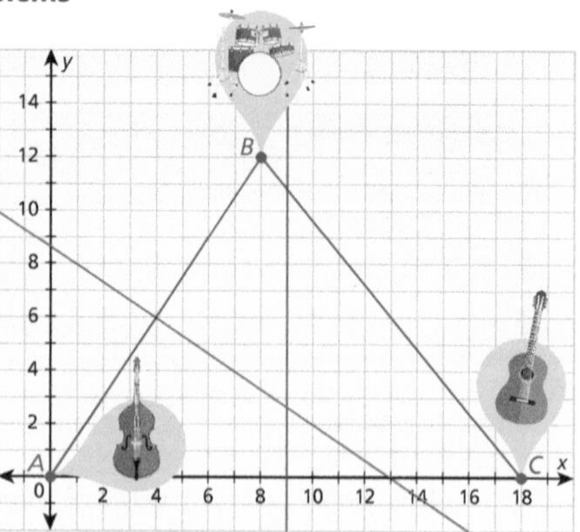

Find the perpendicular bisector of $\overline{AB}$.

midpoint of $\overline{AB} = \left(\frac{0+8}{2}, \frac{0+12}{2}\right) = (4, 6)$

slope of $\overline{AB} = \frac{12-0}{8-0} = \frac{12}{8} = \frac{3}{2}$

The perpendicular bisector of $\overline{AB}$ is $y = -\frac{2}{3}x + \frac{26}{3}$.

A. How was this equation determined?

A, B. See Additional Answers.

Find the perpendicular bisector of $\overline{AC}$.

The midpoint of $\overline{AC} = (9, 0)$.

The perpendicular bisector of $\overline{AC}$ is $x = 9$.

B. How was this equation determined?

Find the intersection of the perpendicular bisectors.

C. What are the coordinates of the circumcenter?

$y = -\frac{2}{3}x + \frac{26}{3}$ and $x = 9$ intersect at $y = -\frac{2}{3}(9) + \frac{26}{3} = \frac{8}{3}$. $\left(9, \frac{8}{3}\right)$

Turn and Talk Compare the processes of circumscribing a triangle with a circle on and off the coordinate plane. See margin.

5 The circumcenter of $\triangle ABC$ is P. Find the length of $\overline{PC}$.

Since P is the circumcenter of the triangle, $PB = PC$.

Solve for x.

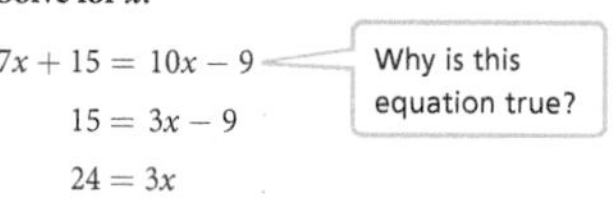

$7x + 15 = 10x - 9$

$15 = 3x - 9$

$24 = 3x$

$8 = x$

Why is this equation true?

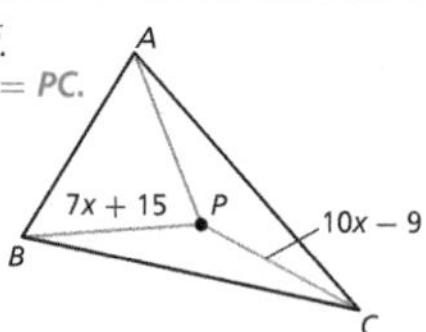

Substitute the value of x into the expression for PC.

$PC = 10(8) - 9 = 80 - 9 = 71$

The length of $\overline{PC}$ is 71.

268

Check Understanding

Copy each triangle. Construct the perpendicular bisectors of each side.

1–5. See Additional Answers.

1.

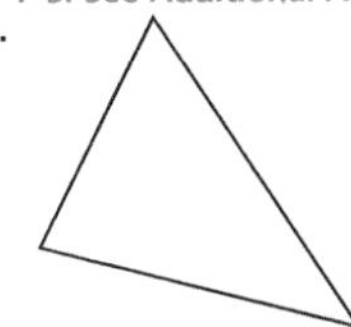

2.

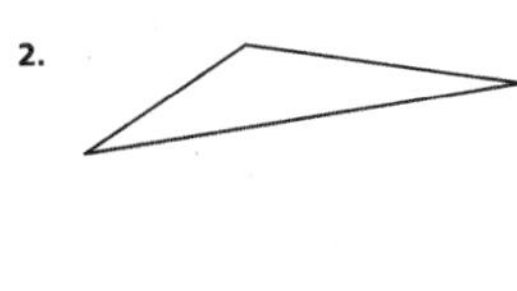

3. (MP) **Critique Reasoning** Alice states that the circumcenter of a triangle must be within the triangle. Nancy says that it is possible for the circumcenter to be outside of the triangle. Who is correct? Explain your reasoning.

4. Sketch a triangle on a coordinate plane with vertices at $(-1, 1)$, $(4, -2)$, and $(3, 3)$. Estimate the coordinates of the circumcenter of the triangle.

5. Point N is the circumcenter of $\triangle QPR$, $QN = 4x - 5$, and $RN = 2x + 3$. Is it possible to find the value of x? If so, explain how.

On Your Own

6. (MP) **Use Structure** Sketch three different right triangles. Find the circumcenter of each triangle. What do you notice about the circumcenter of each right triangle?
The circumcenter of a right triangle is located on the hypotenuse of the triangle.

7. (MP) **Use Structure** Sketch three different obtuse triangles. Find the circumcenter of each triangle. What do you notice about the circumcenter of each obtuse triangle?
The circumcenter of an obtuse triangle is located outside of the triangle.

Copy each triangle. Construct the circumcircle of the triangle.

8–11. See Additional Answers.

8.

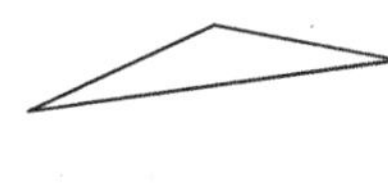

9.

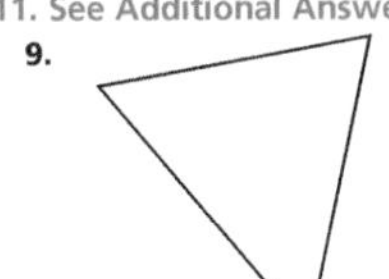

10.

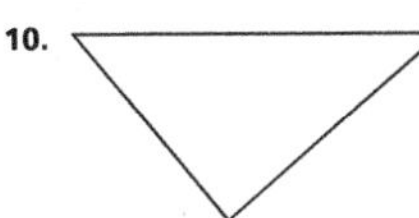

11.

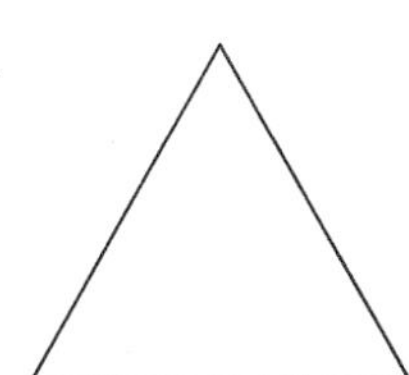

ONLINE

Assign the Digital On Your Own for
- built-in student supports
- Actionable Item Reports
- Standards Analysis Reports

On Your Own

Assignment Guide

The chart below indicates which problems in the On Your Own are associated with each task in the Learn Together. Assign daily homework for tasks completed.

Learn Together Tasks	On Your Own Problems
Task 1, p. 266	Problems 28–30 and 32–34
Task 2, p. 266	Problems 6 and 7
Task 3, p. 267	Problems 6–14, 28, 29, and 31
Task 4, p. 268	Problems 15–23
Task 5, p. 268	Problems 24–27 and 35–38

3 Check Understanding

Formative Assessment

Use formative assessment to determine if your students are successful with this lesson's learning objective.

Students who successfully complete the Check Understanding can continue to the On Your Own practice.

For students who miss 1 problem or more, work in a pulled small group using the Almost There small-group activity on page 265C.

ONLINE

Assign the Digital Check Understanding to determine
- success with the learning objective
- items to review
- grouping and differentiation resources

4 Differentiation Options

Differentiate instruction for all students using small-group activities and math center activities on page 265C.

Reteach

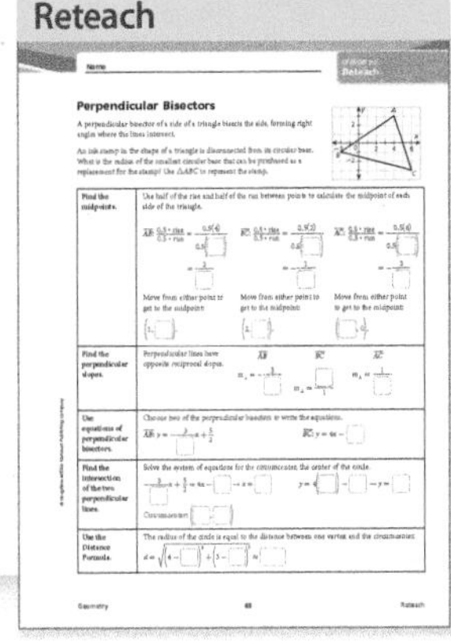

Challenge

Questioning Strategies

Problem 14 The location of the new tower can be found by determining the point of intersection for the perpendicular bisectors of two the sides of the triangle *ABC*. What constructions can you add to the picture that would illustrate that the new tower is the same distance from each of the vertices of the triangle *ABC*? Explain your reasoning. Possible answer: Draw a circle whose center is the circumcenter of the triangle and whose radius is the distance from the center to any of the vertices of the triangle. Since the circle will pass through the three vertices of the triangle, the distance from the new tower to each of the vertices of the triangle is the same.

Watch for Common Errors

Problems 15–18 Students may try to find the circumcenter of each triangle using geometric constructions. Remind students that for triangles shown in the coordinate plane, they should use the coordinates of the vertices of the triangle to find each triangle's circumcenter.

12. (MP) **Construct Arguments** Use a flow proof to prove the Circumcenter Theorem. See Additional Answers.

Given: Lines ℓ, m, and n are the perpendicular bisectors of $\overline{AB}$, $\overline{AC}$, and $\overline{BC}$. P is the intersection of ℓ, m, and n.

Prove: $PA = PB = PC$

13. (MP) **Construct Arguments** Use geometry software to draw a circle. Draw a quadrilateral inside the circle so that all four vertices lie on the circle. What do you notice about the measures of the opposite angles in the quadrilateral?
The measures of opposite angles of the quadrilateral are supplementary.

14. A cell phone company plans to build a new tower that is the same distance from the three existing towers shown in the diagram. Copy the points and show where the new tower should be located.
See Additional Answers.

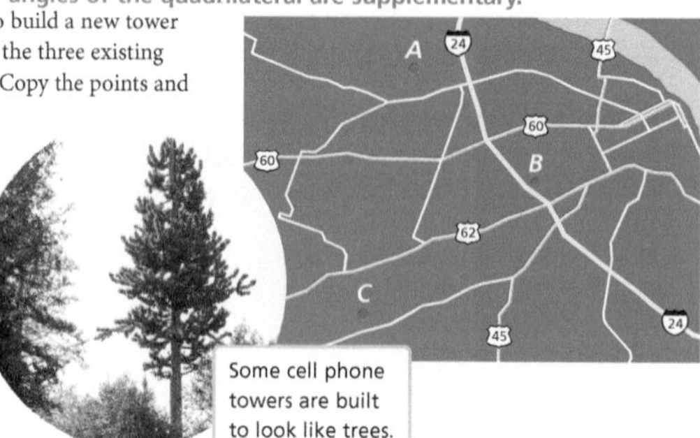

Some cell phone towers are built to look like trees.

©James Hackland/Alamy

Find the circumcenter of the triangle.

15. (2, −3)

16. (−1, 1)

17. (1, 0)

18. (−1, −2)

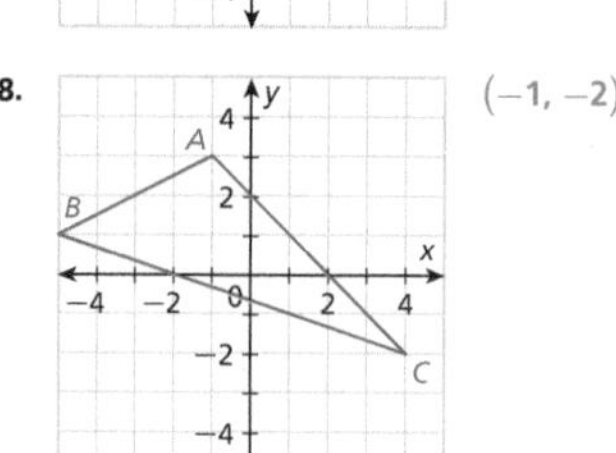

270

19. (MP) **Reason** Suppose you are finding the circumcenter of a triangle on a coordinate plane. Why might you want to find the perpendicular bisector of a horizontal or vertical side of the triangle first? **19, 20. See Additional Answers.**

20. (MP) **Reason** When finding the circumcenter of a triangle on a coordinate plane, only two perpendicular bisectors are needed. Explain why. Then explain how the third perpendicular bisector can be used to check your answer.

The circumcenter P of $\triangle ABC$ on a coordinate plane is given. Find the unknown coordinate of the vertices of the triangle.

21. circumcenter: $P(3, 2)$; vertices: $A(1, 3)$, $B(4, ?)$, $C(5, 1)$ **4 or 0**

22. circumcenter: $P(-1, 3)$; vertices: $A(-4, 0)$, $B(-4, 6)$, and $C(?, 0)$ **2**

23. circumcenter: $P(2.5, 2)$; vertices: $A(7, ?)$, $B(-1, -1)$, and $C(-2, 1)$ **1 or 3**

The circumcenter of $\triangle ABC$ is P. Find the value of x.

24. **$x = 7$**

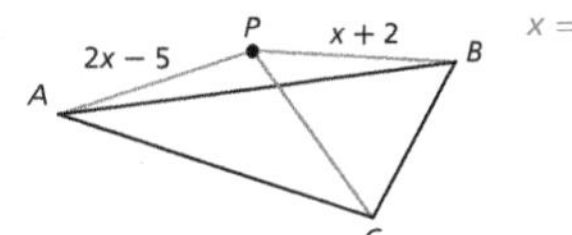

25. **$x = 10$**

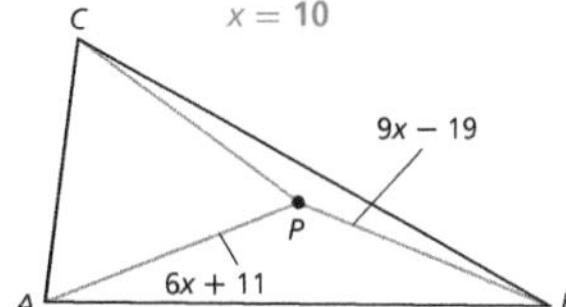

26. **$x = 8$**

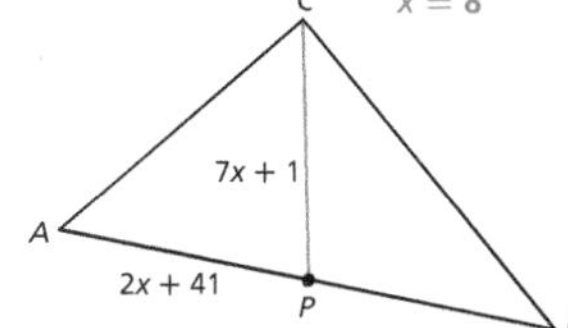

27. **$x = 5$**

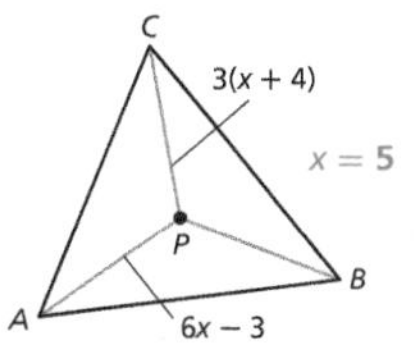

Use the diagram for Problems 28–31. $\overline{DE}$, $\overline{DF}$, and $\overline{DG}$ are the perpendicular bisectors of $\triangle ABC$. Use the given information to find the lengths. Note that the figure is not drawn to scale.

28. Given: $FD = 24$, $DA = 46$, $GC = 32$
Find: AC and DB **$AC = 64$, $DB = 46$**

29. Given: $BD = 26$, $DE = 11$, $BC = 42$
Find: BE and DC **$BE = 21$, $DC = 26$**

30. Given: $BC = 30$, $DC = 17$
Find: DE **$DE = 8$**

31. Name a segment that is a radius of the circumcircle of $\triangle ABC$. **$\overline{DA}$ or $\overline{DB}$ or $\overline{DC}$**

Questioning Strategies

Problem 21 The points A, B and C are all on the circumcircle for the triangle ABC. Determine the coordinates of a point D which is also on the circumcircle. Explain how you found your answer. Possible answer: $(3, 2 + \sqrt{5})$; Each radius of the circumcircle must have the same length. Use the given points and the Distance Formula to show that the length of the radius of the circle is $\sqrt{5}$ units. Point $D(3, 2 + \sqrt{5})$ is $\sqrt{5}$ units from the circumcenter at $(3, 2)$ so it is also on the circumcircle.

Watch for Common Errors

Problem 27 When students are writing and solving an equation to find the value of x, remind students that one of the expressions in their equation will use parentheses, so they should make sure to follow the order or operations to find the correct value for x.

⑤ Wrap-Up

Summarize learning with your class. Consider using the Exit Ticket, Put It in Writing, or I Can scale.

Exit Ticket

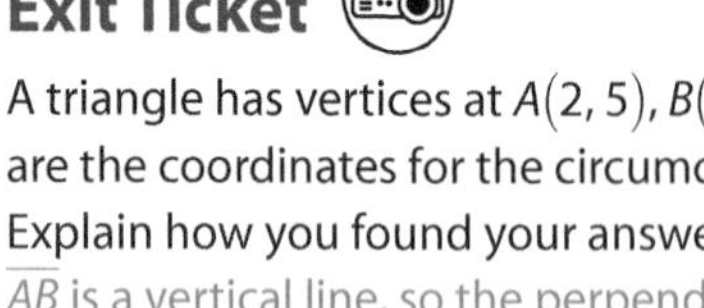

A triangle has vertices at $A(2, 5)$, $B(2, -1)$ and $C(6, 3)$. What are the coordinates for the circumcenter for triangle ABC? Explain how you found your answer. $(3, 2)$; Possible answer: $\overline{AB}$ is a vertical line, so the perpendicular bisector for $\overline{AB}$ will be a horizontal line through the midpoint $(2, 2)$. The perpendicular bisector of $\overline{AB}$ is the line $y = 2$, so the y-coordinate of the circumcenter will be 2. The side $\overline{BC}$ has a slope of 1, so its perpendicular bisector has a slope of -1. The midpoint of $\overline{BC}$ is $(4, 1)$, so the perpendicular bisector will pass through $(4, 1)$. The perpendicular bisector of $\overline{BC}$ is the line $y = -x + 5$. The intersection of the lines is at $x = 3$.

Put It in Writing

Describe the constructions you can make to find the circumcenter of a triangle.

I Can

The scale below can help you and your students understand their progress on a learning goal.

4	I can construct perpendicular bisectors and use the point of concurrency to circumscribe triangles with circles, and I can explain my methods for circumscribing triangles with circles to others.
3	I can construct perpendicular bisectors and use the point of concurrency to circumscribe triangles with circles.
2	I can show that the perpendicular bisectors of a triangle are concurrent lines.
1	I can identify the perpendicular bisector of a side of a triangle.

Spiral Review • Assessment Readiness

These questions will help determine if students have retained information taught in the past and can also prepare them for high-stakes assessments. Here, students must determine the measures of the angles of a triangle **(9.1)**, determine the measures of bisected angles **(1.2)**, and use theorems to show triangle congruence **(8.1–8.4)**.

In the diagram, $\overline{PK}$, $\overline{PL}$, and $\overline{PM}$ are the perpendicular bisectors of sides $\overline{AB}$, $\overline{BC}$, and $\overline{AC}$, respectively. Tell whether the given statement is justified by the figure.

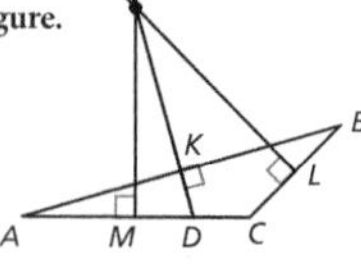

32. $AK = KB$ justified

33. $PA = PB$ justified

34. $PM = PL$ not Justified

The circumcenter of $\triangle ABC$ is P. Find the length of $\overline{PA}$.

35. $PA = 81$

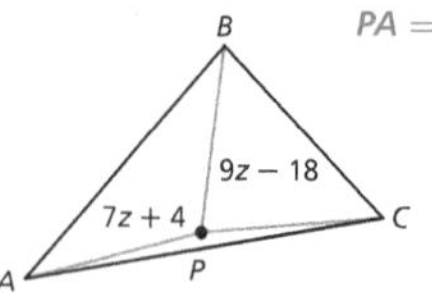

36. $PA = 13$

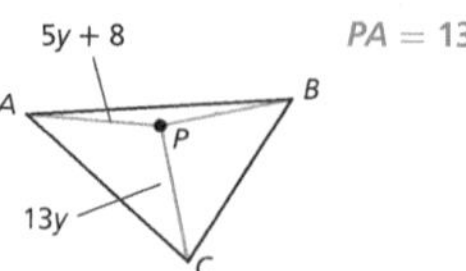

37. $PA = 53$

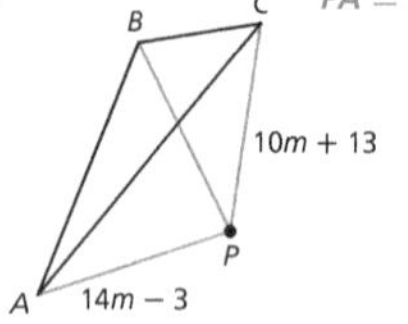

38. $PA = 43$

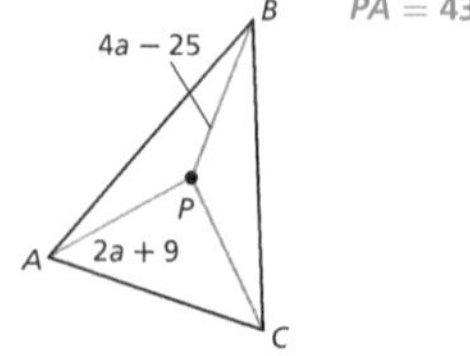

Spiral Review • Assessment Readiness

39. What is the measure of $\angle B$?

Ⓐ 13°
Ⓑ 35°
Ⓒ 60°
Ⓓ 85°

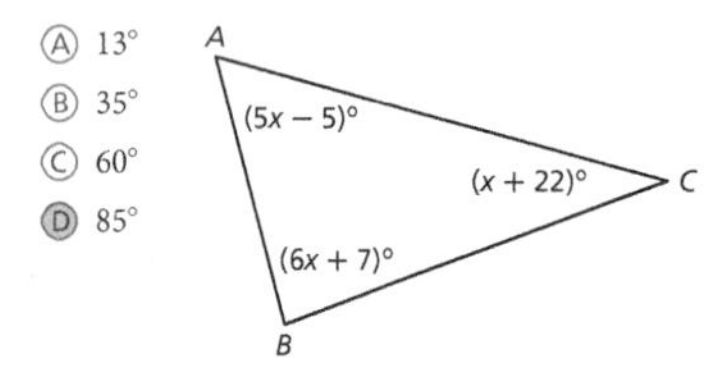

40. The measure of $\angle JKL$ is 88°, and $\angle JKL$ is bisected by $\overrightarrow{KM}$. What is the value of x if $m\angle JKM = (7x - 12)°$?

Ⓐ 8
Ⓑ 14
Ⓒ 44
Ⓓ 56

41. Which theorem can you use to show that $\triangle FGH \cong \triangle RST$?

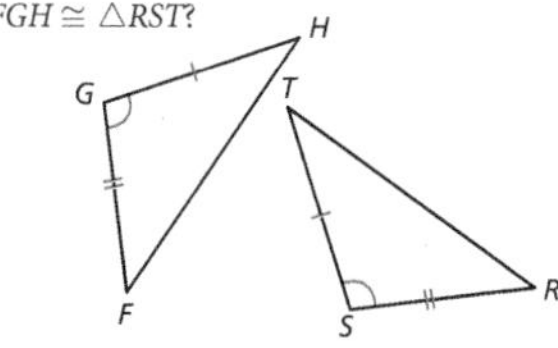

Ⓐ ASA Triangle Congruence
Ⓑ AAS Triangle Congruence
Ⓒ SAS Triangle Congruence
Ⓓ SSS Triangle Congruence

I'm in a Learning Mindset!

What are my strengths when finding and using perpendicular bisectors?

272

Keep Going ▶ Journal and Practice Workbook

Learning Mindset

Challenge Seeking Builds Confidence

Discuss with students the different methods they know for finding a perpendicular bisector, as well as the different kinds of problems where finding a perpendicular bisector can be useful in finding the solution. Students who can identify their strengths with regards to perpendicular bisectors will be more prepared to address the kinds of perpendicular bisector problems which give them difficulty. *What methods for finding a perpendicular bisector are you most confident using? What kinds of problems that involve perpendicular bisectors are you most confident solving?*

9.3 Angle Bisectors in Triangles

LESSON FOCUS AND COHERENCE

Mathematics Standards

- Prove theorems about triangles. *Theorems include: measures of interior angles of a triangle sum to 180°; base angles of isosceles triangles are congruent; the segment joining midpoints of two sides of a triangle is parallel to the third side and half the length; the medians of a triangle meet at a point.*
- Construct the inscribed and circumscribed circles of a triangle, and prove properties of angles for a quadrilateral inscribed in a circle.

Mathematical Practices and Processes 

- Attend to precision.
- Construct viable arguments and critique the reasoning of others.
- Use appropriate tools strategically.

I Can Objective 

I can prove that angle bisectors are concurrent and inscribe circles in triangles.

Learning Objective

Students prove the Angle Bisector Theorem, the converse of the Angle Bisector Theorem, and the Incenter Theorem, and students construct the incenters and inscribed circles of triangles.

Language Objective

Describe the intersection of the angle bisectors of a triangle as a point called the *incenter* that is equidistant from the sides of the triangle and the center of the inscribed circle of a triangle.

Vocabulary

New: angle bisector of a triangle, incenter

Lesson Materials: compass, straightedge, ruler

Mathematical Progressions

Prior Learning	Current Development	Future Connections
Students: • proved theorems about perpendicular bisectors in triangles. **(9.2)**	**Students:** • prove the Angle Bisector Theorem and its converse. • prove that angle bisectors are concurrent. • construct inscribed circles.	**Students:** • will prove that the medians of a triangle meet at a point **(9.4)**

UNPACKING MATH STANDARDS

Construct the inscribed and circumscribed circles of a triangle, and prove properties of angles for a quadrilateral inscribed in a circle.

What it Means to You

Constructing the inscribed circle of a triangle requires two distinct steps. First, students must locate the incenter by constructing the angle bisectors of the triangle. Students should understand that the center of the inscribed circle is the point of intersection of the angle bisectors. Because the angle bisectors are concurrent, students only need to construct two bisectors to find the incenter.

The second step in this construction requires determining the radius of the inscribed circle. Students need to recall that the angle bisectors are equidistant from the sides of the triangle. Therefore, the desired radius is the length of each perpendicular segment from the incenter to a side. Using tools strategically, students construct the perpendicular segment to a side, adjust the width of the compass to reflect its length, and draw the inscribed circle of the triangle.

WARM-UP OPTIONS

ONLINE

PROJECTABLE & PRINTABLE

ACTIVATE PRIOR KNOWLEDGE • Perpendicular Bisectors in Triangles

Use these activities to quickly assess and activate prior knowledge as needed.

Problem of the Day

In this figure, the vertices of $\triangle ABC$ lie on circle O. What is the relationship between points A, B, and C and point O? Explain.
Point O is the center of the circumscribed circle about $\triangle ABC$ and is therefore equidistant from the vertices of the triangle. That means $OA = OB = OC$.

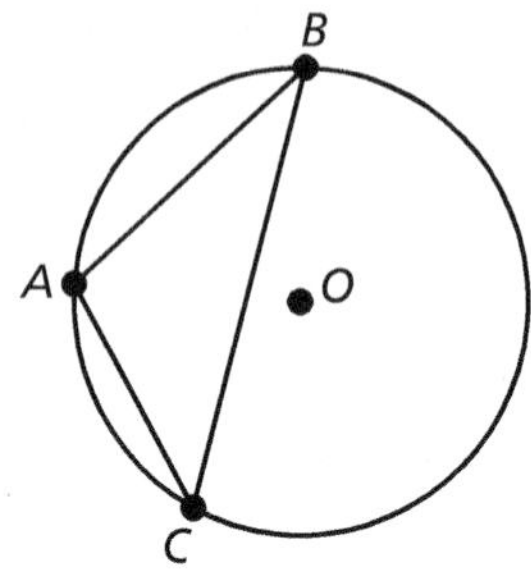

Quick Check for Homework

As part of your daily routine, you may want to display the Teacher Solution Key to have students check their homework.

Make Connections

Based on students' responses to the Problem of the Day, choose one of the following:

 Project the Interactive Reteach, Geometry, Lesson 9.2.

 Complete the Prerequisite Skills Activity:

Have students work in pairs. Each pair of students should have a compass and straightedge. Have one student construct an isosceles right triangle and the other student construct an equilateral triangle. Have them exchange papers and construct the circumcircle of each triangle and describe the location of its circumcenter.

- *Where is the circumcenter for a right triangle?* at the midpoint of the hypotenuse
- *What is the length of the diameter of the circumscribed circle about the right triangle in terms of s, the length of each leg?* $s\sqrt{2}$
- *What types of triangles are formed by the intersection of the radii of the circumcenter and the vertices of the equilateral triangle?* isosceles triangles
- *What congruence theorem can you use to prove that the triangles formed in the circumscribed equilateral triangle are congruent?* SSS

SHARPEN SKILLS

If time permits, use this on-level activity to build fluency and practice basic skills.

Mental Math

Objective: Students use reasoning to express the ratio of the areas of two concentric circles: the circumcircle and the incircle of a triangle.

In this figure, a circumcircle and an incircle have been drawn about and in an equilateral triangle. The circles have the same center at O.

If $OA = r$, and $OB = 2r$, explain why the ratio between the area of the circumscribed circle about the triangle and the area of its incircle is 4:1.

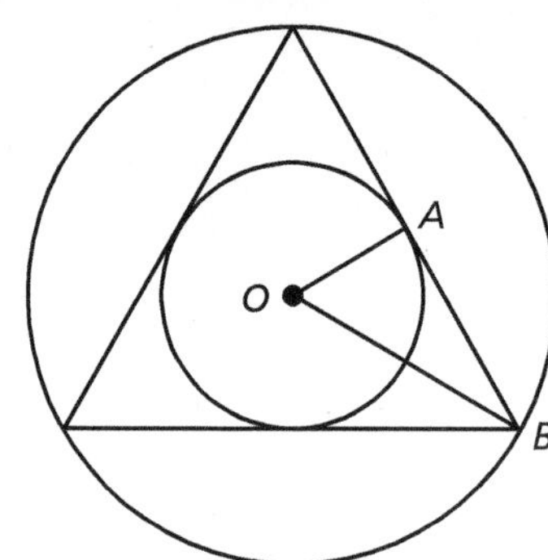

PLAN FOR DIFFERENTIATED INSTRUCTION

Small-Group Options

Use these teacher-guided activities with pulled small groups.

On Track

Materials: compasses and straightedges

Have students use the straightedge to draw a triangle. Have them use their compasses and straightedges to construct two angle bisectors and the inscribed circle of the triangle.

Have students compare their constructions and explain the relationship between each triangle and its incircle.

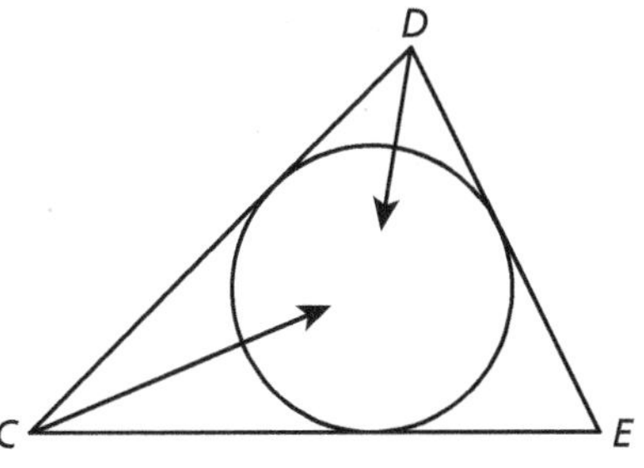

Almost There

Materials: compass, straightedge, and ruler

On a large sheet of paper draw an acute angle, a right angle, and an obtuse angle. Then use compasses and straightedges to construct the angle bisector of each angle.

Then place a point on each bisector and construct the perpendicular segments to each side of the angle. Use a ruler to verify that the pairs of segments are equal.

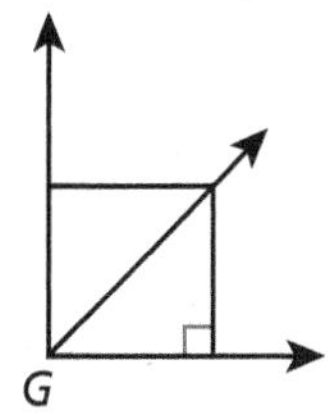

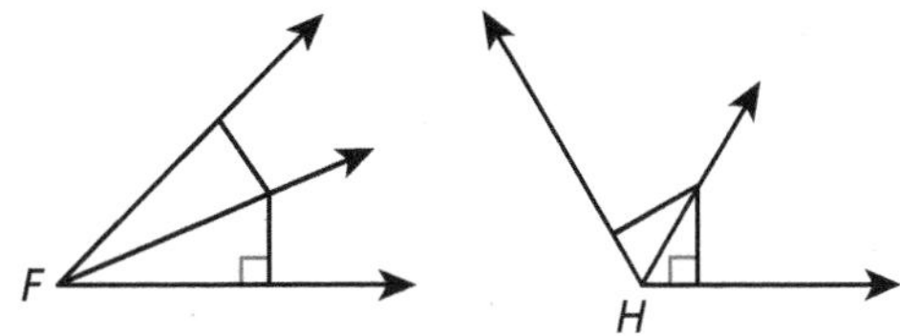

Ready for More

Materials: compass and straightedge

Use compasses and straightedges to construct a square and inscribe a circle in it, such as the figure below. Describe the steps in your construction.

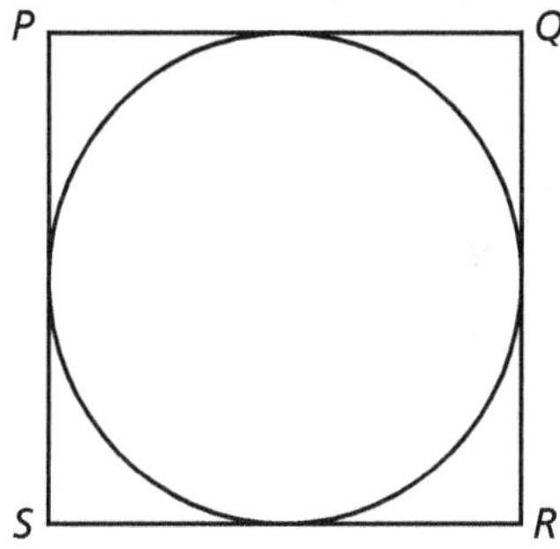

Math Center Options

Use these student self-directed activities at centers or stations. Key: ● Print Resources ● Online Resources

On Track

- ● Interactive Digital Lesson
- ●● Journal and Practice Workbook
- ● Interactive Glossary (printable): **angle bisector of a triangle**, **incenter**
- ●● Standards Practice: Make Formal Geometric Constructions with a Variety of Tools and Methods
- ● Module Performance Task

Almost There

- ● Reteach 9.3 (printable)
- ● Interactive Reteach 9.3

Ready for More

- ● Challenge 9.3 (printable)
- ● Interactive Challenge 9.3

ONLINE 

View data-driven grouping recommendations and assign differentiation resources.

During the *Spark Your Learning,* listen and watch for strategies students use. See samples of student work on this page.

Use a Compass and Ruler

Strategy 1

To find the largest circle, I chose a point inside the triangle that seemed to be the same distance from each side. Then I set the width of the compass so that when I drew a circle it would touch each side of the triangle in only one point.

If students . . . use tools to approximate the location of the center of the largest circle and its radius, they understand that the center has to be in the interior of the triangle and the same distance from each side.

Have these students . . . explain why they thought their strategy would give them the most likely point from which to draw the largest circle. **Ask:**

- Q What appears to be the relationship between the radius of the circle and each side?
- Q What theorem can you recall about the relationship between points on the bisector of an angle and its sides?

Use Trial and Error

Strategy 2

I started by using a compass to draw a circle whose center was inside what might be the center of the triangle. If the circle I drew was too small and I could draw a larger one, I widened the distance between the legs of the compass and drew another circle. I repeated this technique changing the location of the center and radius until by trial and error, I drew the largest circle I could in the interior of the triangle and that touched each side of the triangle in a point.

If students . . . kept experimenting until they seemed to have found the largest circle, they understand that the center is in the interior of the triangle and that the circle must not extend beyond the sides of the triangle.

Activate prior knowledge . . . by having students recall under what circumstance a point can be equidistant from two intersecting lines or segments. **Ask:**

- Q What is the relationship between an angle bisector and the sides of the angle it bisects?
- Q How do you find the minimum distance between a point not on a line and a given line or segment?

COMMON ERROR: Constructs Perpendicular Bisectors

I know that the perpendicular bisectors of the sides of a triangle intersect and that a circle can be drawn about the triangle using that point of intersection as the center. So, I sketched the perpendicular bisectors of two sides and then drew a circle using a distance to one of the vertices as the radius.

If students . . . approximated the center of the largest circle as the point of concurrency of the perpendicular bisectors, they misunderstand the criteria that the points on the inscribed circle must have.

Then intervene . . . by pointing out that the incenter is always equidistant from the points at which the sides of the triangle are tangent to the incircle. **Ask:**

- Q Where is the incenter of a triangle if the triangle is equilateral?
- Q What is the relationship between points on the bisector of an angle and its sides?

9.3

Angle Bisectors in Triangles

I Can prove that angle bisectors are concurrent and inscribe circles in triangles.

Spark Your Learning

A group of friends are making scrapbooks. Maria wants to cut a circle out of a triangular piece of paper.

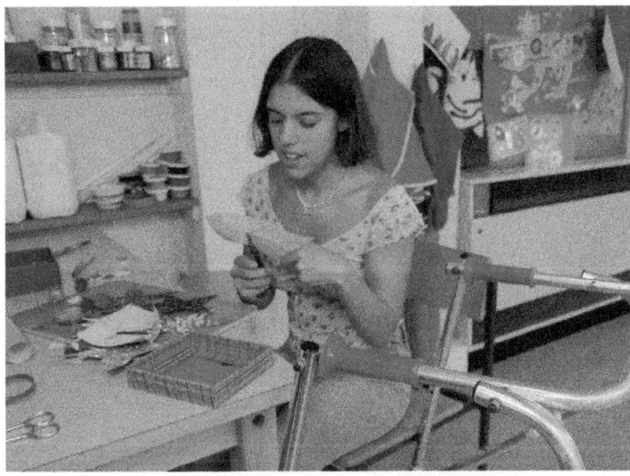

©John Birdsall/age fotostock

Complete Part A as a whole class. Then complete Parts B–D in small groups.

A. What is a geometric question you can ask about this situation? What information would you need to know to answer your question?
A. How can you find the largest circle that can be cut from a triangle?; the dimensions of the triangle

B. To answer your question, what strategy and tool would you use along with all the information you have? What answer do you get?
See Strategies 1 and 2 on the facing page.

C. How should the edge of the circle relate to the sides of the triangle?
Possible answer: The edge of the circle should touch each side of the triangle.

D. How can you check that the center of the circle is in the correct location?
See Additional Answers.

Turn and Talk Do you notice anything about the center of the circle that relates to the sides, vertices, or angles of the triangle? See margin.

 SUPPORT SENSE MAKING • Three Reads

Tell students to read the information in the photo three times and prompt them with a different question each time.

1. What is the information about? The information is about how to determine the center of the largest possible circle that does not have any points outside of the triangle.
2. What are the geometric relationships in this situation? The center of the circle must lie in the interior of the triangle.
3. What are possible questions you could ask about this situation? Possible answer: What is the property of the center of a circle inscribed in a triangle? What are the possible locations of the incenter of a triangle? What is always be true about the location of the center of the largest possible circle inside a triangle?

LESSON 9.3 — Connect Concepts and Skills

1 Spark Your Learning

▶ MOTIVATE 

- Have students look at the photo in their books and read the information contained in the photo. Then complete Part A as a whole-class discussion.
- Give the class the additional information they need to solve the problem. This information is available online as a printable and projectable page in the Teacher Resources.
- Have students work in small groups to complete Parts B–D.

▶ PERSEVERE

If students need support, guide them by asking:

Q **Advancing • Use Tools** Which tool could you use to solve the problem? Why choose that tool and not some other? Students' choices of tools and reasons for choosing them will vary.

Q **Assessing** Is the largest circle you can draw also the circumcircle? No, because except for the vertices of the triangle, a circumcircle lies outside the triangle rather than inside the triangle.

Q **Assessing** What would be the relationship between the center of the largest circle that could be drawn and the triangle? The center would have to be the same distance from each side.

Q **Advancing** What theorem describes the relationship between a segment and points on a line that are equidistant from the endpoints of the segment? the Perpendicular Bisector of a Segment Theorem

Turn and Talk Is the center of the largest circle in a triangle equidistant from the vertices or the sides of the triangle? The center of the circle is the same distance from each side of the triangle.

▶ BUILD SHARED UNDERSTANDING

Select groups of students who used various strategies and tools to share with the class how they solved the problem. As they present their solutions, have each group discuss why they chose a specific strategy and tool.

② Learn Together

Build Understanding

Task 1 (MP) **Attend to Precision** Students use AAS to prove the Angle Bisector Theorem.

CONNECT TO VOCABULARY

Have students use the **Interactive Glossary** to record their understanding of the vocabulary in this task.

Sample Guided Discussion:

Q **In Part A, what is the geometric meaning of the word *distance*?** The distance is the length of the perpendicular segment from R to the ray $\overrightarrow{QP}$.

Q **In Part B, what are the known congruences between corresponding angles and sides in $\triangle XYW$ and $\triangle ZYW$?** $\angle XYW \cong \angle ZYW$, $\angle X \cong \angle Z$, and $WY \cong WY$.

Turn and Talk Have students sketch an isosceles triangle ABC and draw the angle bisector of the vertex angle $\angle A$. Then have them draw the perpendicular bisector to the base $\overline{BC}$ from A. Possible answer: They both divide a part of a triangle into two congruent parts.

Build Understanding

Prove the Angle Bisector Theorem

An **angle bisector of a triangle** is a ray that divides an angle into two congruent angles. You can find the relationship between an angle bisector and the sides of the angle it bisects.

Angle Bisector Theorem

If a point is on the bisector of an angle, then it is equidistant from the two sides of the angle.

If $\overrightarrow{PC}$ bisects $\angle APB$, $\overline{AC} \perp \overrightarrow{PA}$, and $\overline{BC} \perp \overrightarrow{PB}$, then $AC = BC$.

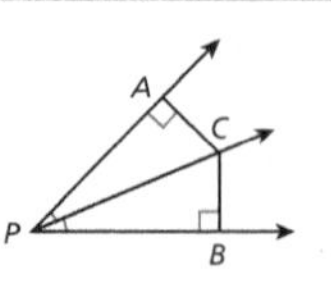

1 Prove the Angle Bisector Theorem.

A, B. See Additional Answers.

A. Many segments can be drawn from a point to a ray. In $\angle PQR$, which segment appears to represent the distance from point R to $\overrightarrow{QP}$? Explain your reasoning.

B. The theorem claims that $WX = WZ$. How can congruent triangles be used to show that $WX = WZ$? How can you show that the triangles are congruent if you know that $\angle XYW \cong \angle ZYW$?

Given: $\overrightarrow{YW}$ bisects $\angle XYZ$, $\overline{WX} \perp \overrightarrow{YX}$, and $\overline{WZ} \perp \overrightarrow{YZ}$.

Prove: $WX = WZ$

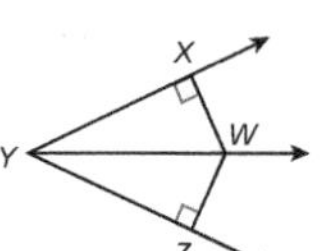

Statements	Reasons
1. $\overrightarrow{YW}$ bisects $\angle XYZ$, $\overline{WX} \perp \overrightarrow{YX}$, and $\overline{WZ} \perp \overrightarrow{YZ}$.	1. Given
2. $\angle XYW \cong \angle ZYW$	2. Definition of angle bisector
3. $\angle WXY$ and $\angle WZY$ are right angles.	3. Definition of perpendicular lines
4. $\angle WXY \cong \angle WZY$	4. All right angles are congruent.
5. $\overline{YW} \cong \overline{YW}$	5. Reflexive Property of Congruence
6. $\triangle YXW \cong \triangle YZW$	6. AAS Triangle Congruence Theorem
7. $\overline{WX} \cong \overline{WZ}$	7. Corresponding parts of congruent figures are congruent.
8. $WX = WZ$	8. Definition of congruent segments

Turn and Talk How is an angle bisector similar to a perpendicular bisector? See margin.

274

LEVELED QUESTIONS

Depth of Knowledge (DOK)	Leveled Questions	What Does This Tell You?
Level 1 **Recall**	Why is $\angle XYW$ congruent to $\angle ZYW$? Both angles are right angles that have measures of 90°.	Students' answers will indicate whether they understand that if the measures of two angles are equal, the angles are congruent.
Level 2 **Basic Application of Skills & Concepts**	What congruence theorem can be used to prove the two triangles congruent? AAS, with $\overline{YW}$ as the common side.	Students' answers will indicate that they understand how to prove triangles congruent if two angles and a side of one triangle are congruent to a pair of corresponding angles and side of another triangle.
Level 3 **Strategic Thinking & Complex Reasoning**	Why is it necessary to prove that $\triangle XYW$ and $\triangle ZYW$ are congruent? Because $\overline{WX}$ and $\overline{WZ}$ are corresponding parts of congruent triangles and therefore, they would be congruent.	Students' answers will indicate that they understand what they need to establish to prove that a given theorem is true.

Prove the Converse of the Angle Bisector Theorem

The converse of the Angle Bisector Theorem is true as well.

Converse of the Angle Bisector Theorem

If a point in the interior of an angle is equidistant from the sides of the angle, then it is on the bisector of the angle.

If $\overline{AC} \perp \overrightarrow{PA}$, $\overline{BC} \perp \overrightarrow{PB}$, and $AC = BC$, then $\overrightarrow{PC}$ bisects $\angle APB$.

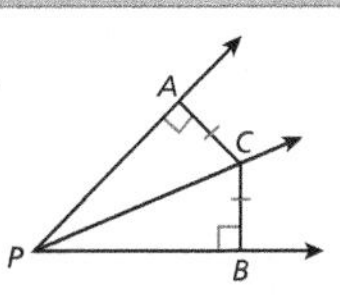

2 Prove the Converse of the Angle Bisector Theorem.

A. In the statement of the Converse of the Angle Bisector Theorem, why is it important to state that the point is in the interior of the angle? See Additional Answers.

B. What do you need to show before you can state that $\overrightarrow{YW}$ bisects $\angle XYZ$? $\angle XYW \cong \angle ZYW$

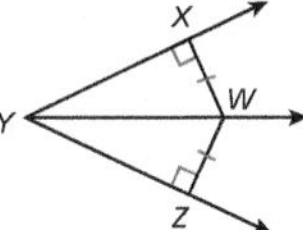

Given: $\overline{WX} \perp \overrightarrow{YX}$, $\overline{WZ} \perp \overrightarrow{YZ}$, and $WX = WZ$.

Prove: $\overrightarrow{YW}$ bisects $\angle XYZ$.

Statements	Reasons
1. $\overline{WX} \perp \overrightarrow{YX}$, $\overline{WZ} \perp \overrightarrow{YZ}$, and $WX = WZ$.	1. Given
2. $\overline{WX} \cong \overline{WZ}$	2. Definition of congruent segments
3. $\angle WXY$ and $\angle WZY$ are right angles.	3. Definition of perpendicular lines
4. $\triangle YXW$ and $\triangle YZW$ are right triangles.	4. Definition of a right triangle
5. $\overline{YW} \cong \overline{YW}$	5. Reflexive Property of Congruence
6. $\triangle YXW \cong \triangle YZW$	6. HL Triangle Congruence Theorem
7. $\angle XYW \cong \angle ZYW$	7. Corresponding parts of congruent figures are congruent.
8. $\overrightarrow{YW}$ bisects $\angle XYZ$.	8. Definition of angle bisector

C. Compare the proofs for the Angle Bisector Theorem and its converse. How are the proofs alike? How are they different?

Turn and Talk Suppose $m\angle XYZ = 68°$ and $m\angle XYW = (3x + 7)°$. Do you have enough information to solve for x? Explain your reasoning. See margin.

Task 2 (MP) **Construct Arguments** Students use congruent triangles to prove the Converse of the Angle Bisector Theorem.

Sample Guided Discussion:

Q **How many perpendicular lines can you draw to a side of a triangle (or side extended) from a point not on the side?** exactly one

Q **What is the relationship between $\angle XYW$ and $\angle ZYW$?** They are adjacent angles with common side $\overline{YW}$.

Turn and Talk Remind students that because $m\angle XYW = (3x + 7)°$ and $m\angle XYW = 34°$, then by the Transitive Property of Equality, $(3x + 7)° = 34°$. yes; Possible answer: By the Converse of the Angle Bisector Theorem, you know that ray $\overrightarrow{MK}$ is an angle bisector of $\angle LMN$. So, $m\angle LMK$ is half of $m\angle LMN$ or $m\angle LMK = \frac{1}{2}(68°) = 34°$. Set $(3x + 7)°$ equal to 34° and solve for x.

PROFICIENCY LEVEL

Beginning

Give students several situations in which either the Angle Bisector Theorem or its converse can be used. Have students verbally identify which theorem should be used.

Intermediate

Have students work in pairs and use a protractor and ruler to draw a right angle. Have them choose a point in the interior of the triangle and draw the perpendicular segment from the point to each side of the angle. Then have them measure the length of each segment and the measures of the two acute angles formed. What appears to be true? Have them discuss their results with others in the class.

Advanced

Have students work individually and use protractors and rulers to draw and label the vertices of a square and its diagonals. Label the point of intersection of the diagonals P. Then have students use protractors and rulers to measure distances and angles and propose three theorems about the diagonals of a square.

Step It Out

Task 3 (MP) **Construct Arguments** Students use the Angle Bisector Theorem and its converse to prove the Incenter Theorem.

CONNECT TO VOCABULARY

Have students use the **Interactive Glossary** to record their understanding of the vocabulary in this task.

Sample Guided Discussion:

Q How many points of intersection are there between two non-parallel lines in the plane? exactly one

Q What is a definition of concurrent lines? three or more lines in a plane that intersect at a single point

Q What is the difference between the incenter of a triangle and its circumcenter? The incenter is the point of concurrence of the angle bisectors of a triangle. The circumcenter is the point of concurrence of the perpendicular bisectors of the sides of the triangle.

Q Can the incenter of a triangle also be the circumcenter of the triangle? Explain. yes, in an equilateral triangle; Each angle bisector is also the perpendicular bisector of the opposite side and the three bisectors intersect at the circumcenter of the triangle.

Step It Out

Prove the Incenter Theorem

Recall that concurrent lines are three or more lines that intersect at one point, called the point of concurrency. The point of concurrency of angle bisectors of a triangle is called the **incenter** of the triangle. You can show that the incenter is equidistant from the sides of the triangle.

Incenter Theorem

The angle bisectors of a triangle intersect at a point that is equidistant from the sides of the triangle.

If $\overline{AP}$, $\overline{BP}$, and $\overline{CP}$ are angle bisectors of $\triangle ABC$, then $PX = PY = PZ$.

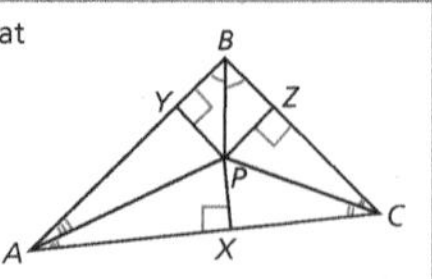

3 Prove the Incenter Theorem.

Given: Angle bisectors of $\angle A$, $\angle B$, and $\angle C$ in $\triangle ABC$.

Prove: The angle bisectors intersect at a point equidistant from $\overline{AB}$, $\overline{BC}$, and $\overline{AC}$.

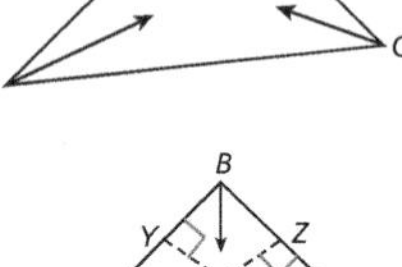

The bisectors of $\angle A$ and $\angle C$ intersect at point P. Perpendicular segments are drawn from point P so that $\overline{PX} \perp \overline{AC}$, $\overline{PY} \perp \overline{AB}$, and $\overline{PZ} \perp \overline{BC}$. So, $PX = PY$ and $PX = PZ$.

A. How do you know that $PX = PY$ and $PX = PZ$?

A, B. See Additional Answers.

Using the Transitive Property of Equality, $PY = PZ$. This means that P lies on the angle bisector of $\angle B$.

B. How do you know that P lies the angle bisector of $\angle B$?

Because $PX = PY = PZ$, point P is equidistant from $\overline{AB}$, $\overline{BC}$, and $\overline{AC}$.

Use Properties of Angle Bisectors

A circle is inscribed in a triangle if each side of the triangle is tangent to the circle. In the figure, circle C is inscribed in $\triangle XYZ$. The center of the circle is the incenter of $\triangle XYZ$.

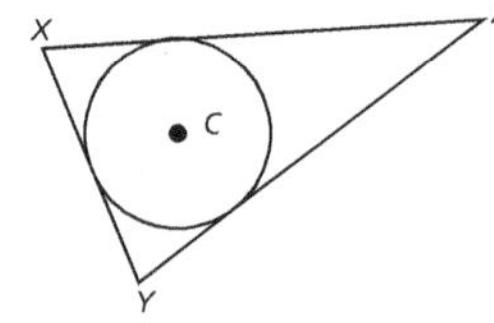

276

4 A circular tent will be set up for a graduation ceremony at a university. The tent will be located on a triangular section of land between a sculpture, a corner of the library building, and a gazebo.

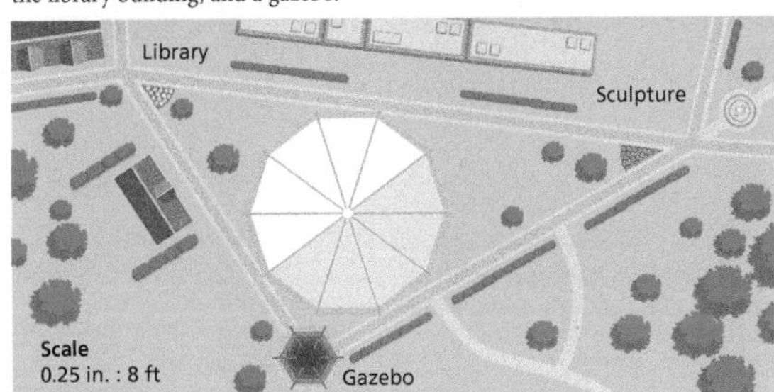

Reggie uses a scale drawing of the area to determine the largest circular tent that can be used. Reggie's steps are shown below. A, B. See Additional Answers.

Step 1 Draw the angle bisector of each angle.

A. How can you construct an angle bisector using a compass and a straightedge?

Step 2 Find the incenter of the triangle. Using a compass, construct an inscribed circle.

B. How is the radius of the circle determined?

Step 3 Use a ruler to measure the radius.

The radius is $\frac{5}{8}$, or 0.625, inch.

Step 4 Use the scale to determine the diameter of the largest possible circular tent that can be used.

$$\frac{0.25 \text{ in.}}{8 \text{ ft}} = \frac{0.625 \text{ in.}}{x \text{ ft}} \quad \rightarrow \quad x = 20$$

The largest circular tent that will fit has a radius of 20 feet.

Turn and Talk Is it possible to use only two angle bisectors to find the incenter? Explain. See margin.

5 Point Z is the incenter of $\triangle PQR$. What is $m\angle QPZ$?

Use the Triangle Sum Theorem to write an equation.

$m\angle QPR + m\angle PRQ + m\angle RQP = 180°$

$m\angle QPR + 2(13°) + 122° = 180°$

$m\angle QPR + 148° = 180°$

$m\angle QPR = 32°$

A, B. See Additional Answers.

A. How do you know that $m\angle PRQ = 2(13°)$?

The measure of $\angle QPZ$ is half the measure of $\angle QPR$: $m\angle QPZ = \frac{1}{2}(32°) = 16°$.

B. How do you know that the measure of $\angle QPZ$ is half the measure of $\angle QPR$?

Task 4 (MP) **Use Tools** Students use compasses and straightedges to locate the incenter of a triangle and determine its radius.

Sample Guided Discussion:

Q What is true about the points on an angle bisector? They are equidistant from the sides of the angle.

Q Why is the incenter the desired center in this example? The incenter is the desired center because it is equidistant from the three sides of the triangle, which is a condition of the given situation.

Q What does it mean to say that each side of a triangle is tangent to the incircle? Each side of the triangle is perpendicular to a radius drawn to the point of contact between the circle and the triangle.

Turn and Talk Point out that the point of intersection between any two angle bisectors in a triangle is equidistant from two pairs of adjacent sides yes; All three angle bisectors intersect at the same point, so you only need to find two to determine the point of intersection.

Task 5 (MP) **Attend to Precision** Students apply the theorems about angle bisectors and the incenter of a triangle to find the measures of angles in a triangle.

Sample Guided Discussion:

Q What is the relationship between the segment from point Z and side $\overline{PR}$ of the triangle? It is perpendicular to it.

Q Could you state that the measures of $\angle QPZ$, $\angle QRZ$, and $\angle PQR$ sum to 180°? no; These angles are not three angles of a triangle.

ONLINE

Assign the Digital On Your Own for

- built-in student supports
- Actionable Item Reports
- Standards Analysis Reports

On Your Own

Assignment Guide

The chart below indicates which problems in the On Your Own are associated with each task in the Learn Together. Assign daily homework for tasks completed.

Learn Together Tasks	On Your Own Problems
Task 1, p. 274	Problem 25
Task 2, p. 275	Problems 9–12
Task 3, p. 276	Problems 8, 21, and 26
Task 4, p. 277	Problems 13–16
Task 5, p. 277	Problems 17–20 and 22–24

Questioning Strategies

Problem 11 Point out the given information in this figure and ask students why the two triangles are congruent. Remind them that although the given information is ASS, which is not a congruence theorem, the two given sides are a leg and the hypotenuse in each triangle. Therefore, the triangles are congruent by the HL Triangle Congruence. Then ask why $\overrightarrow{BQ}$ must be the angle bisector of $\angle ABC$. The angles at vertex B are corresponding parts of congruent right triangles.

Check Understanding

1. $\overline{BD}$ bisects $\angle ABC$. What is the value of x? 53°

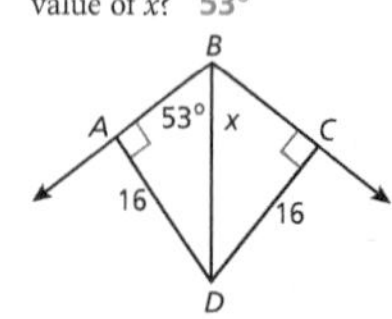

2. For what value of x does D lie on the bisector of $\angle ABC$? 27

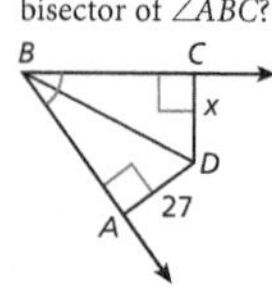

Point G is the incenter of $\triangle KLM$. Tell whether each statement can be determined from the given information.

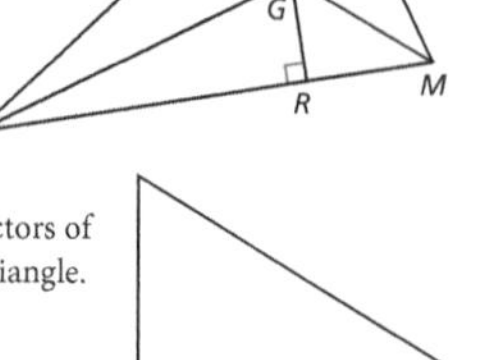

3. $\overline{PG} \cong \overline{QG}$ yes
4. $\overline{LG} \cong \overline{MG}$ no
5. $\angle QMG \cong \angle RMG$ yes
6. $\angle PGK \cong \angle LGQ$ no
7. Copy the triangle. Construct the angle bisectors of each angle. Then label the incenter of the triangle. See Additional Answers.

On Your Own

8. (MP) **Use Tools** Use a geometric drawing tool to draw a triangle. Construct angle bisectors of each angle.
   A. Move the vertices of the triangle. What is true about the angle bisectors? The angle bisectors are concurrent.
   B. Is it possible to move the vertices in such a way that the intersection of the angle bisectors is outside the triangle? Explain. no; The intersection of the angle bisectors is always in the interior of the triangle.

Can you determine that $AQ = CQ$? Explain your reasoning. 9–12. See Additional Answers.

9.

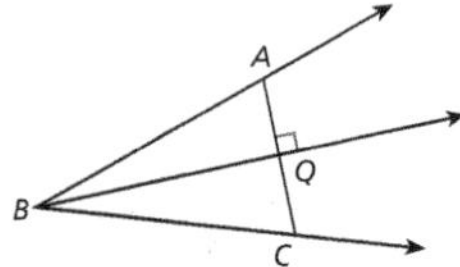

10.

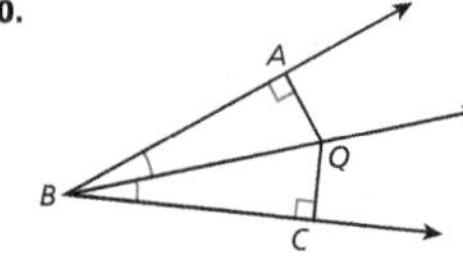

Can you determine that $\overrightarrow{BQ}$ bisects $\angle ABC$? Explain your reasoning.

11.

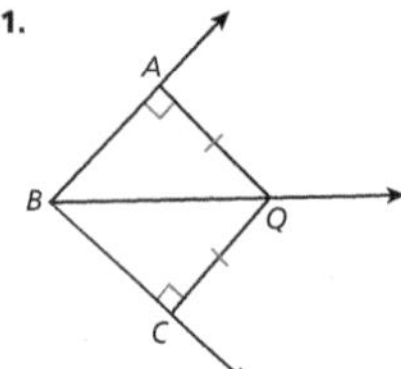

12.

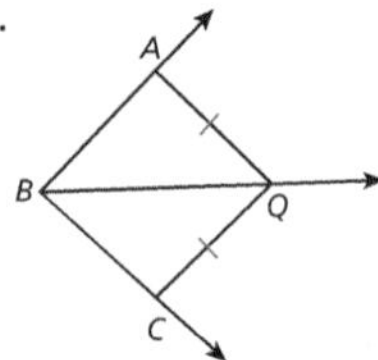

278

3 Check Understanding

Formative Assessment

Use formative assessment to determine if your students are successful with this lesson's learning objective.

Students who successfully complete the Check Understanding can continue to the On Your Own practice.

For students who miss 1 problem or more, work in a pulled small group using the Almost There small-group activity on page 273C.

ONLINE

Assign the Digital Check Understanding to determine

- success with the learning objective
- items to review
- grouping and differentiation resources

4 Differentiation Options

Differentiate instruction for all students using small-group activities and math center activities on page 273C.

Reteach

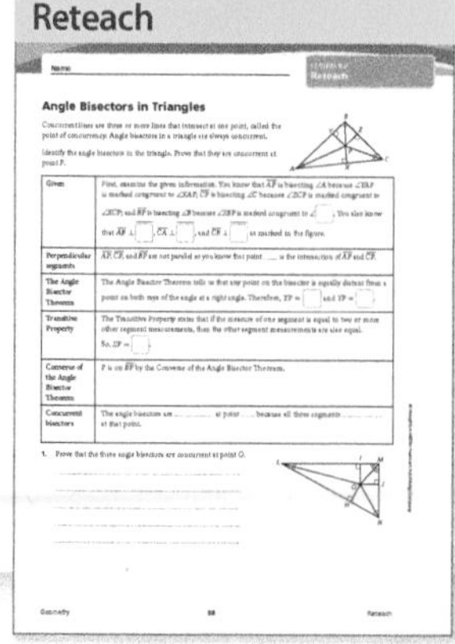

Challenge

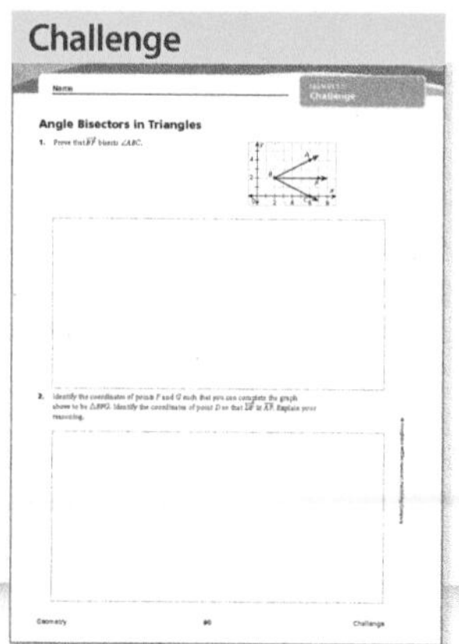

Copy the triangle. Then construct an inscribed circle for each triangle. 13, 14. See Additional Answers.

13.

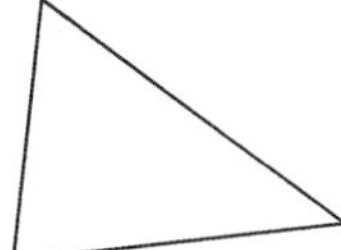

14.

15. Open Ended Draw a triangle. Find the incenter of the triangle. Inscribe a circle in the triangle. Answers will vary.

16. David is installing crown molding. The molding creates a triangular gap with the wall and the ceiling through which a circular tube, called conduit, will be run. The conduit will hold wires and cables. A, B. See Additional Answers.

A. Copy the triangle in the diagram, which shows a cross section of the triangular space created by the molding. Draw an inscribed circle to represent the largest conduit that can be run through this space.

B. Use a ruler and the dimensions given on the diagram to estimate the diameter of the largest conduit that will fit in the space.

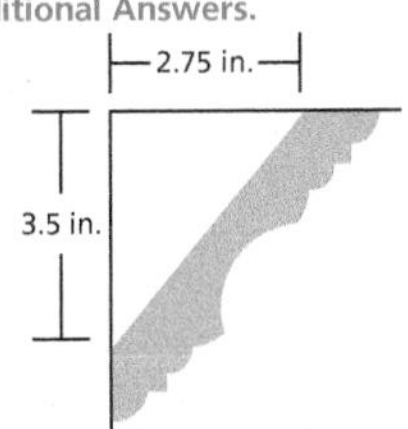

Find each measure.

17. $m\angle ABD$ 50°

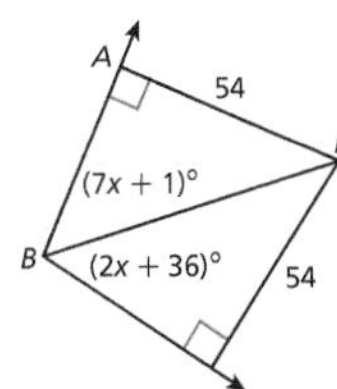

18. DC 19

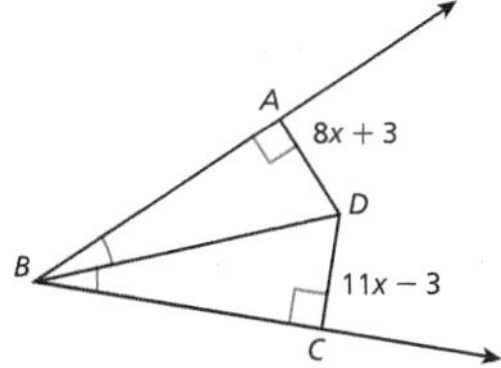

19. WZ 37

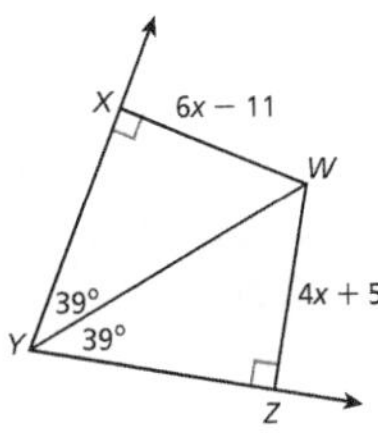

20. $m\angle XYW$ 26°

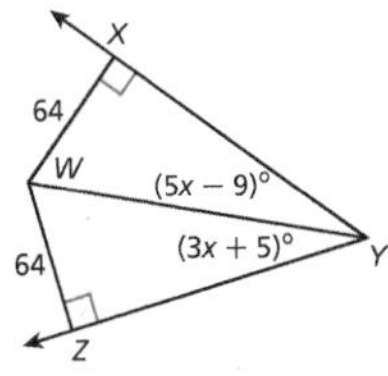

Questioning Strategies

Problem 15 Suggest that students first sketch two or three different types of triangles, such as acute, obtuse, and right triangles, and approximate the location of the incenter for each one. Then ask them if it looks like the incenter in each triangle is the same distance from each side or from each angle of the triangle? By inspection of the different types of triangles, the incenter is not always equidistant from the vertices of each triangle. Rather, it is equidistant from the sides. This distance can be found by drawing the perpendicular segment from the incenter to each side and verifying the answer by using a compass to check that the distances (radii) are equal.

Watch for Common Errors

Problem 20 Students may correctly determine that $\overline{YW}$ is the angle bisector of $\angle XWZ$, and that to find the measure of $\angle XYW$, they can set the two expressions equal and solve for x. However, the value of x they find is not the measure of the angle. Ask students who make this common error if it seems reasonable that the measure of the angle is 7°. Remind them of what they were asked to find and have them substitute 7 into one of the expressions to find the correct answer. Have them also verify that 36° is correct by using substitution to check both expressions.

5 Wrap-Up

Summarize learning with your class. Consider using the Exit Ticket, Put It in Writing, or I Can scale.

Exit Ticket 

In the figure below, circle S is inscribed in $\triangle ABC$ and radii $SP = SQ = SR$. What conclusion can you draw about $\overline{BS}$, $\overline{CS}$, and $\overline{AS}$? They bisect $\angle B$, $\angle C$, and $\angle A$, respectively.

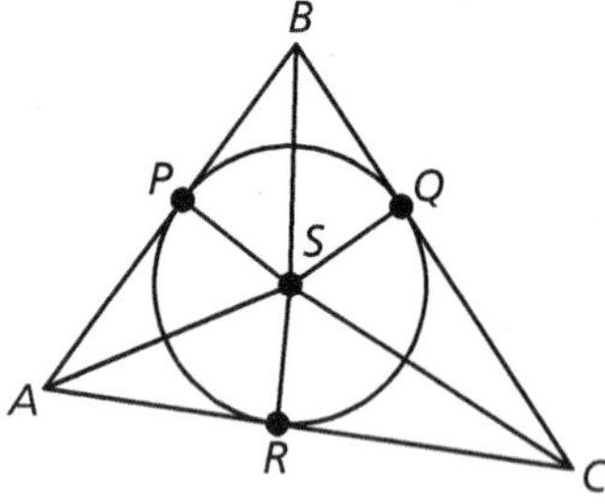

Put It in Writing 

Describe the difference between the incenter of a triangle and the circumcenter of the triangle and how to construct each one.

I Can 

The scale below can help you and your students understand their progress on a learning goal.

4	I can prove that angle bisectors are concurrent and inscribe circles in triangles, and I can explain my proof to others.
3	I can prove that angle bisectors are concurrent and inscribe circles in triangles.
2	I can prove that if a point lies on the bisector of an angle, the point is equidistant from the sides of the angle.
1	I can construct the angle bisector of an angle.

Spiral Review • Assessment Readiness

These questions will help determine if students have retained information taught in the past and can also prepare them for high-stakes assessments. Here, students must use algebra to find the measures of the remote interior angles of a triangle **(9.1)**, find the distances from the circumcenter of a triangle to two vertices **(9.2)**, find a midpoint **(1.4)**, and identify why two triangles are congruent **(8.1–4)**.

Eve is correct; The incenter of a triangle is always within the triangle.

21. (MP) **Critique Reasoning** P and Q are the incenter and circumcenter of the triangle, but not necessarily in that order. Michael states that the incenter of the triangle is P and the circumcenter is Q. Eve states that the incenter of the triangle is Q and the circumcenter is P. Who is correct? Explain your reasoning.

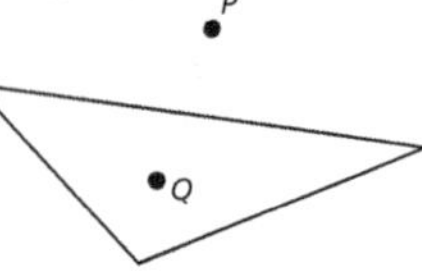

$\overline{AP}$ and $\overline{BP}$ are angle bisectors of $\triangle ABC$. Find each measure.

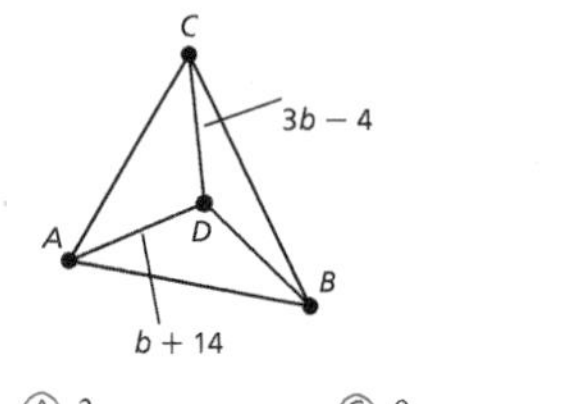

22. distance from P to $\overline{BC}$ 6.4

23. $m\angle PAC$ 47°

24. $m\angle QPB$ 61°

$\overline{AP}$ and $\overline{BP}$ are angle bisectors of $\triangle ABC$. Determine whether each statement is true or false.

25. P lies on the angle bisector of $\angle ACB$. true

26. $\overline{PQ}$ is a perpendicular bisector of $\overline{AB}$. false

A 24° P Q B 24° C

Spiral Review • Assessment Readiness

27. What is the measure of $\angle ABD$?

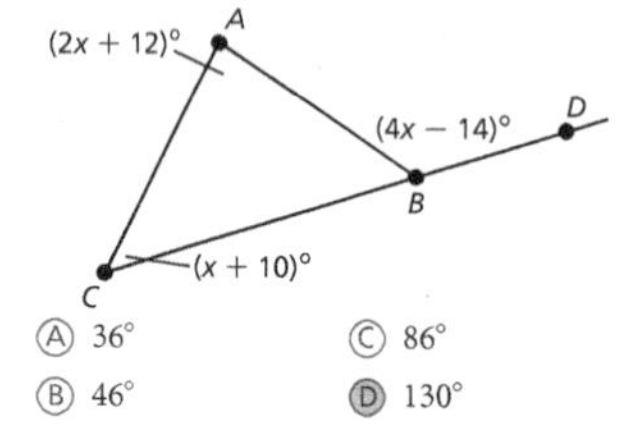

Ⓐ 36° Ⓑ 46° Ⓒ 86° Ⓓ 130°

28. For what value of t is Y the midpoint of $\overline{XZ}$?

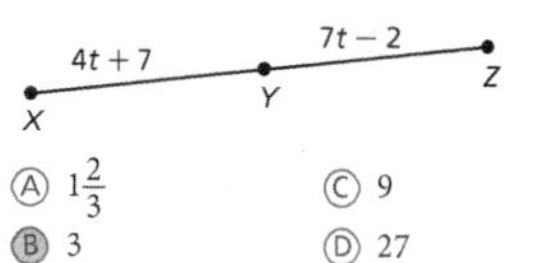

Ⓐ $1\frac{2}{3}$ Ⓑ 3 Ⓒ 9 Ⓓ 27

29. The circumcenter of $\triangle ABC$ is D. What is AD?

C 3b − 4 A D B b + 14

Ⓐ 2 Ⓑ 5 Ⓒ 9 Ⓓ 23

30. Which of the following set of statements does not allow you to conclude that $\triangle ABC \cong \triangle DEF$?

Ⓐ $\angle A \cong \angle D$, $\angle B \cong \angle E$, $\overline{BC} \cong \overline{EF}$

Ⓑ $\angle A \cong \angle D$, $\overline{AC} \cong \overline{DF}$, $\overline{AB} \cong \overline{DE}$

Ⓒ $\overline{AB} \cong \overline{DE}$, $\overline{BC} \cong \overline{EF}$, $\overline{AC} \cong \overline{DF}$

Ⓓ $\angle A \cong \angle D$, $\overline{AB} \cong \overline{DE}$, $\overline{BC} \cong \overline{EF}$

 **I'm in a Learning Mindset!**

How do my time-management skills impact decision-making?

Keep Going ▶ Journal and Practice Workbook

Learning Mindset

Perseverance Checks for Understanding

Start this lesson by checking to see if students recall the definition of an angle bisector and can explain its relationship to the sides of the angle. Then point out that the bisectors of the angles of a triangle always intersect in the interior of the triangle at a common point—a point of concurrency. This point is equidistant from each side of the triangle and is also the center of the incircle that is tangent to each side. *Once students have found the incenter of a triangle, ask them what they also need to find to draw the incircle. That is, how can they determine the distance from a point not on a line to the line? How can they be sure the circle drawn is the incircle? Also, have them compare the properties of the incenter and the properties of the circumcenter of a triangle. How are they alike? different? Does the location of either point depend on the type of triangle?*

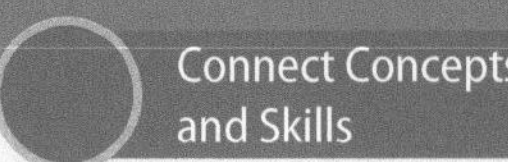

9.4 Medians and Altitudes in Triangles

LESSON FOCUS AND COHERENCE

Mathematics Standards

- Prove theorems about triangles. Theorems include: measures of interior angles of a triangle sum to 180°; base angles of isosceles triangles are congruent; the segment joining midpoints of two sides of a triangle is parallel to the third side and half the length; the medians of a triangle meet at a point.
- Construct the inscribed and circumscribed circles of a triangle, and prove properties of angles for a quadrilateral inscribed in a circle.

Mathematical Practices and Processes 

- Construct viable arguments and critique the reasoning of others.
- Use appropriate tools strategically.

I Can Objective 

I can construct medians and altitudes to find centroids and orthocenters.

Learning Objective

Students will construct medians and altitudes to find centroids and orthocenters.

Language Objective

Discuss the basics of centroids and orthocenters and explain how to find them.

Vocabulary

New: altitude of a triangle, centroid, median of a triangle, orthocenter

Lesson Materials: Straight edge

Mathematical Progressions

Prior Learning	Current Development	Future Connections
Students: • proved theorems about perpendicular bisectors, angle bisectors, medians, and altitudes of triangles. **(9.2 and 9.3)**	**Students:** • find the centroid of a triangle by drawing medians. • use the Centroid Theorem to find the location of a centroid. • draw altitudes of a triangle and use them to find the orthocenter.	**Students:** • will prove theorems about triangle inequalities. **(10.1 and 10.2)** • will prove theorems about quadrilaterals. **(11.1–11.5)**

PROFESSIONAL LEARNING

Using Mathematical Practices and Processes

Use appropriate tools strategically

This lesson provides an opportunity to address this Mathematical Practice Standard, which calls for students to "use appropriate tools." According to the standard, "proficient students are sufficiently familiar with tools appropriate for their grade or course to make sound decisions about when each of these tools might be helpful, recognizing both the insight to be gained and their limitations." This lesson introduces centroids, orthocenters, and the Centroid Theorem. Students learn to use the Centroid Theorem to find lengths of other medians when given algebraic expressions for their distance from the centroid. They also learn to find the orthocenter of a triangle and state whether that point lies inside, outside, or on the triangle.

WARM-UP OPTIONS

ONLINE

PROJECTABLE & PRINTABLE

ACTIVATE PRIOR KNOWLEDGE • Find Medians and Altitudes

Use these activities to quickly assess and activate prior knowledge as needed.

Problem of the Day 

A 10-foot ladder is leaning against a wall and the top of the ladder is 8 feet above the ground. Draw the triangle formed by the wall, the ladder, and the distance between the wall and the bottom of the ladder. Then find both the median and the altitude. What is the length of the altitude? What is the distance between the wall and the point at which the median reaches the ground? Using the Pythagorean Theorem, we find that the distance from the wall to the bottom of the ladder is 6 feet ($6^2 + 8^2 = 10^2$); So the altitude is 8 feet long (the distance from the wall to the top of the ladder), and the distance from the wall to the median is 3 feet.

Quick Check for Homework

As part of your daily routine, you may want to display the Teacher Solution Key to have students check their homework.

Make Connections

Based on students' responses to the Problem of the Day, choose one of the following:

1. Project the Interactive Reteach, Geometry, Lesson 9.4.
2. Complete the Prerequisite Skills Activity:

Have students in groups draw the triangle from the Problem of the Day on graph paper. Ask them to change the triangle from a right triangle to an acute triangle by changing the right angle to a 60° angle. For the original triangle, the angle formed by the ladder with the ground is approximately 53°, and the angle formed by the ladder with the wall is approximately 37°. Increasing the lesser angle to 67° yields a triangle formed by two right triangles next to each other, the one on the left being a 30°, 60°, 90° right triangle, with side lengths x, $\sqrt{3x}$, $2x$. Have students find the point where the median meets the base of the new triangle. (They don't need to know the ratio. The base is approximately 4.6 feet long, so the base of the new triangle is 10.6 feet. The median is at 5.3 feet.)

- *What is the altitude of the new triangle?* The altitude is still 8 feet because the distance from the top vertex to the base hasn't changed.
- *What can you say about how the altitude divides the triangle?* The altitude divides the triangle into two right triangles. The one on the right is the original triangle.
- *About how much longer is the base of the new triangle than the original triangle?* The base is about 4.6 feet longer.

If students continue to struggle, use Tier 2 Skill 8.

SHARPEN SKILLS

If time permits, use this on-level activity to build fluency and practice basic skills.

Quantitative Comparison

Objective: Students make a comparison between two quantities.
Materials: an index card with a triangle showing both the median and the altitude of the triangle

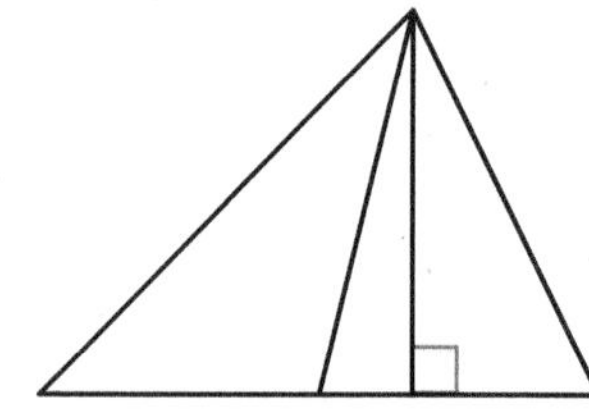

Have students say which line segment is longer, the median or the altitude.

A. The altitude is longer.
B. The median is longer.
C. The two are of equal length.
D. The relationship cannot be determined from the information given.

B; The median is longer.
The altitude is perpendicular to the base.

PLAN FOR DIFFERENTIATED INSTRUCTION

Small-Group Options

Use these teacher-guided activities with pulled small groups.

On Track

Materials: Straight edge

On graph paper, students should draw a square with side length 1. From the midpoint of the top side of the square, they should draw two lines to the bottom two vertices of the square. This will yield two right triangles and an isosceles triangle. Have students then draw the medians of the isosceles triangle and construct the centroid. Ask volunteers to report to the class.

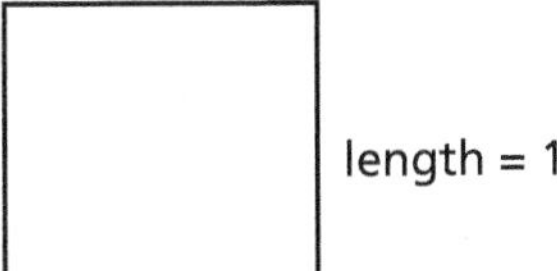

Almost There

Materials: Straight edge

Have students draw an equilateral triangle with a base measuring 6 units. The distance from the midpoint to the top vertex should measure approximately 5.2 units. The right and left legs should both measure 6 units.

Ask students to find the medians of the triangle and draw them. Have volunteers report to the class.

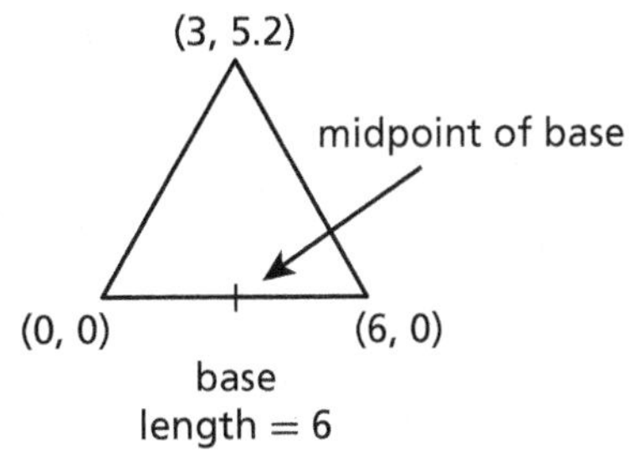

Ready for More

Have students draw an equilateral triangle with a base measuring 6 units. The distance from the midpoint to the top vertex should measure approximately 5.2 units. The right and left legs should both measure 6 units.

Ask students to draw the medians of the triangle and find the centroid. Then have them draw the altitudes and find the orthocenter. Ask volunteers to report to the class on what they have found.

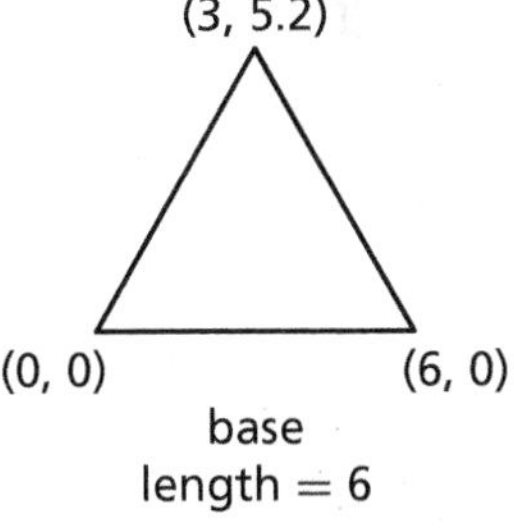

Math Center Options

Use these student self-directed activities at centers or stations.

Key: ● Print Resources ● Online Resources

On Track

- ● Interactive Digital Lesson
- ●● Journal and Practice Workbook
- ● Interactive Glossary (printable): **altitude of a triangle**, **centroid**, **median of a triangle**, **orthocenter**

Almost There

- ● Reteach 9.4 (printable)
- ● Interactive Reteach 9.4
- ● RtI Tier 2 Skill 8: Distance and Midpoint Formulas

Ready for More

- ● Challenge 9.4 (printable)
- ● Interactive Challenge 9.4

View data-driven grouping recommendations and assign differentiation resources.

During the *Spark Your Learning,* listen and watch for strategies students use. See samples of student work on this page.

Use Medians to Find the Centroid

Strategy 1

Find the balancing point of the triangle:
Draw a line segment from the vertex to the midpoint of the opposite leg.
Repeat the procedure with the next vertex, then the third.
See where the lines intersect. This point is the balancing point because the median divides the triangle into two triangles of equal area.
So, the point where the three medians intersect should be the balancing point for the triangle.

If students . . . draw medians to find the centroid, they are employing an efficient method and demonstrating an exemplary understanding of the concept of centroids and medians explained in this lesson.

Have these students . . . explain how they determined that the centroid was the balancing point. **Ask:**

- Q How did you decide to draw medians to find the balancing point?
- Q How did you determine that the intersection of the medians was the balancing point?

Use the Centroid Theorem

Strategy 2

I drew a median from one vertex to the midpoint of the opposite side of the triangle.
I knew that the balancing point must be two-thirds of the distance from the vertex to the midpoint, so I measured that distance. I marked that point as the balancing point. That point is where the sculpture would balance on the pole.

If students . . . use measurements to determine the distance from the vertex to the midpoint, they understand how to use the Centroid Theorem, but may not realize that the intersection of the medians is the centroid.

Activate prior knowledge . . . by having students talk about how to define the median. **Ask:**

- Q How do you find the medians of a triangle?
- Q How does the median divide a triangle?

COMMON ERROR: Uses Incenter

I used the incenter of the triangle because I thought that if a circle is inscribed in the triangle, the medians meet at the center of the incircle.

If students . . . use the incenter or any point other than the centroid as the balancing point, they may not understand that lines such as angle bisectors and altitudes do not divide a triangle into sections of equal area.

Then intervene . . . by pointing out that the incenter does not function as the balancing point of a triangle. **Ask:**

- Q How do you find the areas of two parts of a right triangle divided by the angle bisector?
- Q What type of line segment do you need to divide the triangle into two figures of equal area?

9.4

Medians and Altitudes in Triangles

I Can construct medians and altitudes to find centroids and orthocenters.

Spark Your Learning

A city planner is creating an architectural model that shows the redevelopment plans for an area of the city. The model needs to include a miniature version of the sculpture shown.

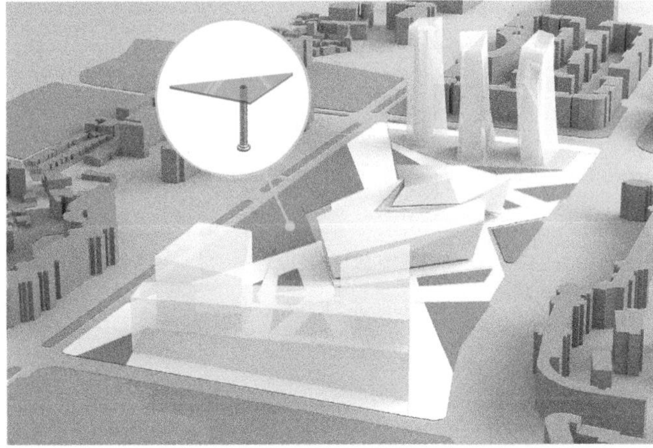

Complete Part A as a whole class. Then complete Parts B–D in small groups.

A. What is a mathematical question you can ask about this sculpture? A. Where should you place the triangle on the pole so that it balances?

B. To answer your question, what strategy and tool would you use along with all the information you have? What answer do you get? See Strategies 1 and 2 on the facing page.

C. Is it easier to balance the triangle on a pole that has a flat top like the one shown or on a pole that comes to a point like the tip of a pencil? Explain. C, D. See Additional Answers.

D. How do you think the shape of the triangle affects the location of the balance point of the triangle?

Turn and Talk Is it important that the triangle has an even thickness and density? See margin.

CULTIVATE CONVERSATION • Information Gap

Ask students questions to help them decide what missing information they need to answer the question "Where should you place the triangle on the pole so that it balances?"

1. Do you have enough information to conclude where to place the triangle on the pole? Explain. no; There are no coordinates or dimensions given for the triangle.
2. Suppose the triangle was not lying level with the ground. Would that affect the chances of balancing the triangle on the pole? If the triangle were not level with the ground, it would tip over unless it was bolted to the pole.
3. Can you determine how to find the balance point? What would you need? yes; I can determine how to find the balance point if I know the dimensions of the triangle.

1 Spark Your Learning

▶ MOTIVATE

- Have students look at the photo in their books and read the information contained in the photo. Then complete Part A as a whole-class discussion.
- Give the class the additional information they need to solve the problem. This information is available online as a printable and projectable page in the Teacher Resources.
- Have students work in small groups to complete Parts B–D.

▶ PERSEVERE

If students need support, guide them by asking:

Q **Advancing • Use Tools** Which tool could you use to solve the problem? Why choose that tool and not some other? Students' choices of tools and reasons for choosing them will vary.

Q **Assessing** To hold up a triangular structure such as the one in the photo, which part would be the safest to attach to a pillar? The safest place would be the point at which the weight of the structure is distributed equally.

Q **Advancing** What factors could contribute to success in balancing the triangle shown in the photo? the weight of the triangle, the size of the pillar, the placement of the pillar, the type of triangle, and its design are a few of the factors

Turn and Talk Help students think about how differences of density can affect the distribution of weight of an object. Possible answer: Yes, it is important that the triangle shape be of uniform thickness and density. If it is not, it will not balance at the theoretical balance point.

▶ BUILD SHARED UNDERSTANDING

Select groups of students who used various strategies and tools to share with the class how they solved the problem. As they present their solutions, have each group discuss why they chose a specific strategy and tool.

② Learn Together

Build Understanding

Task 1 **(MP) Construct Arguments** Explain that students will use a geometric drawing tool to draw a triangle and then draw lines from its vertices to the midpoints of the opposite legs of the triangle. By using the tool to move the vertices of the triangle, students will see how the medians shift.

CONNECT TO VOCABULARY

Have students use the **Interactive Glossary** to record their understanding of the vocabulary in this task.

Sample Guided Discussion:

Q **How does the distance of the centroid from each vertex change when the vertex is moved?** It remains two-thirds of the distance from the vertex to the midpoint of the opposite side.

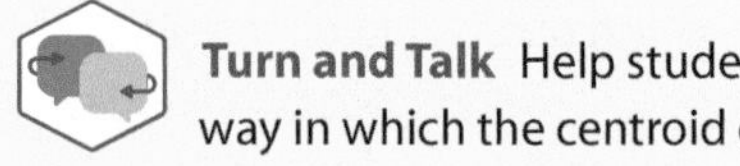

Turn and Talk Help students understand the way in which the centroid of a triangle is related to the three major triangles formed by medians. Possible answer: I think triangles *MPN, MPL,* and *LPN* will have the same areas. I could measure the base and height of each triangle to calculate the area; The centroid, incenter, and circumcenter are generally three different points. They will only coincide if the triangle is equilateral.

Build Understanding

Medians and the Centroid Theorem

A **median of a triangle** is a segment whose endpoints are a vertex of the triangle and the midpoint of the opposite side.

Every triangle has three distinct medians. The medians of a triangle are concurrent. The point of concurrency is called the **centroid** of the triangle.

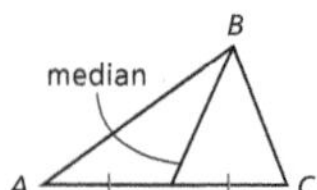

Centroid Theorem

The centroid of a triangle is located $\frac{2}{3}$ of the distance from each vertex to the midpoint of the opposite side.

$AP = \frac{2}{3}AX$ $\quad BP = \frac{2}{3}BY$ $\quad CP = \frac{2}{3}CZ$

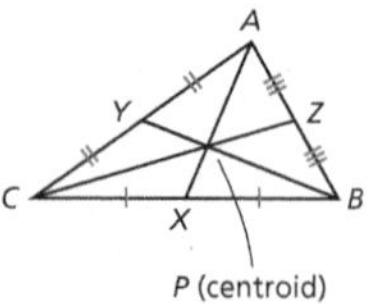

1 Use a geometric drawing tool to draw a triangle. Construct the midpoint of each side of the triangle. Then construct a line segment from each vertex to the midpoint of the opposite side.

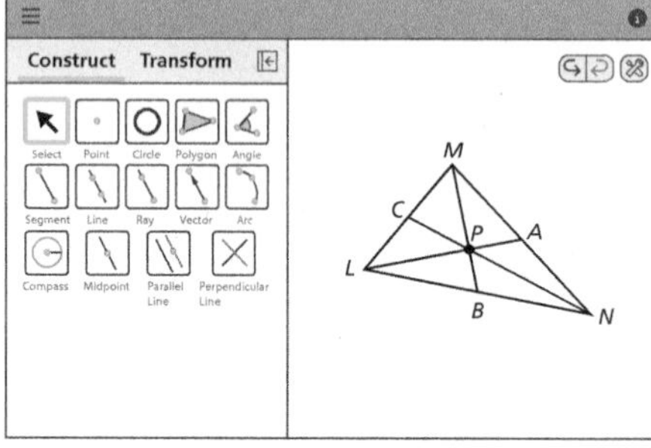

A. Move the vertices of the triangle. What is true about the medians? They appear to intersect at a point.

B. Move the vertices to make acute, obtuse, and right triangles. Describe the location of the centroid in relation to the triangle. The centroid is always inside the triangle.

C. Measure the distance from the centroid to the vertex and the centroid to the midpoint for one segment. What relationship do the lengths have? Is this true for each of the segments? The distance from the vertex to the centroid is twice the distance from the centroid to the midpoint; yes

Turn and Talk

- What do you think might be true about the area of the triangles *MPN, MPL,* and *LPN*? What information would you need to justify this conjecture? See margin.
- How does the centroid of a triangle compare to the circumcenter and incenter?

282

LEVELED QUESTIONS

Depth of Knowledge (DOK)	Leveled Questions	What Does This Tell You?
Level 1 **Recall**	Where is the centroid of a triangle located? The centroid is located at the intersection of the triangle's three medians.	Students' answers will indicate their understanding of what a centroid is.
Level 2 **Basic Application of Skills & Concepts**	If two triangles are similar, what does that say about the distances from the corresponding vertices and the centroids of the respective triangles? The distances would be proportional.	Students' answers will indicate their ability to apply the concepts of medians and centroids to the concept of similar triangles.
Level 3 **Strategic Thinking & Complex Reasoning**	Using the Centroid Theorem, how could you find the third vertex if you were given the centroid and the two other vertices? I could draw the leg between the first two vertices, draw a line from the midpoint of that leg to the centroid, and measure the distance from the centroid to the midpoint of that leg. Then I would multiply that distance by two to find the location of the third vertex.	Students' answers will indicate their ability to use the Centroid Theorem to think strategically about solving a problem involving the centroid and medians of a triangle.

Altitudes and the Orthocenter

An **altitude of a triangle** is a perpendicular segment from a vertex to the opposite side or to a line that contains the opposite side. An altitude can be inside, outside, or on the triangle.

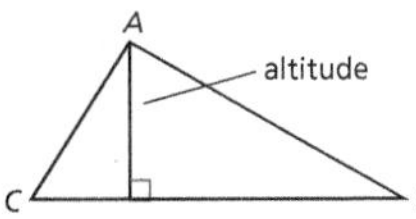

Every triangle has three altitudes. The altitudes of a triangle are concurrent. The point of concurrency is called the **orthocenter** of the triangle. The orthocenter can be inside, outside, or on the triangle.

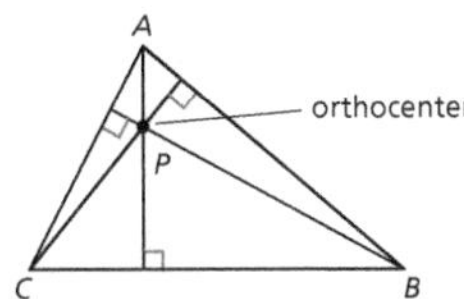

2 Copy $\triangle JKL$. Construct an altitude at each vertex. In some cases, you may have to extend a side of the triangle to construct the altitude, as shown below.

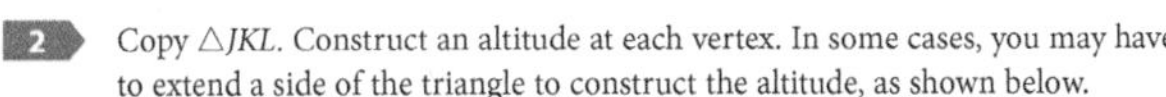

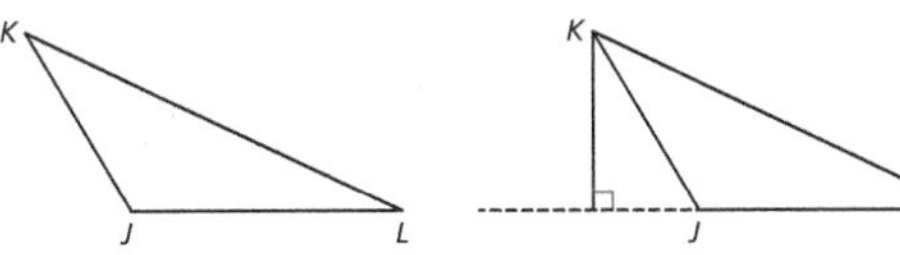

A. Do you think the orthocenter of $\triangle JKL$ will be inside, outside, or on the triangle? outside

B. Extend the altitudes and locate the orthocenter of the triangle. Where is the orthocenter located? The orthocenter lies outside of the triangle.

C. Construct additional triangles and orthocenters. Match the type of triangle on the left with the location of its orthocenter on the right.

Type of Triangle	Location of Orthocenter
A. Acute	**1.** Outside the triangle
B. Right	**2.** Inside the triangle
C. Obtuse	**3.** On the triangle

A. 2, B. 3, C. 1

Turn and Talk Suppose the orthocenter of $\triangle PQR$ occurs at a vertex of the triangle. What type of triangle is $\triangle PQR$? See margin.

Task 2 (MP) **Construct Arguments** Students will draw altitudes for $\triangle JKL$ in order to find the orthocenter. Students will need to think carefully about whether the orthocenter is located inside, outside, or on the triangle.

OPTIMIZE OUTPUT Critique, Correct, and Clarify

Have students work with a partner. Ask them to copy $\triangle JKL$ onto graph paper and find the location of its centroid. Then have them measure the distance from each vertex to the centroid to verify the Centroid Theorem. Ask volunteers to report to the class.

CONNECT TO VOCABULARY

Have students use the **Interactive Glossary** to record their understanding of the vocabulary in this task.

Sample Guided Discussion:

Q What connection do you see between the triangle in Task 2 and the location of the orthocenter? The triangle is obtuse, and the orthocenter is located outside the triangle. An obtuse triangle has its orthocenter outside the triangle.

Q Can you generalize between the type of triangle and the location of the orthocenter? An acute triangle has its orthocenter inside the triangle. A right triangle has its orthocenter on the triangle.

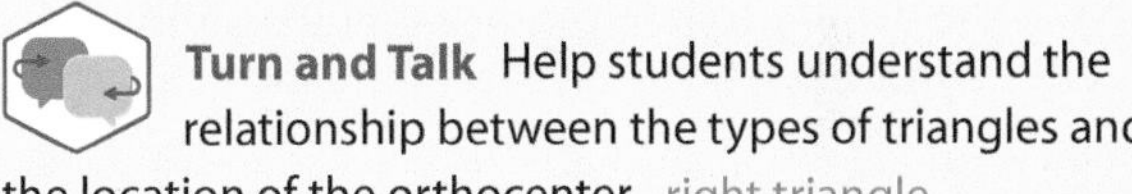

Turn and Talk Help students understand the relationship between the types of triangles and the location of the orthocenter. right triangle

PROFICIENCY LEVEL

Beginning

Ask students where medians intersect. Elicit "at the centroid." Ask where altitudes are concurrent. Elicit "at the orthocenter."

Intermediate

Have students work in pairs. Ask them to use graph paper to draw a right triangle with legs measuring 6, 8, and 10 units. Have them label the centroid and the orthocenter, with one sentence for each explaining where it is located within or on the triangle.

Advanced

Have students work in small groups to write a short paragraph describing the differences in location of the orthocenter of acute, obtuse, and right triangles.

Step It Out

Task 3 **MP** **Use Tools** Encourage students to use all tools at their disposal to accomplish this task.

Sample Guided Discussion:

Q How do you locate the centroid of a triangle? Draw a line from each vertex to the midpoint of the opposite leg of the triangle. The point of intersection of the three lines is the centroid.

Q Why is the centroid important? The centroid represents the center of gravity of an object. The center of gravity of the bike affects how it rides.

Turn and Talk Remind students that adjusting a triangle's vertices will change the position of its centroid. Possible answer: If $\angle DBE$ is smaller and $\angle BDF$ is larger, then $\overline{BC}$ and $\overline{DC}$ will angle down more. In doing so, the intersection of the medians will lower as well.

Task 4 **MP** **Use Tools** Have students use rulers, graph paper, and other tools to complete this task. Explain that they will have to graph triangles using the precise coordinates from the book.

Sample Guided Discussion:

Q What is the relationship between a triangle's altitudes and the orthocenter? The orthocenter is the point at which the altitudes are concurrent.

Q How do you find the slope of a line (or line segment)? The slope is rise over run.

Step It Out

Find the Center of Gravity

3 A bicycle frame consists of two triangles. Describe the location of the center of gravity of each triangle given that $AH = BG = 13.8$ in. and $BF = DE = 18$ in.

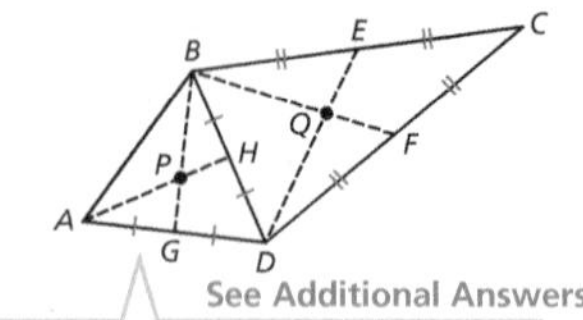

If an object has a consistent density, then its center of gravity is its centroid.

Since point P is the centroid of $\triangle ABD$,
$AP = \frac{2}{3}AH = \frac{2}{3}(13.8) = 9.2$ inches

Since point Q is the centroid of $\triangle BCD$,
$BQ = \frac{2}{3}BF = \frac{2}{3}(18) = 12$ inches

So, the center of gravity of $\triangle ABD$ is 9.2 inches from A and B, and the center of gravity of $\triangle BCD$ is 12 inches from B and D.

See Additional Answers.

Explain why P is the centroid of $\triangle ABD$, and why Q is the centroid $\triangle BCD$.

Turn and Talk Describe how you could modify the frame to lower its center of gravity to increase its stability. See margin.

Find the Orthocenter of a Triangle on the Coordinate Plane

4 Find the orthocenter of $\triangle PQR$ with vertices $P(1, 3)$, $Q(7, 9)$, and $R(9, 3)$.

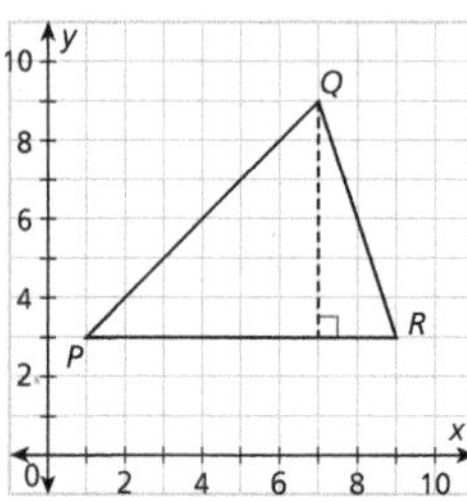

Find the altitude from vertex Q.

Since the side opposite vertex Q is horizontal, the altitude will be vertical.

The altitude is a segment on the vertical line that passes through $(7, 9)$.

Find the altitude from vertex P.

The slope of the side opposite vertex P is -3.

A, B. See Additional Answers.

A. How do you know the slope is -3?

The slope of the line perpendicular to $\overline{QR}$ is $\frac{1}{3}$. Use this information to draw the altitude from P.

B. How do you use this information to draw the altitude?

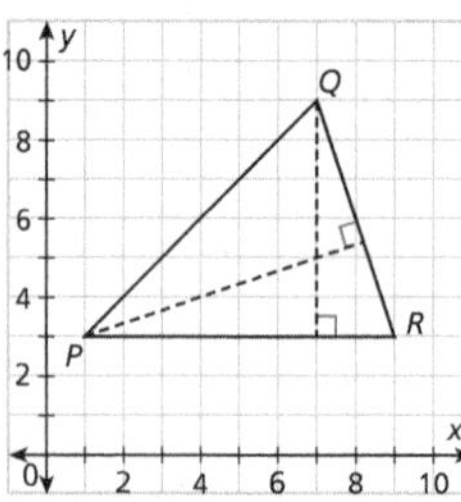

Locate the orthocenter.

The two altitudes intersect at $(7, 5)$.

284

Check Understanding

1. Copy the triangle. Construct the medians and identify the centroid of the triangle.

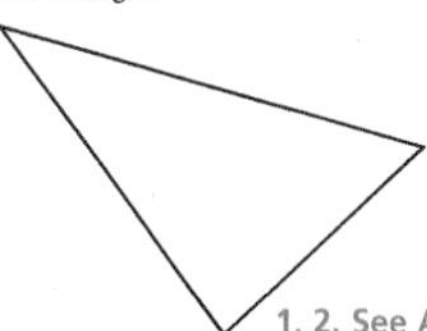

2. Copy the triangle. Construct the altitudes and identify the orthocenter of the triangle.

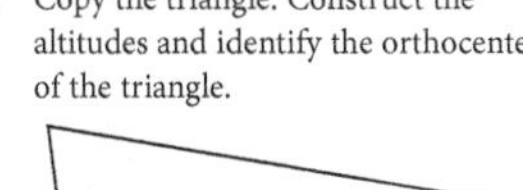

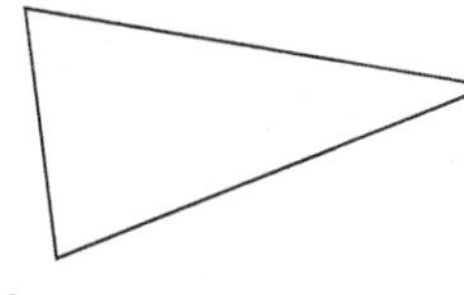

1, 2. See Additional Answers.

3. Suppose the orthocenter of $\triangle PQR$ occurs outside of the triangle. What type of triangle is $\triangle PQR$? obtuse triangle

4. Is it possible for the centroid and the orthocenter a triangle to be the same point? Explain. yes; In an equilateral triangle, the medians intersect in the same point as the altitudes.

5. In $\triangle XYZ$, medians $\overline{XM}$ and $\overline{YN}$ intersect at K. If $XM = 21$, what is XK? 14

6. On a coordinate plane, $\triangle ABC$ has vertices $A(0, 5)$, $B(10, 0)$, and $C(0, 0)$. What are the coordinates of the orthocenter of the triangle? (0, 0)

On Your Own

Copy each triangle and find its centroid.

7.

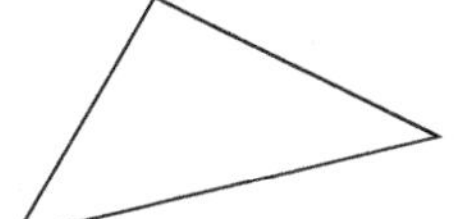

8.

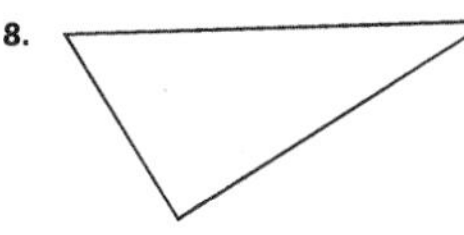

7–9. See Additional Answers.

9. **STEM** In a cable-stayed bridge, cables are used to help support the weight of the bridge. The cables form triangles that are symmetric to each pylon. In the diagram, $\triangle ABC$, $\triangle ADF$, and $\triangle AGH$ are all symmetric about $\overleftrightarrow{AZ}$. Use the diagram to explain why the triangles formed by the cables all have the same centroid.

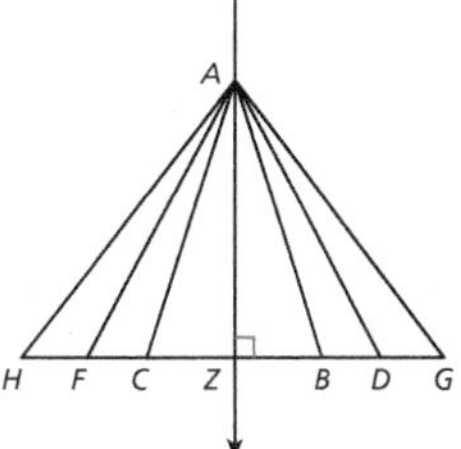

ONLINE

Assign the Digital On Your Own for

- built-in student supports
- Actionable Item Reports
- Standards Analysis Reports

On Your Own

Assignment Guide

The chart below indicates which problems in the On Your Own are associated with each task in the Learn Together. Assign daily homework for tasks completed.

Learn Together Tasks	On Your Own Problems
Task 1, p. 282	Problems 7–11 and 17–40
Task 2, p. 283	Problems 13–16 and 49
Task 3, p. 284	Problem 12
Task 4, p. 284	Problems 41–48

3 Check Understanding

Formative Assessment

Use formative assessment to determine if your students are successful with this lesson's learning objective.

Students who successfully complete the Check Understanding can continue to the On Your Own practice.

For students who miss 1 problem or more, work in a pulled small group using the Almost There small-group activity on page 281C.

ONLINE

Assign the Digital Check Understanding to determine

- success with the learning objective
- items to review
- grouping and differentiation resources

4 Differentiation Options

Differentiate instruction for all students using small-group activities and math center activities on page 281C.

Reteach

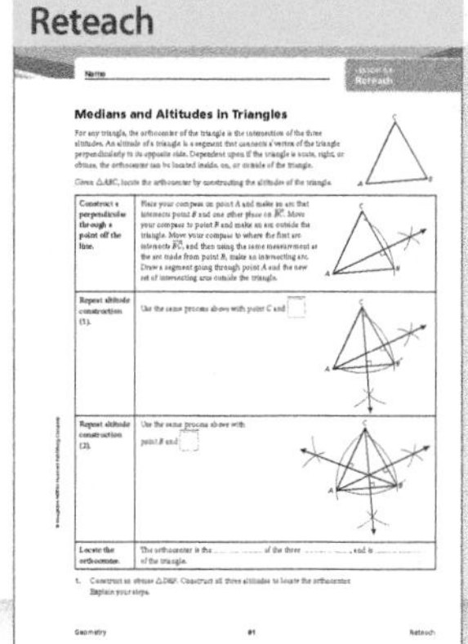

Challenge

Medians and Altitudes in Triangles

Questioning Strategies

Problem 10 Suppose vertex *B* of $\triangle ABC$ was located at (4, 6). Where would the centroid be located? Would your answer fit the Centroid Theorem? The centroid would be located at (4, 4); This fits the Centroid Theorem because the distance from the vertex to the centroid is twice the distance from the centroid to the midpoint of the opposite leg.

10. MP **Reason** The centroid of $\triangle ABC$ is located at (4, 5).

A. Choose one vertex and find the coordinates of the endpoints of the median from that vertex.

B. Show that the Centroid Theorem is true for the median.

10–12. See Additional Answers.

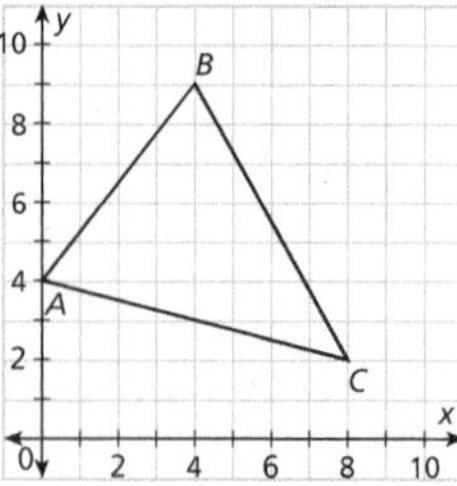

11. MP **Critique Reasoning** Joe draws a triangle and claims that the centroid of his triangle is outside of the triangle. Do you think this is possible? Explain your reasoning.

12. MP **Construct Arguments** Another term for the centroid of a triangle is the balancing point of a triangle. Why do you think this phrase is used to describe the centroid?

Copy each triangle and find its orthocenter. State whether the orthocenter is *inside*, *outside*, or *on* the triangle.

13–16. See Additional Answers.

13.

14.

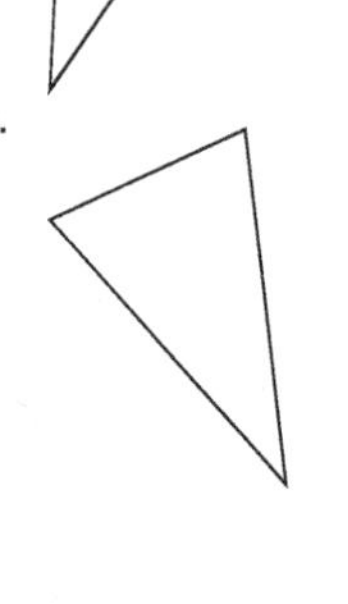

15.

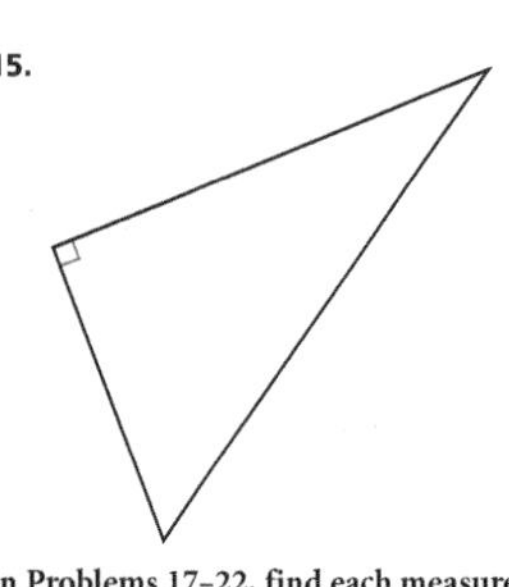

16.

In Problems 17–22, find each measure.

17. AG if $AE = 15$ 10

18. GF if $BF = 24$ 8

19. CD if $CG = 34$ 51

20. GE if $AG = 54$ 27

21. BF if $GF = 12$ 36

22. CG if $DG = 9$ 18

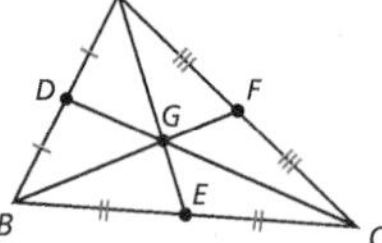

23. MP **Reason** If you are given the distance from a vertex of a triangle to the centroid, can you determine the length of that median? Explain. See Additional Answers.

286

Point G is the centroid of $\triangle ABC$. Find the value of the variable.

24.

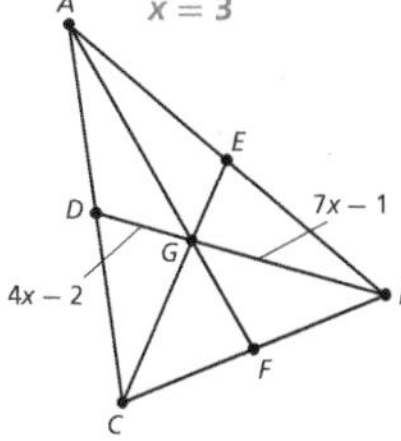

25.

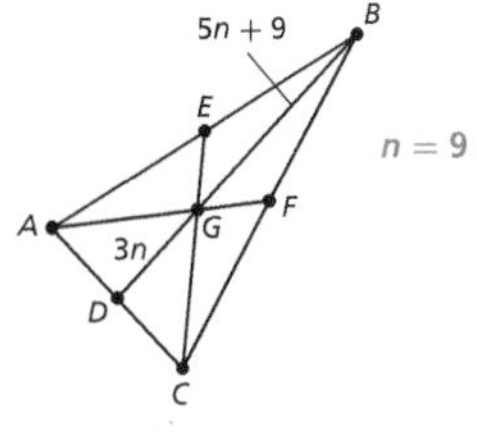

26.

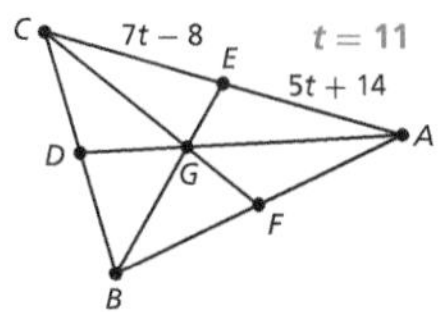

27.

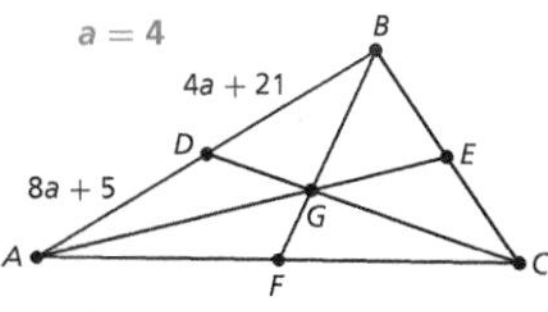

28.

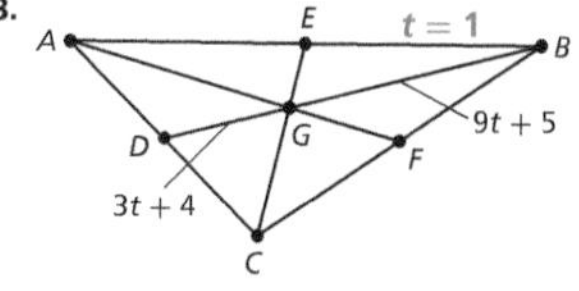

29.

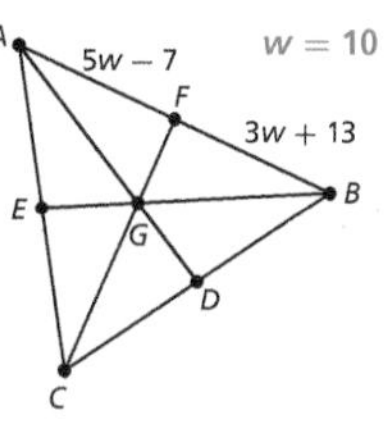

A map of a campground is drawn on a coordinate plane. The coordinates of three cabins are $A(0, 4)$, $B(10, 1)$, and $C(2, -5)$. Find the coordinates of each feature of the campground.

30. the main lodge, which is the midpoint of $\overline{AB}$ (5, 2.5)

31. the fire pit, which is the midpoint of $\overline{AC}$ (1, −0.5)

32. the archery range, which is the midpoint of $\overline{BC}$ (6, −2)

33. the centroid of $\triangle ABC$ (4, 0)

34. Open Ended Draw a triangle on a coordinate plane. Use the medians of the triangle to find the coordinates of the centroid of the triangle. Answers will vary.

Find the centroid of each triangle with the given vertices.

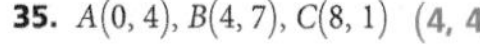

35. $A(0, 4)$, $B(4, 7)$, $C(8, 1)$ (4, 4)

36. $A(-2, 2)$, $B(0, -2)$, $C(5, 0)$ (1, 0)

37. $A(-3, -3)$, $B(4, -4)$, $C(5, 4)$ (2, −1)

38. $A(-5, -4)$, $B(-2, -1)$, $C(-5, -1)$ (−4, −2)

39. $A(-6, -7)$, $B(6, 2)$, $C(0, 5)$ (0, 0)

40. $A(2, 4)$, $B(7, 2)$, $C(3, -3)$ (4, 1)

Watch for Common Errors

Problem 34 Because of confusion around the term centroid, students might tend to think of the distance between the vertex and the centroid as half the distance between the vertex and the midpoint of the opposite side. Explain that the centroid is different from the center of the incircle.

(5) Wrap-Up

Summarize learning with your class. Consider using the Exit Ticket, Put It in Writing, or I Can scale.

Exit Ticket 

Suppose $\triangle ABC$ and $\triangle DEF$ share a centroid, point P. Are the two triangles necessarily similar? What if $\triangle ABC$ is equilateral? Explain. no; Point P, being the intersection of three medians of $\triangle ABC$, for example, could be the intersection of an infinite number of medians, for an infinite number of triangles. The two triangles are not necessarily similar.

Put It in Writing 

Have students write a paragraph describing the differences between the orthocenter and the centroid of any given triangle.

I Can 

The scale below can help you and your students understand their progress on a learning goal.

4	I can construct medians and altitudes to find centroids and orthocenters, and I can explain the procedure to my classmates.
3	I can construct medians and altitudes to find centroids and orthocenters.
2	I can find the orthocenter of a triangle drawn on a coordinate grid.
1	Given two points, I can distinguish between the orthocenter and the centroid of a triangle.

Spiral Review • Assessment Readiness

These questions will help determine if students have retained information taught in the past and can also prepare them for high-stakes assessments. Here, students must find the measure of an angle from algebraic expressions for the other two angles of a triangle **(9.1)**, determine the midpoint of a line segment on a coordinate grid **(9.2)**, and distinguish between an incenter and a circumcenter **(9.3)**.

Find the orthocenter of each triangle with the given vertices.

41. $A(11, 0), B(7, 4), C(5, -2)$ (8, 1)

42. $A(12, 3), B(7, -2), C(1, 3)$ (7, −3)

43. $A(-2, -1), B(2, 3), C(-4, 5)$ (−1, 2)

44. $A(-1, 1), B(3, 6), C(3, 1)$ (3, 1)

45. $A(-1, 0), B(6, 7), C(4, 0)$ (6, −2)

46. $A(5, 6), B(5, -2), C(-2, -1)$ (4, −1)

47. Sergio is creating a pattern that he is going to use to cut out triangular pieces of wood. He draws his pattern on a coordinate plane. What are the coordinates of the orthocenter?
(6, 4)

48. **Open Ended** Draw a triangle on a coordinate plane with an orthocenter that is is outside of the triangle. Draw a triangle on a coordinate plane with an orthocenter that is is on the triangle.
See Additional Answers.

49. (Open Middle™) Using the digits 1 to 9, at most one time each, replace the boxes to create two triangles. One triangle should have an altitude x that is less than 5, and the other triangle should have an altitude y that is greater than 5. Possible answer: 2, 3, 4 and 7, 8, 9

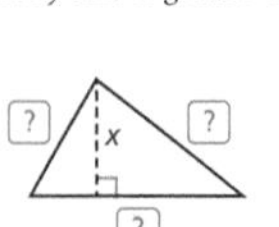
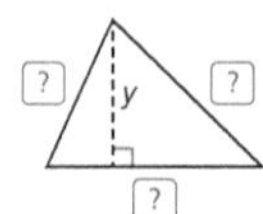

Spiral Review • Assessment Readiness

50. What is the measure of $\angle B$?

Ⓐ 6°
Ⓑ 34°
Ⓒ 73°
Ⓓ 146°

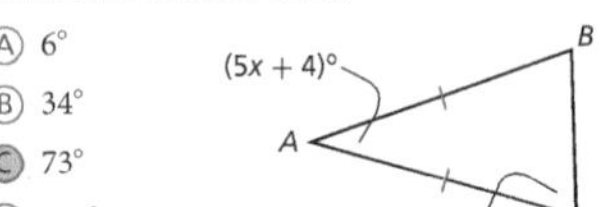

51. What is the midpoint M of $\overline{JK}$?

Ⓐ $M(0, 2)$
Ⓑ $M(1, 2)$
Ⓒ $M(2, 1)$
Ⓓ $M(2, 1.5)$

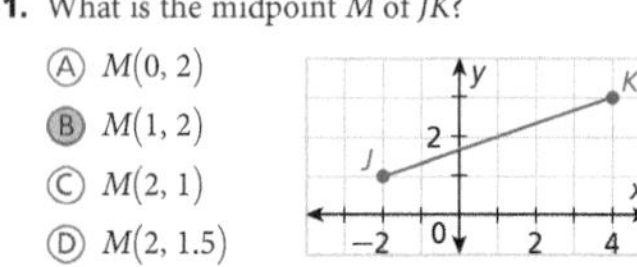

52. Determine whether the statement describes the incenter or circumcenter of a triangle.

Statement	Incenter	Circumcenter
A. This point is equidistant from the sides of the triangle.	?	?
B. This point is equidistant from the vertices of the triangle.	?	?

 **I'm in a Learning Mindset!**

How will I know when my goal is met successfully?

288 Keep Going Journal and Practice Workbook

Learning Mindset

Challenge-seeking Makes Plans to Meet Goals

Once students understand the concepts of the centroid and the orthocenter of a triangle, they may seek new challenges. Given that they can now find the centroid and the orthocenter of a triangle, students may want to find out other important facts about triangles. The important question is how they set goals for themselves and make plans to meet those goals. Ask questions such as the following: *How do you go about building on the concepts and ideas that interest you most? Do you think about specific points to explore? Do you make outlines of the ideas you have learned? Do you organize what you have learned in special notebooks? What are the strategies you have for setting and meeting goals to learn new things and conquer new challenges?*

9.5 The Triangle Midsegment Theorem

LESSON FOCUS AND COHERENCE

Mathematics Standards

- Prove theorems about triangles. Theorems include: measures of interior angles of a triangle sum to 180°; base angles of isosceles triangles are congruent; the segment joining midpoints of two sides of a triangle is parallel to the third side and half the length; the medians of a triangle meet at a point.

Mathematical Practices and Processes 

- Use appropriate tools strategically.
- Attend to precision.
- Construct viable arguments and critique the reasoning of others.

I Can Objective 

I can construct midsegments and prove the Triangle Midsegment Theorem.

Learning Objective

Construct midsegments of a given triangle, prove the Triangle Midsegment Theorem, and apply the theorem to solve for segment lengths and angle measures.

Language Objective

Explain how we can use the Triangle Midsegment Theorem to find the coordinates of the endpoints of a midsegment of a triangle from the coordinates of the vertices of the triangle.

Vocabulary

New: midsegment of a triangle

Lesson Materials: compass, ruler, protractor, straightedge

Mathematical Progressions

Prior Learning	Current Development	Future Connections
Students: • proved theorems about perpendicular bisectors, angle bisectors, medians, and altitudes of triangles. **(9.1–9.4)**	**Students:** • construct midsegments of a triangle. • apply the Triangle Midsegment Theorem to solve for segment lengths and angle measures from given measures. • apply the Triangle Midsegment Theorem to triangles on a coordinate plane.	**Students:** • will prove theorems about triangle inequalities. **(10.1 and 10.2)** • will prove theorems about quadrilaterals. **(11.1–11.5)**

PROFESSIONAL LEARNING

Math Background

An important part of this lesson is understanding the Triangle Midsegment Theorem when it is applied to or verified with a triangle on a coordinate plane. For example, in Task 2 students verify that the midsegment is parallel to a side of the triangle by calculating and comparing their slopes and show with the Distance Formula that the midsegment has length half that of the parallel side of the triangle. In doing this, students connect geometric concepts with fundamental algebraic concepts they learned about in previous classes. The concepts of slope and distance between points come up repeatedly not only in this course, but also in Algebra 2 and beyond. Students can also define midsegments of triangles on coordinate planes by finding the midpoints of sides with the Midpoint Formula. When students engage with these fundamental topics in multiple ways, it helps to deepen their understanding.

WARM-UP OPTIONS

ONLINE

PROJECTABLE & PRINTABLE 

ACTIVATE PRIOR KNOWLEDGE • Locate and Relate Medians and Centroids of Triangles

Use these activities to quickly assess and activate prior knowledge as needed.

Problem of the Day 

Cut out a triangle from a piece of paper and find its centroid by folding the triangle. A midpoint of a side of the triangle can be located by placing one vertex over another and making a fold on the edge connecting those vertices. Then, a median can be constructed by folding or drawing a segment from the midpoint to the opposite vertex. The other two medians can be similarly located. Their intersection is the centroid.

Quick Check for Homework

As part of your daily routine, you may want to display the Teacher Solution Key to have students check their homework.

Make Connections

Based on students' responses to the Problem of the Day, choose one of the following:

1. Project the Interactive Reteach, Geometry, Lesson 9.4.
2. Complete the Prerequisite Skills Activity:

Have students work in pairs. Show the class a triangle like the one below and ask them to find all missing segment lengths.

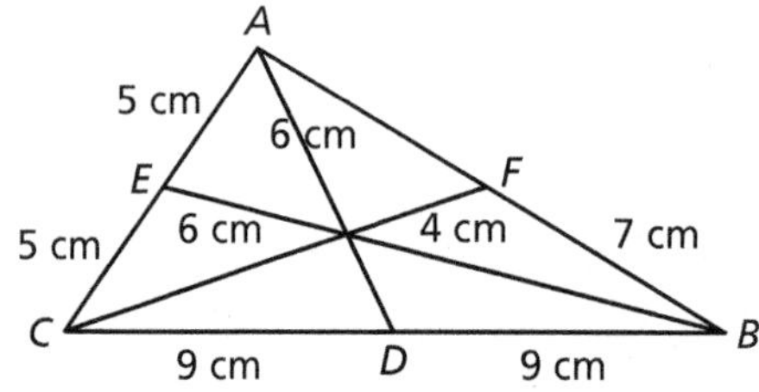

- *How do we define a median of a triangle?* The median is a segment with one endpoint at the midpoint of a side and the other endpoint at the vertex opposite that side.
- *Do the medians of a triangle always intersect?* Yes, their point of concurrence is called the centroid of the triangle.
- *Where on a median is the centroid located?* The centroid is $\frac{2}{3}$ of the way along any given median, measuring from the vertex to the opposite side.

SHARPEN SKILLS

If time permits, use this on-level activity to build fluency and practice basic skills.

Vocabulary Review

Objective: Students demonstrate an understanding of terms related to a transversal.
Materials: Bubble Map (Teacher Resource Masters)

Have students work in pairs. Each pair should work together to build a bubble map for the term "transversal" without looking anything up. Encourage students to work together to recall as much information as possible and to not worry too much about being correct at this point.

When all pairs have had ample time to work, have a class discussion and create a class bubble map on the board.

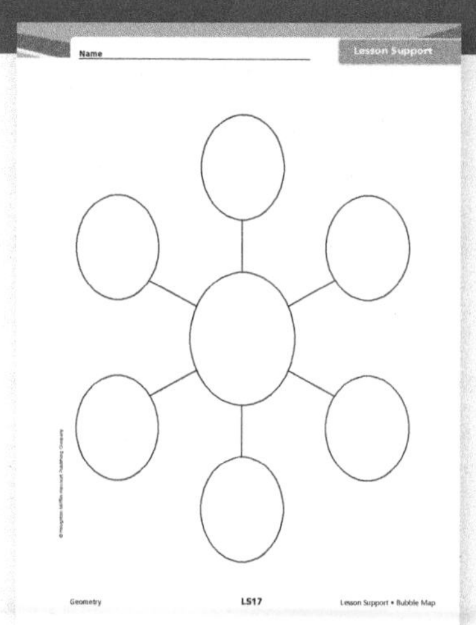

PLAN FOR DIFFERENTIATED INSTRUCTION

Small-Group Options

Use these teacher-guided activities with pulled small groups.

On Track

Materials: index cards

- Give each group a set of index cards. Each card should have the integer coordinates of a point. Groups should draw three cards at a time. If the points are collinear, one card should be put back and replaced with another.
- The three points describe a triangle. Have groups work together to find the endpoints of each midsegment, its length, and its slope.

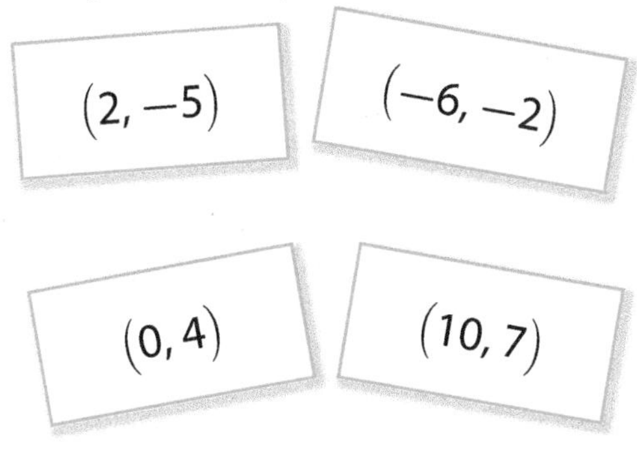

Almost There

Materials: ruler, protractor

Have students work in pairs.

- Have them first cut out a triangle from paper, fold along edges to locate the midpoints of sides, then fold to construct the midsegments of the triangle. Students can then draw the midsegments on the folds of the triangle.
- Ask students to measure the side lengths and interior angles of the smaller triangles and compare.
- Ask students to measure the side lengths and interior angles of the larger triangle and compare to the measurements of the smaller triangles.

Ready for More

Show students a triangle like the one below. Ask:

- Is w a midsegment?
- Do you think segments w and z are parallel?
- How long is w relative to z?
- What if $2x$ and $2y$ were instead $3x$ and $3y$? How long would w be relative to z?
- Can you come up with a general rule that lets us compare the lengths of w and z if $2x$ and $2y$ were replaced with some other multiple of x and y?

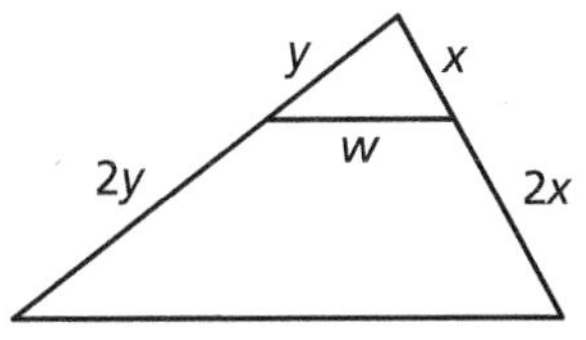

Math Center Options

Use these student self-directed activities at centers or stations. Key: ● Print Resources ● Online Resources

On Track

- ● Interactive Digital Lesson
- ●● Journal and Practice Workbook
- ● Interactive Glossary (printable): **midsegment of a triangle**
- ● Module Performance Task

Almost There

- ● Reteach 9.5 (printable)
- ● Interactive Reteach 9.5

Ready for More

- ● Challenge 9.5 (printable)
- ● Interactive Challenge 9.5
- ● Illustrative Mathematics: Joining Two Midpoints of Sides of a Triangle

ONLINE Ed — View data-driven grouping recommendations and assign differentiation resources.

During the *Spark Your Learning*, listen and watch for strategies students use. See samples of student work on this page.

Use Corresponding Angle Measures

Strategy 1

The corresponding angles have the same measures, so the second floor is parallel to the first floor.

If students . . . use corresponding angle measures to determine that the floors are parallel, they are demonstrating an understanding of how angles can be used to prove lines are parallel from Lesson 3.2.

Have these students . . . explain how they determined their equation and how they solved it. **Ask:**

- Q How did you use the relationship between angles formed when a transversal intersects parallel lines?
- Q Could you have found that the floors were parallel from another pair of angle measures?

Use a Perpendicular

Strategy 2

I drew a segment perpendicular to one floor and found with a protractor that it is perpendicular to the other floor, so the floors are parallel.

If students . . . draw a perpendicular line, they understand that a line perpendicular to one of a pair of parallel lines is perpendicular to the other line but may not understand that this is a consequence of congruent angle pairs.

Activate prior knowledge . . . by having students use the given angle pair. **Ask:**

- Q What do you know about the measures of the angles that the perpendicular segment makes with the floors?
- Q Could you use the same reasoning with acute or obtuse angles?

COMMON ERROR: Applies an Invalid Definition

The floors never touch, so they are parallel.

If students . . . say the floors are parallel since they do not intersect, they do not understand that their condition can only be applied to lines and not segments.

Then intervene . . . by pointing out that this condition applies to lines, not segments with defined endpoints. **Ask:**

- Q Can you draw two line segments that do not intersect but intersect if you make them longer?
- Q How can you define parallel line segments in terms of lines?

9.5

The Triangle Midsegment Theorem

I Can construct midsegments and prove the Triangle Midsegment Theorem.

Spark Your Learning

The A-frame cabins at a camp have a second floor for sleeping.

Complete Part A as a whole class. Then complete Parts B–D in small groups.

A. What is a mathematical question you can ask about this situation? What information would you need to know to answer your question?

A. Is the second floor parallel to the first floor of the cabin?; measures of angles within the triangle

B. What should be true about the angles in the triangles formed by the floors and the roof? See Additional Answers.

C. To answer your question, what strategy and tool would you use along with all the information you have? What answer do you get? See Strategies 1 and 2 on the facing page.

D. What are some factors that determine the location of the second floor in relation to the the first floor? Possible answer: The second floor must be located in a way that leaves enough useable space on both floors.

Turn and Talk What do you notice about the width of the second floor compared to the width of the first floor? See margin.

SUPPORT SENSE-MAKING • Three Reads

Tell students to read the information in the photo three times and prompt them with a different question each time.

1. What is the situation about? The situation is about cabins at a camp. The cabins all have the same triangular design with a second floor for sleeping.
2. What are the quantities in this situation? How are those quantities related? The quantities are the length, width, and height of a cabin, the width of the second floor, and the measures of the angles the walls make with the floors; The length, width, and height would determine the volume of the cabin. The width of the second floor is less than that of the first floor. Angles on opposite ends of each wall seem to have the same measure, and corresponding angles seem to have the same measure.
3. What are possible questions you could ask about this situation? Possible answer: How much room is inside each cabin? How wide is the second floor compared to the first floor? Are the first and second floors parallel?

1 Spark Your Learning

▶ MOTIVATE

- Have students look at the photo in their books and read the information contained in the photo. Then complete Part A as a whole-class discussion.
- Give the class the additional information they need to solve the problem. This information is available online as a printable and projectable page in the Teacher Resources.
- Have students work in small groups to complete Parts B–D.

▶ PERSEVERE

If students need support, guide them by asking:

Q **Advancing • Use Tools** Which tool could you use to solve the problem? Why choose that tool and not some other? Students' choices of tools and reasons for choosing them will vary.

Q **Assessing** If you extended the segments in the plan to lines, should the lines representing the floors intersect? Explain. no; The floors should be parallel, and parallel lines do not intersect.

Q **Assessing** Do the lines representing the floors intersect with a transversal? Explain. yes; The lines representing the walls intersect with the lines representing the floors.

Q **Advancing** What term would you use to describe the pair of labeled angles with respect to the lines representing the floors and the wall on the left? They are a pair of corresponding angles.

Turn and Talk Help students think of different ways to compare the floors' widths. A natural way to compare is by measuring both and finding the difference of their measures, but another way to compare is with a ratio. While students would hopefully notice the second floor is half as wide using the first method, encourage them to think in terms of the second method. The second method can be accomplished by improvising a ruler from another piece of paper by marking the widths of floors with tick marks on the margin. The second floor is half as wide as the first floor.

▶ BUILD SHARED UNDERSTANDING

Select groups of students who used various strategies and tools to share with the class how they solved the problem. As they present their solutions, have each group discuss why they chose a specific strategy and tool.

② Learn Together

Build Understanding

Task 1 (MP) **Use Tools** Students define the midsegment of a triangle and use a compass and a straightedge to construct the midsegments of a triangle. Encourage students to look for patterns in the midsegments and smaller triangles created. Make sure they understand that verification of a few cases does not constitute proof of conjectures.

CONNECT TO VOCABULARY

Have students use the **Interactive Glossary** to record their understanding of the vocabulary in this task.

Sample Guided Discussion:

Q If you didn't have a ruler to measure $\overline{FG}$ and $\overline{PR}$, could you still compare their relative lengths? Yes, I could mark their lengths with tick marks on the margin of another piece of paper and compare.

d Talk Encourage students to sketch triangles of various shapes to test their conjectures. No; The midsegments form a triangle, and it is not possible for the sides of a triangle to intersect in one point; The four triangles appear to be congruent. They have congruent angles and all have sides that are one half of the sides of *ABC*.

Build Understanding

Investigate Midsegments of a Triangle

A **midsegment of a triangle** is a segment that joins the midpoints of two sides of a triangle. Every triangle has three midsegments.

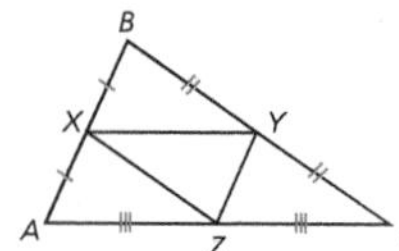

$\overline{XY}$, $\overline{YZ}$, and $\overline{ZX}$ are midsegments of $\triangle ABC$.

1 You can use a compass and straightedge to construct a midsegment of a triangle. Start by finding the midpoint of one side of the triangle.

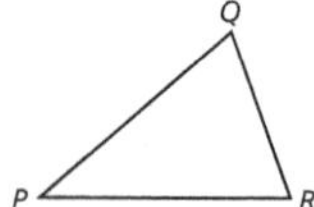

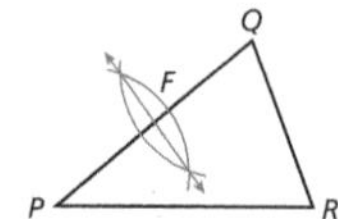

Then find the midpoint of another side, and connect the midpoints.

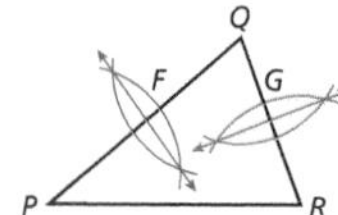

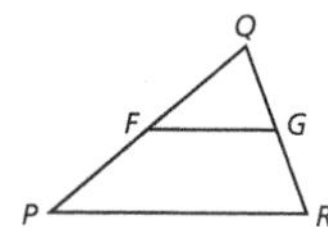

A. Explain why $\overline{FG}$ is a midsegment of $\triangle PQR$. A, B. See Additional Answers.

B. How can you construct the other two midsegments of the triangle?

C. Use a ruler to measure $\overline{FG}$ and $\overline{PR}$. How are the measures of the segments related? *FG* is half of *PR*.

D. Use a protractor to measure $\angle QFG$ and $\angle QPR$. What does this tell you about $\overline{FG}$ and $\overline{PR}$? Explain.
$\overline{FG}$ is parallel to $\overline{PR}$ since the corresponding angles are congruent.

E. Make a conjecture about the midsegment of a triangle. Then construct the other two midsegments in the triangle and test your conjecture. See Additional Answers.

Turn and Talk

- Is it possible for the midsegments of a triangle to intersect in one point? Explain.
- The midsegments divide the original triangle into four smaller triangles. What appears to be true about these triangles? Justify your conjecture. See margin.

290

LEVELED QUESTIONS

Depth of Knowledge (DOK)	Leveled Questions	What Does This Tell You?
Level 1 **Recall**	What kind of segment are you constructing when you find the midpoint of one of the sides? a perpendicular bisector	Students' answers will indicate whether they understand midsegment constructions.
Level 2 **Basic Application of Skills & Concepts**	What kind of transformations does it seem you could do to the top smaller triangle to get the others? I could translate the triangle to get the lower two triangles and rotate and translate to get the middle triangle.	Students' answers will indicate whether they can visualize the result of rigid motions.
Level 3 **Strategic Thinking & Complex Reasoning**	How do you think the areas of the two parts of a triangle on either side of a midsegment compare? One part has three times as much area as the other since there seem to be four congruent triangles with three on one side of the midsegment and one on the other.	Students' answers will indicate whether they understand how the regions of a triangle cut by a midsegment can be further partitioned and can apply their understanding of areas of congruent figures in the plane.

Step It Out

Use the Triangle Midsegment Theorem

Triangle Midsegment Theorem

The segment joining the midpoints of two sides of a triangle is parallel to the third side of the triangle, and its length is half of the length of that side.

2 The walking path connecting Holiday Street to Lakeview Avenue is a midsegment of the triangle formed by the roads. Verify that the midsegment is parallel to a side and half the length of that side.

Lakeview Ave
Meadow Drive
Holiday Street

Compare slopes.

slope of Meadow Dr. $= \frac{9-1}{2-4} = \frac{8}{-2} = -4$

slope of path $= \frac{5-1}{7-8} = \frac{4}{-1} = -4$

The side and midsegment are parallel.

A. How do you know that the side and midsegment of the triangle are parallel?

B. How can you verify that the path is a midsegment of the triangle?

A, B. See Additional Answers.

Compare lengths.

length of Meadow Dr. from $(2, 9)$ to $(4, 1) = \sqrt{(4-2)^2 + (1-9)^2}$
$= \sqrt{68}$
$= 2\sqrt{17}$

length of path $= \sqrt{(8-7)^2 + (1-5)^2} = \sqrt{17}$

The length of the midsegment is half the length of the side of the triangle.

3 What is the value of x?

$4x + 5 = \frac{1}{2}(34)$

$4x + 5 = 17$

$4x = 12$

$x = 3$

Why is this equation true?

See Additional Answers.

Turn and Talk Suppose you know that $m\angle ZMN = 78°$ and $m\angle YXZ = 65°$. What is $m\angle Z$? See margin.

Step It Out

Task 2 (MP) **Attend to Precision** Students are introduced to the Triangle Midsegment Theorem and verify the theorem with a triangle on a coordinate plane. By interpreting the slopes of the segments and the distances between the endpoints of the segments, students can determine and communicate the directions and lengths of segments with greater precision than they could through measuring physical representations with a ruler and protractor.

Sample Guided Discussion:

Q How can we describe the direction of a segment on a coordinate plane? with its slope

Q How can we visualize the Distance Formula when using it to find the length of a segment? The segment is the hypotenuse of a right triangle with vertical and horizontal legs, so we can apply the Pythagorean Theorem to find its length.

Task 3 (MP) **Construct Arguments** Students apply the Triangle Midsegment Theorem to solve for a variable in an expression describing the length of a midsegment.

Sample Guided Discussion:

Q How are the lengths of $\overline{XY}$ and $\overline{NM}$ related? $\overline{NM}$ is half as long as $\overline{XY}$.

Q How can we algebraically describe the relationship between the lengths of $\overline{XY}$ and $\overline{NM}$? We can write $NM = \frac{1}{2}XY$ or $2NM = XY$.

Turn and Talk Help students use the relationships between angles formed by parallel lines and intersecting transversals. If necessary, guide them with prompting questions meant to remind them that they can assume corresponding angles are congruent. They can then use the fact that $m\angle ZMN + m\angle MNZ + m\angle NZM = 180°$. They could also apply the Triangle Angle Sum Theorem to $\triangle XYZ$ to get the same result. 37°

PROFICIENCY LEVEL

Beginning

Show students a diagram of a scalene triangle with a midsegment, for example, $\triangle ABC$ with midsegment $\overline{DE}$ parallel to side $\overline{BC}$. Say, "Segment DE is a midsegment parallel to side BC" while indicating the discussed segments. Construct midpoint F of $\overline{BC}$, and ask students to verbally identify the other midsegments and the sides to which they are parallel.

Intermediate

Give students a coordinate plane with a triangle with midsegment labeled and have them explain in writing their steps in confirming that the midsegment is parallel to a side and half its length.

Advanced

Have students describe in writing the relationships between the sides of a triangle, the midsegments of the triangle, and the angles formed by the sides and midsegments.

ONLINE

Assign the Digital On Your Own for

- built-in student supports
- Actionable Item Reports
- Standards Analysis Reports

On Your Own

Assignment Guide

The chart below indicates which problems in the On Your Own are associated with each task in the Learn Together. Assign daily homework for tasks completed.

Learn Together Tasks	On Your Own Problems
Task 1, p. 290	Problems 6–9
Task 2, p. 291	Problems 18–24
Task 3, p. 291	Problems 10–17 and 25–28

Check Understanding

Copy the triangle. Construct the midsegment parallel to the given side of the triangle. 1, 2. See Additional Answers.

1. $\overline{ST}$ 

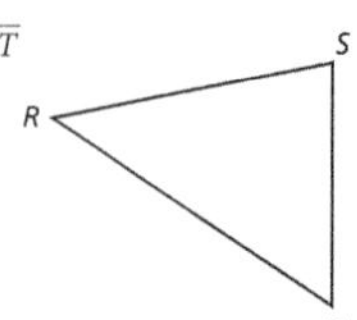

2. $\overline{EF}$

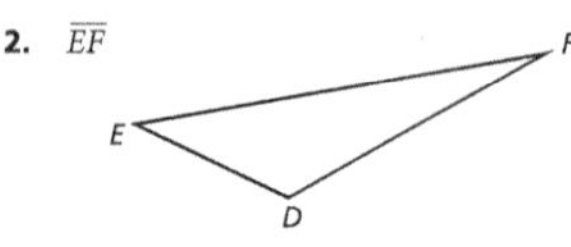

3. A triangle has vertices $F(-2, 3)$, $G(4, 3)$, and $H(2, -1)$. What are the endpoints of the midsegment that is parallel to $\overline{GH}$? (0, 1) and (1, 3)

Find the value of x.

4. $x = 7$

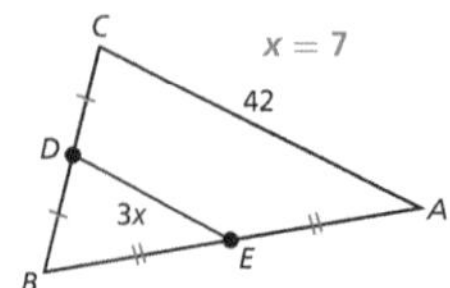

5. $x = 12$

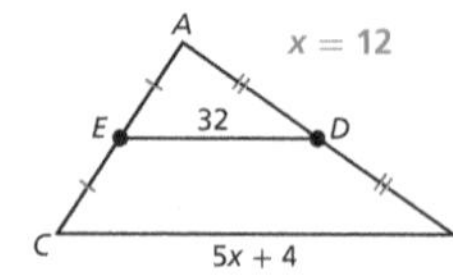

On Your Own

6. MP **Critique Reasoning** Henry claims that $\overline{MN}$ is a midsegment of $\triangle JKL$. Explain why this is not true. *M* and *N* are not the midpoints of $\overline{JK}$ and $\overline{KL}$, respectively.

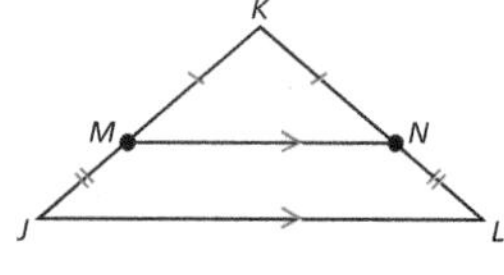

7. MP **Reason** Triangle *PQR* is an isosceles triangle.
   A. Copy the triangle and construct the midsegments of the triangle.
   B. What type of triangle is formed by the midsegments? Explain your reasoning. A, B. See Additional Answers.

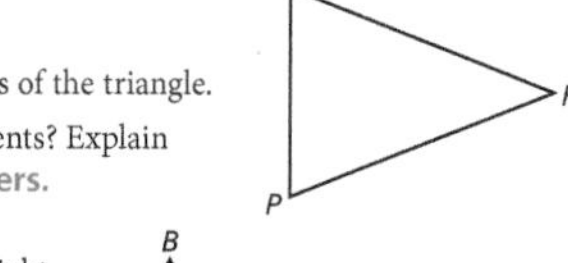

8. An example of a Sierpinski triangle is shown at the right. This triangle is a fractal formed by connecting the midpoints of a large triangle to form a smaller triangle. In the fractal, *AC* is 48. What is *ST*? 24

9. MP **Reason** $\triangle DEF$ is an equilateral triangle. The midsegments of $\triangle DEF$ form $\triangle ABC$. What type of triangle is $\triangle ABC$? Explain your reasoning.
equilateral triangle; The lengths of the midsegments of an equilateral triangle are equal.

292

3 Check Understanding

Formative Assessment

Use formative assessment to determine if your students are successful with this lesson's learning objective.

Students who successfully complete the Check Understanding can continue to the On Your Own practice.

For students who miss 1 problem or more, work in a pulled small group using the Almost There small-group activity on page 289C.

ONLINE

Assign the Digital Check Understanding to determine

- success with the learning objective
- items to review
- grouping and differentiation resources

4 Differentiation Options

Differentiate instruction for all students using small-group activities and math center activities on page 289C.

Reteach

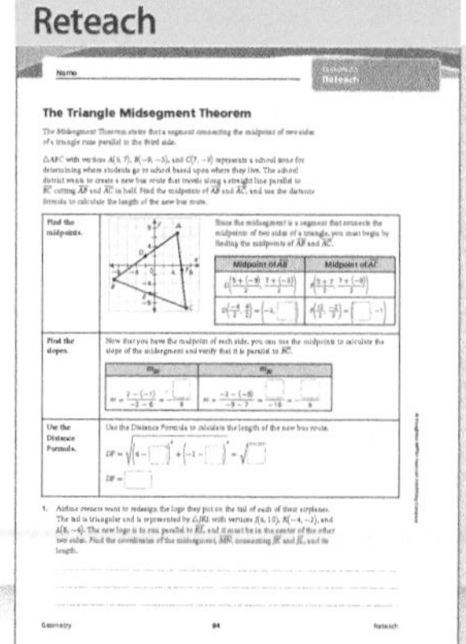

Challenge

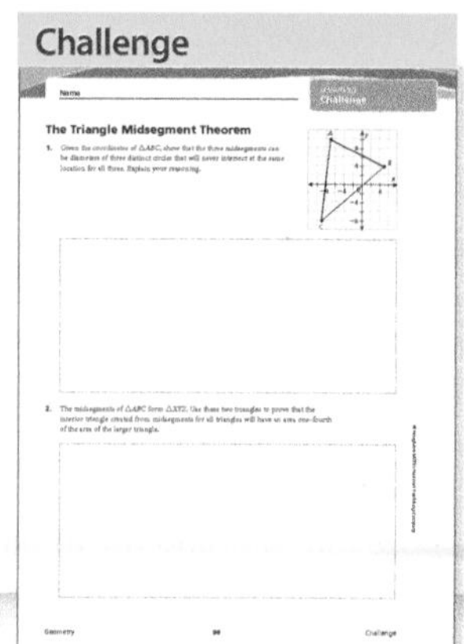

In Problems 10–15, find each measure.

10. CB 45

11. FE 28.5

12. $m\angle DFC$ 64°

13. AE 30

14. $m\angle DFB$ 116°

15. DF 30

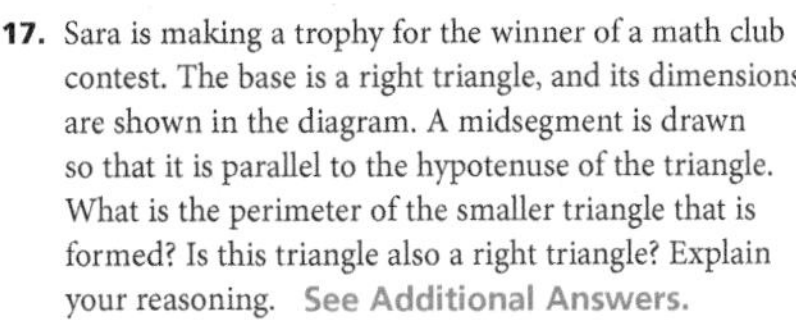

16. MP **Use Repeated Reasoning** $\overline{AB}$ is a midsegment of $\triangle XYZ$. $\overline{CD}$ is a midsegment of $\triangle AYB$. $\overline{EF}$ is a midsegment of $\triangle CYD$.

A. Copy and complete the table.
A. See Additional Answers.

Midsegment	1	2	3
Length	?	?	?

B. If this pattern continues, what will be the length of midsegment 8? 0.25

C. Write an algebraic expression to represent the length of midsegment n. (*Hint:* Relate the pattern to powers of 2.) $64\left(\frac{1}{2}\right)^n = 2^{6-n}$

17. Sara is making a trophy for the winner of a math club contest. The base is a right triangle, and its dimensions are shown in the diagram. A midsegment is drawn so that it is parallel to the hypotenuse of the triangle. What is the perimeter of the smaller triangle that is formed? Is this triangle also a right triangle? Explain your reasoning. See Additional Answers.

32 cm 68 cm 60 cm

18. MP **Construct Arguments** Use coordinates to prove the Triangle Midsegment Theorem.

A. M is the midpoint of $\overline{HJ}$. What are its coordinates? $M(a, b)$

B. N is the midpoint of $\overline{JK}$. What are its coordinates? $N(c + a, b)$

C. Find the slopes of $\overline{MN}$ and $\overline{HK}$. What can you conclude? See Additional Answers.

D. Find MN and HK. What can you conclude? $MN = c$, $HK = 2c$; MN is half of HK.

y, J(2a, 2b), M, N, x, H(0, 0), K(2c, 0)

The coordinates of the vertices of a triangle are given. Find the coordinates of the endpoints of the midsegment parallel to the given side of the triangle.

19. $A(-2, 5)$, $B(4, 9)$, $C(8, 3)$; $\overline{BC}$ (1, 7), (3, 4)

20. $A(1, 1)$, $B(7, -1)$, $C(9, 3)$; $\overline{AB}$ (5, 2), (8, 1)

21. $A(3, -1)$, $B(-3, -3)$, $C(-1, 5)$; $\overline{AC}$ (−2, 1), (0, −2)

22. $A(2, -3)$, $B(-4, -3)$, $C(4, 3)$; $\overline{AB}$ (0, 0), (3, 0)

Watch for Common Errors

Problem 14 Students may correctly identify $\overline{DF}$ and $\overline{AB}$ as parallel segments but confuse $\angle DFB$ and $\angle EBF$ as corresponding angles since the angles have legs going in the same direction. Point out that corresponding angles are in the same position relative to their respective vertices.

Questioning Strategies

Problem 16 Suppose that $XZ = 81$ and segments 1, 2, and 3 are parallel to $\overline{XZ}$ but are constructed so $XA = 2AY$, $AC = 2CY$, and $CE = 2EY$. Make a conjecture about the lengths of segments 1, 2, and 3. Each segment should be one third of its predecessor's length.

Segment	1	2	3
Length	27	9	3

5 Wrap-Up

Summarize learning with your class. Consider using the Exit Ticket, Put It in Writing, or I Can scale.

Exit Ticket

$\triangle CFE \cong \triangle EDB$. Find AD.

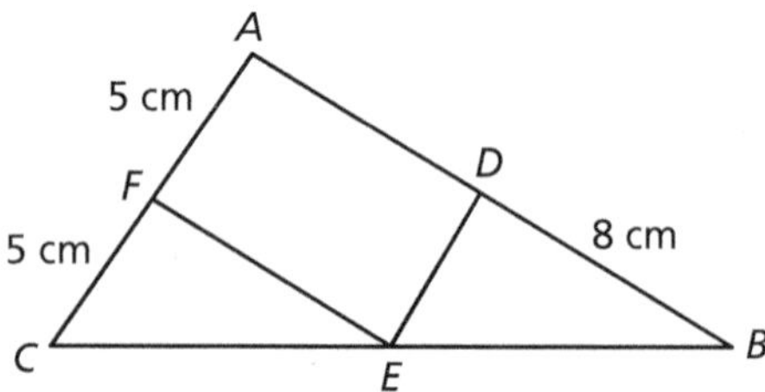

Since $\triangle CFE \cong \triangle EDB$, we know $\overline{CE} \cong \overline{EB}$ by the converse of CPCTC, so $CE = EB$. Then since $AF = FC$, we know FE is a midsegment of $\triangle ABC$ parallel to $\overline{AB}$. Then $FE = \frac{1}{2}AB$, so $AD = DB = 8$ cm.

Put It in Writing 

Explain how to determine the length of a midsegment of a triangle if you know the side lengths of the triangle.

I Can 

The scale below can help you and your students understand their progress on a learning goal.

4	I can construct midsegments and prove the Triangle Midsegment Theorem, and I can explain the process to others.
3	I can construct midsegments and prove the Triangle Midsegment Theorem.
2	I can construct midsegments.
1	I can use the side lengths of a triangle to find the length of a midsegment.

Spiral Review • Assessment Readiness

These questions will help determine if students have retained information taught in the past and can also prepare them for high-stakes assessments. Here, students must locate the centroid along a median **(9.4)**, solve a linear inequality **(Alg1, 2.4)**, and define points of concurrence of a triangle **(9.2–9.4)**.

The vertices of $\triangle ABC$ are $A(3, -4)$, $B(-5, 2)$, and $C(5, 4)$. Verify that each segment is parallel to a side of the triangle and half the length of that side.

23. $\overline{LM}$ with endpoints $L(0, 3)$ and $M(4, 0)$ 23, 24. See Additional Answers.

24. $\overline{MN}$ with endpoints $M(4, 0)$ and $N(-1, -1)$

Find the value of the variable.

25. $v = 5$

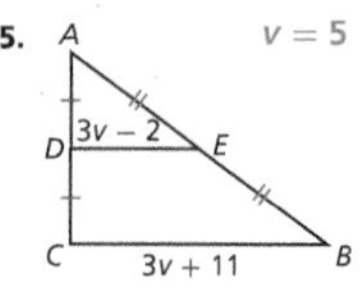

26. $r = 7$

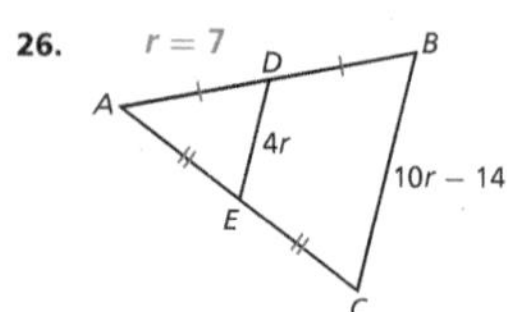

27. $n = 9$

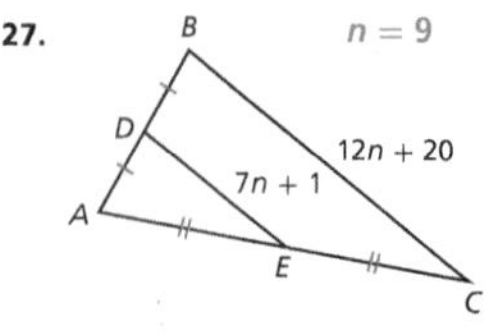

28. $h = 21$

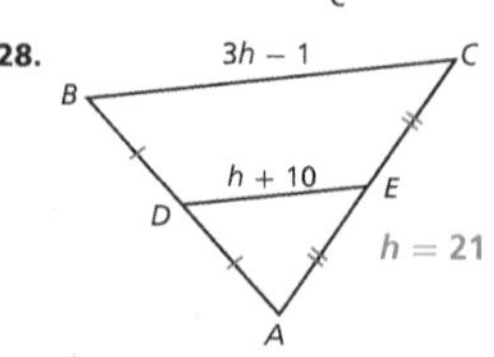

Spiral Review • Assessment Readiness

29. P is the centroid of $\triangle ABC$. What is the value of x?

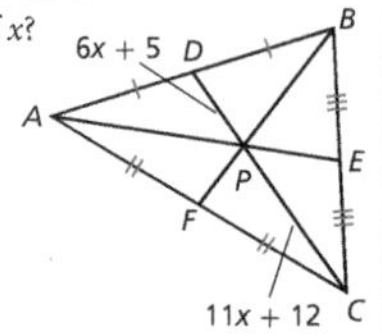

Ⓐ 2
Ⓑ 17
Ⓒ 34
Ⓓ 51

(Figure: 6x + 5, D, B, A, E, P, F, 11x + 12, C)

30. Which of the following values of x make the inequality true? Select all that apply.

$$x + 5 > 17$$

Ⓐ $x = 5$
Ⓑ $x = 11$
Ⓒ $x = 12$
Ⓓ $x = 13$
Ⓔ $x = 17$

31. Match the segments of a triangle with the term for the point of concurrency of the segments.

Segments	Point of Concurrence	
A. altitudes	**1.** centroid	A. 3
B. medians	**2.** circumcenter	B. 1
C. angle bisectors	**3.** orthocenter	C. 4
D. perpendicular bisectors	**4.** incenter	D. 2

 **I'm in a Learning Mindset!**

What is my plan for improving my performance with segments of triangles?

Learning Mindset

Challenge Seeking Sets Achievable Stretch Goals

Point out that achievable short-term goals should be part of any plan to improve performance since they can act as milestones that give students confidence as they progress. Short-term goals can also help students monitor their learning. For example, if a student wants to improve their performance with triangle midsegments, they might plan to first focus on constructing midsegments, then on using the Triangle Midsegment Theorem to solve for missing measures, and then on using the theorem to solve for coordinates of midsegment endpoints. *How does planning achievable short-term goals help give you confidence? How can you include monitoring of your learning as part of your plan?*

Module 9 Review

Triangle Sum Theorem

The sum of the measures of the interior angles of a triangle is 180°.

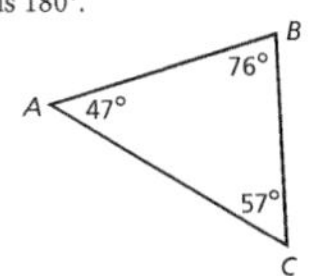

$m\angle A + m\angle B + m\angle C = 47° + 76° + 57° = 180°$

Perpendicular Bisectors

The perpendicular bisectors of a triangle are concurrent. The point of concurrency is called the circumcenter of the triangle.

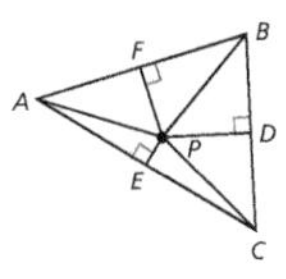

$PD = PE = PF$

Angle Bisectors

The angle bisectors of a triangle are concurrent. The point of concurrency is called the incenter of the triangle.

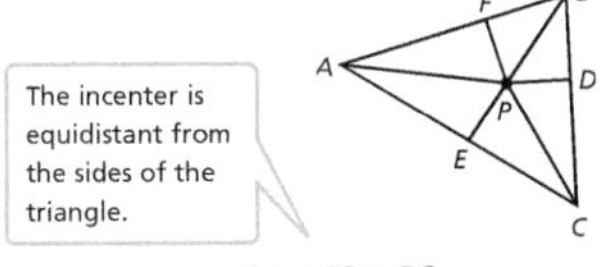

$PA = PB = PC$

Medians

The medians of a triangle are drawn from a vertex to the midpoint of the side opposite the vertex and are concurrent. The point of concurrency is called the centroid of the triangle.

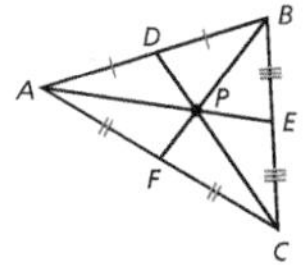

The centroid divides each median into a 2:3 ratio.

$BP = \frac{2}{3}BF,\ AP = \frac{2}{3}AE,\ CP = \frac{2}{3}CD$

Altitudes

The medians of a triangle are drawn from a vertex perpendicular to the side opposite the vertex and are concurrent. The point of concurrency is called the orthocenter of the triangle.

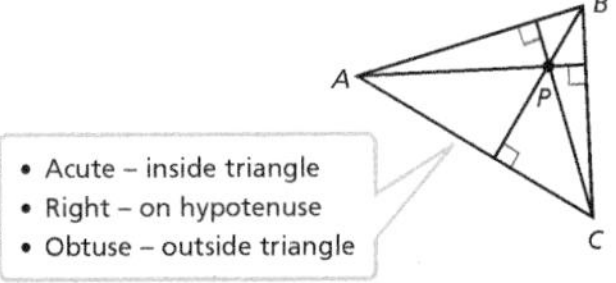

• Acute – inside triangle
• Right – on hypotenuse
• Obtuse – outside triangle

Midsegments

The midsegments are drawn from the midpoints of the sides of the triangle and are parallel to the third side.

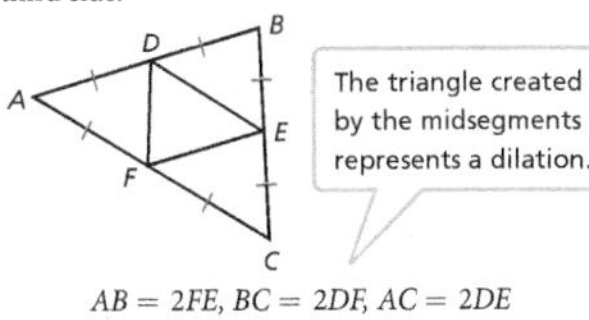

$AB = 2FE,\ BC = 2DF,\ AC = 2DE$

$\overline{AB} \parallel \overline{FE},\ \overline{BC} \parallel \overline{DF},\ \overline{CA} \parallel \overline{ED}$

Module 9 295

Midsegments **How is the perimeter of the triangle formed by the midsegments of a triangle related to the perimeter of the original triangle?** The perimeter of the inner triangle is one-half the perimeter of the enclosing triangle because each side of the inner triangle is one-half the length of its corresponding parallel side in the outer triangle.

ONLINE

Assign the Digital Module Review for
- built-in student supports
- Actionable Item Reports
- Standards Analysis Reports

MODULE 9 REVIEW

Module Review

Use the first page of the Module Review to summarize and connect the main ideas of the module. Use the second page to assess students' understanding of the vocabulary, concepts, and skills presented in the module.

Sample Guided Discussion:

Triangle Sum Theorem **Does the sum of the measures of the interior angles of a triangle depend on the kind of triangle? Explain.** no; In any triangle the sum of the measures of the interior angles is always 180°.

Perpendicular Bisectors **What is the difference between the perpendicular bisector of a side of a triangle and the altitude to that side?** Possible answer: The perpendicular bisector of a side of a triangle is a line that contains the midpoint of the side but may not contain the opposite vertex. The altitude is always a segment from a vertex of an angle perpendicular to its opposite side.

Angle Bisectors **Where is the point of concurrency of the angle bisectors of a triangle located?** In the interior of the triangle—never on a side or in the exterior of the triangle.

Medians **What is the difference between the perpendicular bisector of a side of a triangle and the median to that side?** Possible answer: The perpendicular bisector always contains the midpoint of the side of the triangle, but does not always contain a vertex of the triangle. The endpoints of a median are always a vertex of the triangle and the midpoint of the opposite side.

Altitudes **Can a perpendicular bisector of a side of a triangle also be the altitude to that side?** Yes, the perpendicular bisector of the hypotenuse in a right triangle is also the altitude to the hypotenuse.

Module Review continued

Possible Scoring Guide

Items	Points	Description
1–4	2 each	identifies the correct term(s)
5–8	2 each	correctly finds the measure
9	2	correctly finds the circumcenter
10	2	correctly finds the three angles
11, 12	2 each	correctly determines the coordinates
13	2	correctly justifies why the circumcenter is equidistant from the vertices
14	2	correctly justifies why the incenter of a triangle is equidistant from each side of the triangle
15	2	correctly identifies the transformation whose image is a midsegment
Total points possible = 30 points		

The Unit 5 Test in the Assessment Guide assesses content from Modules 9 and 10.

Vocabulary

Choose the correct term from the box to complete each sentence.

Vocabulary
altitude of a triangle
concurrent
exterior angle
median of a triangle
point of concurrency

1. A(n) __?__ is a segment whose endpoints are a vertex of the triangle and the midpoint of the opposite side. median of a triangle
2. An angle formed by one side of a polygon and an extension of an adjacent side is a(n) __?__. exterior angle
3. A(n) __?__ is a perpendicular segment from a vertex to the line containing the opposite side. altitude of a triangle
4. Three or more lines that intersect at one point are __?__ lines. The point of intersection is the __?__. concurrent; point of concurrency

Concepts and Skills

Determine each measure.

5. $m\angle B$ 54°

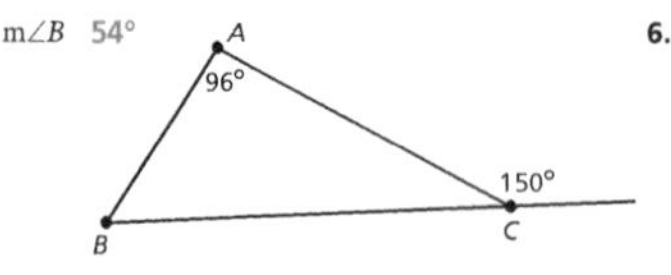

6. AB 62

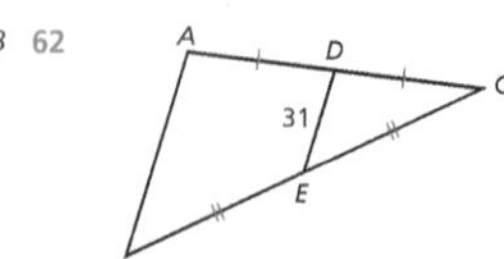

7. PD 9

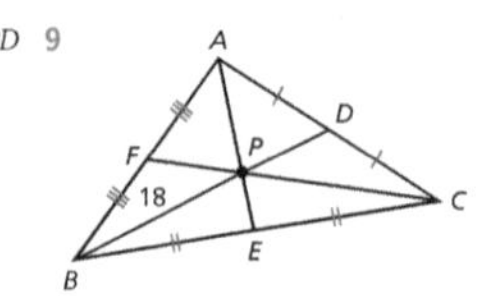

8. $m\angle ABC$ 62°

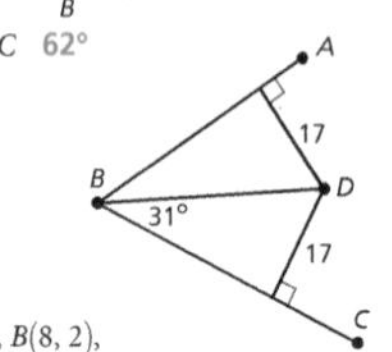

9. Determine the circumcenter of a triangle with vertices at $A(2, 4)$, $B(8, 2)$, and $C(4, -2)$. (4.5, 1.5)
10. A right triangle has an angle that measures 26°. What are the angle measures of all three angles of the triangle? 26°, 64°, 90°
11. The coordinate of the vertices of a triangle are $X(1, 5)$, $Y(9, 3)$, and $Z(3, 1)$. Find the coordinates of the endpoints of the midsegment parallel to $\overline{XY}$. (2, 3) and (6, 2)
12. (MP) **Use Tools** The coordinate of the vertices of a triangle are $J(-2, 4)$, $K(-1, -6)$, and $L(3, 2)$. Find the coordinates of the centroid of the triangle. State what strategy and tool you will use to answer the question, explain your choice, and then find the answer. (0, 0)
13. Explain why the circumcenter is equidistant to each vertex using properties of perpendicular bisectors. 13–15. See Additional Answers.
14. Explain why the incenter is equidistant to each side using properties of angle bisectors.
15. What transformation maps a side of a triangle to the midsegment connecting the other two sides?

296

DATA-DRIVEN INSTRUCTION

Before moving on to the Module Test, use the Module Review results to intervene based on the table below.

Items	Lesson	DOK	Content Focus	Intervention
5	9.1	2	Find the measure of a remote interior angle in a triangle.	Reteach 9.1
6, 11, and 15	9.5	3	Apply the Midsegment Theorem to find the measure of the side opposite the midsegment; find the coordinates of the endpoints of a midsegment.	Reteach 9.5
7 and 12	9.4	2	Determine the coordinates of the centroid of a triangle and find the measure of the length from the centroid to the midpoint of a side.	Reteach 9.4
8	9.3	2	Find the measures of an angle that was bisected.	Reteach 9.3
9 and 13	9.2	3	Find the coordinates of a circumcenter of a triangle; justify why the circumcenter is equidistance from the vertices of a triangle.	Reteach 9.2
10	9.1	2	Find the measure of the three angles in a right triangle.	Reteach 9.1
14	9.3	2	Justify why the circumcenter of a triangle is equidistant from the sides of the triangle.	Reteach 9.3

Module Test

The Module Test is available in alternative versions in your Assessment Guide. All items are presented in standardized test formats.

ONLINE

Assign the Digital Module Test to power actionable reports including

- proficiency by standards
- item analysis

MODULE 9

TEST

Form A

Name

Module 9 • Form A
Module Test

1. The sides of $\triangle FHG$ are tangent to circle P at points M, N, and O as shown in the diagram.

Which statement is true?

Ⓐ The altitudes of the triangle are concurrent at point P.
Ⓑ The angle bisectors of the triangle are concurrent at point P.
Ⓒ The medians of the triangle are concurrent at point P.
Ⓓ The perpendicular bisectors of the triangle are concurrent at point P.

2. The circumcenter of $\triangle HJK$ is point P. Point A lies on $\overline{HJ}$ and $\overline{AP} \perp \overline{HJ}$. If $AP = 7$ in. and $KP = 23.6$ in., what is the length of $\overline{HJ}$?

Ⓐ 22.5 in. **Ⓒ** 45.1 in.
Ⓑ 24.6 in. Ⓓ 47.2 in.

3. The perpendicular bisectors of the sides of $\triangle OPQ$ are concurrent at a point outside of the triangle. What type of triangle is $\triangle OPQ$?

Ⓐ acute Ⓒ right
Ⓑ obtuse Ⓓ equiangular

4. The centroid of $\triangle ABC$ is point P. Point M is the midpoint of $\overline{BC}$, and $MP = 252$ cm. What is the length of $\overline{AP}$?

Ⓐ 126 cm **Ⓒ** 504 cm
Ⓑ 168 cm Ⓓ 756 cm

5. In $\triangle EFG$, $m\angle E = (8x + 5)°$, $m\angle F = (x - 11)°$, and $m\angle G = (2x - 1)°$. What is $m\angle E$ to the nearest degree?

141°

6. The vertices of $\triangle WXY$ are located at $W(-14, 20)$, $X(-2, -4)$, and $Y(10, 4)$. What is the length of the midsegment parallel to $\overline{XY}$?

Ⓐ 4.8 units Ⓒ 9.3 units
Ⓑ 7.2 units Ⓓ 14.4 units

7. In $\triangle ABC$, $m\angle A = \frac{1}{2}m\angle B$ and $m\angle B = 3m\angle C$. What is $m\angle C$ to the nearest degree?

33°

8. The perpendicular bisectors of the sides of $\triangle JKL$ intersect at point P. If $JP = 2x + 5$ meters and $KP = 4x - 13$ meters, what is the length of $\overline{LP}$ to the nearest meter?

23

9. In the figure, D, C, and B are collinear.

What is the measure of $\angle A$ to the nearest degree?

57°

10. In the figure, $\overline{BD}$ bisects $\angle ABC$. $ED = 3x + 5$ cm and $DF = 2(x + 8)$ cm.

If $BF = 150$ cm, what is the length of $\overline{BD}$ to the nearest centimeter?

155

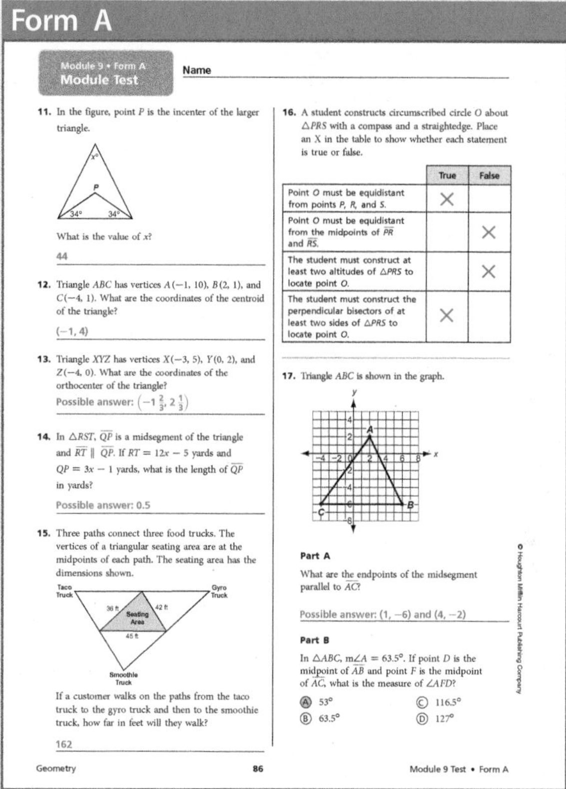

Form A

Module 9 • Form A
Module Test

Name

11. In the figure, point P is the incenter of the larger triangle.

What is the value of x?

44

12. Triangle ABC has vertices $A(-1, 10)$, $B(2, 1)$, and $C(-4, 1)$. What are the coordinates of the centroid of the triangle?

(−1, 4)

13. Triangle XYZ has vertices $X(-3, 5)$, $Y(0, 2)$, and $Z(-4, 0)$. What are the coordinates of the orthocenter of the triangle?

Possible answer: $\left(-1\frac{2}{3}, 2\frac{1}{3}\right)$

14. In $\triangle RST$, $\overline{QP}$ is a midsegment of the triangle and $\overline{RT} \parallel \overline{QP}$. If $RT = 12x - 5$ yards and $QP = 3x - 1$ yards, what is the length of $\overline{QP}$ in yards?

Possible answer: 0.5

15. Three paths connect three food trucks. The vertices of a triangular seating area are at the midpoints of each path. The seating area has the dimensions shown.

If a customer walks on the paths from the taco truck to the gyro truck and then to the smoothie truck, how far in feet will they walk?

162

16. A student constructs circumscribed circle O about $\triangle PRS$ with a compass and a straightedge. Place an X in the table to show whether each statement is true or false.

	True	False
Point O must be equidistant from points P, R, and S.	X	
Point O must be equidistant from the midpoints of $\overline{PR}$ and $\overline{RS}$.		X
The student must construct at least two altitudes of $\triangle PRS$ to locate point O.		X
The student must construct the perpendicular bisectors of at least two sides of $\triangle PRS$ to locate point O.	X	

17. Triangle ABC is shown in the graph.

Part A

What are the endpoints of the midsegment parallel to $\overline{AC}$?

Possible answer: (1, −6) and (4, −2)

Part B

In $\triangle ABC$, $m\angle A = 63.5°$. If point D is the midpoint of $\overline{AB}$ and point F is the midpoint of $\overline{AC}$, what is the measure of $\angle AFD$?

Ⓐ 53° Ⓒ 116.5°
Ⓑ 63.5° Ⓓ 127°

© Houghton Mifflin Harcourt Publishing Company

Geometry 86 Module 9 Test • Form A

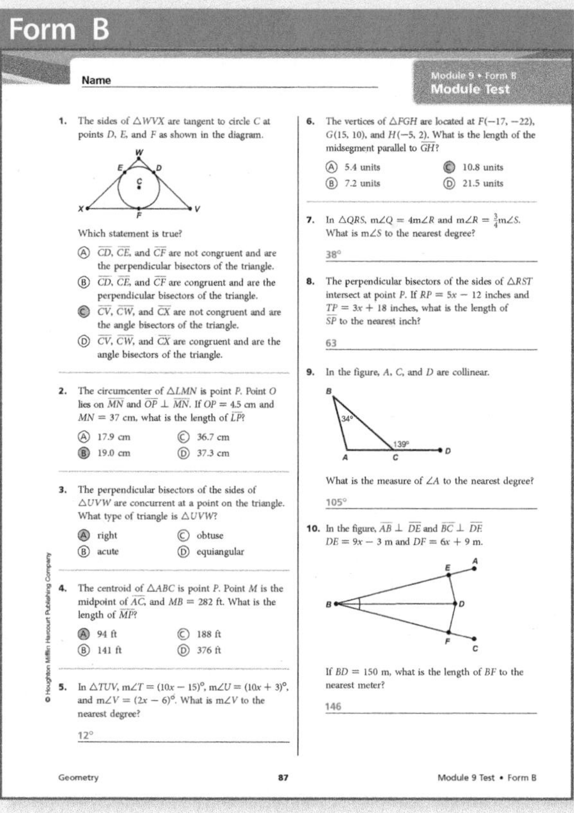

Form B

Name

Module 9 • Form B
Module Test

1. The sides of $\triangle WVX$ are tangent to circle C at points D, E, and F as shown in the diagram.

Which statement is true?

Ⓐ $\overline{CD}$, $\overline{CE}$, and $\overline{CF}$ are not congruent and are the perpendicular bisectors of the triangle.
Ⓑ $\overline{CD}$, $\overline{CE}$, and $\overline{CF}$ are congruent and are the perpendicular bisectors of the triangle.
Ⓒ $\overline{CV}$, $\overline{CW}$, and $\overline{CX}$ are not congruent and are the angle bisectors of the triangle.
Ⓓ $\overline{CV}$, $\overline{CW}$, and $\overline{CX}$ are congruent and are the angle bisectors of the triangle.

2. The circumcenter of $\triangle LMN$ is point P. Point O lies on $\overline{MN}$ and $\overline{OP} \perp \overline{MN}$. If $OP = 4.5$ cm and $MN = 37$ cm, what is the length of $\overline{LP}$?

Ⓐ 17.9 cm Ⓒ 36.7 cm
Ⓑ 19.0 cm Ⓓ 37.3 cm

3. The perpendicular bisectors of the sides of $\triangle UVW$ are concurrent at a point on the triangle. What type of triangle is $\triangle UVW$?

Ⓐ right Ⓒ obtuse
Ⓑ acute Ⓓ equiangular

4. The centroid of $\triangle ABC$ is point P. Point M is the midpoint of $\overline{AC}$, and $MB = 282$ ft. What is the length of $\overline{MP}$?

Ⓐ 94 ft Ⓒ 188 ft
Ⓑ 141 ft Ⓓ 376 ft

5. In $\triangle TUV$, $m\angle T = (10x - 15)°$, $m\angle U = (10x + 3)°$, and $m\angle V = (2x - 6)°$. What is $m\angle V$ to the nearest degree?

12°

6. The vertices of $\triangle FGH$ are located at $F(-17, -22)$, $G(15, 10)$, and $H(-5, 2)$. What is the length of the midsegment parallel to $\overline{GH}$?

Ⓐ 5.4 units **Ⓒ** 10.8 units
Ⓑ 7.2 units Ⓓ 21.5 units

7. In $\triangle QRS$, $m\angle Q = 4m\angle R$ and $m\angle R = \frac{3}{4}m\angle S$. What is $m\angle S$ to the nearest degree?

38°

8. The perpendicular bisectors of the sides of $\triangle RST$ intersect at point P. If $RP = 5x - 12$ inches and $TP = 3x + 18$ inches, what is the length of $\overline{SP}$ to the nearest inch?

63

9. In the figure, A, C, and D are collinear.

What is the measure of $\angle A$ to the nearest degree?

105°

10. In the figure, $\overline{AB} \perp \overline{DE}$ and $\overline{BC} \perp \overline{DF}$. $DE = 9x - 3$ m and $DF = 6x + 9$ m.

If $BD = 150$ m, what is the length of BF to the nearest meter?

146

© Houghton Mifflin Harcourt Publishing Company

Geometry 87 Module 9 Test • Form B

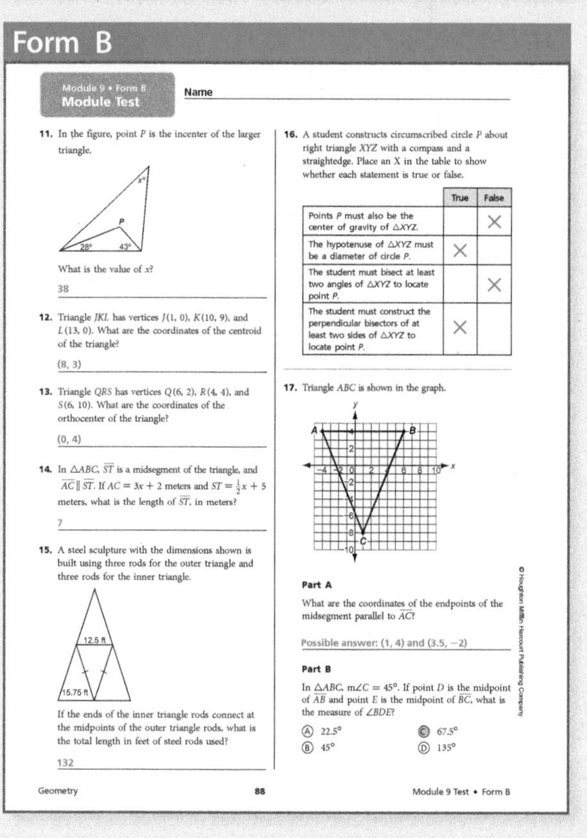

Form B

Module 9 • Form B
Module Test

Name

11. In the figure, point P is the incenter of the larger triangle.

What is the value of x?

38

12. Triangle JKL has vertices $J(1, 0)$, $K(10, 9)$, and $L(13, 0)$. What are the coordinates of the centroid of the triangle?

(8, 3)

13. Triangle QRS has vertices $Q(6, 2)$, $R(4, 4)$, and $S(6, 10)$. What are the coordinates of the orthocenter of the triangle?

(0, 4)

14. In $\triangle ABC$, $\overline{ST}$ is a midsegment of the triangle, and $\overline{AC} \parallel \overline{ST}$. If $AC = 3x + 2$ meters and $ST = \frac{1}{2}x + 5$ meters, what is the length of $\overline{ST}$, in meters?

7

15. A steel sculpture with the dimensions shown is built using three rods for the outer triangle and three rods for the inner triangle.

If the ends of the inner triangle rods connect at the midpoints of the outer triangle rods, what is the total length in feet of steel rods used?

132

16. A student constructs circumscribed circle P about right triangle XYZ with a compass and a straightedge. Place an X in the table to show whether each statement is true or false.

	True	False
Points P must also be the center of gravity of $\triangle XYZ$.		X
The hypotenuse of $\triangle XYZ$ must be a diameter of circle P.	X	
The student must bisect at least two angles of $\triangle XYZ$ to locate point P.		X
The student must construct the perpendicular bisectors of at least two sides of $\triangle XYZ$ to locate point P.	X	

17. Triangle ABC is shown in the graph.

Part A

What are the coordinates of the endpoints of the midsegment parallel to $\overline{AC}$?

Possible answer: (1, 4) and (3.5, −2)

Part B

In $\triangle ABC$, $m\angle C = 45°$. If point D is the midpoint of $\overline{AB}$ and point E is the midpoint of $\overline{BC}$, what is the measure of $\angle BDE$?

Ⓐ 22.5° **Ⓒ** 67.5°
Ⓑ 45° Ⓓ 135°

© Houghton Mifflin Harcourt Publishing Company

Geometry 88 Module 9 Test • Form B

TRIANGLE INEQUALITIES

Introduce and Check for Readiness

- Module Performance Task
- Are You Ready?

Lesson 2.1—2 Days

Inequalities Within a Triangle

Learning Objective: Decide when three lengths can form a triangle, find the possible ranges of side lengths for the third side of a triangle, and order and compare the side lengths and angle measures in a triangle.

Lesson 2.2—2 Days

Inequalities Between Two Triangles

Learning Objective: Apply known information about a triangle(s) in a pair of triangles to determine relative lengths and angle measures.

New Vocabulary: indirect proof

Assessment

- Module 10 Test (Forms A and B)
- Unit 5 Test (Forms A and B)

LEARNING ARC FOCUS

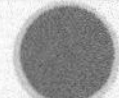 Build Conceptual Understanding
 Connect Concepts and Skills
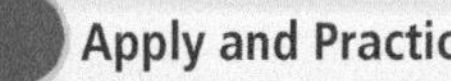 Apply and Practice

TEACHING FOR SUCCESS

TEACHING FOR DEPTH: Triangle Inequalities

Make Connections. Until this module, students have worked with concepts that relate equality relationships within and between triangles. They have learned that if a triangle has at least two congruent sides, it is isosceles. Conversely, if it is isosceles, at least two sides are congruent. Students have also learned that the angles opposite two congruent sides in a triangle are congruent, and conversely, if two angles of a triangle are congruent, the opposite sides are also congruent.

In the two lessons in this module, students consider how inequalities between sides and angles of a triangle are related and how inequalities between corresponding sides and angles of two triangles are related.

Students will begin the module considering the Triangle Inequality Theorem and its converse. The Triangle Inequality Theorem establishes the conditions under which a triangle can exist; that is, the sum of the lengths of any two sides is always greater than the length of the third side. The converse establishes that an inequality between two sides of a triangle results in the same inequality relationship between the angles opposite the sides.

In the second lesson, students prove the Hinge Theorem that relates the inequality relationship between two triangles that have a pair of congruent sides but angles of unequal measure between them. The third sides opposite the angles of unequal measure are unequal in the same order. Also, if two triangles have a pair of congruent sides but third sides of unequal length, then the angle measures included between the sides of unequal length are unequal in the same order.

Mathematical Progressions

Prior Learning	Current Development	Future Connections
Students: • investigated the relationship between an exterior angle and the remote interior angles of a triangle. • used the Triangle Sum Theorem. • wrote and solved inequalities.	**Students:** • learn how to apply inequalities to relate the sides and angles in one triangle. • relate the sides and angles in two triangles.	**Students:** • will investigate segment length and angle measure relationships among lines intersecting circles.

TEACHER TO TEACHER

From the Classroom

Elicit and use evidence of student thinking. My students have more trouble solving inequalities than solving equations in Algebra 1, especially remembering to "change signs" when multiplying or dividing by a negative number. Further, solutions to algebraic inequalities are usually not one number. I've found that this confusion often continues in geometry.

One way I begin introducing the idea of inequality relationships in geometry is to remind my students about the three possibilities that exist between quantities. That is, two quantities are greater than, equal to, or less than each other. I ask students to consider what happens in a triangle if two of its sides have equal length or two of its angles have equal measure. They usually know. What about if the side lengths or the angle measures are unequal? It seems logical to assume that the same "unequalness" also applies. What happens if corresponding side lengths of two triangles are not equal? or corresponding angle measures?

After this informal discussion, I find that it is easy to move into the formality of the inequality theorems that apply within or between triangles. I have students use constructions, either using a compass and straight edge or geometry software, to demonstrate these inequality relationships. I find the latter to be even more helpful for students because it lets them manipulate parts of a triangle or pairs of triangles to see what changes and what stays the same.

LANGUAGE DEVELOPMENT • Planning for Instruction

By giving all students regular exposure to language routines in context, you will provide opportunities for students to **listen**, **speak**, **read**, and **write** about mathematical situations and develop both mathematical language and conceptual understanding at the same time.

Using Language Routines to Develop Understanding

Use the **Professional Cards** for the following routines to plan for effective instruction.

Three Reads Lessons 10.1 and 10.2

Students read a problem three times with a specific focus each time.

1st Read What is the situation about?
2nd Read What are the quantities in the situation?
3rd Read What are the possible mathematical questions that you could ask for the situation?

Information Gap Lesson 10.1

Students recognize when information given in a problem situation is incomplete, and they pose questions and share knowledge with others to discover any missing facts or relationships and work together to solve the problem.

Critique, Correct, and Clarify Lesson 10.2

Students correct the work in a flawed explanation, argument, or solution method and share with a partner and refine the sample work.

Connecting Language to Triangle Inequalities

Watch for students' use of the new term listed below as they explain their reasoning and make connections with new concepts.

Key Academic Vocabulary

Current Development • New Vocabulary	
indirect proof a proof whose assumption is that the conclusion is false and shows that this assumption leads to a contradiction	

Linguistic Note

Listen for students who have difficulty breaking down the verbiage in the theorems in this module. Take time to parse out each theorem, underlining and explaining important words and phrases. The Hinge Theorem, for example, contains 50 words! To help EL students, have them draw sketches of the figures to be analyzed. Then have them read the theorem aloud, stopping after each unfamiliar phrase or term, and point to the relevant parts of the figures they have drawn. Ask students to rephrase the theorem in their own words, again pointing to figures that represent the relationships. Repeat this exercise with each theorem in this module until you are confident that the students understand its meaning and can use it successfully.

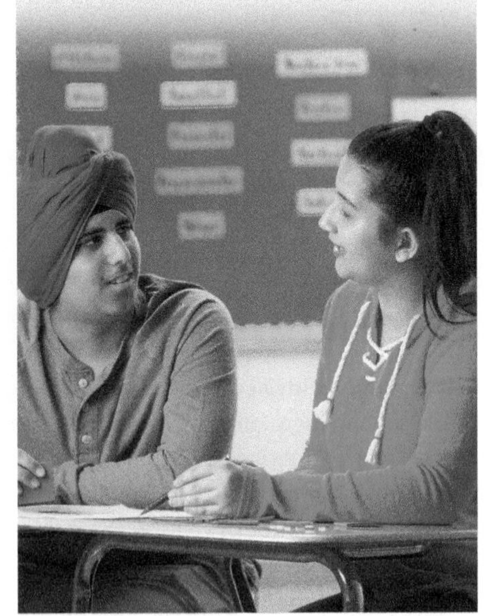

Module 10 Triangle Inequalities

Module Performance Task: Focus on STEM

Robotic Scissor Lift

Triangles are commonly used in robotics to add stability and flexibility to the design. The control system of a robotic scissor lift uses an elastic belt to adjust the height of the lift. A–D. See margin.

A. The lengths of the metal bars, AB and AC, are 8 inches. What is the range of possible lengths of the rubber band? Explain your reasoning.

B. What type of triangle is formed by these components? Explain your reasoning.

C. At one setting, the length of the rubber band is 3 inches. At another setting, the length of the rubber band is 5 inches. At which setting is the measure of $\angle A$ greater? Explain your reasoning.

D. How does the length of the rubber band affect the height of the scissor lift? Explain your reasoning.

Mathematical Connections

Task Part	Prerequisite Skills
Part A	Students should be familiar with the Triangle Inequality Theorem and be able to solve a compound inequality. **(10.1 and Algebra 1, 2.5)**
Part B	Students should be familiar with different types of triangles. **(Grade 7, 9.3)**
Part C	Students should be familiar with the Converse of the Hinge Theorem. **(10.2)**
Part D	Students should be able to rewrite the Pythagorean Theorem and use it to solve for a, the length of a leg in right triangle ABC, where c is the length of the hypotenuse and b is the length of the other leg: $a = \sqrt{c^2 - b^2}$. **(Grade 8, 11.3)**

Robotic Scissor Lift

Overview

A robotic scissor lift is a mechanism of linked supports (arms) that can be used to raise a load to different heights. The supports are in an X-shape, like the blades of a scissor—hence, its name. A robotic scissor lift is controlled by an *actuator*, a pneumatic, hydraulic, or mechanical device that applies a force needed to lift an object. This force depends on the height to which the object needs to be raised and the lengths of the arms of the lift. Robotic scissor lifts have many industrial applications, and models of robotic scissor lifts are popular devices in robotic competitions.

Career Connections

Students interested in robotics have a number of careers to choose from, such as mechanical, electrical, and aerospace engineering, and of course, computer science. Possible careers are also available in sales, marketing and purchasing departments of commercial manufacturers, and users of these devices.

Answers

A. The lengths of the given sides of the triangle are both 8 inches, so by the Triangle Inequality Theorem, the length of the rubber band is greater than 0 inches and less than 16 inches.

B. The lengths of two sides of the triangle are the same, so the triangle is an isosceles triangle.

C. The measure of $\angle A$ is greater when the length of the rubber band is 5 inches. This is true because of the Converse of the Hinge Theorem.

D. As the length of the rubber band decreases, the height of the scissor lift increases. Letting the length of the rubber band be x, then the height is given by $h = \sqrt{8^2 - \left(\frac{x}{2}\right)^2}$.

ONLINE

Assign the Digital Are You Ready? to power actionable reports including

- proficiency by standards
- item analysis

Are You Ready?

Diagnostic Assessment

- Diagnose prerequisite mastery.
- Identify intervention needs.
- Modify or set up leveled groups.

Have students complete the *Are You Ready?* assessment on their own. Items test the prerequisites required to succeed with the new learning in this module.

Solve Two-Step Inequalities Students will use their knowledge of solving simple two-step linear inequalities to apply the Triangle Inequality Theorem and determine possible values for the length of the third side of a triangle.

Draw Triangles with Given Conditions Students will use their knowledge of triangles to determine whether a triangle exists given the measures of its three sides or three angles.

Solve Compound Inequalities Students will apply their previous knowledge of solving compound linear inequalities to determine the range of possible values for the lengths of the third side of a triangle.

ONLINE 

Are You Ready?

Complete these problems to review prior concepts and skills you will need for this module.

Solve Two-Step Inequalities

1–6. See Additional Answers for number lines.

Solve each inequality and graph the solution on a number line.

1. $3x - 5 > 10$ $x > 5$
2. $4z + 7 \leq 23$ $z \leq 4$
3. $7t - 4 \geq -32$ $t \geq -4$
4. $-2b + 4 > 6$ $b < -1$
5. $2x + 6 < 10$ $x < 2$
6. $-5y - 4 \leq 11$ $y \geq -3$

Draw Triangles with Given Conditions

Draw a triangle with the given measurements, if possible.

7, 9. See Additional Answers.

7. Side Lengths: 11, 15, 20
8. Side Lengths: 6, 9, 17 not possible
9. Angle Measures: 50°, 50°, 80°
10. Angle Measures: 30°, 50°, 90° not possible

Solve Compound Inequalities

11–15, 18. See Additional Answers for number lines.

Solve each compound inequality and graph the solution on a number line.

11. $-4 < 3x - 1 < 8$ $-1 < x < 3$
12. $2x + 5 < -1$ OR $2x - 5 \geq 3$ $x < -3$ OR $x \geq 4$
13. $5x + 3 \leq 8$ OR $4x + 1 \geq 17$ $x \leq 1$ OR $x \geq 4$
14. $-2 \leq 3x - 8$ AND $3x - 8 < 16$ $2 \leq x < 8$
15. $1 \leq 2x + 3 \leq 7$ $-1 \leq x \leq 2$
16. $2x - 1 < -7$ AND $-3x + 9 < -3$ no solution
17. $5y - 16 \leq -1$ AND $3y + 4 > 13$ no solution
18. $-4x + 3 > 7$ OR $6x - 15 \geq 3$ $x < -1$ OR $x \geq 3$

Connecting Past and Present Learning

Previously, you learned:

- to write and solve inequalities that model geometric figures,
- to prove theorems about triangles, and
- to find the measure of an angle of a triangle using the other angle measures.

In this module, you will learn:

- to write inequalities comparing the side and angle measures within a triangle,
- to write inequalities comparing the side measures of two triangles, and
- to write inequalities comparing the angle measures of two triangles.

298

DATA-DRIVEN INTERVENTION

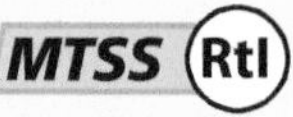

Concept/Skill	Objective	Prior Learning *	Intervene With
Solve Two-Step Inequalities	Solve two-step inequalities and graph the solutions on a number line.	Grade 7, Lesson 8.3	• Tier 3 Skill 5 • Reteach, Grade 7 Lesson 8.3
Draw Triangles with Given Conditions	Draw, if possible, triangles that have given side lengths and/or angle measures.	Grade 7, Lessons 9.2 and 9.3	• Tier 2 Skill 15 • Reteach, Grade 7 Lessons 9.2 and 9.3
Solve Compound Inequalities	Solve compound inequalities involving *and* or *or*, and graph the solutions on a number line.	Algebra 1, Lesson 2.5	• Tier 2 Skill 16 • Reteach, Algebra 1 Lesson 2.5

* Your digital materials include access to resources from Grade 6–Algebra 2. The lessons referenced here contain a variety of resources you can use with students who need support with this content.

10.1 Inequalities Within a Triangle

LESSON FOCUS AND COHERENCE

Mathematics Standards

- Prove theorems about triangles. Theorems include: measures of interior angles of a triangle sum to 180°; base angles of isosceles triangles are congruent; the segment joining the midpoints of two sides of a triangle is parallel to the third side and half the length; the medians of a triangle meet at a point.
- Use geometric shapes, their measures, and their properties to describe objects (e.g., modeling a tree trunk or a human torso as a cylinder).

Mathematical Practices and Processes 

- Use appropriate tools strategically.
- Construct viable arguments and critique the reasoning of others.
- Look for and make use of structure.
- Reason abstractly and quantitatively.

I Can Objective 

I can determine the relative sizes of angles and sides in a triangle.

Learning Objective

Decide when three lengths can form a triangle, find the possible ranges of side lengths for the third side of a triangle, and order and compare the side lengths and angle measures in a triangle.

Language Objective

Explain how to determine when given side lengths describe a triangle.

Lesson Materials: straightedge, compass

Mathematical Progressions

Prior Learning	Current Development	Future Connections
Students: • investigated the relationship between an exterior angle and the remote interior angles of a triangle. **(Gr8, 4.1)** • determined the measure of the third angle in a triangle given information about the other two angle measures. **(9.1)** • wrote and solved inequalities. **(Alg1, 2.4 and 2.5)**	**Students:** • write measurements of a triangle in order by size. • determine whether a triangle can exist given side lengths. • determine possible lengths for the third side of a triangle given the other two lengths.	**Students:** • will investigate segment length and angle measure relationships among lines intersecting circles. **(15.4 and 15.5)**

PROFESSIONAL LEARNING

Visualizing the Math

When finding the possible range of values for a missing side in a triangle given two side lengths, students may find it helpful to organize their work in a three-column table. Once students have solved the three possible inequalities, encourage them to cross out the inequality with the negative value and combine the remaining two into the final solution.

$3 + 5 > x$ $8 > x$	$3 + x > 5$ $x > 2$	~~$5 + x > 3$~~ ~~$x > -2$~~
$2 < x < 8$		

WARM-UP OPTIONS

ONLINE

PROJECTABLE & PRINTABLE

ACTIVATE PRIOR KNOWLEDGE • Find the Measure of the Third Angle in a Triangle

Use these activities to quickly assess and activate prior knowledge as needed.

Problem of the Day

The flooring pattern set in a hotel foyer uses a right triangle-shaped tile with the measurements shown below. Explain the method a flooring company can use to find the other angle measures. Since the entryway flooring is a right triangle, one angle is 90°. The sum of the angles is 180°, so the third angle is 180 − 90 − 32 = 58°.

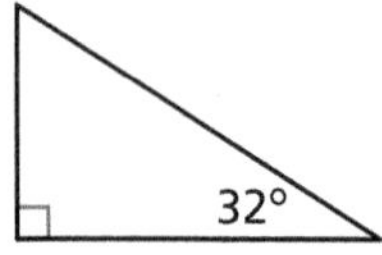

Quick Check for Homework

As part of your daily routine, you may want to display the Teacher Solution Key to have students check their homework.

Make Connections

Based on students' responses to the Problem of the Day, choose one of the following:

1. Project the Interactive Reteach, Geometry, Lesson 9.1.
2. Complete the Prerequisite Skills Activity:

Have students work in pairs to find all the missing angle measures in the diagram.

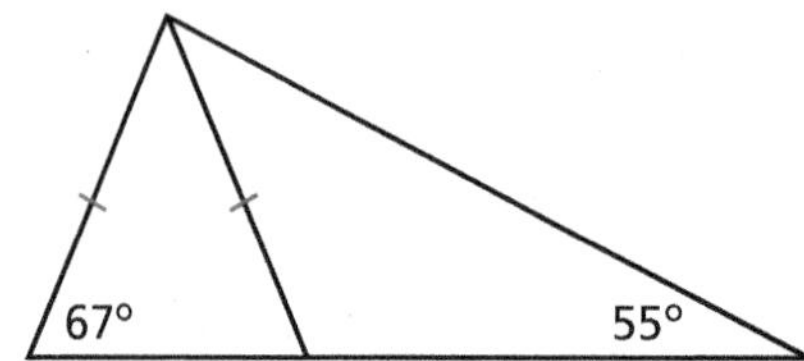

- *The left triangle is what type of triangle and why?* Isosceles because the side markings show that two sides are congruent.
- *What is true about the base angles in an isosceles triangle?* They have the same measure.
- *Two angles that form a straight line form a linear pair. What is the measure of a straight angle?* 180°
- *What is the sum of the three angles in the largest triangle?* 180°
- *What is the missing angle in the left triangle?* 46°
- *What is the other base angle of the right triangle?* 113°
- *What is the top angle of the right triangle?* 12°

If students continue to struggle, use Tier 2 Skill 5.

SHARPEN SKILLS

If time permits, use this on-level activity to build fluency and practice basic skills.

Mental Math

Objective: Students demonstrate they can find the missing angle in a triangle.
Materials: index cards

Draw and label triangles on index cards where two of the three angles are labeled. Have pairs of students take a card and use mental math to determine the missing angle measure.

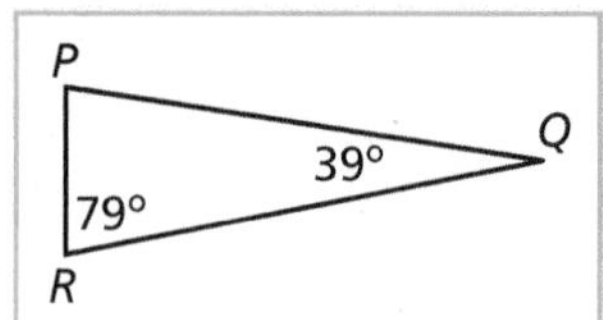

PLAN FOR DIFFERENTIATED INSTRUCTION

Small-Group Options

Use these teacher-guided activities with pulled small groups.

On Track

Materials: index cards, rulers, protractors

Give each student an index card. Ask students to draw a triangle and measure and label the side lengths. Have students trade cards with a partner and order the angles from smallest to largest. Then have students use a protractor to measure the angles and confirm their answers.

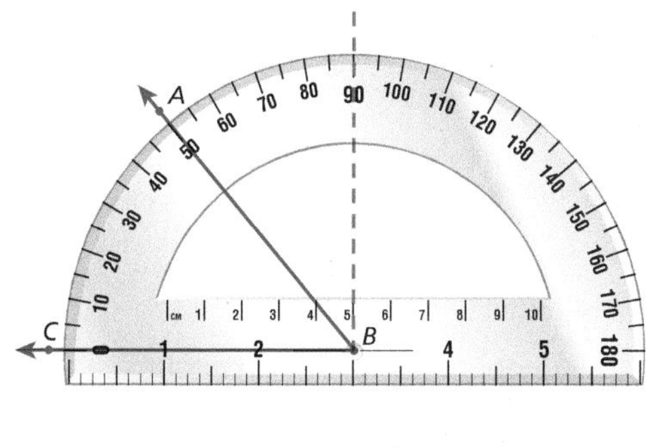

Almost There

Materials: Bubble Map (Teacher Resource Masters)

Complete the top three bubbles of the bubble map with students as shown.

- Have pairs of students work together to draw an example of a triangle in one bubble.
- Then ask students to find the inequality that represents the possible range of side lengths for the unknown side.

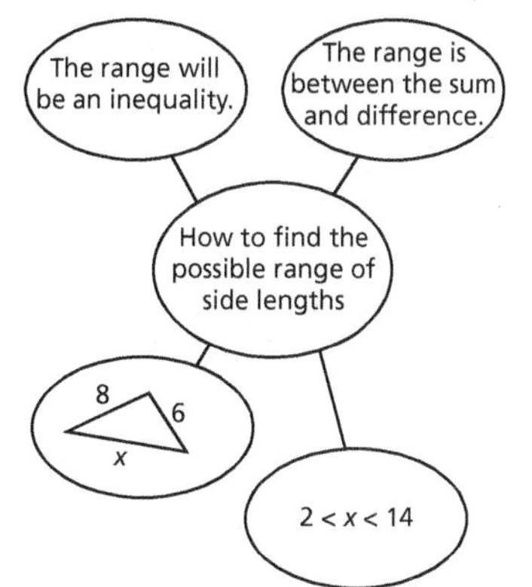

Ready for More

Draw the following triangle on the board.

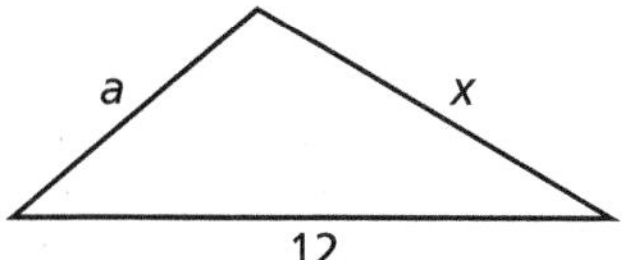

Have students write an inequality to show the range of values for the side x. Then have students discuss the following questions in small groups.

- What must be true about the value of a?
- What happens if a is equal to 12?
- What happens if a is greater than 12?
- What happens if a is less than 12?

Math Center Options

Use these student self-directed activities at centers or stations. Key: ● Print Resources ● Online Resources

On Track

- ● Interactive Digital Lesson
- ●● Journal and Practice Workbook
- ● Module Performance Task

Almost There

- ● Reteach 10.1 (printable)
- ● Interactive Reteach 10.1
- ● RtI Tier 2 Skill 5: Angle Relationships in Triangles
- ● Illustrative Mathematics: Sum of Angles in a Triangle

Ready for More

- ● Challenge 10.1 (printable)
- ● Interactive Challenge 10.1

Unit Project Check students' progress by asking how big or small the third side of a triangle can be in comparison with the other two sides.

View data-driven grouping recommendations and assign differentiation resources.

During the *Spark Your Learning*, listen and watch for strategies students use. See samples of student work on this page.

Use Angle Measures to Compare Relative Sides Strategy 1

In each triangle that had an orange and a purple segment, I compared the side lengths based on the angles given in the diagram. For example, in the triangle with the 60° and 85° angles, the purple segment is longer because it is across from the larger angle. In the triangle with the 70° and 45° angles, the purple segment is longer because it is across from the larger angle. I compared all of the triangles this way and overall, the purple segments are longer.

If students . . . compare angle measures and the sides across from them to solve the problem, they are showing an understanding of the properties of angles in triangles from Grade 8.

Have these students . . . explain how they determined their side comparisons. **Ask:**

- Q How did you know which side was longer in each triangle?
- Q How did you know which path was longer overall?

Measure Path Segments with a Ruler Strategy 2

Some of the segments of the orange and purple paths are the same, so I don't need to measure those. I measured the orange segments and found the sum. Then I measured the purple segments and found the sum. The purple path is longer than the orange path.

If students . . . use a ruler to measure the sides and find the sum, they understand they need to compare the overall length of the two paths but may not understand how to use the angle measures to make these comparisons.

Activate prior knowledge . . . by having students compare the side lengths across from the angles in the triangle with the 60° and 85° angles. **Ask:**

- Q Which side is longer?
- Q How can you use the angle measures to compare side lengths without using a ruler?

COMMON ERROR: Makes General Observations

I traced the orange path with my finger. Then I traced the purple path with my finger. It seems that the paths are close to being the same length because there are some segments of the path that are the same for the orange and purple courses.

If students . . . make general observations, then they aren't seeing how to use the angle measures to compare sides.

Then intervene . . . by pointing out that the side across from a larger angle is longer than a side across from a smaller angle. **Ask:**

- Q When a triangle has an obtuse angle and an acute angle, what is true about the sides across from the two angles?
- Q Find a triangle in the diagram that has one orange and one purple path as two sides. How are the angle and side lengths related?

10.1

Inequalities Within a Triangle

I Can determine the relative sizes of angles and sides in a triangle.

Spark Your Learning

The map shows two paths through a ropes course, one orange and one purple.

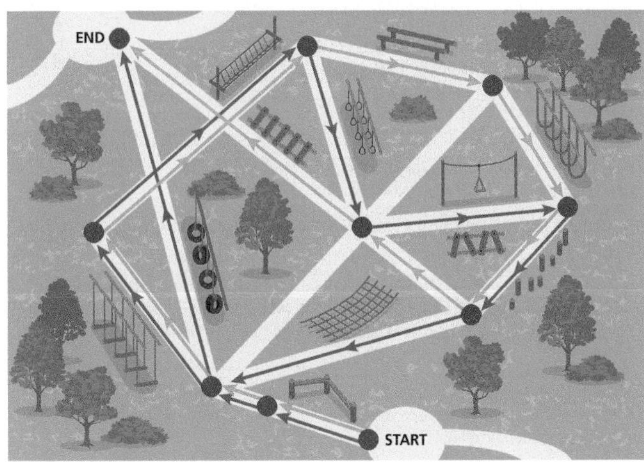

Complete Part A as a whole class. Then complete Parts B–D in small groups.

A. What is a mathematical question you can ask about this situation? What information would you need to know to answer your question? A. Which path is the longest?; the length of each section of the course

B. What variable(s) are involved in this situation? What unit of measurement would you use for each variable? See Additional Answers.

C. To answer your question, what strategy and tool would you use along with all the information you have? What answer do you get? See Strategies 1 and 2 on the facing page.

D. Does your answer make sense in the context of the situation? How do you know? See Additional Answers.

Turn and Talk Create a new path that is longer than any of the given paths. How do you know it is longer? See margin.

CULTIVATE CONVERSATION • Information Gap

Ask students questions to help them decide what information they need to answer the question, "Which path is the longest?"

1. Is there enough information in the picture to determine the length of each path? Explain. no; Without a ruler or some other given measurements, you do not know the measure of the segments in each path.

2. How could knowing the angles measures in the triangles formed by the paths help determine which path is longer? Segments across from greater angle measures would be longer than segments across from smaller angle measures.

3. Would you need know all of the angle measures? Why or why not? no; If you know two angles in a triangle, you can find the third measure because you know the sum of the angles is 180°. You need to know angle measures across from the different paths so you can compare the lengths.

LESSON 10.1

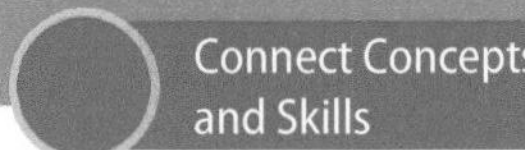

Connect Concepts and Skills

1 Spark Your Learning

▶ MOTIVATE

- Have students look at the illustration in their books and read the information contained in the photo. Then complete Part A as a whole-class discussion.
- Give the class the additional information they need to solve the problem. This information is available online as a printable and projectable page in the Teacher Resources.
- Have students work in small groups to complete Parts B–D.

▶ PERSEVERE

If students need support, guide them by asking:

Q **Advancing • Use Tools** Which tool could you use to solve the problem? Why choose that tool and not some other? Students' choices of tools and reasons for choosing them will vary.

Q **Assessing** What are two types of measurements that would help determine which path is longer? the angle measures and the path lengths

Q **Assessing** If you could measure with a ruler, how would you find the total length of the path? You would measure each section and find the sum of the measurements.

Q **Advancing** How can you use the angle measures to determine the relative lengths of the triangles' sides? An angle measure that is larger would indicate the side opposite the angle is longer than that of a side opposite a smaller angle.

Turn and Talk When you create a new path for the ropes course, have students consider the length of each segment of the path. It may help to label the sides with an angle measure of 30° as x and label the remaining sides as expressions with x, such as $2x$, $2x + 10$, etc. Point out that the paths they make must be along the existing ropes, shown in tan. Answers will vary.

▶ BUILD SHARED UNDERSTANDING

Select groups of students who used various strategies and tools to share with the class how they solved the problem. As they present their solutions, have each group discuss why they chose a specific strategy and tool.

② Learn Together

Build Understanding

Task 1 (MP) **Use Tools** Students use a compass set at a given radius to show possible side lengths in a triangle.

Sample Guided Discussion:

Q What happens when one side of the triangle is formed using the radius of the circle from point B to the point where $\overline{AB}$ intersects the circle? The side of the triangle lies along $\overline{AB}$ and does not form an angle.

Q Why can't the sum of two sides of a triangle be equal to the third side? The two sides arranged so that one of their endpoints touches the endpoint of the other segment would be the same length as the third side. The only way for both of the segment's other endpoints to intersect the endpoints of the third side would be for the two sides to form a straight line that lies directly on top of the third side. If the two sides form an angle, the endpoints of those sides would not intersect the endpoints of the third side, but rather somewhere between the endpoints of the third side.

Turn and Talk Remind students to think about the definition of an isosceles triangle. Is it possible to make another side equal in length to $\overline{AB}$? Is it possible to make two sides of the triangle equal to the radius of the circle?. No, because 1.5 + 1.5 is not greater than 3; Point C would be on $\overline{AB}$.

Build Understanding

Explore Triangle Inequalities

You have learned what combinations of angles are possible in a triangle and the relationships that exist among those angles. A relationship exists among the lengths of the sides of a triangle as well. A–D. See Additional Answers.

1 A. In $\triangle ABC$, $AB = 3$ inches and $BC = 1.5$ inches. To draw $\triangle ABC$, first draw a segment that is 3 inches long with endpoints A and B.

To determine all possible locations for vertex C, draw a circle centered at B with a radius of 1.5 inches.

How many triangles can be formed that have a side length of 3 inches and a side length of 1.5 inches?

B. Is it possible to place vertex C on the circle so that $\triangle ABC$ will have side lengths 1.5 inches, 3 inches, and 4 inches? Explain.

C. Is it possible to place vertex C on the circle so that AC will be greater than the sum of AB and BC? Explain.

D. Choose a placement for vertex C on the circle and draw the segments to form the triangle. Measure the side lengths and the angles. Where is the smallest angle in relation to the smallest side? Where is the largest angle in relation to the largest side?

Turn and Talk Is it possible to place vertex C on the circle so that $\triangle ABC$ will be an isosceles triangle? Explain. See margin.

Use the Triangle Inequality Theorem

The relationship among the lengths of the sides of a triangle has been summarized in the following theorem.

Triangle Inequality Theorem

The sum of the lengths of any two sides of a triangle is greater than the length of the third side.

$AB + BC > AC$
$BC + AC > AB$
$AC + AB > BC$

To be able to form a triangle, each of the three inequalities must be true. So, given three values, you can test to determine if they can be used as side lengths to form a triangle. To show that three values cannot be the side lengths of a triangle, you only need to show that one of the three triangle inequalities is false.

300

LEVELED QUESTIONS

Depth of Knowledge (DOK)	Leveled Questions	What Does This Tell You?
Level 1 **Recall**	What is the sum of the angles in a triangle? 180°	Students' answers will indicate they know the sum of the three angles in a triangle.
Level 2 **Basic Application of Skills & Concepts**	What types of triangles can be formed given the length of two sides, where one given side is 2 times the length of the other? Possible answers: right triangle, obtuse, scalene	Students' answers will indicate they understand that many types of triangles with different side lengths can be formed when given two side lengths.
Level 3 **Strategic Thinking & Complex Reasoning**	Two possible sides of a triangle measure x and $2x$. If a third possible side is equal to $3x$, can the three lengths be used to draw a triangle? No, the sum of the length of two sides of a triangle must be greater than the length of the third side. $x + 2x = 3x$, so the side lengths do not form a triangle.	Students' answers will indicate they understand that a triangle cannot be formed when two sides of the triangle are equal to the third side.

2 In an art class, Disha is designing a triangular picture frame using wooden strips of different lengths. Determine whether each set of wooden strips will form a triangle.

Set 1	Set 2
4 in., 5 in., 7 in.	3 in., 6 in., 11 in.

Compare the sum of each pair of possible side lengths to the third side length.

Set 1	Set 2
$4 + 5 > 7$	$3 + 6 \not> 11$
$5 + 7 > 4$	
$7 + 4 > 5$	

A. Explain why the wooden strips in Set 1 will form a triangle. A, B. See Additional Answers.

B. Why is only one inequality listed for Set 2?

C. What does the Triangle Inequality tell you about Set 2?
A triangle cannot be formed.

Turn and Talk In the same class, Arturo creates another design using three wooden strips with lengths 5 inches, 5 inches, and 5 inches. Does Arturo need to check all three inequalities to determine whether the design will form a triangle? Why? See margin.

3 Find the possible range of values for x in the triangle shown.
A–C. See Additional Answers.

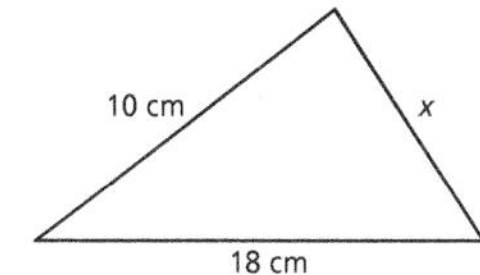

A. What is being compared in each of the inequalities below?

$10 + 18 > x$	$x + 10 > 18$	$x + 18 > 10$
$28 > x$	$x > 8$	$x > -8$

B. What information about the value of x is given in the first two inequalities?

C. Since $18 > 10$, is the information given about x useful? Explain.

Since the information in the third inequality is not useful, the range of values for x can be determined using the first two inequalities.

D. What is the range of values for x? $8 \text{ cm} < x < 28 \text{ cm}$

Turn and Talk Notice the range of values for x is between the sum of the two given side lengths and the difference of the two given side lengths. Is this relationship true for all triangles? Give examples to support your answer. See margin.

Task 2 MP **Construct Arguments** Students use inequalities to explain situations when triangles can be formed.

Sample Guided Discussion:

Q For Set 2, suppose the first inequality you write is $6 + 11 > 3$. Why is this not enough to prove the triangle can be formed? Although this inequality is true, the inequality $3 + 6 > 11$ is not.

Q How many inequalities must be false to prove a triangle cannot be formed given three sides? Only one false inequality proves the triangle cannot be formed.

Turn and Talk Sketch the triangle Arturo creates. All three sides form an equilateral triangle. Write and compare one possible inequality. no; All three inequalities are the same because all three sides are the same.

SUPPORT SENSE-MAKING Three Reads

Have students read the problem three times. Use the questions below for a different focus each read.

1. What is the situation about?
2. What are the quantities in the situation?
3. What are the possible mathematical questions that you could ask for the situation?

Task 3 MP **Use Tools** Students analyze three inequalities to determine the lengths of a third side of a triangle.

Sample Guided Discussion:

Q Why can't the length of the third side be equal to 28 cm? If the third side was equal to 28, that would be equal to the sum of the other two sides which would not form a triangle.

Q In the third inequality, it states that $x > -8$. Why can't x be a negative number? A side length is a positive distance and cannot be less than 0.

Turn and Talk Guide students to consider a triangle with two sides 4 and 7. Let a variable x be the third side. Write the three possible inequalities. Can you find the sum of the two sides? Can you find the difference? Yes, this is true for all triangles; For example, a triangle with sides 4, 7, and x, $3 < x < 11$. A triangle with sides 11, 15, x, $4 < x < 26$.

PROFICIENCY LEVEL

Beginning

Fill in the blanks for the following statements. The ________ side is across from the largest angle. The _________ side is across from the smallest angle.

Intermediate

A triangle has side lengths PQ, QR, and PR. What must be true about the lengths of the sides?

Advanced

Suppose you know the lengths of two sides of a triangle. Explain to a partner why you can find the range of the third side of the triangle with only two inequalities.

Step It Out

Task 4 (MP) **Use Structure** Students will compare the side lengths in order from smallest to largest to determine the angle measures from smallest to largest.

Sample Guided Discussion:

Q Why is the longest side across from the largest angle? A larger angle makes a bigger opening such that the side across must be longer to complete the triangle.

Q What do you know about the measure of $\angle T$? The measure of $\angle T$ is smaller than $\angle R$ and greater than $\angle S$.

Task 5 (MP) **Reason** Students use properties of right triangles to compare angle measures and side lengths.

Sample Guided Discussion:

Q What is another way you can find the measure of $\angle G$? You know the two non-right angles in a right triangle add up to 90°. Subtract 90 − 68 to find the third angle measure.

Q In Part B, how does the Pythagorean Theorem $a^2 + b^2 = c^2$ show the hypotenuse is the longest side of a right triangle? The sum of the squares of the two sides is equal to the square of the third side. This means the side c must be the longest.

Turn and Talk Ask students to name a triangle that that has two congruent angles, i.e. isosceles. In this triangle these angles would both be smaller than the right angle and are both the same measure, so this relationship also applies to the side lengths. The sides would be the same length.

Step It Out

Use Side-Angle Relationships in Triangles

The side-angle realtionships describe how the measures of the angles in a triangle are related to the side lengths of a triangle.

Side-Angle Relationships in Triangles

If one side of a triangle is longer than another side, then the angle opposite the longer side is larger than the angle opposite the shorter side.

$AC > BC$, so $m\angle B > m\angle A$.

If one angle of a triangle is larger than another angle, then the side opposite the larger angle is longer than the side opposite the smaller angle.

$m\angle D > m\angle F$, so $EF > DE$.

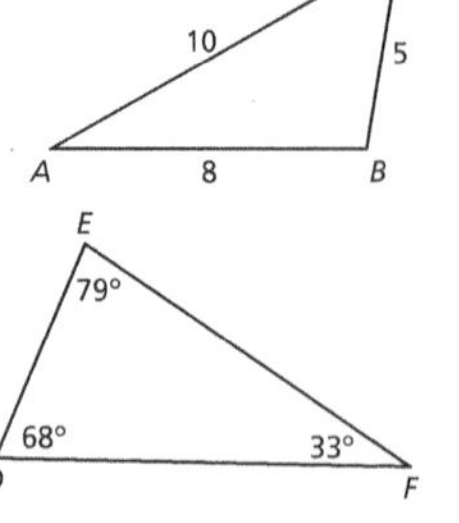

4 List the sides and angles in order from smallest to largest.

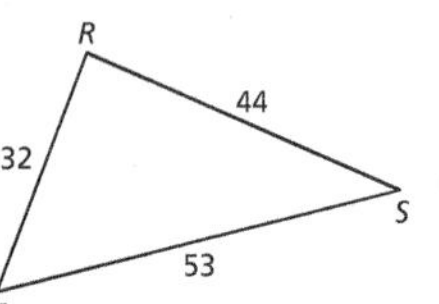

A. Why should the sides be listed in order from smallest to largest first, before listing the angles in order from smallest to largest? See Additional Answers.

Sides from smallest to largest: $\overline{RT}, \overline{RS}, \overline{ST}$

Angles from smallest to largest: $\angle S, \angle T, \angle R$

B. How do you know that $\angle S$ is the smallest angle?

C. How do you know that $\angle R$ is the largest angle?

B. $\angle S$ is opposite the shortest side of the triangle.
C. $\angle R$ is opposite the longest side of the triangle.

5 List the sides and angles in order from smallest to largest.

A. What theorem is used to find the missing angle measure? Triangle Sum Theorem

$m\angle G = 180° - (90° + 68°) = 22°$

Angles from smallest to largest: $\angle G, \angle F, \angle H$

Sides from smallest to largest: $\overline{FH}, \overline{GH}, \overline{FG}$

B. Why must the hypotenuse of a right triangle always be longer than either leg? Since both angles opposite the two legs are acute, the legs are smaller than the hypotenuse.

Turn and Talk Suppose $\angle F$ and $\angle G$ have the same measure. What can you conclude about the sides opposite those angles? See margin.

302

Check Understanding

Use a compass and straightedge to decide whether each set of lengths can form a triangle. Explain your reasoning. 1, 2. See Additional Answers.

1. 7 cm, 9 cm, 18 cm

2. 2 in., 4 in., 5 in.

Determine whether a triangle can be formed with the given side lengths.

3. 3, 9, 11

3. Yes, the sum of the lengths of any two sides is greater than the length of the remaining side.

4. 9, 12, 21 No, $9 + 12 = 21$.

5. A triangle has sides with lengths 15, 23, and x. What is the range of possible values of x? $8 < x < 38$

6. Name each angle in the triangle at the right. For each angle, name the side that is opposite that angle. 6, 7. See Additional Answers.

7. If you know the side lengths of a triangle, how can you determine which angle is the largest and which angle is the smallest?

In Problems 8–10, use the given information about $\triangle ABC$ to list the sides and angles in order from smallest to largest.

8. $m\angle A = 60°$, $m\angle B = 75°$, $m\angle C = 45°$ $\angle C, \angle A, \angle B; \overline{AB}, \overline{BC}, \overline{AC}$

9. $m\angle A = 82°$, $m\angle B = 48°$, $m\angle C = 50°$ $\angle B, \angle C, \angle A; \overline{AC}, \overline{AB}, \overline{BC}$

10. $AB = 14$, $BC = 26$, $AC = 20$ $\overline{AB}, \overline{AC}, \overline{BC}; \angle C, \angle B, \angle A$

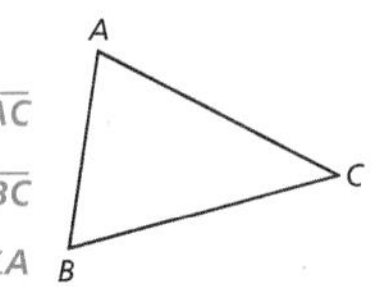

On Your Own

11. The construction below shows two possible triangles that can be formed when $AB = 3$ inches and $BC = 1.5$ inches. Describe what happens to the length of $\overline{AC}$ as point C moves counterclockwise around the circle toward point A.

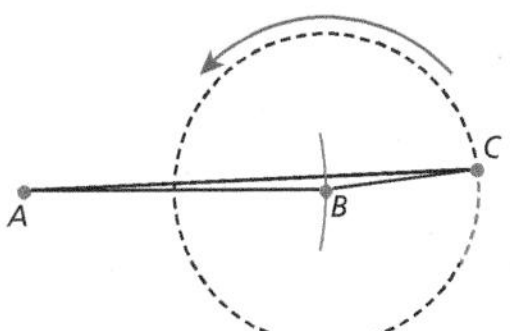

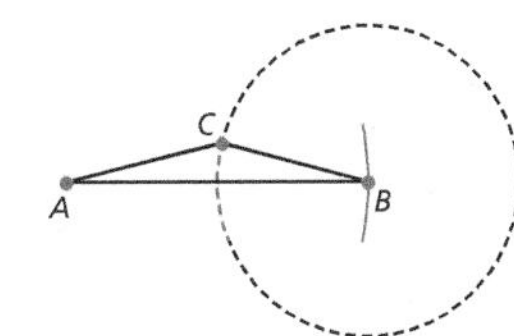

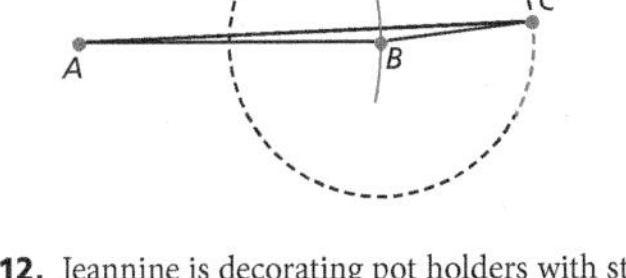

AC decreases as point C moves counterclockwise toward point A.

12. Jeannine is decorating pot holders with strips of fabric. She wants to make triangles using strips of fabric. Can she make a triangle with any of the following sets of strips of fabric? Explain your reasoning. See Additional Answers.

A. 4 inches, 2 inches, and 5 inches

B. 4 inches, 2 inches, and 1 inch no; $2 + 1 < 4$

C. 3 inches, 2 inches, and 5 inches no; $3 + 2 = 5$

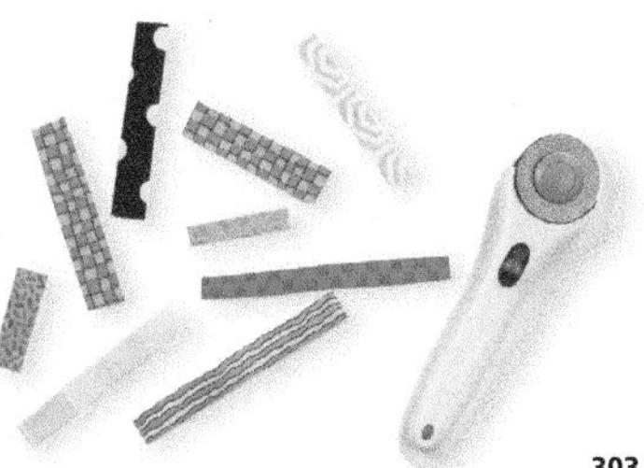

ONLINE

Assign the Digital On Your Own for

- built-in student supports
- Actionable Item Reports
- Standards Analysis Reports

On Your Own

Assignment Guide

The chart below indicates which problems in the On Your Own are associated with each task in the Learn Together. Assign daily homework for tasks completed.

Learn Together Tasks	On Your Own Problems
Task 1, p. 300	Problem 11 and 31
Task 2, p. 301	Problems 12, 19, and 28–30
Task 3, p. 301	Problems 13–18 and 27
Task 4, p. 302	Problems 20–23, 28, and 29
Task 5, p. 302	Problems 24–26

3 Check Understanding

Formative Assessment

Use formative assessment to determine if your students are successful with this lesson's learning objective.

Students who successfully complete the Check Understanding can continue to the On Your Own practice.

For students who miss 1 problem or more, work in a pulled small group using the Almost There small-group activity on page 299C.

ONLINE

Assign the Digital Check Understanding to determine

- success with the learning objective
- items to review
- grouping and differentiation resources

4 Differentiation Options

Differentiate instruction for all students using small-group activities and math center activities on page 299C.

Reteach

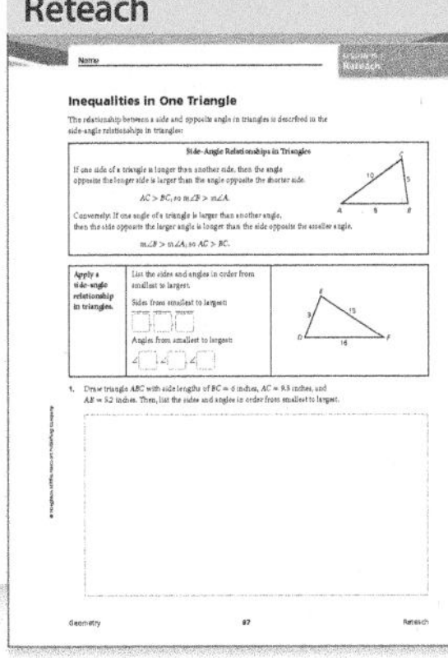
Inequalities in One Triangle

Challenge

Inequalities in One Triangle

Questioning Strategies

Problem 17 There are several ways the triangle could be isosceles. One way is when side $\overline{BC}$ or $\overline{AC}$ is also 8 feet. Another way would be for sides $\overline{BC}$ and $\overline{AC}$ to both have a different length. What does it mean for a triangle to be isosceles and what does that mean for this triangle?

A triangle is isosceles when two of the sides have the same measure. Since the angles are not labeled in this diagram, there is more than one way to make the triangle isosceles.

Use the two given side lengths of a triangle to describe the possible values for x, which represents the third side length.

13. 7, 13 $6 < x < 20$

14. 45, 44 $1 < x < 89$

15. 23, 14 $9 < x < 37$

16. 9, 15 $6 < x < 24$

17. STEM Nakia is an architect designing a house with a peaked roof. She is trying to decide what the limitations are on her design. If AB is 8 feet and $\triangle ABC$ will be isosceles, describe the possible lengths for AC.
$0 < AC < 16$

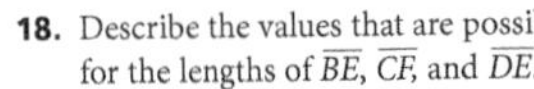

18. Describe the values that are possible for the lengths of $\overline{BE}$, $\overline{CF}$, and $\overline{DE}$.
$0 < BE < 8$;
$1 < CF < 7$, $2 < DE < 10$

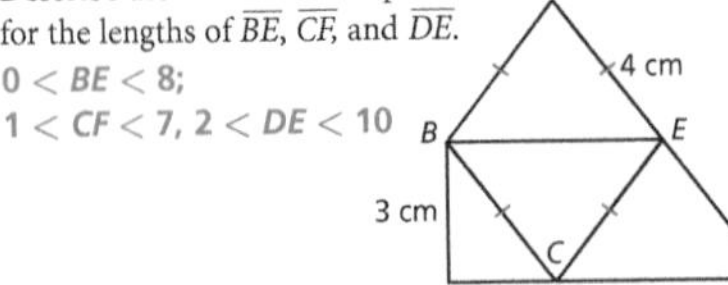

In Problems 19 and 20, use the diagram to prove the Triangle Inequality Theorem.

Given: $\triangle ABC$

Prove: (1) $AB + BC > AC$
(2) $AB + AC > BC$
(3) $AC + BC > AB$

19. If the longest side of $\triangle ABC$ is $\overline{AB}$, why are (1) and (2) true? If $\overline{AB}$ is the longest side, then the length of any other side added to AB will be greater than the length of the third side.

20. Copy and complete the proof to prove (3) $AC + BC > AB$.

Statements	Reasons
1. ?	1. Given
2. Locate D on $\overrightarrow{AC}$ so that $BC = DC$.	2. Ruler Postulate
3. $AC + DC =$?	3. Segment Addition Postulate
4. $\angle 1 \cong \angle 2$	4. ?
5. $m\angle 1 = m\angle 2$	5. ?
6. $m\angle ABD = m\angle 2 +$?	6. Angle Addition Postulate
7. $m\angle ABD > m\angle 2$	7. Comparison Property of Inequality
8. $m\angle ABD > m\angle 1$	8. ?
9. $AD > AB$	9. ?
10. $AC + DC > AB$	10. ?
11. ?	11. Substitution

1. $\triangle ABC$
3. AD
4. Isosceles Triangle Theorem
5. Definition of congruent angles
6. $m\angle 3$
8. Substitution
9. Side-Angle Relationships in Triangles
10. Substitution
11. $AC + BC > AB$

List the sides and angles in order from smallest to largest.

21.

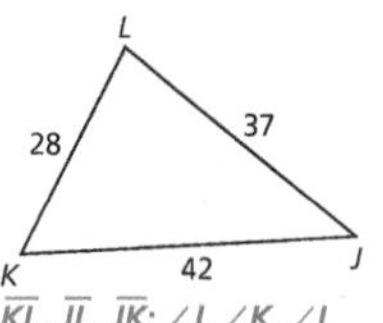

$\overline{KL}$, $\overline{JL}$, $\overline{JK}$; $\angle J$, $\angle K$, $\angle L$

22.

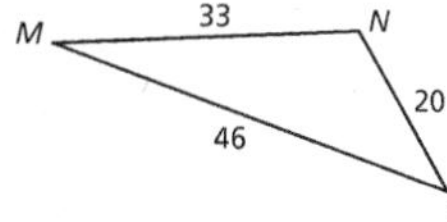

$\overline{NP}$, $\overline{MN}$, $\overline{MP}$; $\angle M$, $\angle P$, $\angle N$

23.

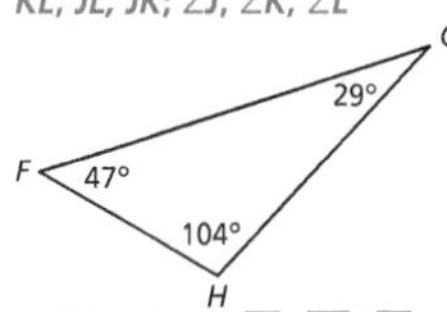

$\angle G$, $\angle F$, $\angle H$; $\overline{FH}$, $\overline{GH}$, $\overline{FG}$

24.

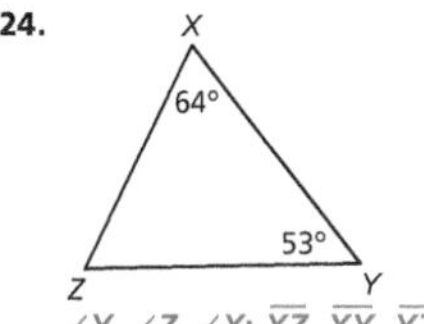

$\angle Y$, $\angle Z$, $\angle X$; $\overline{XZ}$, $\overline{XY}$, $\overline{YZ}$

25. Navigation A large ship is sailing between three small islands. To do so, the ship must sail between two pairs of islands, avoiding sailing between a third pair. The safest route is to avoid the closest pair of islands. Which is the safest route for the ship?

25–27. See Additional Answers.

26. Three cell phone towers form $\triangle PQR$. The measure of $\angle Q$ is 10° less than the measure of $\angle P$. The measure of $\angle R$ is 5° greater than the measure of $\angle Q$. Which two towers are closest together?

27. In any triangle ABC, suppose you know the lengths of $\overline{AB}$ and $\overline{BC}$, and suppose that $AB > BC$. If x is the length of the third side $\overline{AC}$, use the Triangle Inequality Theorem to show that $AB - BC < x < AB + BC$. That is, x must be between the difference and the sum of the other two side lengths.

Prove that the statements are true.

28, 29. See Additional Answers.

28. Given: $\triangle ABC \cong \triangle DEF$

Prove: $d + e > c$

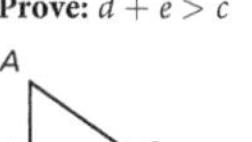

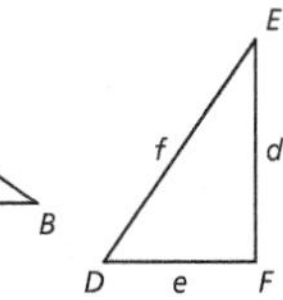

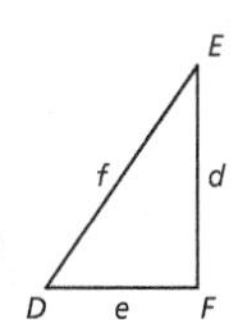

29. Given: $\triangle ABC$

Prove: $AC > AB$

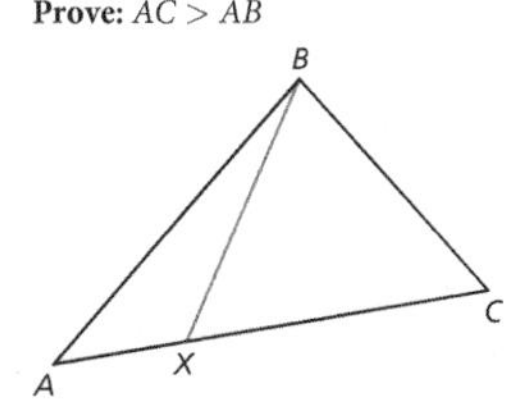

Questioning Strategies

Problem 21 Suppose you wrote the side lengths in order from largest to smallest. How can you use this list to find the angle measures from smallest to largest? The smallest angle will be across from the smallest side. Look for the smallest side length to identify the smallest angle is across from it.

Watch for Common Errors

Problem 26 Students may not know how to label the angles in order to compare them. Encourage students to draw a diagram labeling the measures of the angles with respect to each other. Have students label angle P with a measure of $x°$, then label angle Q, followed by angle R.

5 Wrap-Up

Summarize learning with your class. Consider using the Exit Ticket, Put It in Writing, or I Can scale.

Exit Ticket 

On a map, paths from Brad's house, his school, and the library form a triangle. Brad knows his house is 2 miles from his school. He also knows his house is 1.5 miles from the library. How far apart are the school and library? There are many possible distances between the school and library, but it is greater than 0.5 miles and less than 3.5 miles. If the distance between the school and library is x miles, then the possible values for x are $0.5 < x < 3.5$.

Put It in Writing 

If you are given a set of three possible sides of a triangle, describe how you can tell if the sides form a triangle.

I Can 

The scale below can help you and your students understand their progress on a learning goal.

4	I can determine the relative sizes of angles and sides in a triangle, and I can explain my reasoning to others.
3	I can determine the relative sizes of angles and sides in a triangle.
2	I can describe the relationships between sides and angles in a triangle.
1	I can identify some properties of sides and angles in triangles.

Spiral Review • Assessment Readiness

These questions will help determine if students have retained information taught in the past and can also prepare them for high-stakes assessments. Here, students must find the length of a triangle's midsegment using the Triangle Midsegment Theorem **(9.5)**, find the third angle of a triangle using the Triangle Sum Theorem **(9.1)**, and solve problems involving triangle medians **(9.4)**.

30. Analyzing Cases A hole on a golf course is a dogleg, meaning that it bends in the middle. A golfer will usually start by driving for the bend in the dogleg (from A to B), and then use a second shot to get the ball to the green (from B to C). Sandy believes she may be able to drive the ball far enough to reach the green in one shot, avoiding the bend (from A directly to C). Sandy knows she can accurately drive a distance of 250 yards. Should she attempt to drive for the green on her first shot? Explain. yes; Since AC is less than 250 yd, Sandy has a good chance of reaching the green in one shot.

31. (Open Middle™) Fill in the boxes with possible measures for the three angles such that the measure of angle A is the greatest possible. Use each digit from 1 to 9 at most one time. See Additional Answers.

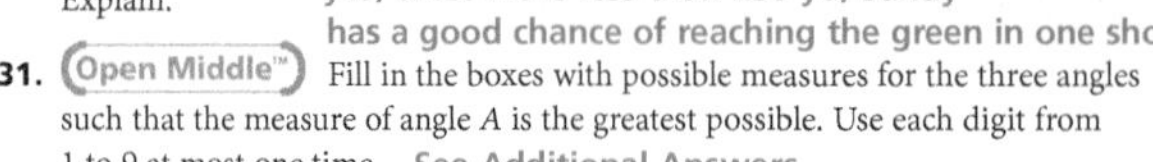

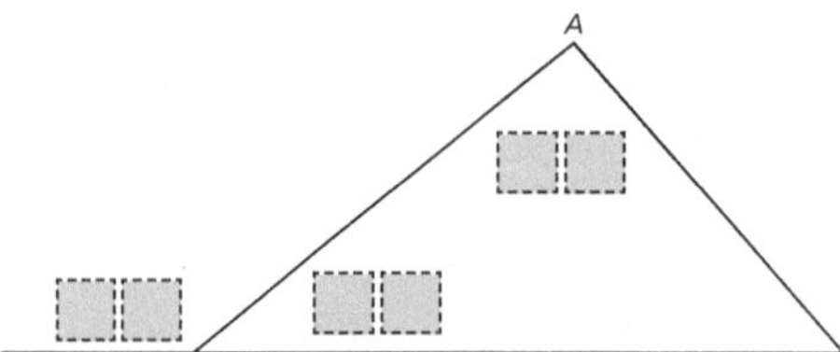

Spiral Review • Assessment Readiness

32. Find the length of $\overline{VW}$.

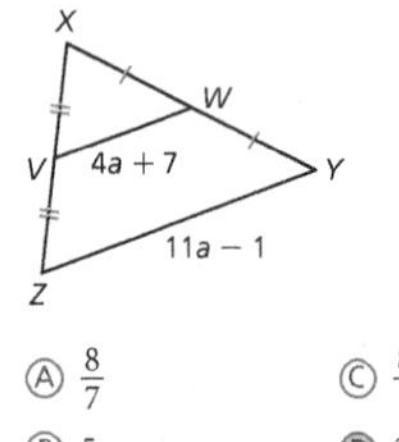

Ⓐ $\frac{8}{7}$ Ⓒ $\frac{81}{7}$

Ⓑ 5 Ⓓ 27

33. A triangle has angles that measure 46° and 76°. What is the measure of the third angle?

Ⓐ 14° Ⓒ 58°

Ⓑ 44° Ⓓ 122°

34. Find GF if $BF = 45$.

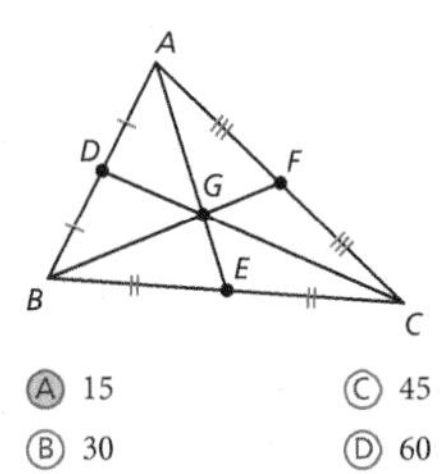

Ⓐ 15 Ⓒ 45

Ⓑ 30 Ⓓ 60

 **I'm in a Learning Mindset!**

Did I have any biases that affected my attempt to understand inequalities in one triangle? How did I address them?

306 Keep Going to Journal and Practice Workbook

Learning Mindset

Resilience Responds to Feedback

Remind students that when they are learning a new skill, they may have ideas about the situation from previous experiences. Remind them that diagrams and drawings can be deceiving and that they should focus more on the labels than the drawings themselves, unless they are told they are drawn to scale. Students should understand that if triangles do not have measurements labeled, it is easy to let your eyes decide which sides shortest or longest. In diagrams involving more than one triangle, encourage students to draw the triangles separately, labeling the known information. Tell students to use the theorems regarding triangle inequalities to make informed decisions. *What information do you know in each problem? What does this information along with the triangle inequality theorems tell you about the side lengths and angle measures? Use your prior knowledge about the different types of triangles and the Triangle Sum Theorem when solving problems.*

10.2 Inequalities Between Two Triangles

LESSON FOCUS AND COHERENCE

Mathematics Standards

- Prove theorems about triangles. *Theorems include: measures of interior angles of a triangle sum to 180°; base angles of isosceles triangles are congruent; the segment joining midpoints of two sides of a triangles is parallel to the third side and half the length; the medians of a triangle meet at a point.*
- Use geometric shapes, their measures, and their properties to describe objects (e.g., modeling a tree trunk or a human torso as a cylinder).

Mathematical Practices and Processes 

- Use appropriate tools strategically.
- Look for and make use of structure.
- Construct viable arguments and critique the reasoning of others.

I Can Objective 

I can determine the relative sizes of angles and sides in two triangles.

Learning Objective

Apply known information about a triangle(s) in a pair of triangles to determine relative lengths and angle measures.

Language Objective

Explain the difference between the Hinge Theorem and the Converse of the Hinge Theorem.

Vocabulary

Review: indirect proof

Lesson Materials: paper strips

Mathematical Progressions

Prior Learning	Current Development	Future Connections
Students: • investigated the relationship between an exterior angle and the remote interior angles of a triangle. **(Gr8, 4.1)** • determined the measure of the third angle in a triangle given information about the other two angle measures. **(9.1)** • wrote and solve inequalities. **(Alg1, 2.4 and 2.5)**	**Students:** • use Hinge Theorem and the Converse of the Hinge Theorem to solve mathematical and real-world problems. • write proofs using properties of triangle theorems. • write indirect proofs using properties of triangle theorems.	**Students:** • will investigate segment length and angle measure relationships among lines intersecting circles. **(Geo, 15.4 and 15.5)**

PROFESSIONAL LEARNING

Math Background

In this lesson, students add to their prior knowledge of inequalities in one triangle by investigating inequalities in two triangles from both an inductive and deductive perspective. The opening activity leads students to make a conjecture about the measure of an angle in a triangle and the resulting length of the side opposite that angle as they move toward an understanding of the Hinge Theorem and its converse.

Student work in Units 1, 2, and 4 regarding writing geometric proofs is revisited in this lesson to prove both theorems, which ultimately strengthens skills they will need to be successful with writing proofs in Unit 6. Developing deep understanding of various theorems regarding triangles, as there are many, is very important as students continue their study of geometry.

WARM-UP OPTIONS

ONLINE

PROJECTABLE & PRINTABLE 

ACTIVATE PRIOR KNOWLEDGE • Find Missing Angle Measure in a Triangle

Use these activities to quickly assess and activate prior knowledge as needed.

Problem of the Day 

The measure of one angle of a triangle is three times the measure of the smallest angle. The third angle is 25 more than the smallest angle. What is the measure of each angle? 31°, 56°, and 93°

Quick Check for Homework

As part of your daily routine, you may want to display the Teacher Solution Key to have students check their homework.

Make Connections

Based on students' responses to the Problem of the Day, choose one of the following:

1. Project the Interactive Reteach, Geometry, Lesson 9.1.
2. Complete the Prerequisite Skills Activity:

Have students work in pairs and tell them that the smallest angle in a triangle is represented by the variable x. Have each student write 2 expressions that would result in a value greater than x, such as $x + 10$ and $3x$. Students should use the three expressions to write an equation and solve for the value of x and for the degree measure of each angle of the triangle.

- *What is the sum of the interior angles of a triangle?* 180°
- *What equation can be written to solve for the value of the smallest angle of the triangle?* x + (expression #1) + (expression #2) = 180°
- *How do you find the measure of each angle once you know the value of x?* Substitute the value of x into the expressions representing the other two sides of the triangle and evaluate those expressions.

If students continue to struggle, use Tier 2 Skill 5.

SHARPEN SKILLS

If time permits, use this on-level activity to build fluency and practice basic skills.

Quantitative Comparison

Objective: Students make a comparison between two quantities.

Write the following problem on the board. Ask students to choose the letter representing the correct answer and to explain their reasoning.

Quantity A	**Quantity B**
the value of x in a triangle with angle measures (degrees) $2x$, $4x$, and $3x - 9$	the value of x in a triangle with angle measures (degrees) $x - 6$, $5x$, and $4x - 24$

A. Quantity A is greater.
B. Quantity B is greater.
C. The two quantities are equal.
D. The relationship cannot be determined from the information given.

The answer is C.

Triangle A:	Triangle B:
$9x - 9 = 180$	$10x - 30 = 180$
$9x = 189$	$10x = 210$
$x = 21$	$x = 21$

PLAN FOR DIFFERENTIATED INSTRUCTION

Small-Group Options

Use these teacher-guided activities with pulled small groups.

On Track

Materials: graphic organizer

Have students work in pairs. Ask them to find the range of possible values for *x* in the figure below, showing their work in the graphic organizer provided.

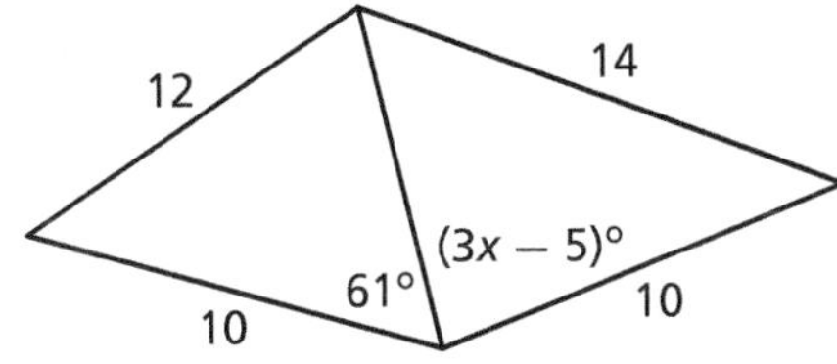

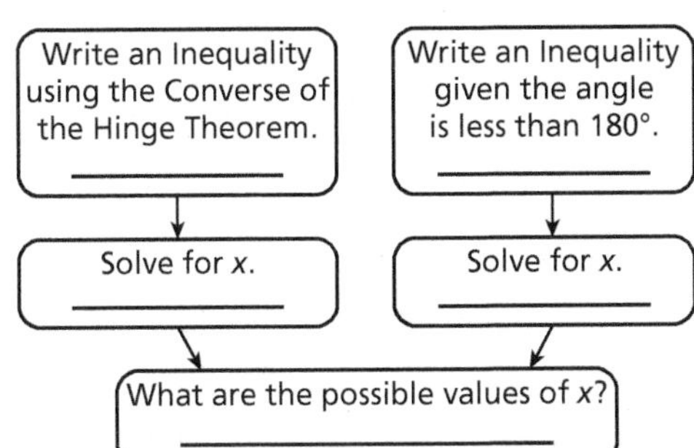

Almost There

Have students work in pairs and ask them to brainstorm a list of examples of hinges in real life and share with the class. Then, ask them to work together to write a response to the questions below, using picture to support their answers

- What time is it when the distance between the tip of a clock's hour hand and the tip of the clock's minute hand is the greatest?
- At which time is the distance between the tip of a clock's hour hand and the tip of the clock's minute hand greater, 3:00 or 3:10?
- At which time is the distance between the tip of a clock's hour hand and the tip of the clock's minute hand smaller, 12:35 or 5:20?

Ready for More

Have students work with a partner to prepare a response to the following question:

- $\triangle DEF$ has median $\overline{EG}$. If $DE < EF$, then $\angle EGD$ is *always, sometimes, or never* greater than $\angle EGF$. Explain your answer.

Math Center Options

Use these student self-directed activities at centers or stations.

Key: ● Print Resources ● Online Resources

On Track

- ● Interactive Digital Lesson
- ●● Journal and Practice Workbook
- ● Interactive Glossary (printable): **indirect proof**
- ● Module Performance Task
- ●● Standards Practice: Prove Congruence Theorems About Triangles

Almost There

- ● Reteach 10.2 (printable)
- ● Interactive Reteach 10.2
- ● RtI Tier 2 Skill 5: Angle Relationships in Triangles

Ready for More

- ● Challenge 10.2 (printable)
- ● Interactive Challenge 10.2

View data-driven grouping recommendations and assign differentiation resources.

During the *Spark Your Learning,* listen and watch for strategies students use. See samples of student work on this page.

Use Deductive Reasoning

Strategy 1

Two of the sections of each option are the same distance. Two of the interior angles measure 43° and 48°. Since 43° is less than 48°, Side 1 is shorter than Side 2, which makes the bottom flight path shorter than the top flight path.

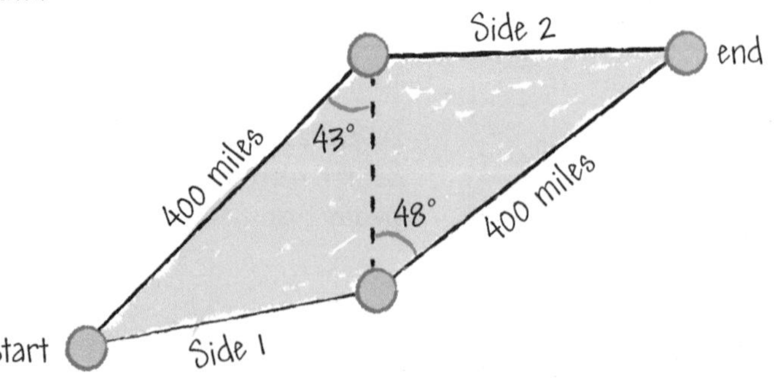

If students . . . use deductive reasoning to determine the relationship between angles in a triangle and sides opposite those angles, they are demonstrating an exemplary understanding of the properties of triangles.

Have these students . . . explain how they determined which path had the shortest distance. **Ask:**

- Q How did you use the given information and the relationship between the sides and angles to come to a conclusion?
- Q What given information was necessary in order to solve the problem?

Use Visual Reasoning

Strategy 2

The first leg of the top option is the same distance as the second leg of the bottom option. By looking at the picture, I can see that the second leg of the top option is longer than the first leg of the bottom option, which makes the top option longer than the bottom option.

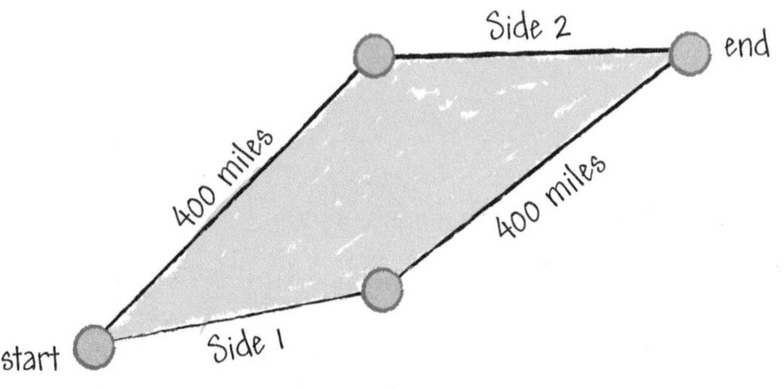

If students . . . use a visual inspection and estimation to solve the problem, they understand how the lengths of the segments shown in the diagram are related but not how to use the given angle measures to verify or confirm that what they see is true.

Activate prior knowledge . . . by having students discuss some important points. **Ask:**

- Q Did you prove mathematically that side 2 is longer than side 1?
- Q How can the angles given in the additional information help you verify that side 2 is longer than side 1?

COMMON ERROR: Uses Triangle Congruency Theorems

Since $\overline{AD} \cong \overline{BC}$, then $\angle ACD \cong \angle BAC$ because angles opposite congruent sides are also congruent. $\overline{AC} \cong \overline{CA}$ because of the Reflexive Property, so $\angle D \cong \angle B$ for the same reason. So, the triangles are congruent by AAS.

If students . . . use triangle congruence theorems to solve the problem, then they do not understand that the same properties that are true within one triangle do not apply two a figure containing two separate triangles.

Then intervene . . . by pointing out that the Isosceles Triangle Theorem does not apply in this situation. **Ask:**

- Q Do the sides labeled with equal distances reside within the same triangle?
- Q Do the angles opposite the two given congruent segments have to be congruent? Can you draw a counterexample?

Inequalities Between Two Triangles

I Can determine the relative sizes of angles and sides in two triangles.

Spark Your Learning

A passenger, flying in a small plane, has two options of flights to get from the starting airport to the destination airport.

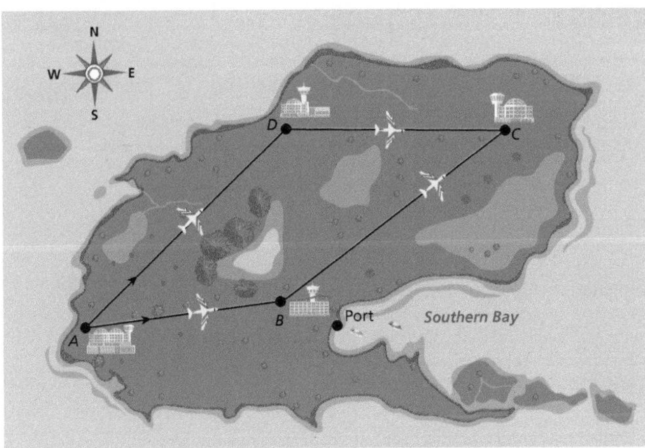

Complete Part A as a whole class. Then complete Parts B–D in small groups.

A. What is a mathematical question you can ask about this situation? What information would you need to know to answer your question? Which option is the shortest?; the distances between the airports

B. What variable(s) are involved in this situation? See Additional Answers.

C. To answer your question, what strategy and tool would you use along with all the information you have? What answer do you get? See Strategies 1 and 2 on the facing page.

D. Does your answer make sense in the context of the situation? How do you know? yes; The path with the shortest segments would be the shortest path.

Turn and Talk Predict how your answer would change if the measure of the angle in the left triangle was 39° and the measure of the angle in the right triangle was 36°. See margin.

SUPPORT SENSE-MAKING • Three Reads

Tell students to read the information in the photo three times and prompt them with a different question each time.

1. What is the situation about? The situation is about the distance of two different flight paths between two points.
2. What are the quantities in this situation? How are those quantities related? The quantities in this situation are the lengths of the line segments that create the flight paths and the angle measure measures inside the triangles created by the two flight paths; The larger the angle, the longer the segment opposite the angle.
3. What are possible questions you could ask about this situation? Possible answer: What would make the two flight paths equal in length? What would make the top flight path longer than the bottom path? What is true about the angles opposite the congruent line segments?

1 Spark Your Learning

▶ MOTIVATE

- Have students look at the illustration in their books and read the information contained in the illustration. Then complete Part A as a whole-class discussion.
- Give the class the additional information they need to solve the problem. This information is available online as a printable and projectable page in the Teacher Resources.
- Have students work in small groups to complete Parts B–D.

▶ PERSEVERE

If students need support, guide them by asking:

Q Advancing • Use Tools Which tool could you use to solve the problem? Why choose that tool and not some other? Students' choices of tools and reasons for choosing them will vary.

Q Assessing What two flight path options exist that get a passenger from the start point to the end point? A passenger can travel 400 miles and then an unknown distance to get to the end point or travel an unknown distance first, followed by a 400-mile distance.

Q Assessing Does knowing that two of the line segments shown are congruent and equivalent to a distance of 400 miles help you solve the problem? Yes, because we only need to compare the lengths of the two unknown segments to see which path is shorter.

Q Advancing What do you know is true about the angles that are opposite the two congruent line segments shown? The angles opposite the two segments that are congruent are also congruent.

Turn and Talk Have students consider what would happen to the angle opposite the given segment length of 400 miles if the number of miles increased or decreased. The top path would be the shortest path.

▶ BUILD SHARED UNDERSTANDING

Select groups of students who used various strategies and tools to share with the class how they solved the problem. As they present their solutions, have each group discuss why they chose a specific strategy and tool.

② Learn Together

Build Understanding

Task 1 (MP) **Use Tools** Students use concrete models to explore and develop a deeper understanding of the relationships that exist among side lengths and angle measures of a triangle.

Sample Guided Discussion:

Q In Part A, what type of angle can you make angle *A* to assure it is the largest angle in $\triangle ABC$? Move the paper strips so that angle *A* is an obtuse angle.

Turn and Talk As students consider their response to the Turn and Talk, encourage them to review their answers in Parts A–E to help them move toward a generalization. The included angle in the first triangle is larger than the included angle in the second triangle.

Task 2 (MP) **Attend to Precision** Students must communicate precisely using clear definitions and accurate symbols when providing explanations about triangle side and angle relationships.

Sample Guided Discussion:

Q Why do the lengths of line segments $\overline{AB}$ and $\overline{AC}$ remain constant? Because line segment $\overline{AC}$ represents the distance from the ground to the top of the ride and line segment $\overline{AB}$ represents the length of the cable between the top of the ride and the gondola.

Build Understanding

Explore Inequalities in Two Triangles

1 Join two paper strips that have different lengths at point *A* to create $\triangle ABC$

A. Move the paper strips so that $\angle A$ is the largest angle in $\triangle ABC$. Which side of the triangle is the longest? $\overline{BC}$

B. Move the paper strips so that $\angle A$ is the smallest angle in $\triangle ABC$. Which side of the triangle is the shortest? $\overline{BC}$

Join another pair of paper strips at point *D* to create $\triangle DEF$. The paper strips should be the same lengths as the paper strips used in $\triangle ABC$.

C. Adjust the paper strips so that $\angle A$ is larger than $\angle D$. How are $\overline{BC}$ and $\overline{EF}$ related? $\overline{BC}$ is longer than $\overline{EF}$.

D. Which corresponding parts in $\triangle ABC$ and $\triangle DEF$ are congruent? Explain. $\overline{AB} \cong \overline{DE}$, $\overline{AC} \cong \overline{DF}$; The lengths of the paper strips do not change.

E. If you are given that two sides of a triangle are congruent to two sides of another triangle, and the included angle in the first triangle is larger than the included angle in the second triangle, what can you conclude about the sides opposite the included angles? See Additional Answers.

Turn and Talk Suppose you are given that two sides of a triangle are congruent to two sides of another triangle, and the third side of the first triangle is longer than the third side of the second triangle. What can you conclude about the angles included between the two congruent sides? Use diagrams of the triangles to support your answer. See margin.

2 Two positions of a swinging circular gondola ride at an amusement park are shown. The triangles show the position of the gondola in relation to its starting point.

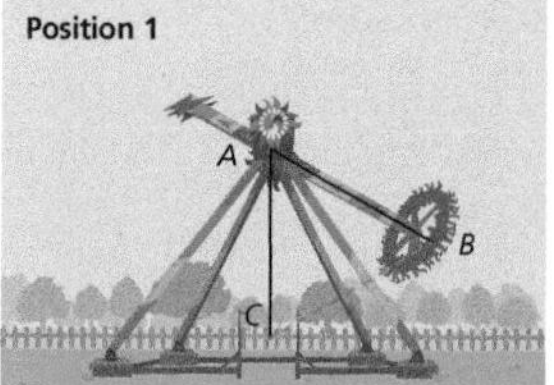

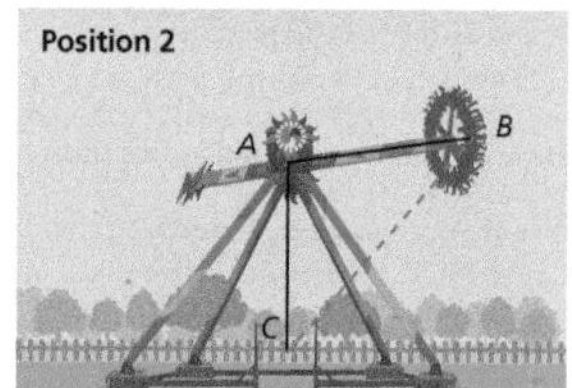

A. Which corresponding sides of the triangles are congruent? $\overline{AB}$ and $\overline{AC}$

B. The distance from *B* to *C* in Position 1 is less than the distance from *B* to *C* in Position 2. What can you conclude about the angles opposite these sides? The angle opposite $\overline{BC}$ in Position 1 is smaller than the angle opposite $\overline{BC}$ in Position 2.

308

LEVELED QUESTIONS

Depth of Knowledge (DOK)	Leveled Questions	What Does This Tell You?
Level 1 **Recall**	Suppose $\triangle ABC \cong \triangle DEF$. What are the corresponding pairs of segments? $\overline{AB}$ and $\overline{DE}$, $\overline{BC}$ and $\overline{EF}$, $\overline{AC}$ and $\overline{DF}$	Students' answers will indicate whether they understand the meaning of corresponding sides.
Level 2 **Basic Application of Skills & Concepts**	Given $\triangle ABC$, what line segment is affected by increasing the measure of angle *A* and how does it change? Increasing the measure of angle *A* will result in line segment $\overline{BC}$ getting longer.	Students' answers will indicate whether they understand the relationship between angles of a triangle and the sides opposite those angles.
Level 3 **Strategic Thinking & Complex Reasoning**	$\triangle ABC$ and $\triangle DEF$ have $\overline{AB} \cong \overline{DE}$ and $\overline{AC} \cong \overline{DF}$. What must be true about the three pairs of corresponding angles if $BC > EF$? $m\angle A > m\angle D$, $m\angle B \cong m\angle E$, and $m\angle C \cong m\angle F$	Students' answers will indicate whether they understand the relationship between angles of a triangle and the sides opposite those angles and if they can interpret given information to develop a logical argument regarding those relationships.

Step It Out

The Hinge Theorem and Its Converse

You have learned how to apply inequalities to relate the sides and angles in one triangle. The following theorems relate the sides and angles in two triangles.

Hinge Theorem

If two sides of one triangle are congruent to two sides of another triangle, and the included angle in the first triangle is larger than the included angle in the second triangle, then the third side of the first triangle is longer than the third side of the second triangle.

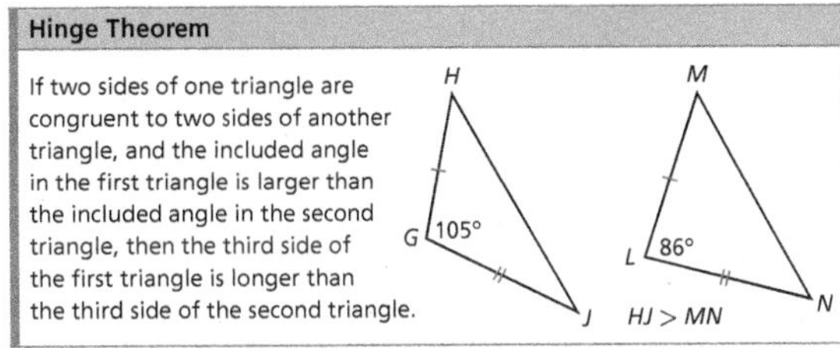

Converse of the Hinge Theorem

If two sides of one triangle are congruent to two sides of another triangle, and the third side of the first triangle is longer than the third side of the second triangle, then the included angle in the first triangle is larger than the included angle in the second triangle.

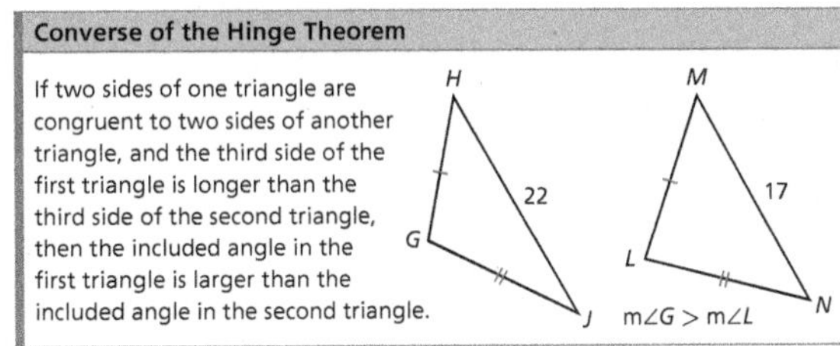

3 Compare DC and BC using the information given in the diagram.
Compare the sides and angles in $\triangle ADC$ and $\triangle ABC$.

$m\angle DAC > m\angle BAC$ $\quad \overline{AD} \cong \overline{AB}$ $\quad \overline{AC} \cong \overline{AC}$

A. How do you know that $m\angle DAC > m\angle BAC$?

B. Why is this congruence statement listed?

A. $32° > 28°$

B, C. See Additional Answers.

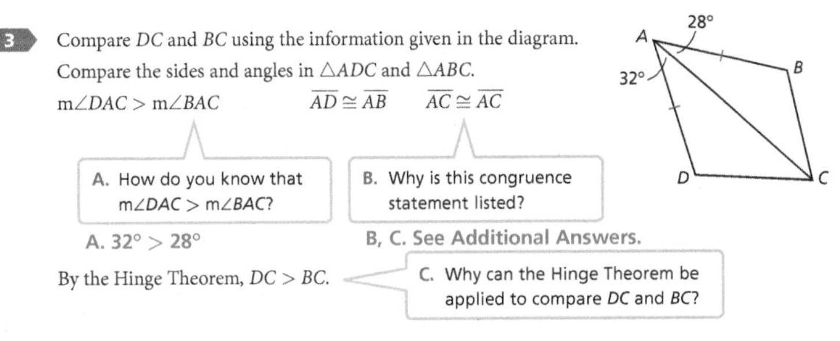

By the Hinge Theorem, $DC > BC$.

C. Why can the Hinge Theorem be applied to compare DC and BC?

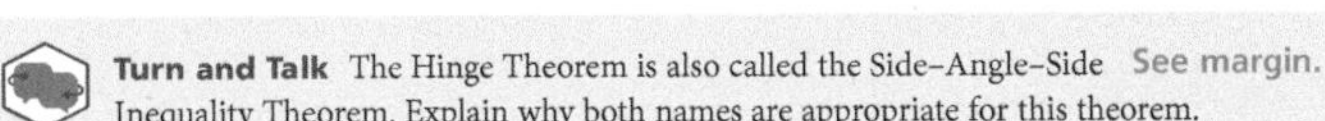

Turn and Talk The Hinge Theorem is also called the Side-Angle-Side Inequality Theorem. Explain why both names are appropriate for this theorem. See margin.

Step It Out

Task 3 (MP) **Use Structure** Students will use what they know about triangles and the Hinge Theorem to compare sides and angles in triangles with a shared side.

Sample Guided Discussion:

Q **In Part B, what property are you using to verify the given statement?** Reflexive Property

Q **How could you change the problem so that the Converse of the Hinge Theorem is used instead of the Hinge Theorem?** The lengths of segments $\overline{DC}$ and $\overline{BC}$ would need to be given, with $\overline{DC}$ longer than $\overline{BC}$, to show that $m\angle DAC > m\angle BAC$.

Turn and Talk Encourage students to think of an actual hinge and how it works. Ask them to explain how two hinges with congruent plates can have a different opening angle and how that affects the size of the opening from one plate to the other. The two sides of the triangle and their included angle can be modeled by a hinge, like on a door. In general, if a hinge is opened with a greater angle than a second hinge, then the distance between the two ends of the first hinge is greater than the distance between the two ends of the second hinge. This relationship can be modeled by an inequality. Additionally, since the Hinge Theorem involves a side, an included angle, and another side the Side-Angle-Side Inequality Theorem is an appropriate name as well.

EL PROFICIENCY LEVEL

Beginning

Have students label two triangles with vertices *ABC* and *DEF*. Then ask them to name the pairs of corresponding angles, the pairs of corresponding sides, and the side opposite each of the 6 angles.

Intermediate

Have students work in groups and provide them with a sheet showing several pairs of isosceles triangles with the included, noncongruent angles labeled with degree measures. Ask them to describe the relationship between the opposite sides for each pair of given angles using short sentences.

Advanced

Have students explain why the perimeter of an isosceles right triangle is greater than an isosceles triangle that whose congruent legs include an 80° vertex angle.

Task 4 **MP** **Use Structure** Students will use what they know about the structure of linear inequalities to solve algebraic problems involving unknown angle measures in triangles.

Sample Guided Discussion:

Q How are the Hinge Theorem and the Converse of the Hinge Theorem different? The Hinge Theorem allows you to compare the lengths of a pair of corresponding sides of two triangles when given the angle measures opposite those sides, while the converse allows you to compare the opposite angles given the lengths of the corresponding sides.

Q In Part B, what types of numbers are excluded from being the measure of an angle? zero and negative numbers

Turn and Talk When creating an expression that meets these criteria, ask students to think about how an expression can be equivalent to a positive value even if the value of the missing variable is negative. Encourage them to write and test some examples. If $m\angle SRT = (2x + 5)°$, then $-2.5 < x < 30$.

Task 5 **MP** **Construct Viable Arguments** Students construct viable arguments whenever they are asked to provide a mathematical proof. A Statement/Reason table allows students to show a coherent development of ideas and definitions.

Sample Guided Discussion:

Q Based on the given information, what two angles do we know are congruent because they are opposite those congruent sides? $\angle NPL \cong \angle PLM$

Q In Part B, why can't we use the Converse of the Hinge Theorem to complete the proof? Because the steps in the proof lead us to the conclusion that a second angle instead is greater than another, rather than one side being greater than a second side.

4 Find the range of possible values for x.

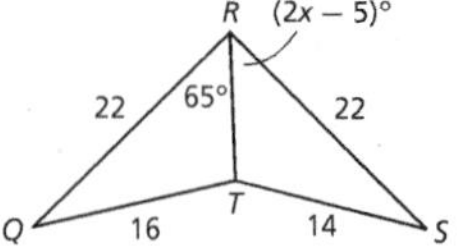

Compare the sides and angles in $\triangle QRT$ and $\triangle SRT$.

The Converse of the Hinge Theorem can be used to conclude that $m\angle SRT < m\angle QRT$.

$2x - 5 < 65$
$2x < 70$
$x < 35$

A. Why can the Converse of the Hinge Theorem be applied to make this conclusion?

Find all positive values for $m\angle SRT$. A, B. See Additional Answers.

$2x - 5 > 0$
$2x > 5$
$x > 2.5$

B. Why is this step necessary?

The range of values for x is $2.5 < x < 35$.

Turn and Talk Write a different variable expression to represent $m\angle SRT$ so that the value of the variable can be negative while the measure of the angle is positive. See margin.

Use the Hinge Theorem in a Proof

5 Use the Hinge Theorem to prove that $LM > NP$.

Given: $\overline{LN} \cong \overline{MP}$

Prove: $LM > NP$

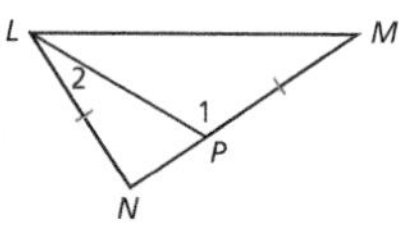

Statements	Reasons
1. $\overline{LN} \cong \overline{MP}$	1. Given
2. $\overline{LP} \cong \overline{LP}$	2. Reflexive Property of Congruence
3. $\angle 1$ is an exterior angle of $\triangle LPN$.	3. Definition of an exterior angle
4. $m\angle 1 = m\angle 2 + m\angle LNP$	4. Exterior Angle Theorem
5. $m\angle 2 + m\angle LNP > m\angle 2$	5. If $a = b + c$, where c is positive, then $a > b$.
6. $m\angle 1 > m\angle 2$	6. Substitution
7. $LM > NP$	7. Hinge Theorem

A, B. See Additional Answers.

A. Explain this property in words rather than in symbols.

B. Why can the Hinge Theorem be applied to make this conclusion?

310

Write an Indirect Proof

In an indirect proof, you begin by assuming that the conclusion is false. Then you show that this assumption leads to a contradiction. This type of proof is also called a proof by contradiction.

How to Write an Indirect Proof
Step 1: Identify the statement to be proven.
Step 2: Assume the opposite (negation) of that statement is true.
Step 3: Use direct reasoning until you reach a contradiction.
Step 4: Conclude that since the assumption is false, the original statement must be true.

6 Prove the Converse of the Hinge Theorem.

Given: $\overline{AB} \cong \overline{DE}$, $\overline{AC} \cong \overline{DF}$, $BC > EF$

Prove: $m\angle A > m\angle D$

A, B. See Additional Answers.

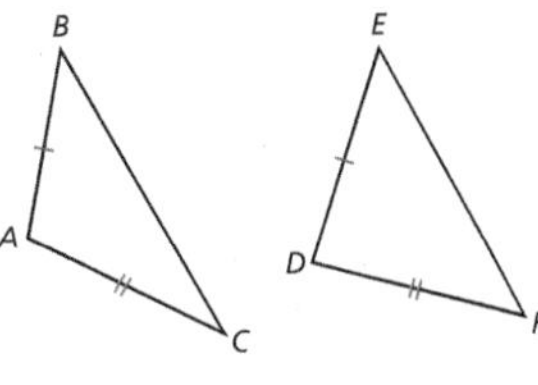

A. Why do you want to make this assumption?

Assume $m\angle A \not> m\angle D$. So, either $m\angle A < m\angle D$ or $m\angle A = m\angle D$.

Check if each case is true.

Case 1

If $m\angle A < m\angle D$, then $BC < EF$ by the Hinge Theorem. This contradicts the given information that $BC > EF$. So, $m\angle A \not< m\angle D$.

B. Why can you use SAS Triangle Congruence to show the triangles are congruent?

Case 2

If $m\angle A = m\angle D$, then $\angle A \cong \angle D$. So $\triangle ABC \cong \triangle DEF$ by SAS Triangle Congruence Theorem. Then $\overline{BC} \cong \overline{EF}$ by CPCTC, and $BC = EF$. This contradicts the given information that $BC > EF$. So, $m\angle A \neq m\angle D$.

The assumption $m\angle A \not> m\angle D$ is false. Therefore $m\angle A > m\angle D$.

Turn and Talk Use the information below to write an indirect proof to show that a right triangle cannot have an obtuse angle. See margin.

Given: $\triangle RST$ is a triangle with right angle $\angle R$.

Prove: $\triangle RST$ does not have an obtuse angle.

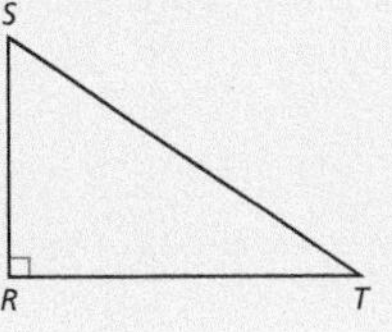

Task 6 (MP) **Construct Viable Arguments** In this task, not only are students required to construct viable arguments, but are doing so indirectly by first assuming a statement is false and then using definitions to contradict this false assumption.

CONNECT TO VOCABULARY

Have students use the **Interactive Glossary** to record their understanding of the vocabulary in this task.

OPTIMIZE OUTPUT Critique, Correct, and Clarify

Have students work with a partner to discuss when it is appropriate to use Hinge Theorem or the Converse of the Hinge Theorem when proving triangle inequalities. Have them each draw diagrams to justify their claims. Encourage students to use vocabulary from the lesson in their discussions.

Sample Guided Discussion:

Q Why does this task involve the Converse of the Hinge Theorem instead of the Hinge Theorem? Because the given information tells us that the length of one side is greater the length of another side, instead of one angle being greater than another angle.

Q Why must we show proof of two contradictions to our assumption instead of just 1? Because if our assumption is that $m\angle A$ is not greater than $m\angle D$, then there are two other possibilities that could be true. Either $m\angle A = m\angle D$ or $m\angle A < m\angle D$.

Turn and Talk Be sure students understand the difference between a direct proof and an indirect proof. Review the similarities and differences. Assume $\triangle RST$ has an obtuse angle at S. Then $m\angle R + m\angle T < 90°$. This contradicts the given information that $\angle R$ is a right angle. So, $\triangle RST$ cannot have an obtuse angle.

ONLINE

Assign the Digital On Your Own for
- built-in student supports
- Actionable Item Reports
- Standards Analysis Reports

On Your Own

Assignment Guide

The chart below indicates which problems in the On Your Own are associated with each task in the Learn Together. Assign daily homework for tasks completed.

Learn Together Tasks	On Your Own Problems
Task 1, p. 308	Problem 25
Task 2, p. 308	Problem 8
Task 3, p. 309	Problems 9–14 and 18–20
Task 4, p. 310	Problems 15–17 and 21
Task 5, p. 310	Problem 22
Task 6, p. 311	Problems 23 and 24

Check Understanding

1. Consider $\triangle ABC$ and $\triangle XYZ$ with $\overline{AB} \cong \overline{XY}$ and $\overline{AC} \cong \overline{XZ}$. If $\angle A$ is smaller than $\angle X$, how are $\overline{BC}$ and $\overline{YZ}$ related? BC is shorter than YZ.
2. Consider $\triangle ABC$ and $\triangle RST$. If $\overline{AC}$ is shorter than $\overline{RT}$, what can you conclude about the angles opposite these sides? $\angle B$ is smaller than $\angle S$.

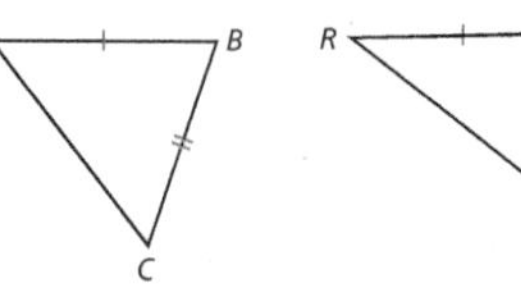

In Problems 3–6, use the triangles shown.

3. Compare $m\angle D$ and $m\angle G$. $m\angle D > m\angle G$
4. Compare BC and HJ. $BC > HJ$
5. Describe the restrictions on x. $7 < x < 23.25$
6. Suppose the Hinge Theorem is being used to prove that EF is greater than HJ. What criteria need to be stated in the proof before the Hinge Theorem can be applied? $\overline{DE} \cong \overline{GH}$, $\overline{DF} \cong \overline{GJ}$, $m\angle D > m\angle G$

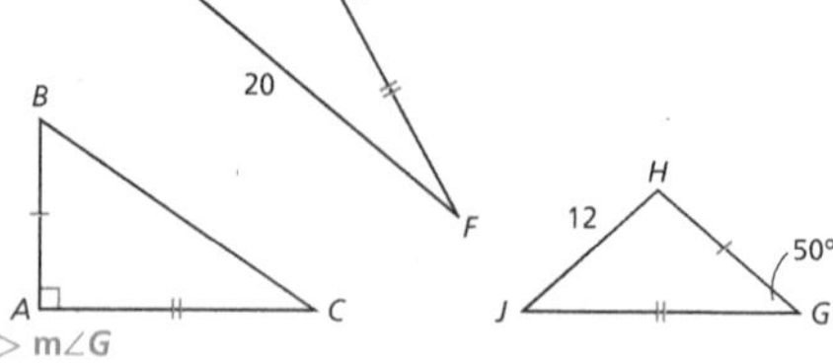

7. Explain why an indirect proof is called proof by contradiction. See Additional Answers.

On Your Own

8. Two positions of an open gate are shown. The triangles show the position of the gate in relation to its closed position. The distance from G to H in Position 1 is less than the distance from G to H in Position 2. What can you conclude about the angles opposite these sides? The angle in Position 1 is smaller than the angle in Position 2.

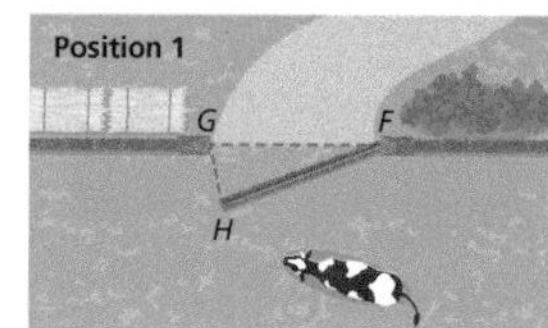

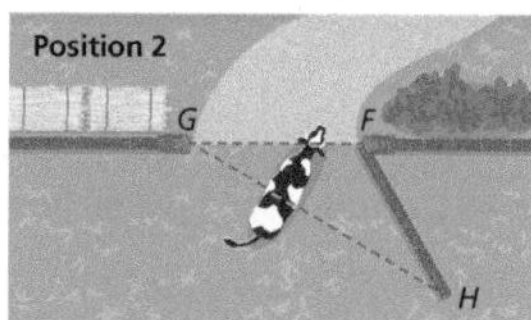

Compare the given measures.

9. ST and VW $ST > VW$
10. AB and BD $AB < BD$
11. VW and CZ $VW > CZ$

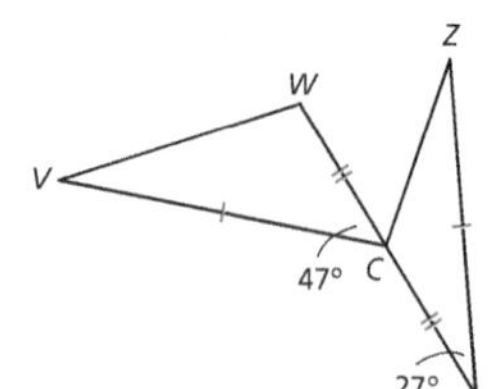

3 Check Understanding

Formative Assessment

Use formative assessment to determine if your students are successful with this lesson's learning objective.

Students who successfully complete the Check Understanding can continue to the On Your Own practice.

For students who miss 1 problem or more, work in a pulled small group using the Almost There small-group activity on page 307C.

ONLINE

Assign the Digital Check Understanding to determine
- success with the learning objective
- items to review
- grouping and differentiation resources

4 Differentiation Options

Differentiate instruction for all students using small-group activities and math center activities on page 307C.

Reteach

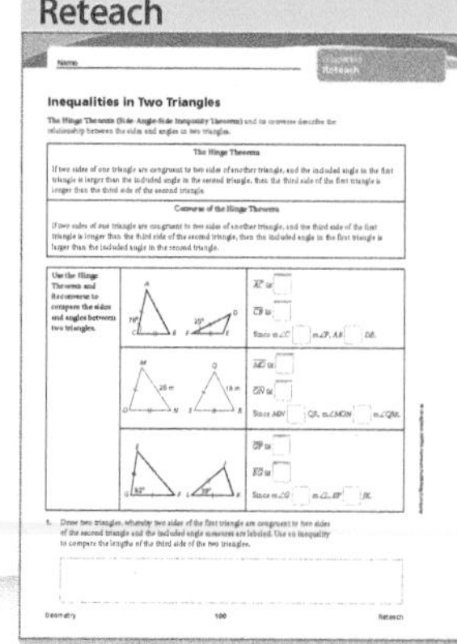

Challenge

Compare the given measures.

12. m$\angle$1 and m$\angle$2 $m\angle 1 < m\angle 2$

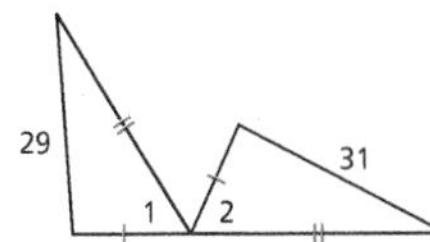

13. m$\angle L$ and m$\angle D$ $m\angle L > m\angle D$

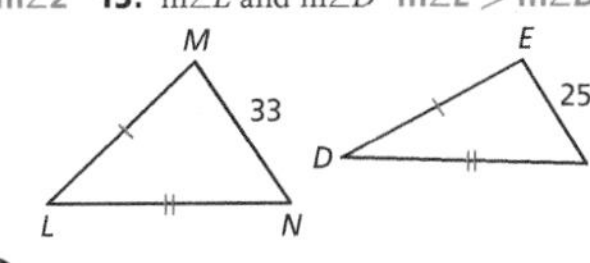

14. m$\angle$1 and m$\angle$2 $m\angle 1 > m\angle 2$

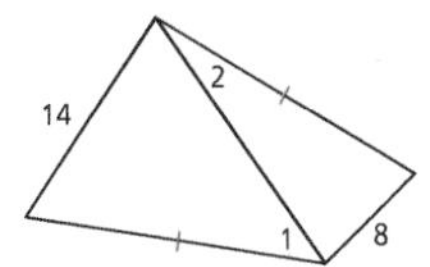

Use the Hinge Theorem and its converse to describe the restrictions on the value of the variable.

15. $2 < x < 10$

B (4x − 8) ft C

40°

30°

A

(2x + 12) ft

D

16. $x > 1$

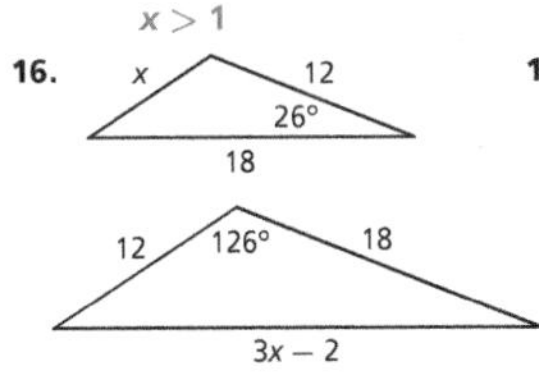

17. $-18 < x < 34$

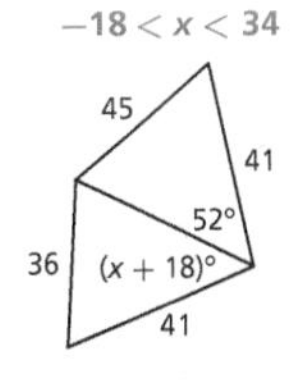

18. Geography The road from A to E and the road from C and D are perpendicular and intersect at point F. Theresa's house is at C, 500 feet from the intersection. Ray's house is at D, 750 feet from the intersection. Ivy's house is at E, 750 feet from the intersection. Zana's house is at A, 500 feet from the intersection, and Tan's house is at B, 750 feet from F. Which distance is longer, the distance from Zana's house to Tan's house or the distance from Theresa's house to Ivy's house? Show how you know.

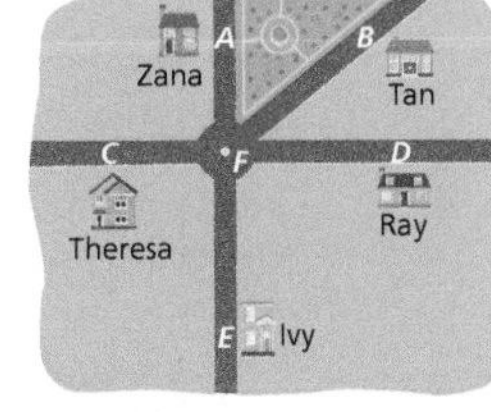

18–20. See Additional Answers.

19. Two pairs of hikers leave the same camp heading in opposite directions. Each travels 2 miles, then changes direction and travels 1.2 miles. The first pair starts due east and then turns 50° toward north. The second pair starts due west and then turns 40° toward south. Which pair is farther from camp? Explain your reasoning.

20. (MP) **Critique Reasoning** Terrence says that $AB > CD$ by the Hinge Theorem. Is he right? Explain why or why not.

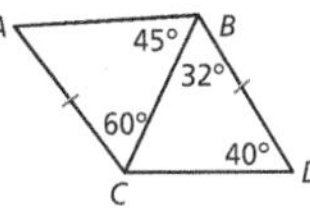

21. The triangles at the right show the paths in a town's public square.

A. Write an inequality for the possible values of x. $\frac{5}{3} < x < \frac{13}{2}$

B. If the triangle with the side length $2(x - 1)$ is equilateral, what is the value of x? 5

C. Using the value of x you found in Part B, what are the lengths of the sides represented by $4x + 3$ and $6x - 10$? See Additional Answers.

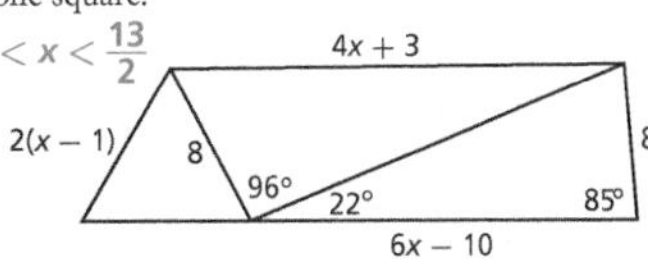

Questioning Strategies

Problem 20 Suppose Terrence said that $AB > CB$. Would Terrence still be able to use the Hinge Theorem to prove this statement? Why or why not? no; The angle opposite $\overline{CB}$ is not the included angle in $\triangle ABC$.

Watch for Common Errors

Problem 21 When working with the relationship between the sides and angles of a triangle, students tend to summarize the theorem to "largest side is opposite largest angle". They sometimes forget that this comparison only works within one triangle. There can be a small obtuse triangle in the same figure as a large acute triangle. Just because the obtuse angle is the largest angle in the figure, does not mean the side opposite of it is the longest among all the segments in the figure, just that it is the longest side in that obtuse triangle.

5 Wrap-Up

Summarize learning with your class. Consider using the Exit Ticket, Put It in Writing, or I Can scale.

Exit Ticket

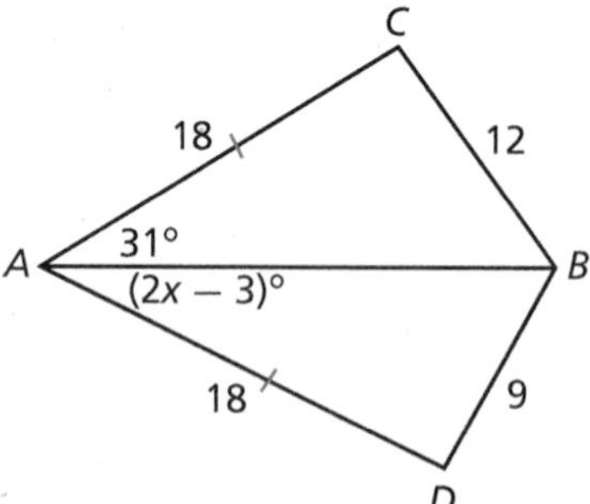

What two inequalities can be used to determine the possible range of values for x? $31 > 2x - 3$; $2x - 3 > 0$

Put It in Writing

Describe the difference between the Hinge Theorem and the Converse of the Hinge Theorem.

I Can

The scale below can help you and your students understand their progress on a learning goal.

4	I can determine the relative sizes of angles and sides in two triangles, and I can explain the process to others.
3	I can determine the relative sizes of angles and sides in two triangles.
2	I can understand the Hinge Theorem and its converse.
1	I can recognize that the measure of an angle in a triangle is related to the length of the opposite side.

Spiral Review • Assessment Readiness

These questions will help determine if students have retained information taught in the past and can also prepare them for high-stakes assessments. Here, students must know how to define the median of a triangle **(9.4)**, understand the Triangle Midsegment Theorem **(9.5)**, apply the Triangle Inequality Theorem **(10.1)**, and apply the properties of two parallel lines cut by a transversal **(3.1)**.

22. Prove that $\angle BCD > m\angle ABC$ using the Converse of the Hinge Theorem. See Additional Answers.
Given: $\overline{AB} \cong \overline{CD}$
Prove: $\angle BCD > m\angle ABC$

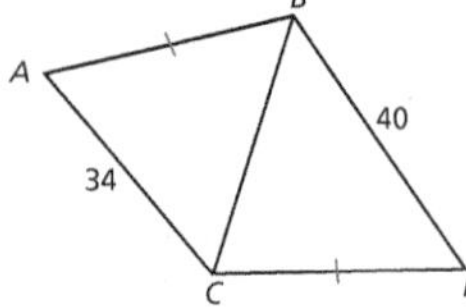

23. Write an indirect proof to prove that two supplementary angles cannot both be obtuse angles.
See Additional Answers.
Given: $\angle 1$ and $\angle 2$ are supplementary.
Prove: $\angle 1$ and $\angle 2$ cannot both be obtuse.

24. (MP) **Critique Reasoning** Tiffany says an indirect proof is less reliable than a direct proof because you have to start by assuming something. Is Tiffany right? Explain why or why not. See Additional Answers.

25. **Open Ended** Reginald is trying to design a stage riser as shown. AC must be equal to CD but the other lengths are flexible. Construct a set of triangles as shown that fit the requirements. Which side is longer, $\overline{AB}$ or $\overline{BD}$?
Answers will vary.

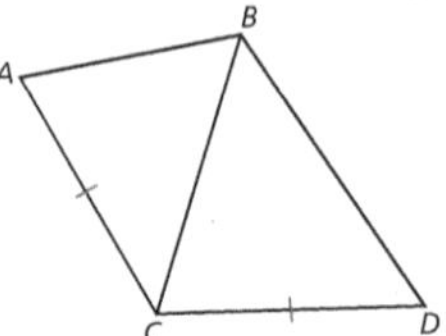

Spiral Review • Assessment Readiness

26. If a line joins a vertex of a triangle to the midpoint of the opposing side, bisecting it, it is the __?__.

Ⓐ altitude
Ⓑ perpendicular bisector
Ⓒ angle bisector
Ⓓ median

27. In the picture shown, $\overline{DE}$ is which of the following?

Ⓐ altitude
Ⓑ median
Ⓒ bisector
Ⓓ midsegment

A, D, E, B, C

28. If $\triangle ABC$ has sides 4 and 11 which of the following are possible measures for the length of the third side? Select all that apply.

Ⓐ 2 Ⓓ 9
Ⓑ 5 Ⓔ 11
Ⓒ 7.5 Ⓕ 15

29. Which angles are congruent to $\angle 1$? Select all that apply.

Ⓐ $\angle 2$ Ⓓ $\angle 5$
Ⓑ $\angle 3$ Ⓔ $\angle 6$
Ⓒ $\angle 4$ Ⓕ $\angle 7$

1 2 4 3 5 6 8 7

I'm in a Learning Mindset!

How did I assert my own needs and viewpoints when learning about inequalities in two triangles?

314 Keep Going Journal and Practice Workbook

Learning Mindset

Resilience Responds to Feedback

Point out that working with theorems involving triangles, especially when using those theorems when writing proofs, can be a task that takes a lot of practice. One way to gain confidence in this area is to work with a partner or in a group so that you can hear different points of view and different strategies that you may not have considered. Being able to listen to the feedback of your teachers and peers, as well as respond to that feedback in a respectful way, allows you to become a student that is able to think at a higher level. It is more important to understand why a theorem is true than it is to be able to memorize that it is. Feedback is one way to attain that understanding. *How does working with a partner help you have a more confident mindset? Do you read the feedback teachers leave you on your graded papers? How does receiving feedback from a peer or a teacher make you feel? Can you see the benefits of receiving constructive feedback?*

Module
10 Review

Side-Angle Relationships

The sum of the interior angles of a triangle is 180°.

To determine the measure of $\angle E$, subtract the other angles from 180°.

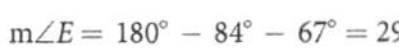

$m\angle E = 180° - 84° - 67° = 29°$

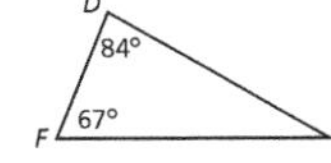

The angles in order least to greatest measure are $\angle D$, $\angle F$, $\angle E$.

If one angle of a triangle is larger than another angle, then the side opposite the larger angle is longer than the side opposite the smaller angle.

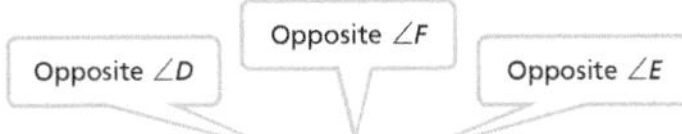

So, the sides in order from least to greatest length are $\overline{EF}$, $\overline{DE}$, $\overline{DF}$.

Triangle Inequality Theorem

Not every combination of sides can be used to form a triangle.

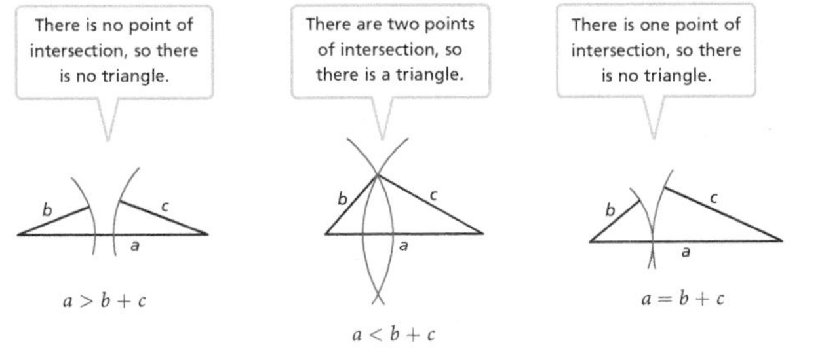

There is no point of intersection, so there is no triangle.
There are two points of intersection, so there is a triangle.
There is one point of intersection, so there is no triangle.
b
c
a
a > b + c
a < b + c
a = b + c

Hinge Theorem

When two triangles have two pairs of congruent sides but the included angles differ, then the sides opposite the included angle differ accordingly.

$m\angle YZX$ is greater than $m\angle WZX$,

so YX is greater than WX.

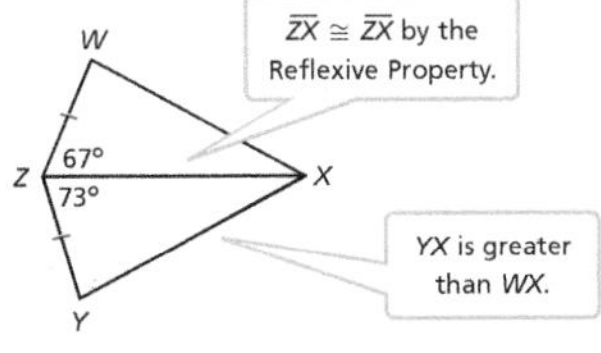

Module 10 315

ONLINE

Assign the Digital Module Review for

- built-in student supports
- Actionable Item Reports
- Standards Analysis Reports

MODULE 10 REVIEW

Module Review

Use the first page of the Module Review to summarize and connect the main ideas of the module. Use the second page to assess students' understanding of the vocabulary, concepts, and skills presented in the module.

Sample Guided Discussion:

Q Side-Angle Relationships **Based on the measures of the angles and sides of $\triangle DEF$, what kind of triangle is $\triangle DEF$? Explain.** an acute scalene triangle; The measure of each angle is less than 90° and the no two sides have the same length.

Q Triangle Inequality Theorem **What is the longest side in a right triangle? Explain using the Triangle Inequality Theorem.** the hypotenuse; It is opposite the right angle, which is the largest angle in a right triangle.

Q Hinge Theorem **What assumption would you make if you wanted to prove the Hinge Theorem indirectly?** Assume that $YX \leq WX$.

Module Review continued

Possible Scoring Guide

Items	Points	Description
1–5	2 each	identifies the correct term
6	2	correctly explains why a triangle cannot be created
7–10	2 each	correctly determines whether a triangle can be formed
11, 12	2 each	correctly uses the Triangle Inequality Theorem to find the range of possible values for the length of the third side
13	1	correctly *BC* and *EF* and explains reasoning
13	2	correctly compares *BC* and *EF* and m∠*B* and m∠*E*, and explains reasoning
14	1	correctly compares m∠*JGI* and m∠*HGI* and explains reasoning
14	2	correctly compares m∠*JGI* and m∠*HGI* and explains reasoning, and correctly writes an indirect proof
Total points possible = 28 points		

The Unit 6 Test in the Assessment Guide assesses content from Modules 9 and 10.

Vocabulary

Choose the correct term from the box to complete each sentence.

Vocabulary
converse
indirect proof
inequality
theorem
triangle

1. A(n) __?__ is a statement that has been proven. theorem
2. The __?__ is a statement formed by exchanging the hypothesis and conclusion of a condition statement. converse
3. A(n) __?__ is a three-sided polygon. triangle
4. A(n) __?__ is a proof in which the statement to be proved is assumed to be false and a contradiction is shown. indirect proof
5. A(n) __?__ is a statement that compares two expressions by using one of the following signs: >, <, ≥, ≤, ≠. inequality

Concepts and Skills

6. Explain why you can not create $\triangle ABC$ such that $AC \geq AB + BC$. See Additional Answers.

Determine whether a triangle can be formed with the given side lengths. Explain your reasoning.

7. 7 in., 2 in., 8 in. See below.
8. 4 cm, 11 cm, 6 cm no; $4 + 6 < 11$
9. 16 m, 10 m, 5 m no; $10 + 5 < 16$
10. 23 ft, 16 ft, 38 ft See below.

Find the range of possible values of *x* using the Triangle Inequality Theorem.

11. 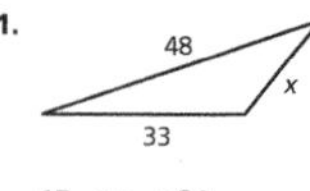

$15 < x < 81$

12. 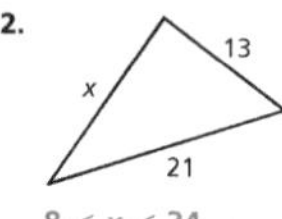

$8 < x < 34$

13. **A.** Compare *BC* and *EF*. Explain your reasoning. $BC < EF$; Hinge Theorem
B. Compare m∠*B* and m∠*E*. Explain your reasoning m∠*E* < m∠*B*; using a scale drawing

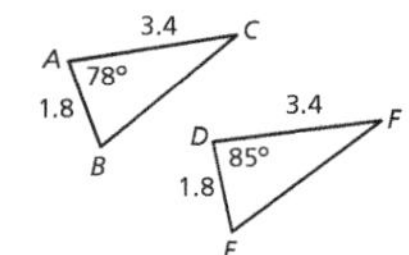

14. **A.** Compare m∠*JGI* and m∠*HGI*. Explain your reasoning. See Additional Answers.
B. (MP) **Use Tools** Write an indirect proof to prove that ∠*JGI* is not congruent to ∠*HGI*. State what strategy and tool you will use to answer the question, explain your choice, and then find the answer. See Additional Answers.

G, J, H, I, 12, 11

7. yes; The sum of any two sides of the triangle is greater than the other side.
10. yes; The sum of any two sides of the triangle is greater than the other side.

316

DATA-DRIVEN INSTRUCTION

Before moving on to the Module Test, use the Module Review results to intervene based on the table below.

Items	Lesson	DOK	Content Focus	Intervention
6	10.1	2	Explain whether a triangle can be formed given an inequality in which one side length is greater than or equal to the sum of the other two side lengths.	Reteach 10.1
7–10	10.1	1	Determine whether a triangle can be created given three side lengths.	Reteach 10.1
11–12	10.1	2	Use the Triangle Inequality Theorem to find the range of possible values for the length of a side or the measure of an angle.	Reteach 10.1
13	10.2	2	Use the Hinge Theorem to compare two noncongruent sides in two triangles and a scale drawing to compare the measures of two corresponding angles.	Reteach 10.2
14	10.2	3	Use the Converse of the Hinge Theorem to compare noncongruent angles in two triangles and prove indirectly that the two included angles between corresponding and congruent sides are also not congruent.	Reteach 10.2

Module Test

The Module Test is available in alternative versions in your Assessment Guide. All items are presented in standardized test formats.

ONLINE

Assign the Digital Module Test to power actionable reports including

- proficiency by standards
- item analysis

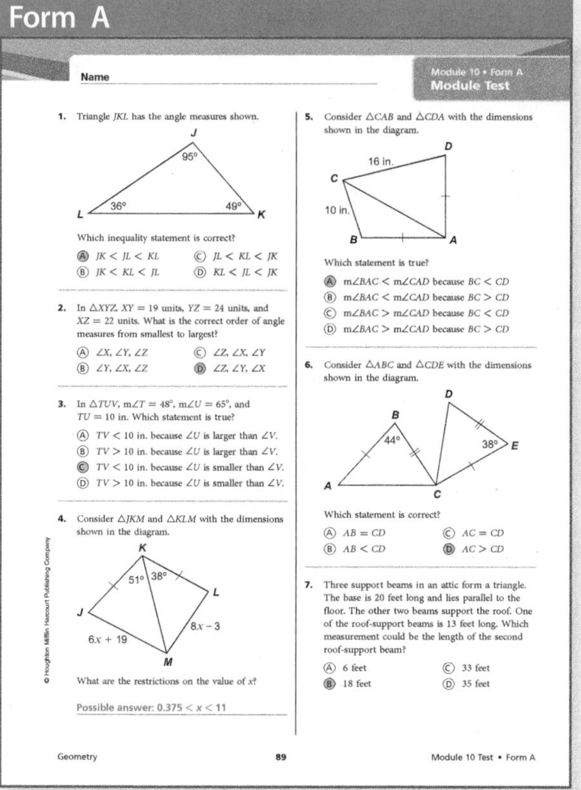

Form A

Module 10 • Form A Module Test

Name

1. Triangle JKL has the angle measures shown.

Which inequality statement is correct?

(A) $JK < JL < KL$ (B) $JK < KL < JL$ (C) $JL < KL < JK$ (D) $KL < JL < JK$

2. In $\triangle XYZ$, $XY = 19$ units, $YZ = 24$ units, and $XZ = 22$ units. What is the correct order of angle measures from smallest to largest?

(A) $\angle X, \angle Y, \angle Z$ (B) $\angle Y, \angle X, \angle Z$ (C) $\angle Z, \angle X, \angle Y$ (D) $\angle Z, \angle Y, \angle X$

3. In $\triangle TUV$, $m\angle T = 48°$, $m\angle U = 65°$, and $TU = 10$ in. Which statement is true?

(A) $TV < 10$ in. because $\angle U$ is larger than $\angle V$.
(B) $TV > 10$ in. because $\angle U$ is larger than $\angle V$.
(C) $TV < 10$ in. because $\angle U$ is smaller than $\angle V$.
(D) $TV > 10$ in. because $\angle U$ is smaller than $\angle V$.

4. Consider $\triangle JKM$ and $\triangle KLM$ with the dimensions shown in the diagram.

What are the restrictions on the value of x?

Possible answer: $0.375 < x < 11$

5. Consider $\triangle CAB$ and $\triangle CDA$ with the dimensions shown in the diagram.

Which statement is true?

(A) $m\angle BAC < m\angle CAD$ because $BC < CD$
(B) $m\angle BAC < m\angle CAD$ because $BC > CD$
(C) $m\angle BAC > m\angle CAD$ because $BC < CD$
(D) $m\angle BAC > m\angle CAD$ because $BC > CD$

6. Consider $\triangle ABC$ and $\triangle CDE$ with the dimensions shown in the diagram.

Which statement is correct?

(A) $AB = CD$ (B) $AB < CD$ (C) $AC = CD$ (D) $AC > CD$

7. Three support beams in an attic form a triangle. The base is 20 feet long and lies parallel to the floor. The other two beams support the roof. One of the roof-support beams is 13 feet long. Which measurement could be the length of the second roof-support beam?

(A) 6 feet (B) 18 feet (C) 33 feet (D) 35 feet

Geometry 89 Module 10 Test • Form A

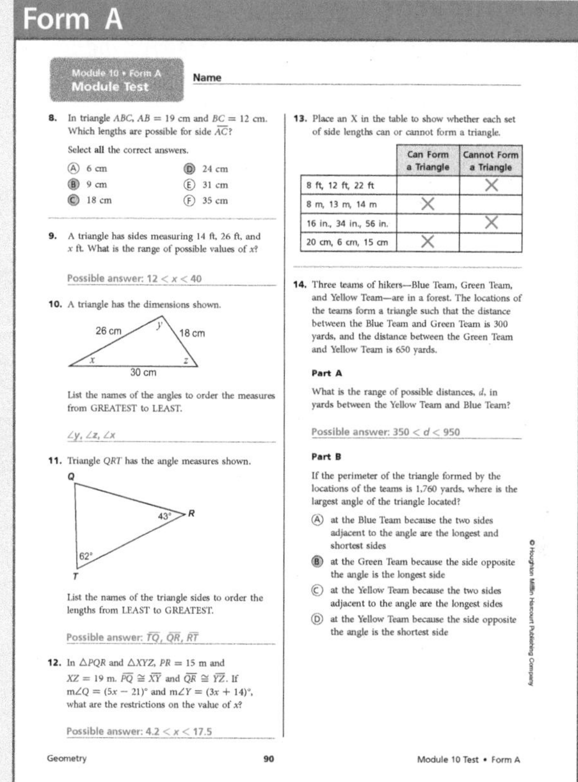

Form A

Module 10 • Form A Module Test

Name

8. In triangle ABC, $AB = 19$ cm and $BC = 12$ cm. Which lengths are possible for side $\overline{AC}$?

Select **all** the correct answers.

(A) 6 cm (B) 9 cm (C) 18 cm (D) 24 cm (E) 31 cm (F) 35 cm

9. A triangle has sides measuring 14 ft, 26 ft, and x ft. What is the range of possible values of x?

Possible answer: $12 < x < 40$

10. A triangle has the dimensions shown.

List the names of the angles to order the measures from GREATEST to LEAST.

$\angle y, \angle z, \angle x$

11. Triangle QRT has the angle measures shown.

List the names of the triangle sides to order the lengths from LEAST to GREATEST.

Possible answer: $\overline{TQ}, \overline{QR}, \overline{RT}$

12. In $\triangle PQR$ and $\triangle XYZ$, $PR = 15$ m and $XZ = 19$ m. $\overline{PQ} \cong \overline{XY}$ and $\overline{QR} \cong \overline{YZ}$. If $m\angle Q = (5x - 21)°$ and $m\angle Y = (3x + 14)°$, what are the restrictions on the value of x?

Possible answer: $4.2 < x < 17.5$

13. Place an X in the table to show whether each set of side lengths can or cannot form a triangle.

	Can Form a Triangle	Cannot Form a Triangle
8 ft, 12 ft, 22 ft		X
8 m, 13 m, 14 m	X	
16 in., 34 in., 56 in.		X
20 cm, 6 cm, 15 cm	X	

14. Three teams of hikers—Blue Team, Green Team, and Yellow Team—are in a forest. The locations of the teams form a triangle such that the distance between the Blue Team and Green Team is 300 yards, and the distance between the Green Team and Yellow Team is 650 yards.

Part A

What is the range of possible distances, d, in yards between the Yellow Team and Blue Team?

Possible answer: $350 < d < 950$

Part B

If the perimeter of the triangle formed by the locations of the teams is 1,760 yards, where is the largest angle of the triangle located?

(A) at the Blue Team because the two sides adjacent to the angle are the longest and shortest sides
(B) at the Green Team because the side opposite the angle is the longest side
(C) at the Yellow Team because the two sides adjacent to the angle are the longest sides
(D) at the Yellow Team because the side opposite the angle is the shortest side

Geometry 90 Module 10 Test • Form A

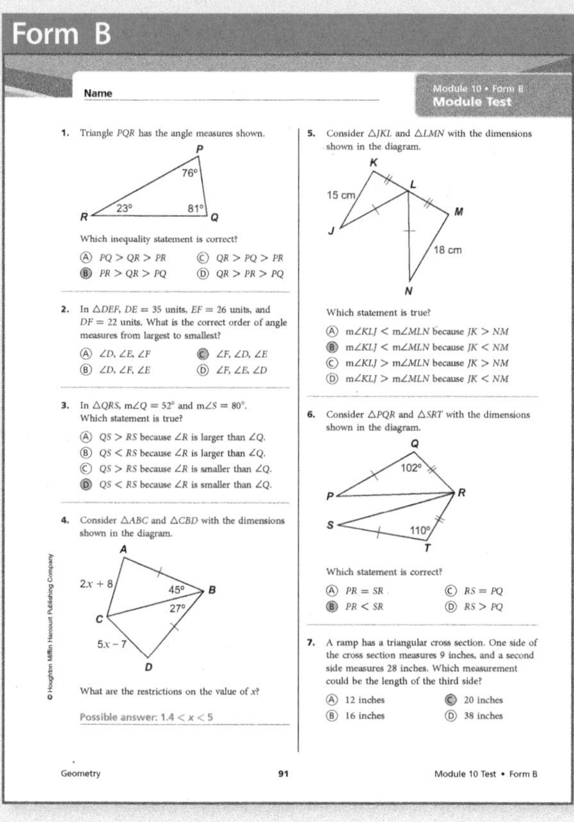

Form B

Module 10 • Form B Module Test

Name

1. Triangle PQR has the angle measures shown.

Which inequality statement is correct?

(A) $PQ > QR > PR$ (B) $PR > QR > PQ$ (C) $QR > PQ > PR$ (D) $QR > PR > PQ$

2. In $\triangle DEF$, $DE = 35$ units, $EF = 26$ units, and $DF = 22$ units. What is the correct order of angle measures from largest to smallest?

(A) $\angle D, \angle E, \angle F$ (B) $\angle D, \angle F, \angle E$ (C) $\angle F, \angle D, \angle E$ (D) $\angle F, \angle E, \angle D$

3. In $\triangle QRS$, $m\angle Q = 52°$ and $m\angle S = 80°$. Which statement is true?

(A) $QS > RS$ because $\angle R$ is larger than $\angle Q$.
(B) $QS < RS$ because $\angle R$ is larger than $\angle Q$.
(C) $QS > RS$ because $\angle R$ is smaller than $\angle Q$.
(D) $QS < RS$ because $\angle R$ is smaller than $\angle Q$.

4. Consider $\triangle ABC$ and $\triangle CBD$ with the dimensions shown in the diagram.

What are the restrictions on the value of x?

Possible answer: $1.4 < x < 5$

5. Consider $\triangle JKL$ and $\triangle LMN$ with the dimensions shown in the diagram.

Which statement is true?

(A) $m\angle KLJ < m\angle MLN$ because $JK > NM$
(B) $m\angle KLJ < m\angle MLN$ because $JK < NM$
(C) $m\angle KLJ > m\angle MLN$ because $JK > NM$
(D) $m\angle KLJ > m\angle MLN$ because $JK < NM$

6. Consider $\triangle PQR$ and $\triangle SRT$ with the dimensions shown in the diagram.

Which statement is correct?

(A) $PR = SR$ (B) $PR < SR$ (C) $RS = PQ$ (D) $RS > PQ$

7. A ramp has a triangular cross section. One side of the cross section measures 9 inches, and a second side measures 28 inches. Which measurement could be the length of the third side?

(A) 12 inches (B) 16 inches (C) 20 inches (D) 38 inches

Geometry 91 Module 10 Test • Form B

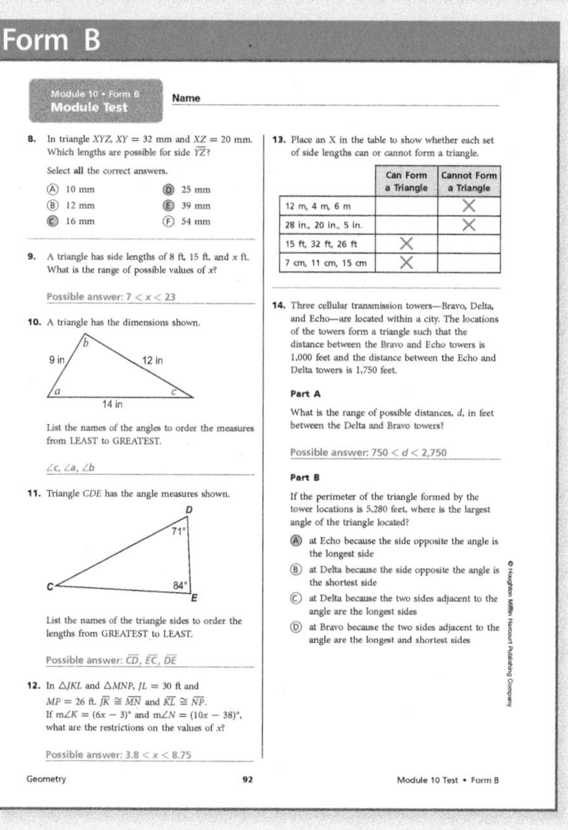

Form B

Module 10 • Form B Module Test

Name

8. In triangle XYZ, $XY = 32$ mm and $XZ = 20$ mm. Which lengths are possible for side $\overline{YZ}$?

Select **all** the correct answers.

(A) 10 mm (B) 12 mm (C) 16 mm (D) 25 mm (E) 39 mm (F) 54 mm

9. A triangle has side lengths of 8 ft, 15 ft, and x ft. What is the range of possible values of x?

Possible answer: $7 < x < 23$

10. A triangle has the dimensions shown.

List the names of the angles to order the measures from LEAST to GREATEST.

$\angle c, \angle a, \angle b$

11. Triangle CDE has the angle measures shown.

List the names of the triangle sides to order the lengths from GREATEST to LEAST.

Possible answer: $\overline{CD}, \overline{EC}, \overline{DE}$

12. In $\triangle JKL$ and $\triangle MNP$, $JL = 30$ ft and $MP = 26$ ft. $\overline{JK} \cong \overline{MN}$ and $\overline{KL} \cong \overline{NP}$. If $m\angle K = (6x - 3)°$ and $m\angle N = (10x - 38)°$, what are the restrictions on the values of x?

Possible answer: $3.8 < x < 8.75$

13. Place an X in the table to show whether each set of side lengths can or cannot form a triangle.

	Can Form a Triangle	Cannot Form a Triangle
12 m, 4 m, 6 m		X
28 in., 20 in., 5 in.		X
15 ft, 32 ft, 26 ft	X	
7 cm, 11 cm, 15 cm	X	

14. Three cellular transmission towers—Bravo, Delta, and Echo—are located within a city. The locations of the towers form a triangle such that the distance between the Bravo and Echo towers is 1,000 feet and the distance between the Echo and Delta towers is 1,750 feet.

Part A

What is the range of possible distances, d, in feet between the Delta and Bravo towers?

Possible answer: $750 < d < 2{,}750$

Part B

If the perimeter of the triangle formed by the tower locations is 5,280 feet, where is the largest angle of the triangle located?

(A) at Echo because the side opposite the angle is the longest side
(B) at Delta because the side opposite the angle is the shortest side
(C) at Delta because the two sides adjacent to the angle are the longest sides
(D) at Bravo because the two sides adjacent to the angle are the longest and shortest sides

Geometry 92 Module 10 Test • Form B

UNIT 6

QUADRILATERALS, POLYGONS, AND TRIANGLE SIMILARITY

Module 11: Quadrilaterals and Polygons

Module 12: Similarity

Digital Animator

- **Say:** *Think about the animations you have seen on the Internet, on TV, in films, and on your tablets and smart phones. These graphic scenarios include 2-D and 3-D computer-generated images in motion created by digital animators. The animators use software to manipulate the characters, colors, settings, and overall compositions you see on various technology platforms.*
- Explain that digital animators use their knowledge of geometry, 3-D modeling graphics, and computer science to produce colorful and dynamic images. Then introduce the STEM task.

STEM Task

Ask students what they know about computer-generated animations.

- Have students describe examples of digital animations they have seen and where they have seen them.
- Have students discuss why digital animations can be appealing and entertaining.
- Have students discuss how and why many advertisers use digital animations to market their products.
- If possible, have students display and describe examples of the best and the worst digital animations they have seen.
- Ask students to compare 2-D and 3-D digital animations and to discuss which type is more effective.

Stem Task Answers

The measure of $\angle C$ is one of the two congruent angles in an isosceles triangle with the central angle as the third angle. Since the third angle must be at least 7°, the maximum possible measure of $\angle C$ is $\frac{180-7}{2} = \frac{173}{2} = 86.5°$

The measure of $\angle C$ is one of the two congruent angles in an isosceles triangle. Since the third angle must be at least $x°$, the maximum possible measure of $\angle C$ is $\left(\frac{180-x}{2}\right)^{\circ}$.

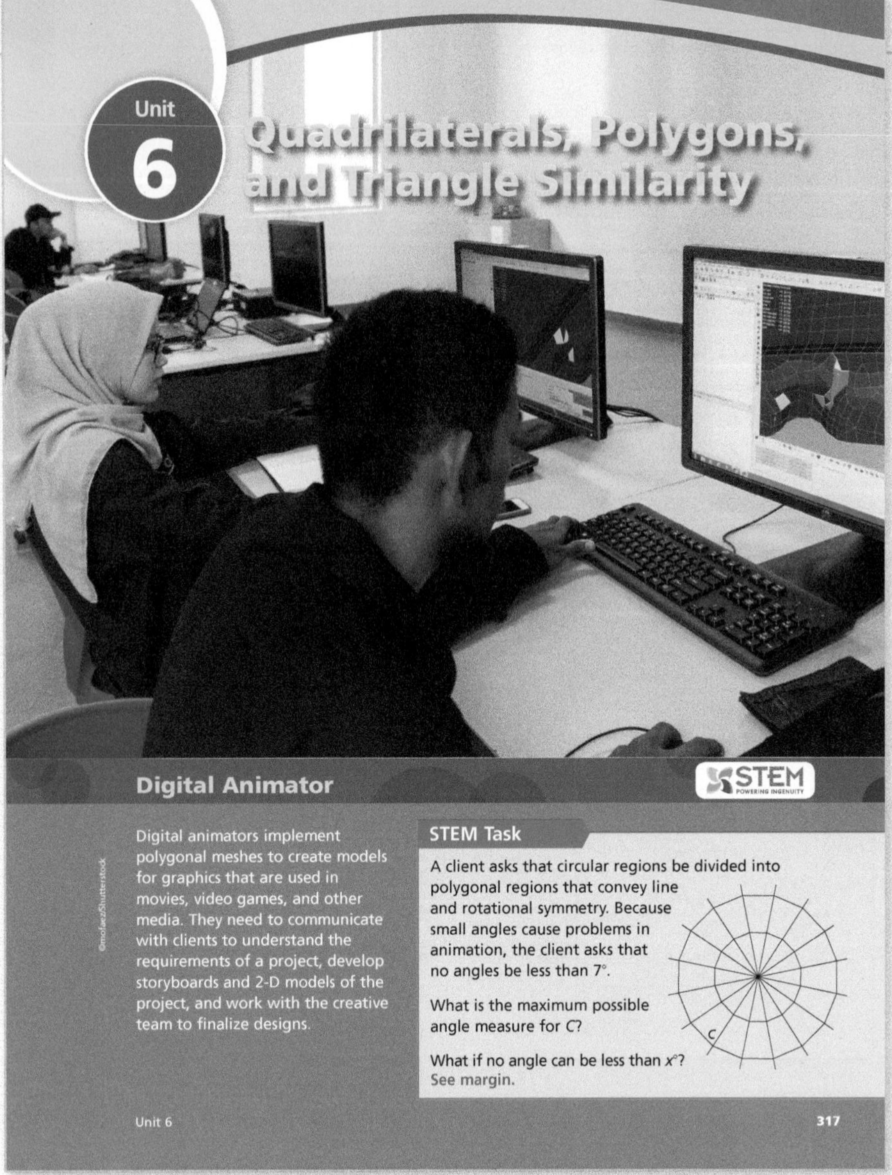

Unit 6
Quadrilaterals, Polygons, and Triangle Similarity
Digital Animator
STEM POWERING INGENUITY
Digital animators implement polygonal meshes to create models for graphics that are used in movies, video games, and other media. They need to communicate with clients to understand the requirements of a project, develop storyboards and 2-D models of the project, and work with the creative team to finalize designs.
STEM Task
A client asks that circular regions be divided into polygonal regions that convey line and rotational symmetry. Because small angles cause problems in animation, the client asks that no angles be less than 7°.
What is the maximum possible angle measure for C?
What if no angle can be less than x°?
See margin.
C
Unit 6
317

Unit 6 Project Meshing Around

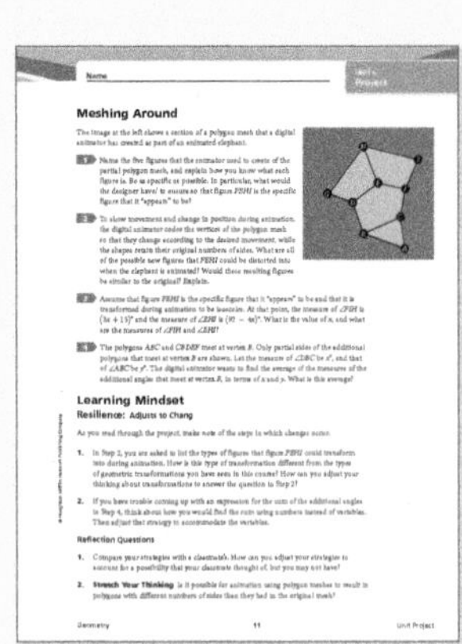

Overview: In this project students use properties of quadrilaterals to describe a section of a polygon mesh created by a digital animator.

Materials: display or poster board for presentations

Assessing Student Performance: Students' presentations should include:

- an accurate description of each of the four shapes in the mesh **(Lesson 11.5)**
- a correct value of x and correct measures of $\angle FIH$ and $\angle EHI$ **(Lesson 11.5)**
- a complete list of possible distorted shapes and an indication that the only way the result would be a similar polygon is if it were still a trapezoid with equal angles and proportional sides **(Lesson 12.1)**
- a correct average for the new angle **(Lesson 12.1 or later)**

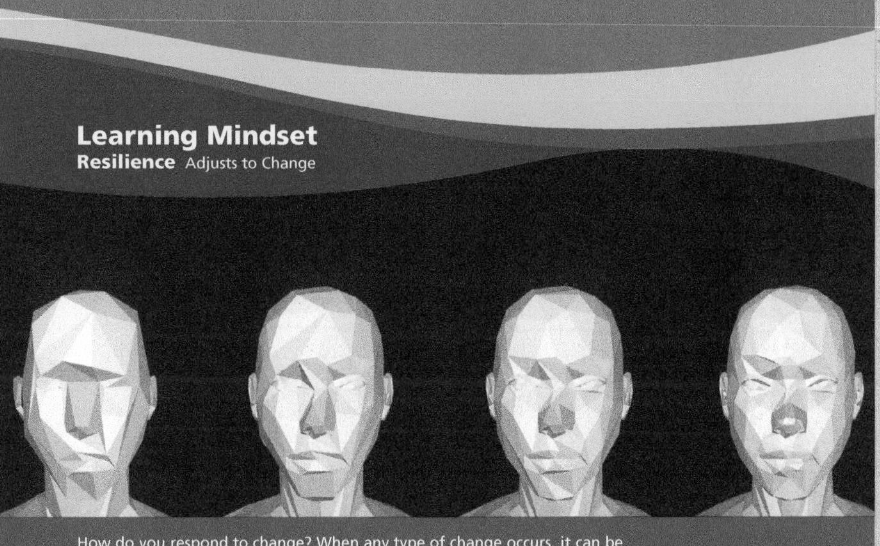

How do you respond to change? When any type of change occurs, it can be uncomfortable. Keep in mind that adapting to change yields growth. Having a plan to adjust to changes can build your confidence as you work through transitional periods. Here are some questions you can ask yourself to help you navigate and adjust to changes:

- Can the situation I am in be changed? Why or why not?
- If the situation can be changed, how am I going to change it? What reasons do I have for wanting the change?
- How am I devoting my energy to change my learning situation? Should I spend my energy on changing the situation or my approach?
- If the situation cannot be changed, how am I going to adapt to the new situation? How can I adjust? How will I grow?
- Who can I ask for advice about change? What questions can I ask?

Reflect

Q **Think about a time when you had to deal with a big change. How did you approach the change? What was your mindset? What was the outcome, and how would you change your approach in the future?**

Q **As a digital animator, how would you adjust when the client changes the requirements of a project? What kind of response would be beneficial? How would this method help you in your career development?**

318

Resilience

Learning Mindset

Adjusts to Change

The learning-mindset focus in this unit is *resilience*, which refers to a person's ability to bounce back after experiencing a change in circumstances. Some students may not perceive a change as a "good thing" and may respond negatively. However, success in learning depends on a flexible mindset that enables a person to push on.

Mindset Beliefs

The ability to adjust to change involves a student's mental, emotional, and physical well-being. Students who are comfortable with the status quo can find it difficult to adjust to change. And learning geometry often requires students to change how they think about mathematics. Much of arithmetic and elementary algebra involves solving equations for x, and in most cases, there is one correct answer. However, in geometry, students are asked to prove statements. There is sometimes no "right" answer or sequence of steps that works. This change in ways of thinking can discourage students who do not appreciate or understand the logical emphasis and rigor introduced into geometry lessons and in later mathematics courses.

Discuss with students how they respond to change. What changes in math have they experienced? How did they deal with such changes? Do they see any benefits to change?

Mindset Behaviors

Encourage students to become more resilient by asking themselves questions about the advantages of change. Pose questions like the following:

When confronted with a change in mathematics:

- Do I turn a blind eye to why it may be a positive thing?
- Do I feel that I am not smart enough to understand what may be required?

When communicating with others:

- Do I readily share my ideas and ask questions if I don't understand something?
- Do I feel threatened or uncomfortable when I am unable to solve a problem or misunderstand what someone says?

When trying to solve a problem in mathematics:

- Do I get stuck and have no idea of what the next steps might be? Do I give up?
- Am I uncomfortable asking others for help?

If students are to respond positively to change, they need to develop a more resilient mindset and convince themselves that change can help them become more successful learners.

What to Watch For

Watch for students who are struggling with adjusting to change in the classroom. Help them become more resilient by

- reassuring them that struggling with change is a common response and
- providing learning strategies that they can use to adjust to an unexpected change.

Watch for students who are upset by change and want to give up. Help them move from a stubborn mindset to a flexible mindset by

- providing a plausible reason for a proposed change and
- providing examples of how change can have positive learning outcomes, such as using technology to understand geometric relationships.

"To improve is to change; to be perfect is to change often."

—Winston S. Churchill

QUADRILATERALS AND POLYGONS

Introduce and Check for Readiness
- Module Performance Task
- Are You Ready?

Lesson 11.1—2 Days

Properties of Parallelograms
Learning Objective: Prove and use properties of parallelograms.
New Vocabulary: diagonal of a polygon, parallelogram

Lesson 11.2—2 Days

Conditions for Parallelograms
Learning Objective: Prove and use conditions for parallelograms.

Lesson 11.3—2 Days

Properties of Rectangles, Rhombuses, and Squares
Learning Objective: Prove and use properties of squares, rectangles, and rhombuses.
New Vocabulary: rectangle, rhombus, square

Lesson 11.4—2 Days

Conditions for Rectangles, Rhombuses, and Squares
Learning Objective: Prove and use conditions for rectangles, rhombuses, and squares.

Lesson 11.5—2 Days

Properties and Conditions for Trapezoids and Kites
Learning Objective: Prove and apply theorems about trapezoids and kites.
New Vocabulary: base angles of a trapezoid, bases of a trapezoid, isosceles trapezoid, kite, legs of a trapezoid, midsegment of a trapezoid, trapezoid

Assessment
- Module 11 Test (Forms A and B)
- Unit 6 Test (Forms A and B)

data checkpoint

LEARNING ARC FOCUS

Build Conceptual Understanding  Connect Concepts and Skills  Apply and Practice

TEACHING FOR SUCCESS

TEACHING FOR DEPTH: Quadrilaterals and Polygons

Meaning of Trapezoids. The inclusive definition of a trapezoid used in this textbook states that a trapezoid is a quadrilateral with **at least one** pair of parallel sides. Parallelograms are trapezoids that have two pairs of parallel sides, and *isosceles* trapezoids have a pair of equal but not parallel sides, called legs.

Using the inclusive definition of a trapezoid, the relationships among the 8 quadrilaterals can be seen on the right side of this tree diagram. Here, parallelograms and their derivatives are also trapezoids.

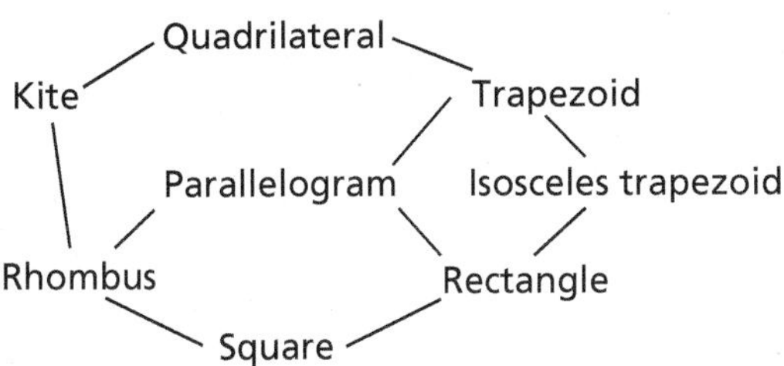

The excusive definition of a trapezoid in this textbook states that a trapezoid is a quadrilateral with **exactly one** pair of parallel sides. The term *exactly* one means "one and only one," which establishes existence and uniqueness.

The exclusive definition of a trapezoid changes the relationships among quadrilaterals as shown in this tree diagram, where the sets of parallelograms and trapezoids are mutually exclusive.

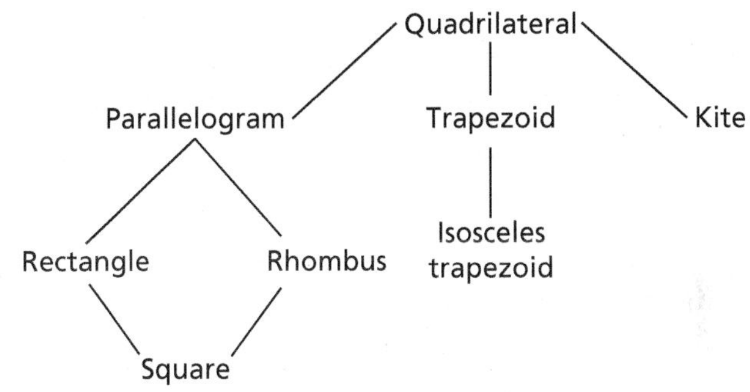

Mathematical Progressions

Prior Learning	Current Development	Future Connections
Students: • proved and used properties of triangles. • proved that lines are parallel. • proved properties of parallelograms.	**Students:** • prove and use properties of and conditions for parallelograms, rectangles, rhombuses, and squares. • prove and apply theorems about trapezoids and kites.	**Students:** • will construct inscribed quadrilaterals and identify properties of their angles.

TEACHER TO TEACHER

From the Classroom

Establish mathematics goals to focus learning. I want my students to be able to categorize quadrilaterals correctly and to list the properties of each type. If my students don't understand the like and unlike properties of these figures—and misapply them when solving problems—they will not reach the math goals set for them.

At the beginning of the module, I have students read the titles of each lesson to anticipate what's coming. The five lessons in the module form a hierarchy of quadrilaterals. Then, at the start of each lesson, students state the learning goal associated with it and discuss how to approach the material so they can be successful.

Throughout this module, I continually engage students in discussions about what they've learned. I then use their responses to plan how to proceed next, always focusing on helping my student achieve the learning goals set for them.

LANGUAGE DEVELOPMENT • Planning for Instruction

By giving all students regular exposure to language routines in context, you will provide opportunities for students to **listen**, **speak**, **read**, and **write** about mathematical situations and develop both mathematical language and conceptual understanding at the same time.

Using Language Routines to Develop Understanding

Use the **Professional Cards** for the following routines to plan for effective instruction.

Co-Craft Questions and Problems Lessons 11.1, 11.2, and 11.5

Students think of natural questions to ask about a given situation or problems similar to a given task and answer the questions they have developed or problems they have created.

Three Reads Lessons 11.1, 11.2, 11.4, and 11.5

Students read a problem three times with a specific focus each time.

1st Read What is the situation about?
2nd Read What are the quantities in the situation?
3rd Read What are the possible mathematical questions that you could ask for the situation?

Information Gap Lesson 11.3

Students recognize when information given to a problem situation is incomplete, and they pose questions and share knowledge with others to discuss any missing facts or relationships and work together to solve the problem.

Critique, Correct, and Clarify Lesson 11.4

Students correct the work in a flawed explanation, argument, or solution method and share with a partner and refine the sample work.

Connecting Language to Quadrilaterals and Polygons

Watch for students' use of the new terms listed below as they explain their reasoning and make connections with new concepts.

Key Academic Vocabulary

Current Development • New Vocabulary

base angles of a trapezoid the pairs of consecutive angles whose common side is a base of the trapezoid

bases of a trapezoid the sides of a trapezoid that are parallel

isosceles trapezoid a trapezoid in which the legs are congruent but not parallel

kite a quadrilateral whose four sides can be grouped into two pairs of consecutive congruent sides

legs of a trapezoid the sides that are not the bases

midsegment of a trapezoid the segment whose endpoints are the midpoints of the legs of a trapezoid

parallelogram a quadrilateral with both pairs of opposite sides parallel

rectangle a parallelogram with four right angles

rhombus a parallelogram with four congruent sides

trapezoid a quadrilateral with at least one pair of parallel sides

Linguistic Note

Listen for how students think about and describe a *trapezoid*. Tell English learners that the word comes from the Greek word *trapeza* that means "table" and *-oeides* that means "shaped." So, the word refers to a table-shaped object. Have them draw different examples of trapezoids and describe what is the same and what is different about each one. Students may also be confused by the two definitions of *trapezoid*. Focus on the inclusive definition used in the textbook. Emphasize the meaning of the phrase *at least one* in the definition. Point out that dogs have at least one pair of legs, but actually have two. Then reinforce the meaning of the phrase by emphasizing that a trapezoid *always* has one pair of parallel sides, but it could have two pairs if the trapezoid is a parallelogram, rectangle, rhombus, or square.

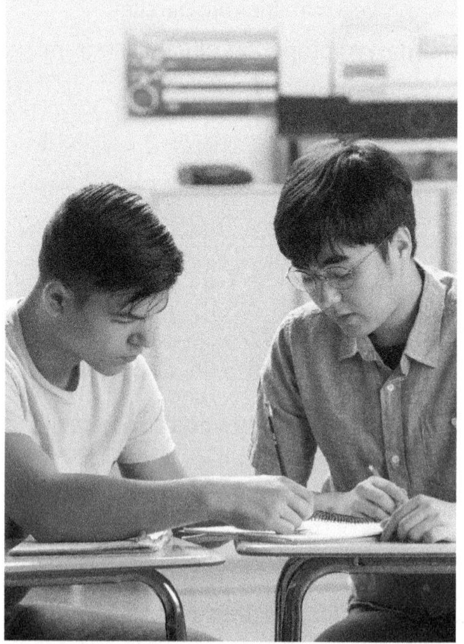

Module

11 Quadrilaterals and Polygons

Module Performance Task: Focus on STEM

4-Link Rear Suspension Systems

A parallel 4-link suspension system is used in both street cars and race cars. This system is designed to keep the rear axle centered while preventing it from rotating. The position of the bars can be changed to improve acceleration to meet the conditions that the car will be driving in. A–D. See margin.

A. For street driving, the top and bottom bars are the same length and the distance between the bolts is equal on both front and rear. What type of quadrilateral is formed? Explain your reasoning.

B. Keeping the bottom bar perpendicular to a segment that passes between the front bolts, what type of quadrilateral is formed? Explain your reasoning.

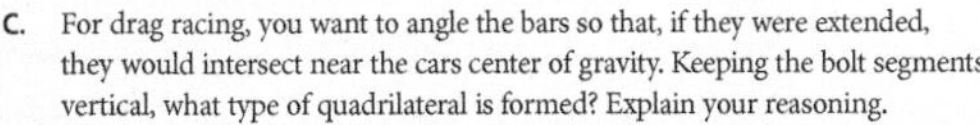

C. For drag racing, you want to angle the bars so that, if they were extended, they would intersect near the cars center of gravity. Keeping the bolt segments vertical, what type of quadrilateral is formed? Explain your reasoning.

D. If the bars were extended, the location of the point of intersection determines the cars acceleration. How can you determine where the point of intersection is located?

Mathematical Connections

Task Part	Prerequisite Skills
Part A	Students should be able to identify a parallelogram. **(11.1 and 11.2)**
Part B	Students should be able to identify a rectangle. **(11.3 and 11.4)**
Part C	Students should be able to identify a trapezoid. **(11.5)**
Part D	Students should be familiar with creating a scale drawing to locate the point of intersection of two segments. **(Grade 7, 1.6)**

4-Link Rear Suspension Systems

Overview

A 4-link rear suspension system is a great way to improve a car's performance and ride quality. It uses two links per side of the axle, which is under the vehicle. The bottom two links keep the axle in place from front to back. The upper two links keep the axle from rotating by keeping the pinion angle as constant as possible. For street driving, the top and bottom bars are the same length, and the distance between the bolts is equal in both the front and rear. The arrangement of the bolts forms the shape of a quadrilateral that has two pairs of equal and opposite sides.

In this module, students will identify types of quadrilaterals. Based on its properties, they will determine whether a quadrilateral is a trapezoid, parallelogram, rectangle, rhombus, square, or kite.

Career Connections

Students with an interest in 4-link rear suspension systems can consider a career as a suspension design engineer or as a mechanical engineer. Each type of engineer is responsible for the initial concept, creating a model of the system, conducting a structural analysis, and validating vehicle production.

Answers

A. Although this appears to be a rectangle, there is not enough information to conclude that it is a rectangle; There is enough information to conclude that the quadrilateral is a parallelogram, but I do not know if the angles are right angles, or if the diagonals are congruent.

B. rectangle; Since one angle of the parallelogram is a right angle, it is a rectangle.

C. trapezoid; I know that one pair of segments are parallel, but the other pair is not, so the quadrilateral formed is a trapezoid.

D. I can use the distance between the bolts and the lengths of the bars to create a scale drawing. Extending the bars in the scale drawing, I can locate the point of intersection.

ONLINE

Assign the Digital Are You Ready? to power actionable reports including

- proficiency by standards
- item analysis

Are You Ready?

Diagnostic Assessment

data checkpoint

- Diagnose prerequisite mastery.
- Identify intervention needs.
- Modify or set up leveled groups.

Have students complete the *Are You Ready?* assessment on their own. Items test the prerequisites required to succeed with the new learning in this module.

Solve Multi-Step Equations Students will apply previous knowledge of solving multi-step equations to find the measures of missing sides and angles in various quadrilaterals.

Congruent Figures Students will apply previous knowledge about triangle congruence criteria to prove properties of types of quadrilaterals: trapezoids, parallelograms, rectangles, rhombuses, squares, and kites.

Parallel Lines Cut by a Transversal Students will apply previous knowledge of the relationships between angles formed when parallel lines are cut by a transversal to prove theorems about various types of quadrilaterals.

ONLINE 

Are You Ready?

Complete these problems to review prior concepts and skills you will need for this module.

Solve Multi-Step Equations

Solve each equation.

1. $5b + 7 = -3b + 39$ $b = 4$
2. $7y + 15 = 10y + 21$ $y = -2$
3. $5(t + 2) = 9t - 14$ $t = 6$
4. $-3m + 1 = 4(m - 5)$ $m = 3$

Congruent Figures

Can you perform a series of transformations on Figure 1 to create Figure 2? Explain your reasoning.

5. 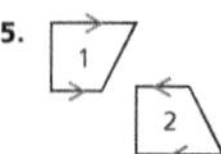

yes; You can translate the figure right, then reflect across a horizontal line.

6. 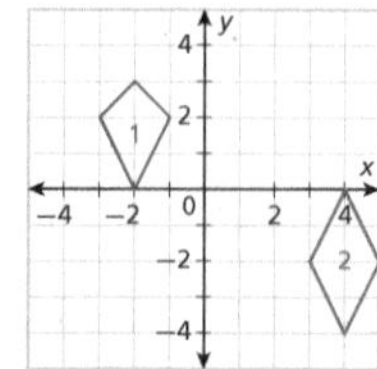

no; There is no sequence of transformations as the figures have non-congruent parts.

Parallel Lines Cut by a Transversal

Find the value of x.

7. 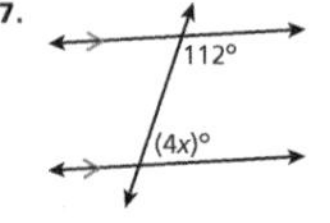

$x = 17$

8. 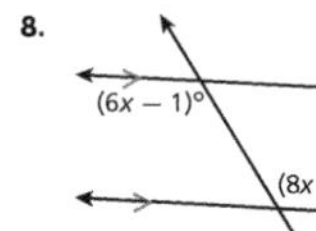

$x = 21$

Connecting Past and Present Learning

Previously, you learned:

- to identify symmetries in figures,
- to determine if figures are congruent using transformations, and
- to use triangle congruence criteria to prove relationships.

In this module, you will learn:

- to classify a quadrilateral as a parallelogram, rectangle, rhombus, square, kite, or trapezoid,
- to use congruence criteria to prove properties of special quadrilaterals, and
- to use the properties of special quadrilaterals to determine unknown measures.

320

DATA-DRIVEN INTERVENTION

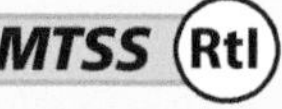

Concept/Skill	Objective	Prior Learning *	Intervene With
Solve Multi-Step Equations	Solve linear equations with grouping symbols and/or the variable on both sides.	Grade 8, Lesson 3.1	• Tier 3 Skill 2 • Reteach, Grade 8 Lesson 3.1
Congruent Figures	Use transformations to determine whether two figures are congruent.	Grade 8, Lesson 1.5	• Tier 2 Skill 14 • Reteach, Grade 8 Lesson 1.5
Parallel Lines Cut by a Transversal	Solve problems about angle measures when parallel lines are cut by a transversal.	Grade 8, Lesson 4.3	• Tier 2 Skill 6 • Reteach, Grade 8 Lesson 4.3

* Your digital materials include access to resources from Grade 6–Algebra 2. The lessons referenced here contain a variety of resources you can use with students who need support with this content.

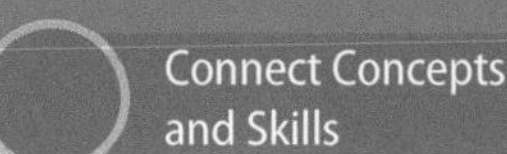

11.1 Properties of Parallelograms

LESSON FOCUS AND COHERENCE

Mathematics Standards

- Prove theorems about parallelograms. *Theorems include: opposite sides are congruent, opposite angles are congruent, the diagonals of a parallelogram bisect each other, and conversely, rectangles are parallelograms with congruent diagonals.*
- Given a rectangle, parallelogram, trapezoid, or regular polygon, describe the rotations and reflections that carry it onto itself.

Mathematical Practices and Processes (MP)

- Reason abstractly and quantitatively.
- Construct viable arguments and critique the reasoning of others.

I Can Objective 

I can prove and use properties of parallelograms.

Learning Objective

Students prove and use properties of parallelograms.

Language Objective

Explain the properties of parallelograms.

Vocabulary

New: **diagonal of a polygon**, **parallelogram**

Lesson Materials: dynamic software, compass, straightedge, straws, protractor

Mathematical Progressions

Prior Learning	Current Development	Future Connections
Students: • proved and used properties of triangles. **(9.1–9.5)** • proved that lines are parallel. **(3.2)**	**Students:** • prove and apply properties of parallelograms.	**Students:** • will prove and use properties of rectangles, rhombuses, and squares. **(11.3)** • will prove and use properties of kites and trapezoids. **(11.5)**

UNPACKING MATH STANDARDS

Prove theorems about parallelograms. *Theorems include: opposite sides are congruent, opposite angles are congruent, the diagonals of a parallelogram bisect each* other, and conversely, rectangles are parallelograms with congruent diagonals.

What It Means to You

Students enrich their understanding of parallelograms by exploring its properties. They use dynamic software to compare sides and angles. They prove theorems and apply them in solving problems to find unknown sides, angles, and diagonals.

In the next lessons they will extend their knowledge of parallelograms and add additional properties to rectangles, squares and rhombi.

WARM-UP OPTIONS

ONLINE

PROJECTABLE & PRINTABLE

ACTIVATE PRIOR KNOWLEDGE • Find the Measure of the Third Angle in a Triangle

Use these activities to quickly assess and activate prior knowledge as needed.

Problem of the Day

The following is true about isosceles triangle *ABC*:

$AB \cong BC$ and $\begin{matrix} m\angle A = (x + 13)^\circ \\ m\angle C = (2x - 17)^\circ \end{matrix}$

What is the measure of angle *B*?

Since the triangle is isosceles, $\angle A \cong \angle C$. Therefore,

$$\begin{aligned} x + 13 &= 2x - 17 \\ x &= 30 \\ m\angle A &= m\angle C = (30 + 13)^\circ = 43^\circ \\ m\angle A + m\angle C + m\angle B &= 180^\circ \\ 43^\circ + 43^\circ + m\angle B &= 180^\circ \\ m\angle B &= 94^\circ \end{aligned}$$

Quick Check for Homework

As part of your daily routine, you may want to display the Teacher Solution Key to have students check their homework

Make Connections

Based on students' responses to the Problem of the Day, choose one of the following:

1. Project the Interactive Reteach, Geometry, Lesson 9.1.

2. Complete the Prerequisite Skills Activity:

Have students work in pairs. Have students classify triangles on the basis of their sides and on the basis of their angles. Have them explain to each other what they know about the angles in right, isosceles, and equilateral triangles.

- *In a right triangle, what is the sum of the angles for which you don't know the measure?* Since the sum of all interior angles is 180° and one of the angles is 90°, the sum of the other two angles is 90°.
- *Can you have a right angle in an isosceles triangle?* Yes, you can. The other angles will be 45° each.
- *What can you tell about a triangle with at least two congruent angles?* It is either isosceles or equilateral.
- *What can you tell about the angles in an obtuse isosceles triangle?* There is one obtuse angle, and the other two angles are congruent with a sum of less than 90°.

If students continue to struggle, use Tier 2 Skill 5.

SHARPEN SKILLS

If time permits, use this on-level activity to build fluency and practice basic skills.

Vocabulary Review

Objective: Students demonstrate understanding of the classification of quadrilaterals.
Materials: paper cut-outs of the following quadrilaterals: a (non-square) rectangle, a square, a (non-square) rhombus, a trapezoid, and a (non-rectangle) parallelogram

Have students place each shape in the Venn diagram according to the definition of the geometric shapes.

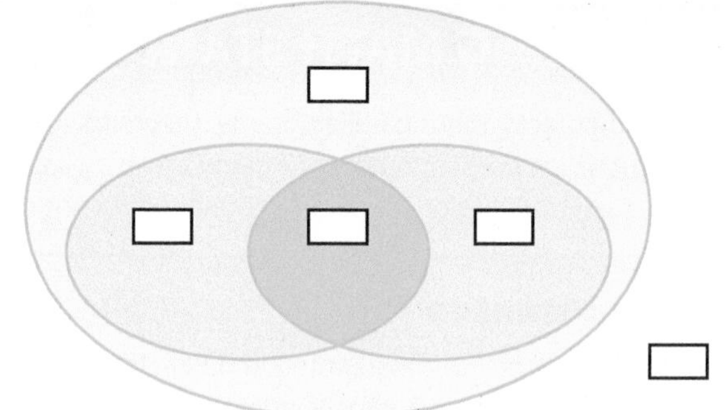

PLAN FOR DIFFERENTIATED INSTRUCTION

Small-Group Options

Use these teacher-guided activities with pulled small groups.

On Track

Materials: dynamic software

Have students work in pairs using the dynamic software.

Have students construct a parallelogram. Then, have them construct the diagonals of the parallelograms and measure the length of the diagonals and half of the diagonals to verify that the diagonals bisect each other.

Almost There

Materials: straws, protractor, ruler

- Students use straws and straw cut-outs to construct parallelograms.
- Then, have students measure the opposite angles of parallelograms with a protractor to verify their congruency.
- Have students cut straws for the diagonals of their parallelograms to verify that the diagonals of parallelograms bisect each other.

Ready for More

Materials: compass and straightedge

Have students work in small groups. Ask students to construct a parallelogram given two diagonals, a and b, such that $a > b$.

Math Center Options

Use these student self-directed activities at centers or stations

Key: ● Print Resources ● Online Resources

On Track

- ● Interactive Digital Lesson
- ●● Journal and Practice Workbook
- ● Interactive Glossary (printable): **diagonal of a polygon**, **parallelogram**
- ● Module Performance Task

Almost There

- ● Reteach 11.1 (printable)
- ● Interactive Reteach 11.1
- ● RtI Tier 2 Skill 5: Angle Relationships in Triangles
- ● Illustrative Mathematics: Is This a Parallelogram?

Ready for More

- ● Challenge 11.1 (printable)
- ● Interactive Challenge 11.1
- ● Illustrative Mathematics: Midpoints of the Sides of a Parallelogram

View data-driven grouping recommendations and assign differentiation resources.

During the *Spark Your Learning*, listen and watch for strategies students use. See samples of student work on this page.

Use a Detailed Narrative

Strategy 1

The fact that the opposite sides are congruent makes me think that this fabric is made of non-square parallelograms and squares. The sides of the squares are equal to the length of either the shorter or the longer sides of the parallelograms.

You can picture the fabric being made up of triangles. You can find congruent triangles, by SSS, within each parallelogram.

If students . . . use a detailed narrative and precise geometric vocabulary to solve the problem, they are showing an exemplary understating of quadrilateral classification from Grade 8.

Have these students . . . explain how they determined that the black shapes are squares. **Ask:**

- Q Can you figure out the angle of the parallelogram from the eight parallelograms that fit in the middle?
- Q Which quadrilaterals have congruent sides?

Use a Short Narrative

Strategy 2

I can tell that these are parallelograms. Their opposite sides are the same.

They also fit in the middle.

I am not sure how I can prove that they are parallelograms.

If students . . . use appropriate vocabulary to describe the situation in the problem, they have a good understanding about the geometric idea inherent in the problem but may lack sufficient knowledge of the theorems needed to prove the validity of their approach.

Activate prior knowledge . . . by having students look at the triangles formed by one of the diagonals in the parallelograms and recall the Triangle Congruence Theorems. **Ask:**

- Q How can you prove that these triangles are congruent knowing that the sides of the shape are congruent?
- Q If we know one of the angles in these triangles, can we use a different Triangle Congruency Theorem?

COMMON ERROR: Uses Imprecise Vocabulary

The fabric is made of squares and parallel figures. Some of them are squares and others are tilted rectangles. That description fits with the additional information that the opposite sides are congruent.

If students . . . use imprecise vocabulary, they need help recalling geometric terminology.

Then intervene . . . by pointing out all of the quadrilaterals that have opposite congruent sides. **Ask:**

- Q How many quadrilaterals do you know that have opposite congruent sides?
- Q If I draw a diagonal, can I look at these triangles and use a Triangle Congruency Theorem based on what I know about the sides?

11.1

Properties of Parallelograms

(I Can) prove and use properties of parallelograms.

Spark Your Learning

Louisa is making a quilt using parallelograms.

Complete Part A as a whole class. Then complete Parts B–D in small groups.

A. What is a geometric question that you can ask about this situation?

A. Possible answer: Which sides and angles of the quadrilaterals are congruent?

B. Does it appear that the quadrilaterals are made from one piece of fabric? How are the fabric quadrilaterals made? See Additional Answers.

C. To answer your question, what strategy and tool would you use along with all the information you have? What answer do you get? See Strategies 1 and 2 on the facing page.

D. Does your answer make sense in the context of the situation? How do you know? yes; Possible answer: In the center of the quilt, the shorter sides of the quadrilaterals align without spaces or overlaps.

Turn and Talk Do your results change if you use the longer diagonal of the quadrilateral to create triangles? See margin.

CULTIVATE CONVERSATION • Co-Craft Questions

If students have difficulty formulating a mathematical question about the situation in the Spark Your Learning, ask them to consider the types of quadrilaterals they know. What are some natural questions to ask about this situation?

Work together to craft the following questions:

- How many types of quadrilaterals do you see?
- How many different-length sides do you see?
- If the black shape has four congruent sides, what could it be?

Then have students think about what additional information, if any, they would need to answer these questions. **Ask:**

- Can you determine if the quadrilaterals are parallelograms? Why or why not?
- Can you determine the measure of the angles of all quadrilaterals? Explain.

1 Spark Your Learning

▶ MOTIVATE

- Have students look at the photo in their books and read the information contained in the photo. Then complete Part A as a whole-class discussion.
- Give the class the additional information they need to solve the problem. This information is available online as a printable and projectable page in the Teacher Resources.
- Have students work in small groups to complete Parts B–D.

▶ PERSEVERE

If students need support, guide them by asking:

Q **Advancing • Use Tools** Which tool could you use to solve the problem? Why choose that tool and not some other? Students' choices of tools and reasons for choosing them will vary.

Q **Assessing** How would you know that in a pattern like this there would be no gaps at a point? There would be no gaps if the sum of the measures of all angles around a point is 360°.

Q **Assessing** What are the measures of the angles of the parallelograms in the center of the fabric? Eight congruent angles add up to 360°. Each acute angle is 45° and the obtuse angle is 135°.

Q **Advancing** What shape is the black quadrilateral? It is a parallelogram with 4 congruent sides. It could be a rhombus or a square. It is a square because its angle makes 360° with the obtuse angles of the parallelograms, $360° - 2(135°) = 90$.

Turn and Talk Point out to students that you can split a parallelogram into two triangles using the two diagonals of the parallelogram. No, the opposite angles and sides are congruent no matter which diagonals you use to form the triangles.

▶ BUILD SHARED UNDERSTANDING

Select groups of students who used various strategies and tools to share with the class how they solved the problem. As they present their solutions, have each group discuss why they chose a specific strategy and tool.

2 Learn Together

Build Understanding

Task 1 **MP** **Attend to Precision** Students use precise vocabulary to find the sum of the measures of the interior angles in a quadrilateral.

Turn and Talk Have students show each other the quadrilaterals they have drawn. Point out that the Triangle Sum Theorem can be applied to any quadrilateral regardless of the measure of its angles. Each angle measures 90°; The sum of the angle measures of the quadrilateral is 360°, so the equation to find the angle measure of one angle is $4x = 360$.

Task 2 **MP** **Construct Arguments** Students use dynamic software, make observations by changing parameters, and reason about properties of parallelograms.

CONNECT TO VOCABULARY

Have students use the **Interactive Glossary** to record their understanding of the vocabulary in this task.

Turn and Talk Point out to students that a rotation preserves distances and angles. Rotate the parallelogram 180°. The angles and sides will align, so you know that they are congruent.

Build Understanding

Explore Quadrilaterals

1 Use a straightedge to draw three different quadrilaterals. Then draw a diagonal from one vertex of each quadrilateral. An example is shown.

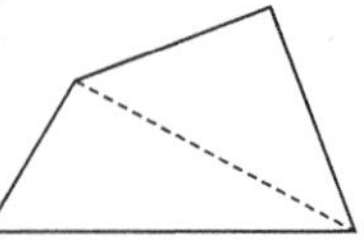

A. How many triangles are formed in each quadrilateral? 2

B. Use what you know about the Triangle Sum Theorem to find the sum of the measures of the interior angles in each quadrilateral. What is the sum? See Additional Answers.

Turn and Talk Suppose you know that all four angles of a quadrilateral are congruent. What is the measure of each angle? Explain how you found your answer. See margin.

Explore Parallelograms

A **parallelogram** is a quadrilateral with both pairs of opposite sides parallel. The symbol ▱ is used to write the name of a parallelogram. In ▱$WXYZ$, $\overline{WZ}$ and $\overline{XY}$ are parallel, and $\overline{WX}$ and $\overline{ZY}$ are parallel.

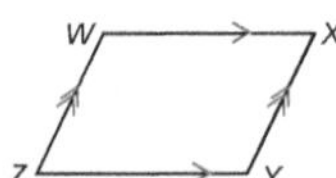

2 Use a geometric drawing tool to construct two pairs of parallel lines. The intersections of the lines create a parallelogram. Label the vertices of the parallelogram A, B, C, and D. Then show the measures of the interior angles and the side lengths of the parallelogram.

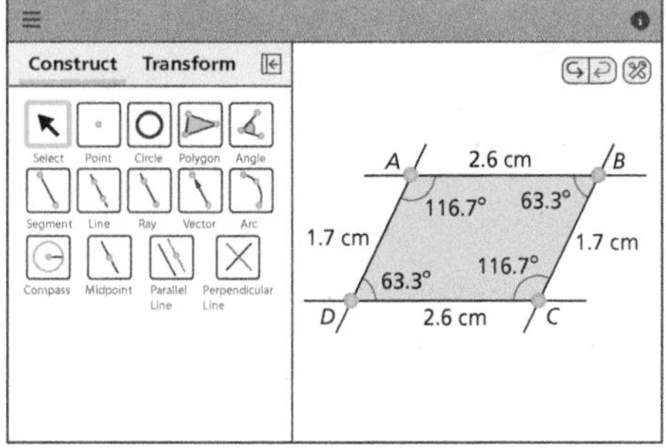

A. What do you notice about the side lengths? The side lengths of opposite sides are equal.

B. What do you notice about the measures of the angles? The angle measures of opposite angles are equal.

C. Drag one of the vertices to change the shape. Does the quadrilateral remain a parallelogram? How do you know? yes; As the sides move, each remains parallel to the side opposite it.

D. Observe the side lengths and the angle measures of the parallelograms formed as you drag one of the vertices. Are your observations from Parts A and B the same for all of the parallelograms? See Additional Answers.

Turn and Talk How could you use rotations to show that your observations in Parts A and B are true? See margin.

322

LEVELED QUESTIONS

Depth of Knowledge (DOK)	Leveled Questions	What Does This Tell You?
Level 1 **Recall**	How can you split this quadrilateral into two triangles? I can connect two of the corners.	Students' answers will indicate whether they recall the meaning of *diagonal*. Remind them there is a difference between a *diagonal* and *diagonally*.
Level 2 **Basic Application of Skills & Concepts**	How can you use the method of partitioning a shape into triangles to find the sum of the measures of the interior angles of a kite? The kite is a quadrilateral, and I will split it along a diagonal to form two triangles.	Students' answers will indicate whether they can understand and apply the Triangle Sum Theorem to find the sum of the measures of the interior angles of any quadrilateral.
Level 3 **Strategic Thinking & Complex Reasoning**	Can you use the Triangle Sum Theorem to find the sum of the interior angles of a pentagon? Possible answer: I can split a pentagon into a quadrilateral and a triangle. The sum would be $360° + 180° = 540°$.	Students' answers will indicate whether they understand the principle of partitioning a quadrilateral into smaller shapes to find the sum of the measures of interior angles and whether they can extend that concept to other polygons.

Theorems about Quadrilaterals and Parallelograms

The observations you made about quadrilaterals and parallelograms can be stated as theorems.

Quadrilateral Sum Theorem

The sum of the angle measures in a quadrilateral is 360°.

Opposite Sides of a Parallelogram Theorem

If a quadrilateral is a parallelogram, then its opposite sides are congruent.

Opposite Angles of a Parallelogram Theorem

If a quadrilateral is a parallelogram, then its opposite angles are congruent.

3 Prove that the opposite angles of a parallelogram are congruent.

Given: *ABCD* is a parallelogram.

Prove: $\angle A \cong \angle C$

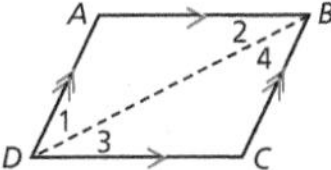

A. How can the two triangles in the parallelogram be used to show that $\angle A \cong \angle C$? See Additional Answers.

Statements	Reasons
1. *ABCD* is a parallelogram.	1. Given
2. Draw $\overline{BD}$.	2. Through any two points, there is exactly one line.
3. $\overline{AB} \parallel \overline{DC}$, $\overline{AD} \parallel \overline{BC}$	3. Definition of parallelogram
4. $\angle 1 \cong \angle 4$, $\angle 2 \cong \angle 3$	4. Alternate Interior Angles Theorem
5. $\overline{DB} \cong \overline{DB}$	5. Reflexive Property of Congruence
6. $\triangle ABD \cong \triangle CDB$	6. ASA Triangle Congruence Theorem
7. $\angle A \cong \angle C$	7. Corresponding parts of congruent figures are congruent.

B. This proof shows only one result of the Opposite Angles of a Parallelogram Theorem. What else needs to be proven? You need to prove that $\angle B \cong \angle D$.

C. Explain how to show that $\angle B \cong \angle D$. See Additional Answers.

Turn and Talk How can the proof shown, in Part A, be used to write a proof for the Opposite Sides of a Parallelogram Theorem? See margin.

Task 3 (MP) **Construct Arguments** Students use the proof of the Opposite Angles of a Parallelogram Theorem to reason about additional results that could be proven in a similar way.

Sample Guided Discussion:

Q **Since we are given that *ABCD* is a parallelogram, what does that tell us?** We know that the opposite sides are parallel.

Q **In Part C, how can we use the other diagonal of the parallelogram in a similar way?** The other diagonal is also a transversal between the two parallel sides and can be used to form two other triangles.

Turn and Talk Remind students that when two triangles are congruent, all of their corresponding parts are also congruent. Change the last step to $\overline{AB} \cong \overline{DC}$ and $\overline{AD} \cong \overline{BC}$ by Corresponding Parts of Congruent Figures are Congruent.

PROFICIENCY LEVEL

Beginning

Have students draw a parallelogram and explain what makes the shape a parallelogram. Have them point to and write the pairs of parallel lines by using the vocabulary of "parallel," "parallelogram," and "quadrilateral."

Intermediate

Have students work in small groups and explain to each other why a rhombus is a parallelogram and what that means for its sides and opposite angles. Then, have them do the same for the trapezoid, i.e., why the trapezoid is not a parallelogram. Have students use sentence starters such as, "The rhombus is a parallelogram, because.... Therefore, its opposite angles are congruent."

Advanced

Have students explain and model why the kite is not a parallelogram although it has one pair of congruent opposite angles.

Step It Out

Task 4 (MP) **Reason** Students explore the properties of the diagonals of parallelograms and reason through theorems to prove that the diagonals bisect each other.

CONNECT TO VOCABULARY

Have students use the **Interactive Glossary** to record their understanding of the vocabulary in this task.

Sample Guided Discussion:

Q **In order to prove congruency of two segments, what might be useful to prove first?** It will be useful to think of the segments as sides of triangles and prove the triangles congruent.

Q **For Part C, can you summarize what you know so far about parallelograms in order to possibly use the information in the proof?** I know that parallelograms have two pairs of parallel and congruent sides. Also, I can think of parallel lines cut by a transversal to look at angles.

Turn and Talk Point out to students that they can go about proving theorems in different ways depending on prior knowledge, that is, as they prove theorems, they can use them to prove other or previous theorems in different ways. You could prove that $\triangle AED \cong \triangle CEB$ using the Alternate Interior Angles Theorem to show that $\angle DAE \cong \angle BCE$ and $\angle ADE \cong \angle EBC$, and the Opposite Sides of a Parallelogram Theorem to show that $\overline{AD} \cong \overline{BC}$.

Step It Out

Prove Diagonals Bisect Each Other

A **diagonal of a polygon** is a segment connecting two nonconsecutive vertices of a polygon.

Diagonals of a Parallelogram Theorem

If a quadrilateral is a parallelogram, then its diagonals bisect each other.

In $ABCD$, $\overline{AE} \cong \overline{CE}$ and $\overline{BE} \cong \overline{DE}$.

4 Prove that the diagonals of a parallelogram bisect each other.

Given: $ABCD$ is a parallelogram.

Prove: $\overline{AE} \cong \overline{CE}$ and $\overline{BE} \cong \overline{DE}$

A. How are $\angle 2$ and $\angle 3$ related? How do you know?
A. $\angle 2$ and $\angle 3$ are congruent; Alternate Interior Angles Theorem

B. How are $\angle 1$ and $\angle 4$ related? How do you know?
B. $\angle 1$ and $\angle 4$ are congruent; Alternate Interior Angles Theorem

Statements	Reasons
1. $ABCD$ is a parallelogram.	1. Given
2. $\overline{AB} \parallel \overline{DC}$	2. Definition of parallelogram
3. $\angle 1 \cong \angle 4$, $\angle 2 \cong \angle 3$	3. Alternate Interior Angles Theorem
4. $\overline{AB} \cong \overline{CD}$	4. Opposite Sides of a Parallelogram Theorem
5. $\triangle ABE \cong \triangle CDE$	5. ASA Triangle Congruence Theorem
6. $\overline{AE} \cong \overline{CE}$ and $\overline{BE} \cong \overline{DE}$	6. Corresponding parts of congruent figures are congruent.

C. Is it possible to prove this theorem using a different triangle congruence theorem? Explain your reasoning. See Additional Answers.

D. Why do you think the Diagonals of a Parallelogram Theorem is presented after the Opposite Sides of a Parallelogram Theorem? See below.

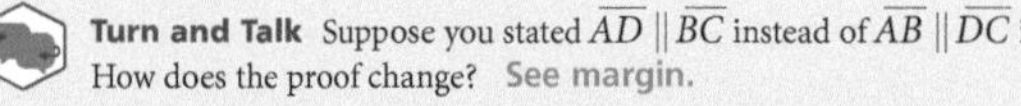

Turn and Talk Suppose you stated $\overline{AD} \parallel \overline{BC}$ instead of $\overline{AB} \parallel \overline{DC}$ in Step 2. How does the proof change? See margin.

D. You need to use the Opposite Sides of a Parallelogram Theorem to show that the triangles are congruent to prove that the diagonals bisect each other.

324

Use Properties of Parallelograms

You can use the properties of parallelograms to find the unknown lengths and unknown angle measures of parallelograms.

5 A guitar has markers on the fret board that are shaped like parallelograms. Find BC and $m\angle B$ in the fret marker shown.

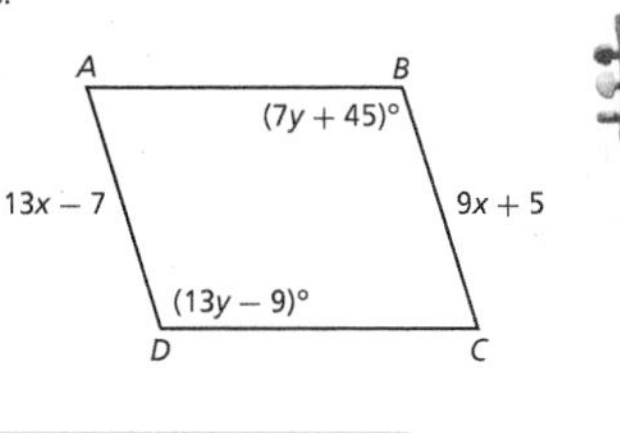

Find *BC*.

To find BC, first find the value of x.

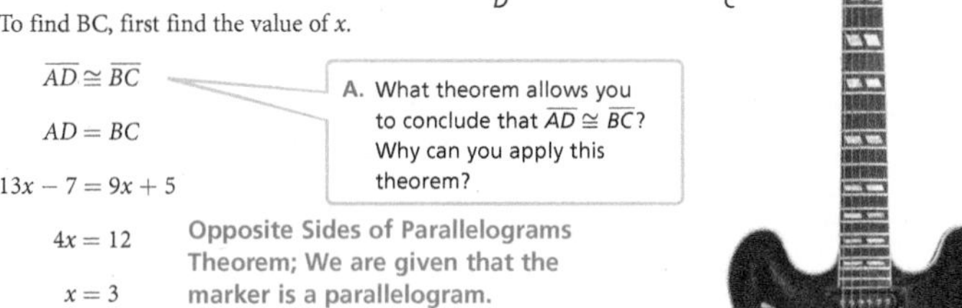

A. What theorem allows you to conclude that $\overline{AD} \cong \overline{BC}$? Why can you apply this theorem?

$\overline{AD} \cong \overline{BC}$

$AD = BC$

$13x - 7 = 9x + 5$

$4x = 12$ Opposite Sides of Parallelograms Theorem; We are given that the marker is a parallelogram.

$x = 3$

Next, substitute the value of x into the expression for BC.

$BC = 9x + 5 = 9(3) + 5 = 32$

The length of $\overline{BC}$ is 32.

Find $m\angle B$.

To find $m\angle B$, first find the value of y.

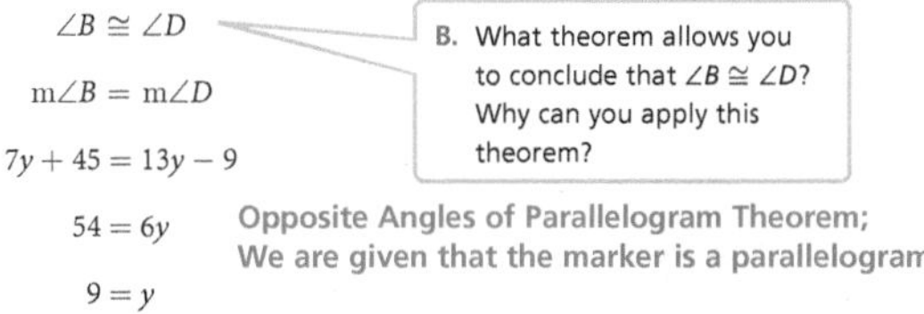

B. What theorem allows you to conclude that $\angle B \cong \angle D$? Why can you apply this theorem?

$\angle B \cong \angle D$

$m\angle B = m\angle D$

$7y + 45 = 13y - 9$

$54 = 6y$ Opposite Angles of Parallelogram Theorem; We are given that the marker is a parallelogram.

$9 = y$

Next, substitute the value of y into the expression for $m\angle B$.
$m\angle B = (7y + 45)^\circ = (7(9) + 45)^\circ = 108^\circ$

The measure of $\angle B$ is 108°.

Turn and Talk Do you have enough information to find AB and $m\angle A$? Explain your reasoning. See margin.

Task 5 (MP) **Reason** Students solve problems wherein they use their knowledge of the properties of parallelograms to find unknown lengths and angle measures.

(EL) **SUPPORT SENSE-MAKING Three Reads**

Have students read the problem three times. Use the questions below for a different focus each read.

1. What is the situation about?
2. What are the quantities in the situation?
3. What are the possible mathematical questions that you could ask for the situation?

Sample Guided Discussion:

Q **Without thinking of finding a concrete answer, just by looking at the picture, what do you know about the parallelogram?** I know that opposite angles are congruent and opposite sides are congruent.

Q **In solving the first equation, you have obtained a value for x. Is this the end of your solution for finding the length of $\overline{BC}$?** No, you still have to substitute the value of x in any one of the expressions given for sides $\overline{AD}$ or $\overline{BC}$.

Turn and Talk Remind students that the sum of the measures of the interior angles is the same for any quadrilateral. There is enough information to find $m\angle A$, but not AB. $m\angle A = m\angle C$, $m\angle B = m\angle D = 108^\circ$, and $m\angle A + m\angle B + m\angle C + m\angle D = 360^\circ$. Solve and find that $m\angle A = 72^\circ$.

ONLINE

Assign the Digital On Your Own for
- built-in student supports
- Actionable Item Reports
- Standards Analysis Reports

On Your Own

Assignment Guide

The chart below indicates which problems in the On Your Own are associated with each task in the Learn Together. Assign daily homework for tasks completed.

Learn Together Tasks	On Your Own Problems
Task 1, p. 322	Problems 9–11
Task 2, p. 322	Problems 16, 17, and 24
Task 3, p. 323	Problems 15, 16, 25, and 26
Task 4, p. 324	Problems 13, 14, and 24
Task 5, p. 325	Problems 10–15 and 18–23

Check Understanding

1. You are given the measures of three of the angles of a quadrilateral. Explain how you can use the Quadrilateral Sum Theorem to find the measure of the fourth angle. See Additional Answers.

$WXYZ$ is a parallelogram. Find each measure.

2. WX 25 cm
3. XY 16 cm
4. $m\angle W$ 108°
5. $m\angle X$ 72°

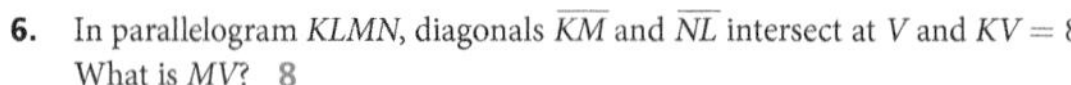

6. In parallelogram $KLMN$, diagonals $\overline{KM}$ and $\overline{NL}$ intersect at V and $KV = 8$. What is MV? 8

For Problems 7 and 8, $ABCD$ is a parallelogram. Find the measure.

7. AD 27

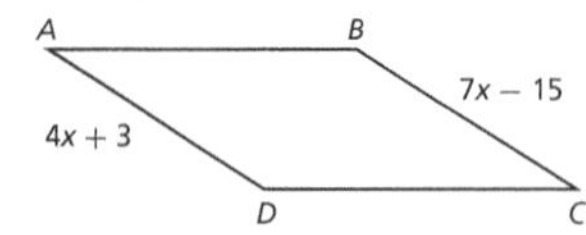

8. $m\angle B$ 75°

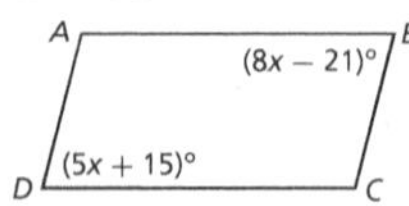

On Your Own

9. (MP) **Reason** Write a paragraph proof of the Quadrilateral Sum Theorem.

Given: Quadrilateral $FGHJ$ See Additional Answers.

Prove: $m\angle F + m\angle G + m\angle H + m\angle J = 360°$

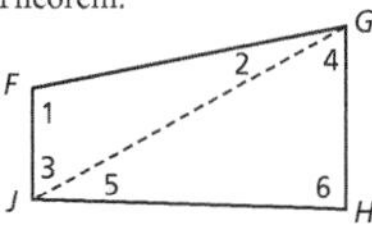

Find the measure of the unknown angle in each quadrilateral.

10.

137°

11.

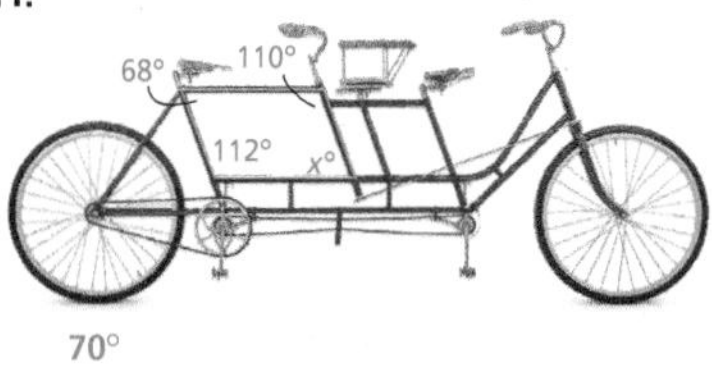

70°

(l) ©Sandra Lass/Shutterstock; (r) ©NamiP/Shutterstock

$QRST$ is a parallelogram. $QR = 2$, $RS = 4.1$, $SP = 2.5$, and $m\angle RST = 76°$. Find each measure.

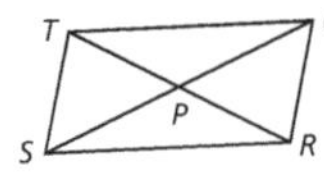

12. QT 4.1
13. PQ 2.5
14. QS 5
15. $m\angle TQR$ 76°

326

3 Check Understanding

Formative Assessment

Use formative assessment to determine if your students are successful with this lesson's learning objective.

Students who successfully complete the Check Understanding can continue to the On Your Own practice.

For students who miss 1 problem or more, work in a pulled small group using the Almost There small-group activity on page 321C.

ONLINE

Assign the Digital Check Understanding to determine
- success with the learning objective
- items to review
- grouping and differentiation resources

4 Differentiation Options

Differentiate instruction for all students using small-group activities and math center activities on page 321C.

Reteach

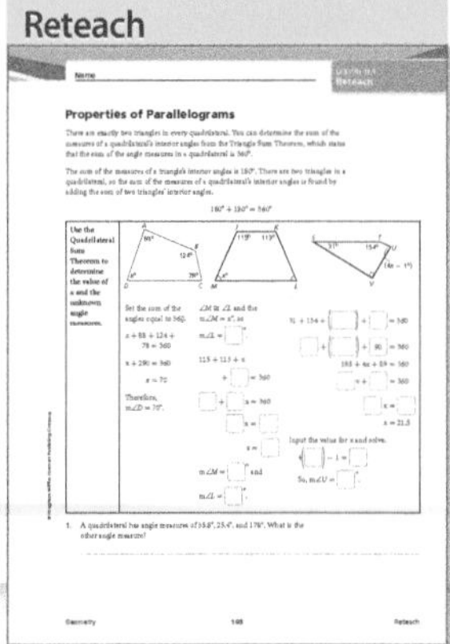

Challenge

16, 17. See Additional Answers.

16. **Construct Arguments** Prove that the opposite sides of the parallelogram are congruent.

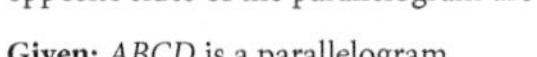

Given: *ABCD* is a parallelogram.

Prove: $\overline{AB} \cong \overline{DC}$, $\overline{AD} \cong \overline{BC}$

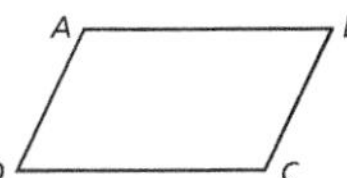

17. **Reason** Prove the Consecutive Angles of a Parallelogram Theorem, which is stated below.

If a quadrilateral is a parallelogram, then its consecutive angles are supplementary.

Given: *ABCD* is a parallelogram.

Prove: $\angle A$ is supplementary to $\angle B$,
$\angle B$ is supplementary to $\angle C$,
$\angle C$ is supplementary to $\angle D$, and
$\angle D$ is supplementary to $\angle A$.

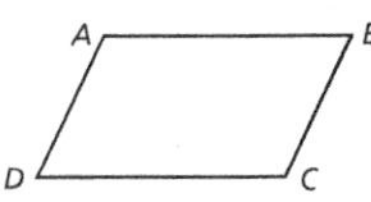

***KLMN* is a parallelogram. Find each measure.**

18. *KN* 30

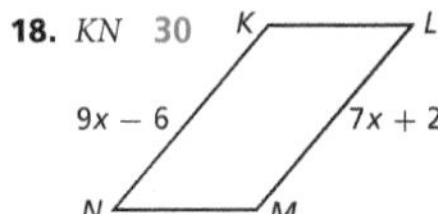

19. m$\angle K$ 117°

K L (7x + 12)° N (4x + 3)° M

20. In parallelogram *EFGH*, diagonals $\overline{EG}$ and $\overline{FH}$ intersect at *I*. $EI = 6s - 10$ and $GI = 3s + 11$. What is *GI*? 32

***WXYZ* is a parallelogram. $WY = 4a - 6$. Find each measure.**

21. *WZ* 43

22. *PY* 27

23. m$\angle PYZ$ 34°

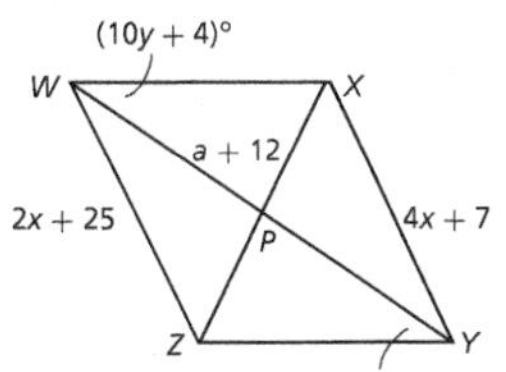

24. Lindsey is designing a stained glass window using parallelograms. She draws a model before starting her work. In her model *BCGF* and *CDHG* are parallelograms, and $\overline{BC} \cong \overline{CD}$. Find all the segments that are congruent to $\overline{KG}$. $\overline{KD}$, $\overline{CJ}$, $\overline{JF}$

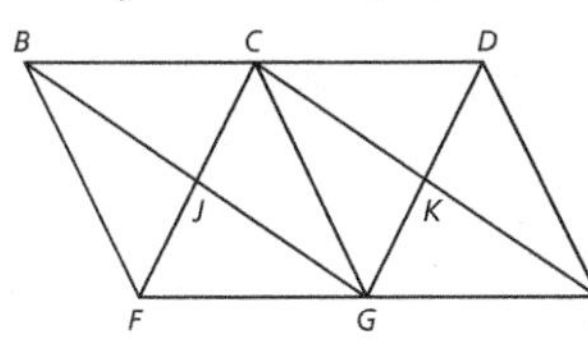

Questioning Strategies

Problem 16 How could you draw a segment to create two triangles? Why would this be helpful? I could draw one of the diagonals; This would be helpful because I could potentially use a triangle congruence theorem to prove the two triangles are congruent, and then identify congruent corresponding sides.

Watch for Common Errors

Problem 20 Point out to students that in problems like that, finding the value of the variable is only part of the solution. The value of the variable then needs to be substituted in an expression to find the length of the side. Furthermore, in this case, students might assume that they are being asked to find the length of the diagonal when in fact they are being asked to find half of the diagonal.

5 Wrap-Up

Summarize learning with your class. Consider using the Exit Ticket, Put It in Writing, or I Can scale.

Exit Ticket 

Half the length of one diagonal of a parallelogram is twice the length of half the length of the other diagonal. The sum of the measures of the lengths of the diagonals is 48 cm. How long is each diagonal?

Let x be half of the shorter diagonal. Then the whole diagonal will be $2x$. Half of the longer diagonal is twice x, so it's $2x$. The longer diagonal is $2(2x)$ or $4x$. The sum of the two diagonals is $2x + 4x = 48$. Solving this equation gives $x = 8$.

The length of the shorter diagonal is 16 cm and the length of the longer is 32 cm.

Put It in Writing 

Explain in writing all the properties of a parallelogram that you now know.

I Can 

The scale below can help you and your students understand their progress on a learning goal.

4	I can prove and use properties of parallelograms, and I can explain my reasoning to others.
3	I can prove and use properties of parallelograms.
2	I understand the proofs about properties of parallelograms, and I can use the properties to solve problems to find unknown sides and angles.
1	I can use the properties of parallelograms to solve problems to find unknown sides and angles.

Spiral Review • Assessment Readiness

These questions will help determine if students have retained information taught in the past and can also prepare them for high-stakes assessments. Here, students must apply the Triangle Inequality Theorem **(10.1),** determine relations between angles and sides (i.e., across longer sides are greater angles) **(9.1),** and review pairs of congruent angles when parallel lines are cut by a transversal **(3.1).**

For Problems 25 and 26, write a two-column proof.

25, 26. See Additional Answers.

25. **Given:** *ABCD* and *AXYZ* are parallelograms.

Prove: $\angle C \cong \angle Y$

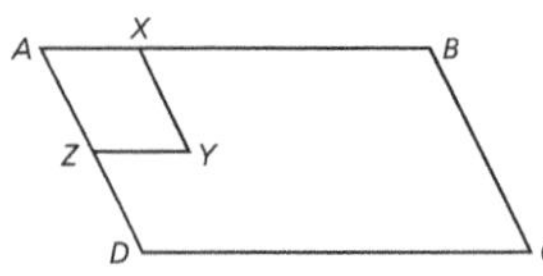

26. **Given:** *CDFE* and *FGIH* are parallelograms, and $EF \cong FG$.

Prove: $\overline{CD} \cong \overline{HI}$

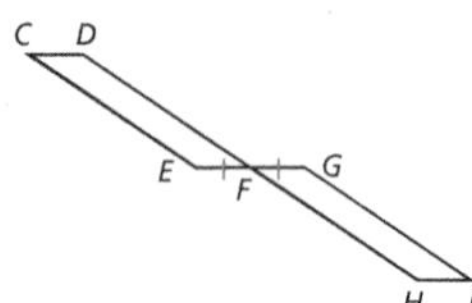

Spiral Review • Assessment Readiness

27. Which of the following sets of numbers can be the side lengths of a triangle? Select all that apply.

Ⓐ 4, 5, 10
Ⓑ 12, 15, 18
Ⓒ 7, 11, 17
Ⓓ 19, 21, 40
Ⓔ 13, 13, 25

28. In $\triangle ABC$, $AB = 5$, $AC = 7$, and $m\angle A = 65°$. In $\triangle XYZ$, $XY = 5$ and $XZ = 7$. If $BC > YZ$, which of the following can be the measure of $\angle X$? Select all that apply.

Ⓐ 25°
Ⓑ 64°
Ⓒ 65°
Ⓓ 66°
Ⓔ 90°
Ⓕ 115°

29. The shaded parallelogram is created from two pairs of parallel lines. Match the statement on the left with its justification on the right.

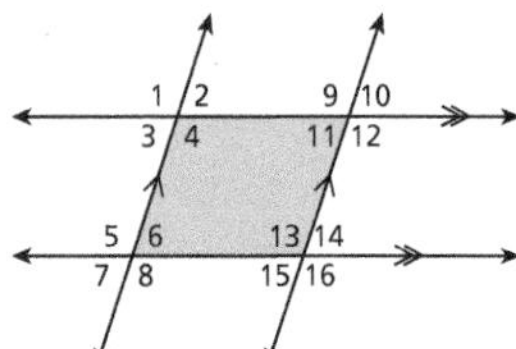

Statement		Justification
A. $\angle 4 \cong \angle 12$	A. 3	**1.** Consecutive Interior Angles Theorem
B. $\angle 11 \cong \angle 14$	B. 4	**2.** Consecutive Exterior Angles Theorem
C. $m\angle 12 + m\angle 14 = 180°$	C. 1	**3.** Corresponding Angles Postulate
D. $m\angle 2 + m\angle 8 = 180°$	D. 2	**4.** Alternate Interior Angles Theorem

I'm in a Learning Mindset!

How am I directing my efforts when my learning environment changes?

328 Keep Going to Journal and Practice Workbook

Learning Mindset

Resilience Adjusts to Change

Point out to students that as they learn more about properties of polygons, the definitions will become richer and more involved. Just like younger kids learn that the shape called a rectangle has right angles, we will add to that definition that the rectangle is in fact a parallelogram and that parallelograms have bisecting diagonals. *Do you feel your understanding of parallelograms has changed or become deeper? Do you feel your understanding has grown?*

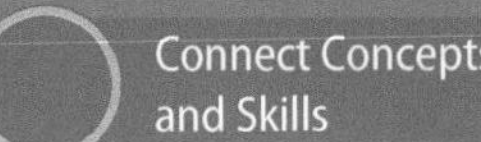

11.2 Conditions for Parallelograms

LESSON FOCUS AND COHERENCE

Mathematics Standards

- Prove theorems about parallelograms. *Theorems include: opposite sides are congruent, opposite angles are congruent, the diagonals of a parallelogram bisect each other, and conversely, rectangles are parallelograms with congruent diagonals.*
- Given a rectangle, parallelogram, trapezoid, or regular polygon, describe the rotations and reflections that carry it onto itself.

Mathematical Practices and Processes 

- Construct viable arguments and critique the reasoning of others.
- Look for and make use of structure.

I Can Objective 

I can prove and use conditions for parallelograms.

Learning Objective

Students prove and use conditions for parallelograms.

Language Objective

Explain the steps needed to prove conditions necessary to prove a quadrilateral is a parallelogram.

Mathematical Progressions

Prior Learning	Current Development	Future Connections
Students: • proved and used properties of triangles. **(9.1–9.5)** • proved that lines are parallel. **(3.2)** • proved properties of parallelograms. **(11.1)**	**Students:** • prove and use conditions for parallelograms.	**Students:** • will prove and use properties of rectangles, rhombuses, and squares. **(11.3)** • will prove and use properties of kites and trapezoids. **(11.5)**

PROFESSIONAL LEARNING

Using Mathematical Practices and Processes

Construct viable arguments and critique the reasoning of others.

This lesson provides an opportunity to address this Mathematical Practice Standard. According to the standard, mathematically proficient students understand and use stated assumptions, definitions, and previously established results in constructing arguments. They justify their conclusions, communicate them to others, and respond to the arguments of others. This lesson provides opportunities for students to analyze constructed arguments, unscramble constructed arguments to put them in logical order, and create their own arguments. The arguments provide criteria to prove that a quadrilateral is a parallelogram. Most of the criteria are converses of Properties of Parallelogram Theorems. Finally, the parallelogram criteria are applied to show that quadrilaterals are parallelograms.

WARM-UP OPTIONS

ONLINE

PROJECTABLE & PRINTABLE

ACTIVATE PRIOR KNOWLEDGE • Properties of Parallelograms

Use these activities to quickly assess and activate prior knowledge as needed.

Problem of the Day

Given: Parallelogram *ABCD* with $SP = 2a + 5$, $RQ = 5a - 1$, $ST = 3b - 3$, and $SQ = 7b - 9$. What are the values of *RQ* and *TQ*?

P S T Q R

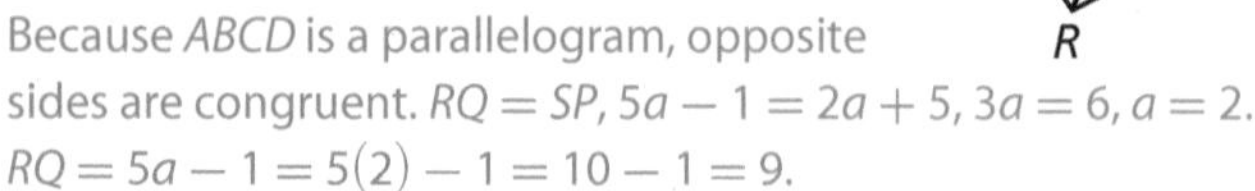

Because *ABCD* is a parallelogram, opposite sides are congruent. $RQ = SP$, $5a - 1 = 2a + 5$, $3a = 6$, $a = 2$. $RQ = 5a - 1 = 5(2) - 1 = 10 - 1 = 9$.

Because *ABCD* is a parallelogram, the diagonals bisect each other. $ST = \frac{1}{2}SQ$. $3b - 3 = \frac{1}{2}(7b - 9)$, $6b - 6 = 7b - 9$, $3 = b$. $TQ = ST = 3b - 3 = 3(3) - 3 = 9 - 3 = 6$.

Quick Check for Homework

As part of your daily routine, you may want to display the Teacher Solution Key to have students check their homework.

Make Connections

Based on students' responses to the Problem of the Day, choose one of the following:

1. Project the Interactive Reteach, Geometry, Lesson 11.1.
2. Complete the Prerequisite Skills Activity:

Have students work in groups of four. Each group creates a list of the properties of parallelograms with a diagram of each property.

- *What do you know about the opposite sides of a parallelogram?* The opposite sides are congruent.
- *What do you know about the opposite angles of a parallelogram?* The opposite angles are congruent.
- *What do you know about the diagonals of a parallelogram?* The diagonals bisect each other.

SHARPEN SKILLS

If time permits, use this on-level activity to build fluency and practice basic skills.

Mental Math

Objective: Students demonstrate their knowledge about the properties of parallelograms.
Materials: index cards

Have students work in pairs. Give each group of students a set of index cards. The cards show a diagram of a parallelogram with a length or an angle measure given and an unknown length or angle measure. One student picks a card and says what the unknown is and states the property of parallelograms that justifies the answer. The other student tells whether the answers are correct.

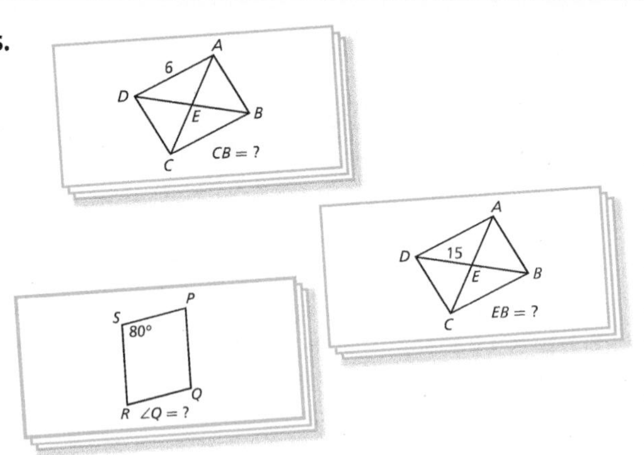

PLAN FOR DIFFERENTIATED INSTRUCTION

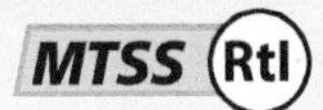

Small-Group Options

Use these teacher-guided activities with pulled small groups.

On Track

Materials: geometry software
Have students working in pairs use geometry software to prove each of the parallelogram criteria.

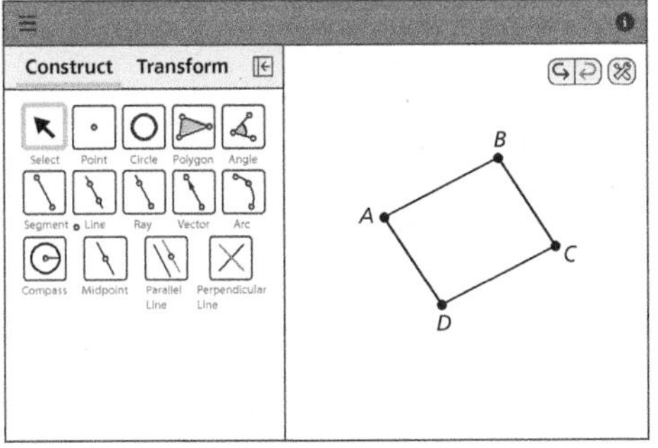

Almost There

Materials: geometry software

Review the properties of parallelograms. Have pairs of students do the following using the geometry software.

- Create a parallelogram.
- Draw both diagonals.
- Rotate one of the triangles formed by the diagonal 180° about the point of intersection of the diagonals.
- Use the geometry software's measure tools to confirm the properties of parallelograms.
- Write the results on paper.

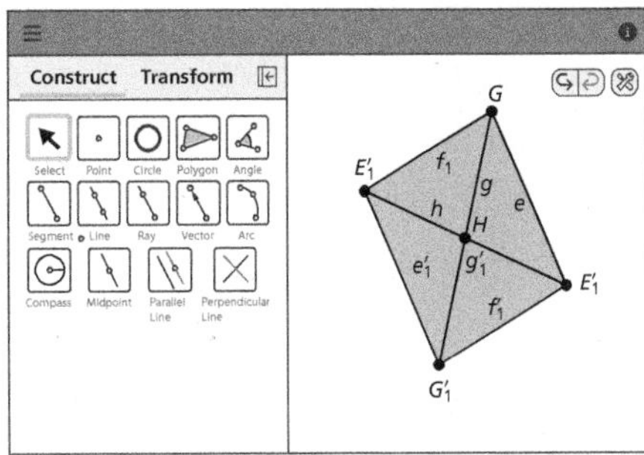

Ready for More

- Each group creates a test, with answer sheets, on using parallelogram criteria to show a quadrilateral is/is not a parallelogram.
- Groups exchange and take the tests.
- Groups verify that the answer sheets are correct.

Math Center Options

Use these student self-directed activities at centers or stations. Key: ● Print Resources ● Online Resources

On Track

- ● Interactive Digital Lesson
- ●● Journal and Practice Workbook
- ● Module Performance Task

Almost There

- ● Reteach 11.2 (printable)
- ● Interactive Reteach 11.2
- ● Illustrative Mathematics: Is this a Parallelogram?

Ready for More

- ● Challenge 11.2 (printable)
- ● Interactive Challenge 11.2
- ● Illustrative Mathematics: Is this a Parallelogram?

ONLINE Ed

View data-driven grouping recommendations and assign differentiation resources.

During the *Spark Your Learning*, listen and watch for strategies students use. See samples of student work on this page.

Use a Property of Parallelograms

Strategy 1

The opposite sides of a parallelogram are congruent. The lengths of the sides do not change as the door opens and closes. The shape is always a parallelogram.

If students . . . use a property of parallelograms to solve the problem, they are employing an efficient method and demonstrating an exemplary understanding of parallelograms from 11.1.

Have these students . . . explain how they determined to use the property and how they solved the problem. **Ask:**

- Q Why did you choose a property of parallelograms to show the hinge is always shaped like a parallelogram?
- Q How did you figure out how the lengths of the sides changed when the door opened and closed?

Use Definition of a Parallelogram

Strategy 2

As the door opens and closes, both pairs of opposite sides of the hinge do not change in length. If a diagonal is drawn, two triangles are created. Each triangle has two adjacent sides: the hinge and the diagonal. The two triangles are congruent by SSS. The angle formed by one side of the hinge and the diagonal is congruent to the angle formed by its opposite side and the diagonal. So, one pair of opposite sides is parallel. The same argument is true for the other pairs of opposite sides. The hinge is always a parallelogram.

If students . . . use the definition of a parallelogram to prove the hinge is always shaped like a parallelogram, they understand parallelograms but their method of proof is cumbersome and awkward.

Activate prior knowledge . . . by having students recall properties of parallelograms. **Ask:**

- Q What is true about both pairs of opposite sides of a parallelogram?
- Q What is true about both pairs of opposite sides of the hinge as it opens and closes?

COMMON ERROR: Uses Wrong Definition

The hinge is always in the shape of a parallelogram because it has four sides.

If students . . . use the wrong definition, then they don't understand parallelograms.

Then intervene . . . by pointing out what a parallelogram is. **Ask:**

- Q Why does the word *parallelogram* have the word *parallel* in it?
- Q Do all quadrilaterals have opposite sides that are parallel?

11.2

Conditions for Parallelograms

I Can prove and use conditions for parallelograms.

Spark Your Learning

A vertical-lift door can be installed in a location where there is not enough space for a swing-out door.

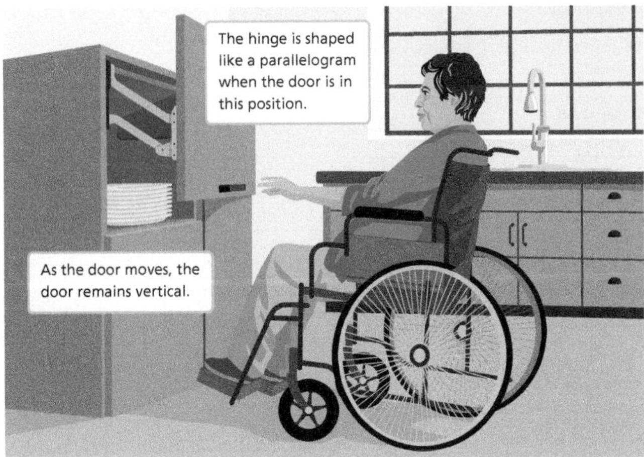

The hinge is shaped like a parallelogram when the door is in this position.
As the door moves, the door remains vertical.

Complete Part A as a whole class. Then complete Parts B–D in small groups.

A. What is a geometric question you can ask about the door hinge in this situation? **A. Is the hinge always shaped like a parallelogram as the door opens and closes?**

B. What information is needed in this situation? **See Additional Answers.**

C. To answer your question, what strategy and tool would you use along with all the information you have? What answer do you get? **See Strategies 1 and 2 on the facing page.**

D. Does your answer make sense in the context of the situation? How do you know? **See Additional Answers.**

Turn and Talk What should be true about the angles and the sides of the hinge if it remains a parallelogram as it moves? **See margin.**

SUPPORT SENSE-MAKING • Three Reads

Tell students to read the information in the photo three times and prompt them with a different question each time.

1. What is the situation about? Possible answer: The situation is about the shape of a hinge of a vertical-lift door as it opens and closes.
2. What are the quantities in this situation? How are those quantities related? The quantities are the lengths of the sides of the hinge and the measures of the angles of the hinge; The sides and angles form a parallelogram.
3. What are possible questions you could ask about this situation? Possible answer: As the door moves up, how does the shape of the hinge change? As the door moves down, how does the shape of the hinge change?

LESSON 11.2 Connect Concepts and Skills

1 Spark Your Learning

▶ MOTIVATE 

- Have students look at the photo in their books and read the information contained in the photo. Then complete Part A as a whole-class discussion.
- Give the class the additional information they need to solve the problem. This information is available online as a printable and projectable page in the Teacher Resources.
- Have students work in small groups to complete Parts B–D.

▶ PERSEVERE

If students need support, guide them by asking:

Q **Advancing • Use Tools** Which tool could you use to solve the problem? Why choose that tool and not some other? Students' choices of tools and reasons for choosing them will vary.

Q **Assessing** How can you tell whether the sides of the hinge form a parallelogram? I could overlay a simplified sketch of the hinge on top of the photo and observe that the sketch has four sides and that the opposite sides are parallel. I could also measure the sides and angles in the sketch. By definition, in the open state this hinge is a parallelogram.

Q **Advancing** How does the shape of the hinge change as the door goes up and down? Explain. Possible answer: As the door opens and closes, the length of the sides of the hinge stays the same; The hinges are attached together in such a way that lengths of the sides stay the same but the angles change.

Turn and Talk Have students recall the definition of a parallelogram. *A parallelogram is a quadrilateral with both pairs of opposite sides parallel.* The opposite sides would be congruent, and the opposite angles would be congruent.

▶ BUILD SHARED UNDERSTANDING

Select groups of students who used various strategies and tools to share with the class how they solved the problem. As they present their solutions, have each group discuss why they chose a specific strategy and tool.

② Learn Together

Build Understanding

Task 1 (MP) **Construct Arguments** Ensure that students understand how the statements and reasons are constructed in the proof of the Converse of the Opposite Angles of a Parallelogram Theorem.

Sample Guided Discussion:

Q **How can you prove that a quadrilateral is a parallelogram?** Possible answer: I can prove a quadrilateral is a parallelogram by using the definition of a parallelogram. That is, I can show that both pairs of opposite sides are parallel.

Q **What properties of quadrilaterals and parallelograms have you learned?** Quadrilateral Sum Theorem, Opposite Sides of a Parallelogram Theorem, Opposite Angles of a Parallelogram Theorem, Diagonals of a Parallelogram Theorem

Build Understanding

Establish Parallelogram Criteria

You can use the definition of a parallelogram to show that a quadrilateral is a parallelogram. To do so, you need to show that both pairs of opposite sides are parallel. However, there are other criteria that guarantee a quadrilateral is a parallelogram.

1 The proof below verifies one of the criteria that can be used to show that a quadrilateral is a parallelogram.

The given information about the quadrilateral is marked on *ABCD*.

Statements	Reasons
1. $\angle A \cong \angle C$ and $\angle B \cong \angle D$	1. Given
2. $m\angle A = m\angle C$ and $m\angle B = m\angle D$	2. Definition of congruent angles
3. $m\angle A + m\angle B + m\angle C + m\angle D = 360°$	3. Quadrilateral Sum Theorem
4. $m\angle A + m\angle B + m\angle A + m\angle B = 360°$ $m\angle A + m\angle D + m\angle A + m\angle D = 360°$	4. Substitution
5. $2m\angle A + 2m\angle B = 360°$ $2m\angle A + 2m\angle D = 360°$	5. Combine like terms.
6. $m\angle A + m\angle B = 180°$ $m\angle A + m\angle D = 180°$	6. Division Property of Equality
7. $\angle A$ and $\angle B$ are supplementary. $\angle A$ and $\angle D$ are supplementary.	7. Definition of supplementary angles
8. $\overline{AD} \parallel \overline{BC}$, $\overline{AB} \parallel \overline{DC}$	8. Converse of the Consecutive Interior Angles Theorem
9. *ABCD* is a parallelogram.	9. Definition of parallelogram

A. What parallelogram criterion is being proved? **If both pairs of opposite angles of a quadrilateral are congruent, then the quadrilateral is a parallelogram.**

B. How is the criterion being proved related to one of the theorems in Lesson 11.1? **It is the converse of the Opposite Angles of a Parallelogram Theorem.**

You have learned theorems about the properties of parallelograms. You can use the converses of these theorems to prove that a quadrilateral is a parallelogram.

Converse of the Opposite Sides of a Parallelogram Theorem

If both pairs of opposite sides of a quadrilateral are congruent, then the quadrilateral is a parallelogram.

If $\overline{AB} \cong \overline{DC}$ and $\overline{AD} \cong \overline{BC}$, then *ABCD* is a parallelogram.

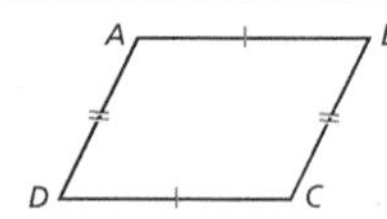

330

LEVELED QUESTIONS

Depth of Knowledge (DOK)	Leveled Questions	What Does This Tell You?
Level 1 **Recall**	What is the sum of the angles of a quadrilateral? 360°	Students' answers will indicate whether they understand the Quadrilateral Sum Theorem.
Level 2 **Basic Application of Skills & Concepts**	In the proof, how does Statement 6 result from using the Quadrilateral Sum Theorem? Possible answer: The sum of the angles of a quadrilateral is 360° (QST). We are given that the pairs of opposite angles are congruent, substituting, and simplifying, the result is Statement 6.	Students' answers will indicate whether they can apply the Quadrilateral Sum Theorem.
Level 3 **Strategic Thinking & Complex Reasoning**	How was the argument in the proof constructed? Write a paragraph to explain. Possible answer: Given that the pairs of opposite angles are congruent, we must show this leads to both pairs of opposite sides being parallel.	Students' answers will indicate whether they can use strategic thinking and complex reasoning to reach a conclusion.

Converse of the Opposite Angles of a Parallelogram Theorem

If both pairs of opposite angles of a quadrilateral are congruent, then the quadrilateral is a parallelogram.

If $\angle A \cong \angle C$ and $\angle B \cong \angle D$, then *ABCD* is a parallelogram.

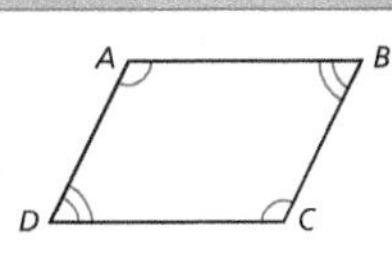

Converse of the Diagonals of a Parallelogram Theorem

If the diagonals of a quadrilateral bisect each other, then the quadrilateral is a parallelogram.

If $\overline{AE} \cong \overline{CE}$ and $\overline{BE} \cong \overline{DE}$, the *ABCD* is a parallelogram.

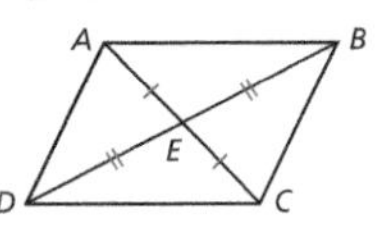

Opposite Sides Criteria for a Parallelogram Theorem

If one pair of opposite sides of a quadrilateral are congruent and parallel, then the quadrilateral is a parallelogram.

If $\overline{AB} \cong \overline{DC}$ and $\overline{AB} \parallel \overline{DC}$, then *ABCD* is a parallelogram.

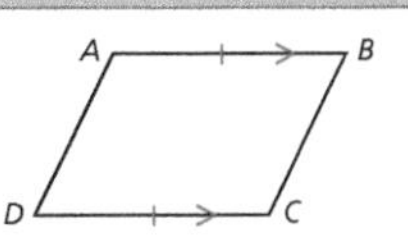

2 The lengths and slopes of two sides of quadrilateral *FGHJ* are shown.

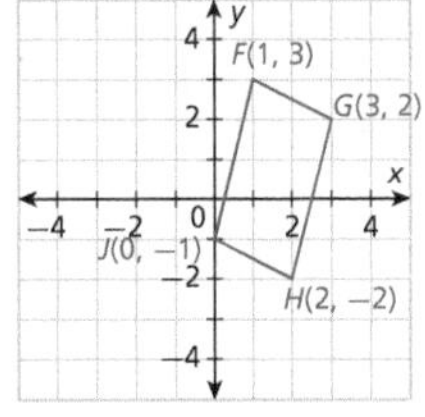

$$GH = \sqrt{(3-2)^2 + (2-(-2))^2} = \sqrt{17}$$

$$FJ = \sqrt{(1-0)^2 + (3-(-1))^2} = \sqrt{17}$$

$$\text{Slope of } \overline{GH} = \frac{2-(-2)}{3-2} = 4$$

$$\text{Slope of } \overline{FJ} = \frac{3-(-1)}{1-0} = 4$$

A. Do you have enough information to conclude that *FGHJ* is a parallelogram? Explain. See below.

B. What additional information is needed to apply the Converse of the Opposite Sides of a Parallelogram Theorem to show that *FGHJ* is a parallelogram?
You would need to know the lengths of $\overline{FG}$ and $\overline{JH}$.

Turn and Talk Use the definition of a parallelogram to show that *FGHJ* is a parallelogram. What do you need to know about the quadrilateral? See margin.

A. yes; *FGHJ* is a parallelogram because you know that one pair of opposite sides are congruent and parallel.

PROFICIENCY LEVEL

Beginning

Pair students. Give each pair a set of index cards with a labeled quadrilateral on one side and a list of pairs of opposite sides and opposite angles on the other side. One student selects a card and shows the quadrilateral to the other. The other student says which sides and angles are opposite.

Intermediate

Pair students. Give each pair blank index cards. Each student draws and labels a quadrilateral on one side and lists pairs of opposite sides and opposite angles on the other side. Students exchange cards and one tells the other what the opposite pairs are using a complete sentence.

Advanced

Pair students. Each student draws and labels two quadrilaterals marking congruent parts. One quadrilateral is a parallelogram and one is not. Students swap cards and discuss why or why not the figure is a parallelogram citing theorems they have just learned.

Task 2 (MP) **Construct Arguments** Students use coordinates to show a special case of the Opposite Sides Criteria for a Parallelogram Theorem.

Sample Guided Discussion:

Q What conditions are necessary to prove a quadrilateral is a parallelogram using the Opposite Sides Criteria for a Parallelogram Theorem? Prove one pair of opposite sides are both parallel and congruent.

Q How do you show two segments are congruent using coordinates? Possible answer: Use the Distance Formula to show they have the same length.

Q How do you show two segments are parallel using coordinates? Possible answer: Show they have the same slope.

Turn and Talk Elicit from students the definition of a parallelogram: A parallelogram is a quadrilateral with both pairs of opposite sides parallel. Have students recall that two segments in a coordinate plane are parallel if their slopes are equal. Possible answer: You can use the coordinates to show that both pairs of opposite sides are parallel by finding the slope of each side.

OPTIMIZE OUTPUT Critique, Correct, and Clarify

Use the logic and steps you just applied for the example shown in Task 2 to the following similar figure. Is *ABCD* a parallelogram? Explain.

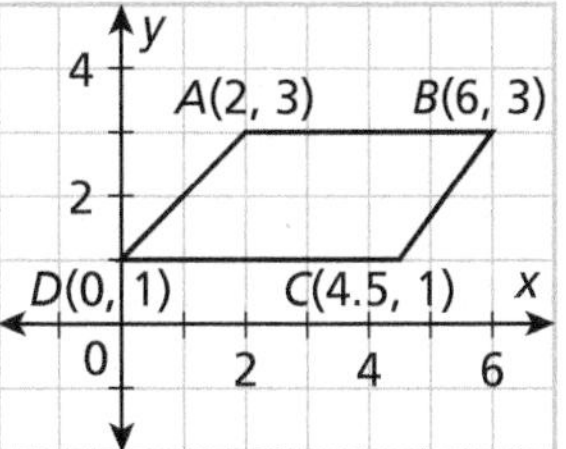

Have students work with a partner to discuss if *ABCD* is a parallelogram. Encourage students to use the properties and conditions of parallelograms in their discussions. Students should revise their answers to Task 2, if necessary, after talking about this new example with their partner.

Step It Out

Task 3 (MP) **Construct Arguments** Students construct an argument to prove the Converse of the Diagonals of a Parallelogram Theorem by unscrambling Statements and Reasons.

Sample Guided Discussion:

Q How can you use the figure? Possible answer: I can use the figure to identify the given information for the first statement of the proof.

Q What should the last statement of the proof be? *ABCD* is a parallelogram.

Task 4 (MP) **Use Structure** Students find values of variables in expressions that result in quadrilaterals that are parallelograms using the theorems in the previous tasks.

Sample Guided Discussion:

Q What is one pair of opposite sides of quadrilateral *KLMN*? $\overline{KL}$ and $\overline{NM}$ are opposite sides.

Q What is the other pair of opposite sides? $\overline{KN}$ and $\overline{LM}$ are opposite sides.

Turn and Talk To review all the methods of proving quadrilaterals are parallelograms, you may want to create a bulletin board with all the methods. Another method would be to divide the class into groups and assign each group one of the methods. Give each group a transparency for them to use to make a presentation to the class. You can prove that a quadrilateral is a parallelogram by showing that both pairs of opposite sides are congruent, both pairs of opposite angles are congruent, the diagonals bisect each other, one pair of opposite sides are congruent and parallel, or that both pairs of opposite sides are parallel.

Step It Out

Prove that a Quadrilateral Is a Parallelogram

3 You are asked to prove that if the diagonals of a quadrilateral bisect each other, then the quadrilateral is a parallelogram. The statements and reasons for the proof are shown, but they have been scrambled.

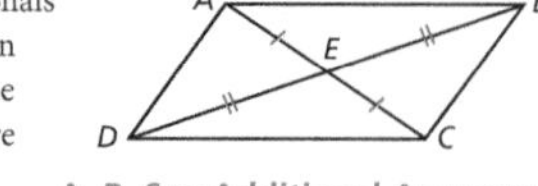

A, B. See Additional Answers.

A. Write the statements in the correct order.

B. Write the reasons in the correct order.

A. Statements	B. Reasons
$\overline{AB} \cong \overline{CD}$, $\overline{AD} \cong \overline{CB}$	SAS Triangle Congruence Theorem
$\angle AEB \cong \angle CED$, $\angle AED \cong \angle CEB$	Given
ABCD is a parallelogram.	Corresponding parts of congruent figures are congruent.
$\overline{AE} \cong \overline{CE}$, $\overline{DE} \cong \overline{BE}$	If both pairs of opposite sides of a quadrilateral are congruent, then it is a parallelogram.
$\triangle AEB \cong \triangle CED$, $\triangle AED \cong \triangle CEB$	Vertical angles are congruent.

Use Parallelogram Criteria

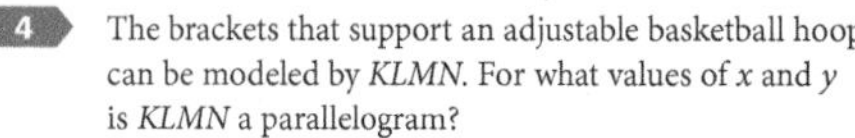

4 The brackets that support an adjustable basketball hoop can be modeled by *KLMN*. For what values of *x* and *y* is *KLMN* a parallelogram?

Find the value of *y*.

$KL = NM$

$7y = 5y + 6$

$2y = 6$

$y = 3$

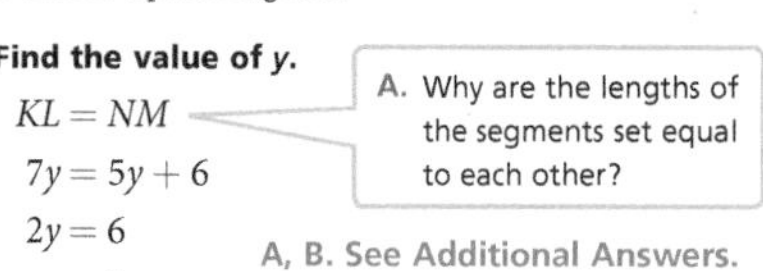

A. Why are the lengths of the segments set equal to each other?

A, B. See Additional Answers.

Find the value of *x*.

$KN = LM$

$x + 4 = 3x - 12$

$16 = 2x$

$8 = x$

When $x = 8$ and $y = 3$, *KLMN* is a parallelogram.

B. Why can this conclusion be made?

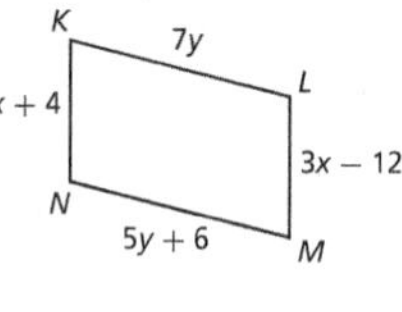

Turn and Talk Summarize the various methods you can use to prove that a quadrilateral is a parallelogram. See margin.

Check Understanding

For Problems 1–4, state the theorem that can be used to show that the quadrilateral is a parallelogram.

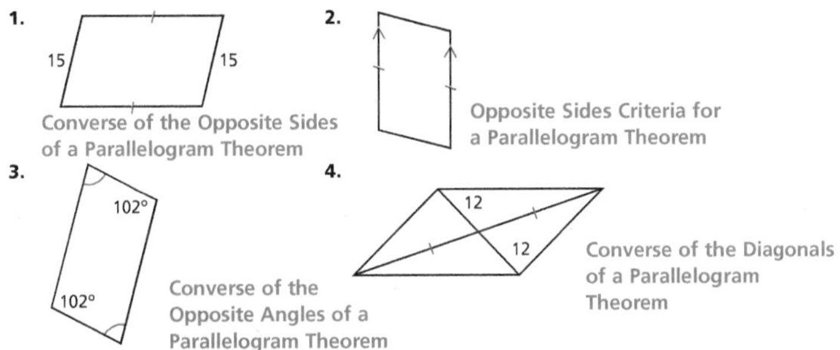

1. Converse of the Opposite Sides of a Parallelogram Theorem

2. Opposite Sides Criteria for a Parallelogram Theorem

3. Converse of the Opposite Angles of a Parallelogram Theorem

4. Converse of the Diagonals of a Parallelogram Theorem

5. The slopes of the sides of $QRST$ are shown below. Can you conclude that $QRST$ is a parallelogram? Explain why or why not.

   Slope of $\overline{QR} = -3$ Slope of $\overline{RS} = \frac{1}{2}$

   Slope of $\overline{ST} = -3$ Slope of $\overline{TQ} = \frac{1}{2}$

   yes; $QRST$ is a parallelogram because opposite sides have the same slope. This means the opposite sides are parallel.

Find the values of the variables that make the quadrilateral a parallelogram.

6.

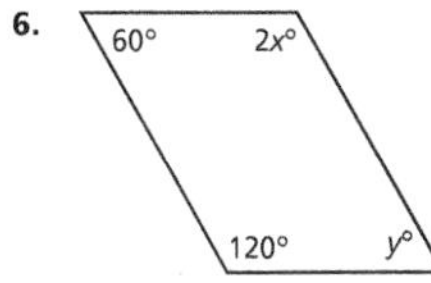

$x = 60$, $y = 60$

7.

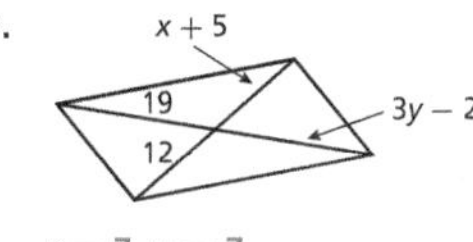

$x = 7$, $y = 7$

On Your Own

8, 9. See Additional Answers.

8. Prove the Converse of Opposite Sides of a Parallelogram Theorem.

   Given: $\overline{AD} \cong \overline{BC}$, $\overline{AB} \cong \overline{DC}$
   Prove: $ABCD$ is a parallelogram.

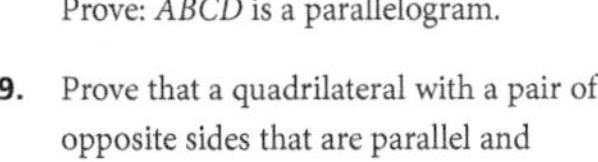

9. Prove that a quadrilateral with a pair of opposite sides that are parallel and congruent is a parallelogram.

   Given: $\overline{AD} \cong \overline{BC}$, $\overline{AD} \parallel \overline{BC}$
   Prove: $ABCD$ is a parallelogram.

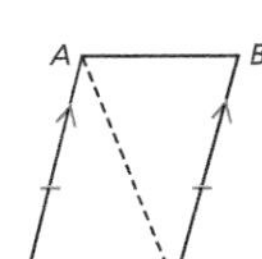

ONLINE

Assign the Digital On Your Own for
- built-in student supports
- Actionable Item Reports
- Standards Analysis Reports

On Your Own

Assignment Guide

The chart below indicates which problems in the On Your Own are associated with each task in the Learn Together. Assign daily homework for tasks completed.

Learn Together Tasks	On Your Own Problems
Task 1, p. 330	Problems 8, 9, and 34
Task 2, p. 331	Problems 20–23
Task 3, p. 332	Problems 28 and 30–33
Task 4, p. 332	Problems 10–19, 24–27, and 29

3 Check Understanding

Formative Assessment

Use formative assessment to determine if your students are successful with this lesson's learning objective.

Students who successfully complete the Check Understanding can continue to the On Your Own practice.

For students who miss 1 problem or more, work in a pulled small group using the Almost There small-group activity on page 329C.

ONLINE

Assign the Digital Check Understanding to determine
- success with the learning objective
- items to review
- grouping and differentiation resources

4 Differentiation Options

Differentiate instruction for all students using small-group activities and math center activities on page 329C.

Reteach

Challenge

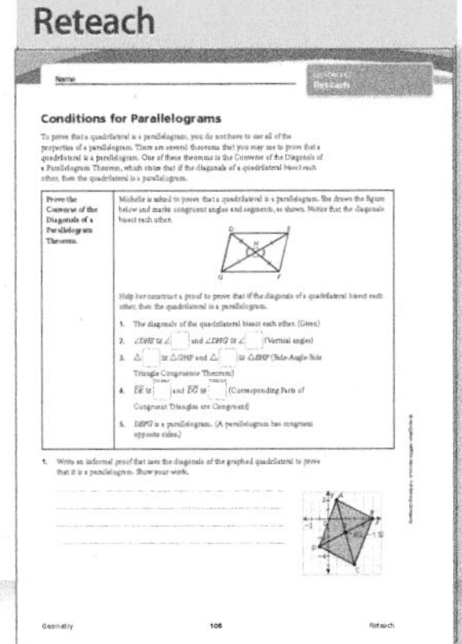

Watch for Common Errors

Problems 14–19 Some students may think that the conditions for Problems 14–19 are cumulative. Tell students that the conditions for each problem are unique.

Questioning Strategies

Problem 20 Ask student to make *ABCD* a parallelogram. What coordinates for point *A* would make *ABCD* a parallelogram? $A(2, 3)$

Determine whether the quadrilateral is a parallelogram. Justify your answer.

10–13. See Additional Answers.

10.

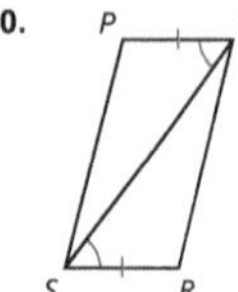

11.

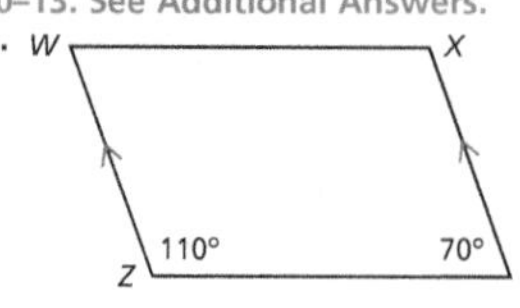

12.

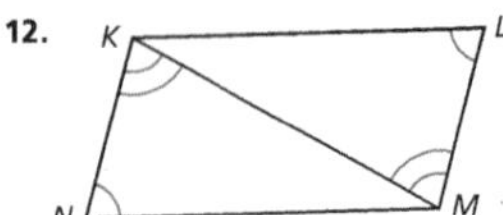

13.

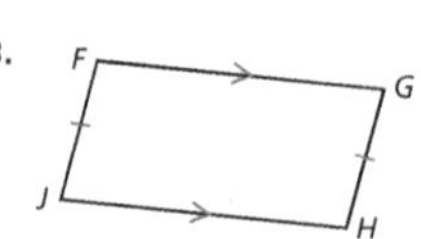

Part of a robotic arm is modeled by *PQRS*. Does each set of given information guarantee that *PQRS* is a parallelogram?

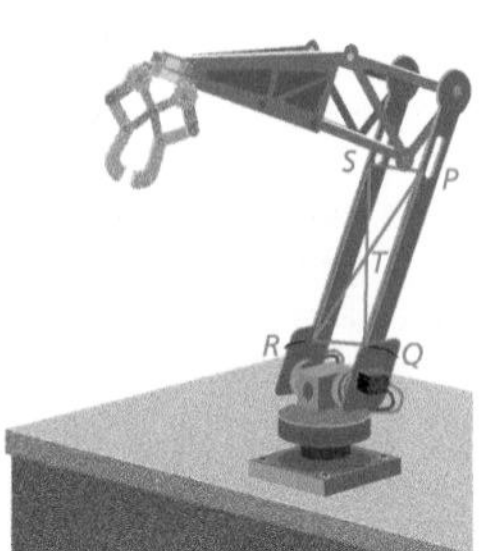

14. $\overline{PS} \cong \overline{PQ}, \overline{RS} \cong \overline{RQ}$ no

15. $\angle PST \cong \angle RQT, \overline{PS} \cong \overline{QR}$ yes

16. $PT = 15, PR = 30, QT = ST = 23$ yes

17. $\angle STR \cong \angle PTQ, \angle PTS \cong \angle QTR$ no

18. $\triangle PRS \cong \triangle RSQ$ no

19. $\triangle STR \cong \triangle QTP$ yes

Determine whether the quadrilateral is a parallelogram. Explain your reasoning.

20–23. See Additional Answers.

20.

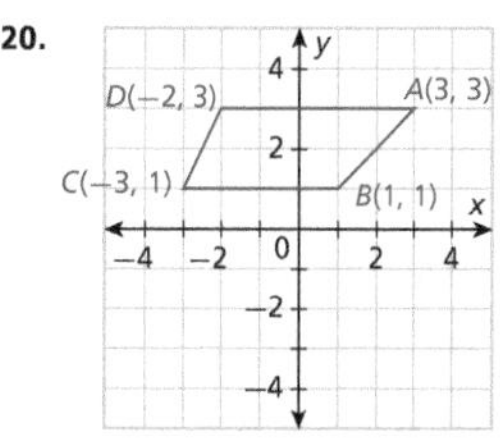

21.

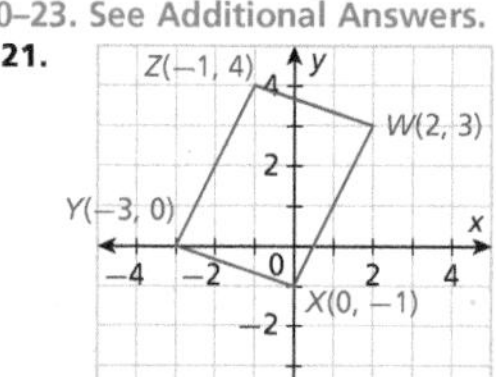

22.

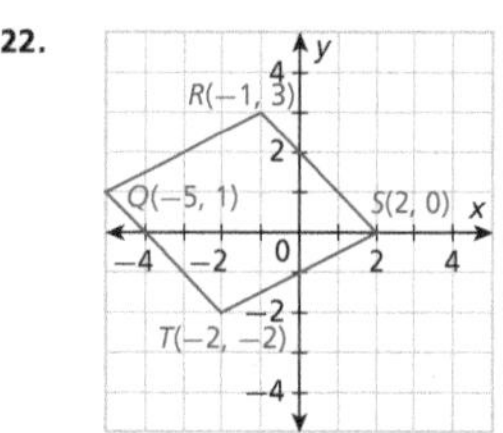

23.

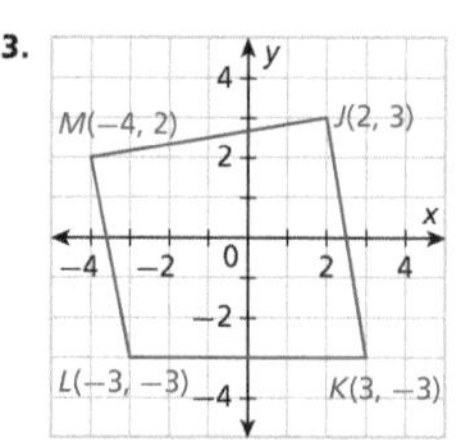

334

Find the values of the variables that make quadrilateral $ABCD$ a parallelogram.

24.

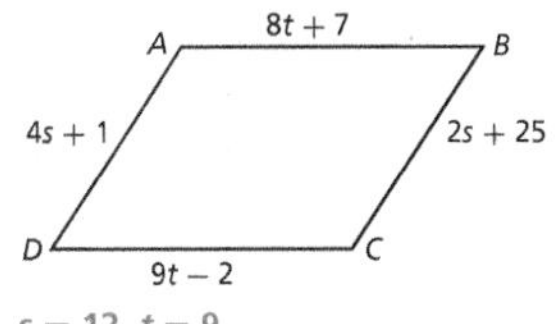

$s = 12, t = 9$

25.

$(10a - 17)°$

$(26b - 13)°$

$(7a + 7)°$

$(20b + 17)°$

$a = 8, b = 5$

26.

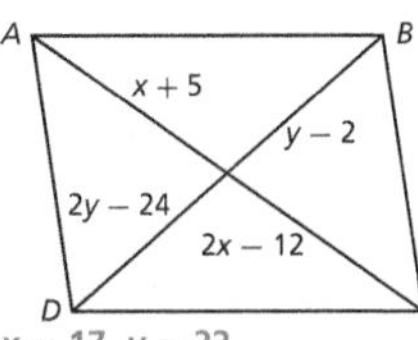

$x = 17, y = 22$

27.

$(7m + 4)°$ $(17n - 3)°$

$(9m)°$ $(11n + 15)°$

$m = 11, n = 6$

28. **Critique Reasoning** Ethan says that if you take a cardboard box apart to flatten it, the resulting shape is always a parallelogram. Jessica disagrees with him. Who is correct? Explain your reasoning. See Additional Answers.

29. Open Ended Draw a quadrilateral that you can prove is a parallelogram using the theorem that if opposite angles are congruent, the quadrilateral is a parallelogram. Find the angle measures. Answers will vary.

30. **Construct Arguments** Explain why you know that a quadrilateral is a parallelogram if adjacent angles are supplementary.

Given: $\angle A$ is supplementary to $\angle B$,
$\angle B$ is supplementary to $\angle C$,
$\angle C$ is supplementary to $\angle D$, and
$\angle D$ is supplementary to $\angle A$.

Prove: $ABCD$ is a parallelogram.

30–32. See Additional Answers.

31. **Construct Arguments** In $\triangle ABC$, M is the midpoint of $\overline{AB}$. Follow the steps below to form a quadrilateral. Is the quadrilateral a parallelogram? Explain your reasoning.

Step 1: Trace the triangle on a piece of tracing paper.
Step 2: Without moving the paper, place your pencil on the midpoint of $\overline{AB}$.
Step 3: Rotate the tracing paper 180°.

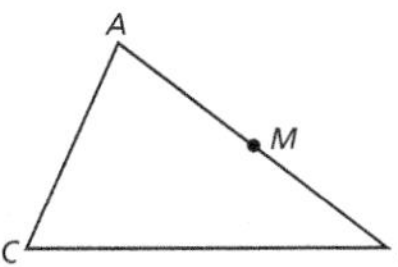

32. Quadrilateral $ABCD$ has vertices $A(0, 6)$, $B(7, 2)$, $C(2, -1)$, and $D(-5, 3)$. Use the Converse of the Diagonals of a Parallelogram Theorem to prove that $ABCD$ is a parallelogram.

Questioning Strategies

Problem 30 Select just 2 of the four conditions as given and still prove $ABCD$ is a parallelogram. Then, select 2 of the four conditions as given for which you cannot prove $ABCD$ is a parallelogram. Possible answer: Given: $\angle A$ is supplementary to $\angle B$ and $\angle B$ is supplementary to $\angle C$. Prove: $ABCD$ is a parallelogram. Proof: If $\angle A$ is supplementary to $\angle B$, then $\overline{AD} \parallel \overline{BC}$ and if $\angle B$ is supplementary to $\angle C$, then $\overline{AB} \parallel \overline{DC}$ by the Converse of the Consecutive Interior Angles Theorem. $ABCD$ is a parallelogram because both pairs of opposite sides are parallel.

Given: $\angle A$ is supplementary to $\angle B$ and $\angle C$ is supplementary to $\angle D$. Prove: $ABCD$ is a parallelogram. If $\angle A$ is supplementary to $\angle B$ and $\angle C$ is supplementary to $\angle D$, you can only prove that $\overline{AD} \parallel \overline{BC}$. There is not enough information for you to prove $ABCD$ is a parallelogram.

Watch for Common Errors

Problem 31 Some students may be confused by how the directions form a quadrilateral. Elicit from students that a quadrilateral is formed by joining two triangles, the original triangle and the triangle on the tracing paper.

5 Wrap-Up

Summarize learning with your class. Consider using the Exit Ticket, Put It in Writing, or I Can scale.

Exit Ticket 

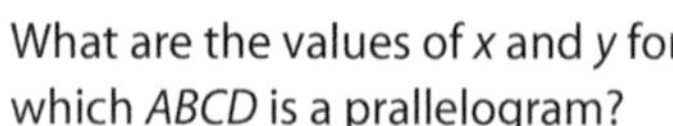

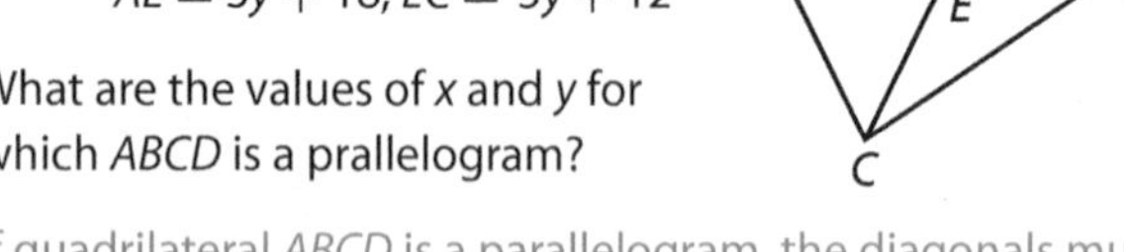

Given: $DE = 5x - 12$, $EB = 2x + 6$,
$AE = 3y + 18$, $EC = 5y + 12$

What are the values of x and y for which $ABCD$ is a prallelogram?

If quadrilateral $ABCD$ is a parallelogram, the diagonals must bisect each other. $DE = EB$, $5x - 12 = 2x + 6$, $3x = 18$, $x = 6$.

$AE = EC$, $3y + 18 = 5y + 12$, $6 = 2y$, $y = 3$.

Put It in Writing 

You are given a quadrilateral $ABCD$. Describe three different sets of measurements you can make to prove the figure is a parallelogram.

I Can 

The scale below can help you and your students understand their progress on a learning goal.

4	I can prove and use conditions for parallelograms, and I can explain my work to others.
3	I can prove and use conditions for parallelograms.
2	I can identify and use one or two conditions for parallelograms.
1	I can determine if some quadrilaterals are parallelograms.

Spiral Review • Assessment Readiness

These questions will help determine if students have retained information taught in the past and can also prepare them for high-stakes assessments. Here, students must determine which sets of numbers can be side lengths of a triangle **(10.1)**, use triangle inequalities to determine measures of angles **(10.2)**, determine the slope of a segment perpendicular to a given segment **(4.2),** and use properties of parallelograms to find the length of a side **(11.1).**

33. Quadrilateral $ABCD$ represents the outdoor area of a dog boarding facility. An inner quadrilateral is formed by fencing that connects the midpoints of the sides of $ABCD$. Explain why quadrilateral $WXYZ$ is a parallelogram.

See Additional Answers.

34. Describe three different ways that you can prove that $ABCD$ is a parallelogram.

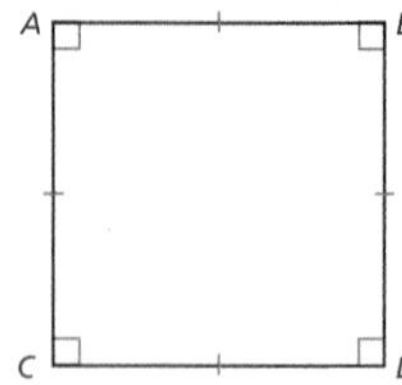

Both pairs of opposite sides are congruent. Both pairs of opposite angles are congruent. You can show that the opposite sides are parallel using the Converse of the Same Side Interior Angles Theorem, then use the fact that a pair of opposite sides is parallel and congruent.

Spiral Review • Assessment Readiness

35. Which of the following sets of numbers can be the side lengths of a triangle? Select all that apply.

Ⓐ 8, 9, 16
Ⓑ 7, 8, 15
Ⓒ 15, 15, 29
Ⓓ 3, 7, 11
Ⓔ 16, 18, 30

36. In $\triangle ABC$, $AB = 11$, $AC = 15$, and $m\angle A = 105°$. In $\triangle XYZ$, $XY = 15$ and $XZ = 17$. If $BC < YZ$, which of the following can be the measure of $\angle X$? Select all that apply.

Ⓐ 75°
Ⓑ 90°
Ⓒ 104°
Ⓓ 105°
Ⓔ 106°
Ⓕ 135°

37. $\overline{MN}$ and $\overline{ST}$ are perpendicular. If the slope of $\overline{MN}$ is $\frac{2}{5}$, what is the slope of $\overline{ST}$?

Ⓐ $-\frac{5}{2}$
Ⓑ $-\frac{2}{5}$
Ⓒ $\frac{2}{5}$
Ⓓ $\frac{5}{2}$

38. $EFGH$ is a parallelogram. Find the length of GH.

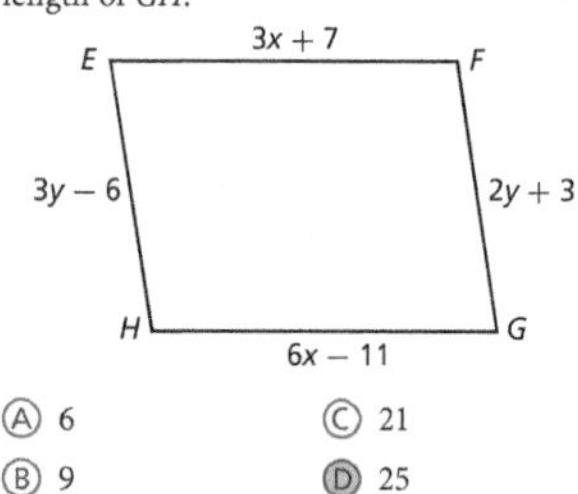

Ⓐ 6
Ⓑ 9
Ⓒ 21
Ⓓ 25

 **I'm in a Learning Mindset!**

How am I devoting my energy to learning what conditions make a quadrilateral a parallelogram?

336 Keep Going Journal and Practice Workbook

Learning Mindset

Resilience Adjust to Changes

Point out that making sure a given step in proving a theorem follows logically from the previous step is essential in all areas of mathematics, not just when proving theorems. Encourage students to develop the habit of always justifying the steps of their proofs. Also, let students know that they just learned conditions for making a quadrilateral a parallelogram in this lesson. They will use these conditions later to prove more involved conjectures. *How does justifying each step when proving a theorem help you have a more confident mindset? When you can't think of a good justification for one of your steps, what might that suggest about your approach to proving the theorem? How well do you know the conditions that make a quadrilateral a parallelogram? Did you devote enough time and energy into learning these conditions? Do you understand the difference between the properties of a parallelogram and the conditions of showing a quadrilateral is a parallelogram?*

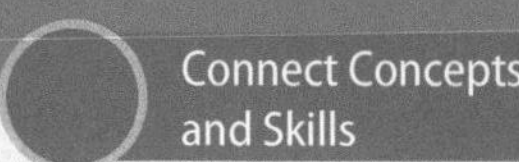

11.3 Properties of Rectangles, Rhombuses, and Squares

LESSON FOCUS AND COHERENCE

Mathematics Standards

- Prove theorems about parallelograms. *Theorems include: opposite sides are congruent, opposite angles are congruent, the diagonals of a parallelogram bisect each other, and conversely, rectangles are parallelograms with congruent diagonals.*

Mathematical Practices and Processes 

- Construct viable arguments and critique the reasoning of others.
- Look for and make use of structure.

I Can Objective 

I can prove and use properties of squares, rectangles, and rhombuses.

Learning Objective

Students will prove theorems about rectangles, rhombuses, and squares, discriminate among them based on their properties, and apply theorems to find measures of sides, diagonals, and interior angles.

Language Objective

Students write and explain the definitions of rectangles, rhombuses, and squares and describe their properties.

Vocabulary

New: **rectangle**, **rhombus**, **square**

Lesson Materials: geometry software

Mathematical Progressions

Prior Learning	Current Development	Future Connections
Students: • proved and used properties of parallelograms. **(11.1 and 11.2)**	**Students:** • prove theorems about rectangles, rhombuses, and squares. • use properties of rectangles, rhombuses, and squares to solve problems.	**Students:** • will prove and use properties of kites and trapezoids. **(11.4)**

PROFESSIONAL LEARNING

Visualizing the Math

Parallelograms are quadrilaterals that include rectangles, rhombuses, and squares. The relationships among them can be represented in a Venn diagram, such as the one on the right. Explain to students that the intersection of the two ovals contains squares—parallelograms that have equal sides and right angles. The set of squares differs only in the lengths of their sides. The set of rhombuses contains squares because they have equal sides. The set of rectangles also contains squares because they have right angles. The region outside of the ovals contains all other parallelograms, those who neither have all four sides congruent nor right angles.

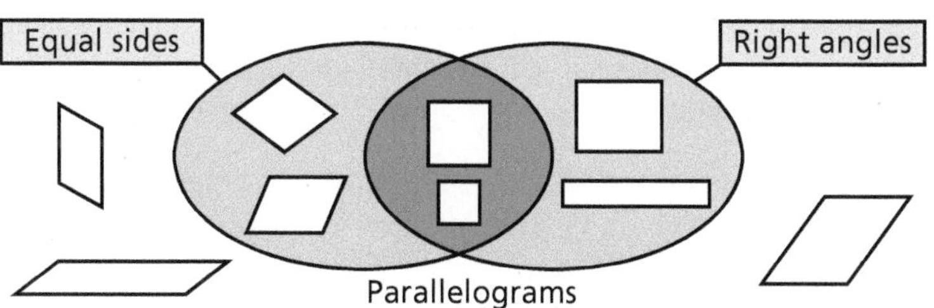

WARM-UP OPTIONS

ONLINE

PROJECTABLE & PRINTABLE

ACTIVATE PRIOR KNOWLEDGE • Properties of Parallelograms

Use these activities to quickly assess and activate prior knowledge as needed.

Problem of the Day

ABCD is a parallelogram, $AE = x + 15$, and $EC = 3 - x$. Find x and AE. $x = -6$ and $AE = 9$

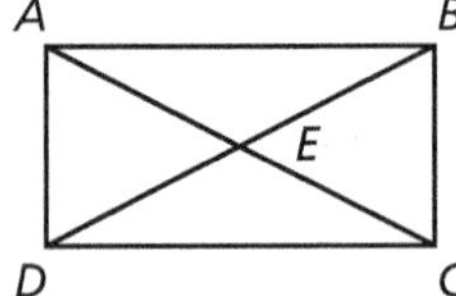

Quick Check for Homework

As part of your daily routine, you may want to display the Teacher Solution Key to have students check their homework.

Make Connections

Based on students' responses to the Problem of the Day, choose one of the following:

1. Project the Interactive Reteach, Geometry, Lesson 11.1.

2. Complete the Prerequisite Skills Activity:

Complete each of the following statements about parallelograms.

- *A parallelogram is a* ________. quadrilateral
- *The opposite sides of a parallelogram are* ________ *and* ________. parallel; congruent
- *The opposite angles in a parallelogram are* ________. congruent
- *The diagonals of a parallelogram* ________. bisect each other

SHARPEN SKILLS

If time permits, use this on-level activity to build fluency and practice basic skills.

Vocabulary Review

Objective: Students complete a graphic organizer about parallelograms.
Materials: Word Description (Teacher Resource Masters), rulers, protractors (optional)

Have students complete each of the regions in the graphic organizer by defining the term *parallelogram*, listing its properties, providing more than two sketches of examples of parallelograms, and more than two sketches of non-examples, such as a convex quadrilateral with no parallel sides, a trapezoid, a circle, or a concave quadrilateral.

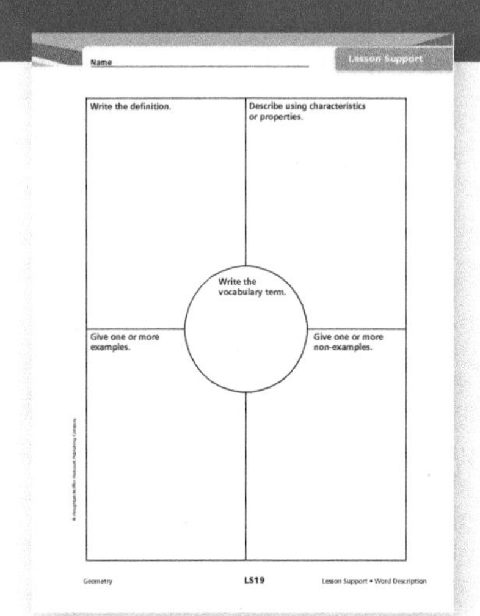

PLAN FOR DIFFERENTIATED INSTRUCTION

MTSS (RtI)

Small-Group Options

Use these teacher-guided activities with pulled small groups.

On Track

MNPO is a rhombus with diagonals $\overline{MO}$ and $\overline{PN}$.

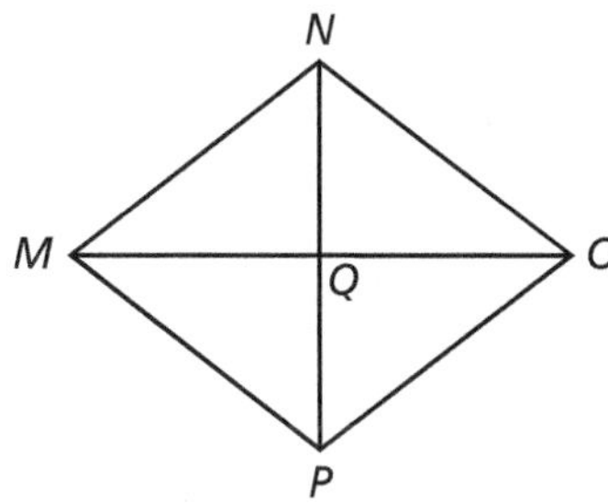

1. What kind of triangle is $\triangle MNO$? Explain.
2. What kind of triangle is $\triangle PQO$? Explain.

Almost There 

Materials: calculator

WXYZ is a rectangle with sides of 5 and 12. *KLMN* is a square whose sides are each 9.

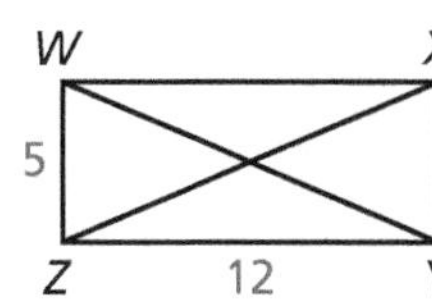

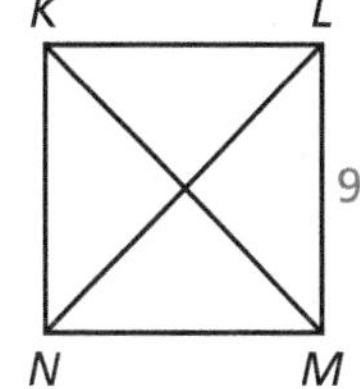

Which quadrilateral has the longer diagonal? Explain.

Ready for More

DEFG is a quadrilateral with two pairs of congruent sides as marked. Prove that the diagonals of *DEFG* are perpendicular.

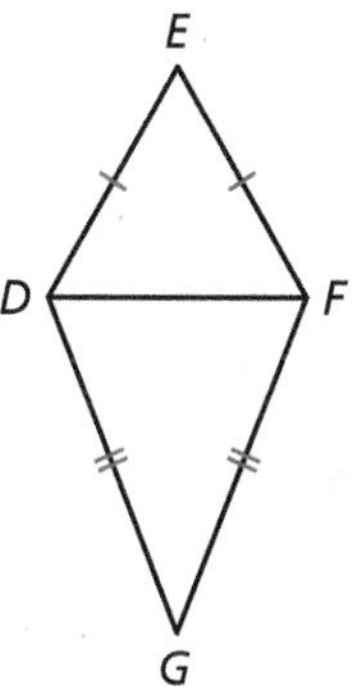

Math Center Options

Use these student self-directed activities at centers or stations.

Key: ● Print Resources ● Online Resources

On Track

- ● Interactive Digital Lesson
- ●● Journal and Practice Workbook
- ● Interactive Glossary (printable): **rectangle**, **rhombus**, **square**
- ● Module Performance Task

Almost There

- ● Reteach 11.3 (printable)
- ● Interactive Reteach 11.3

Ready for More

- ● Challenge 11.3 (printable)
- ● Interactive Challenge 11.3

ONLINE 

View data-driven grouping recommendations and assign differentiation resources.

Spark Your Learning • Student Samples

During the *Spark Your Learning,* listen and watch for strategies students use. See samples of student work on this page.

Use the Small Square and Compare Sides — Strategy 1

The shapes in the tile pattern are rectangles and squares. There are two different-size squares: a small one with sides of 1.5 inches and a larger one with sides of 3 inches because 2 small tiles line up with 1 side of the larger square. The dimensions of the rectangle are 1.5 inches by 3 inches because one of the sides matches exactly with a side of the small square, and the other side matches exactly with the large square.

If students . . . recognized the three unique shapes in this tiling as two squares and a rectangle whose sides are multiples of each other, then have them explain their reasoning.

Have these students . . . explain how they determined the dimensions of each shape. **Ask:**

- Q How did you decide that each side of the large shape is 3?
- Q How did you decide that the lengths of the sides of the other shape is 1.5 inches by 3?

Count Shapes and Match Up Sides — Strategy 2

I noticed that there were 4 congruent large squares, 6 congruent rectangles, and 8 congruent small squares.

- One side of the rectangle matched one side of the small square. So, one dimension of the rectangle was 1.5 inches.
- It took a small square and the 1.5-inch side of a rectangle to match up with a side of the large square. So, each side of the large square was 2 x 1.5 or 3 inches.
- The long side of the rectangle matched up with a side of the large square, so its length was 3 inches.

If students . . . compared a side of the smaller square to find the length of one side of the rectangle, then they understand how the shapes are related, but may not have made the most efficient comparisons.

Activate prior knowledge . . . by having students consider the number of small squares that can form the outer edges of the tiling. **Ask:**

- Q Using the lengths of the sides of the smallest shape, what is the length and width of the tiling in the photograph?
- Q What mathematical types of shapes are these tiles?

COMMON ERROR: Counts Number of Tiles

I counted up 18 shapes and multiplied 18 and 1.5 inches to determine the size of the tiling.

If students . . . thought that the question asked about the number of tiles in the photograph rather than the kinds and sizes of the different shapes in the tiling, then they misread or misunderstood the problem.

Then intervene . . . by pointing out that the tessellation consists of three sets of congruent shapes of different sizes. **Ask:**

- Q How does the length of a side of the largest shape compare to the length of a side of the smallest shape?
- Q How do the unequal lengths of the sides of one of the shapes compare to the lengths of the sides of the smaller and larger shapes?

11.3

Properties of Rectangles, Rhombuses, and Squares

I Can prove and use properties of squares, rectangles, and rhombuses.

Spark Your Learning

Patterns of tiles are commonly used to cover floors and walls.

Complete Part A as a whole class. Then complete Parts B–D in small groups.

A. What is a geometric question you can ask about the shapes that can be used to completely cover a flat surface? How many different shapes and sizes are used to create this pattern?

B. Describe some different tile patterns you have seen. What properties of the patterns could you use to classify different patterns? See Additional Answers.

C. To answer your question, what strategy and tool would you use along with all the information you have? What answer do you get? See Strategies 1 and 2 on the facing page.

D. A shape can tessellate if you can completely cover the plane with congruent copies of it. Do you think that every possible quadrilateral should be able to tessellate? Provide evidence to support your reasoning. See Additional Answers.

Turn and Talk Is the tile pattern shown a tessellation? Explain your reasoning. See margin.

CULTIVATE CONVERSATION • Information Gap

Ask students questions to help them decide what missing information they need to answer the question, "How many different shapes and sizes are used to create this pattern?"

1. Do you have enough information to decide how many different shapes are used in this pattern? Yes, there appear to be three different shapes in this pattern.
2. Do you have information about the sizes of the shapes? No, I need information about the dimensions of the shapes to figure out their sizes.
3. Suppose you know the length of one of the sides of a shape. Would that help you answer the question? Yes, because I could then use that information to determine the lengths of the sides of the other shapes.
4. Is there any more information you would need to answer the question? no

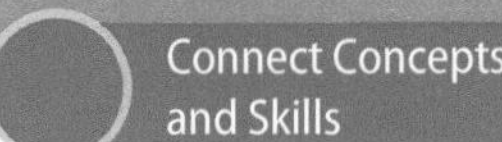

Connect Concepts and Skills

1 Spark Your Learning

▶ MOTIVATE

- Have students look at the photo in their books and read the information contained in the photo. Then complete Part A as a whole-class discussion.
- Give the class the additional information they need to solve the problem. This information is available online as a printable and projectable page in the Teacher Resources.
- Have students work in small groups to complete Parts B–D.

▶ PERSEVERE

If students need support, guide them by asking:

Q **Advancing • Use Tools** Which tool could you use to solve the problem? Why choose that tool and not some other? Students' choices of tools and reasons for choosing them will vary.

Q **Assessing** What appears to be the relationship between the lengths of the sides of the tiles in this picture? Possible answer: The lengths of the sides of the largest tile seem to be twice the lengths of the sides of the smallest tile. The length of a side of the medium-sized tile seems to be the same length as that of the smallest tile, and the length of the other side seems to be twice the length of the sides of the smallest tile.

Q **Assessing** If the length of the smallest tile is s, what seems to be an expression in terms of s for the length and width of the pattern pictured in the photograph? $6s$ by $6s$.

Q **Advancing** What kinds of triangles could be used to tessellate a floor? Possible answer: an equilateral triangle or congruent right triangles

Turn and Talk Remind students that a tessellation is a pattern in which shapes cover a surface without overlapping or leaving any gaps between them. no; A tessellation can cover a plane with congruent copies of the same shape or with a pattern of congruent copies of various shapes. This pattern uses copies of different congruent shapes.

▶ BUILD SHARED UNDERSTANDING

Select groups of students who used various strategies and tools to share with the class how they solved the problem. As they present their solutions, have each group discuss why they chose a specific strategy and tool.

2 Learn Together

Build Understanding

Task 1 MP **Construct Arguments** Students use geometry software to draw a parallelogram and its diagonals and then manipulate it to determine the relationships between the lengths of its diagonals and the measures of the angles between the diagonals and opposite sides.

CONNECT TO VOCABULARY

Have students use the **Interactive Glossary** to record their understanding of the vocabulary in this task.

Sample Guided Discussion:

Q **What is the relationship between the lengths of the two pairs of opposite sides of a parallelogram?** They are equal.

Q **What types of parallelogram have interior angles that are right angles?** a rectangle or a square

Q **In Part B, if the diagonals of the parallelogram are perpendicular and also equal, what type of parallelogram results?** a square

Turn and Talk Remind students that a conjecture is a statement that may or may not be true. The diagonals of a rectangle are the same length. The diagonals of a rhombus are perpendicular.

Build Understanding

Special Parallelograms

There are three types of special parallelograms: rectangles, rhombuses, and squares.

A **rectangle** is a parallelogram with four right angles.

A **rhombus** is a parallelogram with four congruent sides.

A **square** is a parallelogram with four congruent sides and four right angles.

rectangle

rhombus

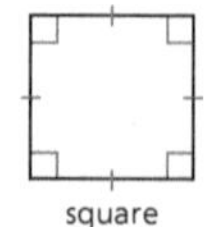

square

A. rectangle; All angles measure 90°.

1 A. Use a geometry drawing tool to draw parallelogram *FGHJ* and its diagonals. Write the side lengths, the measures of the interior angles, the lengths of the diagonals, and the measure of $\angle FKJ$. Move vertex *F* so that the diagonals have the same length. What type of parallelogram results? How do you know?

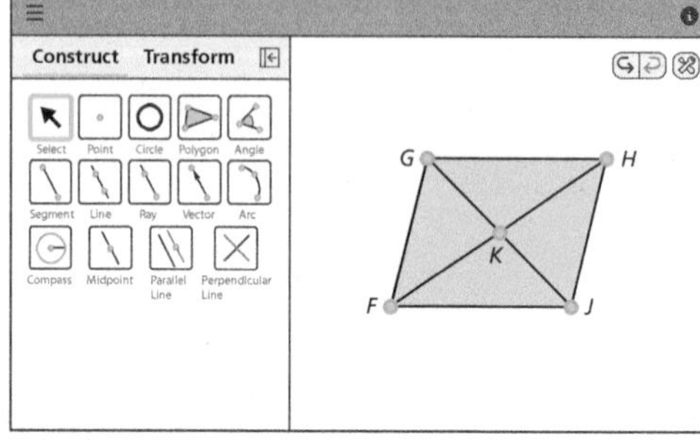

B. Move vertex *F* so that the diagonals are perpendicular. What type of parallelogram results? How do you know? rhombus; All sides have the same length.

Turn and Talk Make conjectures about the diagonals of a rectangle and the diagonals of a rhombus. See margin.

Diagonals of Special Parallelograms

You have learned that the diagonals of a parallelogram bisect each other. The diagonals of rectangles, rhombuses, and squares bisect each other because they are all parallelograms.

Diagonals of a Rectangle Theorem

If a parallelogram is a rectangle, then its diagonals are congruent.

If *ABCD* is a rectangle, then $\overline{AC} \cong \overline{BD}$.

A B D C

338

LEVELED QUESTIONS

Depth of Knowledge (DOK)	Leveled Questions	What Does This Tell You?
Level 1 **Recall**	What properties of a parallelogram are preserved when manipulating it in a geometry software program? It continues to have four sides, where its opposite sides remain parallel, the lengths of its opposite sides remain equal, and the measures of its opposite angles remain equal.	Students' answers indicate whether they understand the definition and basic properties of a parallelogram.
Level 2 **Basic Application of Skills & Concepts**	In parallelogram *ABCD* with diagonals $\overline{AC}$ and $\overline{BD}$ intersecting at *E*, what is the relationship between the diagonals of a parallelogram? They bisect each other.	Students' answers indicate whether they understand that they can use the Triangle Congruence Theorem AAS to prove that the pairs of triangles formed by each diagonal are congruent.
Level 3 **Strategic Thinking & Complex Reasoning**	What type of parallelograms can be inscribed in a circle? a rectangle and a square	Students' answers will indicate that they can reason that unless a parallelogram is a rectangle, at most only three vertices can lie on a circle.

Diagonals of a Rhombus Theorem

If a parallelogram is a rhombus, then its diagonals are perpendicular.

If *ABCD* is a rhombus, then $\overline{AC} \perp \overline{BD}$.

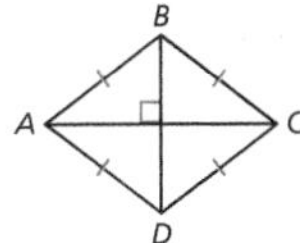

Diagonals of a Square Theorem

If a parallelogram is a square, then its diagonals are congruent and perpendicular.

If *ABCD* is a square, then $\overline{AC} \cong \overline{BD}$ and $\overline{AC} \perp \overline{BD}$.

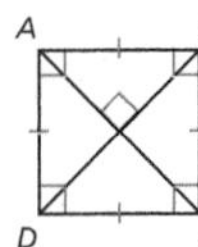

2 Consider the following conditional statements about squares.

- If a quadrilateral is a square, then it is a rectangle.
- If a quadrilateral is a square, then it is a rhombus.

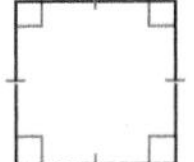

By definition, a square is a parallelogram with four right angles.

By definition, a rectangle is also a parallelogram with four right angles. So, a square is a rectangle, which explains why the first statement is true.

A. Use definitions to explain why the second statement, "If a quadrilateral is a square, then it is a rhombus" is true. See below.

B. Because a square is a rectangle and a rhombus, squares have the same properties as rectangles and rhombuses. Are the properties of the diagonals of a square the same as the properties of the diagonals of a rectangle and a rhombus? Explain. yes; The diagonals of a square are the same length and perpendicular.

Turn and Talk List each type of quadrilateral—parallelogram, rectangle, rhombus, and square—for which the property always applies.

- All sides are congruent. See margin.
- All angles are congruent.
- The diagonals are congruent.
- The diagonals bisect each other.

A. By definition, a square is a parallelogram with four congruent sides. By definition, a rhombus is also a parallelogram with four congruent sides. So, a square is a rhombus.

Task 2 MP **Construct Arguments** Students use the definitions of a square and a rhombus to construct an argument to explain why a square is also a rhombus.

Sample Guided Discussion:

Q **Is the converse of the statement, "If a quadrilateral is a square, then it is a rhombus," also true? Explain.** no; The converse is that a rhombus is also a square. But the angles of a rhombus are not always right angles.

Q **In Part B, what are the properties of the diagonals in a rectangle and in a rhombus?** The diagonals of a rectangle are congruent, and the diagonals of a rhombus are perpendicular.

Turn and Talk Have students draw one or more sketches of each type of quadrilateral to help them decide which ones have the given properties.

- All sides are congruent: square, rhombus
- All angles are congruent: square, rectangle
- The diagonals are congruent: square, rectangle
- The diagonals bisect each other: parallelogram, square, rectangle, rhombus

OPTIMIZE OUTPUT Critique, Correct, and Clarify

Have students work with a partner. On one side of an index card, write the word *rectangle, rhombus,* or *square*. On the other side, write the definition of the parallelogram and draw a sketch of the figure that also includes its diagonals. Then below the figure write the theorem related to the properties of the diagonals in that figure.

PROFICIENCY LEVEL

Beginning

Have students complete the following statement for each figure and read the completed statement aloud:

"A (rectangle, rhombus) always has (four equal sides, four equal angles, equal diagonals)."

Intermediate

Have students use complete sentences to write definitions of *parallelogram*, *rectangle*, *rhombus*, and *square*.

Advanced

Have students write the converse of each of the true statements below and indicate whether the converse is true or false.

- A square is a rectangle.
- A square is a rhombus.
- A parallelogram with perpendicular diagonals is a rhombus.

Step It Out

Task 3 **MP** **Construct Arguments** Students analyze a given proof of the Diagonals of a Rectangle Theorem and use the Distance Formula to calculate the lengths of the diagonals of a rectangle plotted in the coordinate plane.

Sample Guided Discussion:

Q What step in the proof of the Diagonals of a Rectangle Theorem could you delete and still prove that $\overline{AC} \cong \overline{BD}$? Explain. Step 4; It is sufficient to say that the opposite sides of a rectangle are congruent.

Q In Part B, what are the lengths of the legs of right triangles $\triangle ADC$ and $\triangle BCD$? Explain. $(5 - 2)$ and $(7 - 1)$; The sides of the rectangle are vertical and horizontal. Therefore, their lengths equal $(5 - 2)$ and $(7 - 1)$, respectively.

Turn and Talk Point out that one way to show that the diagonals are perpendicular is to draw a rhombus in the coordinate plane and calculate the slopes of its diagonals. Another way is to prove triangles congruent and their corresponding parts congruent. Show that consecutive angles formed by the intersection of the diagonals are congruent and form a linear pair so they are right angles.

Step It Out

Prove Diagonals of a Rectangle Are Congruent

3 Prove that the diagonals of a rectangle are congruent.
Given: *ABCD* is a rectangle.
Prove: $\overline{AC} \cong \overline{BD}$

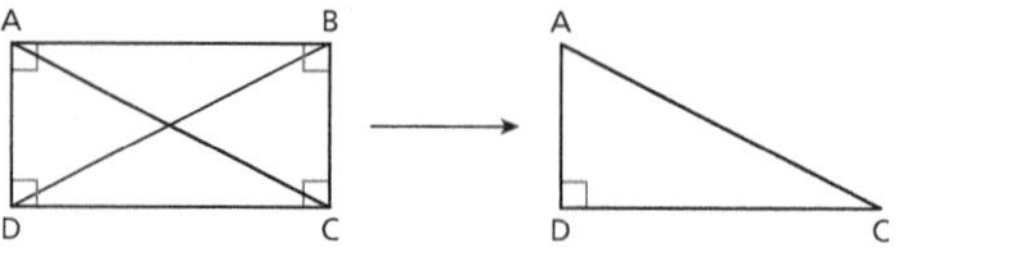

A. Why are both diagrams of the triangles helpful when writing the proof? See Additional Answers.

Statements	Reasons
1. *ABCD* is a rectangle.	1. Given
2. $\angle BCD$ and $\angle ADC$ are right angles.	2. Definition of rectangle
3. $\angle BCD \cong \angle ADC$	3. All right angles are congruent.
4. *ABCD* is a parallelogram.	4. Definition of a rectangle
5. $\overline{AD} \cong \overline{BC}$	5. Opposite Sides of a Parallelogram Theorem
6. $\overline{DC} \cong \overline{DC}$	6. Reflexive Property of Congruence
7. $\triangle ADC \cong \triangle BCD$	7. SAS Triangle Congruence Theorem
8. $\overline{AC} \cong \overline{BD}$	8. Corresponding parts of congruent figures are congruent.

B. The coordinate grid shows rectangle *ABCD*. Verify that the diagonals are congruent.

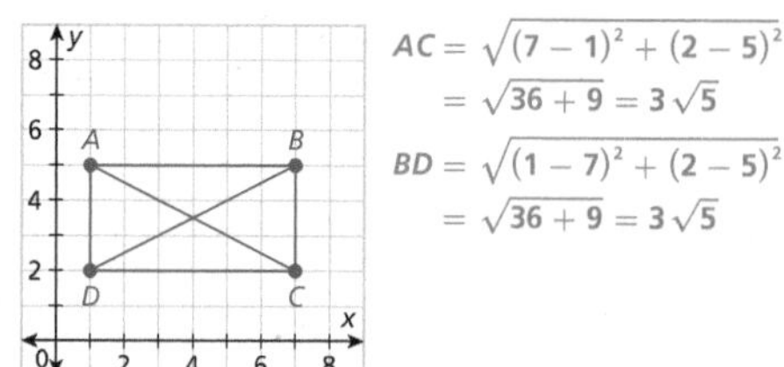

$AC = \sqrt{(7-1)^2 + (2-5)^2}$
$= \sqrt{36 + 9} = 3\sqrt{5}$

$BD = \sqrt{(1-7)^2 + (2-5)^2}$
$= \sqrt{36 + 9} = 3\sqrt{5}$

Turn and Talk How would you show that the diagonals of a rhombus are perpendicular? See margin.

340

Use Properties of Squares, Rectangles, and Rhombuses

You can use the properties of squares, rectangles, and rhombuses to find the measures of sides, diagonals, and angles.

 Cheryl is making a keychain that includes a rhombus trinket. In the rhombus, $m\angle XYZ = (5b - 10)^\circ$. Use rhombus $WXYZ$ to find WZ and $m\angle XYZ$.

W X Y Z V; $9a - 11$ mm; $(5b + 20)^\circ$; $6a + 7$ mm

Find *WZ*.

Substitute the expressions for the lengths WZ and ZY and solve for a.

$WZ = ZY$

$9a - 11 = 6a + 7$

$3a - 11 = 7$

$3a = 18$

$a = 6$

A. Why are you able to set *WZ* and *ZY* equal to each other?

***WXYZ* is a rhombus, so each side length is equal.**

Substitute 6 for a in the expression for WZ and simplify.

$9a - 11 = 9(6) - 11 = 54 - 11 = 43$

So, $WZ = 43$ mm.

Find $m\angle XYZ$.

$m\angle XVY = 90^\circ$, so $5b + 20 = 90$.

Solve for b.

$5b + 20 = 90$

$5b = 70$

$b = 14$

B. How do you know that $m\angle XVY = 90^\circ$?

Substitute 14 for b to find $m\angle XYZ$.

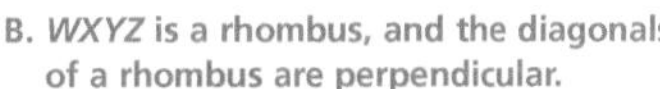

B. *WXYZ* is a rhombus, and the diagonals of a rhombus are perpendicular.

$m\angle XYZ = (5b - 10)^\circ$

$5b - 10 = 5(14) - 10 = 70 - 10 = 60$

$m\angle XYZ = 60^\circ$

Turn and Talk Suppose you were told that quadrilateral $WXYZ$ is a parallelogram but not a rhombus. Is it possible to find WZ and $m\angle XYZ$ using this given information? Explain. See margin.

Task 4 (MP) **Use Structure** Students see an example of how to use the properties of a rhombus and algebra to solve for the lengths of its sides and an angle measure.

Sample Guided Discussion:

Q Why can you also write $ZY = WZ$ to solve this problem? the Reflexive Property of Congruence

Q How can you check that $WZ = 43$ mm? Substitute $a = 6$ into the second expression $6a + 7$.

Q Why is $\triangle XYZ$ an isosceles triangle? In $\triangle XYZ$, $\overline{XY} \cong \overline{ZY}$ because the sides of a rhombus are equal.

Q What is $m\angle XYW$? Explain. $m\angle XYW = \frac{1}{2}m\angle XYZ = 30^\circ$; The diagonals of a rhombus bisect the angles of the rhombus.

Turn and Talk Remind students of the special properties of rhombuses. They are parallelograms with four equal sides and perpendicular diagonals. You cannot find the values with the given information if WXYZ is not a rhombus; You do not know the relationship between the lengths of the sides and you do not know the measure of the angle formed by the intersection of the diagonals.

ONLINE

Assign the Digital On Your Own for
- built-in student supports
- Actionable Item Reports
- Standards Analysis Reports

On Your Own

Assignment Guide

The chart below indicates which problems in the On Your Own are associated with each task in the Learn Together. Assign daily homework for tasks completed.

Learn Together Tasks	On Your Own Problems
Task 1, p. 338	Problems 7–9
Task 2, p. 339	Problems 14–17
Task 3, p. 340	Problems 18–20 and 26
Task 4, p. 341	Problems 5, 6, 10–13, 21–25, 27, and 28

Check Understanding

In Problems 1 and 2, *ABCD* is a parallelogram. Determine whether there is enough information to conclude that the *ABCD* is a rhombus. Explain your reasoning.

1. 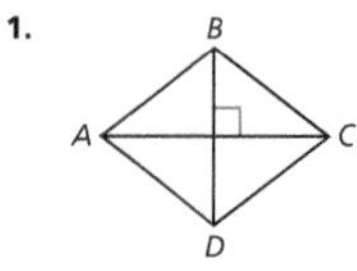

yes; The diagonals are perpendicular, so the shape is a rhombus.

2. 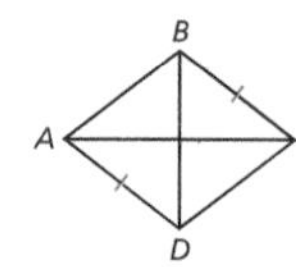

no; You need to know that all sides are congruent to know that the shape is a rhombus.

3. Explain why the following statement is true: If a quadrilateral is a square, then it is a parallelogram. If a quadrilateral is a square, then the opposite sides are parallel, so it is also a parallelogram.

4. Rectangle *LMNQ* has diagonals $\overline{MQ}$ and $\overline{LN}$. Name two right triangles formed by the diagonals. Are the triangles congruent? Explain why or why not. See Additional Answers.

Find the value of the variable. Then give the lengths of the diagonals.

5. *DEFG* is a rectangle, $DF = 5x + 15$, and $GE = 8x - 18$. $x = 11$; $DF = GE = 70$

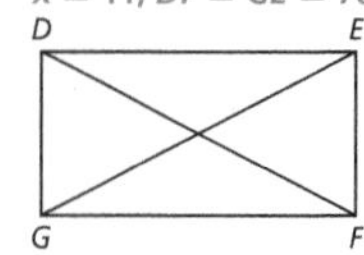

6. *QRST* is a square, $QS = 7n - 12$, and $RT = 3n + 8$. $n = 5$; $QS = RT = 23$

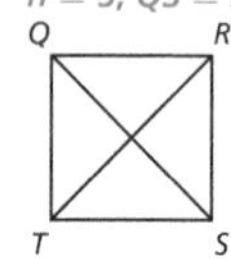

On Your Own

Use geometry software to draw each special parallelogram. Find the lengths of the diagonals and the measures of the angles formed by the intersections of the diagonals. What do you notice about each special parallelogram?

7. rectangle 7–9. See Additional Answers.
8. rhombus
9. square

The photo shows the front door of a house. In rectangle *ABCD*, $AB = 36$ inches, and $BC = 80$ inches. Find each measure.

10. *DC* 36 inches
11. *AD* 80 inches
12. *AC* about 87.7 inches
13. *BD* about 87.7 inches

342

3 Check Understanding

Formative Assessment

Use formative assessment to determine if your students are successful with this lesson's learning objective.

Students who successfully complete the Check Understanding can continue to the On Your Own practice.

For student who miss 1 problem or more, work in a pulled small group using the Almost There small-group activity on page 337C.

ONLINE

Assign the Digital Check Understanding to determine
- success with the learning objective
- items to review
- grouping and differentiation resources

4 Differentiation Options

Differentiate instruction for all students using small-group activities and math center activities on page 337C.

Reteach

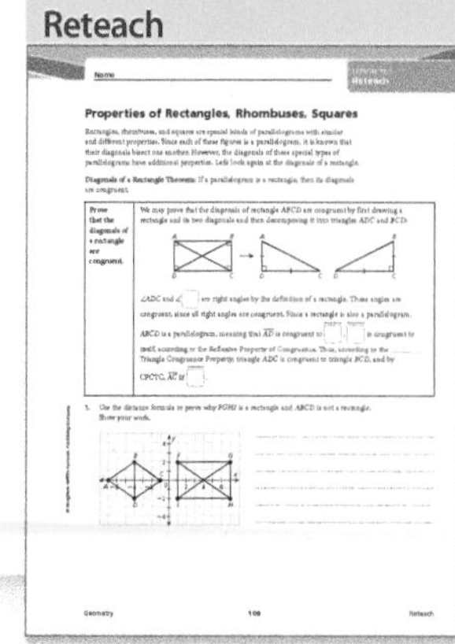
Properties of Rectangles, Rhombuses, Squares

Challenge

Properties of Rectangles, Rhombuses, Squares

List each type of quadrilateral—*parallelogram, rectangle, rhombus,* and *square*—for which the property always applies.

14. The diagonals are perpendicular. rhombus, square

15. Opposite angles are congruent. parallelogram, rectangle, rhombus, square

16. The diagonals are perpendicular and congruent. square

17. All angles are right angles. rectangle, square

In Problems 18 and 19, show that the diagonals of a rhombus are perpendicular.

18. Prove that the diagonals of a rhombus are perpendicular.
Given: *QRST* is a rhombus.
Prove: $\overline{QS} \perp \overline{RT}$ See Additional Answers.

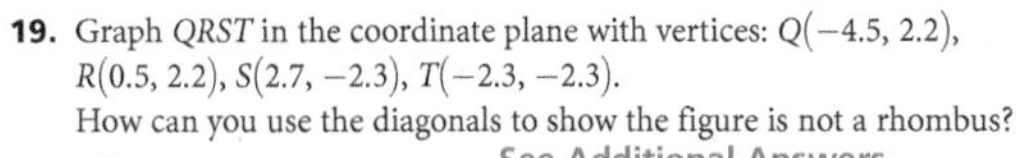

19. Graph *QRST* in the coordinate plane with vertices: $Q(-4.5, 2.2)$, $R(0.5, 2.2)$, $S(2.7, -2.3)$, $T(-2.3, -2.3)$.
How can you use the diagonals to show the figure is not a rhombus?
See Additional Answers.

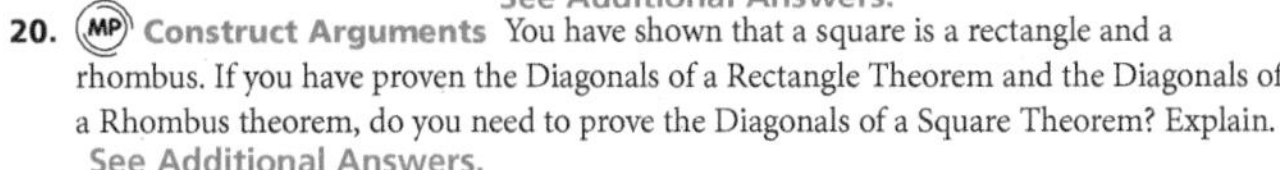

20. (MP) **Construct Arguments** You have shown that a square is a rectangle and a rhombus. If you have proven the Diagonals of a Rectangle Theorem and the Diagonals of a Rhombus theorem, do you need to prove the Diagonals of a Square Theorem? Explain.
See Additional Answers.

Each parallelogram is a rhombus. Find the side length.

21. *BC* 62

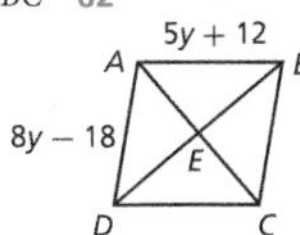

22. *ST* 37

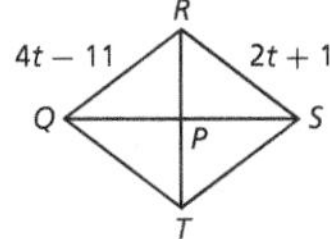

Each parallelogram is a rhombus. Find the angle measure.

23. $m\angle ABC = (10n + 5)°$ 115°

24. $m\angle RQT = (6z - 4)°$ 50°

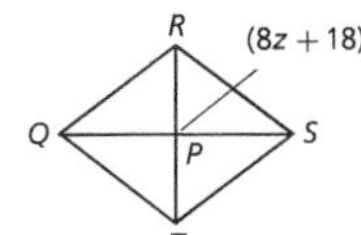

25. Draw a rectangle to represent a field hockey goal. Label the rectangle *ABCD* to show that the distance between the goal posts, $\overline{BC}$, is 2 feet less than twice the distance from the top goal post to the ground. If the perimeter of *ABCD* is 38 feet, what is the length of $\overline{BC}$? 12 ft

Questioning Strategies

Problem 18 One way to prove that $\overline{QS} \perp \overline{RT}$ is to prove triangles congruent and use CPCTC. What is another way to prove that $\overline{QS} \perp \overline{RT}$ without proving triangles congruent? Parallelogram *RSTQ* is a rhombus, so $RS = TS$. Therefore, $\overline{QS}$ is the perpendicular bisector of $\overline{RT}$ because if a point is equidistant from the endpoints of a segment, it lies on the perpendicular bisector of the segment.

Watch for Common Errors

Problem 25 Students may write the incorrect expression for the length of $\overline{BC}$ translating "2 feet less than twice the distance from the top goal post to the ground" as $2 - 2x$, instead of the correct expression $2x - 2$, where x is the vertical distance of $\overline{AB}$.

5 Wrap-Up

Summarize learning with your class. Consider using the Exit Ticket, Put It in Writing, or I Can scale.

Exit Ticket 

Given: $QRST$ is a rhombus.

Prove: $\angle QRT \cong \angle SRT$

$\overline{RQ} \cong \overline{RS}$ because $QRST$ is a rhombus. Diagonals $\overline{RT}$ and $\overline{QS}$ are perpendicular, $\angle RPQ$ and $\angle RPS$ are right angles, and $\overline{RP} \cong \overline{RP}$ by the Reflexive Property of Congruence. Therefore, $\triangle RQP \cong \triangle RSP$ by the HL Congruence Theorem, and $\angle QRT \cong \angle SRT$ because of CPCTC.

Put It in Writing 

Have students write a paragraph explaining the differences among parallelograms, rectangles, squares, and rhombuses.

I Can 

The scale below can help you and your students understand their progress on a learning goal.

4	I can prove and use properties of squares, rectangles, and rhombuses, and I can explain my steps to others.
3	I can prove and use properties of squares, rectangles, and rhombuses.
2	I can define squares, rectangles, and rhombuses as special types of parallelograms.
1	I can prove and use the properties of parallelograms.

Spiral Review • Assessment Readiness

These questions will help determine if students have retained information taught in the past and can also prepare them for high-stakes assessments. Here, students must identify what conditions guarantee that a quadrilateral is a parallelogram **(11.1)**, use the properties of a parallelogram to find the length of a side **(11.2)**, choose the congruence theorem that proves two triangles are congruent **(8.2)**, and apply theorems about angle inequalities in two triangles **(10.2)**.

26. MP **Construct Arguments** Trace the rectangle and fold as shown. Explain how the reflections show that the diagonals of a rectangle are congruent. See Additional Answers.

Step 1	Step 2	Step 3
Trace.	Fold over vertical line.	Fold over horizontal line.
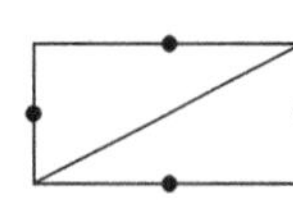	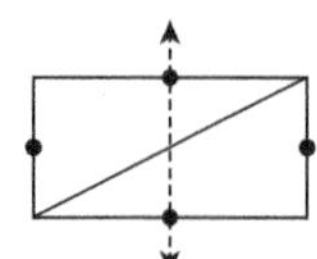	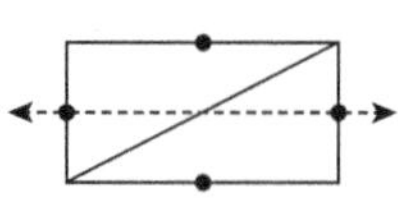

27. **Open Ended** Write a problem that requires solving for a variable to find a length in a special parallelogram. Solve your problem. See Additional Answers.

28. Open Middle™ What is the fewest number of geometric markings needed to demonstrate that a quadrilateral is a rectangle? Give an example.
at least three markings for example: three right angle markings

Spiral Review • Assessment Readiness

29. Which set of information guarantees that quadrilateral $ABCD$ is a parallelogram?

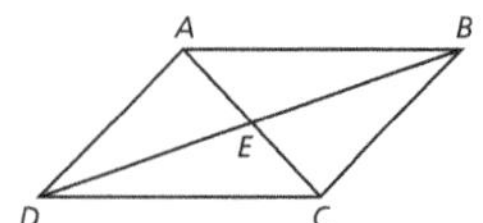

Ⓐ $\overline{AD} \cong \overline{AB}$, $\overline{BC} \cong \overline{DC}$
Ⓑ $\overline{DE} \cong \overline{BE}$, $\overline{AE} \cong \overline{CE}$
Ⓒ $\overline{AE} \cong \overline{BE}$
Ⓓ $\angle ADE \cong \angle DCB$

30. $JKLM$ is a parallelogram with side lengths $JK = 17x + 5$, $KL = 3y - 7$, $LM = 20x - 1$, and $MJ = y + 11$. Find the length of $\overline{MJ}$.

Ⓐ 2 Ⓒ 20
Ⓑ 9 Ⓓ 39

31. Which theorem of congruence can you use to show that the triangles are congruent?

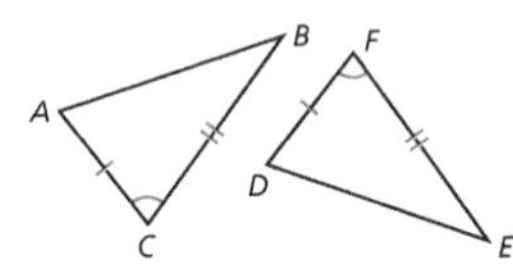

Ⓐ ASA Ⓒ SAS
Ⓑ AAS Ⓓ SSS

32. In $\triangle ABC$, $AB = 21$, $AC = 25$, and $m\angle A = 75°$. In $\triangle XYZ$, $XY = 21$ and $XZ = 25$. If $BC < YZ$, which of the following can be the measure of $\angle X$? Select all that apply.

Ⓐ 15° Ⓓ 76°
Ⓑ 74° Ⓔ 90°
Ⓒ 75° Ⓕ 105°

 **I'm in a Learning Mindset!**

How am I devoting my energy to learning about the properties of rectangles, rhombuses, and squares?

Learning Mindset

Perseverance Checks for Understanding

When introducing students to rectangles, rhombuses, and squares, it is probably helpful to first review the properties of parallelograms. Point out that because each of the new quadrilaterals is a parallelogram, it "inherits" the same set of properties as a parallelogram, plus additional ones. To illustrate this, draw a Venn diagram on the board that has two intersecting ovals or have students draw one. It should show the relationships among rectangles, rhombuses, and squares, with squares in the intersection of the ovals. Remind students that squares have all the properties of each of the other parallelograms. *Keep pointing out the unique properties by asking questions such as these: What type of parallelogram has equal diagonals? What type(s) have perpendicular diagonals? What type(s) have equal angles? equal sides? Is a rectangle a square? Is a square a rectangle? Is a rhombus a rectangle? Is a rhombus a square?*

11.4 Conditions for Rectangles, Rhombuses, and Squares

LESSON FOCUS AND COHERENCE

Mathematics Standards

- Prove theorems about parallelograms. *Theorems include: opposite sides are congruent, opposite angles are congruent, the diagonals of a parallelogram bisect each other and conversely, rectangles are parallelograms with congruent diagonals.*
- Construct an equilateral triangle, a square, and a regular hexagon inscribed in a circle.
- Use coordinates to prove simple geometric theorems algebraically.
- Prove the slope criteria for parallel and perpendicular lines and use them to solve geometric problems (e.g., find the equation of a line parallel or perpendicular to a given line that passes through a given point.)

Mathematical Practices and Processes 

- Look for and make use of structure.
- Construct viable arguments and critique the reasoning of others.

I Can Objective 

I can prove and use conditions for rectangles, rhombuses, and squares.

Learning Objective

Students use the properties of diagonals for rectangles, rhombuses, and squares to identify special parallelograms in the coordinate plane and construct two-step proofs of theorems to prove conditions for rectangles and rhombuses.

Language Objective

Students describe the properties of diagonals for rectangles, rhombuses, squares, and the conditions for rectangles and rhombuses.

Mathematical Progressions

Prior Learning	Current Development	Future Connections
Students: • proved and used properties of triangles. **(9.1–9.5)** • proved properties of parallelograms. **(11.1)**	**Students:** • construct an inscribed quadrilateral within a circle. • identify and prove conditions for rectangles, rhombuses, and squares. • use conditions for squares, rectangles, and rhombuses to prove that quadrilaterals are squares, rectangles, or rhombuses.	**Students:** • will prove and use properties of kites and trapezoids. **(11.5)**

UNPACKING MATH STANDARDS

Construct an equilateral triangle, a square, and a regular hexagon inscribed in a circle.

What it Means to You

In order for students to construct these figures, they need to use theorems and properties about these shapes. For example, they must recognize that all these shapes have congruent sides and congruent angles. Therefore, students must be able to construct congruent lines and angles. When they construct the square inscribed in the circle, they will draw perpendicular bisectors of the diameter, so they will need to know about perpendicular lines. Students benefit from repeated practice until the steps for each construction are memorized.

WARM-UP OPTIONS

ONLINE

PROJECTABLE & PRINTABLE

ACTIVATE PRIOR KNOWLEDGE • Prove and Use Properties of Triangles

Use these activities to quickly assess and activate prior knowledge as needed.

Problem of the Day 

Find the circumcenter of the triangle with vertices $A(3, 2)$, $B(1, 4)$ and $C(5, 4)$. $(3, 4)$

Midpoints: $\overline{AB} = \left(\frac{3+1}{2}, \frac{2+4}{2}\right) = \left(\frac{4}{2}, \frac{6}{2}\right) = (2, 3)$;

$\overline{AC} = \left(\frac{3+5}{2}, \frac{2+4}{2}\right) = \left(\frac{8}{2}, \frac{6}{2}\right) = (4, 3)$

Slopes: $\overline{AB} = \frac{4-2}{1-3} = \frac{2}{-2} = -1$; $\overline{AC} = \frac{4-2}{5-3} = \frac{2}{2} = 1$

Perpendicular slopes: $\overline{AB} = 1$; $\overline{AC} = -1$ Perpendicular bisectors:

$\overline{AB}$:	$\overline{AC}$:
$y = mx + b$	$y = mx + b$
$3 = 1(2) + b$	$3 = -1(4) + b$
$3 = 2 + b$	$3 = -4 + b$
$b = 1$	$b = 7$
$y = x + 1$	$y = -x + 7$

Circumcenter:
$x + 1 = -x + 7$; $2x = 6$; $x = 3$; $y = x + 1$ or $-x + 7$; $y = 4$

Quick Check for Homework

As part of your daily routine, you may want to display the Teacher Solution Key to have students check their homework.

Make Connections

Based on students' responses to the Problem of the Day, choose one of the following:

- Project the Interactive Reteach, Geometry, Lesson 9.2.
- Complete the Prerequisite Skills Activity:

Have students work in pairs. Give each pair of students a piece of graph paper. Have them sketch a triangle, noting the coordinates of the vertices, and then calculate the lengths of the medians.

- *Why is it necessary to calculate the midpoint of each side of the triangle?* It is necessary to calculate the midpoint of each side of the triangle because the median is a segment whose endpoints are a vertex of the triangle and the midpoint of the opposite side.
- *What is the Midpoint Formula?* The Midpoint Formula is $\left(\frac{x_1 + x_2}{2}, \frac{y_1 + y_2}{2}\right)$.
- *What formula is used to determine the length of the median?* The Distance Formula is used to determine the length of the median.
- *What is the Distance Formula?* The Distance Formula is $\sqrt{(x_2 - x_1)^2 + (y_2 - y_1)^2}$.

SHARPEN SKILLS

If time permits, use this on-level activity to build fluency and practice basic skills.

Quantitative Comparison

Students make a comparison between two quantities. Write the following problem on the board. Ask students to choose the letter representing the correct answer and to explain their reasoning.

Quantity A	**Quantity B**
the diagonal of a rectangle with length 5 and width 3	the diagonal of a square with side 4

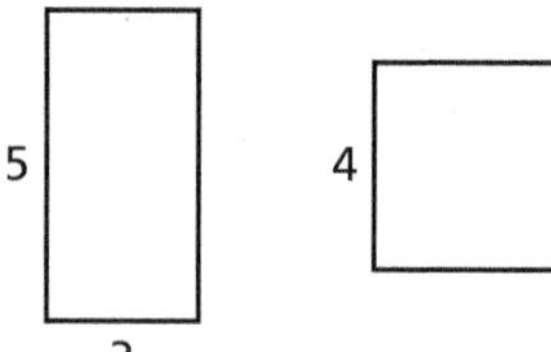

A. Quantity A is greater.
B. Quantity B is greater.
C. The two quantities are equal.
D. The relationship cannot be determined from the information given.
A; Quantity A $= \sqrt{34}$ and Quantity B $= \sqrt{32}$.

PLAN FOR DIFFERENTIATED INSTRUCTION

Small-Group Options

Use these teacher-guided activities with pulled small groups.

On Track

Materials: graph paper

Give students graph paper on which various parallelograms are graphed. Have students pair up and determine if the parallelogram is a rectangle, rhombus, and/or square. Students should list all names that apply.

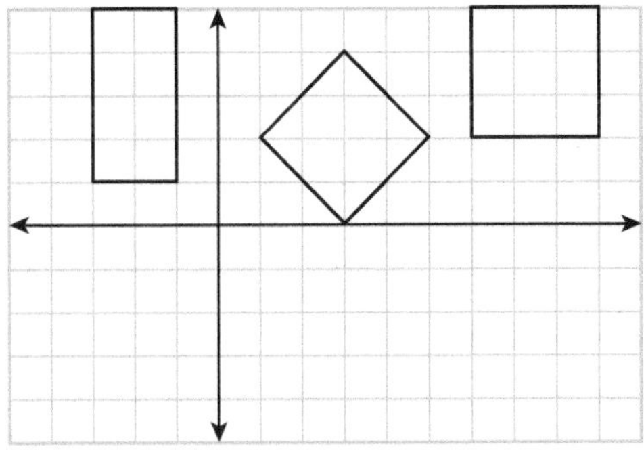

Almost There

Materials: graph paper

Give students graph paper on which various parallelograms are graphed. Have students pair up and calculate the length of the diagonal of each figure and if this is enough information to determine what type of figure each can be classified as.

Ready for More

Have students work in pairs. Provide them with the following statement to prove.

If the diagonals of a parallelogram are perpendicular, then the parallelogram is a rhombus.

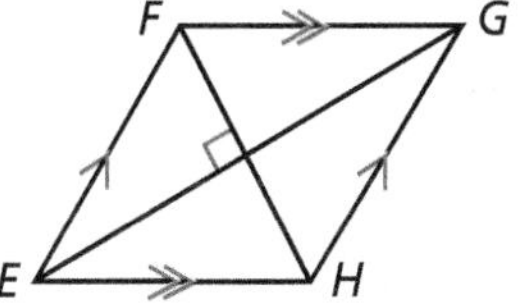

Math Center Options

Use these student self-directed activities at centers or stations. Key: ● Print Resources ● Online Resources

On Track

- ● Interactive Digital Lesson
- ●● Journal and Practice Workbook
- ● Module Performance Task
- ● Illustrative Mathematics: Is this a Parallelogram?

Almost There

- ● Reteach 11.4 (printable)
- ● Interactive 11.4
- ● Illustrative Mathematics: Is this a Rectangle?

Ready for More

- ● Challenge 11.4 (printable)
- ● Interactive Challenge 11.4

View data-driven grouping recommendations and assign differentiation resources.

During the *Spark Your Learning,* listen and watch for strategies students use. See samples of student work on this page.

Use the Definition of Rhombus

Strategy 1

The shapes formed by the folding sides of the accordion mirror are all parallelograms since opposite sides are always congruent. Because all four sides are congruent, *ABCD* is a rhombus.

Also, as it extends, the diagonals intersect at perpendicular angles. This supports the conclusion that *ABCD* is a rhombus.

If students . . . use the definition of rhombus to solve the problem, they are applying properties of geometric shapes from Grade 8 to real-world problems.

Have these students . . . explain how they determined their approach to this problem. **Ask:**

- Q How did you use the definition of a rhombus to solve the problem?
- Q What are the properties of a rhombus that support your answer?

Use Visualization and Elimination

Strategy 2

I know the figure has four equal sides because all sides of the quadrilateral remain congruent as the accordion mirror extends. I don't know the measure of any of the angles, so I can't say that the figure is a square since I don't know that each angle is 90 degrees. It can't be rectangle, trapezoid, or kite because it doesn't look like any of them, so I'm left with one possibility–rhombus. Therefore, *ABCD* is a rhombus.

If students . . . use visualization and elimination, they understand the shapes of different types of quadrilaterals but may have had difficulty making an immediate judgement about the figure being a rhombus.

Activate Prior Knowledge . . . by having students list all possible quadrilaterals. **Ask:**

- Q Which quadrilaterals have four congruent sides?
- Q What do you know about the diagonals of *ABCD*?

COMMON ERROR: Uses Incorrect Reasoning

The figure is a square because it has four congruent sides.

If students . . . use incorrect reasoning about congruent sides to draw conclusions about the figure, then they have an incomplete understanding of the different types of quadrilaterals.

Then intervene . . . by pointing out that some quadrilaterals share characteristics and that they should think about the angles and diagonals of figures before making conclusions. **Ask:**

- Q A quadrilateral with 4 congruent sides could be what two types of figures?
- Q If the interior angles of the figure change, becoming more acute or more obtuse, what could you conclude about *ABCD*?

11.4

Conditions for Rectangles, Rhombuses, and Squares

(I Can) prove and use conditions for rectangles, rhombuses, and squares.

Spark Your Learning

As the accordion mirror extends, the shapes of the quadrilaterals formed on the sides of the extension change.

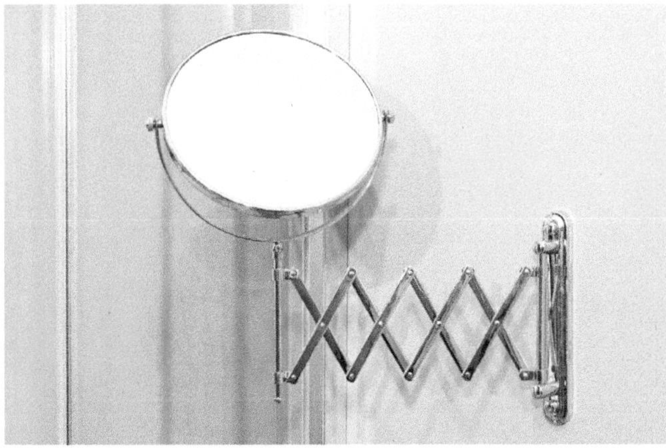

©Baloncici/iStock/Getty Images Plus/Getty Images

Complete Part A as a whole class. Then complete Parts B–D in small groups.

A. What is a mathematical question you can ask about the shapes formed by the folding side-pieces of the mirror? A. What type of quadrilateral is *ABCD* as the mirror extends?

B. How do the side lengths, angles, and diagonals of *ABCD* change as the mirror extends? See Additional Answers.

C. To answer your question, what strategy and tool would you use along with all the information you have? What answer do you get? See Strategies 1 and 2 on the facing page.

D. If you know the length of the full extension, what do you know about the length when the mirror is fulled retracted? See Additional Answers.

Turn and Talk What must be true about the diagonals of *ABCD* as the shape changes? Explain. See margin.

CULTIVATE CONVERSATION • Three Reads

Tell students to read the information in the photo three times and prompt them with a different question each time.

1. What is the situation about? This situation is about identifying the type of quadrilateral formed by the folding side-pieces when an accordion mirror is fully extended.

2. What are the quantities in this situation? How are those quantities related? The quantities in this situation are the length of the sides and the diagonals of the quadrilateral formed by the folding side-pieces when the mirror is fully extended; The sides are congruent, and the diagonals are perpendicular and congruent.

3. What are possible questions you could ask about this situation? Possible answer: When the mirror is fully extended, what is the measure of $\angle BAD$? When is $\angle BAD$ larger: when the mirror is partially extended or fully extended?

1 Spark Your Learning

▶ MOTIVATE

- Have students look at the photo in their books and read the information contained in the photo. Then complete Part A as a whole-class discussion.
- Give the class the additional information they need to solve the problem. This information is available online as a printable and projectable page in the Teacher Resources.
- Have students work in small groups to complete Parts B–D.

▶ PERSEVERE

If students need support, guide them by asking:

Q **Advancing • Use Tools** Which tool could you use to solve the problem? Why choose that tool and not some other? Students' choices of tools and reasons for choosing them will vary.

Q **Assessing** Before the accordion mirror is fully extended, what kind of angles are the left and right angles of the quadrilateral formed on the sides of the mirror? The left and right angles are obtuse angles.

Q **Assessing** Before the accordion mirror is fully extended, which diagonal is longer? The diagonal $\overline{BD}$ is longer.

Q **Advancing** How does the shape of the quadrilateral formed on the sides of the accordion mirror change as the mirror is fully extended? As the accordion mirror is fully extended, the quadrilaterals change from rhombuses to squares.

Turn and Talk Have students sketch a rhombus and square and draw in the diagonals of both. Ask them what kind of angle is formed at the intersection of the diagonals. The diagonals of *ABCD* are perpendicular; This is because the diagonals of a rhombus are perpendicular, and *ABCD* remains a rhombus as the rack extends.

▶ BUILD SHARED UNDERSTANDING

Select groups of students who used various strategies and tools to share with the class how they solved the problem. As they present their solutions, have each group discuss why they chose a specific strategy and tool.

2 Learn Together

Build Understanding

Task 1 (MP) **Use Structure** Students construct circles and draw diameters to inscribe a quadrilateral and reach conclusions about the type of quadrilateral formed by the diameters as diagonals.

Sample Guided Discussion:

Q How do you know the diagonals of the parallelogram you drew bisect each other without measuring them? Each half of each diagonal is the radius of the circle.

Q In Part C, what happens to the angles at the intersections of the diameters as they are drawn further apart until they are perpendicular to each other? The obtuse angles decrease in size until they reach 90°. The acute angles increase in size until they reach 90°

Turn and Talk Encourage students to focus on the congruence and perpendicularity of the diagonals and the way we name parallelograms by their characteristics. If the diagonals are congruent, the parallelogram is a rectangle. If the diagonals are perpendicular, then the parallelogram is a rhombus. If the diagonals are congruent and perpendicular, then the parallelogram is a square.

Build Understanding

Determine Conditions for Special Parallelograms

You can use the definitions of special parallelograms to determine whether a quadrilateral is a rectangle, rhombus, or square. However, there are other conditions that guarantee that a quadrilateral is a rectangle, rhombus, or square.

1 A. Use a compass to construct a circle and label the center. Use a straightedge to draw any two diameters. Then draw a quadrilateral by using a straightedge to connect the endpoints of the diameters. Is the quadrilateral a parallelogram? How do you know? A–E. See Additional Answers.

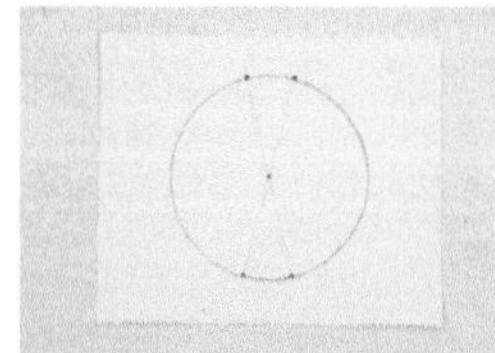

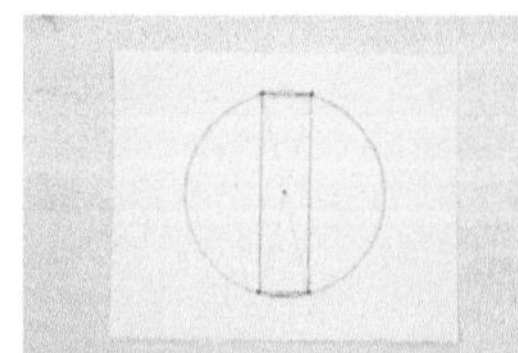

B. Measure the sides and the angles of the quadrilateral. What type of special parallelogram is formed? How do you know?

C. Repeat Part A using different pairs of diameters. Do you form the same type of parallelogram each time?

D. Use a compass to construct another circle and label the center. Use a straightedge to draw one diameter. Then construct the perpendicular bisector of the diameter. Draw a quadrilateral by connecting the endpoints of the segments. What type of special parallelogram is formed? How do you know?

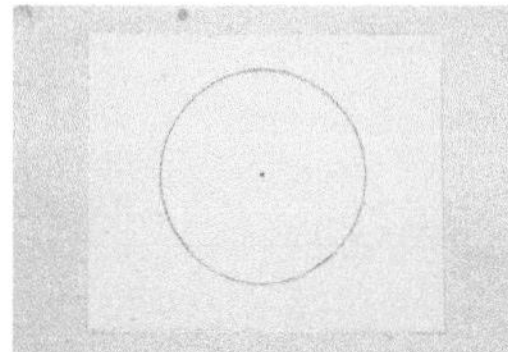

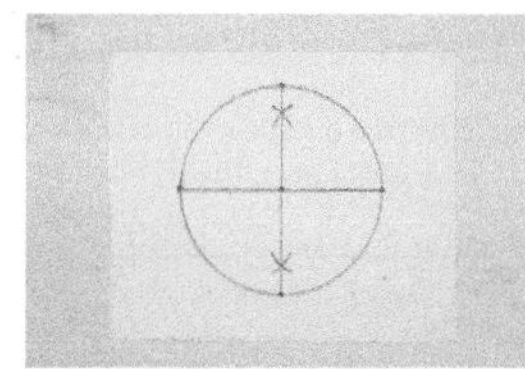

E. Is it possible to draw a rhombus that is not a square so that its vertices lie on a circle and its diagonals intersect at the center of the circle? Explain why or why not.

Turn and Talk Describe how information about the diagonals of a parallelogram can be used to classify the parallelogram as a rectangle, rhombus, or square. See margin.

LEVELED QUESTIONS

Depth of Knowledge (DOK)	Leveled Questions	What Does This Tell You?
Level 1 **Recall**	What parallelograms have congruent diagonals? Rectangles and squares have congruent diagonals.	Students' answers will indicate whether they understand which parallelograms have congruent diagonals.
Level 2 **Basic Application of Skills & Concepts**	Are the diagonals of a parallelogram always perpendicular? Explain. no; The diagonals of a rhombus and a square are perpendicular, but the diagonals of a rectangle are not.	Students' answers will indicate whether they understand which parallelograms have perpendicular diagonals.
Level 3 **Strategic Thinking & Complex Reasoning**	What type of quadrilateral could have diagonals with slopes equal to 2 and $-\frac{1}{2}$? square or rhombus	Students' answers will indicate whether they understand how to use slopes to determine if diagonals are perpendicular and which quadrilaterals have perpendicular diagonals.

Step It Out

Prove Conditions for Rectangles

The theorems below can be used to determine whether a parallelogram is a rectangle.

Theorems: Conditions for Rectangles

If one angle of a parallelogram is a right angle, then the parallelogram is a rectangle.

If the diagonals of a parallelogram are congruent, then the parallelogram is a rectangle.

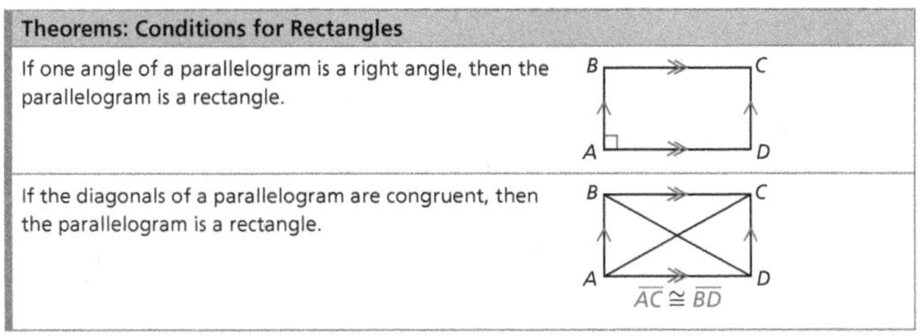

2 Prove that if the diagonals of a parallelogram are congruent, then parallelogram is a rectangle.

Given: $ABCD$ is a parallelogram and $\overline{AC} \cong \overline{BD}$.

Prove: $ABCD$ is a rectangle.

Because the opposite sides of a parallelogram are congruent, $\overline{AB} \cong \overline{CD}$.

It is given that $\overline{AC} \cong \overline{BD}$ and $\overline{AD} \cong \overline{AD}$ by the Reflexive Property of Congruence.

So, $\triangle ABD \cong \triangle DCA$.

A. Which theorem can you use to show that the triangles are congruent? SSS Triangle Congruence Theorem

$\angle BAD \cong \angle CDA$ by CPCTC. These angles are supplementary, so $m\angle BAD + m\angle CDA = 180°$.

B. How do you know that these angles are supplementary? B. See below.

By substitution, $m\angle BAD + m\angle BAD = 180°$.

C. What is $m\angle BAD$? 90°

A similar argument shows that the other angles of $ABCD$ are right angles, so $ABCD$ is a rectangle.

D. Why does this tell you that $ABCD$ is a rectangle? A rectangle is a parallelogram with four right angles.

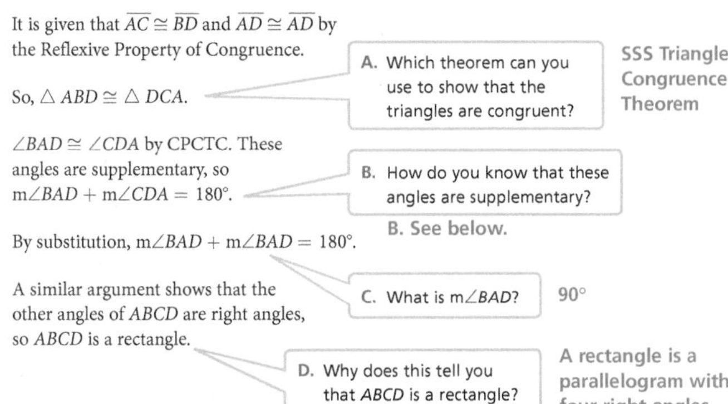

Turn and Talk Suppose the given information did not state that $ABCD$ is a parallelogram. Can it be proven that $ABCD$ is a rectangle? See margin.

B. Since $\overline{AB} \parallel \overline{CD}$ and $\angle BAD$ and $\angle CDA$ are consecutive interior angles of a parallelogram, the angles are supplementary.

Module 11 • Lesson 11.4 347

Step It Out

Task 2 (MP) **Construct Arguments** Students learn to use stated assumptions and previously established rules to prove conditions for rectangles. Students will draw upon their knowledge of congruent and right triangles to complete the proof.

Sample Guided Discussion:

Q What do you need to prove about the angles of a parallelogram to show that it is rectangle? You need to prove that all the angles of the parallelogram are right angles.

Q In Part A, is there another way to prove congruence of the triangles? No, the other ways to show congruence of triangle (SAS, AAS, ASA) require information about angles and HL requires that the triangle be a right triangle. There is no information about the angles or that the triangle is a right triangle.

Turn and Talk Ask students which property of a parallelogram drove the proof. What did they use to establish the congruence of the triangles which lead to the ultimate conclusion? No, you would not know that $\overline{AB} \cong \overline{CD}$.

EL PROFICIENCY LEVEL

Beginning

In the proof for Task 2, students show that $\triangle ABD \cong \triangle DCA$. Have students draw these two triangles and list which sides are congruent and which angles are congruent by CPCTC.

Intermediate

To complete the proof for Task 2, students need to show that other angles of $ABCD$ are right angles. Have students write a proof to show that $\angle CDA$ is a right angle and explain their steps.

Advanced

To complete the proof for Task 2, students need to show that other angles of $ABCD$ are right angles. Have students describe or prove another way to find that all of the angles are right angles.

Task 3 (MP) **Construct Arguments** Students will build a logical progression of statements to prove that *EFGH* is a rhombus. Remind students that the first step of every proof is always the "Given" and the final step of every proof is always the "Prove."

Sample Guided Discussion:

Q What do you need to prove to show that a parallelogram is a rhombus? You need to prove that the four sides are congruent.

Q In the proof, how is the Transitive Property of Congruence used to show $\overline{EH} \cong \overline{GH}$? $\overline{EH} \cong \overline{FE}$; $\overline{FE} \cong \overline{GH}$; $\overline{EH} \cong \overline{GH}$

Turn and Talk Ask students what, besides being perpendicular, do they know about the diagonals of a parallelogram. Remind them they will use the right angle formed by the perpendicular diagonals to create congruent right triangles. You prove that the two right triangles formed by diagonals are congruent using SAS, then the hypotenuses are congruent by CPCTC. The hypotenuses are congruent consecutive sides, so the parallelogram is a rhombus.

Prove Conditions for Rhombuses

The theorems below can be used to determine whether a parallelogram is a rhombus.

Theorems: Conditions for Rhombuses
If one pair of consecutive sides of a parallelogram are congruent, then the parallelogram is a rhombus.
If the diagonals of a parallelogram are perpendicular, then the parallelogram is a rhombus.
If one diagonal of a parallelogram bisects a pair of opposite angles, then the parallelogram is a rhombus.

3 Shown below are the randomly ordered steps for proving that *If one pair of consecutive sides of a parallelogram are congruent, then the parallelogram is a rhombus.*

Given: *EFGH* is a parallelogram and $\overline{EH} \cong \overline{FE}$.

Prove: *EFGH* is a rhombus. A–B. See Additional Answers.

A. Write the statements in the correct order.

B. Write the reasons in the correct order.

A. Statements	B. Reasons
$\overline{EH} \cong \overline{FG}$, $\overline{FE} \cong \overline{GH}$	Transitive Property of Congruence
$\overline{EH} \cong \overline{GH} \cong \overline{FG} \cong \overline{FE}$	Given
EFGH is a rhombus.	Opposite sides of a parallelogram are congruent.
EFGH is a parallelogram and $\overline{EH} \cong \overline{FE}$.	A quadrilateral is a rhombus if and only if it has four congruent sides.

Turn and Talk Describe how you would prove that if the diagonals of a parallelogram are perpendicular, then the parallelogram is a rhombus. See margin.

Apply Conditions for Special Parallelograms

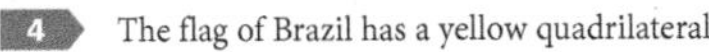

4 The flag of Brazil has a yellow quadrilateral.

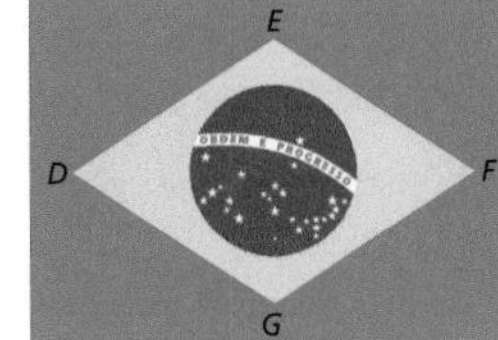

Use the given information to determine whether the conclusion about the flag of Brazil is valid.

Given: $\overline{GE}$ and $\overline{DF}$ bisect each other, and $\overline{GE}$ bisects $\angle DGF$ and $\angle DEF$.

Conclusion: *DEFG* is a rhombus.

Step 1: Determine if *DEFG* is a parallelogram.

A. Why do you need to determine if the quadrilateral is a parallelogram? A. See below.

Because $\overline{GE}$ and $\overline{DF}$ bisect each other, *DEFG* is a parallelogram by the Diagonals of a Parallelogram Theorem.

Step 2: Determine if *DEFG* is a rhombus.

Because $\overline{GE}$ bisects $\angle DGF$ and $\angle DEF$, *DEFG* is a rhombus.

B. Why can this conclusion be made? B. See below.

Identify Special Parallelograms in the Coordinate Plane

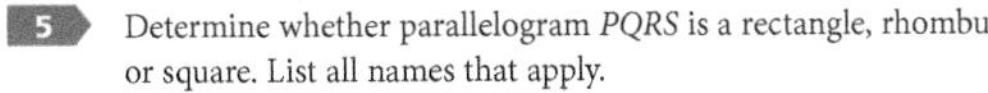

5 Determine whether parallelogram *PQRS* is a rectangle, rhombus, or square. List all names that apply.

Step 1: Determine if *PQRS* is a rectangle.

$PR = \sqrt{(-2-4)^2 + (3-1)^2} = \sqrt{36+4} = \sqrt{40}$

$QS = \sqrt{(0-2)^2 + (5-(-1))^2} = \sqrt{4+36} = \sqrt{40}$

A. Explain why *PQRS* is a rectangle.
PQRS is a rectangle because the diagonals are congruent.

Step 2: Determine if *PQRS* is a rhombus.

Slope of $\overline{PR} = \frac{y_2 - y_1}{x_2 - x_1} = \frac{1-3}{4-(-2)} = -\frac{1}{3}$

Slope of $\overline{QS} = \frac{y_2 - y_1}{x_2 - x_1} = \frac{5-(-1)}{0-2} = -3$

B. Explain why *PQRS* is a not a rhombus. *PQRS* is not a rhombus because the diagonals are not perpendicular. The slopes of the diagonals are not opposite reciprocals.

C. Explain how you know that *PQRS* is a not a square.
A square is a parallelogram that is a rectangle and a rhombus. *PQRS* is not a rhombus, so it cannot be a square.

Turn and Talk Determine the coordinates of a parallelogram that is a rhombus but not a rectangle. Explain how you know. See margin.

A. The theorems for conditions of rhombuses require that the quadrilateral be a parallelogram.

B. It applies the condition of a rhombus that states if one diagonal of a parallelogram bisects a pair of opposite angles, then the parallelogram is a rhombus.

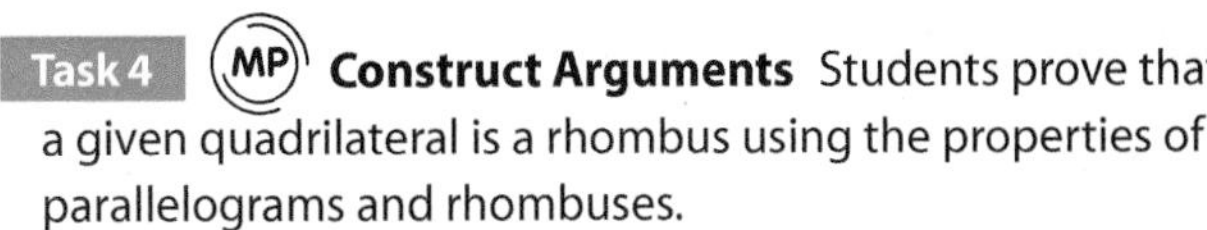

Task 4 (MP) **Construct Arguments** Students prove that a given quadrilateral is a rhombus using the properties of parallelograms and rhombuses.

SUPPORT SENSE-MAKING Three Reads

Have students read the problem three times. Use the questions below for a different focus each read.

1. What is the situation about?
2. What are the quantities in the situation?
3. What are the possible mathematical questions that you could ask for the situation?

Task 5 (MP) **Construct Arguments** Students will determine if a parallelogram on a coordinate plane is a rectangle, rhombus, or square using the Distance Formula to confirm congruent diagonals and the formula for slope to disprove that the diagonals are perpendicular.

Sample Guided Discussion:

Q What must you prove about the diagonals of a parallelogram to show it is a rectangle? You must prove the diagonals are congruent.

Q What is another measure, other than slope, you can use to show that *PQRS* is not a rhombus? Possible answer: You can use the Distance Formula to find the lengths of the four sides of the figure and show that they are not congruent.

Turn and Talk Suggest that students pick an initial ordered pair for the first vertex of the rhombus, draw one side, (noting its coordinates) and then use the symmetry of the rhombus to find the remaining sides. For example, if they start at $(0, 0)$ and their next vertex is at $(m, 1)$, by symmetry the other vertices will be at $(m, -1)$ and $(2m, 0)$. Answers will vary. Possible answer: $A(0, 2)$, $B(3, 4)$, $C(6, 2)$, $D(3, 0)$; The diagonals are not equal. $AC = 6$ and $BD = 4$, so *ABCD* cannot be a rectangle.

ONLINE

Assign the Digital On Your Own for
- built-in student supports
- Actionable Item Reports
- Standards Analysis Reports

On Your Own

Assignment Guide

The chart below indicates which problems in the On Your Own are associated with each task in the Learn Together. Assign daily homework for tasks completed.

Learn Together Tasks	On Your Own Problems
Task 1, p. 346	Problems 7, 8, and 18
Task 2, p. 347	Problems 10, 13, and 17
Task 3, p. 348	Problems 9 and 12
Task 4, p. 349	Problems 11 and 14
Task 5, p. 349	Problems 15, 16, and 19–23

Check Understanding

Copy and complete each theorem.

1. If the diagonals of a parallelogram are perpendicular, then the parallelogram is a __?__. rhombus
2. If one angle of a parallelogram is a right angle, then the parallelogram is a __?__. rectangle
3. If one diagonal of a parallelogram bisects a pair of opposite angles, then the parallelogram is a __?__. rhombus

Each quadrilateral is a parallelogram. Determine whether or not you have enough information to conclude that each parallelogram is a rectangle. Explain your reasoning.

4. 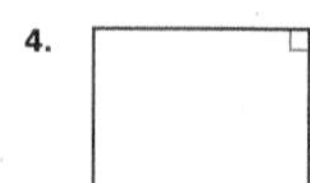

5. 4, 5. See Additional Answers.

6. Parallelogram $WXYZ$ has vertices at $W(-1, 2)$, $X(2, -1)$, $Y(-1, -4)$, $Z(-4, -1)$. Determine whether $WXYZ$ is a rectangle, rhombus, or square. List all names that apply. rectangle, rhombus, square

On Your Own

Each quadrilateral is a parallelogram. Determine whether or not you have enough information to conclude that each parallelogram is a rhombus. Explain your reasoning.
7–10. See Additional Answers.

7. 

8. 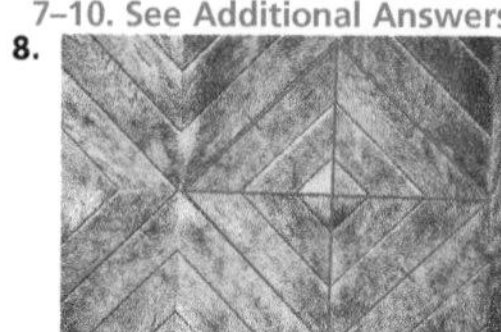

9. Prove that if one diagonal of a parallelogram bisects a pair of opposite angles, then the parallelogram is a rhombus.

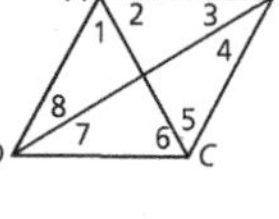

Given: $ABCD$ is a parallelogram.
$\overline{AC}$ bisects $\angle DAB$ and $\angle DCB$.
$\overline{BD}$ bisects $\angle ADC$ and $\angle ABC$.
Prove: $ABCD$ is a rhombus.

10. Prove that if one angle of a parallelogram is a right angle, then the parallelogram is a rectangle.

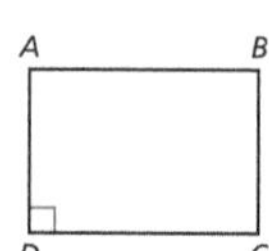

Given: $ABCD$ is a parallelogram. $\angle D$ is a right angle.
Prove: $ABCD$ is a rectangle.

350

3 Check Understanding

Formative Assessment

Use formative assessment to determine if your students are successful with this lesson's learning objective.

Students who successfully complete the Check Understanding can continue to the On Your Own practice.

For students who miss 1 or more problem, work in a pulled small group using the Almost There small-group activity on page 345C.

ONLINE

Assign the Digital Check Understanding to determine
- success with the learning objective
- items to review
- grouping and differentiation resources

4 Differentiation Options

Differentiate instruction for all students using small-group activities and math center activities on page 345C.

Reteach

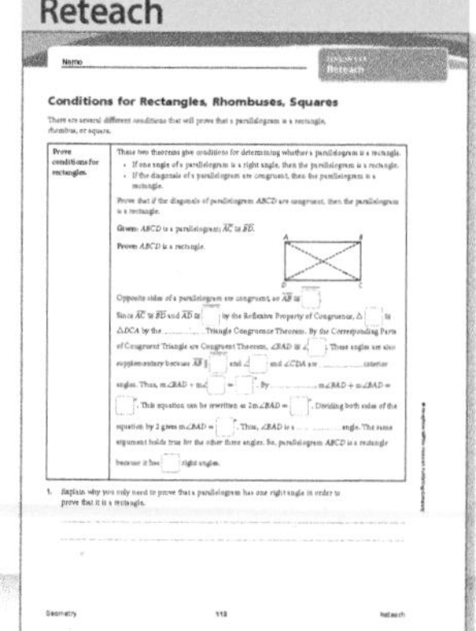

Conditions for Rectangles, Rhombuses, Squares

Challenge

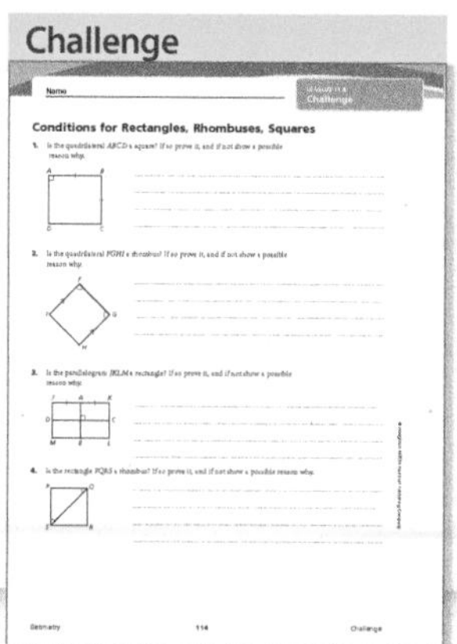

Conditions for Rectangles, Rhombuses, Squares

11. **Critique Reasoning** Lindsey believes she is given enough information to prove that parallelogram *QRST* is a square. Greg disagrees. Who is correct? Explain your reasoning.
11–14. See Additional Answers.

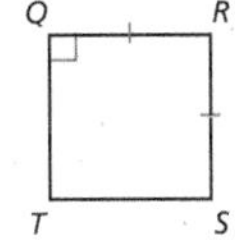

12. Darren has a piece of fabric shaped like *WXYZ*. He makes the following conclusion about the fabric. Determine if the conclusion is valid. If not, tell what additional information is needed to make the conclusion valid.

Given: $\overline{XZ}$ bisects $\angle WXY$ and $\angle WZY$.

Conclusion: *WXYZ* is a rhombus.

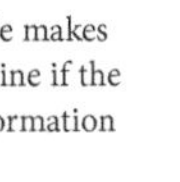

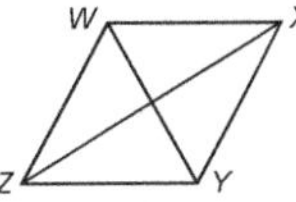

13. A surveyor is mapping a property and determines that the property is a parallelogram. Determine if the conclusion the surveyor makes is valid. If not, tell what additional information is needed to make the conclusion valid.

Given: *KLMN* is a parallelogram. $\overline{KP} \cong \overline{LP}$

Conclusion: *KLMN* is a rectangle.

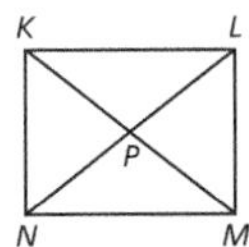

14. A soccer player practices by kicking a ball at a parallelogram drawn on the side of a building. The sides are drawn so that $\overline{PQ} \cong \overline{RS}$ and $\overline{QR} \cong \overline{PS}$. A target is placed at point *Z*, so that *PZ*, *QZ*, *RZ*, and *SZ* are all equal lengths. Why must the parallelogram be a rectangle?

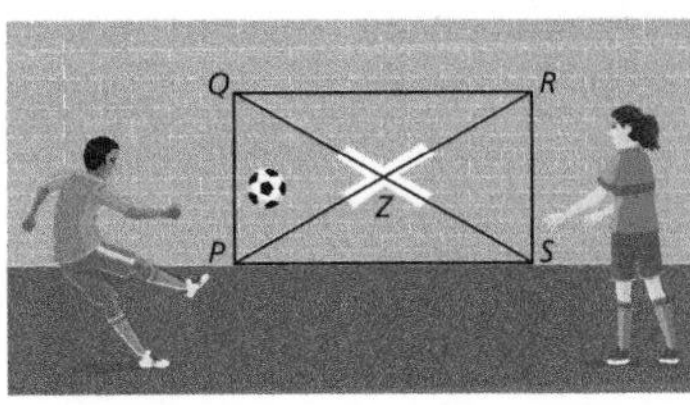

Determine whether the parallelogram is a rectangle, rhombus, or square. List all names that apply.

15.

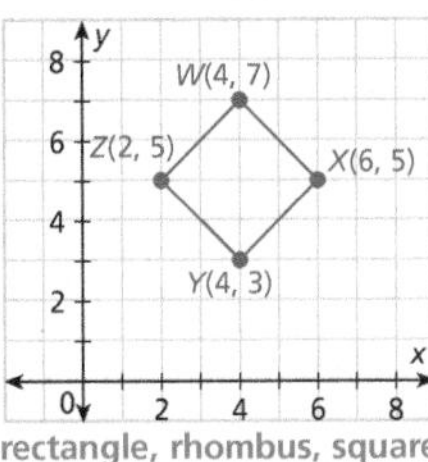

rectangle, rhombus, square

16.

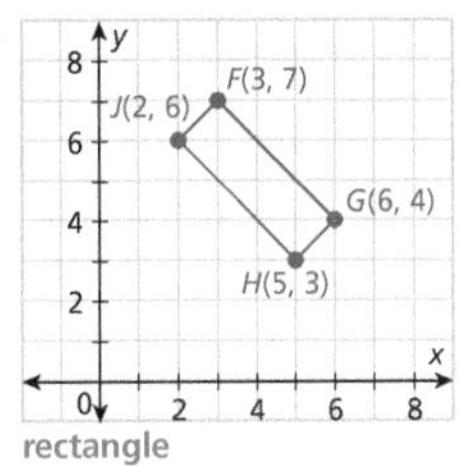

rectangle

17. **Construct Arguments** Use the rotational symmetry of a square to justify the relationships between the diagonals of the square. 17, 18. See Additional Answers.

18. **Attend to Precision** Leon needs to cut a square out of a circular piece of paper. He wants the square to be as large as possible. Sketch how he can make sure that he cuts the largest possible square.

Questioning Strategies

Problem 13 Suppose you are given the additional information that $\angle KPL$ is right. What could you conclude about *KLMN*? You could conclude that *KLMN* is a square because the diagonals are perpendicular.

Watch for Common Errors

Problem 15 Remind students of the hierarchy of quadrilaterals. All squares are rectangles, but not all rectangles are squares. Likewise, all squares are rhombuses, but not all rhombuses are squares. Students may only use one label for the figure. Make sure they understand that figures can have more than one name.

5 Wrap-Up

Summarize learning with your class. Consider using the Exit Ticket, Put It in Writing, or I Can scale.

Exit Ticket 

Determine whether the parallelogram *ABCD* with vertices $A(0, 0)$, $B(4, 5)$, $C(9, 9)$ and $D(5, 4)$ is a rectangle, rhombus, or square. List all names that apply.

rhombus; It is not a rectangle because the diagonals are not congruent. It is not a square because the diagonals are not congruent.

Put It in Writing 

Students should create a graphic organizer that includes the definitions for rectangle, rhombus, and square, the properties of the diagonals of each, and theorems proved in class or for homework about each.

I Can 

The scale below can help you and your students understand their progress on a learning goal.

4	I can prove and use conditions for rectangles, rhombuses, and squares, and I can explain my reasoning to others.
3	I can prove and use conditions for rectangles, rhombuses, and squares.
2	I can state but not prove theorems on conditions for rectangles and for rhombuses.
1	I can identify a rectangle, rhombus, or square based on what I know about its diagonals.

Spiral Review • Assessment Readiness

These questions will help determine if students have retained information taught in the past and can also prepare them for high-stakes assessments. Here, students must state conditions for parallelograms **(11.2)**, use properties of parallelograms to find unknown lengths **(11.1)**, use the Midpoint Formula **(1.1)**, and use the Pythagorean Theorem to find the length of a side of a right triangle. **(Gr8, 11.3)**.

Find the value of x that makes each parallelogram the given type.

19. rhombus 17

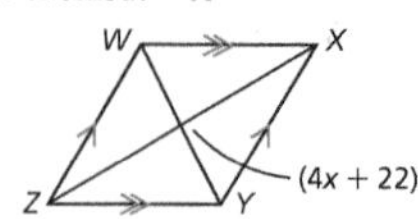

20. square 5

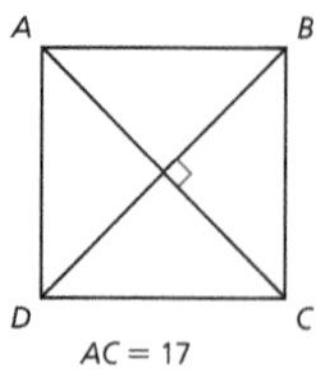

$AC = 17$
$BD = 4x - 3$

21. rhombus 7

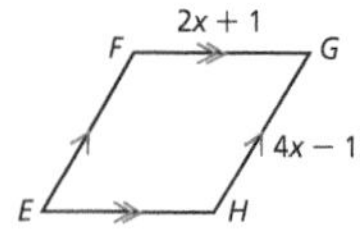

22. rectangle 9

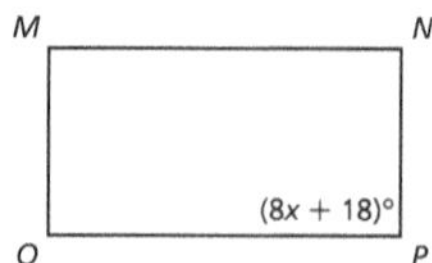

23. Open Middle™ Using the digits 1 to 9, at most one time each, fill in the boxes to create coordinates of the vertices of a square.

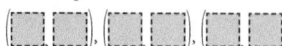

There are several possibilities, including: (1, 6), (3, 5), (4, 7), (2, 8).

Spiral Review • Assessment Readiness

24. Which statement guarantees that *ABCD* is a parallelogram?

Ⓐ Opposite sides $\overline{AB}$ and $\overline{CD}$ are congruent.
Ⓑ Diagonals $\overline{AC}$ and $\overline{BD}$ bisect each other.
Ⓒ $\overline{AB}$ and $\overline{BC}$ are perpendicular.
Ⓓ Opposite sides $\overline{BC}$ and $\overline{AD}$ are parallel.

25. *EFGH* is a parallelogram with $EF = 4x + 11$, $FG = 3y + 15$, $HG = 7x - 1$, and $EH = 5y - 1$. Find the length of $\overline{HG}$.

Ⓐ 4 Ⓒ 27
Ⓑ 8 Ⓓ 39

26. The point $(4, 8)$ is the midpoint of which set of ordered pairs?

Ⓐ $(5, 9), (3, 10)$ Ⓒ $(3, 5), (5, 11)$
Ⓑ $(-4, -8), (8, 4)$ Ⓓ $(2, 6), (6, 2)$

27. Find the length of $\overline{SU}$.

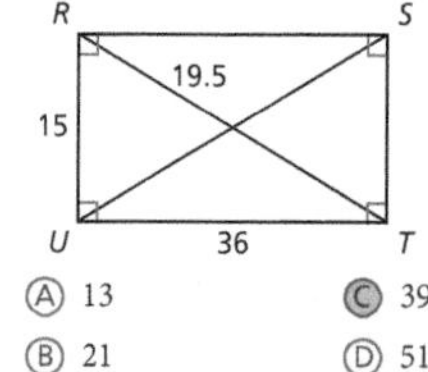

Ⓐ 13 Ⓒ 39
Ⓑ 21 Ⓓ 51

I'm in a Learning Mindset!

How did I evaluate opposing views about conditions for rectangles, rhombuses, and squares?

352 Keep Going Journal and Practice Workbook

Learning Mindset

Resilience Adjusts to Change

Students need to be flexible in their methods for reasoning about quadrilaterals. It is possible for a student to, for example, look at a rectangle with length five times its width, and immediately conclude that it is not a square based on observation. However, they need to be reminded that in this lesson they are making those determinations based on the congruence and/or perpendicularity of the diagonals which may be a new approach to defining quadrilaterals. *Have I used my knowledge of geometry and algebra, including the properties of parallelograms, to complete my proofs? Have I listed a sound mathematical reason to justify each statement in my proof? Have I answered questions using appropriate formulas and shown my work, and not made visual determinations to support my answer?*

11.5 Properties and Conditions for Trapezoids and Kites

LESSON FOCUS AND COHERENCE

Mathematics Standards

- Given a rectangle, parallelogram, trapezoid, or regular polygon, describe the rotations and reflections that carry it onto itself.

Mathematical Practices and Processes 

- Look for and make use of structure.
- Construct viable arguments and critique the reasoning of others.
- Attend to precision.
- Reason abstractly and quantitatively.

I Can Objective 

I can prove and apply theorems about trapezoids and kites.

Learning Objective

Use the exclusive and inclusive definitions of kite and trapezoid to classify quadrilaterals, prove and apply theorems about angle measures and diagonals of kites, prove and apply theorems about conditions on a trapezoid equivalent to it being isosceles, and apply the Trapezoid Midsegment Theorem.

Language Objective

Explain how different types of quadrilaterals can be considered kites or trapezoids.

Vocabulary

New: base angles of a trapezoid, bases of a trapezoid, isosceles trapezoid, kite, legs of a trapezoid, midsegment of a trapezoid, trapezoid

Mathematical Progressions

Prior Learning	Current Development	Future Connections
Students: • proved and used properties of triangles. **(9.1–9.5)** • proved properties of parallelograms. **(11.1)**	**Students:** • classify quadrilaterals under the inclusive and exclusive definitions of kite and trapezoid. • prove and apply theorems about kites. • prove and apply theorems about isosceles trapezoids. • apply the Trapezoid Midsegment Theorem.	**Students:** • will construct inscribed quadrilaterals and identify properties of their angles. **(15.2)**

PROFESSIONAL LEARNING

Using Mathematical Practices and Processes

Reason abstractly and quantitatively.

This lesson provides an opportunity to address this Mathematical Practice Standard. According to the standard, mathematically proficient students demonstrate an ability "to decontextualize—to abstract a given situation and represent it symbolically." In this lesson, students are reintroduced to kites and trapezoids and asked to prove and apply theorems about the figures. As students construct arguments to prove the theorems, they extend their reasoning from making observations of measures of parts of special cases to making more general arguments. For example, in Exercise 27 in the On Your Own exercises, students are asked to prove that the theorem holds for any trapezoid. Doing so requires them to represent geometric objects symbolically.

WARM-UP OPTIONS

ONLINE

PROJECTABLE & PRINTABLE

ACTIVATE PRIOR KNOWLEDGE • Apply Theorems About Parallelograms

Use these activities to quickly assess and activate prior knowledge as needed.

Problem of the Day 

A platform lift has struts that cross to create congruent rhombuses that change shape as the lift is raised and lowered. What are the measures of the angles of each rhombus when one angle measures 50°? A rhombus is a parallelogram, and opposite angles in a parallelogram are congruent. So, two of the angles have a measure of 50°. The sum of the interior angles of each parallelogram is 360°, so the sum of the other two angles' measures is $360° - 2 \cdot 50° = 260°$. Since they are congruent, their measures are each $\frac{260°}{2} = 130°$.

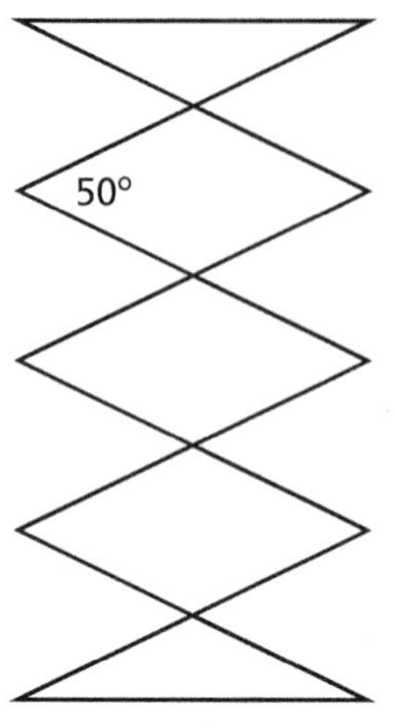

Quick Check for Homework

As part of your daily routine, you may want to display the Teacher Solution Key to have students check their homework.

Make Connections

Based on students' responses to the Problem of the Day, choose one of the following:

1. Project the Interactive Reteach, Geometry, Lesson 11.1.
2. Complete the Prerequisite Skills Activity:

Have students work in small groups. Give each group a set of index cards. Each card should have on it a parallelogram with the lengths of two noncongruent sides and an angle measure indicated. Have students work together to find the missing measures.

- *How are opposite sides of a parallelogram related?* They are congruent.
- *How are opposite angles of a parallelogram related?* They are congruent.
- What do you know about the sum of the angle measures of a parallelogram? It is 360° since a parallelogram is a quadrilateral.

SHARPEN SKILLS

If time permits, use this on-level activity to build fluency and practice basic skills.

Mental Math

Objective: Students identify congruent parts of quadrilaterals.
Materials: index cards

Have students work in small groups. Give each group a set of index cards. Each card should have on it a quadrilateral *ABCD* that is indicated with markings to be a parallelogram, a rectangle, a rhombus, or a square. Have students take turns drawing a card and stating which parts are congruent.

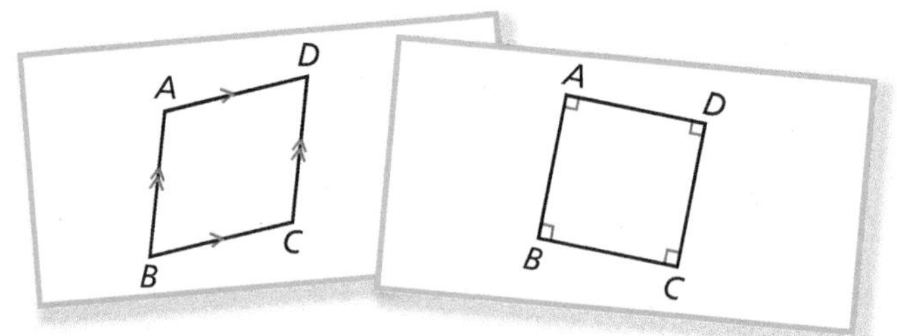

PLAN FOR DIFFERENTIATED INSTRUCTION

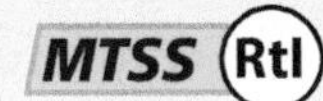

Small-Group Options

Use these teacher-guided activities with pulled small groups.

On Track

Materials: item

Give each group a set of index cards. Each index card should have on it a trapezoid with either:

- legs indicated congruent,
- a pair of base angles indicated congruent, or
- diagonals indicated congruent.

The measure of one angle, one diagonal, and one leg should be given. Have students take turns drawing a card and giving all missing measures.

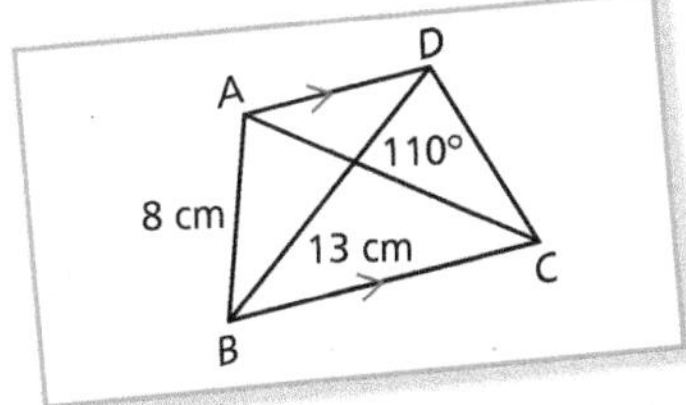

Almost There

Materials: index cards

Give each group a set of index cards. Each index card should have on it a quadrilateral with some sides indicated congruent and/or parallel. Make sure some of the quadrilaterals are squares, rhombuses, rectangles, or parallelograms. Have students take turns drawing a card and telling whether the figure is:

- a kite by the inclusive definition,
- a kite by the exclusive definition,
- a trapezoid by the inclusive definition, or
- a trapezoid by the exclusive definition.

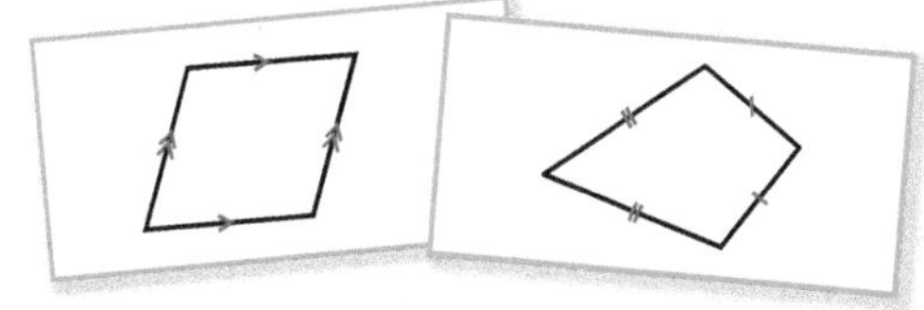

Ready for More

In the Turn and Talk for Task 3, students explained how the measure of every angle of an isosceles trapezoid is determined by the measure of one of the angles. Ask students to work together to make at least two other arguments that do not directly use the fact that the sum of all of the angle measures is 360°.

- If students need prompting, have them construct altitudes in the trapezoid and think about the angle measures of the resulting triangles and rectangle.
- If students need more prompting, have them sketch the diagonals of the trapezoid, think about how they can find pairs of congruent triangles, and then apply CPCTC to draw conclusions.

Math Center Options

Use these student self-directed activities at centers or stations.

Key: ● Print Resources ● Online Resources

On Track

- ● Interactive Digital Lesson
- ●● Journal and Practice Workbook
- ● Interactive Glossary (printable): **base angles of a trapezoid**, **bases of a trapezoid**, **isosceles trapezoid**, **kite**, **legs of a trapezoid**, **midsegment of a trapezoid**, **trapezoid**
- ● Module Performance Task
- ●● Standards Practice: Prove Theorems About Parallelograms

Almost There

- ● Reteach 11.5 (printable)
- ● Interactive Reteach 11.5
- ● Illustrative Mathematics: What is a Trapezoid? (Part 1)

Ready for More

- ● Challenge 11.5 (printable)
- ● Interactive Challenge 11.5

Unit Project Check students' progress by asking to see their equation and by asking for an explanation of which property of trapezoids they are using.

View data-driven grouping recommendations and assign differentiation resources.

During the *Spark Your Learning,* listen and watch for strategies students use. See samples of student work on this page.

Apply CPCTC

Strategy 1

$\triangle GFE$ and $\triangle GHE$ are congruent, so by CPCTC, $\angle F$ and $\angle H$ are congruent.

If students . . . apply CPCTC, they are demonstrating an understanding that they can conclude congruent parts of congruent triangles are congruent from previous triangle work.

Have these students . . . explain how they determined the angles are corresponding and congruent. **Ask:**

- Q How did you know that $\angle F$ and $\angle H$ are corresponding angles?
- Q What allows you to conclude that the angles are congruent?

Sketch and Measure a Reproduction

Strategy 2

I sketched a triangle $\triangle GFE$, traced it, and cut out the traced triangle. I then flipped the triangle and retraced it to draw $\triangle GHE$ since I knew the triangles are congruent. I then measured $\angle F$ and $\angle H$ with a protractor and found that they are congruent.

If students . . . sketch and measure a reproduction of the kite, they understand how the kite is defined with two congruent triangles, but may not understand that they can relate corresponding parts of the triangle without measuring them.

Activate prior knowledge . . . by having students use CPCTC. **Ask:**

- Q Which angle of $\triangle GHE$ corresponds with $\triangle GFE$?
- Q What can you conclude about corresponding parts of congruent triangles?

COMMON ERROR: Incorrectly Identifies Corresponding Angles

$\triangle GFE$ and $\triangle GHE$ are congruent, so by CPCTC, $\angle G$ and $\angle E$ are congruent.

If students . . . incorrectly identify corresponding angles, then they may not understand how to use congruent parts of triangles to identify corresponding parts.

Then intervene . . . by pointing out that $\angle G$ and $\angle E$ are not parts of either triangle, let alone included between corresponding pairs of sides. **Ask:**

- Q Which side of $\triangle GHE$ corresponds with $\overline{GF}$? with $\overline{EF}$?
- Q How can you use pairs of corresponding sides to help identify corresponding angles?

11.5

Properties and Conditions for Trapezoids and Kites

I Can prove and apply theorems about trapezoids and kites.

Spark Your Learning

Anita is flying a kite at the beach.

Complete Part A as a whole class. Then complete Parts B–D in small groups.

A. What is a geometric question you can ask about this situation?
How are the angles of the kite related?

B. What information would you need to know to answer your question?
You need to know how the triangles within the kite are related.

C. To answer your question, what strategy and tool would you use along with all the information you have? What answer do you get?
See Strategies 1 and 2 on the facing page.

D. Use a different method to answer the question. Explain your reasoning.
D. You can show that the triangles are congruent using SSS and then show that the angles are congruent by CPCTC.

Turn and Talk Is it possible that the kite shown is a parallelogram? Explain why or why not. See margin.

CULTIVATE CONVERSATION • Co-Craft Questions

If students have difficulty formulating a mathematical question about the situation in the Spark Your Learning, ask them to consider the shape and parts of the kite. What are some natural questions to ask about this situation?

Work together to craft the following questions:

- What kind of parts does the quadrilateral formed by the kite have?
- How are the parts of the kite related?
- How are the angles of the kite related?

Then have students think about what additional information, if any, they would need to answer these questions. **Ask:**

- Can you determine how the angles of the kite are related from the shape of the kite? Why or why not?
- Can you determine how the angles of the kite are related if you know how some of the side lengths are related? Explain.

1 Spark Your Learning

▶ MOTIVATE

- Have students look at the photo in their books and read the information contained in the photo. Then complete Part A as a whole-class discussion.
- Give the class the additional information they need to solve the problem. This information is available online as a printable and projectable page in the Teacher Resources.
- Have students work in small groups to complete Parts B–D.

▶ PERSEVERE

If students need support, guide them by asking:

Q **Advancing • Use Tools** Which tool could you use to solve the problem? Why choose that tool and not some other? Students' choices of tools and reasons for choosing them will vary.

Q **Assessing** How are $\triangle GFE$ and $\triangle GHE$ related? The triangles are congruent.

Q **Assessing** Which sides of $\triangle GFE$ and $\triangle GHE$ are corresponding? $\overline{GF}$ corresponds with $\overline{GH}$, $\overline{FE}$ corresponds with $\overline{HE}$, and $\overline{EG}$ corresponds with $\overline{EG}$.

Q **Advancing** Can you find a pair of corresponding angles of $\triangle GFE$ and $\triangle GHE$ that are also angles of the kite? $\angle F$ and $\angle H$ are corresponding angles.

Turn and Talk Help students use known facts about parallelograms to deduce whether or not the kite could also be a parallelogram. Remind them to consider the relative positions of congruent sides and angles. If students are stuck and need prompting, ask them to consider the relationship between the diagonals of the kite and what they know about the diagonals of a parallelogram. no; The kite has two pairs of congruent sides and a pair of congruent opposite angles. However, the congruent sides of the kite are adjacent, not opposite as in a parallelogram. Also, the kite does not have parallel opposite sides.

▶ BUILD SHARED UNDERSTANDING

Select groups of students who used various strategies and tools to share with the class how they solved the problem. As they present their solutions, have each group discuss why they chose a specific strategy and tool.

2 Learn Together

Build Understanding

Task 1 (MP) **Use Structure** Students are introduced to the inclusive and exclusive definitions of a kite and trapezoid. Emphasize to students that while figures may have different names, students should be careful not to assume that definitions are mutually distinct. This is especially true when considering inclusive definitions.

CONNECT TO VOCABULARY

Have students use the **Interactive Glossary** to record their understanding of the vocabulary in this task.

Sample Guided Discussion:

Q **In Part D, is it likely that more figures satisfy the exclusive definition of trapezoid, or fewer?** Fewer, since the exclusive definition is more restrictive. This means it is possible for a figure to satisfy the inclusive definition but not the exclusive definition.

Turn and Talk Clarify to students that pairs of sides of a quadrilateral need not be disjoint. For example, given quadrilateral *ABCD*, $\overline{AB}$, $\overline{BC}$ and $\overline{BC}$, $\overline{CD}$ are distinct pairs of consecutive sides even though they share $\overline{BC}$. This is exclusive; By this definition, a rhombus is not a kite. It has 4 pairs of consecutive congruent sides.

Build Understanding

Understand Definitions of Kites and Trapezoids

In this lesson, you will learn about two special quadrilaterals, **kites** and **trapezoids**. Inclusive definitions for kite and trapezoid allow parallelograms to share the properties of kites and trapezoids. Exclusive definitions for kite and trapezoid do not allow parallelograms to share the properties of kites and trapezoids.

Definitions for Kite

Inclusive definition: A kite is a quadrilateral whose four sides can be grouped into two pairs of consecutive congruent sides.

Exclusive definition: A kite is a quadrilateral with two pairs of consecutive congruent sides, but opposite sides are not congruent.

Definitions for Trapezoid

Inclusive definition: A trapezoid is a quadrilateral with at least one pair of parallel sides.

Exclusive definition: A trapezoid is a quadrilateral with exactly one pair of parallel sides.

In this book, kite and trapezoid are defined using the inclusive definitions. That means a rhombus can be classified as a kite and as a trapezoid.

1 Consider the quadrilaterals shown.

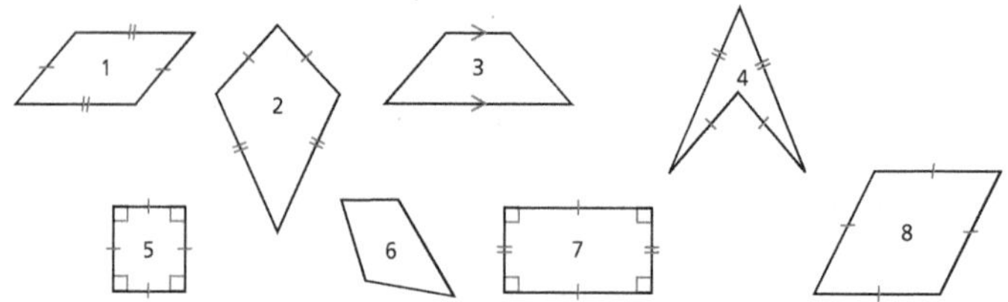

A. Which shape(s) are kites using the inclusive definition? 2, 4, 5, 8

B. Which shape(s) are kites using the exclusive definition? 2, 4

C. Which shape(s) are trapezoids using the inclusive definition? 1, 3, 5, 7, 8

D. Which shape(s) are trapezoids using the exclusive definition? 3

Turn and Talk Another definition for kite is as follows: A kite is a quadrilateral with exactly two pairs of congruent consecutive sides. Is this an inclusive definition or exclusive definition? Explain. See margin.

354

LEVELED QUESTIONS

Depth of Knowledge (DOK)	Leveled Questions	What Does This Tell You?
Level 1 **Recall**	Why is a parallelogram not a trapezoid, using the exclusive definition? A trapezoid has exactly one pair of parallel sides, but a parallelogram has two.	Students' answers will indicate whether they understand the exclusive definition of a trapezoid.
Level 2 **Basic Application of Skills & Concepts**	Is a square a kite using both definitions? a trapezoid using both definitions? A square is a kite and a trapezoid using the inclusive definitions, but it is neither a kite nor a trapezoid using their exclusive definitions.	Students' answers will indicate whether they understand and can apply the inclusive and exclusive definitions of both figures.
Level 3 **Strategic Thinking & Complex Reasoning**	Can you reason and attempt to classify all quadrilaterals that are both trapezoids and kites using their inclusive definitions? I think that every such quadrilateral is a rhombus.	Students' answers will indicate whether they can completely describe the intersection of both sets of figures using their inclusive definitions.

Step It Out

Theorems About Kites

Consider the following theorems.

Theorems About Kites	
Diagonals of a Kite Theorem If a quadrilateral is a kite, then its diagonals are perpendicular. If *ABCD* is a kite, then $\overline{AC} \perp \overline{BD}$.	B, A, C, D
Opposite Angles of a Kite Theorem If a quadrilateral is a kite, then at least one pair of opposite angles are congruent. If *ABCD* is a kite, then $\angle B \cong \angle D$.	B, A, C, D

Shown below are the randomly ordered steps for proving that *If a quadrilateral is a kite, then at least one pair of opposite angles are congruent.*

Given: $\overline{BC} \cong \overline{DC}$ and $\overline{AB} \cong \overline{AD}$.
Prove: $\angle B \cong \angle D$

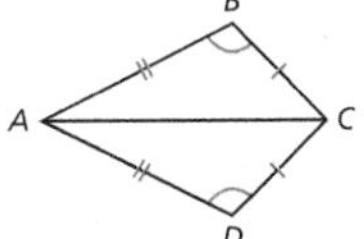

A–C. See Additional Answers.

A. Write the statements in the correct order.

B. Write the reasons in the correct order.

Statements	Reasons
$\triangle ABC \cong \triangle ADC$	Corresponding parts of congruent figures are congruent.
$\overline{AC} \cong \overline{AC}$	Given
$\overline{BC} \cong \overline{DC}$ and $\overline{AB} \cong \overline{AD}$.	SSS Triangle Congruence Theorem
$\angle B \cong \angle D$	Reflexive Property of Congruence

C. Explain how to change the proof to show that $\overline{AC} \perp \overline{BD}$. What general result does your proof verify?

Turn and Talk Using the inclusive definition for kite, what type of quadrilateral is a kite with both pairs of opposite angles congruent? See margin.

PROFICIENCY LEVEL

Beginning

Show students a kite *ABCD* with $\overline{AB} \cong \overline{AD}$ and $\overline{BC} \cong \overline{DC}$. Say, "$\angle B$ is congruent to $\angle D$, and $\overline{AC}$ is perpendicular to $\overline{BD}$." Show students another kite and ask them to verbally identify congruent angles and perpendicular segments.

Intermediate

Have students work in pairs. Have one student sketch a kite with labeled vertices and congruent sides indicated. The other student will then briefly explain which other parts of the kite are congruent and why they believe this is the case.

Advanced

Have students work in small groups. Show students a kite *ABCD* with $\overline{AB} \cong \overline{AD}$ and $\overline{BC} \cong \overline{DC}$. Have groups work together to determine which angles they think are congruent and give a verbal explanation justifying their reasoning. After they have done this, have them work together to build a verbal explanation of why they think the diagonals are perpendicular. Encourage students to be precise in their language and to use correct mathematical terminology.

Step It Out

Task 2 (MP) **Construct Arguments** Students are presented with two theorems about kites and arrange given arguments in a logical order to prove one of the theorems. They then modify the proof to prove the other theorem. When students are proving the Diagonals of a Kite Theorem, make sure they are precise in their arguments. While $\angle BCA$ and $\angle DCA$ are the same angles as two angles of smaller triangles formed when the diagonal $\overline{BD}$ is drawn, students should use the intersection point of the diagonals instead of *A* in referring to the angles when justifying SAS congruence.

Sample Guided Discussion:

Q In Part B, which property justifies a geometric figure being congruent to itself? the Reflexive Property of Congruence

Q In Part C, how can you relate $\angle BCA$ and $\angle DCA$? They are congruent by CPCTC.

Q In Part C, call the intersection of $\overline{AC}$ and $\overline{BD}$ point *E*. How can you relate $\angle BCE$ and $\angle DCE$? They are congruent since $\angle BCA = \angle BCE$ and $\angle DCE = \angle DCE$.

Q In Part C, can you find two pairs of corresponding congruent sides in $\triangle BCE$ and $\triangle DCE$? $\overline{BC} \cong \overline{DC}$ and $\overline{CE} \cong \overline{CE}$.

Turn and Talk As with the Turn and Talk from Task 1, make sure students understand that pairs of sides of a quadrilateral need not be disjoint. In order to better understand the question, students may want to sketch multiple quadrilaterals that vary as much as possible while still satisfying the given conditions. Encourage students to indicate congruent parts on their sketches to assist in identifying the quadrilateral. Remind them that their sketches need not be to scale. rhombus

Task 3 (MP) **Attend to Precision** Students are introduced to theorems involving trapezoids and use one to justify a claim that a trapezoid is isosceles. They then solve for measures of the trapezoid with this result. Make sure students accurately cite the Base Angles of a Trapezoid Theorem when prompted to explain how they know the side lengths in question are equal. Specifically, students should be able to say something equivalent to, "If a trapezoid has one pair of congruent base angles, then the trapezoid is isosceles."

CONNECT TO VOCABULARY

Have students use the **Interactive Glossary** to record their understanding of the vocabulary in this task.

Sample Guided Discussion:

Q What kind of trapezoid do you know *ABCD* is since $\angle A \cong \angle D$? I know that *ABCD* is an isosceles trapezoid.

Q What is an isosceles trapezoid? a trapezoid whose legs are congruent but not parallel

Q Which sides of *ABCD* are congruent? $\overline{AB} \cong \overline{CD}$

Q Can you conclude anything about *AC* and *BD*? I know that $AC = BD$ since the Diagonals of an Isosceles Trapezoid Theorem tells me that diagonals of an isosceles trapezoid are congruent.

Turn and Talk If necessary, remind students that the sum of the interior measures of a quadrilateral is 360° and that they can apply the Isosceles Trapezoid Theorem to conclude that the trapezoid has two pairs of congruent angles. While they may want to begin by considering a special case (for example, a base angle has measure 70°), make sure students are extending their reasoning to the general case. It may be helpful to denote the known angle measure with a variable. You know there are two pairs of congruent angles. You can subtract the known measure from 360° twice, then divide the difference by 2 to find the measure of the other two angles.

Theorems About Trapezoids

An **isosceles trapezoid** is a trapezoid in which the legs are congruent but not parallel. The **bases of a trapezoid** are the sides of the trapezoid that are parallel. The **base angles of a trapezoid** are a pair of consecutive angles whose common side is a base of the trapezoid. The **legs of a trapezoid** are the sides that are not the bases.

Theorems About Trapezoids	
Base Angles of a Trapezoid Theorem If a trapezoid has one pair of congruent base angles, then the trapezoid is isosceles. If *ABCD* is a trapezoid and $\angle C \cong \angle D$, then *ABCD* is an isosceles trapezoid.	A B D C
Isosceles Trapezoid Theorem If a quadrilateral is an isosceles trapezoid, then each pair of base angles are congruent. If *ABCD* is an isosceles trapezoid, then $\angle A \cong \angle B$ and $\angle C \cong \angle D$.	A B D C
Diagonals of an Isosceles Trapezoid Theorem A trapezoid is isosceles if and only if its diagonals are congruent. *ABCD* is an isosceles trapezoid if and only if $\overline{AC} \cong \overline{BD}$.	A B D C $\overline{AC} \cong \overline{BD}$

3 Trapezoids can be seen in the architecture of the Inca civilization. Their buildings contain many trapezoidal doorways and windows. The window shown from an Incan building is a trapezoid with $\angle A \cong \angle D$. What are *AB* and *CD*?

Find the value of *x*.

$8x + 11 = 14x - 7$

$11 = 6x - 7$

$18 = 6x$

$3 = x$

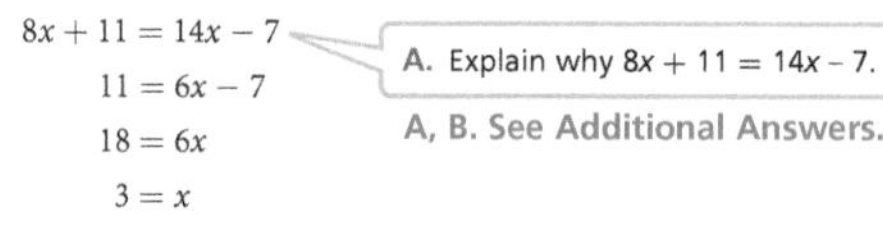

A. Explain why $8x + 11 = 14x - 7$.

A, B. See Additional Answers.

Find *AB* and *CD*.

Substitute $x = 3$ into the expression for one of the legs of the trapezoid.

$8x + 11 = 8(3) + 11 = 24 + 11 = 35$

The length of the legs of the trapezoid are 35 centimeters.

B. Why was the value of *x* substituted in only one of the expressions?

Turn and Talk Explain how you can find all the angle measures of an isosceles trapezoid if you know the measure of one angle. See margin.

The Trapezoid Midsegment Theorem

The **midsegment of a trapezoid** is a segment whose endpoints are the midpoints of the legs of the trapezoid. The Trapezoid Midsegment Theorem below is similar to the Triangle Midsegment Theorem.

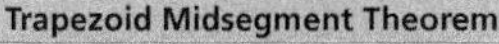

Trapezoid Midsegment Theorem

The midsegment of a trapezoid is parallel to each base, and its length is one-half of the sum of the lengths of the bases.

$\overline{XY} \parallel \overline{BC}, \overline{XY} \parallel \overline{AD}, XY = \frac{1}{2}(AD + BC)$

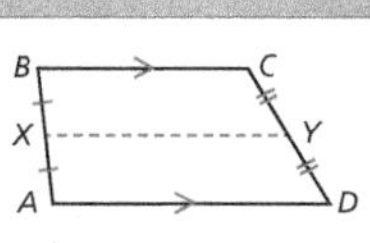

4 Find *DE*.

$MN = \frac{1}{2}(DE + GF)$

$14 = \frac{1}{2}(DE + 20)$

$28 = DE + 20$

$8 = DE$

D, E, M, N, 14, G, F, 20

Why can the Trapezoid Midsegment Theorem be applied?

4. See below.

Transformations and Symmetry in Trapezoids and Kites

5 Consider the types of symmetry for the isosceles trapezoid and the kite shown.

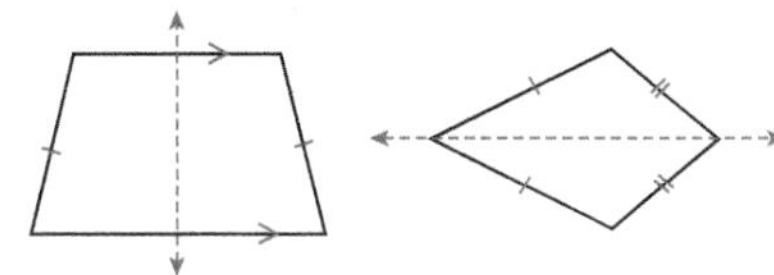

Rotational symmetry?	none	none
Reflection symmetry?	yes, across 1 vertical line	yes, across 1 horizontal line

A. Why do neither of the shapes have rotational symmetry?
Neither shape can be rotated onto itself.

B. Why do both shapes have reflection symmetry?
B. Each shape has sides that have the same length and opposite slope on either side of the axis of reflection.

Turn and Talk Suppose the trapezoid is not isosceles. How does this fact change the symmetry of the trapezoid? See margin.

4. *M* is the midpoint of $\overline{DG}$, and *N* is the midpoint $\overline{EF}$, so $\overline{MN}$ is the midsegment of the trapezoid.

Task 4 MP **Use Structure** Students are introduced to the Trapezoid Midsegment Theorem, then apply it to find the length of a base of a trapezoid. Ask students how the midsegment of a trapezoid is similar to the midsegment of a triangle and how they are different.

CONNECT TO VOCABULARY

Have students use the **Interactive Glossary** to record their understanding of the vocabulary in this task.

SUPPORT SENSE-MAKING Three Reads

Have students read the problem three times. Use the questions below for a different focus each read.

1. What is the situation about?
2. What are the quantities in the situation?
3. What are the possible mathematical questions that you could ask for the situation?

Sample Guided Discussion:

Q **How can you think of the Trapezoid Midsegment Theorem in terms of a mean or average?** The Trapezoid Midsegment Theorem says that the length of the midsegment is the mean, or average, or the lengths of the bases.

Q **What would happen if $MN = 10$ instead?** I would get $DE = 0$, which means the trapezoid is really a triangle.

Task 5 MP **Reason** Students identify transformations that map trapezoids and kites onto themselves. Ask how their conclusions compare if they use inclusive definitions for figures versus exclusive definitions.

Sample Guided Discussion:

Q **Could an isosceles trapezoid have rotational symmetry if we used the inclusive definition?** Yes, if it was a parallelogram it would have 180° symmetry.

Turn and Talk Students could use transparencies or tracing paper to visualize the symmetry of the trapezoid in question, but encourage them to progress to thinking of symmetries in terms of transformations. Ask whether an isosceles trapezoid could somehow "gain" rotational symmetry through losing its isosceles characteristic. Similarly, ask whether the trapezoid could "lose" its reflection symmetry through losing its isosceles characteristic. A trapezoid only has reflection symmetry if it is isosceles.

ONLINE

Assign the Digital On Your Own for

- built-in student supports
- Actionable Item Reports
- Standards Analysis Reports

On Your Own

Assignment Guide

The chart below indicates which problems in the On Your Own are associated with each task in the Learn Together. Assign daily homework for tasks completed.

Learn Together Tasks	On Your Own Problems
Task 1, p. 354	Problems 7, 24, 25, 29–31, 34–36, and 38
Task 2, p. 355	Problems 8, 26, and 27
Task 3, p. 356	Problems 9–12, 19–22, and 37
Task 4, p. 357	Problems 13, 14, 23, 28, 32, and 33
Task 5, p. 357	Problems 15–18

Check Understanding

1. Explain why the words *inclusive* and *exclusive* are appropriate terms to describe the different definitions of kite and trapezoid. See Additional Answers.

In kite *KLMN*, $m\angle KLM = 140°$. Find each measure.

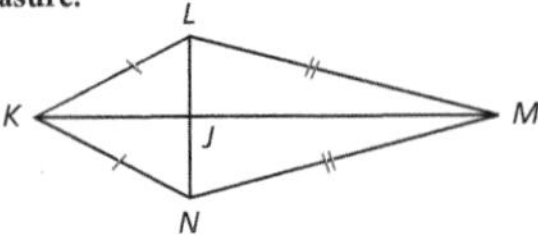

2. $m\angle KNM$ 140°
3. $m\angle KJN$ 90°

Do you have enough information to find the measure? If so, find the measure. If not, explain why.

4. no; You do not know that W is the midpoint of $\overline{SV}$ and X is the midpoint of $\overline{TU}$.

4. WX

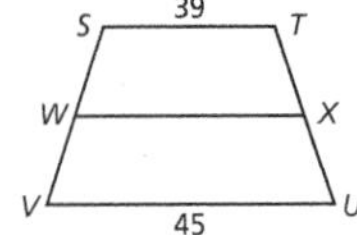

5. AC

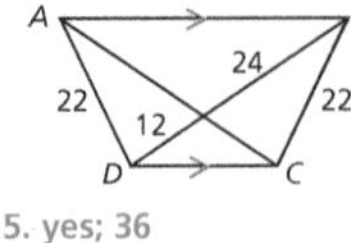

5. yes; 36

6. Is it possible to draw a kite that has no reflection symmetry? Explain why or why not. no; A kite will always be able to be reflected across the diagonal that connects the pairs of opposite angles that are not congruent.

On Your Own

7. **MP Reason** Using the inclusive definition for a trapezoid, are all parallelograms trapezoids? Are all trapezoids parallelograms? Explain.

7–9. See Additional Answers.

8. **Art** In origami, a kite base is used as the starting point for many paper-folded creations. The steps for making a kite base out of a square piece of origami paper are shown. Explain how the paper folding justifies the fact that the diagonals of a kite are perpendicular.

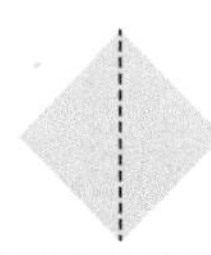

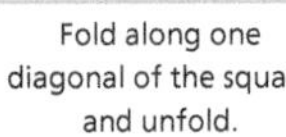

Fold along one diagonal of the square and unfold.

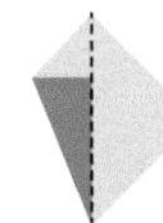

Fold the left side to meet the crease in the center.

Fold the right side to meet the crease in the center.

9. **MP Critique Reasoning** Lexi believes that there is not enough information to determine the measure of $\angle Y$. Ally believes the measure of $\angle Y$ is 117°. Who is correct? Explain your reasoning.

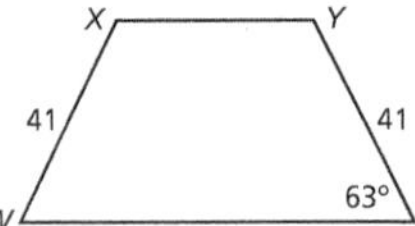

358

3 Check Understanding

Formative Assessment

Use formative assessment to determine if your students are successful with this lesson's learning objective.

Students who successfully complete the Check Understanding can continue to the On Your Own practice.

For students who miss 1 problem or more, work in a pulled small group using the Almost There small-group activity on page 353C.

ONLINE

Assign the Digital Check Understanding to determine

- success with the learning objective
- items to review
- grouping and differentiation resources

4 Differentiation Options

Differentiate instruction for all students using small-group activities and math center activities on page 353C.

Reteach

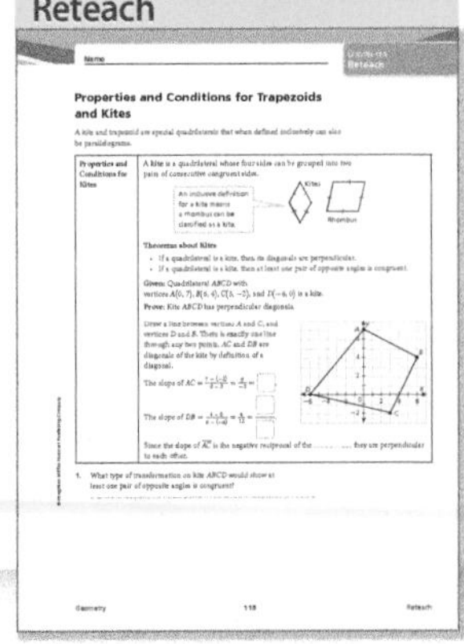

Challenge

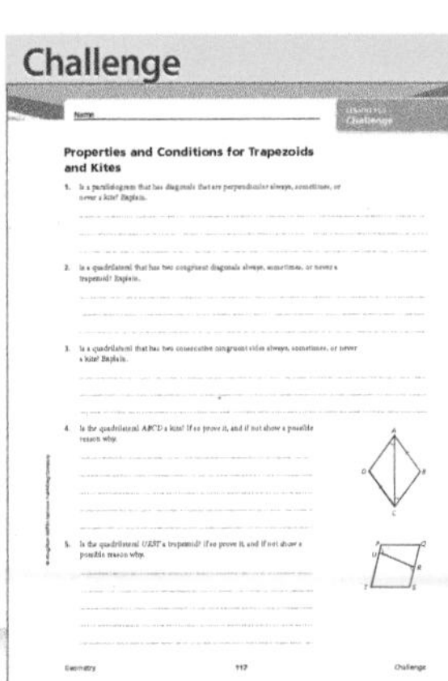

Find the value of x in the isosceles trapezoid. Then find the given measure.

10. JL $x = 7$; $JL = 24$

$KM = 2x + 10$
$JL = 5x - 11$

11. $m\angle K$ $x = 40$ $m\angle K = 135°$

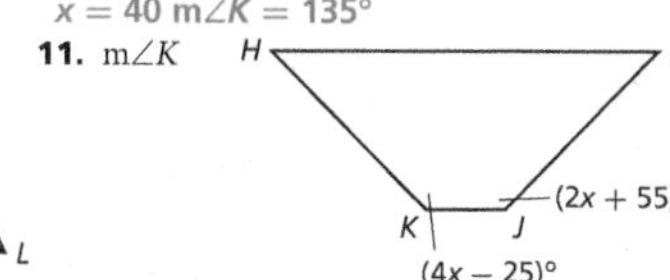

12. STEM An engineer is designing solar panels for a rover that will explore Mars. The solar panel unfolds to an isosceles trapezoid. The length of one leg is $0.1x - 0.5$ meters, and the length of the other leg is $0.3x - 2.1$ meters. What is the length of each of the legs of the solar panel? 0.3 meter

Find the given measure.

13. BC 38

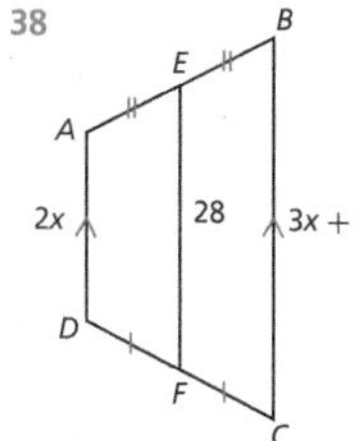

14. RS 12

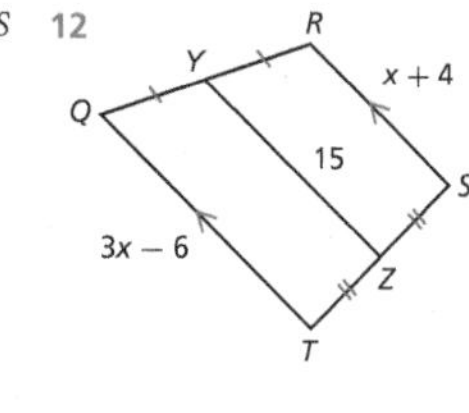

Does the figure have rotational symmetry or reflection symmetry? If so, describe the symmetry. 15–18. See Additional Answers.

15. kite $JKLM$

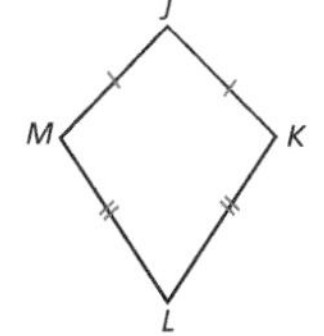

16. trapezoid $ABCD$

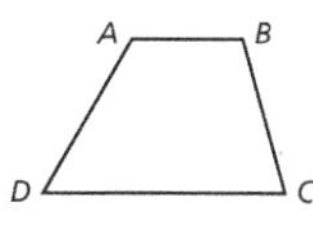

17. trapezoid $WXYZ$

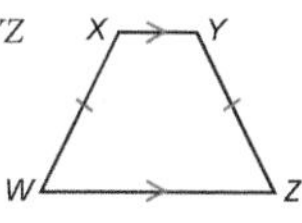

18. kite $QRST$

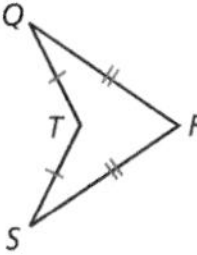

Questioning Strategies

Problem 14 Suppose that $QT = 3x - y$ and $RS = x + y$. Can you solve for each base length? No, I can solve for x.

$$YZ = \frac{1}{2}(QT + RS)$$
$$15 = \frac{1}{2}\left((3x - y) + (x + y)\right)$$
$$15 = \frac{1}{2} \cdot 4x = 2x$$
$$7.5 = x$$

I do not have an equation relating x and y, so I cannot solve for y. Each base length is dependent on both x and y, so I do not have enough information to find the base lengths.

ONLINE

Assign the Digital On Your Own for

- built-in student supports
- Actionable Item Reports
- Standards Analysis Reports

Watch for Common Errors

Problem 21 Students may assume that $\triangle AED$ and $\triangle BEC$ are isosceles triangles as a direct result of *ABCD* being an isosceles trapezoid. While those triangles are in fact isosceles, students would need to make arguments to show that result instead of assuming it from the given information. This would be a less direct proof strategy than the suggested plan for proof, but still valid if carried out rigorously.

In Problems 19 and 20, prove each theorem for trapezoids using the diagram with an auxiliary segment and the plan for the proof.

19–22. See Additional Answers.

19. Prove that if a quadrilateral is an isosceles trapezoid, then each pair of base angles are congruent.

Given: $\overline{BC} \parallel \overline{AD}$, $\overline{AB} \parallel \overline{EC}$, $\overline{AB} \cong \overline{CD}$

Prove: $\angle A \cong \angle D$, $\angle B \cong \angle BCD$

Plan for Proof: Show that $\triangle ECD$ is an isosceles triangle and *ABCE* is a parallelogram. Use these facts to reason that $\angle A \cong \angle D$. The use the Congruent Supplements Theorem to show that $\angle B \cong \angle BCD$.

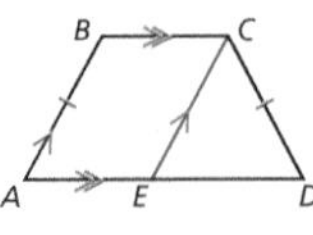

20. Prove that if a trapezoid has one pair of congruent base angles, then the trapezoid is isosceles.

Given: $\overline{GH} \parallel \overline{FJ}$, $\overline{GF} \parallel \overline{HK}$, $\angle F \cong \angle J$

Prove: $\overline{GF} \cong \overline{HJ}$

Plan for Proof: Show that $\triangle KHJ$ is an isosceles triangle. Then reason that $\overline{GF} \cong \overline{HJ}$.

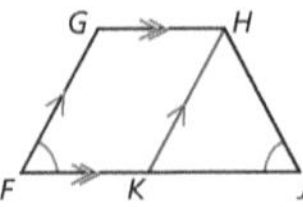

In Problems 21 and 22, prove each part of the third trapezoid theorem: *A trapezoid is isosceles if and only if its diagonals are congruent.* Use the first two trapezoid theorems, which were proven in Problems 19 and 20.

21. Prove that if a trapezoid is isosceles, then its diagonals are congruent, which is the first part of the theorem.

Given: *ABCD* is an isosceles trapezoid with $\overline{BC} \parallel \overline{AD}$ and $\overline{AB} \cong \overline{CD}$.

Prove: $\overline{AC} \cong \overline{DB}$

Plan for Proof: Find a way to show that $\triangle ABD \cong \triangle DCA$. Then reason that $\overline{AC} \cong \overline{DB}$.

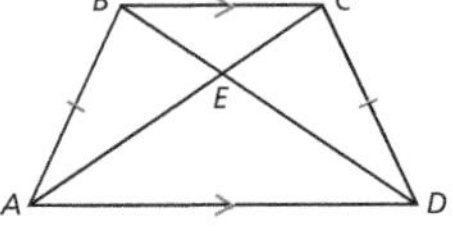

22. Prove that if the diagonals of a trapezoid are congruent, then it is an isosceles trapezoid, which is the second part of the theorem.

Given: $\overline{BC} \parallel \overline{AD}$ and $\overline{AC} \cong \overline{DB}$.

Prove: *ABCD* is an isosceles trapezoid.

Plan for Proof: Show that *BCFE* is a parallelogram. Then show that $\triangle DBE \cong \triangle ACF$. Use information about the congruent triangles to show that $\triangle ABC \cong \triangle DCB$. Then reason that the base angles of the trapezoid are congruent.

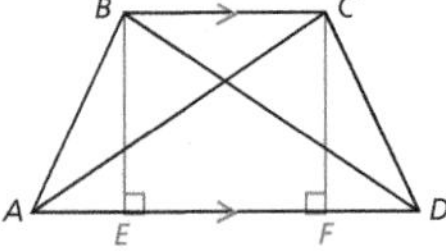

23. Quadrilateral *ABCD* has vertices $A(-3, 4)$, $B(5, 2)$, $C(6, -4)$, and $D(-6, -1)$.

A. Show that *ABCD* is a trapezoid. A–C. See Additional Answers.

B. Find the coordinates of the midsegment of *ABCD* and use them to determine the length and slope of the midsegment.

C. Use the Trapezoid Midsegment Theorem to find the length and slope of the midsegment. Compare your answer to the answer for Part B.

360

Show that the statement about the figure in the coordinate plane is true.

24. *ABCD* is a trapezoid.

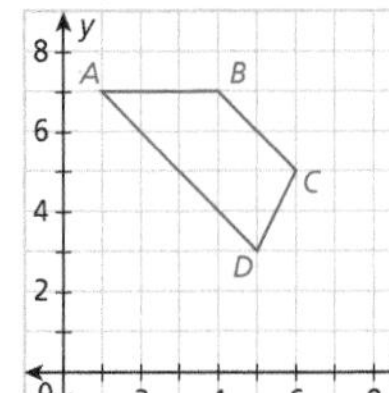

25. *ABCD* is a kite.

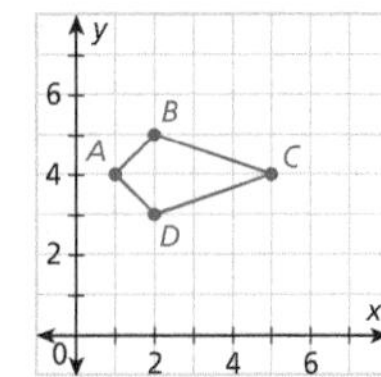

24–28. See Additional Answers.

In Problems 26 and 27, use the diamond shown, which is in the shape of a kite.

26. Find the value of x when $y = 40°$ and $z = 92°$.

27. Write a general formula for finding the value of x for any values of y and z.

28. (MP) **Construct Arguments** Follow the steps below to prove the Trapezoid Midsegment Theorem.

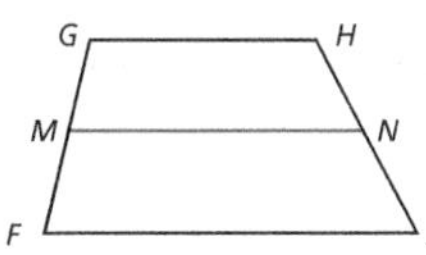

Given: Trapezoid *FGHJ* with midsegment $\overline{MN}$

Prove: $\overline{MN} \parallel \overline{FJ}$, $\overline{MN} \parallel \overline{GH}$, $MN = \frac{1}{2}(FJ + GH)$

A. What does the given information tell you about the diagram?

B. $\overline{GN}$ is drawn to create $\triangle GHN$. Then the triangle is rotated 180° about point N to create $\triangle KJN$ as shown below. Explain how you know that $\overline{GN} \cong \overline{KN}$.

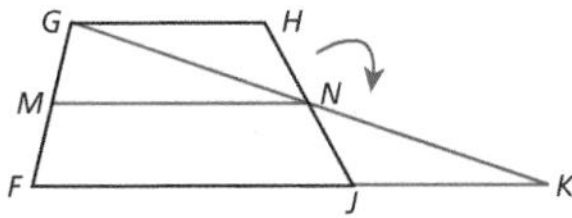

C. Use the fact that N is the midpoint of $\overline{GK}$ to explain why $\overline{MN} \parallel \overline{FK}$. (Hint: Consider how $\overline{MN}$ relates to $\triangle GFK$.)

D. If you know that $\overline{MN} \parallel \overline{FK}$, then explain why $\overline{MN} \parallel \overline{GH}$.

E. How are $\overline{MN}$ and $\overline{FK}$ related? Use this relationship to show that $MN = \frac{1}{2}(FJ + GH)$.

29. Quadrilateral *PQRS* has vertices $P(-1, 3)$, $Q(2, 4)$, $R(5, 3)$, and $S(2, -6)$.

A. Find the side lengths of *PQRS*. A–C. See Additional Answers.

B. What do you notice about the side lengths?

C. Can you conclude that the quadrilateral is a kite? Explain your reasoning.

Questioning Strategies

Problem 24 Suppose you want to know whether *ABCD* is an isosceles trapezoid. What are two ways you could determine this? I know that $AB = 3$, so I could use the Distance Formula to find *CD* and compare it with *AB* to see whether they are equal. Or, I could use the Distance Formula to find *AC* and *BD*. I know *ABCD* is isosceles if and only if $AC = BD$ by the Diagonals of an Isosceles Trapezoid Theorem.

Watch for Common Errors

Problem 28 In Step B, students' eyes might be drawn to $\triangle GMN$, and they might mistakenly conclude that $\triangle KJN \cong \triangle NMG$ due to a translation. Clarify that the arrow in the diagram indicates the rotation of $\triangle GHN$ about point N, so $\triangle KJN \cong \triangle GHN$.

Wrap-Up

Summarize learning with your class. Consider using the Exit Ticket, Put It in Writing, or I Can scale.

Exit Ticket 

If $m\angle ABC = 70°$ and $BD = 22$ cm, find AC, $m\angle BCD$, $m\angle CDA$, and $m\angle DAB$.

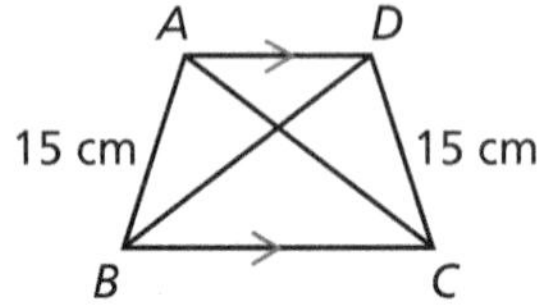

Since $ABCD$ is isosceles, it has two pairs of congruent base angles. So, $m\angle BCD = 70°$ and $360° - 2 \cdot 70° = 220°$, so $m\angle CDA = m\angle DAB = \frac{220°}{2} = 110°$. The diagonals of $ABCD$ are congruent, so $AC = BD = 22$ cm.

Put It in Writing 

Describe three ways you can use a ruler or protractor to determine whether a trapezoid is isosceles.

I Can 

The scale below can help you and your students understand their progress on a learning goal.

4	I can prove and apply theorems about trapezoids and kites, and I can explain my reasoning to others.
3	I can prove and apply theorems about trapezoids and kites.
2	I can explain how different kinds of quadrilaterals can be kites or trapezoids using the inclusive definitions.
1	I can tell if a quadrilateral is a kite or a trapezoid using the exclusive and inclusive definitions.

Spiral Review • Assessment Readiness

These questions will help determine if students have retained information taught in the past and can also prepare them for high-stakes assessments. Here, students must apply conditions for rhombuses **(11.4)**, apply properties of rectangles **(11.3)**, and distinguish between enlargements and reductions of figures **(6.2)**.

In Problems 30–36, determine whether the statement is *always*, *sometimes*, or *never* true. Explain your reasoning. 30–38. See Additional Answers.

30. The bases of a trapezoid are parallel.

31. The legs of a trapezoid are congruent.

32. The midsegment of a trapezoid is parallel to the bases of the trapezoid.

33. The midsegment of a trapezoid is half the length of one of the bases.

34. The opposite sides of a kite are congruent.

35. The adjacent sides of a kite are perpendicular.

36. A trapezoid is a kite.

37. In kite $ABCD$, $\triangle ABD$ and $\triangle CBD$ can be rotated and translated, mapping $\overline{AD}$ onto $\overline{CD}$ and joining the remaining pair of vertices, as shown in the figure. Why does this process produce an isosceles trapezoid?

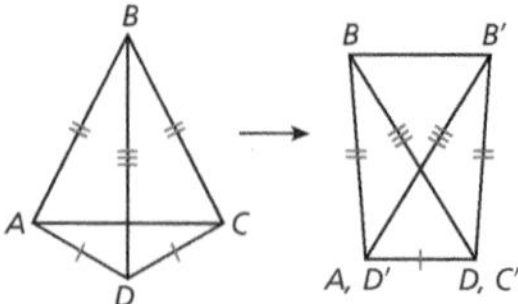

38. Open Middle™ What is the fewest number of geometric markings needed to demonstrate that a quadrilateral is a trapezoid? Explain your reasoning.

Spiral Review • Assessment Readiness

39. Find the value of x that makes $WXYZ$ a rhombus.

Ⓐ 11
Ⓑ 22.5
Ⓒ 90
Ⓓ 180

W X Z Y $(8x + 2)°$

40. $WXYZ$ is a rectangle with intersection of diagonals at point V. Find the length of $\overline{WY}$ if $ZV = 26$.

Ⓐ 13
Ⓑ 26
Ⓒ 39
Ⓓ 52

W X V Z Y

41. For each triangle, identify the dilation as an enlargement or reduction.

	Enlargement	Reduction
A. $(2, 4), (0, 2), (2, 2) \to (3, 6), (0, 3), (3, 3)$	?	?
B. $(6, 2), (-2, 4), (-1, -1) \to (12, 4), (-4, 8), (-2, -2)$	?	?
C. $(-2, 2), (4, 4), (2, -2) \to (-1, 1), (2, 2), (1, -1)$	?	?

 **I'm in a Learning Mindset!**

What did I learn from the mistakes I made with the properties and conditions of trapezoids and kites?

Keep Going to Journal and Practice Workbook

Learning Mindset

Resilience Managing the Learning Process

Point out to students that the best learning happens as a result of mistakes. Remind them that while mistakes can be discouraging, it's important to keep them in perspective. Too strong of an emotional reaction to a mistake precludes rational thought, so the invaluable learning opportunity presented by a mistake might be lost. Encourage students to go back over their work, retrace steps, and ask themselves about the rationale behind each step in an effort to understand where the mistake was made. For example, it is very common to assume the exclusive definitions of figures when working with them—was a mistake the result of assuming that a parallelogram cannot be a trapezoid? *How does going back over your work help you realize where you made a mistake? Why is it important to not get too discouraged when you get an incorrect answer? How does understanding where your work went wrong help keep you in a learning mindset?*

Module

11 Review

Parallelograms

A parallelogram is a quadrilateral with opposite sides parallel.

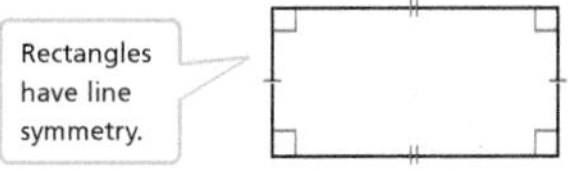
Each diagonal divides the parallelogram into 2 congruent triangles.
There is rotational symmetry.

- Opposite sides are congruent.
- Opposite angles are congruent.
- The diagonals bisect each other.

Rectangles

A rectangle is a parallelogram with right angles.

Rectangles have line symmetry.

- Rectangles have the same properties as parallelograms.
- The diagonals are congruent.

Rhombi

A rhombus is a parallelogram with all sides congruent.

- Rhombi have the same properties as parallelograms.
- The diagonals are perpendicular.

Squares

A square is both a rectangle and a rhombus.

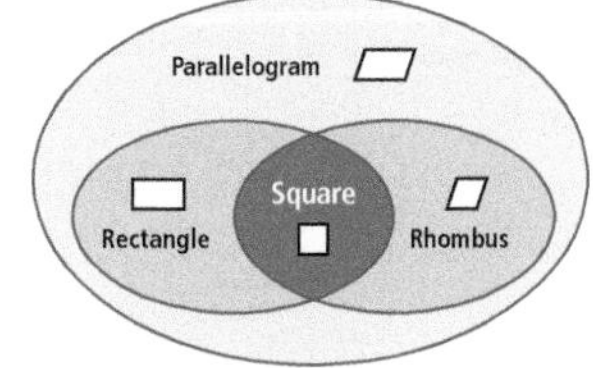
Parallelogram
Square
Rectangle
Rhombus

- Squares have the same properties as parallelograms, rectangles, and rhombi.

Kites

A kite is a quadrilateral with pairs of consecutive congruent sides.

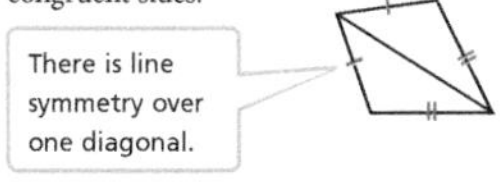
There is line symmetry over one diagonal.

- The diagonals are perpendicular.
- Kites have one pair of opposite angles that are congruent.

Isosceles Trapezoids

A quadrilateral with one pair of parallel sides is a trapezoid. An isosceles trapezoid has congruent legs.

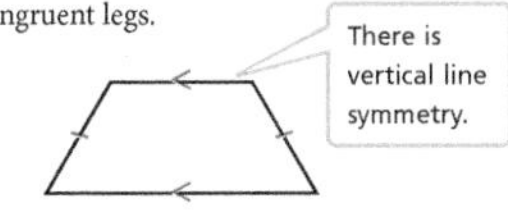
There is vertical line symmetry.

- The diagonals are congruent.
- Each pair of base angles are congruent.

Squares **What are the properties of the diagonals of a square?** The diagonals of a square are congruent and are the perpendicular bisectors of each other.

Kites **Is a square a kite? Explain.** yes; All sides of a square are congruent, so the four sides can be grouped into two pairs of consecutive congruent sides.

Isosceles Trapezoids **Can the bases of an isosceles trapezoid ever be equal? Explain.** Possible answer: yes; If the isosceles trapezoid is a parallelogram, the bases are equal.

ONLINE

Assign the Digital Module Review for

- built-in student supports
- Actionable Item Reports
- Standards Analysis Reports

Module Review

Use the first page of the Module Review to summarize and connect the main ideas of the module. Use the second page to assess students' understanding of the vocabulary, concepts, and skills presented in the module.

Sample Guided Discussion:

Parallelograms **Is a parallelogram a trapezoid? Explain.** Possible answer: yes; Using the inclusive definition, a trapezoid is a quadrilateral with at least one pair of parallel sides. A parallelogram is a quadrilateral with two pairs of parallel sides, so this condition satisfies the definition of a trapezoid.

Rectangles **Do the diagonals of a rectangle bisect the opposite angles? Explain.** Possible answer: no; The adjacent sides of a rectangle are not equal.

The diagonals of a rectangle bisect the opposite angles only if the rectangle is a square.

Rhombi **The points $A(0,0)$, $B(x,y)$, $C(6,0)$ and $D(x,-y)$, where $0 < x < 6$, are the vertices of rhombus *ABCD*. Suppose that *ABCD* is rotated 180° counterclockwise about point *A*, and its image *A′B′C′D′* is reflected across the *y*-axis to form its image *A″B″C″D″*. What kind of quadrilateral is *A″B″C″D″* and what are the coordinates of its vertices? Explain.** *A″B″C″D″* is a rhombus because reflections and rotations are rigid transformations that preserve lengths and angle measures. The vertices are $A''(0,0)$, $B''(x,-y)$, $C''(6,0)$, $D''(x,y)$.

Module Review continued

Possible Scoring Guide

Items	Points	Description
1–4	2 each	identifies the correct term
5–7	2 each	correctly identifies which quadrilaterals always have a given property under a transformation
8–10	2 each	correctly identifies which quadrilaterals always have a given property
11, 12	2 each	correctly proves properties of parallelograms
13–16	2 each	correctly determines measures of sides or angles of given quadrilaterals
17	2	correctly identifies the type(s) of parallelogram given the coordinates of its vertices
18	2	correctly determines the length and width of a rectangle given algebraic expressions for its length and width and its perimeter
Total points possible = 36 points		

The Unit 6 Test in the Assessment Guide assesses content from Modules 11 and 12.

Vocabulary

Choose the correct term from the box to complete each sentence.

Vocabulary
kite
parallelogram
rectangle
rhombus
square

1. A __?__ is a __?__ with four congruent sides. rhombus, parallelogram
2. A __?__ is a quadrilateral whose four sides can be grouped into two pairs of consecutive congruent sides. kite
3. A __?__ is a parallelogram with four right angles. rectangle
4. A __?__ is a parallelogram with four congruent sides and four right angles. square

Concepts and Skills

Consider parallelograms, rectangles, rhombi, squares, kites and trapezoids. Identify which quadrilaterals always have the given property.

5. Rotational Symmetry
   See Additional Answers.
6. Line Symmetry
   kite, rectangle, rhombus, square
7. Both Rotational and Line Symmetry
   rectangle, rhombus, square
8. Opposite sides are congruent
   parallelogram, rectangle, rhombus, square
9. Diagonals are congruent
   rectangle, square
10. Diagonals are perpendicular
    rhombus, square, kite

Complete a proof given that opposite sides in a parallelogram are parallel.

11. Prove that opposite sides are congruent. See Additional Answers.
12. Prove that opposite angles are congruent. See Additional Answers.

Find the measure.

13. AC 40 ($AC = 3x + 7$, $BD = 4x - 4$)
14. HG 28 ($JF = 16$, $MN = 22$)
15. NP 55 ($MN = 6z + 7$, $NP = 8z - 9$)
16. $m\angle C$ 35°

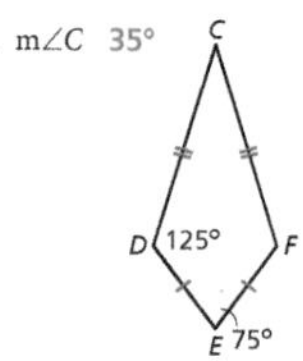

17. Parallelogram $WXYZ$ has vertices at $W(1, 3)$, $X(2, 5)$, $Y(6, 3)$, and $Z(5, 1)$. Determine whether $WXYZ$ is a rectangle, rhombus, or square. List all names that apply. rectangle
18. (MP) Use Tools A rectangular garden has a length of $(4x + 5)$ inches and a width of $(3x - 1)$ inches. Find the length and width of the garden if the perimeter is 36 feet. State what strategy and tool you will use to answer the question, explain your choice, and then find the answer. length 13 feet, width 5 feet

364

DATA-DRIVEN INSTRUCTION

Before moving on to the Module Test, use the Module Review results to intervene based on the table below.

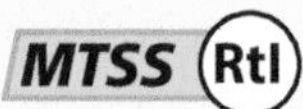

Items	Lesson	DOK	Content Focus	Intervention
5–7	11.5	2	Identify the type of quadrilateral *ABCD* having a given property.	Reteach 11.5
8–10	11.2, 11.4	2	Identify the type of quadrilateral *ABCD* having a given property.	Reteach 11.2, 11.4
11, 12	11.1	2	Prove properties of parallelograms.	Reteach 11.1
13	11.3	2	Find the measures of the sides of a given parallelogram.	Reteach 11.3
14	11.5	2	Find the measures of the bases of a given isosceles trapezoid.	Reteach 11.5
15	11.4, 11.5	2	Find the measures of the sides of a rhombus.	Reteach 11.4, 11.5
16	11.5	2	Find the measures of the angles of a kite.	Reteach 11.5
17	11.4	3	Classify a parallelogram given the coordinates of its vertices.	Reteach 11.4
18	11.3	3	Find missing measures of a rectangle given its perimeter.	Reteach 11.3

Module Test

The Module Test is available in alternative versions in your Assessment Guide. All items are presented in standardized test formats.

ONLINE

Ed

Assign the Digital Module Test to power actionable reports including

- proficiency by standards
- item analysis

Form A

Name

Module 11 • Form A
Module Test

1. In quadrilateral $TUVW$, $m\angle U = 94°$, $m\angle V = 51°$, and $m\angle W = 67°$. What is $m\angle T$?

Ⓐ 32° Ⓑ 51° Ⓒ 129° **Ⓓ 148°**

2. In parallelogram $QRST$, $m\angle R = 63.7°$. What is $m\angle S$?

Ⓐ 26.3° Ⓑ 63.7° **Ⓒ 116.3°** Ⓓ 127.4°

3. In rhombus $JKLM$, $JK = 3x + 9$ meters and $KL = 11x - 7$ meters. What is the perimeter of the rhombus?

Ⓐ 2 meters Ⓑ 8 meters Ⓒ 15 meters **Ⓓ 60 meters**

4. In quadrilateral $DEFG$, $DE = y + 3$ mm, $EF = 4x - 1$ mm, $FG = 2y - 2$ mm, and $DG = 3x + 3$ mm. What value of x makes $DEFG$ a parallelogram?

Ⓐ 4 Ⓑ 5 Ⓒ 8 Ⓓ 15

5. In kite $ABCD$, $\overline{AB} \cong \overline{BC}$ and $\overline{CD} \cong \overline{AD}$. If $AB = 4x + 12$ units and $BC = 6x - 4$ units, what is the value of x?

Ⓐ 4 **Ⓑ 8** Ⓒ 16 Ⓓ 44

6. A diagonal of a rhombus lies on the line $y = 2x + 7$. What is an equation of the line for the second diagonal of the rhombus that passes through the vertex (4, 15)?

Possible answer: $y = -\frac{1}{2}x + 17$

7. In the figure, quadrilateral $BCDE$ is inscribed in circle A.

$\overline{DB}$ and $\overline{CE}$ are diameters of the circle. Which statement proves that quadrilateral $BCDE$ is a rectangle?

Ⓐ Since $\overline{DB}$ and $\overline{CE}$ are diameters of the circle, $\overline{DB} \perp \overline{CE}$. Therefore, $BCDE$ is a rectangle.
Ⓑ Since $\overline{DB}$ and $\overline{CE}$ are diameters of the circle, $\overline{DB} \cong \overline{CE}$. Therefore, $BCDE$ is a rectangle.
Ⓒ Since $\overline{DB}$ and $\overline{CE}$ are diameters of the circle, $\overline{BC} \parallel \overline{DE}$ and $\overline{CD} \parallel \overline{BE}$. Therefore, $BCDE$ is a rectangle.
Ⓓ Since $\overline{DB}$ and $\overline{CE}$ are diameters of the circle, $\overline{DA} \cong \overline{AB}$ and $\overline{CA} \cong \overline{AE}$. Therefore, $BCDE$ is a rectangle.

8. Which of these polygons always map onto themselves when rotated 180° about their center?

Select **all** the correct answers.

Ⓐ square Ⓑ triangle Ⓒ octagon **Ⓓ rhombus** Ⓔ trapezoid **Ⓕ rectangle**

9. In parallelogram $ABCD$, diagonal $\overline{AC}$ and diagonal $\overline{BD}$ intersect at point F. If $AF = 12x - 18.2$ units and $FC = 8x + 11.2$ units, what is the value of x?

7.35

10. In rectangle $DEFG$, $m\angle F = (x + 14)°$. What is the value of x?

76

© Houghton Mifflin Harcourt Publishing Company

Geometry 97 Module 11 Test • Form A

Form A

Module 11 • Form A
Module Test

Name

11. In parallelogram $ABCD$, $m\angle B = 5x + 1$ and $m\angle C = (x - 1)°$. What are $m\angle B$ and $m\angle C$ in degrees?

$m\angle B =$ 151° and $m\angle C =$ 29°

12. A group of paleontologists sections off a plot of land, in the shape of a trapezoid, to excavate for fossils. The plot has the dimensions shown in the diagram.

A 45 m B
D 63 m C

The paleontologists divide the plot into two parts by erecting a straight partition from the midpoint of $\overline{AD}$ to the midpoint of $\overline{BC}$. What is the length in meters of the partition?

54

13. A quadrilateral with vertices $A(-2, -2)$, $B(0, -1)$, $C(5, -3)$, and $D(-3, -7)$ is graphed on the coordinate plane.

Part A

Which statements about the quadrilateral are true?

Select **all** the correct answers.

Ⓐ $\overline{AB} \parallel \overline{DC}$ Ⓑ $\overline{AB} \perp \overline{BC}$ **Ⓒ $\overline{AD} \nparallel \overline{BC}$** Ⓓ $\overline{AB} \cong \overline{DC}$ Ⓔ $\overline{AD} \cong \overline{BC}$ Ⓕ $\overline{AC} \cong \overline{BD}$

Part B

Which type of quadrilateral is $ABCD$?

Ⓐ $ABCD$ is a rectangle because both diagonals are congruent.
Ⓑ $ABCD$ is a rectangle because both pairs of opposite sides are parallel.
Ⓒ $ABCD$ is a trapezoid because only one pair of opposite sides is congruent.
Ⓓ $ABCD$ is a trapezoid because only one pair of opposite sides is parallel.

14. Parallelogram $KLNM$ has the dimensions shown.

K (192 − x) ft L; (120 − y)°; (y − 16)°; M (72 + x) ft N

Part A

What is the length in feet of $\overline{KL}$?

$KL =$ 132

Part B

What is $m\angle N$ in degrees?

$m\angle N =$ 128°

15. A student is asked to construct square $QRST$ inscribed in the circle shown. Two vertices of the square will be at the endpoints of diameter $\overline{QS}$.

Part A

If diagonal $\overline{RT}$ lies on $\overleftrightarrow{RT}$, what is an equation of $\overleftrightarrow{RT}$?

$\overleftrightarrow{RT}$: Possible answer: $y = -\frac{1}{2}x + 3$

Part B

What are the coordinates of the vertices R and T?

Possible answer:

R(−4, 5) and T(4, 1)

© Houghton Mifflin Harcourt Publishing Company

Geometry 98 Module 11 Test • Form A

Form B

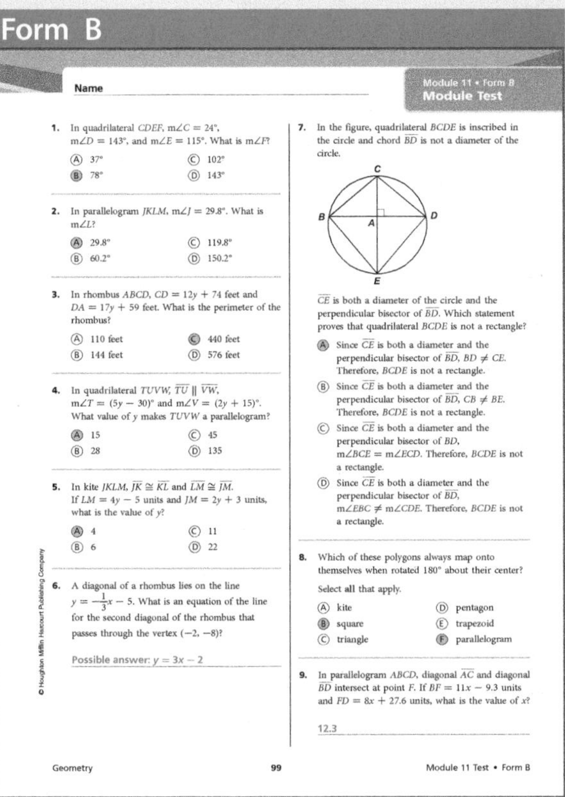

Name

Module 11 • Form B
Module Test

1. In quadrilateral $CDEF$, $m\angle C = 24°$, $m\angle D = 143°$, and $m\angle E = 115°$. What is $m\angle F$?

Ⓐ 37° **Ⓑ 78°** Ⓒ 102° Ⓓ 143°

2. In parallelogram $JKLM$, $m\angle J = 29.8°$. What is $m\angle L$?

Ⓐ 29.8° Ⓑ 60.2° Ⓒ 119.8° Ⓓ 150.2°

3. In rhombus $ABCD$, $CD = 12y + 74$ feet and $DA = 17y + 59$ feet. What is the perimeter of the rhombus?

Ⓐ 110 feet Ⓑ 144 feet **Ⓒ 440 feet** Ⓓ 576 feet

4. In quadrilateral $TUVW$, $\overline{TU} \parallel \overline{VW}$, $m\angle T = (5y - 30)°$ and $m\angle V = (2y + 15)°$. What value of y makes $TUVW$ a parallelogram?

Ⓐ 15 Ⓑ 28 Ⓒ 45 Ⓓ 135

5. In kite $JKLM$, $\overline{JK} \cong \overline{KL}$ and $\overline{LM} \cong \overline{JM}$. If $LM = 4y - 5$ units and $JM = 2y + 3$ units, what is the value of y?

Ⓐ 4 Ⓑ 6 Ⓒ 11 Ⓓ 22

6. A diagonal of a rhombus lies on the line $y = -\frac{1}{3}x - 5$. What is an equation of the line for the second diagonal of the rhombus that passes through the vertex (−2, −8)?

Possible answer: $y = 3x - 2$

7. In the figure, quadrilateral $BCDE$ is inscribed in the circle and chord $\overline{BD}$ is not a diameter of the circle.

$\overline{CE}$ is both a diameter of the circle and the perpendicular bisector of $\overline{BD}$. Which statement proves that quadrilateral $BCDE$ is not a rectangle?

Ⓐ Since $\overline{CE}$ is both a diameter and the perpendicular bisector of $\overline{BD}$, $BD \neq CE$. Therefore, $BCDE$ is not a rectangle.
Ⓑ Since $\overline{CE}$ is both a diameter and the perpendicular bisector of $\overline{BD}$, $CB \neq BE$. Therefore, $BCDE$ is not a rectangle.
Ⓒ Since $\overline{CE}$ is both a diameter and the perpendicular bisector of BD, $m\angle BCE = m\angle ECD$. Therefore, $BCDE$ is not a rectangle.
Ⓓ Since $\overline{CE}$ is both a diameter and the perpendicular bisector of $\overline{BD}$, $m\angle EBC \neq m\angle CDE$. Therefore, $BCDE$ is not a rectangle.

8. Which of these polygons always map onto themselves when rotated 180° about their center?

Select **all** that apply.

Ⓐ kite **Ⓑ square** Ⓒ triangle Ⓓ pentagon Ⓔ trapezoid **Ⓕ parallelogram**

9. In parallelogram $ABCD$, diagonal $\overline{AC}$ and diagonal $\overline{BD}$ intersect at point F. If $BF = 11x - 9.3$ units and $FD = 8x + 27.6$ units, what is the value of x?

12.3

© Houghton Mifflin Harcourt Publishing Company

Geometry 99 Module 11 Test • Form B

Form B

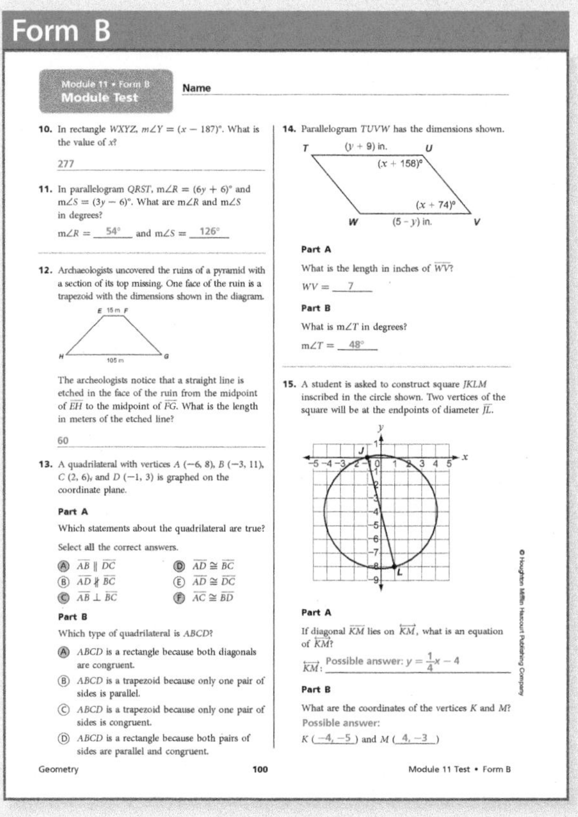

Module 11 • Form B
Module Test

Name

10. In rectangle $WXYZ$, $m\angle Y = (x - 187)°$. What is the value of x?

277

11. In parallelogram $QRST$, $m\angle R = (6y + 6)°$ and $m\angle S = (3y - 6)°$. What are $m\angle R$ and $m\angle S$ in degrees?

$m\angle R =$ 54° and $m\angle S =$ 126°

12. Archaeologists uncovered the ruins of a pyramid with a section of its top missing. One face of the ruin is a trapezoid with the dimensions shown in the diagram.

E 15 m F
H 105 m G

The archeologists notice that a straight line is etched in the face of the ruin from the midpoint of $\overline{EH}$ to the midpoint of $\overline{FG}$. What is the length in meters of the etched line?

60

13. A quadrilateral with vertices $A(-6, 8)$, $B(-3, 11)$, $C(2, 6)$, and $D(-1, 3)$ is graphed on the coordinate plane.

Part A

Which statements about the quadrilateral are true?

Select **all** the correct answers.

Ⓐ $\overline{AB} \parallel \overline{DC}$ Ⓑ $\overline{AD} \nparallel \overline{BC}$ **Ⓒ $\overline{AB} \perp \overline{BC}$** **Ⓓ $\overline{AD} \cong \overline{BC}$** Ⓔ $\overline{AD} \cong \overline{DC}$ **Ⓕ $\overline{AC} \cong \overline{BD}$**

Part B

Which type of quadrilateral is $ABCD$?

Ⓐ $ABCD$ is a rectangle because both diagonals are congruent.
Ⓑ $ABCD$ is a trapezoid because only one pair of sides is parallel.
Ⓒ $ABCD$ is a trapezoid because only one pair of sides is congruent.
Ⓓ $ABCD$ is a rectangle because both pairs of sides are parallel and congruent.

14. Parallelogram $TUVW$ has the dimensions shown.

T (y + 9) in. U; (x + 158)°; (x + 74)°; W (5 − y) in. V

Part A

What is the length in inches of $\overline{WV}$?

$WV =$ 7

Part B

What is $m\angle T$ in degrees?

$m\angle T =$ 48°

15. A student is asked to construct square $JKLM$ inscribed in the circle shown. Two vertices of the square will be at the endpoints of diameter $\overline{JL}$.

Part A

If diagonal $\overline{KM}$ lies on $\overleftrightarrow{KM}$, what is an equation of $\overleftrightarrow{KM}$?

$\overleftrightarrow{KM}$: Possible answer: $y = \frac{1}{4}x - 4$

Part B

What are the coordinates of the vertices K and M?

Possible answer:

K(−4, −5) and M(4, −3)

© Houghton Mifflin Harcourt Publishing Company

Geometry 100 Module 11 Test • Form B

SIMILARITY

Introduce and Check for Readiness

• Module Performance Task • Are You Ready?

Lesson 12.1—2 Days

Use Transformations to Prove Figures are Similar

Learning Objective: Use similarity transformations to prove figures are similar.

New Vocabulary: similar figures, similarity transformation

Lesson 12.2—2 Days

Develop AA Triangle Similarity

Learning Objective: Prove AA, SSS, and SAS Similarity Theorems.

Lesson 12.3—2 Days

Develop and Prove Triangle Proportionality

Learning Objective: Identify and use the connection between parallel lines and proportional segments in triangles.

New Vocabulary: partition

Lesson 12.4—2 Days

Apply Similarity in Right Triangles

Learning Objective: Identify similar right triangles, apply the Geometric Means Theorems, and recognize Pythagorean triples.

New Vocabulary: geometric mean, Pythagorean triple

Assessment

- Module 12 Test (Forms A and B)
- Unit 6 Test (Forms A and B)

LEARNING ARC FOCUS

Build Conceptual Understanding  Connect Concepts and Skills  Apply and Practice

TEACHING FOR SUCCESS

TEACHING FOR DEPTH: Similarity

Meaning of Similar Figures. Two figures are *similar* if their angles are congruent and the ratio between corresponding sides are equal. Similar figures can be mapped onto one another using *dilations*, which are transformations that preserve angle measures but not side measures. Instead, the lengths of corresponding sides depend on the scale factor used in the dilation. If the scale factor is less than 1, the image is smaller than the preimage; if the scale factor is greater than 1, the image is larger than the preimage. In each case, however, the ratio of the lengths between any two corresponding sides is equal to the scale factor.

Basically, similar figures have the same shape, but not necessarily the same size. If two similar figures have the same size, they are congruent. In fact, the symbol for congruence $\cong$ incudes the symbol for similarity $\sim$ and the symbol for equality $=$, indicating that the figures are similar and lengths of corresponding sides are equal.

Similarity plays an important role in measurement. Unknown measures in the real world can be found indirectly. If something is hard to measure, such as the width of a pond, similar triangles may be created from known measures and the proportional relationship between corresponding sides can be used to find the unknown measures.

Similarity is also used in trigonometry, the mathematics of right triangles. The word trigonometry comes from the Greek words meaning "triangle" and "measure." Trigonometric functions can be defined using ratios between the legs and a leg and the hypotenuse of a right triangle. For example, the sine ratio is the ratio between the side opposite one of the acute angles in the triangle and the hypotenuse. For example, if the measure of one of the acute angles in a right triangle is 30°, the sine of 30° always equals $\frac{1}{2}$ regardless of the lengths of the sides of the triangle. Lessons dealing with the trigonometric ratios in right triangles and in all triangles follow in Unit 7.

Mathematical Progressions

Prior Learning	Current Development	Future Connections
Students: • performed dilations and sequences of transformations.	**Students:** • use corresponding parts of similar triangles to solve problems. • prove and use the Triangle Proportionality Theorem. • identify similar right triangles. • develop the AA similarity criterion for triangles.	**Students:** • will explore trigonometry with right triangles. • will explore trigonometry with all triangles.

TEACHER TO TEACHER

From the Classroom

Pose purposeful questions. When I teach mathematics, I try to ask open-ended questions with the purpose of encouraging my students to use reason and make sense of mathematical relationships. For example, to show that two triangles are similar, I ask "What would you do to show the triangles are similar?" Then I ask "Why would you do it that way?" I acknowledge a correct answer, then ask "Did anyone use another method?" "Why did you choose this method?" "Compare the two methods. How are they the same? How are they different?" "Both methods are correct, but which method do you think is better, and why?" This line of questioning makes my students think about how to establish that two triangles are similar and the support their reasoning. Asking good questions makes students think and reason mathematically.

LANGUAGE DEVELOPMENT • Planning for Instruction

By giving all students regular exposure to language routines in context, you will provide opportunities for students to **listen**, **speak**, **read**, and **write** about mathematical situations and develop both mathematical language and conceptual understanding at the same time.

Using Language Routines to Develop Understanding

Use the **Professional Cards** for the following routines to plan for effective instruction.

Co-Craft Questions and Problems Lesson 12.4

Students think of natural questions to ask about a given situation or problems similar to a given task and answer the questions they have developed or problems they have created.

Three Reads Lessons 12.1–12.4

Students read a problem three times with a specific focus each time.

1st Read What is the situation about?
2nd Read What are the quantities in the situation?
3rd Read What are the possible mathematical questions that you could ask for the situation?

Information Gap Lesson 12.2

Students recognize when information given in a problem situation is incomplete, and they pose questions and share knowledge with others to discover any missing facts or relationships and work together to solve the problem.

Critique, Correct, and Clarify Lesson 12.1

Students correct the work in a flawed explanation, argument, or solution method and share with a partner and refine the sample work.

Connecting Language to Similarity

Watch for students' use of the new terms listed below as they explain their reasoning and make connections with new concepts.

Key Academic Vocabulary

Current Development • New Vocabulary

geometric mean the second and third terms in the proportion $\frac{a}{x} = \frac{x}{b}$ where the first and fourth terms are *a* and *b*; equal to the positive value of $\sqrt{ab}$

partition divide a segment into smaller segments

Pythagorean triple a set of three nonzero whole numbers that satisfy the Pythagorean Theorem

similar figures figures that have the same shape but not necessarily the same size

similarity transformation a transformation in which an image has the same shape as the preimage

Linguistic Note

Listen for how students use the word *similar*. It can mean resembling, but not being identical; alike; or in geometry, having the same shape but not necessarily the same size. Have students give examples of how the word is used based on the different definitions. Then have them discuss different meanings of the word, especially the mathematical meaning of the word.

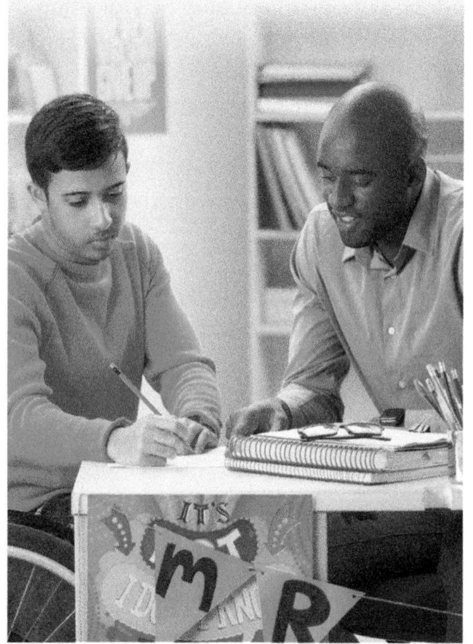

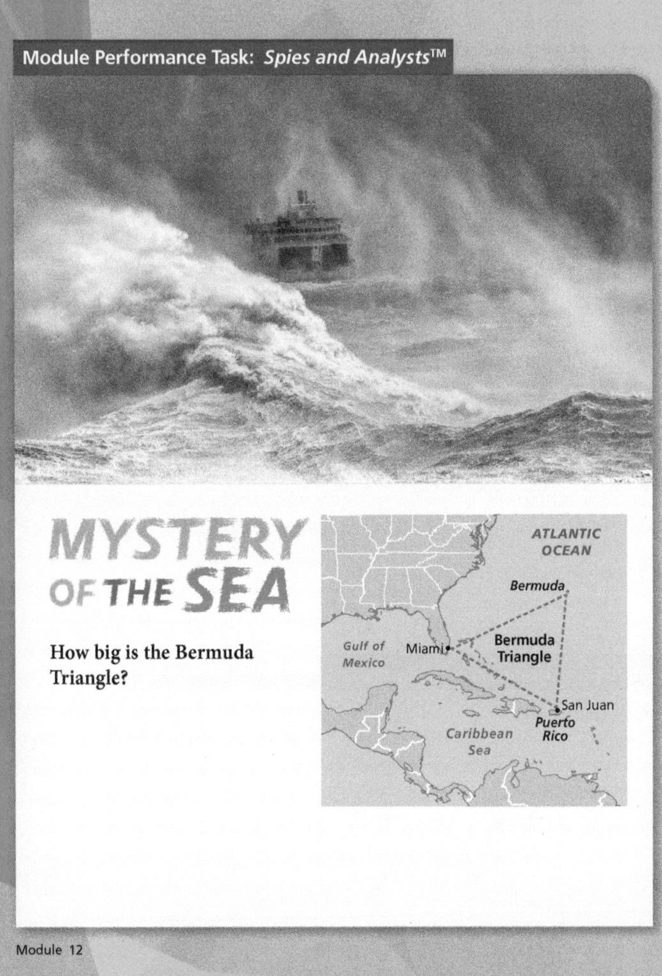

Connections to This Module

One sample solution might include changing the GPS coordinates to (x, y) coordinates.

- Create a triangle similar to the Bermuda Triangle. **(12.1)**
- Use indirect measurement to find the area of the Bermuda Triangle. **(12.2)**

The approximate area of the Bermuda Triangle is about 1 million square miles (2,600,000 square kilometers) depending on the choice of vertices.

MODULE 12

SIMILARITY

MYSTERY OF THE SEA

Overview

To determine "how big" the Bermuda Triangle is, students can make the following assumptions.

- The vertices of the Bermuda Triangle are Miami, Florida, San Juan, Puerta Rico, and Hamilton, Bermuda.
- The area of the triangle indicates "how big" the Bermuda Triangle is.
- The section of Earth that includes the Bermuda triangle lies in a plane.

Be a *Spy*

First, students need to determine the information they need to answer the question. This can include doing the following:

- determine the GPS coordinates of the cities,
- convert the GPS coordinates to x- and y-coordinates.
- graph the coordinates and find the area of the triangle.

Be an *Analyst*

Help students understand that one way to answer the question, "How big is the Bermuda Triangle?" is to ask "What is its area?" Have a map of the area available so that students can use its scale and calculate a reasonably accurate approximation of distances in miles or kilometers between the cities. Then they can find the lengths of a base of the triangle, the altitude to that base, and use the formula $A = \frac{1}{2}bh$, where A is the area measured in square units.

Alternative Approaches

In their analysis, students might:

- use algebraic methods they learn in this module, as shown at the left.
- enclose the Bermuda Triangle in a rectangle made up of four triangular regions—three right triangles and the Bermuda Triangle. The approximate area of the Bermuda Triangle is then the difference between the area of the rectangle minus the sum of the areas of the right triangles.

ONLINE

Assign the Digital Are You Ready? to power actionable reports including

- proficiency by standards
- item analysis

Are You Ready?

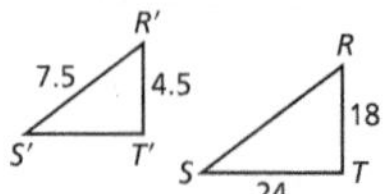

Diagnostic Assessment

- Diagnose prerequisite mastery.
- Identify intervention needs.
- Modify or set up leveled groups.

Have students complete the *Are You Ready?* assessment on their own. Items test the prerequisites required to succeed with the new learning in this module.

Scale Drawings Students will apply previous knowledge of scale drawings to find the lengths of the sides in a scaled right triangle, the scale factor, and the areas of both triangles.

The Pythagorean Theorem and Its Converse Students will apply previous knowledge of the converse of the Pythagorean Theorem to determine whether the lengths of three segments determine a right triangle.

Dilate Figures in the Coordinate Plane Students will apply previous knowledge of dilations to find the coordinates of the vertices of the image of a triangle given its coordinates and a scale factor.

ONLINE 

Are You Ready?

Complete these problems to review concepts and skills you will need for this module.

Scale Drawings

Triangle $R'S'T'$ is a scale of drawing of $\triangle RST$. Answer each question using the figures.

1. What is the length of $\overline{S'T'}$? 6
2. What is the length of $\overline{RS}$? 30
3. What is the scale factor from $\triangle RST$ to $\triangle R'S'T'$? $\frac{1}{4}$
4. Assume $\triangle RST$ and $\triangle R'S'T'$ are right triangles. What is the area of each triangle? 216 square units and 13.5 square units, respectively

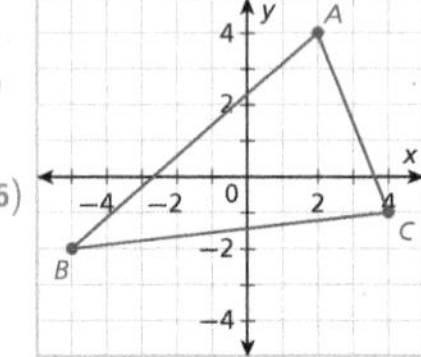

The Pythagorean Theorem and Its Converse

Determine if each triangle is a right triangle using the given side lengths. Explain.

5. $DE = 13.5$, $EF = 18$, $DF = 22.5$ yes; $13.5^2 + 18^2 = 22.5^2$
6. $JL = 7$, $KL = 23$, $JK = 24$ no; $7^2 + 23^2 \neq 24^2$
7. $AB = 6$, $AC = 8$, $BC = 12$ no; $6^2 + 8^2 \neq 12^2$
8. $LM = 9$, $MN = 41$, $LN = 40$ yes; $9^2 + 40^2 = 41^2$

Dilate Figures in the Coordinate Plane

Find the coordinates of $\triangle A'B'C'$ using the given center of dilation and scale factor.

9. Center of dilation: $(0, 0)$
   Scale factor: 2.5
   $A'(5, 10)$, $B'(-12.5, -5)$, $C'(10, -2.5)$
10. Center of dilation: $(2, 4)$
    Scale factor: 40
    $A'(2, 4)$, $B'(-278, -236)$, $C'(82, -196)$

Connecting Past and Present Learning

Previously, you learned:

- to use proportions to perform dilations,
- to develop criteria to prove figures are congruent, and
- to use transformations to prove figures are congruent.

In this module, you will learn:

- to use transformations to prove figures are similar,
- to develop criteria to prove figures are similar, and
- to identify and apply special relationships in similar right triangles.

366

DATA-DRIVEN INTERVENTION

Concept/Skill	Objective	Prior Learning *	Intervene With
Scale Drawings	Compute actual lengths and areas from a scale drawing.	Grade 7, Lesson 1.6	• Tier 3 Skill 4 • Reteach, Grade 7 Lesson 1.6
The Pythagorean Theorem and Its Converse	Use the Pythagorean Theorem and its converse to solve problems.	Grade 8, Lessons 11.1–11.3	• Tier 2 Skill 4 • Reteach, Grade 8 Lessons 11.1–11.3
Dilate Figures in the Coordinate Plane	Find the image of a figure that is dilated in the coordinate plane.	Grade 8, Lesson 2.2	• Tier 2 Skill 12 • Reteach, Grade 8 Lesson 2.2

* Your digital materials include access to resources from Grade 6–Algebra 2. The lessons referenced here contain a variety of resources you can use with students who need support with this content.

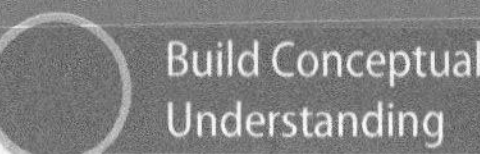

12.1 Use Transformations to Prove Figures are Similar

LESSON FOCUS AND COHERENCE

Mathematics Standards

- A dilation takes a line not passing through the center of the dilation to a parallel line, and leaves a line passing through the center unchanged.
- The dilation of a line segment is longer or shorter in the ratio given by the scale factor.
- Given two figures, use the definition of similarity in terms of similarity transformations to decide if they are similar; explain using similarity transformations the meaning of similarity for triangles as the equality of all corresponding pairs of angles and the proportionality of all corresponding pairs of sides.
- Prove that all circles are similar.

Mathematical Practices and Processes 

- Attend to precision.
- Construct viable arguments and critique the reasoning of others.

I Can Objective 

I can use similarity transformations to prove figures are similar.

Learning Objective

Determine when figures are similar using transformations and comparing corresponding side ratios. Solve problems to prove figures are similar and to find missing values.

Language Objective

Explain how to determine if two figures are similar.

Vocabulary

New: **similar figures**, **similarity transformation**

Lesson Materials: geometric drawing tool

Mathematical Progressions

Prior Learning	Current Development	Future Connections
Students: • performed dilations and sequences of transformations. **(6.1 and 6.2)**	**Students:** • define, identify, and perform dilations. • confirm that figures are similar by finding a sequence of similarity transformations that map one figure onto the other. • prove that all circles are similar. • determine if pairs of figures are similar using sequences of similarity transformations.	**Students:** • will use corresponding parts of similar figures to solve problems. **(12.2)** • will develop the AA similarity criterion for triangles. **(12.3)**

UNPACKING MATH STANDARDS

Given two figures, use the definition of similarity in terms of similarity transformations to decide if they are similar; explain using similarity transformations the meaning of similarity for triangles as the equality of all corresponding pairs of angles and the proportionality of all corresponding pairs of sides.

What it Means to You

Students demonstrate an understanding of similar figures by performing dilations given a scale factor. They can see that when each coordinate is multiplied by the same value, the corresponding side lengths remain proportional.

The emphasis for this standard is on making connections between dilations, scale factors, and similar figures. In future lessons, students will use their understanding of similarity transformations to understand the AA similarity criterion for triangles.

WARM-UP OPTIONS

ONLINE

PROJECTABLE & PRINTABLE 

ACTIVATE PRIOR KNOWLEDGE • Perform Dilations

Use these activities to quickly assess and activate prior knowledge as needed.

Problem of the Day 

Gerald needs to photocopy his driver's license so it is 30% larger than the original. He sets the copier to a scale factor of 1.3 and copies his original license that has dimensions of 3 inches by 2 inches. What are the dimensions of the larger license? 3.9 inches by 2.6 inches

Quick Check for Homework

As part of your daily routine, you may want to display the Teacher Solution Key to have students check their homework.

Make Connections

Based on students' responses to the Problem of the Day, choose one of the following:

 1 Project the Interactive Reteach, Grade 8, Lesson 2.2.

 2 Complete the Prerequisite Skills Activity:

Have students plot the vertices and draw the triangle with the coordinates $A(-3, 2)$, $B(0, -3)$, and $A(3, 1)$. Ask half of the class to dilate the triangle using the scale factor $\frac{1}{3}$ and the other half of the class using the scale factor 2. Ask students to discuss in pairs the effect each scale factor had on the original triangle.

- *How do you use the scale factor to dilate the triangle?* Multiply each x- and y-coordinate by the scale factor.
- *What happens to the triangle when the scale factor is smaller than 1 but greater than 0?* The dilated triangle is smaller than the original.
- What happens to the triangle when the scale factor is greater than 1? The dilated triangle is larger than the original.
- How are the coordinates affected? The coordinates are either divided by 3 or are multiplied by 2.

If students continue to struggle, use Tier 2 Skill 12.

SHARPEN SKILLS

If time permits, use this on-level activity to build fluency and practice basic skills.

Vocabulary Review

Objective: Students demonstrate an understand of similar figures.
Materials: Word Description Template (Teacher Resource Masters)

Have students write "similar figures" in the center box of the word description template. Work together as a class to write a definition and to describe some characteristics of similar figures. Have students work in pairs to draw examples and non-examples of similar figures.

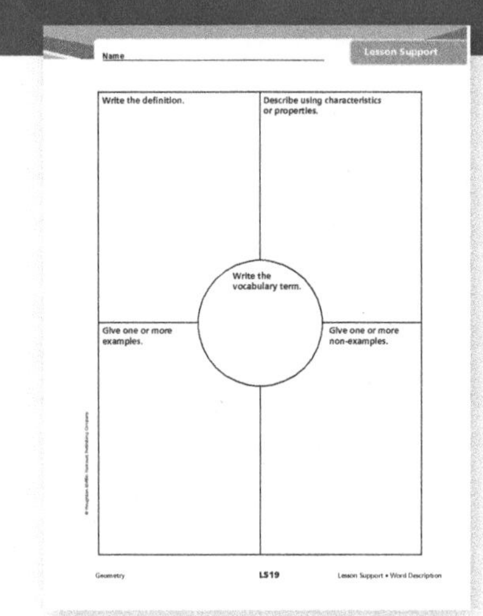
Name
Lesson Support
Write the definition.
Describe using characteristics or properties.
Write the vocabulary term.
Give one or more examples.
Give one or more non-examples.
Geometry LS19 Lesson Support • Word Description

PLAN FOR DIFFERENTIATED INSTRUCTION

Small-Group Options

Use these teacher-guided activities with pulled small groups.

On Track

Materials: graph paper

Have each student draw a triangle or a quadrilateral on graph paper. Ask students to choose a scale factor and dilate their figure. After the dilation, they can also translate, reflect, or rotate their figure to draw the final figure. They should label the coordinates of the original figure, *A*, *B*, *C*, and *D* and the coordinates of the transformed figure, A', B', C', D'.

Collect the students graphs and redistribute them. Using the graph they receive, students should determine the scale factor and any transformations used between the two figures.

Almost There

Draw the labeled rectangles on the board.

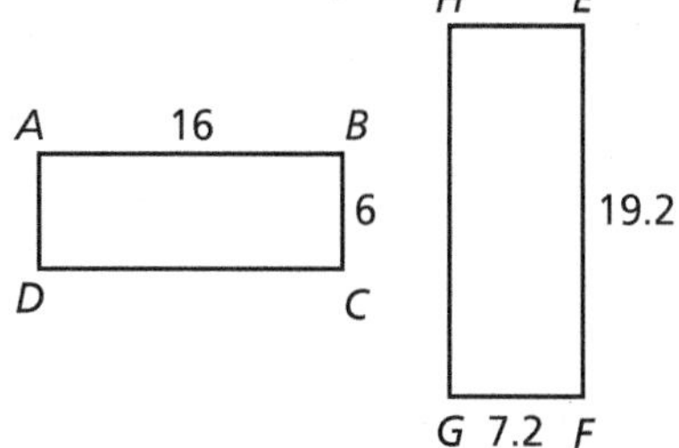

Tell students that the rectangles are similar. Then ask:

- What are the pairs of corresponding angles? If the figures are rectangle, what is true about all of the angles?
- What are the pairs of corresponding sides? How do you know which side in *ABCD* corresponds to the side in *EFGH*?
- What is the scale factor from *ABCD* to *EFGH*?

Ready for More

Materials: graph paper

Write the following sequence of transformations on the board.

$$(x, y) \rightarrow \left(\frac{1}{5}x, \frac{1}{5}y\right) \rightarrow (x - 3, y + 1) \rightarrow (-y, -x)$$

Have pairs of students draw a figure in the coordinate plane and perform the transformations. Then pose the following questions:

- Does the sequence of transformations produce similar figures? Why or why not?
- How can you prove the first and last figure are similar?
- Which of the transformations are rigid and which produce similar figures?

Math Center Options

Use these student self-directed activities at centers or stations.

Key: ● Print Resources ● Online Resources

On Track

- ● Interactive Digital Lesson
- ●● Journal and Practice Workbook
- ● Interactive Glossary (printable): **similar figures**, **similarity transformation**
- ● Module Performance Task
- ● Illustrative Mathematics: Are They Similar?

Almost There

- ● Reteach 12.1 (printable)
- ● Interactive Reteach 12.1
- ● RtI Tier 2 Skill 12: Dilate Figures in the Coordinate Plane
- ● Illustrative Mathematics: Effects of Dilations on Length, Area, and Angles

Ready for More

- ● Challenge 12.1 (printable)
- ● Interactive Challenge 12.1
- ● Illustrative Mathematics: Similar Quadrilaterals

Unit Project Check students' progress by asking them how they know if two figures are similar.

View data-driven grouping recommendations and assign differentiation resources.

During the *Spark Your Learning,* listen and watch for strategies students use. See samples of student work on this page.

Use the Scale Factor to Perform a Dilation

Strategy 1

I know the dimensions of an actual regulation court, and I see how a scale is used for the photograph. I can use the same scale 1 in. = 20 ft to calculate the dimensions of a scale model that is a reduction of the actual regulation court.

If students . . . use the scale factor to solve the problem, they are performing a dilation they learned in an earlier course.

Have these students . . . explain how they used the scale factor to find the new dimensions and why they are allowed to do this. **Ask:**

- Q How did you know the transformation from the photo to the real court is a dilation?
- Q What operation did you use to find the dimensions of the real court?

Make a Table

Strategy 2

I can use the given scale 1 inch to 20 feet to make a table of values so that I can estimate the dimensions of a scale model that is a reduction of the actual regulation court.

Inches	Feet
1	20
2	40
3	60
4	80
5	100
6	120

If students . . . make a table to solve the problem, they understand the relationship between the dimensions in the photo and the dimensions of the real court but may not understand that the dimensions can be multiplied by the scale factor without making a table.

Activate prior knowledge . . . by having students look at the values in the table. **Ask:**

- Q How can you find the number of feet if the inches in the photo are not a whole number?
- Q In general, what operation can you perform on the dimensions in the photograph to find the dimensions of the real court?

COMMON ERROR: Ignores the Scale

I can make a rectangle that is about the same shape and copy the shapes into my drawing.

If students . . . ignore the scale, then they are only comparing the rectangular shape of the photo to the real court.

Then intervene . . . by pointing out that the figures are similar by a given scale factor. **Ask:**

- Q What is true about the side lengths in similar figures?
- Q How can make sure the dimensions of the real court are in the same ratio as the dimensions in the photo?

12.1

Use Transformations to Prove Figures Are Similar

I Can use similarity transformations to prove figures are similar.

Spark Your Learning

The boundaries and lines on basketball courts are drawn to regulatory specifications to ensure the distances are the same.

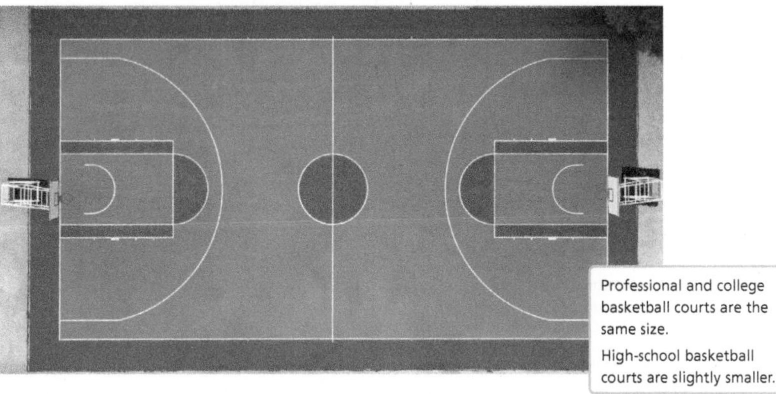

Professional and college basketball courts are the same size.

High-school basketball courts are slightly smaller.

Complete Part A as a whole class. Then complete Parts B–D in small groups.

A. What is a mathematical question you can ask about this basketball court? A. How can I draw the boundaries and lines of a regulation basketball court?

B. Is it possible to use the pictured regulation court to determine the boundaries and lines on an actual court? What additional information do you need? See Additional Answers.

C. To answer your question, what strategy and tool would you use along with all the information you have? What answer do you get? See Strategies 1 and 2 on the facing page.

D. Is a transformation used to create the scale drawing? If so describe the transformation. Possible answer: A transformation called a *dilation* was used with a factor of 0.05 (dividing by 20).

Turn and Talk Consider two congruent polygons. One of the polygons is enlarged by a ratio of 4. What are the relationships between corresponding angles and corresponding sides? See margin.

CULTIVATE CONVERSATION • Three Reads

Tell students to read the information in the photo three times and prompt them with a different question each time.

1. What is the situation about? The situation is about a scale model of a basketball court represented by a photo. The photo shows a basketball court and you want to know how to find the measurements for the actual court.
2. What are the quantities in this situation? How are those quantities related? The quantities are the length and width of the basketball court in the photo and in real life. The dimensions of the photo are related to the real court using the scale 1 inch to 20 feet.
3. What are possible questions you could ask about this situation? Possible answer: Are the angles the same in the photo and in the real court? Can the measurements be found without knowing the scale? How can the scale be used to find the measurements of the real court? Are the photo and the real court similar?

1 Spark Your Learning

MOTIVATE

- Have students look at the photo in their books and read the information contained in the photo. Then complete Part A as a whole-class discussion.
- Give the class the additional information they need to solve the problem. This information is available online as a printable and projectable page in the Teacher Resources.
- Have students work in small groups to complete Parts B–D.

PERSEVERE

If students need support, guide them by asking:

Q **Advancing • Use Tools** Which tool could you use to solve the problem? Why choose that tool and not some other? Students' choices of tools and reasons for choosing them will vary.

Q **Assessing** What measurements do you need to take of the photo in order to help find the measurements of the real court? You should measure the length and width in inches since you know that 1 inch is equal to 20 feet on the real court.

Q **Assessing** How do you know the photo will have similar dimensions to the actual basketball court? A photo shrinks the actual size of the court while keeping the width and length in the same proportion.

Q **Advancing** How can you use the scale to find the lengths for the boundary lines of the court? You can multiply the inches for each dimension by 20 to find the number of feet on the real court.

Turn and Talk Have students consider the angles of the basketball court in the photo and a basketball court in real life. Ask, "Do the right angles change? Think about the side lengths in the photo of the basketball court and a court in real life. How do the side lengths compare?" The corresponding angles in the polygons would stay the same and be congruent. The side lengths would be 4 times as large, or an increase of 400%.

BUILD SHARED UNDERSTANDING

Select groups of students who used various strategies and tools to share with the class how they solved the problem. As they present their solutions, have each group discuss why they chose a specific strategy and tool.

② Learn Together

Build Understanding

Task 1 (MP) **Attend to Precision** Students use drawing tools to measure and compare corresponding side lengths of triangles.

CONNECT TO VOCABULARY

Have students use the **Interactive Glossary** to record their understanding of the vocabulary in this task.

Sample Guided Discussion:

Q In Part B, the dilated triangle is larger, but the corresponding angles are the same. Why is this true? The dilation stretches the sides and makes a larger triangle but does not change the shape or angles of the triangle.

Turn and Talk Point out to students that when the parallel line is constructed in Part C, the parallel line coincides with a segment on the second triangle. Remind students that parallel lines have slopes that are the same or equal. Possible answer: The slopes of the image and preimage of a segment that undergoes a dilation would stay the same; The slopes would remain the same after some other transformations, such as a translation, but the slopes might not be the same after a rotation or reflection, or a composition of transformations.

Build Understanding

Investigate Dilations

You learned a dilation is a transformation that changes the size of a figure but does not change the shape. The scale factor defines the ratio of the dilation. A scale factor greater than 1 enlarges the figure, and a scale factor between 0 and 1 reduces the figure.

A, C. See Additional Answers.

1 Use a geometric drawing tool to construct $\triangle ABC$. Place point D in the interior of the triangle. Then use the dilation command to dilate $\triangle ABC$ by a scale factor of 2 with center of dilation D.

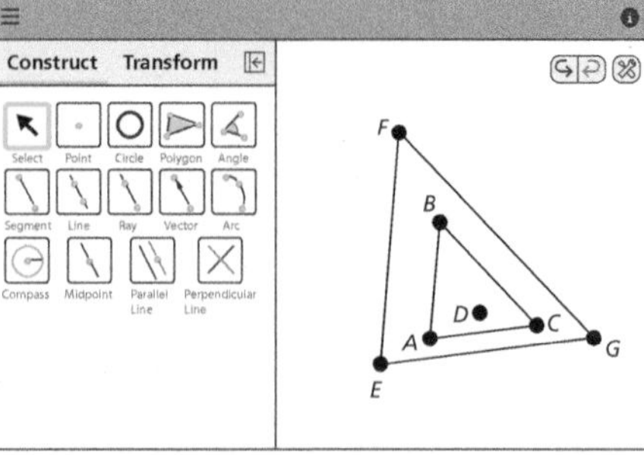

A. Measure the sides of the triangles. How can you use these measures to confirm that $\triangle EFG$ is a dilation of $\triangle ABC$ by a scale factor of 2? Do the sides of $\triangle ABC$ and $\triangle EFG$ have the same relationship as you drag one of the vertices of $\triangle ABC$? Explain.

B. Measure the angles of $\triangle ABC$ and $\triangle EFG$. What do you notice about the measure of each angle?
The corresponding angles of the triangles are congruent.

C. Construct a line parallel to $\overline{BC}$ through a point on $\overline{FG}$. How is this line related to $\overline{FG}$? Does this relationship change as you move one of the vertices of $\triangle ABC$? Does the same relationship exist between other corresponding sides of the image and the preimage?

Turn and Talk Based on your results, what can you conclude about the slopes of the image and preimage of any segment that undergoes a dilation? Does the conclusion apply if the figure undergoes other types of transformations? Explain. See margin.

Find a Sequence of Similarity Transformations

A **similarity transformation** is a transformation in which the image has the same shape as the preimage. If a figure can be mapped to another figure using a sequence of similarity transformations, then the figures are **similar figures**. Two similar figures have the same shape but not necessarily the same size. You can write "$\triangle ABC$ is similar to $\triangle EFG$" as $\triangle ABC \sim \triangle EFG$. Similarity transformations are dilations and rigid motions.

Properties of Similar Figures
Corresponding angles of similar figures are congruent.
Corresponding sides of similar figures are proportional. The ratio of the lengths of two corresponding sides is the scale factor.

368

LEVELED QUESTIONS

Depth of Knowledge (DOK)	Leveled Questions	What Does This Tell You?
Level 1 **Recall**	How do you know if two triangles are similar? They have the same size and the same shape. Corresponding pairs of angles are congruent and corresponding pairs of sides are proportional.	Students' answers will indicate they know the definition of similar triangles and what to look for when determining if triangles are similar.
Level 2 **Basic Application of Skills & Concepts**	What does it mean for corresponding pairs of sides to be proportional in similar figures? The ratio of corresponding sides will always be the same value.	Students' answers will indicate they can compare side lengths for corresponding pairs of sides to see if they have the same ratio.
Level 3 **Strategic Thinking & Complex Reasoning**	Are congruent figures also similar? Explain. Congruent figures will have congruent pairs of corresponding angles and sides. The ratio of corresponding sides will always be the same number and have a scale factor of dilation equal to 1, because they are the same size and shape. So they are also similar.	Students' answers will indicate they know that congruent figures are the similar based on the definition of similar figures.

2 In the figure below, determine whether quadrilateral *ABCD* is similar to *LMNP*.

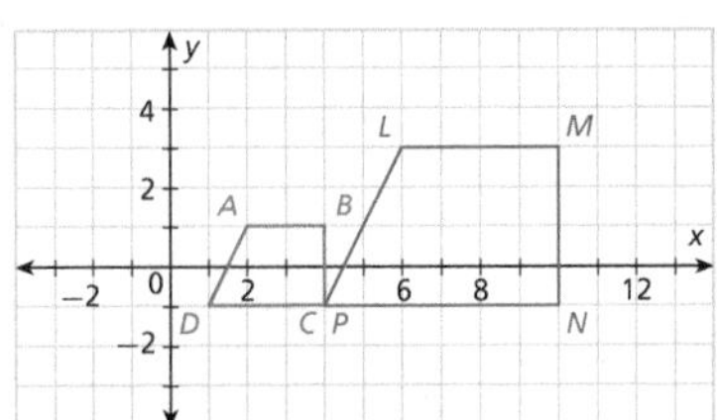

A–F. See Additional Answers.

A. Is there a sequence of similarity transformations that maps *LMNP* onto *ABCD*? If so, would the dilation have a scale factor greater than 1 or between 0 and 1? Explain.

B. Write a ratio between two corresponding sides of *LMNP* and *ABCD* to determine a scale factor *k* of a possible dilation.

C. A transformation of *LMNP* is shown below as *L′M′N′P′*.

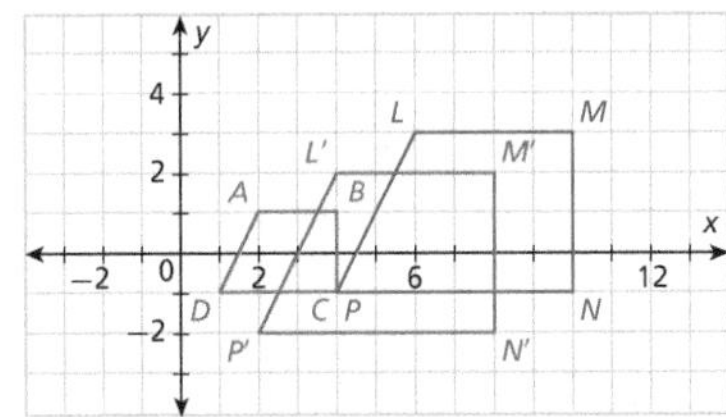

Explain how the coordinates of *LMNP* and *L′M′N′P′* show that a translation maps *LMNP* onto *L′M′N′P′*.

D. The graph shows *ABCD*, which is the image of *L′M′N′P′* after the dilation. What is true about all pairs of corresponding sides? Use examples from the graph to support your answer.

E. A sequence of transformations that maps *LMNP* onto *ABCD* is a translation $(x, y) \to (x-2, y-1)$ followed by a dilation $(x, y) \to \left(\frac{1}{2}x, \frac{1}{2}y\right)$. How do you know *LMNP* and *ABCD* are similar figures? Are there any angles in *ABCD* that must be congruent to $\angle M$ in figure *LMNP*? Explain.

F. Is there only one sequence of similarity transformations that maps *LMNP* to *ABCD*? Explain.

Turn and Talk In the graph above, suppose another quadrilateral *RMNP* has vertex $R(4, 4)$. Would *RMNP* be similar to *ABCD*? Explain. See margin.

Task 2 (MP) **Attend to Precision** Students compare coordinates in a dilation to determine if two figures are similar.

Sample Guided Discussion:

Q What must be true of a scale factor that enlarges a figure? The scale factor must be greater than 1.

Q What must be true of a scale factor that shrinks a figure? The scale factor must be between 0 and 1.

Q In Part C, how do the *x*-coordinates compare for each corresponding point? How do the *y*-coordinates compare for each corresponding point? The *x*-coordinates are 2 less for the translated figure compared to the original figure; The *y*-coordinates are 1 less for the translated figure compared to the original figure.

Turn and Talk Point out to students that when the vertex *L* in *LMNP* changes to *R*, the shape of the quadrilateral changes. Remind them that similar figures have the same shape, even when their sizes are different. Have them compare the corresponding angles in *LMNP* to the angles in *RMNP*. no; Moving one vertex would cause the corresponding angles $\angle R$ and $\angle A$ to be not congruent. So *RMNP* would not be similar to *ABCD*.

EL PROFICIENCY LEVEL

Beginning

Pair students. Give each pair six index cards: three with scaled figures of squares (labeled A–C from smallest to largest), two with scaled circle figures (labeled D–E from smallest to largest), and one with a rectangle (labeled F). Have students pick two cards at random and verbally describe the figures as similar or not similar.

Intermediate

Using the same pairs, ask students to pull out figures A–C. Starting from figure B, have one student explain how figure A relates to figure B. Ask the second student what they know about the scale factor used on B to create A. Then have the pair reverse roles and explain the relationship between figures B and C.

Advanced

Ask students to write a short paragraph to explain why a dilation is not a rigid transformation.

Task 3 (MP) **Construct Arguments** Students prove the Circle Similarity Theorem using transformations.

Sample Guided Discussion:

Q The manufacturer translates and then dilates the circle. Which of these are rigid transformations? The translation is a rigid transformation, but the dilation is not a rigid transformation because the size of the figure changes.

Q In Part D, what do you know about the definition of similar figures that can help you determine if circles are similar? Possible answer: Similar figures have the same shape and corresponding sides that are proportional. The radii of the circles are always in the same ratio.

Turn and Talk Students may think that all shapes of the same type are similar. Give two examples of triangles, one that is a right triangle and one that is an acute triangle. Help students to see that two triangles do not necessarily have corresponding angles that are congruent. Possible answers: There is only one way to draw a shape that fits the description of a circle, and the only variation between any circles is the radius. So, any circle is similar to any other circle. All squares are similar. In the same way as the circle, there is only one way to draw a shape that fits the description of a circle, and the only variation between any square is the side length. But not all rectangles are similar. Though the corresponding angles will be congruent, the corresponding lengths and widths may not have the same scale factor.

Prove All Circles Are Similar

You can use what you know about similarity to prove theorems about circles.

Circle Similarity Theorem
All circles are similar.

3 Prove the Circle Similarity Theorem.

Given: circle C with center C and radius r,
circle D with center D and radius s

Prove: Circle C is similar to circle D.

Suppose a company manufactures vinyl records. To prove similarity, show there is a sequence of similarity transformations that maps circle C, the label, to circle D, the vinyl.

A–E. See Additional Answers.

A. How is the proof generalized by using variables to represents the radii?

B. To place the label on the record, the manufacturer translates circle C so its center coincides with the center of circle D. The image of circle C is circle C'. Is this transformation a similarity transformation? Explain.

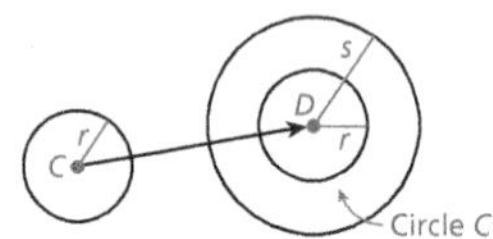

C. Transform circle C' with a dilation centered at point D and a scale factor of $\frac{s}{r}$. The image of circle C' after the dilation has a radius of $\frac{s}{r} \cdot r = s$. How are the images of circle C' after the dilation and circle D related?

D. Why can you conclude the circles are similar?

E. Show that the scale factor also transforms the diameters and circumferences in the same way as the radii.

$\text{Diameter}_D \stackrel{?}{=} \frac{s}{r} \cdot \text{Diameter}_C$

$\text{Circumference}_D \stackrel{?}{=} \frac{s}{r} \cdot \text{Circumference}_C$

Turn and Talk Are all geometric shapes of the same type similar? Explain. See margin.

Step It Out

Apply Properties of Similar Figures

Recall that in similar figures, corresponding angles are congruent and pairs of corresponding sides are proportional. You can use these facts to find unknown angle measures or lengths of corresponding sides.

A pantograph is a tool that allows the user to make a copy of a figure by tracing. The copy can be smaller or larger and will be similar to the original.

4 An artist used a pantograph to make figure $ABCDE$ that is similar to $PQRST$.

Find the values of x and y.

Find the value of *x*.

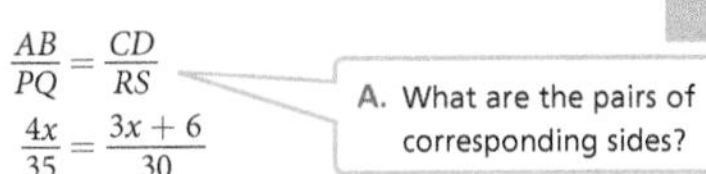

$$\frac{AB}{PQ} = \frac{CD}{RS}$$

$$\frac{4x}{35} = \frac{3x+6}{30}$$

A. What are the pairs of corresponding sides?

$$4x(30) = 35(3x+6)$$
$$120x = 105x + 210$$
$$15x = 210$$
$$x = 14$$

A. Possible answer: The five pairs of corresponding sides are: $\overline{AB}$ and $\overline{PQ}$, $\overline{BC}$ and $\overline{QR}$, $\overline{CD}$ and $\overline{RS}$, $\overline{DE}$ and $\overline{ST}$, and $\overline{EA}$ and $\overline{TP}$.

B. no; These are the only two pairs of corresponding sides that have lengths given.

B. Is it possible to use any other pairs of corresponding sides to write and solve a proportion to solve for x? Explain why or why not.

Find the value of *y*.

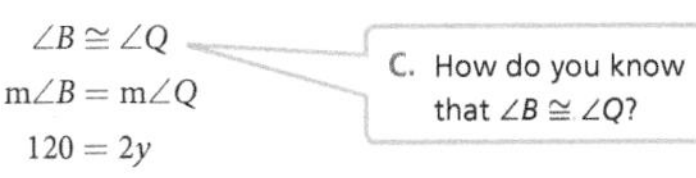

$$\angle B \cong \angle Q$$
$$m\angle B = m\angle Q$$
$$120 = 2y$$
$$y = 60$$

C. How do you know that $\angle B \cong \angle Q$?

Possible answer: The angles are corresponding angles of similar polygons, so they are congruent.

D. What information must you know to determine the scale factor of the dilation that is part of the sequence of similarity transformations that maps $PQRST$ onto $ABCDE$? Do you have enough information to determine the scale factor? If so, what is it? See Additional Answers.

Turn and Talk If two polygons are similar, do the perimeters of the polygons have the same ratio as the corresponding pairs of sides? Why or why not? See margin.

Step It Out

Task 4 **MP** **Construct Arguments** Students use properties of similar figures to find unknown values for side lengths and angle measures and use these values to find the scale factor.

EL **OPTIMIZE OUTPUT** **Critique, Correct, and Clarify**

Provide the following scenario:

A student says the scale factor is the ratio 5:8 for some pairs of sides and 8:5 for other pairs of sides. Is this possible? Why or why not?

Have students work in small groups to identify if the scenario of finding two different scale factors is possible. Ask students to determine what error the student likely made and how it can be avoided.

Sample Guided Discussion:

Q **The polygons have the same shape, but different orientations. How can you use the names *ABCDE* and *PQRST* to determine corresponding angles?** The corresponding angles are in the same place in the names. For example, $\angle C$ is corresponding to $\angle R$ because they are both third in the names.

Q **Why can't you use the measures of corresponding angles to find the scale factor for similar figures?** The corresponding angles in similar figures are congruent. Since the angles are the same measure, comparing the ratio of corresponding angle measures would always be 1.

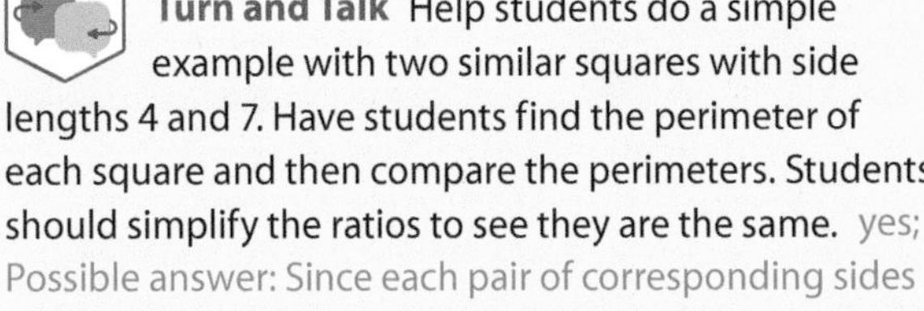

Turn and Talk Help students do a simple example with two similar squares with side lengths 4 and 7. Have students find the perimeter of each square and then compare the perimeters. Students should simplify the ratios to see they are the same. yes; Possible answer: Since each pair of corresponding sides has the same ratio, the sum of the sides, the perimeters, will have the same ratio.

ONLINE

Assign the Digital On Your Own for
- built-in student supports
- Actionable Item Reports
- Standards Analysis Reports

On Your Own

Assignment Guide

The chart below indicates which problems in the On Your Own are associated with each task in the Learn Together. Assign daily homework for tasks completed.

Learn Together Tasks	On Your Own Problems
Task 1, p. 368	Problems 6–8 and 25
Task 2, p. 369	Problems 9–15, and 26
Task 3, p. 370	Problems 16–19
Task 4, p. 371	Problems 20–24

Check Understanding

For Problems 1–3, use the graph where $\triangle PQR$ is a dilation of $\triangle LMN$.

1–5. See Additional Answers.

1. Which side of the image is parallel to $\overline{NL}$? Explain.
2. Which angle in $\triangle PQR$ is congruent to $\angle L$? Explain.
3. Is $\triangle LMN$ similar to $\triangle PQR$? Explain why or why not.
4. Circle P has a radius of 3.6 mm. Circle Q has a radius of 10.8 mm. Is circle Q similar to circle P? If so, what is the scale factor?
5. Figure $ABCD$ is similar to figure $PQRS$. What is the length of $\overline{CD}$?

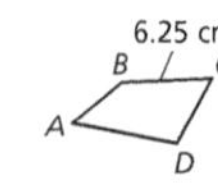

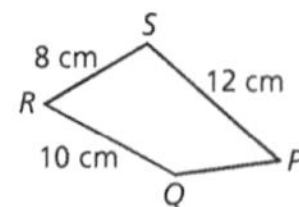

On Your Own

For Problems 6 and 7, determine whether the transformation is a dilation. If so, what is the scale factor? Explain your reasoning. 6–11. See Additional Answers.

6. 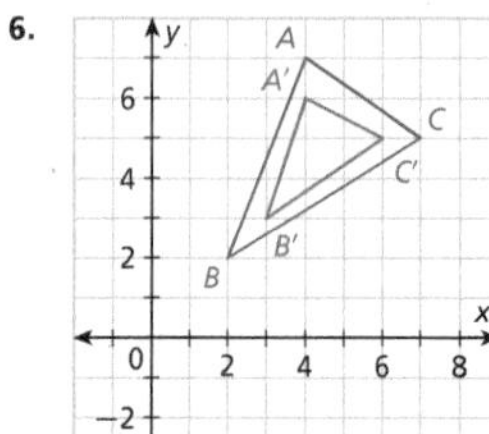

7. 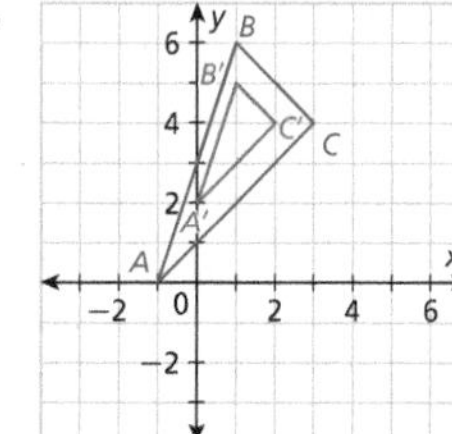

In the diagram, $ABCD \sim FGHJ$.

8. Explain why $ABCD \sim FGHJ$.
9. Is $\angle J \cong \angle D$? Explain.
10. Is $\frac{AB}{BC}$ equal to $\frac{FG}{GH}$? Explain.
11. Is $ABCD \sim LMNP$? Explain why or why not.

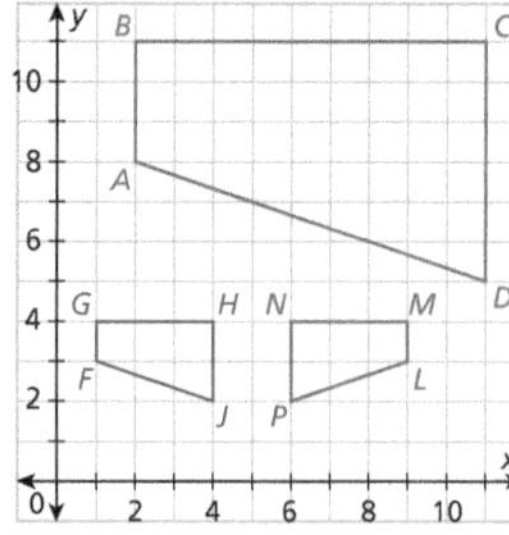

372

data checkpoint

3 Check Understanding

Formative Assessment

Use formative assessment to determine if your students are successful with this lesson's learning objective.

Students who successfully complete the Check Understanding can continue to the On Your Own practice.

For students who miss 1 problem or more, work in a pulled small group using the Almost There small-group activity on page 367C.

ONLINE

Assign the Digital Check Understanding to determine
- success with the learning objective
- items to review
- grouping and differentiation resources

4 Differentiation Options

Differentiate instruction for all students using small-group activities and math center activities on page 367C.

Reteach

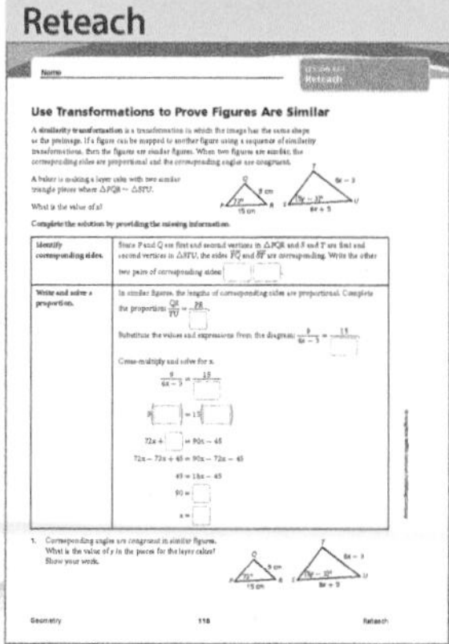

Challenge

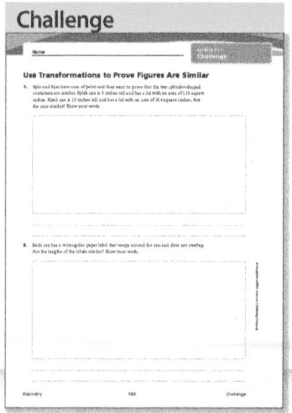

Determine whether each pair of figures is similar using similarity transformations. Explain your reasoning. 12–15. See Additional Answers.

12. *JKLM* to *WXYZ*

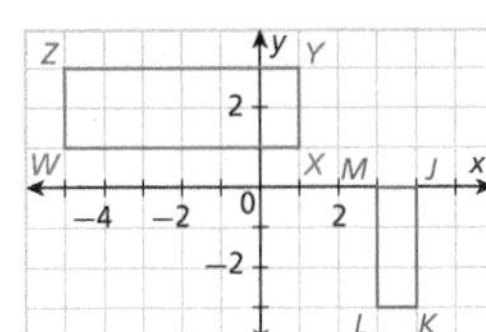

13. $\triangle DEF$ to $\triangle LMN$

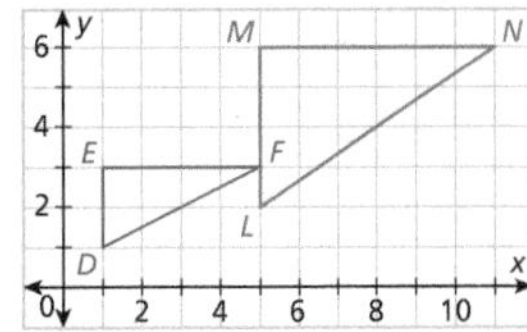

14. *ABCD* to *PQRS*

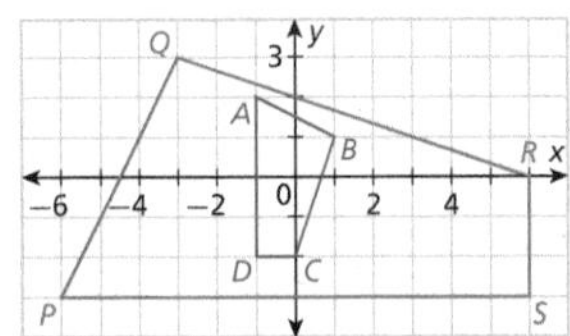

15. *MNPQR* to *ABCDE*

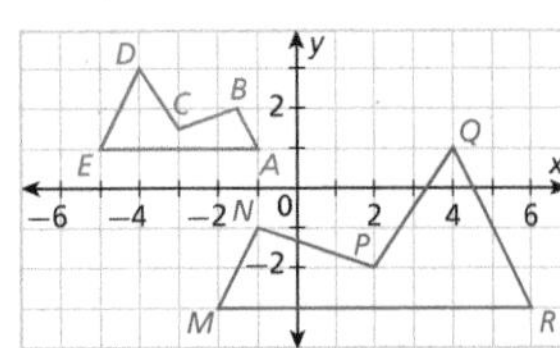

Social Studies **For Problems 16–18, the photo shows the circular terraces at the Moray Ruins in Peru. It is widely believed that the ruins were used as an agricultural laboratory by the Incas. The circular terraces have the same center.**

16–22. See Additional Answers.

16. Are the circles in the terraces similar? Explain your reasoning.

17. Describe the similarity transformation that would take the largest terrace circle onto the smallest. Give an estimate of the scale factor.

18. How many times greater is the circumference of the largest terrace circle than the circumference of the smallest circle?

19. (MP) **Reason** Are congruent circles also similar circles? Explain your reasoning.

In the diagram, $JKLM \sim EFGH$**.**

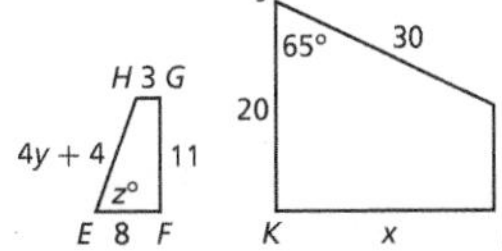

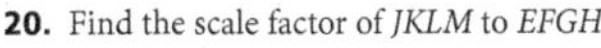

20. Find the scale factor of *JKLM* to *EFGH*.

21. Find the values of *x*, *y*, and *z*.

22. Find the perimeter of each polygon.

23. **Open Ended** Sketch a figure labeled *PQRS* that is larger than and similar to figure *JKLM*. What is the scale factor of *PQRS* to *JKLM*? Check students' sketches.

Questioning Strategies

Problem 17 Suppose the smallest terrace circle needed to be transformed onto the largest terrace circle. Would the scale factor be greater than 1 or between 0 and 1? How do you know? To go from the smaller to the larger, the scale factor would need to be greater than 1; This means the radius of the smaller circle would be multiplied by a number that would increase the value of the radius.

Watch for Common Errors

Problem 20 The two quadrilaterals are similar and the same shape but have different orientation. Some students may try to match the 3 with the 30 because the values are both at the top of figures. Or, they may try to compare 11 and 20 because they are both on the sides of the figures. Remind students to look at the similarity statement $JKLM \sim EFGH$ to determine which sides and angles are corresponding.

(5) Wrap-Up

Summarize learning with your class. Consider using the Exit Ticket, Put It in Writing, or I Can scale.

Exit Ticket 

A family has a circular swimming pool that has a radius of 15 feet. They decide to put a circular pond near the pool that has a radius of 6 feet. What scale factor can the family use to compare the measurements of the pool to the measurements of the pond? The family would multiply the radius of the pool by the scale factor to get the radius of the pond using an equation such as $15 \cdot x = 6$. Solving the equation gives a scale factor of $\frac{2}{5}$.

Put It in Writing 

Write a paragraph describing two ways you can determine if two figures are similar.

I Can 

The scale below can help you and your students understand their progress on a learning goal.

4	I can use similarity transformations to prove figures are similar, and I can explain my reasoning to others.
3	I can use similarity transformations to prove figures are similar.
2	I can solve for missing values when I know figures are similar.
1	I can identify when two figures are similar if I know the side lengths and angle measures.

Spiral Review • Assessment Readiness

These questions will help determine if students have retained information taught in the past and can also prepare them for high-stakes assessments. Here, students must find the value of a variable using the diagonals of a rectangle **(11.3)**, use properties of kites to find congruent angle pairs **(11.5)**, use characteristics of quadrilaterals to identify a specific quadrilateral **(11.4)**, and determine if triangles are similar **(Gr8, 2.3)**.

24–26. See Additional Answers.

24. A. An Olympic-sized pool containing 10 lanes is shown. A shorter-sized pool contains only 8 lanes and is 25 m long by 18.3 m wide. Are the two pools similar in shape? Explain.

B. Financial Literacy It costs about \$10.70 per square meter per year to maintain the Olympic-sized pool. Estimate the total cost per year to maintain a pool that is 75% of the area of the Olympic-sized pool.

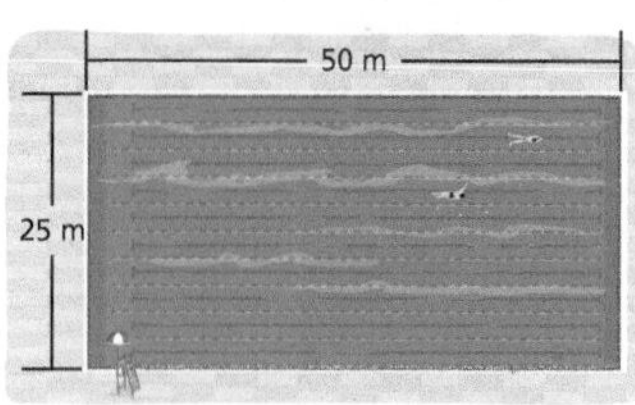

25. (MP) **Critique Reasoning** Michael and Sara are completing the following problem. Michael determines the answer is 3. Sara determines the answer is 2. Which student, if either, is correct? Justify your reasoning.

"A triangle on the coordinate plane is dilated from $\triangle LMN$ to $\triangle L'M'N'$. $\overline{LM}$ lies along the line $y = 2x + 3$, $\overline{MN}$ lies along the line $y = -\frac{1}{3}x + 6$, and $\overline{NL}$ lies along the line $y = \frac{1}{4}x - 4$. The center of dilation is the origin. What is the slope of the line on which $\overline{M'N'}$ lies?" ✓ ✗

26. A figure is transformed by $(x, y) \rightarrow (x + 2, y - 3)$ followed by $(x, y) \rightarrow (x, 3y)$. Does this sequence of transformations produce a pair of similar figures? Explain your reasoning.

Spiral Review • Assessment Readiness

27. In rectangle $ABCD$, $AC = (x + 15)$ ft and $BD = (3x - 10)$ ft. What is the length of $\overline{BD}$?

Ⓐ 2.5 ft
Ⓑ 6.25 ft
Ⓒ 12.5 ft
Ⓓ 27.5 ft

28. Kite $PQRS$ is symmetric across $\overline{PR}$. The diagonals intersect at point T. Which pairs of angles are congruent? Select all that apply.

Ⓐ $\angle PTS \cong \angle PTQ$
Ⓑ $\angle SPQ \cong \angle SRQ$
Ⓒ $\angle PSR \cong \angle PQR$
Ⓓ $\angle PTS \cong \angle RTQ$

29. A quadrilateral is described by the following: two consecutive sides are congruent, two sides are parallel, and an interior angle is 90°. What could be the shape? Select all that apply.

Ⓐ square
Ⓑ rectangle
Ⓒ trapezoid
Ⓓ rhombus
Ⓔ parallelogram
Ⓕ triangle

30. Two triangles have three corresponding congruent angles. Are the triangles similar?

Ⓐ always
Ⓑ never
Ⓒ sometimes
Ⓓ It cannot be determined.

 **I'm in a Learning Mindset!**

How can I improve my approach to solving transformation problems?

374 Keep Going Journal and Practice Workbook

Learning Mindset

Resilience Adjusts to Change

Point out that the definition of similar figures is two figures that have the same size and shape with congruent corresponding pairs of angles and corresponding side lengths that are proportional. . There is often more than one way to find the scale factor between two similar figures or to determine if two figures are similar. Students should recognize that their approach to problems involving similarity does not have to be static. *When your first approach to solving a problem doesn't seem to work, what other options do you have? Could there be more than one way to solve a problem? What have you learned in this lesson that can help you solve problems about similarity transformations?*

12.2 Develop AA Triangle Similarity

LESSON FOCUS AND COHERENCE

Mathematics Standards

- Use the properties of similarity transformations to establish the AA criterion for two triangles to be similar.
- Use congruence and similarity criteria for triangles to solve problems and to prove relationships in geometric figures.

Mathematical Practices and Processes 

- Construct viable arguments and critique the reasoning of others.
- Attend to precision.
- Model with mathematics.

I Can Objective 

I can prove AA, SSS, and SAS Similarity Theorems.

Learning Objective

Prove the AA Triangle Similarity Theorem and use it to find missing dimensions of triangles. Use the SSS and SAS Triangle Similarity Theorems to prove triangles are similar and find missing dimensions.

Language Objective

Explain how to prove the AA Triangle Similarity Theorem.

Vocabulary

Lesson Materials: Cards for Sharpen Skills and Plan for Differentiated Instruction activities

Mathematical Progressions

Prior Learning	Current Development	Future Connections
Students: • performed dilations. **(6.1)** • applied sequences of transformations. **(6.2)**	**Students:** • learn to prove the AA Triangle Similarity Theorem. • apply the AA Triangle Similarity Theorem to show that two triangles are similar. • learn to prove SSS and SAS Triangle Similarity Theorems.	**Students:** • will prove the Triangle Proportionality Theorem. **(12.4)** • will use the Triangle Proportionality Theorem to show the proportionality of given triangles. **(12.4)**

UNPACKING MATH STANDARDS

Use the properties of similarity transformations to establish the AA criterion for two triangles to be similar.

What It Means to You

Students demonstrate an understanding of proving theorems, such as the AA Triangle Similarity Theorem, by using the properties of dilations, for example. Students determine the scale factor for two arbitrary triangles by calculating the proportion of corresponding legs of the triangles.

The emphasis for this standard is understanding the steps in proving theorems and propositions and developing fluency in the use of previously learned concepts to establish and prove the similarity of two triangles whose dimensions are given.

WARM-UP OPTIONS

ONLINE

PROJECTABLE & PRINTABLE 

ACTIVATE PRIOR KNOWLEDGE • Find Scale Factors of Dilated Triangles

Use these activities to quickly assess and activate prior knowledge as needed.

Problem of the Day 

Let $\triangle PQR$ have the coordinates $(0, 0)$, $(0, 5)$, $(12, 0)$, and let $\triangle STU$ have coordinates $(5, 8)$, $(5, 15.5)$, and $(23, 8)$. Have students draw the two triangles, find the length of each leg, and compute the scale factor. The first triangle is a right triangle with lengths 5, 12, and 13. The second triangle has lengths 7.5, 18, and 19.5. The scale factor is $k = \frac{PQ}{ST} = \frac{7.5}{5} = \frac{3}{2}$.

Quick Check for Homework

As part of your daily routine, you may want to display the Teacher Solution Key to have students check their homework.

Make Connections

Based on students' responses to the Problem of the Day, choose one of the following:

1. Project the Interactive Reteach, Geometry, Lesson 6.1.
2. Complete the Prerequisite Skills Activity:

Draw three triangles on the board with the following lengths labeled: 3, 7, 5; 6, 8, 10; 2, 3, 4.

- *Take the first triangle. What would be the lengths of that triangle dilated by a scale factor of 2? a scale factor of 3?* For a scale factor of 2, the first triangle has lengths 6, 14, and 10, the second has lengths of 12, 16, and 20, and the third has lengths of 4, 6, and 8; With a scale factor of 3, the first has lengths of 9, 21, and 15, the second has lengths of 18, 24, 30, and the third has 6, 9, 12.
- *What would be the lengths of those triangles dilated with a scale factor or 0.5?* For a scale factor of 0.5, the first triangle has lengths of 1.5, 3.5, 2.5, the second has 3, 4, 5, and the third has 1, 1.5, and 2.
- *What scale factor would dilate the first triangle to the dimensions 9.3, 21.7, and 15.5?* A scale factor of 3.1 would result in that dilation.
- *What would the dimensions of the other two triangles be with that scale factor?* The dimensions would be 18.6, 24.8, 31 and 6.2, 9.3, and 12.4.

If students continue to struggle, use Tier 2 Skill 12.

SHARPEN SKILLS

If time permits, use this on-level activity to build fluency and practice basic skills.

Mental Math

Objective: Students demonstrate the ability to calculate a scale factor mentally.
Materials: Pair of cards with two triangles, one on each

Have students work in small groups. Give each group a pair of cards. Each card should have a triangle displayed with the lengths labeled, and each group should have both a smaller triangle and a larger one. Have students look at the cards briefly and then turn them over. Then call on volunteers to give the scale factor of the larger triangle.

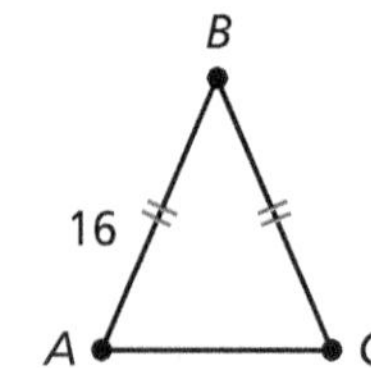

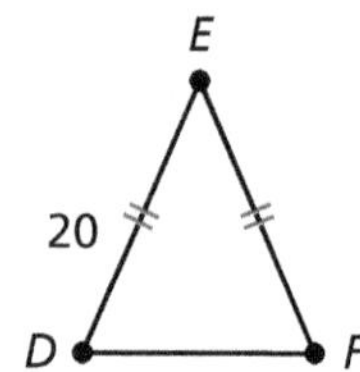

PLAN FOR DIFFERENTIATED INSTRUCTION

Small-Group Options

Use these teacher-guided activities with pulled small groups.

On Track

Materials: Card with triangle and inscribed triangle pictured

Given that the base of the upper triangle joins the midpoints of the legs of the larger triangle, use the AA Triangle Similarity Theorem to prove that the two triangles are similar.

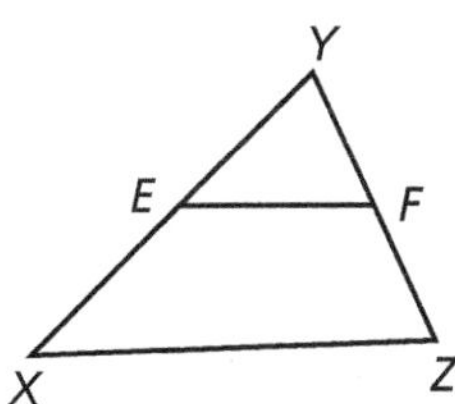

Almost There

Materials: Card with triangle pictured

Have students work in small groups to state why the two triangles shown on the card are not similar. Point out that students cannot judge by the appearance of a figure in an illustration, and ask questions such as the following to provide hints:

- What do the angles shown add up to?
- Which parts of the triangles are congruent?

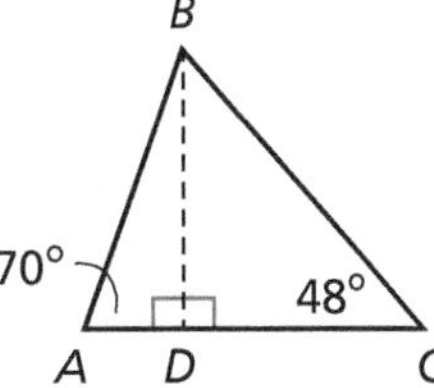

Ready for More

Remind students that any arbitrary parallelogram can be divided into two triangles. Suppose one of those triangles is rigidly dilated by an arbitrary scale factor *k*. Have students work in small groups to write a proof that those two triangles are similar.

Math Center Options

Use these student self-directed activities at centers or stations. **Key: ● Print Resources ● Online Resources**

On Track

- ● Interactive Digital Lesson
- ●● Journal and Practice Workbook
- ●● Standards Practice: Establish the AA Criterion for Two Triangles to be Similar

Almost There

- ● Reteach 12.2 (printable)
- ● Interactive Reteach 12.2
- ● RtI Tier 2 Skill 12: Dilate Figures in the Coordinate Plane

Ready for More

- ● Challenge 12.2 (printable)
- ● Interactive Challenge 12.2

View data-driven grouping recommendations and assign differentiation resources.

During the *Spark Your Learning*, listen and watch for strategies students use. See samples of student work on this page.

Use Proportions to Find the Scale Factor

Strategy 1

I calculated the proportion of the bases of the two triangles in order to find a scale factor:

$k = \frac{\text{base}(\Delta 1)}{\text{base}(\Delta 2)} = \frac{24}{4} = 6$, where Δ1 is the triangle formed by the tree and its shadow and Δ2 is the triangle formed by the man and his shadow.

So, $h = 6k = 36$ ft.

If students . . . use the scale factor proportion to solve the problem, they have mastered one of the key points of this lesson.

Have these students . . . explain how they determined their equation and how they solved it. **Ask:**

- Q How did you use proportions to find the scale factor to solve the SYL problem?

Graph the Triangles

Strategy 2

I wasn't sure how to find h, so I graphed the smaller triangle using the height of the man and the length of his shadow. I then extended the base of the triangle to 24 units and drew a parallel line for the hypotenuse of the larger triangle. I extended that line up to the line on which the vertical leg of the smaller triangle lay. I then extended that leg to meet the hypotenuse. I found that the vertical leg of the larger triangle was 36 units. So $h = 36$.

If students . . . graph the triangles to find the answer, the answer they get will probably be correct (especially since the problem involves integer values), but their method lacks sophistication and will give imprecise or even incorrect answers in more difficult cases.

Activate prior knowledge . . . by having students think about the process of finding a scale factor. **Ask:**

- Q What is the formula for the scale factor?
- Q How do you use the scale factor to find the height of the tree?

COMMON ERROR: Uses the Wrong Proportion

I calculated the proportion of the bases of the two triangles in order to find a scale factor:

$k = \frac{\text{base}(\Delta 1)}{\text{base}(\Delta 2)} = \frac{4}{24} = \frac{1}{6}$, where Δ1 is the triangle formed by the man and his shadow and Δ2 is the triangle formed by the tree and its shadow.

So $h = \frac{k}{6} = 1$ ft.

If students . . . use the reciprocal of the proportion they need, they may understand the concept of the scale factor, but they haven't computed it correctly.

Then intervene . . . by pointing out that the proportion needs to be set up so that the ratio of the bases corresponds to the ratio of the height of the tree to the height of the man. **Ask:**

- Q What is the relationship between the legs of the two triangles?
- Q How can you set up the proportion so that corresponding legs of the triangles are in the numerator and the denominator?

12.2

Develop AA Triangle Similarity

I Can prove AA, SSS, and SAS Similarity Theorems.

Spark Your Learning

An architect is designing a tree house that will have supports reaching to the top of a tree. Both the architect and the tree cast a shadow.

Complete Part A as a whole class. Then complete Parts B–D in small groups.

A. What is a mathematical question you can ask about this situation? How tall is the tree?

B. How can the shadows of the person and the tree be used to help determine the height of the tree? See Additional Answers.

C. To answer your question, what strategy and tool would you use along with all the information you have? What answer do you get? See Strategies 1 and 2 on the facing page.

D. What must be true about the angles of the triangles for you to show that the triangles are similar? See Additional Answers.

 **Turn and Talk** Compare and contrast the two shapes you used to model the scenario. What conclusion can you draw about the relationship between the shapes? See margin.

CULTIVATE CONVERSATION • Information Gap

Ask students questions to help them decide what missing information they need to answer the question "How tall is the tree?"

- Do you have enough information to conclude how tall the tree is? Explain. no; I don't know the height of the tree because, although the man is 6 feet tall, I'm not sure of the relationship between his height and the height of the tree.
- If a dilation transforms all legs of a triangle proportionally, what can you say about the height of the tree? The height of the tree will be proportional to the height of the man.
- If the two heights are proportional, what other information can you use to find that proportion? If the dilation transforms all legs proportionally, then I can use the proportion of two shadows as the proportion of the two heights. That's 6 to 1.

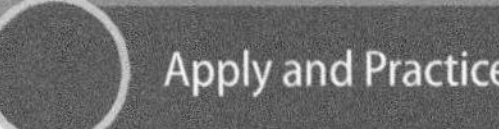

Apply and Practice

1 Spark Your Learning

▶ MOTIVATE 

- Have students look at the illustration in their books and read the information contained in the illustration. Then complete Part A as a whole-class discussion.
- Give students the additional information they need to solve the problem. This information is available online as a printable and projectable page in the Teacher Resources.
- Have students work in small groups to complete Parts B–D.

▶ PERSEVERE

If students need support, guide them by asking:

Q **Advancing • Use Tools** Which tool could you use to solve the problem? Why choose that tool and not some other? Students' choices of tools and reasons for choosing them will vary.

Q **Assessing** What is the ratio of the length of the longer shadow to that of the shorter shadow? The ratio is $\frac{24}{4}$, or $\frac{6}{1}$.

Q **Assessing** If the shadow cast by the tree is on a straight line with the shadow cast by the architect, what do you think is the relationship between the angles formed by the top of the tree and the ground, on the one hand, and the man's head and the ground on the other? I think the angles are congruent.

Q **Advancing** Can you think of a possible relationship between the legs of the two triangles? Maybe the corresponding legs are proportional.

Turn and Talk Make sure students understand that both triangles are right triangles, so there is at least one pair of congruent angles. Remind them of their answer to the second assessing question. Possible answer: The two triangles have different side lengths. The two triangles have at least two pairs of congruent angles.

▶ BUILD SHARED UNDERSTANDING

Select groups of students who used various strategies and tools to share with the class how they solved the problem. As they present their solutions, have each group discuss why they chose a specific strategy and tool.

② Learn Together

Build Understanding

Task 1  **Construct Arguments** Guide students through the process of constructing a basic proof from the information given about two triangles. Students will go through the proof of the AA Triangle Similarity Theorem.

Sample Guided Discussion:

Q What is the relationship between $\triangle JKL$ and $\triangle PQR$? $\triangle PQR$ is larger than $\triangle JKL$. The lengths are not given, but we can assume that the ratio of two corresponding legs of the triangles will be the same as the ratio of two other corresponding legs of the triangles.

Q How does the dilation of $\triangle JKL$ shown on page 370 transform the triangle? It dilates the triangle by a scale factor that is the ratio of two corresponding legs of the two triangles.

Turn and Talk Point out to students that the difference in the names gives them a clue to the principal differences: one is called a similarity theorem and the other is called a congruence theorem. Possible answer: The AA Triangle Similarity Theorem can be used to compare triangles with congruent angles that are different sizes. The ASA Triangle Congruence Theorem is used to compare triangles with both congruent angles and congruent sides. Both are shortcuts to prove similarity or congruence without comparing all parts of the triangles.

Build Understanding

Prove the AA Triangle Similarity Theorem

You have learned previously that two figures are similar when there is a sequence of similarity transformations that maps one figure to the other. You can also show that two triangles are similar by using theorems, such as the Angle-Angle Triangle Similarity Theorem.

Angle-Angle (AA) Triangle Similarity Theorem

If two angles of one triangle are congruent to two angles of another triangle, then the two triangles are similar.

1 Prove the AA Triangle Similarity Theorem.

Given: $\angle J \cong \angle P$ and $\angle K \cong \angle Q$.

Prove: $\triangle JKL \sim \triangle PQR$

To prove that the triangles are similar, find a sequence of similarity transformations that maps $\triangle JKL$ to $\triangle PQR$.

A. Consider a dilation that maps $\triangle JKL$ to $\triangle J'K'L'$. What is the scale factor of the dilation so that $\triangle J'K'L'$ will be congruent to $\triangle PQR$? Why should $\triangle J'K'L'$ be congruent to $\triangle PQR$? A–F. See Additional Answers.

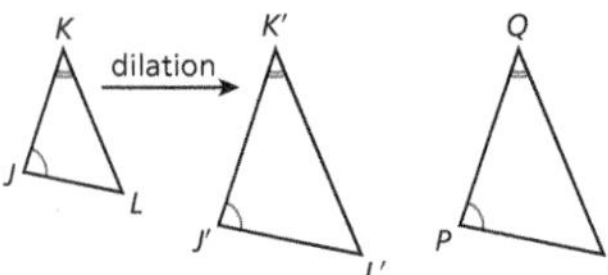

B. Why is $\triangle JKL$ similar to $\triangle J'K'L'$?

C. To complete the proof, you must show $\triangle J'K'L' \cong \triangle PQR$. Because $\triangle JKL \sim \triangle J'K'L'$, $J'K' = JK \cdot$ scale factor. What is the result when you substitute the scale factor from Part A, and simplify?

D. Because $\triangle JKL \sim \triangle J'K'L'$, $\angle J \cong \angle J'$ and $\angle K \cong \angle K'$. Explain why $\angle J' \cong \angle P$ and $\angle K' \cong \angle Q$.

E. Explain why you can conclude that $\triangle J'K'L' \cong \triangle PQR$.

F. Use the results from the steps above to describe how a sequence of similarity transformations maps $\triangle JKL$ to $\triangle PQR$.

 **Turn and Talk** Compare and contrast the AA Triangle Similarity Theorem and the ASA Triangle Congruence Theorem. See margin.

376

LEVELED QUESTIONS

Depth of Knowledge (DOK)	Leveled Questions	What Does This Tell You?
Level 1 **Recall**	What transformation does $\triangle JKL$ undergo to become $\triangle J'K'L'$? $\triangle JKL$ undergoes a dilation.	Students' answers will indicate that they have grasped that the large triangle is a dilation of the first and can identify the transformation as such.
Level 2 **Basic Application of Skills & Concepts**	How are $\triangle JKL$ and $\triangle PQR$ alike and different? $\triangle PQR$ is larger than $\triangle JKL$, but the two triangles are similar because the larger one is a dilation of the smaller one with a constant scale factor.	Students' answers will indicate that they understand the relationship between the two triangles and can identify why they are similar.
Level 3 **Strategic Thinking & Complex Reasoning**	What can you conclude from the fact that $\triangle JKL$ and $\triangle PQR$ are similar triangles? I conclude that all corresponding angles are congruent.	Students' answers will indicate they have understood the concept of similar triangles and that they can explain the two triangles' similarity in terms of the concept of congruent angles.

Step It Out

Apply the AA Triangle Similarity Theorem

You can use the AA Triangle Similarity Theorem to prove triangles are similar and find missing dimensions.

2 During time-lapse photography, a motorized camera slider stabilizer can be used. The camera moves along the slider while it records. Two positions of a camera slider are shown. What is PQ?

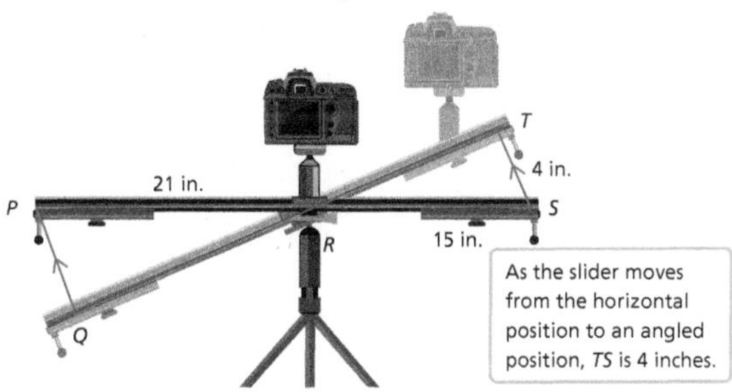
T
4 in.
21 in.
P
S
R
15 in.
Q
As the slider moves from the horizontal position to an angled position, TS is 4 inches.

Determine if $\triangle PQR$ and $\triangle RST$ are similar.

A. How can you use knowledge of parallel lines to prove $\angle P$ and $\angle S$ are congruent and $\angle Q$ and $\angle T$ are congruent? See Additional Answers.

B. Are there any other congruent angles in the triangles? If so, how do you know?
yes; $\angle PRQ$ and $\angle SRT$ are vertical angles.

C. How can you prove the two triangles are similar?
C. Possible answer: The triangles have two congruent angles and are similar by the AA Triangle Similarity Theorem.

Find PQ.

D. Name the corresponding pairs of sides between the two triangles.
$\overline{PR}$ corresponds to $\overline{SR}$, $\overline{PQ}$ corresponds to $\overline{ST}$, and $\overline{QR}$ corresponds to $\overline{TR}$.

E. Why can you use a proportion to find PQ?
E. Possible answer: Because the triangles are similar, the lengths of corresponding sides are proportional.

$$\frac{RP}{PQ} = \frac{RS}{ST}$$

$$\frac{21}{PQ} = \frac{15}{4}$$

$$84 = 15 \cdot PQ$$

$$PQ = 5.6$$

Answer the question.

As the slider moves from the horizontal position to an angled position, PQ is 5.6 inches.

Turn and Talk Is there more than one way to set up a proportion to find PQ? If so, give another proportion that can be used to find PQ. See margin.

PROFICIENCY LEVEL

Beginning

Have students work in pairs to work out the first step—orally or in writing (in two to three words)—in applying the theorem to the example in Task 2. Have volunteers report to the class.

Intermediate

Have students work in pairs to make a list of all the parallel segments in the example in Task 2 and go through the list orally, taking turns checking each one. Have volunteers report to the class.

Advanced

Have students work in groups to explain why they cannot claim that the triangles in Task 2 are congruent. Ask groups to write a paragraph summarizing their thinking. Have volunteers report to the class.

Step It Out

Task 2 (MP) **Attend to Precision** Help students understand the need to be careful with definitions and concepts they have learned up to now that they will need to apply in this task—for example, the concept of vertical angles to show congruence.

Sample Guided Discussion:

Q **What can you say about the angles formed by two intersecting lines?** The two angles formed by the intersection of two lines are known as vertical angles and are congruent.

Q **What do you need to show to prove that two triangles are similar?** We only need to show that two pairs of angles for the two triangles are congruent.

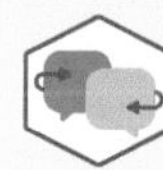

Turn and Talk Explain that proportions can be set up between any two corresponding legs of two similar triangles. yes; Possible answer: You could also write the proportion as $\frac{RP}{RS} = \frac{PQ}{ST}$.

Task 3 **MP** **Attend to Precision** Remind students of the need to be careful with definitions and concepts and to be precise when they apply the SSS and SAS Similarity Theorems.

Sample Guided Discussion:

Q Which side corresponds to the side labeled *x*?
DF corresponds to the side labeled *x*.

Q How would you go about setting up a proportion?
I would think about which sides correspond to each other. I know that the longer sides are going to be the right ones to compare. If the drawing confuses me, I can try two pairs of sides and if the proportions are not equal, I can try the other two pairs of the longer sides.

Turn and Talk Remind students of the details of the ASA Theorem. Possible answer: ASA means two angles are congruent and an included side is proportional, which already meets the criteria of AA Triangle Similarity. An additional ASA theorem is not needed.

Apply the SSS and SAS Triangle Similarity Theorems

In addition to the AA Triangle Similarity Theorem, there are two other theorems that can be used to prove triangle similarity.

Side-Side-Side (SSS) Triangle Similarity Theorem
If the three sides of one triangle are proportional to the corresponding sides of another triangle, then the triangles are similar.

Side-Angle-Side (SAS) Triangle Similarity Theorem
If two sides of one triangle are proportional to the corresponding sides of another triangle and their included angles are congruent, then the triangles are similar.

3 Find the value of x.

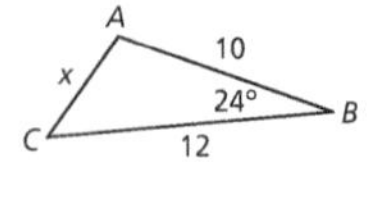

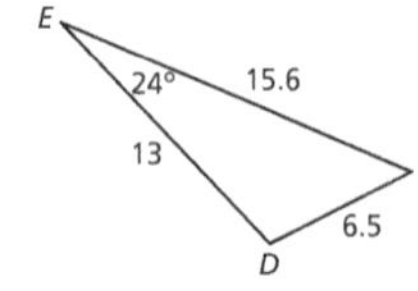

A–D. See Additional Answers.

Determine whether the triangles are similar.

Start by investigating corresponding sides to determine if they are proportional.

$\frac{ED}{AB} = \frac{13}{10} = 1.3$ $\quad \frac{EF}{BC} = \frac{15.6}{12} = 1.3$

A. How do you know which corresponding sides are proportional?

Check for congruent corresponding angles.

$\angle B \cong \angle E$

B. How do you know that the angles are congruent?

C. Why can you conclude that the triangles are similar? Write a similarity statement for the two triangles.

Find the value of *x*.

D. Which solution correctly determines the length of x? Explain.

$\frac{ED}{AB} = \frac{FD}{x}$	$\frac{ED}{CB} = \frac{FD}{x}$
$\frac{13}{10} = \frac{6.5}{x}$	$\frac{13}{12} = \frac{6.5}{x}$
$13x = 65$	$13x = 78$
$x = 5$	$x = 6$

Turn and Talk Why isn't Angle-Side-Angle listed as a way to show two triangles are similar? See margin.

378

Use Indirect Measurement

You can use triangle similarity to find unknown measures in the real world when taking a direct measure may not be possible.

4 A swimmer wants to find the distance across a lake to create a training plan. She wants to use indirect measurement to determine the distance across the lake.

The swimmer stands at point U and identifies a tree directly across the lake from her, and labels it point T. She then turns 90° away from T, walks 200 feet, and marks point V.

She continues walking 300 feet in a straight line from V and marks point W. She then turns another 90° and walks until points T and V align with her location and marks the point as Y. She measures to find she had walked 435 feet from W to Y.

A. Which diagram correctly represents the situation? Diagram 2

Diagram 1

Diagram 2

B. Use each statement and give reasons to prove $\triangle TUV \sim \triangle YWV$.

B, D. See Additional Answers.

$\angle WVY \cong \angle UVT$

$\angle YWV \cong \angle TUV$

$\triangle TUV \sim \triangle YWV$

C. To solve the problem, find TU. Which is the correct solution? $TU = 290$ is correct.

$\frac{TU}{300} = \frac{200}{435}$	$\frac{TU}{435} = \frac{200}{300}$
$435 \cdot TU = 60{,}000$	$300 \cdot TU = 87{,}000$
$TU \approx 138$	$TU = 290$
The distance across the lake is about 138 feet.	The distance across the lake is 290 feet.

D. Does your answer make sense in the context of the problem? Explain.

Turn and Talk In the example, what is another proportion you can write to solve the problem? See margin.

Task 4 (MP) **Model with Mathematics** Explain to students that Task 4 will give them an example of modeling real-world problems with mathematics—in this case, to find an unknown measure with the geometric relationships they have learned.

SUPPORT SENSE-MAKING Three Reads

Have students read the problem three times. Use the questions below for a different focus each read.

1. What is the situation about?
2. What are the quantities in the situation?
3. What are the possible mathematical questions that you could ask for the situation?

Sample Guided Discussion:

Q What are examples of congruent angles in Task 4? $\angle WVX$ is congruent to $\angle TVU$. Also, $\angle TUV$ is congruent to $\angle WVX$. The third pair of corresponding angles is also congruent.

Q Can you give examples of congruent legs of the two triangles? If not, what can you say about the corresponding legs of the two triangles? There are no congruent legs of the two triangles; The triangles do have proportional legs.

Turn and Talk Remind students that taking the reciprocal of both ratios gives an equivalent answer. Answer: $\frac{435}{TU} = \frac{300}{200}$

ONLINE

Assign the Digital On Your Own for

- built-in student supports
- Actionable Item Reports
- Standards Analysis Reports

On Your Own

Assignment Guide

The chart below indicates which problems in the On Your Own are associated with each task in the Learn Together. Assign daily homework for tasks completed.

Learn Together Tasks	On Your Own Problems
Task 1, p. 376	Problem 5
Task 2, p. 377	Problems 6, 7–10, 20, and 21
Task 3, p. 378	Problems 6, 11–15, 17, 19, and 22
Task 4, p. 379	Problems 16 and 18

Check Understanding

1. If $\triangle ABC \sim \triangle LMN$ and $\triangle LMN \sim \triangle XYZ$, is $\triangle ABC \sim \triangle XYZ$? Use the AA Triangle Similarity Theorem to support your answer. See Additional Answers.
2. Two triangular sculptures are similar. The leg of the first sculpture is 4.5 feet, and the base is 7.2 feet. The leg of the second sculpture is 6 feet. What is the length of the base for the second sculpture? 9.6 ft
3. Are the triangles at the right similar? If so, what is x? If not, explain. See Additional Answers.

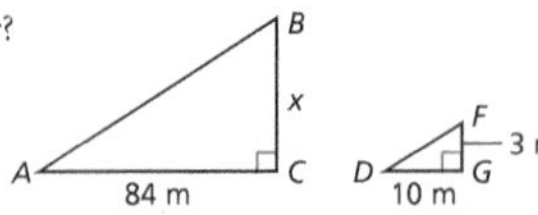

4. Explain how you can create a pair of similar triangles to measure an unknown distance across a canyon. Discuss landmarks, alignment of points, and measurements in your answer. See Additional Answers.

On Your Own

5. Are the two triangular flags similar? Explain. 5–10. See Additional Answers.

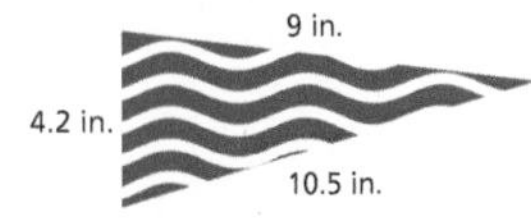

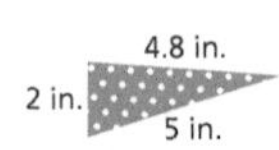

6. Prove the SAS Triangle Similarity Theorem using similarity transformations.

Given: $\frac{DE}{AB} = \frac{DF}{AC}$ and $\angle A \cong \angle D$.

Prove: $\triangle ABC \sim \triangle DEF$

Determine whether each set of triangles is similar. Justify your reasoning.

7. $\triangle ABC$ and $\triangle DEC$

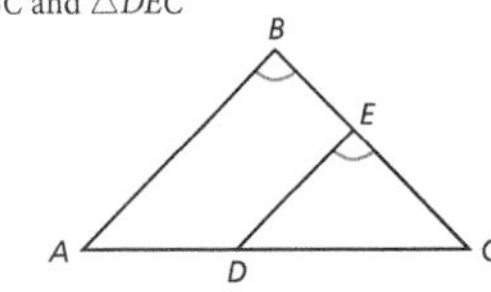

8. $\triangle XYZ$ and $\triangle PQR$

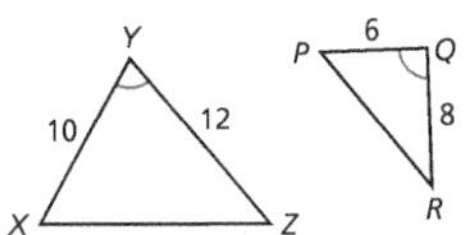

9. $\triangle LMN$ and $\triangle JKN$

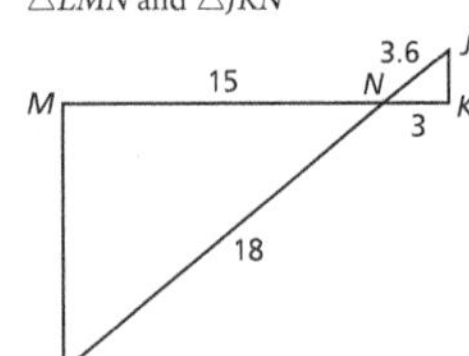

10. $\triangle ABC$ and $\triangle CDB$

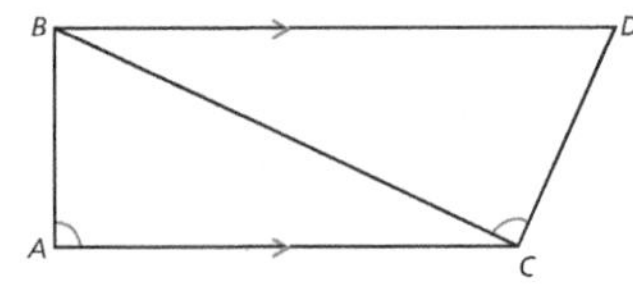

380

3 Check Understanding

Formative Assessment

Use formative assessment to determine if your students are successful with this lesson's learning objective.

Students who successfully complete the Check Understanding can continue to the On Your Own practice.

For students who miss 1 problem or more, work in a pulled small group using the Almost There small-group activity on page 375C.

ONLINE

Assign the Digital Check Understanding to determine

- success with the learning objective
- items to review
- grouping and differentiation resources

4 Differentiation Options

Differentiate instruction for all students using small-group activities and math center activities on page 375C.

Reteach

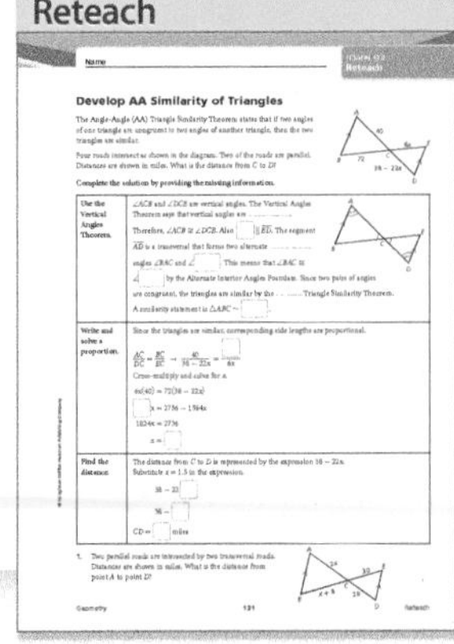

Develop AA Similarity of Triangles

Challenge

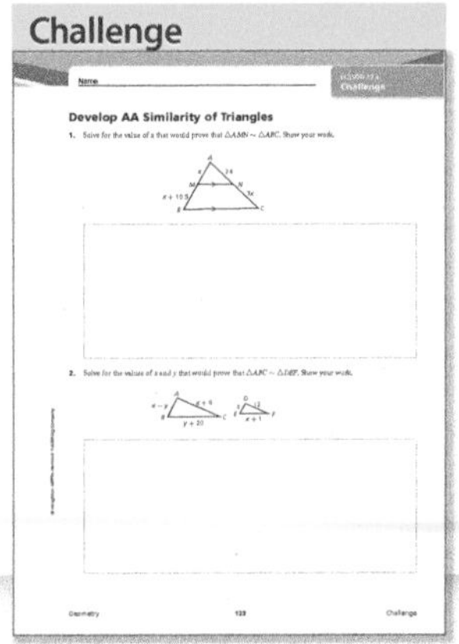

Develop AA Similarity of Triangles

Determine whether each pair of triangles is similar. When possible, find the value of x.

11. $\triangle ABC \sim \triangle ADE$; $x = 72$

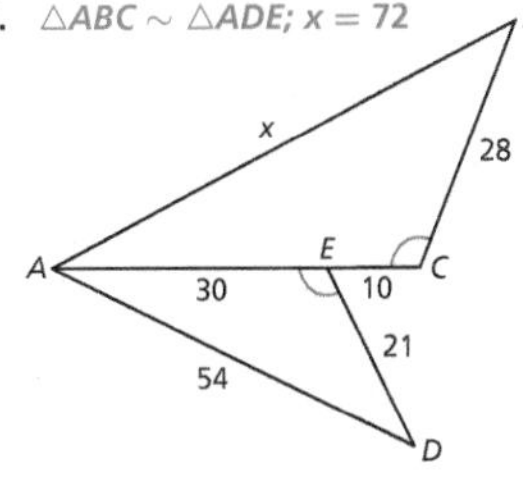

12.

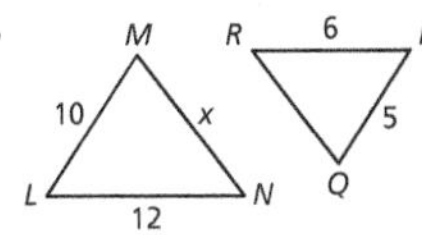

It cannot be determined.

13. $\triangle UTV \sim \triangle USW$; $x = \frac{ac}{b}$

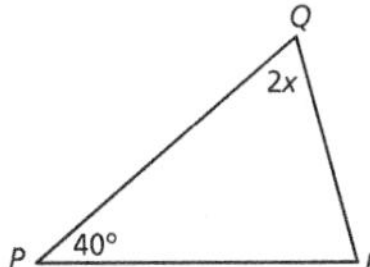

14. It cannot be determined.

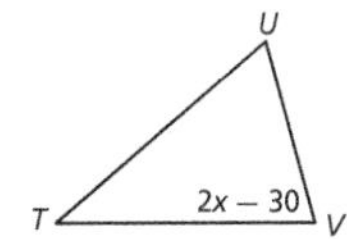

15. Find the value of x that makes $\triangle PQR$ similar to $\triangle TUV$. $x = 42.5$

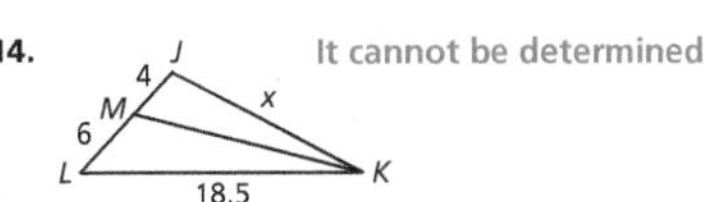

16. STEM In a pinhole camera, light from an object goes through a pinhole to produce an inverted image on its screen. A lighthouse is 21 m from the pinhole of a large pinhole camera. The 12 cm tall image is projected on the screen 30 cm from the pinhole.

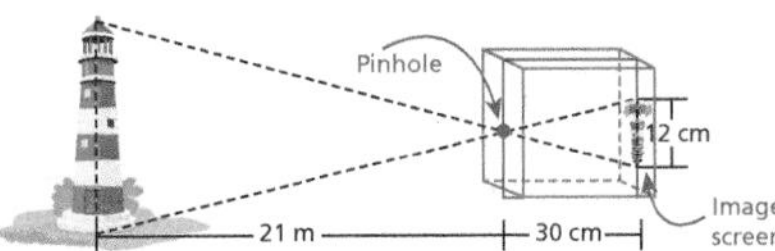

A. How tall is the lighthouse? 8.4 m

B. Suppose the same pinhole camera is moved closer to the lighthouse. If the lighthouse projects a 15 cm tall image, how much closer is the pinhole camera moved? 4.2 m closer

C. Suppose a different pinhole camera is placed 29.4 m from the lighthouse. If the image is 10 cm tall, how far is the camera screen from the pinhole? 35 cm

Questioning Strategies

Problem 11 The ratio of DE to BC is $\frac{3}{4}$. What must be true about two other side lengths for the triangles to be similar? Is it true? The ratio of AE to AC must be $\frac{3}{4}$; It is true because $\frac{AE}{AC} = \frac{30}{30 + 10} = \frac{30}{40} = \frac{3}{4}$

Watch for Common Errors

Problem 12 Students may conclude that the triangles are similar because the side lengths that are given are proportional. However, more than just two side lengths are needed to conclude that the triangles are similar.

5 Wrap-Up

Summarize learning with your class. Consider using the Exit Ticket, Put It in Writing, or I Can scale.

Exit Ticket 

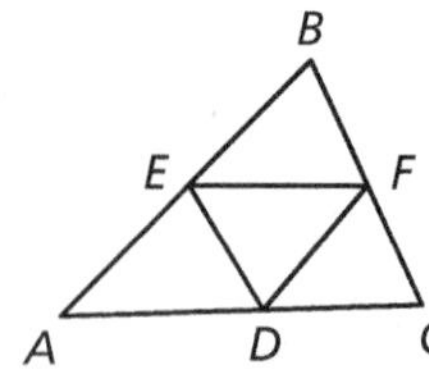

Suppose a triangle, $\triangle ABC$, has an inscribed triangle, $\triangle DEF$, such that the vertices of $\triangle DEF$ are at the midpoints of the legs of $\triangle ABC$. Prove or disprove that the triangles are similar. The triangles are similar. By the Triangle Midsegment Theorem,

$EF = \frac{1}{2}AC$, $DE = \frac{1}{2}BC$, $DF = \frac{1}{2}AB$,

so $\triangle ABC$ has a scale factor of 2, compared to $\triangle DEF$. Since all 3 corresponding sides of the two triangles are proportional, the two triangles are similar by the SSS Triangle Similarity Theorem.

Put It in Writing 

There are five triangles pictured in the illustration above. Which of them, if any, are similar? Which of them are congruent? How do you know?

I Can 

The scale below can help you and your students understand their progress on a learning goal.

4	I can prove the AA, SSS, and SAS Similarity Theorems and explain my proofs to others.
3	I can prove AA, SSS, and SAS Similarity Theorems.
2	I can use given information and do some of the steps in a proof of the similarity of two triangles.
1	I can understand the proof of the similarity of two triangles using the AA Triangle Similarity Theorem.

Spiral Review • Assessment Readiness

These questions will help determine if students have retained information taught in the past and can also prepare them for high-stakes assessments. Here, students must use transformations to prove triangles are similar **(12.1)**, use the Triangle Midsegment Theorem **(9.5)**, and use the properties and conditions for rectangles, rhombuses, squares, trapezoids, and kites **(11.4 and 11.5)**.

17. Prove the SSS Triangle Similarity Theorem using similarity transformations. See Additional Answers.

Given: $\frac{DE}{AB} = \frac{DF}{AC} = \frac{FE}{BC}$

Prove: $\triangle ABC \sim \triangle DEF$

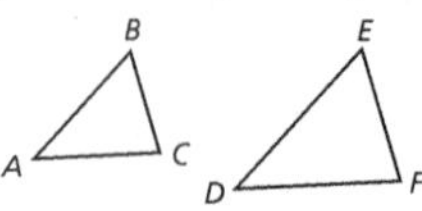

18. (MP) **Attend to Precision** An anthropologist studying an ancient society reaches a river that needs to be crossed. She locates a tree directly across the river at Point A, marks her location as Point B, then turns and walks to Point C. She continues walking to Point D, turns perpendicular to the river, and walks until Points A and C are directly in line with her final location, Point E. What is the width of the river? The river is 50 feet wide.

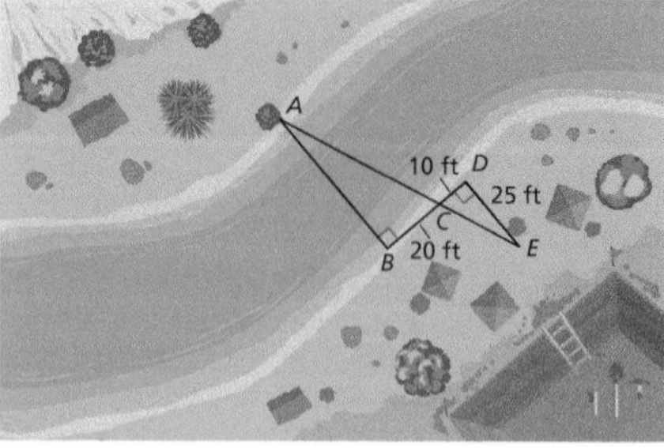

(MP) **Use Repeated Reasoning** **Determine whether each of the following provides enough information to prove $\triangle ABC \sim \triangle DEC$. Justify your answer.**

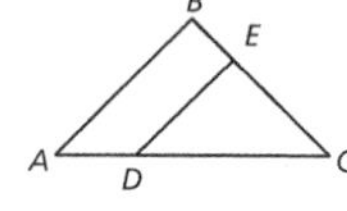

19. $\overline{AB} \parallel \overline{DE}$ yes; AA Similarity

20. $\angle B \cong \angle DEC$ yes; AA Similarity

21. $m\angle DEC = 90°$ no; only one pair of cong. angles

22. $\frac{EC}{EC + EB} = \frac{DC}{DC + AD}$ yes; SAS Similarity

Spiral Review • Assessment Readiness

23. A right triangle with a hypotenuse of 7 cm is dilated so that the hypotenuse is 4 cm. If k is the scale factor, then

Ⓐ $k < -1$
Ⓑ $-1 < k < 0$
Ⓒ $k > 1$
Ⓓ $0 < k < 1$

24. If a quadrilateral is a ____, then its diagonals are perpendicular. Select all that apply.

Ⓐ kite
Ⓑ rectangle
Ⓒ rhombus
Ⓓ trapezoid

25. If $\overline{AB}$ is the midsegment of $\triangle QRP$, what is AB?

Ⓐ $11\frac{1}{3}$
Ⓑ 17
Ⓒ 35
Ⓓ 68

R
34
B
Q
A
P

26. Which shape is not necessarily a parallelogram?

Ⓐ rhombus
Ⓑ trapezoid
Ⓒ square
Ⓓ rectangle

I'm in a Learning Mindset!

When studying similarity in triangles, how do I react when my learning environment changes?

Keep Going Journal and Practice Workbook

Learning Mindset

Resilience Adjusts to Change

Point out that students have to learn to study in a variety of environments and they also have to get used to shifting from one topic to another. It can be disorienting for students to have to focus on a new topic when they have just gotten used to a previous topic. For this reason, it helps them to adjust when you spiral and review previous topics that are related to what's new. You can ask questions such as the following: *What properties of triangles can think of? Can you think of specific concepts that can help you understand the similarity theorems? What is the difference between similarity and congruence? How do you prove propositions in geometry?*

12.3 Develop and Prove Triangle Proportionality

LESSON FOCUS AND COHERENCE

Mathematics Standards

- Prove theorems about triangles. *Theorems include: a line parallel to one side of a triangle divides the other two proportionally, and conversely; the Pythagorean Theorem proved using triangle similarity.*
- Find the point on a directed line segment between two given points that partitions the segment in a given ratio.
- The dilation of a line segment is longer or shorter in the ratio given by the scale factor.

Mathematical Practices and Processes (MP)

- Construct viable arguments and critique the reasoning of others.
- Attend to precision.
- Use appropriate tools strategically.

I Can Objective

I can identify and use the connection between parallel lines and proportional segments in triangles.

Learning Objective

Analyze a proof of the Triangle Proportionality Theorem, apply the theorem to solve for lengths of partitions of triangle sides, apply the converse of the theorem to determine partitions of triangle sides that give a line parallel to another side, and find the point on a directed line segment that partitions the segment in a given ratio.

Language Objective

Explain how the Triangle Proportionality Theorem applies to triangles and how it can be used to solve for unknown lengths.

Vocabulary

New: partition

Lesson Materials: compass, ruler

Mathematical Progressions

Prior Learning	Current Development	Future Connections
Students: • performed dilations and sequences of transformations. **(6.1 and 6.2)**	**Students:** • analyze a proof of the Triangle Proportionality Theorem. • use the theorem to solve for segment lengths. • use the converse of the theorem to find segment lengths resulting in parallel lines. • find the point on a directed line segment that partitions the segment in a given ratio.	**Students:** • will identify similar right triangles. **(13.4)**

PROFESSIONAL LEARNING

Math Background

An important part of this lesson is helping students further build their understanding of proportional relationships between segment lengths. Students have already been exposed to proportional side lengths of figures undergoing dilations, and in this lesson students explore proportions between segment lengths that are not sides of similar figures. For example, in Task 5, students construct a point that partitions a line segment in a given ratio. They do this by applying the Triangle Proportionality Theorem, constructing segments that partition two segments proportionally. In the next lesson, students further explore proportions between side lengths as they investigate geometric means.

WARM-UP OPTIONS

ONLINE

PROJECTABLE & PRINTABLE

ACTIVATE PRIOR KNOWLEDGE • Construct a Dilation

Use these activities to quickly assess and activate prior knowledge as needed.

Problem of the Day

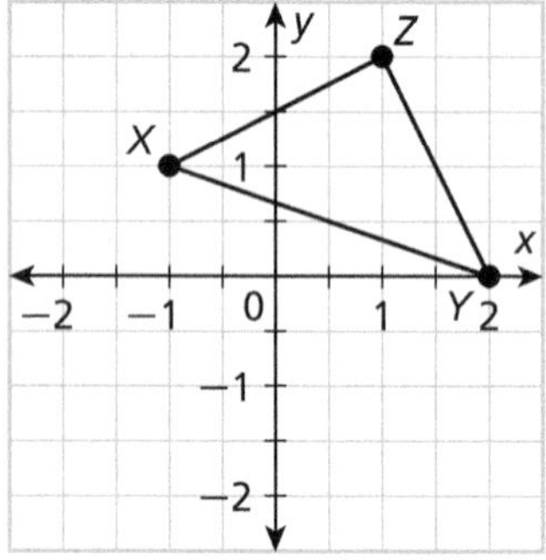

Draw a dilation of $\triangle XYZ$ with a scale factor of $-\frac{1}{2}$ centered at the origin.

Applying the coordinate rule $(x, y) \rightarrow \left(-\frac{1}{2}x, -\frac{1}{2}y\right)$ gives $\triangle X'Y'Z'$ as shown here:

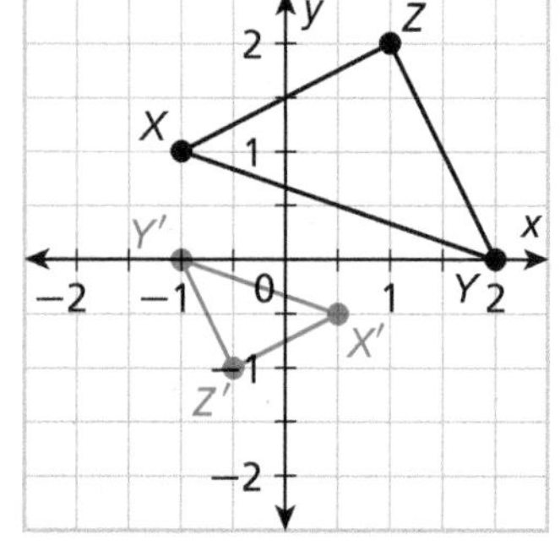

Quick Check for Homework

As part of your daily routine, you may want to display the Teacher Solution Key to have students check their homework.

Make Connections

Based on students' responses to the Problem of the Day, choose one of the following:

1. Project the Interactive Reteach, Geometry, Lesson 6.1.
2. Complete the Prerequisite Skills Activity:

Have students work in small groups. Give each group a set of index cards. Each card should have on it a dilation about the origin of $\triangle XYZ$ from the Problem of the Day. Have groups draw cards and work together to determine the scale factor described on each card.

- *How does the scale factor let you find a coordinate rule for a dilation about the origin?* The coordinate rule is to multiply each coordinate by the scale factor.
- *How can you find the scale factor from a preimage point and its image?* I can divide to find the number the preimage coordinates were multiplied by to change to the image coordinates.
- *How do you know when the scale factor is negative?* When each pair of preimage and image points are on opposite sides of the origin.

If students continue to struggle, use Tier 2 Skill 12.

SHARPEN SKILLS

If time permits, use this on-level activity to build fluency and practice basic skills.

Vocabulary Review

Objective: Students demonstrate an understanding of terms related to similar figures.
Materials: Bubble Map (Teacher Resource Masters)

Have students work in pairs. Each pair should work together to build a bubble map for the term "similar figures" without using reference materials to complete the map. Encourage students to work together to recall as much information as possible and to not worry too much about being correct at this point.

When all pairs have had ample time to work, have a class discussion and create a class bubble map on the board.

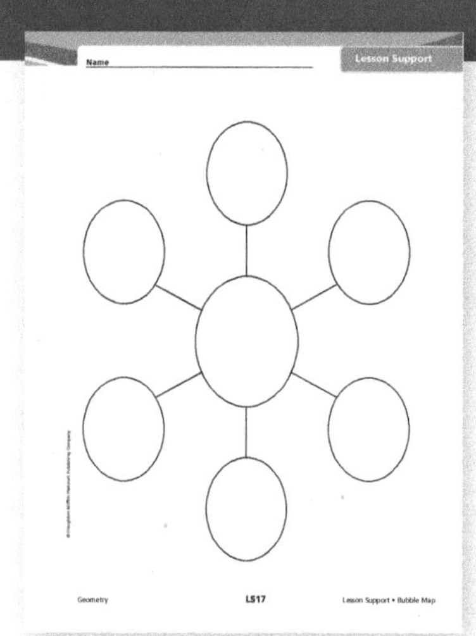

PLAN FOR DIFFERENTIATED INSTRUCTION

Small-Group Options

Use these teacher-guided activities with pulled small groups.

On Track

Materials: calculator

Show students the triangle below.

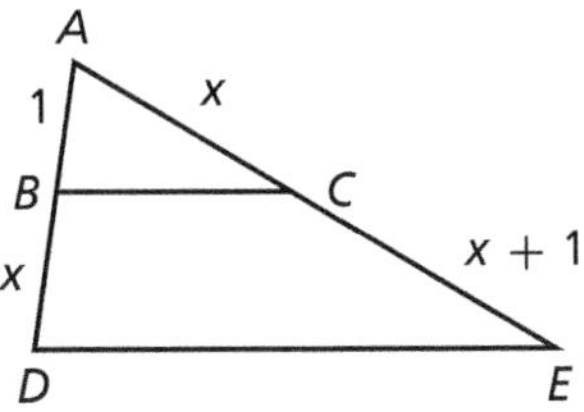

Ask,

- "Is there an integer that you can substitute for x so $\overline{BC} \parallel \overline{DE}$?"
- "How can you solve for x?" (They will have to solve the equation $x^2 = x + 1$, which has solutions $x = \frac{1 \pm \sqrt{5}}{2}$, but only the positive solution makes sense here.)

Almost There

Materials: index cards

Display a triangle like the one shown below.

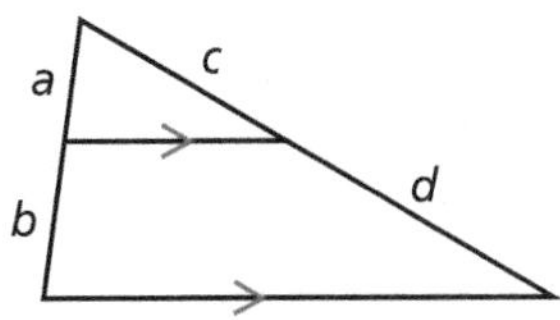

Provide each group with a set of index cards. Four cards in each set should have a, b, c, or d written on them, and the other cards should have positive integers written on them. Have student draw a variable card and an integer card, one pair at at a time, until they have three variable-integer pairs. These pairs indicate the lengths of the segments of the displayed triangle. Have students work together to find the length of the unknown segment. Clarify that the diagram will not be to scale.

Ready for More

Materials: meterstick/yardstick, large sheet of paper, string

Have students do the following to complete a construction of the geometric series $\frac{1}{3} + \frac{1}{9} + \frac{1}{27} + \ldots = \frac{1}{2}$. Draw a very long segment $\overline{AB}$ and ray $\overrightarrow{AC}$ with the meterstick.

- Find points A' and B' on $\overline{AB}$ that partition the segment in a 1 to 2 ratio and a 2 to 1 ratio, respectively.
- Draw a new ray using the same approach to find points A'' and B'' on $\overline{A'B'}$ that partition the segment in a 1 to 2 ratio and a 2 to 1 ratio, respectively.
- Repeat this process as long as possible.

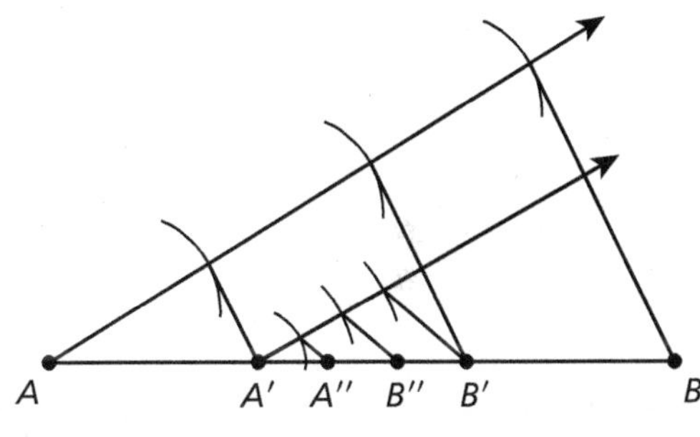

Math Center Options

Use these student self-directed activities at centers or stations. Key: ● Print Resources ● Online Resources

On Track

- ● Interactive Digital Lesson
- ●● Journal and Practice Workbook
- ● Interactive Glossary (printable): **partition**
- ● Module Performance Task

Almost There

- ● Reteach 12.3 (printable)
- ● Interactive Reteach 12.3
- ● RtI Tier 2 Skill 12: Dilate Figures in the Coordinate Plane

Ready for More

- ● Challenge 12.3 (printable)
- ● Interactive Challenge 12.3
- ● Illustrative Mathematics: Finding Triangle Coordinates

ONLINE Ed

View data-driven grouping recommendations and assign differentiation resources.

During the *Spark Your Learning,* listen and watch for strategies students use. See samples of student work on this page.

Use Congruent Corresponding Angles

Strategy 1

The balcony and porch are parallel. I can use the rules about transversals and parallel lines to determine that corresponding angles are congruent. $\triangle ABC$ and $\triangle ADE$ are similar by the AA Triangle Similarity Theorem.

If students . . . identify and use congruent corresponding angles to solve the problem, they are demonstrating an understanding of the AA Triangle Similarity Theorem from Lesson 12.2.

Have these students . . . explain how they determined their strategy and how they carried it through. **Ask:**

- Q How did you decide to use congruent corresponding angles?
- Q Which rules about transversals and parallel lines did you use?

Measure Side Lengths

Strategy 2

I measured the side lengths of $\triangle ABC$ and $\triangle ADE$ with a ruler, divided the side lengths of $\triangle ADE$ by the corresponding side lengths of $\triangle ABC$, and got the same ratio. The side lengths are proportional, so the triangles are similar.

If students . . . measure side lengths of the triangles, they understand that similar triangles have proportional sides, but they may not understand that corresponding angle measures can also be used to determine whether triangles are similar.

Activate prior knowledge . . . by having students apply the AA Triangle Similarity Theorem. **Ask:**

- Q How are the angles of similar triangles related?
- Q How are $\angle B$ and $\angle D$ related?

COMMON ERROR: Compares Incorrect Segment Lengths

I measured side lengths and calculated $\frac{AB}{AD}$, $\frac{AC}{CE}$, and $\frac{BC}{DE}$. $\frac{AB}{BD}$ and $\frac{AC}{CE}$ are not equal to $\frac{BC}{DE}$, so the triangles are not similar.

If students . . . compare incorrect segment lengths, then they may not understand that they are not actually comparing side lengths of the triangles.

Then intervene . . . by pointing out that $\overline{BD}$ and $\overline{CE}$ are not sides of $\triangle ADE$. **Ask:**

- Q What are the names of the sides of $\triangle ADE$?
- Q With which side of $\triangle ADE$ does $\overline{AB}$ correspond?

12.3

Develop and Prove Triangle Proportionality

(I Can) identify and use the connection between parallel lines and proportional segments in triangles.

Spark Your Learning

An A-frame house is styled with a very steep-angled roof that makes the front of the house resemble the letter A.

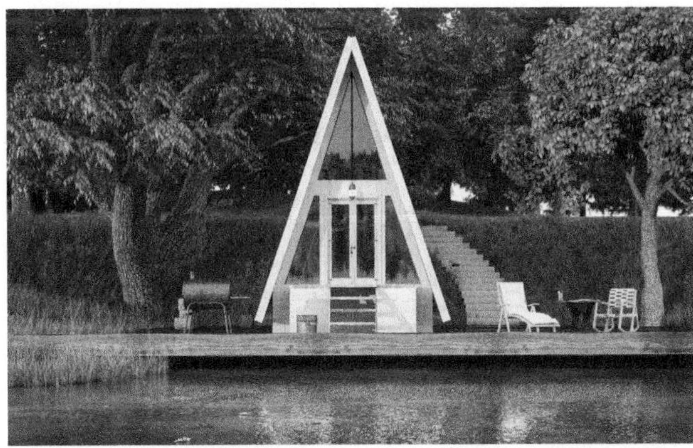

Complete Part A as a whole class. Then complete Parts B–D in small groups.

A. What is a mathematical question you can ask about the design of this house? What information do you need to answer the question?

A. Does the front of the house show similar triangles?; angle and side measurements

B. How many triangles can you identify when looking at the front of the house? 2

C. To answer your question, what strategy and tool would you use along with all the information you have? What answer do you get? See Strategies 1 and 2 on the facing page.

D. How are the corresponding sides of the triangles related? How do you know? See Additional Answers.

Turn and Talk Suppose you are only given the side lengths of the triangles formed. How could you use this information to answer the question? See margin.

SUPPORT SENSE-MAKING • Three Reads

Tell students to read the information in the photo three times and prompt them with a different question each time.

1. What is the situation about? The situation is about a house that looks triangular from the front, with a porch and a balcony going across the front of the house.

2. What are the quantities in this situation? How are those quantities related? The quantities are the lengths of the edges of the roof, the width of the balcony, the width of the porch, and the measures of the angles formed by the roof edges with the deck and balcony; The width of the roof seems to be equal to the lengths of the edges of the roof, and the width of the balcony is less than the other lengths. The measures of the angles seem to be equal.

3. What are possible questions you could ask about this situation? Possible answer: Does the front of the house make an equilateral triangle? Does the balcony cut the front of the house into a trapezoid and another smaller triangle? Is that triangle similar to the bigger triangle formed by the front of the house?

Apply and Practice

1 Spark Your Learning

▶ MOTIVATE

- Have students look at the photo in their books and read the information contained in the photo. Then complete Part A as a whole-class discussion.
- Give the class the additional information they need to solve the problem. This information is available online as a printable and projectable page in the Teacher Resources.
- Have students work in small groups to complete Parts B–D.

▶ PERSEVERE

If students need support, guide them by asking:

Q **Advancing • Use Tools** Which tool could you use to solve the problem? Why choose that tool and not some other? Students' choices of tools and reasons for choosing them will vary.

Q **Assessing** How are the bottom sides of the triangles related? They are parallel.

Q **Assessing** How are the angles of the triangles related? They share an angle at the peak of the roof, and the other corresponding angles are congruent since they are formed by parallel lines cut by transversals.

Q **Advancing** If you made one change to the triangles to ensure that they were not similar, what change could you make? Possible answer: I could change the angle of one base so that they the bases were no longer parallel.

Turn and Talk Remind students that if the side lengths are proportional, then one triangle is a dilation of the other, so there should be a common scale factor by which each side length of the smaller triangle can be multiplied to give the corresponding side length of the larger triangle. Possible answer: If the three pairs of corresponding side lengths are proportional, triangles are similar by the SSS Triangle Similarity Theorem. Because the triangles shown share a common angle, you could use the SAS Triangle Similarity Theorem to show that the triangles are similar.

▶ BUILD SHARED UNDERSTANDING

Select groups of students who used various strategies and tools to share with the class how they solved the problem. As they present their solutions, have each group discuss why they chose a specific strategy and tool.

2 Learn Together

Build Understanding

Task 1 MP **Construct Arguments** Students are introduced to the Triangle Proportionality Theorem and are guided through a proof of the theorem. Make sure students are analyzing and responding to the arguments given in the proof rather than reading passively. Encourage students to work out the given justifications on their own in order to convince themselves of their accuracy.

Sample Guided Discussion:

Q **Why is it incorrect to start the proof with the proportion $\frac{AD}{AB} = \frac{AE}{AC}$?** That proportion is the statement I am trying to prove, so I can't assume that it's true.

Turn and Talk Have students recall and write down the Midsegment Theorem first. Help students understand that one theorem can simply be a special case of another more general theorem. Ask them how they might relate two equal quantities with a ratio. Students may struggle with that idea since they are likely used to ratios comparing only unequal quantities. Possible answer: The Triangle Midsegment Theorem describes a specific case of the Triangle Proportionality Theorem where the ratio of the segments into which each side of the triangle is divided is 1 to 1, or the line divides each side exactly in half.

Build Understanding

The Triangle Proportionality Theorem

The Triangle Proportionality Theorem describes how a line parallel to one of the sides of the triangle divides the two sides that it intersects.

Triangle Proportionality Theorem

If a line parallel to one side of a triangle intersects the other two sides, then it divides those sides proportionally.

If $\overleftrightarrow{DE} \parallel \overline{BC}$, then $\frac{AD}{DB} = \frac{AE}{EC}$.

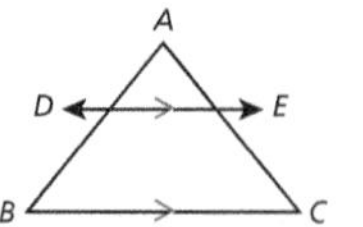

1 Prove the Triangle Proportionality Theorem.

Given: $\overleftrightarrow{DE} \parallel \overline{BC}$

Prove: $\frac{AD}{DB} = \frac{AE}{EC}$

First, show that $\triangle ADE \sim \triangle ABC$.
A–C. See Additional Answers.

A. Explain why $\angle 1 \cong \angle 2$ and $\angle 3 \cong \angle 4$.

B. Is there enough information to conclude that $\triangle ADE \sim \triangle ABC$? Explain.

Next, use the similar triangles to show that $\frac{AD}{DB} = \frac{AE}{EC}$.

$\frac{AB}{AD} = \frac{AC}{AE}$	Corresponding sides are proportional.
$\frac{AD + DB}{AD} = \frac{AE + EC}{AE}$	Segment Addition Postulate
$\frac{AD}{AD} + \frac{DB}{AD} = \frac{AE}{AE} + \frac{EC}{AE}$	Distributive Property of Division
$1 + \frac{DB}{AD} = 1 + \frac{EC}{AE}$	$\frac{a}{a} = 1$
$\frac{DB}{AD} = \frac{EC}{AE}$	Subtract 1 from each side.
$\frac{AD}{DB} = \frac{AE}{EC}$	Take the reciprocal of each side.

C. Suppose the proportion $\frac{AD}{AB} = \frac{AE}{AC}$ is used instead of $\frac{AB}{AD} = \frac{AC}{AE}$. Do you get the same result? Explain.

Turn and Talk Compare the Midsegment Theorem presented earlier to the Triangle Proportionality Theorem. How are they alike? See margin.

384

LEVELED QUESTIONS

Depth of Knowledge (DOK)	Leveled Questions	What Does This Tell You?
Level 1 **Recall**	How many pairs of congruent corresponding angles do you need to find to know that triangles are similar? I need to find at least two pairs.	Students' answers will indicate whether they understand how the triangles in the proof are demonstrated to be similar.
Level 2 **Basic Application of Skills & Concepts**	If $\overleftrightarrow{DE}$ was instead parallel to $\overline{AC}$, write a proportion relating segment lengths. $\frac{BD}{DA} = \frac{BE}{EC}$.	Students' answers will indicate whether they can apply the theorem to a new case.
Level 3 **Strategic Thinking & Complex Reasoning**	Do you think you could apply a similar theorem to a trapezoid? Yes, if a line intersects the legs of a trapezoid and is parallel to the bases, then I think the legs are divided proportionally.	Students' answers will indicate whether they can conjecture and extend the theorem to an analogous result involving a trapezoid.

Step It Out

Apply the Triangle Proportionality Theorem

2 A light sensor is installed above the front door of a building at point G. If the light sensor is moved to point J, how far from the front door will the light from the sensor reach?

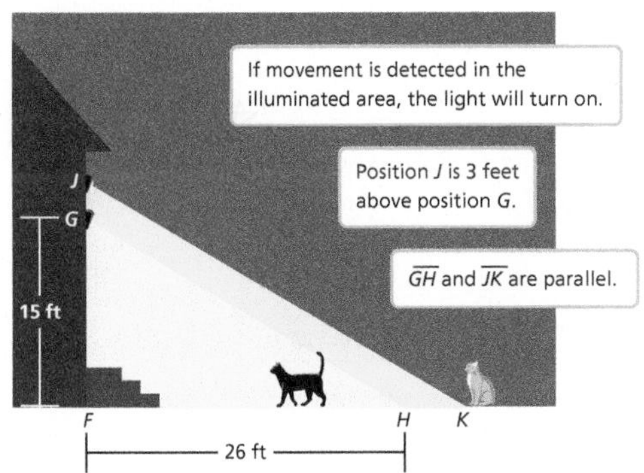
If movement is detected in the illuminated area, the light will turn on.
Position J is 3 feet above position G.
GH and JK are parallel.
J
G
15 ft
F
H
K
26 ft

Find *KH*.

$\frac{JG}{GF} = \frac{KH}{HF}$

$\frac{3}{15} = \frac{KH}{26}$

$78 = 15 \cdot KH$

$KH = 5.2$

A. How do you know this proportion is true?

A. Possible answer: It is given that $\overline{GH}$ is parallel to $\overline{JK}$, and the Triangle Proportionality Theorem confirms the proportion.

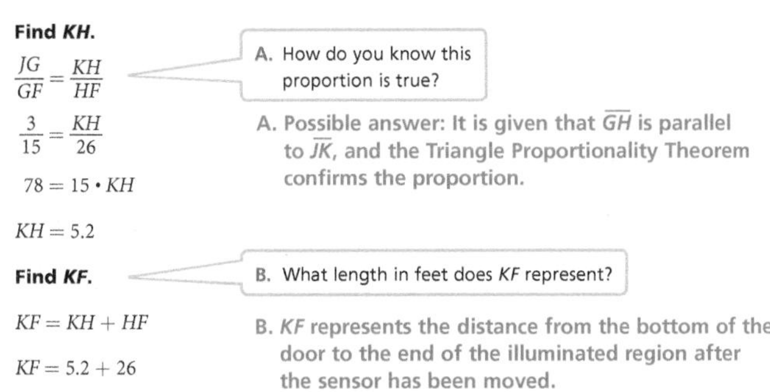

Find *KF*.

B. What length in feet does KF represent?

$KF = KH + HF$

$KF = 5.2 + 26$

$KF = 31.2$

B. KF represents the distance from the bottom of the door to the end of the illuminated region after the sensor has been moved.

The light from the sensor will reach 31.2 feet after it is moved up to point J.

Turn and Talk The sensor is moved to position M in order to reach a horizontal distance of 33 feet from the building. Where is position M located on the wall in relation to position G? See margin.

Step It Out

Task 2 **Attend to Precision** Students follow an application of the Triangle Proportionality Theorem to solve a real-world problem. Make sure students clearly explain their reasoning for each step. For example, in Task A students should specifically mention the fact that $\overline{GH} \parallel \overline{JK}$.

Sample Guided Discussion:

Q In Part A, which triangles are similar? $\triangle JFK \sim \triangle GFH$

Q What is another proportion you could set up for Part A? I could use $\frac{GF}{JG} = \frac{HF}{KH}$.

Q Could you solve this problem without applying Triangle Proportionality Theorem? Yes, I can show that $\triangle JFK \sim \triangle GFH$ and then use the fact that their side lengths are proportional to solve for KF.

Turn and Talk Ask students to sketch a diagram, reminding them the distance described is horizontal and should be along the same line as F, H, and K. Point M is approximately 4 feet above point G.

PROFICIENCY LEVEL

Beginning

Show students $\triangle ADE$ with interior segment $\overline{BC}$ parallel to side $\overline{DE}$. Say, "$\overline{BC}$ is parallel to $\overline{DE}$, so $\frac{AB}{BD}$ is equal to $\frac{AC}{CE}$," then write "$\overline{BC} \parallel \overline{DE}$, so $\frac{AB}{BD} = \frac{AC}{CE}$." Give students another labeled triangle with a labeled interior segment parallel to a side of the triangle, and ask them to write a similar statement.

Intermediate

Show students $\triangle ADE$ with interior segment $\overline{BC}$ parallel to side $\overline{DE}$, and ask them to work together to write two different equality proportions involving segment lengths. Have them explain in words why both equations are true.

Advanced

Show students $\triangle ADE$ with interior segment $\overline{BC}$ parallel to side $\overline{DE}$. Write, "$\overline{BC} \parallel \overline{DE}$, so $\frac{AB}{BD} = \frac{CE}{AC}$." Have students explain what, if anything, is incorrect with the statement in a short paragraph.

Task 3 **MP** **Construct Arguments** Students see how the Converse of the Triangle Proportionality Theorem can be used to verify that a side of a triangle is parallel to an interior segment. Ask students if they believe it is correct to express the Triangle Proportionality Theorem as an "if and only if" statement, and have them justify their reasoning. Point out how if the converse of a theorem is true, then that theorem can be expressed as a biconditional statement.

Sample Guided Discussion:

Q In Part A, what must be true to use the converse of a theorem? The conclusion of the theorem must be true to use its converse.

Q Without doing any calculations, can you use the work done in the task to tell if $\frac{YV}{WY} = \frac{ZX}{WZ}$ is a true statement? It is true since the ratios are reciprocals of the ratios of the proportion in the task.

Task 4 **MP** **Use Tools** Students see how the Triangle Proportionality Theorem can be used to partition a line segment on a coordinate plane. Ask students how two distinct points on a coordinate plane lead naturally to a right triangle, assuming the *x*-and *y*-coordinates are distinct.

CONNECT TO VOCABULARY

Have students use the **Interactive Glossary** to record their understanding of the vocabulary in this task.

Sample Guided Discussion:

Q What is the equation of the horizontal leg of the triangle? $y = -5$

Q What are you doing if you add $\left(\frac{1}{4}\right)12 = 3$ to $y = -5$ to get $y = -2$? I am finding a line parallel to one leg of the triangle that intersects the other leg $\frac{1}{4}$ of the way up.

Q How can you use the line $y = -2$ to find the point P? The line intersects the hypotenuse of the triangle at $(-2, -2)$, which is $\frac{1}{4}$ of the way along the hypotenuse by the Triangle Proportionality Theorem.

The Converse of the Triangle Proportionality Theorem

Converse of the Triangle Proportionality Theorem

If a line divides two sides of a triangle proportionally, then it is parallel to the third side.

If $\frac{AD}{DB} = \frac{AE}{EC}$, then $\overleftrightarrow{DE} \parallel \overline{BC}$.

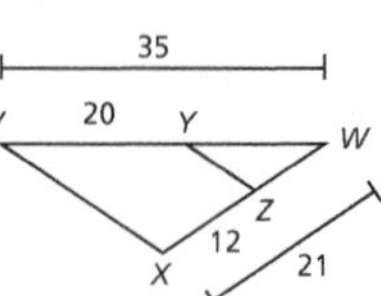

3 In the figure at the right, verify that $\overline{YZ} \parallel \overline{VX}$.

A–C. See Additional Answers.

A. In order to use the Converse of the Triangle Proportionality Theorem to show that $\overline{YZ} \parallel \overline{VX}$, what must be true about the sides of the triangle?

Find *WY*, *YV*, *WZ*, and *ZX*.

$WY = 35 - 20 = 15$ $YV = 20$

$WZ = 21 - 12 = 9$ $ZX = 12$

B. Why do you need to subtract to find *WY* and *WZ*?

Check for proportionality.

$\frac{WY}{YV} \stackrel{?}{=} \frac{WZ}{ZX}$

$\frac{15}{20} \stackrel{?}{=} \frac{9}{12}$

$0.75 = 0.75$ ✓

C. Is $\overline{YZ} \parallel \overline{VX}$? Explain.

Partition Segments

You can apply the Triangle Proportionality Theorem to **partition**, or divide, a line segment into shorter segments with a given ratio.

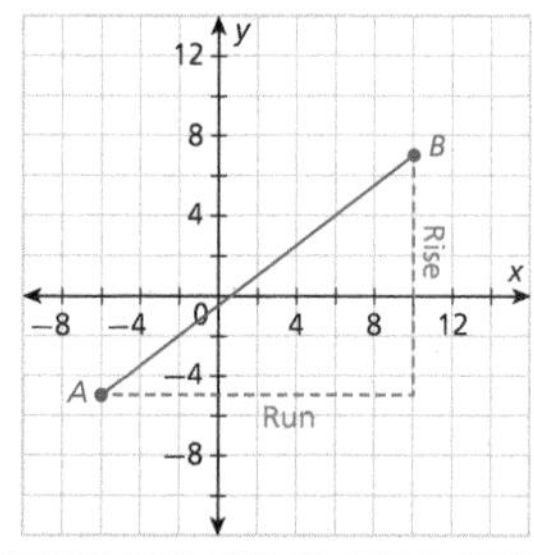

4 Find the coordinates of point P that divides $\overline{AB}$ into a ratio of 1 to 3 from $A(-6, -5)$ to $B(10, 7)$.

Write a ratio to describe the distance of point *P* along the segment from *A* to *B*.

Point P is $\frac{1}{3+1} = \frac{1}{4}$ of the distance from A to B.

Find the rise and the run between points *A* and *B*.

Run (horizontal distance): $10 - (-6) = 16$

Rise (vertical distance): $7 - (-5) = 12$

Find $\frac{1}{4}$ of the rise and run to determine the distance point *P* is from *A*.

$\frac{1}{4}$ of run $= \left(\frac{1}{4}\right)16 = 4$ $\frac{1}{4}$ of rise $= \left(\frac{1}{4}\right)12 = 3$

Point P is 4 units horizontally from Point A and 3 units vertically from Point A.

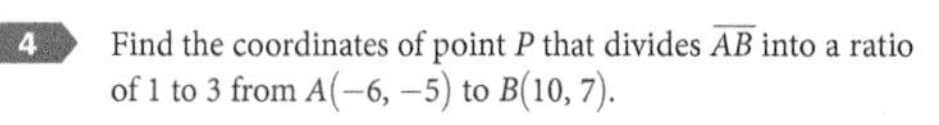

$P(x, y) = (-6 + 4, -5 + 3) = (-2, -2)$

The coordinates of point P are $(-2, -2)$.

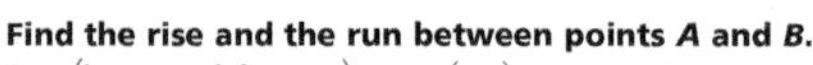

A. What does $\frac{1}{4}$ of the rise represent? $\frac{1}{4}$ of the run represent?

B. Why is $\frac{1}{4}$ of the rise and $\frac{1}{4}$ of the run added to the coordinates of point *A* instead of the coordinates of point *B*?

A, B. See Additional Answers.

386

You can use a construction to partition a segment into several equal parts. Then, you can locate a point that partitions the segment in a given ratio.

For $\overline{AB}$, construct point P that partitions the segment into a ratio of 2 to 3 from point A to point B.

A. Use a ruler to draw $\overrightarrow{AC}$. The exact length or angle is not important, but the construction may be easier if the measure of the angle is about 45° to 60°.
Check students' construction.

B. Placing your compass point on A, draw a small arc through $\overrightarrow{AC}$ and label that point D. The precise distance from A is not important, but for any construction, you will be drawing the number of arcs equal to the total number $a + b$ of portions in the ratio of a to b. Since the construction will divide $\overline{AB}$ in the ratio of 2 to 3, how many total arcs should you draw?
Check students' construction; 5 arcs

C. Using the same compass setting, place the compass point at D and draw a second arc and label the intersection E. Repeat the process 3 more times and label the points of intersection F, G, and H. Why is it necessary to draw an equal number of arcs as the total number of portions in the ratio? See Additional Answers.

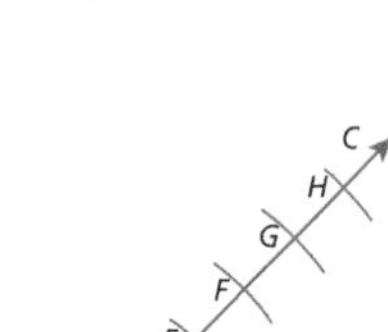

D. Use a straightedge to draw a segment from H to B. Construct an angle congruent to $\angle AHB$ with G as the vertex. Repeat the process for F, E, and D to create 5 segments.
Check students' construction.

E. The construction partitions $\overline{AB}$ into 5 equal parts. Why is point P located as shown?
E, F. See Additional Answers.

F. Why is $\overline{EP}$ parallel to $\overline{HB}$?

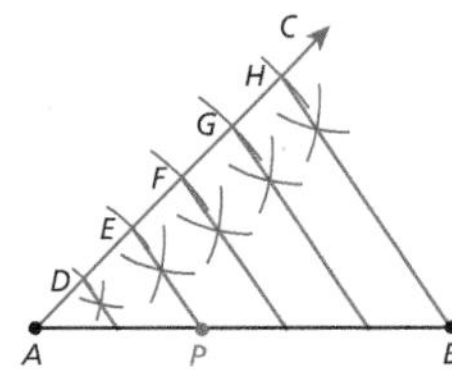

Turn and Talk For the construction in Task 5, explain why this construction works using the Triangle Proportionality Theorem. See margin.

Task 5 (MP) **Use Tools** Students use a compass and ruler to partition a line segment as an application of the Triangle Proportionality Theorem. Make sure students are actively thinking about how this step-by-step construction is an application of the theorem instead of passively following directions. They should understand that a triangle and segments constructed parallel to a side of the triangle are at the heart of this construction.

Sample Guided Discussion:

Q **In Part C, why is it important to keep the compass at the same setting as you draw the five arcs?** A ratio is a comparison of how many equal-sized parts are in two quantities. If I changed the compass setting, then the segments would not be partitioned into equal-sized segments.

Q **In Part D, what is the triangle used in the Triangle Proportionality Theorem?** It is $\triangle ABH$.

Q **If you label the other endpoint of the segment with endpoint F as point I, where is point P along the segment $\overline{AI}$?** It is $\frac{2}{3}$ of the way along $\overline{AI}$.

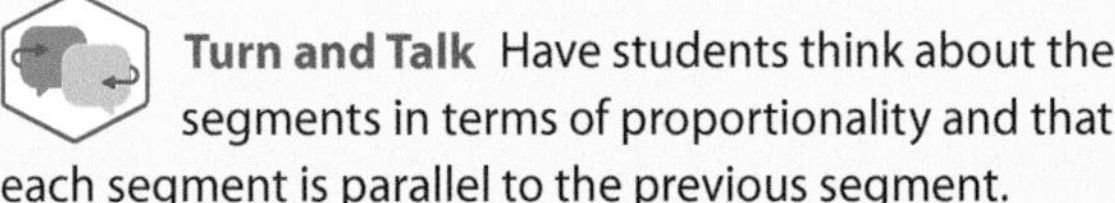

Turn and Talk Have students think about the segments in terms of proportionality and that each segment is parallel to the previous segment.

- The construction ensures that the segments drawn between $\overline{AC}$ and $\overline{AB}$ are parallel by the Converse of the Corresponding Angles Postulate. Because they are parallel, the ratios of the lengths of the divided segment on side $\overline{AH}$ are equal to the ratios of the lengths of the corresponding divided segment on side $\overline{AB}$ by the Triangle Proportionality Theorem. This means that since point E divides $\overline{AH}$ in a ratio of 2 to 3, point P divides $\overline{AB}$ in a ratio of 2 to 3.

PLAN FOR DIFFERENTIATED INSTRUCTION

MTSS 

Small-Group Options

Use these teacher-guided activities with pulled small groups.

On Track

Ask students,

"Can you have a Pythagorean triple *a*, *b*, and *c* where one number is the geometric mean of the other two numbers?"

- Ask which number of the triple the geometric mean could be.
- The two conditions result in a system of equations: $\begin{cases} a^2 + b^2 = c^2 \\ b^2 = ac \end{cases}$.
- Substituting ac for b^2 in the first equation results in $a^2 + ac - c^2 = 0$. Ask students if they think there are any nonzero whole numbers *a* and *c* that satisfy the equation.
- Ask students if there are any nonzero whole numbers satisfying both conditions.

Almost There (RtI)

Materials: index cards

Give each group a set of index cards. Each card should have a right triangle with altitude to its hypotenuse. The legs of the outer right triangle, segments of its hypotenuse, and altitude should be labeled with variables. Make sure triangles on different cards have different orientations and variables.

Have students work together to identify three similar right triangles and proportionally relate their side lengths.

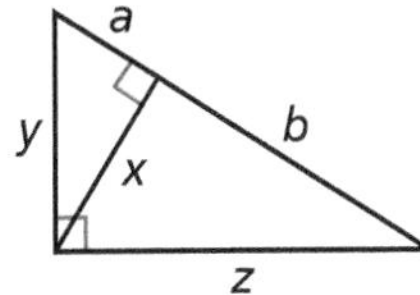

Ready for More

Say, "We know that π is a famous irrational number, and the number φ is famous as well! It's often called the golden mean, and one definition is:

φ is the positive number that is the geometric mean of 1 and $1 + \varphi$."

Ask,

- Can you use the definition of φ to write an expression for φ^2?
- Can you write φ^3 in terms of φ and 1?
- Can you write φ^4 in terms of φ and 1?
- Can you write φ^5 in terms of φ and 1?
- Do you notice a pattern?

Math Center Options

Use these student self-directed activities at centers or stations. Key: ● Print Resources ● Online Resources

On Track

- ● Interactive Digital Lesson
- ●● Journal and Practice Workbook
- ● Interactive Glossary (printable): **geometric mean**, **Pythagorean triple**
- ● Module Performance Task

Standards Practice:

- ●● Prove Similarity Theorems About Triangles
- ●● Prove Relationships Related to Congruence and Similarity of Triangles

Almost There

- ● Reteach 12.4 (printable)
- ● Interactive Reteach 12.4
- ● Illustrative Mathematics: Converse of the Pythagorean Theorem

Ready for More

- ● Challenge 12.4 (printable)
- ● Interactive Challenge 12.4

ONLINE 

View data-driven grouping recommendations and assign differentiation resources.

During the *Spark Your Learning*, listen and watch for strategies students use. See samples of student work on this page.

Relate the Legs of the Triangles Proportionally — Strategy 1

Write and solve a proportion to find the complete distance from the player to the final location of the ball. Subtract 7 to find the distance from the player to the net.

$$\frac{7}{3.5} = \frac{x}{10}$$

$x = 20$

$20 - 7 = 13$ ft

If students . . . proportionally relate the legs of the right triangles, they are demonstrating an understanding of AA Triangle Similarity from Lesson 12.2.

Have these students . . . explain how they determined they could apply a proportion and how they set it up. **Ask:**

- Q How did you know the lengths of the legs of the triangle are proportional?
- Q Are there any other proportions you could have used?

Use the Triangle Proportionality Theorem — Strategy 2

Since $\overline{AB} \parallel \overline{ED}$, I know $\frac{AE}{EC} = \frac{x-7}{7}$, so the Pythagorean Theorem lets me write

$$\frac{\sqrt{x^2+100} - \sqrt{61.25}}{\sqrt{61.25}} = \frac{x-7}{7}.$$

$$7\left(\sqrt{x^2+100} - \sqrt{61.25}\right) = (x-7)\sqrt{61.25}$$

$$7\sqrt{x^2+100} - 7\sqrt{61.25} = x\sqrt{61.25} - 7\sqrt{61.25}$$

$$49(x^2+100) = 61.25x^2$$

$$4900 = 12.25x^2$$

$$400 = x^2$$

$$x = 20 \text{ ft}$$

Then, I can subtract 7 to find the distance from the player to the net: $20 - 7 = 13$ ft.

If students . . . use the Triangle Proportionality Theorem, they understand that a line parallel to a side of the triangle divides sides proportionally, but they may not see that they can relate side lengths of similar triangles to solve the problem more efficiently.

Activate prior knowledge . . . by having students proportionally relate the lengths of the legs of the right triangles. **Ask:**

- Q How are the two triangles related?
- Q How are the lengths of the legs of the triangles related?

COMMON ERROR: Does not Subtract

I can use similar triangles to write and solve a proportion to find the complete distance from the player to the final location of the ball.

$$\frac{7}{3.5} = \frac{x}{10}$$

$x = 20$ ft

If students . . . fail to subtract at the end of the problem, they may not understand that the desired quantity is not the side length of a triangle in the diagram.

Then intervene . . . by pointing out that x is the length of a leg of the larger triangle. **Ask:**

- Q Is the distance from the player to the net the side of a triangle?
- Q How can you get that distance from the side length of triangles?

12.4

Apply Similarity in Right Triangles

I Can identify similar right triangles, apply the Geometric Means Theorems, and recognize Pythagorean triples.

Spark Your Learning

When a tennis player hits the ball using an overhead smash serve, the goal is for the player to hit the ball close to a straight line that passes as close to the net as possible without hitting it.

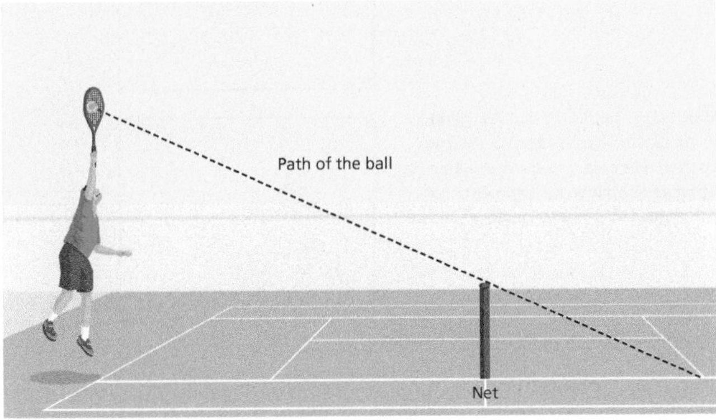

Complete Part A as a whole class. Then complete Parts B–D in small groups.

A. What is a mathematical question you can ask about the location of the player? What is the maximum distance the player can be from the net so the ball still clears the net?

B. Is it possible to find a relationship between any of the triangles shown in the figure using the additional information? Explain. See Additional Answers.

C. To answer your question, what strategy and tool would you use along with all the information you have? What answer do you get? See Strategies 1 and 2 on the facing page.

D. What does your answer mean in the context of the situation? See Additional Answers.

Turn and Talk Suppose the ball is served or hit from a height off the ground equal to the distance from the net to where the ball hits the ground. The total horizontal distance traveled by the ball stays the same. Determine the height of the ball when it is hit in this situation. See margin.

CULTIVATE CONVERSATION • Co-Craft Questions

If students have difficulty formulating a mathematical question about the situation in the Spark Your Learning, ask them to imagine themselves hitting a tennis ball so it barely makes it over the net. What are some natural questions to ask about this situation?

Work together to craft the following questions:

- How far can the ball travel past the net before hitting the ground?
- What is the maximum distance the player can be from the net so the ball still clears the net?

Then have students think about what additional information, if any, they would need to answer these questions. **Ask:**

- Can you determine how far the player can be from the net if you know only the height of the net? Why or why not?
- Can you determine how far the player can be from the net if you know only how far past the net the ball hits the ground? Explain.

1 Spark Your Learning

▶ MOTIVATE

- Have students look at the illustration in their books and read the information contained in the photo. Then complete Part A as a whole-class discussion.
- Give the class the additional information they need to solve the problem. This information is available online as a printable and projectable page in the Teacher Resources.
- Have students work in small groups to complete Parts B–D.

▶ PERSEVERE

If students need support, guide them by asking:

Q **Advancing • Use Tools** Which tool could you use to solve the problem? Why choose that tool and not some other? Students' choices of tools and reasons for choosing them will vary.

Q **Assessing** How many triangles are involved in the situation? two

Q **Assessing** How are the angles of the triangles related? They share two pairs of corresponding congruent angles.

Q **Advancing** How are the corresponding sides of the triangle related? Corresponding sides of the triangle are proportional since the triangles have AA similarity.

Turn and Talk Encourage students to make a sketch to better understand the problem. Help them see that this problem is very similar to the problem in the task, though there are two triangle legs with unknown length instead of one. Using a variable to represent the equal distances in the situation can help students see that they only need to solve for one variable, not two. The height of the ball is $\sqrt{70}$ feet when hit.

$$\frac{x}{3.5} = \frac{20}{x}$$
$$x^2 = 70$$
$$x = \sqrt{70}$$

▶ BUILD SHARED UNDERSTANDING

Select groups of students who used various strategies and tools to share with the class how they solved the problem. As they present their solutions, have each group discuss why they chose a specific strategy and tool.

(2) Learn Together

Build Understanding

Task 1 (MP) **Use Structure** Students fold a piece of paper to construct an altitude to the hypotenuse of a right triangle. Then they investigate the relationship between the angles of the three resulting right triangles and see that all triangles are similar. In doing so, they build an intuitive understanding of the Right Triangle Similarity Theorem. Ask students whether they think the similarity of the triangles was a result of the proportions of the paper, or if there is something about the structure of any right triangle that will give a similar result. Having students use papers with different proportions for this task can help drive this discussion.

Sample Guided Discussion:

Q In Part B, how many pairs of acute angles do you need to match between the triangles? One pair, since I know each have a right angle, so the triangles have AA similarity.

Q In Part E, how can you see if the same result occurs with other right triangles? I can repeat the task with pieces of paper whose length and width are not proportional with the original piece of paper.

Build Understanding

Compare Corresponding Parts of Similar Figures

Right triangles have special features you can use to prove two triangles are similar.

1 Use a ruler to draw a diagonal line on a rectangular piece of paper to form two congruent triangles.

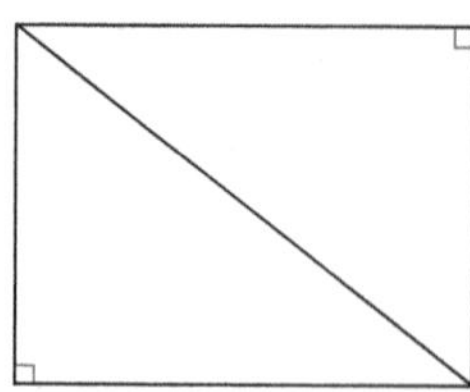

Fold the paper so the line you drew folds back along itself, and make a crease from the bottom left corner to the drawn line, creating an altitude to the hypotenuse of one triangle. Label your triangles as shown.

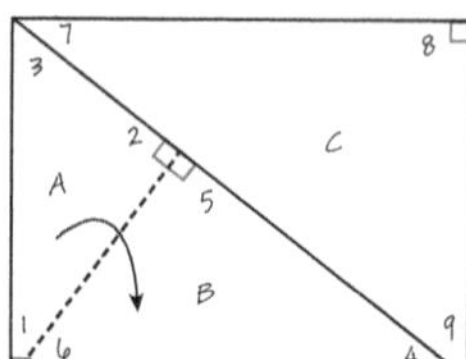

Cut out the three right triangles *A*, *B*, and *C*.

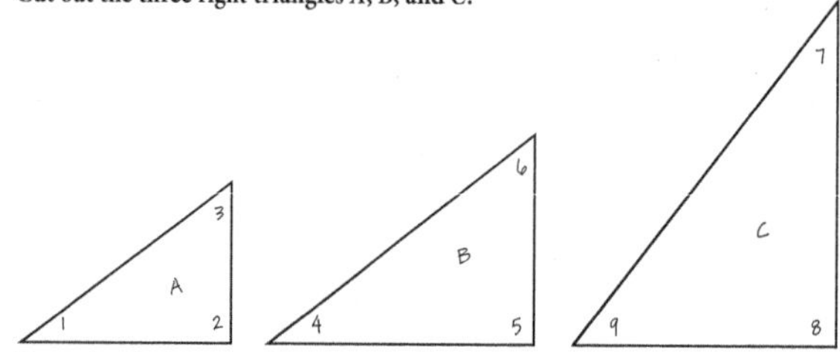

Place triangle *A* on triangle *B*. Move the triangles to match corresponding angles.

A. What do you observe about the angles of triangles *A* and *B*?
$\angle 1 \cong \angle 4$, $\angle 2 \cong \angle 5$, $\angle 3 \cong \angle 6$

B. Are triangles *A* and *B* similar? Why or why not?
yes; They meet the AA Triangle Similarity Criteria.

Move triangle *A* to triangle *C*. Move the triangles to match corresponding angles.

C. What do you observe about the angles of triangles *A* and *C*?
$\angle 1 \cong \angle 9$, $\angle 2 \cong \angle 8$, $\angle 3 \cong \angle 7$

D. What relationship can you identify between triangles *B* and *C*. How do you know? See Additional Answers.

E. In general, what types of figures are created by drawing an altitude to the hypotenuse of a right triangle?
two triangles similar to the original triangle and to themselves

392

LEVELED QUESTIONS

Depth of Knowledge (DOK)	Leveled Questions	What Does This Tell You?
Level 1 **Recall**	What are you constructing with the second fold? I am constructing an altitude to the hypotenuse of one of the triangles I created with the first fold.	Students' answers will indicate whether they understand the construction in the task.
Level 2 **Basic Application of Skills & Concepts**	If you construct an altitude to the hypotenuse of a right triangle that has one acute angle measuring 35°, what will be the measures of the angles of the smallest triangle created? The angles will have measures of 35°, 55°, and 90° since the triangles are similar.	Students' answers will indicate whether they can apply the result of the construction.
Level 3 **Strategic Thinking & Complex Reasoning**	If you extended the second fold in the task all the way to the edge of the paper to create a fourth triangle, how would its side lengths be related to those of the other three triangles? Corresponding sides of all four triangles would be proportional, since all triangles are similar.	Students' answers will indicate whether they can find a corollary to the result by appling relationships between complementary angles or angles created by a transversal intersecting parallel lines.

Step It Out

Identify Properties of Similar Right Triangles

The following theorem describes the figures formed when drawing the altitude to the hypotenuse of a right triangle.

Right Triangle Similarity Theorem
The altitude to the hypotenuse of a right triangle forms two triangles that are similar to each other and to the original triangle.

Segment lengths in the three triangles have a special proportional relationship. In a proportion of the form $\frac{a}{x} = \frac{x}{b}$, two numbers are the same. The x represents the *geometric mean* of a and b. The **geometric mean** of two numbers is the positive square root of their product. The following theorems involve the geometric mean and right triangles.

Geometric Means Theorems		
The length of the altitude to the hypotenuse of a right triangle is the geometric mean of the lengths of the segments of the hypotenuse.	$\frac{x}{h} = \frac{h}{y}$ or $h = \sqrt{xy}$	
The length of the leg of a right triangle is the geometric mean of the lengths of the hypotenuse and the segment of the hypotenuse adjacent to that leg.	$\frac{x}{a} = \frac{a}{c}$ or $a = \sqrt{xc}$ $\frac{y}{b} = \frac{b}{c}$ or $b = \sqrt{yc}$	

 Prove the first Geometric Means Theorem.

Given: Right triangle ABC with altitude $\overline{BD}$

Prove: $\frac{CD}{BD} = \frac{BD}{AD}$

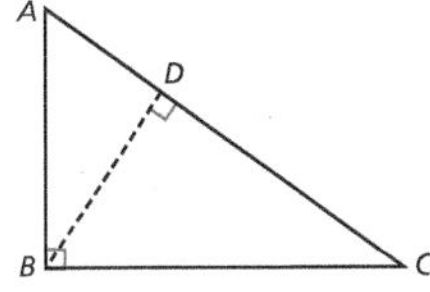

Statements	Reasons
1. Right triangle ABC with altitude $\overline{BD}$	1. Given
2. $\triangle ABD \sim \triangle BCD$	2. The altitude to the hypotenuse of a right triangle forms two triangles that are similar to each other and to the original triangle.
3. $\frac{CD}{BD} = \frac{BD}{AD}$	3. ___?___

A. What property is used for the reason in Step 3 of the proof? Corresponding sides of similar triangles are proportional.

B. Is $BD = \sqrt{CD \cdot AD}$ equivalent to $\frac{CD}{BD} = \frac{BD}{AD}$? Explain why or why not.
See Additional Answers.

Step It Out

Task 2 (MP) **Construct Arguments** Students are given a formal statement of the result they explored in Task 1, and see how it can be used to prove one of the Geometric Mean Theorems. Make sure students organize their thoughts before discussing reasons for the steps in the proof; they can be confused since the same quantity appears in ratios on either side of the proportion. Having them sketch a labeled diagram can help them better make sense of the relationship between the quantities involved.

CONNECT TO VOCABULARY

Have students use the **Interactive Glossary** to record their understanding of the vocabulary in this task.

Sample Guided Discussion:

Q How do you know which of x, y, and h must be the largest? I know that h is the largest since it is the length of the hypotenuse of both of the smaller triangles, and they each have a leg with length x or y.

Q In Part B, does it make sense to take both the positive and negative square roots of an equation when solving for BD? No; it doesn't make sense for BD to be negative, since it is a side length of a triangle.

Q Is $\frac{x}{a} = \frac{a}{c}$ equivalent to $a = \sqrt{xc}$ if a, c, or x are any real numbers with $a, c \neq 0$?
No. $\frac{x}{a} = \frac{a}{c}$ is equivalent to $a^2 = xc$, and a might be negative, so I would need to take the positive and negative square roots of both sides of the equation to get $a = \pm\sqrt{xc}$.

PROFICIENCY LEVEL

Beginning

Show students a right triangle with altitude to the hypotenuse labeled as shown here, say, "x is to h as h is to y," and write the proportion $\frac{x}{h} = \frac{h}{y}$. Then, show them a similar diagram but with different variables and with the outer triangle positioned differently (for example, the hypotenuse is vertical). Have them say a similar statement and write a proportion.

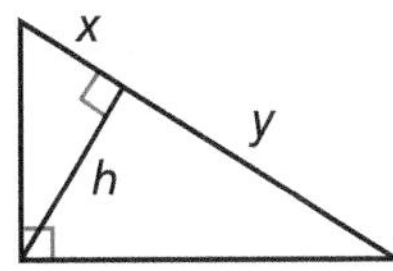

Intermediate

Show students a right triangle with altitude to the hypotenuse labeled as shown here. Have them describe out loud both proportions resulting from the Geometric Means Theorems and write the proportions.

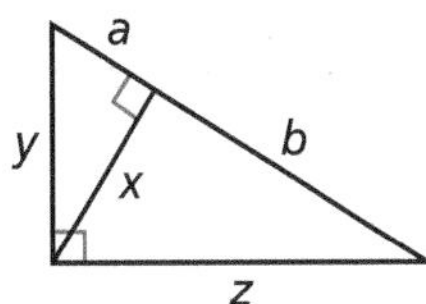

Advanced

Have students give verbal descriptions of both Geometric Means Theorems.

Task 3 (MP) **Attend to Precision** Students see how a Geometric Means Theorem can be applied to find the height of an object. Make sure students understand that the position of the observer is determined by their lines of sight to the top and the bottom of the wall, specifically, so these lines of sight are perpendicular. When students justify the given proportion, make sure they identify the quantities of the proportion as sides of right triangles formed by lines of sight and the wall and make explicit mention of a Geometric Means Theorem.

Sample Guided Discussion:

Q How do you know that the proportion $\frac{5}{14.5} = \frac{14.5}{y}$ must have a solution greater than 14.5? Since the ratios are equivalent and 14.5 is greater than 5, I know that *y* is greater than 14.5.

Q The arithmetic mean of two numbers *a* and *b* is $\frac{a+b}{2}$. Can you think of a way that the arithmetic and geometric means of *a* and *b* are similar? Both means are between *a* and *b*.

Turn and Talk Have students reproduce the given diagram as a sketch and sketch another triangle modeling the proposed scenario on the same diagram. Ask students to compare and contrast the two triangles, making sure to ask questions that draw their attention to the angles. No, a right triangle must be formed to apply the geometric mean theorem.

Task 4 (MP) **Construct Arguments** Students see how a Geometric Means Theorem can be applied to prove the Pythagorean Theorem. While the proof is for the Pythagorean Theorem and not explicitly for the converse, ask students how they might prove the converse. Remind them what it means for algebraic expressions to be equivalent.

Sample Guided Discussion:

Q In Part A, how do you know which Geometric Means Theorem to apply? The first Geometric Means Theorem involves the altitude to the hypotenuse of the triangle, but the proof only involves the legs, the hypotenuse, and segments of the hypotenuse. So, I can apply the second Geometric Means Theorem.

Q Could you use the steps of the proof to prove the Converse of the Pythagorean Theorem? Yes, since every step is equivalent to the previous step, I could write the steps in an order opposite of that in the proof.

Apply a Geometric Means Theorem

3 To find the height of a rock-climbing tower, a climber uses a square of cardboard at eye level to line up the top and bottom of the tower. What is the height of the tower?

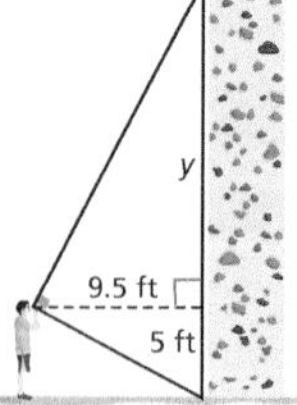

Write a proportion to find the missing portion of the height.

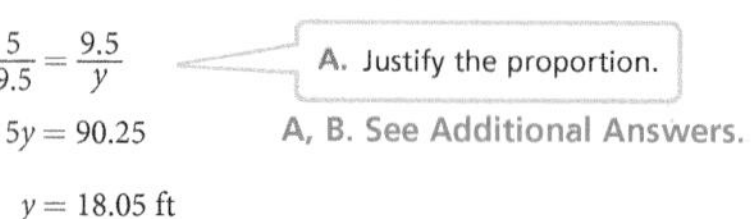

$\frac{5}{9.5} = \frac{9.5}{y}$ — A. Justify the proportion.

$5y = 90.25$ — A, B. See Additional Answers.

$y = 18.05$ ft

Find the total height.

height $= 18.05 + 5 = 23.05$ ft — B. Why do you add 5 to *y* to find the total height?

Turn and Talk A different person with an eye level at 6 ft is finding the height of the tower. Can this person stand at the same distance from the tower and estimate its height with this method? Explain. See margin.

Prove the Pythagorean Theorem

Recall the Pythagorean Theorem and its converse stated below.

Pythagorean Theorem: In a right triangle, the sum of the squares of the lengths of the legs equals the square of the length of the hypotenuse.

Converse of the Pythagorean Theorem: If the square of the length of the longest side of a triangle equals the sum of the squares of the lengths of the other two sides, then the triangle is a right triangle.

There are many proofs of the Pythagorean Theorem. One such proof uses a theorem presented in this lesson.

4 **Given:** $\triangle ABC$ is a right triangle.
Prove: $a^2 + b^2 = c^2$

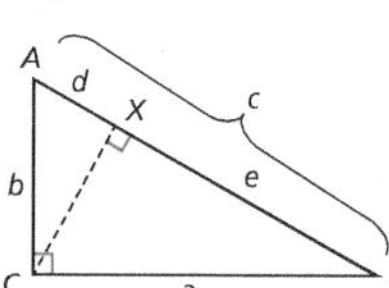

Relate the sides of the triangles.

A. $\overline{CX}$ is the altitude to the hypotenuse of $\triangle ABC$ that divides *c* into segments *d* and *e*. What theorem justifies the following equations?

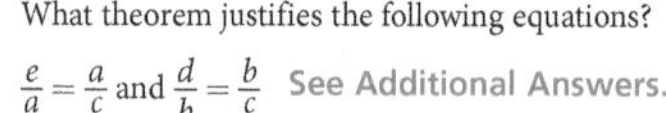
$\frac{e}{a} = \frac{a}{c}$ and $\frac{d}{b} = \frac{b}{c}$ See Additional Answers.

394

Derive the formula.

$\frac{e}{a} = \frac{a}{c}$ and $\frac{d}{b} = \frac{b}{c}$	Original proportions
$a^2 = ec \quad b^2 = cd$	Multiply.
$a^2 + b^2 = ec + cd$	Addition Property of Equality
$a^2 + b^2 = c(e + d)$	Distributive Property
$a^2 + b^2 = c(c)$	Substitute c for $e + d$.
$a^2 + b^2 = c^2$	Multiply.

B. How do you know that $c = e + d$?

B. See Additional Answers.

Use Pythagorean Triples

A **Pythagorean triple** is a set of three nonzero whole numbers a, b, and c that satisfy the equation $a^2 + b^2 = c^2$. Examples of Pythagorean triples are shown in the table.

Pythagorean triple	Used in $a^2 + b^2 = c^2$
3, 4, 5	$3^2 + 4^2 = 5^2$
5, 12, 13	$5^2 + 12^2 = 13^2$
8, 15, 17	$8^2 + 15^2 = 17^2$

5 Determine if the set of three values is a Pythagorean triple.

Set 1	Set 2
7, 24, 25	11, 14, 21
$7^2 + 24^2 \stackrel{?}{=} 25^2$	$11^2 + 14^2 \stackrel{?}{=} 21^2$
$49 + 576 \stackrel{?}{=} 625$	$121 + 196 \stackrel{?}{=} 441$
$625 = 625$	$317 \neq 441$
Both sides of the equation are equivalent, so 7, 24, 25 is a Pythagorean triple.	The sides of the equation are not equivalent, so 11, 14, 21 is not a Pythagorean triple.

A. How is the Converse of the Pythagorean Theorem used to determine whether a set of three numbers is a Pythagorean triple?

A, B. See Additional Answers.

B. Make a sketch of a right triangle that has side lengths 7, 24, and 25. How do the numbers of a Pythagorean triple relate to the side lengths of a right triangle?

C. Explain why 5, 6, $\sqrt{61}$ is not a Pythagorean triple. $\sqrt{61}$ is not a whole number.

Turn and Talk Suppose you multiply each of the numbers of a Pythagorean triple by the same number. Do the resulting numbers also form a Pythagorean triple? If so, give an example. See margin.

Module 12 • Lesson 12.4 395

Task 5 (MP) **Attend to Precision** Students are introduced to Pythagorean triples and see how the Converse of the Pythagorean Theorem can be used to check whether three nonzero whole numbers are a Pythagorean triple. Draw students' attention to the definition's careful specification that the numbers in question are nonzero whole numbers.

CONNECT TO VOCABULARY

Have students use the **Interactive Glossary** to record their understanding of the vocabulary in this task.

SUPPORT SENSE-MAKING Three Reads

Have students read the problem three times. Use the questions below for a different focus each read.

1. What is the situation about?
2. What are the quantities in the situation?
3. What are the possible mathematical questions that you could ask for the situation?

Sample Guided Discussion:

Q If a, b, and c are nonzero whole numbers with c being the largest number, how could you use them to state the Converse of the Pythagorean Theorem? If $a^2 + b^2 = c^2$, then a, b, and c are side lengths of a right triangle.

Q In Part C, can you add 5 and 6 and then square that sum? No, since $5^2 + 6^2 \neq (5 + 6)^2$.

Q Why do you think we say all three numbers of a Pythagorean triple must be nonzero? If I let one of the numbers be zero, then for any whole number b, the set $\{0, b, b\}$ would be a Pythagorean triple.

Turn and Talk Ask students how they can think of this question in terms of right triangles. Ask them how two triangles are related if every side length of one triangle is an integer multiple of a side length of the other triangle. If necessary, follow this by asking students how the angles of similar triangles are related. Yes, any multiple of a Pythagorean Triple is also a Pythagorean Triple; For example, 3-4-5 is a Pythagorean Triple and its product by 2, 6-8-10, is also a Pythagorean Triple.

ONLINE

Assign the Digital On Your Own for
- built-in student supports
- Actionable Item Reports
- Standards Analysis Reports

On Your Own

Assignment Guide

The chart below indicates which problems in the On Your Own are associated with each task in the Learn Together. Assign daily homework for tasks completed.

Learn Together Tasks	On Your Own Problems
Task 1, p. 392	Problems 6–10
Task 2, p. 393	Problem 11
Task 3, p. 394	Problems 12–22 and 32
Task 4, p. 394	Problems 29–31
Task 5, p. 395	Problems 23–28

Check Understanding

1. Braden uses a ruler and protractor to draw a right triangle. He divides that triangle using the altitude to the hypotenuse. How many similar triangles does his drawing contain? Explain. 1–5. See Additional Answers.

In Problems 2 and 3, use the diagram of the right triangle.

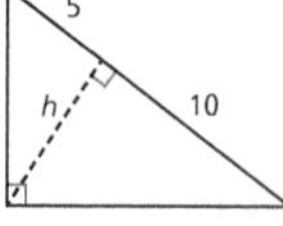

2. In a right triangle, how is the length of the altitude to the hypotenuse related to the lengths of the two segments it creates?

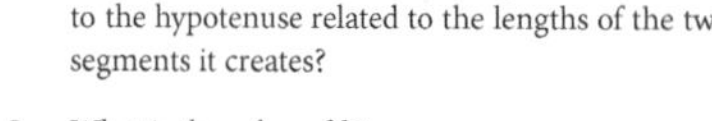

3. What is the value of h?

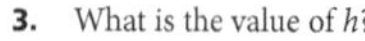

4. Explain how the Geometric Means Theorems are related to the Pythagorean Theorem.

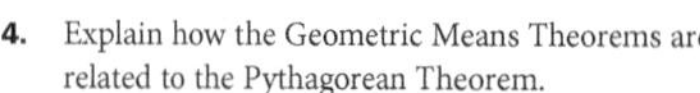

5. Do the values 9, 40, and 41 describe the side lengths and hypotenuse of a right triangle? Explain why or why not.

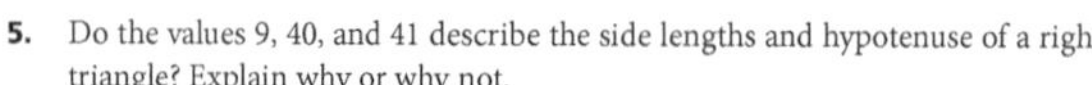

On Your Own

Write a similarity statement comparing the triangles in each figure.

6. 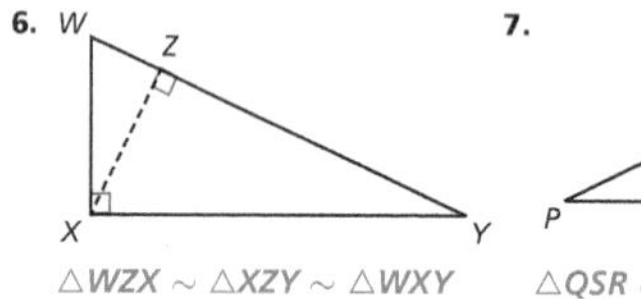

$\triangle WZX \sim \triangle XZY \sim \triangle WXY$

7. 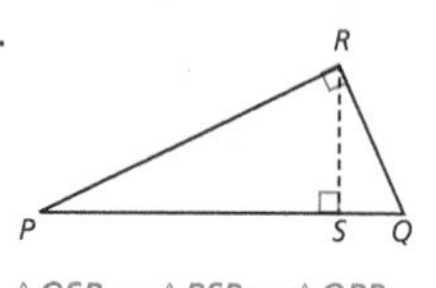

$\triangle QSR \sim \triangle RSP \sim \triangle QRP$

8. 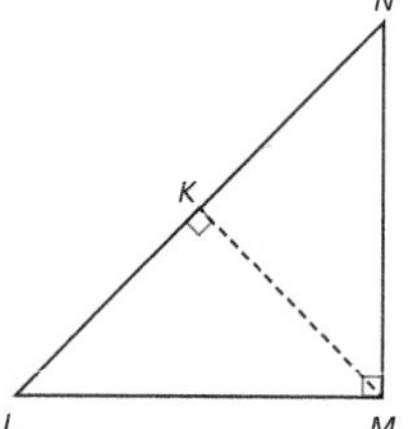

8. $\triangle LKM \sim \triangle MKN \sim \triangle LMN$

9. **Open Ended** Draw a right triangle with an altitude drawn to the hypotenuse and label the vertices of all the triangles formed. Redraw the triangle as three separate triangles, and write a similarity statement that relates the triangles. Why might redrawing the triangles be helpful when writing similarity statements for the triangles? 9–11. See Additional Answers.

10. Prove the Right Triangle Similarity Theorem.
Given: Right triangle ABC with altitude $\overline{BD}$
A. Prove that $\triangle BDC \sim \triangle ABC$.
B. Prove that $\triangle ADB \sim \triangle ABC$.
C. Prove that $\triangle BDC \sim \triangle ADB$.

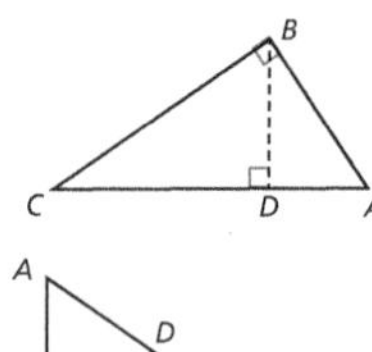

11. Prove the second Geometric Means Theorem.
Given: Right triangle ABC with altitude $\overline{BD}$
Prove: $\frac{AD}{AB} = \frac{AB}{AC}$ and $\frac{DC}{BC} = \frac{BC}{AC}$

A
D
B
C

396

3 Check Understanding

Formative Assessment

Use formative assessment to determine if your students are successful with this lesson's learning objective.

Students who successfully complete the Check Understanding can continue to the On Your Own practice.

For students who miss 1 problem or more, work in a pulled small group using the Almost There small-group activity on page 394.

ONLINE

Assign the Digital Check Understanding to determine
- success with the learning objective
- items to review
- grouping and differentiation resources

4 Differentiation Options

Differentiate instruction for all students using small-group activities and math center activities on page 391C.

Reteach

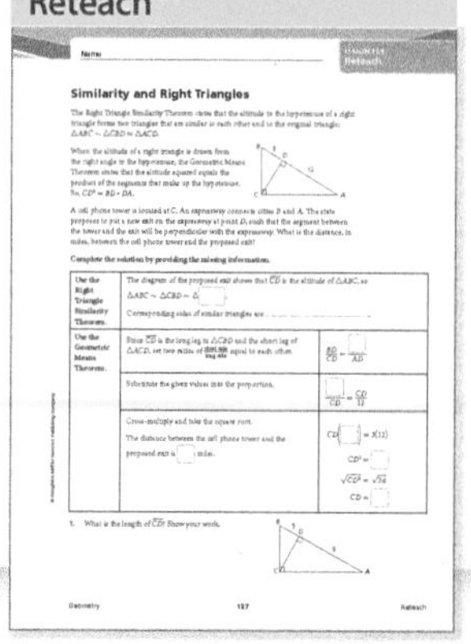

Challenge

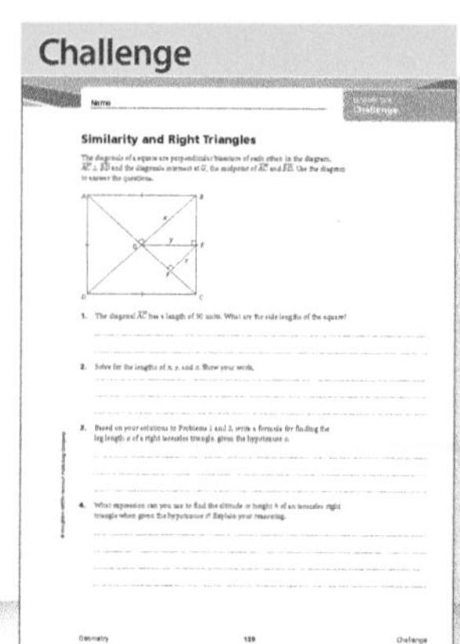

12, 13. See Additional Answers.

For Problems 12–15, find x, y, and z. Write your answer in simplest radical form.

12.

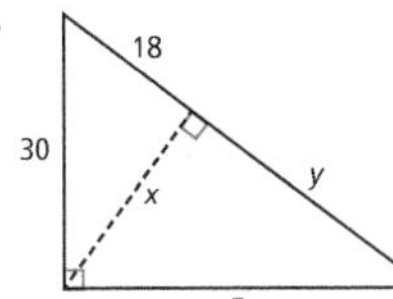

13.

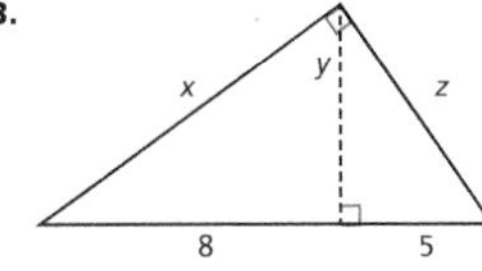

14.

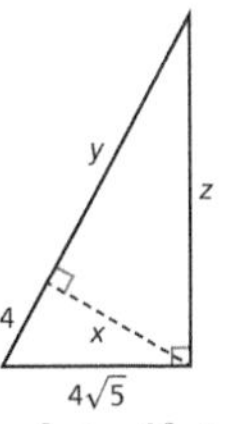

$x = 8$, $y = 16$, $z = 8\sqrt{5}$

15.

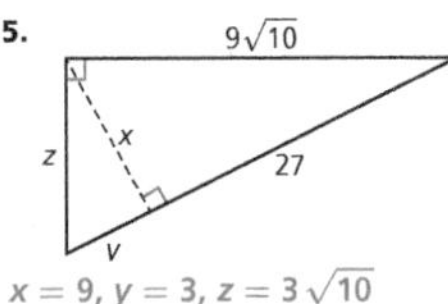

$x = 9$, $y = 3$, $z = 3\sqrt{10}$

Find the length of $\overline{AD}$ using the given segment lengths.

16. $AC = 4$, $AB = 3$ $AD = 2.4$

17. $AC = 80$, $BC = 89$ $AD = \frac{3120}{89} \approx 35.1$

18. $AB = 33$, $BC = 65$ $AD = \frac{1848}{65} \approx 28.4$

19. $BD = 3$, $DC = 12$ $AD = 6$

20. A 9 in. altitude divides the hypotenuse of a right triangle into two segments. One segment is 9 times as long as the other. What are the lengths for the two segments of the hypotenuse? 3 in. and 27 in.

21. STEM One common style of bridge is a truss bridge, where a supporting framework adds strength to the bridge. Some truss members are under a compression load, and some truss members are under a tension load. What is the length of the shortest truss member under compression? $4\sqrt{3}$, or approximately 6.93 feet

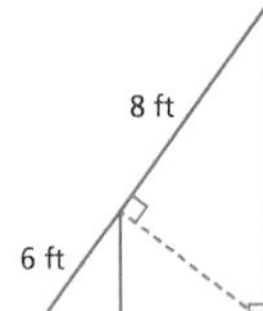

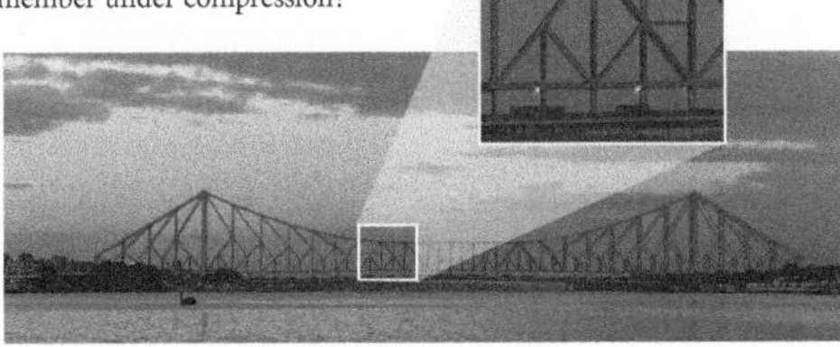

22. (MP) **Repeated Reasoning** An altitude to the hypotenuse of a right triangle divides it into two smaller right triangles. In the smallest triangle, the ratio of the hypotenuse to the longer leg is $\sqrt{2}$ to 1. In the original triangle, what is the ratio of the hypotenuse to the altitude? 2 to 1

Watch for Common Errors

Problem 12 Students may incorrectly relate three side lengths of a right triangle through a proportion. For example, a student might write the proportion $\frac{30}{x} = \frac{x}{18}$. They need to remember that the Geometric Mean Theorems result from similar triangles, which means any proportion must relate side lengths of two triangles.

Questioning Strategies

Problem 20 Suppose the altitude has length a inches and one segment of the hypotenuse is n times as long as the other, where n is a positive whole number. What are the lengths of the segments? We can relate the segment lengths with $\frac{x}{a} = \frac{a}{nx}$, which is equivalent to $nx^2 = a^2$. So,

$$nx^2 = a^2$$
$$x^2 = \frac{a^2}{n}$$
$$x = \sqrt{\frac{a^2}{n}}$$
$$= a\frac{\sqrt{n}}{n} \text{ in.}$$

is the length of one segment, while the other segment has length $nx = n \cdot a\frac{\sqrt{n}}{n} = a\sqrt{n}$ in.

5 Wrap-Up

Summarize learning with your class. Consider using the Exit Ticket, Put It in Writing, or I Can scale.

Exit Ticket

A right triangle with side lengths 5, 12, and 13 cm has an altitude drawn to its hypotenuse, so the hypotenuse is divided into two segments. How long are the segments? Call the shorter segment's length x and the longer segment's length y. Then $\frac{x}{5} = \frac{5}{13}$, so $13x = 25$ and $x = \frac{25}{13} \approx 1.92$ cm, while $\frac{y}{12} = \frac{12}{13}$, so $13y = 144$ and $x = \frac{144}{13} \approx 11.08$ cm.

Put It in Writing

Suppose you draw an altitude to the hypotenuse of a right triangle so the hypotenuse is divided into two segments. How does the length of the altitude compare to the lengths of the segments of the hypotenuse?

I Can

The scale below can help you and your students understand their progress on a learning goal.

4	I can identify similar right triangles, apply the Geometric Means Theorems, and recognize Pythagorean triples, and I can explain my reasoning to others.
3	I can identify similar right triangles, apply the Geometric Means Theorems, and recognize Pythagorean triples.
2	I can apply the Geometric Means Theorems to find side lengths of similar right triangles.
1	I can find similar triangles after drawing an altitude to the hypotenuse of a right triangle.

Spiral Review • Assessment Readiness

These questions will help determine if students have retained information taught in the past and can also prepare them for high-stakes assessments. Here, students must find the coordinates of a point that partitions a line segment in a given ratio **(12.3)**, find a ratio of side lengths of a right triangle **(13.1)**, and recognize and apply AA triangle similarity **(12.2)**.

Tell whether the numbers form a Pythagorean triple. Explain.

23. 4, 6, 10
no; $4^2 + 6^2 \neq 10^2$

24. 5, 12, 13
yes; $5^2 + 12^2 = 13^2$

25. 8, 17, $\sqrt{353}$
no; $\sqrt{353}$ is not a whole number.

26. 13, 24, 60
no; $13^2 + 24^2 \neq 60^2$

27. 24, 32, 40
yes; $24^2 + 32^2 = 40^2$

28. 65, 72, 97
yes; $65^2 + 72^2 = 97^2$

The lengths of the legs of a right triangle are a and b, and the length of the hypotenuse is c. Find the missing side length of the right triangle. Decide if the side lengths form a Pythagorean triple.

29. $a = 6, b = 8, c = ?$
$c = 10$; yes

30. $a = 10, b = ?, c = 26$
$b = 24$; yes

31. $a = ?, b = 32, c = 34$
$a = \sqrt{132} = 2\sqrt{33}$; no

32. (MP) **Attend to Precision** Kira has two plans for the frame of a small doghouse roof. She wants to choose the plan that requires the least amount of lumber.

A. Which plan should she choose? Why?
See Additional Answers.

B. If lumber costs \$2.50 per foot, what is her approximate total lumber cost for the frame?
approximately \$18.25

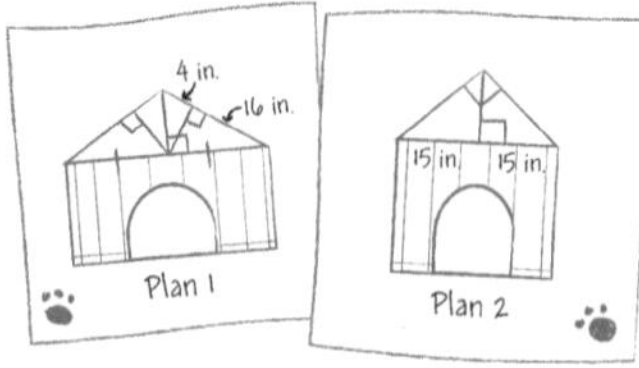

Spiral Review • Assessment Readiness

33. On a coordinate plane, A is at $(0, 0)$, B is at $(10, 4)$, and C is plotted so it divides the line segment from A to B in a 4 to 1 ratio. What are the coordinates of C?

Ⓐ $C(2.5, 1)$
Ⓑ $C(2, 0.8)$
Ⓒ $C(8, 3.2)$
Ⓓ $C(9, 3.6)$

34. In a right triangle with side lengths of 3 m, 4 m, and 5 m, what is the ratio of the length of the leg opposite the smallest interior angle to the length of the leg adjacent to the smallest interior angle?

Ⓐ $\frac{3}{4}$
Ⓑ $\frac{4}{3}$
Ⓒ $\frac{3}{5}$
Ⓓ $\frac{4}{5}$

35. In $\triangle ABC$, $m\angle A = 42°$, $m\angle B = 50°$, $AB = 4$ in., and $AC = 3$ in. In $\triangle XYZ$, $m\angle X = 42°$, $m\angle Y = 50°$, $XY = 14$ in., and $YZ = 9.5$ in. Match the part of each triangle with the correct measure. A. 3; B. 2; C. 1

Triangle part	Measure
A. $\overline{BC}$	**1.** 88°
B. $\overline{XZ}$	**2.** 10.5 in.
C. $\angle BCA$	**3.** 2.7 in.

I'm in a Learning Mindset!

How am I responding to learning new material that builds on my previous understanding?

Keep Going to Journal and Practice Workbook

Learning Mindset

Resilience Adjusts to Change

Students have likely already been told multiple times that learning mathematics is a cumulative process, with new material resting on previously learned material. Encourage students to think deeply about how this actually happens. Point out that previously learned material is not a solid foundation upon which new material is stacked, but more like a fluid set of connections to topics with which new topics are integrated. When students are having an especially hard time with new material, it may well be a sign that previous material was not quite integrated into their understanding of collective topics. This may not mean students have to go all the way back to the beginning, as applying older concepts while newer concepts are being learned often strengthens those "older" connections. For example, a student struggling to apply the Geometric Means Theorems might not have a strong grasp of the proportionality of sides of similar triangles, but applying that concept to this lesson can help strengthen their understanding. *How do you see math: as layers of concepts we stack on top of each other or as a set of connections between different topics? How can applying previous material to new material help you better understand the previous material?*

Module 12 Review

Similarity Transformations

A similarity transformation can use rigid transformations as well as dilations to map a pre-image onto a similar image.

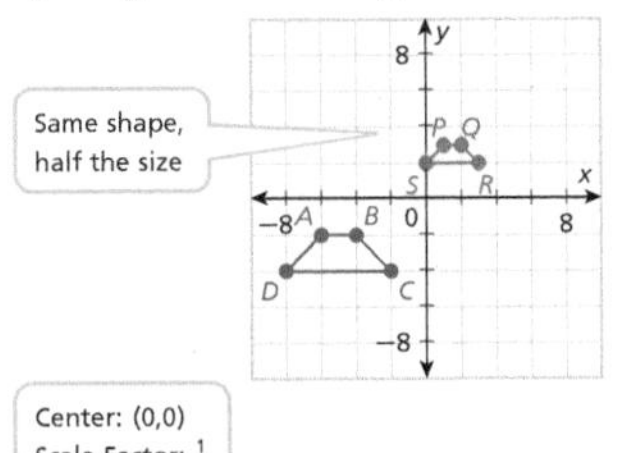

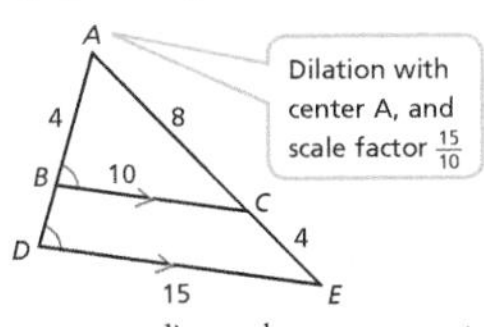

Dilation: $(x, y) \rightarrow \left(\frac{1}{2}x, \frac{1}{2}y\right)$

Translation: $(x, y) \rightarrow (x + 4, y + 4)$

Similarity Theorems

Angle-Angle (AA) Similarity If two angles of one triangle are congruent to two angles of another triangle, then the two triangles are similar.

The third pair of angles are congruent by the Third Angle Theorem.

Side-Angle-Side (SAS) Similarity If two correspnding sides are proportional and the included angles are congruent, then the two triangles are similar.

Side-Side-Side (SSS) Similarity If all three pairs of corresponding sides are proportional, then the two triangles are similar.

Dilate to obtain congruent sides. Then apply SAS or SSS for congruence.

Triangle Proportionality

If a line parallel to one side of a triangle intersects the other two sides, then it divides those sides proportionally.

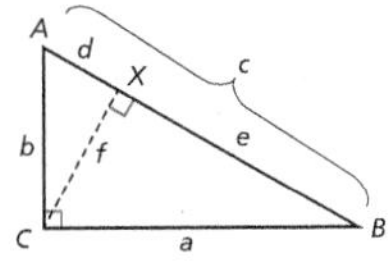

$\overline{DE} \parallel \overline{BC}$, so corresponding angles are congruent. This means that $\angle D \cong \angle B$. $\angle A \cong \angle A$ by the Reflexive Property, so $\triangle ABC \sim \triangle ADE$.

$\frac{DB}{BA} = \frac{EC}{CA} \rightarrow DB = \frac{EC}{CA}(BA) = \frac{4}{8}(4) = 2$

The Pythagorean Theorem

Similarity can be used to prove the Pythagorean Theorem. Take a right triangle and draw in an altitude to the hypotenuse.

All 3 triangles are similar by AA similarity.

$\triangle ABC \sim \triangle ACX \sim \triangle CBX$

Set up equivalent ratios.

$\frac{e}{a} = \frac{a}{c} \rightarrow a^2 = ec$ and $\frac{d}{b} = \frac{b}{c} \rightarrow b^2 = dc$

$\rightarrow a^2 + b^2 = ec + dc = (e + d)c = c^2$

This proves that $a^2 + b^2 = c^2$.

ONLINE

Assign the Digital Module Review for

- built-in student supports
- Actionable Item Reports
- Standards Analysis Reports

MODULE 12 REVIEW

Module Review

Use the first page of the Module Review to summarize and connect the main ideas of the module. Use the second page to assess students' understanding of the vocabulary, concepts, and skills presented in the module.

Sample Guided Discussion:

Q Similarity Transformations Does the order in a sequence of similarity transformations matter? Explain. yes; Possible answer: A translation followed by a rotation and then a dilation will not give the same result as a rotation followed by a translation and then a dilation.

Q Similarity Theorems What is the difference between SSS Congruence and SSS Similarity? Possible answer: SSS Congruence can be thought of as a special case of SSS Similarity where the ratio between the lengths of the corresponding pairs of sides is 1 to 1.

Q Triangle Proportionality What is the ratio between the perimeters of $\triangle ABC$ and $\triangle ADE$? Explain. The ratio between the perimeters of the two triangles is the same as the scale factor, $\frac{15}{10}$. The perimeter of $\triangle ABC$ is $4 + 10 + 8 = 22$, and the perimeter of $\triangle ABC$ is $6 + 15 + 12 = 33$. The ratio $\frac{33}{22} = \frac{3}{2}$, which is equivalent to the scale factor $\frac{15}{10}$.

Q The Pythagorean Theorem The altitude to the hypotenuse of an isosceles right triangle has length r. It partitions the hypotenuse into segments of lengths s and t. What is the relationship of r, s, and t? $r = s = t$

Module Review continued

Possible Scoring Guide

Items	Points	Description
1–4	2 each	identifies the correct term
5	2	correctly compares and contrasts similarity and congruence criteria
6, 7	2 each	correctly describes a sequence of transformations that proves the figures are similar
8	2	correctly finds the measure of the altitude of the hypotenuse
9	2	correctly determines if $\angle LM$ is parallel to $\angle PQ$
10	2	correctly finds the height of the building and states assumptions
11	2	correctly finds the perimeter
12	2	correctly graphs the segment in the coordinate plane and partitions it into two segments in the ratio of 2:3
Total points possible = 24 points		

The Unit 6 Test in the Assessment Guide assesses content from Modules 11 and 12.

Vocabulary

Choose the correct term from the box to complete each sentence.

Vocabulary
dilation
partition
proportional
similar
similarity transformation

1. A __?__ is a transformation that changes the size of a figure but leaves the shape unchanged. dilation
2. When triangles have __?__ corresponding sides and congruent corresponding angles, they are said to be __?__ proportional; similar
3. A __?__ produces an image that is similar to the preimage. similarity transformation
4. To __?__ a line segment is to divide it into subsets. partition

Concepts and Skills

5. Compare and contrast the criteria used to prove two triangles are congruent with the criteria used to prove two triangles are similar. See Additional Answers.

Describe a sequence of transformations that proves each pair of figures are similar.

6.

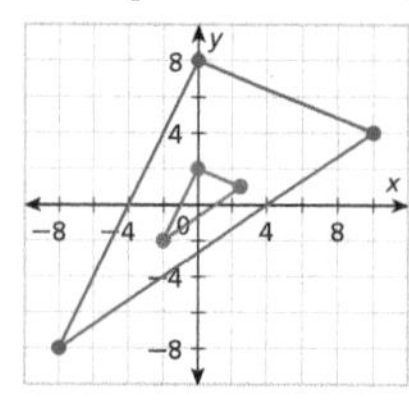

See Additional Answers.

7.

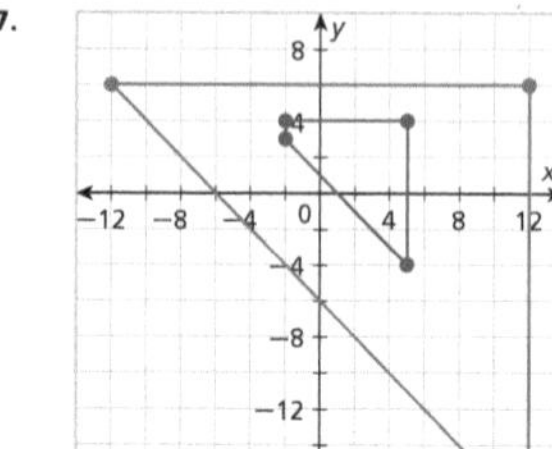

See Additional Answers.

8. Find the value of x.

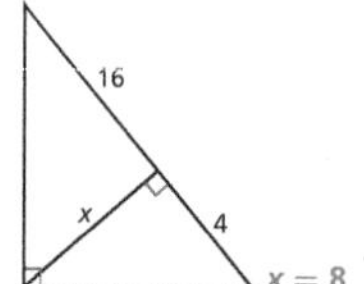

$x = 8$

9. Is $\overline{LM}$ parallel to $\overline{PQ}$? Explain.

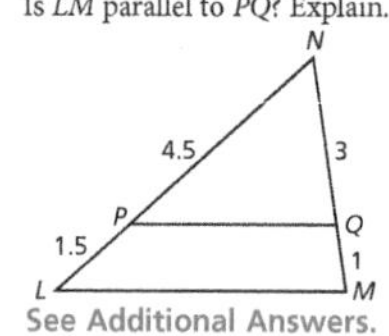

See Additional Answers.

10. (MP) **Use Tools** A 30-foot pole casts a 12-foot shadow. Find the height of a nearby building that casts a shadow of 80 feet. State any assumptions you make. State what strategy and tool you will use to answer the question, explain your choice, and then find the answer. See Additional Answers.

11. $\triangle ABC \sim \triangle XYZ$, where $AB = 18$ cm, $BC = 30$ cm, and $CA = 42$ cm. The longest side of $\triangle XYZ$ is 25.2 cm. What is the perimeter of $\triangle XYZ$? 54 cm

12. Line segment $\overline{AB}$ has endpoints $A(-6, 13)$ and $B(4, -2)$. Graph $\overline{AB}$ and plot point P that partitions the segment in the ratio 2:3 from A to B, and point Q that partitions the segment in the same ratio from B to A. See Additional Answers.

DATA-DRIVEN INSTRUCTION

Before moving on to the Module Test, use the Module Review results to intervene based on the table below.

Items	Lesson	DOK	Content Focus	Intervention
5	12.1	2	Distinguish between congruence and similarity.	Reteach 12.1
6, 7	12.2	3	Use criteria to show figures are similar.	Reteach 12.2
8	12.4	2	Use properties of similar right triangles to find a value.	Reteach 12.4
9	12.3	2	Apply the converse of the Triangle Proportionality Theorem to determine whether segments are parallel.	Reteach 12.3
10	12.2	3	Use indirect measurement to find an unknown measurement.	Reteach 12.2
11	12.1	3	Use properties of similarity to find the perimeter of a triangle.	Reteach 12.1
12	12.3	2	Apply the Triangle Proportionality Theorem to partition a segment in the coordinate plane.	Reteach 12.3

Module Test

The Module Test is available in alternative versions in your Assessment Guide. All items are presented in standardized test formats.

ONLINE

Assign the Digital Module Test to power actionable reports including

- proficiency by standards
- item analysis

Form A

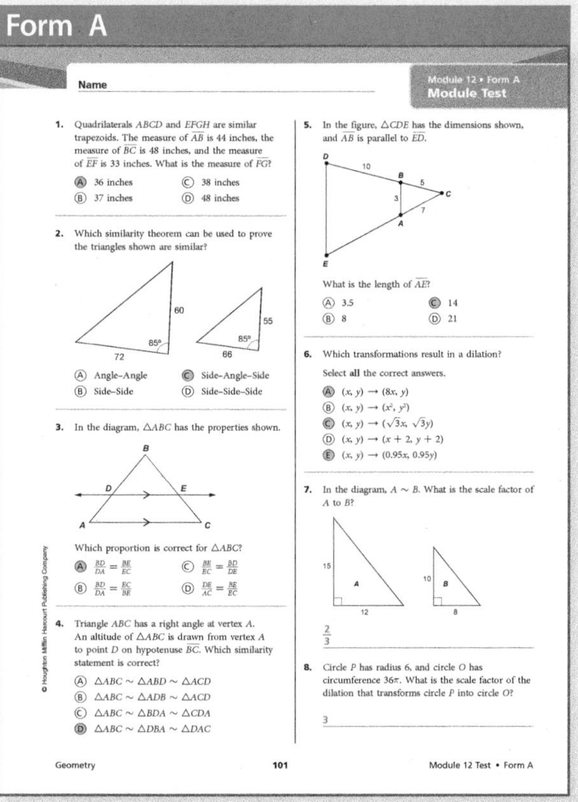

Name

Module 12 • Form A
Module Test

1. Quadrilaterals $ABCD$ and $EFGH$ are similar trapezoids. The measure of $\overline{AB}$ is 44 inches, the measure of $\overline{BC}$ is 48 inches, and the measure of $\overline{EF}$ is 33 inches. What is the measure of $\overline{FG}$?
   Ⓐ 36 inches Ⓒ 38 inches
   Ⓑ 37 inches Ⓓ 48 inches

2. Which similarity theorem can be used to prove the triangles shown are similar?
   Ⓐ Angle-Angle Ⓒ Side-Angle-Side
   Ⓑ Side-Side Ⓓ Side-Side-Side

3. In the diagram, $\triangle ABC$ has the properties shown.
   Which proportion is correct for $\triangle ABC$?
   Ⓐ $\frac{BD}{DA} = \frac{BE}{EC}$ Ⓒ $\frac{BE}{BC} = \frac{BD}{DB}$
   Ⓑ $\frac{BD}{DA} = \frac{BC}{BE}$ Ⓓ $\frac{DE}{AC} = \frac{BE}{EC}$

4. Triangle ABC has a right angle at vertex A. An altitude of $\triangle ABC$ is drawn from vertex A to point D on hypotenuse $\overline{BC}$. Which similarity statement is correct?
   Ⓐ $\triangle ABC \sim \triangle ABD \sim \triangle ACD$
   Ⓑ $\triangle ABC \sim \triangle ADB \sim \triangle ACD$
   Ⓒ $\triangle ABC \sim \triangle BDA \sim \triangle CDA$
   Ⓓ $\triangle ABC \sim \triangle DBA \sim \triangle DAC$

5. In the figure, $\triangle CDE$ has the dimensions shown, and $\overline{AB}$ is parallel to $\overline{ED}$.
   What is the length of $\overline{AE}$?
   Ⓐ 3.5 Ⓒ 14
   Ⓑ 8 Ⓓ 21

6. Which transformations result in a dilation? Select **all** the correct answers.
   Ⓐ $(x, y) \to (8x, y)$
   Ⓑ $(x, y) \to (x^2, y^2)$
   Ⓒ $(x, y) \to (\sqrt{3}x, \sqrt{3}y)$
   Ⓓ $(x, y) \to (x + 2, y + 2)$
   Ⓔ $(x, y) \to (0.95x, 0.95y)$

7. In the diagram, $A \sim B$. What is the scale factor of A to B?
   $\frac{2}{3}$

8. Circle P has radius 6, and circle O has circumference 36π. What is the scale factor of the dilation that transforms circle P into circle O?
   3

© Houghton Mifflin Harcourt Publishing Company

Geometry 101 Module 12 Test • Form A

Form A

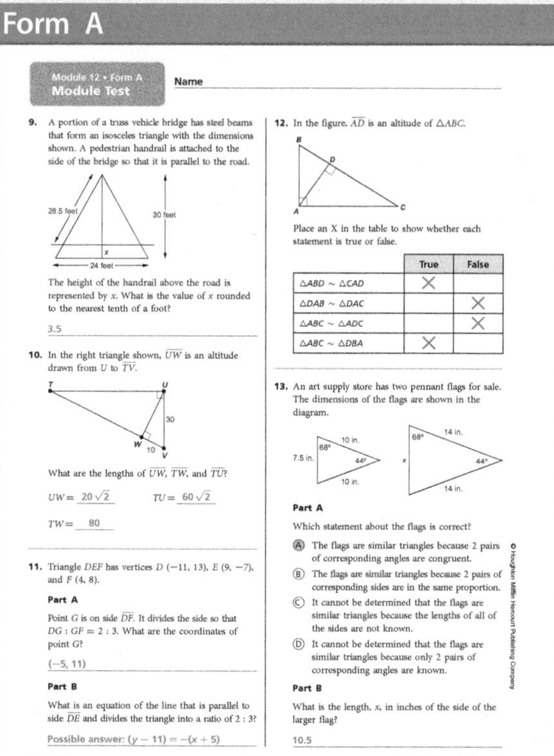

Module 12 • Form A
Module Test

Name

9. A portion of a truss vehicle bridge has steel beams that form an isosceles triangle with the dimensions shown. A pedestrian handrail is attached to the side of the bridge so that it is parallel to the road.
   The height of the handrail above the road is represented by x. What is the value of x rounded to the nearest tenth of a foot?
   3.5

10. In the right triangle shown, $\overline{UW}$ is an altitude drawn from U to $\overline{TV}$.
    What are the lengths of $\overline{UW}$, $\overline{TW}$, and $\overline{TU}$?
    $UW =$ $20\sqrt{2}$ $TU =$ $60\sqrt{2}$
    $TW =$ 80

11. Triangle DEF has vertices $D(-11, 13)$, $E(9, -7)$, and $F(4, 8)$.
    Part A
    Point G is on side $\overline{DF}$. It divides the side so that $DG : GF = 2 : 3$. What are the coordinates of point G?
    $(-5, 11)$
    Part B
    What is an equation of the line that is parallel to side $\overline{DE}$ and divides the triangle into a ratio of 2 : 3?
    Possible answer: $(y - 11) = -(x + 5)$

12. In the figure, $\overline{AD}$ is an altitude of $\triangle ABC$.
    Place an X in the table to show whether each statement is true or false.

	True	False
$\triangle ABD \sim \triangle CAD$	X	
$\triangle DAB \sim \triangle DAC$		X
$\triangle ABC \sim \triangle ADC$		X
$\triangle ABC \sim \triangle DBA$	X	

13. An art supply store has two pennant flags for sale. The dimensions of the flags are shown in the diagram.
    Part A
    Which statement about the flags is correct?
    Ⓐ The flags are similar triangles because 2 pairs of corresponding angles are congruent.
    Ⓑ The flags are similar triangles because 2 pairs of corresponding sides are in the same proportion.
    Ⓒ It cannot be determined that the flags are similar triangles because the lengths of all of the sides are not known.
    Ⓓ It cannot be determined that the flags are similar triangles because only 2 pairs of corresponding angles are known.
    Part B
    What is the length, x, in inches of the side of the larger flag?
    10.5

© Houghton Mifflin Harcourt Publishing Company

Geometry 102 Module 12 Test • Form A

Form B

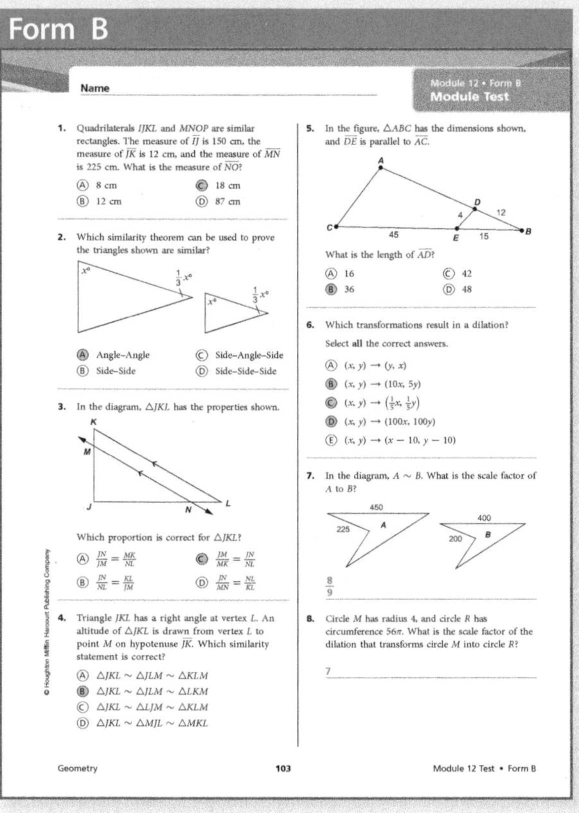

Name

Module 12 • Form B
Module Test

1. Quadrilaterals $IJKL$ and $MNOP$ are similar rectangles. The measure of $\overline{IJ}$ is 150 cm, the measure of $\overline{JK}$ is 12 cm, and the measure of $\overline{MN}$ is 225 cm. What is the measure of $\overline{NO}$?
   Ⓐ 8 cm Ⓒ 18 cm
   Ⓑ 12 cm Ⓓ 87 cm

2. Which similarity theorem can be used to prove the triangles shown are similar?
   Ⓐ Angle-Angle Ⓒ Side-Angle-Side
   Ⓑ Side-Side Ⓓ Side-Side-Side

3. In the diagram, $\triangle JKL$ has the properties shown.
   Which proportion is correct for $\triangle JKL$?
   Ⓐ $\frac{JN}{JM} = \frac{MK}{NL}$ Ⓒ $\frac{JM}{MK} = \frac{JN}{NL}$
   Ⓑ $\frac{JN}{NL} = \frac{KL}{JM}$ Ⓓ $\frac{JN}{MN} = \frac{NL}{KL}$

4. Triangle JKL has a right angle at vertex L. An altitude of $\triangle JKL$ is drawn from vertex L to point M on hypotenuse $\overline{JK}$. Which similarity statement is correct?
   Ⓐ $\triangle JKL \sim \triangle JLM \sim \triangle KLM$
   Ⓑ $\triangle JKL \sim \triangle JLM \sim \triangle LKM$
   Ⓒ $\triangle JKL \sim \triangle LJM \sim \triangle KLM$
   Ⓓ $\triangle JKL \sim \triangle MJL \sim \triangle MKL$

5. In the figure, $\triangle ABC$ has the dimensions shown, and $\overline{DE}$ is parallel to $\overline{AC}$.
   What is the length of $\overline{AD}$?
   Ⓐ 16 Ⓒ 42
   Ⓑ 36 Ⓓ 48

6. Which transformations result in a dilation? Select **all** the correct answers.
   Ⓐ $(x, y) \to (y, x)$
   Ⓑ $(x, y) \to (10x, 5y)$
   Ⓒ $(x, y) \to (\frac{1}{3}x, \frac{1}{3}y)$
   Ⓓ $(x, y) \to (100x, 100y)$
   Ⓔ $(x, y) \to (x - 10, y - 10)$

7. In the diagram, $A \sim B$. What is the scale factor of A to B?
   $\frac{8}{9}$

8. Circle M has radius 4, and circle R has circumference 56π. What is the scale factor of the dilation that transforms circle M into circle R?
   7

© Houghton Mifflin Harcourt Publishing Company

Geometry 103 Module 12 Test • Form B

Form B

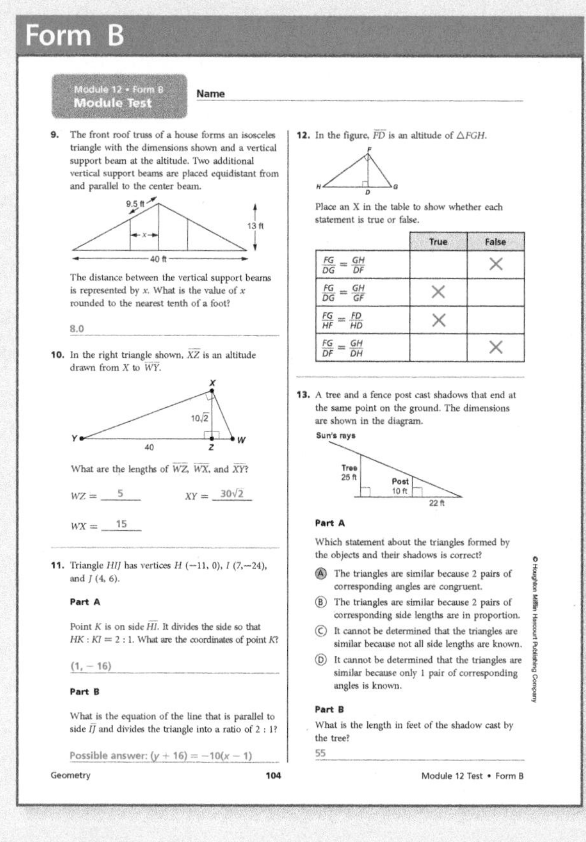

Module 12 • Form B
Module Test

Name

9. The front roof truss of a house forms an isosceles triangle with the dimensions shown and a vertical support beam at the altitude. Two additional vertical support beams are placed equidistant from and parallel to the center beam.
   The distance between the vertical support beams is represented by x. What is the value of x rounded to the nearest tenth of a foot?
   8.0

10. In the right triangle shown, $\overline{XZ}$ is an altitude drawn from X to $\overline{WY}$.
    What are the lengths of $\overline{WZ}$, $\overline{WX}$, and $\overline{XY}$?
    $WZ =$ 5 $XY =$ $30\sqrt{2}$
    $WX =$ 15

11. Triangle HIJ has vertices $H(-11, 0)$, $I(7, -24)$, and $J(4, 6)$.
    Part A
    Point K is on side $\overline{HI}$. It divides the side so that $HK : KI = 2 : 1$. What are the coordinates of point K?
    $(1, -16)$
    Part B
    What is the equation of the line that is parallel to side $\overline{IJ}$ and divides the triangle into a ratio of 2 : 1?
    Possible answer: $(y + 16) = -10(x - 1)$

12. In the figure, $\overline{FD}$ is an altitude of $\triangle FGH$.
    Place an X in the table to show whether each statement is true or false.

	True	False
$\frac{FG}{DG} = \frac{GH}{DF}$		X
$\frac{FG}{DG} = \frac{GH}{GF}$	X	
$\frac{FG}{HF} = \frac{FD}{HD}$	X	
$\frac{FG}{DF} = \frac{GH}{DH}$		X

13. A tree and a fence post cast shadows that end at the same point on the ground. The dimensions are shown in the diagram.
    Part A
    Which statement about the triangles formed by the objects and their shadows is correct?
    Ⓐ The triangles are similar because 2 pairs of corresponding angles are congruent.
    Ⓑ The triangles are similar because 2 pairs of corresponding side lengths are in proportion.
    Ⓒ It cannot be determined that the triangles are similar because not all side lengths are known.
    Ⓓ It cannot be determined that the triangles are similar because only 1 pair of corresponding angles is known.
    Part B
    What is the length in feet of the shadow cast by the tree?
    55

© Houghton Mifflin Harcourt Publishing Company

Geometry 104 Module 12 Test • Form B

Module 13: Trigonometry with Right Triangles

Module 14: Trigonometry with All Triangles

Surveyor STEM

- **Say:** *Have you ever seen a surveyor working along a road or highway? Chances are the surveyor is determining property boundaries and recording information about the shape of Earth's surface to collect information about projects involving engineering, mapmaking, or construction.*
- Explain that surveyors use mathematics and science to measure distances and angles between points on, above, or below Earth's surface. Then introduce the STEM task.

STEM Task

Ask students if they have any previous knowledge of how surveyors do their work.

- Explain that in the field, surveyors use a variety of tools, including Global Positioning System (GPS) technology to locate precise reference points.
- They use handheld GPS units and automated systems known as *robotic total stations* to collect information about the land they are surveying.
- Surveyors also use Geographic Information System (GIS) technology that presents spatial information as maps, reports, and charts. Surveyors use the results to advise clients on where to plan homes, roads, and landfills.
- Another tool that surveyors use is a sight level.
- First, ask students for the distance between the top and bottom sights. The distance shown in the picture is measured in inches, so the distance between sight levels is 4 feet. Have students set up a proportion to find the approximate distance x from the sight rod.

  $\frac{4}{x} = \frac{1}{12}$

- Students can multiply both sides by $12x$ to get $48 = x$, so the approximate distance from the sight rod is 48 feet.

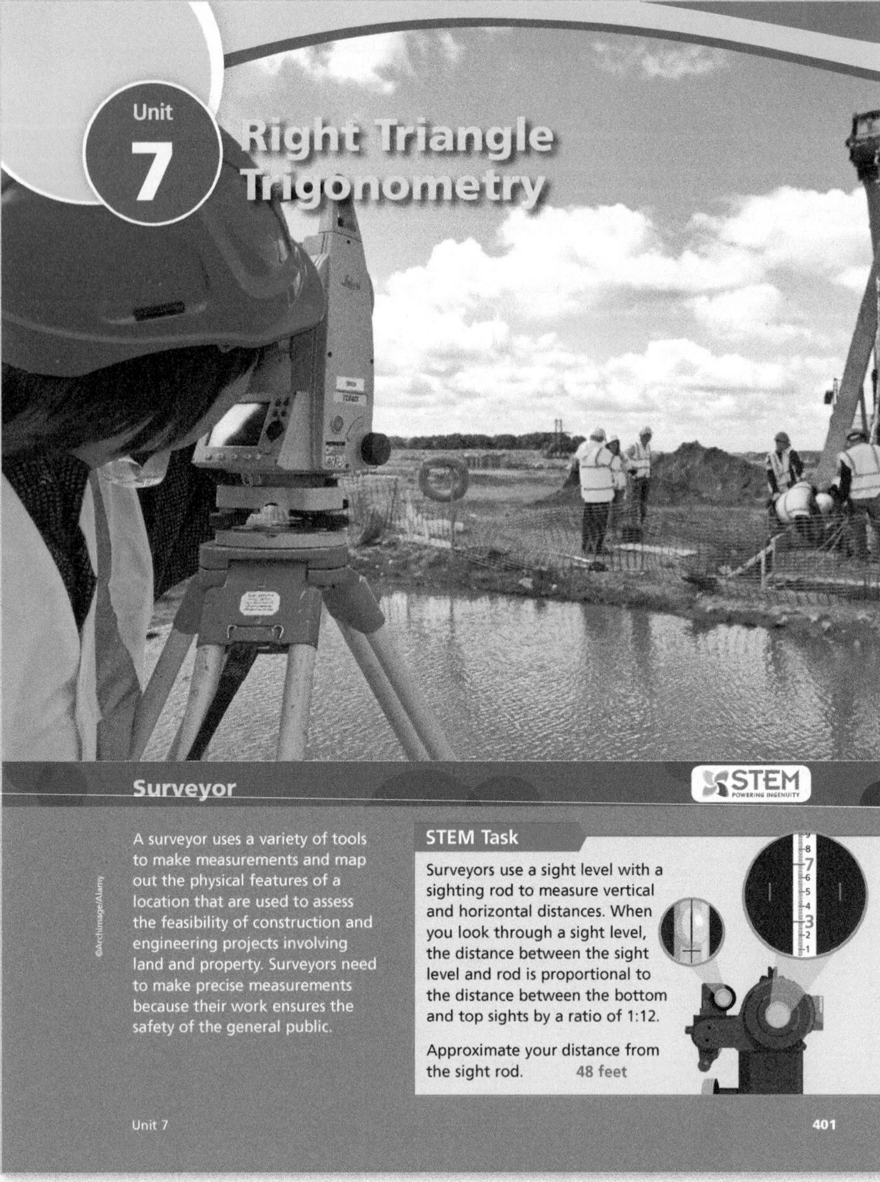

Unit 7
Right Triangle Trigonometry
Surveyor
STEM
A surveyor uses a variety of tools to make measurements and map out the physical features of a location that are used to assess the feasibility of construction and engineering projects involving land and property. Surveyors need to make precise measurements because their work ensures the safety of the general public.
STEM Task
Surveyors use a sight level with a sighting rod to measure vertical and horizontal distances. When you look through a sight level, the distance between the sight level and rod is proportional to the distance between the bottom and top sights by a ratio of 1:12.
Approximate your distance from the sight rod. 48 feet
Unit 7
401

Unit 7 Project Sound Surveying

Overview: In this project students use trigonometry to find the length and height of a bridge being planned by a surveyor.

Materials: paper and pencil or a computer used for word processing

Assessing Student Performance: Students' reports should include:

- a correct length *AB* **(Lesson 13.4 or earlier)**
- a correct length *GH* **(Lesson 13.4 or earlier)**
- a correct height using the tangent ratio **(Lesson 13.4)**
- a correct length *BC* **(Lesson 14.2)**

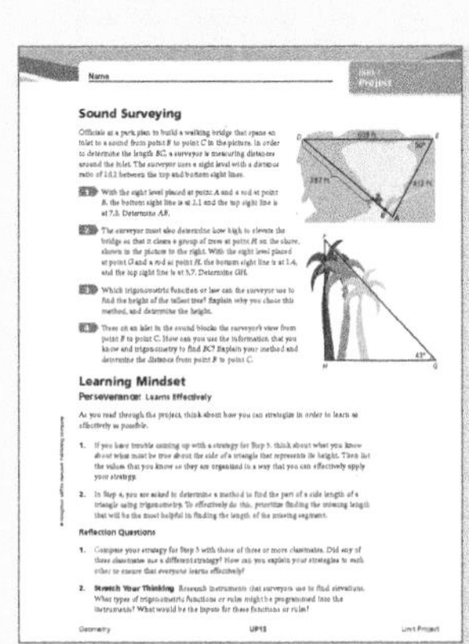

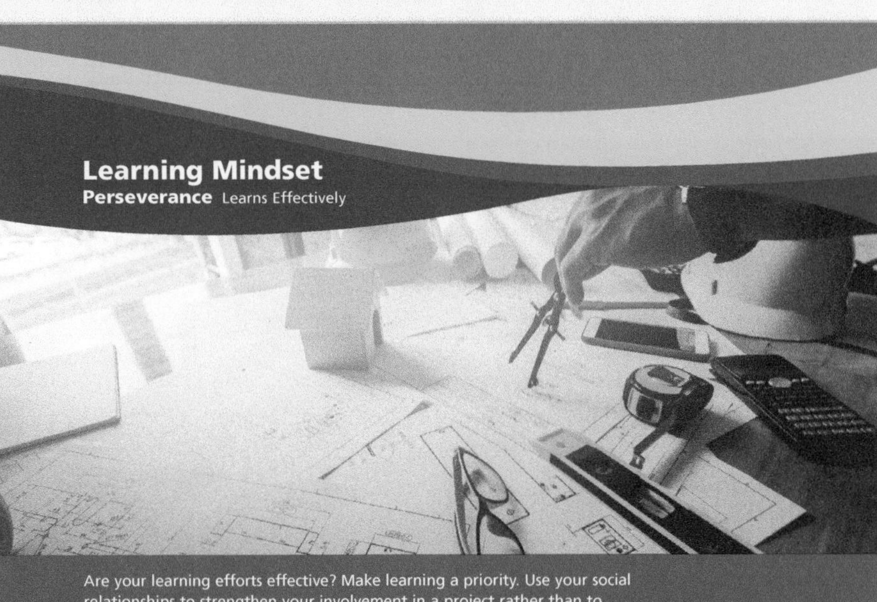

Are your learning efforts effective? Make learning a priority. Use your social relationships to strengthen your involvement in a project rather than to distract from it. Be proactive in getting things done. We all experience failures, but we can recognize our failures or inefficiencies, reflect on them, and then use our experience to improve ourselves. Here are some questions you can ask yourself to monitor how effectively you are learning new material and progressing on a new task:

- How am I prioritizing my learning? What are my greatest strengths? What weaknesses might I need to address? How can I turn my weaknesses into strengths?
- Is a lack of organization affecting my learning outcomes? In what way? What steps can I take to become more organized?
- How do my social relationships affect my academic performance? How can I use these relationships to help me learn more effectively?

©Akira Kaelyn/Shutterstock

Reflect

Q When have you experienced an initial failure that led to growth and learning? How did you benefit from this experience in the long term? How can this experience help you approach new challenges?

Q As a surveyor, you would need to work with a team to make accurate measurements and precise calculations. What steps could you take to better organize your process? How would organization improve the outcome for your team?

402

What to Watch For

Watch for students who are struggling with math and tend to give up. Strategies that may help these students include

- recognizing their frustration,
- asking them how they first approached solving the problem,
- pointing out where they may have made a wrong turn, and
- encouraging them to start over and try a different approach.

Watch for students who want to give up, especially after trying hard to succeed. Help them move to a more effective learning mindset by

- reminding them of past successes,
- encouraging them to be patient and persevere, and
- reminding them that success does not always come easily and is often directly proportional to one's effort.

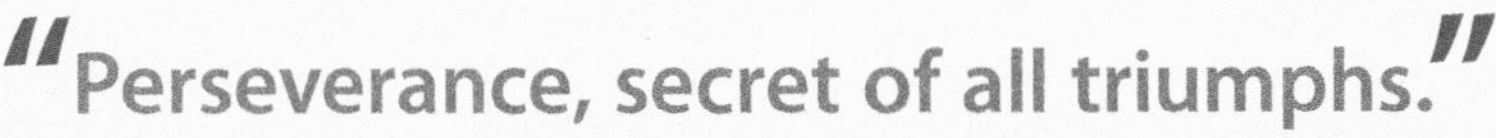

—Victor Hugo

Perseverance
Learning Mindset

Learns Effectively

The learning-mindset focus in this unit is *perseverance*, which refers to a person's ability to learn effectively by continuing to work on solving a problem long after wanting to give up.

Mindset Beliefs

Students who lack perseverance often give up when confronted with a learning challenge—especially in mathematics. They become frustrated and refuse to continue or to seek help. These students do not experience the satisfaction of solving a difficult problem and, instead, may develop resistance to learning.

Students who exhibit perseverance, however, develop ways to overcome stumbling blocks. They start over, perhaps trying a new approach, or carefully analyze the steps in a solution that did not work initially. Students who persevere can usually overcome learning obstacles and develop greater self-confidence for the next learning challenge. They demonstrate a growth mindset and learn effectively.

Discuss with students their own definitions of perseverance. Do they put off taking on a task or challenge because they think they are not up to it? Have they experienced times when sticking with a task has resulted in a positive outcome? Has the perseverance they experienced made them feel more confident about moving on to the next learning challenge?

Mindset Behaviors

Encourage students to learn more effectively by asking themselves questions about how to become more open to sticking with a difficult challenge or task. This self-monitoring can take the form of questions like the following:

When reading a mathematics text:

- Do I understand the mathematics I've learned before in a way that will help me now?
- Do I understand how to apply the skills and concepts well enough to take on this challenge and stick with it?

When attempting to solve a difficult problem:

- Have I read the problem more than once and understood what the situation is about?
- Have I ever seen a simpler version of this problem that I was able to solve?

As students learn to persevere, they will become more effective learners, which will boost their self-confidence. Perseverance takes time and requires applying discrete analytical skills. It also requires trying a new approach when an old one doesn't work. To persevere may require working with others—both peers and teachers—who may provide suggestions that lead to eventual success.

MODULE 13

PLANNING

TRIGONOMETRY WITH RIGHT TRIANGLES

Introduce and Check for Readiness

- Module Performance Task • Are You Ready?

Lesson 13.1—2 Days

Tangent Ratio

Learning Objective: Solve for missing side lengths and angle measures using the tangent, inverse tangent, and properties of similar triangles.

New Vocabulary: inverse tangent, tangent

Lesson 13.2—2 Days

Sine and Cosine Ratios

Learning Objective: Students will use sine and cosine ratios derived from similarity and their inverses to find side lengths and angle measures in right triangles.

New Vocabulary: cosine, inverse cosine, inverse sine, sine, trigonometric ratio

Lesson 13.3—2 Days

Special Right Triangles

Learning Objective: Examine and use trigonometric ratios for special right triangles and use the Pythagorean Theorem to find the side lengths and angle measures of special right triangles.

New Vocabulary: special right triangles

Lesson 13.4—2 Days

Solve Problems Using Trigonometry

Learning Objective: Apply the sine, cosine, and tangent ratios and their inverses to find the areas of triangles and measures of sides and acute angles in right triangles both in the coordinate plane and represented in real-world situations involving angles of elevation and depression.

New Vocabulary: angle of depression, angle of elevation, solve a right triangle

Assessment

- Module 13 Test (Forms A and B)
- Unit 7 Test (Forms A and B)

LEARNING ARC FOCUS

Build Conceptual Understanding  Connect Concepts and Skills  Apply and Practice

TEACHING FOR SUCCESS

TEACHING FOR DEPTH: Trigonometry with Right Triangles

Meaning of Tangent. The word *tangent* has mathematical meanings and a non-mathematical meaning. In the case of the latter, a tangent is a digression, or a change of course. In mathematics, *tangent* comes from the Latin word *tangere* that means "to touch." So one of its mathematical meanings is a line that intersects, or "touches," a curve in exactly one point. If the curve is a circle, the tangent is perpendicular to the radius drawn to that point.

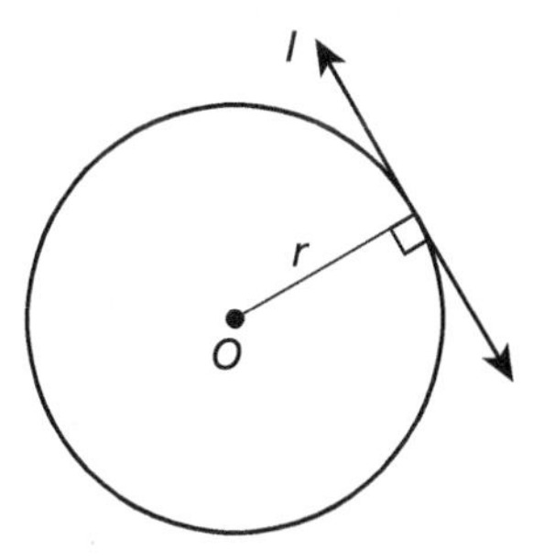

In a second mathematical meaning, *tangent* is defined as the ratio between the length of the side opposite an acute angle in a right triangle and the length of the side adjacent to the angle. It is written as $\tan A = \frac{a}{b}$.

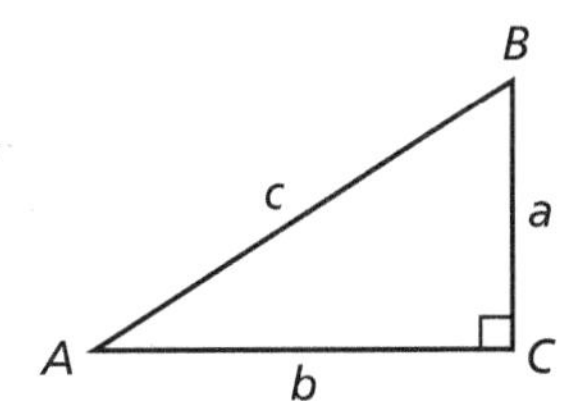

Hold a discussion with your students about the different meanings of the word *tangent*. Some may be familiar with its non-mathematical definition. Ask students to give an example of the non-mathematical usage.

For its mathematical definitions, emphasize that in one case, a tangent is a line. In another case, it's a number related to an angle. Draw a sketch of each usage, such as shown here, and explain that learning about the tangent as a number is the focus of the first lesson in the module.

You may also want to mention that there are several other trigonometric ratios that can be represented in a right triangle, including those that involve the hypotenuse. These also have special names—sine and cosine—and are the subject of the second lesson in this module. Depending on the sophistication of your students, you may also want to point out that the reciprocals of the tangent, sine, and cosine also have names, and that these will be discussed in later courses.

Mathematical Progressions

Prior Learning	Current Development	Future Connections
Students: • used the definition of similarity in terms of similarity transformations to decide if triangles are similar. • used congruence and similarity criteria for triangles to solve problems and to prove relationships in geometric figures.	**Students:** • use trigonometric ratios and their inverses to find side lengths and angle measures in right triangles. • use trigonometric ratios and the Pythagorean Theorem to find the side lengths and angle measures of special right triangles. • solve right triangles in applied problems.	**Students:** • will derive the formula $A = \frac{1}{2}ab \sin C$ for the area of a triangle by drawing an auxiliary line from a vertex perpendicular to the opposite side.

TEACHER TO TEACHER

From the Classroom

Facilitate meaningful mathematical discourse. I enjoy introducing my students to the wonders of the right triangle and all the relationships that hold true in it. My students are usually familiar with the Pythagorean Theorem and its converse, but are less familiar with the trigonometric ratios. Fortunately, there is a mnemonic that I often introduce, albeit reluctantly, to my students to help them remember the tangent, sine, and cosine ratios: SOCATOA. This mnemonic is a way to set up a trigonometric ratio. The letters S, C, and T stand for the three functions, and O and A stand for *opposite* and *adjacent*. I also encourage students to sketch a right triangle to identify the correct ratio for a relevant acute angle and to be able to explain the reasons why they chose that ratio.

LANGUAGE DEVELOPMENT • Planning for instruction

By giving all students regular exposure to language routines in context, you will provide opportunities for students to **listen**, **speak**, **read**, and **write** about mathematical situations and develop both mathematical language and conceptual understanding at the same time.

Using Language Routines to Develop Understanding

Use the **Professional Cards** for the following routines to plan for effective instruction.

Co-Craft Questions and Problems Lesson 13.3

Students think of natural questions to ask about a given situation or problems similar to a given task and answer the questions they have developed or problems they have created.

Three Reads Lessons 13.1, 13.2, and 13.4

Students read a problem three times with a specific focus each time.

1st Read What is the situation about?
2nd Read What are the quantities in the situation?
3rd Read What are the possible mathematical questions that you could ask for the situation?

Information Gap Lesson 13.2

Students recognize when information given in a problem situation is incomplete, and they pose questions and share knowledge with others to discover any missing facts or relationships and work together to solve the problem.

Critique, Correct, and Clarify Lessons 13.2

Students correct the work in a flawed explanation, argument, or solution method and share with a partner and refine the sample work.

Connecting Language to Trigonometry with Right Triangles

Watch for students' use of the new terms listed below as they explain their reasoning and make connections with new concepts.

Key Academic Vocabulary

Current Development • New Vocabulary

cosine in right triangle *ABC* with the right angle at *C*,

$$\cos A = \frac{\text{length of leg adjacent } \angle A}{\text{length of hypotenuse}} = \frac{AC}{AB}$$

inverse cosine in right triangle *ABC* with the right angle at *C*,

$$\sin^{-1}\left(\frac{a}{c}\right) = m\angle A$$

inverse sine in right triangle *ABC* with the right angle at *C*,

$$\cos^{-1}\left(\frac{b}{c}\right) = m\angle A$$

sine in right triangle *ABC* with the right angle at *C*,

$$\sin A = \frac{\text{length of leg opposite } \angle A}{\text{length of hypotenuse}} = \frac{BC}{AB}$$

tangent the ratio of the length of the side opposite an angle in a right triangle to the length of the side adjacent to the angle

trigonometric ratio a ratio of two sides of a right triangle

Linguistic Note

Listen for how students use the vocabulary involved in writing the trigonometric ratios. Before defining them, spend time defining the words *opposite* and *adjacent*, pointing out that *opposite* means "across from" and adjacent means "next to." Use sticky notes or chalk to represent the vertices of a right triangle on the floor and choose three students to stand at each vertex. Give the students lengths of string to form the sides of the triangle. Then point to a student standing at the vertex of an acute angle and ask the other students to indicate the side opposite or adjacent to him or her and the hypotenuse. Then introduce each trigonometric ratio by pointing to the relevant pairs of sides that form the numerators and denominators. Record each ratio on the board using the relevant terms.

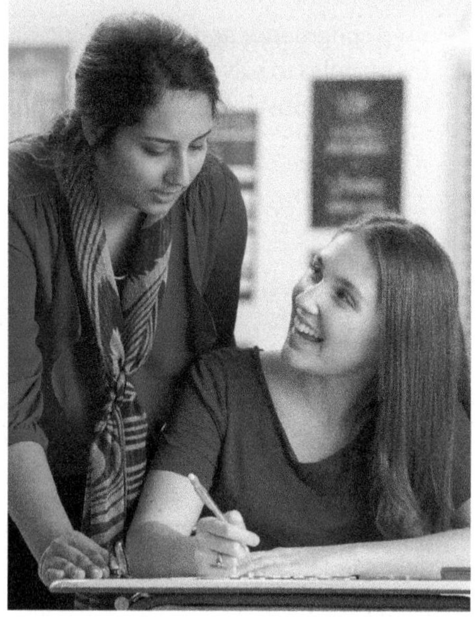

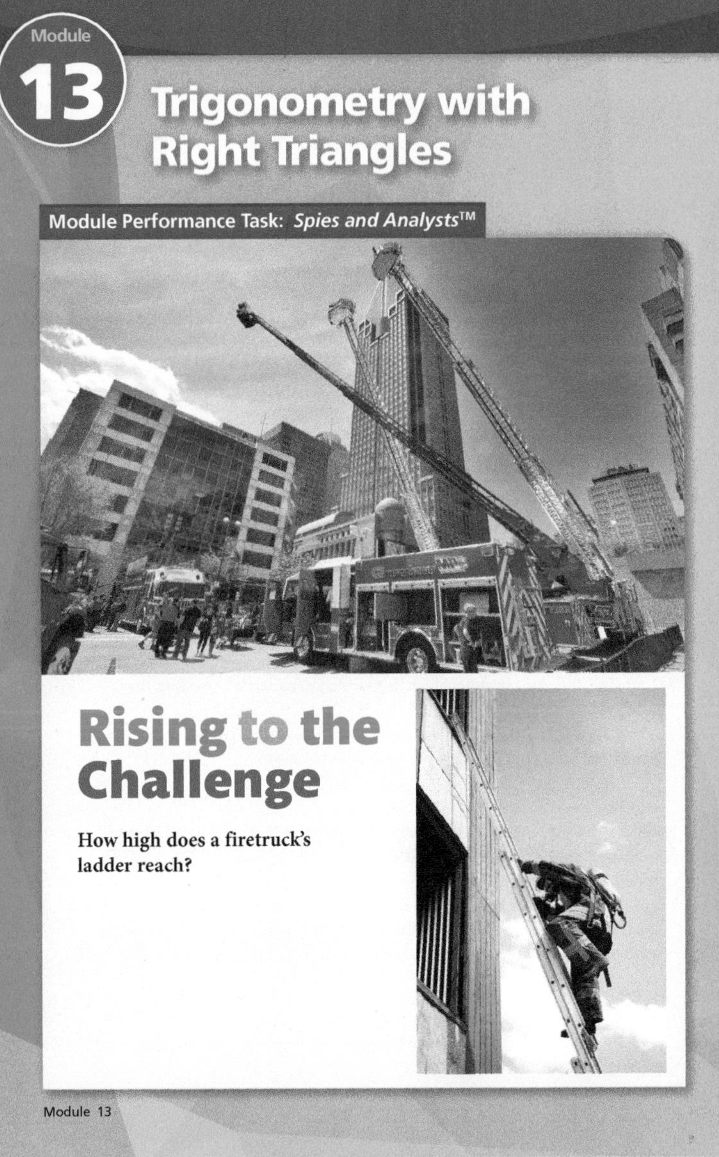

Module
13
Trigonometry with Right Triangles
Module Performance Task: *Spies and Analysts*™
Rising to the Challenge
How high does a firetruck's ladder reach?
Module 13
403

Connections to This Module

One sample solution might involve setting up a trigonometric ratio to first find the measure of the angle at which the ladder can be safely extended from the truck to reach the necessary height of the building. This angle is called the angle of elevation in a right triangle.

Then let h represent how high up the building the ladder must reach, let d represent the horizontal distance of the base of the ladder on the truck to the building, and let l represent the unknown length of the ladder. The variables h and d are the legs of a right triangle, and l is its hypotenuse.

Now let A represent the angle of elevation and use the inverse tangent ratio to determine the angle of elevation of the ladder, $A = \tan^{-1}\left(\frac{h}{d}\right)$.

Once you know the measure of A, you can use the sine or cosine ratio to find l, the length of the ladder: that is, $\sin A = \frac{h}{l}$, and $l = \frac{h}{\sin A}$, or $\cos A = \frac{d}{l}$ and $l = \frac{h}{\cos A}$. **(13.1 and 13.4)**

MODULE 13

TRIGONOMETRY WITH RIGHT TRIANGLES

Rising to the Challenge

Overview

This problem requires students to make assumptions to determine how far up the side of a building a fire truck's ladder can reach. Examples of assumptions:

- The height of a building, the length of a fire truck's ladder, and the horizontal distance of the fire truck and ladder from the building form the sides of a right triangle.
- The angle that the base of the ladder makes with the fire truck must not be too steep for fire fighters to climb.

Be a *Spy*

First, students must determine what information is necessary to know. This should include:

- how far the building is from the ladder.
- how high up the building the ladder must reach.
- what angle the ladder should make with the horizontal.

Students should recognize that the length of the ladder, the height of the side of the building where the ladder must reach, and the distance of the ladder on the truck from the building form a right triangle.

Be an *Analyst*

Help students understand that the fire truck has to be close enough to the building so that the ladder can safely reach the fire when extended. This depends on the angle that the ladder makes with the horizontal.

Alternative Approaches

In their analysis, students might:

- use trigonometric ratios to find the angle of elevation of the ladder and its length, as shown at the left.
- Use the Pythagorean Theorem to solve for the hypotenuse of the triangle represented by the length of the ladder, where the legs are the height of the building to which the ladder can reach and the horizontal distance from the side of the building to the base of the ladder on the truck.

ONLINE

Assign the Digital Are You Ready? to power actionable reports including

- proficiency by standards
- item analysis

Are You Ready?

Diagnostic Assessment

- Diagnose prerequisite mastery.
- Identify intervention needs.
- Modify or set up leveled groups.

Have students complete the *Are You Ready?* assessment on their own. Items test the prerequisites required to succeed with the new learning in this module.

Solve One-Step Equations Students will apply previous knowledge of solving one-step linear equations to determine whether figures are similar using proportionality.

Angle Relationships in Triangles Students will apply their previous understandings about the interior and exterior angles of a triangle to find angle measures.

The Pythagorean Theorem and its Converse Students will apply their previous knowledge of this theorem and its converse to find the measure of a side of a right triangle and to verify whether the lengths of three sides form the sides of a right triangle.

ONLINE 

Are You Ready?

Complete these problems to review prior concepts and skills you will need for this module.

Solve One-Step Equations

Solve each equation.

1. $6 + t = -17$ $t = -23$

2. $7z = 84$ $z = 12$

3. $b - 14 = 16$ $b = 30$

4. $\frac{m}{8} = 11$ $m = 88$

5. $-3x = 12$ $x = -4$

6. $y + 14 = 2$ $y = -12$

7. $\frac{a}{4} = -2$ -8

8. $n - 3 = 1$ $n = 4$

Angle Relationships in Triangles

Determine each value of x.

9.

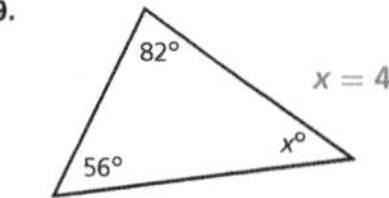

$x = 42$

10.

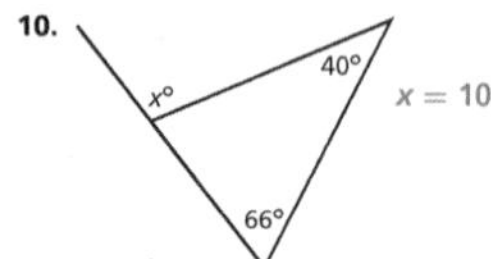

$x = 106$

The Pythagorean Theorem and its Converse

11. A right triangle has a leg that is 11 feet long and hypotenuse that is 22 feet long. How long is the other leg of the triangle? $11\sqrt{3}$ ft

12. A triangle has sides with lengths 8 inches, 15 inches, and 17 inches. Is the triangle a right triangle? Explain how you know. yes; $8^2 + 15^2 = 17^2$

Connecting Past and Present Learning

Previously, you learned:

- to prove if figures are similar using proportionality,
- to use the Pythagorean Theorem to find missing side lengths of right triangles, and
- to use angle relationships in triangles to find the measure of missing angles.

In this module, you will learn:

- to use similarity to define trigonometric ratios of acute angles, and
- to use trigonometric ratios and the Pythagorean Theorem to solve right triangles in applied problems.

404

DATA-DRIVEN INTERVENTION

Concept/Skill	Objective	Prior Learning *	Intervene With
Solve One-Step Equations	Use addition, subtraction, multiplication, or division to solve one-step equations.	Grade 6, Lessons 9.2 and 9.3	• Tier 3 Skill 6 • Reteach, Grade 6 Lessons 9.2 and 9.3
Angle Relationships in Triangles	Use the Triangle Sum Theorem and the Exterior Angle Theorem to solve problems.	Grade 8, Lesson 4.1	• Tier 2 Skill 5 • Reteach, Grade 8 Lesson 4.1
The Pythagorean Theorem and its Converse	Use the Pythagorean Theorem and its converse to solve problems	Grade 8, Lessons 11.1–11.3	• Tier 2 Skill 4 • Reteach, Grade 8 Lessons 11.1–11.3

* Your digital materials include access to resources from Grade 6–Algebra 2. The lessons referenced here contain a variety of resources you can use with students who need support with this content.

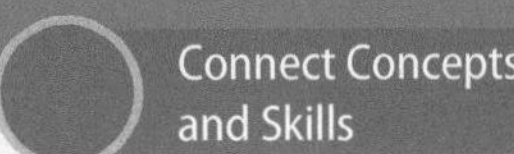

13.1 Tangent Ratio

LESSON FOCUS AND COHERENCE

Mathematics Standards

- Understand that by similarity, side ratios in right triangles are properties of the angles in the triangle, leading to definitions of trigonometric ratios for acute angles.

Mathematical Practices and Processes (MP)

- Use appropriate tools strategically.
- Look for and make use of structure.
- Attend to precision.

I Can Objective 

I can use the tangent ratio and the inverse tangent to find side lengths and angle measures in right triangles.

Learning Objective

Solve for missing side lengths and angle measures using the tangent, inverse tangent, and properties of similar triangles.

Language Objective

Explain which sides of a right triangle are used for the tangent ratio.

Vocabulary

New: **inverse tangent**, **tangent**

Lesson Materials: calculator, geometric drawing tool, tracing paper

Mathematical Progressions

Prior Learning	Current Development	Future Connections
Students: • used the definition of similarity in terms of similarity transformations to decide if two figures are similar; explained using similarity transformations the meaning of similarity for triangles as the equality of all corresponding pairs of angles and the proportionality of all corresponding pairs of sides. **(12.2)** • used similarity criteria for right triangles to solve problems and to prove relationships. **(12.4)**	**Students:** • in any right triangle with a given acute angle measure, understand that by similarity, the ratio of the side opposite the angle to the side adjacent to the angle is constant. • find the tangent of an acute angle in a right triangle. • use the tangent ratio to find lengths in real-world scenarios. • use the inverse tangent to find an angle measure in a right triangle.	**Students:** • will use trigonometric ratios and the Pythagorean Theorem to solve right triangles in applied problems. **(13.4)**

UNPACKING MATH STANDARDS

Understand that by similarity, side ratios in right triangles are properties of the angles in the triangle, leading to definitions of trigonometric ratios for acute angles.

What It Means to You

Students demonstrate an understanding of trigonometric ratios that come from properties of similar right triangles using measurement and comparison of ratios. This builds on previous knowledge that similar triangles have corresponding pairs of proportional sides and congruent pairs of corresponding angles.

The emphasis for this standard is on the measurement of side lengths and the comparison of ratios in similar triangles. This helps students to see how the tangent ratio can be used to find unknown side lengths and how the inverse tangent can be used to find unknown angle measures. In the next lesson, students will extend this knowledge to the sine and cosine ratios for right triangles.

WARM-UP OPTIONS

ONLINE

PROJECTABLE & PRINTABLE 

ACTIVATE PRIOR KNOWLEDGE • Use Properties of Similar Triangles to Solve Problems

Use these activities to quickly assess and activate prior knowledge as needed.

Problem of the Day 

Two ski hills are on mountains that can be modeled by similar triangles. Use a proportion to find the unknown length. $\frac{6}{7.2} = \frac{12}{x}$, $x = 14.4$ km

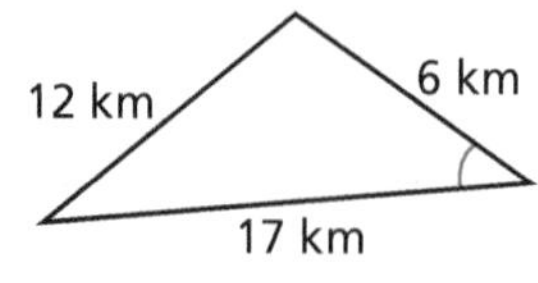

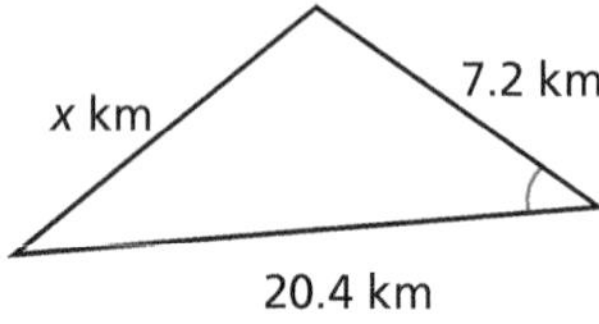

Quick Check for Homework

As part of your daily routine, you may want to display the Teacher Solution Key to have students check their homework.

Make Connections

Based on students' responses to the Problem of the Day, choose one of the following:

 Project the Interactive Reteach, Grade 7, Lesson 1.2.

 Complete the Prerequisite Skills Activity:

Display the figures on the board. Work as a class to answer the questions.

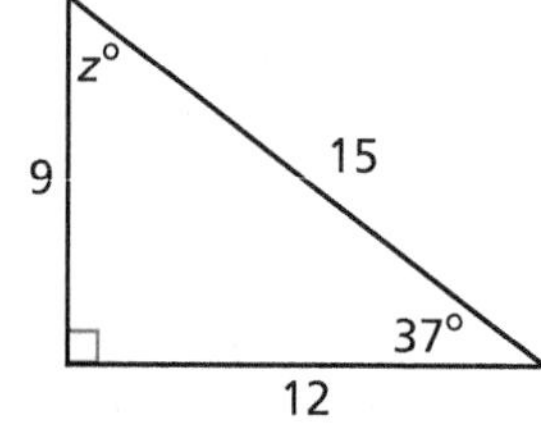

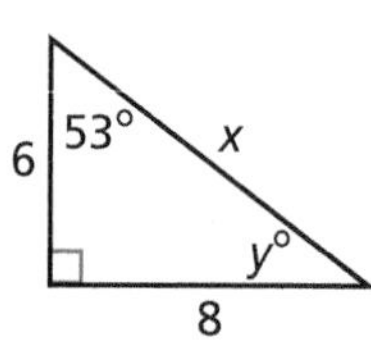

- *What is the sum of angles in a triangle?* 180°
- *What are the missing angle measures y° and z°?* $y° = 37°$, $z° = 53°$
- *How do you know the triangles are similar?* The corresponding pairs of angles are congruent.
- *What proportion can you use to find the value of x? What is x?* $\frac{9}{6} = \frac{15}{x}$ or $\frac{12}{8} = \frac{15}{x}$, $x = 10$

If students continue to struggle, use Tier 3 Skill 9.

SHARPEN SKILLS

If time permits, use this on-level activity to build fluency and practice basic skills.

Vocabulary Review

Objective: Students define the tangent ratio and give examples.
Materials: Bubble Map (Teacher Resource Masters)

Have students use the Bubble Map, writing "Tangent Ratio" in the center circle. In one circle, write when the tangent ratio is used. In another circle, write the tangent ratio related to the sides of the triangle. In the last two circles, have students draw right triangles and give examples of the tangent ratio.

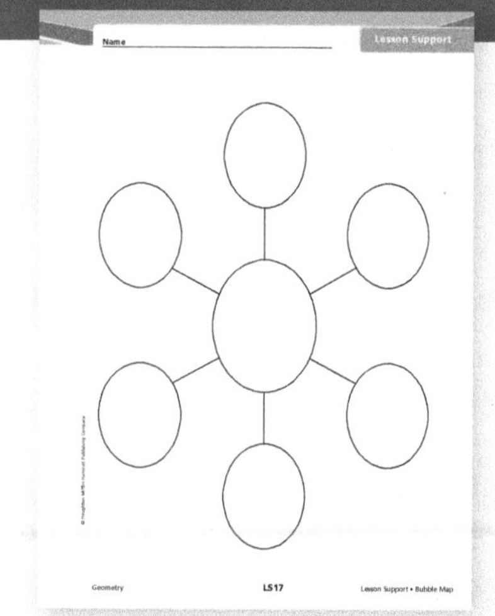

PLAN FOR DIFFERENTIATED INSTRUCTION

Small-Group Options

Use these teacher-guided activities with pulled small groups.

On Track

Materials: index cards

On index cards, draw examples of right triangles, giving enough information to solve for the missing value using the tangent or the inverse tangent. Have students work in pairs to solve the problem. Then have the students switch cards with another group to confirm their answers.

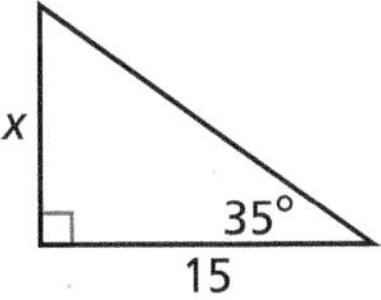

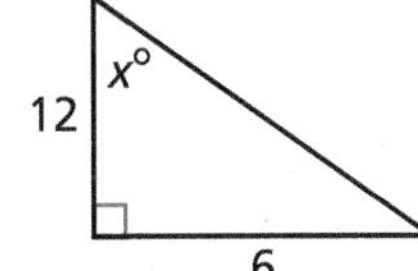

Almost There

Have students sketch a right triangle on a sheet of paper labeling the sides *a*, *b*, and *c*, and the opposite angles *A*, *B*, and *C*. Ask the following questions.

- Which are the acute angles? How do you know?
- Which two sides of the triangle are used in the tangent ratios?
- What is the ratio of tan *A*?
- What is the ratio of tan *B*?
- Why is *c* not used in the ratios?

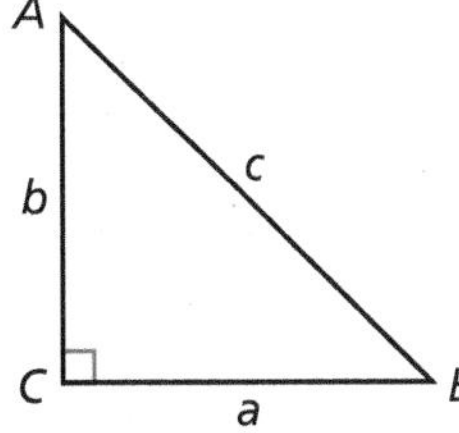

Ready for More

Give the following statement:

The tangent of 90° is undefined.

Have students work in small groups. Ask them to use their knowledge of the tangent ratio to explain why you can only find the tangent ratio for the acute angles in a right triangle, and not the right angle itself.

Math Center Options

Use these student self-directed activities at centers or stations.

Key: ● Print Resources ● Online Resources

On Track

- ● Interactive Digital Lesson
- ●● Journal and Practice Workbook
- ● Interactive Glossary (printable): **inverse tangent**, **tangent**
- ● Module Performance Task

Almost There

- ● Reteach 13.1 (printable)
- ● Interactive Reteach 13.1
- ● RtI Tier 3 Skill 9: Write Equations for Proportional Relationships
- ● Illustrative Mathematics: Creating Similar Triangles

Ready for More

- ● Challenge 13.1 (printable)
- ● Interactive Challenge 13.1

View data-driven grouping recommendations and assign differentiation resources.

During the *Spark Your Learning,* listen and watch for strategies students use. See samples of student work on this page.

Construct a Similar Triangle

Strategy 1

Since similar triangles have proportional side lengths, I drew a right triangle that had the same acute angle measure. I used a ruler to measure the vertical and horizontal distances for the triangle I drew. I wrote a proportion and solved.

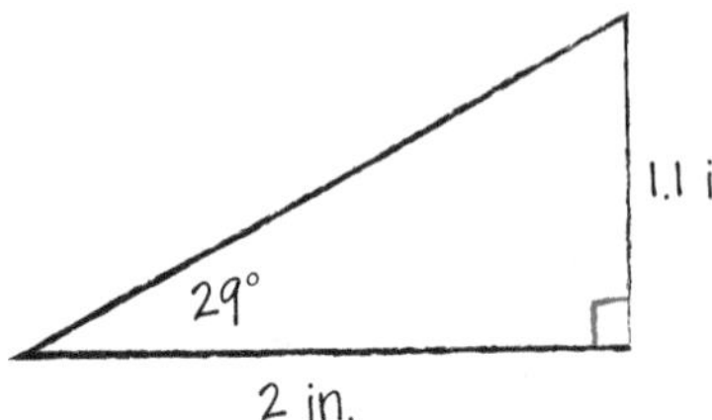

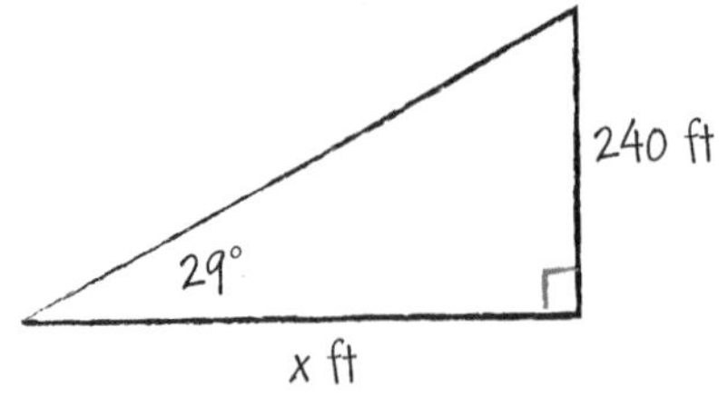

$$\frac{x}{2} = \frac{240}{1.1}$$

$$x \approx 436.4$$

The students are approximately 436.4 feet from the entrance.

If students . . . use similar triangles to solve the problem, they understand proportional relationships from Grade 7.

Have these students . . . explain how they used similar triangles and how they solved the proportion. **Ask:**

- Q How did you use similar triangles to solve the problem?
- Q What measurements did you need to draw the similar triangle?

Measure and Solve a Proportion

Strategy 2

I measured the vertical distance of the triangle in centimeters. Then I measured the horizontal distance in centimeters. Then I wrote and solved a proportion to find the unknown horizontal distance.

$$\frac{\text{horizontal (cm)}}{\text{vertical (cm)}} = \frac{x \text{ ft}}{240 \text{ ft}}$$

$$\frac{5.1 \text{ cm}}{2.8 \text{ cm}} = \frac{x \text{ ft}}{240 \text{ ft}}$$

$$x \approx 437 \text{ ft}$$

If students . . . use a proportion to solve the problem, they know how to use equal ratios to find missing values.

Activate prior knowledge . . . by having students relate their proportion to similar triangles. **Ask:**

- Q How is the triangle that is measured in centimeters related to the triangle in the photo?
- Q What is the relationship between the corresponding side lengths in two similar triangles?

COMMON ERROR: Multiplies the Measurements

The angle measure is 29°, so I multiplied it by 240 ft. These are the only two values in the problem, so the product should be the distance the students are away from the entrance.

$29 \cdot 240 = 6960$ ft

If students . . . multiply the values, then they are guessing about how the numbers are related to the horizontal distance.

Then intervene . . . by pointing out that the product of the angle measure and the vertical distance does not necessarily give the horizontal distance. **Ask:**

- Q What kind of triangle is shown to represent the situation?
- Q How can you draw a triangle that is similar to the one in the picture?

13.1

Tangent Ratio

(I Can) use the tangent ratio and its inverse to find side lengths and angle measures in right triangles.

Spark Your Learning

A group of students see the entrance as they walk toward the Taj Mahal.

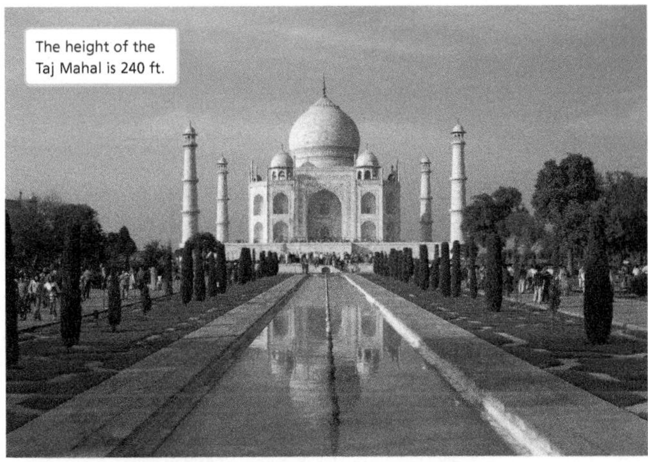

The height of the Taj Mahal is 240 ft.

Complete Part A as a whole class. Then complete Parts B–D in small groups.

A. What is a mathematical question you can ask about this situation? What information would you need to know to answer your question?

A. How far away are the students from the entrance?; a diagram of the Taj Mahal with distances and/or angle measures given

B. What variable(s) are involved in this situation? distances (ft) and/or angle measures

C. To answer your question, what strategy and tool would you use along with all the information you have? What answer do you get? See Strategies 1 and 2 on the facing page.

D. Does your estimate make sense in the context of the situation? How do you know? See Additional Answers.

©Andrei Kazarov/Fotolia

Turn and Talk What if the measure of the angle were a bit smaller? How would this affect the accuracy of your estimate? See margin.

SUPPORT SENSE-MAKING • Three Reads

Tell students to read the information in the photo three times and prompt them with a different question each time.

1 What is the situation about? The situation is about students looking at the Taj Mahal as they approach the entrance. The students know how tall the building is.

2 What are the quantities in this situation? How are the quantities related? The quantities are the height of the building, the distance the students are away from the building, and the angle the ground makes with a line from the students to the top of the building. The quantities are the measures of a right triangle a right triangle.

3 What are possible questions you could ask about this situation? Possible answer: Does the angle change as the students get closer to the building? How can you find the distance from where the students stand to the building? How can you use similar triangles to find the missing distance?

1 Spark Your Learning

▶ MOTIVATE 

- Have students look at the photo in their books and read the information contained in the photo. Then complete Part A as a whole-class discussion.
- Give the class the additional information they need to solve the problem. This information is available online as a printable and projectable page in the Teacher Resources.
- Have students work in small groups to complete Parts B–D.

▶ PERSEVERE

If students need support, guide them by asking:

Q **Advancing • Use Tools** Which tool could you use to solve the problem? Why choose that tool and not some other? Students' choices of tools and reasons for choosing them will vary.

Q **Assessing** How could you find the measures of the triangle shown? You could use a protractor to measure angles and a ruler to measure side lengths.

Q **Assessing** How could you draw a triangle that is similar to the one shown? You would draw a right triangle that has a 29° angle since, similar triangles have the same angle measures.

Q **Advancing** If you had two similar triangles, how could you solve for an unknown length? You can write a proportion and solve for the unknown value.

Turn and Talk Have students try to visualize what happens to the side lengths in a triangle when you adjust the size of the angle. Do the side lengths get larger or smaller? Possible answer: If the given angle was slightly smaller, your estimate would be close but slightly larger than the actual distance.

▶ BUILD SHARED UNDERSTANDING

Select groups of students who used various strategies and tools to share with the class how they solved the problem. As they present their solutions, have each group discuss why they chose a specific strategy and tool.

2 Learn Together

Build Understanding

Task 1 (MP) **Use Tools** Students use tracing paper and rulers to measure and confirm relationships in similar right triangles.

Sample Guided Discussion:

Q What is true about the ratio of corresponding pairs of sides in the two triangles? The ratios of corresponding sides are the same.

Q In Part D, how do you know the triangles are similar? The triangles have corresponding pairs of congruent angles and proportional corresponding side lengths.

Turn and Talk Help students to see that the side opposite the angle is the leg of the right triangle that is not on the ray of the angle and that the adjacent side is the leg next to the angle. Have students check their ratios to make sure they are writing the opposite value in the numerator and the adjacent value in the denominator. Possible answer: The ratios $\frac{\text{opposite}}{\text{adjacent}}$ for $\angle P$ and $\angle X$ are the same and are the reciprocals of the ratios of $\angle R$ and $\angle Z$, respectively.

Build Understanding

Investigate a Ratio in a Right Triangle

In the given right triangle $\triangle ABC$, the sides (or legs) are labeled in reference to $\angle C$. So side $\overline{AB}$ is the opposite leg of $\angle C$, and side $\overline{BC}$ is the adjacent leg to $\angle C$. The side that connects the opposite and adjacent legs is the hypotenuse.

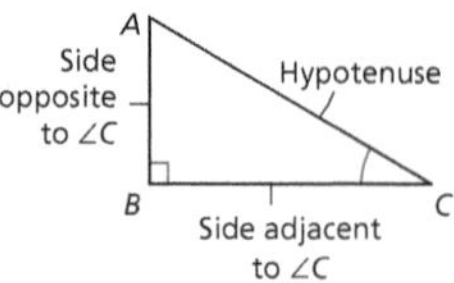

1 Analyze the relationships of opposite and adjacent side lengths of two right triangles that have the same acute angle.

A. Use tracing paper to create a copy of each triangle shown. What are the approximate side lengths, in inches, of $\overline{QR}$ and $\overline{YZ}$? Round to the nearest tenth of an inch. $QR \approx 1.4$ in., $YZ \approx 2.1$ in.

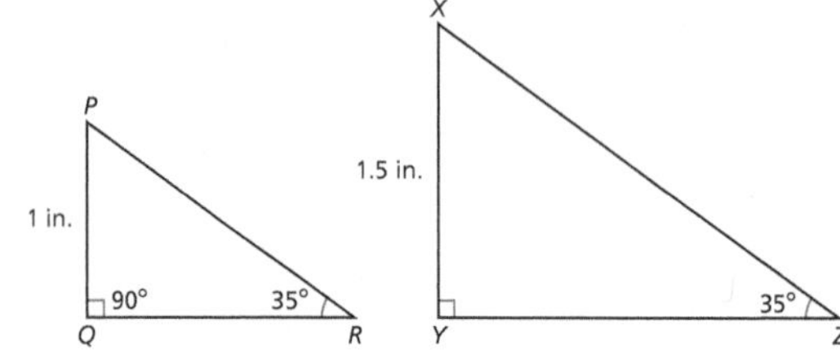

B. Complete the table below. B, C. See Additional Answers.

	Reference angle	Opposite side length (in.)	Adjacent side length (in.)	Ratio of $\frac{\text{opposite}}{\text{adjacent}}$
$\triangle PQR$	$m\angle R = 35°$	$PQ = 1.0$	$QR \approx ?$	$\frac{PQ}{QR} \approx ?$
$\triangle XYZ$	$m\angle Z = 35°$	$XY = 1.5$	$YZ \approx ?$	$\frac{XY}{YZ} \approx ?$

C. Compare the ratios of $\frac{\text{opposite}}{\text{adjacent}}$ of $\angle R$ and $\angle Z$. What do you notice?

D. Compare the triangles. What relationship exists between the triangles? The triangles are similar.

E. If you create a third right triangle with an angle of 35° and its opposite side is 3 in. long, how would your results be similar when comparing it to the triangles above? How would they be different? E, F. See Additional Answers.

F. Why do you think each ratio of the opposite side to the adjacent side between the two right triangles is the same?

Turn and Talk Repeat the process for $\angle P$ and $\angle X$. What do you notice? See margin.

406

LEVELED QUESTIONS

Depth of Knowledge (DOK)	Leveled Questions	What Does This Tell You?
Level 1 **Recall**	What kind of triangle has a 90° angle? a right triangle	Students' answers will indicate whether they know the definition of a right triangle.
Level 2 **Basic Application of Skills & Concepts**	How can you tell which side is opposite and which is adjacent to the reference angle? The opposite side is across from the angle and is a leg. The adjacent side is the leg that forms part of the reference angle.	Students' answers will indicate whether they can find the opposite and adjacent legs in a right triangle with respect to an acute angle, which will help them to write a correct ratio.
Level 3 **Strategic Thinking & Complex Reasoning**	Is a right triangle with a 35° angle similar to an obtuse triangle with a 35° angle? Explain. no; A right triangle will only be similar to other right triangles with a 35° angle because in similar triangles corresponding angles are congruent.	Students' answers will indicate whether they understand that similar triangles must have three pairs of congruent corresponding angles.

Understand Tangent Ratios

The ratio you calculated in Task 1 is called the tangent ratio. The **tangent** of an acute angle in a right triangle is the ratio of the length of the side opposite the angle to the length of the side adjacent to the angle. You can calculate the tangent, abbreviated tan, of any acute angle in a right triangle as shown.

Tangent Ratio

$\tan A = \frac{\text{length of side opposite } \angle A}{\text{length of side adjacent } \angle A} = \frac{a}{b}$

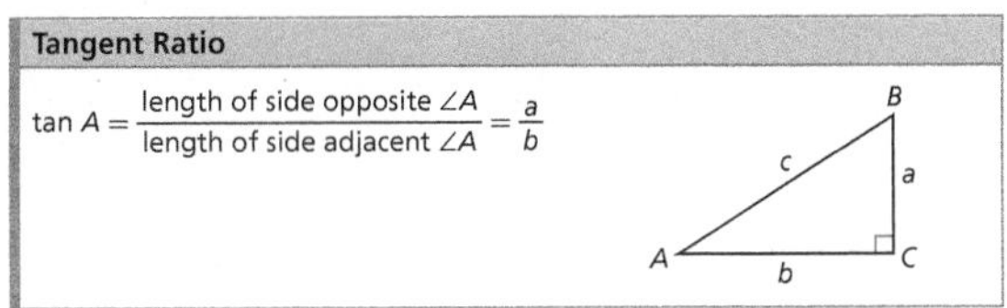

2 Use a geometric drawing tool to create the figures. Write each tangent ratio as a fraction and as a decimal rounded to the nearest hundredth. Then analyze your results.

A. Create the triangle shown. What is tan A?
A, B. See margin.

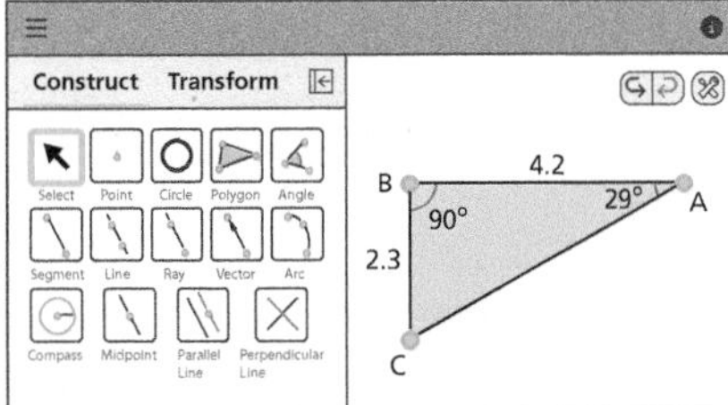

B. Drag point C down so BC is 3. Then measure $\angle A$. What is tan A?

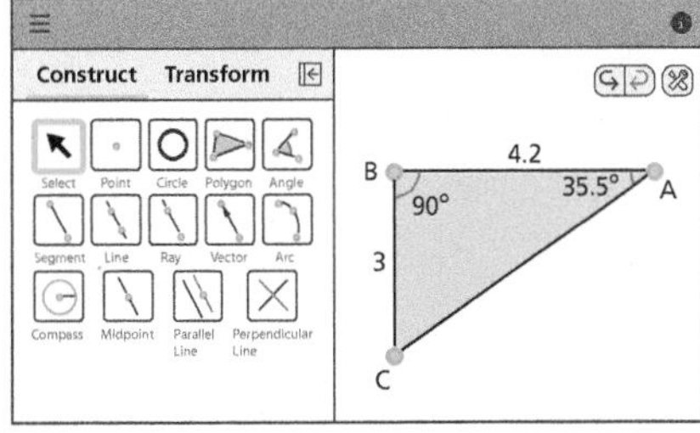

C. What happens to the tangent ratio as the measure of $\angle A$ increases?
The tangent ratio increases as m$\angle A$ increases.

D. Because the triangles are right triangles, the other acute angle, $\angle C$, is complementary to $\angle A$. For each triangle, find the tangent of $\angle C$. What is the relationship between the tangent ratios of complementary angles?
See margin.

Turn and Talk A triangular support is being constructed for a new building. Will the tangent ratio most likely be used to find the opposite or adjacent side? Why do you think this is so? See margin.

PROFICIENCY LEVEL

Beginning

Have students answer the following question: Which two sides of the triangle are compared for the tangent ratio?

Intermediate

Write a sentence describing how to identify the two lengths needed to find the tangent ratio.

Advanced

Have students write several sentences to explain how the tangent ratio is measured. Have them answer this question: Is it measured as a length, in degrees, or with no unit at all? Why? Then have students share their answer with a partner.

Task 2 (MP) **Use Structure** Students create different triangles by changing sides and angles to see how the tangent ratio is affected.

CONNECT TO VOCABULARY

Have students use the **Interactive Glossary** to record their understanding of the vocabulary in this task.

Sample Guided Discussion:

Q Why do you find the tangent ratio of the acute angles and not the right angle? Possible answer: The tangent ratio is the length of the opposite leg divided by the length of the adjacent leg. The side opposite the right angle is the hypotenuse, and the right angle has two adjacent legs.

Q When would the tangent ratio be greater than 1? When the opposite leg is longer than the adjacent leg because the numerator would be greater than the denominator.

Turn and Talk Ask students to brainstorm how they could measure horizontal and vertical distances. Then ask them to think about which might be easier to do. the opposite side; The adjacent side to the angle from the ground is easier to measure, since it is at ground height.

ANSWERS

A. $\tan A = \frac{2.3}{4.2} \approx 0.548$

B. 35.5°; $\tan A = \frac{3}{4.2} \approx 0.714$

D. In Part A, $\tan C = \frac{4.2}{2.3} \approx 1.826$, and in Part B, $\tan C = \frac{4.2}{3} = 1.4$. The tangent ratios of complementary angles are inverses.

Step It Out

Task 3 **(MP) Attend to Precision** Encourage students to wait until the final calculation when rounding the values to get the most accurate answer.

Sample Guided Discussion:

Q **$\overline{AC}$ is also adjacent to the given angle. Why is the side not used in the calculations?** The side $\overline{AC}$ is the hypotenuse, which is not the same as the leg adjacent to the angle.

Q **When solving for x, why do you multiply by x on both sides of the equation?** This is the first step in isolating the variable. Then you can divide by tan 75° to solve for x.

Turn and Talk Point out to students that they do not need to know the measure of $\angle A$ in order to find tan A. Have students apply the definition of the tangent ratio. The opposite leg of $\angle A$ is the adjacent leg of $\angle C$, and the adjacent leg of $\angle A$ is the opposite leg of $\angle C$ so that the tan A is the reciprocal of tan C. You can find tan 25° and then use the inverse key to find the reciprocal. Then you can calculate tan 75° and see that they are the same.

ANSWER

C. Possible answer: If you round the value of the tangent before finishing your calculations, the rounding error will be even greater in the final answer.

Step It Out

Apply Tangent to Find a Length

3 Kathy is looking at the Washington Monument located in Washington, D.C. The monument is 555 feet tall. From where she is standing, the angle made between the ground and the top of the monument is 75°. How far away is she from the monument?

Draw a diagram of the situation.

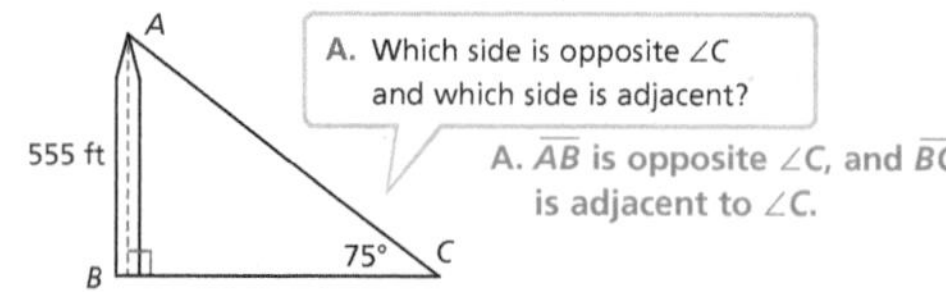

A. $\overline{AB}$ is opposite $\angle C$, and $\overline{BC}$ is adjacent to $\angle C$.

Write the given information.

$m\angle C = 75°$
$AB = 555$ ft
$x = BC$

B. The variable x represents the distance from the Washington Monument.

Write the tangent ratio.

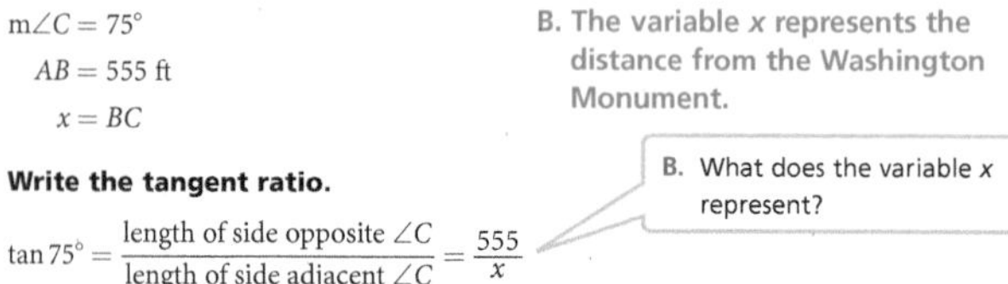

B. What does the variable x represent?

$$\tan 75° = \frac{\text{length of side opposite } \angle C}{\text{length of side adjacent } \angle C} = \frac{555}{x}$$

Solve for the unknown distance. Be sure the calculator is in degree mode.

$\tan 75° = \frac{555}{x}$	Write the ratio using the identified values.
$x \cdot \tan 75° = \frac{555}{x} \cdot x$	Multiply both sides by x.
$x = \frac{555}{\tan 75°}$	Divide both sides by tan 75°.
≈ 149	Simplify. Round to the nearest foot.

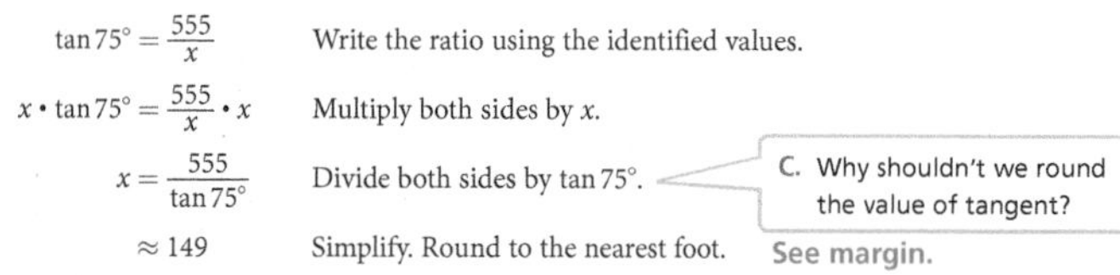

C. Why shouldn't we round the value of tangent?

See margin.

Answer the question.

She is about 149 feet away from the monument.

Turn and Talk How can you use what you know about the tangent of $\angle C$ to find the tangent of $\angle A$? How can you verify this by using a calculator? See margin.

Apply Inverse Tangent to Find an Angle Measure

If you know the length of both legs of a right triangle, but an angle is unknown, you can use the **inverse tangent** to find the missing angle. The inverse tangent, written as $\tan^{-1}$ and read as "the inverse tangent of angle...", gives the acute angle that has a tangent equal to a given value.

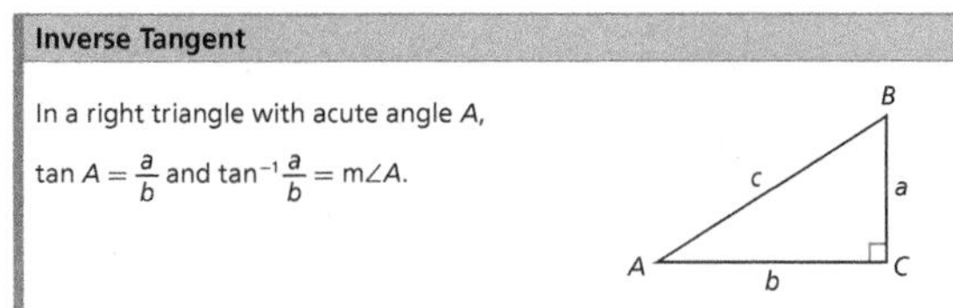

Inverse Tangent

In a right triangle with acute angle A,

$\tan A = \frac{a}{b}$ and $\tan^{-1}\frac{a}{b} = m\angle A$.

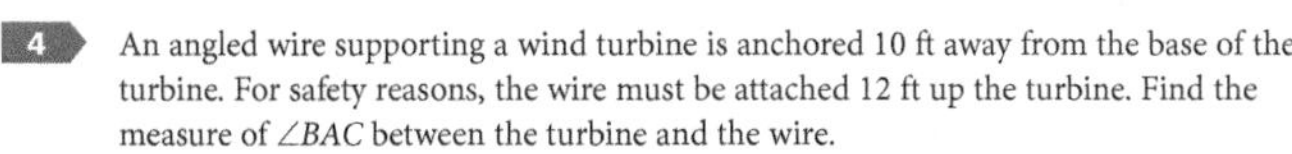

4 An angled wire supporting a wind turbine is anchored 10 ft away from the base of the turbine. For safety reasons, the wire must be attached 12 ft up the turbine. Find the measure of $\angle BAC$ between the turbine and the wire.

Draw and label a diagram.

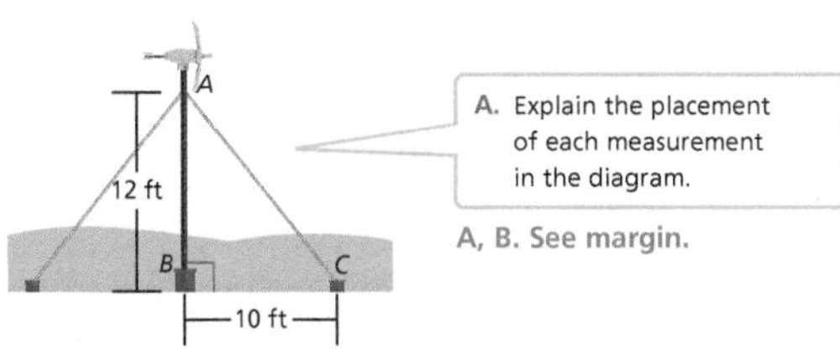

A, B. See margin.

Write the given information.

Length of side opposite $\angle BAC$: 10 ft

Length of side adjacent to $\angle BAC$: 12 ft

B. Why is the opposite side the same as the horizontal side in this problem, while it was the vertical side in Task 3?

Write the tangent ratio.

$\tan BAC = \frac{10}{12} = \frac{5}{6}$

Solve using the inverse tangent.

$m\angle BAC = \tan^{-1}\frac{5}{6}$ Write the inverse tangent equation.

$\approx 40°$ Use the inverse tangent function $(\tan^{-1})$ to evaluate. Round to the nearest degree.

C. How was the inverse tangent equation written from the tangent ratio?

C. The equation was written so the inverse tangent of the given ratio is equal to the measure of the angle.

Answer the question.

The wire makes an angle of about 40° with the turbine.

Turn and Talk Given that $m\angle BAC \approx 40°$, find the measure of the base angle of the triangle without using trigonometry. Verify your answer using the inverse tangent. See margin.

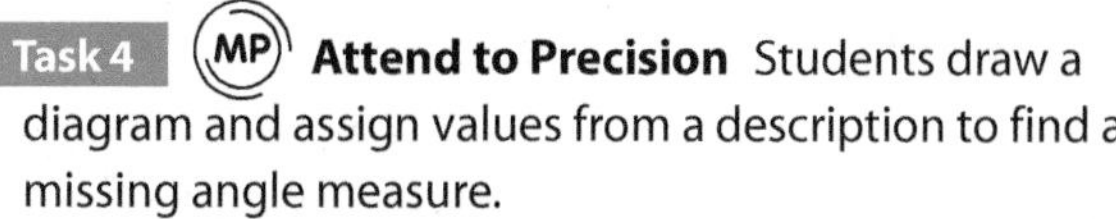

Task 4 (MP) **Attend to Precision** Students draw a diagram and assign values from a description to find a missing angle measure.

CONNECT TO VOCABULARY

Have students use the **Interactive Glossary** to record their understanding of the vocabulary in this task.

Sample Guided Discussion:

Q **The operations addition and subtraction are called inverse operations because one does the inverse of the other. How are the tangent and the inverse tangent related?** Possible answer: Using the tangent finds a ratio of side lengths from an angle measure, while using the inverse tangent finds an angle measure from a ratio of side lengths.

Q **Which angle could be found using the ratio $\frac{12}{10}$?** $\angle C$ because 12 is the length of the leg opposite the angle and 10 is the length of the adjacent leg.

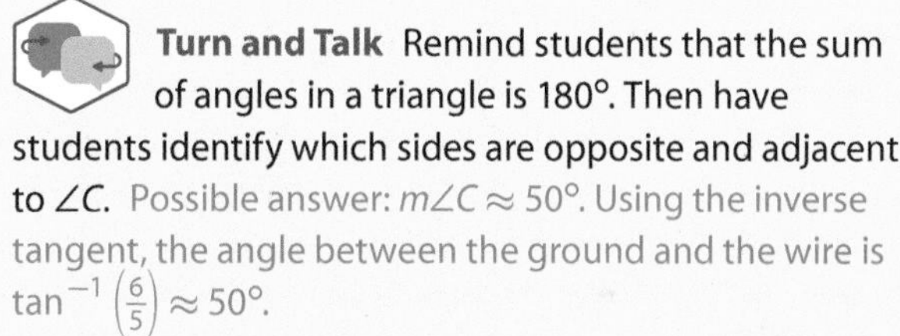

Turn and Talk Remind students that the sum of angles in a triangle is 180°. Then have students identify which sides are opposite and adjacent to $\angle C$. Possible answer: $m\angle C \approx 50°$. Using the inverse tangent, the angle between the ground and the wire is $\tan^{-1}\left(\frac{6}{5}\right) \approx 50°$.

ANSWERS

A. Possible answer: The vertical height of the triangle is 12 ft, because the problem tells us the wire is attached 12 ft up the turbine. The horizontal leg of the triangle is 10 ft, because the problem tells us the wire is anchored 10 ft from the base of the turbine. Angle BAC is the angle between the turbine and wire, so it is placed between the vertical leg and the hypotenuse.

B. Possible answer: The referenced angle is in a different location in the triangle, so the side opposite that angle is the horizontal leg rather than the vertical leg, which is adjacent to the referenced angle.

ONLINE

Assign the Digital On Your Own for
- built-in student supports
- Actionable Item Reports
- Standards Analysis Reports

On Your Own

Assignment Guide

The chart below indicates which problems in the On Your Own are associated with each task in the Learn Together. Assign daily homework for tasks completed.

Learn Together Tasks	On Your Own Problems
Task 1, p. 406	Problems 5–9, and 26
Task 2, p. 407	Problems 22, and 24
Task 3, p. 408	Problems 10–15, and 23
Task 4, p. 409	Problems 16–21, and 25

ANSWERS

5.A. $\tan C \approx \frac{2 \text{ cm}}{4.3 \text{ cm}} \approx 0.47$; $\tan Z \approx \frac{3 \text{ cm}}{6.4 \text{ cm}} \approx 0.47$;
The tangent ratios are about the same.

B. They have the same measure.

C. $\tan^{-1} C \approx \tan^{-1} \frac{2}{4.3} \approx 25°$; $\tan^{-1} Z \approx \tan^{-1} \frac{3}{6.4} \approx 25°$

Check Understanding

1. If two right triangles are similar, what is the relationship between the ratios of the length of the opposite side to the length of the adjacent side for any two corresponding acute angles? The ratios are equal.

2. Find the tangent of $\angle B$.
What is the tangent of $\angle A$?
$\tan \angle B = 2.4$ and $\tan \angle A \approx 0.417$

B, C, A; 10, 26, 24

3. What is the length of $\overline{PR}$? Round to the nearest whole unit. 10

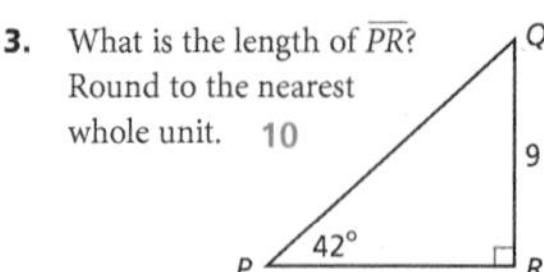

4. A ladder leaning against a wall reaches a vertical height of 14 ft. The base of the ladder is 3.5 ft from the wall. To the nearest degree, what is the angle that the ladder makes with the ground? 76°

On Your Own

5. (MP) **Use Tools** Trace triangles ABC and XYZ. Using a centimeter ruler, measure the lengths of the sides opposite to and adjacent to both $\angle C$ and $\angle Z$.

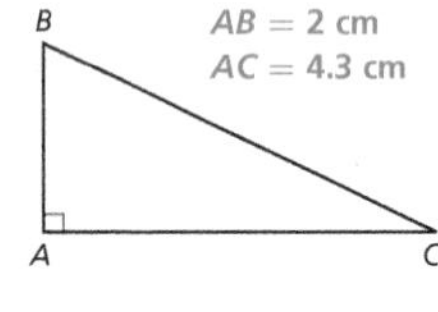

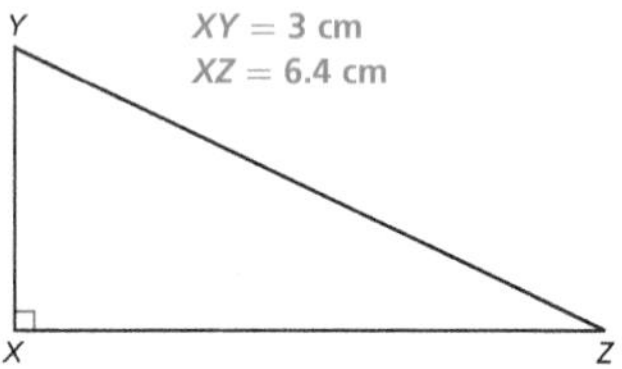

A–C. See margin.

A. Write the tangent ratio for $\angle C$ and $\angle Z$ as a fraction and as a decimal rounded to the nearest hundredth. Then compare the ratios.

B. What can you conclude about $m\angle C$ and $m\angle Z$ without measuring?

C. Use inverse tangent to verify your conclusions about $m\angle C$ and $m\angle Z$.

For each triangle, find the tangent of each given angle.

6. $\angle P$

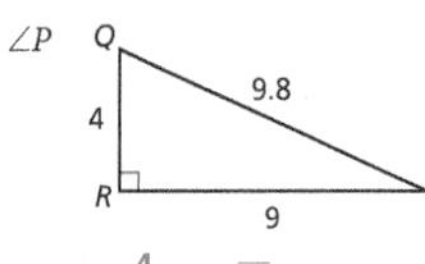

$\tan P = \frac{4}{9} = 0.\overline{44}$

7. $\angle Y$ $\quad \tan Y = \frac{6}{8} = 0.75$

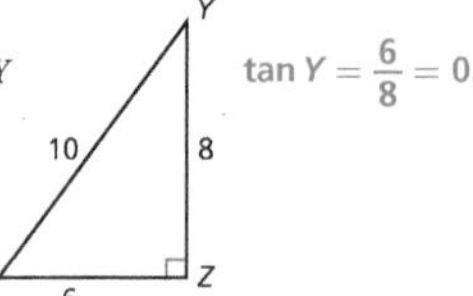

8. $\angle L$ and $\angle M$

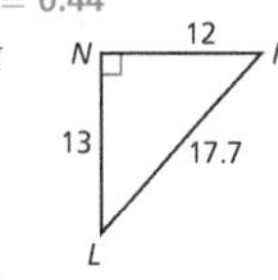

$\tan L = \frac{12}{13} \approx 0.92$; $\tan M = \frac{13}{12} \approx 1.08$

9. $\angle B$ and $\angle C$

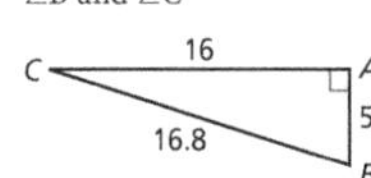

$\tan B = \frac{16}{5} = 3.2$; $\tan C = \frac{5}{16} = 0.3125$

410

3 Check Understanding

Formative Assessment

Use formative assessment to determine if your students are successful with this lesson's learning objective.

Students who successfully complete the Check Understanding can continue to the On Your Own practice.

For students who miss 1 problem or more, work in a pulled small group using the Almost There small-group activity on page 405C.

ONLINE

Assign the Digital Check Understanding to determine
- success with the learning objective
- items to review
- grouping and differentiation resources

4 Differentiation Options

Differentiate instruction for all students using small-group activities and math center activities on page 405C.

Reteach

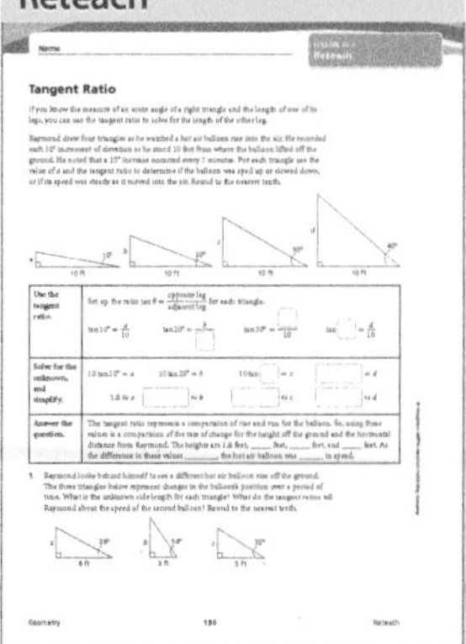

Challenge

For each triangle, find the given side length. Round to the nearest tenth.

10. AB $AB \approx 2.9$

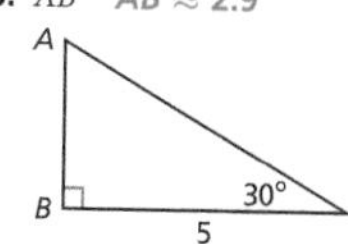

11. PR $PR \approx 10.7$

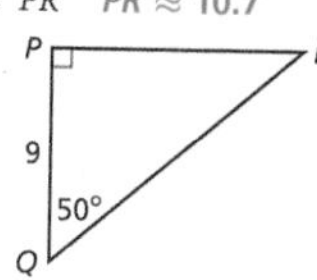

12. MN $MN \approx 30.6$

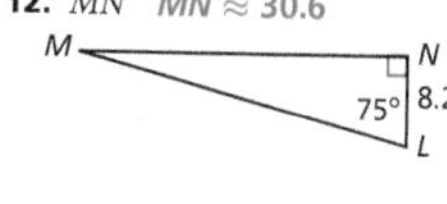

13. ZY $ZY \approx 8.3$

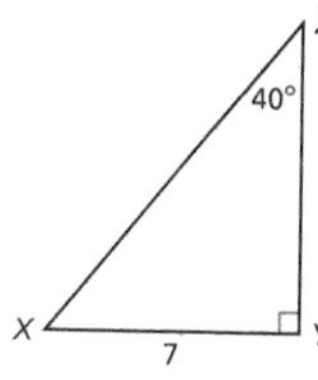

14. AB $AB \approx 18.3$

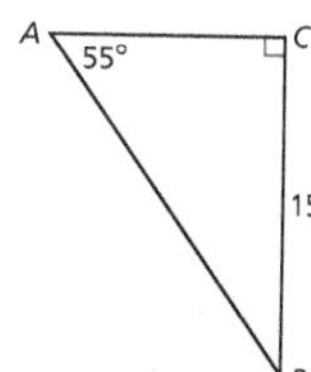

15. YZ $YZ \approx 56.6$

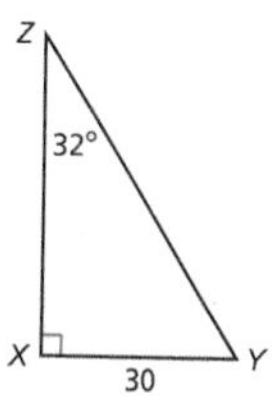

Use a calculator to find the measure of each given angle. Round the value to the nearest tenth of a degree.

16. $m\angle P$ $m\angle P \approx 45°$

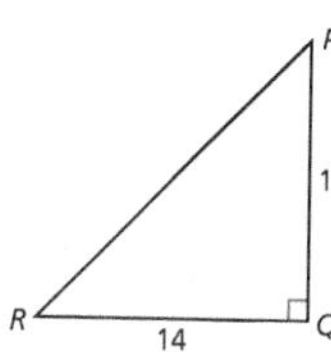

17. $m\angle Q$ $m\angle Q \approx 50.2°$

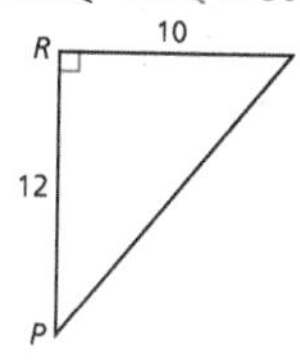

18. $m\angle Z$ $m\angle Z \approx 35°$

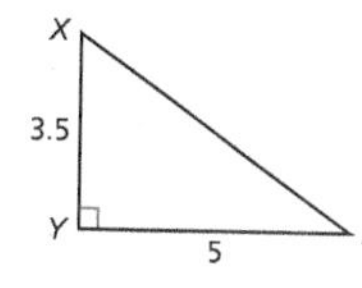

19. $m\angle L$ $m\angle L \approx 68.2°$

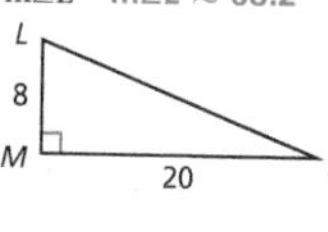

20. $m\angle B$ $m\angle B \approx 50.7°$

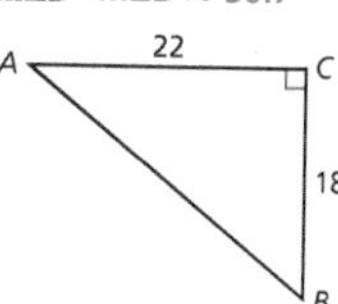

21. $m\angle T$ $m\angle T \approx 62.1°$

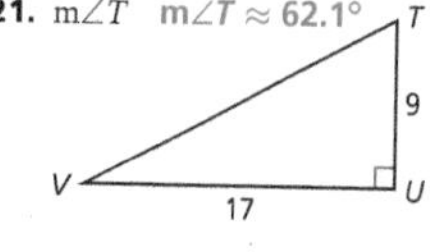

22. A kickball field is being set up by players in a nearby playground.

A. What is the distance between home base where the kicker will kick the ball to the farthest base? Round to the nearest meter. 31 m

B. After kicking the ball, a player ran from home plate to first base then to second base. The player was tagged out and walked straight back to home plate. What is the total distance the player traveled?
The player travels approximately 75 m.

50°
20 m

Questioning Strategies

Problem 10 Suppose you first found that $\angle A$ was equal to 60°. How could you use the tangent ratio to find AB? The opposite side would be 5 and the adjacent side would be the unknown AB. Then you could write the equation $\tan 60° = \frac{5}{AB}$ to solve for the missing side length.

Watch for Common Errors

Problem 20 Sometimes students will write the correct equation, such as $\tan B = \frac{22}{18}$, but then will try to take the tangent of the ratio instead of the inverse tangent. Remind students that if the missing value is a side length, they can use the tangent to solve. When the missing value is an angle measure, they can use the inverse tangent to solve.

⑤ Wrap-Up

Summarize learning with your class. Consider using the Exit Ticket, Put It in Writing, or I Can scale.

Exit Ticket 

A photographer spots an owl in a tree. The photographer is 15 yards from the base of the tree and sees the owl at a 60° angle from the ground. How high in the tree is the owl, to the nearest yard? Write an equation to model the situation and solve. $\tan 60° = \frac{x}{15}$; 26 yards

Put It in Writing 

Ask students to write a small paragraph describing when to use the tangent and when to use the inverse tangent when solving problems.

I Can 

The scale below can help you and your students understand their progress on a learning goal.

4	I can use the tangent ratio and the inverse tangent to find side lengths and angle measures in right triangles, and I can explain my reasoning to others.
3	I can use the tangent ratio and the inverse tangent to find side lengths and angle measures in right triangles.
2	I can write the tangent ratio and find an unknown side length.
1	I can find the opposite and adjacent sides related to a given angle in a right triangle.

Spiral Review • Assessment Readiness

These questions will help determine if students have retained information taught in the past and can also prepare them for high-stakes assessments. Here, students solve problems using proportionality in similar triangles **(12.2)**, find coordinates along a directed line segment **(13.2)**, identify sides in a right triangle **(12.3)**, and find the altitude in a right triangle **(12.4)**.

23. MP **Use Repeated Reasoning** Suppose $\triangle LMN$ and $\triangle OPQ$ are similar right triangles. For $\triangle LMN$, the side opposite $\angle M$ is 10 m, the side opposite $\angle N$ is 24 m, and $NM = 26$ m.

A. What is tan P and tan Q? A. See Additional Answers.

B. What is a possible length of each side of $\triangle OPQ$?
Possible answer: $OP = 12$ m, $PQ = 13$ m; $OQ = 5$ m

24. MP **Attend to Precision** The tangent ratios of the acute angles from three different right triangles are mixed together. Determine which angles belong to the same right triangle.

$\tan\angle 3 = 1.25$ $\tan\angle 4 = 0.6$
$\tan\angle 1 = 0.8$ $\tan\angle 2 = 1.67$
$\tan\angle 5 = 0.25$ $\tan\angle 6 = 4$

$\angle 1$ and $\angle 3$, $\angle 2$ and $\angle 4$, $\angle 5$ and $\angle 6$

25. Find the measure of $\angle C$. Round to the nearest tenth.
$m\angle C \approx 20.5°$

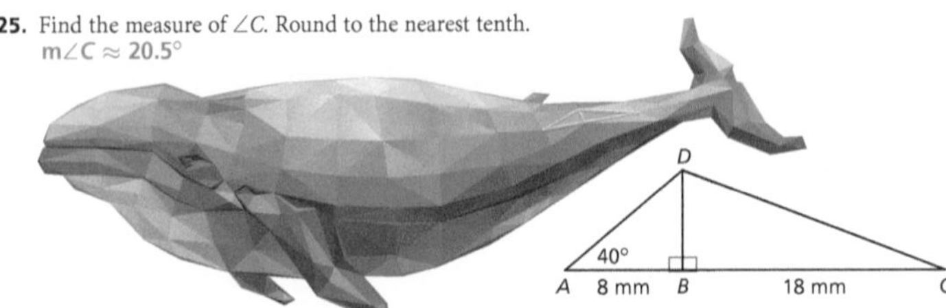

©mr.Zs/Shutterstock

Spiral Review • Assessment Readiness

26. $\triangle ABC$ and $\triangle XYZ$ are similar, and $\triangle ABC$ is a scaled drawing of $\triangle XYZ$. The length of $\overline{AB}$ is 7.2 cm, the length of $\overline{BC}$ is 6 cm, and the length of $\overline{XY}$ is 9.36 km. What is the length of $\overline{YZ}$?

Ⓐ 7.2 km Ⓒ 8.5 km
Ⓑ 7.8 km Ⓓ 12.2 km

27. Segment $\overline{AB}$ has endpoints $A(1, 3)$ and $B(5, 8)$. Point P divides $\overline{AB}$ into a 4 to 1 ratio from A to B. What are the coordinates of P?

Ⓐ $P(4, 6)$ **Ⓒ** $P(4.2, 7)$
Ⓑ $P(2, 4)$ Ⓓ $P(1.8, 4)$

28. In right triangle $\triangle XYZ$, the side opposite $\angle X$ is 6 m long, the side adjacent to $\angle X$ is 8 m long, and the hypotenuse is 10 m long. What is the ratio of the length of the side opposite $\angle X$ to the length of the hypotenuse?

Ⓐ 0.6 Ⓒ 0.8
Ⓑ 0.75 Ⓓ 1.33

29. The altitude to the hypotenuse of a right triangle divides the hypotenuse into a 3 in. segment and a 4 in. segment. What is the approximate length of the altitude?

Ⓐ 3.46 in. Ⓒ 5.5 in.
Ⓑ 5.28 in. Ⓓ 7 in.

 **I'm in a Learning Mindset!**

Does my lack of organization affect my learning outcomes? In what ways?

Keep Going ▶ Journal and Practice Workbook

Learning Mindset

mindset works

Perseverance Learns Effectively

Point out that the tangent ratio can be used to solve for missing lengths and angles in right triangles, but the ratios must be carefully written. Help students to examine each problem to see which sides of the triangle are used to write the tangent ratio and when it is useful to use the tangent or the inverse tangent to solve the problem. Encourage students to look for ways to check their answers. *When should you use the tangent rather than the inverse tangent? How do you know if your ratio is correct? How can you check your answer to make sure it makes sense?*

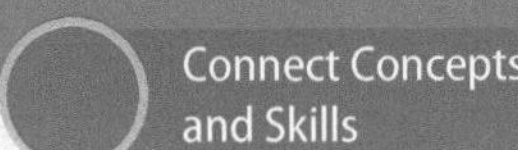

13.2 Sine and Cosine Ratios

LESSON FOCUS AND COHERENCE

Mathematics Standards

- Understand that by similarity, side ratios in right triangles are properties of the angles in the triangle, leading to definitions of trigonometric ratios for acute angles.
- Explain and use the relationship between the sine and cosine of complementary angles.
- Use trigonometric ratios and the Pythagorean Theorem to solve right triangles in applied problems.

Mathematical Practices and Processes 

- Construct viable arguments and critique the reasoning of others.
- Look for and make use of structure.
- Attend to precision.

I Can Objective 

I can use sine and cosine ratios and the inverses to find side lengths and angle measures in right triangles.

Learning Objective

Use sine and cosine ratios and the inverses to find side lengths and angle measures in right triangles.

Language Objective

Use trigonometric terminology appropriately and with understanding.

Vocabulary

New: cosine, inverse cosine, inverse sine, sine, trigonometric ratio

Lesson Materials: calculator, geometric drawing tool

Mathematical Progressions

Prior Learning	Current Development	Future Connections
Students: • explained the similarity of triangles as the equality of all corresponding pairs of angles and the proportionality of all corresponding pairs of sides. **(12.2)** • used similarity criteria for right triangles to solve problems and to prove relationships. **(12.4)**	**Students:** • find the sine and cosine of an acute angle in a right triangle using the definitions of sine and cosine as a ratio. • understand the meanings of inverse sine and inverse cosine and describe how to use them.	**Students:** • will understand and apply the Law of Sines and the Law of Cosines to find unknown measurements in right and non-right triangles in mathematical and real-world problems (e.g., surveying problems, resultant forces). **(14.1 and 14.2)**

PROFESSIONAL LEARNING

Math Background

It might be interesting for students to know that *trigon* is Greek for "triangle" and *metric* is Greek for "measurement." In this lesson, students continue their exploration of trigonometric ratios, namely, the special measurements of a right triangle. In the previous lesson, they understood and used the tangent ratio. Now, they added into their toolbox the sine and cosine ratios. Later on, students will learn about Law of Sines and the Law of Cosines and apply them in a variety of problems to find unknown measurements.

ONLINE

PROJECTABLE & PRINTABLE

WARM-UP OPTIONS

ACTIVATE PRIOR KNOWLEDGE • Set Up Proportions for Sides of Similar Triangles

Use these activities to quickly assess and activate prior knowledge as needed.

Problem of the Day

In triangle $\overline{ABC}$, DE is parallel to $\overline{AC}$. Find the length of $\overline{AD}$.

DE is parallel to AC. Therefore $\angle BDE \cong \angle DAC$ and $\angle BED \cong \angle ECA$.

Triangle ABC is similar to triangle DEB by AA (Angle-Angle Similarity Theorem).

Since the triangles are similar, their sides are proportional.

$$\frac{AB}{DB} = \frac{AC}{DE} \qquad \frac{20 + x}{20} = \frac{24}{14}$$

$$(20 + x)14 = 24(20)$$
$$7(20 + x) = 12(20)$$
$$140 + 7x = 240$$
$$7x = 100$$
$$x = 14.3$$

Quick Check for Homework

As part of your daily routine, you may want to display the Teacher Solution Key to have students check their homework.

Make Connections

Based on students' responses to the Problem of the Day, choose one of the following:

1. Project the Interactive Reteach, Lesson 12.4.

2. Complete the Prerequisite Skills Activity:

Have students work in pairs and use a geometric drawing tool. Ask students to draw a triangle. Then, have them measure the angles and the side lengths of the triangle. After that, have students, dilate or shrink the triangle. Have students measure the angles and lengths of the sides of the image. Have students investigate the ratios of the corresponding sides of the triangles.

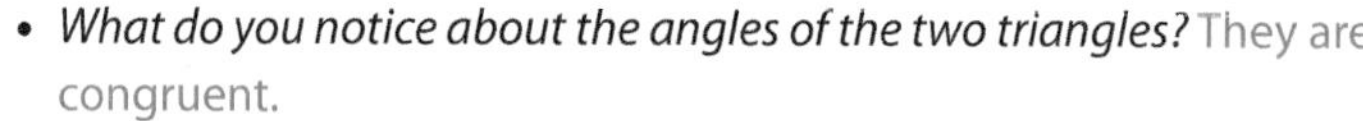

- *What do you notice about the angles of the two triangles?* They are congruent.
- *What do you notice about the ratios of the corresponding sides of the two triangles?* They are proportional.
- *What do we call these two triangles?* similar

If students continue to struggle, use Tier 2 Skill 6.

SHARPEN SKILLS

If time permits, use this on-level activity to build fluency and practice basic skills.

Quantitative Comparison

Objective: Students make a comparison between two quantities.
Write the following problem on the board. Ask students to choose the letter representing the correct answer and to explain their reasoning.

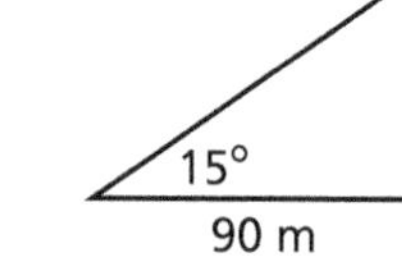

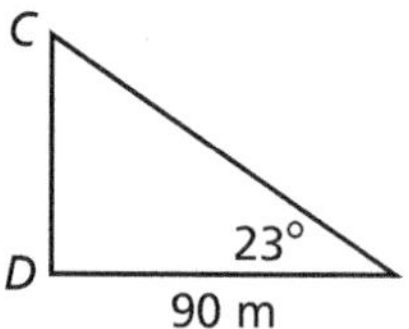

Quantity A	**Quantity B**
Length of $\overline{AB}$	Length of $\overline{CD}$

A. Quantity A is greater.
B. Quantity B is greater. B. Quantity A is 24.3m.
C. The two quantities are equal. Quantity B is 37.8m.
D. The relationship cannot be determined from the information given.

PLAN FOR DIFFERENTIATED INSTRUCTION

Small-Group Options

Use these teacher-guided activities with pulled small groups.

On Track

Materials: straightedge, protractor

Draw an irregular shape to represent a lake. Have students discuss a situation in which they have to measure the lake. They *cannot* measure over the lake, but they can measure around it to discover how long it is indirectly.

If students have difficulty, direct them to draw a right triangle so that the length of the lake lies on the hypotenuse of the triangle.

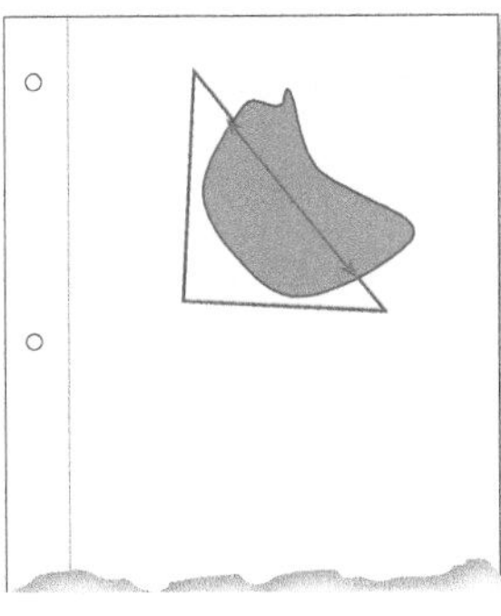

Almost There

Materials: protractor, straws, ruler

Have students construct similar right triangles using straws. Have them use the protractor to make sure the triangles have the same angles. Then they can measure the sides with a ruler to explore sine, cosine, and tangent ratios.

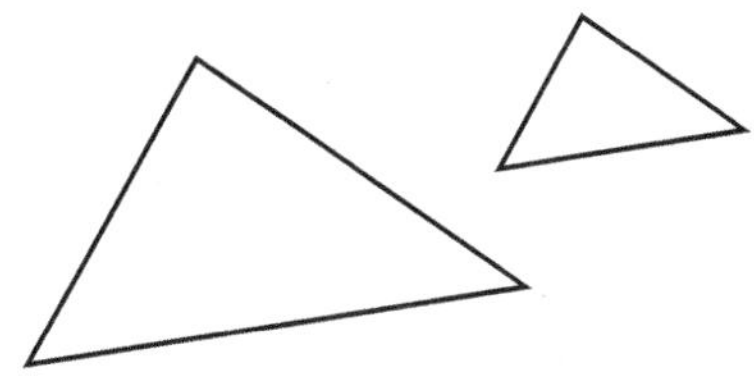

Ready for More

Have students try to complete a table of values of what sine and cosine values will be for angles from 0° to 360° by reasoning about special triangles.

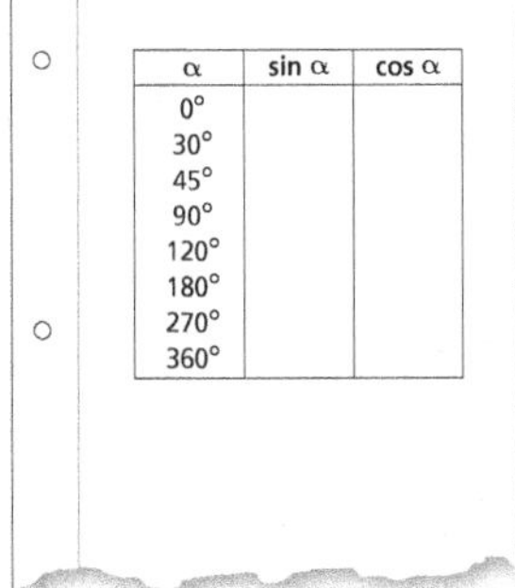

α	$\sin \alpha$	$\cos \alpha$
0°		
30°		
45°		
90°		
120°		
180°		
270°		
360°		

Math Center Options

Use these student self-directed activities at centers or stations.

Key: ● Print Resources ● Online Resources

On Track

- ● Interactive Digital Lesson
- ●● Journal and Practice Workbook
- ● Interactive Glossary (printable): **cosine**, **inverse cosine**, **inverse sine**, **sine**, **trigonometric ratio**
- ● Module Performance Task

Almost There

- ● Reteach 13.2 (printable)
- ● Interactive Reteach 13.2
- ● RtI Tier 2 Skill 6: Parallel Lines Cut by a Transversal

Ready for More

- ● Challenge 13.2 (printable)
- ● Interactive Challenge 13.2
- ● Illustrative Mathematics: Tangent of Acute Angles

ONLINE Ed

View data-driven grouping recommendations and assign differentiation resources.

During the *Spark Your Learning*, listen and watch for strategies students use. See samples of student work on this page.

Use Equation

Strategy 1

$\tan 70° = \frac{x}{2.4}$

$x = 2.4(\tan 70°)$

$x \approx 6.6$ m

$h^2 \approx 6.6^2 + 2.4^2$

$h \approx \sqrt{6.6^2 + 2.4^2}$

≈ 7

The cable is about 7 meters long.

If students . . . use their knowledge of the tangent ratio and the Pythagorean Theorem to solve the problem, they are demonstrating exemplary understanding from Lesson 13.1.

Have these students . . . explain how they decided to use a tangent ratio. **Ask:**

- Q How did you decide to apply the tangent ratio to solve the problem?
- Q How do you know that your answer is reasonable?

Use a Narrative

Strategy 2

This is a two-step problem. First, I found the length of the other leg of the triangle, and then I found the length of the cable.
I used tangent of 70° to equal the ratio of the opposite leg divided by adjacent leg. That gave me 6.6 m.
I used the Pythagorean Theorem and found the length of the cable to be about 7 m.

If students . . . use a narrative to explain their thinking, they show an excellent understanding of the problem but may not be expressing their thinking efficiently or clearly.

Activate prior knowledge . . . by having students use equations to show their calculations. **Ask:**

- Q Can you show me the equation you used to find the length of the unknown leg?
- Q Can you write the equation you used to find the length of the cable?

COMMON ERROR: Finds Partial Solution

$\tan 70° = \frac{x}{2.4}$

$x = 2.4(\tan 70°)$

$x \approx 6.6$ m

The length of the cable is 6.6 m.

If students . . . do not recognize that they have solved only part of the problem, they may not recall all the steps involved in their reasoning.

Then intervene . . . by pointing out that *x* may not represent the length of the cable. **Ask:**

- Q What does your unknown represent?
- Q Now that you know the lengths of two sides of the right triangle, how can you find the length of the third side?

13.2

Sine and Cosine Ratios

I Can use sine and cosine ratios and their inverses to find side lengths and angle measures in right triangles.

Spark Your Learning

Steel cables are sometimes used to anchor a mast to a sailboat's deck.

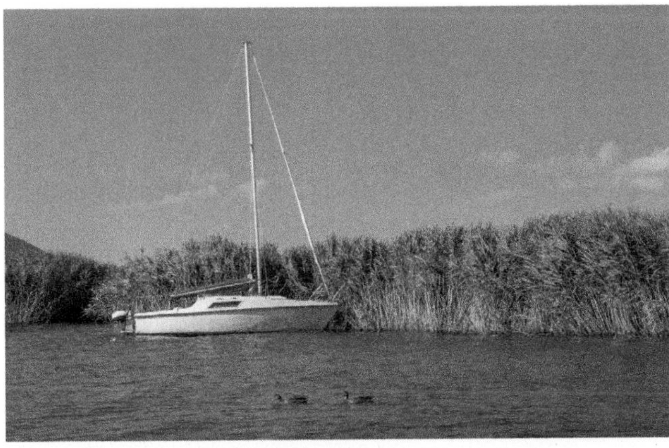

©Plamen Petrov/Shutterstock

Complete Part A as a whole class. Then complete Parts B–D in small groups.

A. What is a mathematical question you can ask about the cable?
What is the length of the cable?

B. What information do you need that can help you answer the question?
In the right triangle, I need the base angle and the base length.

C. To answer your question, what strategy and tool would you use along with all the information you have? What answer do you get?
See Strategies 1 and 2 on the facing page.

D. Does your answer make sense in the context of the situation? How do you know? The answer is reasonable because the hypotenuse is larger than the two sides and also follows the Triangle Inequality Theorem where $7 < 2.4 + 6.6$.

Turn and Talk Predict how your answer would change for each of the following changes in the situation: See margin.

- The cable is connected to the mast at a higher point but anchors to the deck at the same spot.
- The cable is connected to the mast at a lower point but anchors to the deck at the same spot.

CULTIVATE CONVERSATION • Information Gap

Ask students questions to help them decide what missing information they need to answer the question, "What is the length of the cable?"

- Do you have enough information to find the length of the cable? No, I need more information about the triangle.
- What are some possible ways to find the length of the cable? I can use the Pythagorean Theorem, or I can use tangent ratio.
- Suppose you were going to use the tangent ratio. What information will be enough? I will need the lengths of the two legs or I need the length of a leg and the measure of an acute angle.

1 Spark Your Learning

▶ MOTIVATE

- Have students look at the photo in their books and read the information contained in the photo. Then complete Part A as a whole-class discussion.
- Give the class the additional information they need to solve the problem. This information is available online as a printable and projectable page in the Teacher Resources.
- Have students work in small groups to complete Parts B–D.

▶ PERSEVERE

If students need support, guide them by asking:

Q **Advancing • Use Tools** Which tool could you use to solve the problem? Why choose that tool and not some other? Students' choices of tools and reasons for choosing them will vary.

Q **Assessing** If you would like to use a trigonometric ratio, what information do you have? I have an angle measure and the length of the side adjacent an to the angle.

Q **Assessing** What do you need to find out? I need to find the length of the hypotenuse in the right triangle.

Q **Advancing** Does it matter which side and which angle is given (other than the right angle) in order to apply the tangent ratio? Possible answer: I can always find the measure of the other acute angle in a right triangle if the measure of one acute angle is given, but I would need the length of a leg in order apply the tangent ratio.

Turn and Talk Direct students to draw diagrams to better visualize the situations. If the cable is connected at a higher point, it would be longer. If the cable is connected at a lower point, it would be shorter.

▶ BUILD SHARED UNDERSTANDING

Select groups of students who used various strategies and tools to share with the class how they solved the problem. As they present their solutions, have each group discuss why they chose a specific strategy and tool.

② Learn Together

Build Understanding

Task 1 (MP) **Construct Arguments** Students explore ratios of sides in right triangles and reason about the ratios of similar triangles.

Sample Guided Discussion:

Q **What happens to the measure of the sides and the angles of $\triangle AED$ as you move $\overleftrightarrow{ED}$ to the left?**
The length of $\overline{ED}$ becomes shorter, as well as the lengths of the other two sides of the triangle. The angle measures do not change.

Q **In Part E, how do you know you are creating similar triangles as you are moving $\overleftrightarrow{ED}$ to the left?**
As I move $\overleftrightarrow{ED}$, it is still perpendicular to the base of the triangle. It is like two parallel lines cut by two transversals to form two triangles with a shared angle. All angles have the same angle measures, and corresponding side lengths are proportional.

Turn and Talk For students who have difficulty visualizing the situation, have them construct another ray starting at A so that the angle is larger. If $m\angle A$ is increased, ED and AE will increase, while AD will remain the same. The ratio $\frac{ED}{AE}$ increases, and the ratio $\frac{AD}{AE}$ decreases.

Build Understanding

Investigate Ratios in a Right Triangle

Previously, you investigated the ratio between the sides opposite from and adjacent to a given acute angle in a right triangle. There are special ratios between the hypotenuse and the side adjacent to the referenced angle and between the hypotenuse and the side opposite from the referenced angle. The sides in the figure are labeled in relation to $\angle A$.

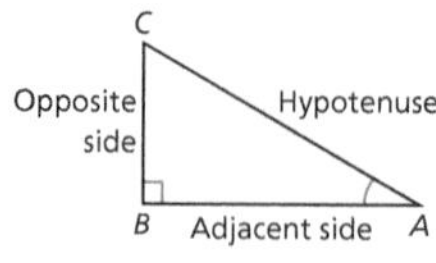

1 Use a geometric drawing tool to construct the figure by following these steps.

Step 1: Draw horizontal ray $\overrightarrow{AB}$.

Step 2: Draw $\overrightarrow{AC}$ so that $\angle A$ is acute.

Step 3: Draw line $\overleftrightarrow{ED}$ so that it is perpendicular to $\overrightarrow{AB}$.

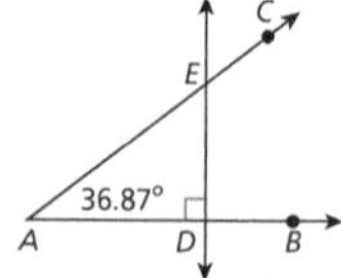

A. Measure each side of $\triangle AED$. What are the lengths of both legs and the hypotenuse? Possible answer: $DE = 1.8$, $AD = 2.4$, $AE = 3$

B. Calculate the ratio of each leg to the hypotenuse given by $\frac{ED}{AE}$ and $\frac{AD}{AE}$. Round to the nearest tenth. The ratio of $\frac{ED}{AE} \approx 0.6$. The ratio of $\frac{AD}{AE} \approx 0.8$.

C. If you slide $\overleftrightarrow{ED}$ horizontally along $\overrightarrow{AB}$, what values within $\triangle AED$ change and what values stay constant? How can you prove the different triangles that are created are similar to each other? C. $m\angle A$ remains constant and the side lengths change; AA Similarity Theorem

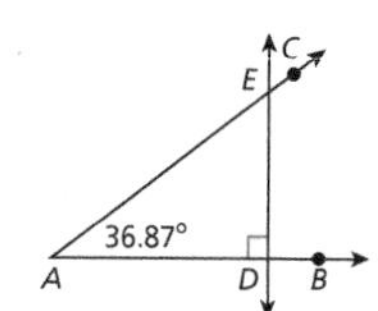

D. As you slide $\overleftrightarrow{ED}$ horizontally along $\overrightarrow{AB}$, what happens to the values of $\frac{ED}{AE}$ and $\frac{AD}{AE}$? The ratios should remain approximately the same.

E. Explain your results from Part D using what you know about similar triangles. Do you think the same results would apply to any similar triangle? Why or why not? See Additional Answers.

Turn and Talk For $\triangle AED$, predict how your answers would change if $m\angle A$ is increased. See margin.

414

LEVELED QUESTIONS

Depth of Knowledge (DOK)	Leveled Questions	What Does This Tell You?
Level 1 **Recall**	How do you check whether two triangles are similar? I will measure their angles and sides to make sure the angles are congruent and the corresponding side lengths are proportional.	Students' answers will indicate whether they understand the meaning of similar triangles.
Level 2 **Basic Application of Skills & Concepts**	How can you use a geometric drawing tool to show that the two triangles obtained by moving $\overleftrightarrow{ED}$ are similar? I can measure the angles and the side lengths to make sure the angles are congruent and the corresponding side lengths are proportional.	Students' answers will demonstrate that they can apply their understanding of the meaning of similar triangles.
Level 3 **Strategic Thinking & Complex Reasoning**	What necessary condition should be preserved when you move $\overleftrightarrow{ED}$ in order for the two triangles to be similar? The slide should still preserve the right angle formed between $\overleftrightarrow{ED}$ and the base.	Students' answers will demonstrate that they can reason about the conditions that need to be fulfilled for two triangles to be similar.

Find the Sine and Cosine of an Angle

The ratios you calculated in Task 1 are called the sine and cosine ratios. The sine ratio is abbreviated as sin and read as "the sine of angle ...". The cosine ratio is abbreviated as cos and read as "the cosine of angle ...".

Trigonometric Ratios

A **trigonometric ratio** is a ratio of two sides of a right triangle. You have already studied the tangent ratio. The two additional trigonometric ratios, sine and cosine, involve the hypotenuse of a right triangle.

sine	$\sin A = \frac{\text{length of leg opposite } \angle A}{\text{length of hypotenuse}} = \frac{BC}{AB}$	$\sin A = \frac{a}{c}$	
cosine	$\cos A = \frac{\text{length of leg adjacent } \angle A}{\text{length of hypotenuse}} = \frac{AC}{AB}$	$\cos A = \frac{b}{c}$	

2 The figure represents a post $(\overline{IE})$ braced by four diagonal parallel supports.

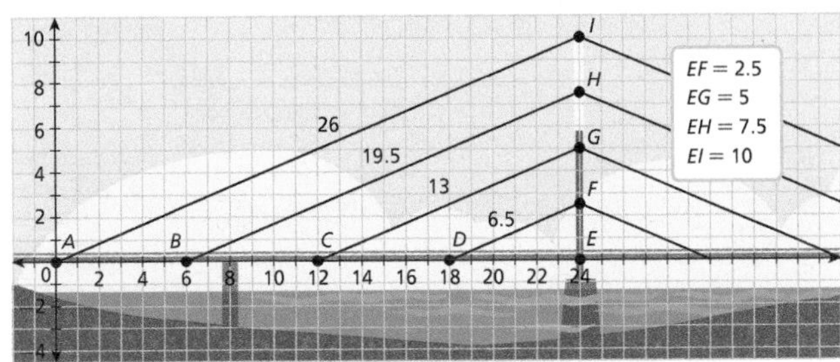

A. In $\triangle DEF$ and $\triangle CEG$, calculate the sine and cosine for $\angle D$, $\angle F$, $\angle C$, and $\angle G$. If necessary, round to the nearest thousandth. See Additional Answers.

B. What do you notice about the sines and cosines you found?
The sine and cosine ratios of corresponding angles are the same.

C. What relationship exists between the sine and cosine ratios of a pair of complementary angles in a right triangle—say $\angle D$ and $\angle F$ or $\angle C$ and $\angle G$? Is this relationship true for any pair of acute angles in a right triangle? Explain.
See Additional Answers.

D. Verify your results from Parts A–C using different pairs of triangles such as $\triangle AEI$ and $\triangle BEH$. Check student's calculations.

Turn and Talk In a right triangle, when one acute angle increases while the adjacent leg does not change, predict how the values change for each of the following: See margin.

- The length of the opposite leg as well as the measure of the complementary angle
- The sine and cosine of the angle as well as the sine and cosine of the complementary angle

Task 2 (MP) **Use Structure** Students learn about the sine and the cosine of an acute angle in a right triangle as ratios. They recall the concept of similarity to show that the ratio for any given angle measure is constant.

CONNECT TO VOCABULARY

Have students use the **Interactive Glossary** to record their understanding of the vocabulary in this task.

Sample Guided Discussion:

Q Look closely at the picture. Can you make a prediction about the sine and cosine ratios? They will be the same because the triangles are similar triangles.

Q In Part D, does it matter which angles you pick? No, the triangles are similar. Therefore, the angles are congruent, so the sine and cosine ratios of the corresponding angles are going to be the same.

Turn and Talk Remind students that in a right triangle, one of the angles is a right angle and the other two angles are acute and complementary. Possible answer: When an acute angle increases while the adjacent side does not change, the opposite side also increases. The measure of the complementary angle will decrease. The cosine ratio will decrease because the denominator of the ratio increases, and the sine ratio will increase. The opposite is true with the sine and cosine ratios of the complementary angle.

Step It Out

Task 3  **Attend to Precision** Students use precise language to justify steps in finding unknown sides and perimeter of a triangle using trigonometric ratios.

Sample Guided Discussion:

Q How do you know whether the side given is the opposite or the adjacent? You look at the angle given. The side across the angle is the opposite. The side attached to the angle is the adjacent.

Q How do you know which side in a right triangle is labeled "c" in the Pythagorean Theorem? The side labeled "c" is the hypotenuse, the side opposite the right angle.

Turn and Talk Point out to students that the question is asked about any angle out of three angles in the right triangle. no; You must be given the measure of either of the two acute angles and any side length. If you were given only the 90° angle and any side length, you would not have enough information to solve the triangle.

Step It Out

Find Side Lengths and Perimeter Using Sine and Cosine

You can use the sine and cosine ratios to find unknown values in a right triangle.

 Find the perimeter of $\triangle ABC$.

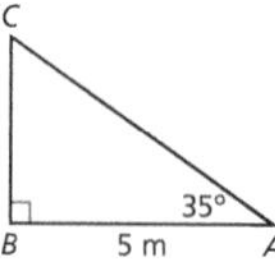

Start by finding the unknown side lengths in the triangle.

Given: $m\angle A$ and its adjacent side length.

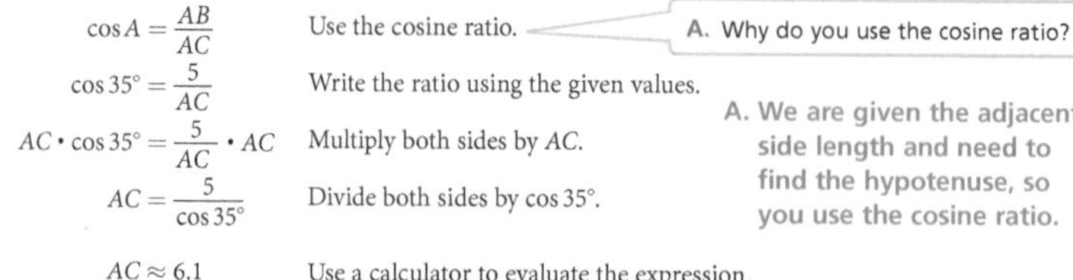

Find AC.

$\cos A = \frac{AB}{AC}$	Use the cosine ratio.
$\cos 35° = \frac{5}{AC}$	Write the ratio using the given values.
$AC \cdot \cos 35° = \frac{5}{AC} \cdot AC$	Multiply both sides by AC.
$AC = \frac{5}{\cos 35°}$	Divide both sides by cos 35°.
$AC \approx 6.1$	Use a calculator to evaluate the expression. Be sure your calculator is in degree mode.

A. Why do you use the cosine ratio?

A. We are given the adjacent side length and need to find the hypotenuse, so you use the cosine ratio.

Find BC.

$AB^2 + BC^2 = AC^2$	Use the Pythagorean Theorem.
$5^2 + BC^2 = 6.1^2$	Substitute.
$25 + BC^2 = 37.21$	Simplify.
$BC^2 = 12.21$	Subtract 25 from both sides.
$BC \approx 3.5$	Take the square root. Round to the nearest tenth.

B. Why can we use the Pythagorean Theorem to find the missing leg?

B. You have two known lengths of a right triangle, so you can solve for the third side.

Calculate the perimeter of $\triangle ABC$.

Perimeter of $\triangle ABC = AB + BC + AC$	Write the perimeter equation.
$= 5 + 3.5 + 6.1$	Substitute the known information.
$= 14.6$	Simplify.

C. It is an approximate value because we rounded $\sqrt{12.21}$ earlier in the solution.

Answer the question.

The perimeter of $\triangle ABC$ is approximately 14.6 m.

C. Why is this an approximate value?

 **Turn and Talk** Can you solve for every side and angle in a right triangle when given any one angle and any one side length? Why or why not? See margin.

416

PROFICIENCY LEVEL

Beginning

Provide students with different right triangles. Have them write the sine and cosine ratios. Encourage them to say the terms: *sine, cosine, opposite, adjacent,* and *hypotenuse* as they write the ratios.

Intermediate

Provide students with different right triangles where a measurement of an angle (other than the right angle) and a side length are provided. Have students explain which trigonometric ratio they would use and why in a short sentence.

Advanced

Provide students with different right triangles where a measurement of an angle (other than the right angle) and a side length are provided. Have students explain and write how they would find the unknown sides and angles.

Find an Angle Measure Using Inverse Sine and Inverse Cosine

You can use the **inverse sine** and the **inverse cosine** to find an unknown angle when given the side lengths of a right triangle. Use a calculator to evaluate the inverse sine, written as $\sin^{-1}$, and inverse cosine, written as $\cos^{-1}$.

inverse sine	$\sin A = \frac{BC}{AB} = \frac{a}{c}$	$\sin^{-1}\left(\frac{a}{c}\right) = m\angle A$	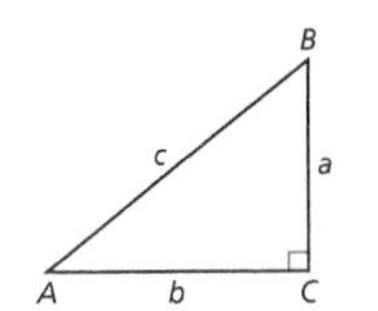
inverse cosine	$\cos A = \frac{AC}{AB} = \frac{b}{c}$	$\cos^{-1}\left(\frac{b}{c}\right) = m\angle A$	

4 A hot air balloon is tethered with cables that run from both sides of the basket to the ground. The tethers allow riders to experience a hot air balloon ride only up to a predetermined height. When the balloon is at the maximum height, what is $m\angle B$? Round to the nearest tenth of a degree.

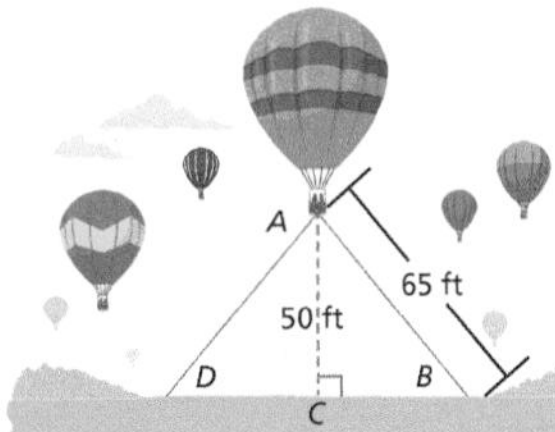

Write a trigonometric ratio for $\angle B$.

We are given the opposite side length and the hypotenuse. Use the sine ratio.

$\sin B = \frac{50}{65}$

A. Why do you use the sine ratio in this situation?

A. We are given the opposite side length and the hypotenuse length, which matches the values needed for sine.

Write the inverse sine ratio.

$\sin B = \frac{50}{65}$

$\sin^{-1}\left(\frac{50}{65}\right) = m\angle B$

B. What does the "−1" represent?

B. It represents the inverse of a trigonometric ratio to find the angle measure.

Use a calculator to solve.

$m\angle B = \sin^{-1}\left(\frac{50}{65}\right)$

$= 50.3°$

C. Possible answer: yes; I know from the Pythagorean Theorem that the height is slightly longer than the base, so the angle opposite the height should be slightly larger than 45°.

Answer the question.

The tether makes a 50.3° angle with the ground.

C. Does the answer make sense in this context?

Turn and Talk Find one or both of the acute angles of a right triangle in a real-world problem given the length of the hypotenuse and one leg length. See margin.

Task 4 (MP) **Attend to Precision** Students use precise language to justify steps in finding angle measures in right triangles using the definitions of sine and cosine ratios.

CONNECT TO VOCABULARY

Have students use the **Interactive Glossary** to record their understanding of the vocabulary in this task.

OPTIMIZE OUTPUT Critique, Correct, and Clarify

Have students work with a partner to discuss which of these solutions is correct. Encourage students to use correct vocabulary terms in these discussions. Students should revise their answers if necessary after talking about the problem with their partner.

Find the measure of $\angle A$.

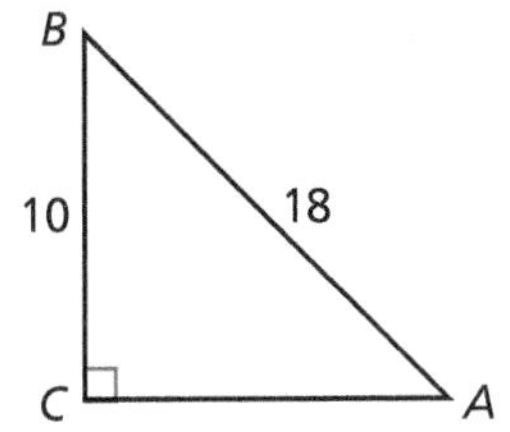

$\sin \angle A = \frac{10}{18}$	$\cos \angle A = \frac{10}{18}$
$\sin \angle A \approx 0.56$	$\cos \angle A \approx 0.56$
$\sin^{-1}(0.56) \approx 33.4°$	$\cos^{-1}(0.56) \approx 56.6°$
$m\angle A \approx 33.4°$	$m\angle A \approx 56.6°$

Sample Guided Discussion:

Q How would you know to use the inverse sine and not sine? You use the inverse sine when the hypotenuse and the opposite side measures are given and you are looking for the measure of the angle.

Q In Part C, how would you know if the answer does not make sense? if the measure of the angle is 90° or greater

Turn and Talk Remind students to first write the trigonometric ratio to make sure they are finding the measure of the corresponding angles. Check students' work.

ONLINE

Assign the Digital On Your Own for

- built-in student supports
- Actionable Item Reports
- Standards Analysis Reports

On Your Own

Assignment Guide

The chart below indicates which problems in the On Your Own are associated with each task in the Learn Together. Assign daily homework for tasks completed.

Learn Together Tasks	On Your Own Problems
Task 1, p. 414	Problems 6 and 7
Task 2, p. 415	Problems 8–13 and 27
Task 3, p. 416	Problems 14–17, 24, and 25–26
Task 4, p. 417	Problems 18–23

ANSWER

3. $\sin \angle A = \frac{18}{30} = \frac{3}{5} = 0.6; \cos \angle A = \frac{24}{30} = \frac{4}{5} = 0.8$

Check Understanding

1. If two right triangles are similar, what is the relationship between the ratios of the lengths of the opposite side to the length of the hypotenuse for any two corresponding acute angles? The ratios are equal.

2. In $\triangle XYZ$, $\angle Y$ is a right angle. What is the relationship between $\angle X$ and $\angle Z$? What is the relationship between $\sin X$ and $\cos Z$? $\angle X$ and $\angle Z$ are complementary; $\sin X = \cos Z$

3. Find the sine and cosine of $\angle A$. See Additional Answers.

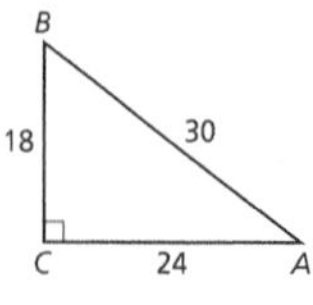

4. What is the length of $\overline{QR}$? Round to the nearest hundredth. 7.45

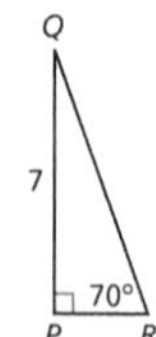

5. An 8-foot long ramp is placed so the end of the ramp is 5 feet from the base of a building. What angle does the ramp make with the ground? Round to the nearest tenth of a degree. 51.3°

On Your Own

Determine whether each pair of right triangles is similar. Justify your reasoning.

6. $\frac{AC}{AB} = \frac{2}{3}, \frac{QR}{PR} = \frac{1.5}{2.5}$

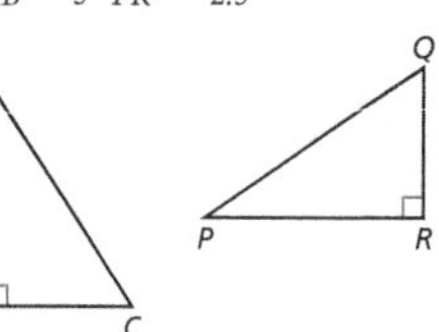

6, 7. See Additional Answers.

7. $\frac{AB}{AC} = \frac{5}{12}, \frac{QR}{PR} = \frac{6}{14.4}$

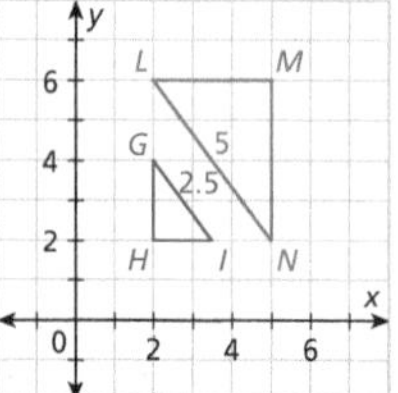

For Problems 8–13, the graph shows image $\triangle LMN$ which is a rotation and dilation of $\triangle GHI$. Write each pair of trigonometric ratios as a fraction and as a decimal.

8–13. See Additional Answers.

8. $\sin G, \cos G$
9. $\sin I, \cos I$
10. $\sin L, \cos L$
11. $\sin N, \cos N$
12. In the figure, how can you prove that the triangles are similar?
13. What is the trigonometric relationship between the complementary angles within each triangle?

418

data checkpoint

3 Check Understanding

Formative Assessment

Use formative assessment to determine if your students are successful with this lesson's learning objective.

Students who successfully complete the Check Understanding can continue to the On Your Own practice.

For students who miss 1 problem or more, work in a pulled small group using the Almost There small-group activity on page 413C.

ONLINE

Assign the Digital Check Understanding to determine

- success with the learning objective
- items to review
- grouping and differentiation resources

4 Differentiation Options

Differentiate instruction for all students using small-group activities and math center activities on page 413C.

Reteach

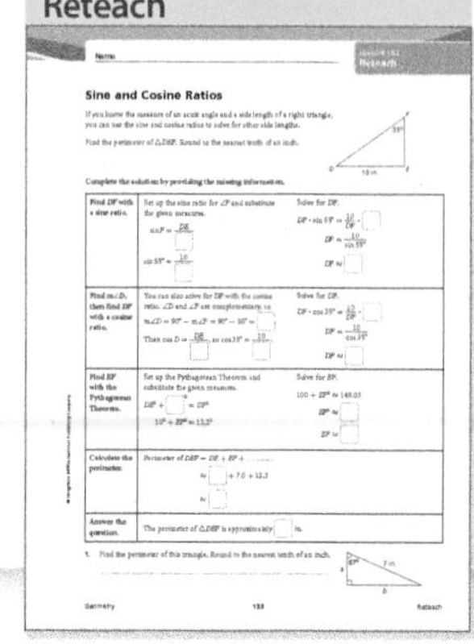

Challenge

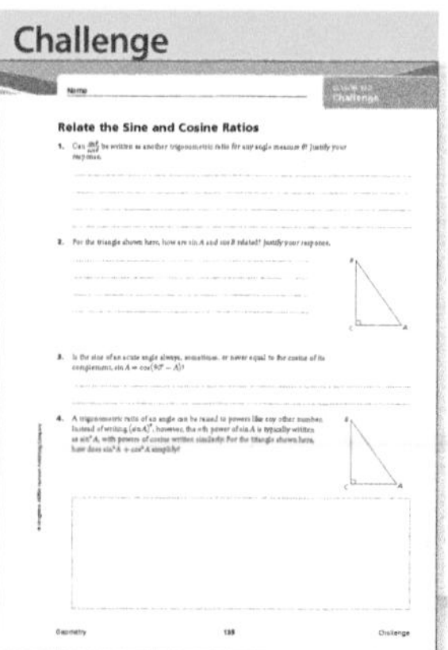

Find each side length. Round to the nearest tenth.

14. *LM* 3.6 in.

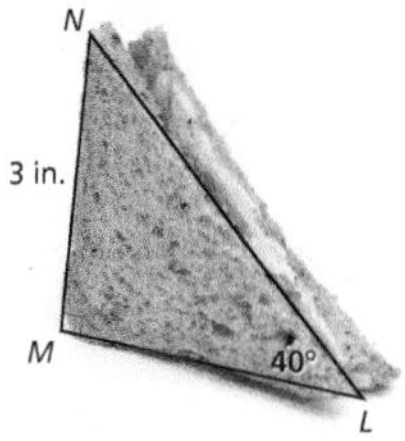

15. *RS* 2.8 in.

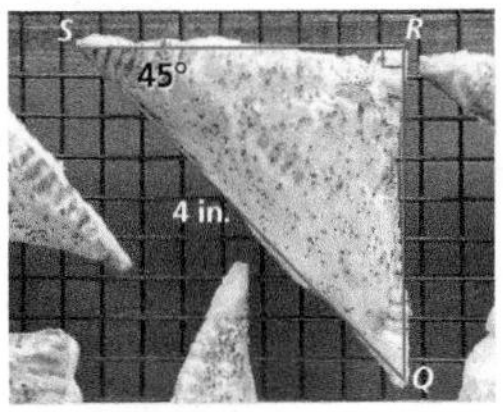

16. *GH* 2.1 in.

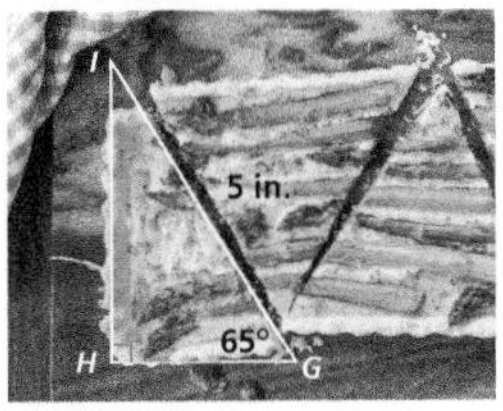

17. *PR* 20.9 in.

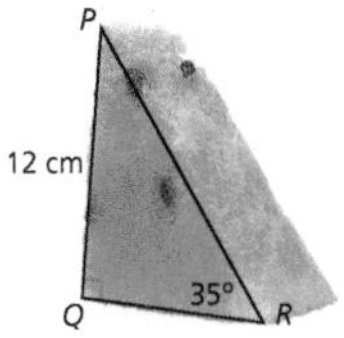

For each triangle, use a calculator to find the measure of the given angle. Round to the nearest tenth of a degree.

18. $\angle K$ $m\angle K \approx 41.8°$

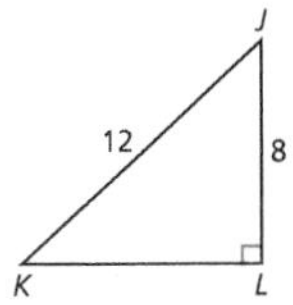

19. $\angle B$ $m\angle B \approx 53.1°$

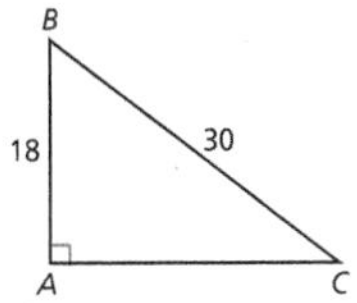

20. $\angle P$ $m\angle P \approx 44.4°$

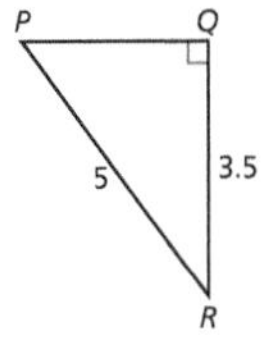

21. $\angle X$ $m\angle X \approx 66.4°$

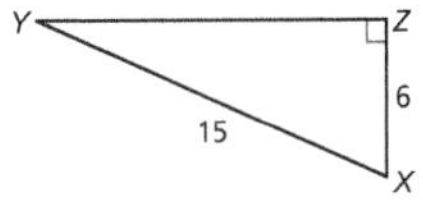

22. $\angle M$ $m\angle M \approx 45.3°$

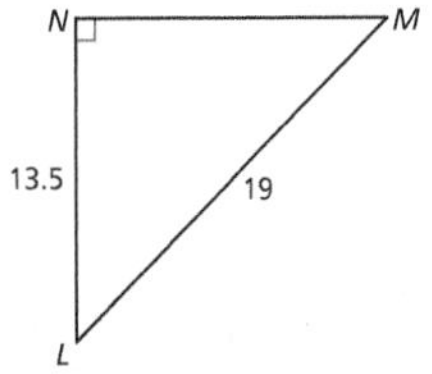

23. $\angle V$ $m\angle V \approx 33.6°$

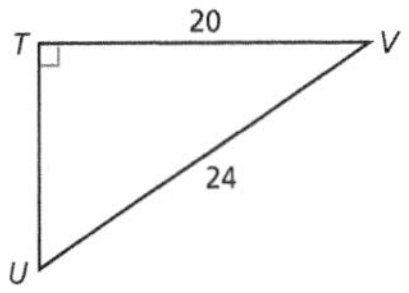

24. (MP) **Reason** Emerson wants to draw a right triangle with acute angles $\angle 1$ and $\angle 2$, where the sine of $\angle 1$ is equal to the cosine of $\angle 2$. How does this constrain the possible measures of the angles in the triangle? See margin.

Questioning Strategies

Problem 24 Let $x°$ be the measure of $\angle 1$. How would you express the measure of $\angle 2$ in terms of x? The measure of $\angle 2$ will be $(90 - x)°$.

Watch for Common Errors

Problem 12 Point out to students that oftentimes triangles are rotated and it might be difficult to figure out the corresponding proportional sides. Suggest to students that they redraw triangles to have the same orientation.

ANSWERS

24. One angle will be 90° so that the triangle is a right triangle. For any pair of acute angles in a right triangle, the sine of one is equal to the cosine of the other, so he can draw any right triangle.

⑤ Wrap-Up

Summarize learning with your class. Consider using the Exit Ticket, Put It in Writing, or I Can scale.

Exit Ticket 

The base of a ski ramp is 12 inches long. The ramp itself is 15 inches long. What is the measure of the angle between the base and the ramp?

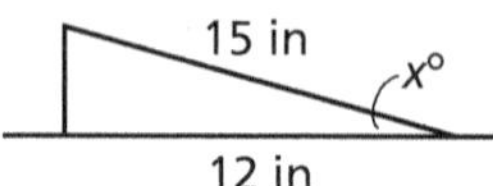

$\cos x° = \frac{12}{15}$; $\cos x° = 0.8$;
$\cos^{-1}(0.8) \approx 36.87°$; $x° \approx 36.87°$

Put It in Writing 

Have students work in pairs. First, each student writes two problems: 1) where cosine needs to be used and 2) where inverse cosine needs to be used. Then, students exchange problems, solve, and discuss their solutions.

I Can 

The scale below can help you and your students understand their progress on a learning goal.

4	I can use sine and cosine ratios and the inverses to find side lengths and angle measures in right triangles, and I can explain my reasoning to others.
3	I can use sine and cosine ratios and the inverses to find side lengths and angle measures in right triangles.
2	I can use sine and cosine ratios to find side lengths in right triangles.
1	I can perform some of the steps needed to find side lengths in right triangles.

Spiral Review • Assessment Readiness

These questions will help determine if students have retained information taught in the past and can also prepare them for high-stakes assessments. Here, students use expressions to represent segments on the basis of similar triangles **(12.3)**, calculate a geometric mean **(12.4)**, use tangent ratios **(13.1)** and use trigonometric ratios to solve real-life problems **(13.3)**.

25. (MP) **Attend to Precision** A farmer is building a set of new corrals. Corrals 1 and 2 share a side that is perpendicular to the barn. The barn serves as one side of each corral.

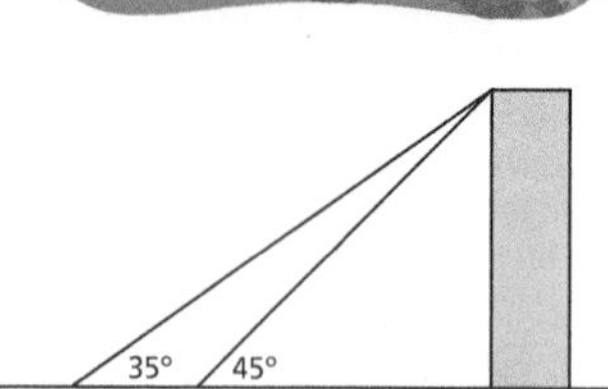

A. To the nearest foot, about how many feet of fence material does the farmer need to build both corrals? 97 ft

B. What is the approximate total area of both corrals? about 760 ft²

26. STEM In Tel Aviv, Israel, it took about a year to build a tower using construction toys. Andrea uses a laser measuring device to find the distance to the top of the tower at an angle of about 35°. If she moves forward so the angle is 45°, the length of the laser beam is reduced by 30 ft.

A. What is the length of each laser beam? Round to the nearest foot. about 159 ft; about 129 ft

B. To the nearest foot, what is the height of the tower? about 91 ft

35° 45°

27. (Open Middle™) Using the digits 1 to 9, at most one time each, fill in the boxes to create two true statements. See Additional Answers.

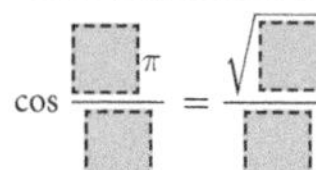

Spiral Review • Assessment Readiness

28. Which expression(s) represent the length of $\overline{AD}$? Select all that apply.

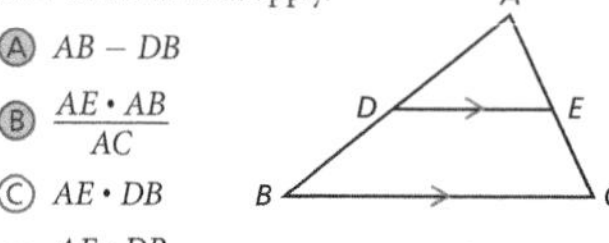

Ⓐ $AB - DB$
Ⓑ $\frac{AE \cdot AB}{AC}$
Ⓒ $AE \cdot DB$
Ⓓ $\frac{AE \cdot DB}{EC}$

29. What is the geometric mean of 5 and 45?

Ⓐ 15
Ⓑ 25
Ⓒ 40
Ⓓ 50

30. If the tangent of $\angle A$ is $\frac{10}{24}$, and the side adjacent to $\angle A$ is 36 cm, what is the length of the side opposite $\angle A$?

Ⓐ 5 cm
Ⓑ 12 cm
Ⓒ 15 cm
Ⓓ 18 cm

31. A skateboard ramp reaches a height of 2 feet and has a horizontal length of 4 feet. What is the angle the ramp makes with the ground?

Ⓐ 63.4°
Ⓑ 26.6°
Ⓒ 30°
Ⓓ 60°

 **I'm in a Learning Mindset!**

How am I prioritizing my learning? What do I do each day to prepare to learn tomorrow?

Keep Going ▶ Journal and Practice Workbook

Learning Mindset

Perseverance Learns Effectively

Point out to students that to learn effectively, they might "box" the important ideas in the lesson or write them on an index card so they can easily refer to the ideas. *What are some of the key ideas in today's lesson? (You may wish to record them on your index card.) Would sketches make any of the key ideas easier to retain?*

13.3 Special Right Triangles

LESSON FOCUS AND COHERENCE

Mathematics Standards

- Understand that by similarity, side ratios in right triangles are properties of the angles in the triangle, leading to definitions of trigonometric ratios for acute angles.
- Explain and use the relationship between the sine and cosine of complementary angles.
- Use trigonometric ratios and the Pythagorean Theorem to solve right triangles in applied problems.

Mathematical Practices and Processes (MP)

- Look for and make use of structure.
- Use appropriate tools strategically.
- Attend to precision.

I Can Objective 

I can use trigonometric ratios and the Pythagorean Theorem to find the side lengths and angle measures of special right triangles.

Learning Objective

Examine and use trigonometric ratios for special right triangles, and use the Pythagorean Theorem to find the side lengths and angle measures of special right triangles.

Language Objective

Explain how to find the ratio of the sides of special right triangles.

Vocabulary

New: special right triangles

Lesson Materials: compass, geometric drawing tool, protractor, ruler

Mathematical Progressions

Prior Learning	Current Development	Future Connections
Students: • proved theorems about triangles. **(9.1)** • explained the similarity of triangles as the equality of all corresponding pairs of angles and the proportionality of all corresponding pairs of sides. **(12.2)** • used similarity criteria for right triangles to solve problems and to prove relationships. **(12.4)**	**Students:** • examine and use trigonometric ratios to find side lengths of special right triangles. • use the Pythagorean Theorem to find the side lengths and angle measures of special right triangles. • use special right triangles to solve real-world problems.	**Students:** • will use trigonometric ratios and the Pythagorean Theorem to solve right triangles in applied problems. **(13.4)**

UNPACKING MATH STANDARDS

Use trigonometric ratios and the Pythagorean Theorem to solve right triangles in applied problems.

What It Means to You

Students demonstrate an understanding of using trigonometric ratios and the Pythagorean Theorem to solve special right triangles and find and use their trigonometric ratios in applied problems. This understanding develops from the previous two lessons, where they were introduced to sine, cosine, and tangent and applied the functions to real-world problems.

In the next lesson, students will use trigonometric ratios, an area formula for a triangle and the Pythagorean Theorem to solve right triangles in applied problems. In the next module, students will use trigonometric ratios and the Pythagorean Theorem to develop the Law of Sines and the Law of Cosines.

WARM-UP OPTIONS

ONLINE

PROJECTABLE & PRINTABLE 

ACTIVATE PRIOR KNOWLEDGE • Similarity and Right Triangles

Use these activities to quickly assess and activate prior knowledge as needed.

Problem of the Day 

Is the set of values 9, 40, and 41 a Pythagorean Triple? Explain.

yes; $9^2 + 40^2 = 81 + 1600 = 1681 = 41^2$

Quick Check for Homework

As part of your daily routine, you may want to display the Teacher Solution Key to have students check their homework.

Make Connections

Based on students' responses to the Problem of the Day, choose one of the following:

1. Project the Interactive Reteach, Geometry, Lesson 12.4.

2. Complete the Prerequisite Skills Activity:

Provide student pairs with the length of one leg of a right triangle. Have them find a second leg and hypotenuse length that would make a Pythagorean Triple.

- *What is the Pythagorean Theorem?* The theorem that describes the relationship between sides of a right triangle as $a^2 + b^2 = c^2$.
- *What is a Pythagorean Triple?* A set of three nonzero whole numbers that satisfy the Pythagorean Theorem.
- *How can you determine two other lengths of the triangle that would result in a Pythagorean Triple?* Choose a value for c that is greater than a, then subtract the squares of a and c to find the value of b^2. Use this method of guess and check to find three values that work with the Pythagorean Theorem.

If students continue to struggle, use Tier 2 Skill 4.

SHARPEN SKILLS

If time permits, use this on-level activity to build fluency and practice basic skills.

Mental Math

Objective: Students determine the missing value of a Pythagorean Triple.
Materials: index cards

Have students work in small groups. Give each group of students a set of index cards. Each card shows two of the three values of a Pythagorean Triple, a, b, c, where a and b are the lengths of the legs and c is the length of the hypotenuse of a right triangle. Have students take turns drawing a card and mentally determining the missing value.

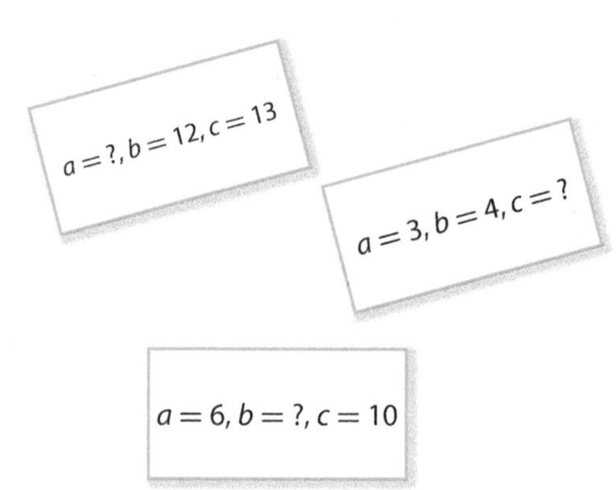

PLAN FOR DIFFERENTIATED INSTRUCTION

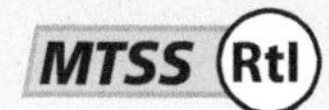

Small-Group Options

Use these teacher-guided activities with pulled small groups.

On Track

Materials: index cards

Create a set of cards with three numbers, most of which are sides of 45°-45°-90° and 30°-60°-90° triangles. Include some non-special Pythagorean Triples and others that include lengths that cannot form triangles.

Divide students into groups. Give each group a set of cards and select one card at a time. Each member of the group classifies the three numbers as forming a special right triangle, a non-special right triangle, a triangle that is not right, or no triangle. The other members of the group checks each member's answer.

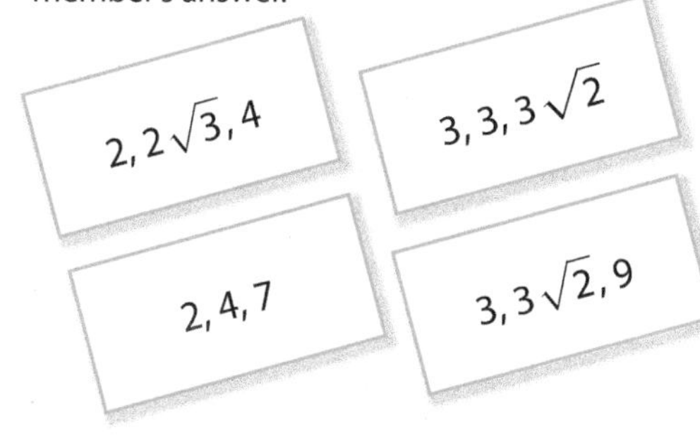

Almost There

Materials: ruler and triangles on a coordinate grid

Provide copies of the following triangles for each student.

Scale: 0.1 □ 0.1

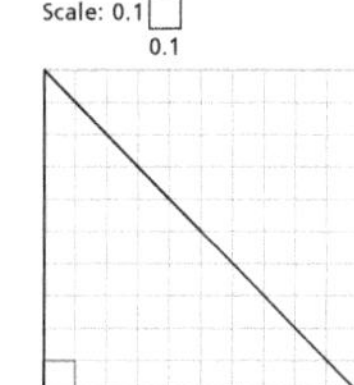

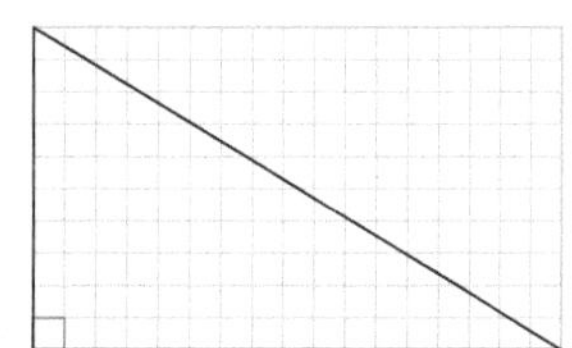

- Explain that the graphs are special right triangles like those in the lesson. Have students write the measure of each angle and label.
- Have students write the length of each side in units by either counting horizontally or vertically or by using a ruler. Remind them to use the scale.
- Have students identify the type of special right triangles by angle measure and the ratio of the sides under the triangle.

Ready for More

Materials: index cards

Divide students into groups of 4. Give each group blank index cards. Have each group do the following:

- On one side of the card, each pair within the group draws a right triangle (to scale) using a Pythagorean Triple, with the side lengths labeled.
- On the other side, write the sine, cosine, and tangent of one of the angles in the following form:

  $\sin A = \frac{3}{5}$ $\quad \sin B = \frac{4}{5}$

  $\cos A = \frac{4}{5}$ $\quad \cos B = \frac{3}{5}$

  $\tan A = \frac{3}{4}$ $\quad \tan B = \frac{4}{3}$
- Exchange cards with the other pair in the group and compare and verify answers.

Cards may be used in other class activities.

Math Center Options

Use these student self-directed activities at centers or stations. **Key: ● Print Resources ● Online Resources**

On Track

- ● Interactive Digital Lesson
- ●● Journal and Practice Workbook
- ● Interactive Glossary (printable): **special right triangles**
- ● Module Performance Task

Almost There

- ● Reteach 13.3 (printable)
- ● Interactive Reteach 13.3
- ● RtI Tier 2 Skill 4: The Pythagorean Theorem and Its Converse
- ● Illustrative Mathematics: Finding the Area of an Equilateral Triangle

Ready for More

- ● Challenge 13.3 (printable)
- ● Interactive Challenge 13.3
- ● Illustrative Mathematics: Constructing Special Angles
- ● Desmos: Pythagorean Triples and Similar Triangles

ONLINE 

View data-driven grouping recommendations and assign differentiation resources.

During the *Spark Your Learning*, listen and watch for strategies students use. See samples of student work on this page.

Use the Pythagorean Theorem

Strategy 1

The height is also 3 feet because this is an isosceles triangle. I can use the Pythagorean Theorem to find the length of the hypotenuse.

$3^2 + 3^2 = c^2$

$9 + 9 = c^2$

$18 = c^2$

$\sqrt{18} = c$

$3\sqrt{2} = c$

The length of the hypotenuse is $3\sqrt{2}$ feet.

If students . . . use the Pythagorean Theorem to solve the problem, they are employing an efficient method and demonstrating an exemplary understanding using the Pythagorean Theorem to solve problems from Grade 8.

Have these students . . . explain how they decided to use the Pythagorean Theorem and how they answered the question. **Ask:**

Q Why did you use the Pythagorean Theorem to find the window dimensions?

Q How did you apply the Pythagorean Theorem to find the window lengths?

Break into Smaller Parts

Strategy 2

I used grid paper and the Pythagorean Theorem to find the window lengths. I made each square 1 square foot. So, the windows on the congruent sides are 3 feet each in length. To find the length of the diagonal window, I broke the diagonal up into the diagonals of 3 smaller triangles. I used the Pythagorean Theorem to find that the hypotenuse of each triangle was $\sqrt{2}$ feet, so the 3 diagonals total $3\sqrt{2}$ feet.

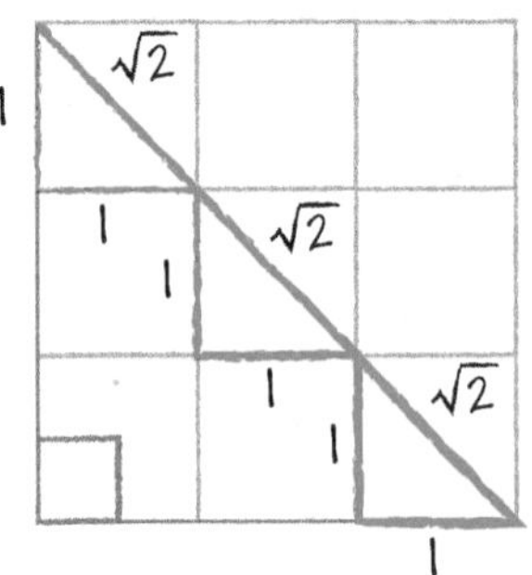

If students . . . used grid paper to solve the problem, they understand how to model the problem and how to apply the Pythagorean Theorem to the graph, but they may not understand that they could apply the Pythagorean Theorem to the window.

Activate prior knowledge . . . by having students apply the Pythagorean Theorem to the window. **Ask:**

Q What is the length of each of the congruent sides?

Q Can you apply the Pythagorean Theorem to the window?

COMMON ERROR: Finds the Perimeter

The lengths of the sides are 3 feet, 3 feet, and $3\sqrt{2}$ feet. So, the length of the window is $6 + 3\sqrt{2}$ feet.

If students . . . find the perimeter, then they do not understand that the question is asking them to find the individual lengths of the window.

Then intervene . . . by pointing out that they need to carefully read the question. **Ask:**

Q What does the question ask you to find?

Q What lengths does the window have? Describe them by name.

13.3

Special Right Triangles

I Can use trigonometric ratios and the Pythagorean Theorem to find the side lengths and angle measures of special right triangles.

Spark Your Learning

Triangles can be used to develop unique features in architectural design.

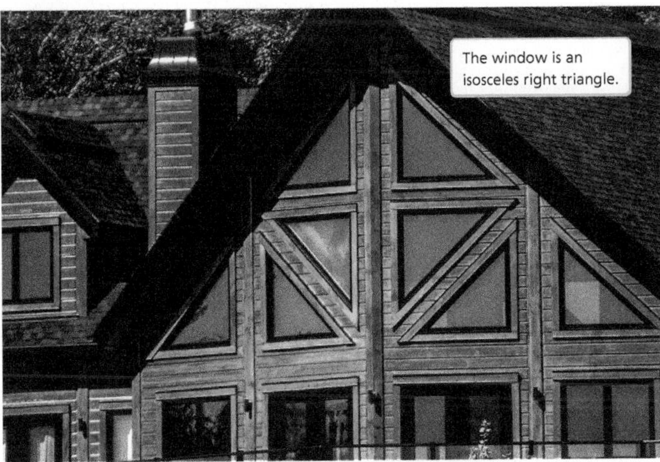
The window is an isosceles right triangle.

©Pascal Guay/Shutterstock

Complete Part A as a whole class. Then complete Parts B–D in small groups.

A. What is a mathematical question you can ask about this situation? What is the length of each side of the window?

B. What previous knowledge do you have that could help you answer the question? I know that the triangle is isosceles, so the leg lengths are the same.

C. To answer your question, what strategy and tool would you use along with all the information you have? What answer do you get? See Strategies 1 and 2 on the facing page.

D. Does your answer make sense in the context of the situation? What trigonometric methods can you use to check your answer? See Additional Answers.

Turn and Talk Suppose the window doubled in height but the angles of the triangle are to be kept the same. How would the ratio of the new base to the new height compare to the ratio of the original base to the original height? See margin.

CULTIVATE CONVERSATION • Co-Craft Questions

If students have difficulty formulating a mathematical question about the situation in the Spark Your Learning, ask them to consider what the shape of the window is. What are some natural questions to ask about this situation?

Work together to craft the following questions:

- What is the shape of the window?
- What are the six parts of a triangle?
- Which parts of the triangles do you know?
- Which parts of the triangles don't you know?

Then have students think about what additional information, if any, they would need to answer these questions. **Ask:**

- Can you determine the lengths of the sides? Why or why not?
- What information do you need to know to find the lengths of the sides of the window?

1 Spark Your Learning

▶ MOTIVATE

- Have students look at the photo in their books and read the information contained in the photo. Then complete Part A as a whole-class discussion.
- Give the class the additional information they need to solve the problem. This information is available online as a printable and projectable page in the Teacher Resources.
- Have students work in small groups to complete Parts B–D.

▶ PERSEVERE

If students need support, guide them by asking:

Q **Advancing • Use Tools** Which tool could you use to solve the problem? Why choose that tool and not some other? Students' choices of tools and reasons for choosing them will vary.

Q **Assessing** What properties does an isosceles triangle have? Possible answer: The legs are congruent, and the base angles are congruent.

Q **Assessing** What are the angle measures of an isosceles right triangle? 45°, 45°, and 90°

Q **Advancing** If the measure of the hypotenuse of the right isosceles triangle is 3 ft, rather than a leg, how could you find the length of the legs? Possible answer: Use the Pythagorean Theorem and the definition of an isosceles triangle. Because the legs are congruent, I can use the variable a for both. $a^2 + a^2 = 3^2$, $2a^2 = 9$, $a^2 = \frac{9}{2}$, so $a = \frac{3\sqrt{2}}{2}$.

Turn and Talk Have students recall that the window is an isosceles right triangle. Possible answer: If both the height and base doubled, the ratio would stay constant even though the values of the numerator and denominator of the ratio fraction doubles.

▶ BUILD SHARED UNDERSTANDING

Select groups of students who used various strategies and tools to share with the class how they solved the problem. As they present their solutions, have each group discuss why they chose a specific strategy and tool.

2 Learn Together

Build Understanding

Task 1 **MP Use Structure** Students use the structure of an isosceles triangle and a right triangle to find the ratio of length of the sides of a 45°-45°-90° triangle.

Sample Guided Discussion:

Q If the hypotenuse of an isosceles right triangle is 2 units, what is the length, in units, of a leg? $\sqrt{2}$

Task 2 **MP Use Tools** Students construct a 30°-60°-90° triangle with side lengths written as algebraic expressions.

Sample Guided Discussion:

Q What is the relationship between an isosceles triangle and its altitude to the base? The altitude to the base in an isosceles triangle lies on the perpendicular bisector of the base.

Turn and Talk The side lengths of a 30°-60°-90° triangle correspond to the proportion 1: $1\sqrt{3}$: 2. The lengths of the legs are 2 and $2\sqrt{3}$, and the length of the hypotenuse is 4.

Build Understanding

Investigate 45°-45°-90° Triangles

1 Discover relationships that always apply to an isosceles right triangle.

A. Construct an isosceles right triangle ABC as shown. Before you begin, what are the measurements of $\angle A$ and $\angle B$? How do you know without measuring? Explain.
See Additional Answers.

B. Label one leg length as x. What is the length of the other leg?
The length of the other leg is x.

C. How can you determine the length of the hypotenuse in terms of x? What is the length? Show your work. See Additional Answers.

D. What is the relationship between the side length and the hypotenuse length in an isosceles right triangle? Is this relationship the same in a 45°-45°-90° triangle? The ratio is $1:1\sqrt{2}$; Yes, because an isosceles right triangle is also a 45°-45°-90° triangle.

E. Given any one side of an isosceles right triangle, can you solve for the other two sides? Why or why not? Yes, even if the hypotenuse is a whole number, you can find the side lengths by dividing the hypotenuse length by $\sqrt{2}$.

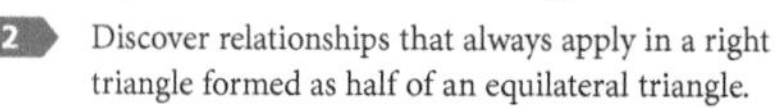

Investigate 30°-60°-90° Triangles

2 Discover relationships that always apply in a right triangle formed as half of an equilateral triangle.

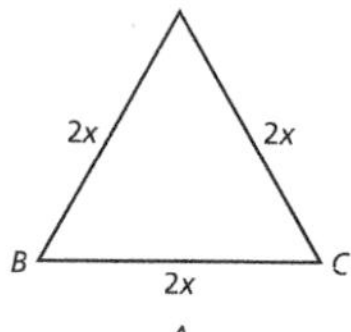

A. Construct an equilateral triangle ABC with side lengths of $2x$ as shown at the right. What are the measures of angles A, B, and C?
All three angles are 60°.

B. Draw an altitude $\overline{AD}$ from point A to $\overline{BC}$ and label the point of intersection D. How does this divide the triangle? Explain. It divides the triangle into two congruent right triangles.

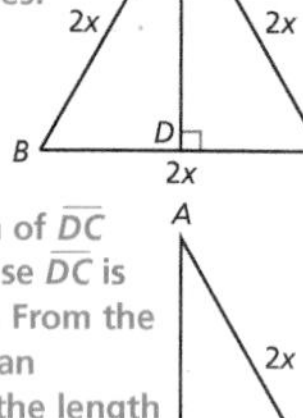

C. Examine one of the triangles created by the altitude as shown at the right. What are the three interior angles?
$m\angle DAC = 30°$, $m\angle ADC = 90°$, $m\angle C = 60°$

D. What is the length of base $\overline{DC}$? How can you find the length of $\overline{AD}$? Show your work. The length of $\overline{DC}$ is x, because $\overline{DC}$ is half of $\overline{BC}$. From the Pythagorean Theorem, the length of $\overline{AD} = x\sqrt{3}$.

E. From Part C, what is a three-term ratio ($f:g:h$, for example) for triangles given by the angle measures you found? Explain. 30°-60°-90° corresponds to $1:1\sqrt{3}:2$.

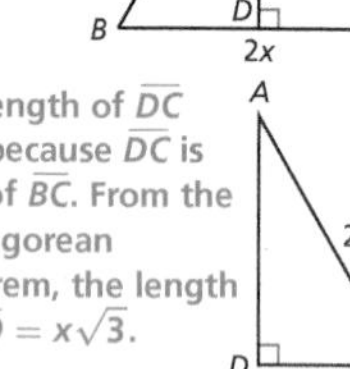

Turn and Talk If the side length of an equilateral triangle is 4, what is the length of each leg and the hypotenuse of one of the 30°-60°-90° triangles created by an altitude to the base? See margin.

422

LEVELED QUESTIONS

Depth of Knowledge (DOK)	Leveled Questions	What Does This Tell You?
Level 1 Recall	What is the Triangle Sum Theorem? The sum of the angle measures of a triangle is 180°.	Students' answers will indicate whether they know the sum of the angle measures of a triangle.
Level 2 Basic Application of Skills & Concepts	In an isosceles right triangle, how can you find the measures of the acute angles? Let $x°$ be the measure of an acute angle. The base angles are congruent and have equal measure. $x° + x° + 90° = 180°$, so solve to get $x = 45°$. Each acute angle measures 45°.	Students' answers will indicate whether they know how to apply the Isosceles Triangle Theorem and the Angle Sum Theorem.
Level 3 Strategic Thinking & Complex Reasoning	If the hypotenuse of an isosceles right triangle has a side that measures $8\sqrt{2}$, what are the possible measures of the other sides of the triangle? If one leg measures $8\sqrt{2}$, then the other leg measures $8\sqrt{2}$, and the hypotenuse is 16. If the hypotenuse measures $8\sqrt{2}$, then each leg measures 8.	Students' answers will indicate whether they know how to apply the ratios of the side lengths of a 45°-45°-90° triangle to find the possible unknown side lengths given a side length.

Step It Out

Find Trigonometric Ratios of Special Right Triangles

The 45°-45°-90° and 30°-60°-90° triangles are known as **special right triangles**.

45°-45°-90° Triangle Theorem

In a right isosceles triangle, the ratio of the angles is 45°:45°:90°, and the ratio of the side lengths is $x:x:x\sqrt{2}$, where $x\sqrt{2}$ is always the hypotenuse.

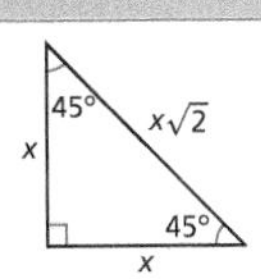

30°-60°-90° Triangle Theorem

In a right triangle where the ratio of the angles is 30°:60°:90°, the ratio of the side lengths is $x:x\sqrt{3}:2x$, where $2x$ is always the hypotenuse.

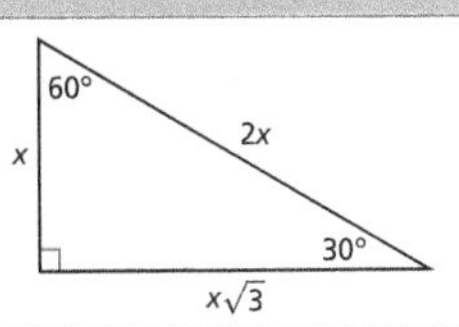

3 Consider the triangles above. Find the trigonometric ratios for 30°, 45°, and 60° when $x = 1$.

Sketch a 45°-45°-90° triangle with a leg length of 1. Then find the trigonometric ratios for either of the 45° angles. A, B. See margin.

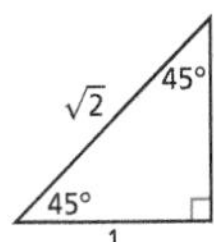

$\sin 45^\circ = \frac{\text{opp}}{\text{hyp}} = \frac{1}{\sqrt{2}} = \frac{\sqrt{2}}{2}$

$\cos 45^\circ = \frac{\text{adj}}{\text{hyp}} = \frac{1}{\sqrt{2}} = \frac{\sqrt{2}}{2}$

$\tan 45^\circ = \frac{\text{opp}}{\text{adj}} = \frac{1}{1} = 1$

A. What is the extra step you need to take to find the sine and cosine ratios?

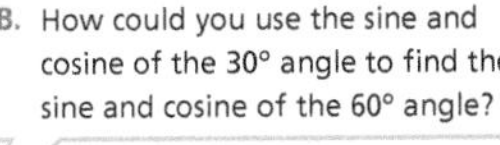

B. How could you use the sine and cosine of the 30° angle to find the sine and cosine of the 60° angle?

Sketch a 30°-60°-90° triangle in which the shortest leg is 1. Then find the trigonometric ratios for the 30° and 60° angles.

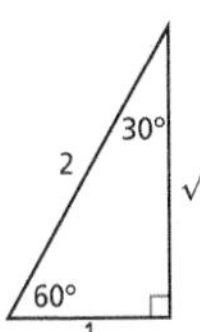

$\sin 30^\circ = \frac{\text{opp}}{\text{hyp}} = \frac{1}{2}$ $\quad\sin 60^\circ = \frac{\text{opp}}{\text{hyp}} = \frac{\sqrt{3}}{2}$

$\cos 30^\circ = \frac{\text{adj}}{\text{hyp}} = \frac{\sqrt{3}}{2}$ $\quad\cos 60^\circ = \frac{\text{adj}}{\text{hyp}} = \frac{1}{2}$

$\tan 30^\circ = \frac{\text{opp}}{\text{adj}} = \frac{1}{\sqrt{3}} = \frac{\sqrt{3}}{3}$ $\quad\tan 60^\circ = \frac{\text{opp}}{\text{adj}} = \frac{\sqrt{3}}{1} = \sqrt{3}$

Turn and Talk In both triangles, what would change if the side length x were not 1? See margin.

Step It Out

Task 3 (MP) **Attend to Precision** Students calculate the sine, cosine, and tangent of the angles in special right triangles and derive the trigonometric ratios for them.

CONNECT TO VOCABULARY

Have students use the **Interactive Glossary** to record their understanding of the vocabulary in this task.

Sample Guided Discussion:

Q Why does $\sin 45^\circ = \cos 45^\circ$? Possible answer: Because the two legs of an isosceles right triangle are congruent.

Q What is the numerical relationship between tan 30° and tan 60°? Possible answer: They are reciprocals.

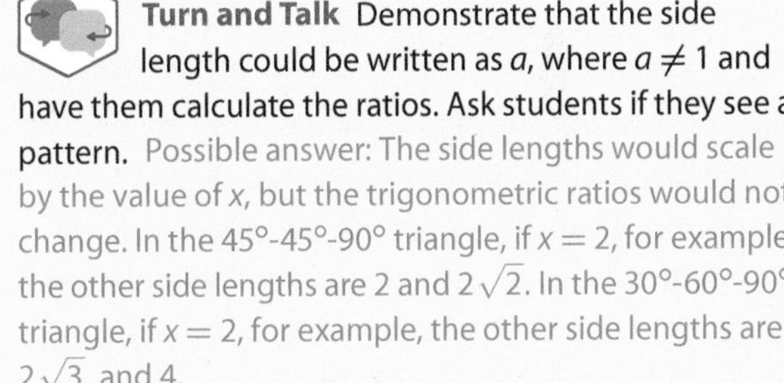

Turn and Talk Demonstrate that the side length could be written as a, where $a \neq 1$ and have them calculate the ratios. Ask students if they see a pattern. Possible answer: The side lengths would scale by the value of x, but the trigonometric ratios would not change. In the 45°-45°-90° triangle, if $x = 2$, for example, the other side lengths are 2 and $2\sqrt{2}$. In the 30°-60°-90° triangle, if $x = 2$, for example, the other side lengths are $2\sqrt{3}$ and 4.

ANSWERS

A. You have to rationalize the denominators since there is a square root there.

B. Possible answer: $\sin 30^\circ = \cos 60^\circ$ and $\cos 30^\circ = \sin 60^\circ$

PROFICIENCY LEVEL

Beginning

Show students sin A, cos A, and tan A. Have student pairs say and write out what the abbreviations stand for and how they are applied to an angle with their partner.

Intermediate

Have students write the abbreviated definition of sin A, cos A, and tan A for a right triangle. Have student pairs draw a special right triangle with an angle labeled A and explain to each other how to find the value of each function using phrases or short sentences.

Advanced

Have each student write, in complete sentences, the definitions of sine, cosine, and tangent. Then have each student draw special right triangles with one leg having length 1 and show how the definitions are applied to each triangle. Students should label all the side and angle measures of their triangles.

Step It Out

Task 4  **Attend to Precision** Students subdivide a hexagon into equilateral triangles and then use a 30°-60°-90° triangle to find the approximate length of a side of Saturn's hexagon with three significant digits.

SUPPORT SENSE-MAKING Three Reads

Have students read the problem three times. Use the questions below for a different focus each read.

1. What is the situation about?
2. What are the quantities in the situation?
3. What are the possible mathematical questions that you could ask for this situation?

Sample Guided Discussion:

Q How do you know that a regular hexagon can be divided into six equilateral triangles? Possible answer: A regular hexagon is a hexagon that is both equiangular and equilateral, so the resulting triangles will also have equal sides and angles.

Q How can you use a 30°-60°-90° triangle when the hexagon was divided into equilateral triangles? If you draw a perpendicular bisector from one side to the opposite angle, this results in the equilateral triangle being divided into two 30°-60°-90° triangles, where the hypotenuse is equal to the length of a side of the hexagon.

Turn and Talk Remind students that the hexagon is divided into six equilateral triangles. Ask "How many 30°-60°-90° triangles are there in six equilateral triangles?" Possible answer: Find the area of one of the 30°-60°-90° triangles. Then multiply that area by 12 because there are six equilateral triangles, or twelve 30°-60°-90° triangles, inside a regular hexagon.

ANSWERS

A. The distance from the center to each vertex is the same, so each triangle is isosceles. The angle at the center measures 60° because the triangles are congruent and $360° \div 6 = 60°$, so the base angles measure 60°. The triangles are equilateral because they are equiangular.

B. You can use a 30°-60°-90° triangle to find half the length of a side of the hexagon because a 30°-60°-90° triangle is half of an equilateral triangle.

C. The longer leg measures 12,500 kilometers because that leg represents half of the shortest distance across the hexagon (25,000 km).

Model Real-World Measurements

From Task 3, you can use trigonometric relationships of special right triangles to find unknown side lengths.

4 Saturn's hexagon is a persisting cloud pattern around the north pole of Saturn. The cloud pattern is roughly a regular hexagon with a vortex at the center of the hexagon with rotating atmospheric gases. The shortest distance across the hexagon is 25,000 km. What is the length of one of the sides of the hexagon?

Sketch a regular hexagon. Then divide the hexagon into six equilateral triangles.

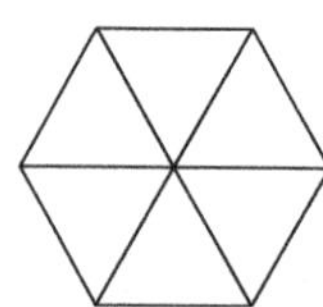

A. Why can you divide the hexagon into six equilateral triangles?

A–C. See margin.

Use a 30°-60°-90° triangle to find the length of half of one of the sides of the hexagon.

B. Why can you use a 30°-60°-90° triangle to find half the length of one side of the hexagon?

The ratio of the shorter leg to the longer leg of the triangle is $x : x\sqrt{3}$.

Write and solve an equation.

Since the length of the longer leg is 12,500 kilometers, set 12,500 equal to $x\sqrt{3}$. Then solve for x. Round the measure to three significant digits since there are three significant digits in 12,500.

$$12{,}500 = x\sqrt{3}$$
$$\frac{12{,}500}{\sqrt{3}} = x$$
$$7220 \approx x$$

C. Why does the triangle show that the longer leg measures 12,500 kilometers?

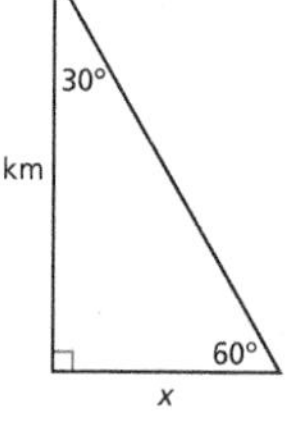

Answer the question.

The length of one of the sides of the hexagon is twice the length of the side of the triangle, so $2x \approx 2(7220) = 14{,}400$.

The length of one side of Saturn's hexagon is approximately 14,400 kilometers.

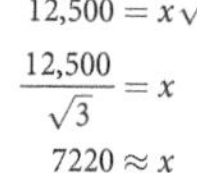

 **Turn and Talk** Without solving, describe how you could find the area of the top view of the cloud cover using trigonometry. See margin.

424

Check Understanding

1. What is the relationship between side lengths in a 45°-45°-90° triangle? What does the ratio mean? The sides have a ratio of $1:1:\sqrt{2}$. The ratio means both leg lengths are the same length, and the hypotenuse is that value multiplied by $\sqrt{2}$.
2. What is the relationship between side lengths in a 30°-60°-90° triangle? What does the ratio mean? The sides have a ratio of $1:1\sqrt{3}:2$. The ratio means the long leg length is the short leg length multiplied by $\sqrt{3}$, and the hypotenuse is the short leg length multiplied by 2.

For Problems 3–5, use the diagram at the right.

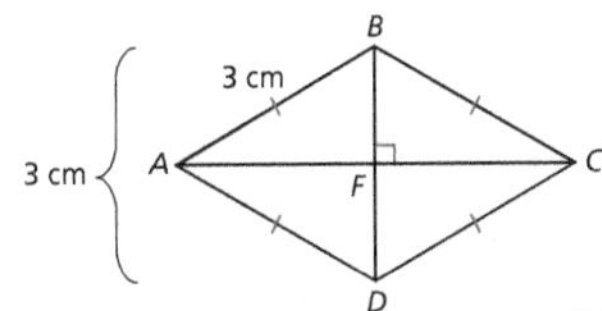

3. What is LN in terms of a? $2a$
4. How is sin N related to cos L? Explain. See Additional Answers.
5. What is MN in terms of a? $a\sqrt{3}$
6. What is the length of $\overline{AC}$ in the rhombus shown? $3\sqrt{3}$ cm

B, 3 cm, 3 cm, A, F, C, D

On Your Own

Determine whether each ratio of side lengths belongs to a 45°-45°-90° triangle, a 30°-60°-90° triangle, or neither.

7. $2:2\sqrt{3}:4$ 30°-60°-90°
8. $5:5:5\sqrt{2}$ 45°-45°-90°
9. $\sqrt{2}:\sqrt{6}:2\sqrt{2}$ 30°-60°-90°
10. $\sqrt{3}:3:2\sqrt{3}$ 30°-60°-90°
11. $2:3:2\sqrt{2}$ neither
12. $4\sqrt{2}:4\sqrt{2}:8$ 45°-45°-90°

Determine whether each triangle is possible. Show your work.

13. yes; $3:3:3\sqrt{2} = x:x:x\sqrt{2}$

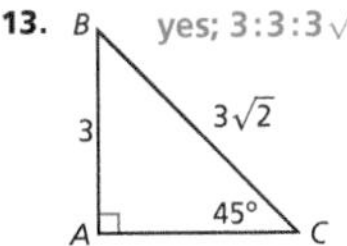

14. yes; $2\sqrt{2}:2\sqrt{6}:4\sqrt{2} = x:x\sqrt{3}:2x$

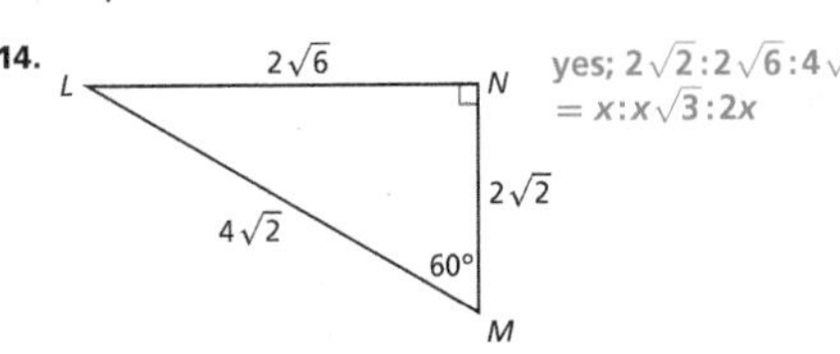

15.

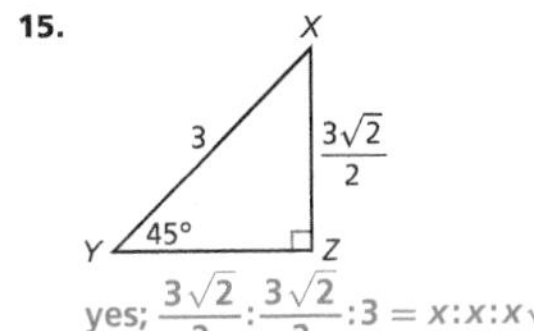

yes; $\frac{3\sqrt{2}}{2}:\frac{3\sqrt{2}}{2}:3 = x:x:x\sqrt{2}$

16. no; The side ratios do not match the 30°-60°-90° criteria.

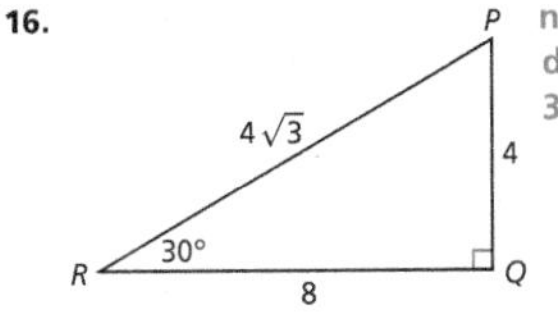

ONLINE

Assign the Digital On Your Own for
- built-in student supports
- Actionable Item Reports
- Standards Analysis Reports

On Your Own

Assignment Guide

The chart below indicates which problems in the On Your Own are associated with each task in the Learn Together. Assign daily homework for tasks completed.

Learn Together Tasks	On Your Own Problems
Task 1, p. 422	Problems 7–16 and 30
Task 2, p. 422	Problems 7–20, 28, 29, and 31–35
Task 3, p. 423	Problems 27 and 37–42
Task 4, p. 424	Problems 21–26, 36, and 43–45

ANSWERS

4. They are congruent ratios. In a right triangle, the sine of one of the acute angles equals the cosine of the other acute angle.

3 Check Understanding

Formative Assessment

Use formative assessment to determine if your students are successful with this lesson's learning objective.

Students who successfully complete the Check Understanding can continue to the On Your Own practice.

For students who miss 1 problem or more, work in a pulled small group using the Almost There small-group activity on page 421C.

ONLINE

Assign the Digital Check Understanding to determine
- success with the learning objective
- items to review
- grouping and differentiation resources

4 Differentiation Options

Differentiate instruction for all students using small-group activities and math center activities on page 421C.

Reteach

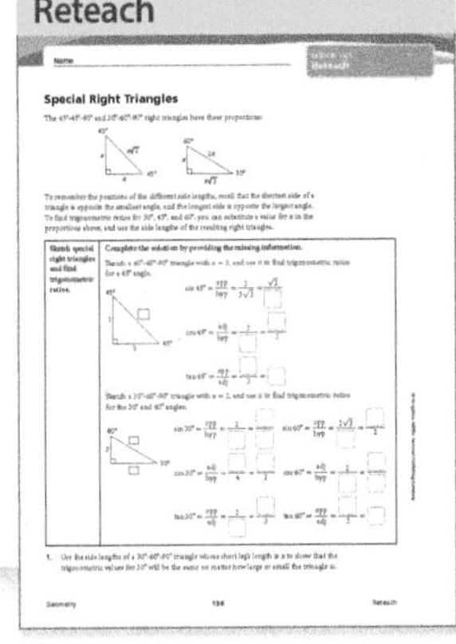

Challenge

Patterns with Special Right Triangles

Questioning Strategies

Problem 17 How can you apply the ratio of the sides and hypotenuse of a special right triangle to find the unknown sides of the triangle? Use proportional reasoning to apply the ratio where $x = 1$ to the known side to find a scale factor for that side. In a 30°-60°-90° triangle the ratio of the sides is $1 : \sqrt{3} : 2$, so the scale factor for the sides is 2.5. The ratio for this triangle's sides is $2.5 : 2.5\sqrt{3} : 5$

Questioning Strategies

Problem 28 If you did not know that the triangle is a special right triangle, what other method can you use to find the value of x that does not involve trigonometric ratios? I can use the Pythagorean Theorem, so $(10x - 12)^2 = x^2 + (x\sqrt{2})^2$. $100x^2 - 240x + 144 = x^2 + 3x^2$; $25x^2 - 60x + 36 = x^2$; $24x^2 - 60x + 36 = 0$; $2x^2 - 5x + 3 = 0$: $(2x - 3)(x - 1) = 0$; $2x - 3 = 0$ or $x - 1 = 0$; $x = \frac{3}{2}$ or $x = 1$. Because the length of the hypotenuse is $10x - 12$, $x = 1$ is not a solution, so $x = \frac{3}{2}$.

Find the unknown side lengths in each triangle. Give an exact answer.

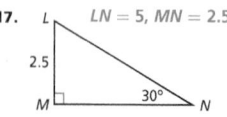

17. $LN = 5$, $MN = 2.5\sqrt{3}$

18. $AC = 4\sqrt{3}$, $CB = 4$

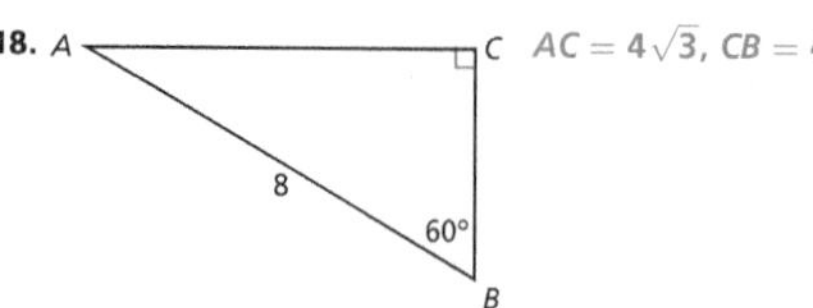

19.

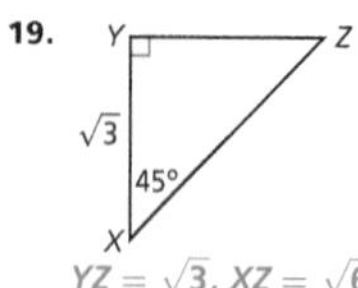

$YZ = \sqrt{3}$, $XZ = \sqrt{6}$

20. $QR = \frac{\sqrt{10}}{2}$, $PR = \frac{\sqrt{10}}{2}$

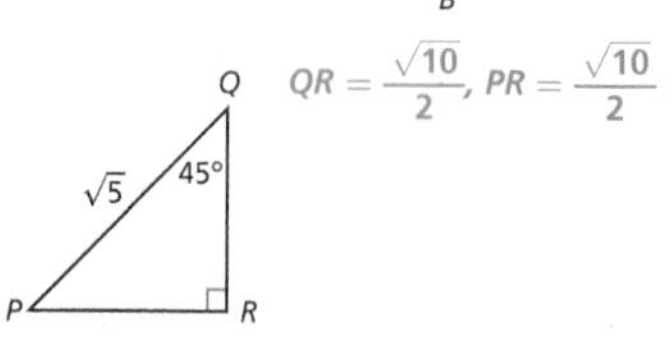

The illustration shows a site where three different triangular areas are being studied by archeologists. Find each measurement.

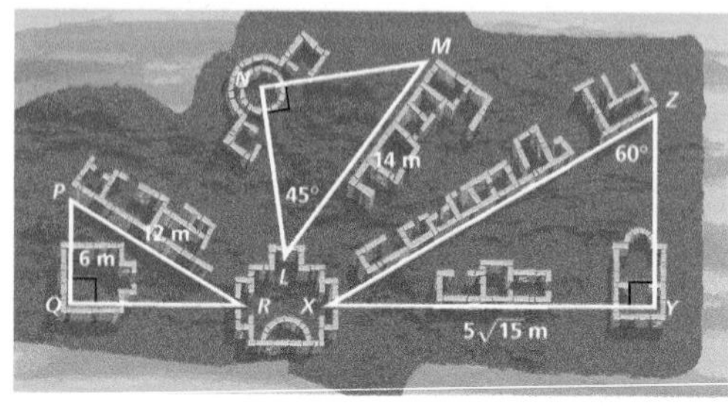

21. QR $6\sqrt{3}$ m

22. m∠P 60°

23. MN $7\sqrt{2}$ m

24. NL $7\sqrt{2}$ m

25. YZ $5\sqrt{5}$ m

26. XZ $10\sqrt{5}$ m

27. Open Ended An acute angle has a sine ratio of $\frac{1}{2}$. Draw a right triangle that meets this criterion given that no side length is either 1 or 2. Check students' work.

Find the value of x in each right triangle.

28. $x = \frac{3}{2}$

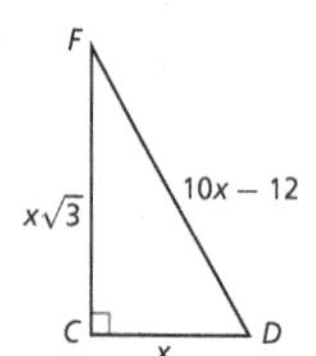

29. $x = \frac{100}{9}$

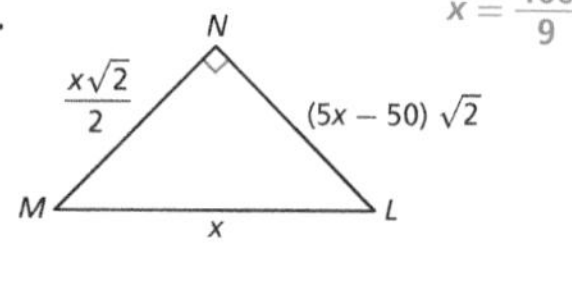

426

30. Is it true that if you know one side length of an isosceles right triangle, then you know all the side lengths? Explain. See Additional Answers.

For Problems 31–35, use the figure below to find the indicated values.

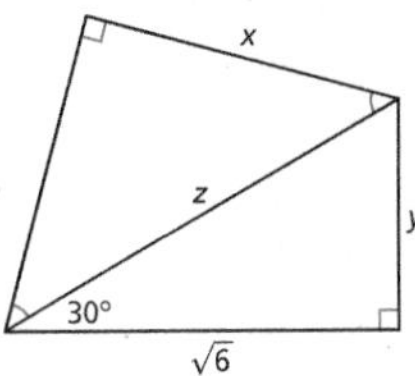

31. y $y = \sqrt{2}$ **32.** z $z = 2\sqrt{2}$ **33.** x $x = 2$

34. the perimeter of the composite figure (excludes z) $4 + \sqrt{6} + \sqrt{2}$

35. the area of the composite figure $2 + \sqrt{3}$

36. (MP) **Construct Arguments** Using your knowledge of right isosceles triangles, prove the length of a diagonal of a square is always the side length multiplied by $\sqrt{2}$. See Additional Answers.

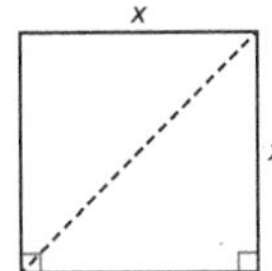

Find the range of acute angle measures x that satisfy each expression.

37. $0 < \cos x < \frac{\sqrt{2}}{2}$ $45° < x < 90°$

38. $\frac{\sqrt{2}}{2} < \sin x < 1$ $45° < x < 90°$

39. $\frac{\sqrt{3}}{2} < \sin x < 1$ $60° < x < 90°$

40. $0 < \tan x < 1$ $0° < x < 45°$

41. $\frac{1}{2} < \cos x < \frac{\sqrt{3}}{2}$ $30° < x < 60°$

42. $1 < \tan x < \sqrt{3}$ $45° < x < 60°$

43. (MP) **Critique Reasoning** Aiden is asked to sketch a right triangle with a 60° acute angle and a hypotenuse length of $24\sqrt{2}$. Identify, explain, and correct his error.

BC should be $12\sqrt{2} \cdot \sqrt{3} = 12\sqrt{6}$.

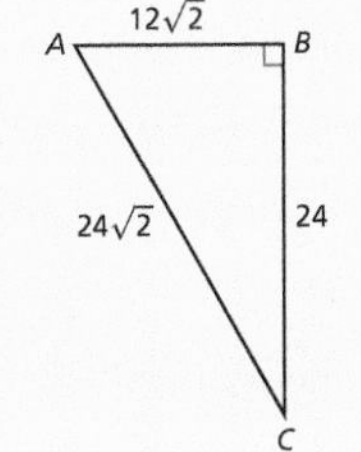

Watch for Common Errors

Problem 34 Even though the problem specifies to exclude z in the perimeter, some students may include it. Explain that the question is asking for the perimeter of the overall figure, not the two triangles.

Watch for Common Errors

Problems 37–42 Some students may be confused by expressions such as cos *x* because in this lesson, *x* has been used to represent a side and the angles have been represented by capital letters. Assure students that *x* does represent a range of acute angle measures in this problem and that expressions such as sin *x*, cos *x*, and tan *x* will be seen often in future math lessons or courses.

ANSWERS

30. yes; If you know one leg length, the other leg length is also equal to this length, and the length of the hypotenuse is this length multiplied by $\sqrt{2}$. If you know the hypotenuse length, then each leg length is this length divided by $\sqrt{2}$.

36. Possible answer: By drawing a diagonal, you create two triangles. Each triangle is an isosceles right triangle because the side lengths are congruent and it contains a right angle. This means the triangle is a 45°-45°-90° triangle. In 45°-45°-9.0° triangles, the the length of the hypotenuse, which is the length of the diagonal, is the side length multiplied by $\sqrt{2}$.

Wrap-Up

Summarize learning with your class. Consider using the Exit Ticket, Put It in Writing, or I Can scale.

Exit Ticket 

The longer leg of a 30°-60°-90° triangle is 4. What are the lengths of the other two sides? Possible answer: The ratio of the sides of a 30°-60°-90° triangle is $x: x\sqrt{3}: 2x$. The longer leg is $x\sqrt{3}$. So, $x\sqrt{3} = 4$. $x = \frac{4}{\sqrt{3}}$. The shorter leg is $x = \frac{4}{\sqrt{3}} \cdot \frac{\sqrt{3}}{\sqrt{3}} = \frac{4\sqrt{3}}{3}$; the hypotenuse is $2x$, so $2x = 2 \cdot \frac{4\sqrt{3}}{3} = \frac{8\sqrt{3}}{3}$.

Put It in Writing 

Explain how to find the ratio of the sides of a 30°-60°-90° triangle starting from an equilateral triangle.

I Can 

The scale below can help you and your students understand their progress on a learning goal.

4	I can use trigonometric ratios and the Pythagorean Theorem to find the side lengths and angle measures of special right triangles, and I can explain my work to others.
3	I can use trigonometric ratios and the Pythagorean Theorem to find the side lengths and angle measures of special right triangles.
2	I can use the special right triangles to solve problems.
1	I know the ratio of the sides in special right triangles.

Spiral Review • Assessment Readiness

These questions will help determine if students have retained information taught in the past and can also prepare them for high-stakes assessments. Here, students use sine to find the height of a right triangle **(13.2)**, use sine to find the base and height of a triangle to determine the area **(13.2)**, and use tangent to determine if statements are true or false **(13.1)**.

44. A rhombus on the Thai door is shown below. The length of $\overline{BD}$ is 4 inches.

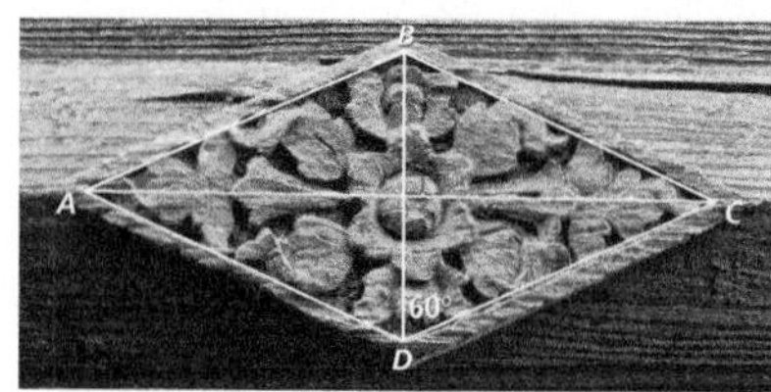

A. Find the perimeter of the rhombus. The perimeter of the rhombus is $4 \cdot 4 = 16$ in.

B. Find the length of $\overline{AC}$. $AC = 2 \cdot 2\sqrt{3} = 4\sqrt{3}$ in.

45. Triangle XYZ is a 30°-60°-90° triangle with a right angle at point X. The longer leg $\overline{XY}$ has vertices $X(2, 2)$ and $Y(7, 2)$. Where is point Z in Quadrant I? $\left(2, 2 + \frac{5\sqrt{3}}{3}\right)$, or approximately $(2, 4.89)$

Spiral Review • Assessment Readiness

46. For the 20-foot wheelchair ramp shown, what is the approximate height h of the ramp? The diagram is not drawn to scale.

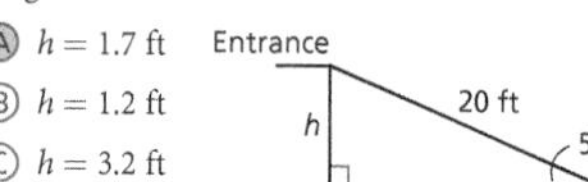

Ⓐ $h = 1.7$ ft (circled)
Ⓑ $h = 1.2$ ft
Ⓒ $h = 3.2$ ft
Ⓓ $h = 3.4$ ft

47. For right triangle ABC, $\sin A = \frac{12}{13}$. Hypotenuse $\overline{AC}$ is 26 in. long. What is the area of $\triangle ABC$?

Ⓐ 30 in²
Ⓑ 60 in²
Ⓒ 90 in²
Ⓓ 120 in² (circled)

48. In right triangle ABC, $m\angle B = 90°$ and $\tan A = \frac{21}{28}$. Determine whether each statement is true or false.

Statement	True	False
A. $\tan C = \frac{28}{21}$	■	?
B. $\tan C = \frac{28}{35}$	?	■
C. $m\angle A = \tan^{-1}\frac{28}{21}$	?	■
D. $m\angle C = \tan^{-1}\frac{28}{21}$	■	?

 **I'm in a Learning Mindset!**

Do I know my own strengths and weaknesses in learning? What is one example of each?

Keep Going to Journal and Practice Workbook

Learning Mindset

Perseverance Learns Effectively

Point out that effective learning is understanding the development of finding side ratios for the special right triangles, not just knowing what the ratios are. Encourage students to make a connection between the side ratios and the evaluation of the trigonometric functions of the special right triangles. Ask students if they think they can effectively use values of trigonometric functions to solve real-world problems. *Are my efforts in learning being effective? How well do I understand how the side ratios of special right triangles was developed? Do I understand the trigonometric ratios? Can I make effective use of them? How does lack of organization affect my learning outcomes? How are my social relationships affecting my academic performance? How do I prioritize my learning? What are my greatest strengths? How can I improve?*

13.4 Solve Problems Using Trigonometry

LESSON FOCUS AND COHERENCE

Mathematics Standards

- Explain and use the relationship between the sine and cosine of complementary angles.
- Use trigonometric ratios and the Pythagorean Theorem to solve right triangles in applied problems.

Mathematical Practices and Processes 

- Look for and make use of structure.
- Attend to precision.
- Model with mathematics.

I Can Objective 

I can use trigonometric ratios, the area formula for a triangle in terms of its side lengths, and the Pythagorean Theorem to solve right triangles in applied problems.

Learning Objective

Apply the sine, cosine, and tangent ratios and the inverses to find the areas of triangles and measures of sides and acute angles in right triangles both in the coordinate plane and represented in real-world situations.

Language Objective

Explain how to derive the area formula $A = \frac{1}{2}ab \sin C$ and how to solve problems involving right triangles and trigonometric ratios.

Vocabulary

New: angle of depression, **angle of elevation**, **solve a right triangle**

Lesson Materials: calculator

Mathematical Progressions

Prior Learning	Current Development	Future Connections
Students: • proved theorems about triangles. **(9.1)** • used similarity criteria for right triangles to solve problems and to prove relationships. **(12.4)**	**Students:** • find the area of a triangle using the formula $A = \frac{1}{2}ab \sin C$, where a and b are the lengths of two sides and $\angle C$ is the included angle. • use sine, cosine, and tangent ratios and their inverses to solve right triangles. • apply the distance formula and trigonometry to solve right triangles on the coordinate plane. • solve real-world problems involving angles of depression and angles of elevation.	**Students:** • will derive the formula $A = \frac{1}{2}ab \sin C$ for the area of a triangle by drawing an auxiliary line from a vertex perpendicular to the opposite side. **(14.1)**

PROFESSIONAL LEARNING

Using Mathematical Practices and Processes

Model with mathematics.

In this lesson, students learn how to use angle measures that describe objects sighted either above or below an observer. Such angles are called *angles of depression*, located outside a right triangle, and *angles of elevation*, in the interior of a right triangle. Explain to students, that a drawing such as this, can help them understand how to apply trigonometric ratios to find unknown measures in these types of problems.

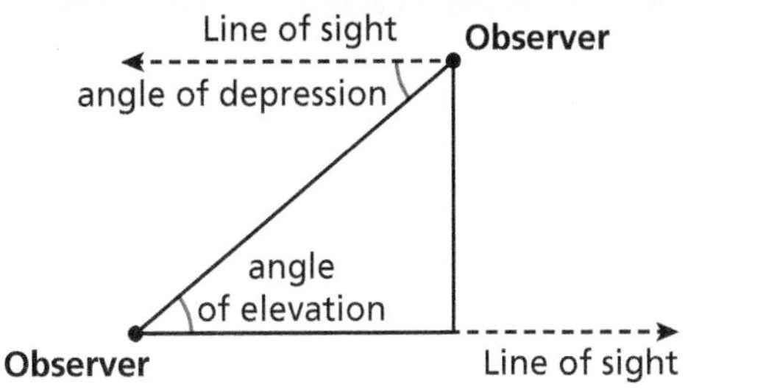

WARM-UP OPTIONS

ONLINE

PROJECTABLE & PRINTABLE

ACTIVATE PRIOR KNOWLEDGE • Use Theorems About Triangles

Use these activities to quickly assess and activate prior knowledge as needed.

Problem of the Day

Triangle MNP is a right triangle, and $\overline{QR} \perp \overline{PN}$. What is MP?

$MP = 9$

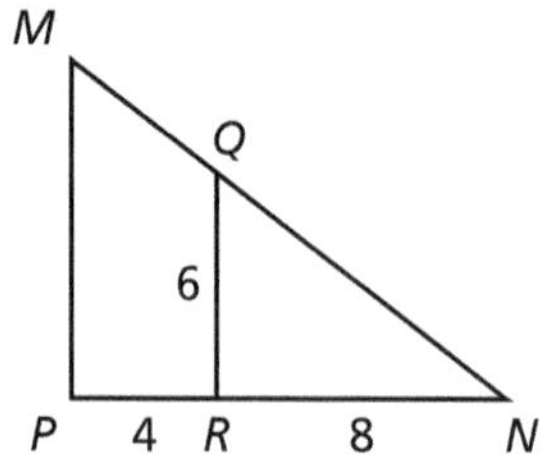

Quick Check for Homework

As part of your daily routine, you may want to display the Teacher Solution Key to have students check their homework.

Make Connections

Based on students' responses to the Problem of the Day, choose one of the following:

1. Project the Interactive Reteach, Lesson 12.3.
2. Complete the Prerequisite Skills Activity:

Have students work in pairs. On an index card, have one student draw a triangle whose dimensions are 3, 4, and 5. Give the card to the other student, and on the reverse side of the card, have that student draw a triangle similar to the 3-4-5 triangle and label its dimensions.

- *What theorem can you use to prove that the two triangles drawn by you and a partner are right triangles?* the Converse of the Pythagorean Theorem
- *What properties do a pair of similar triangles share?* Their corresponding angles are congruent and their corresponding sides are proportional.
- *What theorems can you use to prove that two triangles are similar?* AA and SSS, where the ratios between corresponding sides are equal.
- *Why is the third pair of angles in two similar triangles congruent?* The difference between 180° and the sum of the measures of the two pairs of congruent angles are equal. That is, equals subtracted from equals are equal.

SHARPEN SKILLS

If time permits, use this on-level activity to build fluency and practice basic skills.

Vocabulary Review

Objective: Students use a graphic organizer to review the definitions and properties of congruent and similar triangles.

Have students make a graphic organizer like the one shown. Have them write the definitions of congruent and similar triangle, then list all the theorems that prove triangles either congruent or similar, and add sketches of at least two pairs of examples of each type.

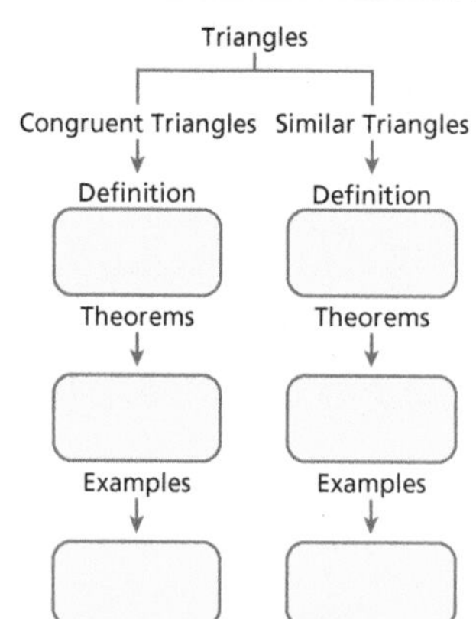

PLAN FOR DIFFERENTIATED INSTRUCTION

Small-Group Options

Use these teacher-guided activities with pulled small groups.

On Track

Ask students to solve the right triangle ABC whose hypotenuse is $AB = 15$ and $m\angle A = 39°$. Round the lengths of the sides to the nearest tenth and the measures of the angles to the nearest degree.

Almost There

Have students show that the area formula of a triangle, $A = \frac{1}{2}bh$, can be written as $A = \frac{1}{2}ab\sin C$.

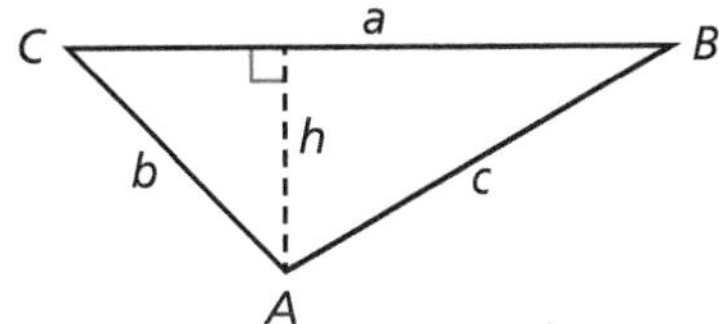

Ready for More

Draw and show students $\angle ABC$. Because the segment with length h is perpendicular to $\overline{BC}$, h is the altitude from vertex A. Ask students to show that $\frac{\sin C}{c} = \frac{\sin B}{b}$.

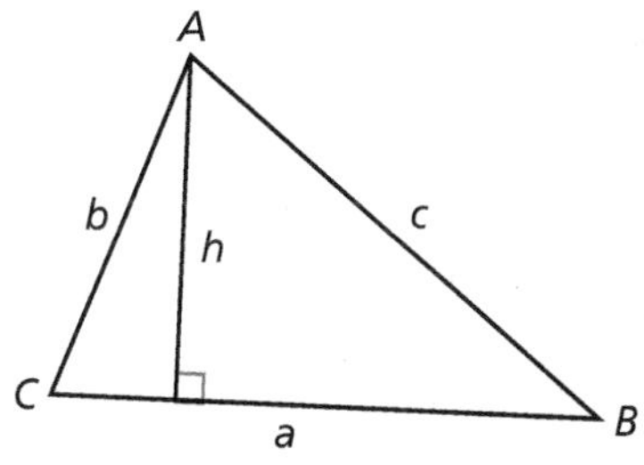

Math Center Options

Use these student self-directed activities at centers or stations. Key: ● Print Resources ● Online Resources

On Track

- ● Interactive Digital Lesson
- ●● Journal and Practice Workbook
- ● Interactive Glossary (printable): **angles of depression**, **angle of elevation**, **solve a right triangle**

Standards Practice:

- ●● Know That Side Ratios in Right Triangles Are Properties of the Angles
- ●● Use the Relationship Between the Sine and Cosine of Complementary Angles
- ●● Solve Right Triangles Using Various Methods

Almost There

- ● Reteach 13.4 (printable)
- ● Interactive Reteach 13.4
- ● Illustrative Mathematics: Are They Similar?

Ready for More

- ● Challenge 13.4 (printable)
- ● Interactive Challenge 13.4
- ● Illustrative Mathematics: Congruent and Similar Triangles
- ● Desmos: Transformation Golf: Rigid Motion

Unit Project Check students' progress by asking what trigonometric ratio they have chosen and asking to see their equation.

View data-driven grouping recommendations and assign differentiation resources.

During the *Spark Your Learning*, listen and watch for strategies students use. See samples of student work on this page.

Use Correct Trigonometric Ratio

Strategy 1

I drew a triangle having sides of 6 and 8 and the included 35° angle. Then I wrote the formula for the area of a triangle $A = \frac{1}{2}bh$ and drew h to the 8-foot base. This gave me a right triangle so I wrote $\sin 35° = \frac{h}{6}$. I solved the equation for h and found $h = 6 \sin 35°$, which is about 3.441. In the formula $A = \frac{1}{2}bh$ I substituted 8 for b, and 3.441 for h to get $A = \frac{1}{2}(8)(3.441)$. I multiplied to find that the area of the highlighted triangle is about 13.8 square feet.

If students . . . use trigonometry to solve the problem, they are applying what they learned in preceding lessons about trigonometric ratios in right triangles.

Have these students . . . explain how they determined their equation and how they solved it. **Ask:**

- Q Why did you use the sine ratio to express the value of h?
- Q Could you have solved this problem by drawing the height to the 6-foot side?

Make Two Right Triangles

Strategy 2

I drew a triangle with sides 6 and 8 with the included angle 35°. I realized by drawing a perpendicular line from the opposite vertex to the 8-foot side, I have created two right triangles. Using the sine ratio, $6 \sin 35°$ is the length of the common leg for both right triangles. Using the cosine ratio, $6 \cos 35°$ is the length of the other leg of one right triangle and $8 - 6 \cos 35°$ is the length of the other leg of the other right triangle. The areas are $\frac{1}{2}(6 \sin 35°)(6 \cos 35°)$ and $\frac{1}{2}(6 \sin 35)(8 - 6 \cos 35°)$, or about 8.5 square feet and about 5.3 square feet. The sum is 13.8 square feet, so the area of the highlighted triangle is about 13.8 square feet.

If students . . . use the trigonometry to find the areas of two right triangles that make up the highlighted triangle, they show a valid but less efficient way to solve the problem.

Activate prior knowledge . . . by having students recognize that they have found the altitude of the larger triangle. **Ask:**

- Q In the area formula of $\frac{1}{2}$ (base)(height) what is the relationship between the base and height?
- Q After drawing the perpendicular line, what two segments represent the base and height of the larger triangle?

COMMON ERROR: Uses the Wrong Trigonometric Ratio

I drew a triangle with sides of 6 and 8 and the included 35° angle. I drew a height h to the 8-foot base and created a right triangle. I then wrote $\cos 35° = \frac{h}{6}$ and solved it for h, which is about 4.9. I used the formula $A = \frac{1}{2}bh$ and substituted 8 for b and 4.9 for h to get $A = \frac{1}{2}(8)(4.9)$. So the area of the highlighted triangle equals 19.6 square feet.

If students . . . choose the incorrect trigonometric ratio to solve for the height, then they are probably confusing the position of the legs associated with each of the trigonometric ratios.

Then intervene . . . by pointing out that the height is opposite the 35° angle, not adjacent to it. **Ask:**

- Q What is the difference between the ratio of the sine and the cosine in terms of the hypotenuse, the opposite side, and adjacent side of an angle?
- Q If the angle was 30°, how can you estimate the height by using a special right triangle? Is that height close to the value you got for the height?

13.4

Solve Problems Using Trigonometry

(I Can) **use trigonometric ratios, the area formula for a triangle in terms of its side lengths, and the Pythagorean Theorem to solve right triangles in applied problems.**

Spark Your Learning

An artist created a sculpture made with several triangles.

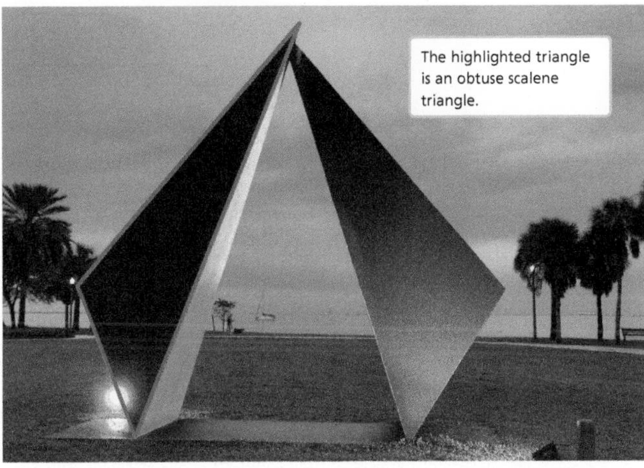
The highlighted triangle is an obtuse scalene triangle.

Complete Part A as a whole class. Then complete Parts B–C in small groups.

A. What is a mathematical question you can ask about the highlighted triangle?
What is the area of the highlighted triangle?

B. What information do you usually have that can help you answer the question?
You are typically given the base and height of the triangle to find the area.

C. To answer your question, what strategy and tool would you use along with all the information you have? What answer do you get?
See Strategies 1 and 2 on the facing page.

Turn and Talk

- How would your solution change if the angle provided is not the included angle?
- Can you determine how it could be possible to use a trigonometric ratio if the given triangle is not a right triangle? **See margin.**

SUPPORT SENSE-MAKING • Three Reads

Ask students to read the information in the photo three times and prompt them with a different question each time.

1. What is the situation about? The situation is about finding the area of the triangular face of a sculpture.
2. What are the quantities in this situation? How are those quantities related? The quantities in this situation are the measures of two sides and the included angle of a triangle that represent one face of the sculpture.
3. What are possible questions you could ask about this situation? Possible answer: What formula can I use to find the area of a triangle? How can I use the information I have to find the values I need to plug into the area formula?

1 Spark Your Learning

▶ MOTIVATE

- Have students look at the photo in their books and read the information contained in the photo. Then complete Part A as a whole-class discussion.
- Give the class the additional information they need to solve the problem. This information is available online as a printable and projectable page in the Teacher Resources.
- Have students work in small groups to complete Parts B–C.

▶ PERSEVERE

If students need support, guide them by asking:

Q **Advancing • Use Tools** Which tool could you use to solve the problem? Why choose that tool and not some other? Students' choices of tools and reasons for choosing them will vary.

Q **Assessing** What is an obtuse scalene triangle? a triangle where one angle is an obtuse angle and in which no two sides are congruent

Q **Assessing** What formula can you use to find the area of a triangle? $A = \frac{1}{2}bh$, where b is the length of a base, and h is the length of the altitude to that base.

Q **Advancing** What auxiliary line could you draw in the triangle to help you answer the question? Possible answer: the altitude (height) to one of the sides

Turn and Talk Remind students of the Area Addition Postulate, which they might use to find the area of the triangular faces of the sculpture.

- Possible answer: I would not have enough information to find a solution.
- Possible answer: I can treat the triangle like a composite figure and break it into triangles where I can identify the bases and heights, find the areas of each triangle, and find the sum of the areas.

▶ BUILD SHARED UNDERSTANDING

Select groups of students who used various strategies and tools to share with the class how they solved the problem. As they present their solutions, have each group discuss why they chose a specific strategy and tool.

2 Learn Together

Build Understanding

Task 1 (MP) **Use Structure** Students recognize that in order to use the familiar area formula for a triangle involving a base and height, they can use trigonometry and express the height using a trigonometric expression.

Sample Guided Discussion:

Q Why can't you use the Pythagorean Theorem to find an expression for h in terms of a, b, and c? Because to use the theorem, you need the lengths of two sides of the right triangle. But, the altitude partitions the base into two segments with no information about the length of either segment.

Q How could you calculate the area of the triangle (and check your answer to Part G) using the cosine ratio? The measure of the other angle in the right triangle is the complement of 38°, which equals 52°. So you can write $\cos 52 = \frac{h}{18}$, and then substitute 25 for a and solve the equation for A like this: $A = \frac{1}{2}(25)(18\cos 52)$.

Turn and Talk Have students draw right $\triangle ABC$ with right angle at C. Label the sides a, b, and c and then write the expression for the sine of C.

- yes; in this case, $\sin C = \sin 90° = 1$, so the formula becomes area $= \frac{1}{2}ab$.
- The area would be $\frac{1}{2}ac\sin B$; the area is one-half the product of the adjacent sides and the sine of the included angle.

Build Understanding

Derive an Area Formula

You can find the area of a triangle without knowing its height using trigonometric ratios.

The diagram shows a triangular pool. Use trigonometry and the lengths shown to find a formula for the area of the triangle.

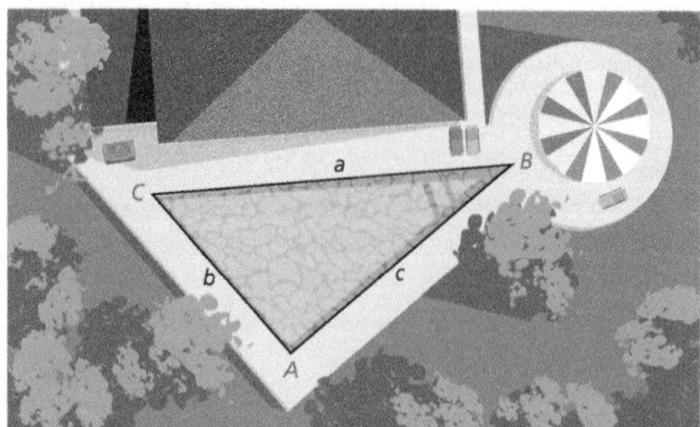

1 A. Redraw the triangle. Then draw an altitude h from vertex A to $\overline{BC}$. Make a conjecture as to why this altitude will help develop the formula. See Additional Answers.

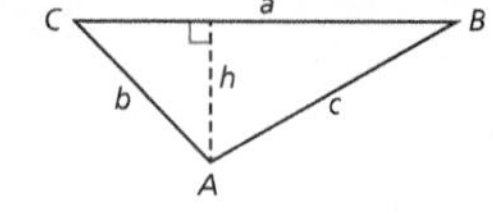

B. Write the equation for the sine of angle C. Then solve the equation for h. $\sin C = \frac{h}{b}$; $h = b\sin C$

C. Recall the general formula for the area of a triangle.

$$\text{area} = \frac{1}{2} \cdot \text{base} \cdot \text{height}$$

Which side of the triangle in the figure represents the base? a

D. Write the formula in terms of the variables from the triangle in the figure. area $= \frac{1}{2}ah$

E. Substitute the expression for h from Part B into the area formula. area $= \frac{1}{2}ab\sin C$

F. What is the relationship between sides a and b and angle C? Would the formula still apply if you know sides a and c and angle B? Why or why not? See Additional Answers.

G. In the winter, a tarp is used to cover the pool. Find the area of the tarp if $a = 25$ feet, $b = 15$ feet, and $m\angle C = 47°$. Round your answer to the nearest hundredth. approximately 137.13 square feet

Turn and Talk

- Does the area formula you found work if $\angle C$ is a right angle? Explain.
- Suppose you used a trigonometric ratio in terms of $\angle B$, h, and a different side length to find the area. How would this change your findings? What does this tell you about the choice of sides and included angle? See margin.

430

LEVELED QUESTIONS

Depth of Knowledge (DOK)	Leveled Questions	What Does This Tell You?
Level 1 **Recall**	What are the sine, cosine, and tangent of acute $\angle A$ in right triangle ABC, in terms of a, b, and c? $\sin A = \frac{a}{c}$, $\cos A = \frac{b}{c}$, $\tan A = \frac{a}{b}$	Students' answers will indicate whether they can identify the trigonometric ratios for an acute angle in a right triangle.
Level 2 **Basic Application of Skills & Concepts**	Why can you substitute the trigonometric expression $b\sin B$ for h in the area formula of $\triangle ABC$? The equation $\sin B = \frac{h}{b}$ solved for h is $h = b\sin B$.	Students' answers will indicate whether they can write a correct trigonometric equation and then solve it for one of the variables, either a side or a trigonometric ratio.
Level 3 **Strategic Thinking & Complex Reasoning**	What is the measure of $\angle B$? Explain your answer. $\tan B = \frac{h}{(a - b\cos C)}$, so $\tan B = \frac{b\sin C}{(a - b\cos C)}$; use inverse tangent with the values to get $m\angle B = 36.6°$.	Students' answers will indicate whether they can correctly evaluate a more complex expression involving a trigonometric ratio.

Step It Out

Solve a Right Triangle

To **solve a right triangle** means to find the length of each side of the triangle and the measure of each interior angle. You can use any of the methods you have learned to solve a right triangle, including trigonometric ratios, inverse trigonometric ratios, and the Pythagorean Theorem.

2 An eagle on top of a 65-foot cliff spots a fish in the water and dives toward it. The angle between the cliff and the eagle's path is 36°. Solve the right triangle that represents this situation. Round each value to the nearest tenth.

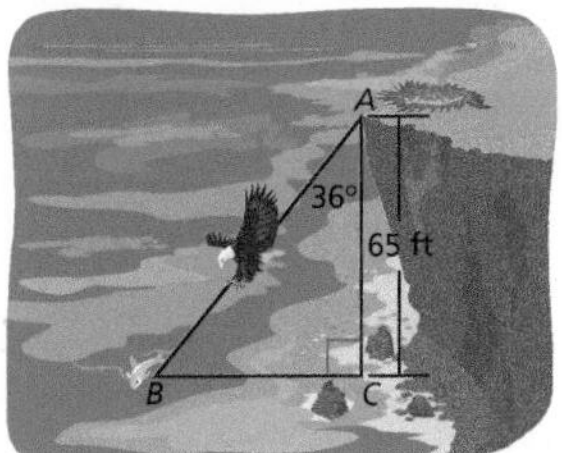

Determine the given information.
You are given an acute angle and an adjacent side length in a right triangle. Choose a trigonometric ratio that uses the given information.

A. What two trigonometric ratios use the side adjacent to the given angle?
Cosine and tangent use the adjacent side.

Find *AB*.

$$\cos 36° = \frac{65}{AB}$$

$$AB\cos 36° = \frac{65}{AB} \cdot AB$$

$$AB = \frac{65}{\cos 36°}$$

$$AB \approx 80.3 \text{ ft}$$

B. Where did each value in this ratio come from?
The acute angle given in the problem is 36°, 65 is the adjacent side length, and *AB* is the length of the hypotenuse.

Find *BC*.

$$\tan 36° = \frac{BC}{65}$$

$$65\tan 36° = BC$$

$$BC \approx 47.2 \text{ ft}$$

C. Why do we start with our given information again and not use the value of *AB*?
The value of *AB* is rounded, so starting with given values gives a more precise answer than using a rounded value in our formula.

Find the measure of *B*.

$$m\angle B = 90° - 36°$$

$$m\angle B = 54°$$

D. What allows us to use this equation to find the measure of angle *B*?
B and *A* are complementary angles.

Answer the question.
Besides the given information, $AB \approx 80.3$ ft, $BC \approx 47.2$ ft, and $m\angle B = 54°$.

Turn and Talk What is the minimum information you could use to solve the right triangle if you had not been given one of the acute angle measures? See margin.

Step It Out

Task 2 (MP) **Attend to Precision** Students solve a right triangle by writing the appropriate trigonometric ratios for a given acute angle in a right triangle and then solve the equations to find the desired lengths.

CONNECT TO VOCABULARY

Have students use the **interactive Glossary** to record their understanding of the vocabulary in this task.

Sample Guided Discussion:

Q How do you decide which are the "opposite" and "adjacent" sides of an acute angle in a right triangle? The opposite side of an angle is not one of the sides of the angle; the adjacent side is one of the sides of the angle.

Q How could you prove that the values of *AB* and *BC* that you found were correct? Explain. by substituting the values of the three sides into the Pythagorean Theorem to see if $AB^2 = BC^2 + AC^2$; Because of rounding, the two sides of the equation may not be exactly equal.

Turn and Talk Remind students that solving a right triangle means finding the measures of each of its sides and angles. If you were not given the measure of one of the angles, you would need any two sides to solve the triangle.

PROFICIENCY LEVEL

Beginning

Draw a right triangle *ABC* with *C* as the vertex of the right angle. Have students copy the statements below and complete each one by writing the terms *adjacent to* and *opposite to* in order to make each statement true.

- $\overline{AB}$ is ___________ angle *A*, and $\overline{BC}$ is ___________ angle *B*.
- $\overline{AB}$ is___________ angle *B*, and $\overline{BC}$ is ___________ angle *A*.

Intermediate

Have students draw a right triangle and label it *MNP*, with *P* as the vertex of the right angle. Have them write the sine, cosine, and tangent of $\angle M$ and $\angle N$ in terms of the sides of the triangle and explain why the ratios are correct.

Advanced

Have students draw an isosceles right triangle with *C* as the vertex of the right angle. Have them explain the relationships between the sine, cosine, and tangent ratios of the acute angles of the triangle.

Task 3 (MP) **Model with Mathematics** Students solve right triangles that model real-world situations involving angles of elevation and angles of depression and see how to use trigonometry to find the lengths of the sides of the triangles.

CONNECT TO VOCABULARY

Have students use the **interactive Glossary** to record their understanding of the vocabulary in this task.

Sample Guided Discussion:

Q In the first step of this task, what reason can you give to justify the statement, "The angle between the vertical and line of sight is $90° - 20° = 70°$"? Because the horizontal and vertical lines are perpendicular to one another and form a 90° angle, the two adjacent angles at that vertex are complementary.

Q What theorem can you use to prove that the angle of depression is congruent to the angle of elevation? If two parallel lines are cut by a transversal, the alternate interior angles are congruent.

Turn and Talk Point out that there are two points of view in this situation. One is a person's view from the boat down toward the whale, and the other is the whale's point of view from beneath the boat. The angle formed by the horizontal line of length x and the 42-foot line of sight from the whale to the boat is an angle of elevation.

ANSWERS

A. The angle of depression is between the horizontal and the line of sight, which is outside the triangle.

B. You are looking for the opposite side length and given the length of the hypotenuse. These are the lengths needed for the sine ratio.

C. You can use the cosine ratio because you know the length of the side adjacent to the 70° angle and the length of the hypotenuse; yes; You can also use the tangent ratio because you know the opposite side length from the previous step, 39.5 ft. However, the opposite side length was rounded so it is more accurate to use the given information and the cosine ratio in this case.

D. yes; You can use the Pythagorean Theorem to find y, although using the method with a trigonometric function will lead to a more precise measurement.

Find an Angle of Elevation (or Depression)

You can use trigonometry to find angles of elevation and depression.

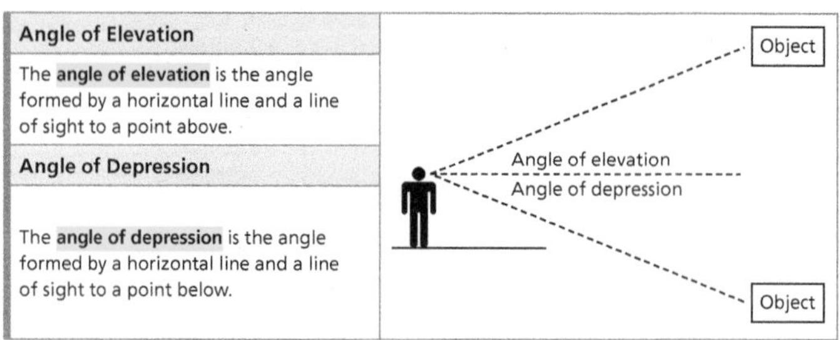

Angle of Elevation
The angle of elevation is the angle formed by a horizontal line and a line of sight to a point above.
Angle of Depression
The angle of depression is the angle formed by a horizontal line and a line of sight to a point below.
Object
Angle of elevation
Angle of depression
Object

3 A boat on the surface of the water detects a whale 42 feet away at an angle of depression of 20°. What is the horizontal distance between the boat and the whale? What is the depth of the whale?

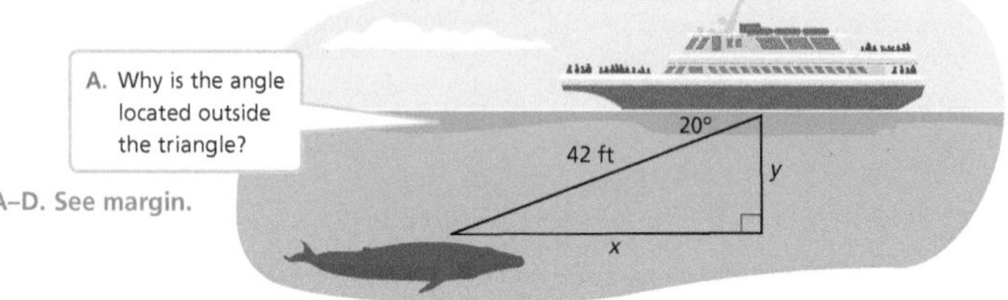

A–D. See margin.

The angle between the vertical and the line of sight is $90° - 20° = 70°$.

Use a trigonometric ratio of the 70° angle to find the horizontal distance.

$\sin 70° = \frac{x}{42}$

$x = 42 \cdot \sin 70°$

$x \approx 39.5$ feet

B. Why would you use the sine ratio?

The horizontal distance is 39.5 feet.

Use a different trigonometric ratio to find depth.

$\cos 70° = \frac{y}{42}$

$y = 42 \cdot \cos 70°$

$y \approx 14.4$ feet

C. Why can you use the cosine ratio? Can you also use the tangent ratio? Explain.

D. Is it possible to use another method to find y after x is determined? Explain.

The depth of the whale is 14.4 feet.

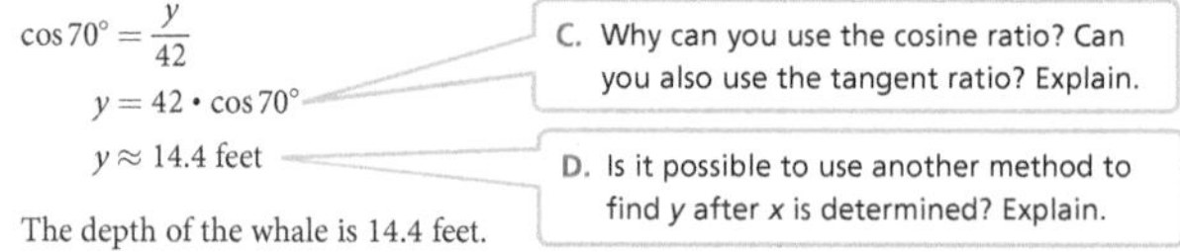

Turn and Talk From Task 3, what angle in the triangle can be considered an angle of elevation? See margin.

432

Solve a Right Triangle in the Coordinate Plane

When a right triangle is in the coordinate plane, you can use the distance formula in addition to trigonometric ratios to solve the triangle.

4 Triangle XYZ has vertices $X(-4, -3)$, $Y(5, -3)$, and $Z(-4, 2)$. Find the side lengths to the nearest tenth and the measure of each angle to the nearest degree.

Plot the coordinates on a graph.

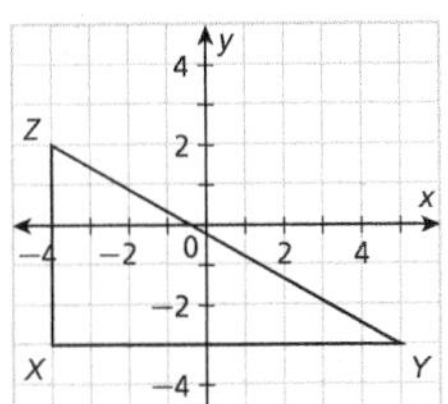

A. We can count the units on the coordinate plane.

Find the side lengths.

$XY = 9$, $XZ = 5$

A. Why is there no calculation needed to find these side lengths?

Use the distance formula to find the length of $\overline{YZ}$.

$YZ = \sqrt{(-4-5)^2 + (2-(-3))^2}$

$YZ = \sqrt{(-9)^2 + (5)^2}$

$YZ = \sqrt{81 + 25} = \sqrt{106} \approx 10.3$

B. One leg is a vertical segment and the other leg is a horizontal segment, which must meet at a 90° angle.

Find the angle measures.

$m\angle X = 90°$

B. How do you know this from the graph?

Use inverse tangent to find $m\angle Y$. C. The tangent ratio uses whole number values that are not rounded and will give a more precise answer.

$\tan Y = \frac{5}{9}$

$\tan^{-1}\left(\frac{5}{9}\right) = m\angle Y$

$m\angle Y \approx 29.1°$

C. Why do you use the inverse tangent to find $m\angle Y$ rather than the other inverse trigonometric functions?

$\angle Z$ and $\angle Y$ are complementary, so $m\angle Z \approx 90° - 29.1° = 60.9°$.

D. Is there only one correct method to solve this triangle? If not, outline a different series of steps you could use to solve this triangle. See margin.

Turn and Talk What would change in the solution method if a right triangle has no horizontal or vertical sides? See margin.

Task 4 (MP) **Attend to Precision** Students graph a right triangle in the coordinate plane given the coordinates of its vertices and use the Distance Formula to determine the measure of a side of the triangle and the measure of an angle using the inverse tangent.

Sample Guided Discussion:

Q How could you use the Distance Formula to find XZ and XY? $XZ = \sqrt{(-4-(-4))^2 + (2-(-3))^2} = 5$

$XY = \sqrt{(-4-5)^2 + ((-3)-(-3))^2} = 9$

Q What trigonometric expression can you use to find the measure of $\angle Z$? $m\angle Z = \tan^{-1}\left(\frac{9}{5}\right)$

Q The measures of the acute angles of $\triangle XYZ$ are nearly equal to 30 degrees and 60 degrees. How could you use those angle measures to verify that YZ equals about 10.3? Possible answer: The measure of the side opposite the 30-degree angle in a 30°-60°-90° triangle is half the length of the hypotenuse. Side $XZ = 5$, so if $\angle Y = 30°$, then YZ would be 10, which is approximately equal to 10.3.

Turn and Talk Remind students that the coordinate plane used to plot points and draw the triangle here is a rectangular coordinate plane, which means that the grid consists of intersecting horizontal and vertical lines. Possible answer: You would have to use the slopes of the lines to prove the triangle was a right triangle. The slopes of the lines must be opposite reciprocals for the lines to be perpendicular.

ANSWERS

D. no; Possible answer: You could use the Pythagorean Theorem to find the length of the hypotenuse and then use sine or cosine to find both angles.

ONLINE

Assign the Digital On Your Own for
- built-in student supports
- Actionable Item Reports
- Standards Analysis Reports

On Your Own

Assignment Guide

The chart below indicates which problems in the On Your Own are associated with each task in the Learn Together. Assign daily homework for tasks completed.

Learn Together Tasks	On Your Own Problems
Task 1, p. 430	Problems 5–8
Task 2, p. 431	Problems 9, 10, and 20
Task 3, p. 432	Problems 11, 12, and 14–19
Task 4, p. 433	Problem 13

Check Understanding

1. In an acute scalene triangle, you are given two adjacent side lengths and the included angle measure. How can you find the area of the triangle? Take one-half the product of the two side lengths and the sine of the included angle.
2. In $\triangle XYZ$, $\angle Y$ is a right angle, $m\angle X = 22°$, and YZ is 5 cm. Solve the triangle. Round to the nearest hundredth. $m\angle Z = 68°$, $XY \approx 12.38$, $ZX \approx 13.35$
3. A tower casts a 24-meter long shadow. The sun is situated where an angle of depression is 67° from an imaginary horizontal line through the top of the tower. How tall is the tower? about 56.5 m
4. Solve the triangle. Round to the nearest hundredth.

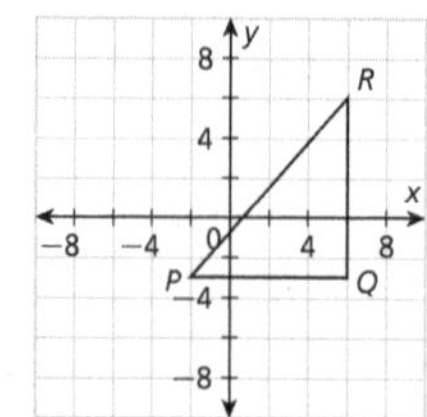

$PQ = 8$, $QR = 9$, $PR \approx 12.04$, $m\angle P = 48.37°$, $m\angle R = 41.63°$

On Your Own

Find the area of each triangle. Round to the nearest tenth.

5. $\triangle PQR$, where $PR = 2.5$ in., $QR = 4$ in., and $m\angle R = 55°$ about 4.1 in^2
6. $\triangle XYZ$, where $XY = 7$ cm, $XZ = 4.2$ cm, and $m\angle X = 68°$ about 13.6 cm^2
7. 5 m, 30°, 5 m — about 6.3 m^2
8.

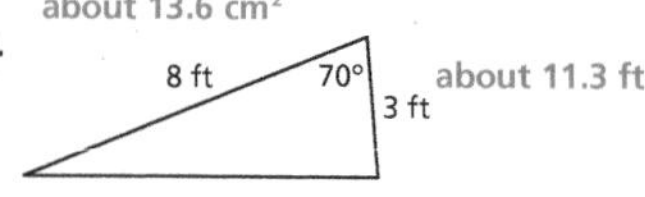

about 11.3 ft^2

Solve each triangle. Round side lengths to the nearest tenth and angles to the nearest degree.

9. $\triangle ABC$, where $\overline{AB} \perp \overline{BC}$, $AC = 5.8$ in., and $m\angle C = 35°$ $AB \approx 3.3$ in., $BC \approx 4.8$ in., $m\angle A = 55°$
10. $\triangle LMN$, where $m\angle M = 90°$, $m\angle L = 24°$, and $LM = 11$ cm $LN \approx 12.0$ cm, $MN \approx 4.9$ cm, $m\angle N = 66°$
11. A 12-foot ladder leans against a tree with an angle of elevation of 75°. Round values to the nearest tenth.
    A. How far up the tree does the ladder go? about 11.6 ft
    B. What is the distance between the tree and the base of the ladder? about 3.1 ft

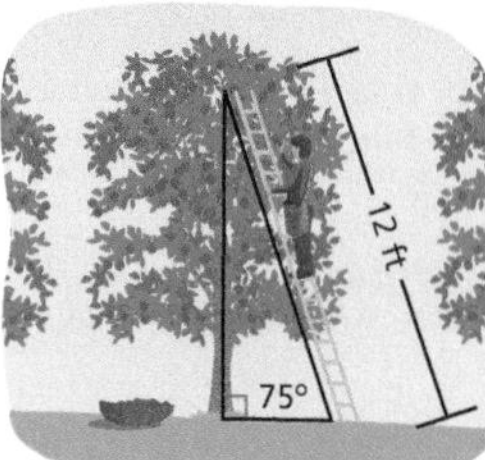

434

③ Check Understanding

Formative Assessment

Use formative assessment to determine if your students are successful with this lesson's learning objective.

Students who successfully complete the Check Understanding can continue to the On Your Own practice.

For students who miss 1 problem or more, work in a pulled small group using the Almost There small-group activity on page 429C.

ONLINE

Assign the Digital Check Understanding to determine
- success with the learning objective
- items to review
- grouping and differentiation resources

④ Differentiation Options

Differentiate instruction for all students using small-group activities and math center activities on page 429C.

Reteach

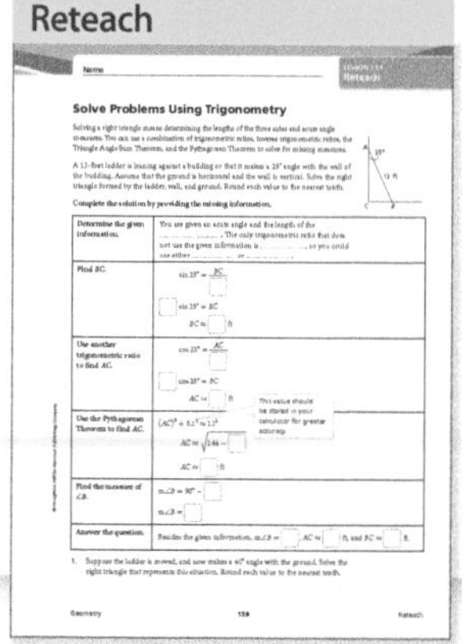

Challenge

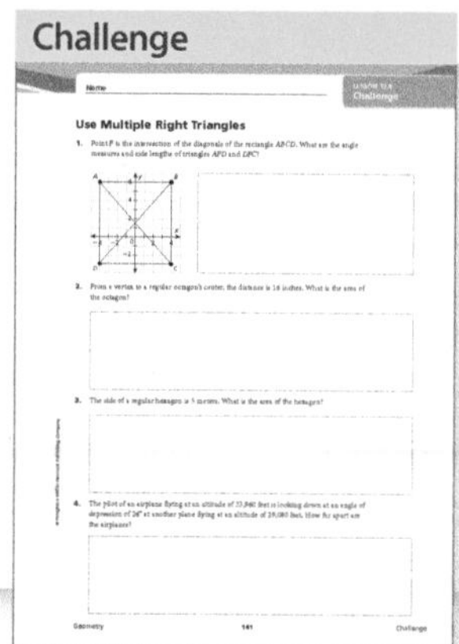

12. The main sail of the sailboat shown at the right has the shape of a right triangle. The boom is 20 feet in length. What is the measure of the angle of elevation of the sail when it is pulled 45 feet up the mast? $m\angle C \approx 66.0°$

13. For Parts A–D, solve each triangle. Round side lengths to the nearest tenth and angles to the nearest degree.
A–D. See margin.

A. $\triangle JKL$: vertices $J(-3, -2)$, $K(6, -2)$, and $L(-3, 7)$

B. $\triangle PQR$: vertices $P(-5, 9)$, $Q(5, 4)$, and $R(-1, 1)$

C.

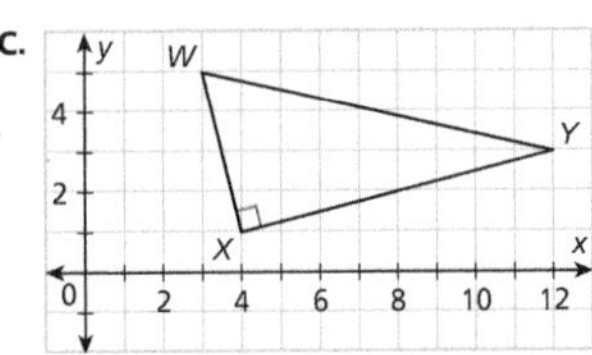

D.

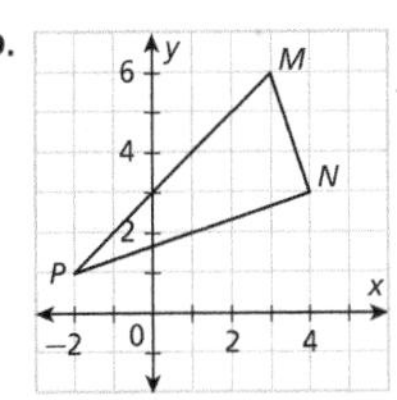

14. **Open Ended** Create a real-world problem involving angle of elevation or depression. Draw and label a diagram, then solve the problem.
Check students' diagrams and calculations.

15. **History** The dimensions of the Great Pyramid of Giza have changed with time. The original height was 146.5 m, and the base distance to the vertical was 115 m. The Egyptians believed that the pyramid would be most pleasing to the eye if the angle of elevation between one of the congruent sides and the ground were $\frac{1}{7}$ of the number of degrees of a full circle. Was the pyramid constructed correctly? Explain. See margin.

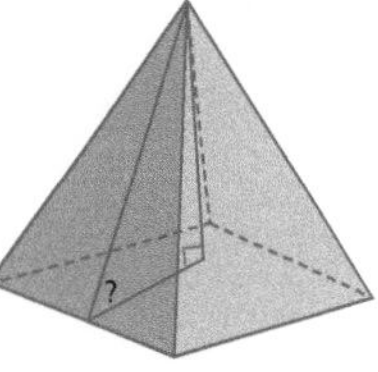

16. **Critique Reasoning** Paige and Bryan try to solve the following problem. Which student solved the problem incorrectly? Explain. See margin.

"A squirrel in a tree looks down toward a dog at a 52° angle of depression from the horizontal. The dog is 6 ft from the base of the tree. How high up the tree is the squirrel?"

Bryan's Work

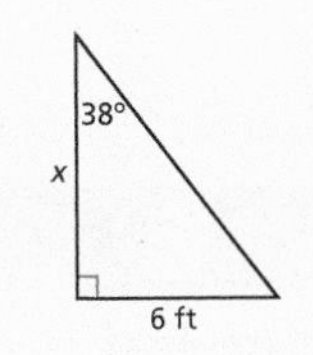

$\tan 38° = \frac{6}{x}$

$x \tan 38° = 6$

$x = \frac{6}{\tan 38°}$

$x \approx 7.7$ ft

Paige's Work

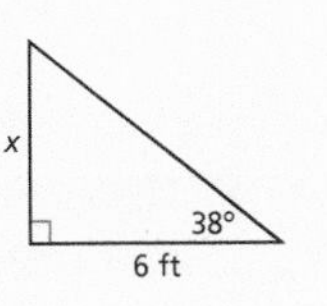

$\tan 38° = \frac{x}{6}$

$x = 6 \tan 38°$

$x \approx 4.7$ ft

Questioning Strategies

Problem 13C Triangle WXY appears to be a right triangle but no sides are horizontal or vertical. How can you verify that it is a right triangle? After verifying that the triangle is right triangle, how can you find the lengths of its sides? How can you find the measures of the angles?

- Find the slopes of the three sides of the triangle and determine whether the slopes of two sides are opposite reciprocals of each.
- To find the sides, substitute the coordinates of the endpoints of each side into the Distance Formula and calculate the lengths of all three sides. Or after finding the lengths of two sides, use the Pythagorean Theorem to find the length of the third side.
- Use trigonometry to find the measure of one of the acute angles and subtract it from 90°.

Watch for Common Errors

Problem 16 Students often confuse which is an angle of depression and which is the angle of elevation in a right triangle problem situation. A side of each angle is horizontal, but as its name suggests, an angle of depression is associated with looking down from the horizontal, and therefore, lies outside a right triangle. An angle of elevation, as its name suggests, is associated with looking upwards and is in the interior of a right triangle. If Paige had drawn the horizontal line through the upper vertex of the triangle, she might have realized that the angle of depression was outside the triangle, and the correct location of the 38° angle had to be adjacent to it.

ANSWERS

13.A. $JK = 9$, $JL = 9$, $LK = 12.7$, $m\angle J = 90°$, $m\angle K = 45°$, $m\angle L = 45°$

13.B. $QR = 6.7$, $PR = 8.9$, $PQ = 11.2$, $m\angle P = 37°$, $m\angle R = 90°$, $m\angle Q = 53°$

13.C. $WX = 4.1$, $WY = 9.2$, $XY = 8.2$, $m\angle W = 63°$, $m\angle X = 90°$, $m\angle Y = 27°$

13.D. $MN = 3.2$, $MP = 7.1$, $NP = 6.3$, $m\angle M = 63°$, $m\angle N = 90°$, $m\angle P = 27°$

15. yes, Since $\frac{1}{7}$ of $360° \approx 51.4°$ and the angle of elevation is $\tan^{-1}\left(\frac{146.5}{115}\right) \approx 51.87°$, the pyramid was built as the Egyptians had planned.

16. Paige; She placed the angle as an angle of elevation. The angle between the tree and the line of sight between the squirrel and the dog is $90° - 52° = 38°$.

5 Wrap-Up

Summarize learning with your class. Consider using the Exit Ticket, Put It in Writing, or I Can scale.

Exit Ticket 

A firefighter's ladder leans against a building just below a window that is 15 feet above the ground. The angle of depression from the viewpoint of a firefighter leaning out the window is 60°. To the nearest tenth, how long is the ladder? How far from the base of the building is the foot of the ladder? about 17.3 feet; about 8.7 feet

Put It in Writing 

Have students work in pairs. Partners draw and label a right triangle on coordinate grid and exchange triangles. Partners write trigonometric ratios for each acute angle, explain how they were determined, and check each other's work.

I Can 

The scale below can help you and your students understand their progress on a learning goal.

4	I can use trigonometric ratios, the area formula for a triangle in terms of its side lengths, and the Pythagorean Theorem to solve right triangles in applied problems, and I can explain my steps.
3	I can use trigonometric ratios, the area formula for a triangle in terms of its side lengths, and the Pythagorean Theorem to solve right triangles in applied problems.
2	I can define the trigonometric ratios for the acute angles in right triangles and use them to solve problems.
1	I can use the Pythagorean Theorem to solve problems in right triangles.

Spiral Review • Assessment Readiness

These questions will help determine if students have retained information taught in the past and can also prepare them for high-stakes assessments. Here, students determine the relationships between the sine and cosine of complementary angles and the ratio of sides in an isosceles right triangle **(13.2 and 13.3)**, and match values to a set of trigonometric expressions **(13.2)**.

17. From the top of a lighthouse, a watchman sites a ship at an angle of depression of 10°. In the same direction, the watchman sees a second ship and measures the angle of depression to the ship as 14.2°. What is the distance between the two ships? about 309 ft

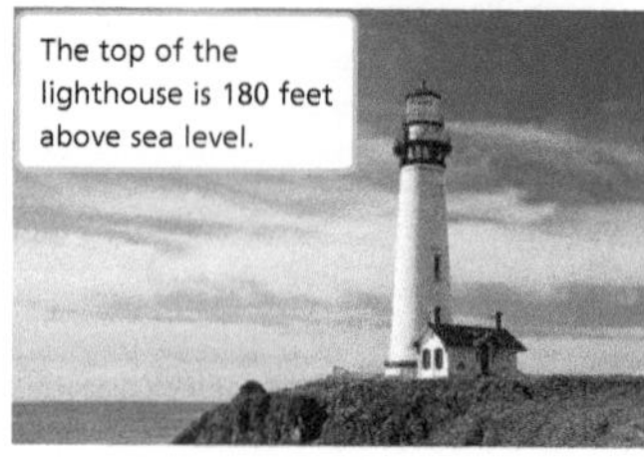

The top of the lighthouse is 180 feet above sea level.

18. MP **Attend to Precision** A tourist 20 miles from the base of a mountain looks up at the peak of the mountain at an angle of elevation of 15°. When the tourist is 10 miles away from the base of the mountain, what is the angle of elevation from his point of view to the peak of the mountain? about 28.2°

19. A man whose eye level is 6 feet off the ground looks up at an angle of elevation of 55° at his friend climbing a rock wall. The man is standing 8 feet from the base of the wall. How high up the rock wall is his friend? about 17.4 ft

20. Open Middle™ Using the digits 1 to 9, at most two times each, fill in the boxes to create two similar right triangles. There are many possibilities, including: 3, 4, 5 and 18, 24, 30.

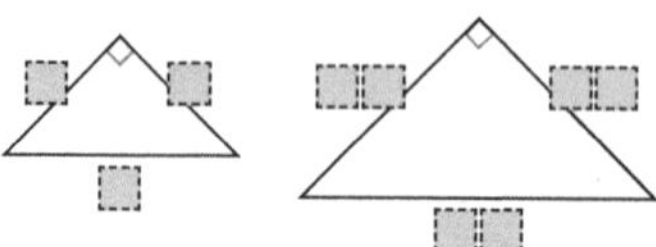

Spiral Review • Assessment Readiness

21. For what values is the statement true? $\sin(x + 20)^\circ = \cos(y - 10)^\circ$

Ⓐ $x = 70, y = 20$
Ⓑ $x = 30, y = 60$
Ⓒ $x = 45, y = 45$
Ⓓ $x = 20, y = 60$

22. For which set of side lengths could the tangent ratio of an angle within a right triangle equal 1?

Ⓐ $1:1\sqrt{3}:2$
Ⓑ $3:3:3\sqrt{2}$
Ⓒ $\sqrt{2}:\sqrt{2}:2\sqrt{2}$
Ⓓ $\sqrt{2}:\frac{\sqrt{2}}{2}:2$

23. Match the trigonometric expression on the left with its approximate decimal value on the right.

Expression	Value
A. sin 38° A. 4	**1.** 0.921
B. cos 38° B. 3	**2.** 0.391
C. sin 67° C. 1	**3.** 0.788
D. cos 67° D. 2	**4.** 0.616

 **I'm in a Learning Mindset!**

How did an initial failure with learning trigonometric ratios improve my learning methods?

Keep Going to Journal and Practice Workbook

Learning Mindset

Perseverance Learns Effectively

Trigonometric ratios may initially be difficult for students. To help students, review similarity with right triangles. Draw a series of nested right triangles on the board. Label the sides of the smallest triangle with a Pythagorean Triples like 3-4-5. Then point out that the corresponding acute angles in each triangle are congruent, and that the ratios between corresponding sides are all equal so the sine, cosine, and tangent of each angle in this set of triangles is always constant.

5 3 4

What are possible lengths of the larger triangles in the figure? What is the sine of the smallest angle in each triangle? What is the cosine? What are the ratios in the largest triangle? What is always true about these values?

Module 13 Review

Trigonometric Ratios

Due to the proportionality of side lengths in similar figures, side ratios in right triangles are properties of the angles.

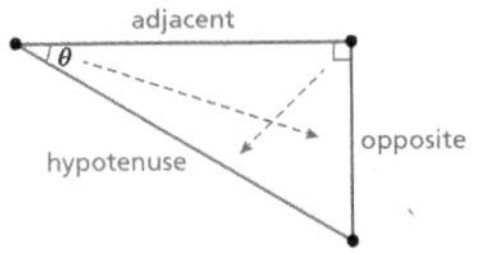

sine	$\sin(\theta) = \frac{\text{opposite}}{\text{hypotenuse}}$
cosine	$\cos(\theta) = \frac{\text{adjacent}}{\text{hypotenuse}}$
tangent	$\tan(\theta) = \frac{\text{opposite}}{\text{adjacent}}$

Missing Side Length

Ava and Nia are hiking up a mountain. The trail has a length of 20,643 feet and rises with a steady angle of elevation of 15 degrees.

At the summit, Ava and Nia wonder how tall the mountain is.

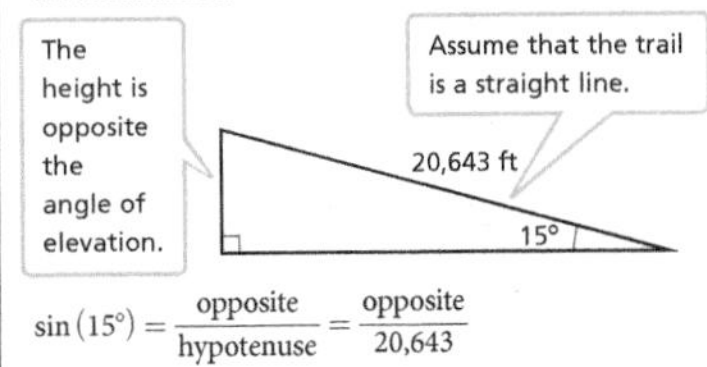

$\sin(15°) = \frac{\text{opposite}}{\text{hypotenuse}} = \frac{\text{opposite}}{20{,}643}$

$\rightarrow \text{opposite} = 20{,}643 \sin(15°) \approx 5343 \text{ ft}$

The height of the mountain is about 5343 feet.

Missing Angle Measure

Ava and Nia want to take the shortest distance back to town. They know that the town lies 5.3 miles north and 6.5 east of their current location.

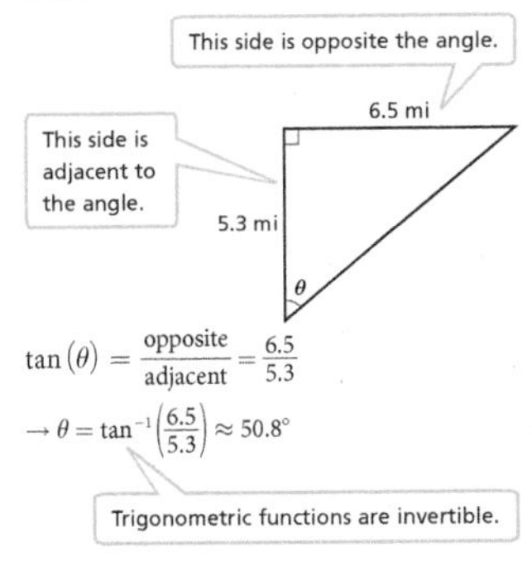

$\tan(\theta) = \frac{\text{opposite}}{\text{adjacent}} = \frac{6.5}{5.3}$

$\rightarrow \theta = \tan^{-1}\left(\frac{6.5}{5.3}\right) \approx 50.8°$

Trigonometric functions are invertible.

Ava and Nia should hike 50.8° east of north.

Special Right Triangles

There are two special right triangles that are helpful to remember.

45-45-90 triangle

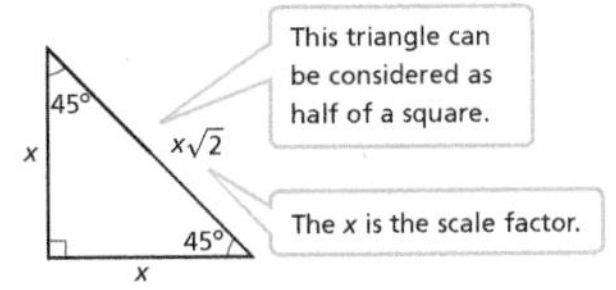

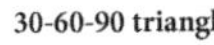

30-60-90 triangle

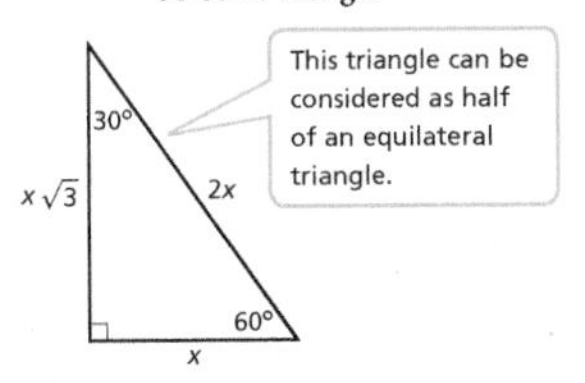

ONLINE

Assign the Digital Module Review for

- built-in student supports
- Actionable Item Reports
- Standards Analysis Reports

MODULE 13 REVIEW

Module Review

Use the first page of the Module Review to summarize and connect the main ideas of the module. Use the second page to assess students' understanding of the vocabulary, concepts, and skills presented in the module.

Sample Guided Discussion:

Q Trigonometric Ratios Prove that $\frac{\sin A}{\cos A} = \tan A$.

$$\frac{\sin A}{\cos A} = \frac{\left(\frac{\text{opposite}}{\text{hypotenuse}}\right)}{\left(\frac{\text{adjacent}}{\text{hypotenuse}}\right)} = \frac{\text{opposite}}{\cancel{\text{hypotenuse}}} \times \frac{\cancel{\text{hypotenuse}}}{\text{adjacent}} = \frac{\text{opposite}}{\text{adjacent}} = \tan A$$

Q Missing Side Length What trigonometric ratio is equal to 15°? cos 75°

Q Missing Angle Measure What is $\tan(90° - \theta)$ in terms of the sides of the given triangle and what is it equal to? $\tan(90° - \theta) = \frac{5.3}{6.5} \approx 39.2°$

Q Special Right Triangles What is the product of sin 60° and sin 45° in simplest radical form? $\sin 60° = \frac{\sqrt{3}}{2}$ and $\sin 45° = \frac{\sqrt{2}}{2}$, so the product is $\sin 60° \cdot \sin 45° = \frac{\sqrt{3}}{2} \cdot \frac{\sqrt{2}}{2} = \frac{\sqrt{6}}{4}$.

Module Review continued

Possible Scoring Guide

Items	Points	Description
1–4	2 each	identifies the correct term
5	1	correctly finds sine, cosine, and tangent for angles of a triangle and determines the measures of the angles
5	2	correctly finds sine, cosine, and tangent for angles of the triangle, determines the measures of the angles, describes the relationship between sine and cosine of complementary angles, and determines the measure of an angle in a similar triangle
6, 7	2 each	correctly solves the right triangle
8	2	correctly uses trigonometry to find the area of the triangle
9	2	correctly uses a trigonometric ratio to find Haylee's distance from the camp
10	2	correctly identifies all expressions equal to $\frac{2}{3}$
11	2	correctly uses a trigonometric ratio to find the height of the kite
Total points possible = 22 points		

The Unit 7 Test in the Assessment Guide assesses content from Modules 13 and 14.

Vocabulary

Choose the correct term from the box to complete each sentence.

Vocabulary
angle of elevation
cosine
sine
tangent

1. In a right triangle, the __?__ is the ratio of the length of the leg opposite an angle to the length of the leg adjacent to an angle. tangent
2. In a right triangle, the __?__ is the ratio of the length of the leg adjacent to an angle to the length of the hypotenuse. cosine
3. The __?__ is the angle between the horizontal and the line of sight when an observer is looking up at an object. angle of elevation
4. In a right triangle, the __?__ is the ratio of the length of the leg opposite an angle to the length of the hypotenuse. sine

Concepts and Skills

5–7. See Additional Answers.

5. $\triangle DEF \sim \triangle ABC$ by a factor of k.
   A. Determine the sine, cosine, and tangent of $\angle A$ and $\angle B$.
   B. Determine $m\angle A$ and $m\angle B$.
   C. Explain the relationship between the sine and cosine of complementary angles and justify your reasoning.
   D. Determine $\sin(m\angle D)$ and justify your reasoning.

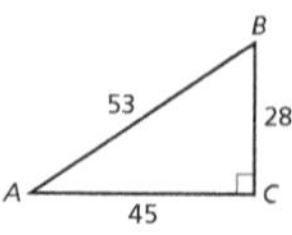

Solve each triangle.

6. 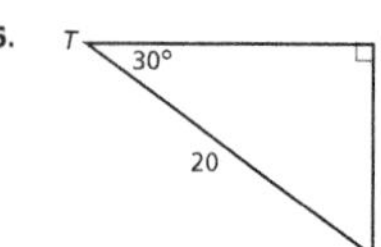

7. 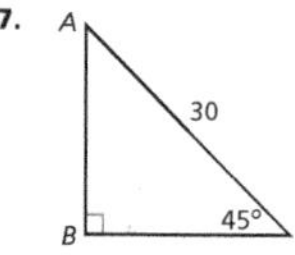

8. Find the area of $\triangle ABC$, with $a = 14$ inches, $b = 9$ inches, and $m\angle C = 42°$. Round your answer to the nearest tenth. 42.2 in²
9. Haylee hikes to the top of a 120-foot vertical cliff. From the top of the cliff, the angle of depression to her campsite is 10°. How far away from the campsite is the base of the cliff? Round to the nearest foot. 681 ft
10. A right triangle has an angle measure x with $\sin(x) = \frac{2}{3}$. Determine if each expression is also equal to $\frac{2}{3}$.
    A. $\cos(x)$ no
    B. $\cos(90° - x)$ yes
    C. $\sin(90° - x)$ no
    D. $\sin(45° - x)$ no
11. (MP) Use Tools Ashley is flying a kite and has let out 150 feet of string. The angle of elevation from where Ashley is holding the string to the kite is 50°. If she is holding the string 4 feet above the ground, how high is the kite above the ground? Round to the nearest foot. State what strategy and tool you will use to answer the question, explain your choice, and then find the answer. 119 feet

438

DATA-DRIVEN INSTRUCTION

Before moving on to the Module Test, use the Module Review results to intervene based on the table below.

Items	Lesson	DOK	Content Focus	Intervention
5	13.1, 13.2	3	Use trigonometry to find relationships between sides and angles in similar triangles.	Reteach 13.1, 13.2
6, 7	13.3	2	Solve special right triangles.	Reteach 13.3
8	13.2	2	Use trigonometry to find the area of a right triangle.	Reteach 13.2
9	13.4	3	Use trigonometry to solve a problem involving an angle of depression.	Reteach 13.4
10	13.2	2	Determine equivalent trigonometric ratios.	Reteach 13.2
11	13.4	3	Use trigonometry to solve a problem involving an angle of elevation.	Reteach 13.4

Module Test

The Module Test is available in alternative versions in your Assessment Guide. All items are presented in standardized test formats.

ONLINE

Ed

Assign the Digital Module Test to power actionable reports including

- proficiency by standards
- item analysis

Form A

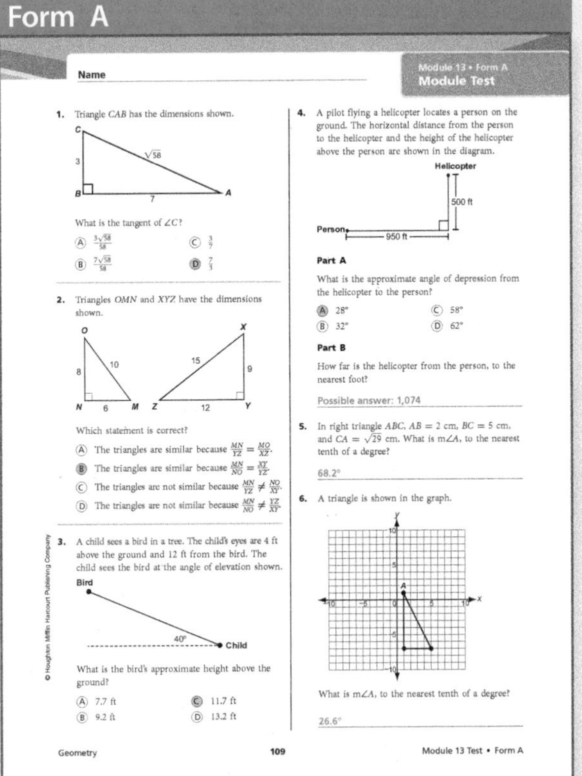

Name

Module 13 • Form A
Module Test

1. Triangle *CAB* has the dimensions shown.

What is the tangent of ∠*C*?

Ⓐ $\frac{3\sqrt{58}}{58}$ Ⓒ $\frac{3}{7}$
Ⓑ $\frac{7\sqrt{58}}{58}$ Ⓓ $\frac{7}{3}$

2. Triangles *OMN* and *XYZ* have the dimensions shown.

Which statement is correct?

Ⓐ The triangles are similar because $\frac{MN}{YZ} = \frac{MO}{XZ}$.
Ⓑ The triangles are similar because $\frac{MN}{NO} = \frac{XY}{YZ}$.
Ⓒ The triangles are not similar because $\frac{MN}{YZ} \neq \frac{NO}{XY}$.
Ⓓ The triangles are not similar because $\frac{MN}{NO} \neq \frac{YZ}{XZ}$.

3. A child sees a bird in a tree. The child's eyes are 4 ft above the ground and 12 ft from the bird. The child sees the bird at the angle of elevation shown.

What is the bird's approximate height above the ground?

Ⓐ 7.7 ft Ⓒ 11.7 ft
Ⓑ 9.2 ft Ⓓ 13.2 ft

4. A pilot flying a helicopter locates a person on the ground. The horizontal distance from the person to the helicopter and the height of the helicopter above the person are shown in the diagram.

Part A

What is the approximate angle of depression from the helicopter to the person?

Ⓐ 28° Ⓒ 58°
Ⓑ 32° Ⓓ 62°

Part B

How far is the helicopter from the person, to the nearest foot?

Possible answer: 1,074

5. In right triangle *ABC*, *AB* = 2 cm, *BC* = 5 cm, and $CA = \sqrt{29}$ cm. What is m∠*A*, to the nearest tenth of a degree?

68.2°

6. A triangle is shown in the graph.

What is m∠*A*, to the nearest tenth of a degree?

26.6°

© Houghton Mifflin Harcourt Publishing Company

Geometry 109 Module 13 Test • Form A

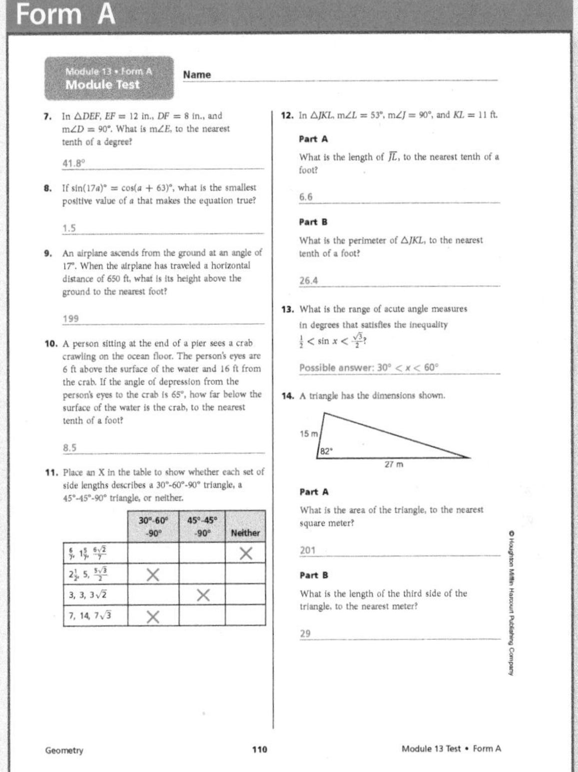

Module 13 • Form A
Module Test

Name

7. In △*DEF*, *EF* = 12 in., *DF* = 8 in., and m∠*D* = 90°. What is m∠*E*, to the nearest tenth of a degree?

41.8°

8. If sin(17*a*)° = cos(*a* + 63)°, what is the smallest positive value of *a* that makes the equation true?

1.5

9. An airplane ascends from the ground at an angle of 17°. When the airplane has traveled a horizontal distance of 650 ft, what is its height above the ground to the nearest foot?

199

10. A person sitting at the end of a pier sees a crab crawling on the ocean floor. The person's eyes are 6 ft above the surface of the water and 16 ft from the crab. If the angle of depression from the person's eyes to the crab is 65°, how far below the surface of the water is the crab, to the nearest tenth of a foot?

8.5

11. Place an X in the table to show whether each set of side lengths describes a 30°-60°-90° triangle, a 45°-45°-90° triangle, or neither.

	30°-60°-90°	45°-45°-90°	Neither
$\frac{6}{7}, 1\frac{5}{7}, \frac{6\sqrt{2}}{7}$			X
$2\frac{1}{2}, 5, \frac{5\sqrt{3}}{2}$	X		
$3, 3, 3\sqrt{2}$		X	
$7, 14, 7\sqrt{3}$	X		

12. In △*JKL*, m∠*L* = 53°, m∠*J* = 90°, and *KL* = 11 ft.

Part A

What is the length of $\overline{JL}$, to the nearest tenth of a foot?

6.6

Part B

What is the perimeter of △*JKL*, to the nearest tenth of a foot?

26.4

13. What is the range of acute angle measures in degrees that satisfies the inequality $\frac{1}{2} < \sin x < \frac{\sqrt{3}}{2}$?

Possible answer: 30° < *x* < 60°

14. A triangle has the dimensions shown.

Part A

What is the area of the triangle, to the nearest square meter?

201

Part B

What is the length of the third side of the triangle, to the nearest meter?

29

© Houghton Mifflin Harcourt Publishing Company

Geometry 110 Module 13 Test • Form A

Form B

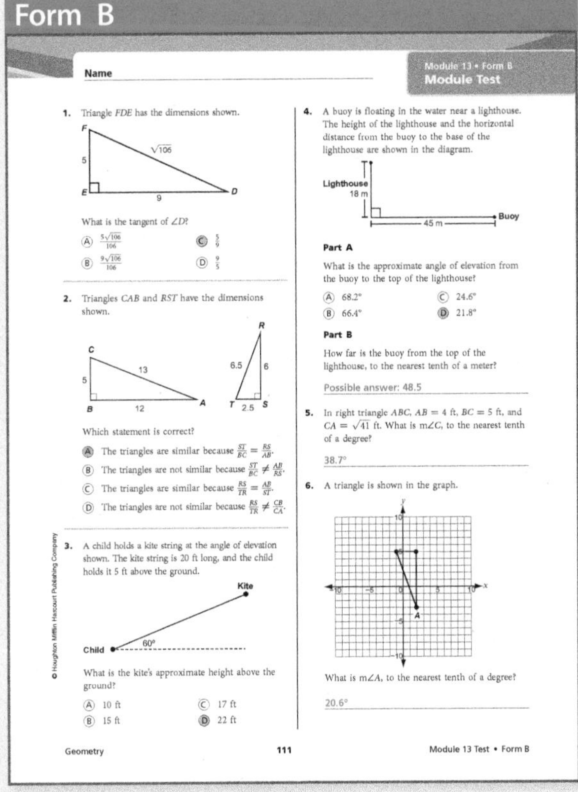

Name

Module 13 • Form B
Module Test

1. Triangle *FDE* has the dimensions shown.

What is the tangent of ∠*D*?

Ⓐ $\frac{5\sqrt{106}}{106}$ Ⓒ $\frac{5}{9}$
Ⓑ $\frac{9\sqrt{106}}{106}$ Ⓓ $\frac{9}{5}$

2. Triangles *CAB* and *RST* have the dimensions shown.

Which statement is correct?

Ⓐ The triangles are similar because $\frac{ST}{BC} = \frac{RS}{AB}$.
Ⓑ The triangles are not similar because $\frac{ST}{BC} \neq \frac{AB}{RS}$.
Ⓒ The triangles are similar because $\frac{RS}{TR} = \frac{AB}{AT}$.
Ⓓ The triangles are not similar because $\frac{RS}{TR} \neq \frac{CB}{CA}$.

3. A child holds a kite string at the angle of elevation shown. The kite string is 20 ft long, and the child holds it 5 ft above the ground.

What is the kite's approximate height above the ground?

Ⓐ 10 ft Ⓒ 17 ft
Ⓑ 15 ft Ⓓ 22 ft

4. A buoy is floating in the water near a lighthouse. The height of the lighthouse and the horizontal distance from the buoy to the base of the lighthouse are shown in the diagram.

Part A

What is the approximate angle of elevation from the buoy to the top of the lighthouse?

Ⓐ 68.2° Ⓒ 24.6°
Ⓑ 66.4° Ⓓ 21.8°

Part B

How far is the buoy from the top of the lighthouse, to the nearest tenth of a meter?

Possible answer: 48.5

5. In right triangle *ABC*, *AB* = 4 ft, *BC* = 5 ft, and $CA = \sqrt{41}$ ft. What is m∠*C*, to the nearest tenth of a degree?

38.7°

6. A triangle is shown in the graph.

What is m∠*A*, to the nearest tenth of a degree?

20.6°

© Houghton Mifflin Harcourt Publishing Company

Geometry 111 Module 13 Test • Form B

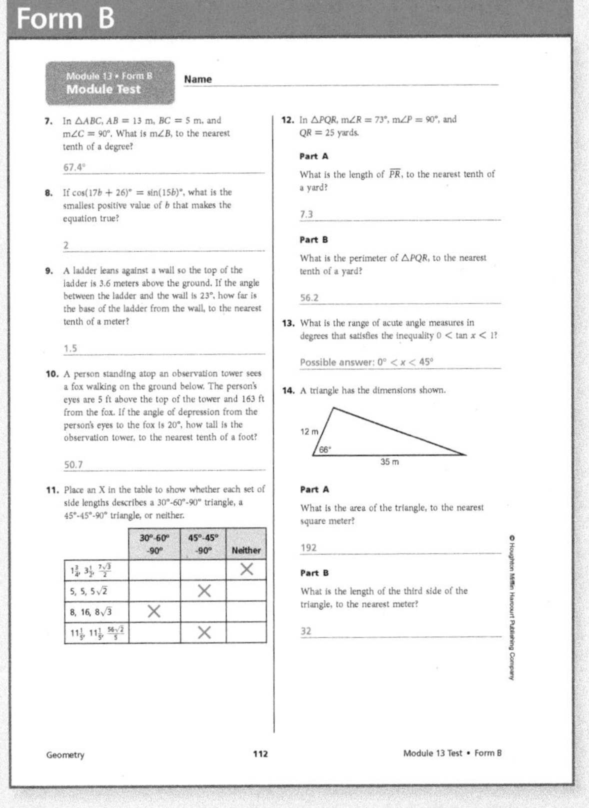

Module 13 • Form B
Module Test

Name

7. In △*ABC*, *AB* = 13 m, *BC* = 5 m, and m∠*C* = 90°. What is m∠*B*, to the nearest tenth of a degree?

67.4°

8. If cos(17*b* + 26)° = sin(15*b*)°, what is the smallest positive value of *b* that makes the equation true?

2

9. A ladder leans against a wall so the top of the ladder is 3.6 meters above the ground. If the angle between the ladder and the wall is 23°, how far is the base of the ladder from the wall, to the nearest tenth of a meter?

1.5

10. A person standing atop an observation tower sees a fox walking on the ground below. The person's eyes are 5 ft above the top of the tower and 163 ft from the fox. If the angle of depression from the person's eyes to the fox is 20°, how tall is the observation tower, to the nearest tenth of a foot?

50.7

11. Place an X in the table to show whether each set of side lengths describes a 30°-60°-90° triangle, a 45°-45°-90° triangle, or neither.

	30°-60°-90°	45°-45°-90°	Neither
$1\frac{3}{4}, 3\frac{1}{2}, \frac{7\sqrt{3}}{2}$			X
$5, 5, 5\sqrt{2}$		X	
$8, 16, 8\sqrt{3}$	X		
$11\frac{1}{5}, 11\frac{1}{5}, \frac{56\sqrt{2}}{5}$		X	

12. In △*PQR*, m∠*R* = 73°, m∠*P* = 90°, and *QR* = 25 yards.

Part A

What is the length of $\overline{PR}$, to the nearest tenth of a yard?

7.3

Part B

What is the perimeter of △*PQR*, to the nearest tenth of a yard?

56.2

13. What is the range of acute angle measures in degrees that satisfies the inequality 0 < tan *x* < 1?

Possible answer: 0° < *x* < 45°

14. A triangle has the dimensions shown.

Part A

What is the area of the triangle, to the nearest square meter?

192

Part B

What is the length of the third side of the triangle, to the nearest meter?

32

© Houghton Mifflin Harcourt Publishing Company

Geometry 112 Module 13 Test • Form B

TRIGONOMETRY WITH ALL TRIANGLES

Introduce and Check for Readiness

- Module Performance Task • Are You Ready?

Lesson 14.1—2 Days

Law of Sines

Learning Objective: Prove the Law of Sines, determine which combinations of given triangle measures are sufficient to apply the Law of Sines, determine how many triangles can exist with an ambiguous (SSA) combination of given measures, and apply the Law of Sines to determine unknown triangle measures.

Lesson 14.2—2 Days

Law of Cosines

Learning Objective: Derive the Law of Cosines, recognize the SAS and SSS cases where the Law of Cosines is applicable and apply it to find unknown measures, derive the area formula $A = \frac{1}{2}ab \sin C$ for a triangle, and find the area of triangles when given an SAS or SSS combination of measures.

Assessment

- Module 14 Test (Forms A and B)
- Unit 7 Test (Forms A and B)

LEARNING ARC FOCUS

 Build Conceptual Understanding  Connect Concepts and Skills  Apply and Practice

TEACHING FOR SUCCESS

TEACHING FOR DEPTH: Trigonometry with All Triangles

Make Connections. The basic conditions for the existence of a triangle depends on the relationship among the lengths of the three sides of the triangle and the sum of the measures of its angles. That is, the sum of the lengths of any two sides is always greater than the length of the third side, and the sum of the measures of the angles is equal to 180°.

Thus, it is possible to construct a unique triangle having the lengths of its three sides (SSS), two angles and a side (ASA), or two sides and the included angle (SAS). But what about two sides and the non-included angle, known as SSA?

In the first lesson in this module that deals with the Law of Sines, students learn about the ambiguous case, where *ambiguous* means *uncertain*. In the case of SSA, where the given angle is not included between the two sides, there can be zero, one, or two triangles.

Whether there are 0, 1, or 2 triangles depends on the lengths h, a, and c as follows:

- If $a < h$, there is no possible triangle.
- If $a = h$, there is 1 right triangle.
- If $h < a < c$, there are 2 possible triangles.
- If $a \geq c$, there is 1 possible triangle.

Mathematical Progressions

Prior Learning	Current Development	Future Connections
Students: • applied the Pythagorean Theorem to determine unknown side lengths in right triangles. • understood that by similarity, side ratios in right triangles are properties of the angles in the triangle, leading to definitions of trigonometric ratios.	**Students:** • use the Law of Sines to find side lengths and angle measures of non-right triangles and solve real-world problems. • use the Law of Cosines to find side lengths and angle measures of non-right triangles and solve real-world problems.	**Students:** • will identify and describe relationships among inscribed angles, radii, and chords. • will construct the inscribed and circumscribed circles of a triangle and prove properties of angles for a quadrilateral inscribed in a circle.

TEACHER TO TEACHER

From the Classroom

Elicit and use evidence of student thinking. My students often make sign errors when using the Law of Cosines. This often happens when the given included angle is obtuse.

At this point in the course, my students have not seen graphs of the trigonometric functions, so I have students draw a right triangle *ABC* with legs along the *x*- and *y*-axes and then reflect the triangle across the *y*-axis. Now they can prove that $\cos\left(180° - A\right) = -\cos A$.

In the figure, students can see that the two triangles are congruent. To find $\cos A$, use the right triangle in Quadrant 1: $\cos A = \frac{x}{r}$. To find $\cos\left(180 - A\right)$, use the right triangle in Quadrant 2: $\left(180 - A\right) = \frac{-x}{r}$. It follows that $\cos\left(180° - A\right) = -\cos A$. Therefore, my students can conclude if an angle is obtuse, its cosine is the opposite of the cosine of its supplement.

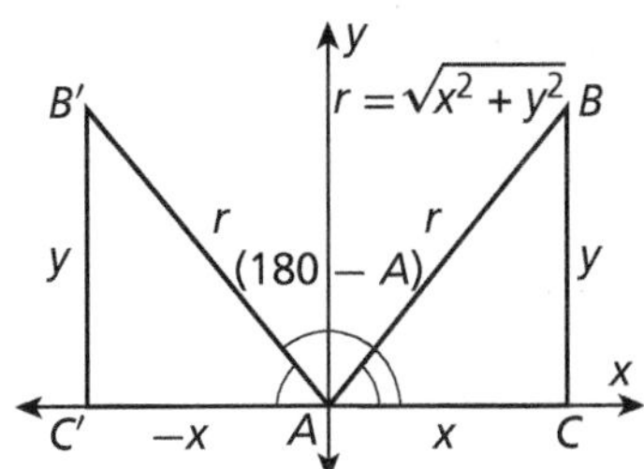

LANGUAGE DEVELOPMENT • Planning for Instruction

By giving all students regular exposure to language routines in context, you will provide opportunities for students to **listen**, **speak**, **read**, and **write** about mathematical situations and develop both mathematical language and conceptual understanding at the same time.

Using Language Routines to Develop Understanding

Use the **Professional Cards** for the following routines to plan for effective instruction.

Co-Craft Questions and Problems Lesson 14.1

Students think of natural questions to ask about a given situation or problems similar to a given task and answer the questions they have developed or problems they have created.

Three Reads Lesson 14.2

Students read a problem three times with a specific focus each time.

1st Read What is the situation about?
2nd Read What are the quantities in the situation?
3rd Read What are the possible mathematical questions that you could ask for the situation?

Information Gap Lesson 14.2

Students recognize when information given in a problem situation is incomplete, and they pose questions and share knowledge with others to discover any missing facts or relationships and work together to solve the problem.

Critique, Correct, and Clarify Lesson 14.1

Students correct the work in a flawed explanation, argument, or solution method and share with a partner and refine the sample work.

Linguistic Note

Listen for how students use the vocabulary necessary to understand the content of the two lessons in this module, the Law of Sines and the Law of Cosines. Write the words *acute, obtuse, opposite, adjacent, hypotenuse, supplementary, complementary, altitude, included, sine, cosine,* and *tangent* on the board. Have students create a table containing three columns whose headings are ***Angle, Segment,*** and ***Number***. Then have the students write each word in the proper category. Point out that *included* can refer to an angle or to a side. When students have completed their tables, have them take turns defining or explaining the meaning of each term.

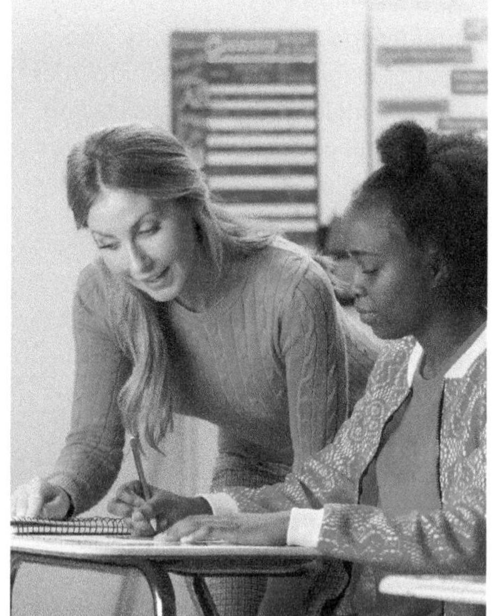

Module
14
Trigonometry with
All Triangles
Module Performance Task: Spies and Analysts™

Towering Over the Competition

How do you measure the height of the world's tallest cookie tower?

Connections to This Module

One sample solution might involve using a surveyor's tool. The tower appears to have the shape of a square pyramid, so begin by using the tool to measure one side of the square base of the tower. Then, using the surveyor's sight tool, you can measure the angle of elevation of the tower; that is, determine the measure of the angle formed between your horizontal line of sight and the top of the tower.

The measure of the angle of elevation and your distance from the midpoint of the base of the tower gives you the necessary information to set up the dimensions of the legs of a right triangle. The height h of the triangle is the perpendicular from the top of the tower to b, the base of the pyramid and equal to one-half of it. If you let A represent the angle of elevation, you can use the tangent ratio to find h.

$\tan A = \frac{h}{\left(\frac{b}{2}\right)}$, where $h = \left(\frac{b}{2}\right) \tan A$. Substituting the values for the tangent of A and $\frac{b}{2}$ and solving for h gives you an approximation of the height of the tower. **(14.2)**

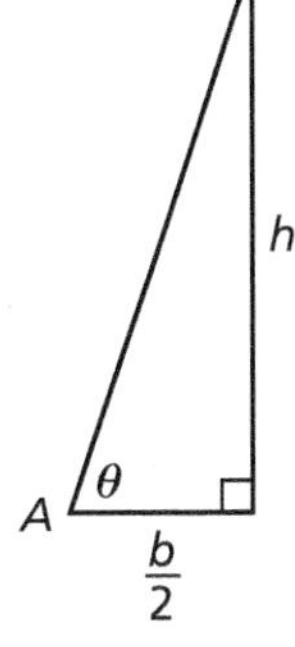

Towering Over the Competition

Overview

This problem requires students to make assumptions to determine the height of a tower made of cookies. Examples of assumptions:

- It is too difficult to get up close to the tower to measure its height directly because it might topple over.
- There is no information about the size or how many cookies were used to build the tower.

Be a *Spy*

First, students must determine what information is necessary to know. This should include:

- identifying the general shape of the tower.
- selecting the proper tools to measure the height from a distance.

Students should recognize that they need information about the length of the base of the tower and the measure of the angle formed by a side of the tower and its base.

Be an *Analyst*

Help students understand that the measurement must be found indirectly because it is not possible to actually get close enough to the tower and use ordinary measuring tapes or sticks.

Alternative Approaches

In their analysis, students might:

- use trigonometric methods they will learn in this module, as shown at the left.
- use lighting to create a shadow of the tower and the shadow of an observer and set up a proportion involving equivalent ratios between h, the unknown height of the tower, and the measured lengths of both shadows and the height of the observer.

ONLINE

Assign the Digital Are You Ready? to power actionable reports including

- proficiency by standards
- item analysis

Are You Ready?

data checkpoint

Diagnostic Assessment

- Diagnose prerequisite mastery.
- Identify intervention needs.
- Modify or set up leveled groups.

Have students complete the *Are You Ready?* assessment on their own. Items test the prerequisites required to succeed with the new learning in this module.

Find Equivalent Ratios Students will apply previous knowledge of finding equivalent ratios to prove and use the Law of Sines in real-world situations.

The Pythagorean Theorem and Its Converse Students will apply previous knowledge of the Pythagorean Theorem and its converse to prove and use the Law of Cosines in real-world situations.

Solve Quadratic Equations by Finding Square Roots Students will apply previous knowledge of solving quadratic equations by finding square roots to use the Law of Cosines in real-world situations.

ONLINE Ed

Are You Ready?

Complete these problems to review prior concepts and skills you will need for this module.

Find Equivalent Ratios

1. Carter walks 14 miles in 4 hours. How long will it take him to walk 24.5 miles? 7 hours
2. A store sells 2 pounds of bananas for \$1.58. How many pounds of bananas does the store sell for \$3.95? 5 pounds
3. Jose earns \$68 for working for 4 hours. How long will it take Jose to earn \$153? 9 hours

The Pythagorean Theorem and Its Converse

4. A right triangle has a leg that is 4 feet long and hypotenuse that is 6 feet long. How long is the other leg of the triangle? $2\sqrt{5}$ feet
5. A triangle has sides with lengths 10 inches, 24 inches, and 26 inches. Is the triangle a right triangle? Explain how you know. yes; $10^2 + 24^2 = 26^2$
6. A triangle has sides with lengths 8 inches, 10 inches, and 13 inches. Is the triangle a right triangle? Explain how you know. no; $8^2 + 10^2 = 164 \neq 13^2$

Solve Quadratic Equations by Finding Square Roots

Solve each equation.

7. $x^2 = 64$ $x = \pm 8$
8. $\frac{1}{2}x^2 + 11 = 19$ $x = \pm 4$
9. $3x^2 - 17 = 55$ $x = \pm 2\sqrt{6}$
10. $2x^2 = 120$ $x = \pm 2\sqrt{15}$
11. $-5x^2 + 20 = 0$ $x = \pm 2$

Connecting Past and Present Learning

Previously, you learned:

- to use properties of similarity to define trigonometric ratios for acute angles, and
- to use trigonometric ratios and the Pythagorean Theorem to solve right triangles in applied problems.

In this module, you will learn:

- to prove the Law of Sines and Cosines, and
- to apply the Law of Sines and Cosines to find unknown measurements in right and non-right triangles.

440

DATA-DRIVEN INTERVENTION

Concept/Skill	Objective	Prior Learning *	Intervene With
Find Equivalent Ratios	Solve real-world problems by finding equivalent ratios.	Grade 6, Lessons 5.5 and 9.3	• Tier 3 Skill 7 • Reteach, Grade 6 Lessons 5.5 and 9.3
The Pythagorean Theorem and Its Converse	Use the Pythagorean Theorem and its converse to solve problems.	Grade 8, Lessons 11.1–11.3	• Tier 2 Skill 4 • Reteach, Grade 8 Lessons 11.1–11.3
Solve Quadratic Equations by Finding Square Roots	Find the solutions of a quadratic equation by taking square roots.	Algebra 1, Lesson 18.1	• Tier 2 Skill 17 • Reteach, Algebra 1 Lesson 18.1

* Your digital materials include access to resources from Grade 6–Algebra 2. The lessons referenced here contain a variety of resources you can use with students who need support with this content.

14.1 Law of Sines

LESSON FOCUS AND COHERENCE

Mathematics Standards

- Prove the Laws of Sines and Cosines and use them to solve problems.
- Understand and apply the Law of Sines and the Law of Cosines to find unknown measurements in right and non-right triangles (e.g., surveying problems, resultant forces).

Mathematical Practices and Processes 

- Look for and make use of structure.
- Use appropriate tools strategically.
- Model with mathematics.

I Can Objective 

I can use the Law of Sines to find side lengths and angle measures of non-right triangles and solve real-world problems.

Learning Objective

Prove the Law of Sines, determine which combinations of given triangle measures are sufficient to apply the Law of Sines, determine how many triangles can exist with an ambiguous (SSA) combination of given measures, and apply the Law of Sines to determine unknown triangle measures.

Language Objective

Explain how more than one triangle can have a given SSA combination of measures.

Lesson Materials: calculator

Mathematical Progressions

Prior Learning	Current Development	Future Connections
Students: • applied the Pythagorean Theorem to determine unknown side lengths in right triangles. **(Gr8, 11.3)** • understood through similarity the definitions of trigonometric ratios for acute angles in right triangles. **(13.2 and 13.4)**	**Students:** • prove the Law of Sines. • determine the measures that are sufficient to apply the Law of Sines. • determine how many triangles can exist with an ambiguous (SSA) combination of given measures. • apply the Law of Sines to determine unknown triangle measures.	**Students:** • will prove, understand, and use the Law of Cosines to solve problem involving finding unknown measurements in triangles. **(14.2)**

PROFESSIONAL LEARNING

Visualizing the Math

The ambiguous (SSA) case of given triangle measures can be especially difficult for students to understand. Suppose $m\angle A = 30°$, $AB = 10$, and $BC = 7$ for $\triangle ABC$. Students need to understand that the relationship between AB, BC, and $h = AB \sin A$ determines how many unique triangles exist that satisfy the given conditions. A physical model can help students better visualize this relationship. Have them use a protractor and ruler to draw $\angle BAC$ and AB, making sure that ray $\overrightarrow{AC}$ is significantly longer than $\overline{AB}$ and point C is not fixed. Students can then measure a length of string representing $\overline{BC}$, hold one end at B, and see how the other end "swings" left and right to define two possible intersection points with $\overrightarrow{AC}$. They can hold the string down on the paper while a partner traces along the string to sketch $\overline{BC}$. Students can then experiment with other lengths of string to better understand the relationship between $m\angle A$, AB, and BC in determining the number of unique triangles ABC.

WARM-UP OPTIONS

ONLINE

PROJECTABLE & PRINTABLE

ACTIVATE PRIOR KNOWLEDGE • Use Trigonometric Ratios to Solve Problems

Use these activities to quickly assess and activate prior knowledge as needed.

Problem of the Day 

Janelle walks 500 ft on a straight road from her home to the store, and the elevation of the store is 20 ft higher than her home. What angle of elevation does the road make with the horizontal? A right triangle with the angle of elevation $\angle A$, the hypotenuse has length 500 ft, and the leg opposite $\angle A$ has length 20 ft, so $\sin A = \frac{20}{500} = 0.04$. The inverse sine of 0.04 is about 2.29°.

Quick Check for Homework

As part of your daily routine, you may want to display the Teacher Solution Key to have students check their homework.

Make Connections

Based on students' responses to the Problem of the Day, choose one of the following:

1. Project the Interactive Reteach, Lesson 13.4.
2. Complete the Prerequisite Skills Activity:

Have students work in pairs or small groups. Give each group a set of index cards. Each card should have a right triangle with two side lengths indicated. Make sure some cards have the lengths of both legs indicated and some have the lengths of the hypotenuse and a leg. Have students draw cards one at a time and find the angle measures of the triangle on the card using trigonometric ratios.

- *How can you relate the lengths of the legs of a right triangle with its acute angles?* I can use the tangent ratio.
- *How can you relate the lengths of a leg and the hypotenuse with the acute angles of a right triangle?* I can use the sine ratio or the cosine ratio.
- *How is the sine of an acute angle in a right triangle related to the cosine of the other acute angle?* The ratios are equivalent.

PROFESSIONAL LEARNING

If time permits, use this on-level activity to build fluency and practice basic skills.

Mental Math

Objective: Students use side lengths of right triangles to estimate their angle measures.

Write the following question on the board.

Estimate the angle measures of an 8-15-17 right triangle and a 20-21-29 right triangle.

If students need prompting, ask them to compare the ratios of the side lengths of the triangles to those of special right triangles. Since the 8-15-17 triangle has the length of its shorter leg roughly half that of the hypotenuse, the proportions of its side lengths are close to that of a 30°-60°-90° triangle, so its acute angles have measures roughly 30° and 60°. The 20-21-29 triangle is nearly isosceles, so its acute angles have measures close to 45°.

PLAN FOR DIFFERENTIATED INSTRUCTION

Small-Group Options

Use these teacher-guided activities with pulled small groups.

On Track

Materials: index cards

Have students work in a group and provide them with a set of index cards. Each card should have an AAS, ASA, or SSA combination of triangle measures. Have students work together to determine how many unique triangles exist with that combination of measures.

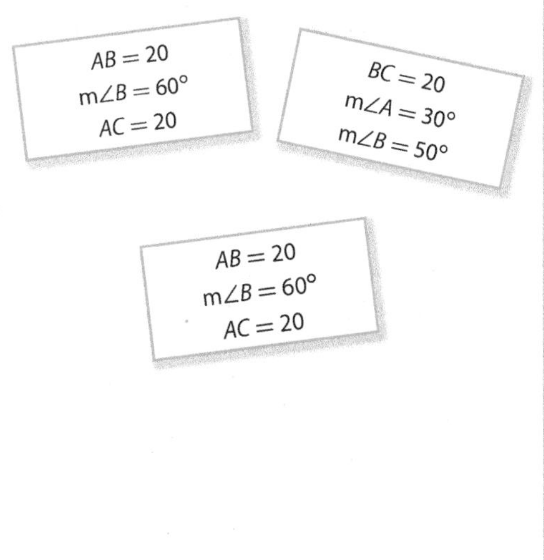

Almost There

Materials: index cards

Have students work in a group and provide them with a set of index cards. Each card should have an SSS, SAS, AAS, ASA, or SSA combination of triangle measures. Have students work together to determine which triangles can be solved with the Law of Sines. If needed, provide students with a reference sheet that describes and explains the law to use when determining their answers.

$AB = 20$
$AC = 15$
$BC = 18$

$XY = 8$
$YZ = 5$
$m\angle Y = 55°$

Ready for More

Show students an acute triangle, $\triangle ABC$, inscribed in a circle with center O and a radius r from O to B, as shown.

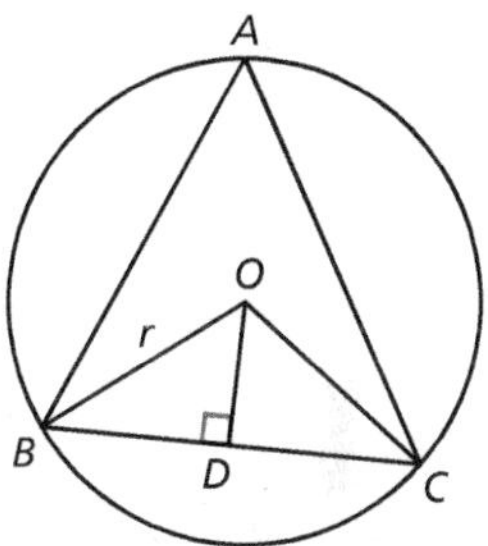

Reason as follows:

- If O is the circumcenter of the triangle, and on the perpendicular bisector of $\overline{BC}$, write an expression for sin BOD in terms of r and BC. $\left(\sin BOD = \frac{BD}{r} = \frac{BC}{2r}\right)$
- Since $\angle A \cong \angle BOD$, write sin A in terms of r and BC. $\left(\sin A = \frac{BC}{2r}\right)$
- Substitute your expression for sin A into $\frac{BC}{\sin A}$. $\left(\frac{BC}{\sin A} = \frac{BC}{BC/2r} = \frac{1}{1/2r} = 2r\right)$
- Use the Law of Sines to express the radius of the circle circumscribed on a triangle. $\left(\frac{a}{\sin A} = \frac{b}{\sin B} = \frac{c}{\sin C} = 2r\right)$

Math Center Options

Use these student self-directed activities at centers or stations. Key: ● Print Resources ● Online Resources

On Track

- ● Interactive Digital Lesson
- ●● Journal and Practice Workbook
- ● Module Performance Task

Almost There

- ● Reteach 14.1 (printable)
- ● Interactive Reteach 14.1
- ● Illustrative Mathematics: Defining Trigonometric Ratios

Ready for More

- ● Challenge 14.1 (printable)
- ● Interactive Challenge 14.1

View data-driven grouping recommendations and assign differentiation resources.

During the *Spark Your Learning*, listen and watch for strategies students use. See samples of student work on this page.

Use Sine Ratios — Strategy 1

I can use the sine ratio in each right triangle to find the length of each suspension rod.

$\sin 16° = \frac{6}{x}$ | $\sin 30° = \frac{6}{x}$
$x \sin 16° = 6$ | $x \sin 30° = 6$
$x = \frac{6}{\sin 16°}$ | $x = \frac{6}{\sin 30°}$
$x \approx 21.8$ | $x = 12$

The lengths of the rods are about 21.8 m and 12 m.

If students . . . use sine ratios to solve the problem, they are demonstrating an exemplary understanding of sine ratios of right triangles from Lesson 13.2.

Have these students . . . explain how they determined their approach and how they implemented it. **Ask:**

- **Q** How did you use the given information and the relationships between the given lengths to set up your equations?
- **Q** How did you solve the equations?

Use Cosine Ratios — Strategy 2

I can subtract $180° - 90° - 16° = 74°$ and $180° - 90° - 30° = 60°$ to find the unknown angles of the right triangles, and then use the cosine ratio in each right triangle to find the length of each suspension rod.

$\cos 74° = \frac{6}{x}$ | $\cos 60° = \frac{6}{x}$
$x \cos 74° = 6$ | $x \cos 60° = 6$
$x = \frac{6}{\cos 74°}$ | $x = \frac{6}{\cos 60°}$
$x \approx 21.8$ | $x = 12$

The lengths of the rods are about 21.8 m and 12 m.

If students . . . use cosine ratios to solve the problem, they understand how side lengths of right triangles are related through the cosine ratio but may not understand that using sine ratios would allow them to solve the problem more directly.

Activate prior knowledge . . . by having students set up sine ratios for each right triangle. **Ask:**

- **Q** Would you say that the 6 m leg of the right triangle is adjacent to the 16° angle, or opposite?
- **Q** Which trigonometric ratio relates the length of the leg opposite from an acute angle to the length of the hypotenuse?

COMMON ERROR: Writes An Incorrect Trigonometric Ratio

I can use the tangent ratio in each right triangle to find the length of each suspension rod.

$\tan 16° = \frac{6}{a}$ | $\tan 30° = \frac{6}{b}$
$a = \frac{6}{\tan 16°} \approx 20.9 \text{ m}$ | $b = \frac{6}{\tan 30°} \approx 10.4 \text{ m}$

The lengths of the rods are about 20.9 m and about 10.4 m.

If students . . . use tangent to find the lengths of the suspension rods, they may not understand that the suspension rods are also the hypotenuse of the triangles which requires the use of sine or cosine.

Then intervene . . . by pointing out the correct definition of the trigonometric ratio used. **Ask:**

- **Q** Which part of each triangle are your trying to find, a leg or the hypotenuse?
- **Q** How do you define the tangent of an acute angle of a right triangle?
- **Q** How would you describe the 6-meter leg and the side with length *a* in terms of the 16° angle?

14.1

Law of Sines

I Can use the Law of Sines to find side lengths and angle measures of non-right triangles and solve real-world problems.

Spark Your Learning

In a tower crane, the mast is the vertical part. The jib is the side of the horizontal section that lifts the weight. The other horizontal section is called the counter jib.

Both the jib and the counter jib are supported by suspension rods.

©Zhao jian kang/Shutterstock

Complete Part A as a whole class. Then complete Parts B–D in small groups.

A. What is a mathematical question you can ask about the suspension rods? What information would you need to know to answer your question? A. What is the length of each suspension rod?; measures of other parts of the crane

B. Is it possible to use right triangles to answer your question? Explain. See Additional Answers.

C. To answer your question, what strategy and tool would you use along with all the information you have? What answer do you get? See Strategies 1 and 2 on the facing page.

D. Does your answer make sense in the context of the situation? How do you know? yes; The shorter suspension rod is across from the smaller angle.

Turn and Talk One way the tower crane can adjust its reach is to lengthen and shorten the jib as well as the height of the tower supporting it. How will these adjustments affect the angles of the triangles made by the mast, jib, counter jib, and the cables supporting them? See margin.

CULTIVATE CONVERSATION • Co-Craft Questions

If students have difficulty formulating a mathematical question about the situation in the Spark Your Learning, ask them to imagine that they were designing or building the crane. What are some natural questions to ask about this situation?

Work together to craft the following questions:

- What are the lengths of the jib and counter jib?
- How are the lengths of the jib and the counter jib related to the lengths of the suspension rods?
- What are the lengths of the suspension rods?

Then have students think about what additional information, if any, they would need to answer these questions. **Ask:**

- Can you determine the lengths of the suspension rods if you know only the lengths of the jib and counter jib? Why or why not?
- Can you determine the lengths of the suspension rods if you know only the angles the rods make with the jib and counter jib? Explain.

1 Spark Your Learning

▶ MOTIVATE

- Have students look at the photo in their books and read the information contained in the photo. Then complete Part A as a whole-class discussion.
- Give the class the additional information they need to solve the problem. This information is available online as a printable and projectable page in the Teacher Resources.
- Have students work in small groups to complete Parts B–D.

▶ PERSEVERE

If students need support, guide them by asking:

Advancing • Use Tools Which tool could you use to solve the problem? Why choose that tool and not some other? Students' choices of tools and reasons for choosing them will vary.

Assessing How can you describe the left triangle made by the jib, mast, and the longer suspension rod? It is a right triangle with its legs formed by the mast and jib, and the suspension rod as its hypotenuse.

Assessing Can you use the Pythagorean Theorem to find the lengths of the suspension rods? no, because I only know the length of one side of each right triangle

Advancing How can you relate the 16° angle, the length of the mast, and the length of the longer suspension rod? The sine of the angle is the ratio of the length of the mast to the length of the suspension rod.

Turn and Talk Ask students to make sketches to show how changing the lengths will affect the angles. Lengthening the jib will make the angle that the cable makes with the jib smaller and the angle the cable makes with the mast larger. Changing the height of the tower does not affect the angles.

▶ BUILD SHARED UNDERSTANDING

Select groups of students who used various strategies and tools to share with the class how they solved the problem. As they present their solutions, have each group discuss why they chose a specific strategy and tool.

② Learn Together

Build Understanding

Task 1 (MP) **Construct Arguments** Students are introduced to the Law of Sines. They go on to derive the law for a non-right triangle by considering the right triangles defined by an altitude, and applying the right triangle definition of sine to angles of the triangle. Make sure students are careful when organizing their arguments, encouraging them to make sketches so they are referring to the intended parts of right triangles.

Sample Guided Discussion:

Q In Part D, why don't you need the additional equation $\frac{\sin A}{a} = \frac{\sin C}{c}$ you could get after drawing an altitude from vertex B? I already know that $\frac{\sin A}{a} = \frac{\sin B}{b}$ and $\frac{\sin B}{b} = \frac{\sin C}{c}$, so I can apply the Transitive Property of Equality to show all three ratios are equivalent.

Turn and Talk Ask students how they derived the equation $\frac{\sin B}{b} = \frac{\sin C}{c}$ in Part B of the task, and how they might modify their steps to get $\frac{b}{\sin B} = \frac{c}{\sin C}$. It is helpful to rewrite the Law of Sines as $\frac{a}{\sin A} = \frac{b}{\sin B} = \frac{c}{\sin C}$ if you are trying to find the length of one of the sides of the triangle. For the triangle in Part A, the sine ratios can be written as $b \sin C = c \sin B$. Then divide both sides of the equation by $\sin C \cdot \sin B$ to get $\frac{b}{\sin B} = \frac{c}{\sin C}$. Similar reasoning can be used for the triangle in Part C to write $\frac{a}{\sin A} = \frac{b}{\sin B}$. Then the Transitive Property of Equality can be used to write $\frac{a}{\sin A} = \frac{b}{\sin B} = \frac{c}{\sin C}$.

Build Understanding

Derive the Law of Sines

Previously, you used trigonometric ratios to find unknown measures in a right triangle. You can also use trigonometric ratios to find unknown measures in a triangle that is not a right triangle. The Law of Sines relates the sines of the angles of any triangle to the lengths of its sides.

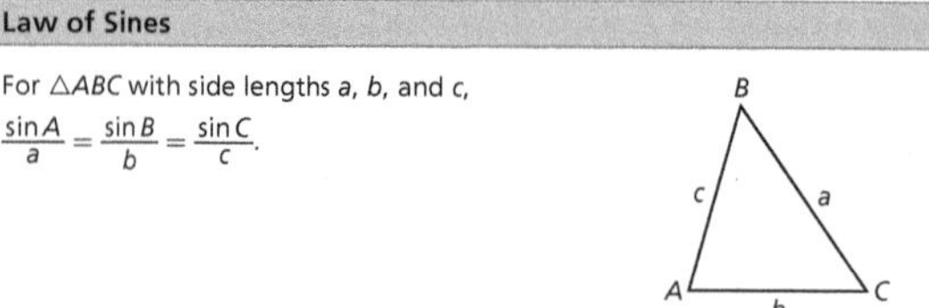

Law of Sines

For $\triangle ABC$ with side lengths a, b, and c,
$\frac{\sin A}{a} = \frac{\sin B}{b} = \frac{\sin C}{c}$.

1 Use sine ratios to derive the Law of Sines.

A. In the triangle shown, h is the altitude from vertex A. Use the right triangles formed by the altitude to find each of the ratios. $\sin C = \frac{h}{b}$; $\sin B = \frac{h}{c}$

$\sin C =$ ___?___

$\sin B =$ ___?___

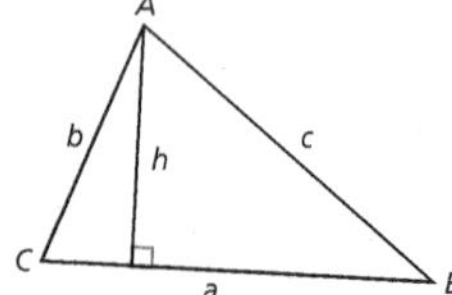

B. See Additional Answers.

B. How can the two equations from Part A be used to write $\frac{\sin B}{b} = \frac{\sin C}{c}$?

C. In the triangle shown, t is the altitude from vertex C. Use the right triangles formed by the altitude to find each of the ratios. $\sin A = \frac{t}{b}$; $\sin B = \frac{t}{a}$

$\sin A =$ ___?___

$\sin B =$ ___?___

Use your ratios to show that $\frac{\sin A}{a} = \frac{\sin B}{b}$.

$(\sin A)b = t$; $(\sin B)a = t$
$(\sin A)b = (\sin B)a$
$\frac{\sin A}{a} = \frac{\sin B}{b}$

D. Use the equations from Parts B and C to explain the following result:

$\frac{\sin A}{a} = \frac{\sin B}{b} = \frac{\sin C}{c}$ See Additional Answers.

Turn and Talk Why would it be useful to rewrite the Law of Sines as $\frac{a}{\sin A} = \frac{b}{\sin B} = \frac{c}{\sin C}$? Explain how it is possible to rewrite the expression this way. See margin.

442

LEVELED QUESTIONS

Depth of Knowledge (DOK)	Leveled Questions	What Does This Tell You?
Level 1 **Recall**	How are the angle and side of the triangle in each ratio related? The side is the only side that doesn't form part of the angle. It is opposite the angle.	Students' answers will indicate whether they understand the relationship between the angle and side described in a ratio of the Law of Sines.
Level 2 **Basic Application of Skills & Concepts**	If $m\angle A = 30°$, $a = 10$ cm, and $m\angle B = 70°$, what is the value of b? b is about 18.79 cm.	Students' answers will indicate whether they can apply the Law of Sines to solve for the side length of a triangle.
Level 3 **Strategic Thinking & Complex Reasoning**	How is the sine ratio you learned about in Module 13 a special case of the Law of Sines? If C is a right angle, then c is the length of the hypotenuse and $\frac{\sin B}{b} = \frac{\sin 90°}{c} = \frac{1}{c}$, so $c \sin B = b$ and $\sin B = \frac{b}{c}$.	Students' answers will indicate whether they can show that the right triangle definition of the sine of an acute angle is a special case of the Law of Sines.

Determine when to Apply the Law of Sines

To find an unknown side length or angle measure in any triangle, you need to know the length of at least one side and two other measures in the triangle. There are five possible combinations of given information that are needed to solve a triangle.

- SSS: All three side lengths are known.
- SAS: Two side lengths and their included angle measure are known.
- SSA: Two side lengths and an angle measure opposite one of them are known.
- ASA: Two angle measures and an included side length are known.
- AAS: Two angle measures and a non-included side length are known.

The Law of Sines cannot be used to find unknown measures for all of the cases. The following task investigates when the Law of Sines can be applied.

2 A. The triangle at the right shows only the lengths of the three sides, so it is an example of the SSS case. The Law of Sines is used to generate the equations below. Explain why it is not possible to apply the Law of Sines to find the value of sin *A*, sin *B*, or sin *C*.

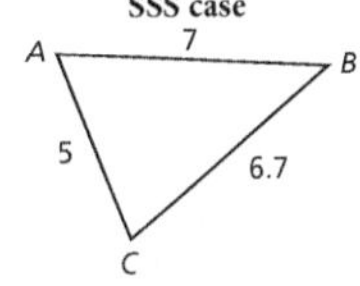

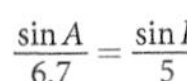

$\frac{\sin A}{6.7} = \frac{\sin B}{5}$ $\qquad$ $\frac{\sin B}{5} = \frac{\sin C}{7}$ $\qquad$ $\frac{\sin A}{6.7} = \frac{\sin C}{7}$

A. See margin.

B. For each triangle below, use the Law of Sines to write three equations that relate the side lengths to the sines of the angles. Can any of the equations be used to determine unknown angle measures or side lengths in the triangle? Explain. B, C. See Additional Answers.

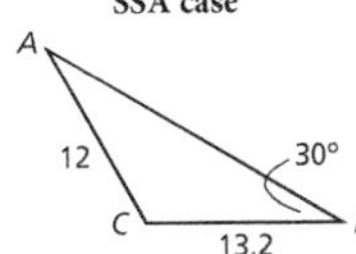

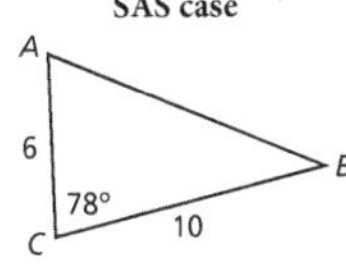

C. For each triangle below, first find m∠*B*. Then use the Law of Sines to write three equations that relate the side lengths to the sines of the angles. Can any of the equations be used to determine unknown side lengths in the triangle? Explain.

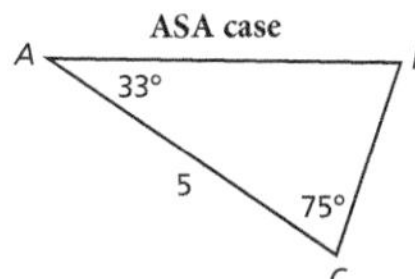

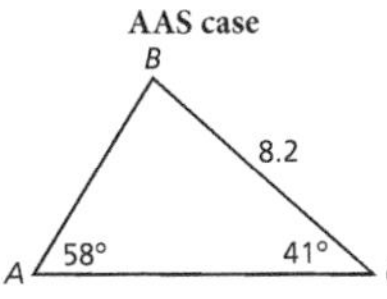

D. Which of the five possible combinations of given information—SSS, SAS, SSA, ASA, AAS—can the Law of Sines be used to find unknown measures? AAS, ASA, and SSA

Turn and Talk How can you figure out a the side lengths of a triangle if you only know the measures of its angles? How do you know? See margin.

Step It Out

Task 2 (MP) **Use Structure** Students explore different cases of triangles with incomplete measures and see which triangles can be solved with the Law of Sines. Emphasize the structure of the proportion defining the Law of Sines, and draw attention to how each ratio involves the sine of an angle and the length of a side opposite the angle.

Sample Guided Discussion:

Q Where is side *a* relative to ∠*A*? Side *b* relative to ∠*B*? The described side is on the opposite side of the triangle from the angle.

Q In Part B, how many values of the variables in a proportion $\frac{w}{x} = \frac{y}{z}$ do you need to know in order to solve the proportion? I would need to know the values of three of the variables.

Q In Part D, how can you think of pairs of opposite angles and sides to tell whether you can solve a triangle with the Law of Sines? In order to use the Law of Sines, I need to know the measures of at least one pair of an angle and side opposite from each other.

Turn and Talk Ask students how two figures with all corresponding angles congruent are related. Is such a pair of figures always congruent? Then have them use this same train of thought to apply to a single triangle with given angle measures. no; Possible answer: A triangle with three known angles can be any size. So, there is no way to determine the side lengths unless at least one side length is given.

PROFICIENCY LEVEL

Beginning

Show students △*ABC* with sides and vertices labeled and write $\frac{\sin A}{a} = \frac{\sin B}{b} = \frac{\sin C}{c}$. Ask students to read each proportion out loud as you point to its related angle.

Intermediate

Show students △*ABC* where $a = 10$ and $m\angle B = 70°$ are indicated. Ask students to identify which other segment length would be needed to illustrate an SSA combination. Then, ask why they choose that segment.

Advanced

Have students explain why the Law of Sines can be used to find unknown measures when an SSA combination of information is given for a triangle but not when an SAS combination is given.

ANSWERS

A. Each individual equation has two unknowns, so each equation cannot be solved for one unknown. When using all three equations to find an unknown angle measure, the result is an identity such as $\sin A = \sin A$.

Task 3 (MP) **Use Tools** Students use a calculator to find sine values of a set of pairwise supplementary angles, and use the values to conjecture about the sine values of supplementary angles. They then see that the inverse sine function does not have obtuse angle measures in its range, and use this to explain why an SSA combination of given measures by itself is ambiguous. Make sure students understand that the angle whose measure is given is not included between the sides with given lengths.

Sample Guided Discussion:

Q In Part A, is the calculator using the right triangle definition of sine that you learned about in Module 13 for all of the angle measures in the table? no; A right triangle cannot have an obtuse angle.

Q In Part B, what do you think is the value of $\sin(180° - x)$ when x is the measure of an angle in the table? It is equal to $\sin(x)$.

Q In Part D, if you knew only that the sine of an angle from the table was 0, would you be able to identify the angle? No, the angle could have measure 0° or 180°, but I would not know which.

Turn and Talk Encourage students to make sketches of the case described, varying the lengths of sides and the measures of angles to see which are possible. Ask them to compare their sketches to the diagrams in the "Ambiguous Case" box on the page. Ask whether they can make any conclusions about the measure of C from the given conditions. $m\angle C$ must be greater than or equal to 90°. If the measure of one angle in a triangle is 90° or greater, then each of the other angles must be acute. So, there is only one possible angle measure for $\angle A$.

ANSWERS

D. When finding an angle measure using the inverse sine operation on a calculator, there are two possible angle measure types, an acute angle or its obtuse supplement.

E. no; In the ASA and AAS cases, two angle measures are known, so there is only one possible measure for the unknown angle.

Explore the SSA Case

When given two side lengths and the angle measure of a non-included angle of a triangle (SSA case), the given measures may determine no triangle, one triangle, or two triangles. This is why the SSA case is called the ambiguous case.

Ambiguous Case

In each diagram, a, b, and $m\angle A$ are given. Side b and $\angle A$ are fixed, and then the possible positions of side a are considered.

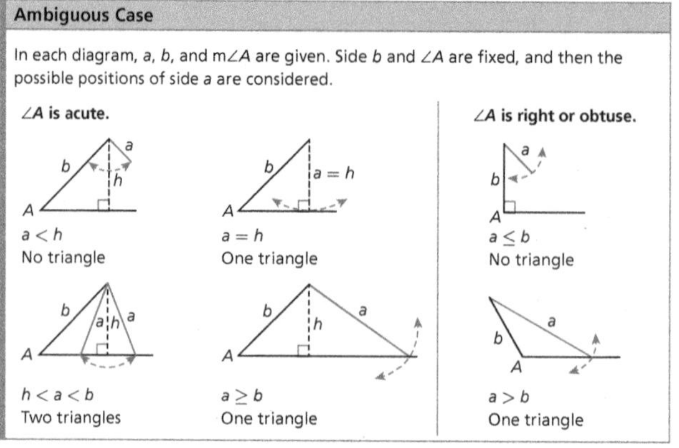

3 A. Copy the table. Complete the table by using a calculator to find the sine of each angle measure in the table. See Additional Answers.

x	0°	30°	45°	90°	135°	150°	180°
$\sin(x)$	?	?	?	?	?	?	?

B. Use the information in your table to make a conjecture about the sine values of supplementary angles. For a pair of supplementary angles, each angle has the same sine value.

C. For each of the $\sin(x)$ values in your table, evaluate the number using the inverse sine function on the calculator. Does the calculator always give the measure of the corresponding acute angle or obtuse angle? Explain. The angle measure given is for the acute angle.

D. For the SSA case, explain what has to be considered when using a calculator to find the measure of an unknown angle.

D, E. See margin.

E. Do the considerations described in Part D apply to the ASA case or the AAS case? Explain why or why not.

Turn and Talk In $\triangle ABC$, suppose you are given AB, AC, and the measure of $\angle C$. For this triangle, there is only one possible measure for $\angle A$. What is true about $\angle C$? Explain your reasoning. See margin.

Step It Out

Apply the Law of Sines

The Law of Sines can be applied when given the following information about a triangle.

- SSA: Two side lengths and an angle measure opposite one of them are known.
- ASA: Two angle measures and an included side length are known.
- AAS: Two angle measures and a non-included side length are known.

4 Two scuba divers at points Y and Z are 15 meters apart when they start to ascend at the same rate to the surface at the angles shown. How far does each scuba diver travel when their paths cross?

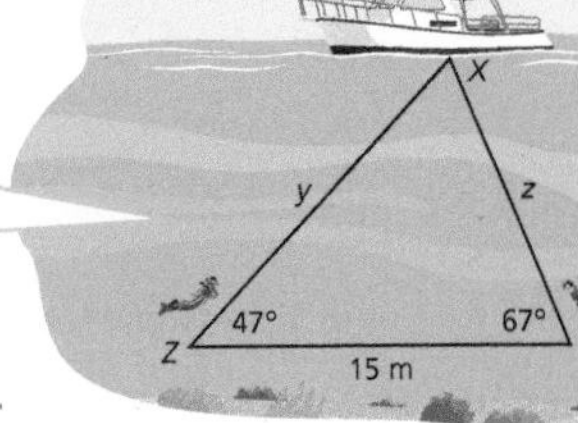

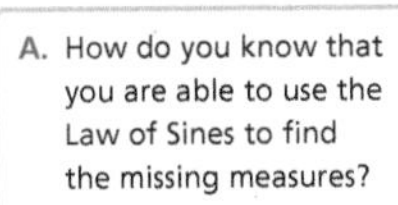

A. How do you know that you are able to use the Law of Sines to find the missing measures?

A. This triangle satisfies the ASA case.

First, find the unknown angle measure.

$m\angle X + m\angle Y + m\angle Z = 180°$	Triangle Sum Theorem
$m\angle X + 67° + 47° = 180°$	Substitute the known angle measures.
$m\angle X = 66°$	Solve for the measure of $\angle X$.

Use the Law of Sines to find the distance traveled by each scuba diver. B. See margin.

B. Explain why $\frac{\sin X}{x}$ is used in both equations.

Find the value of y.		Find the value of z.
$\frac{\sin Y}{y} = \frac{\sin X}{x}$	**Law of Sines**	$\frac{\sin Z}{z} = \frac{\sin X}{x}$
$\frac{\sin 67°}{y} = \frac{\sin 66°}{15}$	**Substitute.**	$\frac{\sin 47°}{z} = \frac{\sin 66°}{15}$
$y = \frac{15 \sin 67°}{\sin 66°}$	**Solve for the unknown.**	$z = \frac{15 \sin 47°}{\sin 66°}$
$y \approx 15.114 \approx 15.1$	**Evaluate.**	$z \approx 12.008 \approx 12.0$

When the paths of the two scuba divers meet, the scuba diver who was at point Z traveled about 15.1 meters. The scuba diver who was at point Y traveled about 12 meters.

Turn and Talk In the triangle in Task 4, can you find the missing angle measure using the Law of Sines instead of using the Triangle Sum Theorem? Explain. See margin.

Task 4 (MP) **Model with Mathematics** Students see how the Law of Sines can be used to solve a real-world problem where two side lengths of a triangle must be found.

Sample Guided Discussion:

Q In Part B, why do you need to know the value of x in order to find the missing side lengths? If I didn't know that side length, both of my proportions from the Law of Sines would have two unknowns.

Q In Part C, how many unknowns would you have in each proportion if you didn't use the Triangle Sum Theorem to find the missing angle measure first? There would be two unknowns in each proportion.

Turn and Talk Ask students where they would need to apply the Law of Sines again to solve for y and z in this new case, or if they can think of a more efficient strategy. Ask how the triangles in the two cases compare by drawing attention to their angle measures. If necessary, ask how the side lengths of similar triangles are related. no; You need to find $m\angle X$ before you can find the other side measures.

ANSWERS

B. The ratio $\frac{\sin X}{x}$ has two known measures, x and $m\angle X$, which can be used to find the unknown value of a variable in the other ratios.

Task 5  **Use Structure** Students see how the Law of Sines can be applied to the SSA case, when two triangles satisfying the given conditions exist. Encourage them to make a sketch of both triangles, imagining $\overline{BC}$ swinging like a door with its hinge at $\angle B$. Have them sketch both triangles with vertices A and B lying on top of each other, so they can better visualize the isosceles triangle with the two copies of $\overline{BC}$ as legs. Point out how the two supplementary cases of $\angle C$ determine the measures of the differing parts of the triangles.

OPTIMIZE OUTPUT Critique, Correct, and Clarify

Have students work with a partner to compare the method shown in the task with an alternate method, where they would begin by setting up $\frac{\sin A}{AB} = \frac{\sin B}{AC}$. Encourage them to use the vocabulary terms *SSA* and *ambiguous case* in their discussions. Students should revise their answers if necessary after discussing the problem with a partner.

Sample Guided Discussion:

Q In Part A, how many triangles would be possible if the value of a was not between h and c? If $a < h < c$, then no triangle is possible; if $h < c < a$, then exactly one triangle is possible.

Q In Part B, what do you notice about the triangle formed by the two possible positions of $\overline{BC}$ and the part of $\overline{AC}$ for the two possible positions of vertex C? It is an isosceles triangle.

Q How can you use the base angles of that isosceles triangle to relate $\angle C_1$ and $\angle C_2$? I know that $\angle C_1$ is one of the base angles of the isosceles triangle, while $\angle C_2$ and the other base angle of the triangle form a linear pair, so they are supplementary. Since the base angles are congruent, $\angle C_1$ and $\angle C_2$ are supplementary.

Turn and Talk Ask students which measure determines the number of triangles possible in the ambiguous case. Remind them that for this triangle, BC is the determining factor. Ask what kind of triangle would result if $\overline{BC}$ was "just barely" long enough to intersect $\overline{AC}$. Help them visualize this case as a right triangle with $\angle C$ a right angle. If the SSA case determines a single triangle, either $\angle C$ is a right angle or $BC > AB$. By the Law of Sines, $\frac{\sin C}{AB} = \frac{\sin A}{BC}$. If $\angle C$ is a right angle, $\sin C = 1$. Replace $\sin C$ with 1 in the equation. Next, to eliminate the fraction, multiply both sides of the equation by $AB \cdot BC$ to get $BC = AB \sin A$. So, to find a length BC that produces one triangle, choose $BC > AB$ or $BC = AB \sin A$.

Apply the Law of Sines to the SSA Case

5 In $\triangle ABC$, $AB = 9$ in., $BC = 6.8$ in., and $m\angle BAC = 42°$. What is AC?

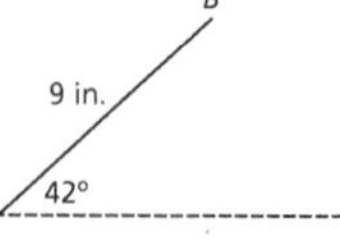

Determine how many triangles are possible.

Find the height.

$\sin 42° = \frac{h}{9}$, so $h = 9 \cdot \sin 42° \approx 6.022 \approx 6$ in.

There are two possible triangles because $h < a < c$.

A. What are the values of h, a, and c?

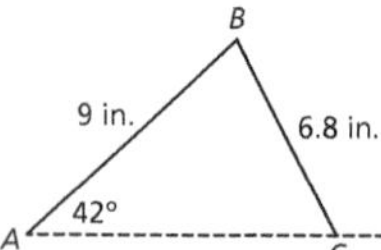

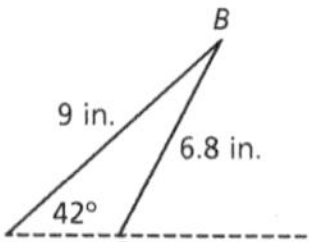

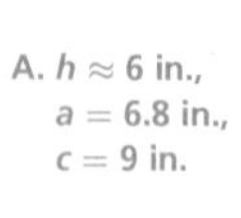

A. $h \approx 6$ in., $a = 6.8$ in., $c = 9$ in.

B. The two angles are supplementary.

Use the Law of Sines to find the possible measures of $\angle C$.

$\frac{\sin 42°}{6.8} = \frac{\sin C}{9}$ Law of Sines

$\sin C = \frac{9 \sin 42°}{6.8}$ Solve for sin C.

B. How are the two angles between 0° and 180° that have this sine value related?

Let $\angle C_1$ be the acute angle with the given sine, and let $\angle C_2$ be the obtuse angle. Use the inverse sine function on a calculator to find $m\angle C_1$. Then find the measure of $\angle C_2$.

$m\angle C_1 = \sin^{-1}\left(\frac{9 \sin 42°}{6.8}\right) \approx 62.327° \approx 62.3°$

$m\angle C_2 \approx 180° - 62.327° \approx 117.673 \approx 117.7°$

Find the measure of $\angle B$ and AC for each triangle.

C. Why is it necessary to find $m\angle B$ in each triangle?

C. See Additional Answers.

When $\angle C$ is acute	When $\angle C$ is obtuse
$m\angle B \approx 180° - 42° - 62.327°$	$m\angle B \approx 180° - 42° - 117.673°$
$\approx 75.673° \approx 75.7°$	$\approx 20.327° \approx 20.3°$
$\frac{\sin 42°}{6.8} \approx \frac{\sin 75.673°}{AC}$	$\frac{\sin 42°}{6.8} \approx \frac{\sin 20.327°}{AC}$
$AC \approx \frac{6.8 \sin 75.673°}{\sin 42°}$	$AC \approx \frac{6.8 \sin 20.327°}{\sin 42°}$
$AC \approx 9.846 \approx 9.8$ in.	$AC \approx 3.530 \approx 3.5$ in.

Turn and Talk Suppose you are given AB, BC, and $m\angle BAC < 90°$ for $\triangle ABC$. How can you find the length BC that will produce only one triangle? See margin.

Check Understanding

1. Use the Law of Sines to write an equation that relates the ratios of each side length to the sine of the angle opposite the side length in $\triangle ABC$. $\frac{\sin(22°)}{2.295} = \frac{\sin(116°)}{5.5} = \frac{\sin(42°)}{4.1}$

2. In $\triangle XYZ$, you are given the measure of $\angle X$, XY and XZ. Can you use the Law of Sines to find YZ? Explain. no; The given information is an SAS case, and the Law of Sines cannot be applied to an SAS case.

3. In $\triangle DEF$, $m\angle D = 38°$, $DE = 3.1$, and $EF = 2.5$. How many triangles are possible with the given measurements? two triangles

Find all the unknown side lengths and angle measures for each. Round your answers to the nearest tenth if necessary.

4. 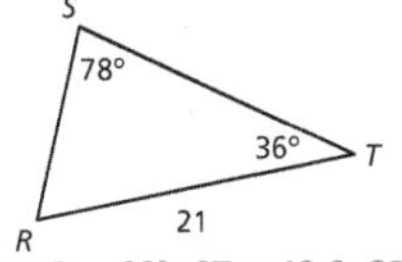

$m\angle R = 66°$; $ST \approx 19.6$; $SR \approx 12.6$

5. $m\angle K = 117°$; $JK \approx 5.3$; $KL \approx 8.7$

Find two angle measures that have the given sine value. Round each angle measure to the nearest tenth of a degree.

6. 0.9191 66.8° and 113.2°

7. 0.0854 4.9° and 175.1°

On Your Own

Match the altitude in $\triangle ABC$ with an equivalent expression.

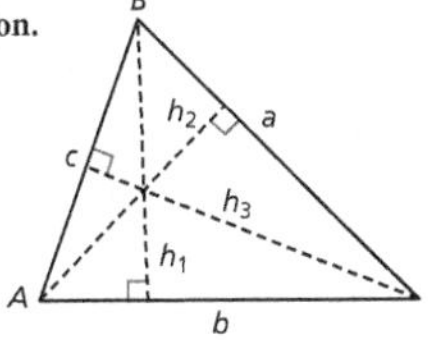

8. h_1 B — **A.** $a \sin B$
9. h_2 C — **B.** $c \sin A$
10. h_3 A — **C.** $b \sin C$

11. (MP) **Reason** Triangle ABC is an obtuse triangle, and h is an altitude of the triangle.
   - **A.** Write a ratio for the sine of $\angle A$. $\sin A = \frac{h}{c}$
   - **B.** Write a ratio for the sine of $\angle BCD$. Use this ratio to find the sine of $\angle C$. Explain your reasoning. B, D. See Additional Answers.
   - **C.** Draw altitude j from vertex C and find the sine of $\angle B$ and the sine of $\angle A$ using this altitude. $\sin B = \frac{j}{a}$, $\sin A = \frac{j}{b}$
   - **D.** Use the sine ratios to derive the Law of Sines for an obtuse triangle.

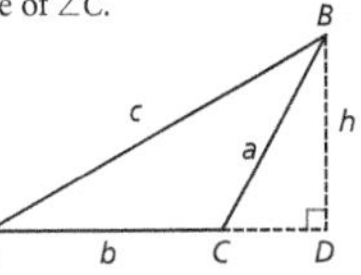

ONLINE

Assign the Digital On Your Own for
- built-in student supports
- Actionable Item Reports
- Standards Analysis Reports

On Your Own

Assignment Guide

The chart below indicates which problems in the On Your Own are associated with each task in the Learn Together. Assign daily homework for tasks completed.

Learn Together Tasks	On Your Own Problems
Task 1, p. 442	Problems 8–11
Task 2, p. 443	Problems 12–15
Task 3, p. 444	Problems 16–19, 29–31, and 41
Task 4, p. 445	Problems 20–26
Task 5, p. 446	Problems 27, 28, and 32–40

3 Check Understanding

Formative Assessment

Use formative assessment to determine if your students are successful with this lesson's learning objective.

Students who successfully complete the Check Understanding can continue to the On Your Own practice.

For student who miss 1 problem or more, work in a pulled small group using the Almost There small-group activity on page 441C.

ONLINE

Assign the Digital Check Understanding to determine
- success with the learning objective
- items to review
- grouping and differentiation resources

4 Differentiation Options

Differentiate instruction for all students using small-group activities and math center activities on page 441C.

Reteach

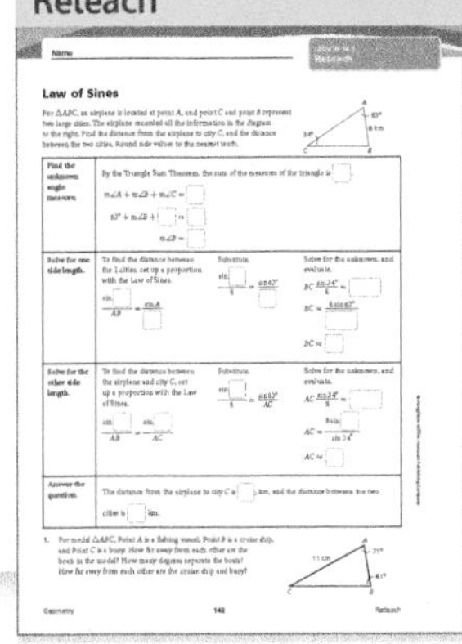

Challenge

Extend the Sine

Questioning Strategies

Problem 20 Suppose that $m\angle A = 117°$ and $m\angle C = 63°$. What do you notice when you apply the Law of Sines to solve for BC? Is this possible? Why do you think you got this strange result? I get $BC = \frac{14 \sin 117°}{\sin 63°} = 14$, so the triangle is isosceles since $AB = BC$. This isn't possible, though, since the base angles $\angle A$ and $\angle C$ aren't congruent. This happened because $\triangle ABC$ does not exist, since $\angle A$ and $\angle C$ are supplementary, violating the Triangle Sum Theorem.

Watch for Common Errors

Problem 13 Students may see that $m\angle A$ is not directly given, and since BC is the only known side length, assume that they do not have enough information to apply the Law of Sines. While the given information is insufficient to directly apply the Law of Sines, students need to understand that it may be possible to find the necessary information. In this case, students can apply the Triangle Sum Theorem to find $m\angle A$ and then apply the Law of Sines.

ANSWERS

12. no; Only two angle measures are given, and this does not satisfy any of the cases where the Law of Sines can be applied.

13. yes; The given measures are an ASA case, and the Law of Sines can be applied to this case.

14. yes; The given measures are an SSA case, and the Law of Sines can be applied to this case.

15. no; The given measures are an SAS case, and the Law of Sines cannot be applied to this case.

Determine whether the distance across each pond can be found using the Law of Sines and the given information. Explain your reasoning. 12–15. See margin.

12. 

13. 

14. 

15. 

The measure of $\angle D$ and side lengths d and f in $\triangle DEF$ are given. Also, the altitude h from vertex E is given. Determine how many triangles are possible for each set of given measures. Explain your reasoning.

16. $m\angle D = 110°$, $d = 5$, $f = 7$, $h = 6.6$ 0 because the side opposite the obtuse angle is not the longest side

17. $m\angle D = 55°$, $d = 8$, $f = 6$, $h = 4.9$ 2 because $\angle D$ is acute and $h < f < d$

18. $m\angle D = 30°$, $d = 7$, $f \approx 7.2$, $h = 7$ 1 because $\angle D$ is acute and $d = h$

19. $m\angle D = 120°$, $d = 11$, $f = 10$, $h = 8.7$ 1 because $\angle D$ is obtuse and $d > f$

Find the unknown measurements using the Law of Sines. Round your answers to the nearest tenth if necessary.

20. 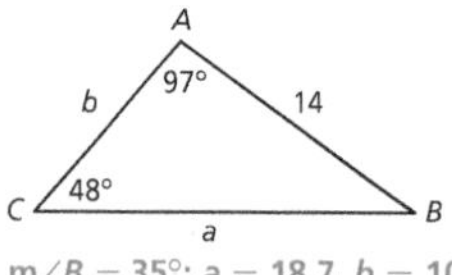

$m\angle B = 35°$; $a = 18.7$, $b = 10.8$

21. 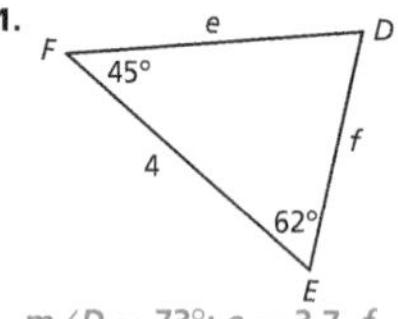

$m\angle D = 73°$; $e = 3.7$, $f = 3.0$

22.

$m\angle J = 131°$; $j = 13.9$, $k = 9.2$

23. 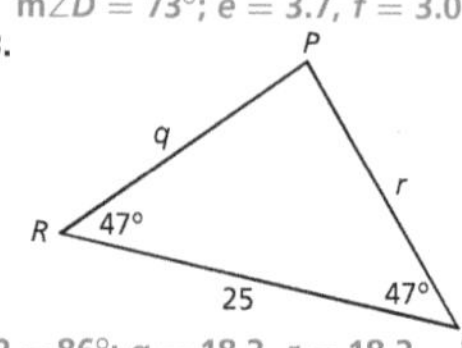

$m\angle P = 86°$; $q = 18.3$, $r = 18.3$

24, 25. See margin.

24. (MP) **Critique Reasoning** Two radio towers that are 65 miles apart track a satellite in orbit. The signal from Tower A forms a 76° angle between the ground and the satellite, and the signal from Tower B forms an 80.5° angle. Rylan says that Tower B is 4.5 miles closer to the satellite than Tower A. Is he correct? Explain.

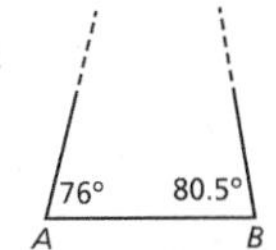

25. For a movie theater screen that is 50 feet wide, a high fidelity audio and visual company suggests that the center of the last row of seats forms the the angles shown with the left side *L* of the screen and the right side *R* of the screen.

A. Use the Law of Sines to find the distance from the center of the last row of seats to both the left side of the screen and to the right side of the screen to the nearest tenth of a foot.

B. Is it possible to find the distances without using the Law of Sines? Explain.

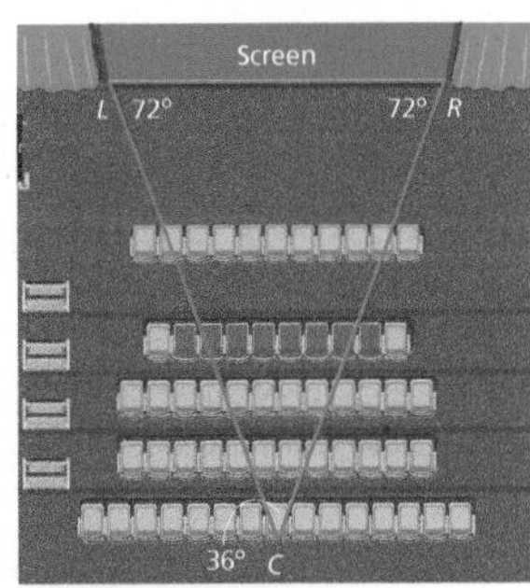

26. Jade, Morgan, and Denae are tossing a flying disc in a park. Their locations form the triangle shown. Jade is 16 feet from Morgan.

A. How far apart are Morgan and Denae? Round to the nearest tenth of a foot. 20.1 ft

B. How far apart are Denae and Jade? Round to the nearest tenth of a foot. 31.8 ft

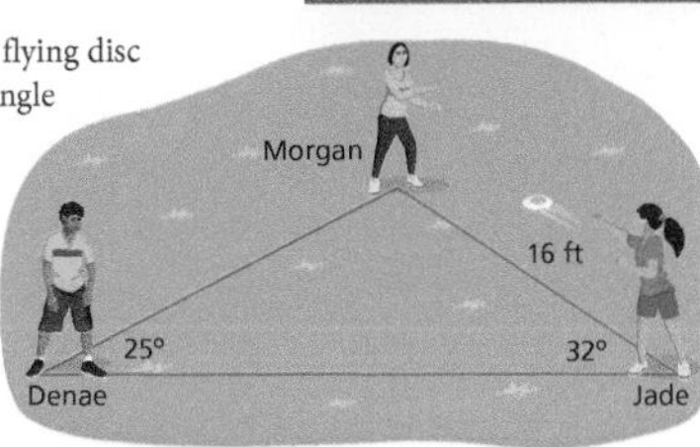

Find the height *h* of each triangle to the nearest tenth.

27. 4.7

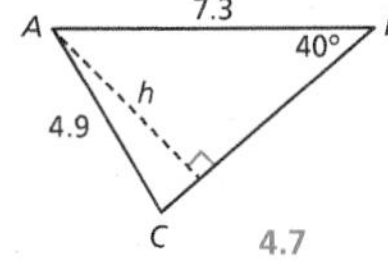

28. 4.9

B, 5.4, h, 9.6, 31°, A, C

Find two angle measures that have the given sine value. Round your answer to the nearest tenth of a degree.

29. 0.6639 41.6°, 138.4° **30.** 0.2504 14.5°, 165.5° **31.** 0.9803 78.6°, 101.4°

Use the given information to find the unknown angle measures and side lengths of $\triangle ABC$, if possible. If more than one triangle is possible, find both sets of measures.

32. $m\angle A = 20°$, $a = 18$, $b = 14$
$m\angle B = 15.4°$, $m\angle C = 144.6°$, $c = 30.5$

33. $m\angle A = 64°$, $a = 13$, $b = 15$
not possible

34. $m\angle A = 36°$, $a = 12$, $b = 18$
See margin.

35. $m\angle A = 106°$, $a = 11$, $b = 20$
not possible

36. $m\angle A = 82°$, $a = 25$, $b = 20$
$m\angle B = 52.4°$, $m\angle C = 45.6°$, $c = 18.0$

37. $m\angle A = 45°$, $a = 27$, $b = 34$
See margin.

Watch for Common Errors

Problem 24 Students may misinterpret the distance from a tower to the satellite as the distance from the base of the tower to a point directly below the satellite. That is, they might solve for the horizontal leg of a right triangle whose vertical leg is an altitude of the triangle defined by the satellite and the bases of the two towers. They need to understand that the distance from a tower to the satellite is the length of a segment drawn from the tower directly to the satellite.

ANSWERS

24. no; Using the Law of Sines, Tower B is 158.2 miles from the satellite, and Tower A is 160.8 miles from the satellite. So Tower B is $160.8 - 158.2 = 2.6$ miles closer.

25A. 80.9 feet for both distances

25B. yes; Because the triangle is isosceles, the height bisects the 50-foot side forming two congruent right triangles. Trigonometric ratios can be used to find the distances in the right triangles.

34. first case: $m\angle B = 61.8°$, $m\angle C = 82.2°$, $c = 20.2$; second case: $m\angle B = 118.2°$, $m\angle C = 25.8°$, $c = 8.9$

37. first case: $m\angle B = 62.9°$, $m\angle C = 72.1°$, $c = 36.3$; second case: $m\angle B = 117.1°$, $m\angle C = 17.9°$, $c = 11.8$

5 Wrap-Up

Summarize learning with your class. Consider using the Exit Ticket, Put It in Writing, or I Can scale.

Exit Ticket 

Tell how many triangles ABC there are with $m\angle A = 35°$, $AC = 10$, and $BC = 8$. If there is more than one triangle, give the missing measures of one of the triangles. This is the SSA case, and $10 \sin 35° \approx 5.74 < 8$, so there are two triangles. $\sin B = \frac{10 \sin 35°}{8} \approx 0.717$, so $m\angle B \approx 45.80°$. Then $m\angle C \approx 180° - 35° - 45.80° \approx 99.20°$. Then, $AB = \frac{8 \sin 99.20°}{\sin 35°} \approx 13.77$.

Put It in Writing 

Explain how you can determine whether you have enough information to use the Law of Sines to find unknown measures of a triangle.

I Can 

The scale below can help you and your students understand their progress on a learning goal.

4	I can use the Law of Sines to find side lengths and angle measures of non-right triangles and solve real-world problems, and I can explain my steps to others.
3	I can use the Law of Sines to find side lengths and angle measures of non-right triangles and solve real-world problems.
2	I can use the Law of Sines to find side lengths and angle measures of a triangle when I already know the measures of two angles and a side length.
1	I can tell for which triangles the Law of Sines will help me find missing side lengths and angle measures.

Spiral Review • Assessment Readiness

These questions will help determine if students have retained information taught in the past and can also prepare them for high-stakes assessments. Here, students must use a trigonometric ratio to solve for an angle measure of a right triangle **(13.4)**, use a trigonometric ratio to solve for a side length of a right triangle **(13.2)**, and identify special right triangles from their side lengths **(13.3)**.

38. Open Ended Write the lengths of two sides and an angle measure that can form two different triangles. Find all dimensions of each triangle.
Answers will vary. Check students' answers.

Find the area of each triangle. Round your answer to the nearest tenth of a square unit.

39. 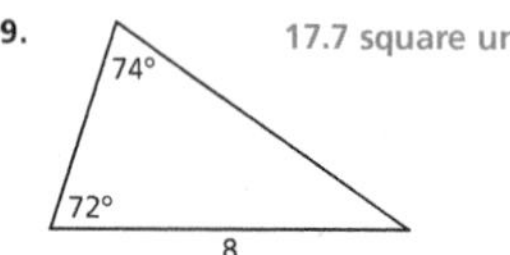

17.7 square units

40. 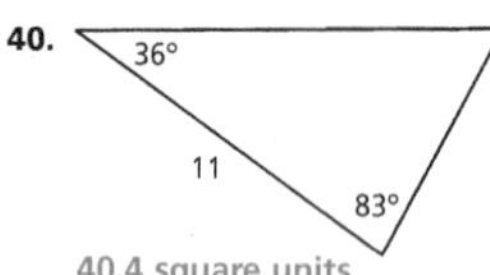

40.4 square units

41. Open Middle™ Using the digits 1 to 9 at most one time each, fill in the boxes three times. For the first set of numbers, fill in the boxes so that exactly two triangles exist. For the second set of numbers, fill in the boxes so that exactly one triangle exists. For the last set, fill in the boxes so no triangle exists. Answers will vary.

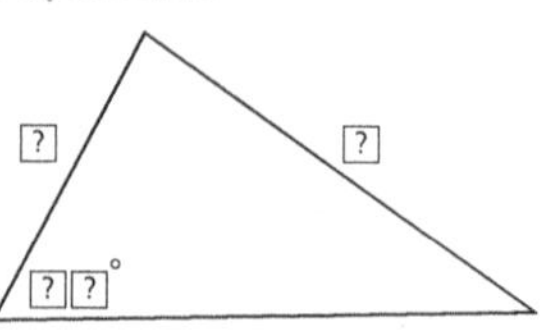

Spiral Review • Assessment Readiness

42. Find $m\angle C$.

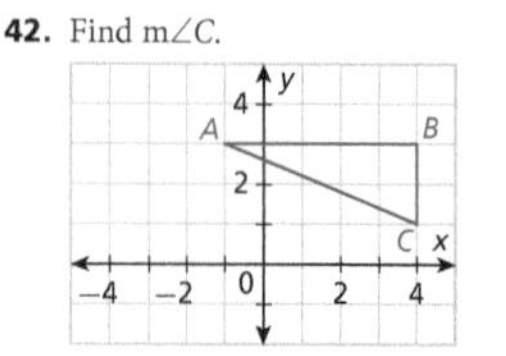

Ⓐ 21.8°
Ⓑ 42.9°
Ⓒ 68.2°
Ⓓ 90°

43. Find x.

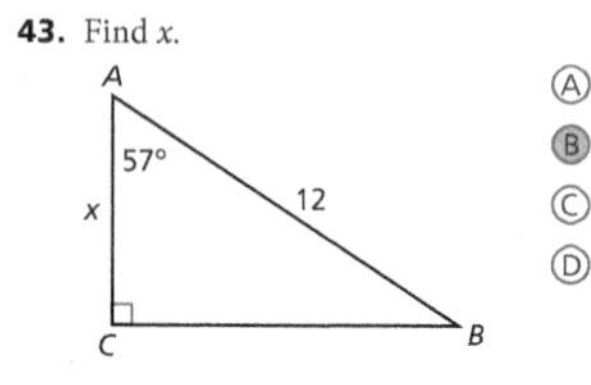

Ⓐ 0.05
Ⓑ 6.5
Ⓒ 10.1
Ⓓ 18.5

44. Determine whether the side lengths of a triangle belong to a 45°-45°-90° triangle, a 30°-60°-90° triangle, or neither.

Side lengths	45°-45°-90° triangle	30°-60°-90° triangle	Neither
A. $3, 3\sqrt{2}, 3$	?	?	?
B. $5, 5\sqrt{3}, 10\sqrt{3}$	?	?	?
C. $2, 4, 2\sqrt{3}$	?	?	?

 **I'm in a Learning Mindset!**

Did procrastination or lack of organization affect my learning outcomes? In what way?

Learning Mindset

Perseverance Learns Effectively

Point out that deep learning takes time and can't be made up through last minute cramming. Procrastination also leads to stress as due dates or assessments approach, and this stress makes it more difficult to focus on learning. Lack of organization can not only cause us to lose assignments or forget dates, but also hurt our ability to make sense of material. It can also can lead to feeling unsettled, which makes it more difficult to think in a logical fashion, which is very important for making sense of mathematical concepts. *How can procrastination raise your stress levels? Do you find it harder to focus when you're feeling stressed or unsettled due to being unorganized?*

14.2 Law of Cosines

LESSON FOCUS AND COHERENCE

Mathematics Standards

- Prove the Laws of Sines and Cosines and use them to solve problems.
- Understand and apply the Law of Sines and the Law of Cosines to find unknown measurements in right and non-right triangles (e.g., surveying problems, resultant forces).
- Derive the formula $A = \frac{1}{2}ab \sin C$ for the area of a triangle by drawing an auxiliary line from a vertex perpendicular to the opposite side.

Mathematical Practices and Processes 

- Look for and make use of structure.
- Look for and express regularity in repeated reasoning.
- Attend to precision.

I Can Objective 

I can use the Law of Cosines to find side lengths and angle measures of non-right triangles and solve real-world problems.

Learning Objective

Derive the Law of Cosines, recognize the SAS and SSS cases where the Law of Cosines is applicable and apply it to find unknown measures, derive the area formula $A = \frac{1}{2}ab \sin C$ for a triangle, and find the area of triangles when given an SAS or SSS combination of measures.

Language Objective

Explain when you can use the Law of Cosines to find an unknown measure in a triangle, and when you can use the Law of Sines.

Lesson Materials: calculator

Mathematical Progressions

Prior Learning	Current Development	Future Connections
Students: • applied the Pythagorean Theorem to determine unknown side lengths in right triangles. **(Gr8, 11.3)** • understood through similarity the definitions of trigonometric ratios for acute angles in right triangles. **(13.2 and 13.4)**	**Students:** • derive the Law of Cosines. • recognize when the Law of Cosines can be applied to solve for unknown measures, and find those measures. • derive the area formula $A = \frac{1}{2}ab \sin C$. • find the areas of non-right triangles given an SAS or SSS combination of measures.	**Students:** • will identify and describe relationships among inscribed angles, radii, and chords. **(15.1–15.3)** • will construct the inscribed and circumscribed circles of a triangle, and prove properties of angles for a quadrilateral inscribed in a circle. **(15.2)**

UNPACKING MATH STANDARDS

Understand and apply the Law of Sines and the Law of Cosines to find unknown measurements in right and non-right triangles (e.g., surveying problems, resultant forces).

What It Means to You

Students demonstrate an ability to judge which relationship to apply when solving for an unknown measure of a triangle and the ability to follow through and solve for the measure. This understanding develops from the previous lesson, where they learned to find unknown measures of triangles using strictly the Law of Sines and more elementary methods.

The emphasis on this standard is a deep understanding of both relationships so students can judiciously determine their steps in solving for all the unknown measures of a triangle. For example, in this lesson students could use the Law of Cosines to solve for an initial unknown measure of a triangle, apply the Law of Sines to find another unknown measure, and then use a more elementary method such as the Triangle Sum Theorem to find the third.

WARM-UP OPTIONS

ONLINE

PROJECTABLE & PRINTABLE

ACTIVATE PRIOR KNOWLEDGE • Apply the Pythagorean Theorem

Use these activities to quickly assess and activate prior knowledge as needed.

Problem of the Day

A surveyor wants to know the distance across Miller's Pond as indicated in the diagram and has measured out the distances as shown. The segment going across Swampy Pond is perpendicular to the 400-foot-long segment. How far is it across Miller's Pond, to the nearest foot?

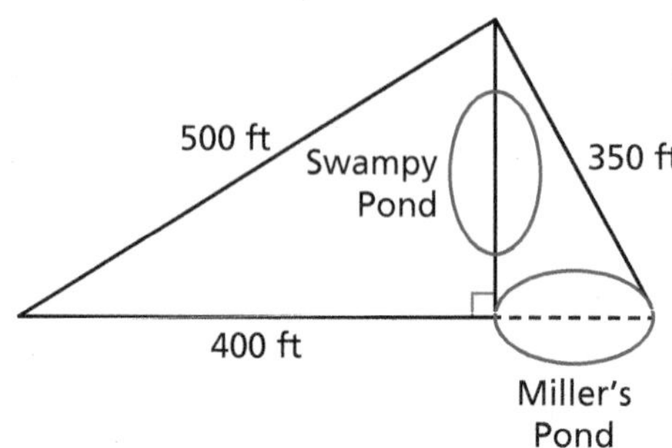

The right triangle with a 500 ft hypotenuse has sides that are a Pythagorean triple, with the short leg across Swampy Pond having length 300 ft. Then calling the distance across Miller's Pond x, we have $x^2 + 300^2 = 350^2$ so $x^2 = 32{,}500$ and $x = \sqrt{32{,}500} \approx 180$ ft.

Quick Check for Homework

As part of your daily routine, you may want to display the Teacher Solution Key to have students check their homework.

Make Connections

Based on students' responses to the Problem of the Day, choose one of the following:

1. Project the Interactive Reteach, Grade 8, Lesson 11.3.
2. Complete the Prerequisite Skills Activity:

Have students work in pairs or small groups. Give each pair a set of index cards; each card having a triangle with an altitude drawn, so there are two right triangles sharing a common leg. The common leg's length should not be labeled, while one right triangle has one side length given and the other triangle has two side lengths given. Students can work together to find the missing side lengths.

- *How can you find the length of the hypotenuse from the lengths of the legs?* I can square the lengths of the legs, add, and then take the square root of the sum.
- *How can you find the length of a leg from the lengths of the hypotenuse and another leg?* I can square the lengths of the hypotenuse and leg, subtract the smaller quantity from the larger quantity, and then take the square root of the difference.

If students continue to struggle, use Tier 2 Skill 4.

SHARPEN SKILLS

If time permits, use this on-level activity to build fluency and practice basic skills.

Quantitative Comparison

Objective: Students make a comparison between two quantities.

Write the following problem on the board. Ask students to choose the letter representing the correct answer and to explain their reasoning.

Quantity A	Quantity B
cos 30°	sin 60°

A. Quantity A is greater.
B. Quantity B is greater.
C. The two quantities are equal.
D. The relationship cannot be determined from the information given.

C; a triangle with side lengths of 1, $\sqrt{3}$, and 2 is a 30°-60°-90° triangle. So, $\cos 30° = \frac{\text{adj}}{\text{hyp}} = \frac{\sqrt{3}}{2}$ and $\sin 60° = \frac{\text{opp}}{\text{hyp}} = \frac{\sqrt{3}}{2}$.

PLAN FOR DIFFERENTIATED INSTRUCTION

Small-Group Options

Use these teacher-guided activities with pulled small groups.

On Track

Materials: index cards

Give each group a set of index cards. Each card should show an angle measure or a positive integer representing a side length. Have students select cards until they have either three side lengths or two side lengths and an angle measure. These three cards will represent measures of a triangle, with the angle included between the sides if applicable. Students can then work together to solve for the unknown measures of the triangle and find its area.

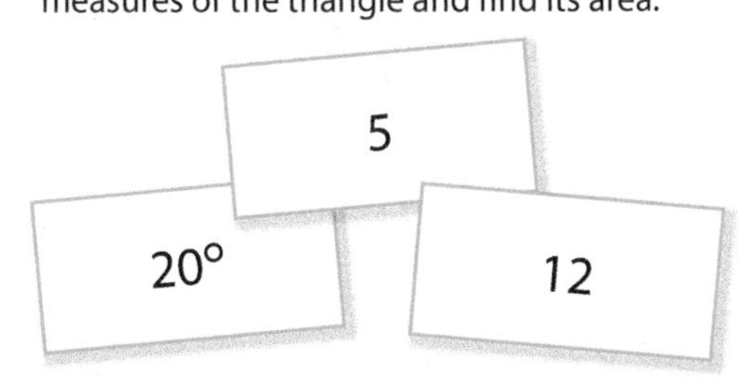

Almost There

Materials: index cards

Give each group a set of index cards. Each card should show an angle measure or a positive integer representing a side length. Have students draw cards until they have two side lengths and an angle measure. These three cards will represent measures of a triangle, with the angle included between the sides. Students can then work together to solve for the unknown measures of the triangle and find its area.

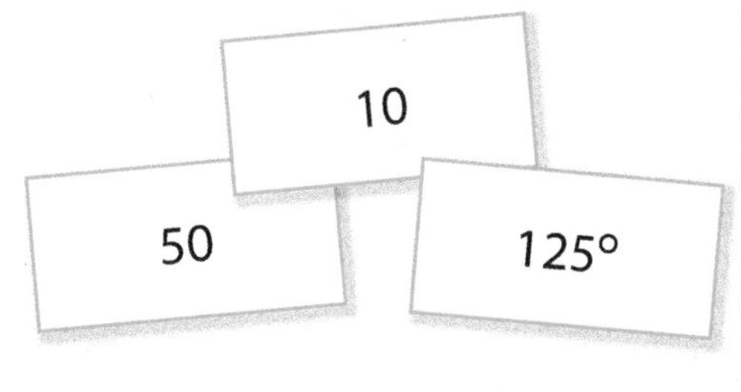

Ready for More

Materials: index cards

Give each group a set of index cards. Each card should show 100°, 50°, 30°, or a positive integer representing a side length. Have students draw three cards at a time, with these three cards representing measures of a triangle. They can then work together to

- determine whether it's possible to find the unknown measures of the triangle,
- determine how different arrangements of the given measures affect the number of possible triangles satisfying those measures (for example, two side lengths and an angle measure in an SAS arrangement or an SSA arrangement), and
- how they would solve for the unknown measures, if possible.

Math Center Options

Use these student self-directed activities at centers or stations. Key: ● Print Resources ● Online Resources

On Track

- ● Interactive Digital Lesson
- ●● Journal and Practice Workbook
- ● Module Performance Task

Standards Practice:

- ●● Derive the Formula $A = \frac{1}{2}ab\sin(C)$ to Find the Area of a Triangle
- ●● Prove and Use the Law of Sines and Law of Cosines
- ●● Use the Law of Sines and Law of Cosines to Solve Triangles

Almost There

- ● Reteach 14.2 (printable)
- ● Interactive Reteach 14.2
- ● RtI Tier 2 Skill 4: The Pythagorean Theorem and Its Converse
- ● Illustrative Mathematics: Finding the Area of an Equilateral Triangle

Ready for More

- ● Challenge 14.2 (printable)
- ● Interactive Challenge 14.2

Unit Project Check students' progress by asking to see their equation that uses the Law of Cosines.

View data-driven grouping recommendations and assign differentiation resources.

During the *Spark Your Learning,* listen and watch for strategies students use. See samples of student work on this page.

Use Cosine Ratios

Strategy 1

I can use the side lengths of the right triangles to write and solve cosine ratios to find the measures of two angles. Then, I can use the Triangle Sum Theorem to find the measure of the third angle. The measure of the angle at the 1st buoy is about 83°, the measure of the angle at the 2nd buoy is about 43°, and the measure of the third angle is 54°.

If students . . . use cosine ratios to solve the problem, they are demonstrating an exemplary understanding of using trigonometric ratios to find acute angle measures of right triangles from Lesson 13.4.

Have these students . . . explain how they determined their ratios and how they solved them. **Ask:**

- Q How did you use the given information and the definition of cosine to write your ratios?
- Q How did you solve the ratios to find the angle measures?

Use Sine Ratios

Strategy 2

I can use the side lengths of one of the right triangles and the Pythagorean Theorem to find the length of the altitude of the triangle. Then I can write and solve sine ratios to find the measures of two angles. Then, I can use the Triangle Sum Theorem to find the measure of the third angle. The measure of the angle at the 1st buoy is about 83°, the measure of the angle at the 2nd buoy is about 43°, and the measure of the third angle is 54°.

If students . . . use sine ratios to solve the problem, they understand that trigonometric ratios can be used to find angle measures of right triangles, but may not know how to choose the ratio that lets them use the given information most efficiently.

Activate Prior Knowledge . . . by having students set up cosine ratios relating the given side lengths of the right triangles. **Ask:**

- Q Is the 26.4 m side of one of the right triangles the opposite leg, the adjacent leg, or the hypotenuse with respect to the angle at Buoy 1?
- Q Which trigonometric ratio relates the 26.4 m side, the 205 m side, and the angle at Buoy 1?

COMMON ERROR: Inverts Cosine Ratio

I can use the side lengths of the right triangles to write and solve cosine ratios to find the measures of two angles. The angle at the 1st buoy has a cosine of $\frac{205}{24.6} \approx 8.33$, but then my calculator gives me an error message when I try to find the inverse cosine of 8.33.

If students . . . invert the cosine ratio, then they may not understand the definition of cosine.

Then intervene . . . by pointing out that since the cosine of an angle cannot be greater than 1, their ratio must be incorrect. **Ask:**

- Q The cosine of an acute angle describes how large its adjacent leg is compared to the hypotenuse. Which side is larger?
- Q If you get a cosine value greater than 1, why might you think you have made a mistake?

14.2

Law of Cosines

(I Can) use the Law of Cosines to find side lengths and angle measures of non-right triangles and solve real-world problems.

Spark Your Learning

In a triathlon with a sprint format, competitors have to swim a course that is 750 meters long. An example of such a course is shown.

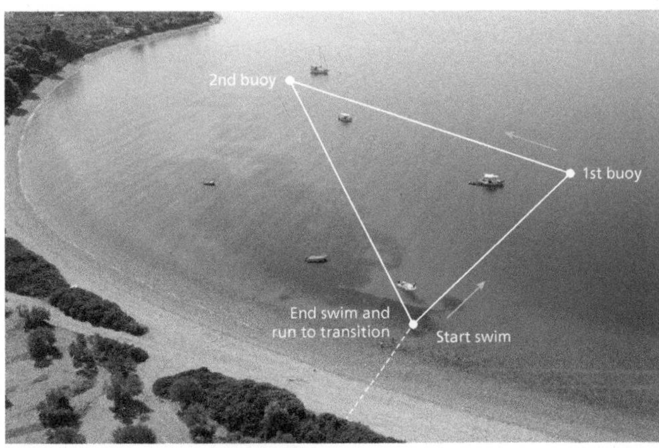

Complete Part A as a whole class. Then complete Parts B–D in small groups.

A. What is a mathematical question you can ask about this situation? What information would you need to know to answer your question?
A. What are the angle measures of the triangular course?; some of the measures in the triangle

B. Is it possible to use right triangle trigonometric ratios to answer your question?
See Additional Answers.

C. To answer your question, what strategy and tool would you use along with all the information you have? What answer do you get?
See Strategies 1 and 2 on the facing page.

D. Does your answer make sense in the context of the situation? How do you know?
yes; The largest angle is opposite the longest side of the triangle.

Turn and Talk How can you use the Law of Sines with the given information to find the unknown angle measures of the triangular course? See margin.

SUPPORT SENSE-MAKING • Information Gap

Ask students questions to help them decide what missing information they need to answer the question, "What are the angle measures of the triangular course?"

Do you have enough information to conclude that the angle measures are related to the side lengths of the triangle? Explain. yes; Since the ratios of the side lengths and the angle measures both determine the shape of the triangle.

Suppose you know the side lengths of the triangle. Can you determine the angle measures? Explain. no; Possible answer: If I knew one angle measure, I could use the Law of Sines to find another angle measure and then the Triangle Sum Theorem to find the third angle. However, I don't know any angle measures.

In addition to the angles, what information might be helpful? Possible answer: I need to know either one angle measure so I could apply the Law of Sines, or I need to know the length of an altitude of the triangle so I could use a cosine ratio to find the measures of both parts of the angle split by the altitude.

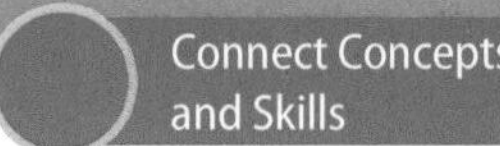
Connect Concepts and Skills

1 Spark Your Learning

▶ MOTIVATE

- Have students look at the photo in their books and read the information contained in the photo. Then complete Part A as a whole-class discussion.
- Give the class the additional information they need to solve the problem. This information is available online as a printable and projectable page in the Teacher Resources.
- Have students work in small groups to complete Parts B–D.

▶ PERSEVERE

If students need support, guide them by asking:

Advancing • Use Tools Which tool could you use to solve the problem? Why choose that tool and not some other? Students' choices of tools and reasons for choosing them will vary.

Assessing Is the triangular course a right triangle? No; the sum of the squares of the shorter sides does not equal the square of the longest side. By the Converse of the Pythagorean Theorem, the course is not a right triangle.

Assessing How would you describe the triangles on either side of the altitude from the start and end of the swim? They are both right triangles.

Advancing How would you name the known sides of the right triangle with an angle at Buoy 1 with respect to that angle? The adjacent leg and the hypotenuse of the triangle.

Turn and Talk Ask students to describe the structure of the Law of Sines, making sure they understand that it is a proportion. Ask how many known quantities they need to have in a proportion in order to be able to solve for any unknown quantities. Ask how many measures they know in any of the ratios relating the sine of an angle with the length of a side in the triangle. no; To use the Law of Sines, you need to know at least one of the angle measures of the triangle.

▶ BUILD SHARED UNDERSTANDING

Select groups of students who used various strategies and tools to share with the class how they solved the problem. As they present their solutions, have each group discuss why they chose a specific strategy and tool.

2 Learn Together

Build Understanding

Task 1 (MP) **Use Structure** Students derive the Law of Cosines with the Pythagorean Theorem. They see how drawing an altitude allows them to divide the triangle into two right triangles and then write an expression for the square of a side of the non-right triangle in terms of side lengths of the right triangles. They then rewrite that expression strictly in terms of measures of the non-right triangle by applying a cosine ratio of one of the right triangles.

Sample Guided Discussion:

Q In Part B, does $(b-x)^2$ equal b^2-x^2? If not, what does it equal? no; $(b-x)^2 = b^2 - 2bx + x^2$

Q How would you derive a similar expression for c^2? I would draw the same altitude from B, set $BD = x$ and $AD = b - x$, and then follow the steps equivalent to those that gave me the expression for a^2

Turn and Talk Ask students about the positions of the sides of a right triangle with respect to the right angle. Help them recall if needed, that the cosine value of a right angle is zero. For $\triangle ABC$, the Law of Cosines states that $c^2 = a^2 + b^2 - 2ab\cos C$. If the measure of $\angle C$ is 90°, we know $\cos 90° = 0$. Then, $c^2 = a^2 + b^2 - 2ab(0)$ or $c^2 = a^2 + b^2$. The resulting equation relating the sides of the right triangle is the Pythagorean Theorem.

Build Understanding

Derive the Law of Cosines

You have learned how to find unknown angle measures and side lengths in a triangle using the Law of Sines. However, the Law of Sines cannot be used to solve triangles with the following combinations of given information:

- SSS: All three side lengths are known.
- SAS: Two side lengths and their included angle measure are known.

For these cases, you must apply a different law, the Law of Cosines.

Law of Cosines

For $\triangle ABC$, the Law of Cosines states that

$a^2 = b^2 + c^2 - 2bc\cos A$
$b^2 = a^2 + c^2 - 2ac\cos B$
$c^2 = a^2 + b^2 - 2ab\cos C$

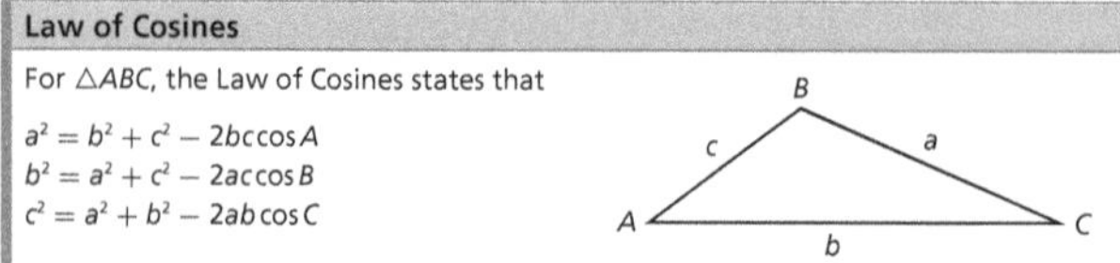

 **1** Use $\triangle ABC$ to derive the Law of Cosines.

A. In the Law of Cosines, the equations contain the cosines of angles. Is it possible to write ratios for cosines of angles in $\triangle ABC$? Explain why altitude $\overline{BD}$ is drawn in the second triangle.
A, B. See Additional Answers.

B. Use the Pythagorean Theorem to write a relationship for the side lengths of $\triangle ABD$ and for the side lengths of $\triangle CBD$.

C. Notice that both equations contain the variables x and h. Since these variables are not in the Law of Cosines, x and h need to be replaced. What does $x^2 + h^2$ equal? Replace $x^2 + h^2$ by this term in the equation for a^2.
c^2; $a^2 = b^2 + c^2 - 2bx$

D. Compare your revised equation for a^2 with the equation for a^2 given in the Law of Cosines. What should x be equivalent to? Explain how to use the ratio for the cosine of $\angle A$ to eliminate x from the equation for a^2.
See Additional Answers.

 **Turn and Talk** Explain why the Pythagorean Theorem can be considered a special case of the Law of Cosines. See margin.

Investigate Cosine Values of Obtuse Angles

You have learned that supplementary angles have the same sine value. In $\triangle ABC$, the sine of $\angle ACB$ is equal to the sine of $\angle ACD$ because they form a linear pair.

The following task investigates how the cosines of supplementary angles are related.

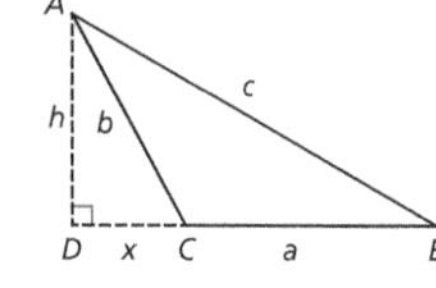

452

LEVELED QUESTIONS

Depth of Knowledge (DOK)	Leveled Questions	What Does This Tell You?
Level 1 **Recall**	Given a, b, and $m\angle C$ for $\triangle ABC$, what expression can you write for c^2? $c^2 = a^2 + b^2 - 2ab\cos C$	Students' answers will indicate whether they can recall the Law of Cosines.
Level 2 **Basic Application of Skills & Concepts**	Which theorem can you apply to help derive the Law of Cosines? I can draw an altitude in the triangle and apply the Pythagorean theorem to both right triangles.	Students' answers will indicate whether they understand the general idea behind the derivation of the Law of Cosines.
Level 3 **Strategic Thinking & Complex Reasoning**	Supposing $\triangle ABC$ is a triangle and you know its side lengths, how can you solve for $m\angle C$ I can use the Law of Cosines to write $\frac{a^2+b^2-c^2}{2ab}$ as an expression for $\cos C$ and then use the inverse cosine button on my calculator.	Students' answers will indicate whether they can recognize the structure of the Law of Cosines and understand how to apply the law to solve for an angle measure of a triangle.

 A. Copy the table. Complete the table by using a calculator to find the cosine of each angle measure in the table. A, B. See Additional Answers.

x	0°	30°	45°	90°	135°	150°	180°
cos(x)	?	?	?	?	?	?	?

B. Use the information in your table to make a conjecture about the cosine values of supplementary angles.

C. For each of the cos(x) values in your table, evaluate the number using the inverse cosine function on the calculator. Does the calculator always give the measure of the corresponding acute angle or obtuse angle? yes

Derive a Formula for the Area of a Triangle

You can use two side lengths of a triangle and their included angle to find the area of any triangle.

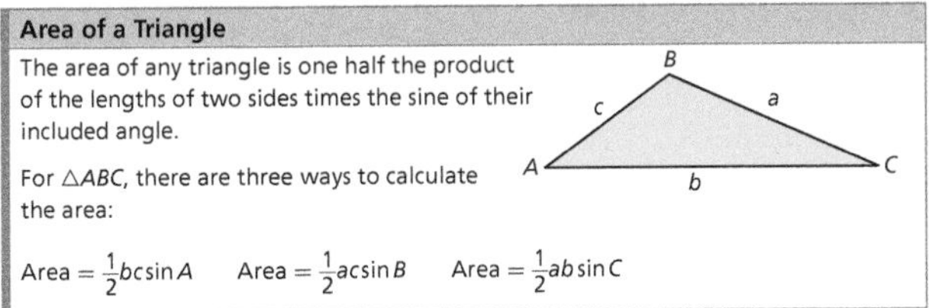

Area of a Triangle
The area of any triangle is one half the product of the lengths of two sides times the sine of their included angle.
For △ABC, there are three ways to calculate the area:
Area = ½bc sin A Area = ½ac sin B Area = ½ab sin C
B
c
a
A
b
C

 Use $\triangle ABC$ to derive a triangle area formula. In $\triangle ABC$, h is the height of the triangle.

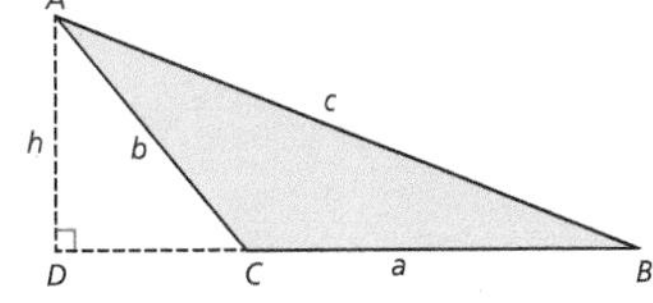

A. Use the formula for the area of a triangle, Area $= \frac{1}{2}bh$, to write the area of $\triangle ABC$ with base a. To generate an area formula that uses only two side lengths and the included angle, how does the formula need to change? A–D. See Additional Answers.

B. Write an equation for the value of h using $\angle ACD$. Use what you know about the value of the sines of supplementary angles to rewrite the expression for h using $\angle C$. Explain your reasoning.

C. Use your equation from Part B to eliminate the variable h from the original equation for the area found in Part A. Explain how you found the equation.

D. Explain how to use the diagram to show that $\frac{1}{2}ab\sin C = \frac{1}{2}bc\sin A$.

 **Turn and Talk** How can you use the equation you found for the area of a triangle in Part C to find the area of a right triangle, with $\angle C$ being a right angle? See margin.

PROFICIENCY LEVEL

Beginning

Have students work in pairs. Show them an obtuse $\triangle ABC$ with sides labeled with lowercase letters opposite the vertices. One student will name either a pair of sides or a side and an angle. The other student assume that the measures of the named parts are known, respond with the side or angle whose measure would allow them to find the area of $\triangle ABC$, and state the expression to find the area of $\triangle ABC$.

Intermediate

Show students an obtuse $\triangle ABC$ with sides labeled with lowercase letters opposite the vertices. Write the expression for the area: $A = \frac{1}{2}ab\sin C$. Have students read the expression aloud, and then have students write *another* expression for the area of $\triangle ABC$.

Advanced

Have students explain how they can find the area of any triangle if they know its side lengths and angle measures.

Task 2 (MP) **Use Repeated Reasoning** Students use a calculator to find cosine values of a pair of supplementary angles. They then inspect the relationship between these cosine values and make a conjecture on a relationship between the cosine values of supplementary angle pairs in general.

Sample Guided Discussion:

- **Q In Part B, what do you notice about the relationship between an angle and the sign of its cosine value?** Acute angles have positive cosine values, while obtuse angles have negative cosine values.
- **Q How does the cosine value of an unknown angle of a triangle give you more information than its sine value?** The cosine value tells me if the angle is acute or obtuse, depending on whether the cosine value is positive or negative. The sine value will be positive no matter the size of the angle.

Task 3 (MP) **Use Repeated Reasoning** Students derive an area formula for a triangle in terms of two sides and the sine of the included angle, using the area formula of a right triangle and the relationship between sine values of supplementary angles. They then extend this reasoning to show a similar formula in terms of another pair of sides and the sine of their included angle gives the same area.

Sample Guided Discussion:

- **Q In Part B, how are the sine values of supplementary angles related?** The sine values are equal.
- **Q In Part D, how will the right triangle letting you write h in terms of $\sin A$ be different from the right triangle you used to derive $\frac{1}{2}ab\sin C$?** The right triangle I used earlier was outside of $\triangle ABC$, but this right triangle will be inside of $\triangle ABC$.

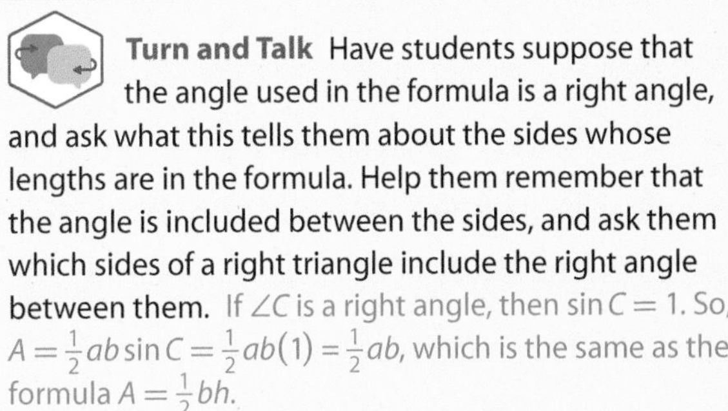

Turn and Talk Have students suppose that the angle used in the formula is a right angle, and ask what this tells them about the sides whose lengths are in the formula. Help them remember that the angle is included between the sides, and ask them which sides of a right triangle include the right angle between them. If $\angle C$ is a right angle, then $\sin C = 1$. So, $A = \frac{1}{2}ab\sin C = \frac{1}{2}ab(1) = \frac{1}{2}ab$, which is the same as the formula $A = \frac{1}{2}bh$.

Step It Out

Task 4 (MP) **Attend to Precision** Students see how the Law of Cosines can be used to find the angle measures of a triangle when all side lengths are given. Encourage them to be precise when explaining their rationales; this can be challenging for this task because there are many closely related, but not equivalent, terms involved, such as "angle," "angle measure," and "cosine of an angle." While "square of a side" can be easily interpreted as "square of a side length," encourage students to use more precise language like the latter so you can get a better sense of their level of comprehension.

Sample Guided Discussion:

- **Q Which angle is the only angle that might be obtuse?** $\angle Y$, since it is opposite the longest side of the triangle and a triangle can only have one obtuse angle
- **Q Why might it not be a good idea to find the measure of $\angle X$ first and then use the Law of Sines to find the measure of $\angle Y$?** $\angle Y$ might be obtuse, so its sine value will not tell me whether it is obtuse or acute.
- **Q How do you know that $\angle Y$ is obtuse?** It has a negative cosine value.

Turn and Talk Point out to students that having three angle measures satisfy the Triangle Sum Theorem does not mean that the angle measures are correct. Help them understand that the angle measures must be related correctly to their opposite side lengths. Ask them if they know of a way the angle measures of any triangle are related to its opposite side lengths, clarifying that a trigonometric ratio may be involved. Use the Law of Sines to write ratios. Then check that the ratios are equivalent.

ANSWERS

B. Possible answer: Law of Sines; There are fewer steps when using the Law of Sines to find the measure of $\angle X$.

C. The largest angle in the triangle, $\angle Y$, is obtuse. So, the two other angles must be acute.

Step It Out

Solve a Triangle Using the Law of Cosines

When using the Law of Cosines to solve a triangle, start by finding the measure of the angle opposite the longest side. Recall that the largest angle in a triangle is opposite the longest side.

- If the largest angle in a triangle is acute, then the other two angles are acute.
- If the largest angle in a triangle is obtuse, then the other two angles are acute.

After using the Law of Cosines to find the measure of one angle, the Law of Sines can be used to find the measure of another angle, but the unknown angle must be acute. This is because there are two angles between 0° and 180° that have the same sine value—an acute angle and its obtuse supplement.

4 In $\triangle XYZ$, $x = 4.7$, $y = 6.8$, and $z = 2.5$. What are the angle measures of the triangle?

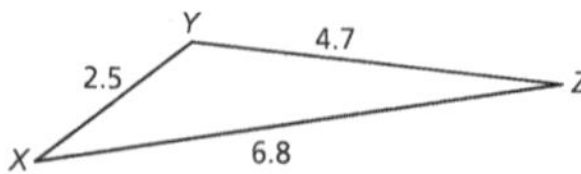

Use the Law of Cosines to find the measure of $\angle Y$.

$$y^2 = x^2 + z^2 - 2xz\cos Y$$
$$6.8^2 = 4.7^2 + 2.5^2 - 2(4.7)(2.5)\cos Y$$
$$46.24 = 22.09 + 6.25 - 23.5\cos Y$$
$$17.9 = -23.5\cos Y$$
$$\frac{17.9}{-23.5} = \cos Y$$
$$m\angle Y = \cos^{-1}\left(\frac{17.9}{-23.5}\right) \approx 139.614° \approx 139.6°$$

A. Why is the measure of $\angle Y$ found first?

A. $\angle Y$ is the largest angle since it is opposite the longest side.

Use the Law of Sines to find the measure of $\angle X$.

$$\frac{\sin X}{x} = \frac{\sin Y}{y}$$
$$\frac{\sin X}{4.7} \approx \frac{\sin 139.614°}{6.8}$$
$$\sin X \approx \frac{4.7 \sin 139.614°}{6.8}$$
$$m\angle X \approx \sin^{-1}\left(\frac{4.7 \sin 139.614}{6.8}\right) \approx 26.605 \approx 26.6°$$

B. Find $m\angle X$ using the Law of Cosines. Do you prefer to use the Law of Sines or the Law of Cosines to find $m\angle X$? Explain.

B, C. See Additional Answers.

C. How do you know that $m\angle X$ is about 26.6°, not 153.4°?

Use the Triangle Sum Theorem to find the measure of $\angle Z$.

$$m\angle X + m\angle Y + m\angle Z = 180°$$
$$26.605° + 139.614° + m\angle Z \approx 180°$$
$$m\angle Z \approx 13.781 \approx 13.8°$$

In $\triangle XYZ$, $m\angle X \approx 26.6$, $m\angle Y \approx 139.6$, and $m\angle Z \approx 13.8°$.

Turn and Talk How can you check your answers without using the Triangle Sum Theorem? See margin.

454

Use the Law of Cosines to Solve a Real-World Problem

5 Tracking devices are used to monitor the elephants in a herd. An ecologist observes an elephant that has fallen behind the herd. Use the ecologist's distance from the herd and from the lone elephant to determine how far the elephant is from the herd.

Apply the Law of Cosines to find the distance between the elephant and the herd.

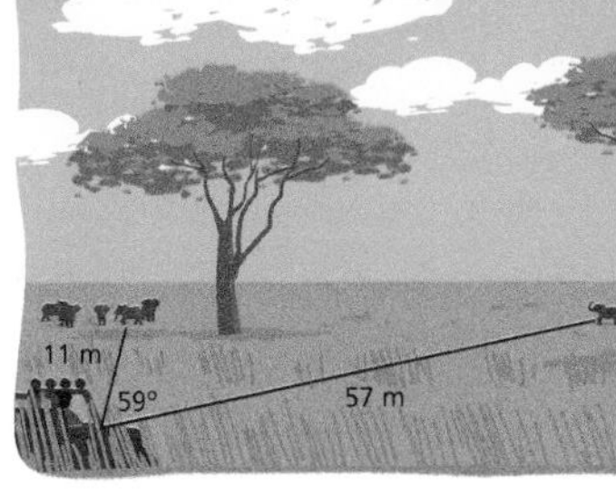

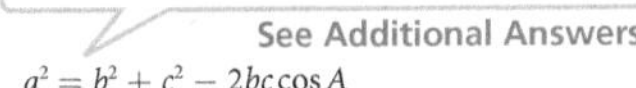

See Additional Answers.

$a^2 = b^2 + c^2 - 2bc\cos A$

$a^2 = 57^2 + 11^2 - 2(57)(11)\cos 59°$

$a^2 = 3249 + 121 - 1254\cos 59°$

$a^2 = 3370 - 1254\cos 59°$

$a = \sqrt{3370 - 1254\cos 59°} \approx 52.193 \approx 52.2$

The elephant is about 52.2 meters from the herd.

Turn and Talk Explain how each of the following will affect the estimate of the elephant's distance from the herd: See margin.

- The distance between the ecologist and the herd is underestimated
- The measure of the angle made by the herd, ecologist, and elephant is smaller

Find the Area of a Triangle

6 A triangular plot of land has the dimensions shown. Find the area of the plot to the nearest square foot.

B, 430 ft, 340 ft, C, A, 588 ft

Find the measure of an angle.

$b^2 = a^2 + c^2 - 2ac\cos B$

$588^2 = 430^2 + 340^2 - 2(430)(340)\cos B$

$45{,}244 = -292{,}400\cos B$

$\frac{45{,}244}{-292{,}400} = \cos B$

$B = \cos^{-1}\left(\frac{45{,}244}{-292{,}400}\right) \approx 98.901° \approx 98.9°$

Why is the measure of an angle needed to find the area?

See Additional Answers.

Find the area of the triangle.

$\text{Area} = \frac{1}{2}ac\sin B$

$\approx \frac{1}{2}(430)(340)\sin 98.901° \approx 72{,}219.669 \approx 72{,}220$

The area of the plot is about 72,200 square feet.

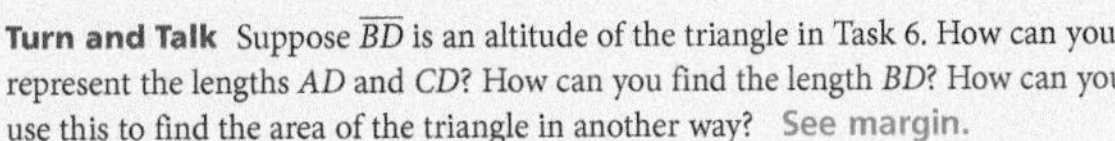

Turn and Talk Suppose $\overline{BD}$ is an altitude of the triangle in Task 6. How can you represent the lengths *AD* and *CD*? How can you find the length *BD*? How can you use this to find the area of the triangle in another way? See margin.

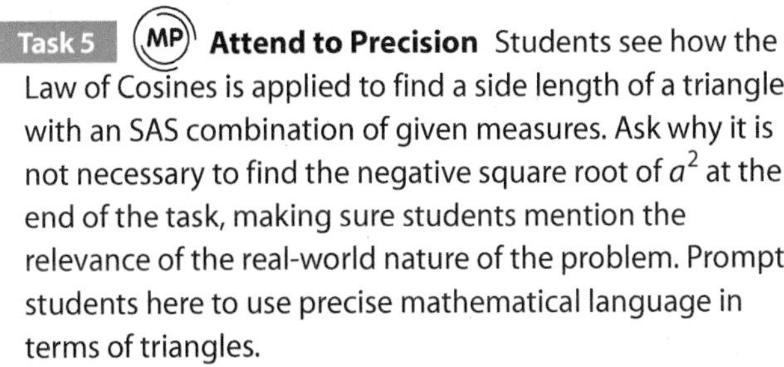

Task 5 **(MP) Attend to Precision** Students see how the Law of Cosines is applied to find a side length of a triangle with an SAS combination of given measures. Ask why it is not necessary to find the negative square root of a^2 at the end of the task, making sure students mention the relevance of the real-world nature of the problem. Prompt students here to use precise mathematical language in terms of triangles.

SUPPORT SENSE-MAKING Three Reads

Have students read the problem three times. Use the questions below for a different focus each read.

1. What is the situation about?
2. What are the quantities in the situation?
3. What are the possible mathematical questions that you could ask for this situation?

Turn and Talk If the distance between the ecologist and herd is underestimated, the estimate of the elephant's distance from the heard will be greater because as you calculate *a* for decreasing values of *c* less than 11, the value of *a* increases. If the measure of the angle made by the herd, ecologist, and elephant is smaller, the estimate will be less because as you calculate *a* for decreasing measures of $\angle A$, the value of *a* decreases.

Task 6 **(MP) Use Structure** Students see how the area of a triangle can be found indirectly from its side lengths.

Turn and Talk Have students think about how they can set up equations using the Pythagorean Theorem. Possible answer: If $\overline{BD}$ is an altitude, then $\triangle ADB$ and $\triangle CDB$ are right triangles, so $AD = \sqrt{AB^2 - BD^2}$ and $CD = \sqrt{BC^2 - BD^2}$. You can solve the equation $\sqrt{AB^2 - BD^2} + \sqrt{BC^2 - BD^2} = AC$ for *BD* since the values of *AB*, *BC*, and *AC* are known. Then the area of the triangle is $\frac{1}{2}(AC)(BD)$.

ONLINE

Assign the Digital On Your Own for
- built-in student supports
- Actionable Item Reports
- Standards Analysis Reports

On Your Own

Assignment Guide

The chart below indicates which problems in the On Your Own are associated with each task in the Learn Together. Assign daily homework for tasks completed.

Learn Together Tasks	On Your Own Problems
Task 1, p. 452	Problems 6 and 13
Task 2, p. 453	Problem 7
Task 3, p. 453	Problems 8, 23, and 27
Task 4, p. 454	Problems 10, 11, 17, 20, and 22
Task 5, p. 455	Problems 9, 12, 14–16, 18, 19, and 21
Task 6, p. 455	Problems 24–26

ANSWERS

2. $\sin K$ is equal to $\sin J$, and $\cos K$ is the opposite of $\cos J$.
3. $m\angle A \approx 71.2°$, $m\angle B \approx 65.4°$, $m\angle C \approx 43.4°$
4. $m\angle A \approx 28.8°$, $m\angle C \approx 24.2°$, $b \approx 33.1$

Check Understanding

1. Draw and label the diagram you would use to derive the form $b^2 = a^2 + c^2 - 2ac\cos B$ from the Law of Cosines. See Additional Answers.
2. Suppose $m\angle K + m\angle J = 180°$. How is the sine of $\angle K$ related to the sine of $\angle J$? How is the cosine of $\angle K$ related to the cosine of $\angle J$? 2–4. See margin.

Solve each triangle. Round intermediate results to three decimal places and final answers to one decimal place.

3. 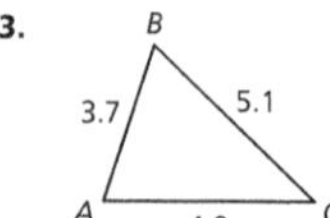

4. 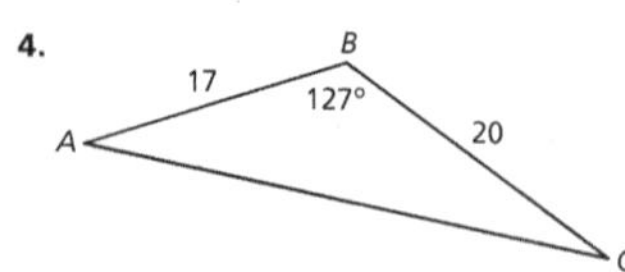

5. John owns a triangular piece of property. The lengths of two sides are 200 feet and 250 feet, and the included angle measures 85°.
   A. What is the length of the third side to the nearest foot? about 306 feet
   B. What is the area of the property to the nearest square foot? about 24,905 square feet

On Your Own

6. MP **Reason** Solve the equation $c^2 = a^2 + b^2 - 2ab\cos C$ from the Law of Cosines for C. When would this form of the equation be useful? See Additional Answers.
7. MP **Reason** Suppose $\angle A$ in $\triangle ABC$ is an obtuse angle. What must be true about the value of x in the equation $\cos^{-1} x = A$? x is negative.
8. MP **Reason** Explain how you can derive a formula for the area of an equilateral triangle with side length s using the formula Area $= \frac{1}{2}bc\sin A$. See Additional Answers.

Solve each triangle. Round intermediate results to three decimal places and final answers to one decimal place. 9–12. See Additional Answers.

9. 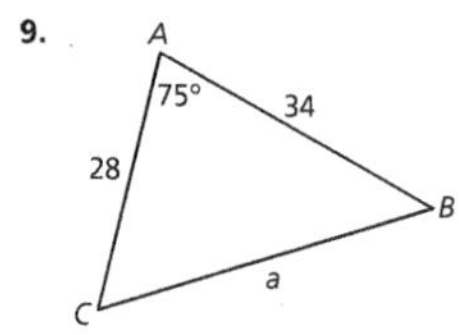

10. 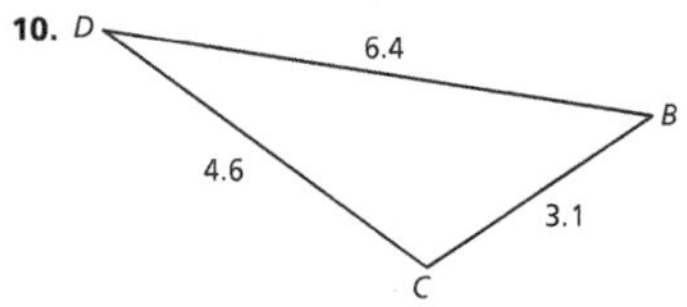

11. 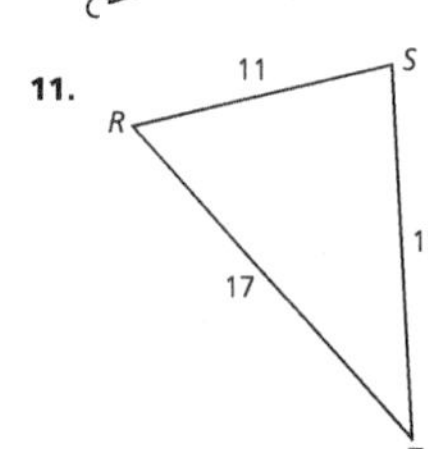

12. 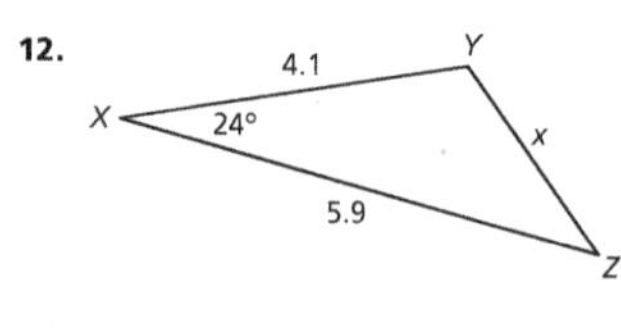

456

3 Check Understanding

Formative Assessment

Use formative assessment to determine if your students are successful with this lesson's learning objective.

Students who successfully complete the Check Understanding can continue to the On Your Own practice.

For students who miss 1 problem or more, work in a pulled small group using the Almost There small-group activity on page 451C.

ONLINE Ed

Assign the Digital Check Understanding to determine
- success with the learning objective
- items to review
- grouping and differentiation resources

4 Differentiation Options

Differentiate instruction for all students using small-group activities and math center activities on page 451C.

Reteach

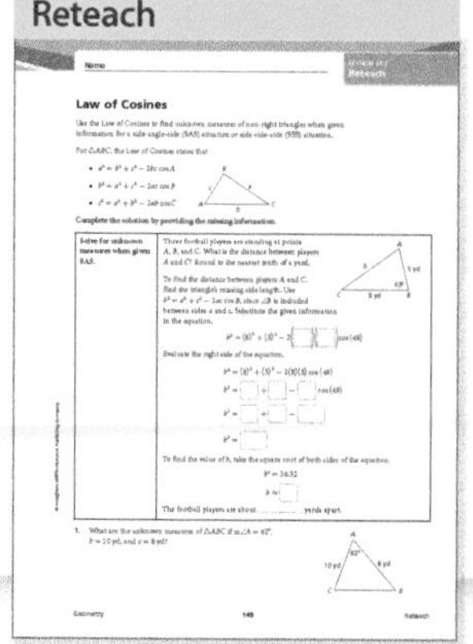

Challenge

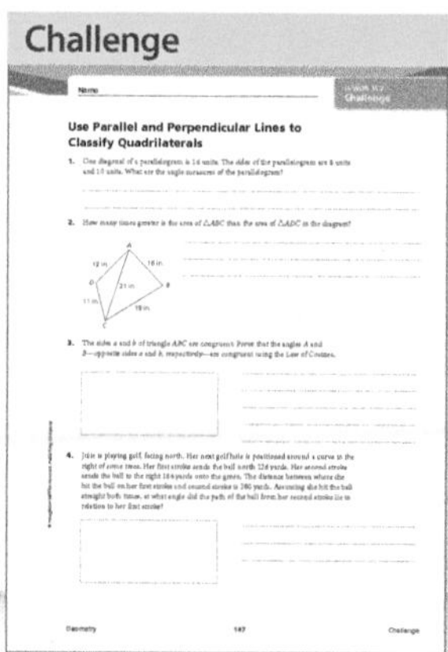

13. Critique Reasoning Melissa believes $\triangle FGH$ can be solved using only the Law of Sines. Sara believes $\triangle FGH$ can be solved using only the Law of Cosines. Who is correct? Explain your reasoning.
See margin.

14. **Open Ended** Draw a triangle. Use a ruler to measure the lengths of two sides, and use a protractor to measure the included angle. Use the Law of Cosines to solve the triangle.
Answers will vary. Check students' work.

15. The distances from home plate to the pitcher's mound and from home plate to first base on a baseball field are shown. Alex is the pitcher, and he stands at point B on the pitcher's mound. A. about 63.7 feet

A. How far does Alex have to throw the ball from the pitcher's mound to reach first base?

B. Alex is facing home plate. Through what angle does he have to turn to face first base?
about 92.8°

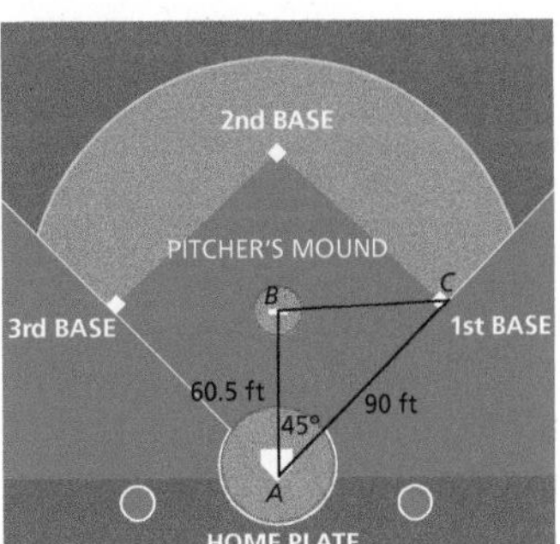

16. **STEM** An engineer is designing a new sidewalk on a college campus that will connect two existing sidewalks. The existing sidewalks start at the same point. One of the sidewalks is 220 yards long. The other is 140 yards long. The angle formed by these sidewalks is 50°. The new sidewalk will connect the ends of the existing sidewalks. How long is the new sidewalk? Round your answer to the nearest yard.
about 169 yards

Use the given measures in $\triangle ABC$ to solve each triangle. Round intermediate results to three decimal places and final answers to one decimal place.

17–23. See margin.

17. $a = 14, b = 9, c = 19$

18. $m\angle B = 62°, a = 5.3, c = 7.6$

19. $m\angle C = 132°, a = 11, b = 8$

20. $a = 6.8, b = 7.2, c = 12.1$

21. $m\angle A = 58°, b = 22, c = 15$

22. $a = 23, b = 16, c = 18$

23. Critique Reasoning Grant wants to find the area of the triangle. Martin says that he cannot find the area because he does not know the height of the triangle. Grant disagrees. Who is correct? Explain your reasoning.

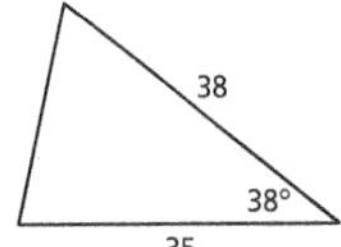

Find the area of each triangle. Round your answer to the nearest tenth of a square unit.

24.

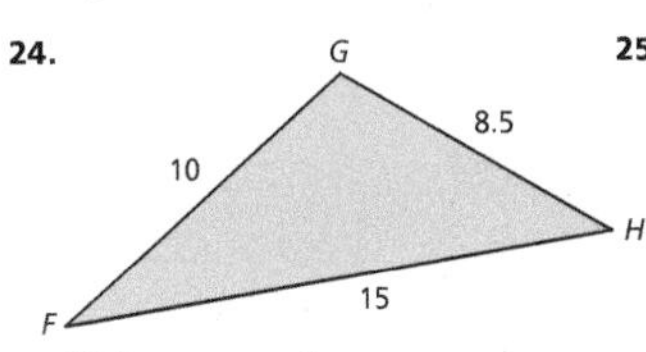

40.4 square units

25.

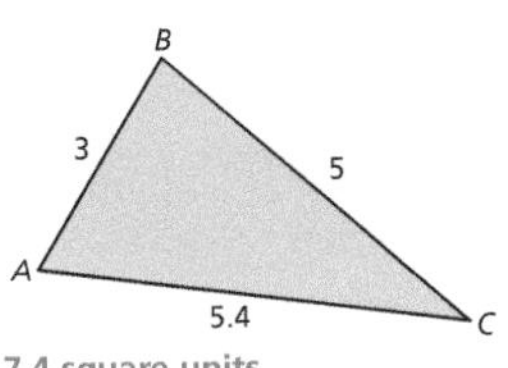

7.4 square units

Questioning Strategies

Problem 25 Suppose that $AB = 3$, $BC = 4$, and $AC = 5$. Can you find the area of $\triangle ABC$ in a more direct way than using the area formula we learned in this lesson? I know that 3, 4, 5 is a Pythagorean triple, so $\triangle ABC$ is a right triangle with area $A = \frac{1}{2}bh = \frac{1}{2}(3)(4) = \frac{1}{2}(12) =$ 6 square units.

Watch for Common Errors

Problem 15 Students might look at the diagram and mistakenly believe that $\angle B$ is a right angle and then deduce that $\triangle ABC$ is isosceles. They need to understand that while $\angle B$ appears to be a right angle, it is actually an obtuse angle. They can confirm this by comparing $\frac{AC}{BC} = \frac{90}{60.5} \approx 1.49$ to $\sqrt{2}$, seeing that the two values are not equal and deducing that the side lengths are not proportional to those of a 45°-45°-90° triangle.

ANSWERS

13. Sara is correct. The given information is a case of side-angle-side, so the Law of Cosines is used to solve the triangle.

17. $m\angle A \approx 44.0°$, $m\angle B \approx 26.5°$, $m\angle C \approx 109.5°$

18. $m\angle A \approx 42.4°$, $m\angle C \approx 75.6°$, $b \approx 6.9$

19. $m\angle A \approx 28.0°$, $m\angle B \approx 20.0°$, $c \approx 17.4$

20. $m\angle A \approx 29.3°$, $m\angle B \approx 31.2°$, $m\angle C \approx 119.6°$

21. $m\angle B \approx 79.7°$, $m\angle C \approx 42.3°$, $a \approx 19.0$

22. $m\angle A \approx 84.9°$, $m\angle B \approx 43.9°$, $m\angle C \approx 51.2°$

23. Grant is correct. Because two sides and the included angle are given, the area formula $A = \frac{1}{2}ac \sin B$ can be used to find the area.

5 Wrap-Up

Summarize learning with your class. Consider using the Exit Ticket, Put It in Writing, or I Can scale.

Exit Ticket 

For $\triangle ABC$, $AB = 10$, $AC = 8$, and $m\angle A = 100°$. Find the other measures of the triangle. $(BC)^2 = 8^2 + 10^2 - 2(8)(10)\cos 100° \approx 191.78$, so $BC \approx 13.85$. Then $\sin B \approx \frac{8 \sin 100°}{13.85} \approx 0.5688$, so $m\angle B \approx 34.67°$, and $m\angle C \approx 180° - 100° - 34.67° = 45.33°$.

Put It in Writing 

Explain how the Pythagorean Theorem is a special case of the Law of Cosines.

I Can 

The scale below can help you and your students understand their progress on a learning goal.

4	I can use the Law of Cosines to find side lengths and angle measures of non-right triangles and solve real-world problems, and I can explain my steps to others.
3	I can use the Law of Cosines to find side lengths and angle measures of non-right triangles and solve real-world problems.
2	I can use the Law of Cosines to find angle measures of a triangle when I know all of its side lengths.
1	I can tell when I can use the Law of Cosines to find missing measures of parts of a triangle.

Spiral Review • Assessment Readiness

These questions will help determine if students have retained information taught in the past and can also prepare them for high-stakes assessments. Here, students must use a trigonometric ratio to find an angle measure of a right triangle **(13.4)**, find the circumference of a circle from its radius **(Gr7, 10.1)**, apply the Law of Sines to find a side length **(14.1)**, and identify special right triangles from their side lengths **(13.3)**.

26. **History** The Historic Triangle in Virginia is a triangle formed by the historic communities of Jamestown, Williamsburg, and Yorktown as shown in the map. Use the distances between the communities to find the area of the Historic Triangle to the nearest tenth of a square mile. 32.2 square miles

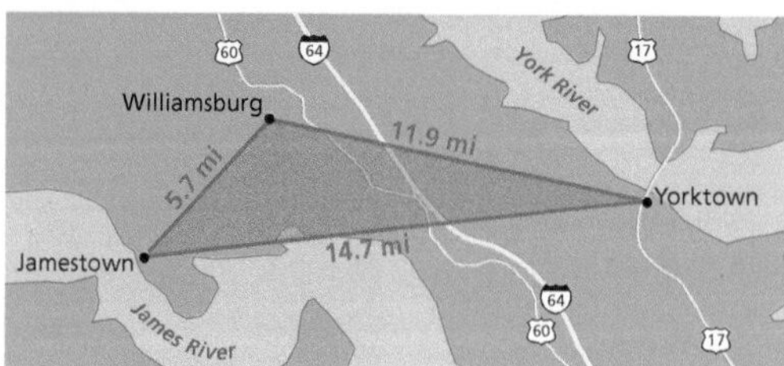

27. (Open Middle™) Using the digits 1 to 9, at most one time each, fill in the boxes to create a triangle with the greatest possible area. side lengths: 8 and 9; included angle: 76°

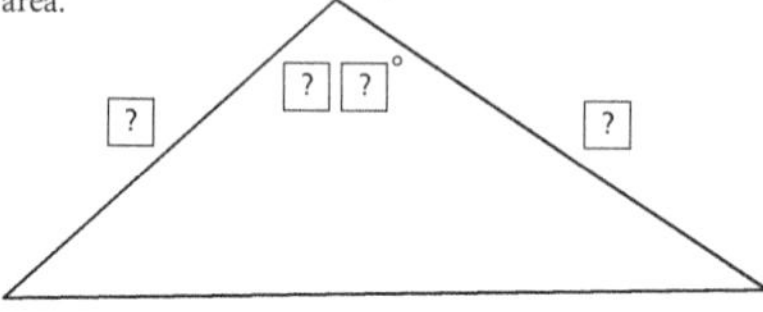

Spiral Review • Assessment Readiness

28. What is $m\angle B$?

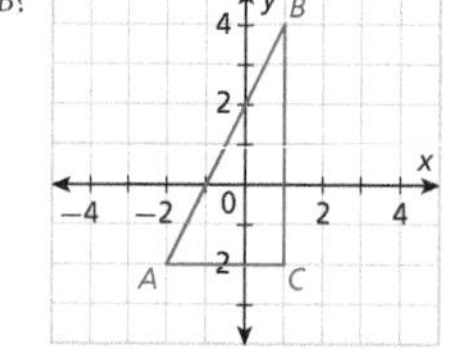

Ⓐ 26.6°
Ⓑ 30°
Ⓒ 63.4°
Ⓓ 90°

29. To the nearest tenth of an inch, what is the circumference of a circle that has a radius of 3.5 inches?

Ⓐ 5.5
Ⓑ 11.0
Ⓒ 22.0
Ⓓ 44.0

30. In $\triangle ABC$, $m\angle A = 71°$, $m\angle C = 68°$, and $b = 14$. What is the value c?

Ⓐ 9.9
Ⓑ 13.7
Ⓒ 19.8
Ⓓ 20.2

31. Which ratios of the side lengths of a triangle belong to a 30°-60°-90° triangle? Select all that apply.

Ⓐ $5 : 5\sqrt{3} : 10$
Ⓑ $3 : 6 : 3\sqrt{2}$
Ⓒ $4\sqrt{3} : 12 : 4$
Ⓓ $8 : 8 : 8\sqrt{3}$
Ⓔ $6 : 12 : 6\sqrt{3}$

I'm in a Learning Mindset!

How am I prioritizing my learning? What are my greatest strengths?

Keep Going ▶ Journal and Practice Workbook

Learning Mindset

Perseverance Learns Effectively

Point out how helpful it can be to prioritize topics when learning them, as the very act requires us to engage with the topics, describe them, and think about how they are connected in order to evaluate their relative importance. This prioritization does not have to be a strict ordering of all topics but can instead consist of a grouping of topics into tiers. It is also important to consider our level of confidence or struggle when it comes to these concepts. Two concepts could roughly have equal importance when viewed strictly by how they fit in a progression from past topics, but if a student is struggling more with one concept than the other they should assign a higher priority to it than the other. *How do you assign importance to different concepts learned in a module? Are some more important in the "big picture" than others? How does already understanding a concept well affect the priority you give it for learning?*

Module

14 Review

Law of Sines

Alicia is surveying a triangular park and takes measurements at two points, A and C.

She needs to find the remaining measurements to complete her report.

$m\angle B = 180° - 38° - 57° = 85°$

Triangle Sum Theorem

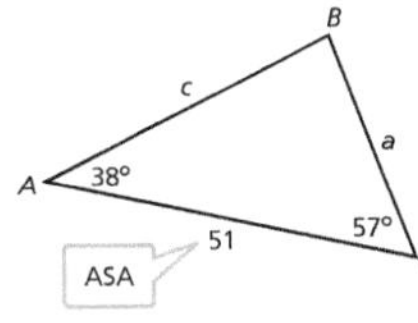

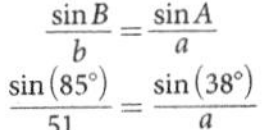

$\frac{\sin B}{b} = \frac{\sin A}{a}$	Law of Sines	$\frac{\sin B}{b} = \frac{\sin C}{c}$
$\frac{\sin(85°)}{51} = \frac{\sin(38°)}{a}$	Substitute.	$\frac{\sin(85°)}{51} = \frac{\sin(57°)}{c}$
$a = \frac{51\sin(38°)}{\sin(85°)}$	Solve for the unknown.	$c = \frac{51\sin(57°)}{\sin(85°)}$
$a \approx 31.5$	Evaluate.	$c \approx 42.9$

Law of Cosines

Alicia must include the triangular service shed that is in the park. She measures the outer dimensions and must determine the angle of each corner.

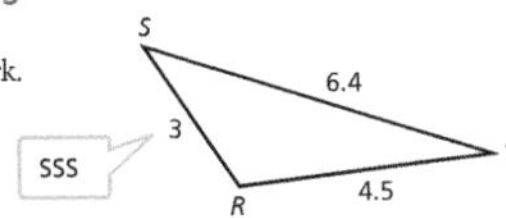

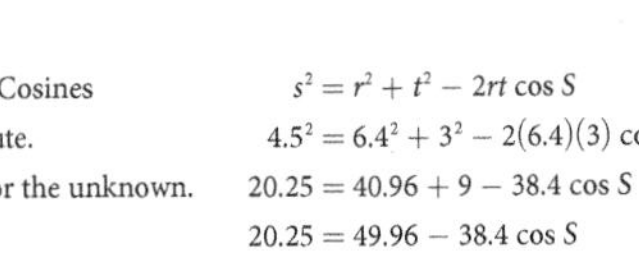

$r^2 = s^2 + t^2 - 2st\cos R$	Law of Cosines	$s^2 = r^2 + t^2 - 2rt\cos S$
$6.4^2 = 4.5^2 + 3^2 - 2(4.5)(3)\cos R$	Substitute.	$4.5^2 = 6.4^2 + 3^2 - 2(6.4)(3)\cos S$
$40.96 = 20.25 + 9 - 27\cos R$	Solve for the unknown.	$20.25 = 40.96 + 9 - 38.4\cos S$
$40.96 = 29.25 - 27\cos R$		$20.25 = 49.96 - 38.4\cos S$
$11.71 = -27\cos R$		$-29.71 = -38.4\cos S$
$\cos R = \frac{11.71}{-27}$		$\cos S = \frac{29.71}{38.4}$
$R = \cos^{-1}\left(\frac{11.71}{-27}\right) \approx 115.7°$	Evaluate.	$S = \cos^{-1}\left(\frac{29.71}{38.4}\right) \approx 39.3°$

Triangle Sum Theorem

$m\angle T = 180° - 115.7° - 39.3° = 25°$

ONLINE

Assign the Digital Module Review for

- built-in student supports
- Actionable Item Reports
- Standards Analysis Reports

Module Review

Use the first page of the Module Review to summarize and connect the main ideas of the module. Use the second page to assess students' understanding of the vocabulary, concepts, and skills presented in the module.

Sample Guided Discussion:

Q Law of Sines What is an alternate way to write the Law of Sines to solve this problem?

Possible answer: $\frac{a}{\sin A} = \frac{b}{\sin B}$ or $\frac{a}{b} = \frac{\sin A}{\sin B}$

Q Law of Cosines How could you rewrite the Law of Cosines to solve for $m\angle R$ directly?

$$r^2 = s^2 + t^2 - 2st\cos R$$

$$2st\cos R = s^2 + t^2 - r^2$$

$$\cos R = \frac{s^2 + t^2 - r^2}{2st}$$

$$R = \cos^{-1}\left(\frac{s^2 + t^2 - r^2}{2st}\right)$$

Module Review continued

Possible Scoring Guide

Items	Points	Description
1–3	2 each	identifies the correct term
4–9	2 each	correctly identifies whether the Law of Sines, Law of Cosines, or neither can be used to solve the triangle
10, 11	2 each	correctly solves the triangle using either the Law of Sines and/or the Law of Cosines
12, 13	2 each	correctly finds the area of a triangle using the Law of Cosines and the formula $A = \frac{1}{2}ab \sin C$
14	2	correctly uses the Law of Cosines to find the third side of a triangle and calculates its perimeter
15	2	correctly explains why SSA is the ambiguous case
16	2	correctly finds the area of the lake
Total points possible = 32 points		

The Unit 7 Test in the Assessment Guide assesses content from Modules 13 and 14.

Vocabulary

Choose the correct term from the box to complete each sentence.

Vocabulary
Cosines
Sines
trigonometric

1. The Law of ___?___ can be used to find the unknown measures of triangle if you know the measure of two angles and a side length. Sines
2. A ___?___ ratio is the ratio of two sides of a right triangle. trigonometric
3. The Law of ___?___ can be used to find the unknown measures of triangle if you know the lengths of the three sides. Cosines

Concepts and Skills

For each case, identify if the Law of Sines, the Law of Cosines, or neither can be applied to solve the triangle.

4. SSS Law of Cosines
5. SSA Law of Sines
6. SAS Law of Cosines
7. AAA neither
8. AAS Law of Sines
9. ASA Law of Sines

Solve each triangle. Round each answer to the nearest tenth.

10. 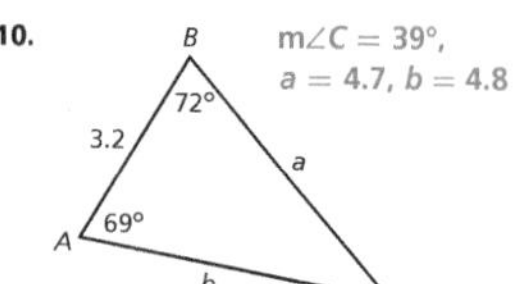

$m\angle C = 39°$, $a = 4.7$, $b = 4.8$

11. 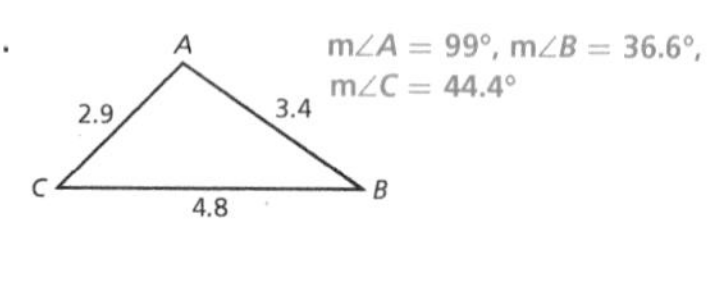

$m\angle A = 99°$, $m\angle B = 36.6°$, $m\angle C = 44.4°$

Find the area of each triangle. Round each answer to the nearest tenth.

12. B, 9, 50°, 9.9, A, C — 34.1 square units

13. 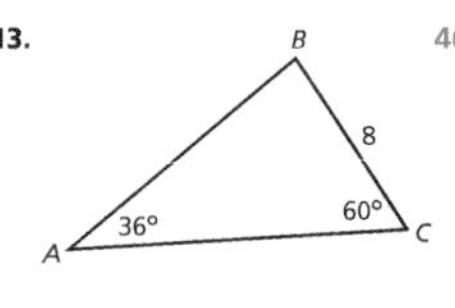

46.9 square units

14. (MP) **Use Tools** Lamar is commissioned to build a fence around a municipal flower garden. The garden is triangle with two sides measuring 7 feet and 10 feet and the included angle measuring 55°. How much fencing does he need to enclose the flower garden? Round your answer to the nearest tenth of a foot. State what strategy and tool you will use to answer the question, explain your choice, and then find the answer. Round your answer to the nearest tenth of a foot. 25.3 ft

15. Felicity walks around a nearly triangular lake and keeps track of each distance. What is the area of the lake? Round your answer to the nearest tenth of a square mile. 6.2 mi²

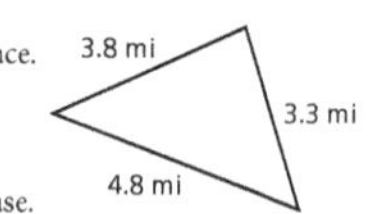

16. Explain why the SSA case of the Law of Sines is known as the ambiguous case. See Additional Answers.

460

DATA-DRIVEN INSTRUCTION

Before moving on to the Module Test, use the Module Review results to intervene based on the table below.

Items	Lesson	DOK	Content Focus	Intervention
4–9	14.1, 14.2	2	Identify whether to use Law of Cosines, Law of Sines, or neither to solve a triangle given the conditions SSS, SAS, ASA, SSA, and AAA.	Reteach 14.1, 14.2
10	14.1, 14.2	2	Use either the Law of Cosines or the Law of Sines to solve a triangle.	Reteach 14.1, 14.2
11	14.2	2	Use the Law of Cosines twice to solve a triangle.	Reteach 14.2
12, 13	14.1, 14.2	2	Use the Law of Cosines and/or the Law of Sines and the formula $A = \frac{1}{2}ab \sin C$ to find the area of a triangle.	Reteach 14.1, 14.2
14	14.2	3	Use the Law of Cosines to find the perimeter of a triangular region.	Reteach 14.2
15	14.1	3	Explain why the condition SSA is the ambiguous case.	Reteach 14.1
16	14.1, 14.2	3	Use the Law of Sines, the Law of Cosines, and the formula $A = \frac{1}{2}ab \sin C$ to find the area of a triangular region.	Reteach 14.1, 14.2

Module Test

The Module Test is available in alternative versions in your Assessment Guide. All items are presented in standardized test formats.

ONLINE

Assign the Digital Module Test to power actionable reports including

- proficiency by standards
- item analysis

Form A

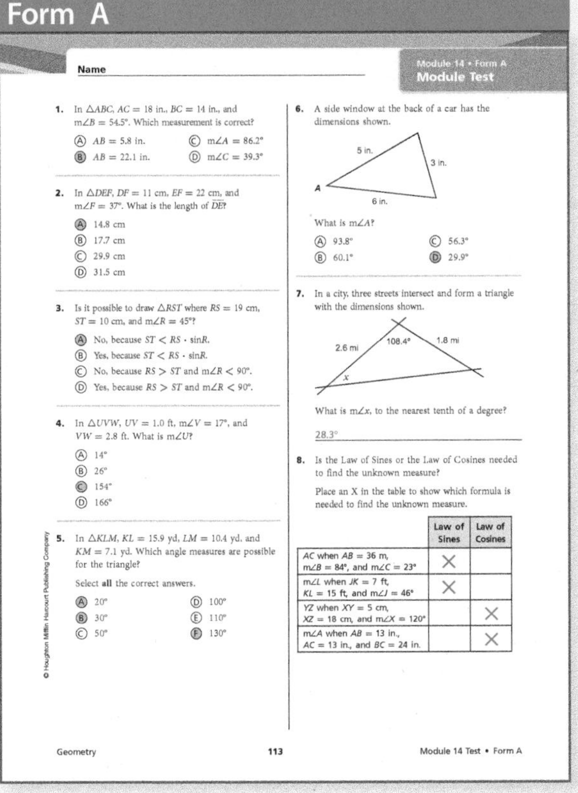

Name

Module 14 • Form A
Module Test

1. In $\triangle ABC$, $AC = 18$ in., $BC = 14$ in., and $m\angle B = 54.5°$. Which measurement is correct?
   - (A) $AB = 5.8$ in.
   - (B) $AB = 22.1$ in.
   - (C) $m\angle A = 86.2°$
   - (D) $m\angle C = 39.3°$

2. In $\triangle DEF$, $DF = 11$ cm, $EF = 22$ cm, and $m\angle F = 37°$. What is the length of $\overline{DE}$?
   - (A) 14.8 cm
   - (B) 17.7 cm
   - (C) 29.9 cm
   - (D) 31.5 cm

3. Is it possible to draw $\triangle RST$ where $RS = 19$ cm, $ST = 10$ cm, and $m\angle R = 45°$?
   - (A) No, because $ST < RS \cdot \sin R$.
   - (B) Yes, because $ST < RS \cdot \sin R$.
   - (C) No, because $RS > ST$ and $m\angle R < 90°$.
   - (D) Yes, because $RS > ST$ and $m\angle R < 90°$.

4. In $\triangle UVW$, $UV = 1.0$ ft, $m\angle V = 17°$, and $VW = 2.8$ ft. What is $m\angle U$?
   - (A) 14°
   - (B) 26°
   - (C) 154°
   - (D) 166°

5. In $\triangle KLM$, $KL = 15.9$ yd, $LM = 10.4$ yd, and $KM = 7.1$ yd. Which angle measures are possible for the triangle?

   Select **all** the correct answers.
   - (A) 20°
   - (B) 30°
   - (C) 50°
   - (D) 100°
   - (E) 110°
   - (F) 130°

6. A side window at the back of a car has the dimensions shown.

   5 in. 3 in. A 6 in.

   What is $m\angle A$?
   - (A) 93.8°
   - (B) 60.1°
   - (C) 56.3°
   - (D) 29.9°

7. In a city, three streets intersect and form a triangle with the dimensions shown.

   2.6 mi 108.4° 1.8 mi x

   What is $m\angle x$, to the nearest tenth of a degree?

   28.3°

8. Is the Law of Sines or the Law of Cosines needed to find the unknown measure?

   Place an X in the table to show which formula is needed to find the unknown measure.

	Law of Sines	Law of Cosines
AC when $AB = 36$ m, $m\angle B = 84°$, and $m\angle C = 23°$	X	
$m\angle L$ when $JK = 7$ ft, $KL = 15$ ft, and $m\angle J = 46°$	X	
YZ when $XY = 5$ cm, $XZ = 18$ cm, and $m\angle X = 120°$		X
$m\angle A$ when $AB = 13$ in., $AC = 13$ in., and $BC = 24$ in.		X

Geometry 113 Module 14 Test • Form A

Form A

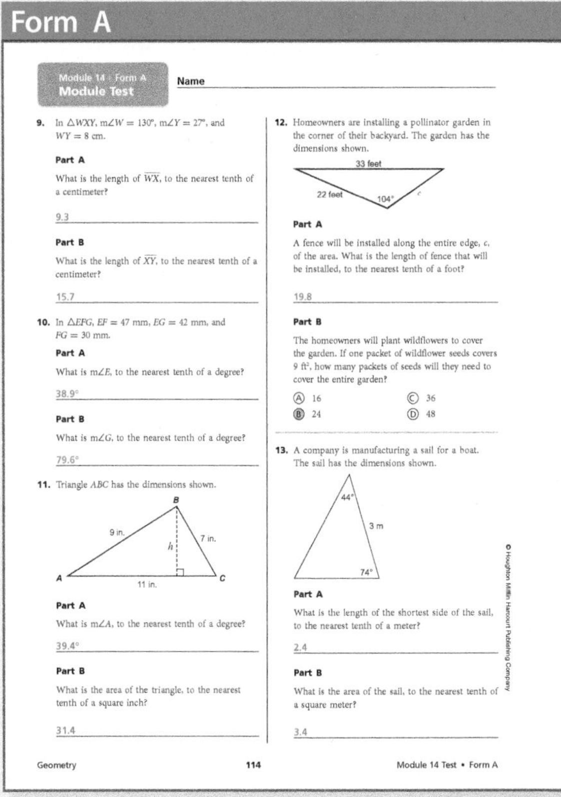

Module 14 • Form A
Module Test

Name

9. In $\triangle WXY$, $m\angle W = 130°$, $m\angle Y = 27°$, and $WY = 8$ cm.

   Part A

   What is the length of $\overline{WX}$, to the nearest tenth of a centimeter?

   9.3

   Part B

   What is the length of $\overline{XY}$, to the nearest tenth of a centimeter?

   15.7

10. In $\triangle EFG$, $EF = 47$ mm, $EG = 42$ mm, and $FG = 30$ mm.

    Part A

    What is $m\angle E$, to the nearest tenth of a degree?

    38.9°

    Part B

    What is $m\angle G$, to the nearest tenth of a degree?

    79.6°

11. Triangle ABC has the dimensions shown.

    B 9 in. 7 in. h A 11 in. C

    Part A

    What is $m\angle A$, to the nearest tenth of a degree?

    39.4°

    Part B

    What is the area of the triangle, to the nearest tenth of a square inch?

    31.4

12. Homeowners are installing a pollinator garden in the corner of their backyard. The garden has the dimensions shown.

    33 feet 22 feet 104° c

    Part A

    A fence will be installed along the entire edge, c, of the area. What is the length of fence that will be installed, to the nearest tenth of a foot?

    19.8

    Part B

    The homeowners will plant wildflowers to cover the garden. If one packet of wildflower seeds covers 9 ft², how many packets of seeds will they need to cover the entire garden?
    - (A) 16
    - (B) 24
    - (C) 36
    - (D) 48

13. A company is manufacturing a sail for a boat. The sail has the dimensions shown.

    44° 3 m 74°

    Part A

    What is the length of the shortest side of the sail, to the nearest tenth of a meter?

    2.4

    Part B

    What is the area of the sail, to the nearest tenth of a square meter?

    3.4

Geometry 114 Module 14 Test • Form A

Form B

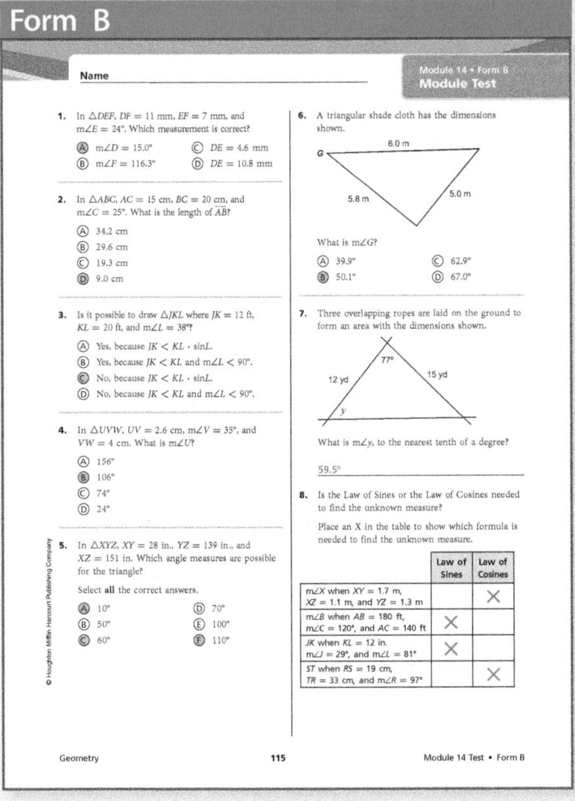

Name

Module 14 • Form B
Module Test

1. In $\triangle DEF$, $DF = 11$ mm, $EF = 7$ mm, and $m\angle E = 24°$. Which measurement is correct?
   - (A) $m\angle D = 15.0°$
   - (B) $m\angle F = 116.3°$
   - (C) $DE = 4.6$ mm
   - (D) $DE = 10.8$ mm

2. In $\triangle ABC$, $AC = 15$ cm, $BC = 20$ cm, and $m\angle C = 25°$. What is the length of $\overline{AB}$?
   - (A) 34.2 cm
   - (B) 29.6 cm
   - (C) 19.3 cm
   - (D) 9.0 cm

3. Is it possible to draw $\triangle JKL$ where $JK = 12$ ft, $KL = 20$ ft, and $m\angle L = 38°$?
   - (A) Yes, because $JK < KL \cdot \sin L$.
   - (B) Yes, because $JK < KL$ and $m\angle L < 90°$.
   - (C) No, because $JK < KL \cdot \sin L$.
   - (D) No, because $JK < KL$ and $m\angle L < 90°$.

4. In $\triangle UVW$, $UV = 2.6$ cm, $m\angle V = 35°$, and $VW = 4$ cm. What is $m\angle U$?
   - (A) 156°
   - (B) 106°
   - (C) 74°
   - (D) 24°

5. In $\triangle XYZ$, $XY = 28$ in., $YZ = 139$ in., and $XZ = 151$ in. Which angle measures are possible for the triangle?

   Select **all** the correct answers.
   - (A) 10°
   - (B) 50°
   - (C) 60°
   - (D) 70°
   - (E) 100°
   - (F) 110°

6. A triangular shade cloth has the dimensions shown.

   6.0 m G 5.8 m 5.0 m

   What is $m\angle G$?
   - (A) 39.9°
   - (B) 50.1°
   - (C) 62.9°
   - (D) 67.0°

7. Three overlapping ropes are laid on the ground to form an area with the dimensions shown.

   77° 12 yd 15 yd y

   What is $m\angle y$, to the nearest tenth of a degree?

   59.5°

8. Is the Law of Sines or the Law of Cosines needed to find the unknown measure?

   Place an X in the table to show which formula is needed to find the unknown measure.

	Law of Sines	Law of Cosines
$m\angle X$ when $XY = 1.7$ m, $XZ = 1.1$ m, and $YZ = 1.3$ m		X
$m\angle B$ when $AB = 180$ ft, $m\angle C = 120°$, and $AC = 140$ ft	X	
JK when $KL = 12$ in. $m\angle J = 29°$, and $m\angle L = 81°$	X	
ST when $RS = 19$ cm, $TR = 33$ cm, and $m\angle R = 97°$		X

Geometry 115 Module 14 Test • Form B

Form B

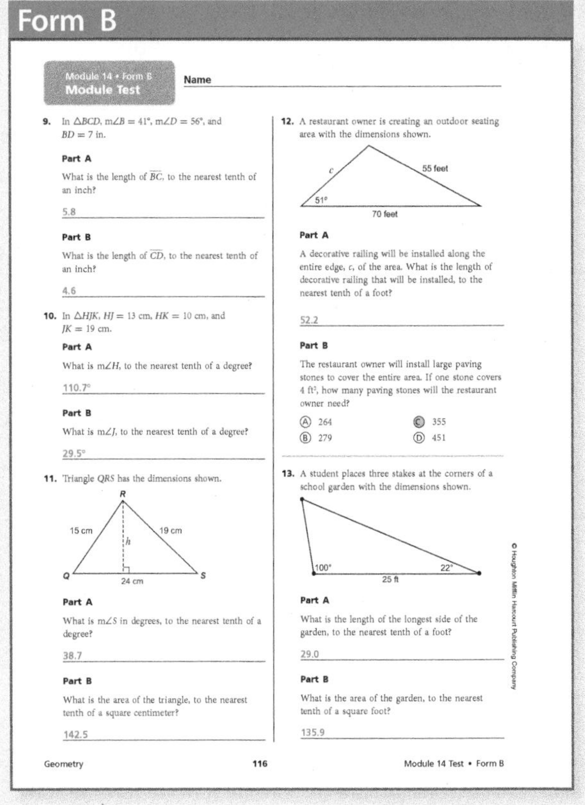

Module 14 • Form B
Module Test

Name

9. In $\triangle BCD$, $m\angle B = 41°$, $m\angle D = 56°$, and $BD = 7$ in.

   Part A

   What is the length of $\overline{BC}$, to the nearest tenth of an inch?

   5.8

   Part B

   What is the length of $\overline{CD}$, to the nearest tenth of an inch?

   4.6

10. In $\triangle HJK$, $HJ = 13$ cm, $HK = 10$ cm, and $JK = 19$ cm.

    Part A

    What is $m\angle H$, to the nearest tenth of a degree?

    110.7°

    Part B

    What is $m\angle J$, to the nearest tenth of a degree?

    29.5°

11. Triangle QRS has the dimensions shown.

    R 15 cm 19 cm h Q 24 cm S

    Part A

    What is $m\angle S$ in degrees, to the nearest tenth of a degree?

    38.7

    Part B

    What is the area of the triangle, to the nearest tenth of a square centimeter?

    142.5

12. A restaurant owner is creating an outdoor seating area with the dimensions shown.

    c 55 feet 51° 70 feet

    Part A

    A decorative railing will be installed along the entire edge, c, of the area. What is the length of decorative railing that will be installed, to the nearest tenth of a foot?

    52.2

    Part B

    The restaurant owner will install large paving stones to cover the entire area. If one stone covers 4 ft², how many paving stones will the restaurant owner need?
    - (A) 264
    - (B) 279
    - (C) 355
    - (D) 451

13. A student places three stakes at the corners of a school garden with the dimensions shown.

    100° 22° 25 ft

    Part A

    What is the length of the longest side of the garden, to the nearest tenth of a foot?

    29.0

    Part B

    What is the area of the garden, to the nearest tenth of a square foot?

    135.9

Geometry 116 Module 14 Test • Form B

Module 15: Angles and Segments in Circles

Module 16: Relationships in Circles

Module 17: Circumference and Area of a Circle

Optical Lens Technician

STEM POWERING INGENUITY

- **Say:** *Optical lens technicians make prescription eyeglasses and contact lenses to match prescriptions provided by opticians, optometrists, and ophthalmologists. Some optical lens technicians also manufacture lenses for optical instruments, such as telescopes and binoculars.*

  Optical lens technicians specialize in lens grinding and lens finishing. In making eyeglasses, a technician starts with a blank lens and follows the specifications in a patient's prescription. The technician then sets up and runs machines that grind and polish the lens. Lens finishers shape and assemble the lens and frame to produce the glasses.

- Explain that optical lens technicians use the lens maker's equation to make thin lenses of particular power from glass that has a given refractive index. This STEM task is focused on using the lens maker's equation to find the effective focal length given a set of known values.

STEM Task

Ask students what they know about how eyeglasses are made. Then present and discuss the lens maker's equation and its parameters. The equation is $\frac{1}{F} = (n-1)\left(\frac{1}{R_1} - \frac{1}{R_2}\right)$, where F is the focal length, n is the refractive index of the glass, and R_1 and R_2 are the radii of curvature of spheres 1 and 2. Students use the equation with $n = 1.5$.

- Ask students what effect R_1 and R_2 have on the equation.
- Why is R_1 positive and R_2 negative? Why is the effective focal length F positive?
- Why is it important to calculate the effective focal length in a lens?

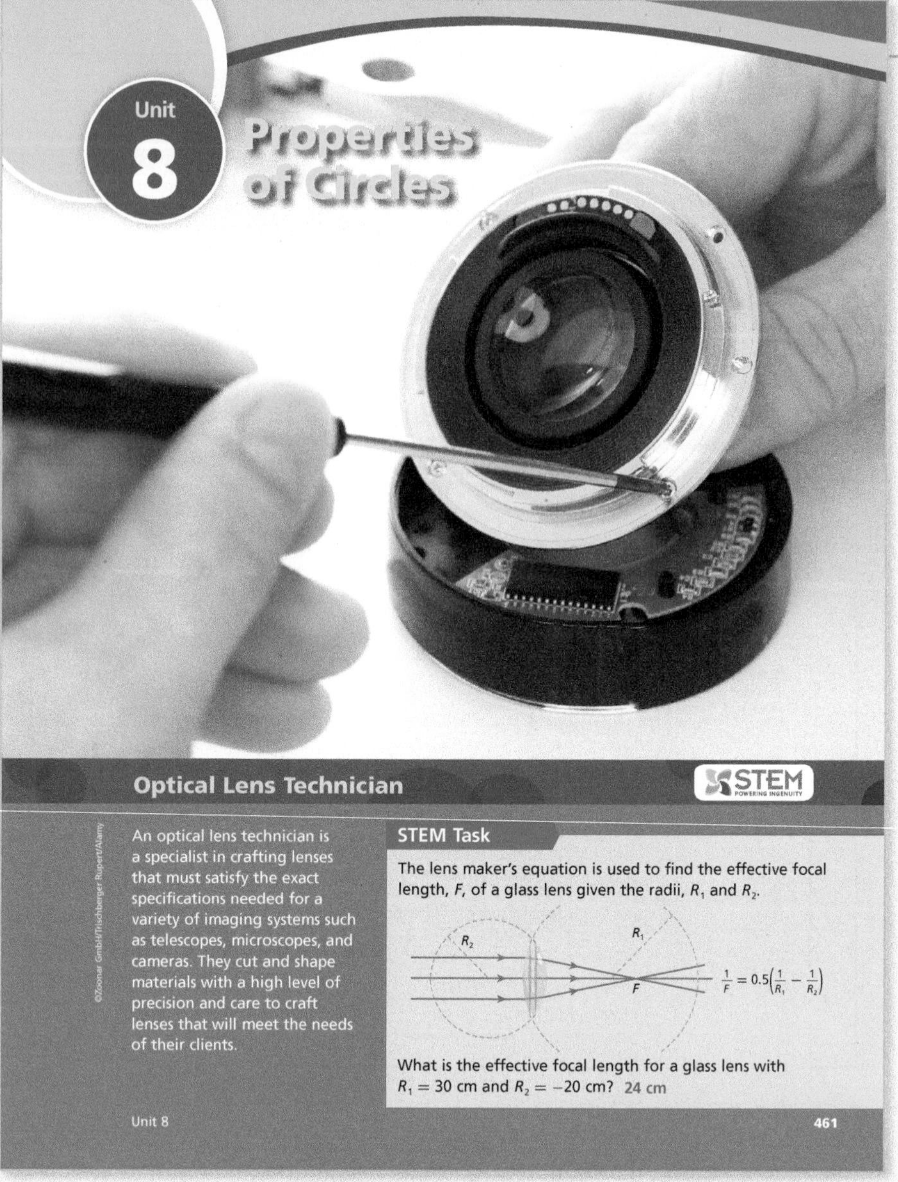

Unit 8 Properties of Circles

Optical Lens Technician

STEM POWERING INGENUITY

©Zoonar GmbH/Tischberger Rupert/Alamy

An optical lens technician is a specialist in crafting lenses that must satisfy the exact specifications needed for a variety of imaging systems such as telescopes, microscopes, and cameras. They cut and shape materials with a high level of precision and care to craft lenses that will meet the needs of their clients.

STEM Task

The lens maker's equation is used to find the effective focal length, F, of a glass lens given the radii, R_1 and R_2.

$\frac{1}{F} = 0.5\left(\frac{1}{R_1} - \frac{1}{R_2}\right)$

What is the effective focal length for a glass lens with $R_1 = 30$ cm and $R_2 = -20$ cm? 24 cm

Unit 8 461

Unit 8 Project Looking Through the Right Lens

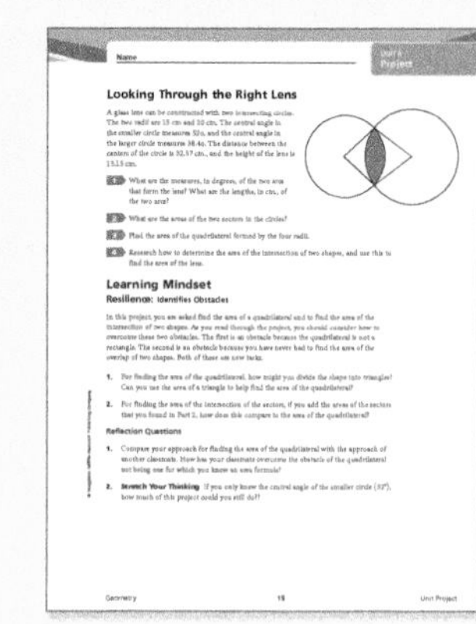

Looking Through the Right Lens

Learning Mindset

Overview: In this project students use arc lengths and sector areas to describe the various properties of an optical lens.

Materials: display or poster board for presentations

Assessing Student Performance: Students' presentations should include:

- two correct arc measures and two correct arc lengths **(Lesson 15.1 and Lesson 17.2)**
- two correct sector areas **(Lesson 17.3)**
- a correct quadrilateral area **(Lesson 17.3 or later)**
- a correct lens area **(Lesson 17.3 or later)**

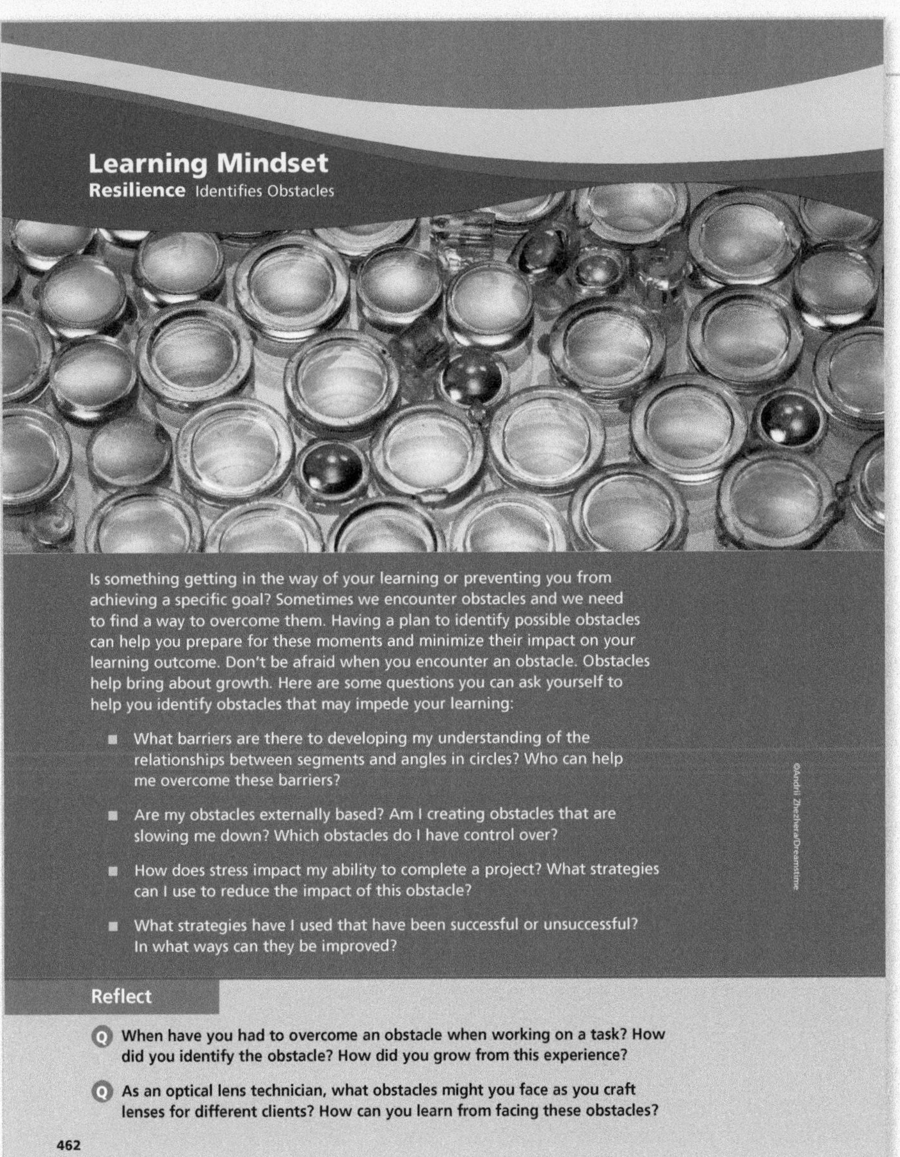

Learning Mindset

Resilience Identifies Obstacles

Is something getting in the way of your learning or preventing you from achieving a specific goal? Sometimes we encounter obstacles and we need to find a way to overcome them. Having a plan to identify possible obstacles can help you prepare for these moments and minimize their impact on your learning outcome. Don't be afraid when you encounter an obstacle. Obstacles help bring about growth. Here are some questions you can ask yourself to help you identify obstacles that may impede your learning:

- What barriers are there to developing my understanding of the relationships between segments and angles in circles? Who can help me overcome these barriers?
- Are my obstacles externally based? Am I creating obstacles that are slowing me down? Which obstacles do I have control over?
- How does stress impact my ability to complete a project? What strategies can I use to reduce the impact of this obstacle?
- What strategies have I used that have been successful or unsuccessful? In what ways can they be improved?

Reflect

- **When have you had to overcome an obstacle when working on a task? How did you identify the obstacle? How did you grow from this experience?**
- **As an optical lens technician, what obstacles might you face as you craft lenses for different clients? How can you learn from facing these obstacles?**

462

What to Watch For

Watch for students who can identify obstacles. Help them become more self-confident by

- evaluating their reactions to dealing with the obstacles,
- asking them how overcoming obstacles in the past has helped them grow, and
- applauding their efforts as they persevere and move forward.

Watch for students who identify and overcome learning obstacles. Encourage a positive attitude to learning by

- acknowledging their determination and perseverance,
- reminding them that becoming self-confident involves meeting obstacles head-on and working to overcome them, and
- encouraging them to mentor other students who lack confidence in overcoming learning obstacles.

"Fall seven times, stand eight."

—Japanese proverb

mindset works

Resilience

Learning Mindset

Identifies Obstacles

The learning-mindset focus in this unit is *resilience*, which refers to a person's ability to adapt and adjust to stressful obstacles or setbacks that must be overcome to succeed.

Mindset Beliefs

Students who are resilient can identify obstacles that interfere with their learning and that prevent them from achieving success. They have a positive attitude that helps them minimize the impact that learning obstacles may present.

Resilient students see obstacles as opportunities to grow. They often exhibit an above-average level of self-confidence and are not likely to step away from a new or difficult situation. They grow by learning from the experience of overcoming a challenge. If they were to confront a similar obstacle in the future, they would be able to cope with it.

Students who can identify and deal with obstacles have the ability to concentrate on what's in front of them, whether it's how to solve a math problem or how to fix something. In mathematics, students who overcome learning obstacles demonstrate resilience and perseverance.

Discuss with students how they react to a learning obstacle. Ask, "Do you analyze the source of an obstacle? Are you confident that you can be successful at overcoming it? What strategies do you use to guarantee success?"

Mindset Behaviors

Encourage students to identify and confront obstacles by asking themselves questions about how they approach learning. This self-monitoring can take the form of questions like the following:

When learning about the geometry of circles:

- What are the possible obstacles to my understanding concepts related to circles?
- How can I overcome these obstacles and be successful?

When identifying a learning obstacle:

- What can I do to get past it?
- What can I learn from this situation?

When confronting an obstacle:

- How confident am I that I can overcome it?
- Why is this obstacle holding me back?

Students who can identify and overcome obstacles grow in confidence. Self-confidence, developed by responding positively to obstacles, will play an important role in their future learning.

ANGLES AND SEGMENTS IN CIRCLES

Introduce and Check for Readiness

- Module Performance Task
- Are You Ready?

Lesson 15.1—2 Days

Central Angles and Inscribed Angles

Learning Objective: Determine the measures of central angles, inscribed angles, and arcs of a circle.
New Vocabulary: adjacent arcs, arc, central angle, chord, inscribed angle, intercepted arc, major arc, minor arc, semicircle
Review Vocabulary: circle, diameter

Lesson 15.2—2 Days

Angles in Inscribed Quadrilaterals

Learning Objective: Use the properties of angles of quadrilaterals inscribed in a circle to prove theorems and solve problems.
New Vocabulary: congruent arcs, congruent circles

Lesson 15.3—2 Days

Tangents and Circumscribed Angles

Learning Objective: Prove theorems about tangents to a circle and use them to solve mathematical and real-world problems.
New Vocabulary: circumscribed angle, exterior of a circle, interior of a circle, point of tangency, tangent of a circle

Lesson 15.4—2 Days

Circles on the Coordinate Plane

Learning Objective: Derive and write the equation of a circle with radius r and center (h, k).

Assessment

- Module 15 Test (Forms A and B)
- Unit 8 Test (Forms A and B)

LEARNING ARC FOCUS

 Build Conceptual Understanding  Connect Concepts and Skills 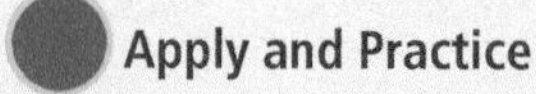 Apply and Practice

TEACHING FOR SUCCESS

TEACHING FOR DEPTH: Angles and Segments in Circles

Make Connections. Students have used the Distance Formula in earlier modules to find distances on the coordinate plane. They have also solved quadratic equations by completing the square and by applying the Quadratic Formula.

In this module, students will use the Distance Formula to derive the equation of a circle with radius r and center (h, k) as $(x-h)^2 + (y-k)^2 = r^2$. Conversely, students also see how completing the square and factoring the quadratic equation $x^2 + ax + y^2 + by = c$ results in the simpler equation $(x-h)^2 + (y-k)^2 = r^2$ which then can be used to identify the coordinates of the center of a circle and its radius.

A circle is an example of a conic section; that is, the 2-D figure formed when a plane intersects a right circular cone so that the plane is parallel to the base of the cone. In later modules, students will learn more about the conic sections and will be able to make the connection between what they've learned in this module about a circle and the general equations of the other conic sections represented by the equation $Ax^2 + Bxy + Cy^2 + Dx + Ey + F = 0$. Students will determine the type of conic section from the values of the coefficients A, B, C, D, E, and F and will learn how to complete the square to rewrite this general equation in a form that is specific to each of the other conic sections.

Mathematical Progressions

Prior Learning	Current Development	Future Connections
Students: • applied the Pythagorean Theorem to determine unknown side lengths in right triangles. • completed the square in a quadratic expression to reveal the maximum or minimum value of the function it defines. • developed definitions of rotations, reflections, and translations in terms of angles, circles, perpendicular lines, parallel lines, and line segments. • wrote coordinate proofs about triangle relationships.	**Students:** • define and determine the measures of central angles, inscribed angles, and arcs of a circle. • use the properties of angles of quadrilaterals inscribed in a circle to prove theorems and solve problems. • prove theorems about tangents to a circle and use them to solve mathematical and real-world problems. • derive and write the equation of a circle with radius r and center (h, k).	**Students:** • will identify and describe relationships among inscribed angles, radii, and chords, including the relationship between central, inscribed, and circumscribed angles. • will derive, using similarity, the fact that the length of the arc intercepted by an angle is proportional to the radius and define the radian measure of the angle as the constant of proportionality. • will derive the formula for the area of a sector.

TEACHER TO TEACHER

From the Classroom

Build procedural fluency from conceptual understanding. I believe that conceptual understanding is an essential foundation for procedural fluency. Students who are fluent in mathematics understand what they read and hear, can think creatively and independently, and can communicate their ideas to others.

In teaching geometry, I want my students to develop procedural fluency as they prove the Inscribed Angle Theorem. I want them to develop the ability to see the need for breaking the proof into three cases. This helps my students build conceptual understanding. I also believe that I am a better teacher when I can provide my students with multiple opportunities to connect procedures with underlying concepts.

LANGUAGE DEVELOPMENT • Planning for Instruction

By giving all students regular exposure to language routines in context, you will provide opportunities for students to **listen**, **speak**, **read**, and **write** about mathematical situations and develop both mathematical language and conceptual understanding at the same time.

Using Language Routines to Develop Understanding

Use the **Professional Cards** for the following routines to plan for effective instruction.

Co-Craft Questions and Problems Lesson 15.1

Students think of natural questions to ask about a given situation or problems similar to a given task and answer the questions they have developed or problems they have created.

Three Reads Lessons 15.1, 15.2, and 15.4

Students read a problem three times with a specific focus each time.

1st Read What is the situation about?
2nd Read What are the quantities in the situation?
3rd Read What are the possible mathematical questions that you could ask for the situation?

Information Gap Lesson 15.3

Students recognize when information given in a problem situation is incomplete, and they pose questions and share knowledge with others to discover any missing facts or relationships and work together to solve the problem.

Critique, Correct, and Clarify Lesson 15.3

Students correct the work in a flawed explanation, argument, or solution method and share with a partner and refine the sample work.

Connecting Language to Angles and Segments in Circles

Watch for students' use of the review and new terms listed below as they explain their reasoning and make connections with new concepts.

Key Academic Vocabulary

Prior Learning and Current Development • Review and New Vocabulary

circle the set of all points in a plane that are equidistant from a given point called the center

diameter a chord of a circle that contains the center of the circle

arc an unbroken part of a circle consisting of two points called endpoints and all points on the circle between them

central angle an angle whose vertex is the center of the circle

chord a segment whose endpoints lie on a circle

circumscribed angle an angle formed by two rays from a common endpoint that are tangent to a circle

inscribed angle an angle whose vertex is on a circle and whose sides are chords of the circle

intercepted arc an arc that consists of its endpoints and all points of the circle between the endpoints

tangent of a circle a line in the same plane as the circle that intersects the circle in exactly one point

Linguistic Note

Listen for students understanding of the word *chord*. In geometry, the word *chord* refers to a segment whose endpoints lie on a circle. In music, a chord is three or more notes that sound harmonious. There is also the homophone, *cord*, which is a string or rope made up of twisted strands or a flexible material made by weaving strands of yarn together. To help English learners learn these words, have them write each word, its definition, and use it in a sentence.

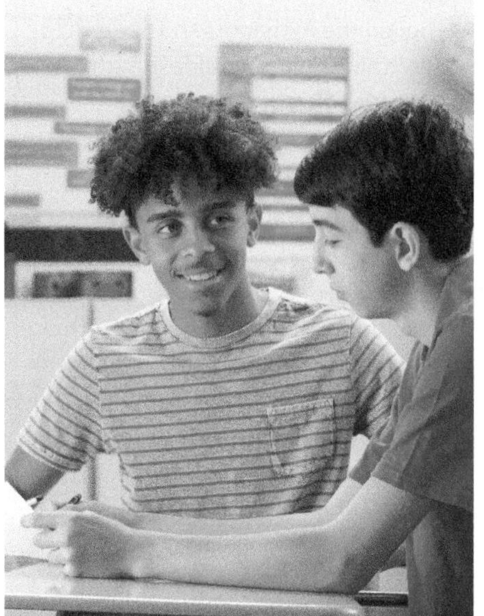

Module 15

Angles and Segments in Circles

Module Performance Task: Focus on STEM

©Mark Nazh/Shutterstock

Virtual Reality

To develop a virtual reality (VR) system, you need to have a geometric understanding of how the eye dynamically receives and composes images.

A horopter is the set of points that are imaged on the corresponding points of the retinas. Objects that lie on the horopter are seen as a single image. In this diagram, the large circle represents the horopter and the two smaller circles represent eyes with congruent radii.

A–C. See margin.

A. What is the relationship between $\angle LAR$ and $\angle LBR$? Explain your reasoning.

B. Prove that $\triangle APL$ and $\triangle BPR$ are similar.

C. If $\overline{LD_L} \cong \overline{RD_R}$ and $\overline{LE_L} \cong \overline{RE_R}$, do you have enough information to show that $\overline{D_LE_L} \cong \overline{D_RE_R}$? Explain your reasoning.

D. What can you conclude about the images on the retinas at $\overset{\frown}{D_LE_L}$ and $\overset{\frown}{D_RE_R}$? Explain your reasoning.
By the Congruent Corresponding Chords Theorem, the arcs are congruent. Therefore, the images on the arcs are congruent.

Module 15 463

Mathematical Connections

Task Part	Prerequisite Skills
Part A	Students can prove inscribed angles are congruent **(15.1)**.
Part B	Students can prove inscribed angles are congruent and triangles are similar by AA **(12.3 and 15.1)**.
Part C	Students can prove triangles are congruent by SAS and use CPCTC **(7.2 and 8.1)**.
Part D	Students can prove that arcs are congruent if their corresponding chords are congruent **(15.2)**.

MODULE 15

Virtual Reality

Overview

Virtual reality (VR) is a 3-D environment created by computer technology. A user can manipulate and explore virtual environments while having a sensation of actually being in that world.

To understand how VR works, it's important to understand how the eyes receive and process images. Although we have two eyes, we do not experience double vision. Fortunately, our brain has the ability to adjust our vision so that we see only a single image rather than a double image. The word *horopter* describes our having single vision. It is defined as the set of points in space that project on corresponding points in the retina of each of our eyes. These retinal images lie at corresponding locations from congruent distances.

Career Connections

Students with an interest in virtual reality can consider careers in software development and design, 3-D modeling, game development, and graphics programming. Less technical VR roles include content producer, content writer, product manager, and quality assurance expert. Students with an interest in the eye and how we see may want to look into a career in optics.

Answers

A. The angles are congruent; $\angle LAR$ and $\angle LBR$ are inscribed angles that subtend the same arc, so they are congruent.

B. $\angle LAR \cong \angle LBR$ because they are inscribed angles that subtend the same arc. $\angle APL \cong \angle BPR$ because they are vertical angles. $\triangle APL \sim \triangle BPR$ by AA Similarity Theorem.

C. yes; Possible answer: We know that $\angle ALP \cong \angle BRP$ because $\triangle ALP \sim \triangle BRP$ or because both angles intercept the same arc. Because vertical angles are congruent, we know $\angle ALP \cong \angle D_LLE_L$ and $\angle BRP \cong \angle D_RRE_R$. Therefore, by the Transitive Property of Congruence, $\angle D_LLE_L \cong \angle D_RRE_R$. We are given that $\overline{LD_L} \cong \overline{RD_R}$ and $\overline{LE_L} \cong \overline{RE_R}$, so $\triangle LD_LE_L \cong \triangle RD_RE_R$ by SAS Triangle Congruence, and $\overline{D_LE_L} \cong \overline{D_RE_R}$ by CPCTC.

ANGLES AND SEGMENTS IN CIRCLES

ONLINE

Assign the Digital Are You Ready? to power actionable reports including

- proficiency by standards
- item analysis

Are You Ready?

Diagnostic Assessment

- Diagnose prerequisite mastery.
- Identify intervention needs.
- Modify or set up leveled groups.

data checkpoint

Have students complete the *Are You Ready?* assessment on their own. Items test the prerequisites required to succeed with the new learning in this module.

Multiply and Divide Rational Numbers Students will apply previous knowledge of multiplication and division of rational numbers to find lengths of segments and measures of angles in circles.

Types of Angle Pairs Students will apply previous knowledge of types of angle pairs to identify relationships among angles in circles and find their measures.

Distance and Midpoint Formulas Students will apply previous knowledge of the Distance and Midpoint Formulas to find lengths of segments and equations of circles graphed in the coordinate plane.

ONLINE 

Are You Ready?

Complete these problems to review prior concepts and skills you will need for this module.

Multiply and Divide Rational Numbers

Simplify each expression.

1. $\frac{5}{4}(-56)$ −70

2. $32 \div \frac{2}{3}$ 48

3. 8.3×0.4 3.32

4. $\frac{9.6}{-1.2}$ −8

5. $\frac{5}{8}(1.2)$ 0.75

6. $6.4 \div \frac{4}{3}$ 4.8

Types of Angle Pairs

Find the measure of each angle.

7. $\angle GED$ 151.4°

8. $\angle CEF$ 56.9°

9. $\angle GEC$ 28.6°

10. $\angle FEA$ 146.9°

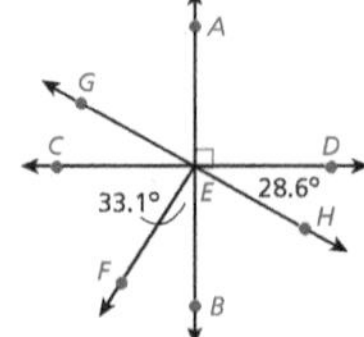

Distance and Midpoint Formulas

11–14. See Additional Answers.

Find the distance and the midpoint of the segment between each set of points.

11. $(-4, -1)$ and $(3, -2)$

12. $(7, -3)$ and $(1, 7)$

13. $(5, 2)$ and $(-2, 4)$

14. $(-6, 3)$ and $(-3, -5)$

Connecting Past and Present Learning

Previously, you learned:

- to locate the circumcenter and incenter of triangles with constructions,
- to find unknown side lengths and angle measures in triangles and quadrilaterals, and
- to use the Pythagorean Theorem to find side lengths of right triangles.

In this module, you will learn:

- to identify relationships between segments and angles in circles,
- to find unknown lengths and angle measures in circles, and
- to write equations of circles on the coordinate plane.

464

DATA-DRIVEN INTERVENTION

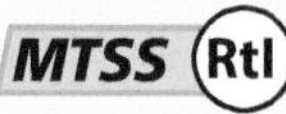

Concept/Skill	Objective	Prior Learning *	Intervene With
Multiply and Divide Rational Numbers	Find products and quotients of rational numbers.	Grade 7, Lesson 5.4	• Tier 3 Skill 8 • Reteach, Grade 7 Lesson 5.4
Types of Angle Pairs	Solve problems involving complementary, supplementary, vertical, and adjacent angles.	Grade 7, Lesson 7.5	• Tier 2 Skill 2 • Reteach, Grade 7 Lesson 7.5
Distance and Midpoint Formulas	Find lengths and midpoints of line segments in the coordinate plane.	Lessons 1.1 and 1.4	• Tier 2 Skill 8 • Reteach, Lessons 1.1 and 1.4

* Your digital materials include access to resources from Grade 6–Algebra 2. The lessons referenced here contain a variety of resources you can use with students who need support with this content.

15.1 Central Angles and Inscribed Angles

LESSON FOCUS AND COHERENCE

Mathematics Standards

- Identify and describe relationships among inscribed angles, radii, and chords. *Include the relationship between central, inscribed, and circumscribed angles; inscribed angles on a diameter are right angles; the radius of a circle is perpendicular to the tangent where the radius intersects the circle.*
- Given a rectangle, parallelogram, trapezoid, or regular polygon, describe the rotations and reflections that carry it onto itself.
- Construct an equilateral triangle, a square, and a regular hexagon inscribed in a circle.

Mathematical Practices and Processes 

- Use appropriate tools strategically.
- Look for and make use of structure.
- Look for and express regularity in repeated reasoning.

I Can Objective 

I can determine the measures of central angles, inscribed angles, and arcs of a circle.

Learning Objective

Identify the relationships between the measures of associated central angles, inscribed angles, and intercepted arcs; apply the Arc Addition Postulate; apply the Inscribed Angle Theorem to solve for unknown arc and angle measures.

Language Objective

Describe the relationships of angles and arcs of circles.

Vocabulary

Review: circle, diameter

New: adjacent arcs, arc, central angle, chord, inscribed angle, intercepted arc, major arc, minor arc, semicircle

Lesson Materials: compass, geometric drawing tool, straightedge

Mathematical Progressions

Prior Learning	Current Development	Future Connections
Students: • explained how the criteria for triangle congruence follow from the definition of congruence in terms of rigid motions. **(8.2)** • given two figures, used the definition of congruence in terms of rigid motions to decide if they are congruent. **(12.1)**	**Students:** • relate the measures of a central angle, its minor arc and major arc, and the measure of an associated inscribed arc. • use the Arc Addition Postulate. • apply the Inscribed Angle Theorem to solve for unknown arc and angle measures.	**Students:** • will identify and describe relationships among inscribed angles, radii, and chords. **(16.1 and 16.2)** • will derive using similarity the fact that the length of the arc intercepted by an angle is proportional to the radius, and derive the formula for the area of a sector. **(17.2 and 17.3)**

PROFESSIONAL LEARNING

Visualizing the Math

Students sometimes have difficulty remembering the relationships between the measures of associated central angles, inscribed angles, and intercepted arcs. For example, they may remember that the measure of one angle is twice the measure of the other but be unsure which measure is greater. It may be helpful for them to tear off a corner of a sheet of paper, place the corner at the vertex of a given angle and an edge along a side of the angle, and fold the paper so the crease is on the other side of the angle. In doing so, they can approximate the given angle with the folded corner and compare this rough copy of the angle to other given angles. Making a rough copy of an inscribed angle in this way and placing the copy on an associated central angle, they can intuit that the measure of the inscribed angle is half that of the central angle.

WARM-UP OPTIONS

ONLINE

PROJECTABLE & PRINTABLE

ACTIVATE PRIOR KNOWLEDGE • Construct a Rotation About a Point

Use these activities to quickly assess and activate prior knowledge as needed.

Problem of the Day 

Jonathon and Mikayla want to rotate a volleyball court on a beach to a different location. $\overline{PQ}$ is an end line of the court, and they want to rotate the end line about point C so the image of point P is at P'. How can they do this while using a long piece of rope as a compass?

They can use the rope as a compass to construct a copy of $\angle PCQ$ with one side along $\overrightarrow{P'C}$. Then, they can use the rope to measure CQ locate point Q' along $\overrightarrow{CQ'}$. They can then connect P' and Q' to get $\overline{P'Q'}$.

Quick Check for Homework

As part of your daily routine, you may want to display the Teacher Solution Key to have students check their homework.

Make Connections

Based on students' responses to the Problem of the Day, choose one of the following:

1. Project the Interactive Reteach, Lesson 5.2.

2. Complete the Prerequisite Skills Activity:

Have students draw a triangle $\triangle ABC$, and plot a point D outside of the triangle. Have them use a ruler and protractor to rotate $\triangle ABC$ counterclockwise by 50° about D.

- *How can you find the ray $\overrightarrow{DA'}$?* *I can draw segment $\overline{DA}$, measure a 50° angle counterclockwise from $\overline{DA}$ with vertex at D, and draw a ray.*
- *How can you determine how far along the ray $\overrightarrow{DA'}$ to plot point A′?* *I can measure DA and then measure that distance along $\overrightarrow{DA}$.*

SHARPEN SKILLS

If time permits, use this on-level activity to build fluency and practice basic skills.

Vocabulary Review

Objective: Students demonstrate an understanding of the definition of a regular polygon.
Materials: Word Definition Map (Teacher Resource Masters)

Have students work in pairs. Each pair should work together to build a word definition map for the term "regular polygon" without looking anything up. Encourage students to work together to recall as much information as possible and to not worry too much about being correct at this point.

When all pairs have had ample time to work, have a class discussion and collaborate to create a word definition map on the board.

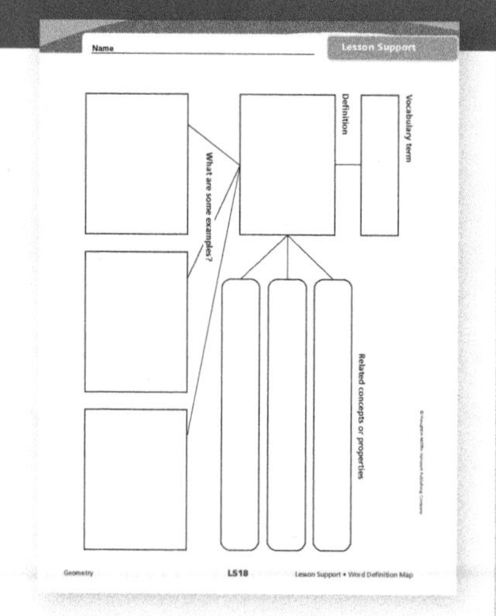

PLAN FOR DIFFERENTIATED INSTRUCTION

Small-Group Options

Use these teacher-guided activities with pulled small groups.

On Track

Materials: index cards

Show students the diagram below:

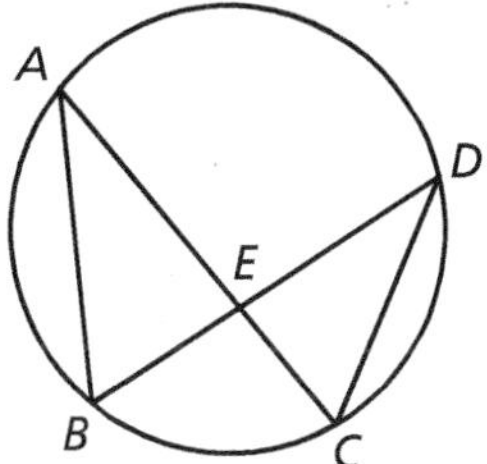

Ask students the following:

- How are $\angle A$, $\angle B$, $\angle C$, and $\angle D$ related?
- How are the two triangles related?
- How are the side lengths of the triangles related?
- What could you conclude if you knew that $AE = DE$?

Almost There

Materials: index cards

Give each group a set of index cards. Each card should have a circle with a central angle and an inscribed angle with common endpoints. Indicate the measure of the intercepted arc, the central angle, or the inscribed angle along with the measure of one of the other arcs. Have students take turns drawing a card and determining the measures of the angles and arcs that are not labeled.

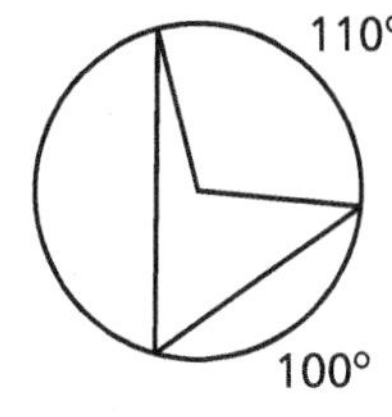

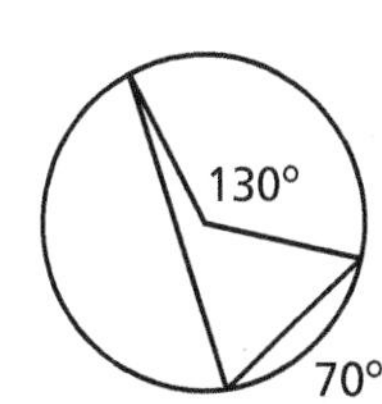

Ready for More

Materials: compass, protractor, ruler

Ask students to draw three circles with radii 2 inches, 4 inches, and 6 inches, respectively. Have them draw a central angle with a measure of 60° in each circle. Ask students,

- What fraction of the circumference of each circle is between the endpoints of the angle? How do you know?
- What is the circumference of each circle?
- What is the length along the minor arc of each central angle?
- What fraction of the circumference of a circle is between the endpoints of a central angle measuring m degrees?
- What do you think is the length along the minor arc of a central angle measuring m degrees in a circle with radius r?

Math Center Options

Use these student self-directed activities at centers or stations. Key: ● Print Resources ● Online Resources

On Track

- ● Interactive Digital Lesson
- ●● Journal and Practice Workbook
- ● Interactive Glossary (printable): **adjacent arcs**, **arc**, **central angle**, **chord**, **circle**, **diameter**, **inscribed angle**, **intercepted arc**, **major arc**, **minor arc**, **semicircle**
- ● Module Performance Task
- ●● Define Geometric Terms That Are Based on Undefined Notions
- ●● Describe Transformations That Carry a Polygon Onto Itself

Almost There

- ● Reteach 15.1 (printable)
- ● Interactive Reteach 15.1
- ● Desmos: ThalEs

Ready for More

- ● Challenge 15.1 (printable)
- ● Interactive Challenge 15.1
- ● Illustrative Mathematics: Right Triangles Inscribed in Circles II

Unit Project Check students' progress by asking what the relationship is between an arc measure and a central angle.

View data-driven grouping recommendations and assign differentiation resources.

During the *Spark Your Learning*, listen and watch for strategies students use. See samples of student work on this page.

Reason with Circumference and Density — Strategy 1

For any row, the length is given by πr because each row is a semicircle. I can then find the total length of all the rows and divide that into seats of equal size to determine the number of people that can be seated in the amphitheater because there can be 1 person for every 2 feet. The number of people that can be seated in the amphitheater is approximately 4650 people.

If students . . . reason with circumference and density, they are demonstrating an understanding of population density from Lesson 1.4.

Have these students . . . explain how they chose to consider the length of a semicircular row and how they determined the density. **Ask:**

- Q How did you know to use the circumference formula to help you find the length of each row?
- Q What was your reasoning to determine the density of people per length of row?

Reason with a Scale Drawing — Strategy 2

I would create a scale drawing of the middle row where 1 foot is represented by 2 millimeters. Then mark off every 2 feet to determine the number of people in the middle row and use that to find the total for all 30 rows. The number of people that can be seated in the amphitheater is approximately 4650 people.

If students . . . reason with a scale drawing, they are demonstrating that they understand how a length can be divided into congruent segments, but they may not realize that their process is inefficient.

Activate prior knowledge . . . by having students determine the length of a semicircular row before drawing it. **Ask:**

- Q How can you determine the length of a semicircular row before you draw it?
- Q How can you determine how many seats would fit along that row if you know the total length of the row and the length of each seat?

COMMON ERROR: Uses Circumference

The circumference of each row is given by $2\pi r$ so I can use that to determine the total length of the rows. The number of people that can be seated in the amphitheater is approximately 4650 people.

If students . . . use the circumference formula, they may not understand that the formula needs to be modified for this situation.

Then intervene . . . by pointing out that each row is a semicircle so the length is only half of the circumference of a corresponding circle. **Ask:**

- Q How can you describe the relationship between the length of each semicircular row and the circumference of a corresponding circle?
- Q How do you think that the formula for circumference should be modified to obtain a formula for the length of a semicircular row?

15.1

Central Angles and Inscribed Angles

I Can determine the measures of central angles, inscribed angles, and arcs of a circle.

Spark Your Learning

An outdoor amphitheater has semicircular rows of seating around a semicircular stage. The tickets for the next performance have been sold out.

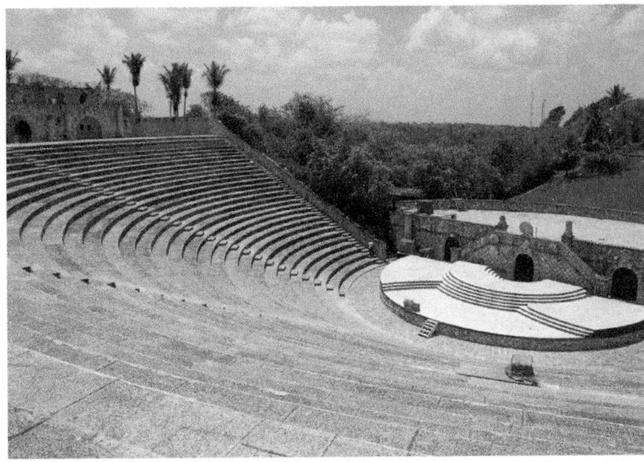

©Lisa S. Engelbrecht/Alamy

Complete Part A as a whole class. Then complete Parts B–C in small groups.

A. What is a mathematical question you can ask about this situation? What information would you need to know to answer your question? *How many people can be seated in the amphitheater?; the number of rows, how the rows increase in length, how much space should be allowed per person.*

B. To answer your question, what strategy and tool would you use along with all the information you have? What answer do you get? *See Strategies 1 and 2 on the facing page.*

C. Does your answer make sense in the context of the situation? How do you know? *yes; on average, each row can seat approximately 155 people and there are 30 rows.*

Turn and Talk How much space would each attendee have if only $\frac{3}{4}$ of all tickets are sold? Explain. *See margin.*

SUPPORT SENSE-MAKING • Three Reads

Tell students to read the information in the photo three times and prompt them with a different question each time.

1. What is the situation about? The situation is about determining the number people that can be seated in the amphitheater.
2. What are the quantities in this situation? How are those quantities related? The quantities are the radius of each row, the distance between each row, the length of each row, the distance between each seat, and the number of people that can be seated in the amphitheater. Because each row is a semicircle, the length of each row can be found by determining the circumference of each circular row and then dividing it in half. The number of people that can be seated in the amphitheater can be found by determining the total length of all rows and then dividing it into seats of equal sizes.
3. What are possible questions you could ask about this situation? Are all the rows and seats equally spaced? What is the distance between each seat? Can people be seated in areas that are not shown in the photo? How big are the aisles? What is the length of each row? How many people can sit in each row?

1 Spark Your Learning

▶ MOTIVATE

- Have students look at the photo in their books and read the information contained in the photo. Then complete Part A as a whole-class discussion.
- Give the class the additional information they need to solve the problem. This information is available online as a printable and projectable page in the Teacher Resources.
- Have students work in small groups to complete Parts B and C.

▶ PERSEVERE

If students need support, guide them by asking:

Q **Advancing • Use Tools** Which tool could you use to solve the problem? Why choose that tool and not some other? Students' choices of tools and reasons for choosing them will vary.

Q **Assessing** How much space will be between people sitting in the same row and between people sitting in adjacent rows? The distance between people sitting in adjacent rows is 3 feet. It is reasonable to assume that each person would also take up approximately 2 feet along the row that they are sitting in.

Q **Assessing** As you move from row to row, how does the length of each row change? The radius of each semicircular row increases by 3 feet from the previous row. Then, the length of each successive row increases by 6π feet.

Q **Advancing** How you can you determine the number of people in the nth row? The radius of the nth row is given by $r_n = 60 + 3(n - 1)$. The number of people in the nth row is then approximately $\frac{2\pi[60 + (3n - 1)]}{3}$.

Turn and Talk Ask students to start by estimating how the amount of space for each attendee would change. If less tickets were sold, would each attendee have more space or less space? If only $\frac{3}{4}$ of the tickets were sold, how would each attendee's amount of space change? They would have more space. The amount of space would increase by 25%.

▶ BUILD SHARED UNDERSTANDING

Select groups of students who used various strategies and tools to share with the class how they solved the problem. As they present their solutions, have each group discuss why they chose a specific strategy and tool.

2 Learn Together

Build Understanding

Task 1  **Use Tools** Students use a geometric drawing tool to construct a circle, an acute inscribed angle, and the corresponding central angle. They then move the vertex of the inscribed angle to explore how the measure of the angle does not change.

CONNECT TO VOCABULARY

Have students use the **Interactive Glossary** to record their understanding of the vocabulary in this task.

Sample Guided Discussion:

Q **In Part A, what seems to determine the measure of the inscribed angle?** the distance between the intersection points of the angle sides and the circle

Q **You moved the vertex of the inscribed angle around the circle in Part A, but why would moving the vertex of the central angle in Part B not make sense?** If the vertex moved to a different location in the circle, the angle would no longer be a central angle.

Turn and Talk Ask students to describe the central angle represented by the diameter. The inscribed angle is a right angle. A central angle formed by a diameter measures 180°, so the inscribed angle will measure half of that.

Build Understanding

Investigate Central Angles and Inscribed Angles

A **circle** is the set of all points in a plane that are equidistant from a given point, called the center of the circle. A **chord** is a segment whose endpoints lie on a circle. A **diameter** is a chord that contains the center of the circle.

A **central angle** of a circle is an angle whose vertex is the center of the circle. An **inscribed angle** is an angle whose vertex is on a circle and whose sides contain chords of the circle.

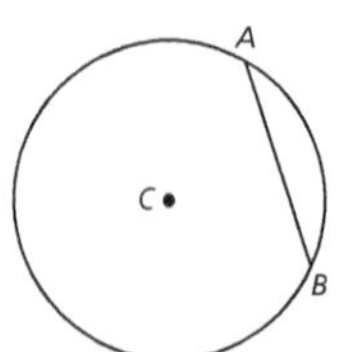

$\overline{AB}$ is a chord.

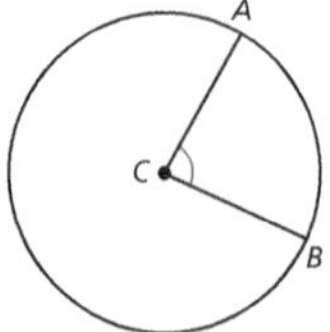

$\angle ACB$ is a central angle.

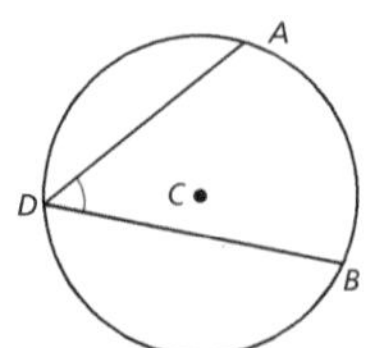

$\angle ADB$ is an inscribed angle.

1 Use a geometric drawing tool to draw a circle. Draw an acute inscribed angle on the circle. Then draw the corresponding central angle, which intersects the sides of the inscribed angle on the circle as shown.

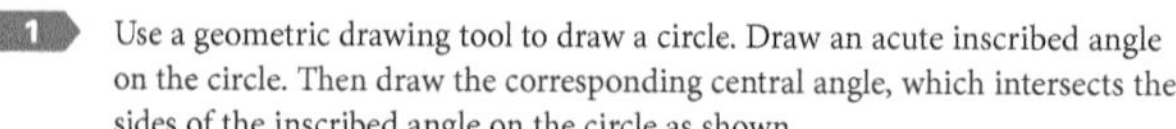

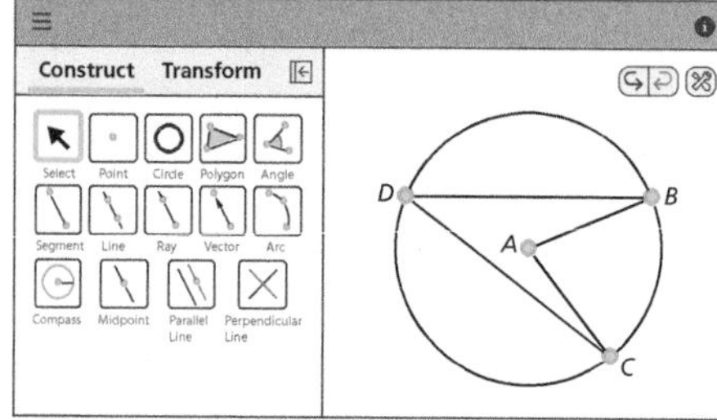

A. Measure the inscribed angle and the central angle. Drag the vertex of the inscribed angle around the circle. What do you notice about the angle measure as the vertex of the inscribed angle moves?

B. Observe the angle measures as you drag one of the points on the circle where the sides of the angles intersect. Make a conjecture about the relationship between the measures of an inscribed angle and its associated central angle. See Additional Answers.

A. The angle measure of the inscribed angle does not change as the vertex moves.

Turn and Talk Suppose $\overline{BC}$ is a diameter of a circle. An inscribed angle is drawn so that it intersects the diameter on the circle at points B and C. Make a conjecture about what type of angle the inscribed angle is. Explain your reasoning. See margin.

466

LEVELED QUESTIONS

Depth of Knowledge (DOK)	Leveled Questions	What Does This Tell You?
Level 1 **Recall**	How are a central angle and an inscribed angle different? A central angle has its vertex at the center of the circle, while an inscribed angle has its vertex on the circle.	Students' answers will indicate whether they can differentiate between central and inscribed angles.
Level 2 **Basic Application of Skills & Concepts**	If a central angle has a measure of 110°, what is the measure of its associated inscribed angle? 55°	Students' answers will indicate whether they can form a conjecture about the relationship between measures of associated central and inscribed angles.
Level 3 **Strategic Thinking & Complex Reasoning**	Can you give two reasons why the measure of an inscribed angle cannot be greater than 180°? The sides of the angle would not contain chords of the circle, and the measure of its associated central angle would be greater than 360°.	Students' answers will indicate whether they can apply the definition of an inscribed angle and relate it to the relationship between measures of associated central and inscribed angles.

Understand Arcs and Arc Measures

An **arc** is an unbroken part of a circle consisting of two points on the circle, called the endpoints, and all the points on the circle between them.

Arc	Measure	Figure
A **minor arc** is an arc of a circle whose points are on or in the interior of a central angle.	The measure of a minor arc is equal to the measure of the central angle. $m\overset{\frown}{AB} = m\angle ACB$	
A **major arc** is an arc of a circle whose points are on or in the exterior of a central angle.	The measure of a major arc is equal to 360° minus the measure of the central angle. $m\overset{\frown}{ADB} = 360° - m\angle ACB$	
A **semicircle** is an arc of a circle whose points lie on a diameter.	The measure of a semicircle is 180°. $m\overset{\frown}{ADB} = 180°$	

Minor arcs are named by their two endpoints. Major arcs and semicircles are named by their two endpoints, and a point on the arc.

2 The minute hand of a circular clock sweeps out an arc as it moves from 9:10 to 9:25.

A. What fraction of a complete rotation did the minute hand travel? $\frac{1}{4}$ of a complete rotation

B. A complete rotation of the minute hand corresponds to 360°. What is the degree measure that the minute hand traveled? What angle does this measure represent? 90°; the central angle

C. What is the measure of the arc formed on the clock as the minute hand moves from 9:10 to 9:25? Why?

C. 90°; The measure of the minor arc equals the measure of its corresponding central angle.

Task 2 (MP) **Use Structure** Students are introduced to major arcs, minor arcs, and arc measure in terms of an associated central angle. They then go on to relate arc measure to a fraction of a complete rotation of a point about the center of the circle. Encourage students to use proportional reasoning to extend their reasoning from this task to a general formula for the measure of the minor arc associated with a fraction of a complete rotation about the center of the circle.

CONNECT TO VOCABULARY

Have students use the **Interactive Glossary** to record their understanding of the vocabulary in this task.

Sample Guided Discussion:

Q Can you set up a proportion for Part A using amounts of time it takes the minute hand to move from one place to another? It takes the minute hand $25 - 10 = 15$ minutes to move from 9:10 to 9:25 and 60 minutes to travel a complete rotation. So, $\frac{15 \text{ min}}{60 \text{ min}} = \frac{x}{1 \text{ rotation}}$ is a proportion where x is how much of a rotation the minute hand travels from 9:10 to 9:25.

Q How can you answer Part A with a central angle formed by the minute hand? The minute hand placed at 9:10 and then at 9:25 makes a central right angle, and 90° is $\frac{1}{4}$ of a complete rotation.

Q Can you use a proportion to find a general rule for the measure m of a minor arc traced by the tip of the minute hand over t minutes? $\frac{t}{60} = \frac{m}{360°}$, so $m = (6t)°$

PROFICIENCY LEVEL

Beginning

Show students a circle with a 50° central angle and associated major and minor arcs indicated. As you point to the parts of the figure, say, "The minor arc has a measure of 50°, and the major arc has a measure of 310°." As you point to the associated minor arc, ask students to identify it by name and give its measure. Then point to the central angle, and ask for the same information.

Intermediate

Have students work in pairs. One student will say a phrase such as, "The measure of a minor arc is 70°," and their partner will respond with, "The measure of the central angle is 70°, and the measure of the major arc is 290°." Partners will then switch roles giving their partner the measure of a hypothetical major arc, minor arc, or central angle.

Advanced

Have students explain in general how the measures of a central angle and its associated minor and major arcs are related.

Task 3 **MP** **Use Repeated Reasoning** Students are introduced to the Arc Addition Postulate and apply it to concentric circles. For an interesting extension, ask students how they might find the measure of an arc formed by multiple adjacent arcs with equal measures.

CONNECT TO VOCABULARY

Have students use the **Interactive Glossary** to record their understanding of the vocabulary in this task.

Sample Guided Discussion:

Q **In Part C, what kind of arc is $\overset{\frown}{CDF}$?** A major arc

Q **What would be the measure of an arc formed by five adjacent arcs if each has a measure equal to $m\overset{\frown}{DC}$?** $5 \cdot m\overset{\frown}{DC} = 5 \cdot 40° = 200°$

Task 4 **MP** **Use Tools** Students use properties of a central angle and its associated minor arc to construct a regular hexagon inscribed in a circle. Draw special attention to the equilateral triangles and ask students about the significance of their angle measures. Make sure students understand that a central angle is not the same thing as its associated minor arc, though they have the same measures and can each be used to define the other.

Sample Guided Discussion:

Q **Why do you think it is important that the six adjacent arcs go completely around the circle?** The endpoints of the arcs will be the vertices of the hexagon, so if they did not go completely around the circle the hexagon would not be closed.

Q **Why do you think it is important to keep the compass set at the radius of the circle?** I need the adjacent arcs to have endpoints that let me construct 6 equilateral triangles with a vertex at the center of the circle.

Turn and Talk Ask students whether they could choose some vertices of a regular hexagon to be the vertices of a triangle. Ask them whether their choice of vertices has an effect on the shape of the triangle. If necessary, remind them that an equilateral triangle is a regular polygon Connect every other point of intersection on the circle instead of every point of intersection.

Adjacent arcs are arcs of the same circle that have a common endpoint. You can add the measures of two adjacent arcs.

Arc Addition Postulate

The measure of an arc formed by two adjacent arcs is the sum of the measures of the two arcs.

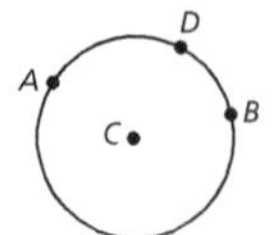

3 A. Two circles with different radii have their centers at A. Explain why $m\overset{\frown}{EB} = 40°$. See margin.

B. What arc measure is being calculated below?
$m\overset{\frown}{DC} + m\overset{\frown}{CF} = 40° + 79° = 119°$ $\overset{\frown}{DF}$

C. What is $m\overset{\frown}{CDF}$? How do you know? See margin.

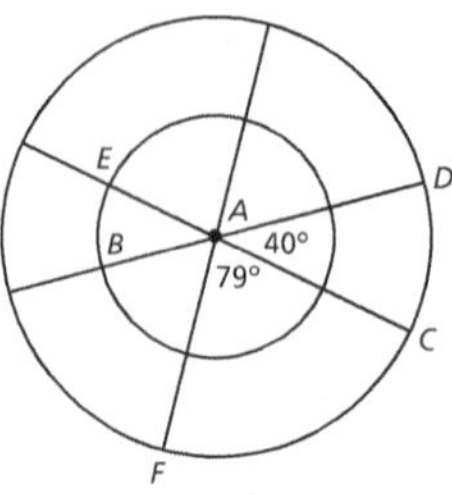

Construct a Regular Hexagon

Recall that a hexagon is a polygon with six sides, and that a regular hexagon is both equilateral and equiangular.

4 With a compass, construct a circle and place point A on the circle. Using the same compass setting, draw an arc from point A that intersects the circle. Label the point of intersection B. Continue this method back to point A, creating points C, D, E, and F. Use a straightedge to draw hexagon $ABCDEF$.

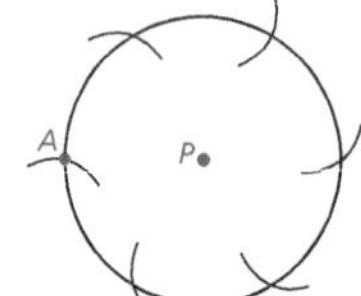

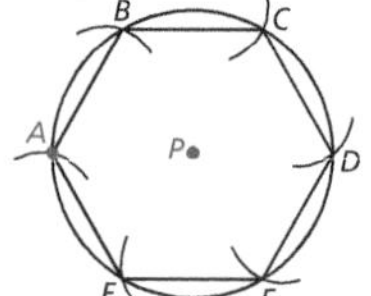

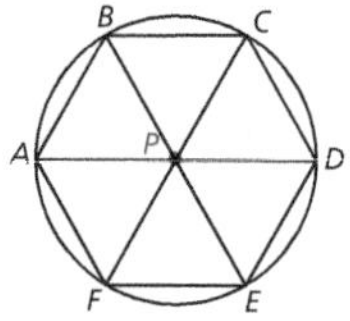

A. Explain why $AB = AP$. How do you know that the sides of the hexagon are congruent? A–E. See Additional Answers.

B. Use a straightedge to draw $\overline{AD}$, $\overline{BE}$, and $\overline{CF}$. Why do these line segments contain center P?

C. Each side of the hexagon along with the two sides connecting it to the center of the circle forms an equiangular triangle. Why?

D. Can you conclude that the hexagon is equiangular? Explain.

E. Why can you conclude that hexagon $ABCDEF$ is a regular polygon?

Turn and Talk How can you use the regular hexagon construction to construct an equilateral triangle inscribed in a circle? See margin.

468

ANSWERS

3.A. Possible answer: Since $\angle DAC$ and $\angle EAB$ are a vertical angle pair, $m\angle EAB = 40°$. The measure of a minor arc of a central angle is equal to the measure of the central angle. So, $m\overset{\frown}{EB} = m\angle EAB = 40°$.

3.C. $m\overset{\frown}{CDF} = 281°$; $\overset{\frown}{CDF}$ is a major arc, so $m\overset{\frown}{CDF} = 360° - m\overset{\frown}{CF} = 281°$.

Step It Out

Prove the Inscribed Angle Theorem

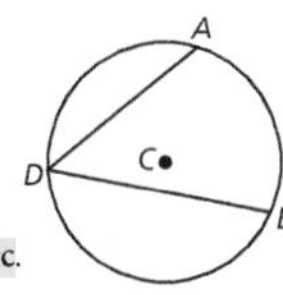

Two chords with a common endpoint form an inscribed angle. The other endpoints and all the points on the circle between them form an **intercepted arc**. $\angle ADB$ is the inscribed angle, and $\overset{\frown}{AB}$ is the intercepted arc.

Inscribed Angle Theorem

The measure of an inscribed angle is equal to half the measure of its intercepted arc.

$m\angle ADB = \frac{1}{2}m\overset{\frown}{AB}$

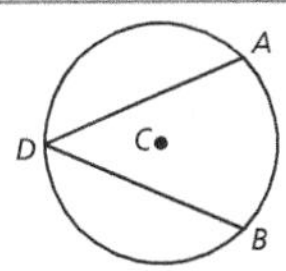

Inscribed Angle of a Diameter Theorem

The endpoints of a diameter lie on the endpoints of an inscribed angle if and only if the inscribed angle is a right angle.

$\overline{AB}$ is a diameter of the circle if and only if $\angle ADB$ is a right angle.

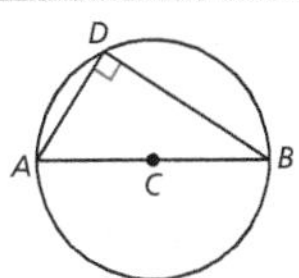

To prove the Inscribed Angle Theorem, the following three cases must be proven:

5 **Case 1:** The center of the circle is on a side of the inscribed angle.

Case 2: The center of the circle is inside the inscribed angle.

Case 3: The center of the circle is outside the inscribed angle.

The proof for Case 1 is shown below.

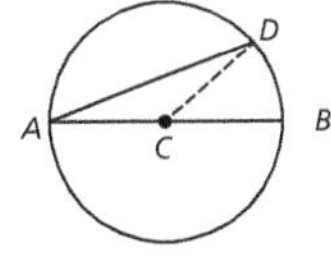

Given: $\angle DAB$ is inscribed in circle C. $\overline{AB}$ contains C.

Prove: $m\angle DAB = \frac{1}{2}m\overset{\frown}{DB}$

A–C. See margin.

Statements	Reasons
1. Draw $\overline{DC}$. $\overline{AC} \cong \overline{DC}$	1. $\overline{AC}$ and $\overline{DC}$ are radii for circle C.
2. $\triangle ADC$ is isosceles.	2. Definition of isosceles triangle
3. $\angle DAB \cong \angle ADC$	3. ? (A. Why are these angles congruent?)
4. $m\angle DAB = m\angle ADC$	4. Congruent angles have equal measures.
5. $m\angle DAB + m\angle ADC = m\angle DCB$ (B. Which angle is the exterior angle?)	5. The measure of an exterior angle equals the sum of the measures of its remote interior angles.
6. $2m\angle DAB = m\angle DCB$	6. $m\angle DAB = m\angle ADC$
7. $m\angle DAB = \frac{1}{2}m\angle DCB$	7. Division Property of Equality
8. $m\angle DAB = \frac{1}{2}m\overset{\frown}{DB}$	8. Substitution (C. Why can you substitute $m\angle DCB$ with $m\overset{\frown}{DB}$?)

Module 15 • Lesson 15.1 469

Step It Out

Task 5 (MP) **Use Structure** Students are introduced to the Inscribed Angle Theorem and the Inscribed Angle of a Diameter Theorem and complete a proof of one case of the Inscribed Angle Theorem. They see how an auxiliary segment can be drawn to give an isosceles triangle, which allows them to apply theorems related to interior and exterior angle measures of the triangle. Encourage students to draw each step of the proof on their own rather than passively reading the steps.

CONNECT TO VOCABULARY

Have students use the **Interactive Glossary** to record their understanding of the vocabulary in this task.

Sample Guided Discussion:

Q **In Part A, how are the angles of an isosceles triangle related?** The base angles are congruent.

Q **In Part C, are $\angle DCB$ and $\overset{\frown}{DB}$ the same geometric object?** No; one is a central angle and the other is an arc, but they have equal measures.

ANSWERS

A. If two sides of a triangle are congruent, then the two angles opposite the sides are congruent.

B. $\angle DCB$ is the exterior angle.

C. The measure of a minor arc is equal to the measure of the central angle so, $m\angle DCB = m\overset{\frown}{DB}$.

Task 6 (MP) **Use Repeated Reasoning** Students see how a right triangle can be inscribed in a circle and repeatedly rotated and inscribed to create a fluid pattern. They apply theorems relating arc and angle measures to determine unknown measures and go on to determine the measures of inscribed angles that give rise to regular designs inscribed in circles.

Sample Guided Discussion:

Q **In Part A, name and describe the angle you can most directly relate to $\widehat{BE}$.** $\angle BDE$ is an inscribed angle with sides through the endpoints of $\widehat{BE}$.

Q **In Part C, what can you conclude about $\overline{BD}$ since it has the center of the circle on it?** $\overline{BD}$ is a diameter of the circle.

Turn and Talk Ask students how they can relate an inscribed angle in a design with an arc. Do they have enough information to conclude that the arcs associated with the inscribed angles are congruent? Ask how they can use the facts that the arcs are consecutive with equal measures to find the sum of their measures. Should the sum of the measures of the arcs be equal to a certain number? How can they determine the measure of an inscribed angle with its associated arc? Design 1: about 25.7°; Design 2: 20°

ANSWERS

B. Determine m$\angle ADB$ using the Inscribed Angle Theorem, then find m$\angle ADE$ using the Angle Addition Postulate. m$\angle ADE = 65°$.

Use Inscribed Angles Theorems

6 Jana creates a circular rainbow art piece by wrapping strings with various colors around pins that are tacked to a board. She needs to know various angle measurements in order to design a fluid pattern.

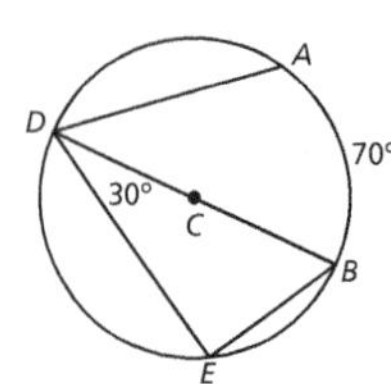

A. Find the measure of $\widehat{BE}$.

$$m\angle BDE = \frac{1}{2}m\widehat{BE}$$
$$30° = \frac{1}{2}m\widehat{BE}$$
$$60° = m\widehat{BE}$$

A. What theorem or postulate can you use to determine m$\widehat{BE}$? What is m$\widehat{BE}$?

A. Inscribed Angle Theorem; 60°

B. Find the measure of $\angle ADE$.

$$m\angle ADE = m\angle BDE + m\angle ADB$$
$$m\angle ADE = 30° + 35°$$
$$m\angle ADE = 65°$$

B. Explain how you can determine m$\angle ADE$. What theorem or postulate did you use? What is m$\angle ADE$?

B. See margin.

C. What type of triangle is $\triangle DBE$? What theorem or postulate did you use to determine this? right triangle; Inscribed Angle of a Diameter Theorem

Turn and Talk In each of the designs below, all of the inscribed angles are congruent. Find the measure of an inscribed angle for each design. See margin.

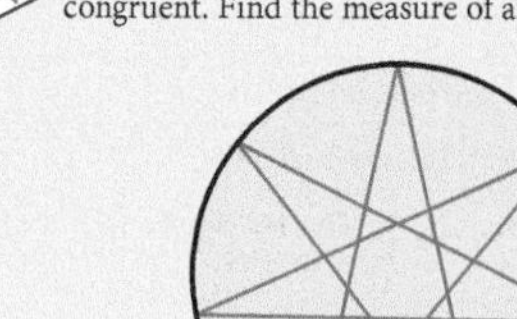

Check Understanding

Identify the chord(s), inscribed angle(s) and central angle(s) in each circle with center C. 1–4. See margin.

1.

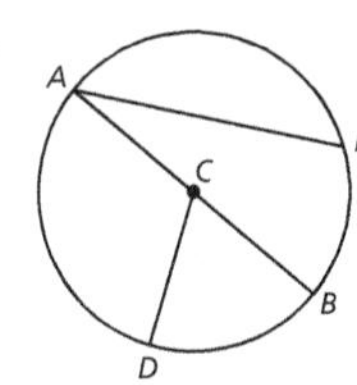

2.

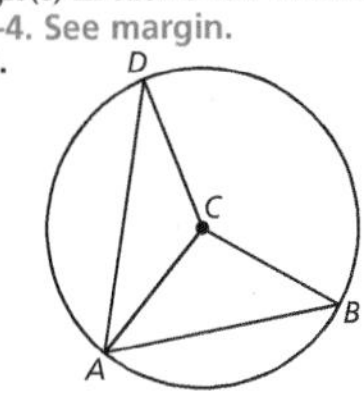

Name a major arc, a minor arc, and a semicircle in each circle with center *C*.

3.

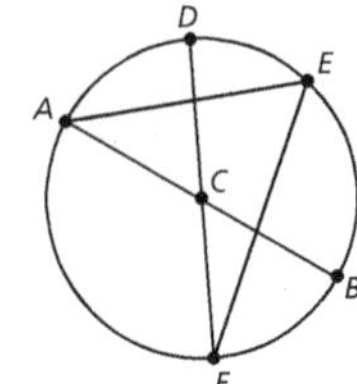

4.

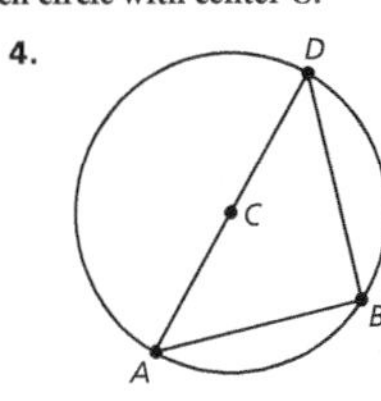

The center of the circle is *A*, and $m\angle EAC = 31°$. Find each measure.

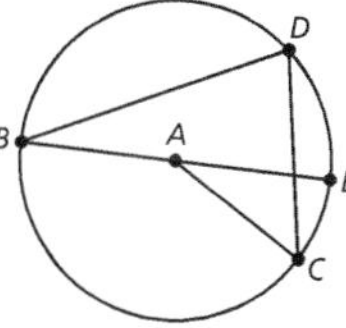

5. $m\overset{\frown}{CE}$ 31°
6. $m\angle BAC$ 149°
7. $m\overset{\frown}{BC}$ 149°
8. $m\angle BDC$ 74.5°

On Your Own

9. **Open Ended** Construct a circle with a central angle and a corresponding inscribed angle. Find the measures of the central angle and the inscribed angle. Answers will vary.

10. The minute hand of a circular clock sweeps out an arc as it moves from 2:35 P.M. to 2:55 P.M. What is the measure of the arc? 120°

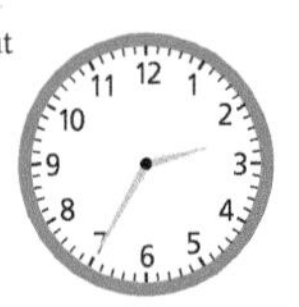

$\overline{RV}$ and $\overline{QT}$ are diameters of circle *P*. Tell whether each arc is a *minor arc*, a *major arc*, or a *semicircle* of circle *P*. Then determine the measure of the arc.

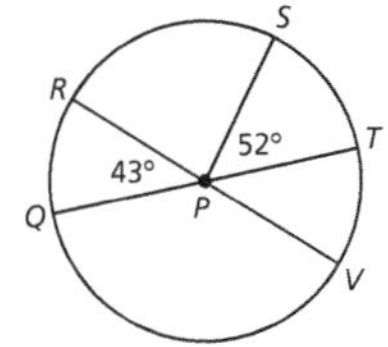

11. $\overset{\frown}{QR}$ 11–14. See Additional Answers.
12. $\overset{\frown}{RS}$
13. $\overset{\frown}{QST}$
14. $\overset{\frown}{QTV}$

ONLINE

Assign the Digital On Your Own for
- built-in student supports
- Actionable Item Reports
- Standards Analysis Reports

On Your Own

Assignment Guide

The chart below indicates which problems in the On Your Own are associated with each task in the Learn Together. Assign daily homework for tasks completed.

Learn Together Tasks	On Your Own Problems
Task 1, p. 466	Problems 9 and 37
Task 2, p. 467	Problems 10–14
Task 3, p. 468	Problems 15, 21 and 22
Task 4, p. 468	Problems 16 and 17
Task 5, p. 469	Problems 18 and 19
Task 6, p. 470	Problems 20 and 23–36

ANSWERS

1. chords: $\overline{AB}$, $\overline{AE}$; inscribed angle $\angle EAB$; central angles: $\angle BCD$, $\angle ACD$, $\angle ACB$
2. chords: $\overline{AD}$, $\overline{AB}$; inscribed angle $\angle BAD$; central angle: $\angle BCA$, $\angle BCD$, $\angle DCA$
3. major arc: $\overset{\frown}{ABF}$; minor arc: $\overset{\frown}{AD}$; semicircle: $\overset{\frown}{ADB}$
4. major arc: $\overset{\frown}{DAB}$; minor arc: $\overset{\frown}{DB}$; semicircle: $\overset{\frown}{DBA}$

3 Check Understanding

Formative Assessment

Use formative assessment to determine if your students are successful with this lesson's learning objective.

Students who successfully complete the Check Understanding can continue to the On Your Own practice.

For students who miss 1 problem or more, work in a pulled small group using the Almost There small-group activity on page 465C.

ONLINE

Assign the Digital Check Understanding to determine
- success with the learning objective
- items to review
- grouping and differentiation resources

4 Differentiation Options

Differentiate instruction for all students using small-group activities and math center activities on page 465C.

Reteach

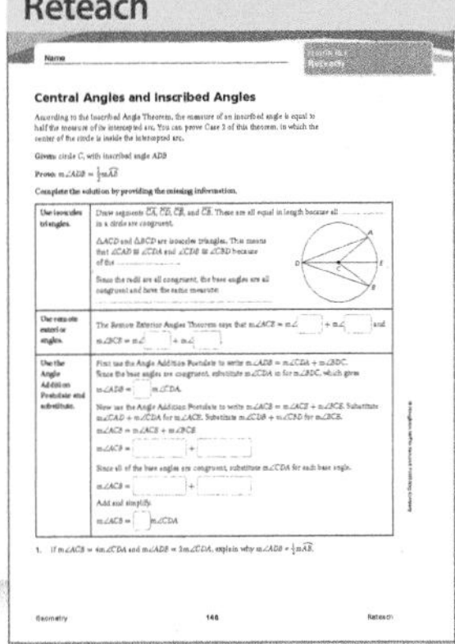

Central Angles and Inscribed Angles

Challenge

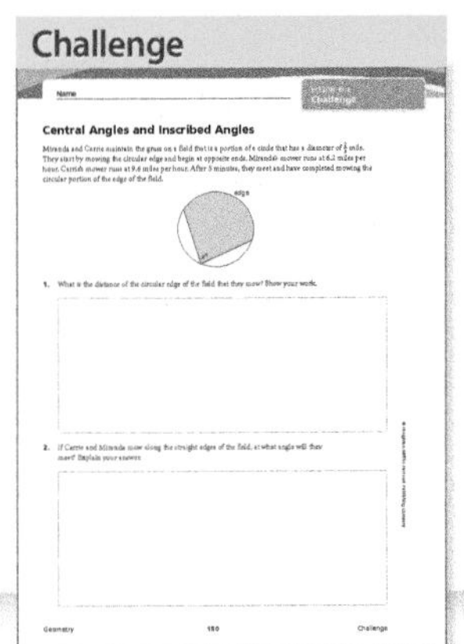

Central Angles and Inscribed Angles

Watch for Common Errors

Problem 15 Students may incorrectly assume that the four points are ordered *A, B, C, D* clockwise or counterclockwise around the circle. While both of those cases are possible, students need to understand that other orderings are possible. For example, the given arc measures make an ordering of *A, B, D, C* possible. There is more than one possible answer for $m\overset{\frown}{DA}$.

Questioning Strategies

Problem 17 How can you use the arcs created by *JKLMN* to determine a reflection that will map the pentagon onto itself? If I reflect the pentagon and circle about the line $\overleftrightarrow{KG}$, the arcs $\overset{\frown}{JK}$ and $\overset{\frown}{NJ}$ are mapped to $\overset{\frown}{LK}$ and $\overset{\frown}{ML}$, respectively, and vice versa. This maps the endpoints *J* and *N* to *L* and *M* and vice versa, and the point *K* to itself.

ANSWER

15. Minor arcs are named by their two endpoints and their measure is equal to the measure of the central angle. Major arcs are named by their two endpoints and a third point on the arc, and their measure is equal to 360° minus the measure of the central angle. The major arc $\overset{\frown}{DCA}$ measures 232° and the only possible value for the measure of minor arc $\overset{\frown}{DA}$ is 128°.

16.B. Divide 360° by the 3 arcs to get 120°.; Each angle in the triangle is an inscribed angle, so the measure of the corresponding arcs can be found by: $2 \times 60° = 120°$.

17.A. $m\overset{\frown}{JK} = m\overset{\frown}{KL} = m\overset{\frown}{LM} = m\overset{\frown}{MN} = m\overset{\frown}{NJ} = 72°$

18. isosceles; $y°$; $2y°$; $2y°$; $(2x + 2y)°$; $(x + y)°$; $m\overset{\frown}{DA}$

19. Possible answer: Construct $\triangle ABC$ and $\triangle BCD$. Show that these triangles are isosceles triangles. Use the fact that the base angles of an isosceles triangle are the same measure, and the fact that the measure of an exterior angle equals the sum of the remote interior angles to show that the measure of the central angle is twice the measure of the inscribed angle. Use the Angle Addition Postulate to subtract the smaller angles from the larger angles.

15. (MP) **Reason** In circle *K*, $m\overset{\frown}{AB} = 58°$, $m\overset{\frown}{BC} = 94°$, and $m\overset{\frown}{CD} = 80°$. Explain why 232° cannot be a possible value for $m\overset{\frown}{DA}$. See margin.

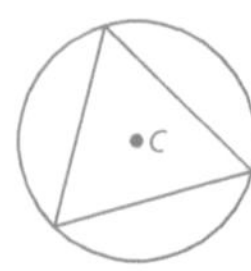

16. **A.** Construct a circle. Then construct an equilateral triangle inside the circle so that its vertices are on the circle.

B. Explain two different ways to determine the measures of the arcs between the vertices. See margin.

17. (MP) **Construct Arguments** Use the regular pentagon *JKLMN* inscribed in circle *G*.

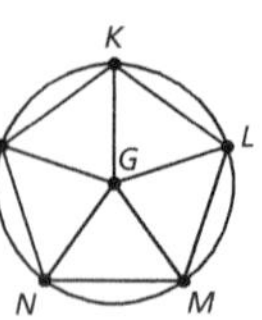

A. What are the measures of $\overset{\frown}{JK}$, $\overset{\frown}{KL}$, $\overset{\frown}{LM}$, $\overset{\frown}{MN}$, and $\overset{\frown}{NJ}$? See margin.

B. Describe the rotational symmetry of pentagon *JKLMN*. rotational symmetry of 72°

C. How can you use the measures of arcs created by an inscribed regular polygon to determine the rotations that will map the polygon onto itself? Divide 360° by the number of arcs.

18. Complete the proof for the second case of the Inscribed Angle Theorem to show that $m\angle DBA = \frac{1}{2}m\overset{\frown}{DA}$. See margin.

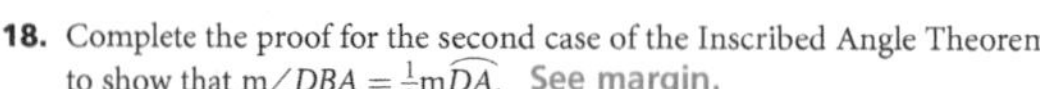

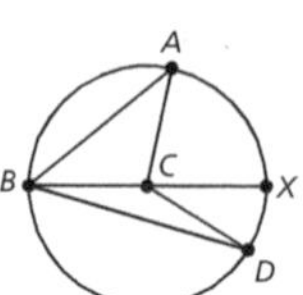

Given: $\angle DBA$ is inscribed in circle *C*, with diameter $\overline{BX}$.

Prove: $m\angle DBA = \frac{1}{2}m\overset{\frown}{DA}$.

Proof: Let $m\angle ABX = x°$ and $m\angle DBX = y°$.

Draw $\overline{AC}$ and $\overline{DC}$.

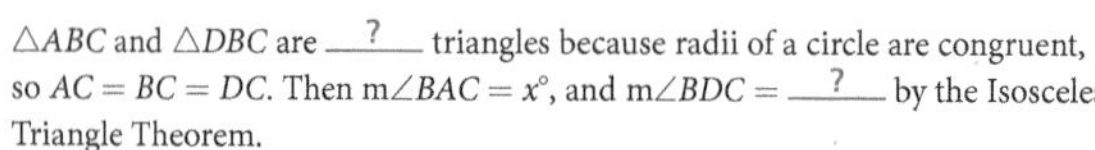

$\triangle ABC$ and $\triangle DBC$ are __?__ triangles because radii of a circle are congruent, so $AC = BC = DC$. Then $m\angle BAC = x°$, and $m\angle BDC =$ __?__ by the Isosceles Triangle Theorem.

$m\angle ACX = 2x°$ and $m\angle DCX =$ __?__ by the Exterior Angle Theorem.

So, $m\angle DCA = (2x + 2y)°$ by the Angle Addition Postulate.

$m\overset{\frown}{AX} = 2x°$ and $m\overset{\frown}{DX} =$ __?__ since the measure of a minor arc equals the measure of its central angle.

$m\overset{\frown}{DA} =$ __?__ by the Arc Addition Postulate.

$m\angle DBA =$ __?__ by the Angle Addition Postulate.

Since $m\overset{\frown}{DA} = (2x + 2y)° = 2(x + y)°$ and $m\angle DBA = (x + y)°$, then $m\angle DBA = \frac{1}{2}$ __?__.

19. Write a plan for how to prove the third case of the Inscribed Angle Theorem.

Given: $\angle DBA$ is inscribed in circle *C*, where *C* is outside $\angle DBA$.

Prove: $m\angle DBA = \frac{1}{2}m\overset{\frown}{DA}$. See margin.

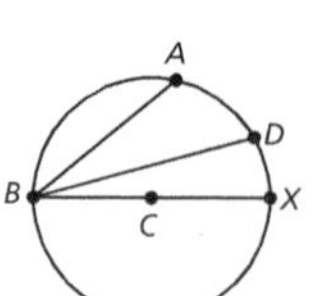

20. A circular garden is divided into three sections with different-colored flowers. The measure of $\angle BCD$ is 21°. The measures of $\overset{\frown}{BC}$ and $\overset{\frown}{DC}$ are equal. What is the measure of each arc? 159°

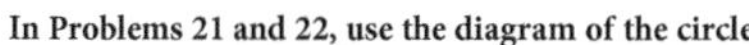

In Problems 21 and 22, use the diagram of the circle.

21. How does the measure of $\angle ABD$ compare to the measure of $\angle ACD$? Explain your reasoning.
21–24. See margin.

22. Points B and D are the endpoints of a diameter of the circle. How are the measures of $\overset{\frown}{AB}$ and $\overset{\frown}{AD}$ related? Explain.

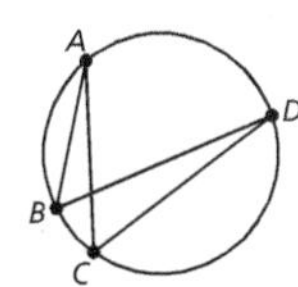

23. A right triangle is inscribed in a circle. How does the hypotenuse relate to the circle? What do you know about the arc formed by the endpoints of the hypotenuse?

24. A carpenter's square is a tool that is used to draw right angles. Suppose you are building a baby toy that has a circle the baby will spin. You need to drill a hole in the center of the circle to attach it to the rest of the toy. Explain how you can use a carpenter's square to find the center of the circle.

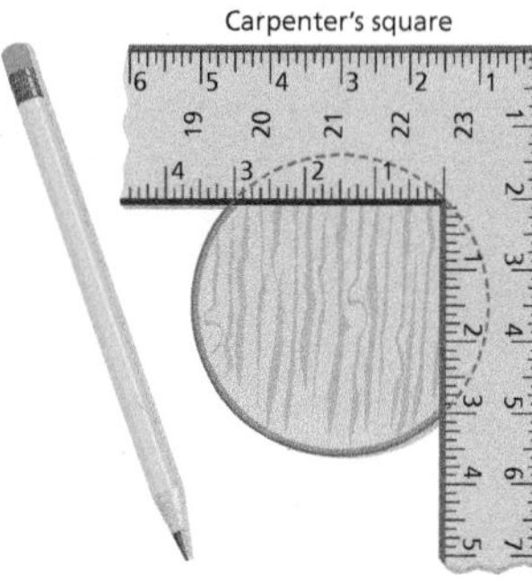

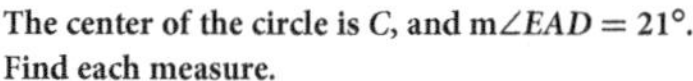

The center of the circle is C, and $m\angle EAD = 21°$. Find each measure.

25. $m\overset{\frown}{ED}$ 42°

26. $m\angle DBE$ 21°

27. $m\overset{\frown}{EB}$ 180°

28. $m\angle BFE$ 90°

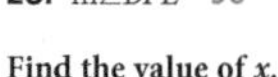

Find the value of x.

29.

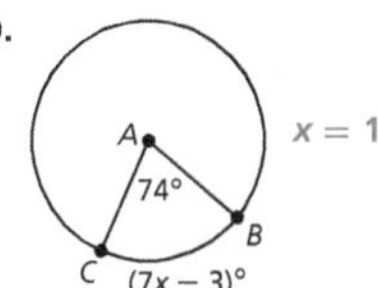

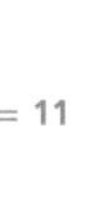

$x = 11$

30.

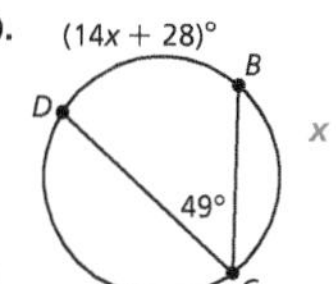

$x = 5$

31.

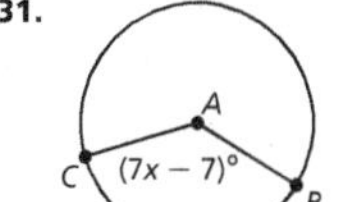
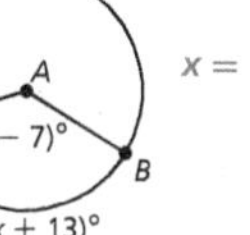

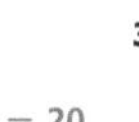

$x = 20$

32.

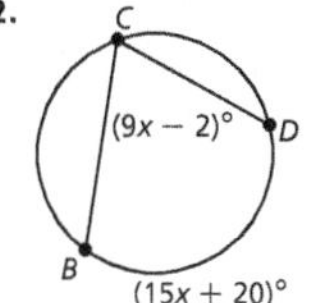

$x = 8$

Watch for Common Errors

Problem 25 Students may use the square to apply the Inscribed Angle of a Diameter Theorem and find a diameter of the circle and assume this is enough to locate the center. While the center is on this diameter, students need to understand that constructing one diameter would be enough for only an approximation of the location of the center. They need to construct another diameter and locate the center at the intersection of the two diameters.

ANSWERS

21. The measures of the angles are equal because the intercepted arc of both angles is the same.

22. Because $\overline{BD}$ is a diameter, then $\overset{\frown}{DAB}$ is a semicircle. So, the sum of $m\overset{\frown}{AB}$ and $m\overset{\frown}{AD}$ will be 180°.

23. The hypotenuse of the triangle is a diameter of the circle. The measure of the arc is 180°.

24. Use the carpenter's square to draw an inscribed right angle. The sides of the angle intersect at the endpoints of a diameter by the Inscribed angle of a diameter theorem. Draw the diameter. Then repeat the process to draw a different diameter. The point of intersection of the two diameters is the center of the circle.

5 Wrap-Up

Summarize learning with your class. Consider using the Exit Ticket, Put It in Writing, or I Can scale.

Exit Ticket 

If $\overparen{AB}$ is the arc intercepted by an inscribed angle $\angle ADB$, $m\angle ADB = 70°$, and $m\overparen{DB} = 100°$, what is $m\overparen{AD}$?

$m\overparen{AB} = 2m\angle ADB = 2 \cdot 70° = 140°$, and

$m\overparen{ADB} = 360° - m\overparen{AB} = 360° - 140° = 220°$.

Since $\overparen{AD}$ and $\overparen{DB}$ are adjacent arcs,

$m\overparen{AD} = m\overparen{ADB} - m\overparen{DB} = 220° - 100° = 120°$.

Put It in Writing 

Have students explain the relationships between inscribed angles and central angles and the relationships among minor, major, and intercepted arcs.

I Can 

The scale below can help you and your students understand their progress on a learning goal.

4	I can determine the measures of central angles, inscribed angles, and arcs of a circle, and I can explain my steps to others.
3	I can determine the measures of central angles, inscribed angles, and arcs of a circle.
2	I can determine the measure of an inscribed angle from its central angle.
1	I can find the measures of a major arc and a minor arc from their central angle.

Spiral Review • Assessment Readiness

These questions will help determine if students have retained information taught in the past and can also prepare them for high-stakes assessments. Here, students must apply the Law of Cosines **(14.2)**, use a trigonometric ratio to solve for an angle of a right triangle **(13.4)**, apply properties of angles of parallelograms **(11.1)**, and apply the Law of Sines **(14.1)**.

33. An inscribed angle with a diameter as a side has a measure of $x°$. If the ratio of $m\overparen{CD}$ to $m\overparen{DB}$ is 1:3, what is $m\overparen{DB}$? What is the value of x? $135°$, $x = 22.5$

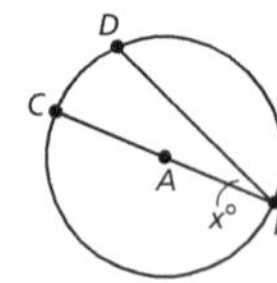

For Problems 34–36, use the two circles which have a common center, point A.

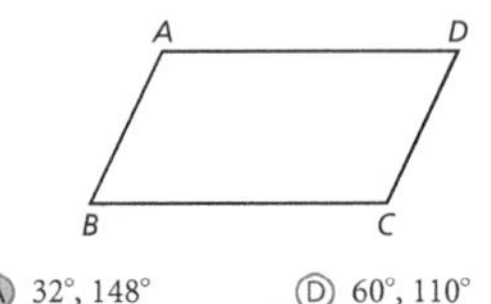

34. List three expressions that are equivalent to $m\angle DAB$.
Possible answer: $m\overparen{DB}$, $m\overparen{CE}$, $2m\angle CFE$, $m\angle CAE$

35. Describe how you could change the location of point F on the circle without changing the value of $m\angle CFE$.
See Additional Answers.

36. Find $m\angle CFE$ if $m\overparen{CFE} = 302°$. $29°$

37.  **Construct Arguments** Inscribe acute, right, and obtuse triangles in a circle. Make a statement about the location of the center of the circle in relation to the triangle. What term have you learned previously that describes this point?
See Additional Answers.

Spiral Review • Assessment Readiness

38. Find the missing length of the triangle.

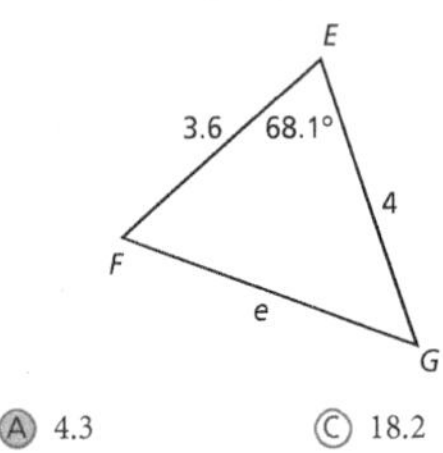

Ⓐ 4.3
Ⓑ 5.4
Ⓒ 18.2
Ⓓ 29.0

39. What is the measure of $\angle A$?

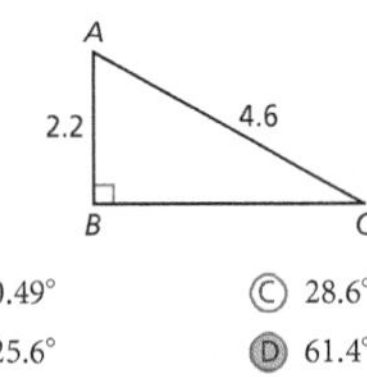

Ⓐ 0.49°
Ⓑ 25.6°
Ⓒ 28.6°
Ⓓ 61.4°

40. $ABCD$ is a parallelogram. Which of the following pairs of angle measures could be the measures of $\angle A$ and $\angle B$? Select all that apply.

Ⓐ 32°, 148°
Ⓑ 63°, 117°
Ⓒ 45°, 45°
Ⓓ 60°, 110°
Ⓔ 90°, 90°
Ⓕ 100°, 100°

41. Find the value of y.

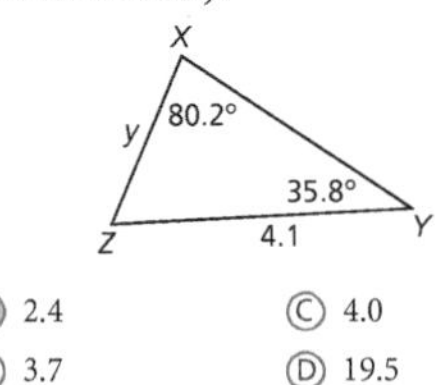

Ⓐ 2.4
Ⓑ 3.7
Ⓒ 4.0
Ⓓ 19.5

I'm in a Learning Mindset!

What barriers are there to understanding central angles and inscribed angles?

474

Keep Going to Journal and Practice Workbook

Learning Mindset

Resilience Identifies Obstacles

Point out how being able to specifically identify obstacles is a critical part of the learning process. If students can articulate aspects of a lesson that were especially troublesome, they are demonstrating an understanding of the various parts of the lesson and how they are related. Articulating problem areas helps students get closer to the root causes of why they might be struggling in those areas. For example, a student who struggles with relating the measures of inscribed angles and the associated central angles might find after some prompting that they do not clearly understand the definitions. After sketching some examples of both types of angles to solidify their understanding of their definitions, they could better determine which angle has the greater measure. *How does trying to identify your specific problem areas help you better understand where you're struggling? How can it help you get closer to the root causes of your struggle?*

15.2 Angles in Inscribed Quadrilaterals

LESSON FOCUS AND COHERENCE

Mathematics Standards

- Identify and describe relationships among inscribed angles, radii, and chords. *Include the relationship between central, inscribed, and circumscribed angles; inscribed angles on a diameter are right angles; the radius of a circle is perpendicular to the tangent where the radius intersects the circle.*
- Construct the inscribed and circumscribed circles of a triangle, and prove properties of angles for a quadrilateral inscribed in a circle.

Mathematical Practices and Processes 

- Use appropriate tools strategically.
- Construct viable arguments and critique the reasoning of others.
- Attend to precision.

I Can Objective 

I can use the properties of angles of quadrilaterals inscribed in a circle to prove theorems and solve problems.

Learning Objective

Students will be able to prove the Inscribed Quadrilateral Theorem and the Congruent Corresponding Chords Theorem and apply them to quadrilaterals inscribed in circles.

Language Objective

Students will be able to explain steps in the proofs of the theorems for quadrilaterals covered in the lesson and the way they're applied to specific examples.

Vocabulary

New: congruent arcs, congruent circles

Lesson Materials: protractor, ruler, compass

Mathematical Progressions

Prior Learning	Current Development	Future Connections
Students: • explained how the criteria for triangle congruence (ASA, SAS, and SSS) follow from the definition of congruence in terms of rigid motions. **(8.2)** • wrote coordinate proofs about parallelograms. **(11.1–11.5)**	**Students:** • prove the Inscribed Quadrilateral Theorem and apply it to specific inscribed quadrilaterals. • prove the Congruent Corresponding Chords Theorem and apply it to circles with inscribed quadrilaterals.	**Students:** • will identify and describe relationships among inscribed angles, radii, and chords. **(16.1 and 16.2)** • will derive using similarity the fact that the length of the arc intercepted by an angle is proportional to the radius, and will define the radian measure of the angle as the constant of proportionality; will derive the formula for the area of a sector. **(17.2)**

PROFESSIONAL LEARNING

Using Mathematical Processes and Practices

Attend to precision.

This lesson provides an opportunity to address this Mathematical Practice Standard. According to the standard, mathematically proficient students "try to communicate precisely with others" and "try to use clear definitions in their discussions with others and in their own reasoning." This lesson introduces theorems involving quadrilaterals inscribed in circles and helps students to understand what conditions permit quadrilaterals to be inscribed in circles. Students learn to prove those theorems and apply them to specific examples. They use principles and theorems previously learned to formulate their proofs. As a result, they have to be careful with the definitions, postulates, and theorems they use.

WARM-UP OPTIONS

ONLINE

PROJECTABLE & PRINTABLE

ACTIVATE PRIOR KNOWLEDGE • Write Coordinate Proofs About Parallelograms

Use these activities to quickly assess and activate prior knowledge as needed.

Problem of the Day 

Prove that the parallelogram ABCD can be divided into two congruent right triangles.

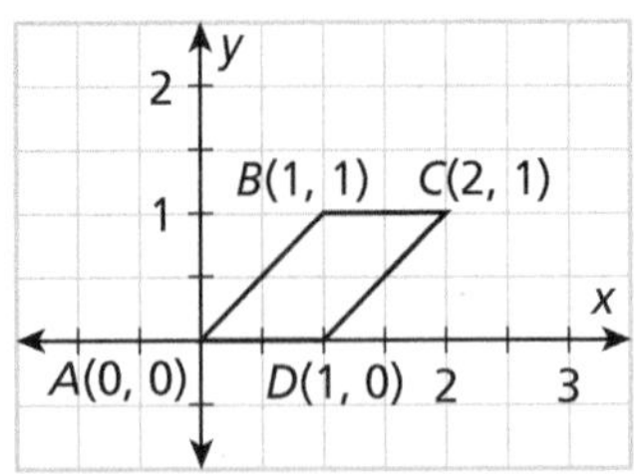

Draw a line from point B to point D. (A line exists between any 2 points.) Note that for both points, $x = 1$, but for B, $y = 1$, whereas for D, $y = 0$. Therefore, by the definition of perpendicularity, $\overline{BD}$ is perpendicular to $\overline{AD}$. By similar reasoning, $\overline{BD}$ is perpendicular to $\overline{BC}$. Because the segments are perpendicular, $m\angle ADB = 90°$ and $m\angle CBD = 90°$ which means $\angle ADB \cong \angle CBD$. Using the distance formula, we can show that $AD = 1$ and $BC = 1$, so $AD \cong BC$. By the SAS Triangle Congruence Theorem, $\triangle ADB \cong \triangle CBD$. So parallelogram $ABCD$ can be divided into two congruent right triangles.

Quick Check for Homework

As part of your daily routine, you may want to display the Teacher Solution Key to have students check their homework.

Make Connections

Based on students' responses to the Problem of the Day, choose one of the following:

1. Project the Interactive Reteach, Lesson 15.1.
2. Complete the Prerequisite Skills Activity:

Have students label the midpoint of $\overline{AC}$ as D and place E on the y-axis such that $\overline{EB} \parallel \overline{AC}$, and then work in groups to prove that the resulting parallelogram $AEBD$ is a rectangle.

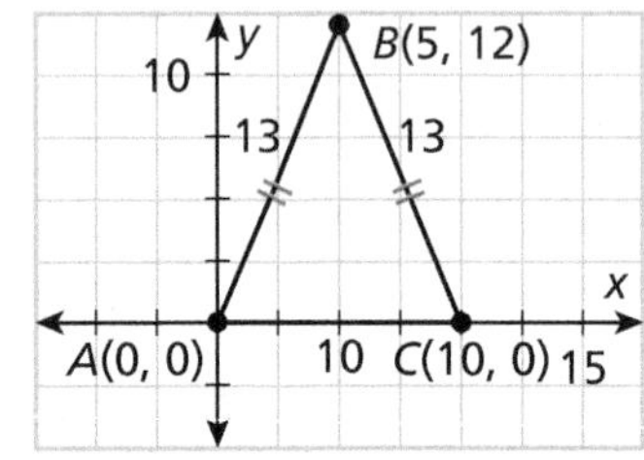

- *Do you see any congruent angles?* There are two congruent right angles on the x-axis. The two other angles at the base of the triangle are congruent because the triangle is isosceles.
- *Do you see any congruent sides?* $\overline{AB} \cong \overline{BC}$ (given), $\overline{BD} \cong \overline{BD}$ by the reflexive property, and $\overline{AD} \cong \overline{DC}$ because they are both of length 5.
- *Which theorems can you think of to prove the two triangles are congruent?* The SAS, AAS, and SSS Triangle Congruence Theorems could prove congruence of the triangles.
- *How can you prove that the parallelogram is a rectangle?* The two top angles are right triangles because $\overline{EB}$ is parallel to the x-axis, and $\overline{AD}$ is on the x-axis, so there are four right angles.

SHARPEN SKILLS

If time permits, use this on-level activity to build fluency and practice basic skills.

Mental Math

Objective: Students calculate lengths of the diagonals of squares and rectangles.

Ask students to mentally compute the length of the diagonal of a square of side length 1 and to share their solution strategies with the class. If no student suggests adding the square of two of the side lengths and taking the square root, suggest this strategy. Ask students to find the diagonals of rectangles with side lengths 3 and 4, 6 and 8, and 5 and 12. Ask students to share their strategies with the class. Ask students to name the theorem that allows them to perform this calculation.

PLAN FOR DIFFERENTIATED INSTRUCTION

Small-Group Options

Use these teacher-guided activities with pulled small groups.

On Track

Have students develop a proof that an isosceles trapezoid can be an inscribed quadrilateral in a circle.

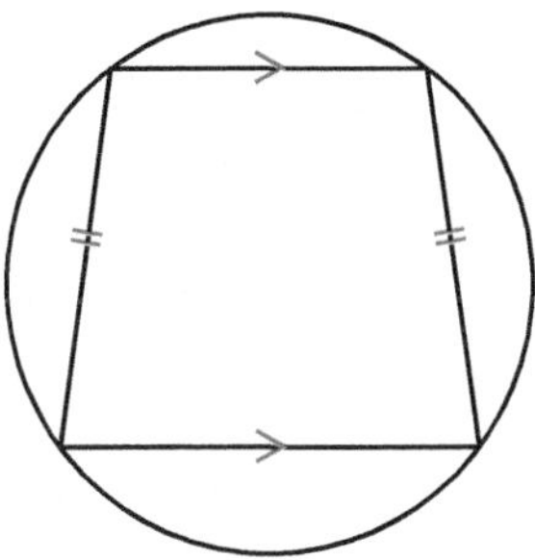

Almost There

Have students think about and write down a computation and short proof showing that a circle of radius 1 can have an inscribed square with side lengths $s = \sqrt{2}$.

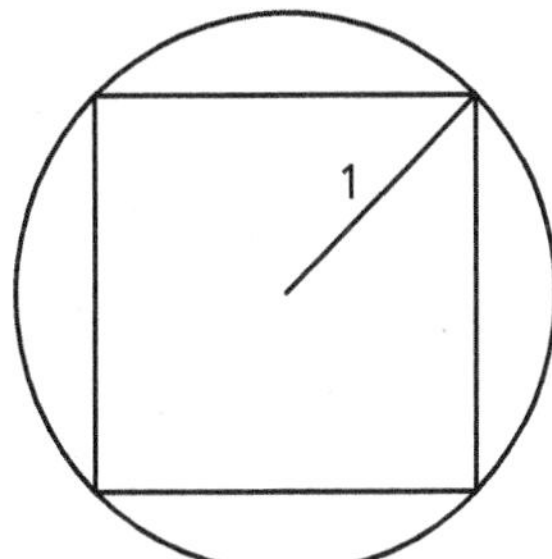

Ready for More

Have students write a proof that a circle can have two inscribed isosceles trapezoids that are congruent and share the same base, with one trapezoid above the other.

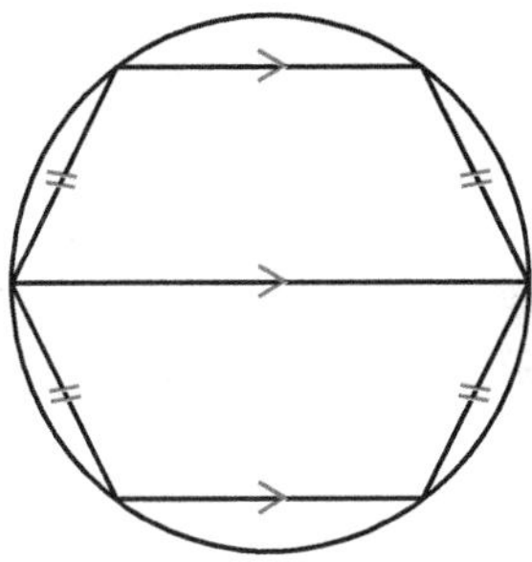

Math Center Options

Use these student self-directed activities at centers or stations.

Key: ● Print Resources ● Online Resources

On Track

- ● Interactive Digital Lesson
- ●● Journal and Practice Workbook
- ● Interactive Glossary (printable): **congruent arcs**, **congruent circles**
- ●● Standards Practice: Prove Angle Properties for an Inscribed Quadrilateral

Almost There

- ● Reteach 15.2 (printable)
- ● Interactive Reteach 15.2

Ready for More

- ● Challenge 15.2 (printable)
- ● Interactive Challenge 15.2

ONLINE Ed

View data-driven grouping recommendations and assign differentiation resources.

During the *Spark Your Learning*, listen and watch for strategies students use. See samples of student work on this page.

Use the Properties of Quadrilaterals
Strategy 1

Given: The diagonals are diameters of the circle. Two adjacent sides of the quadrilateral have side length $s = 15$ cm.

Since the diagonals of the quadrilateral are diameters, I can apply the Inscribed Angle of a Diameter Theorem and conclude that the figure must have right angle vertices.

Since the diagonals are of equal length (by the properties of the diameter of a circle), we have further confirmation that the figure must be a rhombus with right angles. If the angles were not 90°, the diagonals would not be equal. Then they would not be diameters of the circle.

Because the angles must be all right angles and adjacent sides have the same length, the quadrilateral must be a square.

If students . . . use the given information and the properties of quadrilaterals, they have started to understand the way to reason through a proof.

Have these students . . . explain how they determined what information to use and how to use it. **Ask:**

- Q Why did you decide to use the properties of quadrilaterals?
- Q Can you explain how you apply the Inscribed Angle of a Diameter Theorem?

Judge from the Appearance of the Figure
Strategy 2

I looked at the picture. The quadrilateral looked like a square, and the extra information said the adjacent sides are equal. Therefore, I figured that the quadrilateral was a square. This would also fit with the fact that the diagonals are the diameters of a circle.

If students . . . judge by the photo on page 475, they may be at a loss to think of any other way to determine the type of quadrilateral they see.

Activate prior knowledge . . . by having students think about the given information. **Ask:**

- Q Which quadrilaterals have two adjacent sides of equal length?
- Q What does it tell you that the diagonals of the figure are diameters?

COMMON ERROR: Judges Incorrectly from the Photo

I looked at the picture and saw that the two top sides of the quadrilateral looked shorter than the two bottom sides. The given information says that two adjacent sides are the same length, 15 cm.

That rules out rhombuses and rectangles.

That also rules out a trapezoid.

I decided the figure was a kite.

If students . . . use the photo to judge the type of quadrilateral and still guess wrong, they are probably being fooled by the perspective of the photo.

Then intervene . . . by pointing out one cannot judge from the appearance of a figure in a photo or illustration. **Ask:**

- Q Why would the illustration be a poor indicator of the nature of the quadrilateral or any other geometric figure?
- Q What information can you find besides the appearance of the quadrilateral in the illustration?

15.2

Angles in Inscribed Quadrilaterals

I Can use the properties of angles of quadrilaterals inscribed in a circle to prove theorems and solve problems.

Spark Your Learning

The window is in the shape of a circle. A quadrilateral is inside the window.

The four corners of the quadrilateral touch the circle.

©hxdbzxy/Shutterstock

Complete Part A as a whole class. Then complete Parts B–C in small groups.

A. What is a geometric question you can ask about the garden in this photo? What information would you need to know to answer this question?

B. To answer your question, what strategy and tool would you use along with all the information you have? What answer do you get? See Strategies 1 and 2 on the facing page.

C. Does your answer make sense in the context of the situation? How do you know? See Additional Answers.

A. What kind of quadrilateral is inside the circle?; some relationship between the circle and the quadrilateral's diagonals or angles

 **Turn and Talk** How can you use symmetry of a square to construct this design? Explain your process. See margin.

CULTIVATE CONVERSATION • Co-Craft Questions

If students have difficulty formulating a mathematical question about the situation in the Spark Your Learning, ask them to consider what types of quadrilaterals could result from joining four points on a circle. What are some natural questions to ask about this situation?

Work together to craft the following questions:

- What are several notable features of quadrilaterals?
- What is the sum of the angles of a quadrilateral?
- Which quadrilaterals are irregular?

Then have students think about what additional information, if any, they would need to answer these questions. **Ask:**

- Can you determine which of the quadrilaterals you named could be the figure in the photo? Why or why not?
- Can you determine the angles of the quadrilateral in the photo? Explain.

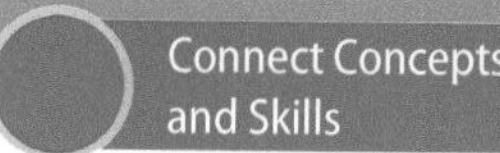
Connect Concepts and Skills

1 Spark Your Learning

MOTIVATE 

- Have students look at the photo in their books and read the information contained in the photo. Then complete Part A as a whole-class discussion.
- Give the class the additional information they need to solve the problem. This information is available online as a printable and projectable page in the Teacher Resources.
- Have students work in small groups to complete Parts B and C.

PERSEVERE

If students need support, guide them by asking:

Advancing • Use Tools Which tool could you use to solve the problem? Why choose that tool and not some other? Students' choices of tools and reasons for choosing them will vary.

Assessing Can absolutely any quadrilateral be inscribed in a circle? No; for example, a rhombus with two angles greater than 90° would have the remaining two angles too small for the vertices to be inscribed.

Assessing What principle would help you determine the type of quadrilateral in the photo? Any chord passing through the center of a circle is a diameter. Therefore, any inscribed quadrilateral must have chords passing through the center that are of equal length. This means that rhombi with angles not equal to 90° are excluded. Using this principle, we can conclude what type of quadrilateral is pictured.

Turn and Talk Remind students that a square has four equal sides and therefore can be divided into two congruent right triangles with side measures equal to x, x, and $x\sqrt{2}$. Since the square can be divided into two right triangles with two legs measuring x, each right triangle can be further divided into two congruent right triangles. A square is symmetric across the diagonal. You can use the diameter to construct a right isosceles triangle to create half of the square. Then you can reflect the triangle across the diameter to create the other half.

BUILD SHARED UNDERSTANDING

Select groups of students who used various strategies and tools to share with the class how they solved the problem. As they present their solutions, have each group discuss why they chose a specific strategy and tool.

② Learn Together

Build Understanding

Task 1 **MP** **Use Tools** Students will use a protractor and a compass to draw inscribed quadrilaterals within a circle and measure their angles.

Sample Guided Discussion:

Q What does the Inscribed Angle of a Diameter Theorem state? It states that the measure of an inscribed angle is half the measure of the central angle. Therefore, since the central angle is a diameter, its measure is 180° and the measure of the inscribed angle is 90°.

Q Would that theorem exclude the possibility of inscribing any quadrilateral other than a rectangle or a square? No, because other quadrilaterals may have diagonals that are not diameters of the circle.

Turn and Talk Have students consider that if the opposite angles are congruent, then why wouldn't the other two angles be congruent as well? Since the sum of the measures of opposite angles in an inscribed quadrilateral is always 180°, the angles must be right angles.

Build Understanding

Investigate Inscribed Quadrilaterals

You have already learned some properties of quadrilaterals. Inscribing them in circles reveals some special ones.

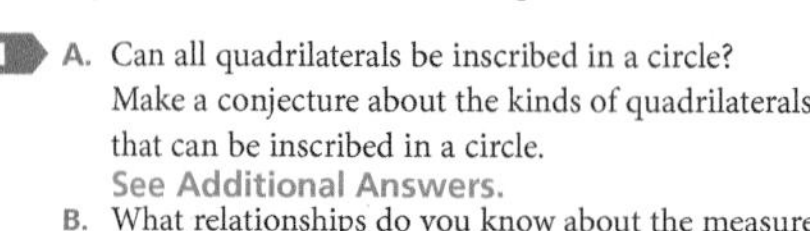

1 A. Can all quadrilaterals be inscribed in a circle? Make a conjecture about the kinds of quadrilaterals that can be inscribed in a circle.
See Additional Answers.

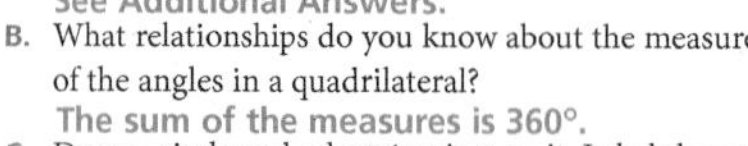

B. What relationships do you know about the measures of the angles in a quadrilateral?
The sum of the measures is 360°.

C. Draw a circle and select 4 points on it. Label the points A, B, C, and D. Draw quadrilateral $ABCD$.
Drawings and angle measures will vary.

D. Measure the angles of the quadrilateral that you inscribed, then find the sums of the measures its adjacent and opposite pairs of angles.
Sums will vary, except $m\angle A + m\angle C = 180°$ and $m\angle B + m\angle D = 180°$.

$m\angle A + m\angle B =$ __?__ $m\angle B + m\angle C =$ __?__

$m\angle A + m\angle C =$ __?__ $m\angle B + m\angle D =$ __?__

$m\angle A + m\angle D =$ __?__ $m\angle C + m\angle D =$ __?__

E. Compare your results with others' in your class. What patterns do you notice about these sums of adjacent and opposite angle pairs? E, F. See Additional Answers.

F. Use what you know about inscribed angles in circles to justify your conjecture. Do you think this pattern is true for all quadrilaterals inscribed in a circle?

G. Does it matter if the center of the circle is inside or outside the inscribed quadrilateral for the relationship between the angles to hold? Explain. See Additional Answers.

Turn and Talk Suppose you have a quadrilateral inscribed in a circle with opposite angles that are congruent. What do you know about those angles? See margin.

Prove the Inscribed Quadrilateral Theorem

The results of your investigation in the previous task can be summarized in the Inscribed Quadrilateral Theorem.

Inscribed Quadrilateral Theorem

If a quadrilateral is inscribed in a circle, then its opposite angles are supplementary.

If A, B, C, and D lie on circle P, then $m\angle A + m\angle C = 180°$ and $m\angle B + m\angle D = 180°$.

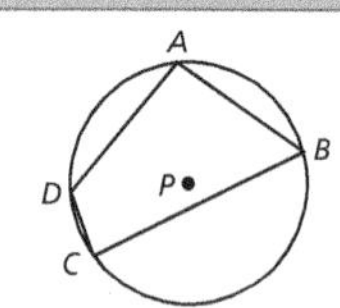

476

LEVELED QUESTIONS

Depth of Knowledge (DOK)	Leveled Questions	What Does This Tell You?
Level 1 **Recall**	What do the opposite angles of an inscribed quadrilateral add up to? They add up to 180°.	Students' answers will indicate whether they recall an important fact.
Level 2 **Basic Application of Skills & Concepts**	How would you find the opposite angle B of an inscribed quadrilateral if you knew the measure of the other angle, A? The opposite angle is equal to 180° minus the measure of angle A.	Students' answers will indicate whether they are able to apply the basic skills.
Level 3 **Strategic Thinking & Complex Reasoning**	Could an isosceles trapezoid be inscribed in a circle? Explain. Could any other trapezoid be inscribed in a circle? Explain why or why not. An isosceles trapezoid could be inscribed in a circle because its opposite angles would have measures adding up to 180°. Other types of trapezoids could not because the measures of their opposite angles would not sum to 180°.	Students' answers will indicate whether they can apply properties of specific types of quadrilaterals to predict whether they can be inscribed in a circle.

The converse of the Inscribed Quadrilateral Theorem is also true.

Converse of the Inscribed Quadrilateral Theorem

If the opposite angles of a quadrilateral are supplementary, then the quadrilateral can be inscribed in a circle.

If $m\angle A + m\angle C = 180°$ and $m\angle B + m\angle D = 180°$, then A, B, C, and D lie on circle P.

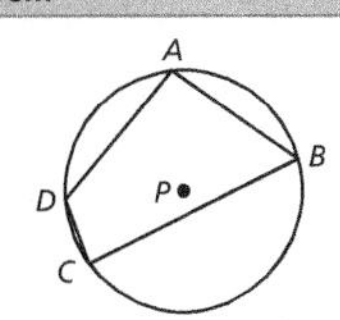

2 Prove the Inscribed Quadrilateral Theorem.

Given: Quadrilateral $ABCD$ is inscribed in circle P.

Prove: $\angle A$ and $\angle C$ are supplementary.
$\angle B$ and $\angle D$ are supplementary.

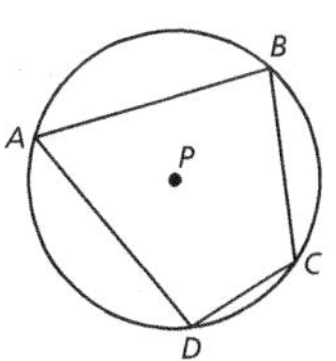

A. What is the sum of the measures of two arcs that together form a complete circle? What is the sum of the measures of the inscribed angles that form these two arcs? **360°; 180°**

Statements	Reasons
1. Quadrilateral $ABCD$ is inscribed in circle P.	1. Given
2. $m\widehat{BCD} + m\widehat{BAD} = 360°$	2. Arc Addition Postulate and definition of a circle
3. $m\widehat{BCD} = 2m\angle A$ $m\widehat{BAD} = 2m\angle C$	3. Inscribed Angle Theorem
4. $2m\angle A + 2m\angle C = 360°$	4. Substitution Property of Equality
5. $2(m\angle A + m\angle C) = 360°$	5. Distributive Property
6. $m\angle A + m\angle C = 180°$	6. Division Property of Equality
7. $\angle A$ and $\angle C$ are supplementary.	7. Definition of supplementary angles

B. The proof shows that $\angle A$ and $\angle C$ are supplementary. Use similar reasoning to show that $\angle B$ and $\angle D$ are supplementary. **See Additional Answers.**

C. Explain how the Inscribed Quadrilateral Theorem can be used to verify the Quadrilateral Sum Theorem presented previously for quadrilaterals that are inscribed in circles. This theorem states that the sum of the measures of the interior angles of a quadrilateral is 360°. **See Additional Answers.**

Turn and Talk Is it possible for a parallelogram to be inscribed in a circle? Explain your reasoning. **See margin.**

Task 2 (MP) **Construct Arguments** Students will use the Arc Addition Postulate and the Inscribed Angle Theorem to prove the Inscribed Quadrilateral Theorem.

Sample Guided Discussion:

Q What does the Arc Addition Postulate state? The Arc Addition Postulate states that the measure of an arc formed by two adjacent arcs of a circle equals the sum of those measures.

Q What differences would there be in the proof that $\angle B$ and $\angle D$ are supplementary? Only the arcs chosen and the angles named would change. The rest of the proof would be the same.

Turn and Talk Remind students of the various types of quadrilaterals that are parallelograms. A parallelogram can be inscribed in a circle if and only if the parallelogram is a rectangle. Rectangles are the only parallelograms that have opposite angles that are supplementary.

PROFICIENCY LEVEL

Beginning

Show students an inscribed quadrilateral $ABCD$. Say, "The measures of arcs $\widehat{ABC}$ and $\widehat{CDA}$ have a sum of 360°. The arcs have inscribed angles $\angle ABC$ and $\angle CDA$." Point at or trace the indicated arcs and angles. Then, have students verbally identify by name another pair of arcs and associated inscribed angles with the same relationship.

Intermediate

Show students an inscribed quadrilateral $ABCD$, and explain that by the Inscribed Angle Theorem, the measure of arc $\widehat{ABC}$ is twice the measure of $\angle D$. Have students state a similar relationship between the measures of another arc and inscribed angle, and give the theorem that justifies the relationship.

Advanced

Have students work in groups to explain what is wrong with this argument: "For an inscribed quadrilateral $ABCD$, $m\widehat{ABC} + m\widehat{BCD} + m\widehat{CDA} + m\widehat{DAB} = 360°$. Since each inscribed angle has half the measure of its intercepted arc, I know that $m\angle ABC + m\angle BCD + m\angle CDA + m\angle DAB = \frac{2}{1}(360°) = 180°$."

Task 3 (MP) **Construct Arguments** Students will use the properties of central angles and minor arcs to prove the Corresponding Chords Theorem and its converse.

CONNECT TO VOCABULARY

Have students use the **Interactive Glossary** to record their understanding of the vocabulary in this task.

Sample Guided Discussion:

Q What is a biconditional statement, and how does it differ from a conditional statement? A biconditional statement is a proposition stated with its converse, both being true. A conditional statement is a proposition without the implication that its converse is true; in many cases, the converse is not true.

Q Would it be possible to use the SAS Triangle Congruence Theorem instead of the SSS Triangle Congruence Theorem? Why or why not? No; given the steps preceding the use of the SSS Triangle Congruence Theorem and the theorems and postulates I know, I would not be able to use the SAS Triangle Congruence Theorem. Nothing in the steps preceding the use of SSS establishes the congruence of any two angles.

Turn and Talk Remind students that central angles have twice the measure of corresponding inscribed angles. Have students therefore draw the proper inference from the fact that the two arcs are congruent and that chords can be drawn from points C and D to points A and B. The central angles are also congruent.

ANSWERS

A. Conditional statement: If two minor arcs in the same circle or congruent circle are congruent, then their corresponding chords are congruent.
Converse: If two chords in the same circle or congruent circles are congruent, then their corresponding minor arcs are congruent.

Prove the Congruent Corresponding Chords Theorem

Two circles are **congruent circles** if they have the same radius. Two arcs are **congruent arcs** if they have the same measure and they are arcs of the same circle or of congruent circles.

Just as corresponding parts of congruent triangles are congruent, you can show that corresponding parts of congruent circles are also congruent.

Congruent Corresponding Chords Theorem

Two minor arcs in the same circle or in congruent circles are congruent if and only if their corresponding chords are congruent.

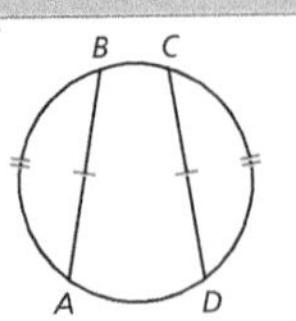

3 A. The Congruent Corresponding Chords Theorem is a biconditional statement. Rewrite the theorem as a conditional statement and its converse.
See margin.

B. Which conditional statement from Part A is proven in the following proof?

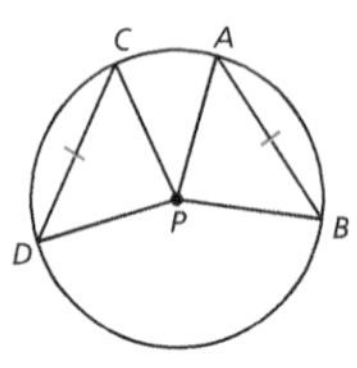

Given: $\overline{AB} \cong \overline{CD}$, P is the center of the circle

Prove: $\overset{\frown}{AB} \cong \overset{\frown}{CD}$

converse of the Congruent Corresponding Chords Theorem

Statements	Reasons
1. $\overline{AB} \cong \overline{CD}$	1. Given
2. Draw $\overline{AP}, \overline{BP}, \overline{CP}, \overline{DP}$	2. Through any two points exists one line.
3. $\overline{AP} \cong \overline{BP} \cong \overline{CP} \cong \overline{DP}$	3. Radii in the same circle are congruent.
4. $\triangle APB \cong \triangle CPD$	4. SSS Triangle Congruence Theorem
5. $\angle APB \cong \angle CPD$	5. Corresponding parts of congruent figures are congruent.
6. $m\angle APB = m\angle CPD$	6. Definition of congruent angles
7. $m\overset{\frown}{AB} = m\angle APB$ $m\overset{\frown}{CD} = m\angle CPD$	7. Definition of minor arc
8. $m\overset{\frown}{AB} = m\overset{\frown}{CD}$	8. Substitution Property of Equality
9. $\overset{\frown}{AB} \cong \overset{\frown}{CD}$	9. Definition of congruent arcs

C. Why can you state that the Congruent Corresponding Chords Theorem is true in congruent circles as well as in the same circle?
If the circles are congruent, then the radii of the circles are congruent.

Turn and Talk What statement can you make about the central angles formed by corresponding congruent chords in a circle? See margin.

Step It Out

Apply the Inscribed Quadrilateral Theorem

You can use the Inscribed Quadrilateral Theorem to find missing angle measures.

 Find the measure of each angle in quadrilateral *WXYZ*.

X (5a + 7)°
(8a − 1)° W
Y
Z (7a + 17)°

Find the value of *a*.

$$m\angle X + m\angle Z = 180^\circ$$
$$(5a + 7)^\circ + (7a + 17)^\circ = 180^\circ$$
$$12a + 24 = 180$$
$$12a = 156$$
$$a = 13$$

A. Why is this a true statement?

A, B. See margin.

Substitute the value of *a* into the expression for each angle measure.

$m\angle W = (8a - 1)^\circ = \big(8(13) - 1\big)^\circ = 103^\circ$

$m\angle X = (5a + 7)^\circ = \big(5(13) + 7\big)^\circ = 72^\circ$

$m\angle Z = (7a + 17)^\circ = \big(7(13) + 17\big)^\circ = 108^\circ$

$m\angle Y =$ ___?___

B. How can you determine $m\angle Y$ even though you are not given an expression for this measure? What is $m\angle Y$?

Construct an Inscribed Square

 Randall is making a tile design. He wants to inscribe a square in a circle as part of his pattern. How can he construct an inscribed square?

To construct a square, start by using a compass to draw a circle and label the center. Use a straightedge to draw a diameter of the circle. Then construct the perpendicular bisector of the diameter. Connect the endpoints of the two diameters to form the square.

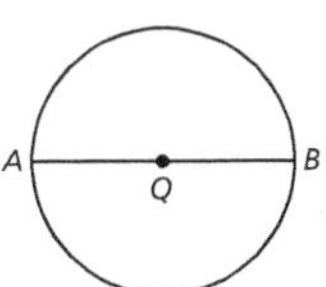

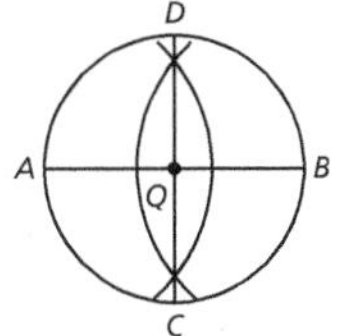

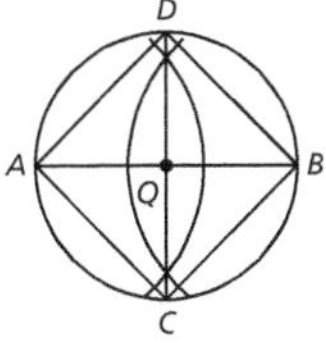

A. For *ADBC* to be a parallelogram, its diagonals must bisect each other. Explain why $\overline{AB}$ bisects $\overline{CD}$. A, B. See margin.

B. For parallelogram *ADBC* to be a square, its diagonals must be congruent and perpendicular. How do you know that *ADBC* is a square?

 **Turn and Talk** How do you know that *ADBC* is a rhombus? See margin.

Step It Out

Task 4 (MP) **Attend to Precision** Students will apply the Inscribed Quadrilateral Theorem to find angle measures of a specific quadrilateral. Students will have to be precise in their application of that theorem.

SUPPORT SENSE-MAKING Three Reads

Have students read the problem three times. Use the questions below for a different focus each read.

1. What is the situation about?
2. What are the quantities in the situation?
3. What are the possible mathematical questions that you could ask for this situation?

Task 5 (MP) **Use Structure** Students will use the Inscribed Quadrilateral Theorem and the properties of central angles to construct an inscribed square in a circle. Students will need to discern patterns and structure that they can use in this process.

Sample Guided Discussion:

Q What can you say about two diameters of a circle? The diameters are of equal length, and they both extend from the circumference through the center to a point at 180° from the opposite end of the diameter.

Q How do you know that the two diameters are perpendicular? The second diameter was constructed by drawing the perpendicular bisector.

 **Turn and Talk** Remind students that every square is a rhombus. The diagonals of *ADBC* are perpendicular.

ANSWERS

4.A. By the Inscribed Quadrilateral Theorem, the opposite angles in an inscribed quadrilateral are supplementary.

4.B. By the Inscribed Quadrilateral Theorem, you know that $\angle W$ and $\angle Y$ are supplementary. So, you can subtract $m\angle W$ from 180° to find $m\angle Y = 77^\circ$.

5.A. $\overline{CD}$ was constructed to perpendicularly bisect $\overline{AB}$ through the center with its endpoints on the same circle, so $\overline{AB}$ also bisects $\overline{CD}$.

5.B. $\overline{AB}$ and $\overline{CD}$ are perpendicular by the definition of a perpendicular bisector. $\overline{AB} \cong \overline{CD}$ because both $\overline{AB}$ and $\overline{CD}$ are diameters of the circle.

ONLINE

Assign the Digital On Your Own for
- built-in student supports
- Actionable Item Reports
- Standards Analysis Reports

On Your Own

Assignment Guide

The chart below indicates which problems in the On Your Own are associated with each task in the Learn Together. Assign daily homework for tasks completed.

Learn Together Tasks	On Your Own Problems
Task 1, p. 476	Problem 11
Task 2, p. 477	Problems 9, 10, and 12
Task 3, p. 478	Problems 14, 15 and 16
Task 4, p. 479	Problems 5–8 and 13
Task 5, p. 479	Problems 17–21

Check Understanding

1. In quadrilateral $ABCD$, $m\angle A = 78°$, and $m\angle B = 116°$. What do the measures of $\angle C$ and $\angle D$ need to be in order for $ABCD$ to be able to be inscribed in a circle? $m\angle C = 102°$, and $m\angle D = 64°$
2. Circles A and B are congruent. List two pairs of congruent arcs. $\widehat{XY} \cong \widehat{FG}$, $\widehat{HJ} \cong \widehat{VW}$

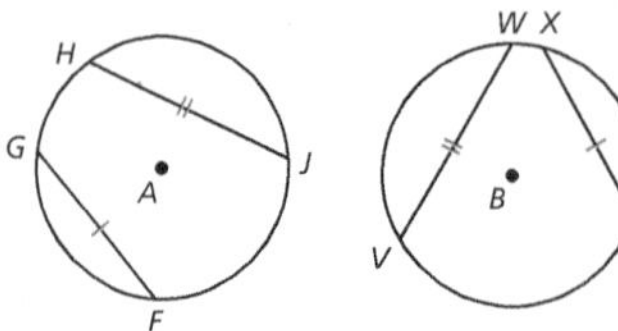

Find the value of x.

3. $x = 15$

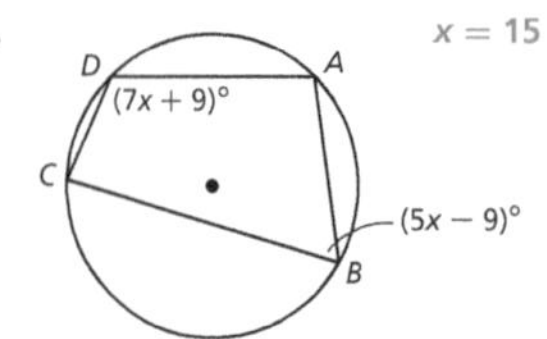

4. $x = 2$

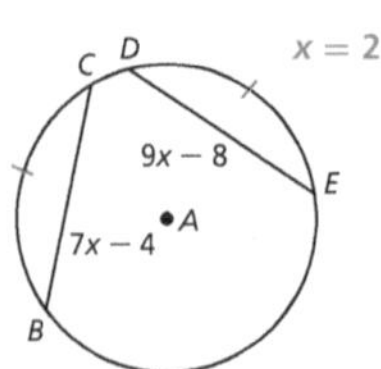

On Your Own

Find the interior angle measures of each inscribed quadrilateral.

5.

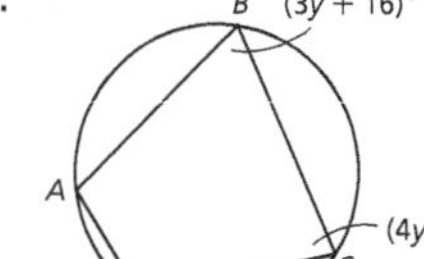

$m\angle A = 102°$, $m\angle B = 70°$, $m\angle C = 78°$, $m\angle D = 110°$

6.

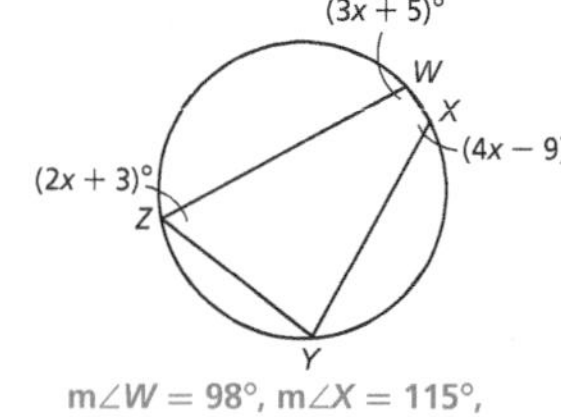

$m\angle W = 98°$, $m\angle X = 115°$, $m\angle Y = 82°$, $m\angle Z = 65°$

7.

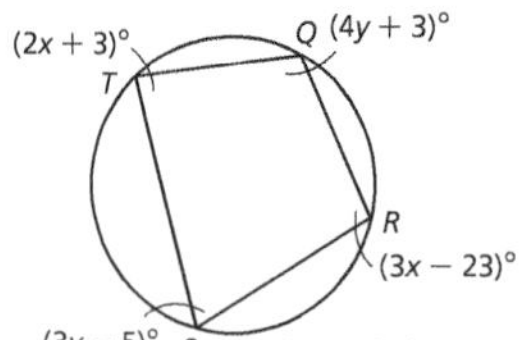

$m\angle Q = 107°$, $m\angle R = 97°$, $m\angle S = 73°$, $m\angle T = 83°$

8.

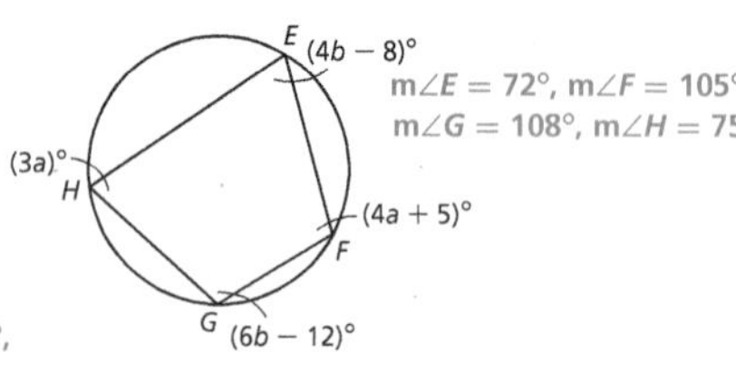

$m\angle E = 72°$, $m\angle F = 105°$, $m\angle G = 108°$, $m\angle H = 75°$

480

3 Check Understanding

Formative Assessment

Use formative assessment to determine if your students are successful with this lesson's learning objective.

Students who successfully complete the Check Understanding can continue to the On Your Own practice.

For students who miss 1 problem or more, work in a pulled small group using the Almost There small-group activity on page 475C.

ONLINE

Assign the Digital Check Understanding to determine
- success with the learning objective
- items to review
- grouping and differentiation resources

4 Differentiation Options

Differentiate instruction for all students using small-group activities and math center activities on page 475C.

Reteach

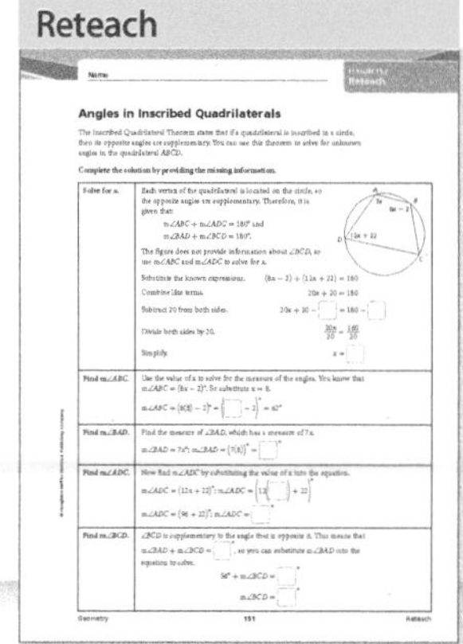

Challenge

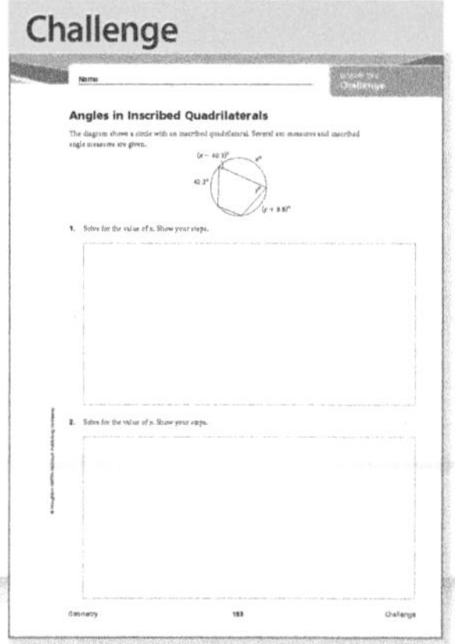

9. **MP Critique Reasoning** Marvin says it is not possible to inscribe a kite in a circle because each pair of opposite angles must be supplementary. Denise disagrees with Marvin. Explain why Denise is correct. See margin.

10. **MP Reason** What must be true about a rhombus that is inscribed in a circle? Explain your reasoning. It must be a square because opposite angles must be congruent and supplementary.

11. Donnie is programming a robot to walk in a quadrilateral-shaped path inscribed within a circular ring. He has the robot start at P toward Q, turn left 75° at Q, and then turn left 100° at R. How many degrees must the robot turn left at S to make it back to the starting point P? 105°

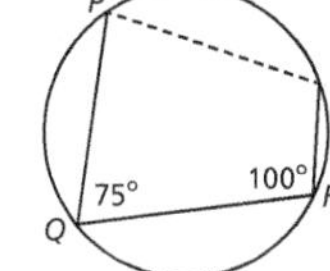

12. $GHJK$ is a quadrilateral inscribed in a circle. Angles H and K are opposite angles that are congruent. Is $\overline{GJ}$ a diameter of the circle? Explain your reasoning. See margin.

13. A company has a logo that contains a square inscribed in a circle. On the sign on the company's building, the circle has a diameter of 12 feet. What is the side length of the square? $6\sqrt{2}$ feet

14. One part of the Congruent Corresponding Chords Theorem was proven in Task 3. Use the statements below to prove the second part of the Congruent Corresponding Chords Theorem. See Additional Answers.

Given: $\overset{\frown}{AB} \cong \overset{\frown}{CD}$, P is the center of the circle.

Prove: $\overline{AB} \cong \overline{CD}$

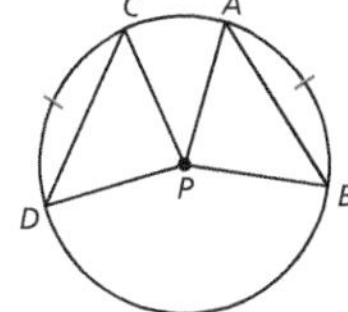

Find the measure of each chord or arc.

15. $\overline{DE}$ 83 

16. $\overset{\frown}{AB}$ 133° 

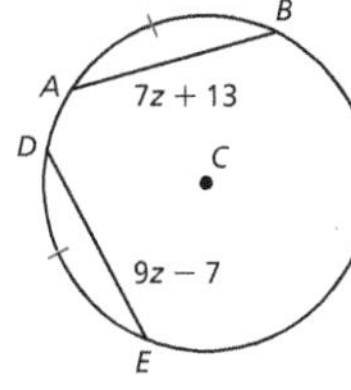

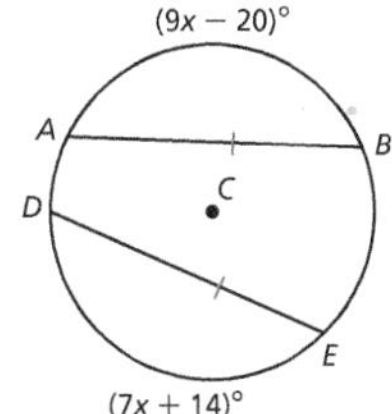

In Problems 17 and 18, tell whether the quadrilateral can be inscribed in a circle. If so, describe a method for doing so using a compass and a straightedge. If not, explain why not.

17. a parallelogram that is not a rectangle 17, 18. See margin.

18. a kite with two right angles

Questioning Strategies

Problem 12 Suppose that, instead of two opposite angles being congruent, two adjacent angles were congruent. How would that change your answer to the question? If adjacent angles were congruent and we didn't know for sure that opposite angles were also congruent, then $PQRS$ could still be a rectangle, but it could also be an isosceles trapezoid, in which case the diagonals would not be diameters of the circle.

Watch for Common Errors

Problem 14 Students might want to duplicate the reasoning in Task 3 in order to prove the converse of the theorem. The converse is proved by using part of the reasoning of the first part of the theorem and building on theorems we already know. Given that $\overset{\frown}{AB} \cong \overset{\frown}{CD}$ and P is the center of the circle, we know that the measures of those arcs are equal. Draw the legs CD and AB (because between any two points there exists a single line). Then we have two triangles, $\triangle CPD$ and $\triangle APB$ with the legs that meet at P congruent because they are radii of the circle. We then note that $m\overset{\frown}{AB} = m\angle AOB$ and $m\overset{\frown}{CD} = m\angle CPD$ by the definition of a minor arc. Therefore, $m\overset{\frown}{AB} = m\overset{\frown}{CD}$ by the substitution property of equality. By the transitivity property of equality, the two angles have equal measures: $m\angle CPD = m\angle APB$. Therefore, by the SAS Triangle Congruence Theorem, $\triangle CPD$ and $\triangle APB$ are congruent, since the radii are congruent. Because the triangles are congruent, we have that $\overline{AB} \cong \overline{CD}$.

ANSWERS

9. Since a kite must have a pair of congruent opposite angles, it is possible to inscribe a kite in a circle as long as the two congruent opposite angles are right angles.

12. yes; The quadrilateral is inscribed in the circle, so opposite angles are supplementary. The only way that supplementary angles are congruent is if they are right angles. By the Inscribed Angle of a Diameter Theorem, if the inscribed angles are right angles, the endpoints of the angle lie on the diameter of a circle.

17. It is not possible to inscribe a parallelogram that is not a rectangle in a circle because the opposite angles must be congruent and supplementary, which means they are right angles.

18. It is possible. Draw a circle. Draw a diameter and a chord perpendicular to the diameter. Connect the end points of the diameter and chord to form a kite.

5 Wrap-Up

Summarize learning with your class. Consider using the Exit Ticket, Put It in Writing, or I Can scale.

Exit Ticket 

Prove that a right trapezoid cannot be an inscribed quadrilateral in a circle. Let $ABCD$ be a right trapezoid and let $\angle BCD \cong \angle CDA$, where the measures of both angles equal 90°. Let the two top vertices, from left to right, be A and B respectively. Then the two other angles add up to 180° but are not congruent. So $m\angle DAB \neq m\angle BCD$. Because $m\angle BCD = 90°$, we know that $m\angle BCD + m\angle CDA = 180°$, whereas $m\angle DAB + m\angle BCD \neq 180°$, so by the Inscribed Quadrilateral Theorem and its converse, $ABCD$ cannot be an inscribed quadrilateral of a circle.

Put It in Writing 

Write several paragraphs explaining how the theorems in this lesson are interrelated. Assume that each theorem given with its converse is a single theorem or biconditional proposition.

I Can 

The scale below can help you and your students understand their progress on a learning goal.

4	I can use the properties of angles of quadrilaterals inscribed in a circle to prove theorems and solve problems, and I can explain my reasoning to others.
3	I can use the properties of angles of quadrilaterals inscribed in a circle to prove theorems and solve problems.
2	I can use the Arc Addition Postulate and the Inscribed Angle Theorem to prove several steps in the Inscribed Quadrilateral Theorem.
1	I can understand parts of the proof of the Inscribed Quadrilateral Theorem.

Spiral Review • Assessment Readiness

These questions will help determine if students have retained information taught in the past and can also prepare them for high-stakes assessments. Here, students use the Law of Sines to find the length of a triangle leg **(14.1)**, use the Law of Cosines to find the length of a triangle leg, and apply the Arc Addition Postulate **(15.1)** to find an arc measure.

19. Chibenashi is making a dream catcher with his little brother. He shows him how to attach two pieces of string that are the same length to the ring. If the two pieces of string share one endpoint and the arc formed by the endpoints of one chord has the measure shown, what is the measure of the angle formed by the pieces of string? 40°

20. Use a compass to draw a circle. Then mark a point on the circle. Inscribe a square in the circle so that one of its vertices is the point marked on the circle. Explain your method. See Additional Answers.

21. (Open Middle™) What is the greatest possible area for a quadrilateral inscribed inside a circle with a circumference of 16π units? 128 square units

Spiral Review • Assessment Readiness

22. Find the value of a.

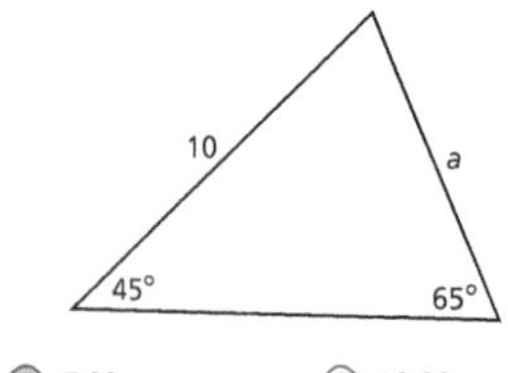

Ⓐ 7.80 Ⓑ 14.22 Ⓒ 12.82 Ⓓ 10.33

23. Find the value of c.

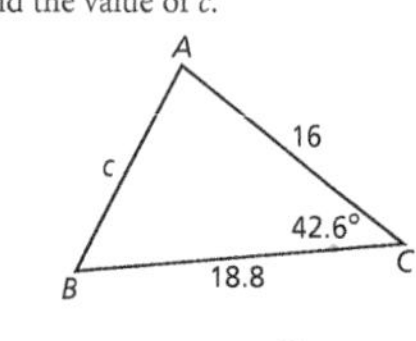

Ⓐ 12.9 Ⓑ 14.2 Ⓒ 16.6 Ⓓ 20.2

24. Use the circle graph to find the measure of $\widehat{DEF}$ to the nearest degree.

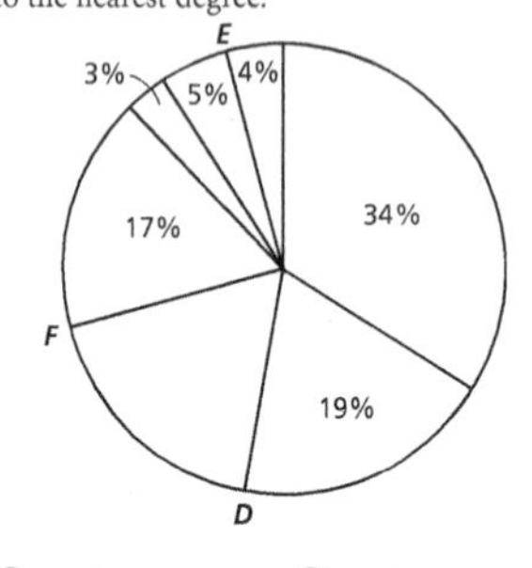

Ⓐ 270° Ⓑ 310° Ⓒ 295° Ⓓ 328°

I'm in a Learning Mindset!

What strategies did I use to overcome barriers when solving problems with angles in inscribed quadrilaterals?

Keep Going Journal and Practice Workbook

Learning Mindset

Resilience Identifies Obstacles

Point out that constructing proofs can sometimes be complicated, but students should always identify what they know and what they don't know. For example, they know the Arc Addition Postulate and the Inscribed Angle Theorem, which they use to prove the Inscribed Quadrilateral Theorem. Each theorem builds on knowledge acquired previously (at some point). *What are the biggest obstacles you face when trying to prove a theorem? Is it the specific arguments to make or the sequence of arguments? What theorems did you know before beginning the lesson that helped you prove parts or the whole of any of the theorems? Which theorems in this lesson helped you prove the next theorems?*

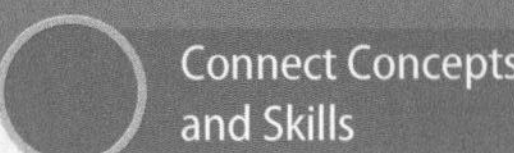

15.3 Tangents and Circumscribed Angles

LESSON FOCUS AND COHERENCE

Mathematics Standards

- Identify and describe relationships among inscribed angles, radii, and chords. *Include the relationship between central, inscribed, and circumscribed angles; inscribed angles on a diameter are right angles; the radius of a circle is perpendicular to the tangent where the radius intersects the circle.*
- Construct a tangent line from a point outside a given circle to the circle.

Mathematical Practices and Processes 

- Use appropriate tools strategically.
- Construct viable arguments and critique the reasoning of others.
- Look for and make use of structure.
- Look for and express regularity in repeated reasoning.
- Attend to precision.

I Can Objective 

I can prove theorems about tangents to a circle and use them to solve mathematical and real-world problems.

Learning Objective

Solve problems where tangent lines form angles with other tangent lines and with the radius of the circle. Prove that two tangent segments that share an exterior point are congruent. Prove that a circumscribed angle is supplementary to its related central angle. Prove that a tangent is perpendicular to the radius it intersects.

Language Objective

Explain how to prove and use the properties of tangents to a circle.

Vocabulary

New: circumscribed angle, **exterior of a circle**, **interior of a circle**, **point of tangency**, **tangent of a circle**

Lesson Materials: compass, geometric drawing tool, straightedge

Mathematical Progressions

Prior Learning	Current Development	Future Connections
Students: • applied the Pythagorean Theorem to determine unknown side lengths in right triangles in real-world and mathematical problems. **(Gr8, 11.3)** • applied the criteria for triangle congruence (ASA, SAS, and SSS). **(8.4)**	**Students:** • use a compass and straightedge to draw a circle and tangent to the circle. • prove theorems involving tangents and radii of circles. • solve a real-world problem involving a tangent to a circle.	**Students:** • will identify and describe relationships among inscribed angles, radii, and chords to include the relationship between central, inscribed, and circumscribed angles. **(16.1 and 16.2)** • will identify the radius of a circle as perpendicular to the tangent where the radius intersects the circle. **(16.1 and 16.2)**

UNPACKING MATH STANDARDS

Identify and describe relationships among inscribed angles, radii, and chords.

What It Means to You

Students first investigate the relationship among inscribed angles, radii, and tangents to a circle by constructing figures using geometric drawing tools and compass and straightedge constructions. This investigation helps students to visualize the relationships that form the theorems.

The emphasis on this standard is on construction and proof, leading to application of the theorems in mathematical and real-world situations. In future lessons, students will extend this knowledge to the relationships of chords, radii, and angles formed within circles.

WARM-UP OPTIONS

ONLINE

PROJECTABLE & PRINTABLE

ACTIVATE PRIOR KNOWLEDGE • Use the Pythagorean Theorem to Solve Problems

Use these activities to quickly assess and activate prior knowledge as needed.

Problem of the Day 

A field manager paints a line of chalk from first base to home plate and then from home plate to third base. The angle formed between the lines of chalk is a right angle. Each line of chalk the manager paints is 90 feet. Use the Pythagorean Theorem to write an equation to find the distance between first base and third base. What is the distance, to the nearest foot?

$90^2 + 90^2 = x^2$, approximately 127 feet

Quick Check for Homework

As part of your daily routine, you may want to display the Teacher Solution Key to have students check their homework.

Make Connections

Based on students' responses to the Problem of the Day, choose one of the following:

1. Project the Interactive Reteach, Grade 8, Lesson 11.3.
2. Complete the Prerequisite Skills Activity:

Draw and label the two triangles on the board. Then ask the following questions.

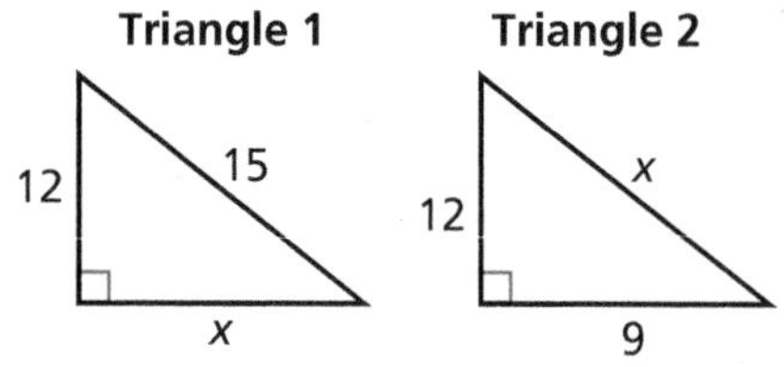

Triangle 1 (legs 12 and x, hypotenuse 15)

Triangle 2 (legs 12 and 9, hypotenuse x)

- *What equations can you write to solve for x in each triangle using the Pythagorean Theorem?* $12^2 + x^2 = 15^2$, $12^2 + 9^2 = x^2$.
- *How are the equations different?* The x variable is in the sum in one equation and alone in the other equation.
- *How can you make sure to write the equations correctly?* You could label the legs and hypotenuse before writing the equation.
- *Solve each equation. What is the value of x in each triangle?* In Triangle 1, $x = 9$. In Triangle 2, $x = 15$.

If students continue to struggle, use Tier 2 Skill 4.

SHARPEN SKILLS

If time permits, use this on-level activity to build fluency and practice basic skills.

Vocabulary Review

Objective: Students show they understand the definition of the term *tangent of a circle* and the properties of a tangent.
Materials: Word Definition Map (Teacher Resource Masters)

Pass out a copy of the Word Definition Map. Have students write the vocabulary term *tangent of a circle* in the box. Work together as a class to formulate a definition and to brainstorm related concepts or properties. Finally, have students work in pairs to sketch, draw, or construct examples of tangent lines and segments with respect to a circle.

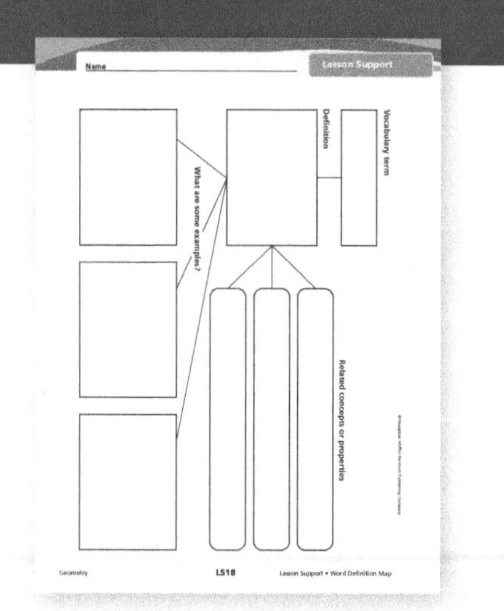

PLAN FOR DIFFERENTIATED INSTRUCTION

Small-Group Options

Use these teacher-guided activities with pulled small groups.

On Track

Display the three figures on the board. Have students copy the figures. Then have students discuss which theorems could be used in each and what information the theorems reveal about the figures.

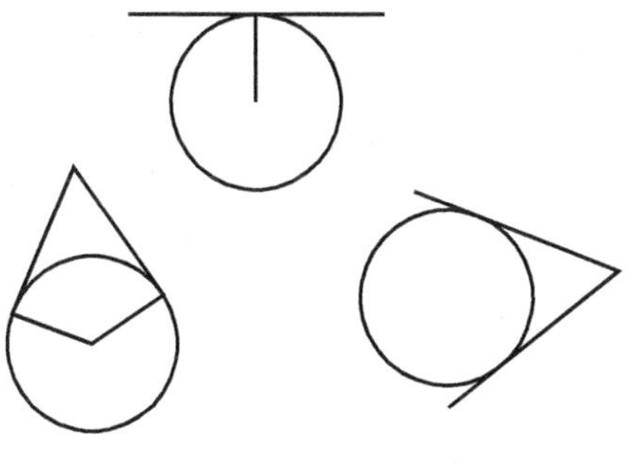

Almost There

Materials: index cards

Make enough cards for students to work in partner groups. Each card should have a circle and a tangent and the sides of a triangle labeled.

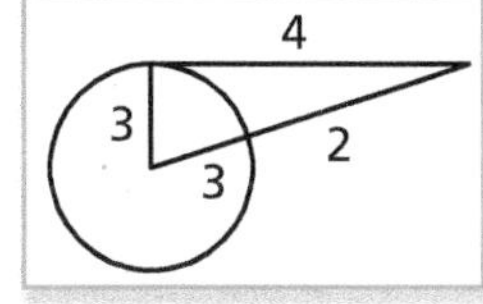

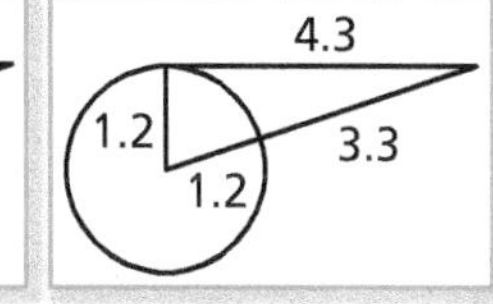

- Have students use the Pythagorean Theorem to write an equation for each of the triangle's sides. For example, write $3^2 + 4^2 = (3 + 2)^2$.
- Are the equations true?
- What does that say about the relationship between a radius and a tangent of a circle?

Ready for More

Materials: compass and straightedge

In the third task of the lesson, students explored the relationship between the central angle and a circumscribed angle. Here, students will prove that two tangents through the endpoints of a diameter cannot form a circumscribed angle.

Have students construct a circle with center *C*. Instruct students to draw a diameter through *C*, labeling the endpoints *P* and *Q*. Next, have students draw a tangent line through each of the points *P* and *Q*.

Next, have students work in pairs to write a proof explaining why the tangent lines cannot form a circumscribed angle.

Math Center Options

Use these student self-directed activities at centers or stations.

Key: ● Print Resources ● Online Resources

On Track

- ● Interactive Digital Lesson
- ●● Journal and Practice Workbook
- ● Interactive Glossary (printable): **circumscribed angle**, **exterior of a circle**, **interior of a circle**, **point of tangency**, **tangent of a circle**
- ● Illustrative Mathematics: Tangent Lines and the Radius of a Circle
- ● Module Performance Task
- ●● Standards Practice: Construct a Tangent Line from a Point Outside a Given Circle to the Circle

Almost There

- ● Reteach 15.3 (printable)
- ● Interactive Reteach 15.3
- ● RtI Tier 2 Skill 4: The Pythagorean Theorem and Its Converse

Ready for More

- ● Challenge 15.3 (printable)
- ● Interactive Challenge 15.3
- ● Illustrative Mathematics: Neglecting the Curvature of the Earth

View data-driven grouping recommendations and assign differentiation resources.

Spark Your Learning • Student Samples

During the *Spark Your Learning*, listen and watch for strategies students use. See samples of student work on this page.

Draw a Right Triangle and Use the Pythagorean Theorem

Strategy 1

I noticed the triangle is a right triangle. I labeled the sides and used the Pythagorean Theorem.

$$?^2 + (4{,}000)^2 = (18{,}560)^2$$
$$?^2 = (18{,}560)^2 - (4{,}000)^2$$
$$? = \sqrt{(18{,}560)^2 - (4{,}000)^2}$$
$$? \approx 18{,}124 \text{ miles}$$

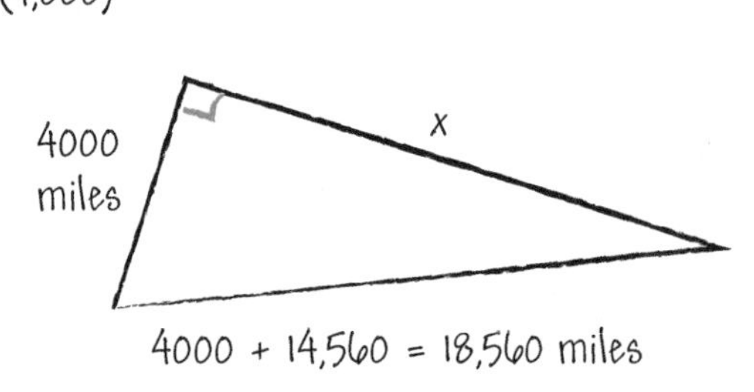

If students . . . use a drawing and an equation to solve the problem, they are using the Pythagorean Theorem from Grade 8.

Have these students . . . explain how they wrote their equation and how they solved it. **Ask:**

- Q How did you know which lengths went in for the variables in the Pythagorean Theorem?
- Q How did you find the length of the hypotenuse?

Use an Equation

Strategy 2

I know the distance from the satellite to the farthest point on Earth is unknown. I also know the radius is 4,000 miles and the distance from the satellite to the center of Earth is 18,560 miles. I think the unknown distance can be found by using the Distance Formula.

$$\sqrt{(4{,}000 - 0)^2 + (x - 0)^2} = 18{,}560$$
$$(4{,}000)^2 + x^2 = (18{,}560)^2$$
$$x^2 = (18{,}560)^2 - (4{,}000)^2$$
$$x = \sqrt{(18{,}560)^2 - (4{,}000)^2}$$
$$x \approx 18{,}124 \text{ miles}$$

If students . . . use the Distance Formula, they understand the quantities are related, but are not making the connection to a right triangle.

Activate prior knowledge . . . by having students draw the right triangle on the diagram. **Ask:**

- Q Which sides of the triangle are the legs, and which side is the hypotenuse?
- Q How can you use the Pythagorean Theorem to solve the problem?

COMMON ERROR: Adds the Radius to the Shortest Distance

I know the shortest distance to Earth is 14,560 miles. So, the longest distance must be 4,000 miles more than 14,560.

14,560 + 4,000 = 18,560 miles

If students . . . add the radius to the shortest distance, they think the longest distance is to the center of Earth.

Then intervene . . . by pointing out that the satellite cannot reach the inside of Earth, but only to the surface of Earth. **Ask:**

- Q What is the longest segment you can draw from the satellite that will touch Earth?
- Q How can you form a right triangle with this longest segment?
- Q What equation can you use to find the missing value in a right triangle?

15.3

Tangents and Circumscribed Angles

(I Can) prove theorems about tangents to a circle and use them to solve mathematical and real-world problems.

Spark Your Learning

A communications satellite is in orbit around Earth.

Complete Part A as a whole class. Then complete Parts B–D in small groups.

A. What is a mathematical question you can ask about this situation? What information would you need to know to answer your question?

B. What quantities are you given in this situation? What unit of measurement is used for each quantity?
B, D. See Additional Answers.

C. To answer your question, what strategy and tool would you use along with all the information you have? What answer do you get?
See Strategies 1 and 2 on the facing page.

D. Does your answer make sense in the context of the situation? How do you know?

A. What is the farthest distance that the satellite's signal needs to travel to a point on the surface of Earth?; distance from the surface of Earth to the satellite, radius of Earth

Turn and Talk How would your answer change if the distance from the surface of Earth to the satellite increased by 10,000 miles? See margin.

CULTIVATE CONVERSATION • Information Gap

Ask students questions to help them decide what missing information they need to answer the question, "What is the farthest distance on Earth from the satellite that the satellite's signal can reach"?

1. Where is the point on Earth that is farthest from the satellite? Explain. If you draw a segment from the satellite tangent to Earth, the point where the segment meets Earth is the farthest point the signal can reach. The satellite signal cannot reach parts of Earth on the far side.

2. When you draw a tangent segment from the satellite to Earth, a radius from the tangent point to the center, and a segment from the center to the satellite, you form a right triangle. What measurements do you need to know? the value of the radius and the distance the satellite is above Earth

3. What is the length of the hypotenuse of the right triangle? the sum of the radius and the distance the satellite is above Earth

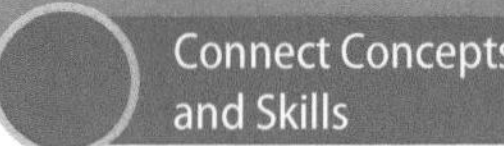

Connect Concepts and Skills

1 Spark Your Learning

▶ MOTIVATE

- Have students look at the illustration in their books and read the information contained in the illustration. Then complete Part A as a whole-class discussion.
- Give the class the additional information they need to solve the problem. This information is available online as a printable and projectable page in the Teacher Resources.
- Have students work in small groups to complete Parts B–D.

▶ PERSEVERE

If students need support, guide them by asking:

Q **Advancing • Use Tools** Which tool could you use to solve the problem? Why choose that tool and not some other? Students' choices of tools and reasons for choosing them will vary.

Q **Assessing** When you draw a radius from the center of Earth to the point where the satellite reaches, what type of angle appears to be formed? a right angle

Q **Assessing** What theorem can help you find a missing length in the right triangle? the Pythagorean Theorem, where c is the hypotenuse

Q **Advancing** In what ways does knowing the length of the radius of Earth help to solve the problem? The length of the radius forms one leg of a right triangle. (That same length also forms part of the hypotenuse.) Then you can use the Pythagorean Theorem to solve for the unknown distance.

Turn and Talk When the distance from the surface of Earth to the satellite increases by 10,000 miles, help students to find the sum of the original distance, the radius, and the additional 10,000 miles. Then have students redraw and label the right triangle in order to use the Pythagorean Theorem.

The farthest distance would be $\sqrt{28{,}560^2 - 4{,}000^2} =$ 28,278.5, or about 28,279 miles.

▶ BUILD SHARED UNDERSTANDING

Select groups of students who used various strategies and tools to share with the class how they solved the problem. As they present their solutions, have each group discuss why they chose a specific strategy and tool.

② Learn Together

Build Understanding

Task 1 (MP) **Use Tools** Students use a geometric drawing tool to investigate the relationship between a tangent and a radius of a circle.

CONNECT TO VOCABULARY

Have students use the **interactive Glossary** to record their understanding of the vocabulary in this task.

Sample Guided Discussion:

Q If the angle is not 90°, what happens to line $\overleftrightarrow{BC}$? The line would intersect the circle and become a secant.

Task 2 (MP) **Construct Arguments** Students write an indirect proof to show a radius is perpendicular to a tangent.

Sample Guided Discussion:

Q In Part A, what is the conclusion you are trying to prove? The radius is perpendicular to the tangent line.

Turn and Talk Remind students that you started the proof by stating the conclusion must be false and showed that this was impossible. Ask students to discuss how a contradiction is used in an indirect proof. It shows that a mathematical impossibility results from the assumption that the assertion is not true. Therefore, the assertion must be true.

Build Understanding

Investigate the Tangent-Radius Theorem

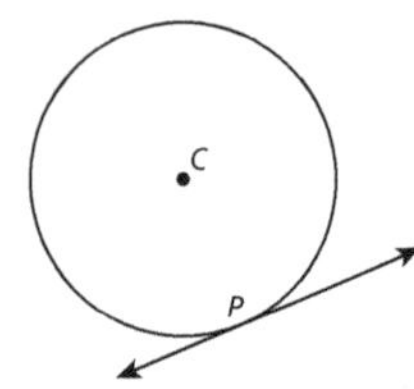

A **tangent of a circle** is a line that is in the same plane as the circle and intersects the circle in exactly one point. The point where the tangent and the circle intersect is called the **point of tangency**. In circle *C*, point *P* is the point of tangency.

The **exterior of a circle** is the set of all points outside a circle. The **interior of a circle** is the set of all points inside a circle. All points on a line tangent to a circle other than the point of tangency are in the exterior of the circle.

1 Use a geometric drawing tool to draw a circle and a point on the circle. Then draw a line tangent to the circle at the point. Place an additional point on the tangent line.

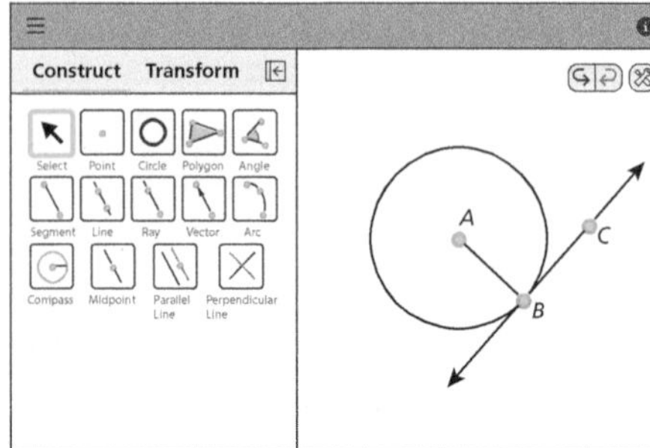

A. What is the measure of $\angle ABC$?
90°

B. Change the location of point *B* on the circle to move the tangent line. What happens to the measure of $\angle ABC$ when you change the location of the point of tangency?
The measure of $\angle ABC$ remains 90°.

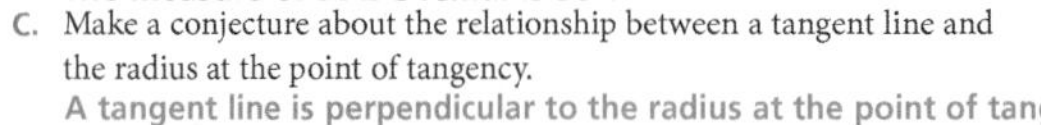

C. Make a conjecture about the relationship between a tangent line and the radius at the point of tangency.
A tangent line is perpendicular to the radius at the point of tangency.

Prove the Tangent-Radius Theorem

The Tangent-Radius Theorem and its converse are both true.

2 An indirect proof is used to prove the Tangent-Radius Theorem. Recall that in an indirect proof, you begin by assuming that the conclusion is false. Then you show that this assumption leads to a contradiction.

Given: Line *m* is tangent to circle *C* at point *P*.

Prove: $\overline{CP} \perp m$

A. What assumption should be made to start an indirect proof?
Assume that $\overline{CP}$ is not perpendicular to line *m*.

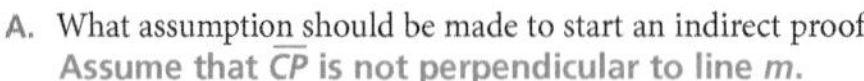

B. What conclusion follows from this assumption?
See margin.

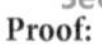

Proof:

If $\overline{CP}$ is not perpendicular to line *m*, then there must be a point *Q* on line *m* such that $\overline{CQ} \perp m$.

If $\overline{CQ}$ is perpendicular to *m*, then $\overline{CP}$ is the hypotenuse for $\triangle CPQ$. So, $CQ < CP$.

484

LEVELED QUESTIONS

Depth of Knowledge (DOK)	Leveled Questions	What Does This Tell You?
Level 1 **Recall**	In how many places does a tangent line intersect a circle? one	Students' answers will indicate whether they know the relationship between a circle and a tangent line.
Level 2 **Basic Application of Skills & Concepts**	What type of angle is formed by a tangent line and the radius of a circle that intersects the tangent line? a right angle	Students' answers will indicate whether they know the relationship between an intersecting tangent and radius of a circle.
Level 3 **Strategic Thinking & Complex Reasoning**	How many right angles could be formed using tangent lines and radii of a circle? infinitely many because there are infinite number of radii that can be drawn in a circle	Students' answers will show they realize that not only one tangent line and radius exists in a circle.

Since line m is a tangent line, it intersects circle C at the point of tangency P, and all other points of line m are in the exterior of the circle. This means point Q is in the exterior of the circle.

You can conclude that $CP < CQ$ because $\overline{CP}$ is a radius of circle C.

$CP < CQ$ contradicts $CQ < CP$. Thus, $\overline{CP} \perp m$.

C. Explain how exterior point Q leads to a contradiction of $CQ < CP$. C, D. See margin.

D. What does this contradiction mean?

Turn and Talk How does an indirect proof demonstrate that an assertion must be true? See margin.

Prove the Circumscribed Angle Theorem

A **circumscribed angle** is an angle formed by two rays from a common endpoint that are tangent to a circle.

Circumscribed Angle Theorem
A circumscribed angle of a circle and its associated central angle are supplementary.

3 Prove the Circumscribed Angle Theorem.

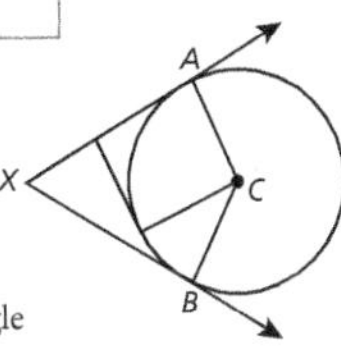

Given: $\angle AXB$ is a circumscribed angle of circle C.

Prove: $\angle AXB$ and $\angle ACB$ are supplementary.

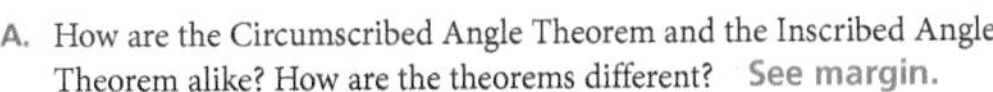

A. How are the Circumscribed Angle Theorem and the Inscribed Angle Theorem alike? How are the theorems different? See margin.

Statements	Reasons
1. $\overline{XA}$ and $\overline{XB}$ are tangents to circle C at points A and B.	1. Definition of circumscribed angle
2. $\angle CAX$ and $\angle CBX$ are right angles.	2. Tangent-Radius Theorem
3. $m\angle AXB + m\angle CAX + m\angle ACB + m\angle CBX = 360°$	3. Sum of measures of interior angles of a quadrilateral is 360°.
4. $m\angle AXB + 90° + m\angle ACB + 90° = 360°$	4. Substitution Property of Equality
5. $m\angle AXB + m\angle ACB = 180°$	5. Subtraction Property of Equality
6. $\angle AXB$ and $\angle ACB$ are supplementary.	6. Definition of supplementary angles

B. Is it possible for $ACBX$ to be a parallelogram? If so, what type? yes; a square

Turn and Talk Suppose that points A and B are endpoints of a diameter. Do the lines tangent to the circle at A and B form a circumscribed angle? How do you know? See margin.

PROFICIENCY LEVEL

Beginning
Have students sketch a circle, radius, and tangent line intersecting the radius. Ask students to label the radius, tangent line, and center of the circle.

Intermediate
Have students explain to a partner how a radius and a tangent line that intersect are related using a complete sentence.

Advanced
Ask students to write a paragraph explaining how they can use the Tangent-Radius Theorem to help prove the Circumscribed Angle Theorem.

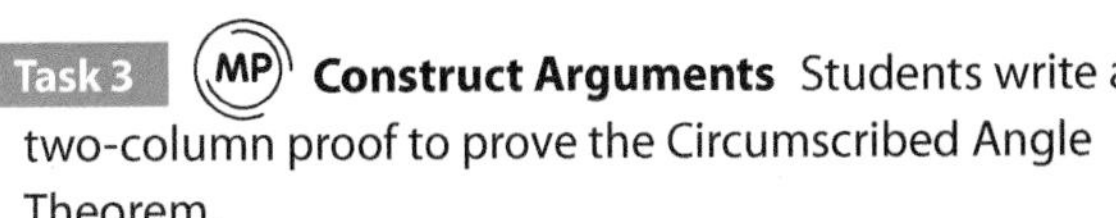

Task 3 (MP) **Construct Arguments** Students write a two-column proof to prove the Circumscribed Angle Theorem.

CONNECT TO VOCABULARY

Have students use the **interactive Glossary** to record their understanding of the vocabulary in this task.

Sample Guided Discussion:

Q What is an inscribed angle? an angle whose vertex is on the circle

Q In Step 2 of the proof, how do you know the angles are right angles? They are formed by a radius of the circle and a tangent and are right angles by the Tangent-Radius Theorem.

Q In Part B, what happens when the central angle $\angle ACB$ is a right angle? The circumscribed angle must also be a right angle because 3 of the 4 angles in the quadrilateral already are right angles.

Turn and Talk Encourage students to sketch a diagram of the situation. Ask students to identify what type of angles are formed by the tangent lines and how they are related. no; Two lines perpendicular to the same line are parallel, so they do not intersect at a point.

ANSWERS

TASK 2

2.B. The assumption $CP \not\perp m$ leads to the conclusion that there must be a point Q on line m such that $\overline{CQ} \perp m$, which means $\overline{CP}$ would be the hypotenuse of $\triangle CPQ$ and $CP > CQ$.

2.C. Since Q is in the exterior of circle C, $\overline{CQ}$ is longer than any radius of circle C. P is on circle C, so $\overline{CP}$ is a radius of circle C. Thus, $CP < CQ$, which is a contradiction of $CQ < CP$.

2.D. The assumption $\overline{CP} \not\perp m$ is false, and $\overline{CP} \perp m$ must be true.

TASK 3

3.A. Both theorems relate an angle to a central angle of a circle. The Inscribed Angle Theorem relates an angle within the circle to a central angle while the Circumscribed Angle Theorem relates an angle outside the circle to a central angle.

Step It Out

Task 4 (MP) **Use Structure** Encourage students to be consistent in the structure of their constructions and reassure them that the resulting figures have the same properties, regardless of the exact measurements.

Sample Guided Discussion:

Q If two students draw different lengths for the segment $\overline{CX}$, will this affect whether the tangents are perpendicular to the radius? Explain. no; The basic construction is the same—creating two tangent lines to the circle. A tangent line is perpendicular to the radius of a circle by the Tangent-Radius Theorem.

Q In Part B, how do you know if a segment drawn is a tangent? A tangent will intersect the circle in only one place and will not enter the interior of the circle.

Turn and Talk Encourage students to draw a situation where M might be on the inside of the circle. Have students look at Step 1 to see how they can change where point X and point M will be. Point M can be inside or outside of circle C, depending on how close X is to the circle.

ANSWERS

C. right angle; $\overline{CX}$ is a diameter of circle M, so the inscribed angle $\angle CAX$ is a right angle by the Inscribed Angle of a Diameter Theorem.

Step It Out

Construct a Tangent to a Circle

4 Use a compass and a straightedge to construct tangents to a circle.

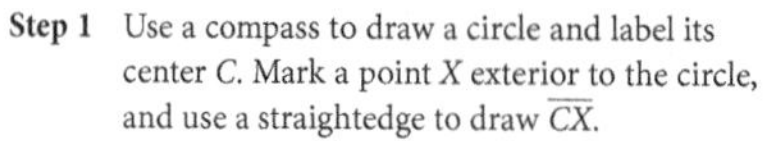

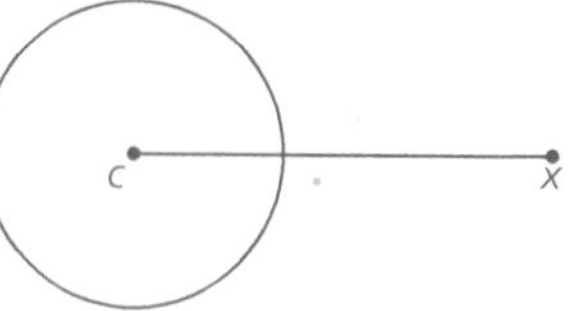

Step 1 Use a compass to draw a circle and label its center C. Mark a point X exterior to the circle, and use a straightedge to draw $\overline{CX}$.

Step 2 Bisect $\overline{CX}$ to find its midpoint. Label the midpoint M. Then use a compass to construct a circle centered at the point M that contains point C.

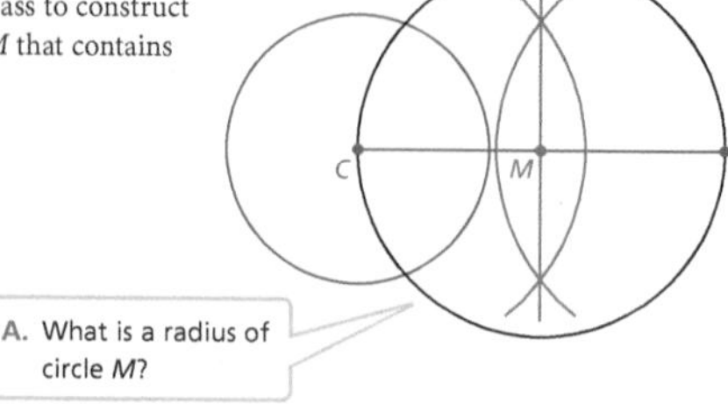

A. What is a radius of circle M?

A. $\overline{MC}$ or $\overline{MX}$

Step 3 From point X, use a straightedge to draw the tangents.

B. What are the points of tangency? How are the points of tangency determined?

B. A and B; The points of tangency are where the two circles intersect.

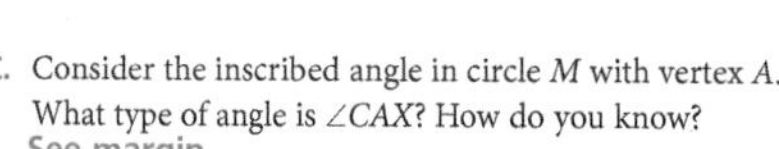

C. Consider the inscribed angle in circle M with vertex A. What type of angle is $\angle CAX$? How do you know? See margin.

D. How do you know that $\overrightarrow{XA}$ and $\overrightarrow{XB}$ are tangents to circle C?
$\overrightarrow{XA}$ and $\overrightarrow{XB}$ are each perpendicular to radii $\overline{CA}$ and $\overline{CB}$ of circle C, so they are tangents to circle C by the Converse of the Tangent-Radius Theorem.

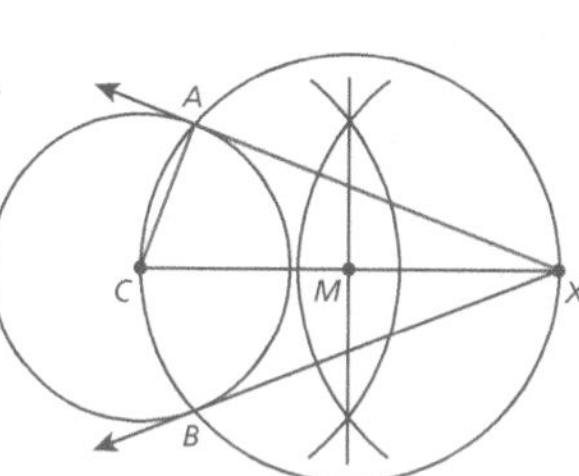

Turn and Talk Does point M have to be in the exterior of circle C for $\overrightarrow{XA}$ and $\overrightarrow{XB}$ to be tangent lines? Explain your reasoning. See margin.

486

Prove the Two-Tangent Theorem

You can use the Tangent-Radius Theorem to prove the Two-Tangent Theorem.

Two-Tangent Theorem

If two segments from the same exterior point are tangent to a circle, then the segments are congruent.

If $\overline{XA}$ and $\overline{XB}$ are tangents, then $\overline{XA} \cong \overline{XB}$.

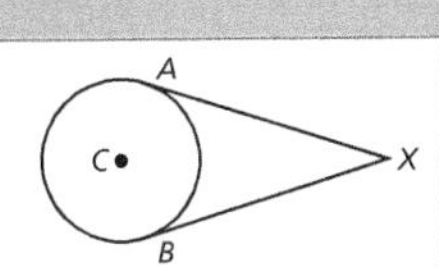

5 Prove the Two-Tangent Theorem.

Given: $\overline{XA}$ and $\overline{XB}$ are tangent to circle C.

Prove: $\overline{XA} \cong \overline{XB}$

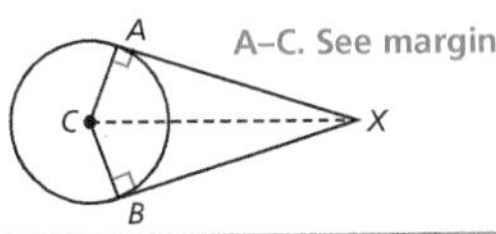

A–C. See margin.

Statements	Reasons
1. $\overline{XA}$ and $\overline{XB}$ are tangent to circle C.	1. Given
2. $\overline{XA} \perp \overline{CA}$; $\overline{XB} \perp \overline{CB}$	2. __?__
3. $\angle A$ and $\angle B$ are right angles.	3. Definition of perpendicular lines
4. $\triangle XAC$ and $\triangle XBC$ are right triangles.	4. Definition of a right triangle
5. $\overline{CA} \cong \overline{CB}$	5. __?__
6. $\overline{CX} \cong \overline{CX}$	6. Reflexive Property of Congruence
7. $\triangle XAC \cong \triangle XBC$	7. __?__
8. $\overline{XA} \cong \overline{XB}$	8. Corresponding parts of congruent figures are congruent.

A. What theorem justifies $\overline{XA} \perp \overline{CA}$ and $\overline{XB} \perp \overline{CB}$?

B. Why is $\overline{CA} \cong \overline{CB}$?

C. What theorem justifies $\triangle XAC \cong \triangle XBC$?

Apply the Two-Tangent Theorem

6 A bicycle is hung from the ceiling of an apartment using two wheel clips. One clip is attached to the ceiling, and the other is attached to the wall. The ceiling is tangent to the wheel at point J. The wall is tangent to the wheel at point L. The ceiling and the wall meet at point K. What are KJ and KL?

Find the value of y.

$KJ = KL$

$y + 20 = 4y - 7$

$27 = 3y$ A, B. See margin.

$9 = y$

A. How do you know that $KJ = KL$?

Substitute 9 for y in the expression $y + 20$.

$KJ = y + 20$

$= 9 + 20$

$= 29$ cm

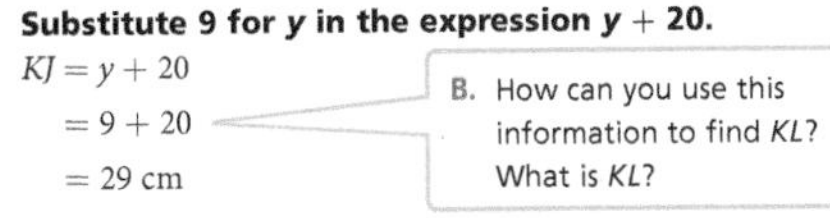

B. How can you use this information to find KL? What is KL?

Task 5 (MP) **Use Repeated Reasoning** Remind students to use theorems and definitions they already know when writing the proof of a new theorem.

SUPPORT SENSE-MAKING Critique, Correct, and Clarify

Have students consider the following alternative explanation of the proof. Then have pairs of students discuss the validity of this option. Ask for volunteers to explain their reasoning. In the proof, a student states that $\overline{CX} \cong \overline{CX}$ first before proving that $\overline{CA} \cong \overline{CB}$. Then the student uses SAS to prove the triangles are congruent.

Sample Guided Discussion:

- **Q Why can't you use the Two-Tangent Theorem to say that $\overline{XA} \cong \overline{XB}$?** This is the theorem you are trying to prove. You cannot use the theorem to prove itself.
- **Q Why are all radii in a circle congruent?** A circle is defined as the set of points that are equidistant from a given center point.

Task 6 (MP) **Attend to Precision** Students solve an algebraic equation and use the solution to find measurements in a real-world situation.

Sample Guided Discussion:

- **Q Why is 9 not the length of KL or KJ?** The value of 9 is only the value of the variable, not the value of the expression or length in the diagram.
- **Q Why could you also substitute 9 in for the other expression, $4y - 7$?** The two segments are equal in length.

ANSWERS

5.A. Tangent-Radius Theorem

5.B. $\overline{CA}$ and $\overline{CB}$ are radii of the same circle, so they are congruent.

5.C. HL Triangle Congruence Theorem

6.A. Two-Tangent Theorem

6.B. You know that $KJ = KL$, so $KL = 29$ cm.

ONLINE

Assign the Digital On Your Own for

- built-in student supports
- Actionable Item Reports
- Standards Analysis Reports

On Your Own

Assignment Guide

The chart below indicates which problems in the On Your Own are associated with each task in the Learn Together. Assign daily homework for tasks completed.

Learn Together Tasks	On Your Own Problems
Task 1, p. 484	Problem 6
Task 2, p. 484	Problems 7, 9, 11, and 12
Task 3, p. 485	Problems 8, 10, 14, 17, 23, and 25
Task 4, p. 486	Problems 13 and 17
Task 5, p. 487	Problem 18
Task 6, p. 487	Problems 19–22, and 24

ANSWERS

8. always, by the Circumscribed Angle Theorem
9. sometimes, if $\angle GHJ$ is a right angle
10. sometimes, if $\angle GHJ$ and $\angle GFJ$ are right angles
11. always, by the Tangent-Radius Theorem
12. sometimes, if $\angle GHJ$ is a right angle

Check Understanding

1. In the figure, $\overline{BC}$ is tangent to circle A at point B. What is $m\angle ACB$? Explain your reasoning. See Additional Answers.

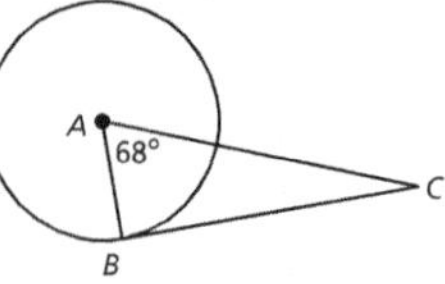

In each circle, A and B are points of tangency. Find the measure of $\angle BCA$.

2. 132°

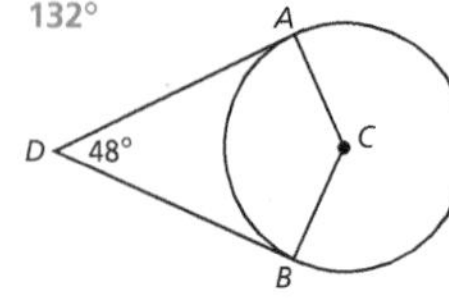

3. 75°

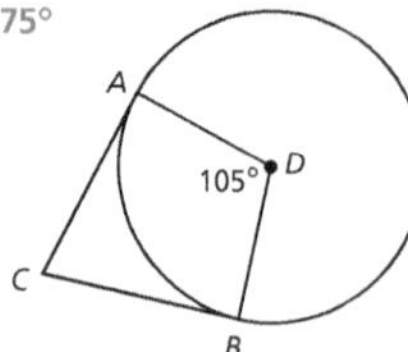

$\overline{AC}$ and $\overline{BC}$ are tangent to circle D. Find the value of x.

4. $x = 3$

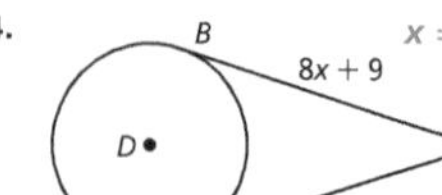

5. $x = 10$

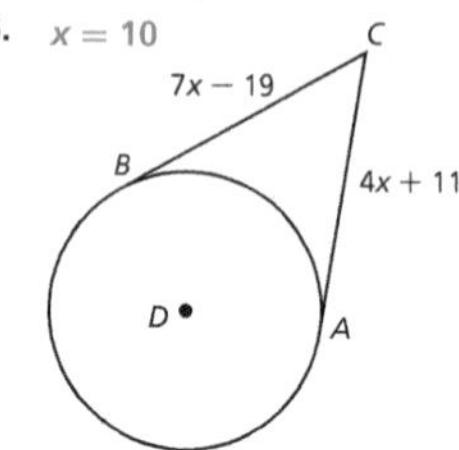

On Your Own

6. (MP) **Use Tools** Use a compass and a straightedge to create a diagram that demonstrates the Tangent-Radius Theorem. 6, 7. See Additional Answers.

7. (MP) **Reason** Prove the Converse of the Tangent-Radius Theorem.

Given: Line t is in the plane of circle C, A is a point of circle C, and $\overline{CA} \perp t$.

Prove: t is tangent to circle C at point A.

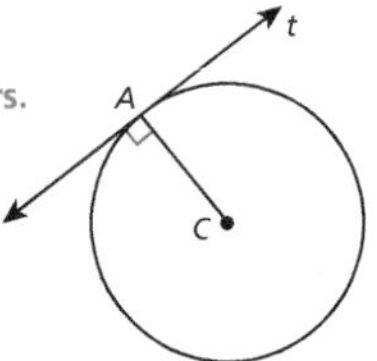

In circle F, G and J are points of tangency. Determine whether each statement in Problems 8–12 is *always* or *sometimes* true. Explain your reasoning.

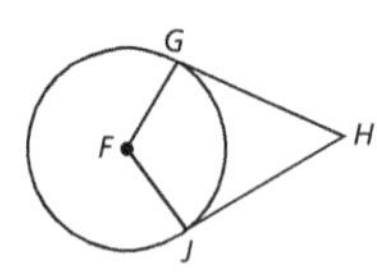

8. $\angle GFJ$ and $\angle GHJ$ are supplementary. 8–12. See margin.
9. $\angle HJF$ and $\angle GHJ$ are supplementary.
10. $\angle GHJ \cong \angle GFJ$
11. $\angle HGF \cong \angle HJF$
12. $\angle FGH \cong \angle GHJ$

488

3 Check Understanding

Formative Assessment

Use formative assessment to determine if your students are successful with this lesson's learning objective.

Students who successfully complete the Check Understanding can continue to the On Your Own practice.

For students who miss 1 problem or more, work in a pulled small group using the Almost There small-group activity on page 483C.

ONLINE

Assign the Digital Check Understanding to determine

- success with the learning objective
- items to review
- grouping and differentiation resources

4 Differentiation Options

Differentiate instruction for all students using small-group activities and math center activities on page 483C.

Reteach **Challenge**

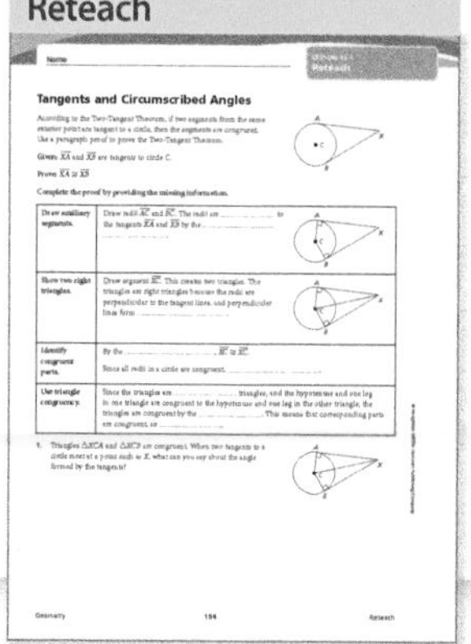

13. (MP) **Use Tools** A park in a town has a statue and a circular fountain. The town has built two sidewalks from the statue to the fountain that pass by the edges of the fountain. Use a compass and a straightedge to create a model of this situation. 13, 14. See margin.

14. Samantha is designing a logo with a circle and a circumscribed angle. The measure of the central angle is twice the measure of the circumscribed angle. What is the measure of each angle?

In each circle, A and B are points of tangency. Find the measures of the inscribed angle and the circumscribed angle.

15.

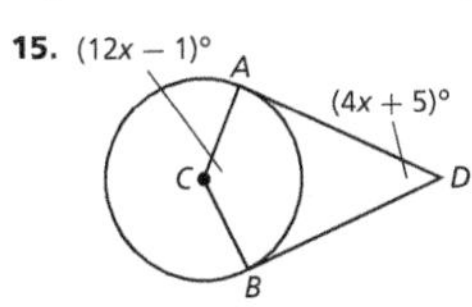

$m\angle ACB = 131^\circ$; $m\angle ADB = 49^\circ$

16. $(4x - 6)^\circ$ A, C, D, B $(6x - 14)^\circ$

$m\angle ACB = 74^\circ$; $m\angle ADB = 106^\circ$

17. **Health and Fitness** Rachel is standing at the center of a circle on a basketball court, and she has the ball. Derek and Ayush are standing on the circle. A line connecting Derek and Lia is tangent to the circle, and a line connecting Ayush and Lia is tangent to the circle. The ball can be passed along the blue path or the red path. Does one path cover more distance than the other? Explain your reasoning. 17, 18. See margin.

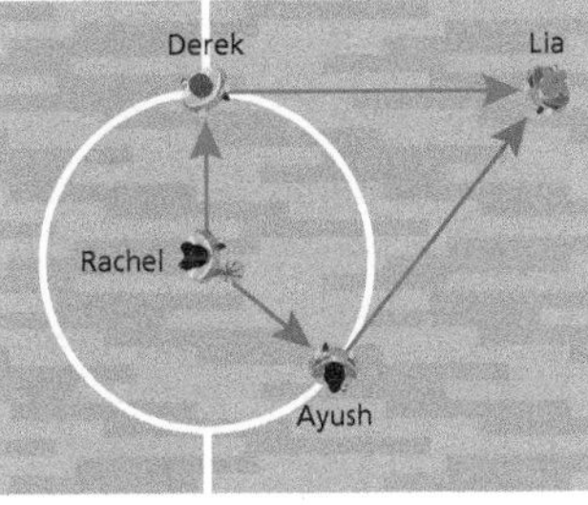

18. Given a circle with a diameter $\overline{CD}$, is it possible to construct tangents to points C and D from an external point P? If so, make a construction. If not, explain why it is not possible.

The segments in each figure are tangent to the circle at the points shown. Find each length.

19. A, $5r + 7$, D, C, $7r - 9$, B

Each length is 47 units.

20. B, $11z + 7$, D, C, $14z - 2$, A

Each length is 40 units.

21.

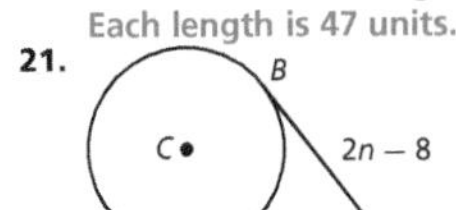

Each length is 24 units.

22. B, C, $8c - 1$, A, $10c - 15$, D

Each length is 55 units.

Watch for Common Errors

Problem 16 Point out to students that the angles across from each other in a quadrilateral may look congruent in a diagram but are not necessarily congruent. Remind students to use the Circumscribed Angle Theorem to write an equation where the angles across from each other add up to 180°.

Questioning Strategies

Problem 17 Suppose you knew the lengths of the segments in the diagram. Which segments would you add together? Which pairs of segments do you know are equal? Why are the pairs of segments equal? You would add a radius with a tangent segment; You know the two radii are equal in length because all radii in a circle are congruent; You know the tangents are equal in length because of the Two-Tangent Theorem.

ANSWERS

13.

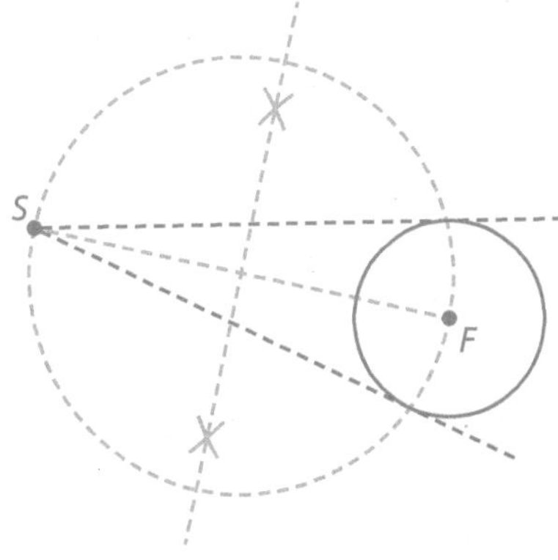

14. The central angle measure is 120°. The circumscribed angle measure is 60°.

17. no; The paths are the same length. The two radii of the circle are the same length and the two tangent lines are the same length by the Two-Tangent Theorem.

18. It is not possible. If it were possible, $\triangle CDP$ would contain two right angles, one each where the ends of the diameter intersect the tangents. This contradicts the Triangle Sum Theorem.

5 Wrap-Up

Summarize learning with your class. Consider using the Exit Ticket, Put It in Writing, or I Can scale.

Exit Ticket 

A sign for an ice cream store has a shape shown below where $\overline{SV}$ is tangent to the circle at point V and $\overline{SC}$ is tangent to the circle at point C.

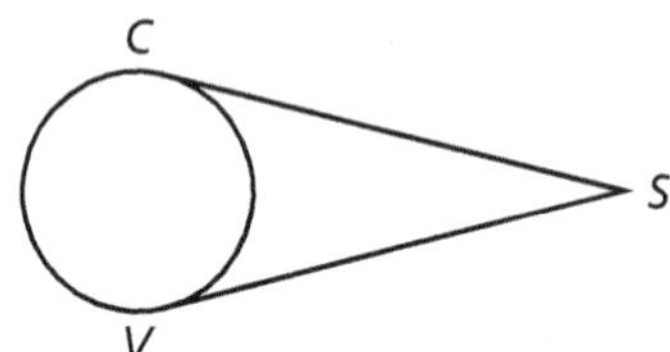

How do you know that $SV = SC$? The Two-Tangents Theorem says that two tangents to a circle from the same exterior point are congruent. Congruent segments have the same length by definition.

Put It in Writing 

Choose one of the theorems from this lesson. Write the theorem in your own words, and draw a diagram to support your explanation.

I Can 

The scale below can help you and your students understand their progress on a learning goal.

4	I can prove theorems about tangents to a circle and use them to solve mathematical and real-world problems, and I can explain my reasoning to others.
3	I can prove theorems about tangents to a circle and use them to solve problems.
2	I can use the theorems about tangents to a circle to solve problems.
1	I can identify tangents, radii, and circumscribed angles related to a triangle.

Spiral Review • Assessment Readiness

These questions will help determine if students have retained information taught in the past and can also prepare them for high-stakes assessments. Here, students use the Law of Cosines **(14.2)**, find the measures of arcs and angles in a circle **(15.1)**, and find the measures of inscribed angles **(15.2)**.

23. In circle A, $m\angle DCE = 48°$. Find $m\widehat{DBE}$. 228°

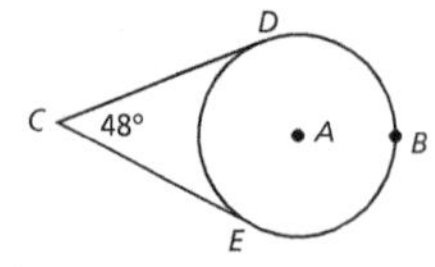

24. MP **Critique Reasoning** Given $CD = 41$, Rebecca says that $AC = 40$. Linda says that there is not enough information given to find AC. Who is correct? Explain your reasoning. Rebecca is correct. She used the Pythagorean Theorem to find that $CB = 40$. Then she used the Two-Tangent Theorem to find $AC = 40$.

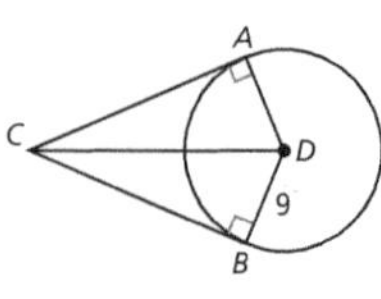

25. In the life preserver hanging from a rope, $\overline{BC}$ is tangent to circle A at B, $\overline{DC}$ is tangent to circle A at D, and $m\angle C = 58°$. Use the figure to find $m\angle EAH$ and $m\angle BAE$.
$m\angle EAH = 122°$, $m\angle BAE = 58°$

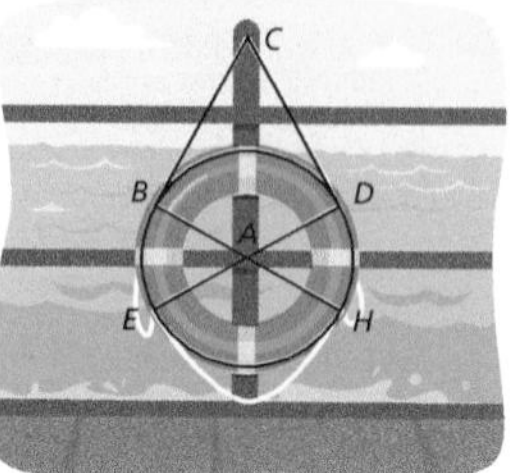

Spiral Review • Assessment Readiness

26. Find the value of a.

- (A) 3.8
- (B) 14.7
- (C) 40.6
- (D) 1648

27. Find $m\angle D$.

- (A) 69°
- (B) 82°
- (C) 98°
- (D) 111°

28. Match each item with its measurement.

Circle part	Measurement
A. $\angle CDB$	**1.** 50°
B. $\widehat{CDB}$	**2.** 100°
C. $\widehat{BC}$	**3.** 260°

$(10x - 20)°$ $(6x + 28)°$ A 1; B 3; C 2

 **I'm in a Learning Mindset!**

What barriers are there to finding tangents and circumscribed angles?

Keep Going to Journal and Practice Workbook

Learning Mindset

Resilience Identifies Obstacles

Point out that sometimes a pair of segments or angles may look congruent or not congruent in a diagram, but the opposite may be true. Remind students to identify the type of segment, line, angle, or figure represented in a diagram and to choose the appropriate theorem to apply. *How do you know when a tangent is present in a diagram? What theorems can be applied to tangents with respect to angles or lengths? When two segments appear to be congruent in a diagram, ask yourself how you know they are congruent and support your answer with a theorem.*

15.4 Circles on the Coordinate Plane

LESSON FOCUS AND COHERENCE

Mathematics Standards

- Derive the equation of a circle of given center and radius using the Pythagorean Theorem; complete the square to find the center and radius of a circle given by an equation.
- Use coordinates to prove simple geometric theorems algebraically.

Mathematical Practices and Processes 

- Look for and make use of structure.
- Attend to precision.
- Construct viable arguments and critique the reasoning of others.

I Can Objective 

I can derive and write the equation of a circle with radius r and center (h, k).

Learning Objective

Students will write an equation of a circle given its radius and the coordinates of its center, complete the square to rewrite an equation of a circle so its center and radius can be easily identified, and determine whether a given point lies on a circle given the equation of the circle.

Language Objective

Students will be able to explain the process of writing the equation of the circle given its radius and the coordinates of its center

Lesson Materials: compass, geometric drawing tool, straightedge, graph paper

Mathematical Progressions

Prior Learning	Current Development	Future Connections
Students: • applied the Pythagorean Theorem to determine unknown side lengths in right triangles in real-world and mathematical problems in two and three dimensions. **(Gr8, 11.1–11.3)** • completed the square in a quadratic expression to reveal the maximum or minimum value of the function it defines. **(Alg1, 18.2)** • wrote coordinate proofs about triangle relationships. **(9.4 and 9.5)**	**Students:** • understand that a circle is the set of all points on the coordinate plane that are a fixed distance from the center. • derive the equation of a circle using the Pythagorean Theorem. • given a center and radius use the formula to write an equation of a circle. • given an equation of a circle, find its center and radius. • prove or disprove that a point lies on a given circle.	**Students:** • will identify and describe relationships among inscribed angles, radii, and chords to include the relationship between central, inscribed, and circumscribed angles. **(16.1 and 16.2)** • will identify inscribed angles on a diameter as right angles. **(16.1 and 16.2)** • will identify the radius of a circle as perpendicular to the tangent where the radius intersects the circle. **(16.1 and 16.2)**

PROFESSIONAL LEARNING

Math Background

Many students look at mathematics with a silo approach, clearly delineating their algebra and geometry skills. Writing an equation of a circle calls for students to integrate these skills. The algebra needed to complete the square to rewrite an equation of a circle so its center and radius can be easily identified or to prove or disprove that a point lies on a given circle are examples of the blending of these branches of mathematics. The skills that students use to write equations of circles will be transferred to their later learning of writing equations for other conics such as parabolas, ellipses, and hyperbolas. Therefore, it is important that students begin to appreciate the connections between algebra and geometry and see how they often overlap.

WARM-UP OPTIONS

ONLINE

PROJECTABLE & PRINTABLE 

ACTIVATE PRIOR KNOWLEDGE • Applied the Pythagorean Theorem

Use these activities to quickly assess and activate prior knowledge as needed.

Problem of the Day 

The hypotenuse of a right triangle measures 50 feet, and one of the legs measures 14 feet. What is the measure of the other leg?

$\text{leg} = \sqrt{(\text{hyp})^2 - (\text{leg})^2}; = \sqrt{(50)^2 - (14)^2};$

$= \sqrt{2500 - 196} = \sqrt{2304} = 48;$ 48 feet

Quick Check for Homework

As part of your daily routine, you may want to display the Teacher Solution Key to have students check their homework.

Make Connections

Based on students' responses to the Problem of the Day, choose one of the following:

 Project the Interactive Reteach, Grade 8, Lesson 11.3.

 Complete the Prerequisite Skills Activity:

Have students work in pairs. Each student writes down a different positive real number. Students then use each number as the length of the legs of a right triangle and calculate its hypotenuse, and then use the larger number as the length of the hypotenuse and the smaller number as the length of the leg, and calculate the length of the other leg.

- *What is the Pythagorean Theorem?* $(\text{hyp})^2 = (\text{leg})^2 + (\text{leg})^2$
- *Why doesn't it matter in which order you add the squares of the legs?* Addition is commutative. $a + b = b + a$
- *Why must you use the larger of the two numbers as the hypotenuse?* The hypotenuse is the longest side of a right triangle.
- *How can you solve the Pythagorean Theorem for the length of a leg?* $\text{leg} = \sqrt{(\text{hyp})^2 - (\text{leg})^2}$

If students continue to struggle, use Tier 2 Skill 4.

SHARPEN SKILLS

If time permits, use this on-level activity to build fluency and practice basic skills.

Quantitative Comparison

Students make a comparison between two quantities.

Write the following problem on the board. Ask students to choose the letter representing the correct answer and to explain their reasoning.

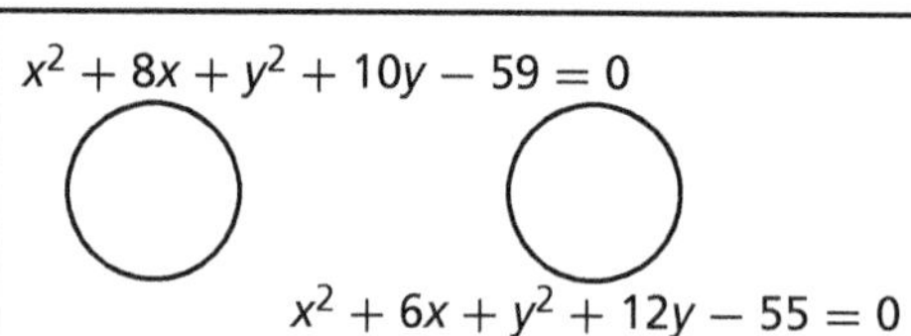

Quantity A	**Quantity B**
the radius of circle given by the equation $x^2 + 8x + y^2 + 10y - 59 = 0$	the radius of the circle given by the equation $x^2 + 6x + y^2 + 12y - 55 = 0$

A. Quantity A is greater.

B. Quantity B is greater. C, The radius of each circle is 10.

C. The two quantities are equal. $(x + 4)^2 + (y + 5)^2 = 10^2; (x + 3)^2 + (y + 6)^2 = 10^2$

D. The relationship cannot be determined from the information given.

PLAN FOR DIFFERENTIATED INSTRUCTION

Small-Group Options

Use these teacher-guided activities with pulled small groups.

On Track

Materials: index cards

Give each pair of students a card with an ordered pair, a positive number, and a second ordered pair such as

$(3, 4)$, 6 and $(9,12)$ or

$(-5, -2)$, 12 and $(2, 4)$.

Have students pair up and write the equation of the circle with the first ordered pair as the center of the circle and the positive number as the radius of the circle. Then have students prove or disprove if the second ordered pair lies on the circle.

$(3, 4) = (h, k)$; $6 = r$
$(9, 12)$ lies on circle?

$(-5,-2) = (h, k)$; $12 = r$
$(2, 4)$ lies on circle?

Almost There

Materials: graph paper

Write the equation of a circle, $(x-h)^2 + (y-k)^2 = r^2$,

the ordered pair $(2, 5)$, and $r = 6$ on the board.

- Have students label the ordered pair (h, k).
- Check that students have plugged in correct values into the formulas.
- Have students sketch the circle on graph paper.

Ready for More

What are the center and radius of the circle whose equation is

$x^2 - 12x + y^2 - 8y + 3 = 0$?

- The point $(-1, y)$ lies on the circle. What is the value of y?
- What are the coordinates of three other points that lie on the circle?

$x^2 - 12x + y^2 - 8y + 3 = 0$

Math Center Options

Use these student self-directed activities at centers or stations. Key: ● Print Resources ● Online Resources

On Track

- ● Interactive Digital Lesson
- ●● Journal and Practice Workbook
- ● Module Performance Task

Standards Practice:

- ●● Find and Use the Equation of a Circle
- ●● Use Coordinates to Prove Simple Geometric Theorems Algebraically

Almost There

- ● Reteach 15.4 (printable)
- ● Interactive Reteach 15.4
- ● RtI Tier 2 Skill 4: The Pythagorean Theorem and its Converse
- ● Illustrative Mathematics: Explaining the equation for a circle

Ready for More

- ● Challenge 15.4 (printable)
- ● Interactive Challenge 15.4
- ● Illustrative Mathematics: Slopes and Circles
- ● Desmos: Circle Patterns

ONLINE 

View data-driven grouping recommendations and assign differentiation resources.

During the *Spark Your Learning,* listen and watch for strategies students use. See samples of student work on this page.

Use the Diameter to Find the Radius

Strategy 1

I know that the radius is half the diameter, so because the diameter is 120 feet, the radius is 60 feet. I can use the equation $x^2 + y^2 = r^2 = 60^2 = 3600$ to algebraically describe the red path with respect to the center of the island.

If students . . . use the diameter to find the radius to solve the problem, they are using properties of circles from Grade 7.

Have these students . . . explain how they determined their equation and how they solved it. **Ask:**

- Q How did you use the diameter to find the radius to solve the Spark Your Learning problem?
- Q How does the equation $x^2 + y^2 = r^2 = 60^2 = 3600$ describe the red path with respect to the center of the island?

Use the Pythagorean Theorem

Strategy 2

I know the diameter is 120 feet, so I know the radius is 60 feet. I drew a radius and let it be the hypotenuse of a right triangle and I observed the legs measured x and y. I discovered that it didn't matter which radius I drew—the legs always measured x and y. So, I figured out that $x^2 + y^2 = r^2 = 60^2 = 3600$

If students . . . use the Pythagorean Theorem to solve the problem, they understand how to draw auxiliary lines to create right triangles and use relationships created by those right triangles, but are using a less efficient method to solve the problem.

Activate prior knowledge . . . by having students state the Pythagorean Theorem. **Ask:**

- Q How can you use the Distance Formula to determine the length of the legs of the right triangle?
- Q Why are the lengths of the legs of the right triangle always x and y, regardless of which radius is used to represent the hypotenuse ?

COMMON ERROR: Calculates the Radius Incorrectly

I know that the radius is twice the diameter, so the radius is 240 feet. So I can use the equation $x^2 + y^2 = r^2 = 240^2 = 57,600$.

If students . . . calculate the radius incorrectly, then they don't fully understand the definitions of radius and diameter.

Then intervene . . . by pointing out that they should draw a circle with a diameter of 120 feet. **Ask:**

- Q How do you draw the radius of the circle?
- Q What can you conclude if you draw radii that cover the diameter you've drawn?

15.4

Circles on the Coordinate Plane

I Can derive and write the equation of a circle with radius r and center (h, k).

Spark Your Learning

A traffic circle is an intersection where traffic moves in one direction around a central circular island.

©Darjus Urbanovic/Shutterstock

Complete Part A as a whole class. Then complete Parts B–D in small groups.

A. What is a mathematical question you can ask about this situation? What information would you need to know to answer your question? A. How can I algebraically describe the road with respect to the center of the island?; the radius of the traffic circle

B. What variable(s) are involved in this situation? What unit of measurement would you use for each variable? See Additional Answers.

C. To answer your question, what strategy and tool would you use along with all the information you have? What answer do you get? See Strategies 1 and 2 on the facing page.

D. Does your answer make sense in the context of the situation? How do you know? yes; A traffic circle is a circle, so an equation of a circle would describe the path around the traffic circle algebraically.

Turn and Talk A second traffic circle is constructed at a different location. The diameter of this traffic circle is 250 feet. Compare an equation that would describe this traffic circle to the equation that describes the original circle. See margin.

CULTIVATE CONVERSATION • Three Reads

After each read students answer a different question.

1. What is the situation about? The question calls for algebraically describing the red path of the traffic circle with respect to the center of the island.
2. What are the quantities in this situation? How are those quantities related? The quantities are x, the position to the east or west of the center, y, the position to the north or south of the center, and r, the radius of the traffic circle. They are related by the equation $x^2 + y^2 = r^2$.
3. What are possible questions you could ask about this situation?
   How does knowing the distance around the traffic circle, or circumference, provide enough information to determine its radius?

   What are the coordinates of any one point on a traffic circle with a radius of 60 feet?

   Would knowing the area of the traffic circle provide enough information to determine its radius?

1 Spark Your Learning

▶ MOTIVATE

- Have students look at the photo in their books and read the information contained in the photo. Then complete Part A as a whole-class discussion.
- Give the class the additional information they need to solve the problem. This information is available online as a printable and projectable page in the Teacher Resources.
- Have students work in small groups to complete Parts B–D.

▶ PERSEVERE

If students need support, guide them by asking:

Q **Advancing • Use Tools** Which tool could you use to solve the problem? Why choose that tool and not some other? Students' choices of tools and reasons for choosing them will vary.

Q **Assessing** Why is the position north or south of the center of the circle represented by the variable y? The position north and south of the center of the circle would represent points on the y-axis.

Q **Assessing** How are the radius and the diameter of a circle related? The radius equals half the diameter.

Q **Advancing** Assuming the center of a circle corresponds to the origin of the coordinate plane, how is the Pythagorean Theorem used to develop the equation that describes a circle? Let (x, y) be any point on the circumference of the circle. The radius, r, is the distance from the origin to (x, y). Draw a right triangle whose hypotenuse is the radius of the circle, and whose legs are, by virtue of (x, y), x units long and y units high. By the Pythagorean Theorem, $leg^2 + leg^2 = hyp^2$ or $x^2 + y^2 = r^2$.

Turn and Talk Ask students which of the variables in the equation $x^2 + y^2 = r^2$ vary based on the position of the car traveling around the traffic circle and which variable remains constant. The side of the equations with the variables would remain the same. The side of the equations with the square of the radius of the circle would change from 3,600 to 15,625.

▶ BUILD SHARED UNDERSTANDING

Select groups of students who used various strategies and tools to share how they solved the problem. As they present their solutions, have each group discuss why they chose a specific strategy and tool.

② Learn Together

Build Understanding

Task 1 (MP) **Use Structure** Students use structure by recognizing the significance of the radius of the circle, using the strategy of drawing auxiliary lines in the figure. The proof hinges on the right triangle formed by drawing a horizontal line through C and a vertical line through P.

Sample Guided Discussion:

Q How can P be selected so that it is impossible to draw triangle CPA? If P is selected either to the extreme left or right of C or directly above or below C, it is impossible to draw triangle CPA.

Turn and Talk Ask students to write down the general equation for a circle with center $C(h, k)$ and radius r. Ask them to define the variables in their equation. Since the circle in this question has its center C at the origin, what do they know about the values of h and k for this circle? Finally, have them substitute those values into the general equation and simplify.
$x^2 + y^2 = r^2$

Build Understanding

Derive the Equation of a Circle

In the previous lessons, you have worked with circles. In this lesson, you will investigate circles on the coordinate plane and learn how to write an equation of a circle. Recall that a circle is the set of all points on the coordinate plane that are a fixed distance from the center (h, k).

1 Consider the circle on the coordinate plane that has its center at $C(h, k)$ and a radius r. Let P be any point on the circle with coordinates (x, y).

Draw a horizontal line through C and a vertical line through P and label their intersection A, as shown.

A–F. See Additional Answers.

A. If P can be any point on the circle, what does that mean about the maximum and minimum distances between C and A?

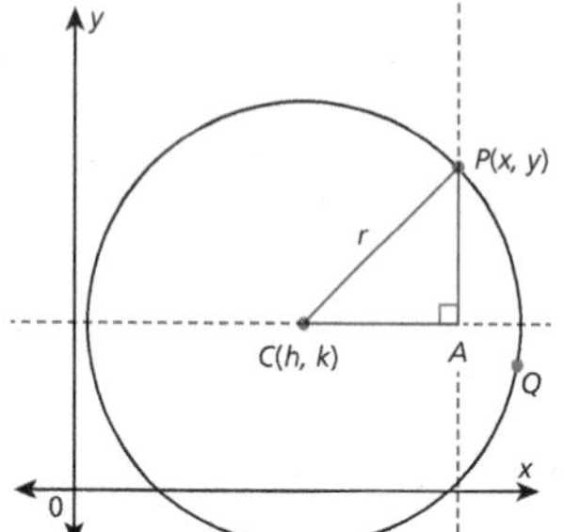

B. What are the coordinates of point A? Explain how you found them.

Right triangle $\triangle PCA$ has a hypotenuse of length r.

Its leg $\overline{CA}$ has length $|x - h|$.

Its leg $\overline{PA}$ has length $|y - k|$.

C. Why are absolute values needed to express the lengths of $\overline{CA}$ and $\overline{PA}$?

Now apply the Pythagorean Theorem to write a relationship between the side lengths of $\triangle PCA$ and its hypotenuse.

$(x - h)^2 + (y - k)^2 = r^2$

D. Why are absolute values now no longer needed for the lengths of $\overline{CA}$ and $\overline{PA}$?

This is the equation for a circle with center $C(h, k)$ and radius r.

Take the square root of both sides of this equation.

$r = \sqrt{(x - h)^2 + (y - k)^2}$

E. Are these relationships true for values of x that are less than h, and for values of y that are less than k? Explain your reasoning.

This distance formula expresses the distance r between the center of the circle $C(h, k)$ and a point on its circumference $P(x, y)$.

Q is another point on the circle. Through it, draw a vertical line. Where that intersects the horizontal line through C, label the point of intersection B.

F. Compare and contrast this new $\triangle QCB$ with $\triangle PCA$. What will always be true about any triangle constructed in this way, provided that the point chosen on the circle does not lie directly right, left, above, or below the circle's center?

Turn and Talk Suppose a circle has its center C at the origin. What is the equation of the circle in this case? See margin.

LEVELED QUESTIONS

Depth of Knowledge (DOK)	Leveled Questions	What Does This Tell You?
Level 1 **Recall**	What is the set of all points on the coordinate plane that are a fixed distance from the center (h, k)? a circle	Students' answers will indicate whether they understand the definition of a circle.
Level 2 **Basic Application of Skills & Concepts**	What are the values of h and k if the center of the circle is at the origin? Both values are 0.	Students' answers will indicate whether they understand the origin of the coordinate plane.
Level 3 **Strategic Thinking & Complex Reasoning**	How can you use the Distance Formula to derive the equation of a circle? Let $C(h, k)$ be the center and $P(h, k)$ be any point on the circle. The distance between C and P is r. $D = \sqrt{(x_2 - h_1)^2 + (y_2 - k_1)^2}$; $D = \sqrt{(x - h)^2 + (y - k)^2}$; $D^2 = r^2 = (x - h)^2 + (y - k)^2$	Students' answers will indicate whether they can derive the equation of a circle using an alternative proof.

Step It Out

Write an Equation of a Circle

You can write the equation of any circle on a coordinate plane if you know its radius and the coordinates of its center.

Equation of a Circle
The equation of a circle with center (h, k) and radius r is $(x - h)^2 + (y - k)^2 = r^2$.

Fairy rings are growths that naturally take the shape of a circle. They begin when a fungal spore lands in a spot and begins to grow underground evenly in all directions, creating a circular organism. Mushroom caps then develop at the edges of the organism.

The center of a fairy ring is located in a field 7 feet east and 5 feet north from a stone trail marker. The ring has a radius of 3 feet.

Write an equation that represents the fairy ring on a coordinate plane with respect to the stone trail marker.

Identify the center of the circle and the radius.

The center of the circle is (7, 5), and the radius is 3 feet, as shown in the graph.

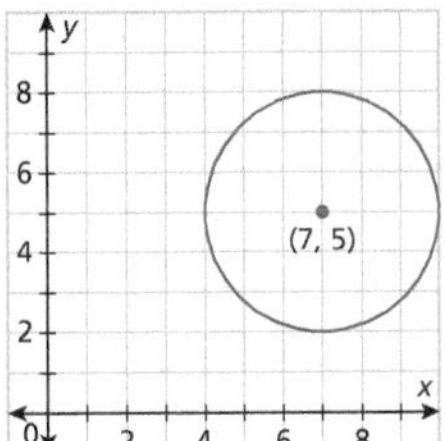

A. Explain how the center of the circle was determined.

A. The center of the circle is the distance east and north of the stone trail marker.

B. What does the origin of the graph represent?

B. The origin represents the location of the stone trail marker.

Write an equation for the circle.

Substitute the values for the center of the circle and radius into $(x - h)^2 + (y - k)^2 = r^2$.

$$(x - 7)^2 + (y - 5)^2 = 3^2 = 9$$

C. Another fairy ring is the same size but is located 4 feet west and 5 feet south of the stone trail marker. What is an equation that represents this fairy ring? $(x + 4)^2 + (y + 5)^2 = 9$

Turn and Talk What does the transformation $(x, y) \rightarrow (x - h, y - k)$ represent in this context? See margin.

PROFICIENCY LEVEL

Beginning

Show students the equation $(x - 12)^2 + (y + 3)^2 = 25$ and say, "The graph of this equation is a circle with center $(12, -3)$ and radius 5." Then, show them the equation $(x + 5)^2 + (y - 1)^2 = 36$, ask what kind of shape the graph has, and ask for its center and radius. Encourage students to use the terms *center* and *radius* when they respond.

Intermediate

Have students work in pairs to compare the given equation and the equation of a circle and explain what is wrong with this student's reasoning about the graph of $(x - 2)^2 + (y + 3)^2 = 16$ by pointing out the mistakes in the sentence concerning the center and radius of the circle.

"The graph is a circle with center at $(-2, 3)$ and radius 16."

Advanced

Have students give a general explanation of how to determine the center and radius of a circle from its equation and how to write the equation of a circle when given its center and radius.

Step It Out

Task 2 (MP) **Attend to Precision** Students learn to use concrete references such as a drawing to write the equation of a circle. Remind students that the drawing helps them visualize and contextualize the word problem.

Sample Guided Discussion:

Q How do you know the point (4, 5) lies on the circle? (4, 5) satisfied the equation of the circle with center (7, 5) and radius 3.

$(h, k) = (7, 5)$; $(x, y) = (4, 5)$; $r = 3$

$(x - h)^2 + (y - k)^2 = r^2$;

$(4 - 7)^2 + (5 - 5)^2 = (-3)^2 + (0)^2 = 9 = 3^2 = r^2$

Q In Part C, suppose the fairy ring is located 5 feet north of the stone trail marker instead of south. What is the equation of the circle that this new fairy ring represents? $(x + 4)^2 + (y - 5)^2 = 9$

Turn and Talk Encourage students to sketch a circle with center (h, k) and perform the transformation so they can see what is happening. It represents translating the center of the circle with center (h, k) to the origin.

Task 3 (MP) **Use Structure** Students will see that a rather complicated mathematical equation of a circle $x^2 + ax + y^2 + by = c$ can be rewritten in a form so that the center and radius of the circle are easily discerned. By completing the square to rewrite the equation, they are recognizing that there are multiple ways to express the same concept.

Sample Guided Discussion:

Q What is the purpose of putting empty spaces on both sides of the equation? The empty spaces on the left are used when you take $\frac{1}{2}$ of the coefficients of the x and y terms, square those amounts, and add them to complete the square. The empty space on the right is used to remind you to add the same amount so that the equation remains balanced.

Q In Part A, how can you use verbalization to confirm what the final equation of the circle will be? To complete the x term, ask yourself what is $\frac{1}{2}$ of negative 4. The answer is negative 2, so the completed square should look like x minus 2. To complete the y term, ask yourself what is $\frac{1}{2}$ of positive 12. The answer is positive 6, so the completed square should look like y plus 6.

Turn and Talk With this new equation, we will be adding to a positive number on the right side of the equation. Ask students to consider what will be the final result of adding to a positive number. Possible answer: I predict that $x^2 - 4x + y^2 + 12y = 24$ will also have its center at $(2, -6)$ but have a larger radius.

ANSWERS

A. The terms $\left(\frac{-4}{2}\right)^2$ and $\left(\frac{12}{2}\right)^2$ were added to the right side of the equation in order to balance the equation after completing the squares on the left side.

B. The equation was rewritten so that $+6$ would be in the form of $-k$ in $(x - h)^2 + (y - k)^2 = r^2$.

C. Expand $(x - 2)^2 + (y + 6)^2 = 16$ and rewrite it as an equivalent equation with the same form as $x^2 - 4x + y^2 + 12y = -24$. If the equations are the same, then the two equations are equivalent.

D. The coefficients a and b in the first equation can be used to find the center of the circle because $h = -\frac{a}{2}$ and $k = -\frac{b}{2}$. The value of the constant c affects the radius, but does not affect the coordinates of the center of the circle.

Find the Center and Radius of a Circle

The equation of a circle can also take the form $x^2 + ax + y^2 + by = c$. You can rewrite such an equation so that is in the form $(x - h)^2 + (y - k)^2 = r^2$ by completing the square. Then you can identify the center and radius of the circle.

3 Find the center and radius of the circle with equation $x^2 - 4x + y^2 + 12y = -24$. Then graph the circle.

Rewrite the equation in the form $(x - h)^2 + (y - k)^2 = r^2$ by completing the square twice.

$x^2 - 4x + y^2 + 12y = -24$	Original equation
$x^2 - 4x + (\ )^2 + y^2 + 12y + (\ )^2 = -24 + (\ )^2 + (\ )^2$	Add spaces.
$x^2 - 4x + \left(\frac{-4}{2}\right)^2 + y^2 + 12y + \left(\frac{12}{2}\right)^2 = -24 + \left(\frac{-4}{2}\right)^2 + \left(\frac{12}{2}\right)^2$	
$x^2 - 4x + 4 + y^2 + 12y + 36 = -24 + 4 + 36$	Simplify.
$(x - 2)^2 + (y + 6)^2 = 16$	Factor.

A. What expressions were added to both sides of the equation? Why?

A–D. See margin.

Identify *h*, *k*, and *r*.

$(x - 2)^2 + (y + 6)^2 = 16$

$(x - 2)^2 + [y - (-6)]^2 = 16$

So, $h = 2$, $k = -6$, and $r = 4$.

The center of the circle is $(2, -6)$, and the radius is 4.

B. What is the benefit of rewriting the second squared term in this equation?

C. How can you check that $(x - 2)^2 + (y + 6)^2 = 16$ is equivalent to $x^2 - 4x + y^2 + 12y = -24$?

Graph the circle.

Locate the center of the circle on the coordinate plane. Place the point of your compass at the center, open the compass to the radius, and then draw the circle.

D. It is much easier to visualize this graph after completing the squares of the original equation to rewrite it as the equation of a circle. But, either equation can give you some information. For the equation in the form $x^2 + ax + y^2 + by = c$, how do the coefficients a and b and the constant c relate to the coordinates (h, k) of the center of the circle?

Turn and Talk Change the sign of the constant on the right side of the original equation so that it becomes $x^2 - 4x + y^2 + 12y = 24$. Predict how this will affect the graph. Then graph it, and discuss your predictions and your results. See margin.

Write a Coordinate Proof

You can use a coordinate proof to show whether or not a given point lies on a given circle on the coordinate plane.

4 Does $(\sqrt{5}, 2)$ lie on a circle that is centered at the origin and contains the point $(0, -3)$?

Graph the circle.

Plot points at the origin and $(0, -3)$ to help you draw the circle.

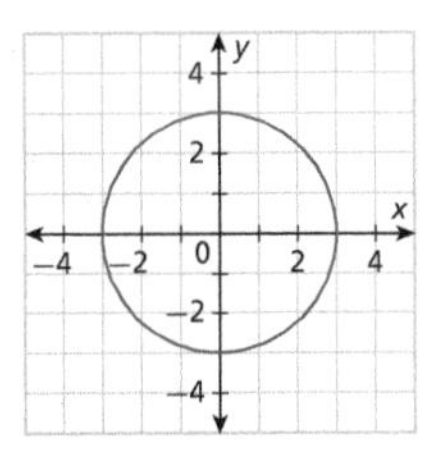

A. Estimate the location of $(\sqrt{5}, 2)$ on the graph. Can you tell that it lies on the circle by inspection?

A. Possible answer: I estimate that it is near the circle in Quadrant I, but by inspection I cannot be certain that it lies on the circle.

Determine the radius.

The radius of the circle is 3.

B. How do you know that the radius is 3?

B. You know that the circle is centered at the origin and contains the point $(0, -3)$. The distance from the origin to $(0, -3)$ is 3.

Write an equation for the circle.

Use the radius and the coordinates of the center to write the equation of the circle.

$(x - h)^2 + (y - k)^2 = r^2$

$(x - 0)^2 + (y - 0)^2 = 3^2$

$x^2 + y^2 = 9$

C. How do you determine the values for h and k?

C. The value of h is the x-coordinate of the center of the circle, and the value of k is the y-coordinate of the center of the circle.

Substitute the given point in the equation.

Use the equation of the circle to check if the point $(\sqrt{5}, 2)$ lies on the circle.

$x^2 + y^2 = 9$

$(\sqrt{5})^2 + (2)^2 \stackrel{?}{=} 9$

$5 + 4 \stackrel{?}{=} 9$

$9 \stackrel{?}{=} 9$

D. Does $(\sqrt{5}, 2)$ lie on the circle? Explain.

D. yes; The point lies on the circle because substituting the values of x and y makes the equation true.

Turn and Talk Find another point that lies on the circle. Explain how you know that the point lies on the circle. See margin.

Task 4  **Construct Arguments or Critique Reasoning** Students will use the previously derived equation of a circle to test whether a given point lies on the circle. They will use a concrete reference, a sketch of a circle, to visually determine the plausibility of the result and verify their conclusion by an algebraic check.

Sample Guided Discussion:

Q Why can you use the simpler equation $x^2 + y^2 = r^2$ for this problem? You can use the simpler equation because the center of the circle is at the origin, and (h, k) is $(0, 0)$.

Q In Part B, how can you determine the radius if you were given that the circle was centered at the origin and contained the point $(\sqrt{7}, \sqrt{2})$? The distance between (0, 0) and the point is the radius of the circle, so you would use the Distance Formula.

Turn and Talk Encourage students to look at the sketch of the circle to determine points whose coordinates are precisely known. Sample answer: $(\sqrt{5}, -2)$; $(\sqrt{5})^2 + (-2)^2 = 5 + 4 = 9$. The points (0, 3) and $(0, -3)$ are also points that would be logical to check.

ONLINE

Assign the Digital On Your Own for

- built-in student supports
- Actionable Item Reports
- Standards Analysis Reports

On Your Own

Assignment Guide

The chart below indicates which problems in the On Your Own are associated with each task in the Learn Together. Assign daily homework for tasks completed.

Learn Together Tasks	On Your Own Problems
Task 1, p. 492	Problem 8
Task 2, p. 493	Problems 9–15, 18, 20–24, and 34
Task 3, p. 494	Problems 16, 17, and 25–31A
Task 4, p. 495	Problems 19, and 31B–33

ANSWERS

8. You can draw a right triangle with the end points of the hypotenuse being the given points. You can find the length of the legs of the right triangle by subtracting the *x*- and *y*-coordinates of the given points. You can then use the Pythagorean Theorem to find the length of the hypotenuse. This is the radius of the circle. Then you can substitute the radius and the coordinates of the center into the equation of a circle. The equation of this circle is $(x-2)^2+(y+3)^2=25$.

Check Understanding

1. What are the coordinates of the center of the circle represented by the equation $x^2+y^2=25$? $(0, 0)$

Write the equation of the circle with the given center and radius.

2. center: $(-2, 7)$; radius: 2 $(x+2)^2+(y-7)^2=4$
3. center: $(-1, -4)$; radius: 8 $(x+1)^2+(y+4)^2=64$

Find the center and radius of the circle with the given equation.

4. $x^2-14x+y^2-2y+41=0$ center: $(7, 1)$; radius: 3
5. $x^2-8x+y^2+10y=59$ center: $(4, -5)$; radius: 10
6. Prove or disprove that the point $(7, \sqrt{29})$ lies on the circle that is centered at the origin and passes through the point $(0, -6)$. $(7, \sqrt{29})$ does not lie on the circle. $7^2+(\sqrt{29})^2=49+29=78\neq 6^2$
7. Prove or disprove that the point $(\sqrt{21}, 2)$ lies on the circle that is centered at the origin and passes through the point $(5, 0)$. $(\sqrt{21}, 2)$ does lie on the circle. $(\sqrt{21})^2+2^2=21+4=25=5^2$

On Your Own

8. 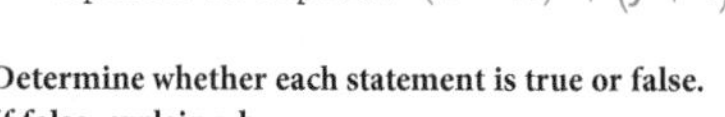 Reason Describe how to use the Pythagorean Theorem to write the equation for a circle that is centered at $(2, -3)$ and passes through $(5, 1)$. What is the equation of the circle? See margin.

Write the equation of the circle with the given center and radius.

9. center: $(5, -1)$; radius: 1 $(x-5)^2+(y+1)^2=1$
10. center: $(6, 4)$; radius: 4 $(x-6)^2+(y-4)^2=16$
11. center: $(-2, -7)$; radius: $\sqrt{11}$ $(x+2)^2+(y+7)^2=11$
12. center: $(-3, 5)$; radius: $\sqrt{29}$ $(x+3)^2+(y-5)^2=29$
13. center: $(0, -4)$; radius: 9 $x^2+(y+4)^2=81$
14. center: $(7, 0)$; radius: $\sqrt{15}$ $(x-7)^2+y^2=15$
15. Chloe overlaid a coordinate plane on a photograph of a crop circle. The location of the center of the crop circle is at $(11, -4)$, and the radius of the circle is 3 units. Write an equation for a circle that represents the crop circle. $(x-11)^2+(y+4)^2=9$

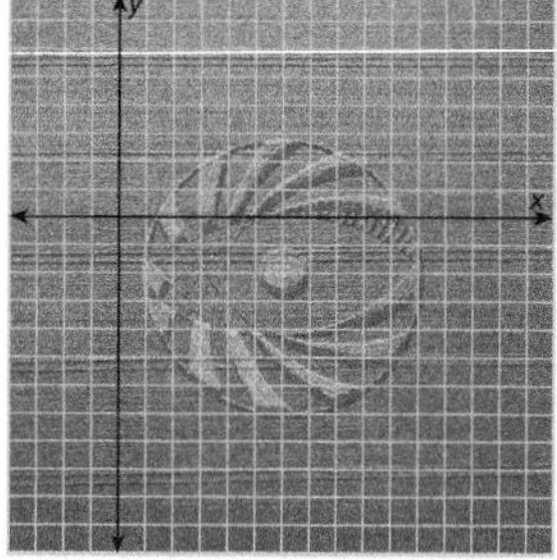

Determine whether each statement is true or false. If false, explain why.

16. The circle $(x-4)^2+(y+1)^2=17$ has a radius of 17. false; The radius is $\sqrt{17}$.
17. The center of the circle $(x+5)^2+(y-2)^2=25$ lies in the second quadrant. true
18. The equation of a circle centered at $(-3, 7)$ with a diameter of 8 is $(x+3)^2+(y-7)^2=64$. false; If the diameter is 8, the radius is 4. The equation is $(x+3)^2+(y-7)^2=16$.
19. The circle $(x+1)^2+(y-8)^2=25$ passes through the point $(-1, 3)$. true

496

3 Check Understanding

Formative Assessment

Use formative assessment to determine if your students are successful with this lesson's learning objective.

Students who successfully complete the Check Understanding can continue to the On Your Own practice.

For students who miss 1 problem or more, work in a pulled small group using the Almost There small-group activity on page 485C.

ONLINE

Assign the Digital Check Understanding to determine

- success with the learning objective
- items to review
- grouping and differentiation resources

4 Differentiation Options

Differentiate instruction for all students using small-group activities and math center activities on page 485C.

Reteach

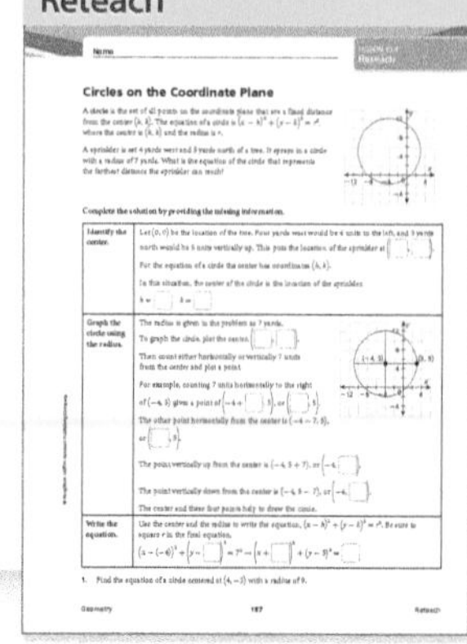

Challenge

20. **Open Ended** Write an equation for your own circle. Identify the center and radius of your circle. Answers will vary.

Write the equation of each circle.

21. 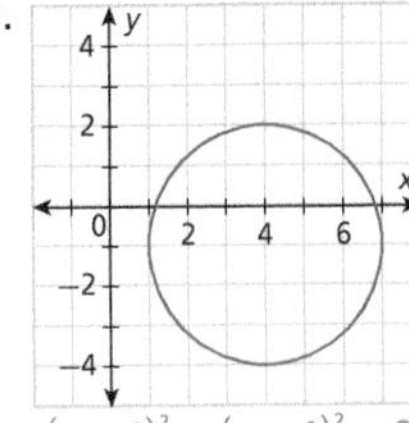

$(x-4)^2+(y+1)^2=9$

22. 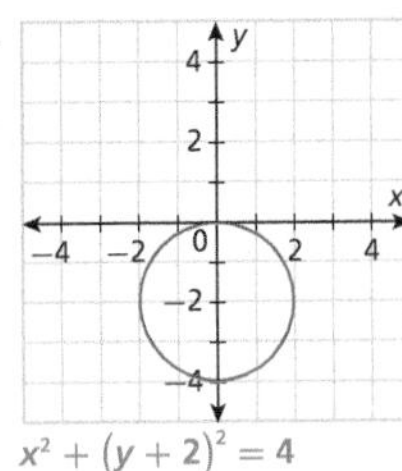

$x^2+(y+2)^2=4$

23. 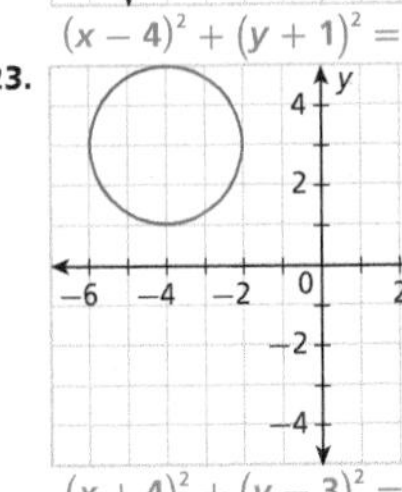

$(x+4)^2+(y-3)^2=4$

24. $(x+1)^2+(y+1)^2=16$

Find the center and radius of the circle with the given equation.

25. $x^2+6x+y^2-18y+86=0$
center: $(-3, 9)$; radius: 2

26. $x^2-14x+y^2-2y=-25$
center: $(7, 1)$; radius: 5

27. $x^2+8x+y^2+10y+8=0$
center: $(-4, -5)$; radius: $\sqrt{33}$

28. $x^2-2x+y^2+2y-20=0$
center: $(1, -1)$; radius: $\sqrt{22}$

29. $x^2+y^2+10y=-18$
center: $(0, -5)$; radius: $\sqrt{7}$

30. $x^2-12x+y^2=5$
center: $(6, 0)$; radius: $\sqrt{41}$

31. **STEM** An engineer is designing a Ferris wheel that has a diameter of 100 feet.

A. If the center of the Ferris wheel is at the origin of a coordinate plane, what is the equation of the Ferris wheel? $x^2+y^2=2500$

B. Is it possible for one of the cars of the Ferris wheel to be attached to the wheel at the point $(14, 48)$? Explain your reasoning.
yes; $14^2+48^2=196+2304=2500=50^2$

Watch for Common Errors

Problem 25 Students are anxious to start completing the square and sometimes forget the first crucial step of moving any constants that appear on the left-hand side of the equation to the right-hand side. Remind students that the left-hand side must contain only terms with variables before they can begin the process for completing the square.

Questioning Strategies

Problem 31 Suppose the diameter of the Ferris wheel is doubled. If the coordinates of the point (14, 48) are also doubled, is it possible for one of the Ferris wheel cars to be attached at that point? Explain. yes; The doubled diameter is 200, so the radius is 100. The doubled coordinates are (28, 96).

$28^2+96^2=784+9{,}216=10{,}000=100^2$

5 Wrap-Up

Summarize learning with your class. Consider using the Exit Ticket, Put It in Writing, or I Can scale.

Exit Ticket 

Find the center and radius of the circle given by the equation $x^2 + 8x + y^2 - 12y = 29$. center $(-4, 6)$, radius $= 9$

$x^2 + 8x + (\ ) + y^2 - 12y + (\ ) = 29 + (\ ) + (\ );$

$x^2 + 8x + 16 + y^2 - 12y + 36 = 29 + 16 + 36;$

$(x + 4)^2 + (y - 6)^2 = 81 = 9^2$

$(x - (-4))^2 + (y - 6)^2 = 9^2$

Put It in Writing 

A house, located at (0, 0) has a nearby circular garden. You are to write and solve a word problem that gives the location of the garden in terms of direction (N/S/E/W) from the house, and the radius of the garden (don't forget units) and asks for the equation of the circle that represents the garden.

I Can 

The scale below can help you and your students understand their progress on a learning goal.

4	I can derive and write the equation of a circle with radius r and center (h, k), and I can explain my steps to others.
3	I can derive and write the equation of a circle with radius r and center (h, k).
2	I can determine if a point lies on a circle when given the equation of the circle in the form $(x - h)^2 + (y - k)^2 = r^2$.
1	I can identify the center and radius of a circle given an equation in the form $(x - h)^2 + (y - k)^2 = r^2$.

Spiral Review • Assessment Readiness

These questions will help determine if students have retained information taught in the past and can also prepare them for high-stakes assessments. Here, students apply the inscribed quadrilateral theorem **(15.2)**, apply the two-tangent theorem **(15.3)**, and classify arcs as major, minor or semicircle **(15.1)**.

32. (MP) **Critique Reasoning** Eve says that the point $(\sqrt{3}, \sqrt{7})$ lies on a circle that is centered at the origin and passes through the point $(-4, 0)$. Brandy says that the point does not lie on the circle. Who is correct? Explain your reasoning. See Additional Answers.

33. Prove or disprove that the point $(-3\sqrt{3}, 3)$ lies on the circle that is centered at the origin and passes through the point $(0, -6)$. $(-3\sqrt{3}, 3)$ does lie on the circle. $(-3\sqrt{3})^2 + 3^2 = 27 + 9 = 36 = 6^2$

34. (Open Middle™) Using the digits 1 to 9, at most one time each, fill in the boxes to create the equation of a circle and a point on the circle. See Additional Answers.

$(x - \square)^2 + (y - \square)^2 = \square^2$ with a point on the circle $(\square, \square)$

Spiral Review • Assessment Readiness

35. Find m∠B.

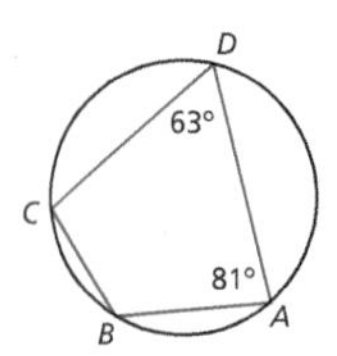

Ⓐ 63°
Ⓑ 81°
Ⓒ 99°
Ⓓ 117°

36. Find XZ.

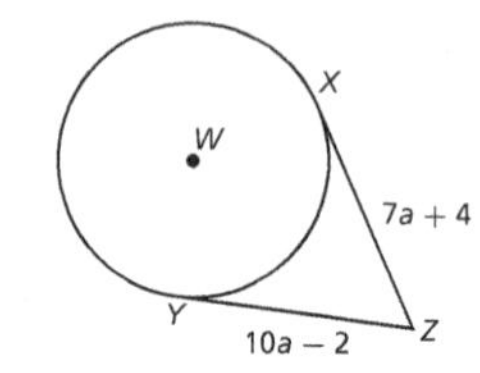

Ⓐ 2
Ⓑ 14
Ⓒ 18
Ⓓ 20

37. Classify each arc of circle C as a major arc, a minor arc, or a semicircle.

Circle part	Major arc	Minor arc	Semicircle
A. $\overset{\frown}{ADB}$	?	?	**?**
B. $\overset{\frown}{AE}$	?	**?**	?
C. $\overset{\frown}{ABE}$	**?**	?	?
D. $\overset{\frown}{AD}$	?	**?**	?

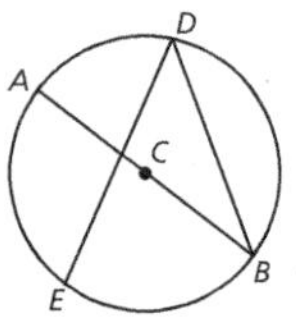

I'm in a Learning Mindset!

What strategy did you use to overcome barriers to understanding a circle on a coordinate plane?

498

Keep Going ▶ Journal and Practice Workbook

Learning Mindset

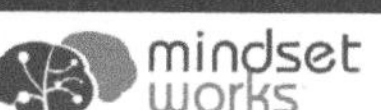

Resilience Identifies Obstacles

Students may need help recognizing barriers they need to overcome in order to be successful at writing and solving equations of circles. For example, one student may need to practice completing the square until that skill is mastered and can be handily applied to writing an equation of a circle in the form $(x - h)^2 + (y - k)^2 = r^2$. Another student may need to brush up on the algebra skills involved in checking if a point lies on a circle. Identifying and mastering these prerequisite skills will help students achieve success with their current learning. *Is there a skill I learned before that I need to practice so I can correctly apply it to what I'm learning now? Do I struggle solving problems at the same point in the process? For example, do I struggle with signs when writing the equation of the circle when the coordinates of the center are negative numbers? Can I figure out what is causing me to struggle? What can I do to improve my skills for that topic?*

Module

15 Review

Central Angles and Inscribed Angles

Two angles are inscribed in ⊙A so that they subtend the same arc, which measures 110°.

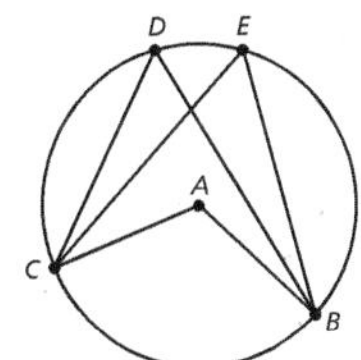

The central angle $\angle CAB$ has the same angle measure as the arc it subtends.

Each inscribed angle is equal to half the measure of the central angle it subtends.

$$m\angle CDB = m\angle CEB = \frac{m\angle CAB}{2} = \frac{110°}{2} = 55°$$

Inscribed Quadrilaterals

A quadrilateral, $WXYZ$, is inscribed in ⊙O.

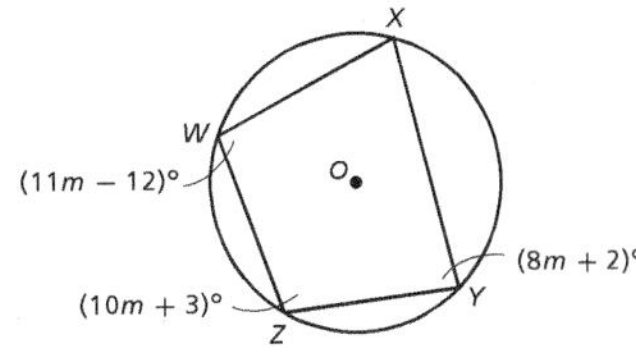

$\angle W$ and $\angle Y$ are supplementary by the Inscribed Quadrilateral Theorem.

$$m\angle W + m\angle Y = 180°$$
$$11m - 12 + 8m + 2 = 180$$
$$19m - 10 = 180$$
$$19m = 190$$
$$m = 10$$

$$m\angle W = (11(10) - 12)° = (110 - 12)° = 98°$$
$$m\angle Y = (8(10) + 2)° = (80 + 2)° = 82°$$
$$m\angle Z = (10(10) + 3)° = (100 + 3)° = 103°$$
$$m\angle X = 180° - 103° = 77°$$

Tangents

⊙O has a radius of 3. Tangent segments are drawn from A to ⊙O.

The two tangent segments are congruent.

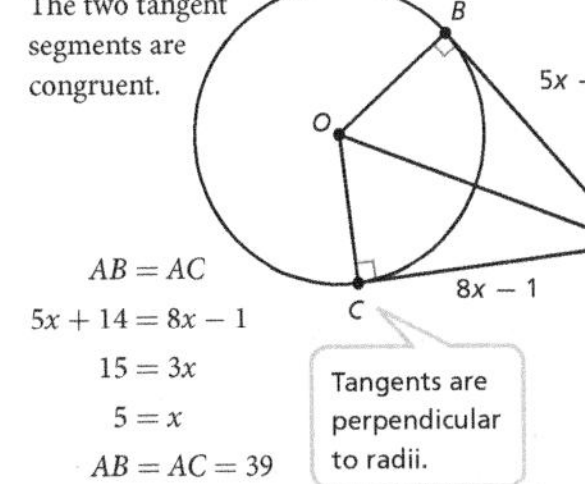

$$AB = AC$$
$$5x + 14 = 8x - 1$$
$$15 = 3x$$
$$5 = x$$
$$AB = AC = 39$$

Equation of a Circle

Given a point on the circle $B\ (x, y)$, the horizontal distance to the center is $(x - (-1)) = x + 1$ while the vertical distance is $(y - 2)$.

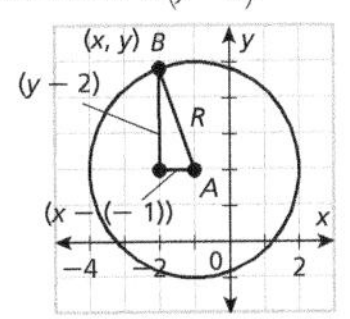

By the Pythagorean Theorem, the points on the circle must satisfy the equation $(x + 1)^2 + (y - 2)^2 = R^2 = 3^2 = 9$.

ONLINE

Assign the Digital Module Review for

- built-in student supports
- Actionable Item Reports
- Standards Analysis Reports

MODULE 15

REVIEW

Module Review

Use the first page of the Module Review to summarize and connect the main ideas of the module. Use the second page to assess students' understanding of the vocabulary, concepts, and skills presented in the module.

Sample Guided Discussion:

Q Central Angles and Inscribed Angles If diameter $\overline{CF}$ is drawn, what is $m\angle CFB$? Explain. $m\angle CFB = \frac{1}{2}m\angle CAB = 55°$; $\angle CFB$ is an inscribed angle and equal to one-half the measure of the central angle.

Q Inscribed Quadrilaterals Explain why quadrilateral *WXYZ* is not a trapezoid. Possible answer: For *WXYZ* to be a trapezoid, two sides must be parallel. This means that each pair of adjacent angles, like $\angle W$ and $\angle Z$ or $\angle W$ and $\angle X$, would have to be supplementary because they are pairs of interior angles on the same side of a transversal. But the sum of the measures of each pair of angles is not equal to 180°, so $\overline{WX}$ and $\overline{ZY}$ are not parallel, and $\overline{WZ}$ and $\overline{XY}$ are also not parallel, and quadrilateral *WXYZ* is not a trapezoid.

Q Tangent Explain how to find $m\angle BC$. Possible answer: In right triangle *ABO*, $\tan^{-1}(BAO) = \tan^{-1}\left(\frac{3}{39}\right) \approx 4.4°$. So, $m\angle BAO \approx 4.4°$. $\angle BAO \cong \angle CAO$ because they are corresponding parts of congruent triangles, and $m\angle BAC \approx 8.8°$. By the Circumscribed Angle Theorem, $m\angle BOC \approx (180 - 8.8)° \approx 171.2°$ The measures of a central angle and its arc are equal, so $m\angle BC \approx 171.2°$.

Q Equation of a circle What is the general form of the equation of the circle with center (−1, 2)? Expand the equation $(x + 1)^2 + (y - 2)^2 = 9 \Rightarrow x^2 + 2x + y^2 - 4y = 4$

Module Review continued

Possible Scoring Guide

Items	Points	Description
1–4	2 each	identifies the correct term(s)
5–7	2 each	correctly determines the measures of central and inscribed angles
8–13	2 each	correctly determines the lengths of tangents to the circle and the measures of circumscribed angles
14, 15	2 each	correctly determines the coordinates of the center of the circle and its radius from an equation of the circle
16	2	correctly solves the algebraic equation to find the measures of angles inscribed along the diameter
17	2	correctly applies the Inscribed Quadrilateral Theorem to find the value of *x* in expressions for measures of opposite angles in the quadrilateral
Total points possible = 34 points		

The Unit 8 Test in the Assessment Guide assesses content from Modules 15, 16, and 17.

Vocabulary

Choose the correct term from the box to complete each sentence.

Vocabulary
central angle
chords
circle
inscribed angle
radii
tangent

1. A(n) __?__ of a circle is a line that lies in the same plane as the circle and intersects the circle at exactly one point. tangent
2. The set of points in a plane that are a fixed distance from a given point is a(n) __?__. circle
3. An angle whose vertex is the center of a circle and whose sides contain __?__ of the circle is a(n) __?__. radii; central angle
4. An angle whose vertex is on a circle and whose sides contain __?__ of the circle is a(n) __?__. chords; inscribed angle

Concepts and Skills

5–13. See Additional Answers.

Given that $m\angle ABC = 38°$, determine each angle measure and justify your reasoning.

5. $m\angle ACB$ 6. $m\angle CAB$ 7. $m\angle CDB$

Given that $\odot E$ has a radius of 5 and that $\overline{GF}$ and $\overline{GH}$ are tangent to $\odot E$, determine each measure and justify your reasoning.

8. $m\angle EFG$ 9. FG 10. EG
11. $m\angle FEG$ 12. $m\angle HEF$ 13. $m\angle FGH$

17x + 3

21x − 5

Find the center and radius each circle.

14. $(x+3)^2 + (y-4)^2 = 36$. center: $(-3, 4)$, radius: 6
15. $x^2 + y^2 - 10y + 23 = 0$ center: $(0, 5)$, radius: $\sqrt{2}$
16. In $\odot S$, an angle is inscribed along a diameter. Given that $m\angle PQR = (2x - 2)$ and $m\angle QRP = (3x + 12)$, determine the measures of all three angles in $\triangle PQR$. See Additional Answers.

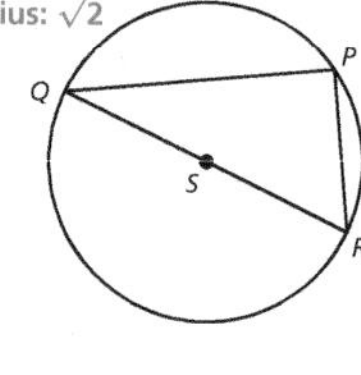

17. (MP) **Use Tools** Quadrilateral *WXYZ* is inscribed in a circle. Determine the value of *x*. State what strategy and tool you will use to answer the question, explain your choice, and then find the answer. $x = 27$

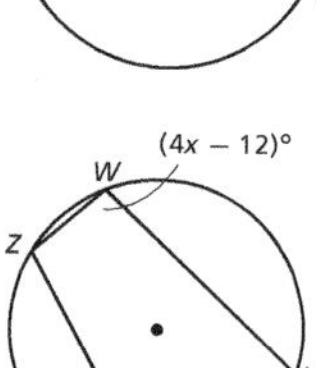

500

DATA-DRIVEN INSTRUCTION

Before moving on to the Module Test, use the Module Review results to intervene based on the table below.

MTSS (RtI)

Items	Lesson	DOK	Content Focus	Intervention
5–7	15.1	2	Determine the measures of central angles and inscribed angles.	Reteach 15.1
8–13	15.3	2	Determine the lengths and angle measures of segments and angles of tangents and circumscribed angles.	Reteach 15.3
14, 15	15.4	2	Determine the center and radius from the equation of a circle.	Reteach 15.4
16	15.1	2	Determine the measures of inscribed angles.	Reteach 15.1
17	15.2	2	Determine a value of a variable associated with angles inscribed in a quadrilateral.	Reteach 15.2

Module Test

data checkpoint

The Module Test is available in alternative versions in your Assessment Guide. All items are presented in standardized test formats.

ONLINE

Assign the Digital Module Test to power actionable reports including

- proficiency by standards
- item analysis

Form A

Name ____________________

Module 15 • Form A
Module Test

1. In circle C, $\overset{\frown}{AB}$ measures 62°. What is the measure of the inscribed angle that intercepts $\overset{\frown}{AB}$?

Ⓐ 31° Ⓑ 62° Ⓒ 118° Ⓓ 124°

2. In circle A, $m\angle CBD = 38°$.

What is the measure of $\overset{\frown}{CBD}$?

Ⓐ 38° Ⓑ 76° Ⓒ 284° Ⓓ 322°

3. Quadrilateral $ABCD$ is inscribed in a circle. If $m\angle A = 56°$, what is $m\angle C$?

Ⓐ 34° Ⓑ 56° Ⓒ 124° Ⓓ 304°

4. What is an equation of a circle with center $(-7, 4)$ and a radius of 12 units?

Ⓐ $(x-7)^2+(y+4)^2=12$
Ⓑ $(x+7)^2+(y-4)^2=12$
Ⓒ $(x-7)^2+(y+4)^2=144$
Ⓓ $(x+7)^2+(y-4)^2=144$

5. Circle P has center $(-12, -2)$ and passes through point $(-10, 2)$. Which point also lies on circle P?

Ⓐ $(-14, -6)$ Ⓑ $(-12, 0)$ Ⓒ $(10, -2)$ Ⓓ $(16, 4)$

6. The minute hand of a circular clock sweeps an arc as it moves from 1:22 p.m. to 2:05 p.m. in the same day. What is the measure of the arc in degrees?

258°

7. In circle C, the inscribed angle DEF measures $(4x+17)°$ and $\overset{\frown}{DF}$ measures $(9x+25)°$. What is $m\angle DCF$ in degrees?

106°

8. In circle O, $\overline{AB}$ and $\overline{CD}$ are chords.

If $m\overset{\frown}{AB} = (21x+7)°$ and $m\overset{\frown}{CD} = (15x+27)°$, what is the measure of $\overset{\frown}{AB}$ in degrees?

77°

9. Angle BAC is circumscribed about circle P. If $BA = \frac{2}{3}x$ m and $AC = x - 5$ m, what is the length of $\overline{BA}$ in meters?

10

10. Quadrilateral $ABCD$ is inscribed in a circle. If $m\angle B$ is four times $m\angle D$, what is $m\angle B$ in degrees?

144°

11. Regular octagon $ABCDEFGH$ is inscribed in a circle. What is the measure of $\overset{\frown}{CD}$ in degrees?

45°

12. What is an equation of the circle with center $(3, -10)$ and a radius of 11 units?

Possible answer: $(x-3)^2+(y+10)^2=121$

13. Consider the equation $x^2+16x+y^2+2y=-29$. What is the radius of the circle in units?

6

Geometry 121 Module 15 Test • Form A

Form A

Module 15 • Form A
Module Test

Name ____________________

14. A circle has the equation $x^2+6x+y^2-8y=0$.

Graph the circle.

15. In the figure, $\overline{QS}$ is tangent to circle T at point S.

Part A

If $m\angle Q = 27°$, what is the measure of $\overset{\frown}{RS}$ in degrees?

63°

Part B

If $QS = 8$ ft and $TR = 2$ ft, what is the length of $\overline{QT}$ in feet?

$2\sqrt{17}$

16. Circle O and circle M intersect at points A and B. $\overline{OP}$ is a diameter of circle M. If the radius of circle O is 7 in. and $AP = 24$ in., what is the diameter of circle M in inches?

25

17. Kite $ABCD$ is inscribed in circle O. Vertices B and D are endpoints of a diameter of the circle and $m\angle BDC = 27°$.

Part A

What is the measure of $\overset{\frown}{AB}$ in degrees?

54°

Part B

What is the measure of $\overset{\frown}{AD}$ in degrees?

126°

18. In the diagram, a ball rests on the floor with a board leaning against it. The lengths shown represent the distances from points of tangency to where the board meets the floor.

(14y + 8)°
(5x − 11.25) cm
(2y)°
(2x + 13.5) cm

Part A

What is the measure of the angle between the board and the floor?

Ⓐ 10.75° Ⓑ 11.75° Ⓒ 21.50° Ⓓ 158.50°

Part B

If the diameter of the ball is 11 cm, what is the distance from the center of the ball to where the board meets the floor, to the nearest tenth of a centimeter?

30.5

Geometry 122 Module 15 Test • Form A

Form B

Name ____________________

Module 15 • Form B
Module Test

1. In circle O, inscribed angle ABC measures 84°. What is the measure of $\overset{\frown}{AC}$?

Ⓐ 42° Ⓑ 84° Ⓒ 96° Ⓓ 168°

2. In circle O, $m\angle RPQ = 268°$.

What is the measure of $\angle RPQ$?

Ⓐ 46° Ⓑ 92° Ⓒ 184° Ⓓ 268°

3. Quadrilateral $PQRS$ is inscribed in a circle. If $m\angle Q = 98°$, what $m\angle S$?

Ⓐ 82° Ⓑ 98° Ⓒ 172° Ⓓ 262°

4. What is an equation of a circle with center $(9, -8)$ and a radius of 5 units?

Ⓐ $(x-9)^2+(y+8)^2=5$
Ⓑ $(x+9)^2+(y-8)^2=5$
Ⓒ $(x-9)^2+(y+8)^2=25$
Ⓓ $(x+9)^2+(y-8)^2=25$

5. Circle O has center $(1, -7)$ and passes through point $(5, -6)$. Which point also lies on circle O?

Ⓐ $(-3, 6)$ Ⓑ $(-2, 3)$ Ⓒ $(2, -1)$ Ⓓ $(5, -8)$

6. The minute hand of a circular clock sweeps an arc as it moves from 10:49 a.m. to 11:17 a.m. in the same day. What is the measure of the arc in degrees?

168°

7. In circle M, the central angle LMN measures $(14x-48)°$ and inscribed angle LPN measures $(5x+12)°$. What is the measure of $\overset{\frown}{LN}$ in degrees?

204°

8. In circle C, $\overline{QR}$ and $\overline{RS}$ are chords.

If $m\overset{\frown}{QR} = (46x+14)°$ and $m\overset{\frown}{RS} = (60x-21)°$, what is the measure of $\overset{\frown}{RS}$ in degrees?

129°

9. Angle QRS is circumscribed about circle O. If $QR = 12x$ ft and $RS = \frac{4x+5}{2}$ ft, what is the length of $\overline{RS}$ in feet?

3

10. Quadrilateral $JKLM$ is inscribed in a circle. If $m\angle L$ is five times $m\angle J$, what is $m\angle J$ in degrees?

30°

11. Regular nonagon $ABCDEFGHI$ is inscribed in a circle. What is the measure of $\overset{\frown}{FG}$ in degrees?

40°

12. What is an equation of the circle with center $(5, 13)$ and a radius of 9 units?

Possible answer: $(x-5)^2+(y-13)^2=81$

13. Consider the equation $x^2-24x+y^2+36y=-243$. What is the radius of the circle in units?

15

Geometry 123 Module 15 Test • Form B

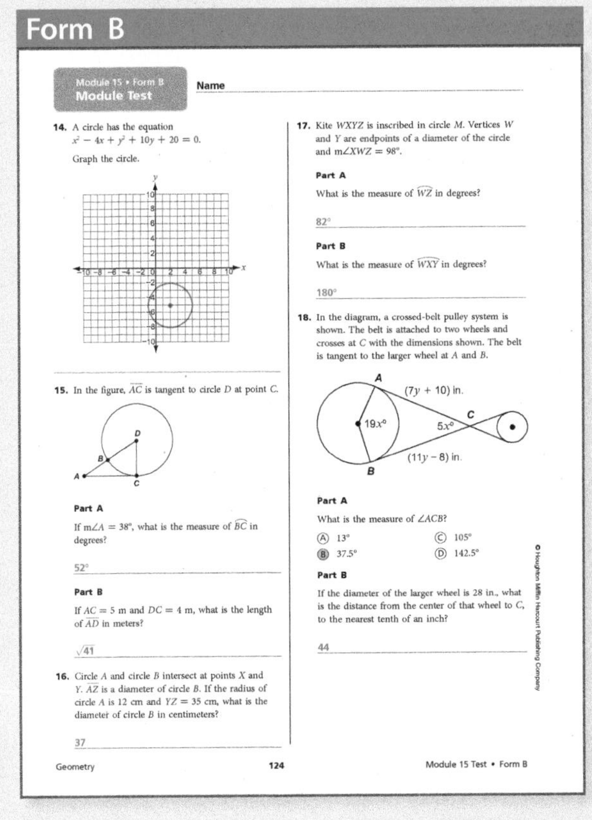

Form B

Module 15 • Form B
Module Test

Name ____________________

14. A circle has the equation $x^2-4x+y^2+10y+20=0$.

Graph the circle.

15. In the figure, $\overline{AC}$ is tangent to circle D at point C.

Part A

If $m\angle A = 38°$, what is the measure of $\overset{\frown}{BC}$ in degrees?

52°

Part B

If $AC = 5$ m and $DC = 4$ m, what is the length of $\overline{AD}$ in meters?

$\sqrt{41}$

16. Circle A and circle B intersect at points X and Y. $\overline{AZ}$ is a diameter of circle B. If the radius of circle A is 12 cm and $YZ = 35$ cm, what is the diameter of circle B in centimeters?

37

17. Kite $WXYZ$ is inscribed in circle M. Vertices W and Y are endpoints of a diameter of the circle and $m\angle XWZ = 98°$.

Part A

What is the measure of $\overset{\frown}{WZ}$ in degrees?

82°

Part B

What is the measure of $\overset{\frown}{WXY}$ in degrees?

180°

18. In the diagram, a crossed-belt pulley system is shown. The belt is attached to two wheels and crosses at C with the dimensions shown. The belt is tangent to the larger wheel at A and B.

A
(7y + 10) in.
C
19x°
5x°
(11y − 8) in.
B

Part A

What is the measure of $\angle ACB$?

Ⓐ 13° Ⓑ 37.5° Ⓒ 105° Ⓓ 142.5°

Part B

If the diameter of the larger wheel is 28 in., what is the distance from the center of that wheel to C, to the nearest tenth of an inch?

44

Geometry 124 Module 15 Test • Form B

MODULE 16

PLANNING

RELATIONSHIPS IN CIRCLES

Introduce and Check for Readiness

- Module Performance Task
- Are You Ready?

Lesson 16.1—2 Days

Segment Relationships in Circles

Learning Objective: Use proportional relationships in circles to prove the Chord-Chord, Secant-Secant, and Secant-Tangent Product Theorems. Apply the theorems to solve for segment lengths in mathematical and real-world problems.

New Vocabulary: external secant segment, secant, secant segment, tangent segment

Lesson 16.2—2 Days

Angle Relationships in Circles

Learning Objective: Determine the relationships that exist between secants, tangents, and chords in a circle and the angles and arcs formed by them. Prove and use theorems about these relationships to solve mathematical and real-world problems.

Assessment

- Module 16 Test (Forms A and B)
- Unit 8 Test (Forms A and B)

LEARNING ARC FOCUS

 Build Conceptual Understanding

 Connect Concepts and Skills

 Apply and Practice

TEACHING FOR SUCCESS

TEACHING FOR DEPTH: Relationships in Circles

Represent and Explain. When finding products involving chords and finding products involving secants, students should understand the differences and similarities in the relationships. Consider the following diagram:

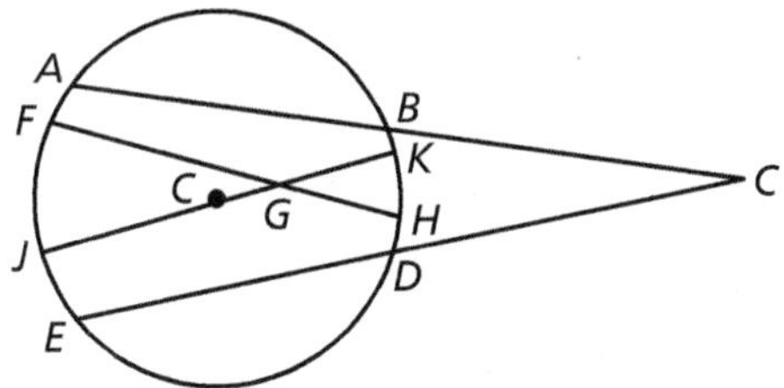

For the Chord-Chord Product Theorem, students find the products of the lengths of parts—that is, $FG \cdot GH$ (lengths of parts of the same chord $\overline{FH}$) and $JG \cdot GK$ (lengths of parts of the same chord $\overline{JK}$).

For the Secant-Secant Product Theorem, students find the products of the lengths of a whole and a part—that is, $AC \cdot BC$ (lengths of all of the secant segment $\overline{AC}$ and the external part of $\overline{AC}$) and $EC \cdot DC$ (lengths of all of the secant segment $\overline{EC}$ and the external part of $\overline{EC}$).

Have students observe that the intersecting chords all share a common point *G* and that the intersecting secant segments and the external segments all share a common endpoint *C*.

Mathematical Progressions

Prior Learning	Current Development	Future Connections
Students: • proved theorems about lines, angles, and triangles. • used congruence and similarity criteria of triangles to solve problems and to prove relationships. • identified relationships among inscribed angles, radii, and chords of circles.	**Students:** • prove theorems about the relationships of chords, secants, and tangents of circles. • use segment and angle relationships in circles to solve mathematical and real-world problems. • prove theorems about angle relationships of circles.	**Students:** • justify the formulas for the circumference and area of a circle.

TEACHER TO TEACHER

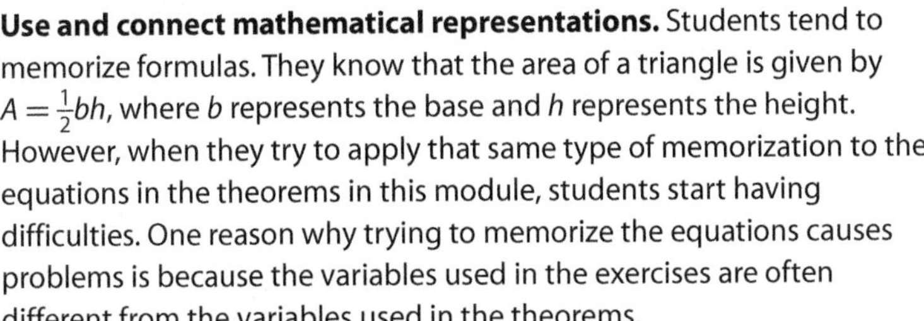

From the Classroom

Use and connect mathematical representations. Students tend to memorize formulas. They know that the area of a triangle is given by $A = \frac{1}{2}bh$, where *b* represents the base and *h* represents the height. However, when they try to apply that same type of memorization to the equations in the theorems in this module, students start having difficulties. One reason why trying to memorize the equations causes problems is because the variables used in the exercises are often different from the variables used in the theorems.

Therefore, I have my students make explicit connections between the illustrations, the equations, and the verbal descriptions in each theorem in this module. For example, consider the Secant-Secant Product Theorem (on page 505). I ask my students to circle the phrase "intersect in the exterior of a circle" in red, and likewise to circle the point *T* in the diagram in red. Then I have them highlight the phrase "one secant segment," $\overline{AT}$ in the diagram, and *AT* in the equation yellow. Next, I have them underline the phrase "its external segment" the first time it appears in the theorem, $\overline{BT}$ in the diagram, and *BT* in the equation blue. Similarly, I have them use a color to connect "the other secant segment" with $\overline{CT}$ and *CT* and the second occurrence of "its external segment" with $\overline{DT}$ and *DT*.

Using this process helps my students internalize the parts of the theorems and apply them to other problems that do not use the same variables.

LANGUAGE DEVELOPMENT • Planning for Instruction

By giving all students regular exposure to language routines in context, you will provide opportunities for students to **listen**, **speak**, **read**, and **write** about mathematical situations and develop both mathematical language and conceptual understanding at the same time.

Using Language Routines to Develop Understanding

Use the **Professional Cards** for the following routines to plan for effective instruction.

Co-Craft Questions Lesson 16.2

Students think of natural questions to ask about a given situation or problems similar to a given task and answer the questions they have developed or problems they have created.

Three Reads Lesson 16.2

Students read a problem three times with a specific focus each time.

1st Read What is the situation about?
2nd Read What are the quantities in the situation?
3rd Read What are the possible mathematical questions that we could ask for the situation?

Information Gap Lesson 16.1

Students recognize when information given in a problem situation is incomplete, and they pose questions and share knowledge with others to discover any missing facts or relationships and work together to solve the problem.

Critique, Correct, and Clarify Lesson 16.1

Students correct the work in a flawed explanation, argument, or solution method and share with a partner and refine the sample work.

Connecting Language to Relationships in Circles

Watch for students' use of the new terms listed below as they explain their reasoning and make connections with new concepts.

Key Academic Vocabulary

Current Development • New Vocabulary

external secant segment a secant segment that lies in the exterior of a circle with one point on the circle

secant a line that intersects a circle at exactly two points

secant segment a segment of a secant line with at least one endpoint on the circle

tangent segment a segment of a tangent with one endpoint on the circle

Linguistic Note

Listen for students who do not distinguish the differences between *secant, secant segment*, and *external secant segment*.

Have a visual representation in the classroom of a circle with each of the above terms drawn and labeled on the circle. When students are writing or speaking about the module content, ask them to use precise mathematical vocabulary to describe their work and refer to the visual model as needed to correct any errors.

Module 16

Relationships in Circles

Module Performance Task: Focus on STEM

Planetary Exploration

A space probe, S, approaches an unknown planet and takes measurements at different points.

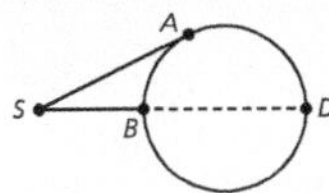

$\overline{SA}$ is tangent to the sphere at A.
$\overline{BD}$ is a diameter of the sphere.
Time for a radar signal to travel:

- from S to A and back to S: 0.0494 second
- from S to B and back to S: 0.0317 second

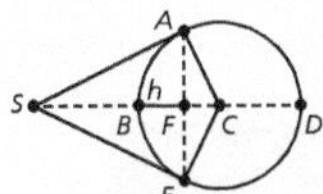

$\overline{SE}$ is tangent to the sphere at E.

$m\angle ASE = 52°$

A–D. See margin.

A. Radar signals travel at the speed of light, approximately 186,000 miles per second. Use this information to determine the shortest distance to the surface, SB, and the distance to the horizon, SA. Round your answers to the nearest mile.

B. Approximate the radius of the planet to the nearest mile. Justify your reasoning.

C. The surface area of a sphere is given by the formula $A_{sphere} = 4\pi r^2$. Use this formula to approximate the surface area of the planet to the nearest square mile.

D. The portion of the planet that is viewable from S is a spherical cap with height h. The surface area of a spherical cap is given by the formula $A_{cap} = 2\pi rh$. What percentage of the planet's surface area is visible from the space probe? Justify your reasoning.

©Dotted Yeti/Shutterstock

Module 16 501

Mathematical Connections

Task Part	Prerequisite Skills
Part A	Students should be able to apply the formula for the distance traveled at a constant rate, $d = rt$ **(Algebra 1, 2.2)**.
Part B	Students should understand and apply the Secant-Tangent Product Theorem **(16.1)**.
Part C	Students should understand the concept of surface area **(Grade 6, 13.1)** and be able to evaluate a formula **(Grade 6, 8.2)**.
Part D	Students should be able to apply the Tangent-Secant Exterior Angle Measure Theorem **(16.2)** and understand the relationship between radii and chords **(16.1)**. They also need to understand and apply the cosine ratio in a right triangle **(13.2)**.

Planetary Exploration

Overview

As space probes orbit planets, they take a variety of measurements. Using distance measurements, angle measurements, and theorems about circle relationships, scientists can calculate other quantities to learn more about the planet.

Career Connections

Suggest that students research a career as an aerospace engineer, astronomer, or astrophysicist.

Answers

A. $SA \approx 4594$ mi; $SB \approx 2948$ mi

B. 2106 miles; Using the Secant-Tangent Product Theorem, $4594^2 = 2948(2948 + d)$, where d is the diameter, so $d \approx 4211$ and $r \approx 2106$.

C. about 55,734,819 mi^2

D. about 28%; by the Tangent-Secant Exterior Angle Measure Theorem: $52° = \frac{1}{2}\left(\left(360° - m\widehat{AE}\right) - m\widehat{AE}\right)$, so $m\widehat{AE} = 128°$ and $m\angle ACE = 128°$. $CA = CE$ (both radii), so by the Converse of the $\perp$ Bisector Theorem, $\overline{FC} \perp \overline{AE}$ and $m\angle AFC = 90°$. By HL, $\triangle ACF \cong \triangle ECF$, and $m\angle ACF = \frac{1}{2} m\angle ACE = 64°$. Applying trigonometric ratios to $\triangle ACF$, $\cos 64° = \frac{r-h}{r}$, so $h = r - r\cos 64° \approx 1183$. $A_{cap} = 2\pi rh \approx 2\pi(2106)(1183) \approx 15{,}653{,}915$ mi^2; $15{,}653{,}915 \div 55{,}734{,}819 \approx 0.28$.

ONLINE

Assign the Digital Are You Ready? to power actionable reports including

- proficiency by standards
- item analysis

Are You Ready?

Diagnostic Assessment

- Diagnose prerequisite mastery.
- Identify intervention needs.
- Modify or set up leveled groups.

Have students complete the *Are You Ready?* assessment on their own. Items test the prerequisites required to succeed with the new learning in this module.

Solve Multi-Step Equations Students will apply their previous knowledge of solving multi-step equations to finding segment and angle measures in circles by solving multi-step equations.

Solve Quadratic Equations by Finding Square Roots Students will extend their ability to solve quadratic equations using square roots to solve problems using the Secant Tangent Product Theorem.

Solve Quadratic Equations Using the Quadratic Formula Students will build upon their skill of using the Quadratic Formula in solving equations to solve real-world quadratic problems involving circles.

ONLINE

Are You Ready?

Complete these problems to review prior concepts and skills you will need for this module.

Solve Multi-Step Equations

Solve each equation.

1. $5x - 6 = 64$ $x = 14$
2. $7 = 7(1 - 5n)$ $n = 0$
3. $-5p = -2(8 + 6p)$ $p = -\frac{16}{7}$
4. $-12(-3a - 3) = 8(5a + 12)$ $a = -15$
5. $7(w + 3) = 5 - 4(5 - 2w)$ $w = 36$
6. $2(3 - c) + 6c = 8 - 3(2c - 5)$ $c = \frac{17}{10}$

Solve Quadratic Equations by Finding Square Roots

Solve each equation by finding square roots.

7. $x^2 = 49$ $x = \pm 7$
8. $3x^2 = 48$ $x = \pm 4$
9. $25x^2 - 5 = 95$ $x = \pm 2$
10. $9x^2 + 7 = 23$ $x = \pm\frac{4}{3}$
11. $40 - x^2 = 3x^2$ $x = \pm\sqrt{10}$
12. $8x^2 - 7 = 35$ $x = \pm\frac{\sqrt{21}}{2}$

Solve Quadratic Equations Using the Quadratic Formula

Solve each equation using the Quadratic Formula.

13. $x^2 - 5x + 6 = 0$ $\{2, 3\}$
14. $2x^2 + 5x - 3 = -5$ $\left\{-\frac{1}{2}, -2\right\}$
15. $2x^2 - 14 = -3x$ $\left\{-\frac{7}{2}, 2\right\}$
16. $-3x^2 + 4 = 3x - 2x^2$ $\{-4, 1\}$
17. $3x^2 - x = x^2 + 1$ $\left\{-\frac{1}{2}, 1\right\}$
18. $-x^2 - 3 = 4 - 2x^2$ $\{-\sqrt{7}, \sqrt{7}\}$
19. $x^2 - 4 = 2x$ $\{1 - \sqrt{5}, 1 + \sqrt{5}\}$
20. $3x^2 - 4 = 2x - 1$ $\left\{\frac{1 - \sqrt{10}}{3}, \frac{1 + \sqrt{10}}{3}\right\}$

Connecting Past and Present Learning

Previously, you learned:

- to solve multi-step linear and quadratic equations,
- to prove theorems about lines and angles, and
- to prove relationships among inscribed angles, radii, and chords.

In this module, you will learn:

- to prove relationships formed by segments in circles and
- to use angle and segment relationships in circles to solve real-world problems.

502

DATA-DRIVEN INTERVENTION

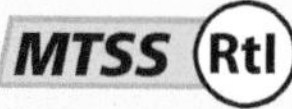

Concept/Skill	Objective	Prior Learning *	Intervene With
Solve Multi-Step Equations	Solve linear equations with grouping symbols and/or variable on both sides.	Grade 8, Lesson 3.1	• Tier 3 Skill 2 • Reteach, Grade 8 Lesson 3.1
Solve Quadratic Equations by Finding Square Roots	Find the solutions of a quadratic equation by taking square roots.	Algebra 1, Lesson 18.1	• Tier 2 Skill 17 • Reteach, Algebra 1 Lesson 18.1
Solve Quadratic Equations Using the Quadratic Formula	Find the solutions of a quadratic equation by using the Quadratic Formula.	Algebra 1, Lesson 18.3	• Tier 2 Skill 18 • Reteach, Algebra 1 Lesson 18.3

* Your digital materials include access to resources from Grade 6–Algebra 2. The lessons referenced here contain a variety of resources you can use with students who need support with this content.

16.1 Segment Relationships in Circles

LESSON FOCUS AND COHERENCE

Mathematics Standards

- Identify and describe relationships among inscribed angles, radii, and chords. *Include the relationship between central, inscribed, and circumscribed angles; inscribed angles on a diameter are right angles; the radius of a circle is perpendicular to the tangent where the radius intersects the circle.*

Mathematical Practices and Processes 

- Construct viable arguments and critique the reasoning of others.
- Model with mathematics.

I Can Objective 

I can use segment relationships in circles to solve mathematical and real-world problems.

Learning Objective

Use proportional relationships in circles to prove the Chord-Chord, Secant-Secant, and Secant-Tangent Product Theorems. Apply the theorems to solve for segment lengths in mathematical and real-world problems.

Language Objective

Explain how to derive the equation of the Chord-Chord Product Theorem from the relationship of similar triangles.

Vocabulary

New: **external secant segment**, **secant**, **secant segment**, **tangent segment**

Lesson Materials: geometric drawing tool

Mathematical Progressions

Prior Learning	Current Development	Future Connections
Students: • proved theorems about lines and angles. **(2.4, 3.2, and 3.3)** • used similarity transformations to establish the AA Triangle Similarity Theorem for two triangles to be similar. **(12.2)** • used similarity criteria for triangles to solve problems and to prove relationships in geometric figures. **(12.1–12.4)**	**Students:** • prove the Chord-Chord, Secant-Secant, and Secant-Tangent Product Theorems. • solve for segment lengths of chords, secants, and tangents of circles in purely geometric and real-world problems.	**Students:** • justify the formulas for the circumference and area of a circle. **(17.1)**

PROFESSIONAL LEARNING

Using Mathematical Practices and Processes

Model with mathematics.

This lesson provides an opportunity to address this Mathematical Practice. According to the standard, mathematically proficient students "are comfortable making assumptions and approximations to simplify a complicated situation." Circles are commonly encountered in geometry due to their basic definition that can be applied in situations involving equidistance. When used to model physical objects, circles are rarely a true representation of the object in question. However, their simplicity makes them attractive to use as an idealized approximation. Students will learn that approximating a physical object, such as the mouth of a volcano, as a circle allows them to approximate a distance, such as the distance across the volcano, as a diameter of the circle by using theorems relating the lengths of chords in a circle.

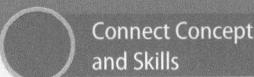

WARM-UP OPTIONS

ONLINE

PROJECTABLE & PRINTABLE 

ACTIVATE PRIOR KNOWLEDGE • Identify Similar Triangles and Solve for Side Lengths

Use these activities to quickly assess and activate prior knowledge as needed.

Problem of the Day 

When a picture is projected by a certain lens, it is turned upside down and made larger. If the height of the picture is $AB = 5$ in., what is CD, the height of the projected image? Assume that $\overline{AB} \parallel \overline{CD}$.

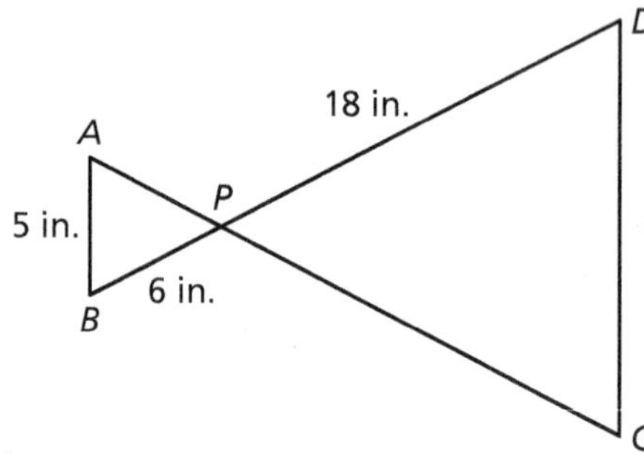

Since $\overline{AB} \parallel \overline{CD}$ and $\angle A$, $\angle C$ is a pair of alternate interior angles, $\angle A \cong \angle C$. Then $\angle APB$ and $\angle CPD$ are a pair of vertical angles, so $\triangle APB \sim \triangle CPD$ by AA similarity. So, $\frac{BP}{DP} = \frac{AB}{CD}$, and $CD = 3 \cdot AB = 15$ in.

Quick Check for Homework

As part of your daily routine, you may want to display the Teacher Solution Key to have students check their homework.

Make Connections

Based on students' responses to the Problem of the Day, choose one of the following:

 Project the Interactive Reteach, Lesson 12.2.

 Complete the Prerequisite Skills Activity:

Have students work in pairs. Show them a diagram with two parallel lines intersected by a pair of transversals, with the transversals intersecting between the parallel lines. Make sure all intersections are labeled. Have students work together to identify all pairs of congruent angles, write a triangle similarity statement, and write all proportions relating corresponding side lengths of the triangles.

- *What do you know about a pair of angles on opposite sides of an intersection of two lines?* They are a vertical pair, and are congruent.
- *What kind of congruent angle pairs are formed when two parallel lines are intersected by a transversal?* Possible answers: corresponding angles, alternate interior angles, alternate exterior angles
- *What do you need to know about their angles to conclude that a pair of triangles are similar?* At least two pairs of corresponding angles are congruent.
- *How can you write proportions relating side lengths of similar triangles?* by matching corresponding sides

SHARPEN SKILLS

If time permits, use this on-level activity to build fluency and practice basic skills.

Mental Math

Objective: Students estimate solutions to proportion problems.
Materials: index cards

Have students work in small groups. Give each group a set of index cards. Each card should have an equation declaring A, B, C, or D equal to an integer in the range from 1 to 15. Have students take turns drawing cards until they have three values in the proportion $\frac{A}{B} = \frac{C}{D}$. The have them estimate the fourth value.

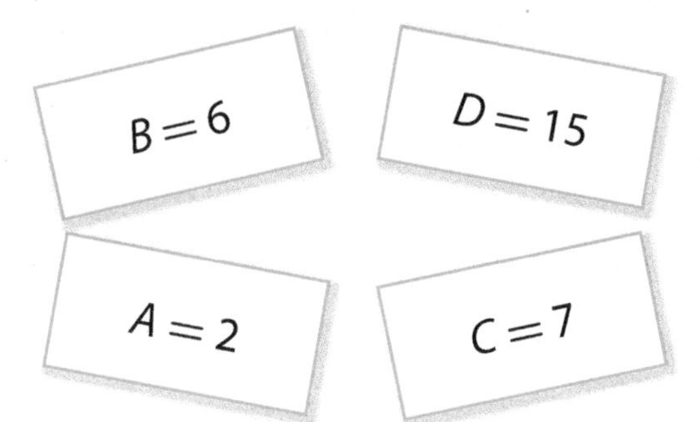

PLAN FOR DIFFERENTIATED INSTRUCTION

Small-Group Options

Use these teacher-guided activities with pulled small groups.

On Track

Materials: index cards

Give each group a set of index cards. Each card should have an integer in the range from 1 to 10. Have students draw cards three at a time. The numbers on the cards represent the lengths of three of the four segments of two intersecting chords. Have students work together to do the following:

- Make a sketch of a circle with the chords inside.
- Solve for the missing segment length.
- Revise their sketch to more accurately show the scale (sketches need not be very accurate; just encourage students to try to visualize how the chords might be arranged).

Almost There

Materials: index cards, yarn

Give each group a set of index cards and a sheet of paper with a circle on it. The set of cards should consist of six cards, with one of the terms *chord*, *secant segment*, and *tangent segment* appearing on each card, using each term twice. Have students take turns drawing cards two at a time and using pieces of yarn to represent on the paper an intersecting pair of the described objects.

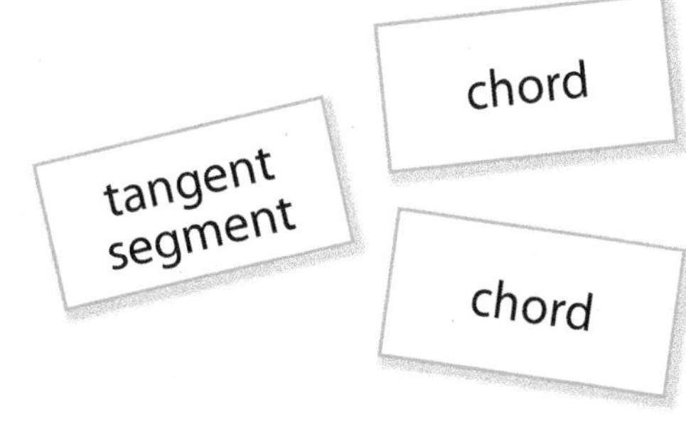

Ready for More

Show students the following statement:

"A secant segment and a tangent segment intersect in the exterior of a circle. The length of the tangent segment is x, the length of the secant segment is 4, and the length of the external part of the secant segment is x. What is the value of x?"

Ask students the following questions:

- Which theorem might you apply here?
- What equation would you write?
- What are the solutions of the equation?
- Do the solutions make sense if you try to sketch them?

Math Center Options

Use these student self-directed activities at centers or stations. Key: ● Print Resources ● Online Resources

On Track

- ● Interactive Digital Lesson
- ●● Journal and Practice Workbook
- ● Interactive Glossary (printable): **external secant segment**, **secant**, **secant segment**, **tangent segment**
- ● Module Performance Task

Almost There

- ● Reteach 16.1 (printable)
- ● Interactive Reteach 16.1

Ready for More

- ● Challenge 16.1 (printable)
- ● Interactive Challenge 16.1

View data-driven grouping recommendations and assign differentiation resources.

During the *Spark Your Learning*, listen and watch for strategies students use. See samples of student work on this page.

Recognize and Use Similar Triangles

Strategy 1

The triangles are similar because vertical angles are congruent and another pair of angles in the triangle are labeled as congruent since they both intercept the same arc on the circle. Therefore, I can use AA Triangle Similarity to prove the triangles are similar. Since the triangles are similar, corresponding sides are proportional. The lengths of the shorter segments of the chords are proportional to the lengths of the larger segments of the chords.

If students . . . recognize that similar triangle relationships can be used to solve the problem, they are demonstrating an exemplary understanding of the AA Triangle Similarity Theorem and triangle proportionality from Lessons 12.2 and 12.3.

Have these students . . . explain how they determined the relationship between the triangles and how they used that relationship. **Ask:**

- Q How did you use the given information and the relationship between the angles of the triangles to conclude that the triangles are similar?
- Q How did you conclude that the segments of the chords are proportional?

Measure and Compare Chord Segment Lengths

Strategy 2

I can use a ruler to measure the lengths of the segments of the chords in the photo in my textbook, and then compare the lengths. I found that the lengths of the smaller segments of the chords are proportional to the lengths of the larger segments of the chords.

If students . . . directly measure the lengths of the segments of the chords to solve the problem, they understand how segment lengths can be proportional, but may not recognize that similar triangles can be used to deduce this relationship.

Have these students . . . relate the angles of the two triangles and then reason about side lengths of similar triangles. **Ask:**

- Q What do you notice about the angles of the two triangles?
- Q What can you conclude about side lengths of similar triangles?

COMMON ERROR: Writes an Incorrect Proportion

The triangles are similar because vertical angles are congruent and another pair of angles in the triangle is labeled as congruent since the angles both intercept the same arc on the circle. Therefore, I can use AA Triangle Similarity to conclude the triangles are similar. Since the triangles are similar, I can conclude that $\frac{AB}{BD} = \frac{BC}{BE}$.

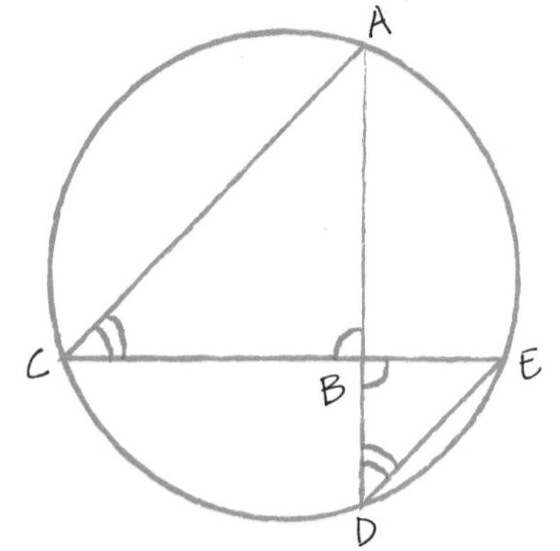

If students . . . write an incorrect proportion relating the chord segment lengths, then they may not understand that a proportion describing side lengths of similar triangles relates corresponding side lengths.

Then intervene . . . by pointing out that students should pair corresponding sides of the triangles, and then write a proportion relating their lengths. **Ask:**

- Q Since $\overline{AB}$ is the longer leg of $\triangle ABC$, which side of the other triangle corresponds with it?
- Q How can you change your proportion so it relates corresponding side lengths?

16.1

Segment Relationships in Circles

I Can use segment relationships in circles to solve mathematical and real-world problems.

Spark Your Learning

A photographer is cropping a photograph into a circle. The photographer uses two chords to compose the image inside the circle.

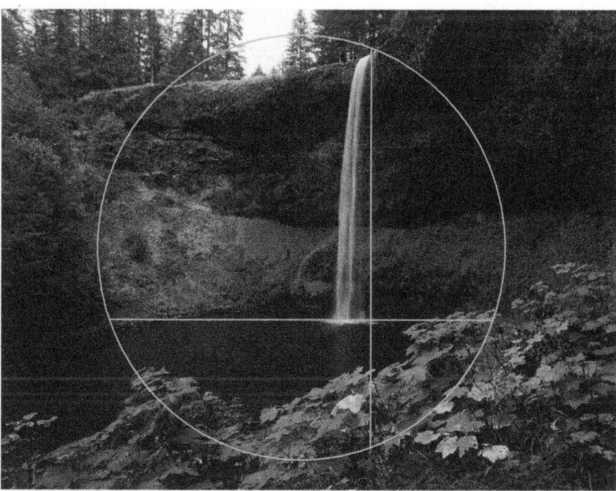

Complete Part A as a whole class. Then complete Parts B–D in small groups.

A. What is a mathematical question you can ask about this situation? What information would you need to know to answer your question?
A, B. See Additional Answers.

B. What are the shapes involved in this situation? What theorem in geometry could help you identify the relationship between these shapes?

C. To answer your question, what strategy and tool would you use along with all the information you have? What answer do you get?
See Strategies 1 and 2 on the facing page.

D. Does your answer make sense in the context of the situation? How do you know? yes; Corresponding sides of similar triangles are proportional.

Turn and Talk Would changing the intersection point of the chords affect the result of their relationship? Explain. See margin.

SUPPORT SENSE-MAKING • Information Gap

Ask students questions to help them decide what missing information they need to answer the question, "How are the lengths of the segments of the intersecting chords related?"

Do you have enough information to conclude that the lengths of the chords are related? Explain. yes; Since the chords intersect, their lengths are related by the lengths of their segments on either side of the intersection point.

Suppose you know that the chords are perpendicular. Can you determine how the lengths of the chords are related? Explain. no; Possible explanation: If I know that the chords are perpendicular, then I know that I can draw two right triangles by connecting the endpoints of the intersecting chords, but I don't know how the side lengths of the triangles are related.

In addition to the measure of the angle made by the chords, what information might be helpful? Possible answer: If I know that the triangles made by connecting the endpoints of the chords are similar, I can use corresponding sides of similar triangles to tell how the lengths of the segments of the intersecting chords are related.

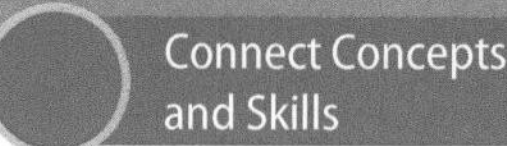

Connect Concepts and Skills

1 Spark Your Learning

▶ MOTIVATE

- Have students look at the photo in their books and read the information contained in the photo. Then complete Part A as a whole-class discussion.
- Give the class the additional information they need to solve the problem. This information is available online as a printable and projectable page in the Teacher Resources.
- Have students work in small groups to complete Parts B–D.

▶ PERSEVERE

If students need support, guide them by asking:

Q **Advancing • Use Tools** Which tool could you use to solve the problem? Why choose that tool and not some other? Students' choices of tools and reasons for choosing them will vary.

Q **Assessing** How are the angles of the triangles related? One pair of angles can be shown to be congruent since they intercept the same arc and another pair is congruent since the angles are vertical angles.

Q **Assessing** What do the angles tell you about the triangles? The triangles are similar because they have two pairs of congruent angles.

Q **Advancing** How are the segments of the chords related? Their lengths are proportional since they are corresponding sides of similar triangles.

Turn and Talk Help students understand that the actual values of any segment lengths or angle measures are not important. Instead, ask them questions drawing their attention to the reasons why the two triangles are similar. Specifically, they should realize that one pair of corresponding angles is a vertical pair and the other is a pair of inscribed angles intercepting the same arc. no; There would still be a pair of vertical angles and a pair of inscribed angles intercepting the same arc, so the two triangles would still be similar.

▶ BUILD SHARED UNDERSTANDING

Select groups of students who used various strategies and tools to share with the class how they solved the problem. As they present their solutions, have each group discuss why they chose a specific strategy and tool.

② Learn Together

Build Understanding

Task 1 (MP) **Construct Arguments** Students are introduced to the Chord-Chord Product Theorem and are then guided through a proof. In a formalization of the Spark Your Learning, students can apply the similarity of triangles formed by the intersecting chords to write a statement equivalent to the proportionality of the lengths of the chord segments.

Sample Guided Discussion:

Q **In Part B, how are the measures of an inscribed angle and its intercepted arc related?** The measure of the angle is half the measure of the arc.

Q **In Part D, how is $DP \cdot BP$ related to the ratios $\frac{AP}{DP}$ and $\frac{CP}{BP}$ when you treat them as fractions?** It is a common denominator of the ratios.

Turn and Talk Make sure that students understand the angle relationships between $\triangle APC$ and $\triangle DPB$ that led to $\triangle APC \sim \triangle DPB$. Ask them if similar angle relationships exist between $\triangle APD$ and $\triangle CPA$. no; You cannot establish a relationship between the two triangles formed, so you cannot establish a relationship between their sides.

Build Understanding

Prove the Chord-Chord Product Theorem

Recall that a chord is a segment whose endpoints lie on a circle. The following theorem describes a relationship among the four segments formed when two chords intersect in a circle.

Chord-Chord Product Theorem

If two chords intersect inside a circle, then the products of the lengths of the segments of the chords are equal.

$AE \cdot EB = CE \cdot ED$

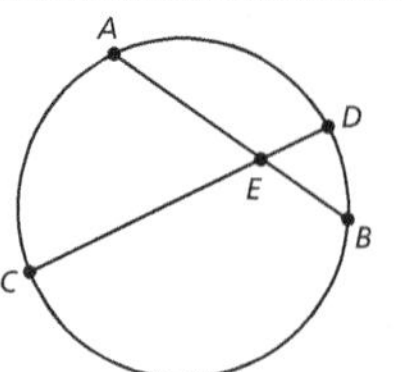

1 Prove the Chord-Chord Product Theorem.

Given: Chords $\overline{AB}$ and $\overline{CD}$ intersect in the interior of the circle. $\overline{AC}$ and $\overline{DB}$ are drawn to create triangles.

Prove: $AP \cdot BP = DP \cdot CP$

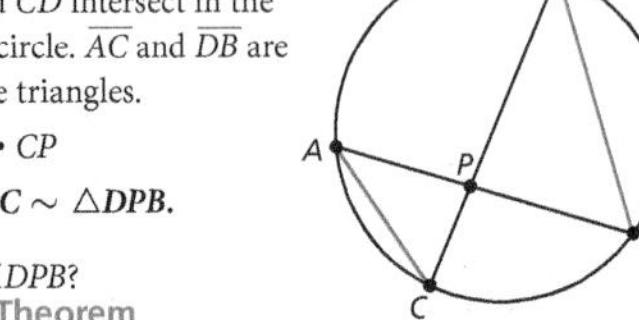

First, show that $\triangle APC \sim \triangle DPB$.

A. Why is $\angle APC \cong \angle DPB$?
Vertical Angles Theorem

B. Name at least one other pair of corresponding congruent angles shown in the triangles. Explain your reasoning.
$\angle CAB \cong \angle CDB$ or $\angle ACD \cong \angle DBA$; the angles intercept the same arc.

C. Explain why $\triangle APC \sim \triangle DPB$.
AA Triangle Similarity

Now, use the similar triangles to show that $AP \cdot BP = DP \cdot CP$.

D. Give a reason for each step in the calculation.

$$\frac{AP}{DP} = \frac{CP}{BP}$$

$$DP \cdot BP \cdot \frac{AP}{DP} = DP \cdot BP \cdot \frac{CP}{BP}$$

$$AP \cdot BP = DP \cdot CP$$

D, E. See Additional Answers.

E. Suppose $\overline{AD}$ and $\overline{CB}$ are drawn instead of $\overline{AC}$ and $\overline{DB}$ to create triangles $\triangle DAP$ and $\triangle BCP$. Do you get the same result? Explain your reasoning.

Turn and Talk Consider the triangles *APD* and *CPA* formed by connecting *A* to *D* and *C* to *A*. Can the Theorem be proven using these triangles? See margin.

504

LEVELED QUESTIONS

Depth of Knowledge (DOK)	Leveled Questions	What Does This Tell You?
Level 1 **Recall**	If $\overline{WX}$ and $\overline{YZ}$ are chords of a circle and intersect at point *P*, can you write an equation relating the lengths of the four segments of the chords? $WP \cdot PX = YP \cdot PZ$ (or any equivalent equation)	Students' answers will indicate whether they can recall and state the result of the theorem.
Level 2 **Basic Application of Skills & Concepts**	If, as in the proof, you create two triangles on opposite sides of the intersection point of intersecting chords, how can you find pairs of congruent corresponding angles between the triangles? One angle pair is a vertical pair, and I can pair other angles since they are inscribed angles intercepting the same arc.	Students' answers will indicate whether they understand the angle relationships leading to the triangle similarity applied in the proof.
Level 3 **Strategic Thinking & Complex Reasoning**	If chords $\overline{AB}$, $\overline{CD}$, and $\overline{EF}$ all intersect at point *P*, how are the lengths of the six segments of the chords related? $AP \cdot PB = CP \cdot PD = EP \cdot PF$	Students' answers will indicate whether they can extend the theorem by transitivity to a case where three chords intersect at the same point.

Investigate Segment Relationships in Circles

A **tangent segment** is a segment of a tangent with one endpoint on the circle, such as $\overline{AB}$.

A **secant** of a circle is a line that intersects a circle at exactly two points, such as $\overleftrightarrow{DB}$. A **secant segment**, such as $\overline{DB}$, is a segment of a secant line with at least one endpoint on the circle. A secant segment that lies in the exterior of a circle with one point on the circle is called an **external secant segment**.

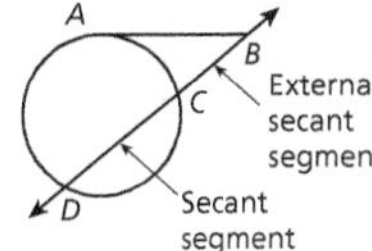

There is a special relationship in the lengths of secant segments drawn from the same point outside a circle.

Secant-Secant Product Theorem

If two secants intersect in the exterior of a circle, then the product of the lengths of one secant segment and its external segment is equal to the product of the lengths of the other secant segment and its external segment.

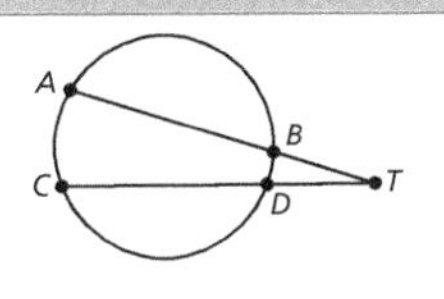

$AT \cdot BT = CT \cdot DT$

Investigate the Secant-Secant Product Theorem using a geometric drawing tool.

Draw a circle with two secants that intersect in the exterior of the circle as shown.

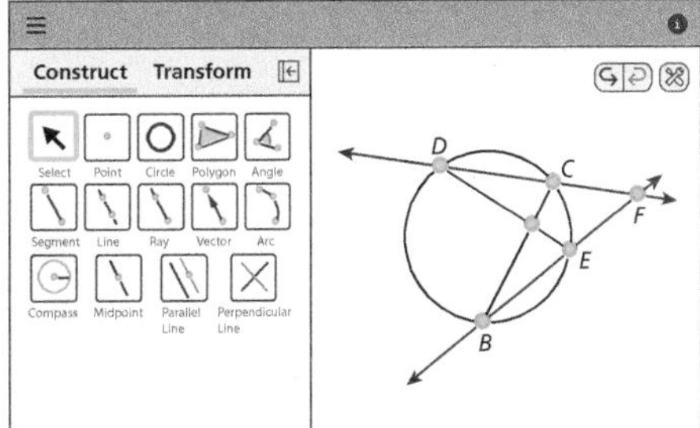

A. Use the Secant-Secant Product Theorem to write an equation that relates the lengths of the secant segments. Does this relationship remain true as points are moved around the circle? $DF \cdot CF = BF \cdot EF$; yes

B. Measure $\angle CDE$ and $\angle EBC$. What happens to the angle measures as you move point D closer to point C along the circle?
The measures of the angles remain equal as point D moves closer to point C.

C. Explain why $\triangle FDE$ and $\triangle FBC$ are similar triangles. How does this relationship explain the result of the Secant-Secant Product Theorem?
See margin.

Turn and Talk Move point D along the circle until it coincides with point C, and name the new point D. Name a pair of similar triangles. Write an equation that relates DF, BF, and EF. Make a conjecture about the relationship between a tangent and a secant drawn from the same point outside the circle. See margin.

Task 2 (MP) **Model with Mathematics** Students are introduced to the Secant-Secant Product Theorem, and investigate the theorem with a geometric drawing tool. In doing so, they go on to identify similar triangles whose corresponding side lengths imply the conclusion of the theorem. Students then use the drawing tool to explore the relationship between an external secant segment and a tangent segment, and conjecture a result that they will later see presented as the Secant-Tangent Product Theorem.

CONNECT TO VOCABULARY

Have students use the **Interactive Glossary** to record their understanding of the vocabulary in this task.

Sample Guided Discussion:

Q In Part A, how is each product in the equation of the Secant-Secant Product Theorem similar to that of the Chord-Chord Product Theorem? The factors of each product are segments with a common endpoint.

Q How are the segments described on one side of the equation of the Secant-Secant Product Theorem different from those of the Chord-Chord Product Theorem? Possible answer: In the Secant-Secant Product Theorem, one segment is part of the other, while in the Chord-Chord Product Theorem, the segments are separate.

Q In Part C, how are $\angle FDE$ and $\angle FBC$ related in terms of the circle? They are inscribed angles that intercept the same arc.

Turn and Talk Make sure students understand that as they drag point D as specified, the external secant segment $\overline{DF}$ becomes a tangent segment. Students who do not fully understand the theorem may think that one side of the equation is zero since $CD = 0$; make sure they understand that CD was not a factor in the original equation, but rather CF. Possible answer: $\triangle FBD \sim \triangle FDE$; $FB \cdot FE = FD \cdot FD$; If a tangent and a secant intersect in the exterior of the circle, then the product of the lengths of the secant segment and its external segment is equal to the square of the tangent segment.

PROFICIENCY LEVEL

Beginning
Show students a diagram with a circle, a secant segment with associated external secant segment, and a tangent segment sharing an endpoint with the external secant segment. Identify each segment, writing the term describing it and saying the term out loud. Then, show students a similar diagram and have them identify the three segments as you did.

Intermediate
Show students a diagram with a circle, two secants intersecting in the exterior of the circle, and two intersecting chords connecting the endpoints of the secant segments. Have them work together to write short sentences identifying pairs of similar triangles.

Advanced
Have students explain in a paragraph how they can make an external secant segment "become" a tangent segment by moving a point.

ANSWERS

2.C. $\angle FDE$ and $\angle FBC$ are congruent inscribed angles. $\angle F \cong \angle F$ by the Reflexive Property of Congruence. Then, $\triangle FDE$ is similar to $\triangle FBC$ by AA Similarity Theorem. Corresponding sides of the triangles are proportional. So, $\frac{FD}{FB} = \frac{FE}{FC}$, which means $FD \cdot FC = FB \cdot FE$.

Step It Out

Task 3 (MP) **Construct Arguments** Students are introduced to the Secant-Tangent Product Theorem, and see how it is applied to solve for the length of a tangent segment. They invoke the theorem to justify an equation relating the square of the length of the tangent segment to the product of the secant segment and associated external secant segment.

OPTIMIZE OUTPUT Critique, Correct, and Clarify

Have students work with a partner to discuss the approach in the text, compare it with the following approach, and determine whether it is also correct or incorrect: "$\overline{FJ}$ is perpendicular to $\overline{JH}$, and $FJ = 7$, so I can use the Pythagorean Theorem to solve for x."

Encourage students to use the vocabulary terms *secant segment* and *tangent segment* in their discussions. Students should revise their answers, if necessary, after talking with their partners.

Sample Guided Discussion:

- **Q Why don't you take the negative square root of 120 as well?** Since x is a side length, it has to be positive.
- **Q What do you call a quantity calculated the way you found x?** a geometric mean

Task 4 (MP) **Model with Mathematics** Students see how the Chord-Chord Product Theorem can be applied to solve a real-world problem. In the problem, the diameter of a volcano's mouth cannot be measured directly, so the length of another chord and a segment of the diameter are used to calculate the diameter.

Sample Guided Discussion:

- **Q In Part B, what kind of segments are $\overline{AC}$ and $\overline{BD}$?** They are two chords.
- **Q In Part C, how is the diameter related to the segment lengths in Part B?** The diameter is $AX + XC$.

Turn and Talk Draw students' attention to the congruent segments in the diagram. Possible answer: $\overline{AB}$ and $\overline{DC}$; $\triangle DXC \sim \triangle AXB$; $\angle DXC \cong \angle AXB$ by the Vertical Angles Theorem, and $\angle CDB \cong \angle CAB$ because they intersect the same arc. So, by the AA Similarity Theorem, $\triangle DXC \sim \triangle AXB$.

Step It Out

Use the Secant-Tangent Product Theorem

There is a special relationship in the lengths of a secant segment and a tangent that intersect outside a circle.

Secant-Tangent Product Theorem

If a secant and a tangent intersect in the exterior of a circle, then the product of the lengths of the secant segment and its external segment is equal to the square of the length of the tangent segment.

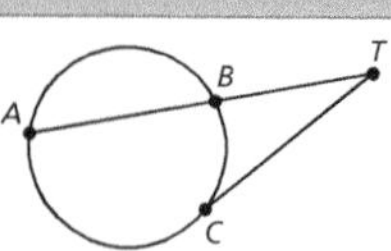

$$AT \cdot BT = CT^2$$

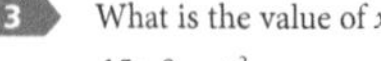

3 What is the value of x?

$15 \cdot 8 = x^2$

$120 = x^2$ — Explain why this equation is true. See margin.

$\sqrt{120} = x$

$11 \approx x$

F, 7, G, 8, H, x, J

Apply Segment Relationships

4 Molokini is an island in the Pacific Ocean. It is the remains of the mouth of an extinct volcano. A geologist uses information about the distance BD to estimate the diameter of the mouth of the volcano. Assuming the mouth of the volcano is a circle, what is the length of diameter $\overline{AC}$? A. lengths XC, DX, and XB

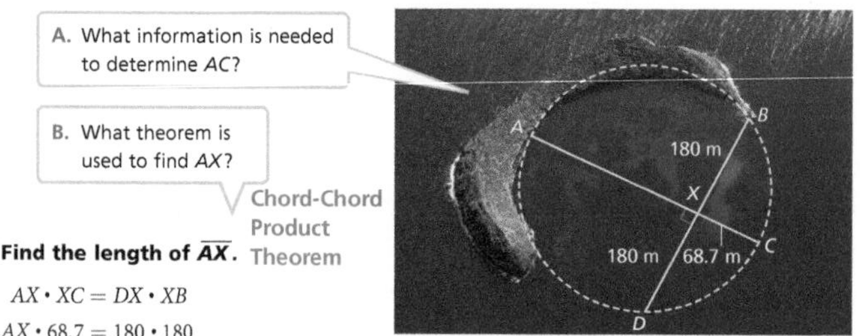

Chord-Chord Product Theorem

Find the length of $\overline{AX}$.

$AX \cdot XC = DX \cdot XB$

$AX \cdot 68.7 = 180 \cdot 180$

$AX \cdot 68.7 = 32{,}400$

$AX \approx 471.6$ m

Find the length of $\overline{AC}$.

The diameter is about 540.3 meters. — C. How was the diameter determined? $AX + XC = 471.6 + 68.7 = 540.3$

Turn and Talk What segments can be drawn in the diagram to create two similar triangles? Write a similarity statement for the two triangles. Then justify the statement. See margin.

506

ANSWERS

3. $\overline{FH}$ is a secant segment of the circle and its length is 15. $\overline{GH}$ is the external segment of $\overline{FH}$ and its length is 8. $\overline{JH}$ is the tangent segment of the circle and its length is x. So, by the Secant-Tangent Product Theorem, the equation $15 \cdot 8 = x^2$ is true.

Check Understanding

1. Name two segments that can be drawn to create similar triangles in the diagram. What is a similarity statement that justifies the equation $XZ \cdot YZ = UZ \cdot VZ$?
   1, 2. See margin.
2. Two secants intersect in the exterior of a circle and form four segments. What is the relationship between these four segments?

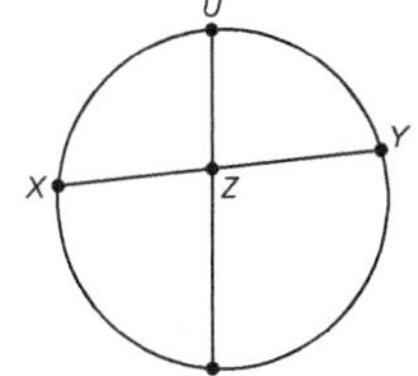

In Problems 3 and 4, find the value of *x*.

3. 3. 7.8

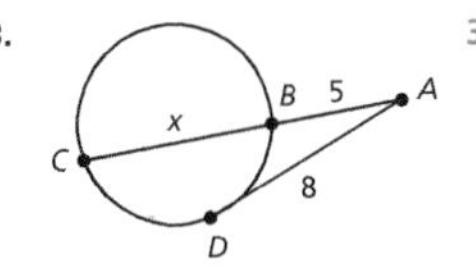

4. 4. 10

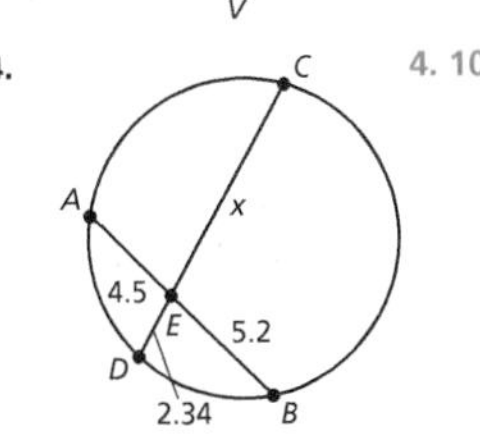

5. Chords $\overline{AB}$ and $\overline{CD}$ intersect inside a circle at point T. The chord $AB = 14$ centimeters, $AT = 5$ centimeters, and $CT = 6$ centimeters. What is the length CD?
   $CD = 13.5$ centimeters

On Your Own

6. Two chords intersect inside a circle and form four segments. What is the relationship between these four segments? 6, 7. See Additional Answers.
7. (MP) **Construct an Argument** The circle in the diagram has radius c. Use the diagram and the Chord-Chord Product Theorem to prove the Pythagorean Theorem.

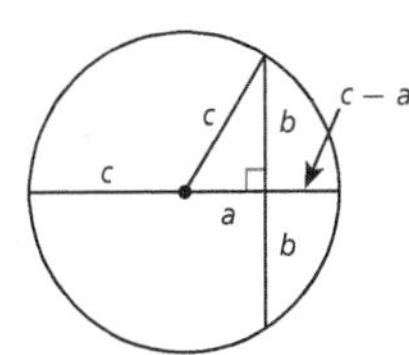

Use a geometric drawing tool to construct a circle and two chords $\overline{MN}$ and $\overline{PQ}$ that intersect inside the circle at *C*.

8. What do you know about the relationship between the lengths CM, CN, CP, and CQ?
   $CM \cdot CN = CP \cdot CQ$
9. Drag the points around the circle to change the lengths CM, CN, CP, and CQ. Does the relationship between the lengths CM, CN, CP, and CQ change?
   9, 10. See Additional Answers.
10. Is there a place in the circle where you could place point C so that the lengths CM, CN, CP, and CQ are all equal?

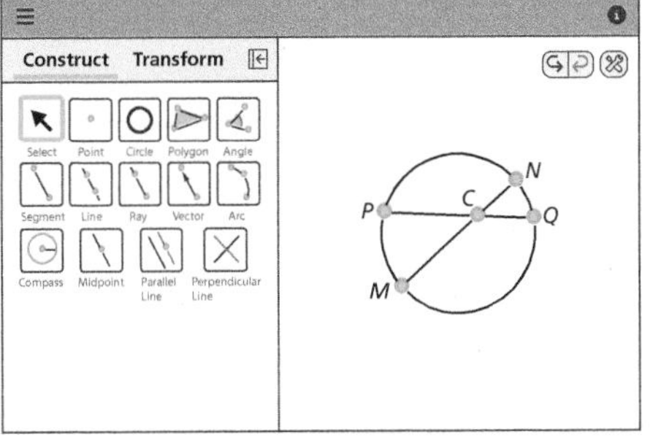

ONLINE

Assign the Digital On Your Own for
- built-in student supports
- Actionable Item Reports
- Standards Analysis Reports

On Your Own

Assignment Guide

The chart below indicates which problems in the On Your Own are associated with each task in the Learn Together. Assign daily homework for tasks completed.

Learn Together Tasks	On Your Own Problems
Task 1, p. 504	Problems 6–10, 16, 19–21, 23, and 31
Task 2, p. 505	Problems 11–13, 16, 17, 24, 25, and 31
Task 3, p. 506	Problems 14–16, 18, 26, 27, and 31
Task 4, p. 506	Problems 22, 28, 29, and 30

ANSWERS

1. $\overline{XU}$ and $\overline{VY}$ can be drawn to create $\triangle XZU$ and $\triangle VZY$, or $\overline{XV}$ and $\overline{UY}$ can be drawn to create $\triangle XZV$ and $\triangle UZY$; $\triangle XZU \sim \triangle VZY$ or $\triangle XZV \sim \triangle UZY$.
2. The product of the lengths of one secant segment and its external segment equals the product of the lengths of the other secant segment and its external segment.

(3) Check Understanding

Formative Assessment

Use formative assessment to determine if your students are successful with this lesson's learning objective.

Students who successfully complete the Check Understanding can continue to the On Your Own practice.

For students who miss 1 problem or more, work in a pulled small group using the Almost There small-group activity on page 503C.

ONLINE

Assign the Digital Check Understanding to determine
- success with the learning objective
- items to review
- grouping and differentiation resources

(4) Differentiation Options

Differentiate instruction for all students using small-group activities and math center activities on page 503C.

Reteach

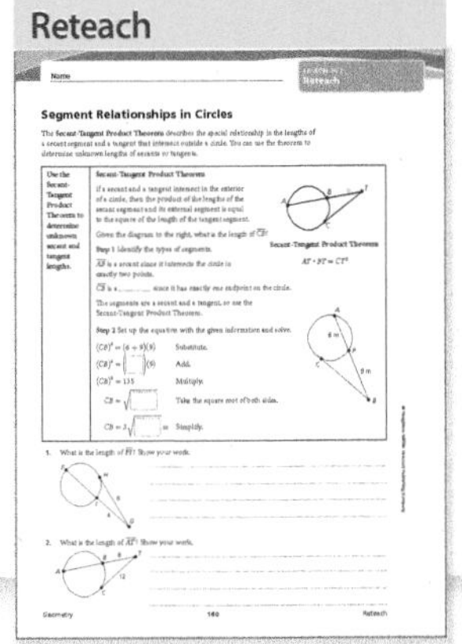

Challenge

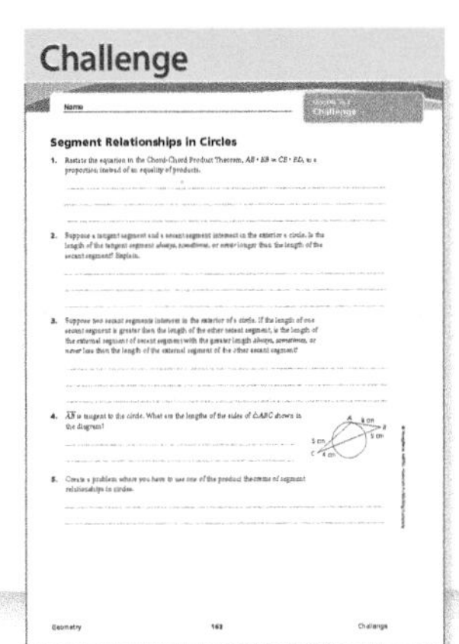

Questioning Strategies

Problem 15 Suppose another tangent intersects the given tangent and the secant so they are all concurrent at a point in the exterior of the circle. How are the lengths of the tangent segments related? The square of the length of each tangent segment is equal to the product of the secant segment and its external secant segment, so the squares of the tangent segment lengths are equal. Therefore, the lengths of the tangent segments are equal.

11. Prove the Secant-Secant Product Theorem.

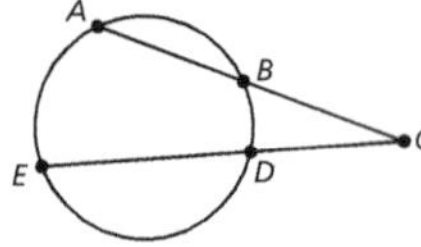

Given: $\overline{AC}$ and $\overline{EC}$ are secant segments.

Prove: $AC \cdot BC = EC \cdot DC$

See Additional Answers.

12. Can you apply the Secant-Secant Product Theorem to two secant segments that intersect on the circle? Explain. no; The Secant-Secant Product Theorem is based on the secants intersecting outside the circle.

13. Two secant segments intersect outside the circle. Their internal secant segments are congruent. Can you conclude that the secant segments are congruent? Explain. 13, 14. See Additional Answers.

14. Prove the Secant-Tangent Product Theorem.

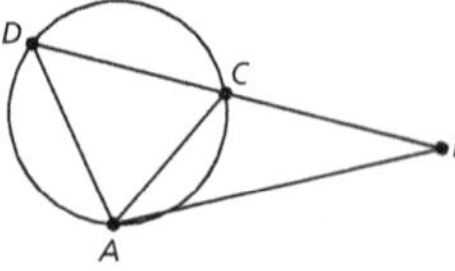

Given: $\overline{CB}$ is the external segment to secant segment $\overline{DB}$. Secant $\overline{DB}$ and tangent $\overline{AB}$ intersect in the exterior of the circle. $\overline{DA}$ and $\overline{AC}$ are drawn to create triangles.

Prove: $AB^2 = BC \cdot DB$

15. A secant and a tangent intersect in the exterior of a circle and form two secant segments and a tangent segment. What is the relationship between these three segments? The product of the lengths of the secant segment and its exterior segment equals the square of the tangent segment.

16. What theorem is the basis for each of the product theorems in this lesson showing how segments in circles are related? Explain why. 16–18. See Additional Answers.

17. Two secant segments intersect outside the circle where each of their corresponding external secant segments is equal, as shown in the diagram. What must be true about the relationship between the two secants? Justify your response.

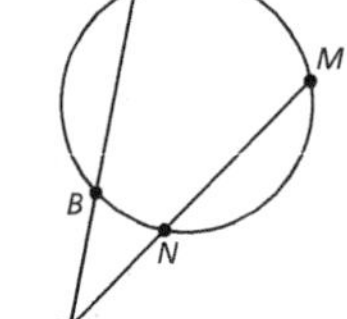

18. Can a tangent segment and a secant segment formed by a secant and a tangent intersecting in the exterior of a circle ever be equal? Explain.

For Problems 19–21, use the construction described below.

(MP) **Use Tools** Construct a perpendicular bisector of a chord.

Step 1: Use a compass to draw a circle. Mark the center.

Step 2: Use a straightedge to draw a chord.

Step 3: To construct the perpendicular bisector of the chord, draw two arcs with equal diameter, one with each endpoint as a center. Then connect the intersections of the arcs.

19. Does a diameter of the circle lie on the perpendicular bisector? Explain.
yes; The perpendicular bisector passes through the center of the circle.

20. Draw a different chord in the circle and construct its perpendicular bisector. Do you have the same result?
yes; The perpendicular bisector passes through the center of the circle.

21. Make a conjecture about the perpendicular bisector of a chord.
The perpendicular bisector of a chord is also a diameter of the circle.

508

22. Find the diameter of a circular plate from an archeological dig. The length of the chord $\overline{LM}$ is 14 inches. $\overline{AB}$ is the perpendicular bisector of $\overline{LM}$.

16.25 inches

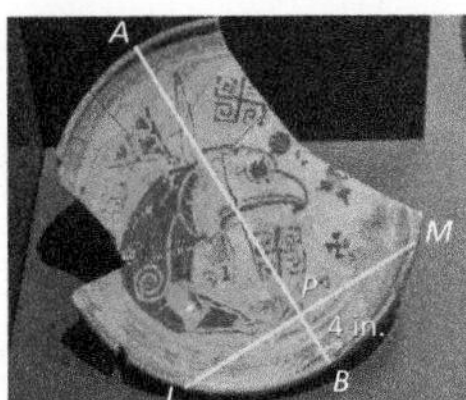

For Problems 23–26, find the length of the indicated segments.

23. AB and CD

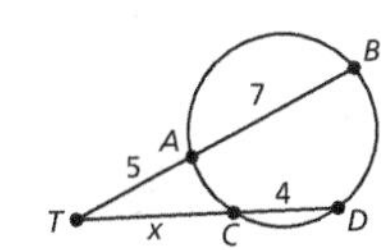

$AB = 12$ and $CD = 9$

24. TC, TB, and TD

$TC = 6$, $TB = 12$, and $TD = 10$

25. CD, TD, and TB

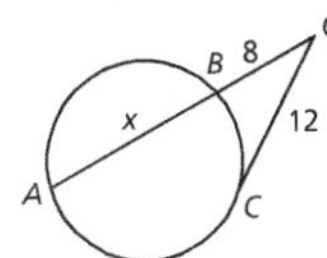

$CD = 23.25$, $TD = 31.25$, and $TB = 25$

26. AB, AQ, and CQ

$AB = 10$, $AQ = 18$, $CQ = 12$

27. A secant and a tangent intersect in the exterior of a circle. The tangent segment is equal to the internal segment that is part of the secant. The external secant segment is 6 inches. What are the lengths of the tangent segment and secant segment to the nearest tenth of an inch? The tangent segment is about 9.7 inches, and the secant segment is about 15.7 inches.

28. A communication satellite's orbit is 6400 miles above the Earth. Its geographic range is the furthest distance it can reach on Earth at this orbit. This distance, shown by $\overline{SP}$, is a tangent segment. Find the geographic range of the satellite. 9600 miles

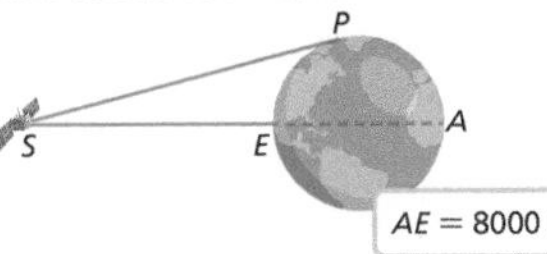

29. **STEM** A graphic arts designer is creating a logo for a company. The diagram shown is an outline of the logo that shows how much space the designer has for the company's slogan and name. According to the diagram, how many pixels are available for the company's name and for the company's slogan?
See margin.

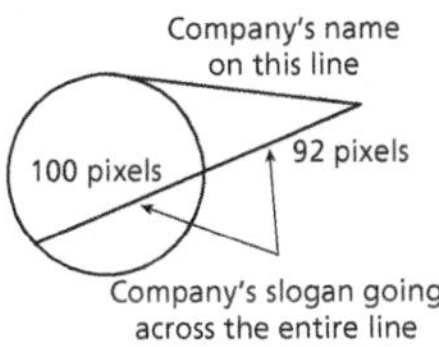

Questioning Strategies

Problem 22 Suppose that $\overline{AB}$ bisects $\overline{LM}$ with $PB = 4$ inches, but the two chords are not perpendicular. Is it still possible to find the diameter of the plate with the information given? Explain your answer. no; Because $\overline{AB}$ is a diameter of the circle if and only if it is a perpendicular bisector of $\overline{LM}$, unless $\overline{LM}$ is also a diameter. I know that the two chords are not both diameters since two diameters intersect at the center of a circle and LP and PB are not equal, so they cannot both be radii of the circle. Neither chord is a diameter, so I do not have enough information to find the diameter of the circle.

Watch for Common Errors

Problem 24 Students may confuse the structure of the equation in the Secant-Secant Product Theorem with that of the Chord-Chord Product Theorem and write $4x = 7 \cdot 5$. They need to understand that both sides of the equation should be the product of an exterior secant segment length and the length of the entire secant segment, not its portion in the interior of the circle.

ANSWERS

29. Approximately 133 pixels are available for the company's name, and 192 pixels are available for the company's slogan.

5 Wrap-Up

Summarize learning with your class. Consider using the Exit Ticket, Put It in Writing, or I Can scale.

Exit Ticket 

A chord $\overline{AB}$ of a circle O has a length of 10 cm, and is intersected at its midpoint P by a diameter $\overline{CD}$ of the circle, so $CP = 10$ cm. What is the diameter of the circle? Apply the Chord-Chord Product Theorem to get:

$AP \cdot PB = CP \cdot PD$

$5 \cdot 5 = 10 \cdot PD$

$2.5 = PD$

The length of $\overline{CD}$ is then $CD = CP + PD = 10 + 2.5 = 12.5$ cm.

Put It in Writing 

Explain how you can move a point on a secant until it becomes a tangent to change the equation of the Secant-Secant Product Theorem to the equation of the Secant-Tangent Product Theorem.

I Can 

The scale below can help you and your students understand their progress on a learning goal.

4	I can use segment relationships in circles to solve mathematical and real-world problems, and I can explain to others how to prove the relationships.
3	I can use segment relationships in circles to solve mathematical and real-world problems.
2	I can find the length of a tangent segment when I know the length of a secant segment and external secant segment sharing an endpoint with the tangent segment.
1	I can identify chords, tangent segments, secant segments, and external secant segments.

Spiral Review • Assessment Readiness

These questions will help determine if students have retained information taught in the past and can also prepare them for high-stakes assessments. Here, students must find angle measures of an inscribed quadrilateral **(15.2)**, find the length of a tangent segment **(15.3)**, complete the square to identify the radius of a circle **(15.4)**, and find the measure of an angle **(1.2)**.

30. Lisa and Martha are in a park. The diagram shows how far each of them has to walk in order to reach a picnic table where they will eat lunch. How far is Lisa from the picnic table?
$20 + \left(-10 + 10\sqrt{17}\right) \approx 51.2$ feet

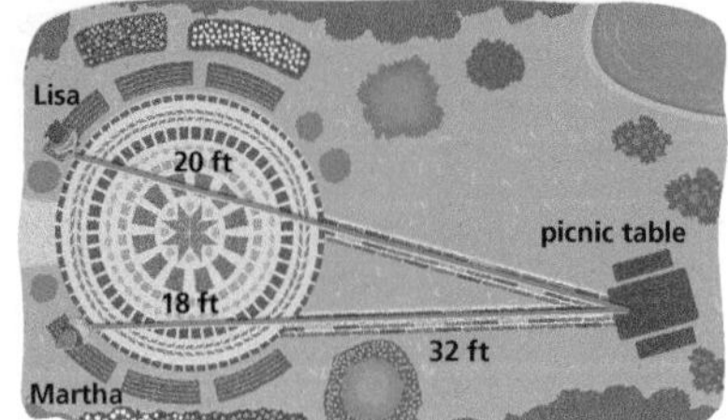

31. A student claims that all of the theorems in this lesson can be written as "When two lines intersect each other and intersect the same circle, the product of the distances between each intersection point of a line with the circle and the point where the lines intersect will be the same for both lines." Do you agree? Explain your reasoning. See Additional Answers.

Spiral Review • Assessment Readiness

32. Which angle measure is correct for the inscribed quadrilateral?

Ⓐ $m\angle B = 130°$

Ⓑ $m\angle C = 100°$

Ⓒ $m\angle A = 50°$

Ⓓ $m\angle D = 40°$

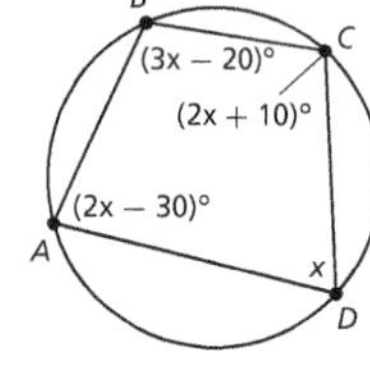

33. A Geostationary Operational Environmental Satellite is above Earth 26,199.5 miles from the center of the Earth. Earth's radius is approximately 3960 miles. What is the approximate distance from the satellite to Earth's horizon (which can be represented by a tangent segment)?

Ⓐ 25,898 mi
Ⓑ 26,497 mi
Ⓒ 702,095,400 mi
Ⓓ 909,565,281 mi

34. Which statement is true for the equation of the circle $x^2 + y^2 - 8x + 18y + 61 = 0$?

Ⓐ center $(-4, 9)$

Ⓑ center $(9, -4)$

Ⓒ radius 6 units

Ⓓ radius 12 units

35. In the diagram, $m\widehat{AC} = 34°$. What is the measure of angle ABC?

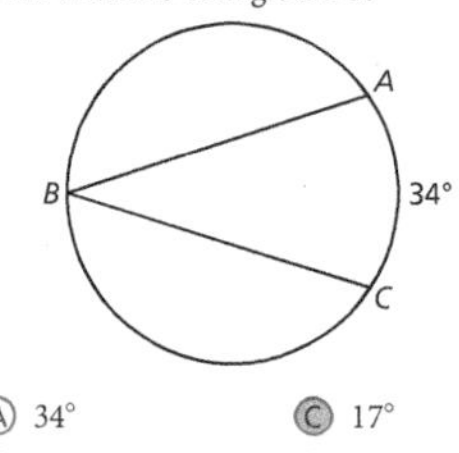

Ⓐ 34°
Ⓑ 76°
Ⓒ 17°
Ⓓ 6°

I'm in a Learning Mindset!

How am I responding to segment relationships in circles instead of segment relationships of common polygons, such as rectangles?

Keep Going to Journal and Practice Workbook

Learning Mindset

Resilience Adjusts to Change

Point out that while segment relationships in circles may have some parallels with segment relationships in polygons, the two concepts are markedly different for several reasons, the main reason being that circles are not composed of segments. Ask students which types of polygons have segment relationships that are easier to relate to than those of circles. After some thought, they may come to the understanding that more similarities can be found between a circle and a polygon when the polygon is regular, and even more as the number of sides of the polygon increases. Drawing parallels like these can help students deal with change and new topics more easily. *How does finding similarities between new topics and old topics help you better adjust to change? How does being able to list similarities and differences help you better understand both topics?*

16.2 Angle Relationships in Circles

LESSON FOCUS AND COHERENCE

Mathematics Standards

- Identify and describe relationships among inscribed angles, radii, and chords. Include the relationship between central, inscribed, and circumscribed angles; inscribed angles on a diameter are right angles; the radius of a circle is perpendicular to the tangent where the radius intersects the circle.

Mathematical Practices and Processes 

- Construct viable arguments and critique the reasoning of others.
- Look for and express regularity in repeated reasoning.

I Can Objective 

I can use angle relationships in circles to solve mathematical and real-world problems.

Learning Objective

Determine the relationships that exist between secants, tangents, and chords in a circle and the angles and arcs formed by them. Prove and use theorems about these relationships to solve mathematical and real-world problems.

Learning Objective

Describe the differences in the relationships of arcs in circles when the intersecting secants, tangents, and chords are inside, on, or outside the circle.

Mathematical Progressions

Prior Learning	Current Development	Future Connections
Students: • proved theorems about lines and angles. **(2.4, 3.2, and 3.3)** • proved theorems about triangles. **(9.1)** • identified and described relationships among inscribed angles, radii, and chords. **(15.1 and 15.3)**	**Students:** • prove the Intersecting Chords Angle Measure Theorem. • prove the Tangent-Secant Interior Angle Measure Theorem. • prove the Tangent-Secant Exterior Angle Measure Theorem, specifically the case where two secants intersect outside the circle. • use the relationships between angle measures in circles to solve mathematical and real-world problems.	**Students:** • justify the formulas for the circumference and area of a circle. **(17.1)**

PROFESSIONAL LEARNING

Visualizing the Math

Students may find it helpful to label the types of lines in a diagram. This can help them to identify which theorem to use. Students can also highlight or color the arcs and determine if the angle formed by the lines is inside or outside the circle.

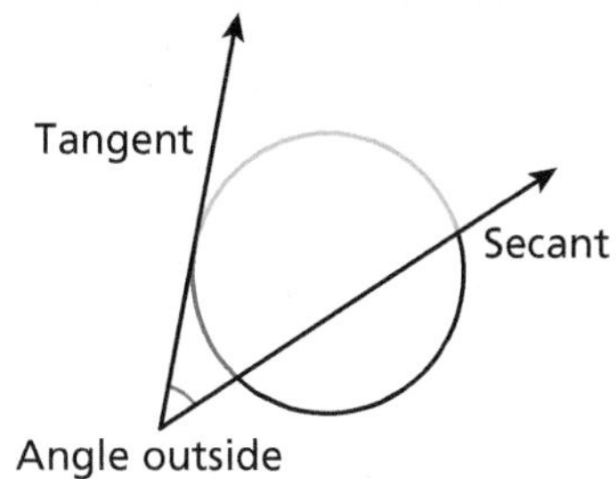

WARM-UP OPTIONS

ONLINE

PROJECTABLE & PRINTABLE 

ACTIVATE PRIOR KNOWLEDGE • Find the Measure of Central and Inscribed Angles

Use these activities to quickly assess and activate prior knowledge as needed.

Problem of the Day 

The logo for a sports team has two angles within a circle with center at point C. Both angles intercept an 87° arc. What are the measures of $\angle 1$ and $\angle 2$? $\angle 1$ is a central angle and will have the same measure as the intercepted arc, 87°. $\angle 2$ is an inscribed angle and will have a measure half that of the intercepted arc: $m\angle 2 = \frac{1}{2}(87) = 43.5°$

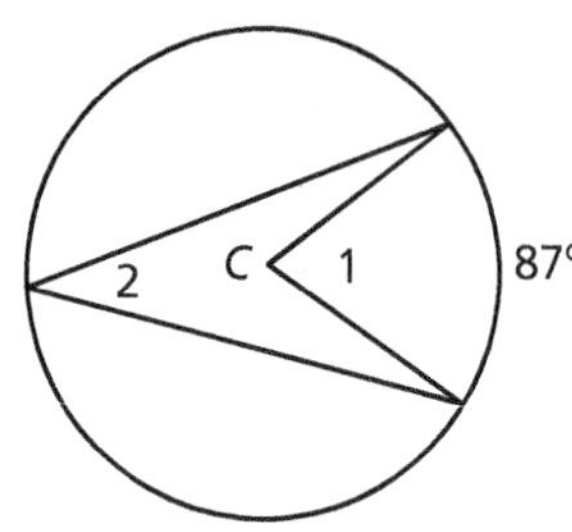

Quick Check for Homework

As part of your daily routine, you may want to display the Teacher Solution Key to have students check their homework.

Make Connections

Based on students' responses to the Problem of the Day, choose one of the following:

1. Project the Interactive Reteach, Geometry, Lesson 15.1.
2. Complete the Prerequisite Skills Activity:

Project the two statements:

A central angle has the same measure as its intercepted arc.

An inscribed angle has $\frac{1}{2}$ the measure of its intercepted arc.

Then have students work in pairs to draw examples of each statement.

- *How can you tell the difference between a central angle and an inscribed angle?* The vertex of a central angle is on the center of the circle. An inscribed angle has a vertex on the circle.
- *What is an intercepted arc?* The part of a circle between the points where the rays of an angle intersect with the circle.
- *If you know the measure of an inscribed angle, how can you find the measure of the intercepted arc?* You can multiply the measure of the angle by 2.

SHARPEN SKILLS

If time permits, use this on-level activity to build fluency and practice basic skills.

Vocabulary Review

Objective: Students compare types of angles in a circle.
Materials: Bubble Map (Teacher Resource Masters)

Have students write "Angles in Circles" in the center of the Bubble Map. Then have students write each type of angle in the surrounding circles. Work with students to sketch an example of each angle.

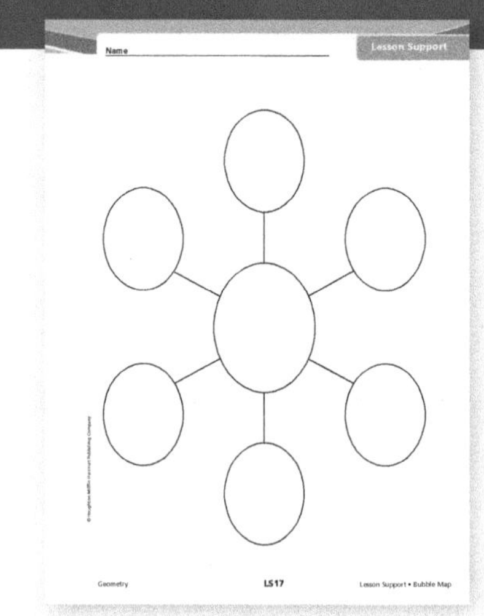

PLAN FOR DIFFERENTIATED INSTRUCTION

Small-Group Options

Use these teacher-guided activities with pulled small groups.

On Track

Materials: index cards

Pass out an index card to each student.

Have students draw a circle problem that could be solved using one of the theorems from the lesson. On the front of the card, draw the problem. On the back of the card, write the solution and the theorem used. Then have students trade cards with a classmate and solve the new problem, checking their answer on the back.

Intersecting Chords Angle Measure Theorem

Almost There

Draw the diagram on the board.

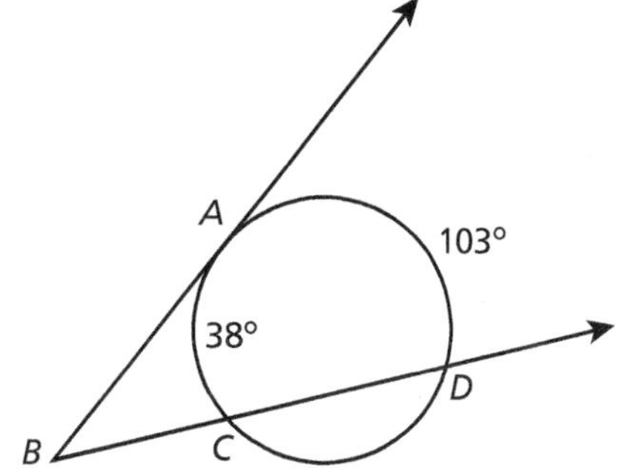

Then ask the following questions to help students solve for the measure of $\angle ABC$.

- What kind of lines form the angle?
- Is the vertex of the angle inside, outside, or on the circle?
- Which theorem can you use to relate the arcs and the angles?
- Solve for the measure of $\angle ABC$.

Ready for More

Give students the diagram, stating that $\triangle RST$ is isosceles. What is the measure of $\angle S$ in terms of a and b?

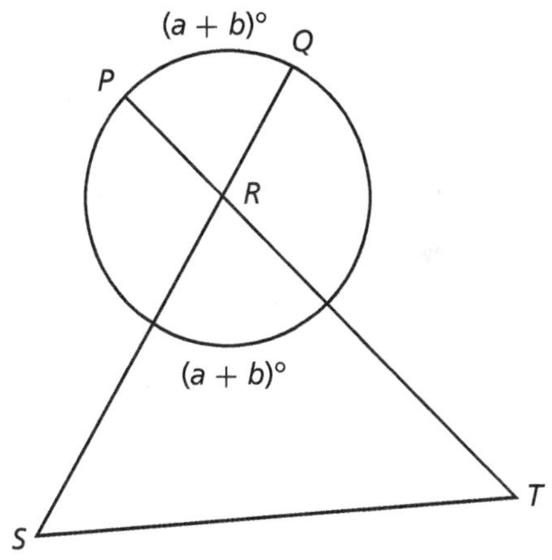

Math Center Options

Use these student self-directed activities at centers or stations. Key: ● Print Resources ● Online Resources

On Track

- ● Interactive Digital Lesson
- ●● Journal and Practice Workbook
- ● Module Performance Task
- ●● Standards Practice: Identify and Describe Relationships Among Parts of a Circle

Almost There

- ● Reteach 16.2 (printable)
- ● Interactive Reteach 16.2
- ● Illustrative Mathematics: Right Triangles Inscribed in Circles I

Ready for More

- ● Challenge 16.2 (printable)
- ● Interactive Challenge 16.2

ONLINE 

View data-driven grouping recommendations and assign differentiation resources.

During the Spark *Your Learning*, listen and watch for strategies students use. See samples of student work on this page.

Use Trigonometry to Solve

Strategy 1

I drew a right triangle $\triangle ABC$. I know the central angle $\angle BCA$ has a measure equal to its intercepted arc. Since $BC = 1080$ and $AC = 253{,}645$, I can use cosine to find the central angle measurement.

$$\cos(m\angle BCA) = \frac{1080}{253{,}645}$$

The inverse cosine gives a central angle of about 89.76 degrees, which means the intercepted arc has the same measure. The two triangles are congruent. This means I can multiply this measurement by 2 to find the total arc of about 179.52 degrees.

If students ... use trigonometry to solve the problem, they are using trigonometric ratios from Lesson 13.2.

Have these students ... explain how they determined their cosine ratio equation and how they solved the equation. **Ask:**

- Q How did you use the cosine ratio to relate the sides in the triangle?
- Q How did you know the triangles are congruent?

Use Repeated Trigonometry

Strategy 2

I drew two right triangles. Then I calculated the central angles using the cosine ratio. I know that $AC = 252{,}565 + 1080 = 253{,}645$.

$$\cos(m\angle BCA) = \frac{1080}{253{,}645} \qquad \cos(m\angle DCA) = \frac{1080}{253{,}645}$$

$$m\angle BCA = \cos^{-1}\left(\frac{1080}{253{,}645}\right) \qquad m\angle DCA = \cos^{-1}\left(\frac{1080}{253{,}645}\right)$$

$$m\angle BCA \approx 89.76^\circ \qquad m\angle DCA \approx 89.76^\circ$$

Then, I know the central angle has the same measure of its intercepted arc, so I added $89.76 + 89.76$ to get a total arc measure of about 179.52°.

If students ... use repeated trigonometry to solve the problem, they understand how to use the cosine ratio to find a missing angle measure, but they may not realize they can perform the calculation once and then multiply by 2.

Activate prior knowledge ... by having students find corresponding sides and angles in the two triangles. **Ask:**

- Q Which pairs of corresponding angles and sides are congruent?
- Q How do you know the triangles are congruent?
- Q If the triangles are congruent, do you have to do the calculation twice? Why or why not?

COMMON ERROR: Forgets to Add Radius and Double the Answer

I found the measure of the central angle by using the cosine ratio equation.

$$\cos(m\angle BCA) = \frac{1080}{252{,}565}$$

$$m\angle BCA = \cos^{-1}\left(\frac{1080}{252{,}565}\right)$$

$$m\angle BCA \approx 89.75^\circ$$

If students ... use the smaller value, they are forgetting to use the total length of $\overline{AC}$ and are only using a measurement for half the arc $\widehat{BGD}$.

Then intervene ... by pointing out that the hypotenuse of the triangle is the sum of two segments. **Ask:**

- Q How can you find the length of $\overline{AC}$?
- Q What central angle is related to the arc $\widehat{BGD}$?

16.2

Angle Relationships in Circles

I Can use angle relationships in circles to solve mathematical and real-world problems.

Spark Your Learning

John is observing the moon with a telescope. The two tangents drawn from Earth to the moon show the part of the surface of the moon John can see through the telescope.

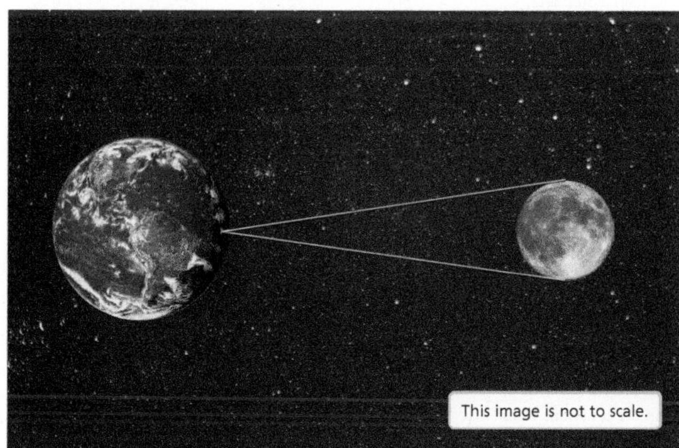
This image is not to scale.

(bg) ©gutesk7/Shutterstock; (l) ©Stocktrek Images, Inc/Brand X Pictures/Getty Images; (r) ©Tristan3D/Shutterstock

Complete Part A as a whole class. Then complete Parts B–D in small groups.

A. What is a mathematical question you can ask about this situation? What information would you need to know to answer your question?

B. What shape can represent the situation? How can you use parts of the shape to solve the problem? B, D. See Additional Answers.

C. To answer your question, what strategy and tool would you use along with all the information you have? What answer do you get? See Strategies 1 and 2 on the facing page.

D. Does your answer make sense in the context of the situation? How do you know?

A. What is the measure of the arc of the moon that can be seen from Earth?; the radius of the moon, the distance from Earth to the moon, and a measure of corresponding angles for triangles ABC and ADC

Turn and Talk What happens to the measure of the arc of the moon that can be seen from Earth with the same points of tangency when the moon is closer to Earth, such as at lunar perigee when they are 221,559 miles apart? See margin.

CULTIVATE CONVERSATION • Co-Craft Questions

If students have difficulty formulating a mathematical question about the situation in the Spark Your Learning, ask them to consider what point on the moon the tangent lines touch. What are some natural questions to ask about this situation?

Work together to craft the following questions:

- When a line or line segment touches a circle in only one point, what kind of segment or line is it?
- What angle is formed when a tangent intersects a radius of a circle?
- What could the distance between the two points where the tangents touch the moon represent?

Then have students think about what additional information, if any, they would need to answer these questions. **Ask:**

- Can you draw a right triangle? Why or why not?
- Can you determine the measure of the central angle based on the measurements given? Explain.

1 Spark Your Learning

▶ MOTIVATE

- Have students look at the photo in their books and read the information contained in the photo. Then complete Part A as a whole-class discussion.
- Give the class the additional information they need to solve the problem. This information is available online as a printable and projectable page in the Teacher Resources.
- Have students work in small groups to complete Parts B–D.

▶ PERSEVERE

If students need support, guide them by asking:

Q **Advancing • Use Tools** Which tool could you use to solve the problem? Why choose that tool and not some other? Students' choices of tools and reasons for choosing them will vary.

Q **Assessing** What is the relationship between the tangent $\overline{AB}$ and a radius of the moon? They are perpendicular segments and form a right angle where they intersect.

Q **Assessing** How is the measurement of an inscribed angle related to the measurement of its intercepted arc? The measurements are the same in degrees.

Q **Advancing** If you draw a right triangle, how can you use a trigonometric ratio to find the measure of one of the acute angles? You know the adjacent side and the hypotenuse for the acute angle which is a central angle. You can use a cosine ratio to help find the angle measure.

Turn and Talk Have students decide which value in their equation changes when the moon is closer to the Earth. Encourage students to recalculate the angle measure and compare it to the original answer.

$221{,}559 + 1080 = 222{,}639$;

$\cos\theta = \frac{\text{adj}}{\text{hyp}}$, so $\theta = \cos^{-1}\left(\frac{1080}{222{,}639}\right) \approx 89.72°$.

The measure of the arc decreases to about 179.44°.

▶ BUILD SHARED UNDERSTANDING

Select groups of students who used various strategies and tools to share with the class how they solved the problem. As they present their solutions, have each group discuss why they chose a specific strategy and tool.

② Learn Together

Build Understanding

Task 1 (MP) **Construct Arguments** Students use properties of angles formed by intersecting chords to prove the Intersecting Chords Angle Measure Theorem.

Sample Guided Discussion:

Q In Part A, how are $\angle AEC$ and $\angle BEC$ related? They are supplementary because they form a linear pair. This means that $\angle AEC$ must be supplementary to the other two angles $\angle EBC$ and $\angle ECB$ in the triangle because the sum of the three angles in the triangle is 180°.

Q In Part D, how does the measure of $\angle AEC$ compare to the measure of $\overset{\frown}{AC}$ if point E is the center of the circle? The measures would be equal because $\angle AEC$ would be a central angle.

Turn and Talk Encourage students to sketch a picture using $\overline{AD}$ instead of $\overline{CB}$. Then have students identify the relationships between the measures of angles in the new diagram. yes; The measure of $\angle AEC$ would be equal to the sum of the measures of $\angle BAD$ and $\angle CDA$.

Build Understanding

Prove the Intersecting Chords Angle Measure Theorem

The following theorem describes a relationship among the measures of angles and arcs formed when two lines intersect inside a circle.

Intersecting Chords Angle Measure Theorem

If two secants or chords intersect in the interior of a circle, then the measure of each angle formed is half the sum of the measures of its intercepted arcs.

Chords $\overline{AB}$ and $\overline{CD}$ intersect at E.

$$m\angle AEC = \frac{1}{2}\left(m\overset{\frown}{AC} + m\overset{\frown}{DB}\right)$$

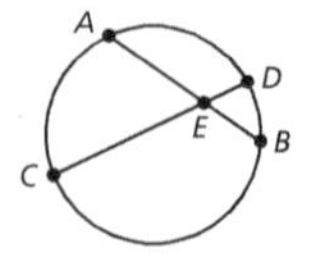

Prove the Intersecting Chords Angle Measure Theorem.

Given: Chords $\overline{AB}$ and $\overline{CD}$ intersect at E.

Prove: $m\angle AEC = \frac{1}{2}\left(m\overset{\frown}{AC} + m\overset{\frown}{DB}\right)$

A. How is the measure of $\angle AEC$ related to the measures of the angles in $\triangle CEB$? Write an equation for this relationship. What theorem justifies this equation?
A–D. See Additional Answers.

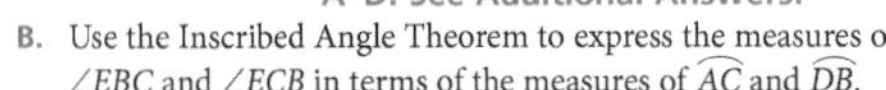

B. Use the Inscribed Angle Theorem to express the measures of $\angle EBC$ and $\angle ECB$ in terms of the measures of $\overset{\frown}{AC}$ and $\overset{\frown}{DB}$.

C. How can the equations from Parts A and B help you write $m\angle AEC = \frac{1}{2}\left(m\overset{\frown}{AC} + m\overset{\frown}{DB}\right)$.

D. Describe a special case of intersecting chords in which $m\angle AEC = m\overset{\frown}{DB}$.

Turn and Talk Could the theorem be proved using $\overline{AD}$ instead of $\overline{CB}$? Explain. See margin.

Explore the Tangent-Secant Interior Angle Measure Theorem

The following theorem describes the relationship between the measures of angles and the measures of arcs formed when two lines intersect on a circle.

Tangent-Secant Interior Angle Measure Theorem

If a tangent and a secant (or a chord) intersect on a circle at the point of tangency, then the measure of the angle formed is half the measure of the intercepted arc.

Tangent $\overrightarrow{BC}$ and secant $\overrightarrow{BA}$ intersect at B.

$$m\angle ABC = \frac{1}{2}m\overset{\frown}{AB}$$

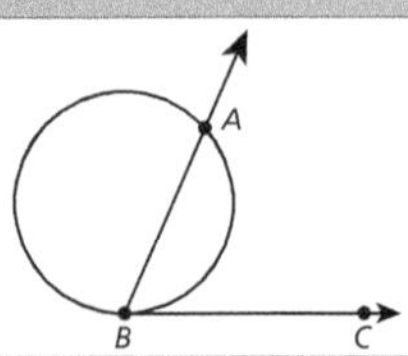

512

LEVELED QUESTIONS

Depth of Knowledge (DOK)	Leveled Questions	What Does This Tell You?
Level 1 **Recall**	What is the Exterior Angle Theorem for a triangle? The measure of an exterior angle is equal to the sum of the measures of the nonadjacent angles in the triangle.	Students' answers will indicate whether they know the relationship between an exterior angle of a triangle and the remote interior angles.
Level 2 **Basic Application of Skills & Concepts**	How is an inscribed angle related to its intercepted arc? The inscribed angle has a measure that is half the measure of its intercepted arc.	Students' answers will indicate whether they know how to compare the measures of inscribed angles and their intercepted arcs.
Level 3 **Strategic Thinking & Complex Reasoning**	How does the proof change if the chords intersect in the center of the circle? The angles with E as a vertex are central angles. The intercepted arcs will all have the same measure as the central angles.	Students' answers will indicate whether they understand how the Intersecting Chords Angle Measure Theorem applies to all possibilities of intersecting chords.

2 Consider what happens to the segments and angles as C approaches B in the circles.

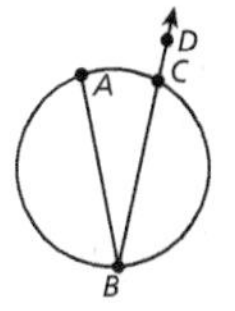

Circle 1

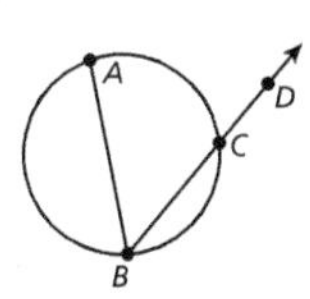

Circle 2

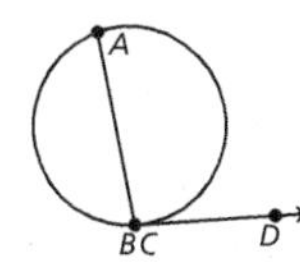

Circle 3

A. Describe the location of point C in relation to point B as you move from the first to the third circle. A–C. See margin.

B. What do you know about the measure of the intercepted arc of an inscribed angle?

C. How does the measure of the intercepted arc of an inscribed angle relate to the Tangent-Secant Interior Angle Measure Theorem?

Prove the Tangent-Secant Interior Angle Measure Theorem

There are three cases of the Tangent-Secant Interior Angle Measure Theorem.

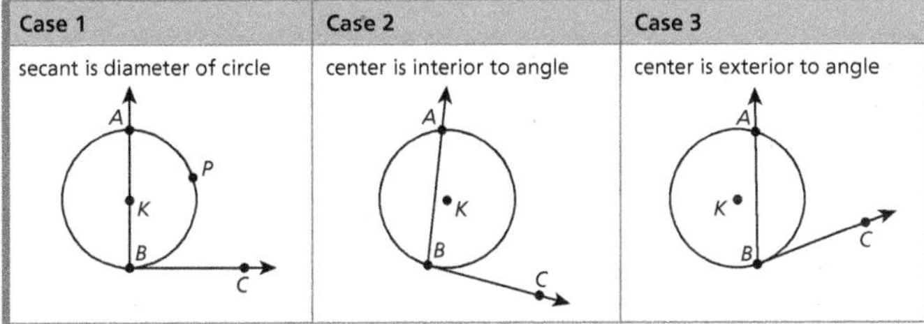

Case 1	Case 2	Case 3
secant is diameter of circle	center is interior to angle	center is exterior to angle

3 Prove Case 1 of the Tangent-Secant Interior Angle Measure Theorem.

Given: Secant $\overrightarrow{BC}$ tangent to circle K at point B and diameter $\overline{AB}$.

Prove: $m\angle ABC = \frac{1}{2}m\widehat{APB}$

First, find the measure of $\angle ABC$.

A. What is $\overline{BK}$ for the circle? Explain. A–D. See Additional Answers.

B. Why is $\overline{BC} \perp \overline{BK}$? What is $m\angle KBC$? What is $m\angle ABC$? Explain.

Now, find the measure of $\widehat{APB}$.

C. Why is $m\widehat{APB} = 180°$?

Finally, show you have enough information to prove Case 1.

D. How does $m\angle ABC$ compare to $m\widehat{APB}$?

Turn and Talk If you use the Angle Addition Postulate to prove the two other cases, what postulate will you have to use for the arcs? Arc Addition Postulate

Task 2 MP **Use Repeated Reasoning** Students examine several cases to investigate the angles formed by a chord and a secant or tangent.

Sample Guided Discussion:

Q **How are the segments $\overline{BD}$ different in Circles 1 and 2 compared to Circle 3?** In the first two circles they are secants, and in the last circle it is a tangent.

Q **If $\overline{AB}$ went through the center of the circle, what would be true about the angle formed with the tangent $\overline{BD}$ in Circle 3?** It would be 90° and would intercept a semi-circle of 180°.

Task 3 MP **Construct Arguments** Students use theorems and properties about circles to prove the relationship between the angle formed by a tangent and a diameter and the intercepted arc.

Sample Guided Discussion:

Q **What is the measure of a straight angle?** 180°

Q **What kind of angle is $\angle AKB$, formed by two radii in the circle?** a straight angle

Q **What is the measure of the arc formed by a semi-circle?** 180°

Turn and Talk Remind students that the Angle Addition Postulate says the sum of the measures of two angles is equal to the measure of the larger angle. Ask students to brainstorm how to find the measure of a larger arc if the measures of two smaller arcs that form the larger arc are known.

PROFICIENCY LEVEL

Beginning

Have students draw a circle, secant, tangent, and inscribed angle and label each part of their drawing clearly.

Intermediate

Ask students to write one sentence explaining how the measure of an inscribed angle is related to the measure of its intercepted arc. Then have students read their sentences to a partner.

Advanced

Ask students to write the Tangent-Secant Interior Angle Measure Theorem in their own words, using a diagram to support their explanation.

ANSWERS

2.A. Point C moves to the same location as point B, turning secant $\overline{BC}$ into tangent $\overline{CD}$.

2.B. It is twice the measure of the inscribed angle.

2.C. The Inscribed Angle Theorem states that the measure of an inscribed angle is equal to half the measure of its intercepted arc. The Tangent-Secant Interior Angle Measure Theorem describes an angle formed by a secant and a tangent intersecting at the point of tangency. This angle formed by a tangent and a secant is compared to an inscribed angle formed by two secants (or chords). Therefore, the measure of the intercepted arc in both cases is double the measure of the angle that intercepts it.

Step It Out

Task 4 **MP** **Construct Arguments** Encourage students to talk through the proof before putting the statements in order to prove the Tangent-Secant Exterior Angle Measure Theorem.

Sample Guided Discussion:

Q **Does the Exterior Angle Theorem involve a sum or a difference of angle measures?** The sum of the measures of the remote interior angles is equal to the measure of the exterior angle.

Q **In the last two steps of the proof, how could you check to see that the equations are equivalent?** You could distribute the $\frac{1}{2}$ to both terms in the difference in the parentheses.

Turn and Talk Have students look at the diagram of the secant-tangent case for the theorem. What extra lines would have to be drawn in order to use the Exterior Angles Theorem? yes; You will draw an auxiliary segment so that you can use the Exterior Angles Theorem and the Inscribed Angle Theorem. This allows you to introduce half arc measures into the equation from the Tangent-Secant Exterior Angle Measure Theorem.

ANSWERS

2. $m\angle CAE + m\angle BEA = m\angle CBE$
3. $m\angle CAE = m\angle CBE - m\angle BEA$
4. $m\angle BEA = \frac{1}{2}m\widehat{BD}$ and $m\angle CBE = \frac{1}{2}m\widehat{CE}$
5. $m\angle CAE = \frac{1}{2}m\widehat{CE} - \frac{1}{2}m\widehat{BD}$
6. $m\angle CAE = \frac{1}{2}(m\widehat{CE} - m\widehat{BD})$

Step It Out

Prove the Tangent-Secant Exterior Angle Measure Theorem

The following theorem relates the measures of angles and arcs formed when two lines intersect outside of a circle.

Tangent-Secant Exterior Angle Measure Theorem

If a tangent and a secant, two tangents, or two secants intersect in the exterior of a circle, then the measure of the angle formed is half the difference of the measures of its intercepted arcs.

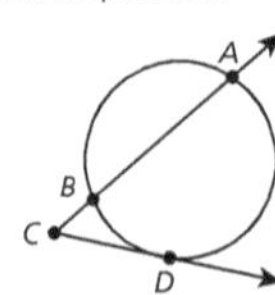

$m\angle ACD = \frac{1}{2}(m\widehat{AD} - m\widehat{BD})$

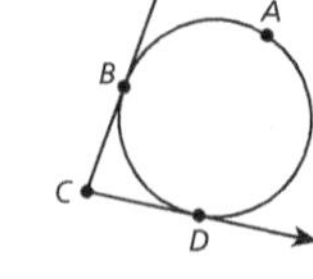

$m\angle BCD = \frac{1}{2}(m\widehat{BAD} - m\widehat{BD})$

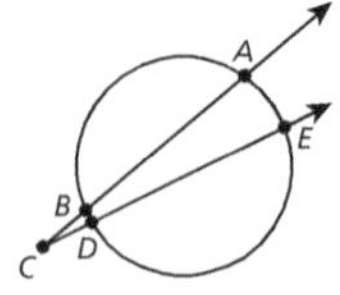

$m\angle ACE = \frac{1}{2}(m\widehat{AE} - m\widehat{BD})$

 **4** The statements in the left column of the proof of the Tangent-Secant Exterior Angle Measure Theorem for two secants are scrambled below.

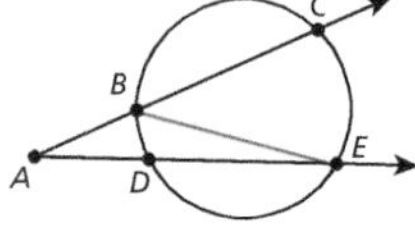

Given: Secants $\overrightarrow{AC}$ and $\overrightarrow{AE}$

Prove: $m\angle CAE = \frac{1}{2}(m\widehat{CE} - m\widehat{BD})$

Put the proof statements in the correct sequence. See margin.

Statements	Reasons
1. Secants $\overrightarrow{AC}$ and $\overrightarrow{AE}$	1. Given.
$m\angle CAE = \frac{1}{2}m\widehat{CE} - \frac{1}{2}m\widehat{BD}$	2. Exterior Angle Theorem
$m\angle CAE = m\angle CBE - m\angle BEA$	3. Subtraction Property of Equality
$m\angle CAE + m\angle BEA = m\angle CBE$	4. Inscribed Angle Theorem
$m\angle BEA = \frac{1}{2}m\widehat{BD}$ and $m\angle CBE = \frac{1}{2}m\widehat{CE}$	5. Substitution.
$m\angle CAE = \frac{1}{2}(m\widehat{CE} - m\widehat{BD})$	6. Distribution.

 **Turn and Talk** Can the same reasoning that was used to prove the Tangent-Secant Exterior Angle Measure Theorem for two secants be used to prove the case with the secant and the tangent? Explain why or why not. See margin.

514

Apply Angle Relationships in Circles

5 Find $x°$, the measure of $\angle DTC$.

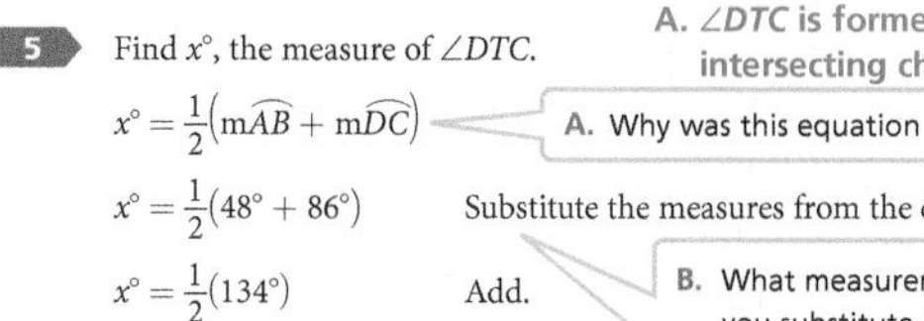

$x° = \frac{1}{2}(m\overset{\frown}{AB} + m\overset{\frown}{DC})$ — A. Why was this equation chosen?

A. $\angle DTC$ is formed from two intersecting chords.

$x° = \frac{1}{2}(48° + 86°)$ Substitute the measures from the diagram.

$x° = \frac{1}{2}(134°)$ Add.

$x° = 67°$ Multiply.

B. What measurements do you substitute into the equation? Explain.

C. Which theorem is used for the problem, and what is the measure of $\angle DTC$?

B, C. See margin.

6 **Archaeology** Stonehenge is a circular arrangement of massive stones near Salisbury, England. The diagram represents a viewer at V that observes the monument from a point where two of the stones A and B are aligned with stones at the endpoints of a diameter of the circular shape. If $m\overset{\frown}{AB} = 48°$, then what is $m\angle AVB$?

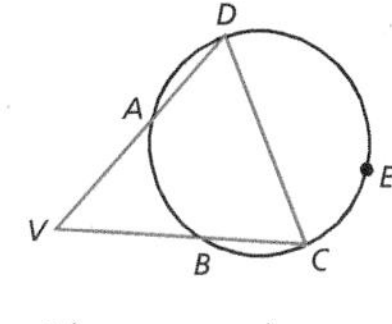

A. Angle AVB is formed by two secant segments that intersect in the exterior of the circle.

$m\angle AVB = \frac{1}{2}(m\overset{\frown}{DC} - m\overset{\frown}{AB})$ — A. Why was this equation chosen?

$= \frac{1}{2}(180° - 48°)$ Substitute the given measures.

B. Why 180°?

B. Since $\overline{DC}$ is the diameter of the circle, the arc formed by the endpoints of $\overline{DC}$ is half the circle or 180°.

$= \frac{1}{2}(132°)$ Subtract.

$= 66°$ Multiply.

C. Which theorem is used for the problem, and what is the measure of $\angle AVB$?

C. Tangent-Secant Exterior Angle Measure Theorem; 66°

Turn and Talk If $\overset{\frown}{AB}$ was formed with two tangent lines instead of two secant lines, does $m\angle AVB$ change? Explain why or why not. See margin.

ANSWERS

5.B. Degree measures of the labeled arcs. The formula uses these values to find the measure of the angle.

5.C. Intersecting Chords Angle Measure Theorem; 67°

Task 5 (MP) **Construct Arguments** Have students explain why the equation can be used when two chords intersect within a circle and give a theorem to support their answer.

Sample Guided Discussion:

Q **How do you know to use the sum and not the difference in the equation?** The chords intersect and the angle is formed inside the circle.

Q **How can you check that $m\angle DTC = 67°$?** You can substitute the value back into the equation $x = \frac{1}{2}(48 + 86)$.

Task 6 (MP) **Construct Arguments** Students examine the structure of a stone arrangement and decide why a given equation is appropriate to solve for a missing measure, justifying their work with a theorem.

SUPPORT SENSE-MAKING Three Reads

Have students read the problem three times. Use the questions below for a different focus each read.

1. What is the situation about?
2. What are the quantities in the situation?
3. What are some possible mathematical questions that you could ask for this situation?

Sample Guided Discussion:

Q **What type of lines form $\angle AVB$?** two secants

Q **How do you know $\overline{DC}$ goes through the center of the circle?** It is a diameter, which is given information in the problem.

Turn and Talk Have students sketch a drawing of what it means to use tangent lines instead of secants. Then point out the arcs that are used in the different cases in the Tangent-Secant Exterior Angle Measure Theorem. yes; It changes because you would use the arcs formed where the tangents meet the circle.

$$m\angle AVB = \frac{1}{2}(m\overset{\frown}{AEB} - m\overset{\frown}{AB})$$
$$= \frac{1}{2}((360° - 48°) - 48°)$$
$$= \frac{1}{2}(312° - 48°)$$
$$= 132°$$

Assign the Digital On Your Own for

- built-in student supports
- Actionable Item Reports
- Standards Analysis Reports

On Your Own

Assignment Guide

The chart below indicates which problems in the On Your Own are associated with each task in the Learn Together. Assign daily homework for tasks completed.

Learn Together Tasks	On Your Own Problems
Task 1, p. 512	Problems 5–7
Task 2, p. 513	Problem 9
Task 3, p. 513	Problem 8
Task 4, p. 514	Problems 10 and 11
Task 5, p. 515	Problems 12–15 and 19
Task 6, p. 515	Problems 16–18

ANSWERS

1. The measure of an angle whose vertex is inside the circle is half the sum of its intercepted arcs.
2. The measure of an angle whose vertex is on the circle is half the measure of its intercepted arc because the angle is an inscribed angle.
3. Each ray of an inscribed angle intersects the circle. So, the angle intersects the circle in three places once along each ray and at its vertex, but a tangent-secant interior angle intersects the circle once along its secant and at its vertex.

Check Understanding

1. Explain the relationship between angles formed by lines intersecting inside a circle and the intercepted arcs. 1–3. See margin.
2. What is the relationship between an angle whose vertex is on the circle and its intercepted arc? Explain.
3. How are tangent-secant interior angles different from inscribed angles?
4. Use a theorem and the diagram shown to find the measures of $\widehat{EB}$ and $\angle ABC$. $m\widehat{EB} = 84°$; $m\angle ABC = 78°$

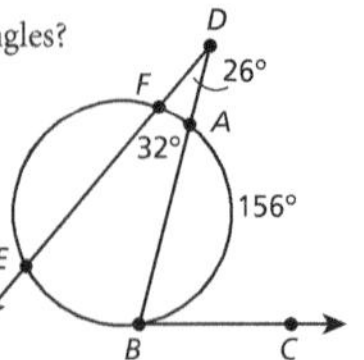

On Your Own

5. Is the equation in the Intersecting Chords Angle Measure Theorem an equation for finding an average? Explain. See Additional Answers.

In Problems 6 and 7, use the diagram and given information. Given: Chords $\overline{AB}$ and $\overline{CD}$ intersect at the center of the circle, point E. 6–10. See Additional Answers.

6. (MP) **Construct Arguments** Show that $m\angle AEC = \frac{1}{2}(m\widehat{CA} + m\widehat{BD})$ using the diagram and a two-column proof.
7. How does your proof for Problem 6 use different geometric relationships and theorems than the proof in Task 1?

8. Prove the Tangent-Secant Interior Angle Measure Theorem, where the center of the circle is exterior to $\angle ABC$

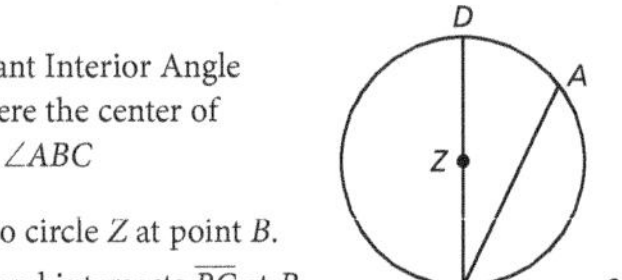

Given: $\overline{BC}$ is tangent to circle Z at point B. $\overline{AB}$ is a secant and intersects $\overline{BC}$ at B.

Prove: $m\angle ABC = \frac{1}{2}m\widehat{AB}$

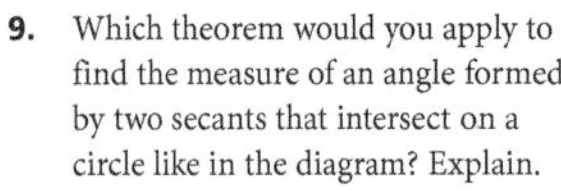

9. Which theorem would you apply to find the measure of an angle formed by two secants that intersect on a circle like in the diagram? Explain.

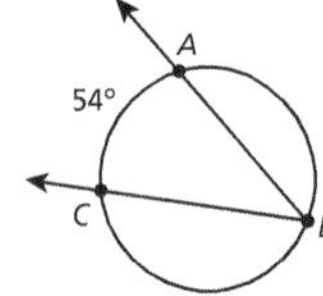

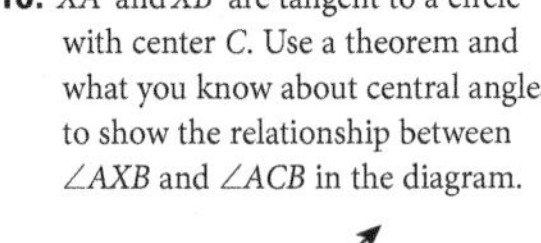

10. $\overrightarrow{XA}$ and $\overrightarrow{XB}$ are tangent to a circle with center C. Use a theorem and what you know about central angles to show the relationship between $\angle AXB$ and $\angle ACB$ in the diagram.

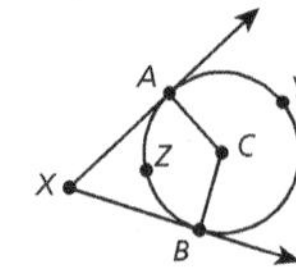

516

3 Check Understanding

Formative Assessment

Use formative assessment to determine if your students are successful with this lesson's learning objective.

Students who successfully complete the Check Understanding can continue to the On Your Own practice.

For students who miss 1 problem or more, work in a pulled small group using the Almost There small-group activity on page 511C.

ONLINE

Assign the Digital Check Understanding to determine

- success with the learning objective
- items to review
- grouping and differentiation resources

4 Differentiation Options

Differentiate instruction for all students using small-group activities and math center activities on page 511C.

Reteach

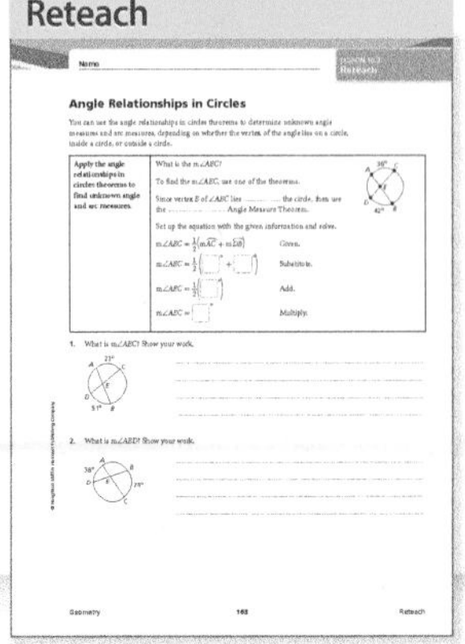

Challenge

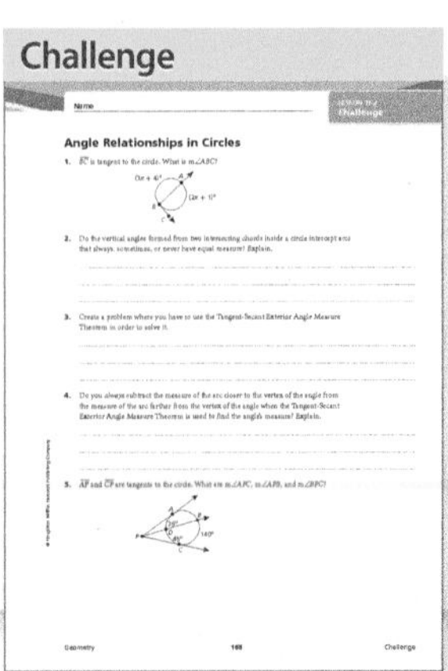

11. Prove the Tangent-Secant Exterior Angle Measure Theorem for a secant and tangent, and then compare the proof for two tangents. See Additional Answers.

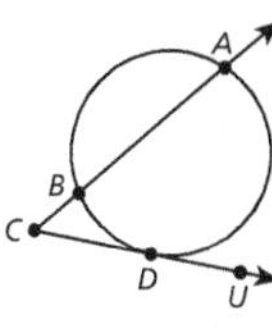

A. Write a proof for a secant and tangent that intersect in the exterior of the circle.

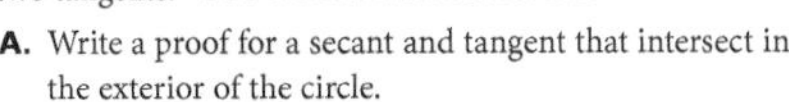

Given: Tangent $\overline{CD}$ and secant $\overrightarrow{CA}$ intersect at C in the exterior of the circle.

Prove: $m\angle ACD = \frac{1}{2}(m\widehat{AD} - m\widehat{BD})$

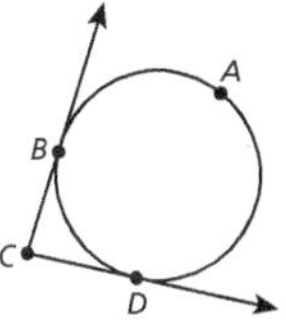

B. Suppose two tangents intersect outside the circle. Describe how you can modify the proof in part A to prove that $m\angle BCD = \frac{1}{2}(m\widehat{BAD} - m\widehat{BD})$

Find the indicated measure.

12. $m\widehat{AEB}$ 146°

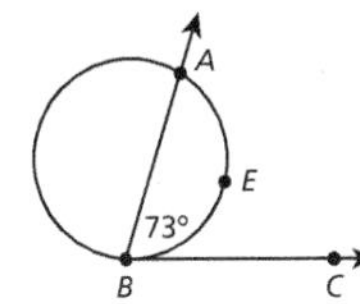

13. $m\angle ACD$ 61.5°

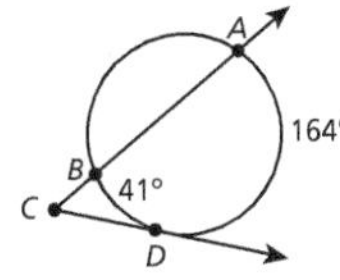

14. $m\angle AEC$ 43°

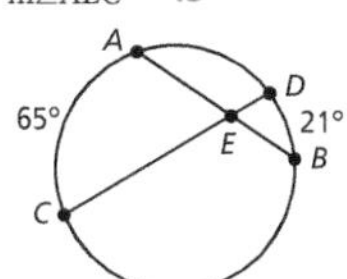

15. $m\angle ACE$ 9°

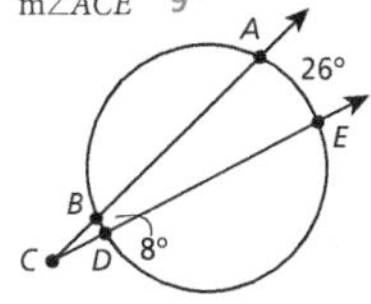

16. Sean designed the necklace shown. It is a circular stone held by a triangular wire. The measure of the angle at vertex C where the triangle attaches to a chain is 72°. The base of the wire crosses the stone across its center. What is the measure of $\widehat{BD}$? 36°

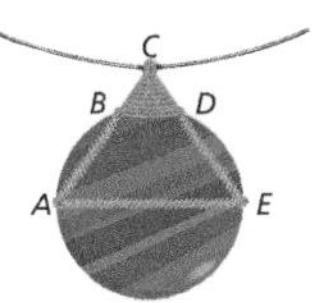

17. The outline of a symmetrical logo is shown in the diagram. It is an X, W, and O layered on top of each other. $\overrightarrow{BA}$ and $\overrightarrow{DE}$ are tangent to the circle. The difference between $m\widehat{BD}$ and $m\widehat{FG}$ is 84°. What is $m\widehat{FG}$ and $m\angle ABG$?

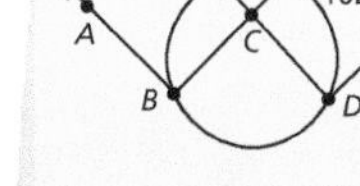

$m\widehat{FG} = 36°$ and $m\angle ABG = 69°$

Questioning Strategies

Problem 12 Make sure students understand what the 73° refers to in the diagram. Does the measure given refer to the angle or to the arc measure? How does this affect the values that you use in your equation?

The measure refers to the angle that will go on the left side of the equation, $m\angle ABC = \frac{1}{2}m\widehat{AEB}$. You can write the equation $73 = \frac{1}{2}m\widehat{AEB}$ and solve for the measure of the arc.

Watch for Common Errors

Problem 15 Point out that some theorems in this lesson use addition and some use subtraction. Have students identify that the secants in this problem are intersecting outside of the circle. Students may use addition in their equation by mistake, but encouraging them to double check the theorems and how they relate to the diagrams may help avoid this error.

5 Wrap-Up

Summarize learning with your class. Consider using the Exit Ticket, Put It in Writing, or I Can scale.

Exit Ticket 

An artist makes a design out of metal. The artist measures the two arcs as shown. Does the design intersect in the center of the circle? How do you know? yes; To find the unknown angle measure in the middle, use the equation $x° = \frac{1}{2}(45° + 45°)$ where $x°$ is the unknown angle measure. Solving this gives that the unknown angle measure is half of 90, or 45°. Since the unknown angle measure matches the arc measure, the angle is a central angle whose vertex is located at the center of the circle.

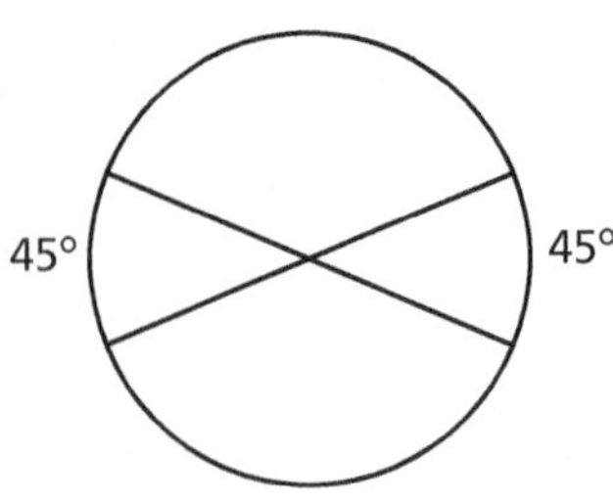

Put It in Writing 

Have students write a summary of the different theorems in this lesson that relate angle and arc measures in a circle.

I Can 

The scale below can help you and your students understand their progress on a learning goal.

4	I can use angle relationships in circles to solve mathematical and real-world problems, and I can explain my reasoning to others.
3	I can use angle relationships in circles to solve mathematical and real-world problems.
2	I can use equations that relate angles and arcs in circles.
1	I can identify tangents, secants, and chords in circles.

Spiral Review • Assessment Readiness

These questions will help determine if students have retained information taught in the past and can also prepare them for high-stakes assessments. Here, students use the property that a tangent is perpendicular to a radius to solve a real-world problem **(15.3)**, find the center and radius from an equation of a circle **(15.4)**, find the length of a tangent segment given the length of a secant **(16.1)**, and use the circumference of a circle to find its radius **(Gr7, 10.1)**.

18. STEM The superior oblique and inferior rectus are two muscles that help control eye movement. They intersect behind the eye to create $\angle ACB$. If $m\widehat{AEB} = 200°$ what is $m\angle ACB$? 20°

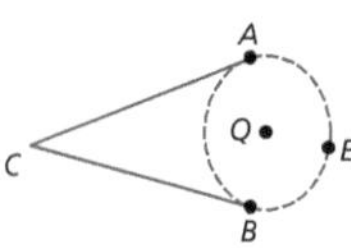

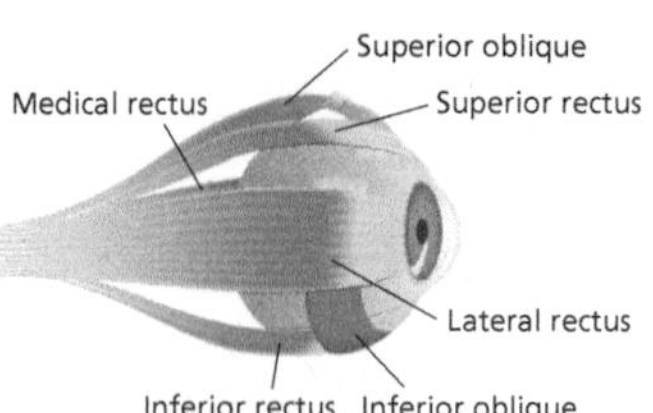

19. Open Middle™ Using the digits 0 to 9, at most one time each, fill in the boxes so that the central angle, the inscribed angle, and the angle formed by two tangent segments have the correct relationship. The diagram is not shown to scale. 76°, 38°, 142°

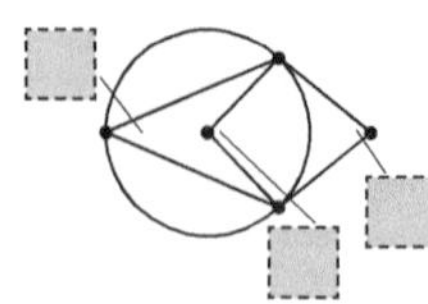

Spiral Review • Assessment Readiness

20. During an orbit of Saturn, the Cassini satellite was 1,680 miles from Saturn. The radius of Saturn is approximately 33,780 miles where Cassini was directly above Saturn. What is the approximate distance from the satellite to Saturn's horizon (which can be represented by a tangent segment)?

Ⓐ 10,785 mi
Ⓑ 33,738 mi
Ⓒ 33,911 mi
Ⓓ 48,974 mi

21. Which statement is true for the equation of the circle $x^2 + y^2 + 12x + 4y + 24 = 0$?

Ⓐ radius 16
Ⓑ radius 8
Ⓒ center (2, 6)
Ⓓ center (−6, −2)

22. Given the circle, the secant, and the tangent shown in the diagram, what is the length of the tangent segment to the nearest tenth?

Ⓐ 6.7
Ⓑ 8.4
Ⓒ 11.2
Ⓓ 20.1

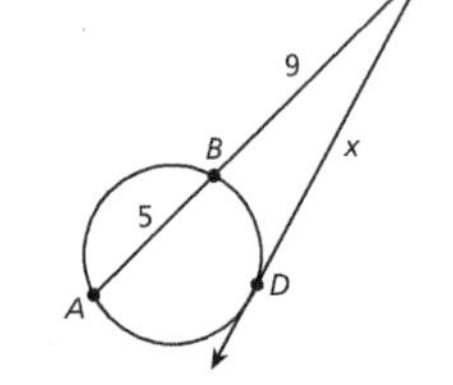

23. A decorative circular plate has a circumference of 33.9 inches. What is the plate's radius to the nearest tenth of an inch?

Ⓐ 3.3 inches
Ⓑ 5.4 inches
Ⓒ 10.8 inches
Ⓓ 17 inches

I'm in a Learning Mindset!

How am I responding to angle relationships in circles instead of segment relationships in circles?

Keep Going ▶ Journal and Practice Workbook

Learning Mindset

Resilience Adjusts to Change

In the previous lesson, students learned about the relationships between the lengths of tangents and secants. Now students are using the tangents and secants to find angle and arc measures. There are a lot of theorems, and students may feel confused because there are many different situations. Encourage students to examine each diagram and carefully choose which theorem applies. *How can you tell which theorem is best to use? What is the difference between a tangent and a secant? What is the difference between an angle measure and an arc measure? What can you do to help you remember which theorem to use?*

Module

16 Review

Segment Relationships in Circles

Jasmine plants a circular garden. She places steppingstones along $\overline{AC}$ and $\overline{BD}$.

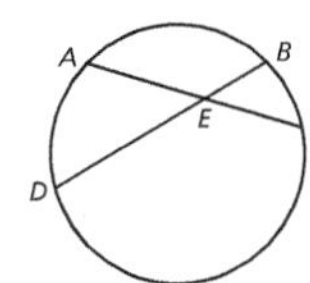

Given $AE = 11$ feet, $EC = 9$ feet, and $EB = 7$ feet, you can use the Chord-Chord Product Theorem to find DE.

$$AE \cdot EC = DE \cdot EB$$
$$11 \cdot 9 = DE \cdot 7$$
$$99 = DE \cdot 7$$
$$14.14 \approx DE$$

DE is approximately 14 feet.

Jasmine's cousin also plants a circular garden.

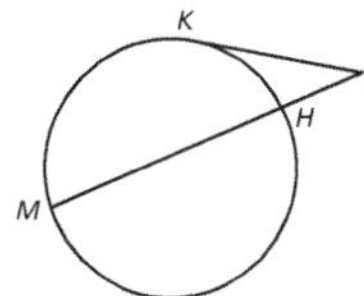

He builds a tool shed at point J. Given $HM = 18$ feet and $JH = 4$ feet, you can use the Secant-Tangent Product Theorem to find the distance from point K to the tool shed.

$$JH \cdot JM = JK^2$$
$$4 \cdot (4 + 18) = JK^2$$
$$4 \cdot 22 = JK^2$$
$$88 = JK^2$$
$$9.38 \approx JK$$

The tool shed is approximately 9.4 feet from point K.

Angle Relationships in Circles

Jasmine is planning to place a statue at E. Given $m\overset{\frown}{AD} = 15°$ and $m\overset{\frown}{BC} = 9.8°$, you can use the Intersecting Chords Angle Measure Theorem to find $m\angle AED$.

$$m\angle AED = \frac{1}{2}\left(m\overset{\frown}{AD} + m\overset{\frown}{BC}\right)$$
$$m\angle AED = \frac{1}{2}(15° + 9.8°)$$
$$m\angle AED = \frac{1}{2}(24.8°)$$
$$m\angle AED = 12.4°$$

Jasmine's cousin places a spotlight on the corner of his tool shed. Given $m\overset{\frown}{MK} = 135°$ and $m\angle KJM = 24°$, you can use the Tangent-Secant Exterior Angle Measure Theorem to find the amount of the garden's edge illuminated by the spotlight.

$$m\angle KJM = \frac{1}{2}\left(m\overset{\frown}{MK} - m\overset{\frown}{HK}\right)$$
$$24° = \frac{1}{2}\left(135° - m\overset{\frown}{HK}\right)$$
$$48° = 135° - m\overset{\frown}{HK}$$
$$m\overset{\frown}{HK} = 87°$$

Approximately 87° of the garden's edge is illuminated by the spotlight.

ONLINE

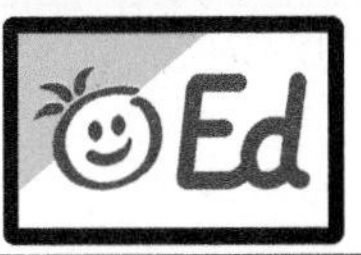

Assign the Digital Module Review for

- built-in student supports
- Actionable Item Reports
- Standards Analysis Reports

Module Review

Use the first page of the Module Review to summarize and connect the main ideas of the module. Use the second page to assess students' understanding of the vocabulary, concepts, and skills presented in the module.

Sample Guided Discussion:

Q Segment Relationships in Circles When finding the distance to the tool shed, why is 4 used as both a factor and an addend? In applying the Secant-Tangent Product Theorem, you find the product of JH and JM. Because JH is 4 feet, 4 is a factor in the equation $JH \cdot JM = JK^2$. The length of $\overline{JM}$ is not given, but $JM = JH = HM$, which is why 4 is also used as an addend in the equation.

Q Angle Relationships in Circles When applying the Tangent-Secant Exterior Angle Measure Theorem, how do you know what arc length you need to find? Possible answer: The Tangent-Secant Exterior Angle Measure Theorem describes the relationship between the exterior angle created by the intersection of a tangent and a secant of the circle and the arcs formed by the tangent and secant. In this problem, tangent $\overline{JK}$ and secant $\overline{JM}$ form $\overset{\frown}{MK}$ and $\overset{\frown}{HK}$. The measure of $\overset{\frown}{MK}$ is given, so you need to find the measure of $\overset{\frown}{HK}$.

Module Review continued

Possible Scoring Guide

Items	Points	Description
1–3	2 each	identifies the correct term(s)
4	1	correctly identifies the theorem
4	2	correctly identifies the theorem, and correctly finds the length of the chord
5	1	correctly states who has farther to travel and by how much
5	2	correctly states who has farther to travel and by how much, and correctly states who has farther to swim and by how much
6	2	correctly calculates the horizontal distance
7	2	correctly calculates $m\widehat{EC}$
8	2	correctly calculates $m\angle S$
9	2	calculates the length of the fountain wall that needs to be replaced correctly and provides an appropriate explanation
Total points possible = 18 points		

The Unit 8 Test in the Assessment Guide assesses content from Modules 15–17.

Vocabulary

Choose the correct term from the box to complete each sentence.

Vocabulary
chord
external secant segment
secant
tangent
tangent segment

1. A(n) __?__ is a line that intersects a circle at two points. A segment of this line that is outside of the circle with only one endpoint on the circle is a(n) __?__. secant; external secant segment
2. A(n) __?__ is a segment whose endpoints lie on a circle. chord
3. A(n) __?__ is a line that is in the same plane as a circle and intersects the circle in exactly one point. A segment of this line that has one endpoint on the circle is a(n) __?__. tangent; tangent segment

Concepts and Skills

4. **A.** What theorem can be used to find x? A. Chord-Chord Product Theorem
   B. What is BD? 4.9

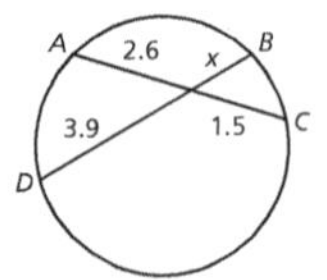

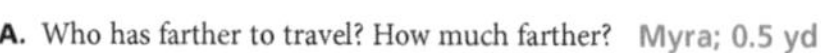

5. Myra and Leonard are sitting on the edge of a circular pool and batting a beach ball back and forth. A sudden gust of wind blows the ball across the pool, and it comes to rest at point B. Myra and Leonard swim across the pool, hop over the side, and walk to reach the ball.
   A. Who has farther to travel? How much farther? Myra; 0.5 yd
   B. Who has farther to swim? How much farther? Myra; about 0.7 yd

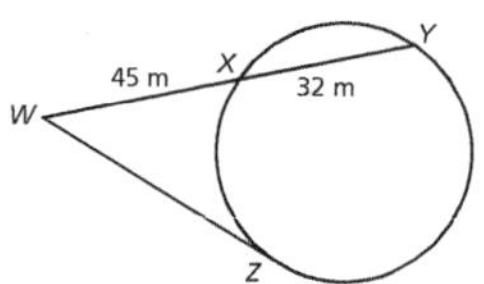

6. (MP) **Use Tools** Divers are exploring the areas surrounding a shipwreck. The points X, Y, and Z mark spots at the circular bottom of the lake where the divers locate artifacts. What is the horizontal distance from the artifact located at point Z and basecamp, W? Round to the nearest meter. State what strategy and tool you will use to answer the question, explain your choice, and then find the answer. 59 m

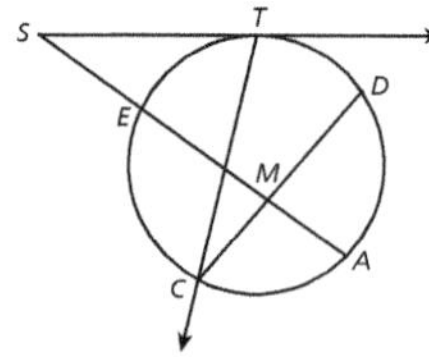

A fountain with a circumference of 300 feet has several points of interest.

7. Given $m\angle DMA = 84°$ and $m\widehat{DA} = 72°$, find $m\widehat{EC}$. 96°
8. Given $m\widehat{ET} = 85°$ and $m\widehat{AT} = 155°$, find $m\angle S$. 35°
9. The wall of the fountain along $\widehat{TC}$ needs to be repaired. Given $m\angle STC = 69°$, what length of the fountain wall needs to be repaired? Explain your reasoning. See Additional Answers.

520

DATA-DRIVEN INSTRUCTION

Before moving on to the Module Test, use the Module Review results to intervene based on the table below.

MTSS (RtI)

Items	Lesson	DOK	Content Focus	Intervention
4	16.1	2	Use the Chord-Chord Product Theorem to calculate the chord.	Reteach 16.1
5	16.1	3	Use the Secant-Secant Product Theorem to solve a real-world problem.	Reteach 16.1
6	16.1	3	Use the Secant-Tangent Product Theorem to solve a real-world problem.	Reteach 16.1
7	16.2	2	Use the Intersecting Chords Angle Measure Theorem to find the measure of an arc.	Reteach 16.2
8	16.2	2	Use the Tangent-Secant Exterior Angle Measure Theorem to find the measure of an angle.	Reteach 16.2
9	16.2	3	Use the Tangent-Secant Product Theorem to solve a real-world problem.	Reteach 16.2

Module Test

The Module Test is available in alternative versions in your Assessment Guide. All items are presented in standardized test formats.

ONLINE

Ed

Assign the Digital Module Test to power actionable reports including

- proficiency by standards
- item analysis

Form A

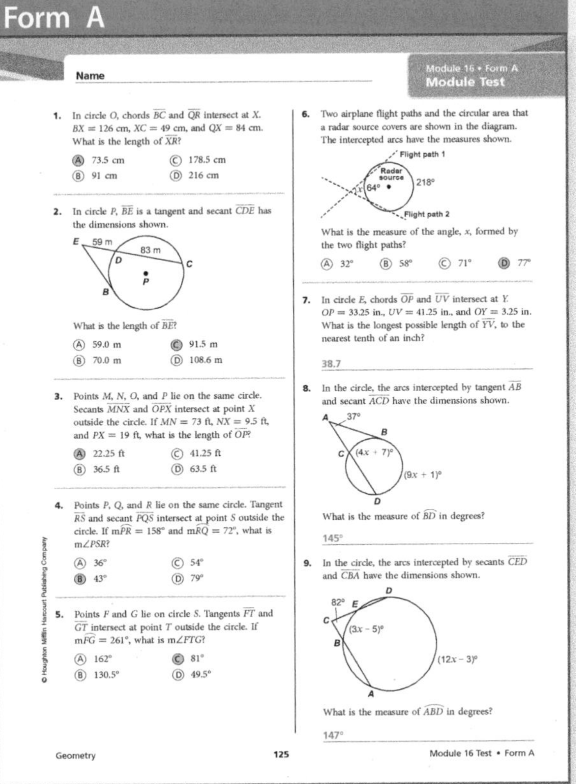

Module 16 • Form A
Module Test

Name

1. In circle O, chords $\overline{BC}$ and $\overline{QR}$ intersect at X. $BX = 126$ cm, $XC = 49$ cm, and $QX = 84$ cm. What is the length of $\overline{XR}$?
Ⓐ 73.5 cm Ⓒ 178.5 cm
Ⓑ 91 cm Ⓓ 216 cm

2. In circle P, $\overline{BE}$ is a tangent and secant $\overline{CDE}$ has the dimensions shown.
What is the length of $\overline{BE}$?
Ⓐ 59.0 m Ⓒ 91.5 m
Ⓑ 70.0 m Ⓓ 108.6 m

3. Points M, N, O, and P lie on the same circle. Secants $\overline{MNX}$ and $\overline{OPX}$ intersect at point X outside the circle. If $MN = 73$ ft, $NX = 9.5$ ft, and $PX = 19$ ft, what is the length of $\overline{OP}$?
Ⓐ 22.25 ft Ⓒ 41.25 ft
Ⓑ 36.5 ft Ⓓ 63.5 ft

4. Points P, Q, and R lie on the same circle. Tangent $\overline{RS}$ and secant $\overline{PQS}$ intersect at point S outside the circle. If $m\widehat{PR} = 158°$ and $m\widehat{RQ} = 72°$, what is $m\angle PSR$?
Ⓐ 36° Ⓒ 54°
Ⓑ 43° Ⓓ 79°

5. Points F and G lie on circle S. Tangents $\overline{FT}$ and $\overline{GT}$ intersect at point T outside the circle. If $m\widehat{FG} = 261°$, what is $m\angle FTG$?
Ⓐ 162° Ⓒ 81°
Ⓑ 130.5° Ⓓ 49.5°

6. Two airplane flight paths and the circular area that a radar source covers are shown in the diagram. The intercepted arcs have the measures shown.
What is the measure of the angle, x, formed by the two flight paths?
Ⓐ 32° Ⓑ 58° Ⓒ 71° Ⓓ 77°

7. In circle E, chords $\overline{OP}$ and $\overline{UV}$ intersect at Y. $OP = 33.25$ in., $UV = 41.25$ in., and $OY = 3.25$ in. What is the longest possible length of $\overline{YV}$, to the nearest tenth of an inch?
38.7

8. In the circle, the arcs intercepted by tangent $\overline{AB}$ and secant $\overline{ACD}$ have the dimensions shown.
What is the measure of $\widehat{BD}$ in degrees?
145°

9. In the circle, the arcs intercepted by secants $\overline{CED}$ and $\overline{CBA}$ have the dimensions shown.
What is the measure of $\widehat{ABD}$ in degrees?
147°

Geometry 125 Module 16 Test • Form A

© Houghton Mifflin Harcourt Publishing Company

Form A

Module 16 • Form A
Module Test

Name

10. Points J and K lie on circle O. $\overline{KL}$ is tangent to the circle. If $m\widehat{JK} = 239°$, what is the smallest possible measure of $\angle JKL$ in degrees?
Possible answer: 60.5°

11. Points A, B, and C lie on the same circle. Tangent $\overline{AD}$ and secant $\overline{CBD}$ intersect at point D outside the circle. If $CB = 0.7$ in. and $BD = 1.2$ in., what is the length of $\overline{AD}$, to the nearest tenth of an inch?
1.5

12. In the circle, $\overline{PS}$ and $\overline{PT}$ are secants with the dimensions shown.
What is the length of $\overline{RT}$, to the nearest tenth of an inch?
22.2

13. In the circle, chords $\overline{FOH}$ and $\overline{GOK}$ have the dimensions shown.
What is OG in centimeters and $m\angle GOH$ in degrees?
$OG =$ 16.5
$m\angle GOH =$ 102.5°

14. A satellite orbiting Earth is at point A in the diagram. $\overline{AB}$ and $\overline{BC}$ are tangent to Earth's surface.
If $m\angle BAC = 80°$, what is the measure of the arc not visible to the satellite, to the nearest degree?
260°

15. Five exercise stations in a park are connected by a series of walking paths, including one circular path. The exercise stations are at points A, B, C, D, and E in the diagram.

Part A
If the path from the entrance to exercise station A is tangent to the circular path, what is the shortest distance between exercise stations C and D, to the nearest foot?
70

Part B
What is the length of the path from exercise station A to E, to the nearest foot?
64

Geometry 126 Module 16 Test • Form A

© Houghton Mifflin Harcourt Publishing Company

Form B

Module 16 • Form B
Module Test

Name

1. In circle P, chords $\overline{DE}$ and $\overline{JK}$ intersect at X. $DX = 86$ in., $XE = 86$ in., and $JX = 40$ in. What is the length of $\overline{XK}$?
Ⓐ 40 in. Ⓒ 184.9 in.
Ⓑ 132 in. Ⓓ 329.8 in.

2. In circle Q, $\overline{FI}$ is a tangent and secant $\overline{IHG}$ has the dimensions shown.
What is the length of $\overline{GH}$?
Ⓐ 7.3 ft Ⓒ 10.9 ft
Ⓑ 9.2 ft Ⓓ 16.8 ft

3. Points Q, R, S, and T lie on the same circle. Secants $\overline{QRZ}$ and $\overline{STZ}$ intersect at point Z outside the circle. If $RZ = 18.5$ m, $ST = 49$ m, and $TZ = 37$ m, what is the length of $\overline{QR}$?
Ⓐ 67.5 m Ⓒ 153.5 m
Ⓑ 98 m Ⓓ 172 m

4. Points K, L, and M lie on the same circle. Tangent $\overline{MN}$ and secant $\overline{KLN}$ intersect at point N outside the circle. If $m\widehat{KM} = 204°$ and $m\widehat{LM} = 92°$, what is $m\angle MNK$?
Ⓐ 44° Ⓒ 56°
Ⓑ 46° Ⓓ 78°

5. Points H and I lie on circle O. Tangents $\overline{HP}$ and $\overline{IP}$ intersect at point P outside the circle. If $m\widehat{HI} = 129°$, what is $m\angle HPI$?
Ⓐ 115.5° Ⓒ 64.5°
Ⓑ 102° Ⓓ 51°

6. A ribbon is attached to a circular mirror that hangs from a hook on the wall as shown in the diagram. The intercepted arcs have the measures shown.
What is the measure of the angle, x, formed by the two parts of the ribbon at the hook?
Ⓐ 34° Ⓑ 39° Ⓒ 56° Ⓓ 73°

7. In circle F, chords $\overline{MN}$ and $\overline{UT}$ intersect at Y. $MN = 53.25$ cm, $MY = 36$ cm, and $UT = 52$ cm. What is the shortest possible length of $\overline{YT}$, to the nearest tenth of a centimeter?
18.6

8. In the circle, the arcs intercepted by secant $\overline{LKM}$ and tangent $\overline{MJ}$ have the dimensions shown.
What is the measure of $\widehat{JK}$ in degrees?
106°

9. In the circle, the arcs intercepted by secants $\overline{HKJ}$ and $\overline{HGF}$ have the dimensions shown.
What is the measure of $\widehat{FGJ}$ in degrees?
267°

Geometry 127 Module 16 Test • Form B

© Houghton Mifflin Harcourt Publishing Company

Form B

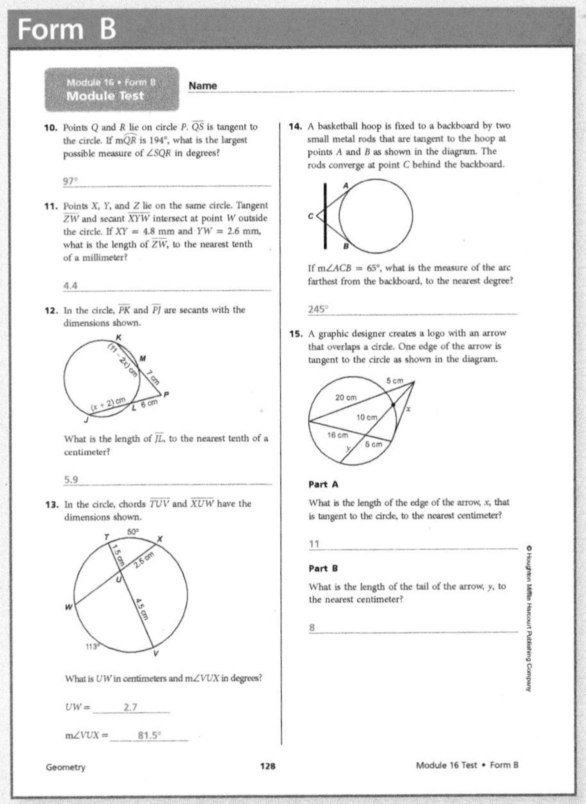

Module 16 • Form B
Module Test

Name

10. Points Q and R lie on circle P. $\overline{QS}$ is tangent to the circle. If $m\widehat{QR} = 194°$, what is the largest possible measure of $\angle SQR$ in degrees?
97°

11. Points X, Y, and Z lie on the same circle. Tangent $\overline{ZW}$ and secant $\overline{XYW}$ intersect at point W outside the circle. If $XY = 4.8$ mm and $YW = 2.6$ mm, what is the length of $\overline{ZW}$, to the nearest tenth of a millimeter?
4.4

12. In the circle, $\overline{PK}$ and $\overline{PJ}$ are secants with the dimensions shown.
What is the length of $\overline{JL}$, to the nearest tenth of a centimeter?
5.9

13. In the circle, chords $\overline{TUV}$ and $\overline{XUW}$ have the dimensions shown.
What is UW in centimeters and $m\angle VUX$ in degrees?
$UW =$ 2.7
$m\angle VUX =$ 81.5°

14. A basketball hoop is fixed to a backboard by two small metal rods that are tangent to the hoop at points A and B as shown in the diagram. The rods converge at point C behind the backboard.
If $m\angle ACB = 65°$, what is the measure of the arc farthest from the backboard, to the nearest degree?
245°

15. A graphic designer creates a logo with an arrow that overlaps a circle. One edge of the arrow is tangent to the circle as shown in the diagram.

Part A
What is the length of the edge of the arrow, x, that is tangent to the circle, to the nearest centimeter?
11

Part B
What is the length of the tail of the arrow, y, to the nearest centimeter?
8

Geometry 128 Module 16 Test • Form B

© Houghton Mifflin Harcourt Publishing Company

CIRCUMFERENCE AND AREA OF A CIRCLE

Introduce and Check for Readiness

- Module Performance Task • Are You Ready?

Lesson 17.1—2 Days

Measure Circumference and Area of a Circle

Learning Objective: Justify and use the formulas for the circumference and area of a circle to solve real-world and mathematical problems.

New Vocabulary: limit

Review Vocabulary: circumference

Lesson 17.2—2 Days

Measure Arc Length and Use Radians

Learning Objective: Use the arc length formula and apply it to real-world problems, and convert between degree and radian measure.

New Vocabulary: arc length, concentric circles, radian measure

Review Vocabulary: arc

Lesson 17.3—2 Days

Measure Sector Area

Learning Objective: Derive the formula for the area of a sector of a circle and use that formula to compute the area of sectors of circles having different central angles and radii.

New Vocabulary: sector

Assessment

- Module 17 Test (Forms A and B)
- Unit 8 Test (Forms A and B)

LEARNING ARC FOCUS

Build Conceptual Understanding  Connect Concepts and Skills  Apply and Practice

TEACHING FOR SUCCESS

TEACHING FOR DEPTH: Circumference and Area of a Circle

Meaning of Radian measure. Students have measured angles using degrees in their previous math courses. In this module, students develop another, method for measuring angles.

The concept of the radian measure of an angle relates the measure of a central angle to the arc length of the related intercepted arc. A radian is the measure of a central angle that intercepts an arc having length equal to the length of the radius of the circle. An angle that measures 1 radian is about 57.3°.

For a circle with radius of 1, or 1 radian, the radian measure of an angle and the length of the intercepted arc for the angle is the same. For example, consider a central angle of 60° in a circle with radius 1. The arc length is $\frac{60°}{360°}$, or $\frac{1}{6}$, of the circumference of the circle. So, the length of the arc is $\frac{1}{6} \cdot 2\pi \cdot 1$ radians, or $\frac{\pi}{3}$ radians. In radians, the angle measure of a 60° angle is $\frac{\pi}{3}$.

As students continue their mathematical studies, they will use radian measure when they study the equations and graphs of trigonometric functions.

Mathematical Progressions

Prior Learning	Current Development	Future Connections
Students: • used the formulas for the circumference and area of a circle to solve problems. • used trigonometric ratios and the Pythagorean Theorem to solve right triangles in applied problems. • proved that all circles are similar. • identified relationships among inscribed angles, radii, and chords of circles.	**Students:** • justify the formulas for the circumference and area of a circle. • use the formulas for the circumference and area of a circle to solve mathematical and real-world problems. • derive and use the formula for arc length and area of a sector to solve mathematical and real-world problems. • convert between degree and radian measure.	**Students:** • will use geometric shapes, their measures, and their properties to describe three-dimensional objects. • will find the volumes and surface areas of cylinders, pyramids, cones, and spheres. • will use the formulas for volume and surface area to solve problems. • will explain how the unit circle in the coordinate plane can be used to extend the trigonometric functions to all real numbers.

From the Classroom

Support productive struggle in learning mathematics. In this module, students are asked to justify and derive several formulas. These skills are quite different from learning to apply formulas and often cause my students to struggle.

However, I feel that the struggle they experience is crucial for them to develop the reasoning and analytical skills they will need to be successful in later courses and in solving real-world problems.

To support their struggle, I use the following process for each task that asks students to justify or derive a formula. First, I ask the students to read through the task individually and write down any questions they have. I circulate throughout the room as students record their questions and then place them in small groups. Within their groups, students work through the task, discuss, and agree on their responses as a group.

I continue to circulate during the group work, listening to the conversations and offering help as needed. Then, I choose one or two groups to share with the class how they were able to justify or derive the formula. This method of starting with individual reading, continuing on to group discussion and problem solving, and finally, sharing as a class really helps my students work productively through their struggles with the content.

LANGUAGE DEVELOPMENT • Planning for Instruction

By giving all students regular exposure to language routines in context, you will provide opportunities for students to **listen**, **speak**, **read**, and **write** about mathematical situations and develop both mathematical language and conceptual understanding at the same time.

Using Language Routines to Develop Understanding

Use the **Professional Cards** for the following routines to plan for effective instruction.

Co-Craft Questions   Lesson 17.2

Students think of natural questions to ask about a given situation or problems similar to a given task and answer the questions they have developed or problems they have created.

Three Reads   Lessons 17.1 and 17.2

Students read a problem three times with a specific focus each time.

1st Read What is the situation about?
2nd Read What are the quantities in the situation?
3rd Read What are the possible mathematical questions that we could ask for the situation?

Information Gap   Lesson 17.3

Students recognize when information given in a problem situation is incomplete, and they pose questions and share knowledge with others to discover any missing facts or relationships and work together to solve the problem.

Critique, Correct, and Clarify   Lesson 17.1

Students correct the work in a flawed explanation, argument, or solution method and share with a partner and refine the sample work.

Connecting Language to Circumference and Area of a Circle

Watch for students' use of the review and new terms listed below as they explain their reasoning and make connections with new concepts.

Key Academic Vocabulary

Prior Learning and Current Development • Review and New Vocabulary

arc an unbroken part of a circle consisting of two points on the circle, called the endpoints, and all the points on the circle between them

circumference the distance around a circle

arc length the distance along an arc, measured in linear units, such as centimeters

concentric circles coplanar circles that have the same center

limit a value that the output of a function approaches as the input increases or decreases without bound or approaches a given value

radian measure the ratio of the length of a circular arc to the radius of the arc

sector a portion of a circle bounded by two radii and their intercepted arc

Linguistic Note

Listen for the terms *radians* and *radian measure*. Although students may understand that *radian measure* is the ratio of a circular arc to the radius of the arc, they may have some difficulty understanding that a *radian* is a unit of measure similar to a degree.

For Spanish-speaking English learners, it may be helpful to tell them that radian/radián are English/Spanish cognates.

It may also be helpful to display an image of 1 radian, which is much larger than 1 degree. While a circle measures 360°, it measures only 2π, or just over 6 radians.

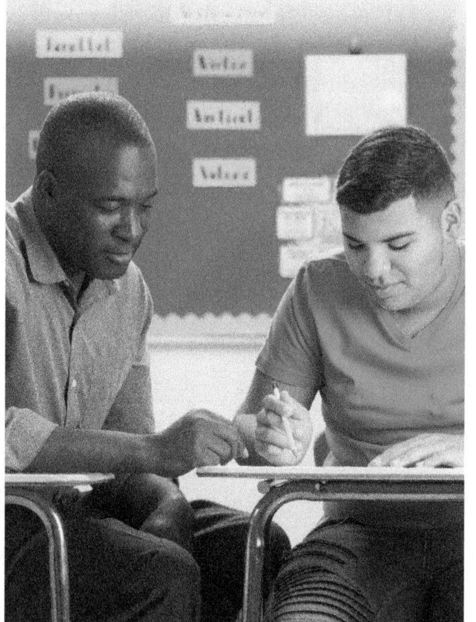

Module

17 Circumference and Area of a Circle

Module Performance Task: Focus on STEM

The Coriolis Effect

Earth rotates as an airplane flies from Buffalo, New York, to Quito, Ecuador. The radius of Earth is 3959 miles and the speed of the plane is 525 miles per hour.

A–D. See margin.

Buffalo, New York
latitude: 43° N
longitude: 78° W

Quito, Ecuador
latitude: 0°
longitude: 78° W

A. What is the distance between Buffalo and Quito? How long does it take the plane to fly this distance?

B. While the plane is in flight, Earth continues to rotate. How far around the circumference of Earth will Quito have moved while the plane is in flight? Has the rotation of Earth also moved the plane? Explain your reasoning.

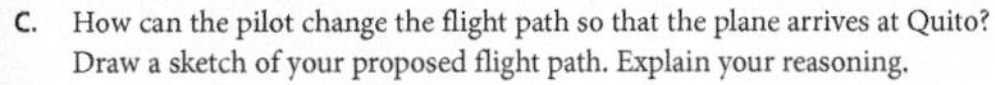

C. How can the pilot change the flight path so that the plane arrives at Quito? Draw a sketch of your proposed flight path. Explain your reasoning.

D. How would this situation change if the plane were flying north from a location in the Southern Hemisphere to Quito? Draw a sketch of your proposed flight path. Compare and contrast this flight path with the flight path from Buffalo.

Mathematical Connections

Task Part	Prerequisite Skills
Part A	Students should be able to apply the arc length formula **(17.2)**, and the formula for the distance traveled at a constant rate, $d = rt$ **(Algebra 1, 2.2).**
Part B	Students should understand that the equator represents the circumference of Earth **(17.1).** They should be able to set up and solve a proportion **(Grade 7, 1.5)**. Students should also be able to apply the arc length formula **(17.2).**
Parts C and D	Students should understand and apply the concept of rotation **(5.2).**

The Coriolis Effect

Overview

The Coriolis effect is the apparent deflection of an object in a rotating system. For example, as Earth rotates, a plane in flight along a north-south line appears not to fly in a straight line with regard to Earth. In the Northern Hemisphere, objects moving in air along a north-south line appear to be deflected to the right, while in the Southern Hemisphere, they appear to be deflected to the left. The Coriolis effect also explains why storms swirl counterclockwise in the Northern Hemisphere and clockwise in the Southern Hemisphere.

Career Connections

Suggest that students research what it would be like to have a career as a meteorologist or oceanographer. Astrophysicists apply the Coriolis effect when studying sunspots or other phenomena on other planets.

Answers

A. about 2971 miles; about 5.66 hours

B. about 5866 miles; yes, but to a lesser extent; the plane is in Earth's atmosphere, which is also rotating, but the plane moves against the force of the rotating atmosphere.

C. The pilot could steer the plane to its left in a southeasterly direction to compensate for the effect of the rotation. The flight path should be a curved line starting from the top and curving down and to its left.

D. If the plane were in the Southern Hemisphere, the plane would appear to deflect to the left, so the pilot would need to steer to the right. The flight path should be a curved line starting from the bottom and curving up and to the right. In both cases, Quito has moved east, but east is to the right when coming from south to north and to the left when coming from north to south.

ONLINE

Assign the Digital Are You Ready? to power actionable reports including

- proficiency by standards
- item analysis

Are You Ready?

Diagnostic Assessment

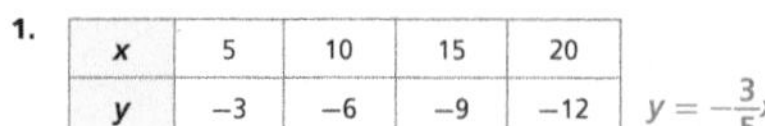

- Diagnose prerequisite mastery.
- Identify intervention needs.
- Modify or set up leveled groups.

Have students complete the *Are You Ready?* assessment on their own. Items test the prerequisites required to succeed with the new learning in this module.

Write Equations for Proportional Relationships Students will apply their previous knowledge of writing equations involving proportional relationships to understand that the length of an arc on a circle intercepted by an angle is proportional to the radius of the circle.

Circumferences of Circles Students will build upon their knowledge of the radius, diameter, and circumference of a circle to define the radian measure of an angle as the ratio of the length of the intercepted arc to the length of the radius of the circle.

Areas of Circles Students will extend their understanding of the area of a circle to discover the formula for the area of a sector of a circle.

ONLINE 

Are You Ready?

Complete these problems to review prior concepts and skills you will need for this module.

Write Equations for Proportional Relationships

Write an equation for each proportional relationship.

1.

x	5	10	15	20
y	−3	−6	−9	−12

$y = -\frac{3}{5}x$

2. One inch is equal to 2.54 centimeters. Write an equation that can be used to convert n inches to c centimeters. $c = 2.54n$

Circumferences of Circles

Use Figures 1 and 2 for Problems 3, 4, 7, and 8. Round to the nearest hundredth.

3. What is the circumference of Figure 1? 20.11 cm
4. What is the circumference of Figure 2? 50.27 in.
5. What is the circumference of a circle with diameter 4 feet? 12.57 ft
6. What is the circumference of a circle with radius 2.5 meters? 15.71 m

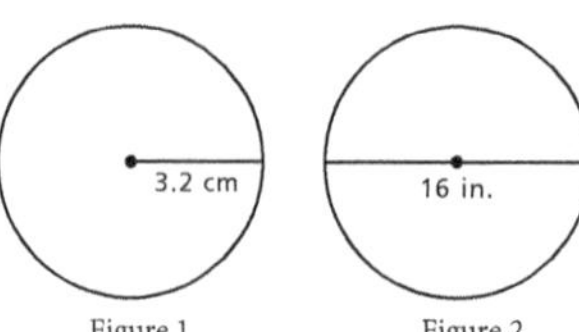

Figure 1 Figure 2

Areas of Circles

7. What is the area of Figure 1? 32.17 cm²
8. What is the area of Figure 2? 201.06 in²
9. What is the area of a circle with diameter 6 feet? 28.27 ft²
10. What is the area of a circle with radius 1.8 meters? 10.18 m²

Connecting Past and Present Learning

Previously, you learned:

- to use proportional relationships to determine side lengths of similar figures,
- to derive the relationship between the circumference and the radius of a circle, and
- to use similarity criteria to solve applied problems.

In this module, you will learn:

- to derive the fact that the length of an arc in a circle intercepted by an angle is proportional to the radius of the circle,
- to define the radian measure of an angle as the constant of proportionality given by the ratio of the length of the subtended arc and the length of the radius, and
- to give an informal argument to develop the formula for the area of a sector.

522

DATA-DRIVEN INTERVENTION

Concept/Skill	Objective	Prior Learning *	Intervene With
Write Equations for Proportional Relationships	Write an equation for a proportional relationship represented by a table or in words.	Grade 7, Lessons 1.2 and 1.5	• Tier 3 Skill 9 • Reteach, Grade 7 Lessons 1.2 and 1.5
Circumferences of Circles	Find the circumference of a circle given the circle's radius or diameter.	Grade 7, Lesson 10.1	• Tier 2 Skill 19 • Reteach, Grade 7 Lesson 10.1
Areas of Circles	Find the area of a circle given the circle's radius or diameter.	Grade 7, Lesson 10.2	• Tier 2 Skill 20 • Reteach, Grade 7 Lesson 10.2

* Your digital materials include access to resources from Grade 6–Algebra 2. The lessons referenced here contain a variety of resources you can use with students who need support with this content.

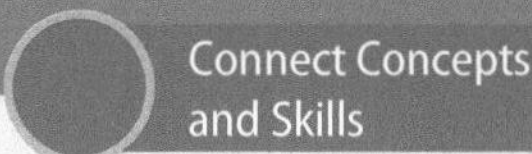

17.1 Measure Circumference and Area of a Circle

LESSON FOCUS AND COHERENCE

Mathematics Standards

- Give an informal argument for the formulas for the circumference of a circle, area of a circle, volume of a cylinder, pyramid, and cone. *Use dissection arguments, Cavalieri's principle, and informal limit arguments.*

Mathematical Practices and Processes 

- Look for and make use of structure.
- Attend to precision.
- Model with mathematics.

I Can Objective 

I can justify and use the circumference and area of a circle formulas to solve real-world problems.

Learning Objective

Justify and use the formulas for the circumference and area of a circle to solve real-world and mathematical problems.

Language Objective

Explain two different ways to find the circumference of a circle when given the measure of its radius.

Vocabulary

Review: circumference

New: limit

Lesson Materials: spreadsheet

Mathematical Progressions

Prior Learning	Current Development	Future Connections
Students: • knew the formulas for circumference and area of a circle and used them to solve problems. **(Gr7, 10.1 and 10.2)** • identified and described relationships among inscribed angles, radii, and chords. **(15.1–15.3)** • used trigonometric ratios and the Pythagorean Theorem to solve right triangles in applied problems. **(13.2–13.4)**	**Students:** • justify formulas for area and circumference. • use formulas for circumference and area of a circle to solve mathematical and real-world problems. • use polygons inscribed in circles to understand formulas for circumference and area of a circle.	**Students:** • will apply geometric methods to solve design problems. **(18.2–18.4 and 19.1–19.3)** • will give an informal argument for the formulas for the volume of a cylinder and cone. **(19.1–19.3)** • will use volume formulas for cylinders, cones, and spheres to solve problems. **(19.1–19.3)**

UNPACKING MATH STANDARDS

Give an informal argument for the formulas for the circumference of a circle, area of a circle, … *Use… informal limit arguments.*

What It Means to You

Students produce informal arguments for circumference and areas of a circle and approach this standard with prior knowledge of how to calculate volume, perimeter, circumference, and area. At this point, students are expected to know many area formulas, but have not experienced the derivation or proof of the formula. Producing informal arguments for circumference and area formulas helps prepare students for later lessons when they will produce arguments for formulas involving three-dimensional solids.

WARM-UP OPTIONS

ONLINE

PROJECTABLE & PRINTABLE

ACTIVATE PRIOR KNOWLEDGE • Use Trigonometric Ratios to Solve Right Triangles

Use these activities to quickly assess and activate prior knowledge as needed.

Problem of the Day 

A telephone pole that is 40 feet high has a support cable that runs from the top of the pole to the ground. The cable meets the ground at a 34° angle. How long, to the nearest foot, is the cable?

72 feet; $x = \frac{40}{\sin(34°)} \approx 72$

Quick Check for Homework

As part of your daily routine, you may want to display the Teacher Solution Key to have students check their homework.

Make Connections

Based on students' responses to the Problem of the Day, choose one of the following:

1. Project the Interactive Reteach, Lesson 13.2.
2. Complete the Prerequisite Skills Activity:

Have students work in pairs. Have one student draw a right triangle, labeling one side and one angle with dimensions, and a second side with the variable *x*. The other student uses trigonometric ratios to solve for the value of *x*. Then students switch roles to create and solve a different right triangle.

- *Based on the position of the given angle measure, which sides are the opposite side, the adjacent side, and the hypotenuse?* Answers will vary. For example, in triangle *ABC*, where angle *C* is the right angle, if angle *B* contains the angle measure, then side *AC* is opposite, side *BC* is adjacent, and side *AB* is the hypotenuse.
- *If the opposite side is labeled with a dimension and the hypotenuse is labeled with x, then what trigonometric function will you use to solve the problem?* sine
- *If the adjacent side is labeled with a dimension and the opposite side is labeled with x, then what trigonometric function will you use to solve the problem?* tangent
- *How do you know you are looking for a side dimension and not an angle measure?* Because an angle and two sides are labeled instead of two sides and no angle.

SHARPEN SKILLS

If time permits, use this on-level activity to build fluency and practice basic skills.

Quantitative Comparison

Objective: Students make a comparison between two quantities.
Write the following problem on the board. Ask students to choose the letter representing the correct answer and to explain their reasoning.

Quantity A	Quantity B
Value of *x* in $\triangle ABC$ when	Value of *x* in $\triangle DEF$ when
$m\angle C = 90°$, $m\angle B = 30°$, $AC = 55$, $AB = x$	$m\angle F = 90°$, $m\angle D = 60°$, $DF = 55$, $DE = x$

C; Quantity A is $x = \frac{55}{\sin(30°)} = 110$, and Quantity B is $x = \frac{55}{\cos(60°)} = 110$.

A. Quantity A is greater.
B. Quantity B is greater.
C. The two quantities are equal.
D. The relationship cannot be determined from the information given.

PLAN FOR DIFFERENTIATED INSTRUCTION

Small-Group Options

Use these teacher-guided activities with pulled small groups.

On Track

Materials: index cards

Have students work in pairs. Give each pair a set of index cards with either an area or circumference written on it. Ask students to work on finding the area if given the circumference or finding the circumference if given a circle's area.

$A = 78.4\,\text{cm}^2$

$C = 47.1$ cm

Almost There

Materials: index cards and worksheet

Have students work in pairs and give each a set of index cards and a worksheet. The number of rows in the worksheet should match the number of cards in the set. Each card should display a circle with either the radius or diameter drawn and labeled with its dimension. Ask students to fill in the chart with the information in each column, including the answer for both circumference and area, rounded to the nearest tenth.

Circle	r	d	$C = \pi \times d$	$A = \pi \times r^2$
8	4	8	$C = \pi \times 8$ $= 25.1$	$A = \pi \times 4^2$ $= 50.3$

Ready for More

Display an isosceles right triangle

that has a hypotenuse of $3\sqrt{2}$.

and is inscribed in one quarter of a circle. Ask students to find both the area and the circumference of the circle in terms of π.

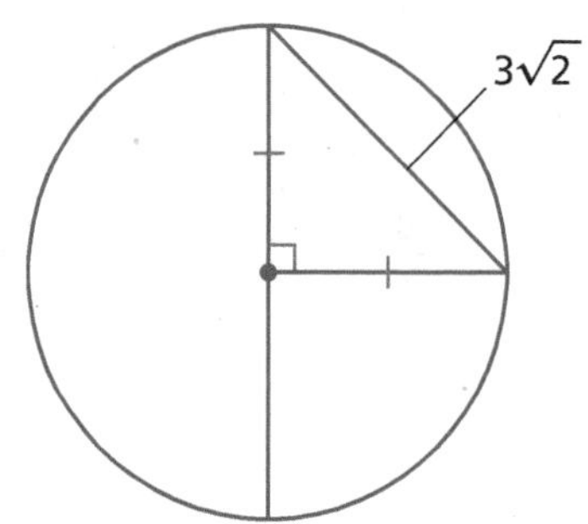

Math Center Options

Use these student self-directed activities at centers or stations. Key: ● Print Resources ● Online Resources

On Track

- ● Interactive Digital Lesson
- ●● Journal and Practice Workbook
- ● Interactive Glossary (printable): circumference, limit
- ● Module Performance Task

Almost There

- ● Reteach 17.1 (printable)
- ● Interactive Reteach 17.1
- ● Illustrative Mathematics: Measuring the Area of a Circle, Circumference of a Circle (Gr7)

Ready for More

- ● Challenge 17.1 (printable)
- ● Interactive Challenge 17.1
- ● Illustrative Mathematics: Circumference of a Circle, Area of a Circle

View data-driven grouping recommendations and assign differentiation resources.

Spark Your Learning • Student Samples

During the *Spark Your Learning*, listen and watch for strategies students use. See samples of student work on this page.

Use Circumference Formula

Strategy 1

It looks like 8 line segments with the same width as the bricks surrounding the cover can be placed across end-to-end through the center of the circle, which means the diameter of the circle is equal to 8 small bricks.

Length of diameter: $8 \times 3 = 24$ inches

$c = \pi \times d$

$c = \pi \times 24$

$c \approx 75.4$ inches

If students . . . use the formula for circumference, they understand how the diameter of a circle can be used to find its circumference, and can apply the circumference formula to find the distance around a circle from Grade 7.

Have these students . . . explain how they determined their estimate. **Ask:**

- Q How did you find the additional information that helped to solve the SYL problem?
- Q How did you use your estimated diameter length to find the circumference?

Use Multiplication

Strategy 2

There are 25 bricks surrounding the cover and each brick is 3 inches wide. Therefore, the total distance around the cover can be estimated by multiplying 25 by 3.

$25 \times 3 = 75$

Since circumference is the distance around a circle, the estimated circumference in this situation is 75 inches.

If students . . . use multiplication to solve the problem, they are employing an efficient method of finding the estimate of the circumference of the cover as well as displaying an exemplary understanding of estimating techniques from previous grade levels.

Activate prior knowledge . . . by having students explain how they determined their equation and how they solved it. **Ask:**

- Q Can you find the circumference of a circle using a formula?
- Q How can you use the additional information given to estimate the diameter of the maintenance hole cover?

COMMON ERROR: Uses Counting

There are 25 bricks that surround the cover, so the circumference is 25 inches.

If students . . . count the number of bricks that surround the cover and state that the number they arrive at is the circumference, then although they understand the meaning of circumference, they neglected to take into consideration the width of each brick.

Then intervene . . . by pointing out the additional information again. **Ask:**

- Q Is each brick that surrounds the cover 1 inch in length?
- Q What part of the given additional information did you not consider?

17.1

Measure Circumference and Area of a Circle

I Can justify and use the circumference and area of a circle formulas to solve real-world problems.

Spark Your Learning

Maintenance hole covers allow workers to access underground systems such as water lines and maintenance tunnels.

Complete Part A as a whole class. Then complete Parts B–D in small groups.

A. What is a mathematical question you can ask about the cover? What information would you need to know about the cover to precisely answer your question? What is the circumference of the cover?; the radius or diameter of the cover

B. What estimates can you make using information in the photo? B, D. See Additional Answers.

C. To answer your question, what strategy and tool would you use along with all the information you have? What answer do you get? See Strategies 1 and 2 on the facing page.

D. How can you determine if your estimates are reasonable?

Turn and Talk Suppose the diameter doubled in length.

- How would you adjust your estimate?
- How would your new estimate relate to your original estimate? See margin.

SUPPORT SENSE-MAKING • Three Reads

Tell students to read the information in the photo three times and prompt them with a different question each time.

1. What is the situation about? The situation is about the distance around a circular maintenance hole cover.
2. What are the quantities in this situation? How are those quantities related? The quantities are the width and number of bricks surrounding the cover and the distance around the cover.; The distance around the cover is approximately the product of the number of bricks and the width of a brick.
3. What are possible questions you could ask about this situation? Possible answer: What is the diameter of the cover? What is the circumference of the cover?

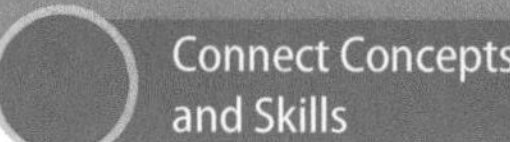

Connect Concepts and Skills

1 Spark Your Learning

▶ MOTIVATE

- Have students look at the photo in their books and read the information contained in the photo. Then complete Part A as a whole-class discussion.
- Give the class the additional information they need to solve the problem. This information is available online as a printable and projectable page in the Teacher Resources.
- Have students work in small groups to complete Parts B–D.

▶ PERSEVERE

If students need support, guide them by asking:

Q **Advancing • Use Tools** Which tool could you use to solve the problem? Why choose that tool and not some other? Students' choices of tools and reasons for choosing them will vary.

Q **Assessing** What do the radius and the diameter of a circle have in common? Both contain the center point of the circle.

Q **Assessing** How are the diameter and the radius of a circle related? The diameter of a circle is twice the radius.

Q **Advancing** Why doesn't it matter which side of the cover you use to estimate the length of the diameter? because a circle contains an infinite number of diameters

Turn and Talk When predicting how the estimate would be adjusted, suggest students create an example to explore what happens to the circumference of a circle when its diameter is doubled. If the diameter doubled, my estimate would be larger. I would guess that my estimate should double as well because the new circle would be twice as big as the original circle.

▶ BUILD SHARED UNDERSTANDING

Select groups of students who used various strategies and tools to share with the class how they solved the problem. As they present their solutions, have each group discuss why they chose a specific strategy and tool.

② Learn Together

Build Understanding

Task 1 (MP) **Use Structure** Students look closely at a given diagram to determine a pattern, to combine numbers and expressions, and to shift their perspective from a numerical example to a generalization.

CONNECT TO VOCABULARY

Have students use the **Interactive Glossary** to record their understanding of the vocabulary in this task.

Sample Guided Discussion:

Q What happens to the perimeter of the inscribed *n*-gon every time you add a side? The perimeter gets larger.

Q In Part B, what are the three trigonometric ratios and which one applies in finding the value of *x*? cos = adjacent ÷ hypotenuse, sin = opposite ÷ hypotenuse, and tan = opposite ÷ adjacent; Since the side with length of 1 is the hypotenuse, and the side labeled *x* is the opposite side, then use sin to find the value of *x*.

Turn and Talk Review with students the definition of π. Then students can use the definition to verify that no matter what the diameter of the circle is, the ratio is always equal to π. same; Possible answer: I expect the ratio of the circumference to the diameter of any circle to be $\frac{\pi d}{d}$, or π.

Build Understanding

Justify the Formula for the Circumference of a Circle

The **circumference** of a circle is the distance around the circle. The ratio of the circumference of any circle to its diameter is a constant that is defined as pi (π), which is the irrational number 3.14159. . . . Throughout this book, you should use a calculator when performing calculations with π.

Circumference of a Circle

The circumference C of a circle is $C = \pi d$ or $C = 2\pi r$, where d is the diameter of the circle and r is the radius of the circle.

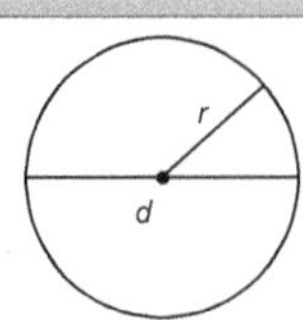

1 A. The diagrams show regular *n*-gons inscribed in circles with radius 1. Consider the perimeter of each *n*-gon. As *n* increases, what value should the perimeters approach? Explain your reasoning. See Additional Answers.

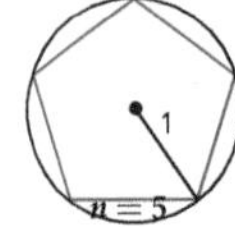

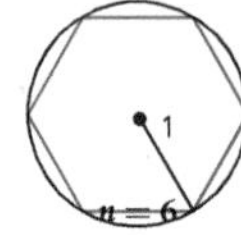

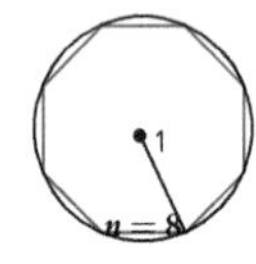

B. Consider the circle with an inscribed regular hexagon. The hexagon can be divided into six congruent triangles. Each triangle has a vertex angle measuring $\frac{360°}{6} = 60°$ and legs with length 1. Then each triangle can be divided into two right triangles. Use trigonometric ratios to write an expression for the value of *x*. Then use the value of *x* to write the perimeter of the hexagon. $x = \sin 30°$; perimeter $= 12x = 12\sin 30°$

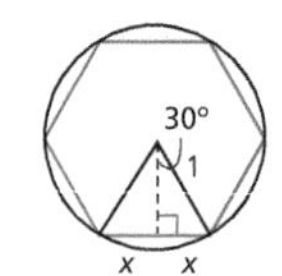

C. Consider an inscribed regular polygon with *n* sides. The triangle shown is from a regular *n*-gon inscribed in a circle with radius 1. What is an expression for the value of *x*? Use this expression to write a formula for the perimeter of the *n*-gon. C, D. See Additional Answers.

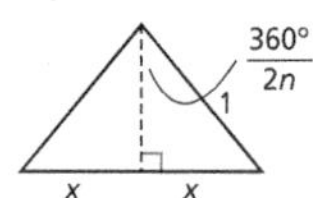

D. Use the formula in Part C to find the perimeters of regular *n*-gons inscribed in circles with radius 1 for large values of *n*. What value do the perimeters approach as *n* gets very large? How does this justify the formula $C = 2\pi r$?

Turn and Talk Suppose you know the diameter and circumference of two different-sized circles. Would you expect the ratio of the circumference to the diameter of the circles to be the same or to be different? Explain your reasoning. See margin.

524

LEVELED QUESTIONS

Depth of Knowledge (DOK)	Leveled Questions	What Does This Tell You?
Level 1 **Recall**	How can the irrational number π be described as a ratio? π is the ratio of the circumference of a circle to its diameter.	Students' answers will indicate whether they understand how the value of π is calculated.
Level 2 **Basic Application of Skills & Concepts**	How can you find the diameter of a circle if you are given its circumference? Divide the circumference by π to find the diameter.	Students' answers will indicate whether they understand the relationship between the circumference of a circle and its diameter.
Level 3 **Strategic Thinking & Complex Reasoning**	When a polygon is inscribed in a circle, how can you use trigonometry to find the perimeter of that polygon? Trigonometry can be used to find the length of half of the side of the polygon, and that measure can be used to find the distance around the polygon.	Students' answers will indicate whether they understand the properties of a circle and how right triangle trigonometry applies to this task.

Justify the Formula for the Area of a Circle

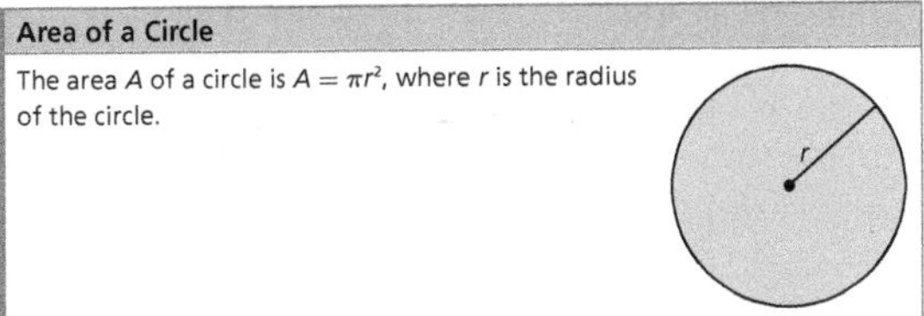

Area of a Circle
The area A of a circle is $A = \pi r^2$, where r is the radius of the circle.
r

In mathematics, a **limit** is a value that the output of a function approaches as the input increases or decreases without bound or approaches a given value. From the diagram in the previous task, it appears that the perimeters of the inscribed n-gons approach the limit of the circumference of the circle as n increases.

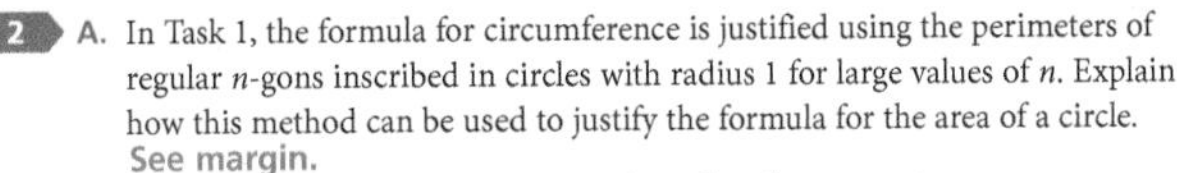

2 A. In Task 1, the formula for circumference is justified using the perimeters of regular n-gons inscribed in circles with radius 1 for large values of n. Explain how this method can be used to justify the formula for the area of a circle. See margin.

B. A regular n-gon is inscribed in a circle with radius 1. The n-gon is divided into n congruent triangles, one of which is shown. What is the area of the triangle in terms of h and x? Use this expression to write the area of the n-gon.
Area of triangle $= xh$; Area of n-gon $= nxh$

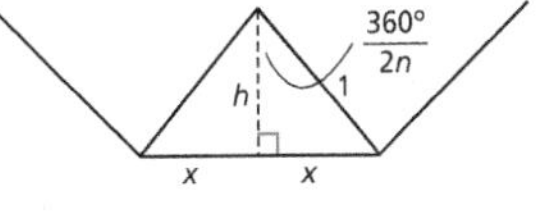

C. Use trigonometric ratios to write expressions for h and x in terms of n. Then use the expressions to write the area of the n-gon in terms of n. See margin.

D. Suppose a regular n-gon is inscribed in a circle with radius 1. Use a spreadsheet to find the area of the n-gon as n increases to very large numbers. What value does the area approach for larger values of n? 3.14, or π

B2 | fx | =A2*SIN(RADIANS(360/(2*A2)))*COS(RADIANS(360/(2*A2)))

	A	B	C	D
1	n	Area of n-gon		
2	50	3.1333308		
3	100			
4	150			
5	200			

E. Explain how the result from Part D justifies the formula for the area of a circle. See margin.

Turn and Talk Is it possible for n to be a large enough value such that the area of the n-gon is equal to the area of the circle? Why or why not? See margin.

PROFICIENCY LEVEL

Beginning

Write the equation for circumference of a circle, $C = \pi d$. Say, "The circumference of a circle is found by multiplying pi by the value of the diameter." Then say, "A circle has a radius of 10". Ask students to write a phrase in words that describes what must be multiplied to find the circumference of this circle.

Intermediate

Display two circles, one with a radius drawn and labeled 5 inches, one with a diameter drawn and labeled 8 inches. Have students work in pairs and ask them to describe what needs to be multiplied to find both the circumference and the area of each circle.

Advanced

Have students explain in a paragraph how to find the measure of the radius of a circle if given the circle's circumference.

Task 2 (MP) **Use Structure** Students make use of structure by exploring a polygon with n sides inscribed in a circle in order to estimate measurements, such as area. Students decompose the n-gon into congruent triangles and use prior knowledge to gain a deeper understanding of limits.

CONNECT TO VOCABULARY

Have students use the **Interactive Glossary** to record their understanding of the vocabulary in this task.

Sample Guided Discussion:

Q How do you calculate the area of a triangle? Area $= \frac{1}{2}bh$

Q Why is the angle indicated in the picture represented by the expression $\frac{360}{2n}$? Because a circle contains 360°, each central angle created by the n congruent triangles of the n-gon has a measure of $\frac{360}{n}$. Since each of those central angles are bisected into two congruent angles, each is represented by $\frac{360}{2n}$.

Q In Part D, What term can you use to describe your answer? limit

Turn and Talk As students consider this question, discuss and review why a circle is not considered a polygon. This should help them realize that since a circle is not an enclosed figure made up of line segments, that no matter how large the value of n gets, the inscribed polygon will never become the circle. no; The area of the n-gon gets very close to the area of the circle, but it will always be less than the area of the circle.

ANSWERS

A. The areas of regular n-gons inscribed in circles of radius 1 should approach the area of the circle as n increases.

C. $x = \sin\left(\frac{360°}{2n}\right)$, $h = \cos\left(\frac{360°}{2n}\right)$; Area of n-gon $= n\sin\left(\frac{360°}{2n}\right)\cos\left(\frac{360°}{2n}\right)$

E. For large values of n, the n-gon resembles a circle. So, the area of the n-gon should approximate the area of the circle with radius 1, which is $\pi(1)^2 \approx 3.14$.

Task 3 (MP) **Attend to Precision** Students will communicate using precise language, vocabulary, and mathematical symbols when they write inequality statements to relate the circumferences of inscribed and circumscribed shapes.

OPTIMIZE OUTPUT Critique, Correct, and Clarify

Suppose a student's answer to Part A is $P_1 \leq \pi \leq P_2$. Have students work with a partner to discuss why this inequality would be incorrect. Encourage students to use terms such as *limit*, *inscribed*, and *circumscribed* in their discussions. Ask students to share their thoughts with the class.

Sample Guided Discussion:

Q Suppose both regular hexagons were inscribed in the circle. How would the inequality statement you wrote in Part A change? $P_1 = P_2 < \pi$

Q In Part C, why is the segment labeled with the dimension of 1 the height of the congruent triangle in the smaller hexagon instead of the length of the triangle's side, as it is in the larger hexagon? When the hexagon is inscribed, a side of the hexagon is a chord inside the circle, so the segment labeled with 1 is the radius of the circle. When the hexagon is circumscribed, a side of the hexagon is a tangent outside the circle, so the segment labeled with 1 is an apothem in the hexagon.

Turn and Talk Encourage students to think of a measure other than 1 for the radius, where this alternate definition would also be true. If students mention that an *r*-value of −1 would also make the area equal to pi, remind them that segment length can never be negative. The ratio of the area of a circle with radius *r* to the area of a circle with radius 1 is. The area of a circle with radius 1 is π square units, so the area formula is $A = r^2 \cdot \text{(area of circle with radius 1)} = \pi r^2$.

Justify π

As you have learned, π (pi) represents the ratio of the circumference of any circle to its diameter.

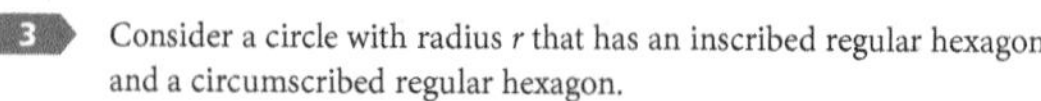

3 Consider a circle with radius *r* that has an inscribed regular hexagon and a circumscribed regular hexagon.

A. Suppose P_1 is the perimeter of the smaller hexagon, P_2 is the perimeter of the larger hexagon, and *C* is the circumference of the circle. This relationship can be represented as shown.

$$P_1 < C < P_2$$

Explain how to rewrite the inequality in terms of π.

___?___ $< \pi <$ ___?___ See margin.

B. What would be true about the inequalities in Part A if regular polygons with larger numbers of sides were used? The values $\frac{P_1}{d}$ and $\frac{P_2}{d}$ would both approach π.

C. Consider a circle with radius 1 that has an inscribed regular *n*-gon and a circumscribed regular *n*-gon. The *n*-gons are divided into *n* congruent triangles, and one triangle from each *n*-gon is shown below. Explain how the formulas for the perimeters of the *n*-gons were derived.
C, D. See margin.

Inscribed *n*-gon

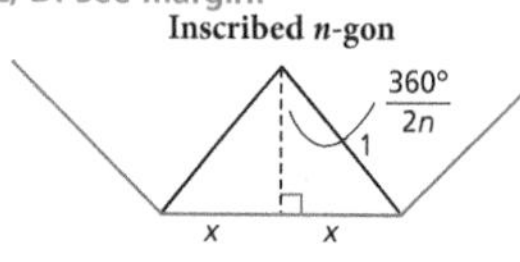

$P_1 = 2n\sin\left(\frac{360°}{2n}\right)$

Circumscribed *n*-gon

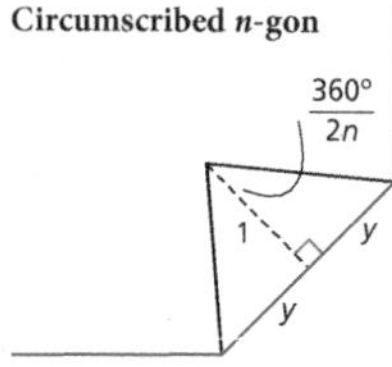

$P_2 = 2n\tan\left(\frac{360°}{2n}\right)$

D. Rewrite the inequality in terms of π from Part A using the formulas from Part C and a radius of 1. Then use a spreadsheet to calculate a lower bound for π, an upper bound for π, and the average of the two bounds for different *n*-gons. What do you observe?

B2 | fx = AVERAGE(B2:C2)

	A	B	C	D
1	*n*	Lower Bound	Upper Bound	Average
2	50	3.1395260	3.1457334	3.1426297
3	100			
4	150			
5	200			

Turn and Talk Explain why an alternate definition of π is the area of a circle with radius 1. See margin.

526

ANSWERS

A. $\frac{P_1}{d} < \pi < \frac{P_2}{d}$; Because $C = \pi d$, divide each part of the inequality by *d* to isolate π in the center.

C. $P_1 = 2nx$ and $x = \sin\left(\frac{360°}{2n}\right)$, so $P_1 = 2n\sin\left(\frac{360°}{2n}\right)$. $P_2 = 2ny$ and $y = \tan\left(\frac{360°}{2n}\right)$, so $P_2 = 2n\tan\left(\frac{360°}{2n}\right)$.

D. $\frac{2n\sin\left(\frac{360°}{2n}\right)}{d} < \pi < \frac{2n\tan\left(\frac{360°}{2n}\right)}{d}$; As *n* increases, the lower and upper bounds average to approximate π.

Step It Out

Apply the Circumference Formula

4 A new traffic circle will have an outer circumference of $\frac{1}{4}$ mile. What should the radius of the traffic circle be to the nearest foot? Use the fact that there are 5280 feet in 1 mile.

A. The final answer needs to be in feet, and integers are easier to work with than fractions.

Determine the circumference in feet.

$\frac{1}{4}\text{ mi} \cdot \frac{5280\text{ ft}}{1\text{ mi}} = 1320\text{ ft}$

A. Why would you choose to convert the circumference from miles to feet?

Use the circumference formula.

$C = 2\pi r$

B. Why is this version of the formula used?

$1320 = 2\pi r$

$r = \frac{1320}{2\pi} \approx 210\text{ ft}$

B. The radius needs to be found, and the formula $C = 2\pi r$ can be used to find the value of r directly.

Apply the Area Formula

5 The plan for a splash pad shows a concrete area with circular nonslip surfaces. The diameter d for each circle is given. The nonslip surfaces cost \$11 per square foot. What is an estimate for the cost for the nonslip surfaces?

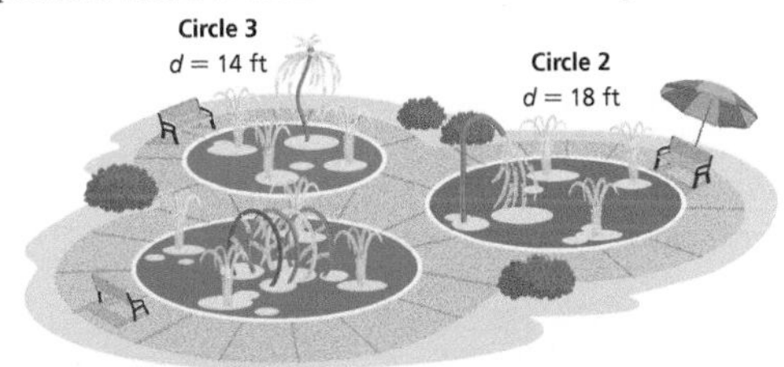

Find the area of each circle.

Area of Circle 1 $= \pi r^2 = \pi \cdot 10^2 = 100\pi$

Area of Circle 2 $= \pi r^2 = \pi \cdot 9^2 = 81\pi$

Area of Circle 3 $= \pi r^2 = \pi \cdot 7^2 = 49\pi$

A. Why are the answers left in terms of π?

Find the total area.

$100\pi + 81\pi + 49\pi = 230\pi \approx 723\text{ ft}^2$

A, B. See margin.

Estimate the cost.

$723 \times 11 = \$7953$

B. What are the units in the cost equation?

Turn and Talk Suppose the cost of the circular surfaces should be close to but not exceed \$7500. How can the design change to meet the budget? See margin.

Step It Out

Task 4 MP **Attend to Precision** Students attend to precision when applying the definitions they have learned and in making decisions as to what formulas and units are appropriate in the context of a problem.

Sample Guided Discussion:

Q Suppose you did not convert the outer circumference, given in miles, to feet. How would your work change and what would be the length of the radius in miles?

$\frac{1}{4} = \pi d$

$\frac{0.25}{\pi} = d$

$d \approx 0.08$

$r \approx 0.08 \div 2 \approx 0.04$

The radius is about 4 hundredths of a mile.

Task 5 MP **Model with Mathematics** Students use visual representations to identify important quantities in a practical situation and apply the mathematics they know to solve problems in everyday life.

Sample Guided Discussion:

Q Suppose someone estimated the cost to be only \$1,793. What mistake was made in the calculation of the cost? The circumference of each circle was found instead of the area of each circle.

Turn and Talk Point out that when trying to change the dimensions so that the cost does not exceed \$7,500, there are a variety of answers that are appropriate. Modeling allows students to adjust dimensions slightly to fit the context of the problem. Encourage students to find more than one possibility. Possible answer: Change the diameter of circle 2 to 16 feet. Then the total area will be about 669 square feet and cost \$7,359.

ANSWERS

A. Using an approximate value for π early on in calculations will introduce a rounding error. Calculations should be completed in terms of π until a numeric answer is needed.

B. square feet $\times$ dollars per square foot $=$ dollars

ONLINE

Assign the Digital On Your Own for

- built-in student supports
- Actionable Item Reports
- Standards Analysis Reports

On Your Own

Assignment Guide

The chart below indicates which problems in the On Your Own are associated with each task in the Learn Together. Assign daily homework for tasks completed.

Learn Together Tasks	On Your Own Problems
Task 1, p. 524	Problems 6 and 8a
Task 2, p. 525	Problems 8b and 8c
Task 3, p. 526	Problem 9
Task 4, p. 527	Problems 7, 10–15, 20–23, and 28
Task 5, p. 527	Problems 7, 16–19, and 24–27

Check Understanding

1. A regular 20-gon is inscribed in a circle with radius 2 units. A regular 30-gon is inscribed in another circle with radius 3. Which inscribed polygon has a perimeter closer to the circumference of the circle in which it is inscribed? Explain. 1–3. See Additional Answers.
2. How do you justify and use the formula for the area of a circle?
3. Is it possible to draw a circle whose ratio of circumference to diameter is not π? Explain why or why not.

A pair of sunglasses has circular lenses that each have a diameter of 1.8 inches.

4. Determine the length of wire that frames each lens. Round to the nearest tenth. about 5.7 in.
5. A polarized film is embedded in each lens to filter sunlight. How many square inches of polarized film is needed for both lenses? Round to the nearest tenth of a square inch. 5.1 in^2

On Your Own

6. **Reason** A regular n-gon is inscribed in a circle with radius r. Is there any value of n that will produce an n-gon with a perimeter of $2\pi r$? Explain why or why not. 6–9. See Additional Answers.
7. **Critique Reasoning** Disha states that the circumference and the area of the larger circle are both double the circumference and the area of the smaller circle. Is she correct? Explain why or why not.

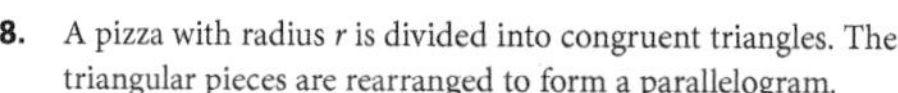

8. A pizza with radius r is divided into congruent triangles. The triangular pieces are rearranged to form a parallelogram.
   - **A.** How are the base and the height of the parallelogram related to the circle?
   - **B.** How can the parallelogram be used to justify the formula for the area of a circle?
   - **C.** Suppose the circle is divided into 16 congruent triangles. Will the new parallelogram formed by these pieces be a better estimate for the area of the circle? Explain.

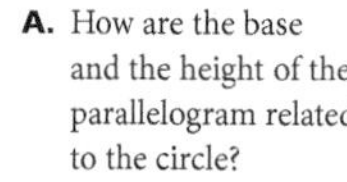

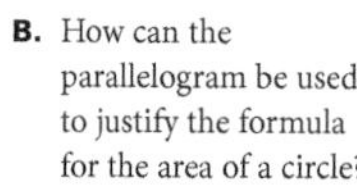

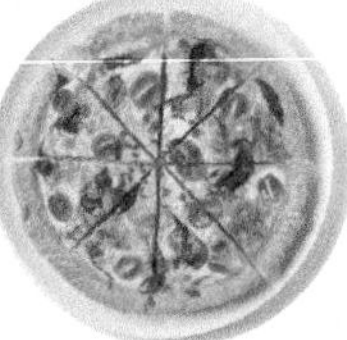

9. **Reason** Refer back to Task 3. If you only inscribed a regular n-gon in the circle, is that enough to find an approximate value of π? Explain.

Find the circumference of each circle with the given radius r or diameter d. Round answers to the nearest hundredth.

10. A circle with $r = 4$ in. 25.13 in.
11. A circle with $d = 7.5$ m 23.56 m
12. A circle with $r = 16.2$ mm 101.79 mm
13. A circle with $d = 15$ cm 47.12 cm

528

3 Check Understanding

Formative Assessment

Use formative assessment to determine if your students are successful with this lesson's learning objective.

Students who successfully complete the Check Understanding can continue to the On Your Own practice.

For students who miss 1 problem or more, work in a pulled small group using the Almost There small-group activity on page 523C.

ONLINE

Assign the Digital Check Understanding to determine

- success with the learning objective
- items to review
- grouping and differentiation resources

4 Differentiation Options

Differentiate instruction for all students using small-group activities and math center activities on page 523C.

Reteach

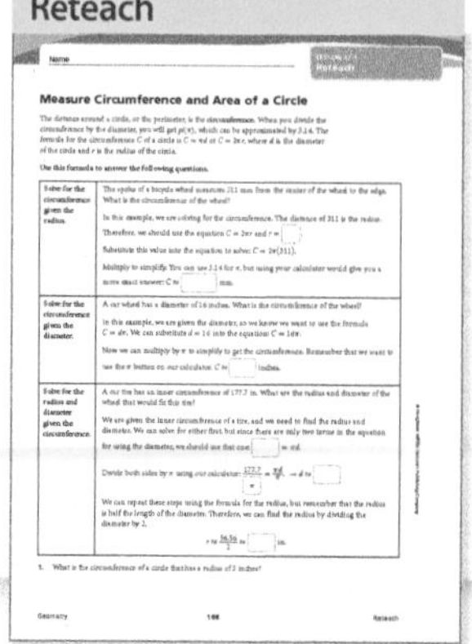

Challenge

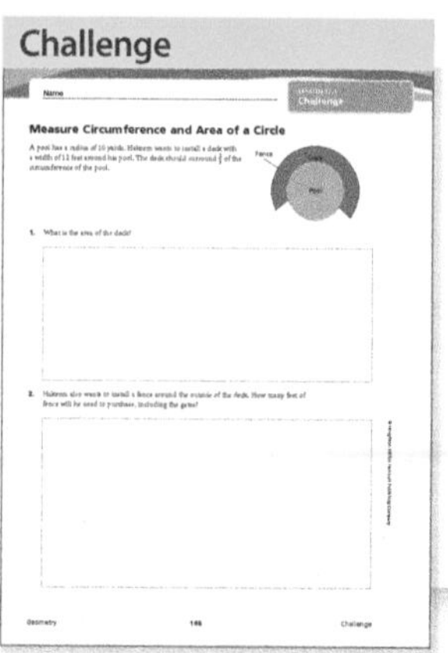

14. A circular horse pen is used to provide rehabilitation exercise for a horse recovering from an injury.

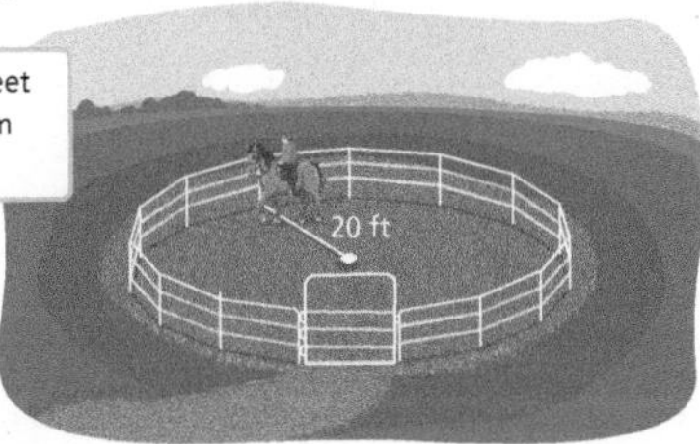
The horse trots at 15 feet per second 20 feet from the center of the pen.
20 ft

A. To the nearest foot, what is the total distance the horse travels during one lap around the pen? 126 feet

B. How many seconds does it take to complete 5 laps around the pen? 42 seconds

C. Suppose the horse trots around the pen 18 feet from the center at the same speed. How much less time would it take the horse to complete 5 laps? about 4.2 seconds

15. **STEM** A trundle wheel is used by a surveyor to measure distances by rolling it on the ground and counting the number of revolutions.

A. A trundle wheel has a diameter of 12.5 inches. To the nearest tenth of an inch, how much ground is covered with every rotation of the wheel? 39.3 inches

B. The trundle wheel measures a distance of 78.6 feet. How many rotations did the wheel make while measuring this distance? 24 rotations

C. Suppose a trundle wheel is designed so one revolution measures 1 meter. What is the radius of the wheel in centimeters? Round to the nearest centimeter. 16 cm

Find the area of each circle with radius r or diameter d. Round answers to the nearest tenth.

16. $r = 7$ mm 153.9 mm^2

17. $d = 22$ yd 380.1 yd^2

18. $d = 37$ cm 1075.2 cm^2

19. $r = 5.3$ ft 88.2 ft^2

Find the circumference of the circle with the given area A. Round answers to the nearest hundredth.

20. $A = 121\pi$ in^2 69.12 in.

21. $A = 49\pi$ cm^2 43.98 cm

22. $A = 400\pi$ mm^2 125.66 mm

23. $A = 16\pi$ ft^2 25.13 ft

In Problems 24 and 25, use the figure which shows two circles with the same center.

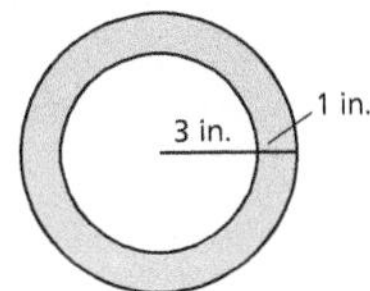
3 in.
1 in.

24. What percent of the circle's area is within the shaded region? about 43.75%

25. Suppose the radius of the white circle changes so the shaded area represents half of the total area. What is the new radius of the white circle? $\sqrt{8}$, or $2\sqrt{2} \approx 2.83$ in.

Questioning Strategies

Problem 15 Suppose a student arrives at an answer of 2 rotations for Part B. What mistake was made? The student may not have converted 78.6 feet into inches before dividing by 39.3.

Problem 25 Suppose you are asked for the percent of the circle's area that is not shaded. How would your work and answer change? I would divide 9π by 16π instead of 7π by 16π and get 56.25% instead of 43.75%.

Watch for Common Errors

Problem 20 A problem like this involves attention to precision and detail because it involves using a given area to determine the value of a missing parameter, which may or may not be the parameter needed to solve the problem. It is a good idea for students to first write out both formulas, without substituting in any given dimensions. Then, after deciding which formula must be used, substitute in the given information, and solve for the missing variable (in this case, the radius of the circle). Next, inspecting the formula for circumference before making any substitutions will help eliminate error when solving the problem.

5 Wrap-Up

Summarize learning with your class. Consider using the Exit Ticket, Put It in Writing, or I Can scale.

Exit Ticket 

The diameter of a circular table is 48 inches and will be covered with a circular table cloth. The table cloth will hang over the edge of the table 18 inches. What is the area of the table cloth to the nearest square foot? 38 square feet; The diameter of the table cloth is $48 + 18 + 18 = 84$ inches and $84 \div 12 = 7$ feet, which makes the radius 3.5 feet. $A = \pi(3.5)^2 \approx 38$ square feet

Put It in Writing 

Describe the difference between area and circumference of a circle, referring to both their definitions and the formulas.

I Can 

The scale below can help you and your students understand their progress on a learning goal.

4	I can justify and use the circumference and area of a circle formulas to solve real-world problems, and I can explain my steps to others.
3	I can justify and use the circumference and area of a circle formulas to solve real-world problems.
2	I can use the circumference and area of a circle formulas to solve mathematical problems.
1	I can write the formulas for area and circumference of a circle.

Spiral Review • Assessment Readiness

These questions will help determine if students have retained information taught in the past and can also prepare them for high-stakes assessments. Here, students understand the relationships of segments and angles of circles **(16.1 and 16.2)**, how to find the measure of a central angle in a circle **(15.1)**, and are able to interpret the equation of a circle **(15.4)**.

---

26. MP **Reason** A pizzeria charges \$10.50 for a medium cheese pizza and \$12.50 for a large cheese pizza. The medium pizza has a diameter of 12 inches, while the large pizza has a diameter of 14 inches. Which pizza is a better buy? Explain.
See Additional Answers.

27. A weather advisory states that the orange circular region with an average diameter of 72 miles is sustaining tropical storm force winds. The diameter of the circular region is increasing at a rate of 5 miles per hour. At this rate, how long will it take for the circular region to cover 6650 square miles?
about 4 hours

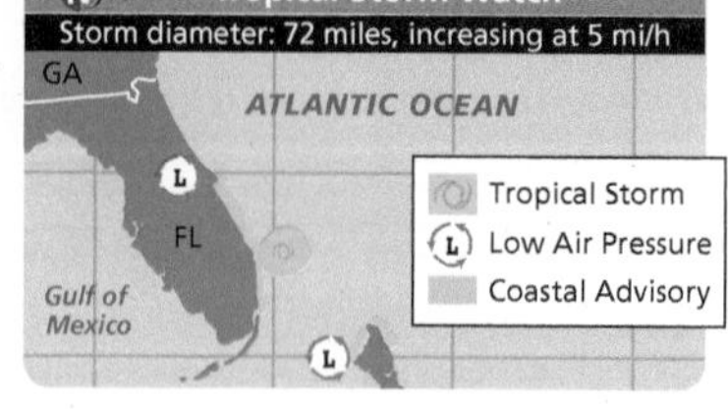

28. **STEM** An engineer designs a circular gasket with a manufacturing tolerance of ± 1.5 mm, which means that any dimension of a gasket produced in the factory must be within the range of 1.5 mm bigger than or smaller than the drawing specifications to pass quality control. If the outer diameter of the gasket is labeled as 50 mm on the engineering drawing, what is the acceptable range of the circumference of the part?
from 48.5π mm to 51.5π mm, or from 152.4 mm to 161.8 mm

Spiral Review • Assessment Readiness

29. Which of the following are true statements? Select all that apply.

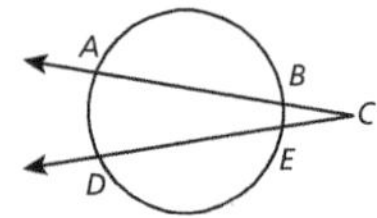

Ⓐ $BC = \frac{DE \cdot EC}{AC}$

Ⓑ $m\angle C = m\widehat{AD}$

Ⓒ $m\angle C \leq 180°$

Ⓓ $EC = \frac{AC \cdot BC}{DC}$

Ⓔ $m\angle C = AD \cdot DE$

Ⓕ $m\widehat{AD} = 2 \cdot m\widehat{BE}$

30. Diameters of a circle are drawn to divide the circle into congruent parts as shown. What is the value of x?

Ⓐ 22.5°

Ⓑ 30°

Ⓒ 45°

Ⓓ 60°

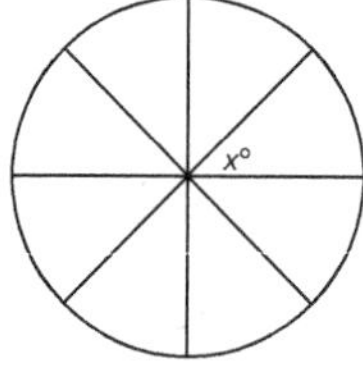

31. Match the equation of a circle on the left to its description on the right.

Equation	Description	
A. $(x-2)^2 + (y+1)^2 = 9$	**1.** center at $(1, -2)$, radius of $\sqrt{3}$ units	A. 4
B. $(x-1)^2 + (y+2)^2 = 3$	**2.** center at $(-2, 1)$, radius of 3 units	B. 1
C. $(x+1)^2 + (y-2)^2 = 1$	**3.** center at $(-1, 2)$, radius of 1 unit	C. 3
D. $(x+2)^2 + (y-1)^2 = 9$	**4.** center at $(2, -1)$, radius of 3 units	D. 2

I'm in a Learning Mindset!

What strategy did I use to overcome difficulties with applying previous knowledge in new ways to justify the circumference and area formulas?

530 Keep Going Journal and Practice Workbook

Learning Mindset

Resilience Identifies Obstacles

Point out that looking at mathematics theoretically, in a general sense, can result in obstacles that are more challenging to overcome, but can also help develop a much deeper understanding of mathematical concepts. Encourage students to work together to brainstorm prior math knowledge that can help strengthen an understanding of how formulas are derived. Confirm that the obstacles experienced in this lesson will aid them in the lessons to come on sectors, arcs, and formulas for volume and surface area. *What have you learned in the past about right triangles? What obstacles did you experience in Tasks 1, 2, and 3, and how did you overcome those? Was it helpful to work in groups or did you prefer to work independently? Was there a time where you left a task still feeling confused? Reflect on that experience and apply what you have learned.*

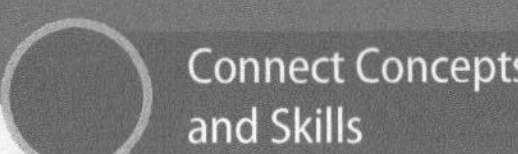

17.2 Measure Arc Length and Use Radians

LESSON FOCUS AND COHERENCE

Mathematics Standards

- Derive using similarity the fact that the length of the arc intercepted by an angle is proportional to the radius, and define the radian measure of the angle as the constant of proportionality; derive the formula for the area of a sector.

Mathematical Practices and Processes 

- Look for and make use of structure.
- Look for and express regularity in repeated reasoning.
- Model with mathematics.
- Attend to precision.

I Can Objective 

I can use similarity of circles to find arc length.

Learning Objective

Use the arc length formula and apply it to real-world problems, and convert between degree and radian measure.

Language Objective

Explain how to derive the arc length formula using similarity of circles.

Vocabulary

Review: **arc**

New: **arc length**, **concentric circles**, **radian measure**

Mathematical Progressions

Prior Learning	Current Development	Future Connections
Students: • knew the formulas for the area and circumference of a circle and used them to solve problems. **(Gr7, 10.1)** • identified and described relationships among inscribed angles, radii, and chords, including the relationship between central, inscribed, and circumscribed angles. **(15.1–15.3)** • proved that all circles are similar. **(12.1)**	**Students:** • understand that the ratio of the length of an arc intercepted by an angle to the circumference of the circle is equal to the ratio of the measure of the central angle to 360°. • use similarity of circles to derive the formula for arc length. • use the arc length formula to solve real-world problems. • convert between radian measure and degree measure.	**Students:** • will derive the formula for the area of a sector. **(15.3)** • will explain how the unit circle in the coordinate plane enables the extension of trigonometric functions to all real numbers, interpreted as radian measures of angles traversed counterclockwise around the unit circle. **(Alg2, 15.1)**

PROFESSIONAL LEARNING

Math Background

Arc lengths are denoted by the letter s. This designation comes from the Latin word *spatium*, which means "length" or "size."

Roger Cotes, who worked with Sir Isaac Newton, first described radian measure around 1714, but he did not have a name for it. During the late 1760s and early 1770s, mathematicians debated the terms *rad*, *radial*, and *radian* before eventually settling on *radian*.

WARM-UP OPTIONS

ONLINE

PROJECTABLE & PRINTABLE

ACTIVATE PRIOR KNOWLEDGE • Area and Circumference Formulas

Use these activities to quickly assess and activate prior knowledge as needed.

Problem of the Day 

A circle has a diameter of 12 inches. Find its area and circumference.

$d = 12 \rightarrow r = 6$

$A = \pi r^2 = \pi(6)^2 = 36\pi$ square inches

$C = 2\pi r = 2\pi(6) = 12\pi$ inches

Quick Check for Homework

As part of your daily routine, you may want to display the Teacher Solution Key to have students check their homework.

Make Connections

Based on students' responses to the Problem of the Day, choose one of the following:

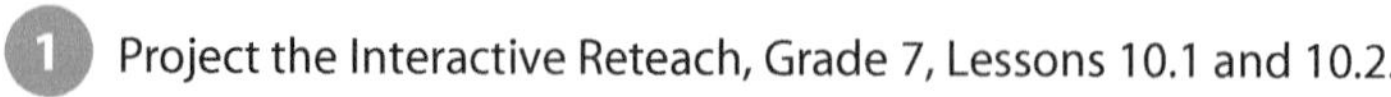

1. Project the Interactive Reteach, Grade 7, Lessons 10.1 and 10.2.
2. Complete the Prerequisite Skills Activity:

Have students work in pairs. Give each pair a positive even integer, multiplied by π, such as 14π. The first student will use this number to represent the area of the circle, and find its radius, diameter, and circumference. The second student will use this number to represent the circumference of the circle, and find its radius, diameter, and area. Students will check each other's work.

- *What is the relationship between the radius and the diameter of a circle?* $d = 2r; r = \frac{1}{2}d$
- *What is the formula for the area of a circle?* $A = \pi r^2$
- *What is the formula for the circumference of a circle?* $C = \pi d = 2\pi r$
- *Why is it easier to solve the problem if the circumference is an even multiple of π rather than an odd multiple?* It is easier to solve because, to find the radius, the circumference is divided by 2π. The 2's will cancel out.

If students continue to struggle, use Tier 2 Skills 19 and 20.

SHARPEN SKILLS

If time permits, use this on-level activity to build fluency and practice basic skills.

Mental Math

Objective: Students match fractional parts of a circle and measures of central angles
Materials: index cards

Have students work in small groups. Give each group of students index cards. The cards should show fractions whose denominators are factors of 360, such as $\frac{1}{2}, \frac{1}{3}, \frac{1}{4}, \frac{1}{6}, \frac{1}{12}, \frac{2}{3}$, and so on, and central angles that are factors of 360°, such as 180°, 120°, 90°, 60°, 30°, 240°, and so on. Have students take turns drawing one card and mentally calculating the central angle represented by the fraction or the fraction represented by the central angle.

$\frac{1}{2}$	$\frac{1}{6}$

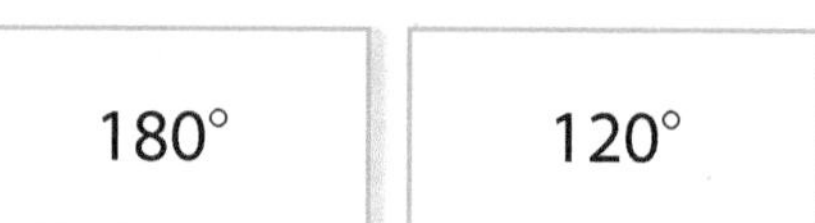

PLAN FOR DIFFERENTIATED INSTRUCTION

Small-Group Options

Use these teacher-guided activities with pulled small groups.

On Track

Materials: index cards

Give each student a card on which is drawn a circle with a central angle in degrees and a radius. Have students calculate the arc length formed by the angle and the radian measure of the angle.

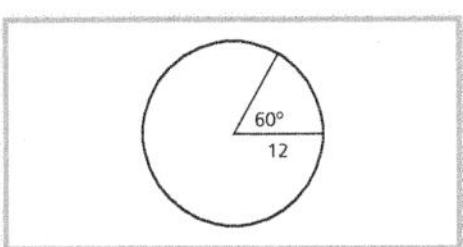

Almost There

Draw a circle with a central angle of 60 degrees and a radius of 12 and write the arc length formula.

- Have students identify the variables in the formula.
- Check that students have substituted correct values into the formula.
- Have students calculate the arc length.

Ready for More

A circle with a radius of 12 has a central angle that forms an arc length of 6π. What is the arc length of a circle with a radius of 4 with the same central angle?

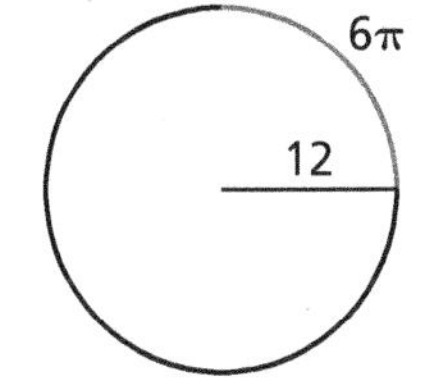

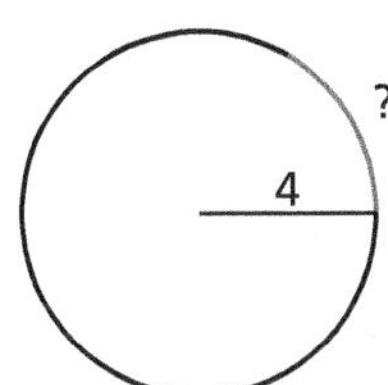

Math Center Options

Use these student self-directed activities at centers or stations. **Key: ● Print Resources ● Online Resources**

On Track

- ● Interactive Digital Lesson
- ●● Journal and Practice Workbook
- ● Interactive Glossary (printable): **arc**, **arc length**, **concentric circles**, **radian measure**
- ● Module Performance Task

Almost There

- ● Reteach 17.2 (printable)
- ● Interactive Reteach 17.2
- ● RtI Tier 2 Skill 19: Circumferences of Circles
- ● RtI Tier 2 Skill 20: Areas of Circles
- ● Illustrative Mathematics: Orbiting Satellite

Ready for More

- ● Challenge 17.2 (printable)
- ● Interactive Challenge 17.2
- ● Illustrative Mathematics: Two Wheels and a Belt

Unit Project Check students' progress by asking how to use the circumference of a circle and the central angle to determine arc length.

ONLINE 

View data-driven grouping recommendations and assign differentiation resources.

During the *Spark Your Learning,* listen and watch for strategies students use. See samples of student work on this page.

Use the Radius and Fractional Part of the Circle

Strategy 1

Because the angle measured 290°, I knew it represented $\frac{290°}{360°}$ of the circle. The arc, therefore, would represent a portion of the circumference or a full circle. $C = 2\pi r = 2\pi(4)$ for the inner circle and $C = 2\pi r = 2\pi(5)$ for the outer circle.

$\frac{290°}{360°} \cdot 8\pi \approx 20.2$ ft and $\frac{290°}{360°} \cdot 10\pi \approx 25.3$ ft

If students ... use the radius and fractional part of the circle to solve the problem, they are using properties of circles from Grade 7.

Have these students ... explain how they calculated the fractional part of the circle, and why they decided to use the formula for circumference that uses the radius, as opposed to the formula for circumference that uses diameter, to solve the problem. **Ask:**

- Q How did you use the radius to find the circumference to solve the problem?
- Q Why would you expect the length of the arcs to be greater than half the circumference of the circles?

Use the Diameter and Percentages

Strategy 2

I doubled the radius of both circles to find the diameters. Then, I applied the formula $C = \pi d$ to calculate the circumference of each circle. I found what percent 290° represents of the entire circumference, and I multiplied that number by the circumference to determine the arc length.

$r = 4 \rightarrow d = 8;\ r = 5 \rightarrow d = 10$

$C = \pi d = 8\pi,\ 10\pi$

$\frac{290°}{360°} = \frac{x\%}{100\%}$

$x = 80.6\%$

$80.6\% \cdot 8\pi \approx 20.3$ ft

$80.6\% \cdot 10\pi \approx 25.3$ ft

If students ... use diameter and percentages to solve the problem, they understand how to use percentages, but they are not being as efficient since they can use the circumference formula with radius nor do they need to find the percent.

Activate prior knowledge ... by having students explain the formula for circumference using the radius. **Ask:**

- Q What is the quicker way to find the circumference given the radius?
- Q Why is it not necessary to change $\frac{230°}{360°}$ to a percent?

COMMON ERROR: Use the Formula for Area

$C = \pi r^2;\ C = \pi(4)^2 = 16\pi$ for the inner circle

$\frac{290°}{360°} \cdot 16\pi \approx 40.5$ ft for the inner arc

$C = \pi r^2;\ C = \pi(5)^2 = 25\pi$ for the outer circle

$\frac{290°}{360°} \cdot 25\pi \approx 63.3$ ft for the outer arc

If students ... use the formula for area of a circle to calculate the circumference, then they are confusing the formulas for area and circumference.

Then intervene ... by pointing out that the units their answer would be for area is ft^2 but the unit for length is ft. **Ask:**

- Q What is the difference between the circumference of a circle and the area of a circle?
- Q What two formulas are used to calculate the circumference of a circle?

17.2

Measure Arc Length and Use Radians

I Can use similarity of circles to find arc length.

Spark Your Learning

Portions of circles called arcs are often used in architectural design to add visual interest.

Complete Part A as a whole class. Then complete Parts B–D in small groups.

A. What is a mathematical question you can ask about the border of this entrance? What information would you need to know to answer your question? What is the length of each arc?; the radius of each arc

B. How can you determine which arc to analyze? Are both the same size? See Additional Answers.

C. To answer your question, what strategy and tool would you use along with all the information you have? What answer do you get? See Strategies 1 and 2 on the facing page.

D. Does the answer make sense in the context of the situation? How do you know? yes; By comparing radii to arc lengths, the answers seem reasonable. Also, the length of the inner arc is less than the length of the outer arc.

Turn and Talk Consider the following changes to the intercepted arc. See margin.

- How would your answer change if the intercepted arc were 180°, or exactly half the circle? How would the size of the arc relate to the circumference?
- How would your answer change if the intercepted arc were 270°? How would the size of the arc relate to the circumference?

CULTIVATE CONVERSATION • Co-Craft Questions

If students have difficulty formulating a mathematical question about the situation in the Spark Your Learning, ask them to consider that they are measuring a curved length. What are some natural questions to ask about this situation?

Work together to craft the following questions:

- Why is it difficult to measure the length of a curve with a ruler?
- What is the name given to the distance around a circle?
- How is this distance calculated?

Then have students think about what additional information, if any, they would need to answer these questions. **Ask:**

- Can you determine the circumference of the circle? Why or why not?
- Can you determine the proportion of the circumference represented by the arc length? Explain.

1 Spark Your Learning

▶ MOTIVATE

- Have students look at the photo in their books and read the information contained in the photo. Then complete Part A as a whole-class discussion.
- Give the class the additional information they need to solve the problem. This information is available online as a printable and projectable page in the Teacher Resources.
- Have students work in small groups to complete Parts B–D.

▶ PERSEVERE

If students need support, guide them by asking:

Q **Advancing • Use Tools** Which tool could you use to solve the problem? Why choose that tool and not some other? Students' choices of tools and reasons for choosing them will vary.

Q **Assessing** What is the formula for circumference of a circle when the radius is known? $C = 2\pi r$

Q **Assessing** How many degrees are in a circle? 360°

Q **Advancing** Given the angle that the arc length spans, what fractional part of the circumference is the arc length?
The arc length equals $\frac{\text{angle}}{360^\circ} \cdot$ circumference

Turn and Talk Ask students what fractional part of a circle is represented by 270°. Remind students that the angle and arc length are directly proportional and will vary, but that the radius of the circle remains constant throughout the calculations.

The length of the arc would be about 12.5 feet for the inner border and 15.7 feet for the outer border, or half the circumference of each full circle.

The length of the arc would be about 18.8 feet for the inner border and 23.6 feet for the outer border, or three-fourths of the circumference of each full circle.

▶ BUILD SHARED UNDERSTANDING

Select groups of students who used various strategies and tools to share with the class how they solved the problem. As they present their solutions, have each group discuss why they chose a specific strategy and tool.

② Learn Together

Build Understanding

Task 1 (MP) **Use Structure** Encourage students to use the structure of the table to complete it. Remind students of the pattern—the last column of the table is equal to the product of the second and fourth columns.

CONNECT TO VOCABULARY

Have students use the **Interactive Glossary** to record their understanding of the vocabulary in this task.

Sample Guided Discussion:

Q **Given a circle with radius *r*, how does the length of the arc vary as the arc measure in degrees increases?** The arc length also increases. Arc length is directly proportional to the arc measure.

Q **In Part D, suppose the radii of both circles were the same. How would the arc lengths of 60° and 120° angles be related?** The arc length of the 60° angle is half the arc length of the 120° (or the arc length of the 120° is twice the arc length of the 60°).

Turn and Talk Ask students what formula is used to calculate the circumference of a circle if the diameter is known. Possible answer: The arc length is equal to the ratio of the arc measure to 360° times the product of π and the diameter of the circle.

Build Understanding

Derive the Formula for Arc Length

An **arc** is an unbroken part of a circle consisting of two points, called endpoints, and all the points on the circle between them. The distance along an arc, measured in linear units such as centimeters, is called the **arc length**. You can apply the fact that there are 360° in a circle to find the length of an arc given the radius of the circle.

1 The table shows information about two circles.

Circle	$C = 2\pi r$	$m\overset{\frown}{AB}$	Fraction of circle	Length of $\overset{\frown}{AB}$
Circle P with radius 10, points A and B, right angle at P	$C = 2 \cdot \pi \cdot 10$ $= 20\pi$	90°	$\frac{1}{4}$	$\frac{1}{4}(20\pi) = 5\pi$
Circle P with radius 7, diameter AB	$C = 2 \cdot \pi \cdot 7$ $= 14\pi$	180°	$\frac{1}{2}$	$\frac{1}{2}(14\pi) = 7\pi$

A–F. See Additional Answers.

A. How can you find the fraction of the circle that each arc represents?

B. Describe how the arc length in the last column is calculated.

C. What units are used to measure arc length? Explain.

D. Create a similar table for the circles shown.

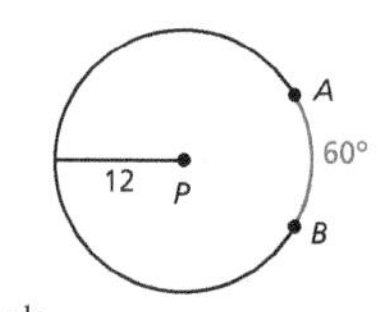

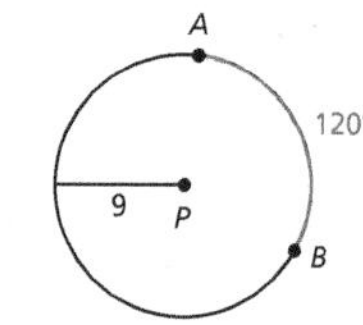

E. Suppose $m\overset{\frown}{AB} = x°$. What fraction of the circumference is contained within the arc?

F. Using the same reasoning, what formula can you write to determine the arc length s for any arc measuring $x°$ with radius r?

Turn and Talk Explain in your own words how the arc length relates to the diameter in terms of the arc measure and 360°. See margin.

532

LEVELED QUESTIONS

Depth of Knowledge (DOK)	Leveled Questions	What Does This Tell You?
Level 1 **Recall**	What formula is used to find the fraction of a circle represented by an arc of $x°$? $\frac{x°}{360°}$	Students' answers will indicate whether they understand that a circle contains 360° and an arc of $x°$ represents a portion of that circle.
Level 2 **Basic Application of Skills & Concepts**	What is the length of an arc in a circle with radius of 3 and arc measure of 150°? $s = \frac{150°}{360°} \cdot 2\pi(3) = \frac{5}{2}\pi$	Students' answers will indicate whether they can calculate the length of an arc in a circle given the radius and arc measure.
Level 3 **Strategic Thinking & Complex Reasoning**	What is the radius of a circle if an angle of 45° has an arc length of 2π? $r = 8$; $2\pi = \frac{45}{360} \cdot 2\pi r$; $1 = \frac{1}{8}r$; $8 = r$	Students' answers will indicate whether they can manipulate the formula for determining the arc length to solve for indicated unknowns.

Derive an Expression for Radian Measure

Angles are commonly measured using degrees, and there are 360° in a circle. Another measure that can be used for angles is *radian measure*, which is defined using the relationship between the length of an arc intercepted by a central angle in a circle and the radius of the circle.

Radian measure can be investigated using **concentric circles**, which are coplanar circles that have the same center. The radar screen shown uses concentric circles to mark distances from a specific point.

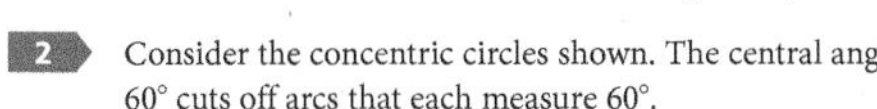

2 Consider the concentric circles shown. The central angle of 60° cuts off arcs that each measure 60°.

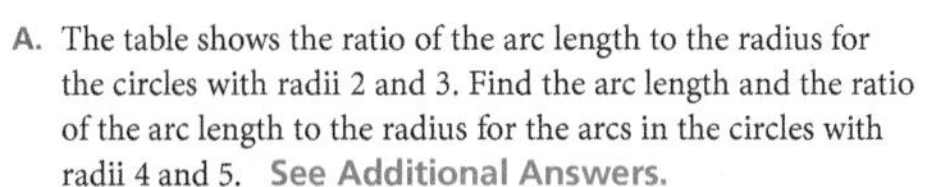

A. The table shows the ratio of the arc length to the radius for the circles with radii 2 and 3. Find the arc length and the ratio of the arc length to the radius for the arcs in the circles with radii 4 and 5. See Additional Answers.

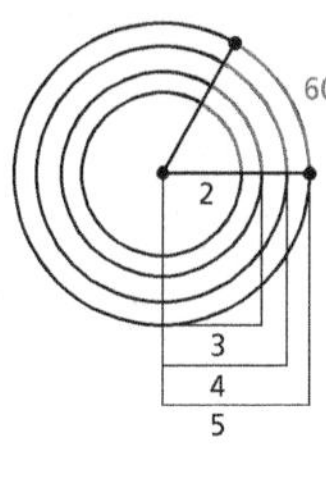

Radius r	Arc length	$\frac{\text{arc length}}{\text{radius}}$
2	$\frac{60°}{360°} \cdot 2\pi(2) = \frac{2}{3}\pi$	$\frac{\frac{2}{3}\pi}{2} = \frac{\pi}{3}$
3	$\frac{60°}{360°} \cdot 2\pi(3) = \pi$	$\frac{\pi}{3}$

B. What do you notice about the ratio of the arc length to the radius of a circle for an arc defined by a central angle of 60°? B–E. See margin.

C. Suppose the concentric circles have a central angle of 90° instead of 60°. What should be true about the ratios of the corresponding arc lengths and radii? Explain.

D. As you have learned, all circles are similar. Use what you know about the ratios of corresponding parts of similar figures to explain the relationship between the arc lengths and the radii in the concentric circles.

E. The expression $s = \frac{x°}{360°} \cdot 2\pi r$ gives the arc length s for an arc intercepted by a central angle of $x°$ in a circle with radius r. When $x°$ is fixed, the expression $\frac{x°}{360°} \cdot 2\pi$ is the constant of proportionality, and this constant is defined as the **radian measure** of the fixed angle. How is the radian measure of an angle related to the ratio of an arc's length to the radius of the circle? Explain.

Turn and Talk Write an ordered pair of the form (radius, arc length) for each circle in Part A, using a decimal to the nearest tenth for each arc length. Should the points lie on a line? If so, what should be the slope of the line? See margin.

PROFICIENCY LEVEL

Beginning

Give students a drawing of a circle and the formula $s = \frac{x°}{360°} \cdot 2\pi r$, and ask them to label the circle with the variables of the formula.

Intermediate

Give students a drawing of a circle with a central angle of 150°, and ask them to explain what formula they will need to convert degrees to radians and what their final answer is after they have applied the formula.

Advanced

Give students a drawing of a circle with a central angle of 40°, and ask them to explain how to calculate the radian measure of the angle. Ask what conjectures they could make about the radian measures of central angles of 80°, 120°, and 160° and to explain their reasoning.

Task 2 (MP) **Use Repeated Reasoning** Remind students to look for shortcuts when doing repeated calculations. After they have completed several calculations with specific values of r, suggest that they replace the specific values of the radius with the variable r and confirm the pattern they have found.

CONNECT TO VOCABULARY

Have students use the **Interactive Glossary** to record their understanding of the vocabulary in this task.

Sample Guided Discussion:

Q What does it mean if two circles are similar? Circles are similar if the ratios of corresponding measures of the circle are equal to the same scale factor.

Q In Part C, suppose the concentric circles had a central angle of 45°. What should be true about the ratios of the corresponding arc lengths and radii? The ratio of the arc length to the radius should be equal for all arcs with a central angle of 45°. For any value of r, that ratio would be $\frac{\frac{45°}{360°} \cdot 2\pi r}{r} = \frac{\pi}{4}$.

Turn and Talk Remind students that as the radius increases, the arc length increases in the same ratio. (2, 2.09), (3, 3.14), (4, 4.19), (5, 5.24); yes; The slope should be $\frac{\pi}{3}$ or 1.05.

ANSWERS

2.B. The ratio of the arc length to the radius is equal for all arcs with a central angle of 60°.

2.C. The ratio of the arc length to the radius should also be equal for all arcs with a central angle of 90°. For any value of r, that ratio would be $\frac{\frac{90°}{360°} \cdot 2\pi r}{r} = \frac{\pi}{2}$.

2.D. Corresponding parts of similar figures are proportional. So, the ratio of the length of an arc formed by the same angle to the radius is equal for all circles.

2.E. They are the same; The ratio of arc length to radius is $\frac{\text{arc length}}{\text{radius}} = \frac{\frac{x°}{360°} \cdot 2\pi r}{r} = \frac{x°}{360°} \cdot 2\pi$, which is the same as the constant of proportionality.

Step It Out

Task 3  **Model with Mathematics** Encourage students to realize that they can apply mathematics to problems arising in everyday life. Remind them to draw a diagram to help them to identify the important quantities of the problem and to apply what they know—the formula for arc length—to solve the problem.

Sample Guided Discussion:

Q What values are needed to apply the arc length formula? The radius of the circle and the arc measure in degrees are needed to apply the arc length formula.

Q In Part B, how can you use the diameter to find the arc length? $s = \frac{120°}{360°} \cdot \pi d = \frac{120°}{360°} \cdot \pi(23) = \frac{23\pi}{3}$

Turn and Talk Ask students to compare the number of minutes that pass from 1:32 to 1:52 with the number of minutes that pass from 10:05 to 10:25. no; The minute hand still travels 20 minutes and sweeps an arc measure of 120°. So, the distance traveled around the clock face is the same.

ANSWERS

3.A. There are 60 minutes in a complete rotation of the minute hand. Since the fraction of 20 minutes in 60 minutes is $\frac{20}{60} = \frac{1}{3}$, progressing 20 minutes is $\frac{1}{3}$ of a complete rotation.

3.B. The arc is $\frac{1}{3}$ of a complete rotation, so the arc is $\frac{1}{3}$ of the circumference. So, $\frac{1}{3}C = \frac{1}{3}\pi d = \frac{23}{3}\pi \approx 24.1$ feet.

Step It Out

Apply the Formula for Arc Length

Arc Length

The arc length s of an arc with measure $x°$ and radius r is given by the formula $s = \frac{x°}{360°} \cdot 2\pi r$.

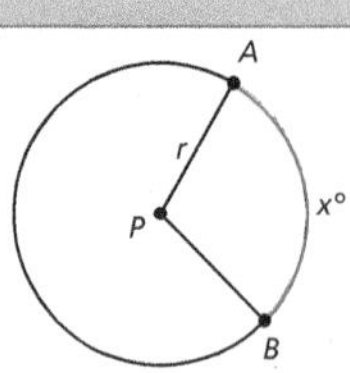

3 The Palace of Westminster in London, England, has a tower with a circular clock face set in an iron frame. The tip of the minute hand sweeps across the iron frame as the time changes. How far does the tip of the minute hand travel as the clock progresses from 10:05 to 10:25? Round your answer to the nearest tenth.

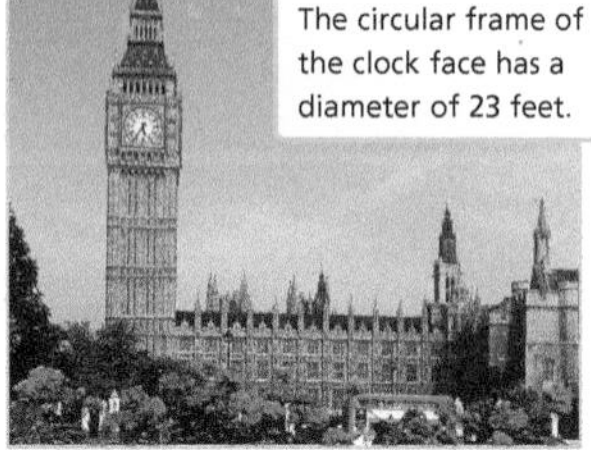

The circular frame of the clock face has a diameter of 23 feet.

Find the radius.

The radius is half of the diameter.

$r = \frac{23}{2} = 11.5$ ft

Find the arc measure.

The minute hand sweeps an arc over 20 minutes, which is $\frac{1}{3}$ of a complete rotation. So, the measure of the arc is $\frac{1}{3} \cdot 360° = 120°$.

A. Why is 20 minutes $\frac{1}{3}$ of a complete rotation of the minute hand?

Find the arc length. A, B. See margin.

$$s = \frac{x°}{360°} \cdot 2\pi r$$
$$= \frac{120°}{360°} \cdot 2\pi(11.5)$$
$$= \frac{23}{3}\pi$$
$$\approx 24.1 \text{ ft}$$

B. What is another way to find the arc length given that the arc is $\frac{1}{3}$ of a complete rotation of the minute hand?

The minute hand of the clock travels about 24.1 feet as the clock progresses from 10:05 to 10:25.

 **Turn and Talk** Would the results change if you were asked to find the distance as the clock progressed from 1:32 to 1:52? Why or why not? See margin.

534

Convert Between Radian Measure and Degree Measure

As you learned in Task 2, the radian measure of an angle that measures $x°$ is $\frac{x°}{360°} \cdot 2\pi$. Because there are 360° in a circle, there are 2π radians in a circle. That means $360° = 2\pi$ radians and $180° = \pi$ radians. This proportional relationship can be used to convert between radians and degrees.

Convert Degrees to Radians	Convert Radians to Degrees
Multiply the degree measure by $\frac{\pi \text{ radians}}{180°}$ and simplify the fraction.	Multiply the radian measure by $\frac{180°}{\pi \text{ radians}}$ and simplify the fraction.

Radian measures are usually given in terms of π. When giving an angle measure in radians, the unit "radians" is often not included. For example, $\frac{\pi}{2}$ radians can be written as $\frac{\pi}{2}$.

Working with radian measures will be useful in future coursework in math and physics that involves trigonometry functions and circular and angular motion.

4 Convert 270° to radian measure.

$$270° = 270°\left(\frac{\pi \text{ radians}}{180°}\right)$$

$$= \frac{270° \cdot \pi \text{ radians}}{180°}$$

$$= \frac{3\pi}{2} \text{ radians}$$

Why are there no degree measures in the final answer?

See margin.

Turn and Talk Give answers in radian measure. See margin.

- What is the sum of the measures of the angles in a triangle?
- What is the sum of the measures of the angles in a quadrilateral?

5 Convert $\frac{\pi}{6}$ to degree measure.

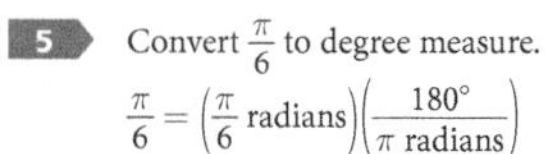

$$\frac{\pi}{6} = \left(\frac{\pi}{6} \text{ radians}\right)\left(\frac{180°}{\pi \text{ radians}}\right)$$

$$= \frac{180°}{6}$$

$$= 30°$$

How do you know that the degree measure of $\frac{\pi}{6}$ will be less than 180°?

The degree measure of π radians is 180°. A fraction of π radians, such as $\frac{\pi}{6}$, will have a degree measure that is a fraction of 180°.

Turn and Talk Use the fact that $180° = \pi$ radians to set up and solve a proportion to find the degree measure of $\frac{\pi}{6}$ radians. See margin.

Task 4 **MP** **Attend to Precision** Students sometimes confuse which conversion relationship to use. Remind them that when the final answer is in radians, the degrees must cancel out during the calculation.

EL SUPPORT SENSE-MAKING Three Reads

Have students read the problem three times. Use the questions below for a different focus each read.

1. What is the situation about?
2. What are the quantities in the situation?
3. What are the possible mathematical questions that you could ask for this situation?

Turn and Talk Have students answer the question in degrees, and then convert the degrees to radians. π; 2π

Task 5 **MP** **Attend to Precision** Students sometimes confuse which conversion relationship to use. Remind them that when the final answer is in degrees, the π's must cancel out during the calculation.

Sample Guided Discussion:

Q **What are the units of the term multiplied by radians to convert it to degree measure?** $\frac{180°}{\pi \text{ radians}}$

Q **How can you check that your answer is correct?** Convert your answer in degrees back to radians and make sure it equals what you started with.

Turn and Talk Remind students that the units in a proportion need to be consistent. A ratio with units $\frac{\text{degrees}}{\text{radians}}$ must be set equal to another ratio with units $\frac{\text{degrees}}{\text{radians}}$.

Let x = degree measure of $\frac{\pi}{6}$. So, $\frac{180°}{\pi} = \frac{x}{\frac{\pi}{6}}$ and $x = 30°$.

ANSWERS

4. The unit of degrees is in the numerator and the denominator, so they cancel when divided.

ONLINE

Assign the Digital On Your Own for
- built-in student supports
- Actionable Item Reports
- Standards Analysis Reports

On Your Own

Assignment Guide

The chart below indicates which problems in the On Your Own are associated with each task in the Learn Together. Assign daily homework for tasks completed.

Learn Together Tasks	On Your Own Problems
Task 1, p. 532	Problems 10–15 and 31
Task 2, p. 533	Problems 17 and 18
Task 3, p. 534	Problems 16, 18, 20, 26, 31, and 32
Task 4, p. 535	Problems 21–25
Task 5, p. 535	Problems 27–30

ANSWERS

1. The measure of an arc is the measure of its central angle while the length of an arc is the distance of an arc along a circle. An arc's measure is in degrees, and its length is in linear units.

Check Understanding

1. Describe the difference between the measure of an arc and the length of an arc. See margin.

Answer each question about the formula for the arc length s of an arc $s = \frac{x°}{360°} \cdot 2\pi r$.

2. What part of the circle does the expression $2\pi r$ represent? circumference
3. What does $x°$ represent? arc measure
4. What does $\frac{x°}{360°}$ represent? the fraction of the circle represented by the arc
5. For a fixed angle that measures $x°$, what does $\frac{x°}{360°} \cdot 2\pi$ represent? the radian measure of the angle

Find the length of $\widehat{AB}$ using the given measure of $\widehat{AB}$ in a circle with a radius of 5 meters. Round to the nearest hundredth.

6. $m\widehat{AB} = 135°$ 11.78 m
7. $m\widehat{AB} = 80°$ 6.98 m

Complete each statement with the equivalent angle measure.

8. $120° =$ __?__ radians $\frac{2\pi}{3}$
9. __?__ $= \frac{\pi}{5}$ radians 36°

On Your Own

Tell whether each statement is *always*, *sometimes*, or *never* true. Explain. 10–12. See Additional Answers.

10. Two arcs with the same measure have the same arc length.
11. The length of the arc of a circle is greater than the circumference of the circle.
12. In a circle, two arcs with the same length have the same measure.

Find each length of $\widehat{AB}$. Round to the nearest hundredth.

13.
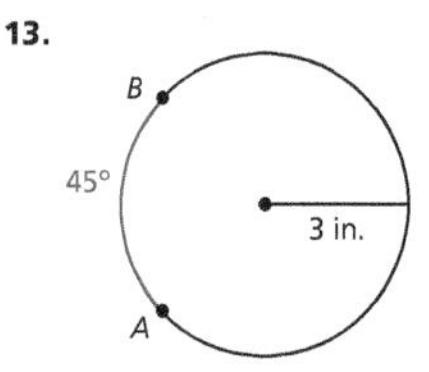

2.36 in.

14.
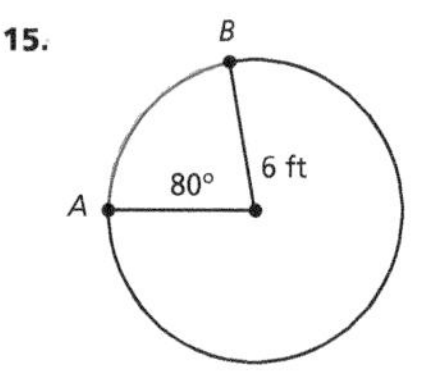

22.34 cm

15.

8.38 ft

16. **MUSIC** Metronomes provide a tempo for musicians by producing clicks at a set interval. Each swing of the inverted pendulum produces a click. With each swing, the pendulum sweeps through an angle of 70°. What distance does the tip of the pendulum travel on each swing? Round to the nearest centimeter. 22 cm

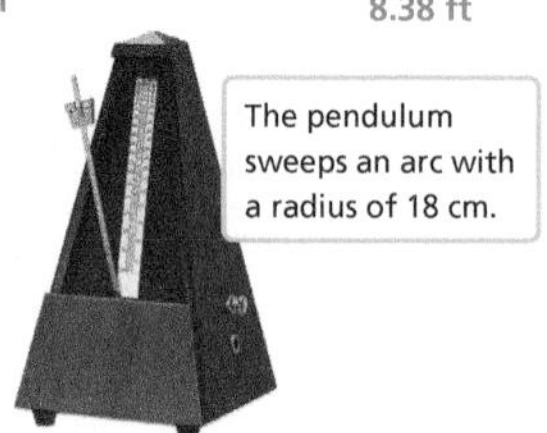

The pendulum sweeps an arc with a radius of 18 cm.

3 Check Understanding

Formative Assessment

Use formative assessment to determine if your students are successful with this lesson's learning objective.

Students who successfully complete the Check Understanding can continue to the On Your Own practice.

For students who miss 1 problem or more, work in a pulled small group using the Almost There small-group activity on page 531C.

ONLINE

Assign the Digital Check Understanding to determine
- success with the learning objective
- items to review
- grouping and differentiation resources

4 Differentiation Options

Differentiate instruction for all students using small-group activities and math center activities on page 531C.

Reteach

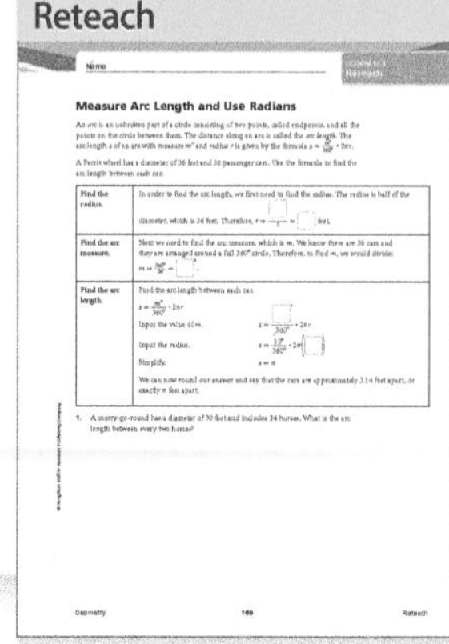

Challenge

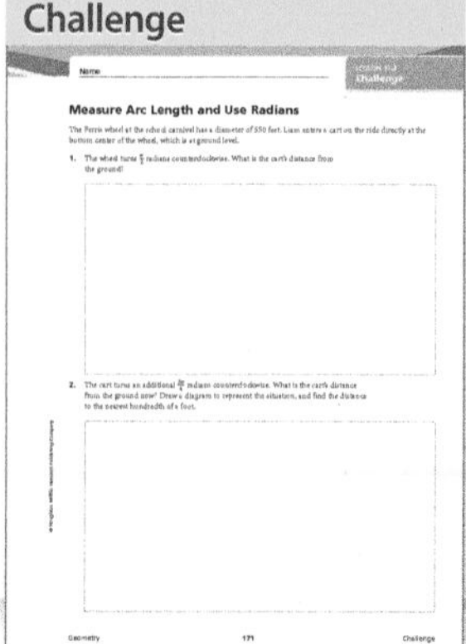

In Problems 17 and 18, two concentric circles on a pocket watch are shown. Find the unknown measure. Round to the nearest hundredth.

17. length of $\widehat{AB} = 3.87$ cm, length of $\widehat{CD} = 6.9$ cm, $PA = 1.85$ cm, $PD =$ __?__

3.30 cm

18. length of $\widehat{AB} =$ __?__, length of $\widehat{CD} = 8.38$ cm, $PA = 1.7$ cm, $PD = 3$ cm

4.75 cm

19. Snow skis have curved edges. The curve is determined by the sidecut radius, which is the radius of a very large circle. The ski shown has an effective edge of 150.1 cm, and the measure of $\widehat{AB}$ is 6.14°. What is its sidecut radius in meters? 14 m

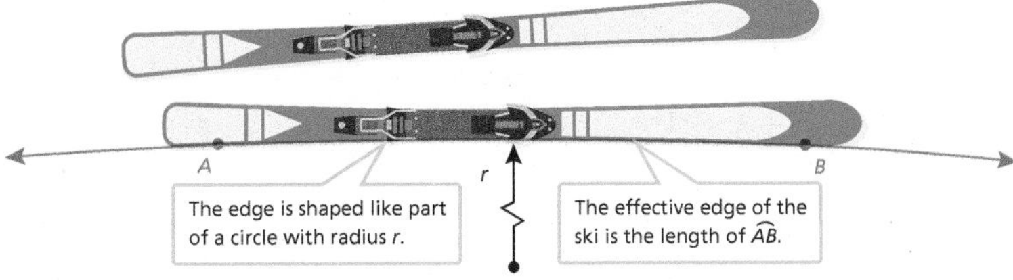
A
r
B
The edge is shaped like part of a circle with radius r.
The effective edge of the ski is the length of $\widehat{AB}$.

20. Belt drives transmit power between drive shafts in equipment. In the belt drive shown, the smaller pulley is being driven by a power source and is delivering power to the larger pulley through the belt. The difference in size is used to change the speed at which the larger pulley turns. For each full turn of the smaller pulley, through what angle does the larger pulley turn? Round to the nearest tenth. 150°

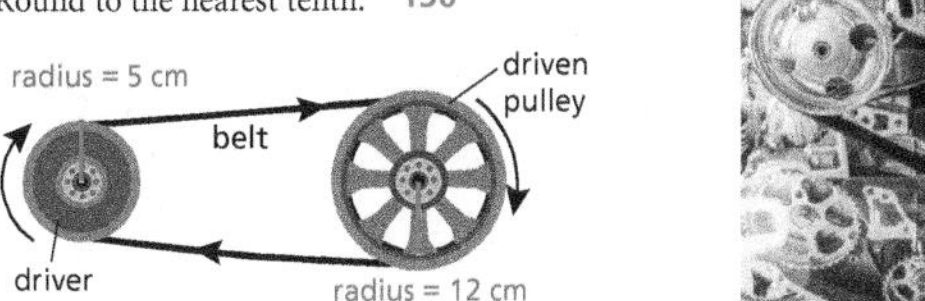
radius = 5 cm
belt
driven pulley
driver
radius = 12 cm

Convert each angle from degree measure to radian measure.

21. 15° $\frac{\pi}{12}$ radians **22.** 55° $\frac{11\pi}{36}$ radians **23.** 110° $\frac{11\pi}{18}$ radians **24.** 200° $\frac{10\pi}{9}$ radians

25. STEM The angular velocity ω of an object gives the radians traveled per unit of time. This can be applied to objects traveling a circular path. To find the linear velocity when given the angular velocity, use the formula $v = r\omega$, where v is the linear velocity, r is the radius, and ω is the angular velocity. If a racecar on a circular track with a radius of 500 ft travels an arc of 30° per second, what is the linear velocity of the car in feet per second? 261.8 feet per second

Questioning Strategies

Problem 20 Suppose the radius of the larger circle is 18 cm. For each full turn of the smaller pulley, through what angle does the larger pulley turn?
100°; $10\pi = \frac{x°}{360°} \cdot 36\pi$; $x = 100°$

Watch for Common Errors

Problem 19 Students need to watch the units in this problem. The effective edge is given in centimeters, but the problem asks for the radius in meters. Students need to convert centimeters to meters, either before beginning the problem so that their answer will be in meters or after working the problem in centimeters.

5 Wrap-Up

Summarize learning with your class. Consider using the Exit Ticket, Put It in Writing, or I Can scale.

Exit Ticket 

A jogger runs a distance of 12π feet around a circular track that has a radius of 18 feet. Through what angle did he run from his starting point to his ending point? Express your answer in both degrees and radians.

$120°, \frac{2}{3}\pi$ radians; $12\pi = \frac{x°}{360°} \cdot 2\pi(18); x° = 120°;$

$120° \cdot \frac{\pi \text{ radians}}{180°} = \frac{2}{3}\pi$ radians

Put It in Writing 

Students are to write a letter to a student who will be taking this class next year describing what they have learned on arc length and radian measure. The letter should include one paragraph on the formulas for calculating arc length and one paragraph on converting between degrees and radians.

I Can 

The scale below can help you and your students understand their progress on a learning goal.

4	I can use similarity of circles to find arc length, and I can explain my steps to others.
3	I can use similarity of circles to find arc length.
2	I can convert between radian and degree measures.
1	I can find the fraction of a circle represented by a central angle.

Spiral Review • Assessment Readiness

These questions will help determine if students have retained information taught in the past and can also prepare them for high-stakes assessments. Here, students use the secant-tangent product theorem **(16.1)**, apply the area of a circle formula **(17.1)**, use the tangent-secant exterior angle measure theorem **(16.2)**, and calculate the area of a circle in preparation for work with sectors **(17.1)**.

26. MP **Reason** In circle Q, the length of $\overset{\frown}{CD}$ is equal to its radius.

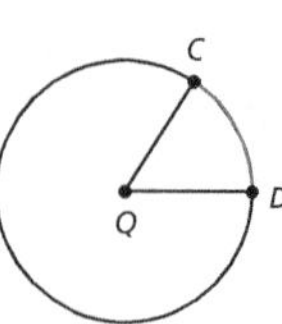

A. How many times does $\overset{\frown}{CD}$ fit around the circle? Leave your answer in terms of π. 2π

B. Is it possible that the terms *radians* and *radius* are related to each other? Use your observations to explain your answer. See Additional Answers.

Convert each angle from radian measure to degrees.

27. $\frac{3\pi}{2}$ 270° **28.** $\frac{\pi}{6}$ 30° **29.** $\frac{7\pi}{8}$ 157.5° **30.** $\frac{7\pi}{9}$ 140°

31. MP **Use Repeated Reasoning** Complete the table to find the angle measure in radians for the benchmark angles. See Additional Answers.

Benchmark Angles									
Degree measure	0°	30°	45°	60°	90°	120°	135°	150°	180°
Radian measure	?	?	?	?	?	?	?	?	?

32. Open Middle™ Using the digits 1 to 9, at most one time each, fill in the boxes so that the radius and angle measure result in the arc length measure.

$r =$  cm $\theta =$  ° $\overset{\frown}{AB} =$  π cm

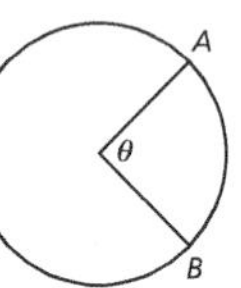

See Additional Answers.

Spiral Review • Assessment Readiness

33. Find the value of x.

Ⓐ 8
Ⓑ 11
Ⓒ 12
Ⓓ 16

34. A circle has a circumference of 25 feet. What is the approximate area of the circle?

Ⓐ 13.2 ft^2
Ⓑ 25.6 ft^2
Ⓒ 36.2 ft^2
Ⓓ 49.7 ft^2

35. What is the measure of $\angle ABC$?

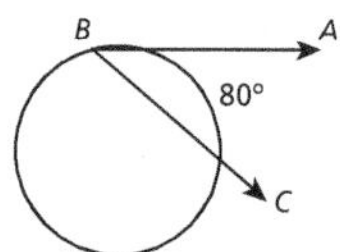

Ⓐ 20°
Ⓑ 40°
Ⓒ 80°
Ⓓ 160°

36. A circle has a radius of 6 inches. What is the area of the circle?

Ⓐ 9π in^2
Ⓑ 20π in^2
Ⓒ 27π in^2
Ⓓ 36π in^2

 **I'm in a Learning Mindset!**

What barriers are there to understanding how radian measure and degree measure relate to each other?

Keep Going ▶ Journal and Practice Workbook

Learning Mindset

Resilience Identifies Obstacles

Students may need help recognizing barriers they need to overcome in order to be successful at measuring arc length and converting between degree and radian measures. For example, one student may need to brush up on multiplying fractions and reducing fractions to lowest terms while another student may need practice working with units. Identifying and mastering these prerequisite skills will help students achieve success with their current learning. *Are there concepts that I need to review so I can use them in this section? Can I use my calculator to help me perform operations like fraction multiplication or simplification to lowest terms? Do I always check my units to make sure they make sense in terms of the problem?*

17.3 Measure Sector Area

LESSON FOCUS AND COHERENCE

Mathematics Standards

- Derive using similarity the fact that the length of the arc intercepted by an angle is proportional to the radius, and define the radian measure of the angle as the constant of proportionality; derive the formula for the area of a sector.

Mathematical Practices and Processes 

- Use appropriate tools strategically.
- Look for and make use of structure.
- Model with mathematics.

I Can Objective 

I can use sector area to solve real-world problems.

Learning Objective

Derive the formula for the area of a sector of a circle and use that formula to compute the area of sectors of circles having different central angles and radii.

Language Objective

Explain using correct terminology how to find the area of a sector of a circle.

Vocabulary

New: sector

Mathematical Progressions

Prior Learning	Current Development	Future Connections
Students: • identified the formulas for the area and circumference of a circle and used them to solve problems; gave an informal derivation of the relationship between the circumference and area of a circle. **(Gr7, 10.1 and 10.2)** • proved that all circles are similar. **(12.1)**	**Students:** • understand the derivation of the formula for the area of a sector of a circle. • use the formula for sector area to find the area of sectors of circles of varying radii and central angles.	**Students:** • will explain how the unit circle in the coordinate plane enables the extension of trigonometric functions to all real numbers, interpreted as radian measures of angles traversed counterclockwise around the unit circle. **(Alg2, 15.1)**

UNPACKING MATH STANDARDS

Derive using similarity the fact that the length of the arc intercepted by an angle is proportional to the radius, and define the radian measure of the angle as the constant of proportionality; derive the formula for the area of a sector.

What It Means to You

Students will find arc lengths and the areas of sectors of circles. They will derive and understand the derivation of the formula for the arc length of a circle. Their understanding of these concepts develops from previous lessons, in which arc length was related to inscribed and circumscribed angles. Using the ratio of central angles to 360°, for example, students will learn the relationship between the area of a circle and the area of a sector of a circle.

The emphasis for this standard is on computing sector area and understanding the derivation of the formula. In the next lesson, students will develop fluency with three-dimensional figures.

WARM-UP OPTIONS

ONLINE

PROJECTABLE & PRINTABLE

ACTIVATE PRIOR KNOWLEDGE • Use the Formula for Area of a Circle to Solve Problems

Use these activities to quickly assess and activate prior knowledge as needed.

Problem of the Day 

Ron wants to cut his grass by mowing in a circle moving outward. The outer boundary of his yard is 200 ft from the wall of his house. If you assume that he is mowing in perfect circles, how much of his lawn will he have cut before he has to stop mowing in a circle? Since the distance from the wall of his house to the end of his yard is 200 ft, the radius is 100 ft (half the diameter of the circle, which would be tangent to both the wall and the edge of his yard). Therefore, the area of the circle he will have cut is $A = \pi r^2 = \pi \cdot 100^2 = 10{,}000\pi \approx 31{,}400$ square feet, using the approximation $\pi \approx 3.14$.

Quick Check for Homework

As part of your daily routine, you may want to display the Teacher Solution Key to have students check their homework.

Make Connections

Based on students' responses to the Problem of the Day, choose one of the following:

1. Project the Interactive Reteach, 17.1.
2. Complete the Prerequisite Skills Activity:

Find the radii of four concentric circles if the largest has a diameter of 28 cm and each successive circle has a diameter 4 cm less than the one larger than it. Then square each radius and multiply each square radius by π (not the approximation) to give the area of each circle.

- *What can you tell us about the diameter of a circle?* The diameter of a circle is twice the radius.
- *What are the diameters of the three smaller circles?* The diameters are 24, 20, and 16, in descending order.
- *Given the diameters, what are the radii?* The radii are 14, 12, 10, and 8.
- *What are the squares of those radii?* The squares of the radii are 196, 144, 100, and 64. So the four circles' areas are 196π, 144π, 100π, and 64π respectively.

SHARPEN SKILLS

If time permits, use this on-level activity to build fluency and practice basic skills.

Quantitative Comparison

Objective: Students demonstrate an understanding of the area of a circle.

Write the following problem on the board. Ask students to choose the letter representing the correct letter and to explain their reasoning.

Two circles, C_1 and C_2, have diameters of 10 cm and 6 cm respectively.

Quantity A	Quantity B
The area of C_2	The difference of the areas of C_1 and C_2

A. Quantity A is greater.
B. Quantity B is greater.
C. The two quantities are equal.
D. The answer cannot be determined from the information given.

B. $Area(C_1) = 25\pi$; $Area(C_2) = 9\pi$, so the difference between the two areas is 16π.

PLAN FOR DIFFERENTIATED INSTRUCTION

Small-Group Options

Use these teacher-guided activities with pulled small groups.

On Track

Have students work in small groups to find the central angles of two circles, one whose sector has an area of $\frac{9}{8}\pi$ and a radius of 3, and the other whose sector has an area of $\frac{4}{3}\pi$ and a radius of 4. Ask students to draw the circles with the angle measures and radii indicated. Have volunteers share their drawings with the whole class.

Almost There

Have students work in groups to draw circles and sectors with the following radii and central angles: 1, 90°; 2, 60°; 3, 30°. Ask them to choose one of them and compute the sector area.

Have volunteers share their results with the whole class.

Ready for More

Have students work in groups to find the total area of a group of sectors that represent the areas of a semicircular fan opened so that the right end of the fan is open at 45° and the left end is open at 135°, with each arc length equal to 10°. The length of each end of the fan is 6 inches.

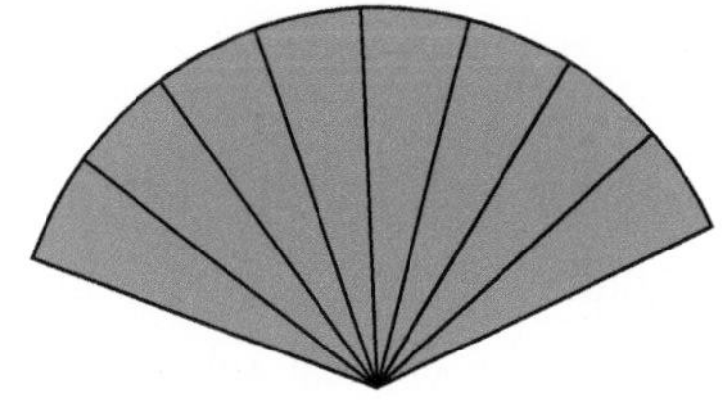

Math Center Options

Use these student self-directed activities at centers or stations. Key: ● Print Resources ● Online Resources

On Track

- ● Interactive Digital Lesson
- ●● Journal and Practice Workbook
- ● Interactive Glossary (printable): **sector**
- ●● Standards Practice: Define Radian and Solve Problems Involving Sectors of a Circle

Almost There

- ● Reteach 17.3 (printable)
- ● Interactive Reteach 17.3

Ready for More

- ● Challenge 17.3 (printable)
- ● Interactive Challenge 17.3

Unit Project Check students' progress by asking how to use the area of a circle and the central angle to determine the area of a sector.

View data-driven grouping recommendations and assign differentiation resources.

During the *Spark Your Learning,* listen and watch for strategies students use. See samples of student work on this page.

Use the Central Angle and Radius — Strategy 1

The central angle is 60°, and the radius is 350 feet.
Then I found the total area of the circle to be
$A = \pi r^2 = \pi \cdot 350^2 = 122{,}500\pi$ square feet.
And $60° = \frac{1}{6}360°$.
Then I divided the area of the whole circle by 6 to get the area of the portion of the neighborhood, which is
$A = \frac{1}{6}(122{,}500)\pi \cong 64{,}141$ square feet.

If students . . . use the formula for the area of a circle, they are on the right track in understanding the concept of sector area.

Have these students . . . explain how they determined the inputs to their equation and how they solved it. **Ask:**

- Q How did you decide to divide the area formula by 4?
- Q How did you decide to use the area formula?

Graph the Figure — Strategy 2

I used the fact that the central angle is 60°, so I drew two line segments of length 2 inches at a 60° angle. I got out my protractor and drew an arc from the vertical side to the horizontal side. The radius is 350 feet. So I measured the legs of the arc in my graph and figured that I needed to scale up my drawing by a factor of 175 feet per inch.
Then I measured the area of the figure in the drawing and multiplied by 175^2 to a get a number of about 64,000. So I would say that the area of the sector is about 64,000 square feet.

If students . . . decide to make a visual representation, they may not recognize that the area of the portion is a fraction of the area of a circle.

Activate prior knowledge . . . by having students talk about the area of a circle. **Ask:**

- Q Could you think of a way to find the answer by using the formula for the area of a circle?
- Q Since the portion of the neighborhood is only part of a circle, how would you use the formula for the area of a circle to make it describe the area of the portion?

COMMON ERROR: Use the Wrong Formula

I used the facts that the central angle is 60° and the radius is 350 feet.
So I calculated the total area of the circle:
$A = 2\pi r = 2\pi(350) \approx 2199$ square feet.
Then I divided the total area by 6 and found the answer to be about 367 square feet.

If students . . . use the formula for the circumference of a circle, they may not have a good understanding of the formulas for area and circumference of a circle.

Then intervene . . . by pointing out that area is measured in square units and circumference is measured in units. **Ask:**

- Q If radius is measured in feet, what would the area of the circle be measured in?
- Q Do you think the formula for the area of a circle would have the radius squared?

17.3

Measure Sector Area

(I Can) use sector area to solve real-world problems.

Spark Your Learning

The neighborhood is divided by concentric circular streets and streets along radii.

©Hans Blossey/imageBROKER/Alamy

Complete Part A as a whole class. Then complete Parts B–D in small groups.

A. What is a mathematical question you can ask about the neighborhood? What information would you need to know to answer your question?

B. How does the shape of one portion of the neighborhood compare to the shape of a circle with the same radius?
The shape is a part of a whole circle.

C. To answer your question, what strategy and tool would you use along with all the information you have? What answer do you get?
See Strategies 1 and 2 on the facing page.

D. Does the answer make sense in the context of the situation? How do you know? See Additional Answers.

A. What is the area of one portion of the neighborhood?; the measure of the central angle and radius

Turn and Talk How would the area of one portion of the neighborhood relate to the area of the complete circle if the central angle were 180°, or exactly half the circle? See margin.

CULTIVATE CONVERSATION • Information Gap

Ask students questions to help them decide what missing information they need to answer the question "What is the area covered by one portion of the neighborhood?"

1. Do you have enough information to determine the area of the portion? Explain. yes; I know the radius and the measure the central angle.
2. Suppose you knew the radius was 150 feet. What would the area of one portion be then? The area would be $A = \frac{22{,}500\pi}{6} = 3750\pi$ square feet.
3. Suppose you knew the radius was 300 feet. Then what would the area of one portion be? The area would be $A = \frac{90{,}000\pi}{6} = 15{,}000\pi$ square feet.

1 Spark Your Learning

▶ MOTIVATE 

- Have students look at the photo in their books and read the information contained in the photo. Then complete Part A as a whole-class discussion.
- Give the class the additional information they need to solve the problem. This information is also available online as a printable and projectable page in the Teacher Resources.
- Have students work in small groups to complete Parts B–D.

▶ PERSEVERE

If students need support, guide them by asking:

Q **Advancing • Use Tools** Which tool could you use to solve the problem? Why choose that tool and not some other? Students' choices of tools and reasons for choosing them will vary.

Q **Assessing** How would you begin to think about the area of a sector of a circle? Possible answer: A circle has area $A = \pi r^2$, so I would try to estimate the radius and calculate the total area of the circle. Then I would multiply that by my estimate of the portion of the circle the neighborhood represents.

Q **Advancing** How would you think about calculating the area if you had no idea of the length of the radius? I would set the radius equal to 1, which would make the area of the total circle equal to π. Then I would multiply that area by my estimate of the fraction of the circle that the portion of the neighborhood represents, say, $\frac{1}{c}$, for some integer c. Then that gives me an area of $\frac{\pi}{c}$. When I get more information about the length of the radius (or if I can better estimate the radius), I will multiply $\frac{\pi}{c}$ by the square of the approximate radius to compute the area of the portion.

Turn and Talk Remind students that a circle is 360°. The area of one portion would be half the area of the complete circle, because 180° is half of 360°.

▶ BUILD SHARED UNDERSTANDING

Select groups of students who used various strategies and tools to share with the class how they solved the problem. As they present their solutions, have each group discuss why they chose a specific strategy and tool.

2 Learn Together

Build Understanding

Task 1 (MP) **Use Tools** Students will use proportional reasoning by using the ratio of the central angle of a sector to 360 to derive the formula for the area of a sector. Students will take simple ratios such as $\frac{1}{2}, \frac{1}{3}, \frac{1}{4},$ and so on. They may also construct circles and measure the radii and angles of different arcs and central angles.

CONNECT TO VOCABULARY

Have students use the **Interactive Glossary** to record their understanding of the vocabulary in this task.

Sample Guided Discussion:

Q In the formula you just found, what does the fraction $\frac{1}{6}$ represent? In the formula, I got the fraction $\frac{1}{6}$ by dividing the central angle by 360°: $\frac{1}{6} = \frac{60°}{360°}$.

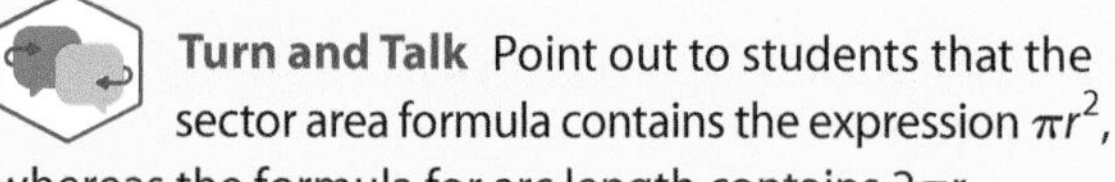

Turn and Talk Point out to students that the sector area formula contains the expression πr^2, whereas the formula for arc length contains $2\pi r$. Possible answer: They are the same because the arc length and sector area can be found by multiplying the fraction of the circle by the arc measure. They are different because the fraction is multiplied by the circumference to find the arc length, and the fraction is multiplied by the area to find the sector area.

Build Understanding

Derive the Formula for Sector Area

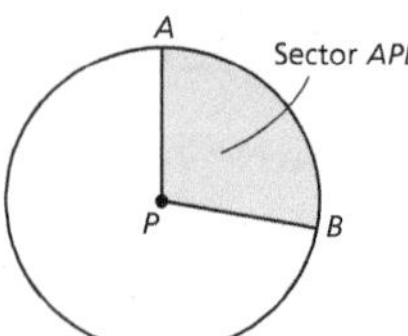

A **sector** of a circle is a portion of the circle bounded by two radii and their intercepted arc. A sector is named using the ends of each radius and the center of the circle. You have learned how to find the area of a circle. You can use proportional reasoning to find the area of a sector of a circle.

1 Find the area of each sector.

A. Find the area of the circle at the right. Leave your answer in terms of π. Suppose a sector represents $\frac{1}{5}$ of the area of the circle. What is the area of the sector? 100π cm²; 20π cm²

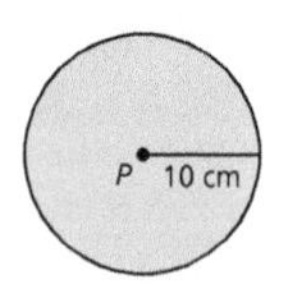

B. The table shows the areas of two sectors of a circle with radius 10 cm. How is the fraction of the circle represented by the sector calculated? How is the fraction used to find the area of the sector? B, C. See Additional Answers.

Circle	Central angle	Fraction of circle	Area of whole circle	Area of sector *APB*
A, 90°, P, 10 cm, B	90°	$\frac{90°}{360°} = \frac{1}{4}$	100π cm²	$\frac{1}{4}(100\pi) = 25\pi$ cm²
P, A, B, 20 cm	180°	$\frac{180°}{360°} = \frac{1}{2}$	100π cm²	$\frac{1}{2}(100\pi) = 50\pi$ cm²

C. Create a similar table for the circles shown.

D. Suppose $m\angle APB = x°$. What fraction of the circle's area is represented by the sector? $\frac{x°}{360°}$ of the area is contained within the sector.

E. The area A is the area of a sector of a circle with central angle $x°$ and radius r. Write a proportion that relates A, $x°$, the total area of the circle, and the number of degrees in a circle. Then solve the proportion for A. See Additional Answers.

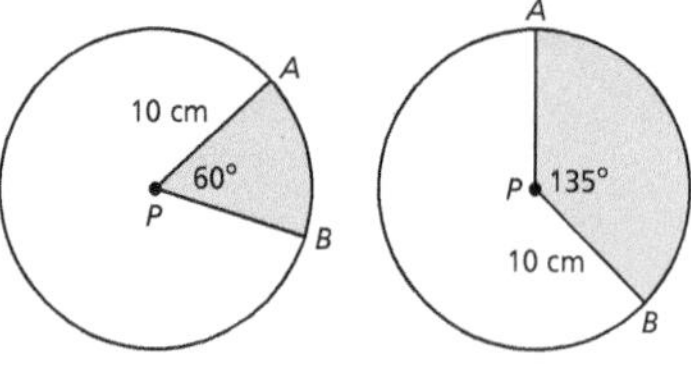

Turn and Talk Compare the formulas for finding arc length and sector area. How are they the same? How are they different? See margin.

540

LEVELED QUESTIONS

Depth of Knowledge (DOK)	Leveled Questions	What Does This Tell You?
Level 1 **Recall**	What is a sector bounded by? two radii and an intercepted arc	Students' answers will indicate whether they recall and can recite the boundaries of a sector of a circle.
Level 2 **Basic Application of Skills & Concepts**	For the 90° sector shown in Task 1, what is the difference between the arc length and the sector area if $r = 3$? The arc length is *Arc length* $= \frac{1}{4}2\pi r = \frac{1}{4}2\pi(3) = \frac{3\pi}{2}$, but the area of the sector is $A = \frac{1}{4}\pi r^2 = \frac{1}{4}\pi(3^2) = \frac{9\pi}{4}$.	Students' answers will indicate whether they have mastered the formulas for arc length and the area of a sector of a circle and are able to apply them to a routine problem.
Level 3 **Strategic Thinking & Complex Reasoning**	Could the arc length and the sector area ever be equal? Explain. They could be the same if $2r = r^2$, which is the case when $r = 2$.	Students' answers will indicate whether they are able to analyze the difference between arc length and sector area and draw a logical conclusion from that difference.

Step It Out

Use the Formula for Sector Area

The proportional reasoning used to find the area of a sector in the previous task can be generalized by the following formula.

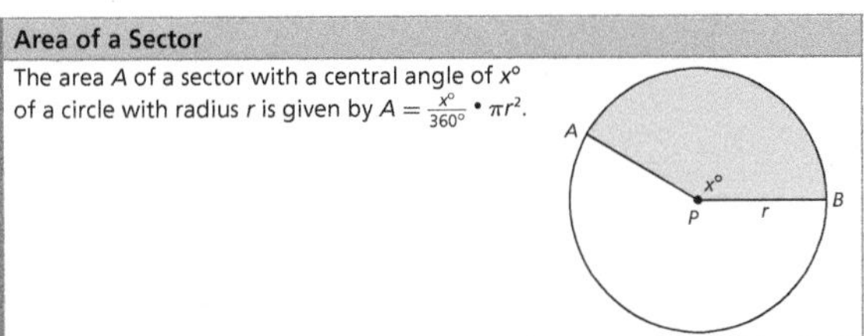

Area of a Sector
The area A of a sector with a central angle of $x°$ of a circle with radius r is given by $A = \frac{x°}{360°} \cdot \pi r^2$.
A
$x°$
P
r
B

2 The bristles of a fan brush are a sector of a circle. Find the area of the bristles.

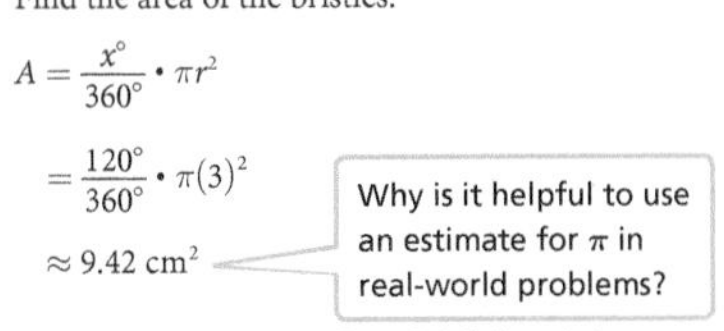

$\approx 9.42 \text{ cm}^2$

Why is it helpful to use an estimate for π in real-world problems?

120°
3 cm

See Additional Answers.

3 One part of a pizza is topped with only green peppers, and the other part is topped with only mushrooms. The part topped with mushrooms has an area of $40\pi \text{ in}^2$. What is the radius of the pizza?

A, B. See Additional Answers.

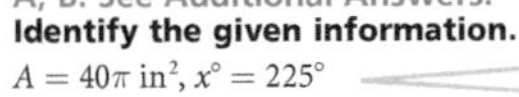

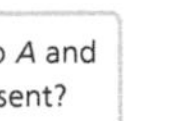

A. What do A and $x°$ represent?

225°
X
Y

Use the formula for area of a sector.

$A = \frac{x°}{360°} \cdot \pi r^2$	Write the formula.
$40\pi = \frac{225°}{360°} \cdot \pi r^2$	Substitute the given values.
$\frac{360°}{225°} \cdot 40\pi = \frac{225°}{360°} \cdot \pi r^2 \cdot \frac{360°}{225°}$	Multiply to clear the fraction.
$64\pi = \pi r^2$	Simplify.
$64 = r^2$	Divide both sides by π.
$8 = r$	B. How can you check the answer?

Turn and Talk Why can your answer contain either π or an approximated decimal? Explain a situation where you would use each method. What does it mean if a problem asks for either an exact or approximate answer? See margin.

PROFICIENCY LEVEL

Beginning

Have students work in pairs to name the variables in the formula for the area of a sector of a circle.

Intermediate

Have students write a few short sentences or phrases describing the process of finding the area of a sector of a circle. Have volunteers read their sentences to the class.

Advanced

Have students work in pairs, taking turns reading the steps in the computation for Task 3: one student reads a step from the book and the other supplies the language for the next step from memory. The one not looking at the book writes down what the other says, does the next step on paper, and reads it to her partner. Pairs go through the computation twice, switching roles.

Step It Out

Task 2 (MP) **Use Structure** Students will apply the formula for the area of a sector of a circle to a specific problem. Students will need to recognize and use the pattern in the problem and understand its significance.

Sample Guided Discussion:

- **Q What does the sector represent? Is it part of an actual circle?** Even if the bristles of the fan do not make up a sector of a perfect circle, we can apply the sector area formula to compute an approximate area of the fan.
- **Q When do we want to use π instead of an estimate for π?** We prefer to use π when we need a precise number that can be irrational. If we need a rational number and π will not cancel out, we would prefer to use an estimate for π.

Task 3 (MP) **Model with Mathematics** Students will apply the formula they know for the area of a sector of a circle to solve a real-world problem that may present complications for some learners.

Sample Guided Discussion:

- **Q How do you solve a problem for an unknown that is not the variable on the left-hand side of an equation and is part of an expression on the right-hand side?** Solving such an equation requires some basic algebraic manipulation: when the given information is substituted for the other variables in the equation, the equation can be rewritten by isolating the unknown.
- **Q When can we justify using only the positive square root of a variable x, as opposed to $-\sqrt{x}$?** We can use the positive square root of a variable x (i.e., $\sqrt{x}$, rather than $-\sqrt{x}$) when the variable or unknown in question takes no negative values, such as the radius of a circle.

Turn and Talk Point out that in the computation for Task 3, π cancels out. Your answer is exact and more precise by leaving the answer in terms of π. If you round when using a calculator for π, your answer is an approximate answer. An approximate answer is more useful in a real-world problem. An exact answer is more useful in a complex mathematical calculation. If a problem asks for an exact answer, leave the answer in terms of π. If a problem asks for an approximate answer, round by using a calculator for π in your calculation.

ONLINE

Assign the Digital On Your Own for
- built-in student supports
- Actionable Item Reports
- Standards Analysis Reports

On Your Own

Assignment Guide

The chart below indicates which problems in the On Your Own are associated with each task in the Learn Together. Assign daily homework for tasks completed.

Learn Together Tasks	On Your Own Problems
Task 1, p. 540	Problems 6–8
Task 2, p. 541	Problems 9–11, 16–19, and 23
Task 3, p. 541	Problems 12–15, 20–22, 24–28

ANSWERS

1. Write a measure of the central angle over. Then multiply by the area of the entire circle.
6. Possible answer: Find the fraction of the circle contained by the sector by dividing the given central angle by 360°. Then multiply the area of the circle by the fraction to find the area of the sector.
7. Check students' work. Students should show two sectors with one sector having a central angle whose measure is twice the measure of the central angle of the other sector. Any radius is acceptable.
8. disagree; The original area is $\frac{x^\circ}{360^\circ} \cdot \pi r^2$ and the new area is $\frac{x^\circ}{360^\circ} \cdot \pi(2r)^2$ or $4 \cdot \frac{m^\circ}{360^\circ} \cdot \pi r^2$. So, the area is 4 times the original area.

Check Understanding

1. Describe the process you can use to find the area of a sector of a circle when you know the area of the circle and the central angle of the sector. See margin.
2. If the angle measure of a sector is 210°, what fraction of the circle is the sector? $\frac{7}{12}$

In Problems 3 and 4, find the area of each shaded sector. Give your answer in terms of π.

3. 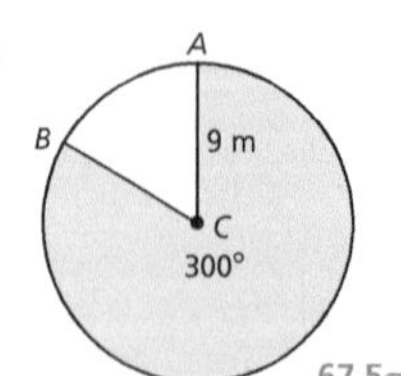

67.5π m²

4. 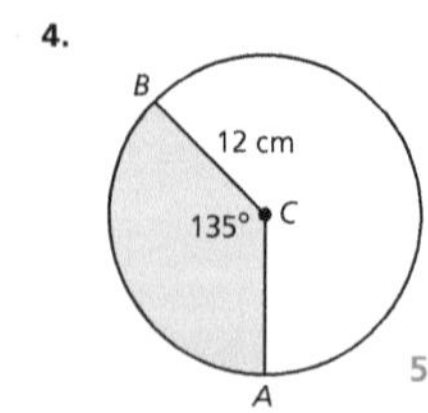

54π cm²

5. A circular stained-glass window is divided into 10 congruent wedge-shaped panes that meet at the center of the circle. The radius of the window is 2 ft. What is the area of one pane? Round the answer to the nearest hundredth. 1.26 ft²

On Your Own

6–8. See margin.

6. (MP) **Reason** Suppose you are given the angle measure of a sector and the radius of the circle, but you have forgotten the formula for finding the area of a sector. Explain how you can use proportional reasoning to find the area of the sector.
7. **Open Ended** Two congruent circles each have a sector. The area of one sector is twice the area of the other sector. Draw and label two different scenarios that meet the criteria, including the central angle of each sector and the radius of the circles.
8. (MP) **Critique Reasoning** Parisa says that when you double the radius of a sector while keeping the central angle constant, you double the area of the sector. Do you agree or disagree? Explain.

Find the exact area of each sector.

9. 10. 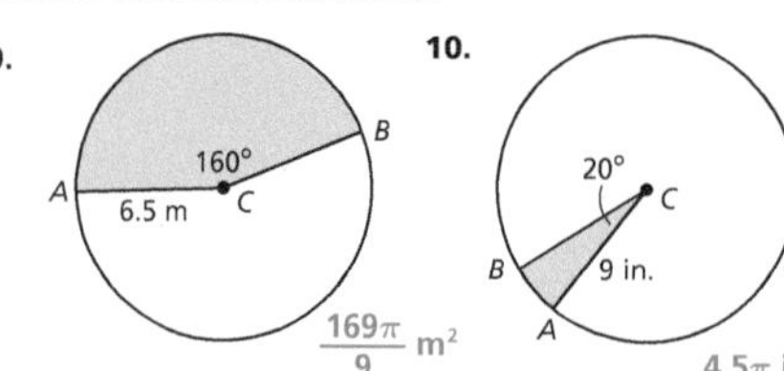

9. $\frac{169\pi}{9}$ m²
10. 4.5π in²

11. 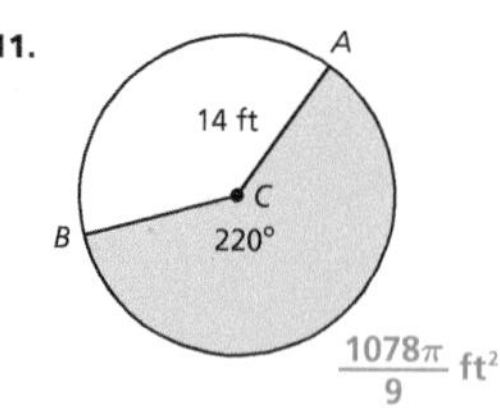

$\frac{1078\pi}{9}$ ft²

Find the approximate area of each sector with the given measure x° in a circle with diameter d or radius r. Round the answer to the nearest hundredth.

12. $x^\circ = 25^\circ$, $d = 13$ mm 9.22 mm²
13. $x^\circ = 73^\circ$, $r = 8.2$ ft 42.83 ft²
14. $x^\circ = 205^\circ$, $r = 4$ in. 28.62 in²
15. $x^\circ = 315^\circ$, $d = 27$ m 500.99 m²

542

3 Check Understanding

Formative Assessment

Use formative assessment to determine if your students are successful with this lesson's learning objective.

Students who successfully complete the Check Understanding can continue to the On Your Own practice.

For students who miss 1 problem or more, work in a pulled small group using the Almost There small-group activity on page 539C.

ONLINE

Assign the Digital Check Understanding to determine
- success with the learning objective
- items to review
- grouping and differentiation resources

4 Differentiation Options

Differentiate instruction for all students using small-group activities and math center activities on page 539C.

Reteach

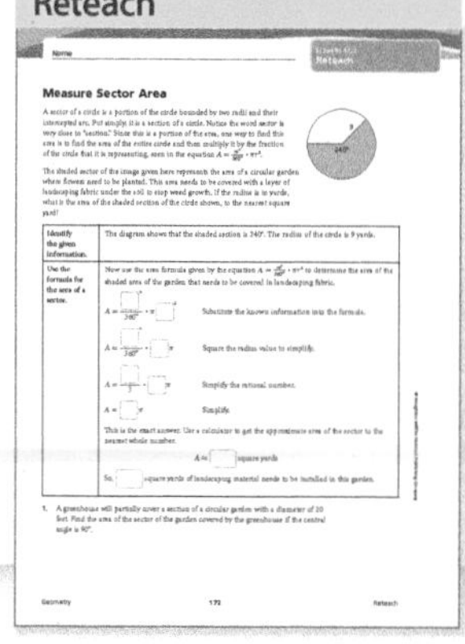

Challenge

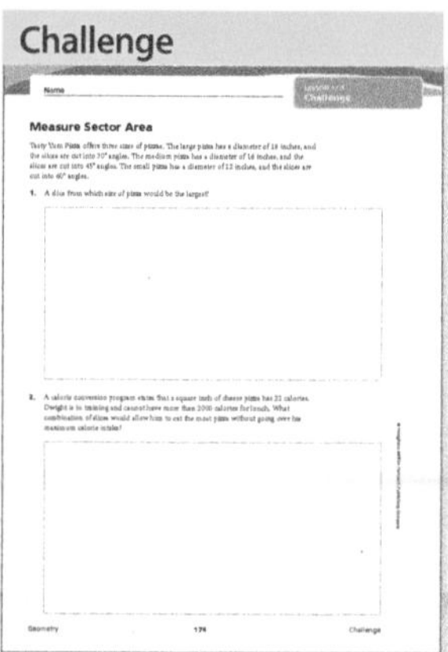

16. Sundar makes the logo shown for a florist. It consists of a rectangle with two sectors on top of the rectangle and 3 sectors below.

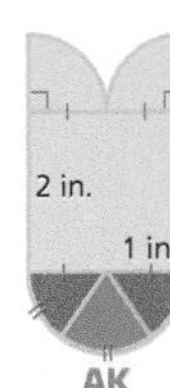

A. Estimate the area of the logo to the nearest hundredth. 7.14 in^2

B. One ounce of ink in Sundar's printer can cover 625 square inches. About how many times can the logo be printed at the given size to use 1 ounce of ink? about 87 times

17. A semicircular window is mounted over a rectangular entryway. The diameter of the window is 48 inches. To the nearest square inch, what is the area of the window? 905 in^2

Find each area occupied by the sprouts between the concentric circles. Round the answer to the nearest hundredth.

18.

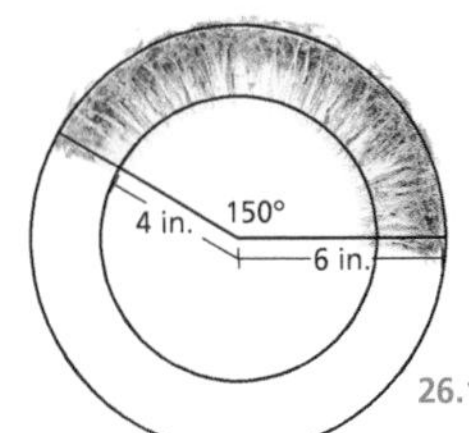

26.18 in^2

19.

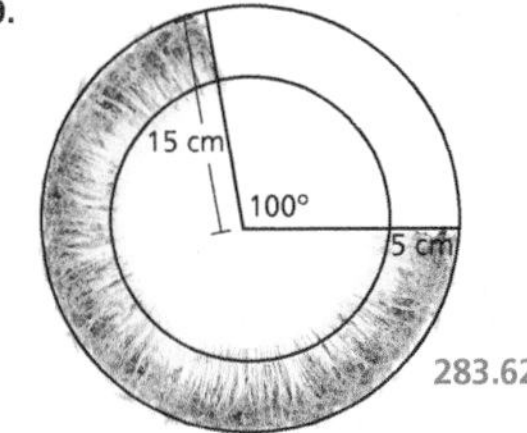

283.62 cm^2

20. (MP) Model with Mathematics Sectors from two different circles are outlined on a construction site. The radius of sector A is 10 feet, and the radius of sector B is 12 feet. The area of sector B is twice as much as the area of sector A. The central angle of sector A is 14° less than the central angle of sector B.

A. Using the area of a sector formula, write an equation that could be used to solve for the central angle of one sector. (Hint: Use the given information to write the equation in terms of just one variable.) $2\left(\frac{x° - 14°}{360°} \cdot 10^2 \cdot \pi\right) = \frac{x°}{360°} \cdot 12^2 \cdot \pi$

B. What are the measures of the central angles of sector A and of sector B? sector A: 36°; sector B: 50°

C. What are the areas of sector A and of sector B? Round to the nearest tenth. sector A: $10\pi \approx 31.4 \text{ ft}^2$; sector B: $20\pi \approx 62.8 \text{ ft}^2$

21. A farmer uses an irrigation system that covers an equal distance from the center of a crop. This irrigation method produces circular regions of crops. An irrigation system has a radius of 1300 feet and covers a sector of 20° every 45 seconds. What is the rate of coverage of the system in square feet per second? Round to the nearest tenth. $6554.7 \text{ ft}^2/\text{sec}$

22. The spokes on the rim of a bicycle wheel are equally spaced. Each spoke is connected from the center of the wheel to the rim of the wheel. There are 24 spokes connected to the rim.

A. What is the angle formed between each spoke? 15°

B. If each spoke is 14 inches long, what is the area of the sector formed between any two consecutive spokes? Round to the nearest hundredth. 25.66 in^2

C. Suppose the area of the sector formed by two consecutive spokes is 6π square feet. What is the length of each spoke? 12 inches

Questioning Strategies

Problem 17 Suppose the window forms only a quarter of a complete circle, with the radii spreading out at 45° angles with respect to the rectangular entryway, and suppose the window has the same diameter—48 inches. What would be the area of the window in this new case? The radius is half the diameter, 24 inches. So we use the formula $A = \pi r^2$ to obtain the total area of the circle, which is $A = 576\pi$. A quarter of that area is $\frac{576}{4}\pi = 144\pi \approx 452.16$, using $\pi \approx 3.14$. So rounding to the nearest inch, we get 452 square inches.

Watch for Common Errors

Problem 19 Students often think they will save time by focusing on the minor arc between the two radii, but here they should look at the big picture and compute the area occupied by the sprouts directly. Subtract the area of the smaller circle from that of the larger circle, then multiply that value by the fraction of total area represented by the major arc $\left(\frac{13}{18}\right)$.

Point out to students that they should wait almost until the end of their calculations to substitute 3.14 for π, so that they can then do the final multiplication using that approximation.

5 Wrap-Up

Summarize learning with your class. Consider using the Exit Ticket, Put It in Writing, or I Can scale.

Exit Ticket 

A couple would like a three-tiered wedding cake such that the base cake has a diameter of 16 inches, the middle cake has a diameter of 12 inches, and the top cake has a diameter of 8 inches. If they cut 30° pieces from each cake, what will be the area of the top of each piece from each cake? The top of the base cake has an area $A = 64\pi \text{ in}^2$. The top of the middle cake has an area $A = 36\pi \text{ in}^2$, and the top of the top cake has an area $A = 16\pi \text{ in}^2$. Since $\frac{30°}{360°} = \frac{1}{12}$, multiply the area of each top by $\frac{1}{12}$ to find the area of the top of each piece. So, $\frac{1}{12}(64\pi) = \frac{16}{3}\pi \text{ in}^2$, $\frac{1}{12}(36\pi) = 3\pi \text{ in}^2$, and $\frac{1}{12}(16\pi) = \frac{4}{3}\pi \text{ in}^2$.

Put It in Writing 

Have students write a paragraph or two, complete with computations, giving several steps in the derivation of the formula for the area of a sector of a circle.

I Can 

The scale below can help you and your students understand their progress on a learning goal.

4	I can use sector area to solve problems, and I can derive the formula for sector area and explain it to others.
3	I can use sector area to solve real-world problems.
2	I can find the area of a sector of a circle.
1	I can understand the steps in finding the area of a sector of a circle.

Spiral Review • Assessment Readiness

These questions will help determine if students have retained information taught in the past and can also prepare them for high-stakes assessments. Here, students use angle relationships to find an unknown angle measure **(16.2)**, convert radians into degrees **(17.2)**, find the equivalent of the expression for the area of a circle **(17.1)**, and understand the relationship between two-dimensional and three-dimensional figures **(Gr7, 11.1)**.

---

23. (MP) **Financial Literacy** A college student creates a table to track monthly budgeting. If the student creates a pie chart with a radius of 6 cm to represent the monthly budget, what is the area of each sector? Round to the nearest tenth. See Additional Answers.

Category	% of Budget
Rent	25%
Bills	10%
Food	15%
Savings	20%
Other	30%

Estimate the radius r of each sector with the given measure $x°$ in a circle with the given area. Round the answer to the nearest whole unit.

24. $x° = 50°$, area $= 85.5 \text{ m}^2$ about 14 m

25. $x° = 85°$, area $= 107 \text{ ft}^2$ about 12 ft

26. $x° = 210°$, area $= 148 \text{ cm}^2$ about 9 cm

27. $x° = 315°$, area $= 1718 \text{ in}^2$ about 25 in.

28. The area of the shaded sector is 51.3 square feet. What is an estimate for the radius of the circle? Round the answer to the nearest foot. about 7 feet

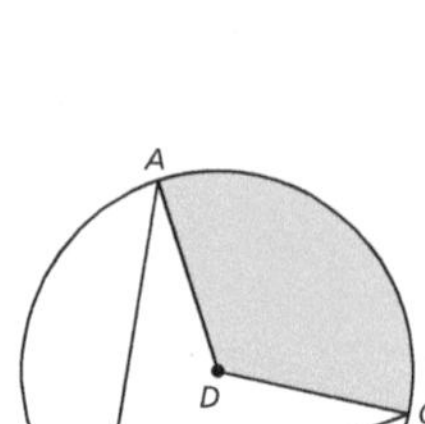

Spiral Review • Assessment Readiness

29. What is the value of x?

212°
$x°$

Ⓐ 64
Ⓑ 32
Ⓒ 128
Ⓓ 74

30. What is the degree measure of 2π radians?

Ⓐ 180°
Ⓑ 270°
Ⓒ 360°
Ⓓ 540°

31. Which expressions describe the area of a circle with radius r or diameter d? Select all that apply.

Ⓐ $r(\pi r)$
Ⓑ $\frac{360°}{360°} \cdot \pi r^2$
Ⓒ πd^2
Ⓓ $2\pi r$
Ⓔ $\pi\left(\frac{d}{2}\right)^2$
Ⓕ $x \cdot \pi r^2$

32. If a cube is sliced parallel to a face, what is the shape of the surface of the slice?

Ⓐ rectangle
Ⓑ square
Ⓒ circle
Ⓓ triangle

 I'm in a Learning Mindset!

How did procrastination impact my ability to perform my best today?

Learning Mindset

Resilience Identifies Obstacles

Point out that computations can sometimes seem like obstacles, and students should try to realize that when they find a computation especially challenging, they can ask for help. Students need to find the strength to bounce back from the frustration of not being able to solve a problem. Sometimes a student just needs to get up, move around a bit, and then go back to the problem. Physical activity can be very helpful. *What do you do when you feel frustrated in your attempts to solve a math problem? Do you ask for help from your classmates, tutors, or the teacher? When you work on a problem, do you get up and move around? Do you find that physical activity can help you get out of a rut? What can you do to avoid losing confidence completely?*

Module

17 Review

Circumference and Area

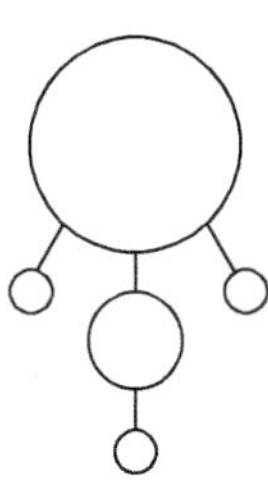

Jenna draws five circles to model a dream catcher she saw at a local craft fair.

- The large circle has a radius of 5 inches.
- The medium circle has a diameter of 4.5 inches.
- The small circles each have a radius of 1 inch.

Jenna calculates the circumference of the large circle to determine how much ribbon she would need to cover it.

To find the circumference of the large circle, use the formula for the circumference of a circle with $r = 5$.

$C = 2\pi r = 2\pi(5) \approx 31.4$ in.

Jenna wonders whether the area covered by the medium circle will be large enough to cover a mark on the wall.

To find the area of the medium circle, use the formula for the area of a circle with $r = 2.25$.

$A = \pi r^2 = \pi(2.25)^2 \approx 15.9\ \text{in}^2$

Arc Length

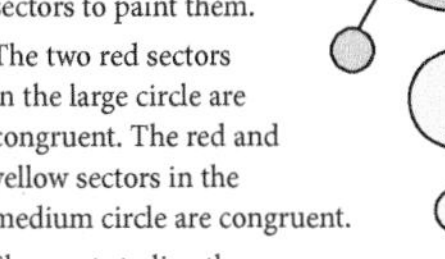

Jenna decides to paint the dream catcher on her wall. She divides the large and medium circles symmetrically into sectors to paint them.

The two red sectors in the large circle are congruent. The red and yellow sectors in the medium circle are congruent.

She wants to line the arc formed by the red sector in the medium circle with patterned duct tape.

To find the length of the arc formed by the red sector in the medium circle, use the formula for arc length with $x° = 140°$ and $r = 2.25$.

$s = \frac{x°}{360°} \cdot 2\pi r = \left(\frac{140°}{360°}\right) \cdot 2\pi(2.25) \approx 5.5$ in.

Area of a Sector

Jenna estimates she only has about one-fourth of a bottle of blue paint left and that an entire bottle can cover up to 16 square feet.

To find the area of the blue sector in the large circle, use the formula for the area of a sector with $x° = 166°$ and $r = 5$.

$A = \frac{x°}{360°} \cdot \pi r^2 = \frac{166°}{360°} \cdot \pi(5)^2 \approx 36.2\ \text{in}^2$

Find the area of the blue sector in the medium circle.

$A = \frac{80°}{360°} \cdot \pi(2.25)^2 \approx 3.5\ \text{in}^2$

Find the area of the small blue circle.

$A = \pi(1)^2 \approx 3.1\ \text{in}^2$

The area Jenna wants to cover with blue paint is the sum.

$36.2 + 3.5 + 3.1 = 42.8\ \text{in}^2$

Because $\frac{1}{4} \cdot 16\ \text{ft}^2 \cdot \frac{144\ \text{in}^2}{1\ \text{ft}^2} = 576\ \text{in}^2$ is greater than 43 in², she won't need more blue paint.

ONLINE

Assign the Digital Module Review for

- built-in student supports
- Actionable Item Reports
- Standards Analysis Reports

Module Review

Use the first page of the Module Review to summarize and connect the main ideas of the module. Use the second page to assess students' understanding of the vocabulary, concepts, and skills presented in the module.

Sample Guided Discussion:

Q Circumference and Area Is the total area of the three small circles greater than the area of the medium circle? Justify your answer. no; The area of each small circle is $A = \pi r^2 = \pi(1)^2 \approx 3.14$ square inches. So, the total area of the three small circles is $3(3.14) \approx 9.42$ square inches. The total area of the small circles is less than the area of the medium circle because $9.42 < 15.9$.

Q Arc Length How can you determine that the arc length of the red sector in the medium circle measures 140°? The central angle of the blue sector measures 80° and the red and yellow sectors are congruent. So, subtract 80° from 360° and divide by 2 to find the measure of the central angle of the red sector, 140°. The measure of the arc length is equal to the measure of the central angle that forms the arc.

Q Area of a Sector What is the significance of calculating $\frac{1}{4} \cdot 16\ \text{ft}^2\ \frac{144\text{in.}^2}{1\ \text{ft}^2}$? Is there another way you could have determined whether Jenna needed more paint? The first part of the calculation $\frac{1}{4} \cdot 16\ \text{ft}^2\ \frac{144\text{in.}^2}{1\ \text{ft}^2}$ is completed to determine the number of square feet Jenna can cover with blue paint. The second part of the calculation is completed to convert that area from square feet to square inches so that it can be compared to the area of the blue sectors. Rather than converting the area covered by blue paint to square inches, you could have converted the area of the blue sectors to square feet to determine whether Jenna needed more blue paint.

Module Review continued

Possible Scoring Guide

Items	Points	Description
1–4	2 each	identifies the correct term
5	2	correctly calculates the area of a circle
6	2	correctly calculates the circumference of a circle
7	2	correctly identifies the range of seats that can be weatherproofed
8	2	correctly converts the angle from degrees to radian measure
9	2	correctly converts the angle from radian measure to degrees
10, 11	2 each	correctly approximates the area of the sector and the corresponding arc length
12	2	correctly calculates the distance covered by the security camera
13	2	correctly calculates the unpainted area of the sector
14	2	correctly calculates the area of the window that is tinted blue
Total points possible = 28 points		

The Unit 8 Test in the Assessment Guide assesses content from Modules 15–17.

Vocabulary

Choose the correct term from the box to complete each sentence.

Vocabulary
arc
concentric
radian measure
sector

1. __?__ is the constant of proportionality that relates the radius to arc length for a fixed value m. radian measure
2. A(n) __?__ is an unbroken part of a circle consisting of two endpoints and all the points on the circle between them. arc
3. A portion of a circle bounded by two radii and their intercepted arc is a(n) __?__. sector
4. Circles that share a center are called __?__ circles. concentric

Concepts and Skills

5. Find the area of a circle with a circumference of 18 inches. Round to the nearest tenth. $A \approx 25.8$ in^2
6. Find the circumference of a circle with an area of 25π m^2. $C = 31.42$ m
7. Andrew sells seats made from tree stumps. He only uses stumps with a diameter between 1.5 feet and 2 feet. Andrew weatherproofs the top of the seats by sealing them with lacquer. A can of lacquer will cover 200 square feet. How many seats can Andrew weatherproof if he has only one can of lacquer left? between 64 and 113 seats

Convert each angle from degrees to radian measure or from radian measure to degrees.

8. 80° $\frac{4\pi}{9}$
9. $\frac{5\pi}{6}$ 150°

Approximate each shaded sector area and corresponding arc length. Round to the nearest hundredth.

10. 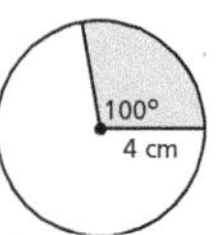

$A \approx 13.96$ cm^2
$s \approx 6.98$ cm;

11. 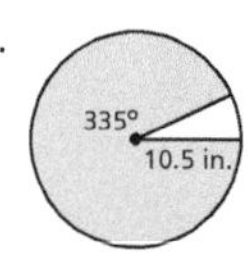

$A \approx 322.31$ in^2
$s \approx 61.39$ in.;

12. A security camera is attached to the corner of a store building. The camera is guaranteed to see with a radius of 100 feet. It rotates around an arc measuring 245°. What distance around the perimeter of the building does the security camera cover? Round to the nearest foot. 428 ft
13. (MP) **Use Tools** Josiah is designing a mural for a contest. The mural consists of a circle with radius 3 feet, in which he sketches a central angle of 65°. Josiah intends to paint the sector purple but paints only one-fourth of the sector before he runs out of purple paint. What is the unpainted area of the sector? Round to the nearest tenth of a square foot. State what strategy and tool you will use to answer the question, explain your choice, and then find the answer. 3.8 ft^2
14. A window is in the shape of a circle and divided into 12 equal-sized sectors. Five of the sectors are tinted blue. The diameter of the window is 50 inches. What is the area of the window that is tinted blue? Round to the nearest square inch. 818 in^2

546

DATA-DRIVEN INSTRUCTION

Before moving on to the Module Test, use the Module Review results to intervene based on the table below.

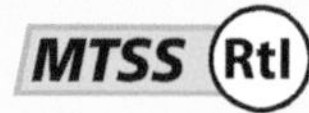

Items	Lesson	DOK	Content Focus	Intervention
5	17.1	2	Find the area of a circle.	Reteach 17.1
6	17.1	2	Find the circumference of a circle.	Reteach 17.1
7	17.1	3	Solve a real-world area problem.	Reteach 17.1
8	17.2	2	Convert an angle from degrees to radian measure.	Reteach 17.2
9	17.2	2	Convert an angle from radian measure to degrees.	Reteach 17.2
10, 11	17.3	2	Approximate the area of sectors and the corresponding arc lengths.	Reteach 17.3
12–14	17.3	3	Solve a real-world problem involving sectors.	Reteach 17.3

Module Test

The Module Test is available in alternative versions in your Assessment Guide. All items are presented in standardized test formats.

ONLINE

Assign the Digital Module Test to power actionable reports including

- proficiency by standards
- item analysis

MODULE 17

TEST

Form A

Name

Module 17 • Form A
Module Test

1. A circle has an area of 289π cm². What is the circumference of the circle?
   Ⓐ 17 cm Ⓑ 34 cm Ⓒ 106.8 cm Ⓓ 144.5 cm

2. A circle has a circumference of approximately 25 in. and a diameter of 8 in. What is the relationship between the circumference, C, and the diameter, d, of the circle?
   Ⓐ $\frac{C}{d} = \pi$ Ⓑ $\frac{C}{2d} = \pi$ Ⓒ $\frac{C}{\pi} = 2d$ Ⓓ $\frac{d}{C} = \pi$

3. The diameter of a tree increases at a rate of 0.9 in. per year. If the initial diameter of this tree is 7.5 in., how many years will it be before the area of its cross-section is 107.5 in.²?
   Ⓐ 5 years Ⓑ 9 years Ⓒ 11 years Ⓓ 13 years

4. Which angle measure is equivalent to 330°?
   Ⓐ $\frac{11}{6\pi}$ radians Ⓑ $\frac{11}{3\pi}$ radians Ⓒ $\frac{11\pi}{6}$ radians Ⓓ $\frac{11\pi}{3}$ radians

5. Which angle measure is equivalent to $\frac{19\pi}{45}$ radians?
   Ⓐ 38° Ⓑ 76° Ⓒ 114° Ⓓ 152°

6. Circle O has a radius of 7 inches. What is the approximate area of the sector subtended by a central angle measuring $\frac{9\pi}{16}$ radians?
   Ⓐ 12.4 in.² Ⓑ 43.3 in.² Ⓒ 77.7 in.² Ⓓ 87.1 in.²

7. Two concentric circles are shown in the diagram. The shaded region is 36% of the total area of the outer circle that has the radius shown.

   58 cm

   What is the radius of the inner circle, to the nearest tenth of a centimeter?
   46.4

8. A circle has a circumference of 16 inches. What is the exact area of the circle in square inches?
   $\frac{64}{\pi}$

9. What is the degree measure of an angle that measures $\frac{8\pi}{3}$ radians?
   480°

10. What is the radian measure of an angle that measures 900°?
    5π

11. Points A and B lie on a circle with a radius of 6 inches. If the measure of $\widehat{AB}$ is 78°, what is the length of $\widehat{AB}$, to the nearest tenth of an inch?
    8.2

12. A sector of a circle with radius 8 cm has an angle measure of 120°. What is the exact area of the sector in square centimeters?
    $\frac{64\pi}{3}$

Form A

Module 17 • Form A
Module Test

Name

13. A circle has a radius of 34 cm. A sector of the circle has an area of 1,618.4 cm². What is the arc length of the sector, to the nearest tenth of a centimeter?
    95.2

14. Points Q, R, and S lie on the same circle. If $m\widehat{QRS} = 210°$ and the length of $\widehat{QRS}$ is 45 ft, what is the area of the sector subtended by the central angle of $\widehat{QRS}$, to the nearest tenth of a square foot?
    276.2

15. Points K and L lie on circle Q, with the angle measure and arc length shown.

    Q, 35°, L, K, 8.5 ft

    Part A
    What is the equivalent m∠KQL in radians?
    $\frac{7\pi}{36}$

    Part B
    What is the radius of circle Q, to the nearest tenth of a foot?
    13.9

16. A circle graph shows the results of a survey about how students get to school. The radius of the circle graph is 4 inches. If the sector representing students who ride the bus has a central angle of 150°, what is the area of this sector, to the nearest tenth of a square inch?
    20.9

17. A circular parachute has a diameter of 24 feet. What is the approximate area of the parachute, to the nearest square foot?
    452

18. The center of a pendulum hangs 7.5 feet from its pivot point. If the pendulum swings through 72°, how far does the center of the pendulum travel, to the nearest tenth of a foot?
    9.4

19. Points E and F lie on circle P with the dimensions shown.

    F, 42 cm, 130°, P, E

    Part A
    What is the exact arc length of $\widehat{EF}$ in centimeters?
    $\frac{91\pi}{3}$

    Part B
    What is the exact area of the shaded region in square centimeters?
    $1{,}127\pi$

Form B

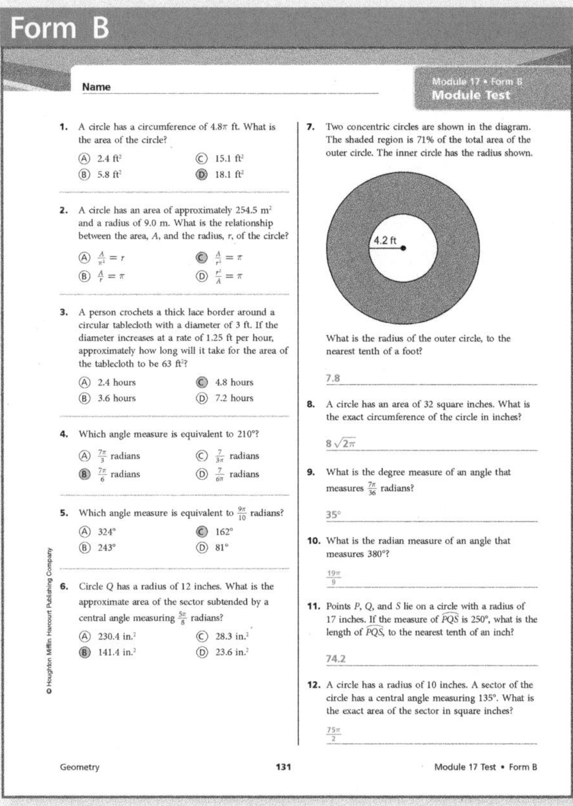

Name

Module 17 • Form B
Module Test

1. A circle has a circumference of 4.8π ft. What is the area of the circle?
   Ⓐ 2.4 ft² Ⓑ 5.8 ft² Ⓒ 15.1 ft² Ⓓ 18.1 ft²

2. A circle has an area of approximately 254.5 m² and a radius of 9.0 m. What is the relationship between the area, A, and the radius, r, of the circle?
   Ⓐ $\frac{A}{\pi^2} = r$ Ⓑ $\frac{A}{r} = \pi$ Ⓒ $\frac{A}{r^2} = \pi$ Ⓓ $\frac{r^2}{A} = \pi$

3. A person crochets a thick lace border around a circular tablecloth with a diameter of 3 ft. If the diameter increases at a rate of 1.25 ft per hour, approximately how long will it take for the area of the tablecloth to be 63 ft²?
   Ⓐ 2.4 hours Ⓑ 3.6 hours Ⓒ 4.8 hours Ⓓ 7.2 hours

4. Which angle measure is equivalent to 210°?
   Ⓐ $\frac{7\pi}{3}$ radians Ⓑ $\frac{7\pi}{6}$ radians Ⓒ $\frac{7}{3\pi}$ radians Ⓓ $\frac{7}{6\pi}$ radians

5. Which angle measure is equivalent to $\frac{9\pi}{10}$ radians?
   Ⓐ 324° Ⓑ 243° Ⓒ 162° Ⓓ 81°

6. Circle Q has a radius of 12 inches. What is the approximate area of the sector subtended by a central angle measuring $\frac{5\pi}{8}$ radians?
   Ⓐ 230.4 in.² Ⓑ 141.4 in.² Ⓒ 28.3 in.² Ⓓ 23.6 in.²

7. Two concentric circles are shown in the diagram. The shaded region is 71% of the total area of the outer circle. The inner circle has the radius shown.

   4.2 ft

   What is the radius of the outer circle, to the nearest tenth of a foot?
   7.8

8. A circle has an area of 32 square inches. What is the exact circumference of the circle in inches?
   $8\sqrt{2\pi}$

9. What is the degree measure of an angle that measures $\frac{7\pi}{36}$ radians?
   35°

10. What is the radian measure of an angle that measures 380°?
    $\frac{19\pi}{9}$

11. Points P, Q, and S lie on a circle with a radius of 17 inches. If the measure of $\widehat{PQS}$ is 250°, what is the length of $\widehat{PQS}$, to the nearest tenth of an inch?
    74.2

12. A circle has a radius of 10 inches. A sector of the circle has a central angle measuring 135°. What is the exact area of the sector in square inches?
    $\frac{75\pi}{2}$

© Houghton Mifflin Harcourt Publishing Company

Geometry 131 Module 17 Test • Form B

Form B

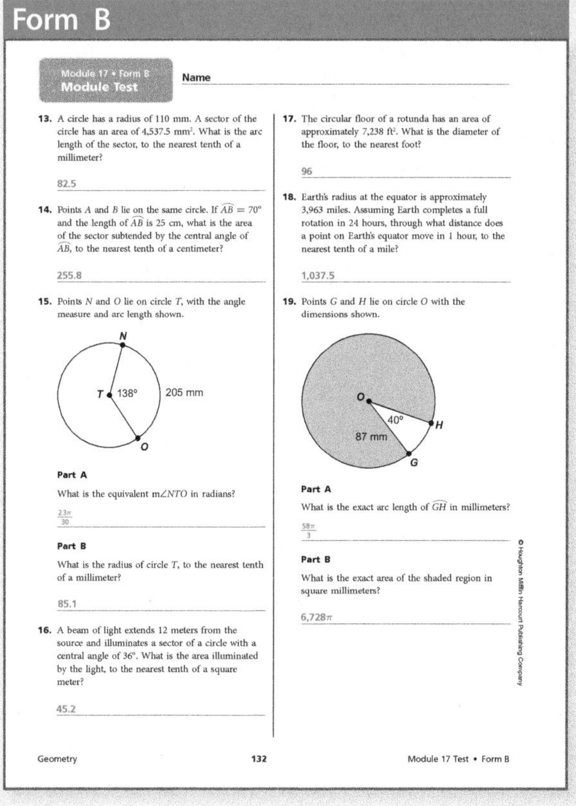

Module 17 • Form B
Module Test

Name

13. A circle has a radius of 110 mm. A sector of the circle has an area of 4,537.5 mm². What is the arc length of the sector, to the nearest tenth of a millimeter?
    82.5

14. Points A and B lie on the same circle. If $\widehat{AB} = 70°$ and the length of $\widehat{AB}$ is 25 cm, what is the area of the sector subtended by the central angle of $\widehat{AB}$, to the nearest tenth of a centimeter?
    255.8

15. Points N and O lie on circle T, with the angle measure and arc length shown.

    N, T, 138°, 205 mm, O

    Part A
    What is the equivalent m∠NTO in radians?
    $\frac{23\pi}{30}$

    Part B
    What is the radius of circle T, to the nearest tenth of a millimeter?
    85.1

16. A beam of light extends 12 meters from the source and illuminates a sector of a circle with a central angle of 36°. What is the area illuminated by the light, to the nearest tenth of a square meter?
    45.2

17. The circular floor of a rotunda has an area of approximately 7,238 ft². What is the diameter of the floor, to the nearest foot?
    96

18. Earth's radius at the equator is approximately 3,963 miles. Assuming Earth completes a full rotation in 24 hours, through what distance does a point on Earth's equator move in 1 hour, to the nearest tenth of a mile?
    1,037.5

19. Points G and H lie on circle O with the dimensions shown.

    O, 40°, H, 87 mm, G

    Part A
    What is the exact arc length of $\widehat{GH}$ in millimeters?
    $\frac{58\pi}{3}$

    Part B
    What is the exact area of the shaded region in square millimeters?
    $6{,}728\pi$

© Houghton Mifflin Harcourt Publishing Company

Geometry 132 Module 17 Test • Form B

Module 18: Surface Area

Module 19: Volume

Naval Architect

STEM POWERING INGENUITY

- **Say:** *Think about the different types of ships that cross the oceans. Naval architects are responsible for designing those ships and for keeping those ships running safely.*
- Explain that naval architects apply math and engineering to design ships and other marine vessels. Then introduce the STEM task.

STEM Task

Ask students about their previous knowledge of the terms *volume* and *rectangular prism*.

- Elicit from students that the standardized cargo container is a rectangular prism.
- In the problem shown, students will need to convert units in the height before calculating the volume.

  Have students start with the dimensions of one standardized cargo container (TEU). Since there are 12 inches in a foot, the height of 8 feet 6 inches is the same as 8.5 feet. The volume of a container in the shape of a rectangular prism is the product of its length, width, and height.

  $V = \ell \cdot w \cdot h = (20)(8)(8.5) = 1360 \text{ ft}^3$

  Since the cargo container ship can hold 21,413 TEUs, the total available volume is the product of the volume of one TEU by the number of TEUs.

  $1360 \cdot 21{,}413 = 29{,}121{,}680 \text{ ft}^3$

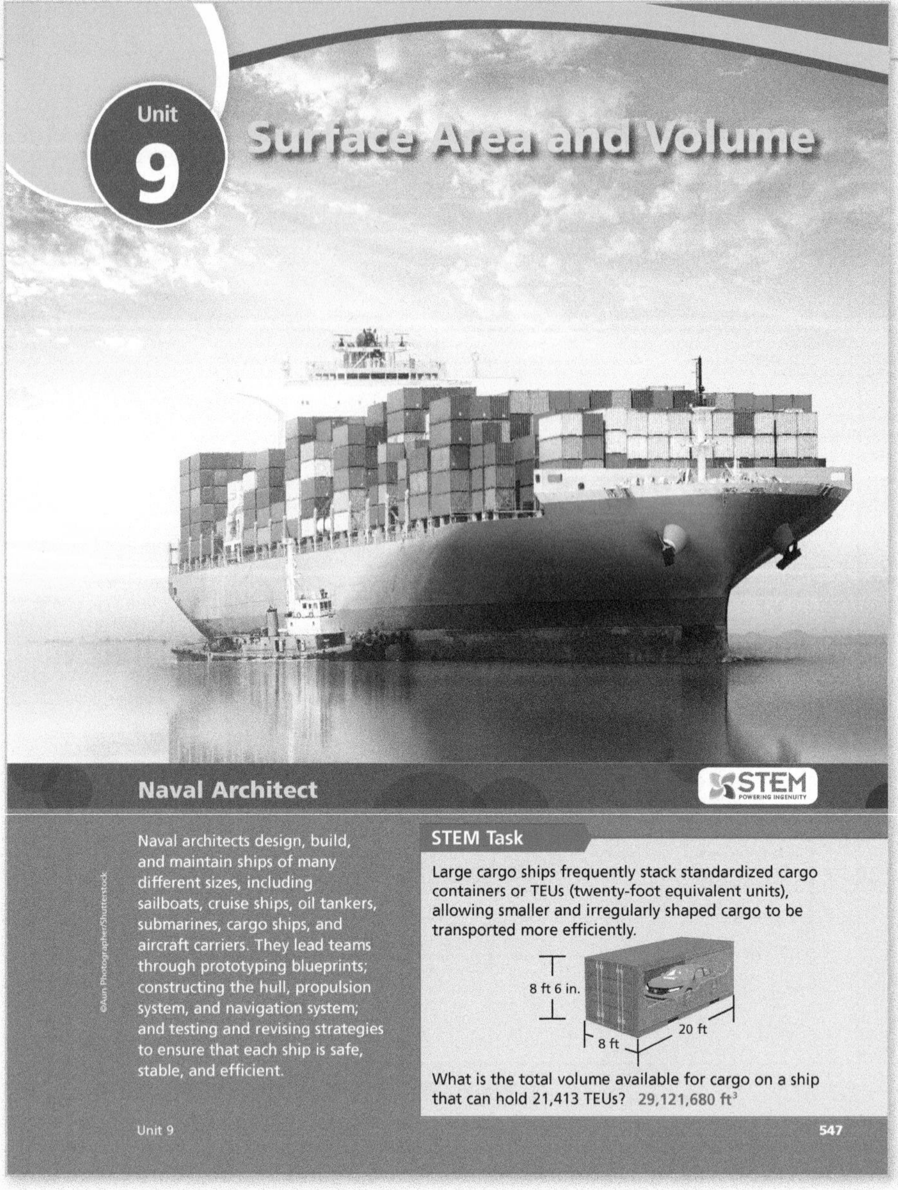

Unit 9

Surface Area and Volume

Naval Architect

STEM POWERING INGENUITY

Naval architects design, build, and maintain ships of many different sizes, including sailboats, cruise ships, oil tankers, submarines, cargo ships, and aircraft carriers. They lead teams through prototyping blueprints; constructing the hull, propulsion system, and navigation system; and testing and revising strategies to ensure that each ship is safe, stable, and efficient.

©Aun Photographer/Shutterstock

STEM Task

Large cargo ships frequently stack standardized cargo containers or TEUs (twenty-foot equivalent units), allowing smaller and irregularly shaped cargo to be transported more efficiently.

What is the total volume available for cargo on a ship that can hold 21,413 TEUs? $29{,}121{,}680 \text{ ft}^3$

Unit 9 547

Unit 9 Project Tremendous Tanks

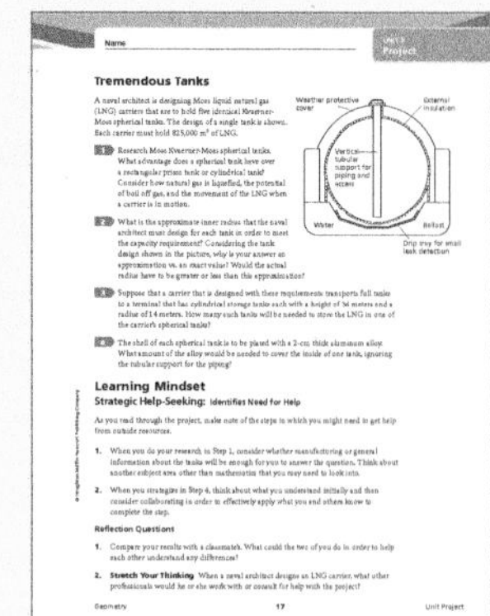

Overview: In this project students find the dimensions, surface area, and volume of Kvaerner-Moss spherical tanks.

Materials: display or poster board for presentations (even numbered)

Materials: paper and pencil or a computer used for word processing (odd numbered)

Assessing Student Performance: Students' presentations/reports should include:

- an explanation of why a spherical tank is preferable to a cylindrical tank or a tank in the shape of a rectangular prism **(Lesson 18.2 or earlier)**
- a correct surface area for the cylindrical transportation tanks **(Lesson 18.2)**
- an accurate estimate for the inner radius of the tank and an indication that the actual radius would need to be greater than this in order to fit the liquid **(Lesson 19.3)**
- a correct volume for the alloy plate **(Lesson 19.3)**

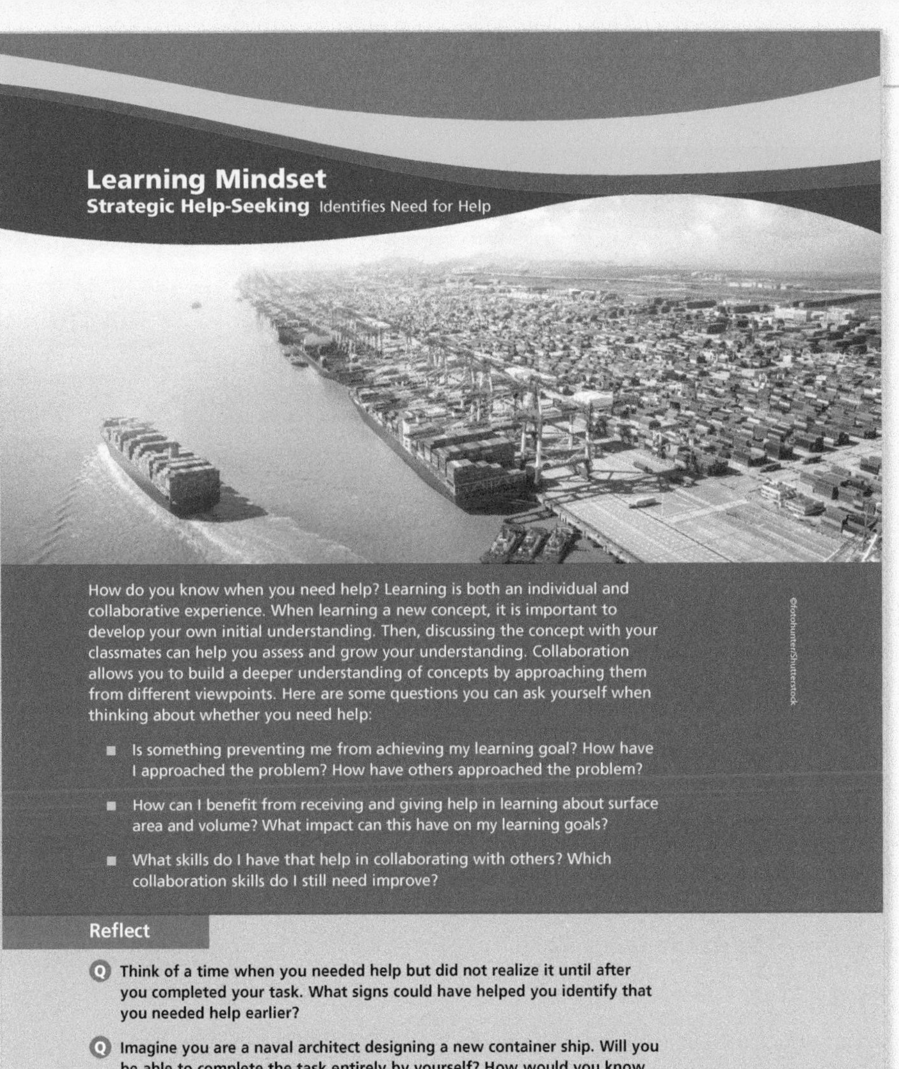
Learning Mindset
Strategic Help-Seeking Identifies Need for Help

©fotohunter/Shutterstock

How do you know when you need help? Learning is both an individual and collaborative experience. When learning a new concept, it is important to develop your own initial understanding. Then, discussing the concept with your classmates can help you assess and grow your understanding. Collaboration allows you to build a deeper understanding of concepts by approaching them from different viewpoints. Here are some questions you can ask yourself when thinking about whether you need help:

- Is something preventing me from achieving my learning goal? How have I approached the problem? How have others approached the problem?
- How can I benefit from receiving and giving help in learning about surface area and volume? What impact can this have on my learning goals?
- What skills do I have that help in collaborating with others? Which collaboration skills do I still need improve?

Reflect

Q **Think of a time when you needed help but did not realize it until after you completed your task. What signs could have helped you identify that you needed help earlier?**

Q **Imagine you are a naval architect designing a new container ship. Will you be able to complete the task entirely by yourself? How would you know when you need to ask for help from others? How could it be beneficial to seek help with some parts of the task?**

548

What to Watch For

Watch for students who never ask for help. Help them become more strategic about seeking help by

- making them feel comfortable asking for help when they need it,
- encouraging all students when they ask for help, and
- encouraging collaboration between students in groups of different sizes.

Watch for students who ask for help before trying to work on their own. Help them move from a fixed mindset to a learning mindset by

- reminding them that developing a problem-solving plan can help them to get started and
- assuring them that asking for help is okay, but that struggling to solve the problem first can lead to better learning.

"Sometimes it takes more courage to ask for help than to act alone."

—Ken Petti, children's book author

Strategic Help-Seeking
Learning Mindset

Identifies Need for Help
The learning-mindset focus in this unit is *strategic help-seeking*, which refers to a person's ability to know when they need help. In math, students often help each other as they are learning.

Mindset Beliefs
Students who react to difficulties by shutting out all outside influences and yet are unable to make progress are indicating a belief that they must solve all problems by themselves and cannot gain insight from others. These students are demonstrating a fixed mindset.

While it is good for students to work alone and experience some struggles as they are learning new material, a person who uses strategic help-seeking will look for help from others when they are stuck. In math, this involves sharing strategies for problem solving and identifying tools that can help students to advance in their learning goals. Students who ask for help and offer help to others are demonstrating evidence of a growth mindset.

Discuss with students their own definitions of help and collaboration. Do they believe that there are times when it is most useful to ask for help from others, to collaborate, or to offer assistance to others who are struggling? Use the difference in expectation between their work on quizzes and tests and their work during initial learning experiences as examples.

Mindset Behaviors
Encourage students to identify when they need help and when others need help. Some questions that can suggest whether it is a good time to ask for or to offer help include the following:

When solving a mathematical problem:

- Have I considered multiple methods for solving the problem?
- Am I making progress but getting stuck?
- Have I solved the problem, or have I found a solution that cannot be true?

When observing others solving problems:

- Are they making repeated mistakes that show misunderstanding?
- Are they getting frustrated by their lack of progress?

As students ask for help from others and offer help to others, both the student who is asking and the student who is offering will improve their knowledge and skills. When offering help, it is important not to just tell the direct way to solve a problem, but to lead others to a greater understanding through asking questions and suggesting strategies that may be more efficient. Students will still find themselves getting stuck from time to time, but the productive struggle will lead them to greater learning when they ask for help. Later, after understanding the material themselves, students can offer help to others.

SURFACE AREA

Introduce and Check for Readiness

- Module Performance Task • Are You Ready?

Lesson 18.1—2 Days

Three-Dimensional Figures

Learning Objective: Identify and classify three-dimensional solids by name, identify solids of rotation with plane figures rotated about axes, and identify cross sections of solids in planes parallel and not parallel to bases.

New Vocabulary: **cone, cross section, cylinder, oblique solid, prism, pyramid, right solid, solid of rotation, sphere**

Lesson 18.2—2 Days

Surface Areas of Prisms and Cylinders

Learning Objective: Develop the formulas for the surface areas of right prisms and right cylinders, and use the formulas to solve mathematical problems. Apply surface area to population density problems in the real world.

New Vocabulary: **lateral area, net, population density, surface area**

Review Vocabulary: **composite figure**

Lesson 18.3—2 Days

Surface Areas of Pyramids and Cones

Learning Objective: Use formulas for the surface areas of pyramids and cones to solve real-world problems.

New Vocabulary: **regular pyramid, right cone, slant height**

Lesson 18.4—2 Days

Surface Areas of Spheres

Learning Objective: Find the surface area of a sphere, and use the formula to find the surface areas of hemispheres and composite figures in real-world problems.

New Vocabulary: **hemisphere**

Assessment

- Module 18 Test (Forms A and B)
- Unit 9 Test (Forms A and B)

LEARNING ARC FOCUS

 Build Conceptual Understanding  Connect Concepts and Skills  Apply and Practice

TEACHING FOR SUCCESS

TEACHING FOR DEPTH: Surface Area

Making Connections. It is important for students to understand that a formula for the surface area of a three-dimensional figure is simply based on the formulas for the two-dimensional figures that make up the three-dimensional figure. For example, the surface of a square pyramid is four congruent triangles and one square.

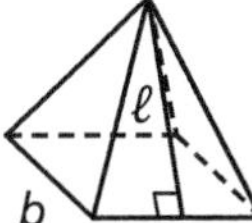

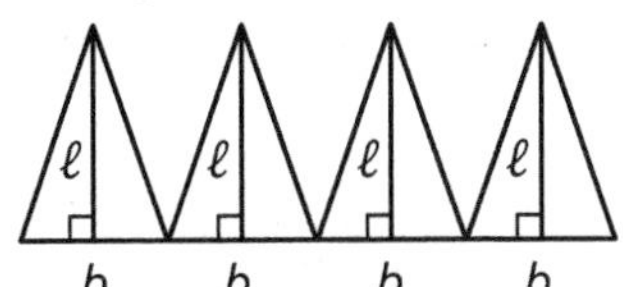

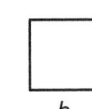

$$\begin{aligned}\text{Surface Area} &= \text{Lateral area} + \text{Base area}\\ &= 4 \cdot \frac{1}{2}b\ell + b^2\\ &= \frac{1}{2} \cdot 4b\ell + b^2\\ &= \frac{1}{2}P\ell + b^2\end{aligned}$$

Students' fluency with writing formulas and finding surface area for three-dimensional figures and composite figures will increase as they are able to visualize the two-dimensional parts of a figure.

Mathematical Progressions

Prior Learning	Current Development	Future Connections
Students: • represented three-dimensional figures using nets. • used nets to find the surface area of three-dimensional figures. • solved mathematical and real-world problems involving area, volume, and surface area.	**Students:** • identify solids of rotation. • identify cross sections of solids. • use nets to develop formulas for surface area. • use formulas for the surface areas of solids to solve problems. • find surface areas of hemispheres and composite figures.	**Students:** • will translate between the geometric description and the equation for a conic section. • for a function that models a relationship between two quantities, will interpret key features of graphs and tables in terms of the quantities, and sketch the graphs showing key features given a verbal description of the relationship.

TEACHER TO TEACHER

From the Classroom

Implement tasks that promote reasoning and problem solving. Given the opportunity, students can really develop their reasoning skills when they develop the surface area formula for a right cone and when they investigate the formula for the surface area of a sphere.

When developing the surface area formula for a right cone, I either let my students create their own cones out of construction paper or provide them with premade cones that they can unfold and take apart to help them with their investigations. I have discovered that allowing them to spend time with physical manipulatives helps them to reason through the formula better than when they only see a diagram.

When investigating the formula for the surface area of a sphere, my more advanced students attempt to use algebra to express the height of the cylinder that contains the sphere in terms of the radius and become stuck when they are unable to do so. I encourage them to take a closer look at the diagram. Eventually, they conclude that $h = 2r$.

Giving students the opportunity to make these connections on their own increases their ability to reason and problem solve through more complicated scenarios.

LANGUAGE DEVELOPMENT • Planning for Instruction

By giving all students regular exposure to language routines in context, you will provide opportunities for students to **listen**, **speak**, **read**, and **write** about mathematical situations and develop both mathematical language and conceptual understanding at the same time.

Using Language Routines to Develop Understanding

Use the **Professional Cards** for the following routines to plan for effective instruction.

Co-Craft Questions Lesson 18.1

Students think of natural questions to ask about a given situation or problems similar to a given task and answer the questions they have developed or problems they have created.

Three Reads Lessons 18.2–18.4

Students read a problem three times with a specific focus each time.

1st Read What is the situation about?
2nd Read What are the quantities in the situation?
3rd Read What are the possible mathematical questions that we could ask for the situation?

Information Gap Lesson 18.3

Students recognize when information given in a problem situation is incomplete, and they pose questions and share knowledge with others to discover any missing facts or relationships and work together to solve the problem.

Critique, Correct, and Clarify Lesson 18.2

Students correct the work in a flawed explanation, argument, or solution method and share with a partner and refine the sample work.

Connecting Language to Surface Area

Watch for students' use of the new terms listed below as they explain their reasoning and make connections with new concepts.

Key Academic Vocabulary

Current Development • New Vocabulary

cross section the intersection of a three-dimensional figure and a plane

hemisphere half of a sphere

lateral area the sum of the areas of the lateral faces of a three-dimensional figure

net a diagram of a three-dimensional figure arranged in such a way that the diagram can be folded to form the three-dimensional figure

oblique solid a solid in which the axis or the lateral edges are *not* perpendicular to the base(s)

population density the number of organisms of a particular type per square unit of area

right solid a solid in which the axis or the lateral edges are perpendicular to the base(s)

slant height the height of each lateral face of a regular pyramid

solid of rotation a solid that is formed by rotating a shape about an axis

surface area the total area of all faces and curved surfaces of a three-dimensional figure

Linguistic Note

Listen for vocabulary involved with calculating surface area, such as *net* and *slant height*.

The term *net* has multiple meanings. Ensure that students understand that in this context, a net is a diagram of a three-dimensional figure that can be folded to form the figure and not a bag made of thread or a person's earnings once deductions have been subtracted.

Help students clearly distinguish between the height and the *slant height* of a pyramid by displaying a diagram with both terms labeled in color.

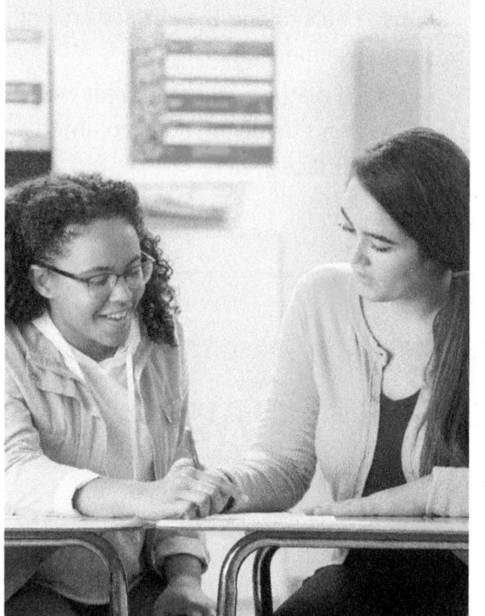

Module 18 Surface Area
Module Performance Task: Spies and Analysts™
On the SPOT
How can you describe the area of a sunspot?
(t) ©solarseven/Shutterstock; (b) ©NASA Goddard Space Flight Center
Module 18
549

Connections to This Module

One sample solution might involve students measuring the approximate dimensions of the sunspot in the image, measuring the diameter of the sun in the image, and then using these measurements to compute the area of the sunspot as a percent of the surface area of the sun.

- The surface area of the sun is computed using the formula for the surface area of a sphere: $A = 4\pi r^2$. **(18.4)**
- The sunspot can be approximated as a rectangle, and its area can be approximated using the area formula for a rectangle: $A = \ell w$.
- The area of the sunspot can be expressed as a percent of the total surface area of the sun.

On the SPOT

Overview

This problem requires students to ask questions in order to make decisions about how to approach the problem. Examples of questions:

- How do you measure the dimensions of the irregularly shaped sunspot? Do you only include the larger section, or do you include the smaller dots next to the larger section?
- Does the curvature of the sun affect the area of the sunspot? Is the effect of the curvature small enough that it can be disregarded?
- How can the answer be expressed in a way that is comprehensible?

Be a *Spy*

First, students must determine what information is necessary to know. This should include:

- the dimensions of the sunspot in the image.
- the diameter of the sun in the image.
- the formula for the surface area of a sphere.

Be an *Analyst*

Help students understand that the size of the sunspot in the image is meaningful only if they compare it to the surface area of the sun. Students will also need to determine how to present their answer in a way that can be understood; square miles or square kilometers may be too small to make the area comprehensible.

Alternative Approaches

In their analysis, students might:

- compute the area of the sunspot as a percent of the surface area of the sun.
- use reference sources to find the actual diameter of the sun, compute the actual surface area of the sun, and then set up and solve a proportion to estimate the area of the sunspot.
- use nonstandard units to express the area.
- express the area of the sunspot as a fraction or percent of the visible surface of the sun.

ONLINE

Assign the Digital Are You Ready? to power actionable reports including

- proficiency by standards
- item analysis

Are You Ready?

data checkpoint

Diagnostic Assessment

- Diagnose prerequisite mastery.
- Identify intervention needs.
- Modify or set up leveled groups.

Have students complete the *Are You Ready?* assessment on their own. Items test the prerequisites required to succeed with the new learning in this module.

Nets and Surface Area Students will apply their previous knowledge of finding surface area using nets to discover the formulas for the surface areas of three-dimensional figures.

Areas of Composite Figures Students will build upon their knowledge of finding the area of composite figures to find the density of real-world objects using area and volume.

Cross Sections of Solids Students will extend their knowledge of the cross sections of three-dimensional figures to describe real-world objects using geometric shapes and their properties.

ONLINE 

Are You Ready?

Complete these problems to review prior concepts and skills you will need for this module.

Nets and Surface Area

Find the surface area of the three-dimensional figure represented by the given net. Round to the nearest hundredth.

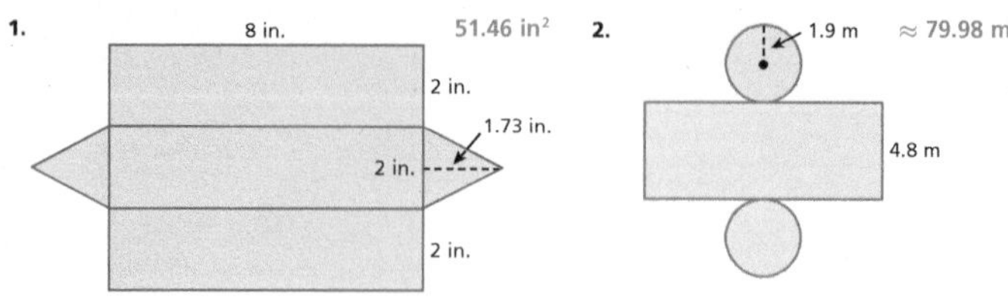

Areas of Composite Figures

Find the area of each composite figure. Round to the nearest hundredth.

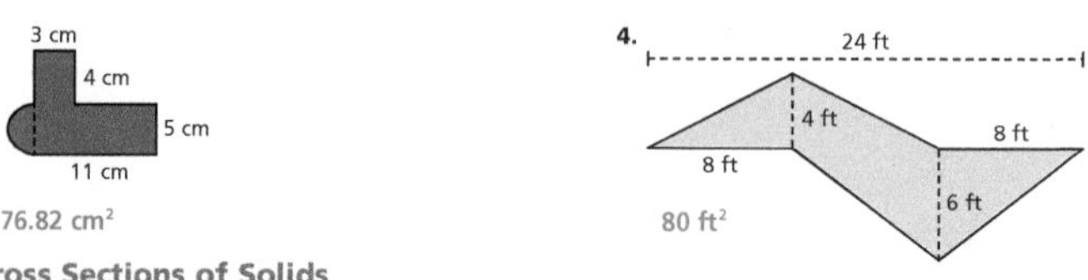

Cross Sections of Solids

Name the shape of each two-dimensional cross section.

5. A cylinder is sliced parallel to its base. circle
6. A square pyramid is sliced perpendicular to its base. triangle

Connecting Past and Present Learning

Previously, you learned:

- to create nets to model the surface areas of three-dimensional figures,
- to solve problems involving areas of figures composed of polygons, and
- to describe cross-sections that result from slicing three-dimensional solids.

In this module, you will learn:

- to model real-world objects using geometric shapes and their properties,
- to use formulas to find the surface areas of three-dimensional figures, and
- to apply concepts of density based on area and volume in modeling situations.

550

DATA-DRIVEN INTERVENTION

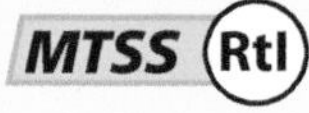

Concept/Skill	Objective	Prior Learning *	Intervene With
Nets and Surface Area	Use nets made up of rectangles and/or triangles to find surface areas of prisms and pyramids.	Grade 6, Lesson 13.1	• Tier 3 Skill 10 • Reteach, Grade 6 Lesson 13.1
Areas of Composite Figures	Find areas of figures composed of triangles and quadrilaterals.	Grade 7, Lesson 10.4	• Tier 2 Skill 1 • Reteach, Grade 7 Lesson 10.4
Cross Sections of Solids	Describe the two-dimensional figures that result from slicing three-dimensional figures.	Grade 7, Lessons 10.3 and 11.1	• Tier 2 Skill 21 • Reteach, Grade 7 Lessons 10.3 and 11.1

* Your digital materials include access to resources from Grade 6–Algebra 2. The lessons referenced here contain a variety of resources you can use with students who need support with this content.

18.1 Three-Dimensional Figures

LESSON FOCUS AND COHERENCE

Mathematics Standards

- Identify the shapes of two-dimensional cross-sections of three-dimensional objects, and identify three-dimensional objects generated by rotations of two-dimensional objects.
- Use geometric shapes, their measures, and their properties to describe objects (e.g., modeling a tree trunk or a human torso as a cylinder).
- Apply geometric methods to solve design problems (e.g., designing an object or structure to satisfy physical constraints or minimize cost; working with typographic grid systems based on ratios).

Mathematical Practices and Processes 

- Look for and make use of structure.
- Model with mathematics.

I Can Objective 

I can identify the characteristics of three-dimensional figures and represent them using drawings.

Learning Objective

Identify and classify three-dimensional solids by name, identify solids of rotation with plane figures rotated about axes, and identify cross sections of solids in planes parallel and not parallel to bases.

Language Objective

Explain when a cross section of a solid is a transformation of a base of the solid.

Vocabulary

New: cone, cross section, cylinder, oblique solid, prism, pyramid, right solid, solid of rotation, sphere

Mathematical Progressions

Prior Learning	Current Development	Future Connections
Students: • developed definitions of rotations, reflections, and translations in terms of angles, circles, perpendicular lines, parallel lines, and line segments. **(5.2)** • given a geometric figure and a rotation, reflection, or translation, drew the transformed figure. **(5.2)**	**Students:** • identify and classify three-dimensional solids by name. • identify solids of rotation, and recognize them as the result of plane figures rotated about an axis. • identify cross sections of solids in planes parallel or not parallel to a base of the solid, when applicable.	**Students:** • will translate between the geometric description and the equation for a conic section. **(4th-Year Course)**

UNPACKING MATH STANDARDS

Identify the shapes of two-dimensional cross-sections of three-dimensional objects, and identify three-dimensional objects generated by rotations of two-dimensional objects.

What It Means to You

The emphasis for this standard is understanding how a three-dimensional solid can be related to a two-dimensional object, with the two-dimensional object as a "slice" of the solid. When the two-dimensional object is a cross section, students associate it with a slice through the solid. When the three-dimensional object is generated by a rotation of the two-dimensional object, students can associate the two-dimensional object with a slice from the outer surface of the solid to a central axis.

This visualization allows students to better understand measurable aspects of three-dimensional objects such as surface area and volume. Over the next two modules, students will develop an understanding of the formulas for surface area and volume of the solids introduced in this lesson.

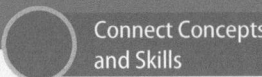

WARM-UP OPTIONS

ONLINE

PROJECTABLE & PRINTABLE

ACTIVATE PRIOR KNOWLEDGE • Identify a Transformation from a Preimage and Image

Use these activities to quickly assess and activate prior knowledge as needed.

Problem of the Day

A slide in a projector and its larger projected image on a screen are represented by the quadrilaterals shown here. What transformation maps the slide to its projected image?

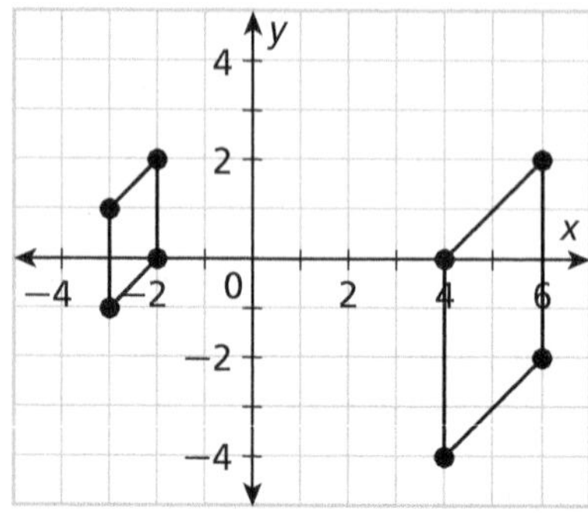

Possible answer: A dilation with a scale factor of 2.

Quick Check for Homework

As part of your daily routine, you may want to display the Teacher Solution Key to have students check their homework.

Make Connections

Based on students' responses to the Problem of the Day, choose one of the following:

1. Project the Interactive Reteach, Lesson 6.1.
2. Complete the Prerequisite Skills Activity:

Have students work in small groups. Give each group a set of index cards. Each card should have a triangle or quadrilateral and its image under a single transformation on the same coordinate plane. Have students work together to determine the transformation that maps the preimage to the image.

- *How can you tell if the transformation is a dilation?* The preimage and image are similar but not congruent.
- *How can you tell if the transformation is a rotation?* The orientations of the preimage and image are the same, and pairs of corresponding points all define congruent angles from a center.
- *How can you tell if the transformation is a translation?* Every pair of corresponding points is connected with the same vector.
- *How can you tell if the transformation is a reflection?* The preimage and image are congruent, but their orientations are reversed.

If students continue to struggle, use Tier 2 Skills 9–12.

SHARPEN SKILLS

If time permits, use this on-level activity to build fluency and practice basic skills.

Vocabulary Review

Objective: Students demonstrate an understanding of terms related to a transformation.
Materials: Bubble Map (Teacher Resource Masters)

Have students work in pairs. Each pair should work together to build a bubble map for the term "transformation" without looking anything up. Encourage students to work together to recall as much information as possible and to not worry too much about being correct at this point.

When all pairs have had ample time to work, have a class discussion and create a class bubble map on the board.

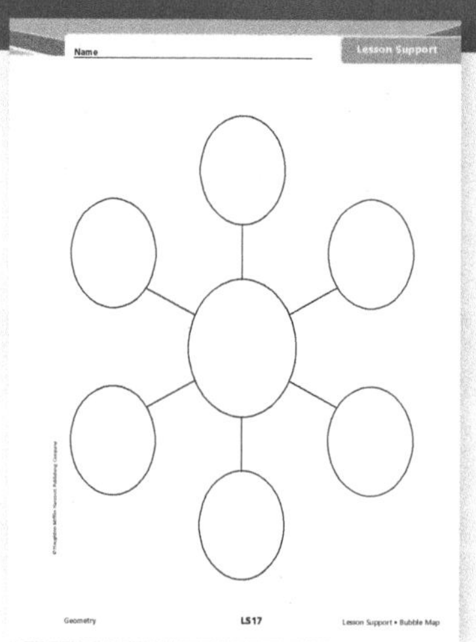

PLAN FOR DIFFERENTIATED INSTRUCTION

Small-Group Options

Use these teacher-guided activities with pulled small groups.

On Track

Materials: index cards

Give each group two sets of index cards: one set with the names of solids introduced in this section, limiting pyramids and prisms to square bases; another set of four cards with one of the following statements:

- parallel to a base
- not parallel to a base but not intersecting a base
- intersects one base
- intersects both bases

Have students take turns drawing one card from each set. They will then identify the cross section defined by the intersection of the solid with the plane described on the other card, or explain why such an intersection is not possible.

Almost There

Materials: 1-inch-thick foam board, glue

Assign each group one of the solids introduced in this lesson, limiting pyramids and prisms to rectangular bases. Do not assign a sphere. Assume each solid has a height of 10 inches. Have each group work together on the following tasks:

- Identify the cross section of their solid in a plane parallel to a base.
- Choose dimensions of the base (limit them to no more than 10 inches).
- Find the dimensions of cross sections parallel to the base at 1-inch intervals.
- Draw the base and each cross section on foam board and cut them out.
- Qlue together the foam board cutouts to create a rough representation of the solid.

Ready for More

Materials: 1-inch-thick foam board, compass, glue, graphing calculator

Each group will work together to build a rough model of a 10-inch diameter sphere with approximating layers of foam board.

When students need prompting, ask:

- What is the equation of a circle with center at the origin and diameter 10?
- How can you graph the upper half of the circle on your calculator?
- How can you use graphs of parallel cross sections with smaller and smaller radii to build your sphere?

Math Center Options

Use these student self-directed activities at centers or stations.

Key: ● Print Resources ● Online Resources

On Track

- ● Interactive Digital Lesson
- ●● Journal and Practice Workbook
- ● Interactive Glossary (printable): **cone**, **cross section**, **cylinder**, **oblique solid**, **prism**, **pyramid**, **right solid**, **solid of rotation**, **sphere**
- ● Module Performance Task
- ●● Standards Practice: Identify Cross-Sections and 3-D Objects Generated by Rotations

Almost There

- ● Reteach 18.1 (printable)
- ● Interactive Reteach 18.1
- ● RtI Tier 2 Skills 9–12: Translate, Rotate, Reflect, and Dilate Figures in the Coordinate Plane
- ● Illustrative Mathematics: Cube Ninjas!

Ready for More

- ● Challenge 18.1 (printable)
- ● Interactive Challenge 18.1
- ● Illustrative Mathematics: Tennis Balls in a Can

View data-driven grouping recommendations and assign differentiation resources.

During the *Spark Your Learning*, listen and watch for strategies students use. See samples of student work on this page.

Use a Rotation About an Axis

Strategy 1

The shape is formed by spinning around a central axis, the center of the ride, with lights sweeping out the ringed pattern.

If students . . . use a rotation about an axis to solve the problem, they are demonstrating an understanding of rotations from Lesson 5.2.

Have these students . . . explain how they reasoned about the motion of the lights of the ride. **Ask:**

- Q How did you use the positions of the lights and the motion of the ride to solve the problem?
- Q What kind of transformation did you use?

Make a Sketch

Strategy 2

I made a sketch of what the ride might look like at rest, with only a few lights. Then I sketched the positions of the lights at other positions as the ride turns. The lights eventually traced out a ringed pattern as I drew more and more of them.

If students . . . make a sketch to solve the problem, then they intuitively understand the time-lapse photograph as the result of multiple transformations, but they may not know how to explicitly describe it as such.

Activate prior knowledge . . . by having students consider the effects of repeated rotations of the lights about an axis. **Ask:**

- Q What kind of transformation is happening with the lights?
- Q Where is the center of the rotation?

COMMON ERROR: Assume Photo Shows the Ride at an Instant

The shape looks like a ring because that is how the ride always looks. The lights are in a circle around the ride.

If students . . . assume the photo shows the ride at an instant, then they may not understand that the time-lapse photograph is a combination of many images over an interval of time.

Then intervene . . . by pointing out that the image should be viewed as a combination of many images over an interval of time. **Ask:**

- Q What kind of picture would a camera produce if the photo was taken over many seconds instead of instantly?
- Q What would it look like if you took many photographs of a moving object over a period of time and laid them on top of each other?

18.1

Three-Dimensional Figures

I Can identify the characteristics of three-dimensional figures and represent them using drawings.

Spark Your Learning

A photographer takes a long-exposure photo of a carnival ride.

©Tami Freed/Alamy

Complete Part A as a whole class. Then complete Parts B and C in small groups.

A. What is a mathematical question you can ask about this situation? What information would you need to know to answer your question?

A. Why does the shape look like a ring?; the shape of the ride at rest and how the ride moves

B. To answer your question, what strategy and tool would you use along with all the information you have? What answer do you get?
See Strategies 1 and 2 on the facing page.

C. What shape do the outermost lights appear to form? How do you know?
The outermost lights appear to form a cylinder. The shape has two circular bases.

Turn and Talk What kinds of solid shapes can be generated this way? Are there shapes that cannot be generated? See margin.

CULTIVATE CONVERSATION • Co-Craft Questions

If students have difficulty formulating a mathematical question about the situation in the Spark Your Learning, ask them to visualize what the ride might look like at rest, and then how its appearance changes as it rotates at an increasing rate. What are some natural questions to ask about this situation?

Work together to craft the following questions:

- What does the ride look like when it is at rest?
- What kind of symmetry does the shape of the moving ride have?
- Why does the shape look like a ring?

Then have students think about what additional information, if any, they would need to answer these questions. **Ask:**

- Can you determine why the shape looks like a ring if you know only what the ride looks like at rest? Why or why not?
- Can you determine why the shape looks like a ring if you know only what kind of lights it has? Why or why not?

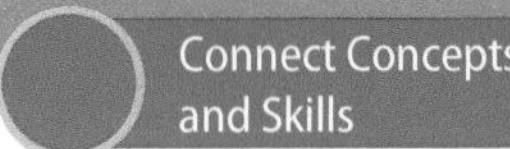

Connect Concepts and Skills

1 Spark Your Learning

MOTIVATE

- Have students look at the photo in their books and read the information contained in the photo. Then complete Part A as a whole-class discussion.
- Give the class the additional information they need to solve the problem. This information is available online as a printable and projectable page in the Teacher Resources.
- Have students work in small groups to complete Parts B and C.

PERSEVERE

If students need support, guide them by asking:

Q **Advancing • Use Tools** Which tool could you use to solve the problem? Why choose that tool and not some other? Students' choices of tools and reasons for choosing them will vary.

Q **Assessing** What do you know about the lights when the ride is at rest? There are only a few lights at regular heights.

Q **Assessing** How do the lights move as the ride moves? They spin around the center of the ride.

Q **Advancing** Can you think of the photograph as the result of repeated transformations, and what does that mean in terms of the images made? The circular light patterns are the images of the lights being rotated repeatedly about the central axis.

Turn and Talk Ask students about similarities between shapes that could be generated this way, drawing attention to the edges of the shapes and whether they are straight or curved. Remind them, if necessary, that the ring in the photograph is the result of rotations about a central axis, and that any rotation of a point involves moving that point along a circle. Only shapes with a circular cross section can be generated by rotating a line or point. A cube is a shape that cannot be generated this way.

BUILD SHARED UNDERSTANDING

Select groups of students who used various strategies and tools to share with the class how they solved the problem. As they present their solutions, have each group discuss why they chose a specific strategy and tool.

2 Learn Together

Build Understanding

Task 1 MP **Use Structure** After being introduced to definitions of various solids, students are presented with various scenarios involving a plane figure being rotated about an axis and identify each result as one of the previously defined solids. Students then classify the solids defined before the task by whether they are solids of rotation.

CONNECT TO VOCABULARY

Have students use the **Interactive Glossary** to record their understanding of the vocabulary in this task.

Sample Guided Discussion:

Q How can you model the rotation in Part B? I can tape one edge of a rectangular piece of paper or an index card to a pencil and then spin the pencil.

Q Is there a shape you could cut out of paper, tape an edge to a pencil, and then spin the pencil to model a sphere? a semicircle

Turn and Talk Encourage students to visualize regions on other sides of the axis sweeping out their own solid parts. It may be helpful to have them cut out a figure, tape to a pencil, and spin the pencil. If the shape crosses the axis of rotation, the solid of rotation will have two parts. It does not matter if the axis of rotation is a line of symmetry for the shape.

Build Understanding

Identify Solids of Rotation

Right solids have an axis or lateral edges that are perpendicular to their base(s). **Oblique solids** have an axis or lateral edges that are not perpendicular to their base(s).

A **prism** has two parallel congruent polygonal bases connected by lateral faces. Prisms are named by the shapes of their bases. A **cylinder** has two parallel congruent circular bases connected by a curved lateral surface. Its axis connects the centers of the bases.

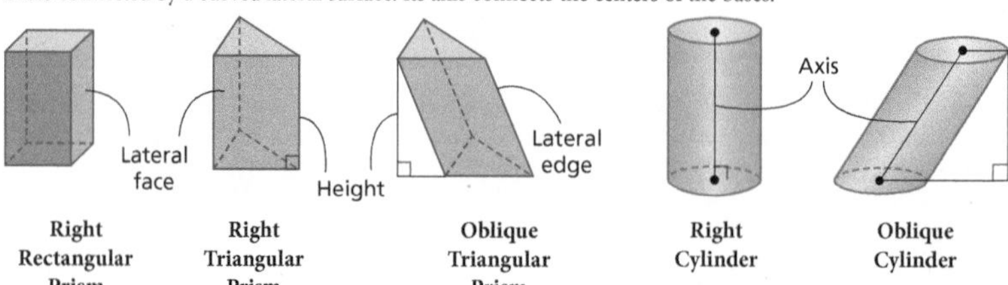

A **pyramid** has a polygonal base with triangular faces that meet at a vertex. A **cone** has a circular base and a curved surface that connects the edge of the circular base to its vertex. A **sphere** is the locus of points that are a fixed distance from its center.

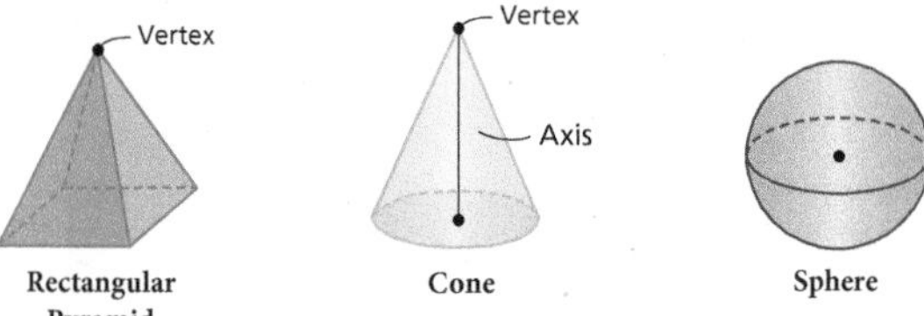

1 A. Suppose a right triangle is rotated about a line that contains one of its legs. This line is called the axis of rotation. You can model the rotation by taping a right triangular piece of paper to a pencil and spinning the pencil. What solid is formed by the rotation? cone

B. Suppose a rectangle is rotated about a line that contains one of its sides. What solid is formed by the rotation? cylinder

C. When a polygon is rotated about a line containing one of its sides, what shape is swept out in space by a point on another side of the polygon? circle

D. A solid that is formed by rotating a shape about an axis is a **solid of rotation**. Which of the eight solids above are solids of rotation? right cylinders, cones, and spheres

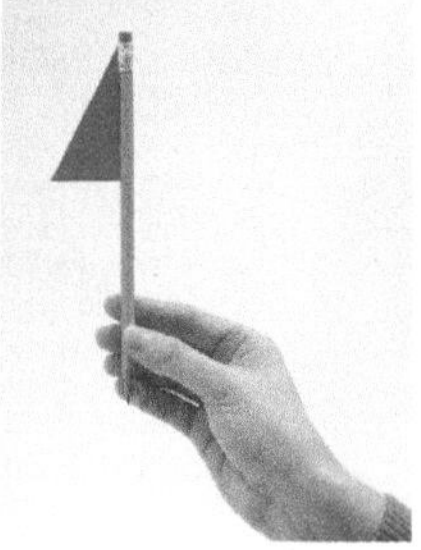

Turn and Talk What happens if the rotated shape crosses the axis of rotation? Does it matter whether the axis of rotation is a line of symmetry for the shape? See margin.

LEVELED QUESTIONS

Depth of Knowledge (DOK)	Leveled Questions	What Does This Tell You?
Level 1 **Recall**	How would you classify a solid with a pentagonal base and triangular faces that connect the base to a vertex? The solid is a pyramid.	Students' answers will indicate whether they can recall one of the given definitions of solids and apply it to classify a given solid.
Level 2 **Basic Application of Skills & Concepts**	What makes a pyramid different from a cone? A cone has a circular base, while a pyramid has a polygonal base; and a cone has a curved surface that connects the base to its vertex, while a pyramid's base is connected to its vertex by triangular faces. Also, a cone is a solid of rotation, while a pyramid is not.	Students' answers will indicate whether they can contrast two types of solids by their definitions.
Level 3 **Strategic Thinking & Complex Reasoning**	Is it possible for a solid of rotation to have more than one axis? Give an example. Yes, a sphere has infinitely many axes described by all lines through its center.	Students' answers will indicate whether they can visualize the construction of a solid through rotations about multiple axes.

Cross Sections of Solids

A **cross section** is the intersection of a three-dimensional figure and a plane.

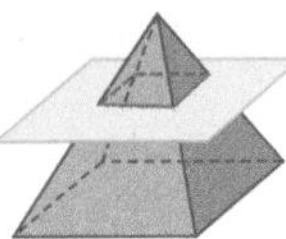

The cross sections of three-dimensional figures can be simple figures, such as triangles, rectangles, or circles.

2 A. In each figure shown, a plane parallel to the base intersects the solid. Describe the cross section of each figure. Compare the cross sections of the figures. What can you conclude? A–D. See margin.

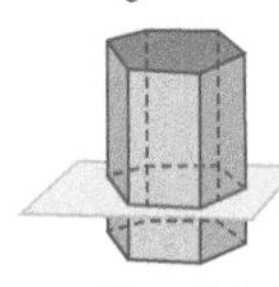
Figure 1

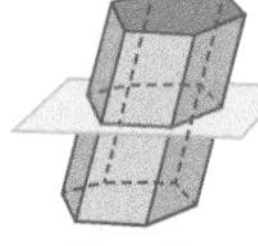
Figure 2

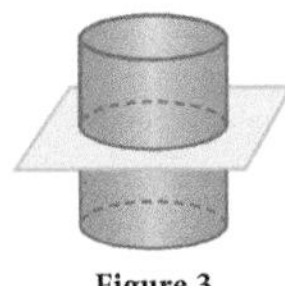
Figure 3

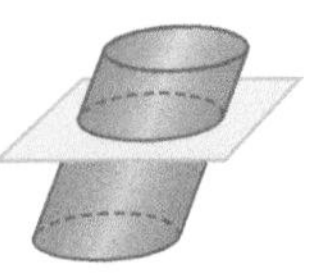
Figure 4

B. In each figure shown, a plane parallel to the base intersects the solid. Describe the cross section of each figure. What type of transformation maps the shape of the base to the cross section in each figure?

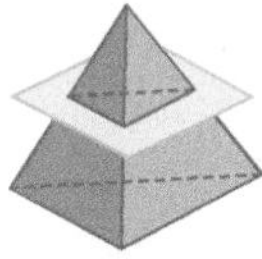
Figure 5

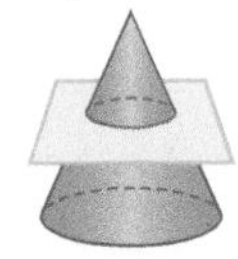
Figure 6

C. Suppose a rectangle is drawn on the coordinate plane. Describe how to use transformations of the rectangle to generate a rectangular pyramid.

D. Consider the different intersections of a plane and a cylinder as shown. What is true about the cross sections formed when a plane intersects a cylinder along its lateral surface? What is true about the cross sections formed when a plane intersects a cylinder through its two bases? See Additional Answers.

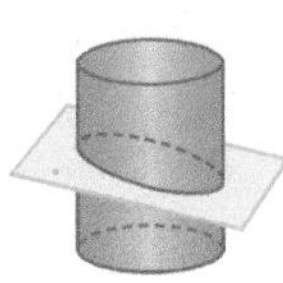
Figure 7

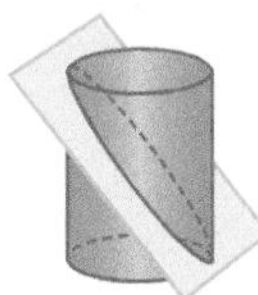
Figure 8

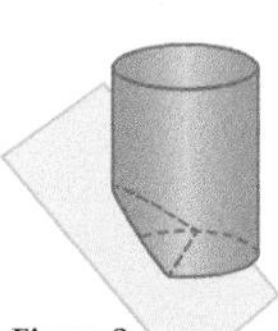
Figure 9

E. What is true about all cross sections of a sphere? All cross sections of a sphere are circles.

Turn and Talk Determine whether it is possible to create a cross section that is a triangle in a prism, cylinder, pyramid, cone, and sphere. Explain your reasoning. See margin.

Task 2 (MP) **Use Structure** Students describe cross sections of solids in planes parallel to a base and how some solids can be generated through transformations of a base. Students then go on to identify cross sections in planes that are not parallel to a base, with some planes not intersecting any base and others intersecting one or two bases. Finally, they identify cross sections of a sphere.

CONNECT TO VOCABULARY

Have students use the **Interactive Glossary** to record their understanding of the vocabulary in this task.

Sample Guided Discussion:

Q **In Part A, how can you visualize a cross section of a solid by imagining you were "chopping" it parallel to the base?** I can imagine that I'm chopping the solid along the plane, and then look at the shape of the slice.

Q **In Part B, can you think of the cross section as the result of a rigid motion of the base?** no; since the cross section is smaller than the base

Q **In Part C, how can you think of building the pyramid with very thin layers or slices?** I can think of stacking very thin rectangular slices that are all similar, and getting smaller at a constant rate as I stack upward from the base.

Turn and Talk Remind students that a cross section of a solid does not necessarily lie in a plane parallel to a base of the solid. Ask them how the vertices of a cross section relate to edges of the solid. It is possible to create a cross section that is a triangle for all of the solids except the cylinder and the sphere. The cross section of a rectangular prism can be a triangle if the cross section is diagonal, through one of the vertices of the prism. The cross section of a pyramid or cone is a triangle when the cross section goes through the vertex and the base.

PROFICIENCY LEVEL

Beginning
Have students work in pairs. One student will say the name of a solid, and the other student will say the name of the transformation mapping a base of the solid to a cross section in a plane parallel to the base, or say that the given solid does not have a base. Have students take turns quizzing each other.

Intermediate
Show students a diagram with a square pyramid and say, "This is a square pyramid." Then sketch a representation of a plane parallel to a base that intersects the pyramid. Draw a square, indicating how it comes from the intersection of the plane with the pyramid, and say, "The cross section is a square." Do the same for a cone, but have students identify the solid and cross section.

Advanced
Have students work in small groups to craft an explanation of how they could construct a cone through transformations of its base.

ANSWERS

2.A. Figure 1: hexagon; Figure 2: hexagon; Figure 3: circle; Figure 4: circle. The cross section of two solids with bases that are the same size are the same, even if one of the solids is oblique.

2.B. Figure 5: triangle; Figure 6: circle; a dilation; the cross section is similar to the base, just smaller.

2.C. You can generate a rectangular pyramid using dilations. Each rectangle is a reduction of the rectangle before it. Stack each reduction above the previous one until the top most rectangle becomes a point.

Step It Out

Task 3 (MP) **Model with Mathematics** Students are shown an example of a mineral that naturally forms cubes, and see how the area of a certain cross section can be found. Ask them how the area calculation might change if the cross section is in another plane.

Sample Guided Discussion:

Q **In Part A, where on the cube are the sides of the cross section rectangle?** One pair of opposite sides are on a pair of vertical edges of the cube, and the other pair are diagonals of two of the cube's faces.

Q **What kind of triangle does a side of the cross section form with upper edges of the cube?** An isosceles right triangle, or a 45°-45°-90° triangle.

Turn and Talk Ask students to visualize a plane perpendicular to a pair of faces of the cube and through the center of these faces. Ask them to imagine how the cross section described by the intersection of the plane with the cube varies as the plane is rotated about an axis through the centers of the two faces. Is this cross section the same size as the cross section shown in the photograph? Cut the cube through the midpoints of opposite sides of one of the faces, perpendicular to the face. The resulting cross section is a square with an area of $3 \times 3 = 9\text{ cm}^2$.

Task 4 (MP) **Model with Mathematics** Students see that a baluster for a staircase is produced through a technique called "turning" on an instrument called a lathe. They are asked to match finished balusters with their profile drawings. In doing so, students see that a baluster is a model of a solid of rotation.

Sample Guided Discussion:

Q **Why do we say that a baluster is "turned" on a lathe when it is being made?** The lathe makes the baluster from a piece of wood by turning, or rotating, the wood while cutting it.

Q **How does somebody use the profile drawing to make a baluster with the lathe?** The profile drawing tells how deeply to cut the wood relative to its axis, so as the piece of wood rotates about its axis, the operator of the lathe or a computer raises or lowers the blade of the lathe to match the profile drawing.

Q **What kind of solid is the turned part of each baluster?** a solid of rotation

Step It Out

Apply Cross Sections

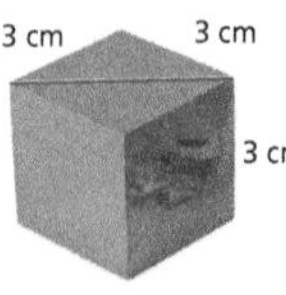

3 One form of the mineral pyrite naturally forms shiny metallic cubes. A cube of pyrite is cut in half along the diagonal of one of its faces and perpendicular to that face. What is the area of the newly exposed cross section?

The cross section is a rectangle.

A. Why is the cross section a rectangle but not a square?

Use the Pythagorean Theorem to find the length of the rectangle.

A, B. See margin.

$a^2 + b^2 = c^2$

$3^2 + 3^2 = c^2$

$18 = c^2$

$3\sqrt{2} = c$

B. Why is the width of the rectangle 3 cm?

Use the length of the rectangle and its width, 3 cm, to find the area of the rectangle.

$A = \ell w = (3\sqrt{2})(3) = 9\sqrt{2} \approx 12.7\text{ cm}^2$

Turn and Talk Describe a different way to cut the pyrite cube in half so that the area of the cross section formed is less than the area of the cross section found in the task. See margin.

Model a Real-World Solid

4 A carpenter needs to replace several turned balusters to repair a staircase. The carpenter uses a contour gauge to draw the profile of each existing baluster and delivers the profile drawings shown to the shop to get the balusters turned on a lathe. Match the profile drawings with the balusters.

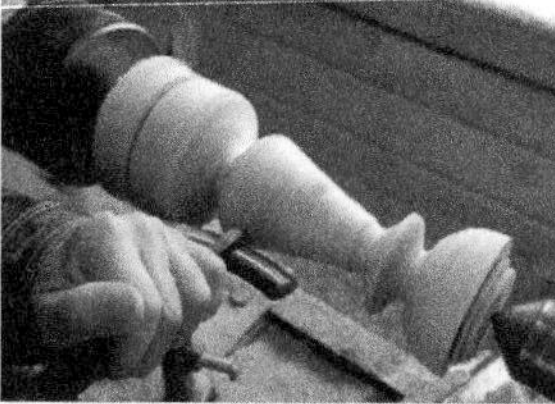

(t) ©MarcelClemens/Shutterstock; (b) ©losch/Shutterstock

A. A. 3

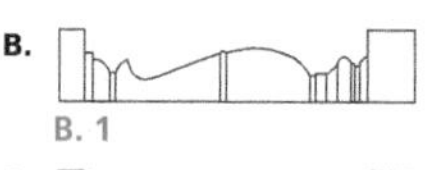

B. B. 1

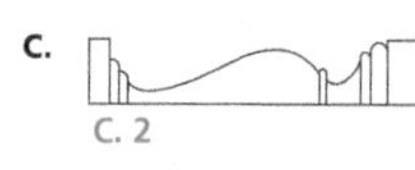

C. C. 2

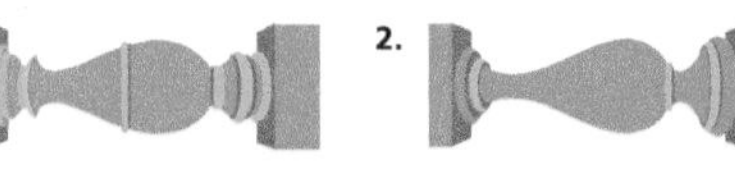

1.

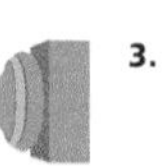

2.

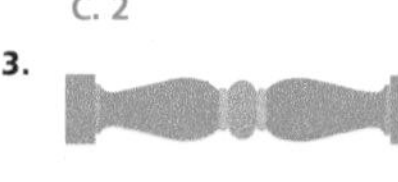

3.

554

ANSWERS

Task 3

A. The diagonal of the square face represents the length of the rectangular cross section, and the diagonal is longer than the sides of the square face.

B. In the rectangular cross section of the cube, the width of the rectangle is the height of the cube.

Check Understanding

1. What solid will be created by rotating the shape about the given axis of rotation? cylinder

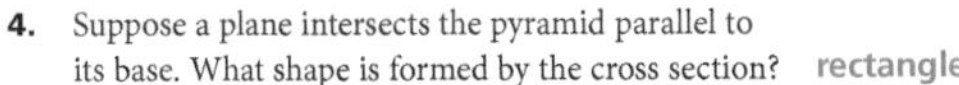

In Problems 2–4, use the figure of the rectangular pyramid. The plane is perpendicular to the base of the pyramid.

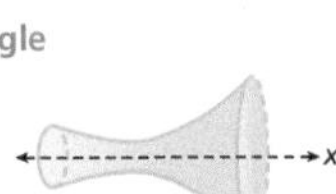

2. Describe the cross section shown. triangle
3. Find the area of the cross section. 36 in^2
4. Suppose a plane intersects the pyramid parallel to its base. What shape is formed by the cross section? rectangle
5. Raheem used a drafting application on a computer to generate the solid shown. He created the solid by rotating a two-dimensional shape around a horizontal axis of rotation. Draw the shape that was rotated. See Additional Answers.

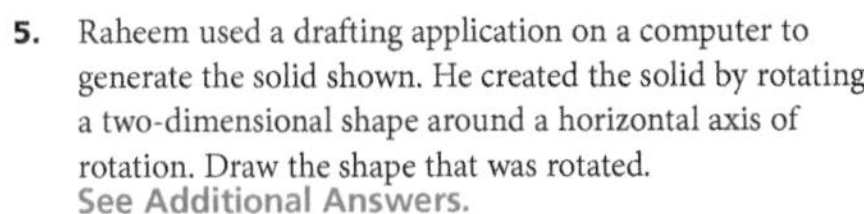

On Your Own

6. **Critique Reasoning** Kyle believes that rotating the given shape about the axis of rotation will produce a sphere. Juan believes the shape that will be formed is round in certain aspects, but not a sphere. Who is correct? Explain your reasoning. See Additional Answers.

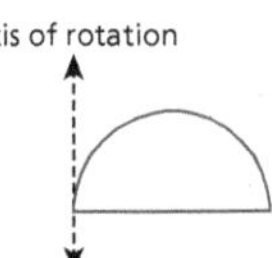

Sketch the figure created by rotating each given shape about the axis. 7–9. See Additional Answers.

7. 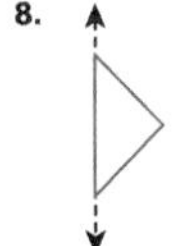

8.

9. 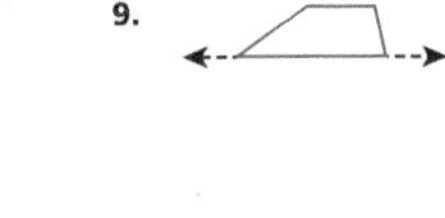

Describe each cross section.

10. 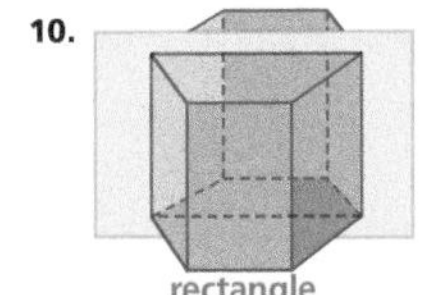 rectangle

11. 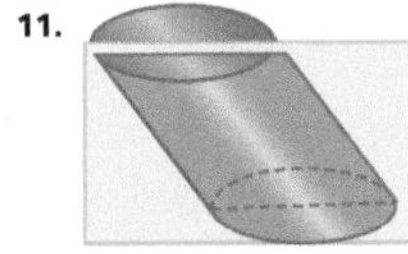 parallelogram

12. 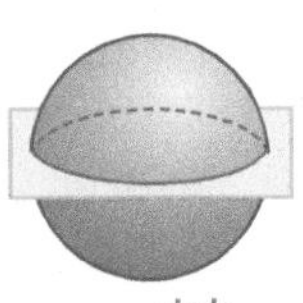 circle

13. A plane intersects a prism and forms a cross section that is congruent to its base. What is true about the plane? The plane is parallel to the base.
14. A plane intersects a prism perpendicular to the base and forms a cross section that is congruent to its base. What is true about the prism? The prism is a cube.
15. **Open Ended** Sketch the intersection of a solid and a plane with a cross section that is a triangle. Check students' sketches.

ONLINE

Assign the Digital On Your Own for
- built-in student supports
- Actionable Item Reports
- Standards Analysis Reports

On Your Own

Assignment Guide

The chart below indicates which problems in the On Your Own are associated with each task in the Learn Together. Assign daily homework for tasks completed.

Learn Together Tasks	On Your Own Problems
Task 1, p. 552	Problems 6–9
Task 2, p. 553	Problems 10–15
Task 3, p. 554	Problems 16 and 17
Task 4, p. 554	Problems 18–20

Watch for Common Errors

Problem 8 Students might see the straight sides and vertices of the triangle and think that the solid has planar faces. They need to understand that a solid of rotation has circular cross sections, so the solid will have curved surfaces.

Questioning Strategies

Problem 13 Suppose a solid has a base, and every cross section in a plane parallel to the base is congruent to the base. What is true about the solid? It has two bases.

ANSWER

6. Juan is correct. If the semicircle was rotated about a horizontal line of rotation, it would be a sphere.

3 Check Understanding

Formative Assessment

Use formative assessment to determine if your students are successful with this lesson's learning objective.

Students who successfully complete the Check Understanding can continue to the On Your Own practice.

For students who miss 1 problem or more, work in a pulled small group using the Almost There small-group activity on page 551C.

ONLINE

Assign the Digital Check Understanding to determine
- success with the learning objective
- items to review
- grouping and differentiation resources

4 Differentiation Options

Differentiate instruction for all students using small-group activities and math center activities on page 551C.

Reteach

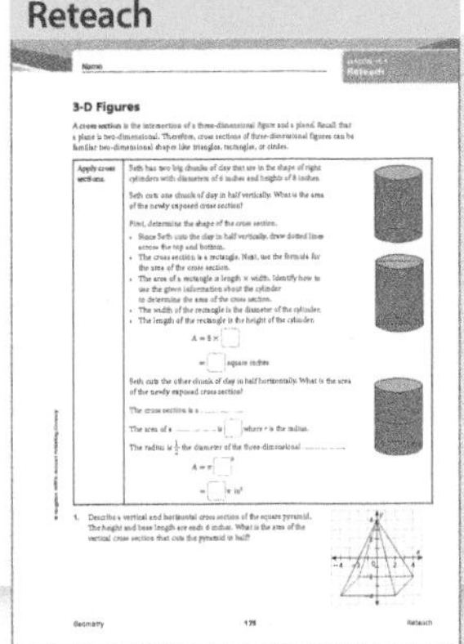

Challenge

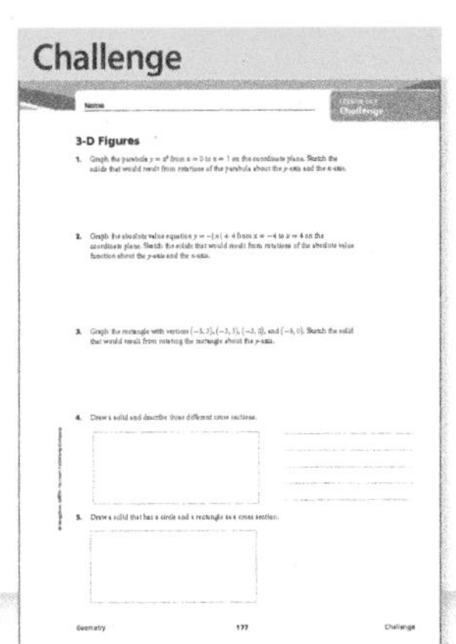

5 Wrap-Up

Summarize learning with your class. Consider using the Exit Ticket, Put It in Writing, or I Can scale.

Exit Ticket 

Suppose a cube has edge length 4 cm, and a plane intersects the cube at a vertex on the top face and at the opposite vertex on the same face. What is the area of the cross section?

The cross section is a $4\sqrt{2} \times 4$ rectangle, so its area is $4(4\sqrt{2}) = 4 \cdot 4(\sqrt{2}) = 16 \cdot \sqrt{2} = 16\sqrt{2}$ cm^2.

Put It in Writing 

Explain how cross sections in planes parallel to the base of a prism and a pyramid are similar and how they are different.

I Can 

The scale below can help you and your students understand their progress on a learning goal.

4	I can identify the characteristics of three-dimensional figures and represent them using drawings, and I can explain my steps to others.
3	I can identify the characteristics of three-dimensional figures and represent them using drawings.
2	I can describe a cross section of cylinder solid of rotation when the plane of the cross section intersects a base, and when it does not.
1	I can identify three-dimensional figures by name.

Spiral Review • Assessment Readiness

These questions will help determine if students have retained information taught in the past and can also prepare them for high-stakes assessments. Here, students must find the circumference of a circle from its radius **(17.1)**, find the area of a rectangle **(G3)**, convert a radian measure to a degree measure **(17.2)**, and find the area of a sector of a circle **(17.3)**.

An art teacher uses a piece of yarn to slice through a block of clay. Find the area of the cross section created by each slice.

16. 6 in., 6 in., 4 in., 10 in.

24 in²

17.

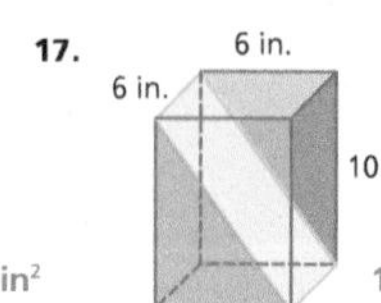

$12\sqrt{34}$ in²

Match each shape with its solid of rotation.

18. C

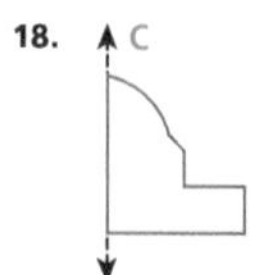

19. A

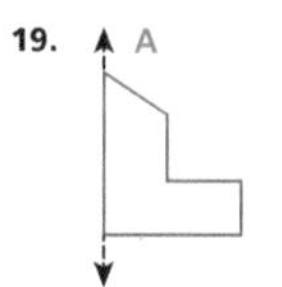

20. B

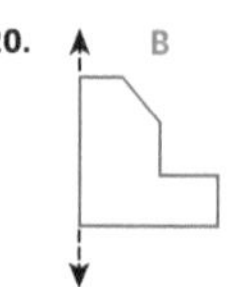

A.

B.

C.

Spiral Review • Assessment Readiness

21. Find the circumference of a circle that has a radius of 4.

Ⓐ 2π
Ⓑ 4π
Ⓒ 8π
Ⓓ 16π

22. What is the area of a rectangle with a height of 3 inches and a length of 4π inches?

Ⓐ π in²
Ⓑ 6π in²
Ⓒ 7π in²
Ⓓ 12π in²

23. What is $\frac{2\pi}{3}$ in degrees?

Ⓐ 30°
Ⓑ 60°
Ⓒ 120°
Ⓓ 240°

24. A circular game piece 1.5 inches in diameter is divided into sectors. Each sector has a central angle of 60°. What is the area of one sector of the game piece?

Ⓐ 0.29 in²
Ⓑ 1.18 in²
Ⓒ 2.36 in²
Ⓓ 7.07 in²

 **I'm in a Learning Mindset!**

Was collaboration an effective tool for describing cross sections of three-dimensional figures? Explain.

556

Keep Going Journal and Practice Workbook

Learning Mindset

Strategic Help-Seeking Identifies Need for Help

Point out that collaboration can be helpful in learning since participating in discussions can give a student a better idea of their level of understanding. If a student found others' explanations of cross sections of solids confusing or found that others were not understanding their explanations of cross sections of various solids, it may be that the student's understanding is not yet fully formed. This can give the student an idea of where they might need additional help. *How can working with others give you a better idea of your level of understanding? How can participating in conversations give you clues about where you might need additional help?*

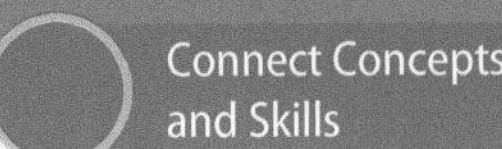

18.2 Surface Areas of Prisms and Cylinders

LESSON FOCUS AND COHERENCE

Mathematics Standards

- Use geometric shapes, their measures, and their properties to describe objects (e.g., modeling a tree trunk or a human torso as a cylinder).

Mathematical Practices and Processes (MP)

- Look for and make use of structure.
- Model with mathematics.
- Attend to precision.

I Can Objective

I can find the surface area of a prism or cylinder.

Learning Objective

Develop the formulas for the surface areas of right prisms and right cylinders and use the formulas to solve mathematical problems. Apply surface area to population density problems in the real world.

Language Objective

Describe how the net of a figure relates to the formula for surface area.

Vocabulary

Review: composite figure

New: lateral area, net, population density, surface area

Mathematical Progressions

Prior Learning	Current Development	Future Connections
Students: • represented three-dimensional figures using nets made up of rectangles and triangle, and used the nets to find the surface area of these figures; applied these techniques in the context of solving real-world and mathematical problems. **(Gr6, 18.1)** • knew the formulas for the area and circumference of a circle and used them to solve problems. **(Gr7, 17.1)** • solved real-world and mathematical problems involving two-dimensional objects composed of circles, triangles, rectangles, and rhombuses. **(Gr7, 11.4)**	**Students:** • use the net of a right prism to develop a general formula for the surface area of a right prism. • use the net of a right cylinder to develop a general formula for the surface area of a right cylinder. • use the formulas for surface area of right prisms and cylinders to find the surface area of a composite figure. • apply the formula for surface area of a prism to solve a real-world problem.	**Students:** • for a function that models a relationship between two quantities, will interpret key features of graphs and tables in terms of the quantities, and sketch graphs showing key features given a verbal description of the relationship. Key features include: intercepts; intervals where the function is increasing, decreasing, positive, or negative; relative maximums and minimums; symmetries; end behavior; and periodicity. **(4th-Year Course)**

PROFESSIONAL LEARNING

Visualizing the Math

Even when students have the formula for surface area of a prism or cylinder, they may still find it helpful to draw the net. For the right cylinder, have students highlight the circumference of the circular base and the length of the rectangle that forms that lateral surface area. This can help students see why the circumference $2\pi r$ is part of the total surface area formula.

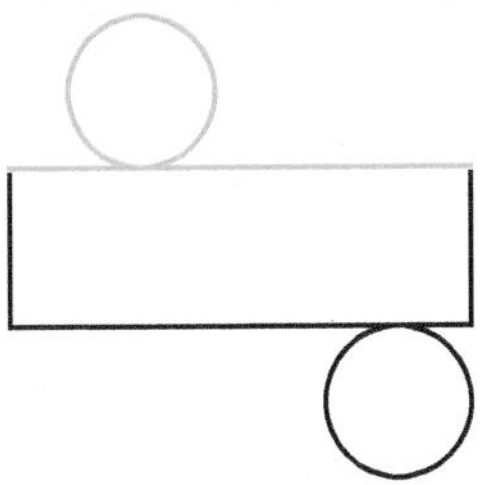

WARM-UP OPTIONS

ONLINE

PROJECTABLE & PRINTABLE

ACTIVATE PRIOR KNOWLEDGE • Solve Problems Using Surface Area and Volume

Use these activities to quickly assess and activate prior knowledge as needed.

Problem of the Day

To mail a box, the postal service calculates the cost based on the surface area and the volume. If the box is a cube with side length 8 inches, what is the surface area and the volume?

$SA = 384 \text{ in}^2$, $V = 512 \text{ in}^3$

Quick Check for Homework

As part of your daily routine, you may want to display the Teacher Solution Key to have students check their homework.

Make Connections

Based on students' responses to the Problem of the Day, choose one of the following:

1. Project the Interactive Reteach, Grade 7, Lesson 11.4.
2. Complete the Prerequisite Skills Activity:

Let a 4 in. by 6 in. rectangle be the base of the rectangular prism with height 2 inches. What are the surface area and volume of the prism?

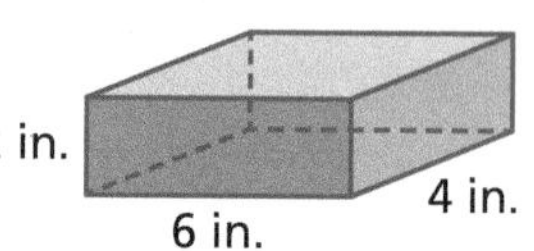

- *How is the perimeter of the base related to the surface area?* To find the lateral area, you can multiply the height, 2, by the perimeter of the base.
- *How do you find the volume of a rectangular prism?* Multiply the length by the width by the height.

If students continue to struggle, use Tier 3 Skill 11.

SHARPEN SKILLS

If time permits, use this on-level activity to build fluency and practice basic skills.

Mental Math

Objective: Students find the surface area and volume of a cube when given a side length.
Materials: index cards

Have students work in pairs. Pass out a card to each pair. Tell students that they will find the surface area or the volume of a cube when given a side length. Have the students work together to mentally compute the measurement described on the card. Each card will have a side length of a cube and the words *surface area* or *volume*.

3 cm surface area	**5 in.** volume

PLAN FOR DIFFERENTIATED INSTRUCTION

Small-Group Options

Use these teacher-guided activities with pulled small groups.

On Track

Have students work in small groups to answer the question, using calculations to support their reasoning.

The two rectangular prisms have the same volume. Does this mean they have the same surface area?

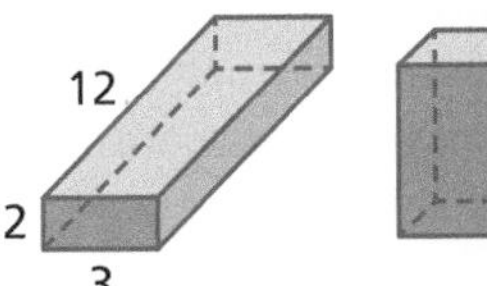

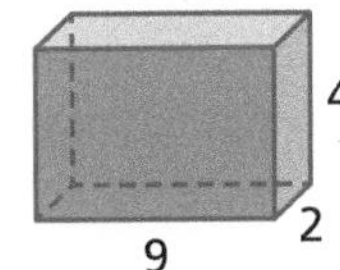

Almost There

Materials: tissue box or other rectangular prism and canned food or other cylinder

- Have students discuss the formula for the surface area of a rectangular prism and a cylinder, using the box and can to visualize each part of the formula.
- Place the can on the tissue box and discuss how to find the surface area of the composite figure. Which part of the box is covered by the can? Which part of the can is covered by the box? How does this affect the total surface area?

Ready for More

Have students work in small groups to solve the problem below.

- A gingerbread house sits on a table and has a cube base with side lengths 8 inches. The distance from the table to the peak of the roof is 16 inches.
- What is the surface area of the house? Do not include the bottom of the house.
- The house should be covered with 10 jellybeans per square inch. How many jellybeans are needed to cover the whole house?

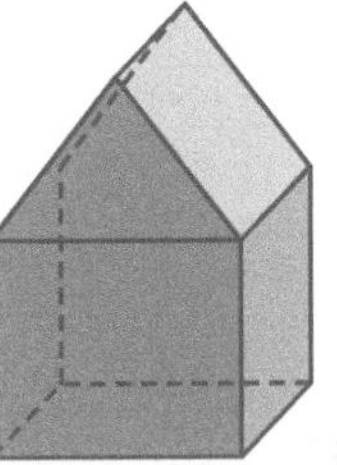

Math Center Options

Use these student self-directed activities at centers or stations. Key: ● Print Resources ● Online Resources

On Track

- ● Interactive Digital Lesson
- ●● Journal and Practice Workbook
- ● Interactive Glossary (printable): **composite figure**, **lateral area**, **net**, **population density**, **surface area**
- ● Module Performance Task

Almost There

- ● Reteach 18.2 (printable)
- ● Interactive Reteach 18.2
- ● RtI Tier 3 Skill 11: Volumes of Right Rectangular Prisms
- ● Illustrative Mathematics: Nets for Pyramids and Prisms

Ready for More

- ● Challenge 18.2 (printable)
- ● Interactive Challenge 18.2

Unit Project Check students' progress by asking to see their setup for the surface area of the cylinder.

View data-driven grouping recommendations and assign differentiation resources.

During the *Spark Your Learning,* listen and watch for strategies students use. See samples of student work on this page.

Use a Net to Find the Area

Strategy 1

I labeled the length and width of each rectangle in the unfolded box. Then I multiplied the length by the width to find each area. There are 2 of each rectangle, so I multiplied by 2 and then found the sum.

$2(7.5)(7) = 105$

$2(7.5)(5) = 75$

$2(7)(5) = 70$

The total surface area is $105 + 75 + 70 = 250$ in^2.

If students . . . use area calculations to solve the problem, they are using a net to find the areas of the sides of the rectangle to find the total surface area from Grade 6.

Have these students . . . explain how they determined their calculations and how they found the surface area. **Ask:**

- Q How did you use a net to solve the problem?
- Q Why did you multiply each area by 2?

Label the Net

Strategy 2

I labeled the unfolded box and found the area of each section. Then, I added the areas together to find the total area.

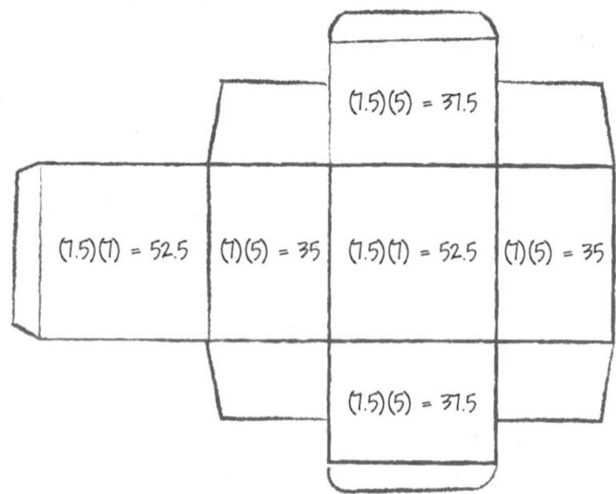

$52.5 + 52.5 + 35 + 35 + 37.5 + 37.5 = 250$ in^2

If students . . . label each part of the net, they understand that there are six rectangles on the box that make up the total surface area, but they did not realize that some of the areas were duplicates.

Activate prior knowledge . . . by having students identify which sides of the box are congruent rectangles. **Ask:**

- Q What is true about the top and bottom of the box?
- Q How could you simplify your calculation?

COMMON ERROR: Use the Visible Sides of the Box

I multiplied the length and width for each side and found the sum.

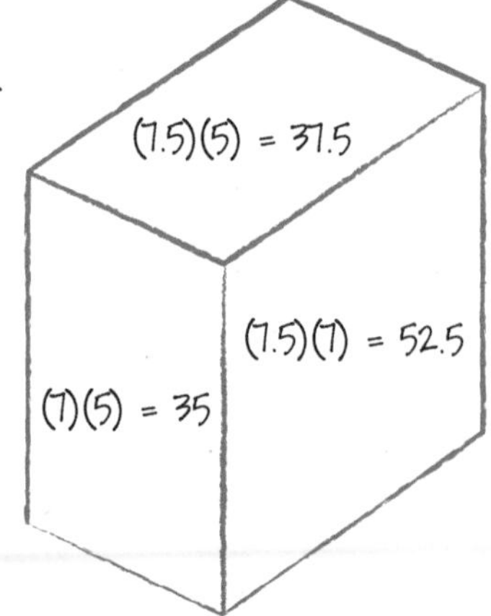

$35 + 52.5 + 37.5 = 125$ in^2

If students . . . use the built box to find the surface area, they are only using the visible sides of the box.

Then intervene . . . by pointing out that in the picture of the built box, three sides are not visible. **Ask:**

- Q How can you use the unfolded box to find the surface area?
- Q What does the unfolded box tell you about the number of sides in the built box?

18.2

Surface Areas of Prisms and Cylinders

I Can find the surface area of a prism or cylinder.

Spark Your Learning

Caleb is folding gift boxes and wrapping gifts.

Complete Part A as a whole class. Then complete Parts B–D in small groups.

A. What is a mathematical question you can ask about this situation? What information would you need to know to answer your question? A. How could I use the unfolded box to find the surface area of the built box?; the lengths and widths of each section of the box

B. What variable(s) are involved in this situation? What unit of measurement would you use for each variable? See Additional Answers.

C. To answer your question, what strategy and tool would you use along with all the information you have? What answer do you get? See Strategies 1 and 2 on the facing page.

D. Does your answer make sense in the context of the situation? How do you know? See Additional Answers.

Turn and Talk Make a copy of the image of the unfolded box. Shade the sections that are not included in the surface area of the built box. How does the surface area of the built box compare to the area of the unfolded box? See margin.

SUPPORT SENSE-MAKING • Three Reads

Tell students to read the information in the photo three times and prompt them with a different question each time.

1. What is the situation about? The situation is about using an unfolded box to help find the surface area of a built box.
2. What are the quantities in this situation? How are those quantities related? The quantities are the dimensions of the box which include the length and width of each side of the box. The length and width of each side can be multiplied together to find the area of each side.
3. What are possible questions you could ask about this situation? When the box is unfolded, which parts are used to find the surface area of the built box? Are the lengths and widths the same for each side? How can I use the given dimensions to label the diagram of the unfolded box?

1 Spark Your Learning

▶ MOTIVATE

- Have students look at the photo in their books and read the information contained in the photo. Then complete Part A as a whole-class discussion.
- Give the class the additional information they need to solve the problem. This information is available online as a printable and projectable page in the Teacher Resources.
- Have students work in small groups to complete Parts B–D.

▶ PERSEVERE

If students need support, guide them by asking:

Q **Advancing • Use Tools** Which tool could you use to solve the problem? Why choose that tool and not some other? Students' choices of tools and reasons for choosing them will vary.

Q **Assessing** How many sides make up the box? six

Q **Assessing** Does each side of the box have the same area? Explain. not unless the dimensions are all the same; For example, if the length, width, and height are all the same value, the sides would have the same area.

Q **Advancing** When you unfold the box, which parts of the diagram are not part of the surface area? the tabs and inserts that help secure the box together

Turn and Talk Ask students to identify the shape of each side of the box. The sides are rectangles and students can use this information to identify which parts of the unfolded box are part of the surface area for the built box. The unfolded box would have a larger surface area.

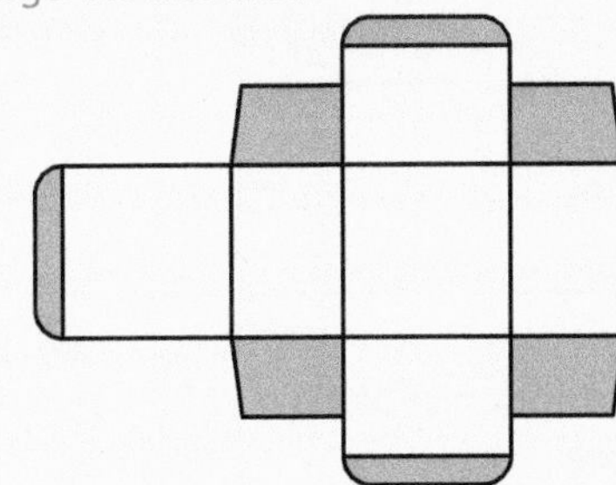

▶ BUILD SHARED UNDERSTANDING

Select groups of students who used various strategies and tools to share with the class how they solved the problem. As they present their solutions, have each group discuss why they chose a specific strategy and tool.

2 Learn Together

Build Understanding

Task 1 (MP) **Use Structure** Students compare the net and surface area formula for a right rectangular prism to make conjectures about the surface area for an oblique prism.

CONNECT TO VOCABULARY

Have students use the **Interactive Glossary** to record their understanding of the vocabulary in this task.

Sample Guided Discussion:

Q **When writing the formula in Part B, how many terms are there in your formula?** There are six, $lw + lw + hl + hl + wh + wh$. You could also multiply each of three terms by 2, $2lw + 2hl + 2wh$.

Q **How are the shapes in the nets of the rectangular prism and the oblique prism different?** The net of the rectangular prism has rectangles, and the oblique prism has parallelograms and rectangles.

Turn and Talk Show students a rectangular prism, such as a tissue box. Set the box on a table and point out that the lateral surface area is the part of the prism that is not on the table or on the top of the box. $L = 2hl + 2hw$; Find the area of the base of the prism. Because the base is a regular polygon, each lateral face will have the same dimensions. The surface area is twice the area of the base plus the area of a lateral face times the number of lateral faces.

Build Understanding

Develop a Surface Area Formula for a Right Prism

Surface area is the total area of all faces and curved surfaces of a three-dimensional figure. The **lateral area** of a prism is the sum of the areas of the lateral faces.

A **net** is a diagram of a three-dimensional figure arranged in such a way that the diagram can be folded to form the three-dimensional figure.

1 Consider the net of the right rectangular prism.

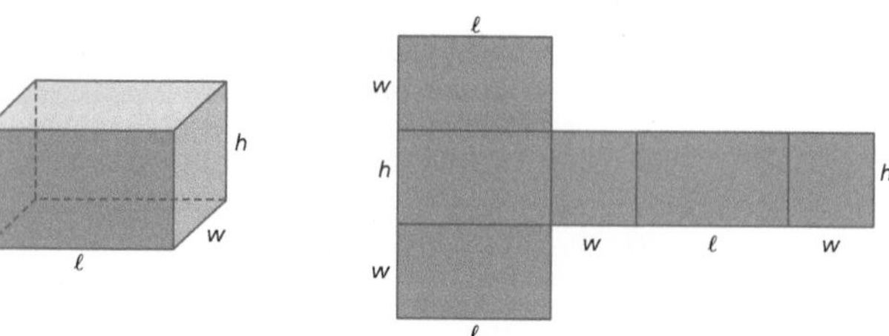

A. Describe the surface area of the prism represented by the net.
A. The surface area is equal to the sum of the areas of each rectangle in the net of the prism.

B. Use your description of the surface area of the prism to write a formula for the surface area of a right rectangular prism.
$SA = 2\ell w + 2\ell h + 2wh$

Consider the net of the oblique rectangular prism shown.

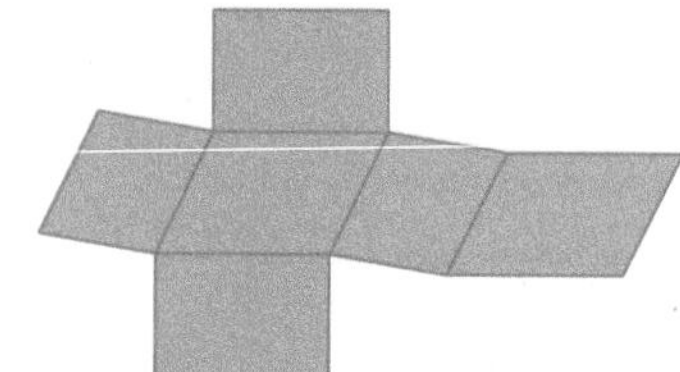

C. Draw a sketch of the oblique prism that you would get if you folded the net.
C–E. See Additional Answers.

D. Describe how the surface area of the prism represented by the net is the same as the right rectangular prism. How is it different?

E. Is it possible to write a general formula for the surface area of an oblique prism using the dimensions of the net? Explain your reasoning.

Turn and Talk

- For the prism in Parts A and B, if $w\ell$ represents the area of the bases, what is the lateral area of the prism that does not include base areas?
- Describe how to find the surface area of a prism with a base that is a regular polygon but not a square. See margin.

558

LEVELED QUESTIONS

Depth of Knowledge (DOK)	Leveled Questions	What Does This Tell You?
Level 1 **Recall**	How do you find the area of a rectangle and a parallelogram? The area of a rectangle is length times width. The area of a parallelogram is the length multiplied by the height.	Students' answers will indicate whether they know the formula for the area of a rectangle and a parallelogram.
Level 2 **Basic Application of Skills & Concepts**	How does the total surface area of a prism differ from the lateral surface area? The lateral surface area does not include the area of the bases, and the total surface area includes all surfaces.	Students' answers will indicate whether they can identify the bases in a prism and distinguish between total and lateral surface area.
Level 3 **Strategic Thinking & Complex Reasoning**	How do the bases in a right rectangular prism differ from the bases in an oblique prism? The bases in a right prism are parallel and directly above and below each other. The bases in an oblique prism are parallel, but not exactly in line above and below.	Students' answers will indicate whether they can compare oblique and right prisms. They can explain differences and similarities between the two types of prisms.

Develop a Surface Area Formula for a Right Cylinder

You can also use a net of a right cylinder to develop a formula for the surface area of a right cylinder.

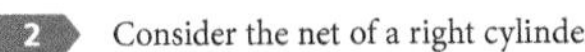

2 Consider the net of a right cylinder.

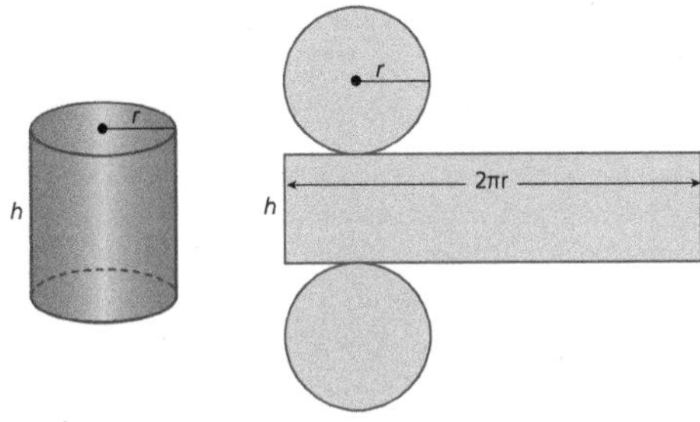

A. What shapes make up the surface of the cylinder? a rectangle and two circles

B. Describe the surface area of the cylinder represented by the net. The surface area is equal to the area of each circle plus the area of the rectangle.

C. Why is the length of the rectangle represented by the expression $2\pi r$? Explain your reasoning. See Additional Answers.

D. Use your description of the surface area of the cylinder to write a formula for the surface area of a right cylinder. $SA = 2\pi rh + 2\pi r^2$

Consider an oblique cylinder.

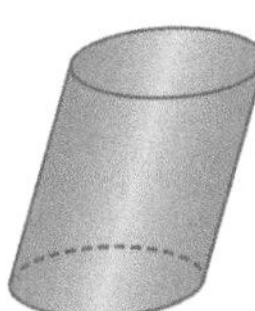

E. Would the lateral area of the oblique cylinder be represented by a rectangle? The lateral area would be represented by a parallelogram instead of a rectangle.

F. Can you write a general formula for the surface area of an oblique cylinder? Explain your reasoning. See Additional Answers.

Turn and Talk Suppose you squeeze a right cylinder so that the shape of each base is an ellipse or oval shape, not a circle. What would you need to know in order to calculate the surface area? See margin.

Task 2 (MP) **Use Structure** Students compare the shapes of the figures that make up the surface area of a cylinder to develop a formula.

Sample Guided Discussion:

Q What is the formula for the area of a circle? $A = \pi r^2$

Q In Part C, how is the circular base related to the length of the rectangle? The length of the rectangle is the same as the distance around the circle.

Turn and Talk Remind students that for the cylinder with circular bases, they needed to know the circumference of the circle and the area of the circles. Ask students how the measurements change if the circle becomes an ellipse or oval. You would need to know the area of the ellipse and the distance around the ellipse.

ANSWERS

2.C. The length of the rectangle is equal to the circumference of the circular base, because the rectangle wraps around the entire circle.

2.F. You cannot write a general formula for the surface area using the dimensions of the cylinder. You need to know the height of the cylinder perpendicular to a base, not the side lengths of the lateral face.

PROFICIENCY LEVEL

Beginning

Ask students to describe the shapes that make up the surface area of a cylinder. Listen for correct use of terminology and assist as needed.

Intermediate

Have student pairs write the formula for a right cylinder and then write what it means in words.

Advanced

Ask each student to write a paragraph describing the differences between how to find the surface area for a right cylinder and an oblique cylinder.

Step It Out

Task 3 (MP) **Model with Mathematics** Encourage students to identify the three-dimensional shapes that model the small house and to count the number of sides that make up the surface area.

CONNECT TO VOCABULARY

Have students use the **Interactive Glossary** to record their understanding of the vocabulary in this task.

Sample Guided Discussion:

Q How many sides of the rectangular prism and triangular prism are present in the surface area of the house? There are four sides in the triangular prism and four sides in the rectangular prism.

Q In Part B, how could you write an expression for the surface area of the house without using subtraction? You would use the formula for the rectangular prism and triangular prism without the (16)(12) parts.

Turn and Talk Help students to identify which value changes in the diagram and in the formula. Ask students to consider if the formula itself changes, or only the substituted value. You would substitute 6.5 for h in the formula for the surface area of the rectangular prism instead of 6.

ANSWERS

3.A. $2\left(\frac{1}{2}(4)(12)\right)$ represents the triangular bases. $2(7.2)(16)$ represents the two top faces. $(16)(12)$ represents the bottom face.

3.B. The top of the rectangular prism and the bottom of the triangular prism are not on the exterior of the figure.

Step It Out

Find the Surface Area of a Composite Figure

Surface Area of a Right Prism

The surface area of a right prism is the sum of the lateral area L and the area of the two bases $2B$.

$S = L + 2B$

Surface Area of a Right Cylinder

The surface area of a right cylinder is the sum of the lateral area L and the area of the two bases $2B$.

$S = L + 2B$

Recall that a **composite figure** is a three-dimensional figure made up of prisms, cones, pyramids, cylinders, and other simple three-dimensional figures. You can find the surface area of a composite figure by finding the surface areas of the individual figures that make up the composite figure.

3 Find the surface area of the small house. The three dimensional figures that make up the house are a rectangular prism and a triangular prism.

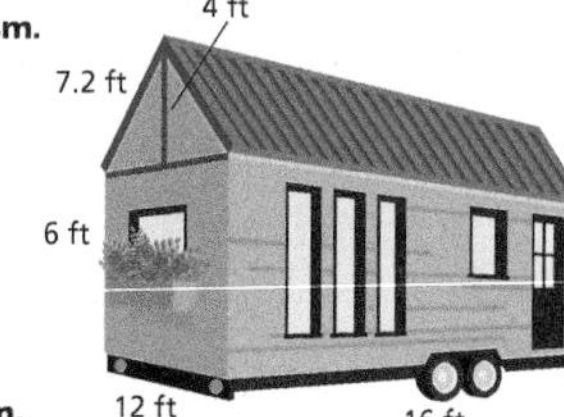

First, find the surface area of the rectangular prism.

$S = 2\ell w + 2\ell h + 2wh$

$S = 2(16)(12) + 2(16)(6) + 2(12)(6)$

$S = 384 + 192 + 144$

$S = 720 \text{ ft}^2$

Then, find the surface area of the triangular prism.

$S = 2\left(\frac{1}{2}(4)(12)\right) + 2(7.2)(16) + (16)(12)$

$S = 48 + 230.4 + 192$

$S = 470.4 \text{ ft}^2$

A. What does each term in this equation represent in the triangular prism?

A, B. See margin.

Find the surface area of the composite figure.

The surface area is the sum of the areas of all surfaces on the exterior of the figure.

$S = 720 + 470.4 - 2(16)(12)$

$S = 806.4 \text{ ft}^2$

B. Why is 2(16)(12) subtracted from this equation?

Turn and Talk Another model of this small house has a rectangular prism that is 6.5 feet high. How would the equation for the surface area change? See margin.

Apply a Surface Area Formula

Surface area can be used to answer many real-world questions. One use of surface area is in population density problems. **Population density** is the number of organisms of a particular type per square unit of area.

4 A treasure chest in an aquarium is covered with algae. If the algae covered the entire treasure chest, the population would be about 30,000,000 cells. What would be the population density of the algae?

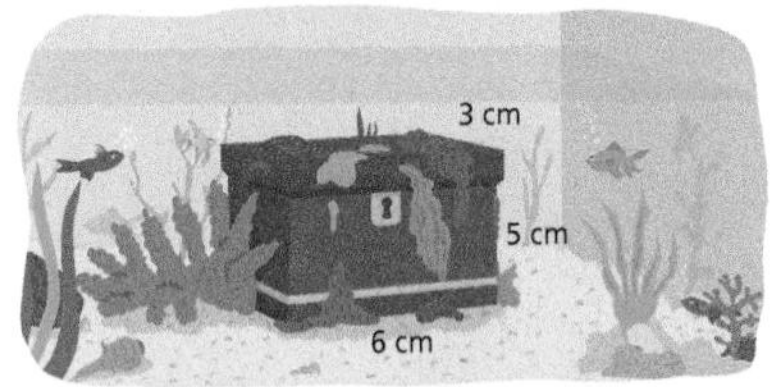

First, find the surface area of the rectangular prism.

$S = 2\ell w + 2\ell h + 2wh$

$S = 2(6)(5) + 2(6)(3) + 2(5)(3)$

$S = 60 + 36 + 30$

$S = 126$

The surface area of the treasure chest is 126 square centimeters.

Answer the question.

To find the population density of the algae on the entire treasure chest, divide the total population of algae by the surface area of the treasure chest. A–D. See margin.

A. Why do you divide the total population by the surface area?

$$\text{Population Density} = \frac{30{,}000{,}000}{126}$$

$$\approx 238{,}000$$

B. What do the units mean in this situation?

If 30,000,000 cells of algae covered the entire chest, the population density would be about 238,000 algae cells per square centimeter.

C. How many algae would you expect to be on the top of an algae-covered treasure chest? Explain your reasoning.

D. What would the population density of the algae be if 30,000,000 algae cells covered all except the bottom of the treasure chest?

Turn and Talk Suppose the treasure chest had dimensions of 12 cm, 10 cm, and 6 cm, with the same population of algae. How would the population density of this chest compare to the population density of the original treasure chest? Explain. See margin.

ANSWERS

4.A. Per definition of population density, you divide to determine the average number of organisms (or algae cells) per square unit of area (or square centimeter of the treasure chest).

4.B. The units say that this number represents the number of cells in square centimeter of the surface of the treasure chest.

4.C. about 4,284,000 cells; Multiply the population density of the algae by 18 cm^2, the area of the top of the treasure chest.

4.D. about 277,800 algae cells per square centimeter

Task 4 (MP) **Attend to Precision** Students make precise calculations to find surface area and population density.

CONNECT TO VOCABULARY

Have students use the **Interactive Glossary** to record their understanding of the vocabulary in this task.

OPTIMIZE OUTPUT Critique, Correct, and Clarify

Have students consider the two expressions below.

$$\frac{30{,}000{,}000}{126} \qquad \frac{126}{30{,}000{,}000}$$

Have students use the task to assign units to the numbers in the numerators and denominators. What are the units for each expression? Which expression correctly describes the population density? Explain. The first expression gives units of cells per square centimeter. The second expression gives units of square centimeters per cell. The first expression is correct because it gives the population per square inch. Population should be in the numerator when finding population density.

Sample Guided Discussion:

Q **In Part C, how can you find the area of only the top of the treasure chest?** Multiply the dimensions of the rectangle that makes up the top of the treasure chest.

Q **How do the answers in Parts C and D change if the population density is kept at 238,095 and not rounded to 238,000?** The answers will be slightly larger.

Turn and Talk If students have difficulty visualizing that this is a larger chest, have them draw and label a picture. Then have students calculate the total surface area and then the population density. The population density would decrease because the new treasure chest has a greater surface area.

ONLINE

Assign the Digital On Your Own for
- built-in student supports
- Actionable Item Reports
- Standards Analysis Reports

On Your Own

Assignment Guide

The chart below indicates which problems in the On Your Own are associated with each task in the Learn Together. Assign daily homework for tasks completed.

Learn Together Tasks	On Your Own Problems
Task 1, p. 558	Problems 7–10, 18, 20, 21, and 22
Task 2, p. 559	Problems 6, 11–13, 16, 17, 19, and 23
Task 3, p. 560	Problem 14
Task 4, p. 561	Problem 15

ANSWER

6.A. no; He is not correct, because the vase does not have a top, so the area of the circle should only be used once when calculating the surface area.

Check Understanding

Find the surface area. Round your answer to the nearest tenth, if necessary.

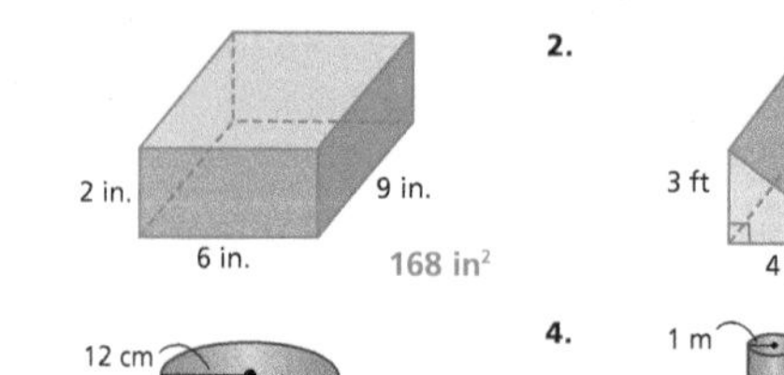

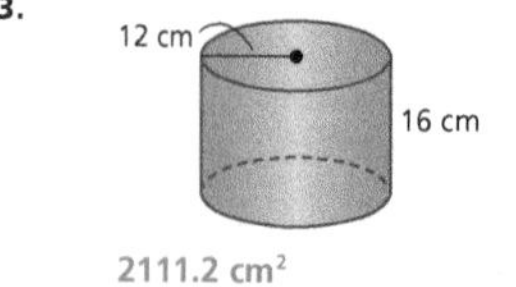

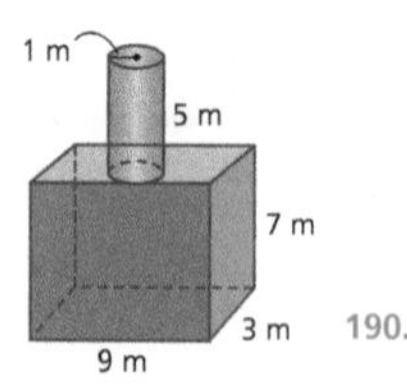

5. Raul is wrapping a gift that is a box shaped like a rectangular prism. The box is 14 inches long, 8 inches wide, and 3 inches tall. How much wrapping paper does he need to cover the box? at least 356 in²

On Your Own

6. Dante has a vase that is shaped as shown.

A. (MP) **Critique Reasoning** He says that the surface area of the vase is about 50.3 square inches. Is he correct? Explain your reasoning. See margin.

B. If he wants to redecorate the lateral part of the vase only, what is the lateral area of the vase? Round to the nearest square inch about 44 in²

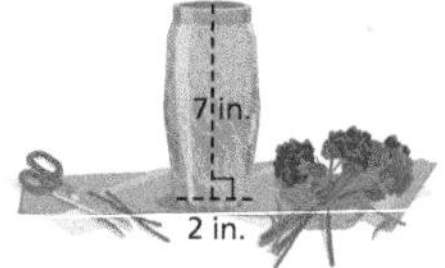

Find the surface area. Round your answer to the nearest tenth, if necessary.

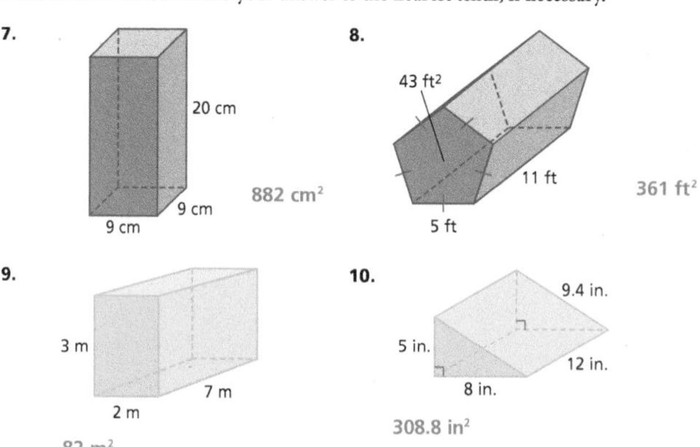

562

3 Check Understanding

Formative Assessment

Use formative assessment to determine if your students are successful with this lesson's learning objective.

Students who successfully complete the Check Understanding can continue to the On Your Own practice.

For students who miss 1 problem or more, work in a pulled small group using the Almost There small-group activity on page 557C.

ONLINE

Assign the Digital Check Understanding to determine
- success with the learning objective
- items to review
- grouping and differentiation resources

4 Differentiation Options

Differentiate instruction for all students using small-group activities and math center activities on page 557C.

Reteach

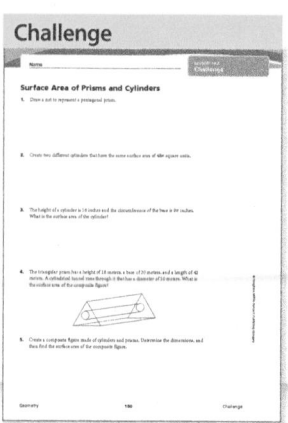

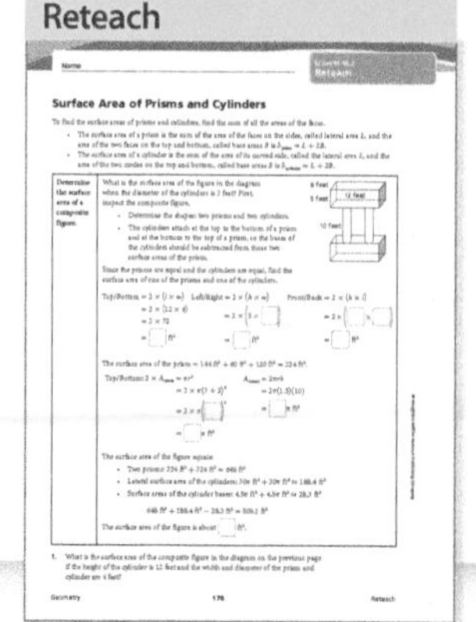

Challenge

Surface Area of Prisms and Cylinders

Find the surface area of each figure. Leave your answer in terms of π.

11. **12.** **13.** **14.**

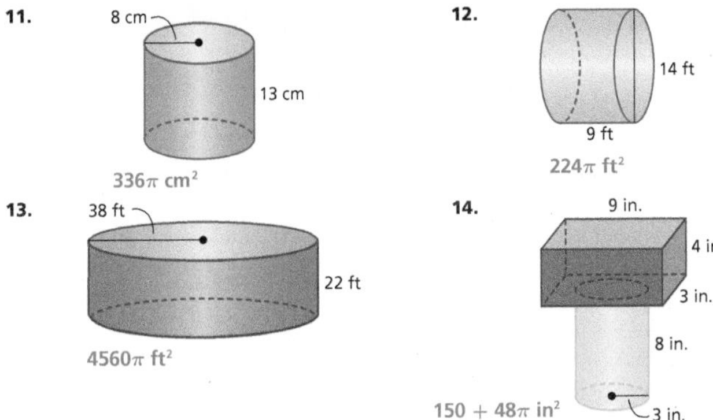

11. 336π cm²

12. 224π ft²

13. 4560π ft²

14. $150 + 48\pi$ in²

15. STEM A scientist is conducting an experiment with bacteria. The growth medium is shaped like a rectangular prism as shown.

A. What is the entire surface area of the block?

B. What is the population density of the bacteria on the block after 24 hours?

C. What change in the surface area would decrease the population density?

A–C. See margin.

After 24 hours, 63,000 bacteria have colonized on the block.

3 in. 3 in. 4 in.

16. Doug is building a decoration and wants to add a cylindrical candle to it. He finds a candle that has a diameter of 3 inches and a surface area of about 61.3 square inches. He determines that he needs to cut two inches off the bottom of the candle in order for it to fit in the decoration. He cuts the candle parallel to the base. What is the surface area of the candle after cutting the bottom off? about 42.4 square inches

17. (MP) **Reason** Erica draws and labels the rectangle and wants to know what the surface area would be if it is rotated about the line. Does she have enough information to calculate the surface area? Explain your reasoning. If she has enough information, what is the surface area?

17, 18. See margin.

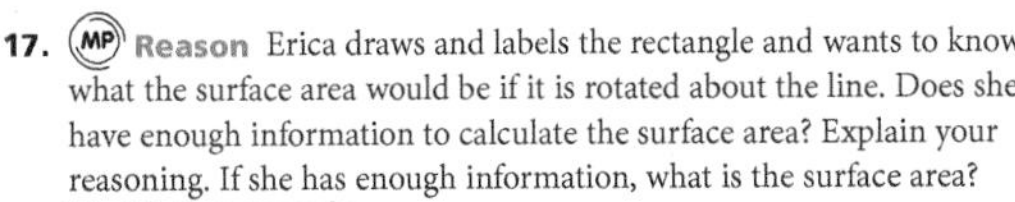

18. A company is designing the boxes that they will use to ship their products. The company wants to choose the box design that will require the least amount of cardboard. Which box design should the company use? Explain your reasoning.

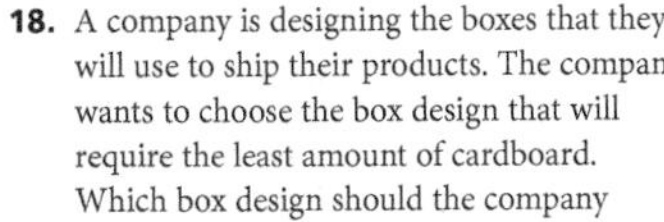

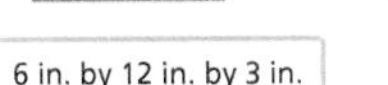

Questioning Strategies

Problem 14 Suppose the two three-dimensional figures were not attached. There would be no surfaces that are shared by the two figures. However, when they are attached, they share a surface. What is the shape formed where the two figures are attached? How does this affect the formula for the total surface area? The two figures are attached with one circular base of the cylinder. This means when you find the surface area of the rectangular prism, you have to subtract the area of one circle. Similarly, when you find the surface area of the cylinder, you have to subtract the area of one circle.

Watch for Common Errors

Problem 16 Students may try to find original height by assuming the given surface area is the lateral surface area. Have students write the formula for total surface area of a cylinder, $SA = 2\pi rh + 2\pi r^2$, and substitute the information they know about the candle before Doug adjusts the size.

ANSWERS

15.A. 66 in^2

15.B. approximately 955 bacteria per square inch.

15.C. an increase in surface area

17. The figure after it is rotated is a cylinder, and because she knows the height and radius of the cylinder, she can calculate the surface area. The surface area is about 1508 square feet.

18. Box A; The surface area of Box A 252 square inches and the surface area of Box B is 264 square inches, so Box A uses the least amount of cardboard.

5 Wrap-Up

Summarize learning with your class. Consider using the Exit Ticket, Put It in Writing, or I Can scale.

Exit Ticket 

A cylindrical can of pears has a label that exactly covers the lateral surface area. If the height of the can is 18 cm and the radius is 4 cm, how much paper is used to make the label? Round to the nearest square centimeter.

$2\pi(4)(18) \approx 452 \text{ cm}^2$

Put It in Writing 

Explain in words what each part of the formula for the surface area of a right cylinder means.

I Can 

The scale below can help you and your students understand their progress on a learning goal.

4	I can find the surface area of a prism or cylinder, and I can explain the process to others.
3	I can find the surface area of a prism or cylinder.
2	I can write the formula for the surface area of a prism or cylinder and identify the dimensions to use in the formula.
1	I can identify if the figure is a prism or cylinder and determine which formula to use.

Spiral Review • Assessment Readiness

These questions will help determine if students have retained information taught in the past and can also prepare them for high-stakes assessments. Here, students find the area of a sector of a circle **(17.3)**, identify a cross section of a three-dimensional figure **(18.1)**, and convert from degree to radian measure **(17.2)**.

19. Find the height of a right cylinder with a radius of 24 centimeters and a surface area of 1632π square centimeters. 10 centimeters

20. Find the surface area of a hexagonal prism with regular hexagons as the bases. The area of the base is 210 square inches. Each side of the base is 9 inches, and the height of the prism is 6 inches. 744 square inches

21. **Geology** Volcanic activity near basalt rock formations sometimes results in naturally occurring columns. The columns shown are slowly eroding because of exposure to rain and other elements. Suppose the bases of the columns that are labeled can be modeled by regular hexagons. The area of the base of the taller labeled one is approximately 1.5 square feet. Each side of its base is 9 inches.

A. Draw a net that represents the taller hexagonal column. Then shade the area of the net that appears to be exposed to the elements. See Additional Answers.

B. From Part A, what is the surface area that is exposed to the elements? Possible answer: 11.25 square feet

22. Find the width of a right rectangular prism with a length of 16 inches, a height of 4 inches, and a surface area of 368 square inches. 6 inches

23. Open Middle™ What is the greatest possible volume for a cylinder with a surface area of 24π square units? (The formula for the volume of a cylinder is $V = \pi r^2 h$.) 16π cubic units

Spiral Review • Assessment Readiness

24. Find the area of the sector.

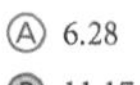

Ⓐ 6.28
Ⓑ 11.17
Ⓒ 12.57
Ⓓ 50.27

25. Which of the following could be a vertical or horizontal cross section of a cylinder? Select all that apply.

Ⓐ rectangle
Ⓑ triangle
Ⓒ trapezoid
Ⓓ circle
Ⓔ cone
Ⓕ pyramid

26. Match the angle measure on the left with its equivalent radian measure on the right.

A. 135°	**1.** $\frac{2\pi}{3}$	A. 4; B. 1; C. 3; D. 2
B. 120°	**2.** $\frac{\pi}{3}$	
C. 45°	**3.** $\frac{\pi}{4}$	
D. 60°	**4.** $\frac{3\pi}{4}$	

I'm in a Learning Mindset!

What skills do I have that benefit collaboration? Which collaboration skills still need improvement?

564 Keep Going ▶ Journal and Practice Workbook

Learning Mindset

Strategic Help-Seeking Identifies Need for Help

The formulas for the surface area of a prism and a cylinder have several parts. The area of the base is determined by which type of figure is represented, and this also affects the shaped formed by the lateral surface area. Encourage students to ask for help when they are unsure what shapes are formed or do not know what the dimensions represent. *What is the shape of the base? How does the shape of the base affect the lateral surface area? How can drawing a net help you confirm the formula or calculations for the surface area?*

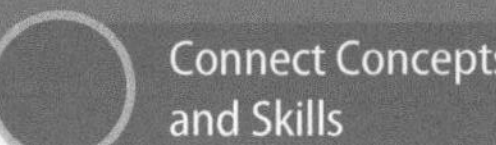

18.3 Surface Areas of Pyramids and Cones

LESSON FOCUS AND COHERENCE

Mathematics Standards

- Use geometric shapes, their measures, and their properties to describe objects.
- Apply concepts of density based on area and volume in modeling situations.
- Apply geometric methods to solve design problems.

Mathematical Practices and Processes 

- Look for and make use of structure.
- Model with mathematics.
- Attend to precision.

I Can Objective 

I can use formulas for the surface area of pyramids and cones to solve real-world problems.

Learning Objective

Use formulas for the surface area of pyramids and cones to solve real-world problems.

Language Objective

Explain why a pyramid that is not regular does not have a slant height.

Vocabulary

New: regular pyramid, slant height, right cone

Mathematical Progressions

Prior Learning	Current Development	Future Connections
Students: • used the nets to find the surface area of three-dimensional figures. **(Gr6, 18.1)** • knew the formulas for the area and circumference of a circle and used them to solve problems. **(Gr7, 17.1)** • solved real-world and mathematical problems involving area of two- dimensional objects composed of circles triangles, rectangles, and rhombuses. **(Gr7, 11.4)**	**Students:** • develop formulas for surface area of regular pyramids and right cones. • calculate the surface area of composite solids. • apply surface area formulas to solve real-world problems.	**Students:** • for a function that models the relationship between two quantities, will interpret key features of graphs and tables in terms of the quantities, and will sketch graphs showing key features given a verbal description of the relationship, including: intercepts; intervals where the function is increasing, decreasing, positive; or negative; relative maximums and minimums; symmetries; end behavior; and periodicity. **(4th-Year Course)**

UNPACKING MATH STANDARDS

Use geometric shapes, their measures, and their properties to describe objects.

What It Means to You

Students derive and apply formulas for surface area of pyramids and cones to solve mathematical and real-world problems, as well as deconstruct composite solids into component shapes to determine total surface area of those composite solids. At this point, students are expected to know the surface area formulas for prisms and cylinders, so they can use this prior knowledge in conjunction with the new formulas. Experience with applying these formulas will help students prepare for work in later lessons, when surface area of a sphere and the volume of three-dimensional solids are is introduced.

WARM-UP OPTIONS

ONLINE

PROJECTABLE & PRINTABLE

ACTIVATE PRIOR KNOWLEDGE • Apply Area Formula

Use these activities to quickly assess and activate prior knowledge as needed.

Problem of the Day 

A pizza shop sells an 8-inch-diameter pizza for $5.00 and a 10-inch-diameter pizza for $10. Which offer gives you more pizza for your money? Area of 8-inch pizza $= \pi(4)^2 \approx 50.26$ in^2; Area of 10 inch pizza $= \pi(5)^2 \approx 78.54$ in^2; 50.26 ÷ \$8 ≈ 6.28 square inches of pizza per dollar spent; 78.54 ÷ \$10 ≈ 7.85 square inches of pizza per dollar spent. So, the 10-inch pizza is the better buy.

Quick Check for Homework

As part of your daily routine, you may want to display the Teacher Solution Key to have students check their homework.

Make Connections

Based on students' responses to the Problem of the Day, choose one of the following:

1. Project the Interactive Reteach, Grade 7, Lesson 10.2.
2. Complete the Prerequisite Skills Activity:

Have students work in pairs and display the following information:

radius: 4 in.; diameter: 6 in.; area: 201.06 in^2

Ask students to find the values of the diameter and area when $r = 4$, the radius and area when $d = 6$, and the radius and diameter when $A = 201.06$.

- *When given the radius, how can you find the diameter and the area of the circle?* Find the diameter by multiplying the radius by 2 and the area by multiplying π by r^2.
- *How can you determine the radius and the area when given the diameter of the circle?* Divide the diameter by 2 to find the radius, and then multiply π by r^2 to find the area.
- *How can you use the given area to determine the radius and diameter of the circle?* Divide the given area by π to find the value of r^2, and then take the square root of r^2 to find the measure of the radius. Multiply that value by 2 to find the measure of the diameter.

If students continue to struggle, use Tier 2 Skill 20.

SHARPEN SKILLS

If time permits, use this on-level activity to build fluency and practice basic skills.

Quantitative Comparison

Objective: Students make a comparison between two quantities.

Write the following problem on the board. Ask students to choose the letter representing the correct answer and to explain their reasoning.

Quantity A	**Quantity B**
Surface area of a right cone with a diameter of 8 and a slant height of 4	Surface area of a regular pyramid with a base edge length of 8 and a slant height of 4

A. Quantity A is greater.
B. Quantity B is greater.
C. The two quantities are equal.
D. The relationship cannot be determined from the information given.

B;
Quantity A: $\pi(4)^2 + \pi(4)(4) = 100.53$ units^2
Quantity B: $8^2 + \frac{1}{2}(32)(4) = 128$ units^2

PLAN FOR DIFFERENTIATED INSTRUCTION

Small-Group Options

Use these teacher-guided activities with pulled small groups.

On Track

Materials: index cards

Have students work in pairs and provide each pair with an index card that gives the dimensions of either a right cone or a regular pyramid. For cones, give the radius of the base and the height of the cone. For pyramids, list the edge length of the base and slant height. Ask pairs to draw a diagram of the solid described on their card on a sheet of chart paper. Underneath the diagram, have students list the steps to finding the surface area of the solid. Have students share their work with the class.

Almost There

Materials: index cards

Have students work in pairs and provide each pair with a set of index cards. Each index card shows a net of either a right cone or a square pyramid, with all necessary dimensions labeled. Have students work together to find the surface area of each solid. Display the formulas for the surface area of each on the flip side of each index card.

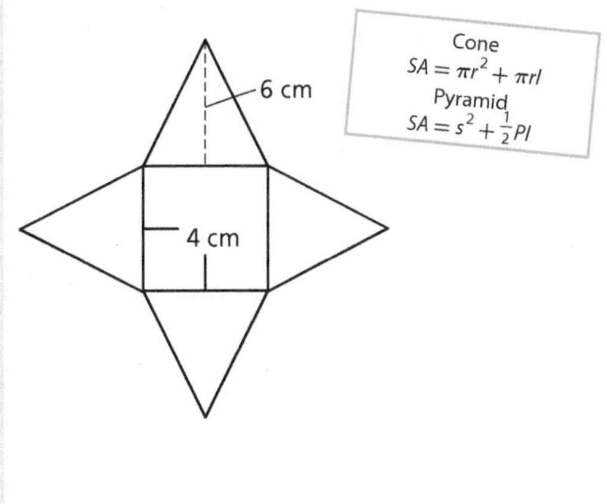

6 cm
4 cm
Cone
$SA = \pi r^2 + \pi rl$
Pyramid
$SA = s^2 + \frac{1}{2}Pl$

Ready for More

Provide students with the following information:

- A right cone has a radius of 4 inches.
- A square pyramid and the cone have the same slant height of 5 inches.
- Both solids have the same surface area.

Ask students to find the length of a base edge of the pyramid.

Math Center Options

Use these student self-directed activities at centers or stations. Key: ● Print Resources ● Online Resources

On Track

- ● Interactive Digital Lesson
- ●● Journal and Practice Workbook
- ● Interactive Glossary (printable): **regular pyramid**, **slant height**, **right cone**
- ● Module Performance Task

Almost There

- ● Reteach 18.3 (printable)
- ● Interactive Reteach 18.3
- ● RtI Tier 2 Skill 20: Area of Circles
- ● Illustrative Mathematics: Nets for Pyramids and Prisms

Ready for More

- ● Challenge 18.3 (printable)
- ● Interactive Challenge 18.3

View data-driven grouping recommendations and assign differentiation resources.

During the *Spark Your Learning,* listen and watch for strategies students use. See samples of student work on this page.

Use Area Formula for Triangle — Strategy 1

There are 8 triangular sections of the roof.

$A = \frac{1}{2}bh$

$A = \frac{1}{2}(60)(82) = 2460$

$8 \times 2460 = 19{,}680$ square inches

If students . . . use the formula for the area of a triangle to solve the problem, they are displaying an exemplary understanding of the concept of area from Grade 6 and are able to recognize that three-dimensional solids can be decomposed into two-dimensional shapes.

Have these students . . . explain how they determined which formula they should apply to solve this problem. **Ask:**

- Q Why was the additional information given necessary to solve the problem?
- Q What is different about the roof of the gazebo and a two-dimensional triangle?

Use Area Formula for Parallelogram — Strategy 2

There are 8 triangles that have a base of 60 and a height of 82 that make up the octagonal gazebo. If I put two triangles together, I can make a parallelogram with base *b* and height *l*. There will be 4 parallelograms.

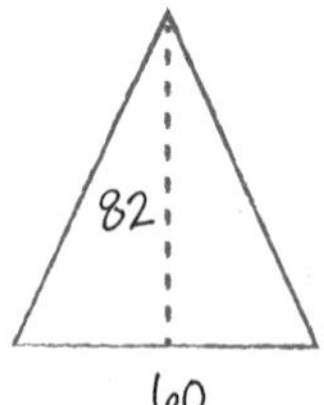

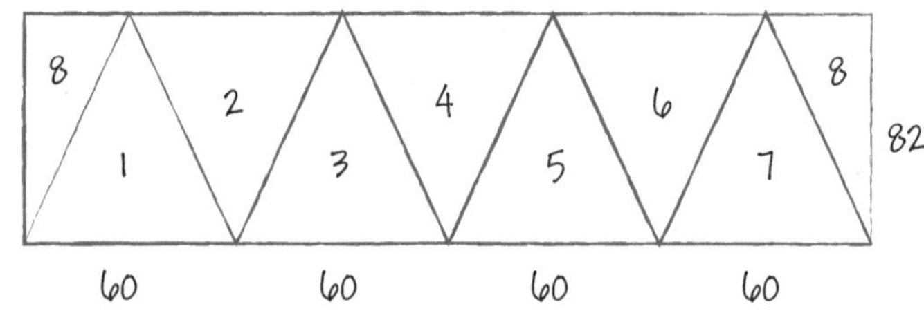

Area of a parallelogram is $b \times h = 60 \times 82 = 4920$ square inches.

There are 4 parallelograms, so the area is $4 \times 4920 = 19{,}680$ square inches.

If students . . . use the area formula for a parallelogram, they understand how to decompose a solid to create a two-dimensional shape of which they know how to find the area, like a parallelogram, but may not know the formula for finding the area of a triangle.

Activate prior knowledge . . . by having students derive the formula for the area of a triangle. **Ask:**

- Q If you draw a diagonal in a parallelogram, what two congruent shapes do you see?
- Q What do you need to do to the area formula of a parallelogram to find the area of one of the congruent triangles?

COMMON ERROR: Find Circumference

The base of one triangle is 60 inches long. An octagon has 8 sides. $8 \times 60 = 480$, so there are 480 square inches of roof that need to be shingled.

If students . . . find the circumference of the octagon, then they do not understand that area is the number of square units that cover a shape and that circumference is the number of units that surround a shape.

Then intervene . . . by reminding students of the context of the problem. **Ask:**

- Q What is Geno doing to the roof?
- Q Does the distance around the outer edge of the roof tell you the number of square feet that covers the roof?

18.3 Surface Areas of Pyramids and Cones

I Can use formulas for the surface area of pyramids and cones to solve real-world problems.

Spark Your Learning

Geno is shingling the roof of an octagonal gazebo.

Complete Part A as a whole class. Then complete Parts B–D in small groups.

A. What is a mathematical question you can ask about this situation? What information would you need to know to answer your question? A. How much material is required to shingle the roof of the gazebo?; the dimensions of the roof

B. What variable(s) are involved in this situation? What unit of measurement would you use for each variable? See Additional Answers.

C. To answer your question, what strategy and tool would you use along with all the information you have? What answer do you get? See Strategies 1 and 2 on the facing page.

D. Does your answer make sense in the context of the situation? How do you know? See Additional Answers.

Turn and Talk How would your method to find the area change if the gazebo had 6 sides instead of 8 sides? See margin.

CULTIVATE CONVERSATION • Information Gap

Ask students questions to help them decide what information they need to answer the question, "How much material is required to shingle the roof of the gazebo?"

1. Do you have enough information to determine the area of the octagon-shaped roof? Explain. no; I do not have a formula for finding the area of an octagon.
2. Is the roof of the gazebo made up of shapes whose area can be found using a formula? Explain. yes; The roof is made up of eight congruent triangles.
3. What information do you need to determine the area of each triangle? I need to know the length of the height and the length of the base of each triangle.

Connect Concepts and Skills

1 Spark Your Learning

▶ MOTIVATE

- Have students look at the photo in their books and read the information contained in the photo. Then complete Part A as a whole-class discussion.
- Give the class the additional information they need to solve the problem. This information is available online as a printable and projectable page in the Teacher Resources.
- Have students work in small groups to complete Parts B–D.

▶ PERSEVERE

If students need support, guide them by asking:

Advancing • Use Tools Which tool could you use to solve the problem? Why choose that tool and not some other? Students' choices of tools and reasons for choosing them will vary.

Assessing What shapes make up the roof of the gazebo? eight congruent triangles

Assessing How do you find the area of a triangle? Multiply the base by the height and divide the result by 2.

Advancing Since you do not have a formula for finding the area of an octagon, what is another way you can find the area of the roof of the gazebo? Find the sum of the areas of the triangles.

Turn and Talk Before students predict how the method would change, discuss finding the area of irregular figures that are made up of different shapes that they do know the area of. This will allow them to understand that the area of any shape or solid can be found. You would still find the area of each triangular section of the roof, but you would multiply that area by 6 instead of 8 to find the total area.

▶ BUILD SHARED UNDERSTANDING

Select groups of students who used various strategies and tools to share with the class how they solved the problem. As they present their solutions, have each group discuss why they chose a specific strategy and tool.

2 Learn Together

Build Understanding

Task 1 MP **Use Structure** Students use structure when they break apart a solid and explore the different faces of it as shapes with which they are familiar and of which they have prior knowledge.

CONNECT TO VOCABULARY

Have students use the **Interactive Glossary** to record their understanding of the vocabulary in this task.

Sample Guided Discussion:

Q How would you describe the height of the pyramid? The height of the pyramid is the distance between the very tip of the pyramid and the center of its base.

Q In Part D, why does the second term of the general equation have a coefficient of 2? Because the formula for area of one triangle is (base × height) ÷ 2, but since there are 4 congruent triangles, we multiply the area of 1 triangle by 4, and 4 ÷ 2 is 2.

Turn and Talk Before students think about the shapes that make up the faces of a cone, discuss the differences between a pyramid and a cone in terms of the bases. The net of a cone is a circle and a sector of a circle.

Build Understanding

Develop a Surface Area Formula for a Regular Pyramid

A **regular pyramid** is a pyramid whose base is a regular polygon and whose lateral faces are congruent isosceles triangles. The **slant height** ℓ of a regular pyramid is the height of each lateral face. Three examples of regular pyramids are shown.

Square Pyramid

4 congruent lateral faces

Triangular Pyramid

3 congruent lateral faces

Hexagonal Pyramid

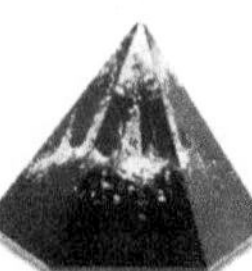

6 congruent lateral faces

1 A. How is the slant height different from the height of a pyramid?
A–C. See Additional Answers.

Consider the regular square pyramid as shown.

B. Describe the base and the lateral faces of the pyramid. What are the dimensions of the base and a lateral face?

C. Use the description of the base and lateral faces of the square pyramid to draw a net of the pyramid.

D. Suppose the side length s of the pyramid is 7 centimeters and the slant height ℓ is 5 centimeters. Use your net to calculate the surface area of the pyramid. Then write a general formula for the surface area of a square pyramid.
119 square centimeters; $A = s^2 + 2s\ell$

Consider a regular hexagonal pyramid as shown.

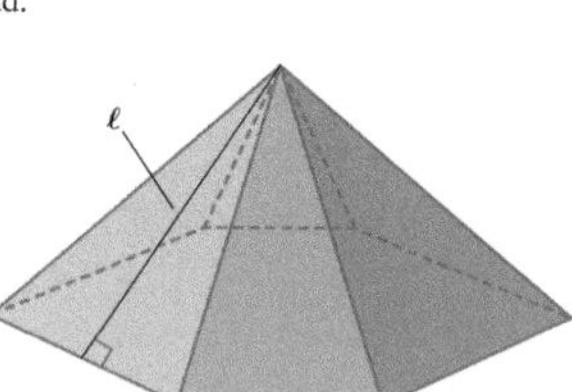

E. Can you generalize the formula you wrote for the surface area of a square pyramid to find the surface area of a hexagonal pyramid? If so, write the formula. If not, what additional information do you need?
E, F. See Additional Answers.

F. Describe a general formula for the surface area of any regular pyramid. Do you think this formula would apply to the surface area of a cone?

Turn and Talk What shapes make up the net of a cone? See margin.

LEVELED QUESTIONS

Depth of Knowledge (DOK)	Leveled Questions	What Does This Tell You?
Level 1 **Recall**	What shapes make up the faces of a regular pyramid? a square and four congruent triangles	Students' answers will indicate whether they can recognize a three-dimensional solid and the shapes of which it is composed.
Level 2 **Basic Application of Skills & Concepts**	What is the area of a triangle that has a base that is 4.5 inches long and a height that is twice as long as the base? $\frac{4.5 \times 9}{2} = 2025 \text{ in}^2$	Students' answers will indicate whether they are able to apply the formula for area of a triangle to solve problems.
Level 3 **Strategic Thinking & Complex Reasoning**	Justify why the formula for the surface area of a regular pyramid is $A = s^2 + 2sl$. The area of the square base is $s \times s$, or s^2. The area of one triangular face is $\frac{sl}{2}$, so the area of 4 congruent triangular faces is $4 \times \frac{sl}{2}$, or $2sl$. So the area of the pyramid is $s^2 + 2sl$.	Students' answers will indicate whether they understand how the formula for surface area of a regular pyramid is derived.

Develop a Surface Area Formula for a Right Cone

A **right cone** is a cone whose axis is perpendicular to its base. The slant height ℓ is the length from the base to the vertex along the lateral edge.

2 Consider the glass-blown right cone displayed.

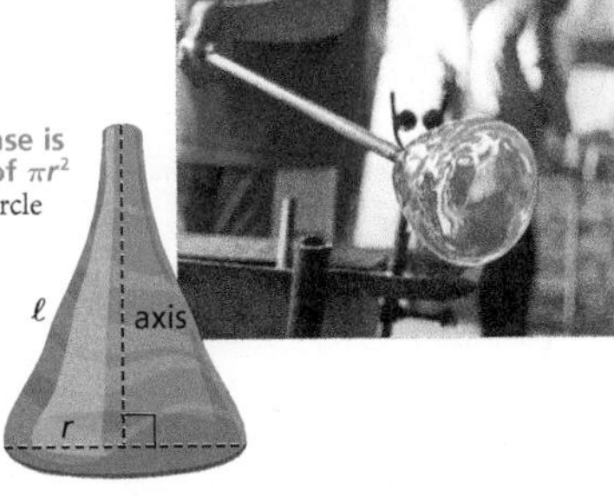

A. Describe the base of the cone. What does r represent in the figure? What attributes of the base can you determine using r? The base is a circle; radius; circumference of $2\pi r$; area of πr^2

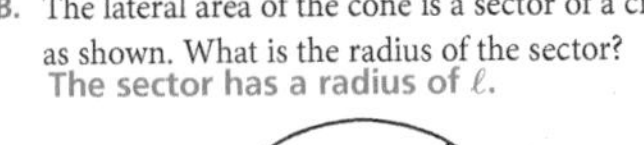

B. The lateral area of the cone is a sector of a circle as shown. What is the radius of the sector? The sector has a radius of ℓ.

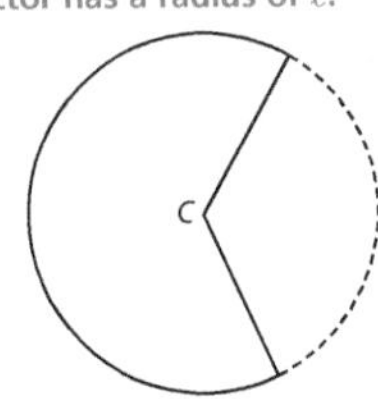

C. From Part B, what is the arc length of the sector? How do you know? See margin.

D. Mackenzie and Cole both draw nets to represent the glass-blown cone. Which is the correct net you would use to find the surface area of the cone? the net drawn by Cole

Mackenzie's Drawing **Cole's Drawing**

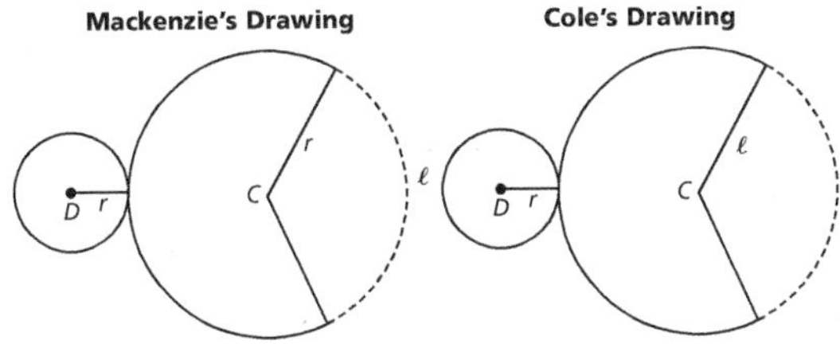

E. Suppose the radius of the cone is 2 inches and the slant height is 6 inches. Use the correct net from Part D to find the surface area of the cone. See margin.

F. Can you write a general formula for the surface area of a right cone? If so, write the formula. If not, what additional information do you need? yes; $S = \pi r^2 + \pi r\ell$

G. How are the surface area formulas of a cone and a pyramid similar? How are they different? See margin.

Turn and Talk How are cones and pyramids related? See margin.

PROFICIENCY LEVEL

Beginning

Display a right cone and say, "This solid is a cone, and its net has two faces. One is a circle and the other is a sector." Then display a regular pyramid and ask students to name the faces of its net verbally and in writing.

Intermediate

Display a cone that shows the measures of the radius of the base and the slant height. Ask students to explain how to find the area of the base and the area of the lateral face using the formula for each.

Advanced

Have students work in groups and provide them with a set of index cards. Each card shows a cone with the measures of the radius and slant height labeled. Ask students to take turns explaining how to calculate the surface area of the cone using its formula.

Task 2 (MP) **Use Structure** Students make use of structure by exploring the lateral faces of a cone. At first, they see a three-dimensional solid, and may find it difficult to visualize the lateral face of a cone. This lesson helps them decompose a cone into a base and a lateral face.

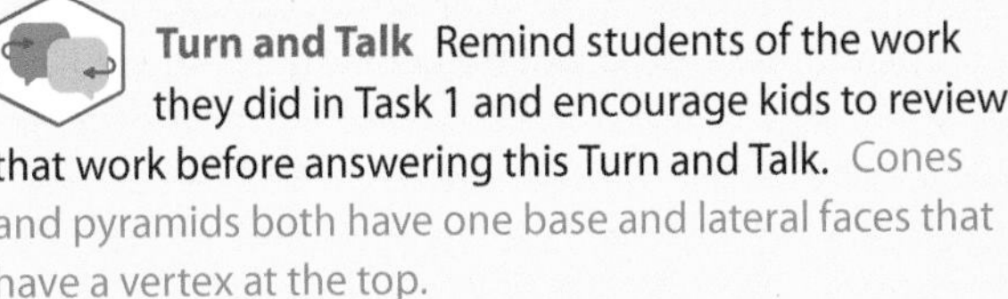

CONNECT TO VOCABULARY

Have students use the **Interactive Glossary** to record their understanding of the vocabulary in this task.

Sample Guided Discussion:

Q **In Part B, when you open up the cone and lay it out flat, where does the circumference of the circular base end up?** The circumference of the base ends up being the major arc of the lateral face, or sector.

Q **In Part D, why is the arc that Mackenzie labeled with l incorrect?** Because if you roll up Mackenzie's net, turning it back into a cone, the arc that she labeled with l disappears and does not exist.

Turn and Talk Remind students of the work they did in Task 1 and encourage kids to review that work before answering this Turn and Talk. Cones and pyramids both have one base and lateral faces that have a vertex at the top.

ANSWERS

2.C. The arc length of the sector is $2\pi r$. The arc length of the sector is equal to the circumference of the base.

2.E. Area of base is $\pi r^2 = 4\pi$. Area of the sector is $\pi rl = 12\pi$. Surface area is $4\pi + 12\pi = 16\pi \approx 50.3$ in.^{2}.

2.G. The surface areas are similar because each is the sum of the lateral area and the area of the base. The shapes of the bases are different, so the actual surface area formulas are different.

Step It Out

Task 3 (MP) **Model with Mathematics** Students model with mathematics when they identify the important quantities described in the task, relate those quantities using formulas that they already know, and analyze those relationships mathematically to solve the problem in the task.

Sample Guided Discussion:

Q How should the right trapezoid shown be sliced so that two shapes you recognize are created? Slice the shape 80 mm up from the bottom of the trapezoid.

Q In Part A, is there a name for the solid that is created by spinning the right trapezoid about an axis? No, the triangular section above the 80 mm point creates a cone when spun, and the rectangular section below the 80 mm point creates a cylinder when spun, but the figure created by spinning the entire trapezoid does not have a specific name.

Q What do the base, altitude, and hypotenuse of the triangular section of the trapezoid represent in the cone that is created when the triangle is spun? The base of the triangle becomes the radius of the base of the cone, the altitude of the triangle acts as the axis about which the triangle is spun, and the hypotenuse becomes the slant height of the cone.

Q In Part C, why is approximating pi necessary in any problem where you are finding surface area of a cylinder or a cone? Pi must be approximated because it is an irrational number, so the decimal never ends and never repeats.

ANSWERS

3.A. rectangle and triangle; cone and cylinder

3.B. The base of the cone is not on the exterior of the composite figure.

3.C. to not introduce inaccuracy when approximating

Step It Out

Find the Surface Area of a Composite Figure

Surface Area of a Regular Pyramid

The lateral area of a regular pyramid with perimeter P and slant height ℓ is $L = \frac{1}{2}P\ell$.

The surface area of a regular pyramid with lateral area L and base area B is $S = L + B$, or $S = \frac{1}{2}P\ell + B$.

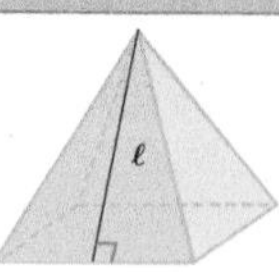

Surface Area of a Right Cone

The lateral area of a right cone with a radius r and slant height ℓ is $L = \pi r\ell$.

The surface area of a right cone with a lateral area L and base area B is $S = L + B$, or $S = \pi r\ell + \pi r^2$.

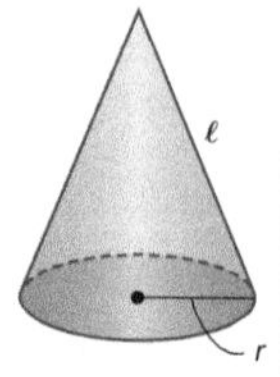

You learned previously how to find the surface area of composite figures made up of prisms and cylinders. Now you can find the surface area of composite figures with pyramids and cones as well.

3 A crayon can be modeled by rotating the two-dimensional shape shown about an axis. Find the surface area of the crayon.

A–C. See margin.

A. What simple shapes make up this shape? What figures are formed when this is rotated about an axis?

Find the lateral area of the cone, the lateral area of the cylinder, and the area of the base of the cylinder.

The height of the cone is $92 - 80 = 12$.

The slant height of the cone is $\ell = \sqrt{4^2 + 12^2} = 4\sqrt{10}$.

The lateral area of the cone is $\pi r\ell = \pi(4)(4\sqrt{10}) \approx 50.6\pi$.

The lateral area of the cylinder is $2\pi rh = 2\pi(4)(80) = 640\pi$.

The area of the base of the cylinder is $\pi r^2 = \pi(4^2) = 16\pi$.

B. Why do you not need to find the area of the base of the cone?

Answer the question.

C. Why is approximating the value of π done only at the last step?

The surface area of the crayon is $50.6\pi + 640\pi + 16\pi = 706.6\pi$, or about 2220 square millimeters.

568

Apply a Surface Area Formula

Recall that the population density is the number of organisms of a particular type per square unit of area. Surface density is similar in that it is the number of objects per square unit of area. You can find the surface density by dividing the number of objects by the area of the solid.

4 Evan is decorating a crystal square pyramid paperweight. He has a bottle of 250 gold flakes to cover the paperweight. He covers the pyramid with glue and rolls it into the flakes, making sure that all flakes in the bottle stick to the surface of the pyramid. What is the surface density—the number of gold flakes per square centimeter?

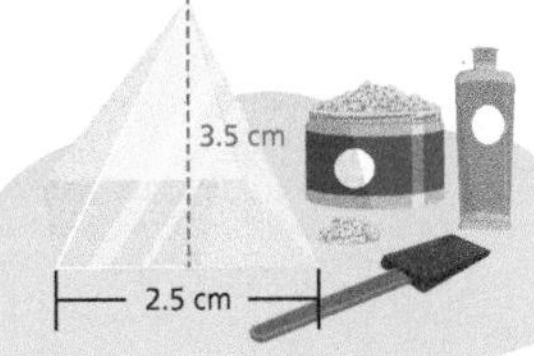

Find the surface area of the pyramid.

$S = \frac{1}{2}P\ell + B$

$S = \frac{1}{2}(10)(3.5) + 2.5^2$

$S = 17.5 + 6.25 = 23.75$

The surface area is 23.75 square centimeters.

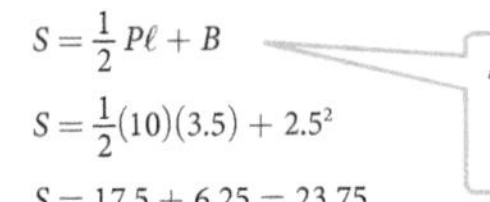

A. What do each of the variables in the formula mean in terms of the crystal pyramid?

A. See margin.

Calculate the density.

$D = \frac{250}{23.75} \approx 11$

B. What are the units of the density?

B. flakes per square centimeter

Answer the question.

Evan uses about 11 flakes per square centimeter to cover the paperweight.

C. How would the calculation of the density be different if there were no flakes added to the bottom of the paperweight? What would the density be in this situation? Instead of using the surface area, you would use the lateral area of the paperweight; about 14 flakes per square centimeter

D. How many flakes would be on one of the triangular faces of the pyramid? Explain how you found your answer.
48 flakes; You multiply the area of a triangular face by the density.

E. When Evan is rolling the paperweight in the flakes, suppose he knocks some of the flakes on the floor and loses them. How would the density of the flakes that end up on the paperweight compare to the calculated population density? Explain. See margin.

Turn and Talk

- How would the density change if the bottle of flakes contained 100 flakes instead?
- If you know the surface density is 20 flakes per square centimeter, and you use the bottle of 250 flakes to cover the surface of the square pyramid, estimate the area of the base of the pyramid if the slant height is 2 centimeters. Show your work. See margin.

Task 4 MP **Attend to Precision** Students must attend to precision when they apply clear definitions and formulas to a real-world situation, when stating the variables described in the problem, and when choosing the appropriate numbers that should be substituted in for those variables.

SUPPORT SENSE-MAKING Three Reads

Have students read the problem three times. Use the questions below for a different focus each read.

1. What is the situation about?
2. What are the quantities in the situation?
3. What are the possible mathematical questions that you could ask for this situation?

Sample Guided Discussion:

Q While solving for the surface area of the paper weight, why is the number 10 substituted in place of P instead of 2.5? Because there are 4 triangles that have a base of length 2.5, and $2.5 \times 4 = 10$.

Q In Part C, how does the work change from the work shown in Part B? In Part C, you divide 250 by the lateral area of the pyramid, 17.5, instead of dividing by the total surface area of 23.75.

Turn and Talk Discuss with students the concept of working backward and how that strategy may apply here. When solving for a variable, evaluating an expression for given values of the variables may not be enough. Sometimes an equation must be solved in order to find the missing variable.

$D = \frac{100}{23.75} \approx 4$ flakes per square centimeter

$20 = \frac{250}{S}; S = 12.5 \text{ cm}^2$

$S = \frac{1}{2}P\ell + B$

$S = 12.5, P = 4\sqrt{B}, \ell = 2$

$12.5 = \frac{1}{2}(4\sqrt{B})(2) + B$

$12.5 = 4\sqrt{B} + B$

Using a graphing calculator or table, $B \approx 4.25 \text{ cm}^2$.

ANSWERS

4.A. P is the perimeter of the base of the pyramid, ℓ is the slant height of the pyramid, which is also the height of a triangular face, and B is the area of the base of the pyramid.

4.E. The density would be less because the number of flakes will be less than 250 and the surface area is still 23.75. So, the ratio will decrease.

ONLINE

Assign the Digital On Your Own for
- built-in student supports
- Actionable Item Reports
- Standards Analysis Reports

On Your Own

Assignment Guide

The chart below indicates which problems in the On Your Own are associated with each task in the Learn Together. Assign daily homework for tasks completed.

Learn Together Tasks	On Your Own Problems
Task 1, p. 566	Problems 6–12, 19, 21, 23A, 24, 25, and 26
Task 2, p. 567	Problems 3–15, 22, 23B, and 26
Task 3, p. 568	Problems 16–18
Task 4, p. 569	Problem 20

ANSWER

5. The surface area is about 125.7 in^2. The density is about $\frac{2000}{125.7} \approx 16$; 16 flakes per square inch.

Check Understanding

Find the surface area of each figure. All pyramids have a regular base. Round to the nearest tenth, if necessary.

1. 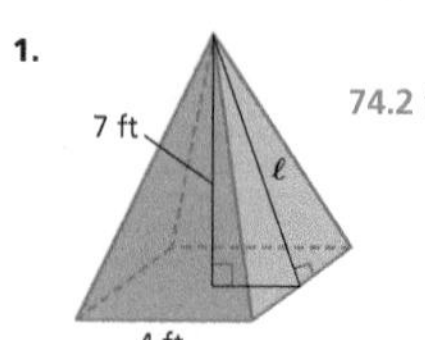

74.2 ft^2

2. 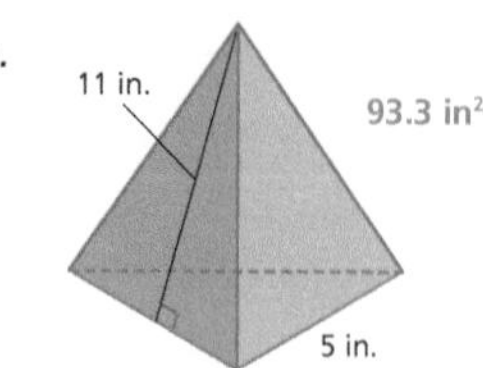

93.3 in^2

3. 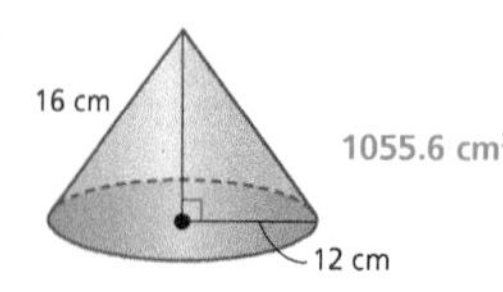

1055.6 cm^2

4. 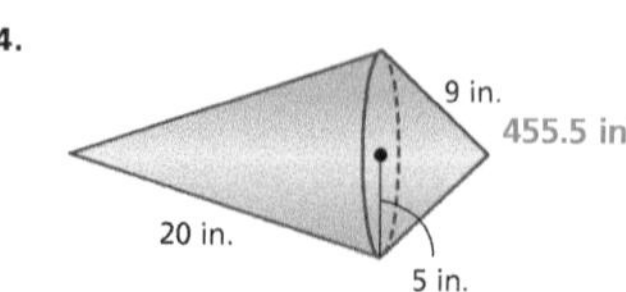

455.5 in^2

5. Jake wants to decorate a ceramic right cone. The base radius of the cone is 4 inches, and the slant height is 6 inches tall. There are 2000 golden metallic flakes inside a jar. If he uses the entire jar to cover the cone, what is the surface density of the flakes on the cone? See margin.

On Your Own

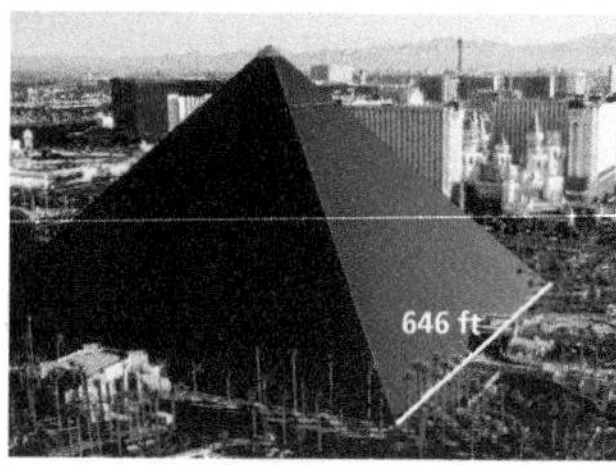

©Pegaz/Alamy

6. The Luxor Hotel in Las Vegas, Nevada is a square pyramid with a height of 350 ft.
   A. What is the slant height of one of its faces? Round to the nearest foot. 476 ft
   B. What is the lateral area of the pyramid? 614,992 ft^2
   C. (MP) **Model with Mathematics** A model of the hotel uses a scale of 1 in^2 = 1000 ft^2. To the nearest square inch, what is the surface area of the model? about 1032 in^2

Find the surface area of each regular pyramid. Round to the nearest tenth, if necessary.

7. 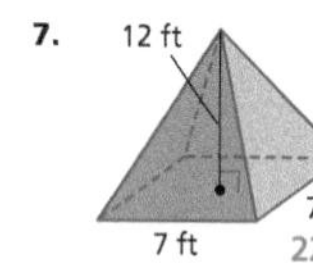

224 ft^2

8. 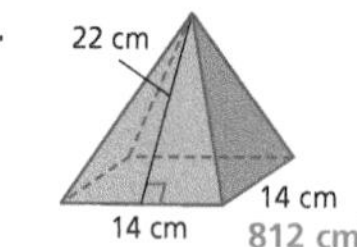

812 cm^2

9. 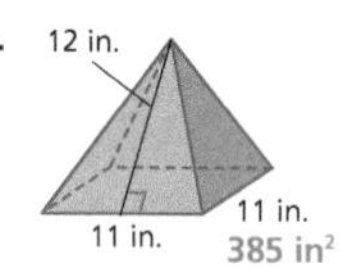

385 in^2

10. 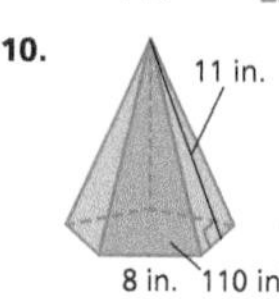

330 in^2

11. 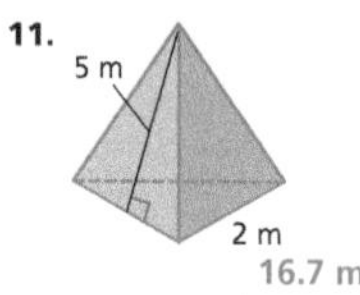

16.7 m^2

12. 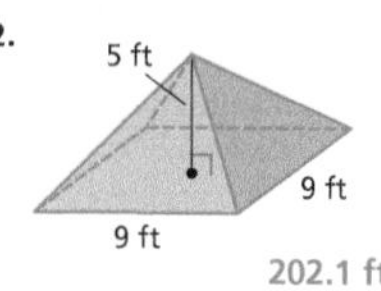

202.1 ft^2

570

3 Check Understanding

Formative Assessment

Use formative assessment to determine if your students are successful with this lesson's learning objective.

Students who successfully complete the Check Understanding can continue to the On Your Own practice.

For students who miss 1 problem or more, work in pulled small groups using the Almost There small-group activity on page 565C.

ONLINE

Assign the Digital Check Understanding to determine
- success with the learning objective
- items to review
- grouping and differentiation resources

4 Differentiation Options

Differentiate instruction for all students using small-group activities and math center activities on page 565C.

Reteach

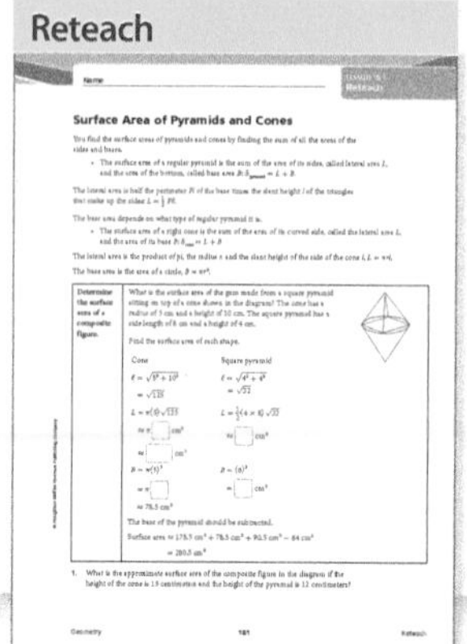

Challenge

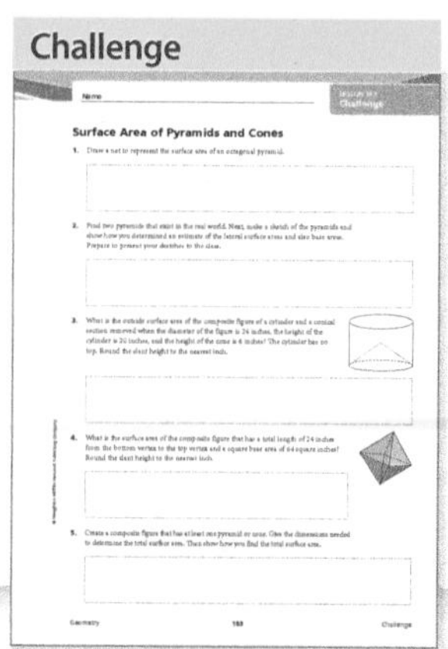

Find the surface area of each right cone. Leave your answer in terms of π.

13.

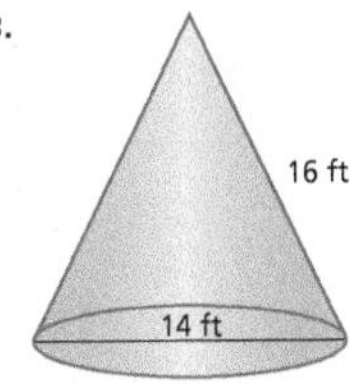

161π ft^2

14.

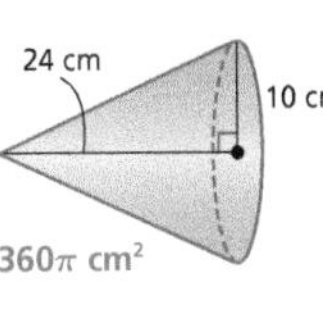

360π cm^2

15.

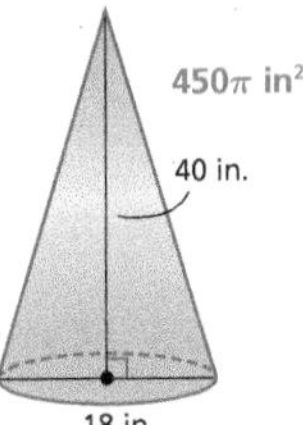

450π in^2

Find the surface area of each figure. If necessary, round to the nearest tenth.

16.

12 ft
5 ft
12 ft
12 ft

605.3 ft^2

17.

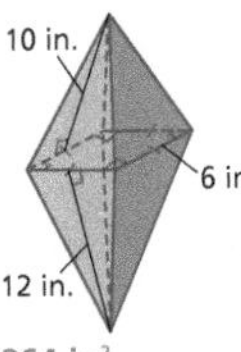

264 in^2

18.

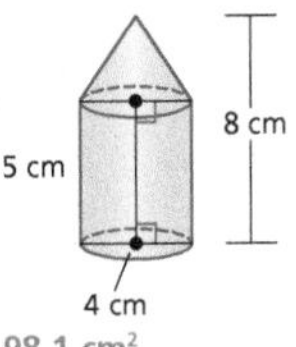

98.1 cm^2

19. (MP) **Reason** A regular pyramid has a surface area of 212 square feet. What would the surface area of the figure be if you multiplied each dimension of the pyramid by 2? Explain your reasoning. See margin.

20. Suppose a mooring buoy shown in the diagram is attached to an anchor and is pulled underwater by a chain malfunction. While underwater, the buoy collected about 1 million algae cells on its surface. What is the approximate population density of the algae on the buoy?

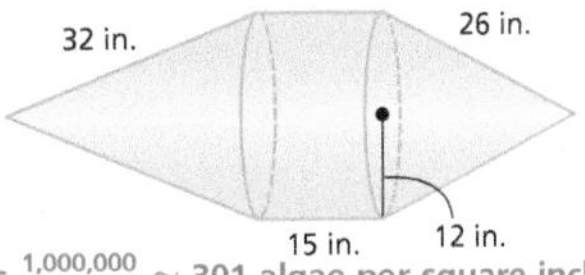

With a surface area of 1056π in^2, the density is $\frac{1{,}000{,}000}{1056\pi} \approx 301$ algae per square inch.

21. A square pyramid has a surface area of 272 square feet, and one of the side lengths of the square base is 8 feet. What is the slant height of the pyramid?
13 ft

22. A cone has a surface area of 1650 square centimeters, and the radius of the base is 15 centimeters. What is the slant height of the cone? Round to the nearest tenth.
20.0 cm

23. (MP) **Reason** An ice tray makes ice "cubes" that are square pyramids with height of 1.5 centimeters and with each side of the base 2.5 centimeters.

A. Find the surface area of the ice cube to the nearest tenth of a centimeter. 16 cm^2

B. Suppose an ice cube tray makes ice cubes in the shape of a right cone which has the same surface area as the pyramid. The diameter of the base is 3 cm. Find the height of the cone to the nearest tenth of a centimeter. 1.2 cm

Watch for Common Errors

Problem 14 Be sure students are careful when choosing the values for each variable in the equation for surface area of a cone. In this problem, the given radius is 10 and the given height is 24. Many students will substitute 24 into the formula for the slant height, *l*. Encourage students to make a list of the given information, and then a second list of the information needed. The Pythagorean Theorem is needed here to find the slant height of the cone.

Problem 19 Students tend to think that the change taking place to the dimensions of a solid results in the same change to its surface area or volume. Have students find the surface area of a pyramid with dimensions of their choice for *s* and *l*. Then have them double those dimensions and find the surface area again. This will help them discover that the surface area gets four times larger.

Questioning Strategies

Problem 21 Suppose the solid in this problem was a right cone instead of a regular pyramid with a radius of 8 feet and a surface area of 272 square feet. How would the slant height change? Justify your answer.

The slant height would be 2.8 feet.

$$S = \pi r^2 + \pi r l$$

$$272 = (8)^2\pi + \frac{1}{2}(8)l$$

$$272 = 64\pi + 8\pi l$$

$$272 = 8\pi(8 + l)$$

$$\frac{272}{8\pi} = \frac{8\pi(8 + l)}{8\pi}$$

$$10.8 = 8 + l$$

$$l = 2.8$$

ANSWER

19. 848 ft^2; If each dimension is multiplied by 2, when finding the surface area, each product would be $2 \times 2 = 4$ times as great as in the original figure, so the surface area would be 4 times as great.

5 Wrap-Up

Summarize learning with your class. Consider using the Exit Ticket, Put It in Writing, or I Can scale.

Exit Ticket 

The roof of a building is being shingled. Its shape is a square pyramid with no base. The altitude of the roof is 6 feet and the slant height measures 21 feet. How many square feet will be covered with shingles, to the nearest square foot? The segment, s, that joins the slant height and the altitude is $\frac{1}{2}$ of a side of the base of the square pyramid. The slant height, the altitude and s form a right triangle.

$6^2 + s^2 = 21^2; s^2 = 21^2 - 6^2; s^2 = 405; s = \sqrt{405} \approx 20.1$

A side of the square base is about 2×20.1 or 40.2 feet.
$S = \frac{1}{2}Pl = \frac{1}{2}(4 \times 40.2)(21) = 1688.4$ square feet

Put It in Writing

Describe the difference in the process for calculating surface area of a regular pyramid and surface area of a right cone.

I Can 

The scale below can help you and your students understand their progress on a learning goal.

4	I can use formulas for the surface area of pyramids and cones to solve real-world problems, and I can explain my steps to others.
3	I can use formulas for the surface area of pyramids and cones to solve real-world problems.
2	I can use formulas for the surface area of pyramids and cones to solve mathematical problems.
1	I can write formulas for the surface area of right cones and regular pyramids.

Spiral Review • Assessment Readiness

These questions will help determine if students have retained information taught in the past and can also prepare them for high-stakes assessments. Here, students calculate surface area of a cylinder **(18.2)**, recognize the shapes of cross sections for geometric solids **(18.1)**, calculate area of a sector **(17.3)**, and apply the area formula for a circle to find missing dimensions **(17.1)**.

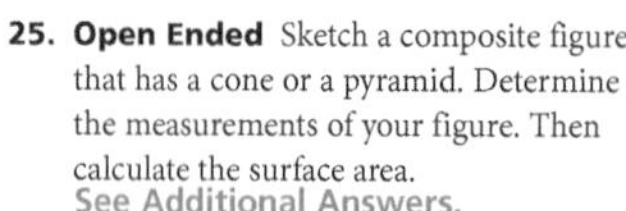

24. Suppose the greenhouse shown has two seven-sided pyramids with corresponding faces connected by trapezoids. Each of the seven trapezoids has an area of 88 ft^2. Find the amount of glass, in square feet, needed for the exterior surface area of the greenhouse. 1624 ft^2

25. Open Ended Sketch a composite figure that has a cone or a pyramid. Determine the measurements of your figure. Then calculate the surface area.
See Additional Answers.

26. Emily is designing a terrarium for a contest at her school as shown. The terrarium has a base that is a rectangular prism and a top that is a square pyramid. What is the surface area of the terrarium?
125 in^2

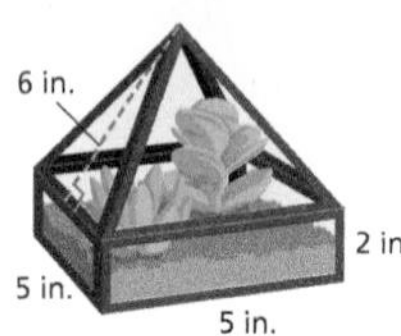

Spiral Review • Assessment Readiness

27. Find the surface area, in square feet, of the cylinder.
(A) 18π
(B) 54π
(C) 63π
(D) 72π

3 ft
9 ft

28. Which of the following could be a vertical or horizontal cross section of a cone? Select all that apply.
(A) rectangle
(B) triangle
(C) square
(D) circle
(E) trapezoid
(F) parallelogram

29. Find the area of the sector.
(A) 113.1
(B) 226.2
(C) 339.3
(D) 452.4

12
90°

30. Find the diameter of a circle with an area of 615.44 cm^2.
(A) 14 cm
(B) 28 cm
(C) 196 cm
(D) 392 cm

I'm in a Learning Mindset!

How did I benefit by giving help with finding the surface area of a composite figure? What impact did it have on my learning outcome?

Keep Going Journal and Practice Workbook

Learning Mindset

Strategic Help-Seeking Identifies Need for Help

Point out that calculating the surface area of a composite figure can be complicated because typically, not all faces of each solid are included in the total surface area. So, although decomposing the given figure into solids of which we can find the surface area is important, students need to realize which faces they should be considering when doing this. This lesson provides ample opportunity for students to recognize when they need support, utilize classmates to get that support, and give and receive feedback as either the receiver or provider of that support. These opportunities create a stronger depth of knowledge for students, as well as allow students to practice collaborating with others in a positive way. *What tasks in this lesson did you find you needed the most support on? Was there a time when a classmate depended on you for some extra support when solving a problem? Did that experience of helping another student understand a concept result in a stronger understanding of the concept for you?*

18.4 Surface Areas of Spheres

LESSON FOCUS AND COHERENCE

Mathematics Standards

- Use geometric shapes, their measures, and their properties to describe objects (e.g., modeling a tree trunk or a human torso as a cylinder).
- Apply concepts of density based on area and volume in modeling situations (e.g., persons per square mile, BTUs per cubic foot).
- Apply geometric methods to solve design problems (e.g., designing an object or structure to satisfy physical constraints or minimize cost; working with typographic grid systems based on ratios).

Mathematical Practices and Processes 

- Look for and make use of structure.
- Attend to precision.
- Model with mathematics.

I Can Objective 

I can use the formula for the surface area of a sphere to calculate the surface areas of composite figures.

Learning Objective

Find the surface area of a sphere and use the formula to find the surface area of hemispheres and composite figures in real-world problems.

Language Objective

Explain how the lateral area of a cylinder (containing a sphere of the same radius that intersects each base of the cylinder in exactly one point) and the surface area of the sphere are related.

Vocabulary

New: hemisphere

Mathematical Progressions

Prior Learning	Current Development	Future Connections
Students: • learned the formulas for area and circumference of a circle and used them to solve problems. **(Gr7, 17.1)** • solved real-world and mathematical problems involving area of two-dimensional objects composed of circles, triangles, rectangles, and rhombuses. **(Gr7, 11.4)**	**Students:** • understand how the surface area of a sphere is related to the lateral area of a cylinder. • use surface area formulas to find the surface area of spheres and hemispheres. • apply the formula for surface area of a sphere to solve a real-world problem. • use surface area formulas to find the surface area of a composite figure.	**Students:** • for a function that models a relationship between two quantities, will interpret key features of graphs and tables in terms of the quantities, and sketch the graphs showing key features given a verbal description of the relationship. **(4th- Year Course)**

PROFESSIONAL LEARNING

Using Mathematical Practices and Processes

Look for and make use of structure.

This lesson provides an opportunity to address this Mathematical Practice Standard. According to the standard, mathematically proficient students "see complicated things, such as some algebraic expressions, as single objects or as composed of several objects." In this lesson, students use the formula for surface area of a sphere as one part of finding the surface area of a composite figure. Students view the surface area of a real-world object, a silo, as being composed of the lateral area of a cylinder and the surface area of a spherical part of a hemisphere. Breaking up a problem into small, manageable pieces is a good starting point for solving the problem.

ONLINE

PROJECTABLE & PRINTABLE 

WARM-UP OPTIONS

ACTIVATE PRIOR KNOWLEDGE • Find the Volume of a Cube

Use these activities to quickly assess and activate prior knowledge as needed.

Problem of the Day 

What is the volume of a cube with a side that measures 5 inches?

$V = s^3 = 5^3 = 125 \text{ in}^3$

Quick Check for Homework

As part of your daily routine, you may want to display the Teacher Solution Key to have students check their homework.

Make Connections

Based on students' responses to the Problem of the Day, choose one of the following:

1. Project the Interactive Reteach, Grade 8, Lesson 13.4.
2. Complete the Prerequisite Skills Activity:

Have students work in pairs. Give each pair of students a paper on which is drawn a cube with a side that measures 4 inches.

- *What is a cube?* A cube is a prism with six square faces.
- *What is the formula for the volume of a prism?* $V = lwh$
- *What do you know about the length, width, and height of a cube?* They are equal.
- *If you substitute s for the length, width, and height of the cube into the formula for the volume of a prism, what is the formula for the volume of a cube with edge 4 inches?* $V = s^3 = 4^3 = 64 \text{ in}^3$

If students continue to struggle, use Tier 3 Skill 11.

SHARPEN SKILLS

If time permits, use this on-level activity to build fluency and practice basic skills.

Vocabulary Review

Objective: Students demonstrate their understanding of the term *hemisphere.*
Materials: Word Definition Map (Teacher Resource Masters)

Have students create a word definition map for the term *hemisphere.*

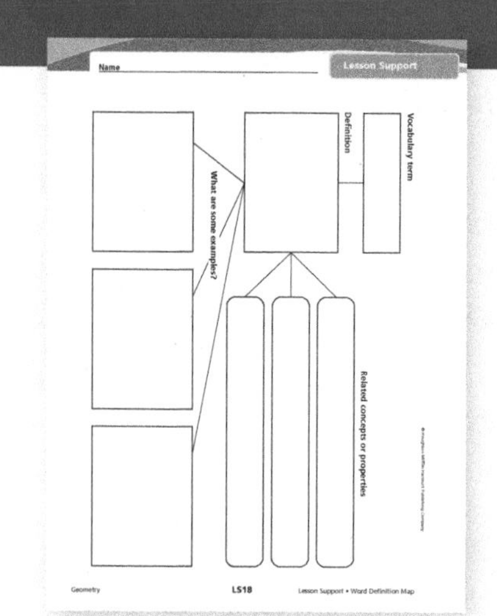

PLAN FOR DIFFERENTIATED INSTRUCTION

Small-Group Options

Use these teacher-guided activities with pulled small groups.

On Track

Materials: index cards

Provide each student with a card that has either the diameter or height of a cylinder. Have students work in pairs to calculate the surface area of the spheres that fit inside the cylinder and have the same radius.

$d = 11$ cm

$h = 7$ cm

Almost There

Materials: oranges

Provide each student with an orange and have them draw four circles, each with the diameter of their orange, on a piece of paper. Have students peel their oranges and place the peel pieces on the circles, keeping space between pieces minimal. Have students calculate the area of each circle. Ask them to explain how the combined area of the circles is the same as the surface area of the sphere. Then have them explain how they could have used a single measure, the radius of the sphere or circle, to determine the surface area of the orange.

Ready for More

Materials: balls and/or round foods of different sizes (oranges, grapefruits, cherries, tomatoes), tape measure

Have students measure the circumference of each object and calculate its radius and surface area.

Math Center Options

Use these student self-directed activities at centers or stations. Key: ● Print Resources ● Online Resources

On Track

- ● Interactive Digital Lesson
- ●● Journal and Practice Workbook
- ● Interactive Glossary (printable): **hemisphere**
- ● Module Performance Task
- ● Illustrative Mathematics: How Thick Is a Soda Can? Variation I

Almost There

- ● Reteach 18.4 (printable)
- ● Interactive Reteach 18.4
- ● RtI Tier 3 Skill 11: Use a formula to find the volume of a right rectangular prism.

Ready for More

- ● Challenge 18.4 (printable)
- ● Interactive Challenge 18.4
- ● Illustrative Mathematics: How Thick Is a Soda Can? Variation II

ONLINE Ed

View data-driven grouping recommendations and assign differentiation resources.

During the *Spark Your Learning*, listen and watch for strategies students use. See samples of student work on this page.

Approximate the Area as Four Circles

Strategy 1

The area of each strip of leather is about two circles, with a radius of 1.45 inches. Because there are two strips, the surface area of the leather is made of about four circles.

$S = 4(\pi r^2) = 4\pi(1.45)^2 \approx 26.42 \text{ in}^2$

If students . . . approximated the area as four circles to solve the problem, they are showing an understanding of the formula for the area of a circle from Grade 7.

Have these students . . . explain how they determined the equation to use and how they understood that each leather strip was approximately two circles. **Ask:**

- Q How did your two-dimensional sketch of the leather covering the baseball help you solve the problem?
- Q How did you use approximation to solve the problem?

Use a Sketch

Strategy 2

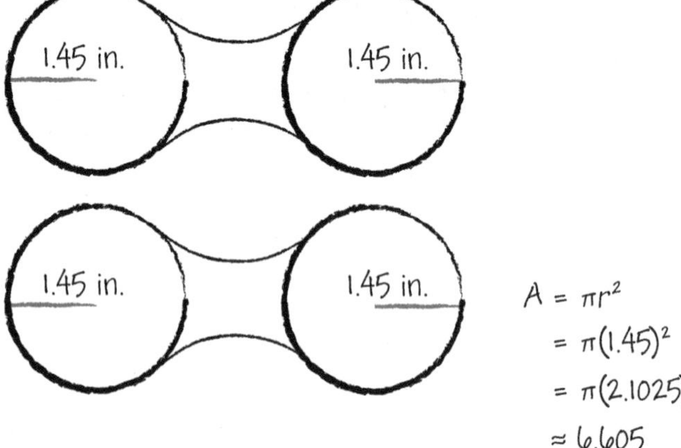

$A = \pi r^2$
$= \pi(1.45)^2$
$= \pi(2.1025)$
≈ 6.605

There are about 4 complete circles, so the area of the leather is about $4(6.605)$ or $S \approx 26.42$ square inches.

If students . . . use a sketch of four circles to solve the problem, they understand how to approximate an area of an irregular figure using figures they are familiar with. However, they failed to initially multiply by four.

Activate prior knowledge . . . by having students identify the formula for the area of a circle and ask how they could use that information without drawing a sketch. **Ask:**

- Q How can you approximate the surface area using circles without sketching the leather strips?
- Q How is the area of a circle related to its radius?

COMMON ERROR: Use the Wrong Formula for Area

$A_{circle} = 2\pi r$; $A = 2\pi(1.45) \approx 9.1 \text{ in}^2$
$4(9.1) = 36.4 \text{ in}^2$

If students . . . use the formula for circumference of a circle, then they are confusing area and circumference.

Then intervene . . . by pointing out that the units of area in this problem will be square inches, so the formula they use must also have a value that is squared, or r^2. **Ask:**

- Q What does the circumference of a circle measure?
- Q What is the formula for the area of a circle?

18.4

Surface Areas of Spheres

I Can use the formula for the surface area of a sphere to calculate the surface areas of composite figures.

Spark Your Learning

The baseballs used in Major League Baseball games must be covered in two strips of white leather, tightly stitched together.

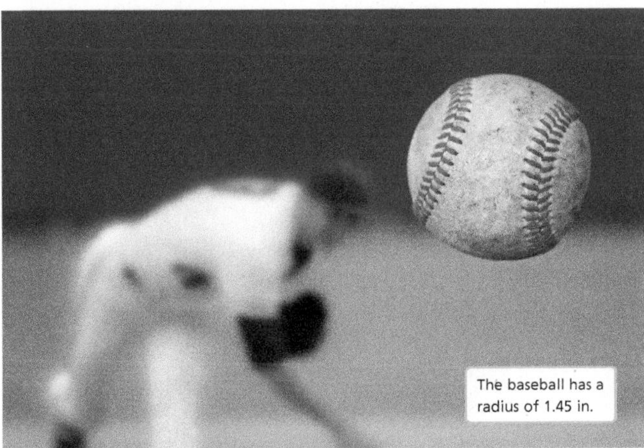

The baseball has a radius of 1.45 in.

©Brocreative/Shutterstock

Complete Part A as a whole class. Then complete Parts B–D in small groups.

A. What is a mathematical question you can ask about this situation? What information would you need to know to answer your question? A. What is a good estimate of the surface area of the ball?; a visual of the surface of the baseball cover as well as the dimensions

B. Suppose you could take the leather covering off of the baseball and lay the pieces flat. Make a sketch of what the pieces would look like. See Additional Answers.

C. To answer your question, what strategy and tool would you use along with all the information you have? What answer do you get? See Strategies 1 and 2 on the facing page.

D. Does your answer make sense in the context of the situation? How do you know? See Additional Answers.

Turn and Talk In Major League Baseball, the rules state that the circumference of a baseball must be 9 inches or greater but not exceed 9.25 inches. What are the minimum and maximum surface areas that a baseball can have? See margin.

CULTIVATE CONVERSATION • Three Reads

After each read, students answer a different question.

1 What is the situation about? The situation is about finding the surface area of a baseball, which is a sphere.

2 What are the quantities in this situation? How are those quantities related? The quantities in this situation are the radius and surface area of the baseball. In this situation, they are related because it appears that the area of four circles is about equal to the surface area of the baseball.

3 What are possible questions you could ask about this situation? How could I find the surface area of the baseball given its diameter? How could I find the surface area of the baseball given its circumference?

1 Spark Your Learning

▶ MOTIVATE 

- Have students look at the photo in their books and read the information contained in the photo. Then complete Part A as a whole-class discussion.
- Give the class the additional information they need to solve the problem. This information is available online as a printable and projectable page in the Teacher Resources.
- Have students work in small groups to complete Parts B–D.

▶ PERSEVERE

If students need support, guide them by asking:

Q **Advancing • Use Tools** Which tool could you use to solve the problem? Why choose that tool and not some other? Students' choices of tools and reasons for choosing them will vary.

Q **Assessing** What is the surface area of an object? Surface area is the total area of all faces and curved surfaces of a three-dimensional figure.

Q **Assessing** What unit of measure will be used to represent the surface area of the baseball? The unit will be in^2 because the radius is measured in inches, and area is always represented in square units.

Q **Advancing** If you know the area of a circle, how could you use your calculations for the surface area of the baseball leather to write the formula for the surface area of all spheres? I know that the area of a circle is $A = \pi r^2$. The baseball leather used, approximately four circles to make up the surface area, so the formula for surface area of a sphere is $A = 4\pi r^2$.

Turn and Talk Remind students that they will need to perform two calculations: one for minimum and one for maximum. The minimum surface area is about 25.8 in^2 and the maximum surface area is about 27.2 in^2.

▶ BUILD SHARED UNDERSTANDING

Select groups of students who used various strategies and tools to share with the class how they solved the problem. As they present their solutions, have each group discuss why they chose a specific strategy and tool.

② Learn Together

Build Understanding

Task 1 (MP) **Use Structure** Students use the formula for surface area of a sphere. Since it is composed of the lateral area of the cylinder, they then write the height of the cylinder in terms of the radius.

CONNECT TO VOCABULARY

Have students use the **Interactive Glossary** to record their understanding of the vocabulary in this task.

Sample Guided Discussion:

Q **What is lateral area?** Lateral area is the sum of the areas of the faces of a prism or pyramid or the area of the surface of a cylinder or cone.

Q **In Part B, how is the radius related to the height?** The radius is half of the height of the cylinder, $r = \frac{1}{2}h$.

Turn and Talk Encourage students to sketch the cross sections of the sphere. The cross sections would be circles stacked concentrically, where each one, starting from the middle and moving upward, is a little bit smaller than the previous one. Each circle stacked on top of the middle circle would be mirrored by a circle stacked beneath it.

Build Understanding

Investigate the Formula for the Surface Area of a Sphere

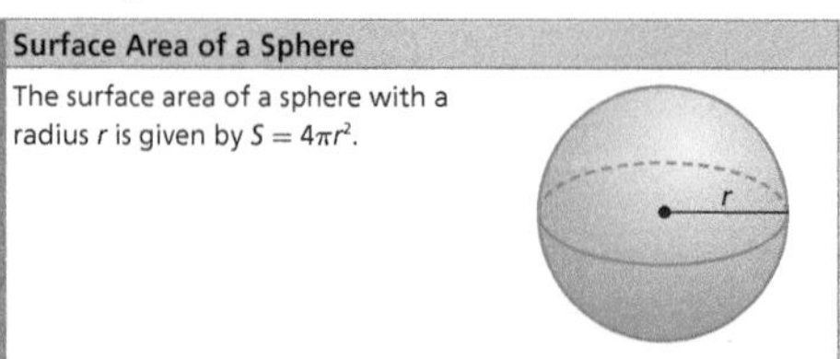

A **hemisphere** is half of a sphere. The surface area of a hemisphere is half the surface area of the related sphere plus the area of its circular base.

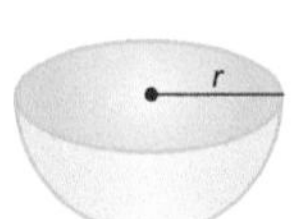

1 A cylinder and a sphere have the same radius r. The sphere fits inside the cylinder so that the sphere intersects each base of the cylinder in exactly one point.

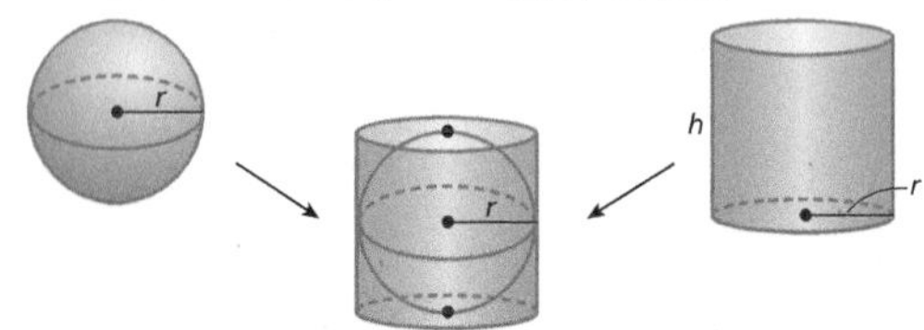

A. Write the lateral area of the cylinder in terms of the radius r and height h of the cylinder. $2\pi rh$

B. What is the height of the cylinder in terms of the radius r? $2r$

C. Write the lateral area of the cylinder in terms of r only. How is the lateral area of the cylinder related to the surface area of the sphere? $4\pi r^2$; They are the same.

D. A hemisphere with radius r fits inside a cylinder with radius r and height r. One base of the cylinder is the circular surface of the hemisphere, and the hemisphere intersects the other base in one point. Is the surface area of the hemisphere half the surface area of the cylinder? Explain. See Additional Answers.

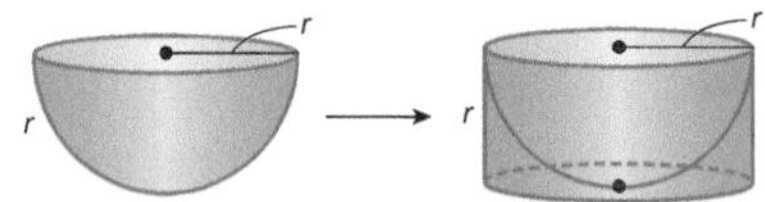

Turn and Talk Suppose you wanted to construct a solid that approximates the shape of a sphere by stacking cross sections on top of each other. Describe the cross sections that you could use. See margin.

574

LEVELED QUESTIONS

Depth of Knowledge (DOK)	Leveled Questions	What Does This Tell You?
Level 1 **Recall**	What is a hemisphere? A hemisphere is half of a sphere.	Students' answers will indicate whether they understand the definition of hemisphere.
Level 2 **Basic Application of Skills & Concepts**	What is the surface area of a sphere with radius of 5 inches? $S = 4\pi r^2 = 4\pi(5)^2 = 4\pi(25) = 100\pi$ in^2	Students' answers will indicate whether they can apply the formula for surface area of a sphere.
Level 3 **Strategic Thinking & Complex Reasoning**	What is the radius of a sphere that has the same surface area as a cylinder with a radius of 1 inch and a height of 7 inches? 2 inches Cylinder: $S = 2\pi r^2 + 2\pi rh = 2\pi(1)^2 + 2\pi(1)(7) = 16\pi$ Sphere: $S = 16\pi = 4\pi r^2$; Solve for r to get $r = 2$.	Students' answers will indicate whether they can apply the formula for the surface area of a sphere to find its radius when given the measurements of a cylinder with an equal surface area.

Step It Out

Use the Surface Area of a Sphere

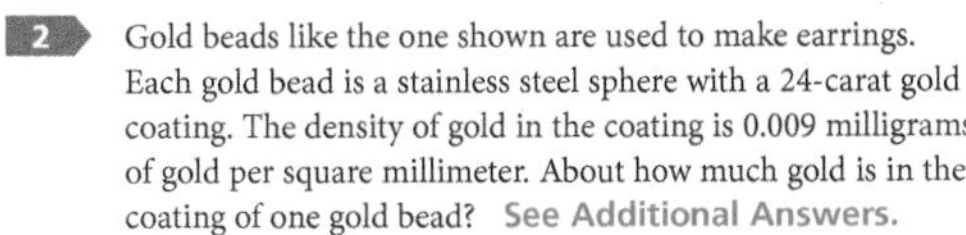

2 Gold beads like the one shown are used to make earrings. Each gold bead is a stainless steel sphere with a 24-carat gold coating. The density of gold in the coating is 0.009 milligrams of gold per square millimeter. About how much gold is in the coating of one gold bead? See Additional Answers.

The radius of each bead is 6 mm.

Surface area of sphere $= 4\pi r^2 = 4\pi(6)^2 \approx 452 \text{ mm}^2$

Total amount of gold $\approx 0.009 \text{ mg/mm}^2 \cdot 452 \text{ mm}^2 \approx 4.068 \text{ mg}$

How do the units of 0.009 and 452 produce the units of the total amount of gold?

One gold bead is covered with about 4.1 milligrams of gold.

Find the Surface Area of a Composite Figure

Recall that you can find the surface area of composite figures by using appropriate formulas to find the areas of the different surfaces of the figure.

3 Find the surface area of the silo to the nearest square foot.

Determine the different surfaces of the silo.
The surface area of the silo is composed of the lateral area of a cylinder and the surface area of the spherical part of a hemisphere.

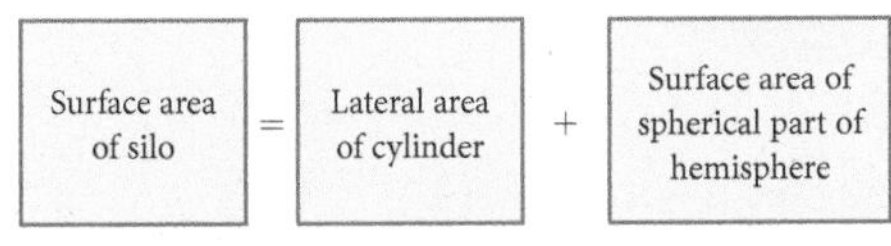

Surface area of silo = Lateral area of cylinder + Surface area of spherical part of hemisphere

Find the lateral area of the cylindrical part of the silo.

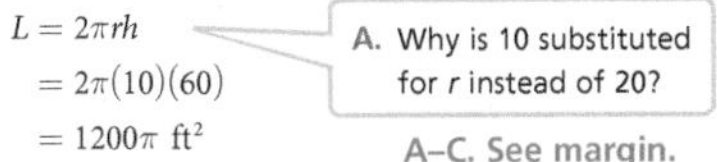

$L = 2\pi rh$
$= 2\pi(10)(60)$
$= 1200\pi \text{ ft}^2$

A. Why is 10 substituted for r instead of 20?

A–C. See margin.

Find the surface area of the spherical part of the silo. Then answer the question.

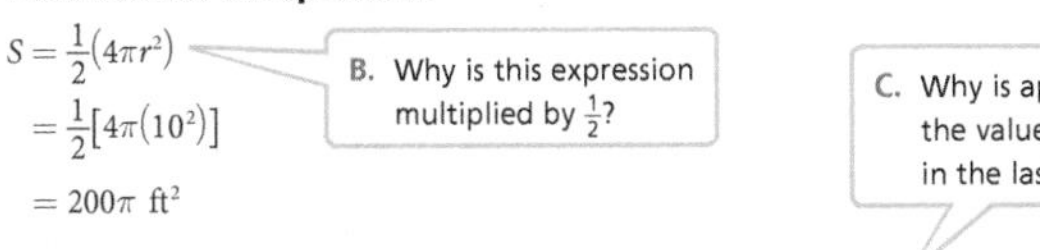

$S = \frac{1}{2}(4\pi r^2)$
$= \frac{1}{2}[4\pi(10^2)]$
$= 200\pi \text{ ft}^2$

B. Why is this expression multiplied by $\frac{1}{2}$?

C. Why is approximating the value of π done in the last step?

The surface area of the silo $= 1200\pi + 200\pi \approx 4398$ square feet.

Turn and Talk How could you find the total amount of material needed to create the silo if you are given the thickness of the silo? What are the units of this amount? See margin.

PROFICIENCY LEVEL

Beginning
Have students write the formula for surface area of a sphere and then name each variable and say what it represents.

Intermediate
Have students calculate the surface area of a sphere with radius 10 feet and then recount each step of their calculations with a partner.

Advanced
Have students calculate the surface area of spheres and hemispheres with radii of 1, 2, 4, and 8 inches, recording their calculations in a table. Have them describe in a paragraph any patterns they notice in their answers.

Step It Out

Task 2 (MP) **Attend to Precision** Students calculate the surface area of a sphere, expressed in square millimeters, to determine the density of the sphere expressed in milligrams.

Sample Guided Discussion:

Q What does the 24-karat gold coating around the bead represent? The 24-karat gold coating represents the surface area of the bead.

Q Why is the surface area of the bead multiplied by the density of the gold? The two quantities are multiplied to find the amount of gold in the coating.

$$\text{mm}^2 \cdot \frac{\text{milligrams}}{\text{mm}^2} = \text{milligrams}$$

Task 3 (MP) **Model with Mathematics** Students use formulas for the lateral area of a cylinder and the surface area of the spherical part of a hemisphere to find the surface area of a grain silo.

Sample Guided Discussion:

Q What is the radius of the silo? 10 ft; The radius is half the diameter.

Q Why do you need to find the lateral surface area of the cylinder instead of the total surface area? The total surface area would include both bases, but the bases are not part of the silo, as the top is made up of a hemisphere and the bottom is made up of the ground.

Turn and Talk Encourage students to think about what they already know and how the thickness of the material would affect the surface area of the silo. Multiply the surface area by the thickness. The units would be cubic feet.

ANSWERS

Task 3

A. The diameter of the base of the cylinder is 20 feet, so the radius in the formula is 10 which represents 10 feet.

B. A hemisphere is half of a sphere, so the surface area of the hemisphere is half of the surface area of a sphere with the same radius.

C. You want to approximate as few times as possible, so you have the most accurate answer.

ONLINE

Assign the Digital On Your Own for
- built-in student supports
- Actionable Item Reports
- Standards Analysis Reports

On Your Own

Assignment Guide

The chart below indicates which problems in the On Your Own are associated with each task in the Learn Together. Assign daily homework for tasks completed.

Learn Together Tasks	On Your Own Problems
Task 1, p. 574	Problems 7–13, 22, and 23
Task 2, p. 575	Problems 14, 16, 17, 20 and 21
Task 3, p. 575	Problems 15, 18, and 19

ANSWER

11. yes; The lateral surface area of the cylinder is greater than the surface area of the sphere.

Check Understanding

1. Explain how two-dimensional shapes can be used to represent the surface area of a sphere that has a radius of 8 centimeters. 4 circles with radius 8 cm can be drawn.

Find the surface area of each sphere. Leave answers in terms of π.

2. 784π cm^2

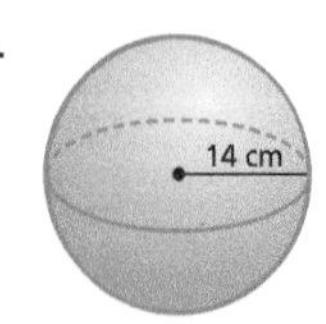

3. 484π m^2

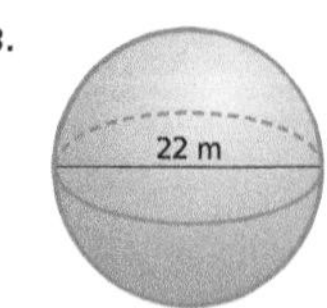

4. A polystyrene sphere representing the moon has a radius of 18 inches. The sphere is coated with extra fine glitter. If 1 ounce of glitter covers 900 square inches, how many ounces of glitter are needed to cover the sphere? Round your answer to the nearest tenth of an ounce. 4.5 ounces

Find the surface area of each composite figure. Round answers to the nearest tenth.

5.

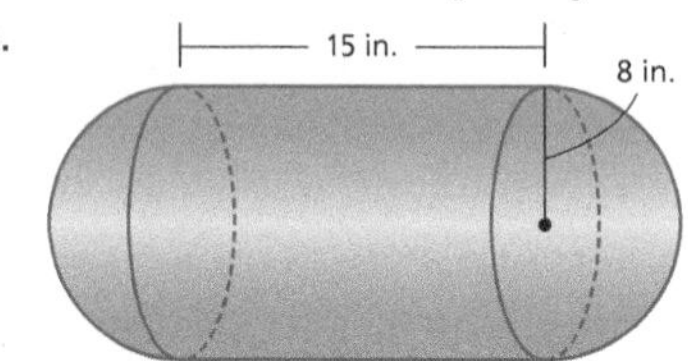

1558.2 in^2

6.

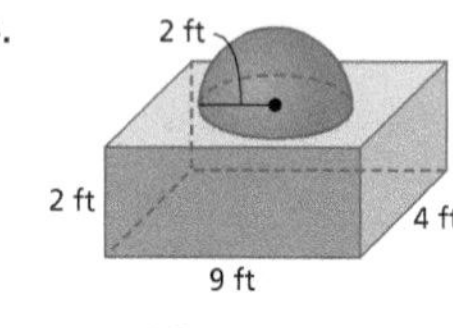

136.6 ft^2

On Your Own

Find the surface area of each figure. Leave answers in terms of π.

7. 64π ft^2

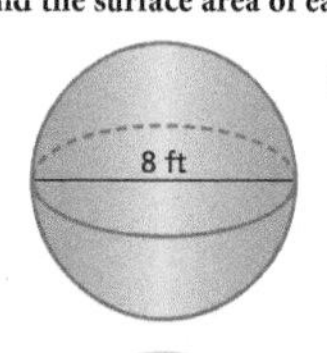

8. 324π in^2

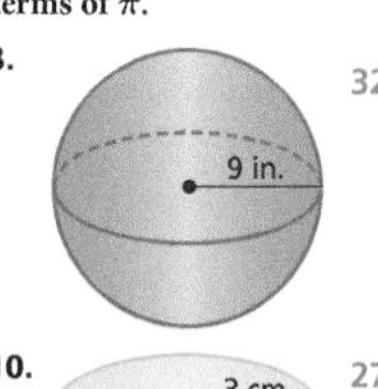

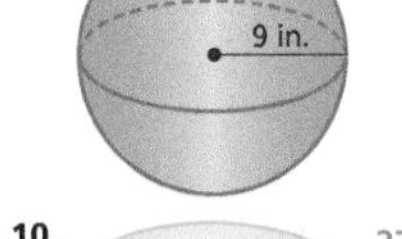

9. 432π yd^2

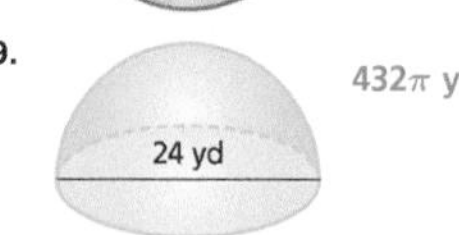

10. 27π cm^2

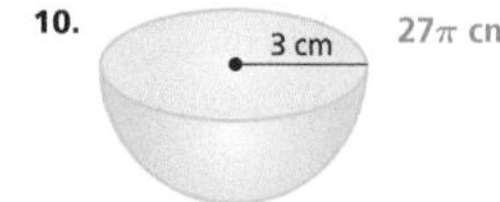

11. The figure shows how a long piece of rope was wrapped around the cylinder to completely cover its lateral surface. Then the rope is taken from the cylinder and wrapped around the sphere. Will the rope completely cover the surface of the sphere? Why or why not?
See margin.

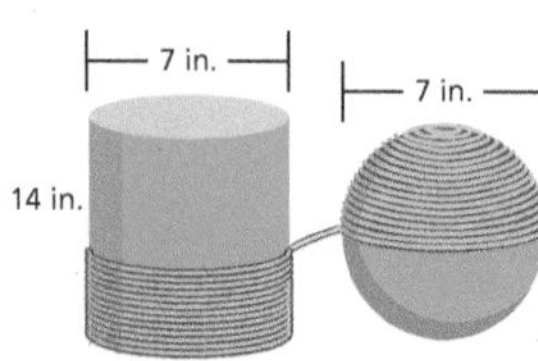

576

③ Check Understanding

Formative Assessment

Use formative assessment to determine if your students are successful with this lesson's learning objective.

Students who successfully complete the Check Understanding can continue to the On Your Own practice.

For students who miss 1 problem or more, work in a pulled small group using the Almost There small-group activity on page 573C.

ONLINE

Assign the Digital Check Understanding to determine
- success with the learning objective
- items to review
- grouping and differentiation resources

④ Differentiation Options

Differentiate instruction for all students using small-group activities and math center activities on page 573C.

Reteach

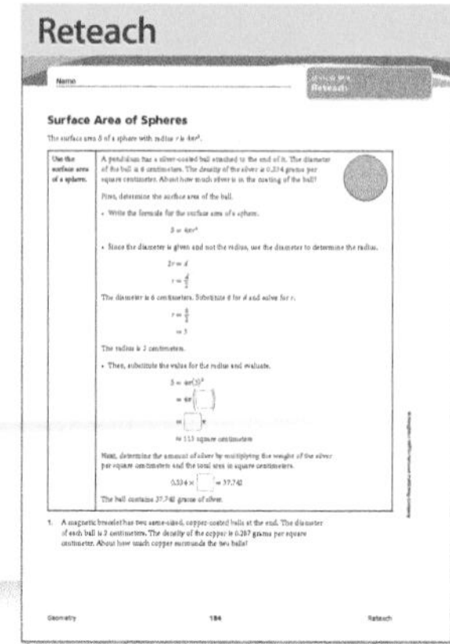

Challenge

12. **Open Ended** Draw a sphere and label the radius of the sphere. Find the surface area of your sphere. Answers will vary. Check students' drawings.

13. (MP) **Critique Reasoning** Glen and Nancy both attempt to calculate the surface area of the closed hemisphere shown. The results are shown below. Who is correct? Explain your reasoning. See margin.

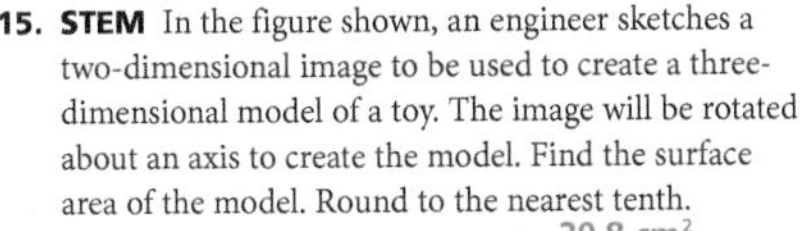

Nancy's Calculations

$S = \frac{1}{2}(4\pi r^2) + \pi r^2$

$= \frac{1}{2}[4\pi(7^2)] + \pi(7)^2$

$= 147\pi \text{ in}^2$

Glen's Calculations

$S = \frac{1}{2}(4\pi r^2)$

$= \frac{1}{2}[4\pi(7^2)]$

$= 98\pi \text{ in}^2$

14. If two pieces of ice have the same volume, the one with the greater surface area will melt faster. One piece of ice shown is a sphere, and the other piece is a cube. Given that the pieces of ice have approximately the same volume, which will melt faster? Explain. See margin.

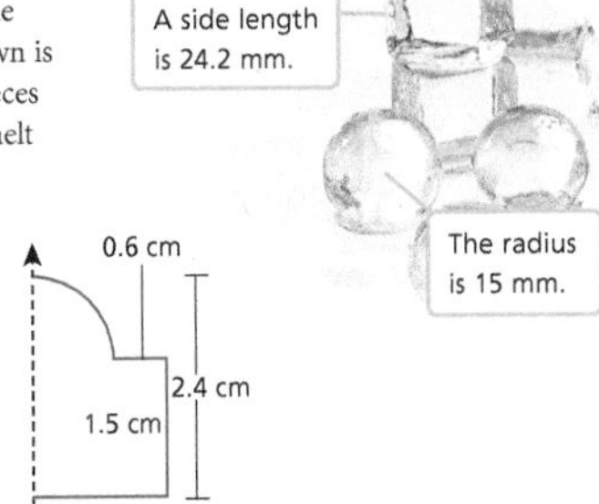

15. **STEM** In the figure shown, an engineer sketches a two-dimensional image to be used to create a three-dimensional model of a toy. The image will be rotated about an axis to create the model. Find the surface area of the model. Round to the nearest tenth. 30.8 cm²

0.6 cm
2.4 cm
1.5 cm
1.5 cm

16. (MP) **Reason** What happens to the surface area of a sphere if you triple its radius? Explain your reasoning. See margin.

17. **Art** The fountain shown is called a floating sphere fountain because a person can easily spin a granite sphere weighing thousands of pounds. This is due to the force provided by the water in the socket that the sphere rests in. 362 in²

A. In one such fountain, 5% of the surface of the sphere is in contact with the water in the socket. How many square inches of the sphere is in contact with the water?

B. The water in the socket applies a force of 15 pounds per square inch to the sphere. How much force is applied to the sphere? 5430 pounds

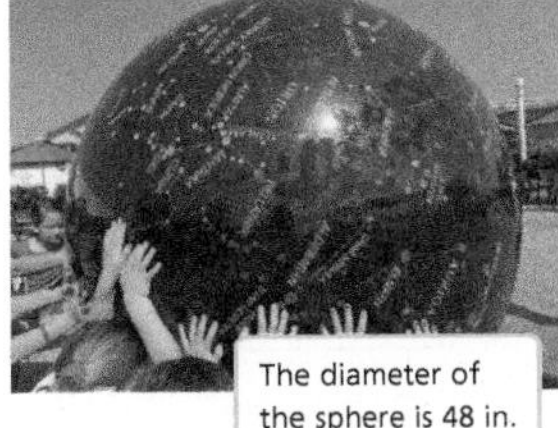

Find the surface area of each composite figure. Round answers to the nearest tenth.

18.

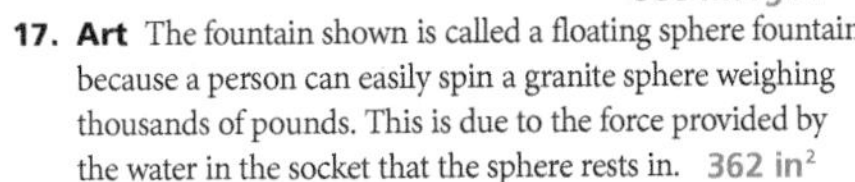

279.6 cm²

19.

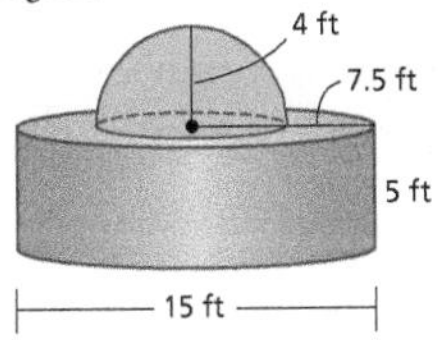

639.3 ft²

Watch for Common Errors

Problem 13 Students may recall that the surface area of a hemisphere is found using the formula $S = 3\pi r^2$, but this is not obvious in either calculation. Have students simplify the first line in each calculation to determine which accurately uses the formula.

Questioning Strategies

Problem 16 Suppose the radius is reduced to half its current measure. What happens to the surface area of the sphere? It is divided by 4.

$S = 4\pi r^2; r_{new} = \frac{1}{2}r;$

$S = 4\pi\left(\frac{1}{2}r\right)^2 = 4\pi\left(\frac{1}{4}r^2\right) = \pi r^2$

ANSWERS

13. Nancy is correct. Because the hemisphere is a closed hemisphere, the area of the circular base is included in the surface area. Glen found the surface area of the curved surface, but Nancy found the sum of the curved surface plus the circular base of the closed hemisphere.

14. The surface area of the cube is about 3514 mm². The surface area of the sphere is about 2827 mm². The cube will melt faster.

16. The surface area is multiplied by 9. Because the radius is multiplied by 3, when you square the radius, the 3 is squared, so the original surface area is multiplied by 9.

⑤ Wrap-Up

Summarize learning with your class. Consider using the Exit Ticket, Put It in Writing, or I Can scale.

Exit Ticket 

Sketch a sphere with a diameter of 8 feet. What is its surface area?

$d = 8 \text{ ft} \rightarrow r = 4 \text{ ft};$

$S = 4\pi r^2 = 4\pi(4)^2 = 4\pi(16) = 64\pi \text{ ft}^2$

Put It in Writing 

Have students research the radii of each of the planets and calculate their surface areas.

I Can 

The scale below can help you and your students understand their progress on a learning goal.

4	I can use the formula for the surface area of a sphere to calculate the surface areas of composite figures, and I can explain my calculations to others.
3	I can use the formula for the surface area of a sphere to calculate the surface areas of composite figures.
2	I can calculate the surface area of a sphere by using the formula for surface area of a sphere.
1	I can define the variables in the formula for the surface area of a sphere.

Spiral Review • Assessment Readiness

These questions will help determine if students have retained information taught in the past and can also prepare them for high-stakes assessments. Here, students apply the surface area formula for a right cone **(18.3)**, identify solids of rotation **(18.1)**, and connect the surface area formula for a right cylinder to two-dimensional figures **(18.2)**.

20. Find the radius of the globe shown. If you want to purchase a globe with a radius 50% larger, what is the approximate surface area of the new globe 6 in.; about 1018 in²

21. A dog has a ball that is 2.7 in. wide. A scientist estimates that the population of bacteria on the surface of the ball is about 41,000 cells. What is the approximate population density of bacteria on the ball? 1790 bacteria per square inch

The globe has a surface area of about 452.4 in².

22. A sphere has a surface area of about 452.2 square meters. What is the radius of the sphere? Round to the nearest tenth. 6 m

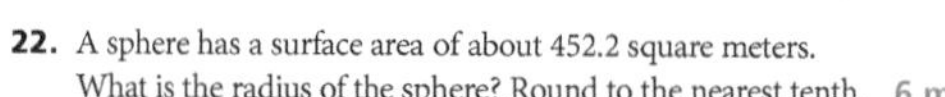

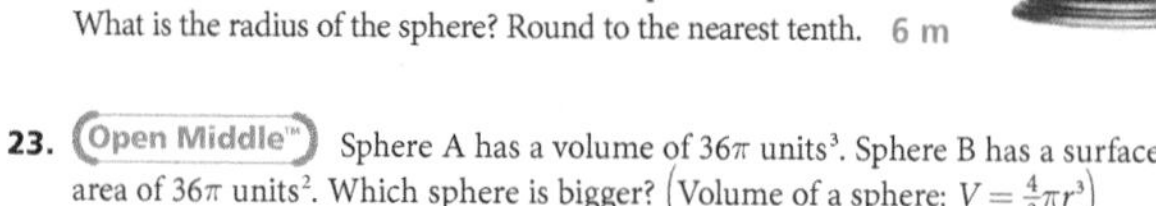

23. Open Middle™ Sphere A has a volume of 36π units³. Sphere B has a surface area of 36π units². Which sphere is bigger? (Volume of a sphere: $V = \frac{4}{3}\pi r^3$)
Spheres A and B are congruent, so neither is bigger.

Spiral Review • Assessment Readiness

24. What is the surface area of the cone?

Ⓐ 85π in²
Ⓑ 90π in²
Ⓒ 220π in²
Ⓓ 230π in²

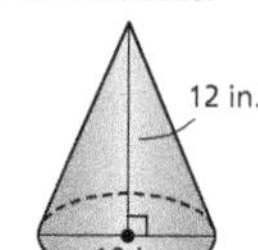

25. __?__ have an axis or lateral edges that are perpendicular to its base(s). Select all that apply.

Ⓐ Spheres
Ⓑ Oblique solids
Ⓒ Right solids
Ⓓ Cones

26. Consider the net of a right cylinder. Can you use the given shape to show the surface of the cylinder?

Shape	Yes	No
A. rectangle	**?**	?
B. circle	**?**	?
C. triangle	?	**?**
D. trapezoid	?	**?**

 I'm in a Learning Mindset!

How did I benefit by giving help with finding surface area of a composite figure? What impact did it have on my learning outcome?

Keep Going Journal and Practice Workbook

Learning Mindset

Strategic Help-Seeking Identifies Need for Help

Encourage students to help each other, when appropriate, while solving problems. Remind students that it is often useful to break a complex problem, such as finding the surface area of a composite figure, into smaller, manageable pieces. Suggest that students who are struggling with solving for the surface area of a composite figure draw separate sketches of each component of the figure and find the individual surface areas. *How can you help another student find the surface area of a composite figure? How can you break the task into smaller, easier-to-solve parts, and then put all the work together to solve the big-picture problem? How can sketching out each part of the composite figure help in solving the big-picture problem?*

Module 18

Review

Cross Sections

A cross section is the intersection of a three-dimensional figure and a plane.

A cross section that is made parallel to the base of a prism or pyramid is similar to the base.

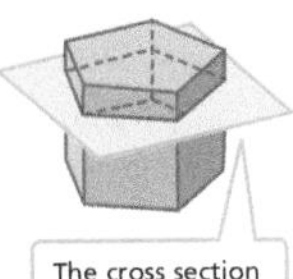

The cross section is a pentagon.

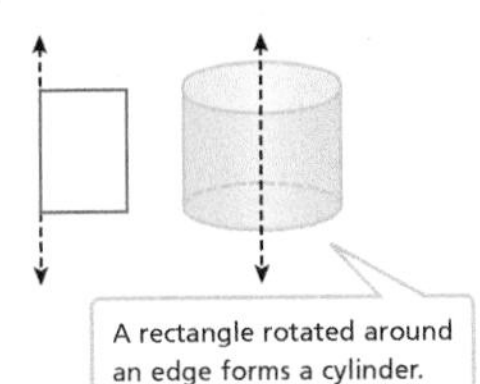

A rectangle rotated around an edge forms a cylinder.

A two-dimensional shape and an axis of rotation can determine a three-dimensional figure.

Surface Area of Prisms and Cylinders

The surface area of right prisms and right cylinders is $S = L + 2B$, where L is the lateral area and B is the area of a base.

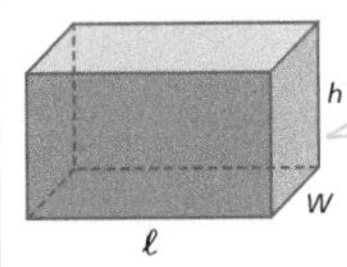

You can also find the area of each of the six faces of a prism and add them together.

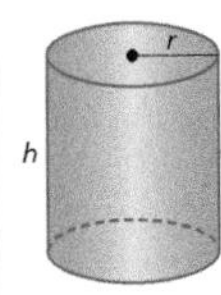

The lateral area of a cylinder is $L = 2\pi rh$. The area of the base of a cylinder is $B = \pi r^2$.

Surface Area of Pyramids and Cones

The surface area of regular pyramids and right cones is $S = L + B$, where L is the lateral area and B is the area of the base.

For a regular pyramid, the lateral area is half the product of the perimeter of the base and the slant height, or $L = \frac{1}{2}P\ell$.

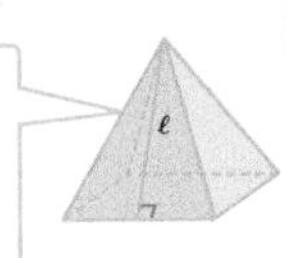

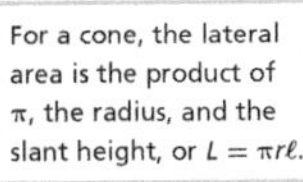

For a cone, the lateral area is the product of π, the radius, and the slant height, or $L = \pi r\ell$.

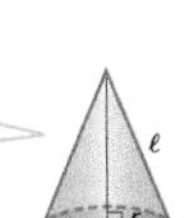

Surface Area of Spheres

The surface area of a sphere is $S = 4\pi r^2$.

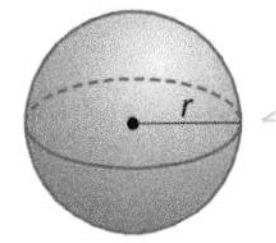

The curved part of a hemisphere has half the surface area of a sphere.

One base of the cylinder is not exposed. Omit this area.

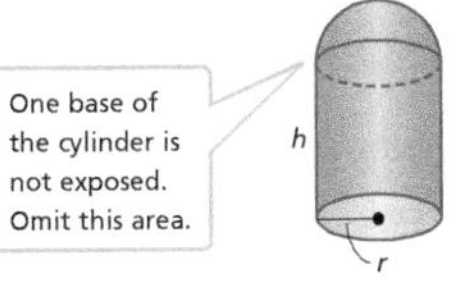

ONLINE

Assign the Digital Module Review for

- built-in student supports
- Actionable Item Reports
- Standards Analysis Reports

MODULE 18 REVIEW

Module Review

Use the first page of the Module Review to summarize and connect the main ideas of the module. Use the second page to assess students' understanding of the vocabulary, concepts, and skills presented in the module.

Sample Guided Discussion:

Q Cross Sections What is the cross section of a cylinder cut parallel to its base? a circle

Q Surface Area of Prisms and Cylinders Why is the lateral area of a cylinder equal to $2\pi rh$? The length of the lateral surface of the cylinder is the same as the circumference of the circular base. So, the length of the lateral surface is $2\pi r$. To get the lateral area of the cylinder, this length must be multiplied by the height h.

Q Surface Area of Pyramids and Cones Why is the product of the perimeter of the base and the slant height multiplied by $\frac{1}{2}$ to find the lateral area of a regular pyramid? The area of one triangular side of a regular pyramid is $\frac{1}{2}b\ell$, where b is the base of the triangle and 1 is the height of the triangle. However, because this is a regular pyramid, $b = \frac{1}{2}P$ and the lateral area is the area of all four triangular sides. So, $L = 4\left(\frac{1}{2}b\ell\right) = 2b\ell = 2\left(\frac{1}{4}P\right)\ell = \frac{1}{2}P\ell$.

Q Surface Area of Spheres How could you find the surface area of the composite figure made up of the cylinder and the hemisphere? To find the surface area from the cylinder, calculate $2\pi rh + \pi r^2$ because only one base of the cylinder is exposed. To find the surface area of the hemisphere, calculate $2\pi r^2$ because the surface area of a hemisphere is half that of a sphere. Add the two expressions to find the surface area of the composite figure, $2\pi rh + \pi r^2 + 2\pi r^2$.

Module Review continued

Possible Scoring Guide

Items	Points	Description
1–5	2 each	identifies the correct term
6	2	correctly identifies the shape of the cross section
7	2	correctly identifies the solid created
8, 9	2 each	correctly calculates the surface area of the figure
10	2	correctly compares the surface area of a cone and pyramid
11	2	correctly calculates the density of sequins on the sphere
Total points possible = 22 points		

The Unit 9 Test in the Assessment Guide assesses content from Modules 18 and 19.

Vocabulary

Choose the correct term from the box to complete each sentence.

Vocabulary
hemisphere
lateral area
oblique solid
population density
regular pyramid

1. The _?_ of a prism is the sum of the areas of the lateral faces. lateral area
2. Half a sphere is a(n) _?_. hemisphere
3. A(n) _?_ is a solid that has an axis or sides that are not perpendicular to its base. oblique solid
4. A solid whose base is a regular polygon and whose five lateral faces are congruent isosceles triangles is a(n) _?_. regular pyramid
5. The number of organisms of a particular type per square unit of area is the _?_. population density

Concepts and Skills

6. Describe the cross section. circle

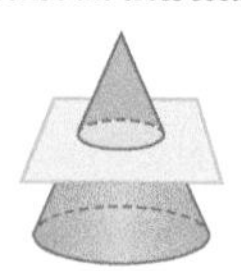

7. What solid will be created by rotating the shape about the given axis of rotation? sphere

Find the surface area of each figure. Round your answers to the nearest tenth, if necessary.

8.

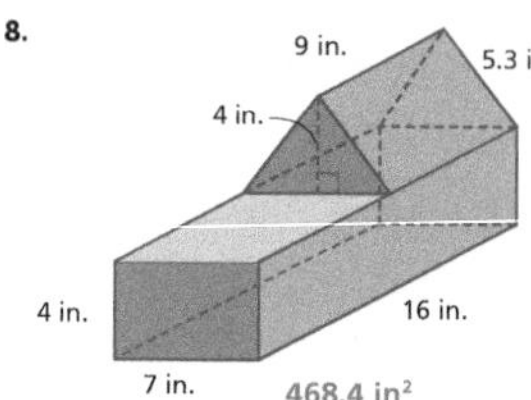

468.4 in²

9.

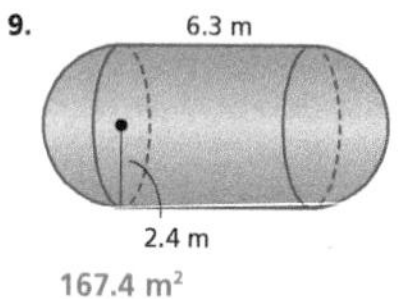

167.4 m²

10. Travis uses a 3D printer to create a regular square pyramid with a base length of 6 centimeters and a slant height of 6 centimeters. Rhonda uses the printer to create a cone with a radius of 4 centimeters and a slant height of 5 centimeters. Whose figure has a smaller surface area? Explain. Travis's figure; The surface area of the pyramid is 108 cm², and the surface area of the cone is about 113 cm².

11. MP **Use Tools** Nadia is making a scale model of a disco ball. She has a plastic-foam sphere with a radius of 5.2 inches. She has a package that contains 1400 sequins. She covers the sphere with glue and rolls it in the sequins, making sure that all the sequins stick to the sphere. Find the density of sequins on the sphere. State what strategy and tool you will use to answer the question, explain your choice, and then find the answer. approximately 4 sequins per square inch

580

DATA-DRIVEN INSTRUCTION

Before moving on to the Module Test, use the Module Review results to intervene based on the table below.

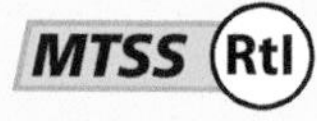

Items	Lesson	DOK	Content Focus	Intervention
6	18.1	2	Describe the cross section formed by slicing a three-dimensional figure.	Reteach 18.1
7	18.1	2	Identify the solid of rotation.	Reteach 18.1
8, 9	18.2	2	Calculate the surface area of a composite figure.	Reteach 18.2
10	18.3	3	Use the formulas for the surface area of a cone and the surface area of a square pyramid to solve a real-world problem.	Reteach 18.3
11	18.4	3	Use the formula for the surface area of a sphere to solve a problem involving density.	Reteach 18.4

Module Test

The Module Test is available in alternative versions in your Assessment Guide. All items are presented in standardized test formats.

ONLINE

Assign the Digital Module Test to power actionable reports including

- proficiency by standards
- item analysis

Form A

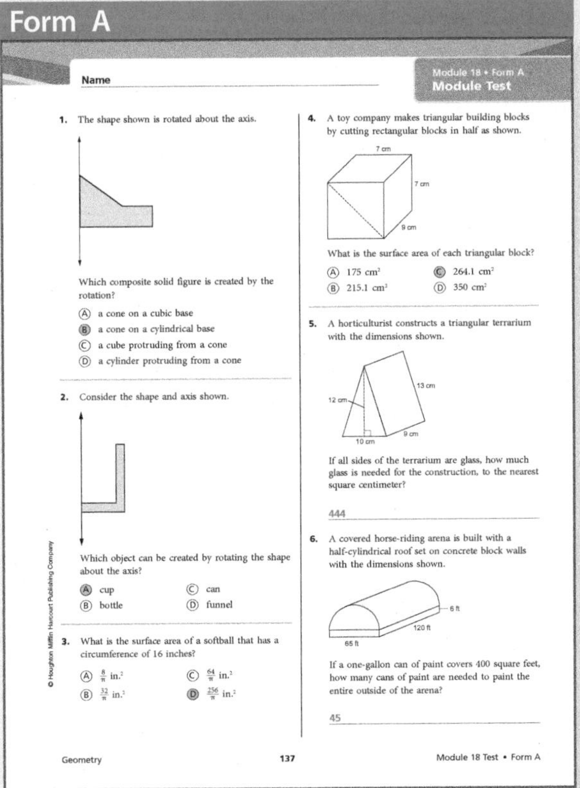

Name

Module 18 • Form A
Module Test

1. The shape shown is rotated about the axis.

Which composite solid figure is created by the rotation?

Ⓐ a cone on a cubic base
Ⓑ a cone on a cylindrical base
Ⓒ a cube protruding from a cone
Ⓓ a cylinder protruding from a cone

2. Consider the shape and axis shown.

Which object can be created by rotating the shape about the axis?

Ⓐ cup Ⓒ can
Ⓑ bottle Ⓓ funnel

3. What is the surface area of a softball that has a circumference of 16 inches?

Ⓐ $\frac{8}{\pi}$ in.^{2} Ⓒ $\frac{64}{\pi}$ in.^{2}
Ⓑ $\frac{32}{\pi}$ in.^{2} **Ⓓ $\frac{256}{\pi}$ in.^{2}**

4. A toy company makes triangular building blocks by cutting rectangular blocks in half as shown.

What is the surface area of each triangular block?

Ⓐ 175 cm^2 **Ⓒ 264.1 cm^2**
Ⓑ 215.1 cm^2 Ⓓ 350 cm^2

5. A horticulturist constructs a triangular terrarium with the dimensions shown.

If all sides of the terrarium are glass, how much glass is needed for the construction, to the nearest square centimeter?

444

6. A covered horse-riding arena is built with a half-cylindrical roof set on concrete block walls with the dimensions shown.

If a one-gallon can of paint covers 400 square feet, how many cans of paint are needed to paint the entire outside of the arena?

45

Form A

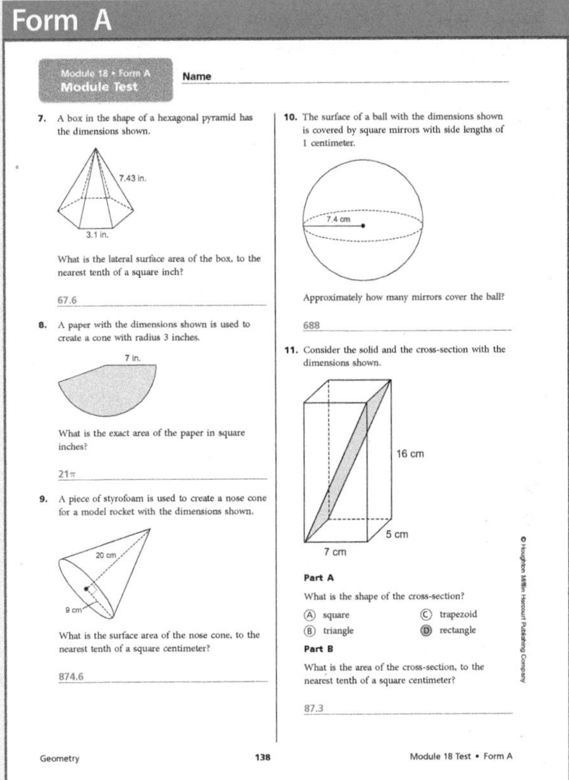

Module 18 • Form A
Module Test

Name

7. A box in the shape of a hexagonal pyramid has the dimensions shown.

What is the lateral surface area of the box, to the nearest tenth of a square inch?

67.6

8. A paper with the dimensions shown is used to create a cone with radius 3 inches.

What is the exact area of the paper in square inches?

21π

9. A piece of styrofoam is used to create a nose cone for a model rocket with the dimensions shown.

What is the surface area of the nose cone, to the nearest tenth of a square centimeter?

874.6

10. The surface of a ball with the dimensions shown is covered by square mirrors with side lengths of 1 centimeter.

Approximately how many mirrors cover the ball?

688

11. Consider the solid and the cross-section with the dimensions shown.

Part A

What is the shape of the cross-section?

Ⓐ square Ⓒ trapezoid
Ⓑ triangle **Ⓓ rectangle**

Part B

What is the area of the cross-section, to the nearest tenth of a square centimeter?

87.3

Form B

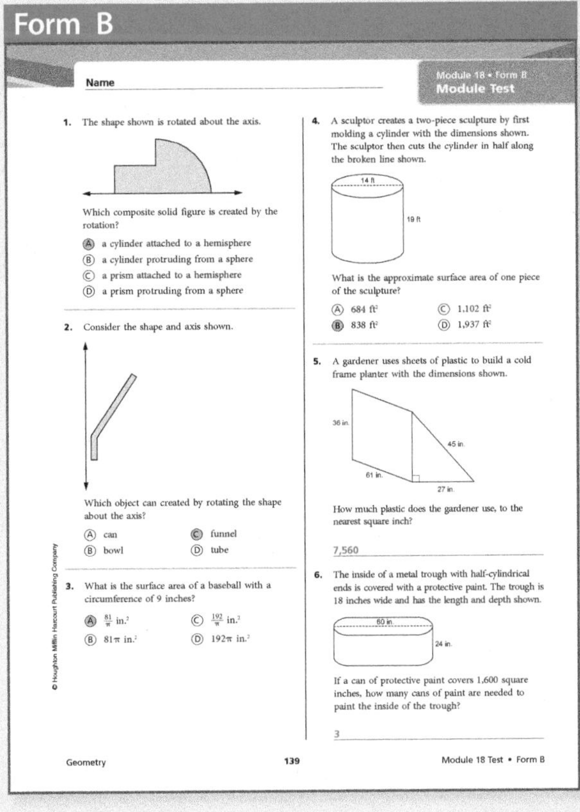

Name

Module 18 • Form B
Module Test

1. The shape shown is rotated about the axis.

Which composite solid figure is created by the rotation?

Ⓐ a cylinder attached to a hemisphere
Ⓑ a cylinder protruding from a sphere
Ⓒ a prism attached to a hemisphere
Ⓓ a prism protruding from a sphere

2. Consider the shape and axis shown.

Which object can created by rotating the shape about the axis?

Ⓐ can **Ⓒ funnel**
Ⓑ bowl Ⓓ tube

3. What is the surface area of a baseball with a circumference of 9 inches?

Ⓐ $\frac{81}{\pi}$ in.^{2} Ⓒ $\frac{192}{\pi}$ in.^{2}
Ⓑ 81π in.^{2} Ⓓ 192π in.^{2}

4. A sculptor creates a two-piece sculpture by first molding a cylinder with the dimensions shown. The sculptor then cuts the cylinder in half along the broken line shown.

What is the approximate surface area of one piece of the sculpture?

Ⓐ 684 ft^2 Ⓒ 1,102 ft^2
Ⓑ 838 ft^2 Ⓓ 1,937 ft^2

5. A gardener uses sheets of plastic to build a cold frame planter with the dimensions shown.

How much plastic does the gardener use, to the nearest square inch?

7,560

6. The inside of a metal trough with half-cylindrical ends is covered with a protective paint. The trough is 18 inches wide and has the length and depth shown.

If a can of protective paint covers 1,600 square inches, how many cans of paint are needed to paint the inside of the trough?

3

Form B

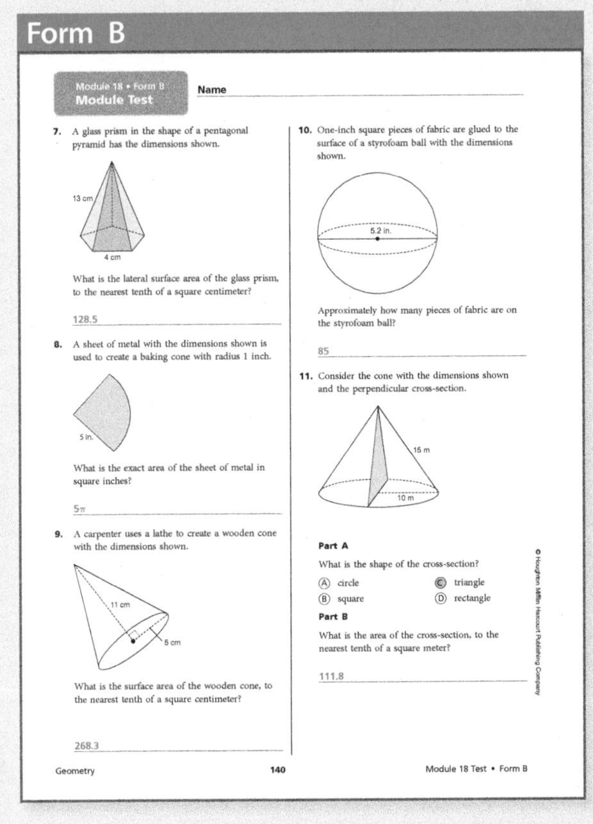

Module 18 • Form B
Module Test

Name

7. A glass prism in the shape of a pentagonal pyramid has the dimensions shown.

What is the lateral surface area of the glass prism, to the nearest tenth of a square centimeter?

128.5

8. A sheet of metal with the dimensions shown is used to create a baking cone with radius 1 inch.

What is the exact area of the sheet of metal in square inches?

5π

9. A carpenter uses a lathe to create a wooden cone with the dimensions shown.

What is the surface area of the wooden cone, to the nearest tenth of a square centimeter?

268.3

10. One-inch square pieces of fabric are glued to the surface of a styrofoam ball with the dimensions shown.

Approximately how many pieces of fabric are on the styrofoam ball?

85

11. Consider the cone with the dimensions shown and the perpendicular cross-section.

Part A

What is the shape of the cross-section?

Ⓐ circle **Ⓒ triangle**
Ⓑ square Ⓓ rectangle

Part B

What is the area of the cross-section, to the nearest tenth of a square meter?

111.8

VOLUME

Introduce and Check for Readiness

- Module Performance Task
- Are You Ready?

Lesson 19.1—2 Days

Volumes of Prisms and Cylinders

Learning Objective: Develop, relate, and use formulas for the volumes of prisms and cylinders. Use algebraic models and a graphing calculator to maximize the volumes of rectangular prisms.

Review Vocabulary: volume

Lesson 19.2—2 Days

Volumes of Pyramids and Cones

Learning Objective: Derive formulas for the volume of a cone and a pyramid, and relate them to the formulas for the volume of cylinders and prisms. Solve real-world and mathematical problems by finding the volumes of pyramids, cones, and composite figures.

Lesson 19.3—2 Days

Volumes of Spheres

Learning Objective: Derive and use a formula for the volume of a sphere. Use the volume formula to solve real-world problems including calculating capacity dimensions. Calculate the volume of composite figures involving hemispheres and other known solids.

Assessment

- Module 19 Test (Forms A and B)
- Unit 9 Test (Forms A and B)

LEARNING ARC FOCUS

Build Conceptual Understanding  Connect Concepts and Skills  Apply and Practice

TEACHING FOR SUCCESS

TEACHING FOR DEPTH: Volume

Make Connections. Students have worked in previous grade levels to find areas of two-dimensional shapes, including areas of composite figures. They have identified equivalent algebraic expressions and have also found volumes of rectangular prisms.

Now students will learn that the volume of many three-dimensional figures can be expressed as the product of the area of the base and the height. Students will recognize that $V = Bh$ and $V = \pi r^2 h$ are equivalent equations used to express the volume of a cylinder since $B = \pi r^2$. They will apply their knowledge of volume formulas to calculate the volumes of composite figures and to solve real-world problems.

In the future, some students will learn how to use integration to derive volume formulas and to calculate the volumes of irregular solids.

Mathematical Progressions

Prior Learning	Current Development	Future Connections
Students: • used formulas for the volumes of cones, cylinders, prisms, pyramids, and spheres to solve real-world and mathematical problems. • solved real-world and mathematical problems involving the areas of two-dimensional objects composed of known shapes. • solved real-world and mathematical problems involving the surface area of three-dimensional objects.	**Students:** • develop, relate, and apply formulas for the volumes of right and oblique prisms and cylinders. • algebraically model the volumes of rectangular prisms and use a graphing calculator to find the maximum volume and the associated dimensions. • apply a formula for volume to solve a real-world problem involving density. • find the volume of composite figures involving spheres. • estimate volume in a real-world situation.	**Students:** • will interpret key features for a function that models a relationship between two quantities. • will sketch graphs showing key features of a function given a verbal description of the relationship.

TEACHER TO TEACHER

From the Classroom

Pose purposeful questions. Purposeful questions asked during tasks help students really make sense of the content. Although some students prefer to memorize volume formulas, those who develop an understanding of where the formulas come from tend to have an easier time applying them in real-world situations.

I begin the module with a class discussion about the different types of three-dimensional solids. I ask, "How can these different solids have the same formula for volume? Why is the volume the same for right solids and for oblique solids with the same height and base?"

As we continue through the module, I ask, "Why is it important to work through the steps of developing formulas for cones and spheres? How does working through this process help you make sense of the formulas?"

Before applying a volume formula to find the density an object, I determine whether students have any previous experience with the term *density*. If not, I ask them to brainstorm what the word means and then to look it up to confirm or refute their ideas.

I feel this type of questioning really encourages my students to think about the processes involved in learning rather than simply focusing on the results.

LANGUAGE DEVELOPMENT • Planning for Instruction

By giving all students regular exposure to language routines in context, you will provide opportunities for students to **listen**, **speak**, **read**, and **write** about mathematical situations and develop both mathematical language and conceptual understanding at the same time.

Using Language Routines to Develop Understanding

Use the **Professional Cards** for the following routines to plan for effective instruction.

Co-Craft Questions Lesson 19.1

Students think of natural questions to ask about a given situation or about problems similar to a given task and answer the questions they have developed or problems they have created.

Three Reads Lessons 19.1 and 19.2

Students read a problem three times with a specific focus each time.

1st Read What is the situation about?
2nd Read What are the quantities in the situation?
3rd Read What are the possible mathematical questions that we could ask for the situation?

Information Gap Lesson 19.3

Students recognize when information given in a problem situation is incomplete, and they pose questions and share knowledge with others to discover any missing facts or relationships and work together to solve the problem.

Critique, Correct, and Clarify Lesson 19.3

Students correct the work in a flawed explanation, argument, or solution method and share with a partner and refine the sample work.

Connecting Language to Volume

Watch for students' use of the review term listed below as they explain their reasoning and make connections with new concepts.

Key Academic Vocabulary

Prior Learning • Review Vocabulary

volume the number of non-overlapping cubic units contained within a three-dimensional figure

Linguistic Note

Listen for students who do not distinguish the difference between *prisms* and *pyramids*. Students who confuse the two terms may also be confusing the terms *faces* and *bases*. Remind students that the base(s) of both prisms and pyramids can be any shape, but that the faces of prisms are rectangles and the faces of pyramids are triangles. Emphasize that a prism has two bases whereas a pyramid has only one. It may also be helpful to note for Spanish-speaking English learners that prism and pyramid are English/Spanish cognates: prisma and pirámide.

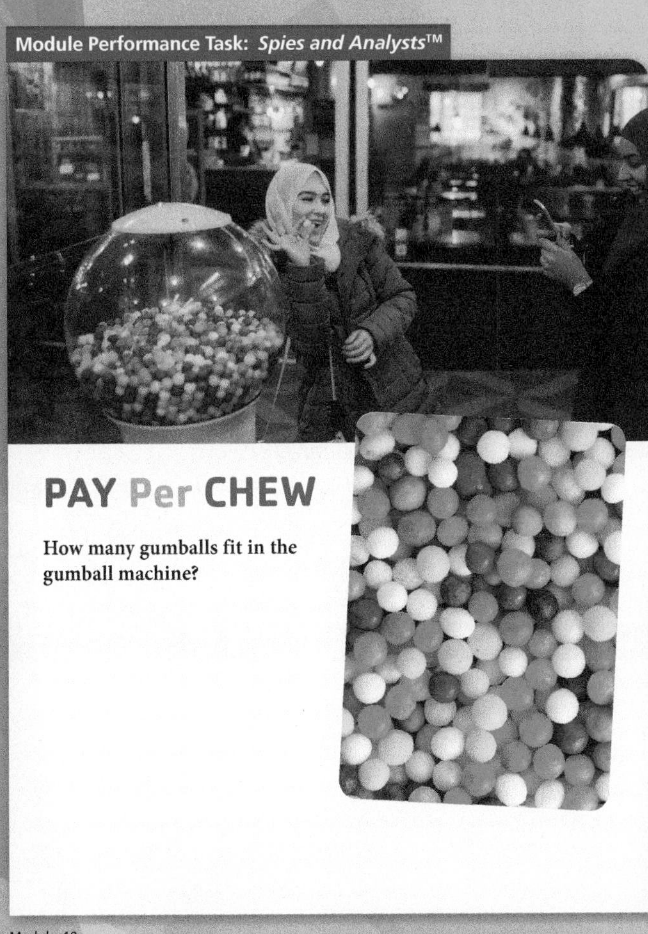

Module Performance Task: *Spies and Analysts*™
PAY Per CHEW
How many gumballs fit in the gumball machine?
(t) ©Maskot/Getty Images; (b) ©peeyun/Shutterstock
Module 19
581

PAY Per CHEW

Overview

This problem requires students to make assumptions to determine how to compute the number of gumballs the machine can hold. Examples of assumptions:

- The container that holds the gumballs can be modeled as a sphere.
- The container can be completely filled with gumballs.
- Each gumball can be modeled as a sphere.
- There are gaps of empty space between gumballs that must be accounted for.

Be a *Spy*

First, students must determine what information is necessary to know. This should include:

- the diameter or radius of the gumball container.
- the diameter or radius of a gumball.
- the percentage or fraction of space in the container that will actually contain gumballs.
- the formula for the volume of a sphere.

Be an *Analyst*

Make sure students understand that both the gumball container and the gumballs are three-dimensional shapes. Help students understand that when filling a larger sphere with smaller spheres, there are gaps between the smaller spheres.

Alternative Approaches

In their analysis, students might:

- fill containers with small spheres, such as ping pong balls, in order to better estimate the portion of space in the gumball container that the gumballs take up.
- use estimating and counting strategies to attempt to count the number of gumballs in the container in the picture in order to check the reasonableness of their answer.
- express their solution as a formula based on the radii of the gumball container and a gumball.
- consider how their solution would change if the gumball container were another shape, such as a cylinder.

Connections to This Module

One sample solution might involve students estimating the radii based on the picture and what they know about the size of gumballs and estimating the portion of space in the container that is occupied by the gumballs.

- Students apply the formula for the volume of a sphere, $V = \frac{4}{3}\pi r^3$, to find the volume of the gumball container and the volume of a gumball **(19.3)**.
- Students divide the volume of the container by the volume of a gumball to get the number of gumballs that would fit in the container with no gaps **(Grade 6, 9.3)**.
- Students use their understanding of three-dimensional figures to visualize the gaps between the gumballs **(18.1)**. They estimate of the portion of space inside the gumball machine, and they use the estimate to adjust their answer.

ONLINE

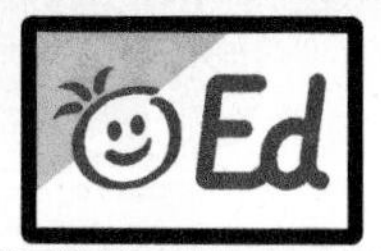

Assign the Digital Are You Ready? to power actionable reports including

- proficiency by standards
- item analysis

Are You Ready?

Diagnostic Assessment

data checkpoint

- Diagnose prerequisite mastery.
- Identify intervention needs.
- Modify or set up leveled groups.

Have students complete the *Are You Ready?* assessment on their own. Items test the prerequisites required to succeed with the new learning in this module.

Areas of Triangles Students will apply their previous knowledge of finding the area of a triangle to give an informal argument for the formula for the volume of a pyramid.

Volumes of Right Rectangular Prisms Students will build upon their skill of using a formula to find the volume of a right rectangular prism by applying Cavalieri's Principle in an informal argument to determine formulas for the volumes of solid figures.

Areas of Circles Students will apply their previous knowledge of finding the area of a circle to solve real-world problems by calculating the volumes of cylinders, cones, and spheres.

ONLINE 

Are You Ready?

Complete these problems to review prior concepts and skills you will need for this module.

Areas of Triangles

Calculate the area of each triangle.

1. 27 cm²

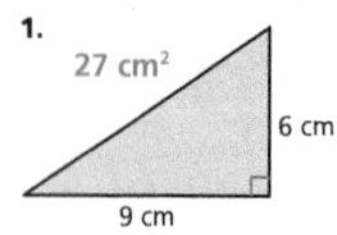

2. 48 in²

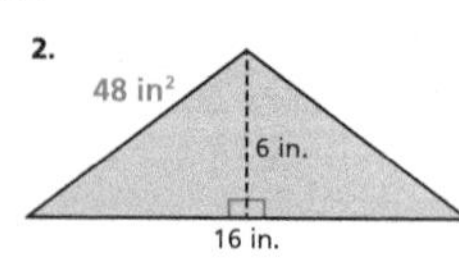

3. 0.5 m²

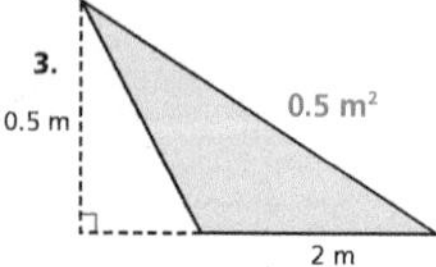

Volumes of Right Rectangular Prisms

Calculate the volume of each right rectangular prism.

4. 24 ft³

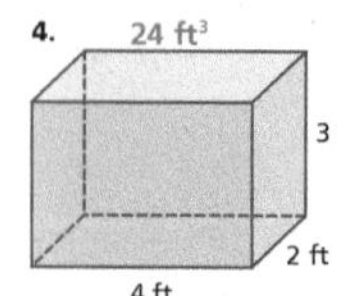

5. 16.25 in³

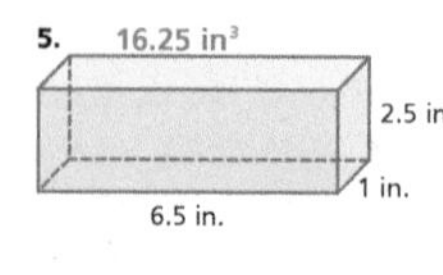

6. 945 cm³

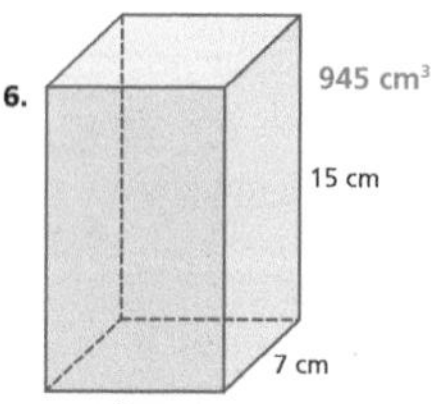

Areas of Circles

Calculate the area of each circle. Leave answers in terms of π.

7. 49π in²

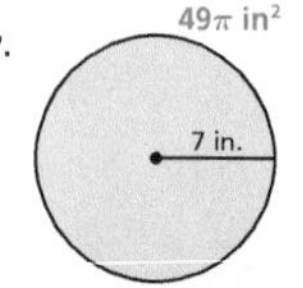

8. 0.5625π mi²

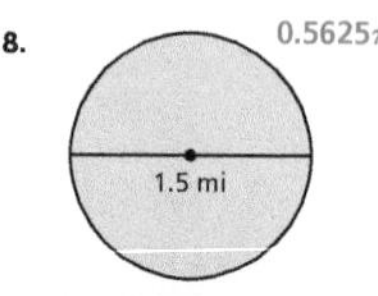

9. 18.49π m²

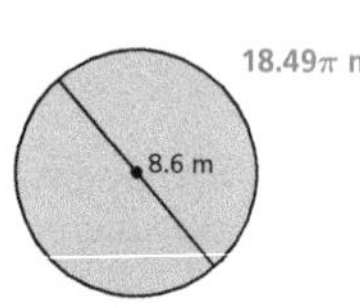

Connecting Past and Present Learning

Previously, you learned:

- to construct arguments for the circumference and area of a circle,
- to identify the shapes of two-dimensional cross sections of three-dimensional objects, and
- to apply surface area to solve design problems.

In this module, you will learn:

- to give an informal argument for the formulas for the volume of prisms and cylinders,
- to give an informal argument using Cavalieri's principle for the formulas for the volume of solid figures, and
- to use volume formulas for cylinders, pyramids, cones, and spheres to solve problems.

582

DATA-DRIVEN INTERVENTION

MTSS 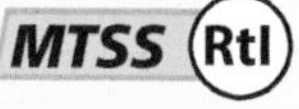

Concept/Skill	Objective	Prior Learning *	Intervene With
Areas of Triangles	Find the area of a triangle given the triangle's base and height.	Grade 6, Lesson 12.2	• Tier 3 Skill 1 • Reteach, Grade 6 Lesson 12.2
Volumes of Right Rectangular Prisms	Use a formula to find the volume of a right rectangular prism.	Grade 6, Lesson 13.2	• Tier 3 Skill 11 • Reteach, Grade 6 Lesson 13.2
Areas of Circles	Find the area of a circle given the circle's radius or diameter.	Grade 7, Lesson 10.2	• Tier 2 Skill 20 • Reteach, Grade 7 Lesson 10.2

* Your digital materials include access to resources from Grade 6–Algebra 2. The lessons referenced here contain a variety of resources you can use with students who need support with this content.

19.1 Volumes of Prisms and Cylinders

LESSON FOCUS AND COHERENCE

Mathematics Standards

- Give an informal argument for the formulas for the circumference of a circle, area of a circle, volume of a cylinder, pyramid, and cone. Use dissection arguments, Cavalieri's Principle, and informal limit arguments.
- Use volume formulas for cylinders, pyramids, cones, and spheres to solve problems.
- Apply geometric methods to solve design problems (e.g., designing an object or structure to satisfy physical constraints or minimize cost; working with typographic grid systems based on ratios).

Mathematical Practices and Processes (MP)

- Look for and make use of structure.
- Reason abstractly and quantitatively.
- Look for and express regularity in repeated reasoning.
- Use appropriate tools strategically.
- Model with mathematics.

I Can Objective 

I can develop and use formulas for the volume of a prism and a cylinder.

Learning Objective

Develop, relate, and use formulas for the volumes of prisms and cylinders. Use algebraic models and a graphing calculator to maximize the volumes of rectangular prisms.

Language Objective

Explain how to find the volume of an oblique prism or cylinder.

Vocabulary

Review: volume

Lesson Materials: graphing calculator

Mathematical Progressions

Prior Learning	Current Development	Future Connections
Students: • solved real-world and mathematical problems involving area, volume and surface area of two- and three-dimensional objects. **(Gr7, 11.4)** • applied the formulas for the volumes of cones, cylinders, and spheres in solving real-world and mathematical problems. **(Gr8, 13.1 and 13.4)**	**Students:** • develop, relate, and apply formulas for right and oblique prisms and cylinders. • develop and apply Cavalieri's Principle for volumes of oblique prisms and cylinders. • algebraically model the volumes of rectangular prisms and use a graphing calculator to find the maximum volume and the dimensions associated with that volume.	**Students:** • for a function that models a relationship between two quantities, will interpret key features of graphs and tables in terms of the quantities, and sketch graphs showing key features given a verbal description of the relationship. **(4th-year course)**

UNPACKING MATH STANDARDS

Apply geometric methods to solve design problems (e.g., designing an object or structure to satisfy physical constraints…).

What It Means to You

Students model physical objects as cylinders or prisms, and use appropriate formulas to model the volumes of the objects. For example, a box made by folding a piece of cardboard has physical constraints due to the dimensions of the cardboard, so students write algebraic expressions in the same variable for the dimensions of the box modeled as a rectangular prism, find a volume function, and then use a graphing calculator to determine the maximum volume of such a box.

Students also solve for a missing dimension of cylinders and prisms corresponding to constraints on the other dimensions and a required volume.

WARM-UP OPTIONS

ONLINE

PROJECTABLE & PRINTABLE

ACTIVATE PRIOR KNOWLEDGE • Apply the Relationship Between Area and Circumference of a Circle

Use these activities to quickly assess and activate prior knowledge as needed.

Problem of the Day

Mark has a 10-foot-long rope and a 20-foot-long rope that he is going to use to make two circles. How do the areas of the two circles compare?

The radii of the circles are

$C = 2\pi r$; $10 = 2\pi r$; $\frac{10}{2\pi} = \frac{5}{\pi} = r$ and

$C = 2\pi r$; $20 = 2\pi r$; $\frac{20}{2\pi} = \frac{10}{\pi} = r$, so their areas are

$A = \pi r^2$; $A = \pi\left(\frac{5}{\pi}\right)^2 = \frac{25}{\pi}$ square feet and

$A = \pi r^2$; $A = \pi\left(\frac{10}{\pi}\right)^2 = \frac{100}{\pi}$ square feet. The area of the larger circle is 4 times the area of the smaller circle.

Quick Check for Homework

As part of your daily routine, you may want to display the Teacher Solution Key to have students check their homework.

Make Connections

Based on students' responses to the Problem of the Day, choose one of the following:

1. Project the Interactive Reteach, Grade 7, Lessons 10.1 and 10.2.
2. Complete the Prerequisite Skills Activity:

Have students work in pairs. One student will give the other student either the diameter, circumference, or area of a hypothetical circle, and the other student will find the remaining two measures of the circle.

- *How are the diameter and the circumference of a circle related?* The circumference is the diameter multiplied by pi, $C = \pi d$.
- *How are the diameter and the area of a circle related?* One half of the diameter is the radius, $r = \frac{d}{2}$, and the area is the radius squared multiplied by pi, $A = \pi r^2$.
- *How can you find radius of a circle given the circumference or the area of the circle?* The circumference is found by $C = 2\pi r$, so the radius is $r = \frac{C}{2\pi}$. The area is found by $A = \pi r^2$, so the radius is $r = \sqrt{\frac{A}{\pi}}$.

If students continue to struggle, use Tier 2 Skills 19 and 20.

SHARPEN SKILLS

If time permits, use this on-level activity to build fluency and practice basic skills.

Vocabulary Review

Objective: Students demonstrate an understanding of the term prism.
Materials: Word Description (Teacher Resource Masters)

Have students work in pairs. Each pair should work together to build a word description graphic organizer for the term prism without looking anything up. Encourage students to work together to recall as much information as possible and to not worry too much about being correct at this point.

When all pairs have had ample time to work, have a class discussion and create a class word description on the board.

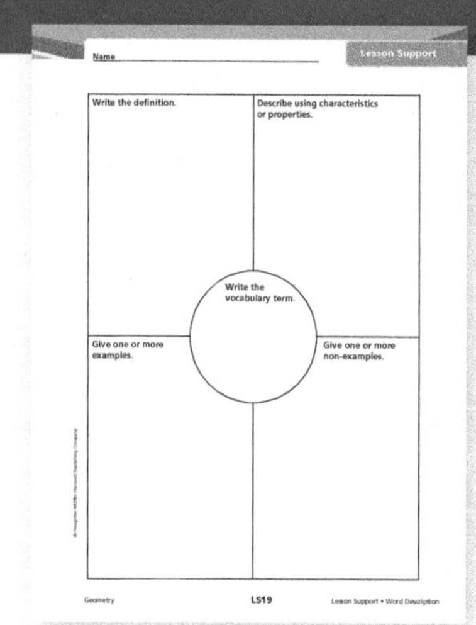

Name

Lesson Support

Write the definition.

Describe using characteristics or properties.

Write the vocabulary term.

Give one or more examples.

Give one or more non-examples.

Geometry LS19 Lesson Support • Word Description

PLAN FOR DIFFERENTIATED INSTRUCTION

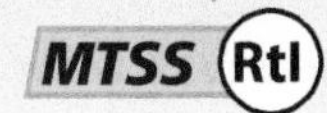

Small-Group Options

Use these teacher-guided activities with pulled small groups.

On Track

Materials: index cards

Give each group two sets of index cards. One set should have integers from 1 to 20, and the other set should be three cards with the terms *radius*, *height*, and *volume*. For each turn, students draw a pair of cards from each set, and treat the drawn integers as the values of the attributes of a cylinder named on the other two cards. If *volume* is drawn, then it must be assigned to the larger integer. Have students work together to solve for the third attribute of the cylinder.

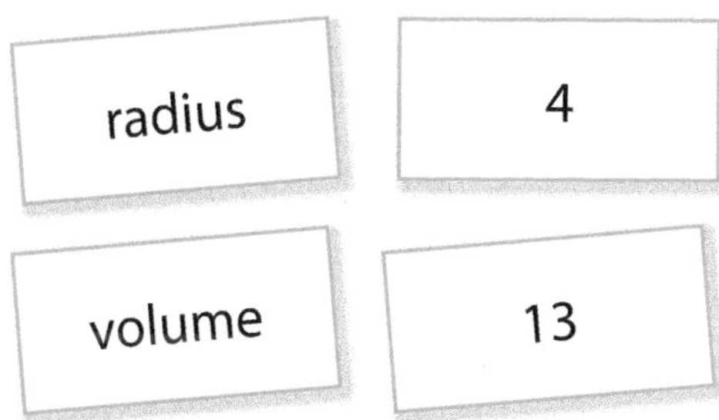

Almost There

Materials: interlocking cubes

Have students build a 3 by 4 array of cubes. Ask, "What is the volume of this rectangular prism?" Then, have them build an identical array of cubes and create a larger rectangular prism by stacking the two arrays. Ask students to calculate the volume of the larger prism and to explain their reasoning. Then do the following:

- Have them build a 2 by 4 array of cubes. Ask, "How can you recreate the large prism from earlier using this smaller array of cubes? How is this volume calculated?"
- Ask, "Is there another prism you can repeatedly stack to build the large prism? How is the volume calculated?"

Ready for More

Say, "Suppose two prisms are similar, with the dimensions of one prism *c* times as large as the other, where *c* is a positive integer." Then ask the following:

- How do the base areas and lateral areas of the prisms compare?
- How do the surface areas of the prisms compare?
- How do the volumes of the prisms compare?
- Do you think this relationship holds true for other figures such as pyramids, cones, and spheres?

Math Center Options

Use these student self-directed activities at centers or stations.

Key: ● Print Resources ● Online Resources

On Track

- ● Interactive Digital Lesson
- ●● Journal and Practice Workbook
- ● Interactive Glossary (printable): **volume**
- ● Module Performance Task

Almost There

- ● Reteach 19.1 (printable)
- ● Interactive Reteach 19.1
- ● RtI Tier 2 Skills 19 and 20: Circumferences and Areas of Circles
- ● Illustrative Mathematics: Volume formulas for cylinders and prisms

Ready for More

- ● Challenge 19.1 (printable)
- ● Interactive Challenge 19.1
- ● Illustrative Mathematics: Centerpiece

View data-driven grouping recommendations and assign differentiation resources.

During the *Spark Your Learning,* listen and watch for strategies students use. See samples of student work on this page.

Multiply the Volume of One Container

Strategy 1

The volume of a circular container is 46 in^3. The volume for the stack of 4 containers is: $V = 4(46)$

$= 184$ in^3

Replace the "4" with "n" in the above equation to create a formula for finding the volume of n containers: $V = n(46)$.

If students . . . find the volume of one container and multiply it by the number of containers to solve the problem, they are demonstrating a recognition that volume is additive from Grade 5 and are extending repeated addition of the same volume to multiplication.

Have these students . . . explain how they determined the formula to use and how they applied it. **Ask:**

- Q How did you reason about the relationship between the volume of a group of containers and the volume of each individual container?
- Q How did you know that you could multiply instead of repeatedly adding the individual volumes?

Add the Volumes of the Containers

Strategy 2

The volume of a circular container is 46 in^3. So, the volume of 4 stacked containers is:

$V = 46 + 46 + 46 + 46$

$= 184$ in^3

If students . . . add the volumes of the containers, they understand that volume is additive, but may not recognize that they can also find the volume using multiplication.

Activate Prior Knowledge . . . by having students multiply the volume of a single container. **Ask:**

- Q How do the volumes of the individual containers compare?
- Q What is a more efficient way to repeatedly add the same quantity?

COMMON ERROR: Misunderstands the Given Volume as the Total Volume

The total volume of the circular container is 46 in^3. Each container has a volume of $\frac{46}{4} = 11.5$ in^3.

If students . . . do not recognize the given volume was for an individual container in the stack, then they may not realize that they need to find the volume of 4 containers of that size.

Then intervene . . . by pointing out that the stack is formed from individual containers, and each container has a volume of 46 cubic inches. **Ask:**

- Q For which of the pieces are you given the volume, the stack or 1 container?
- Q How do the volumes of the individual containers relate to the volume of the stack?

19.1

Volumes of Prisms and Cylinders

I Can develop and use formulas for the volume of a prism and a cylinder.

Spark Your Learning

Circular containers can be stacked to save storage space. The stacked containers are in the shape of a cylinder.

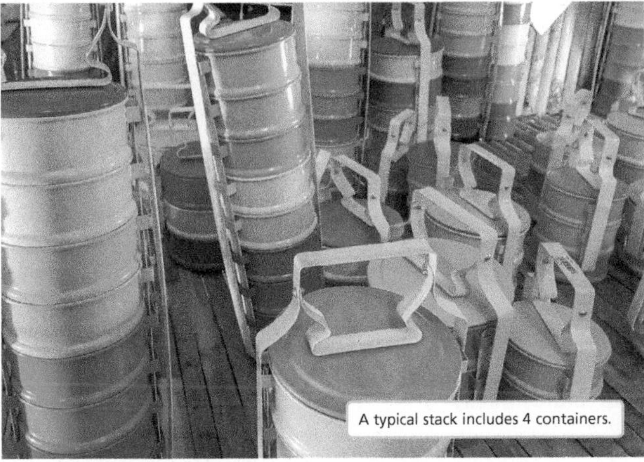

A typical stack includes 4 containers.

Complete Part A as a whole class. Then complete Parts B–D in small groups.

A. What is a mathematical question you can ask about this situation? What information would you need to know to answer your question? A. What is the volume of a typical stack of containers?; the volume of one container and the number of containers in the stack

B. What variables are involved in this situation? What unit of measurement would you use for each variable? B, D. See Additional Answers.

C. To answer your question, what strategy and tool would you use along with all the information you have? What answer do you get? See Strategies 1 and 2 on the facing page.

D. Does your answer make sense in the context of the situation? How do you know?

Turn and Talk How does finding the volume of one layer of a cylinder help you find the volume of the entire cylinder? See margin.

CULTIVATE CONVERSATION • Co-Craft Questions

If students have difficulty formulating a mathematical question about the situation in the Spark Your Learning, ask them to imagine themselves using containers like these to efficiently store ingredients. What are some natural questions to ask about this situation?

Work together to craft the following questions:

- How large is each container?
- How much of an ingredient can a container hold?
- What is the volume of this stack of containers?

Then have students think about what additional information, if any, they would need to answer these questions. **Ask:**

- Can you determine the volume of this stack of containers if you know only how many containers are in the stack? Why or why not?
- Can you determine the volume of this stack of containers if you know only the height of each container? Explain.

1 Spark Your Learning

MOTIVATE

- Have students look at the photo in their books and read the information contained in the photo. Then complete Part A as a whole-class discussion.
- Give the class the additional information they need to solve the problem. This information is available online as a printable and projectable page in the Teacher Resources.
- Have students work in small groups to complete Parts B–D.

PERSEVERE

If students need support, guide them by asking:

Q **Advancing • Use Tools** Which tool could you use to solve the problem? Why choose that tool and not some other? Students' choices of tools and reasons for choosing them will vary.

Q **Assessing** Are the containers all the same size? Yes, they are identical in size.

Q **Assessing** What mathematical term describes the amount of an ingredient a container can hold? volume

Q **Advancing** How is the volume of the stack related to the volumes of the individual container? The volume of the stack is the sum of the volumes of the individual containers, or 6 times the volume of 1 container.

Turn and Talk Help students understand that volume is an additive measure, so students can find the volume of the stack by multiplying the volume of an individual container. If you find the volume of one layer, you can multiply that volume by the total number of layers to find the volume of the entire cylindrical stack.

BUILD SHARED UNDERSTANDING

Select groups of students who used various strategies and tools to share with the class how they solved the problem. As they present their solutions, have each group discuss why they chose a specific strategy and tool.

2 Learn Together

Build Understanding

Task 1 (MP) **Use Structure** Students visualize a rectangular prism as a stack of identical layers to derive a formula for the volume of the prism. They use similar reasoning to derive a formula for the volume of the right cylinder, and then generalize both formulas as they give a rule for the volume of a right prism or right cylinder.

Sample Guided Discussion:

Q In Part E, how were your approaches for Parts C and D similar? In both parts, I multiplied the area of the base of the solid by its height to give a formula for the volume.

Task 2 (MP) **Reason** Students model a right cylinder as a stack of nickels and shift the nickels to create stacks with different shapes and equal heights. They see that the volume of the stack is invariant as the shape changes. In doing so, they build an intuitive understanding of Cavalieri's Principle.

Sample Guided Discussion:

Q How does the volume of the cylinder change as you shift the stack? Explain. It does not change, because the base area and height remain the same.

Turn and Talk Ask students whether their general reasoning about volume would change here, or if this is simply an extension of Parts A–C. If you add 3 more nickels, the volume would increase by about $3 \cdot 689 = 2067$ cubic millimeters.

Build Understanding

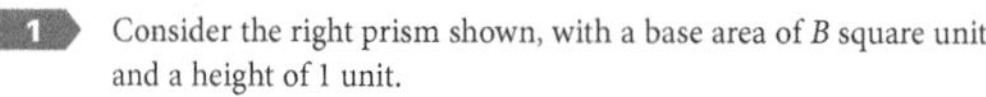

Develop a Basic Volume Formula

The **volume** of a three-dimensional figure is the number of nonoverlapping cubic units contained within the figure.

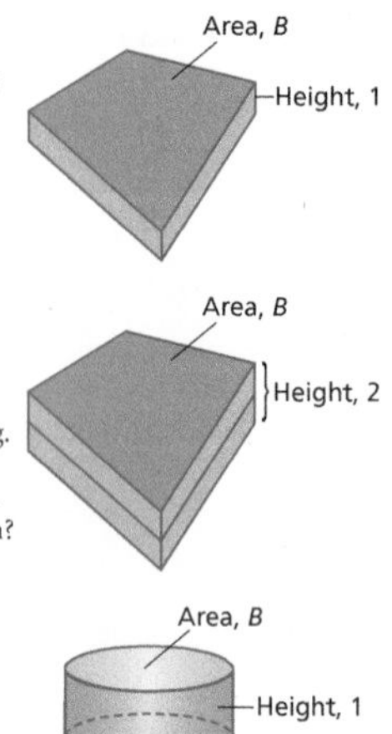

1 Consider the right prism shown, with a base area of B square units and a height of 1 unit.

A. Each side of a unit cube is 1 unit. How many unit cubes could you place in the prism? What is the volume of this prism? A, B. See Additional Answers.

Suppose you stack another identical prism on top of this prism.

B. What is the volume of the stacked prism? Explain your reasoning.

C. Suppose you continue adding prisms on top of each other until you have h prisms stacked. What is the volume of the full prism? The volume is hB cubic units.

D. Can you use the same method to find the volume of a three-dimensional figure if the base is a circle with an area of B square units and a height of 1 unit? What is the volume of a right cylinder made by stacking four of these cylinders? D, E. See Additional Answers.

E. What is a general rule you can use to find the volume of a right prism or a right cylinder? Write an equation for this rule.

Justify the Procedure for Finding Volumes of Oblique Prisms and Cylinders

You can use the same formula you found for the volume of a prism and a cylinder to find the volume of an oblique prism and an oblique cylinder.

2 Consider a stack of ten nickels.

A. The diameter of a nickel is 21.21 millimeters, and the height is 1.95 millimeters. What is the volume of a nickel? What is the volume of this stack of nickels? Write your answer to the nearest cubic millimeter. A–E. See Additional Answers.

B. How are these stacks of nickels similar to the previous stack of nickels? How are they different?

C. What is the volume of each stack of nickels? Explain your reasoning.

D. Use what you found to make a general statement about any stack of 10 nickels.

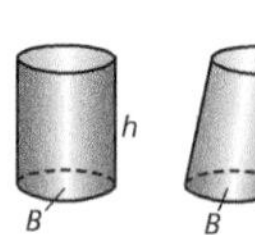
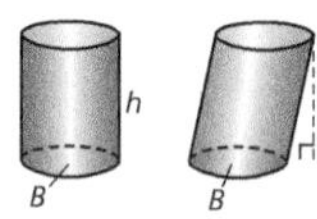

E. The volume of the right cylinder shown is 80π cubic inches. Can you determine the volume of the oblique cylinder? Explain.

Turn and Talk How does the volume of the stack of nickels change if you add three more nickels? See margin.

584

LEVELED QUESTIONS

Depth of Knowledge (DOK)	Leveled Questions	What Does This Tell You?
Level 1 **Recall**	What is the volume of a right rectangular prism with base area B square units and height h units? $V = Bh$	Students' answers will indicate whether they can recall the volume formula for a right rectangular prism.
Level 2 **Basic Application of Skills & Concepts**	What is the volume of a right cylinder with base area of 50 square inches and height 3 inches? $V = Bh = (50)(3) = 150 \text{ in}^2$	Students' answers will indicate whether they can apply the volume formula for a right cylinder with a given base area.
Level 3 **Strategic Thinking & Complex Reasoning**	You drill a circular hole through a right prism with base area B and height h, with the drill perpendicular to both bases. If the hole has a base area of B and is completely contained in both bases of the prism, what is the volume of the resulting solid? $V = B_{prism}h - B_{cylinder}h$	Students' answers will indicate whether they can use the volume formula of a right prism and a right cylinder to find the volume of a solid resulting from removing a smaller solid from another.

Investigate the Volume of a Solid Formed by Rotation

In Tasks 1 and 2, you investigated the formulas for the volumes of prisms and cylinders with vertical edges and those with slanted edges. Cavalieri's Principle guarantees that those oblique solids have the same volumes as their corresponding right solids.

Volume of a Prism or Cylinder	Cavalieri's Principle
The general formula for the volume of a prism or a cylinder is $V = Bh$, where B is the area of the base of the figure and h is the height of the figure.	If two solids have the same height and the same cross-sectional area at every level, then the two solids have the same volume.

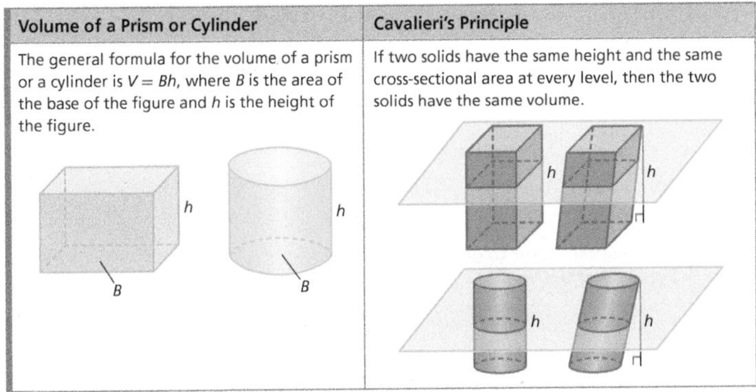

Recall that you have explored revolving two-dimensional shapes about an axis or a line to form three-dimensional figures. You have also found the surface area of these three-dimensional figures.

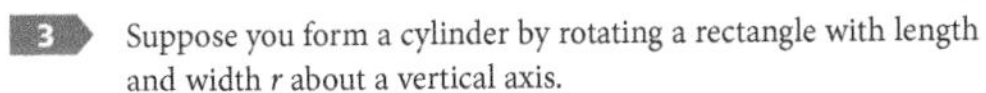

3 Suppose you form a cylinder by rotating a rectangle with length h and width r about a vertical axis.

A. What are the dimensions of the cylinder shown?
The height is h, and the radius is r.

B. Approximate the volume of the cylinder by cutting the cylinder into a series of triangular prisms. Let the short base edge of each triangular prism be 1 unit and the height of the triangular base be equal to the radius of the base of the cylinder. How many of these triangular prisms can the cylinder be divided into? Explain your reasoning. B–D. See margin.

C. What is the volume of each triangular prism?

D. Show how you can use the volume of the triangular prism and the number of triangular prisms formed to estimate the volume of the cylinder.

E. Compare the volume estimate you found in part D to the volume of a cylinder found by using the formula $V = Bh$. What do you notice?
$V = Bh = (\pi r^2)h$; The volumes are the same.

Turn and Talk Suppose you approximated the volume of the cylinder in Task 3 by dividing the cylinder into triangular prisms with a base of 2 units. How would this affect the resulting volume formula? See margin.

PROFICIENCY LEVEL

Beginning

Show students a diagram of a right circular cylinder with height 10 units, radius 4 units, and a triangular prism as in the task with short edge length 1 unit. Identify the cylinder, the prism, and their measurable aspects mentioned out loud while clearly indicating them. Then, show students a similar diagram with different measurements and have them name the measurable aspect when you say its value.

Intermediate

Have students work in pairs. One student will give the other student a positive integer representing the length of a short base edge of the triangular prism as in the task, and the other will briefly explain how to find the volume of the prism. Then, students will switch roles.

Advanced

Have students explain why they can use triangular prisms with any short base edge length, as long as they use congruent triangular prisms, to approximate the volume of a cylinder.

Task 3 (MP) **Use Repeated Reasoning** Students consider a right cylinder as a solid of rotation, and approximate the cylinder as a series of congruent isosceles triangular prisms with a common edge on the axis of the cylinder. They compare the sum of the volumes of the triangular prisms to the volume of the cylinder itself and see that the two quantities are equal.

Sample Guided Discussion:

Q In Part B, how can you use a transformation to visualize the cylinder being approximated by these prisms? I can start with one prism and repeatedly rotate it about the axis of the cylinder.

Q In Part B, is it possible to go around the cylinder exactly once with a whole number of these prisms? Explain. no; The circumference of the cylinder is $C = 2\pi r$, and the short base edge length of the triangle cannot divide C evenly since it is a rational number and C is irrational due to pi.

Q In Part C, how can you find the base area B of one triangular prism? The area for a triangle is $A = \frac{1}{2}bh$. In this prism the triangular base has base of 1, which is the short base edge, and a height that is equal to the radius of the cylinder, r. So, for this triangular base the area can be written as $B = \frac{1}{2}(1)r$.

Turn and Talk Have students examine the cylinder from the task and the method used to find the volume. Would you change the way you calculate the volume? The base area of the prisms would be doubled, but the number of prisms the cylinder is divided into would be half as many, so the volume formula would remain the same.

ANSWERS

B. $2\pi r$ triangular prisms; The circumference of the base of the cylinder is $2\pi r$ and the base edge of each triangular face of each triangular prism is 1 unit.

C. $V = \frac{1}{2}r(1)(h) = \frac{1}{2}rh$ cubic units

D. Multiply the number of triangular prisms by the volume of one triangular prism; $V = (2\pi r)\left(\frac{1}{2}rh\right) = \pi r^2 h$ cubic units

Step It Out

Task 4 **(MP) Use Tools** Students interpret a model of an open-top box in the shape of a rectangular prism and a function that represents its volume, in terms of the dimensions of the piece of cardboard from which the box is constructed. Then they use a graphing calculator to graph the function and determine the dimensions of the box with the maximum volume.

Sample Guided Discussion:

Q In Part A, where does the $(10 - 2x)$ factor come from? The cardboard is 10 inches long and the squares cut out from the corners have a side length of x, so a length of x is removed twice from the length of the cardboard. When the cardboard is folded, $(10 - 2x)$ is the length of the box.

Q In Part C, how can you interpret the coordinates of any point on the graph of $y = V(x)$? The x-coordinate is the length of the squares cut from each corner of the cardboard, and the y-coordinate is the volume of the resulting box.

Task 5 **(MP) Model with Mathematics** Students are presented with a scenario where a drinking vessel is made by cutting out a cylinder from a log approximated as a cylinder. They see how the height of the vessel is determined from the desired thickness of the walls of the vessel and the amount of water it needs to hold.

SUPPORT SENSE-MAKING **Three Reads**

Have students read the problem three times. Use the questions below for a different focus each read.

1. What is the situation about?
2. What are the quantities in the situation?
3. What are the possible mathematical questions that you could ask for this situation?

Sample Guided Discussion:

Q In Part B, what are the units of the right-hand side of the equation? Since the radius and height are both measured in inches, the units are $V = (\text{inches})^2\,(\text{inches})$

Turn and Talk Help students apply the reasoning used in the task to this variation of the problem. Ask them if the decreased thickness of the vessel results in a larger or smaller radius of the portion to be filled. Since the radius of the filled portion is larger, how should the height change if the volume is invariant? about 4.4 inches

Step It Out

Maximize the Volume of a Rectangular Prism

4 A manufacturer is building a box with no top. The box is constructed from a piece of cardboard that is 10 inches long and 15 inches wide by cutting out the corners and folding up the sides. The manufacturer wants to build a box with maximum volume.
A–C. See Additional Answers.

Write an equation for the volume of the box.

$V = x(10 - 2x)(15 - 2x)$

A. How was this equation developed?

B. What values of x are valid in this situation? Explain your reasoning.

Use a graphing calculator to graph the function for the volume of the box.

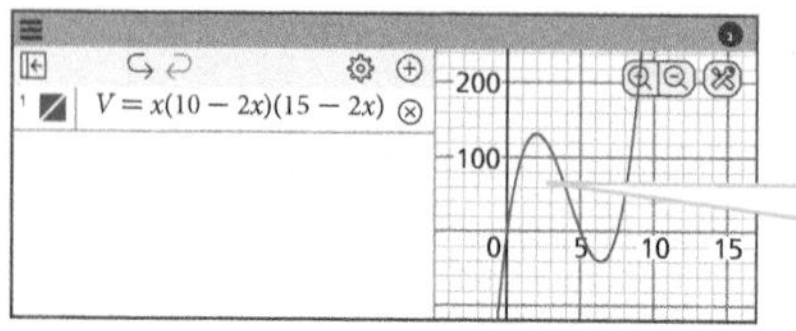

C. Explain why a box with approximate dimensions of height 2 inches, width 6 inches, and length 11 inches will give the maximum volume. What is that volume?

©Oleg Dudko/Dreamstime

Estimate Volume in a Real-World Scenario

5 A craftsperson wants to make a wooden drinking vessel for wild animals that holds approximately 3 gallons of water (1 gallon ≈ 231 cubic inches) by cutting a portion of a log and hollowing it out. One inch of wood will be retained on the bottom and lateral part (side) of the vessel. What should be the height of the vessel?

The drinking vessel will be a cylinder, so use the formula $V = \pi r^2 h$.

To find the height, substitute the values of the radius and volume in the formula for the volume and solve for h. The radius of the drinking vessel is 7 inches.

A. Why is the radius 7 inches?

The volume of the drinking vessel is 3 gallons, or about 693 cubic inches.

$$V = \pi r^2 h$$
$$693 = \pi(7^2)h$$
$$693 = 49\pi h$$
$$\frac{693}{49\pi} = h$$
$$h \approx 4.5$$

B. Why is 693 cubic inches used for the volume instead of 3 gallons? A–C. See margin.

C. Explain why the height of the vessel is 1 inch more than the height of the water.

So, the drinking vessel will be about 5.5 inches high.

Turn and Talk What would be the height of the drinking vessel if the amount of wood retained on the bottom and side were 0.5 inch? See margin.

586

ANSWERS

5A. 1 inch of the log will be left on the outside of the drinking vessel, so the radius of the vessel is 1 inch less than the radius of the log.

5B. The dimensions of the drinking vessel are given in inches, so you would want the volume in cubic inches, instead of gallons, when performing calculations.

5C. 4.5 inches is the height of the area that will be filled with water. There is a one-inch bottom to the vessel, so its height should be about 5.5 inches.

Check Understanding

1. Explain how you can find the volume of a prism or cylinder.
Multiply the area of the base of the figure by the height of the figure.

Find the volume of each figure. Round your answer to the nearest hundredth.

2. 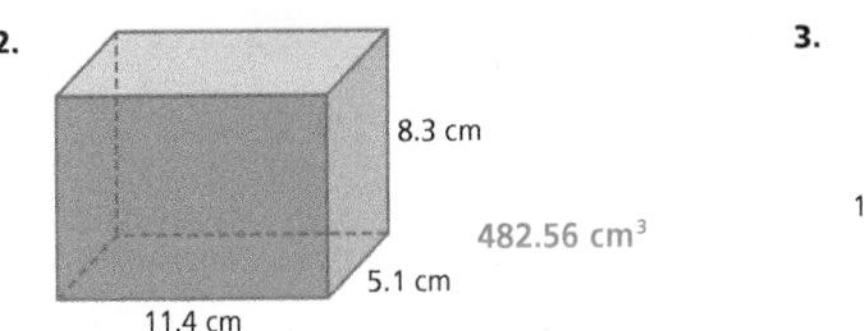

482.56 cm^3

3.
9.2 in.
14 in.
3722.66 in^3

4. The diameter of a dime is 17.91 millimeters, and the height is 1.35 millimeters. What is the volume of a dime? What is the volume of an oblique cylinder formed by stacking 50 dimes? Round your answers to the nearest hundredth. See margin.

5. A rectangle with a height of 6 inches and a width of 6 inches is rotated around an axis along one side. Find the volume of the solid that is created by this rotation. Round your answer to the nearest hundredth. 678.58 in^3

6. Find the maximum volume of a box with side lengths of x, $24 - 2x$, and $16 - 2x$. What are the approximate dimensions that produce the maximum volume and what is that volume? Round all answers to the nearest hundredth. See margin.

7. Mike has a shoe box that is 14.75 inches long, 10 inches wide, and 5.5 inches tall. What is the volume of his shoe box? 811.25 in^3

On Your Own

8. (MP) **Critique Reasoning** Francisco believes that the volume of an oblique cylinder with a height of 3 inches and a radius of 1 inch is the same as the volume of a right cylinder with a height of 3 inches and a radius of 1 inch. Is Francisco correct? Explain your reasoning. See Additional Answers.

9. A rectangle is rotated about a line containing one of its long sides. The rectangle has a length of 5 inches and a width of 2 inches. Find the volume of the solid that is created by this rotation. What type of solid is created? 62.83 in^3; right cylinder

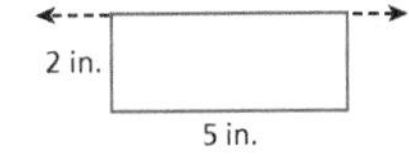

10. (MP) **Reason** Deanna is studying two cylindrical blocks. The radius of the base of one block is 6 inches. The diameter of the base of the other block is 10 inches. Each block has a height of 12 inches. Without calculating, identify which block has the greater volume. Explain your reasoning. See Additional Answers.

11. Ramon has a cylindrical can of chili with the dimensions shown. How much chili can fit in the can? Round your answer to the nearest hundredth. 35.26 in^3

ONLINE

Assign the Digital On Your Own for

- built-in student supports
- Actionable Item Reports
- Standards Analysis Reports

On Your Own

Assignment Guide

The chart below indicates which problems in the On Your Own are associated with each task in the Learn Together. Assign daily homework for tasks completed.

Learn Together Tasks	On Your Own Problems
Task 1, p. 584	Problems 10, 12–15, 29, 30, and 33
Task 2, p. 584	Problems 8, 16–19, and 24
Task 3, p. 585	Problem 9
Task 4, p. 586	Problems 20, 21, and 27
Task 5, p. 586	Problems 11, 22, 23, 25, 26, 28, 31, and 32

ANSWERS

4. Volume of one dime: about 340.11 mm^3; volume of 50 dimes: about 17,005.32 mm^3

6. Maximum volume is about 540.83 cubic units. Dimensions of the box are about 3.14 units by 17.72 units by 9.72 units.

8. Francisco is correct; By Cavalieri's Principle, if two solids have the same height and the same cross-sectional area at every level, then the two solids have the same volume.

3 Check Understanding

Formative Assessment

Use formative assessment to determine if your students are successful with this lesson's learning objective.

Students who successfully complete the Check Understanding can continue to the On Your Own practice.

For students who miss 1 problem or more, work in a pulled small group using the Almost There small-group activity on page 583C.

ONLINE

Assign the Digital Check Understanding to determine

- success with the learning objective
- items to review
- grouping and differentiation resources

4 Differentiation Options

Differentiate instruction for all students using small-group activities and math center activities on page 583C.

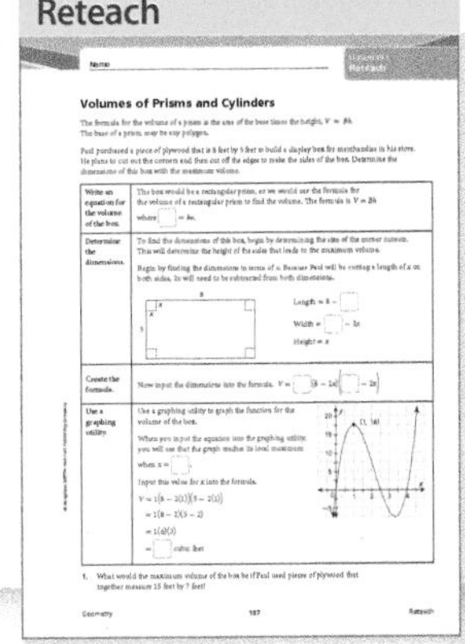

Questioning Strategies

Problem 17 Suppose that the prism has base dimensions as shown, but now the height is unknown. The slanted edges have length 12 in., are perpendicular to the 6.8-in. edges, and form 70° angles with the 7.2-in. edges. What is the volume of the prism? Hint: Use a trigonometric ratio to find the height. The height of the prism can be found by calculating $h = 12 \sin 70°$ in. The volume is then $V = (7.2)(6.8)12 \sin 70° \approx 552.1$ in^3.

Watch for Common Errors

Problem 22 Students may incorrectly multiply the volume of the water by its mass when finding the density of the water. They need to understand that density is the ratio of unit of mass per unit of volume (grams per cubic centimeter). Thinking of a very "dense" object as having a large mass squeezed inside a small volume can help them remember the relationship between the three measures.

ANSWERS

20. Maximum volume is about 262.68 in^2. Dimensions of box are about 2.43 in. by 7.14 in. by 15.14 in.

21. Maximum volume is about 788.28 in^2. Dimensions of box are about 3.38 in. by 9.24 in. by 25.24 in.

22. Volume is 20,000 cm^3; Density is 1 gram per cubic centimeter.

Find the volume of each prism or cylinder. Round your answer to the nearest hundredth.

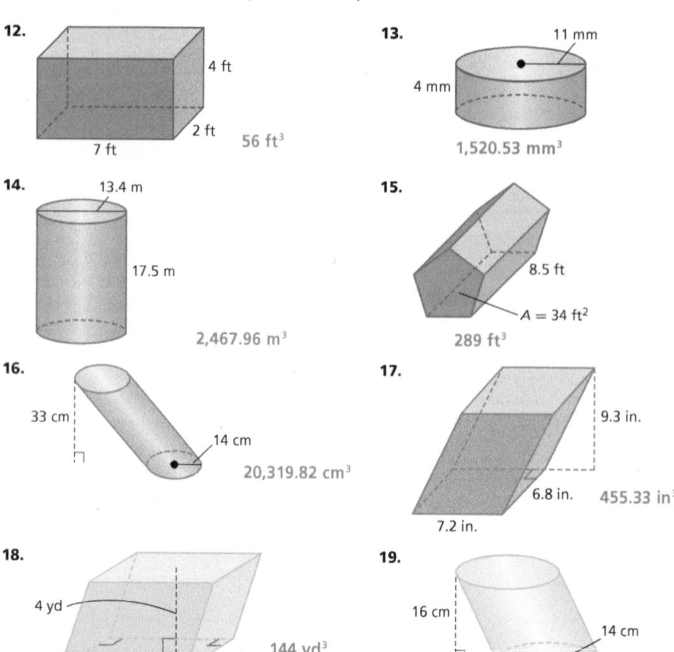

Find the maximum volume of each box and the dimensions that produce the maximum volume. Round all answers to the nearest hundredth. 20–22. See margins.

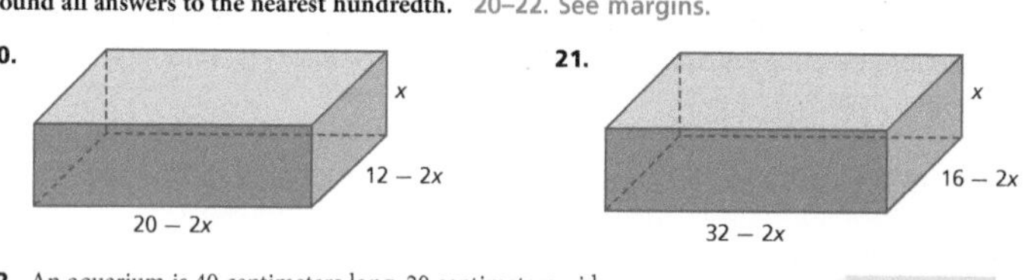

22. An aquarium is 40 centimeters long, 20 centimeters wide, and 25 centimeters high. When filled with water, the mass of the water in the aquarium is 20 kilograms. What is the volume of the aquarium? What is the density of the water in the aquarium in grams per cubic centimeter?

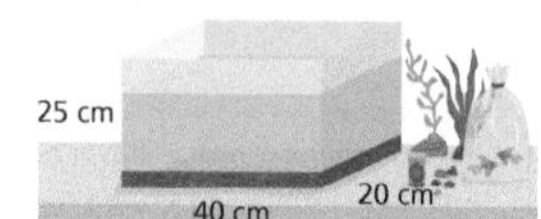

588

Find the volume of each composite figure. Round your answer to the nearest hundredth.

23.

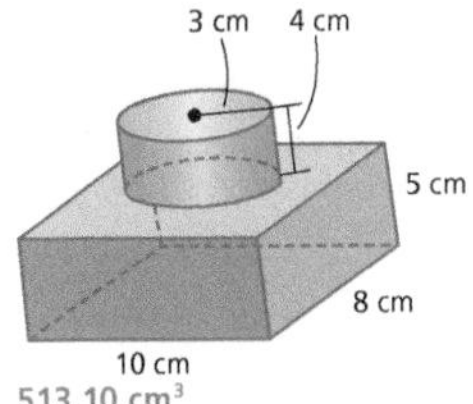

513.10 cm^3

24.

1165.45 m^3

25.

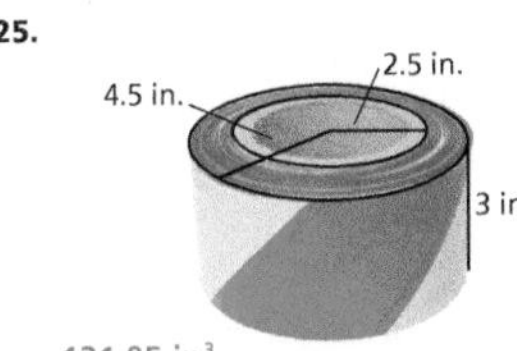

131.95 in^3

26.

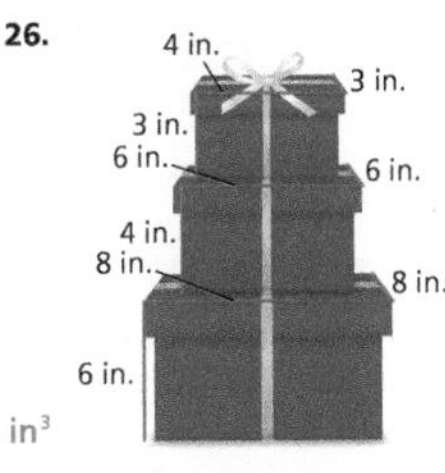

564 in^3

27. Dwight is digging a hole for a post for a mailbox. The post is 4 inches wide and 4 inches long. The post must be buried between 12 inches and 18 inches deep in the ground. What is the minimum and maximum volume of the hole that Dwight must dig? minimum: 192 in^3, maximum: 288 in^3

Find the missing dimension of each figure. Round your answer to the nearest tenth.

28. $V = 706.86$ ft^3 5 feet

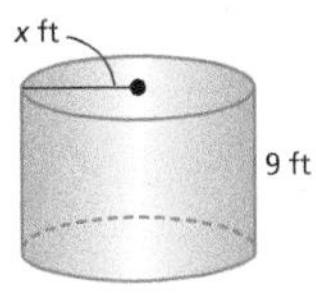

29. $V = 436.05$ in^3 10.2 inches

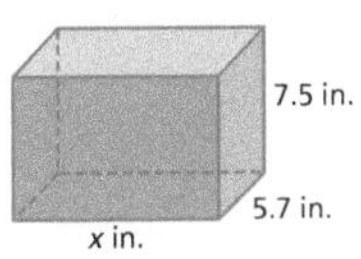

30. Cylinder A and cylinder B are similar solids. What is the volume of cylinder B? 1187.5 m^3

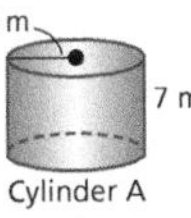

Cylinder A

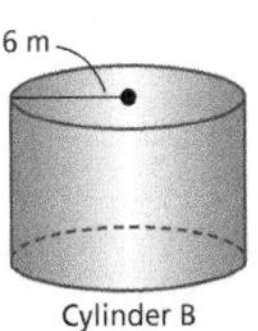

Cylinder B

31. **STEM** An engineer designs an oil filter. The filter is a cylinder with a base diameter of 3.5 inches. The filter must fit in a space that is 5 inches long by 5 inches wide by 5 inches tall. What is the greatest volume the filter can have and still fit in the space? about 48.11 in^3

Questioning Strategies

Problem 30 Suppose that the radius and height of cylinder A are r and h and the radius of cylinder B is $2r$, with the cylinders still similar. How do the volumes of the cylinders compare? For the figures to still be similar, the height must also increase by a scale factor of 2. The volume of cylinder B is $V = \pi(2r)^2(2h) = \pi(4r^2)(2h) = 8\pi r^2 h$, which is 8 times the volume of cylinder A.

5 Wrap-Up

Summarize learning with your class. Consider using the Exit Ticket, Put It in Writing, or I Can scale.

Exit Ticket 

Find the volume of an oblique cylinder with diameter 3 m and height 6 m. $V = \pi r^2 h = \pi(1.5)^2(6) \approx 42.4 \text{ m}^3$

Put It in Writing 

Explain how the formulas for the volume of a right rectangular prism and the volume of a right cylinder are special cases of the same rule.

I Can 

The scale below can help you and your students understand their progress on a learning goal.

4	I can develop and use formulas for the volume of a prism and a cylinder, and I can explain the process to others.
3	I can develop and use formulas for the volume of a prism and a cylinder.
2	I can use the formulas for the volume of a prism and a cylinder.
1	I can use the formula for the volume of a rectangular prism.

Spiral Review • Assessment Readiness

These questions will help determine if students have retained information taught in the past and can also prepare them for high-stakes assessments. Here, students must identify expressions equivalent to the surface area of a cylinder **(18.2)**, use a trigonometric ratio to find a side length of a triangle **(13.2)**, find the surface area of a cone **(18.3)**, and find the surface area of a sphere **(18.4)**.

32. A landscaping company has 3 cubic yards of mulch to deliver to a house. The homeowner has a box where they want the mulch stored. The box is a prism with a base that is 1.5 yards wide and 1.5 yards long. What is the least height the box can be to hold all of the mulch? 1.3 yards tall

33. Open Middle™ Using the digits 1 to 9 at most two times each, fill in the boxes for the length of the radius and height such that it creates a true statement where the volume of a cone plus the volume of a sphere equals the volume of a cylinder. See Additional Answers.

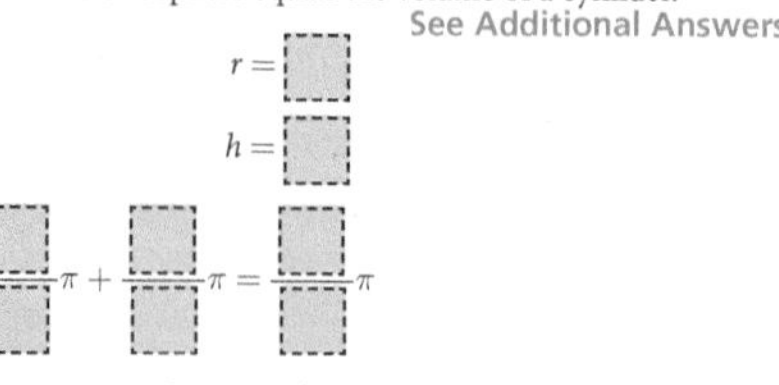

Spiral Review • Assessment Readiness

34. Which of the following expressions could be used to find the surface area of the cylinder? Select all that apply.

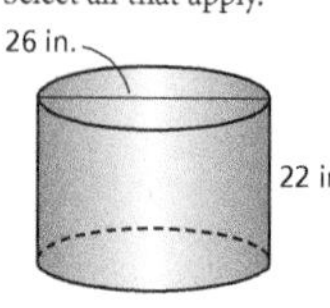

Ⓐ $\pi r^2 h$
Ⓑ $L + 2\pi r^2$
Ⓒ $L + 2B$
Ⓓ $2\pi r^2$
Ⓔ Bh
Ⓕ $2L + B$

35. Find the length of $\overline{BC}$.

Ⓐ 2.64
Ⓑ 3.26
Ⓒ 3.4
Ⓓ 4.2

A, 4.2, 39°, C, B

36. Find the surface area of the cone.

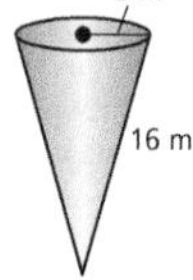

Ⓐ $85\pi \text{ m}^2$
Ⓑ $105\pi \text{ m}^2$
Ⓒ $261\pi \text{ m}^2$
Ⓓ $281\pi \text{ m}^2$

37. Find the surface area of the sphere.

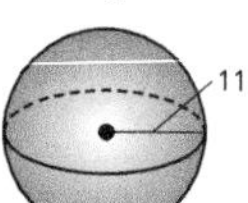

Ⓐ $11\pi \text{ mm}^2$
Ⓑ $121\pi \text{ mm}^2$
Ⓒ $242\pi \text{ mm}^2$
Ⓓ $484\pi \text{ mm}^2$

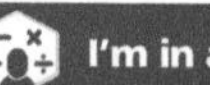 **I'm in a Learning Mindset!**

What skills do I have that benefit collaboration? Which collaboration skills still need improvement?

590

Keep Going ▶ Journal and Practice Workbook

Learning Mindset

Strategic Help-Seeking **Identifies Need for Help**

Point out how working with others to derive and make sense of the volume formulas for prisms and cylinders can be helpful for identifying strengths and areas for improvement of collaboration skills. Being able to explain to others how the two formulas are special cases of the same general concept can be a sign that a student has strong communication skills. Struggling to find the right clarifying questions for explained concepts can be a sign that a student may need work on their listening/understanding skills. *How does being able to help others understand how two topics are related tell you that you have strong communication skills? How might others not understanding your explanations tell you that your communication skills might need work? Is being able to ask the right clarifying questions a sign of strong collaboration skills?*

19.2 Volumes of Pyramids and Cones

LESSON FOCUS AND COHERENCE

Mathematics Standards

- Use volume formulas for cylinders, pyramids, cones, and spheres to solve problems.
- Give an informal argument for the formulas for the circumference of a circle, area of a circle, and volume of a cylinder, pyramid, and cone. Use dissection arguments, Cavalieri's principle, and informal limit arguments.
- Apply concepts of density based on area and volume in modeling situations (e.g., person per square mile, BTUs per cubic foot).
- Use geometric shapes, their measures, and their properties to describe objects (e.g. modeling a tree trunk or a human torso as a cylinder).

Mathematical Practices and Processes 

- Look for and make use of structure.
- Construct viable arguments and critique the reasoning of others.
- Model with mathematics.
- Attend to precision.

I Can Objective 

I can show the relationship between the volume formulas for pyramids and cones.

Learning Objective

Derive formulas for the volume of a cone and a pyramid, relating the formulas to the volumes of cylinders and prisms; and solve real-world and mathematical problems by finding the volumes of pyramids, cones, and composite figures.

Language Objective

Describe how the volume formulas for a cone and pyramid are derived from the volume formulas of prisms and cylinders.

Lesson Materials: graphing calculator

Mathematical Progressions

Prior Learning	Current Development	Future Connections
Students: • solved real-world and mathematical problems involving area, volume, and surface area of two- and three-dimensional objects. **(Gr7, 11.4)** • knew the formulas for the volumes of cones, right prisms cylinders, and spheres and used them to solve real-world and mathematical problems. **(Gr8, 13.2 and 13.4)**	**Students:** • develop the formula for the volume of a pyramid by recognizing that the pyramid occupies $\frac{1}{3}$ of the volume of a prism. • develop the formula for the volume of a cone based on the volume of a pyramid and the concept of circle area as a limit of the area of an inscribed polygon. • apply the formula for volume to solve a real-world problem involving density.	**Students:** • for a function that models a relationship between two quantities, will interpret key features of graphs and tables in terms of the quantities, and sketch graphs showing key features given a verbal description of the relationship. Key features will include: intercepts; intervals where the function is increasing, decreasing, positive, or negative; relative maximums and minimums; symmetries; end behavior; and periodicity. **(4th year course)**

PROFESSIONAL LEARNING

Math Background

An important goal of this lesson is to help students see the relationship between a cone and a cylinder and between a pyramid and a prism, and how this relates to the derivations of the formulas for volume. Students are building on prior knowledge that the volume of a cone is one-third the volume of a cylinder from Grade 8. This contributes to their understanding of the mathematical formulas for both the volume of a cone and the volume of a pyramid. They can also use this knowledge to see how, as the sides of the base of a pyramid increase, the base eventually becomes close to the shape of a circle, which is the base of a cone.

ONLINE

PROJECTABLE & PRINTABLE 

WARM-UP OPTIONS

ACTIVATE PRIOR KNOWLEDGE • Find the Area and Circumference of a Circle

Use these activities to quickly assess and activate prior knowledge as needed.

Problem of the Day 

A circular clock face has a diameter of 12 inches. The minute hand extends to the very edge of the clock. How much area does the minute hand cover in one hour? What is the distance the tip of the minute hand travels in one hour?

Area covered: $A = \pi(6)^2 \approx 113$ in^2

Distance traveled: $C = \pi(12) \approx 38$ in.

Quick Check for Homework

As part of your daily routine, you may want to display the Teacher Solution Key to have students check their homework.

Make Connections

Based on students' responses to the Problem of the Day, choose one of the following:

1. Project the Interactive Reteach, Grade 7, Lesson 10.2.
2. Complete the Prerequisite Skills Activity:

Have students work in pairs. Students should draw a circle and label the radius with a value. Have students calculate the circumference and the area showing their work. Then ask students to trade circles and work with another pair to check the solutions.

- *What is the formula for the circumference of a circle?* $C = 2\pi r$
- *What is the formula for the area of a circle?* $A = \pi r^2$
- *What do you need to do to find the area if you know the diameter of a circle?* Divide the diameter by 2.
- *How do the units for circumference compare with the units for area?* Circumference is a length and will be a single unit like centimeters. Area is a space and will be a unit squared, like square centimeters.

If students continue to struggle, use Tier 2 Skills 19 and 20.

SHARPEN SKILLS

If time permits, use this on-level activity to build fluency and practice basic skills.

Quantitative Comparison

Objective: Students make a comparison between two quantities.
Write the following problem on the board. Ask students to choose the letter representing the correct answer and to explain their reasoning.

Quantity A

Figure A

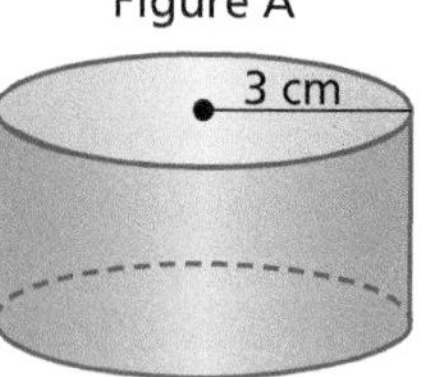

Height = 10 cm

Quantity B

Figure B

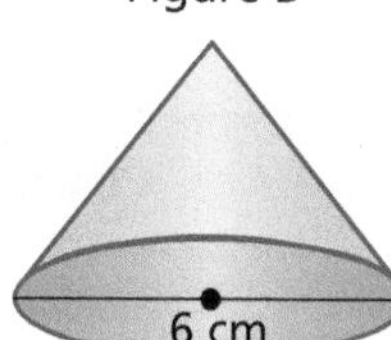

Height = 10 cm

A. Quantity A is greater.
B. Quantity B is greater.
C. The two quantities are equal.
D. The relationship cannot be determined from the information given.

A; The volume of Figure A is three times greater because it is 90π cm^3 and Figure B is 30π cm^3.

PLAN FOR DIFFERENTIATED INSTRUCTION

Small-Group Options

Use these teacher-guided activities with pulled small groups.

On Track

Materials: index cards

Make sets of index cards where one gives the dimensions of a pyramid and one gives the same dimensions for a prism. Pass out one card to each student. Have students find the volume of the figure on their card. Then have them find the student with the card that has the same dimensions and discuss the question below.

How does the volume of the pyramid compare to the volume of the prism with the same dimensions?

Pyramid
length = 5 cm
width = 6 cm
height = 2 cm

Prism
length = 5 cm
width = 6 cm
height = 2 cm

Almost There

Materials: number cubes

Have students work in pairs and have them sketch a rectangular pyramid. Then have students roll a number cube three times, using the three numbers as dimensions for the pyramid.

- What is the formula to find the volume of the pyramid?
- What is the volume of the pyramid?
- If the dimensions are listed in inches, what are the units on the volume of the pyramid?
- If you were to find the volume of a prism with the same dimensions, what would you predict the volume to be?

Ready for More

Consider a cone with radius a and height $2a + 4$.

- What is the volume of the cone?
- If a cylinder with the same dimensions as the cone has a volume of 198π in^3, what are the dimensions of the cylinder and cone?
- What do you think will happen to the volume of the cone if the radius is doubled?
- Find the volume of the cone with a radius $2a$ and compare the volume with the original cone.
- What will happen to the volume if the radius of the cylinder is also doubled?

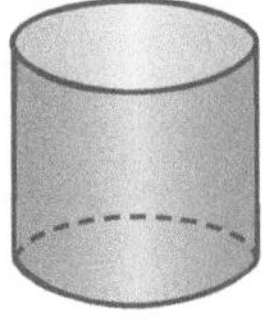

Math Center Options

Use these student self-directed activities at centers or stations. Key: ● Print Resources ● Online Resources

On Track

- ● Interactive Digital Lesson
- ●● Journal and Practice Workbook
- ● Module Performance Task
- ● Illustrative Mathematics: Doctor's Appointment

Almost There

- ● Reteach 19.2 (printable)
- ● Interactive Reteach 19.2
- ● RtI Tier 2 Skill 19: Circumferences of Circles
- ● RtI Tier 2 Skill 20: Areas of Circles
- ● Illustrative Mathematics: The Great Egyptian Pyramids

Ready for More

- ● Challenge 19.2 (printable)
- ● Interactive Challenge 19.2
- ● Illustrative Mathematics: Volume Estimation

View data-driven grouping recommendations and assign differentiation resources.

During the *Spark Your Learning*, listen and watch for strategies students use. See samples of student work on this page.

Compare to the Volume of a Prism

Strategy 1

I know that a pyramid takes up one-third the volume of a prism with the same dimensions. So, a prism with the same dimensions would have a volume 3 times as great as the pyramid.

$V = 147 \cdot 230 \cdot 230$

$V = 7{,}776{,}300 \text{ m}^3$

$V = \frac{147 \cdot 230 \cdot 230}{3}$

$V = 2{,}592{,}100 \text{ m}^3$

If students . . . base the calculations on previous knowledge, they are making connections to the volume of a prism from Grade 8.

Have these students . . . explain how they determined their calculations for the volume of the pyramid. **Ask:**

- Q How did you use the volume of a prism to find the volume of the pyramid?
- Q How did you know to divide the standard calculation for volume of $V = Bh$?

Draw a Diagram

Strategy 2

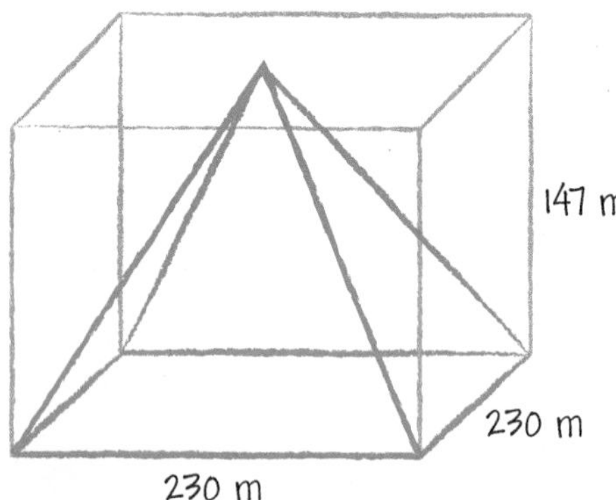

The volume of the prism is $V = 147 \cdot 230 \cdot 230 = 7{,}776{,}300 \text{ m}^3$.

The pyramid appears to take up about one third of the space of the prism, so the volume of the pyramid is $V = \frac{147 \cdot 230 \cdot 230}{3} = 2{,}592{,}100 \text{ m}^3$.

If students . . . draw a diagram of the pyramid within a prism of the same dimensions, they understand that volume of the pyramid is less than the volume of the prism, but they may be unsure of the exact relationship between a prism and a pyramid.

Activate prior knowledge . . . by having students think about how they could have visualized the relationship between the volume of the prism and the pyramid without drawing a diagram. **Ask:**

- Q What do you think is the formula for the volume of a pyramid?
- Q How could you have visualized the relationship between a pyramid and the prism without drawing a diagram?

COMMON ERROR: Finds the Volume of a Prism

I multiplied the length by the width of the square base.

$230 \cdot 230 = 52{,}900$

Then I multiplied by the height to find the volume in cubic meters.

$52{,}900 \cdot 147 = 7{,}776{,}300 \text{ m}^3$

If students . . . multiply the three dimensions, then they know how to find the volume of a rectangular prism, but may not understand how a prism relates to a pyramid of the same dimensions.

Then intervene . . . by pointing out that the pyramid could fit inside a prism with the same dimensions. **Ask:**

- Q How many pyramids could you fit inside a prism with the same dimensions?
- Q Why does the pyramid not have the same volume as a prism with the same dimensions?

19.2

Volumes of Pyramids and Cones

I Can show the relationship between the volume formulas for pyramids and cones.

Spark Your Learning

Brooke is researching the pyramids of ancient Egypt. The Great Pyramid of Giza is the largest one.

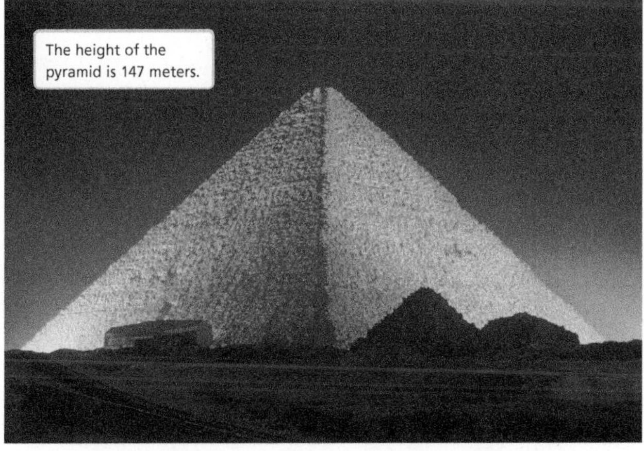

The height of the pyramid is 147 meters.

Complete Part A as a whole class. Then complete Parts B–D in small groups.

A. What is a mathematical question you can ask about this situation? What information would you need to know to answer your question? A. What is the volume of the Great Pyramid of Giza?; the dimensions of the base of the pyramid and the height

B. A pyramid is related to what other three-dimensional figure? About what fraction of the space inside that figure does a pyramid take up—more than half as much space, half as much space, or less than half as much space? Give an approximate fraction and explain your reasoning.
B, C. See Additional Answers.

C. What variable(s) are involved in this situation about the Giza Pyramid? What unit of measurement would you use for each variable?

D. To answer your question, what strategy and tool would you use along with all the information you have? What answer do you get?
See Strategies 1 and 2 on the facing page.

Turn and Talk What would it mean to find the volume of one of the pyramids at Giza? What would you need to know to find this volume? See margin.

EL SUPPORT SENSE-MAKING • Three Reads

Tell students to read the information in the photo three times and prompt them with a different question each time.

1. What is the situation about? The situation is about finding the volume of the Great Pyramid of Giza.
2. What are the quantities in this situation? How are those quantities related? The quantities are the height of the pyramid and the length and width of the base. One-third of the product of the length, width, and height gives the volume of the pyramid.
3. What are possible questions you could ask about this situation? Does it matter which order you multiply the dimensions in? Can you divide by three or multiply by $\frac{1}{3}$ once you know the product of the length, width, and height? Do all pyramids have a base that is a square?

1 Spark Your Learning

▶ MOTIVATE

- Have students look at the photo in their books and read the information contained in the photo. Then complete Part A as a whole-class discussion.
- Give the class the additional information they need to solve the problem. This information is available online as a printable and projectable page in the Teacher Resources.
- Have students work in small groups to complete Parts B–D.

▶ PERSEVERE

If students need support, guide them by asking:

Q **Advancing • Use Tools** Which tool could you use to solve the problem? Why choose that tool and not some other? Students' choices of tools and reasons for choosing them will vary.

Q **Assessing** How does the volume of the pyramid compare with the volume of a rectangular prism with the same dimensions? The pyramid will have a smaller volume.

Q **Assessing** How would you find the volume of a rectangular prism with the same dimensions? Multiply the length by the width by the height.

Q **Advancing** How could you use the volume of the rectangular prism to find the volume of the pyramid? You could find the volume of the prism, $l \cdot w \cdot h$, and divide it by 3 since it is one-third the volume.

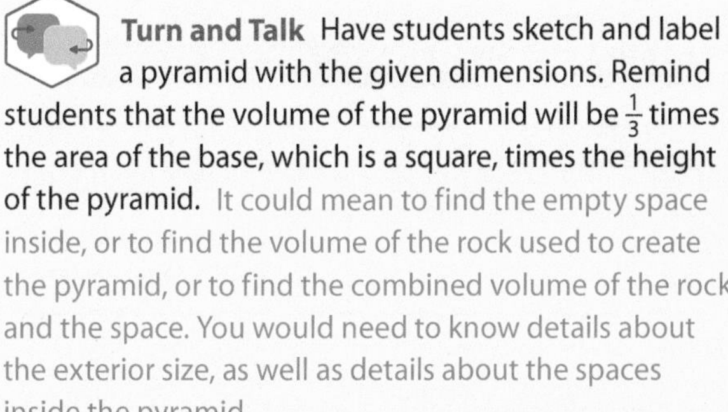

Turn and Talk Have students sketch and label a pyramid with the given dimensions. Remind students that the volume of the pyramid will be $\frac{1}{3}$ times the area of the base, which is a square, times the height of the pyramid. It could mean to find the empty space inside, or to find the volume of the rock used to create the pyramid, or to find the combined volume of the rock and the space. You would need to know details about the exterior size, as well as details about the spaces inside the pyramid.

▶ BUILD SHARED UNDERSTANDING

Select groups of students who used various strategies and tools to share with the class how they solved the problem. As they present their solutions, have each group discuss why they chose a specific strategy and tool.

② Learn Together

Build Understanding

Task 1 **(MP) Use Structure** Students look at how one triangular prism has the same volume as three triangular pyramids with the same base and height as the prism.

Sample Guided Discussion:

Q Do the three pyramids in Part B overlap? Explain. no, They have different bases and take up different space in the prism.

Q In Part C, how do you know that $\triangle EHC \cong \triangle CFE$? You can use SSS because the opposite sides of a rectangle are congruent and the diagonal $\overline{CE}$ is congruent to itself.

Turn and Talk Have students compare the shape of the base of a triangular pyramid to the base of a rectangular pyramid. Then ask students to compare the shapes of the bases of a triangular and a rectangular prism. Ask students to think about how many rectangular pyramids would fit into a rectangular prism with the same dimensions. This is true of all pyramids, not just triangular pyramids.

ANSWERS

B. If you remove *G-HCD* from the prism, and then slice the remaining solid on the plane *GEC*, you get pyramids *G-EHC* and *G-CFE*. The three pyramids do not overlap and contain the entire prism, so the sum of their volumes is equal to the volume of the prism.

Build Understanding

Develop a Formula for the Volume of a Pyramid

The following postulate is helpful when developing a formula for the volume of a pyramid.

Postulate
Pyramids that have equal base areas and equal heights have equal volumes.

1 Consider a triangular pyramid with vertex *G* directly over vertex *D* of the base *HCD*. This triangular pyramid can be thought of as part of triangular prism with $\triangle EFG \cong \triangle HCD$. Let the area of the base of the pyramid be *B* and let $GD = h$.

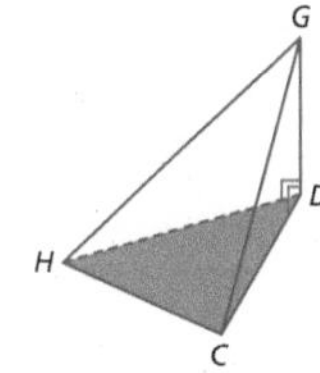

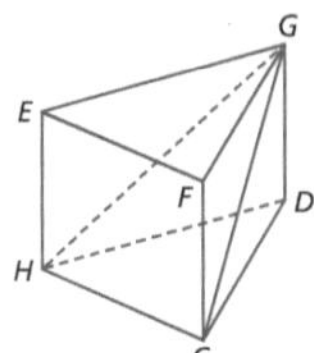

A. What is the volume of the triangular prism in terms of *B* and *h*? $V = Bh$

B. Draw $\overline{EC}$. Consider the three pyramids: *G-EHC*, *G-CFE*, and *G-HCD*. Explain why the sum of the volumes of these pyramids is equal to the volume of the prism. B–D. See Additional Answers

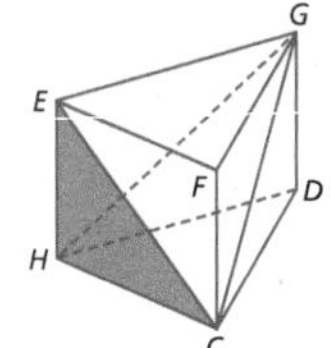

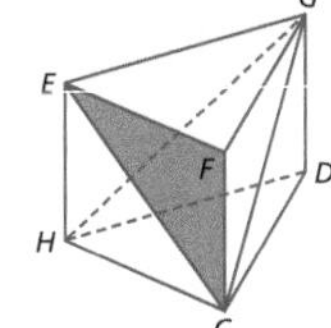

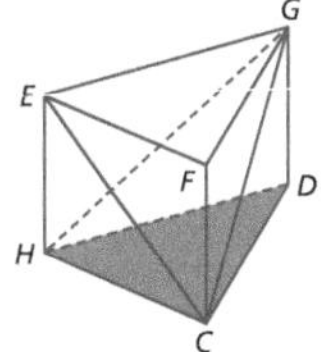

C. $\overline{EC}$ is the diagonal of a rectangle, so $\triangle EHC \cong \triangle CFE$. Explain why pyramids *G-EHC* and *G-CFE* have the same volume. Explain why pyramids *G-HCD* and *C-EFG* have the same volume.

D. *G-CFE* and *C-EFG* are two names for the same pyramid, so you have shown that the three pyramids that form the triangular prism have the same volume. Compare the volume of pyramid *G-HCD* to the volume of the triangular prism. Write the volume of pyramid *G-HCD* in terms of *B* and *h*.

Turn and Talk If the bases of a pyramid and a prism are congruent, is it always true, sometimes true, or never true that the volume of the pyramid is $\frac{1}{3}$ the volume of the prism? See margin.

592

LEVELED QUESTIONS

Depth of Knowledge (DOK)	Leveled Questions	What Does This Tell You?
Level 1 **Recall**	What is a triangular prism? A prism with two parallel bases that are triangular in shape. The lateral faces are rectangles.	Students' answers will indicate whether they know the definition and shape of a triangular prism.
Level 2 **Basic Application of Skills & Concepts**	How does the volume of a triangular prism compare to the volume of a triangular pyramid with the same dimensions? The volume of the prism is three times as great as the volume of the pyramid.	Students' answers will indicate whether they know how the volumes of a pyramid and prism are related, which will help them understand the formula for the volume of a pyramid.
Level 3 **Strategic Thinking & Complex Reasoning**	How is the volume of a cone and cylinder of the same dimensions related to the volume of a pyramid and prism with the same dimensions? A cone is one-third the volume of a cylinder, which is the same relationship as between a pyramid and prism.	Students' answers will indicate whether they are making connections to prior knowledge and understand that a cone is one-third the volume of a cylinder with the same dimensions.

Develop a Formula for the Volume of a Cone

Volume of a Pyramid

The volume V of a pyramid with base area B and height h is given by $V = \frac{1}{3}Bh$.

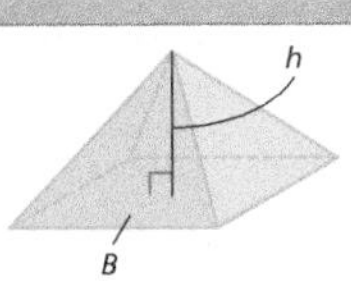

You can approximate the volume of a cone by finding the volume of inscribed pyramids.

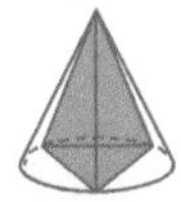

Base of inscribed pyramid has 3 sides

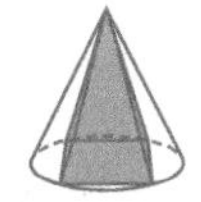

Base of inscribed pyramid has 4 sides

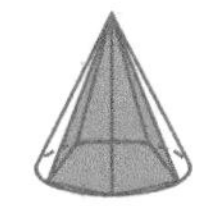

Base of inscribed pyramid has 5 sides

A. The shape of the base gets closer to a circle as the number of sides increases.

2 A. Consider pyramids with regular polygon bases: equilateral triangles, squares, regular pentagons, and so on. As the number of sides increases, what shape does the base begin to resemble?

B. The base of the pyramid is inscribed in the base of the cone and is a regular n-gon. Let P be the center of the cone's base, let r be the radius of the cone's base, and let h be the height of the cone.

B, C. See margin.

Construct $\overline{PM}$ from P to the midpoint M of $\overline{AB}$. How do you know that $\angle 1 \cong \angle 2$? What do you know about $\angle AMP$ and $\angle BMP$?

C. There are n triangles congruent to $\triangle APB$ in the n-gon, so $m\angle APB = \frac{360°}{n}$ and $m\angle 1 = \frac{180°}{n}$. Using the fact that $\sin(\angle 1) = \frac{x}{r}$ and $\cos(\angle 1) = \frac{y}{r}$, you know that $x = r\sin\left(\frac{180°}{n}\right)$ and $y = r\cos\left(\frac{180°}{n}\right)$. Using these equations for x and y, and the formula for the area of a triangle, write an equation for the area of the n-gon.

D. The expression for the area of the base of the pyramid includes the expression $n\sin\left(\frac{180°}{n}\right) \cdot \cos\left(\frac{180°}{n}\right)$. Use a calculator to discover what happens to this expression as n gets larger. The value gets closer and closer to π.

- Enter the expression $n\sin\left(\frac{180°}{n}\right) \cdot \cos\left(\frac{180°}{n}\right)$ as Y_1, using x for n.
- Go to the **TABLE SETUP** menu and have the table start at 3, then view the table for the function and scroll down.

What happens to the expression as n gets very large?

E. Use your answer from Part D. What happens to the expression for the volume of the inscribed pyramid as n increases?

The expression for the volume of the inscribed pyramid gets closer to $\frac{1}{3}\pi r^2 h$.

PROFICIENCY LEVEL

Beginning

Have students draw a square pyramid, labeling the drawing with the words *base*, *length*, *width*, and *height*.

Intermediate

Have students describe to a partner why a pyramid has $\frac{1}{3}$ as part of the volume formula.

Advanced

Ask students to write a paragraph explaining how the formulas for the volume of a prism and a pyramid are related.

Task 2 MP **Construct Arguments** Students use triangles and angles to show that as the base of a pyramid inscribed in a cone approaches a circle, the volume gets close to $\frac{1}{3}\pi r^2 h$, or the volume of the cone.

Sample Guided Discussion:

Q What theorem can you use to prove $\triangle APM \cong \triangle BPM$? SSS; $\overline{PM} \cong \overline{PM}$ by the reflexive property, $\overline{AM} \cong \overline{MB}$ by definition of midpoint, and $\overline{PA} \cong \overline{PB}$ are both radii.

Q In Part C, why does the expression for $m\angle APB$ use 360° and $m\angle 1$ use 180°? There are 360° in a circle, and the angle is divided in 2 to give $m\angle 1$, so it uses 180°.

ANSWERS

B. $\triangle APM \cong \triangle BPM$ by SSS Triangle Congruence. So, $\angle 1 \cong \angle 2$ by CPCTC. $\angle AMP$ and $\angle BMP$ are congruent by CPCTC and form a linear pair, so they are right angles.

C. $A = n\left[\frac{1}{2}(2xy)\right] = nxy = n\left[r\sin\left(\frac{180°}{n}\right) \cdot r\cos\left(\frac{180°}{n}\right)\right] = r^2 n\left[\sin\left(\frac{180°}{n}\right) \cdot \cos\left(\frac{180°}{n}\right)\right]$

Step It Out

Task 3 (MP) **Model with Mathematics** Students use the formulas for the volumes of a cone and cylinder when finding the volume of a real-world figure.

Sample Guided Discussion:

Q How is the diameter of a circle related to its radius? The diameter is two times the length of the radius.

Q In Part B, why do you have to use two formulas to find the total volume? The figure is a composite figure made up of a cone and a cylinder. You need to use the formula for a cone for the volume of the cone and a different formula for the volume of the cylinder.

Turn and Talk Have students compare the formula for the volume of a cylinder, $\pi r^2 h$, and the volume of a cone, $\frac{1}{3}\pi r^2 h$. They can substitute values in for r and h to help see the relationship. The volume of the cylinder always is always greater than the volume of the cone. The volume of the cylinder is $\pi r^2 h$ and the volume of the cone is $\frac{1}{3}\pi r^2 h$, and $\pi r^2 h$ is always greater than $\frac{1}{3}\pi r^2 h$.

Step It Out

Model a Real-World Structure to Estimate Volume

Volume of a Cone

The volume of a cone with base radius r and base area $B = \pi r^2$ and height h is given by $V = \frac{1}{3}Bh$ or $V = \frac{1}{3}\pi r^2 h$.

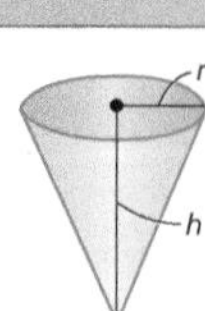

3 An ice cream shop has a large decorative cone next to its front door. What is the volume of the cone?

©incamerastock/Alamy

Use the formula $V = \frac{1}{3}\pi r^2 h$ for the volume of the cone and substitute the values of the height and the radius.

$V = \frac{1}{3}\pi r^2 h$

$V = \frac{1}{3}\pi(1.5^2)(4.5)$

A. Why is 1.5 substituted for r?

$V = 3.375\pi$

A. 3 ft is shown as the diameter, not the radius. So, divide 3 by 2 to find the radius.

$V \approx 10.6$

The volume of the cone is about 10.6 cubic feet.

A second cone with the same dimensions is added upside down on a cylinder with a diameter of 3 feet and a height of 2 feet. Find the volume of the composite figure.

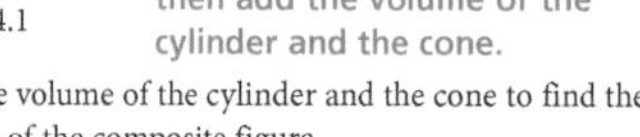

Find the volume of the cylinder.

$V = \pi r^2 h$

$V = \pi(1.5^2)(2)$

B. How can you find the volume of the composite figure?

$V = 4.5\pi$

B. Find the volume of the cylinder, then add the volume of the cylinder and the cone.

$V \approx 14.1$

Add the volume of the cylinder and the cone to find the volume of the composite figure.

The volume of the composite figure is about $10.6 + 14.1 = 24.7$ cubic feet.

Turn and Talk Given a cylinder and a cone with the same height and radius, is the volume of the cylinder always, sometimes, or never greater than the volume of the cone? See margin.

594

Apply a Volume Formula to Find Density of a Real-World Object

A crystal is cut into a shape formed by two square pyramids joined at the base.

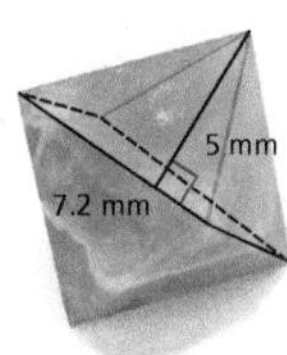

The mass of the crystal is 3 grams. Find the density of the crystal in grams per cubic millimeter.

A. How do the units inform the process of finding the density?

A. In this case, density must be written in grams per cubic millimeter, so we can conclude that density is mass per unit of volume.

Find the volume of one of the pyramids.

$V = \frac{1}{3}Bh$

$V = \frac{1}{3}(7.2^2)(5)$

B. Why is 7.2^2 substituted for B?

$V = 86.4$

The volume of a pyramid is 86.4 cubic millimeters.

B. B represents the area of the base of the pyramid. The base of the pyramid is a square, so its area is the square of the side length, 7.2 mm.

Find the volume of the crystal.

The volume of the crystal is $2(86.4) = 172.8$ cubic millimeters.

C. Why is the volume of the crystal two times the volume of the pyramid?

Find the density.

$\text{density} = \frac{\text{mass}}{\text{volume}}$

$d = \frac{3}{172.8} \approx 0.0174$

The density is about 0.0174 gram per cubic millimeter.

C. The crystal is made of two identical pyramids, so we must multiply the volume of the pyramid by 2 to get the volume of the entire crystal.

D. What is the density of the crystal if the mass of the crystal is 5 grams? about 0.0289 gram per cubic millimeter

Turn and Talk Suppose a different crystal has the same edge length and mass, but double the height. How would that affect the density? See margin.

©Reika/Shutterstock

Task 4 (MP) **Attend to Precision** Students find the volume of a composite figure and use the value to find the density of a crystal.

Sample Guided Discussion:

Q How can you use what you know about squares to find the base area of the pyramid? A square has congruent sides, so the area of the base of the pyramid is equal to the side squared.

Q If you wanted to find the volume of the entire figure at once, how could you adjust the formula for the volume of a pyramid? You can multiply it by 2; $V = 2 \cdot \frac{1}{3}Bh$

Turn and Talk Have students think about what happens to a quotient when the numerator becomes larger and the denominator stays the same. **Should the density become a larger or a smaller number if the mass increases?** The density would be half as much.

ONLINE

Assign the Digital On Your Own for
- built-in student supports
- Actionable Item Reports
- Standards Analysis Reports

On Your Own

Assignment Guide

The chart below indicates which problems in the On Your Own are associated with each task in the Learn Together. Assign daily homework for tasks completed.

Learn Together Tasks	On Your Own Problems
Task 1, p. 592	Problems 8–14, 24–27, 32, 39, and 41
Task 2, p. 593	Problems 7, 15–18, 28–30, and 33
Task 3, p. 594	Problems 19–22, 31, 34, 36, 37, 38 and 40
Task 4, p. 595	Problems 23, and 35

ANSWERS

7. They are both correct. The volume of a cone with the same base as a cylinder is $\frac{1}{3}$ of the volume of the cylinder. The volume of the cylinder is 3 times the volume of the cone.

9. When finding the volume of the composite figure, you subtract the volume of the pyramid from the volume of the prism. When finding the surface area, you add the lateral area to the surface area of all sides except the top of the prism.

Check Understanding

1. How does the volume of a pyramid compare to the volume of a prism with the same base and height as the pyramid? The volume of the pyramid is $\frac{1}{3}$ of the volume of the prism.

Find the volume of each pyramid. Round your answer to the nearest tenth.

2. 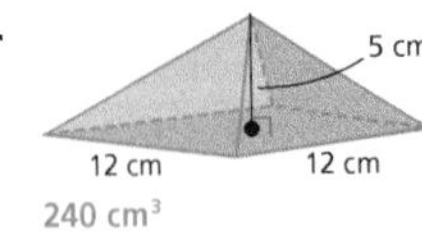

240 cm^3

3. 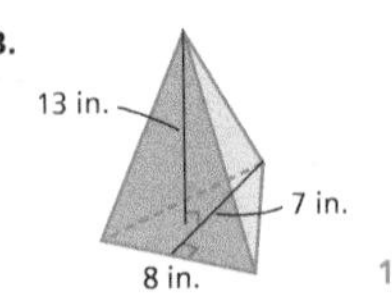

121.3 in^3

Find the volume of each cone. Round your answer to the nearest tenth.

4. 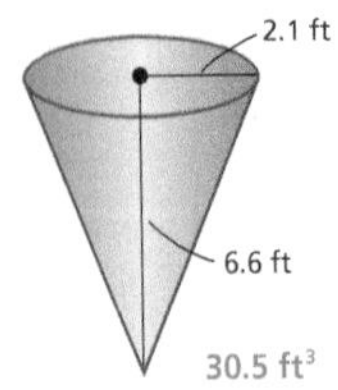

30.5 ft^3

5. 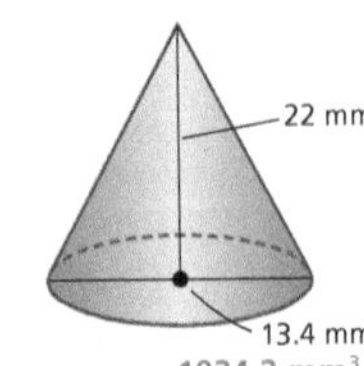

1034.2 mm^3

6. Julian is making candles. Each has the same volume. The dimensions of the candles are shown. If he has 200 cubic centimeters of wax, how many candles can he make? 5 candles

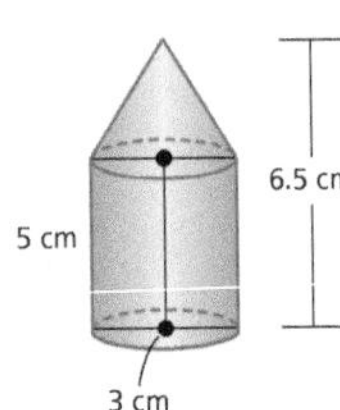

On Your Own

7. MP **Critique Reasoning** Malcolm states that it takes 3 cones to fill a cylinder with the same base. Angus states that it takes $\frac{1}{3}$ of a cylinder to fill a cone with the same base. Who is correct? Explain your reasoning. See margin.

8. MP **Reason** Dana knows the volume of a pyramid. How can she find the volume of a prism with the same base and the same height as the pyramid? She can multiply the volume of the pyramid by 3 to find the volume of the prism.

9. MP **Reason** How is finding the volume of a composite figure with a pyramid removed different from finding the surface area of a composite figure with a pyramid removed? See Additional Answers.

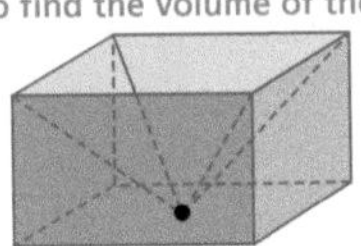

10. Find the volume of a hexagonal pyramid with a base area of 147 square meters and a height of 4 meters. 196 m^3

596

3 Check Understanding

Formative Assessment

Use formative assessment to determine if your students are successful with this lesson's learning objective.

Students who successfully complete the Check Understanding can continue to the On Your Own practice.

For students who miss 1 problem or more, work in a pulled small group using the Almost There small-group activity on page 591C.

ONLINE

Assign the Digital Check Understanding to determine
- success with the learning objective
- items to review
- grouping and differentiation resources

4 Differentiation Options

Differentiate instruction for all students using small-group activities and math center activities on page 591C.

Reteach

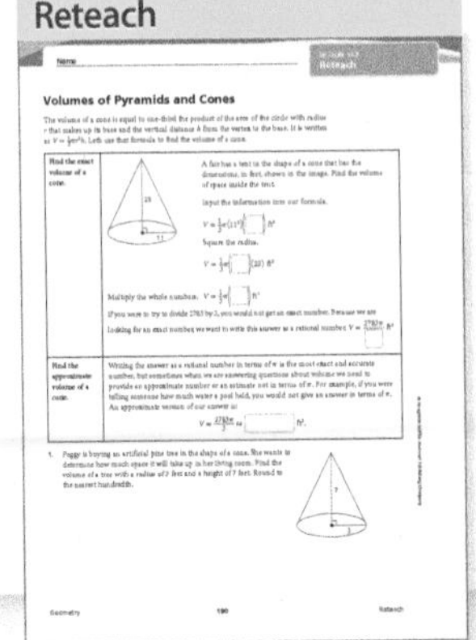

Challenge

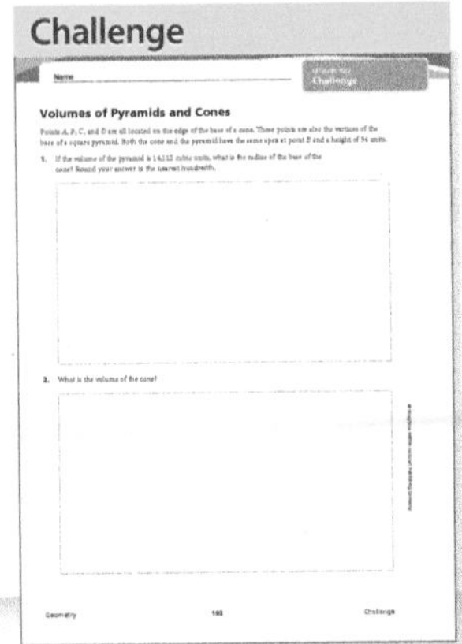

Find the volume of each pyramid. Round your answer to the nearest tenth.

11. 750 ft^3

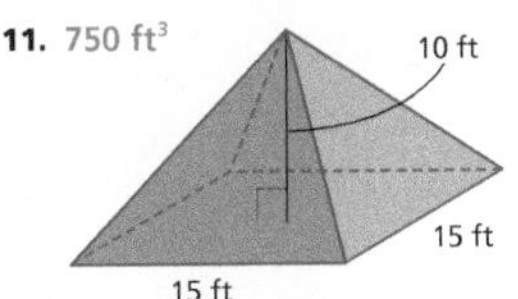

12. 47.5 cm^3

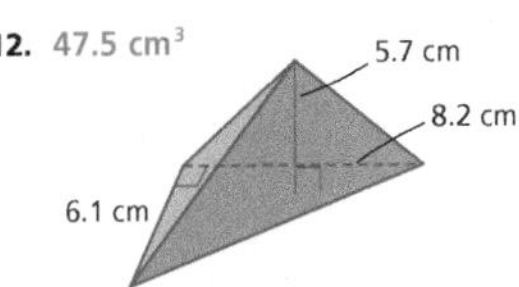

13. 128 in^3

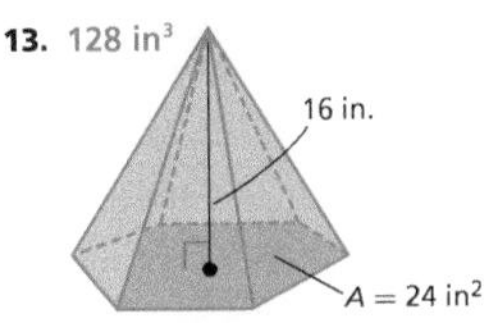

14. 210.9 m^3

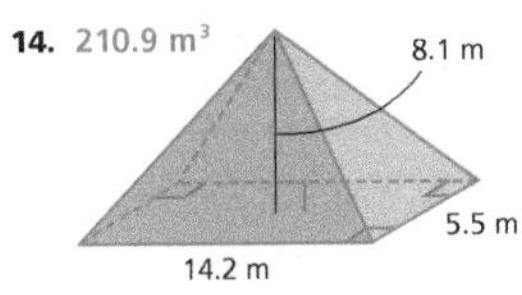

Find the volume of each cone. Round your answer to the nearest tenth.

15.

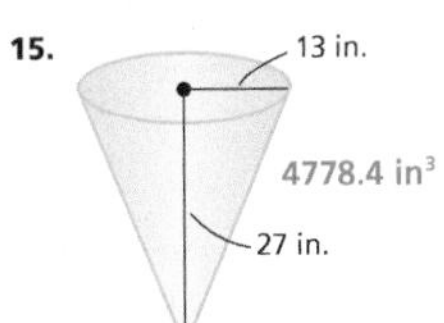

16.

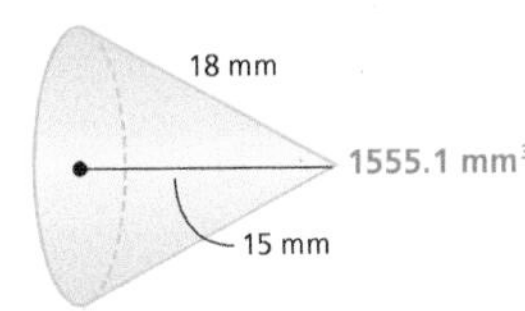

17.

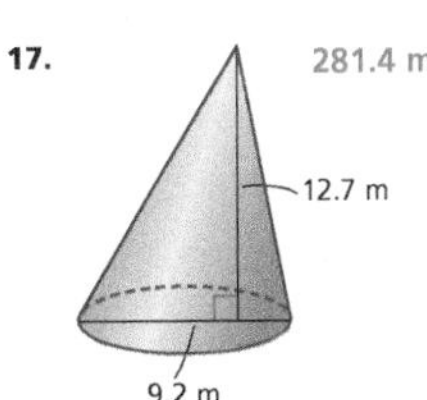

18.

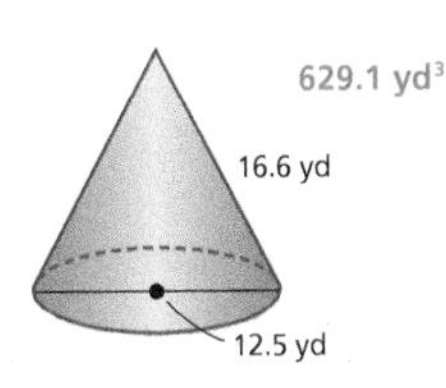

Find the volume of each composite figure. Round your answer to the nearest tenth.

19. 1436.8 ft^3

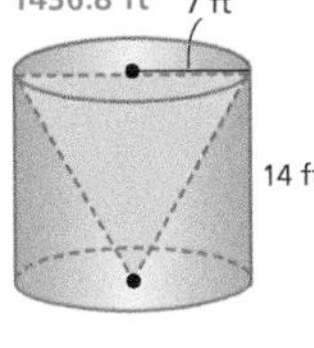

20. 522.7 in^3

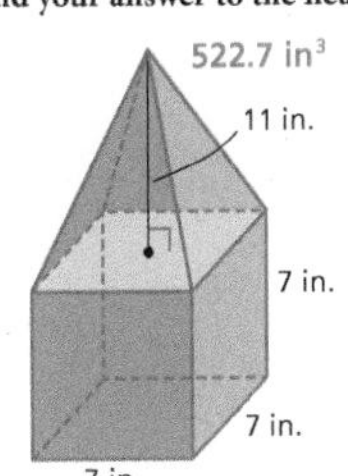

Questioning Strategies

Problem 19 Suppose the cone was added onto the cylinder, not taken out of it. How would you find the volume of the composite figure if the cone was placed on top of the cylinder? How does this differ from the actual problem? If the cone was on top of the cylinder, you would find the volume of the cone and the volume of the cylinder and add them together. In this picture, you can find the volume of the cylinder and subtract the volume of the cone from it.

Watch for Common Errors

Problem 21 Students may not connect that the label of 16 feet at the top of the photo is the diameter. Remind students to find the radius of the circular bases and not to use the diameter to determine the area of the base.

Questioning Strategies

Problem 26 Suppose you know the length and width of the base and you know the height of the pyramid. How would you find the volume? How can you use this to find the missing height when you know the volume? You would find the volume using the formula $V = \frac{1}{3} l \cdot w \cdot h$. Since you know the volume, substitute the values in for V, l, and w and solve for h, $79.7 = \frac{1}{3}(9.5)(3.7)(h)$.

21. **STEM** A water tower is built on top of a building. The dimensions of the water tower are shown.
    A. What is the volume of the water tower? About 3351 ft^3
    B. Assume the entire volume of the water tower is filled with water. The weight of the water in the tower is about 209,193 pounds. What is the density of the water in the water tower? about 62.4 pounds per cubic foot

22. A garage is constructed as a rectangular prism with a roof that is a pyramid. The rectangular prism is 42 feet long, 20 feet wide, and 9 feet tall. The pyramid is 3 feet tall. What is the volume of the garage? 8400 ft^3

23. Hope is creating sand art. She is using a pyramid-shaped mold. The base of the mold is shaped like a square and has a side length of 8 inches. The height of the mold is 8 inches.
    A. What is the volume of sand that the mold will hold? Round your answer to the nearest tenth of a cubic inch. about 170.7 in^3
    B. When the mold is full, the sand weighs 9.43 pounds. What is the density of the sand in the mold? Round your answer to three decimal places. about 0.055 pounds per cubic inch

24. **Open Ended** Draw a square pyramid and label its dimensions. Find the volume of your pyramid. Answers will vary.

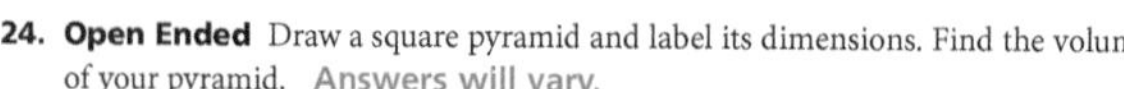

Find the missing dimension of each figure. Round your answer to the nearest tenth.

25. $V = 66.7\ ft^3$ Base lengths: 5 ft

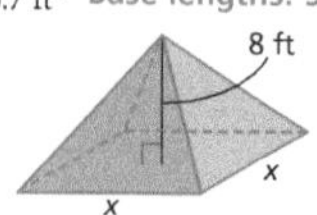

26. $V = 79.7\ cm^3$ Height: 6.8 cm

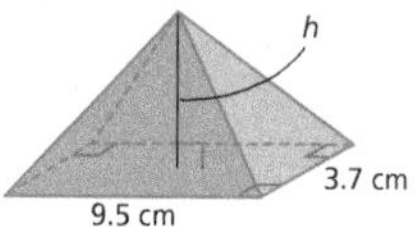

27. $V = 481.7\ in^3$ Height: 17 in.

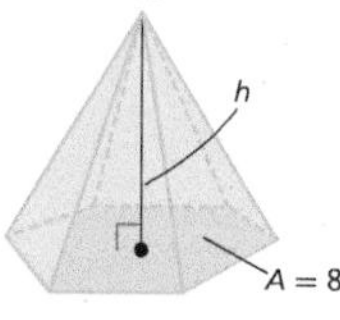

28. $V = 15{,}500\ mm^3$ Height: 41 mm

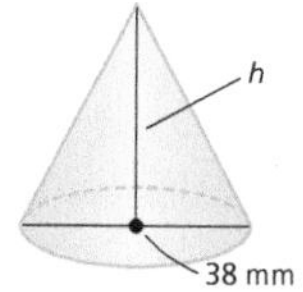

29. $V = 14.3\ ft^3$ Radius: 1.8 ft

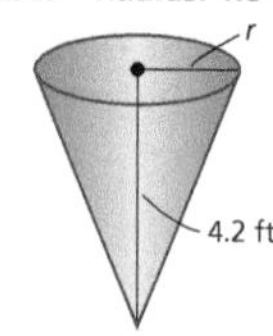

30. $V = 3468\ cm^3$ Diameter: 24 cm

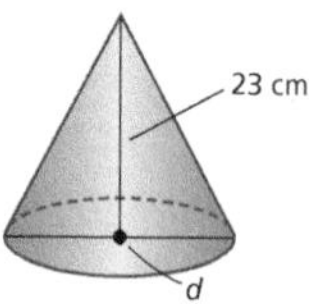

©Stuart Monk/Shutterstock

598

31. Dan has a plastic cup that is shaped as shown.

1.5 in.

8 in.

A. How much plastic is needed to make his cup? 37.7 in^3

B. The cup cost Dan \$4. What is the cost per cubic inch of plastic? about \$0.11 per cubic inch

C. Dan buys another cup with the same radius but a height of 6 inches. Using your result in Part B, what should Dan pay for this cup? about \$3.11

32. (MP) **Reason** A prism has a square base with sides of 4 inches and a height of 8 inches. Without calculating the volumes, find the height of a pyramid with the same base and the same volume as the prism. Explain your reasoning. See margin.

33. Consider the volume of a cone. How does the volume of the cone change if the height of the cone is doubled? How does the volume of the cone change if the radius of the cone is doubled? See margin.

34. A composite figure is formed by a rectangular prism with two square pyramids on top of it. What is its volume? Round your answer to the nearest tenth of a cubic inch. 597.3 in^3

35. Randall needs 50 pounds of sand. He is scooping it with a cone-shaped tool whose dimensions are shown. The density of the sand is 0.055 pound per cubic inch. How many times does Randall need to fill his tool to get the amount of sand he needs? 7 times

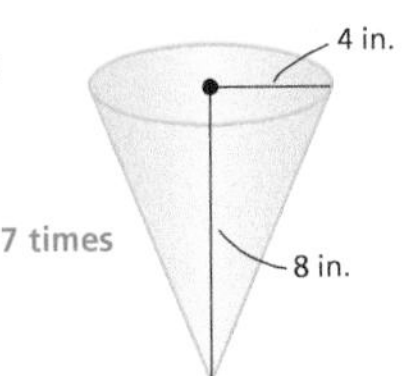

Find the volume of each composite figure. Round your answer to the nearest tenth.

36.

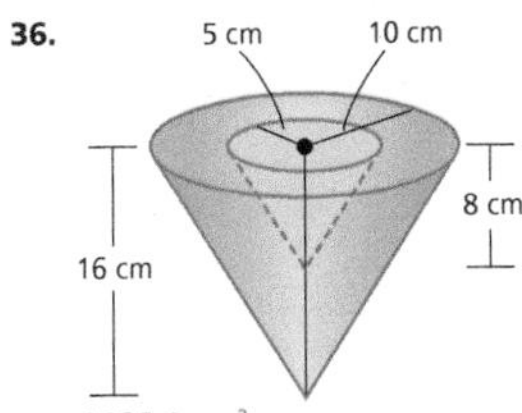

1466.1 cm^3

37.

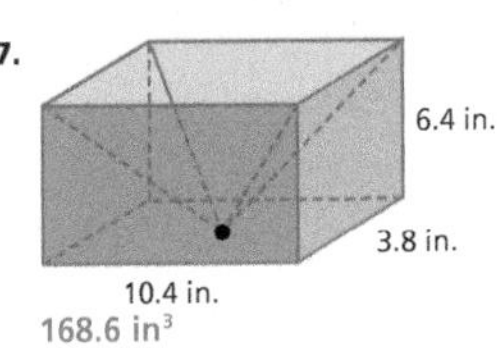

168.6 in^3

38. A company is building pyramid-shaped paperweights. The paperweights will ship in a box with the dimensions shown. The paperweights will be surrounded by one-half inch of packaging to keep them from breaking. What is the maximum volume of a paperweight that can be shipped in the box? 45 in^3

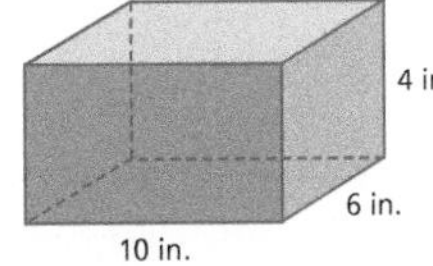

39. A square pyramid has a volume of 132 cubic inches. If the height of the pyramid is 11 inches, what is the length of a side of the base? 6 inches

Watch for Common Errors

Problem 31 In Part A, students may be confused about what the shape of the cup is. They may think it is just a cone or only a cylinder, but it is a cylinder with a cone removed, so the liquid will take shape of the cone when the cup is filled. Have students describe in words what composite shape makes up the cup. For example, students may say the cup is a cone subtracted out of a cylinder. Then work with students to write an expression that represents the volume of the cup.

Questioning Strategies

Problem 35 How much sand can be scooped with the cone-shaped tool? What ratio can you write to determine the density of the sand per scoop? First, find the volume of the cone:

$$V = \frac{1}{3}\pi r^2 h$$
$$= \frac{1}{3}\pi(4)^2(8)$$
$$= \frac{1}{3}\pi(16)(8)$$
$$\approx 134.04 \text{ in}^3; \frac{0.055 \text{ lb}}{\text{in}^3}$$

ANSWERS

32. $h = 24$ in; The volume of a pyramid with the same base and height as the prism is $\frac{1}{3}$ the volume of the prism. For the pyramid to have the same volume as the cylinder, the height of the pyramid must be 3 times the height of the prism.

33. If the height of the cone is doubled, the volume of the cone is doubled; If the radius of the cone is doubled, the volume of the cone is four times as great.

⑤ Wrap-Up

Summarize learning with your class. Consider using the Exit Ticket, Put It in Writing, or I Can scale.

Exit Ticket 

Thomas carves a solid wooden cone to put atop a cylindrical bird feeder, so that the circumferences of the bases are the same. Both figures have a diameter of 11.5 inches and a height of 3 inches. What is the total volume of the figures, to the nearest cubic inch?

$V_{cone} = \frac{1}{3}\pi(5.75)^2(3)$, $V \approx 104 \text{ in}^3$

$V_{cylinder} \approx (3)104 \approx 312 \text{ in}^3$

$V_{total} \approx 104 + 312 \approx 416 \text{ in}^3$

Put It in Writing 

Describe how the formula for the volume of a cone is related to the formula for the volume of a cylinder.

I Can 

The scale below can help you and your students understand their progress on a learning goal.

4	I can show the relationship between the volume formulas for pyramids and cones, and I can explain my reasoning to others.
3	I can show the relationship between the volume formulas for pyramids and cones.
2	I can use the formulas to find the volume of a pyramid or a cone.
1	I can identify whether the shape is a pyramid or a cone and determine some of the dimensions.

Spiral Review • Assessment Readiness

These questions will help determine if students have retained information taught in the past and can also prepare them for high-stakes assessments. Here, students find the area of a circle **(17.1)**, find the volume of a cylinder **(19.1)**, and identify and calculate measurements of a sphere **(18.4)**.

40. Social Studies Some pastoral nomads of central Asia live in portable structures adapted to a mobile life on the windy Great Steppe. A *ger*, the Mongolian word for "home," consists of a cylindrical frame under a conical crown. This one is 4 meters in diameter. Its cylindrical frame is 40 cm taller than its crown. If this *ger* has a total volume of 8π cubic meters, how tall is its crown? 1.2 m

41. Open Middle™ Using the digits 1 to 9, at most one time each, fill in the boxes for a square pyramid's height and base edge length so that the volume has the greatest possible value.

Base edge = ☐

Height = ☐

Volume = ☐☐☐

Base edge = 9; Height = 8; Volume = 216

Spiral Review • Assessment Readiness

42. Find the area of the circle.

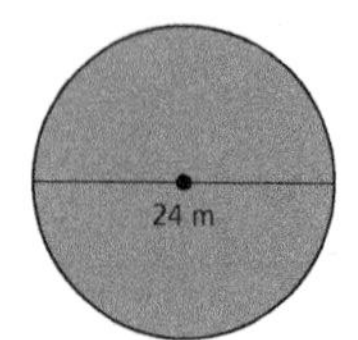

Ⓐ $24\pi \text{ m}^2$
Ⓑ $48\pi \text{ m}^2$
Ⓒ $144\pi \text{ m}^2$
Ⓓ $576\pi \text{ m}^2$

43. Find the volume of the cylinder.

5 ft
8 ft

Ⓐ 40 ft^3
Ⓑ 125.6 ft^3
Ⓒ 200 ft^3
Ⓓ 628 ft^3

44. Match the quantity with what it measures.

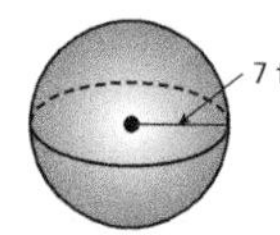

Quantity	Measurement
A. 14π ft	**1.** area of maximum cross-section
B. $\frac{1372\pi}{3} \text{ ft}^3$	**2.** volume
C. $49\pi \text{ ft}^2$	**3.** circumference of maximum cross-section
D. $196\pi \text{ ft}^2$	**4.** surface area

A. 3
B. 2
C. 1
D. 4

 **I'm in a Learning Mindset!**

Was collaboration an effective tool for finding volumes of cylinders and pyramids? Explain.

Keep Going ▶ Journal and Practice Workbook

Learning Mindset

Strategic Help Seeking **Identifies Need for Help**

Point out that sometimes the formulas can be difficult to recall when finding the volumes of cones, pyramids, cylinders, and prisms. Encourage students to collaborate, helping each other to find ways to remember where the formulas come from. Remembering the connection between cones and cylinders and between pyramids and prisms can solidify the expressions for the formulas. *What could you do to help remember how a cone and cylinder are related? What could you do to help remember how a pyramid and prism are related? How are pyramids related to cones? What basic formulas do you need to know in order to find the volume formulas for cones and pyramids?*

19.3 Volumes of Spheres

LESSON FOCUS AND COHERENCE

Mathematics Standards

- Use volume formulas for cylinders, pyramids, cones, and spheres to solve problems.
- Use geometric shapes, their measures, and their properties to describe objects (e.g., modeling a tree trunk or a human torso as a cylinder).
- Apply concepts of density based on area and volume in modeling situations (e.g., persons per square mile, BTUs per cubic foot).

Mathematical Practices and Processes (MP)

- Look for and make use of structure.
- Attend to precision.
- Model with mathematics.

I Can Objective 

I can use the formula for the volume of a sphere to calculate the volumes of composite figures.

Learning Objective

Derive and use a formula for the volume of a sphere. Use the volume formula to solve real-world problems including calculating capacity and dimensions. Calculate the volume of composite figures involving hemispheres and other known solids.

Language Objective

Explain how the volume of a sphere and a hemisphere are related to each other and to a cylinder.

Mathematical Progressions

Prior Learning	Current Development	Future Connections
Students: • solved real-world and mathematical problems involving area, volume, and surface area of two- and three-dimensional objects. **(Gr7, 11.4)** • used the formulas for volumes of cones, cylinders, and spheres to solve real-world and mathematical problems. **(Gr8, 13.2 and 13.4)**	**Students:** • develop the formula for the volume of a sphere. • use the formula for volume of a sphere to solve real-world problems. • estimate volume in a real-world situation. • find the volume of composite figures involving spheres.	**Students:** • for a function that models a relationship between two quantities, will interpret key features and sketch graphs showing key features given a verbal description. Key features will include: intercepts, intervals where the function is increasing, decreasing, positive, or negative; relative maximums and minimums; symmetries; and end behavior. **(4th year course)**

PROFESSIONAL LEARNING

Visualizing the Math

Students may find it helpful to see an example supporting the volume formula for a sphere using two cones that have the same radius as the sphere. Demonstrate that filling the cone twice with water and dumping the water into the sphere will fill the sphere to the top. Then the volume formula for a sphere is twice the volume formula for a cone. Be sure to point out that the two radii of the cones, when used to form a diameter, are the same as the diameter of the sphere.

$$V_{\text{sphere}} = 2\left(\frac{1}{3}\pi r^2 h\right) = \frac{2}{3}\pi r^2(2r) = \frac{2}{3}\cdot 2\cdot \pi\cdot r^2\cdot r = \frac{4}{3}\pi r^3$$

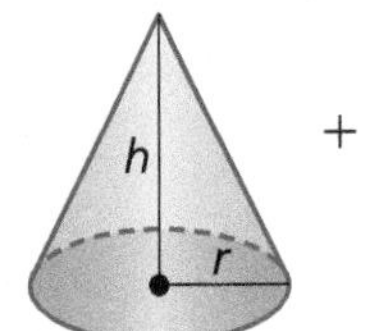

+
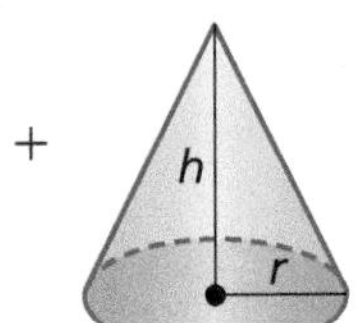

=
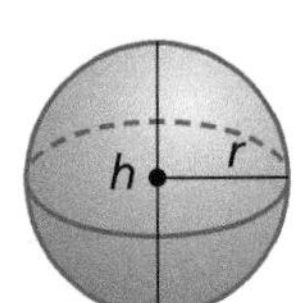

WARM-UP OPTIONS

ONLINE

ACTIVATE PRIOR KNOWLEDGE • Use Surface Area and Volume Formulas

Use these activities to quickly assess and activate prior knowledge as needed.

Problem of the Day 

A fish tank that is 24 inches long, 6 inches deep, and 13.5 inches high is filled so that the water level is at 80% capacity. How many cubic inches of water are in the tank? What is the surface area of the tank if the top is open?

Volume $= 1{,}555.2 \text{ in}^3$; $0.80(24 \times 6 \times 13.5)$

Surface area $= 954 \text{ in}^2$;

base: (24×6)

front and back: $(2 \times 24 \times 13.5)$

left and right: $(2 \times 6 \times 13.5)$

$(24 \times 6) + (2 \times 24 \times 13.5) + (2 \times 6 \times 13.5)$

$144 + 648 + 162 = 954 \text{ in}^2$

Quick Check for Homework

As part of your daily routine, you may want to display the Teacher Solution Key to have students check their homework.

Make Connections

Based on students' responses to the Problem of the Day, choose one of the following:

Project the Interactive Reteach, Grade 7, Lesson 11.4.

Complete the Prerequisite Skills Activity:

Have students work in pairs and give each pair a volume, such as 725 cubic units. Ask students to find three dimensions of a rectangular prism or cube that are appropriate for each volume. Dimensions should be whole numbers greater than 1. Then, have students use the dimensions to find the prism's surface area.

- *What numbers is 725 divisible by?* 25
- *What is 725 divided by 25?* 29
- *What are the factors of 29?* 29 and 1
- *What other number is 725 divisible by?* 5
- *Can a dimension of 5 be used to find two other whole number dimensions? Explain.* yes; $725 \div 5 = 145$, and $145 \div 5 = 29$, so three possible dimensions are 5, 5, and 29.

If students continue to struggle, use Tier 3 Skill 11.

SHARPEN SKILLS

If time permits, use this on-level activity to build fluency and practice basic skills.

Mental Math

Objective: Students estimate volumes using mental math.
Materials: index cards

Have students work in groups and give each group a set of 4 index cards. Students draw a rectangular prism or cube on each card and label the dimensions. Students then take turns calculating or estimating the volume of each figure using mental math only.

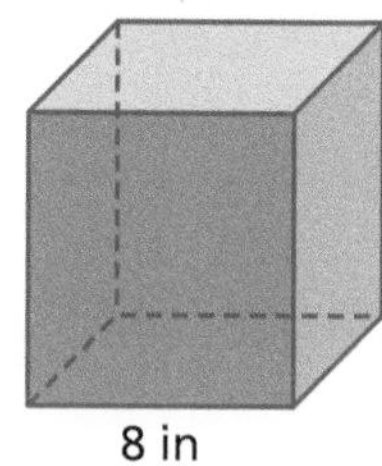

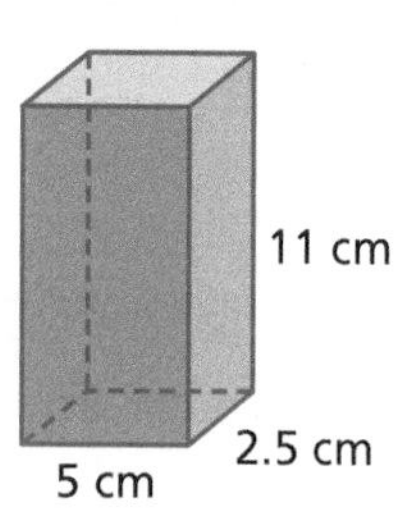

PLAN FOR DIFFERENTIATED INSTRUCTION

Small-Group Options

Use these teacher-guided activities with pulled small groups.

On Track

Display the following information:

- diameter of a gumball = 1 inch
- diameter of hollow center = $\frac{1}{8}$ inch
- 1 case of gumballs costs \$51.00
- there are 850 gumballs in 1 case

Have students estimate the cost of gum, per cubic inch, to the nearest cent.

Almost There

Materials: index cards

Have students work in pairs and give each pair a set of index cards that each show the volume formula for a solid with the value of any dimensions (i.e. edge length, height, or radius) substituted into the formula. Have students do the following:

- state the name of the solid
- name and state the values of the solid's dimensions (radius, diameter, height, edge length, etc.)
- find the volume of the solid

Ready for More

Display the following information:

$V = 18{,}816.6$ units^3

Have students find the dimensions of a cube and a sphere by applying the solid's volume formula and the given volume. Then, have students find the surface area of each solid.

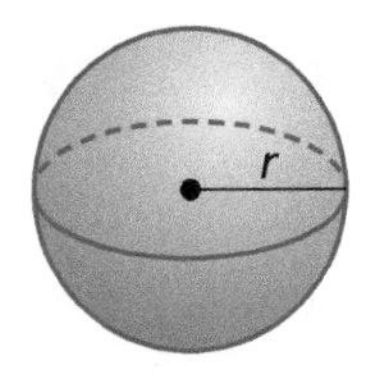

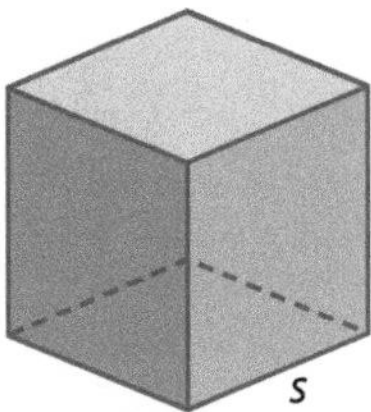

Math Center Options

Use these student self-directed activities at centers or stations.

Key: ● Print Resources ● Online Resources

On Track

- ● Interactive Digital Lesson
- ●● Journal and Practice Workbook
- ● Module Performance Task

Standards Practice:

- ●● Give Informal Arguments for Area, Circumference, and Volume Formulas
- ●● Use Volume Formulas for Cylinders, Pyramids, Cones, and Spheres
- ●● Use Properties of Geometric Shapes to Describe Objects
- ●● Apply Concepts of Density in Modeling Situations
- ●● Apply Geometric Methods to Solve Design Problems

Almost There

- ● Reteach 19.3 (printable)
- ● Interactive Reteach 19.3
- ● RtI Tier 3 Skill 11: Volumes of Right Rectangular Prisms
- ● Illustrative Mathematics: Doctor's Appointment

Ready for More

- ● Challenge 19.3 (printable)
- ● Interactive Challenge 19.3
- ● Illustrative Mathematics: Centerpiece

Unit Project Check students' progress by asking to see their equation for the volume of the sphere and to see their setup for calculating the difference of two spheres.

View data-driven grouping recommendations and assign differentiation resources.

During the *Spark Your Learning,* listen and watch for strategies students use. See samples of student work on this page.

Use a Cylinder to Model

Strategy 1

I can model one tomato using cylinders. One tomato has a diameter of 1 inch, or radius of $\frac{1}{2}$ inch, and a height of 1 inch.

Volume of tomato $= \pi\left(\frac{1}{2}\right)^2(1) \approx 0.785 \text{ in}^3$

Volume of container $= 3.25 \times 3.25 \times 2.875 \approx 30.37 \text{ in}^3$

$\frac{30.37}{0.785} \approx 39$ tomatoes

If students ... use the formula for volume of a cylinder to estimate the volume of 1 tomato, they are displaying a strong understanding of how to use models to make estimates of measures they do not yet know how to calculate using a formula.

Have these students ... explain how they determined their model and equation. **Ask:**

- Q How did you decide on your dimensions?
- Q Is there another solid you could have used to model the tomato?

Use Estimation Techniques

Strategy 2

Since the container is square, the dimension of 2.875 must be the height of the container. When I count all of the tomatoes that I see in one container, I count 13, so each layer of tomatoes contains about 13 tomatoes, and there are about 2.875 layers of tomatoes.

$2.875 \times 13 \approx 38$ tomatoes

If students ... use the height of the container and the number of tomatoes shown visible in the container to estimate the number of tomatoes, they have a solid understanding of estimation techniques learned in previous grades, but may not be making the connection between the information given in the problem and the volume formulas they have learned in previous lessons.

Activate prior knowledge ... by having students make a list of all of the volume formulas they have learned so far. **Ask:**

- Q What geometric solids do you already know the volume formula for?
- Q Which of those solids can be used to model the tomato in this situation?

COMMON ERROR: Uses Surface Area Formula

Container size $= 3.25 \times 3.25 \times 2.875 \approx 30.37$

1 tomato $= 4\pi r^2 = 4\pi\left(\frac{1}{2}\right)^2 = 4\left(\frac{1}{4}\right)\pi = \pi$

≈ 3.14

$\frac{30.37}{3.14} \approx 10$ tomatoes

If students ... use the formula for surface area of a sphere, they may not have a strong understanding of the difference between the capacity and the area of the surface of an object.

Then intervene ... by pointing out that since we are looking for the number of tomatoes that fill a container, the amount of space 1 tomato takes up must be considered. **Ask:**

- Q What did you do to determine the amount of space inside the container?
- Q What does the formula you used tell you about the tomato?

19.3

Volumes of Spheres

I Can use the formula for the volume of a sphere to calculate the volumes of composite figures.

Spark Your Learning

Cherry tomatoes are often sold in pint baskets.

©David Kay/Shutterstock

Complete Part A as a whole class. Then complete Parts B–D in small groups.

A. What is a mathematical question you can ask about this situation? What information would you need to know to answer your question? A. How many cherry tomatoes can fit in a pint basket?; the dimensions of a cherry tomato

B. What variable(s) are involved in this situation? What unit of measurement would you use for each variable? B, D. See Additional Answers.

C. To answer your question, what strategy and tool would you use along with all the information you have? What answer do you get? See Strategies 1 and 2 on the facing page.

D. Does your answer make sense in the context of the situation? How do you know?

Turn and Talk Suppose the cherry tomatoes of another variety have, on average, the same volume as the ones above, but are shaped more like eggs than spheres. How could you decide whether more or fewer would fit in the same size basket? See margin.

Module 19 • Lesson 19.3 601

CULTIVATE CONVERSATION • Information Gap

Ask students questions to help them decide what information they need to answer the question, "How many cherry tomatoes can fit in a pint basket?"

Do you have enough information to determine the number of cubic inches that the pint container holds? Yes, I know the dimensions of the pint container, so I can find its volume.

What do you need to do to determine how many tomatoes will fit in a container? Divide the total volume of the container by the volume of 1 tomato.

What dimension do you need to determine the volume of each tomato? I need to know the radius of 1 tomato.

1 Spark Your Learning

▶ MOTIVATE 

- Have students look at the photo in their books and read the information contained in the photo. Then complete Part A as a whole-class discussion.
- Give the class the additional information they need to solve the problem. This information is available online as a printable and projectable page in the Teacher Resources.
- Have students work in small groups to complete Parts B–D.

▶ PERSEVERE

If students need support, guide them by asking:

Q **Advancing • Use Tools** Which tool could you use to solve the problem? Why choose that tool and not some other? Students' choices of tools and reasons for choosing them will vary.

Q **Assessing** What solids could you use to model the tomatoes and the basket? Possible answers: cylinder or sphere and rectangular prism

Q **Assessing** What information can the dimensions of the baskets allow you to find? They allow you to calculate the volume of each basket.

Q **Advancing** What parameters do you think would be involved in writing a formula that calculates the volume of a tomato, and why? Possible answer: Since a tomato is roughly spherical, which is circular, the formula will have π in it. A sphere's volume is a fraction of a cylinder's volume, so there will be a fraction involved as well. And since the cross section of a sphere is a circle, the formula will involve the radius of that circle as well.

Turn and Talk Students should think about how the change in the shape of the tomato affects the figure they used to represent the original tomato. Possible answer: I could estimate the volume of the new variety of cherry tomato using a cylinder that has a greater height and a lesser radius.

▶ BUILD SHARED UNDERSTANDING

Select groups of students who used various strategies and tools to share with the class how they solved the problem. As they present their solutions, have each group discuss why they chose a specific strategy and tool.

2 Learn Together

Build Understanding

Task 1 **MP** **Use Structure** Students compare the volume of a hemisphere to that of a cylinder with a cone removed from its interior. They visualize the relationship between the dimensions and volumes to derive a formula for the volume of a sphere.

Sample Guided Discussion:

Q **In Part A, why is it necessary to call the radius of the cross-section R when the radius of the base of the hemisphere is called r?** The radius of the cross section of the hemisphere is not equal to the radius of the hemisphere's base, so we must use two different variables.

Q **How do you know that the height, h, of the cylinder with the cone removed is equal to r?** The height of the cylinder is the same as the radius of the hemisphere, which is represented by the variable r.

Turn and Talk Have students discuss what would happen to the volume of the cylinder if the radius and height doubled. Let r be the radius of the initial sphere. Let $2r$ be the radius of the enlarged sphere. When you do the calculations, you find that the volume increases by 8, or 2^3.

Build Understanding

Develop a Formula for the Volume of a Sphere

1 To develop a formula for the volume of a sphere, compare one of its hemispheres to a cylinder with the same height and radius from which a cone has been removed.

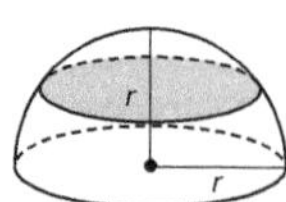

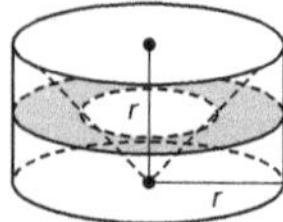

A. Use the Pythagorean Theorem to find the area of a cross section of the hemisphere. A, B. See Additional Answers.

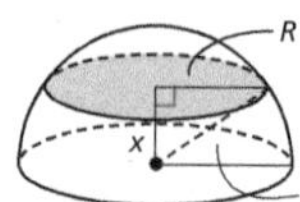

B. A cross section of the cylinder with the cross section of the cone removed is a ring. How can you find the area of the ring? What is the area of the ring?

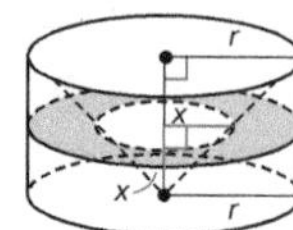

C. What do you notice about the area of the cross section of the hemisphere and the area of the cross section of the cylinder with the cone removed? The areas are equal.

D. Find the volume of the cylinder with the cone removed. Explain the steps you use and your reasoning. D, E. See Additional Answers.

E. How can you use Cavalieri's Principle to find the volume of the hemisphere? Use this result to write a formula for the volume of a sphere with radius r.

F. Find the volume of the sphere using your formula from Part E. Leave your answer in terms of π. 972π in^3

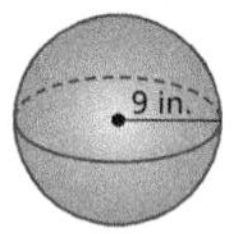

Turn and Talk How can you determine the increase in volume if the radius of a sphere doubles? See margin.

602

LEVELED QUESTIONS

Depth of Knowledge (DOK)	Leveled Questions	What Does This Tell You?
Level 1 **Recall**	What three-dimensional solid has a volume that is equal to $\frac{1}{3}\pi r^2 h$? a cone	Students' answers will indicate whether they can recall formulas for solids they have previously explored.
Level 2 **Basic Application of Skills & Concepts**	In terms of π, what is the volume of a cylinder that has a base with a diameter of 29 and a height of 22? $V = \pi(14.5)^2(22) \approx 4625.5\pi$ cubic units	Students' answers will indicate whether they can apply the formula for volume of a cylinder.
Level 3 **Strategic Thinking & Complex Reasoning**	A cone and a cylinder has the same height and base radius. If the radius of the cylinder is doubled, what is the ratio of the volume of the original cone to the new cylinder? The original ratio is 1:3. When the radius is doubled the volume is quadrupled, so the new ratio is 1:12.	Students' answers will indicate whether they know how to apply changes in dimensions to changes in the volume when comparing volumes of cylinders and cones.

Step It Out

Use a Volume Formula to Solve a Real-World Problem

Volume of a Sphere

The volume of a sphere with radius r is given by $V = \frac{4}{3}\pi r^3$.

 BTU (British Thermal Units) is a common unit in the United States for measuring the amount of heat energy or heat capacity used.

The spherical propane storage tank shown is full of propane. What is the heating capacity of the propane in BTUs? Use the fact that 1 cubic foot of propane contains 2516 BTUs.

The radius of the tank is 30 feet.

A. What steps do you need to take to answer this question?
See margin.

Find the volume of propane.

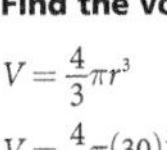

$V = \frac{4}{3}\pi r^3$

$V = \frac{4}{3}\pi (30)^3$

$V = 36{,}000\pi$

$V \approx 113{,}097$

The volume of the propane is about 113,097 cubic feet.

B. What information in the problem statement tells you that you need to find the volume?

B. The problem gives the conversion of BTUs to cubic feet, a unit that measures volume.

Find the number of BTUs.

$113{,}097 \cdot 2516 = 284{,}552{,}052$

C. What are the units of the conversion factor? C. BTUs per cubic foot.

The heating capacity of the propane is about 284,552,052 BTUs.

D. What is the heating capacity of a full spherical propane storage tank with a radius of 2 feet?
about 84,312 BTUs

 **Turn and Talk** How would the process to find the heating capacity change if you were given the diameter instead of the radius of the propane storage tank? See margin.

Step It Out

Task 2 (MP) **Attend to Precision** Students not only calculate a quantity, but then convert the quantity to a different unit of measure using a given rate. Remind students to show the steps and label all answers in those steps with units.

OPTIMIZE OUTPUT Critique, Correct, and Clarify

Have students work with a partner to discuss what would happen if a student made the mistake of using the formula $V = \frac{4}{3}\pi r^2$ to find the volume of the tank instead of $V = \frac{4}{3}\pi r^3$. Have them share why using the wrong formula in this task does not change the final numeric value, but why in most cases, the conclusions drawn by making this mistake would be inaccurate. Encourage students to use mathematical vocabulary in their explanations.

Sample Guided Discussion:

Q How can you use a proportion to prove that the number of BTUs is found by multiplying the volume by 2,516? You can use a proportion that sets the given conversion rate equal to the ratio of the tank's volume to its capacity in BTUs: $\frac{1\text{ ft}^3}{2516\text{ BTUs}} = \frac{113{,}097\text{ ft}^3}{x\text{ BTUs}}$. To solve this, you multiply 113,097 by 2516.

 **Turn and Talk** Students should parallel the process of first finding the volume with the process of finding the area of a circle when given the diameter. Divide the diameter by 2 to find the radius, and then find the heating capacity as before.

ANSWERS

A. Find the volume of the propane. Then multiply the volume by the number of BTUs per cubic foot.

PROFICIENCY LEVEL

Beginning

Write the formula $V = \frac{1}{3}\pi (2)^2(6)$ and say, "This formula is used to find the volume of a cone that has a base with a radius of 2, a diameter of 4, and a height of 6." Then, write the formula $V = \frac{4}{3}\pi (5)^3$ and ask students to describe the name of the solid and the measure of its radius, diameter, and height.

Intermediate

Have students briefly explain the relationship between the volume of a sphere and the volume of a cone with the same height and a base congruent to a cross section through the center of the sphere.

Advanced

Have students write a paragraph to explain why the formula for volume of a cone contains the fraction $\frac{1}{3}$ and r^2 but the formula for volume of a sphere contains the fraction $\frac{4}{3}$ and r^3.

Task 3  **Model with Mathematics** Students use a sphere to estimate the volume of the canopy of a tree and investigate the changes in dimensions and volume as the radius or volume changes.

Sample Guided Discussion:

Q Does finding the volume of the canopy involve evaluating an expression or solving an equation? Finding the volume involves evaluating the expression $\frac{4}{3}\pi r^3$ by substituting known dimensions into the expression.

Q In Part B, how is the process when solving for the diameter of a canopy with a volume of 5,000 cubic feet different than the process used to find the volume in Part A? The process is different because you have to substitute 5,000 in place of the volume in the formula and solve for the value of the radius instead of evaluating an expression for the value of the volume.

Estimate Volume in a Real-World Situation

You can use the volume formula to estimate the volume of a real-world object that is close to, but not exactly, a sphere.

3 The canopy of the tree is nearly spherical. The diameter of the canopy is 25 feet. Estimate the volume of the canopy.

You can use the formula for the volume of a sphere to estimate the volume.

$$V = \frac{4}{3}\pi r^3$$

Substitute the value of the radius of the canopy and simplify.

$$V = \frac{4}{3}\pi r^3$$

$$V = \frac{4}{3}\pi(12.5)^3$$

$$V = \frac{4}{3}\pi(1953.125)$$

$$V \approx 8181.2$$

A. Why is 12.5 substituted for r instead of 25?

A. 25 is the diameter of the canopy. The formula for the volume of a sphere uses the radius, which is half of the diameter.

The volume of the canopy of the tree is about 8181 cubic feet.

Suppose a landscaping company trims the tree so that the canopy is still spherical but has a new volume of 5000 cubic feet. What is the diameter of the canopy after trimming?

$$V = \frac{4}{3}\pi r^3$$

$$5000 = \frac{4}{3}\pi r^3$$

$$1193.66 \approx r^3$$

$$10.6 \approx r$$

B. Why is this formula used when it doesn't involve the diameter of the tree?

B. You can multiply the radius found using this formula by 2 to find the diameter.

C. How is the equation with r^3 solved for r?

C. Use the definition of the cube root to solve for r.

The radius of the canopy is 10.6 feet after trimming, so the diameter is 21.2 feet.

When the diameter of the canopy grows to 24 feet, it will be trimmed again. How many cubic feet of the canopy must be trimmed so the canopy's diameter is cut back to 21.2 feet?

$$V = \frac{4}{3}\pi r^3$$

$$V = \frac{4}{3}\pi(12)^3$$

$$V = \frac{4}{3}\pi(1728)$$

$$V \approx 7238.2$$

D. What does this value represent?

D. The volume of the tree when the diameter of the canopy reaches 24 feet.

E. How is this value determined?

About 2238 cubic feet of the canopy must be trimmed for the diameter to be 21.2 feet.

E. The value is determined by finding the difference between the volume of the canopy with a 21.2-foot diameter and the volume with a 24-foot diameter.

Find the Volume of a Composite Figure

You can use the formula for the volume of a sphere to help find the volume of composite figures that contain hemispheres.

4 Find the volume of the composite figure.

A. What figures make up the composite figure?

A. cone and hemisphere

Find the volume of the cone.

$V = \frac{1}{3}\pi r^2 h$

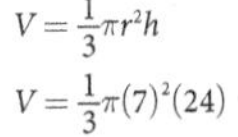

$V = \frac{1}{3}\pi(7)^2(24)$

B. How do you know that the radius of the base of the cone is 7 cm?

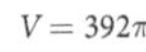

$V = 392\pi$

B–D. See margin.

$V \approx 1231.5$

The volume of the cone is approximately 1,231.5 cubic centimeters.

Find the volume of the hemisphere.

$V = \frac{2}{3}\pi r^3$

$V = \frac{2}{3}\pi(7)^3$

C. Why is the formula for the volume of a hemisphere $V = \frac{2}{3}\pi r^3$?

$V \approx 718.4$

The volume of the hemisphere is approximately 718.4 cubic centimeters.

Find the volume of the composite figure.

Add the volumes of the cone and the hemisphere.

$V = 1231.5 + 718.4 = 1949.9$

The volume of the composite figure is approximately 1950 cubic centimeters.

The composite figure below shows a hemisphere on top of a cube.

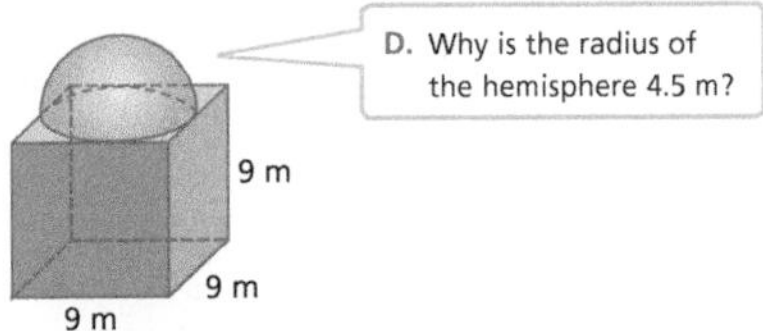

D. Why is the radius of the hemisphere 4.5 m?

Turn and Talk How would you find the volume of a composite figure? See margin.

Task 4 (MP) **Use Structure** Students explore composite figures made up of hemispheres, cones, and cubes, and find the volume of these composite figures.

Sample Guided Discussion:

Q What is the volume of the composite figure comprised of the hemisphere and cone in terms of π?

620.7π cubic inches

Q In Part D, what is the relationship between the edge length of the cube and the radius of the hemisphere?

The edge length of the cube is equivalent to the diameter of the hemisphere, so the radius is half the edge length, or 4.5 meters.

Turn and Talk Have students consider how they might estimate the volume of real-life objects using a model made from known figures.

Possible answer: Find the volume of each part of the composite figure and add the volumes.

ONLINE

Assign the **Modeling Success Extension** for this lesson.

Student and teacher resources available for download.

ANSWERS

B. The cone and the hemisphere have the same base. The radius of the hemisphere is 7 cm, so the radius of the cone is 7 cm.

C. The volume of a sphere is $V = \frac{4}{3}\pi r^3$, and a hemisphere is half of a sphere, so $V = \frac{1}{2}\left(\frac{4}{3}\pi r^3\right) = \frac{2}{3}\pi r^3$.

D. The base of the hemisphere sits on top of the cube. The length of a side of the cube equals the length of the diameter of the circular base of the hemisphere so the radius of the circular base is one-half the diameter.

ONLINE

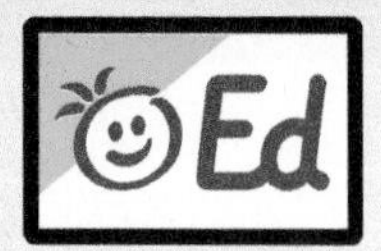

Assign the Digital On Your Own for

- built-in student supports
- Actionable Item Reports
- Standards Analysis Reports

On Your Own

Assignment Guide

The chart below indicates which problems in the On Your Own are associated with each task in the Learn Together. Assign daily homework for tasks completed.

Learn Together Tasks	On Your Own Problems
Task 1, p. 602	Problems 7 and 20
Task 2, p. 603	Problems 9–12 and 21
Task 3, p. 604	Problems 8, 14–17, and 22–24
Task 4, p. 605	Problems 13 and 18–19

Check Understanding

Find the volume of each sphere. Round your answer to the nearest tenth.

1. 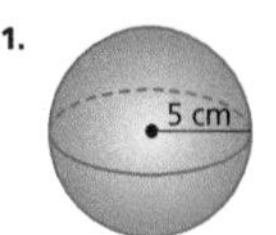

523.6 cm^3

2. 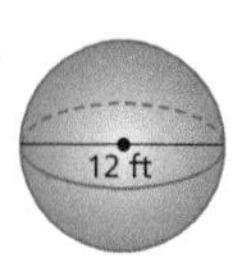

904.8 ft^3

3. Su is making a model of Earth out of papier mâché. The diameter of her model is 13 inches. What is the volume of her model? Round your answer to the nearest tenth of a cubic inch. 1150.3 in^3

4. A chickpea is approximately spherical. Chickpeas have a diameter of about 9 millimeters. Estimate the volume of a chickpea. Round your answer to the nearest tenth of a cubic millimeter. about 381.7 mm^3

Find the volume of each composite figure. Round your answer to the nearest tenth.

5. 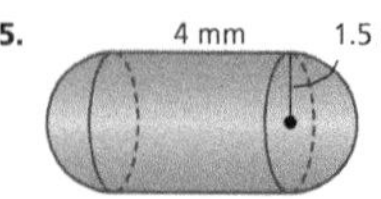

42.4 mm^3

6. 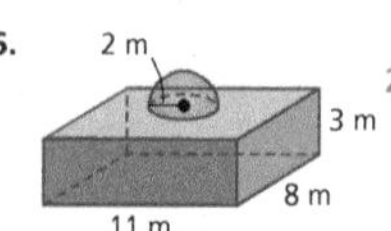

280.8 m^3

On Your Own

7. (MP) **Critique Reasoning** Dawn states that if you double the radius of a sphere, you double the volume of the sphere. Is she correct? Explain your reasoning. Dawn is incorrect; If you double the radius of a sphere, the volume is 8 times as large.

8. A trout lays eggs that are approximately spherical. The eggs have a diameter of about 3.5 millimeters. Estimate the volume of one egg. Round your answer to the nearest tenth of a cubic millimeter. about 22.5 cubic millimeters

Find the volume of each sphere. Leave the answer in terms of π.

9. 

$\frac{32}{3}\pi \text{ m}^3$

10. 

$36\pi \text{ in}^3$

606

3 Check Understanding

Formative Assessment

Use formative assessment to determine if your students are successful with this lesson's learning objective.

Students who successfully complete the Check Understanding can continue to the On Your Own practice.

For students who miss 1 problem or more, work in a pulled small group using the Almost There small-group activity on page 601C.

ONLINE

Assign the Digital Check Understanding to determine

- success with the learning objective
- items to review
- grouping and differentiation resources

4 Differentiation Options

Differentiate instruction for all students using small-group activities and math center activities on page 601C.

Reteach

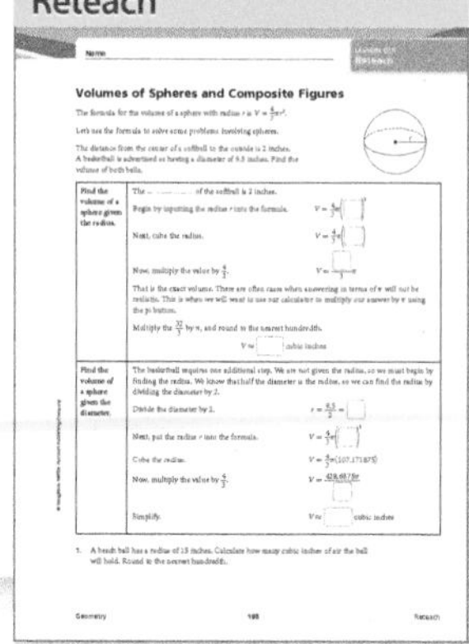

Volumes of Spheres and Composite Figures

Challenge

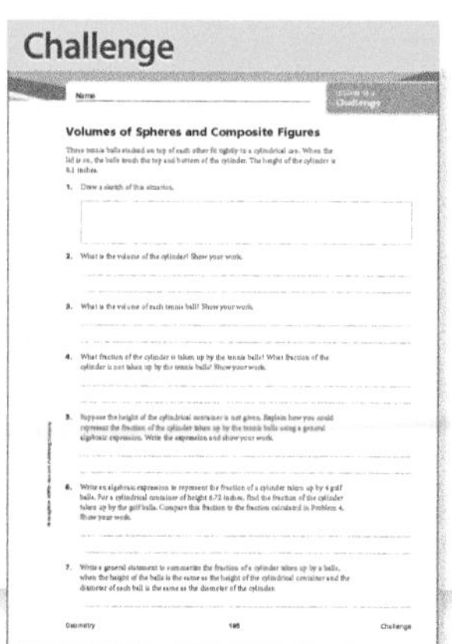

Volumes of Spheres and Composite Figures

Find the volume of each hemisphere. Leave the answer in terms of π.

11. 

$486\pi\ \text{mm}^3$

12. 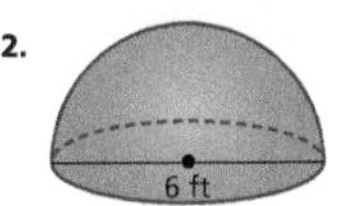

$18\pi\ \text{ft}^3$

13. A homeowner installs a propane tank with the dimensions shown. The tank is then filled with propane. What is the heating capacity of the propane in BTUs? Round your answer to the nearest whole number. (1 cubic foot of propane contains 2516 BTUs.) 73,773 BTUs

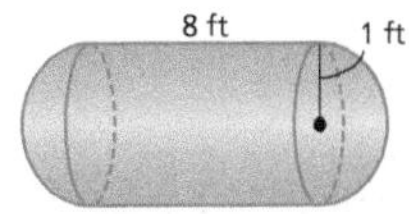

14. A landscaper is trimming shrubs. Each shrub is nearly spherical, with a diameter of 4 feet. Estimate the volume of a shrub. Round your answer to the nearest tenth of cubic foot. $33.5\ \text{ft}^3$

The diameter of each shrub is about 4 feet.

15. A sphere has a volume of 697 cubic meters. What is the radius of the sphere? Round your answer to the nearest tenth of a meter. 5.5 m

16. A sphere has a volume of 2482 cubic inches. What is the diameter of the sphere? Round your answer to the nearest tenth of an inch. 16.8 in.

17. A basketball has a circumference of 29.5 inches. What is the volume of the basketball? Round your answer to the nearest cubic inch. $434\ \text{in}^3$

Find the volume of each composite figure. Round your answer to the nearest tenth.

18. 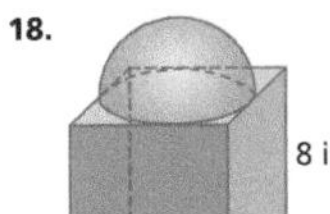
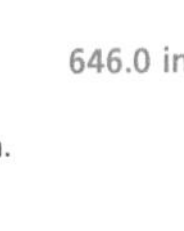

$646.0\ \text{in}^3$

19. 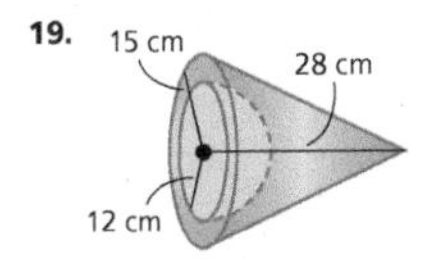

$2978.2\ \text{cm}^3$

20. Jenna has a scoop that is shaped like a hemisphere. She is filling the scoop with water and is dumping the water in a bucket that is a cylinder with the dimensions shown. How many scoops of water will it take to fill the bucket? 20 scoops

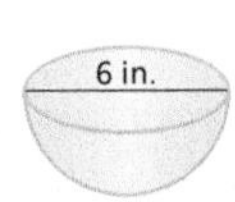

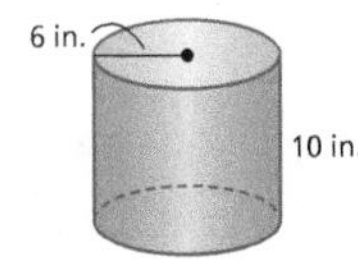

21. A cube has a side length of 14 centimeters. What is the volume of the largest sphere that can fit inside of the cube? Leave your answer in terms of π. $\frac{1372}{3}\pi\ \text{cm}^3$

Questioning Strategies

Problem 11 Suppose the radius of the hemisphere is tripled. What effect does this have on the volume of the hemisphere? The volume of the hemisphere becomes 27 times larger.

Problem 18 Suppose the hemisphere on top of the cube were rotated 180° and placed inside the cube. How would the process for finding the volume change? The volume would be found by subtracting the volume of the hemisphere from the volume of the cube instead of adding the two volumes to find the answer.

Watch for Common Errors

Problem 13 Be sure students understand that in order to find the heating capacity of a solid, you must know the volume of the solid first. Students may try and solve for the heating capacity without realizing that this problem is a two-step process. You can clear up this confusion by asking them to concentrate on the units of the conversion ratio given, 1 cubic foot of propane contains 2516 BTUs. The units are cubic feet and BTUs. So to find the heating capacity, you must have a number of cubic feet to convert to BTUs.

Problem 19 The most common mistake made when finding the volume of a cone is students forget to divide by three. The first step students should take when beginning a problem involving volume, in this case volume of a composite solid, is to make a decision as to whether or not the shapes involved are prisms or pyramids/cones. Once they have decided what solids are involved, they should write the formulas for each.

5 Wrap-Up

Summarize learning with your class. Consider using the Exit Ticket, Put It in Writing, or I Can scale.

Exit Ticket 

A basketball has a diameter of 9.5 inches and a softball has a diameter of 3.8 inches. How much greater is the volume of the basketball, in terms of π? Volume of basketball $= \frac{4}{3}\pi(9.5)^3 = 1143.2\pi$ cubic inches, volume of softball $= \frac{4}{3}\pi(3.8)^3 = 73.2\pi$ cubic inches, $1143.2\pi - 73.2\pi = 1070\pi$ cubic inches; The volume of the basketball is 1070π cubic inches greater than the volume of the softball.

Put It in Writing 

Describe the difference between finding the sum of the volumes of a cone and a sphere with equal radii, and finding the volume of a composite solid made up of the same cone with the sphere placed on top of it.

I Can 

The scale below can help you and your students understand their progress on a learning goal.

4	I can use the formula for the volume of a sphere to calculate the volumes of composite figures and explain my reasoning to others.
3	I can use the formula for the volume of a sphere to calculate the volumes of composite figures.
2	I can identify the formula for the volume of a sphere.
1	I can identify and name the characteristics of a sphere and its volume.

Spiral Review • Assessment Readiness

These questions will help determine if students have retained information taught in the past and can also prepare them for high-stakes assessments. Here, students calculate the volume of a square pyramid **(19.2)**, calculate the volume of a cylinder **(19.1)**, and classify triangles by their angles and sides **(9.1)**.

22. A bead is formed by drilling a cylindrical hole with a 2-millimeter diameter through a sphere with a 6-millimeter diameter. Estimate the volume of the bead to the nearest cubic millimeter. 94 mm^3

23. A cube-shaped box has sides 24 centimeters long. A sphere with a diameter of 24 centimeters is placed inside the box. How much empty space is in the box? Round your answer to the nearest cubic centimeter. 6586 cm^3

24. A golf ball is nearly spherical. The volume of a golf ball is about 2.48 cubic inches.

A. Estimate the diameter of a golf ball. Round your answer to the nearest hundredth of an inch. 1.68 inches

B. A golf ball weighs about 1.62 ounces. Find the density of the golf ball. Round your answer to the nearest hundredth. about 0.65 ounce per cubic inch

Spiral Review • Assessment Readiness

25. Find the volume of the pyramid.

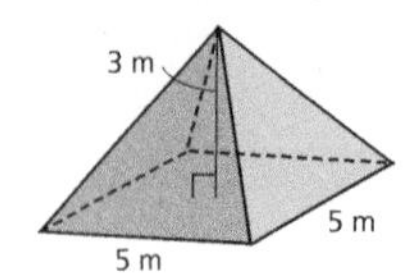

Ⓐ 13 m^3
Ⓑ 25 m^3
Ⓒ 55 m^3
Ⓓ 75 m^3

26. Find the volume of the cylinder

Ⓐ $37\pi\text{ in}^3$
Ⓑ $55\pi\text{ in}^3$
Ⓒ $110\pi\text{ in}^3$
Ⓓ $275\pi\text{ in}^3$

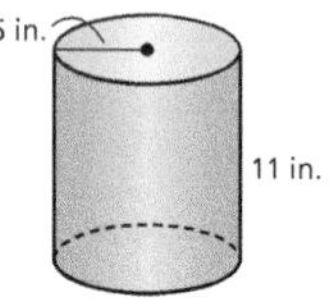

27. Classify each triangle given the following characteristics.

Characteristics	Isosceles	Equilateral	Scalene
Each vertex angle measures 60°.	?	?	?
Has side lengths of 3–3–5.	?	?	?
Has side lengths of 3–5–7.	?	?	?

I'm in a Learning Mindset!

How did I benefit by giving help with finding the volume of a sphere? What impact did it have on my learning outcome?

Keep Going ▶ Journal and Practice Workbook

Learning Mindset

Strategic-Help Seeking Identifies Need for Help

This lesson provides several examples where students will realize that at some point in the process of solving a problem, they may need assistance, or perhaps may be needed to offer support to a classmate. Willingness to lend support to a peer or ask questions to clear up your own confusion will lead students toward having stronger communication skills, as well as attaining a deeper understanding of mathematical content. *What tasks in this lesson did you find you needed the most support on? Was there a time when a classmate depended on you for some extra support when solving a problem? Did that experience of helping another student understand a concept result in a stronger understanding of the concept for you?*

Module 19

Review

Volume of Prisms and Cylinders

The volume of a prism and a cylinder is $V = Bh$, where B is the area of the base and h is the height.

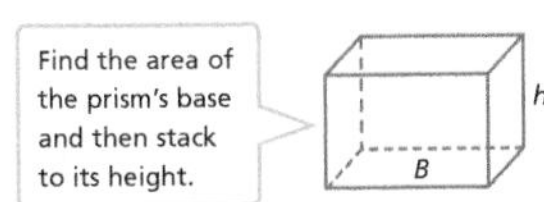

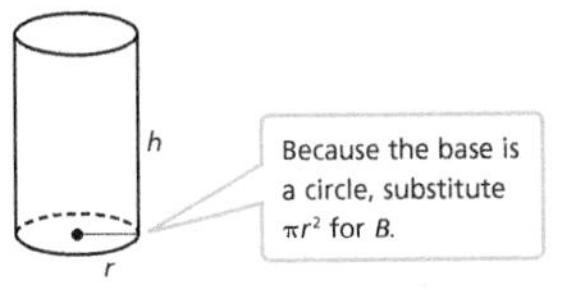

h
r
Because the base is a circle, substitute πr^2 for B.

Volume of Pyramids and Cones

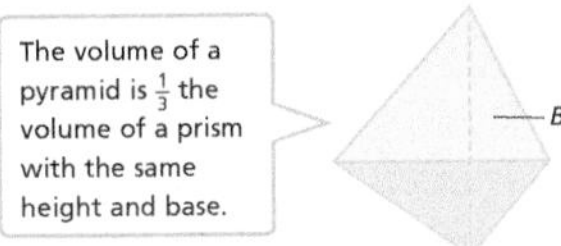

The volume of a pyramid is $\frac{1}{3}$ the volume of a prism with the same height and base.
B

The volume of a pyramid and a cone is $V = \frac{1}{3}Bh$ with base area B and height h.

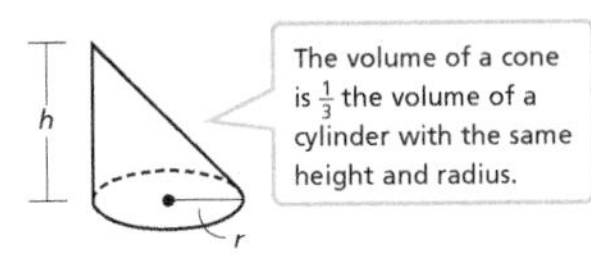

h
r
The volume of a cone is $\frac{1}{3}$ the volume of a cylinder with the same height and radius.

Volume of Spheres

The volume of a sphere with radius r is given by $V = \frac{4}{3}\pi r^3$.

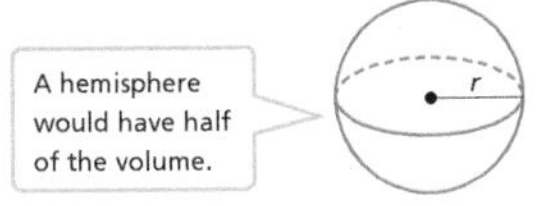

The radius of a hamster ball is 7 inches. The amount of space a hamster would have to run around in is $V = \frac{4}{3}\pi(7)^3 \approx 1437$ cubic inches.

Volume of Composite Figures

Decompose a complex figure into simpler parts. The volume of the complex figure is the sum of the volumes of the parts. Consider a hamster cage with a climbing tube.

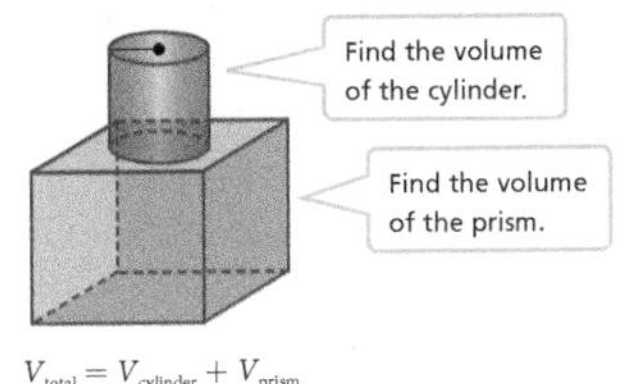

$V_{total} = V_{cylinder} + V_{prism}$

ONLINE

Assign the Digital Module Review for

- built-in student supports
- Actionable Item Reports
- Standards Analysis Reports

Module Review

Use the first page of the Module Review to summarize and connect the main ideas of the module. Use the second page to assess students' understanding of the vocabulary, concepts, and skills presented in the module.

Sample Guided Discussion:

Q Volume of Prisms and Cylinders How is finding the area of the prism's base and then stacking the height equal to the base times the height? Repeated addition is the same as multiplication. So, if the area of the base is repeatedly added while the base is stacked until the height is reached, that is the same as multiplying the area of the base by the height.

Q Volume of Pyramids and Cones Suppose the triangular pyramid and the cone have same height and volume. How can you find the radius of the cone if the base area of the pyramid is 5 square units? The pyramid and the cone have volume equal to one-third the product of the base area and the height. Since they have the same volume and height, the base areas are the same, so $\pi r^2 = 5$, or $r = \sqrt{\frac{5}{\pi}}$.

Q Volume of Spheres What is the formula for the volume of a hemisphere? What is the volume of a hemisphere with a radius of 7 inches? Because a hemisphere is half of a sphere, the volume is $\frac{2}{3}\pi r^3$. The volume of a hemisphere with a radius of 7 inches is $\frac{2}{3}\pi r^3 = \frac{2}{3}\pi(7)^3 \approx 718$ cubic inches.

Q Volume of Composite Figures How can you calculate the volume of the composite figure? Possible answer: Separate the composite figure into a cylinder and a prism. Use $\pi r^2 h$ to find the volume of the cylinder. Use Bh to find the volume of the prism. Then add the two volumes together.

Module Review continued

Possible Scoring Guide

Items	Points	Description
1–4	2 each	identifies the correct term
5	2	correctly determines the greater volume without making calculations
6, 7	2 each	correctly identifies the missing dimension
8	2	correctly finds the density of the water
9, 10	2 each	correctly finds the volume of the composite figure
11	2	correctly finds the amount of space left in the bowl
Total points possible = 22 points		

The Unit 9 Test in the Assessment Guide assesses content from Modules 18 and 19.

Vocabulary

Choose the correct term from the box to complete each sentence.

Vocabulary
cylinder
cone
prism
pyramid

1. A three-dimensional figure with a polygonal base and triangular sides that meet at a point is a __?__. pyramid
2. A three-dimensional figure with two circular bases and one curved side is a __?__. cylinder
3. A three-dimensional figure with two congruent polygonal bases and rectangular sides is a __?__. prism
4. A three-dimensional figure with a circular base and a curved lateral surface that connects the base to the vertex is a __?__. cone

Concepts and Skills

5. Melanie has stacked multiple decks of cards on a slant with a height of 5 inches. The length of each card is 3.5 inches. Jorge has also created a straight stack of cards with a height of 5 inches and the length of each card is 4.5 inches. All cards have the same width. Without making calculations, whose stack of cards has a greater volume? Explain your reasoning. See Additional Answers.

Find the missing dimension of each figure. Round your answer to the nearest whole number.

6. $V = 7{,}100 \text{ in}^3$ 43 in.

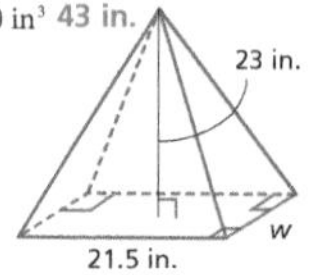

7. $V = 450 \text{ cm}^3$ 11 cm

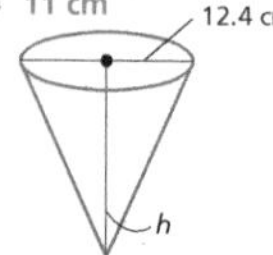

8. (MP) **Use Tools** A cylindrical tank of emergency water has a radius of 12.6 meters and a height of 10 meters. The mass of the water is 4,972,630 kilograms. What is the density of the water? Round to the nearest whole number. State what strategy and tool you will use to answer the question, explain your choice, and then find the answer. 997 kg/m³

Find the volume of each composite figure. Round your answer to the nearest whole number.

9.

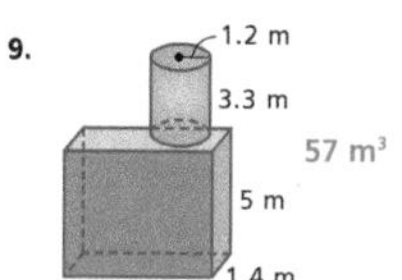

57 m³

10.

1006 yd³

11. A melon ball with a circumference of 7.85 centimeters is placed in a cylindrical bowl with a diameter of 12 centimeters and a height of 7 centimeters. How much space is left in the bowl? Round your answer to the nearest tenth. approximately 783.5 cm³

610

DATA-DRIVEN INSTRUCTION

Before moving on to the Module Test, use the Module Review results to intervene based on the table below.

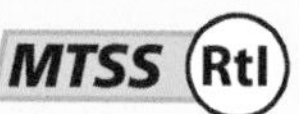

Items	Lesson	DOK	Content Focus	Intervention
5	19.1	3	Compare volumes of prisms without calculations.	Reteach 19.1
6, 7	19.2	2	Calculate the missing dimension of a pyramid and a cone.	Reteach 19.2
8	19.1, 19.2	3	Use mass and volume to find the density of a cylinder.	Reteach 19.1, 19.2
9, 10	19.1–19.3	2	Find the volume of composite figures.	Reteach 19.1–19.3
11	19.3	3	Apply the formula for the volume of a sphere to solve a real-world problem.	Reteach 19.3

Module Test

The Module Test is available in alternative versions in your Assessment Guide. All items are presented in standardized test formats.

ONLINE

Ed

Assign the Digital Module Test to power actionable reports including

- proficiency by standards
- item analysis

MODULE 19

TEST

Form A

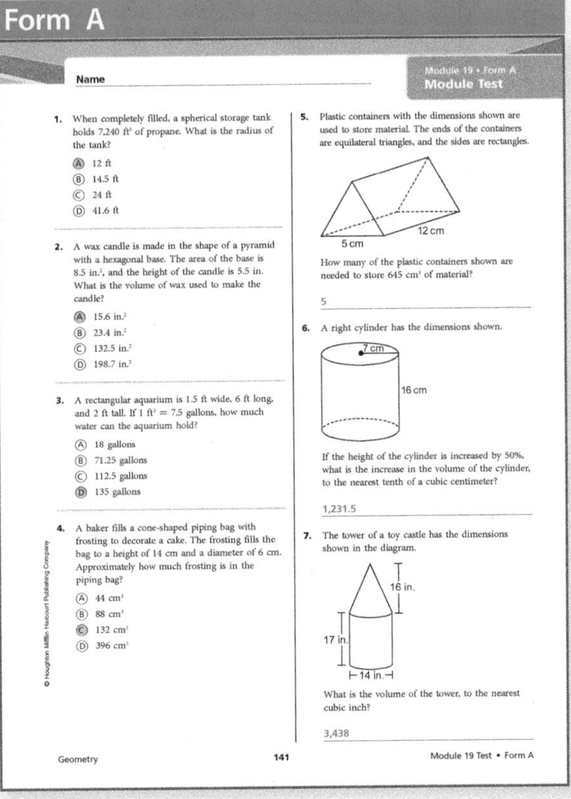

Name

Module 19 • Form A
Module Test

1. When completely filled, a spherical storage tank holds 7,240 ft³ of propane. What is the radius of the tank?
   Ⓐ 12 ft
   Ⓑ 14.5 ft
   Ⓒ 24 ft
   Ⓓ 41.6 ft

2. A wax candle is made in the shape of a pyramid with a hexagonal base. The area of the base is 8.5 in.², and the height of the candle is 5.5 in. What is the volume of wax used to make the candle?
   Ⓐ 15.6 in.³
   Ⓑ 23.4 in.³
   Ⓒ 132.5 in.³
   Ⓓ 198.7 in.³

3. A rectangular aquarium is 1.5 ft wide, 6 ft long, and 2 ft tall. If 1 ft³ = 7.5 gallons, how much water can the aquarium hold?
   Ⓐ 18 gallons
   Ⓑ 71.25 gallons
   Ⓒ 112.5 gallons
   Ⓓ 135 gallons

4. A baker fills a cone-shaped piping bag with frosting to decorate a cake. The frosting fills the bag to a height of 14 cm and a diameter of 6 cm. Approximately how much frosting is in the piping bag?
   Ⓐ 44 cm³
   Ⓑ 88 cm³
   Ⓒ 132 cm³
   Ⓓ 396 cm³

5. Plastic containers with the dimensions shown are used to store material. The ends of the containers are equilateral triangles, and the sides are rectangles.

   12 cm
   5 cm

   How many of the plastic containers shown are needed to store 645 cm³ of material?

   5

6. A right cylinder has the dimensions shown.

   7 cm
   16 cm

   If the height of the cylinder is increased by 50%, what is the increase in the volume of the cylinder, to the nearest tenth of a cubic centimeter?

   1,231.5

7. The tower of a toy castle has the dimensions shown in the diagram.

   16 in.
   17 in.
   14 in.

   What is the volume of the tower, to the nearest cubic inch?

   3,438

© Houghton Mifflin Harcourt Publishing Company

Geometry 141 Module 19 Test • Form A

Form A

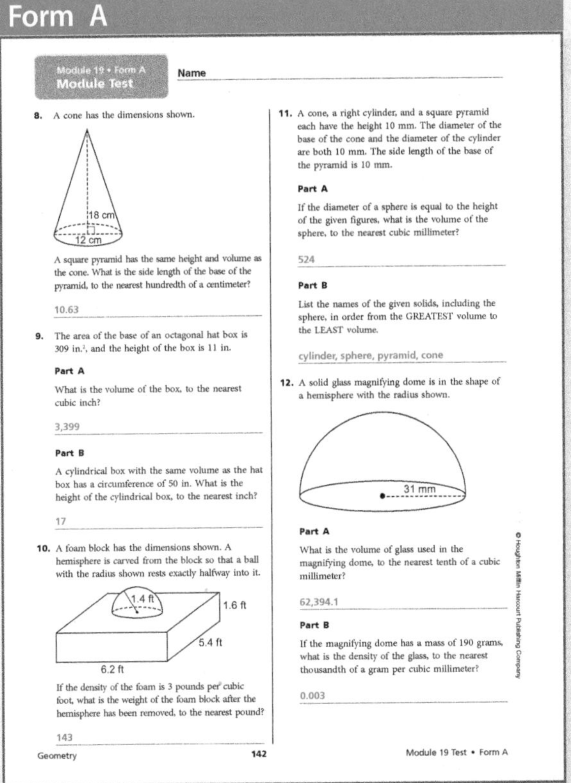

Module 19 • Form A
Module Test

Name

8. A cone has the dimensions shown.

   18 cm
   12 cm

   A square pyramid has the same height and volume as the cone. What is the side length of the base of the pyramid, to the nearest hundredth of a centimeter?

   10.63

9. The area of the base of an octagonal hat box is 309 in.², and the height of the box is 11 in.

   Part A

   What is the volume of the box, to the nearest cubic inch?

   3,399

   Part B

   A cylindrical box with the same volume as the hat box has a circumference of 50 in. What is the height of the cylindrical box, to the nearest inch?

   17

10. A foam block has the dimensions shown. A hemisphere is carved from the block so that a ball with the radius shown rests exactly halfway into it.

    1.4 ft
    1.6 ft
    5.4 ft
    6.2 ft

    If the density of the foam is 3 pounds per cubic foot, what is the weight of the foam block after the hemisphere has been removed, to the nearest pound?

    143

11. A cone, a right cylinder, and a square pyramid each have the height 10 mm. The diameter of the base of the cone and the diameter of the cylinder are both 10 mm. The side length of the base of the pyramid is 10 mm.

    Part A

    If the diameter of a sphere is equal to the height of the given figures, what is the volume of the sphere, to the nearest cubic millimeter?

    524

    Part B

    List the names of the given solids, including the sphere, in order from the GREATEST volume to the LEAST volume.

    cylinder, sphere, pyramid, cone

12. A solid glass magnifying dome is in the shape of a hemisphere with the radius shown.

    31 mm

    Part A

    What is the volume of glass used in the magnifying dome, to the nearest tenth of a cubic millimeter?

    62,394.1

    Part B

    If the magnifying dome has a mass of 190 grams, what is the density of the glass, to the nearest thousandth of a gram per cubic millimeter?

    0.003

© Houghton Mifflin Harcourt Publishing Company

Geometry 142 Module 19 Test • Form A

Form B

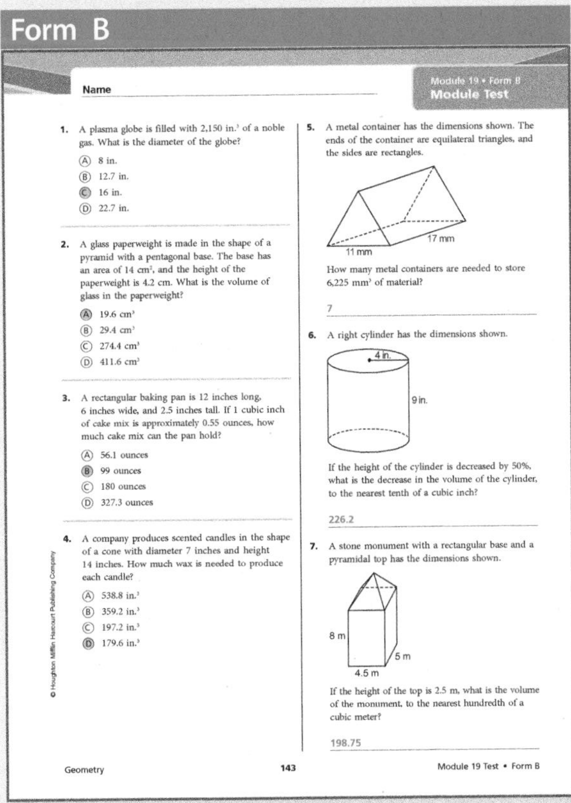

Name

Module 19 • Form B
Module Test

1. A plasma globe is filled with 2,150 in.³ of a noble gas. What is the diameter of the globe?
   Ⓐ 8 in.
   Ⓑ 12.7 in.
   Ⓒ 16 in.
   Ⓓ 22.7 in.

2. A glass paperweight is made in the shape of a pyramid with a pentagonal base. The base has an area of 14 cm², and the height of the paperweight is 4.2 cm. What is the volume of glass in the paperweight?
   Ⓐ 19.6 cm³
   Ⓑ 29.4 cm³
   Ⓒ 274.4 cm³
   Ⓓ 411.6 cm³

3. A rectangular baking pan is 12 inches long, 6 inches wide, and 2.5 inches tall. If 1 cubic inch of cake mix is approximately 0.55 ounces, how much cake mix can the pan hold?
   Ⓐ 56.1 ounces
   Ⓑ 99 ounces
   Ⓒ 180 ounces
   Ⓓ 327.3 ounces

4. A company produces scented candles in the shape of a cone with diameter 7 inches and height 14 inches. How much wax is needed to produce each candle?
   Ⓐ 538.8 in.³
   Ⓑ 359.2 in.³
   Ⓒ 197.2 in.³
   Ⓓ 179.6 in.³

5. A metal container has the dimensions shown. The ends of the container are equilateral triangles, and the sides are rectangles.

   17 mm
   11 mm

   How many metal containers are needed to store 6,225 mm³ of material?

   7

6. A right cylinder has the dimensions shown.

   4 in.
   9 in.

   If the height of the cylinder is decreased by 50%, what is the decrease in the volume of the cylinder, to the nearest tenth of a cubic inch?

   226.2

7. A stone monument with a rectangular base and a pyramidal top has the dimensions shown.

   8 m
   5 m
   4.5 m

   If the height of the top is 2.5 m, what is the volume of the monument, to the nearest hundredth of a cubic meter?

   198.75

© Houghton Mifflin Harcourt Publishing Company

Geometry 143 Module 19 Test • Form B

Form B

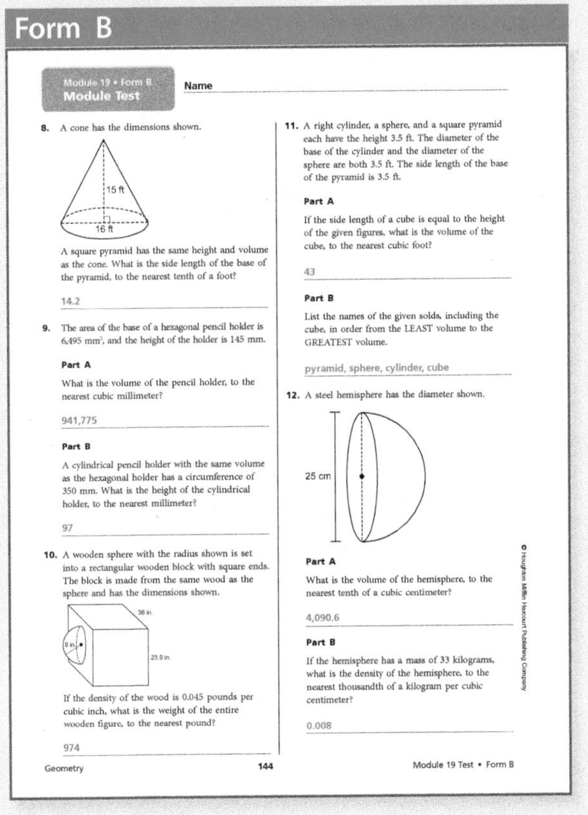

Module 19 • Form B
Module Test

Name

8. A cone has the dimensions shown.

   15 ft
   16 ft

   A square pyramid has the same height and volume as the cone. What is the side length of the base of the pyramid, to the nearest tenth of a foot?

   14.2

9. The area of the base of a hexagonal pencil holder is 6,495 mm², and the height of the holder is 145 mm.

   Part A

   What is the volume of the pencil holder, to the nearest cubic millimeter?

   941,775

   Part B

   A cylindrical pencil holder with the same volume as the hexagonal holder has a circumference of 350 mm. What is the height of the cylindrical holder, to the nearest millimeter?

   97

10. A wooden sphere with the radius shown is set into a rectangular wooden block with square ends. The block is made from the same wood as the sphere and has the dimensions shown.

    If the density of the wood is 0.045 pounds per cubic inch, what is the weight of the entire wooden figure, to the nearest pound?

    974

11. A right cylinder, a sphere, and a square pyramid each have the height 3.5 ft. The diameter of the base of the cylinder and the diameter of the sphere are both 3.5 ft. The side length of the base of the pyramid is 3.5 ft.

    Part A

    If the side length of a cube is equal to the height of the given figures, what is the volume of the cube, to the nearest cubic foot?

    43

    Part B

    List the names of the given solids, including the cube, in order from the LEAST volume to the GREATEST volume.

    pyramid, sphere, cylinder, cube

12. A steel hemisphere has the diameter shown.

    25 cm

    Part A

    What is the volume of the hemisphere, to the nearest tenth of a cubic centimeter?

    4,090.6

    Part B

    If the hemisphere has a mass of 33 kilograms, what is the density of the hemisphere, to the nearest thousandth of a kilogram per cubic centimeter?

    0.008

© Houghton Mifflin Harcourt Publishing Company

Geometry 144 Module 19 Test • Form B

UNIT 10

PROBABILITY

Module 20: Probability of Multiple Events

Module 21: Conditional Probability and Independence of Events

Genetic Counselor STEM

- **Say:** *Individuals or families visit with genetic counselors when they need information or have concerns about the risk of inherited conditions. Genetic counselors may perform DNA tests and analyze the results. They help patients to understand the results and to make informed decisions.*
- Explain that genetic counselors need an understanding of probability and statistics when analyzing test results. Then introduce the STEM task.

STEM Task

Ask students if they have any previous knowledge of genetics and heredity.

- Explain that parents pass characteristics and traits on to their children. Inherited traits are passed through DNA.
- The table shown is called a Punnett square. Punnett squares are used to determine the probability of a child having an inherited trait. Ask students if the Punnett square resembles a type of table with which they are familiar. Elicit from students that the Punnett square is similar to a two-way frequency table.
- Students should understand that there is an equal likelihood that a child will receive X^B or X^b from the mother and X^B or Y from the father.
- Have students think about the Punnett square as a two-way relative frequency table and have them calculate the probability for the outcomes X^BX^B, X^BX^b, X^BY, and X^bY. Ask students what the sum of the probabilities should be.
- Encourage students to complete the Punnett square as a two-way relative frequency table.
- The chance that the child will be colorblind, which corresponds to the bottom right cell in the relative frequency table, is $\frac{1}{2}\left(\frac{1}{2}\right) = \frac{1}{4}$.

		Carrier Mother	
		X^B	X^b
Not Colorblind Father	X^B	$\frac{1}{4}$	$\frac{1}{4}$
	Y	$\frac{1}{4}$	$\frac{1}{4}$

Unit 10 Probability

Genetic Counselor STEM

Genetic counselors help people understand the facts of their genetic makeup. They analyze genetic information to identify risks for passing on certain medical conditions. They discuss the risks, benefits, and limitations of genetic testing options for individuals and families, and write detailed reports of test results.

STEM Task

Colorblindness is the decreased ability to see differences in color. The most common cause is genes on the X chromosome. A person will not be colorblind if they have at least one gene for unaffected sight.

B = Unaffected Sight
b = Colorblindness

		Carrier Mother	
		X^B	X^b
Not Colorblind Father	X^B	X^BX^B	X^BX^b
	Y	X^BY	X^bY

What is the chance that a child of these two parents will be colorblind? See margin.

Unit 10 611

Unit 10 Project It's Probably Genetic

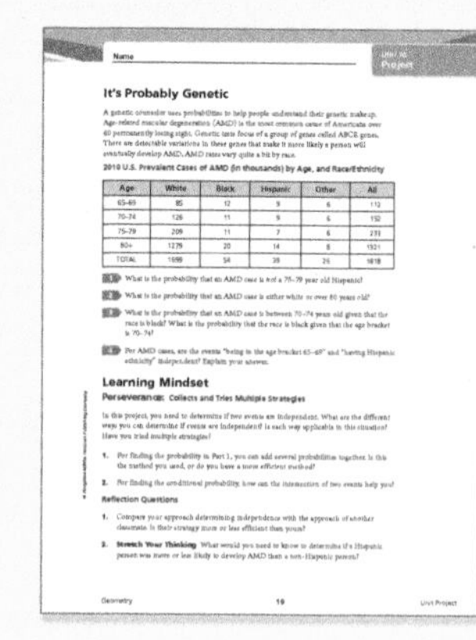

It's Probably Genetic

Learning Mindset

Overview: In this project students use a two-way table to calculate the probability of having a genetic ocular disease for various age and race groups.

Materials: display or poster board for presentations

Assessing Student Performance: Students' presentations should include:

- a correct probability for the complement of the event **(Lesson 20.1)**
- a correct probability for the inclusive events **(Lesson 20.2)**
- two correct conditional probabilities **(Lesson 21.1)**
- a correct explanation for why the two events are dependent **(Lesson 21.2)**

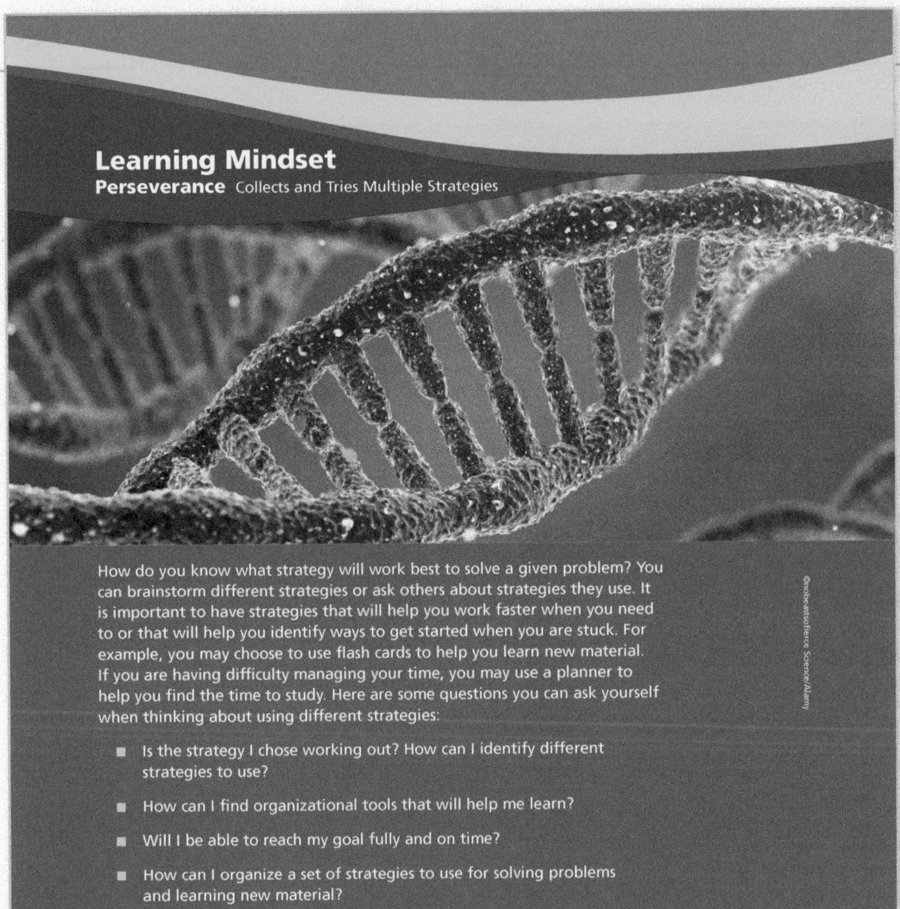

Learning Mindset

Perseverance Collects and Tries Multiple Strategies

How do you know what strategy will work best to solve a given problem? You can brainstorm different strategies or ask others about strategies they use. It is important to have strategies that will help you work faster when you need to or that will help you identify ways to get started when you are stuck. For example, you may choose to use flash cards to help you learn new material. If you are having difficulty managing your time, you may use a planner to help you find the time to study. Here are some questions you can ask yourself when thinking about using different strategies:

- Is the strategy I chose working out? How can I identify different strategies to use?
- How can I find organizational tools that will help me learn?
- Will I be able to reach my goal fully and on time?
- How can I organize a set of strategies to use for solving problems and learning new material?

Reflect

Q **Think of a time when you needed to persevere. What strategies did you use to help you persevere? How did you find or choose those strategies?**

Q **Imagine that you are a genetic counselor. What if your presentation style doesn't work for a client? How can you work with the client to develop a successful strategy? Why is it important to have different communication strategies?**

612

Perseverance

Learning Mindset

Collects and Tries Multiple Strategies

The learning-mindset focus in this unit is *perseverance*, which refers to a continued effort to achieve something despite difficulties, failure, or opposition. In math, students must persevere when mastery of concepts does not come easily.

Mindset Beliefs

Students may solve the same problem in different ways. Generally, there are many strategies that can be applied to solve a math problem. Students should realize that there is no single strategy to solve a type of problem. Students who continually use the same solution strategy demonstrate a fixed mindset.

Students should recognize that by learning and mastering multiple strategies, they can find and apply a more efficient strategy to solve a problem, even if their usual strategy works. Students who are willing to try another strategy or approach demonstrate a growth mindset.

Discuss with students their own definitions of perseverance. Have students discuss situations in which they had recognized that a strategy they were using was not working and how they found another strategy to use. Have students discuss and compare different general strategies they use to help organize their work and manage their time.

Mindset Behaviors

Encourage students to persevere by asking themselves questions about the various strategies they use to solve problems and stay organized. This self-monitoring can take the form of questions like the following:

When completely stuck:

- Why is my strategy not helping solve this problem?
- What other strategies can I use to solve this problem?
- How are my classmates approaching this problem?

When optimizing efficiency:

- Which strategy is the most effective if I have to solve a set of similar problems?
- Is there a strategy I could use to increase my efficiency?

When time is of the essence:

- How can I best use my time to solve this problem?
- Could another solution strategy help me solve the problem more quickly?

As students persevere in collecting solution and organizational strategies, they will become more effective learners.

What to Watch For

Watch for students who are quickly giving up after their first attempt to solve a problem. Encourage them by

- having them think about similar problems they have solved,
- asking them to describe other strategies that can be used to solve the problem, and
- reminding them that success often happens after numerous attempts and failures.

Watch for students who repeatedly try the same strategy. Help them move from a fixed mindset to a learning mindset by

- asking them to analyze why their strategy is not working,
- having them explore other strategies, and
- reminding them that they should look at the problem in another way.

"For every failure, there's an alternative course of action. You just have to find it."

—Mary Kay Ash, businesswoman

MODULE 20

PLANNING

PROBABILITY OF MULTIPLE EVENTS

Introduce and Check for Readiness

- Module Performance Task • Are You Ready?

Lesson 20.1—2 Days

Probability and Set Theory

Learning Objective: Describe sets and their relationships, including the universal set and complement, calculate theoretical probabilities and outcomes of events, and use the complement of an event to calculate probability.

New Vocabulary: **complement of a set**, **element**, **empty set**, **intersection**, **set**, **subset**, **union**, **universal set**

Review Vocabulary: **complement of an event**, **event**, **outcome**, **probability experiment**, **sample space**, **theoretical probability**, **trial**

Lesson 20.2—2 Days

Disjoint and Overlapping Events

Learning Objective: Recognize disjoint and overlapping events. Devise the Addition Rule for the probability of the union of overlapping events. Calculate probabilities of unions of events.

New Vocabulary: **disjoint events**, **mutually exclusive events**, **overlapping events**

Assessment

- Module 20 Test (Forms A and B)
- Unit 10 Test (Forms A and B)

LEARNING ARC FOCUS

 Build Conceptual Understanding

 Connect Concepts and Skills

 Apply and Practice

TEACHING FOR SUCCESS

TEACHING FOR DEPTH: Probability of Multiple Events

Meaning of Set Theory. Set theory is the study of sets and their properties. A set is a collection of objects, known as elements. In the study of set theory, students learn about subsets, unions, intersections, and complements of sets. Students learn new symbols and mathematical representations to describe set relationships. These concepts are fundamental to calculating probabilities of multiple events.

By representing sets with Venn diagrams, students will be able to visualize why the sum of the probability of an event and the probability of its complement is 1. They will also be able to see why the probability of mutually exclusive events is calculated differently than the probability of overlapping events.

It is important for students to be able to distinguish between mutually exclusive and overlapping events given a verbal description. This allows them to determine whether they can simply add the probabilities of the two events or whether they also need to subtract the probability of the intersection of the events from that sum.

Mathematical Progressions

Prior Learning

Students:

- expressed the likelihood of an event as a number between 0 and 1.
- used positive and negative numbers to represent quantities in real-world contexts.
- summarized categorical data in two-way frequency tables.
- interpreted relative frequencies in the context of the data.

Current Development

Students:

- understand how elements and subsets relate to a larger universal set.
- describe and represent sets, their characteristics, and their relationships.
- calculate theoretical probabilities.
- devise the Addition Rule for the probability of the union of overlapping events.
- calculate probabilities of unions of events.

Future Connections

Students:

- will describe events as subsets of a sample space using characteristics of the outcomes, or as unions, intersections, or complements of other events.
- will understand that two events are independent if the probability of the events occurring together is the product of their probabilities.
- will understand the conditional probability of an event occurring given that another event has already occurred.

TEACHER TO TEACHER

From the Classroom

Establish mathematics goals to focus learning. Some students prefer to learn fixed algorithms for solving problems. That mindset may work for them in other math courses, but it proves to be less successful in higher-level classes. For example, in Algebra I, I can teach students how to graph a line by plotting the y-intercept and using the slope to locate another point on the line. They can then complete several similar problems using the exact same method.

However, in statistics, students may not be able to apply the same algorithm to solve a group of problems. Rather, they may need to analyze each problem and understand which method or tool they can use to solve it. Thus, I like to establish goals in the study of statistics that will help my students understand the need to analyze each problem before they determine the appropriate solution path. Consider these possibilities:

- I can use set notation to describe relationships between sets.
- I understand the relationship between an event and its complement, and I can use that relationship to solve real-world problems.
- I understand the difference between mutually exclusive and overlapping events, and I can calculate their probabilities accordingly.

I display each goal to the class at the beginning of the applicable lesson. I have students discuss in pairs what they think the lesson will be about. I gather informal assessments throughout the lesson and check in with students to determine whether they feel they are making progress toward their goals. I use an exit ticket to help me determine where to continue the lesson in the next class period.

LANGUAGE DEVELOPMENT • Planning for Instruction

By giving all students regular exposure to language routines in context, you will provide opportunities for students to **listen**, **speak**, **read**, and **write** about mathematical situations and develop both mathematical language and conceptual understanding at the same time.

Using Language Routines to Develop Understanding

Use the **Professional Cards** for the following routines to plan for effective instruction.

Three Reads Lessons 20.1 and 20.2

Students read a problem three times with a specific focus each time.

1st Read What is the situation about?
2nd Read What are the quantities in the situation?
3rd Read What are the possible mathematical questions that we could ask for the situation?

Information Gap Lesson 20.2

Students recognize when information given in a problem situation is incomplete, and they pose questions and share knowledge with others to discover any missing facts or relationships and work together to solve the problem.

Critique, Correct, and Clarify Lesson 20.1

Students correct the work in a flawed explanation, argument, or solution method and share with a partner and refine the sample work.

Connecting Language to Probability of Multiple Events

Watch for students' use of the review and new terms listed below as they explain their reasoning and make connections with new concepts.

Key Academic Vocabulary

Prior Learning and Current Development • Review and New Vocabulary

complement of an event for a given event, all the outcomes that are not part of the event; The sum of the probabilities of an event and its complement must equal 1

disjoint when two events that cannot occur in the same trial at the same time; another word for *mutually exclusive*

element each object in a set

empty set a set with no elements, denoted by ∅ or {}

intersection of two sets the set of elements that belong to both sets

mutually exclusive two events that cannot occur in the same trial at the same time; another term for *disjoint*

set a collection of distinct objects

subset of set *A* a set in which all elements of the set are also contained within set *A*

union of two sets the set of elements that belong to one or both sets

universal set the set of all elements in a particular context

Linguistic Note

Listen for the terms *complements* and *elements*. Point out that while compliments are nice things we say about other people, they are different from *complements* in mathematics. Also review the mathematical definition of *elements* so that English learners understand the term is neither referring to the natural elements of earth, water, air, and fire nor the particles of matter composed of atoms they may be learning about in science class.

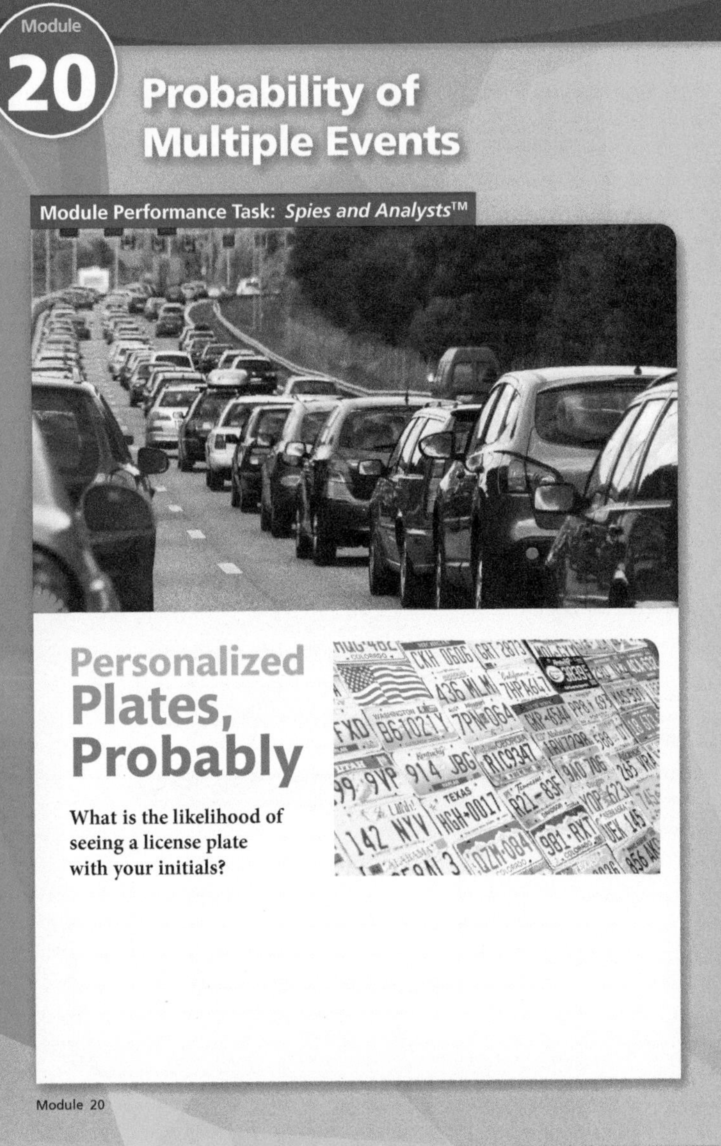

Module
20
Probability of Multiple Events
Module Performance Task: Spies and Analysts™
Personalized Plates, Probably
What is the likelihood of seeing a license plate with your initials?
Module 20
613

Connections to This Module

One sample solution might use the license plate number pattern for a specific state. For example, the license plate number may consist of three letters followed by four digits, with any letters and digits allowed and equally likely.

- Students define a sample space of all possible license plates and determine the number of outcomes in the sample space **(20.1)**.
- Students define an event that a license plate shows their initials and determine the number of outcomes in the event **(20.1)**.
- Students determine the probability that a license plate has their initials **(20.1)**.

Personalized Plates, Probably

Overview

This problem requires students to make assumptions about how license plate numbers are assigned and what combinations of letters or numbers might not be used. Examples of assumptions:

- A license plate number consists of three letters followed by four digits.
- Any license plate number is equally likely.
- Letters and digits can be repeated in a license plate number.

Be a *Spy*

First, students must determine what information is necessary to know. This should include:

- the pattern for license plate numbers in a selected state.
- any letters or digits or combinations of letters and digits that are not assigned on license plates.

Be an *Analyst*

Make sure students understand that, for example, a license plate with two letters will have $26 \cdot 26 = 676$ possible combinations, because there are 26 choices for each letter. It may help students to start with simple cases, such as license plates with one letter and two digits, to determine how many possible plates there are. They can then work up to the license plate pattern they are using.

Alternative Approaches

In their analysis, students might:

- account for some combinations of letters not being used.
- only consider the letter portion of the license plates when counting outcomes.
- find the probability of a license plate with their two-letter initials in any position in the letter sequence of the license plate.

ONLINE

Assign the Digital Are You Ready? to power actionable reports including

- proficiency by standards
- item analysis

Are You Ready?

Diagnostic Assessment

- Diagnose prerequisite mastery.
- Identify intervention needs.
- Modify or set up leveled groups.

Have students complete the *Are You Ready?* assessment on their own. Items test the prerequisites required to succeed with the new learning in this module.

Write Decimals and Fractions as Percents Students will apply their previous knowledge of expressing decimals or fractions as percents to calculate the probabilities of mutually exclusive events.

Probability of Simple Events Students will build upon their skill in finding the theoretical probability of simple events to use two-way tables as a sample space to determine conditional probabilities.

Probability of Compound Events Involving *Or* Students will apply their previous knowledge of finding the theoretical probability of a compound event involving *or* to describe events as subsets of a sample space, including complements of other events.

ONLINE

Are You Ready?

Complete these problems to review prior concepts and skills you will need for this module.

Write Decimals and Fractions as Percents

Express each rational number as a percent.

1. 1.2 120%
2. 0.065 6.5%
3. $\frac{7}{10}$ 70%
4. $\frac{24}{5}$ 480%
5. $\frac{5}{8}$ 62.5%
6. $\frac{133}{400}$ 33.25%

Probability of Simple Events

Calculate each theoretical probability.

7. What is the probability of rolling a number greater than 4 on a number cube? $\frac{1}{3}$
8. What is the probability of drawing a silver marble from a bag that contains 4 blue marbles, 7 silver marbles, and 5 green marbles? $\frac{7}{16}$
9. A high school debate team has 5 seniors, 4 juniors, 4 sophomores, and 3 freshmen. What is the probability of randomly selecting a junior from the debate team? $\frac{1}{4}$

Probability of Compound Events Involving *Or*

Calculate each probability.

10. The letters of the word MATHEMATICS are written on slips of paper and placed in a bowl. What is the probability of drawing an M or a vowel? $\frac{6}{11}$
11. Andrew has several pairs of socks in a drawer: 6 are black, 4 are gray, 3 are blue, and 8 are white. He chooses a pair of socks without looking. What is the probability Andrew chooses a black pair or a gray pair of socks? $\frac{10}{21}$

Connecting Past and Present Learning

Previously, you learned:

- to apply concepts of density based on area and volume in modeling situations,
- to describe events as subsets of a sample space using characteristics of the outcomes, and
- to summarize categorical data in two-way frequency tables.

In this module, you will learn:

- to describe events as subsets of a sample space, including unions and complements of other events,
- to calculate the probabilities of mutually exclusive events, and
- to use two-way tables as a sample space to approximate conditional probabilities.

614

DATA-DRIVEN INTERVENTION

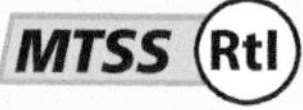

Concept/Skill	Objective	Prior Learning *	Intervene With
Write Decimals and Fractions as Percents	Express a decimal or fraction as a percent.	Grade 6, Lesson 7.1	• Tier 3 Skill 12 • Reteach, Grade 6 Lesson 7.1
Probability of Simple Events	Find the theoretical probability of a simple event.	Grade 7, Lesson 15.1	• Tier 2 Skill 22 • Reteach, Grade 7 Lesson 15.1
Probability of Compound Events Involving *Or*	Find the theoretical probability of a compound event involving *or*.	Grade 7, Lesson 15.2	• Tier 2 Skill 23 • Reteach, Grade 7 Lesson 15.2

* Your digital materials include access to resources from Grade 6–Algebra 2. The lessons referenced here contain a variety of resources you can use with students who need support with this content.

20.1 Probability and Set Theory

LESSON FOCUS AND COHERENCE

Mathematics Standards

- Describe events as subsets of a sample space (the set of outcomes) using characteristics (or categories) of the outcomes, or as unions, intersections, or complements of other events ("or," "and," "not").

Mathematical Practices and Processes 

- Look for and make use of structure.
- Look for and express regularity in repeated reasoning.
- Reason abstractly and quantitatively.
- Attend to precision.

I Can Objective 

I can use sets and their relationships to understand and calculate probabilities.

Learning Objective

Describe sets and their relationships, including the universal set and complement, calculate theoretical probabilities and outcomes of events, and use the complement of an event to calculate probability.

Language Objective

Use precise language to describe sets and calculate probabilities.

Vocabulary

Review: **complement of an event**, **event**, **outcome**, **probability experiment**, **sample space**, **theoretical probability**, **trial**

New: **complement of a set**, **element**, **empty set**, **intersection**, **set**, **subset**, **union**, **universal set**

Mathematical Progressions

Prior Learning	Current Development	Future Connections
Students: • expressed the likelihood of an event as a number between 0 and 1. **(Gr.7, 14.1–14.4)** • used positive and negative numbers to represent quantities in real-world contexts, explaining the meaning of 0 in each situation. **(Gr6, 1.1)**	**Students:** • understand how elements and subsets relate to the universal set of which they are a part. • describe and represent sets, their characteristics, and their relationships. • calculate theoretical probabilities. • use the complement of an event to calculate probability in a real-world scenario.	**Students:** • will understand conditional probability and interpret independence using conditional probability. **(21.1; Alg2, 18.1, 18.2)** • will understand that two events *A* and *B* are independent if the probability of *A* and *B* occurring together is the product of their probabilities. **(21.2; Alg2, 18.2)**

UNPACKING MATH STANDARDS

Describe events as subsets of a sample space (the set of outcomes) using characteristics (or categories) of the outcomes, or as unions, intersections, or complements of other events ("or," "and," "not").

What It Means to You

Students demonstrate an understanding of sets and their characteristics. They visualize the elements using Venn diagrams and describe how they relate to each other and to the universal set. Students discuss situations involving intersection and union of events. They understand that "intersection" or a "repeated event" implies "and" and that "union" implies "or."

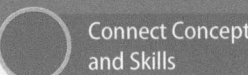

WARM-UP OPTIONS

ONLINE

PROJECTABLE & PRINTABLE 

ACTIVATE PRIOR KNOWLEDGE • Express the Likelihood of an Event as a Number Between 0 and 1

Use these activities to quickly assess and activate prior knowledge as needed.

Problem of the Day 

A spinner is labeled with numbers 1 to 10, where each number occurs only once and is given an equal portion of the spinner. Use fractions to express the probability of:

a) spinning an even or an odd number?

b) spinning a number that is a square?

c) spinning a prime or a composite number?

d) not spinning a one-digit number?

a) 1, b) $\frac{3}{10}$, c) $\frac{9}{10}$, d) $\frac{1}{10}$

Quick Check for Homework

As part of your daily routine, you may want to display the Teacher Solution Key to have students check their homework.

Make Connections

Based on students' responses to the Problem of the Day, choose one of the following:

1. Project the Interactive Reteach, Grade 7, Lesson 14.1.
2. Complete the Prerequisite Skills Activity:

Have students work independently to recall that probability can be expressed by percentage, decimal, or fraction. Ask them to consider the following: Alex has a coin and a bag of marbles containing 3 blue and 2 red marbles.

- *What is the probability of getting a tail if Alex flips the coin?* It is 50% or 0.5 or $\frac{1}{2}$.
- *What is the probability of Alex picking a red marble from the bag?* It is $\frac{2}{5}$ or 0.4 or 40%.
- *What is the equivalent of 100% as a fraction?* It is 1 or $\frac{100}{100}$.
- *Why would someone use fractions rather than percent to express probability?* Possible answer: Sometimes the probability cannot easily be written as a percent because it is a large decimal, so it is much more convenient to use fractions.

SHARPEN SKILLS

If time permits, use this on-level activity to build fluency and practice basic skills.

Vocabulary Review

Objective: Students demonstrate understanding of a tree diagram and the terms *sample space*, *events*, and *theoretical probability*.
Materials: index cards

Appetizers
Chicken Fingers
Mozzarella Sticks
Main Course
Pasta
Tacos
Burger
Dessert
Chocolate Cake
Ice Cream

Have students work in small groups. Provide them with index cards labeled with food selections (an Appetizer card, Main Course card, and Dessert card). Create a tree diagram listing all possible ways to have one appetizer, one main course, and one dessert.

Have students discuss the number of events in the sample space if they were to select one item from each card. Then have students think of the theoretical probability of getting any one of the choices.

PLAN FOR DIFFERENTIATED INSTRUCTION

Small-Group Options

Use these teacher-guided activities with pulled small groups.

On Track

Materials: standard deck of 52 cards

Select four cards from the deck: A, K, Q, and J.

Ask students to determine the following from the four cards:

- the sample space
- *P*(K)
- *P*(Q or J)
- *P*(not A)
- *P*(10)

Then have students use the complement to find the probability of selecting any of those cards from the entire deck of 52 cards.

Almost There

Materials: cubes (or marbles) in two colors, bags

Place 8 of one color and 14 of the other color cubes in a bag. Ask students to determine the following:

- the universal set
- the sample space of picking any two cubes together
- the theoretical probability of picking one particular color
- the theoretical probability of not picking one particular color

Have students do an experiment picking a color cube 20 times and then compare the results to the theoretical probability.

Ready for More

Materials: standard deck of 52 cards

Provide small groups with a deck of cards. Then have them find probabilities:

- What is the probability of drawing a 7?
- What is the probability of drawing a black 7?
- What is the probability of drawing the queen of hearts and then the king of diamonds?
- What is the probability of drawing a 10 or a club?

Have groups write two of their own questions involving the probability of multiple outcomes, such as "What is the probability of choosing a king, a jack, and a 5?" Then swap questions with another group and solve.

Math Center Options

Use these student self-directed activities at centers or stations. Key: ● Print Resources ● Online Resources

On Track

- ● Interactive Digital Lesson
- ●● Journal and Practice Workbook
- ● Interactive Glossary (printable): **complement of an event**, **event**, **outcome**, **probability experiment**, **sample space**, **theoretical probability**, **trial**, **complement of a set**, **element**, **empty set**, **intersection**, **set**, **subset**, **union**, **universal set**
- ● Module Performance Task

Almost There

- ● Reteach 20.1 (printable)
- ● Interactive Reteach 20.1
- ● Illustrative Mathematics: Describing Events
- ● Desmos: Chance Experiments

Ready for More

- ● Challenge 20.1 (printable)
- ● Interactive Challenge 20.1

Unit Project Check students' progress by asking how they plan to use the complement of an event to find the desired probability.

View data-driven grouping recommendations and assign differentiation resources.

Spark Your Learning • Student Samples

During the *Spark Your Learning*, listen and watch for strategies students use. See samples of student work on this page.

Organize Data

Strategy 1

There are 170 dogs with straight hair: $42 + 21 + 77 + 30 = 170$

There are 74 dogs with curly hair: $12 + 32 + 18 + 12 = 74$

There are 244 total dogs: $170 + 74 = 244$

Dogs with straight hair: $\frac{170}{244} = \frac{85}{122}$

Dogs with curly hair: $\frac{74}{244} = \frac{37}{122}$

If students . . . organize the data to find the sum of each type of hair and the total number of dogs to write their ratios, they are doing an exemplary job and showing understanding of how to find ratios from Grade 7.

Have these students . . . explain how they used the data shown in the table to write their ratios. **Ask:**

- Q How did you determine which categories to use in the ratios?
- Q What made you construct the ratios this way?

Extend the Table

Strategy 2

	Dogs	Ratio
Straight	170	$\frac{170}{244}$
Curly	74	$\frac{74}{244}$
Total	244	

If students . . . construct a table to extend the table from the problem, they are demonstrating a good understanding of the problem, but they may not realize there may be a more efficient method without having to construct a table.

Activate prior knowledge . . . by having students explain another method they could use without drawing the table. **Ask:**

- Q How did you find the total number of dogs per coat style?
- Q Could you have found the total number of dogs per coat style and the ratios without using a table?

COMMON ERROR: Uses the Wrong Ratio

There are 170 dogs with straight hair and 74 with curly hair. The ratio is $\frac{74}{170}$.

If students . . . find the correct number of dogs per coat style but write a ratio that compares the two coat types rather than each coat type to the total number of dogs, they may need help understanding the question.

Then intervene . . . by pointing out that the question asks about the ratios of straight-haired dogs to the total number of dogs brought to grooming. **Ask:**

- Q How can you write a ratio of straight-haired dogs to the total number of dogs?
- Q What is the difference in the meanings of the ratio you wrote and the ratio you should write?

20.1

Probability and Set Theory

I Can use sets and their relationships to understand and calculate probabilities.

Spark Your Learning

Dogs can be categorized into seven groups—terrier, toy, working, sporting, nonsporting, hound, and herding. Each group will have different types of coats.

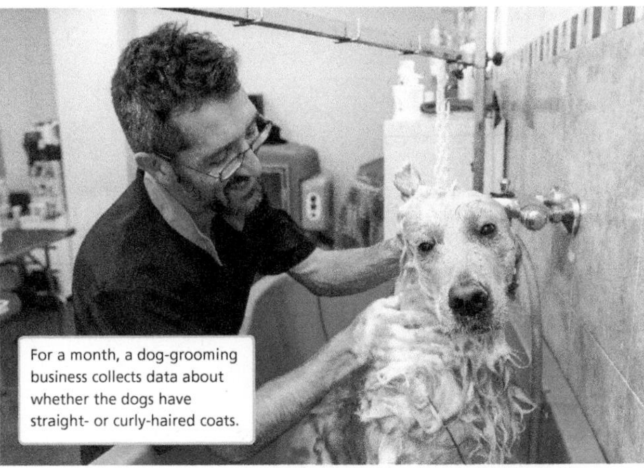

For a month, a dog-grooming business collects data about whether the dogs have straight- or curly-haired coats.

©Sam Wordley/Shutterstock

Complete Part A as a whole class. Then complete Parts B–D in small groups.

A. What is a mathematical question you can ask about this situation? What information would you need to know to answer your question? See Additional Answers.

B. What variable(s) are involved in this situation? The variables are the groups of dogs and whether they have curly or straight hair.

C. To answer your question, what strategy and tool would you use along with all the information you have? What answer do you get? See Strategies 1 and 2 on the facing page.

D. Does your answer make sense in the context of the situation? How do you know? See Additional Answers.

Turn and Talk During the month, what ratio of groomings for large dogs were for large dogs with straight hair? See margin.

SUPPORT SENSE-MAKING • Three Reads

Tell students to read the information in the photo and the table three times and prompt them with a different question each time.

1. What is the situation about? This is about a dog-grooming business and the number of dogs with straight or curly hair brought in a month.
2. What are the quantities in this situation? How are the quantities related? The number of different-sized dogs, some with straight hair and some curly hair. We have to relate the number of types of dogs to total number of dogs.
3. What are the possible questions you could ask about the situation? How many dogs were counted? How many dogs have straight hair? How many dogs have curly hair? How many dogs are large? How many large dogs have straight hair?

1 Spark Your Learning

▶ MOTIVATE

- Have students look at the photo in their books and read the information contained in the photo. Then complete Part A as a whole-class discussion.
- Give the class the additional information they need to solve the problem. This information is available online as a printable and projectable page in the Teacher Resources.
- Have students work in small groups to complete Parts B–D.

▶ PERSEVERE

If students need support, guide them by asking:

Q **Advancing • Use Tools** Which tool could you use to solve the problem? Why choose that tool and not some other? Students' choices of tools and reasons for choosing them will vary.

Q **Assessing** How could you determine the number of dogs the business saw during the month it collected data? Add up all of the dogs in both the straight and curly hair categories.

Q **Assessing** Do you need to consider the different groups of dogs to find the numbers of dogs with straight or curly hair? Explain. no; Two dogs in the same group could have different types of coats.

Q **Advancing** How could the business use the data it collects? Possible answer: If the business determines that more dogs with straight hair—or more dogs with curly hair—are brought in for grooming, it could advertise accordingly and better train its technicians to groom those dogs.

Turn and Talk Have students identify the parts of the ratio they need to know in order to answer the question. 30 out of 42 or 5 out of 7

▶ BUILD SHARED UNDERSTANDING

Select groups of students who used various strategies and tools to share with the class how they solved the problem. As they present their solutions, have each group discuss why they chose a specific strategy and tool.

2 Learn Together

Build Understanding

Task 1 (MP) **Use Structure** Students construct a Venn diagram and reason about its structure to illustrate a universal set, subsets, intersection, union, and complement of sets.

CONNECT TO VOCABULARY

Have students use the **Interactive Glossary** to record their understanding of the vocabulary in this task.

Sample Guided Discussion:

Q How would you determine the number of elements in the universal set if you only have the Venn diagram to show the information? I would add the elements of all sets including "others."

Q What does it mean to look at *A* and *B* instead of *A* or *B*? *A* and *B* means an element that belongs in both sets or in the intersection. *A* or *B* means that all elements are included in the sets, or the union.

Q In Part E, how can you determine the complement of *A*? Add everything in the universal set that is not included in *A*.

Build Understanding

Work with Sets

A **set** is a collection of distinct objects. Each object in a set is called an **element** of the set. A set is often denoted by writing the elements in braces. A set with no elements is called the **empty set**, denoted by ∅ or {}. The set of all elements in a particular context is called the **universal set**.

To denote the number of elements in a set, use the letter n. For example, given a set A, $n(A)$ denotes the number of elements in set A.

For example, consider the natural numbers less than 10.

Set	Set Notation	Number of Elements
the universal set	$U = \{1, 2, 3, 4, 5, 6, 7, 8, 9\}$	$n(U) = 9$
the set of even numbers	$A = \{2, 4, 6, 8\}$	$n(A) = 4$
the set of multiples of 3	$B = \{3, 6, 9\}$	$n(B) = 3$
the set of multiples of 12	$C = \{\ \}$	$n(C) = 0$

Term	Definition	Symbol	Venn Diagram
subset	A **subset** is a set that is contained entirely within another set. Set *A* is a subset of set *B* if every element of *A* is contained in *B*.	$A \subset B$	U, B, A
intersection	The **intersection** of sets *A* and *B* is the set of all elements that are in both set *A* and set *B*.	$A \cap B$	U, A, B
union	The **union** of sets *A* and *B* is the set of all elements that are in set *A* or set *B*, including elements that are in both sets.	$A \cup B$	U, A, B
complement	The **complement** of set *A* is the set of all elements in the universal set *U* that are not in set *A*.	A^c	U, A

616

LEVELED QUESTIONS

Depth of Knowledge (DOK)	Leveled Questions	What Does This Tell You?
Level 1 **Recall**	What is the meaning of the universal set in any set of data? The universal set is all elements in a set of information or a particular context.	Students' answers will indicate whether they understand and can apply the meaning of a universal set.
Level 2 **Basic Application of Skills & Concepts**	In Task 1, what can you tell about the intersection of *A* and *B* in the Venn diagram? The intersection contains parrots that have a blue and yellow body and a green crown.	Students' answers will indicate whether they understand the relationship between sets and the intersection of those sets of data.
Level 3 **Strategic Thinking & Complex Reasoning**	Is it possible for a subset to not have a complement in another subset or to not intersect with another subset.? Explain. yes; Two subsets do not to have objects that are not in one subset but in the other (complement) nor do two subsets have to have objects in common (intersection).	Students' answers will indicate whether they understand how sets are related and how that relationship can differ between different contexts.

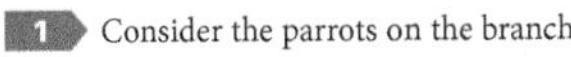

 1 Consider the parrots on the branch.

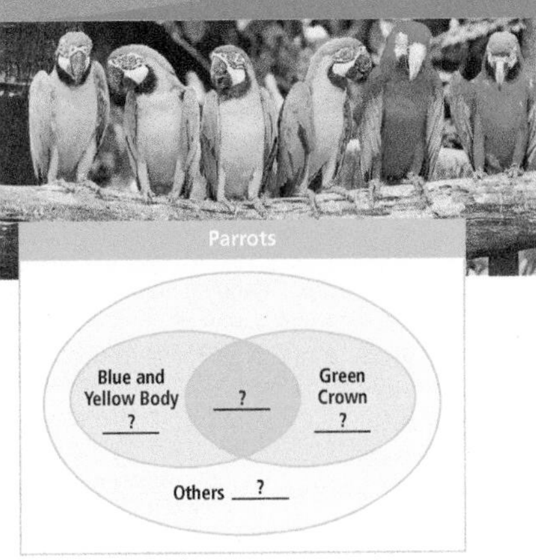

A. What is the universal set shown in the photo? How many elements are in the universal set?
A–C. See Additional Answers.

B. Describe a subset of the universal set.

C. Let event A be the set of all parrots with blue and yellow body color. Let event B be the set of all parrots with green color on the top of their head, called the crown. Copy and complete the Venn diagram.

D. How many elements are in each set?
$n(A) = 4$; $n(B) = 5$

E. How many elements are in a union of A and B? Intersection of A and B? Complement of A?
$n(A \text{ or } B) = 5$; $n(A \text{ and } B) = 4$; $n(A^c) = 2$

Parrots

Blue and Yellow Body ?
?
Green Crown ?
Others ?

Calculate Theoretical Probabilities

A **probability experiment** is an activity involving chance. Each repetition of the experiment is called a **trial**, and each possible result of the experiment is called an **outcome**. A set of outcomes is known as an **event**, and the set of all possible outcomes is called a **sample space**.

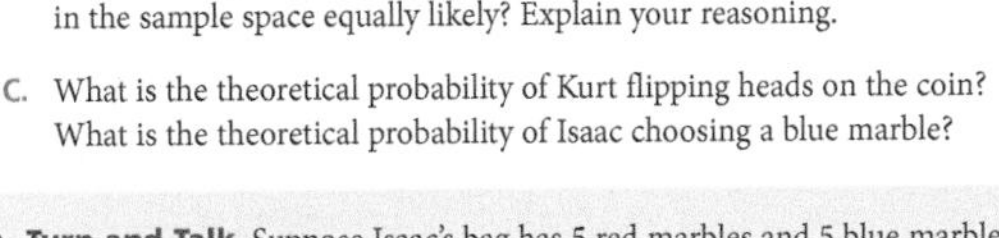

Probability measures how likely an event is to occur. The probability of an event occurring is a number between 0 and 1.

If all of the outcomes of a probability experiment are equally likely to occur, the **theoretical probability** of an event A in the sample space S is

$$P(A) = \frac{\text{number of outcomes in the event}}{\text{number of outcomes in the sample space}} = \frac{n(A)}{n(S)}.$$

 2 Kurt and Isaac are performing probability experiments. Kurt flips a coin. Isaac randomly chooses a marble from a bag of 4 red marbles and 6 blue marbles.
A–C. See margin.

A. Identify the trial, an event, a possible outcome of the event, and the sample space for each of the probability experiments.

B. Within each probability experiment, are all of the possible outcomes in the sample space equally likely? Explain your reasoning.

C. What is the theoretical probability of Kurt flipping heads on the coin? What is the theoretical probability of Isaac choosing a blue marble?

Turn and Talk Suppose Isaac's bag has 5 red marbles and 5 blue marbles. Are all the outcomes in the sample space equally likely? Explain. See margin.

Task 2 (MP) **Use Repeated Reasoning** Students consider a couple of situations where they have to calculate the theoretical probability.

CONNECT TO VOCABULARY

Have students use the **Interactive Glossary** to record their understanding of the vocabulary in this task.

Sample Guided Discussion:

Q In Part A, what is the difference between an event and a sample space? Possible answer: An event is a set of outcomes, but the sample space is the set of all outcomes.

Q If the theoretical probability for flipping heads is 50%, does that mean that if you flip a coin 10 times, 5 of those times you will result in heads? Theoretical probability describes the chances of what you may get but that is not necessarily what you will get in an experiment with multiple trials.

Turn and Talk Point out to students that they might want to think about comparing the likelihood of choosing a blue marble with choosing a red marble. The events are equally likely. There are 5 ways to choose red and 5 ways to choose blue.

ANSWERS

2 A. Kurt: trial: flipping a coin; event: flipping a coin and getting a result of heads; outcome: getting a result of heads; sample space: heads, tails
Isaac: trial: choosing a marble; event: choosing any red marble. outcome: choosing a red marble; sample space: 4 red marbles, 6 blue marbles

B. For Kurt, outcomes are equally likely. Each side of the coin is just as likely to be landed on. For Isaac, the outcomes are not equally likely because each marble has the same probability, but there are 4 ways to choose a red marble, and 6 ways to choose a blue marble.

C. $P(\text{heads}) = \frac{1}{2}$; $P(\text{blue marble}) = \frac{6}{10}$

PROFICIENCY LEVEL

Beginning
Have students work in small groups. Give each group a standard deck of 52 cards. Ask students to state the number of outcomes in the sample space (52 cards). Then have each student select a card and state the theoretical probability of pulling that card from the deck.

Intermediate
Have students consider the class and record some characteristic of their classmates, such as eye color. Then have them write sentences describing the theoretical probability of selecting a student with a certain characteristic, such as blue eyes.

Advanced
Have students write on slips of paper the numbers 1 to 10. Then ask them to describe the theoretical probability of picking a prime number. Have students perform 5 trials and write a paragraph discussing whether the experiment matched their prediction.

Step It Out

Task 3 (MP) **Reason** Students determine the complement of an event and the advantages in using the complement to find the probability of an event.

CONNECT TO VOCABULARY

Have students use the **Interactive Glossary** to record their understanding of the vocabulary in this task.

Sample Guided Discussion:

Q How do you determine the complement of an event? The complement is any outcome that is not part of the probability of an event. In this case, rolling a 1 would not be part of the event for which the probability is being determined.

Q In the expressions for $P(A)$, what does 1 represent? 1 is the sample space. It is the whole, or 100% of the possible outcomes.

Turn and Talk Point out to students that choosing the complement should make finding the probability easier, because there are only 2 possibilities for the complement. Event A is rolling a 1, 2, 3, or 4, so A^C is rolling a 5 or 6. $P(A) = 1 - \frac{2}{6} = \frac{4}{6} = \frac{2}{3}$

ANSWERS

A. The only number that he can roll that is not greater than 1 is 1, so the complement of rolling a number greater than 1 is rolling a 1.

B. There is only one outcome in the complement of the event.

Step It Out

Use the Complement of an Event

You may have noticed that the probability of an event occurring and the probability of the event not occurring (the probability of the complement of the event) have a sum of 1. This relationship can be used when it is more convenient to calculate the probability of the **complement of an event** than the probability of the event. Use a superscript c to denote the complement of set.

Probabilities of an Event and Its Complement	
$P(A) + P(A^c) = 1$	The sum of the probability of an event and the probability of its complement is 1.
$P(A) = 1 - P(A^c)$	The probability of an event is 1 minus the probability of its complement.
$P(A^c) = 1 - P(A)$	The probability of the complement of an event is 1 minus the probability of the event.

Use the complement to calculate the probabilities.

3 Lamar rolls a six-sided number cube once. What is the probability that he rolls a number greater than 1?

Let A^c be that he rolls a 1.

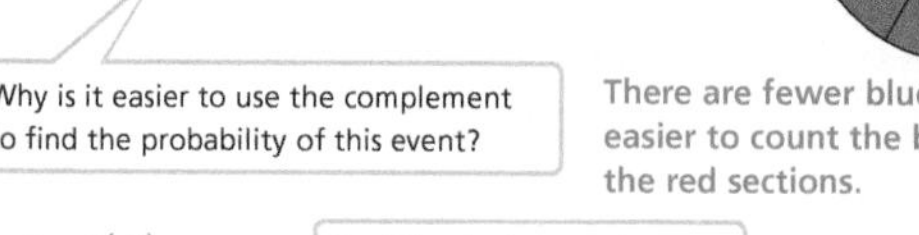
A. Why is rolling a 1 the complement of rolling a number greater than 1?

A, B. See margin.

$P(A) = 1 - P(A^c)$

$P(A) = 1 - \frac{1}{6} = \frac{5}{6}$

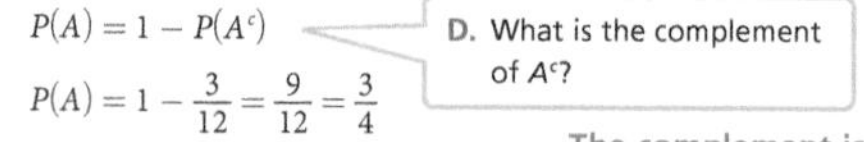
B. Why is it easier to use the complement to find the probability of this event?

The probability of rolling a number greater than 1 is $\frac{5}{6}$.

Erica spins the spinner shown once. The spinner is divided into 12 equal sections. Find the probability that the pointer lands on red.

Let A^c be the event that the pointer lands on blue.

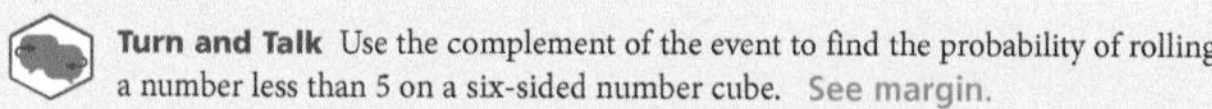
C. Why is it easier to use the complement to find the probability of this event?

There are fewer blue sections so it is easier to count the blue sections than the red sections.

$P(A) = 1 - P(A^c)$

$P(A) = 1 - \frac{3}{12} = \frac{9}{12} = \frac{3}{4}$

D. What is the complement of A^c?

The probability of spinning red is $\frac{3}{4}$.

The complement is the event that the pointer lands on red.

Turn and Talk Use the complement of the event to find the probability of rolling a number less than 5 on a six-sided number cube. See margin.

Calculate Probability in a Real-World Scenario

4 Veronica attends a conference where there is a possibility that she can win a prize as described. What is the probability that Veronica does not win a prize?

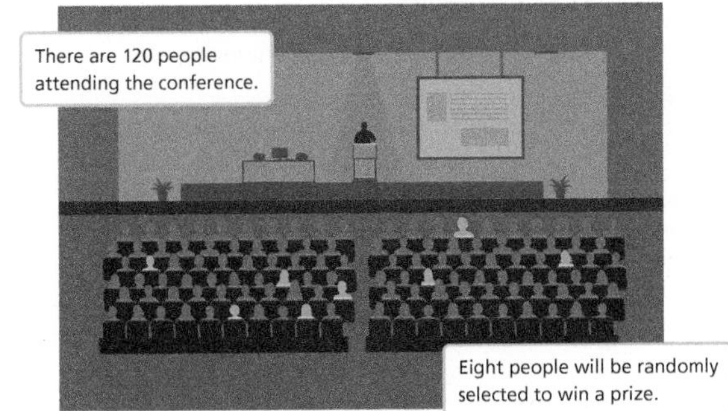

There are 120 people attending the conference.
Eight people will be randomly selected to win a prize.

To find the probability that Veronica does not win a prize, find the probability of the complement of the event that she wins a prize.

A. Why is it easier to use the complement to find the probability of this event?
The information given is the complement of the event.

Let A be the event that Veronica wins a prize.

$$P(A) = \frac{\text{attendees that win a prize}}{\text{total number of attendees}} = \frac{8}{120}$$
$$= \frac{1}{15}$$

B. What is the sample space?
120 attendees

Find the probability that Veronica does not win a prize.

$$P(A^c) = 1 - P(A)$$
$$= 1 - \frac{1}{15}$$
$$= \frac{14}{15}$$

C. What does A^c represent?
Event A is winning a prize, so the complement of A is not winning a prize.

Answer the question.

The probability that Veronica does not win a prize is $\frac{14}{15}$.

Turn and Talk

- The next day, Veronica attends a conference that has 30 fewer attendees. Ten people are selected at random to win a prize. Find the probability that she wins a prize. Then find the probability that she does not win a prize.
- Does Veronica have a better chance to win a prize on the first or the second day?

See margin.

Task 4 (MP) **Attend to Precision** Students define and use the complement of an event to calculate a probability in a real-world scenario.

OPTIMIZE OUTPUT Critique, Correct, and Clarify

Have students work with a partner and discuss which solution makes sense. Encourage students to explain their thinking.

Cameron bought bagels for the project group. He bought 6 plain bagels, 8 poppy seed bagels, and 10 cinnamon bagels. Dana prefers poppy seed bagels. What is the probability that she will not get a poppy seed bagel?

Solution A	Solution B
Let A be the event that she doesn't get one.	Let A be the event that she gets one.
$P(A) = \frac{8}{24} = \frac{1}{3}$	$P(A) = \frac{8}{24} = \frac{1}{3}$ $P(A^c) = 1 - P(A)$ $= 1 - \frac{1}{3} = \frac{2}{3}$

Solution B

Sample Guided Discussion:

Q If Veronica is only 1 person, why do you not calculate the probability as 1 out of 120 instead of 8 out of 120? There are 8 people who can win a prize, so Veronica has 8 chances out of 120.

Q In Part B, why isn't 15 the sample space? There are 120 people, so the sample space is 120. The denominator of the reduced ratio, 15, is the total number of parts in the fraction in terms of probability, but not the total number of outcomes in the sample space.

Turn and Talk Students should compare the ratio of the first day to the ratio of winners to attendees on the second day. Let B be the event of Veronica winning a prize on the second day. The number of people attending the conference that day is $120 - 30 = 90$.

$$P(B) = \frac{\text{number of people winning a prize}}{\text{number of people attending conference}}$$
$$= \frac{10}{90} = \frac{1}{9}$$
$$P(B^C) = 1 - P(B)$$
$$P(B^C) = 1 - \frac{1}{9}$$
$$P(B^C) = \frac{8}{9}$$

Veronica has a better chance to win on the second day because the probability is greater; $\frac{1}{9} > \frac{1}{15}$

ONLINE

Assign the Digital On Your Own for
- built-in student supports
- Actionable Item Reports
- Standards Analysis Reports

On Your Own

Assignment Guide

The chart below indicates which problems in the On Your Own are associated with each task in the Learn Together. Assign daily homework for tasks completed.

Learn Together Tasks	On Your Own Problems
Task 1, p. 617	Problems 7–16 and 30–33
Task 2, p. 617	Problems 5, 6, 17–22, 24–29, 31, and 40–41
Task 3, p. 618	Problems 23 and 38–39
Task 4, p. 619	Problems 34–37

Watch for Common Errors

Problem 5B Students often misunderstand the language of logic, namely in this problem, "a red or black." Contrary to common meaning where one is to choose when hearing "or," here it is meant as an inclusive term. Point out to students that what is meant is that either one or the other would be fine, so we need to include both subsets in calculating the probability.

Check Understanding

1. Let A be the set of natural number multiples of 3 that are less than 25. Let B be the set of even natural numbers less than 25. Find $A \cap B$, the intersection of the sets. $A \cap B = \{6, 12, 18, 24\}$

2. Cameron has a bag with 8 green marbles and 7 yellow marbles. If Cameron chooses a marble at random, what is the probability that he chooses a yellow marble? $\frac{7}{15}$

3. Lindsey is writing a report about one of the 50 states. She chooses a state at random. What is the probability that the state she chooses isn't Ohio, Oregon, or Oklahoma? $\frac{47}{50}$

4. Nancy's school is selecting 4 students from each class at random to attend a presentation. There are 28 students in Nancy's class. What is the probability that Nancy will be selected to attend the presentation? What is the probability that she will not be selected? $\frac{1}{7}, \frac{6}{7}$

On Your Own

5. A store marks certain items with special colored tags. Suppose you choose an item at random. Find the probability of each event.

A. the item has a white tag $\frac{1}{4}$

B. the item has a red or black tag $\frac{3}{5}$

There are 8 red tags, 3 blue tags, 5 white tags and 4 black tags.

6. (MP) **Critique Reasoning** Zack says that the probability of rolling a number less than 4 on a six-sided number cube is $\frac{4}{6}$. Valeria says that the probability is $\frac{3}{6}$. Who is correct? Explain.
Valeria; There are 3 numbers on a six-sided number cube that are less than 4, namely 1, 2, and 3.

Let A be the set of factors of 10, B be the set of even natural numbers less than or equal to 10, C be the set of odd natural numbers less than or equal to 10, and D be the set of factors of 9. The universal set is the set of natural numbers less than or equal to 10.

7. Write sets A–D using set notation.

8. Is $D \subset C$? Explain.

7, 8. See Additional Answers.

9. What is $A \cap B$? $A \cap B = \{2, 10\}$

10. What is $A \cap C$? $A \cap C = \{1, 5\}$

11. What is $A \cup B$? $A \cup B = \{1, 2, 4, 5, 6, 8, 10\}$

12. What is $A \cup C$? $A \cup C = \{1, 2, 3, 5, 7, 9, 10\}$

13. What is $B \cap C$? $B \cap C = \{\}$

14. What is $A \cup D$? $A \cup D = \{1, 2, 3, 5, 9, 10\}$

15. What is C^c? $C^c = \{2, 4, 6, 8, 10\}$

16. What is D^c? $D^c = \{2, 4, 5, 6, 7, 8, 10\}$

A set of 12 cards is numbered 1 to 12. A card is chosen at random. Event A is choosing a card greater than 4. Event B is choosing an even number. Calculate each probability.

17. $P(A)$ $\frac{2}{3}$

18. $P(B)$ $\frac{1}{2}$

19. $P(A \cap B)$ $\frac{1}{3}$

20. $P(A \cup B)$ $\frac{5}{6}$

21. $P(A^c)$ $\frac{1}{3}$

22. $P(B^c)$ $\frac{1}{2}$

620

data checkpoint

3 Check Understanding

Formative Assessment

Use formative assessment to determine if your students are successful with this lesson's learning objective.

Students who successfully complete the Check Understanding can continue to the On Your Own practice.

For students who miss 1 problem or more, work in a pulled small group using the Almost There small-group activity on page 615C.

ONLINE

Assign the Digital Check Understanding to determine
- success with the learning objective
- items to review
- grouping and differentiation resources

4 Differentiation Options

Differentiate instruction for all students using small-group activities and math center activities on page 615C.

Reteach

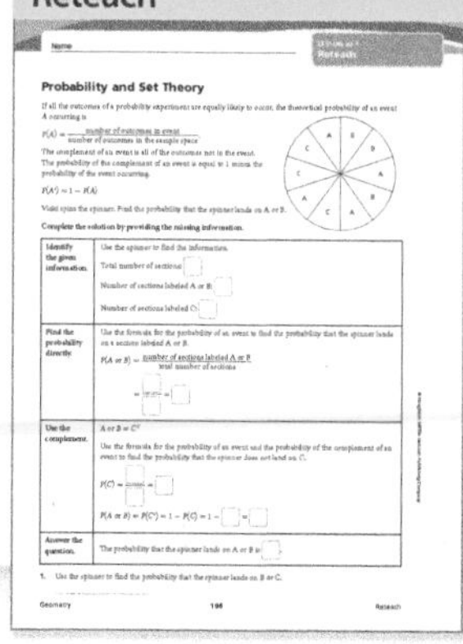

Challenge

23. Kiera has a bag with 20 chips in it. There are 3 red chips and 4 yellow chips. The rest of the chips are blue. If she chooses a chip at random from the bag, use the complement of the event to find each probability.

A. She chooses a blue chip. $\frac{13}{20}$

B. She chooses a blue or red chip. $\frac{4}{5}$

C. She does not choose a red chip. $\frac{17}{20}$

The spinner is divided into 8 equal sections. Suppose you spin the spinner once. Find the probability of the pointer landing on each number described.

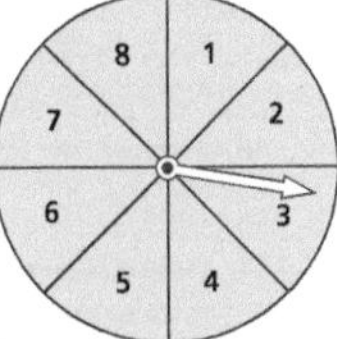

24. odd $\frac{1}{2}$

25. less than 3 $\frac{1}{4}$

26. greater than or equal to 4 $\frac{5}{8}$

27. not less than 8 $\frac{1}{8}$

28. greater than 10 0

29. less than 9 1

30. (MP) **Model with Mathematics** A car dealership has 10 SUVs on sale. There are 4 black SUVs, 5 silver SUVs, and 1 tan SUV. Use set notation to represent the universal set in this situation. {black, black, black, black, silver, silver, silver, silver, silver, tan}

31. Anita randomly chooses a marble from a bag. The bag contains 6 red marbles, 4 blue marbles, and 5 green marbles.

A. What are the possible outcomes of the event "not choosing a blue marble"? {red, green}

B. What is the probability of not choosing a blue marble? $\frac{11}{15}$

Social Studies **In a recent survey, parents and teens were asked about their cellphone use. Some of the results are shown in the table.**

Cellphone Habits		
Parents (%)	**Teens (%)**	**Said that they...**
36	54	spend too much time on their cellphone.
26	44	check their phone as soon as they wake up.
14	30	feel their teen/parent is distracted when having in-person conversations.
15	8	lose focus at work/school because they are checking their cellphone.

32. Describe the sample space in the study. the total number of parents and teens surveyed

33. What is an event in the study? Possible answer: A parent checks their phone as soon as they wake up.

34. What is the probability that a parent surveyed does not lose focus at work/school because they are checking their cellphone? 85%

35. What is the probability that a teen surveyed does not feel their parent is distracted when having in-person conversations? 70%

36. What is the probability that a teen surveyed spends too much time on their cellphone? 54%

37. What is the probability that a parent surveyed spends too much time on their cellphone? 36%

Questioning Strategies

Problem 23B Suppose you had to find the probability of choosing a yellow chip. How could you use the complement of this event to find the probability of choosing a blue or red chip? There are only 3 colors of chips in the bag. If the probability of choosing a yellow chip is 4 out of 20, then the complement of this event (choosing a red or blue chip) is $P(A^C) = 1 - \frac{4}{20}$.

$$P(A^C) = 1 - P(A)$$

$$P(A^C) = 1 - \frac{4}{20}$$

$$P(A^C) = \frac{16}{20}$$

Watch for Common Errors

Problem 27 Students may not include 8 in their probability calculation, because they assume "not less than 8" does not include 8. Point out that they need to read the statement carefully and think about what numbers are included in "not less than 8."

5 Wrap-Up

Summarize learning with your class. Consider using the Exit Ticket, Put It in Writing, or I Can scale.

Exit Ticket

You have a deck of cards. You are wondering what the theoretical probability is of *not* getting a 2, 3, 4, or 5.

A. What is the universal set? 52 cards in the deck.

B. What are the possible outcomes of the event "not getting a 2, 3, 4, or 5"? {6, 7, 8, 9, 10, J, Q, K, A}.

C. Use the complement to find the probability.
The complement is getting 2, 3, 4, or 5. That is 16 cards out of 52, because there are 4 of each card (1 per suit).
$1 - \frac{16}{52} = \frac{36}{52} = \frac{9}{13}$

Put It in Writing 

Have students explain, in writing, why it is sometimes easier to use the complement to calculate the theoretical probability of an event.

I Can

The scale below can help you and your students understand their progress on a learning goal.

4	I can use sets and their relationships to understand and calculate probability, and I can explain my reasoning to others.
3	I can use sets and their relationships to understand and calculate probabilities.
2	I can understand and calculate probabilities of simple events.
1	I can perform some of the steps needed to calculate probabilities of simple events.

Spiral Review • Assessment Readiness

These questions will help determine if students have retained information taught in the past and can also prepare them for high-stakes assessments. Here, students calculate the volume of a cone **(19.1)**, calculate the volume of a triangular prism **(19.2)**, interpret a frequency table **(Gr8, 9.3)**, and calculate the volume of a sphere **(19.3)**.

38. MP **Reason** Greg buys a ticket for a raffle. On the ticket, it states that the probability of winning a prize is $\frac{1}{25}$. He wants to know what the probability is that he will not win a prize. Explain how he can determine this probability. See Additional Answers.

Observe the jar of marbles at the right.

39. The jar has the marbles shown. What is the probability that a randomly selected marble is not yellow? Use the complement to find the probability. $\frac{4}{5}$

40. Suppose one marble of each color is removed from the jar. What is the probability that a randomly selected marble is red? $\frac{2}{7}$

41. Jerome is going to randomly select one of his friends to go to an amusement park with him. The list shows the names of his friends. What is the probability that he chooses a name that does not start with a J, A, or D? $\frac{2}{5}$

Ana	Calvin	Deandre
Donna	Jorge	Jake
Kim	Bruce	Phu
Jasmine		

Spiral Review • Assessment Readiness

42. Find the volume of the cone.

Ⓐ 42π cm³
Ⓑ 294π cm³
Ⓒ 882π cm³
Ⓓ 1176π cm³

7 cm
18 cm

43. Find the volume of the prism.

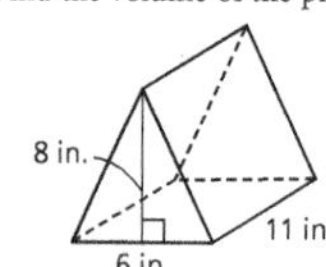

Ⓐ 25 in³
Ⓑ 132 in³
Ⓒ 264 in³
Ⓓ 528 in³

44. Use the two-way table to find the number of people in the east side of a street who swim.

	East	West	Total
Swims	28	24	52
Does not swim	17	19	36
Total	45	43	88

Ⓐ 17
Ⓑ 24
Ⓒ 28
Ⓓ 45

45. A ball bearing has a diameter of 6.5 mm. What is its volume? Round to the nearest whole.

Ⓐ 144 mm³
Ⓑ 575 mm³
Ⓒ 863 mm³
Ⓓ 1150 mm³

6.5 mm

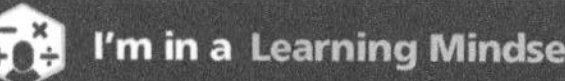

I'm in a Learning Mindset!

Did I manage my time effectively while I was calculating probabilities? What steps did I take to manage my time?

622

Keep Going Journal and Practice Workbook

Learning Mindset

Perseverance Collects and Tries Multiple Strategies

Point out that using the complement to calculate probability is useful when calculating the probability of an event with a large number of outcomes, but it certainly does not negate other strategies. Encourage students to use different ways to calculate probability and compare the process. *Which strategy is more efficient? Do you find it useful to use two different strategies to decide which one is more efficient?*

20.2 Disjoint and Overlapping Events

LESSON FOCUS AND COHERENCE

Mathematics Standards

- Construct and interpret two-way frequency tables of data when two categories are associated with each object being classified. Use the two-way table as a sample space to decide if events are independent and to approximate conditional probabilities.
- Apply the Addition Rule, $P(A \text{ or } B) = P(A) + P(B) - P(A \text{ and } B)$, and interpret the answer in terms of the model.

Mathematical Practices and Processes 

- Look for and make use of structure.
- Attend to precision.
- Model with mathematics.

I Can Objective 

I can calculate probabilities for both disjoint and overlapping events.

Learning Objective

Recognize disjoint and overlapping events and understand the Addition Rule to find the probability of the union of overlapping events.

Language Objective

Explain in everyday language what it means for events to be mutually exclusive.

Vocabulary

New: disjoint events, mutually exclusive events, overlapping events

Mathematical Progressions

Prior Learning	Current Development	Future Connections
Students: • summarized categorical data for two categories in two-way frequency tables. **(Alg1, 21.1)**	**Students:** • recognize disjoint and overlapping events. • devise the Addition Rule for the probability of the union of overlapping events. • calculate probabilities of unions of events.	**Students:** • will understand conditional probability and interpret independence using conditional probability. **(21.1; Alg2, 18.1 and 18.2)** • will understand that two events *A* and *B* are independent if the probability of *A* and *B* occurring together is the product of their probabilities. **(21.2; Alg2, 18.2)**

PROFESSIONAL LEARNING

Math Background

An important goal of this lesson is to help students understand the concept of overlapping events. For example, in Task 2 they see that the intersection of the events "the result is an odd number" and "the result is a factor of 6" is nonempty, so it is possible for both events to occur at the same time. Students go on to find the probability of the intersection of the events and use the probability in finding the probability of the union of the events. This concept is important in probability, as events often overlap and the probability of the intersection of overlapping events is involved in finding other probabilities. Later in the course, students will find conditional probabilities, which involves dividing the probability of the intersection of events by the probability of one of the events.

WARM-UP OPTIONS

ONLINE

PROJECTABLE & PRINTABLE

ACTIVATE PRIOR KNOWLEDGE • Calculate Theoretical Probabilities

Use these activities to quickly assess and activate prior knowledge as needed.

Problem of the Day 

Mia is getting her lunch for the day ready and reaches into a cabinet to get a can of soup. There are 6 cans of soup in the cabinet, and each can is a different kind. If Mia cannot see into the cabinet, what is the probability that she chooses vegetable beef soup or lentil soup? $\frac{2}{6} = \frac{1}{3}$

Quick Check for Homework

As part of your daily routine, you may want to display the Teacher Solution Key to have students check their homework.

Make Connections

Based on students' responses to the Problem of the Day, choose one of the following:

 Project the Interactive Reteach, Lesson 20.1.

 Complete the Prerequisite Skills Activity:

Have students work in pairs. Show the class a diagram of a spinner with equal-sized sectors labeled with integers 1 through 12. Have students imagine the spinner being spun. They can then take turns giving each other a simple event, such as "The result is an even number," "The result is less than 5," or "The result is a factor of 8," while their partner responds with the probability of the event.

- *What is the sample space?* all of the integers in the range 1 through 12
- *What is the set of outcomes in the event?* Possible answer: the integers 2, 4, 6, 8, 10, and 12
- *How do you find the probability of the event?* Divide the size of the event by the size of the sample space.

If students continue to struggle, use Tier 2 Skill 22.

SHARPEN SKILLS

If time permits, use this on-level activity to build fluency and practice basic skills.

Quantitative Comparison

Objective: Students make a comparison between two quantities.
Write the following problem on the board. Ask students to choose the letter representing the correct answer and to explain their reasoning.

Quantity A	Quantity B
The probability that a randomly chosen integer in the range from 1 to 12 is a factor of 7.	The probability that a randomly chosen integer in the range from 1 to 100 is a multiple of 5.

A. Quantity A is greater.
B. Quantity B is greater. B; Quantity A is $\frac{2}{12} = \frac{1}{6}$, while Quantity B is $\frac{20}{100} = \frac{1}{5}$.
C. The two quantities are equal.
D. The relationship cannot be determined from the information given.

PLAN FOR DIFFERENTIATED INSTRUCTION

Small-Group Options

Use these teacher-guided activities with pulled small groups.

On Track

Materials: index cards

Give each group a set of outcome cards with each card describing an event, such as "The number is less than 5," "The number is odd," or "The number is a factor of 10." Have students do the following:

- Select a pair of event cards and assume each event refers to a randomly chosen integer in the range from 1 to 20.
- Decide whether the events are disjoint or overlapping, and explain.
- Find the probability of the union of the events.

Almost There

Materials: index cards

Give each group two sets of cards. One set is 12 outcome cards, numbered from 1 to 12. The statements on the cards for the other set describe events such as "The number is less than 5," "The number is odd," or "The number is a factor of 10." Have students do the following:

- Select a pair of event cards and place them side by side.
- Place any outcome cards in both events between the event cards.
- Place any outcome cards in one event, but not in the other, under the event card.
- Decide whether the events are disjoint or overlapping, and explain.

Ready for More

Materials: index cards

Give each group two sets of cards. One set is 12 outcome cards, numbered from 1 to 12. The statements on the cards for the other set events such as "The number is less than 5," "The number is odd," or "The number is a factor of 10." Have students to do the following:

- Select an event card and place all outcome cards that apply below the card as the sample space.
- Select another event card, and move any cards from the sample space that apply below the new event.
- Divide the number of outcome cards in the new event by the original size of the sample space, and describe what this probability represents.

Math Center Options

Use these student self-directed activities at centers or stations.

Key: ● Print Resources ● Online Resources

On Track

- ● Interactive Digital Lesson
- ●● Journal and Practice Workbook
- ● Interactive Glossary (printable): **disjoint events**, **mutually exclusive events**, **overlapping events**
- ● Module Performance Task

Standards Practice:

- ●● Construct and Use Two-Way Frequency Tables
- ●● Apply the Addition Rule of Probability

Almost There

- ● Reteach 20.2 (printable)
- ● Interactive Reteach 20.2
- ● RtI Tier 2 Skill 22: Probability of Simple Events
- ● Illustrative Mathematics: Describing Events

Ready for More

- ● Challenge 20.2 (printable)
- ● Interactive Challenge 20.2
- ● Illustrative Mathematics: The Titanic I

Unit Project Check students' progress by asking how the probability of the union of two events relates to the intersection of the two events.

View data-driven grouping recommendations and assign differentiation resources.

Spark Your Learning • Student Samples

During the *Spark Your Learning*, listen and watch for strategies students use. See samples of student work on this page.

Use the Number of Favorable Outcomes

Strategy 1

Using the table, I can see that there are 32 pieces of fruit, and 8 of them are green, so the probability is $\frac{8}{32} = \frac{1}{4}$.

If students . . . find the ratio of the number of favorable outcomes to the size of the sample space to solve the problem, they are demonstrating an understanding of theoretical probability from Lesson 20.1.

Have these students . . . explain how they determined their approach and how they solved it. **Ask:**

- Q How did you use the relationship between the pieces of fruit and the concept of theoretical probability?
- Q What did you assume about the likelihoods of selecting different pieces of fruit?

Use a Complement

Strategy 2

Using the table, I can see that there are 32 pieces of fruit, and $4 + 8 + 12 = 24$ of the pieces are a color other than green, so the probability of selecting a color other than green is $\frac{24}{32} = \frac{3}{4}$. This means the probability of selecting a green piece of fruit is $1 - \frac{3}{4} = \frac{1}{4}$.

If students . . . use a complement to solve the problem, then they understand theoretical probability but may not realize that finding this probability directly would be more efficient.

Activate prior knowledge . . . by having students solve for the probability directly. **Ask:**

- Q What would you describe as the set of favorable outcomes?
- Q How many favorable outcomes are there?

COMMON ERROR: Misinterpret the Number of Favorable Outcomes

Reading across the top row of the table, I see there are 6 green apples, so the probability is $\frac{6}{32} = \frac{3}{16}$.

If students . . . misinterpret the number of favorable outcomes, then they may not understand that the set of favorable outcomes is the union of two types of fruit.

Then intervene . . . by pointing out that there is another type of fruit that might be green. **Ask:**

- Q Did the problem specifically ask for the probability of selecting a green apple?
- Q Is there another type of fruit that might be green?

20.2

Disjoint and Overlapping Events

I Can calculate probabilities for both disjoint and overlapping events.

Spark Your Learning

Fruits can be classified in many ways, such as the type of seeds they contain, the type of core they have, or what type of climate they grow in. You can also categorize fruits by the color of their surfaces.

Complete Part A as a whole class. Then complete Parts B–D in small groups.

A. What is a mathematical question you can ask about this situation? What information would you need to know to answer your question? A. What is the probability of selecting a green piece of fruit at random?; the total number of fruits and the colors of the fruits

B. What variable(s) are involved in this situation? See Additional Answers.

C. To answer your question, what strategy and tool would you use along with all the information you have? What answer do you get? See Strategies 1 and 2 on the facing page.

D. Does your answer make sense in the context of the situation? How do you know? yes; Possible answer: About 25% of the picture is green.

Turn and Talk What are two groups of fruit characteristics that have items in common? See margin.

1 Spark Your Learning

▶ MOTIVATE

- Have students look at the photo in their books and read the information contained in the photo. Then complete Part A as a whole-class discussion.
- Give the class the additional information they need to solve the problem. This information is available online as a printable and projectable page in the Teacher Resources.
- Have students work in small groups to complete Parts B–D.

▶ PERSEVERE

If students need support, guide them by asking:

Q **Advancing • Use Tools** Which tool could you use to solve the problem? Why choose that tool and not some other? Students' choices of tools and reasons for choosing them will vary.

Q **Assessing** How many pieces of green fruit are there? There are 6 green apples and 2 green pears, so there are $6 + 2 = 8$ green pieces of fruit.

Q **Assessing** How does the likelihood of selecting one specific piece of fruit compare to the likelihood of selecting a different specific piece of fruit? The likelihoods are equal.

Q **Advancing** How can you describe the likelihood of selecting a green piece of fruit? I can divide the number of pieces of green fruit by the total number of pieces of fruit.

Turn and Talk Encourage students to think about intersecting rows and columns in the table shown on the projectable. Ask them how they can think of a green apple as a piece of fruit that is in two categories. It may be helpful to refer to an apple as "a fruit that is an apple" and to point out that the set of apples includes both apples that are green and apples that are red. fruits that are apples and fruits that are green

SUPPORT SENSE-MAKING • Information Gap

Ask students questions to help them decide what missing information they need to answer the question, "What is the probability of selecting a green piece of fruit at random?"

1. Do you have enough information to conclude that there is such a probability? Explain. yes; Since there is an equal likelihood of selecting any particular piece of fruit, there is a theoretical probability of selecting a piece of fruit in any subset of the sample space.
2. Suppose you knew how many green apples there are. Can you determine the probability? Explain. no; Possible explanation: If I knew the total number of pieces of fruit, I could solve for the probability of selecting a green apple, but that is not quite the probability we are looking for.
3. In addition to the total number of pieces of fruit, what information might be helpful? Possible answer: I also need to know how many green pieces of fruit there are.

▶ BUILD SHARED UNDERSTANDING

Select groups of students who used various strategies and tools to share with the class how they solved the problem. As they present their solutions, have each group discuss why they chose a specific strategy and tool.

② Learn Together

Build Understanding

Task 1 (MP) **Use Structure** Students recognize disjoint, or mutually exclusive, events in the given scenario. They then go on to deduce the probability of the union of two such events.

CONNECT TO VOCABULARY

Have students use the **Interactive Glossary** to record their understanding of the vocabulary in this task.

Sample Guided Discussion:

Q **In Part C, how would you get the set of marbles that are orange and purple from the set of orange marbles and the set of purple marbles?** It is the intersection of the sets of orange marbles and purple marbles.

Turn and Talk For the first part, ask students to explain if there should be any overlap in the circles representing the sets of marbles of various colors. Check students' diagrams. Diagrams should show no overlaps; $\frac{8}{15}$; You can count the number of green marbles and purple marbles and then divide by the total number of marbles, or you can add the probabilities of choosing a green marble and of choosing a purple marble.

Build Understanding

Recognize Disjoint Events

Two events are **disjoint**, or **mutually exclusive**, if they cannot occur in the same trial at the same time.

If you flip a coin, it cannot have a result of both heads and tails in the same trial. If you roll a number cube, you cannot roll a number that is both even and odd. These events are disjoint.

The Venn diagram shows two events A and B that are disjoint. Observe that the circles do not intersect.

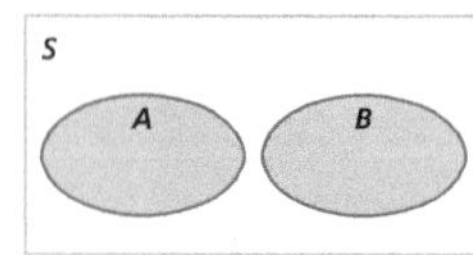

1 Haylee randomly chooses a marble from the jar of marbles.

A. What is the probability that she chose an orange marble? $\frac{7}{15}$

B. What is the probability that she chose a purple marble? $\frac{1}{5}$

C. What is the probability that she chose a marble that is orange and purple? Explain your reasoning. The probability is 0. There are no marbles that are orange and purple. It is not possible to choose an orange and purple marble on the same trial, so these events are disjoint.

D. What is the probability that Haylee will randomly choose a marble that is orange or purple? Explain how you found your answer. $\frac{2}{3}$; You can count the number of orange marbles and purple marbles.

E. What relationship do you notice about the probability of choosing an orange marble, the probability of choosing a purple marble, and the probability of choosing an orange marble or a purple marble? See Additional Answers.

F. Use the relationship you found in Part E to write an equation to determine the probability of randomly choosing an orange marble or a green marble from the jar of marbles. $\frac{7}{15}+\frac{5}{15}=\frac{12}{15}=\frac{4}{5}$

Turn and Talk

- For Part D, sketch a Venn diagram that would be helpful in finding the probability.
- What is the probability that Haylee chooses a green marble or a purple marble? How can you calculate this probability in two different ways? See margin.

624

LEVELED QUESTIONS

Depth of Knowledge (DOK)	Leveled Questions	What Does This Tell You?
Level 1 **Recall**	What does it mean when we say events are disjoint? They cannot occur in the same trial at the same time.	Students' answers will indicate whether they understand what is meant by *disjoint events.*
Level 2 **Basic Application of Skills & Concepts**	Can you give another pair of disjoint events in the scenario from the task? Possible answer: A marble is purple, and a marble is green.	Students' answers will indicate whether they can identify another pair of disjoint events in the scenario from the task.
Level 3 **Strategic Thinking & Complex Reasoning**	Can you give an example of a pair of real-world events that are not disjoint? Possible answer: It rained on a particular day, and it snowed on a particular day.	Students' answers will indicate whether they can devise an example of a pair of events with a non-empty intersection.

Recognize Overlapping Events

Two or more events are **overlapping events** if they share one or more outcomes in common.

If you roll a six-sided number cube, you can roll a number that is both even and a multiple of 3. If you randomly choose a car from a lot, you can choose a car that is blue and is a sedan. These are overlapping events.

The Venn diagram shows two events A and B that are overlapping. Notice that the circles intersect.

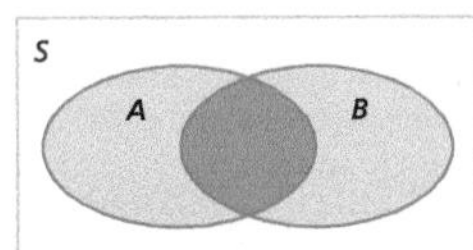

2 Dylan spins the spinner once.

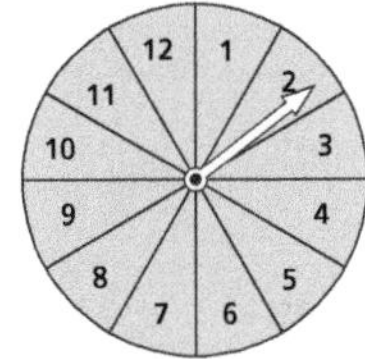

A. What is the probability that the result is an odd number? $\frac{1}{2}$

B. What is the probability that the result is a factor of 6? $\frac{1}{3}$

It is possible that the result is both an odd number and a factor of 6, so these events are not disjoint. They are overlapping events.

C. Sketch a Venn diagram that represents the overlapping events. What is the probability that the result is an odd number or a factor of 6? See Additional Answers.

D. Compare the probability that the result is an odd number or a factor of 6 to the sum of the individual event probabilities. What do you notice?
D, E. See margin.

E. Use your response in Part D to find the probability that the number is even or a multiple of 5. Write an equation that shows how you determined this probability.

Turn and Talk See margin.
- What does *and* represent when describing the probability of multiple events?
- What does *or* represent when describing the probability of multiple events?

Task 2 (MP) **Attend to Precision** Students recognize overlapping events in a scenario involving a spinner with integers 1–12. They then go on to deduce the probability of the union of two such events, using a Venn diagram to help visualize the relationship between the sets of favorable outcomes of the events. Make sure students understand that the word "or" in this context does not mean the "exclusive or."

CONNECT TO VOCABULARY

Have students use the **Interactive Glossary** to record their understanding of the vocabulary in this task.

Sample Guided Discussion:

Q In Part C, how are the numbers that are an odd number and a factor of 6 related to the set of odd numbers and the set of factors of 6? They form the intersection of the two sets.

Q In Part D, why do you not just add the number of spaces with an odd number and the number of spaces with a factor of 6? Since the spaces with 1 and 3 fall in both of these sets, I would be adding them twice.

Turn and Talk Remind students that imposing an "and" statement to two nonequivalent conditions results in a more exclusive condition, while an "or" statement results in a more inclusive condition. Encourage them to sketch a Venn diagram to better visualize this. When "and" is used, it is talking about the intersection of the events. When "or" is used, it is talking about the union of the events.

PROFICIENCY LEVEL

Beginning

Show students a Venn diagram representing overlapping events A and B. Indicate the intersection of the events and say, "This represents A and B." Indicate all three portions of the union of the events and say, "This represents A or B." Give students a new Venn diagram depicting a new overlapping set of events C and D, and have them say similar statements while indicating the relevant portions of the diagram.

Intermediate

Have students work in pairs, and give each pair a diagram of a spinner with integers 1–10. One student will quiz the other with a pair of statements such as "The number is even" and "The number is greater than four." The other student will explain how to find the union of the two statements and their probability.

Advanced

Have students write a paragraph to explain how to find the probability of the union of overlapping events.

ANSWERS

D. The probabilities are not equal. You need to subtract the probability of spinning an odd number and a factor of 6 from the sum of the individual event probabilities to get the same result as the probability of spinning an odd number or a factor of 6.

$P(A \text{ or } B) = P(A) + P(B) - P(A \text{ and } B)$

E. $\frac{6}{12} + \frac{2}{12} - \frac{1}{12} = \frac{7}{12}$

Step It Out

Task 3 (MP) **Attend to Precision** Students see how the Addition Rule for the probability of overlapping events is applied to an example. Make sure students are precise when explaining why the events in the example are not disjoint. Instead of making a general statement, they should make explicit mention of an outcome that is an outcome of both events. While students will go on to find all outcomes in the intersection of the events when they find the probability of their union, they only need to show that the intersection is nonempty in order to prove that the events are not disjoint.

Sample Guided Discussion:

Q **In Part A, what do you need to find in order to show that two events are not disjoint?** I need to find one case that is an outcome of both events.

Q **In Part B, what kind of set is the intersection of disjoint sets?** It is the empty set.

Turn and Talk If necessary, prompt students by asking them what it means for two events to be disjoint. Ask them for the probability of an outcome that cannot happen. It may be helpful to connect this question with the task more directly and ask students again about Part B. Two events are disjoint if the probability of both events occurring at the same time is 0. If the probability is 0, then it is impossible for both events to occur, so the events are disjoint.

ANSWERS

C. If you count the number of outcomes in each event, then the outcomes that are in both events are counted twice. You have to subtract that number so they are only counted once.

Step It Out

Find the Probability of Disjoint Events

The probability of disjoint events can be determined by adding the probabilities of the individual events.

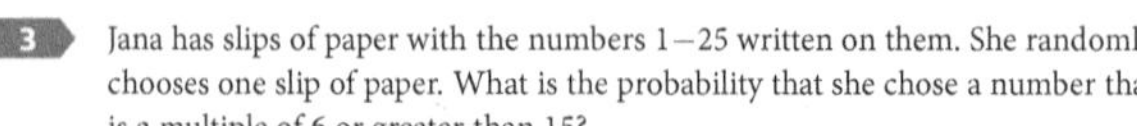

Disjoint Events

If A and B are disjoint events, then $P(A \text{ or } B) = P(A) + P(B)$.

The probability of overlapping events can be determined by adding the probabilities of the individual events and subtracting the probability of both events occurring.

The Addition Rule

$P(A \text{ or } B) = P(A) + P(B) - P(A \text{ and } B)$.

3 Jana has slips of paper with the numbers 1–25 written on them. She randomly chooses one slip of paper. What is the probability that she chose a number that is a multiple of 6 or greater than 15?

Let A be the event that the number is a multiple of 6. Let B be the event that the number is greater than 15.

A. Are A and B disjoint events? Explain your reasoning.
no; The probability of A and B is not 0.

Make a table to organize the different probabilities.

Probability	Value
$P(A)$	$\frac{4}{25}$
$P(B)$	$\frac{10}{25} = \frac{2}{5}$
$P(A \text{ and } B)$	$\frac{2}{25}$

B. If A and B are disjoint, then $P(A \text{ and } B)$ is 0, and the equation simplifies to $P(A \text{ or } B) = P(A) + P(B)$.

B. Explain why you can use the Addition Rule even if A and B were disjoint.

Use the Addition Rule to find the probability. See margin.

$$P(A \text{ or } B) = P(A) + P(B) - P(A \text{ and } B)$$
$$= \frac{4}{25} + \frac{10}{25} - \frac{2}{25} = \frac{12}{25}$$

C. Explain why you subtract the probability of A and B in terms of the number of outcomes.

Answer the question.

The probability that she chose a number that is a multiple of 6 or greater than 15 is $\frac{12}{25}$.

Turn and Talk How can you use probability to determine whether or not two events are disjoint? Explain. See margin.

Find a Probability from a Two-Way Table of Data

You can use a two-way table to determine the probabilities of events.

4 Alyssa took a survey of some students in her school. She asked them what their favorite season is. The results of the survey are shown in the table.

Favorite Season	Year in High School				
	Freshman	Sophomore	Junior	Senior	Total
Spring	14	22	19	21	76
Summer	56	55	46	63	220
Fall	12	3	24	7	46
Winter	8	11	0	10	29
Total	90	91	89	101	371

What is the probability that a randomly selected student is a junior or their favorite season is winter?

Find each of three probabilities—the student is a junior, their favorite season is winter, and they are both a junior and their favorite season is winter.

A, B. See margin.

Let A be the event that a student is a junior.
Let B be the event that a student said winter.

$P(A) = \frac{89}{371}$; $P(B) = \frac{29}{371}$; $P(A \text{ and } B) = \frac{0}{371} = 0$

A. How did you use the table to find the values in the numerator and dominator of these probabilities?

Use the Addition Rule to find $P(A \text{ or } B)$.

$P(A \text{ or } B) = P(A) + P(B) - P(A \text{ and } B)$

$= \frac{89}{371} + \frac{29}{371} - 0 = \frac{118}{371}$

B. Do you have to use the Addition Rule to find this probability? Explain.

The probability that the student is a junior or their favorite season is winter is $\frac{118}{371}$.

What is the probability that a randomly selected student is a sophomore or their favorite season is summer?

$P(\text{sophomore}) = \frac{91}{371}$; $P(\text{summer}) = \frac{220}{371}$; $P(\text{sophomore and summer}) = \frac{55}{371}$

$P(A \text{ or } B) = P(A) + P(B) - P(A \text{ and } B)$

$= \frac{91}{371} + \frac{220}{371} - \frac{55}{371} = \frac{256}{371}$

C. Do you have to use the Addition Rule to find this probability? Explain.

The probability that the student is a sophomore or their favorite season is summer is $\frac{256}{371}$.

C. yes; Because the probability of both is not 0, the events are overlapping.

Turn and Talk What does a 0 in a cell of the two-way table indicate to you? See margin.

Task 4 (MP) **Model with Mathematics** Students see that a two-way frequency table can be used to find the probabilities of unions of two disjoint events and of two overlapping events. They see how such a representation of data explicitly indicates the cardinality of intersection of events, so events can be determined to be disjoint or overlapping simply by inspecting the frequency shown in the appropriate cell.

SUPPORT SENSE-MAKING Three Reads

Have students read the problem three times. Use the questions below for a different focus each read.

1. What is the situation about?
2. What are the quantities in the situation?
3. What are the possible mathematical questions that you could ask for this situation?

Sample Guided Discussion:

Q In Part A, how can you find $P(A \text{ and } B)$? I can look for the cell in the "Winter" row and "Junior" column and divide its number by the number in the Total cell of the table.

Q Does "a randomly selected student is a sophomore or their favorite season is summer" mean a student can meet both conditions? yes

Turn and Talk Remind students that the number in a cell indicates how many favorable outcomes there were for that particular event. If necessary, prompt them with a question such as "What does it mean when there are zero outcomes in two events?" The events in the row and in the column are disjoint.

ANSWERS

A. The denominator of 371 is the total number of students surveyed. The total number of juniors is 89, the total number of students who said winter is 29, and the number of juniors who said winter is 0.

B. no; Because the probability of both is 0, the events are disjoint. You can just add the probabilities of the two events.

ONLINE

Assign the Digital On Your Own for
- built-in student supports
- Actionable Item Reports
- Standards Analysis Reports

On Your Own

Assignment Guide

The chart below indicates which problems in the On Your Own are associated with each task in the Learn Together. Assign daily homework for tasks completed.

Learn Together Tasks	On Your Own Problems
Task 1, p. 624	Problems 7, 10, and 11
Task 2, p. 625	Problems 8, 9, and 12
Task 3, p. 626	Problems 13, 20–24, and 26
Task 4, p. 627	Problems 14–19 and 25

Check Understanding

Determine if the events are disjoint or overlapping.

1. You roll a six-sided number cube once and the result is less than 4 and a factor of 4. overlapping events
2. You roll a six-sided number cube once and the result is odd or greater than 5. disjoint events

Find the probability of each event.

3. Juanita has a bag that contains 5 red marbles, 9 yellow marbles, and 6 green marbles. She randomly chooses a yellow marble or a green marble. $\frac{3}{4}$
4. Jun spins a spinner with equal sized sections labeled 1–10. The result is a number that is a multiple of 3 or a factor of 6. $\frac{1}{2}$

Use the two-way table to find each probability.

		Enjoys Cooking		
		Yes	No	Total
Tries New Foods	Yes	4	18	22
	No	13	23	36
	Total	17	41	58

5. $P(\text{enjoys cooking or tries new foods})$ $\frac{35}{58}$
6. $P(\text{does not enjoy cooking or does not try new foods})$ $\frac{54}{58}$

On Your Own

7. **Critique Reasoning** Mario is rolling a six-sided number cube. He states that rolling an even number and rolling less than 2 are disjoint events. Is he correct? Explain your reasoning. Mario is correct; It is not possible to roll an even number less than 2, so the events are disjoint.

A bag contains slips of paper with the numbers 1–20 on them. A slip of paper is selected at random. Find each probability.

8. $P(\text{less than 7 or even})$ $\frac{13}{20}$
9. $P(\text{multiple of 4 or less than 10})$ $\frac{3}{5}$
10. $P(\text{even or factor of 15})$ $\frac{7}{10}$
11. $P(\text{greater than 14 or} \leq 2)$ $\frac{2}{5}$

12. A bag contains 26 tiles, each with a different letter of the alphabet written on it. Consider the letters a, e, i, o, or u as vowels. You choose a tile without looking. What is the probability of each event?

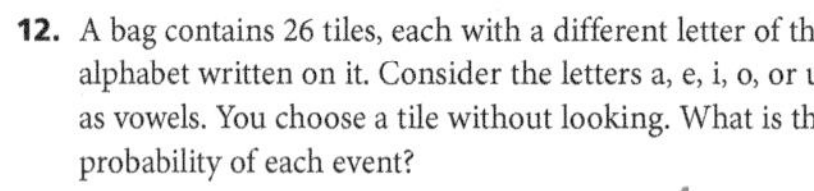

   A. a vowel or a letter in the word MATH $\frac{4}{13}$
   B. a consonant or a letter not in the word MATH $\frac{25}{26}$

13. **Open Ended** Write two events that are disjoint. Write two events that are overlapping. Find the probability of your events. Check students' answers.

628

Check Understanding

Formative Assessment

Use formative assessment to determine if your students are successful with this lesson's learning objective.

Students who successfully complete the Check Understanding can continue to the On Your Own practice.

For students who miss 1 problem or more, work in a pulled small group using the Almost There small-group activity on page 623C.

ONLINE

Assign the Digital Check Understanding to determine
- success with the learning objective
- items to review
- grouping and differentiation resources

Differentiation Options

Differentiate instruction for all students using small-group activities and math center activities on page 623C.

Reteach

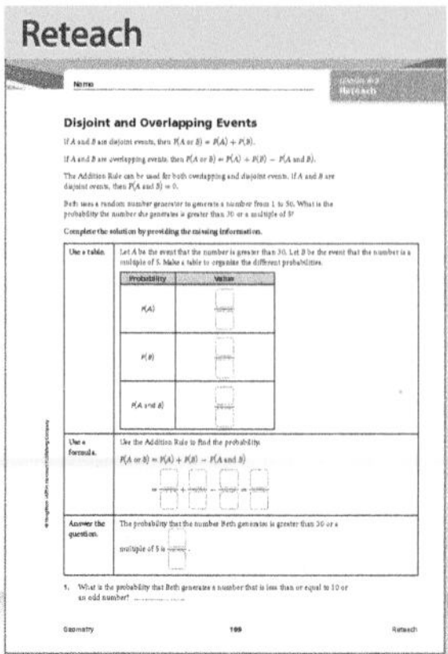

Challenge

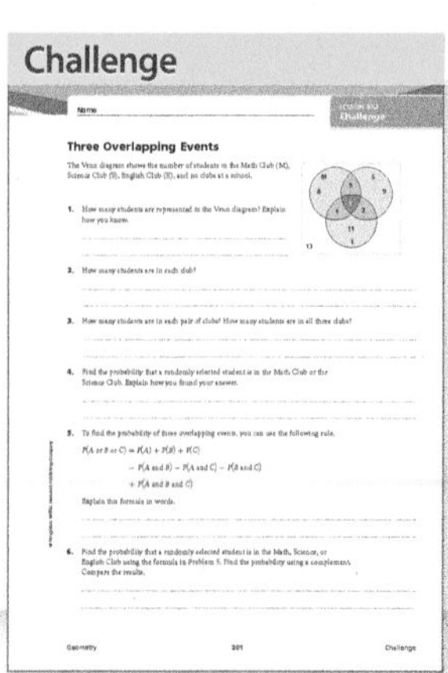

The table shows the results of a survey asking residents of three towns if they had attended college. You randomly choose one person from the survey. Find the probability of each event.

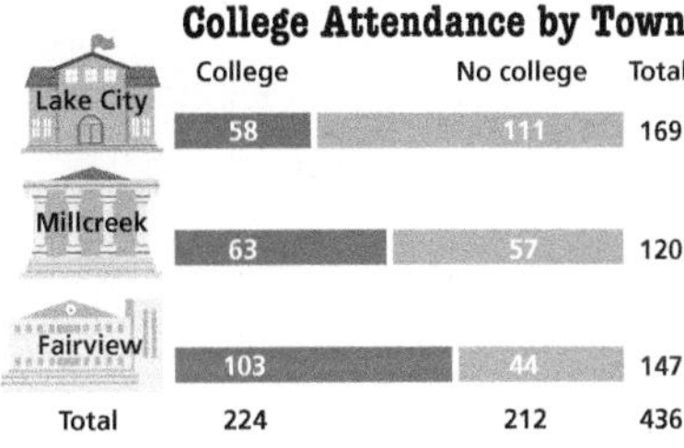

College Attendance by Town

	College	No college	Total
Lake City	58	111	169
Millcreek	63	57	120
Fairview	103	44	147
Total	224	212	436

14. The person is a resident of Lake City. $\frac{169}{436}$

15. The person attended college. $\frac{56}{109}$

16. The person is a resident of Millcreek and attended college. $\frac{63}{436}$

17. The person is a resident of Fairview or attended college. $\frac{67}{109}$

18. The person is a resident of Lake City or did not attend college. $\frac{135}{218}$

19. The person attended college or is not a resident of Millcreek. $\frac{379}{436}$

STEM The table shows partial climate data in Honolulu, Hawaii throughout the year. If a meteorologist chooses a month at random to analyze the climate, find the probability that she chooses a month with the given data.

	Average Climate Data—Honolulu, Hawaii											
	Jan	Feb	Mar	Apr	May	Jun	Jul	Aug	Sep	Oct	Nov	Dec
High (°F)	80	80	81	83	85	87	88	89	89	87	84	81
Low (°F)	66	66	68	69	71	73	74	75	74	73	71	68
Sunshine (h)	227	202	250	255	276	280	293	290	279	257	221	211
Precipitation (in.)	2.3	2.0	2.0	0.6	0.6	0.3	0.5	0.6	0.7	1.9	2.4	3.2

20. The high temperature is above 80 °F and below 87 °F. $\frac{5}{12}$

21. The low temperature is below 70 °F or the precipitation is above 2.0 inches. $\frac{1}{2}$

22. The number of hours of sunshine is 280 h or the precipitation is below 0.5 in. $\frac{1}{12}$

23. The high temperature is 80 °F or the number of hours of sunshine is above 225 h. $\frac{5}{6}$

Questioning Strategies

Problem 17 What is the probability that the person is a resident of Fairview or is a resident of Millcreek or attended college? $\frac{147}{436}+\frac{120}{436}+\frac{224}{436}-\frac{103}{436}-\frac{63}{436}=\frac{325}{436}$

Watch for Common Errors

Problem 22 Students may incorrectly interpret the "or" in the statement as the "exclusive or" and not include January and December as outcomes of the event since those months had both a low temperature below 70 °F and precipitation above 2.0 inches. They need to understand that unless otherwise specified, "or" should be interpreted as the "inclusive or."

5 Wrap-Up

Summarize learning with your class. Consider using the Exit Ticket, Put It in Writing, or I Can scale.

Exit Ticket 

Find the probability that a card you draw from a standard 52-card deck is a red card or an ace. $\frac{1}{2}+\frac{4}{52}-\frac{2}{52}=\frac{28}{52}=\frac{7}{13}$

Put It in Writing 

Explain why the Addition Rule $P(A \text{ or } B) = P(A) + P(B) - P(A \text{ and } B)$ can still be used when A and B are disjoint events.

I Can 

The scale below can help you and your students understand their progress on a learning goal.

4	I can calculate probabilities for both disjoint and overlapping events, and I can explain my steps to others.
3	I can calculate probabilities for both disjoint and overlapping events.
2	I can use a two-way frequency table to tell whether two events are disjoint or overlapping.
1	I can explain what are disjoint events and what are overlapping events.

Spiral Review • Assessment Readiness

These questions will help determine if students have retained information taught in the past and can also prepare them for high-stakes assessments. Here, students must multiply two rational numbers **(Gr7, 5.2)**, find the volume of a sphere **(19.3)**, and determine theoretical probabilities of simple events **(20.1)**.

24. Jessica surveys middle school and high school students and asks if they play video games. Construct a two-way table showing her results. For a randomly selected student, find the probability of each event.

A. a middle school student or does not play video games A–C. See Additional Answers.

B. a high school student or plays video games

C. any student who plays video games or does not play video games

54 out of 60 middle school students and 48 out of 55 high school students play video games.

25. MP **Reason** Explain why it makes sense to subtract the probability of events A and B from the sum of the probabilities of A and B for overlapping events when using the Addition Rule. See Additional Answers.

26. Last year, Michael painted 17 portraits, drew 11 sketches, and made 7 sculptures. This year, he painted 21 portraits and made 14 sculptures. He chooses a piece of artwork at random. Find the probability of each event.

A. chooses a sketch or the piece of artwork was made this year $\frac{23}{35}$

B. chooses a portrait or the piece of artwork was made last year $\frac{4}{5}$

Spiral Review • Assessment Readiness

27. Find $\frac{2}{3} \times \frac{9}{16}$.

(A) $\frac{3}{8}$ (B) $\frac{11}{48}$ (C) $\frac{11}{19}$ (D) $\frac{59}{48}$

28. Find the volume of the sphere.

(A) 4608π in^3
(B) 9216π in^3
(C) $18{,}432\pi$ in^3
(D) $55{,}296\pi$ in^3

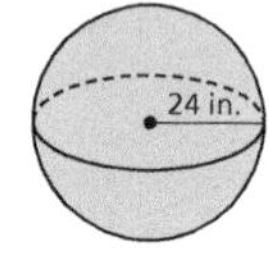

29. A standard deck of 52 cards has 13 cards (2, 3, 4, 5, 6, 7, 8, 9, 10, jack, queen, king, ace) in each of 4 suits (hearts, clubs, diamonds, spades). Suppose you select a card at random. Match the event on the left with its probability on the right.

Event	Probability
A. select a 2, 3, or 4	**1.** $\frac{3}{13}$
B. select an ace	**2.** $\frac{6}{13}$
C. do not select a card from 2 through 8	**3.** $\frac{1}{13}$
D. do not select a diamond	**4.** $\frac{3}{4}$

A. 1
B. 3
C. 2
D. 4

 **I'm in a Learning Mindset!**

How effective is making a table to find the probability of overlapping events?

Keep Going to Journal and Practice Workbook

Learning Mindset

Perseverance Collects and Tries Multiple Strategies

Point out how trying multiple strategies is an invaluable tool for learning. Students not only find more efficient ways to determine probabilities of overlapping events but also deepen their understanding when they compare the relative efficiencies of making a two-way frequency table to finding theoretical probabilities directly and applying the Addition Rule. Ask which factors they might consider when deciding whether to use a table, such as the relative complexity of the different events. *How can making a careful decision about which method to use to solve a problem help make your work more efficient? How does it help you develop a deeper understanding of the material?*

Module

20 Review

Events

Students are picking natural numbers less than or equal to 20. Set D contains multiples of 6. Set E contains multiples of 3.

$D = \{6, 12, 18\}$
$E = \{3, 6, 9, 12, 15, 18\}$

Subset
$D \subset E$

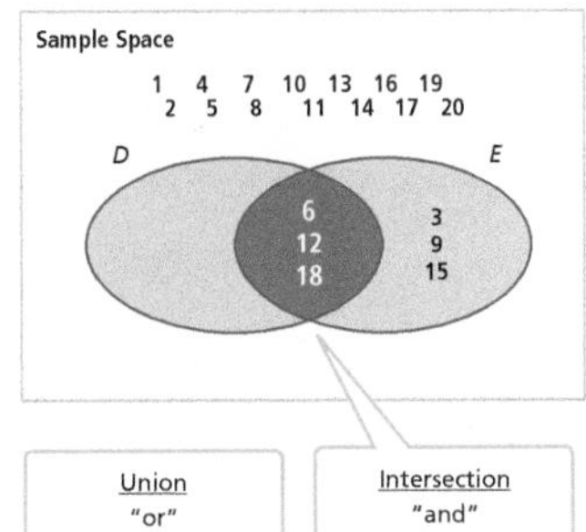

Union
"or"
$D \cup E =$
$\{3, 6, 9, 12, 15, 18\}$

Intersection
"and"
$D \cap E = \{6, 12, 18\}$

Theoretical Probability

The theoretical probability of an event D in the sample space S is

$$P(D) = \frac{\text{number of outcomes in the event}}{\text{number of outcomes in the sample space}}.$$

$P(D) = \frac{3}{20}$ — probability of picking a multiple of 6

$P(E) = \frac{6}{20}$ — probability of picking a multiple of 3

An event and its complement contain all elements of a sample space, with no overlapping elements.

Complement
$D^c = \{1, 2, 3, 4, 5, 7, 8, 9, 10, 11, 13, 14, 15, 16, 17, 19, 20\}$
$E^c = \{1, 2, 4, 5, 7, 8, 10, 11, 13, 14, 16, 17, 19, 20\}$

The sum of the probability of picking a number from set D and the probability of its complement is 1.

$P(D) + P(D^c) = 1$
$\frac{3}{20} + \frac{17}{20} = 1$

Mutually Exclusive Events

Students can also pick from the set of odd numbers, F. Students cannot pick a multiple of 6 and an odd number in a single pick. Thus, the events D and F are mutually exclusive events.

$P(D \cup F) = P(D) + P(F)$
$= \frac{3}{20} + \frac{10}{20}$
$= \frac{13}{20}$

Take the sum of the probabilities of each event.

Overlapping Events

$P(D \cup E) = P(D) + P(E) - P(D \cap E)$
$= \frac{3}{20} + \frac{6}{20} - \frac{3}{20}$
$= \frac{6}{20} = \frac{3}{10}$

6, 12 and 18 would be counted twice because they are in both sets.

The probability of overlapping events can be found by adding the probabilities of picking a number from set D and set E individually and subtracting the probability of both events occurring.

Module 20 631

ONLINE

Assign the Digital Module Review for
- built-in student supports
- Actionable Item Reports
- Standards Analysis Reports

MODULE 20 REVIEW

Module Review

Use the first page of the Module Review to summarize and connect the main ideas of the module. Use the second page to assess students' understanding of the vocabulary, concepts, and skills presented in the module.

Sample Guided Discussion:

Q Events What do the numbers that are outside of the circles in the Venn diagram represent? They are the natural numbers that are less than or equal to 20, but they do not fit the requirements for set D or set E.

Q Theoretical Probability What is the sum of the probability of picking a number from set E and the probability of its complement? Explain. 1; Because an event and its complement contain all of the elements of a sample space without overlapping, the sum of the probabilities of an event and its complement must equal 1.

$$P(E) + P(E^C) = 1$$

$$\frac{6}{20} + \frac{14}{20} = 1$$

Q Mutually Exclusive Events Why are the probabilities of each event added together for mutually exclusive events without subtracting the probability of both events occurring at the same time? Both events cannot occur in the same trial of the experiment, so the probability of both events occurring at the same time is 0. So, subtraction is not needed, and the probabilities of each event are added.

Q Overlapping Events What is the probability of the event of picking from set E or set F? Explain. $\frac{13}{20}$; $P(E) + P(F) - P(E \cap F) =$

$$\frac{6}{20} + \frac{10}{20} - \frac{3}{20} = \frac{13}{20}$$

Module Review continued

Possible Scoring Guide

Items	Points	Description
1–5	2 each	identifies the correct term(s)
6	1	correctly determines the probabilities of landing on green and of landing on yellow
6	2	correctly determines the three probabilities
7	1	correctly determines and explains the probabilities of choosing a shirt that is not blue and of choosing a shirt that is red or white
7	2	correctly determines and explains the probabilities of choosing a shirt that is not blue, of choosing a shirt that is red or white, and of choosing a shirt that is red and white
8	1	correctly determines the probabilities of choosing a piece of fruit and of choosing a refrigerated item
8	2	correctly determines the probabilities, as well as explains how to use the probabilities to determine the probability of choosing a piece of fruit or a refrigerated item
Total points possible = 16 points		

The Unit 10 Test in the Assessment Guide assesses content from Modules 20 and 21.

Vocabulary

Choose the correct term from the box to complete each sentence.

Vocabulary
elements
mutually exclusive
outcomes
overlapping events
sample space
subset
union

1. Two events are __?__ if they cannot both occur in the same trial of an experiment. mutually exclusive

2. A __?__ is a set that is contained entirely within another set. subset

3. The set of all possible __?__ is called a __?__. outcomes; sample space

4. The __?__ of sets *F* and *G* is the set of all __?__ that are in set *F* or set *G*. union; elements

5. Two or more events are __?__ if they have one or more outcomes in common. overlapping events

Concepts and Skills

6. (MP) Use Tools Denny and his friends are using a spinner. State what strategy and tool you will use to answer the questions, explain your choice, and then find the answers.

A. What is the probability that the spinner lands on green? $\frac{3}{8}$

B. What is the probability that the spinner lands on yellow? $\frac{1}{4}$

C. What is the probability that the spinner lands on red or blue? $\frac{3}{8}$

7. Elaine is choosing a shirt for school. She has 3 red shirts, 8 blue shirts, 4 green shirts, and 6 white shirts. See Additional Answers.

A. What is the probability that a shirt is not blue? Explain your reasoning. $\frac{13}{21}$

B. What is the probability that a shirt is red or white? Explain your reasoning. $\frac{9}{21}$

C. What is the probability that a shirt is red and white? Explain your reasoning. 0

8. The table shows the food in the school refrigerator and freezer. Keith cannot decide what he wants to eat, so he randomly chooses an item.

		Food			
		Vegetable	Fruit	Meat	Total
Location	Refrigerated	7	8	2	17
	Frozen	11	3	5	19
	Total	18	11	7	36

A. Determine the probability that the item is a piece of fruit. $\frac{11}{36}$

B. Determine the probability that the item is refrigerated. $\frac{17}{36}$

C. Determine the probability that the item is a piece of fruit and refrigerated. $\frac{2}{9}$

D. Explain how to use these probabilities to determine the probability that the item is a piece of fruit or refrigerated. See Additional Answers.

632

DATA-DRIVEN INSTRUCTION

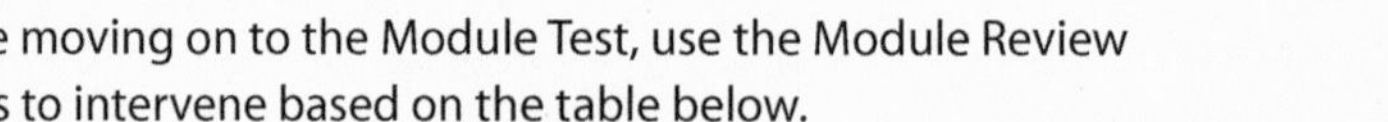
Before moving on to the Module Test, use the Module Review results to intervene based on the table below.

Items	Lesson	DOK	Content Focus	Intervention
6	20.1, 20.2	2	Use a spinner to determine probabilities.	Reteach 20.1, 20.2
7	20.1, 20.2	2	Use a list to determine probabilities.	Reteach 20.1, 20.2
8	20.1, 20.2	2	Use a table to determine probabilities.	Reteach 20.1, 20.2

Module Test

The Module Test is available in alternative versions in your Assessment Guide. All items are presented in standardized test formats.

ONLINE

Ed

Assign the Digital Module Test to power actionable reports including

- proficiency by standards
- item analysis

Form A

Module 20 • Form A
Module Test

Name ____________

1. $P(A) = 0.1$ and $P(B) = 0.7$. If events A and B are disjoint, what is $P(A \text{ or } B)$?
   (A) 0.07 (B) 0.2 (C) 0.73 (D) 0.8

2. A fair number cube is six-sided with the numbers 1–6. What is the intersection of favorable outcomes for the events "roll an even number with the number cube" and "roll a number greater than 4 with the number cube"?
   (A) 6 (B) 5, 6 (C) 2, 4, 6 (D) 2, 4, 5, 6

3. A fair number cube is six-sided with the numbers 1–6. What is the theoretical probability of rolling a number greater than 4 with the number cube?
   (A) $\frac{1}{6}$ (B) $\frac{1}{3}$ (C) $\frac{1}{2}$ (D) $\frac{2}{3}$

4. A bag contains cubes of different colors. If the probability of drawing a red cube is $\frac{1}{6}$ and the probability of drawing a green cube is $\frac{1}{3}$, what is the probability of drawing a red or a green cube?
   (A) $\frac{1}{18}$ (B) $\frac{1}{9}$ (C) $\frac{1}{6}$ (D) $\frac{1}{2}$

5. There are 3 sophomores, 12 juniors, and 9 seniors on the debate team. A student is selected at random to lead a discussion about the next debate topic. What is the theoretical probability that the student is a senior?
   (A) $\frac{3}{8}$ (B) $\frac{3}{7}$ (C) $\frac{1}{2}$ (D) $\frac{5}{8}$

6. The probability that event Z occurs is 41.3%. What is the probability that event Z does not occur?

   Possible answer: 58.7%

7. $P(E) = 0.05$ and $P(F) = 0.01$. If $P(E \text{ and } F) = 0.005$, what is $P(E \text{ or } F)$?

   Possible answer: 0.055

8. The two-way table describes office buildings in a business district based on if there are less than 10, 10 to 20, or more than 20 companies leasing offices in the building. The buildings are also described based on whether they were constructed before 2000. The two-way table shows the results.

	Before 2000	2000 to Present	Total
Less than 10	8	5	13
10 to 20	19	3	22
More than 20	4	10	14
Total	31	18	49

   What is the probability that a randomly selected office building has more than 20 companies that lease offices in the building or was constructed before 2000?

   Possible answer: $\frac{22}{49}$

9. Set A represents the letters in the word HYDROMAGNETIC. Set B represents the letters in the word BIRTHPLACES. How many elements are in the union of the two sets?

   17

10. A student takes a multiple-choice test. With 30 seconds left to complete the test, the student randomly selects answers for the last two questions. Each question has answer choices A, B, C, and D, and both questions have a correct answer of B. What is the theoretical probability that the student does not guess either answer correctly?

    Possible answer: $\frac{9}{16}$

© Houghton Mifflin Harcourt Publishing Company

Geometry 149 Module 20 Test • Form A

Form A

Module 20 • Form A
Module Test

Name ____________

11. Two fair number cubes that are six-sided each have the numbers 1–6. What is the theoretical probability of tossing both cubes and not having the number 1 come up at least once?

    Possible answer: $\frac{25}{36}$

12. The numbers 1–10 are written on separate sheets of paper, which are then folded and put into a jar. What is the probability of drawing a number that is even or composite?

    Possible answer: $\frac{3}{5}$

13. A company produces backpacks in different colors. The available colors are blue, red, yellow, green, brown, gold, and pink. The sets below show the different color backpacks produced in three of the company's factories.

    Z: {blue, red, green, brown}

    P: {blue, red, yellow, green, pink}

    U: {red, green, gold}

    Place an X in the table to show whether each statement is true or false.

	True	False
Set U is the subset of set P.		X
The complement of set P is {brown, gold}.	X	
The intersection of sets Z and P is {blue, red, green}.	X	
The union of sets Z and P is {blue, red, yellow, green, brown, pink}.	X	
The intersection of sets Z, P, and U is {blue, red, yellow, green, brown, gold, pink}.		X

14. Do the following events overlap, or are they disjoint?

    Place an X in the table to show whether the events overlap or are disjoint.

	Overlap	Disjoint
A number is even and a number is odd.		X
A coin lands on heads and a coin lands on tails.		X
An animal has brown fur and an animal is a dog.	X	
A word begins with a vowel and a word ends with a vowel.	X	

15. A pet shop owner surveys 42 customers about whether they own a dog or a cat. The results are recorded in the Venn diagram.

    Type of Pets Owned

    Dog Cat

    17 5 12

    8

    Part A

    What is the probability that a randomly selected customer owns both a dog and a cat?

    Possible answer: $\frac{5}{42}$

    Part B

    What is the probability that a randomly selected customer owns a dog or a cat?

    Possible answer: $\frac{4}{7}$

© Houghton Mifflin Harcourt Publishing Company

Geometry 150 Module 20 Test • Form A

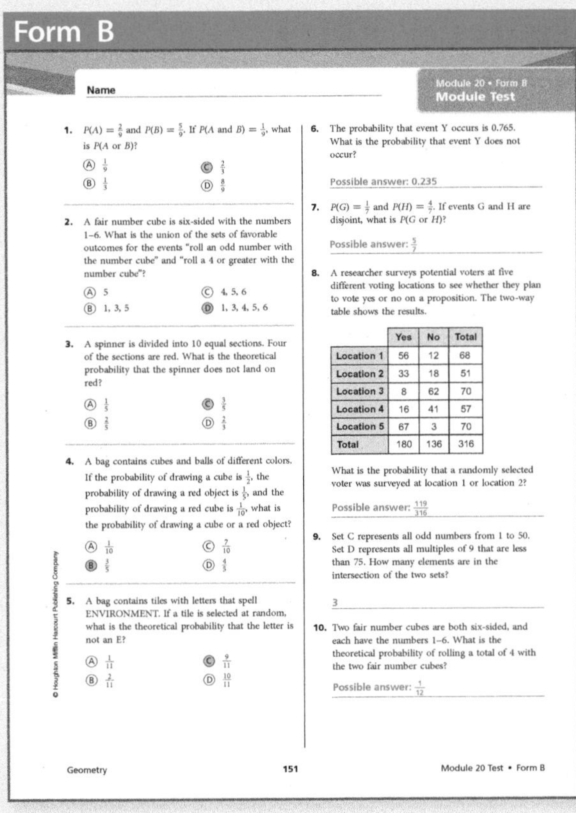

Form B

Module 20 • Form B
Module Test

Name ____________

1. $P(A) = \frac{2}{9}$ and $P(B) = \frac{5}{9}$. If $P(A \text{ and } B) = \frac{1}{9}$, what is $P(A \text{ or } B)$?
   (A) $\frac{1}{9}$ (B) $\frac{1}{3}$ (C) $\frac{2}{3}$ (D) $\frac{8}{9}$

2. A fair number cube is six-sided with the numbers 1–6. What is the union of the sets of favorable outcomes for the events "roll an odd number with the number cube" and "roll a 4 or greater with the number cube"?
   (A) 5 (B) 1, 3, 5 (C) 4, 5, 6 (D) 1, 3, 4, 5, 6

3. A spinner is divided into 10 equal sections. Four of the sections are red. What is the theoretical probability that the spinner does not land on red?
   (A) $\frac{1}{5}$ (B) $\frac{2}{5}$ (C) $\frac{3}{5}$ (D) $\frac{2}{3}$

4. A bag contains cubes and balls of different colors. If the probability of drawing a cube is $\frac{1}{2}$, the probability of drawing a red object is $\frac{1}{5}$, and the probability of drawing a red cube is $\frac{1}{10}$, what is the probability of drawing a cube or a red object?
   (A) $\frac{1}{10}$ (B) $\frac{3}{5}$ (C) $\frac{7}{10}$ (D) $\frac{4}{5}$

5. A bag contains tiles with letters that spell ENVIRONMENT. If a tile is selected at random, what is the theoretical probability that the letter is not an E?
   (A) $\frac{1}{11}$ (B) $\frac{2}{11}$ (C) $\frac{9}{11}$ (D) $\frac{10}{11}$

6. The probability that event Y occurs is 0.765. What is the probability that event Y does not occur?

   Possible answer: 0.235

7. $P(G) = \frac{1}{7}$ and $P(H) = \frac{4}{7}$. If events G and H are disjoint, what is $P(G \text{ or } H)$?

   Possible answer: $\frac{5}{7}$

8. A researcher surveys potential voters at five different voting locations to see whether they plan to vote yes or no on a proposition. The two-way table shows the results.

	Yes	No	Total
Location 1	56	12	68
Location 2	33	18	51
Location 3	8	62	70
Location 4	16	41	57
Location 5	67	3	70
Total	180	136	316

   What is the probability that a randomly selected voter was surveyed at location 1 or location 2?

   Possible answer: $\frac{119}{316}$

9. Set C represents all odd numbers from 1 to 50. Set D represents all multiples of 9 that are less than 75. How many elements are in the intersection of the two sets?

   3

10. Two fair number cubes are both six-sided, and each have the numbers 1–6. What is the theoretical probability of rolling a total of 4 with the two fair number cubes?

    Possible answer: $\frac{1}{12}$

© Houghton Mifflin Harcourt Publishing Company

Geometry 151 Module 20 Test • Form B

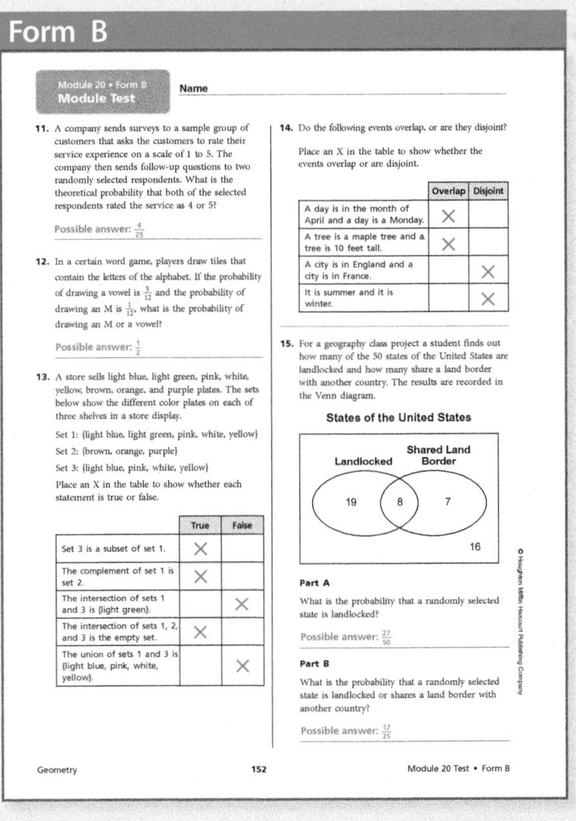

Form B

Module 20 • Form B
Module Test

Name ____________

11. A company sends surveys to a sample group of customers that asks the customers to rate their service experience on a scale of 1 to 5. The company then sends follow-up questions to two randomly selected respondents. What is the theoretical probability that both of the selected respondents rated the service as 4 or 5?

    Possible answer: $\frac{4}{25}$

12. In a certain word game, players draw tiles that contain the letters of the alphabet. If the probability of drawing a vowel is $\frac{5}{12}$ and the probability of drawing an M is $\frac{1}{12}$, what is the probability of drawing an M or a vowel?

    Possible answer: $\frac{1}{2}$

13. A store sells light blue, light green, pink, white, yellow, brown, orange, and purple plates. The sets below show the different color plates on each of three shelves in a store display.

    Set 1: {light blue, light green, pink, white, yellow}

    Set 2: {brown, orange, purple}

    Set 3: {light blue, pink, white, yellow}

    Place an X in the table to show whether each statement is true or false.

	True	False
Set 3 is a subset of set 1.	X	
The complement of set 1 is set 2.	X	
The intersection of sets 1 and 3 is {light green}.		X
The intersection of sets 1, 2, and 3 is the empty set.	X	
The union of sets 1 and 3 is {light blue, pink, white, yellow}.		X

14. Do the following events overlap, or are they disjoint?

    Place an X in the table to show whether the events overlap or are disjoint.

	Overlap	Disjoint
A day is in the month of April and a day is a Monday.	X	
A tree is a maple tree and a tree is 10 feet tall.	X	
A city is in England and a city is in France.		X
It is summer and it is winter.		X

15. For a geography class project a student finds out how many of the 50 states of the United States are landlocked and how many share a land border with another country. The results are recorded in the Venn diagram.

    States of the United States

    Landlocked Shared Land Border

    19 8 7

    16

    Part A

    What is the probability that a randomly selected state is landlocked?

    Possible answer: $\frac{27}{50}$

    Part B

    What is the probability that a randomly selected state is landlocked or shares a land border with another country?

    Possible answer: $\frac{17}{25}$

© Houghton Mifflin Harcourt Publishing Company

Geometry 152 Module 20 Test • Form B

MODULE 21

PLANNING

CONDITIONAL PROBABILITY AND INDEPENDENCE OF EVENTS

Introduce and Check for Readiness

- Module Performance Task • Are You Ready?

Lesson 21.1—2 Days

Conditional Probability

Learning Objective: Calculate conditional probability and use it to solve real-world problems.

New Vocabulary: conditional probability, relative frequency

Lesson 21.2—2 Days

Independent Events

Learning Objective: Interpret independence and its definition in terms of conditional probability. Derive and apply the Multiplication Rule for the probability of the intersection of independent events and use the rule to test the independence of events. Use the concept of independence to solve real-world problems.

New Vocabulary: independent events

Lesson 21.3—2 Days

Dependent Events

Learning Objective: Determine whether two events are dependent and find their probabilities.

New Vocabulary: dependent events

Assessment

- Module 21 Test (Forms A and B)
- Unit 10 Test (Forms A and B)

LEARNING ARC FOCUS

Build Conceptual Understanding  Connect Concepts and Skills  Apply and Practice

TEACHING FOR SUCCESS

TEACHING FOR DEPTH: Conditional Probability and Independence of Events

Represent and Explain. Students intuitively understand that the probability of randomly selecting marbles of a certain color from a bag is independent if each marble is returned before another is drawn and dependent if the selected marble is not returned.

Understanding how to determine independence algebraically may be difficult for some students. Consider these two methods.

Method 1: Events A and B are independent if and only if $P(A \mid B) = P(A)$.

Method 2: Events A and B are independent if and only if $P(A \cap B) = P(A) \cdot P(B)$.

Both methods are equally valid. Often the method chosen to solve a problem depends on the given information. Method 1 states that the probability of A given B is equal to the probability of A because the occurrence of event B does not affect the occurrence of event A.

Method 2 uses the same argument but is interpreted slightly differently. For intersections, students usually apply the Multiplication Rule: $P(A \cap B) = P(A) \cdot P(B \mid A)$. But, in the case of independent events, $P(B \mid A) = P(B)$, as expressed by Method 1. Therefore, students can use substitution to rewrite $P(A \cap B) = P(A) \cdot P(B \mid A)$ as $P(A \cap B) = P(A) \cdot P(B)$.

Mathematical Progressions

Prior Learning	Current Development	Future Connections
Students: • summarized and interpreted categorical data in two-way frequency tables. • interpreted relative frequencies in the context of the data. • used two-way frequency tables to determine if events are independent and to determine approximate conditional probabilities.	**Students:** • find conditional probabilities from a two-way table. • derive and apply the Conditional Probability Formula. • use the Multiplication Rule to find the probability of events. • determine whether two events are independent.	**Students:** • will describe events as subsets of a sample space using characteristics of the outcomes, or as unions, intersections, or complements of other events. • will understand that two events are independent if the probability of the events occurring together is the product of their probabilities. • will understand the conditional probability of an event occurring given that another event has already occurred.

TEACHER TO TEACHER

From the Classroom

Build procedural fluency from conceptual understanding. This module consists of many rules with symbolic notation that is still relatively new to students. I have discovered that the best way for students to develop procedural fluency in calculating probabilities is to develop their conceptual understanding of the symbols and formulas.

One way to help students develop their conceptual understanding is through the use of flashcards. One side should show a symbolic rule, such as $P(B \mid A) = \frac{P(A \cap B)}{P(A)}$, and the other side should contain a verbal description of the rule: The probably of B given A is the probability of both A and B occurring divided by the probability of A.

Another way is to have students work in pairs. For a given problem, one student chooses a formula, and the other student either agrees by explaining the formula in the context of the problem or provides another formula to use with explanation. Pairs discuss until they reach agreement and then switch roles for the next problem.

By continually verbalizing the notation and the rules and by explaining why a given formula will work in a problem situation, students develop the conceptual understanding they need to fluently solve problems.

LANGUAGE DEVELOPMENT • PLANNING FOR INSTRUCTION

By giving all students regular exposure to language routines in context, you will provide opportunities for students to **listen**, **speak**, **read**, and **write** about mathematical situations and develop both mathematical language and conceptual understanding at the same time.

Using Language Routines to Develop Understanding

Use the **Professional Cards** for the following routines to plan for effective instruction.

Co-Craft Questions Lesson 21.1

Students think of natural questions to ask about a given situation or problems similar to a given task and answer the questions they have developed or problems they have created.

Three Reads Lessons 21.1 and 21.2

Students read a problem three times with a specific focus each time.

1st Read What is the situation about?
2nd Read What are the quantities in the situation?
3rd Read What are the possible mathematical questions that we could ask for the situation?

Information Gap Lesson 21.3

Students recognize when information given in a problem situation is incomplete, and they pose questions and share knowledge with others to discover any missing facts or relationships and work together to solve the problem.

Critique, Correct, and Clarify Lesson 21.2

Students correct the work in a flawed explanation, argument, or solution method and share with a partner and refine the sample work.

Connecting Language to Conditional Probability and Independence of Events

Watch for students' use of the new terms listed below as they explain their reasoning and make connections with new concepts.

Key Academic Vocabulary

Current Development • New Vocabulary

conditional probability given two events *A* and *B*, the probability that event *B* occurs given that event *A* has already occurred

dependent event an event affected by occurrence of another event

independent event an event that is not affected by occurrence of another event

relative frequency the frequency of one outcome divided by the frequency of all outcomes

Linguistic Note

Listen for students who relate *conditional probability* to conditionals they learned about in English grammar. In grammar, conditionals are known as "if clauses," so students may make an immediate connection. However, those "if clauses" in grammar describe what *might* happen or something that might have happened but *didn't*. Both understandings are different from conditional probability, which gives the probability of one event given that another event has *already* occurred.

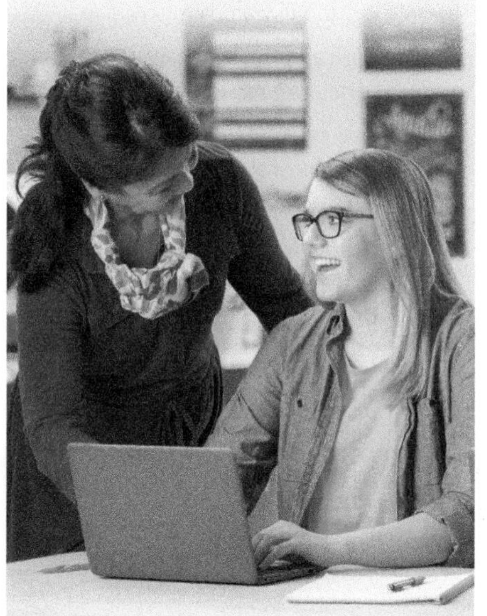

Module 21

Conditional Probability and Independence of Events

Module Performance Task: *Spies and Analysts*™

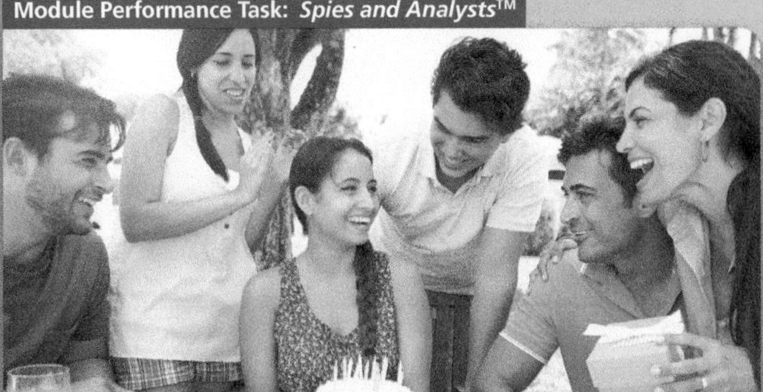

Birthday BASH!

What are the chances that two people in this group have the same birthday?

NOVEMBER				
Mon.	Tues.	Wed.	Thur.	Fri.
31	1	2	3	4
6	7	8 John and Bobby's Birthday!	9	10
13	14	15	16	17
20	21	22	23	24

Connections to This Module

One sample solution involves students using complementary events to find the probability.

- Students recognize that for this problem, it will be simpler to find the probability of the complementary event: that no students share a birthday **(20.1)**.
- Students realize that the first student can have any birthday. They define event *A* as the event that the second student has a birthday different than the first student, event *B* as the event that the third student has a birthday different than the first two students, and so on **(21.3)**.
- Students find the probability that no students share a birthday as $\frac{364}{365} \cdot \frac{363}{365} \cdot \frac{362}{365} \cdot \ldots \cdot \frac{365 - n + 1}{365}$ **(21.3)**.
- Students subtract the probability found in the previous step from 1 to find the answer **(20.1)**.

Birthday BASH!

Overview

This problem requires students to make assumptions about the birthdays of the people in the group. Examples of assumptions:

- There are no twins, triplets, etc. in the group.
- No one has a birthday of February 29.
- All possible birthdays are equally likely.

Be a *Spy*

First, students should determine what information is necessary to know. This should include:

- the number of people in the group.
- the number of different birthdays possible.

Be an *Analyst*

Help students understand that the events of different people's birthdays being any particular day are independent. It may help students to start with a simpler problem, such as finding the probability that both people in a group of two share a birthday. Suggest that they expand the problem to find the probability that at least two out of three share a birthday, then at least two out of four, and so on. Students will realize that the problem becomes increasingly more complex as the group size increases, and they may decide to find the probability of the complement instead. As students define events to find probabilities, discuss whether the events are independent or dependent.

Alternative Approaches

In their analysis, students might:

- survey groups of students about their birthdays to find the experimental probability that two students share a birthday.
- find the probability that two or more students share a birthday in groups of different sizes, making a table to organize their results.
- find a generalized formula for the probability that at least two people share a birthday in a group of *n* people.

ONLINE

Assign the Digital Are You Ready? to power actionable reports including

- proficiency by standards
- item analysis

Are You Ready?

Diagnostic Assessment

- Diagnose prerequisite mastery.
- Identify intervention needs.
- Modify or set up leveled groups.

Have students complete the *Are You Ready?* assessment on their own. Items test the prerequisites required to succeed with the new learning in this module.

Multiply and Divide Rational Numbers Students will apply previous knowledge of finding products and quotients of rational numbers to find and interpret conditional probabilities.

Probability of Simple Events Students will build upon their skill in finding the theoretical probability of a simple event to understand and determine independence between two events.

Probability of Compound Events Involving *And* Students will apply their previous knowledge of finding the theoretical probability of a compound event involving *and* to recognize and explain the concepts of conditional probability and independence in everyday situations using everyday language.

ONLINE 

Are You Ready?

Complete these problems to review prior concepts and skills you will need for this module.

Multiply and Divide Rational Numbers

Simplify each expression.

1. 5.03×14.1 70.923
2. $125.248 \div 30.4$ 4.12
3. $\frac{2}{3} \div \frac{7}{9}$ $\frac{6}{7}$
4. $\frac{5}{12} \times \frac{4}{15}$ $\frac{1}{9}$

Probability of Simple Events

Calculate each theoretical probability.

5. A spinner has five congruent sections numbered 1 through 5. What is the probability of spinning a 2? 20%
6. A bowl contains 15 strawberries, 32 red raspberries, 27 black raspberries, and 24 blackberries. What is the probability of randomly selecting a red raspberry? 32.65%

Probability of Compound Events Involving *And*

Calculate each probability.

7. Laiken can choose one sandwich (ham and cheese, egg salad, peanut butter and jelly) and one side (apple, carrots, pretzels) for her lunch. If she chooses each item at random, what is the probability she chooses the egg salad sandwich and the apple? 11.11%
8. On a field trip, students can choose one free item from each of three tables. At the first table, students can choose a pen or a keychain. At the second table, they can choose a notepad, a bottle of water, or a coupon. At the third table, they can choose a T-shirt or a hat. If Trevor chooses an item from each table at random, what is the probability he chooses a pen, a notepad, and a hat? 8.33%

Connecting Past and Present Learning

Previously, you learned:

- to describe events as subsets of a sample space,
- to calculate probabilities of disjoint and overlapping events, and
- to represent data on two quantitative variables and describe the association between the variables.

In this module, you will learn:

- to find and interpret conditional probabilities,
- to understand and determine independence between two events, and
- to recognize and explain the concepts of conditional probability and independence in everyday language and everyday situations.

634

DATA-DRIVEN INTERVENTION

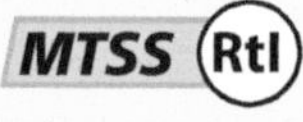

Concept/Skill	Objective	Prior Learning *	Intervene With
Multiply and Divide Rational Numbers	Find products and quotients of rational numbers.	Grade 7, Lesson 5.4	• Tier 3 Skill 8 • Reteach, Grade 7 Lesson 5.4
Probability of Simple Events	Find the theoretical probability of a simple event.	Grade 7, Lesson 15.1	• Tier 2 Skill 22 • Reteach, Grade 7 Lesson 15.1
Probability of Compound Events Involving *And*	Find the theoretical probability of a compound event involving *and.*	Grade 7, Lesson 15.2	• Tier 2 Skill 24 • Reteach, Grade 7 Lesson 15.2

* Your digital materials include access to resources from Grade 6–Algebra 2. The lessons referenced here contain a variety of resources you can use with students who need support with this content.

21.1 Conditional Probability

LESSON FOCUS AND COHERENCE

Mathematics Standards

- Describe events as subsets of a sample space (the set of outcomes) using characteristics (or categories) of the outcomes, or as unions, intersections, or complements of other events ("or," "and," "not").
- Understand the conditional probability of *A* given *B* as (*A* and *B*)/*P*(*B*), and interpret independence of *A* and *B* as saying that the conditional probability of *A* given *B* is the same as the probability of *A*, and the conditional probability of *B* given *A* is the same as the probability of *B*.
- Find the conditional probability of *A* given *B* as the fraction of *B*'s outcomes that also belong to *A*, and interpret the answer in terms of the model.

Mathematical Practices and Processes 

- Look for and make use of structure.
- Attend to precision.
- Look for and express regularity in repeated reasoning.

I Can Objective 

I can calculate conditional probability and use it to solve real-world problems.

Learning Objective

Calculate conditional probability and use it to solve real-world problems.

Language Objective

Explain in speaking or in writing the meaning of conditional probability and how it can be calculated.

Vocabulary

New: conditional probability, relative frequency

Mathematical Progressions

Prior Learning	Current Development	Future Connections
Students: • summarized categorical data for two categories in two-way frequency tables. **(Alg1, 20.1)** • described events as subsets of a sample space using characteristics of the outcomes, or as unions, intersections, or complements of other events. **(20.1)**	**Students:** • find conditional probabilities from a two-way table. • derive the Conditional Probability Formula. • use the Conditional Probability Formula to solve a real-world problem.	**Students:** • will understand conditional probability and interpret independence using conditional probability. **(Alg2, 18.1 and 18.2)** • will understand that two events *A* and *B* are independent if the probability of *A* and *B* occurring together is the product of their probabilities. **(Alg2, 18.2)**

UNPACKING MATH STANDARDS

Find the conditional probability of *A* given *B* as the fraction of *B*'s outcomes that also belong to *A*, and interpret the answer in terms of the model.

What It Means to You

The word *conditional* in *conditional probability* means that the probability of an event depends on a particular circumstance. That is, it represents the likelihood of the occurrence of a particular event *B* occurring after a prior event *A* has occurred. The sample space of the event is therefore restricted.

WARM-UP OPTIONS

ONLINE

PROJECTABLE & PRINTABLE

ACTIVATE PRIOR KNOWLEDGE • Theoretical Probability of Outcomes in a Sample Space

Use these activities to quickly assess and activate prior knowledge as needed.

Problem of the Day

To the nearest whole number percent, what is the theoretical probability that a number chosen randomly from the set $\{0, 1, 2, 3, \ldots, 9\}$ is an odd number? 56%

Quick Check for Homework

As part of your daily routine, you may want to display the Teacher Solution Key to have students check their homework.

Make Connections

Based on students' responses to the Problem of the Day, choose one of the following:

1. Project the Interactive Reteach, Lesson 20.1.
2. Complete the Prerequisite Skills Activity:

Connie tosses a balanced six-sided number cube. What is the theoretical probability that, in the first trial, a number greater than 4 will turn up?

- *What is the sample space for this trial?* the numbers 1 to 6
- *If* A *represents the set of numbers greater than 4 when tossing the number cube, what are the theoretical outcomes in this trial?* 5 and 6
- *What is* P(A)? $\frac{1}{3}$ or about 33%
- *What is the probability of the complement of set* A? $\frac{2}{3}$ or about 67%
- *What is the probability that Connie tosses a 7?* 0

If students continue to struggle, use Tier 2 Skill 22.

SHARPEN SKILLS

If time permits, use this on-level activity to build fluency and practice basic skills.

Mental Math

Objective: Students demonstrate how to use the Conditional Probability Formula to calculate a conditional probability of two independent events.

Have pairs of students write the two-way table shown on the right, which gives the hypothetical results of a genetics experiment involving fruit flies. The table represents the relative frequency of two events *A* and *B*, where *A* is the sex of the fruit fly, and *B* is its eye color. Have students take turns asking one another questions such as, "What is the probability, written as a percent, that red-eyed fruit flies were males?"

	Eye Color		
	Red	White	Total
Male	0.50	0.25	0.75
Female	0.25	0.05	0.25
Total	0.70	0.30	1.00

PLAN FOR DIFFERENTIATED INSTRUCTION

Small-Group Options

Use these teacher-guided activities with pulled small groups.

On Track

Pose the following question to students:

Given that $P(A) = 0.3$, $P(B) = 0.6$, and $P(A \cap B) = 0.15$, what are $P(A \mid B)$ and $P(B \mid A)$? $P(A \mid B) = 0.25$, $P(B \mid A) = 0.5$

Almost There

A student spins a wheel divided into 12 equal sections. Each section has a number.

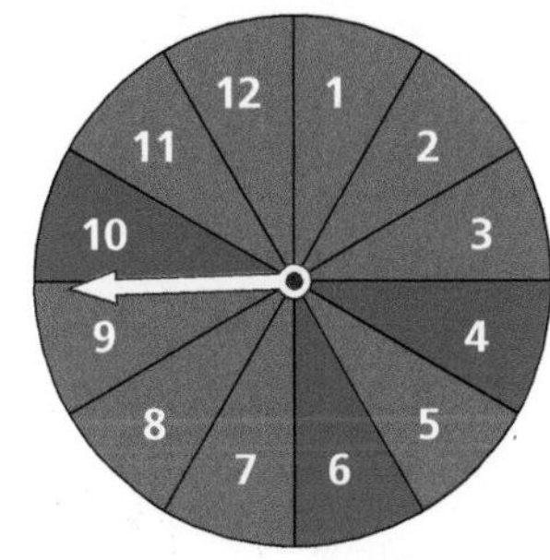

- What is the probability of landing on a blue section? $\frac{1}{4} \approx 25\%$
- What is the probability of landing on a number greater than 5? $\frac{7}{12} \approx 58\%$
- What is the probability of A and B? $\frac{7}{48} \approx 15\%$

Ready for More

The definition of independent events A and B is $P(A \cap B) = P(A)P(B)$, where neither $P(A)$ nor $P(B) = 0$. Use the Conditional Probability Formula to prove that $P(A \mid B) = P(A)$.

1. By the Conditional Probability Theorem, $P(A \mid B) = \dfrac{P(A \cap B)}{P(B)}$
2. Since the events are independent, then by definition, $P(A \cap B) = P(A)P(B)$
3. By substitution, $P(A \mid B) = \dfrac{P(A)P(B)}{P(B)} = P(A)$

Math Center Options

Use these student self-directed activities at centers or stations. Key: ● Print Resources ● Online Resources

On Track

- ● Interactive Digital Lesson
- ●● Journal and Practice Workbook
- ● Interactive Glossary (printable): **conditional probability**, **relative frequency**
- ● Module Performance Task

Standards Practice:

- ●● Describe Events Using Set Language
- ●● Understand and Apply Conditional Probability Concepts
- ●● Find Conditional Probabilities

Almost There

- ● Reteach 21.1 (printable)
- ● Interactive Reteach 21.1
- ● RtI Tier 2 Skill 22 Probability of Compound Events
- ● Desmos: Chance Experiments

Ready for More

- ● Challenge 21.1 (printable)
- ● Interactive Challenge 21.1

Unit Project Check students' progress by asking them to state the formula for conditional probability.

View data-driven grouping recommendations and assign differentiation resources.

Spark Your Learning • Student Samples

During the *Spark Your Learning*, listen and watch for strategies students use. See samples of student work on this page.

Use Relative Frequency Data

Strategy 1

I looked at the frequency table and saw that 100 of the hatched eggs were blue dragons. Of these, only 25 had wings. So, the probability of winged blue dragons was $\frac{25}{100}$, or 25%.

If students . . . use relative frequency data to solve the problem, they are applying what they know about probability from Lessons 20.1 and 20.2.

Have these students . . . explain how they determined the correct probability. **Ask:**

- Q How many hatched eggs were blue?
- Q How many had wings?

Calculate Relative Frequencies

Strategy 2

I calculated the relative frequencies of each color dragon that had wings by dividing the frequencies in the Wings column of the table by the total number of that color dragon in the game. This gave me the probability of winged red, winged green, and winged blue dragons that hatched. I circled the answer for winged blue dragons and wrote it as 25% percent.

	Wings	No wings
Red	0.09	0.91
Green	0.33	0.66
Blue	0.25 (circled)	0.75

If students . . . converted the relative frequencies to percentages, they are applying what they know about decimal-percent equivalencies but they did not use the most efficient strategy.

Activate prior knowledge . . . by having students explain how they determined each percent. **Ask:**

- Q What set notation can you use to represent the relationship between set *A* (color) and set *B* (winged) in the circled cell?
- Q What does the ratio $\frac{25}{57}$ represent?

COMMON ERROR: Use Incorrect Values

Of the 262 eggs that hatched, I saw that 25 were blue dragons with wings. So, the probability of winged blue dragons was $\frac{25}{262}$, or about 10%.

If students . . . chose the incorrect sample space, then they may not understand how to read the table.

Then intervene . . . by pointing out that 262 included all of the eggs in the game, not just winged blue dragons. **Ask:**

- Q How many blue dragons were hatched in the game?
- Q How many dragons that hatched were both blue and winged?

21.1

Conditional Probability

I Can calculate conditional probability and use it to solve real-world problems.

Spark Your Learning

Miguel and Bryn play a video game. In the game, each player collects and hatches dragon eggs.

Dragons can be red, green, or blue.
Some dragons have wings. Some dragons do not.

Complete Part A as a whole class. Then complete Parts B–D in small groups.

A. What is a mathematical question you can ask about this situation? What information would you need to know to answer your question?

A. If you hatch a blue dragon, what is the probability that the dragon will have wings?; how many eggs are in the game, how many contain a blue dragon, and how many of the blue dragons have wings

B. What variables are involved in this situation?
B, D. See Additional Answers.

C. To answer your question, what strategy and tool would you use along with all the information you have? What answer do you get?
See Strategies 1 and 2 on the facing page.

D. Does your answer make sense in the context of the situation? How do you know?

Turn and Talk Predict how your answer would change for each of the following changes in the situation: See margin.

- The game designer adds 10 more eggs that contain red dragons.
- The game designer adds 20 more eggs that contain blue dragons with no wings.
- The game designer adds 2 eggs that contain blue, winged dragons.

CULTIVATE CONVERSATION • Co-Craft Questions

If students have difficulty formulating a mathematical question about the situation in the Spark Your Learning, ask them to consider expressing other probabilities. What are some natural questions to ask about this situation?

Work together to craft the following questions:

- What is the probability that a hatched dragon will have wings? will not have wings?
- What is the probability that a hatched dragon will not be blue?
- What is the probability that a hatched dragon will be red? green?

Then have students think about what additional information, if any, they would need to answer these questions. **Ask:**

- Can you determine whether all of the eggs in a game are hatched? Why or why not?
- If there were no red dragons in this game, can you determine the probability that a hatched dragon is blue and has wings? Explain.

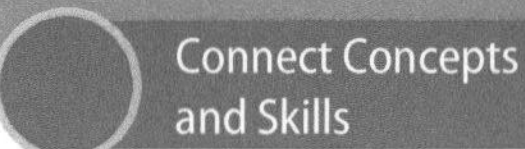

Connect Concepts and Skills

1 Spark Your Learning

▶ MOTIVATE

- Have students look at the illustration in their books and read the information contained in the table. Then complete Part A as a whole-class discussion.
- Give the class the additional information they need to solve the problem. This information is available online as a printable and projectable page in the Teacher Resources.
- Have students work in small groups to complete Parts B–D.

▶ PERSEVERE

If students need support, guide them by asking:

Q **Advancing • Use Tools** Which tool could you use to solve the problem? Why choose that tool and not some other? Students' choices of tools and reasons for choosing them will vary.

Q **Assessing** What is the total number of eggs in this game? 272

Q **Assessing** Of the total number of eggs in the game, how many are blue? 100

Q **Advancing** What is the probability of hatching a blue dragon? $\frac{100}{272} \approx 37\%$

Q **Advancing** How many of the hatched eggs were blue and had wings? 25

Turn and Talk Point out that the number of hatched blue dragons that have wings is the intersection of two sets: the set of blue dragons and the set of winged dragons. Together, these two outcomes make up an event. no change; probability decreases; probability increases

▶ BUILD SHARED UNDERSTANDING

Select groups of students who used various strategies and tools to share with the class how they solved the problem. As they present their solutions, have each group discuss why they chose a specific strategy and tool.

② Learn Together

Build Understanding

Task 1 (MP) **Use Structure** Students examine the entries in a two-way frequency table and use the variables A and B to write the conditional probabilities $P(A \mid B)$ and $P(B \mid A)$.

CONNECT TO VOCABULARY

Have students use the **Interactive Glossary** to record their understanding of the vocabulary in this task.

Task 2 (MP) **Attend to Precision** Students analyze the entries in a two-way frequency table and choose the correct values that can be used to express the conditional probabilities for the given outcomes in an experiment.

Sample Guided Discussion:

Q What are $P(A)$ and $P(B)$? $P(A) = \frac{55}{100}$, $P(B) = \frac{62}{100}$

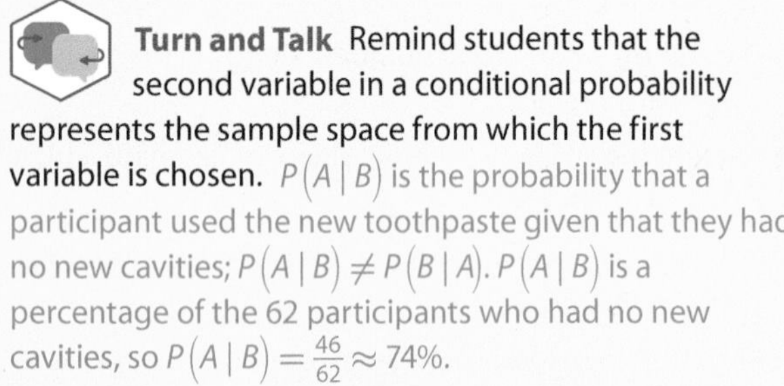

Turn and Talk Remind students that the second variable in a conditional probability represents the sample space from which the first variable is chosen. $P(A \mid B)$ is the probability that a participant used the new toothpaste given that they had no new cavities; $P(A \mid B) \neq P(B \mid A)$. $P(A \mid B)$ is a percentage of the 62 participants who had no new cavities, so $P(A \mid B) = \frac{46}{62} \approx 74\%$.

Build Understanding

Find Conditional Probabilities from a Two-Way Frequency Table

For two events A and B, the probability that event B occurs given that event A has already occurred is the **conditional probability** of B given A and is written $P(B \mid A)$.

1 Suppose 30 people choose a playing card at random from a standard deck and then return the card to the deck before the next person chooses. The two-way frequency table shows how many times each outcome occurred.

	Face card	Other card	Total
Red card	6	3	9
Black card	10	11	21
Total	16	14	30

A. Describe $P(\text{face} \mid \text{red})$ in words. the probability of selecting a face card given that it is a red card

B. Which parts of the table would you use to find $P(\text{red})$ and $P(\text{face} \mid \text{red})$? What is $P(\text{face} \mid \text{red})$? See Additional Answers.

C. Which parts of the table would you use to find $P(\text{red} \mid \text{face})$? What is $P(\text{red} \mid \text{face})$?

C. $P(\text{red} \mid \text{face})$: row 1 column 1 and row 3 column 1; $P(\text{red} \mid \text{face}) = \frac{6}{16} = \frac{3}{8}$, or 37.5%

D. What is a difference between finding $P(\text{face} \mid \text{red})$ and finding $P(\text{red} \mid \text{face})$? See Additional Answers.

2 In an experiment, one group of participants uses a new toothpaste and another group uses their old toothpaste. After a year, both groups are examined to see if they have any new cavities. The two-way frequency table shows the results.

	No new cavities	At least 1 new cavity	Total
New toothpaste	46	9	55
Old toothpaste	16	29	45
Total	62	38	100

A. What do the values 16, 45, and 100 represent in this situation? See Additional Answers.

B. Let A be the event that a person uses the new toothpaste. Let B be the event that a person has no new cavities. Which values from the table would you use to find $P(B \mid A)$? 46 and 55

C. What is $P(B \mid A)$? $P(B \mid A) = \frac{46}{55} \approx 84\%$

D. Did you need to use any values from the second row to find $P(B \mid A)$? Explain why or why not. See Additional Answers.

Turn and Talk What does $P(A \mid B)$ represent in the toothpaste experiment? Is $P(A \mid B)$ equivalent to $P(B \mid A)$? Explain your reasoning. See margin.

636

LEVELED QUESTIONS

Depth of Knowledge (DOK)	Leveled Questions	What Does This Tell You?
Level 1 **Recall**	What is a conditional probability? Given event A and event B, it is the probability that event B occurs after event A has occurred.	Students' answers will indicate whether they understand the meaning of a conditional probability.
Level 2 **Basic Application of Skills & Concepts**	How can you calculate the probability of an event reported in a frequency table? It is the ratio between the frequency of the event and the total number of events.	Students' answers will indicate whether they understand how to calculate the probability of an event as either a decimal or a percent.
Level 3 **Strategic Thinking & Complex Reasoning**	How can you represent A, the set of red face cards, and B, the set of black face cards, using set notation? $A \cap B$ and $A^C \cap B$, where A^C is the complement of A.	Students' answers will indicate whether they understand how two events can be represented using set notation, where $A \cap B$ is the intersection of two sets, and A' is the complement of set A.

Derive the Conditional Probability Formula

3 Recall the experiment in Task 1 where 30 people chose a card from a standard deck. Let event A be choosing a red card. Let event B be choosing a face card.

You can use set notation to describe each cell of the two-way frequency table. The first cell is the intersection of event A and event B. The value 6 in that cell tells you that 6 people chose a card that is both a red card and a face card.

	Face card	Other card	Total
Red card	$A \cap B$	?	A
Black card	?	?	A^C
Total	?	?	?

A. Copy and complete the table using set notation to describe the remaining cells. See Additional Answers.

B. Do any of the cells represent a union? Explain your thinking.

B. yes; The cells that show totals represent unions. For example, $B = (A \cap B) \cup (A^C \cap B)$.

C. What is the value of $A \cap B^C$? What does the value of $A \cap B^C$ represent in this situation? What is $P(A \cap B^C)$?

C. 3; three people chose a red card that is not a face card; $\frac{1}{10}$, or 10%

D. What does $P(B^C \mid A)$ represent in this situation? How does this compare to $P(A \cap B^C)$? D–F. See Additional Answers.

Relative frequency is the frequency of one outcome divided by the frequency of all outcomes. To get the relative frequency table for the toothpaste experiment in Task 2, divide each value in the table by the total number of participants in the experiment.

E. Copy and complete the relative frequency table for the experiment in Task 2.

	No new cavities	At least 1 new cavity	Total
New toothpaste	$\frac{46}{100} = 0.46$	?	?
Old toothpaste	?	?	?
Total	?	?	?

F. Recall that event A is that a participant used the new toothpaste and event B is that a participant has no new cavities. What is $P(A)$? $P(A \cap B)$?

G. Which values from the table would you use to find $P(B \mid A)$? 0.46 and 0.55

H. What is $P(B \mid A)$? Use set notation to show how you found your answer.

$$P(B \mid A) = \frac{0.46}{0.55} = \frac{P(A \cap B)}{P(A)} \approx 84\%$$

Turn and Talk What is the difference between frequency and relative frequency? What does each represent? See margin.

PROFICIENCY LEVEL

Beginning

Use the completed table in this task and have students complete the following statements for the toothpaste experiment.

- Sets A and B are __________ of the experiment.
- The number 46 is the ______________ of A.
- $\frac{46}{100}$ is the ______________ of A.

Intermediate

Have students describe in words the expressions $P(B \mid A)$ and $P(A \mid B)$, where A and B are the variables when choosing cards from a standard deck.

Advanced

Have students write a paragraph to explain the differences between the frequency of an event and the relative frequency of an event. Give examples of each from the toothpaste experiment.

Task 3 (MP) **Use Repeated Reasoning** Students refer to previous tasks and use frequency tables and set theory to express the probabilities and conditional probabilities of given events.

CONNECT TO VOCABULARY

Have students use the **Interactive Glossary** to record their understanding of the vocabulary in this task.

Sample Guided Discussion:

Q What does the expression $(A \cap B)$ represent if A represents choosing a red card and B represents choosing a face card? Possible answer: It represents the intersection of the set of playing cards that are red and are face cards. For example, the red jack of hearts would be in this set, but the black jack of clubs would not be.

Q If A represents the set of red cards, what does the expression A^C represent? the set of cards that are not red; The C stands for the complement of A.

Q In Part C, what does the expression $A \cap B^C$ represent? the intersection of set A and the complement of set B; that is, the intersection of the set of red cards and the complement of B, cards that are not face cards

Q In Part C, what does the expression $P(A \cap B^C)$ represent? Give an example using cards. the probability of A and B; a red card that is not a face card

Q In Part H, what is an example of an outcome in $A \cap B$? Possible answer: a participant who used the new toothpaste and who had no cavities

Q In Part H, what does the expression $P(B \mid A)$ represent in words? Possible answer: the number of participants who had no cavities, given that they were using the new toothpaste

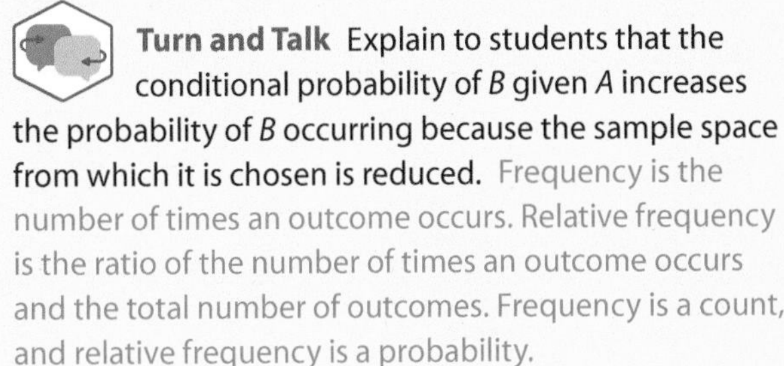

Turn and Talk Explain to students that the conditional probability of B given A increases the probability of B occurring because the sample space from which it is chosen is reduced. Frequency is the number of times an outcome occurs. Relative frequency is the ratio of the number of times an outcome occurs and the total number of outcomes. Frequency is a count, and relative frequency is a probability.

Step It Out

Task 4 (MP) **Attend to Precision** Students synthesize what they have learned about conditional probability to match given probability expressions to values in a frequency table.

Q Describe in words the expressions $P(A)$ and $P(B)$. the probability of a patient in the study who did not exercise before bed, and the probability of patients in the study who slept well, respectively

Q What is $P(A) + P(B)$ expressed as a whole-number percent? Show your work. $\frac{22}{50} + \frac{17}{50} = \frac{39}{50} = 78\%$

Q What is $P(A^C)$? Explain. $P(A^C)$ is the complement of $P(A)$ and is equal to $\frac{28}{50} = 56\%$.

Turn and Talk Remind students of the difference between $P(A \cap B)$ and $P(A \mid B)$ and that, under most circumstances, $P(A \mid B) \neq P(B \mid A)$. $P(A \cap B)$ is the probability that a patient did not exercise before bed and slept well. $P(B \mid A)$ is the probability that a patient who did not exercise before bed did sleep well. $P(A \mid B)$ is the probability a patient who slept well did not exercise before bed.

Step It Out

Use the Conditional Probability Formula

The previous task explores the relationship between the probability of an event given that a second event has already occurred, the probability of the intersection of the two events, and the probability of the event that has already happened. The equation that describes this relationship is the conditional probability formula.

Conditional Probability Formula

For events A and B, the conditional probability of B given A is the probability of the intersection of A and B divided by the probability of A.

$$P(B \mid A) = \frac{P(A \cap B)}{P(A)}$$

4 Dr. Lin studies a group of patients with insomnia. She has some of the patients do exercise like yoga before they go to bed. The other patients do not exercise before bed. In the morning, patients record how well they had slept.

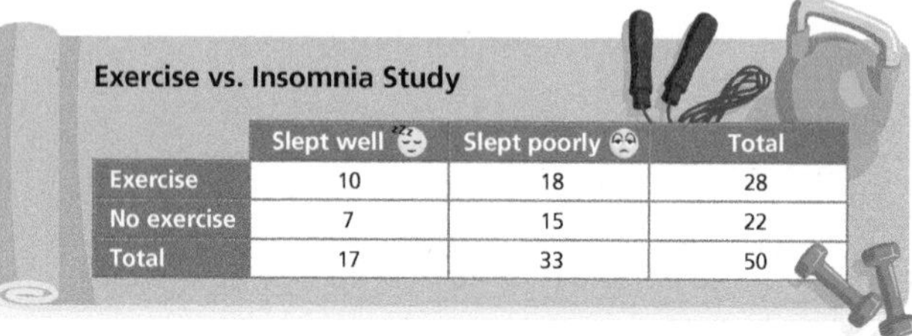

Exercise vs. Insomnia Study

	Slept well	Slept poorly	Total
Exercise	10	18	28
No exercise	7	15	22
Total	17	33	50

Let A be the event that a patient did not exercise before bed and B be the event that a patient slept well. Match each probability with its value.

Probability		Value
A. $P(A)$	5	1. $\frac{7}{50}$
B. $P(B)$	4	2. $\frac{7}{22}$
C. $P(A \cap B)$	1	3. $\frac{7}{17}$
D. $P(B \mid A)$	2	4. $\frac{17}{50}$
E. $P(A \mid B)$	3	5. $\frac{11}{25}$

Turn and Talk What do $P(A \cap B)$, $P(B \mid A)$, and $P(A \mid B)$ represent in the study? See margin.

Use the Conditional Probability Formula to Solve a Real-World Problem

Quality control is the process of checking a sample of a factory's output for defects. A high percentage of defects shows that there is a problem with a machine or process in the factory.

Only 2.8% of ListenUp! earbuds made are defective.

 **5** Julia buys a pair of ListenUp! earbuds, but they are defective. What is the probability that Julia's earbuds were made on a Friday?

ListenUp! Earbuds Manufacturing Defects in Q3

	M	T	W	Th	F	Total
Earbuds manufactured	6000	6000	6000	6000	6000	30,000
Defective earbuds (average based on daily samples)	90	72	80	99	510	851
Percent defective earbuds (average for the quarter)	0.3%	0.2%	0.3%	0.3%	1.7%	2.8%

Describe each event.

Let A be the event that a pair of earbuds is defective. Let B be the event that a pair of earbuds is made on a Friday.

Identify the given probabilities.

$P(A) = 0.028$

$P(A \cap B) = 0.017$

A. What does $P(A \cap B)$ represent in this situation?

A. $P(A \cap B)$ is the probability that a pair of earbuds is defective and is made on a Friday.

Solve.

$P(B \mid A) = \dfrac{P(A \cap B)}{P(A)}$ ____?____

B. What is the justification for this step?

$= \dfrac{0.017}{0.028}$ Substitute.

≈ 0.607 Divide.

B. The equation is the formula for conditional probability. The problem asks for the conditional probability of a pair of earbuds being made on a Friday given that the earbuds are defective.

Answer the question.

The probability that Julia's earbuds were made on a Friday is about 61%.

Turn and Talk Suppose Julia's earbuds were not defective. See margin.

- Do you have enough information to find the probability that her earbuds were made on a Friday?
- If yes, what is the probability? If not, what additional information do you need?

Task 5 (MP) **Attend to Precision** Students use the given information and the Conditional Probability Formula $P(B \mid A) = \frac{P(A \cap B)}{P(A)}$, $P(A) \neq 0$ in a real-world situation to find the probability that a defective earbud was made on a Friday.

Sample Questions

Q Based on the advertisement for ListenUp! earbuds, what is the probability that no earbuds are defective? Explain. Possible answer: The probability of being defective is 2.8%, so the probability of not being defective is its complement, or (100% − 2.8%) = 97.2%.

Q Based on the answer to this problem, what is the probability that Julia's earbuds were not made on a Friday? Explain. about 39%; The probability that they were made on a Friday was about 61%, so the difference between 100% and 61% is 39%.

Turn and Talk Remind students that the B in this situation is the event that occurs after A occurs. That is, we are looking at the probability of the manufacture of defective earbuds on a Friday, not at the probability of all manufactured earbuds. yes; The probability that the earbuds were made on Friday given that they are NOT defective is about 18.8%:

$$\frac{P(A^c \cap B)}{P(A^c)} = \frac{\frac{6000 - 510}{30{,}000}}{\frac{30{,}000 - 851}{30{,}000}} \approx \frac{0.183}{0.972}, \text{ or about } 0.188$$

SUPPORT SENSE-MAKING Three Reads

Have students read the problem three times. Use the questions below for a different focus each read.

1. What is the situation about?
2. What are the quantities in the situation?
3. What are the possible mathematical questions that you could ask for this situation?

ONLINE

Assign the Digital On Your Own for
- built-in student supports
- Actionable Item Reports
- Standards Analysis Reports

On Your Own

Assignment Guide

The chart below indicates which problems in the On Your Own are associated with each task in the Learn Together. Assign daily homework for tasks completed.

Learn Together Tasks	On Your Own Problems
Task 1, p. 636	Problems 6–8, 10, and 20
Task 2, p. 636	Problems 6–8, 10–16, and 20
Task 3, p. 637	Problem 9
Task 4, p. 638	Problems 9, 20, and 21
Task 5, p. 639	Problems 17–20

ANSWERS

6. Row 1: 0.60, the probability that a spin is red; Row 2: 0.17, the probability that a spin is not red and a toss is a multiple of 3; Row 3: 0.70, the probability that a toss is not a multiple of 3

9. the probability that a person who spins red does not toss a multiple of 3; 78%

Check Understanding

Use the table for Problems 1–3. A computer assigns a randomly selected letter of the alphabet to 100 people. The table shows the results. Assume the letter *Y* is a vowel.

	First letter of name	Not first letter	Total
Vowel	5	21	26
Consonant	9	65	74
Total	14	86	100

1. What is $P(\text{first letter} \mid \text{consonant})$? Round to the nearest whole percent. approximately 12%
2. Let A be the event that a person receives the first letter of their name. Let B be the event that a person receives a vowel. What is $P(B \mid A^C)$? Round to the nearest whole percent. approximately 24%
3. Create a relative frequency table for the two-way frequency table. See Additional Answers.
4. Ivan tosses an icosahedron, numbered from 1 to 20. Let event A be tossing a number less than or equal to 7. Let event B be tossing an odd number. Find $P(A)$, $P(A \cap B)$, and $P(B \mid A)$ to the nearest whole percent. 35%; 20%; 57%
5. Darren and Paula are dishwashers at a restaurant. On a randomly selected night, there is a 0.64% probability that Darren breaks a plate, and a 0.4% probability that Darren and Paula each break a plate. What is the probability that on a randomly selected night, Paula will break a plate given that Darren has also broken a plate? 62.5%

On Your Own

On each turn in a game, a player spins a spinner and tosses a 6-sided number cube. The spinner has same-sized sectors colored red, blue, red, green, red, white, and red. Use the table below that shows the relative frequencies for one game.

	Multiple of 3	Not a multiple of 3	Total
Red	0.13	0.47	?
Not red	?	0.23	0.40
Total	0.30	?	1.00

6. What is the missing value in each row? What does the value represent? See margin.
7. Let A be the event that a player spins a red sector, and let B be the event that a player does not toss a multiple of 3. What is $P(A)$ to the nearest whole percent? 60%
8. What is $P(A \cap B)$ to the nearest whole percent? 47%
9. What does $P(B \mid A)$ represent? What is $P(B \mid A)$ to the nearest percent? See margin.
10. (MP) **Use Structure** Is it possible to use the table to find how many turns are taken in the game? Explain. no; The values in the table are probabilities, not frequencies.

640

data checkpoint

3 Check Understanding

Formative Assessment

Use formative assessment to determine if your students are successful with this lesson's learning objective.

Students who successfully complete the Check Understanding can continue to the On Your Own practice.

For students who miss 1 problem or more, work in a pulled small group using the Almost There small-group activity on page 635C.

ONLINE

Assign the Digital Check Understanding to determine
- success with the learning objective
- items to review
- grouping and differentiation resources

4 Differentiation Options

Differentiate instruction for all students using small-group activities and math center activities on page 635C.

Reteach

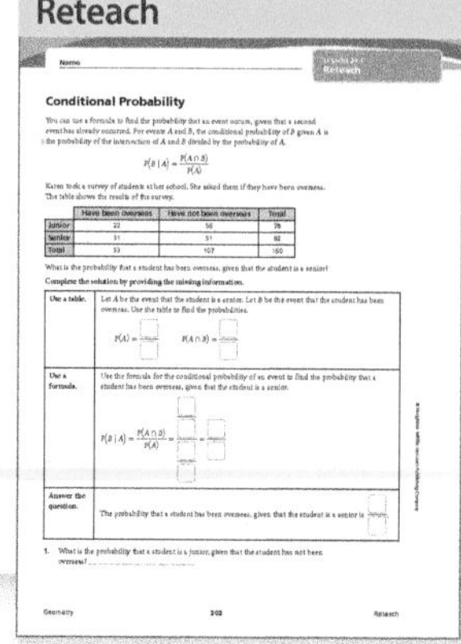

Challenge

Conditional Probability with Multiple Events

You have 26 hand-carved wooden letters—one for each letter in the alphabet. You choose a letter at random. Let event *A* represent choosing a letter in the range V–Z. Let event *B* represent choosing a consonant (assume Y is a vowel). Let event *C* represent choosing a letter in the range N–W. Write each probability as a fraction.

11. $P(A \mid B)$ $\frac{1}{5}$

12. $P(B \mid C)$ $\frac{4}{5}$

13. $P(C \mid A)$ $\frac{2}{5}$

14. $P(A \mid C)$ $\frac{1}{5}$

15. $P(B \mid A)$ $\frac{4}{5}$

16. $P(C \mid B)$ $\frac{2}{5}$

17. A survey asks 1000 people to choose their favorite movie from a list of comedy and drama movies. Overall, 420 of the people in the survey choose a comedy. For Parts A–D, find each probability to the nearest percent.

200 people choose a drama and are less than 21 years old.

170 people choose a comedy and are at least 21 years old.

A. chooses a drama given that the person is at least 21 years old $\frac{380}{550} \approx 69\%$

B. chooses a comedy given that the person is under 21 years old $\frac{250}{450} \approx 56\%$

C. is under 21 years old given that the person chooses a comedy $\frac{250}{420} \approx 60\%$

D. is at least 21 years old given that the person chooses a drama $\frac{380}{580} \approx 66\%$

18. In a carnival game, Carl randomly pulls two ducks from a box of yellow and green ducks. To win, he must pull a yellow duck first, then a green duck. To the nearest whole percent, what is Carl's probability of winning if he pulls a yellow duck on his first draw? 25%

19. (MP) **Attend to Precision** A company has a website that sells only jeans and T-shirts. During a recent month, 61% of customers buy jeans and T-shirts, while 83% of customers buy jeans. What is the probability that a customer who buys jeans also buys a T-shirt? Round to the nearest percent. 73%

P(yellow first) = 76%
P(yellow first and green second) = 19%

Questioning Strategies

Problem 17 What two-way table could you use to help you answer this question? The two variables are type of movie, comedy or drama, and age, less than or equal to or great than 21. The given frequencies are inserted in the table.

	Age < 21	Age > 21	Total
Comedy	?	170	420
Drama	200	?	?
Totals	?	?	1000

Watch for Common Errors

Problem 18 Students may confuse the expressions $P(\text{green second} \mid \text{yellow first})$ and $P(\text{green second} \cap \text{yellow first})$ and think that the latter is the solution of this problem. Point out the difference between these two expressions and remind students of the Conditional Probability Formula that enables them to find $P(\text{green second} \mid \text{yellow first})$. That is,

$$P(\text{green second} \mid \text{yellow first}) = \frac{P(\text{green second} \cap \text{yellow first})}{P(\text{yellow first})} = \frac{0.19}{0.76} = 0.25, \text{ or } 25\%.$$

5 Wrap-Up

Summarize learning with your class. Consider using the Exit Ticket, Put It in Writing, or I Can scale.

Exit Ticket 

65% of the boys in your class like football, and 15% of them like football and soccer. What percentage of the students who like football also like soccer? This is a conditional probability situation that you can solve using the Conditional Probability Formula.

$$P(soccer \mid football) = \frac{P(soccer \cap football)}{P(football)}$$

$$= \frac{0.15}{0.65} \approx 23\%$$

Put It in Writing 

If you are given two events *A* and *B*, explain the differences among $P(A)$, $P(B)$, and $P(A \mid B)$.

I Can 

The scale below can help you and your students understand their progress on a learning goal.

4	I can calculate conditional probability and use it to solve real-world problems, and I can explain my steps to others.
3	I can calculate conditional probability and use it to solve real-world problems.
2	I can use a frequency table to represent the number of outcomes in an event and calculate the probability of each event.
1	I can calculate the probability of an event.

Spiral Review • Assessment Readiness

These questions will help determine if students have retained information taught in the past and can also prepare them for high-stakes assessments. Here, students calculate the probability of whole numbers less than 50 that are divisible by 4 **(20.1)**, calculate the probability of an odd number or multiple of 3 from the set of factors of 30 **(20.1)**, and determine whether three given sets are disjoint or overlapping **(20.2)**.

20. (MP) **Critique Arguments** Mark surveys 100 cell phone owners about whether their phone is new (less than 1 year old) or old, and whether they had problems with their phone.

	No problems	Problems (one or more)	Total
New phone	46	9	55
Old phone	16	29	45
Total	62	38	100

Mark calculates $P(\text{problems} \mid \text{new cell phone}) = \frac{9}{38}$, or about 24%.

A. What is Mark's error? A, B. See Additional Answers.

B. What is the correct probability to the nearest whole percent?

21. (Open Middle™) Using the digits 1 to 9, at most one time each, fill in the boxes to make a true statement. See Additional Answers.

There is a bag of ☐ marbles with ☐ blue marbles and ☐ white marbles.
The probability of picking a white marble first and picking a blue marble second (without replacement) is ☐☐%. Round to the nearest whole percent.

Spiral Review • Assessment Readiness

22. A number from 1 to 50 is chosen at random. What is the probability that the number is not divisible by 4?

Ⓐ $\frac{4}{25}$ Ⓒ $\frac{14}{25}$
Ⓑ $\frac{6}{25}$ Ⓓ $\frac{19}{25}$

23. If you randomly choose a factor of 30, what is the probability of choosing an odd number or a multiple of 3?

Ⓐ $\frac{1}{4}$ Ⓒ $\frac{3}{4}$
Ⓑ $\frac{1}{2}$ Ⓓ $\frac{7}{8}$

24. Let event *A* be that a puppy weighs more than 6 pounds, event *B* be that the puppy weighs no more than 8 pounds, and event *C* be that the puppy weighs 5 pounds or less. For each pair of events, identify whether the events are disjoint or overlapping.

Events	Disjoint	Overlapping
A. *A* and *B*	?	?
B. *B* and *C*	?	?
C. *C* and *A*	?	?

 **I'm in a Learning Mindset!**

How effective was using the conditional probability formula in finding probabilities?

642 Keep Going Journal and Practice Workbook

Learning Mindset 

Perseverance Checks for Understanding

Check that students understand the convenience of using the Conditional Probability Formula to calculate the probability of events *A* and *B*; $P(B \mid A) = \frac{P(A \cap B)}{P(A)}$, $PI(AI) \neq 0$. Point out that they can also rewrite the formula and use it to calculate the probability of the intersection of the two sets or the probability of the first event. If students still have difficulty understanding this concept, remind them that it is also helpful to use tables to calculate these probabilities. *Check students' understanding by having them use words to explain each probability expression. Encourage them to create tables that display frequencies or relative frequencies of events. Ask them to explain the use of set notation to represent the events and their probabilities. In particular, have students recall what a probability statement means and how, like estimation, it is often a useful way to understand real-world situations.*

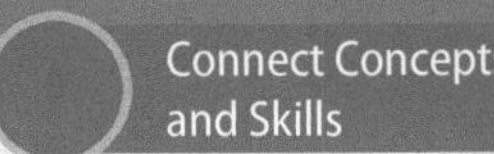

21.2 Independent Events

LESSON FOCUS AND COHERENCE

Mathematics Standards

- Understand that two events *A* and *B* are independent if the probability of *A* and *B* occurring together is the product of their probabilities, and use this characterization to determine if they are independent.
- Recognize and explain the concepts of conditional probability and independence in everyday language and everyday situations.

Mathematical Practices and Processes

- Reason abstractly and quantitatively.
- Construct viable arguments and critique the reasoning of others.
- Attend to precision.

I Can Objective

I can determine whether two events are independent and find their probabilities.

Learning Objective

Interpret independence and its definition in terms of conditional probability. Derive and apply the Multiplication Rule for the probability of the intersection of independent events and use the rule to test for independence of events. Use the concept of independence to solve real-world problems.

Language Objective

Explain what it means for two events to be independent in terms of a conditional probability.

Vocabulary

New: independent events

Mathematical Progressions

Prior Learning	Current Development	Future Connections
Students: • summarized categorical data for two categories in two-way frequency tables. **(Alg1, 21.1)** • described events as subsets of a sample space using characteristics of the outcomes, or as unions, intersections, or complements of other events. **(20.1)**	**Students:** • interpret independence and its definition in terms of conditional probability. • derive and apply the Multiplication Rule for the probability of the intersection of independent events, and use the rule to test for independence of events. • use the concept of independence to solve real-world problems.	**Students:** • will understand conditional probability and interpret independence using conditional probability. **(Alg2, 18.1 and 18.2)** • will understand that two events *A* and *B* are independent if the probability of *A* and *B* occurring together is the product of their probabilities. **(Alg2, 18.2)**

PROFESSIONAL LEARNING

Using Mathematical Practices and Processes

Construct viable arguments and critique the reasoning of others.

This lesson introduces the concept of independent events and defines independent events *A* and *B* as those for which $P(A) = P(A \cap B)$ and $P(B) = P(B \cap A)$. Soon thereafter, students are introduced to the Multiplication Rule for independent events, $P(A \cap B) = P(A) \cdot P(B)$.

A point of emphasis of the lesson is using given data or probabilities to determine whether events are independent, giving ample opportunity to address the Mathematical Practice Standard through discussion. While students build understanding through an intuitive definition of independence, they apply formal definitions as justification in concluding that two events are independent.

WARM-UP OPTIONS

ONLINE

PROJECTABLE & PRINTABLE 

ACTIVATE PRIOR KNOWLEDGE • Construct a Two-Way Frequency Table

Use these activities to quickly assess and activate prior knowledge as needed.

Problem of the Day 

Two hundred customers at a pet supply store were surveyed and asked whether they own a dog or a cat. There were 110 cat owners, 120 dog owners, and 20 customers who own neither a dog nor a cat. Construct and complete a two-way frequency table describing how many customers own or do not own either kind of animal.

	Own a dog	Do not own a dog	Total
Own a cat	50	60	110
Do not own a cat	70	20	90
Total	120	80	200

Quick Check for Homework

As part of your daily routine, you may want to display the Teacher Solution Key to have students check their homework.

Make Connections

Based on students' responses to the Problem of the Day, choose one of the following:

1. Project the Interactive Reteach, Lesson 20.2.
2. Complete the Prerequisite Skills Activity:

Have students work in pairs to construct and complete a two-way frequency table from this data:

Over a span of 100 days, it rained on 40 days, snowed on 60 days, and both rained and snowed on 20 days.

- *How can you find the number that should go in a cell in the Total column?* I can add the two other numbers in the cell's row, if I know both of their values.
- *If you know a value in the Total row and another value above it, how can you find the other value in that column?* I can subtract the known value from the value in the Total cell.

SHARPEN SKILLS

If time permits, use this on-level activity to build fluency and practice basic skills.

Vocabulary Review

Objective: Students demonstrate an understanding of terms related to the term "conditional probability."
Materials: Bubble Map (Teacher Resource Masters)

Have students work in pairs. Each pair should work together to build a bubble map for the term "conditional probability" without looking anything up. Encourage students to work together to recall as much information as possible and to not worry too much about being correct at this point.

When all pairs have had ample time to work, have a class discussion and create a class bubble map on the board.

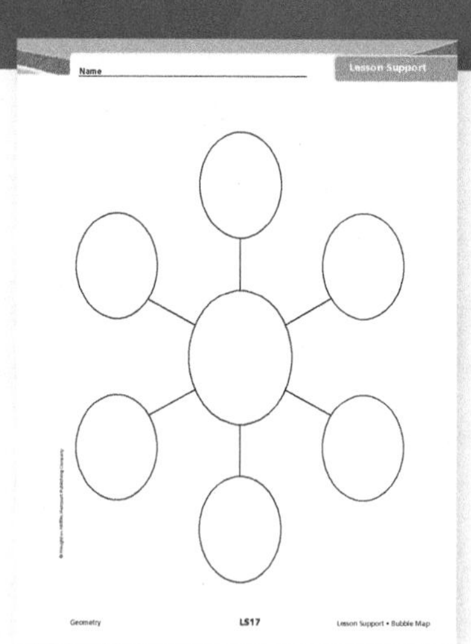

PLAN FOR DIFFERENTIATED INSTRUCTION

Small-Group Options

Use these teacher-guided activities with pulled small groups.

On Track

Materials: index cards

Give each group a set of index cards. Each card will have on it one of the numbers 1, 2, 3, 4, or 6. For each number, there are two cards with that number. Have students select four cards at a time and do the following:

- Assign the largest value to $n(S)$, the size of the sample space, and the smallest value to $n(A \cap B)$, the size of the intersection of events A and B. The other two values are assigned to $n(A)$ and $n(B)$.
- Calculate and compare $P(A)$ to $P(A \mid B)$ and $P(B)$ to $P(B \mid A)$.
- Calculate and compare $P(A \cap B)$ to $P(A) \cdot P(B)$.
- State whether or not A and B are independent.

Almost There

Materials: index cards

Give each group a set of index cards. Each card will have on it one of the numbers 1, 2, 3, 4, or 6. For each number, there are two cards with that number. Have students select four cards at a time and do the following:

- Assign the largest value to $n(S)$, the size of the sample space, and the smallest value to $n(A \cap B)$, the size of the intersection of events A and B. The other two values are assigned to $n(A)$ and $n(B)$.
- Construct and complete a two-way frequency table for events A and B.
- Calculate and compare $P(A)$ to $P(A \mid B)$ and $P(B)$ to $P(B \mid A)$, and state whether or not A and B are independent.

Ready for More

Materials: index cards

Give each group a set of index cards. Each card will have on it an integer in the range from 1 to 10. Have students draw one card at a time and do the following:

- Assign the value drawn to $n(S)$, the size of the sample space for two events A and B.
- Determine all integer triples for $n(A)$, $n(B)$, and $n(A \cap B)$ that would make A and B independent events, or explain why there are no such positive integers.

Math Center Options

Use these student self-directed activities at centers or stations.

Key: ● Print Resources ● Online Resources

On Track

- ● Interactive Digital Lesson
- ●● Journal and Practice Workbook
- ● Interactive Glossary (printable): **independent events**
- ● Module Performance Task
- ● Illustrative Mathematics: The Titanic 2
- ●● Standards Practice: Understand and Use the Definition of Independent Events

Almost There

- ● Reteach 21.2 (printable)
- ● Interactive Reteach 21.2
- ● Illustrative Mathematics: Cards and Independence

Ready for More

- ● Challenge 21.2 (printable)
- ● Interactive Challenge 21.2
- ● Illustrative Mathematics: The Titanic 3

Unit Project Check students' progress by asking what method they plan to use to determine if two events are independent.

View data-driven grouping recommendations and assign differentiation resources.

During the *Spark Your Learning,* listen and watch for strategies students use. See samples of student work on this page.

Use a Conditional Probability

Strategy 1

I can find the conditional probability that a student gets 60 minutes or more of aerobic activity each day given that he or she owns a dog by dividing 19 by 87: $\frac{19}{87} \approx 22\%$. Next, I can find the probability of students that are active: $\frac{38}{245} \approx 16\%$. Then I can compare the two probabilities. The answer to the question is yes.

If students . . . use a conditional probability to solve the problem, then they are demonstrating an understanding of conditional probability from Lesson 21.1.

Have these students . . . explain how they found the conditional probability and how they used it to solve the problem. **Ask:**

- Q How did you use the given data to find the conditional probability?
- Q How did you compare the probabilities to come to a conclusion?

Use Complements

Strategy 2

I can use the given data to make a two-way frequency table:

	Dog	No dog	Total
Active	19	19	38
Not active	68	139	207
Total	87	158	245

The probability that a student is not active each day given that they do not own a dog is $\frac{139}{158} \approx 88\%$ while the probability that a student is not active is $\frac{207}{245} \approx 84\%$. Not owning a dog means a student is less likely to be active, so the answer to the question is yes.

If students . . . use complements of events related to the given data, they may understand conditional probabilities but may not realize that the question could be answered more directly by finding probabilities related to the given data.

Activate prior knowledge . . . by having students find probabilities of events directly related to the given data. **Ask:**

- Q What is the probability that a student owns a dog?
- Q How can you find the probability that a student is active, given that they own a dog?

COMMON ERROR: Assumes Students Are Active

Since 38 students are active and 19 of them own a dog, a student who owns a dog has a $\frac{19}{38} = 50\%$ probability of being active. Students are equally likely to be active if they own a dog as they would be if they did not own a dog.

If students . . . assume students are active, then they may not understand which event or condition is to be assumed.

Then intervene . . . by pointing out that a student owning a dog should be assumed, and then a conditional probability should be found. **Ask:**

- Q How does the question in the problem begin?
- Q What should you assume about a student in the question?

21.2

Independent Events

(I Can) determine whether two events are independent and find their probabilities.

Spark Your Learning

Tasha surveyed 245 students at her school.

Eighty-seven students own a dog.
Thirty-eight students get 60 minutes or more of aerobic activity each day.

©H. Mark Weidman Photography/Alamy

Complete Part A as a whole class. Then complete Parts B–D in small groups.

A. What is a mathematical question you can ask about this situation? What information would you need to know to answer your question?

B. What probabilities are involved in this situation? How can you use these probabilities to answer your question?
B, D. See Additional Answers.

C. To answer your question, what strategy and tool would you use along with all the information you have? What answer do you get?
See Strategies 1 and 2 on the facing page.

D. Does your answer make sense in the context of the situation? How do you know?

A. Is a student who owns a dog more likely to be more active?; the number of students who own a dog and get 60 minutes or more of aerobic activity each day

Turn and Talk Predict how the answer to your question would change if the following numbers changed as described:

- The number of students who own dogs and get 60 or more minutes of aerobic activity each day is 15.
- The number of students who own dogs is 125.
- The number of students who get 60 or more minutes of exercise is 54. See margin.

SUPPORT SENSE-MAKING • Three Reads

Tell students to read the information in the photo three times and prompt them with a different question each time.

1. What is the situation about? The situation is about students at a school. Tasha asked students whether they get 60 minutes or more of exercise each day, and whether they own a dog.
2. What are the quantities in this situation? How are those quantities related? The quantities are the number of students surveyed, the number of students who are active, the number of students who own a dog, and the number of students who are active and own a dog. The number surveyed is the largest, the number of dog owners second largest, the number of active students third largest, and the number of students who are both must be the smallest since such a student has to satisfy both conditions.
3. What are possible questions you could ask about this situation? How many students who own a dog aren't active? How many students who aren't active also own a dog? Is a student who owns a dog more likely to be active?

1 Spark Your Learning

▶ MOTIVATE

- Have students look at the photo in their books and read the information contained in the photo. Then complete Part A as a whole-class discussion.
- Give the class the additional information they need to solve the problem. This information is available online as a printable and projectable page in the Teacher Resources.
- Have students work in small groups to complete Parts B–D.

▶ PERSEVERE

If students need support, guide them by asking:

Q **Advancing • Use Tools** Which tool could you use to solve the problem? Why choose that tool and not some other? Students' choices of tools and reasons for choosing them will vary.

Q **Assessing** What is the relationship between the four sets of students described? The set of active dog owners is the intersection of the set of active students and the set of dog owners, and all three sets are contained in the set of surveyed students.

Q **Advancing** How can you determine whether owning a dog makes a student more likely to be active? I can find the conditional probability that dog owners are active, and compare it to the probability that a student is active.

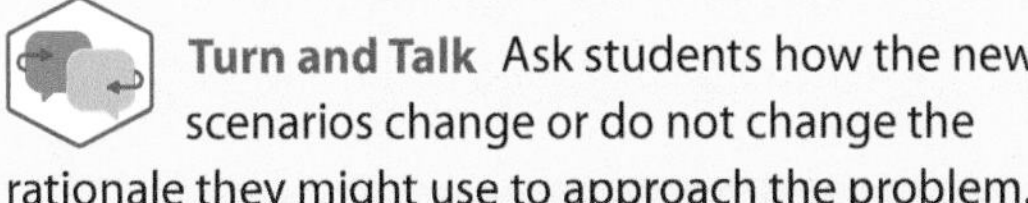

Turn and Talk Ask students how the new scenarios change or do not change the rationale they might use to approach the problem.
Possible answer: The conditional probability that a student gets 60 minutes or more of aerobic activity each day given that he or she owns a dog is $\frac{15}{125} = 12\%$. The probability that a student gets 60 or more minutes of exercise is $\frac{54}{245} \approx 22\%$. A student who owns a dog is not more likely to be active.

▶ BUILD SHARED UNDERSTANDING

Select groups of students who used various strategies and tools to share with the class how they solved the problem. As they present their solutions, have each group discuss why they chose a specific strategy and tool.

2 Learn Together

Build Understanding

Task 1 MP **Reason** Students are introduced to the definition of independent events and reason about the definition when it is applied to an example.

CONNECT TO VOCABULARY

Have students use the **Interactive Glossary** to record their understanding of the vocabulary in this task.

Turn and Talk Have students look at their calculations in the task. $P(A \mid B) = P(A)$ and $P(B \mid A) = P(B)$

Task 2 MP **Construct Arguments** Students derive a formula for the probability of the intersection of independent events.

Turn and Talk Encourage students to think about the steps in Part D in reverse. Given $P(A \cap B) = P(A) \cdot P(B)$, I can divide each side by $P(A)$, so $\frac{P(A \cap B)}{P(A)} = P(B)$. Using the definition of conditional probability, substitute $P(B \mid A)$ for $\frac{P(A \cap B)}{P(A)}$, so $P(B \mid A) = P(B)$. Events A and B are independent.

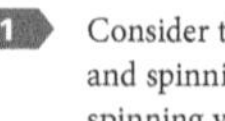

Build Understanding

Understand Independence of Events

Two events are called **independent events** when the occurrence of one event does not influence the occurrence of the other.

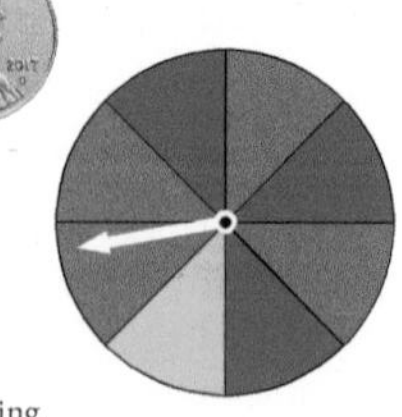

1 Consider the two simple events of flipping a coin and spinning a spinner where the result is heads and spinning yellow. Any outcome from flipping the coin does not influence the outcome of spinning the spinner, so the events are independent. A–C. See Additional Answers.

A. Suppose that you want to use probability to show that the two events are independent. How should the probability of spinning yellow compare to the conditional probability of spinning yellow given that the result of the coin flip is heads? Explain your reasoning.

B. How should the probability of flipping heads compare to the conditional probability of flipping heads given that yellow is spun? Explain your reasoning.

C. How can you calculate the conditional probabilities? Do the results agree with your expectations?

Turn and Talk If A and B are independent events, what must be true about the conditional probabilities $P(A \mid B)$ and $P(B \mid A)$? See margin.

Derive the Formula for the Probability of the Intersection of Independent Events

Events A and B are independent if and only if $P(A \mid B) = P(A)$ and $P(B \mid A) = P(B)$.

2 From Task 1, suppose you spin the spinner twice. Let R be the event that you spin red first and Y be the event that you spin yellow second.

A. Why are events R and Y independent? The conditional probability of event Y given that event R has occurred is equal to the probability of event Y independently.

B. What are $P(R)$, $P(Y)$, and $P(R \cap Y)$? $P(R) = \frac{1}{2}$, $P(Y) = \frac{1}{8}$, $P(R \cap Y) = \frac{1}{16}$

C. Find the product of $P(R)$ and $P(Y)$. How does it compare to $P(R \cap Y)$? C, D. See Additional Answers.

D. Recall from the definition of conditional probability that for any events A and B, $P(A \mid B) = \frac{P(A \cap B)}{P(B)}$. Suppose events A and B are independent so that $P(A \mid B) = P(A)$. Use the definition of conditional probability to show that $P(A \cap B) = P(A) \cdot P(B)$.

Turn and Talk Suppose that for two events A and B, $P(A \cap B) = P(A) \cdot P(B)$. How can you show that events A and B are independent? See margin.

644

LEVELED QUESTIONS

Depth of Knowledge (DOK)	Leveled Questions	What Does This Tell You?
Level 1 **Recall**	What does it mean when we say that two events are independent? The outcome of one event does not affect the outcome of the other event.	Students' answers will indicate whether they can recall and interpret the definition of independent events.
Level 2 **Basic Application of Skills & Concepts**	If you roll a pair of dice and say A = "The first die comes up even" and B = "The second die comes up odd," are A and B independent events? yes	Students' answers will indicate whether they can recognize independent events.
Level 3 **Strategic Thinking & Complex Reasoning**	If A = "A coin lands tails," B = "A die comes up odd," and C = "The spinner from Task 1 stops on blue," can you write a formula for the probability of all three events occurring? $P(A) \cdot P(B) \cdot P(C)$	Students' answers will indicate whether they can extend the formula for the probability of the intersection of independent events to the probability of the intersection of three independent events.

Step It Out

Find the Probability of Independent Events

Your findings from Task 2 give the following result.

Probability of Independent Events
Events A and B are independent if and only if $P(A \cap B) = P(A) \cdot P(B)$.

 3 A bag contains 12 marbles as shown. You select a marble, return it, and then select another marble. What is the probability that you select a red marble first and a blue marble second?

Let R be the event of selecting a red marble first and B be the event of selecting a blue marble second.

So, $P(R) = \frac{1}{3}$, and $P(B) = \frac{1}{2}$.

A. How do you find $P(R)$ and $P(B)$?

A, B. See Additional Answers.

The events are independent, so multiply their probabilities to find $P(R \cap B)$.

$P(R \cap B) = P(R) \cdot P(B) = \frac{1}{3} \cdot \frac{1}{2} = \frac{1}{6}$

B. Why are the two events independent?

Show that Two Real-World Events are Independent

To show that two events A and B are independent, you can either show that $P(A \mid B) = P(A)$ or show that $P(A \cap B) = P(A) \cdot P(B)$.

 4 A local theater wants to determine if the event that a youth ticket is sold is independent of the event that the ticket sold is for a musical.

Let Y be the event that a youth ticket is sold and M be the event that the ticket sold is for a musical.

	Musical	Non-musical	Total
Youth	36	48	84
Adult	84	112	196
Total	120	160	280

From the table, $P(Y) = \frac{84}{280} = 30\%$, and $P(Y \mid M) = \frac{36}{120} = 30\%$. Since $P(Y) = P(Y \mid M)$, the events are independent.

A. How do you use data in the table to find $P(Y)$ and $P(Y \mid M)$?

A, B. See Additional Answers.

Likewise, you can find $P(M)$ and $P(Y \cap M)$ to show that $P(Y \cap M) = P(Y) \cdot P(M)$.

B. How do you show that $P(Y \cap M) = P(Y) \cdot P(M)$ using the data in the table?

Step It Out

Task 3 (MP) **Attend to Precision** Students see how probabilities of independent events are determined. Encourage them to be precise in their explanation of why the two events in the task are independent and refer to conditional probabilities of the events.

Sample Guided Discussion:

Q In Part B, how can you use probabilities to determine whether events are independent? I can compare the probability of an event to its conditional probability assuming the other event has occurred. If they are equal, the events are independent.

Task 4 (MP) **Construct Arguments** Students see how empirical probabilities can be used to determine whether two real-world events are independent. They see how data from a two-way frequency table can be used to apply the definition of independent events. Students then see how another argument can be made by checking whether the probabilities of the events and their intersections satisfy a condition of independent events.

OPTIMIZE OUTPUT Critique, Correct, and Clarify

Have students work with a partner to discuss the approaches in Parts A and B of Task 4 and which is correct and which is incorrect. Encourage students to use the vocabulary terms *independent events*, *intersection*, and *conditional probability* in their discussions. Students should revise their answers, if necessary, after talking with their partner.

Sample Guided Discussion:

Q In Part A, why is 120 the denominator of $P(Y \mid M)$ instead of 280? For this conditional probability, I am assuming that the ticket sold is for a musical.

Q In Part B, are you applying the definition of independent events? No, I am checking to see whether the probabilities of the events and their intersection satisfy a condition that is true for any pair of independent events.

PROFICIENCY LEVEL

Beginning

Show students a set of index cards, with 5 having a red mark on one side, 3 cards having a blue mark, and 2 green. Write "5 red," "3 blue," and "2 green," and turn the cards facedown and mix them up. Say, "The probability that I draw a blue card is $\frac{3}{10}$." Ask them a different color, and have them give the probability that they would draw that color card.

Intermediate

Have students work in pairs. Give each pair three index cards with $P(A)$, $P(B)$, or $P(A \cap B)$ written on them. Have students take turns drawing a pair of cards and explaining to their partner how to calculate the third probability, making sure to begin their explanation with, "If A and B are independent events, then. . . ."

Advanced

Have students work in small groups and craft an explanation of how $P(A)$, $P(B)$, $P(A \mid B)$, and $P(A \cap B)$ are related when A and B are independent events.

ONLINE

Assign the Digital On Your Own for

- built-in student supports
- Actionable Item Reports
- Standards Analysis Reports

On Your Own

Assignment Guide

The chart below indicates which problems in the On Your Own are associated with each task in the Learn Together. Assign daily homework for tasks completed.

Learn Together Tasks	On Your Own Problems
Task 1, p. 644	Problems 5–9 and 15
Task 2, p. 644	Problems 10 and 16
Task 3, p. 645	Problems 11 and 17
Task 4, p. 645	Problems 12–14

ANSWERS

5. $P(G) = \frac{1}{3}$ and $P(G \mid Y) = \frac{1}{3}$, so $P(G) = P(G \mid Y)$, and the events are independent.

6. $P(G) = \frac{1}{3}$ and $P(G \mid R) = \frac{1}{3}$, so $P(G) = P(G \mid R)$, and the events are independent.

9. $P(Y) = \frac{2}{5}$ and $P(Y \mid R) = 0$, so $P(Y) \neq P(Y \mid R)$, and the events are not independent.

Check Understanding

1. Give an example of two simple independent events. Use conditional probability to explain why the events are independent. 1, 2. See Additional Answers.

2. For two standard 6-sided number cubes, you roll the green number cube and the red number cube. Let A be the event that you roll an even number on the green number cube and B be the event that you roll a 4 or 5 on the red number cube. Show that events A and B are independent by finding $P(A)$, $P(B)$, $P(A \mid B)$, and $P(B \mid A)$.

3. A box has 20 tiles with shapes etched on them. Three tiles have triangles, 12 tiles have squares, and the rest have circles. You randomly select a tile, put it back in the box, and then randomly select again. Let T be the probability that you select a tile with a triangle first and C be the probability that you select a tile with a circle second. What is $P(T \cap C)$? $P(T \cap C) = \frac{3}{80}$

4. A bookstore owner surveyed customers on whether they regularly read science fiction and mystery books. The two-way frequency table shows the survey results.

	Reads mystery books	Does not read mystery books	Total
Reads science fiction books	45	54	99
Does not read science fiction books	35	42	77
Total	80	96	176

Let M be the event that the customer reads mystery books. Let S be the event that the customer reads science fiction books. Determine if M and S are independent in two ways. Show the following: A, B. See Additional Answers.

A. $P(M \mid S) = P(M)$ **B.** $P(M \cap S) = P(M) \cdot P(S)$

On Your Own

For Problems 5–9, a bag contains 10 green, 12 yellow, and 8 red marbles. You randomly select a marble, return it, and then randomly select another marble. Let G be the event that you select a green marble first, Y be the event that you select a yellow marble second, and R be the event that you select a red marble second.

5. Explain why G and Y are independent events. 5, 6. See margin.
6. Explain why R and G are independent events.
7. Find $P(G \cap Y)$. $\frac{2}{15}$
8. Find $P(G \cap R)$. $\frac{4}{45}$
9. Explain why Y and R are not independent events. See margin.

646

data checkpoint

3 Check Understanding

Formative Assessment

Use formative assessment to determine if your students are successful with this lesson's learning objective.

Students who successfully complete the Check Understanding can continue to the On Your Own practice.

For students who miss 1 problem or more, work in a pulled small group using the Almost There small-group activity on page 643C.

ONLINE

Assign the Digital Check Understanding to determine

- success with the learning objective
- items to review
- grouping and differentiation resources

4 Differentiation Options

Differentiate instruction for all students using small-group activities and math center activities on page 643C.

Reteach

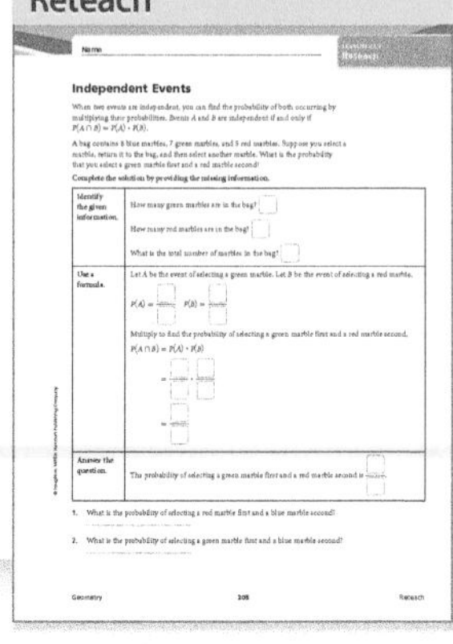

Challenge

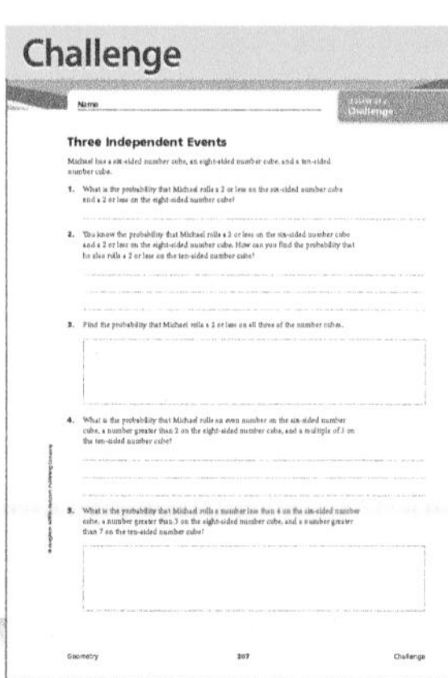

10. **Use Structure** Events M and N are independent. Show that $P(M \cap N \mid N) = P(M)$. See margin.

11. **STEM** Gregor Mendel was an early pioneer in using probability in genetics. He discovered that various traits in pea plants, such as flower color, were inherited independently. As shown, each parent has purple flowers and can donate either B or b to the new plant. A pea plant can have purple flowers (BB or Bb) or white flowers (bb) depending on what it inherits from the parents. If the events of inheriting either B or b from each parent are independent and equally likely, what is the probability that the new pea plant inherits b from each parent and has white flowers? $\frac{1}{4}$

		Pollen ♂ B	Pollen ♂ b
Pistil ♀	B	BB	Bb
	b	Bb	bb

For Problems 12–14, determine if the events are independent.

12. A restaurant manager wants to know if the event that a customer orders dessert at dinner is independent of the event that the customer is dining on the weekend. independent

	Weekend diners	Weekday diners	Total
Dessert	81	36	117
No dessert	459	204	663
Total	540	240	780

13. Town officials are considering a property tax increase to finance a new park. They want to know if the event that a person has children is independent of the event that person supports the tax. not independent

	Supports tax	Does not support tax	Total
Has children	100	40	140
Does not have children	20	60	80
Total	120	100	220

14. The school cafeteria takes orders for a field trip. Students have a choice to order a salad or a sandwich and water or juice. Determine if the event that a student orders a sandwich is independent of the event that the student orders water. independent

	Water	Juice	Total
Salad	27	12	39
Sandwich	54	24	78
Total	81	36	117

Questioning Strategies

Problem 11 When a parent pea plant donates B or b to an offspring plant, it can still donate the same gene to another plant. Suppose two pea plants share the same parent plants as shown in the diagram. What is the probability that both new plants have purple flowers? $\left(\frac{3}{4}\right)\left(\frac{3}{4}\right) = \frac{9}{16}$

Watch for Common Errors

Problem 14 Given events S = "orders a sandwich" and W = "orders water," students might compare the probabilities $P(S \mid W) = \frac{54}{81}$ and $P(W) = \frac{81}{117}$. They need to understand that $P(S \mid W)$ should be compared with $P(S)$, or $P(W \mid S)$ with $P(W)$.

ANSWER

10. $P(M \cap N \mid N) = \frac{P(M \cap N \cap N)}{P(N)} = \frac{P(M \cap N)}{P(N)}$
$= \frac{P(M) \cdot P(N)}{P(N)} = P(M)$

5 Wrap-Up

Summarize learning with your class. Consider using the Exit Ticket, Put It in Writing, or I Can scale.

Exit Ticket 

What is the probability that you draw a card with any suit but diamonds from a standard 52-deck of cards, put the card back, and then draw a king? $\left(\frac{3}{4}\right)\left(\frac{1}{3}\right) = \frac{3}{52}$

Put It in Writing 

Explain how you can use the probabilities of two events and the probability of their intersection to determine whether the events are independent.

I Can 

The scale below can help you and your students understand their progress on a learning goal.

4	I can determine whether two events are independent and find their probabilities, and I can explain my steps to others.
3	I can determine whether two events are independent and find their probabilities.
2	I can use probabilities to determine whether two events are independent.
1	I can explain what independent events are.

Spiral Review • Assessment Readiness

These questions will help determine if students have retained information taught in the past and can also prepare them for high-stakes assessments. Here, students must determine the probability of the intersection of two events **(20.1)**, determine the probability of the union of two events **(20.2)**, and determine conditional probabilities **(21.1)**.

15. MP **Reason** You randomly select a sock from the laundry basket shown. Let D be the event that the sock has dots, R be the event that the sock has some red in its pattern, and S be the event that the sock has stripes. 15–17. See Additional Answers.

A. Are events D and R independent events? Explain.

B. Are events D and S independent events? Explain.

Dotted socks: 2 red and 2 blue
Striped socks: 4 red and 4 blue

16. Explain why $P(B \mid A) = P(B)$ is true if $P(A \mid B) = P(A)$ is true.

17. Open Middle™ Using the digits 1 to 9, at most one time each, fill in the boxes to make a true statement.

Rolling a sum of ☐ on two ☐-sided dice is the same probability as rolling a sum of ☐ on two ☐-sided dice.

Spiral Review • Assessment Readiness

18. Suppose you select a number at random from 1 to 20. Let A be the event of selecting a multiple of 3. Let B be the event of selecting a number greater than 12. What is $P(A \cap B)$?

Ⓐ $\frac{1}{20}$ Ⓒ $\frac{3}{20}$
Ⓑ $\frac{1}{10}$ Ⓓ $\frac{1}{5}$

19. You roll two 4-sided dice, each numbered from 1 to 4. What is the probability that the two numbers rolled are the same or that the sum of the two numbers is odd?

Ⓐ $\frac{1}{4}$ Ⓒ $\frac{5}{8}$
Ⓑ $\frac{1}{2}$ Ⓓ $\frac{3}{4}$

20. You randomly select a card from a standard 52-card deck of playing cards. Let event A be selecting a king, queen, or jack, event B be selecting a red card, and event C be selecting a club. Match the probability on the left with its value on the right.

A. $P(A \mid C)$	**1.** 0	A. 2
B. $P(B \mid A)$	**2.** $\frac{3}{13}$	B. 4
C. $P(B \mid C)$	**3.** $\frac{1}{4}$	C. 1
D. $P(C \mid A)$	**4.** $\frac{1}{2}$	D. 3

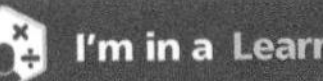

I'm in a Learning Mindset!

Which strategy works best to determine whether two events are independent when I am given information in a two-way frequency table?

648

Learning Mindset

Perseverance Collects and Tries Multiple Strategies

Point out how having multiple tools at their disposal can be invaluable for students when they are working with events and trying to determine whether they are independent. Remind them that while testing probabilities against the definition of independent events is an attractive default method, data given in a two-way frequency table may make it more efficient to check whether the product of the probabilities of the events are equal to the probability of their intersection. Testing mathematical objects with conditions can often be a much quicker way to classify them than checking definitions. *How can having more than one way to answer a question make your work more efficient? How does knowing a condition that independent events always meet give you another way to test whether events are independent?*

21.3 Dependent Events

LESSON FOCUS AND COHERENCE

Mathematics Standards

- Recognize and explain the concepts of conditional probability and independence in everyday language and everyday situations.
- Apply the general Multiplication Rule in a uniform probability model, $P(A \text{ and } B) = P(A)P(B \mid A) = P(B)P(A \mid B)$, and interpret the answer in terms of the model.

Mathematical Practices and Processes 

- Construct viable arguments and critique the reasoning of others.
- Attend to precision.
- Look for and express regularity in repeated reasoning.

I Can Objective 

I can determine whether two events are dependent and find their probabilities.

Learning Objective

Determine whether two events are dependent and find their probabilities.

Language Objective

Explain how to determine whether two events are dependent and find their probabilities.

Vocabulary

New: dependent events

Mathematical Progressions

Prior Learning	Current Development	Future Connections
Students: • summarized categorical data for two categories in two-way frequency tables. Interpreted relative frequencies in the context of the data. **(Alg1, 21.1)** • used the two-way table as a sample space to decide if events are independent and to approximate conditional probabilities. **(20.2)**	**Students:** • determine whether two events are independent. • use the Multiplication Rule to find the probability of two dependent events. • extend the Multiplication Rule to apply to three or more dependent events and give the rule for three events.	**Students:** • will understand that two events *A* and *B* are independent if the probability of *A* and *B* occurring together is the product of their probabilities and use this characterization to determine if they are independent. **(Alg2 18.2)** • will recognize and explain the concepts of conditional probability and independence in everyday language and everyday situations. **(Alg2, 18.1–18.3)**

UNPACKING MATH STANDARDS

Recognize and explain the concepts of conditional probability and independence in everyday language and everyday situations.

What It Means to You

Students are introduced to conditional probability, which is used to illustrate the concept of independence, in a previous lesson. For events *A* and *B*, the conditional probability of *B* given *A* is given by $P(B \mid A) = \frac{P(A \cap B)}{P(A)}$. The next lesson uses conditional probability to define independence. Two events are called independent events when the occurrence of one event does not influence the occurrence of the other. Events *A* and *B* are independent if and only if $P(A \mid B) = P(A)$ and $P(B \mid A) = P(B)$. Independent events can also be defined in terms of probability. Events *A* and *B* are independent if and only if $P(A \cap B) = P(A) \cdot P(B)$.

This lesson defines dependent events. Events are dependent events if the occurrence of one event affects the probability of the other. When events are dependent, the Multiplication Rule can be used to find the probabilities. For dependent event *A* and *B*, $P(A \cap B) = P(A) \cdot P(B \mid A)$, where $P(B \mid A)$ is the conditional probability of event *B* given that event *A* has occurred.

WARM-UP OPTIONS

ONLINE

PROJECTABLE & PRINTABLE

ACTIVATE PRIOR KNOWLEDGE • Probability and Set Theory

Use these activities to quickly assess and activate prior knowledge as needed.

Problem of the Day 

A set of 10 cards is numbered 1 to 10. A card is chosen at random. Event A is choosing a card greater than 5. Event B is choosing a prime number. What is $P(A \cap B)$? Explain.
$A = \{6, 7, 8, 9, 10\}$, $B = \{2, 3, 5, 7\}$, $A \cap B = \{7\}$, $P(A \cap B) = \frac{1}{10}$

Quick Check for Homework

As part of your daily routine, you may want to display the Teacher Solution Key to have students check their homework.

Make Connections

Based on students' responses to the Problem of the Day, choose one of the following:

1. Project the Interactive Reteach, Geometry, Lesson 20.1.
2. Complete the Prerequisite Skills Activity:

Show students the following sets: universal set $U = \{$natural numbers less than or equal to 25$\}$; $A = \{$prime numbers$\}$, $B = \{$prime factors of 12$\}$, $C = \{$prime factors of 15$\}$, $D = \{$prime factors of 20$\}$. The elements of U are placed on cards and a card is selected at random.

- *What is* P(A), P(B), *and* P(A ∩ B)? $P(A) = \frac{9}{25}$, $P(B) = \frac{2}{25}$, and $P(A \cap B) = \frac{2}{25}$
- *What is* P(C), P(D), *and* P(C ∩ D)? $P(C) = \frac{2}{25}$, $P(D) = \frac{2}{25}$, and $P(C \cap D) = \frac{1}{25}$

If students continue to struggle, use Tier 2 Skill 24.

SHARPEN SKILLS

If time permits, use this on-level activity to build fluency and practice basic skills.

Quantitative Comparison

Objective: Students make a comparison between two quantities.

Write the following problem on the board. Ask students to choose the letter representing the correct answer and to explain their reasoning.

Quantity A	**Quantity B**
P(rolling an even number)	*P*(rolling a number less than 4)

A. Quantity A is greater.
B. Quantity B is greater.
C. The two quantities are equal. C; Quantity A is $\frac{1}{2}$. Quantity B is $\frac{1}{2}$. So, the two quantities are equal.
D. The relationship cannot be determined from the given information.

PLAN FOR DIFFERENTIATED INSTRUCTION

Small-Group Options

Use these teacher-guided activities with pulled small groups.

On Track

Materials: blank tables and index cards

Divide students into groups of 4. Give each group a sheet of paper with blank tables like the one shown and a set of index cards with numbers in a frequency table. Students pick a card and enter the numbers in the table and, in pairs, determine whether events *A* and *B* are independent or dependent. The two pairs discuss their results.

	A	Not *A*	Total
B			
Not *B*			
Total			

Almost There

Provide students with the following experiment: There are 7 marbles in a bag: 4 green and 3 yellow. Two marbles are selected at random.

- Lead students to calculate $P(G \mid G)$, $P(Y \mid G)$, $P(G \mid Y)$, and $P(Y \mid Y)$ with replacement.
- Lead students to calculate $P(G \mid G)$, $P(Y \mid G)$, $P(G \mid Y)$, and $P(Y \mid Y)$ without replacement.
- Distinguish between independent and dependent events.

Ready for More

Materials: index cards

Divide students into groups of 4. Students fill in two blank tables like the one shown.

	A	Not *A*	Total
B			
Not *B*			
Total			

Each pair of students fills in the table to create one with independent events and one with dependent events.

Students check each other's cards.

Math Center Options

Use these student self-directed activities at centers or stations. Key: ● Print Resources ● Online Resources

On Track

- ● Interactive Digital Lesson
- ●● Journal and Practice Workbook
- ● Interactive Glossary (printable): **dependent events**
- ● Module Performance Task

Standards Practice:

- ●● Explain Concepts of Conditional Probability and Independence
- ●● Apply the Multiplication Rule of Probability

Almost There

- ● Reteach 21.3 (printable)
- ● Interactive Reteach 21.3
- ● RtI Tier 2 Skill 24: Probability of Compound Events Involving *And*
- ● Illustrative Mathematics: Describing Events

Ready for More

- ● Challenge 21.3 (printable)
- ● Interactive Challenge 21.3
- ● Illustrative Mathematics: Describing Events
- ● Desmos: Chance Experiments

ONLINE Ed — View data-driven grouping recommendations and assign differentiation resources.

During the *Spark Your Learning,* listen and watch for strategies students use. See samples of student work on this page.

Use a Two-Way Table and $P(A \mid B) = P(A)$ — Strategy 1

	Raining	Not raining	Total
Drive time of 30 minutes or less	10	55	65
Drive time of more than 30 minutes	20	15	35
Total	30	70	100

From the table, $P(R \mid D) = \frac{2}{13}$ and $P(R) = \frac{3}{10}$. Because $P(R \mid D) \neq P(R)$, the events are dependent.

If students . . . use "If $P(A \mid B) = P(A)$, then A and B are independent" to solve the problem, they are employing an efficient method and demonstrating an exemplary understanding applying rules to solve problems.

Have these students . . . explain how they decided to use the rule "If $P(A \mid B) = P(A)$, then A and B are independent" and how they answered the question. **Ask:**

- Q Why did you use the rule with conditional probability?
- Q How did you apply the rule?

Use a Two-Way Table and $P(A \cap B) = P(A) \times P(B)$ — Strategy 2

	Raining	Not raining	Total
Drive time of 30 minutes or less	10	55	65
Drive time of more than 30 minutes	20	15	35
Total	30	70	100

From the table, $P(R) = \frac{3}{10}$, $P(D) = \frac{13}{20}$ and $P(R \cap D) = \frac{1}{10}$.

$P(R) \bullet P(D) = \frac{3}{10} \bullet \frac{13}{20} = \frac{39}{200}$. Because $P(R \cap D) \neq P(R) \bullet P(D)$, the events are dependent.

If students . . . used "If $P(A \cap B) = P(A) \bullet P(B)$, then A and B are independent" to solve the problem, they are also employing a correct method and demonstrating an exemplary understanding of applying rules to solve problems but are using a less efficient method.

Activate prior knowledge . . . by having students consider an alternate method. **Ask:**

- Q Is there another rule that could be used?
- Q Which rule is more efficient?

COMMON ERROR: Misunderstand Two-Way Tables

	Raining	Not raining	Total
Drive time of 30 minutes or less	10	55	65
Drive time of more than 30 minutes	20	15	35
Total	30	70	100

I made a table to show the probabilities.

If students . . . can make a two-way table but do not understand how to read it, they will have trouble identifying dependent or independent events.

Then intervene . . . by having students create a two-way table. **Ask:**

- Q What do the numbers in each cell represent?
- Q What probabilities can I find using these numbers?

21.3

Dependent Events

I Can determine whether two events are dependent and find their probabilities.

Spark Your Learning

Evanston, Illinois, is a suburb north of Chicago. Many residents of Evanston drive to and from Chicago every weekday for work.

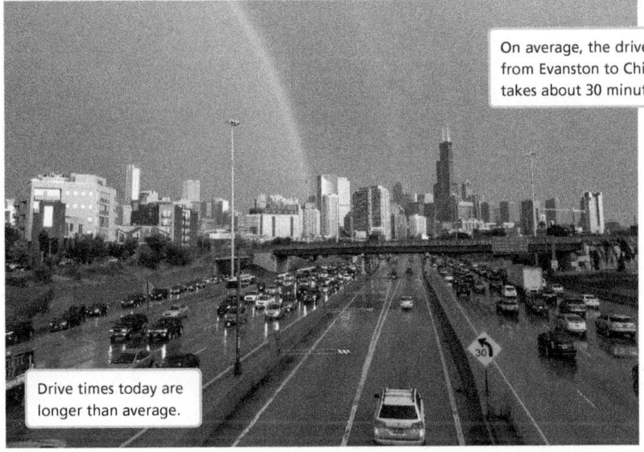

On average, the drive from Evanston to Chicago takes about 30 minutes.
Drive times today are longer than average.

©Antwon McMullen/Shutterstock

Complete Part A as a whole class. Then complete Parts B–D in small groups.

A. What is a mathematical question you could ask about this situation? What information would you need to know to answer your question? A. Are drive times and rain independent events?; data on drive times in clear weather and in the rain

B. What events could you define in this situation? See Additional Answers.

C. To answer your question, what strategy and tool would you use along with all the information you have? What answer do you get? See Strategies 1 and 2 on the facing page.

D. Does your answer make sense in the context of this situation? How do you know? yes; It makes sense that it takes longer to drive a certain distance when it is raining, so drive time depends (in part) on whether it is raining.

Turn and Talk What does it mean if two events are not independent? What other situations can you think of where two events might not be independent? See margin.

CULTIVATE CONVERSATION • Information Gap

Ask students questions to help them decide what missing information they need to answer the question "Are drive time and raining independent events?"

1. Do you have enough information to conclude that drive time and raining are independent events? Explain. Possible answer: I'm not sure. I have a lot of information but don't know if it is enough.
2. How can you organize the given information to make sense of it? Explain. Possible answer: I could put the information in a two-way table.
3. How will organizing the information help you solve the problem? Possible answer: The organized information will help me to find how the probabilities of the events relate to each other.

1 Spark Your Learning

▶ MOTIVATE

- Have students look at the photo in their books and read the information contained in the photo. Then complete Part A as a whole-class discussion.
- Give the class the additional information they need to solve the problem. This information is available online as a printable and projectable page in the Teacher Resources.
- Have students work in small groups to complete Parts B–D.

▶ PERSEVERE

If students need support, guide them by asking:

Q **Advancing • Use Tools** Which tool could you use to solve the problem? Why choose that tool and not some other? Students' choices of tools and reasons for choosing them will vary.

Q **Assessing** What is an event? Possible answer: A set of outcomes is known as an event.

Q **Assessing** What events apply to this situation? Possible answer: drive times and rain

Q **Advancing** How can you set up a two-way table to display the given information? Possible answer: Make the row heads "Raining" and "Not raining" and the column heads "Drive time of 30 minutes or less" and "Drive time of more than 30 minutes."

Turn and Talk Have students recall the meaning of independent events and give examples. Possible answer: If two events are not independent, whether or not one event occurs influences the probability of the other event. Other examples might be wait times in grocery store lines and time of day, or a customer's choice of which product to buy and its price.

▶ BUILD SHARED UNDERSTANDING

Select groups of students who used various strategies and tools to share with the class how they solved the problem. As they present their solutions, have each group discuss why they chose a specific strategy and tool.

2 Learn Together

Build Understanding

Task 1 MP **Construct Arguments** Students use two-way tables to construct an argument about computing probabilities of dependent events.

CONNECT TO VOCABULARY

Have students use the **Interactive Glossary** to record their understanding of the vocabulary in this task.

Sample Guided Discussion:

Q What is the probability of selecting a red marble and then selecting a blue marble?
$P(RB) = \frac{3}{5} \cdot \frac{2}{4} = \frac{6}{20} = \frac{3}{10}$

Q What is the probability of selecting a blue marble and then selecting a red marble?
$P(BR) = \frac{2}{5} \cdot \frac{3}{4} = \frac{6}{20} = \frac{3}{10}$

Turn and Talk Point out to students that the entries in the table in the problem are frequencies, but the entries in the table in Part C are probabilities. Both products are the products of the probabilities of events Y and A, but $P(Y) \cdot P(A \mid Y)$ uses a subset of event A to only consider the probability of A if Y has occurred.

Build Understanding

Derive a Formula for the Probability of the Occurrence of Two Dependent Events

Sometimes the probability of an event depends on the occurrence of another event. For example, suppose you have 3 red marbles and 2 blue marbles as shown in the diagram. You choose a marble at random. When you select another marble at random, the probability of selecting blue can be either $\frac{2}{4}$ or $\frac{1}{4}$. Events are **dependent events** if the occurrence of one event affects the probability of the other.

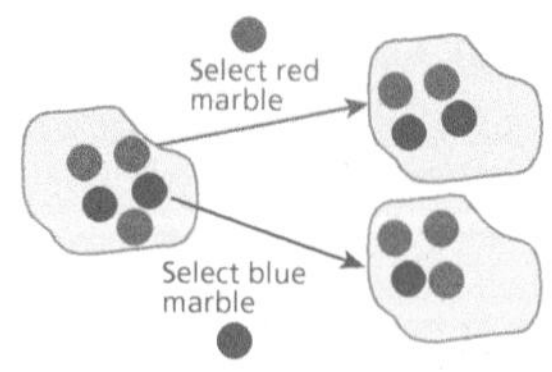

You have learned two ways to test whether events A and B are independent:

1. If $P(A \mid B) = P(A)$, then A and B are independent.
2. If $P(A \cap B) = P(A) \cdot P(B)$, then A and B are independent.

If either of these tests fails, then the two events are dependent.

1 The two-way frequency table shows the numbers of people in two age categories who went to different restaurants at a mall. Let Y be the event that a person is 20 years old or younger. Let A be the event that the person ate at the casual dining restaurant.

	Casual dining	Family style	Total
20 years old or younger	25	15	40
Over 20 years old	30	30	60
Total	55	45	100

A. Find $P(Y)$, $P(A)$, and $P(Y \cap A)$ as fractions. Are events Y and A independent or dependent events? A–D. See Additional Answers.

B. What are the conditional probabilities $P(Y \mid A)$ and $P(A \mid Y)$?

C. Copy and complete the multiplication table using the probabilities you computed.

×	$P(Y)$	$P(A)$
$P(Y \mid A)$	?	?
$P(A \mid Y)$	?	?

D. Do any of the products equal $P(Y \cap A)$? What does this suggest about computing the probability of dependent events?

Turn and Talk How are the products $P(Y) \cdot P(A \mid Y)$ and $P(Y) \cdot P(A)$ alike and how are they different? See margin.

650

LEVELED QUESTIONS

Depth of Knowledge (DOK)	Leveled Questions	What Does This Tell You?
Level 1 **Recall**	How do you find the probability that a person is 20 years old or younger? Divide the number of people 20 years old or younger by the total number of people.	Students' answers will indicate whether they know how to find a probability.
Level 2 **Basic Application of Skills & Concepts**	How can the table be used to determine $P(Y \cap A)$? Possible answer: The number of people, who are 20 years old or younger and who ate at the casual is 25. The total number of people is 100. $P(Y \cap A) = \frac{25}{100} = \frac{1}{4}$.	Students' answers will indicate whether they know how to interpret the two-way table to find a probability.
Level 3 **Strategic Thinking & Complex Reasoning**	How can you change the table so that instead of giving the frequency of each cell, it gives the probability of each cell? Possible answer: Divide each cell by the total in the cell located in the last row, last column.	Students' answers will indicate whether they know how to apply their knowledge of probability and the structure of the table to present the data in a different way.

Step It Out

Find the Probability of Two Dependent Events

When events are dependent, you can use the Multiplication Rule to find the probability of the events occurring.

Multiplication Rule
$P(A \cap B) = P(A) \cdot P(B \mid A)$ where $P(B \mid A)$ is the conditional probability of event B given that event A has occurred.

 2 A board game includes tiles with circles and squares. On each turn, players place all the tiles in a bag, select a tile, then select a tile again without replacing the first tile.

Find the probability that the player selects a circle and then a square.

Let C be the event that the first tile is a circle. Let S be the event that the second tile is a square. There are 3 circles and 3 squares.

After the first tile is removed, the probability of event S depends on whether event C occurred, so S and C are dependent.

A–C. See margin.

$P(C) = \frac{3}{6} = \frac{1}{2}$

$P(S \mid C) = \frac{3}{5}$

A. Why is $P(S \mid C)$ equal to $\frac{3}{5}$ and not $\frac{3}{6}$ or $\frac{2}{5}$?

Apply the Multiplication Rule.

$P(C \cap S) = P(C) \cdot P(S \mid C) = \frac{1}{2} \cdot \frac{3}{5} = \frac{3}{10}$

Find the probability that the player selects a circle followed by another circle.

Let C1 be the event that the first tile is a circle.

Let C2 be the event that the second tile is a circle.

B. Why are C1 and C2 dependent events?

$P(C1) = \frac{3}{6} = \frac{1}{2}$

$P(C2 \mid C1) = \frac{2}{5}$

C. Why are $P(C2 \mid C1)$ and $P(S \mid C)$ not equal?

Apply the Multiplication Rule.

$P(C1 \cap C2) = P(C1) \cdot P(C2 \mid C1) = \frac{1}{2} \cdot \frac{2}{5} = \frac{1}{5}$

 **Turn and Talk** How would the probabilities be different if the player put the first tile back in the bag before drawing the second tile? See margin.

PROFICIENCY LEVEL

Beginning

Write $P(A \cap B)$, $P(A)$, and $P(B \mid A)$. Have student pairs say and write out the words for which each term stands.

Intermediate

Have students write the definitions of independent and dependent events and give an example of each. Have them exchange examples with a partner. Partners can identify whether each example is independent or dependent and briefly explain why.

Advanced

Have students use complete sentences to write a definition of independent and dependent events and a definition of independent and dependent that are not mathematical. Have partners discuss their definitions with each other and produce an example of how each is used.

Step It Out

Task 2 (MP) **Attend to Precision** Students attend to precision when they find probabilities by applying the Multiplication Rule.

Sample Guided Discussion:

Q How is this board game the same as the marble example before Task 1? Possible answer: In both cases, you pick an item and do not replace it.

Q How is it different? Possible answer: The number of the two items is different.

 **Turn and Talk** Discuss the difference between drawing with replacement and drawing without replacement. If the player put the first tile back, there would again be 3 squares and 3 circles for the second draw, so the events C and S would be independent.

ANSWERS

A. $P(S|C)$ means the probability of event S given that event C occurred. If C occurred, there are 5 tiles in the bag: 3 squares and 2 circles. The probability that the second tile is a square is $\frac{3}{5}$.

B. The number of each type of tile remaining in the bag depends on what the first tile was. The probability of C2 would be $\frac{3}{5}$ if the first tile is a square, and $\frac{2}{5}$ if the first tile is a circle.

C. Both probabilities are conditional on a circle being the first tile, which means there are 2 circles and 3 squares left. $P(S|C)$ is the probability of taking a square, so it is $\frac{3}{5}$. $P(C2|C1)$ is the probability of taking a circle, so it is $\frac{2}{5}$.

Step It Out

Task 3 (MP) **Use Repeated Reasoning** Students use repeated reasoning to extend the Multiplication Rule to three or more events.

Sample Guided Discussion:

Q What is the complement of an event? Possible answer: all outcomes that are not the event

Q How do you find the probability of the complement of an event? Possible answer: Subtract the probability of the event from 1.

ANSWERS

A. For each event, whether or not the event before it occurs affects the probability of that event because it changes the number of correct and incorrect passwords remaining on the list to be tried.

B. Think of the complement of *one of the first three passwords is correct* as *it is not true that one of the first three passwords is correct.* If it is not true that one of the first three passwords is correct, then none of the first three passwords are correct.

Find the Probability of Three or More Dependent Events

The Multiplication Rule can be extended to three or more dependent events. For dependent events A, B, and C, $P(A \cap B \cap C) = P(A) \cdot P(B \mid A) \cdot P(C \mid A \cap B)$.

3 Suppose you keep your passwords for secure websites on an encrypted list. The list contains five passwords, and you have forgotten which one to use for your bank website. You try the passwords one at a time, without repeating any.

What is the probability that the third password you try is correct?

Let $W1$ be the event that the first password you try is incorrect: $P(W1) = \frac{4}{5}$.

Let $W2$ be the event that the second password you try is incorrect: $P(W2 \mid W1) = \frac{3}{4}$.

Let R be the event that the third password you try is correct: $P(R \mid W1 \cap W2) = \frac{1}{3}$

A. Why are $W1$, $W2$, and R dependent events?

Apply the Multiplication Rule. Then answer the question.

$$P(W1 \cap W2 \cap R) = P(W1) \cdot P(W2 \mid W1) \cdot P(R \mid W1 \cap W2)$$
$$= \frac{4}{5} \cdot \frac{3}{4} \cdot \frac{1}{3} = \frac{1}{5}$$

A, B. See margin.

The probability that the third password you try is correct is $\frac{1}{5}$.

What is the probability that any one of the first three passwords is correct?

Find the probability of the complement: *none of the first three passwords are correct.*

Let $W1$, $W2$, and $W3$ be the events that the first, second, and third passwords are incorrect. From above, $P(W1) = \frac{4}{5}$, and $P(W2 \mid W1) = \frac{3}{4}$. On the third try, three passwords remain, so $P(W3 \mid W1 \cap W2) = \frac{2}{3}$.

B. Why is *none of the first three passwords are correct* the complement of *one of the first three passwords is correct*?

Apply the Multiplication Rule. Then answer the question.

$$P(W1 \cap W2 \cap W3) = P(W1) \cdot P(W2 \mid W1) \cdot P(W3 \mid W1 \cap W2)$$
$$= \frac{4}{5} \cdot \frac{3}{4} \cdot \frac{2}{3} = \frac{2}{5}$$

This is the probability that none of the first three passwords are correct. To find the probability that one of the first three passwords is correct, subtract from 1. So, $1 - \frac{2}{5} = \frac{3}{5}$.

The probability that one of the first three passwords is correct is $\frac{3}{5}$.

Check Understanding

1. The two-way frequency table shows the numbers of 9th and 10th graders who are on the track team and the robotics team. $P(N) = \frac{3}{5}$, $P(R) = \frac{3}{10}$, $P(N \cap R) = \frac{1}{10}$; dependent

	9th graders	10th graders	Total
On track team	50	20	70
On robotics team	10	20	30
Total	60	40	100

Let N be the event that a person is in 9th grade. Let R be the event that the person is on the robotics team. Find $P(N)$, $P(R)$, and $P(N \cap R)$. Are events N and R independent or dependent events?

There are 5 orange bumper cars and 3 green bumper cars that are being tested for safety for a ride at an amusement park. Two bumper cars are tested at random, one at a time, without retesting the same car.

2. Find the probability that both cars are orange. $\frac{5}{14}$
3. Find the probability that the first car is green and the second is orange. $\frac{15}{56}$

There are 6 chemical elements represented by Li, Na, Mg, Rb, Cs, and Fr that are written on separate pieces of paper. You randomly choose 3 elements, one at a time, without replacement.

4. Find the probability that the third element is Na. $\frac{1}{6}$
5. Find the probability that the first element is Fr, the second element is Rb, and the third element is Li. $\frac{1}{120}$

On Your Own

6. A survey at a high school asked students and teachers whether they read books for entertainment. The two-way frequency table shows the results of the survey. Let S be the event that a person is a student. Let N be the event that the person does not read for entertainment.

	Teacher	Student	Total
Reads books for entertainment	28	57	85
Does not read books for entertainment	2	113	115
Total	30	170	200

A. Find $P(S)$, $P(N)$, and $P(S \cap N)$ as fractions. Are events S and N independent or dependent events? $P(S) = \frac{17}{20}$, $P(N) = \frac{23}{40}$, $P(S \cap N) = \frac{113}{200}$; dependent

B. What are the conditional probabilities $P(S \mid N)$ and $P(N \mid S)$? $P(S \mid N) = \frac{113}{115}$, $P(N \mid S) = \frac{113}{170}$

C. Which probabilities can you multiply to find $P(S \cap N)$? $P(S) \cdot P(N \mid S)$ or $P(N) \cdot P(S \mid N)$

ONLINE

Assign the Digital On Your Own for
- built-in student supports
- Actionable Item Reports
- Standards Analysis Reports

On Your Own

Assignment Guide

The chart below indicates which problems in the On Your Own are associated with each task in the Learn Together. Assign daily homework for tasks completed.

Learn Together Tasks	On Your Own Problems
Task 1, p. 650	Problems 6, 7, 11, and 12
Task 2, p. 651	Problems 8–10, 13, and 19
Task 3, p. 652	Problems 10 and 14–18

 Check Understanding

Formative Assessment

Use formative assessment to determine if your students are successful with this lesson's learning objective.

Students who successfully complete the Check Understanding can continue to the On Your Own practice.

For students who miss 1 problem or more, work in a pulled small group using the Almost There small-group activity on page 649C.

ONLINE

Assign the Digital Check Understanding to determine
- success with the learning objective
- items to review
- grouping and differentiation resources

4 Differentiation Options

Differentiate instruction for all students using small-group activities and math center activities on page 649C.

Reteach

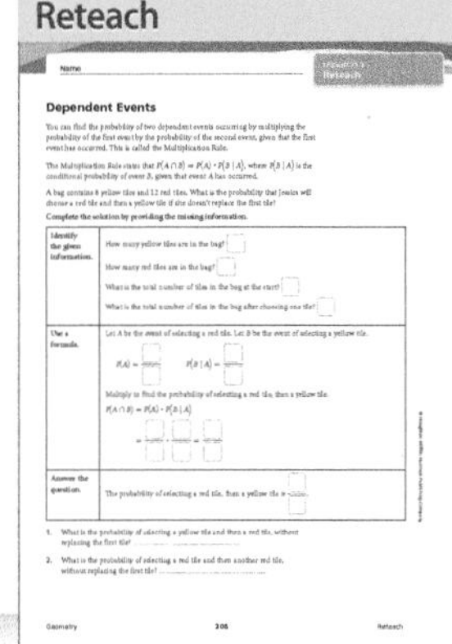

Challenge

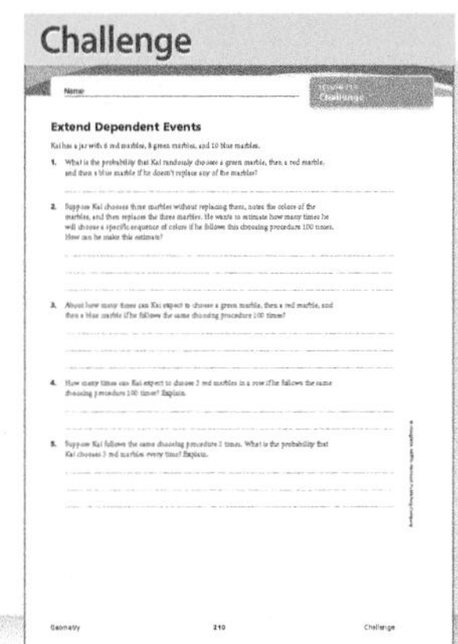

Watch for Common Errors

Problem 9 Some students may not understand what Part B is asking. Point out that they are finding the probability that the two members selected to be interviewed are stage hands. Another concern is that if students did not answer Part A correctly, they will not be able to answer Part B.

Questioning Strategies

Problem 10 If most students answered Part B incorrectly, you may want to solve Part A with them. Then ask them to answer Part B, making sure they understand what Part B is asking. yes; For any specific dog, the probability that it will be walked first is $\frac{1}{4}$. The solution to Part A was not specific to Skipper; the same probability would hold for any specific dog to be walked second. The probability that any specific dog is walked third is the product of the probabilities that it will not be walked first, will not be walked second given that it was not walked first, and will be walked third given that it was not walked first or second: $\frac{3}{4} \cdot \frac{2}{3} \cdot \frac{1}{2} = \frac{1}{4}$. After three dogs have been walked, there is only one choice for the dog to be walked fourth. The probability that any specific dog will be walked fourth is the product of the probabilities that it will not be walked first, will not be walked second given that it was not walked first, will not be walked third given that it was not walked first or second, and will be walked fourth given that it was not walked first, second, or third: $\frac{3}{4} \cdot \frac{2}{3} \cdot \frac{1}{2} \cdot 1 = \frac{1}{4}$.

ANSWERS

7A. dependent; $P(D) = \frac{3}{10}$, $P(A) = \frac{11}{25}$, $P(D \cap A) = \frac{6}{25}$; Since $P(D) \cdot P(A) \neq P(D \cap A)$, the events are dependent.

7B. $P(D|A) = \frac{6}{11}$, $P(A|D) = \frac{4}{5}$; Multiply $P(A)$ and $P(D|A)$, or multiply $P(D)$ and $P(A|D)$ to find $P(D \cap A)$.

10. yes; For any specific dog, the probability that it will be walked first is $\frac{1}{4}$. The solution to Part A was not specific to Skipper; the same probability would hold for any specific dog to be walked second. The probability that any specific dog is walked third is the product of the probabilities that it will not be walked first, will not be walked second given that it was not walked first, and will be walked third given that it was not walked first or second: $\frac{3}{4} \cdot \frac{2}{3} \cdot \frac{1}{2} = \frac{1}{4}$. After three dogs have been walked, there is only one choice for the dog to walk fourth. The probability that any specific dog will be walked fourth is the product of the probabilities that it will not be walked first, will not be walked second given that it was not walked first, will not be walked third given that it was not walked first or second, and will be walked fourth given that it was not walked first, second, or third: $\frac{3}{4} \cdot \frac{2}{3} \cdot \frac{1}{2} \cdot 1 = \frac{1}{4}$.

---

7. Geography The two-way frequency table shows the number of states by area and population density.

	Population density less than 200 people per square mile	Population density greater than or equal to 200 people per square mile	Total
Land area less than 50,000 square miles	10	12	22
Land area greater than or equal to 50,000 square miles	25	3	28
Total	35	15	50

A. Let D be the event that a state has a population density greater than or equal to 200 people per square mile. Let A be the event that a state has an area less than 50,000 square miles. Are events D and A independent or dependent? Explain how you know. A, B. See margin.

B. Find $P(D \mid A)$ and $P(A \mid D)$. How can you use these probabilities to find $P(A \cap D)$?

8. A box of cereal bars contains 4 blueberry bars and 4 apple bars. Jason reaches into the box and takes a cereal bar. Then Amanda reaches into the box and takes a bar. What is the probability that both Jason and Amanda take a blueberry bar? Let $B1$ be the event that Jason takes a blueberry bar and $B2$ be the event that Amanda takes a blueberry bar.

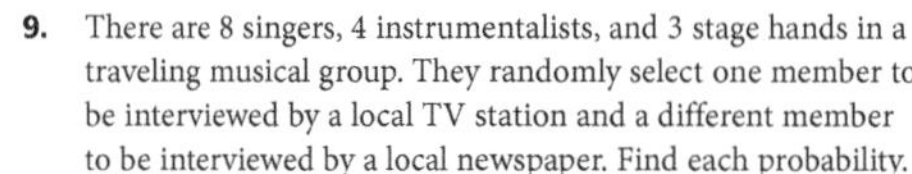

9. There are 8 singers, 4 instrumentalists, and 3 stage hands in a traveling musical group. They randomly select one member to be interviewed by a local TV station and a different member to be interviewed by a local newspaper. Find each probability.

A. the member interviewed by the TV station is a singer and the member interviewed by the newspaper is an instrumentalist $\frac{16}{105}$

B. both members are stage hands $\frac{1}{35}$

10. Andrew is a volunteer dog walker at an animal shelter. He has been asked to walk Rex, Lulu, Spot, and Skipper today, one at a time, in any order he chooses.

A. If he randomly chooses the order, what is the probability that Skipper will be second? $\frac{1}{4}$

B. Is the probability that any specific dog will be walked in any specific order position the same as the probability you found in Part A? Justify your answer. See margin.

Andrew walks four dogs today in any order he chooses.

654

11. **Reason** If $P(A) = P(B \cap A)$, what can you conclude? Give an example of events where $P(A) = P(B \cap A)$. 11–13. See margin.

12. **Open Ended** In your own words, describe the difference between independent and dependent events. Give an example of two events that are independent and two events that are dependent.

13. **Critique Reasoning** Robert solved the problem shown at the right. His work is shown below. Is his answer correct? Explain your answer.

> There are 4 girls (Ana, Charise, Greta, and Jane) and 3 boys (Ethan, Pablo, and Tom) in the Spanish club. If they randomly select a student to be president and a different student to be secretary, what is the probability that a girl will be the president and Greta will be the secretary?

Let A be the event that a girl is president. Let B be the event that Greta is secretary. There are 4 girls out of 7 total members, so $P(A) = \frac{4}{7}$. If a girl is selected as president, there are 6 members left, and Greta is one of them, so $P(B|A) = \frac{1}{6}$. $P(A \cap B) = P(A) \cdot P(B|A) = \frac{4}{7} \cdot \frac{1}{6} = \frac{2}{21}$.

14. You randomly select three tiles, one at a time without replacement, from a bag containing the tiles shown.

   A. Let B be the event that the first tile is B, G be the event that the second tile is G, and A be the event that the third tile is A. What is $P(B \cap G \cap A)$? $\frac{1}{120}$

   B. Let V be the event that the first tile is a vowel, $C1$ be the event that the second tile is a consonant, and $C2$ be the event that the third tile is a consonant. What is $P(V \cap C1 \cap C2)$? $\frac{1}{5}$

15. Ann, Ben, Chandra, Diondre, Emma, Franklin, and Gina will be giving presentations in English class today. The teacher randomly selects the order in which the students will give their presentations. A–C. See margin.

   A. Show how to find the probability that Emma will be one of the first three students by computing the probability of the complement and subtracting from 1.

   B. Show how to find the probability that Emma will be one of the first three students by adding the probabilities that (1) Emma will be first, (2) Emma will be second, and (3) Emma will be third.

   C. Which method do you prefer? Why?

16. A basketball coach randomly assigns players to Team A or Team B for a practice game. She writes "A" on 10 slips of paper and "B" on 10 slips of paper and then places the slips in a bag. The players take turns removing one slip of paper to be assigned to a team. What is the probability that the first three players will be assigned to the same team? $\frac{4}{19}$

Questioning Strategies

Problem 11 Ask how the answer would change if the hypothesis were $P(B) = P(B \cap A)$. Possible answer: You can conclude that $P(A \mid B) = 1$, which means that event A always occurs if event B occurs.

Watch for Common Errors

Problem 12 Students may try to describe independent and dependent events using criteria that determines which they are. The description should not include any equations.

ANSWERS

11. You can conclude that $P(B|A) = 1$, which means that event B always occurs if event A occurs. For example, you roll a number cube. Let event A be rolling a 4, and let event B be rolling an even number. If you roll a 4, you also roll an even number.

12. For independent events, you do not need to know whether one occurred to determine the probability of the other. For dependent events, you do need to know whether one occurred to determine the probability of the other. Independent events: flipping a coin and pulling a tile from a bag of tiles; dependent events: pulling a tile from a bag of tiles, then pulling a second tile from a bag of tiles

13. no; Event A should be the event that a girl other than Greta is president. Since the president and secretary are different, Greta can only be the secretary if she is not the president. If event A is the event that a girl other than Greta is president, and event B is the event that Greta is secretary, then $P(A) = \frac{3}{7}$, $P(B|A) = \frac{1}{6}$, and $P(A \cap B) = \frac{1}{14}$.

15A. Let $N1$ be the event that Emma is not first, $N2$ be the event that Emma is not second, and $N3$ be the event that Emma is not third. $P(N1) = \frac{6}{7}$, $P(N2|N1) = \frac{5}{6}$, $P(N3|N1 \cap N2) = \frac{4}{5}$, $P(N1 \cap N2 \cap N3) = \frac{6}{7} \cdot \frac{5}{6} \cdot \frac{4}{5} = \frac{4}{7}$. The probability that Emma will be one of the first three students is $1 - \frac{4}{7} = \frac{3}{7}$.

15B. Let F be the event that Emma is first: $P(F) = \frac{1}{7}$. Let $N1$ be the event that Emma is not first and S be the event that Emma is second: $P(N1) = \frac{6}{7}$, $P(S|N1) = \frac{1}{6}$, $P(N1 \cap S) = \frac{1}{7}$. Let $N1$ be the event that Emma is not first, $N2$ be the event that Emma is not second, and T be the event that Emma is third: $P(N1) = \frac{6}{7}$, $P(N2|N1) = \frac{5}{6}$, $P(T|N1 \cap N2) = \frac{1}{5}$, $P(N1 \cap N2 \cap T) = \frac{1}{7}$. Probability that Emma will be one of the first three: $\frac{1}{7} + \frac{1}{7} + \frac{1}{7} = \frac{3}{7}$

15C. Answers will vary.

5 Wrap-Up

Summarize learning with your class. Consider using the Exit Ticket, Put It in Writing, or I Can scale.

Exit Ticket 

A set of cards each has one of the 26 letters of the alphabet on them: 5 vowels and 21 consonants. Padma selects a card and gives it to Brian. Padma selects another card and gives it to Brian. What is the probability that the two cards given to Brian were vowels? $P(VV) = \frac{5}{26} \cdot \frac{4}{25} = \frac{20}{650} = \frac{2}{65}$

Put It in Writing 

Explain how selecting two marbles from a bag of marbles with replacement is different than selecting two marbles without replacement.

I Can 

The scale below can help you and your students understand their progress on a learning goal.

4	I can determine whether two events are dependent and find their probabilities, and I can explain my steps to others.
3	I can determine whether two events are dependent and find their probabilities.
2	I can determine whether two events are dependent.
1	I can find probabilities of simple events.

Spiral Review • Assessment Readiness

These questions will help determine if students have retained information taught in the past and can also prepare them for high-stakes assessments. Here, students determine whether events are independent **(21.1)**, determine mutually exclusive events **(20.2)**, and determine conditional probabilities **(21.1)**.

17. A bag contains 3 red marbles, 3 green marbles, and 2 blue marbles. You remove 3 marbles, one at a time, without replacement. What is the probability that the first marble is blue, the second marble is green, and the third marble is not red? $\frac{3}{56}$

18. (MP) **Construct Arguments** Given that $P(A \cap B \cap C) = P((A \cap B) \cap C)$, use the Multiplication Rule for two dependent events to prove the Multiplication Rule for three dependent events. 18, 19. See Additional Answers.

19. (Open Middle™) Using the digits 1 to 9, at most one time each, fill in the boxes to find the conditions with the greatest probability.

There is a bag of ☐ marbles with ☐ blue marbles and ☐ white marbles.

The probability of getting 1 white marble and 1 blue marble without replacement (rounding to the nearest percent) is ☐☐%.

Spiral Review • Assessment Readiness

20. If $P(A) = \frac{1}{2}$ and $P(B) = \frac{1}{6}$, which probability shows that A and B are independent?

Ⓐ $P(B \mid A) = \frac{2}{3}$
Ⓑ $P(A \mid B) = \frac{1}{6}$
Ⓒ $P(A \cap B) = \frac{1}{12}$
Ⓓ $P(A \cap B) = \frac{1}{3}$

21. You randomly pull a card from a deck of playing cards. Which events are mutually exclusive? Select all that apply.

Ⓐ pulling a 7 and pulling a heart
Ⓑ pulling a 5 and pulling a face card
Ⓒ pulling a red card and pulling a number card
Ⓓ pulling a club and pulling a number card divisible by 5
Ⓔ pulling a heart and pulling a spade
Ⓕ pulling a black card and pulling a face card.

22. You roll a 10-sided number cube with the numbers 2 through 11. Let A be the event that the number is even. Let B be the event that the number is prime. Let C be the event that the number is divisible by 4. Match the probability on the left with its value on the right.

Probability	Value	
A. $P(A \mid B)$	**1.** $\frac{2}{5}$	A. 4
B. $P(B \mid C)$	**2.** 1	B. 3
C. $P(C \mid A)$	**3.** 0	C. 1
D. $P(A \mid C)$	**4.** $\frac{1}{5}$	D. 2

 **I'm in a Learning Mindset!**

What different strategies can I use to find the probability of dependent events?

Keep Going Journal and Practice Workbook

Learning Mindset

Perseverance Collects and Tries Multiple Strategies

Point out that effective learning is achieved by collecting and applying multiple strategies. Strategies may include Spark Your Learning, Build Understanding, Step It Out, Turn and Talk, Check Understanding, and On Your Own practice. Ask students if they think that they effectively made use of these strategies in learning about dependent events. *Are my efforts in learning being effective? Did Spark Your Learning make me activate my prior knowledge of independent events and make me curious about dependent events? Am I an active participant in Turn and Talk? How well does Turn and Talk help me understand dependent events? Do I complete the task in Build Understanding and Step It Out? How well do I understand those tasks? How did I do in Check Understanding? Do I practice enough? How does lack of organization affect my learning outcomes? What steps can I take to manage my time better?*

Module 21 Review

Conditional Probability

At Lakeview High, juniors and seniors must join one of three academies.

	Juniors	Seniors	Total
Engineering	103	142	245
Medical	173	228	401
Arts	91	137	228
Total	367	507	874

To find the probability that a student is in the medical academy given that the student is a senior, P(medical | senior), divide the number of seniors who are in the medical academy by the total number of seniors: $\frac{228}{507} \approx 45.0\%$. To find P(junior | arts), divide the number of juniors in the arts academy by the total number of students in the arts academy: $\frac{91}{228} \approx 39.9\%$.

Independent Events

Events A and B are independent if and only if $P(A \cap B) = P(A) \cdot P(B)$.

the probability that a student is a junior and in the engineering academy

$P(J \cap E) \stackrel{?}{=} P(J) \cdot P(E)$

$\frac{103}{874} \stackrel{?}{=} \frac{367}{874} \cdot \frac{245}{874}$

$0.118 = 0.118$

Being a junior and being in the engineering academy are independent.

At a university near Lakeview High, three-fifths of all students live on campus. Living on campus, C, and a student's major are independent events. From Lakeview's senior class, 78 students have been accepted into the university's engineering program.

the probability that a student in the senior class will major in engineering and live on campus

$P(E \cap C) = P(E) \cdot P(C)$

$= \frac{78}{507} \cdot \frac{3}{5} \approx 9.2\%$

Dependent Events

The university has a large summer internship program for high school students considering careers in the medical field. The university only accepts 12% of the students who apply. The funding sources for the program require that exactly 25% of the accepted students are high school juniors.

the probability that a student who applies is accepted and is a junior

$P(A \cap J) = P(A) \cdot P(J \mid A)$

$= (0.12) \cdot (0.25)$

$= 0.03$, or 3%

Module 21 657

ONLINE

Assign the Digital Module Review for

- built-in student supports
- Actionable Item Reports
- Standards Analysis Reports

MODULE 21 REVIEW

Module Review

Use the first page of the Module Review to summarize and connect the main ideas of the module. Use the second page to assess students' understanding of the vocabulary, concepts, and skills presented in the module.

Sample Guided Discussion:

Q Conditional Probability What group could be represented by the probability $\frac{142}{245} \approx 57.96\%$? Explain. The probability represents the number of seniors who are in the engineering academy because 142 is the number of seniors and the total number of engineering students is 245.

Q Independent Events Are the events that a student in the senior class and that a student is in the medical academy considered independent or dependent? Explain. $P(\text{senior} \cap \text{medical}) = \frac{228}{874} \approx 26.1\%$ and $P(\text{senior}) \cdot P(\text{medical}) = \frac{507}{874} \cdot \frac{401}{874} \approx 26.6\%$. The probabilities are close enough that the events can be considered independent.

Q Dependent Events Why are the events that a student who applies is accepted to the summer internship program and that a student who applies is a junior dependent? The probability that an accepted student is a junior is fixed at 25%, but the probability that a student that applies is a junior is not necessarily 25%, so $P(\text{junior}) \neq P(\text{junior} \mid \text{accepted})$. The events are dependent.

Module Review continued

Possible Scoring Guide

Items	Points	Description
1–4	2 each	identifies the correct term
5–8	2 each	correctly finds the probabilities based on the spinner
9	2	correctly determines an independent event
10	1	correctly explains what it means for an event to be independent or dependent and explains how the events are independent or dependent
10	2	correctly explains what it means for an event to be independent or dependent, explains how the events are independent or dependent, and determines the probability of the events
11	2	correctly finds the probability that the first marble is yellow and the next two marbles are red
Total points possible = 22 points		

The Unit 10 Test in the Assessment Guide assesses content from Modules 20 and 21.

Vocabulary

Choose the correct term from the box to complete each sentence.

Vocabulary
conditional probability dependent events independent events relative frequency

1. Two events for which the occurrence of one event does not influence the occurrence of the other are __?__. independent events
2. For two events A and B, the probability that event B occurs given that event A has already occurred is the __?__ of B given A and is written $P(B \mid A)$. conditional probability
3. __?__ is the frequency of one outcome divided by the frequency of all outcomes. relative frequency
4. Events are __?__ if the occurrence of one event affects the probability of the other. dependent events

Concepts and Skills

You create a game that uses the spinner shown. Find each probability.

5. $P(\text{blue} \mid \text{even})$ $\frac{1}{4}$
6. $P(\text{red} \mid \geq 7)$ 0
7. $P(< 6 \mid \text{yellow})$ $\frac{2}{3}$
8. $P(\text{blue} \mid \text{odd})$ $\frac{1}{3}$
9. $P(\text{odd} \mid \text{green})$ 1
10. $P(\text{green} \mid \text{even})$ 0

11. Students attending a summer camp get to choose between climbing a rock wall or practicing archery. Determine whether the event that a student chooses archery is independent of the event that the student is 10–12 years old. Show your work.
11, 12. See Additional Answers.

	Rock wall	Archery	Total
7–9 years	45	32	77
10–12 years	38	27	65
Total	83	59	142

12. A student polled her class. Let M be the event that a student has more than two siblings and C be the event that the student takes a drafting class.
    - **A.** What would it mean for M and C to be dependent or independent in this context?
    - **B.** Determine $P(M \cap C)$.
    - **C.** Are M and C independent? Explain your reasoning.

	Drafting class	No drafting class	Total
≤ 2 siblings	109	53	162
> 2 siblings	16	31	47
Total	125	84	209

13. (MP) **Use Tools** There are 12 yellow, 21 blue, and 17 red marbles in a bag. You take three marbles out of the bag, one at a time, without replacement. What is the probability that the first marble is yellow and the next two marbles are red? State what strategy and tool you will use to answer the question, explain your choice, and then find the answer. $\frac{34}{1225} \approx 0.028$

658

DATA-DRIVEN INSTRUCTION

Before moving on to the Module Test, use the Module Review results to intervene based on the table below.

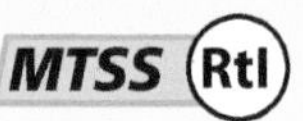

Items	Lesson	DOK	Content Focus	Intervention
5–8	21.1	2	Find probabilities.	Reteach 21.1
9	21.2	3	Determine if an event is independent in a real-world context.	Reteach 21.2
10	21.2, 21.3	3	Explain and calculate events that are independent or dependent in a real-world context.	Reteach 21.2, 21.3
11	21.3	3	Calculate the probability of dependent events in a real-world context.	Reteach 21.3

Module Test

data checkpoint

The Module Test is available in alternative versions in your Assessment Guide. All items are presented in standardized test formats.

ONLINE

Ed

Assign the Digital Module Test to power actionable reports including

- proficiency by standards
- item analysis

MODULE 21

TEST

Form A

Name

Module 21 • Form A
Module Test

1. The probability of event A occurring is 0.25. The probability of event B occurring is 0.60. If the probability of both events A and B occurring is 0.15, are events A and B independent?

   (A) Yes, because $P(A) \cdot P(B) = P(A \text{ and } B)$.
   (B) Yes, because $P(A) \cdot P(B) \neq P(A \text{ and } B)$.
   (C) No, because $P(A) \cdot P(B) = P(A \text{ and } B)$.
   (D) No, because $P(A) \cdot P(B) \neq P(A \text{ and } B)$.

2. Events A and B are independent. If $P(A) = 0.32$ and $P(B) = 0.63$, what is the approximate probability of events A and B occurring together?

   (A) 0.20 (C) 0.51
   (B) 0.31 (D) 0.95

3. For events C and D, the probability that event C will occur is 38%, and the probability that event D occurs if event C occurs is 17%. What is the probability that both events will occur?

   (A) 5.5% (C) 21%
   (B) 6.5% (D) 55%

4. A bicycle lock has three dials. Each dial has positions labeled with the digits 0 through 9. The locks are given randomly generated three-digit combinations when they are manufactured. What is the probability that a lock combination has three even digits?

   (A) 0.5 (C) 0.083
   (B) 0.27 **(D) 0.125**

5. If it rains, there is a 20% chance that the bus will be late. The weather forecast says that there is a 60% chance of rain today. What is the probability that the bus will be late today?

   (A) 80% (C) 33%
   (B) 40% **(D) 12%**

6. A student reports the following statement:

   The probability that a tropical storm develops into a hurricane, given that the storm is in the North Atlantic Ocean, is greater than the probability that a tropical storm develops into a hurricane, given that the storm is in the South Atlantic Ocean.

   Which statement correctly interprets the conditional probability described?

   (A) Tropical storms in the North Atlantic Ocean are less likely to develop into hurricanes than tropical storms in the South Atlantic Ocean.
   (B) Tropical storms in the North Atlantic Ocean are more likely to develop into hurricanes than tropical storms in the South Atlantic Ocean.
   (C) Tropical storms that develop into hurricanes are less likely to be in the North Atlantic Ocean than tropical storms that don't develop into hurricanes.
   (D) Tropical storms that develop into hurricanes are more likely to be in the North Atlantic Ocean than tropical storms that don't develop into hurricanes.

7. Students at a school were asked to write down whether they kick a ball with their left or right foot and whether or not their favorite color is a primary color. The results are in the two-way table.

	Primary	Not Primary	Total
Left	10	10	20
Right	140	40	180
Total	150	50	200

   Let event L be that a student kicks with the left foot and event N be that a student's favorite color is not a primary color. Are L and N independent?

   (A) Yes, because $P(L) \cdot P(N) = P(L \text{ and } N)$.
   (B) Yes, because $P(L) \cdot P(N) \neq P(L \text{ and } N)$.
   (C) No, because $P(L) \cdot P(N) = P(L \text{ and } N)$.
   (D) No, because $P(L) \cdot P(N) \neq P(L \text{ and } N)$.

Form A

Module 21 • Form A
Module Test

Name

8. If $P(A) = \frac{7}{10}$, $P(B) = \frac{2}{3}$, and $P(A \text{ and } B) = \frac{1}{2}$, what is the probability of B given A?

   Possible answer: $\frac{5}{7}$

9. For events C and D, the probability that event C will occur is 19% and the probability that event D occurs if event C occurs is 42%. What is the probability that both events will occur?

   Possible answer: 7.98%

10. A student has a deck of cards that consists of 13 red cards, 13 blue cards, 13 green cards, and 13 yellow cards. The student draws a card from the deck and then draws two more cards at random. What is the probability that the second and third cards are the same color as the first?

    Possible answer: $\frac{22}{425}$

11. A college preparation company holds a meeting for high school juniors. Each student at the meeting is given a ticket for a raffle with one prize of a free one-on-one college counseling session. Five chairs at the meeting have an envelope under them that contains a coupon for a free test-prep book. There are 40 students at the meeting. What is the probability that a student wins both the one-on-one college counseling session and a test-prep book?

    Possible answer: $\frac{1}{320}$

12. At a picnic, a basket of utensils holds 14 forks, 5 knives, and 6 spoons. A guest takes two utensils without looking. If the first utensil is a fork, what is the probability that the second utensil is a spoon?

    Possible answer: 0.25

13. At a high school, 18% of seniors study Spanish and 12% of seniors study Spanish and French. What is the probability that a randomly selected senior who studies Spanish also studies French?

    Possible answer: $\frac{2}{3}$

14. A bag of marbles has 8 blue marbles and 8 yellow marbles.

    Place an X in the table to show whether each pair of events is independent or dependent.

	Independent	Dependent
Drawing a blue marble, then drawing a yellow marble after replacing the first marble.	X	
Drawing a blue marble, then drawing a yellow marble without replacing the first marble.		X
Drawing a blue marble, then drawing another blue marble without replacing the first marble.		X

15. A survey asked 100 high school students if they played basketball on a regular basis and if they planned to watch a nationally televised basketball game tonight. The results are shown in the table.

	Plans to Watch	Does Not Plan to Watch	Total
Plays Basketball	14	13	27
Does Not Play Basketball	26	47	73
Total	40	60	100

    Part A

    What is the probability that a randomly selected student who plays basketball regularly plans to watch the game tonight, to the nearest percent?

    Possible answer: 52%

    Part B

    What is the probability that a randomly selected student who does not plan to watch the game tonight does not play basketball regularly, to the nearest percent?

    Possible answer: 78%

Form B

Name

Module 21 • Form B
Module Test

1. The probability of event J occurring is 0.35. The probability of event K occurring is 0.40. If the probability of both events J and K occurring is 0.09, are events J and K independent?

   (A) Yes, because $P(J) \cdot P(K) = P(J \text{ and } K)$.
   (B) Yes, because $P(J) \cdot P(K) \neq P(J \text{ and } K)$.
   (C) No, because $P(J) \cdot P(K) = P(J \text{ and } K)$.
   (D) No, because $P(J) \cdot P(K) \neq P(J \text{ and } K)$.

2. Events L and M are independent. If $P(L) = 0.82$ and $P(M) = 0.60$, what is the approximate probability of events L and M occurring together?

   (A) 0.22 (C) 0.73
   (B) 0.49 (D) 0.82

3. For events A and B, the probability that event A will occur is 69%, and the probability that event B occurs if event A occurs is 28%. What is the probability that both events will occur?

   (A) 10% (C) 41%
   (B) 19% (D) 97%

4. Two friends are playing a card game. Each friend has a deck of cards that consists of 52 cards: 26 cards are yellow and lettered from A to Z; 26 cards are blue and lettered from A to Z. If the friends pick one card from their respective deck of cards without looking, what is the probability that both friends pick a yellow T card?

   (A) 3.8% **(C) 0.037%**
   (B) 3.9% (D) 0.038%

5. The Sharks are playing in the baseball league championship. The Tigers have a 50% chance of playing in the championship game. If the Sharks play the Tigers, the Sharks have a 20% chance of winning. What is the probability that the Sharks will play the Tigers and win the championship?

   (A) 10% (C) 40%
   (B) 30% (D) 70%

6. A student reports the following statement:

   The probability that a dog has two different color eyes, given the dog is a beagle, is greater than the probability that a dog has two different color eyes, given the dog is a poodle.

   Which statement correctly interprets the conditional probability described?

   (A) Dogs with two different color eyes are less likely to be beagles than poodles.
   (B) Dogs with two different color eyes are more likely to be beagles than poodles.
   (C) Dogs that are beagles are less likely than dogs that are poodles to have two different color eyes.
   (D) Dogs that are beagles are more likely than dogs that are poodles to have two different color eyes.

7. A high school received a grant to make improvements to its gymnasium. Students at the high school from each grade voted on whether to spend the grant money on a new scoreboard or on new bleachers. The results are shown in the two-way table.

	New Scoreboard	New Bleachers	Total
9th	45	9	54
10th	55	11	66
11th	40	8	48
12th	30	6	36
Total	170	34	204

   Let event N be that a student is a 9th grader. Let event B be that a student voted to get new bleachers. Are events N and B independent?

   (A) Yes, because $P(N) \cdot P(B) = P(N \text{ and } B)$.
   (B) Yes, because $P(N) \cdot P(B) \neq P(N \text{ and } B)$.
   (C) No, because $P(N) \cdot P(B) = P(N \text{ and } B)$.
   (D) No, because $P(N) \cdot P(B) \neq P(N \text{ and } B)$.

Form B

Module 21 • Form B
Module Test

Name

8. If $P(A) = \frac{1}{2}$, $P(B) = \frac{7}{20}$, and $P(A \text{ and } B) = \frac{3}{20}$, what is the probability of A given B?

   Possible answer: $\frac{3}{7}$

9. For events C and D, the probability that event C will occur is 56%, and the probability that event D occurs if event C occurs is 32%. What is the probability that both events will occur?

   Possible answer: 17.92

10. There are 15 students on the chess team: 2 freshmen, 3 sophomores, 4 juniors, and 6 seniors. Three of the students are randomly selected to represent the team at a tournament. What is the probability that all three students are seniors?

    Possible answer: $\frac{4}{91}$

11. A student creates a playlist of 50 songs. The playlist contains 6 songs by the student's favorite artist. Each time the student shuffles the playlist, there is an equal chance that any one of the 50 songs will play first. The student shuffles the playlist twice on Monday. What is the probability that a song by the student's favorite artist will be the first song played both times?

    Possible answer: $\frac{9}{625}$

12. A cooler contains an assortment of juice boxes: 8 apple, 6 grape, 7 lemonade, and 5 fruit punch. If someone takes two juice boxes from the cooler without looking and the first juice box is grape, what is the probability that the second juice box is lemonade?

    Possible answer: $\frac{7}{25}$

13. At a high school, 16% of seniors study chemistry and 12% of seniors study physics and chemistry. What is the probability that a randomly selected student who studies chemistry also studies physics?

    Possible answer: 0.75

14. A bag of marbles contains 6 red marbles, 6 black marbles, and 6 green marbles.

    Place an X in the table to show whether each pair of events is independent or dependent.

	Independent	Dependent
Drawing a black marble, then drawing a green marble without replacing the first marble.		X
Drawing a red marble, then drawing a green marble after replacing the first marble.	X	
Drawing a green marble, then drawing another green marble after replacing the first marble.	X	

15. A survey asked 100 tea drinkers if they drink tea with milk and if they drink tea with sugar. The results are shown in the table below.

	Sugar	No Sugar	Total
Milk	18	28	46
No Milk	39	15	54
Total	57	43	100

    Part A

    What is the probability that a randomly selected tea drinker who drinks tea with no milk drinks tea with sugar, to the nearest percent?

    Possible answer: 72%

    Part B

    What is the probability that a randomly selected tea drinker who drinks tea with no sugar drinks tea with milk, to the nearest percent?

    Possible answer: 65%

A

English	Spanish	Examples
acute angle An angle that measures greater than 0° and less than 90°.	**ángulo agudo** Ángulo que mide más de 0° y menos de 90°.	
adjacent angles Two angles in the same plane with a common vertex and a common side, but no common interior points.	**ángulos adyacentes** Dos ángulos en el mismo plano que tienen un vértice y un lado común pero no comparten puntos internos.	$\angle 1$ and $\angle 2$ are adjacent angles.
adjacent arcs Two arcs of the same circle that intersect at exactly one point, which is a shared endpoint.	**arcos adyacentes** Dos arcos del mismo círculo que se cruzan en un punto exacto, que es un extremo compartido.	$\overset{\frown}{RS}$ and $\overset{\frown}{ST}$ are adjacent arcs.
alternate exterior angles For two lines intersected by a transversal, a pair of angles that lie on opposite sides of the transversal and outside the intersected lines.	**ángulos alternos externos** Dadas dos líneas cortadas por una transversal, par de ángulos no adyacentes ubicados en los lados opuestos de la transversal y fuera de las otras dos líneas.	$\angle 4$ and $\angle 5$ are alternate exterior angles.
alternate interior angles For two lines intersected by a transversal, a pair of nonadjacent angles that lie on opposite sides of the transversal and between the intersected lines.	**ángulos alternos internos** Dadas dos líneas cortadas por una transversal, par de ángulos no adyacentes ubicados en los lados opuestos de la transversal y entre las líneas cortadas.	$\angle 3$ and $\angle 6$ are alternate interior angles.
altitude of a triangle A perpendicular segment from a vertex to the line containing the opposite side.	**altura de un triángulo** Segmento perpendicular que se extiende desde un vértice hasta la línea que forma el lado opuesto.	
angle bisector A ray that divides an angle into two congruent angles.	**bisectriz de un ángulo** Rayo que divide un ángulo en dos ángulos congruentes.	$\overrightarrow{JK}$ is an angle bisector of $\angle LJM$.

English	Spanish	Examples
angle of depression The angle formed by a horizontal line and a line of sight to a point below.	**ángulo de depresión** Ángulo formado por una línea horizontal y una línea de visión dirigida a un punto ubicado por debajo de la primera.	
angle of elevation The angle formed by a horizontal line and a line of sight to a point above.	**ángulo de elevación** Ángulo formado por una línea horizontal y una línea de visión dirigida a un punto ubicado por encima de la primera.	
angle of rotation The angle through which a figure is turned when a rotation is applied.	**ángulo de rotación** Ángulo a través del cual una figura gira cuando se aplica una rotación.	
angle of rotational symmetry The smallest angle through which a figure with rotational symmetry can be rotated to coincide with itself.	**ángulo de simetría de rotación** El ángulo más pequeño alrededor del cual se puede rotar una figura con simetría de rotación para que coincida consigo misma.	
arc An unbroken part of a circle consisting of two points on the circle, called the endpoints, and all the points on the circle between them.	**arco** Parte continua de una circunferencia formada por dos puntos de la circunferencia denominados extremos y todos los puntos de la circunferencia comprendidos entre éstos.	
arc length The distance along an arc measured in linear units.	**longitud de arco** Distancia a lo largo de un arco medida en unidades lineales.	
area The surface contained within the boundaries of a two-dimensional object such as a triangle, rectangle, or trapezoid.	**área** Superficie comprendida dentro de los límites de un objeto bidimensional, como un triángulo, un rectángulo o un trapecio.	

English	Spanish	Examples
auxiliary line A line drawn in a figure to aid in a proof.	**línea auxiliar** Línea dibujada en una figura como ayuda en una demostración.	Auxiliary line; B; ℓ; 4; 2; 5; 1; 3; A; C
axiom *See* postulate.	**axioma** *Ver* postulado.	
axis of a cone The segment with endpoints at the vertex and the center of the base.	**eje de un cono** Segmento cuyos extremos se encuentran en el vértice y en el centro de la base.	Axis
axis of a cylinder The segment with endpoints at the centers of the two bases.	**eje de un cilindro** Segmentos cuyos extremos se encuentran en los centros de las dos bases.	Axis

B

English	Spanish	Examples
base angle of a trapezoid One of a pair of consecutive angles whose common side is a base of the trapezoid.	**ángulo base de un trapecio** Uno de los dos ángulos consecutivos cuyo lado en común es la base del trapecio.	Base angles; B; C; A; D; Base angles
base angle of an isosceles triangle One of the two angles that have the base of the triangle as a side.	**ángulo base de un triángulo isósceles** Uno de los dos ángulos que tienen como lado la base del triángulo.	A; Base angle; Base angle; B; C
base of a geometric figure A side of a polygon; a face of a three-dimensional figure, by which the figure is measured or classified.	**base de una figura geométrica** Lado de un polígono; cara de una figura tridimensional por la cual se mide o clasifica la figura.	h; b; Bases
bases of a trapezoid Two sides of a trapezoid that are parallel.	**bases de un trapecio** Los dos lados de un trapecio que son paralelos.	B; C; A; D
between Given three points A, B, and C, B is between A and C if and only if all three of the points lie on the same line and $AB + BC = AC$.	**entre** Dados tres puntos A, B y C, B está entre A y C si y sólo si los tres puntos se encuentran en la misma línea y $AB + BC = AC$.	A; B; C

Glossary/Glosario

English	Spanish	Examples
biconditional statement A statement that can be written in the form "p if and only if q," where p is the hypothesis and q is the conclusion.	**enunciado bicondicional** Enunciado que puede expresarse en la forma "p si y sólo si q", donde p es la hipótesis y q es la conclusión.	A figure is a triangle if and only if it is a three-sided polygon.
bisect To divide a figure into two congruent parts.	**trazar una bisectriz** Dividir una figura en dos partes congruentes.	L K J M $\overrightarrow{JK}$ bisects $\angle LJM$.

C

English	Spanish	Examples
center of dilation The fixed point in the plane that does not change when the dilation is applied.	**centro de dilatación** Punto fijo en el plano que no cambia cuando se aplica la dilatación.	center
center of rotation The point around which a figure is rotated.	**centro de rotación** Punto alrededor del cual rota una figura.	90° 90° Center 90° 90°
central angle of a circle An angle with measure less than or equal to 180° whose vertex is the center of a circle.	**ángulo central de un círculo** Ángulo con medida inferior o igual a 180° cuyo vértice es el centro de un círculo.	
centroid of a triangle The point of concurrency of the three medians of a triangle. Also known as the *center of gravity.*	**centroide de un triángulo** Punto donde se encuentran las tres medianas de un triángulo. También conocido como *centro de gravedad.*	B X P Y A Z C The centroid is P.
chord A segment whose endpoints lie on a circle.	**cuerda** Segmento cuyos extremos se encuentran en un círculo.	A Chord B

English	Spanish	Examples
circle The set of points in a plane that are a fixed distance from a given point called the center of the circle.	**círculo** Conjunto de puntos en un plano que se encuentran a una distancia fija de un punto determinado denominado centro del círculo.	
circumcenter of a triangle The point of concurrency of the three perpendicular bisectors of a triangle.	**circuncentro de un triángulo** Punto donde se cortan las tres mediatrices de un triángulo.	B, P, A, C The circumcenter is *P*.
circumcircle *See* circumscribed circle.	**circuncírculo** *Ver* círculo circunscrito.	
circumference The distance around a circle.	**circunferencia** Distancia alrededor del círculo.	Circumference
circumscribed angle An angle formed by two rays from a common endpoint that are tangent to a circle	**ángulo circunscrito** Ángulo formado por dos semirrectas tangentes a un círculo que parten desde un extremo común.	A, X, C, B
circumscribed circle A circle that intersects all vertices of a polygon and intersects no other points of the polygon.	**círculo circunscrito** Círculo que interseca todos los vértices de un polígono y no interseca otros puntos del polígono.	
circumscribed polygon A polygon in which each side is tangent to a circle drawn in the interior of the polygon.	**polígono circunscrito** Todos los lados del polígono son tangentes al círculo.	
collinear Points that lie on the same line.	**colineal** Puntos que se encuentran sobre la misma línea.	K, L, M *K*, *L*, and *M* are collinear points.
complement of an event All outcomes in the sample space that are not in an event *E*, denoted E^C.	**complemento de un suceso** Todos los resultados en el espacio muestral que no están en el suceso *E* y se expresan E^C.	In the experiment of rolling a number cube, the complement of rolling a 3 is rolling a 1, 2, 4, 5, or 6.

English	Spanish	Examples
complement of set *A* The set of all elements in the universal set U that are not in set A.	**complemento del conjunto A** Conjunto de todos los elementos del conjunto universal U que no están en el conjunto A.	
complementary angles Two angles whose measures have a sum of 90°.	**ángulos complementarios** Dos ángulos cuyas medidas suman 90°.	The complement of a 53° angle is a 37° angle.
component form The form of a vector that lists the vertical and horizontal change from the initial point to the terminal point.	**forma de componente** Forma de un vector que muestra el cambio horizontal y vertical desde el punto inicial hasta el punto terminal.	The component form of $\overrightarrow{CD}$ is $\langle 2, 3 \rangle$.
composite figure A plane figure made up of triangles, rectangles, trapezoids, circles, and other simple shapes, or a three-dimensional figure made up of prisms, cones, pyramids, cylinders, and other simple three-dimensional figures.	**figura compuesta** Figura plana compuesta por triángulos, rectángulos, trapecios, círculos y otras figuras simples, o figura tridimensional compuesta por prismas, conos, pirámides, cilindros y otras figuras tridimensionales simples.	
composition of transformations A transformation that directly maps a preimage to the final image after each image is used as a preimage in a sequence of two or more transformations.	**composición de transformaciones** Transformación que establece una correspondencia directa entre una imagen original y la imagen final tras usar cada imagen como imagen original en una sucesión de dos o más transformaciones.	
compression A nonrigid transformation that changes the shape of a figure in one direction by a factor greater than 0 and less than 1.	**compresión** Transformación no rígida que cambia la forma de una figura en una dirección por un factor mayor que 0 pero menor que 1.	vertical compression
concentric circles Coplanar circles that have the same center.	**círculos concéntricos** Círculos coplanares que tienen el mismo centro.	

English	Spanish	Examples
conclusion The part of a conditional statement following the word *then*.	**conclusión** Parte de un enunciado condicional que sigue a la palabra *entonces*.	If $x + 1 = 5$, then $\underbrace{x = 4}_{\text{Conclusion}}$.
concurrent lines Three or more lines that intersect at one point.	**líneas concurrentes** Tres o más líneas que se cortan en un punto.	
conditional probability The probability of event B, given that event A has already occurred or is certain to occur, denoted $P(B \mid A)$; used to find probability of dependent events.	**probabilidad condicional** Probabilidad del suceso B, dado que el suceso A ya ha ocurrido o es seguro que ocurrirá, expresada como $P(B \mid A)$; se utiliza para calcular la probabilidad de sucesos dependientes.	
conditional statement A statement that can be written in the form "if p, then q," where p is the hypothesis and q is the conclusion.	**enunciado condicional** Enunciado que se puede expresar como "si p, entonces q", donde p es la hipótesis y q es la conclusión.	If $\underbrace{x + 1 = 5}_{\text{Hypothesis}}$, then $\underbrace{x = 4}_{\text{Conclusion}}$.
cone A three-dimensional figure with a circular base and a curved lateral surface that connects the base to a point called the vertex.	**cono** Figura tridimensional con una base circular y una superficie lateral curva que conecta la base con un punto denominado vértice.	
congruence transformation *See* rigid motion.	**transformación de congruencia** *Ver* movimiento rígido.	
congruent Having the same size and shape, denoted by $\cong$.	**congruente** Que tiene el mismo tamaño y la misma forma, expresado por $\cong$.	Q R P S $\overline{PQ} \cong \overline{SR}$
congruent arcs Two arcs that have the same measure and are arcs of the same circle or of congruent circles.	**arcos congruentes** Dos arcos que tienen la misma medida y que son arcos del mismo círculo o de círculos congruentes.	B C A D
congruent circles Two circles that have the same radius.	**círculos congruentes** Dos círculos que tienen el mismo radio.	
congruent Two figures whose corresponding sides and angles are congruent. One figure can be obtained from the other by a sequence of rigid motions.	**polígonos congruentes** Dos figuras cuyos lados y ángulos correspondientes son congruentes. Una figura se puede obtener de la otra mediante una sucesión de movimientos rígidos.	P Q S R Z W Y X

English	Spanish	Examples
conjecture A statement that is believed to be true.	**conjetura** Enunciado que se supone verdadero.	A sequence begins with the terms 2, 4, 6, 8, 10. A reasonable conjecture is that the next term in the sequence is 12.
consecutive exterior angles For two lines intersected by a transversal, a pair of angles that lie on the same side of the transversal and outside the intersected lines.	**ángulos externos consecutivos** Dadas dos líneas cortadas por una transversal, el par de ángulos ubicados en el mismo lado de la transversal y del lado externo de las otras líneas cortadas.	1 2 5 6 3 4 7 8 $\angle 1$ and $\angle 4$ are consecutive exterior angles.
consecutive interior angles For two lines intersected by a transversal, a pair of angles that lie on the same side of the transversal and between the intersected lines.	**ángulos internos consecutivos** Dadas dos líneas cortadas por una transversal, el par de ángulos ubicados en el mismo lado de la transversal y entre las líneas cortadas.	1 2 5 6 3 4 7 8 $\angle 2$ and $\angle 3$ are consecutive interior angles.
contrapositive The statement formed by both exchanging and negating the hypothesis and conclusion of a conditional statement.	**contrarrecíproco** Enunciado que se forma al intercambiar y negar la hipótesis y la conclusión de un enunciado condicional.	Statement: If $n + 1 = 3$, then $n = 2$. Contrapositive: If $n \neq 2$, then $n + 1 \neq 3$.
converse The statement formed by exchanging the hypothesis and conclusion of a conditional statement.	**recíproco** Enunciado que se forma intercambiando la hipótesis y la conclusión de un enunciado condicional.	Statement: If $n + 1 = 3$, then $n = 2$. Converse: If $n = 2$, then $n + 1 = 3$.
coordinate proof A style of proof that uses coordinate geometry and algebra.	**prueba de coordenadas** Tipo de demostración que utiliza geometría de coordenadas y álgebra.	
coplanar Figures that lie in the same plane.	**coplanar** Figuras que se encuentran en el mismo plano.	
corollary A theorem whose proof follows directly from another theorem.	**corolario** Teorema cuya demostración proviene directamente de otro teorema.	
corresponding angles of lines intersected by a transversal For two lines intersected by a transversal, a pair of angles that lie on the same side of the transversal and on the same sides of the intersected lines.	**ángulos correspondientes de líneas cortadas por una transversal** Dadas dos líneas cortadas por una transversal, el par de ángulos ubicados en el mismo lado de la transversal y en los mismos lados de las otras dos líneas.	1 2 5 6 3 4 7 8 $\angle 1$ and $\angle 3$ are corresponding angles.

English	Spanish	Examples
corresponding angles of polygons Angles in the same position in two different polygons that have the same number of angles.	**ángulos correspondientes de los polígonos** Ángulos que tienen la misma posición en dos polígonos diferentes que tienen el mismo número de ángulos.	A B C D E F $\angle A$ and $\angle D$ are corresponding angles.
corresponding sides of polygons Sides in the same position in two different polygons that have the same number of sides.	**lados correspondientes de los polígonos** Lados que tienen la misma posición en dos polígonos diferentes que tienen el mismo número de lados.	A B C D E F $\overline{AB}$ and $\overline{DE}$ are corresponding sides.
cosine In a right triangle, the cosine of $\angle A$ is the ratio of the length of the leg adjacent to $\angle A$ to the length of the hypotenuse. It is the reciprocal of the secant function.	**coseno** En un triángulo rectángulo, el coseno del ángulo A es la razón entre la longitud del cateto adyacente al ángulo A y la longitud de la hipotenusa. Es la inversa de la función secante.	hypotenuse A adjacent $\cos A = \frac{\text{adjacent}}{\text{hypotenuse}} = \frac{1}{\sec A}$
counterexample An example that proves that a conjecture or statement is false.	**contraejemplo** Ejemplo que demuestra que una conjetura o enunciado es falso.	
CPCTC An abbreviation for "Corresponding Parts of Congruent Triangles are Congruent," which can be used as a justification in a proof after two triangles are proven congruent.	**PCTCC** Abreviatura que significa "Las partes correspondientes de los triángulos congruentes son congruentes", que se puede utilizar para justificar una demostración después de demostrar que dos triángulos son congruentes (CPCTC, por sus siglas en inglés).	
cross section The intersection of a three-dimensional figure and a plane.	**sección transversal** Intersección de una figura tridimensional y un plano.	
cylinder A three-dimensional figure with two parallel congruent circular bases and a curved lateral surface that connects the bases.	**cilindro** Figura tridimensional con dos bases circulares congruentes y paralelas y una superficie lateral curva que conecta las bases.	

D

English	Spanish	Examples
deductive reasoning The process of using logic to draw conclusions.	**razonamiento deductivo** Proceso en el que se utiliza la lógica para sacar conclusiones.	

English	Spanish	Examples
definition A statement that describes a mathematical object and can be written as a true biconditional statement.	**definición** Enunciado que describe un objeto matemático y se puede expresar como un enunciado bicondicional verdadero.	
density The amount of matter that an object has in a given unit of volume. The density of an object is calculated by dividing its mass by its volume.	**densidad** La cantidad de materia que tiene un objeto en una unidad de volumen determinada. La densidad de un objeto se calcula dividiendo su masa entre su volumen.	$\text{density} = \frac{\text{mass}}{\text{volume}}$
dependent events Events for which the occurrence or nonoccurrence of one event affects the probability of the other event.	**sucesos dependientes** Dos sucesos son dependientes si el hecho de que uno de ellos se cumpla o no afecta la probabilidad del otro.	From a bag containing 3 red marbles and 2 blue marbles, drawing a red marble, and then drawing a blue marble without replacing the first marble.
diagonal of a polygon A segment connecting two nonconsecutive vertices of a polygon.	**diagonal de un polígono** Segmento que conecta dos vértices no consecutivos de un polígono.	A B C D E Diagonal
diameter A segment that has endpoints on the circle and that passes through the center of the circle; also, the length of that segment.	**diámetro** Segmento que atraviesa el centro de un círculo y cuyos extremos están sobre la circunferencia; longitud de dicho segmento.	
dilation A transformation in which the lines connecting every point P with its preimage P' all intersect at a point C known as the center of dilation, and $\frac{CP'}{CP}$ is the same for every point P; a transformation that changes the size of a figure but not its shape.	**dilatación** Transformación en la cual las líneas que conectan cada punto P con su imagen original P' se cruzan en un punto C conocido como centro de dilatación, y $\frac{CP'}{CP}$ es igual para cada punto P; transformación que cambia el tamaño de una figura pero no su forma.	C D E F D′ E′ F′
disjoint Two events are disjoint, or mutually exclusive, if they cannot occur in the same trial at the same time.	**disjunto** Dos sucesos son disjuntos o mutuamente excluyentes si no pueden ocurrir en la misma prueba al mismo tiempo.	
distance between two points The absolute value of the difference of the coordinates of the points. The length of the shortest line segment that can connect two points.	**distancia entre dos puntos** Valor absoluto de la diferencia entre las coordenadas de los puntos. La longitud del segmento de recta más corto que puede conectar dos puntos.	A B a b $AB = \lvert a - b\rvert = \lvert b - a\rvert$

English	Spanish	Examples
distance from a point to a line The length of the perpendicular segment from the point to the line.	**distancia desde un punto hasta una línea** Longitud del segmento perpendicular desde el punto hasta la línea.	P A C The distance from P to $\overleftrightarrow{AC}$ is 5 units.

E

English	Spanish	Examples
element of a set An item in a set.	**elemento de un conjunto** Componente de un conjunto.	4 is an element of the set of even numbers. $4 \in \{\text{even numbers}\}$
empty set A set with no elements.	**conjunto vacío** Conjunto sin elementos.	The solution set of $\lvert x \rvert < 0$ is the empty set, $\{\}$, or $\varnothing$.
endpoint A point at an end of a segment or the starting point of a ray.	**extremo** Punto en el final de un segmento o punto de inicio de un rayo.	A B D
event An outcome or set of outcomes in a probability experiment.	**suceso** Resultado o conjunto de resultados en un experimento de probabilidad.	In the experiement of rolling a number cube, the event "an odd number" consists of the outcomes 1, 3, 5.
exterior of a circle The set of all points outside a circle.	**exterior de un círculo** Conjunto de todos los puntos que se encuentran fuera de un círculo.	Exterior
exterior angle of a polygon An angle formed by one side of a polygon and the extension of an adjacent side.	**ángulo externo de un polígono** Ángulo formado por un lado de un polígono y la prolongación del lado adyacente.	2 1 3 4 $\angle 4$ is an exterior angle.
external secant segment A segment of a secant that lies in the exterior of the circle with one endpoint on the circle.	**segmento secante externo** Segmento de una secante que se encuentra en el exterior del círculo y tiene un extremo sobre el círculo.	P N M Q J K $\overline{NM}$ is an external secant segment.

F

English	Spanish	Examples
flow proof A proof format that uses boxes and arrows to show the structure of a logical argument.	**demostración de flujo** Formato de demostración que utiliza recuadros y flechas para mostrar la estructura de un argumento lógico.	

G

English	Spanish	Examples
geometric mean For positive numbers a and b, the positive number x such that $\frac{a}{x} = \frac{x}{b}$. In a geometric sequence, a term that comes between two given nonconsecutive terms of the sequence.	**media geométrica** Dados los números positivos a y b, el número positivo x tal que $\frac{a}{x} = \frac{x}{b}$. En una sucesión geométrica, un término que está entre dos términos no consecutivos dados de la sucesión.	$\frac{a}{x} = \frac{x}{b}$ $x^2 = ab$ $x = \sqrt{ab}$
glide reflection A composition of a translation and a reflection across a line parallel to the translation vector.	**deslizamiento con inversión** Composición de una traslación y una reflexión sobre una línea paralela al vector de traslación.	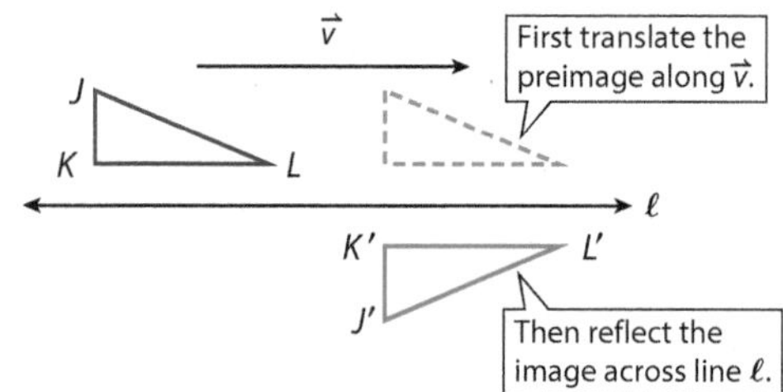

H

English	Spanish	Examples
hemisphere Half of a sphere.	**hemisferio** Mitad de una esfera.	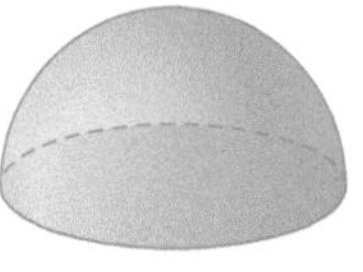
hypothesis The part of a conditional statement following the word *if*.	**hipótesis** La parte de un enunciado condicional que sigue a la palabra *si*.	If $\underbrace{x + 1 = 5}_{\text{Hypothesis}}$, then $x = 4$.

I

English	Spanish	Examples
image A shape that results from a transformation of a figure known as the preimage.	**imagen** Forma resultante de la transformación de una figura conocida como imagen original.	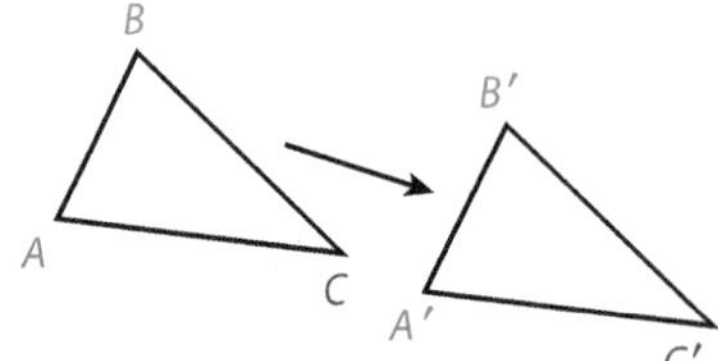

English	Spanish	Examples
incenter of a triangle The point of concurrency of the three angle bisectors of a triangle.	**incentro de un triángulo** Punto donde se encuentran las tres bisectrices de los ángulos de un triángulo.	G, K, M, P, H, L, J — *P* is the incenter.
incircle *See* inscribed circle.	**incírculo** *Ver* círculo inscrito.	
included angle The angle formed by two adjacent sides of a polygon.	**ángulo incluido** Ángulo formado por dos lados adyacentes de un polígono.	A, B, C — $\angle B$ is the included angle between $\overline{AB}$ and $\overline{BC}$.
included side The common side connecting two consecutive angles of a polygon.	**lado incluido** Lado común que conecta dos ángulos consecutivos de un polígono.	Q, P, R — $\overline{PQ}$ is the included side between $\angle P$ and $\angle Q$.
independent events Events for which the occurrence or nonoccurrence of one event does not affect the probability of the other event.	**sucesos independientes** Dos sucesos son independientes si el hecho de que se produzca o no uno de ellos no afecta la probabilidad del otro suceso.	From a bag containing 3 red marbles and 2 blue marbles, drawing a red marble, replacing it, and then drawing a blue marble.
indirect proof A proof in which the statement to be proved is assumed to be false and a contradiction is shown.	**demostración indirecta** Prueba en la que se supone que el enunciado a demostrar es falso y se muestra una contradicción.	
inductive reasoning The process of reasoning that a rule or statement is true because specific cases are true.	**razonamiento inductivo** Proceso de razonamiento por el que se determina que una regla o enunciado son verdaderos porque ciertos casos específicos son verdaderos.	
initial point of a vector The starting point of a vector.	**punto inicial de un vector** Punto donde comienza un vector.	B, $\vec{v}$, A, Initial point
inscribed angle An angle whose vertex is on a circle and whose sides contain chords of the circle.	**ángulo inscrito** Ángulo cuyo vértice se encuentra sobre un círculo y cuyos lados contienen cuerdas del círculo.	D, E, F

English	Spanish	Examples
inscribed circle A circle drawn inside a polygon so that each side of the polygon is tangent to the circle.	**círculo inscrito** Círculo trazado dentro de un polígono de tal forma que cada lado del polígono es tangente al círculo.	
inscribed polygon A polygon drawn inside a circle so that every vertex of the polygon lies on the circle.	**polígono inscrito** Polígono trazado dentro de un círculo de tal forma que todos los vértices se encuentran sobre el círculo.	
intercepted arc An arc that consists of endpoints that lie on the sides of an inscribed angle and all the points of the circle between the endpoints.	**arco abarcado** Arco cuyos extremos se encuentran en los lados de un ángulo inscrito y consta de todos los puntos del círculo ubicados entre dichos extremos.	D E F $\overset{\frown}{DF}$ is the intercepted arc.
interior angle An angle formed by two sides of a polygon with a common vertex.	**ángulo interno** Ángulo formado por dos lados de un polígono con un vértice común.	1 $\angle 1$ is an interior angle.
interior of a circle The set of all points inside a circle.	**interior de un círculo** Conjunto de todos los puntos que se encuentran dentro de un círculo.	Interior
intersection of two sets The set of all elements that are in both set A and set B.	**intersección de dos conjuntos** Conjunto de todos los elementos compartidos por los conjuntos A y B.	
inverse The statement formed by negating the hypothesis and conclusion of a conditional statement.	**inverso** Enunciado formado al negar la hipótesis y la conclusión de un enunciado condicional.	Statement: If $n + 1 = 3$, then $n = 2$. Inverse: If $n + 1 \neq 3$, then $n \neq 2$.
inverse cosine The trigonometric function used to find the measure of an angle whose cosine ratio is known.	**coseno inverso** La función trigonométrica utilizada para hallar la medida de un ángulo cuya razón coseno es conocida.	If $\cos A = x$, then $\cos^{-1} x = m\angle A$.
inverse sine The trigonometric function used to find the measure of an angle whose sine ratio is known.	**seno inverso** La función trigonométrica utilizada para hallar la medida de un ángulo cuya razón seno es conocida.	If $\sin A = x$, then $\sin^{-1} x = m\angle A$.

English	Spanish	Examples
inverse tangent The trigonometric function used to find the measure of an angle whose tangent ratio is known.	**tangente inversa** La función trigonométrica utilizada para hallar la medida de un ángulo cuya razón tangente es conocida.	If $\tan A = x$, then $\tan^{-1} x = m\angle A$.
isometry *See* rigid motion.	**isometría** *Ver* movimiento rígido.	Reflections, translations, and rotations are all examples of isometries.
isosceles trapezoid A trapezoid in which the legs are congruent but not parallel.	**trapecio isósceles** Trapecio cuyos lados no paralelos son congruentes.	B C A D
isosceles triangle A triangle with at least two congruent sides.	**triángulo isósceles** Triángulo que tiene al menos dos lados congruentes.	

K

English	Spanish	Examples
kite A quadrilateral whose four sides can be grouped into two pairs of consecutive congruent sides.	**cometa o papalote** Cuadrilátero cuyos cuatro lados se pueden agrupar en dos pares de lados congruentes consecutivos.	B A C D Kite *ABCD*

L

English	Spanish	Examples
lateral area The sum of the areas of the lateral faces of a prism or pyramid, or the area of the lateral surface of a cylinder or cone.	**área lateral** Suma de las áreas de las caras laterales de un prisma o pirámide, o área de la superficie lateral de un cilindro o cono.	12 cm 6 cm 8 cm Lateral area $= 2(6)(12) + 2(8)(12)$ $= 336 \text{ cm}^2$
leg of an isosceles triangle One of the two congruent sides of an isosceles triangle.	**cateto de un triángulo isósceles** Uno de los dos lados congruentes del triángulo isósceles.	leg leg
legs of a trapezoid The sides of a trapezoid that are not the bases.	**catetos de un trapecio** Los lados del trapecio que no son las bases.	B C Leg Leg A D

English	Spanish	Examples
length The distance between the two endpoints of a segment.	**longitud** Distancia entre los dos extremos de un segmento.	$AB = \|a - b\| = \|b - a\|$
limit A value that the output of a function approaches as the input increases or decreases without bound or as the input approaches a given value.	**límite** Valor al que se aproxima el valor de salida de una función a medida que el valor de entrada aumenta o disminuye sin límite o se aproxima a un valor determinado.	
line An undefined term in geometry, a line is a straight path that has no thickness and extends forever in one dimension.	**línea** Término indefinido en geometría; una línea es un trazo recto que no tiene grosor y se extiende infinitamente en una dimensión.	
line of reflection A line over which a figure is reflected. The line of reflection is the perpendicular bisector of each segment joining each point and its image.	**línea de reflexión** Línea sobre la cual se refleja una figura. La línea de reflexión es la mediatriz de todos los segmentos que unen cada punto y su imagen.	m is the line of reflection.
line of symmetry A line that divides a plane figure into two congruent reflected halves.	**eje de simetría** Línea que divide una figura plana en dos mitades reflejas congruentes.	
line segment *See* segment of a line.	**segmento** *Ver* segmento de recta.	
line symmetry A property of a figure that means it can be reflected across a line so that the image coincides with the preimage.	**simetría axial** Propiedad de una figura que implica que puede reflejarse sobre una línea de forma tal que la imagen coincida con la imagen original.	
linear pair A pair of adjacent angles whose noncommon sides are opposite rays.	**par lineal** Par de ángulos adyacentes cuyos lados no comunes son rayos opuestos.	$\angle 3$ and $\angle 4$ form a linear pair.

M

English	Spanish	Examples
major arc An arc of a circle whose points are on or in the exterior of a central angle.	**arco mayor** Arco de un círculo cuyos puntos están sobre un ángulo central o en su exterior.	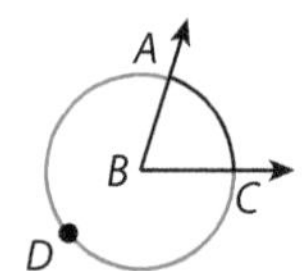 $\overparen{ADC}$ is a major arc of the circle.

English	Spanish	Examples
mapping An operation that matches each point in a plane with another point in the plane, and which is often used to describe an operation used to match a figure (the preimage) to another figure (the image).	**correspondencia** Operación que establece una correlación entre cada elemento de un conjunto con otro elemento, su imagen, en el mismo conjunto.	
median of a triangle A segment whose endpoints are a vertex of the triangle and the midpoint of the opposite side.	**mediana de un triángulo** Segmento cuyos extremos son un vértice del triángulo y el punto medio del lado opuesto.	
midpoint The point that divides a segment into two congruent segments.	**punto medio** Punto que divide un segmento en dos segmentos congruentes.	B is the midpoint of $\overline{AC}$.
midsegment of a trapezoid The segment whose endpoints are the midpoints of the legs of the trapezoid.	**segmento medio de un trapecio** Segmento cuyos extremos son los puntos medios de los catetos del trapecio.	
midsegment of a triangle A segment that joins the midpoints of two sides of the triangle.	**segmento medio de un triángulo** Segmento que une los puntos medios de dos lados del triángulo.	
minor arc An arc of a circle whose points are on or in the interior of a central angle.	**arco menor** Arco de un círculo cuyos puntos están sobre un ángulo central o en su interior.	$\widehat{AC}$ is a minor arc of the circle.
mutually exclusive events Two events are mutually exclusive if they cannot both occur in the same trial of an experiment.	**sucesos mutuamente excluyentes** Dos sucesos son mutuamente excluyentes si ambos no pueden ocurrir en la misma prueba de un experimento.	In the experiment of rolling a number cube, rolling a 3 and rolling an even number are mutually exclusive events.

N

English	Spanish	Examples
net A diagram of the faces of a three-dimensional figure arranged in such a way that the diagram can be folded to form the three-dimensional figure.	**plantilla** Diagrama de las caras de una figura tridimensional que se puede plegar para formar la figura tridimensional.	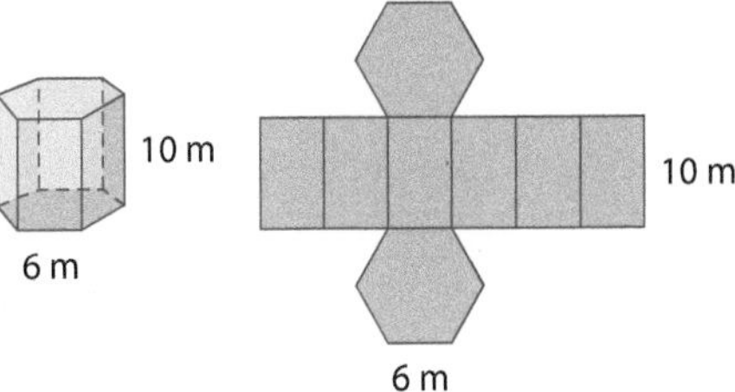

English	Spanish	Examples
n-gon An n-sided polygon.	**n-ágono** Polígono de n lados.	
nonpolygon A two-dimensional geometric object that does not meet the definition of a polygon.	**no polígono** Objeto geométrico bidimensional que no cumple con la definición de un polígono.	

O

English	Spanish	Examples
oblique solid A three-dimensional figure that has an axis or lateral edge that is not perpendicular to its base(s).	**sólido oblicuo** Figura tridimensional que tiene un eje o arista lateral que no es perpendicular a su(s) base(s).	
obtuse angle An angle that measures greater than 90° and less than 180°.	**ángulo obtuso** Ángulo que mide más de 90° y menos de 180°.	
opposite rays Two rays that have a common endpoint and form a line.	**rayos opuestos** Dos rayos que tienen un extremo común y forman una línea.	F E G $\overrightarrow{EF}$ and $\overrightarrow{EG}$ are opposite rays.
orthocenter of a triangle The point of concurrency of the three altitudes of a triangle.	**ortocentro de un triángulo** Punto de intersección de las tres alturas de un triángulo.	P is the orthocenter.
outcome A possible result of a probability experiment.	**resultado** Resultado posible de un experimento de probabilidad.	In the experiment of rolling a number cube, the possible outcomes are 1, 2, 3, 4, 5, and 6.
overlapping events Events that have one or more outcomes in common. Also called inclusive events.	**sucesos superpuestos** Sucesos que tienen uno o más resultados en común. También se denominan sucesos inclusivos.	Rolling an even number and rolling a prime number on a number cube are overlapping events because they both contain the outcome rolling a 2.

P

English	Spanish	Examples
parallel lines Lines in the same plane that do not intersect.	**líneas paralelas** Líneas rectas en el mismo plano que no se cruzan.	r s $r \parallel s$

English	Spanish	Examples
parallel planes Planes that do not intersect.	**planos paralelos** Planos que no se cruzan.	A B E F C D G H Plane *AEF* and plane *CGH* are parallel planes.
parallelogram A quadrilateral with two pairs of parallel sides.	**paralelogramo** Cuadrilátero con dos pares de lados paralelos.	
partition a segment To divide a line segment into shorter segments with a given ratio.	**dividir un segmento** Fraccionar un segmento de recta en segmentos más cortos de acuerdo con una razón determinada.	
perpendicular Intersecting to form 90° angles, denoted by $\perp$.	**perpendicular** Que se cruza para formar ángulos de 90°, expresado por $\perp$.	n m $m \perp n$
perpendicular bisector A line perpendicular to a segment or to a side of a triangle at the segment's midpoint.	**mediatriz** Línea perpendicular a un segmento o a un lado de un triángulo, trazada en el punto medio del segmento.	ℓ A B ℓ is the perpendicular bisector of $\overline{AB}$.
perpendicular lines Lines that intersect to form 90° angles.	**líneas perpendiculares** Líneas que se cruzan para formar ángulos de 90°.	n m $m \perp n$
pi The ratio of the circumference of a circle to its diameter, denoted by the Greek letter π (pi). The value of π is irrational, often approximated by 3.14 or $\frac{22}{7}$.	**pi** Razón entre la circunferencia de un círculo y su diámetro, expresado por la letra griega π (pi). El valor de π es irracional y por lo general se aproxima a 3.14 ó $\frac{22}{7}$.	If a circle has a diameter of 5 inches and a circumference of C inches, then $\frac{C}{5} = \pi$, or $C = 5\pi$ inches, or about 15.7 inches.
plane An undefined term in geometry, it is a flat surface that has no thickness and extends forever in two dimensions.	**plano** Término indefinido en geometría; un plano es una superficie plana que no tiene grosor y se extiende infinitamente en dos dimensiones.	A C ℛ B plane *R* or plane *ABC*

English	Spanish	Examples
point An undefined term in geometry, it names a location and has no size. A point has no dimension.	**punto** Término indefinido de la geometría que denomina una ubicación y no tiene tamaño. Un punto no tiene dimensión.	P • point P
point of concurrency A point where three or more lines coincide.	**punto de concurrencia** Punto donde se cruzan tres o más líneas.	B
point of tangency The point of intersection of a circle or sphere with a tangent line or plane.	**punto de tangencia** Punto de intersección de un círculo o esfera con una línea o plano tangente.	m Tangent C Point of tangency
polygon A closed plane figure formed by three or more segments such that each segment intersects exactly two other segments only at their endpoints and no two segments with a common endpoint are collinear.	**polígono** Figura plana cerrada formada por tres o más segmentos tal que cada segmento se cruza únicamente con otros dos segmentos sólo en sus extremos y ningún segmento con un extremo común a otro es colineal con éste.	
polyhedron A closed three-dimensional figure formed by four or more polygons that intersect only at their edges.	**poliedro** Figura tridimensional cerrada formada por cuatro o más polígonos que se cruzan sólo en sus aristas.	
population density The number of organisms of a particular type per square unit of area.	**densidad de población** Número de organismos de un tipo particular por unidad cuadrada de área.	
postulate A statement that is accepted as true without proof. Also called an *axiom*.	**postulado** Enunciado que se acepta como verdadero sin demostración. También denominado *axioma*.	
preimage The original figure in a transformation.	**imagen original** Figura original en una transformación.	B, A, C → B′, A′, C′
prism A polyhedron formed by two parallel congruent polygonal bases connected by lateral faces that are parallelograms.	**prisma** Poliedro formado por dos bases poligonales congruentes y paralelas conectadas por caras laterales que son paralelogramos.	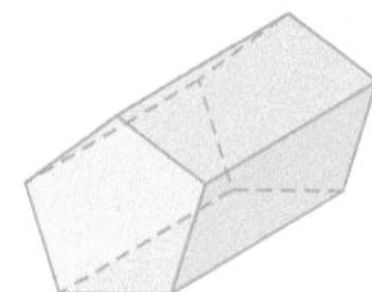

English	Spanish	Examples
probability experiment An activity that has a defined set of possible outcomes and which can be repeated.	**experimento de probabilidad** Actividad que tiene un conjunto definido de resultados posibles que pueden repetirse.	
proof An argument that uses logic to show that a conclusion is true.	**demostración** Argumento que se vale de la lógica para probar que una conclusión es verdadera.	
proof by contradiction *See* indirect proof.	**demostración por contradicción** *Ver* demostración indirecta.	
pyramid A polyhedron formed by a polygonal base and triangular lateral faces that meet at a common vertex.	**pirámide** Poliedro formado por una base poligonal y caras laterales triangulares que se encuentran en un vértice común.	
Pythagorean triple A set of three nonzero whole numbers a, b, and c such that $a^2 + b^2 = c^2$.	**Tripleta de Pitágoras** Conjunto de tres números enteros distintos de cero a, b y c tal que $a^2 + b^2 = c^2$.	$\{3, 4, 5\}$ $3^2 + 4^2 = 5^2$

Q

English	Spanish	Examples
quadrilateral A four-sided polygon.	**cuadrilátero** Polígono de cuatro lados.	

R

English	Spanish	Examples
radial symmetry *See* rotational symmetry.	**simetría radial** *Ver* simetría de rotación.	
radian A unit of angle measure based on arc length. In a circle of radius r, if a central angle has a measure of 1 radian, then the length of the intercepted arc is r units. 2π radians $= 360°$ 1 radian $\approx 57°$	**radián** Unidad de medida de un ángulo basada en la longitud del arco. En un círculo de radio r, si un ángulo central mide 1 radián, entonces la longitud del arco abarcado es r unidades. 2π radians $= 360°$ 1 radian $\approx 57°$	r r O $\theta = 1$ radian
ray A part of a line that starts at an endpoint and extends forever in one direction.	**rayo** Parte de una línea que comienza en un extremo y se extiende infinitamente en una dirección.	D
rectangle A parallelogram with four right angles.	**rectángulo** Paralelogramo con cuatro ángulos rectos.	

English	Spanish	Examples
reflection A transformation across a line, called the line of reflection, such that the line of reflection is the perpendicular bisector of each segment joining each point and its image.	**reflexión** Transformación sobre una línea, denominada la línea de reflexión. La línea de reflexión es la mediatriz de cada segmento que une un punto con su imagen.	B, B′, A, C, C′, A′
reflection symmetry *See* line symmetry.	**simetría de reflexión** *Ver* simetría axial.	
reflex angle An angle with measure greater than 180° and less than 360°.	**ángulo reflejo** Ángulo cuya medida es mayor que 180° y menor que 360°.	
regular polygon A polygon that is both equilateral and equiangular.	**polígono regular** Polígono equilátero de ángulos iguales.	
regular pyramid A pyramid whose base is a regular polygon and whose lateral faces are congruent isosceles triangles.	**pirámide regular** Pirámide cuya base es un polígono regular y cuyas caras laterales son triángulos isósceles congruentes.	
relative frequency The ratio of the frequency of one outcome to the frequency of all outcomes.	**frecuencia relativa** Razón entre la frecuencia de un resultado y la frecuencia de todos los resultados.	
remote interior angle An interior angle of a polygon that is not adjacent to the exterior angle.	**ángulo interno remoto** Ángulo interno de un polígono que no es adyacente al ángulo externo.	2, 1, 3, 4 — The remote interior angles of ∠4 are ∠1 and ∠2.
rhombus A parallelogram with four congruent sides.	**rombo** Paralelogramo con cuatro lados congruentes.	
right angle An angle that measures 90°.	**ángulo recto** Ángulo que mide 90°.	
right cone A cone whose axis is perpendicular to its base.	**cono recto** Cono cuyo eje es perpendicular a su base.	Axis

English	Spanish	Examples
right solid A three-dimensional figure that has an axis or lateral edge that is perpendicular to its base(s).	**sólido recto** Figura tridimensional que tiene un eje o arista lateral que es perpendicular a su(s) base(s).	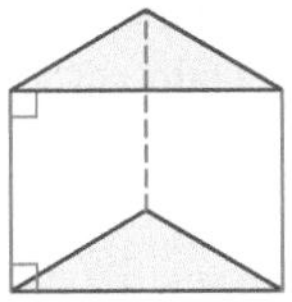
rigid motion A transformation that does not change the size or shape of a figure.	**movimiento rígido** Transformación que no cambia el tamaño ni la forma de una figura.	Reflections, translations, and rotations are all examples of rigid motions.
rigid transformation *See* rigid motion.	**transformación rígida** *Ver* movimiento rígido.	
rotation A transformation about a point *P*, also known as the center of rotation, such that each point and its image are the same distance from *P*. All of the angles with vertex *P* formed by a point and its image are congruent.	**rotación** Transformación sobre un punto *P*, también conocido como el centro de rotación, tal que cada punto y su imagen estén a la misma distancia de *P*. Todos los ángulos con vértice *P* formados por un punto y su imagen son congruentes.	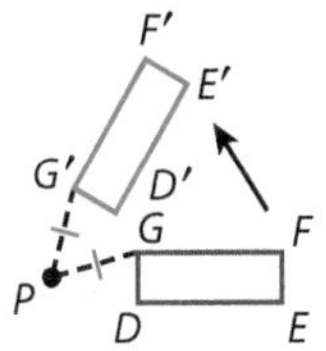 F′ E′ G′ D′ G F P D E
rotational symmetry A property of a figure that can be rotated about a point by an angle less than 360° so that the image coincides with the preimage.	**simetría de rotación** Propiedad de una figura que puede rotarse alrededor de un punto en un ángulo menor de 360° de forma tal que la imagen coincide con la imagen original.	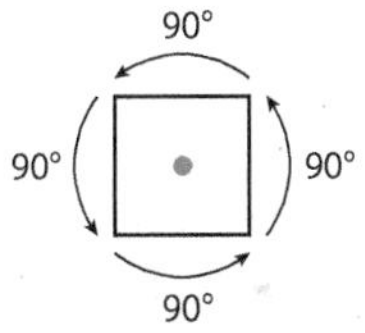 90° 90° 90° 90° Order of rotational symmetry: 4

S

English	Spanish	Examples
sample space The set of all possible outcomes of a probability experiment.	**espacio muestral** Conjunto de todos los resultados posibles de un experimento de probabilidad.	In the experiment of rolling a number cube, the sample space is {1, 2, 3, 4, 5, 6}.
scale factor The multiplier used on each dimension to change one figure into a similar figure.	**factor de escala** El multiplicador utilizado en cada dimensión para transformar una figura en una figura semejante.	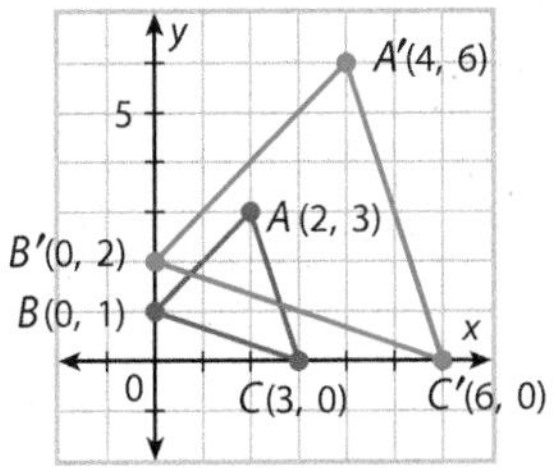 y A′(4, 6) 5 A(2, 3) B′(0, 2) B(0, 1) x 0 C(3, 0) C′(6, 0) Scale factor: 2
secant of a circle A line that intersects a circle at two points.	**secante de un círculo** Línea que corta un círculo en dos puntos.	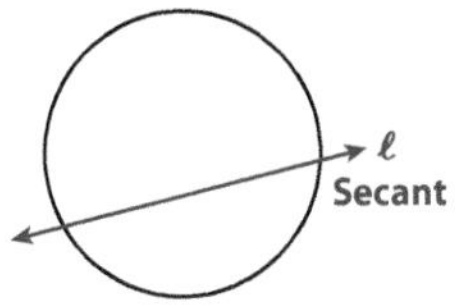 ℓ Secant

English	Spanish	Examples
secant segment A segment of a secant with at least one endpoint on the circle.	**segmento secante** Segmento de una secante que tiene al menos un extremo sobre el círculo.	$\overline{NM}$ is an external secant segment. $\overline{JK}$ is an internal secant segment.
sector of a circle A region inside a circle bounded by two radii of the circle and their intercepted arc.	**sector de un círculo** Región dentro de un círculo delimitado por dos radios del círculo y por su arco abarcado.	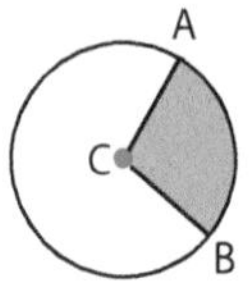
segment bisector A line, ray, or segment that divides a segment into two congruent segments.	**bisectriz de un segmento** Línea, rayo o segmento que divide un segmento en dos segmentos congruentes.	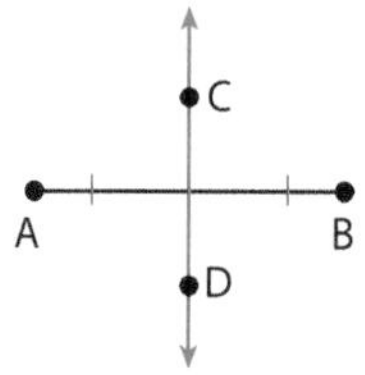
segment of a circle A region inside a circle bounded by a chord and an arc.	**segmento de un círculo** Región dentro de un círculo delimitada por una cuerda y un arco.	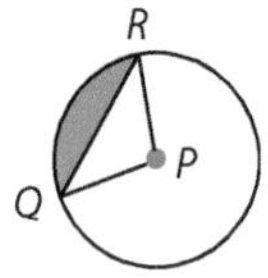
segment of a line A part of a line consisting of two endpoints and all points between them.	**segmento de una línea** Parte de una línea que consiste en dos extremos y todos los puntos entre éstos.	
semicircle An arc of a circle whose endpoints lie on a diameter.	**semicírculo** Arco de un círculo cuyos extremos se encuentran sobre un diámetro.	
set A collection of items called elements.	**conjunto** Grupo de componentes denominados elementos.	$\{1, 2, 3\}$
similar Two figures are similar if they have the same shape but not necessarily the same size.	**semejantes** Dos figuras con la misma forma pero no necesariamente del mismo tamaño.	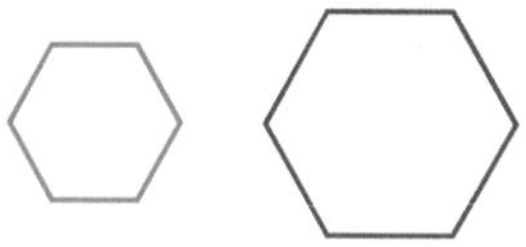
similar figures If a figure can be mapped to another figure using a sequence of similarity transformations, the figures are similar.	**figuras semejantes** Si se puede establecer una correspondencia entre una figura y otra usando una sucesión de transformaciones de semejanza, las figuras son semejantes.	

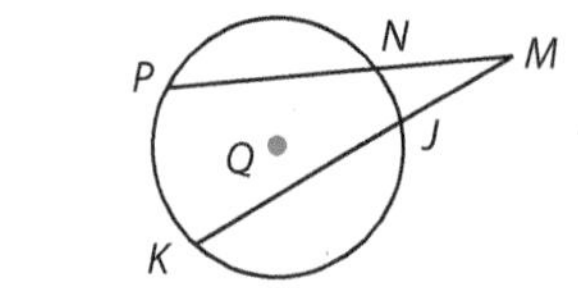

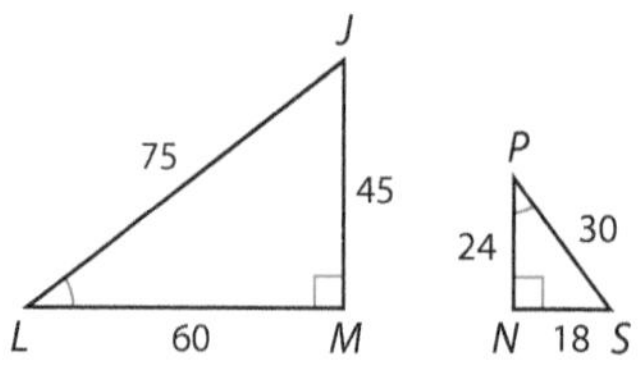

English	Spanish	Examples
similarity transformation A transformation that produces similar figures.	**transformación de semejanza** Una transformación que resulta en figuras semejantes.	Dilations are similarity transformations.
sine In a right triangle, the ratio of the length of the leg opposite $\angle A$ to the length of the hypotenuse.	**seno** En un triángulo rectángulo, razón entre la longitud del cateto opuesto a $\angle A$ y la longitud de la hipotenusa.	opposite; hypotenuse; A $\sin A = \frac{\text{opposite}}{\text{hypotenuse}}$
skew lines Lines that are not coplanar.	**líneas oblicuas** Líneas que no son coplanares.	A, B, E, F, C, D, G, H $\overleftrightarrow{AE}$ *and* $\overleftrightarrow{CD}$ *are skew lines.*
slant height The height of each lateral face of a right pyramid or the lateral surface of a right cone.	**altura inclinada** Altura de cada cara lateral de una pirámide recta o superficie lateral de un cono recto.	ℓ, s, s, s, s The slant height is ℓ.
slide *See* translation.	**deslizamiento** *Ver* traslación.	
slope A measure of the steepness of a line. If (x_1, y_1) and (x_2, y_2) are any two points on the line, the slope of the line, denoted m, is represented by the equation $m = \frac{y_2 - y_1}{x_2 - x_1}$.	**pendiente** Medida de la inclinación de una línea. Dados dos puntos (x_1, y_1) y (x_2, y_2) en una línea, la pendiente de la línea, denominada m, se representa con la ecuación $m = \frac{y_2 - y_1}{x_2 - x_1}$.	
solid A three-dimensional figure.	**cuerpo geométrico** Figura tridimensional.	
solid of rotation A solid that is formed by rotating a two-dimensional shape about an axis.	**sólido de rotación** Cuerpo geométrico que se forma al rotar una figura bidimensional sobre un eje.	

English	Spanish	Examples
solving a triangle Using given measures to find unknown angle measures or side lengths of a triangle.	**resolución de un triángulo** Utilizar medidas dadas para hallar las medidas desconocidas de los ángulos o las longitudes de los lados de un triángulo.	
special right triangle A $45°-45°-90°$ triangle or a $30°-60°-90°$ triangle.	**triángulo rectángulo especial** Triángulo de $45°-45°-90°$ o triángulo de $30°-60°-90°$.	A, 2s, 60°, s, B, 30°, C, $s\sqrt{3}$; B, $\ell\sqrt{2}$, 45°, ℓ, A, 45°, C, ℓ
sphere The set of points in space that are a fixed distance from a given point called the center of the sphere.	**esfera** Conjunto de puntos en el espacio que se encuentran a una distancia fija de un punto determinado denominado centro de la esfera.	
square A parallelogram with four congruent sides and four right angles.	**cuadrado** Paralelogramo con cuatro lados congruentes y cuatro ángulos rectos.	
straight angle A 180° angle.	**ángulo llano** Ángulo que mide 180°.	
stretch A nonrigid transformation that changes the shape of a figure in one direction by a factor greater than 1.	**estiramiento** Transformación no rígida que cambia la forma de una figura en una dirección por un factor mayor que 1.	horizontal stretch
subset A set that is contained entirely within another set. Set *B* is a subset of set *A* if every element of *B* is contained in *A*, denoted $B \subset A$.	**subconjunto** Conjunto que se encuentra dentro de otro conjunto. El conjunto *B* es un subconjunto del conjunto *A* si todos los elementos de *B* son elementos de *A*; se expresa $B \subset A$.	The set of integers is a subset of the set of rational numbers.
supplementary angles Two angles whose measures have a sum of 180°.	**ángulos suplementarios** Dos ángulos cuyas medidas suman 180°.	4, 3; $\angle 3$ and $\angle 4$ are supplementary angles.

English	Spanish	Examples
surface area The total area of all faces and curved surfaces of a three-dimensional figure.	**área total** Área total de todas las caras y superficies curvas de una figura tridimensional.	12 cm 6 cm 8 cm Surface area $= 2(8)(12) + 2(8)(6) + 2(12)(6) = 432 \text{ cm}^2$
symmetry A property of a figure that can be mapped to itself using a rigid motion such as a reflection or rotation.	**simetría** Propiedad de una figura que puede establecer una correspondencia consigo misma mediante un movimiento rígido, como la reflexión o la rotación.	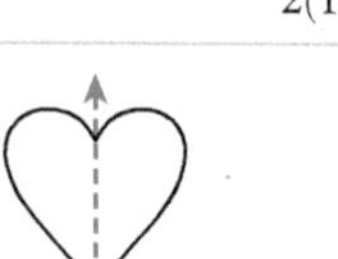

T

English	Spanish	Examples
tangent of an angle In a right triangle, the ratio of the length of the leg opposite $\angle A$ to the length of the leg adjacent to $\angle A$.	**tangente de un ángulo** En un triángulo rectángulo, razón entre la longitud del cateto opuesto a $\angle A$ y la longitud del cateto adyacente a $\angle A$.	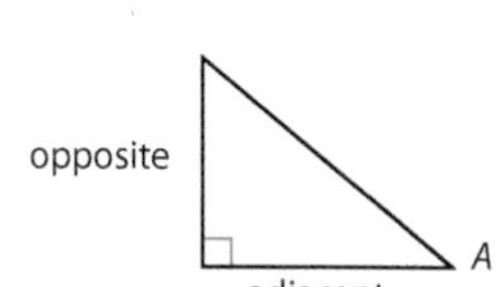 $\tan A = \frac{\text{opposite}}{\text{adjacent}}$
tangent segment A segment of a tangent with one endpoint on the circle.	**segmento tangente** Segmento de una tangente con un extremo en el círculo.	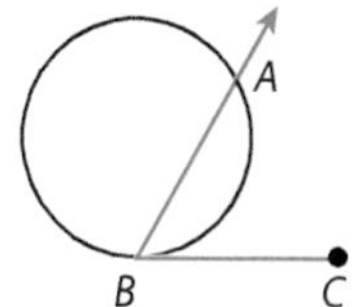 $\overline{BC}$ is a tangent segment.
tangent of a circle A line that is in the same plane as a circle and intersects the circle at exactly one point.	**tangente de un círculo** Línea que se encuentra en el mismo plano que un círculo y lo cruza únicamente en un punto.	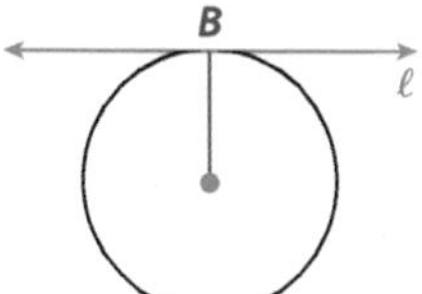
terminal point of a vector The endpoint of a vector.	**punto terminal de un vector** Extremo de un vector.	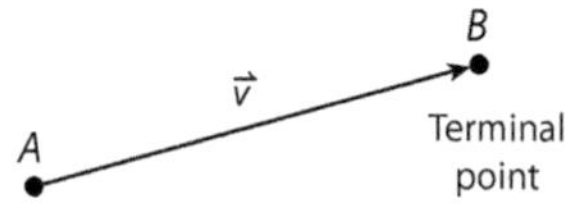
theorem A statement that has been proven.	**teorema** Enunciado que ha sido demostrado.	

English	Spanish	Examples
theoretical probability The ratio of the number of equally-likely outcomes in an event to the total number of possible outcomes.	**probabilidad teórica** Razón entre el número de resultados igualmente probables de un suceso y el número total de resultados posibles.	In the experiment of rolling a number cube, the theoretical probability of rolling an odd number is $\frac{3}{6} = \frac{1}{2}$.
transformation A function that changes the position, size, or shape of a figure or graph.	**transformación** Función que cambia la posición, tamaño o forma de una figura o gráfica.	Preimage, Image; $\triangle ABC \rightarrow \triangle A'B'C'$
translation A transformation that shifts or slides every point of a figure or graph the same distance in the same direction.	**traslación** Transformación en la que todos los puntos de una figura o gráfica se mueven la misma distancia en la misma dirección.	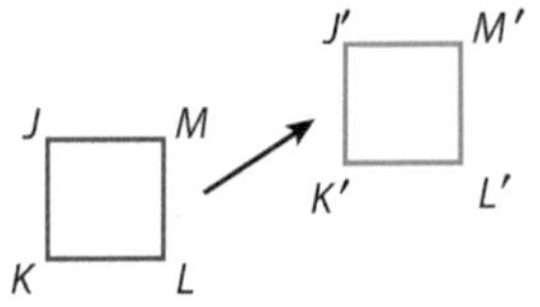 J, M, K, L → J′, M′, K′, L′
transversal A line that intersects two or more coplanar lines at two different points.	**transversal** Línea que corta dos o más líneas coplanares en dos puntos diferentes.	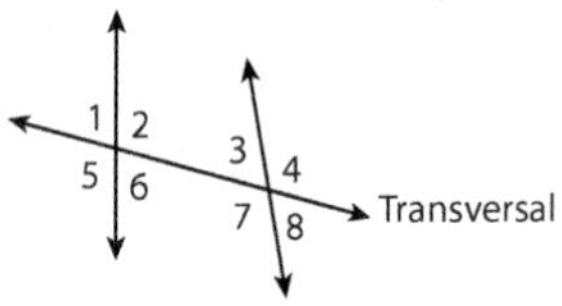 1, 2, 5, 6, 3, 4, 7, 8; Transversal
trapezoid A quadrilateral with at least one pair of parallel sides.	**trapecio** Cuadrilátero con al menos un par de lados paralelos.	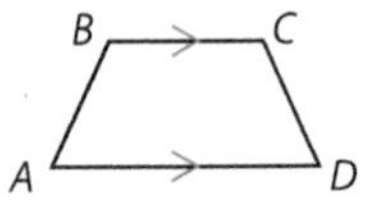 B, C, A, D
trial In probability, a single repetition or observation of an experiment.	**prueba** En probabilidad, una sola repetición u observación de un experimento.	In the experiment of rolling a number cube, each roll is one trial.
trigonometric ratio A ratio of two sides of a right triangle.	**razón trigonométrica** Razón entre dos lados de un triángulo rectángulo.	A, B, C; a, b, c; $\sin A = \frac{a}{c}$; $\cos A = \frac{b}{c}$; $\tan A = \frac{a}{b}$

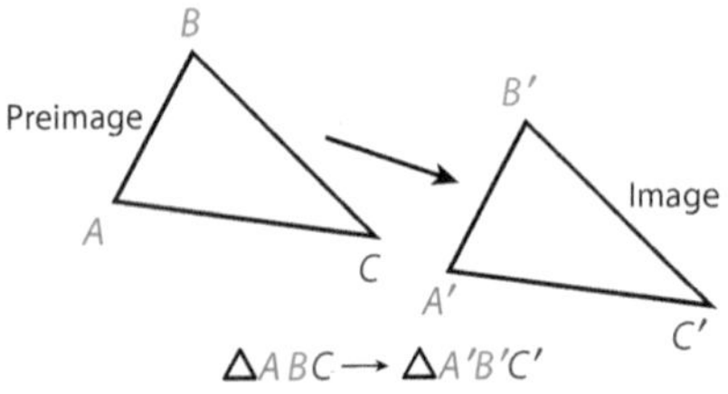

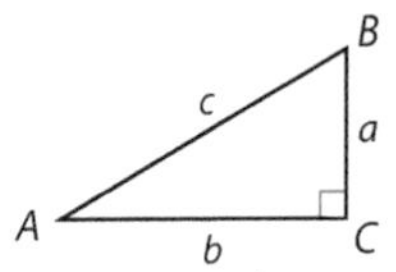

U

English	Spanish	Examples
undefined term A basic figure that is not defined in terms of other figures. The undefined terms in geometry are *point*, *line*, and *plane*.	**término indefinido** Figura básica que no está definida en función de otras figuras. Los términos indefinidos en geometría son el punto, la línea y el plano.	

English	Spanish	Examples
union of two sets The set of all elements that are in either set, denoted by ∪.	**unión de dos conjuntos** Conjunto de todos los elementos que se encuentran en ambos conjuntos, expresado por ∪.	$A = \{1, 2, 3, 4\}$ $B = \{1, 3, 5, 7, 9\}$ $A \cup B = \{1, 2, 3, 4, 5, 7, 9\}$
universal set The set of all elements in a particular context.	**conjunto universal** Conjunto de todos los elementos de un contexto determinado.	

V

English	Spanish	Examples
vector A quantity that has both magnitude and direction.	**vector** Cantidad que tiene magnitud y dirección.	$\vec{u}$
vertex of an angle The common endpoint of the sides of an angle.	**vértice de un ángulo** Extremo común de los lados del ángulo.	C, A, B *A* is the vertex of ∠*CAB*.
vertex of a cone The point opposite the base of a cone.	**vértice de un cono** Punto opuesto a la base del cono.	Vertex
vertex of a polygon The intersection of two sides of a polygon.	**vértice de un polígono** La intersección de dos lados del polígono.	A, B, C, D, E, Vertex *A*, *B*, *C*, *D*, and *E* are vertices of the polygon.
vertex of a pyramid The point opposite the base of a pyramid.	**vértice de una pirámide** Punto opuesto a la base de la pirámide.	Vertex
vertex of a three-dimensional figure The point that is the intersection of three or more faces of a figure.	**vértice de una figura tridimensional** Punto que representa la intersección de tres o más caras de la figura.	Vertex

English	Spanish	Examples
vertex of a triangle The intersection of two sides of a triangle.	**vértice de un triángulo** Intersección de dos lados del triángulo.	A, B, and C are vertices of $\triangle ABC$.
vertical angles The nonadjacent angles formed by two intersecting lines.	**ángulos opuestos por el vértice** Ángulos no adyacentes formados por dos rectas que se cruzan.	$\angle 1$ and $\angle 3$ are vertical angles. $\angle 2$ and $\angle 4$ are vertical angles.
volume The number of nonoverlapping unit cubes of a given size that will exactly fill the interior of a three-dimensional figure.	**volumen** Cantidad de cubos unitarios no superpuestos de un determinado tamaño que llenan exactamente el interior de una figura tridimensional.	4 ft, 3 ft, 12 ft Volume $= (3)(4)(12) = 144$ ft^3

A

B

C

Index

D

Teacher Edition references are in italics; *Teacher Edition: Planning and Pacing Guide* references begin with PG.

N

O

P

Index

Index

Index

U

V

W

X

Y

Tables of Measures, Symbols, and Formulas

LENGTH

1 meter (m) = 1000 millimeters (mm)	1 inch = 2.54 centimeters
1 meter = 100 centimeters (cm)	1 foot (ft) = 12 inches (in.)
1 meter ≈ 39.37 inches	1 yard (yd) = 3 feet
1 kilometer (km) = 1000 meters	1 mile (mi) = 1760 yards
1 kilometer ≈ 0.62 mile	1 mile = 5280 feet
	1 mile ≈ 1.609 kilometers

CAPACITY

1 liter (L) = 1000 milliliters (mL)	1 cup (c) = 8 fluid ounces (fl oz)
1 liter = 1000 cubic centimeters	1 pint (pt) = 2 cups
1 liter ≈ 0.264 gallon	1 quart (qt) = 2 pints
1 kiloliter (kL) = 1000 liters	1 gallon (gal) = 4 quarts
	1 gallon ≈ 3.785 liters

MASS/WEIGHT

1 gram (g) = 1000 milligrams (mg)	1 pound (lb) = 16 ounces (oz)
1 kilogram (kg) = 1000 grams	1 pound ≈ 0.454 kilogram
1 kilogram ≈ 2.2 pounds	1 ton = 2000 pounds

TIME

1 minute (min) = 60 seconds (s)	1 year (yr) = about 52 weeks
1 hour (h) = 60 minutes	1 year = 12 months (mo)
1 day = 24 hours	1 year = 365 days
1 week = 7 days	1 decade = 10 years

SYMBOLS

$=$	is equal to	x^2	x squared
$\neq$	is not equal to	x^3	x cubed
$\approx$	is approximately equal to	$\lvert x \rvert$	absolute value of x
$>$	is greater than	$\frac{1}{x}$	reciprocal of x ($x \neq 0$)
$<$	is less than	$\sqrt{x}$	square root of x
$\geq$	is greater than or equal to	$\sqrt[3]{x}$	cube root of x
$\leq$	is less than or equal to	x_n	x sub n ($n = 0, 1, 2, \ldots$)

FORMULAS

Perimeter and Circumference

Polygon	$P =$ sum of the lengths of sides
Rectangle	$P = 2\ell + 2w$
Square	$P = 4s$
Circle	$C = \pi d$ or $C = 2\pi r$

Area

Rectangle	$A = \ell w$
Parallelogram	$A = bh$
Triangle	$A = \frac{1}{2}bh$
Trapezoid	$A = \frac{1}{2}h(b_1 + b_2)$
Square	$A = s^2$
Circle	$A = \pi r^2$

Volume

Right Prism	$V = \ell wh$ or $V = Bh$
Cube	$V = s^3$
Pyramid	$V = \frac{1}{3}Bh$
Cylinder	$V = \pi r^2 h$
Cone	$V = \frac{1}{3}\pi r^2 h$
Sphere	$V = \frac{4}{3}\pi r^3$

Surface Area

Right Prism	$S = Ph + 2B$
Cube	$S = 6s^2$
Square Pyramid	$S = \frac{1}{2}P\ell + B$

Pythagorean Theorem

$a^2 + b^2 = c^2$

UNIT 1

MODULE 1, LESSON 1.1

Spark Your Learning

A. Mathematical terms that are suggested by the photo are point, line segment, and arc. A point is an exact location. In the picture, a point is the exact location where the wire connects to the middle section. A line segment goes between two points and is not curved. In the picture, it is represented by the wires that suspend the bridge. An arc is a curve or part of a circle. In the picture, it is represented by the curved bridge.

C. Possible answer: A line in English can mean many things: lines on your face, the length of a rope, wire etc. A line in math has no thickness, extends in one dimension, and goes on forever.

D. Possible answer: Yes it does; It shows me that words can have multiple definitions and you need to interpret those definitions differently depending if you are talking about math or English.

Build Understanding

Task 1 **A.** Possible answer:

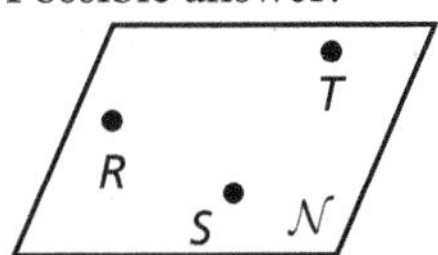

plane $\mathcal{N}$, plane RST; 3 points; No, the order of the points does not matter.

B. Possible answer:

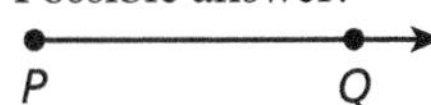

ray PQ or $\overrightarrow{PQ}$; 2 points; Yes, you have to name the endpoint first.

C. Possible answer: You can name the segment using the letters in either order, but to name the ray you have to start with the endpoint.

Task 2 **B.** AB and MN are equal; The same arc was used to draw them.

C. To add $\overline{CD}$ to $\overline{MN}$, open the compass to length CD, place the compass on point N, and draw an arc; The resulting segment is equal to the sum of the lengths of $\overline{CD}$ and $\overline{MN}$ because the same arc was used to draw each respective part.

D. the same; Since addition is commutative, it will not matter which segment you copy first.

Task 3 **D.** I could use a ruler to check that the lengths of each segment are equal.

Step it Out

Task 4 **A.** These points all lie on the same horizontal line, so they have the same y-coordinate.

B. These points all lie on the same vertical line, so they have the same x-coordinate.

C. yes; The intersection is the midpoint of $\overline{AB}$ because it is 4 units away from each of the endpoints.

Check Understanding

1. Possible answer: The image should have arrows on either side, with an A and B on the line, and should be labeled $\overleftrightarrow{AB}$.

2. Possible answer: Use a ruler to measure the lengths of the segments. Add the measures. Draw a segment with length equal to the sum.

3. Possible answer: To find the midpoint of a segment open the compass to $\frac{2}{3}$ the length of the line segment. Place the compass on one of the endpoints and draw an arc above and below the segment. Switch to the other endpoint and draw an arc above and below. Use a straightedge to draw a line through the two points of intersection of the arcs. This line passes through the midpoint of the segment.

On Your Own

15. no; Since $AB + BC = AC$, we can write the equation $(7x - 1) + (4x + 6) = 38$. I can add on the left side to get $11x + 5 = 38$. Subtract 5 from both sides and divide by 11 to find that $x = 3$. This means that $AB = 7(3) - 1 = 20$ cm and $BC = 4(3) + 6 = 18$ cm. B is not the midpoint because $AB \neq BC$.

30. Possible answer: Three points are $(11, 6)$, $(4, 13)$, and $(-3, 6)$. The respective midpoints are $\left(\frac{15}{2}, 6\right)$, $\left(4, \frac{19}{2}\right)$, and $\left(\frac{1}{2}, 6\right)$.

32. no; Possible explanation: Let (a, b) represent one of the points. Then the other point would have to be $(-a, -b)$ for the midpoint to be the origin. The signs of the corresponding coordinates must be the same for the points to be in the same quadrant, but (a, b) and $(-a, -b)$ have coordinates with the same signs only when $a = 0$ and $b = 0$.

33. Possible answer: He did not use the Midpoint Formula correctly. Instead of subtracting the values in the numerator, he should have added them. The correct answer should be $(-3, 3)$.

35. $(1, 2)$, $(1, 10)$, $(-3, 6)$, $(5, 6)$; Possible answer: I found the points by adding and subtracting 4 units from either the y-axis or the x-axis. The midpoint is 2 units from each side, therefore the other endpoint is 2 units away from that point.

MODULE 1, LESSON 1.2

Spark Your Learning

B. Possible answer: The most room for error would occur if the ball was directly in front of the goal. The closer the ball is to the goal, the greater the angle between the posts of the goal and the ball, which gives more room for error.

Additional Answers

D. Using precise language ensures that my thoughts are conveyed as clearly as possible, whether for my own use or when communicating with others.

Build Understanding

Task 1 **A.** Rays that form an angle:

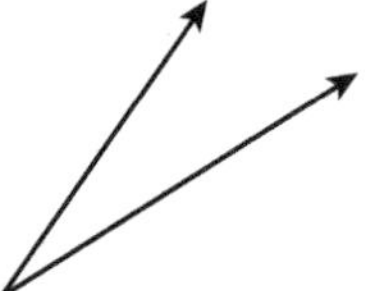

Rays that do not form an angle:

B. $\angle ABD$, $\angle DBA$; $\angle ABC$, $\angle CBA$; $\angle CBD$, $\angle DBC$

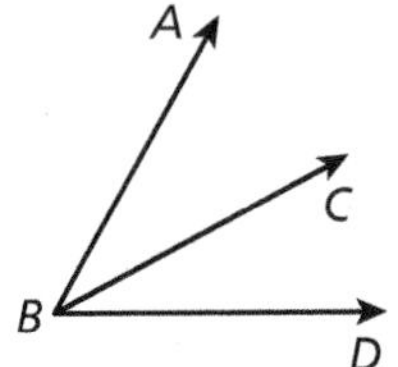

C.

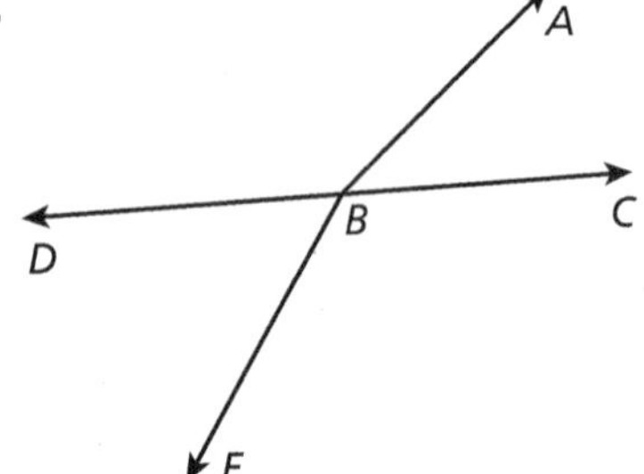

Task 2 **A.** acute angles: $\angle ABG$, $\angle KHC$, $\angle DBC$, $\angle ABF$

right angles: $\angle LHC$

obtuse angles: $\angle ABD$, $\angle GBD$, $\angle LHK$

straight angles: $\angle DBF$, $\angle ABC$

B. $m\angle ABG = 30°$

$m\angle ABD = 150°$

$m\angle GBD = 120°$

$m\angle DBF = 176°$

$m\angle KHC = 25°$

$m\angle DBC = 30°$

$m\angle LHC = 90°$

$m\angle ABF = 26°$

$m\angle LHK = 115°$

$m\angle ABC = 180°$

C. Possible answer: It can be useful to estimate an angle while solving a problem to make sure you are on the correct path.

Check Understanding

3.

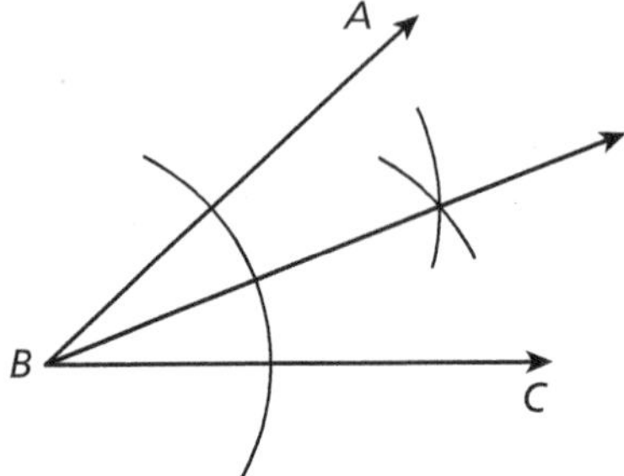

On Your Own

8. Possible answer:

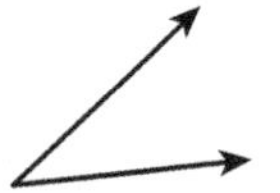

9. Possible answer:

10. Possible answer:

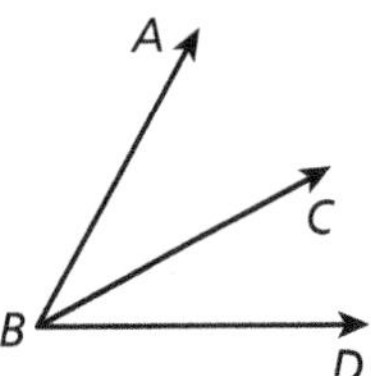

11. Possible answer:

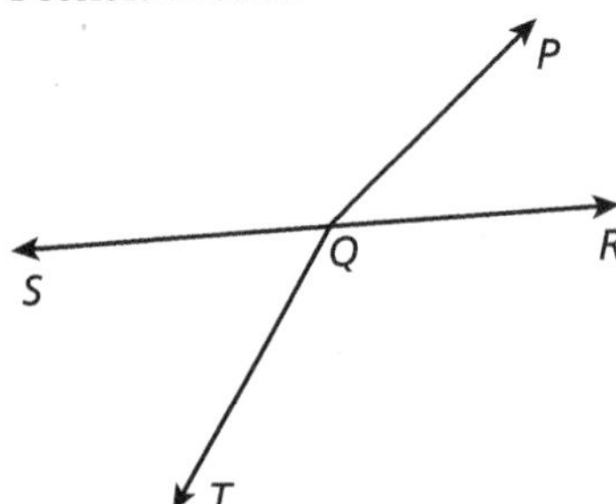

Other angles: $\angle PQS$ and $\angle RQT$

12. Possible answer:

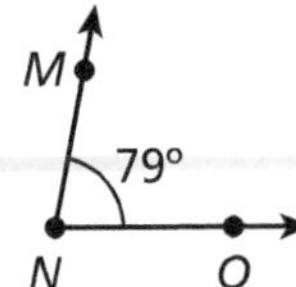

13. Possible answer:

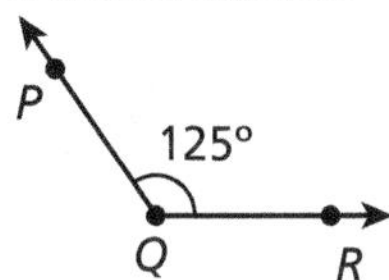

14. Possible answer:

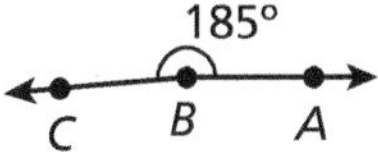

15. Possible answer:

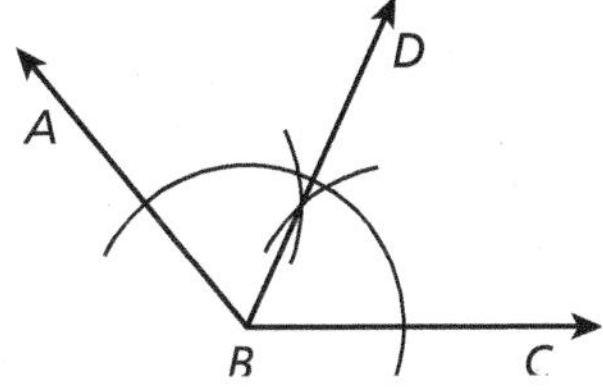

16. Possible answer:

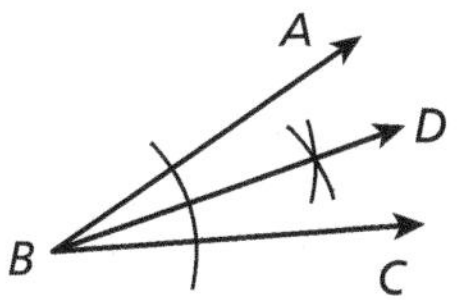

31. Possible answer:

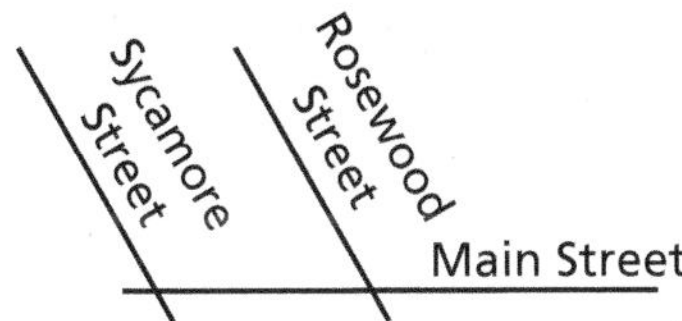

MODULE 1, LESSON 1.3

Spark Your Learning

B. Possible answer: The two boroughs can be modeled by geometric shapes. Someone could approximate the area of both boroughs by using basic shapes to represent the islands.

D. Possible answer: The greater the population density, the likelier the region will feel crowded, maybe have traffic congestion, etc.; This would be an important comparison if you were considering commuting to a location, opening a new business, cost of land, etc.

Build Understanding

Task 1 **B.** **a.** The figure is formed by three or more line segments such that each segment intersects exactly two other segments only at their endpoints. No two segments with a common endpoint are collinear.

b. The figure has a curved side.

c. The figure contains segments that intersect each other in places other than their endpoints.

d. The figure is formed by three or more line segments such that each segment intersects exactly two other segments only at their endpoints. No two segments with a common endpoint are collinear.

e. The figure contains segments that intersect each other in places other than their endpoints.

Task 2 **A.**

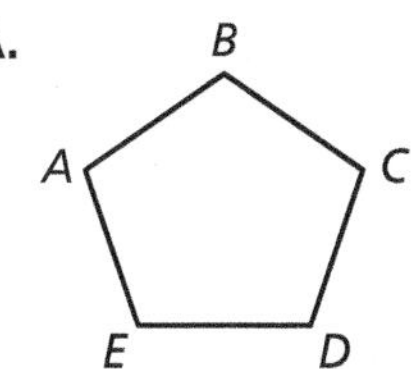

B.

C.

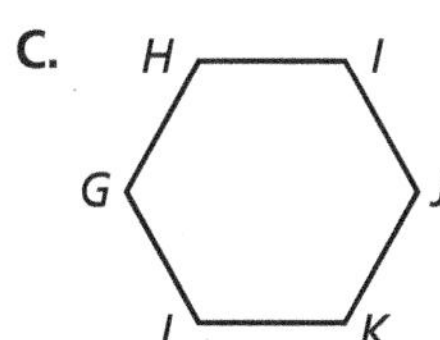

Step it Out

Task 3 **A.** Step 1: Mark the center and draw a circle using the compass.

Step 2: Draw a line across the diameter; mark the two intersection points with the circle.

Step 3: Using each new intersection point as a new center, draw two circles going through the original center point. Mark the intersection points with the original circle.

Step 4: Use the ruler to connect the points. You have six congruent segments forming a regular hexagon.

B. This construction creates 6 points that are evenly spaced around the circle, which allows you to draw a regular hexagon inscribed in the circle; No, you do not achieve a regular polygon if the points are not evenly spaced.

C. They show each side is congruent.

Task 4 **A.** Possible answer: two triangles and one rectangle

B. Triangle A: $A = 0.5bh = 0.5(4.5)(4) = 9$

Rectangle B: $A = lw = (7)(6.5) = 45.5$

Triangle C: $A = 0.5(5)(4.5) = 11.25$

Each grid square is 200 by 200, so has area 200^2 square feet. Use the Area Addition Postulate.

Total Area $= (9 + 45.5 + 11.25)(200^2) = (65.75)(200^2) =$ 2,630,000 ft^2

C. Possible answer: The given solution may be more accurate because the chosen shapes fit the island better.

D. The units are birds per square foot. Square feet is a unit for area, so the population density is found by dividing the number of endangered birds by the area.

E. The population density of endangered birds is approximately $\frac{1058}{2{,}630{,}000} \approx 0.0004$ bird per square foot.

Task 5 A. The length AB is c, which is the side length of the square.

B. The side length of the smaller central square is the length of the longer leg, a, reduced by the length of the shorter leg, b, or $a - b$.

C. Multiply each term of the first factor with each term of the second factor to get $a(a) + a(-b) + a(-b) + (-b)(-b)$, which simplifies to $a^2 - 2ab + b^2$.

Check Understanding

1. Check students' work. Sample answer: a square, a circle, a polygon with overlapping lines
2. 7-gon or heptagon, *PQRSTUV;* It is not regular because the sides aren't shown to be congruent.
3. 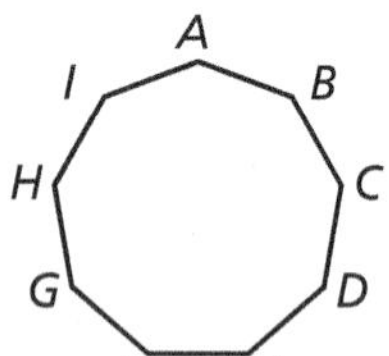

9-gon or nonagon

On Your Own

5. Possible answer: a circle; It contains a curve rather than only straight lines.
6. no; It is a hexagon but not a regular hexagon. All sides and angles are not congruent.
7. No, because both bases must be perpendicular to the height; Jason could divide the yard into a triangle and trapezoid and use the Area Addition Postulate.
13. Possible answers: *PQRST; QPTSR; STPQR*
14. Possible answers: *CDEF; DEFC; CFED*
15. Possible answers: *DEFGHI; GHIDEF; HGFEDI*
16. 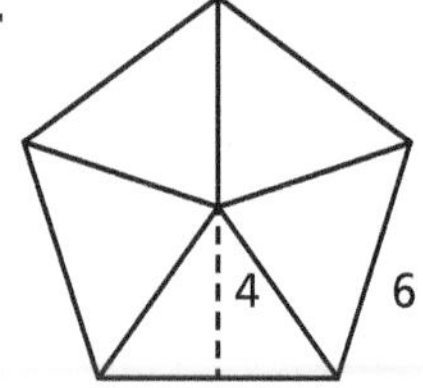

19. A. Mr. Edwards's class has a greater population density. The population density of Ms. Chang's class is approximately 0.06 student per square foot. The population density of Mr. Edwards's class is approximately 0.11 student per square foot.

20. Black bears: $\frac{2500}{9000} = 0.28$ black bears per square mile
Caribou: $\frac{2200}{9000} = 0.24$ caribou per square mile
Dall sheep: $\frac{1700}{9000} = 0.19$ dall sheep per square mile
Grizzly bears: $\frac{600}{9000} = 0.07$ grizzly bears per square mile
Moose: $\frac{1500}{9000} = 0.17$ moose per square mile
Wolves: $\frac{65}{9000} = 0.007$ wolves per square mile
Possible answer: The animals with a lower population density are animals that prey on the animals with greater population density. The animals with a greater population density may live in larger groups than those with a lower population density.

MODULE 1, LESSON 1.4

Spark Your Learning

B. Possible Answer: if the regions are divided into equal population or equal square mileage

D. by using the area formulas for squares, triangles, or other simple shapes

Build Understanding

Task 1 A. I can find the absolute value of the difference between the x-coordinates to find the difference between P and R; I can find the absolute value of the difference between the y-coordinates to find the difference between Q and S. The horizontal distance is 10 units and the vertical distance is 4 units.

B. I can divide the parallelogram along the x-axis into two non-overlapping triangles.

$A = \frac{1}{2}bh + \frac{1}{2}bh$

$A = (0.5)(10)(2) + (0.5)(10)(2)$

$A = 20 \text{ units}^2$

C. The area of the new parallelogram is 20 units^2. The areas of the two parallelograms are the same.

Possible graph:

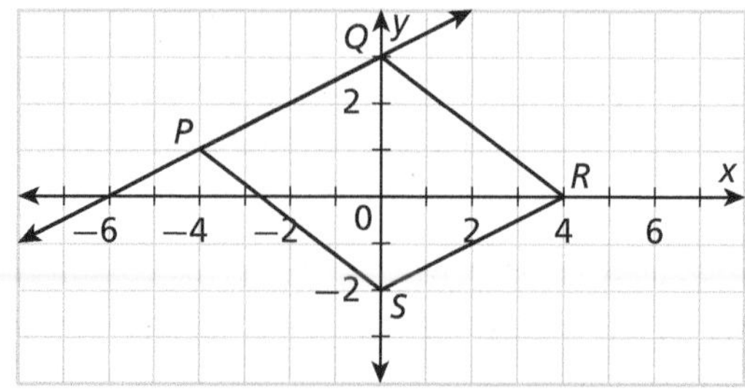

D. yes; If you move any one side along the line that contains it, the area will remain the same.

Possible graph:

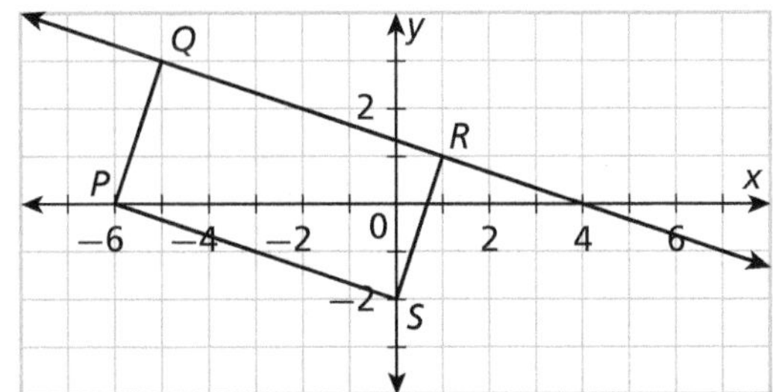

E. Possible answer: I used a different strategy to calculate the area using two triangles and a rectangle; This orientation involves using more shapes to calculate the area, so there are more opportunities to make a mistake.

Step it Out

Task 3 **A.** Possible answer: trapezoid, triangle, and square

B. Check students' work.

C. Possible answer: I used more shapes that more closely match the outline, so my solution is more precise; There is more than one way to estimate the area, but the solutions should all provide answers that are reasonably close.

D. Possible answer: I could divide the area into more/smaller shapes to make the area estimate more accurate; The challenge is choosing and calculating the areas of more and more shapes. The increase in accuracy may not be worth the extra time to make the calculations depending on the needed precision.

Check Understanding

3. Both students are correct; The Distance Formula is derived from the Pythagorean Theorem, so both methods are correct.

On Your Own

5. yes; The parallelograms share a base, and the sides opposite the base lie on the same line, so the parallelograms have the same height and therefore the same area.

6. no; The parallelograms share a base, but the sides opposite the base lie on different lines, so the parallelograms have different heights and therefore different areas.

7. Move one side along the line that contains it. The further you move the side, the more "stretched" the figure becomes. The perimeter will increase while the area remains the same.

14. no; He made an error in step 1. He wrote $\sqrt{\left(-3-(-4)\right)^2+(1-7)^2}$ instead of $\sqrt{(-3-1)^2+(-4-7)^2}$. The correct length of the segment is approximately 11.7 units.

22. No, the trail is too short; Using the Distance Formula for each segment, the trail is approximately 9.33 km and is too short to qualify for a 10K.

24. Check students' work for choice of geometric shape(s) to represent orange trees; area is about 250,000 ft^2

MODULE 2, ARE YOU READY?

1.

$4x+7=39$	Given
$4x+7-7=39-7$	Subtraction Property of Equality
$4x=32$	Simplify.
$\frac{4x}{4}=\frac{32}{4}$	Division Property of Equality
$x=8$	Simplify.

2.

$\frac{5}{8}t=2\frac{1}{2}$	Given
$\frac{5}{8}t=\frac{5}{2}$	Rewrite mixed number as improper fraction.
$\frac{8}{5}\left(\frac{5}{8}t\right)=\frac{8}{5}\left(\frac{5}{2}\right)$	Multiplication Property of Equality
$t=4$	Simplify.

3.

$6m-11=2m+13$	Given
$6m-11-2m=2m+13-2m$	Subtraction Property of Equality
$4m-11=13$	Simplify.
$4m-11+11=13+11$	Addition Property of Equality
$4m=24$	Simplify.
$\frac{4m}{4}=\frac{24}{4}$	Division Property of Equality
$m=6$	Simplify.

4.

$0.4(c-2)=-1.6$	Given
$0.4c-0.8=-1.6$	Distributive Property
$0.4c-0.8+0.8=-1.6+0.8$	Addition Property of Equality
$0.4c=-0.8$	Simplify.
$\frac{0.4c}{0.4}=\frac{-0.8}{0.4}$	Division Property of Equality
$c=-2$	Simplify.

MODULE 2, LESSON 2.1

Spark Your Learning

B. If you use fireworks, then you will be fined \$300.

If you fish, then you will be fined \$100.

If you litter, then you will be fined \$100.

If you smoke, then you will be fined \$100.

When the "if" part of the statement states the rule that is broken and the "then" part of the statement states the correct amount of the fine, the statement is true. When the "if" part of the statement states the amount of the fine and the "then" part states the rule that is broken, the statement may or may not be true.

D. If you do not use fireworks, then you will not be fined \$300; Basically, if you do not break the rule, you will not be fined.

Build Understanding

Task 1 **A.** Check students' sketches.

B. It is not possible to find a counterexample of the statement.

C. The statement is always true.

D. Check students' sketches.

E. It is not possible to find a counterexample of the statement.

F. The statement is always true.

G. Check students' sketches.

H. It is not possible to find a counterexample of each statement.

I. The statements are always true.

Task 2 **A.** hypothesis—the product of two real numbers is a negative number; conclusion—exactly one of the factors is a negative number

B. If the product of two real numbers is a negative number, then exactly one of the factors is a negative number.

C. If exactly one of the factors is a negative number, then the product of two real numbers is a negative number. This is a not a true statement. One factor could be a negative number and the other factor could be zero.

D. If the product of two real numbers is not a negative number, then the number of factors that are negative numbers is not one. This is a not a true statement. One factor could be a negative number and the other factor could be zero.

E. If the number of factors that are negative numbers is not one, then the product of two real numbers is not a negative number. This is a true statement.

Step It Out

Task 3 **A.** No, either statement can be the hypothesis or conclusion in a biconditional statement because the statement will be true either way.

B. A point divides a segment into two congruent segments if and only if the point is the midpoint of the line segment.

C. An angle is an acute angle if and only if its measure is between 0 and 90 degrees.

Check Understanding

1. Sketches will vary. It is not possible to draw a counterexample.

2. Statement: If yesterday was Monday, then today is Tuesday; Converse: If today is Tuesday, then yesterday was Monday. Inverse: If yesterday was not Monday, then today is not Tuesday. Contrapositive: If today is not Tuesday, then yesterday was not Monday.

3. An angle is a right angle if and only if it measures 90 degrees.

4. If your pizza isn't made right, then your next one is free.

On Your Own

5. Sketches will vary. It is not possible to find a counterexample.

6. Sketches will vary. It is not possible to find a counterexample.

10. Statement: If $t = 5$, then $2t + 3 = 13$. Converse: If $2t + 3 = 13$, then $t = 5$. Inverse: If $t \neq 5$, then $2t + 3 \neq 13$. Contrapositive: If $2t + 3 \neq 13$, then $t \neq 5$.

11. If the dog performs the trick, then it gets a treat. Converse: If the dog gets a treat, then it performed a trick. Inverse: If the dog does not perform a trick, then it did not get a treat. Contrapositive: If the dog does not get a treat, then it did not perform a trick.

12. Statement: If a triangle is an isosceles triangle, then it has two sides with the same length. Converse: If a triangle has two sides with the same length, then the triangle is an isosceles triangle. Inverse: If a triangle is not an isosceles triangle, then it does not have two sides with the same length. Contrapositive: If a triangle does not have two sides with the same length, then it is not an isosceles triangle.

13. A triangle is a scalene triangle if and only if it has three sides with different lengths.

14. A rectangle is a square if and only if it has four sides that are the same length.

15. Two lines are perpendicular if and only if they intersect at a 90° angle.

16. Statement: If two numbers are positive, then the sum is positive; Converse: If the sum of two number is positive, then the two numbers are positive; The converse is not true. For example, $8 + (-3) = 5$.

20. A point is the y-intercept of a line if and only if the point is where the line intersects the y-axis.

21. Statement: If the job wasn't done by us, then it wasn't done right; Converse: If the job wasn't done right, then it wasn't done by us. Inverse: If the job was done by us, then it was done right. Contrapositive: If the job was done right, then it was done by us.

22. Converse: If you attended the concert, then you bought a ticket to the concert. Inverse: If you did not buy a ticket to the concert, then you did not attend the concert. Contrapositive: If you did not attend the concert, then you did not buy a ticket to the concert; The converse is not true; You could have been given a ticket, or won a ticket.

MODULE 2, LESSON 2.2

Spark Your Learning

C. Answers will vary. Felicia can tell how well she prepared for the fair by assessing her overall profit, if she ran out of an item, if she had an abundance of items, and customer feedback and comparing to previous experiences; Felica might learn which items are in demand and which booths attracted the most customers in preparation for the next fair.

Build Understanding

Task 1 **A.** The first, third, and fifth reviews use inductive reasoning; Possible answer: They use specific cases to draw their conclusion.

The second and fourth reviews use deductive reasoning; Possible answer: They both draw a conclusion using facts about the phone.

B. No, the type of reasoning does not determine the type of review; For example, the first review is a positive review using inductive reasoning, but the third review is a negative review using inductive reasoning.

C. Possible answers: Everyone in my family uses this phone and loves it. This is inductive because it uses the specific cases of the people in the family.

D. Possible answers: I recommend this phone because it was the best-selling phone last year; This is deductive because it is based on the fact that the phone was the best-selling phone last year.

Task 2 **A.** Addition Property of Equality; This step allows you to collect the variable terms on one side of the equal symbol.

B. Subtraction Property of Equality; This step allows you to collect the constant term on one side of the equal symbol.

C. Division Property of Equality; This step is done so that the coefficient of the variable term is 1.

Step It Out

Task 3 **A.** Addition Property of Equality; The same term is being added to both sides.

B. Multiplication Property of Equality; The same term is being multiplied on both sides of the equation.

Check Understanding

1. The conclusion is based on observing four numbers.

2. deductive reasoning; This proof would show that the formula is true for all cases.

On Your Own

4. The conclusion is based on a limited number of observations.

5. The conclusion is based on observing three numbers.

6. The conclusion is based on observing the shirt Justin wears for four days.

7. The conclusion is based on observing the scores of 7 soccer games.

8. The conclusion is based on observing the sums of three pairs of positive numbers.

9. The conclusion is based on observing the pattern in 3 different images.

10. inductive reasoning; The conclusion is based on observing Jose going to the store several times.

11. deductive reasoning; The definition of a right angle is used to draw the conclusion.

12. deductive reasoning; The definition of a mammal is used to draw the conclusion.

13. inductive reasoning; The conclusion is based on observing Cindy's six rolls of a number cube.

14. deductive reasoning; The definition of a scalene triangle is used to draw the conclusion.

15. inductive reasoning; The conclusion is based on observing Tammy's 3 bowling scores.

16. **A.** Possible answer: The pattern could continue with two diamonds, then three of each shape, and then four of each shape; inductive reasoning

B. Possible answer: The pattern could continue with 1 diamond, then one circle, then one diamond, then two circles, etc.

38.

$y - 23 = 7$	Given
$y - 23 + 23 = 7 + 23$	Addition Property of Equality
$y = 30$	Simplify.

39.

$n + 14 = 19$	Given
$n + 14 - 14 = 19 - 14$	Subtraction Property of Equality
$n = 5$	Simplify.

40.

$3a + 7 = 16$	Given
$3a + 7 - 7 = 16 - 7$	Subtraction Property of Equality
$3a = 9$	Simplify.
$\frac{3a}{3} = \frac{9}{3}$	Division Property of Equality
$a = 3$	Simplify.

41.

$2r - 11 = 1$	Given
$2r - 11 + 11 = 1 + 11$	Addition Property of Equality
$2r = 12$	Simplify.
$\frac{2r}{2} = \frac{12}{2}$	Division Property of Equality
$r = 6$	Simplify.

42. Given: $6b + 7 = b + 2$

Prove: $b = -1$

Statements	Reasons
$6b + 7 = b + 2$	Given
$6b + 7 - b = b + 2 - b$	Subtraction Property of Equality
$5b + 7 = 2$	Simplify.
$5b + 7 - 7 = 2 - 7$	Subtraction Property of Equality
$5b = -5$	Simplify.
$\frac{5b}{5} = \frac{-5}{5}$	Division Property of Equality
$b = -1$	Simplify.

43. Given: $9k - 5 = 7k + 3$

Prove: $k = 4$

Statements	Reasons
$9k - 5 = 7k + 3$	Given
$9k - 5 - 7k = 7k + 3 - 7k$	Subtraction Property of Equality
$2k - 5 = 3$	Simplify.
$2k - 5 + 5 = 3 + 5$	Addition Property of Equality
$2k = 8$	Simplify.
$\frac{2k}{2} = \frac{8}{2}$	Division Property of Equality
$k = 4$	Simplify.

48. Jerome knows that the sum of the lengths of the two parts of a segment are equal to the length of the entire segment using the Segment Addition Postulate. He then subtracted 9 from 23 to find the unknown length.

MODULE 2, LESSON 2.3

Spark Your Learning

B. the distances between locations along Brianna and Carl's way home from school and the total distance from the school to their home; Segment Addition Postulate.

D. yes; The sum of the lengths of the segments must be equal to the length of the segment representing the walk from school to home.

Build Understanding

Task 1 **A.** The structure is the same in some places, so some of the rope segments should be congruent.

B. Use the Symmetric Property. If $\overline{WX} \cong \overline{TV}$, then $\overline{TV} \cong \overline{WX}$.

C. Use the Transitive Property. If $\overline{GH} \cong \overline{JK}$ and $\overline{JK} \cong \overline{LM}$, then $\overline{GH} \cong \overline{LM}$.

D. The Reflexive Property of Congruence tells you that every rope segment is congruent to itself.

Step It Out

Task 3 **A.** Segment Addition Postulate

B. Subtraction Property of Equality

C. Possible answer: Use $AB = 12$ in Step 1 and $BC = 12$ in Step 5 to know that $AB = BC$. The definition of congruent segments states that two segments are equal if they have the same length.

Task 4 **A.** *ABED* is a square.

$\overline{AB} \cong \overline{BC}$

$AB = DE$

$\overline{AB} \cong \overline{DE}$

$\overline{DE} \cong \overline{BC}$

B. Given

Given

All sides of a square are equal length.

Definition of congruent segments

Transitive Property of Congruent Segments

Task 5 **A.** Possible answer: You want to know the value of x that will make the segments congruent. If the segments are congruent, their lengths are equal, so you can set the lengths equal to each other.

B. $5x - 17 + 17 = 53 + 17$

$5x = 70$

$\frac{5x}{5} = \frac{70}{5}$

$x = 14$

Check Understanding

2. $\overline{AB}, \overline{AC}; \overline{DB}, \overline{CE}; \overline{DF}, \overline{EG}$

4.

Statements	Reasons
1. $\overline{DB} \cong \overline{CE}$	**1.** Given
2. $DB = CE$	**2.** Definition of congruent segments
3. $DE = DB + BC + CE$	**3.** Segment Addition Postulate
4. $DE = DB + BC + DB$	**4.** Substitution
5. $DE = FG$	**5.** Opposite sides of a rectangle have equal length.
6. $FG = DB + BC + DB$	**6.** Substitution
7. $FG = 3BC$	**7.** Given
8. $3BC = DB + BC + DB$	**8.** Substitution
9. $BC = DB$	**9.** Simplify.
10. $\overline{BC} \cong \overline{DB}$	**10.** Definition of congruent segments

On Your Own

20. first group: $\overline{AB}$, $\overline{FG}$, and $\overline{DC}$; second group: $\overline{AD}$ and $\overline{BC}$; third group: $\overline{AE}, \overline{EB}, \overline{FI}$, and $\overline{IG}$; fourth group: $\overline{AF}, \overline{EI}$, and $\overline{BG}$; fifth group: $\overline{FD}$ and $\overline{GC}$

21. first group: $\overline{XY}, \overline{YZ}$, and $\overline{XZ}$; second group: $\overline{YW}$ and $\overline{WZ}$

22. yes; $RS + ST = RT$, so $25 + ST = 50$. Solve for ST to find $ST = 25$. Since, $RS = ST$, $\overline{RS}$ is congruent to $\overline{ST}$.

23.

Statements	Reasons
1. Triangle *DEF* is an equilateral triangle. $\overline{FG} \cong \overline{DE}$	**1.** Given
2. $\overline{DE} \cong \overline{EF}$	**2.** Definition of an equilateral triangle
3. $\overline{EF} \cong \overline{FG}$	**3.** Transitive Property of Segment Congruence

24.

Statements	Reasons
1. $AB = CD$	**1.** Given
2. $AB + BC = CD + BC$	**2.** Addition Property of Equality
3. $AB + BC = AC$	**3.** Segment Addition Postulate
4. $CD + BC = BD$	**4.** Segment Addition Postulate
5. $AC = BD$	**5.** Transitive Property of Equality

35.

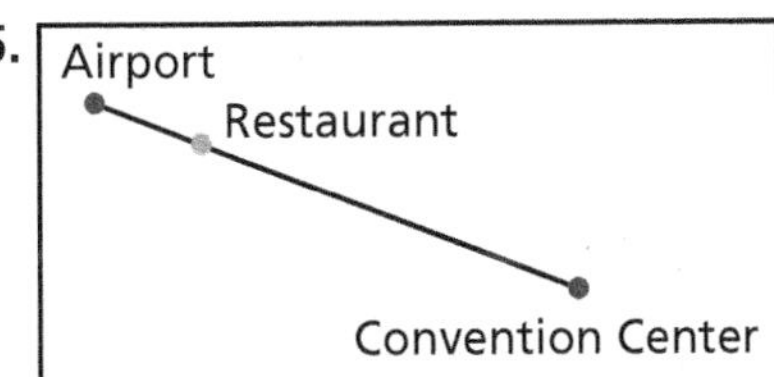

MODULE 2, LESSON 2.4

Spark Your Learning

B. We can assume that the scissor legs do not bend; The scissor legs can be modeled with line segments, angles and triangles; We know that the line segments are all congruent because the scissor legs intersect at their midpoint, so we can, more specifically, use isosceles triangles.

D. yes; My answer makes sense because it is reasonable to assume that the legs are rigid, and so the angles have to change in order for the height of the platform to change.

Build Understanding

Task 1 **A.** Transitive Property of Congruence

B. Possible answer: Showing that each measure is equal to the same value allows you to use the Transitive Property in the next step.

angles have the same measure, they are congruent angles.

D. Possible answer: You need to write the angle measures to show that the angles have equal measures. Once you know the measures are equal, you know the angles are congruent.

Step It Out

Task 2 **A.** Angle Addition Postulate

B. Subtraction Property of Equality

C. Possible answer: The steps to solve this problem are specific to this problem. They cannot be used for every angle addition problem.

Task 3 **A.** Definition of supplementary angles

B. Transitive Property

C. Definition of congruent angles

Task 4 **A.** Definition of linear pair

B. Angle Addition Postulate

Task 5 **A.** Transitive Property

B. Subtraction Property of Equality

Check Understanding

1. yes; Possible answer: $\angle 1$ is a vertical angle with the right angle, so it is a right angle. $\angle 2$ forms a linear pair with $\angle 1$, so $\angle 2$ is a right angle because $90° + 90° = 180°$. Because $\angle 1$ and $\angle 2$ have the same measure, they are congruent.

On Your Own

17.

Statements	Reasons
1. $\angle DEF \cong \angle FEB$	**1.** Given
2. $m\angle DEF = m\angle FEB$	**2.** Definition of congruent angles
3. $\angle AEC \cong \angle DEB$	**3.** Vertical Angles Theorem
4. $m\angle AEC = m\angle DEB$	**4.** Definition of congruent angles
5. $m\angle DEB = m\angle DEF + m\angle FEB$	**5.** Angle Addition Postulate
6. $m\angle AEC = m\angle DEF + m\angle FEB$	**6.** Transitive Property
7. $m\angle AEC = m\angle FEB + m\angle FEB$	**7.** Substitution
8. $m\angle AEC = 2m\angle FEB$	**8.** Combine like terms.

18. If two angles are supplementary, then they form a linear pair; This is not a true statement. Supplementary angles do not have to be adjacent (they do not need to have a common side or noncommon sides that are opposite rays).

19.

Statements	Reasons
1. $\angle 1 \cong \angle 2$	**1.** Given
2. $\angle 2$ is a complement of $\angle 3$.	**2.** Given
3. $m\angle 2 + m\angle 3 = 90°$	**3.** Definition of complementary angles
4. $m\angle 1 = m\angle 2$	**4.** Definition of congruent angles
5. $m\angle 1 + m\angle 3 = 90°$	**5.** Substitution
6. $\angle 1$ is a complement of $\angle 3$.	**6.** Definition of complementary angles

30. $m\angle 1 = 65°$ because of the Linear Pairs Theorem; $m\angle 2 = 115°$ because of the Vertical Angle Theorem; $m\angle 3 = 65°$ because of the Linear Pairs Theorem.

34. $m\angle 2 = m\angle 3 = 76°$ by the Linear Pairs Theorem; $m\angle 1 = 40°$ by the Vertical Angles Theorem and the Angle Addition Postulate.

37. Possible answer: When one of the angles is bisected, the bisecting ray divides the rhombus into two congruent isosceles triangles.

MODULE 2 REVIEW

Concepts and Skills

5. An angle is an obtuse angle if and only if its measure is greater than 90° and less than 180°.

6. A polygon is a pentagon if and only if it has five sides.

7. If it is Monday, then it is a weekday. Converse: If it is a weekday, then it is Monday. Inverse: If it is not Monday, then it is not a weekday. Contrapositive: If it is not a weekday, then it is not Monday.

UNIT 2

MODULE 3, LESSON 3.1

Build Understanding

Task 1 **A.** Check students' diagrams. Possible answer:

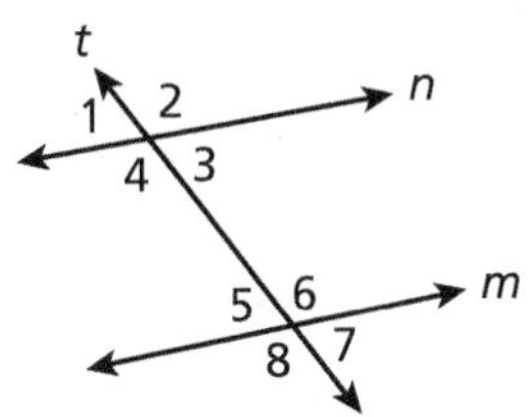

B. Possible answers using the diagram given for Part A: corresponding angles: ∠1 and ∠5; alternate exterior angles: ∠1 and ∠7; alternate interior angles: ∠3 and ∠5; consecutive exterior angles: ∠1 and ∠8; consecutive interior angles: ∠4 and ∠5

C. yes; yes; Every pair of vertical angles is congruent. Every linear pair of angles is supplementary. This is true any time two lines intersect, where the relationship is not affected by the other line.

Step It Out

Task 4 m∠1 = 70°; Possible answer: Consecutive Interior Angles Theorem

Check Understanding

1. Possible answers: corresponding angles: ∠1 and ∠5, alternate exterior angles: ∠1 and ∠7, alternate interior angles: ∠3 and ∠5, consecutive exterior angles: ∠1 and ∠8, consecutive interior angles: ∠3 and ∠6
2. ∠1, ∠5, ∠7
3. By the Consecutive Angles Theorem, m∠4 + m∠5 = 180°. As a linear pair, m∠4 + m∠3 = 180°. Therefore m∠4 + m∠3 = m∠4 + m∠5, so m∠3 = m∠5.
4. $x = 12$

On Your Own

5. *See below.*
6. ∠4, ∠6, ∠8
7. ∠2, ∠4, ∠6, ∠8
8. Corresponding Angles Postulate
9. Consecutive Exterior Angles Theorem
10. Alternate Interior Angles Theorem
11. Consecutive Interior Angles Theorem
12. Answers will vary.
13. m∠1 = 80°, m∠2 = 100°
14.

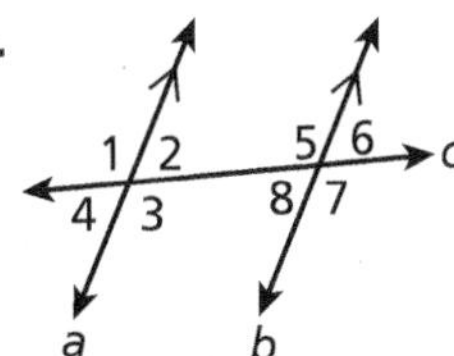

Statements	Reasons
1. Line *a* is parallel to line *b*.	1. Given
2. ∠1 is supplementary to ∠2.	2. Linear Pairs Theorem
3. ∠2 ≅ ∠6	3. Corresponding Angles Postulate
4. ∠1 is supplementary to ∠6.	4. Substitution Property

15.

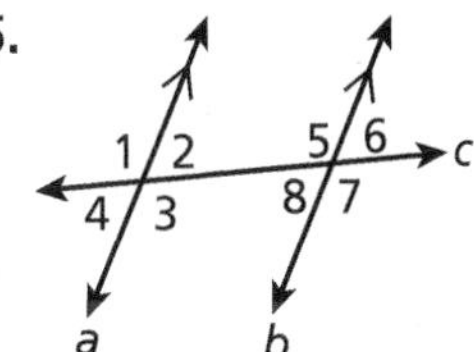

Statements	Reasons
1. Line *a* is parallel to line *b*.	1. Given
2. ∠5 is supplementary to ∠2.	2. Consecutive Interior Angles Postulate
3. ∠1 is supplementary to ∠2.	3. Linear Pairs Theorem
4. ∠1 ≅ ∠5	4. Congruent Supplements Theorem

5.

Corresponding angles	Consecutive interior angles	Consecutive exterior angles	Alternate interior angles	Alternate exterior angles
∠1 and ∠5 ∠2 and ∠6 ∠3 and ∠7 ∠4 and ∠8	∠3 and ∠6 ∠4 and ∠5	∠1 and ∠8 ∠2 and ∠7	∠3 and ∠5 ∠4 and ∠6	∠1 and ∠7 ∠2 and ∠8

16. $x = 16$

17. $x = 11$

18. $x = 27$

19. $x = 13$

20. $x = 7$

21. $x = 21$

22.

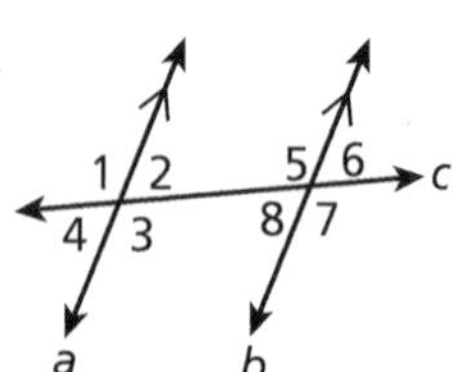

Statements	Reasons
1. Line *a* is parallel to line *b*.	**1.** Given
2. $\angle 1 \cong \angle 3$	**2.** Vertical Angles Theorem
3. $\angle 3 \cong \angle 7$	**3.** Corresponding Angles Postulate
4. $\angle 1 \cong \angle 7$	**4.** Transitive Property of Congruence

23. A. $\angle 2$

B. Corresponding Angles Postulate

C. $m\angle 1 = 105°$, $m\angle 2 = 75°$, $m\angle 3 = 105°$

24.

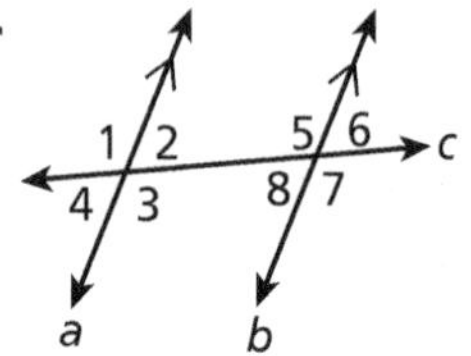

Statements	Reasons
1. Line *a* is parallel to line *b*.	**1.** Given
2. $\angle 5$ is supplementary to $\angle 6$.	**2.** Linear Pairs Theorem
3. $\angle 2 \cong \angle 6$	**3.** Corresponding Angles Postulate
4. $\angle 2$ is supplementary to $\angle 5$.	**4.** Congruent Supplements Theorem

25. $m\angle 3 = m\angle 9 = 125°$

26. $m\angle 12 = 59°$, $m\angle 15 = 121°$

27.

Statements	Reasons
1. Line *a* is parallel to line *b*. Line *x* is parallel to line *y*.	**1.** Given
2. $\angle 2 \cong \angle 6$	**2.** Corresponding Angles Postulate
3. $\angle 6$ is supplementary to $\angle 7$.	**3.** Linear Pairs Theorem
4. $\angle 7 \cong \angle 13$	**4.** Alternate Interior Angles Theorem
5. $\angle 6$ is supplementary to $\angle 13$.	**5.** Congruent Supplements Theorem
6. $\angle 2$ is supplementary to $\angle 13$.	**6.** Congruent Supplements Theorem

MODULE 3, LESSON 3.2

Spark Your Learning

B. The illusion of curved lines is not present when looking at a small area of the illusion.

D. Possible answer: Using a picture of the illusion straight on, you could use a ruler and measure the distance between two lines at various places along the lines. Equal measures will show that the lines are the same distance apart, so they are parallel.

Build Understanding

Task 2 **B.** Possible answer: You could state that the angle that is a vertical angle of $\angle 1$ and $\angle 2$ are congruent alternate interior angles, and use the Converse of the Alternate Interior Angles Theorem.

Step It Out

Task 4 **A.** $m\angle DAB = 72°$, $m\angle EBC = 109°$, $m\angle FCB = 71°$; Given: $m\angle DAB = 72°$, $m\angle EBC = 109°$, $m\angle FCB = 71°$

Task 5 **A.** Vertical Angles Theorem

B. Converse of the Consecutive Interior Angles Theorem

C. Converse of the Corresponding Angles Postulate

D. Transitive Property of Parallel Lines

Check Understanding

1.

One line can be drawn through P that is parallel to k by the Parallel Postulate.

On Your Own

8. yes; Converse of the Alternate Exterior Angles Theorem

9. yes; Converse of the Consecutive Interior Angles Theorem

16.

Statements	Reasons
1. $\angle 1 \cong \angle 7$	1. Given
2. $\angle 5 \cong \angle 7$	2. Vertical Angles Theorem
3. $\angle 1 \cong \angle 5$	3. Transitive Property of Congruence
4. $a \parallel b$	4. Converse of Corresponding Angles Postulate

17.

Statements	Reasons
1. $\angle 4$ and $\angle 5$ are supplementary.	1. Given
2. $\angle 5$ and $\angle 8$ are supplementary.	2. Linear Pairs Theorem
3. $\angle 4 \cong \angle 8$	3. Congruent Supplements Theorem
4. $a \parallel b$	4. Converse of Corresponding Angles Postulate

18. Each lane is parallel to the lane next to it. By the Transitive Property of Parallel Lines, the left side of the pool is parallel to the right side of the pool.

19. Possible answer: The corresponding angles are all equal.

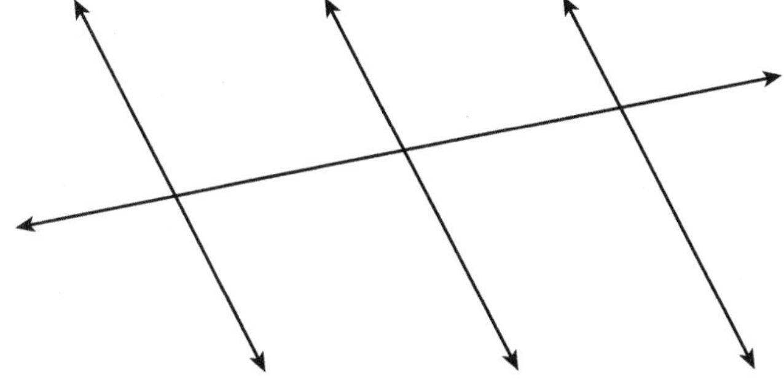

20.

Statements	Reasons
1. $\angle 1$ and $\angle 2$ are supplementary.	1. Given
2. $x \parallel y$	2. Converse of Consecutive Exterior Angles Theorem
3. $\angle 3 \cong \angle 4$	3. Given
4. $y \parallel z$	4. Converse of Alternate Interior Angles Theorem
5. $x \parallel z$	5. Transitive Property of Parallel Lines

21. A. $b \parallel c$; The measure of the vertical angle of the 45° angle measures 45°. 45° and 135° angles are supplements. So, the lines are parallel by the Converse of Consecutive Interior Angles Theorem.

B. Line a is not parallel with the other lines; Because 46° and 135° are not supplementary or congruent, the lines are not parallel by any of the converses of the parallel lines theorems.

MODULE 3, LESSON 3.3

Spark Your Learning

D. Possible answer: relationships between intersecting paths and the relationship between a path and a vertex; The paths that intersect are perpendicular because of congruent angles. The path that connects a vertex to the opposite side is also perpendicular to the opposite side because of right angles.

Build Understanding

Task 2 **E.** Because $AD = BD$, then $\overline{CD}$ bisects $\overline{AB}$. $\overline{CD}$ is also perpendicular to $\overline{AB}$, so C lies on the perpendicular bisector of $\overline{AB}$.

Step It Out

Task 3 **A.** Since the arc radii are the same, the distances from X to each intersection and from Y to each intersection must be equal. So, the intersections lie on the perpendicular bisector of $\overline{XY}$ by the Converse of the Perpendicular Bisector Theorem.

B. The midpoint is the point that is equal distance from each endpoint. The perpendicular bisector is all points that are equal distance from each end point. This means that the midpoint is on the perpendicular bisector.

Task 4 **B.** Point P is equidistant from points A and B because the arcs have the same radius. The point at the intersection of the drawn arcs is also equidistant from points A and B. Therefore, by the Converse of the Perpendicular Bisector Theorem, the line drawn is the perpendicular bisector of the segment.

Check Understanding

1. When you fold the paper in half to match up the endpoints of $\overline{RS}$, the fold is perpendicular to the segment, and it is halfway between the endpoints of the segment. So, the fold is a perpendicular bisector.

2. Since $AD = CD$, D lies on the perpendicular bisector of $\overline{AC}$. Since $AB = CB$, B also lies on the perpendicular bisector of $\overline{AC}$. $\overline{BD}$ is the perpendicular bisector of $\overline{AC}$ because there is one line that passes through the two points. Thus, $\overline{BD} \perp \overline{AC}$.

On Your Own

5. No, you need to know PR.

6. Yes, the distances from the endpoints are the same for both sides of the triangle. Every point on the segment is equal distance from the endpoints.

7.

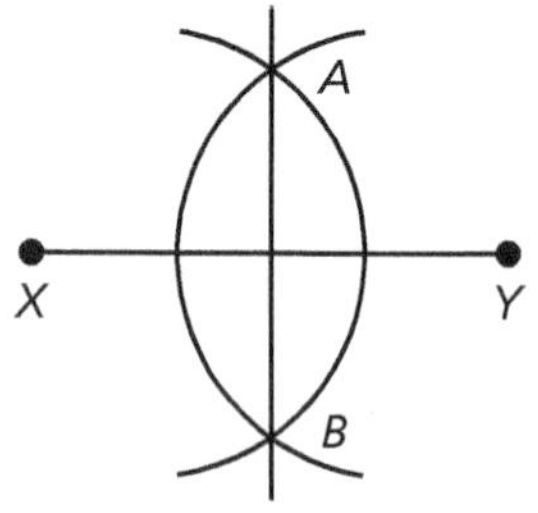

13.

Statements	Reasons
1. $AC = BC$ $AD = BD$	**1.** Given
2. C lies on the perpendicular bisector of $\overline{AB}$.	**2.** Converse of the Perpendicular Bisector Theorem
3. D lies on the perpendicular bisector of $\overline{AB}$.	**3.** Converse of the Perpendicular Bisector Theorem
4. $\overline{AB} \perp \overline{CD}$	**4.** Definition of perpendicular bisector

18. A.

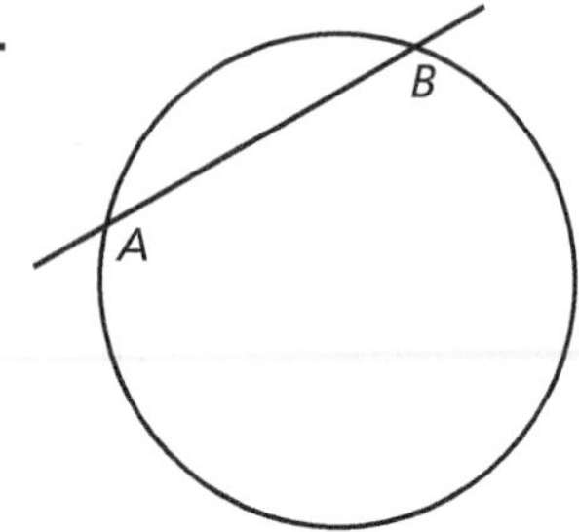

B.

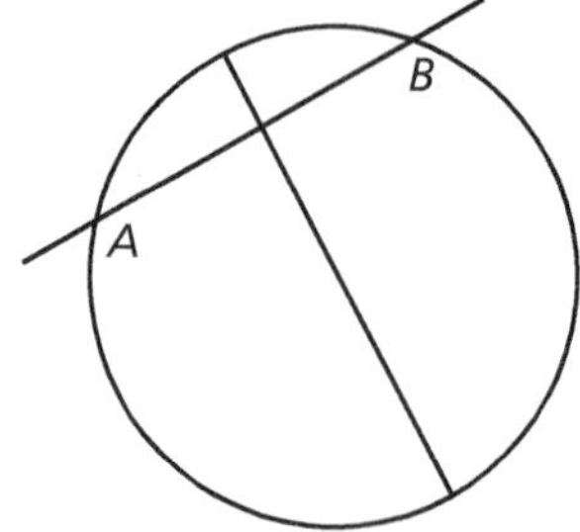

Every point on the perpendicular bisector is equal distance from the endpoints of $\overline{AB}$. The center of the circle has to be on the perpendicular bisector because the center is equal distance from all points on the circle. A segment that intersects a circle at two points and passes through the center is a diameter of the circle.

C.

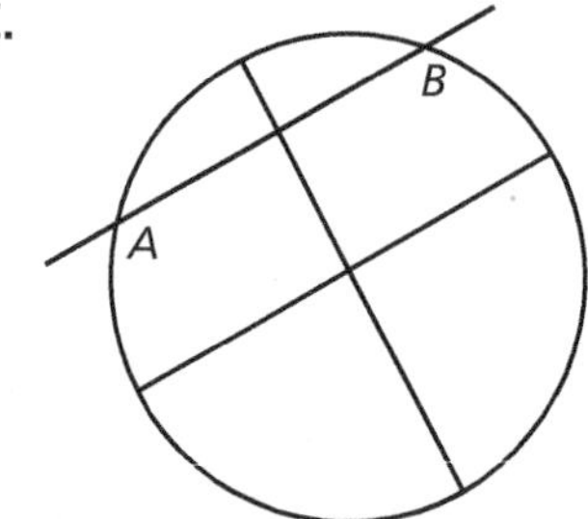

The perpendicular bisector constructed at this step is also a diameter. The point where two diameters intersect is the center of the circle.

D. It is not possible. By definition, a chord is a segment whose endpoints are on a circle. Every point that is on the perpendicular bisector of the chord is the same distance from the endpoints of the chord. In particular, the center is the same distance from the endpoints of the chord because the center is the same distance from all points on the circle. So, the center of the circle will be on the perpendicular bisector.

19.

Statements	Reasons
1. $a \perp t$ and $a \parallel b$	**1.** Given
2. $m\angle 1 = 90°$	**2.** Definition of perpendicular lines
3. $m\angle 1 = m\angle 2$	**3.** Corresponding Angles Postulate
4. $m\angle 2 = 90°$	**4.** Transitive Property of Equality
5. $b \perp t$	**5.** Definition of perpendicular lines

25. By the Converse of the Perpendicular Bisector Theorem, you know that the oxygen atom is on the perpendicular bisector.

26.

Statements	Reasons
1. $m \perp p$ and $n \perp p$	1. Given
2. $m\angle 1 = 90°$, $m\angle 2 = 90°$	2. Definition of perpendicular lines
3. $m \parallel n$	3. Converse of Corresponding Angles Postulate

27. $w = 5$, $v = 7.2$; The linear pair formed are congruent angles, so the lines are perpendicular. $(15w + 15)° = 90°$, so $w = 5$. $\frac{25v}{2} = 90$, so $v = 7.2$.

28. The valve pistons are lines that are perpendicular to the same line (the lead pipe), so they form right angles with the same line. By the Converse of the Corresponding Angles Postulate, all the congruent right angles mean the valve pistons are parallel to each other.

MODULE 3 REVIEW

Concepts and Skills

4. Possible answer: $\angle 1$ and $\angle 5$; congruent
5. Possible answer: $\angle 3$ and $\angle 5$; supplementary
6. Possible answer: $\angle 4$ and $\angle 5$; congruent
7. Possible answer: $\angle 1$ and $\angle 8$; congruent
9. Line a and line b are parallel. The measures of the vertical angle of 68° and 112° are supplements. So, the lines are parallel by the Converse of the Consecutive Interior Angles Theorem. Line b and line c are not parallel because the consecutive exterior angles 67° and 112° are not supplementary. Line a and line c are not parallel because the alternate exterior angles 68° and 67° are not congruent.

MODULE 4, LESSON 4.1

Spark Your Learning

B. Possible answer: I know parallel lines have the same slope; The slope-intercept form of a line could be helpful.

Build Understanding

Task 1 **A.** no; no; As long as the numerator of the slope formula finds the difference of the y-coordinates in the same order as the difference of the x-coordinates in the denominator, the slopes will be the same.

D. Possible answer: I think the third line would also have the same slope. The new line is parallel to the two given lines.

Task 2 **A.** The coordinates are needed so that they can be substituted into the slope equation.

C. Segments $\overline{BC}$ and $\overline{DA}$ are parallel. Since the shape has only one pair of parallel sides, it is not a parallelogram.

D. no; The segments do not have to be the same length or near each other on the coordinate plane to be parallel.

Step It Out

Task 3 **A.** Possible answer: The values of a and b represent the translation of the x- and y-coordinates of line p to line q.

Task 5 **A.** Both equations describe horizontal lines, and all horizontal lines are parallel.

B. Both equations describe vertical lines, and all vertical lines have an undefined slope.

Check Understanding

1. no; Possible answer: Parallel lines must not intersect. This would create the same line twice, which intersects itself infinitely many times.

3. Yes. Line u has a slope of $\frac{(y_2 - y_1)}{(x_2 - x_1)}$. Line v has a slope of $\frac{(y_2 - 1) - (y_1 - 1)}{(x_2 + 2) - (x_1 + 2)} = \frac{(y_2 - y_1)}{(x_2 - x_1)}$. Since the two lines have the same slope, the lines are parallel.

On Your Own

10.

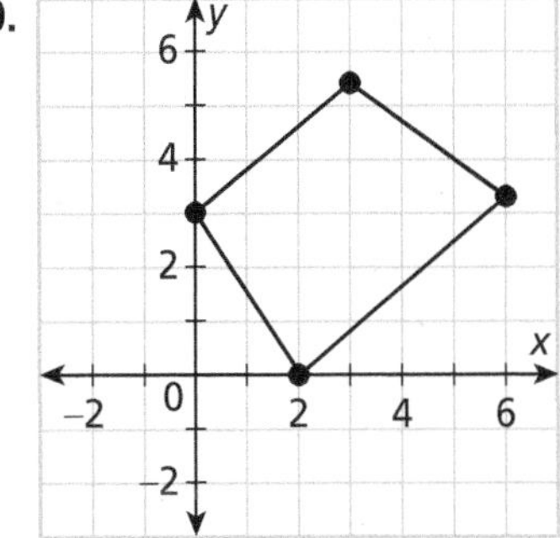

11.

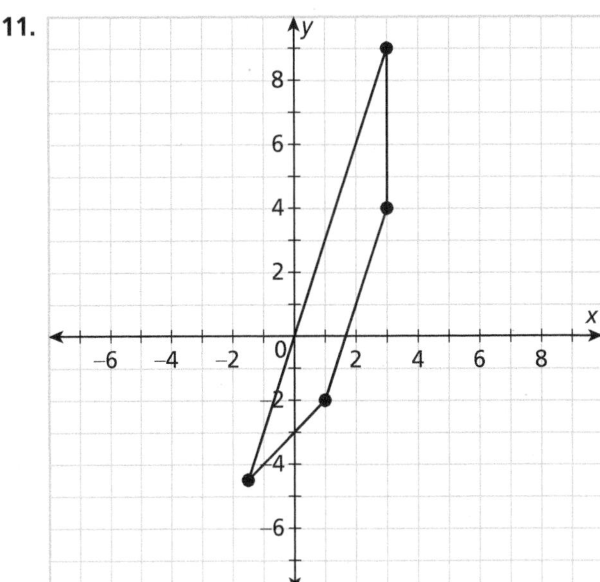

16. Yes, she graphs two pairs of parallel segments that meet at each endpoint; One pair of segments is vertical (undefined slope), and the other is horizontal (slope of zero). This creates a parallelogram.

34. $y = -\frac{3}{4}x + 9.25$; for $x = -1$ and $x = 7$

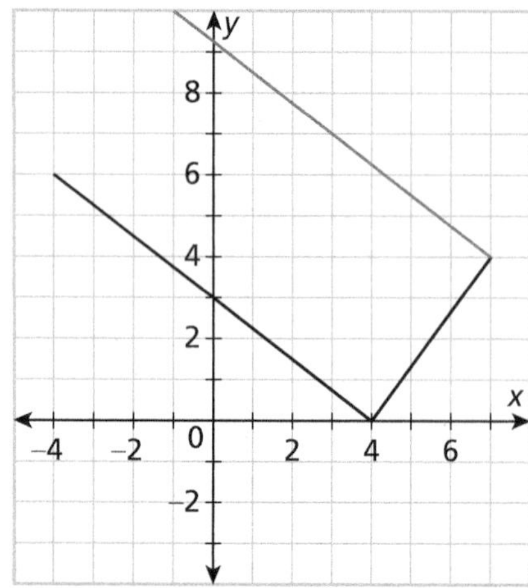

35. A. no; The two lines do not have the same slope.

B. $y = -\frac{1}{3}x + 3$

C. The two lines would be parallel.

36. false; Vertical lines are parallel and do not have a real number slope.

37. light rail: $y = -0.25x + 107.5$; community walking path: $y = -0.25x + 160$

38. Possible answers: $y = \frac{1}{2}x + 5$ and $4x - 8y = 7$

MODULE 4, LESSON 4.2

Spark Your Learning

B. The two intersecting lines are perpendicular. The two lines representing the four blades meet to form four congruent angles. Choose any two adjacent angles. Since any two congruent angles that form a linear pair are right angles, repeating this four times shows that all four angles are right angles. Thus, the lines are perpendicular.

C. no; A six-bladed windmill could be represented by 3 intersecting lines. The adjacent angles formed would be congruent, but they would not form linear pairs. So, the lines would not be perpendicular.

Build Understanding

Task 1 A.

Name	Slope
$\overleftrightarrow{AD}$	4
$\overline{AB}$	$-\frac{1}{4}$
$\overline{BC}$	4
$\overline{CD}$	$-\frac{1}{4}$

B. The slopes of $\overleftrightarrow{AD}$ and $\overline{BC}$ are the same; The slopes of $\overline{AB}$ and $\overline{CD}$ are the same; The slopes have the opposite sign and are reciprocals.

C. Possible answer: I could put the equation in slope-intercept form, and then check whether or not the slopes are opposite reciprocals.

D. The lines $y = 3x - 0.5$ and $x + 3y = 6$ are perpendicular.

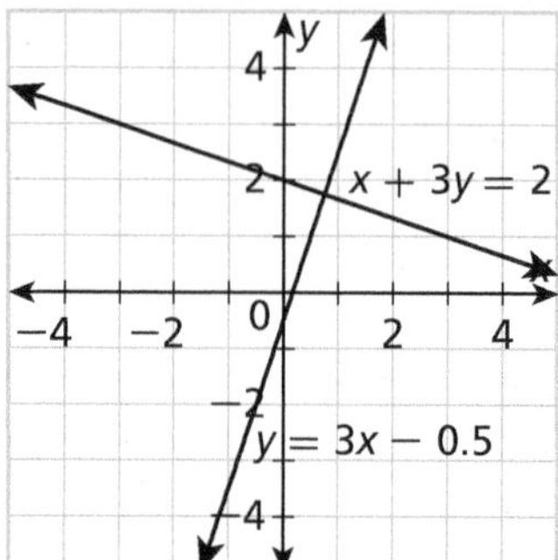

Task 2 A. Possible answer: In a rotation, line segments go to line segments, lines go to lines, and so on. All segments and angles are congruent before and after a rotation, just in a different position.

B. yes; The two slopes must be opposite reciprocals. So, one must be positive and one must be negative.

C. no; The slope of a vertical line is undefined, so it can't be multiplied by anything or have an opposite reciprocal.

Step It Out

Task 3 A. $m = \frac{y_2 - y_1}{x_2 - x_1}$

B. 4 is the opposite reciprocal of $-\frac{1}{4}$.

C. Possible answer: You substitute the values you've found for both the slope and y-intercept back into the equation, using x and y as the variables, to create a general equation that describes any point on the line.

Task 4 A. $(5, -3)$

B. Possible answer: For any vertical line passing through point (a, b), $x = a$.

Task 5 A. Possible answer: I counted the rise and run using the figure.

B. Possible answer: The slope is needed to substitute into the slope-intercept form of a line to create the equation.

C. Possible answer: I would need to use the slope and the coordinates of the given point in the slope-intercept form of a line to find the unknown y-intercept.

On Your Own

20. A. no; Possible answer: No two line segments have slopes that have a product of -1.

B. $\overline{AB}$: $y = -4x + 5$; $\perp : y = \frac{1}{4}x - 29$

C. $\overline{CD}$: $y = \frac{1}{3}x + 2$; $\perp : y = -3x - 37$

D. $\overline{EF}$: $y = 5x + 38$; $\perp : y = -\frac{1}{5}x + 3$

E. Possible answer: $(0, -10)$

21. no; You need two sets of parallel lines, perpendicular to each other, to create a rectangle.

22. A. Steps 2 and 3; He uses $-\frac{4}{3}$ for the slope instead of $\frac{4}{3}$.

B. $y = mx + b$

$6 = \frac{4}{3} -2 + b$

$b = \frac{26}{3}$

$y = mx + b$

$y = \frac{4}{3}x + \frac{26}{3}$

23. Student may or may not include art. Possible answer:

Given: The slopes of two lines are opposite reciprocals.

Prove: The lines are perpendicular.

If the two slopes are opposite reciprocals, I can name them $\frac{a}{b}$ and $-\frac{b}{a}$. I can draw two lines on the coordinate plane to represent these slopes. The distances a and b create right triangles with each line and represent the rise and run of each line. A rotation of 90° maps one triangle to the other. Therefore, the angle between the two lines is 90°. The lines are perpendicular.

26. A. slope of current trail $= -\frac{1}{3}$, slope of perpendicular path $= 3$

D. yes; The side path to the waterfall must have a slope of 3 to be parallel to the path to the tree. Any path with a slope of 3 is perpendicular to the trail with slope $-\frac{1}{3}$.

27. Possible answer: $y = \frac{1}{2}x + 5$ and $8x + 4y = 7$

MODULE 4, LESSON 4.3

Spark Your Learning

B. Possible answer: a ruler, counting the grid units and estimating, string to model the trail

Build Understanding

Task 1 **A.** The leg $\overline{PR}$ is a horizontal line segment because both endpoints have the same y-coordinate. The leg $\overline{QR}$ is a vertical line segment because both endpoints have the same x-coordinate. A horizontal line and vertical line create a 90° angle at the intersection point. Thus, the triangle is a right triangle.

B. yes; By squaring the difference, you get a positive result, even if the difference is negative.

Task 2 **A.** $y_1 + \frac{y_2 - y_1}{2}$

$\frac{2y_1}{2} + \frac{y_2 - y_1}{2}$

$\frac{2y_1 + y_2 - y_1}{2}$

$\frac{y_1 + y_2}{2}$

B. yes; Possible answer: If the two x-coordinates or the two y-coordinates are the same, then the line is vertical or horizontal, respectively, and there is no need to find that coordinate of the midpoint. That coordinate of the midpoint would be the same number as the x-coordinate or y-coordinate, respectively.

C. Yes, you could find the x-coordinate using that method as well; Possible answer: M is exactly halfway between A and B, so adding the horizontal distance to the x-coordinate of A or subtracting the horizontal distance from the x-coordinate of B will result in the same x-coordinate for M.

Step It Out

Task 3 **A.** $10\sqrt{5}$

B. Substituting the x-coordinates into the equation gives the corresponding y-coordinates. You need both coordinates of a point for the Distance Formula.

C. no; It does not matter in which order you add them. Addition is commutative.

Task 4 **A.** The problem states the anchor point is along the same line but 30 miles away.

B. Possible answer: An answer of "about 54.1 NM" may give a better mental picture of the distance than "the square root of 2925". Approximations help understand how the answer applies to the real world.

Check Understanding

1. Possible answer: The Distance Formula is derived from the Pythagorean Theorem where coordinate expressions have been substituted for the lengths of the legs of a right triangle.

2. Possible answer: The midpoint is halfway between both the x-coordinates and both the y-coordinates. To find the halfway point between two coordinates, add them and divide by two, or average them.

On Your Own

6. Possible answer: If you allow either to be equal, you don't end up with a right triangle with which to use the Pythagorean Theorem. You would have either a horizontal or vertical line or a point.

7. Possible answer: If you subtract x_1 from x_2 and divide by 2, you will find the horizontal distance between x_1 and the midpoint, not the actual x-coordinate of the midpoint, which is the horizontal distance between the origin and the midpoint.

8. They are not congruent. AB: $9\sqrt{5}$; CD: $\sqrt{346}$

9. They are congruent. AB: 25; CD: 25

10. They are congruent. AB: $2\sqrt{5}$; CD: $2\sqrt{5}$

11. They are not congruent. AB: $\sqrt{109}$; CD: $\sqrt{85}$

12. They are congruent. AB: $2\sqrt{37}$; CD: $2\sqrt{37}$

13. They are congruent. AB: $3\sqrt{2}$; CD: $3\sqrt{2}$

14. They are not congruent. AB: $\sqrt{34}$; CD: $\sqrt{37}$

15. They are not congruent. AB: $\sqrt{85}$; CD: $\sqrt{74}$

16. distance: $3\sqrt{5}$; midpoint: $\left(-\frac{1}{2}, 1\right)$

17. distance: $4\sqrt{2}$; midpoint: $(3, 2)$

18. distance: $\sqrt{53}$; midpoint: $(1.5, -1)$

24. A. $26 = \sqrt{(10-0)^2 + (y_2 - 0)^2}$

$26 = \sqrt{100 + (y_2)^2}$

$26^2 = 100 + (y_2)^2$

$576 = (y_2)^2$

$y_2 = 24$

MODULE 4 REVIEW

Concepts and Skills

5. Accept any answer with $m = 4$ as parallel and $m = -\frac{1}{4}$ as perpendicular.

6. Accept any answer with $m = 3$ as parallel and $m = -\frac{1}{3}$ as perpendicular.

7. Accept any answer with $m = -2$ as parallel and $m = \frac{1}{2}$ as perpendicular.

8. Accept any answer with $m = -\frac{4}{5}$ as parallel and $m = \frac{5}{4}$ as perpendicular.

9. Accept any answer with $m = \frac{3}{4}$ as parallel and $m = -\frac{4}{3}$ as perpendicular.

10. Accept any answer with $m = 3$ as parallel and $m = -\frac{1}{3}$ as perpendicular.

11. Parallel: $y = -\frac{2}{3}x - \frac{22}{3}$

Perpendicular: $y = \frac{3}{2}x - 3$

12. Parallel: $y = \frac{1}{4}x + \frac{7}{4}$

Perpendicular: $y = -4x + 23$

13. Parallel: $y = -4x + 5$

Perpendicular: $y = \frac{1}{4}x + \frac{3}{4}$

14. Parallel: $y = \frac{1}{3}x + \frac{10}{3}$

Perpendicular: $y = -3x - 10$

16. A. by construction

B. Pythagorean Theorem

C. Substitution

D. Square Root Property of Equality

17. no; The segment along $y = 4x + 9$ from $x = 0$ to $x = 5$ has the endpoints $(0, 9)$ and $(5, 29)$. Using the Distance Formula, the length of the segment is $\sqrt{425}$. The segment along $y = -0.5x - 2$ from $x = -3$ to $x = 1$ has the endpoints $(-3, -0.5)$ and $(1, -2.5)$. Using the Distance Formula, the length of the segment is $\sqrt{20}$. The segments do not have the same length, so they are not congruent.

18. It is an isosceles triangle. Two of the sides are 5 units long and one side is $2\sqrt{5}$ units long.

UNIT 3

MODULE 5, LESSON 5.1

Spark Your Learning

B. The placement of a subsequent tile must match the available shape resulting from the placement of the original tile.

D. Since the pattern is being used on a surface in a kitchen, it is important that there are no large gaps. A gap can leave space where things spilled in a kitchen could collect.

Build Understanding

Task 1 **A.** Corresponding sides of the triangles have the same length. Translating an image preserves the lengths.

B. Corresponding angles of the triangles have the same measure. Translating an image preserves the measure of the angles.

C. Yes, the corresponding sides of the preimage and image are parallel; Possible answer: A line can be drawn connecting corresponding vertices. The measures of the corresponding angles formed by this transversal and the corresponding sides are equal, so the sides are parallel.

D. The distance between the corresponding vertices is the length of vector v.

Task 2 **A.** The direction of the vector is up and to the right, so the quadrilateral will be translated up and to the right.

B. parallel; I can match the segments with the lines on the lined paper. Since the lines on the paper are parallel, the segments are parallel.

C. Each vertex in the image was found by copying the distance XY. So, the distances between corresponding vertices are equal to XY.

Step It Out

Task 4 **A.** The first graph goes with vector $\langle 6, -3 \rangle$. The second graph goes with vector $\langle 6, -4 \rangle$; The vector in the first graph shows a translation 6 units to the right and 3 units down. The vector in the second graph shows a translation 6 units to the right and 4 units down.

Check Understanding

2.

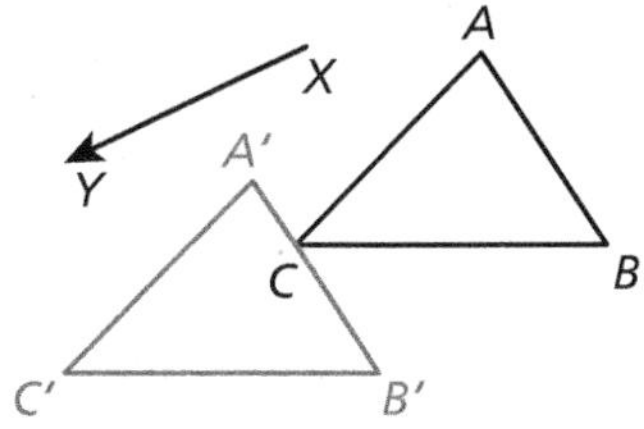

3. $(x, y) \rightarrow (x - 4, y - 2)$

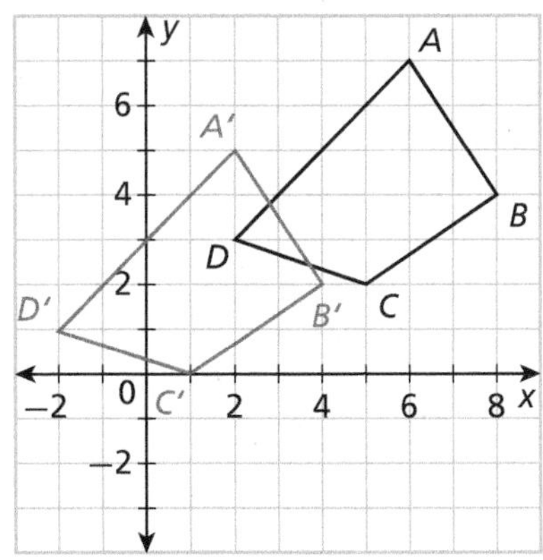

On Your Own

10.

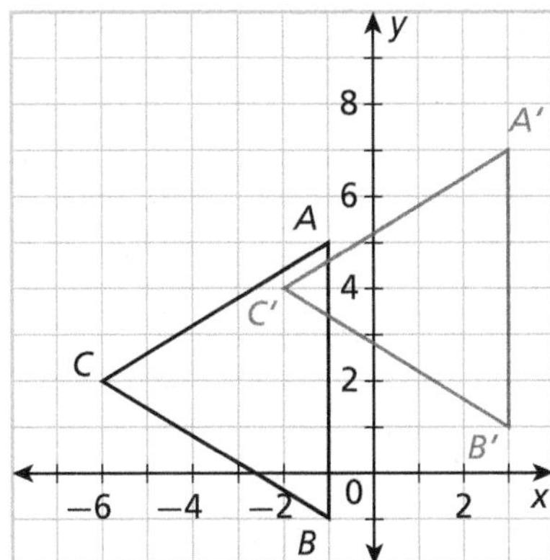

11.

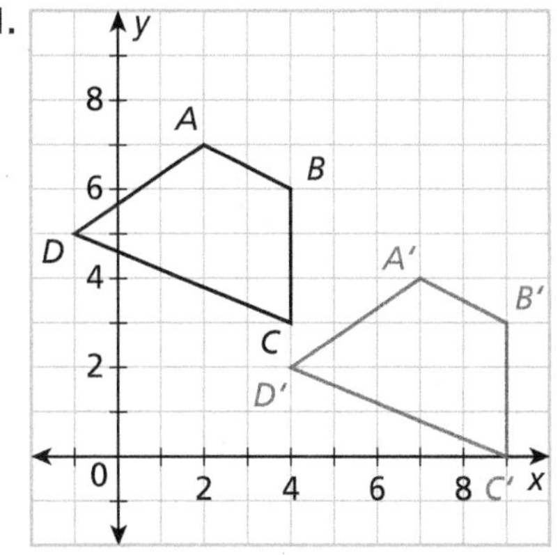

12.

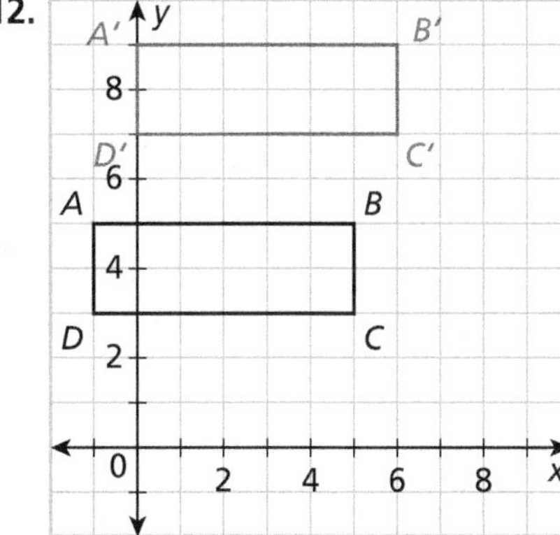

19. $A \rightarrow B$: $\langle 0, -10 \rangle$;
$A \rightarrow C$: $\langle 10, -5 \rangle$;
$A \rightarrow D$: $\langle 10, -15 \rangle$

20. yes; no; Graph C is incorrect; Possible answer: A translation by $\langle 2, 2 \rangle$ maps the points in graph A onto the points in graph B. However, a translation cannot map points in either graph A or B to graph C, so the points in graph C are in a different position than those in graphs A and B.

23. Possible answer: The x-coordinate translates in a positive direction, not a negative one. The correct vector should be $\langle 3, -3 \rangle$.

MODULE 5, LESSON 5.2

Spark Your Learning

B. A person's location in the world may affect how that person views the world. This perspective may be more appropriate for someone who lives in the southern hemisphere.

Additional Answers

Build Understanding

Task 1 **A.** yes; yes; Possible answer: The triangles are congruent, and the vertices are both labeled clockwise. So, this rotation is a rigid motion, and it preserves orientation.

B. These angles are angles of rotation. Since every point on the preimage is rotated using the angle of rotation to find the corresponding image point, all of the angles are congruent.

Task 2 **C.** Vertices A and A' lie on the same circle, B and B' lie on the same circle, and C and C' lie on the same circle.

D. $\triangle ABC$ can be copied on the tracing paper. Without moving the paper, place a pencil tip on point P and turn the paper. The copy of $\triangle ABC$ should align with $\triangle A'B'C'$.

E. Draw a segment connecting point P and point A. Use a protractor to measure the reference angle, then draw an angle with the same measure with segment $\overline{PA}$ as one side of the angle. Use a ruler to find PA. Then use this distance to mark A' along the new segment so that $PA = PA'$. Repeat for other vertices. Connect A', B', and C' to form $\triangle A'B'C'$.

Step It Out

Task 4 graph B; In order for the image to be rotated 180° about the origin, the coordinates should follow this pattern: $(x, y) \rightarrow (-x, -y)$.

Check Understanding

1.

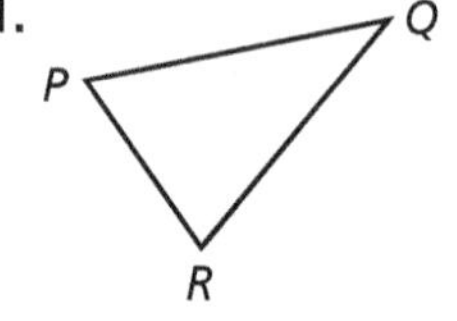

M

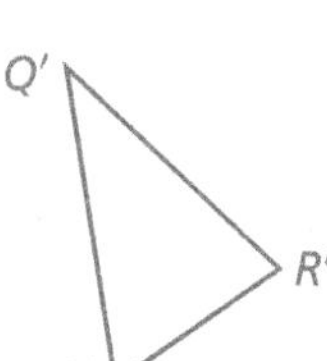

4.

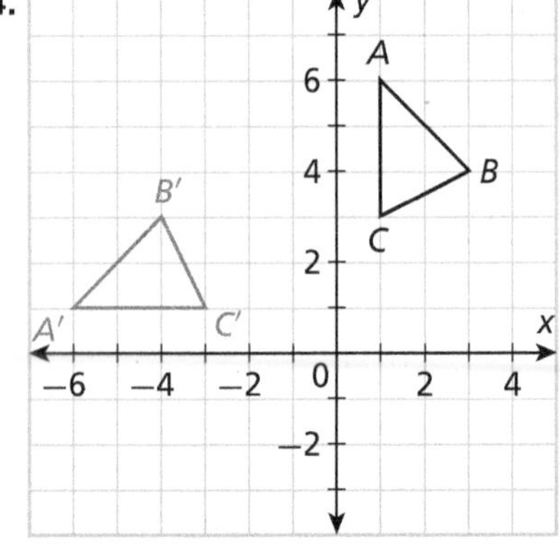

5.

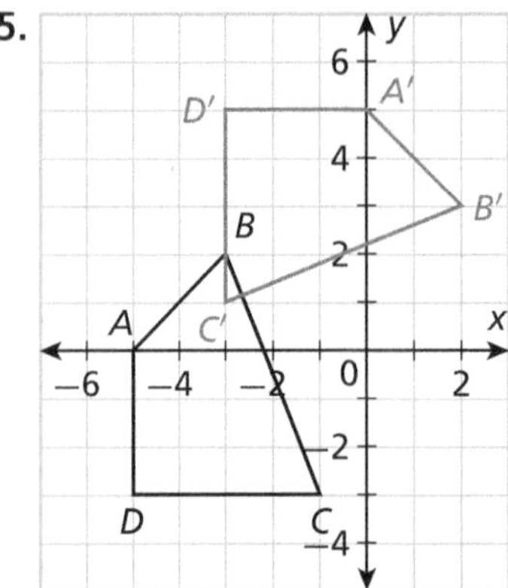

On Your Own

7. always; Since rotations are rigid motions, the preimage and image will always be congruent.

8. always; The distance from the point of rotation to a point on the preimage is always equal to the distance from the point of rotation to the corresponding point on the image.

9. sometimes; $\angle GHF$ and $\angle HFG$ are angles in the same triangle. They could be congruent. Whether they are congruent or not is not determined by a rotation.

10.

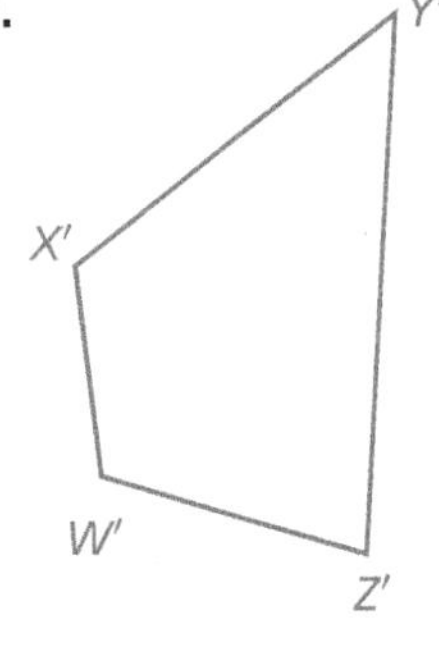

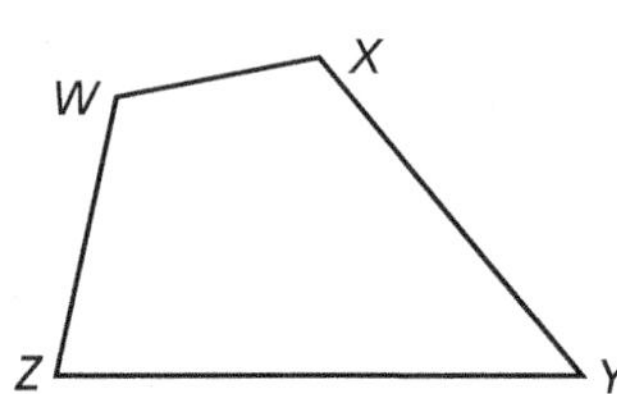

Q

11. no; possible answer: Rocco located vertex L' correctly, but he then translated the triangle instead of rotating it.

12. Angle of rotation: 60°

A B

D C

E

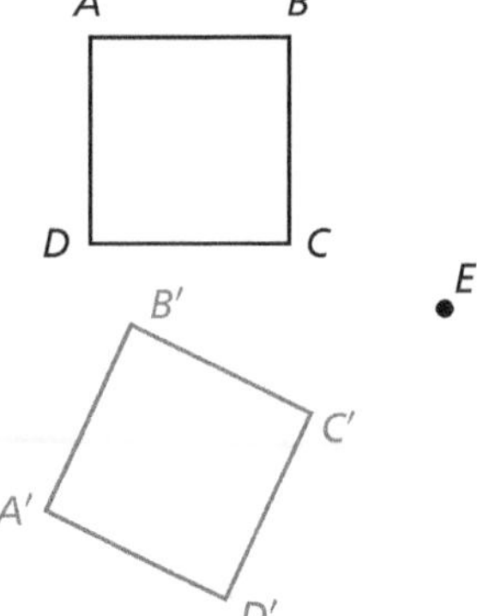

13. Angle of rotation: 20°

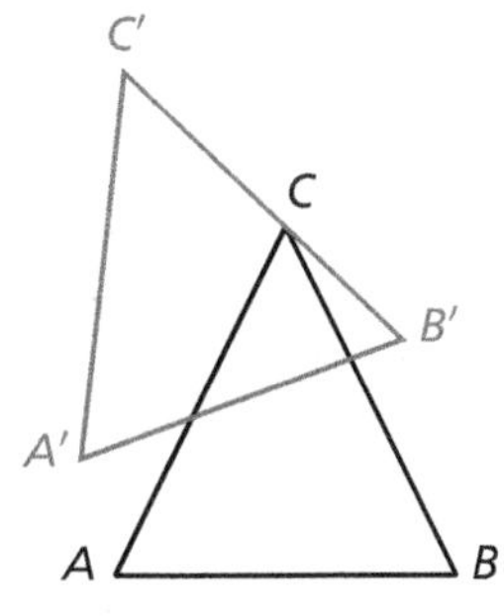

14. no; possible answer: Although all points rotate through the same angle, points closer to the center of rotation move a shorter distance than points farther from the center of rotation.

15.

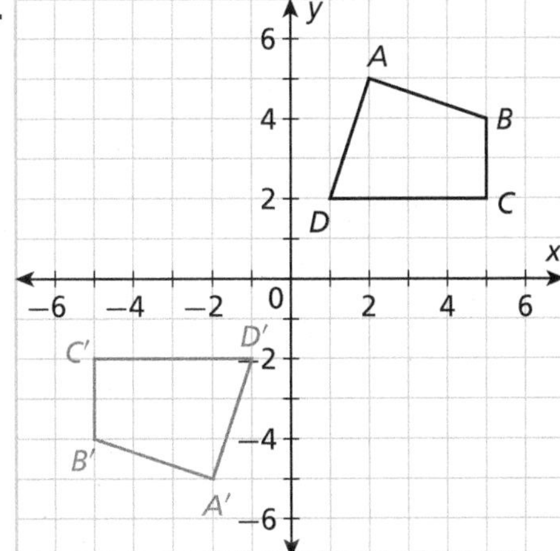

16.

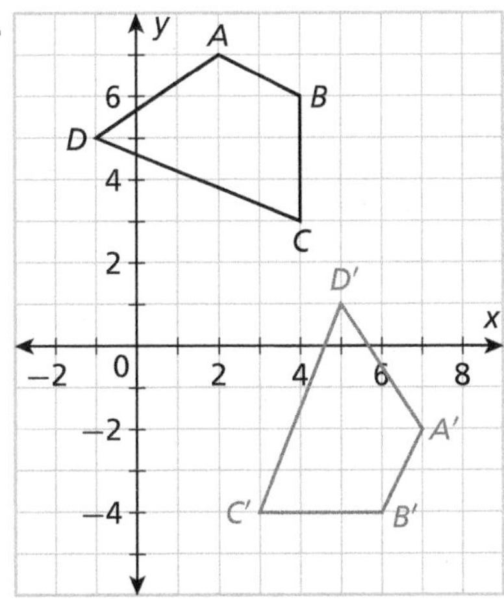

17.

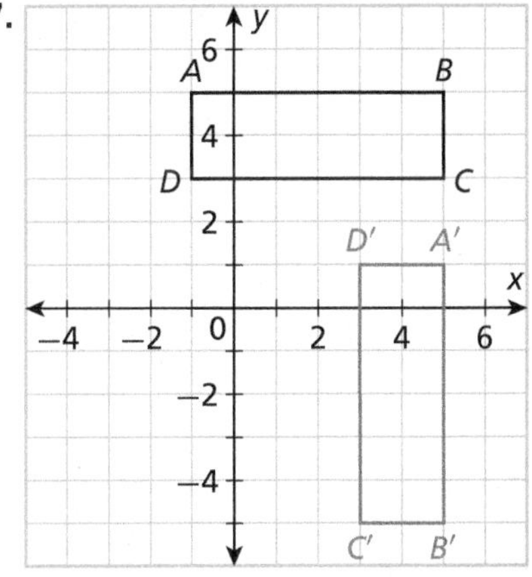

18. positive; Possible answer: The rule for a 270° rotation in a coordinate plane is $(x, y) \rightarrow (y, -x)$. So, the x-coordinate of a point on the image is the y-coordinate the corresponding point on the preimage. Since the y-coordinates of the vertices of the preimage are positive, the x-coordinates of the image will be positive.

MODULE 5, LESSON 5.3

Spark Your Learning

B. Possible answer: The image was created by placing blots of paint on a piece of paper. Then the paper was folded in half and pressed so that the same image is applied to both halves of the paper.

D. Possible answer: Translations and rotations preserve orientation. The halves of this image have different orientations.

Build Understanding

Task 1 **A.** yes; no; Possible answer: The vertices in *DEFG* follow a clockwise pattern, while the vertices in *D'E'F'G* follow a counterclockwise pattern. So, the reflection did not preserve the orientation.

Task 2 **A.** Line p, the line of reflection, is the perpendicular bisector of the segment connecting points A and A'.

B. yes; possible explanation: $\triangle A'B'C'$ is a reflection of $\triangle ABC$, and reflections are rigid motions. So $\triangle ABC \cong \triangle A'B'C'$.

C. yes; The order in which the vertices are reflected does not affect the resulting image.

Step It Out

Task 3 Correct order of steps:

D. Locate corresponding points on the preimage and image.

A. Connect A to A'.

C. Use a compass to make marks to determine the perpendicular bisector.

B. A straightedge is used to connect the intersecting points to create the perpendicular bisector and the line of reflection.

Task 4 B; Possible answer: To reflect across the line $y = -x$, apply the rule $(x, y) \rightarrow (-y, -x)$. The preimage $\triangle ABC$ has the vertices $A(-4, -3)$, $B(-2, -1)$, and $C(-1, -5)$. So, the vertices of the image $\triangle A'B'C'$ are $A(3, 4)$, $B(1, 2)$, and $C(5, 1)$, which is graph B.

Check Understanding

1. A.

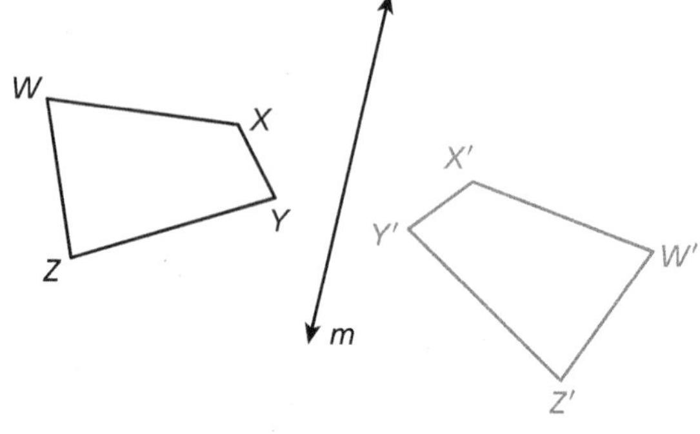

B. The segments should be perpendicular to line m, and the segments should be bisected by line m.

2.

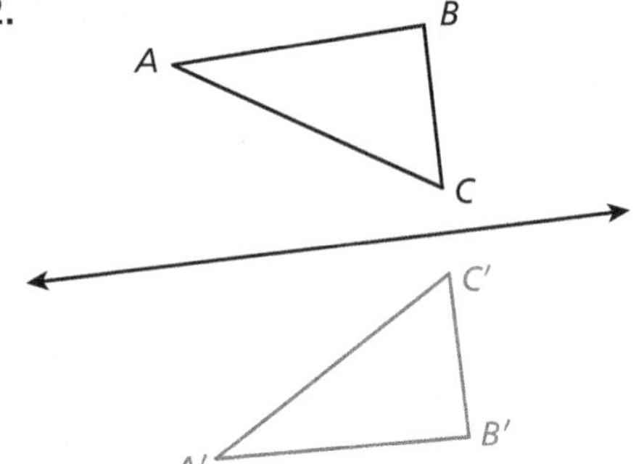

3.

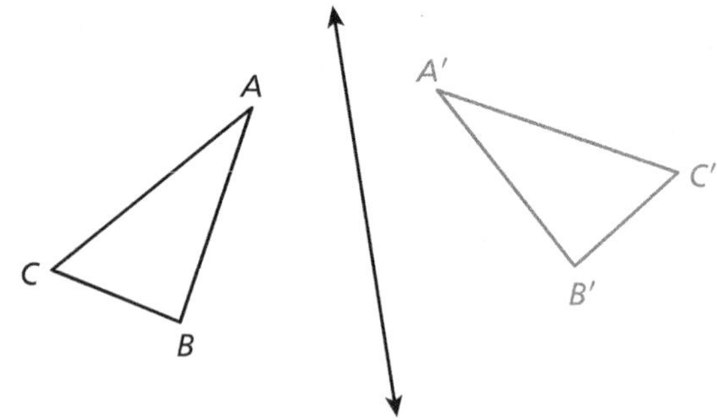

4. $P'(0, -7)$, $Q'(4, -6)$, $R'(2, -3)$, $S'(-1, -3)$

5. $P'(7, 0)$, $Q'(6, 4)$, $R'(3, 2)$, $S'(3, -1)$

On Your Own

7.

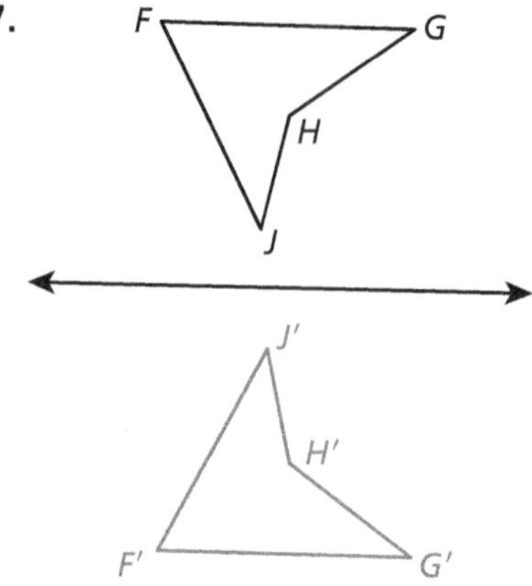

8.

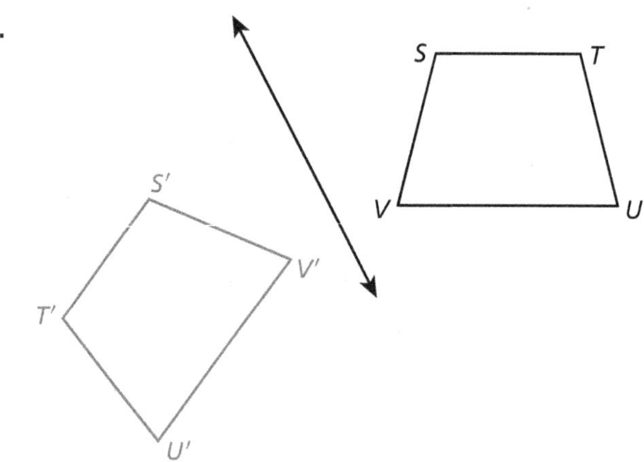

9. A–B.

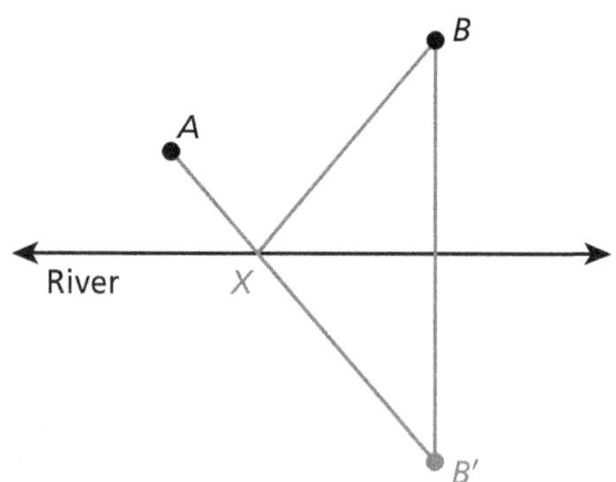

C. AB'; Since AB' is the shortest distance between A and B', and $AB' = AX + B'X$, $AX + B'X$ is the shortest distance between A and B'. Since $BX = B'X$, $AX + BX$ is the shortest distance between A and B.

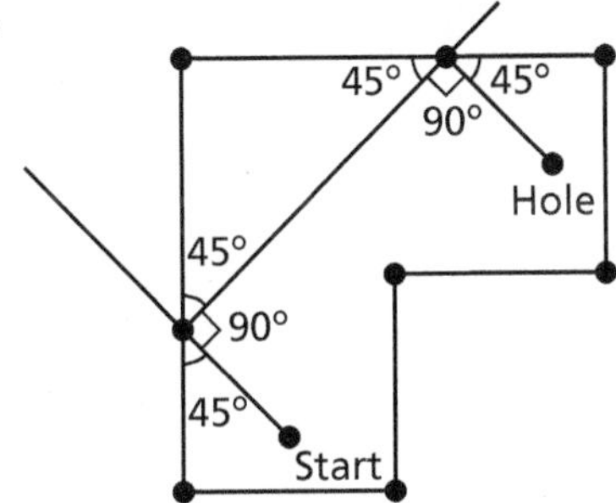

11.

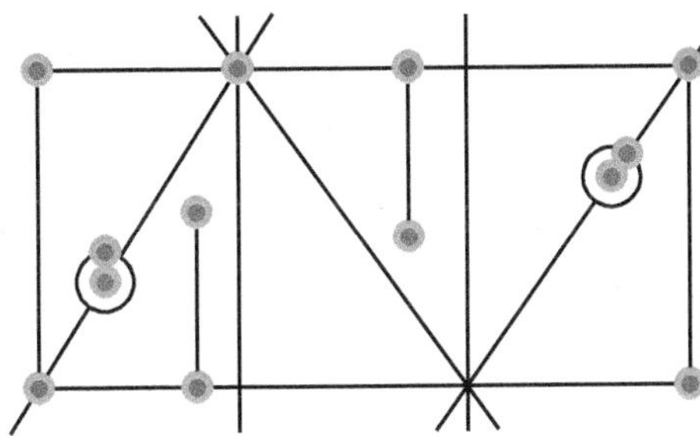

12.

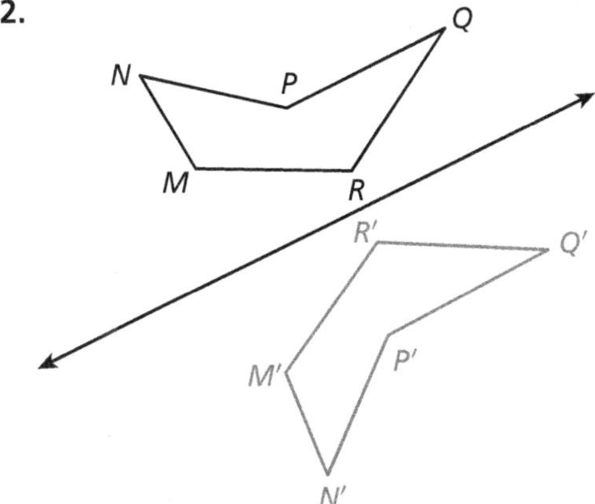

13.

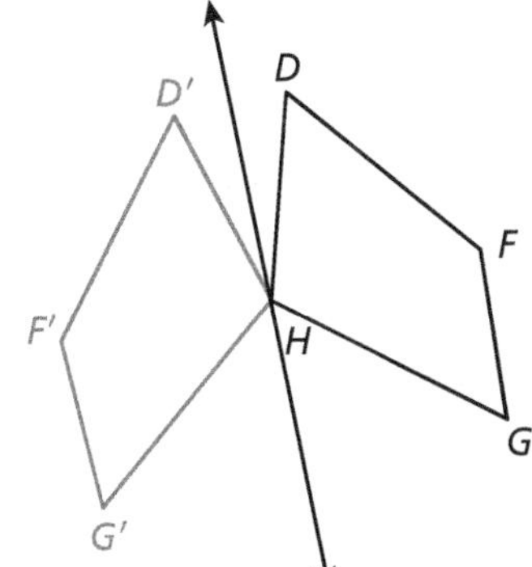

14. $A'\left(-\frac{5}{2}, -\frac{5}{2}\right)$, $B'(0, -4)$, $C'\left(-3, -\frac{1}{2}\right)$

15. $A'(1, 1)$, $B'(-2, 3)$, $C'(2, -2)$

16. rotate 180° counterclockwise about the origin

17. Reflect $\triangle ABC$ across the x-axis. Then reflect the image across the y-axis.

18. The coordinates of the image after the reflection are $A'(-4, 5)$, $B'(-1, 1)$, $C'(1, 4)$; Reflect the image across $y = -x$ to map to $\triangle PQR$.

19. A. $y = \frac{1}{2}x + 3$

B. Since $\overline{AA'}$ is perpendicular to the line of reflection, the slope of $\overline{AA'}$ is the opposite reciprocal of the slope of the line of reflection; -2

20. Possible answer: The student reflected the image across $y = 0$ instead of $y = -2$. The student should have translated the image up two units, reflected the image across $y = 0$, and then translated the image down two units; The correct image is $A'(-2, -10)$, $B'(5, -9)$ and $C'(5, -7)$.

MODULE 5, LESSON 5.4

Spark Your Learning

C. Possible answer: I see shapes that are rotated by several different angle measures. I see reflections if I were to fold it in half.

D. Possible answer: To create art like this, I would need a compass, protractor, tracing paper, pencils, pens, and fine-tip markers to color art.

Build Understanding

Task 1 **A.** Each figure has 1 line of symmetry; Check students' drawings.

B. The triangle has 3 lines of symmetry, and the rectangle has two lines of symmetry; Check students' drawings.

C. Possible answer: The line of symmetry will pass through a vertex and through a point on the side opposite the vertex.

Step it Out

B. The angle of rotational symmetry for a regular polygon is 360° divided by the number of lines of symmetry: $\frac{360°}{6} = 60°$.

Task 4 **A.** Since point E lies on the x-axis, its image after a reflection across the x-axis is the same point.

E. Possible answer: The portion of the final image in quadrants I and III have a different orientation than the original image in quadrant II. A rotation preserves orientation, so a rotation cannot be used to create an image that has the given reflective symmetry.

On Your Own

7. yes;

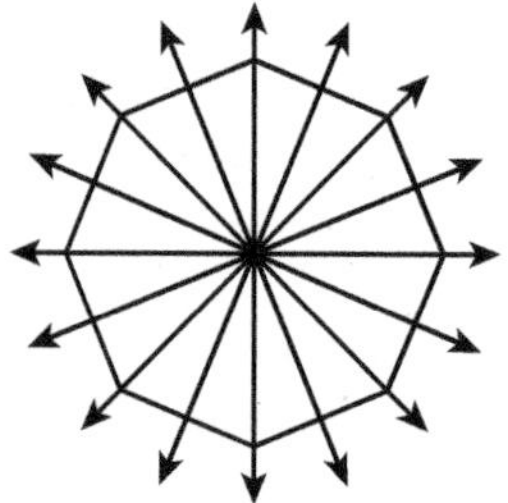

9. yes;

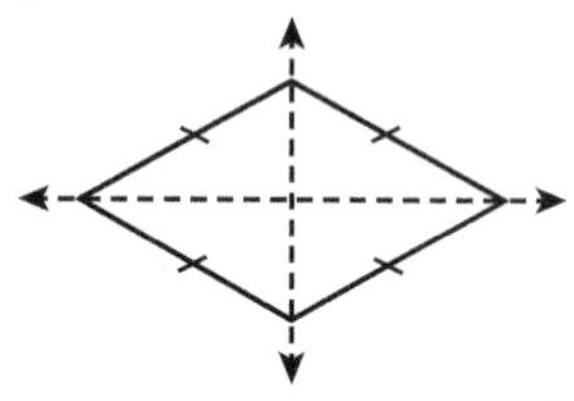

10. Possible answer: The diagonal of a rectangle is not a line of symmetry because adjacent sides do not have the same length.

14. Possible answer: An equilateral triangle can rotate onto itself three times, so its angle of rotational symmetry is 120°, not 60°.

15. Possible answer: It must be formed so that all six points are congruent.

16. Possible answer: A hexagon has point symmetry. Any regular polygon with an even number of sides has point symmetry.

17. Possible answer:

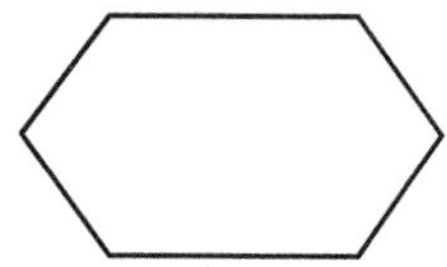

18. Possible answer: Both shapes have 2 lines of symmetry and an angle of rotation of 180°.

27. A.

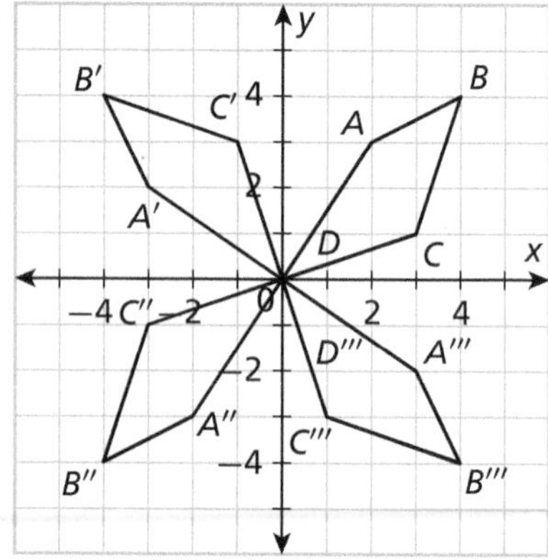

C. yes; Possible answer: The image of the figure in quadrant III can be found by reflecting *ABCD* over the *y*-axis and then reflecting its image over the *x*-axis. Note that reflecting the image just once would cause the image to “point” the wrong direction so the points would not map onto each other. So, two reflections would make the image map onto itself.

MODULE 5 REVIEW

Concepts and Skills

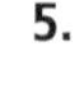

5.

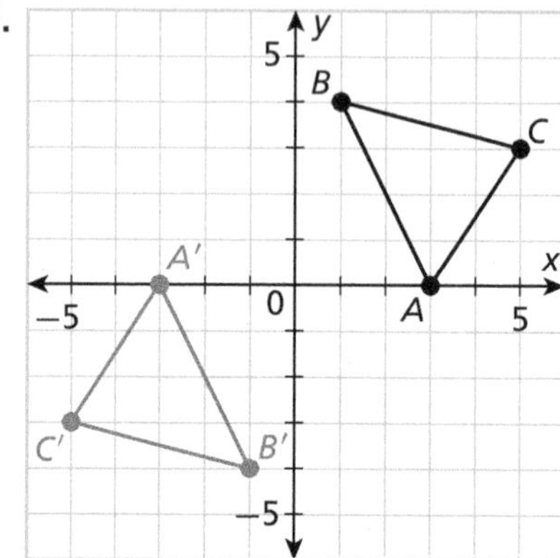

6.

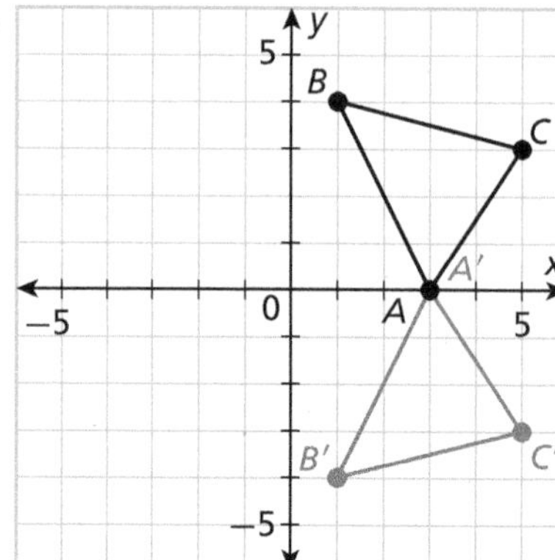

7.

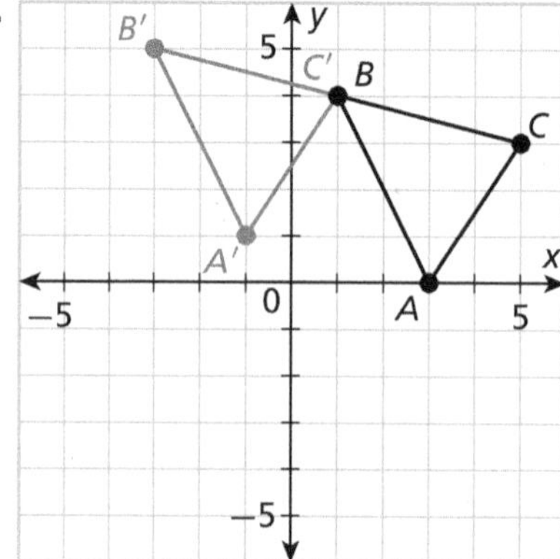

8.

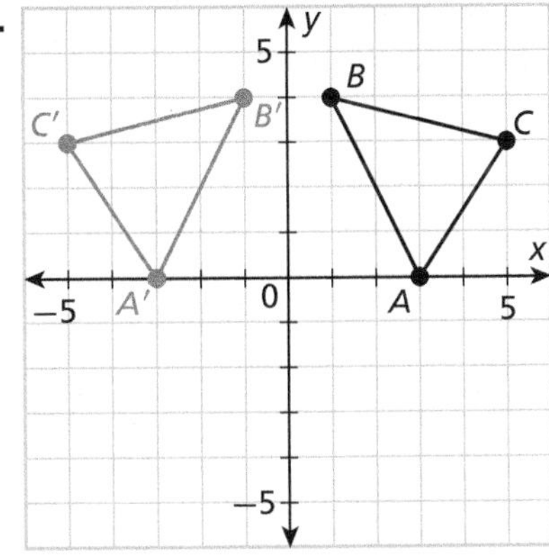

MODULE 6 PERFORMANCE TASK

B. Object beyond C

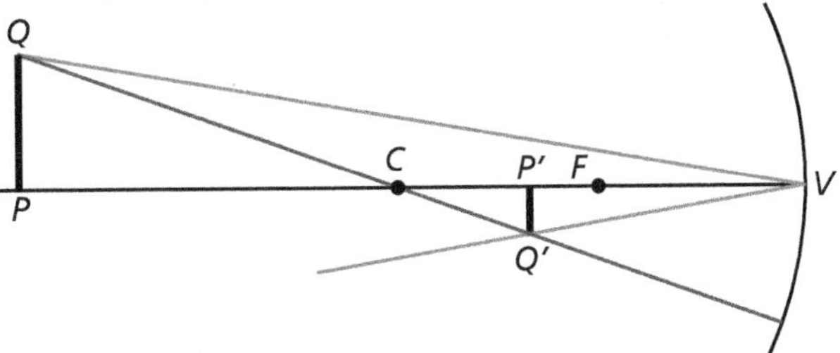

Object at C

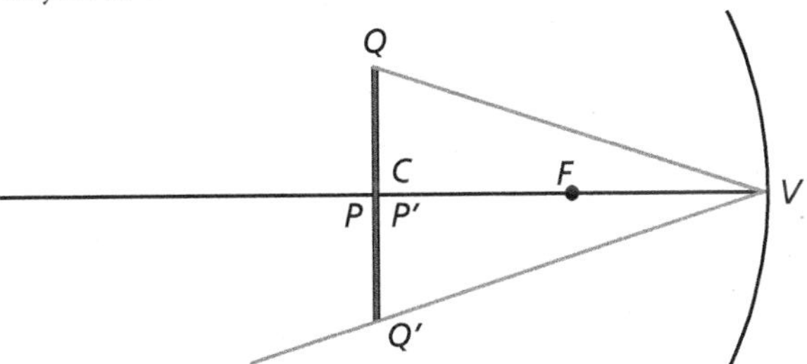

Object between C and F

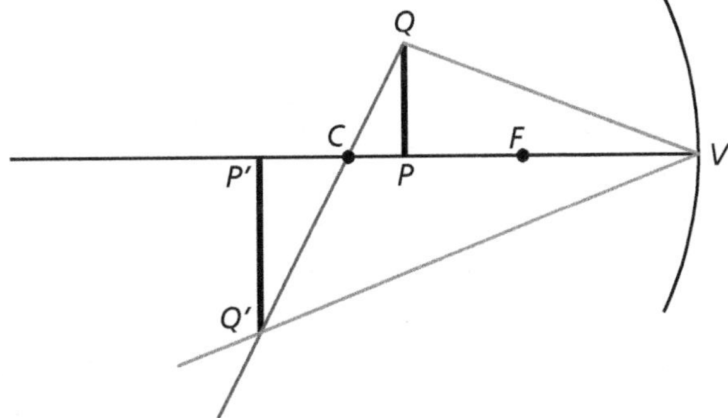

Object at F

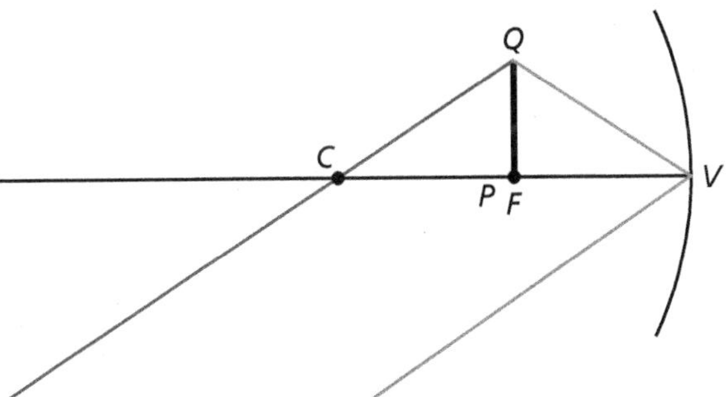

Object between F and V

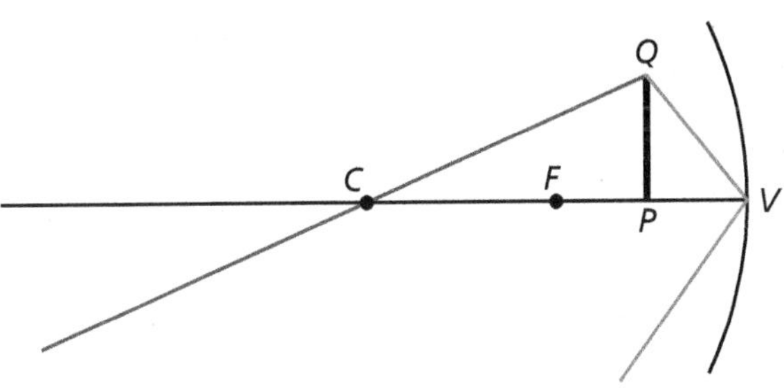

MODULE 6, LESSON 6.1

Spark Your Learning

C. The Subjects Still Studied display accurately shows the relative popularity of the subjects. The Favorite Subject in School display seems to over-represent the more common subjects because the area of the larger triangles is so much bigger than that of the smaller ones.

Build Understanding

Task 1 **A.** $A'(-4, 3)$, $B'(0, 6)$, $C'(0, 3)$; The image is translated left and up.

B. $A'(-6, 6)$, $B'(6, 15)$, $C'(6, 6)$; The side lengths increased and the angles stayed the same; This is not a rigid transformation because the size changed.

C. $A'(-1, 1)$, $B'(1, 2.5)$, $C'(1, 1)$; The side lengths decreased and the angles stayed the same; This is not a rigid transformation because the size changed.

D. $A'(-2, 4)$, $B'(2, 10)$, $C'(2, 4)$; Some side lengths increased, and some angles changed. This is not a rigid transformation because the size changed. The image looks stretched.

Task 2 **C.** All three ratios are equal. Each side of the image is twice as long as the corresponding side of the preimage.

D. yes; Check the measures of the corresponding angles. For example, if $m\angle CPR = m\angle CP'R'$, then $\overline{PR} \parallel \overline{P'R'}$.

Step It Out

Task 3 **A.** If you draw rays $\overrightarrow{AA'}$, $\overrightarrow{BB'}$, and $\overrightarrow{CC'}$, the point where the rays intersect is the center of dilation.

B. The scale factor is $\frac{1}{3}$.

Check Understanding

2.

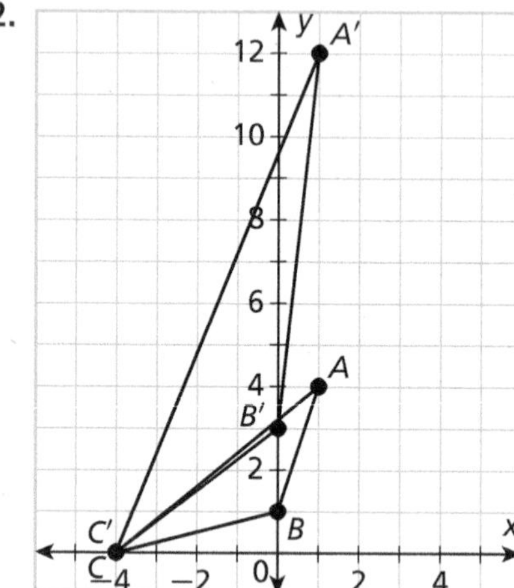

3.

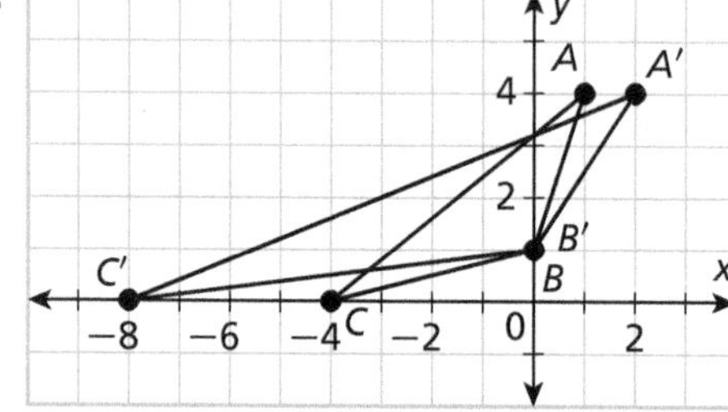

On Your Own

7. A. Maya can compare the coordinates of the vertices of the triangles.

B. The scales of the graphs are different, which makes it hard to determine what happened just by looking at the images.

C. A horizontal stretch.

15. vertical compression with scale factor of $\frac{1}{2}$ and translation to the right 2 units

16. reflection across the x-axis and translated up 1 unit

17. translation to the right 1 unit, down 2 units, and dilated by scale factor of 2

20. no; Dilations preserve angle measure, so Joshua must have made an error in calculating the coordinates of the image.

22. A. $(p, q) \rightarrow (p - a, q - b)$

B. $(p - a, q - b) \rightarrow (k(p - a), k(q - b))$

C. $(k(p - a), k(q - b)) \rightarrow (k(p - a) + a, k(q - b) + b)$

D. See students' drawings. Possible answer: Triangle ABC has vertices $A(5, 3)$, $B(7, 8)$, and $C(9, 3)$. The vertices of the image $A'B'C'$ after a dilation by a scale factor of 2 with center $(1, 5)$ is:

$A'(2(5 - 1) + 1, 2(3 - 5) + 5) = A'(9, 1)$

$B'(2(7 - 1) + 1, 2(8 - 5) + 5) = B'(13, 11)$

$C'(2(9 - 1) + 1, 2(3 - 5) + 5) = C'(17, 1)$

23. no; Possible answer: The lanes are different sizes, and translations preserve the size of the figure.

24. Kevin cannot use the same rule for each dilation. The ratios between each pair of heights is not the same.

25. Possible answer: Image vertices: $(-2, -7)$, $(0, 5)$, and $(-6, -5)$; Dilation point: $(4, 1)$; Scale factor: 2

MODULE 6, LESSON 6.2

Spark Your Learning

B. translation and reflection; A translation is responsible for tracks seen in different locations. A reflection is responsible for tracks that are mirror images of each other.

D. yes; The two tracks are the same. They are reflections of each other but displaced.

Build Understanding

Task 1 **A.** translation down 1 unit and right 2 units; translation up 2 units and right 1 one unit; no; Changing the order does not affect the final image.

B. no; Two translations can always be described as the sum of the changes to the x-value right or left and the sum of the changes to the y-value up or down; ABC can be translated to $A''B''C''$ by the transformation $(x, y) \rightarrow (x + 2 + 1, y - 1 + 2)$ or $(x, y) \rightarrow (x + 3, y + 1)$.

C. rotation 90° clockwise about the origin; reflection across the x-axis; yes; If you reflect first, the rotation cannot map the figure to the same location as if you rotate first.

D. order does not matter: translation/translation, reflection/reflection, rotation/rotation (using the same center); order does or may matter: translation/reflection, translation/rotation, reflection/rotation

E. Possible answer: When the transformations are not the same type, such as a reflection and a translation, the order of transformations may affect the final result.

Task 2 **A.** translation right 2 units; dilation with center at the origin and a scale factor of $\frac{1}{2}$; Either translate left 1 unit and dilate with center at the origin and scale factor of 2, or dilate with center at the origin and scale factor 2 and then translate left 2 units.

B. The scale factors would only affect the x-coordinates.

C. Use the reciprocal of the original scale factor as the new scale factor.

D. Sample answer: Reflect across the line $x = 5$, then dilate with center at the origin and scale factor $\frac{1}{2}$, or rotate 180° about $(5, 0)$, then dilate with center at the origin and scale factor $\frac{1}{2}$.

E. translation right 1 unit and a reflection across the x-axis; yes; You get the previous polygon.

Step It Out

Task 3 **A.** The orientation changed which means that a reflection must have been used; reflect across the y-axis: $(x, y) \rightarrow (-x, y)$

B. dilation; the origin; scale factor of $\frac{1}{2}$; $(x, y) \rightarrow \left(\frac{1}{2}x, \frac{1}{2}y\right)$

Step It Out

Task 4 **B.** $(x, y) \rightarrow (2x + 4, 2y - 3)$ because the translation was not doubled.

Task 5 **B.** Since the reflection only affects the y-coordinates and the translation only affects the x-coordinates, the order does not matter because neither transformation affects the other.

Check Understanding

2. Translate left 2 units and down 2 units, rotate 90° counterclockwise about the origin, stretch vertically by a factor of $\frac{3}{2}$, and horizontally stretch by a factor of $\frac{4}{3}$.

3. rotate 90° counterclockwise about the origin, translate right 2 units and up 2 units

On Your Own

6. rotate 90 degrees clockwise, translate left 3 units

7. The order of the transformations on $BCDE$ does matter.

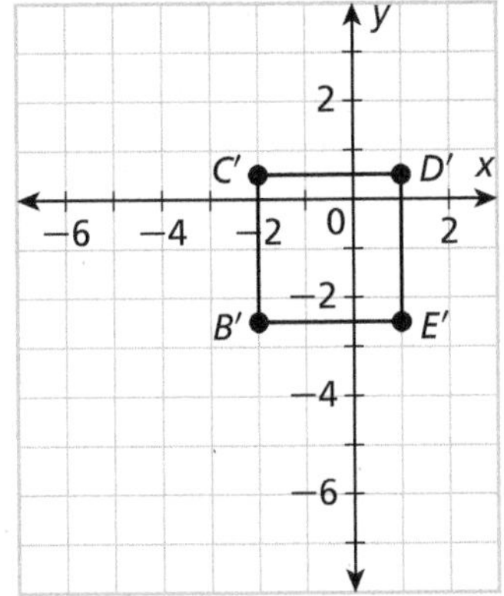

8. The order of the transformations on $BCDE$ does not matter.

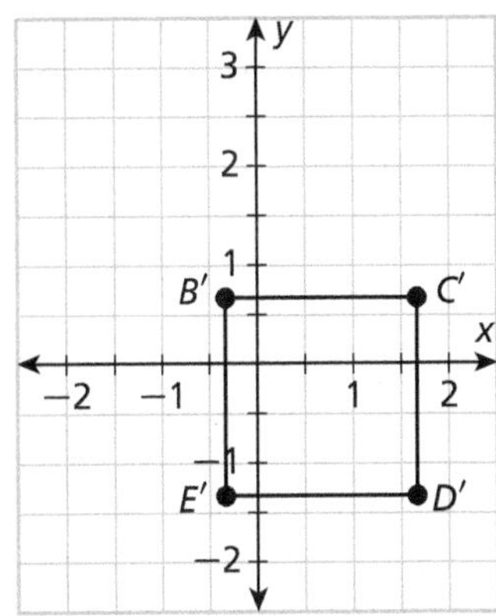

9. The prediction is incorrect. The image changes depending on whether the scale factor will be inside or outside of the parentheses.

10. $(x, y) \rightarrow (-x, -y)$

11. $(x, y) \rightarrow (-x + 5, y - 6)$

13. $A'(6, -18)$, $B'(-14, -2)$, $C'(-14, -18)$

14. $A'(4, 6)$, $B'(0, -9)$, $C'(4, -9)$

15. Sample answer: $(x, y) \rightarrow (-x, y)$ then $(x, y) \rightarrow (x, -y)$; $(x, y) \rightarrow (-x, y)$ then $(x, y) \rightarrow (x, -y)$

16. $(x, y) \rightarrow (-x, y)$ then $(x, y) \rightarrow (x, y + 6)$; $(x, y) \rightarrow (-x, y)$ then $(x, y) \rightarrow (x, y - 6)$

18. $(x, y) \rightarrow \left(-\frac{1}{3}x, -\frac{1}{3}y\right)$

19. $(x, y) \rightarrow (-y + 1, -x - 4)$

20. $(x, y) \rightarrow (-4x - 12, 4y + 20)$

22. Translate up 2 units and right 1 unit; translate down 2 units; reflect across the center line of the character

23. A. isosceles; Two of the sides have the same length.

B. $A'\left(-\frac{5}{2}, \frac{1}{2}\right)$, $B'\left(-\frac{3}{2}, -3\right)$, $C'\left(-\frac{1}{2}, \frac{1}{2}\right)$

C. yes; $\triangle A'B'C'$ is an isosceles triangle.

D. Possible answer: translation down 1 unit, vertical compression by a factor of $\frac{2\sqrt{3}}{7}$. The image will be an equilateral triangle.

24. A. Possible answer: Sequence 1: Reflect figure A across the x-axis: $(x, y) \rightarrow (x, -y)$. Then translate the image: $(x, y) \rightarrow (x + 6, y + 8)$. Sequence 2: Rotate figure A 180° about the point $(8, 4)$.

B. yes; The reflection in Sequence 1 changes the orientation of the figure. The rotation in Sequence 2 does not change the orientation of the figure.

MODULE 6 REVIEW

Concepts and Skills

5. Draw ray CA and ray CB. Copy the length CA 3 times along ray CA to locate A'. Copy the length CB 3 times along ray CB to locate B'. C remains where it is as it is the center of dilation.

6. **A.** Possible answer. It is a dilation because corresponding angles are congruent.

UNIT 4

MODULE 7, LESSON 7.1

Spark Your Learning

B. Possible answer: I could measure the side lengths and angles and compare them.

Build Understanding

Task 1 **C.** Counterclockwise rotation about point D by $m\angle EAB$; $\overline{AB} \cong \overline{DE}$, so the image of point B must be point E.

Step It Out

Task 5 **A.** It is given that the figures are congruent, so the corresponding angles will be congruent by CPCFC.

B. It is given that the figures are congruent, so the corresponding sides will be congruent by CPCFC.

Check Understanding

2. A. The figures are congruent if and only if $PQRS$ can be mapped to $TUVW$ by a sequence of rigid motions.

B. 90 degree clockwise rotation, translation

C. $\overline{PQ} \cong \overline{TU}$, $\overline{QR} \cong \overline{UV}$, $\overline{RS} \cong \overline{VW}$, $\overline{SP} \cong \overline{WT}$, $\angle P \cong \angle T$, $\angle Q \cong \angle U$, $\angle R \cong \angle V$, $\angle S \cong \angle W$

On Your Own

13. not congruent; Not all pairs of corresponding parts are congruent.

14. congruent; All pairs of corresponding sides and angles are congruent.

17. A. The design is created from 16 congruent triangles. Each quarter of the design consists of 4 of the triangles joined to form a square.

B. Possible answer: There are many ways to transform the triangle with base $\overline{AB}$ to the position of the triangle with base $\overline{CD}$. One way is to translate the triangle with base $\overline{AB}$ so that point B is at point C, then rotate it 90° counterclockwise about B, and then translate it to the right.

C. $\overline{CD} \cong \overline{AB}$ because corresponding parts of congruent figures are congruent, so, by the definition of congruent segments, the sides have equal lengths.

MODULE 7, LESSON 7.2

Build Understanding

Task 1 **B.** $ABCD \cong EFGH$; It can be easy to get confused and map the figures so that the corresponding parts do not match if the figure has symmetry or congruent sides.

C. $\overline{AB} \cong \overline{EF}$, $\overline{BC} \cong \overline{FG}$, $\overline{CD} \cong \overline{GH}$, $\overline{DA} \cong \overline{HE}$; All of the corresponding side lengths are equal; Corresponding parts of congruent figures are congruent.

D. $\angle A \cong \angle E$, $\angle B \cong \angle F$, $\angle C \cong \angle G$, $\angle D \cong \angle H$; All of the corresponding angle measures are equal; Corresponding parts of congruent figures are congruent.

Step It Out

Task 3 **B.** If two lines are cut by a transversal and the alternate interior angles are congruent, then the lines are parallel.

Check Understanding

1. The figures are congruent because a rotation and then a translation maps one figure to the other.

2.

Statements	Reasons
1. $PQTU \cong QRST$	1. Given
2. $\overline{PQ} \cong \overline{QR}$	2. CPCFC
3. Q is the midpoint of $\overline{PR}$.	3. Definition of midpoint
4. $\overline{QT}$ bisects $\overline{PR}$.	4. Definition of segment bisector

3.

Statements	Reasons
$\triangle AFG \cong \triangle CAB$	Given
$\triangle DCG \cong \triangle CAB$	Given
$\triangle AFG \cong \triangle DCG$	Transitive Property of Congruence
$\overline{FG} \cong \overline{CG}$	CPCFC

On Your Own

4.

Statements	Reasons
$\triangle ABC \cong \triangle ADC$	1. Given
$\angle BAC \cong \angle DAC$	2. Corr. parts of $\cong$ fig. are $\cong$.
$\angle BCA \cong \angle DCA$	3. Corr. parts of $\cong$ fig. are $\cong$.
$\overline{AC}$ bisects $\angle BAD$ and $\overline{AC}$ bisects $\angle BCD$.	4. Definition of angle bisector

5. She is right; Since the corresponding sides of congruent triangles are congruent, the sum of the total lengths of the sides (perimeter) will be the same for both triangles.

6. no; If $\triangle ABC$ is a right triangle, one of its angles is a right angle. Since corresponding parts of congruent figures are congruent, one of the angles of $\triangle DEF$ must also be a right angle, which means $\triangle DEF$ must be a right triangle.

7. A. and 7. B.

Translate *ABCD* so that *A* corresponds with *X*.	Definition of translation
Rotate *ABCD* counterclockwise about *A* by $m\angle BAW$ so that $\overline{AB}$ corresponds with $\overline{XW}$.	Definition of rotation
Reflect *ABCD* over $\overline{XW}$ so that $\overline{BC}$ corresponds with $\overline{WZ}$.	Definition of reflection
$ABCD \cong XWZY$	Definition of congruence
$\overline{AD} \cong \overline{XY}$	CPCFC

8.

Statements	Reasons
$\triangle SVT \cong \triangle SWT$	**1.** Given
$\angle VST \cong \angle WST$	**2.** CPCFC
$\overline{ST}$ bisects $\angle VSW$.	**3.** Definition of angle bisector

9. The triangles *XYZ* and *PQR* are congruent because all corresponding sides are marked as congruent along with 2 of the pairs of corresponding angles. By the Third Angles Theorem, the third pair of corresponding angles are also congruent, so the triangles are congruent; $w = 27$.

10. The triangles *EFG* and *HJK* are congruent because all corresponding sides and angles are marked as congruent; $y = 5$

11. $\angle T \cong \angle W$, but $\overline{TU}$ is not congruent to $\overline{WX}$ or $\overline{WV}$ (by counting). So, the triangles are not congruent.

12. $\overline{AB} \cong \overline{DE}$, $\overline{BC} \cong \overline{EF}$, and $\overline{AC} \cong \overline{DF}$ by the Distance Formula. It is given that $\angle A \cong \angle D$ and $\angle B \cong \angle E$. So, by the Third Angles Theorem$\angle C \cong \angle F$. All corresponding sides and all corresponding angles are congruent, so the triangles are congruent by the converse of CPCTC. $m\angle C = 56.3°$

13. The correct puzzle piece will be light blue in color, will have two indentations on opposite sides, and will have two protrusions on opposite sides. The two puzzle pieces A and C could fit in the empty space.

14. A. $m\angle D = 6(y + 2)° = (6(9) + 2)° = 56°$.

B. $AB = 3x + 8 = 3(4) + 8 = 12 + 8 = 20$ in.

15.

Statements	Reasons
$\triangle STU \cong \triangle VTU$	Given
$\overline{ST} \cong \overline{SV}$	Given
$\overline{ST} \cong \overline{VT}$	CPCFC
$\overline{SV} \cong \overline{VT}$	Transitive Property of Congruence
$\triangle STV$ is equilateral.	Definition of equilateral triangle

16.

Translate $\triangle PQR$ so that *P* corresponds with *N*.	Definition of translation
Rotate *PQR* clockwise about *P* by $m\angle RPO$ so that $\overline{PR}$ corresponds with $\overline{NO}$.	Definition of rotation
Reflect $\triangle PQR$ over $\overline{PR}$.	Definition of reflection
$\triangle MNO \cong \triangle QPR$	Definition of congruence

17. A. First, rotate A 90° clockwise about the origin. Then, reflect A over the line $x = 3$.

B. no; Translate C down 3 and left 3 (so that both B and C have a vertex at the origin). Then, rotate 180° about the origin.

MODULE 7, LESSON 7.3

Spark Your Learning

B. The hearts are identical and are evenly spaced around the edge of the pie.

D. yes; There are 12 identical hearts around the edge of the pie top, so the pie can be cut into 12, 6, 4, 3, and 2 slices (the divisors of 12 other than 1). The greatest number of identical slices is 12.

Build Understanding

Task 1 **A.** You can map *ABCD* to *EFGH* with a translation right 11 units and down 4 units; They are congruent because there is a sequence of rigid transformations that maps one to the other.

B. You can map *JKLM* to *NPQR* with a reflection over the *y*-axis and a translation down; They are congruent because there is a sequence of rigid transformations that maps one to the other.

C. There is no sequence of rigid transformations that maps *STUV* to *WXYZ*; They are not congruent; There is not a rigid transformation that maps one to the other.

Step It Out

Task 2 **A.** The *y*-coordinates all change sign, which suggests a reflection across the *x*-axis.

Check Understanding

1. false; The figures do not have the same orientation, so the sequence of transformations must include a reflection.

2. Map *PQRST* to *DEFGH* with a rotation of 180° around the origin. The coordinate notation for the rotation is $(x, y) \rightarrow (-x, -y)$.

On Your Own

4. Map *RSTU* to *WXYZ* with a reflection across the *y*-axis, followed by a translation; reflection: $(x, y) \rightarrow (-x, y)$, translation: $(x, y) \rightarrow (x + 1, y - 4)$

5. Map $\triangle ABC$ to $\triangle DEF$ with a rotation of 180° around the origin, followed by a translation; rotation: $(x, y) \rightarrow (-x, -y)$, translation: $(x, y) \rightarrow (x + 2, y + 6)$

6. Map *DEFGH* to *PQRST* with a reflection across the *y*-axis, followed by a vertical translation; reflection: $(x, y) \rightarrow (-x, y)$, translation: $(x, y) \rightarrow (x, y - 8)$.

7. Map $\triangle CDE$ to $\triangle WXY$ with a rotation of 180° around the origin, followed by a horizontal translation; rotation: $(x, y) \rightarrow (-x, -y)$, translation: $(x, y) \rightarrow (x - 2, y)$

8. You can map $\triangle CDE$ to $\triangle JKL$ by reflecting $\triangle CDE$ over the *x*-axis, followed by a horizontal translation. So, the two figures are congruent; reflection: $(x, y) \rightarrow (x, -y)$, translation: $(x, y) \rightarrow (x + 8, y)$

9. You can map *WXYZ* to *DEFG* with a reflection across the *x*-axis, followed by a horizontal translation. So, the two figures are congruent; reflection: $(x, y) \rightarrow (x, -y)$, translation: $(x, y) \rightarrow (x + 10, y)$

10. You can map *ABCDE* to *PQRST* with a translation. So, the figures are congruent; translation: $(x, y) \rightarrow (x - 2, y - 7)$.

11. There is no sequence of rigid transformations that will map one figure onto the other. So, they are not congruent.

12. 120° and 240° rotations about the center of the logo

13. 45°, 90°, 135°, 180°, 225°, 270°, and 315° rotations about the center of the logo.

15. Map *PQRSTU* to *ABCDEF* with a translation; The coordinate notation for the translation is $(x, y) \rightarrow (x + 6, y + 10)$.

16. Map $\triangle DEF$ to $\triangle KLM$ with a rotation of 180° about the origin, followed by a horizontal translation; Rotation: $(x, y) \rightarrow (-x, -y)$, translation: $(x, y) \rightarrow (x - 4, y)$

MODULE 7 REVIEW

5. $\angle KJL \cong \angle QPR$, $\angle JKL \cong \angle PQR$, $\angle JLK \cong \angle PRQ$, $\overline{JK} \cong \overline{PQ}$, $\overline{KL} \cong \overline{QR}$, $\overline{LJ} \cong \overline{RP}$

7. A. Possible answer: Rotate the triangle 90° counterclockwise about the origin, reflect it across the x-axis, and translate it left 4 units and up 2 units.

B. Possible answer: Translate the triangle right 4 units and down 2 units, reflect it across the x-axis, and rotate it 90° clockwise about the origin.

C. Possible answer: Rotate the triangle 90° about the origin, reflect it across the *x*-axis, and translate it right 2 units and down 4 units.

8.

Statements	Reasons
$\triangle ABC \cong \triangle ADC$	Given
$\overline{BC} \cong \overline{CD}$	CPCTC
$BC = CD$	Definition of congruent segments
$BD = BC + CD$	Segment Addition Postulate
$BD = 2BC$	Substitution

9. C. $\overline{PR}$ and $\overline{RP}$; the Symmetric Property of Congruence

D.

Statements	Reasons
$\overline{PQ} \cong \overline{SR}$	Given
$\overline{PS} \cong \overline{RQ}$	Given
$\overline{PR} \cong \overline{RP}$	Symmetric Property of Congruence
$\triangle PQR \cong \triangle RSP$	CPCTC

MODULE 8 PERFORMANCE TASK

A. Check students' work. Students should draw an equilateral triangle between three circle centers, where each side is 38 mm. Then they should divide the equilateral triangle into two right triangles to find the "horizontal" distance between two CCRs.

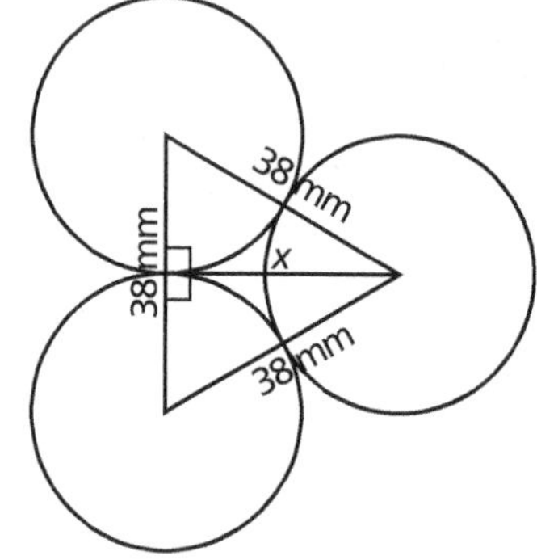

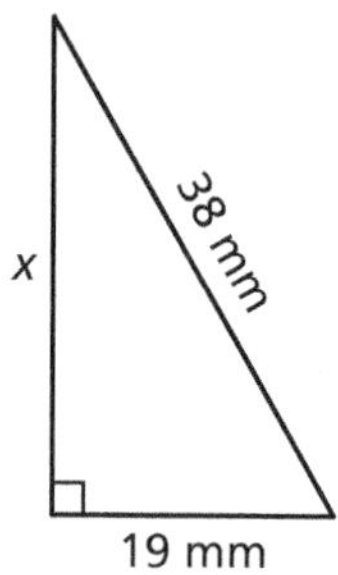

$$19^2 + x^2 = 38^2$$
$$361 + x^2 = 1444$$
$$x^2 = 1083$$
$$x = 19\sqrt{3}$$

Total surface area of 12 × 9 array is approximately $12(38) \cdot 9(19\sqrt{3}) = 4104 \cdot 19\sqrt{3} \approx 135{,}058 \text{ mm}^2$.

Possible answer: I used congruence to determine that the triangle created between the centers of three circles is an equilateral triangle and therefore has three interior angles of 60°. When I divide the triangle in half, I have two congruent right triangles and can use the Pythagorean Theorem on either to find the horizontal distance between two CCRs.

B. If the new diameter is 100 mm, the same process applies from the previous step, substituting 100 for 38. An equilateral triangle is created between three circles, then divided into two right triangles with hypotenuses of 100mm.

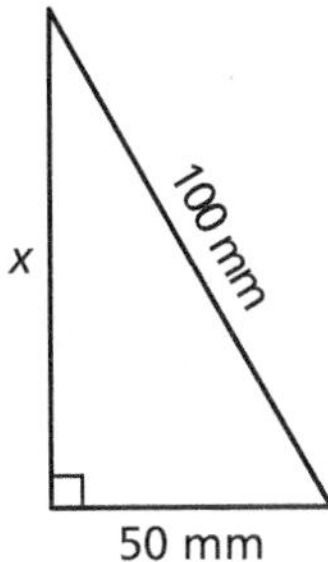

$$50^2 + x^2 = 100^2$$
$$2500 + x^2 = 10000$$
$$x^2 = 7500$$
$$x = 50\sqrt{3}$$

Total surface area of 12×9 array of 100 mm diameter CCR is approximately $12(100) \cdot 9(50\sqrt{3}) = 10,800 \cdot 50\sqrt{3} \approx 935,307\text{ mm}^2$.

The surface area of the new array is about 7 times larger than the surface area of the old array.

MODULE 8, LESSON 8.1

Spark Your Learning

C. Possible answer: The corresponding angles of both triangles are congruent; The corresponding sides of both triangles are congruent.

D. Possible answer: There may be rules such that when some corresponding parts of a two triangles are congruent, all corresponding parts are congruent.

Build Understanding

Task 1 **A–C.**

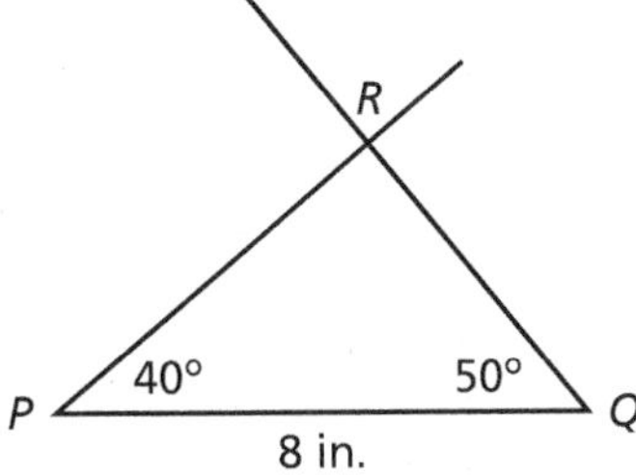

Task 4 **C.** You would need either a given side length or a mark noting the two sides are congruent for the two sides included between the congruent angles to be congruent.

Check Understanding

1. Possible answer: I can prove two pairs of corresponding angles and the included sides between are congruent.

4. You must have two pairs of corresponding angles and a congruent included side between them to use ASA Triangle Congruence to show two triangles are congruent.

On Your Own

6.

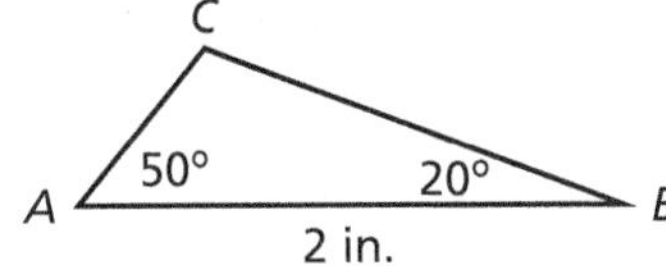

yes; Possible answer: The two triangles must be congruent, because they meet the ASA Triangle Congruence criteria. Any congruent triangles have a series of rigid transformations that map one to the other.

7. First use a translation that maps M to Y. Then use a rotation about point Y with $\angle LMX$ as the angle of rotation.

8. no; We do not know if the included sides are congruent.

9. yes; We can find the missing side using the Pythagorean Theorem. The triangles are congruent by the ASA Triangle Congruence Theorem.

10. yes; Since two segments are parallel, we know the alternate interior angles are congruent. Side XY is shared and therefore congruent.

11. no; We are not given two congruent angles and a congruent side.

12. Possible answer: We would need to know that $\overline{JK}$ and $\overline{HL}$ are parallel.

13. **1.** $\overline{WX}$ bisects $\angle YWZ$ and $\angle YXZ$: Given

4. $XW \cong XW$: Reflexive Property of Congruence

5. $\triangle YWX \cong \triangle ZWX$: ASA Triangle Congruence Theorem

14.

Statements	Reasons
1. $\overline{LM}$ is parallel to $\overline{PQ}$.	1. Given
2. $\angle MLQ \cong \angle PQL$	2. Alternate Interior Angles Theorem
3. $\angle MQL \cong \angle PLQ$	3. Alternate Interior Angles Theorem
4. $\overline{LQ} \cong \overline{LQ}$	4. Reflexive Property of Congruence
5. $\triangle LMQ \cong \triangle QPL$	5. ASA Triangle Congruence Theorem

15.

Statements	Reasons
1. $\overline{BD} \perp \overline{AC}$	**1.** Given
2. $\angle BDA \cong \angle BDC$	**2.** Definition of perpendicular lines
3. D is the midpoint of $\overline{AC}$.	**3.** Given
4. $\overline{AD} \cong \overline{CD}$	**4.** Definition of midpoint
5. $\angle A \cong \angle C$	**5.** Given
6. $\triangle ADB \cong \triangle CDB$	**6.** ASA Triangle Congruence Theorem

16.

Statements	Reasons
1. $\angle T \cong \angle P$	**1.** Given
2. $\overline{SQ}$ bisects $\overline{PT}$.	**2.** Given
3. $\overline{PR} \cong \overline{RT}$	**3.** Definition of segment bisector
4. $\angle PRQ \cong \angle SRT$	**4.** Vertical Angles Theorem
5. $\triangle SRT \cong \triangle QRP$	**5.** ASA Triangle Congruence Theorem

18. no; Perpendicularity gives one pair of corresponding angles congruent [all right angles are congruent], and one pair of corresponding sides are congruent ($\overline{LN} \cong \overline{LN}$ [Reflexive Property of Congruence]), but no other information is known.

19. no; One pair of corresponding angles are congruent, and one pair of corresponding sides are congruent $(\overline{LN} \cong \overline{LN})$, but no other information is known.

20. There is not enough information to prove congruence using the ASA Triangle Congruence Theorem; Two pairs of sides are congruent ($\overline{MN} \cong \overline{KN}$ [Definition of segment bisector] and $\overline{LN} \cong \overline{LN}$) and the corresponding right angles are congruent, but those pairs do not fit ASA.

21. yes; Since two pairs of angles ($\angle MNL \cong \angle KNL$ [Given] and $\angle MLN \cong \angle KLN$ [Definition of angle bisector] and the included side $(\overline{LN} \cong \overline{LN})$ are known, the triangles are congruent using the ASA Triangle Congruence Theorem.

22. A. Since one pair of congruent, corresponding angles is given, one pair of congruent corresponding sides is given, and the vertical angles are congruent, $\triangle MNR \cong \triangle PQR$ by the ASA Triangle Congruence Theorem.

B. Given: $\angle MNR \cong \angle PQR$, $\overline{NR} \cong \overline{QR}$

Prove: $\overline{MN} \cong \overline{PQ}$

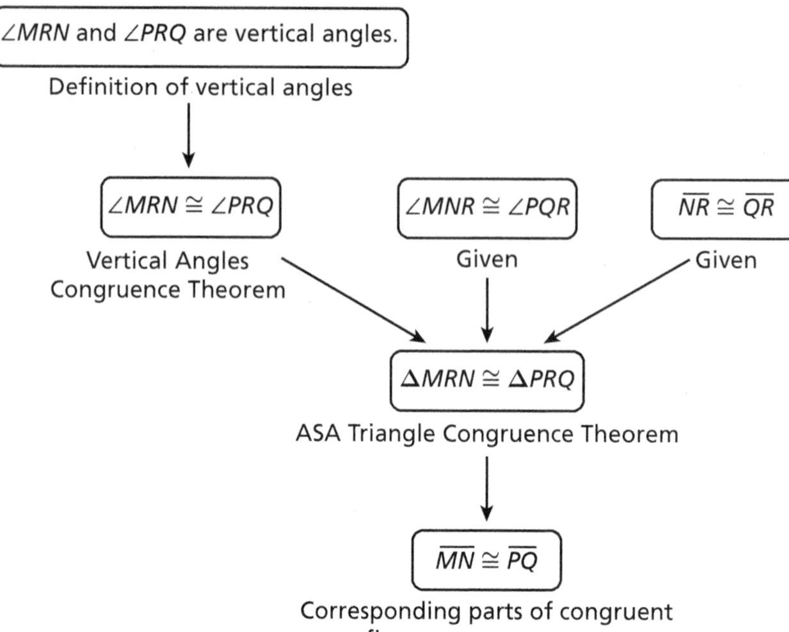

MODULE 8, LESSON 8.2

Spark Your Learning

B. Possible answer: I know all corresponding parts of congruent triangles are congruent. I know triangles with two congruent pairs of angles and a pair of congruent included sides are congruent. I know only one triangle can be created when given two side lengths and the angle between those two sides.

D. Possible answer: Any triangles with two pairs of congruent sides and a pair of congruent included angles are congruent.

Build Understanding

Task 1 **A.** Two triangles are possible;

B. No, there is only one way to complete the triangle; Possible answer: The distance between the endpoints of the two already-drawn lines is set by the given length and the included angle. Only one possible line goes between the two endpoints.

Task 3 **A.** The sides are congruent because they are the same length.

B. No, the angles are not the same degree measurement.

C. no; The two corresponding sides and an included angle are provided, but the corresponding included angles are not congruent.

Check Understanding

2. yes; The triangles are congruent by SAS criteria, and any congruent triangles can be mapped to each other using rigid transformations.

3. To use the ASA Triangle Congruence Theorem, you must prove two pairs of angles and the pair of included sides are congruent. To use the SAS Triangle Congruence Theorem, you must prove two pairs of sides and the pair of included angles are congruent.

On Your Own

5. yes; Possible answer: The triangles have congruent 3 inch sides, a congruent included angle, and a congruent shared side. They are congruent by the SAS Triangle Congruence Theorem.

6. no; Possible answer: The corresponding included angles are not congruent.

7. yes; Possible answer: They have congruent 20 mm sides, a congruent included angle, and a congruent shared side. They are congruent by the SAS Triangle Congruence Theorem.

8. Not enough information is given. I would need to know more about the unlabeled side and angles of each triangle.

9. yes; The triangles have two pairs of congruent sides and pair of included congruent angles, so you can prove the triangles are congruent by the SAS Triangle Congruence Theorem.

10. Check students' work for a triangle in which two congruent sides are twice the length of the noncongruent side (for example, 4-8-8). The two congruent angles in the triangle should be marked with the same number of tick marks, and the noncongruent angle should be marked differently than the two congruent angles.

11. Zach left a step out of his proof. He has only proved one pair of congruent sides and one pair of congruent angles. He could prove $\angle A$ and $\angle E$ are congruent by a transversal through parallel lines to complete the proof.

12.

Statements	Reasons
1. *LMNOP* is a regular pentagon.	1. Given
2. $MN = LP$, $NO = PO$	2. Definition of regular polygon
3. $\overline{MN} \cong \overline{LP}$, $\overline{NO} \cong \overline{PO}$	3. Definition of congruence
4. $\angle P = \angle N$	4. Definition of regular polygon
5. $\angle P \cong \angle N$	5. Definition of congruence
6. $\triangle LPO \cong \triangle MNO$	6. SAS Triangle Congruence Theorem
7. $\overline{LO} \cong \overline{MO}$	7. Corresponding parts of congruent figures are congruent.

13.

Statements	Reasons
1. $\overline{PQ} \parallel \overline{RS}$	1. Given
2. $PQ = RS$	2. Given
3. $\angle PQS \cong \angle RSQ$	3. Alternate Interior Angles Theorem
4. $\overline{SQ} \cong \overline{SQ}$	4. Reflexive Property of Congruence
5. $\triangle PQS \cong \triangle RSQ$	5. SAS Triangle Congruence Theorem

14.

Statements	Reasons
1. $\overline{AC} \cong \overline{DE}$	1. Given
2. $\overline{BC} \cong \overline{BD}$	2. Given
3. $\angle BCD \cong \angle BDC$	3. Given
4. $\angle ACB \cong \angle EDB$	4. Congruent Supplements Theorem
6. $\triangle ABC \cong \triangle EBD$	6. SAS Triangle Congruence Theorem

17. no; Possible answer: There is not enough information given to use the SAS Congruence Theorem; She can't use the ASA Congruence Theorem because we don't know if the 9 in. side is the included side. No theorem can prove congruence.

18. A. Possible answer: The figure shows that the triangles have a pair of congruent right angles. I can count the units along each leg and determine the triangles have two pairs of congruent legs. The triangles are congruent by the SAS Triangle Congruence Theorem.

B. Possible answer: A clockwise rotation of 180° about the origin will map triangle *XYZ* to triangle *ABC*.

19. There is not enough information to prove congruency; Possible answer: The student would need to know the actual length of either the longest side or shortest side of Triangle 1.

20. yes; They both construct a triangle with two sides of lengths 3 inches and 5 inches with an included angle of 35°. By SAS, the triangles are congruent.

22. Possible answer: If I knew one other angle, I could prove whether or not the triangles were congruent.

23. Let *D* be the point where the angle bisector intersects $\overline{BC}$. We know that $\overline{AB} \cong \overline{AC}$ by the definition of congruent segments and the given information. Since *D* is on the bisector of $\angle A$, $\angle CAD \cong \angle BAD$. Also, $\overline{AD} \cong \overline{AD}$. Using this information, $\triangle ACD \cong \triangle ABD$ by SAS.

24. Since $\overline{FH}$ and $\overline{GJ}$ bisect each other and are the same length $\overline{FE} \cong \overline{EH} \cong \overline{GE} \cong \overline{EJ}$. By the Vertical Angles Theorem, $\angle GEF \cong \angle HEJ$. Two sides from each triangle and the included angles are congruent, therefore $\triangle GEF \cong \triangle HEJ$ by the SAS Triangle Congruence Theorem. Since all corresponding parts of congruent figures are congruent, $\overline{FG} \cong \overline{JH}$ and $\overline{FJ} \cong \overline{GH}$.

Additional Answers

MODULE 8, LESSON 8.3

Spark Your Learning

C. Possible answer: This construction uses side measurements only, while the previous constructions used angles and side measurements.

D. Possible answer: If a series of rigid transformations maps the new triangle to the given triangle, the two triangles are congruent.

Build Understanding

Task 1 **D.** Possible answer: I know the result is a triangle with side lengths of 2, 3, and 4 inches because the intersection of the arcs indicates where the side lengths for a triangle with those measurements would intersect.

E. Yes, the corresponding angles and sides are all congruent.

F. The triangles are congruent; Possible answer: A series of transformations maps the triangle to the ones drawn in the previous parts.

Task 2 **B.** Yes; $\overline{LM} \cong \overline{PQ}$ and $\overline{MN} \cong \overline{QR}$. If M lies along the intersection of those segments, and the equal segments are interchangeable by the Substitution Property of Equality, then Q must be the same point as M; Yes, the transformations proved the triangles are congruent.

Task 3 **A.** $BC = 12$ m, $DF = 13$ m; The two triangles are right triangles because we know two sides are perpendicular in each triangle.

C. Possible answer: The SSS Triangle Congruence Theorem was used because there was more information about the sides of the triangles than the angles; yes; You can also use the SAS Triangle Congruence Theorem since you can determine the missing side using the Pythagorean Theorem and then use the right angle as the included angle. Since two pairs of congruent angles were not provided, you cannot use the ASA Triangle Congruence Theorem.

Step It Out

Task 4 **A.** Possible answer: We are given information about all three sides but none of the angles.

B. Possible answer: If the triangles are congruent, all corresponding parts must be congruent. We can set the expressions for the two side lengths equal to each other because they must be congruent.

Check Understanding

1. no; If two triangles have 3 congruent side lengths, the triangles must be congruent.

2. Translate $\triangle ABC$ along $\overrightarrow{AD}$ to map A to D. Rotate $\triangle ABC$ counterclockwise about D with an angle of rotation of $\angle CDF$ to map C to F. Reflect $\triangle ABC$ across $\overline{DF}$ to map B to E.

3. You would need to know enough information about the other two sides of each triangle to prove all three corresponding sides are congruent.

On Your Own

5. Yes, they are congruent.

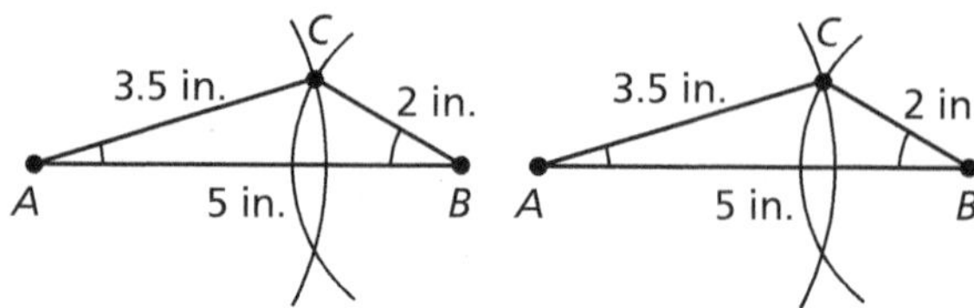

6. Yes, they are congruent.

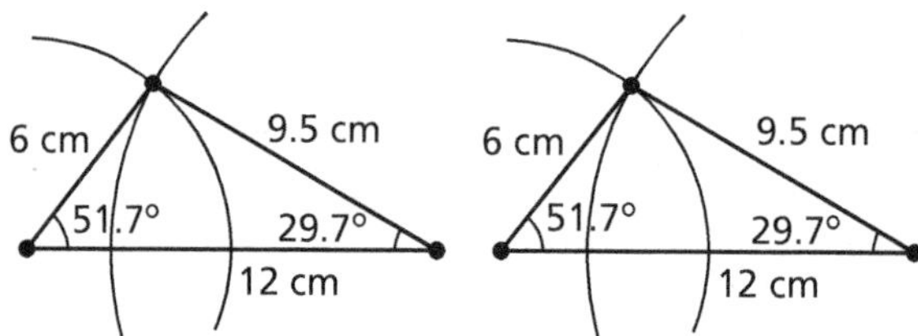

7. congruent; SSS Triangle Congruence Theorem; Possible answer: The triangles have two sides marked as congruent and share a third side, which must be congruent.

8. Not enough information is given about the side lengths, angles, or both.

9. congruent; ASA Triangle Congruence; There is a congruent pair of angles, a pair of congruent included sides, and a congruent pair of vertical angles.

10. congruent; Possible answer: SAS Triangle Congruence; There is a pair of congruent sides, a pair of included congruent alternate interior angles, and a shared congruent side.

11. not congruent; Possible answer: There are 4 side lengths shown. Two congruent triangles have a maximum of 3 different side lengths.

12. Not enough information is given; Information about the side lengths is needed as well.

13. Possible answer: Translate $\triangle ABC$ along $\overrightarrow{AD}$ to map A to D. Rotate $\triangle ABC$ clockwise about D with an angle of rotation of $\angle CDF$ to map $\triangle ABC$ to $\triangle DEF$.

15. $x = 4$

16. $x = 1.5$

17. $x = 4.5, y = 9$ or $x = 3, y = 6$

18. $x = 5, y = 7$

19. $x = 12, y = 3$ or $x = 18, y = 2$

20. $x = 2, y = 2$

21. no; Possible answer: Many triangles have the same angle measurements but different side lengths. For example, the triangles could have side lengths of 2-3-4 and 4-6-8 and have congruent corresponding angles.

23. ;

2 in. 2 in.

2 in.

Possible answer: I drew a 2-inch segment, then set my compass to 2 inches, put the compass on each endpoint, and drew an arc. I put a point where the arcs intersected and drew a triangle using the point as the third vertex.

24.

Statements	Reasons
1. $\overline{AC} \cong \overline{AB}$, $\overline{AD} \cong \overline{AE}$	**1.** Given
2. $AC = AB$, $AD = AE$	**2.** Definition of congruence
3. $AC = AD + DC$, $AB = AE + EB$	**3.** Segment Addition Postulate
4. $AD + DC = AE + EB$	**4.** Transitive Property of Equality
5. $DC = EB$	**5.** Subtraction Property of Equality
6. $\overline{DC} \cong \overline{EB}$	**6.** Definition of congruence
7. $\overline{BD} \cong \overline{CE}$	**7.** Given
8. $\overline{BC} \cong \overline{CB}$	**8.** Reflexive Property of Congruence
9. $\triangle BCD \cong \triangle CBE$	**9.** SSS Triangle Congruence Theorem

25. Create two triangles with the center of the circle, the satellite, and the signal distance points as the vertices. The length of the sides from the center to the signal distance points is r. The shared side from the center to the satellite is $35{,}700 + r$. Both triangles have a corresponding included angle of $90° - 9° = 81°$, so the triangles are congruent by the SAS Triangle Congruence Theorem. Therefore all corresponding sides are congruent, and the signal is proven to travel the same distance in both directions.

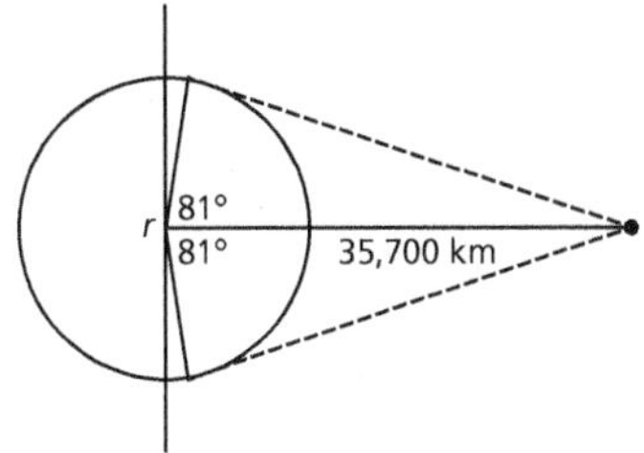

26.

O
1.5 in.

Possible answer: I drew a 1.5-inch diameter segment, placed the point of my compass at the midpoint, and drew a circle connecting the two endpoints. Using the same compass setting, I placed the compass where each diameter endpoint intersected the circle, and drew two more circles. I placed a point at each of the four locations where those circles intersected my original circle and used a ruler to connect the points to make a regular hexagon.

27. no; You cannot prove two triangles are congruent using only information about the angles.

28. *See below.*

28. The triangles are congruent, because $\overline{PQ} \cong \overline{XY}$, $\overline{QR} \cong \overline{YZ}$, and $\overline{RP} \cong \overline{ZX}$.

$P(-8, 2)$, $Q(-4, 6)$, $R(4, 4)$

$X(-7, -4)$, $Y(-3, -8)$, $Z(5, -6)$

$PQ = \sqrt{(x_2 - x_1)^2 + (y_2 - y_1)^2}$

$PQ = \sqrt{(-4 - (-8))^2 + (6 - 2)^2}$

$PQ = \sqrt{32} = 4\sqrt{2}$

$XY = \sqrt{(x_2 - x_1)^2 + (y_2 - y_1)^2}$

$XY = \sqrt{(-3 - (-7))^2 + (-8 - (-4))^2}$

$XY = \sqrt{32} = 4\sqrt{2}$

$QR = \sqrt{(x_2 - x_1)^2 + (y_2 - y_1)^2}$

$QR = \sqrt{(-4 - 4)^2 + (6 - 4)^2}$

$QR = \sqrt{68} = 2\sqrt{17}$

$YZ = \sqrt{(x_2 - x_1)^2 + (y_2 - y_1)^2}$

$YZ = \sqrt{(-3 - 5)^2 + (-8 - (-6))^2}$

$YZ = \sqrt{68} = 2\sqrt{17}$

$RP = \sqrt{(x_2 - x_1)^2 + (y_2 - y_1)^2}$

$RP = \sqrt{(-8 - 4)^2 + (2 - 4)^2}$

$RP = \sqrt{148} = 2\sqrt{37}$

$ZX = \sqrt{(x_2 - x_1)^2 + (y_2 - y_1)^2}$

$ZX = \sqrt{(-7 - 5)^2 + (-4 - (-6))^2}$

$ZX = \sqrt{148} = 2\sqrt{37}$

Additional Answers

MODULE 8, LESSON 8.4

Spark Your Learning

B. I know the SAS, ASA, and SSS Triangle Congruence Theorems. I know that if two triangles are congruent, you can map one to the other with rigid transformations.

Build Understanding

Task 1 **C.** Possible answer: We have previously proven that if two angles in a triangle are congruent to two angles in another triangle, the third angles in each triangle must be congruent; We have proved if two triangles have two pairs of congruent angles and a pair of congruent included sides, the triangles are congruent.

D. There is no AAA triangle congruence criteria. An infinite number of triangles with the same three angle values are possible.

Task 2 **C.** Possible answer: Two triangles are possible because there are two ways in which you can arrange side lengths of 2 inches and 3 inches with a 30 degree angle and create a triangle.

D. No, it is not possible to use SSA criteria to prove two triangles are congruent; We have just created two triangles that share SSA criteria but are not congruent.

Step it Out

Task 3 **B.** Possible answer: We can write this equation because we have shown $c^2 = z^2$, which allows us to set the two Pythagorean Theorem equations equal to each other; We can make the substitution in step two because we have already shown $b^2 = y^2$.

Task 4 **A.** By the Vertical Angles Theorem, all vertical angles (like the ones formed by 20th Street and North Broadway) are congruent.

B. The sides we know are congruent are along 20th Street. Those are not included sides between the angles we know are congruent, so the AAS Triangle Congruence Theorem should be used to determine congruence.

Task 5 **A.** The given information tells us that Welton Street and 21st Street are perpendicular. By the definition of perpendicular lines, the angles formed at the intersection are right angles.

B. In a right triangle, the side opposite the right angle is the hypotenuse.

C. The side along Welton Street is shared by both triangles; It is congruent to itself by the Reflexive Property of Congruence.

D. 2 right angles, 2 congruent sides, and 2 congruent hypotenuses

Check Understanding

1. Possible answer: In the AAS Triangle Congruence Theorem, the congruent side is not included between the two congruent angles. In the ASA Triangle Congruence Theorem, the congruent side is included between the congruent angles. Both are shortcuts to prove two triangles are congruent without proving all six parts are congruent.

2. Possible answer: This is the SSA ambiguous case, so there are two triangles.

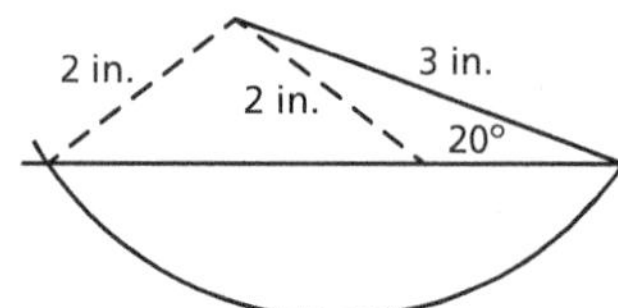

3. Possible answer: When the triangle is a right triangle, SSA criteria applies because the sides and hypotenuse have a defined relationship already. Since the sides and hypotenuse must meet the Pythagorean criteria, two right triangles that meet the SSA criteria will be congruent.

5. yes; We are given two right triangles with congruent hypotenuses and a shared congruent side. The triangles are congruent by the HL Triangle Congruence Theorem.

On Your Own

14.

Statements	Reasons
1. $\overline{JK} \parallel \overline{MN}$	1. Given
2. $\overline{JN}$ bisects $\overline{KM}$.	2. Given
3. $\angle MNL \cong \angle KJL$	3. Alternate Interior Angles
4. $\overline{ML} \cong \overline{KL}$	4. Definition of segment bisector
5. $\angle MLN \cong \angle KLJ$	5. Vertical Angles Theorem
6. $\triangle JKL \cong \triangle NML$	6. AAS Triangle Congruence Theorem

15.

Statements	Reasons
1. $\angle A \cong \angle C$	1. Given
2. $\overline{DB}$ bisects $\angle ABC$	2. Given
3. $\angle ABD \cong \angle CBD$	3. Definition of angle bisector
4. $\overline{DB} \cong \overline{DB}$	4. Reflexive Property of Congruence
5. $\triangle ABD \cong \triangle CBD$	5. AAS Triangle Congruence Theorem

16.

Statements	Reasons
1. $\overline{PT} \perp \overline{SQ}$, $\overline{ST} \cong \overline{PQ}$	1. Given
2. R is the midpoint of $\overline{PT}$.	2. Given
3. $\angle SRT = 90°$, $\angle PRQ = 90°$	3. Definition of perpendicular lines
4. $\overline{PR} \cong \overline{RT}$	4. Definition of segment bisector
6. $\triangle PQR \cong \triangle TSR$	6. HL Triangle Congruence Theorem

17.

Statements	Reasons
1. $\angle E = 90°$, $\angle C = 90°$	1. Given
2. $\overline{AE} \cong \overline{BC}$	2. Given
4. $\overline{AB} \cong \overline{AB}$	4. Reflexive Property of Congruence
6. $\triangle ABC \cong \triangle BAE$	6. HL Triangle Congruence Theorem

18. Given: $\angle DAB \cong \angle DCB$ and $\angle DBA \cong \angle DBC$.

Prove: $\triangle DAB \cong \triangle DCB$

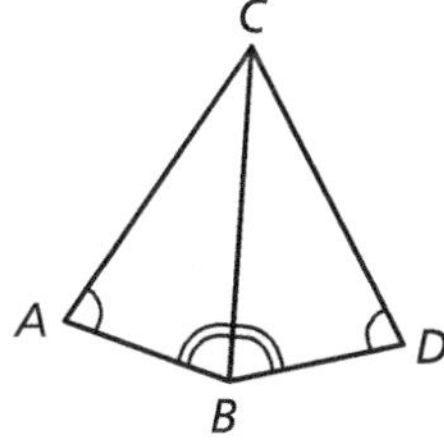

Because of the given congruent angles and the shared side $\overline{BC}$, the triangles are congruent by the AAS Triangle Congruence Theorem.

19. The king post is 10 ft tall, and the bottom chord is 25 ft long.

20. Yes; Two right triangles that share a hypotenuse create a rectangle, and will have two pairs of legs that meet at 90° angles. Since opposite sides are parallel, we know the included angles between the legs and the shared hypotenuse are pairs of congruent alternate interior angle. Congruency can be proven in many ways, but ASA Triangle Congruence is one option.

21. 32 m; The two triangles are right triangles. By using the Pythagorean Theorem, I can find $YW = 8$. The sum of the sides is $10 + 10 + 8 + 8 = 36$ m.

MODULE 8 REVIEW

6. ASA, AAS, SAS, SSS, HL; Examples will vary. See students' drawings.

9. There is no SSA Triangle Congruence Theorem because it is possible to create more than one triangle with these specifications.

12.

$\angle A \cong \angle F$, $\overline{AB} \cong \overline{FE}$, $\overline{AC} \cong \overline{FD}$	Given
$\triangle ABC \cong \triangle FED$	SAS

13.

$\angle HGI \cong \angle JGI$, $\angle HIG \cong \angle JIG$	Given
$\overline{GI} \cong \overline{GI}$	Reflexive Property of Congruence
$\triangle HGI \cong \triangle JGI$	ASA

14.

$\overline{PQ} \cong \overline{RQ}$; $m\angle PSQ = 90°$	Given
$m\angle PSQ + m\angle RSQ = 180°$	Linear Pair Postulate
$m\angle RSQ = 180 - m\angle PSQ = 90°$	Subtraction Property of Equality
$\angle PSQ \cong \angle RSQ$	Definition of congruence
$\overline{SQ} \cong \overline{SQ}$	Reflexive property of congruence
$\triangle PSQ \cong \triangle RSQ$	HL

15. Possible Answer

$\overline{KO} \cong \overline{MO}$; $\overline{KL} \parallel \overline{NM}$	Given
$\angle OKL \cong \angle OMN$; $\angle OLK \cong \angle ONM$	Alternate Interior Angles
$\triangle KOL \cong \triangle MON$	AAS

UNIT 5

UNIT 5 STEM TASK

Possible answer: Potentially, because 40% of 2315 is 926. At 926, the biomass is about 800, which is greater than 783 for a mild bloom. It would be better than it currently is, but it still produces a mild bloom.

MODULE 9, LESSON 9.1

Build Understanding

Task 1 **A.** Possible answer: The auxiliary line is drawn to create angles 4 and 5, which relates them to angles 1, 2, and 3; By the Angle Addition Postulate, $m\angle 4 + m\angle 2 + m\angle 5 = 180°$. Also, $m\angle 1 = m\angle 4$ and $m\angle 3 = m\angle 5$ because they are alternate interior angles formed by a transversal and parallel lines.

Additional Answers

Task 2 **A.** $\angle 4$ forms a linear pair with $\angle 3$; $\angle 3$ and $\angle 4$ are supplementary because of the Linear Pair Theorem.

B. The Triangle Sum Theorem is used to show that the measure of the angles in the triangle is equal to the sum of the measures of $\angle 3$ and $\angle 4$.

Task 3 **A.** Since $\overline{AD}$ bisects $\angle A$, $\angle DAB \cong \angle DAC$. $\overline{AD} \cong \overline{AD}$ by the Reflexive Property of Congruence, and $\overline{AB} \cong \overline{AC}$ is given. So, the triangles are congruent by the SAS Triangle Congruence Theorem.

B. If D is the midpoint of $\overline{BC}$, then $\overline{BD} \cong \overline{CD}$ by the definition of a midpoint. $\overline{AD} \cong \overline{AD}$ by the Reflexive Property of Congruence, and $\overline{AB} \cong \overline{AC}$ is given. So $\triangle ADB \cong \triangle ADC$ by the SSS Triangle Congruence Theorem.

On Your Own

6. no; The sum of the measure of the two obtuse angles would be greater than 180°, and the total sum of the measures of the interior angles of a triangle is 180°.

7. yes; The exterior angle of the right angle in a right triangle is a right angle.

8. yes; The angle that measures 100° cannot be a base angle because then the sum of the measures of the base angles would be 200°. So, the vertex angle measures 100°, and the measure of each base angle is half of $180° - 100° = 80°$.

9.

Statements	Reasons
1. $\angle B \cong \angle C$	1. Given
2. Draw $\overline{AD}$ so that it bisects $\angle A$.	2. An angle has one angle bisector.
3. $\angle DAB \cong \angle DAC$	3. Definition of angle bisector
4. $\overline{AD} \cong \overline{AD}$	4. Reflexive Property of Congruence
5. $\triangle ADB \cong \triangle ADC$	5. AAS Triangle Congruence Theorem
6. $\overline{AB} \cong \overline{AC}$	6. Corresponding parts of congruent triangles are congruent.

10. $x = 60$

11. $x = 59$

18. Jessica; She finds the measure of the exterior angle using the Exterior Angle Theorem. Tristan found the measure of the third angle of the triangle.

19. Answers will vary. Sample answer: The entrance to a tent is shaped like an isosceles triangle. One of the congruent angles has a measure of 57°. What are the measures of the other two angles?

22. 240 ft; By the Angle Addition Postulate, $m\angle ATB = 70° - 35° = 35°$. By the Alternate Interior Angles Theorem, $m\angle BAT = 35°$. So, $\angle ATB \cong \angle BAT$ by the definition of congruence. $\overline{BA} \cong \overline{BT}$ by the Converse of the Isosceles Triangle Theorem.

23. no; The Exterior Angle Theorem only involves the relationship between the sum of the measures of the remote interior angles and that of the remote exterior angle.

33. Possible answer: The sum of the congruent base angles will be an even number. The measure of the vertex angle will be an even number since 180° minus an even number will be an even number.

34. Possible answer: $m\angle KAB = 137°$, $m\angle ABC = 42°$, $m\angle BCA = 95°$

MODULE 9, LESSON 9.2

Spark Your Learning

B. Possible answer: The locations of the food trucks can be represented with a map, and then I can look for a central location.

Build Understanding

Task 2 **A.** Possible answer: Line l and line m are perpendicular bisectors. The sides of a triangle are not parallel, so the perpendicular bisectors are not parallel. Since the lines are not parallel, they intersect at one point.

B. The lines are concurrent because point P lies on lines l, m, and n, which satisfies the definition of concurrent lines.

Task 3 **A.** Definition of perpendicular bisector, Definition of perpendicular bisector, and Transitive Property of Equality; no; For a triangle, if two of the perpendicular bisectors go through a point, the third one also passes through that point.

Step It Out

Task 4 **A.** First, find the slope of the line perpendicular to a line that has a slope of $\frac{3}{2}$. The slope is $-\frac{2}{3}$. Next, find the y-intercept of the line that passes through $(4, 6)$ with a slope of $-\frac{2}{3}$. The y-intercept is $\frac{26}{3}$. So, the equation of the perpendicular bisector of $\overline{AB}$ is $y = -\frac{2}{3}x + \frac{26}{3}$.

B. The perpendicular bisector of $\overline{AC}$ is a vertical line through the point $(9, 0)$. The equation of this line is $x = 9$.

Check Understanding

1.

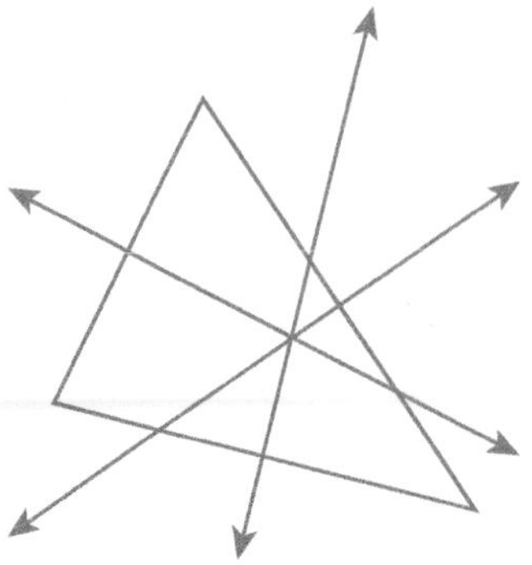

2.

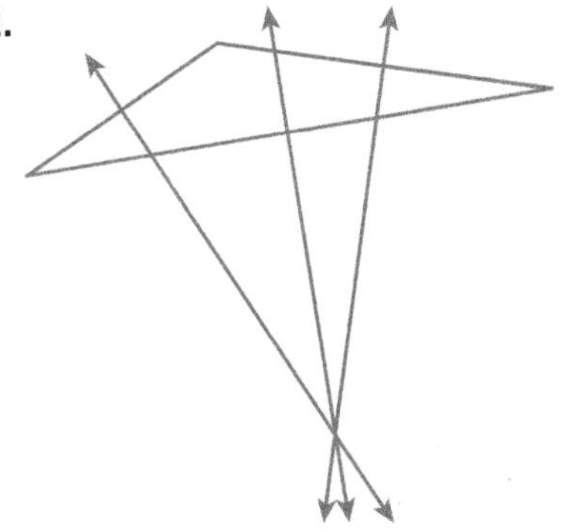

3. Nancy is correct; If a triangle is an obtuse triangle, the circumcenter is located outside of the triangle. If a triangle is a right triangle, the circumcenter is located on the triangle.

4.

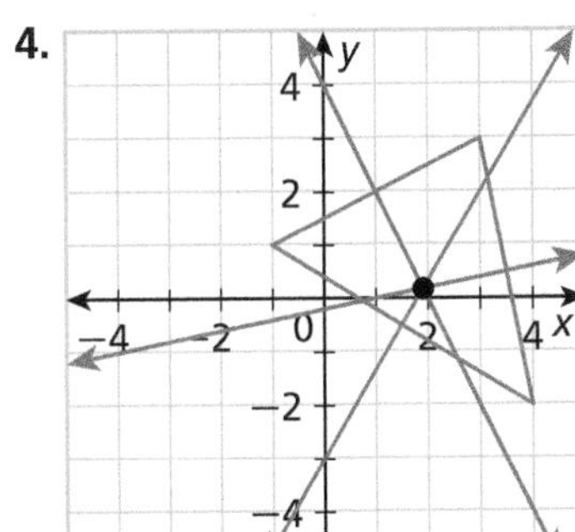

The x-coordinate of the circumcenter is slightly less than 2, and the y coordinate is slightly greater than 0.

5. yes; Point N is equidistant from each vertex of the triangle, so $QN = RN$. To find the value of x, set the algebraic expressions equal to each other and solve for x.

On Your Own

8.

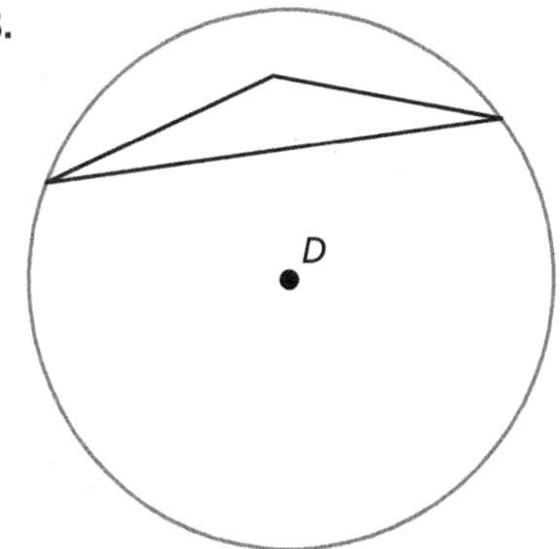

9.

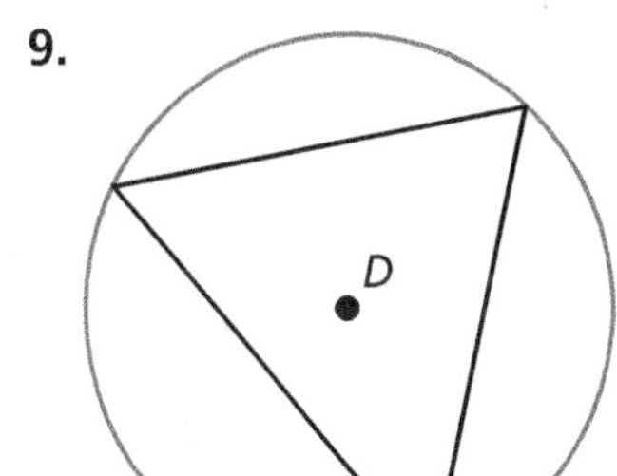

10.

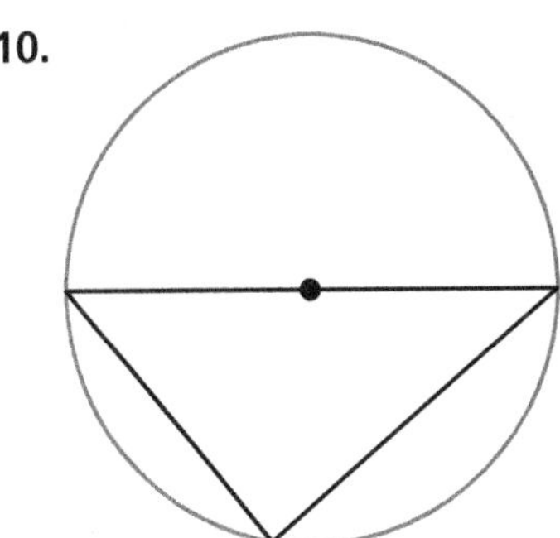

11.

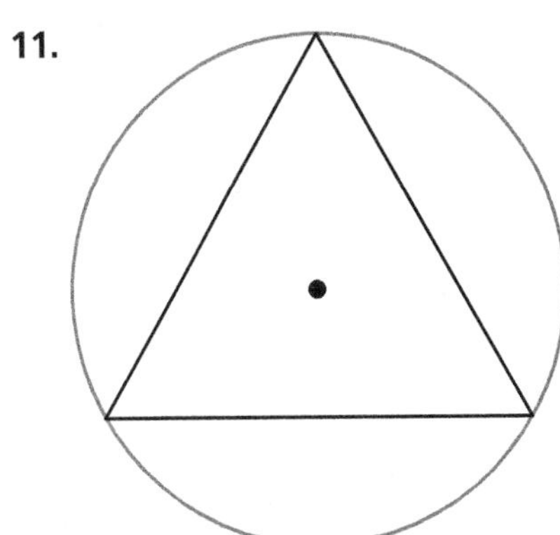

12.

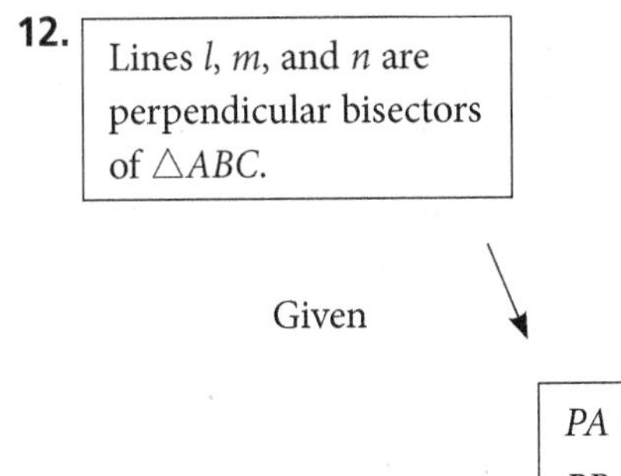

Lines l, m, and n are perpendicular bisectors of $\triangle ABC$.

Given

P be the intersection of l, m, and n.

Given

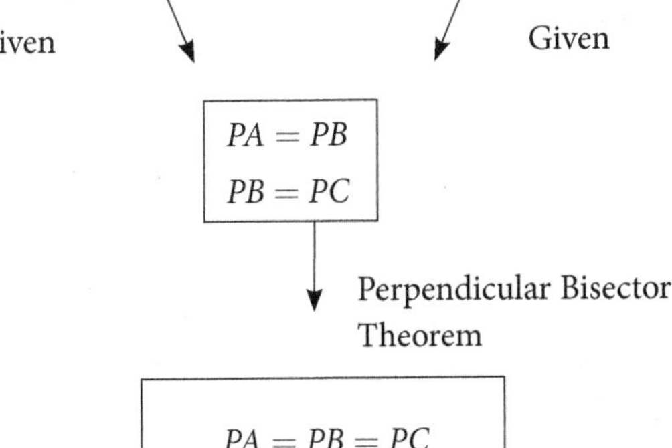

$PA = PB$
$PB = PC$

Perpendicular Bisector Theorem

$PA = PB = PC$

Transitive Property of Equality

14.

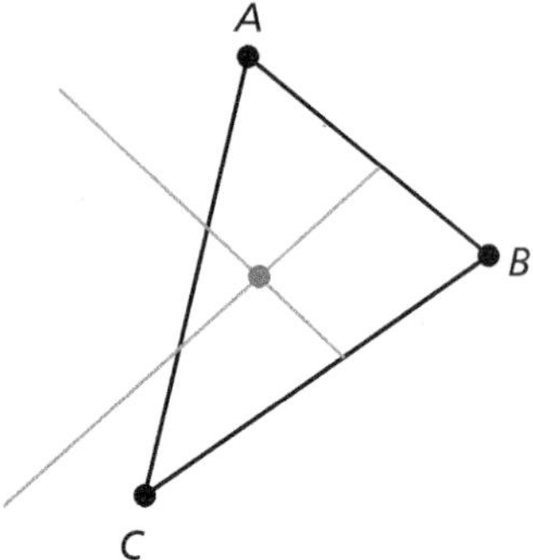

19. It is fairly easy to find an equation for a line perpendicular to a horizontal or vertical line through a given point.

20. Since all three perpendicular bisectors intersect in the same point, you need only two lines to determine the point of intersection; After the circumcenter has been determined using two of the perpendicular bisectors, the third perpendicular bisector can be drawn. It should pass through the point given as the circumcenter.

MODULE 9, LESSON 9.3

Spark Your Learning

D. Possible answer: You can use a compass to draw a circle and see if the circle touches each side of the triangle.

Build Understanding

Task 1 **A.** $\overline{RY}$; The distance between a point and a line is the shortest distance from the point to any point on the line. The segment that represents the distance is perpendicular to the line. In the diagram, $\overline{RY}$ appears to be perpendicular to $\overrightarrow{QP}$.

B. Show that $\triangle YXW \cong \triangle YZW$, and then use CPCFC to show $WX = WZ$; The AAS Triangle Congruence Theorem can be used since we know that $\angle XYW \cong \angle ZYW$, $\angle WXY \cong \angle WZY$ (congruent right angles), and $\overline{YW} \cong \overline{YW}$ (shared side).

Task 2 **A.** Possible answer: If a point lies on the exterior of the angle, it is not necessarily possible to draw perpendicular segments to each ray. The distance to each ray is not necessarily defined.

C. Possible answer: In both proofs, it is shown that $\triangle YXW \cong \triangle YZW$, and then congruent parts of congruent figures is used to state the desired result; In the proof for the Angle Bisector Theorem, the congruent angles are used with AAS Triangle Congruence Theorem to show that $WX = WZ$. In the proof for the converse, the congruent segments are used with HL Triangle Congruence Theorem to show that $\angle XYW \cong \angle ZYW$.

Step It Out

Task 3 **A.** By the Angle Bisector Theorem, a point on the angle bisector is equidistant from the two sides of the angle.

B. By the Converse of the Angle Bisector Theorem, a point equidistant from the sides of an angle lies on the angle bisector.

Task 4 **A.** Draw an arc that intersects both sides of the angle. At each point of intersection, place the compass point and draw an arc. The point of intersection of these arcs lies on the angle bisector. Use a straightedge to connect this point and the vertex of the angle.

B. The radius of the circle is the distance between the incenter of the triangle and each side of the triangle.

Task 5 **A.** Since Z is the incenter, it lies on the angle bisector of $\angle PRQ$. So, $\angle PRZ$ is congruent to $\angle ZRQ$.

B. Since Z is the incenter, it lies on the angle bisector of $\angle QPR$. So, $\angle QPZ$ is congruent to $\angle ZPR$.

Check Understanding

7.

incenter

On Your Own

9. no; Possible answer: For $AQ = CQ$, point Q needs to be on the angle bisector of $\angle ABC$. There is no information given about $\angle ABC$.

10. yes; Possible answer: $\angle ABQ \cong \angle CBQ$, so Q lies on the angle bisector. By the Angle Bisector Theorem, $AQ = CQ$.

11. yes; Possible answer: Since $\overline{AQ} \perp \overline{AB}$, $\overline{CQ} \perp \overline{CB}$, and $AQ = CQ$, the Converse to the Angle Bisector Theorem can be applied to conclude that $\overrightarrow{BQ}$ bisects $\angle ABC$.

12. no; Possible answer: For $\overrightarrow{BQ}$ to bisect $\angle ABC$, $\overline{AQ} \perp \overline{AB}$ and $\overline{CQ} \perp \overline{CB}$. However, this information is not given.

13.

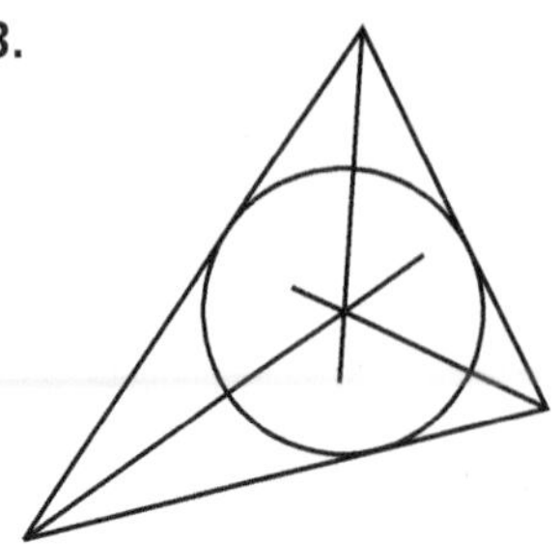

14.

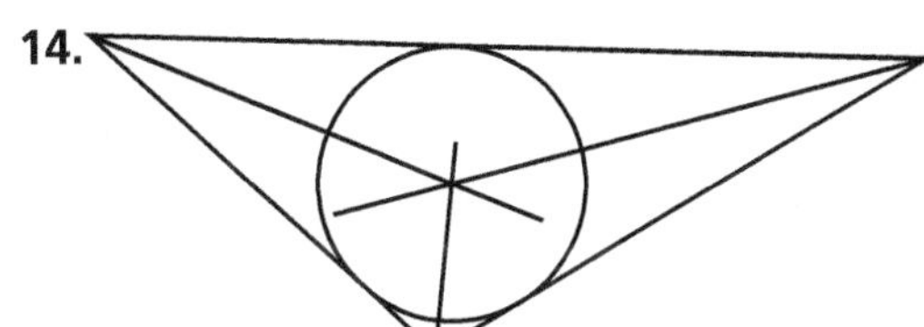

16. A.

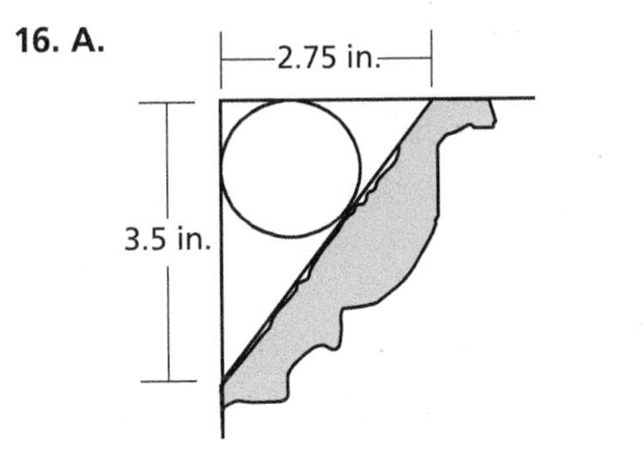

B. Answers will vary, but should be about 1.75 inches.

MODULE 9, LESSON 9.4

Spark Your Learning

C. a pole that has a flat top; If the pole has a tip like a pencil, then the pole has to be placed at the exact point of balance. A pole with a flat top offers some room for error.

D. If the triangle is an equilateral triangle, the point where it will balance will be close to what we think of as the center of the triangle. If it is a scalene triangle, the triangle will be lightest near the angle with the least measure. So, its balance point will be closer to the angles with the larger measures.

Step It Out

Task 3 The centroid is the intersection point of the three medians of a triangle. The intersection point can be found using any two medians. So, P is the centroid of $\triangle ABD$ and Q is the centroid of $\triangle BCD$ because these points are the intersection points of the medians of the triangles.

Task 4 **A.** Calculate the slope using $Q(7, 9)$ and $R(9, 3)$: $m = m = \frac{3-9}{9-7} = \frac{-6}{2} = -3$.

B. Start at $P(1, 3)$. Move 1 unit up, 3 units to the right, and plot a point. Continue plotting points to side $\overline{QR}$.

Check Understanding

1.

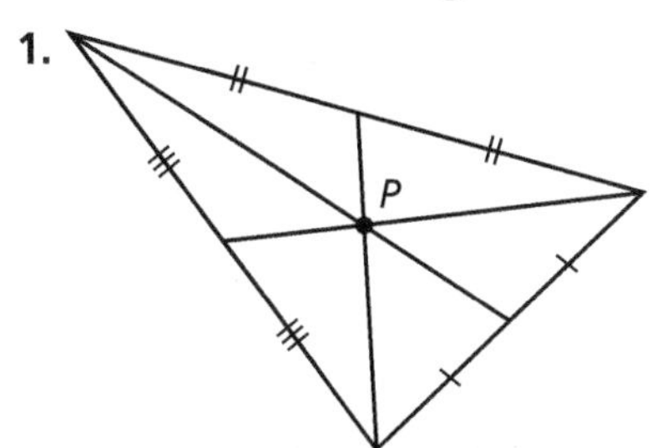

2.

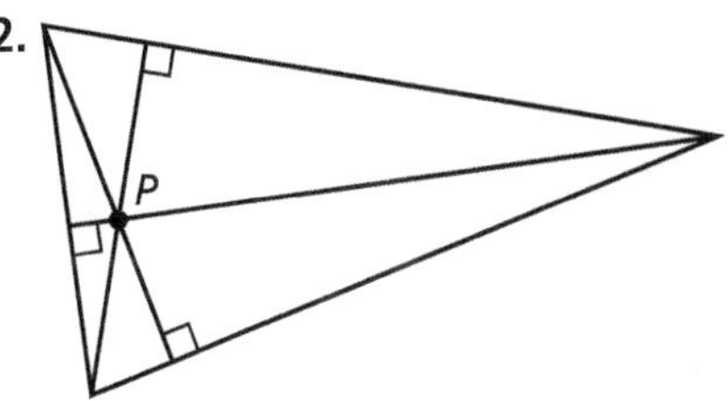

On Your Own

7.

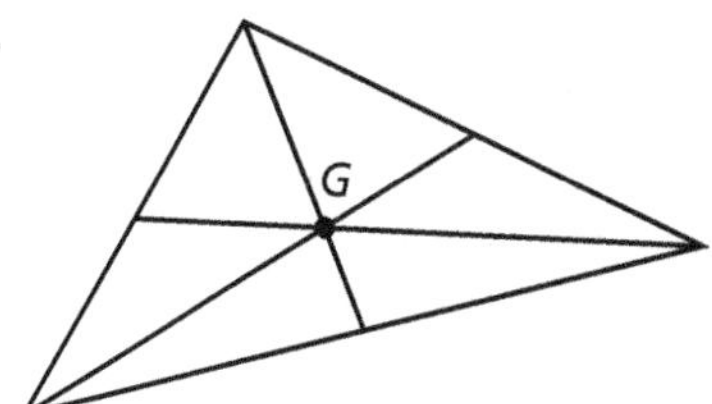

8.

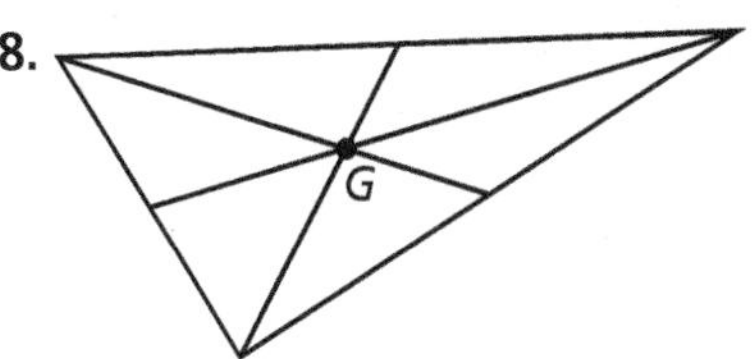

9. Possible answer: In each triangle, the side opposite vertex A is point Z. So, each triangle has $\overline{AZ}$ as a median. By the Centroid Theorem, the centroid is located $\frac{2}{3}$ down $\overline{AZ}$ from A.

10. A. Possible answer: The median from vertex B has endpoints $(4, 9)$ and $(4, 3)$.

B. Possible answer: The distance from $(4, 9)$ to $(4, 3)$ is 6. The distance from $(4, 9)$ to the centroid $(4, 5)$ is 4, which is $\frac{2}{3}$ of 6. The distance from $(4, 3)$ to the centroid $(4, 5)$ is 2, which is $\frac{1}{3}$ of 6.

11. no; The medians of a triangle are inside the triangle, so the intersection of the medians has to be inside the triangle.

12. Possible answer: If the triangle were to be cut out of a thick piece of paper or thin sheet metal, the centroid would be the location where you could balance the triangle.

13. inside

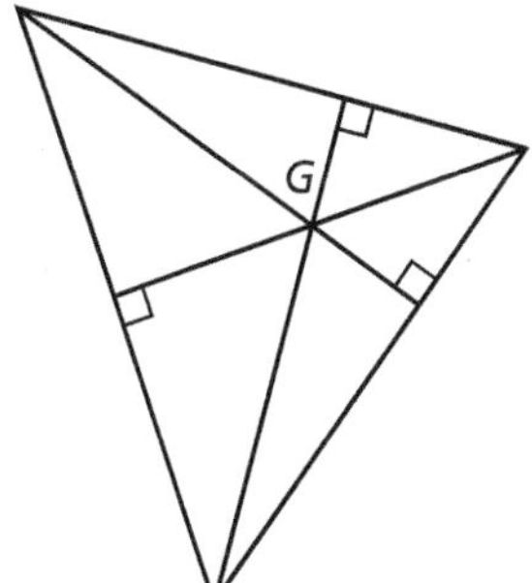

14. outside

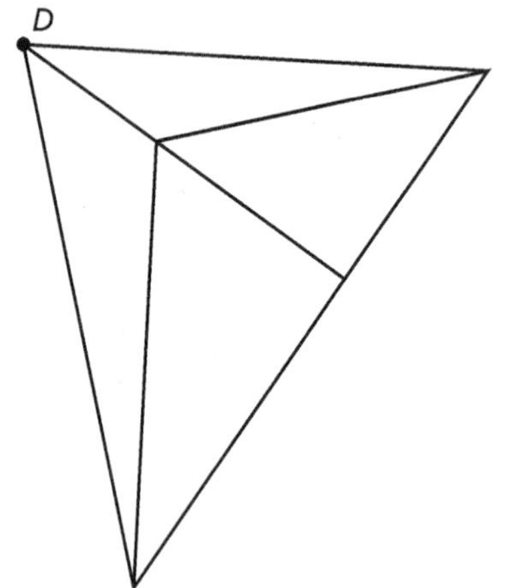

15. on

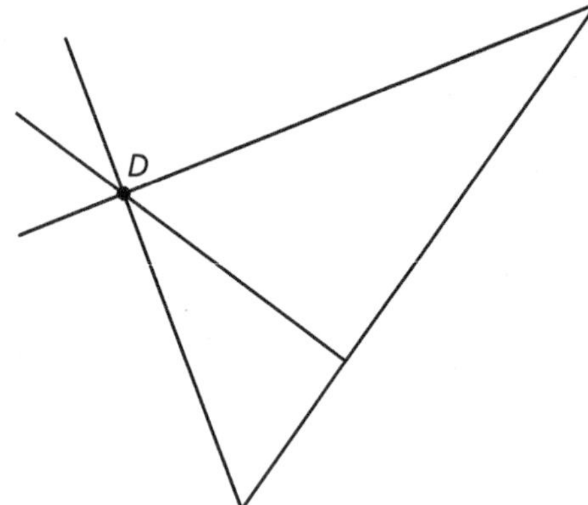

16. inside

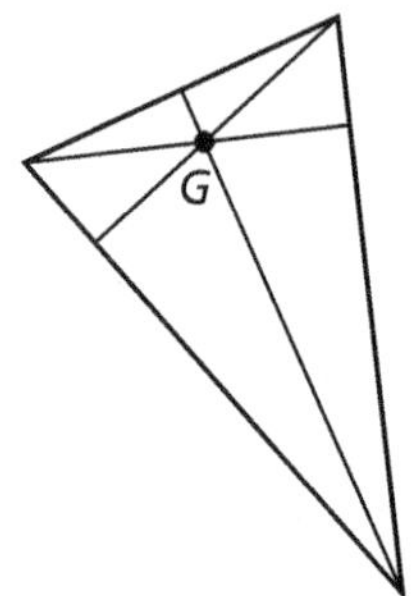

23. yes; The distance you are given is $\frac{2}{3}$ of the length of the median. You can find the length of the median by multiplying the given length by $\frac{3}{2}$.

48. Triangles will vary. Check students' drawings. A triangle with an orthocenter outside of the triangle must be an obtuse triangle; A triangle with an orthocenter on the triangle must be a right triangle.

MODULE 9, LESSON 9.5

Spark Your Learning

B. If the first floor and second floor are parallel, then the corresponding angles should be congruent.

Build Understanding

Task 1 **A.** Point F is the midpoint of $\overline{PQ}$, and point G is the midpoint of $\overline{QR}$. A segment that joins the midpoints of two sides of a triangle is a midsegment of the triangle.

B. Construct the midpoint of $\overline{PR}$. Then draw a segment that connects the midpoint to F and a segment that connects the midpoint to G.

E. A midsegment of a triangle is parallel to the third side of the triangle and is half as long as the third side.

Step It Out

Task 2 **A.** Segments on a coordinate plane that have the same slope are parallel.

B. Find the coordinates of the midpoints of the two sides. The end points of the path should coincide with the midpoints.

Task 3 The congruent marks on the diagram show that M is the midpoint of $\overline{YZ}$, and N is the midpoint of $\overline{XZ}$. So, $\overline{MN}$ is a midsegment of the triangle. By the Midsegment Theorem, $MN = \frac{1}{2}YX$.

Check Understanding

1.

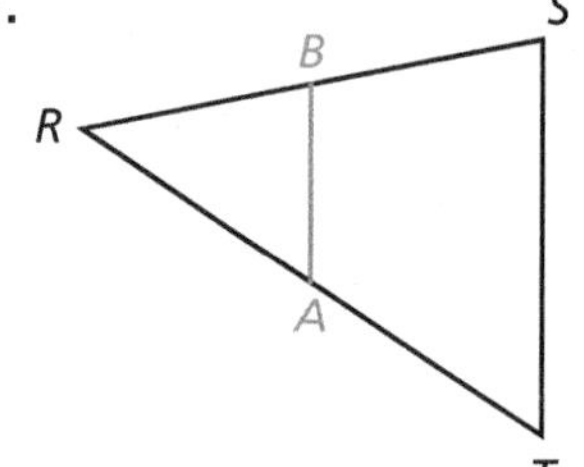

2.

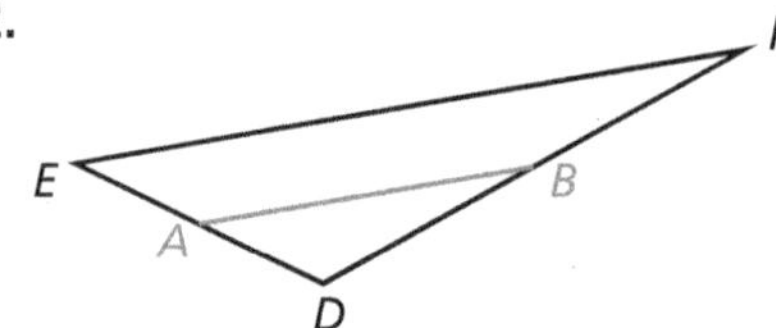

On Your Own

7. A.

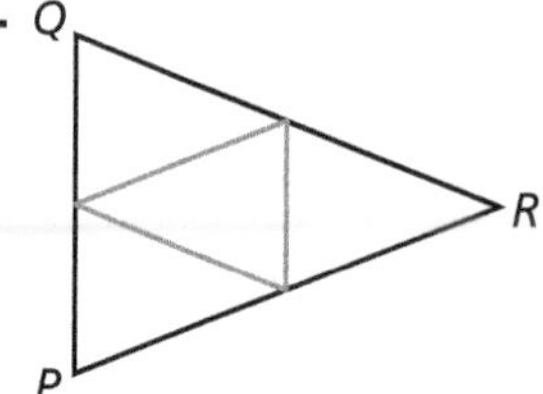

B. isosceles triangle; Two of the midsegments will be congruent since their lengths are half the length of the congruent sides.

16. A.

Midsegment	1	2	3
Length	32	16	8

17. 80 cm; yes; The new side is parallel to the hypotenuse, so it is also a right triangle.

18. C. Both segments have a slope of 0; $\overline{MN} \parallel \overline{HK}$

23. Slope of $\overline{LM} = -\frac{3}{4}$, and slope of $\overline{AB} = -\frac{3}{4}$. So, $\overline{LM} \parallel \overline{AB}$; $LM = 5$ and $AB = 10$. So, $LM = \frac{1}{2}AB$.

24. Slope of $\overline{LM} = \frac{1}{5}$, and slope of $\overline{BC} = \frac{1}{5}$. So, $\overline{LM} \parallel \overline{BC}$; $NM = \sqrt{26}$ and $BC = 2\sqrt{26}$. So, $NM = \frac{1}{2}BC$.

MODULE 9 REVIEW

13. Points on a perpendicular bisector are equidistant from the endpoints of the segment. The perpendicular bisectors used to create the circumcenter use the vertices of the triangle as the endpoints of each segments, thus the point of intersection (concurrency) is equidistant to each vertex.

14. Points on an angle bisector are equidistant from the rays of the angle. The angle bisectors used to create the incenter use the sides of the triangle as the rays, thus the point of intersection (concurrency) is equidistant to each side.

15. A dilation centered at the opposite vertex with a scale factor of $\frac{1}{2}$ causes midsegments to be parallel to the third side.

MODULE 10, ARE YOU READY?

1. 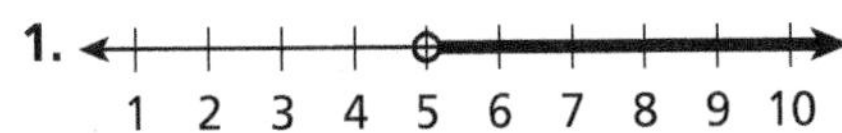

2.
1 2 3 4 5 6 7 8 9 10

3.
−5 −4 −3 −2 −1 0 1 2 3 4 5

4.
−5 −4 −3 −2 −1 0 1 2 3 4 5

5. 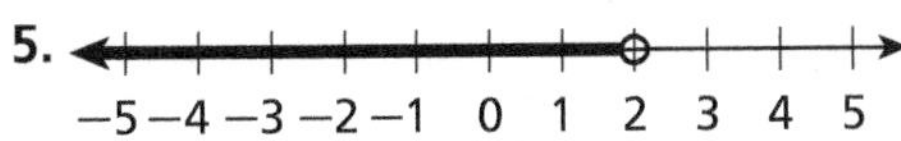

6. 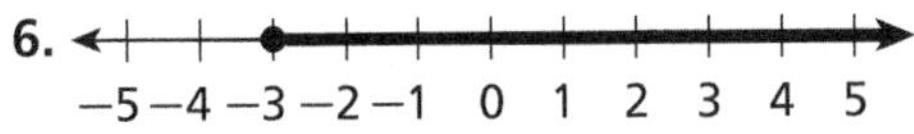

7. 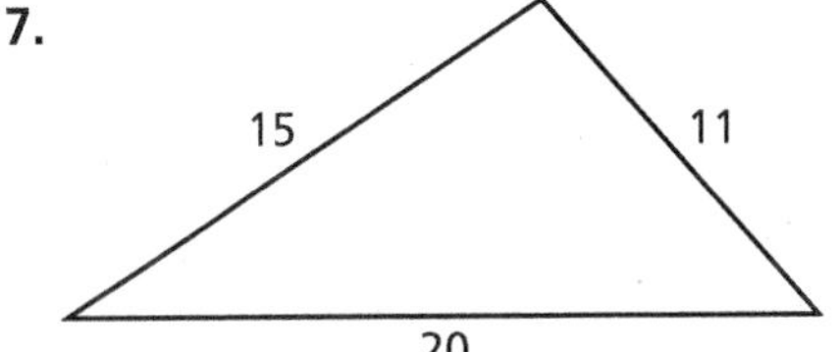

9. 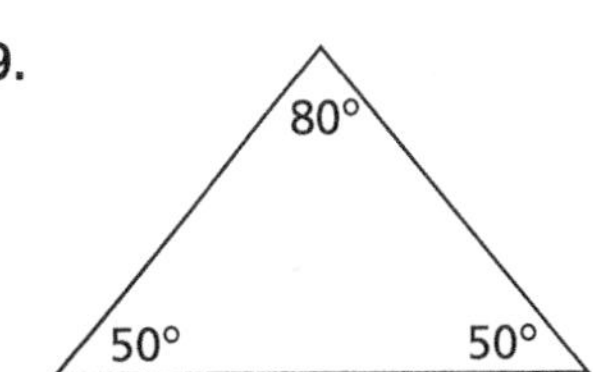

11. 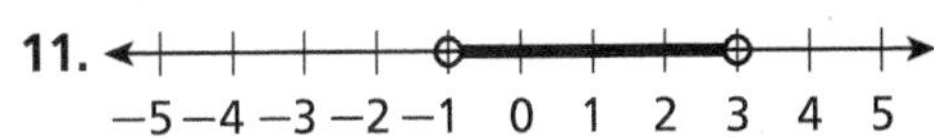

12. 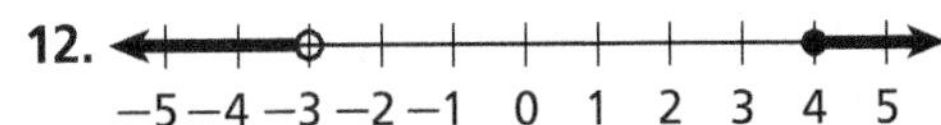

13. 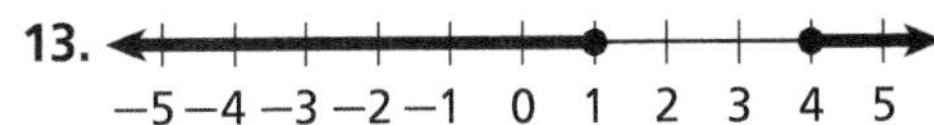

14. 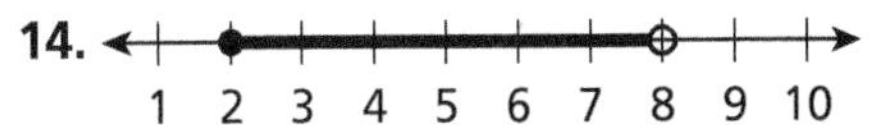

15.
−5 −4 −3 −2 −1 0 1 2 3 4 5

18. 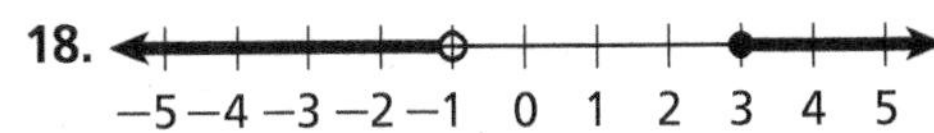

MODULE 10, LESSON 10.1

Spark Your Learning

B. the measures of the angles formed by each section of the course; degrees

D. yes; You could use a ruler and measure each section and then add the sections to verify the longest path.

Build Understanding

Task 1 **A.** There are an infinite number of triangles that can be formed. Vertex C can be anywhere on the circle except on $\overline{AB}$ or on the line passing through $\overline{AB}$.

B. yes; Since the sum of 1.5 inches and 3 inches is greater than 4 inches, a triangle with side lengths 1.5 inches, 3 inches, and 4 inches is possible.

C. no; The longest $\overline{AC}$ can be is $AB + BC$, which is a straight angle and does not form a triangle.

D. The smallest angle is opposite the smallest side; The largest angle is opposite the largest side.

Task 2 **A.** The sum of each pair of sides is greater than the third side.

B. Since three inequalities must be true, if one inequality does not satisfy the Triangle Inequality Theorem, there is no need to check any others.

Task 3 **A.** The sum of the lengths of two sides of the triangle is compared to the length of the third side.

B. the smallest value of x and the largest value of x

C. no; Any positive number plus 18 is greater than 10, and x cannot be a negative number.

Step It Out

Task 4 **A.** Because you are given the side lengths, ordering them first will help determine the order of the angles opposite the given sides.

Check Understanding

1. no; If the base is 18 cm, compass arcs of lengths 7 cm and 9 cm from each end of the base do not intersect.

2. yes; If the base is 5 in., compass arcs of lengths 2 in. and 4 in. from each end of the base have two intersections, each forming a triangle.

6. $\overline{EF}$ is opposite $\angle D$. $\overline{DF}$ is opposite $\angle E$. $\overline{DE}$ is opposite $\angle F$.

7. The largest angle will be opposite longest side. The smallest angle will be opposite shortest side.

On Your Own

12. A. yes; The sum of any two lengths is greater than the third length, so a triangle can be formed with these lengths.

25. $m\angle Z < m\angle X < m\angle Y$, so $XY < YZ < XZ$. Therefore, the safest route is to avoid sailing between the islands at X and Y.

26. The towers at P and R are the closest together.

27. $AB + BC > x$

$AB + x > BC$

$x > BC - AB$

$BC + x > AB$

$x > AB - BC$

Since $AB > BC$, $BC - AB < 0$, so the second inequality is not relevant. Combining the first and last inequalities gives $AB - BC < x < AB + BC$.

28.

Statements	Reasons
1. $\triangle ABC \cong \triangle DEF$	1. Given
2. $\overline{BC} \cong \overline{EF}$, $\overline{AC} \cong \overline{DF}$	2. CPCTC
3. $a = d$, $b = e$	3. Definition of congruent segments
4. $a + b > c$	4. Triangle Inequality Theorem
5. $d + e > c$	5. Substitution

29.

Statements	Reasons
1. $\triangle ABC$	1. Given
2. Draw $\overline{BX}$ so that $m\angle XBC = m\angle XCB$	2. By construction
3. $XB = XC$	3. Sides opposite angles of equal measure are equal in length.
4. $AX + XB = AX + XC$	4. Addition Property of Equality
5. $AC = AX + XC$	5. Segment Addition Postulate
6. $AC = AX + XB$	6. Substitution
7. $AX + XB > AB$	7. The sum of the lengths of any two sides of a triangle is greater than the length of the remaining side.
8. $AC > AB$	8. Substitution

31.

A
7 5
9 6
8 4

MODULE 10, LESSON 10.2

Spark Your Learning

B. The variables are the lengths of the line segments that represent the distances between airports.

Build Understanding

Task 1 **E.** The side opposite the greater included angle is longer. Therefore, the side in the first triangle is longer than the side in the second triangle.

Step It Out

Task 3 **B.** To use the Hinge Theorem, you need to have two congruent sides and the included angle. $\overline{AC}$ is the second congruent side in each triangle.

C. You have two sides in $\triangle ADC$ congruent to two sides in $\triangle ABC$, and you know the measures of the included angles for both triangles.

Task 4 **A.** You have two congruent sides of two triangles and an included angle, and you know the lengths of the sides opposite the included angle. You know that $ST < QT$, so $m\angle SRT < m\angle QRT$.

B. $m\angle SRT$ is a measure of an angle of a triangle, so it must be greater than zero.

Task 5 **A.** Possible explanation: If one of the numbers in a sum of two numbers is removed, the remaining number will be less than the original sum. Here, the sum of the two angle measures will be greater than the measure of either individual angle.

B. You know that two corresponding sides of the triangles are congruent, and you know that the included angle for $\triangle LPM$ is greater than the included angle for $\triangle LPN$.

Task 6 **A.** Assume the opposite of what you want to prove to write an indirect proof.

B. You are given that $\overline{AB} \cong \overline{DE}$ and $\overline{AC} \cong \overline{DF}$, and you are assuming that $\angle A \cong \angle D$.

Check Understanding

7. You assume the opposite of what you want to prove, and then find a contradiction to the given information to show that the opposite is not true.

On Your Own

18. The distance from Teresa's house to Ivy's house is longer; $\overline{CF} \cong \overline{AF}$ and $\overline{BF} \cong \overline{EF}$. $\angle CFE$ is a right angle and $\angle AFB$ is acute, so $m\angle CFE > m\angle AFB$. So, $CE > AB$.

19. Pair 1; Both sides of the triangles created by their paths are congruent, but the included angle is greater for Pair 1, so the distance from camp is greater.

20. Yes, he is right; The two triangles have two congruent sides and the included angle opposite $\overline{AB}$ has a greater measure than the included angle opposite $\overline{CD}$.

21. C. The length of the side represented by $4x + 3$ is 23, and the length of the side represented by $6x - 10$ is 20.

22.

Statements	Reasons
1. $\overline{AB} \cong \overline{CD}$	1. Given
2. $\overline{BC} \cong \overline{BC}$	2. Reflexive Property of Congruence
3. $BD > AC$	3. Given
4. $m\angle BCD > m\angle ABC$	4. Converse of the Hinge Theorem

23. Assume that both $\angle 1$ and $\angle 2$ can be obtuse. If $\angle 1$ and $\angle 2$ are both obtuse, then $m\angle 1 > 90°$ and $m\angle 2 > 90°$, so $m\angle 1 + m\angle 2 > 180°$. But $\angle 1$ and $\angle 2$ are supplementary, so $m\angle 1 + m\angle 2 = 180°$. So, $\angle 1$ and $\angle 2$ cannot both be obtuse.

24. She is incorrect; You make an assumption at the beginning of the proof but you prove your assumption is incorrect by the end of the proof.

MODULE 10 REVIEW

Concepts and Skills

6. The sides AB and BC must have enough length to reach the distance from point A to point C. If AC is greater than the sum of their lengths, then they will be unable to close the triangle. If they are exactly equal, then they will be able to close the distance but both will be collinear with side AC.

14. A. $m\angle JGI > m\angle HGI$; Converse of the Hinge Theorem

B. Assume that $\angle JGI \cong \angle HGI$. Given that $\overline{JG} \cong \overline{HG}$ and $\overline{GI} \cong \overline{GI}$ by the Reflexive Property of Congruence, $\triangle JGI \cong \triangle HGI$ by SAS. So $\overline{JI} \cong \overline{HI}$, contradicting $JI = 12 \neq 11 = HI$. Therefore, the assumption is false and $\angle JGI$ is not congruent to $\angle HGI$.

UNIT 6

MODULE 11, LESSON 11.1

Spark Your Learning

B. no; It appears that the quadrilaterals are made from one of five different pieces of fabric (orange, green, blue, pink, and black).

Build Understanding

Task 1 **B.** The sum of the angles of the quadrilateral is 360° because the sum of the angles of each triangle is 180° by the Triangle Sum Theorem.

Task 2 **D.** Yes, all parallelograms have side lengths of opposite sides that are equal and angle measures of opposite angles that are equal.

Task 3 **A.** You can show that the triangles are congruent, so the corresponding parts of the triangle are congruent.

C. In step 2, you would draw $\overline{AC}$ instead of $\overline{BD}$. Label the angles on opposite sides of $\overline{AC}$ as done previously, then complete the proof as was done for $\angle A \cong \angle C$.

Task 4 **C.** yes; You can use the fact that $\angle AEB \cong \angle DEC$ by the Vertical Angle Theorem, and use $\overline{AB} \cong \overline{DC}$ and $\angle ABD \cong \angle CDB$ to prove that the triangles are congruent using AAS Triangle Congruence Theorem.

Check Understanding

1. From Quadrilateral Sum Theorem, you know the sum of the angles is 360°. Add the three given angle measures and subtract the sum from 360° to find the fourth angle measure.

On Your Own

9. Draw diagonal $\overline{GJ}$ to form two triangles. By the Triangle Sum Theorem, you know that $m\angle 1 + m\angle 2 + m\angle 3 = 180°$ and $m\angle 4 + m\angle 5 + m\angle 6 = 180°$. Add these two equations to find $m\angle 1 + m\angle 2 + m\angle 3 + m\angle 4 + m\angle 5 + m\angle 6 = 360°$. Use Associative Property to rewrite the equation as $m\angle 1 + m\angle 2 + m\angle 4 + m\angle 3 + m\angle 5 + m\angle 6 = 360°$. For the Angle Addition Postulate, $m\angle J = m\angle 3 + m\angle 5$ and $m\angle G = m\angle 2 + m\angle 4$. Substitute to find $m\angle F + m\angle G + m\angle H + m\angle J = 360°$.

16.

Statements	Reasons
1. *ABCD* is a parallelogram.	1. Given
2. Draw $\overline{AC}$.	2. Through any two points, there is exactly one line.
3. $\overline{AB} \parallel \overline{DC}$, $\overline{AD} \parallel \overline{BC}$	3. Definition of parallelogram
4. $\angle BAC \cong \angle DCA$ $\angle DAC \cong \angle BCA$	4. Alternate Interior Angles Theorem
5. $\overline{AC} \cong \overline{AC}$	5. Reflexive Property of Congruence
6. $\triangle ACB \cong \triangle CAD$	6. ASA Triangle Congruence Theorem
7. $\overline{AB} \cong \overline{DC}$, $\overline{AD} \cong \overline{BC}$	7. Corresponding parts of congruent triangles are congruent.

17.

Statements	Reasons
1. *ABCD* is a parallelogram	1. Given
2. $\overline{AB} \parallel \overline{DC}$, $\overline{AD} \parallel \overline{BC}$	2. Definition of parallelogram
3. $\angle A$ is supplementary to $\angle B$, $\angle B$ is supplementary to $\angle C$, $\angle C$ is supplementary to $\angle D$, and $\angle D$ is supplementary to $\angle A$.	3. Consecutive Interior Angles Theorem

25.

Statements	Reasons
1. *ABCD* and *AXYZ* are parallelograms.	1. Given
2. $\angle A \cong \angle Y$; $\angle A \cong \angle C$	2. Opposite angles of parallelograms are congruent.
3. $\angle Y \cong \angle C$	3. Transitive Property of Congruence

26.

Statements	Reasons
1. *GDFE* and *FGIH* are parallelograms.	1. Given
2. $\overline{EF} \cong \overline{FG}$	2. Given
3. $\overline{EF} \cong \overline{CD}$, $\overline{FG} \cong \overline{HI}$	3. Opposite sides of parallelograms are congruent.
4. $\overline{CD} \cong \overline{FG}$	4. Transitive Property of Congruence
5. $\overline{CD} \cong \overline{HI}$	5. Transitive Property of Congruence

MODULE 11, LESSON 11.2

Spark Your Learning

B. The information needed is the measures of the angles formed by the hinges and the door.

D. yes; The lengths of the sides do not change as the shape is transformed. The shape still has two pairs of opposite sides that are parallel.

Step It Out

Task 3 **A.1.** $\overline{AE} \cong \overline{CE}$, $\overline{DE} \cong \overline{BE}$

2. $\angle AEB \cong \angle CED$, $\angle AED \cong \angle CEB$
3. $\triangle AEB \cong \triangle CED$, $\triangle AED \cong \triangle CEB$
4. $\overline{AB} \cong \overline{CD}$, $\overline{AD} \cong \overline{CB}$
5. *ABCD* is a parallelogram.

B.1. Given

2. Vertical angles are congruent.
3. SAS Triangle Congruence Theorem
4. Corresponding parts of congruent figures are congruent.
5. If both pairs of opposite sides of a quadrilateral are congruent, then it is a parallelogram.

Task 4 **A.** If the quadrilateral is a parallelogram, then opposite sides are the same length. So to find the values that make *KLMN* a parallelogram, set the opposite sides equal to each other and solve for the variables.

B. When $x = 8$, $KN = LM = 12$. When $y = 3$, $KL = NM = 21$. Opposite sides are equal, so *KLMN* is a parallelogram.

On Your Own

8.

Statements	Reasons
1. $\overline{AD} \cong \overline{BC}$, $\overline{AB} \cong \overline{DC}$	1. Given
2. Draw $\overline{AC}$.	2. Through any two points, there is exactly one line.
3. $\overline{AC} \cong \overline{AC}$	3. Reflexive Property of Congruence
4. $\triangle DAC \cong \triangle BCA$	4. SSS Triangle Congruence Theorem
5. $\angle BAC \cong \angle DCA$, $\angle BCA \cong \angle DAC$	5. Corresponding parts of congruent triangles are congruent.
6. $\overline{AD} \parallel \overline{BC}$, $\overline{AB} \parallel \overline{DC}$	6. Converse of the Alternate Interior Angles Theorem
7. *ABCD* is a parallelogram.	7. Definition of parallelogram

9.

Statements	Reasons
1. $\overline{AD} \cong \overline{BC}$, $\overline{AD} \parallel \overline{BC}$	1. Given
2. Draw $\overline{AC}$.	2. Through any two points, there is exactly one line.
3. $\overline{AC} \cong \overline{AC}$	3. Reflexive Property of Congruence
4. $\angle DAC \cong \angle BCA$	4. Alternate Interior Angles Theorem
5. $\triangle DAC \cong \triangle BCA$	5. SAS Triangle Congruence Theorem
6. $\overline{AB} \cong \overline{DC}$	6. Corresponding parts of congruent triangles are congruent.
7. *ABCD* is a parallelogram.	7. If both pairs of opposite sides of a quadrilateral are congruent, then the quadrilateral is a parallelogram.

10. yes; the alternate interior angles are congruent, so the opposite sides are parallel. By SAS and CPCTC, the opposite sides are congruent.

11. no; You do not know any information about the angles opposite the given angle measures.

12. yes; The third pair of angles in the triangles are congruent, so the opposite angles are congruent.

13. no; A pair of opposite sides is congruent, but the other pair of opposite sides is parallel.

20. no; *DC* is not equal to *AB*.

21. yes; Each pair of opposite sides has the same length.

22. yes; Each pair of opposite sides has the same slope, so they are parallel.

23. no; *MJ* is not equal to *KL*.

28. Ethan is correct; The shape will always be a parallelogram because both pairs of opposite sides will always be congruent.

30.

Statements	Reasons
1. $\angle A$ is supplementary to $\angle B$, $\angle B$ is supplementary to $\angle C$, $\angle C$ is supplementary to $\angle D$, $\angle D$ is supplementary to $\angle A$.	1. Given
2. $AB \parallel DC$, $AD \parallel BC$	2. Converse of the Consecutive Interior Angles Theorem
3. *ABCD* is a parallelogram.	3. Definition of parallelogram

31. The triangles are congruent because a rotation does not change the size or shape of the triangle. So, the opposite sides of the quadrilateral are congruent because they are corresponding parts of congruent triangles. So, the quadrilateral is a parallelogram.

32. The diagonals are $\overline{AC}$ and $\overline{BD}$. The midpoint of $\overline{AC}$ is $\left(\frac{0+2}{2}, \frac{6+(-1)}{2}\right)$ or $\left(1, \frac{5}{2}\right)$. The midpoint of $\overline{BD}$ is $\left(\frac{7+(-5)}{2}, \frac{2+3}{2}\right)$ or $\left(1, \frac{5}{2}\right)$. Since the diagonals have the same midpoint, they bisect each other. By the Converse of the Diagonals of a Parallelogram Theorem, *ABCD* is a parallelogram.

33. Draw $\overline{AC}$. $\overline{ZW}$ is the midsegment of $\triangle ADC$, so $\overline{ZW} \parallel \overline{AC}$ and $ZW = \frac{1}{2}AC$. Also, $\overline{YX}$ is the midsegment of $\triangle BAC$, so $\overline{XY} \parallel \overline{AC}$ and $XY = \frac{1}{2}AC$. So, $\overline{XY} \parallel \overline{ZW}$ and $\overline{XY} \cong \overline{ZW}$, so $WXYZ$ is a parallelogram.

MODULE 11, LESSON 11.3

Spark Your Learning

B. Possible answer: You can describe a tile pattern using the shapes, the sizes of shapes, and the direction of the shapes.

D. All quadrilaterals can tessellate; The sum of each angle of the quadrilateral will meet at a point in the tessellation and the sum of the angle measures is 360°.

Step It Out

Task 3 **A.** You can see the rectangle as a whole, and you can also see how the individual triangles are formed and how they have a side in common.

Check Understanding

4. $\triangle MNQ$ and $\triangle LQN$; The triangles are congruent; The legs of the triangles are the sides of the rectangle and the hypotenuses are the diagonals of the rectangle. The triangles are congruent by SSS.

On Your Own

7. Answers will vary. The lengths of the diagonals will be equal. The diagonals of a rectangle have the same length.

8. Answers will vary. The angles formed by the diagonals will all measure 90°. The diagonals of a rhombus are perpendicular.

9. Answers will vary. The lengths of the diagonals will be equal and the angles formed by the diagonals will all measure 90°. The diagonals of a square have the same length and are perpendicular.

18. *See below.*

19.

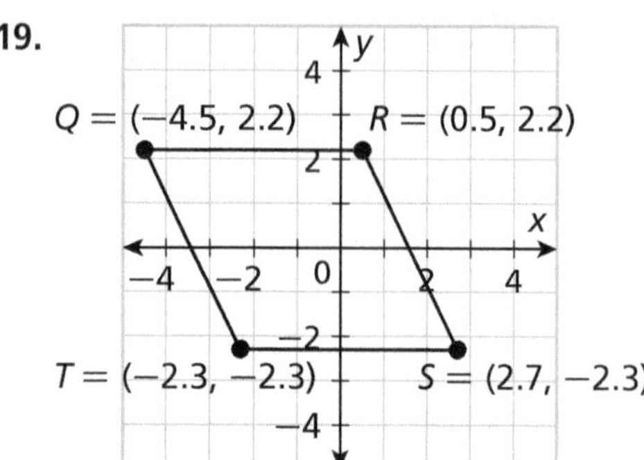

If the diagonals are perpendicular, their slopes will be opposite reciprocals.

slope of $\overline{RT} = \dfrac{2.2 - (-2.3)}{0.5 - (-2.3)} = \dfrac{4.5}{2.8} \approx 1.607...$

slope of $\overline{QS} = \dfrac{-2.3 - 2.2}{2.7 - (-4.5)} = \dfrac{-4.5}{7.2} = -\dfrac{5}{8} = -0.625$

The slopes of the diagonals are close to, but not opposite reciprocals (their product is $-1.004...$), so the diagonals are not perpendicular and the figure is not a rhombus.

20. no; A square is a rectangle, so the Diagonals of a Rectangle Theorem is true for a square. A square is also a rhombus, so the Diagonals of a Rhombus Theorem is true for a square as well.

26. The reflections show that each half of the other diagonal is the same length as half of the diagonal that is drawn. So, the diagonals have the same length.

18.

Statements	Reasons
1. $QRST$ is a rhombus.	1. Given
2. $\overline{QR} \cong \overline{RS}$	2. Definition of rhombus
3. $\overline{QP} \cong \overline{SP}$	3. If a quadrilateral is a parallelogram, the diagonals bisect each other.
4. $\overline{RP} \cong \overline{RP}$	4. Reflexive Property of Congruence
5. $\triangle QRP \cong \triangle SRP$	5. SSS Triangle Congruence Theorem
6. $\angle RPQ \cong \angle RPS$	6. CPCTC
7. $m\angle RPQ + m\angle RPS = 180°$	7. Linear Pairs Theorem
8. $m\angle RPQ + m\angle RPQ = 180°$	8. Substitution Property of Equality
9. $2m\angle RPQ = 180°$	9. Combine like terms.
10. $m\angle RPQ = 90°$	10. Division Property of Equality
11. $\overline{QS} \perp \overline{RT}$	11. Definition of perpendicular segments

27. Answers will vary. Possible answer:

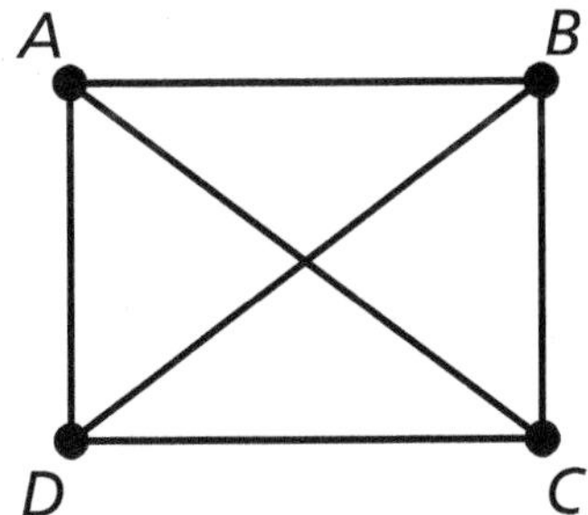

ABCD is rectangle, $AC = 4x + 1$, and $BD = 5x - 6$. Find *BD*; $BD = 29$

MODULE 11, LESSON 11.4

Spark Your Learning

B. The lengths of the side-pieces do not change. The measures of $\angle A$ and $\angle C$ decrease as the mirrored is extended is opened. The measures of $\angle B$ and $\angle D$ increase as the mirror is extended. As the mirror is extended, the length of the vertical diagonal decreases and the length of the horizontal diagonal increases. Throughout the mirror's range of motion, those diagonals intersect at a perpendicular angle.

D. You know that the retracted length is less than the length of the full extension.

Build Understanding

Task 1 **A.** yes; The quadrilateral is a parallelogram because the diagonals bisect each other. Each half of each diagonal is the radius of the circle, which makes them the same length.

B. rectangle; The diagonals of the parallelogram are congruent.

C. Yes, each parallelogram formed is a rectangle because the diagonals are always congruent.

D. square; The diagonals are congruent and perpendicular.

E. no; In order for all four vertices of a rhombus to lie on the circle, its perpendicular diagonals would have to be the same length (diameters of the circle).

Step It Out

Task 3 **A.** *EFGH* is a parallelogram and $\overline{EH} \cong \overline{FE}$.

$\overline{EH} \cong \overline{FG}$, $\overline{FE} \cong \overline{GH}$

$\overline{EH} \cong \overline{GH} \cong \overline{FG} \cong \overline{FE}$

EFGH is a rhombus.

B. Given

Opposite sides of a parallelogram are congruent.

Transitive Property of Congruence

A quadrilateral is a rhombus if and only if it has four congruent sides.

Check Understanding

4. yes; A parallelogram with one right angle is a rectangle.

5. no; Although this parallelogram looks like a square, you are only given enough information to determine that it is a rhombus.

On Your Own

7. yes; A parallelogram with consecutive congruent sides is a rhombus.

8. no; Although the diagonals appear to be perpendicular, you are not given that information.

9. *See below.*

9.

Statements	Reasons
1. *ABCD* is a parallelogram. $\overline{AC}$ bisects $\angle DAB$, and $\angle DCB$. $\overline{BD}$ bisects $\angle ADC$, and $\angle ABC$.	**1.** Given
2. $\angle 1 \cong \angle 2, \angle 3 \cong \angle 4, \angle 5 \cong \angle 6, \angle 7 \cong \angle 8$	**2.** Definition of angle bisector
3. $\overline{AB} \cong \overline{DC}$, $\overline{AD} \cong \overline{BC}$	**3.** Opposite sides of a parallelogram are congruent.
4. $\overline{BD} \cong \overline{BD}$, $\overline{AC} \cong \overline{AC}$	**4.** Reflexive Property of Congruence
5. $\triangle ABC \cong \triangle CDA$, $\triangle ADB \cong \triangle CBD$	**5.** SSS Triangle Congruence Theorem
6. $\angle 1 \cong \angle 5$, $\angle 2 \cong \angle 6$, $\angle 3 \cong \angle 7$, $\angle 4 \cong \angle 8$	**6.** CPCTC
7. $\angle 1 \cong \angle 6$, $\angle 2 \cong \angle 5$, $\angle 3 \cong \angle 8$, $\angle 4 \cong \angle 7$	**7.** Transitive Property of Congruence
8. $\overline{AB} \cong \overline{AD}$, $\overline{DC} \cong \overline{BC}$, $\overline{AB} \cong \overline{BC}$, $\overline{DC} \cong \overline{AD}$	**8.** Converse of Isosceles Triangle Theorem
9. $\overline{AB} \cong \overline{DC} \cong \overline{AD} \cong \overline{BC}$	**9.** Transitive Property of Congruence
10. *ABCD* is a rhombus.	**10.** Definition of rhombus

10. *See below.*

11. Lindsey is correct; You are given that the parallelogram has one right angle, so QRST is a rectangle. You are also given that consecutive sides are congruent, so QRST is a rhombus. Because QRST is a rectangle and a rhombus, it is a square.

12. The conclusion is not valid; You need to know that *WXYZ* is a parallelogram.

13. The conclusion is valid; Because *KLMN* is a parallelogram, the diagonals bisect each other. It was further given that the diagonals are congruent. Therefore, the parallelogram is a rectangle.

14. Because *PZ, QZ, RZ,* and *SZ* are equal in length, *QS* and *PR* are also equal length. Because the diagonals are equal, the parallelogram is a rectangle.

17. A rotation of 90° will map each diagonal onto the other, so it shows that the diagonals are congruent. Because this rotation maps one side of an angle formed by the diagonals onto the other side of the angle, it shows that the measure of the angle is 90°. So, the diagonals are perpendicular.

18.

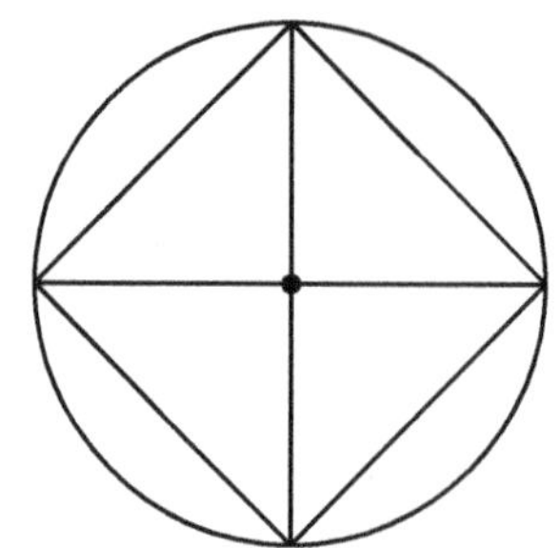

MODULE 11, LESSON 11.5

Step It Out

Task 2 **A.** $\overline{BC} \cong \overline{DC}$ and $\overline{AB} \cong \overline{AD}$. $\overline{AC} \cong \overline{AC}$, $\triangle ABC \cong \triangle ADC$, $\angle B \cong \angle D$

B. Given, Reflexive Property of Congruence, SSS Triangle Congruence Theorem, Corresponding parts of congruent figures are congruent.

C. After showing that $\triangle ABC \cong \triangle ADC$, you can show that the two triangles in $\triangle BCD$ are congruent, so the angles formed by the intersection of both diagonals are congruent and form a linear pair. Therefore they are right angles, and $\overline{AC} \perp \overline{BD}$; The diagonals of a kite are perpendicular.

10.

Statements	Reasons
1. *ABCD* is a parallelogram. $\angle D$ is a right angle.	1. Given
2. $m\angle D = 90°$	2. Definition of right angle
3. $\angle D \cong \angle B$, $\angle A \cong \angle C$	3. Opposite angles of a parallelogram are congruent.
4. $m\angle B = 90°$	4. Definition of congruent angles
5. $\angle B$ is a right angle.	5. Definition of right angle
6. $m\angle A + m\angle B + m\angle C + m\angle D = 360°$	6. Polygon Angle Sum Theorem
7. $m\angle A + 90° + m\angle C + 90° = 360°$	7. Substitution Property of Equality
8. $m\angle A + m\angle C = 180°$	8. Subtraction Property of Equality
9. $m\angle A + m\angle A = 180°$	9. Substitution Property of Equality
10. $2m\angle A = 180°$	10. Combine like terms.
11. $m\angle A = 90°$	11. Division Property of Equality
12. $m\angle C = 90°$	12. Transitive Property of Equality
13. $\angle A$ and $\angle C$ are right angles.	13. Definition of right angle
14. *ABCD* is a rectangle.	14. Definition of a rectangle

Task 3 **A.** If a trapezoid has one pair of congruent base angles, then the trapezoid is isosceles. Because the trapezoid is isosceles, its legs have the same length.

B. The trapezoid is isosceles, so each leg is the same length. You only need to evaluate one expression.

Check Understanding

1. The inclusive definitions include more shapes as kites and trapezoids. The exclusive definitions exclude those additional shapes.

On Your Own

7. yes; A parallelogram has two pairs of opposite parallel sides. The inclusive definition of a trapezoid only requires one pair of opposite parallel sides; no; A trapezoid with noncongruent bases is not a parallelogram.

8. The corners of the paper are right angles and they line up on the vertical diagonal when folded. They horizontally align with each other and form a straight line with the horizontal diagonal.

9. Lexi is correct; Even though the figure appears to be an isosceles trapezoid, the bases are not labeled parallel, so you cannot assume they are parallel.

15. There is reflection symmetry across the vertical diagonal. There is no rotational symmetry.

16. There is no reflection symmetry or rotational symmetry.

17. There is reflection symmetry across the vertical line through the midpoints of the bases. There is no rotational symmetry.

18. There is reflection symmetry across the horizontal diagonal. There is no rotational symmetry.

19. Because $\overline{BC} \parallel \overline{AD}$ and $\overline{AB} \parallel \overline{EC}$, $ABCE$ is a parallelogram. So, $\overline{AB} \cong \overline{EC}$. Using the Transitive Property of Congruence, $\overline{CE} \cong \overline{CD}$, so $\triangle ECD$ is an isosceles triangle and $\angle D \cong \angle CED$. By the Corresponding Angles Postulate, $\angle A \cong \angle CED$, therefore $\angle A \cong \angle D$ by the Transitive Property of Congruence. By the Consecutive Interior Angles Theorem, $\angle A$ and $\angle B$ are supplementary and $\angle D$ and $\angle BCD$ are supplementary. $\angle B$ and $\angle BCD$ are congruent by Congruent Supplements Theorem.

20. It is given that $\overline{GH} \parallel \overline{FJ}$. By the Parallel Postulate, $\overline{HK}$ can be drawn parallel to $\overline{GF}$ so that $\overline{HK}$ intersects $\overline{FJ}$ at K. By the Corresponding Angles Postulate, $\angle F \cong \angle HKJ$. It is given that $\angle F \cong \angle J$, so $\angle HKJ \cong \angle J$ by the Transitive Property of Congruence. By the inverse of the Converse of Isosceles Triangle Theorem, $\overline{HJ} \cong \overline{HK}$. By definition, $FGHK$ is a parallelogram, so $\overline{GF} \cong \overline{HK}$. By the Transitive Property of Congruence, $\overline{GF} \cong \overline{HJ}$.

21. It is given that $\overline{AB} \cong \overline{CD}$. The base angles of an isosceles triangle are congruent, so $\angle BAD \cong \angle CDA$. By the Reflexive Property of Congruence, $\overline{AD} \cong \overline{AD}$. By SAS Triangle Congruence Theorem, $\triangle ABD \cong \triangle DCA$. By CPCTC, $\overline{AC} \cong \overline{DB}$.

22. *See next page.*

23. A. slope of $\overline{AB} = \frac{2-4}{5-(-3)} = -\frac{1}{4}$, slope of $CD = \frac{-4-(-1)}{6-(-6)} = -\frac{1}{4}$. Since one pair of opposite sides of the quadrilateral are parallel, the quadrilateral is a trapezoid.

B. midpoint of $\overline{AD}$: $\left(\frac{-3+(-6)}{2}, \frac{4+(-1)}{2}\right) = \left(-\frac{9}{2}, \frac{3}{2}\right)$; midpoint of $\overline{BC}$: $\left(\frac{5+6}{2}, \frac{2+(-4)}{2}\right) = \left(\frac{11}{2}, -1\right)$. The midsegment has endpoints of $\left(-\frac{9}{2}, \frac{3}{2}\right)$ and $\left(\frac{11}{2}, -1\right)$. Using the Distance Formula, the length of the midsegment is $5\frac{\sqrt{17}}{2}$. The slope is $-\frac{1}{4}$.

C. Based on the Trapezoid Midsegment Theorem, the slope of the midsegment is equal to the slope of the bases, $-\frac{1}{4}$. The lengths of the bases are $AB = 2\sqrt{17}$ and $CD = 3\sqrt{17}$. Using the Trapezoid Midsegment Theorem, the length of the midsegment is the average of the lengths of the bases, $\frac{5\sqrt{17}}{2}$. This is the same as the result found in Part B.

24. slope $\overline{AD} = \frac{3-7}{5-1} = -1$; slope $\overline{BC} = \frac{5-7}{6-4} = -1$; The slopes are equal, so one set of opposite sides are parallel. Since the quadrilateral has one pair of parallel sides, $ABCD$ is a trapezoid.

25. $AB = \sqrt{2}$; $AD = \sqrt{2}$; $BC = \sqrt{10}$; $CD = \sqrt{10}$;

The two pairs of consecutive sides are congruent, so $ABCD$ is a kite.

26. 114°

27. $x = 180° - \frac{y+z}{2}$

28. A. $\overline{GM} \cong \overline{MF}$ and $\overline{HN} \cong \overline{NJ}$.

B. $\angle HGN \cong \angle NKJ$ by Alternate Interior Angles. $\overline{GH} \cong \overline{JK}$ by construction, so $\triangle NHG \cong \triangle NJK$ by SAS Triangle Congruence. So $\overline{GN} \cong \overline{KN}$ by CPCTC.

C. $\overline{MN} \parallel \overline{FK}$ because $\overline{MN}$ is the midsegment of $\triangle GFK$.

D. $\overline{GH} \parallel \overline{FK}$, so $\overline{MN} \parallel \overline{GH}$ because if two segments are parallel to a third segment, then they are parallel.

E. $MN = \frac{1}{2} FK$ and $FK = FJ + GH$, so $MN = \frac{1}{2}(FJ + GH)$.

29. A. $PQ = \sqrt{10}$, $QR = \sqrt{10}$, $RS = 3\sqrt{10}$, $PS = 3\sqrt{10}$

B. $PQ = QR$ and $RS = PS$

C. yes; By definition, a kite has two pairs of congruent consecutive sides. In this case, $\overline{PQ}$ and $\overline{QR}$ are consecutive sides with the same length, and $\overline{RS}$ and $\overline{PS}$ are consecutive sides with the same length.

30. always; definition of a trapezoid

31. sometimes; If the trapezoid is isosceles, the legs will be congruent.

32. always; definition of midsegment of a trapezoid

33. never; One of the bases would have to be 0 for this to be true.

34. sometimes; If the kite is a rhombus or square, opposite sides are congruent.

35. sometimes; If the kite is a square, adjacent sides are perpendicular.

36. sometimes; If the trapezoid is a rhombus, then it is a kite.

Additional Answers

22.

Statements	Reasons
1. Draw $\overline{BE} \perp \overline{AD}$ and $\overline{CF} \perp \overline{AD}$.	**1.** There is only one line through a given point perpendicular to a given line.
2. $\overline{BE} \parallel \overline{CF}$	**2.** Two lines perpendicular to the same line are parallel.
3. $\overline{BC} \parallel \overline{AD}$	**3.** Given
4. *BCFE* is a parallelogram.	**4.** Definition of parallelogram
5. $\overline{BE} \cong \overline{CF}$	**5.** If a quadrilateral is a parallelogram, then its opposite sides are congruent.
6. $\overline{AC} \cong \overline{DB}$	**6.** Given
7. $\angle BED$ and $\angle CFA$ are right angles	**7.** Definition of perpendicular lines
8. $\triangle BED \cong \triangle CFA$	**8.** HL Triangle Congruence Theorem
9. $\angle BDE \cong \angle CAF$	**9.** CPCTC
10. $\angle CBD \cong \angle BDE$, $\angle BCA \cong \angle CAF$	**10.** Alternate Interior Angles Theorem
11. $\angle CBD \cong \angle BCA$	**11.** Transitive Property of Congruence
12. $BC \cong BC$	**12.** Reflexive Property of Congruence
13. $\triangle ABC \cong \triangle DCB$	**13.** SAS Triangle Congruence Theorem
14. $\angle BAC \cong \angle CDB$	**14.** CPCTC
15. $\angle BAD \cong \angle CDA$	**15.** Angle Addition Postulate
16. *ABCD* is isosceles.	**16.** If a trapezoid has one pair of base angles congruent, then the trapezoid is isosceles.

37. $\overline{AB} \cong \overline{CB}$, $\angle BAD \cong \angle BCD$, and $\overline{AD} \parallel \overline{BB'}$. So, *ABB'D* is an isosceles trapezoid.

38. Two markings are needed to demonstrate that a quadrilateral is a trapezoid; The pair of parallel markings on opposite sides are needed.

MODULE 11 REVIEW

Concepts and Skills

5. parallelogram, rectangle, rhombus, square

11. Possible answer: Let the parallelogram be *PQRS*, labeled circularly. Draw diagonal $\overline{PR}$.

$\overline{PQ} \parallel \overline{RS}$; $\overline{QR} \parallel \overline{SP}$	Given
$\angle QPR \cong \angle SRP$; $\angle QRP \cong \angle SPR$	Alternate Interior Angles Theorem
$\overline{PR} \cong \overline{RP}$	Symmetric Property of Congruence
$\triangle QPR \cong \triangle SRP$	ASA
$\overline{PS} \cong \overline{RQ}$; $\overline{QP} \cong \overline{SR}$	CPCTC

12. Possible answer: Let the parallelogram be *PQRS*, labeled circularly. Draw diagonal $\overline{PR}$.

$\overline{PQ} \parallel \overline{RS}$; $\overline{QR} \parallel \overline{SP}$	Given
$\angle QPR \cong \angle SRP$; $\angle QRP \cong \angle SPR$	Alternate Interior Angles Theorem
$PR \cong RP$	Symmetric Property of Congruence
$\triangle QPR \cong \triangle SRP$	ASA
$\angle PQR \cong \angle RSP$	CPCTC

Repeat reasoning with diagonal $\overline{QS}$ to prove $\angle QPS \cong \angle SRQ$.

MODULE 12, LESSON 12.1

Spark Your Learning

B. Possible answer: I can estimate the dimensions of the court, but I cannot determine the dimensions of an actual regulation court; The scale of the picture compared to the actual court is needed.

Build Understanding

Task 1 **A.** Measuring the sides and comparing the ratios of corresponding sides confirms that the triangles are similar with a scale factor of 2; As you move one vertex of $\triangle ABC$, the corresponding vertex of the dilated image moves to maintain the relationship.

C. Possible answer: The only line parallel to $\overline{BC}$ through $\overline{FG}$ is $\overline{FG}$; This does not change by dragging a vertex of $\triangle ABC$; The corresponding sides of these two triangles with the same center of dilation remain parallel.

Task 2 **A.** Possible answer: Yes, there is a sequence of transformations to map *LMNP* onto *ABCD*: translation then dilation; Since *ABCD* is smaller than *LMNP*, the scale factor would be between 0 and 1.

B. $k = \frac{BC}{MN} = \frac{2}{4} = \frac{1}{2}$; The scale factor is $\frac{1}{2}$.

C. Possible answer: The transformation of *LMNP* to *L'M'N'P'* is a translation. The change in the coordinates is the same for all four points. $(x, y) \rightarrow (x - 2, y - 1)$

D. Possible answer: The lengths of the corresponding sides are all reduced by the scale factor, $\frac{1}{2}$; The sides of *ABCD* are half the lengths of the corresponding sides of *LMNP*.

E. Possible answer: The transformations (translation and dilation) are similarity transformations and do not change the angles of the figures; $\angle B$ in *ABCD* must be congruent to $\angle M$ in *LMNP*.

F. Possible answer: no; For example, the dilation could happen first, then the translation. The transformations would be $(x, y) \rightarrow \left(\frac{1}{2}x, \frac{1}{2}y\right)$, then $(x, y) \rightarrow \left(x - 1, y - \frac{1}{2}\right)$

Task 3 **A.** Possible answer: Using variables for the lengths of the radii means the proof applies to all values, not specific values in just two circles.

B. Possible answer: yes; There is no change in the shape of the figure. It is still a circle.

C. Possible answer: The circles coincide. They have the same radius and the same center.

D. Possible answer: Every circle can be mapped onto another circle by translating it so that the two centers correspond and then dilating it by the appropriate scale factor.

E. Yes, the diameter and circumference of a circle are transformed under a dilation in the same way as the radius.

$$\text{Diameter}_D \stackrel{?}{=} \frac{s}{r} \cdot \text{Diameter}_C$$

$$2s \stackrel{?}{=} \frac{s}{r} \cdot 2r$$

$$2s = 2s \checkmark$$

$$\text{Circumference}_D \stackrel{?}{=} \frac{s}{r} \cdot \text{Circumference}_C$$

$$2\pi s \stackrel{?}{=} \frac{s}{r} \cdot 2\pi r$$

$$2\pi s = 2\pi s \checkmark$$

Step It Out

Task 4 **D.** Possible answer: The lengths of a pair of corresponding sides are needed to find the scale factor; yes; Since $AB = 4x$ and $x = 14$, $AB = 56$. The scale factor is the ratio of *PQ* to *AB*, or 35:56 = 5:8.

Check Understanding

1. $\overline{RP}$; Possible answer: The two sides are corresponding parts after a dilation with no other transformations, so they are parallel.

2. $\angle P$; Possible answer: Corresponding angles in similar polygons are congruent.

3. Yes, the triangles are similar; Possible answer: Corresponding sides have the same scale factor. Corresponding angles are congruent.

4. Yes, all circles are similar; Possible answers: The scale factor is 3 or $\frac{1}{3}$.

5. $CD = 5$ cm

$$\frac{BC}{QR} = \frac{CD}{RS}$$

$$\frac{6.25}{10} = \frac{x}{8}$$

$$10x = 8(6.25)$$

$$10x = 50$$

$$x = 5$$

On Your Own

6. not a dilation; Possible answer: The slope of $\overline{AB}$ is $\frac{5}{2}$ and the slope of $\overline{A'B'}$ is $\frac{6}{2}$, or 3. The slopes of corresponding sides are not equal, so the triangles are not similar.

7. a dilation; Possible answer: The slopes of $\overline{AB}$ and $\overline{A'B'}$ are equal to 3. The slopes of $\overline{BC}$ and $\overline{B'C'}$ are equal to -1. The slopes of $\overline{CA}$ and $\overline{C'A'}$ are equal to 1. So, $\triangle A'B'C'$ is a dilation of $\triangle ABC$. The scale factor is $\frac{A'B'}{AB} = \frac{B'C'}{BC} = \frac{C'A'}{CA} = \frac{1}{2}$.

8. Possible answer: Corresponding angles are congruent since each pair of corresponding sides is parallel. The lengths of each pair of corresponding sides have the same scale factor, 3 or $\frac{1}{3}$. So $ABCD \sim FGHJ$.

9. yes; Possible answer: The sides of both angles have the same slopes, so the angles are congruent.

10. yes; Possible answer: Using the lengths of the sides, $\frac{AB}{BC} = \frac{3}{9} = \frac{1}{3}$ and $\frac{FG}{GH} = \frac{1}{3}$. So $\frac{AB}{BC} = \frac{FG}{GH}$.

11. yes; Possible answer: Since $FGHJ \cong LMNP$ and $ABCD \sim FGHJ$, corresponding sides have the same scale factor and corresponding angles are congruent. So $ABCD \sim LMNP$.

12. yes; Possible answer: The transformations are dilation, rotation, and translation.

13. no; Possible answer: The ratios of corresponding side lengths are not equal, so the triangles are not similar. Therefore, $\triangle LMN$ cannot be obtained from a sequence of similarity transformations performed on $\triangle DEF$.

14. yes; Possible answer: The transformations are dilation and rotation.

15. yes; Possible answer: The transformations are dilation, reflection, and translation.

16. yes; Possible answer: All circles are similar.

17. Possible answer: The circumference of the largest circle is about 4 times as big as the circumference of the smallest circle. The transformation will have a scale factor of $\frac{32}{120} \approx \frac{1}{4}$.

18. $\frac{120}{32} = 3.75$ times as great

19. Possible answer: Congruent circles, or any congruent figures, are similar with a scale factor of 1.

20. scale factor $= \frac{EF}{JK} = \frac{8}{20} = \frac{2}{5} = 0.4$

21. $\frac{11}{x} = \frac{8}{20}$; $220 = 8x$; $x = \frac{220}{8} = 27.5$

$\frac{4y+4}{30} = \frac{8}{20}$; $80y + 80 = 240$; $80y = 160$; $y = 2$

$z = 65$

22. perimeter of $EFGH = 3 + 11 + 8 + 12 = 34$

Since the polygons are similar with scale factor 2.5, the perimeter of $JKLM = 2.5\times$ perimeter of $EFGH$, so $2.5 \times 34 = 85$.

24. A. No, the pools are not similar; Possible answer: There is not a constant scale factor between corresponding sides: $\frac{50}{25} \neq \frac{25}{18.3}$.

B. The area of the Olympic-sized pool is $50 \cdot 25$, or 1250 square meters. 75% of 1250 is 937.5 square meters. $937.5 \cdot \$10.7 \approx$ \$10,000 per year

25. Neither student is correct; The slope of $\overline{M'N'}$ is $-\frac{1}{3}$; Possible answer: After a dilation, corresponding sides are parallel, so $\overline{MN}$ and $\overline{M'N'}$ would have the same slope.

26. no; Possible answer: The first transformation is a translation, but the second transformation is not a dilation because it only changes the scale in one direction.

MODULE 12, LESSON 12.2

Spark Your Learning

B. Possible answer: I can create triangles using the heights and shadows as legs. Then I can estimate whether the relationship of the triangles appear to be a dilation from one triangle to the other.

D. Possible answer: The corresponding angles of the triangles must be congruent.

Build Understanding

Task 1 **A.** $k = \frac{PQ}{JK}$; Possible answer: $\triangle J'K'L'$ must be congruent to $\triangle PQR$ so that the similarity transformations from $\triangle JKL$ to $\triangle J'K'L'$ will be the same as those from $\triangle JKL$ to $\triangle PQR$.

B. Possible answer: I know that any triangle which undergoes only a dilation creates a similar image and preimage.

$J'K' = \text{scale factor} \cdot JK$

C. $J'K' = \frac{PQ}{JK} \cdot JK$

$J'K' = PQ$

D. Possible answer: It is given that $\angle J \cong \angle P$ and $\angle K \cong \angle Q$. Because $\angle J \cong \angle J'$ and $\angle K \cong \angle K'$, by the Transitive Property of Congruence, $\angle J' \cong \angle P$ and $\angle K' \cong \angle Q$.

E. Possible answer: $J'K' = PQ$, $\angle J' \cong \angle P$, and $\angle K' \cong \angle Q$. Therefore $\triangle J'K'L' \cong \triangle PQR$ by the ASA Triangle Congruence Theorem.

F. Possible answer: A dilation transformed $\triangle JKL$ to $\triangle J'K'L'$, and because $\triangle J'K'L' \cong \triangle PQR$, there is a sequence of rigid transformations that maps $\triangle J'K'L'$ to $\triangle PQR$. This means there is a sequence of similarity transformations that maps $\triangle JKL$ to $\triangle PQR$.

Step It Out

Task 2 **A.** Possible answer: Segments PQ and TS are parallel and are crossed by two different transversals. Both $\angle P \cong \angle S$ and $\angle Q \cong \angle T$ by the Alternate Interior Angles Theorem.

Task 3 **A.** Possible answer: I can orient both triangles in the same manner so they are easier to compare. Then I can write ratios to compare the lengths. The ratios are the same, so the sides are proportional.

B. Possible answer: The two angles are congruent because they are each labeled as 24°.

C. Possible answer: The triangles are similar by the SAS Triangle Similarity Theorem; $\triangle ABC \sim \triangle DEF$

D. The correct solution is $x = 5$; This solution sets up the proportion correctly.

Task 4 **B.** Possible answer:

$\angle WVY \cong \angle UVT$ because vertical angles are congruent.

$\angle YWV \cong \angle TUV$ because it is given that both measure 90°.

$\triangle TUV \sim \triangle YWV$ by the AA Triangle Similarity Theorem.

D. Yes, because segment WY is longer than segment WV in the top triangle, then corresponding segment TU should be longer than segment VU in the bottom triangle, and $290 > 200$, but 138 would not be.

Check Understanding

1. yes; Possible answer: Similar triangles have congruent angles. If $\triangle ABC \sim \triangle LMN$ and $\triangle LMN \sim \triangle XYZ$, all three triangles have corresponding congruent angles, so by the AA Triangle Similarity Theorem, $\triangle ABC \sim \triangle XYZ$.

3. no; The SAS Triangle Similarity Theorem cannot be applied because we do not know BC. The AA Triangle Similarity Theorem cannot be applied because we do not know that two pairs of corresponding angles are congruent.

4. Possible answer: You can identify a landmark directly across the canyon from your location. Turn at a right angle, walk a certain distance, and mark a point. Continue walking a certain distance, turn away at a right angle, and walk until the landmark across the canyon aligns with the marked point. Mark your current location, and create similar triangles using the marked points. Using the measurements of the segments, you can find the unknown distance across the canyon to the landmark.

On Your Own

5. no; $\frac{4.2}{2} = \frac{10.5}{5} = 2.1, \frac{9}{4.8} \neq 2.1$

6. Possible answer: Dilate $\triangle ABC$ using the scale factor $k = \frac{DE}{AB}$. The image of $\triangle ABC$ is $\triangle A'B'C'$. We know that $\angle A \cong \angle A'$ and it is given that $\angle A \cong \angle D$, so by the Transitive Property of Congruence $\angle D \cong \angle A'$. The length of $\overline{A'B'}$ is the length of $\overline{AB}$ multiplied by the scale factor k. So $A'B' = k \cdot AB = \frac{DE}{AB} \cdot AB = DE$. By the same logic, $A'C' = k \cdot AC = \frac{DF}{AC} \cdot AC = DF$. Therefore $\triangle A'B'C' \cong \triangle DEF$ by the SAS Triangle Congruence Theorem. Because a dilation maps $\triangle ABC$ to $\triangle A'B'C'$ and is followed by a sequence of rigid motions that maps $\triangle A'B'C'$ to $\triangle DEF$, this shows there is a sequence of similarity transformations that maps $\triangle ABC$ to $\triangle DEF$. So, $\triangle ABC \sim \triangle DEF$.

7. yes; The triangles share $\angle C$, and $\angle B \cong \angle E$. The triangles are similar by the AA Triangle Similarity Theorem.

8. no; The corresponding sides are not proportional.

9. yes; The triangles have proportional sides and an included congruent angle by the Vertical Angles Theorem. The triangles are similar by the SAS Triangle Similarity Theorem.

10. yes; The triangles have a given pair of congruent angles, and $\angle BCA \cong \angle CBD$ by the Alternate Interior Angles Theorem. The triangles are similar by the AA Triangle Similarity Theorem.

17. Possible answer: Dilate $\triangle ABC$ using the scale factor $k = \frac{DE}{AB}$. The image of $\triangle ABC$ is $\triangle A'B'C'$. The length of $\overline{A'B'}$ is the length of $\overline{AB}$ multiplied by the scale factor k. So $A'B' = k \cdot AB = \frac{DE}{AB} \cdot AB = DE$. By the same logic, $A'C' = k \cdot AC = \frac{DF}{AC} \cdot AC = DF$, and $C'B' = k \cdot CB = \frac{FE}{CB} \cdot CB = FE$. Therefore $\triangle A'B'C' \cong \triangle DEF$ by the SSS Triangle Congruence Theorem. Because a dilation maps $\triangle ABC$ to $\triangle A'B'C'$ and is followed by a sequence of rigid motions that maps $\triangle A'B'C'$ to $\triangle DEF$, this shows there is a sequence of similarity transformations that maps $\triangle ABC$ to $\triangle DEF$. So, $\triangle ABC \sim \triangle DEF$.

MODULE 12, LESSON 12.3

Spark Your Learning

D. Possible answer: The corresponding sides are proportional; They are similar triangles.

Build Understanding

Task 1 **A.** Possible answer: Angles 1 and 2 are corresponding angles where a transversal crosses two parallel lines, therefore they are congruent. Angles 3 and 4 are also corresponding angles where a transversal crosses two parallel lines, so they are congruent as well.

B. Possible answer: yes; The triangles are similar by the AA Triangle Similarity Theorem.

C. Possible answer: Yes, but not by using the same steps; The segment addition postulate now applies to the denominator instead of the numerator, so you must take the reciprocal earlier in the proof to move it to the numerator, or use different algebraic manipulation to achieve the same result.

Step It Out

Task 3 **A.** Possible answer: The sides of the triangle must be divided proportionally.

B. From the Segment Addition Postulate, $WV = WY + VY$, so $WY = WV - VY$. Similarly, $WX = WZ + XZ$, so $WZ = WX - XZ$.

C. yes; Possible answer: The ratio of the lengths of each pair of segments is equal, so we know the triangle is divided proportionally. The criterion is met to state $\overline{YZ} \parallel \overline{VX}$ by the Converse of the Triangle Proportionality Theorem.

Task 4 **A.** Possible answer: $\frac{1}{4}$ of the rise represents the distance P is from A in the y direction. $\frac{1}{4}$ of the run represents the distance P is from A in the x direction.

B. Possible answer: It is added to A because we want $\frac{1}{4}$ the distance from A going towards B.

Task 5 **C.** Possible answer: The arcs help you divide the line segment into 5 equal portions. The ratio is 2 to 3, so you need a total of 5 portions in order to place point P where 2 portions are on one side and 3 are on the other.

E. Point P is located where there are 2 partitions from point A and 3 partitions from point B.

F. Possible answer: $\angle AEP$ was constructed to be congruent to $\angle AHB$. These are congruent correspondig angles, so $\overline{EP} \parallel \overline{HB}$.

Check Understanding

1. Possible answer: The line that intersects two sides of a triangle must be parallel to the third side.

3. yes; $\frac{3}{4} = \frac{1.5}{2}$, so the segments are parallel by the Triangle Proportionality Theorem.

5. yes; Possible answer: If you divide the segment in the ratio 2:3, you divide the segment into $2 + 3 = 5$ portions, and place a point at 2 portions from A and 3 portions from B. A point placed at the end of 2 out of 5 portions is $\frac{2}{5}$ the distance from A to B.

On Your Own

6. Because $\overline{LM} \parallel \overline{QR}$, we know $\angle PQR \cong \angle PLM$ and $\angle PRQ \cong \angle PML$ by the Corresponding Angles Postulate. $\triangle PQR$ is similar to $\triangle PLM$ by the AA Triangle Similarity Theorem. Similar triangles have proportional sides, so $\frac{PQ}{PL} = \frac{PR}{PM}$. Use the Segment Addition Postulate to rewrite $\frac{PQ}{PL} = \frac{PR}{PM}$ as $\frac{PL+LQ}{PL} = \frac{PM+MR}{PM}$. Use the Distributive Property of Division to rewrite that statement as $\frac{PL}{PL} + \frac{LQ}{PL} = \frac{PM}{PM} + \frac{MR}{PM}$. Because any nonzero number divided by itself is 1, rewrite the equation as $1 + \frac{LQ}{PL} = 1 + \frac{MR}{PM}$. Subtracting 1 from each side leaves you with $\frac{LQ}{PL} = \frac{MR}{PM}$. Taking the reciprocal of both sides results in $\frac{PL}{LQ} = \frac{PM}{MR}$.

9.

Statements	Reasons
1. $\frac{AD}{DB} = \frac{AE}{EC}$	1. Given
2. $\frac{DB}{AD} = \frac{EC}{AE}$	2. Take the reciprocal of each side. (Multiplication/Division Properties of Equality)
3. $1 + \frac{DB}{AD} = 1 + \frac{EC}{AE}$	3. Addition Property of Equality
4. $\frac{AD}{AD} + \frac{DB}{AD} = \frac{AE}{AE} + \frac{EC}{AE}$	4. $\frac{a}{a} = 1$
5. $\frac{AD + DB}{AD} = \frac{AE + EC}{AE}$	5. $\frac{a+b}{c} = \frac{a}{c} + \frac{b}{c}$
6. $\frac{AB}{AD} = \frac{AC}{AE}$	6. Segment Addition Postulate
7. $\angle A \cong \angle A$	7. Reflexive Property
8. $\triangle ABC \sim \triangle ADE$	8. AA Triangle Similarity
9. $\angle ADE \cong \angle ABC$	9. Corresponding angles in similar triangles are congruent.
10. $\overline{DE} \parallel \overline{BC}$	10. Converse of Corresponding Angles Postulate

15. 2. Two points; 3. Triangle Proportionality Theorem; 4. $\frac{UA}{AZ} = \frac{VX}{XZ}$; 5. Transitive

16.

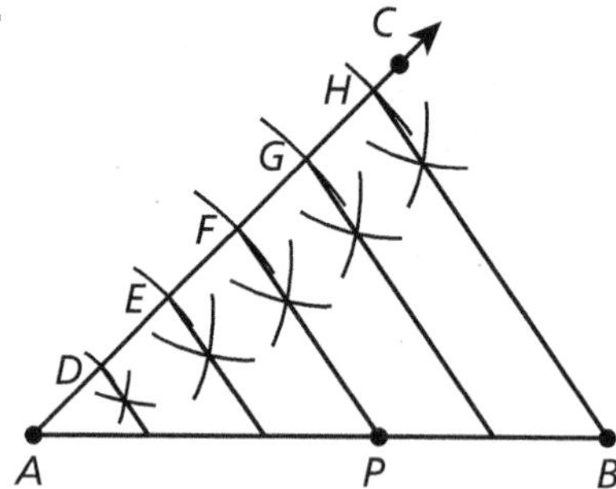

17.

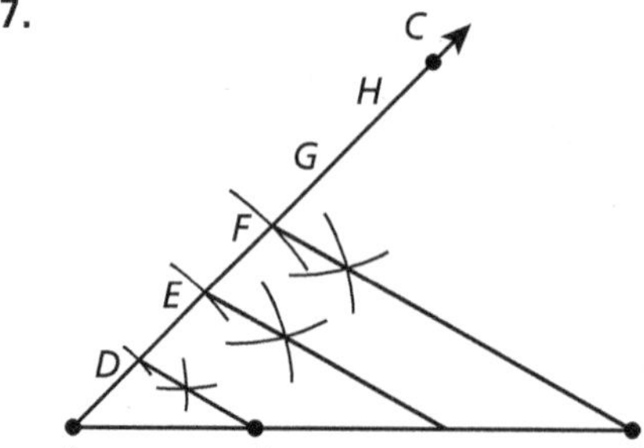

18. Check students' work. Students may describe locations in various ways.

14.4 miles from Harper City exit to Westhaven exit; 9.6 miles from Westhaven exit to Hudsonville exit; 7.2 miles from Harper City exit to the first new sign; 4.8 miles from Westhaven to second new sign

MODULE 12, LESSON 12.4

Spark Your Learning

B. yes; Possible answers: I can tell the triangles are similar because they meet the AA Triangle Similarity Criteria.

D. Possible answer: If the player hits the ball when 13 ft from the net, the ball will hit the net. The player must hit the ball at a distance of less than 13 ft from the net to clear the net.

Build Understanding

Task 1 **D.** They are similar triangles; If $\angle 1 \sim \angle 4$ and $\angle 1 \sim \angle 9$, $\angle 4 \sim \angle 9$ by the Transitive Property. The same applies for each pair of angles between triangles B and C.

Step It Out

Task 2 **B.** yes; $\frac{CD}{BD} = \frac{BD}{AD}$ is equivalent to $CD \cdot AD = BD^2$ after multiplying both sides by $BD \cdot AD$. Use the definition of square root to get $BD = \pm\sqrt{CD \cdot AD}$. Reject the negative square root because BD is a length. So, $BD = \sqrt{CD \cdot AD}$.

Task 3 **A.** The length of the altitude to the hypotenuse of a right triangle is the geometric mean of the lengths of the segments of the hypotenuse.

B. The eye level of the climber is 5 feet, so you must add 5 to the calculated height to find the total height.

Task 4 **A.** The Geometric Means Theorem that states, "The length of the leg of a right triangle is the geometric mean of the lengths of the hypotenuse and the segment of the hypotenuse adjacent to that leg."

B. Possible answer: The Segment Addition Postulate states the length of any segment is equal to the sum of the lengths of its parts.

Task 5 **A.** If the side lengths and hypotenuse satisfy the Pythagorean Theorem, then the triangle is a right triangle. The set of numbers is a Pythagorean triple.

B. The two shorter lengths are side lengths and the longest length is always the hypotenuse.

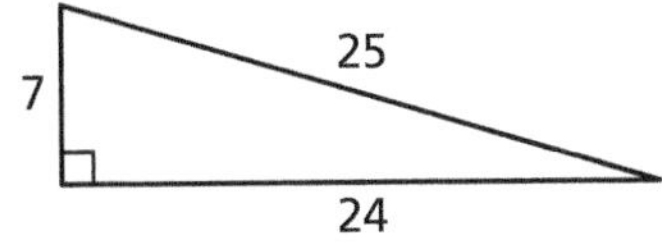

Check Understanding

1. His drawing contains 3 similar triangles; Drawing an altitude to the hypotenuse of a right triangle creates two smaller similar triangles that are also similar to the original triangle.

2. The length of the altitude is the geometric mean of the lengths of the two segments it creates.

3. $\frac{5}{h} = \frac{h}{10}$

$h^2 = 50$

$h = 5\sqrt{2}$

4. Possible answer: The Geometric Means Theorems provide proportions that apply to a right triangle. By manipulating the proportions using algebra and applying the Segment Addition Postulate, you can rearrange the proportions from the Geometric Means Theorems into the Pythagorean Theorem.

5. yes; They satisfy the Pythagorean Theorem.

$9^2 + 40^2 \stackrel{?}{=} 41^2$

$81 + 1600 \stackrel{?}{=} 1681$

$1681 \stackrel{?}{=} 1681$

On Your Own

9.

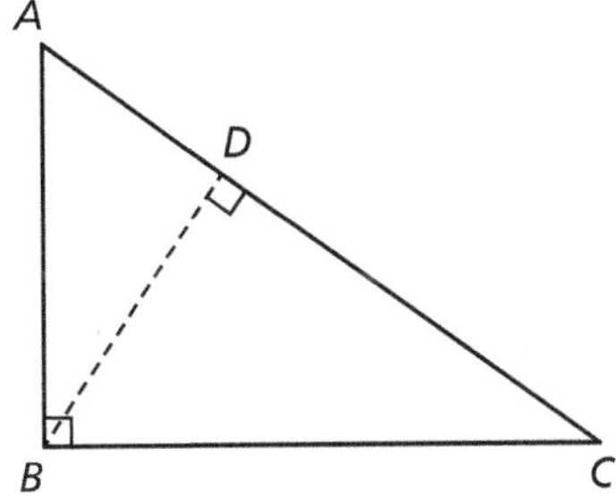

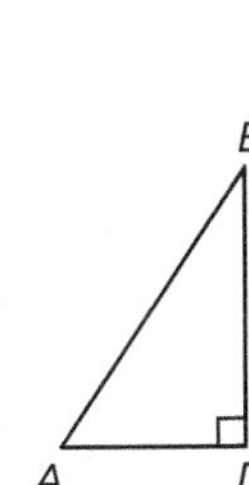

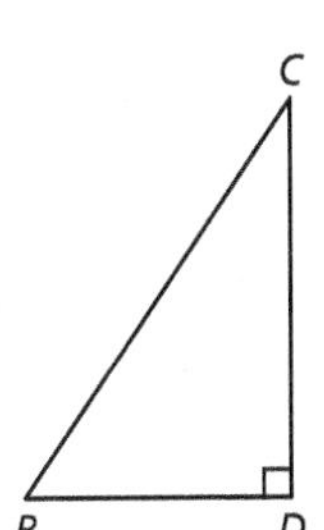

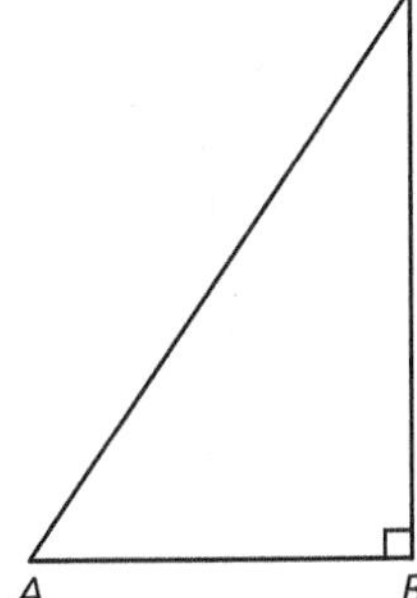

$\triangle ABC \sim \triangle ADB \sim \triangle BDC$; Redrawing the triangles helps to ensure you orient them correctly to write a correct similarity statement.

10. A.

Statements	Reasons
1. Right triangle *ABC* with altitude $\overline{BD}$.	1. Given
2. $\angle BDC$ is a right angle.	2. Definition of an altitude
3. $\angle BDC \cong \angle ABC$	3. All right angles are congruent.
4. $\angle BCD \cong \angle ACB$	4. Reflexive Property of Angle Congruence
5. $\triangle BDC \sim \triangle ABC$	5. AA Triangle Similarity Theorem

B.

Statements	Reasons
1. Right triangle *ABC* with altitude $\overline{BD}$	1. Given
2. $\angle ADB$ is a right angle.	2. Definition of an altitude
3. $\angle ADB \cong \angle ABC$	3. All right angles are congruent.
4. $\angle DAB \cong \angle BAC$	4. Reflexive Property of Angle Congruence
5. $\triangle ADB \sim \triangle ABC$	5. AA Triangle Similarity Theorem

C.

Statements	Reasons
1. Right triangle *ABC* with altitude $\overline{BD}$	1. Given
2. $\triangle BDC \sim \triangle ABC$ and $\triangle ADB \sim \triangle ABC$	2. Proven in Parts A and B
3. $\angle DBC \cong \angle BAC$ and $\angle DAB \cong \angle BAC$	3. Corresponding angles in similar triangles are congruent.
4. $\angle DBC \cong \angle DAB$	4. Transitive Property of Angle Congruence
5. $\angle BDC$ and $\angle ADB$ are right angles.	5. Definition of an altitude
6. $\angle BDC \cong \angle ADB$	6. All right angles are congruent.
7. $\triangle BCD \sim \triangle ABD$	7. AA Triangle Similarity Theorem

11.

Statements	Reasons
1. Right triangle *ABC* with altitude $\overline{BD}$	1. Given
2. $\triangle ABC \sim \triangle ADB \sim \triangle BDC$	2. Right Triangle Similarity Theorem
3. $\frac{AD}{AB} = \frac{AB}{AC}$	3. Corresponding sides of similar triangles are proportional.
4. $\frac{DC}{BC} = \frac{BC}{AC}$	4. Corresponding sides of similar triangles are proportional.

12. $x = 24, y = 32, z = 40$

13. $x = 2\sqrt{26}, y = 2\sqrt{10}, z = \sqrt{65}$

32. A. She should choose Plan 2; Plan 1 $\approx$ 100.7 in., Plan 2 $\approx$ 87.4 in.

MODULE 12 REVIEW

Concepts and Skills

5. Possible answer: The criteria for congruence are stricter than the criteria for similarity as congruence is a special case of similarity. In this way, the criteria for similarity and congruence are alike because the criteria for similarity are contained within the criteria for congruence.

6. dilation with center at $(0,0)$ and a scale factor of $\frac{1}{4}$

7. Possible answer: dilation with center at $(0,0)$ and a scale factor of 3, reflect over line $y = x$, translate down 9 units

9. yes; By using the Converse of the Triangle Proportionality Theorem, $\frac{4.5}{1.5} = \frac{3}{1}$.

10. 200 ft; Assume that the ground is horizontal and both the pole and the building are standing vertically.

12. $P(-2, 7)$; $Q(0, 4)$

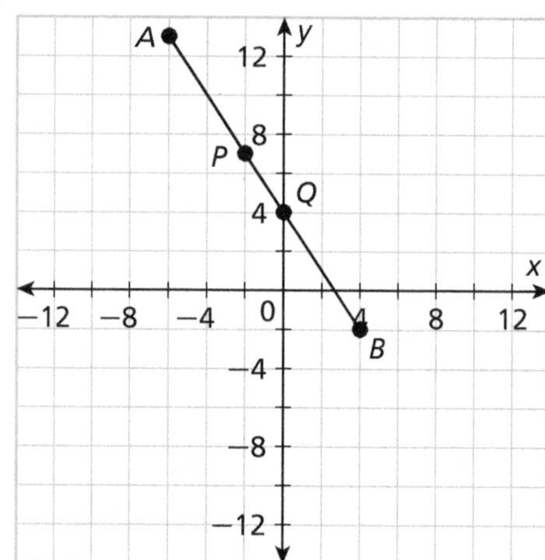

UNIT 7

MODULE 13, LESSON 13.1

Spark Your Learning

D. Possible answer: yes; Based on the side-angle relationships in the triangle, the larger acute angle has a longer opposite leg, and the smaller acute angle has a shorter opposite leg.

Build Understanding

Task 1 **B.**

	Reference angle	Opposite side length (in.)	Adjacent side length (in.)	Ratio of $\frac{\text{opposite}}{\text{adjacent}}$
$\triangle PQR$	$m\angle R = 35°$	$PQ = 1.0$	1.4	0.7
$\triangle XYZ$	$m\angle Z = 35°$	$XY = 1.5$	2.1	0.7

C. The ratios of the oppose side to the adjacent side between the two triangles are about the same.

E. The ratio would be very close because the triangle is similar to the given triangles.

F. Possible answer: The ratios are the same because corresponding side lengths between similar triangles are proportional.

On Your Own

23. A. $\tan P = \frac{10}{24} \approx 0.417$; $\tan Q = \frac{24}{10} = 2.4$

MODULE 13, LESSON 13.2

Build Understanding

Task 1 **E.** Similar triangles have congruent corresponding angles and proportional sides. The ratios between leg length and hypotenuse would be approximately the same as long as the measures of the angles stay the same.

Task 2 **A.** $\sin D \approx 0.385$ $\sin F \approx 0.923$ $\sin C \approx 0.385$ $\sin G \approx 0.923$

$\cos D \approx 0.923$ $\cos F \approx 0.385$ $\cos C \approx 0.923$ $\cos G \approx 0.385$

C. $\sin D = \cos F$, $\sin C = \cos G$; This relationship between complementary angles is always true because the leg opposite one acute angle is adjacent to the other one.

Check Understanding

3. $\sin \angle A = \frac{18}{30} = \frac{3}{5} = 0.6$; $\cos \angle A = \frac{24}{30} = \frac{4}{5} = 0.8$

On Your Own

6. no; Given that $\frac{AC}{AB} = \frac{2}{3}$ and $\frac{QR}{PR} = \frac{1.5}{2.5}$, the triangles are not similar because $\frac{AC}{AB} \neq \frac{QR}{PR}$.

7. yes; Given that $\frac{AB}{AC} = \frac{5}{12}$ and $\frac{QR}{PR} = \frac{6}{14.4}$, the triangles are similar because $\frac{AB}{AC} = \frac{QR}{PR}$.

8. $\sin G = \frac{1.5}{2.5} = 0.6$, $\cos G = \frac{2}{2.5} = 0.8$

9. $\sin I = \frac{2}{2.5} = 0.8$, $\cos I = \frac{1.5}{2.5} = 0.6$

10. $\sin L = \frac{4}{5} = 0.8$, $\cos L = \frac{3}{5} = 0.6$

11. $\sin N = \frac{3}{5} = 0.6$, $\cos N = \frac{4}{5} = 0.8$

12. Since the trigonometric ratios are the same for corresponding angles, the corresponding angles have the same measure. So, the triangles are similar.

13. For any pair of complementary angles, the sine ratio of one angle is equal to the cosine ratio of its complementary angle, and vice versa. For example, $\sin G = \cos I$ and $\sin L = \cos N$.

27. There are many possible answers. Possible answers:

$\cos\frac{1\pi}{6} = \frac{\sqrt{3}}{2}$; $\cos\frac{5\pi}{3} = \frac{\sqrt{1}}{2}$; $\cos\frac{7\pi}{3} = \frac{\sqrt{1}}{2}$

MODULE 13, LESSON 13.3

Spark Your Learning

D. Possible answer: yes; I could use sine, cosine or tangent ratios by finding the inverse of one of the acute angles and see if it matches the ratio of side lengths involved.

Build Understanding

Task 1 **A.** $m\angle A = m\angle B = 45°$; Because all three angles must have a sum of 180°, the two acute angles must be congruent, and $\frac{180° - 90°}{2} = 45°$.

C. Use the Pythagorean Theorem.

$$a^2 + b^2 = c^2$$
$$AC^2 + CB^2 = AB^2$$
$$x^2 + x^2 = AB^2$$
$$2x^2 = AB^2$$
$$\sqrt{2x^2} = AB$$
$$AB = x\sqrt{2}$$

Check Understanding

4. They are congruent ratios. In a right triangle, the sine of one of the acute angles equals the cosine of the other acute angle.

On Your Own

30. yes; If you know one leg length, the other leg length is also equal to this length, and the length of the hypotenuse is this length multiplied by $\sqrt{2}$. If you know the hypotenuse length, then each leg length is this length divided by $\sqrt{2}$.

36. Possible answer: By drawing a diagonal, you create two triangles. Each triangle is an isosceles right triangle because the side lengths are congruent and it contains a right angle. This means the triangle is a 45°-45°-90° triangle. In 45°-45°-90° triangles, the the length of the hypotenuse, which is the length of the diagonal, is the side length multiplied by $\sqrt{2}$.

MODULE 13, LESSON 13.4

Build Understanding

Task 1 **A.** The altitude h creates two right triangles. You can then use a trigonometric ratio of an angle in terms of h and other side lengths to possibly develop an area formula.

F. The sides a and b are adjacent, and angle C is the included angle; yes; Two adjacent sides and an included angle are still provided.

MODULE 13 REVIEW

Concepts and Skills

5. A. $\sin A = \frac{28}{53}$, $\cos A = \frac{45}{53}$, $\tan A = \frac{28}{45}$; $\sin B = \frac{45}{53}$, $\cos B = \frac{28}{53}$, $\tan B = \frac{45}{28}$

B. $m\angle A \approx 32°$; $m\angle B \approx 58°$

C. The sine and cosine of complementary angles are identical. They refer to the same ratio of sides, except they are labeled according to a different reference angle.

D. $\sin D = \sin A = \frac{28}{53}$. The sines of corresponding angles are equivalent because the scale factor will cancel out: $\sin D = \frac{28k}{53k} = \frac{28}{53}$.

6. $m\angle R = 60°$, $IR = 10$, $TI = 10\sqrt{3}$

7. $m\angle A = 45°$, $AB = BC = 15\sqrt{2}$

MODULE 14, LESSON 14.1

Spark Your Learning

B. yes; The large triangle formed by the suspension rods can be divided into two right triangles at the mast. Then right triangle trigonometric ratios could be used to find side lengths.

Build Understanding

Task 1 **B.** Solve each equation for h and use the Transitive Property of Equality to write $b \sin C = c \sin B$. Then divide both sides of the equation by bc to get $\frac{\sin B}{b} = \frac{\sin C}{c}$.

D. The equations in Part C can be used to write $\frac{\sin A}{a} = \frac{\sin B}{b}$. The Transitive Property of Equality can be used to write $\frac{\sin A}{a} = \frac{\sin B}{b}$ and $\frac{\sin B}{b} = \frac{\sin C}{c}$ as $\frac{\sin A}{a} = \frac{\sin B}{b} = \frac{\sin C}{c}$.

Task 2 **B.** For SSA case: $\frac{\sin A}{13.2} = \frac{\sin 30°}{12}$, $\frac{\sin 30°}{12} = \frac{\sin C}{c}$, $\frac{\sin A}{13.2} = \frac{\sin C}{c}$. The equation $\frac{\sin A}{13.2} = \frac{\sin 30°}{12}$ can be use to find the measure of angle A. For SAS case: $\frac{\sin A}{10} = \frac{\sin B}{6}$, $\frac{\sin B}{6} = \frac{\sin 78°}{c}$, $\frac{\sin A}{10} = \frac{\sin 78°}{c}$. None of the equations can be used to find unknown angle measures or side lengths since each equation has two unknowns.

C. For the ASA case, $m\angle B = 72°$, $\frac{\sin 33°}{a} = \frac{\sin 72°}{5}$, $\frac{\sin 72°}{5} = \frac{\sin 75°}{c}$, $\frac{\sin 33°}{a} = \frac{\sin 75°}{c}$. The equations $\frac{\sin 33°}{a} = \frac{\sin 72°}{5}$ and $\frac{\sin 72°}{5} = \frac{\sin 75°}{c}$ can both be used to find unknown side lengths. For the AAS case, $m\angle B = 81°$, $\frac{\sin 58°}{8.2} = \frac{\sin 81°}{b}$, $\frac{\sin 81°}{b} = \frac{\sin 41°}{c}$, $\frac{\sin 58°}{8.2} = \frac{\sin 41°}{c}$. The equations $\frac{\sin 58°}{8.2} = \frac{\sin 81°}{b}$ and $\frac{\sin 58°}{8.2} = \frac{\sin 41°}{c}$ can be used to find unknown side lengths.

Task 3 **A.**

x	0°	30°	45°	90°	135°	150°	180°
$\sin(x)$	0	0.5	0.707	1	0.707	0.5	0

D. When finding an angle measure using the inverse sine operation on a calculator, there are two possible angle measure types, an acute angle or its obtuse supplement.

E. no; In the ASA and AAS cases, two angle measures are known, so there is only one possible measure for the unknown angle.

Step It Out

Task 5 **C.** If the measure of $m\angle B$ is unknown, then the equations written using the Law of Sines cannot be solved since AC is also unknown.

On Your Own

11. B. $\sin \angle BCD = \frac{h}{a}$; $\sin C = \frac{h}{a}$; $\angle BCD$ and $\angle C$ are supplementary, so they have the same sine value.

D. Using the sine ratios from Parts A and B, $c \sin A = a \sin C$, or $\frac{\sin A}{a} = \frac{\sin C}{c}$. Using the ratios from Part C, $b \sin A = a \sin B$, or $\frac{\sin A}{a} = \frac{\sin B}{b}$. The Transitive Property of Equality can be used to write $\frac{\sin A}{a} = \frac{\sin B}{b} = \frac{\sin C}{c}$.

MODULE 14, LESSON 14.2

Spark Your Learning

B. yes, but I would need to form right triangles by drawing an altitude in the triangle

Build Understanding

Task 1 **A.** no; The altitude drawn in the second triangle creates two right triangles. A right triangle is needed to write a ratio for the cosine of an angle.

B. $\triangle ABD$: $x^2 + h^2 = c^2$; $\triangle CBD$: $h^2 + b^2 - 2bx + x^2 = a^2$

D. $c \cos A$; In $\triangle ABD$, $\cos A = \frac{x}{c}$, which can be written as $c \cos A = x$. Then substitute $c \cos A$ for x in the equation for a^2.

Task 2 **A.**

x	0°	30°	45°	90°	135°	150°	180°
cos(x)	1	0.8660	0.7071	0	−0.7071	−0.8660	−1

B. For two supplementary angles, the cosine of one angle is the opposite of the cosine of its supplement.

Task 3 **A.** $A = \frac{1}{2}ah$; The height h must be replaced with measures that involve a side adjacent to side a and the included angle.

B. $h = b \sin(\angle ACD)$; $h = b \sin C$; Supplementary angles have the same sine value, so sine of $\angle ACD$ equals the sine of $\angle C$.

C. $A = \frac{1}{2}ab \sin C$; Substitute $b \sin C$ for h in the equation $A = \frac{1}{2}ah$.

D. Draw an altitude from vertex C to side c. Use the same reasoning in Parts A–C, but use side c as the base of the triangle and use the sine ratio for $\angle A$ to write an expression for the height of the triangle.

Step It Out

Task 4 **B.** Possible answer: Law of Sines; There are fewer steps when using the Law of Sines to find the measure of $\angle X$.

C. The largest angle in the triangle, $\angle Y$, is obtuse. So, the two other angles must be acute.

Task 5 The variable a is the distance between the elephant and the herd, b is the distance from the ecologist to the elephant, and c is the distance from the ecologist to the herd.

Task 6 The height of the triangle is not given, so the area formula $A = \frac{1}{2}bh$ cannot be used. However, the area formula $A = \frac{1}{2}ac \sin B$ can be applied. To use this fomula, at least one angle measure is needed.

Check Understanding

1.

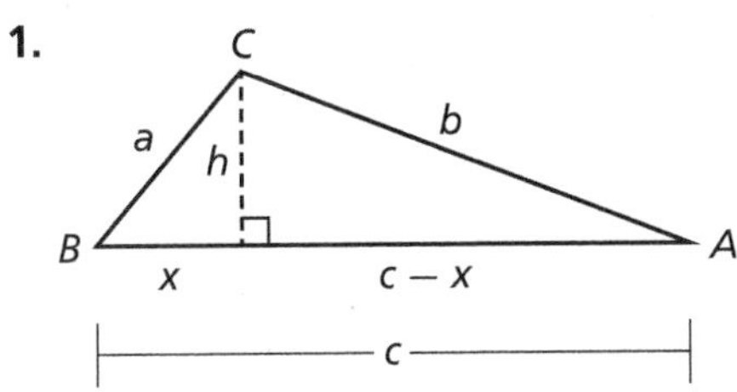

On Your Own

6. $C = \cos^{-1}\left(\frac{c^2 - a^2 - b^2}{-2ab}\right)$; This form is useful when finding an unknown angle measure when given the three side lengths of a triangle.

8. For an equilateral triangle with side s and angle measures of 60°, $A = \frac{1}{2} \cdot s \cdot s \cdot \sin 60°$. The area can be rewritten as $A = \frac{1}{2}s^2 \frac{\sqrt{3}}{2} = \frac{\sqrt{3}}{4}s^2$.

9. $a \approx 38.0$, $m\angle B \approx 45.3°$, $m\angle C \approx 59.7°$

10. $m\angle B \approx 42.2°$, $m\angle C \approx 110.9°$, $m\angle D \approx 26.9°$

11. $m\angle R \approx 60.4°$, $m\angle S \approx 80°$, $m\angle T \approx 39.6°$

12. $x \approx 2.7$, $m\angle Y \approx 117.3°$, $m\angle Z \approx 38.7°$

MODULE 14 REVIEW

Concepts and Skills

16. It is known as the ambiguous case because it may be possible to form 0, 1, or 2 different triangles when you know SSA.

UNIT 8

MODULE 15, ARE YOU READY?

11. distance $= 5\sqrt{2}$, midpoint $= \left(-\frac{1}{2}, -\frac{3}{2}\right)$

12. distance $= 2\sqrt{34}$, midpoint $= (4, 2)$

13. distance $= \sqrt{53}$, midpoint $= \left(\frac{3}{2}, 3\right)$

14. distance $= \sqrt{73}$, midpoint $= \left(-\frac{9}{2}, -1\right)$

MODULE 15, LESSON 15.1

Build Understanding

Task 1 **B.** The measure of the inscribed angle is half of the measure of the associated central angle.

Task 4 **A.** Possible answer: $\overline{AB}$ represents a copy of the radius of circle P, and $\overline{AP}$ is a radius of circle P; All of the sides of the hexagon represent copies of the radius of circle P, so they are congruent.

B. Possible answer: The six arcs created on the circle have the same measure, so each arc measures 60°. $\widehat{ABD}$ is a semicircle since $m\widehat{ABD} = m\widehat{AB} + m\widehat{BC} + m\widehat{CD} = 180°$. $\overline{AD}$ is a diameter since the endpoints lie on the endpoints of the semicircle, so $\overline{AD}$ must contain the center P. The same can be said of $\overline{BE}$ and $\overline{CF}$.

C. Possible answer: Each line segment from the vertices of the hexagon to the center of the circle are radii since $\overline{AD}$, $\overline{BE}$, and $\overline{CF}$ are diameters. You know the measures of the sides of the hexagon equal the measure of the radius from part A, so each of the six triangles formed are equilateral. Equilateral triangles are also equiangular.

D. yes; Possible answer: Each vertex angle of the hexagon is the sum of two adjacent angles of the triangles. Each angle in the triangles is 60°, so each vertex angle of the hexagon is $60° + 60° = 120°$.

E. The side congruence was proven in Part A and the angle congruence was proven in Part D.

On Your Own

11. minor arc; 43°

12. minor arc; 85°

13. semicircle; 180°

14. major arc; 223°

35. *F* can be anywhere on $\widehat{CFE}$. As long as the *C* and *E* are in the same location, the angle measure will remain the same.

37. For an acute triangle, the center of the circle will be inside the triangle. For a right triangle, the center of the circle is on the hypotenuse of the triangle. For an obtuse triangle, the center of the circle is outside the triangle. The term *circumcenter* describes this point.

MODULE 15, LESSON 15.2

Spark Your Learning

C. Possible answer: yes; My answer makes sense because only a square satisfies all the requirements provided by both the photo and the additional information.

Build Understanding

Task 1 **A.** Not all quadrilaterals can be inscribed in a circle. Only those with opposite angles that are supplementary can be inscribed in a circle.

E. The sum of the measures of opposite angle pairs is 180°. There appears to be no pattern for the sum of the measures of adjacent angle pairs.

F. Inscribed angles are half the measure of the arc. If the entire circle measures 360 degrees, opposite inscribed angles must sum to half that measure, or 180 degrees. Yes, this pattern is true for all quadrilaterals inscribed in a circle.

G. no; It does not matter if the center of the circle is inside the inscribed quadrilateral. The circumcenter of a shape does not need to be inside of a shape.

Task 2 **B.**

Statements	Reasons
1. $m\widehat{ADC} + m\widehat{ABC} = 360°$	**1.** Arc Addition Postulate and definition of a circle
2. $m\widehat{ADC} = 2m\angle B$ $m\widehat{ABC} = 2m\angle D$	**2.** Inscribed Angle Theorem
3. $2m\angle B + 2m\angle D = 360°$	**3.** Substitution Property of Equality
4. $2(m\angle B + m\angle D) = 360°$	**4.** Distributive Property
5. $m\angle B + m\angle D = 180°$	**5.** Division Property of Equality
6. $\angle B$ and $\angle D$ are supplementary.	**6.** Definition of supplementary angles

C. For quadrilaterals inscribed in circles, opposite angles are supplementary, so the sum of their measures is 180°. There are two pairs of opposite angles, so the sum of the measures of the two pairs is $180° + 180° = 360°$.

On Your Own

14.

Statements	Reasons
1. $\widehat{AB} \cong \widehat{CD}$	**1.** Given
2. Draw $\overline{AP}$, $\overline{BP}$, $\overline{CP}$, $\overline{DP}$	**2.** Through any two points exists one line
3. $\overline{AP} \cong \overline{BP} \cong \overline{CP} \cong \overline{DP}$	**3.** Radii in the same circle are congruent.
4. $m\widehat{AB} = m\widehat{CD}$	**4.** Definition of congruent arcs
5. $m\widehat{AB} = m\angle APB$ $m\widehat{CD} = m\angle CPD$	**5.** Definition of measure of a minor arc
6. $m\angle APB = m\angle CPD$	**6.** Substitution Property of Equality
7. $\angle APB \cong \angle CPD$	**7.** Definition of congruent angles
8. $\triangle APB \cong \triangle CPD$	**8.** SAS Triangle Congruence Theorem
9. $\overline{AB} \cong \overline{CD}$	**9.** Corresponding Parts of Congruent Triangles Are Congruent

20. Use the point you drew as an endpoint of a diameter of the circle by going through the center. Draw a perpendicular bisector of the diameter. Then connect the endpoints of the diameters.

MODULE 15, LESSON 15.3

Spark Your Learning

B. The quantities are the distance from the satellite to the surface of Earth and the distance the satellite's signal can travel to the surface of Earth. The unit of measurement used is miles.

D. This answer makes sense. The length of the hypotenuse of the right triangle is longer than the farthest distance from the satellite to the surface of Earth.

Check Understanding

1. 22°;$\angle ABC$ is a right angle by the Tangent-Radius Theorem so, $\triangle ABC$ is a right triangle. $m\angle ACB$ can be found by subtracting the $m\angle BAC$ from 90°.

On Your Own

6.

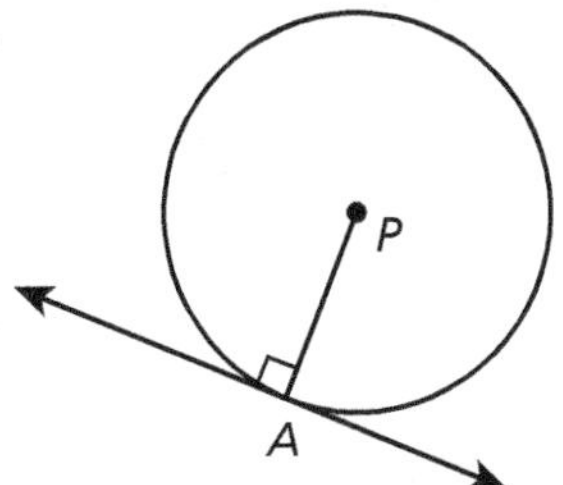

7. Let B be any point on t other than A.

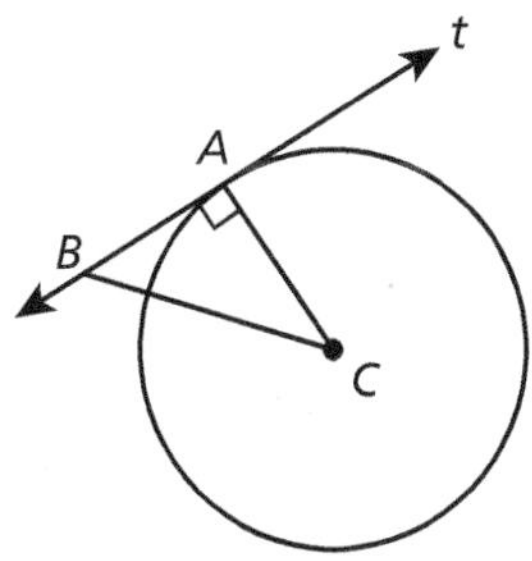

Then $\triangle CAB$ is a right triangle with hypotenuse $\overline{CB}$. Therefore, $CB > CA$ since the hypotenuse is the longest side of a right triangle. Since $\overline{CA}$ is a radius, point B must be in the exterior of circle C. So, A is the only point of line t that is on circle C. Since line t intersects circle C at exactly one point, line t is tangent to the circle at A.

MODULE 15, LESSON 15.4

Spark Your Learning

B. The variables are the position to the east or west of the center of the circle and the position north and south of the center of the circle. The units of measurement would be feet.

Build Understanding

Task 1

A. $\overline{CA}$ may range in length from r (if P is chosen directly to the right or left of C) to 0 (if it is chosen directly above or below C, which would make A and C coincident).

B. The coordinates of A are (x, k). The x-coordinate of A is the same as that of P, and the y-coordinate of A is the same as that of C.

C. The absolute value of the differences is used because distances are nonnegative.

D. The absolute values are no longer needed because squaring the differences produces nonnegative values.

E. yes; because squaring these differences makes them nonnegative

F. $\triangle QCB$ differs from $\triangle PCA$ in that it is on the opposite side of $\overleftrightarrow{AC}$ and has a different base and height. Both $\triangle PCA$ and $\triangle QCB$ have right angles opposite a radius of the circle, as would any triangle constructed in this way.

On Your Own

32. Brandy is correct. The circle has a radius of 4, so the equation of the circle is $x^2 + y^2 = 16$. If you substitute for x and y, you find $\left(\sqrt{3}\right)^2 + \left(\sqrt{7}\right)^2 = 3 + 7 = 10$, which is not equal to 16.

34. Possible answer:
$(x-1)^2 + (y-2)^2 = 5^2$ with point on the circle $(4, 6)$
$(x-1)^2 + (y-3)^2 = 5^2$ with point on the circle $(4, 7)$

MODULE 15 REVIEW

Concepts and Skills

5. The measure of $\angle ACB$ is 38° because the triangle is isosceles.

6. The measure of $\angle CAB$ is 104° by the Triangle Sum Theorem.

7. The measure of $\angle CDB$ is 52° by the Inscribed Angle Theorem.

8. The measure of $\angle EFG$ is 90° because tangents are perpendicular to radii.

9. FG is equal to 37. The two tangent segments are congruent, so $x = 2$.

10. Using the Pythagorean Theorem, EG is approximately equal to 37.3.

11. The measure of $\angle FEG$ is approximately 82.3°. This can be found using the inverse tangent.

12. The measure of $\angle HEF$ is 164.6° by the Angle Addition Postulate.

13. The measure of $\angle FGH$ is 15.4° by the Triangle Sum Theorem and Angle Addition Postulate.

16. $m\angle QPR = 90°$, $m\angle PQR = 30°$, $m\angle QRP = 60°$

MODULE 16, LESSON 16.1

Spark Your Learning

A. How are the lengths of the segments of the intersecting chords related?; I might need to know if the chords are perpendicular. I might need to know that if I form two triangles by connecting the endpoints of the chords, that the triangles will be similar.

B. There is a circle, two chords, and a point of intersection of the chords that looks perpendicular. You could use theorems about the similarity of triangles by connecting the endpoints of the chords to form triangles.

Build Understanding

Task 1 **D.** Corresponding sides of similar triangles are proportional; Multiplication Property of Equality; Identity Property of Division

E. yes; $\angle DAB$ and $\angle BCD$ are congruent inscribed angles that intercept the same arc. $\angle APD$ and $\angle BPC$ are congruent vertical angles. $\triangle DAP \sim \triangle BCP$ by AA Triangle Similarity. $\frac{DP}{BP} = \frac{AP}{CP}$ because corresponding sides of similar triangles are proportional. $DP \cdot CP = AP \cdot BP$ by the Multiplication Property of Equality and Identity Property of Division.

On Your Own

6. The product of the lengths of the segments of one chord equals the product of the lengths of the segments of the other chord.

7. The diagram shows that the diameter bisects the chord. Since the diameter is itself a chord, the diagram shows the intersection of two chords. The segments of one chord have lengths b and b, and the segments of the other chord have lengths $c + a$ and $c - a$. By the Chord–Chord Product Theorem

$$b^2 = (c + a)(c - a)$$
$$b^2 = c^2 - a^2$$
$$a^2 + b^2 = c^2$$

9. no; The Chord–Chord Product rule still applies, so the relationship would stay the same.

10. If you drag C to the center of the circle, then all four segments would be equal because they would be the radius of the circle.

11.

Statements	Reasons
1. Draw segments $\overline{AD}$ and $\overline{EB}$.	1. There is one line through any two points.
2. $\angle CAD \cong \angle CEB$	2. Tangent Chord Theorem
3. $\angle C \cong \angle C$	3. Reflexive Property
4. $\triangle CAD \sim \triangle CEB$	4. AA Triangle Similarity Theorem
5. $\frac{AC}{EC} = \frac{DC}{BC}$	5. Definition of similar figures
6. $AC \cdot BC = EC \cdot DC$	6. Cross Products Property

13. yes; Let $\overline{AB}$ be the internal segment and $\overline{BC}$ be the external segment that form secant $\overline{AC}$. Let $\overline{DE}$ be the internal segment and $\overline{EC}$ be the external segment that form secant $\overline{DC}$. Since the two internal secant segments are given as congruent, their lengths are equal, so $AB = DE$.

Use the Secant-Secant Product Theorem.
$AC \cdot BC = DC \cdot EC$; $(AB + BC) \cdot BC = (DE + EC) \cdot EC$.

For this statement to be true, BC and EC (the lengths of the external segments) must be equal. Since the lengths of the internal segments and the external segments are equal, the secant segments must also be equal.

14.

Statements	Reasons
1. $\overline{CB}$ is the external segment to secant segment $\overline{DB}$. Secant $\overline{DB}$ and tangent $\overline{AB}$ intersect in the exterior of the circle. $\overline{DA}$ and $\overline{AC}$ are drawn to create triangles.	**1.** Given
2. $m\angle CAB = m\angle ADB$	**2.** Tangent Chord Theorem
3. $m\angle ABC = m\angle DBA$	**3.** Common angles
4. $\triangle CAB \sim \triangle ADB$	**4.** AA Similarity Theorem
5. $\frac{BC}{AB} = \frac{AB}{DB}$	**5.** Definition of similar figures
6. $AB^2 = BC \cdot DB$	**6.** Cross Products Property

16. AA Similarity Theorem; by connecting points, you can form triangles with congruent angles. These similar triangles can be used to create proportions using specific segment lengths.

17. The secant segments must also be equal since $(AB + BC) \cdot BC = (MN + NC) \cdot NC$, and $BC = NC$. Then you can rewrite the product as follows.

$(AB + NC) \cdot NC = (MN + NC) \cdot NC$	Substitute NC for BC
$AB \cdot NC + NC \cdot NC = MN \cdot NC + NC \cdot NC$	Distributive Property
$AB \cdot NC = MN \cdot NC$	Subtract $NC \cdot NC$ from both sides
$AB = MN$	Divide both sides by NC

18. no; A secant will always be longer than a tangent because the secant intersects the circle at two points, while a tangent intersects the circle at a single point.

31. Possible answer: yes; When the intersection is inside the circle, this is just a restatement of the theorem. When the intersection is outside the circle, then the distance from one intersection with the circle to the intersection of the lines is the whole secant segment and the other is the external secant segment. When one of the intersecting lines is tangent to the circle, you can think of the two points of intersection as being the same point, so the lengths are the same.

MODULE 16, LESSON 16.2

Spark Your Learning

B. Two congruent right triangles represent the situation. I can use trigonometry to find the measures of angles *BCG* and *DCG*. Then add the measure of these angles to determine the measure of angle *BCD*. The measure of angle *BCD* is the measure of arc *BGD* because *BCD* is a central angle.

D. This makes sense because the measure of the arc is close to 180° which is the measure of the arc representing half the circumference of the moon.

Build Understanding

Task 1 **A.** $m\angle AEC = m\angle ABC + m\angle DCB$ by the Exterior Angle Theorem.

B. You can use the Inscribed Angles Theorem to write the following.

$m\angle ABC = \frac{1}{2}m\overset{\frown}{AC}$

$m\angle DCB = \frac{1}{2}m\overset{\frown}{DB}$

C. Since $m\angle AEC = m\angle ABC + m\angle DCB$, then by substitution, $m\angle AEC = \frac{1}{2}m\overset{\frown}{AC} + \frac{1}{2}m\overset{\frown}{DB}$. Factoring, you have $m\angle AEC = \frac{1}{2}\left(m\overset{\frown}{AC} + m\overset{\frown}{DB}\right)$.

D. In order for the measure of angle *AEC* to equal the measure of the arc *DB*, *E* would have to be the center of the circle. Then you could use the Central Angles Theorem that central angles equal the arc they intersect.

By the Central Angles Theorem. $m\overset{\frown}{AC} = m\overset{\frown}{DB}$. Substituting you have the following.

$m\angle AEC = \frac{1}{2}\left(m\overset{\frown}{AC} + m\overset{\frown}{DB}\right)$

$m\angle AEC = \frac{1}{2}\left(m\overset{\frown}{DB} + m\overset{\frown}{DB}\right)$

$m\angle AEC = \frac{1}{2}\left(2m\overset{\frown}{DB}\right)$

$m\angle AEC = m\overset{\frown}{DB}$

Task 3 **A.** radius; Since $\overline{AB}$ is a diameter that contains center *K* and *B* is on the circle $\overline{BK}$ is a radius for the circle.

B. $\overline{BC}$ is tangent to the circle at point *B*; $m\angle KBC = 90°$; 90°; because *K* lies on the same line as $\overline{AB}$, which is a diameter.

C. $m\overset{\frown}{APB} = 180°$ because $\overline{AB}$ is a diameter, and the measure of central angle $\angle AKB$ is 180°

D. The measure of right angle *ABC* is one-half the measure of arc *APB*: $m\angle ABC = 90°$ and $m\overset{\frown}{APB} = 180°$

On Your Own

5. yes; You are finding the average of the lengths of the intercepted arcs because multiplying the sum of the arcs by one-half is the same as adding the two arcs and dividing by 2.

6.

Statement	Reason
1. $m\overset{\frown}{CA} = 80°$, $\overline{CD}$ and $\overline{AB}$ intersect at center *E*.	1. Given
2. $m\angle AEC = 80°$	2. A central angle's measure equals the measure of the arc it intercepts.
3. $m\angle BED = 80°$	3. Vertical Angles Theorem
4. $m\overset{\frown}{BD} = 80°$	4. A central angle's measure equals the measure of the arc it intercepts.
5. $m\angle AEC = \frac{1}{2}(80° + 80°)$ $= 80°$	5. Rewrite as an equivalent expression.
6. $m\angle AEC = \frac{1}{2}\left(m\overset{\frown}{CA} + m\overset{\frown}{BD}\right)$	6. Substitution

7. In the proof in Task 1, the Exterior Angle Theorem and inscribed angles were used. In this proof, central angles and verticals angles can be used instead since the chords intersect at the center of the circle.

8. Draw radii $\overline{ZA}$ and $\overline{ZB}$ so $\triangle AZB$ is isosceles. Thus $\angle ZAB \cong \angle ZBA$.

For $\triangle AZB$, $m\angle ZBA + m\angle ZAB + m\angle AZB = 180$.

By substitution, $2m\angle ZBA + m\angle AZB = 180$

$2m\angle ZBA = 180 - m\angle AZB$

$m\angle ZBA = 90 - \frac{1}{2}m\angle AZB$

$\angle ZBC$ is a right angle because a tangent is perpendicular to the radius of a circle.

$m\angle ZBA + m\angle ABC = 90°$

$m\angle ABC = 90° - m\angle ZBA$

By substitution, $m\angle ABC = 90° - (90 - \frac{1}{2}m\angle AZB)$.

Simplify to get, $m\angle ABC = \frac{1}{2}m\angle AZB$.

The measure of angle *AZB* equals the measure of arc *AB* because angle *AZB* is a central angle, so $m\angle ABC = \frac{1}{2}mAB$ by substitution.

9. Tangent-Secant Interior Angle Measure because the angle that is formed is an inscribed angle. The measure of an inscribed angle is half the arc it intercepts.

10. $m\angle AXB + m\angle ACB = \frac{1}{2}\left(m\widehat{AYB} - m\widehat{AB}\right) + m\widehat{AB}$

$= \frac{1}{2}\left(m\widehat{AYB} + m\widehat{AB}\right)$

$= \frac{1}{2}(360°) = 180°$

11. A.

Statement	Reason
1. $\overrightarrow{CA}$ is a secant, and $\overline{CD}$ is a tangent.	1. Given
2. Draw $\overline{DA}$.	2. There is exactly one line through any two points.
3. $m\angle DCA + m\angle CAD = m\angle UDA$	3. Exterior Angle Theorem
4. $m\angle DCA = m\angle UDA - m\angle CAD$	4. Subtract $m\angle CAD$ from both sides.
5. $m\angle UDA = \frac{1}{2}m\widehat{AD}$ $m\angle CAD = \frac{1}{2}m\widehat{BD}$	5. Inscribed Angle Theorem
6. $m\angle DCA = \frac{1}{2}m\widehat{AD} - \frac{1}{2}m\widehat{BD}$	6. Substitution
7. $m\angle DCA = \frac{1}{2}\left(m\widehat{AD} - m\widehat{BD}\right)$	7. Distributive Property

B. The segment that is drawn connects to the points of tangency, but you still use the Exterior Angles Theorem and substitute the inscribed angles.

MODULE 16 REVIEW

Concepts and Skills

9. 115 ft; $69° = \frac{1}{2}\widehat{TC}$ so $\widehat{TC} = 138°$. Find the fraction of the circumference that $\widehat{TC}$ covers and multiply that by the circumference of the fountain. $\frac{138°}{360°} = \frac{23}{60}$ and $\frac{23}{60}(300) = 115$

MODULE 17, LESSON 17.1

Spark Your Learning

B. Possible answer: I can estimate the diameter by using the width of the brick shown near the cover. I can estimate the circumference by considering the width of the bricks that surround the manhole.

D. Possible answer: I could use the estimated diameter to calculate the circumference of the circle using the circumference formula.

Build Understanding

Task 1 A. As n increases, the perimeters should approach the circumference of the circle, $2 \cdot \pi \cdot 1 \approx 6.28$. Possible answer: For greater values of n, then the perimeters of the n-gons will get closer to the circumference of the circle because the shapes of the n-gons will get closer to the shape of the circle.

C. $x = \sin\left(\frac{360°}{2n}\right)$; perimeter $= n \cdot 2x = 2n\sin\left(\frac{360°}{2n}\right)$

D. Possible answer: When $n = 100$, perimeter $= 2(100)\sin\left(\frac{360°}{2(100)}\right) \approx 6.28$; The perimeters approach 6.28, or 2π; For large values of n, the n-gon resembles a circle. So, the perimeter of the n-gon should approximate the circumference of the circle with radius 1, which is $2\pi \approx 6.28$.

Check Understanding

1. The 30-gon has a circumference closer to the circumference of the circle in which it is inscribed, because as n increases, the perimeter more closely approximates the circumference of the circle.

2. You can justify the formula for the area of a circle by finding the area for an n-gon inscribed within the circle and identifying that the area gets closer to πr^2 as n increases. To use the formula, find the radius, square it, and then multiply by π.

3. no; The circumference C of a circle with diameter d is $C = \pi d$. Since $\frac{C}{d} = \pi$, the ratio of the circumference to diameter of any circle will always be π.

On Your Own

6. no; The perimeter of the inscribed n-gon gets very close to the circumference of the circle, but it will always be less than the circumference of the circle.

7. no; The circumference of the larger circle (20π) is double the circumference of the smaller circle (10π). However, the area of the larger circle (100π) is four times larger than the area of the smaller circle (25π).

8. A. The height is close to the radius of the circle, and the base is about half the circumference of the circle.

B. The area of the parallelogram is almost equal to the area of the circle. The area of the parallelogram can be approximated using a base of πr and a height of r.

Area $= \pi r \cdot r = \pi r^2$

C. yes; The smaller the pieces of the circle, the more the arranged pieces resemble the parallelogram.

9. Possible answer: no; If you only use an inscribed regular n-gon, you can generate a sequence of approximations of the value of π, but you wouldn't know the accuracy of those approximations because you have nothing against which to compare them. The inscribed regular n-gon will provide just a lower bound on the value of π. When finding an approximate value of π, you should both inscribe a regular n-gon in the circle and circumscribe a regular n-gon about the circle to get both a lower bound and an upper bound on the value of π.

26. The large pizza is a better value because it has a lower cost per square inch. \$0.08 per square inch is less than \$0.09 per square inch.

MODULE 17, LESSON 17.2

Spark Your Learning

B. Possible answer: You can analyze either arc. Since you know the radius of each, you can calculate additional information about both rings. No, they are not the same size.

Build Understanding

Task 1 **A.** I can divide each arc measure in degrees by 360, because there are 360 degrees in a circle.

B. The arc length is equal to the product of the circumference of the circle and the fraction of the circle that the arc represents.

C. the same units as the circumference; The arc is a part of the circle, so its length will have the same units as the circumference. Both arc length and circumference are linear measures.

D.

$C = 2\pi r$	$m\widehat{AB}$	Fraction of circle	Length of $\widehat{AB}$
$C = 2 \cdot \pi \cdot 12 = 24\pi$	60°	$\frac{1}{6}$	$\frac{1}{6}(24\pi) = 4\pi$
$C = 2 \cdot \pi \cdot 9 = 18\pi$	120°	$\frac{1}{3}$	$\frac{1}{3}(18\pi) = 6\pi$

E. $\frac{x}{360}$ of the circumference is contained within the arc

F. $s = \frac{x^\circ}{360^\circ} \cdot 2\pi r$

Task 2 **A.**

Radius r	Arc length	$\frac{\text{arc length}}{\text{radius}}$
4	$\frac{60^\circ}{360^\circ} \cdot 2\pi(4) = \frac{4}{3}\pi$	$\frac{\frac{4}{3}\pi}{4} = \frac{\pi}{3}$
5	$\frac{60^\circ}{360^\circ} \cdot 2\pi(5) = \frac{5}{3}\pi$	$\frac{\frac{5}{3}\pi}{5} = \frac{\pi}{3}$

On Your Own

10. sometimes; If the arcs have the same radii, then they will have the same length. However, if the arcs have different radii, then they will have different lengths.

11. never; An arc is part of a circle, so its length will be less than the circumference.

12. always; The two arcs have the same radii since they are in the same circle. If two arcs have the same radii and the same measure, then they have the same length.

26. B. yes; There are 360° in a circle and there are 2π radians in a circle. The radius will fit around a circle 2π times.

31. *See below.*

32. There are multiple answers, including the following:

$r = 6$ cm, $\theta = 30^\circ$, $\widehat{AB} = 1\pi$ cm

$r = 9$ cm, $\theta = 60^\circ$, $\widehat{AB} = 3\pi$ cm

MODULE 17, LESSON 17.3

Spark Your Learning

D. yes; Since 60° is $\frac{1}{6}$ of 360°, and the formula for the area of a complete circle is $A = \pi r^2$, the area for one portion of the neighborhood could be found by using the formula $A = \frac{1}{6}\pi r^2$.

31.

Benchmark Angles									
Degree measure	0°	30°	45°	60°	90°	120°	135°	150°	180°
Radian measure	0	$\frac{\pi}{6}$	$\frac{\pi}{4}$	$\frac{\pi}{3}$	$\frac{\pi}{2}$	$\frac{2\pi}{3}$	$\frac{3\pi}{4}$	$\frac{5\pi}{6}$	π

Build Understanding

Task 1 **B.** Divide the central angle by 360°; The total area is multiplied by the fraction.

C.

Central angle	Fraction of circle	Area of whole circle	Area of sector AOB
60°	$\frac{60°}{360°} = \frac{1}{6}$	100π cm^2	$\frac{1}{6}(100\pi) = \frac{50}{3}\pi$ cm^2
135°	$\frac{135°}{360°} = \frac{3}{8}$	100π cm^2	$\frac{3}{8}(100\pi) = \frac{75}{2}\pi$ cm^2

E. Possible answer: $\frac{A}{\pi r^2} = \frac{x°}{360°}$; $A = \frac{x°}{360°} \cdot \pi r^2$

Step It Out

Task 2 It is helpful to us an approximation for π because it is easier to visualize 8.64 square centimeters than $\frac{11}{4}\pi$ square centimeters.

Task 3 **A.** A is the area of the pizza that is topped with only mushrooms, and $x°$ is the angle measure of the sector of the pizza that is topped with only mushrooms.

B. Substitute 225° for $x°$ and 8 in. for r in the formula for the area of the sector. The resulting sector area should be 40π in^2.

On Your Own

23.

Category	Central angle	Area (cm^2)
Rent	90°	28.3
Bills	36°	11.3
Food	54°	17
Savings	72°	22.6
Other	108°	33.9

UNIT 9

MODULE 18, LESSON 18.1

Build Understanding

Task 2 **D.** Any cross section that intersects a cylinder along its lateral surface will be a curved shape. When a plane intersects a cylinder through its two bases, it creates a four-sided figure with two curved sides opposite each other; the other two sides are straight, and run parallel to each other. When the plane intersects a cylinder through only one of its bases it creates a figure with a curved side and a straight side. If the plane intersects the cylinder perpendicularly to its bases, it creates a cross section of a rectangle.

Check Understanding

5.

On Your Own

6. Juan is correct. If the semicircle was rotated about a horizontal line of rotation, it would be a sphere.

7.

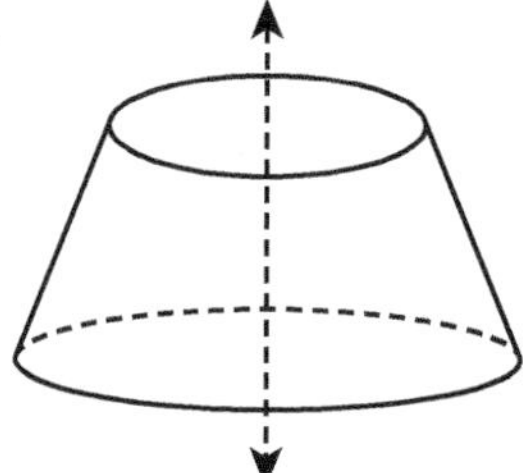

8.

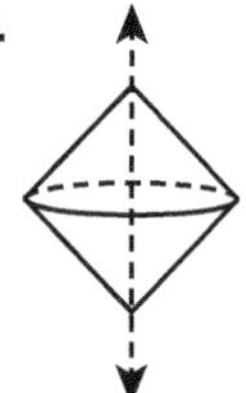

9.

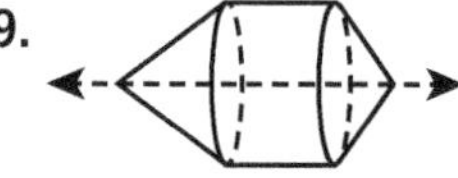

MODULE 18, LESSON 18.2

Spark Your Learning

B. The variables are the lengths and widths of each section of the box. The units of measure for the box are inches.

D. The answer makes sense because the unfolded box shows all of the surfaces of the completed box.

Build Understanding

Task 1 **C.**

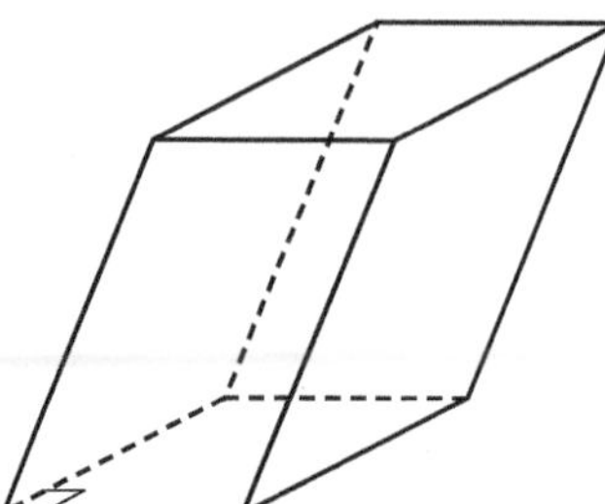

D. Like the right rectangular prism, the surface area of the oblique prism is the sum of the areas of all the faces of the prism. It is different because each face isn't a rectangle. Some of the faces are parallelograms.

E. You cannot write a general formula for the surface area using the dimensions of the net. You need to know the height perpendicular to the base instead of the slant height.

On Your Own

21. A. Possible answer:

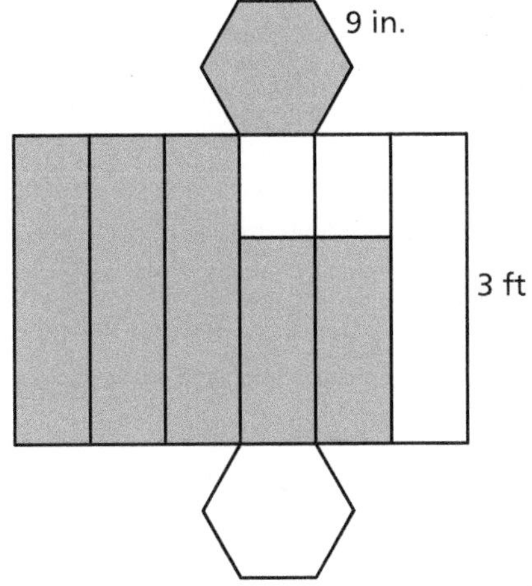

MODULE 18, LESSON 18.3

Spark Your Learning

B. The variables are the length of each side of the octagon and the height of each triangle that makes up the pyramid; Units can be meters, centimeters, inches, or feet, but inches are used in this example.

D. The area 19,680 in^2 is an underestimate because shingles overlap on a roof. The total material you need is greater than 19,680 square inches.

Build Understanding

Task 1 **A.** The slant height is the altitude of a lateral face, while the height of a pyramid is the distance between its vertex and its base.

B. The base is a square and the lateral faces are isosceles triangles. There are 4 congruent lateral faces. The square has a side length of s units. The triangles have a base of s units and a height of ℓ units.

C.

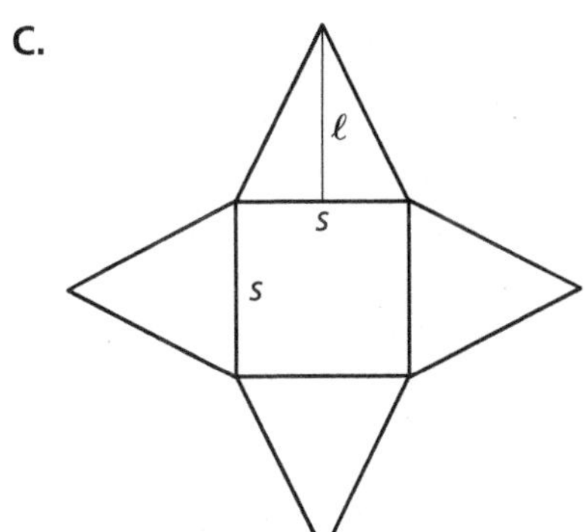

E. You need to replace the s^2 term with the area of the hexagonal base, and you need to replace the $2s\ell$ term with $6\left(\frac{1}{2}s\ell\right) = 3s\ell$. You need to be able to calculate the area of the hexagon, so you need more information about the hexagon to write a formula.

F. The surface area S for any regular pyramid is the area of the base plus the lateral area of the pyramid. This would also apply to a cone. The surface area of a cone is also the area of the base plus the lateral area of the cone. However, the lateral area of a regular pyramid is the sum of the areas of triangles, while the lateral area of a cone is not the sum of polygonal areas.

On Your Own

25. Possible answer:

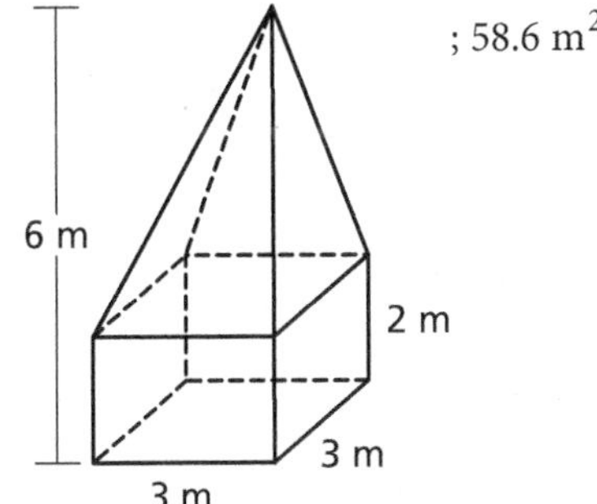

; 58.6 m^2

MODULE 18, LESSON 18.4

Spark Your Learning

B. Possible sketch:

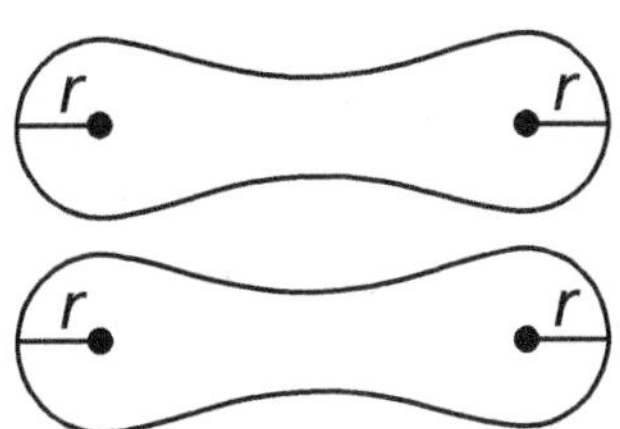

D. The answer makes sense because it is 4 times the area of a circle with a radius of 1.45 inches.

Build Understanding

Task 1 **D.** no; The surface area of the cylinder is $4\pi r^2$. The surface area of the hemisphere is the sum of half the surface area of the entire sphere and the area of the circular base. So, $\frac{1}{2}(4\pi r^2) + \pi r^2 = 3\pi r^2$, and $3\pi r^2 \neq 4\pi r^2$

Step It Out

Task 2 The density, given as the amount of gold per unit of area (mg/mm^2), is multiplied by the area (mm^2) to produce the amount of gold (mg) because $mg/mm^2 \bullet mm^2 = mg$.

MODULE 19, LESSON 19.1

Spark Your Learning

B. The variables are the volume of a container and the number of containers in the stack. The units used are cubic inches.

D. Possible answer: The answer makes sense in the context of the situation. The answer is in the correct unit of measurement and is less than 10 times the volume of one container.

Build Understanding

Task 1 **A.** Since the area of the base of a unit cube is 1 square unit, you could fit B unit cubes on the base of the prism. These cubes would fit perfectly inside the prism since the height of each cube is 1 unit and the height of the prism is 1 unit; The volume of the prism is B cubic units.

B. The volume is $2B$ cubic units; You add a prism with a volume of B cubic units to another prism with a volume of B cubic units. $B + B = 2B$

D. yes; $4B$ cubic units

E. To find the volume of a prism or cylinder, multiply the area of the base by the height of the figure; $V = Bh$

Task 2 **A.** The volume of one nickel is about 689 cubic millimeters; The volume of a stack of nickels is about 6,890 cubic millimeters.

B. They all have the same number of nickels; They are not aligned vertically.

C. The volume of each stack of nickels is about 6,890 cubic millimeters; The volume of each stack is the same as the volume of the original stack because they contain the same number of nickels. The position of the nickels does not affect the volume.

D. Any stack of 10 nickels will have the same volume as the stack of nickels that is aligned vertically. The volume is about 6,890 cubic millimeters.

E. yes; The volume does not change, so the volume of the oblique cylinder is also 80π cubic inches. Both cylinders have the same height and same base area, so their volumes are the same.

Step It Out

Task 4 **A.** Multiply the dimensions of the base of the prism to find the area of the base, then multiply the area of the base by the height of the prism.

B. $0 < x < 5$; x must be greater than 0 for volume to exist. x cannot be greater than 5 because $10 - 2x$ must be greater than 0 for volume to exist.

C. The graph shows that, for the constraints from Part B, a maximum occurs at about $x = 2$. So, height $= 2$, width $= 10 - 2(2) = 6$, and length $= 15 - 2(2) = 11$; The maximum volume is about $2 \bullet 6 \bullet 11$, or about 132 in^3.

On Your Own

8. Francisco is correct. By Cavalieri's Principle, if two solids have the same height and the same cross-sectional area at every level, then the two solids have the same volume.

10. The block with a radius of 6 inches has a greater volume. The radius of the other block is 5 inches. The heights are the same, so the block with a greater radius has the greater volume.

33. There are many possible answers, but they should be equivalent to the following:

$r = 1$

$h = 2$

$\frac{2}{3}\pi + \frac{4}{3}\pi = \frac{8}{4}\pi$

MODULE 19, LESSON 19.2

Spark Your Learning

B. Possible answer: a rectangular prism; A pyramid takes up less than half the space of a prism; approximately $\frac{1}{3}$ the space

C. The variables are the length and width of the base of the pyramid and the height of the pyramid; The units of measure are meters.

Build Understanding

Task 1 **B.** If you remove G-HCD from the prism, and then slice the remaining solid on the plane GEC, you get pyramids G-EHC and G-CFE. The three pyramids do not overlap and contain the entire prism, so the sum of their volumes is equal to the volume of the prism.

C. $\overline{GC}$ is the diagonal of a rectangle, so $\triangle GCD \cong \triangle GFC$. The bases of each pair of pyramids are congruent and they have the same height. By the postulate, the pairs of pyramids have the same volume.

D. The volume of the pyramid is one-third of the volume of the triangular prism. An equation for the volume of the pyramid is $V = \frac{1}{3}Bh$.

9. When finding the volume of the composite figure, you subtract the volume of the pyramid from the volume of the prism. When finding the surface area, you add the lateral area to the surface area of all sides except the top of the prism.

MODULE 19, LESSON 19.3

Spark Your Learning

B. The variables are the dimensions of a cherry tomato and the dimensions of the pint basket.

D. yes; If you multiply the volume of a cherry tomato by the number of tomatoes, it equals the volume of the basket.

Build Understanding

Task 1 **A.** Using the Pythagorean Theorem, $x^2 + R^2 = r^2$. Solving for R, $R = \sqrt{r^2 - x^2}$. Each cross section is a circle, so the area of a cross section is $A = \pi R^2 = \pi\left(\sqrt{r^2 - x^2}\right)^2 = \pi\left(r^2 - x^2\right)$.

B. Subtract the area of the inner circle of the ring from the area of the outer circle. The area of the ring is $A = \pi r^2 - \pi x^2 = \pi r^2 - x^2$.

D. $V = \frac{2}{3}\pi r^3$; Subtract the volume of the cone from the volume of the cylinder.

$$V = \left(\pi r^2 \bullet r\right) - \left(\frac{1}{3}\pi r^2 \bullet r\right)$$
$$= \pi r^3 - \frac{1}{3}\pi r^3$$
$$= \frac{2}{3}\pi r^3$$

E. Cavalieri's Principle states that two figures with the same height and same cross-sectional area at every level have the same volume. So the hemisphere has the same volume as the cylinder with the cone removed. $V = \frac{2}{3}\pi r^3$; $V = \frac{4}{3}\pi r^3$

MODULE 19 REVIEW

Concepts and Skills

5. Jorge's stack of cards has a greater volume. The heights of the stacks are the same. The lengths of Jorge's cards are longer, and the widths are the same, so the area of the base of his stack is greater. Thus, the volume of Jorge's stack of cards is greater by Cavalieri's principle.

UNIT 10

MODULE 20, LESSON 20.1

Spark Your Learning

A. What ratio of the dogs that are brought in for grooming have straight hair? curly hair?; the number of dogs that have curly hair and the number of dogs that have straight hair

D. yes; Since you know that there are 244 groomings and $170 + 74 = 244$, the ratios are reasonable.

Build Understanding

Task 1 **A.** The universal set is all birds on the branch. There are 6 elements in the set.

B. Possible answer: birds with white beaks; There are 2 elements in the subset.

C.

Parrots

Blue and Yellow Body 0

4

Green Crown 1

Others 1

On Your Own

7. $A = \{1, 2, 5, 10\}$, $B = \{2, 4, 6, 8, 10\}$, $C = \{1, 3, 5, 7, 9\}$, $D = \{1, 3, 9\}$

8. yes; because every element of D is in C and there is at least one element in C that is not in D

38. He can find the probability that he does not win a prize by finding the complement of the probability of winning a prize. The sum of an event and its complement is 1, so the probability of not winning a prize is $1 - \frac{1}{25} = \frac{24}{25}$.

MODULE 20, LESSON 20.2

Spark Your Learning

B. The variables involved are the characteristics of the fruit (color) and the numbers of fruits. In this case, the kind of fruits (oranges, apples, bananas, pineapples) is not a variable.

Build Understanding

Task 1 **E.** The probability of choosing an orange or purple marble is equal to the sum of the probability of choosing an orange marble and the probability of choosing a purple marble.

Task 2 **C.**

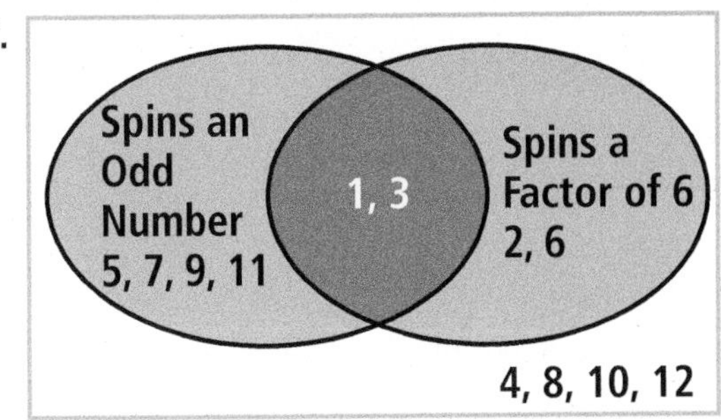

$\frac{2}{3}$

On Your Own

24.

		Plays Video Games		
		Yes	No	Total
School	Middle school	54	6	60
	High school	48	7	55
	Total	102	13	115

A. $\frac{67}{115}$

B. $\frac{109}{115}$

C. $\frac{115}{115} = 1$

25. When adding the probabilities of each event, the outcomes that are in both events are counted twice, so you need to subtract the probability of both events. Then they are only counted once.

MODULE 20 REVIEW

Concepts and Skills

7. A. The probability of choosing a shirt that is not blue is the complement of the probability of choosing a shirt that is blue. $1 - \frac{8}{21} = \frac{13}{21}$

B. The shirts can only be one color, so they are mutually exclusive events. Add the probability of each individual event occurring to get the union. $\frac{3+6}{21} = \frac{9}{21} = \frac{3}{7}$

C. In this situation, a shirt can be red *or* white, but it cannot be red *and* white. There are no elements in the intersection of the two events.

8. D. The events "chooses a piece of fruit" and "chooses an item that is refrigerated" are not mutually exclusive. Therefore, use the Addition Rule.

$$P(A \cup B) = P(A) + P(B) - P(A \cap B)$$
$$= \frac{11}{36} + \frac{17}{36} - \frac{8}{36} = \frac{20}{36} = \frac{5}{9}$$

There are 8 items that are refrigerated fruits. These items should not be counted twice, which is why $\frac{8}{36}$ must be subtracted from the sum of *P*(chooses a fruit) and *P*(refrigerated).

MODULE 21, LESSON 21.1

Spark Your Learning

B. color of the dragon; number of dragons of each color having wings and no wings

D. yes; for a blue dragon, there are two possibilities: wings or no wings. Since the number of blue dragons without wings (75) is much greater than the number with wings (25), I would expect the probability to be less than 50%.

Building Understanding

Task 1 **B.** *P*(red): row 1 column 3 and row 3 column 3, $P(\text{face} \mid \text{red})$: row 1 column 1 and row 1 column 3; $P(\text{face} \mid \text{red}) = \frac{6}{9} = \frac{2}{3}$, or approximately 66.7%

D. Possible answer: The difference is the denominator. The denominator represents the total number of occurrences of the given event. There 9 total occurrences of red cards and 16 total occurrences of face cards.

Task 2 **A.** The value 16 represents the number of participants who used their old toothpaste and had no new cavities. The value 45 represents the number of participants who used their old toothpaste. The value 100 represents the number of participants in the experiment.

D. no; All the data in the second row refers to participants who used their old toothpaste. $P(B|A)$ is the probability that someone had no new cavities given that they used the new toothpaste.

Task 3 **A.**

	Face card	Other card	Total
Red card	$A \cap B$	$A \cap B^C$	A
Black card	$A^C \cap B$	$A^C \cap B^C$	A^C
Total	B	B^C	U

D. the probability that a person chooses a card that is not a face card given that the card is red; Both probabilities describe selecting a red, non-face card. The difference is the denominator. $P(B^C \mid A)$ describes selecting a non-face card from just the red cards, and $P(A \cap B^C)$ describes selecting a red, non-face card from all the cards.

E.

	No new cavities	At least 1 new cavity	Total
New toothpaste	$\frac{46}{100} = 0.46$	$\frac{9}{100} = 0.09$	$\frac{55}{100} = 0.55$
Old toothpaste	$\frac{16}{100} = 0.16$	$\frac{29}{100} = 0.29$	$\frac{45}{100} = 0.45$
Total	$\frac{62}{100} = 0.62$	$\frac{38}{100} = 0.38$	$\frac{100}{100} = 1.00$

F. $P(A) = \frac{55}{100}$, or 55%; $P(A \cap B) = \frac{46}{100}$, or 46%

Check Understanding

3.

	First letter of name	Not first letter	Total
Vowel	0.05	0.21	0.26
Consonant	0.09	0.65	0.74
Total	0.14	0.86	1.00

On Your Own

20. A. Mark incorrectly calculated the conditional probability of those who have a new phone among those who have problems with their phone $(9 \div 38)$. He should have calculated the conditional probability of those who have problems with their phone among those who have new phones $(9 \div 55)$.

B. $$P(\text{problem} \mid \text{new phone}) = \frac{P(\text{problem} \cap \text{new phone})}{P(\text{new phone})} = \frac{9}{55} \approx 0.16 = 16\%$$

21. Possible answer: 9, 5, 4, 2, 8

MODULE 21, LESSON 21.2

Spark Your Learning

B. Possible answer: the probability that a student owns a dog, the probability that a student gets 60 minutes or more of aerobic activity each day, the probability that a student owns a dog and gets 60 minutes or more of aerobic activity each day

D. Possible answer: yes; It makes sense that a student who owns a dog is more likely to be active than a student who does not. A student who owns a dog most likely has to take the dog for walks on a regular basis.

Building Understanding

Task 1 **A.** The probabilities should be the same since flipping the coin with a result of heads should not affect the occurrence of spinning yellow.

B. The probabilities should be the same since spinning yellow should not affect the occurrence of flipping the coin with a result of heads.

C. Possible answer: I can use the conditional probability formula:

$$P(Y \mid H) = \frac{P(Y \cap H)}{P(H)} = \frac{\frac{1}{16}}{\frac{1}{2}} = \frac{1}{8} = P(Y) \text{ and}$$

$$P(H \mid Y) = \frac{P(H \cap Y)}{P(Y)} = \frac{\frac{1}{16}}{\frac{1}{8}} = \frac{1}{2} = P(H).$$

Yes, the results agree with my thinking.

Task 2 **C.** $P(R) \cdot P(Y) = \frac{1}{16}$; The product of $P(R)$ and $P(Y)$ is equal to $P(R \cap Y)$.

D. $$P(A \mid B) = \frac{P(A \cap B)}{P(B)}$$

Since $P(A \mid B) = P(A)$, substitute $P(A)$ for $P(A \mid B)$ in the definition.

$$P(A) = \frac{P(A \cap B)}{P(B)}$$

Multiply each side by $P(B)$.

$P(A) \cdot P(B) = P(A \cap B)$

Step It Out

Task 3 **A.** There are 4 red marbles, 6 blue marbles, and a total 12 marbles for each selection. $P(R) = \frac{4}{12} = \frac{1}{3}$ and $P(B) = \frac{6}{12} = \frac{1}{2}$.

B. $P(B \mid R) = \frac{6}{12} = \frac{1}{2}$. Since $P(B \mid R) = P(B)$, events R and B are independent.

Task 4 **A.** Eighty–four youth tickets were sold from a total of 280 tickets sold, so I divide 84 by 280 to find $P(Y)$. There were 36 youth tickets for a musical and a total of 120 ticket for a musical, so I divide 36 by 120 to find $P(Y \mid M)$.

B. To find $P(Y \cap M)$, divide 36 by 280, so $P(Y \cap M) = \frac{36}{280} = \frac{9}{70}$.
To find $P(Y)$, to divide 84 by 280, so $P(Y) = \frac{84}{280} = \frac{3}{10}$.
To find $P(M)$, divide 120 by 280, so $P(M) = \frac{120}{280} = \frac{3}{7}$.
$P(Y) \cdot P(M) = \frac{3}{10} \cdot \frac{3}{7} = \frac{9}{70}$, so $P(Y \cap M) = P(Y) \cdot P(M)$.

Check Understanding

1. Possible answer: flip a coin and get tails and roll a 3 or 6 on a number cube; Let A be the event of flipping a coin and getting tails and B be the event of rolling a 3 or 6. $P(A) = \frac{1}{2}$ and $P(B) = \frac{1}{3}$. Using the conditional probability formula, $P(A \mid B) = \frac{P(A \cap B)}{P(B)} = \frac{\frac{2}{12}}{\frac{1}{3}} = \frac{1}{2}$ and $P(B \mid A) = \frac{P(A \cap B)}{P(B)} = \frac{\frac{2}{12}}{\frac{1}{2}} = \frac{1}{3}$. Since $P(A) = P(A \mid B)$ and $P(B) = P(B \mid A)$, the two events are independent.

2. $P(A) = \frac{1}{2}$, $P(B) = \frac{1}{3}$, $P(A \mid B) = \frac{1}{2}$, and $P(B \mid A) = \frac{1}{3}$, so $P(A \mid B) = P(A)$ and $P(B \mid A) = P(B)$. The events are independent.

4. A. $P(M \mid S) = \frac{45}{99} = \frac{5}{11}$ and $P(M) = \frac{80}{176} = \frac{5}{11}$, so $P(M \mid S) = P(M)$. M and S are independent.

B. $P(M \cap S) = \frac{45}{176}$, $P(M) = \frac{80}{176} = \frac{5}{11}$, $P(S) = \frac{99}{176} = \frac{9}{16}$, and $P(M) \cdot P(S) = \frac{5}{11} \cdot \frac{9}{16} = \frac{45}{176}$, so $P(M \cap S) = P(M) \cdot P(S)$. M and S are independent.

On Your Own

15. A. yes; $P(D) = \frac{4}{12} = \frac{1}{3}$, $P(R) = \frac{6}{12} = \frac{1}{2}$, $P(D \cap R) = \frac{2}{12} = \frac{1}{6}$, so $P(D \cap R) = P(D) \cdot P(R)$. Since $P(D \cap R) = P(D) \cdot P(R)$, D and R are independent.

B. no; $P(D) = \frac{4}{12} = \frac{1}{3}$, $P(S) = \frac{8}{12} = \frac{2}{3}$, $P(D \cap S) = \frac{0}{12} = 0$, so $P(D \cap S) \neq P(D) \cdot P(S)$.

16. Since $P(A \mid B) = P(A)$, use the definition of conditional probability to get $\frac{P(A \cap B)}{P(B)} = P(A)$. Multiply both sides by $P(B)$ to get $P(A \cap B) = P(A) \cdot P(B)$. Then divide both sides by $P(A)$ to get $\frac{P(A \cap B)}{P(A)} = P(B)$. Since $P(B \mid A) = \frac{P(A \cap B)}{P(A)}$ by definition of conditional probability, $P(B \mid A) = P(B)$.

17. Possible answer: Rolling a sum of [2] on two [4]-sided dice is the same probability as rolling a sum of [5] on two [8]-sided dice.

MODULE 21, LESSON 21.3

Spark Your Learning

B. Possible answer: D is the event that the drive time from Evanston to Chicago takes 30 minutes or less, R is the event that it is raining.

Build Understanding

Task 1 **A.** $P(Y) = \frac{2}{5}$, $P(A) = \frac{11}{20}$, $P(A \cap Y) = \frac{1}{4}$; $\frac{2}{5} \cdot \frac{11}{20} = \frac{11}{50}$; The events are dependent.

B. $P(Y \mid A) = \frac{5}{11}$, $P(A \mid Y) = \frac{5}{8}$

C.

×	P(Y)	P(A)
P(Y|A)	$\frac{2}{11}$	$\frac{1}{4}$
P(A|Y)	$\frac{1}{4}$	$\frac{11}{32}$

D. Yes, $P(A) \cdot P(Y \mid A)$ and $P(Y) \cdot P(A \mid Y)$ are both equal to $P(A \cap Y)$; If events A and B are dependent, you can compute $P(A \cap B)$ as either the product of $P(A)$ and $P(B \mid A)$ or the product of $P(B)$ and $P(A \mid B)$.

On Your Own

18.
$$P(A \cap B \cap C) = P\big((A \cap B) \cap C\big)$$
$$= P(A \cap B) \cdot P(C \mid A \cap B)$$
$$= P(A) \cdot P(B \mid A) \cdot P(C \mid A \cap B)$$

19. 7 marbles, 3 blue, 4 white, 29% OR 7 marbles, 4 blue, 3 white, 29%

MODULE 21 REVIEW

Concepts and Skills

11. Let T be the event that the student is 10–12 years old and A be the event that the student chooses archery.

$$P(T) \stackrel{?}{=} P(T \mid A)$$

$$PT \stackrel{?}{=} \frac{P(T \cap A)}{P(A)}$$

$$\frac{65}{142} \stackrel{?}{=} \frac{\frac{27}{142}}{\frac{59}{142}}$$

$$\frac{65}{142} \stackrel{?}{=} \frac{27}{142} \cdot \frac{142}{59}$$

$$\frac{65}{142} \stackrel{?}{=} \frac{27}{59}$$

$$0.458 \approx 0.458$$

The event that the student is 10–12 years old and the event that the student chooses archery are independent.

12. A. Possible answer: The fact that the events "the student has more than two siblings" and "the student takes a drafting class" are dependent means that a student with two or more siblings is more or less likely to take a drafting class than any student. The fact the events are independent would mean that having more than 2 siblings has no effect on the probability that a student would take a drafting class.

B. $P(M \cap C) = \frac{16}{209} \approx 0.077$

C. The events are dependent.

Two events are independent if and only if $P(M \cap C) = P(M) \cdot P(C)$.

$$P(M \cap C) = \frac{16}{209} \approx 0.08$$

$$P(M) \cdot P(C) = \frac{47}{209} \cdot \frac{125}{209} \approx 0.13$$

$$P(M \cap C) \approx 0.08 \neq 0.13 \approx P(M) \cdot P(C)$$